NEWTON'S TELECOM DICTIONARY

11th Edition

The Official Dictionary of Telecommunications
Computer Telephony, Data Communications
Voice Processing, Internet Telephony
Windows 95 & NT Communications
LAN and WAN Networking

#1 SELLER 140,000 SOLD

by Harry Newton

NEWTON'S TELECOM DICTIONARY

A Flatiron Publishing, Inc. Book
copyright © 1996 Harry Newton
published by Flatiron Publishing, Inc.

All rights reserved under International and Pan-American Copyright conventions, including the right to reproduce this book or portions thereof in any form whatsoever. Published in the United States by Flatiron Publishing, Inc., New York.

12 West 21 Street
New York, NY 10010
212-691-8215 Fax 212-691-1191
1-800-999-0345
1-800-LIBRARY
Email harrynewton@mcimail.com

ISBN # 0-936648-87-2

July, 1996

Manufactured in the United States of America

Eleventh Edition
Cover design by Mara Seinfeld
Printed at Command Web, New Jersey

THE LOGIC OF MY DICTIONARY
by Harry Newton

First, an apology. I've drastically cut the size of the type in this dictionary. For 10 editions, I've believed in "readability," i.e. BIG TYPE. Suddenly, my dictionary blossomed to 1,400 pages, 5 1/2 lbs and was costing $5.50 to ship domestically and $35+ internationally. It was more to ship overseas than to buy. Totally ridiculous. Skip the long story. We spent days trying different type faces, different type sizes, different leading (space between lines) and different spacing between letters. This is the most readable we could create and keep the weight down. It ain't perfect. For that I apologize.

I wrote this book for those of us trying to keep up. This is becoming increasingly difficult because of our industry's growth. I'm adding and changing words every single day. Some days I add as many as 50 words. I'm averaging 100 new words a week. Some words take an hour to define. To find the new words, I subscribe to every telecommunications, networking, PC and business magazine I can find. My Monday pile of magazines and books is about 24 inches high. Fortunately, many of my readers kindly send me new words, point out my mistakes and make suggestions. Please feel free to contact me on harrynewton@mcimail.com.

No other industry is changing as fast as or growing as big as telecommunications. After health and defense comes telecom. It's big. And as it changes our lives, it will get even bigger. There are seven driving forces:

1. The Technology. The main excitement centers on digital signal processors, fiber optics and new algorithms in speech recognition, compression, etc. And the continuing speedup in microprocessors (as reflected in Moore's Law) is making possible new telecom concepts like switches inside a PC, fast ATM, 100 meg Ethernet, etc. But the reality is that technology in telecom is improving in cost-performance far faster than it is in computing. Hence, the computer industry's new obsession with telecom, networking and voice.

2. New Standards. Barely 25 years ago, AT&T was the primary telecom standards setter in North America. It created the 1200 baud standard for modem. That was essentially its last. Since then the company has dropped out of standards-setting, leaving that task to a hodge-podge of volunteer bodies (from the ATM Forum to the ECTF, to the MVIP, to the SCSA, to the ITU-T, to ANSI, to the Internet Engineering Task Force) and to companies (from Hayes to Microsoft to Intel to Dialogic). There's been an explosion of new standards (and they're still coming) — defining everything from telecom operating systems, to buses that carry voice inside PCs, to high-speed lines, to new telecom "building block" software. Standards lead to growth. It is now mid 1996. As I listen to my phone calls and hear the industry "buzz," I can feel the excitement in hundreds of engineering labs frenetically creating new telecom products and services. We're about to see telecom product out in the next few years based on these new standards that will positively blow your brains.

3. New Government Awareness That Competition Is Better Than Monopoly. The telecom industry has historically been closed — closed in architecture, closed to new suppliers, closed to new entrants. No longer. Europe, Japan, Australia, Russia, China — everywhere — governments are waking to one realization. Telecom is infrastructure. Companies and business go where infrastructure is strongest. You can't make modern infrastructure when you have one bloated, sluggish, government-run phone company. And you can't limit your entire country's telecom purchases to two or three hand-chosen suppliers (as each country in Europe has historically done). By the end of this century, European telecom should be as open as North America is today. That's another 300 million people hungry for decent, cheap communications. Everywhere, telecommunications is going from a totally-closed, government-controlled, highly-regulated, monopolistic, bureaucratic, anti-progress, anti-customer industry to an exploding free-for-all.

4. New Startups. The hottest place for venture capital is telecommunications. Smart engineers are leaving big, sluggish telecom manufacturers in droves and starting new companies. There are hundreds of "new telecom" millionaires already. There will be thousands by the end of the decade. Nothing can match the enthusiasm of an engineer obsessed with a great idea that he thinks he will make himself a millionaire. I'm enthusiastically behind each and every one of them. I publish three monthly telecom magazines — Call Center, Computer Telephony and TELECONNECT. I run one annual trade show — Computer Telephony Expo. I welcome these people and their products with open arms to my editorial pages and my speaking platforms.

5. New Uses. Telecommunications is becoming computer telephony. This is the name of the new industry which adds computer intelligence and computer power to the making and receiving of phone calls. As you add intelligence and power, you create brand new products — ones we can't even dream about today. An example: Who would have imagined the 1992-1995 phenomenon which produced the democratization and commercialization of the Internet? I believe that Internet happening is as important to the dissemination of knowledge as the invention of the Gutenberg Press was in 1453.

6. Telecom's Basically Incredible Economics. No industry in the world — neither software nor petroleum — can match the incredible profitability of the zero-marginal cost telephone industry. If I call from New York to California, that call costs my supplier — AT&T, MCI, Sprint or whoever — essentially nothing. Nobody to this day knows how many conversations or how much information you can put down a single strand of fiber. They just keep putting more and more down it. It's truly amazing.

7. The skyrocketing demand for telecom. Children want high-speed Internet access. Business people want "everywhere" computing at a speed that matches what they have in their office. People would like to see who they're talking to. Everyone wants access to instant information — whether in words, in images or in video. Everyone wants to work at home. It's called call it SOHO, as in Small Office Home Office. It's brand-new, and one of telecom's fastest-growing markets. And all of us want the best quality and widest range of digital home telecom services — from movies to three-dimensional tele-gaming.

Writing telecom dictionary updates every six months is exhausting, exhilarating and very time-consuming. My children tell me, "Daddy, get a life!" I apologize to Claire and Michael for seeing so much of Daddy's back and Daddy's monitors.

MY DICTIONARY

This is a dictionary to work every day with. Salespeople tell me they include the definitions in proposals to customers. Novices tell me they love it because it cuts through the clutter. Users explain telecom things to their boss with my definitions. Management uses it to understand telecom technicalities. Lawyers even use it in court. Often they call me as an "expert" witness. It probably pays better than being a juror. But it's probably as boring and as abusive. I always decline, though I'm always flattered.

You can give my dictionary to your users, to your customers, to your boss. You can even give it to your kids to let them understand what you do. They'll then understand why you, too, have no life.

Most technical dictionaries define terms tersely, often in other technical terms. As a result they leave you more confused. This dictionary is different, deliberately so. My definitions tell you what the term is, how it works, how you use it, what its benefits are, what its negatives are. I tell you how it fits into the greater scheme of things, and occasionally some warnings or checklists.

I don't claim my dictionary is comprehensive. But each edition gets bigger and better. I add new terms. I re-work and I update definitions. I'm always looking for new ones. That's an invitation. Send me your product names, your service descriptions, your in-house glossaries, your corrections.

The best way to get me is through harrynewton@mcimail.com

HOW TO USE MY DICTIONARY My definitions are in a combination ASCII / alphabetical order. Definitions containing non-letters — like @, #, / and numbers — are in ASCII order. Definitions with real letters are in alphabetical order. I use ASCII because it gives alleged order to hyphens, periods, forward slashes, numbers, etc. Here is the order of the more common characters you'll find in this dictionary:

Blank Space	= ASCII 32	5	= ASCII 53
!	= ASCII 33	6	= ASCII 54
#	= ASCII 35	7	= ASCII 55
& (Ampersand)	= ASCII 38	8	= ASCII 56
- (Hyphen or dash)	= ASCII 45	9	= ASCII 57
. (Period)	= ASCII 46	: (colon)	= ASCII 58
/ (Forward slash)	= ASCII 47	; (semi colon)	= ASCII 59
0 (zero)	= ASCII 48	A (capital A)	= ASCII 65
1	= ASCII 49	Capital letters to	ASCII 90
2	= ASCII 50	\ (back slash)	= ASCII 92
3	= ASCII 51	Lower case letters	
4	= ASCII 52	start with a	= ASCII 97

ON SPELLING

My dictionary conforms to American spelling. To convert American spelling to British and Canadian spelling typically requires adding a second "L" in words like signaling and dialing (they're American) and changing "Z" in words like analyze to analyse. Center in American is Center. In Britain, Australia and Canada, it's Centre. This dictionary contains more British, Australian and European words than my previous editions — a result of several overseas lecture tours.

ON STYLE

All high-tech industries make up new words by joining words together. They typically start by putting two words next to each other. Later, they join them with a hyphen. Then, with age and familiarity, the hyphen tends to disappear. An example: Kinder garten. Kinder-garten. and now Kindergarten.

Sometimes it's just a matter of personal choice. Some people spell database as one word. Some as two, i.e. data base. I prefer it as one, since it has acquired its own logic by now. Sometimes it's a matter of how it looks. I prefer T1 (T-one) as T-1, simply because T-1 is easier to recognize on paper. I define co-location as co-location. Websters spells it collocation, with two Ls, one more than mine. I think mine is more logical.

Plurals give trouble. The plural of PBX is PBXs, not PBX's. The plural of PC is PCs, not PC's, despite what the New York Times says. The Wall Street Journal and all the major computer magazines agree with me. So there! In this dictionary, I spell out the numbers one through nine. Above nine, the numbers are written as arabic numerals, i.e. 10, 11, 12, etc. In short, there are no rights or wrongs in the spelling business, except that my dictionary is now the correct way of spelling telecom words, because it's the biggest seller (by far) and because lawyers use it in court. (God help the justice system.)

THANK YOUs

A big "Thank You" to the dozens of people and dozens of companies who helped. I bet I left some of you out. If so, I apologize. Among the manufacturers, special thanks to Amdahl, Anixter, Aspect Telecommunications, AT&T, Bellcore, Dialogic, Ecos Electronics, General Cable, Micom, MCI, NEC, Newbridge Networks, New York Telephone (now Nynex), Northern Telecom (now Nortel), Racal Data, Ricoh, Sigma Designs, Sharp and Teknekron. They'll recognize some of their words in this dictionary. Among the magazines I borrowed (or stole), the best were PC Magazine and our own TELECONNECT, Call Center and IMAGING Magazines. Special thank yous also to internetworking expert, Tad Witkowicz at CrossComm, Marlboro MA.; Ken Guy of Micom, Simi Valley (near LA); Michael Marcus, president of Able Communications, an excellent interconnect company based in Scarsdale NY, Frank Derfler of PC Magazine; Chris Gahan of 3Com; Jeff Deneen of the Norstar Division of Northern Telecom in Nashville; Stephen Doster of Telco Research in Nashville; bugging expert Jim Ross of Ross Engineering, Adamstown, MD; wiring experts John and Carl Siemon of The Siemon Company, Watertown CT; the fine folk at Novell, the local area network company in Provo, UT; to Jim Gordon and Parker Ladd at TCS Communications, Nashville, TN, the people who do workforce management software for automatic call distributors; to Judy Marterie and the electricity wiring, grounding and test experts at Ecos Electronics Corporation in Oak Park, Il; to Brian Newman of MCI, who understands wireless; to John Perri of SoftCom, NYC; to John Taylor of GammaLink, a Sunnyvale, CA company which produces beautiful fax products; to Charles Fitzgerald at Microsoft and Herman D'Hooge at Intel who jointly helped created Windows Telephony; to Charles Fitzgerald, again, who allowed me to steal all the words Microsoft makes up (you'd be surprised how many); to Bill Flanagan who's written fine books on T-1 and voice and data networking; to Jane Laino of Corporate Communications Consultants, NYC; to Henry Baird of Seattle consultants Baird & Associates; to Sharon O'Brien formerly of Hayes Microcomputer Products in Norcross (Atlanta); to Howard Bubb, John Landau, Jim Shinn and Nick Zwick at leading voice processing component manufacturer, Dialogic Corporation of Parsippany, NJ; to Al Wokas of San Jose, CA; to Alison Golan of networking company, Interphase Corporation in Dallas, which allowed me to steal some of the definitions from their excellent booklet, "A Hitchhiker's Guide to Internetworking Terms and Acronyms;" to Ian Angus at the Angus TeleManagement Group in Ajax, Ontario, who embarrassed me into expanding my Canadian coverage.

I'm very grateful to The ATM Forum of Mountain View, CA (www.atmforum.com) for allowing me to use many of their definitions from their really well-done ATM Forum Glossary.

At Ziff Davis, which publishes this dictionary on CD-ROM discs (for your copy, call 800-827-7889 or 212-503-4400 or in England 44-344-714-441), I'm grateful to Denis Haskin. Their new CD-ROM division is called Information Access Corp, Medford, MA. In my own office, I'm very grateful to Muriel Fullam, Nick Morley, Christine Kern, Rose Bodin, Mara Leonardi and Jennifer Cooper-Farrow. Without all these wonderful people, this dictionary wouldn't be as good as it's actually turning out. If I sound surprised, you're right. It's now the largest-selling telecommunications dictionary in the world. If I've left any definitions out, or if some of my definitions are unclear, contact me on harrynewton@mcimail.com. I wrote this dictionary on a series of ever-newer, ever-faster Toshiba laptops (very reliable machines) using The Semware Editor, a very beautiful text editor, which Sammy Mitchell of Marietta, GA wrote, and which I wholeheartedly recommend. The Toshiba laptop for this dictionary was the Tecra 700CT. It's a 120 Mhz Pentium machine with 48 Meg of RAM and an 1.4 gigabyte removable hard disk. It also has a removable floppy disk which can be replaced with a CD-ROM. Nora Long was very helpful in fixing the logic and catching stuff. Jennifer and Christine Kern (ne. Fullam) typeset it on a Power PC Macintosh using QuarkXpress and Adobe Illustrator. Mara Leonardi did the cover. It's gorgeous, as usual.

I welcome comments, corrections, suggestions, improvements.

Harry Newton
12 West 21 Street
New York, NY 10010
212-691-8215 Fax 212-691-1191
Electronic mail HarryNewton@MCIMail.com

July, 1996

NEWTON'S TELECOM DICTIONARY

NUMBERS

" Double quotation marks. Typically used to signify something your computer should print (to screen, disk or paper), as in PRINT "Thanks for being a good guy."
Some programs allow you to use single quotation marks interchangeably with double ones. Some programs don't. Try one or the other if in doubt.
The character on the bottom right of your touchtone keypad, which is also typically above the 3 on your computer keyboard. It's commonly called the pound sign, but it's also called the number sign, the crosshatch sign, the tic-tack-toe sign, the enter key, the octothorpe (also spelled octathorp) and the hash. Musicians call the # sign a "sharp."
On some phones the # key represents an "Enter" key like the Enter key on a computer. On some phones it represents "NO." And on others it represents "YES." MCI, AT&T and some other long distance companies use it as the key for making another long distance credit card call without having to redial. Hold down the # key for at least two seconds before the person at the other end has hung up, you'll get a dial tone, punch in your phone number and you can make another long distance call — without having to punch in your credit card number again. (This service is often called Call Reorigination or just plain Next Call.)
The # key is used in the paging industry — national and local. When you dial a phone number which represents someone's beeper, you will typically hear a double beep. At that point you punch in your phone number, ending it with a #. At that point, the machine hangs up on you and sends out the numbers you punched in to the pager you just dialed.
Many digital phone pagers allow people to send actual text messages to pagers. Many use the # sign to signal the use of certain digits (c, f, i, l, o, s, v, y) as well as to signal the end of the transmittal.
& The "and" sign. Its real name is an "ampersand."
* The star sign. It's often used to represent a wild card or a joker. For example, the command
ERASE JOHN.*
will erase all the files on your disk beginning with JOHN, e.g. JOHN.TXT, JOHN.NEW, JOHN.OLD, JOHN.BAK, etc.
IEX in Richardson * key as a cancel character for its Call Valet application.
*57 The North American universal dialing code which you touchtone in immediately after receiving a harassing, obscene or annoying phone call. By touchtoning that number in, you have alerted your central office to "tag" that phone call. Should a law enforcement agency get involved in investigating your annoying calls, they would be able to go into your records and find the phone number from which the annoying call was made. See TRAP and TRACE.
*67 Dial *67 before you make a call in North and the person you call won't see your CallerID.
++ In C programming, the expression i++ means use the current variable i and add 1 to it. In more contemporary usage, ++ has come to mean an expansion, an improvement, an upgrading, etc. So, C++ is an improved version of the C programming language. MAE-East++ is an expansion of MAE-West located nearby. See MAE-East.
- The dash. The minus sign. Often we take two words and join them with a dash into a new word. As the word becomes more and more common, we remove the dash and the double word now becomes a fully-fledged single word.
/ The forward slash. Lotus made it famous.
@ The character typically above the 2 on your keyboard. It's called the "at sign." In English, its biggest use would be two apples @ 50 cents each equals $1 total." But in computerese, its big use is in electronic mail addressing. For example, Bill Gates' e-mail address is billg@Microsoft.com. Mine is harrynewton@mcimail.com. The @ is also used in spreadsheets and computer publishing.
**** The backslash. Used for designating directories on your MS-DOS machine. This dictionary is located in
C:\WORK\DICTIONA
That means it's in the "dictiona" subdirectory of the "work" directory.
^ The character typically above the 6 on your keyboard. It was originally a circumflex. In computer language it became the symbol that was written to represent the Control (Ctrl) key. It's also called the "hat."
~ This character is a tilde. It tells you how to pronounce the n in senor. According to William Safire, it's a Spanish word from the Latin term for a tiny diacritical mark used to change the phonetic value of a letter.
0345 Numbers A British Telecom LinkLine service in England where the caller is charged at the local rate irrespective of the distance of the call. The subscriber pays installation and rental charges in addition to a charge for each call.
0800 Numbers 1. A British Telecom LinkLine service in England where the caller is not charged for the call. Similar to the North American 800 IN-WATS service.
2. Ericsson has 0800-type service on cellular systems it's put in. Ericsson describes it as a "network-oriented service — based on time of day, day of week, or special day — which allows calls to be redirected to other numbers.
0891 And 0898 Numbers A British Telecom Premium rate service in England where the caller is charged at a premium rate for the call. The calls are normally made to receive information or a service. The service provides revenue for the information provider who receives part of the call charge.
0839 And 0881 Numbers Mercury's premium rate service numbers in England.
1+ Pronounced "One plus." In North America, dialing 1 as the first digit has come to signal to your local phone company that the phone number you are dialing is long distance, i.e. is designed to reach a long distance number in the United States, Canada,

or several of the Caribbean islands, (including Bermuda, Puerto Rico, the Virgin Islands, Barbados and the Dominican Republic). The number 1 will typically be followed by an area code and then seven digits. For example, to dial me from outside New York City, you would dial 1-212-691-8215. To reach other international countries, from the United States, you dial the international access code "011." This "1+" will work with your local phone company, signaling it that you want to reach another local area code. (In New York City, calling from Manhattan to Brooklyn means dialing 1+718+the seven digit Brooklyn number.) It also will work with the long distance company you have pre subscribed to, through the process known as equal access. To reach another long distance carrier and route calls over their network, you will need to dial 1-0XXX and then the area code and number.

1.544 MBPS The speed of a North American T-1 circuit. See T-1.

1/4/80 The strange starting date embedded in the original IBM PC.

10-baggers Venture capital jargon for companies returning 1000% on their investment.

10-Net An original local area network invented by a company called Fox Research, Dayton, OH. 10-Net is a baseband, Ethernet CSMA/CD peer-to-peer LAN running on one twisted pair at one megabit per second. It is easy to install and has many advantages. It's also slow. Very slow. See ETHERNET.

10 Base T See 10Base-T.

10 Base T See 100Base

100 Test Line A Northern Telecom switching term. The 100 test line, also known as a quiet or balanced termination, is used for noise and loss measurements. The S100 provides a quiet termination for noise measurements only. In this 100 family, there is the T100, S100 and N100 tests. The N100, a more recent version of the test, also includes a milliwatt test (i.e., a 102 test line) and therefore can be used for far-to-near loss measurements. The T100 is used when the equipment at the terminating office is unknown. When the T100 test line is performed, a two-second time-out is introduced to detect the presence or absence of a milliwatt tone. If the T100 test detects the milliwatt tone, it executes the N100 version of the test; otherwise, the S100 version is initiated. If the version of the distant office test line is known, then that version of the test line can be performed directly, and thus the two-second delay per trunk of the T100 test line test is eliminated.

The 101 test line is used to establish two-way communications between the test position and any trunk incoming to the system. The connection to the 101 test line is established through the switching network.

The 102 test line, also known as the milliwatt line, applies a 1004 Hz test tone towards the originating office to facilitate simple one-way or automatic transmission loss measurements. The test tone is applied for a timed duration of nine-seconds during which answer (off-hook) signal is provided. Then an on-hook signal followed by a quiet termination is transmitted to the originating end until the connection is released by the originating end.

The T103 is used for the overall testing of supervisory and signaling features on intertoll trunks. The test is performed to the far end to check overall supervisory and signaling features of the trunk. If the test fails or if a false tone signal is detected, the test is abandoned and the condition indicated.

The T104 test is used for two-way transmission loss measurements, far-to-near noise measurements and near-to-far noise checks. Normally used in testing toll trunks.

When a 105 test line at a far-end office is called and seized, timing functions are initiated and an off-hook supervisory signal and test progress tone are returned to the originating office. If the responder is idle, the test line is connected to the responder and test progress tone is removed. Transmission tests are then initiated.

The T108 test provides far-end loop-around terminations to which a near-end echo suppression measuring set is connected for the purpose of testing echo suppressors.

100Base-Fx 100-Mbps Ethernet implementation over fiber. The MAC layer is compatible with the 802.3 MAC layer. See 10Base-F.

100Base-T A proposed standard for a 100 megabit-per-second local area network. The standard has the backing of the Fast Ethernet Alliance of 40 companies, including Intel, Synoptics, 3Com, Sun and Digital Equipment. The concept is that 100Base-T would be completely compatible with today's common 10Base-T 10 megabit-per-second, except that it would be ten times as fast. You could use 100Base-T LANs to connect 10Base-T LANs, for example, thus expanding the number of people on the LAN and increasing the overall speed. See 100VG-ANYLAN for the competing 100 megabit LAN standard.

100Base-T2 A LAN that transmits data at 100 megabits per second over copper cabling. It is a half-duplex version of 100Base-T that uses two pairs of category 3, 4 or 5 UTP cable. Officially called 802.3y.

100Base-T4 100-Mbps Ethernet implementation using four-pair category 3,4, or 5 cabling. The MAC layer is compatible with the 802.3 MAC layer.

100Base-TX 100-Mbps Ethernet implementation over Category 5 and Type 1 cabling. The MAC layer is compatible with the 802.3 MAC layer.

100VG-AnyLAN A proposed standard for a 100 megabit-per-second local area network. The 10VG-AnyLAN is a joint Hewlett Packard-AT&T proposal. It would use all four pairs using Category 5 cabling in the 10Base-T twisted pair wiring scheme to transmit or receive, rather than today's present system of using one pair to transmit and one pair to receive. The 5B6B coding used in 100VG splits the signal across the four wire pairs at 25 Mhz each. 100VG is based on quartet signaling and demand priority protocol. It retains 802.3 frame format and bridging with 10-Mbps CSMA/CD. See 100BASE-T for the competing 100 megabit LAN standard.

10Base-2 IEEE 802.3 standard for baseband Ethernet at 10 Mbps in baseband form over coaxial cable to a maximum distance of 185 meters. Also known as THIN ETHERNET or THINWIRE ETHERNET.

10Base-5 A transmission medium specified by IEEE 802.3 that carries information at 10 Mbps in baseband form using 50-ohm coaxial cable and accommodates 500-meter repeaterless runs. 10Base-5 was specified by the original Ethernet standards and is sometimes called ThickWire Ethernet.

10Base-F Standard for Fiber optic Active and Passive Star based Ethernet segments. Described in IEEE 802.1j-1993 (not in an 802.3 supplement, as you might expect). 10Base-F includes the 10Base-FL standards.

10Base-FB Part of the new IEEE 802.3 10BASE-F specification, "Synchronous Ethernet" which is a special-purpose link that links repeaters and allows the limit on segments and repeaters to be enlarged. It is not used to connect user stations. See 10Base-F. 10Base-FL A part of the IEEE Base-F specification that covers Ethernet over fiber. It is interoperable with FOIRL. See 10Base-F.

10Base-T An Ethernet local area network which works on twisted pair wiring that looks and feels remarkably like telephone cabling. In fact, 10Base-T was really invented to run on telephone cable. 10Base-T Ethernet local area networks work on home runs in which the wire from each workstation snakes directly to 10Base-T hub (like the wiring of a phone system). 10Base-T cards which fit inside PCs typically cost the same as those for Ethernet running on coaxial cable. The advantages are twofold — namely if one machine crashes, it doesn't bring down the whole network (coax Ethernet LANs are typically one long line, looping from one machine to another); and secondly, a 10Base-T Ethernet network is easier to manage because the 10Base-T hubs often come with sophisticated management software. Though 10Base-T is designed to work on "normal" telephone lines, there are multiple categories of wiring. See Category 1, 2, etc. If in doubt, pick a higher level. Sometimes old phone cabling will work. Mostly, it won't. Be safe. Put in new cabling. Put in decent new cabling. See 802.3 10Base-T.

10Broad36 An IEEE 802.3 network specification employing a broadband transmission scheme using thick coax cable and running at 10 Mbits/sec.

10XXX Calling An access code that is dialed, in addition to the telephone number, to connect with a long distance carrier. Each carrier has a unique three digit code represented here by XXX. 1-0-XXX calling allows customers to use long distance carriers other than one they subscribe to. AT&T's code is 1-0-288 (as in 1-0-ATT). MCI's is 1-0-222. Sprint's is 1-0-333. At some point in the future, that will change to 1-0-XXXX.

10XXXX Calling See 10XXX.

110-type Connecting Block The part of a 110-type cross connect, developed by AT&T, that terminates twisted-pair wiring and can be used with either jumper wires or patch cords to establish circuit connections.

110-type Cross Connect A compact cross connect, developed by AT&T, that can be arranged for use with either jumper wires or patch cords. Jumper wires, used for more permanent circuits, must be cut down to make circuit connections. Patch cords allow ease of circuit administration for frequently rearranged circuits. The 110-type cross connect also provides straightforward labeling methods to identify circuits.

119 Japan's equivalent of the United States' emergency 911 number.

1394 An IEEE data transport bus that supports up to 63 nodes per bus, and up to 1023 buses. The bus can be tree, daisy chained or any combination. It supports both asynchronous and isochronous data. 1394 is a complimentary technology with higher bandwidth (and associated cost) than Universal Serial Bus. Intel told me in summer of 1996 that it was supporting USB for most devices that attach to PC up through audio and video conferencing. Intel told that they are "supporting IEEE 1394 as the prefered interface for higher bandwidth applications such as high quality digital video editing, and connection to new digital consumer electronics equipment. We expect that 1394 will show up in low volumes in 97, and ramp into high volumes in 99." See USB.

144-line Weighting In telephone systems, a noise weighting used in a noise measuring set to measure noise on a line that would be terminated by an instrument with a No 144-receiver, or a similar instrument.

1455 Johann Gutenberg, a goldsmith from Mainz, Germany, prints his Mazarin Bible, which is believed to be the first book printed with movable type, i.e. printed on his famous Gutenberg Press. It took Gutenberg two years to compose the type for his first bible. But once he had done that he could print multiple copies. Before Gutenberg, all books were copied by hand. Monks usually did the copying. They seldom managed to make more than one book a year. The Gutenberg press was a major advance. Before Gutenberg, there were only 30,000 books on the continent of Europe. By the year 1500, there were nine million. Some people (including me) have likened the invention of the Internet to the Gutenberg Press. Johann Gutenberg lived from 1397 to 1468.

16-bit An adjective that describes systems and software that handle information in words that are 2 bytes (16 bits) wide.

16-bit Computer A computer that uses a central processing unit (CPU) with a 16-bit data bus and processes two bytes (16 bits) of information at a time. The IBM Personal Computer AT, introduced in 1984, was the first true 16-bit PC.

16-CAP An ATM term. Carrierless Amplitude/Phase Modulation with 16 constellation points: The modulation technique used in the 51.84 Mb mid-range Physical Layer Specification for Category 3 Unshielded Twisted-Pair (UTP-3).

16450/8250A Found in most current PCs, these older UART chips use a 1-byte buffer that must be serviced immediately by the CPU. If not, interrupt overruns will result. See 16550 and UART.

16550 An enhanced version of the original National Semiconductor 16xxx series UART, which sits in and controls the flow of information into and out of virtually every PC serial port in the world. The older version contains only a one-byte buffer. This can slow down the transmission of high-speed data especially when you're using a multitasking program, like Windows. The "solution" is to get a serial card or port containing the 16550. This chip contains two 16-byte FIFO buffers, one each for incoming and outgoing data. Also new is the 16550's level-sensitive interrupt-triggering mechanism, which controls the amount of incoming data the buffer can store before generating on interrupt request. Together, these features help reduce your CPU's interrupt overhead and thus speed up your communications. See 16450/8250A and UART.

1791 April 27, Samuel Finley Breese Morse born.
1793 Semaphore invented.
1833 Analytical engine by Charles Babbage.
1836 Elisha Gray born. He invented the telephone at around the same time Alexander Graham Bell did. But Bell got his phone patented before Gray.
1837 Telegraphy by Samuel F. B. Morse.
1840 Samuel Morse patents the telegraph.
1843 1. First commercial test of Morse's telegraph. The US Government paid for a telegraph line between Baltimore and

NEWTON'S TELECOM DICTIONARY

Washington, D.C. It worked. 2. First successful fax machine patented by Scottish inventor, Alexander Bain. His "Recording Telegraph" worked over a telegraph line, using electromagnetically controlled pendulums for both a driving mechanism and timing. At the sending end, a style swept across a block of metal type, providing a voltage to be applied to a similar stylus at the receiving end, reproducing an arc of the image on a block holding a paper saturated with electrolytic solution which discolored when an electric current was applied through it. The blocks at both ends were lowered a fraction of an inch after each pendulum sweep until the image was completed. Bain's device transmitted strictly black and white images.

1844 Samuel Morse send sends first public telegraph message.
1845 First rotary printing press by Richard M. Hoe.
1865 First commercial fax service started by Giovanni Casselli, using his "Pantelegraph" machine, with a circuit between Paris and Lyon, which was later extended to other cities.
1866 First experimental wireless by Mahlon Loomis.
1867 Typewriter invented by Christopher L. Sholes.
1874 April 25, 1874, Guglielmo Marconi born in Bologna, Italy.
1876 Braving a hostile ocean, the men of the Faraday, a steam-driven ship with three masts, laid the first transatlantic cable between Ireland and America. The cable was made by Siemens. It could carry 22 messages at one time. And it carried the world into a new era of communications.
1876 Telephone patent issued to Alexander Graham Bell; first long distance call transmitted by Bell.
1877 1. First telephone in a private home. First telephone in New York City. 2. Phonograph invented by Thomas Edison.
1878 The New Haven Telephone Company publishes the first telephone directory. It has 75 pages and consists of one page with 50 listings. In 1996, some 6,200 telephone directories will be published in the United States and will generate about $10 billion in advertising revenues.
1880 Alexander Graham Bell develops the photophone which uses sunlight to carry messages. It was never commercially produced.
1881 First long distance line, Boston to Providence.
1887 AT&T (American Telephone & Telegraph Co.) starts business.
1889 1. A. B. Strowger invents the telephone switch, dial telephone. 2. Punch card tabulating machine invented by Herman Hollerith.
1891 Undersea telephone cable, England to France.
1895 Guglielmo Marconi of Italy invents wireless telegraph.
1899 Magnetic voice recorder by Vlademar Poulsen.
1906 Motion picture sound by Eugene Augustin Lauste.
1911 Multiplying and dividing calculating machine by Jay R. Monroe.
1915 First transcontinental phone call in USA.
1917 VHF Transatlantic radio by Guglielmo Marconi.
1920 First commercial AM radio broadcast in the U.S. KDKA, Pittsburgh.
1921 Facsimile technology (Wirephoto) from Western Union.
1922 First dial exchange in New York City — PE-6 from PEnnsylvania 6.
1924 1. Thomas J. Watson renames Computing-Tabulating-Recording (CTR) the International Business Machines Corporation. 2. The work of Herbert Ives at Bell Labs on the photoelectric effect leads to the first demonstration of the transmission of pictures over telephone lines.
1925 1. IBM begins selling punch-card machinery in Japan. 2. Frank B. Jewett becomes the first president of Bell Labs.
1926 1. AT&T Bell Labs invents sound motion pictures. 2. First public test of trans-atlantic radiotelephone service — between New York and London.
1927 1. April 7, 1929. First public demonstration of long distance TV transmission. Moving black and white pictures were sent over telephone wires between Secretary of Commerce Herbert Hoover in Washington DC and AT&T executives in New York. They went at 18 frames per second. Further development of this technology led to the creation of TV. 2. Harold S. Black's negative feedback amplifier cuts distortion in long distance telephony. Black is at Bell Labs.
1929 1. Coaxial cable invented in Bell Telephone Laboratories; Herbert Hoover first president to have phone installed on his desk. 2. AT&T Bell Laboratories and Western Electric introduced the Sound Newsreel Camera. It used an AT&T "Light Valve" to record sound directly on the film as it passed through the camera. It was the first single system sound camera.
1933 1. Karl G. Jansky at Bell Labs discovers radio waves from the Milky Way. His discovery leads to the science of radio astronomy. 2. Bell Labs transmits first stereo sound, a symphony concert, over phone lines from Philadelphia to Washington. 3. FM radio invented by Edwin H. Armstrong.
1936 First TV broadcast by the BBC in Great Britain.
193rd Bit The frame bit for a T-1 frame.
1938 Xerography invented by Chester Carlson.
1942 Harry Newton born Sydney, Australia on June 10.
1944 Electronic Numerical Integrator and Computor (spelled with an O). Early computer, built in 1944.
1946 Dr. Robert N. Metcalfe, co-inventor of Ethernet, born Brooklyn, New York on April 7.
1947 Transistor invented at AT&T's Bell Labs in New Jersey. Inventors were John Bardeen, Walter Brattain and William Shockley. In 1956 they shared the Nobel Prize in Physics for creating the transistor.
1948 1. May 11: Birth of the International Communications Association, among the larger groups of telecommunications users in North America. 2. Claude E. Shannon announces the discovery of information theory, the cornerstone of current understanding of the communication process. Shannon is a Bell Labs.

1951 1. First direct distance calling. Phone users can dial long distance without an operator's assistance. 2. Sony unveils the first transistor radio.
1954 The solar cell invented by Gerald L. Pearson, Daryl M. Chapin and Calvin S. Fuller at Bell Labs. Also, the year William Shockley left Bell Labs to pursue the commercial opportunities offered by his invention of the transistor.
1955 Bill Gates, founder and chairman of Microsoft, born October 28, 1955.
1956 1. First transatlantic repeatered telephone cable. 2. First modem was invented by AT&T Bell Laboratories, according to AT&T. 3. Videotape recorder invented by Ampex.
1957 1. The U.S.S.R. launched Sputnik on October 4, 1957. It embarrassed the U.S. Government into a frenzy of space investments, culminating in the U.S. being the first country to have people walk on the moon. 2. Dr. Gordon E. Moore and Dr. Robert N. Noyce leave Fairchild Semiconductor and form Intel Corporation. Intel was founded with less than $5 million in start-up monies using a two-page business plan written by legendary venture capitalist, Arthur Rock.
1958 Integrated circuit invented by Jack S. Kilby at Texas Instruments.
1960 1. First test of an electronic switch. 2. MITI creates the Japan Electronic Computer Corporation to promote its domestic computer industry. 3. Laser invented by Theodore Maiman of the U.S. Laser stands for Light Amplification by the Stimulated Emission of Radiation.
1962 1. LEDs — Light Emitting Diodes — invented. 2. First commercial communications satellite (Telstar). Owned by AT&T. 3. August 31, 1962 President Kennedy signed Communications Satellite Act. 4. Semiconductor laser invented. 5. Ross Perot forms EDS with a reputed $1,000. In 1984, he sold it to General Motors for $2.6 billion. In 1999, GM decided to spin it off in a deal worth $22.2 billion.
1963 1. Touch Tone service introduced. 2. Audio cassette tape introduced by Philips. 3. C. Kumar N. Patel at Bell Labs develops the carbon dioxide laser now used around the world as a cutting tool in surgery and industry.
1964 1. Prototype of the first video phone made by the Bell System shown at The World's Fair in Queens, New York City. Pictures were black and white and the technology was very expensive. It was called the Picturephone. 2. IBM showed the first word processor.
1965 1. PDP-8 minicomputer introduced by Digital Equipment Corporation. 2. First trial offers for reversing telephone charges.
1966 October, 1966 the Electronic Industries Association issues its first fax standard: the EIA Standard RS-328, Message Facsimile Equipment for Operation on Switched Voice Facilities Using Data Communications Equipment. The Group 1 standard, as it later became known, made possible the more generalized business use of fax. Transmission was analog and it took four to six minutes to send a page.
1967 1. First 800 call made in the United States. 2. Electronic handheld calculator introduced by Texas Instruments.
1968 Fiber optics for communications invented by Robert Maurer.
1969 1. Ken Thompson and Dennis Ritchie, computer scientists at AT&T Bell Laboratories, create the Unix software operating system. 2. ARPANET introduced by the Advanced Research Projects Agency of the U.S. Defense Department. 3. Traffic Service Position System replaces traditional cord switchboards. The system automates many operator functions for the first time.
1970 1. Optical fiber for long-range communications developed. 2. Relational database invented by Dr. E. F. "Ted" Codd at IBM. 3. Floppy disk invented by IBM. 4. Gilbert Chin creates a new type of magnetic alloy now used in most telephone handset speakers.
1971 Ted Hoff at Intel invents the microprocessor — a single chip that contained most of the logic elements used to make a computer. Intel's twin innovations with the device were to put most of the transistors that make up a computer's logic circuits on to a single chip and to make that chip programmable. Here, for the first time, according to Robert X. Cringely's book "Accidental Empires," was a programmable device to which a clever engineer could add a few memory chips and a support chip or two and turn it into a real computer you could hold in your hands. Intel's first microprocessor, the 4004, was released in November, 1971. See 4004. See also MICROPROCESSOR.
1972 First commercial video game (Pong) introduced by Nolan Bushnell at Atari.
1973 1. Computerized Axial Tomography (CAT Scan) invented by Allan Cormack and Godfrey N. Hounsfield. 2. Ethernet invented by Dr. Robert N. Metcalfe on May 22, 1973 at the Xerox Palo Alto Research Center (PARC). Dave Boggs (Dr. David R. Boggs) was the co-inventor. Metcalfe and Boggs (in that order) were the authors of THE Ethernet paper, published July 76 in CACM. CACM is the Communications of the ACM. ACM is the Association for Computing Machinery. 3. Vinton Cert, computer scientist, invents the basic design of the Internet - the intermediate level gateways (now called routers), the global address space and the concept of end/end acknowledgement. 4. Gerhard Sessler and James E. West of Bell Labs receive a patent for their unidirectional microphone that improves hands-free telephone conversations.
1974 1. AT&T introduces Picturephone, a two-way color videoconferencing service at 12 locations around the country. Businesses rented meeting rooms equipped with the technology. 2. Hewlett-Packard introduces the first programmable pocket calculator, the HP-65. 3. Structured Query Language (SQL) invented by Don Chamberlain and colleagues at IBM Research.
1975 1. MITS Altair 8800 personal computer kit from MITS. The Altair was the first commercially available personal computer kit. It was on the cover of the Popular Electronics January 1975 issue. 2. Live TV satellite feed. Ali-Frazier fight from HBO (Home Box Office). 3. Sony introduces Betamax, which didn't do as well as the Video Home System (VHS) introduced later by Matsushita/JVC.
1976 1. First digital electronic central office switch installed. 2. Apple Computer founded in a Cupertino, CA garage by Steve Jobs and Steve Wozniak.
1977 1. First lightwave system installed and begins operation. It's under the streets of Chicago. 2. Interactive cable system (Qube) installed by Warner Cable. 3. Commodore PET was among the hot PCs of 1977.
1978 Bell Labs invents cellular technology. CCITT comes out with Group 2 recommendation on fax. Intel introduces the

NEWTON'S TELECOM DICTIONARY

8086 chip, with 29,000 transistors and processing 16 bits of data at one time. A variation of this chip, the 8088, introduced in 1980, caught IBM's eye and IBM used it in its first PC.

1979 1. Chapter 11 Federal bankruptcy provision introduced. Chapter 11 is reorganization. Chapter 7 is liquidation. 2. CompuServe Information Service goes on-line. 3. Gordon Matthews invents corporate voice mail. See VMX.

1980 1. CCITT comes out with Group 3 recommendation on fax. Group 3 machines are much faster than Group 2 or 1. With Group 3 machines, after an initial 15-second handshake that is not repeated, they can send an average page of text in 30 seconds or less. 2. Supreme Court of the United States rules that patents for software can be issued.

1981 1. IBM PC debuts. 2. First portable computer, by Osborne. 3. National electronic phone directory (minitel) starts in France.

1982 January 8, 1982 the consent decree to break up AT&T into seven regional holding companies and what was left (long distance and manufacturing) is announced. The divestiture takes place two years later on January 1, 1984.
March 3, 1982, the FCC formally approved the startup of cellular phone services. The FCC indicated that it would accept applications for licenses in the top 30 markets 90 days after procedures were published and for smaller markets, 180 days after publication. The FCC subsequently gave one license in each market to the local phone company (the "wireline") and one for a competitor (the "non-wireline") carrier.
October 21, 1982, the FCC awards the first construction permit for a cellular radio license to AT&T's Ameritech subsidiary (this was prior to the AT&T breakup).

1983 1. Novell introduces its first network local area network software called NetWare. It was originally introduced to allow a handful of personal computers to share a single hard disk, which at that stage was a costly and scarce resource. As hard disks became more available, the product evolved to allow the sharing of printers and file servers. 2. Nintendo introduces Famicom, a computer turned video game. 3. Cellular radio in the United States gets its first subscriber. 4. Sony introduces the Camcorder. 5. October 13, 1983, Ameritech turns on its new cellular radio system in Chicago, the first in the nation. 6. IBM introduces the PC XT, the first IBM PC to contain a hard disk. 7. Bill Gates of Microsoft announces Windows at November's Comdex.

1984 January 2, 1984. The breakup of the Bell System. AT&T gave up its local operating phone companies, which got formed into seven, roughly equal holding companies. In turn, AT&T got the Justice Department off its back for an antitrust suit and got the right to get into industries other than telecommunications. Its chosen industry was the computer industry for which it felt it had unique skills.
January 24, 1984 Apple Computer Inc.'s Steve Jobs introduces the first Macintosh computer. It was the machine that changed the world of PC computing. Mr. Jobs often described the little machine as "insanely great."
March, 1984, Motorola introduced the DynaTAC 8000X, the first portable cellular phone. It listed for $3,995 and it weighed two pounds.
Ken Oshman sells Rolm to IBM for $1.26 billion. It was not one of IBM's better investments. Rolm is now part of Siemens, which understands telecommunications.
Prodigy Information Service, a service of IBM and Sears, starts.

1985 1. CD-ROM introduced by Philips and Sony. 2. Steve Jobs driven from Apple Computer by John Sculley. 3. John Sculley, head of Apple Computer, licenses the copyrights protecting the "look and feel" of the Apple Macintosh operating system to Microsoft in order to get Microsoft to write more applications for the Mac. The license allowed Microsoft to launch its hugely successful Windows operating system — and to defend itself against a lawsuit brought by Apple alleging Windows was so similar to the Mac that it violated Apple's copyrights.

1986 Novell's SFT NetWare, first fault tolerant local area network operating system.

1987 October, 1987, the one-millionth cellular subscriber signs up for service in America. George Forrester Colony of Forrester Research is believed to have coined the term "client-server" computing. See CLIENT-SERVER.

1988 1. First transatlantic optical fiber cable. 2. Robert E. Allen takes over as CEO of AT&T.

1989 1. Fiber to the home field trial, Cerritos, CA. 2. Novell releases NetWare 3.0, the first 32-bit network operating system for Intel 80386/486-based servers. 3. 1989: Panasonic's household-size video phone with moving color images debuts in Tokyo.

1990 1. Demonstration of 2,000 kilometer link using optical amplifiers without repeaters. 2. MVIP formed and first product shipped.

1991 1. AT&T buys NCR for a gigantic $7.4 billion and soon renames it AT&T Global Information Solutions. Later AT&T took hundreds of millions of dollars in restructuring and other charges related to the fact that NCR lost pots of money after AT&T bought it. Part of the problem, according to analysts, was that AT&T bought NCR right at the time NCR was making the transition from traditional mainframe computers to so-called massively parallel computers powered by collections of small, cheaper processors run in tandem. NCR also got hit by a decline in its traditional cash register business as low-margin PCs came in. The skinny around the industry at the time AT&T bought NCR was that AT&T bought NCR to disguise the fact that its own computer operations at that time were losing so much money. And the senior management of AT&T at that time wanted to retire with the glories of booming long distance revenues and not lousy computer results. 2. Motorola introduces the lightest cellular phone yet, the MicroTAC Lite for about $1,000 retail. 3. The Electronic Industries Association approves and publishes on July 9, 1991, the Commercial Building Telecommunications Wiring Standard, the most important wiring standard ever published in the history of telecommunications. 4. Scott Hinton at Bell Labs heads a team that builds the first photonic switching fabric, bringing light-based switching technology in telecommunications networks closer to reality.

1992 1. AT&T introduces VideoPhone 2500 marketed as the first home-model color video phone which works on normal dial up analog phone lines. It meets cool reception because of poor image quality and its high price, namely $1,500. 2. Microsoft Windows 3.1 and IBM's OS/2 2.0 operating systems introduced. Windows NT (32-bit operating system) debuts in beta form. 3. Wang files for Chapter 11. 4. MCI introduces VideoPhone for normal dial-up analog phone lines. It retails for $750. It is not compatible with the AT&T phone. 5. The cellular industry signs its ten millionth subscriber on November 23, 1992. At least that's what the press releases said. Some carriers claimed that at year end, cellular subscribers in the United States had actually hit 11 million, way ahead of all predictions. 6. Apple, EO and others introduce the PDA, the Personal Digital Assistant.

1993 FCC announces its intention to auction off a chunk of spectrum larger than that used in 1993 for cellular radio. The new airwaves will be used for new types of wireless communications, including portable digital communications devices from phones to laptops, palmtops and PDAs equipped to receive and transmit data of all types, including faxes and video.
February 25, McCaw Cellular announces North America's first all-digital cellular service, in Orlando, Florida.
March 17. The Clinton administration urges Congress to eschew comparative hearings and institute a lottery for awarding new radio spectrum. See PCS.
August. AT&T agrees to buy McCaw Cellular for $12.6 billion. The idea is to help get AT&T back into local phone service.
October. Bell Atlantic sets $21.7 billion merger with TCI, cable TV giant. The assumption is that cable tv and telephone networks are "converging" into an information highway for transporting video, voice and data. The deal later fell apart after the FCC cut cable TV rates and TCI's profitability fell part.
December 31. Thomas J. Watson dies, age 79.
AT&T introduces the AT&T EO Personal Communicator 440, based on the Bell Labs-developed Hobbit microprocessor. This hand-held device combines the features of pen-based personal computers, telephones and fax machines. The device is later withdrawn from the market because of poor demand.
1994 1. GO-MVIP formed. Trade association for developers and manufacturers of MVIP computer telephony products. 2. AT&T pays $12.6 billion for McGaw Cellular Communications Inc. Robert E. Allen is CEO of AT&T.
1995 IBM buys Lotus for $3.5 billion, the main attraction being Lotus Notes. One of the key attractions of Lotus Notes is that it saves on phone bills.
August 24, Windows 95 finally ships. It contains heavy computer telephony features, including TAPI, VoiceView, fax on demand and binary file transfers using
August. Disney agrees to buy Capital Cities/ABC for $19 billion. The idea of the merger was, according to the New York Times, that "biggest is best." the idea was to combine the most profitable TV network with a name-brand family entertainment empire.
August. CBS accepts Westinghouse's $5.4 billion takeover offer. According to the New York Times, the idea of the takeover is that "even an ailing Big 3 TV network is worth owning, if it comes with the collection of radio and TV stations reaching one-third of U.S. households."
September. Time Warner agrees to buy Turner Broadcasting for $7.5 billion.
September 20, 1995, AT&T announces it will split itself into three companies — long distance, equipment manufacturing and computers. Wall Street applauds the decision and in one day lifts the price of AT&T's stock by 10%, or about $6 1/2 billion. Meantime, AT&T announces that it will substantially reduce the size of its failed computer activities, which were called AT&T Global Information Solutions. See 1991.
December 8. Digital Versatile Disk (DVD) is announced. DVD is a specification announced by nine companies for a new type of digital videodisk, similar to CD-ROMs but able to store far more music, video or data in a common format. DVDs will be 5 inches in diameter and will be able to store 4.7 gigabytes on each side, equivalent to 133 minutes of motion picture and sound, or enough to hold most feature-length movies. The companies announcing DVD were Philips, Toshiba, Matsushita Electric Industrial, Sony, Time Warner, Pioneer Electronic, the JVC unit of Matsushita, Hitachi and Mitsubishi Electric.
1996 February. US West signs $5.3 billion deal for Continental Cablevision. The assumption, according to the New York Times, is that it will now have a wire into many homes, by dominating local phone service in 14 states and reaching 16.3 million cable subscribers.
April. SBC Communications (the name for the holding company owning Southwestern Bell Telephone) buys Pacific Telesis (the holding company for Pacific Bell) for $16.7 billion. The assumption of this merger, according to the New York Times, is that regional telephone companies can become national players by combining 30 million phone lines in 5 states with potential coast-to-coast cellular market of 80 million people.
April. Bell Atlantic buys NYNEX (the holding company for New York Telephone and New England Telephone) for $22.1 billion. The new company will be called Bell Atlantic.

1A AT&T's first generation of standardized KEY TELEPHONE SYSTEM equipment based on a variety of interconnected phone-line-powered relays. Prior to 1A1, key systems were often patched together from a variety of non-standard parts, with varying wiring schemes, making repairs and upgrades very difficult.

1A1 AT&T's second generation of standardized KEY TELEPHONE SYSTEM equipment. Unlike the phone-line powered 1A1, it used commercial AC power for added features such as illuminated buttons to indicate line status.

1A2 AT&T's third generation of standardized KEY TELEPHONE SYSTEMS. It was distinctive for its use of plug-in circuit cards, making it much easier to add features or diagnose and cure problems.

1A3 A cute term for an historic TIE electronic key system that provided advanced features, but was priced competitively with 1A2 electromechanical key systems.

1Base-5 A transmission medium specified by IEEE 802.3 that carries 1.0 Mbps in baseband form using twisted-pair conductors and accommodates 500 meter repeaterless runs.

1FB One Flat rate Business phone line. A phone line you pay a single monthly charge for and you may make as many local phone calls as you wish for free during that month. See also 1MB.

1MB One Message rate Business phone line. A phone line you pay a single monthly charge for. That charge typically allows you to make a small number of local calls for free. But that each additional local call will cost you. That cost may be by the minute, by the distance, or just by the call. See 1FB.

1PSS Packet Switching System. The AT&T Western Electric 1PSS is a high-capacity, X.25 packet switch.

2-way Trunk A trunk that can be seized at either end.

2-wire Facility A 2-wire facility is characterized by supporting transmission in two directions simultaneously, where the only method of separating the two signals is by the propagation directions. Impedance mismatches cause signal energy passing in each direction to mix with the signal passing in the opposite direction. See 4-WIRE FACILITY.

23B+D An easy way of saying the ISDN Primary Rate Interface circuit. 23B+D has 23 64 Kbps (kilobits per second) paths for carrying voice, data, video or other information and one 64 Kbps channel for carrying out-of-band signaling information. ISDN PRI bears a remarkable similarity to today's T-1 line, except that T-1 can carry 24 voice channels. In ISDN 23B+D, the one D channel is out-of-band signaling. In T-1, signaling is handled in-band using robbed bit signaling. Increasingly, 23B+D is the preferred way of getting T-1 service since the out of band signaling is richer (delivers more information — like ANI and DNIS) and is more reliable than the in-band signaling on the older T-1. One good thing about PRI: You can now organize with your phone company to deliver the signaling for a bunch of ISDN PRI cards on one D channel. Thus your first line would have 23 voice channels, for example. Your second would have 24 voice channels, etc. Several of the more modern voice cards will accept the signaling for up to eight ISDN PRI channels on the D channel of the first one. See ISDN PRI, ROBBED BIT SIGNALING and T-1.

24-bit Mode The standard addressing mode of Apple Macintosh's System 6 operating system, where only 24 bits are used to designate addresses. Limits address space to 16MB (2 to the 24th power), of which only 8MB is normally available for application memory. This mode is also used under System 7 (the Mac's more modern operating system) if 32-bit addressing is turned off. See 32-BIT.

24-bit Video Adapter A color video adapter that can display more than 16 million colors simultaneously. With a 24-bit video card and monitor, a PC can display photographic-quality images.

2500 Set The "normal" single-line touchtone desk telephone. It has replaced the rotary dial 500 set in most — but definitely not all — areas of the United States and Canada. No one seems to know why the addition of a "2" in front of a model number came to denote touchtone in the old Bell System.

2600 Tone Until the late 1960s, America's telephone network was run 100% by AT&T and used 100% in-band signaling. For in-band signaling to work there needs to be a way to figure when a channel is NOT being used. You can't have nothing on the line, because that "nothing" might be a pause in the conversation. So, in the old days, AT&T put a tone on its vacant long distance lines, those between its switching offices. That tone was 2600 Hertz. If its switching offices heard a 2600 Hz, it knew that that line was not being used. At one point in the 1960s, a breakfast cereal included a small promotion in its cereal boxes. It was a toy whistle. When you blew the whistle, it let out a precise 2,600 Hz tone. If you blew that whistle into the mouthpiece of a telephone after dialing any long distance number, it terminated the call as far as the AT&T long distance phone system knew, while still allowing the connection to remain open. If you dialed an 800 number, blew the whistle and then touchtoned in a series of tones (called MF — multi-frequency — tones) you could make long distance and international calls for free. The man who discovered the whistle was called John Draper and he picked up the handle of Cap'n Crunch in the nether world of the late 1960s phone phreaks. Since then, in-band signaling has been replaced by out-of-band signaling, the newest incarnation being called Signaling System 7. See 2600 CAPTAIN CRUNCH, MULTI-FREQUENCY SIGNALING and SIGNALING SYSTEM 7.

2780 A batch standard used to communicate with IBM mainframes or compatible systems.

2B+D A shortened way of saying ISDN's Basic Rate Interface interface, namely two bearer channels and one data channel. A single ISDN circuit divided into two 64 Kbps digital channels for voice or data and one 16Kbps channel for low speed data (up to 9,600 baud) and signaling. Either or both of the 64 Kbps channels may be used for voice or data. In ISDN 2B+D is known as the Basic Rate Interface. In ISDN, 2B+D is carried on one or two pairs of wires (depending on the interface) — the same wire pairs that today bring a single voice circuit into your home or office. See ISDN.

2B1Q Two Binary, One Quaternary. An ISDN line encoding technique which uses two bits to represent four variations in amplitude and polarity.

2W Two-Wire. See 2-WIRE FACILITY.

3:2 Pull-down A method for overcoming the incompatibility of film and video frame rates when converting or transferring film (shot at 24 frames per second) to video (shot at 30 frames per second).

3172 IBM's network controller. It connects to the mainframe channel on one end and the LAN media (Ethernet, Token Ring, FDDI) on the other.

3174 IBM's cluster controller. It connects to terminals and other I/O devices on one end, and a mainframe channel on the other.

32-bit An adjective that describes hardware or software that manages data, program code, and program address information in 32-bit-wide words. What is the significance of 32-bit? With 32-bit memory, each program can address up to 4 gigabytes (2 to the 32nd power) of memory. This is in contrast to Windows 3.x where programs are limited to 16 MB of memory. Possibly more significant than the amount of memory that is available to a 32-bit application is how that memory is accessed. Under Windows 3.x, memory is accessed by using two 16-bit values that are combined to form a 24-bit memory address. (24-bits is the size of the memory addressing path of the Intel 80286. The 80286 is the architecture that Windows 3.x was designed for.) The first 16-bit value (selector) is used to determine a base address. The second 16-bit value (offset) indicates the offset from the base address. One of the side effects of this architecture is that the maximum size of a single chunk of memory is 64 KB. Windows 95 and Windows NT are 32-bit operating systems. Windows 95 and Windows NT developers can address memory with a single 32-bit value. Such an addressing scheme allows developers to view memory as one flat, linear space with no artificial limits on the size of a single segment. No longer are programmers concerned about selectors and offsets and the 64 KB segment limit. Also, Windows 95 and Windows NT take full advantage of the protection features of the Intel 80386 microprocessor. 32-bit applications are given their own protected address space which tends to prevent applications from inadvertently overwriting each other.

32-bit Addressing See 32-BIT.

32-bit Computer A computer that uses a central processing unit (CPU) with a 32-bit data bus and central processing unit (CPU) which processes four bytes (32 bits) of information at a time. Personal computers advertised as 32-bit machines — such as Macintosh SE, and PCs based on the 80386X microprocessor — aren't true 32-bit computers. These computers use microprocessors (such as the Motorola 68000 and Intel 80386SX) that can process four bytes at a time internally, but the external data bus is only 16 bits wide. 32-bit microprocessors, such as the Intel 80386DX, the Pentium and the Motorola 68030, use a true 32-bit external data bus and can use 32-bit peripherals.

32-megabyte Barrier Versions of MS-DOS prior to 4.0 had a built-in limit on the size of a disk partition. Using the original design parameters of DOS, we can show how we got to the maximum size disk partition. One word, 16 bits, is defined for DOS to access sectors within its hard disk partition. A single 16-bit binary word can represent values from zero through 65,535. This limits the partition's total sector count to 65,536. Hard disk sectors are 512 bytes long. 512 bytes times 65,536 sectors = 33,554,432. Since there are 1,048,576 bytes in each megabyte, the maximum size partition calculates to 32 megabytes. MS-DOS 4.0 and 5.X can now create partitions of and read in one go a disk of up to 512 megabytes.

3270 IBM class of terminals (or printers) used in SNA networks.

3270 Gateway An electronic link which uses 3270 terminals to handle data communications between PCs and IBM mainframes.

3270SNA A specific variation of IBM's System Network Architecture for controlling communications between a 3270 terminal connected to an IBM mainframe.

3274 IBM series of Control Units or Cluster Controllers provide a control interface between host computers and clusters of 3270 compatible terminals.

327X Belonging to IBM's 3270 collection of data communications terminals.

3299 A communications device for an IBM mainframe computer.

347x The type of fixed function computer terminals used with IBM mainframe computers.

3745 IBM's communications controllers, often called front-end processors. 3745 devices channel-attach to the mainframe and support connections to LANs and other FEPS.

3780 a batch protocol used to communicate with an IBM mainframe or compatible system.

386 See 80386.

386 Enhanced Mode An operating mode of Microsoft Windows that takes full advantage of advanced technical capabilities such as multitasking, virtual memory, and protected mode. To use 386 Enhanced mode, your computer must have an 80386 or higher microprocessor and at least 2M of random-access memory (RAM).

386SLC The IBM 386SLC is IBM's improved version of Intel's 80386SX microprocessor chip. IBM designed it with Intel's help, so it's compatible with the Intel family of 8088, 80286, 80386, 80486 chips.

3DGF 3-D Geometry File. A platform independent format for exchanging 3-D geometry data among applications. Developed by Macromind.

4-wire Facility A 4-wire facility supports transmission in two directions, but isolates the signals by frequency division, time division, space division, or other techniques that enable reflections to occur without causing the signals to mix together. A facility is also called 4-wire if its interfaces to other equipment meet this 4-wire criteria (even if 2-wire facilities are used internally), as long as crosstalk between the two transmission directions, as measured at the interface, is negligible. See 2-WIRE FACILITY.

4.9% No regional Bell Operating Company is presently allowed to own more than 4.9% of the stock of a telecommunications manufacturing company. See DIVESTITURE.

4004 The world's first general-purpose microprocessor (computer on a chip). The 4004 was made by Intel, was 4-bit, was released on November 15, 1971 and contained 2,300 transistors. It executed 60,000 instructions per second. The tiny 4004 had as much computing power as the first electronic computer, ENIAC, which filled 3,000 cubic feet with 18,000 vacuum tubes when it was built in 1946. The 4004 found a home in desktop calculators, traffic lights and electronic scales. Despite its power, its 4-bit structure was too small to process all the bits of data at one time to handle all the letters of the alphabet. It was followed by the 8-bit 8008. See also 1971 and 1978.

41449 AT&T's specifications for its ISDN PRI (Primary Rate Interface). It is different to the ANSI standard T1.607.

419A A famous old Bell System tool which many installers found very convenient to hold a diminishing marijuana cigarette (called a 'roach') in the 1960s.

42A An early terminal block. The Model 42A is a plastic mounting base about two inches square with four screws and a cover. Before modular connections became widespread, the 42A was used to connect a phone's line cord to the wire inside a wall or running around the baseboard. Adapters, such as the No. 725A made by AT&T and Suttle Apparatus, can be used to convert a 42A into a 4-conductor modular jack. See also TERMINAL BLOCK.

46-49 In North America, most cordless phones operate within the band 46-49 MHz. That band contains only 10 channels and is horribly overcrowded. Recently, the FCC authorized a new frequency range — 905-928 MHz — for use by, amongst other things, cordless phones. The 900 Mhz contains 50 channels.

486 A shortened name for Intel's family of 80486 chips, the successor to and continuation of the line of chips that started with the 8088 and grew into the 8086, the 80286 and the 80386. The 80486 is a major step forward in speed, complexity and capability. In simple terms, the 486 is a combination microprocessor, floating point math coprocessor, memory cache controller and 8K of RAM cache, all in one chip. The 486 chip contains 1.2 million transistors and is capable of 41 MIPS operating at 50 MHz. See 80486.

486DX The full-powered 80486 from Intel. See 80486.

486SL A power-saving version of Intel's '486 chip, which it introduced in the fall of 1992 and targeted at manufacturers of laptops and notebooks. The chip runs at 3.3 volts, compared with 5.0 volts for its previous power-saving chip, the '386SL.

486SLC A name given to clones of 80486 chips. These clones are made by companies other than Intel, the manufacturer of the original chip. These clones are often not identical in power with the Intel chip. The Cyrix 486SLC, for example, is really a hybrid of the 386 and 486 chips. The Cyrix chip is intended to offer notebook manufacturers an easy way to upgrade machines originally designed to run on Intel 386SX processors. Cyrix built its 486SLC chip to fit the Intel 386SX sockets already in place in many popular notebook designs, thus requiring only minimal changes to the BIOS and motherboard. As with the Intel 386SX, the Cyrix processor has a 32-bit internal data path but communicates with the rest of the computer over a 16-bit data path. However, it uses the 486SX instruction set and boasts an on-board cache, although only 1K. The 486SLC's hardware multiplier, although not as fast as a true math coprocessor, can significantly speed some graphics calculations. Like Intel's 386SL chip, the Cyrix 486SLC operates at a low 3.3 volt. This is designed to extend battery life in portables. Despite its 486 name, though, it's probably more appropriate to compare the performance of 486SLC notebooks to machines powered by 386SX and 386SL chips, not those with Intel 486-class processors. Given their different design goals, it's not fair to expect the 486SLC to match the speed of the 486SX.

486SX The Intel 486SX processor is essentially a full-powered 486DX with its math coprocessor disconnected. Unlike the 386 line, in which the SX version used a 16-bit external path, Intel created the 486SX by simply disconnecting the 486DX's math coprocessor. The 486SX shares the DX's full 32-bit data path and 8K cache.

4A The last generation of "telco-quality" add-on speakerphones, with separately-housed microphone and speaker; made by both Western Electric (AT&T) and Precision Components, Inc.

4GL Fourth Generation Language.

4WL-WDM Four Wavelength Wave Division Multiplexing, also called Quad-WDM. MCI announced this technology in the Spring of 1996 as a method of allowing a single fiber to accommodate four light signals instead of one, by routing them at different wavelengths through the use of narrow-band wave division multiplexing equipment. The technology allowed MCI to transmit four times the amount of traffic along existing fiber. At that time MCI's backbone network operated at 2.5 gigabits per second (2.5 billion bits) over a single strand of fiber optic glass. Using Quad-WDM the same fiber's capacity will rise to 10 gigabits — enough capacity to carry 64,500 simultaneous transmissions over one single strand of fiber.

500 Set The old rotary dial telephone deskset. The touchtone version was called a 2500 set.

5250 IBM class of terminals for midrange (System 3x and AS/400) environments.

5250 Gateway An electronic link which uses 5250 terminals to handle communications between PCs and IBM mini-computers.

56 Kbps A 64 Kbps digital circuit with 8 Kbps used for signaling. Sometimes called Switched 56, DDS or ADN. Each of the carriers have their own name for this service. The phone companies are obsoleting this service in favor of the more modern ISDN BRI, which has two 64 Kbps circuits and one 16 Kbps packet service.

5ESS A digital central office switching system made by AT&T. It is typically used as an "end-office," serving local subscribers.

5XB 5 X-Bar central office equipment

64 Bit Architecture The wide data path over which instructions (words composed of bits) are moved to and from Intel's i860's internal registers and memory. Most conventional mainframes use a word length of 32 bits. Intel's i860 RISC-based microprocessor line, introduced in 1989, incorporates many firsts, including more than 1 million transistors on a single chip. The line currently comprises the i860 XR, with 1.2 million transistors, and the i860 XP, with 2.5 million transistors. See also 8-BIT, 16-BIT and 32-BIT.

64-cap An ATM term. Carrierless Amplitude/Phase Modulation with 64 constellation points.

64 Kbps 64,000 bits per second. The standard speed for V.35 interface, DDS service, and also the effective top speed of a robbed-bit 64 Kbps channel. A 64 kbps circuit (DS0). "Clear Channel" is 64 kbps where entire bandwidth is used. See also ISDN.

66 Block The most common type of connecting block used to terminate and cross-connect twisted-pair cables. It was invented by Western Electric eons ago and has stood the test of time. It's still being installed. Its main claims to fame: Simplicity, speed, economy of space. You don't need to strip your cable of its plastic insulation covering. You simply lay each single conductor down inside the 66 block's two metal teeth and punch the conductor down with a special tool, called a punch-down tool. As you punch it down, the cable descends between the two metal teeth, which remove its plastic insulation (it's called insulation displacement) and the cable is cut. The installation is then neat and secure.

66-type Connecting Block A type of connecting block used to terminate twisted-pair cables. All wires are manually cut down with a special tool to terminate or connect them. See 66 BLOCK.

66-type Cross Connect A cross connect made up of the 66-type connecting blocks and jumper wires for administering circuits. All wires, including jumper wires, must be cut down (or punched down) and seated with a special tool. See 66 BLOCK.

6611 IBM's multi protocol router, which supports APPN in addition to TCP/IP, DECnet, AppleTalk, IPX, NetBIOS, and other protocols.

6800 A Motorola microprocessor used in early Macintoshes, Mac Plus and SE, plus recent low cost units such as the Classic, Portable and PowerBook 100. The 6800 does not support virtual memory or 32-bit addressing.

68020 A Motorola microprocessor used in the Macintosh II and LC. On the Mac II there is a socket to hold a coprocessor called a PMMU which enables the Mac II to use virtual memory. There is no socket on the LC motherboard, so the processor must be upgraded in order to use virtual memory.

68030 A Motorola microprocessor used in the SE/30, PowerBook 140/145 and 170, and Modular Macs, except the LC, Mac II, and Quadras. It supports virtual memory without any additional hardware.

68040 A Motorola microprocessor used in Quadras. It supports virtual memory without any additional hardware. Has built-in floating point math capabilities (an optional coprocessor on the 68030).

7-bit ASCII The standard code for text in which a byte (eight bits) holds the seven ASCII digits that define the character plus one bit for parity.

700 Service An "area code" you dial, as in 1-700-XXX-XXXX. This "area code" has been reserved for the long distance companies to do so as they will. So dial 1-700- and then certain "office codes" (the next three digits), you'll get AT&T and a service it calls Easyreach. It will give a service allowing your calls to follow you. Dial other 700 office codes, get MCI and you'll be able to make intralata "long distance" phone calls, such as those from Manhattan to Westchester County — for presumably cheaper than with the local phone company. 700 service is still evolving. Each carrier has the right to create whatever service it wants with its 700 numbers.

8.3 1. Under the MS-DOS naming structure, a file's name can be eight letters in front of the period and three after it, e.g. LAZARUS8.TXT. 2. 8.3 minutes. The time it takes for light to travel from the Sun to the Earth.

8-bit Computer A computer that uses a central processing unit (CPU) with an 8-bit data bus and that processes one byte (8 bits) of information at a time. The first microprocessors used in personal computers, such as the MOS Technology 6502, Intel 8080, and Zilog Z-80, were installed in 8-bit computers such as the Apple II, the MSAI 8080, and the Commodore 64.

800 Portability 800 Portability refers to the fact that you can take your 800 number to any long distance carrier. A case example, we had 1-800-LIBRARY. For many years, that number was serviced by AT&T. When portability came along, we were able to change it from AT&T to MCI and still keep 1-800-LIBRARY. 800 Portability is provided by a series of complex databases the local phone companies, under FCC mandate, have built. 800 Portability started on May 1, 1993.

800 Service Eight-hundred service. A generic and common (and not trademarked) term for AT&T's, MCI's, US Sprint's and the Bell operating companies' IN-WATS service. All these IN-WATS services have 800 and 888 as their area code. Dialing an 800-number is free to the person making the call. The call is billed to the person or company being called. The telephone company suppliers of 800 services use various ways to configure and bill their 800-services. One way: you can buy an 800 line which will ring on your normal phone line. You'll only pay per call, but you won't receive any incoming call if you're making an outgoing one. You can even terminate an 800 number on your cellular phone or your home number. You might pay a flat monthly rate plus "so-much" (i.e. timed usage) per call. That timed usage may include some calculation for the distance the incoming call traveled. 800-Service is now available for calls from Canada and many countries overseas Europe, though many of those companies pay a normal toll call to reach an American "toll-free" 800-number.

800 Service works like this: You're somewhere in North America. You dial 1-800 or 1-888 and seven digits. Your local central office sees the "1" and recognizes the call as long distance. It ships that call to a bigger central office (or perhaps processes the call itself). At that central office it's processed, a machine will recognize the 800 or 888 "area code" and examine the next six digits. Those six digits will tell which long distance carrier to ship the call to locally. That carrier then holds the call, while it sends a data communications query to a computer containing a big "translation" database. That computer will look up the 800 number and send back information on how the carrier should route the call. There are many ways the carrier can route the call. The simplest is to say "Dial the following number, i.e. 212-691-8215 and connect the call. Or it might say Connect it to your dedicated T-1 number 12345 in New York City.

As a real-life example, Flatiron Publishing, publishers of this book, has an 800 number, namely 800-LIBRARY (or 800-542-7279). When you call that number, MCI routes that number to the first available circuit on the dedicated T-1 line which we have rented from MCI's POP (Point Of Presence) to our New York City office.

Because 800 long distance service is essentially a database lookup and translation telephone service, there are endless "800 services" you can create. You can change the routing instructions based on time of day, day of week, number calling etc. See EIGHT HUNDRED SERVICE for more, especially all the features you can now get on 800 service.

In May of 1993 the FCC has mandated that all 800 numbers will become "portable." That means that customers can take their 800 telephone number from one long distance company to another, and still keep the same number.

800 Valuflex Service A Nynex service which lets you make and receive regular phone calls and 800-number calls from local areas on your normal business phone lines. There are big advantages here. You don't have to rent additional phone lines. You don't have to expand your existing phone system — or buy a new one (if getting extra lines means you'd grow out). And you can combine your incoming 800 lines with features you can get on business lines — like call forwarding, conferencing, etc. Have your 800 calls come into your office during the day. Have them call forwarded at night to your home. 800 Valuflex is aimed at smaller business.

8008 Intel's 8-bit 8008 microprocessor was introduced in April 1972. It contained 2300 transistors. Its typical used was in dumb terminals, general calculators and bottling machines.

80186 The 80286 was Intel's second generation 8086 chip. But between the 8086 and the 80286, there was a chip called the 80186, which was used in the famous Tandy 2000, a powerful — but not fully IBM-compatible — PC, that was used as the in-store PC in all Radio Shacks for several years, and achieved a bit of a cult following. Michael Marcus has two of them in my closet, if someone would like to buy them.

802 IEEE committee on Local Area Networks (LANs) standards. See 802 STANDARDS.

802 Standards The 802 Standards are a set of standards for LAN communications developed through the IEEE's Project 802. The 802 Standards segment the data link layer into two sublayers:

1. A Medium Access Control (MAC) layer that includes specific methods for gaining access to the LAN (Local Area Network). These methods — such as Ethernet's random access method and Token Ring's token procedure — are in the 802.3, 802.5 and 802.6 standards.

2. A Logical Link Control (LLC) Layer, described in 802.2 standard, that provides for connection establishment, data transfer, and connection termination services. LLC specifies three types of communications links:

• An Unacknowledged connectionless Link, where the sending and receiving devices do not set up a connection before transmitting. Instead, messages are sent on a "best try" basis and there is no provision for error detection, error recovery, and message sequencing. This type of link is best suited for applications where the higher layer protocols can provide the error correction and functions, or where the loss of broadcast messages is not critical.

• A Connection-mode Link, where a connection between message source and destination is established prior to transmission. This type of link works best in applications, such as file transfer, where large amounts of data are being transmitted at one time.

• An Acknowledged-connectionless Link that, as its name indicates, provides for acknowledgement of messages without bur-

NEWTON'S TELECOM DICTIONARY

dening the receiving devices with maintaining a connection. For this reason, it is most often used for applications where a central processor communicates with a large number of devices with limited processing capabilities.

802.1 IEEE standard for overall architecture of LANs and internetworking.
802.11 IEEE standard for wireless LANs.
802.1B IEEE standard for network management.
802.1D IEEE standard for inter-LAN bridges (specifically between 802.3, 802.4, and 802.5 networks). Works at the MAC level.
802.2 A format used for frames of data sent on Ethernet, token ring and several other types of local area networks. Now the format favored by Novell for NetWare 4.x LANs over the 802.3.
802.3 IEEE standard for carrier sense multiple access with collision detection (CSMA/CD). A physical layer standard specifying a LAN with a CSMA/CD access method on a bus topology. Ethernet and Starlan both follow the 802.3 standard. Typically they transmit at 10 megabits per second (Mbps). The theoretical limit of Ethernet, measured in 64 byte packets, is 14,800 packets per second (PPS). By comparison, Token Ring is 30,000 and FDDI is 170,000.
802.3 1Base5 IEEE standard for baseband Ethernet at 1 Mbps over twisted pair wire to a maximum distance of 500 meters. Also called Starlan.
802.3 10Base2 IEEE standard for baseband Ethernet at 10 Mbps over coaxial cable to a maximum distance of 185 meters. Also called "Cheapernet" or "Thin Ethernet."
802.3 10Base5 IEEE standard for baseband Ethernet at 10 Mbps over fax coaxial cable to a maximum distance of 500 meters.
802.3 10Broad36 IEEE standard for broadband Ethernet at 10 Mbps over broadband cable to a maximum distance of 3600 meters.
802.3 10Base-T 10Base-T is an IEEE standard for operating Ethernet local area networks (LANs) on twisted-pair cabling using the home run method of wiring (exactly the same as a phone system uses) and a wiring hub that contains electronics performing similar functions to a central telephone switch. The full name for the standard is IEEE 802.3 10Base-T. The 10Base-T standard, issued in the fall of 1990, defined the requirements for sending information at 10 million bits per second on ordinary unshielded twisted-pair cabling. The 10Base-T standard defines various aspects of running Ethernet on twisted-pair cabling such as:
• Connector types (typically eight-pin RJ-45),
• Pin connections (1 and 2 for transmit, 3 and 6 for receive),
• Voltage levels (2.2 volts to 2.8 volts peak), and
• Noise immunity requirements to filter outside interference from telephone lines or other electronic equipment.
Ethernet is the most popular LAN in the world. Ethernet running on loop coaxial cable — typically called thin Ethernet or thinnet — is the most popular way of running Ethernet local area networks. Loop networks suffer from the major problem that one cut in the cable can destroy the complete network. 10Base-T is a much more reliable — though more expensive — way of connecting LANs, since it requires electronics at the center of the home run. As I write this, the most common form of 10Base-T electronics is a small box joining about 12 workstations together. To get more on the LAN, you simply daisy chain the boxes together. The boxes are unbelievably reliably. They're easy to install and they often come with LAN management software, which gives you statistics on who's using the network, for how long, what the performance is, and what potential problems might crop up, etc. The cable 10Base-T networks use to connect between their central electronics and their attached workstations is typically standard twisted pair phone wiring, which is a lot easier to install than coaxial cable. 10Base-T networks are now becoming most popular and are being installed at faster rate than old-style loop coaxial wired LANs.
802.4 IEEE physical layer standard specifying a LAN with a token-passing access method on a bus topology. It is typically used with Manufacturing Automation Protocol (MAP) LANs. MAP was developed by General Motors. Typical transmission speed is 10 megabits per second.
802.5 IEEE physical layer standard specifying a LAN with a token-passing access method on a ring topology using unshielded twisted pair. Used by IBM's Token Ring hardware. Typical transmission speed is 4 or 16 megabits per second.
802.6 IEEE standard for MANs (Metropolitan Area Networks). Formerly known as QPSX (Queued Packet and Synchronous Exchange), now known as DQDB (Distributed Queue DOUBLE Bus.
802.7 IEEE technical advisory group on broadband LANs.
802.8 IEEE technical advisory group for fiber-optic LANs.
802.9 IEEE technical advisory on ISLAN, which stands for Integrated Services LAN. ISLAN is Isoethernet with switched or packetized voice on an Ethernet LAN, which is 10 megabits per seconds of Ethernet (used for data) plus six megabits per second of ISDN B channels, which gives you 96 B ISDN channels plus a D channel. You can used the B channels for voice.
802.X The Institute of Electrical and Electronics Engineers (IEEE) committee that developed a set of standards describing the cabling, electrical topology, physical topology, and access scheme of network products; in other words, the 802.X standards define the physical and data-link layers of LAN architectures. Or, in simple language, the set of IEEE standards for the definition of LAN protocols. IEEE 802.3 is the work of an 802 subcommittee that describes the cabling and signaling for a system nearly identical to classic Ethernet. IEEE 802.5 comes from another subcommittee and similarly describes IBM's Token-Ring architecture.
80286 In February, 1982 Intel unveiled its second generation (thus the 2) 8086 microprocessor chip — the 16-bit 80286. The 80286 featured 134,000 transistors. The 80286 (computer on a chip) drives the IBM AT. It is a true 16-bit processor. This chip will run at 6 MHz to 16 MHz. At 10 MHz it runs 1.5 mips. The 80286 has a 24-bit address bus and can address 16 MB of memory directly (which is two (2) raised to the 24th power). In late 1991 the chip cost $8.
80386DX Introduced in October, 1985, the 80386 is the third generation of the 8086 family (hence the 3). The Intel microprocessor (computer on a chip) which drives the IBM Model 80 and many others. It contains 275,000 transistors and is a 32-bit processor. It is typically clocked at 16 MHz, but there are now versions of it that will run at 33 MHz. Its 32-bit processor

NEWTON'S TELECOM DICTIONARY

means is it can process data in 32-bit chunks, rather than the 16-bit chunks of the 80286 family. The 80386 DX comes in 20 MHz (4 MIPS), 25 MHz (6 MIPS) and 33 MHz (8 MIPS). The principal architectural advantage over its predecessors is the 386 CPU's superior memory management ability, which improves performance by enabling the CPU to work with memory segments larger than 64K and to have access to more than one megabyte of memory (both limitations of the 286). The 386 can address up to four gigabytes (billion bytes) of physical memory and 64 terabytes of virtual memory at a time. It can address the four million byte of memory because it has a 32-bit address bus which can address 2 to the 32 bytes or over 4 billion bytes! Virtual memory is an area of hard disk that is treated by the processor as part of main memory. The Intel 386 DX has a 32-bit external data bus. This 80386's 32-bit bus requires a larger bus — e.g. IBM's Micro Channel or EISA, the Extended Industry Standard Architecture. EISA is the independent computer industry's alternative to IBM's Micro Channel data bus architecture which IBM uses in its high end PS/2 line of desktop computers. EISA, like Micro Channel, is a 32-bit channel. But, unlike IBM's Micro Channel, plug-in boards which work inside the XT and AT-series of IBM and IBM clone desktop computers will work inside EISA machines. They won't work in the Micro Channel machines. There are now several versions of the 80386. The most important one is the 80386DX which is the "genuine" 386 and features 25 times the performance of the original IBM PC. Another important 386 is the 80386SX, also called the 80386sx. Intel calls its 386SX microprocessor "the entry level member of the Intel 386 family." It is a genuine 80386 chip, except that its external data bus is only 16 bits (i.e. the size of the normal AT bus), not 32 bits. It is capable of sustained execution of 2.5 to 3.5 million instructions per second. A third 386 chip is the 80386SL introduced in October 1990 designed specially for laptop and notebook portable personal computers. The 386SL contains a 386 CPU that has been re-engineered for low-power operation. Surrounding the CPU are a memory controller and a complete set of input/output (I/O) controllers like those used in most PCs. By combining these two components with a few external devices, OEMs (Original Equipment Manufacturers) can build a complete, portable 386 PC with the same functionality as a 386 desktop with just five chips — nine, including memory. As a result, systems based on the 80386SL will consume one-third less board space and all components should fit easily on a 3" x 5" motherboard. PCs based on the 386SL SuperSet (Intel's name for its chip set) will also feature 50% longer battery life. The 386 SL will also let notebook users use "auto resume" which lets you shut your laptop off in the middle of a program, turn it on later and find yourself at exactly the place you were when you shut off, i.e. no rebooting is necessary. The 386SL contains 85,000 transistors — more than three times as many as the 386SX.
For history's sake: In 1984 the typical motherboard of a PC required about 170 components plus memory chips. By 1987 the number of components had dropped to 70 excluding memory. With Intel's 386SL, a design engineer can now implement a complete PC AT compatible in 10 components plus memory. By 1995 that number had dropped to under 5.

80386SL Intel introduced the 80386SL chip in October 1990. Intel designed the chip specially for laptop and notebook portable personal computers. The 386SL contains a 386 microprocessor-compatible CPU that has been re-engineered for low-power operation. Surrounding the CPU are a memory controller and a complete set of input/output (I/O) controllers like those used in most PCs. By combining these two components with a few external devices, OEMs (Original Equipment Manufacturers) can build a complete, portable 386 PC with the same functionality as a 386 desktop with just five chips — nine, including memory. As a result, systems based on the 80386SL will consume one-third less board space and all components should fit easily on a 3" x 5" motherboard. PCs based on the 386SL SuperSet (Intel's name for its chip set) will also feature 50% longer battery life. The 386SL will also let notebook users use "auto resume" which lets you shut your laptop off in the middle of a program, turn it on later and find yourself at exactly the place you were when you shut off, i.e. no rebooting is necessary. The 386SL contains 855,000 transistors. In September 1991, Intel introduced a 25 MHz version of the 80386SL, which will run at 5.3 mips (million instructions per second). The 386SL contains 85,000 transistors — more than three times as many as the 386SX.

80386SX Intel calls its 80386SX microprocessor "the entry level member of the Intel 386 family." It is a genuine 80386 chip, except that its external data bus is only 16 bits (i.e. the size of the normal AT bus), not 32 bits. It is capable of sustained execution of 2.5 to 3.5 million instructions per second. The 80386SX contains 275,000 transistors. See also 80386DX.

80387 The Intel 80387 SX is a high performance floating point coprocessor designed to be used in 80385 SX microprocessor-based computers. It's basically a special purpose microprocessor designed to help the main "computer on a chip," the 80387 with mathematical calculations — like recalculating spreadsheets. You can install this chip in virtually all 80386SX based computers. But having the chip is only useful if you run software that can take advantage of it. Not all can.

80486DX Introduced on June 5, 1991, The Intel 80486DX microprocessor (computer on a chip) sports 1,185,000 transistors. It has a clock speed of 20 MHz to 50 MHz. It sports 50 times the performance of the original IBM PC. The Intel 486DX running at 50 MHz will process 41 million instructions per second (MIPS). As of early November, 1991, an 80486DX running at 50 MHz costs $644 in lots of a thousand. See also 486.

80486SX The Intel 80486 microprocessor without a math coprocessor. The 80486SX is a cheaper version of the 80486 chip.

80586 An Intel microprocessor containing more than three million transistors, introduced in April, 1993 and called the Pentium. It is 80% faster than the fastest 486. It is capable of executing 112 million instructions per second.

80686 An upcoming Intel microprocessor containing more than seven million transistors. It will be capable of executing 175 million instructions per second.

80786 An upcoming Intel microprocessor containing more than 20 million transistors. It will be capable of executing 250 million instructions per second.

8080 Intel brought out the 8-bit 8080 microprocessor in April, 1974. It contained 5,000 transistors and was 10-times more powerful than the 8008. The 8080 was, probably the "first" microprocessor. When it was invented, memory for computers was very expensive. The 8080 could directly address 64 thousand bytes of information. This limit of 64K was the direct result of the fact that the CPU chip had only 16 address lines. The address lines are a set of wires coming out of the CPU which allow the CPU to indicate what item of memory it wants to read or write. In most computers, the size of the "pieces" are bytes, or 8-bit characters. These 16 wires are called the Address Bus. The voltages on the 16 address lines are interpreted as a binary number (with the first pin representing the 1's place, the second pin representing the 2's place, the third pin representing the 4's place,

13

etc. The resultant number is the Address addressable by the CPU (Central Processing Unit). The number of distinct patterns of 16 things, each of which can have two values, is 2 raised to the 16th power, or 65,536. This number is 64 times the quantity "1K" which is 1,024. See also 8086 which was the Intel chip which following the 8080.

8086 On June 8, 1978, Intel introduced the 8086, a 16-bit microprocessor with 29,000 transistors and ten times the performance of the 8080. The 8086, or more precisely its 16-bit counterpart, the 8088, became the brain of IBM's first personal computer. The 8086 chip is an upgraded version of the basic 8080 architecture. It is basically a doubling of a small portion of the internal workings of the 8086's address circuitry. Intel basically duplicated the address register (the transistors that hold the pattern of bits to place on the address bus), slid it left four bits, and added some simple circuitry to add it arithmetically to the "old" address register. This is called the Intel Segment Register. By making this simple kludge to the 8080, Intel created the 8086 and 8088 microprocessors which now effectively had twenty address lines and could therefore address two to the 20th locations, or a little over one million bytes. See also 8088 and 80286.

8088 Introduced in June 1979 by Intel, this is the 16-bit chip which drove the original IBM PC and the subsequent IBM XT. It is identical to the 8080 except that external bus width is 8-bit (the 8080 is 16). The 8088 has a clock speed of 4.77 MHz at 0.33 mips and 8 MHz at 0.75 mips. It had 29,000 transistors. It cost between $3 and $4 in late 1991. See also V20, which is a Japanese NEC version of the 8088. It's about 10% to 20% faster than the 8088. You can pull an 8088 out of your old PC (e.g. the AT&T 6300) and replace it with this chip. You'll get about a 20% (and noticeable) increase in speed. We have. It works fine. Some key systems use this chip.

8088-2 A speeded-up version of the 8088 microprocessor. It has a clock speed of 8 mhz. See also the V20-2 which is the Japanese version of the 8088-2. It's 10% to 20% faster than the 8088-2.

82596 The 82596 is an intelligent, 16-/32-bit local area network coprocessor from Intel. The 82596 implements the CSMA/CD access method and can be configured to support all existing 802.3 standards. Coupled with the 82503 Dual Serial Transceiver, the 82596 provides the optimal Ethernet connection to Intel1386 and Intel486 client PCs and servers. The board space required for an 82596/82503 motherboard implementation is less than six square inches. provides full Ethernet bandwidth performance while allowing the CPU to work independently. An on-board four-channel DMA controller along with an intelligent micro machine automatically manages memory structures and provide command chaining and autonomous block transfers while two large independent FIFOs accommodate long bus latencies and provide programmable thresholds.

888 Service North America is running out of 800 numbers. So it is adopting a new prefix — 888. At the time of writing, it looked like the first 888 number would come in around April, 1996. That 888 prefix will have all the characteristics of today's 800 Service. See 800 SERVICE.

9-track A standard for 1/2" magnetic tape designed for data storage. Its nine tracks hold a byte (eight bits) plus one parity bit in a row across the tape width-wise.

900 Service A generic and common (and not trademarked) term for AT&T's, MCI's, Sprint's and other long distance companies' 900 services. All these services have "900" as their "area code." Dialing a 900-number is free to the company or person receiving the call, but costs money to the person making the call. Here's the story: 900 service was introduced as the industry's "information service" area code. You'd dial 1-900-WEATHER, for example, and punch in some touchtones in response to prompts and you could hear the weather in Sydney, Australia or Paris, France, wherever you might be planning your next vacation. For this service, you'd be charged perhaps 75 to 95 cents a minute. And you'd get the bill as part of your normal monthly phone bill. That was the original idea. Then some people got the idea that 900 would make a wonderful porn number and they started advertising "Call 900-666-3333 and speak with Diana. She really wants you." And they started charging $5 a minute. When huge 900-call bills started appearing on people's bills, there was an outcry from many subscribers who wouldn't pay the bills. Some children called on their parents' phones. Employees made calls from work and the company's accountants went nuts. So the industry retreated from 900 porn. Then someone thought — "Why not sell things through an 900 number?" We could sell a set of ginzu knives for just calling this 900 number. No messing with credit cards or checks. The bill goes straight on your phone bill. At about the same time someone thought that 900 numbers would be great for running sweepstakes. "Call up, register your name for a free trip with a racing car team to the Australian Indianapolis 500. Three lucky people will be chosen. The call will cost you only $2.75." So 900 services became a new type of gambling.

The long distance companies providing 900 services reacted predictably to some of the newer services. They clamped down on who they would sign up, which service and/or product you could, or could not sell. And, rather than charging "a piece of the action" as they did in the beginning, the long distance companies began to charge for them as if they were normal long distance calls: charge a set-up fee, a fee for carrying the call, a fee for collecting the money and a fee for the possibility of bad debts. There are variations on these themes.

The 900-service business is rife with stories of people who are alleged to have made millions overnight with innovative 900 numbers. Clearly enormous monies have been made — especially in the beginning when there was novelty to 900 calls. The prognosis is that the 900 number business will grow, that it will mature and that the North American public will wake up to its various scams and discover real value in many of its services. For example, one of the author's "genuine value" and favorite 900 services is fax-back. Dial a 900 number, punch in some touchtone digits, hang up and within seconds your fax machine begins to churn out useful information.

In the summer of 1991, AT&T issued guidelines for its EXPRESS900 service. Those guidelines included:

• The predominant purpose of the calls does not include Entertainment, Children's Programming, Credit/Loan Information, Fulfillment, Political Fundraising, Games of Chance, Postcard Sweepstakes, Job Lines and Personal Lines;

• Every program must have a Preamble and Caller Grace Period, with notification to callers of the opportunity to hang up before charging begins.

• Sponsors may not route calls to any telecommunications equipment or arrangements which allow charging to begin before the caller realizes any value on the call, e.g., Automatic Call Distribution (ACD) with call queuing, or Caller Hold.

NEWTON'S TELECOM DICTIONARY

9001 ISO 9001 is a rigorous international quality standard covering a company's design, development, production, installation and service procedures. Compliance with the standard is of increasing significance for vendors trading in international markets, in particular in Europe where ISO 9001 registration is widely recognized as an indication of the integrity of a supplier's quality processes. ISO is the International Standards Organization in Paris.

905-928 MHz A relatively new frequency range for use by, amongst other things, cordless phones. Most present cordless phones operate within the band 46-49 MHz. The 900 contains 50 channels for cordless telephone transmission. The 46-49 MHz band contains only 10.

911 Service 911 is an emergency reporting system whereby a caller call dial a common number — 911 — for all emergency services. The caller will be answered at a common answering location which will figure the nature of the emergency and dispatch the proper response teams. The first 911 service came on line in 1968. Here are the reasons why 911 benefits a community: Only one number for all emergency services. It's an easy number to remember. It's an easy number to dial. It's great for travelers and new residents. Calls are received by trained personnel. See also E-911, which stands for Enhanced 911 service and typically includes ANI (Automatic Number Identification) and ALI (Automatic Location Information). 911 service is sometimes called B-911 which stands for basic 911 service. See also B-911 and E-911.

9145 A common term in the southern part of America for a customer service representative (i.e. a salesman) of the local telephone company.

958 Dial 958 in New York City and a computer run by Nynex New York will read you the phone number you're calling from. This is very useful. Imagine having a jack on your wall. You've lost track of which phone line it's connected to. Dial 958 and bingo! You know. Other phone companies have similar services but they have different numbers.

9751 CBX A popular PBX from Rolm that ranges in size from 80 to 40,000 ports.

999 Great Britain's equivalent of the United States emergency number 911.

**** Filenames or other resource names that begin with the string \\ mean that they exist on a remote computer. They are called UNC (as in Universal Naming Convention) Names.

NEWTON'S TELECOM DICTIONARY

A 1. Abbreviation for AMP or AMPERE, a unit of electric current. See AMPERE.
2. The local wireline cellular carrier. In one of its less intelligent decisions, the FCC decided to issue two cellular franchises in each cif the United States. They gave one to the local phone company and one to a competitor. This duopoly has naturally meant little real competition. And one day, the FCC will issue other licenses and the price of cellular phone calls will drop dramatically. Meantime, the "A" carrier on your cellular phone is the local wireline carrier, i.e. the local phone company (or that was the company who had the license originally). And "B" is the other one.

A & A1 Control leads that come from 1A2 key telephone sets to operate features like flashing of lights to indicate on hold, line ringing, etc.

A & B Bits Bits used in digital environments to convey signaling information. A bit value of one generally corresponds to loop current flowing in an analog environment. A bit zero corresponds to no loop current, i.e. to no connection. Other signals are made by changing bit values; for example a flash-hook is set by briefly setting the A bit to zero.

A & B Leads Additional leads used typically with a channel bank two-wire E&M interface to certain types of PBXs (also used to return talk battery to the PBX).

A & B Signaling Procedure used in most T-1 transmission links where one bit, robbed from each of the 24 subchannels in every sixth frame, is used for carrying dialing and controlling information. A type of in-band signaling used in T-1 transmission. A and B signaling reduces the available user bandwidth from 1.544 Mbps to 1.536 Mbps.

A Battery Another term for TALK BATTERY.

A Block Cellular licenses received from the FCC with no initial association to a telephone company. Also referred to as non-wireline.

A Law The PCM coding and companding standard used in Europe. See A LAW ENCODING.

A Law Encoding The method of encoding sampled audio waveforms used in the 2048K bit(s) 30 channel PCM primary system, widely used outside North America.

A Link Access Link. A CCS/SS7 signaling link used to connect a Signaling End Point (SEP) and its home pair of STPs (Signal Transfer Points).

A Port Refers to the port in an FDDI topology which connects the incoming primary ring and the outgoing secondary ring of the FDDI dual ring. This port is part of the dual attachment station or a dual attachment concentrator.

A-B Rolls A technique by which audio/video information is played back from two videotape machines rolled sequentially, often for the purpose of dubbing the sequential information onto a third tape, usually a composite master.

A-B Test Direct comparison of the sound/picture quality of two pieces of audio/TV equipment by playing one, then the other.

A/B Switch 1. A switch that allows manual or remote switching between one input and two outputs. See A/B SWITCH BOX.
2. A feature found on all new cellular telephones permitting the user to select either the "A" (non-wireline) carrier or the "B" (wireline) carrier when roaming away from home.

A/B Switch Box A device used to switch one input between two devices, such as printers, modems, plotters, mice, phone lines, etc. An example of how you use such a box: You plug one phone line into the "C" (for Common) jack. You plug a fax machine into the "A" jack. You plug a modem into the "B" jack. By turning the switch, you can use one phone line for either a modem or a fax. A/B switch boxes comes in many many flavors, including also serial and parallel port versions. There are also A/B/C switches that switch among three devices, e.g. a fax, a modem and a phone. There is also a Crossover Switch that connect two inputs and two outputs. In one position the switch might connect input A with output D and input B with output C. In the other position, it might connect input A with output C and input B with output D.

A/D Analog to Digital conversion.

A/D Converter Analog to Digital converter, or digitizer. It is a device which converts analog signals (such as sound or voice from microphone), to digital data so that the signal can be processed by digital circuit such as a digital signal processor.

A/UX An alternate operating system for the Macintosh based on UNIX. A/UX has its own 32-bit addressing mode.

A20 Line A control line on the Intel 80386 microprocessor that allows MS-DOS and an extended memory manager to create the High Memory Area, or HMA. Only one program can claim control over the A20 at a time.

AA Automated Attendant. A device which answers callers with a digital recording, and allows callers to route themselves to an extension.

AABS A new software feature being introduced in the telephone industry. It allows collect and third-number billed calls to be placed by the caller using voice synthesis of an operator's voice plus a digital recording of the caller's voice without using a human operator. It is automated in much the same way as calling card services have been automated.

AAC-1 ATM Access Concentrator for T-1/E-1 networks from ADC Kentrox uses cell multiplexing to concentrate multiple protocols onto a single T-1 or E-1 access facility to the ATM or SMDS network, including SMDS, DXI, ATM DXI, SRI and Frame Relay.

AAC-3 ATM Access Concentrator for T-3/E-3 networks from ADC Kentrox uses cell multiplexing to concentrate multiple protocols onto a T-3 or E-3 access facility to the ATM or SMDS network (including SMDS, DXI, ATM DXI, SRI and Frame Relay).

AAL ATM Adaptation Layer. An ATM term. ATM Adaptation Layer: The standards layer that allows multiple applications to have data converted to and from the ATM call. A protocol used that translates higher layer services into the size and format of an ATM

NEWTON'S TELECOM DICTIONARY

cell. ATM Adaptation Layer sits above ATM and converts non-ATM bit streams — end user data — into ATM cells. The AAL is the protocol used on top of ATM to support higher-layer service requirements. For data communications services, the AAL defines a segmentation/reassembly protocol for mapping large data packets into the 48-octet payload field of an ATM cell. See the next the five definitions.

AAL-1 ATM Adaptation Layer Type 1: AAL functions in support of constant bit rate, time-dependent traffic such as voice and video.

AAL-2 ATM Adaptation Layer Type 2: This AAL is still undefined by the International Standards bodies. It is a placeholder for variable bit rate video transmission.

AAL-3/4 ATM Adaptation Layer Type 3/4: AAL functions in support of variable bit rate, delay-tolerant data traffic requiring some sequencing and/or error detection support. Originally two AAL types, i.e. connection-oriented and connectionless, which have been combined.

AAL-5 ATM Adaptation Layer Type 5: AAL functions in support of variable bit rate, delay-tolerant connection-oriented data traffic requiring minimal sequencing or error detection support.

AAL Connection An ATM term. Association established by the AAL between two or more next higher layer entities.

AAP Administrative and Accounting Package. The name of a feature on an InteCom PBX which provides call accounting reports relating to usage and cost.

AAR Automatic Alternate Routing.

AARP AppleTalk Address Resolution Protocol.

ABAM A designation for 22 gauge, 110 ohm, plastic insulated, twisted pair Western Electric cable normally used in central offices.

Abandoned Call The nontechnical explanation is: A call that is answered, but disconnected before any conversation happens. The technical explanation is: A call which has been offered into a communications network or telephone system, but which is terminated by the person originating the call before it is answered by the person being called. Follow this sequence for an explanation: You call an airline. You hear ringing. Their phone rings. A machine, called an automatic call distributor, answers the call, plays you some dumb message, "Please don't hang up. A real human will answer eventually." You, the caller, are put on Eternity Hold. You get bored waiting and hang up before a live operator answers. You have just abandoned your phone call. Thus the term, Abandoned Call. Information about abandonealls is useful for planning how many people (also called operators, agents or telephone attendants) an airline should employ on what days and during what times of the day. Thus you can organize to get the right percentage of incoming calls answered within the "right" amount of time and give your callers (i.e. your customers) the service they deserve (or the service you think they deserve, or, the service they think they deserve).

Abandoned Call Cost The amount of revenue lost because of abandoned calls. This is calculated based on the number of calls, your estimate of the percentage abandoning, and your estimate of the revenue per call. It's an impossible number to calculate since many callers do, in fact, call back and place their orders on another later call.

Abbreviated Address Calling A calling method that allows the user to employ an address having fewer characters. The destination's assigned device addresses these characters when initiating a connection. May also be called abbreviated dialing when specifically used in connection with telephone systems.

Abbreviated Dialing A feature of phone systems which allows users to place calls by dialing typically one or two digits. The system translates the abbreviated number into the destination number. See also SPEED DIALING and PREDICTIVE DIALING.

ABC 1. Automatic Bill Calling — a method of billing for payphone calls. Changed in 1982 to Calling Card service. 2. Automated Business Connection.

ABCD Signaling Bits These are bits robbed from bytes in each DS-0 or T-1 channel in particular subframes and used to carry in band all status information such as E&M signaling states.

ABDN Attendant Blocking of Directory Number.

ABEND ABnormal END. When Novell's NetWare local area operating system detects a serious problem, such as a hardware or software failure, the system issues an abend (abnormal end) message. The abend stops the file server. And you're dead in the water. Abend is also called ABortive END. Also called system crash. Almost always bad news. Usually caused by input or data presented to a computer which is beyond its ability to cope. If an abend happens in a single-task program (like MS-DOS), the machine will cease to take input ("lock up") and must be restarted ("re-booted"). Abends can be caused by a variety of factors, such as poorly functioning Netware Loadable Modules (NLMs), power problems and heavy network traffic. They are the bane of a NetWare operating manager's existence. Multitasking operating systems (like UNIX) allow other programs to continue running while only stopping the one causing trouble. See also VREPAIR.

Ablation Optical memory data writing techniques in which a laser burns holes (or pits) into thin metal film.

ABM 1. Asynchronous Balanced Mode. A service of the data link level (Logical Link Control) in IBM's token-passing ring. ABM operates at the Systems Network Architecture data link control level and allows devices to send data link commands at any time. 2. Anyone But Microsoft

ABN ABNormal alarm status.

Abort To stop doing something. Often to get out of a software program. Also, to discontinue sending or receiving a message.

Abort Sequence A series of 12 to 18 1-bits appearing at the end of an AppleTalk LLAP frame. The sequence delineates the end of the frame.

Above 890 Decision The 1959 FCC decision which allowed companies to build their own private microwave communications systems. The decision resulted from AT&T Long Lines' reluctance to provide companies with long distance service to remote places, such as oil wells, gas pipelines, power stations and paper plants. The decision got its name because the FCC allowed privately-owned microwave systems using radio frequencies "above 890" megahertz — which are naturally called "microwave." See also MICROWAVE and ENTELEC.

Above The Line Expenses incurred by telephone company that are charged to the ratepayer by being allowed in the company's rate-base.

17

NEWTON'S TELECOM DICTIONARY

ABR 1. Available Bate. Traffic defined by the ATM Forum. ABR is an ATM layer service category for which the limiting ATM layer transfer characteristics provided by the network may change subsequent to connection establishment. A flow control mechanism is specified which supports several types of feedback to control the source rate in response to changing ATM layer transfer characteristics. It is expected that an end-system that adapts its traffic in accordance with the feedback will experience a low cell loss ratio and obtain a fair share of the available bandwidth according to a network specific allocation policy. Cell delay variation is not controlled in this service, although admitted cells are not delayed unnecessarily.
2. AutoBaud Rate detect. A process by which a receiving data device determines the speed, code level, and stop bits of incoming data by examining the first character — usually a preselected sign-on character (often a carriage return). ABR allows the receiving device to accept data from a variety of transmitting devices operating at different speeds without needing to establish data rates in advance.
Abrasion Resistance Ability of material or cable to resist surface wear.
Abrupt Close Close of a connection on a network without any attempt to prevent any loss of data.
ABS Alternate Billing Services. These are IN (Intelligent Network) services that allow subscribers to charge a call to a number or telephone other than the one they are using. For example, by using a charge card, credit card or personal identification number.
Absent-subscriber Service A service offered by local telephone companies to subscribers who will be away. A live operator or a machine intercepts the calls and delivers a message. When you come back, you get your old number. But in the meantime, while you're away, you pay less money per month than you would for normal phone service.
Absolute Delay The time interval or phase difference between transmission and reception of a signal.
Absolute Gain 1. The ratio, usually expressed in decibels, between the signal level at the output of a device and its input under specified operating conditions; e.g., no-load gain, full-load gain, small signal gain.
2. Of an antenna, for a given direction: the ratio, usually expressed in decibels, of the power that would be required at the input of an ideal isotropic radiator to the power actually supplied to the given antenna so that the radiation intensity in the far-field region in the given direction would be the same. If no direction is quoted, that corresponding to maximum radiation is assumed.
Absolute Zero A temperature about 460 degrees below zero in Fahrenheit.
Absorption Attenuation caused by dissipation of energy. See ABSORPTION LOSS.
Absorption Loss The attenuation of an optical signal within a fiber optic transmission system and specified in terms of dB/km.
ABT Advanced Broadcast Television.
AC Alternating Current. Typically refers to the 120 volt electricity delivered by your local power utility to the three-pin power outlet in your wall. Called "alternating current" because the polarity of the current alternates between plus and minus, 60 times a second. The other form of electricity is DC, or direct current, in which the polarity of the current stays constant. Direct current, for example, is what comes from batteries. Outside North America, electricity typically alternates at 50 times a second — which is neither better nor worse, just different. In North America, standard 120 volt AC may be also be referred to as 110 volts, 115 volts, 117 volts or 125 volts. Con Edison, the electricity supplier to New York City, told me that they are only obliged to deliver voltage to 120 volts plus or minus 10%. This means your outlet may deliver anywhere from 118 volts to 132 volts before your power company will get concerned. But you probably should. Telephone and computer are sensitive to voltage fluctuations — some more to high voltages; some more to low voes. My suggestion: If your stuff is valuable, protect it with a voltage regulator. Or better, power it with stable DC current, which you've converted from fluctuating AC power
Your AC electrical circuit consists of two supply conductors — hot and neutral. There is also a "load." That is the term for the device you're running. The hot, energized or live conductor is ungrounded and delivers energy to the load. The hot conductor is connected to the fuse or circuit breaker at the main service entrance. The neutral or common conductor is grounded and completes the circuit from the load back to the utility transformer. The load is any electric or electronic appliance or gadget plugged into the AC electrical outlet. It completes the circuit from the transformer through the hot conductor, to the load, through the neutral conductor and back to the utility transformer. Standard 120 volt circuits also include an equipment ground conductor. This equipment grounding conductor provides an intended path for fault current and is never intended to be a part of the load circuit. The equipment ground serves three very important purposes.
1. It maintains metal appliance cases at zero volts, thus protecting people who touch the cases from receiving an electrical shock.
2. It provides an intentional fault path of low impedance path for current flow when the hot conductor contacts equipment cases (ground fault). This current causes the fuse or circuit breaker to open the circuit to protect people from electric shock.
3. Any electronic equipment (not electrical) uses the equipment ground as a zero volt reference for logic circuits to provide proper equipment performance.
See AC POWER, GROUNDING, BATTERY and SURGE ARRESTOR.
AC Power Phone systems typically run on AC, Alternating Current. Except for small systems, phone systems typically need their own dedicated AC power line. This line should be "cleaned" with a power conditioner, voltage regulator. It also should be protected with a surge arrestor. If possible, the phone system should also be backed by a battery-based power pack, called UPS for Uninterruptible Power Supply. The most reliable battery backup is lead acid, the same technology as used in your car. Phone systems consume more power as they process more phone calls. For example, a PBX brochure says that at minimum capacity, it needs less power than eight 100-watt light bulbs. But that at its maximum duplex capacity, it needs the same power as 26 100-watt light bulbs — or 2600 watts. See AC, GROUND and GROUNDING.
AC To DC Converter An electronic device which converts alternating current (AC) to direct current (DC). Most phone systems, computers and consumer electronic devices (from answering machines to TVs) run on DC. Most phone systems have an AC to DC converter in them. Hint: it's probably buried in the power supply.
AC-DC Ringing A type of telephone signaling that uses both AC and DC components — alternating current to operate a ringer and direct current to aid the relay action that stops the ringing when the called telephone is answered.
ACA Automatic Circuit Assurance.

NEWTON'S TELECOM DICTIONARY

ACADEMMET A network within Russia which connects universities.
Academic Computing Research Facility Network ACRFNET. A network connecting various research units such as colleges and research and development laboratories in the U.S.
ACAT Additional Cooperative Acceptance Testing. A method of testing switched access service that provides a telephone company technician at the central office and a carrier's technician at its location, with suitable test equipment to perform the required tests. ACAT may, for example, consist of the following tests:
Impulse Noise Phase Jitter Signal-to-C-Notched Noise Ratio Intermodulation (Nonlinear) Distortion Frequency Shift (Offset) Envelope Delay Distortion Dial Pulse Percent Break
ACB 1. A Northern Telecom term for Architecture Control Board.
2. Annoyance Call Bureau.
3. Automatic Call Back.
ACC Automatic Cack Calling.
Accelerated Aging A test performed on material or cable meant to duplicate long term environmental conditions in a short period of time.
Accelerated Depreciation A method which allows greater depreciation charges in the early years of an asset's life and progressively smaller ones later on. The total amount of depreciation charged is still equal to 100% of the asset's value. By taking the charges early on in the asset's life, you get the time value of money, i.e. depreciation charged today (and tax saved today) is worth more than the same amount of depreciation charged (or tax saved) tomorrow.
Accelerated Graphics Port A new Intel product which bypasses a common traffic jam between the Pentium and graphics chips.
Accelerator 1. A chemical additive which hastens a chemical reaction under specific conditions. A term used in the telecommunications cable industry.
2. In a Windows program, an accelerator is a keystroke that dispatches a message to a program, invoking one of its functions. For example, Alt-F4 tells the current Windows application of Windows itself to quit.
Accelerator Board A board added onto a personal computer's main board and designed to increase the PC's performance in writing to screen or disk, etc. See also ACCELERATOR CARD.
Accelerator Card An Apple term. An accelerator card is an add-on product that upgrades the CPU of a Macintosh to a higher speed or more powerful generation of processor. An accelerator card is usually a "daughter board" that clips onto the original CPU or is inserted into the socket that held the original CPU. You need an accelerator card to use virtual memory or enhanced 24-bit addressing on LC, Mac Plus, SE and Classic. See ACCELERATOR BOARD.
Acceptable Use Policy AUP. Many transit networks have policies which restrict the use to which the network may be put. A well known example is NSFNET's (National Science Foundation Network) AUP which does not allow commercial use. Enforcement of AUP which does not allow commercial use. Enforcement of AUPs varies with the network.
Acceptance Acceptance refers to the amount of time within which a buyer has to decide whether an item is acceptable. Different from a warranty, it applies more to the equipment's appearance and to whether or not it is what the buyer ordered. Acceptance periods usually range from two to four weeks, but are determined on a case-by-case basis. It's commonly used in the secondary telecom equipment marketplace. See ACCEPTANCE TEST.
Acceptance Angle The angle over which the core of an optical fiber accepts incoming light. It's usually measured from the fiber axis.
Acceptance Test The final test of a new telephone system. If the system passes the test — i.e. it meets all specifications laid down in the sales contract — and is working well, then, and only then, will the customer finish paying for it. See also ACCEPTANCE.
Access As a verb, to dial into a feature, a circuit or a piece of equipment. As a noun, the point at which you enter a circuit or other communications facility. In data processing, access means to retrieve information from, or store data in memory or mass storage.
Access Bus Access Bus is correctly spelled Access.bus. It is a 100KBps bus currently being implemented as part of the Video Electronics Standards Association's Display Data Channel for controlling PC monitors. Access Bus has four pins per connector. See also USB and FIREWIRE.
Access Charge After the FCC deregulated the long distance phone business, it declared that all users having access to United States public switched long distance networks would pay an Access Charge. The charge will differ for residential and business users and may also differ between Centrex and PBX users. It varies from $1 to $6 per month. The charge is theoretically, to make up for the "subsidies" once paid by AT&T Long Lines (now called AT&T) to the local operating telephone companies — Bell and independent. A more formal definition is — A surce levied per the of Federal Regulations. Title 47, part 69, on each line or circuit that has the ability to access or be accessed by a public exchange network. The fee is paid by landline and cellular subscribers, as is a federal telephone excise tax.
Access Code A series of digits or characters which must be dialed, typed or entered in some way to get use of something. That "something" might be the programming of a telephone system, a long distance company, an electronic mail service, a private corporate network, a mainframe computer, a local area network. Once the user dials the main number for the service he must then enter his assigned Access Code to get permission use the system. An Access Code becomes an Account Code when it is used for identifying the caller and doing the billing. Access Code may also mean the digit, or digits, a user must dial to be connected to an outgoing trunk. For example, the user picks up his phone and dials "9" for a local line, dials "8" for long distance, dials "76" for the tie line to Chicago, etc. In programming a phone system such as Northern Telecom's Norstar, there are Access Codes to begin Startup, Configuration programming, and Administration programming.
Access Control A technique used to define or restrict the rights of individuals or application programs to obtain data from, or place data into, a storage device.
Access Control List ACL. Most network security systems operate by allowing selective use of services. An Access Control

NEWTON'S TELECOM DICTIONARY

List is the usual means by which access to, and denial of, services is controlled. It is simply a list of the services available, each with a list of the computers permitted to use the service.

Access Control Method Sets of rules distinguishing different local area network technologies. It regulates each workstation's physical access to the transmission medium (normally cable), directs traffic around the network and determines the order in which nodes gain access so that each user gets service. Examples are token passing, used by Token Ring, ARCnet, FDDI and Ethernet's CSMA/CD.

Access Controls An electronic messaging term. Controls that enable a system to restrict access to a directory entry or mailbox either inclusively or exclusively.

Access Control System A system designed to provide secure access to services, resources, or data; for computers, telephone switches or LANs.

Access Coordination An MCI definition. The process of ordering, installing, and maintaining the local access channel for MCI customers.

Access Coupler A device placed between two fiber optic ends to allow signals to be withdrawn from or entered into one of the fibers.

Access Customer Name Abbreviation ACNA. A three-character abbreviation assigned to each InterExchange Carrier (IXC) and listed in the Local Exchange Routing Guide (LERG).

Access Event Information with a logical content that the functional user and the Network Access FE (Functional Entity) exchange. Definition from Bellcore.

Access Group All terminals or phones have identical rights to make use of the computer, the network or the data PBX.

Access Level Used interchangeably with ACCESS CODE. "Level" in dialing tends to mean a number.

Access Line A telephone line reaching from the telephone company central office to a point usually on your premises. Beyond this point the wire is considered inside wiring.

Access Link The local access connection between a customer's premises and a carrier's POP (Point Of Presence). The POP is the carrier's switching central office. That carrier might be a local or long distance carrier.

Access Method The technique or the program code in a computer operating system that provides input/output services. By concentrating the control instruction sequences in a common sub-routine, the programmer's task of producing a program is simplified. The access method typically carries with it an implied data and/or file structure with logically similar devices sharing access methods. The twas coined, along with data set, by IBM in the 1964 introduction of the System/360 family. It provides a logical, rather than physical, set of references. Early communications access methods were primitive; recently they have gained enough sophistication to be very useful to programmers. Communications access methods have always required large amounts of main memory. In a medium size system supporting a few dozen terminals of dissimilar types, 80K to 100K bytes of storage is not an unusual requirement. The IEEE's 802 standards 802.3 through 802.9 specify access methods for LANs and MANs. See ACCESS METHODS.

Access Methods Techniques and rules for figuring which of several communications devices — e.g. computers — will be the next to use a shared transmission medium. This term relates especially to Local Area Networks (LANs). Access method is one of the main methods used to distinguish between LAN hardware. How a LAN governs users' physical (electrical) access to the cable significantly affects its features and performance. Examples of access methods are token passing (Arcnet and Token Ring) and Carrier Sense Multiple Access with Collision Detection (CSMA/CD) (Ethernet). See ACCESS METHOD.

Access Protection Refers to the process of protecting a local loop from network outages and failures. Access protection can take many forms, such as purchasing two geographically diverse local facilities, adding protection switches to the ends of geographically diverse local loops, or buying service from a local access provider which offers a survivable ring-based architecture to automatically route around network failures.

Access Service Request ASR. The ASR is used by the Carrier to request the ICSC to provide Special Access or Switched Access as specified in the various Access Services Tariffs.

Access Signaling A term which Northern Telecom's Norstar telephones use to indicate their ability to access a remote system (such as a Centrex or a PBX), or dial a number on an alternate carrier by means of Access Signaling (also referred to as "End-to-End" Signaling).

Access Switch Feeder node to Enterprise Network Switches that perform multiprotocol bridge/routing and support a wide range of serial-link (e.g., SDLC BSC, asynchronous) attached devices.

Access Tandem A Local Exchange Carrier switching system that provides a concentration and distribution function for originating or terminating traffic between end offices and a Carrier's location. In short, a type of local phone company central office specifically designed to provide equal access for all long distance carriers in that area. The access tandem provides the interexchange carrier with access to more than one end office within the LATA. More than one access tandem may be needed to provide access to all end offices within LATA.

Access Time There are many definitions of access time: 1. In a telecommunications system, the elapsed time between the start of an access attempt and successful access. Note: Access time values are measured only on access attempts that result in successful access.

2. In a computer, the time interval between the instant at which an instruction control unit initiates a call for data and the instant at which delivery of the data is completed.

3. The time interval between the instant at which storage of data is requested and the instant at which storage is started.

4. In magnetic disk devices, the time for the access arm to reach the desired track and for the rotation of the disk to bring the required sector under the read-write mechanism.

5. The amount of time that lapses between a request for information from memory and the delivery of the information; usually stated in nanoseconds (ns). When accessing data from a disk, access time includes only the time the disk heads take to settle

20

down reaching the correct track (seek time); the time the heads take to settle down after reaching the correct track (seek time); and the time required for the correct sector to moveer the head (latency). Disk access times range between 9ms (fast) and 100 ms (slow).

Access Toolkit Northern Telecom makes a telephone system for up to 100 users called Norstar. The phone system is "open," meaning you can program it. To be able to program it, you need the Norstar Access Toolkit, which you may or may not have to pay for. The Norstar Access Toolkit is a basically set of C Language routines which allows programmers to build applications on an industry standard MS-DOS Personal Computer which is joined to the Norstar with a Norstar interface card. The Toolkit provides access to the D channel messages broadcast to each device on the Norstar. The PC interfaces with the D channel via a PC Interface Card which is inserted into a standard PC-AT slot (the standard ISA bus).

Through the use of the interface card and a set of TSR (Terminate and Stay Resident) drivers, PC applications can insert messages in the D channel and read messages from it which can be received by any other device on the system. Applications on the PC can also be accessed by other devices on the system. In other words, a telephone set can use a feature code which addresses an application on the PC and have access to special information on the PC. Conversely, the application on the PC can monitor a telephone on the system and control how it functions under certain sets of predefined situations.

The basic relationship established between the Norstar telephone sets and the PC is one of server and terminals. Each telephone (terminal) has access to information on the PC (server) and vice versa. No longer are users limited to the set of features and functions which have been defined by Northern Telecom, they can have a phone system which meets their needs. This means you can program a Norstar to act like an automatic call distributor or a predictive dialer or a contact management phone system, or whatever your heart desires. So long as you can program it, or buy the program from someone else who's already done it. As I wrote this in the winter of 1992, shrink-wrapped software packages for Norstar were beginning to appear, as they have appeared for over 10-years for PCs.

Access Unit 1. AU. An electronic messaging term, used for implementing value-added services such as fax, Telex, and Physical Delivery via X.400.
2. In the token ring LAN community, an access unit is a wiring concentrator.

Access bus See ACCESS BUS.

ACCOLC Access Overload Class. A term used in the cellular phone business to allow the cellular system some way of choosing which calls to complete based on some sort of priority. Originally, when the Federal government began designing cellular systems, the government intended to give certain emergency vehicles (such as police, ambulances, and fire departments) codes in their cellular phones that would allow them priority over other subscribers to communicate during emergencies. There is no standard in use within the United States at this time.

Account Code (Voluntary or Enforced) A code assigned to a customer, a project, a department, a division — whatever. Typically, a person dialing a long distance phone call must enter that code so the computer can bill the cost of that call at the end of the month or designated time period. Many service companies, such as law offices, engineering firms, advertising agencies, etc. use account codes to bill their clients. Some account codes are very complicated. They include the client's number and the number of the particular project. The Account Code then includes Client and Matter number. These long codes can tax many call accounting systems, even some very sophisticated ones.

Account-A-Call Pioneers in providing call accounting services since 1972. Account-A-Call works with telecommunications management to assist in cost allocation and analysis of telephone expenses by offering a series of reports and consulting services. One of the country's biggest service bureaus. Based in Pasadena, CA.

Accountant Som who fig your numbers, then numbers your figures and then sends you a bill. See also ECONOMIST.

Accounting Rate A price used between long distance companies to "balance up" what they owe each other. For example, if AT&T sends France Telecom one million minutes of calls, but France sends only 500,000 minutes back to AT&T, then AT&T will have to pay France Telecom for the imbalance. If the accounting rate between France and the United States is $1 per minute, then AT&T will pay France Telecom $250,000 for its work in completing the 500,000 extra calls. AT&T pays France Telecom only half the cost because AT&T does half the work itself.

Accounting Servers A local area network costs money to set up and run. Thus it makes sense to charge for usage on it. In Novell's NetWare LANs, the network supervisor sets up accounting through a program called SYSCON. When this happens, the current file server automatically begins to charge for services. The supervisor can authorize other network services (print servers, database servers, or gateways) to charge for services, or can revoke a server's right to charge.

ACCS Automatic Calling Card Service.

ACCUCALL A PBX integration utility for voice processing from Rhetorex, Campbell, CA.

ACCUNET A family of digital long distance transmission services from AT&T. Services are leased, switched or high speed. One, called Accunet Packet Service, is a packet switching digital network. Another one is called Accunet Reserved 1.5 Service, which AT&T describes as a general purpose, common user, modal-based digital service using dual 1.544 Mbps terrestrial and satellite facilities. Accunet Spectrum of Digital Services (ASDS) is AT&T's leased line (also called private line) digital service at 56 Kbps. MCI and Sprint have similar services. It is available in N x 56/64 Kbps, for N = 1, 2, 4, 6, 8, 12.

Accuracy Absence of error. The extent to which a transmission or mathematical computation is error-free. There are obvious ways of measuring accuracy, such as the percentage of accurate information received compared to the total transmitted.

AC/DC Ringing A common way of signaling a telephone. An alternating current (AC) rings the phone bell and a direct current (DC) is used to work a relay to stop ringing when the called person answers.

ACD See the next seven definitions, AUTOMATIC CALL DISTRIBUTOR and ACIS.

ACD Application Bridge Refers to the link between an ACD and a database of information resident on a user's data system. It allows the ACD to communicate with a data system and gain access to a database of call processing information such as Data Directed Call Routing.

ACD Application-based Call Routing In addition to the traditional methods of routing and tracking calls by trunk and agent group, the latest ACDs route and track calls by application. An application is a type of call, e.g. sales vs. service. Tracking calls in this manner allows accurately reported calls especially when they are overflowed to different agent groups.
ACD Call Back Messaging This ACD capability allows callers to leave messages for agents rather than wait for a live agent. It helps to balance agent workloads between peak and off-peak hours. In specific applications, it offers callers the option of waiting on hold. A good example is someone who only wishes to receive a catalog. Rather than wait while people place extensive orders, they leave their name and address as a message for later follow-up by an agent. This makes things simpler for them and speeds up service to those wanting to place orders.
ACD Caller Directed Call Routing Sometimes referred to as an auto attendant capability within the industry, this ACD capability allows callers to direct themselves to the appropriate agent group without the an operator. The caller responds to prompts (Press 1 for sales, Press 2 for service) and is automatically routed to the designated agent group.
ACD Conditional Routing The ability of an ACD to monitor various parametershin the system and call center and to intelligently route calls based on that information. Parameters include volume levels of calls in queue, the number of agents available in designated overflow agent groups, or the length of the longest call. Calls are routed on a conditional basis. "If the number of calls in queue for agent group #1 exceeds 25 and there are at least 4 agents available in agent group #2, then route the call to agent group #2.
ACD Data Directed Call Routing A capability whereby an ACD can automatically process calls based on data provided by a database of information resident in a separate data system. For example, a caller inputs an account number via touch tone phone. The number is sent to a data system holding a database of information on customers. The number is identified, validated and the call is distributed automatically based on the specific account type (VIP vs. regular business subscriber, as an example).
ACD Intelligent Call Processing The ability of the latest ACDs to intelligently route calls based on information provided by the caller, a database on callers and system parameters within the ACD such as call volumes within agent groups and number of agents available.
ACD-DN A Northern Telecom term for an Automatic Call Distribution-Directory Number (ACD-DN), which refers to the queue where incoming calls wait until they are answered. Calls are answered in order in which they entered the queue.
ACF Advanced Communication Function. A family of software products used by IBM allowing its computers to communicate.
ACF/NCP Advanced Communication Function/ Network Control Program. In host-based IBM SNA networks, ACF/NCP is the control software running on a communications controller that supports the operation of the SNA backbone network.
ACF/VTAM Advanced Communication Function/Virtual Terminal Access Method. In host-based IBM SNA networks the ACF/VTAM is the control software running on a host computer that allows the host to communicate with terminals on an SNA network.
ACFG Short for AutoConFiGuration. The Plug and Play BIOS extensions, now turning up on PCs, are also known as the ACFG BIOS extensions.
ACIS Automatic Customer/Caller Identification. This is a feature of many sophisticated ACD systems. ACIS allows the capture of incoming network identification digits such as DID or DNIS and interprets them to identify the call type or caller. With greater information, such as ANI, this data can identify a calling subscriber number. This is also possible by employing a voice response device to request an inbound caller to identify themselves with a unique code. This could be a phone number, a subscriber number or some other identifying factor. This data can be used to route the call, inform the agent of the call type and even pre-stage the first data screen associated with this call type automatically. See ANI.
ACK In data communications, ACK is a character transmitted by the receiver of data to ACKnowledge a signal, information or packet received from the sender. See also ACKNOWLEDGMENT.
ACK Ahead A variation of the XMODEM protocol that speeds up file transmission across error-free links. See XMODEM.
ACK1 Bisync acknowledgment for odd-numbered message.
Acknowledgment In data communications, the transmission of acknowledgment (ACK) characters from the receiving device to the sending device indicates the data sent has been received correctly.
ACL 1. Access Control List. A roster of users and groups of users, along with their rights.
2. Applications Connectivity Link. Siemens' protocol for linking its PBX to an external computer and having that computer control the movement of calls within a Siemens PBX. See also OPEN APPLICATION INTERFACE.
ACM 1. An ATM term. Address Complete Message. One of the ISUP call set-up messages. A message sent in the backward direction indicating that all the address signals required for routing the call to the called party have been received. See .
2. Association for Computing Machinery. www.acm.org
3. Automatic Call Manager. The integration of both inbound call distribution and automated outbound call placement from a list of phone contacts to be made from a database. Telemarketing and collections applications are targets for this type of system.
ACO 1. Additional Call Offering
2. Alarm Cut Off
ACONET A research network in Austria.
Acoustic Coupler An acoustic modem. A modem designed to transfer data to the telephone network acoustically (i.e by sound), rather than electronically. An acoustic coupler looks like the reverse of a telephone handset and is typically made of rubber. The data communications link is achieved through acoustic (sound) signals rather than through direct electrical connection. It is attached to the computer or data terminal through an RS-232-C connector. To work the acoustic coupler, start the computer's communications program, dial the distant computer on a single line telephone with a normal (e.g. old-fashioned) handset. When the distant computer answers with a higher pitched "carrier tone," you place the telephone handset in the acoustic coupler and transmit data. Since the data is transmitted by sound between the handset and the acoustic coupler (and vice versa), the quality isn't always reliable. You can usually transmit up to 300 baud. People use acoustic couplers when they're short of

time or cannot physically connect their modem electrically, e.g. they're using a payphone without an RJ-11 jack. (There are precious few.)

Acoustic Model In automatic speech recognition, an acoustic model models acoustic behavior of words by gluing together models of smaller units, such as phonemes. (Sorry for the definition of the word model with the word model. But it's actually the best way of defining this term. HN)

Acoustic (or Air) Suspension A loudspeaker system that uses an air-tight sealed enclosure.

Acoustics That branch of science pertaining to the transmission of sound. The qualities of an enclosed space describing how sound is transmitted, e.g. its clarity.

Acousto-optic The interactions between acoustic waves and light in a solid medium. Acoustic waves can be made to modulate, deflect, and focus light waves by causing a variation in the refractive index of the medium. See also FIBER OPTICS.

Acquisition 1. In satellite communications, the process of locking tracking equipment on a signal from a communications satellite.
2. The process of achieving synchronization.
3. In servo systems, the process of entering the boundary conditions that will allow the loop to capture the signal and achieve lock-on. See also phase-locked loop.

Acquisition Time 1. In a communication system, the amount of time required to attain synchronization.
2. In satellite control communications, the time required for locking tracking equipment on a signal from a communications satellite. See also satellite.

ACP Activity Concentration Point.

ACR 1. Attenuation to Crosstalk Ratio: One of the factors that limits the distance a signal may be sent through a given media. ACR is the ratio of the power of the received signal, attenuated by the media, over the power of the NEXT crosstalk from the local transmitter, usually expressed in decibels (db). To achieve a desired bit error rate, the received signal power must usually be several times larger than the NEXT power or plus several db. Increasing the marginal ACR may decrease the bit error rate.
2. An ATM term. Allowed Cell Rate: An ABR service parameter, ACR is the current rate in cells/sec at which a source is allowed to send.

ACRFNET Academic Computing Research Facility Network. A network connecting various research units such as colleges and research and development laboratories in the U.S.

Acrobat A standardized way of viewing a file without needing the associated software. For example, you run Word 6.0 or QuarkXpress, make a pretty desktop published document, replete with diagrams, photos and drgs. Now you want to send the file to someone to view it in all its glory. Simple. Convert the file to an Acrobat file (which has a .PDF extension) and modem it or send it on disk. The receiving person will run an Acrobat viewer program and see your beautiful work. They won't be able to change your work. But they will be able to see it. Acrobat is from Adobe, the Los Altos, CA company which produces PostScript. Acrobat has three benefits: The Acrobat viewing program is very cheap. You can use Acrobat to view any virtually any Windows software created file. Third, an Acrobat file can be up to 75% smaller than the original file in its native form, i.e. the original Word or QuarkXpress file.

Acronym A pronounceable artificial word formed from the first (or first few) letters of each word or group of words. For example, BASIC, the Beginner's All-purpose Symbolic Instruction Code, or COBOL, the computer language, which comes from COmmon Business Oriented Language.

ACS 1. Automatic Call Sequencer. A rudimentary automatic call distributor. See AUTOMATIC CALL SEQUENCER.
2. Advanced Communication System. An old name for AT&T's data communications/data processing service, later called Net 1000. In late 1986, after 10 years in birth, AT&T finally buried ACS because it offered too little in the face of what — by then — had become cheap, powerful desktop microcomputers and under $200 1200 baud modems.

ACSE Association Control Service Element. An application-level protocol. The method used in OSI for establishing a connection between two applications.

ACT Applied Computer Telephony is Hewlett Packard's program that is a strategy and set of open architecture commands and interfaces for integrating voice and database technologies. The idea is that with ACT a call will arrive at the telephone simultaneously with the database record of the caller. And such call and database record can be transferred simultaneously to an expert, a supervisor, etc. ACT works on both HP 3000 and HP 9000 computers. ACT essentially controls the telephone call movement within PBXs it connects to. See also OPEN APPLICATION INTERFACE.

ACTA America's Carriers Telecommunications Association, a Casselberry, FL organization founded in 1985 by 15 small long distance companies wishing to create an association in which the members controlled the direction of the organization. (That's their words.) "The focus established was to provide national representation before legislative and regulatory bodies, while continuing to improve industry business relations." There are now more than 165 members, which include MCI and Sprint, but not AT&T. 407-332-9382

ACTAS Alliance of Computer-Based Telephony Application Suppliers — a part of the North American Telecommunications Association (NATA), which has now changed its name to MMTA. ACTAS's mission, according to ACTAS, is to deliver the benefits of computer-based telephone applications to the broadest possible range of customers. The high costs of customization and marketing currently relegate computer-based telephony applications to Fortune 100-level companies. ACTAS will work to lower the threshold for delivering the benefits of these applications, which include integrated voice and database processing systems automation and customer service, to the general business market. See MMTA.

ACTGA Attendant Control of Trunk Group Access. A complicated term for a simple concept, namely that your operator completes long distance calls. A primitive form of toll control.

ACTIUS Association of Computer Telephone Integration Users and Suppliers. A British organization based in Orpington, Kent. 011-44-689-873-333.

NEWTON'S TELECOM DICTIONARY

Action Media Board Intel i750 based PC board set that performs realtime compression and full-screen playback.

Actionmedia An imaging term. DVI Technology's product family, introduced in 1990 and consisting of single-board delivery and single board capture capability for AT or Micro Channel architecture buses (introduced with IBM)g. ActionMedia 750 ADP, etc.

Activated Return Capacity A cable TV term. The capability of transmitting signals from a subscriber or user premises to the cable headend.

Activation A one time initial connection fee to get cellular phone service. As competition has intensified, so more and more carriers are dropping or severely reducing their activation fees. They do this in order to attract more new subscribers.

Activation Fee Fee for the initial connection to the cellular system.

ActiveX New software code from Microsoft which allows a developer to add move things to an otherwise static Web page. ActiveX is positioned by Microsoft as a competitive move against Java. See JAVA.

Active Call An term which Hayes defines in its Hayes ISDN System adapter manual. An active call is a voice call to which you are connected that is not on hold.

Active Circuits An MCI definition. MCI circuits for which are there is a completed "install order" and a "completed date."

Active Contract One you must sign. See CONTRACT.

Active Coupler A fiber optic coupler that includes a receiver and one or more transmitters. It regenerates (thus "active") input signals and sends them through output fibers, instead of passively dividing input light.

Active Hub A device used to amplify transmission signals in certain local area network topologies. You can use an active hub to add workstations to a network or to lengthen the cable distance between workstations and the file saver.

Active Line A voice or data communications channel currently in use.

Active Matrix Liquid Crystal Display A technique of making liquid crystal displays for computers in which each of the screen's pixels — the tiny elements that make up a picture — is controlled by its own transistor. Active matrix LCD display technique uses a transistor for each monochrome or each red, green and blue pixel. It provides sharp contrast, speedier screen refresh and doesn't lose your cursor when you move it fast (also knowing as submarining). Some active matrix CD screens are as fast as normal glass CRTs.

Active Medium The material in fiber optic transmission, such as crystal, gas, glass, liquid or semiconductor, which actually "lases." It's also called laser medium, lasing medium, or active material.

Active Open Used in TCP to request connection with another node.

Active Pixel Region On a computer display, the area of the screen used for actual display of pixel information.

Active (recurring) Customer An MCI definition. An MCI customer who was installed and not canceled as of the first day of the month. The system determines the customer to be recurring if the install date does not equal the current billing month.

Active Splicing Aligning the ends of two optical fibers with the aim of minimizing the splice loss.

Active Terminator A terminator that can compensate for variations in the terminator power supplied by the host adapter through means of a built-in voltage regulator.

Active Video Lines All video lines not occurring in the horizontal and vertical blanking intervals. In other words, the lines conveying the video and audio signals.

Active Vocabulary A phrase used in voice recognition to mean a group of words which a recognizer has been trained to understand and is attempting to understand at a given time. It is a subset of the total vocabulary of the recognizer.

Active Window A Windows term. The active windows is the window in which the user is currently working. An active window is typically at the top of the window order and is distinguished by the color of its title bar, typically dark blue.

Active/passive Device On a local area network, a device that supplies current for the loop is considered active. Such a device is s Token Ring MAU (Multistation Access Unit). A device which does not supply current is considered passive.

Activity Concentration Point ACP. A location on a telecommunications network where there is high communications traffic, includvoice, data, document distribution and teleconferencing. Generally, there will be some switching equipment present at the ACP.

Activity Factor A decimal fraction which represents the percentage of speech on a voice channel versus those periods of (non-talking) silence on that channel. Most voice channels carry actual speech 30% to 40% of the total available time. This represents an activity factor of 0.3 to 0.4.

Activity Report A report printed by a facsimile machine which lists all transmissions and receptions — their time, date, and number of documents; the remote unit type, diagnostic codes; and machine identification.

ACTS 1. Association of Competitive Telecommunications Suppliers. Trade association of telephone equipment dealers in Canada.
2. Automatic Coin Telephone Service includes a telephone company central office that can complete all types of payphone calls automatically without an operator. Recorded announcements are used to convey instructions to the customer.

ACU Automatic Calling Unit. Also an 801 ACU. A telephone company-provided device instructed by a computer to place a call on behalf of the computer. The call is then connected to a telephone company-provided Data Set. Anyone other than an IBM shop would simply buy a Hayes or Hayes-compatible modem, and not bother with the trouble and expense of an ACU.

ADA 1. Average Delay to Abandon. Average time a caller is held in queue before they get frustrated and decide to hang up.
2. A high level computer language which the Department of Defense has been trying to foist on its suppliers and thus, make a standard. ADA is named for British mathematician Ada Lovelace, known at the time as Lady Lovelace. She was the girlfriend of Charles Babbage, the inventor of the computer.

ADACC Automatic Directory Assistance Call Completion.

ADAD Automatic Dialing and Announcing Device. Device which automatically places calls and connects them to a recording or agent. A Canadian term for an automatic dialer.

Adapter A device used to connect a terminal to some circuit or channel so it will be compatible with the system to which it is attached. An adapter converts one type of jack or plug to another, for example, from old 4-prong telephone jacks to new modu-

24

lar. An adapter may also combine various items, such as putting three plugs in one jack.

Adapter Card A printed circuit card installed inside of a computer. It takes data from memory and transmits it over cable to connected devices such as a modem, or printer.

Adapter Segment A name sometimes used for the upper memory area of a PC, at hexadecimal addresses A000 through EFFF (640K to 1024K).

Adaptive Antenna Array An antenna array in which the received signal is continually monitored in respect of interference (usually adjacent or co-channel). Its directional characteristics are then automatically adjusted to null out the interference. Such a concept often employs computer control of a planar type antenna.

Adaptive Channel Allocation A method of multiplexing wherein the information-handling capacities of channels are not predetermined but are assigned on demand.

Adaptive Communication Any communication system, or portion thereof, that automatically uses feedback information obtained from the system itself or from the signals carried by the system to modify dynamically one or more of the system operational parameters to improve system performance or to resist degradation.

Adaptive Compression Data compression software that continuously analyzes and compensates its algorithm (technique), depending on the type and content of the data and the storage medium.

Adaptive Differential Pulse Code Modulation See ADPCM.

Adaptive Equalization An electronic technique that allows a modem to continuously analyze and compensate for variations in the quality of a telephone line.

Adaptive Interframe Transform Coding A class of compression algorithms commonly used in video codecs to reduce the data transmission rate.

Adaptive Logically Provided Services An AT&T term for services that adapt to an end customer's needs, on demand, in any combination of voice, data or image by the dynamic allocation of network resources (bandwidth, switching, bridging, feature processing, etc.) under sophisticated software logic control.

Adaptive Pulse Code Modulation A way of encoding analog voice signals into digital signals by adaptively predicting future encodings by looking at the immediate past. The adaptive part reduces the number of bits per second that another rival and more common method called PCM (Pulse Code Modulation) requires to encode voice. Adaptive PCM is not common because, even though it reduces the number of bits required to encode voice, the electronics to do it are expensive. See PULSE CODE MODULATION.

Adaptive Routing A method of routing packets of data or data messages in which the system's intelligence selects the best path. This path might change with altered traffic patterns or link failures.

ADAS Automated Directory Assistance Service. A service from Northern Telecom which automates the greeting and inquiry portion of the directory assistance call. With ADAS, directory assistance callers are greeted by the automated system and asked to state the name of the city and the listing they are seeking. They are then connected with an operator. The ADAS service knocks a few seconds off each directory assistance call.

ADC Analog-to-Digital Converter.

ADCCP Advanced Data Communications Control Procedures, A bit-oriented ANSI-standard communications protocol. It is a link-layer protocol. ADCCP is ANSI's version of SDLC/HDLC.

ADCU Association of Data Communications Users.

Add-in Card An expansion board that fits into the computer's slots and is used to expand the system's memory or extend the operation of another device.

Add-on 1. A telephone system feature which allows connecting a third telephone to an existing conversation. This "add-on" feature is initiated by the originator of the call. The feature is also known as "Three-Way Calling."
2. Hardware, often referred to as peripheral equipment, that is added to a system to improve its performance, add memory or increase its capabilities. Voice mail, Automated Attendant and Call Detail Recording Equipment are examples of PBX add-on devices. AT&T and some other manufacturers call them APPLICATIONS PROCESSORS.

Add-on Conference A PBX feature. Almost always used in conjunction with another feature called consultation hold, this feature allows an extension user to add a third person to an existing two-person conversation. The user places an existing central office call or internal call on Hold, and obtains system dial tone. The user can then call another internal extension or an outside party. After speaking with the "consulted" party, the originating phone reactivates the initiating command (typically a button push) and creates a three-party conference with the call previously placed on Hold.

Add-on Conference — Intercom Only Allows a telephone user to add someone else to an existing intercom (within-the-same office) conversation.

Add-on Data Module Plug-in circuit cards which allow a PBX to send and receive analog (voice) and digital (data) signals.

Added Bit A bit delivered to the intended destination user in addition to intended user information bits and delivered overhead bit. An added bit might be used to round out the number of bits to some error checking scheme, for example.

Additional Cooperative Acceptance Testing ACAT. A method of testing switched access service that provides a telephone company technician at the central office and a carrier's technician at its location, with suitable test equipment to perform the required tests. ACAT may, for example, consist of the following tests:
Impulse Noise Phase Jitter Signal-to-C-Notched Noise Ratio Intermodulation (Nonlinear) Distortion Frequency Shift (Offset) Envelope Delay Distortion Dial Pulse Percent Break

Additional Period Billing periods charged after initial, first or minimum period on a call. Usually, long distance toll/DDD has a one-minute initial period at premium rate; subsequent "additional" minutes (period) are billed at a lower rate. Additional period billing increments vary by long distance company.

Additive Primaries By definition, three primary colors result when light is viewed directly as opposed to being reflected:

red, green and blue (RGB). According to the tri-stimulus theory of color perception, all other colors can be adequately approximated by blending some mixture of these three lights together. This theory is harnessed in color television and video communications. It doesn't work so well in color printing where special colors are often printed separately.

ADDMD Administrative Directory Management Domain. A X.500 directory management domain run by a PTT (Posts, Telegraph, and Telephone administration) or other public network provider.

Address Characters identifying the recipient or originator of transmitted data. An address is the destination of a message sent through a communications system. A telephone number is considered the address of the called person. In computer terms, an address is a set of numbers that uniquely identifies something — a workstation on a LAN, a location in computer memory, a packet of data traveling through a network. IEEE 802.3 and 802.5 recommend having a unique address for each device worldwide. An address may also denote the position of data in computer memory or the data packet itself while in transit through a network.

Address Complete Message ACM. A CCS/SS7 signaling message that contains call-status information. This message is sent prior to the called customer going off-hook.

Address Field In data transmission, the sequence of bits immediately following the opening flag of a frame identifying the secondary station sending, or designated to receive, the frame.

Address Filtering A way of deciding which data packets are allowed through a device. The decision is based on the source and destination MAC (Media Access Control, the lower part of ISO layer two) addresses of the data packet.

Address Mask An electronic messaging term. A bit mask used to select bits from a network address (e.g. Internet) for subnet addressing. The mask is 32 bits long and selects the network portion of the address and one or more bits of the local portion. Sometimes called sub-net mask.

Address Prefix An ATM term. A string of 0 or more bits up to a maximum of 152 bits that is the lead portion of one or more ATM addresses.

Address Resolution 1. An internetworking term. A discovery process used when, as in LAN protocols such as TCP/IP and IBM NetBIOS, only the Network Layer address is known and the MAC address is needed to enable delivery to the correct device. The originating end station sends broadcast packets with the device's NLA to all nodes on the LAN; the end station with the specified NLA address responds with a unicast packet, addressed to the originating end station, and containing the MAC address. See ADDRESS RESOLUTION PROTOCOL.
2. An ATM term. Address Resolution is the procedure by which a client associates a LAN destination with the ATM address of another client or the BUS.

Address Resolution Protocol The Internet protocol used to map dynamic Internet addresses to physical (hardware) addresses on local area networks. Limited to networks that support hardware broadcasts.

Address Signaling Signals either the end user's telephone or the central office switching equipment that a call is coming in.

Address Signals Address signals provide information concerning the desired destination of the call. This is usually the dialed digits of the called telephone number or access codes. Typical types of address signals are DP (Dial Pulse), DTMF, and MF.

Address Space The amount of memory a PC can use directly is called its address space. MS-DOS can directly access 1024K of memory (one megabyte). A protected mode control program like Microsoft Windows 3.x or OS/2 can directly address up to 16 megabytes of memory. Here is a definition of address space, as supplied by the Personal Computer Memory Card International Association (PCMCIA) as address space applies to PCMCIA cards: "An address space is a collection of registers and storage locations contained on a PC Card which are distinguished from each other by the value of the Address Lines applied to the Card. There are three, separate, address spaces possible for a card. These are the Common Memory space, the Attribute Memory space and the I/O space."

Addressable Programming A cable TV (CATV) industry term. A subscriber orders a movie or sports event. He does that calling a phone number (generally an 800 number). A computer answers, grabs the calling number, confirms the request, then hangs up. The computer passes the request onto the cable company's computer, which checks the calling phone number against its accounting records. If the subscriber has good credit, the cable company sends a coded message down its cable network to the caller's set-top cable box/converter. The message temporarily enables that particular converter to descramble the channel offering the desired program.

Addressability 1. In computer graphics, the number of addressable points on a display surface or in storage.
2. In micrographics, the number of addressable points, within a specified film frame, written as follows: the number of addressable horizontal points by the number of addressable vertical points, for example, 3000 by 4000.
3. A cable TV term. The capability of controlling the operation of cable subscriber set-top converters by sending commands from a central computer. Such addressability is absolutely require for a cable system to offer pay-per-view services.

Addressable Point In computer graphics, any point of a device that can be addressed. See ADDRESSABILITY.

Addressee The intended recipient of a message.

Addressing Refers to the way that the operating system knows where to find a specific piece of information or software in the application memory. Every memory location has an address.

ADF Automatic Document Feeder

ADH Average Delay to Handle. Average time a caller to an automatic call distributor waits before being connected to an agent.

Adherence A term used in telephone call centers to connote whether the people working in the center are doing what they're meant to be doing. Are they at work? Are they on break? Are they answering the phone? Are they at lunch? All these activities are scheduled by workforce management software. If they're in line, the workers is "in adherence." If not, they're "out of adherence." See ADHERENCE MONITORING.

Adherence Monitoring Adherence monitoring means comparing real-time data coming out of an ACD with forecast call volumes, forecast service levels and forecast workforce employment levels. The idea is to see if the people, the calls and the system are working as forecast. This a measure of how well your forecasting work. You need to know how well it works since it's your forecasting on which you base your employment. See ADHERENCE.

Adjacent Channel Interference When two or more carrier channels are placed too close together, they interfere with each other and mess up each other's conversations.

Adjacency An ATM term. The relationship between two communicating neighboring peer nodes.

Adjacent MTA An MTA (Message Transfer Agent) that directly connects to another MTA. A Message Transfer Agent operated by a public service provider or PTT (Post, Telegraph, and Telephone administration), or a client MTA.

Adjacent Signaling Points Two CCS/SS7 signaling points that are directly interconnected by signaling links.

Adjunct 1. Network system in the Advanced Intelligent Network Release 1 architecture that contains SLEE (Service Logic Execution Environment) functionality, and that communicates with an Advanced Intelligent Network Release 1 Switching System in processing AIN Release 1 calls. Definition from Bellcore. See also ADJUNCT PROCESSOR.
2. An auxiliary device connected to the ISDN set, such as a speakerphone, headset adapter, or an analog interface.

Adjunct Key System A system installed behind a PBX or a Centrex. Such key system provides the users with several more features than the PBX or Centrex. Not a common term today.

Adjunct Processor A computer outside a telephone switching system that "talks" to the switch and gives it switching commands. An adjunct processor might include a database of customers and their recent buying activities. If the database shows that a customer lives in Indiana, the call from the customer might be switched to the group of agents handling Indiana customers. Adjunct processors might be concerned with energy management, building security etc.

Adjunct System Application Interface See ASAI.

Adjusted Ring Length When a segment of Token Ring (in practice a dual ring) trunk cable fails, a function known as the Wrap connects the main path to the backup path. In the worse case — the longest path — would occur if the shortest trunk cable segment failed, so ARL is calculated during network design to ensure the network will always work.

ADK Application Definable Keys

ADM An AT&T term for an Add/Drop Multiplexer. A network element that can add and drop standard DSn or SONET signals from a line signal.

ADMD Administration Management Domain. An X.400 Message Handling System public carrier. Examples include MCImail and ATTmail in the U.S., British Telecom's Gold400mail in the U.K. The ADMDs in all countries worldwide together provide the X.400 backbone.

Admin Administration.

Adminisphere The rarefied organizational layers beginning just above the rank and file. Decisions that fall from the adminisphere are often profoundly inappropriate or irrelevant to the problems they were designed to solve. This definition from Wired Magazine.

Administrable Service Provider An SCSA definition. A service provider which supports administrable services (for example, SCSA Call Router).

Administration A term used by the telephone industry to program features into a phone system. On a Northern Telecom Norstar system, administration includes making settings on 1. System speed dial; 2. Names on phones; 3. Time and date; 4. Restrictions; 5. Overrides; 6. Permissions; 7. Night Service and 8. Passwords.

Administration By Telephone The capability for the system administrator to perform most routine system administrative functions remotely from any Touch Tone pad. Such functions include mailbox maintenance (e.g. create, delete, set password, set class of service, etc.) and disk maintenance.

Administration Directory Management Domain Administrative Directory Management Domain. A X.500 directory management domain run by a PTT (Posts, Telegraph, and Telephone administration) or other public network provider.

Administration Sub-system Part of AT&T's premises distribution system that distributes hardware components for the addition or rearrangement of circuits.

Administrative Alerts A Window NT term. Administrative alerts relate to server and resource use; they warn about problems in areas such as security and access, user sessions, server shutdown because of power loss (when UPS is available), directory replication, and printing. When a computer generates an administrative alert, a message is sent to a predefined list of users and computers.

Administrative Domain 1. AD. A group of hosts and networks operated and managed by a single organization. An Internet term.
2. An ATM term. A collection of managed entities grouped for administrative reasons.

Administrative Management Domain An X.400 electronic mail term: A network domain maintained by a telecommunications carrier.

Administrative Point A location at which communication circuits are administered, i.e. rearranged or rerouted, by means of cross connections, interconnections, or information outlets.

Administrative Service Logic Program ASLP. The SLP responsible for managing the feature interactions between Advanced Intelligent Network AIN Release 1 features resident on a single SLEE (Service Logic Execution Environment).

Administrative Subsystem That part of a premises distribution system where circuits can be rearranged or rerouted. It includes cross connect hardware, and jacks used as information outlets.

Administrator The individual responsible for managing the local area network (LAN). This person configures the network, maintains the network's shared resources and security, assigns passwords and privileges, and helps users.

Administrivia A silly term for administrative tasks, most often related to the maintenance of mailing lists, digests, news gateways, etc. An Internet term.

ADN Advanced Digital Network. ADN is Pacific Bell of California's low-cost leased 56-Kbps digital service. ADN is available for intraLATA calls.

NEWTON'S TELECOM DICTIONARY

ADP 1. Apple Desktop Bus. A synchronous serial bus allowing connection of the Mac keyboard, mouse and other items to the CPU. A Mac keyboard or mouse is called an ADB device. Contrast with peripherals, which attach through the SCSI interface. 2. Automatic Data Processing. The same as DP, data processing. 3. The name of a company which processes my pay check.

ADPCM Adaptive Differential Pulse Code Modulation. A speech coding method which uses fewer bits than the traditional PCM (Pulse Code Modulation). ADPCM calculates the difference between two consecutive speech samples in standard PCM coded telecom voice signals. This calculation is encoded using an adaptive filter and therefore, is transmitted at a lower rate than the standard 64 Kbps technique. Typically, ADPCM allows an analog voice conversation to be carried within a 32Kbit digital channel; 3 or 4 bits are used to describe each sample, which represents the difference between two adjacent samples. Sampling is done 8,000 times a second. ADPCM, which many voice processing makers use, allows encoding of voice signals in half the space PCM allows. In short, ADPCM is a reduced bit rate variant of PCM audio encoding (see also PCM).

ADRMP (pronounced add-rump) AutoDialing Recorded Message Player. A device that calls a bunch of telephone numbers and upon connection will play a message to the answering person. ADRMPs are used for lead solicitation and message delivery. They are often unpopular due to their indiscriminate dialing pattern and random message playing.

ADS AudioGram Delivery Services.

ADSI Analog Display Services Interface. ADSI is a Bellcore standard defining a protocol on the flow of information between something (a switch, a server, a voice mail system, a service bureau) and a subscriber's telephone, PC, data terminal or other communicating device with a screen. The simple idea of ADSI is to add words to, and therefore a modicum of simplicity of use to a system that usually uses only touchtones. Imagine a normal voice mail system. You call it. It answers with a voice menu. Push 1 to listen to your messages, 2 to erase them, 3 to store them, 4 to forward them, etc. It's confusing. You have to remember which is which. ADSI is designed to solve that. It's designed to send to your phone's screen the choices in words that you're hearing. You then have the choice of responding to what you hear or what you see. Your response is the same — a touchtone button. ADSI's signaling is DTMF and standard Bell 202 modem signals from the service to your 202-modem equipped phone. From the phone to the service it's only touchtone. With ADSI, you don't hear the modem signaling because every time the service gets ready to send you information, it first sends a "mute" tone. ADSI works on every phone line in the world.

For ADSI to work visually, you'll need a special ADSI-equipped phone or a piece of ADSI software in your PC. The nice feature of ADSI is that the standard is so flexible, it can work on cheap phones with a small display and more expensive phones with a bigger display and on a PC with a real big display. These three diagrams show a little of the basic concepts behind ADSI — Information Page Mapping and Softkey Mapping.

28

NEWTON'S TELECOM DICTIONARY

Information Page			Softkey definer	Label text	Return String
	Softkey tuple acts as index into softkey table		1		No Return
			2	ERA	DTMF 2
			3	EXIT	Hang-up
4 New / old	10:23A Oct 10 1xNew	2211101	4	DIAL	Line No.
o1) John Jones	11:30P Oct 9 1xNew	4671114	5	UND	Line No.
02) Bob Davis	10:23P Oct 9 3xNew	4171114	6	NUM	Line No.
03) Leslie Donald	11:03A Oct 9 1xNew	4671111	7	REM	Line No.
04) Liz Folsy	8:30A Oct 8 2xOld	1671111	8		Goto Line
05) M. Marcisewski		4671110	9		Goto Line

Softkey information is taken from the softkey table to define the softkeys and scroll keys

ADSL Asymmetrical Digital Subscriber Line. Bellcore's term for one-way T-1 (1.544 million bits per second) to the home over the plain old, single twisted pair wiring already going to homes. ADSL is designed to carry video to the home. ADSL is one-way video with control signals returning from the home at 16 Kbps. Those speeds were the original implementation of ADSL. Recently there has been talk of ADSL running at 6 million downstream and 640K upstream. ADSL, like ISDN, uses adaptive digital filtering, which is a way of adjusting itself to overcome noise and other problems on the line.
According to Northern Telecom, initial ADSL field trails and business cases have focused on ADSL's potential for Video on Demand service, in competition with cable pay-per-view and neighborhood video rental stores. But ADSL offers a wide range of other applications, including education and healthcare. Once telephone companies are able to deliver megabits to the home, Northern Telecom expects an explosion in potential applications including work-at-home access to corporate LANs, interactive services such as home shopping and home banking and even multi-party video gaming, interactive travelogues, and remote medical diagnosis. Multimedia retrieval will also become possible, enabling the home user to browse through libraries of text, audio, and image data — or simply subscribe to CD-quality music services. In the field of education, ADSL could make it possible to provide a low-cost "scholar's workstation" — little more than a keyboard, mouse, and screen — to every student, providing access to unlimited computer processing resources from their home. For a more modern version of ADSL, see DMT, which stands for Discrete Multi-Tone. See also ADSL Forum and VDSL.
ADSL Forum The ADSL Forum is an industry association formed to promote the ADSL concept and to facilitate the development of ADSL system architectures and protocols for major ADSL applications. www.sbexpos.com/sbexpos/associations/adsl/home.html See also ADSL and VDSL.
ADSTAR Automated Document STorage And Retrieval.
ADT Abstract Data Type.
ADTF ACR Decrease Time Factor: This is the time permitted between sending RM-cells before the rate is decreased to ICR (Initial Cell Rate). The ADTF range is .01 to 10.23 sec. with granularity of 10 ms.
ADU Asynchronous Data Unit.
Adobe Systems The name of the company which produces PostScript. See POSTSCRIPT and OUTLINE FONT. Adobe Systems got its name from the creek than ran past its founder's home in Los Altos, California. See ACROBAT.
Advance Replacement Warranty A warranty service whereby the dealer sends the customer a replacement component before the customer returns the defective product. This not only accelerates the replacement time, but also helps the buyer if the component is vital. When you buy vital telecom gear, it's good to check that your equipment has an Advance Replacement Warranty or Guarantee.
Advanced 800 Services AT&T's name for a family of 800 In-WATS services that include time of day and area code routing, single number 800 service (for both interstate and intrastate), automatic number identification, etc. See also 800 SERVICE and ENHANCED 800 SERVICES.
Advanced Branch Exchange ABX. An uncommon term meaning a private branch exchange (PBX) with advanced features normally including the ability to handle both voice and data in some sort of integrated way.
Advanced Communication System ACS. The old name for a proposed packet switched network from AT&T. The service was called Net 1000. AT&T finally killed the service in late 1986. For a bigger explanation, see ACS.
Advanced Digital Network ADN is Pacific Bell of California's low-cost leased 56-Kbps digital service. ADN is available for intraLATA calls.
Advanced Intelligent Network AIN. The local Bell telephone companies' architecture for the 1990s and beyond. See AIN for a much fuller explanation.
Advanced Interactive Executive AIX. An IBM version of UNIX. AIX runs on PS/2 computers, IBM workstations, mini-computers, and mainframes.

Advanced Interactive Video AIV. Interactive videodisc format and system using LV-ROM, a method of storing analog video, digital audio, and digital data on a single videodisc. The system was developed by Philips UK, the British Broadcasting Corporation, Acorn Computer, and Logica Ltd. Most prominent application was the BBC's Domesday Project.

Advanced Private Line Termination An AT&T term which means the PBX user gets access to all the services of an Enhanced Private Switched Communications Services (EPCS) network. It also works when it is associated with AT&T's Common Control Switching Arrangement (CCSA) network.

Advancenet An Ethernet-based local area network from Hewlett Packard, Palo Alto, CA. See ETHERNET.

Advantis A partnership between IBM and Sears Roebuck to provide advanced voice and data network services to users worldwide. As of writing, it's not exactly clear what this new company will do.

Advisory Tones Signals such as dial tone, busy, ringing, fast-busy, call-waiting, camp-on and all the other tones your telephone system uses to tell you that something is happening or about to happen.

Advocacy A General Magic term. Advocacy is the process for getting developers and information providers to commit to developing for Magic Cap devices and Telescript.

AEB Analog Expansion Bus. The analog voice processing bus designed by Dialogic which allows multiple cards to route audio signals within a PC. It is used to interface DTI/124 and D/4x voice response component boards which fit in an AT-expansion slot of a PC. See also PEB and SCSA which are more modern digital expansion buses.

AECS Plan Aeronautical Emergency Communications System Plan. The AECS Plan provides for the operation of aeronautical communications stations on a voluntary, organized basis to provide the President and the Federal Government, as well as heads of state and local governments, or their designated representatives, and the aeronautical industry, with a means of communicating during an emergency.

AEMIS Automatic Electronic Management Information System. This was the first computerized UCD/ACD reporting system introduced by AT&T under CO UCD (Uniform Call Distribution). This package was updated to become the PRO 150/500 system for UCD management on the Dimension PBX/UCD. AEMIS was the successor to the FADS or Force Administration Data System. It was an electro-mechanical system of peg counters and different colored busy lamp fields used to note trunk and position status.

Aerial Cable Cables strung outside and overhead. They're called aerial even though they only hang from poles or buildings. Some aerial cable hangs by its own strength. Some is supported by steel wire above it. Stringing aerial cable is cheaper than burying it, though buried cable lasts longer.

Aerial Distribution Method A method of running cables through the air, typically pole-to-pole. The old fashioned way. Some phone companies say aerial cable is more reliable than underground. Certainly, it's cheaper to fix or add to. It just looks less appetizing.

Aeronautical Radio Inc. ARINC. The organization that coordinates the design and management of telecommunications systems for the airline industry. It's one of the largest buyers of telecommunications services and equipment in the world.

Aerospace Air force publicists coined the term "aerospace" to convince everyone that space was the business of those who fly in the air. According to the Economist Magazine, the "aerospace industry" was quickly accepted into the language, perhaps because President Eisenhower's alternative, the "military industrial complex," sounded rather more sinister. After the Apollo program, which ended in 1972, the "space" in aerospace often seemed like a syllable tacked on to make building airplanes sound grander. But the growth in satellite use in the 1980s made space a respectable business in its own right. In America as of writing in the fall of 1991, the annual sales of space hardware are now bigger than those of civilian aircraft.

AET Application Entity Title. The authoritative name of an OSI application entity, usually a Distinguished Name from the Directory.

AFACTS Automatic FACilities Test System. AFACTS is a Rolm CBX feature. An automatic testing system for identifying faulty tie and central office trunks. AFACTS can pinpoint faulty trunks and generate exception and summary reports.

AFAIK As Far As I Know.

AFCEA Armed Forces Communications and Electronics Association. An organization of military communications personnel and suppliers who fulfill the specialized needs of government and military communications. They run an big convention each year in Washington in May-June. 703-631-4693.

AFE See ANALOG FRONT END.

AFI An ATM term. Authority and Format Identifier: This identifier is part of the network level address header.

Affiliated Sales Agency ASA. A term for a company which resells the service of a phone company. Typically, the phone company pays the ASA a commission. Sometimes the commission is so large that it blurs the thinking of the ASA into recommending to its customers telecom products and services they would be better without.

AFIPS American Federation of Information Processing Societies. A national, highly-respected organization formed by data processing societies to keep abreast of advances in the field. AFIPS organizes one of the biggest trade shows in the data processing industry — the NCC (National Computer Conference).

AFNOR Acronym for Association Francais Normal. France's national standards-setting organization.

AFP AppleTalk File Protocol. Apple's network protocol, used to provide access between file servers and clients in an AppleShare network. AFP is also used by Novell's products for the Macintosh.

AFT Automatic Fine Tuning; SEE AFC.

After-call Wrap-up The time an employee spends completing a transaction after the call has been disconnected. Sometimes it's a few seconds. Sometimes it can be minutes. Depends on what the caller wants.

AGC Automatic Gain Control. There are two electronic ways you can control the recording of something — Manual or Automatic Gain Control (AGC). AGC is an electronic circuit in tape recorders, speakerphones, and other voice devices which is used to maintain volume. AGC is not always a brilliant idea since it will attempt to produce a constant volume level, that is, it will try to equalize all sounds — the volume of your voice, and, when you stop talking, the circuit static and/or general room noise which you do not want amplified. Never record a seminar or speech using AGC. The recording will be decidedly amateurish. Manual Gain Control means there is record volume control and is thus, preferred in professional applications.

NEWTON'S TELECOM DICTIONARY

AGCOMNET US Department of Agriculture's voice and data communications network.

Aged Packet A data packet which has exceeded its maximum predefined node visit count or time in the network.

Agent 1. This term comes from the huge telephone call-in reservation centers which the airlines, hotels and car rental services run. An agent is the person who answers your call, takes your order or answers your question. Agents are also called Telephone Sales Representatives or Communicators. The term "agent" was first used in the airline business. It came from gate or counter ticket agent.
2. An "Agent" is the person or persons you have legally authorized to order your telephone service and equipment from telephone companies.
3. In the computer programming sense of the word, an agent acts on behalf of another person or thing, with delegated authority. The agent's goals are those of the entity that created it. An agent is an active object with a mission, but agents are abstractions that can be implemented in any way, whereas an object has a formal definition.
Business Week in its February 14, 1994 issue wrote, "It's what computer scientists call an 'agent' — a kind of software program that's powerful and autonomous enough to do what all goods robots should: help the harried humans by carrying out tedious, time-consuming, and complex tasks. Software agents just now emerging from the research labs can scan data banks by the dozen, schedule meetings, tidy up electronic in-boxes, and handle a growing list of clerical jobs."
4. Windows 95 Resource Kit defined agent slightly differently. It said that an agent was software that runs on a client computer for use by administrative software running on a server. Agents are typically used to support administrative actions, such as detecting system information or running services.

Agent Logon/Logoff A call center term. The agent begins their day by punching some buttons on their phone. This indicates to the automatic call distributor that they are now ready to take calls. Later in the day, they punch some other buttons and indicate to the ACD that they are now ready to stop working. This is called logoff.

Agent Sign On/Sign Off A feature which allows any ACD agent to occupy any position in the ACD without losing his or her personal identity. Statistics are collected and consolidated about this agent and calls are routed to this agent no matter where he sits or how many positions he may occupy at one time.

Aggregate Bandwidth The total bandwidth of channel carrying a multiplexed bit stream.

Aggregate Rate The sum of the channel data rates for a given application.

Aggregation 1. An AMA (Automatic Message Accounting) function that accumulates AMA data, resulting in a less than detailed AMA record. Definition from Bellcore.
2. An ATM term. Token A number assigned to an outside link by the border nodes at the ends of the outside link. The same number is associated with all uplinks and induced uplinks associated with the outside link. In the parent and all higher-level peer group, all uplinks with the same aggregation token are aggregated.

Aggregation Device A specialized ISDN terminal adapter that can aggregate, or bond, the two B channels "on the fly" into a single higher-speed connection. some aggregation devices also include an Ethernet bridge, i.e. a connection to a local area network.

Aggregator A new breed of long distance reseller. An aggregator is essentially a sales agent for a long distance company. Here's how it works: The aggregator goes to a long distance company and says "May I sell your long distance service at a discount?" The long distance company says Yes! The aggregator hits the street and sells cut-rate long distance service to any and everyone. The long distance provider installs the service and bills it. The aggregator makes his profit by charging a fixed monthly service fee, a percentage of savings or some other arrangement. The key to it: The end user saves some money because his calls are "aggregated" with those of ALL the customers of the aggregator and the long distance company extends a bulk savings to the aggregator. Here's what TELECONNECT Magazine wrote about aggregators under the headline, "Aggregator Warning."

"Aggregators are companies which buy long distance wholesale and sell it retail. Aggregators exist because AT&T decided it wanted to win back long distance business it had lost. AT&T sliced its rates, liberalized its bulk billing rules and encouraged those consultants who had recommended their clients switch to MCI and US Sprint to become aggregators. These consultant-turned aggregators simply solicit anyone's long distance business and add it to their collection. AT&T sends their end-user a bill and the consultant-turned-aggregator a commission check.

"Should you — as an end-user — consider buying your long distance from an aggregator? The simple answer is YES? AT&T's discounts are so deep it's not uncommon for a company using AT&T today directly to switch to billing through an aggregator and save 20% to 25% — with nothing of substance happening. They still get their bills from AT&T and they still place and receive calls on AT&T as they had been doing. No wires are touched. No routing is changed.

"What about the pitfalls? There are some: First, don't buy long distance that isn't billed directly by the long distance carrier providing the service. If the aggregator does the billing, there's too much opportunity for "mischief," says Dick Kuehn, Cleveland consultant. "There's opportunity for doing things like increasing each of your calls by 30 seconds. And because a user has no answer supervision on his call detail records, it's very hard for the user to figure his exact timing." The problem, says Dick, is there's no way for a user to verify his own bill. Dick says "Carriers are honest. Resellers (aggregators that bill) are open to question."

"TELECONNECT also believes you probably shouldn't deal with an aggregator who bills you a percentage of "savings." This is also open to abuse. There are so many rates, so many changes monthly, so many options that it's virtually impossible for the user to figure out what he would have paid had he not gone with the aggregator. The calculation is too open to abuse. TELECONNECT's feeling: pay a flat service fee.

"P.S. Imagine buying an automobile tire. The local garage carries two options: a Bridgestone for $30 and a Pirelli for $100. You opt for the Bridgestone at $30. You've thus saved $70. The garage proprietor splits your "savings" with you and charges you $35. By late 1992, the panoply of companies in the long distance business — not only aggregators — had expanded dramatically. And confusion between companies and what they did became rife. All, of course, purport to save you money on your long distance bills. And many do. Here's a simple explanation of the major categories:

CARRIER. Owns most of its circuits. Has own sales force and possibly independent sales agents. Best examples: AT&T, MCI, Allnet and Sprint.

NEWTON'S TELECOM DICTIONARY

TRADITIONAL RESELLER. Rents/leases most circuits or buys bulk time from carrier. Resells under own brand name, has published prices, sends own bills. Appears to be (and for all practical purposes is) same as the carriers.
AGGREGATOR. "Sponsor" who buys carrier's (typically AT&T) multi-location 800 or outbound service; enrolls other businesses as sites; volume discounts for all based on total calling at all sites. End user is still the carrier's, not the aggregator's. The carrier typically does the billing.
REBILLER: (Also called "Switchless Reseller"). Buys service as multi-location customer from carrier. Signs up individual sites (just like aggregator). Generates own end-user bills. No switch or network, but does sales, customer service, billing for long distance calls. Sometimes the rebiller's bills are more detailed than the bills you get directly from the carrier.
SALES AGENTS: Businesses or groups who are not direct employees of carrier, but who receive sales commissions from carrier. Customers belong to carrier and carrier does billing.
OTHER THIRD-PARTY MARKETERS. Buying co-ops, user groups, long distance brokers, pyramid (legal) marketing systems, shared tenant providers, Centrex aggregators, affinity groups (like college alumni and church congregation groups).
Aging The change in properties of a material with time under specific conditions.
AGTK Application Generator ToolKit. A set of tools that are used to implement and modify a voice-processing application. It includes software to create the script and packages for the creation and editing of prompts. See APPLICATION GENERATOR.
AHD Audio High Density. System of digital audio recording on grooveless discs, employing an electronically-guided capacitance pickup.
AHFG ATM-attached Host Functional Group: The group of functions performed by an ATM-attached host that is participating in the MPOA service.
AHR Abbreviation for ampere hour, measurement of battery power: how much current may be drawn for an hour. Important specification for portable computers, cellular phones, etc.
AHT Average Handle Time. The amount of time an employee is occupied with an incoming call. This is the sum of talk time and after-call-work time.
AHT Distribution Average Handle Time Distribution. A set of factors (either 48 or 96) for each day of the week that defines the typical distribution of average handle times throughout the day. Each factor measure how far AHT in the half or quarter hour deviates from the AHT for day as a whole.
Ahoy See HELLO.
AI Artificial Intelligence. Perhaps the next phase of computing. The present forms of AI in computer software are called Expert or Knowledge Based systems.
Ai An ATM term. Signaling ID assigned by Exchange A.
AIA 1. An SCSA term. Application Interface Adapter. A software module that converts the function calls issued by a client application, via the API, into standard SCSA messages. See SCSA MESSAGE PROTOCOL.
2. Automatic Internal Administration.
AICC Automatic Incoming Call Connection. A Rolm term for connecting an incoming call to the person's phone, without requiring them to press any keys.
AIDS A Trojan Horse software program (a virus) which caused extensive damage in December 1989.
AIFF Audio Interchange File Format.
AIIM Association for Information and Image Management.
AIM Amplitude Intensity Modulation.
AIMS An Acronym for Auto Indexing Mass Storage. Indicates the AIMS Specification which is a standard card interface for storing large data such as image and multimedia files.
AIMUX ATM Inverse Multiplexing: A device that allows multiple T1 or E1 communications facilities to be combined into a single broadband facility for the transmission of ATM cells.
AIN Advanced Intelligent Network. A term promoted by Bellcore (Bell Communications Research Inc.), adopted by Bellcore's owners, the regional Bell holding companies, and by AT&T and virtually every other phone company to indicate the architecture of their networks for the 1990s and beyond. While every phone company has a different interpretation of what their AIN is, there seems to be two consistent threads. First, the network can affect (i.e. change) the routing of calls within it from moment to moment based on some criteria other than the normal, old-time criteria of simply finding a path through the network for the call. Second, the originator or the ultimate receiver of the call can somehow inject intelligence into the network and affect the flow of his call (either outbound or inbound). The concept of AIN is simple. Before calls are sent to their final destination, the network queries a database. "What should I do at this very moment with this phone call?" Depending on the response, depends the disposition of the call. That database may belong to the phone company. Or it may belong to the customer. It makes no difference, so long as they're connected. And various carriers (phone companies) have proposed and implemented various ways of joining these databases. Initial AIN services tend to be focused on inbound 800 toll-free calls. Although no two phone companies seem to have the same idea as to what an Advanced Intelligent Network is, (some call it just an Intelligent Network), it generally includes three basic elements:
1. Signal Control Points. SCPs. Computers that hold databases in which customer-specific information used by the network to route calls is stored.
2. Signal Switching Points. SSPs. Digital telephone switches, which can talk to SCPs and ask them for customer-specific instructions as to how the call should be completed.
3. Signal Transfer Points. STPs. Packet switches that shuttle messages between SSPs and SCPs.
All three communicate via out of band signaling, typically using Signaling System 7 (SS7) protocol. The AIN has increased in complexity, as carriers have added voice response equipment that can prompt callers to enter further instructions as to how they'd like their call handled. Despite the differences between AIN networks, all work fundamentally the same, according to Mark

Langner of TeleChoice, Verona, NJ: The SS7 identifies that a call requires intelligent network processing. The SSP creates a query to find out how this call should be handled. The query is passed via out-of-band signaling through STPs to an SCP. That interprets the query based on the criteria in its database and information provided by the SSP. Once the SSP has determined how the calls is to be handled, it returns a message through STPs to the SSP. This message instructs the SSP how the call should be handled in the network. According to Langner, the number of actions that could take place at the SCP are truly infinite. The call could be translated into a different number for completion. It could be routed to a user's private network for on-net handling. It could be sent to a voice response unit in the carrier network, where a message is played to the caller. Or it could even be blocked, preventing completion of the call.

Ericsson has done focus groups on Mobile Intelligent Network Services. Among the new IN (Intelligent Network) services, Ericsson identified: * Enhanced number translation services functions * Enhanced screening services, i.e. selective call diversion * Selective forwarding of calls * Location-dependent call forwarding * Improvements to voice announcements * Services to support fixed and mobile integration, i.e. personal communications services, PCS and universal personal telecommunications, UPT, and * Enhanced billing.

See IN, NCD, SCP, SiteRP, SS7, SSP, STP and the AIN definitions below.

AIN Release 0 Advanced Intelligent Network Release defined by individual Bell Operating Companies for initial deployment in 1991, or so. See AIN.

AIN Release 0.0 Advanced Intelligent Network Release based on Ameritech specifications with input from Bellcore and some vendors. Contains three trigger detection points. Deployed in 1992 (US) and end of 1993 (Canada). First service for this architecture was "Switch Redirect" for Bell Atlantic (for switch or line failure.) See AIN.

AIN Release 0.1 Advanced Intelligent Network Release provides for some additional functionality and more extensions to Rel 0.0. Contains 5 trigger detection points. Promised deployment by NT and AT&T by mid 1994. See AIN.

AIN Rel 1.0 Advanced Intelligent Network Release target architecture for AIN. Researched by US West 1991. Contains 32 trigger detection points. Too difficult/complicated for switch vendors to implement right now. (Hence Rel 0.0 & 0.1). See AIN.

AIN Release 1 Logical Resources For Bell Operating Companies, the logical network resources configured and updated to provide Advanced Intelligent Network Release 1 subscriber services (e.g., SLP and trigger data). Definition from Bellcore. See AIN.

AIN Release 1 Switching System An access tandem, local tandem or end office that contains an ASC (Advanced Intelligent Network Release 1 Switch Capabilities) functional group. Definition from Bellcore. See AIN.

AIN Release 2 An Advanced Intelligent Network Release for initial deployment in 1995, involving from AIN Release 1 and supporting an expanded range of information networking services from the Bell operating telephone companies. Definition from Bellcore. See AIN.

AIN Switch Capabilities ASC. A functional group residing in an Advanced Intelligent Network Release 1 Switching System that contains the Network Access, Service Switching, Information Management, Service Assistance and Operations FEs (Functional Entities). Definition from Bellcore. See AIN.

AINTCC Automated INTercept Call Completion. A new feature of Northern Telecom's central offices. The AINTCC feature provides options for

- connecting a caller automatically to an intercepted number after hearing an announcement, or
- connecting a caller to an intercepted number without an announcement.

Not using an announcement makes the number change transparent to the caller. The called (intercepted) party then has the option of informing the caller of the number change.

AIOD Automatic Identification of Outward Dialing is the ability of the telephone system to know the specific extension placing a call. It's used as part of the process of recording the detail of each telephone call for billback and cost control purposes. See AIOD LEADS and CALL ACCOUNTING SYSTEM.

AIOD Leads Terminal equipment leads used solely to transmit automatic identified outward dialing (AIOD) data from a PBX to the public switched telephone network or to switched service networks (e.g., EPSDS), so that a vendor can provide a detailed monthly bill identifying long-distance usage by individual PBX extensions, tie-trunks, or the attendant.

AIR An ATM term. Additive Increase Rate: An ABR service parameter, AIR controls the rate at which the cell transmission rate increases.

Air Conditioning In the Department of Defense, air conditioning is a synonym for the term "environmental control," which is the process of simultaneously controlling the temperature, relative humidity, air cleanliness, and air motion in a space to meet the requirements of the occupants, a process, or equipment.

AIRF Additive Increase Rate Factor: Refer to AIR.

Air Space Coaxial Cable One in which air is the essential dielectric material. A spirally wound synthetic filament of spacer may be used to center the conductor.

Air Time Time spent talking on a cellular network to calculate billing. See also AIRTIME.

Airbrush A computer imaging term. A fine-mist paint tool used to create halos, fog, clouds, and similar effects. Most paint programs let you control the size and shape of the application area. Some packages provide a transparency adjustment that determines the density of the applied paint.

Aircraft Earth Station A mobile Earth station in the aeronautical mobile-satellite service located on board an aircraft.

Airline Mileage The monthly charge for many leased circuits is billed on the basis of "airline mileage" between the two points. Though it sounds as if it's the distance a crow would fly directly between the two points, when in reality, it is the distance in mileage between two Rate Centers whose position is laid down according to industry standards, originally created by AT&T. The entire U.S. is divided by a vertical and horizontal grid. The coordinates — vertical and horizontal — of each rate center are defined and applied to a square root formula which yields the distance between the two points. Think back to school. There's a right-angled triangle. At the top is one Rate Center. At the side is the other Rate Center. The horizontal is the horizontal coordinate. The vertical is the vertical coordinate. The formula is simple: Square the vertical distance. Square the horizontal distance. Add the two together. Then take their square root. That will give you the distance across the hypotenuse — the side opposite the right angle in the triangle. Thus, your "airline" mileage. For sample V and H city coordinates and the formula on how to calculate airline mileage, see V & H under the letter V.

Airline Miles See AIRLINE MILEAGE.

Airtime Actual time spent talking on a cellular telephone. Most cellular carriers bill their customers based on how many minutes of airtime they use each month. Whether the calls are incoming or outgoing makes no difference, the customer is still billed. Whether the calls are going to a toll-number or a toll-free 800 number also makes no difference. The customer racks up airtime and he pays. The more minutes of time spent talking on the phone, the higher the bill. Airtime charges during peak periods of the day in North America vary from 25 to 80 cents per minute. Most carriers offer a discount on these rates for off-peak usage. Some carriers offer a discount on these rates if the customer pays a higher minimum usage charge each month.

AIS 1. Alarm Indication Signal. Formerly called a "Blue Alarm" or "Blue Signal." A signal that replaces the normal traffic signal when a maintenance alarm indication has been activated. An AIS is a signal transmitted downstream informing that an upstream failure has been detected.
In ATM, an alarm indication signal is an all ones signal sent down or up stream by a device when it detects an error condition or receives an error condition or receives an error notification from another unit in the transmission path.
2. Automatic Intercept System.

AIX Advanced Interactive eXecutive: IBM's implementation of UNIX. The Open Software Foundation (OSF) based its first operating system (OSF-1) on AIX. The next revision of the OSF operation system (OSF-2) will also be based on AIX with a Mach kernel (Mach was developed by Carnegie Mellon University).

AIW Application Implementer's Workshop. A group of vendors working with IBM to develop software and hardware consistent IBM's Advanced Peer-to-Peer Networking protocol.

AKO Bisync acknowledgment for even-numbered message.

Aka Denwa Japanese for a red telephone. Some coin phones in Japan are red and are often known as aka denwas.

ALAP AppleTalk Link Access Protocol. In an AppleTalk network, this link access-layer (or data link-layer) protocol governs packet transmission on LocalTalk.

Alarm Display Attendant console indicators show the status (i.e. what's happening) in the telephone system. There are usually two types of displays — minor and major. Minor displays may be something as "minor" as a "hung" trunk, i.e. one that didn't hang up when the person speaking on it hung up. They can often be remedied by turning the PBX off, counting to ten, and then turning it on. (Before you do, check it will load itself.) Major problems — such as a blown line card in the PBX, one console out or half the trunks out — often require a service call and are often covered under the Emergency Conditions section of telephone service contracts.

Alarm Indication Signal AIS. A signal that replaces the normal traffic signal when a maintenance alarm indication has been activated.

ALBO Automatic Line BuildOut. ALBO is a means of automatic cable equalization used in T-1 span-line interface equipment.

ALC 1. Automatic Level Control.
2. See AUTOMATIC LIGHT CONTROL.

ALE 1. Approvals Liaison Engineer. This engineer acts on your behalf to asses design and component changes to your BABT approved products. BABT is the British Approvals Board for Telecommunications.
2. See AUTOMATIC LINK ESTABLISHMENT

Alerter In an AT&T ISDN phone, the alerter takes the place of a mechanical ringer. The alerter has eight ringing signals, eight personalized ringing patterns, and the ability to play music.

Alerter Service A Window NT term. Notifies selected users and computers of administrative alerts that occur on a computer. Used by the Server and other services. Requires the Messenger service.

Alerting A signal sent to a customer, PBX or switching system to indicate an incoming call. A common form is the signal that rings a bell in the telephone set. Others signals can trigger such devices as whistles, gongs and chimes.

Alerting Signal A ringing signal put on subscriber access lines to indicate an incoming call. Bellcore defines an alerting signals more broadly; thus: "Alerting signals (for example, ringing, receiver off-hook) are transmitted over the loop to notify the customer of some activity on the line."

ALEX Software which provides Internet users with a transparent read capability of remote files at anonymous FTP sites.

Algorithm A prescribed finite set of well defined rules or processes for the solution of a problem in a finite number of steps. In normal English, it is the mathematical formula for an operation, such as computing the check digits on packets of data that travel via packet switched networks.

ALI Automatic Location Information. ALI is a feature of E-911 (Enhanced 911) systems. ALI is provided to agents answering E-911 calls. It may include information such as name, phone number, address, nearest cross street and special pre-existing conditions (i.e. hazardous materials). On some systems it may also provide the appropriate emergency service address for the particular address. ALI is retrieved from a computer database. The database may be held on site or at a remote location and may be maintained by the local phone company (or its parent) or another agency.

Alias 1. A feature of the Apple Macintosh System 7 allowing the user to create a file that points to the original file. When you click on an alias, the original application is launched. Aliases can work across a network; so you can access a program residing on a file server or a Mac that runs System 7 file sharing.
2. An assumed name under which users of an electronic bulletin board may post messages. For example, Jane Smith may post as "Marketing Group." The system usually provides a list of aliases and the names of the users to which they belong. Some BBS packages allow anonymous message posting.

3. Unwanted signals generated during the A-to-D (Analog to Digital) conversion process. This is typically caused by a sampling rate that is too low to faithfully represent the original analog signal in digital form. Typically, a rate that is less than half the highest frequency to be sampled.
4. A nickname for a domain or host computer on the Internet.

Aliasing Distortion in a video signal. It shows up in different ways depending on the type of aliasing in question. When the sampling rate interferes with the frequency of program material the aliasing takes the form of artifact frequencies that are known as sidebands. Spectral aliasing is caused by interference between two frequencies such the luminance and chrominance signals. It appears as herringbone patterns, wavy lines where straight lines should be and lack of color fidelity. Temporal aliasing is caused when information is lost between line or field scans. It appears when a video camera is focused on a CRT and the lack of scanning synchronization produces a very annoying flickering on the screen of the receiving device.

Aliasing Noise A distortion component that is created when frequencies present in a sampled signal are greater that one-half the sample rate. See ANTI-ALIASING FILTER.

Aligned Bundle A bundle of optical fibers in which the relative spatial coordinates of each fiber are the same at the two ends of the bundle. Also called "Coherent Bundle."

ALIT Automatic Line Insulation Testing.

ALJ An Administrative Law Judge appointed by a State Commission to review a Commission docket, such as a rate case or incentive regulation proposal, and to make recommendations to the Commissioners.

All Call Paging With this feature, a user can broadcast an announcement — a page — to someone through the speakers of all the telephones on the system and, possibly, any external loudspeakers. If you want instant fame, ask your secretary to call all the airports in the country and page you. Mike Todd, the movie mogul, used to have this secretary perform this wonderful task. Mr. Todd gave gigantic egos a whole new meaning.

All Channel Tuning Ability of a television set to receive all assigned channels. VHF and UHF, channels 2 through 83.

All Dielectric Cable Cable made entirely of dielectric (insulating) materials without any metal conductors.

All Number Calling Once upon a time, the first two digits of telephone exchanges sort of corresponded to their location. For example, MU-8 meant Murray Hill 8 in Murray Hill, Manhattan, New York City. Then the phone company started running out of letters, so it went to All Number Calling. The All Number Calling provides a theoretical maximum of 792 central office exchange (NNX) codes per area code (NPA). This is derived on the basis of 800 NXX code combinations (8x10x10) leaving out eight special service combinations, including 411, 611, 911.

All Trunks Busy When a user tries to make an outside call through a telephone system and receives a "fast" busy signal (twice as many signals as a normal busy in the same amount of time), he is usually experiencing the joy of All Trunks Busy. No trunks are available to handle that call. The trunks are all being used at that time for other calls or are out of service. These days, many long distance companies are replacing a "fast" busy signal with a recording that might say something like, "I'm sorry. All circuits are busy. Please try your call later."

Allen, Robert Chairman of the board of AT&T since 1988. In 1980, he headed a task force to look into AT&T's future. The recommendation of the task force: keep equipment manufacturing at all costs.

Alliance Teleconferencing Service A direct-dial AT&T service which allows you to teleconference as few as three and as many as 59 telephones around the world. You can dial the numbers to call or you can schedule a meeting, called "Meet-Me Service." If you want to set the conference up, you dial 0-700-456-1000, then punch in the number of locations, then put in the numbers, ending each with a #. End your dialing with another # — to add yourself to the teleconference.

Alloy A combination of two or more metals that forms a new or different metal with specific or desirable qualities.

ALM 1. AppWare Loadable Module. A visual computer telephony applications generator that works on Novell's NetWare. An ALM works by tying into Novell NetWare's NLMs. See APPWARE.
2. Automated Loan Machine. Like an ATM (Automated Teller Machine), an ALM sits in the wall of a building or inside a building on a wall. However, instead of giving money you own, an ALM dispenses money in the form of an instant loan. One of the leading ALM manufacturing companies is Affinity Technologies of Columbia, South Carolina. Alan Fishman of Columbia Financial Partners contributed this definition. Mr. Fishman is a leading New York City venture capitalist, who helped Affinity get started. Mr. Fishman's company is Columbia Financial Partners.
3. Airline Miles. The method used to calculate the distance (for pricing purposes) of the point-to-point long distance lines in long distance telephone networks. See AIRLINE MILEAGE.

Aloha A method of data transmission in which the device transmits whenever it wants to. If it gets an acknowledgement from the device it's trying to reach, it continues to transmit. If not (as in the case of a collision with someone else trying to transmit simultaneously), it starts all over again. The ALOHA method get its name from a dying satellite that was donated to university researchers in the Pacific. It was used to transmit data by satellites among South Sea islands, especially Hawaii. The ALOHA "method" — called "transmit at will" — was invented because the users were short of funds to develop more sophisticated data transmission protocols, and they had a free satellite, which typically had more bandwidth than they had stuff to send. See ALOHANET.

Alohanet An experimental form of frequency modulation radio network developed by the University of Hawaii. Alohanet is implemented by creating transmission frames containing data, control information, and source and destination addresses which are broadcast for reception by the destination receiver and ignored by all others. Actually, Alohanet is an early version of Ethernet, the local area network technique. See ALOHA.

Alpeth Aluminum-polyethylene primary covering known as the sheath for aerial cable.

Alpha 1. Only alphabetic characters.
2. A family of microprocessors from Digital Equipment Corporation first introduced in February 1992. The first chip introduced was the 21064, described by DEC as a .75 micron CMOS-based super-scalar, super-pipelined processor using dual instruction issue and a 150 MHz cycle time. By relying on its 64-bit RISC architecture, the Alpha can deliver 400 million instructions per second.

3. The first version of hardware or software. It typically has so many bugs you only let your employees play with it. A beta is the next version. It's a pre-release version and selected customers (and the press) become your guinea pigs. After beta, and when the bugs are removed, comes "general availability." That's when the product is finally available for buying by the general public.

Alpha Channel The upper 8 bits of the 32-bit data path in some 24-bit graphics adapters. The alpha channel is used by some software for controlling the color information in the lower 24 bits.

Alpha Geek The most knowledgeable, technically proficient person in an office or work group. "Ask Harry, he's the alpha geek around here."

Alpha Test Testing of product prototypes performed by the vendor. When testing is done by potential users, it's called a beta test.

Alphabetic Only alphabetic characters. See also ALPHANUMERIC.

Alphanumeric A set of characters that contains both letters and numbers — either individually or in combination. Numeric is 12345. Alphabetic is ABCDEF: Alphanumeric is 1A4F6HH8. American and Australian zip codes are numeric. Canadian and English postal codes are alphanumeric. No one knows why.

Alphanumeric Display A display on a phone or console showing calling phone number, called number, trunk number, type of call, class of service and perhaps, some other characteristics of the call. It may also contain instructions as to how to move the call around, set up a conference call, etc. The display may be liquid crystal or light emitting diode. Typically, it's liquid crystal.

Alphanumeric Memory A cellular radio feature that allows you to store names with auto-dial phone numbers.

Altair Ethernet Motorola's name for its wireless local area network, which transmits at the very high frequency of 18 to 18 megahertz. Altair users need to fill out a small, one-page FCC application in order to use the system.

ALTEL Association of Long distance TELephone companies. A trade association composed of alternative (to AT&T) long distance carriers and resellers of long distance services.

Alternate Answering Position Usually refers to a second receptionist's desk which has a telephone switchboard or console functioning like the main one. Also refers to when the main receptionist is away from his/her desk, or is very busy taking calls, the telephone system automatically sends the calls to another console or to a phone that will be answered.

Alternate Buffer In a data communications device, the section of memory set aside for the transmission or receipt of data after the primary buffer is full. This helps the device control the flow of data so transmission is not interrupted due to lack of space for the incoming or outgoing data.

Alternate Lock Code A three-digit lock code to be used with the partial lock feature in some cellular phones.

Alternate Recipient An electronic messaging term. In X.400 terms, a user or distribution list that a recipient MTA (Message Transfer Agent) delivers a message to (if allowed) when the message cannot be sent to the preferred recipient.

Alternate Route A second or subsequent choice path between two exchanges, usually consisting of two or more circuit groups in tandem. Sometimes called "alternative route" or "second-choice route."

Alternate Routing A mechanism that supports the use of a new path after an attempt to set up a connection along a previously selected path fails. It's a feature used in network design and also in PBXs. For example, with PBXs it's a feature used with long distance calls that permits the telephone system (typically a PBX) to send calls over different (alternate) phone lines. It might do this because of congestion of the primary phone lines the calls would normally be sent over. Alternate routing is often confused with LEAST COST ROUTING, in which the telephone system chooses the least expensive way (available at that time) to route that call. Least Cost Routing typically works with so-called "look-up" tables in the memory of the PBX. These tables are put into the PBX by the user. The PBX does not automatically know how to route each call. It must be told by the user. That "telling" might be as simple as saying "all 312 area codes will go via the AT&T FX line." Or it might be as complex as actually listing which exchanges in the 312 area code go by which method. LEAST COST ROUTING tells the calls to go over the lines which are perceived by the user to be the least cost way of getting the call from point A to point B. Alternate routing happens when the least cost routes get congested and alternate routes (typically more expensive) are found from the look-up tables in the PBX's memory.

Alternate Use The ability to switch communications facilities from one type of service to another, i.e., voice to data, etc.

Alternate Voice Data AVD. An older service which is a single transmission facility which can be used for either voice or data (up to 9600 bps). Arrangement includes a manually operated switch (on each end) to allow customers to alternately connect the line to their modem or PBX.

Alternating Current See AC.

Alternative Non-Traffic Sensitive Cost-Recovery Plans New charges proposed by the regional Bell holding companies to supplement subscriber line charges. In short, another charge on the subscriber with an interesting, though dubious, justification. They have not been fully implemented.

Alternative Regulatory Framework ARF.

Alternator A machine which generates electricity which is alternating current. See AC.

ALTS Alternative Access Providers to the local telephone network i.e., Teleport.

ALU Arithmetic Logic Unit. The part of the CPU (Central Processing Unit) that performs the arithmetic and logical operations. See MICROPROCESSOR.

Alvyn Aluminum-polyethylene, the sheath used for riser cable where a flame retardant sheath is required.

AM See Amplitude Modulation.

AMA Automatic Message Accounting. AMA is the phone company's name for Call Detail Recording or Station Message Detail Recording (SMDR). See AMA TAPE.

AMA Tape A telephone company machine-readable magnetic tape which contains the customer's long distance calling and billing data for a given month.

AMA Teleprocessing System AMATPS. The primary method for delivery of AMA data from the network to billing systems. The current AMATPS architecture consists of an AMA Transmitter (AMAT) and a collector. Definition from Bellcore.

AMATPS AMA TeleProcessing System.
Ambient Lighting The general level of illumination throughout a room or area.
Ambient Noise The level of noise present all the time. There is always noise, unless you're in an anechoic chamber. When measured with a sound level meter, it is usually measured in decibels above a reference pressure level of 0.00002 pascal in SI units, or 0.00002 dyne per square centimeter in cgs units.
AMD Air Moving Device. IBM-speak for a fan.
America Online An on-line computer service you call with your PC or Macintosh which provides e-mail, forums, software downloads, news, weather, sports, financial information, conferences, on-line gaming, an encyclopedia, and other features, to its subscribers.
American Bell, Inc. The old name for the unregulated telephone equipment supply subsidiary of American Telephone & Telegraph. American Bell Inc. had its name changed to AT&T Information Systems. At that point it was no longer separately incorporated. It became a division of AT&T. It has been reorganized many times. When it was American Bell, it was only selling telecommunications products and services to end users. When it become AT&T Information Systems, it sold AT&T phone systems and AT&T computer systems. It then merged with AT&T Long Lines, which was then called AT&T Communications and is now called simply AT&T. By the time you read this, it will probably be called something else. Sadly, old gadgetry, knick-knacks and momentos bearing the name American Bell, Inc. have no marketable value as antiques or examples of American folk art. See also LUCENT TECHNOLOGIES.
American National Standards Institute See ANSI.
American Standard Code For Information Interchange ASCII. The standard 7-bit code for transferring information asynchronously on local and long distance telecommunications lines. The ASCII code enables you to represent 128 separate numbers, letters, and control characters. By using an eighth bit — as in extended ASCII or IBM's EBCDIC — you can represent 256 different characters. ASCII often uses an eighth bit as a parity check or a way of encoding word processing symbols, not as a way of broadening the number of characters and symbols which it can represent. See also ASCII.
American Wire Gauge AWG. Standard measuring gauge for non-ferrous conductors (i.e. non-iron and non-steel). AWG covers copper, aluminum, and other conductors. Gauge is a measure of the diameter of the conductor. See AWG for a bigger explanation.
Ameritech Ameritech Corp is one of the Regional Bell operating companies formed as a result of the AT&T Divestiture. Ameritech covers five states and includes the operating telephone companies: Illinois Bell, Indiana Bell, Michigan Bell, Ohio Bell, and Wisconsin Bell. It also includes some other subsidiaries, which fit into two classifications — administrative (centralized buying, real estate, etc.) or entrepreneurial (cellular mobile radio, venture capital, etc.).
AMI Alternate Mark Inversion. The line-coding format in T-1 transmission systems whereby successive ones (marks) are alternately inverted (sent with polarity opposite that of the preceding mark). Here's AT&T definition: A line code that employs a ternary signal to convey binary digits, in which successive binary ones are represented by signal elements that are normally of alternating, positive and negative polarity but equal in amplitude, and in which binary zeros are represented by signal elements that have zero amplitude. This is an AT&T definition.
AMIS See AUDIO MESSAGING INTERCHANGE SPECIFICATION. A standard for networking voice mail systems.
AML Anolog Microwave Link
Ampacity The maximum current an insulated wire or cable can safely carry without excluding either the insulation or jacket materials limitations.
Amperage Rating The amperage which may be safely applied to a circuit, service or equipment. See also AMPERE.
Ampere The unit of measurement of electric current or the flow of electrons. One volt of potential across a one ohm impedance causes a current flow of one ampere. AMP is the abbreviation for ampere. It is mathematically equal to watts divided by volts. Note that in the electrical context, WATTS is spelled with two "Ts." In telecommunications, WATS, meaning Wide Area Telecommunications Service, is spelled with only one "T."
Ampere-hour Unit Measurement of battery capacity, determined by multiplying the current delivered by the time it is delivered for. See AMPERE.
Amphenol Connector Amphenol is a manufacturer of electrical and electronic connectors. They make many different models, many of which are compatible with products made by other companies. Their most famous connector is the 25-pair connector used on 1A2 key telephones and for connecting cables to many electronic key systems and PBXs. The telephone companies call the 25-pair Amphenol connector used as a demarcation point the RJ-21X. The RJ-21X connector is made by other companies including 3M, AMP and TRW. People in the phone business often call non-Amphenol-made 25-pair connectors, amphenol connectors.
Amplified Handset An amplified handset is the best phone gadget you can buy. You use it to crank up the volume of incoming calls (and in some cases the volume of outgoing calls) and save yourself enormous amounts of money on callbacks. "We have a bad line. I'll call you back." There are three types of amplified handsets: 1. The handset with a built-in amplifier. These devices suck their power from the phone line and since the phone line doesn't have much power, you won't have much amplification. I'm not overly impressed with amplified handsets. 2. The handset with amplifying circuits built into the phone. Ditto for our comments about power. 3. The handset with the little external box amplifier which is powered by either AC or by several batteries, typically AA alkalines. Such an external amplifier will produce much greater amplification. This is the type we prefer at TELECONNECT Magazine.
Amplifier When telephone conversations travel through a medium, such as a copper wire, they encounter resistance and thus become weaker and more difficult to hear. An amplifier is an electrical device which strengthens the signal. Unfortunately, amplifiers in analog circuits also strengthen noise and other extraneous garbage on the line. Amplifiers are used in all telephone systems, analog and digital. But in digital systems, signals are regenerated and then amplified. As a result, noise is much less prevalent and less likely to be amplified in digital systems.

NEWTON'S TELECOM DICTIONARY

Amplitude The distance between high or low points of a waveform or signal. Also referred to as the wave "height." See AMPLITUDE MODULATION.

Amplitude Distortion The difference between the output wave shape and the input wave shape.

Amplitude Equalizer A corrective network that is designed to modify the amplitude characteristics of a circuit or system over a desired frequency range. Such devices may be fixed, manually adjustable, or automatic.

Amplitude Modulation Also called AM, it's a method of adding information to an electronic signal in which the signal is varied by its height to impose information on it. "Modulation" is the term given to imposing information on an electrical signal. The information being carried causes the amplitude (height of the sine wave) to vary. In the case of LANs, the change in the signal is registered by the receiving device as a 1 or a 0. A combination of these conveys different information, such as words, numbers or punctuation marks. The principal forms of Amplitude Modulation are
QDM: Double-band Amplitude Modulation QAM: Quadrature Amplitude Modulation SSB: Single-sideband Modulation VSB: Vestigial Sideband Modulation

AMPS Advanced Mobile Phone Service. It's another word for analog cellular radio. Advanced Mobile Phone Service is the term used by AT&T's Bell Laboratories (prior to the break-up in 1984) to refer to its cellular technology. The AMPS standard has been the foundation for the industry in the United States, although it has been slightly modified in recent years. "AMPS-compatible" means equipment designed to work with most analog cellular telephones. Digital cellular is called D-AMPS.

AMS 1. Account Management System.
2. Attendant Management System. An NEC term. With the NEAX2400 IMS, the AMS is an on-screen, dynamic Open Applications Interface (OAI) computer application that emulates and enhances attendant console capabilities. A typical AMS workstation combines the NEC HDAC console and headset with a color display and standard keyboard. The computer is equipped with the UNIX System V operating system, the NEC Applications Manager support platform, and a comprehensive package of software components. Communication between the HDAC and the AMS computer software is made possible by the OAI connection between the computer and the NEAX2400 IMS (ICS).

ANA Assigned Night Answer.

Analog Comes from the word "analogous," which means "similar to." In telephone transmission, the signal being transmitted — voice, video, or image — is "analogous" to the original signal. In other words, if you speak into a microphone and see your voice on an oscilloscope and you take the same voice as it is transmitted on the phone line and ran that signal into the oscilloscope, the two signals would look essentially the same. The only difference is that the electrically transmitted signal (the one over the phone line) is at a higher frequency. In correct English usage, "analog" is meaningless as a word by itself. But in telecommunications, analog means telephone transmission and/or switching which is not digital. Outside the telecom industry, analog is often called linear and covers the physical world of time, temperature, pressure, sound, which are represented by time-variant electrical characteristics, such as frequency and voltage. See ANALOG TRANSMISSION.

Analog Bridge A circuit which allows a normal two-person voice conversation to be extended to include a third person without degrading the quality of the call.

Analog Cellular The current standard for cellular communications.

Analog Channel A channel which transmits in analog waveforms. See ANALOG.

Analog Computer A computer that performs its tasks by measuring continuous physical variables — pressure, voltage, flow — and manipulating these variables to produce a solution, which is then converted into a numerical equivalent. Analog computers are largely used as special purpose machines in scientific or technical applications. The earliest analog computers were purely mechanical devices with levers, cogs, cams, etc., representing the data or operator values. Modern analog computers typically employ electrical parameters such as voltage, resistance, or current to represent the quantities being manipulated.

Analog Digital Converter An A/D Converter. Pronounced: "A to D Converter." A device which converts an analog signal to a digital signal.

Analog Facsimile Facsimile which can transmit and receive grey shadings — not just black and white. It is called analog because of its ability to transmit what appear to be continuous shades of grey. "Analog" facsimile is usually transmitted digitally.

Analog Front End The part of the fax machine that converts between the digitally modulated signal and the analog signal used on the telephone line.

Analog Loop-back A method of testing modems and data terminals by disconnecting the device from the telephone line and looping a signal out through the device's transmit side and in through its receive side. The test tells if the trouble is with the telephone line or with the modem.

Analog Monitor A computer screen that uses an analog signal, a smoothly varying value of current or voltage that varies continuously. VGA, SVGA and Macintosh models are examples of analog monitors. Most computer screens are analog. Most analog monitors are designed to accept input signals at a precise frequency. Higher frequencies are needed to carry higher-resolution images to the monitor. For this reason, multiscanning monitors have been developed that automatically adjust themselves to the incoming frequency. See also ANALOG and DIGITAL MONITOR.

Analog Recording System of recording in which music is converted into electrical impulses that form "patterns" in the grooves of phonograph record masters or in the oxide particles of master tapes representing (or analogous to) musical waveforms.

Analog Switch Telephone switching equipment that switches signals without changing the analog form. The major form of analog switching is circuit switching.

Analog Synchronization A synchronization control system in which the relationship between the actual phase error between clocks and the error signal device is a continuous function over a given range.

Analog Terminal Adapter ATA. A Northern Telecom telephone device permits analog devices (for example, faxes, answering machines, and single line telephones) to connect to the Northern Telecom Norstar telephone system.

Analog Transmission A way of sending signals — voice, video, data — in which the transmitted signal is analogous to the original signal. In other words, if you spoke into a microphone and saw your voice on an oscilloscope and you took the same voice as it was transmitted on the phone line and threw that signal onto the oscilloscope, the two signals would look essentially the same. The only difference would be that the electrically transmitted signal would be at a higher frequency.

Analog Video Signals represented by an infinite number of smooth transitions between video levels. TV signals are analog. By contrast, a digital video signal assigns a finite set of levels. Because computer signals are digital, analog video must be converted into a digital form before it can be shown on a computer screen.

Analog Wireless The dominant radio transmission standard in the United States; also called AMPS.

Analogue An English/European way of spelling ANALOG, which is the correct North American spelling. See ANALOG.

Anamorphic Unequally scaled in vertical and horizontal dimensions.

ANC All Number Calling.

Ancestor Node An ATM term. A logical group node that has a direct parent relationship to a given node (i.e., it is the parent of that node, or the parent's parent.)

AND Automatic Network Dialing.

Anechoic Chamber A perfectly quiet room. A room in which sound or radio waves do not reflect off the walls. An anechoic chamber is the only place in which a speakerphone will work perfectly. The more a room resembles an anechoic chamber — i.e. lots of drapes, plush carpet, etc. — the better a speakerphone will work.

Angle Bracket The term for these two brackets < and >. These two brackets have major use in the HTML language. See HTML.

Angle Modulation Modulation in which phase angle or frequency of a sine wave carrier is varied.

Angle Of Deviation In fiber optics, the net angular deflection experienced by a light ray after one or more refractions or reflections. The term is generally used in reference to prisms, assuming air interfaces. The angle of deviation is then the angle between the incident ray and the emergent ray.

Angle of Incidence The angle between an incident ray and the normal to a reflecting or refracting surface.

Angled End An optical fiber whose end is deliberately polished at an angle to reduce reflections.

Angstrom A unit of length in optical measurement.

Angular Misalignment Loss The optical power loss caused by angular deviation from the optimum alignment of source to optical fiber — fiber-to-fiber, or fiber-to-detector.

ANI Automatic Number Identification. A phone call arrives at your home or office. Somewhere in that phone call is a series of digits which tell you the phone number of the phone calling you. These digits may arrive in analog or digital form. They may arrive as touchtone digits inside the phone call or in a digital form on the same circuit or on a separate circuit. You will need some equipment to decipher the digits AND to do "something" with them. That "something" might be throwing them into a database and bringing your customer's record up on a screen in front of your telephone agent as he answers the call. "Good morning, Mr. Smith." Some large users say they could save as much as 30 seconds on the average IN-WATS call if they knew the phone number of the person calling them. They would avoid asking regular customers for routine identification information since it would all be there in the database. ANI is touted as one of ISDN's most compelling advantages — but it is really an advantage of Signaling System 7 (and therefore distinct from ISDN) and you don't need ISDN to get ANI. In the US, there are various types of "ANI." There's the ANI you get from a long distance phone company, which may arrive over the D channel of an ISDN PRI circuit or on a dedicated single line before the first ring. In contrast, the signaling for Caller ID, as delivered by a local phone company, is delivered between the first and second rings. In Canada, caller ID for both local and long distance is delivered in the same technical way — between the first and second rings. In the US, there are no accepted standards, as yet. In November of 1995, local phone companies in the US are scheduled to deliver both local and long distance ANI exactly as they do today in Canada — between the first and second rings. Thus normal dial-up users, who subscribe to caller ID (and usually pay a few extra dollars a month for the privilege) will be able to figure who's calling before, or as they pick up the incoming call. At one stage, ANI was not available in many states. But those restrictions are disappearing. There are some people who believe ANI is long distance and delivered by long distance phone companies; and Caller ID is local and delivered by local phone companies. And these same people believe the technologies of delivery are different. In reality, ANI and Caller ID are rapidly becoming synonymous. See CALLER ID, CLASS, COMMON CHANNEL SIGNALING, DNIS, ISDN and ISUP.

Anima Someone who communicates with you telepathically.

Animation The process of displaying a sequential series of still images to achieve a motion effect.

Anisochronous Pertaining to transmission in which the time interval separating any two significant instants in sequential signals is not necessarily related to the time interval separating any other two significant instants. Isochronous and anisochronous are characteristics, while synchronous and asynchronous are relationships.

Anisotropic Pertaining to a material whose electrical or optical properties vary with the direction of propagation or with different polarizations of a traveling wave.

ANM ANswer Message. The fourth of the ISUP call set-up messages. A message sent in the backward direction indicating that the call has been answered. See ISUP and COMMON CHANNEL SIGNALING.

Anneal The act of using heat to soften a metal such as copper, making it less brittle.

Annex A The first of the frame relay standard extensions, Annex A outlines provisions for a Local Management Interface (LMI) between customer premises equipment and the frame relay network for the purpose of querying network status information

Annex D The second frame relay standard extension dealing with the communication and signaling between customer premises equipment and frame relay network equipment for the purpose of querying network status information.

Announcement Service Allows a phone user to hear a recording when he dials a certain phone number or extension. These days, announcement services are provided increasingly by totally solid-state digital announcers. These gadgets are more reliable, deliver a clearer message and last much longer than analog tape-based machines (like answering machines), which use recording tape.

NEWTON'S TELECOM DICTIONARY

Announcement System A arrangement for providing information by means of recorded announcements.

Annoyance Call Bureau The department in your local phone company which you call when you need help with annoying or harassing phone calls you are receiving. The Bureau will recommend you file a report with your local phone company. And then it may apply TRAP and TRACE equipment and techniques to try to locate the source of your annoying phone calls. The Annoyance Call Bureau is the stepchild of the phone industry, which means it is typically underfunded. See TRAP and TRACE.

Annual Percentage Rate APR. A percentage calculation of the finance charge portion of financing contact.

Annular Ring An indicator (or ring) around the circumference of the coaxial cable every so many feet — often 2.5 meters (8.2 feet) — to indicate a point where transceivers are to be connected. Same as transceiver attachment mark.

Annunciator Original name for the indicator on magnetic switchboards which indicates the particular line that is calling the exchange. Now it is simply a light, a bell or a device that tells you something. That something might be the ringing of a phone or it might be a problem that you're having with some piece of remote equipment. A communicating annunciator is a sophisticated device that is connected to a phone line and gets on that line (dial-up or leased) to let you know that something is broken.

Anonymous Call Rejection ACR. A service some local phone companies are providing their subscribers. It allows subscribers to automatically stop certain calls from ringing their phone. The calls stopped are "restricted," namely they would be displayed as "P" or "Private" on a subscriber's Caller ID device, meaning that the calling person did not send you his calling number. The person who makes such a call would hear, "We're sorry. The party you have reached is not accepting private calls. To make your call, hang up, dial *82 or 1182 on a rotary phone and re-dial." The caller will be able to reach you only by re-dialing without restricting display of his or her number.

Anonymous FTP A way of logging in anonymously to distant hosts on the Internet and often freeware (free software) from the Internet. With an implementation of the FTP protocol, users can get public domain software from Internet sites, using the word "anonymous" for a login ID, and their userid@hostname.domain as the password. A database called Archie contains a list of what is available from anonymous FTP sites, and can be reached at "archie.mcgill.ca" and at "archie.sura.net." See FTP.

Anonymous Telephone Number A telephone number that should not be displayed or voiced back to the called party. Such a designation is stored in switch memory and is included in signaling information sent to the terminating switch for interSPCS calls.

ANS Answer. (What else?)

ANSI American National Standards Institute. A standards-setting, non-government organization, which develops and publishes standards for "voluntary" use in the United States. The British have their own equivalent of ANSI. It's called the BSI, British Standards Institute. Standards set by national organizations are accepted by vendors in that country. ANSI is located at 1430 Broadway, New York NY 10018 212-642-4900. They put out a biweekly newsletter called "ANSI Standards in Action." See also CCITT, ECMA, IEEE, and ISO. See ANSI CHARACTER SET. www.ansi.org

In a press release, ANSI described itself as "a private non-profit membership organization that coordinates the U.S. voluntary standards system, bringing together interests from the private and public sectors to develop voluntary standards for a wide array of U.S. industries. ANSI is the official U.S. member body to the world's leading standards bodies — the International Organization for Standardization and the International Electronic Commission via the U.S. National Committee. The Institute's membership includes approximately 1,300 national and international companies, 30 government agencies, 20 institutions and 250 professional, technical, trade, labor and consumer organizations."

ANSI T1.110-1987 Signaling system number 7 (SS7) - General Information.

ANSI T1.111-1988 Signaling system number 7 (SS7) - Message Transfer Part (MTP)

ANSI T1.112-1988 Signaling System number 7 (SS7) - Signaling Connection Control Part (SCCP)

ANSI T1.113-1988 Signaling System 7 (SS7) - Integrated Services Digital Network (ISDN) user part.

ANSI T1.114-1988 Signaling System 7 (SS7) - Transaction Capability Application Part (TCAP).

ANSI T1.206 Digital Exchanges and PBXs - Digital circuit loopback test lines.

ANSI T1.301 ANSI ADPCM standard.

ANSI T1.401-1988 Interface between carriers and customer installations - Analog voice grade switched access lines using loop-start and ground-start signaling.

ANSI T1.501-1988 Network performance - Tandem encoding limits for 32 Kbit/s Adaptive Differential Pulse-Code Modulation (ADPCM).

ANSI T1.601-1988 Integrated Services Digital Network (ISDN) - Basic access interface for use on metallic loops for application on the network side of the NT (Layer 1 specification).

ANSI T1.605-1989 Integrated Services Digital Network (ISDN) - Basic access interface for S and T reference points (Layer 1 specification).

ANSI T1.Q1 ANSI's standard for telecommunications network performance standards, switched exchange access network transmission performance standard exchange carrier-to-interexchange carrier standards.

ANSI T1X9.4 ANSI's Sonet standard.

ANSI X3T9.5 A committee sponsored by the American National Standards Institute (ANSI) that is responsible for a variety of system interconnection standards. The committee has produced draft standards for high-speed coaxial cable bus and fiber optic ring local networks.

ANSI X3T9.5 TPDDI Twisted-Pair Distributed Data Interface (TPDDI) is a new technology that allows users to run the FDDI standard 100 Mbps transmission speed over twisted-pair wiring. Unshielded twisted-pair has been tested for distances over 50 meters (164 feet). TPDDI is designed to help users make an earlier transition to 100 Mbps at the workstation.

ANSI Character Set The American National Standards Institute 8-bit character set. It contains 256 characters.

ANSI.SYS A device driver is used by MS-DOS to control output to the screen. Some (very few) MS-DOS programs require you to have the statement DEVICE=ANSI.SYS in your CONFIG.SYS file. It provides the "buttons" that these programs then push to make certain things happen on the screen. Windows has made ANSI.SYS obsolete.

ANSI.SYS Terminal Emulation ANSI.SYS is a popular terminal emulation used on many bulletin board systems. ANSI X3.64 control sequences are recognized to provide visual attributes including color display and full support of the IBM PC graphics character set. Some editors, chiefly those in communications packages such as Smartcom, are configured especially to create ANSI.SYS documents.

Answer Back A signal or tone sent by a receiving equipment or data set to the sending station to indicate that it is ready to accept transmission. Or a signal or tone sent to acknowledge receipt of a transmission. See ANSWER SUPERVISION.

Answer Back Supervision Another word for answer supervision. See ANSWER SUPERVISION.

Answer Call The name of a Bell Atlantic service. Here are Bell Atlantic's words. Answer Call is an answering machine without the machine. This automated messaging service answers your calls right through your touch-tone phone — 24 hours a day — even when you're on the phone. And since it's on Bell Atlantic network, there's no equipment to buy...nothing to turn to...no wires to connect... and no maintenance. By simply dialing a private passcode, you can listen to your messages, replay them or even change your greeting. What's more, Answer Call gives you the option of providing your employees (who share one line) with up to eight private "mailboxes" to receive and retrieve their own messages.

Answer Detect The use of a digital signal processing technique to determine the presence of voice energy on a telephone line. It is used with call (answer) supervision, to identify an answered line. It's beginning to be used with computerized dialing equipment as it eliminates the need for a telephone representative to constantly monitor call set-up progress on each telephone line in the event a call is answered. See ANSWER SUPERVISION and ANSWER SIGNAL.

Answer Message ANM. A CCS/SS7 signaling message that informs the signaling points involved in a telephone call that the call has been answered and that call charging should start.

Answer Mode When a modem is set by the user to receive data, it is in Answer Mode. In any conversation involving two computers, two terminals or one computer and one terminal, one side of the conversation must always be in Answer Mode. Putting a modem/computer in answer mode is sometimes done through software and sometimes through hardware, i.e. a switch on the side of the machine. You cannot run a data communications "conversation" if both sending and receiving equipment are in "Answer Mode." Computers — mainframe and mini — which receive a lot of phone calls are typically put in "Answer Mode." The terminals or computers calling them are typically in transmit mode.

Answer Signal A supervisory signal, usually in the form of a closed loop, returned from the called telephone to the originating switch when the called party answers. This signal stops the ringback signal from being returned to the caller.

Answer Supervision Follow this scenario: I call you long distance. My central office must know when you answer your phone so my central office can start billing me for the call. It works like this: when you, the called party, answer your phone, your central office sends a signal back to my central office (the originating CO). This tells my central office to start billing me for the call. This signal is called Answer Supervision. Before the Divestiture of the Bell System in early 1984, most of the nation's long distance companies — with the exception of AT&T Communications — did not receive Answer Supervision. They did not know precisely when the called party answered. So they started their billing cycle after some time — 20 or 30 seconds after the caller completed dialing. These long distance companies presumed that after this time, some one will have answered and the call will be in progress and can then be timed and billed. Without Answer Supervision, their billing of calls is inaccurate. They may bill for calls which didn't occur. And you may pay more for calls which did occur.

With the Divestiture of the Bell System, and the introduction of Equal Access, the local phone companies have been told by the FCC that they must provide accurate answer supervision to all long distance phone. And with that answer supervision, the pricing of your long distance calls should be accurate. Not all long distance companies, however, choose to buy answer supervision (it costs a little more). And thus your long distance calls may still be billed inaccurately.

Check. If you are "accessing" your preferred long distance carrier by dialing a seven digit local number, then dialing your number and your account code, your carrier is probably not receiving Answer Supervision and the timing and billing of your long distance calls may be inaccurate. Check this out. Remember: just because your town has equal access doesn't mean your preferred long distance phone company has opted for it because it is expensive or for some other reason.

Virtually no hotels have answer supervision. So they start billing you arbitrarily. Some start billing you after three rings. Some after four. When you check out, carefully check your phone bill. You will, in most instances, find you have been billed for many uncompleted calls. Tell your family to pick up the phone quickly when you're out of town and may be calling them. Don't let the phone ring too many times as you're likely to be billed for the dubious pleasure of listening to ringing signals.

"Answer supervision" is getting better, however, as the electronics of "listening" to sounds on phone lines get better. Electronics are now available to do — to a 95% accuracy — what we as humans do — to a 100% accuracy — namely distinguish between a normal ringing sound, a fast busy sound and a person or fax machine answering the phone and saying "Hello." These electronics are getting better and less expensive, by the month. See ANSWER SUPERVISION-LINE SIDE.

Answer Supervision-line Side Answer Supervision-Line Side is a service I first read about in a US West publication, which describes it as "providing an electrical signal that is passed back to the originating end of a switched connection. This signal indicates that the called line has gone off-hook. This service offering has applicability for record start and end, announcement start and end, dialtone reorigination prevention, call progress sequence indications, and other uses. This service offering may be used by terminal equipment (PBX, pay telephone, call diverter, etc) connected to the calling line to determine that the call has been answered."

Answerback In data communications, answerback is a response programmed into a data terminal to identify itself when polled by a remote computer or terminal. This response is usually in reply to a Control-E (ASCII Character 5, Inquiry), which is known on the Telex and TWX networks as a "Who Are You?" character, or "WRU." The Answerback allows a remote computer to verify it has dialed correctly (usually on the Telex or TWX networks) by matching the Answerback received with the Answerback expected.

Answering Tone The tone an asynchronous modem will transmit when it answers the phone. The tone indicates that it is willing to accept data.

NEWTON'S TELECOM DICTIONARY

Ant Farm Gigantic multiscreen movie theater complex with glass facade, often found near American malls. Also called multiplexes or gigaplexes.

ANTC Advanced Networking Test Center. An FDDI interoperability testing center established in 1990.

Antediluvian Before the flood. Very old fashioned.

Antenna A device for transmitting, receiving or transmitting and receiving signals. Antennas come in all shapes and sizes. Their shape depends on the frequency signal they're receiving. Electrical signals with frequencies higher on the spectrum, for example, are shorter and more directional. As they get higher on the spectrum, they look more like light. These must be focused and thus, require antennas which are shaped like the mirror reflector of a focusing flashlight. This parabolic shape focuses the broad beam (of the bulb or the electrical signal) into a narrow, focused beam. The weaker the received signal, the bigger the antenna must be. Antennas come in many varieties and have cute names, like parabola, caresgrain, helix, lens and horn.

Antenna Gain The ratio, usually expressed in decibels, of the power required at the input of a loss-free reference antenna to the power supplied to the input of the given antenna to produce, in a given direction, the same field strength, or the same irradiance, at the same distance. When not specified otherwise, the gain refers to the direction of maximum radiation. The gain may be considered for a specified polarization.

Antenna Lobe A picture showing an antenna's radiation pattern. A more technical explanation: A three-dimensional radiation pattern of a directional antenna bounded by one or more cones of nulls (regions of diminished intensity).

Antenna Matching The process of adjusting impedance so that the input impedance of an antenna equals or approximates the characteristic impedance of its transmission line over a specified range of frequencies. The impedance of either the transmission line, or the antenna, or both, may be adjusted to effect the match.

Anthropomorphism The process of giving human qualities to inanimate objects. For example, getting a file cabinet to talk about what's inside it, or getting a modem to explain how to do communications.

Anti Aliasing 1. A filter (normally low pass) that band limits an input signal before sampling to prevent aliasing noise. See ALIASING NOSE.
2. A computer imaging term. A blending effect that smooths sharp contrasts between two regions of different colors. Properly done, this eliminates the jagged edges of text or colored objects. Used in voice processing, anti-aliasing usually refers to the process of removing spurious frequencies from waveforms produced by converting digital signals back to analog.

Anti Curl A feature marketed by manufacturers of slimy paper fax machines (i.e. thermal paper). As the paper emerges from the fax machine, "anti-curl" simply sends the paper through a path which causes it to bend slightly in the opposite direction to which it was rolled over the roll. This bending purports to make the paper less curly when it emerges. It works to an extent. Virtually all slimy fax machines now have the "feature," though most don't advertise it.

Anti Reflection Coating A thin, dielectric or metallic film (or several such films) applied to an optical surface to reduce its reflectance and thereby increase the transmittance of the optical fiber. The ideal value of the refractive index of a single layer film is the square root of the product of the refractive indices on either side of the film, the ideal optical thickness being one quarter of a wavelength.

Anti Static A material, such as packing material, that is treated to prevent the build-up of static electricity. The static charges gradually dissipate instead of building up a sudden discharge.

Anti Stuffing A mechanical flap in a coin phone which prevents the blocking by paper or other material of coin chutes. An anti-stuffing flap is meant to assure that you, the user, get your money back after you've tried to make a call but didn't get through.

Anti Tinkling Device Leahtta Welch of Tampa Florida

Anti Viral Programs Programs which scan disks looking for the tell-tale signatures of computer viruses.

Antialiasing See ANTI ALIASING.

Anticipointment Raising people's levels of anticipation and then disappointing them. A definition contributed by Gerald Taylor, president, of MCI.

Antistatic A material, such as packing material, that is treated to prevent the build-up of static electricity. The static charges gradually dissipate instead of building up a sudden discharge.

AnyLAN A proposed high speed local area network which Hewlett-Packard Co. and International Business Machines Corp. have created the specifications for. According to HP, AnyLAN will allow work groups operating token ring and ethernet lans to swap many more data-intensive applications at much greater speeds than at present. AnyLAN, which was announced in the early fall of 1993, is proposed to increase the speed and capacity of LANs sixfold to tenfold. HP and IBM announced they would give the technology of AnyLAN to any competitor free of charge, in order to establish it as a standard and expand the size of the total market.

Anynet/MVS IBM product name for the ACF/VTAM feature that implements IBM's "Networking Blueprint" technology on hosts and OS/2 workstations and permits SNA LU 6.2 applications to work over TCP/IP or TCP/IP-oriented sockets applications to run over SNA.

Anywhere The name of an ISDN service from Bell Atlantic and BellSouth. Nynex thought up the name, but delayed so long implementing the service that either Northern Telecom or AT&T (one of the two ISDN equipment vendors to Nynex) took the idea to Bell Atlantic. And Bell Atlantic announced the service very quickly. This irritated Nynex's senior executives, especially its EVP at the time, Arnie Eckelman, who delayed introducing Nynex's own ISDN Anywhere for reasons that were unclear to both Northern and AT&T.

Anywhere Fix The ability of a Global Positioning System (GPS) receiver to start position calculations without being given an approximate location and approximate time. See GPS.

AoC Advice of Charge. A wireless telecommunications term. A supplementary service provided to a customer under GSM, Global System for Mobile Communications.

AOL A Windows-based on-line computer service you call with your PC which provides e-mail, forums, software downloads, news, weather, sports, financial information, conferences, on-line gaming, an encyclopedia, and other features, to its subscribers. Competitors include CompuServe and the Microsoft Network.

AOS Alternate Operator Services. Today there are many Operator Services Providers not owned by the Bell Telephone Companies or AT&T. The AOS industry is dropping the descriptive term "alternate" and communicating that they be known as OSPs. AOS was coined by AT&T. See AOSP and OPERATOR SERVICE PROVIDERS.

AOSP Alternate Operator Service Provider. A new breed of long distance phone company. It handles operator-assisted calls, in particular Credit Card, Collect, Third Party Billed and Person to Person. Phone calls provided by OSP companies are typically fare more expensive than phone calls provided by "normal" long distance companies, i.e. those which have their own long distance networks and which you see advertised on TV. You normally encounter an OSP only when you're making a phone call from a hotel or hospital phone, or a privately-owned payphone. It's a good idea to ask the operator the cost of your call before you make it.

AOSSVR Auxiliary Operator Services System Voice Response.

AOW Asia and Oceania Workshop. One of the three regional OSI Implementors' Workshops.

AP See ADD-ON or APPLICATIONS PROCESSOR. AP is an AT&T word for a piece of equipment which hangs off the side of their PBX and makes it do more things, like voice mail.

APA All Points Addressable (APA) method of host graphics implementation which uses vertical and horizontal pixel coordinates to create a more graphic image. An SNA definition.

APAD Asynchronous Packet Assembler/Disassembler.

APC Adaptive Predictive Coding. A narrowband analog-to-digital conversion technique employing a one-level or multilevel sampling system in which the value of the signal at each sample time is adaptively predicted to be a linear function of the past values of the quantized signals. APC is related to linear predictive coding (LPC) in that both use adaptive predictors. However, APC uses fewer prediction coefficients, thus requiring a higher bit rate than LPC.

APCC The American Public Communications Council, which is part of the North American Telecommunications Association (NATA).

APD Avalanche PhotoDiode. A diode that, when hit by light, increases its electrical conductivity by a multiplication effect. APDs are used in lightwave receivers because the APDs have great sensitivity to weakened light signals (i.e. those which have traveled long distances over fiber). APDs are designed to take advantage of avalanche multiplication of photocurrent.

Aperiodic Antenna An antenna designed to have an approximately constant input impedance over a wide range of frequencies; e.g., terminated rhombic antennas and wave antennas.

Aperture For a parabolic reflector or a horn antenna, aperture is the dimension of the open mouth and represents a surface over which it is possible to calculate the radiation pattern. For a series of n stacked transmitting elements such as dipoles or slots, the vertical aperture is usually defined as n times the element spacing in wavelengths.

Aperture Distortion In facsimile, the distortions in resolution, density, and shape of the recorded image caused by the shape and finite size of the scanning and recording apertures or spots.

Aperture Grille A type of monitor screen made up of thin vertical wires. Said to be less susceptible to doming than iron shadow mask.

API An Application Programming Interface is software that an application program uses to request and carry out lower-level services performed by the computer's or a telephone system's operating system. For Windows, the API also helps applications manage windows, menus, icons, and other GUI elements. In short, an API is a "hook" into software. An API is a set of standard software interrupts, calls, and data formats that application programs use to initiate contact with network services, mainframe communications programs, telephone equipment or program-to-program communications. For example, applications use APIs to call services that transport data across a network. Standardization of APIs at various layers of a communications protocol stack provides a uniform way to write applications. NetBIOS is an early example of a network API. Applications use APIs to call services that transport data across a network.

API_connection An ATM term. Native ATM Application Program Interface Connection: API_connection is a relationship between an API_endpoint and other ATM devices that has the following characteristics:
- Data communication may occur between the API_endpoint and the other ATM devices comprising the API_connection
- Each API_connection may occur over a duration of time only once; the same set of communicating ATM devices may form a new connection after a prior connection is released
- The API_connection may be presently active (able to transfer data), or merely anticipated for the future

APL Automatic Program Load in telecom. In data processing, it's a popular programming language.

APLT Advanced Private Line Termination. Provides the PBX user with access to all the services of an associated enhanced private switched communications services (EPSCS) network. it also functions when associated with a common control switching arrangement (CCSA) network. See Advanced Private Line Termination.

APM 1. Average Positions Manned, the average number of ACD positions manned during the reporting period for a particular group. 2. Advanced Power Management. A specification sponsored by Intel and Microsoft to extend the life of batteries in battery-powered computers. The idea of the specification if for the application programs, the system BIOS and the hardware to work together to reduce power consumption. An APM-compliant BIOS provides built-in power management services to the operating system. The operating system passes calls and information between the BIOS and the application programs. It also arbitrates power management calls in a multi-tasking environment (such as Windows) and identifies power-saving opportunities not apparent to applications. The application software communicates power-saving data via predefined APM interfaces. Only a very few programs presently conform to APM.

Apocalypse, Four Horsemen Of The four horsemen of the Apocalypse were War, Plague, Famine and Death.

Apogee The point on a satellite orbit that is most distant from the center of the gravitational field of the Earth.

Apologize To lay the foundation for a future offense.

APPC Advanced Program-To-Program Communications. In SNA, the architectural component that allows sessions between peer-level application transaction programs. The LUs (Logical Units) that communicate during these sessions are known as LU type 6.2. APPC is an IBM protocol analogous to the OSI model's session layer: it sets up the necessary conditions that enable application programs to send data to each other through the network.

NEWTON'S TELECOM DICTIONARY

APPC/PC An IBM product that implements APPC on a PC.
Appearance Usually refers to a private branch exchange line or extension which is on (i.e. "appears") on a multi-button key telephone. For example, extension 445 appears on three key systems.
Appearance Test Point The point at which a circuit may be measured by test equipment.
Append To add the contents of a list, or file, to those of another.
APPGEN A shortened form of the words APPLications GENerator.
Apple Computer, Inc., Cupertino, CA. Manufacturer of personal computers. Heavy penetration in the graphics/desktop publishing business. Apple was formed on April Fool's Day, 1976, by Steve Wozniak and Steve Jobs, aided greatly by Mike Markkula.
Apple Desktop Bus The interface on a Mac where non-peripheral devices, such as the keyboard, attaches. A Mac keyboard or mouse is called an ADB device. Contrast with peripherals, which attach through the SCSI interface. See also USB, which is a new bus for use on PCs but fulfilling essentially the same function as the Apple Desktop Bus.
Apple Desktop Interface ADI. A set of user-interface guidelines, developed by Apple Computer and published by Addison-Wesley, intended to ensure that the appearance and operation of all Macintosh applications are similar.
Apple Menu The Apple icon in the upper left hand corner of the Apple Macintosh screen. The Apple menu contains aliases, control panels, the chooser and other desk accessories.
Apple Pie Both an American icon, and the name chosen for Apple Computer's Personal Interactive Electronics (PIE) division, chartered with extending the company into new growth areas such as Personal Digital Assistants (PDAs), e.g. the Apple Newton. The PIE division includes Apple Online Services, Newton and Telecommunications group, publishing activities, and ScriptX-based multimedia PDA development.
Apple Remote Access ARA is Apple Computer's dial-in client software for Macintosh users allowing remote access to Apple and third party servers.
Apple URP Apple Update Routing Protocol. The network routing protocol developed by Apple for use with Appletalk.
AppleShare Apple Computer's local area network. It uses AppleTalk protocols. AppleShare is Apple system software that allows sharing of files and network services via a file server in the Apple Macintosh environment. See APPLETALK.
Applet A small application that performs a specific task, such as the Cardfile and Calculator in Microsoft Windows. See also JAVA.
AppleTalk Apple Computer's proprietary networking protocol for linking Macintosh computers and peripherals, especially printers. This protocol is independent of what network it is layered on. Current implementations exist for LocalTalk (230.4 Kbps) and EtherTalk (10Mbps).
AppleTalk Zone and Device Filtering Provides an additional level of security for AppleTalk networks. On AppleTalk networks, network managers can selectively hide or show devices and/or zones to ARA clients. See ARA.
Application A software program that carries out some useful task. Database managers, spreadsheets, communications packages, graphics programs and word processors are all applications.
Application Binary Interface ABI. The rules by which software code is written to operate specific computer hardware. Application software, written to conform to an ABI, is able to be run on a wide variety of system platforms that use the computer hardware for which the ABI is designed.
Application Bridge Aspect Telecommunications' ACD to host computer link. Originally it ran only over R2-232 serial connections, but it now runs over Ethernet, using the TCP/IP link protocol. See also OPEN APPLICATION INTERFACE.
Application Class An SCSA term. A group of client applications that perform similar services, such as voice messaging or fax-back services.
Application Equipment Module AEM. A Northern Telecom term for a device within the Meridian 1 Universal Equipment Module that supports Meridian Link Modules. The Meridian Link Module (MLM) is an Application Module, specially configured to support the Meridian Link interface to host computers.
Application For Service A standard telephone company order form that includes pertinent billing, technical and other descriptive information which enables the company to provide communications network service to the customer and its authorized users.
Application Framework This usually means a class library with a fundamental base class for defining a complete program. The framework provides at least some of the facilities through which a program interfaces with the user, such as menus and windows, in a style that is internally consistent and abstracted from the specific environment for which it has been developed.
This is an explanation I received from Borland. I don't quite understand it, yet. An application framework is an object-oriented class library that integrates user-interface building blocks, fundamental data structures, and support for object-oriented input and output. It defines an application's standard user interface and behavior so that the programmer can concentrate on implementing the specifics of the application. An application framework allows developers to reuse the abstract design of an entire application by modeling each major component of an applications as an abstract class.
Application Generator AG. A program to generate actual programming code. An applications generator will let you produce software quickly, but it will not allow you the flexibility had you programmed it from scratch. Voice processing "applications generators," despite the name, often do not generate programming code. Instead they are self-contained environments which allow a user to define and execute applications. They are more commonly called applications generator, since one generator can define and execute many applications. See APPLICATIONS GENERATOR for a longer explanation.
Application Module A Northern Telecom term for a computer that can be attached to a Northern Telecom phone system and add intelligence and programmability to the phone system. Often, the AM will be a computer conforming to open standards, such as DOS or Windows, or it may be VME-based.
Application Module Link AML. A Northern Telecom internal and proprietary link that connects the Meridian 1 (via EDSI or MSDL port) to the Meridian Link Module.

44

NEWTON'S TELECOM DICTIONARY

Application Program A computer software program designed for a specific job, such as word processing, accounting, spreadsheet, etc.

Application Program Interface API. A set of formalized software calls and routines that can be referenced by an application program to access underlying network services.

Application Programming Interface API. A set of functions and values used by one program to communicate with another program or with an operating system. See API for a better explanation.

Application Profile As SCSA term. A description of the kinds of resources and services required by a client application (or an application class). An application profile is defined once for an instance of an application; then system services such as the SCR will be able to fulfill the needs of the application without the application having to state its needs explicitly.

Application Service Element ASE. A messaging term. A module or portion of a protocol in the application layer 7 of the OSI (Open Systems Interconnection) protocol stack. Several ASEs are usually combined to form a complete protocol, e.g., the X.400 P1 protocol which consists of the MTSE (Message Transfer Service Element), and the RTSE (Reliable Transfer Service Element).

Application Software Interface ASI. The Application Software Interface is a product of the Application Software Interface Expert Working Group of the ISDN Implementor's Workshop. The Interface focuses on the definition of a common application interface for accessing and administering ISDN services provided by hardware commonly referred to in the vendor community as Network Adapters (NAs) and responds to the applications requirements generated by the ISDN Users Workshop (IUW). The characteristics of this Application Interface shall be
* Portable across the broadest range of system architectures;
* Extensible (their words, not mine)
* Abstracted beyond ISDN to facilitate interworking;
* Defined in terms of services and facilities consistent with OSI layer interface standards.
According to Application Software Interface Group, the primary goal of the ASI is to provide a consistent set of application software interface services and application software interface implementation agreement(s) in order that an ISDN application may operate across a broad range of ISDN vendor products and platforms. The application software interface implementation agreements will be referenced by (and tested against) the IUW (ISDN Users Workshop) generated applications. It is anticipated that the vendor companies involved in the development of these implementation agreements will build products for the ISDN user marketplace which conform to them. ASI Implementation Agreements are likely to become a US Government Federal Information Processing Standard (FIPS).

Applications Engineering Applications engineering is the process of analyzing your telephone network to find products and services that will reduce your monthly bill without sacrificing network quality. It can be as simple as calling the telephone company to convert a particular service to a Rate Stabilization Plan (RSP). In many instances, the use of applications engineering concepts will increase the quality of your network. For example, putting DIDs onto a T1 will save you money and provide your network with a digital backbone. Unfortunately, most applications engineering is done by the telephone company or by their sales agents. Their main goal is not to save you money, but rather to sell telephone company products. Therefore, they are unlikely to advise you of all the hidden costs of converting to a particular service. A true application engineer will provide you with a complete cost analysis that includes all the conversion costs, and provides you with the "break-even date." The break-even date is the date that your monthly saving offsets the initial conversion cost of the service. It is often used synonymously with the term break-even point.

Applications Generator An Application generator (AG) is a software tool that, in response to your input, writes code a computer can understand. In simple terms, it is software that writes software. Applications generators have three major benefits: 1. They save time. You can write software faster. 2. They are perfect for quickly demonstrating an application. 3. They can often be used by non-programmers. Applications generators have two disadvantages. 1. The code they produce is often not as efficient as the code produced by a good programmer. 2. They are often limited in what they can produce. Applications generators tend to be either general purpose tools or very specific tools, providing support for specific applications, such as connecting voice response units to mainframe databases, voice messaging system development, audiotex system development, etc. There are simple AGs. There are complex AGs. There are general purpose AGs. There are specialized AGs. There are character-based AGs. There are GUI-based AGS. In researching AGs to write computer telephony and interactive voice response applications, I found three different levels of AG packages. First, there are the sort of non-generator generators. They don't really create new software, but they allow you to tweak existing application blocks. There's no compiling and they're pretty simple to use (though they often lack database and host connectivity). Then there are the pretty GUI forms-based app gens. They usually entail building a call-flow picture, using either pretty icons or easy to understand templates. When you're done filling in all the blanks, you compile it and actually "generate" new software. They're very cute. Finally, there's the script level language of a company like Parity Software, San Francisco. Real programmers dig this. They often feel it gives them a lot more power and flexibility. For very complex apps (with T-1/ISDN, ANI, host connection, speech recognition, multimedia capabilities, etc.) you'll probably need the power and flexibility of a script language. Most of the better GUI application generators let you drop down to a script-level language (and C too).

Applications Layer The seventh and highest layer of the Open Systems Interconnection (OSI) data communications model of the International Standards Organization (ISO). It supplies functions to applications or nodes allowing them to communicate with other applications or nodes. File transfer and electronic mail work at this layer. See OSI MODEL.

Applications Partner An Applications Partner is AT&T's new name for an outside company which will write software to work on AT&T phone systems, such as the Merlin, Legend and the Definitely. AT&T is setting up an Applications Partner Program to work with companies to help them develop programs and distribute their products. See also DESKTOP CONNECTION.

Applications Processor A special purpose computer which attaches to a telephone system and allows it (and the people using it) to perform different "applications," such as voice mail, electronic mail or packet switching. We think AT&T invented the term. See also ADD-ON.

45

Applique Circuit components added to an existing system to provide additional or alternate functions. Some carrier telephone equipment designed for ringdown manual operation can be modified with applique to allow for use between points having dial equipment.

APPN Advanced Peer-to-Peer Networking (APPN) is, according to its creator IBM, a leading-edge distributed networking feature IBM has added to its Systems Network Architecture (SNA). It provides optimized routing of communications between devices. In addition to simplifying the addition of workstations and systems to a network and enabling users to send data and messages to each other faster, APPN is designed to support efficient and transparent sharing of applications in a distributed computing environment. Because APPN permits direct communication between users anywhere on a network, it facilitates the development of client/server computing, in which workstation users anywhere on a network can share processing power, applications and data without regard to where the information is located. Workstations on an APPN network are dynamically defined so they can be relocated easily on the network without extensive re-programming. APPN also allows remote workstations to communicate with each other, without intervention by a central computer. Also, IBM's Advanced Peer-to-Peer Networking software.

APPN End Node An APPN end node is the final destination of user data and cannot function as an intermediate node in an APPN network and cannot perform routing functions. See APPN.

Approved Ground Grounds that meet the requirements of the NEC (National Electrical Code), such as building steel, concrete-encased electrodes, ground rings, and other devices. See AC and GROUNDING.

AppServer A SCSA term. AppServer defines the software environment that enables voice processing applications to run on any computing platform. AppServer sits on a PC equipped with call processing hardware and allows a remotely hosted application to control the call processing hardware.

APR Annual Percentage Rate. A percentage calculation of the finance charge portion of financing contact.

APS Automatic Protection Switching. A switching device which automatically initiates the switching from a primary circuit to a secondary (usually geographically diverse) circuit when the error rate on the primary line exceeds a set threshold.

AR Automatic Recall.

ARA AppleTalk Remote Access. Provides an asynchronous AppleTalk connection to another Macintosh and its network services through a modem. A remote user using ARA can log on to a remote server and mount the volume on his desktop as if he were connected locally.

ARAM Audio grade DRAM. DRAMS are low cost integrated circuits that are widely use in consumer electronic's products to store digital data.

Arbitrated Loop A Fiber Channel topology where nodes are linked together in a closed loop. An arbitrated loop can connect to fabric via an FL_Port.

Archie An Internet term. Located on several computers around the country, Archie is a kind of superdirectory to the files on the internet. If you're looking for a file or even a particular topic, Archie provides its specific location. Veronica, Jughead and WAIS (Wide Area Information Servers) are other tools for searching the huge libraries of information on the Internet. Some companies, such as Hayes, make Archie software which give you a menu driven interface that lets you browse through the various Archie servers on the Internet as though browsing through card catalogs of remote libraries.

Archival Readable (and sometimes writable) media. Archival media have defined minimum life-spans over which the information will remain stable (i.e. accurate without degradation).

Archive Bit A Window NT term. Backup programs use the archive bit to mark files after backing them up, using the normal or incremental backup types.

Archive Server An email-based file transfer facility offered by some computers on Internet.

Architect One who drafts a plan of your office and then plans a draft of your money.

Architectural Freedom An AT&T term for flexibility in locating functions, such as control, storage or processing of information, at any site in or around a network, such as customer premises, central offices or regional service bureaus. Architectural freedom also means the ability to distribute functions among combinations of locations and have them interrelate through a high-throughput, low-delay, transparent network. See also ARCHITECTURE.

Architecture The architecture of a system refers to how it is designed and how the components of the system are connected to, and operate with, each other. It covers voice, video, data and text. Architecture also includes the ability of the system to carry narrow, medium and broadband signals. It also includes the ability of the system to grow "seamlessly" (i.e. without too many large jumps in price).

Architecture Police An individual or group within a company that makes sure software and hardware development follows established corporate guidelines. The architecture police tend to rein in creative development efforts.

Archive A backup of a file. An archived file may contain backup copies of programs and files in use or data and materials no longer in use, but perhaps needed for historical or tax purposes. Archive files are kept on paper, on microfilm, on disk, on floppies, etc. They may be kept in compressed or uncompressed form. See ARCHIVER.

Archiver A software program for compressing files. If you compress files, you will save on communications charges, since you will be able to transmit those files faster as they're now smaller. My favorite MS-DOS archiver, also called file compression utility is Phil Katz's PKZIP.EXE and PKUNZIP.EXE. You can cut a database by as much as 90% and a word processed file by maybe 30% by using PKZIP. How much you can cut is determined by how much fluff is in the file. PKZIP is the most widely-used archive and compression utility today. You can recognized "zipped" files because their extension is always ZIP. There are other compression programs out there which you will recognize by these extensions, ARC, AR7, ARJ, LZH, PAK and ZOO.

Archiving Files This is a process where the information contained in an active computer file is made ready for storing in a non-active file, perhaps in off-line or near-line storage. Typically when files are archived, they are compressed to reduce their size. To restore the file to its original size requires a process known as unarchiving. See also ARCHIVER.

ARCNET Attached Resource Computer NETwork. One of the earliest and most popular local area networks. A 2.5M-bits-per-second LAN that uses a modified token-passing protocol. Developed by Datapoint, San Antonio, TX, Arcnet interface cards are now manufactured by many vendors, including Standard Microsystems and Pure Data, Ltd. Arcnet has lost popularity in recent years to Ethernet (IEEE 802.3) and Token Ring (IEEE 802.5).

ARDIS A public data communications wireless network that allows people carrying handheld devices to send and receive short data messages. Such messages might be from sheriff standing in the street searching his department's data base for unpaid parking tickets. ARDIS is jointly owned by Motorola and IBM. It is an outgrowth of a network originally created for IBM service technicians. A competitor to Ardis is RAM Mobile Data.

ARE An ATM term. All Routes Explorer: A specific frame initiated by a source which is sent on all possible routes in Source Route Bridging.

Area Code A three-digit code designating a "toll" center in the United States and Canada. Until January, 1995 the first digit of an area code was any number from 2 through 9. The second digit was always a "1" or "0." In January 1995, North America (i.e. the US and Canada) adopted the North American Numbering Plan (NANP) and second digits could be any number. This dramatically increased the number of possible area codes — from 152 to 792 and the number of phone numbers to more than six billion. For a full explanation, see NORTH AMERICAN NUMBERING PLAN. For a full listing of area codes, see NORTH AMERICAN AREA CODES.

Area Code Expansion The new North American Numbering Plan (NANP) allowed basically any three numbers to become an area code. This exploded the number of area codes now possible. Some manufacturers of phone equipment, e.g. Rockwell, choose to call this happening "Area Code Expansion." They claimed that their switch would accommodate all future permutations and combinations of area codes.

Area Code Restriction The ability of the telephone equipment (or its ancillary devices) to selectively deny calls to specific (but not all) area codes. Area code restriction is often confused with "0/1" (zero/one) restriction which denies calls to all area codes by sampling the first and second dialed digits (is it a 0 or 1?) and thus, identifying and blocking an attempt at making a toll call. For a full listing of area codes, see NORTH AMERICAN AREA CODES.

Arena A term from a ISV called Integrated North Coast. Here's their explanation: North Coast's Arena Applications allow one computer to work with a NEAX PBX more efficiently than several computers. The Arena is a common platform which allows for the sharing of information among programs. This allows several functions; such as voice mail, call accounting, and IVR to sit on one computer and operate on one common platform. It is easiest to understand the Arena when described in theatrical terms. The Arena platform itself is a "stage" shared by a number of programs (Arena Actors). When information is received by one actor, it is "spoken" into the Arena. This information is then shared by all other actors. Information such as SMDR or CDR, even credit card numbers (literally hundreds of types of information), can be acted upon. This creates the ability to: cost calls, file CDR, call a bank and act as an ATM, open or close a relay, etc. Actors can be added or removed to change the functionally of the Arena.

ARENIX Arena for UNIX. A platform developed by North Coast Logic. This UNIX-based application takes advantage of UNIX multi-tasking and, using a centralized location for sharing data, allows a variety of telecommunications solutions to operate on the same computer. The platform allows programs to offer a true single point of entry — ensuring that information such as names and extension numbers only need to be entered once. Arenix Applications (Call Accounting, Voice Messaging, Fax, Facilities Management), which share the Arenix Platform, are used to provide additional functionality to the NEC NEAX 2400 IMS/ICS and NEAX 2000 IVS.

ARF Alternative Regulatory Framework.

Argument Separator In spreadsheet programs and programming languages, a comma or other punctuation mark that sets off one argument from another in a command or statement. The argument separator is essential in commands that require more than one argument. Without the separator, the program can't tell one argument from another.

Ariane The name of a family of rockets used, amongst other things, for sending communications satellites into space. Ariane is a product of the European Space Agency, the equivalent of the U.S.'s NASA.

ARINC Aeronautical Radio INC. The collective organization that coordinates the design and management of telecommunications systems for the airline industry. It's one of the largest buyers of telecommunications services and equipment in the world. In its own words, "ARINC develops and operates communications and information processing systems for the aviation and travel industries and provides systems engineering and integration solutions to government and industry. Founded in 1929 to provide reliable and efficient radio communications for the airlines, ARINC is a $280 million company headquartered Annapolis, MD with over 2,000 employees worldwide." www.arinc.com

Arithmetic Logic Unit ALU. The part of the CPU (Central Processing Unit) that performs the arithmetic and logical operations. See MICROPROCESSOR.

Arithmetic Operation The process that results in a mathematically correct solution during the execution of an arithmetic statement or the evaluation of an arithmetic expression.

Arithmetic Register A register (i.e. short-term storage location) that holds the operands or the results of operations such as arithmetic operations, logic operations, and shifts.

Arithmetic Unit The part of a computing system which contains the circuits that perform the arithmetic operations. See also ALU.

ARAB Attendant Release Loop. A feature of the PBX console. See RELEASE.

ARM Asynchronous Response Mode. A communication mode involving one primary station and at least one secondary station, where either the primary or one of the secondaries can initiate transmission.

Armageddon The fabled battlefield where God's heavenly forces are to defeat the demon-led forces of evil. The final battle.

Armor Mechanical protection usually accomplished by a metallic layer of tape, braid or served wires or by a combination of jute, steel tapes or wires applied over a cable sheath for additional protection. It is normally found only over the outer sheath. Armor is used mostly on cables lying on lake or river bottoms or on the shore ends of oceans. See ARMORED CABLE.

NEWTON'S TELECOM DICTIONARY

Armored Cable A stainless steel handset cord which is meant to resist vandalism. Typically used on a coin phone, most stainless steel handset cords are too short. This is said to be because they were first ordered for use in prisons, where guards wanted to be certain they would not be used by the prisoners as hanging devices. Thus, they requested Western Electric to make them too short for such a use. Whether there is any validity to this story is dubious, however, it is part of telephone industry folk history and therefore, worth preserving.

ARP Address Resolution Protocol. A protocol within the Transmission Control Protocol/Internet Protocol (TCP/IP) suite that "maps" IP addresses to Ethernet addresses. TCP/IP requires ARP for use with Ethernet.

ARPA Advanced Research Projects Agency of the U.S. Department of Defense. (The whole DOD annual telecommunications bill exceeds $1 billion.) Much of the country's early work on packet switching was done at ARPA. At one stage it was called DARPA, which stands for Defense Advanced Research Projects Agency. ARPA was the US government agency that funded research and experimentation with the ARPANET and later the Internet. The group within DARPA responsible for the ARPANET is ISTO (Information Systems Techniques Office), formerly IPTO (Information Processing Techniques Office). See also DARPA INTERNET. DARPA has changed its name to ARPA and back again. It's hard to keep up.

ARPANET Advanced Research Projects Agency NETwork. A Department of Defense data network, developed by ARPA, which ties together many users and computers in universities, government and businesses. ARPANET has been the forerunner of many developments in commercial data communications, including packet switching, which was first tested on a large scale on this network. ARPANet is a predecessor of the Internet. It was started in 1969 with funds from the Defense Department's Advanced Projects Research Agency.

ARQ Automatic Retransmission reQuest. The standard method of checking transmitted data, used on virtually all high-speed data communications systems. The sender encodes an error-detection field based on the contents of the message. The receiver recalculates the check field and compares it with that received. If they match, an "ACK" (acknowledgment) is transmitted to the sender. If they don't match, a "NAK" (negative acknowledgment) is returned, and the sender retransmits the message. Note: this method of error correction assumes the sender temporarily or permanently stores the data it has sent. Otherwise, it couldn't possibly retransmit the data. No error detection scheme in data transmission is foolproof. This one is no exception.

Array 1. The description of a location of points by coordinates. A 2-D array is described with x,y coordinates. A 3-D array is described with x,y,z coordinates.
2. A named, ordered collection of data elements that have identical attributes; or an ordered collection of identical structures.

Array Antenna Take a bunch of directional antennas. Aim them at the same transmitting source. Join them together. Presto, you now have a very powerful giant antenna. Array antennas are used for picking up weak signals. They are often used in astronomical and defense communications systems.

Array Connector A connector for use with ribbon fiber cable that joins 12 fibers simultaneously. A fan-out array design can be used to connect ribbon fiber cables to non-ribbon cables.

Array Processor A processor capable of executing instructions in which the operands may be arrays rather than data elements.

Arrestor A device used to protect telephone equipment from lightning, electrical storms, etc. An arrestor is typically gas filled so when lightning strikes, the gas ionizes and, bingo, a low resistance to the ground that drains the damaging high voltage elements of the lightning away.

ARS Automatic Route Selection, also called Least Cost Routing. A way that your phone system automatically chooses the least expensive way of making the call that it is presented with. That least expensive way may be a tie line or a WATS line, etc. It may even be dial-up. See LEAST COST ROUTING and ALTERNATE ROUTING.

Article An Internet term. An article is a USENET conversation element. It is a computer file that contains a question or piece of information made available to the USENET community by posting to a newsgroup.

Artifacts Distortions in a video signal. Unintended, unwanted visual aberrations in a video image.

Artificial Intelligence In 1950, Alan Turing, a British mathematician, challenged scientists to create a machine that could trick people into thinking it was one of them. And this for long was THE classic definition of artificial intelligence. One way to trick people is to have the computer make typing mistakes, like real humans do. The real challenge these days with artificial intelligence, now more commonly called "expert systems," is not to recreate people but to recognize the uniqueness of machine intelligence and learn to work with it in intelligent, useful ways.

Artificial Line Interface In T-1 transmission, refers to the ability of a piece of transmission equipment to attenuate its output level to meet the required loop loss of 15-22.5 dB normally switch selectable between 0,7.5, and dB.

ARTS A Rockwell ACD term. Audio Real Time Status. From any touchtone phone, you can call into a Rockwell Spectrum ACD, enter a password, and obtain real time statistics...e.g. average speed of answer, # call in queue, activity by agent group. Neat feature.

ARU Audio Response Unit.

AS Autonomous Systems. Part of the internet layer that routers use to relate to network connectivity and packet addressing. The router checks the network address and only routes on the host address if the source and destination are on the same network.

As Is A term used in the secondary telecom equipment business. "As is" is equipment that is bought or sold with no stated or implied warranties. You should expect any condition from good to bad, from complete to incomplete. Buy As Is equipment at your own risk.

As Is Tested or As Is Working A term used in the secondary telecom equipment business. One step up from "as is" condition. The product has been tested. It works and is complete, unless otherwise specified. Buyer should test upon receipt. There is no warranty beyond receipt. Seller is guaranteeing the product will work upon arrival. After that, the buyer is responsible for any problems.

AS&C Alarm Surveillance and Control

AS/400 IBM's mid-range mini-computer. AS/400 stands for Application System/400. IBM has a product called CallPath/400 which allows AS/400 computers to link to PBXs from the leading manufacturers.

48

NEWTON'S TELECOM DICTIONARY

ASA 1. Average Speed of Answer. How long the average caller has to wait before they speak to an agent. The time can vary, even over the course of one day, due to call volumes and staff levels. An important measure of service quality.
2. Affiliated Sales Agency. A term for a company which resells the service of a phone company. Typically, the phone company pays the ASA a commission. Sometimes the commission is so large that it blurs the thinking of the ASA into recommending to its customers telecom products and services they would be better without.

ASAI AT&T's Adjunct Switch Application Interface. A long detailed set of technical specifications for having an outside computer control an AT&T PBX. See also APPLICATIONS PARTNER, OPEN APPLICATION INTERFACE, PSAI and SCAI.

ASC AIN Switch Capabilities. See AIN.

ASCII Pronounced: as'-kee. American Standard Code for Information Interchange. It's the most popular coding method used by small computers for converting letters, numbers, punctuation and control codes into digital form. (Computers can only understands zeros or ones.) Once defined, ASCII characters can be recognized and understood by other computers and by communications devices. ASCII represents characters, numbers, punctuation marks or signals in seven on-off bits. A capital "C", for example, is 1000011, while a "3" is 0110011. As a seven-bit code, and since each bit can only be a "one" or a "zero," ASCII can represent 128 "things," i.e. 2 x 2 x 2 x 2 x 2 x 2 x 2 which equals 128. ASCII is the code virtually every personal computer in the world encodes "things," including IBM, Apple and Radio Shack/Tandy. This compatible encoding (it was developed by ANSI — the American National Standards Institute) allows virtually all personal computers to talk to each other, if they use a compatible modem, or null modem cable and transmit and receive at the same speed. There are variations of ASCII. (Nothing is totally standard anymore.) The most important variation — one originally from IBM — is called Extended ASCII. It codes characters into eight bits (or one byte) and uses those ASCII characters above 127 to represent foreign language letters, and other useful symbols, such as those to draw boxes. But at 127 and below, extended 8-bit ASCII is identical to standard 7-bit ASCII. The CCITT (now called the ITU-T) calls ASCII International Telegraph Alphabet 5.
The other major method of encoding is IBM's EBCDIC (pronounced ebb'-si-dick). It's largely used on IBM and IBM-compatible mainframe computers (but not their PCs, which use ASCII and extended ASCII.) EBCDIC is an eight-bit encoding scheme, thus allowing up to 256 "things" to be encoded, i.e. 2 x 2 x 2 x 2 x 2 x 2 x 2 x 2 = 256. EBCDIC codes letters, characters and punctuation marks in a totally different way than ASCII. For ASCII files to be read by an IBM mainframe (one that reads EBCDIC), those ASCII files must be translated into EBCDIC by one of the many translation programs available. See also ASCII EDITOR, BAUDOT, EBCDIC, EXTENDED GRAPHICS CHARACTER SET, MORSE CODE and UNICODE.

ASCII Editor An ASCII editor (also called a "text," "DOS" or "non-document mode" editor) does NOT use extended ASCII and printer [ESCAPE] codes, which are used by word processor to create advanced features such as bold, italic, underlining, and super/subscript printing effects; and fancy formatting such as automatic paragraph reformat, pagination, hyphenation, footers, headers, and margins. I initially wrote this dictionary using an ASCII editor called ZEdit, which is a customized version of QEdit, undoubtedly the best editor ever written. Then, the author QEdit, produced a new and more powerful editor, called The Semware Editor. And I'm now using it to write this edition. Since an ASCII editor can't do so much, why would anyone use one? Well, its strength is in the lack of those very things a word processor has, which clutter it and slow it down! Here are my benefits:
1. It's lightning fast. No word processor can match an ASCII editor's speed at loading itself, loading files, finding things in files, etc.
2. A file produced by an ASCII editor can be read and edited by any word processor (absolutely any). Thus it's the universal word processing file. A WordPerfect file typically can't be read by WordStar and vice versa. The reason is that every word processor uses different high-level codes for the same features (underlining, bolding, etc.) There is no consistency among word processors as to how they encode their text so they can tell printers to do bolding, etc.
3. An ASCII editor is better to type programming languages, such as EDLIN (for batch files), BASIC, FORTRAN, PASCAL, etc. If QEDIT used extended ASCII and printer codes, it could not be used by these programs...for each program interprets these "high level" codes differently from another program. An ASCII editor types straight, "vanilla" text...nothing fancy about it.

ASCII File An ASCII file consists solely of ASCII 127 and below ASCII characters that are visible. You create an ASCII file using a simple editor, also called an ASCII editor. An ASCII file is also called a text file. See ASCII.

ASCII-To-Fax Conversion Allows the transfer of a word-processed file directly to your fax board so it can be faxed without being scanned from a hard copy print-out. Documents faxed with ASCII-TO-FAX conversion come out much cleaner at the other end, since the scanning process always degrades the image.

ASDS Accunet Spectrum of Digital Services. AT&T's leased line (also called private line) digital service at 56 Kbps. MCI and Sprint have similar services. It is available in N x 56/64 Kbps, for N = 1, 2, 4, 6, 8, 12. The 56/64 Kbps POP-POP service (between long distance carrier central offices) costs the same as an analog line.

ASE A messaging term. Application Service Element. A module or portion of a protocol in the application layer 7 of the OSI (Open Systems Interconnection) protocol stack. Several ASEs are usually combined to form a complete protocol, e.g., the X.400 P1 protocol which consists of the MTSE (Message Transfer Service Element), and the RTSE (Reliable Transfer Service Element).

ASH Ardire-Stratigakis-Hayduk, a synchronous compression algorithm that is said to offer four times throughput on a typical synchronous channel. It can be used in bridges, routers, ISDN and modems. Transcend of Cleveland, OH said at one point that it was the exclusive licensor of ASH.

ASI 1. Advanced Services Implementation.
2. Application Software Interface. An important ISDN term. See APPLICATION SOFTWARE INTERFACE.
3. Adapter Support Interface. The driver specification developed by IBM for networking over IEEE 802.5 Token-Rings.

ASIC Application Specific Integrated Circuit. This is a chip that has been built for a specific application. Manufacturers use it to consolidate many chips into a single package, reducing system board size and power consumption. Many video boards and modems use ASICs. ASICs span programmable array logic (PAL) devices, electrically programmable logic devices (EPLDs), field programmable logic devices (FPGAs), gate arrays, standard cell-based devices, and full custom, designed from scratch ICs. See also ASSP.

49

NEWTON'S TELECOM DICTIONARY

ASIC Chip Application Specific Integrated Circuit Chip. A fancy name for microprocessor chips which do specific tasks. For example, an ASIC chip might be responsible for a graphics display.
ASL Adaptive Speed Leveling. A US Robotics term for adjusting the transmission speed of a modem up or down, depending on the conditions on the line. US Robotics says it can adjust speed in 2 or 3 seconds after detecting changed line conditions. It requires like modems on either end of the transmission.
ASN Abstract Syntax Notation.
ASN.1 Abstract Syntax Notation One. LAN "grammar," with rules and symbols, that is used to describe and define protocols and programming languages. ASN.1 is the OSI standard language to describe data types.
ASP 1. A Northern Telecom term for Attached Support Processor.
2. Adjunct Service Point. An intelligent-network feature that resides at the intelligent peripheral equipment and responds to service logic interpreter requests for service processing. See also AIN.. 3. Administrable Service Provider. A SCSA term.
3. Abstract Service Primitive. An ATM term. An implementation-independent description of an interaction between a service-user and a service-provider at a particular service boundary, as defined by Open Systems Interconnection (OSI).
Aspect Ratio The ratio of the width of a rectangular image to its height. The aspect ratio of NTSC and PAL TV is four units of width to every three units of height. This is expressed as 4 x 3 aspect ratio. A 35mm frame measures 12 x 24 mm, which means it has two units of width to one unit of height. It is different in size from a TV screen. This is why the side parts of movies are chopped off on TV. For VGA and Indeo video technology, the aspect ratio is 4:3 yielding 160X120, 320X240 and 640X480 sizes.
Aspect Telecommunications The name of a very successful automatic call distributor manufacturer based in San Jose. Allegedly, the name of the company came from a conglomeration of the words Applications Specific Telecommunications.
ASPI ASPI stands for Advanced SCSI Programming Interface set, which is software primitives and data structures which allow software using the ASPI interface to be SCSI host adapter-independent. SCSI stands for Small Computer System Interface. (Pronounced Scuzzie.) ASPI is software that acts as a liaison between SCSI device drivers (the software that drives the SCSI devices) and the interface card (also known as the host adapter). Whenever a new device is added to a computer system, a software program called a "driver" must tell the computer how to talk to the new device. Instead of forcing vendors to write drivers for every host adapter, ASPI lets them write a driver to ASPI standards, supposedly guaranteeing that the device the driver controls will work with all ASPI-compatible host adapters.
The idea behind ASPI is to create a "black box" software interface - one which allows programmers to create software without having to know anything about the details of the SCSI interface hardware used in your computer. With ASPI, it's possible to write programs that can be used with any SCSI-based device used on a computer system that supports ASPI. While things are not always 100% perfect in all cases, ASPI greatly reduces potential compatibility problems for you, the user.
How does ASPI work? Essentially, there are two parts to an ASPI implementation. First, there's the ASPI "manager" which is a device driver supplied by the hardware manufacturer, and the ASPI software application. It's important to note that without an ASPI manager, ASPI compatibility is not possible. It's the manager that creates the standard ASPI-compatibility layer between the SCSI host adapter hardware and the ASPI-compatibility application. The manager is very hardware-specific, and is almost always supplied by the manufacturer of your SCSI host adapter.
ASQ Automated Status Query.
ASR 1. Automatic Speech Recognition. See INTERACTIVE VOICE RESPONSE.
2. Automatic Send-Receive teletype or telex machine. Such a machine, if left on and loaded with paper, will receive incoming messages and print them, even when nobody is present. See also AUTOMATIC SEND RECEIVE.
Assembler A program which translates an assembly programming language into the code of ones and zeros used by computers. See also ASSEMBLY LANGUAGE.
Assembly Pertaining to the translation of a program from symbolic language into machine code. See ASSEMBLY LANGUAGE.
Assembly Language A computer language for writing software. It is a language which is converted by programs called compilers or interpreters into machine language programs which consist of only 1s and 0s and which a computer can understand. Even though an assembly language consists of recognizable nemonics and meaningful words, it's not easy to program in. It is referred to as a "low-level language". Assembly language programs run faster than high-level language programs, such as BASIC, COBOL or FORTRAN, which are much easier to learn and program in. Choosing a programming language is typically a tradeoff of ease for speed.
Asserted A signal is asserted when it is in the state which is indicated by the name of the signal. Opposite of Negated.
Assignation A secret romantic rendezvous. The invitation to an assignation doesn't work if she doesn't know the meaning of the word. Are you listening Jane Laino?
Assigned Cell An ATM term. Cell that provides a service to an upper layer entity or ATM Layer Management entity (ATM-entity).
Assigned Frequency The center of the assigned frequency band assigned to a station.
Assigned Frequency Band The frequency band within which the emission of a station is authorized; the width of the band equals the necessary bandwidth plus twice the absolute value of the frequency tolerance. Where space stations are concerned, the assigned frequency band includes twice the maximum Doppler shift that may occur in relation to any point of the Earth's surface.
Assigned Night Answer ANA. After business hours or when you place your phone system on "Night Answer," this feature sends calls from specified trunks to designated extensions or departments. You may use this feature to send calls directly to modems, or to emergency numbers, or even to outside home numbers.
Assigned Plant Concept A pair is dedicated from the central office to the subscriber home and maintained at that address, even when idle. See REASSIGNMENT.
Assignment A call center term. The process of assigning individual employees to specific schedules in a Master File or Daily Workfile. Master File assignment can be done either manually or automatically (based on employee schedule preference and seniority).

50

NEWTON'S TELECOM DICTIONARY

Associate A verb used in Windows by File Manager. You associate a three character extension with an application. This tells File Manager that, when you click twice on the file, File Manager will know which application to launch. For example, you may tell Windows that the .QXD extension is associated with QuarkXpress. When you click on a QXD file, File Manager will launch Quark and load that particular file.

Associated Common-Channel Signaling A form of common-channel signaling in which the signaling channel is associated with a specific trunk group and terminates at the same pair of switches as the trunk group. The signal channel is usually transmitted by the same facilities as the trunk group.

Associated Voice Port AVP. A Northern Telecom term. A voice port is used to pass signaling and supervision information to establish a connection for the data port. The term is also used in reference to the configuration of modems and voice grade switched circuits connected to the voice-port-data-port pair.

Association A relationship between two connection segments that share a common Leg O (i.e., a common subscriber is in control of connection segments). Definition from Bellcore.

Association Control Service Element ACSE. The International Standards Organization's Open Systems Interconnect (OSI) application layer services used, for example, in Manufacturing Automation Protocol V3.0 (MAP).

ASSP Application Specific Standard Product is an integrated circuit that performs functions for a single application (e.g., keyboard controller). ASSP is a more precise term for a device often referred to as an ASIC.

AST Automatic Scheduled Testing. A method of testing switched access service (Feature Groups B, C, and D) where the customer provides remote office test lines and 105 test lines with associated responders or their functions' equivalent; consists of monthly loss and C-message noise tests and annual balance test.

ASTM American Society for Testing and Materials, a non-profit industry-wide organization which publishes standards, methods of test, recommended practices, definitions and other related material.

ASU Application-Specific Unit.

Asymmetric Faster in one direction than the other. The faster direction is called the main channel and the slower is the back channel.

Asymmetrical Compression Techniques where the decompression process is not the reverse of the compression process. Asymmetrical compression is more computer-intensive on the compression side so that the decompression of video images can be easily performed at the desktop or in applications where sophisticated codecs are not cost effective. In short, any compression technique that requires a lot of processing on the compression end, but little processing to decompress the image. Used in CD-ROM creation, where time and costs can be incurred on the production end, but playback must be inexpensive and easy. See ASYN.

Asymmetrical Modem A type of modem which uses most of the available bandwidth for transmission and only a small part for reception.

Asymmetrical Modulation A duplex transmission technique which splits the communications channel into one high speed channel and one slower channel. During a call under asymmetrical modulation, the modem with the greatest amount of data to transmit is allocated the high speed channel. The modem with less data is allocated the slow, or back channel. The modems dynamically reverse the channels during a call if the volume of data transfer changes.

Asymmetrical Multiprocessing A relatively simple implementation of multiprocessing in which the operating system kernel runs on one dedicated CPU and assigns tasks as they come in to other "slave processors." It is also known as "master/slave" processing.

Asymmetrical PVC This terms refers to a PVC (Private Virtual Circuit) which supports simplex, or asymmetrical, assignments of committed information rate in each direction of transmission. A PVC transmission path is duplex, meaning that there must be a communications path in each direction between the two points being connected. However with an asymmetrical PVC, the network capacity in each direction does not necessarily have to be equal.

Asyn Greek prefix meaning "not."

Asynchronous See ASYNCHRONOUS TRANSMISSION.

Asynchronous Balanced Mode ABM. Used in the IBM Token Ring's Logical Link Control (LLC), ABM operates at the SNA data link control and allows devices on a Token Ring to send data link commands at any time and to initiate responses independently.

Asynchronous Completion A Versit definition. A domain issues a service request and need not wait for it to complete. If the application waits for this completion, this is known as synchronous, but if it is sent off to to another system entity and the domain goes on to other activities before the service request completes (and the system later sends a message to the domain announcing the service's completion), that completion is known as Asynchronous.

Asynchronous Gateway A routing device used for dial-up services such as modem communications.

Asynchronous Request An SCSA term. A request where the client does not wait for completion of the request, but does intend to accept results later. Contrast with synchronous request.

Asynchronous Terminal A terminal which uses asynchronous transmissions. See ASYNCHRONOUS TRANSMISSION.

Asynchronous Time Division Multiplexing A multiplexing technique in which a transmission capability is organized in a priori unassigned time slots. The time slots are assigned to cells upon request of each application's instantaneous real need.

Asynchronous Transfer Mode ATM is the technology selected by the Consultative Committee on International Telephone & Telegraph (CCITT) International standards organization in 1988 (now called the ITU-T) to realize a Broadband Integrated Services Digital Network (B-ISDN). It is a fast, cell-switched technology based on a fixed-length 53-byte cell. All broadband transmissions (whether audio, data, imaging or video) are divided into a series of cells and routed across an ATM network consisting of links connected by ATM switches. Each ATM link comprises a constant stream of ATM cell slots into which transmissions are placed or left idle, if unused. The most significant benefit of ATM is its uniform handling of services, allowing one network to meet the needs of many broadband services. ATM accomplishes this because its cell-switching technology combines the best advantages of both circuit-switching (for constant bit rate services such as voice and image) and packet-switching (for variable bit rate services such as data and full motion video) technologies. The result is the bandwidth guarantee of circuit switching combined with the high efficiency of packet switching. For a longer explanation, see ATM.

Asynchronous Transmission Literally, not synchronous. A method of data transmission which allows characters to be sent at irregular intervals by preceding each character with a start bit, and following it with a stop bit. It is the method most small computers (especially PCs) use to communicate with each other and with mainframes today. In every form of data transmission, every letter, number or punctuation mark is transmitted digitally as "ons" or "offs." These characters are also represented as "zeros" and "ones" (See ASCII). The problem in data transmission is to define when the letter, the number or the punctuation mark begins. Without knowing when it begins, the receiving computer or terminal won't be able to figure out what the transmission means.

One way to do this is by using some form of clocking signal. At a precise time, the transmission starts, etc. This is called SYNCHRONOUS TRANSMISSION. In ASYNCHRONOUS transmission there's no clocking signal. The receiving terminal or computer knows what's what because each letter, number or punctuation mark begins with a start bit and ends with a stop bit. Transmission of data is called SYNCHRONOUS if the exact sending or receiving of each bit is determined before it is transmitted or received. It is called ASYNCHRONOUS if the timing of the transmission is not determined by the timing of a previous character.

ASYNCHRONOUS is used in lower speed transmission and by less expensive computer transmission systems. Large systems and computer networks typically use more sophisticated methods of transmission, such as SYNCHRONOUS or BISYNCHRONOUS, because of the large overhead penalty of 20% in ASYNCHRONOUS transmission. This is caused by adding one start bit and one stop bit to an eight bit word — thus 2 bits out of ten.

The second problem with large transfers is error checking. The user sitting in front of his own screen checks his ASYNCHRONOUS transmission by looking at the screen and re-typing his mistakes. This is impractical for transferring long files at high speed if there is not a person in attendance.

In SYNCHRONOUS transmission start and stop bits are not used. According to the book Understanding Data Communications, characters are sent in groups called blocks with special synchronization characters placed at the beginning of the block and within it to ensure that enough 0 to 1 or 1 to 0 transitions occur for the receiver clock to remain accurate. Error checking is done automatically on the entire block. If any errors occur, then the entire block is retransmitted. This technique also carries an overhead penalty (nothing is free), but the overhead is far less than 20% for blocks or more than a few dozen characters.

AT 1. Access Tandem.

2. Advanced Technology. Refers to a 16 bit Personal Computer architecture using the 80X86 processor family which formed the basis for the ISA Bus as found in the first IBM PC.

3. AudioTex. See AUDIOTEX.

AT Bus The electrical channel used by the IBM AT and compatible computers to connect the computer's motherboard and peripheral devices, such as memory boards, video controllers, PC card modems, bus mouse boards, hard and floppy disk controllers and serial/parallel input/output devices. The AT bus supports 16 bits of data in one slug, whereas the original IBM PC supported only 8 bits (and was called the ISA bus for Industry Standard Architecture). These days there are much faster "buses," including the EISA, MCA (MicroChannel Architecture), Local Bus, PCI, VESA, etc.

AT Command Set Also known as the Hayes Standard AT Command Set. A language that enables PC communications software to get an asynchronous and "Hayes-compatible modem" to do what you want it to do. So called "AT" because all the commands begin with "AT," which is short for ATtention. The most common commands include ATDT (touchtone a number), ATA (manually answer the phone), ATZ (reset modem — it will answer OK), ATS0=0 (disable auto-answer), and ATH (hang up the phone).

To avoid having yourself knocked off your data call by the beep that comes in on the phone company's call waiting, put the following line in your modem setup: ATS10=20. That will increase your S10 register to two seconds. This register sets the time between loss of carrier (caused by the 1.5 second call waiting signal) and internal modem disconnect. Factory default on most modems is 1.4 seconds — just perfect to be cut off by the wall waiting tone! (Dumb.)

If you have to dial through several phone systems, waiting for dial tone on the way and/or going through fax/modem switches, you may consider a dial stream that looks like ATDT 1-800-433-9800 [W]212-989-4675 [W]22, where [W] means (in some software programs) "Wait for any key. When you get it, touchtone out the next digits." In other software programs — pure Hayes command — W means wait for second dialtone.

If [W] doesn't work for you, then change X3 in your setup line to X1; change your computer's dialed number to 9; and dial your distant computer with your phone. When you hear the modem at the other end answer, tell your computer's software to dial 9. It will dial 9, hear the modem tone at the other end and connect as though it had dialed it all by itself. X1 tells your modem to dial (or touchtone) immediately — without waiting for dial tone.

You can use several AT commands on one line. You only need AT before the first one. Some modems require commands typed in capital letters. When your dialing fails and you can't figure why, get out of your communications software program and start again. Or in total desperation, turn your computer and modem completely off and start again. The word "Hayes" comes from the manufacturer of the modems called Hayes Microcomputer, Norcross, GA, the creator of the command set. Not all Hayes compatible modems are. See also HAYES COMMAND SET.

At Local Mode One of the command modes available on the ISDN set. It is used for compatibility with existing communications packages for analog modems or for data-only application programs. See AT COMMAND SET.

At Work Pronounced "At Work." Microsoft's office equipment architecture announced on June 9, 1993. Microsoft's idea is to put a set of software building blocks into both office machines and PC products, including:

* Desktop and network-connected printers. * Digital monochrome and color copiers. * Telephones and voice messaging systems. * Fax machines and PC fax products. * Handheld systems. * Hybrid combinations of the above.

According to Microsoft, the Microsoft At Work architecture focuses on creating digital connections between machines (i.e. the ones above) to allow information to flow freely throughout the workplace. The Microsoft At Work software architecture consists of several technology components that serve as building blocks to enable these connections. Only one of the components, desktop software, will reside on PCs. The rest will be incorporated into other types of office devices (the ones above), making these products easier to use, compatible with one another and compatible with Microsoft Windows-based PCs. The components, according to Microsoft, are:

* Microsoft At Work operating system. A real-time, pre-emotive, multitasking operating system that is designed to specifically address the requirements of the office automation and communication industries. The new operating system supports Windows-compatible application programming interfaces (APIs) where appropriate for the device.

* Microsoft At Work communications. Will provide the connectivity between Microsoft At Work-based devices and PCs. It will support the secure transmission of original digital documents, and it is compatible with the Windows Messaging API and the Windows Telephony API of the Windows Open Services Architecture (WOSA).

* Microsoft At Work rendering. Will make the transmission of digital documents, with formatting and fonts intact, very fast and, consequently, cost-effective; will ensure that a document sent to any of these devices will produce high-quality output, referred to as "What You Print Is What You Fax Is What You Copy Is What You See."

* Microsoft At Work graphical user interface. Will make all devices very easy to use and will make sophisticated features accessible; will provide useful feedback to users. Leveraging Microsoft's experience in the Windows user interface, Microsoft At Work-based products will use very simple graphical user interfaces designed for people who are not computer users.

* Microsoft At Work desktop software for Windows-based PCs. Will provide Windows-based PC applications the ability to control, access and exchange information with any product based on Microsoft At Work. Desktop software is the one piece of the Microsoft At Work architecture that will reside on PCs.

See also FAX AT WORK, VOICE SERVER, WINDOWS, WINDOWS TELEPHONY and WOSA.

AT&T American Telephone and Telegraph Company. Not much telegraph, plenty of telephone and some computing. AT&T, once the largest company in the world, is now 75% smaller as a result of the divestiture of its operating telephone companies on December 31, 1983. At one stage, it had two main divisions, AT&T Communications (the long distance phone company) and AT&T Technologies (equipment). These were merged to form one AT&T with several divisions. Then we rewrote this definition. Then they reorganized. We rewrote this definition so many times we finally gave up. Suffice, AT&T is American Telephone and Telegraph Company. It provides long distance phone service, makes and sells phones and phone switches (from small to very large), is in the cellular phone business (through its acquisition of McCaw Cellular) and makes and sells computers (in 1991, it bought NCR). It has also increasingly widespread investments in entrepreneurial "information age" startups. See also AT&T BELL LABS.

AT&T Alliance A direct-dial AT&T service which allows you to teleconference as few as three and as many as 59 telephones around the world. You can dial the numbers to call or you can schedule a meeting, called "Meet-Me Service." If you want to set the conference up, you dial 0-700-456-1000, then punch in the number of locations, then put in the numbers, ending each with a #. End your dialing with another # — to add yourself to the teleconference.

AT&T Applications Partner A software developer who develops applications for AT&T hardware and software. See also DESKTOP CONNECTION.

AT&T Bell Labs The research and development arm of AT&T. It used to be known simply as Bell Laboratories. It is one of the most important research laboratory in the United States. Bell Labs is credited with inventing many electronics things including the transistor. AT&T Bell Labs is now part of a new company called Lucent Technologies.

AT&T Communications AT&T-Comm or AT&T-C. The part of AT&T that was at one stage responsible for long distance services. It used to be called AT&T Long Lines, which we liked better. AT&T Communications is now the largest part of AT&T.

AT&T Credit Corporation AT&T's financing arm, which provides customized leases for telephone equipment including the MERLIN II, MERLIN Plus, Legend, SPIRIT and Definity Systems, as well as for other equipment, such as AT&T FAX Machines and computers.

AT&T Fax Attendant System A voice mail-like product providing fax store-and-forward broadcast, personal fax mailboxes for confidential reception, remote retrieval and plain paper fax. The FAX Attendant also adds Fax-on-Demand through a automated attendant/fax response. The AT&T FAX Attendant System can be integrated with AT&T's AUDIX Voice Power Voice Mail system to provide a single interface for retrieving messages and combined notification.

AT&T GBCS AT&T Global Business Communications Systems. The new name for the merger of two divisions into one division now responsible for all the business telephone system sales of AT&T worldwide. The two divisions that were merged into GBCS were called General Business Systems (office systems 80 lines and smaller) and Business Communications Systems (systems bigger than 80 lines). The merger happened during the summer of 1992. AT&T GBCS is now part of Lucent Technologies.

AT&T Global Information Solutions A new name for NCR (originally National Cash Register). AT&T bought NCR in 1991. The new name was announced in January 1994. In early 1996, it was changed back to NCR.

AT&T Information Systems AT&T-IS used to be one of AT&T's sales arms. It used to be responsible for (or used to be responsible for) sales of certain equipment, including telephone systems and computers, to both end-users and some selected distributors. AT&T-IS used to be responsible for selling a service called Net 1000, which was an end-to-end data communications/data processing service, whose exact service is hard to define and has now been killed.

AT&T Interexchange Company ATTIX, a suggested name for the long distance part of AT&T, which, at the time was called AT&T Long Lines. It was then called AT&T Communications. The reason the name ATTIX was rejected is that it reminded someone at AT&T of the word "drug addicts." No one has explained logically why AT&T ever needed to change the name "Long Lines," which most people in the industry related to very warmly. Including us.

AT&T Long Lines The long distance part of AT&T. It was called that for over 100 years. Then they changed the name to just AT&T. Long Lines remains a favorite for old timers.

NEWTON'S TELECOM DICTIONARY

AT&T Mail A way of sending electronic ASCII messages, binary files and faxes over phone lines and paper mail (through the US Postal Service) from your PC instantly to other people who have AT&T Mail Mailboxes, to anyone on an electronic mail system connected to AT&T Mail (like Internet, MCI Mail, CompuServe, etc.) to anyone who has a fax machine anywhere in the world or to anyone who receives US Mail (i.e. all of us).

AT&T Network Services This is an AT&T Technologies division which sells equipment primarily to Bell Operating companies, independent phone companies, overseas phone companies and large, Fortune 500-type users. It's doing very well.

AT&T Technologies Formerly called Western Electric, it is the equipment manufacturing arm of AT&T. No one knows why this company changed its name, since Western Electric had a wonderful reputation and is remembered with great fondness. It has an excellent reputation for high quality products and is still used as a brand name on some AT&T Technologies' products. AT&T Technologies is now called Lucent Technologies.

AT&T WorldWorx AT&T WorldWorx Solutions is a family of AT&T products and services that provide simultaneous, real-time interactive voice, video and data communications among two or more people. According to AT&T, WorldWorx allows teams of people at locations all over the world to work together and see each other on their desktop computers, while sharing files and data.

ATA 1. American Telemarketing Association.
2. Analog Terminal Adapter. A device for a Northern Telecom Norstar phone system that lets it use analog devices, for example FAX, answering machines, modems and single line phones, behind the Norstar's central telephone unit (its KSU). Before you buy the analog terminal adapter, check that its speed is fast enough for you. In mid-1995, it was constrained to 9,600 bps, or 14,400 bps if the phone line was clear.
3. AT Attachment. Refers to the interface and protocol used to access a hard disk on AT compatible computers. Disk drives adhering to the ATA protocol are commonly referred to as IDE interfaced drives for PC compatible computers. The ATA specification is fully backward compatible with the ST-506 standard it superseded. IDE drives are sometimes referred to as ATA drives or AT bus drives. The newer ATA-2 specification defines the EIDE interface, which improves upon the IDE standard. See ATA2, IDE and ENHANCED IDE.

ATA2 The second generation AT attachment specification for IDE devices that defines faster transfer speeds and LBA (Logical Block Address) sector-locating method. See ATA, IDE and ENHANCED IDE.

ATA Document The latest draft of the ANSI X3.T9 subcommittee AT Attachment document.

ATA Registers These registers are accessed by a host to implement the ATA protocol for transferring data, control and status information to and from the PC Card. They are defined in the ATA Document. These registers include the Cylinder High, Cylinder Low, Sector Number, Sector Count, DriveHead, Drive Address, Device Control, Error, Feature, Status and Data registers. The I/O and memory address decoding options for these registers are defined within this specification.

ATAPI Attachment Packet Interface specification does for CD-ROM and tape drives why ATA-2 does for hard drives. It defines device-side characteristics for an IDE-connected peripheral. The benefits of having a single interface for the most common non-disk storage device in the desktop world, the CD-ROM are obvious. For the manufacturer, there is no need to add a separate controller card for the CD-ROM. For the end-user it means no more fussing with interrupts, cards and proprietary driver software. ATAPI essentially adapts the established SCSI command set to the IDE interface.

ATB All Trunks Busy. One measure which your phone company or phone systems might give you of telephone traffic in and out of your office. See ALL TRUNKS BUSY.

ATD 1. Asynchronous Time Division.
2. ATtention Dial the phone. The first three letters in the most frequently-used command in the Hayes command set for asynchronous modems — typically those used with microcomputers.

ATIS Alliance for Telecommunications Industry Solutions, a trade group based in Washington, D.C. It was originally called The Exchange Carriers Standards Association.

ATM 1. Automated Teller Machine. The street corner banking machine which is usually hooked up to a central computer through leased local lines and a multiplexed data network.
2. Asynchronous Transfer Mode. Very high speed transmission technology. ATM is a high bandwidth, low-delay, packet-like switching and multiplexing technique. Usable capacity is segmented into 53-byte fixed-size cells, consisting of header and information fields, allocated to services on demand. The ITU has selected ATM as the basis for the future broadband network in view of its flexibility and suitability for both transmission and switching. Each ATM cell contains a 48 octet payload field and a 5-octet Header which identifies the cells virtual circuit. The fixed length cells require lower processing overhead and allow higher transmission speeds than traditional packet switching methods. ATM allocates bandwidth on demand, making it suitable for high-speed connection of voice, data, and video services. ATM services will be available at access speeds up to 622 Mbps.
Here's a full explanation: Conventional networks carry data in a synchronous manner. Because empty slots are circulating even when the link is not needed, network capacity is wasted. The ATM concept which has been developed for use in broadband metropolitan area networks (MAN) and optical fiber based systems is supported by both CCITT (now called the ITU-T) and ANSI standards, can also be interfaced to SONET (Synchronous Optical Network). ATM automatically adjusts the network capacity to meet the system needs and can handle data, voice, video and television signals. These are transferred in a sequence of fixed length data units called cells. Common standards definitions are provided for both private and public networks so that ATM systems can be interfaced to either or both. ATM is therefore a wideband, low delay, packet-like switching and multiplexing concept that allows flexible use of the transmission bandwidth and capable of working at data rates as high as 622.08 Mbit/s. Each data packet consists of five bytes of header field plus 48 bytes for user data. The header contains data that identifies with related cell, a logical address that identifies the routing, forward error correction (FEC) bits, plus bit for priority handling and network management functions. FEC applies only to the header as it is assumed that the network medium will not degrade the error rate below an acceptable level. All the cells of a virtual connection (VC) follow the same path through the network that was determined during call set-up. There are no fixed time slots in the system so that any user can access the transmission medium whenever an

empty cell is available. ATM is capable of operating at bit rates of 155.52 and 622.08 million bits per second and the cell stream is continuous and without gaps. The position of the cells associated with a particular VC is random and depends upon the activity of the network. Cells produced by different streams to the ATM multiplex are stored in queues awaiting cell assignment. Since a call is only accepted when the necessary bandwidth is available, there is a probability of queue overflow. Cell loss due to this forms one ATM impairment. However, this can be minimized through the use of statistical multiplexers. Bit errors in the header which are beyond the FEC capability can lead to misrouting. See ATM FORUM and ATM FORM UNI V3.0.

ATM Address Defined in the UNI Specification as 3 formats, each having 20 bytes in length including country, area and end-system identifiers.

ATM Forum, The An industry organization with some 800 members, co-founded by N.E.T. and three other leading networking companies, which focuses on speeding the development, standardization and deployment of ATM (Asynchronous Transfer Mode) products. It has been remarkably successful. The ATM Forum is based in Mountain View, CA. Their phone number is 415-949-6700. See ATM. www.atmforum.com

ATM Forum UNI V3.0 The ATM Forum UNI V3.0 implementation agreement is based on a subset of the ITU-TS broadband access signaling protocol standards. Additions to this subset have been made where necessary to support early deployment and interoperability of ATM equipment. The procedures and protocol defined in the agreement apply to both public and private UNIs. Moreover, since the protocol is symmetrical, it also applies in the configuration ATM-end-point to ATM-end-point. See ATM and ATM FORUM.

ATMARP ATM Address Resolution Protocol.

ATM Layer Link A section of an ATM Layer connection between two adjacent active ATM Layer entities (ATM-entities).

ATM Link A virtual path link (VPL) or a virtual channel link (VCL).

ATM Peer-to-Peer Connection A virtual channel connection (VCC) or a virtual path connection (VPC).

ATM Traffic Descriptor A generic list of traffic parameters that can be used to capture the intrinsic traffic characteristics of a requested ATM connection.

ATM User-User Connection An association established by the ATM Layer to support communication between two or more ATM service users (i.e., between two or more next higher entities or between two or more ATM-entities). The communications over an ATM Layer connection may be either bidirectional or unidirectional. The same Virtual Channel Identifier (VCI) issued for both directions of a connection at an interface.

ATS Abstract Test Suite: A set of abstract test cases for testing a particular protocol. An "executable" test suite may be derived from an abstract test suite.

ATSC Advanced Television System Committee. www.atsc.org

ATT Automatic Toll Ticketing. A system which telephone companies use to automatically keep call detail records including calling number, number called, time of day and length of call. The phone company uses this information, together with the cost of phone calls, to generate an invoice to its customers.

Attach A command that assigns a connection number to a workstation and attaches the workstation to the LOGIN directory on the default (or specified) file server. As many as 100 workstations can be attached to a file server running NetWare v2.2. When loaded the NetWare shell (workstation file NETx.COM) automatically attaches your workstation to the nearest file server. You can also specify in SHELL.CFG which server you prefer to attach to.

Attach Terminal To assign a terminal for exclusive use by the application program. Contrast with Detach Terminal.

Attachment The process of attaching a file to an electronic message.

Attack Time The time interval between the instant that a signal at the input of a device or circuit exceeds the activation threshold of the device or circuit, and the instant that the device or circuit reacts in a specified manner, or to a specified degree, to the input. The term often implies a protective action such as the provided by a clipper (peak limiter) or compressor, but may be used to describe the action of a device such as a vox (Voice Operated circuit), where the action is not protective.

Attempt Trying to make a telephone call. Also defined as a call offered to a telecommunications system, regardless of whether it is completed.

Attendant The "operator" of a phone system console. Typically, the first person to answer an incoming call. That person usually directs incoming calls to the proper person or department. That person may also assign outgoing lines or trunks to people requesting them. Few companies spend any time training their attendants. They should. There are two types of things attendants should be trained for: 1. Manners, including the correct way to keep people waiting and to screen incoming calls, and 2. The structure of the company. If a caller asks for some help, the attendant should know which department or person might be responsible for providing that help. Increasingly in North America, company phone systems are answered by devices called "automated attendants." They allow a caller to punch in the extension he wants or the department he wants. See AUTOMATED ATTENDANT.

Attendant Busy Lamp Field Lamps, lights or LEDs that show whether a PBX or key system extension is busy or not. These days, many attendant busy lamp fields are being incorporated into CRT displays. We hope more will do this as many lamp-based attendant busy lamp fields are difficult to read.

Attendant Call Waiting Indication An unusual feature on a PBX console. The call waiting button on the attendant console lights to indicate a predetermined number of calls in queue. The light flashes when a second (programmable) threshold is reached.

Attendant Camp-On If the extension is busy, the attendant or operator can place the call in a queue behind the call already in progress. When the call is over, the "camped-on" call will automatically ring the extension.

Attendant Conference PBX feature that allows the attendant (or operator) to establish a conference connection between central office trunks and internal phones.

Attendant Console An attendant console is the larger, specialized telephone set used by the operator or attendant to answer incoming calls and send those calls to the proper extension. Consoles are becoming more sophisticated these days in several ways. Operators need to punch fewer buttons to move calls around, while the information they present to the attendant is more useful for keeping tabs on calls and letting people know what's happening. Many consoles are acquiring TV screens that report the status of each extension, who's speaking, where the call is going, and whether there are problems, such as broken lines or trunks, etc. anywhere on the system. Some of the more modern screens will allow the operator to send messages around the company that can alert someone as to who's calling before he/she picks up the phone. You can also easily program switches through consoles with CRT (also called TV) screens. In the old days you needed to punch in complex codes. Now you can respond to "Yes/No" decisions on a screen with lots of explanatory words and help menus.

Attendant Control of Trunk Group Access The telephone operator or attendant controls the users' access to trunks for making local and/or long distance calls. This may reduce long distance call abuse.

Attendant Direct Station Select This feature gives an operator the ability to reach an extension by simply pushing one button. In direct station select, every extension has its own button. Direct Station Select usually comes with some form of Attendant Busy Lamp Field which shows whether the extensions are busy. Some attendants like direct station select. Others don't, preferring to simply punch in 345, instead of hunting for the button which corresponds to extension 345. The best consoles these days are using some form of easy-to-read screen prompts.

Attendant Exclusion A PBX feature which stops the attendant from listening in on a phone call once she or he has passed the call to the correct extension.

Attendant Forced Release An attendant-activated (pushbutton) facility that will automatically "disconnect" all parties on a given circuit when that circuit is "entered" by the attendant.

Attendant Incoming Call Control A PBX feature which diverts incoming trunk calls automatically to a predetermined phone after a predesignated period of time or number of rings.

Attendant Key Pad Allows the attendant to perform all functions using a standard touch tone key pad on the console or adjacent to it.

Attendant Locked Loop Operation PBX feature which allows the attendant at a console to retain supervision or recall capability of any particular call which has been processed.

Attendant Lockout This feature denies an attendant the ability to re-enter a phone call unless specifically recalled by that PBX extension.

Attendant Loop Transfer Allows the attendant to transfer any call to another attendant.

Attendant Monitor A special attendant circuit which allows "listening in" on all circuits with the console handset/headset transmitter deactivated.

Attendant Override A feature that allows an attendant to enter a busy trunk connection and key the trunk number within the PBX. A warning tone will be heard by the connected parties, after which they connected parties and the attendant will be in a three-way connection.

Attendant Position Where a telephone operator sits to answer calls and send them on to the people in the company. This is usually in front of a telephone system with buttons, toggle switches, etc. that facilitate this process.

Attendant Recall When a phone call has been transferred to a telephone extension and not answered, this telephone system feature sends the call back to the attendant. Sometimes the call will return to a special part of the attendant console which will indicate to the attendant that it is a "returned" call. It's a good idea to pay attention to the speed of recall back to the operator. People hate to be extended into endless ringing. Think of calling a hotel and how aggravating it is to wait until the call comes back to an operator after she/he extended it to the room...and it rings and rings.

Attendant Recall On Trunk Hold The system will recall the attendant if a trunk placed on hold is not re-entered within a predetermined time.

Attendant Transfer of Incoming Calls A PBX and Centrex feature. A telephone extension is talking on a line but that person wants to transfer the call to someone else. The person hits their hookswitch a couple of times. (The hookswitch is the toggle switch the handset depresses when you replace it.) This flashing of the hookswitch signals the attendant to join the call. The person asks the attendant "to please transfer this call." The attendant then transfers the call to the new extension. This feature is totally inefficient as it's hard to reach the attendant, who's always busy, etc. etc. All newer phones can transfer both incoming and outgoing calls automatically by just flashing the hookswitch, dialing the extension number and hanging up.

Attended A telephone system having an attendant or receptionist whose primary job is to answer all incoming calls. Many smaller systems, such as key systems, are not centrally "attended." The phone is simply answered by whoever is near. A non-attended phone system should be set up so anyone can answer an incoming call. Some systems, such as most key systems, come this way from the factory. Others, such as PBXs, have to be specially set up. Some systems can be set up so an attendant will get first shot at answering the incoming call, but then, after a couple of rings, anyone else can answer the call (perhaps a loud "night" bell will ring). This is the system we use at TELECONNECT Magazine.

Attended Mode Imagine a communications situation where your computer is connected over a phone line to another user on another computer and you are uploading and downloading files. Attended mode refers to a situation where both users manually enter the commands required to send or receive a file concurrently, usually while conversing over the phone. Compare this to UNATTENDED MODE.

Attention Key A key or combination of keys on a computer or terminal which signals the main computer to stop its present task and wait for a new command. The ESCape key is often the Attention Key. In Crosstalk, it's Control A.

Attenuate To decrease electrical current, voltage or power in communicating channel. Refers to audio, radio or carrier frequencies. See ATTENUATION.

Attenuation The decrease in power of a signal, light beam, or lightwave, either absolutely or as a fraction of a reference value. The decrease usually occurs as a result of absorption, reflection, diffusion, scattering, deflection or dispersion from an original level and usually not as a result of geometric spreading, i.e., the inverse square of the distance effect. Optical fibers have been classified as high-loss (over 100 dB/km), medium- loss (20 to 100 dB/km), and low-loss (less than 20 dB/km). In other words, attenuation is the loss of volume during transmission. The received signal is lower in volume than the transmitted signal due to losses in the transmission medium (such as that caused by resistance in the cable). Attenuation is measured in decibels. It is the opposite of Gain. Some electrical components are listed as "with attenuation" which means they will compensate for irregular electrical supply (e.g. surges). See GAIN.

Attenuation Coefficient The rate at which average power decreases with distance.

Attenuation Equalizer Any device inserted in a transmission line or amplifier circuit to improve the shape of its frequency response.

Attenuator A device to reduce signal amplitude by a known amount without introducing distortion.

ATTIX See AT&T INTEREXCHANGE COMPANY.

ATTND Attendant. (What else?)

Atto Atto means one quintillion, which is 10 to the power of 18. See also FEMTO, which is 10 to the power 15.

Attribute The form of information items provided by the X.500 Directory Service. The directory information base consists of entries, each containing one or more attributes. Each attribute consists of a type identifier together with one or more values. See ATTRIBUTES.

Attributes Information about a MS-DOS file that indicates whether the file is read-only, hidden, or system, and whether it has been changed since it was last backed up. You can assign attributes to a file using the ATTRIB command. You can identify a file as read-only (meaning others can't change it, but can read it) and/or as a file you want to archive when using the BACKUP, RESTORE, and XCOPY commands. The command to make a file read only is typically

ATTRIB +R filename

By using the ATTRIB command to make a file "read only," you also make it impossible to erase the file from your disk. If you want to remove the "read only" protection, i.e. make the file "read and write, the command is

ATTRIB -R filename

ATUG Australian Telecommunications Users Group, based in Milsons Point, Sydney, Australia.

ATV Advanced TV. Refers to any system of distributing television programming that generally results in better video and audio quality than that offered by the NTSC 525-line standard. This group of techniques is based on digital signal processing and transmission. HDTV (High Definition TV) can be considered one type of ATV. Although ATV systems are collectively considered to offer better quality than the NTSC signal, they can carry multiple pictures of lower-quality and can also support the cancellation of artifacts in ordinary NTSC signals.

ATX Audiotex. Interactive voice response systems that deliver information or entertainment to general telephone callers, i.e. anyone with a phone. Audiotex services are typically widely advertised. They include everything from sex to the weather. See also AUDIOTEX.

AuC Authentication Center. A functional piece of the HLR (Home Location Register) used to authenticate the user of mobile station equipment, and to protect each subscriber or subscription number from unauthorized access. The AuC performs secret, mathematical computations to verify the authenticity of the mobile station to complete the call.

Audible Indication Control Three fancy words for the ability to turn up or down the bell or beeper on your PBX attendant console.

Audible Ring A sound sent from the called party's switch to inform the calling party that the called line is being rung. A long explanation for a bell or buzzer that tells you it's for you.

Audible Ringing Tone The information tone sent back to the calling telephone subscriber as an indication that the called line is being rung.

Audible Tones Audible tones are the sounds provided by the network or an attached switch to inform callers of the status of the line or of an event. Audible tones most frequently encountered in computer telephony applications include: ringing, busy tones (called party busy and network busy), SIT tones (Special Information Tones), and special tones used by computer telephony systems such as the "record at the beep" (which is usually a 1,000 HZ tone).

Audio Sound you hear which may be converted to electrical signals for transmission. A human being who hasn't had his ears blown by listening to a Sony Walkman or a ghetto blaster can hear sounds from about 15 to 20,000 hertz.

Audio Frequencies Those frequencies which the human ear can detect (usually in the range of 20 to 20,000 hertz). Only those from 300 to 3,000 hertz are transmitted through the phone, which is why the phone doesn't sound "Hi-fi."

Audio Frequency The band of frequencies (approximately 30 hertz to 20,000 hertz) that can be heard by the healthy human ear.

Audio Interchange File Format (AIFF) A sound file format common to the Macintosh platform used to store digital audio data. This file format also supports position and loop point markers used by audio recording/editing software. Can be compressed into the AIFF-C format.

Audio Menu Options spoken by a voice processing system. The user can choose what he wants done by simply choosing a menu option by hitting a touchtone on his phone or speaking a word or two. Computer or voice processing software can be organized in two basic ways — menu-driven and non-menu driven. Menu-driven programs are easier for users to use, but they can only present as many options as can be reasonably spoken in a few seconds. Audio menus are typically played to callers in automated attendant/voice messaging, voice response and transaction processing applications. See also MENU and PROMPTS.

Audio Messaging Interchange Specification AMIS. Issued in February 1990, AMIS is a series of standards aimed at addressing the problem of how voice messaging systems produced by different vendors can network or inter-network. Before AMIS, systems from different vendors could not exchange voice messages. AMIS deals only with the interaction between two systems for the purpose of exchange voice messages. It does not describe the user interface to a voice messaging system, specify how to implement AMIS in a particular systems or limit the features a vendor may implement.

AMIS is really two specifications. One, called AMIS-Digital, is based on completely digital interaction between two voice messaging systems. All the control information and the voice message itself, is conveyed between systems in digital form. By contrast, the AMIS-Analog specification calls for the use of DTMF tones to convey control information and transmission of the message itself is in analog form. AMIS was discussed in detail in the October 1990 issue of Business Communications Review, a monthly magazine out of Hinsdale, IL. AMIS specifications are available from Alison Caughman 404-355-7785.

Audio Response Unit A device which translates computer output into spoken voice. Let's say you dial a computer and it said "If you want the weather in Chicago, push 123, then it would give you the weather. But that weather would be "spoken" by an audio response unit. Here's a slightly more technical explanation: An audio response unit is a device that provides synthesized voice responses to dual-tone multi-frequency signaling input. These devices process calls based on the caller's input, information received from a host data base, and information carried with the incoming call (e.g., time of day). ARUs are used to increase the number of information calls handled and to provide consistent quality in information retrieval. See also AUDIOTEX and INTERACTIVE VOICE RESPONSE.

Audio Track The section of a videodisc or tape which contains the sound signal that accompanies the video signal. Systems with two separate audio tracks (most videodiscs) can offer either stereo sound or two independent soundtracks.

Audiographics The technology which allows sound and visual images to be transmitted simultaneously. According to AT&T, audiographics generally refers to single frame or slow frame visual images as opposed to continuous frame image transmission (e.g. television). Audiographic transmission is often used to teach or train people in remote locations from an educational institution or business training center, saving travel and housing expense.

Audiotex A generic term for interactive voice response equipment and services. Audiotex is to voice what on-line data processing is to data terminals. The idea is you call a phone number. A machine answers, presenting you with several options, "Push 1 for information on Plays, Push 2 for information on movies, Push 3 for information on Museums." If you push 2, the machine may come back, "Push 1 for movies on the south side of town, Push 2 for movies on the north side of town, etc." See also INFORMATION CENTER MAILBOX.

Audiotext A different, and less preferred, spelling of AUDIOTEX. See AUDIOTEX.

Audio/visual Multimedia Services AMS is an ATM term. It specifies service requirements and defines application requirements and application program interfaces (APIs) for broadcast video, videoconferencing, and multimedia traffic. AMS is being developed by the ATM Forum's Service Aspects and Applications (SAA) working group. An important debate in the SAA concerns how MPEG-2 applications will travel over ATM (asynchronous transfer mode). Early developers chose to carry MPEG-2 over ATM adaptation layer 1 (AAL 1); others found AAL 5 a more workable solution. Recently, some have suggested coming up with a new video-only AAL using the still-undefined AAL 2.

Auditory Pattern Recognition Auditory pattern recognition is the ability to recognize spoken words.

Audit To conduct an independent review and examination of system records and activities in order to test the adequacy and effectiveness of data security and data integrity procedures, to ensure compliance with established policy and operational procedures, and to recommend any necessary changes.

Audit File On some systems, each time a billing file is generated, an audit file is created to record the details of the generation process.

Audit Trail A record of all the events that occur when users request and use specific resources. An audit trail gives you the ability to trace who did what and who was responsible for what. An audit trail is a chronological record of system activities that is sufficient to enable the reconstruction, review, and examination of the sequence of environments and activities surrounding or leading to an operation, a procedure or an event in a transaction from its inception to final results. Audit trail may apply to information in an automated information system, to the routing of messages in a communications system, or to material exchange transactions, such as in financial audit trails. Audit trails are great for finding out what activity caused the disaster, the network to crash, etc. Audit trails are great for tracing crooks, i.e. unauthorized break-ins.

Auditing 1. An AT&T ISDN term for the monitoring of the status of either the network or the ISDN set.
2. Tracking activities of users by record and selected types of events in the security log of a server or a workstation.

Audix Voice Power A voice processing application, part of Integrated Solution III (IS-III), that provides Automated Attendant, call answering, voice-mail, message drop, and information services for use with the AT&T Merlin telephone system.

Auger A type of drill bit typically used to make large, deep holes for passing wire or cable through wood.

Augmentation Of Bandwidth Bandwidth augmentation is the ability to add another communications channel to an already existing communications channel.

AUI Autonomous Unit Interface or Attachment Unit Interface. Most commonly used in reference to the 15 pin D type connector and cables used to connect single and multiple channel equipment to an Ethernet transceiver.

AUP Acceptable Use Policy. The term used to refer to the restrictions placed on use of a network; usually refers to restrictions on use for commercial purposes.

Aural Relating to the sense of hearing.

AUSTPAC A packet-switched network implemented in Australia.

Authenticate To establish, usually by challenge and response, that a transmission attempt is authorized and valid. To verify the identity of a user, device, or other entity in a computer system, or to verify the integrity of data that have been stored, transmitted, or otherwise exposed to possible unauthorized modification. A challenge given by voice or electrical means to attest to the authenticity of a message or transmission.

Authentication The process whereby a user or information source proves they are who they claim to be. In other words, the process of determining the identify of a user attempting to access a system. Authentication codes are exchanged between two computers that use SMTP. I believed that defining authentication was pretty simple until I came across the following definition in a book on cryptography. Here it is: Authentication is any technique enabling the receiver to automatically identify and reject messages that have been altered deliberately or by channel errors. Also, can be used to provide positive identification of the sender of a message. Although secret (symmetric) key algorithms may be used for some authentication without the prior sharing of any secrets between the messaging parties. See AUTHENTICATION TOKEN.

Authentication Center See AuC.
Authentication Token A portable device used for authenticating a user. Authentication tokens operate by challenge/response, time-based code sequences, or other techniques. An example is Security Dynamics Technologies Inc.'s SecurID card. See AUTHENTICATION.
Auto Discovery Auto discovery is what takes place when a new device is added to a network. That device (e.g., an ethernet board, a modem, a router) communicates to the network management system that it is online and defines its key characteristics.
Autochanger A jukebox-style optical media system that permits the storage and playback multiple discs. Autochangers are available for both 5-,8, and 12-inch optical discs.
Authoring Authoring is the process of using multimedia applications to create multimedia materials for others to view. Multimedia authoring uses many tools, from the more familiar text editor or desktop publishing application, to tools for capturing and manipulating video images or editing audio files. Authors might include specialized creators of training, sales, or corporate applications such as insurance claims processing. Or, they might be creators of everyday business communications like voice-annotated email. Over time, everyone involved in business communications will probably have some level of multimedia authoring capability.
Authoring System Software which helps developers design interactive courseware easily, without heavy computer programming. A specialized, high-level, plain-English computer language which permits non-programmers to perform the programming function of courseware development. The program logic and program content are combined. See AUTHORING.
Authorized Agent Also called Authorized Sales Agent. A term chosen by some of the Bell operating companies and many of the cellular phone companies to refer to companies which sell their network services on commission. Some of these companies have specific industry knowledge and have written specialized software. The idea is to work with businesses to arm them with the absolute best package of telecommunications hardware, software and services.
Authorized Bandwidth The necessary bandwidth required for transmission and reception of intelligence. This definition does not include allowance for transmitter drift or Doppler shift.
Authorized Dealer See DEALER.
Authorized Frequency A frequency that is allocated and assigned by an authority to a specific user for a specific purpose.
Authorization Think of charging things on your MasterCard, Visa, or American Express card. If the store cannot authorize the amount of your purchase, your Visa card will not allow you to make the purchase. Authorization is needed for many long distance calls, especially those made using credit cards, telephone company calling card, etc. Authorization is done by the operator's computer checking with the remote validation database service. See BVA, BVS and VALIDATION.
Authorization Code A code in numbers and/or letters employed by a user to gain access to a system or service. If you are making a call out on a restricted line, the PBX will ask you for an authorization code. If you have one, your call will go through. If not, your call will be denied (i.e. not go through). Authorization codes come in various flavors. Some can be used for making long distance calls. Some can be used also for international calls, etc. See AUTHORIZED USER.
Authorized User A person, firm, corporation or any other legal entity authorized by the provider of the service to use the service being provided.
Auto Adjust An AT&T ISDN term. The automatic setting of the baud rate. The baud rate is automatically set through the use of the AT portion of the command line. The speed of the computer is determined from the A and the parity is determined from the combination of the A and T. For this reason, it is not necessary to set speed and parity parameters for the ISDN set.
Auto Answer The capability of a phone, a terminal, a video phone, a modem or a computer to answer an incoming call and to set up a connection without anyone actually doing anything to physically answer the call. I have auto answer on my video phone. When it rings, my phone answers and I see who's calling. It's nice.
Auto Attendant A shortened name for an automated attendant, a device which answers a company's phones, encourages you to touchtone in the extension you want, and rings that extension. If that extension doesn't answer, it may send the call to voice mail or back to the attendant. It may also allow you to punch in digits and hear information, e.g. the company's hours of business, addresses of local branches, etc. See also AUTOMATED ATTENDANT.
Auto Baud Automatic speed recognition. The ability of a device to adapt to the data rate of a companion device at the other end of the link.
Auto Baud Detect See AUTO BAUD.
Auto Busy Redial A feature of a phone or phone system where the phone has the ability to keep trying a busy number until answered. The circuit actually recognizes the busy tone, hangs up, and dials again. One of the greatest time-savers ever invented.
Auto Call Automatic Calling; a machine feature that allows a transmission control unit or a station to automatically initiate access to (i.e. dial) a remote system over a switched line.
Auto Dial A feature of phone systems and modems which allows them to dial a long phone number (usually long distance) by punching fewer buttons than there are numbers to dial. One button auto dial on electronic phones is very common these days. Most communications software programs will allow you to auto dial a string of 35 to 40 digits, which you may need if you're dialing through a complex network.
Auto Dial Auto Answer A modem feature. Auto Dial lets you dial a phone number through your modem, using your personal computer or data terminal keyboard. Auto Answer permits the modem to automatically answer the incoming call without anybody having to be there.
Auto Dialer See AUTOMATIC DIALER.
Auto Fax Tone Also called CNG, or Calling Tone. This tone is the sound produced by virtually all Group 3 fax machines when they dial another fax machine. CNG is a medium pitch tone (1100 Hz) that lasts 1/2 second and repeats every 3 1/2 seconds. A FAX machine will produce CNG for about 45 seconds after its dials. See also CNG.

NEWTON'S TELECOM DICTIONARY

Auto Line Feed An instruction in a communications program which causes the program to perform a Line Fee (LF) when you hit a carriage return or the "Enter" key.

Auto Partition A feature of 10BaseT. When 32 consecutive collisions are sensed by a port in a hub or concentrator from its attached workstation or network segment, or when a packet that far exceeds the maximum allowable length is received, the port stops forwarding packets. The port continues to monitor traffic and will automatically begin normal packet forwarding when the first correct packet is received.

Auto Recognition A term used in file conversion in which your conversion software figures out by itself in what form the original file was — WordPerfect 5.0, Word 6.0, Wordstar 5.5 etc. See also AUTO STYLING.

Auto Selection Tool An imaging term. A tool that selects an entire area within a specified range of color values around a selected pixel.

Auto Sensing See AUTO STYLING.

Auto Start A standby electrical power system that starts up when the normal supply of commercial power fails.

Auto Stream An AT&T ISDN term. The method of data flow in which both channels between the ISDN set and the application are in use simultaneously.

Auto Styling Auto styling is a term we found in a database conversion software program. What it means is that the program looks at the data in a field and determines from that data if the field is a numeric, character or memo, etc. The problem with auto styling is that it's frequently wrong. For example, it might check one field, find all numbers and decide it's a numeric field. Such a field might be a zip code, which actually is normally a character field. One reason why you might want you zip code to be a character field is that character fields are set left. Numeric fields are set right. (They line up at the decimal point.) Another name for auto styling is auto sensing.

Autoanswer A feature of a telephone which automatically answers incoming calls without the user of the phone lifting a handset or otherwise answering the call. Modems and fax machines also autoanswer. In North America, it's spelled AUTO ANSWER. In Britain, it's spelled AUTOANSWER.

Autoattendant See AUTOMATED ATTENDANT.

Autobauding The process by which the terminal software determines the line speed on a dial-up line.

AUTODIN The worldwide data communications network of the U.S. Department of Defense. Acronym for "AUTOmatic DIgital Network."

Autodial Button An Autodial button on a phone provides one-touch dialing of outside numbers, intercom numbers, or feature codes.

Autoexec.bat The first file executed after a PC is started (i.e. cold or warm booted). AUTOmatic EXECute BATch file contains a list of MS-DOS commands which are executed by your PC once your machine is turned on. Essentially, once you've started your MS-DOS computer, your asks "OK, you're got me started now. What would you like me to do?" Your AUTOEXEC.BAT contains these answers. Your AUTOEXEC.BAT file allows you to tailor the MS-DOS operating system to your needs, for example, setting the PATH command. The AUTOEXEC.BAT file must be installed in the root directory of your boot disk — either hard or floppy disk (the place your computer looks for the initial MS-DOS commands).

Automated Attendant A device which is connected to a PBX. When a call comes in, this device answers it and says something like, "Thanks for calling the ABC Company. If you know the extension number you'd like, pushbutton that extension now and you'll be transferred. If you don't know your extension, pushbutton "0" (zero) and the live operator will come on. Or, wait a few seconds and the operator will come on anyway." Sometimes the automated attendant might give you other options, such as, "dial 3" for a directory of last names and dial 4 for a directory of first names. Automated attendants are also connected also to voice mail systems ("I'm not here. Leave a message for me."). Some people react well to automated attendants. Others don't. A good rule to remember is before you spring an automated attendant on your people/customers/subscribers, etc., let them know. Train them a little. Ease them into it. They'll probably react more favorably than if it comes as a complete surprise. The first impression is rarely forgotten, so try to make it a good experience for the caller. See also DIAL BY NAME.

Automated Coin Toll Service ACTS. In the old days, operators handled routine toll calls by counting the sound of coins hitting the box, checking prices, putting calls through, figuring and collecting overtime charges, etc. ACTS does all this automatically. It figures charges, tells those charges by digitized computerized voice to the customer, counts the coins as they are deposited and then sets up the call.

Automated Intercept Call Completion AINTCC. A new feature of Northern Telecom's central offices. The AINTCC feature provides options for
- connecting a caller automatically to an intercepted number after hearing an announcement, or
- connecting a caller to an intercepted number without an announcement.
Not using an announcement makes the number change transparent to the caller. The called (intercepted) party then has the option of informing the caller of the number change.

Automated Maritime Telecommunications System An automatic, integrated and interconnected maritime communications system serving ship stations on specified inland and coastal waters of the United States.

Automated Radio A radio with the capability for automatically controlled operation by electronic devices that requires little or no operator intervention.

Automated Tactical Command And Control System A command and control system or part thereof which manipulates the movement of information from source to user without intervention. Automated execution of a decision without human intervention is not mandatory.

Automated Voice Response Systems AVRS. Devices which automatically answer calls. They may simply inform the caller that the call is in a queue and will be answered soon; alternatively they can prompt the caller to use voice commands, or touchtones to seek more information.

Automatic Address Discovery This refers to the process by which a network device can poll other network devices to discover the network addresses which each device supports. Automatic address discovery makes the set up and on-going maintenance of complex internetworks much simpler than if all address updates were performed manually.

Automatic Button Restoration When the telephone handset of a multi-line instrument (typically a 1A2 multi-line key set) is placed back in its cradle, the line button being used automatically "pops" back up. Conversely, when a user picks up the handset, he must always push down a line button to make a call. Most phones with this feature can be disabled, so the buttons stay down when the handset sits on the cradle. A twist of a single screw inside the instrument will usually solve the aggravation of the automatic button restoration. Some people like automatic button restoration because it saves a user from accidentally barging into someone else's call. This was a much greater problem with 1A2 key systems. It no longer is a problem with most electronic key systems since they usually extend the user automatic privacy once they get on a call so no one else can barge in, even if they want to.

Automatic Call Distributor ACD. A specialized phone system designed originally for handling many incoming calls, now increasingly used by companies also making outgoing calls. You receive and make lots of phone calls typically to customers. You need an ACD. Once used only by airlines, rent-a-car companies, mail order companies, hotels, etc., it is now used by any company that has many incoming calls (e.g. order taking, dispatching of service technicians, taxis, railroads, help desks answering technical questions, etc.). There are very few large companies today that don't have at least one ACD. Many smaller companies, like the company that publishes this dictionary, also has one.

An ACD performs four functions. 1. It will recognize and answer an incoming call. 2. It will look in its database for instructions on what to do with that call. 3. Based on these instructions, it will send the call to a recording that "somebody will be with you soon, please don't hang up!" or to a voice response unit (VRU). 4. It will send the call to an agent as soon as that operator has completed his/her previous call, and/or the caller has heard the canned message.

The term Automatic Call Distributor comes from distributing the incoming calls in some logical pattern to a group of operators. That pattern might be Uniform (to distribute the work uniformly) or it may be Top-down (the same agents in the same order get the calls and are kept busy. The ones on the top are kept busier than the ones on the bottom). Or it may be Specialty Routing, where the calls are routed to answerers who are most likely to be able to help the caller the most. Distributing calls logically is the function most people associate with an ACD, though it's not the most important.

The management information which the ACD produces is much more valuable. This information is of three sorts: 1. The arrival of incoming calls (when, how many, which lines, from where, etc.) 2. How many callers were put on hold, asked to wait and didn't. This is called information on ABANDONED CALLS. This information is very important for staffing, buying lines from the phone company, figuring what level of service to provide to the customer and what different levels of service (how long for people to answer the phone) might cost. And 3. Information on the origination of the call. That information will typically include ANI (Automatic Number Identification — picking up the calling number and DNIS (Direct Number Identification Service) picking up the called number. Knowing the ANI allows the ACD and its associated computer to look up the caller's record and thus offer the caller much faster service. Knowing the DNIS may allow the ACD to route the caller to particular agent or keep track of the success of various advertising campaigns. Ad agencies will routinely run the same ad in different towns using different 800 phone numbers. Picking up which number was called identifies which TV station the ad ran on.

The seven definitions that follow the definition "ACD" show some of the features which newer ACDs have. See also 800 SERVICE, ACD and AUTOMATIC CALL SEQUENCER.

Automatic Call Intercept A feature of a Rolm ACD. This feature automatically forwards calls to an attendant if the dialed number is not installed or out of order. It can also intercept an attempted trunk call that is in violation of a Class of Service restriction. Automatic Call Intercept will also recall the attendant after a predetermined period of offshoot inactivity (e.g. flash or hold).

Automatic Call Rescheduling When a call is unsuccessful, either no reply, or busy, the system will automatically dial the number again after a pre-determined time.

Automatic Call Sequencer A device for handling incoming calls. Typically it performs three functions. 1. It answers an incoming call, gives the caller a message, and puts them on "Hold." 2. It signals the agent (the person who will answer the call) which call on which line to answer. Typically, the call which it signals to be answered is the call which has been on "hold" the longest. 3. It provides management information, such as how many abandoned calls there were, how long the longest person was kept on hold, how long the average "on hold" was, etc.

There are three types of devices which handle incoming calls. The least expensive is the Automatic Call Sequencer which is traditionally used with key systems. It differs from Uniform Call Distributors (UCDs) and Automatic Call Distributors (ACDs) in that it has no internal switching mechanism and does not affect the call in any way. It simply recommends which call should be picked up and keeps statistical information on the progress of calls. A more expensive type of device is the UCD.

The most full-featured and expensive is the ACD. Distinctions between ACDs and UCDs and/or PBXs with features called UCDs and ACDs are blurring as UCDs get more sophisticated. The main difference, as we understand it, is that a UCD offers fewer options for routing an incoming call and answering calls in any particular order. ACDs typically produce the most detailed management information reports. One company also makes something called an Electronic Call Distributor. It is essentially an automatic call distributor.

Automatic Callback When a caller dials another internal extension and finds it busy, the caller dials some digits on his phone or presses a special "automatic callback" button. When the person he's calling hangs up, the phone system rings his number and the number of the original caller and the phone system automatically connects the two together. This feature saves a lot of time by automatically retrying the call until the extension is free. See also CAMP ON. Wouldn't it be nice if they had this feature on long distance calls?

Automatic Calling Unit ACU. A device that places a telephone call on behalf of a computer.

Automatic Circuit Assurance ACA is a PBX feature that helps you find bad trunks. The PBX keeps records of calls of very

NEWTON'S TELECOM DICTIONARY

short and very long duration. If these calls exceed a certain parameter, the attendant is notified. The logic is that a lot of very short calls or one very long call may suggest that a trunk is hung, broken or out of order. The attendant can then physically dial into that trunk and check it.

Automatic Cover Letter In a fax transmission, an automatic cover letter allows the user to automatically attach a cover letter to the document being sent. This is especially convenient when sending material directly from your PC.

Automatic Dialer Or Autodialer A device which allows the user to dial pre-programmed telephone numbers by pushing one or two buttons. Sometimes referred to as a "repertory" dialer. Dialers can be bought as a separate device and added to a phone, however, today most telephone sets are outfitted with autodialers. There are four basic measures of an automatic dialer's efficiency. 1. What's the longest number it will dial automatically? This is important because using some of AT&T's long distance competitors requires dialing lots of numbers, with lots of pauses. 2. How many numbers will it dial? Some people like to have a dialer which dials hundreds of numbers. Others like a small one, just for their most frequently called numbers. 3. Will the dialer recognize dial tone? This is important because using a long distance company or dialing through a PBX requires one to recognize consecutive dial tones. 4. Can you "chain" dial? In other words, can you hit one speed dial button after another and have the machine dial through a complex network and throw in authorization codes, etc.?

Automatic Dialing See SPEED DIALING.

Automatic Directory Propagation In electronic mail, automatic directory propagation is the ability to update addresses automatically in one domain after manually entering address changes in another domain, whether on the same LAN or another LAN connected by a gateway. In general, automatic directory propagation can be peer-to-peer, where changes in any post office are sent to all other post offices, or master-to-slave, where changes in the master post office are sent to the slaves, but changes in the slave post office do not go to the master.

Automatic Equalization The process of compensating for distortion of data communications signals over an analog circuit.

Automatic Exchange A term for a central office which automatically and electronically switches calls between subscribers without using an operator. Not a common term.

Automatic Facilities Test System AFACTS is a Rolm CBX feature. It is an automatic testing system for identifying faulty tie and central office trunks. AFACTS can pinpoint faulty trunks and generate exception and summary reports.

Automatic Fallback A modem's ability to negotiate an appropriate data rate with the modem on the other end of the link, depending on line quality. For example, if two 2400 baud modems can not pass data at 2400 baud, they would "fall back" to 1200 baud automatically in order to transmit data without excessive errors.

Automatic Frequency Control A circuit in a radio receiver which automatically brings the tuning units of the set into resonance with a wave which is partially tuned in.

Automatic Gain This is an electronic circuit which automatically increases the volume when someone is speaking quietly and drops it when someone is speaking loudly. The idea is to keep the transmitted signal even. Most tape recorders, for example, have automatic gain circuits. This allows them to pick up voices of people in a room, even though the volume of each person's conversation arriving at the tape recorder is different. The problem with automatic gain circuits is they're always looking for something to amplify. Such that when it's quiet (and meant to be) the automatic gain circuit will also try and amplify the ambient noise in the room — to keep the sound level constant. All professionally recorded tapes are done on tape reorders with manual volume controls.

Automatic Hold — Station or Intercom When a user is having a conversation and receives another call, he may press the button to answer that new call. The call he was on originally is automatically put on hold.

Automatic Identified Outward Dialing AIOD. The toll calls placed by all extensions on the telephone systems are automatically recorded. This information allows bills to be sent, long distance lines to be chosen, etc. Some central offices, for example, can provide an itemized breakdown of charges (including individual charges for toll calls) for calls made by each CPE telephone extension. See CALL ACCOUNTING, CALL DETAIL RECORDING, SMDR and AIOD.

Automatic Intercept Center AIC. A Bellcore definition: "The centrally located set of equipment that is part of an Automatic Intercept System (AIS) that automatically advises the calling customer, by means of their recorded or electronically assembled announcements, of the prevailing situation that prevents completion of connection to the called number."

Automatic Level Control ALC. A control system that adjusts the incoming signal to a predetermined level. Somewhat similar to automatic gain control. See AUTOMATIC GAIN CONTROL.

Automatic Light Control ALC. Vidicon camera control which automatically adjusts the target voltage to compensate for variations in light levels. See also AUTOMATIC GAIN.

Automatic Line Hold A PBX feature. As long as a phone does not go "on-hook," activation of various line pushbuttons will automatically place the first line on hold without the use of special "hold" button.

Automatic Link Establishment ALE. The capability of an HF radio station to contact, or initiate a circuit, between itself and another specified radio station, without operator assistance and usually under computer control. ALE techniques include automatic signaling, selective calling, and automatic handshaking. Other automatic techniques that are related to ALE are channel scanning and selection, Link Quality Analysis (LQA), polling, sounding, message store and forward, address protection, and anti-spoofing.

Automatic Message Accounting AMA. 1. Service mark of AT&T. A system which records and documents billing information for long distance calls made by a corporate subscriber. The calls must be made to billable destinations to be documented. The information recorded is generally either 20% of all information (CCSA networks) or 100% of the information (all others).
2. The network functionality that measures, collects, formats, and outputs subscriber network-usage data to upstream billing OSs and other OSs (Operations Systems). Definition from Bellcore.

Automatic Message Link The name of a Bell Atlantic service, which allows you to forward your calls directly to your company message-taking system so your customers can always leave messages.

Automatic Message Switching A technique of sending messages to their appropriate destination through information contained in the message itself — typically in its "address."

Automatic Network Dialing A Rolm feature which is said to simplify on-network calling among multiple business locations by providing single-step dialing for private networks and a single numbering/dialing plan.

Automatic Network Restoral Automatic network restoral is a term which reflects the ability of a network to restore service rapidly and automatically following a catastrophic failure, such as that of a network cable cut. The result is higher network availability and reliability.

Automatic Number Identification ANI. Being able to recognize the phone number of the person calling you. You must have equipment at your office. And the network must have the ability to send the calling number to you. For a much longer explanation, see ANI, CALLER ID, CLASS, ISDN and System Signaling 7.

Automatic Overflow To DDD Toll calls jump to expensive direct distance dialed calls, when all lower cost FX, WATS lines, etc. are busy.

Automatic Phone Relocation The Automatic Phone Relocation feature now available on some phone systems allows a telephone to retain its personal and system programming when it is reconnected to another physical location.

Automatic Privacy When someone is speaking on a phone line or on an intercom, this feature ensures no one else can accidentally or deliberately butt into that conversation. If you did, however, want somebody else to come into the conversation (for example, someone to provide some additional information), there's usually another feature called Privacy Release. By pushing this button on the phone, other people who have the same extension button or intercom button on their phones can then push their buttons and join the conversation. Or you can bring them into your conversation by dialing them in. Most modern key systems come with Automatic Privacy. Many people don't like it, especially those who live in small offices. Some newer phone systems are coming standard without it. And you have to program it in, if you want it.

Automatic Program Load APL is a PBX feature that allows it to load its own software into RAM from a local device such as a hard disk or a floppy disk. All this takes place automatically without human intervention. APL is an important feature since it often determines how fast a PBX can get back into service after some sort of failure — usually a failure in commercial power.

Automatic Protection Switching Switching architecture designed for SONET to perform error protection and network management from any point on the signal path.

Automatic Queuing Queuing is exactly as it sounds. Something you want is being used. So you get placed in line for that device. There are two types of queuing — automatic and manual. Manual is when you're put in queue by a person, for example an operator. Automatic is when you're put in queue by a machine, for example a PBX aided by its software.

Automatic Recall 1. A central office feature which gives telephone subscribers the ability to automatically redial their last incoming call — without actually knowing that number. On some central offices it is now possible for the calling person to block the ability of someone they've called to automatically call them back.
2. A PBX feature which returns a call to the PBX attendant (or alerts the attendant) if a call extended to a telephone is not answered within a pre-set period of time. The most logical time is three rings, or 18 seconds. This feature allows the attendant to give the caller some information, take a message or connect the caller to someone else. Most hotel switches have this feature. And when the call doesn't get answered, the switch sends it back to the operator. The sad thing is that the hotel operator is usually so busy, he/she keeps you waiting another 20 or 30 seconds, irritating you.

Automatic Recovery Your telephone system dies — typically because its power is cut off. Once the power comes on, instructions in the machine direct it to reload its software so that within minutes the system can be back and running normally. Those "instructions" are normally not affected by power drops.

Automatic Replication Also known as data replication. It refers to the process of automatically duplicating and updating data in multiple computers on a network. The word "automatically" in this case means that the process of duplication is handled automatically by the software responsible for the data. The idea is that if one or several of the machines do down, the company using the data will still have reliable delivery of data.

Automatic Rerouting This refers to the process by which an intelligent voice or data network can automatically route a call, or virtual circuit, around a network failure. With frame relay, PVCs represent a fixed path through the network. However, in the event of a network failure along the primary path over which the PVC is routed, the PVC will be automatically routed to a secondary network path until the primary path is physically restored.

Automatic Restart 1. The mechanism whereby, after a power failure and then power resumes, a process automatically restarts. Restart is from the exact point of interruption.
2. System facilities that allow restart from the point of departure after system failure.

Automatic Ringdown Tie Trunk A direct path signaling facility to a distant phone. Signaling happens automatically when you lift the receiver on either phone. See also MANUAL RINGDOWN TIE TRUNK.

Automatic Rollback A feature of the Transaction Tracking System (TTS) that returns a database on a Novell NetWare local area network to its original state. When a network running under TTS fails during a transaction, the database is "rolled back" to its most recent complete state. This prevents the database from being corrupted by the incomplete transaction.

Automatic Route Selection Your phone system automatically chooses the least cost way of sending a long distance call. See LEAST COST ROUTING and ALTERNATE ROUTING.

Automatic Scheduled Testing AST. A method of testing switched access service (Feature Groups B, C, and D) where the customer provides remote office test lines and 105 test lines with associated responders or their functions' equivalent; consists of monthly loss and C-message noise tests and annual balance test.

Automatic Secure Voice Communications Network AUTOSEVOCOM. A worldwide, switched, secure voice network developed to fulfill DoD long-haul, secure voice needs.

Automatic Send/Receive ASR. A data device in which the transmitting thing is different from the receiving part, thus enabling the device to receive calls and transmit them simultaneously. See ASR for a different definition. We're not sure precisely which one is right. Both could be.

Automatic Sequential Connection A service feature provided by a data service to connect automatically, in a predetermined sequence, the terminals at each of a set of specified addresses to a single terminal at a specified address.

Automatic Speed Matching The ability of an asynchronous modem to automatically determine whether it is expected to communicate at 300, 1200 or 2400 bps.

Automatic Time-Out On Uncompleted Call A PBX feature. If a phone stays "off-hook" without dialing for a predetermined time interval, or stays connected to a busy signal longer than the predetermined time interval, the intercom switching equipment will automatically connect this phone to intercept.

Automatic Toll Ticketing A system which makes a record of the calling phone number, the called number, the time of day, the length of the call, etc. and then generates an instant phone bill for that call. Often used in hotel/motels.

Automatic Traffic Overload Protection ATOP. A Rolm feature defined as a dynamic form of line-load control, which automatically denies a dial tone during those periods when the Rolm CBX may become overloaded. One wonders why someone would create this feature.

Automatic Voice Network AUTOVON. The principal long-haul, unsecure (meaning it's not secure) voice communications network within the Defense Communications System.

Automatic Volume Control A circuit in a radio receiver; automatically maintains various received transmissions at approximately the same volume.

Automatic Wakeup The capability for the user to schedule a wake-up call to a predetermined telephone number, either one time or daily.

Automatic Wakeup Service The guest or the operator dials into a machine which records a request for a guest wakeup call the following morning. The auto wakeup machine is a glorified, programmable auto-dial answering machine. The machine is said to save hotels money and make wakeup calls more reliable, and certainly more anonymous.

AUTONET A packet-switched network implemented in the U.S. by ADP, Ann Arbor, MI 48106.

Autonomous System A collection of routers under a single administrative authority using a common Interior Gateway Protocol for routing packets.

Autonomous System Number ASN. A unique number assigned by the InterNIC that identifies an autonomous system in the Internet. ASNs are used by routing protocols (like BGP — Border Gateway Protocol) to uniquely define an autonomous system.

Autoscaling A drawing feature that automatically adjusts the axis units of a graph to the minimum and maximum numerical values of a set of data.

Autosearch See RECORDER.

Autoseed An United Kingdom definition. The process of selecting records from a database using pre-defined criteria and allocating the records to outbound or mailing campaigns.

AUTOSEVOCOM The AUTOmatic SEcure VOice COMmunications system of the U.S. Department of Defense. A worldwide, switched, secure voice network developed to fulfill DoD long-haul, secure voice needs.

Autotimed Recall When a user places a call on hold and forgets about it, Autotimed Recall will ring that user or the receptionist after a predetermined time. That time is usually programmable. It shouldn't be longer than 30 seconds, otherwise your customers, sitting endlessly on your eternity hold, will go nuts and go elsewhere.

Autotimed Transfer This telephone system feature switches unanswered incoming calls to a backup answering position after a predetermined (usually adjustable) interval of time.

Autotype Protocol Hayes Microcomputer definition for a file transfer protocol which allows the user to automatically "type" a disk file, the clipboard or the contents of Smartcom Editor in either plain text (ASCII) or ANSI.SYS format to a remote computer. Pacing, send lines and await character echo options are provided. If necessary, character set mapping translates between different code pages and systems (Macintosh or Windows text files, for example).

AUTOVON The AUTOmatic VOice Network. The principle long-haul voice communications network within the Defense Communications System. The system includes conferencing and secure voice communications (scrambling), among other features.

Auxiliary Equipment See also PERIPHERAL DEVICE or APPLICATIONS PROCESSOR.

Auxiliary Equipment Access The ability of a telephone system to interface with (i.e. talk to) auxiliary equipment such as a paging system or dial dictation system.

Auxiliary Line A telephone trunk in addition to the main number you rent from the phone company. Phone systems are often equipped for calls to hunt from a busy main number to one or more auxiliary lines (Incoming Service Group, or ISG). For example, The Telecom Library's main number is 212-691-8215. But The Telecom Library also has 8216, 8217, 8218 and several unmarked or coded trunks. These are auxiliary lines and they don't receive their own billing or listing from the phone company. Sometimes, people have single line private lines which "appear" on their phone and no one else's. Sometimes they call these auxiliary lines. Sometimes these are called private lines.

Auxiliary Network Address In IBM parlance, in ACF/VTM, any network address except the main network address, assigned to a logical unit which is capable of having parallel sessions.

Auxiliary Power An alternate source of electric power, serving as backup for the primary power at the station main bus or prescribed sub-bus. An off-line unit provides electrical isolation between the primary power and the critical technical load; an on-line unit does not. These are government definitions: A Class A power source is primary power source; i.e., a source that assures an essentially continuous supply of power. Types of auxiliary power service include: Class B: a standby power plant to cover extended outages (days); Class C: a quick-start (10 to 60 seconds) unit to cover short-term outages (hours); Class D: an uninterruptible (no-break) unit using stored energy to provide continuous power within specified voltage and frequency tolerances.

NEWTON'S TELECOM DICTIONARY

Auxiliary Ringer This is a separate external telephone ringer or bell. It can be programmed to ring when a line or a telephone, or both ring; or, when Night Service is turned on.

Auxiliary Service Trunk Groups A category of trunk groups that provides selected services for customer or operators and terminates on announcement systems, switchboards, or desks. Examples include Directory Assistance, Intercept, Public Announcement, Repair Service, Time, and Weather.

Auxiliary Storage A mass storage device capable of holding a larger amount of information than the main memory (i.e. RAM) of the computer or telephone system, but with slower access time. For example — magnetic tape, floppy disks, etc.

AUU ATM User-to-User

Available In automatic call distribution language, an agent state, between calls, when an agent, having finished the previous transaction, returns to accept the next inbound caller. See also AVAILABILITY.

Available State According to Bellcore, an available circuit state occurs when all of the following are true:
1. The Bit Error Ratio (BER) is better than 1 in 10 to the nth power for a specific number of consecutive observation periods of fixed duration.
2. Block Error Ratio (BLER) is better than 1 in 10 to the nth power under the same conditions.
3. There are a specific number of consecutive observation periods of fixed duration without a severely errored unit time.

Availability The amount of time a computer or a telephone system is available for processing transactions or telephone calls. Here's a more technical definition: The ratio of the total time a functional unit is capable of being used during a given interval to the length of the interval; e.g., if the unit is capable of being used for 100 hours in a week, the availability is 100/168. Contrast this with the term Reliability, which is different. See RELIABILITY.

Availability Reports Availability reports show how often and for how long nodes, links or paths were unavailable due to outages between specified dates. They can be used to monitor network reliability and to calculate rebates for users.

Avalanche Photo Diode APD. A fiber optic transmission device. A light detector that generates an output current many times the light energy striking its face. A photodiode that shows gain in its output power compared to the optical power that it receives through avalanche multiplication (signal gain) of the current that flows through a photosensitive device. This type of diode is used in receivers requiring high light sensitivity. See APD.

Avalanching The process by which an electrical signal is multiplied within a device by electron impact ionization.

Avatar An image that represents an individual's cyber presence, i.e. on the Internet. That image will be a cartoon or caricature, but not a photograph of the person.

AVC Automatic Volume Control. In radio it maintains constant sound level despite undesired differences in strength of incoming signal.

AVD Alternative Voice Data or Alternating Voice Data. AVD used to mean a voice circuit that would also handle data. See AVD Circuits. Now AVD means it's a normal switched analog phone line connected to which is a AVD modem, which lets you transfer bursts of fax, data and images between your voice conversation. The protocol mostly used is VoiceView. See AVD CIRCUITS and DSVD.

AVD Circuits Alternate Voice Data Circuits. Telephone lines which have been electrically treated to handle both voice and data signals. Typically used on leased overseas circuits to save money. See AVD.

Average Call Duration Divide the total number of minutes of conversation by the number of conversations. Bingo, that's your average call duration.

Average Customer Arrival Rate Represents the number of entities (humans, packets, calls, etc.) reaching a queuing system in a unit of time. This average is denoted by the Greek letter lambda. One would prefer to know, if possible, the full distribution of the calls arriving.

Average Delay The delay between the time a call is answered by the ACD and the time it is answered by a person. This typically includes time for an initial recorded announcement plus time spent waiting in queue. Average delay can be chosen as the criterion for measuring service quality.

Average Handle Time AHT. The period of time an employee is occupied with an incoming call. This is the sum of talk time and after-call-work time.

Average Holding Time The sum of the lengths (in minutes or seconds) of all phone calls during the busiest hour of the day divided by the number of calls. There are two definitions. The one above refers to average speaking time (it's the more common one). There's a second definition for "average holding time." This refers to how long each call was on hold, and thus not speaking. This second definition is typically found in the automatic call distribution business (ACD). Check before you do your calculations.

Average Latency The time required for a disk to rotate one-half revolution.

Average Pulse Density In T-1 bipolar transmissions, refers to the number of "1" pulses per "0" conditions and is usually tied to a maximum number of "0"s in a row (i.e., FCC Part 68 requires 12.5% pulse density and no more than 80 consecutive "0"s where as AT&T Pub 62411 uses a formula and no more than 15 consecutive "0"s).

Average Speed Of Answer ASA. How many seconds it takes an operator on average to answer a call.

Average Transfer Delay Average time between the arrival of a packet to a station interface and its complete delivery to the destination station.

AVI Audio Video Interleaved. File format for digital video and audio under Windows. Use the "Media Player," which comes with Windows, to play AVI files. The AVI file format is cross-platform compatible, allowing .AVI video files to be played under Windows and other operating systems.

AVK Audio Video Kernel. DVI (Digital Video) system software designed to play motion video and audio across hardware and operating system environments.

AVOX A code name for Northern Telecom's "dumb" (i.e. programmable) switch which it intends to release in the North America some time in mid 1992.

Avoidable Costs A wonderful concept used by the regulated telephone industry. It refers to those costs which would be avoided (i.e. not incurred) if the service were not offered. Examples of costs to avoid are maintenance, taxes, labor, and other direct costs. The concept of Avoidable Costs is to allow the phone industry the justification to price a competitive service very low.
AVP A Northern Telecom word for Associated Voice Port.
AVP-100 Introduced in early April 1992, AT&T Microelectronics' three-chip set that supports Px64, MPEG and JPEG coding and compression schemes. Its support of these standards for full-motion video and audio, stored motion video and still image compression is expected to reduce the cost of video codec equipment and stimulate development of non-proprietary codecs. The chip set performs 25 billion operations per second most of which is motion estimation. AT&T Microelectronics is now called Lucent Technologies, Microelectronics group.
AVP-L A Northern Telecom word for Associated Voice Port and Line.
AVRS Automated Voice Response System. See IVR and VOICE RESPONSE SYSTEM.
AVSS Audio-Video Support System. DVI (Digital Video) system software for DOS. It plays motion video and audio.
AWC Area-Wide Centrex.
AWG American Wire Gauge. Standard measuring gauge for non-ferrous conductors (i.e. non-iron and non-steel). AWG covers copper, aluminum, and other conductors. Gauge is a measure of the diameter of the conductor. Also known as the Brown and Sharpe (B&S) Wire Gauge. A method of measuring the thickness of cable. The AWG numbering system is backwards: The higher the AWG number the thinner the wire. For example, heavy industrial wiring may be No. 2. Homes are typically wired with No. 12 or No. 14. Telephone systems typically use No. 22, No. 24 or No. 26. The thicker the wire, the more current it can carry further. The thicker the wire, the less resistance the current will encounter and therefore the longer it will travel. You need thicker phone cabling when your phones are further away. Some vendors save money by installing systems with thin wire. Make sure you specify.
AX.25 An amateur radio implementation of the X.25 protocol. Used by some private VANs (Value Added Networks) to avoid PTT (post, Telephone, and Telegraph Administration) monopolies (and thus high prices) on X.25 transmission and switching.
Axial Ratio Of a wave having elliptical polarization, the ratio of the major axis to the minor axis of the ellipse described by the tip of the electric field vector.
Axial Ray A ray that travels along the axis of an optical fiber.
Axial Slab Interferometry Synonym for Slab Interferometry.
Axis The center of an optical fiber.
Azimuth The horizontal angle which the radiating lobe of an antenna makes in angular degrees, in a clockwise direction, from a north-south line in the northern hemisphere. In the southern hemisphere, the reference is the south-north line. Azimuth actually involves a lot more than antennas. For example, it covers the alignment of a recording head in a tape recorder.

B The local wireline cellular carrier. In one of its less intelligent decisions, the Federal Communications Commission decided to issue two cellular franchises in each city of the United States. They gave one to the local phone company and one to a competitor. This duopoly has naturally meant little real price competition. And perhaps one day, the FCC will issue other licenses and the price of cellular phone calls will drop dramatically. Meantime, the "A" carrier on your cellular phone is the local wireline carrier. And the "B" is the other one. A new form of cellular carrier is about to be born. It's called PCS, Personal Communications Services. See PCS.

B Battery A section of a phone system power supply that provides unfiltered Direct Current for operating relays and various other components. Typically 20 volts. See A BATTERY.

B Channel A "bearer" channel is a fundamental component of ISDN interfaces. It carries 64,000 bits per seconds in both directions, is circuit switched and is able to carry either voice or data. Whether it does or not depends on how your local telephone company has tariffed its ISDN service. See BASIC RATE INTERFACE (BRI) and ISDN.

B Connector A commonly-used wire-splicing device consisting of a flexible plastic sleeve over a toothed metal cylinder that bites through insulation when crimped with pliers or special tool. It is about one inch long and can hold three or four wires. A gel-filled version is available for installation in damp or humid areas. B connectors are also known as chiclets, beans, beanies, and rodent rubbers.

B Frame Bi-directional or B frames (often called B pictures) refer to part of the MPEG video compression process whereby both past and future pictures / frames are used as references. B frames typically produce the most compression.

B Link Bridge Link. A CCS/SS7 signaling link used to connect STP (Signal Transfer Point) pairs that perform work at the same functional level. These links are arranged in sets of four (called quads).

B Port The port which connects the outgoing primary ring and the incoming secondary ring of the FDDI dual ring. This port is part of a dual attached station or concentrator. See FDDI.

B-911 Basic 911. B-911 is a centralized emergency reporting system which may have many features but which does NOT provide ALI (Automatic Location Information) to the 911 operator. In most cases, it does not provide ANI (Automatic Number Identification) either. B-911 provides a common emergency response number and relies on Emergency Hold and Forced Disconnect to maintain effective service.

B-Block Cellular licenses received from the FCC with an initial association to a telephone company. Also referred to as wireline.

B-CDMA Broadband Code Division Multiple Access.

B-CRYPT A symmetric cryptographic algorithm designed by British Telecom.

B-DCS Broadband Digital Cross-connect System. B-DCS is a generic term for an electronic digital cross-connect system capable of cross-connecting signals at or above the DS3 rate.

B-ICI B-ISDN Inter-Carrier Interface: An ATM Forum defined specification for the interface between public ATM networks to support user services across multiple public carriers.

B-ICI SAAL B-ICI Signaling ATM Adaptation Layer: A signaling layer that permits the transfer of connection control signaling and ensures reliable delivery of the protocol message. The SAAL is divided into a Service Specific part and a Common part (AAL5).

B-ISDN Broadband ISDN. Normal ISDN tends to assume all "conversations" are in chunks of 64 Kbps. ITU-T recommendation I.113 [45] defines Broadband as "a service or system requiring transmission channels capable of supporting (transmission) rates greater than the primary rate." Thus, broadband ISDN is a new concept in information transfer, although exactly what it is isn't clear yet. It could assume information comes in chunks as large as 150-million bits per second. In another ITU-T recommendation (I.121 [47]), the ITU-T presents an overview of what it sees as B-ISDN capabilities:
"B-ISDN supports switched, semipermanent and permanent point-to-point and point-to-multipoint connections and provides on demand, reserved and permanent services. Connections in B-ISDN support both circuit mode and packet mode services of a mono- and/or multi-media type and of connectionless or connection-oriented nature and in a bidirectional or unidirectional configuration. A B-ISDN will contain intelligent capabilities for the purpose of providing advanced service characteristics, supporting powerful operation and maintenance tools, network control and management." CCITT (now called ITU) is to decide on an international standard for B-ISDN by 1996.
Bellcore says that "National and international standards bodies have made the Asynchronous Transfer Mode (ATM) the target solution for providing the flexibility required by B-ISDN. ATM provides a common platform capable of supporting both broadband and narrowband services ... The physical layer-transmission standard for B-ISDN is the Synchronous Optical Network (SONET), also known as the Synchronous Digital Hierarchy (SDH). See N-ISDN and SONET."

B-LLI Broadband Lower Layer Information: This is a Q.2931 information element that identifies a layer 2 and a layer 3 protocol used by the application.

B-TE Broadband Terminal Equipment: An equipment category for B-ISDN which includes terminal adapters and terminals.

B3ZS Bipolar 3-Zero Substitution.

B8ZS Binary 8 Zero Substitution. A technique used to accommodate the ones density requirement for digital T-carrier facilities in the public network, while allowing 64 Kbps clear data per channel. Rather than inserting a one for every seven consecutive zeros, B8ZS inserts two violations of the bipolar line encoding technique for digital transmission links.

NEWTON'S TELECOM DICTIONARY

Babble Just what it sounds like — crosstalk from several interfering communications circuits or channels.

Babbling Tributary "A station that continuously transmits meaningless messages," as defined by John McNamara, of DEC and author of "Local Area Networks, an introduction to the technology." Some people might argue this was another word for Harry Newton, the author who didn't know when to stop and wanted to make this dictionary the most comprehensive telecom dictionary ever.

BABT British Approvals Board for Telecommunications. You need their approval before you can sell telecom equipment in Great Britain.

Babyphone Feature allowing calls to an off-hook telephone to listen to room noises, for example, to check if a baby is crying.

Back Board A piece of plywood mounted on a wall. Phone equipment is mounted on the plywood. It is more efficient to first mount phone equipment on plywood in the service bay, test it out while it's convenient and diagnostic tools are handy. Then take the phone equipment and the back board (which typically consists of the KSU, power supply and 66-blocks) and install them on the customer's premises. This "pre-installation" makes enormous sense — economically and reliably. Sadly, few installation companies do it.

Back End Database server functions and procedures for manipulating data on a network.

Back End Processor A server is often called a back end and a workstation is often called the front end. On a LAN (Local Area Network), A back-end processor runs on a server. It is responsible for preserving data integrity and handles most of the processor-intensive work, such as data storage and manipulation.

Back Haul Back haul is a verb. A communications channel is back hauling when it takes traffic beyond its destination and back. There are many reasons it might do this. The first is that it may be cheaper to go that route instead of going directly. You might, for example, have a full-time private line from New York Dallas. You might find it cheaper to reach Nashville by going to Dallas first, then dialing back to Nashville. The economics of backhauling may change from one moment to another as the line to Dallas is empty, close to full or full. Another reason for backhauling is that you may do it to accommodate changes in your calling or staffing patterns. You may have an automatic call distributor in Omaha and one in Chicago. A call from New York may come into your Omaha ACD, but when it gets there you may discover that there are no agents available to handle the call. So it may now make sense to back haul the call to the Chicago ACD, where an agent is available.

Back Porch The portion of a video signal that contains the color burst information. It occurs between the end of the horizontal synch pulse and the beginning of active video.

Back Projection When the projection is placed behind a screen (as it is in television and various video conferencing applications where the image is displayed on a monitor or a fabric screen) it is described as a back projection system. In these systems the viewer sees the image via the transmission of light as opposed to reflection used in front projection systems. Audiences generally prefer back projection systems since they seem brighter.

Back To Back Channel Bank The connection of voice frequency and signaling leads between channel banks to allow dropping (i.e. removing) and inserting (i.e. adding) of channels.

Back To Back Connection A connection between the output of a transmitting device and the input of an associated receiving device. When used for equipment measurements or testing purposes, this eliminates the effects of the transmission channel or medium.

Back Up Server A program or device that copies files so at least two up-to-date copies always exist.

Backbone The backbone is the part of the communications network which carries the heaviest traffic. The backbone is also that part of a network which joins LANs together — either inside a building or across a city or the country. LANs are connected to the backbone via bridges and/or routers and the backbone serves as a communications highway for LAN-to-LAN traffic. The backbone is one basis for design of the overall network service. The backbone may be the more permanent part of the network. A backbone in a LAN, a WAN, or a combination of both dedicated to providing connectivity between subnetworks in an enterprise-wide network.

Backbone Closet The closet in a building where the backbone cable is terminated and cross connected to either horizontal distribution cable or other backbone cable.

Backbone Network The part of a communications facility that connects primary nodes; a primary shared communications path that serves multiple users via multiplexing at designated jumping-off points. A transmission facility, or arrangement of such facilities, designed to connect lower speed channels or clusters of dispersed users or devices.

Backbone Subsystem See RISER SUBSYSTEM.

Backbone To Horizontal Cross-Connect BHC. Point of interconnection between backbone wiring and horizontal wiring.

Backbone Wiring The physical/electrical interconnections between telecommunications closets and equipment rooms. Cross-Connect hardware and cabling in the Main and Intermediate Cross-Connects are considered part of the backbone wiring.

Backcharging A phone fraud term. Backcharging is starting the clock on a phone call at the time a customer contacts the long-distant phone service provider — not when the person being called answers the phone — which is what it should be.

Backfeed Pull A method used to pull cable into a conduit or a duct liner when the cable is long or when placing cable into controlled environmental vaults, central offices, or under streets. With this method, the cable pays off its reel at an intermediate manhole and is first pulled in one direction. The remaining cable is then removed from the reel, laid on the ground, and then pulled in the opposite direction.

Backfile Conversion The process of scanning in, indexing and storing a large backlog of paper or microform documents in preparation of an imaging system. Because of the time-consuming and specialized nature of the task, it is generally performed by a service bureau.

Backfilling To designate memory on an expanded memory card and make it available for use as conventional memory.

Background See BACKGROUND PROCESSING.

Background Communication Data communication, such as downloading a file from a bulletin board, that takes place in the background while the user concentrates on another application (e.g. a spreadsheet) in the foreground.

68

Background Music This feature allows music to be played through speakers in the ceiling and/or through speakers in each telephone, throughout the office, or office-by-office, or selectively. Background music is typically played through paging speakers, but it can also be played through the speakers of speakerphones. In fact, the two — paging and background music — often go hand-in-hand. When you want to page someone, the music turns off automatically and comes back on when the paging is over. The same thing happens on airplanes. Background music is said to motivate workers, often into shutting it off.

Background Noise The noise you hear when nothing else is being transmitted. See also WHITE NOISE.

Background Processing The automatic execution of lower priority computer programs when higher priority programs are not using the computer's resources. A higher priority task would be completing calls. A lower priority task would be running diagnostics. Some PBXs have this feature. Some insist on running their diagnostics even though they are choked with calls. The smarter ones tone down their diagnostics when they get busier, which makes sense.

Background Program A low priority program operating automatically when a higher priority (foreground) program is not using the computer system's resources.

Background Task A secondary job performed while the user is performing a primary task. For example, many network servers will carry out the duties of the network (like controlling who is talking to whom) in the background, while at the same time the user is running his own foreground application (like word processing). See also BACKGROUND PROCESSING.

Backhaul See BACK HAUL.

Backoff When a device attempts to transmit data and it finds trouble, the sending device must try again. It may not try again immediately. It may "back off" for a little time so the trouble on the line can be cleared. This happens with LANs. For example, an earlier attempt to transmit may have resulted in a collision in a CSMA/CD (Carrier Sense Multiple Access/Collision Detection) Local Area Network (LAN). So the device "backs off," waits a little and then tries again. How long it waits is determined by preset protocols.

Backoff Algorithm The formula built into a contention local area network used after collision by the media access controller to determine when to try again to get back onto the LAN. See also BACKOFF.

Backplane The high-speed communications line to which individual components, especially slide-in cards, are connected. For example, all the extensions of a PBX are connected to line cards (circuit boards) which slide into the PBX's cage. At the rear of the PBX cage, there are several connectors. Each of these connectors is wired to the PBX's backplane, also called a backplane bus. This backplane bus is typically running at a very high speed, since it carries many conversations, address information and considerable signaling. These days, the backplane bus is typically a time division multiplexed line — somewhat like a train with many cars, each representing a time slice of another conversation (data, voice, video or image). The backplane's capacity determines the overall capacity of the switch.

Backplane Bus See BACKPLANE.

Backpressure Propagation effects in a communications network of hop-by-hop flow control to upstream nodes.

Backscattering 1. In fiber optics, the scattering of light into a direction opposite to the original one.
2. Radio wave propagation in which the direction of the incident and scattered waves resolved along a reference direction (usually horizontal) are oppositely directed. A signal received by backscattering is often referred to as "backscatter."

Backslash Also called a virgule, the backslash key achieved fame because Microsoft used it to bring distinguish between subdirectories in MS-DOS. This is a backslash: \

Backup A copy of computer data on an external storage medium, such as a floppy disk or tape. Computers and telephone systems (which are computers) are unreliable. They glitch and lose data for all sorts of unusual and impossible-to-predict reasons. Thus the necessity for backups. The theory is when (not "if") a glitch will occur and the PBX's database will disappear off the face of the earth. If this happens, you have a backup and you simply retrieve the back-up file, load it up and, presto, you're back live. Only information changed (since the backup was made) is lost. Backups save time in restoring the system after a loss. Most modern PBXs work with a database and other extensive customized instructions the user loads in. Most PBX users forget to make and keep backups of their PBX data. They expect their vendor to make backups, but he rarely does. This carelessness costs weeks of aggravation, as the PBX's database and instruction set is manually (and painfully) put back together.
The method by which backups are maintained is also important. The medium should clearly be reliable, i.e. the best quality magnetic medium. The method of backing up is also important. For example, a streaming tape backup is less reliable than a file-by-file backup. In a streaming backup, the backup medium simply captures the original data one bit after another in one long stream. In a file-by-file backup, the data moves over in logical segments — command files, data files, etc. Streaming backups will work if their data is placed back on the same precise device from which they were originally taken. But, if they are placed on a different device (even though the same model number, etc.), they may barf because the tape assumes bad sectors are in the same place. This will probably not be true. Streaming tape backup devices are less expensive to buy and much faster to use. Avoid them.

Backup Domain Controller A server in a network domain that keeps and uses a copy by a computer without interrupting its current or primary task. For Windows NT Server domains, refers to a computer that receives a copy of the domain's security policy and domain database, and authenticates network logons.

Backup Link A resilient (fault tolerant) link which is not used until the primary link fails.

Backup Ring The token ring cabling between MAUs or CAUs consists of the main ring and the backup ring. The data is normally transmitted on the main ring, but if an error occurs, the data can be transmitted on the backup ring until the main ring is repaired. In MAU networks the switching is done manually while in CAU (or CAM) networks it is done automatically.

Backward Channel In data transmission, a secondary channel whose direction of transmission is constrained to be opposite to that of the primary (or forward) channel. The direction of transmission in the backward channel is restricted by the control interchange circuit that controls the direction of transmission in the primary channel. The channel of a data circuit that passes data in a direction opposite to that of its associated forward channel. The backward channel is usually used for transmission of supervisory, acknowledgement, or error-control signals. The direction of flow of these signals is opposite to that in which information is being transferred. The bandwidth of this channel is usually less than that of the forward channel; i.e., the information channel.

Backward Learning Routing algorithm based on assumed symmetric network conditions. Source node assumes best route to given destination is via neighbor node that was on best route from destination to source.

Backward Recovery The reconstruction of an earlier version of a file by using a newer version of data recorded in a journal.

Backward Signal A signal sent in the direction from the called to the calling station, or from the original communications sink to the original communications source. The backward signal is usually sent in the backward channel and consists of supervisory, acknowledgement, or error control elements.

Backward Supervision The use of supervisory sequences from a secondary to a primary station.

Bad Block A defective unit on a storage medium that software cannot read or write.

Bad Line Key When the PBX attendant encounters a bad trunk, he/she pushes this bad line button on the console, automatically flagging the trunk for later checking and repair.

Bad Line Reporting Automatically reports a poor connection without interrupting the current call.

Bad Sectors Defective areas on a hard or soft disk. The MS-DOS FORMAT command locks out bad sectors so they are never used. Other operating systems have similar commands. See also HOT FIX.

BAF Billing format of the 0122 structure code defined by the Bellcore Automatic Message Accounting Format (BAF) Requirements TR#030#NWT-001100. This format identifies paths according to the resource they terminate on.

Baffle A partition used with a loud speaker to prevent air vibrations from the back of the diaphragm from cancelling out the vibrations from the front of the diaphragm. Particularly valuable in the reproduction of bass notes.

Bag Phone A slang expression for a transportable cellular phone whose characteristics are 3 WATT output, heavy weight (for a portable), and a bag with a handle. Bag phones are not designed for carrying around. They are designed to carry from one place to another and used at that place for serious conversations, including possibly faxing and modemming. Bag phones' big "plus" is that they give off five times as much power as a handheld cellular phone. This makes them useful for semipermanent "installation" in places like construction sites, etc. They are as powerful as a car phone.

BAIC Barring of All Incoming Calls. A wireless telecommunications term. A supplementary service provided under GSM (Global System for Mobile Communications).

Bakelite An obsolete insulating material of the phenolic (synthetic resin) group. Jewelry made of bakelite is now particularly prized.

Balance To equalize load or current between parts or elements of a telephone line or circuit. Balancing helps get the best out of a phone line. In more technical terms, balancing a line is to adjust the impedance of circuits and balance networks to achieve specified return loss objectives at junctions of two-wire and four-wire circuits.

Balanced Circuit Telephone circuit in which the two conductors are electrically balanced to each other and to the ground. A balanced electrical interface generally allows data to be transmitted over longer distances than does an unbalanced circuit. See BALANCE.

Balanced Configuration Point-to-point network configuration in HDLC with two combined stations.

Balanced Electrical Interface An electrical interface on which each circuit consists of a separate pair of wires. A balanced electrical interface generally allows data to be transmitted over longer distances than does an unbalanced electrical interface.

Balanced Line A transmission line which has two conductors and a ground. When the voltages of the two conductors are equal in strength but opposite in direction, then you have a balanced line.

Balanced Mode Transmission Data transmission with information conveyed by differences in voltages on two circuits to minimize effects of induced voltages.

Balanced Modulator An amplitude modulating circuit that suppresses the carrier signal, producing an output consisting only of upper and lower sidebands.

Balanced Return Loss A measure of the effectiveness with which a balanced network simulates the impedance of the two-wire circuit at a hybrid coil. More generally, a measure of the degree of balance between two impedances connected to two conjugate sides of a hybrid set, network, or junction.

Balanced To Ground In a two-conductor circuit, a balanced-to-ground condition exists where the impedance-to-ground on one wire equals the impedance-to-ground on the other. This is the preferred condition for decent data communications.

Balanced Transmission Line A line having conductors with equal resistance per unit length and equal capacitance and inductance between each conductor and ground. Coaxial cable, for example, is configured easily as a balanced transmission system by the use of resistance-to-ground terminators.

Balancing Network A network used in a set ending a four-wire circuit to match the impedance of the two-wire circuit. 2. Sometimes employed as a synonym for balun.

Balcony A little platform up a telephone pole where people can work or sleep safely.

Ballast Device that modifies incoming voltage and current to provide the circuit conditions necessary to start and operate electric discharge lamps, e.g. fluorescent bulbs.

Balun BALanced/UNbalanced. An impedance matching transformer. Baluns are small, passive devices that convert the impedance of coaxial cable so that its signal can run on twisted-pair wiring. They are used often so that IBM 3270-type terminals, which traditionally require coaxial cable connection to their host computer, can run off twisted-pair. Works for some types of protocols and not for others. There is often some performance degradation with baluns. And the signal cannot run as far on twisted wire as it can on coaxial cable.

Band 1. Originally referred to AT&T's WATS Bands. AT&T WATS service was organized into circles of increasing distance from the caller. Each circle or BAND (also called SERVICE AREA), cost more per minute. But within each service area, each call costs the same per minute, even though the distances the calls travel might be different. There were typically six interstate bands covering the US and several intrastate bands (depending on how large the state is) which a customer can buy. The word "band" was

NEWTON'S TELECOM DICTIONARY

invented by AT&T Communications (originally known as AT&T Long Lines) when it introduced WATS service. Recently it changed the word "band" to "Service Area." Nobody knows why. See POSTALIZED and WATS.
2. Band can also to the range of frequencies between two defined limits. For example, the band of frequencies able to be heard by the human ear ranges between 30 to 25,000 hertz. The ear can hear a band (or more correctly, a bandwidth) of about 25,000 hertz.
Band Elimination Filter BEF. A filter that has a single continuous attenuation band, with neither the upper nor lower cut-off frequencies being zero or infinite.
Band Splitter A multiplexer designed to split the available frequency band into several smaller channels. A band splitter can use time division or frequency division multiplexing.
Band-Stop Filter A device that bars passage of frequencies within its designed range(s), and allows passage of higher or lower frequencies, or both.
Band, Citizens One of two bands used for low power radio transmissions in the United States — either 26.965 to 27.225 megahertz or 462.55 to 469.95 megahertz. Citizens band radio is not allowed in many countries, even some civilized countries. In some countries they use different frequencies. CB radios, in the United States, are limited by FCC rule to four WATTS of power, which gives each CB radio a range of several miles. Some naughty people boost their CBs with external power. The author of this dictionary has actually spoken to Australia while driving on the Santa Monica Freeway in Los Angeles. See also CB.
Band, Frequency The frequencies between the upper and lower bands. See also BAND. Here is the accepted explanation of "bands:"

BAND	AMERICAN	
Below 300 Hertz	— ELF —	Extremely low frequency
300—3,000 Hertz	— ILF —	Infra Low Frequency
3—30 kHz	— VLF —	Very Low Frequency
30—300 kHz	— LF —	Low Frequency
300—3,000 kHz	— MF —	Medium Frequency
3—30 MHz	— HF —	High Frequency
30—300 MHz	— VHF —	Very High Frequency
300—3,000 MHz	— UHF —	Ultra High Frequency
3—30GHz	— SHF —	Super High Frequency
30—300GHz	— EHF —	Extremely High Frequency
300—3,000 GHz	— THF —	Tremendously High Frequency
	European	
P	0.2-1.0 Ghz	0.2-0.375 Ghz
L	1-2 Ghz	0.375-1.5 Ghz
S	2-4 Ghz	1.5-3.75 Ghz
C	4-8 Ghz	3.75-6 Ghz
X	8-12.5 Ghz	6-11.5 Ghz
J	-	11.5-18 Ghz
Ku	12.5-18 Ghz	-
K	18-26.5 Ghz	18-30 Ghz
Ka	26.5-40 Ghz	-
Q	-	30-47 Ghz

Banded Memory In a PostScript printer, virtual printer memory is a part of memory that stores font information. The memory in PostScript printers is divided into banded memory and virtual memory. Banded memory contains graphics and page-layout information needed to print your documents. Virtual memory contains any font information that is sent to your printer either when you print a document or when you download fonts.
Banded Rate A price range for regulated telephone service that has a minimum floor and maximum ceiling. The minimum covers the cost of service; the maximum is the rate filed in the price list.
Bandmarking A continuous circumferential band applied to an insulated conductor at regular intervals for identification.
Bandpass The range of frequencies that a channel will transmit (i.e. pass through) without excessive attenuation.
Bandpass Filter A device which transmits a band of frequencies and blocks or absorbs all other frequencies not in the specified band. Often used in frequency division multiplexing to separate one conversation from many.
Bandpass Limiter A device that imposes hard limiting on a signal and contains a filter that suppresses the unwanted products of the limiting process.
Bandwidth 1. In telecommunications, bandwidth is the width of a communications channel. In analog communications, bandwidth is typically measured in Hertz — cycles per second. In digital communications, bandwidth is typically measured in bits per second (not bytes per second). A voice conversation in analog format is typically 3,000 hertz. In digital communications, encoded in PCM, it's 64,000 bits per second. Do confuse bandwidth with band. Let's say we're running a communications

device in the 12 GHz band. What's its bandwidth? That's the space it's occupying. Let's say it's occupying from 12 GHz to 12.1 GHz. This means that it's occupying the space from 12,000,000,000 to 12,100,000,000. This means its bandwidth is one million cycles or one megahertz. Affiliated terms are narrow bandwidth and wideband circuit. These are not precise terms, except that wideband circuits are wider than narrow ones.
2. The capacity to move information. A person who can master hardware, software, manufacturing and marketing — and plays the oboe or some other musical instrument — is "high bandwidth." The term is believed to have originated in Redmond, WA in the headquarters of Microsoft. People there who are super-intelligent and have generally broad capabilities are said to have "high bandwidth."
3. Microsoft jargon for schedule. For example, "I have a bandwidth problem" means that I have an overloaded schedule.

Bandwidth Augmentation Bandwidth augmentation is the ability to add another communications channel to an already existing communications channel.

Bandwidth Compression A technique to reduce the bandwidth needed to transmit a given amount of information. Bandwidth compression is used typically in "picture type" transmissions — such as facsimile, imaging or video-conferencing. For example, early facsimile machines scanned each bit of the document to be sent and sent a YES or NO (if there was material in that spot or not). More modern machines simply skip over all the blank spaces and transmit a message to the receiving facsimile machine when to start printing dots again. A facsimile "picture" is made up of tiny dots, similar to printing photos in a magazine. Today, bandwidth compression is used to transmit voice, video and data. There are many techniques, few of which are standard. The key, of course, is that if you're going to compress a "conversation" at one end, you must "de-compress" it at the other end. Thus, in every bandwidth compressed conversation there must be two sets of equipment, one at each end. And they better be compatible.

Bandwidth Limited Operation The condition prevailing when the system bandwidth, rather than the amplitude (or power) of the received signal, limits performance. The condition is reached when the system distorts the shape of the signal waveform beyond specified limits. For linear systems, bandwidth-limited operation is equivalent to distortion-limited operation.

Bandwidth On Demand Just what it sounds like. You want two 56 Kbps circuits this moment for a videoconference. No problem. Use one of the newer pieces of telecommunications equipment and "dial up" the bandwidth you need. An example of such a piece of equipment is an inverse multiplexer. Uses for bandwidth on demand include video conferencing, LAN interconnection and disaster recovery. Bandwidth on demand is typically done only with digital circuits (they're easier to combine). Bandwidth on demand is typically carved out of a T-1 circuit, which is permanently connected to the customer's premises from a long distance carrier's central office, also called a POP — Point of Presence.

Bang An exclamation point (!) used in a Unix-to-Unix Copy Program (UUCP) electronic mail address. People who are on AT&T Mail often give you their mail address as "Bang Their Name." My AT&T Mail address is Bang HarryNewton, i.e. !HarryNewton.

Bang Path A series of UUCP nodes mail will pass through to reach a remote user. Node names are separated by exclamation marks nicknamed "bangs." The first node in the path must be on the local system, the second node must be linked to the first, and so on. To reach user 1 on sys2 if your computer's address is sys1 you would use the following address:
sys1! sys2! sys3! user1

Bank Switching A way of expanding memory beyond an operating system's or microprocessor's address limitations by switching rapidly between two banks of memory. IN MS-DOS, a 64K bank of memory between 640K and one megabyte is set aside. When more money is needed, the bank, or page, is switched with a 64K page of free memory. This is repeated with additional 64K pages of memory. When the computer requires data or program instructions not now in memory, expanded memory software finds the bank containing the data and switches it with the current bank of memory. Although effective, bank switching results in memory access times that are slower than true, extended memory.

BAOC Barring of All Outgoing Calls. A wireless telecommunications term. A supplementary service provided under GSM (Global System for Mobile Communications).

Bar Code A bunch of lines of varying width printed on something. The bar code is designed to be read optically by some data capturing device. Bar codes are turning up on letters. They are read by image scanning devices in the post office and allegedly help the mails move faster. Bar codes are on most things you buy now in supermarkets. By scanning those bar codes at the checkout counter, the supermarket knows what's being sold and not being sold. And presumably the supermarket, or its computer, can order supplies to keep the supermarket stocked with goods that are selling and not re-order those which aren't.

Barbie, Jane The electronic "Voice With A Smile" on many telephone company intercept recordings. Ms. Barbie does her work for Electronic Communications, Inc. of Atlanta, GA.

Bare Wire An electrical conductor having no covering or insulation.

Barge-In Interrupting a call in progress.

Barge-Out Leaving a call in progress without notice.

Barium Ferrite A type of magnetic particle used in some recording media including floptical diskettes. See also FERRITE, HARD FERRITE and SOFT FERRITE.

Barrel An imaging term. Distortion that swells an image in the middle, narrows it at the top and the bottom.

Barrel Connector This connector is a cylindrical (barrel-shaped) connector used to splice together two lengths of thick Ethernet coaxial cable.

Barrel Contact A term in cabling. A barrel contact is an insulation displacement type contact consisting of a slotted tube that cuts the insulation when the wire is inserted.

Barrel Distortion When a screen is distorted — with the top, bottom and sides pushing outwards (like a beer barrel) — the screen is said to be suffering barrel distortion.

Barton Enos Barton once said he was "disgusted" when told that it would be possible to send conversation along a wire. He later co-founded (with Elisha Gray) the Western Electric Company, which became AT&T's manufacturing subsidiary and was once the largest electrical equipment manufacturer in the US. In addition to phones, the company made sewing machines, typewriters, movie sound equipment, radio station gear, radar systems and guided missile parts. See also GRAYBAR.

Base Address The first address in a series of addresses in memory, often used to describe the beginning of a network interface card's I/O space.

Base Amount A call center term. One of the historical patterns; what the monthly call volume would be if there were no long-term trend or seasonal fluctuation- in other words, the average number of calls per month.

Base Load In trunk forecasting, an amount of telephone traffic measuring during a certain defined time. See BASE PERIOD.

Base Memory What many people refer to as the first 640 kilobytes of memory in an MS-DOS PC.

Base Period In trunk forecasting, a time span of consecutive study during which a base load is determined.

Base Schedules A call center term. A fixed set of pre-existing schedules that you can use as a starting point in scheduling. And new schedules created are in addition to the base schedules.

Base Staff A call center term. The minimum number of people, or "bodies in chairs," required to handle the workload in a given period. The actual required number of staff is always greater than base the staff, because of various human factors such as the need for breaks and time off.

Base Station A wireless term. A base station is the fixed device a mobile radio transceiver (transmitter/receiver) talks to, to talk to a person or to get to the landline phone network, public or private.

Base-1 Service An MCI definition. MCI Service providing customers network management capabilities by allowing them to operate and administer their own telecommunications usage.

Baseband A form of modulation in which signals are pulsed directly on the transmission medium without frequency division. Local area networks as a rule, fall into two categories — broadband and baseband. The simpler, cheaper and less sophisticated of the two is baseband. In baseband LANs, the entire bandwidth (capacity) of the LAN cable is used to transmit a single digital signal. In broadband networks, the capacity of the cable is divided into many channels, which can transmit many simultaneous signals. While a baseband channel can only transmit one signal and that signal is usually digital, a broadband LAN can transmit video, voice and data simultaneously by splitting the signals on that cable using frequency division multiplexing. The electronics of a baseband LAN are simpler than a broadband LAN. The digital signals from the sending devices are put directly onto the cable without modulation of any kind. Only one signal is transmitted at a time. Multiple "simultaneous" transmissions can be achieved by a technique called time division multiplexing (see multiplexing). In contrast, broadband networks (which typically run on coaxial cable) need more complex electronics to decipher and pick off the various signals they transmit. Attached devices on a broadband network require modems to transmit. Attached devices to baseband networks do not.

Baseband LANs typically work with one high speed channel, which all the attached devices — printers, computers, databases — share. They share it by using it in turns — for example, passing a "token" to the next device. That token entitles the device with the token to transmit. IBM's LAN is a token ring passing local area network. Another way of sharing the baseband LAN is that each device, when it is ready to transmit, simply transmits into the channel and waits for a reply. If it doesn't receive a reply, it retransmits. Thus there are two main network or baseband access control schemes — Token Ring Passing and CSMA/CD. See also CSMA/CD, BROADBAND, ETHERNET and LOCAL AREA NETWORKS.

Baseband Modem A modem which does not apply a complex modulation scheme to the data before transmission, but which applies the digital input (or a simple transformation of it) to the transmission channel. This technique can only be used when a very wide bandwidth is available. It and only operates over short distances where signal amplification is not necessary. Sometimes called a limited distance or short-haul modem.

Baseband Signaling Transmission of a digital or analog signal at its original frequencies, i.e., a signal in its original form, not changed by modulation.

Baseboard A term used in voice processing/computer telephony to mean a printed circuit board without any daughterboards attached.

Baseboard Raceway A floor distribution method in which metal or wood channels, containing cables, run along the baseboards of the building. The front panel of the baseboard channel is removable, and outlets may be placed at any point along the channel.

Baseline 1. The line from which a graph is drawn. The base line is the X axis on vertically oriented graphs, the Y axis on horizontal bar graphs, or the line representing zero if the data contains both positive and negative numbers.

2. The imaginary line extending through a font and representing the line on which characters are aligned for printing. In conventional, alphanumeric fonts, the baseline is usually defined as the imaginary line touching the bottom of uppercase characters.

Baseline Sequential JPEG The most popular of the JPEG modes which employs the lossy DCT (Discrete Cosine Transform) to compress image data as well as lossless processes based on variations of DPCM (Differential Pulse Code Modulation). The "baseline" system represents a minimum capability that must be present in all Sequential JPEG decoder systems. In this mode image components are compressed either individually or in groups. A single scan pass completely codes a component or group of components.

BASIC Beginners All-purpose Symbolic Instruction Code. A programming language. BASIC is an easy language to learn. It's worth learning. Not all "Basics" are the same. A sort of de facto standard is Microsoft Basic. If you're wish to learn some programming, we recommend you begin with the latest version of Microsoft's Visual Basic, a Windows version of Basic. We've tried it. We're very impressed with it.

Basic Budget Service An inexpensive local phone service often restricted to people with limited incomes. It may or may not include any outgoing calls. It may include only a few outgoing calls.

Basic Call A call between two users that does not require Advanced Intelligent Network Release 1 features (e.g. a POTS call). Definition from Bellcore. See AIN.

Basic Call State Model BCSM. An abstraction of the ASC call processing activities for a basic two-party call. The BCSM is split into Originating and Terminating BCSMs.

Basic Exchange Telecommunications Radio Service BETRS. A service that can extend telephone service to rural areas by replacing the local loop with radio communications, sharing the UHF and VHF common carrier and private radio frequencies.

Basic Mode Link Control Control of data links by use of the control characters of the 7-bit character set for information processing interchange as given in ISO Standard 646-1983 and ITU-T Recommendation V.3-1972.

Basic Rate The default rate paid by a telephone subscriber who does not elect any other calling plan. Basic rate can apply to phone service from either a local phone company or a long distance phone company. Basic rate for local service can be a flat monthly rate or a monthly rate plus charges for each call — on length of call, time of day and distance sensitive charging. A basic rate for long distance or local toll service will be length of call, time of day and distance sensitive charging.

Basic Rate Interface BRI. There are two "interfaces" in ISDN: BRI and PRI. In BRI, you get two bearer B-channels at 64 kilobits per second and a data D-channel at 16 kilobits per second. The bearer B-channels are designed for PCM voice, video conferencing, group 4 facsimile machines, or whatever you can squeeze into 64,000 bits per second full duplex. The data D-channel is for bringing in information about incoming calls and taking out information about outgoing calls. It is also for access to slow-speed data networks, like videotex, packet switched networks, etc. One BRI standard is the "U" interface, which uses two wires. Another BRI standard is the "T" interface which uses four wires. See ISDN for a much fuller explanation.

Basic Service A telephone company service limited to providing local switching and transmission. Basic Service does not include equipment. The term Basic Service is unclear and varies between telephone companies and data communications service providers.

Basic Service Elements BSEs. An Open Network Architecture term. BSEs are services which value-added companies could get from their phone company. BSEs are optional basic network functions that are not required for an ESP to have a BSA, but when combined with BSEs can offer additional features and services. Most BSEs allow an ESP to offer enhanced services to their customers. BSEs fall into four general categories: Switching, where call routing, call management and processing are required; Signaling, for applications like remote alarm monitoring and meter reading; Transmission, where dedicated bandwidth or bit rate is allocated to a customer application; and Network Management, where a customer is given the ability to monitor network performance and reallocate certain capabilities. The selection of available BSEs is an ongoing process, with new arrangements being developed. ANI, Audiotext "Dial-It" Services, and Message Waiting Notification are all examples of BSEs. See AIN and OPEN NETWORK ARCHITECTURE.

Basic Serving Arrangement BSA. An old term defining the relationship of an enhanced service provider (value added provider) to the phone company providing the line/s. Under ONA (Open Network Architecture), a BSA is the basic interconnection access arrangement which offers a customer access to the public network (i.e. the normal switch phone service) and provides for the selection of available Basic Service Elements. It includes an ESP (Enhanced Service Provider) access link, the features and functions associated with that access link at the central office serving the ESP and/or other offices, and the transport (dedicated or switched) within the network that completes the connection from the ESP to the central office serving its customers or to capabilities associated with the customer's complementary network services. Each component may have a number of categories of network characteristics. Within these categories of network characteristics are alternatives from among which the customer must choose. Examples of BSA components are ESP access link, transport and/or usage. See OPEN NETWORK ARCHITECTURE.

Basic Telephony The lowest level of service in Windows Telephony Services is called Basic Telephony and provides a guaranteed set of functions that corresponds to "Plain Old Telephone Service" (POTS - only make calls and receive calls). See WINDOWS TELEPHONY SERVICES.

BASR Buffered Automatic Send/Receive.

Batch File A text file for personal computers with the extension .BAT that contains MS-DOS commands. When you type the name of the file, DOS carries out the commands in the file. Many computer programs are installed on your hard disk or loaded using batch files, as are many LAN operating systems. Batch files simplify life. If used wisely, you can save yourself typing things over and over again. It also means you don't have to remember all the commands to do something, just the first one. Batch files can be used for converting DOS batch files into .EXE or .COM files (this makes them run faster), to load other programs, or to change the parameters of your computer. Anything that can be done from DOS (and some things that can't) can be done from a batch file. Norton Utilities has an enhanced version of DOS's Batch file.

Batch Job File See BATCH FILE.

Batch Processing There are two basic types of data processing. One is batch processing. Also called deferred time processing and off-line processing. Batch processing occurs where everything relating to one complete job — such as preparing this week's payroll — is bunched together and transmitted for processing (locally, in the same building or long distance, across the country), usually by the same computer and under the same application program. Batch processing does not permit interaction between the program and the user once the program has been read (i.e. fed into) the computer. In batch processing with telecommunications (i.e. sending the task to be done over the phone line), network response time is not critical, since no one is sitting in front of a screen waiting for a response. On the other hand, accuracy of communications is very critical, since no one is sitting in front of a screen checking entries and responses.

The second type of processing is called INTERACTIVE or REAL TIME processing. Under this method of processing, a user sends in transactions and awaits a response from the distant computer before continuing. In this case, response time on the data communications facility is critical. Seconds count, especially if a customer is sitting at the other end of a voice call awaiting information on whether there's space on that airline flight, for example. See BATCH FILE.

Batteries A, B and C In ancient times, when vacuum tubes were still in use, the A battery (often about 6 volts) heated the filament or cathode to boil off electrons, the B battery (usually several hundred volts), positive with respect to the cathode, sucked the negative electrons to the plate, while the C battery, a few volts negative with respect to the cathode, tended to repel the electrons back toward the cathode. By varying a small signal voltage riding on the C battery, the flow of electrons to the plate could be controlled to generate a much larger voltage.

Battery All telephone systems work on DC (direct current). DC power is what you use to talk on. Often the DC power is called

NEWTON'S TELECOM DICTIONARY

"talking battery." Most key systems and many PBXs plug directly into an AC on the wall, but that AC power is converted by a built-in power supply to the DC power the phone system needs. All central offices (public exchanges) used rechargeable lead acid batteries to drive them. These batteries perform several functions: 1. They provide the necessary power. 2. They serve as a filter to smooth out fluctuations in the commercial power and remove the "noise" that power often carries. 3. They provide necessary backup power should commercial power stop, as in a "blackout" or should it get very weak, as in a "brownout."
In short, "battery" is the term used to reference the DC power source of a telephone system. Often called "Talking Battery." 2. Storage battery used with central office switching systems and PBXs serving locations which cannot tolerate outages. Batteries serve the following purposes: Act as a filter across the generator or power rectifier output to smooth out the current and reduce noise; provide a cushion against periodic overloads exceeding the generator/rectifier capacity; supply emergency power for a limited time in event of commercial power failure. See also AC, AC POWER, BATTERY RESERVE and CENTRAL OFFICE BATTERY.

Battery Backup A battery which provides power to your phone system when the main AC power fails especially during blackouts and brownouts. Hospitals, brokerage companies, airlines and hotel reservation services must have battery backup because of the integral importance of their phone systems to their business.

Battery Eliminator A device which has a rectifier and (hopefully) a filter. This device will convert AC power into the correct DC voltages necessary to drive a telephone system. Such a battery eliminator, or power supply, should deliver "clean" power, i.e. with little "noise" and of low impedance.

Battery Reserve The capability of the fully charged battery cells to carry the central office power load imposed when commercial power fails and the primary power source (generators/rectifiers) is out of service. Properly described in terms of the number of hours the batteries can furnish operating power for dependent CO apparatus for a demand equal to that on the CO during its busy hour. A busy-hour reserve of eight hours is typical for a telephone office battery plant.

Battle Faxes A cycle of vitriolic faxes exchanged between clients and lawyers, fighting lovers, etc. "Here's the latest round of battle faxes with my record company." Wired Magazine defined this term.

Baud Rate A measure of transmission speed over an analog phone line — i.e. a common POTS line. (POTS stands for Plain Old Telephone Service). Imagine that you want to send digital information (say from your computer) over a POTS phone line. You buy a modem. A modem is a device for converting digital on-off signals, which your computer speaks, to the analog, sine-wave signals your phone line "speaks." For your modem to put data on your phone line means it must send out an analog sine wave (called the carrier signal) and change that carrier signal in concert with the data it's sending. Baud rate measures the number of number of changes per second in that analog sine wave signal. According to Bell Labs, the most changes you can get out of a 3 KHz (3000 cycles per second) voice channel (which is what all voice channels are) is theoretically twice the bandwidth, or 6,000 baud.
Baud rate is often confused with bits per second, which is a transfer rate measuring exactly how many bits of computer data per second can be sent over a telephone line. You can get more data per second — i.e. more bits per second — on a voice channel than you can change the signal. You do this through the magic of coding techniques, such as phase shift keying. Advanced coding techniques mean that more than one bit can be placed on a baud, so to speak. To take a common example, a 9,600 bit per second modem is, in reality, a 2,400 baud modem with advanced coding such that four bits are impressed on each baud. The continuing development of newer and newer modems point to increasingly advanced coding techniques, bringing higher and higher bit per second speeds. My latest modem, for example, is 28,800 bits per second.

Baudot Code The code set used in Telex Transmission, named for French telegrapher Emile Baudot (1845-1903) who invented it. Also known by the CCITT approved name, International Telegraph Alphabet 2. The Baudot code has only five bits, meaning that only 32 separate and distinct characters are possible from this code, i.e. 2 x 2 x 2 x 2 x 2 equals 32. By having one character called Letters (usually marked LTRS on the keyboard) which means "all the characters that follow are alphabetic characters," and having one other key called Figures (marked FIGS), meaning "all characters that follow are numerals or punctuation characters," the Baudot character set can represent 52 (26 x 2) printing characters. The characters "space," "carriage return," "line feed" and "blank" mean the same in either FIGS or LTRS. TDD devices (Telecommunications Devices for the Deaf) use the Baudot method of communications to communicate with distant TDD devices over phone lines. See also ASCII, EBCDIC, MORSE CODE and UNICODE, which are other ways of encoding characters into the ones or zeros needed by computers.

Bay A telephone industry term for the space between the vertical panels or mounting strips ("rails") of the rack. One rack may contain several bays. A bay is another place you put equipment.

BBG Basic Business Group.

BBG-I ISDN Basic Business Group.

BBN Bolt, Beranek, and Newman, Inc. The Cambridge, MA company responsible for development, operation and monitoring of the ARPANET, and later, the Internet core gateway system, the CSNET Coordination and Information Center (CIC), and NSFnet Network Service Center (NNSC).

BBS Bulletin Board System. Another term for an electronic bulletin board. Typically a PC, modem/s and communications bulletin board software attached to the end of one or more phone lines. Callers can call the BBS, read messages and download public domain software. The person who operates a BBS is called a system operator, commonly shortened to SYSOP (pronounced "sis-op"). See also ELECTRONIC BULLETIN BOARD.

BCC 1. Bellcore Client Company. What Bellcore calls its owners — the seven regional Bell operating companies and their operating phone company subsidiaries.
2. Block Check Character. A control character appended to blocks in character-oriented protocols and used for figuring if the block was received in error. BCC is especially used in longitudinal and cyclic redundancy checking. As a packet (or in IBM jargon, a frame) is assembled for transmission, the bits are passed through an algorithm to come up with a BCC. When the packet is received at the other end, the receiving computer also runs the same algorithm. Both machines should come up with the same BCC. If they do, the transmission is correct and the receiving computer sends an ACK — a positive acknowledgement. If

75

they don't, an error has occurred during transmission, and they don't have the same bits in the packet. The receiver transmits a signal (a NAK, for negative acknowledgment) that an error has occurred, and the sender retransmits the packet. This process goes on until the BCC checks.

BCD Binary Coded Decimal. A system of binary numbers where each digit of a number is represented by four bits. See BINARY. In ATM, binary code decimal is a form of coding of each octet within a cell where each bit has one of two allowable states, 1 or 0.

BCH Bose, Chaudhuri and Hocquenghem error correction code. Named after the three guys who invented it.

BCM 1. Bit Compression Multiplexer.
2. Basic Call Model (a term from the Bellcore discussion of Advanced Intelligent Networks). See AIN.

BCOB An ATM term. Broadband Connection Oriented Bearer: Information in the SETUP message that indicates the type of service requested by the calling user.

BCOB-A Bearer Class A: Indicated by ATM end user in SETUP message for connection-oriented, constant bit rate service. The network may perform internetworking based on AAL information element (IE).

BCOB-C Bearer Class C: Indicated by ATM end user in SETUP message for connection-oriented, variable bit rate service. The network may perform internetworking based on AAl information element (IE).

BCOB-X Bearer Class X: Indicated by ATM end user in SETUP message for ATM transport service where AAL, traffic type and timing requirements are transparent to the network.

BCR Business Communications Review, a good magazine based in Hinsdale, IL. To get a sub, call 1-800-BCR-1234.

BCS Batch Change Supplement.

BDLC Burroughs Data Link Control, a bit oriented protocol.

Beacon A token ring frame sent by an adapter indicating that it has detected a serious ring problem, such as a broken cable or a MAU. An adapter sending such frames is said to be beaconing. See BEACONING.

Beacon Initiative A "plan and rationale to help bring the information highway to Canadian businesses and consumers. The central theme of the initiative is an open collaborative effort with all interested players to bring enhanced interactive data, image and video services to Canadians." That's what the press release of April 5, 1994 said. Companies in the Beacon Initiative include BC Telephone, AGT Limited, SaskTel, Manitoba Telephone System, Bell Canada. NBTel, Maritime Tel & Tel. Island Telephone, Newfoundland Telephone and Stentor Communications.

Beaconing Token Ring process to recover the network when any attached station has sensed that the ring is inoperable due to a hard error. Stations can withdraw themselves from the ring if necessary. A station detecting a ring failure upstream transmits (beacons) a special MAC frame used to isolate the location of the error using beacon transmit and beacon repeat modes. See BEACON.

Beam A way to exchange information between Newton users. Beaming uses the small built-in infrared unit at the top of the Newton MessagePad to send anything that's on one MessagePad to another MessagePad, or to a Sharp Wizard. This can be done across a conference table.

Beam Diameter The distance between the diametrically opposed points on a plane perpendicular to the beam axis at which the irradiance is a specified fraction of the beam's peak irradiance. The term is most commonly applied to beams that are circular or nearly circular in cross-section.

Beam Divergence As through the air telecom signal go further, they diverge. Beam divergence measures that divergence.

Beamsplitter A device for dividing an optical beam into two or more separate beams, often a partially reflecting mirror.

Beans Telco slang for B-connectors, which are about the size and shape of common beans. See B CONNECTOR.

Beat Frequency An old radio term: The frequency resulting when an oscillation of one frequency is "beat" or heterodyned against an oscillation of different frequency. The figure given is normally in cycles per second.

Beat Reception An old radio term. The resultant audible frequency when two sources of unequal undamped electrical oscillations of constant amplitude act simultaneously in the same circuit. See BEAT FREQUENCY.

Beating The phenomenon in which two or more periodic quantities having slightly different frequencies produce a resultant having periodic variations in amplitude.

BECKON See BECN.

BECN An ATM term. Backward Explicit Congestion Notification. A resource Management (RM) cell type generated by the network or the destination, indicating congestion or approaching congestion for the traffic flowing in the direction opposite. This bit notifies the user that congestion-avoidance procedures should be initiated for traffic in the opposite direction of the received frame. It indicates that the frames that the user transmits on this logical connection may encounter congested resources. In other words, slow down, you move too fast. Your bits may not get through or get delayed.

Bed Of Nails Cord A description of a type of alligator clip that attaches to the end of a craft test set, also called a butte set.

BEF Band Elimination Filter. A filter that has a single continuous attenuation band, with neither the upper nor lower cut-off frequencies being zero or infinite.

Behind Switch Mode An AT&T Merlin term. A mode of operation in which the communications system control unit is connected to (is "behind") another communications system. In other words, running Merlin behind Centrex.

Bel A relative measurement, denoting a factor of ten change. Rarely used in practice; most measurements are in decibels (0.1 bel).

Belden A major manufacturer of communications cable. Belden is such a high quality manufacturer and has set many cabling standards that some many other manufacturers follow their specs.

Bell A bell in a telephone instrument rings when a 20 Hz signal of about 90 volts AC is applied to the subscriber loop. In contrast, the normal voltage applied to a subscriber loop and used for speaking and listening is 48 volts DC.

Bell 103 AT&T specification for a modem providing asynchronous originate/answer transmission at speeds up to 300 bits per second (300 baud). This is the most common standard for modems running with personal computers. Every dial service in the U.S. adheres to this standard. The second most common standard is Bell 212. See the next few definitions.

Bell 201 AT&T specification for a modem providing synchronous data transmission at 2,400 bits per second.

Bell 202 AT&T specification for a modem providing asynchronous data transmission at speeds up to 1,800 bits per second. Requires four-wire line for full duplex transmission.
Bell 208 AT&T specification for a modem providing synchronous data transmission at speeds up to 4,800 bits per second.
Bell 209 AT&T specification for a modem providing synchronous data transmission at speeds up to 9,600 bits per second.
Bell 212 AT&T specification for a modem providing full duplex asynchronous or synchronous data transmission at speeds up to 1,200 bits per second on the voice dial-up phone network. This is the second most common standard for modems running with personal computers. The first is Bell 103.
Bell 43401 Bell Publication defining requirements for transmission over telco-supplied circuits that have dc continuity (that are metallic).
Bell Atlantic Formed as a holding company after the AT&T Divestiture. Includes Bell of Pennsylvania, C&P Telephone Companies of D.C., Maryland, Virginia, and West Virginia, Diamond State Telephone (MD), New Jersey Bell and several Bell Atlantic business activities. Check out Bell Atlantic's logo. It has a wonderful, small wave in the "A" in Atlantic. Very charming. Very subtle. See BELL OPERATING COMPANIES.
Bell Communications Research Bellcore. Formed at Divestiture to provide certain centralized services to the seven Regional Bell Holding Companies (RHCs) and their operating company subsidiaries. Also serves as a coordinating point for national security and emergency preparedness communications matters of the federal government. Bellcore does not work on customer premise equipment or other areas of potential competition among its owners — the seven Regional Bell Operating companies. Bellcore also works on standardizing methods by which customers of long distance companies will reach their favorite long distance companies. At time of writing, around 8,000 people worked at Bellcore, mostly in northern New Jersey. Bellcore has the unenviable task of trying to service the needs of the seven competitors, which own it. Its annual budget is around $1 billion, paid for by the seven Bell regional operating companies. It does work on ISDN and common channel signaling standards. See also BELLCORE.
Bell Compatible A term sometimes applied to modems. A modem is said to be "Bell compatible" if it conforms to the technical specifications set forth by AT&T for the various devices, such as Bell 212.
Bell Operating Company BOC. A BOC is one the 22 regulated telephone companies of the former Bell System, which was broken apart (the Divestiture of the Bell System) at midnight on December 31, 1983. At Divestiture, the Bell operating companies were grouped into seven Regional Holding Companies (RHCs). According to the terms of the Divestiture Agreement between the Federal Courts, the Federal Government and AT&T, the divested companies must limit their activities to local telephone services, directory service, customer premise equipment, cellular radio and any other ventures as the Federal Court may approve from time to time. BOCs are specifically limited from manufacturing equipment and from providing long distance service. See also REGIONAL BELL OPERATING COMPANY.
The following definition of a Bell Operating Company comes from the Telecommunications Act of 1996: The term `Bell operating company'
(A) means any of the following companies: Bell Telephone Company of Nevada, Illinois Bell Telephone Company, Indiana Bell Telephone Company, Incorporated, Michigan Bell Telephone Company, New England Telephone and Telegraph Company, New Jersey Bell Telephone Company, New York Telephone Company, U S West Communications Company, South Central Bell Telephone Company, Southern Bell Telephone and Telegraph Company, Southwestern Bell Telephone Company, The Bell Telephone Company of Pennsylvania, The Chesapeake and Potomac Telephone Company, The Chesapeake and Potomac Telephone Company of Maryland, The Chesapeake and Potomac Telephone Company of Virginia, The Chesapeake and Potomac Telephone Company of West Virginia, The Diamond State Telephone Company, The Ohio Bell Telephone Company, The Pacific Telephone and Telegraph Company, or Wisconsin Telephone Company; and
(B) includes any successor or assign of any such company that provides wireline telephone exchange service; but
(C) does not include an affiliate of any such company, other than an affiliate described in subparagraph (A) or (B).
Bell Speak A term coined by Michael Marcus for insider jargon spoken by "real" telephone people — those who practiced pre-divestiture. Such old jargon is usually incomprehensible to anyone in today's telephone industry who is younger than 46.
Bell System The entire AT&T organization prior to when it was broken up — at the end of 1984. The Bell System included Bell Labs, Long Lines, Western Electric and the 23 Bell operating companies.
Bellboy A public paging system run by local Bell phone companies. The name survived for many years and only drew criticism in the middle to late seventies with the rise of the Women's Liberation movement. It's now not used. Shucks.
Bellcore Bell Communications Research. Formed at Divestiture to provide certain centralized services to the seven Regional Bell Holding Companies (RBOCs) and their operating company subsidiaries. Also serves as a coordinating point for national security and emergency preparedness, and communications matters of the federal government. Bellcore does not work on customer premise equipment (e.g. the telephone set) or other areas of potential competition between its owners — the seven Regional Bell Operating Companies. It is a key player in the design of AIN — the Advanced Intelligent Network. You'll find many AIN definitions in this dictionary. You can acquire Bellcore documents from Bellcore — Document Registrar, 445 South Street, Room 2J-125, P.O. Box 1910, Morristown, NJ 07962-1910. Fax 201-829-5692. See also BELL COMMUNICATIONS RESEARCH. www.bellcore.com
Bellcore AMA Format BAF. The standard data format for AMA (Automatic Message Accounting) data to be delivered to the Revenue Accounting Office (RAO) in Advanced Intelligent Network Release 1.
Bellcore Multi-Vendor Interactions MVI. The process for coordinating the efforts of Bellcore, the Bell Operating telephone companies and vendors to address technical issues associated with Advanced Intelligent Network.
Bellman-Ford-Algorithm Shortest-path routing algorithm that figures on number of hops in a route to find shortest-path spanning tree.
BellSouth Corporation The largest regional Bell holding company formed at the Divestiture of AT&T. Includes Southern

Bell and South Central Bell and several other BellSouth businesses. See BELL OPERATING COMPANY.

Below-the-line Expenses incurred that are charged to shareholders of regulated operating telephone companies, not ratepayers.

Benchmark A standardized task to test the capabilities of various devices against each other for such measures as speed.

Bend Loss A form of increased attenuation caused by allowing high-order signals to radiate from the side of the fiber. The two common types of bend losses are those occurring when the fiber is curved and microbends caused by small distortions of the fiber imposed by poor cabling techniques.

Bending Radius The smallest bend which may be put into a cable under a stated pulling force. Bending radius affects size of bends in conduit, trays or ducts. It affects pulley size. It affects the size of openings at pull boxes where loops may form. Very critical in all aspects of cable laying, especially with under-carpet cabling and fiber optic cable.

Bentogram A syntax for encoding Versit's vCard in a binary encoding. Bentograms are based on Benton, the OpenDoc standard interchange format.

BEP Back End Processor.

BER 1. Bit Error Rate. The ratio of error bits to the total number of bits transmitted. If the BER gets too high, it might be worth while to go to a slower baud rate. Otherwise, you would spend more time retransmitting bad packets than getting good ones through. The theory is that the faster the speed of data transmission the more likelihood of error. This is not always so. But if you are getting lots of errors, the first — and easiest — step is to drop the transmission speed. Bit Error Rate is thus a measure of transmission quality. It is generally shown as a negative exponent, (e.g., 10 to the minus 7 which means one out of 10,000,000 bits are in error.) See also BIT ERROR RATE.

2. A LAN term, which means Basic Encoding Rules. Rules for encoding data units described in ASN.1. (Abstract Syntax Notation One. LAN "grammar," with rules and symbols, that is used to describe and define protocols and programming languages. ASN.1 is the OSI standard language to describe data types.) BER is sometimes incorrectly lumped under the term ASN.1, which properly refers only to the abstract syntax description language, not the encoding technique.

BERT Bit Error Rate Test, or Tester. A known pattern of bits is transmitted, and errors received are counted to figure the BER. The idea is to measure the quality of data transmission. The bit error rate is the ratio of received bits that are in error, relative to the number of bits received. Usually expressed in a power of 10. Sometimes called Block Error Rate Tester.

Bernoulli Daniel Bernoulli was the 18th century Swiss mathematician who first expressed the principle of fluid dynamics — the basis of Bernoulli "boxes" also called disk cartridges. A Bernoulli box, which is a mass storage device, uses both floppy and hard disk technologies. Bernoulli disks are removable. They physically look like large floppy disks.

Best Path An internetworking term: The optimal route between source and destination end stations through a wide area network. Determined through routing protocols such as RIP and OSPF, best path can be based on lowest delay, cost or other criteria.

Beta 1. Refers to the final stages of development and testing before a product is released to market. "Alpha" is the term used when a product is in preliminary development. "Her baby is in beta," according to Peter Lewis of the New York Times, means she is expecting soon. In the software industry, beta has been known to last a year or more. Microsoft's Bill Gates has given the word "beta" a whole new meaning by having as many as 500,000 "beta testers" for his Windows 95 operating system. At this level, beta testing is no longer testing, it's marketing. And it's positively brilliant. See BETA TEST.

2. Business Equipment Trade Association (UK).

Beta Site A place a beta test is conducted. See BETA and BETA TEST.

Beta Test Typically the last step in the testing of a product before it is officially released. A beta test is often conducted with customers in their offices. Some customers pay for the equipment or software they get under a beta test; some don't. Some beta tests stay in (if they work). Some don't. Most products don't work when they're first introduced. So beta tests are a good idea. Unfortunately, most manufacturers don't do sufficient beta testing. They want to get their product to market before the competition does. This often means we now have two or three new products on the market, none of which work reliably or do exactly what they're meant to do. Our rule: always wait several months after a product is introduced before buying it. By then the major bugs will have been fixed. The test before the beta test is called the Alpha. It isn't that common. See BETA.

Betazoid A planet in the second Star Trek TV series, inhabited by Betazoids, beings with great powers of empathy and telepathy.

BETRS Basic Exchange Telecommunications Radio Service. A service that can extend telephone service to rural areas by replacing the local loop with radio communications, sharing the UHF and VHF common carrier and private radio frequencies.

Bezek The name of the Israeli long and long distance phone company.

Bezel The metal or plastic part — in short, the frame — that surrounds a cathode ray tube — a "boob" tube.

BEZS Bandwidth Efficient Zero Suppression. N.E.T.'s patented T-1 zero suppression technique; maintains Bell specifications for T-1 pulse density without creating errors in end-user data; uses a 32 kbps overhead channel.

BFT Binary File Transfer. BFT is a method of routing digital files using facsimile protocols instead of traditional modem file transfer protocols. See BINARY FILE TRANSFER for a fuller explanation.

BFV Bipolar violations: The digital data format consists of pulses of opposite polarity. No two consecutive pulses should be the same polarity; if two are detected in a row, the term is a violation, which is also a warning flag.

BGE-I ISDN Business Group Elements.

BGID Business Group ID.

BGP Border Gateway Protocol is a Transmission Control Protocol/Internet Protocol (TCP/IP) routing protocol for interdomain routing in large networks. It is used in the Internet and is an alternative to EGP (Exterior Gateway Protocol).

BH Bandwidth Hog. A term defined by Philip Elmer-DeWitt, technology editor of Time Magazine in 1994, who spearheaded the launch of Time Online, the first fully electronic national magazine. He defined BH "as a person who uses the online medium like a bullhorn and attracts like-minded people who then rove in a pack, filling them with up with screeds." (Screed is a long discourse or essay.)

NEWTON'S TELECOM DICTIONARY

BHANG Broadband High Layer Information: This is a Q.2931 information element that identifies an application (or session layer protocol of an application).
BHC Backbone to Horizontal Cross-connect. Point of interconnection between backbone wiring and horizontal wiring.
Bi A Latin prefix meaning twice.
Bias 1. A systemic deviation of a value from a reference value.
2. The amount by which the average of a set of values departs from a reference value.
3. An electrical, mechanical, magnetic, or other force field applied to a device to establish a reference level to operate the device.
4. Effect on telegraph signals produced by the electrical characteristics of the terminal equipment.
Bias Distortion Distortion affecting a two-condition (binary) coding in which all the significant intervals corresponding to one of the two significant conditions have uniformly longer or shorter durations than the corresponding theoretical durations. The magnitude of the distortion is expressed in percent of a perfect unit pulse length.
Bias Generator A CBX printed circuit card that generates a signal that reduces idle channel noise for all coders installed in the CBX.
Bias Potential The potential impressed on the grid of a vacuum tube to cause it to operate at the desired part of its characteristic curve.
Bib Signaling ID assigned by Exchange B.
BICEP An ATM term. Bit Interleaved Parity: A method used at the PHY layer to monitor the error performance of the link. A check bit or word is sent in the link overhead covering the previous block or frame. Bit errors in the payload will be detected and may be reported as maintenance information.
Biconic Fiber Optic Connector developed by AT&T.
Biconical Antenna An antenna consisting of two conical conductors having a common axis and vertex. Excitation occurs at the common vertex. If one of the cones is flattened into a plane, the antenna is called a discone.
BICSI Building Industry Consulting Service International, a professional organization. For those who acquire certain requisite education and experience by BICSI, the association makes them a RCDD, Registered Communication Distribution Designer. BICSI is at 813-397-1991.
Bidirectional Bus A bus that may carry information in either direction but not in both simultaneously.
Bidirectional Couplers Fiber optic couplers that operate in the same way regardless of the direction light passes through them.
Bidirectional Line Switched Ring Commonly referred to as BLSR, bi-directional line switched ring is a method of SONET transport in which half of the working network is sent counter-clockwise over one fiber and the other half is sent clockwise over another fiber. BLSR offers bandwidth use advantages for distributed traffic in single-ring architectures.
Bidirectional Printing A typewriter always prints from left to right. So did the early computer printers. The newer computer printers will print from left to right, drop down a line, then print from right to left. This increases the printer's speed.
Bifurcated Routing Routing that may split one traffic flow among multiple routes.
Big-endian A networking term. A format for storage or transmission of binary data in which the most significant bit (or byte) comes first. The reverse convention is called little-endian.
Bigamy The only crime in which two rites make a wrong.
Bilateral Synchronization A synchronization control system between exchanges A and B in which the clock at exchange A controls the received data at exchange B and the clock at exchange B controls the received data at exchange A. Normally implemented by deriving the receive timing from the incoming bit stream.
Bildschirmtext German word for interactive videotex. The German Bundespost likes this service. But the German version isn't as successful as the French because the French gave away the videotex terminals. And the Germans didn't. Also, the French really encourage videotex entrepreneurs by giving a real piece of the action — 60% of the collected revenues.
Bill Of Materials A list of specific types and amounts of direct materials expected to be used to produce a given job or quantity of output.
Billboard Electronic sales pitches that come up on your computer screen at any time.
Billboard Antenna A broadside antenna array with flat reflectors.
Billed Number Screening You (at home or your business) establish who can and cannot charge a call to your phone by making an agreement with your local telephone company to screen your calls. (e.g. Refusal of all collect call requests.)
Billed Telephone Number BTN. The primary telephone number used for billing regardless of the number of telephone lines associated with that number.
Billibit Someone's absolutely awful term for one billion bits. Also (and better) called a gigabit.
Billing Increment The increments of time in which the phone company (long distance or local) bills. Some services are measured and billed in one minute increments. Others are measured and billed in six or ten second increments. The billing increment is a major competitive weapon between long distance companies. Short billing increments become important to you, as a user, when your average calls are very short — for example, if you're making a lot of very short data calls (say for credit card authorizations). Being billed for a lot of six second calls is a lot cheaper than being billed for a lot of one minute calls.
Billing Media Converter A Billing Media Converter, as made by the Cook division of Northern Telecom, provides a means of transporting Automatic Message Accounting (AMA) data from DMS-10 central offices to regional accounting offices with the physical transfer of magnetic tapes. The BMC is polled.
Billing Telephone Number A Billing Telephone Number is the number to which calls to given location are billed. It is the seven-digit number with the area code followed by an alphanumeric code assigned by the local telephone company (e.g. NPA-NXX-XXXX).
Billing Validation Service See BVA and BVS
Billion In North America, a billion is a thousand million. In many countries overseas, a billion is a million million.

NEWTON'S TELECOM DICTIONARY

Binaries Binary, machine readable forms of programs which have been compiled or assembled. As opposed to source language forms of programs.

Binary Where only two values or states are possible for a particular condition, such as "ON" or "OFF" or "One" or "Zero." Binary is the way digital computers function because they can only represent things as "ON" or "OFF." This binary system contrasts with the "normal" way we write numbers — i.e. decimal. In decimal, every time you push the number one position to the left, it means you increase it by ten. For example, 100 is ten times the number 10. Computers don't work this way. They work with binary notation. Every time you push the number one position to the right it means you double it. In binary, only two digits are used — the "0" (zero) and the "1" (the one). If you write the number 10101 in binary, and you want to figure it in decimal as we know it, here's how you do it. 1 is one thing; Zero x 2 = zero; 1 times 2 x 2 = 4; 0 x 2 x 2 x 2 = 0; 1 x 2 x 2 x 2 x 2 = 16. Therefore the total 10101 in binary = 1 + 0 + 4 + 0 + 16 = 21 in decimal.

Binary notation differs slightly from notation used in ASCII or EBCDIC. In ASCII and EBCDIC, the binary values are used for coding of individual characters or keys or symbols on keyboards or in computers. So each string of seven (as in ASCII) or eight (as in EBCDIC) ones and zeros is a unique value — but not a mathematical one.

ASCII uses a seven bit coding scheme. Thus, the maximum number of different things you can code using seven bits is 128, i.e. 2 x 2 x 2 x 2 x 2 x 2 x 2 = 128. The maximum number represented by a byte (8 bits) or the IBM EBCDIC coding system is 256. i.e. 2 x 2 x 2 x 2 x 2 x 2 x 2 x 2 = 256. See BINARY CODED DECIMAL, BINARY FILE and BINARY TRANSFER.

Binary Code A code in which every element has only one of two possible values, which may be the presence or absence of a pulse, a pulse, a one or a zero, or high or a low condition for a voltage or current.

Binary Coded Decimal BCD. A system of binary numbering that uses a 4-bit code to represent each decimal digit from 0 to 9 and multiple 4-bit patterns for higher numbers. The decimal numbers 0 to 9 are represented by the four-bit binary numbers from 0000 to 1010.

Binary Digit A number in the binary system of notation.

Binary File A file containing information that is in machine-readable form. It is an application. Or it can only be read by an application. See BINARY TRANSFER.

Binary File Transfer BFT. The transmission of binary files, software, documents, images, video and electronic data exchange information between communicating devices, including PCs, fax devices, etc. When binary files are transferred via telecommunications, they must not be changed in any way during the transfer otherwise they will be destroyed. Many electronic mail services, such as CompuServe or MCI Mail, have ways of "attaching" binary files to electronic mail, such that the binary file will not be affected or changed in any way during transmission. But that only works between MCI users or between CompuServe users. The Internet uses a different system. It's called MIME, which stands for Multipurpose Internet Mail Extension. What it does is it uses software to encode the binary file into an ASCII file. It transmits the ASCII file. At the other end, that file is decoded into its original binary format. MIME uses many software techniques for doing this. But all have the same effect of not changing the original binary file. See MIME.

Binary Notation Any notation that uses two different characters, usually the binary digits 0 and 1.

Binary Symmetric Channel A channel designed so that the probability of changing binary bits in one direction is the same as the probability of changing them back to the correct state.

Binary Synchronous Communications BISYNC or BSC. 1. In data transmission the synchronization of the transmitted characters by timing signals. The timing elements at the sending and receiving terminal define where one character ends and another begins. There are no start or stop elements in this form of transmission.

2. Also a uniform discipline or protocol for synchronized transmission of binary coded data using a set of control characters and control character sequences.

Binary To Decimal Conversion Conversion from base 2 to base 10. See BINARY.

Binary Transfer See BINARY FILE TRANSFER and MIME.

Bind A request to activate a session between two logical units (LUs).

Binder A helically applied colored thread, yarn, or plastic ribbon used to confine and sometimes separate and identify groups of fibers or wires in a cable. Binders are usually used for holding assembled cable components in place.

Binder Group A group of wires within a large cable that can be distinguished from other groups because it is wrapped with colored threads. Normal telephone color-coding provides for only 25 pairs of wire, so binder groups allow multiple pairs of the same color wire to be in one cable. A 50-pair cable has blue and orange binder groups; a 75-pair cable has blue, orange, and green groups. Since several wire pairs have the same color markings, installers must be careful when stripping cable insulation so they do not destroy the binder threads.

Bindery A Novell NetWare database containing definitions for entities such as users, groups, and workgroups. The bindery contains three components: objects, properties, and property data sets. Objects represent any physical or logical entity, including users, user groups, file servers, print servers, or any other entity given a name. Properties are the characteristics of each bindery object, including passwords, account restrictions, account balances, internetwork addresses, list of authorized clients, workgroups, and group members. Property data sets are values assigned to entities bindery properties.

Binding A Windows 95 definition. Binding is a process that establishes the communication channel between a protocol driver and a network adapter driver.

Binding Post A screw with a small nut. You take your wires and join them together on a binding post by wrapping them together around the screw and then tightening the nut on them. These days, 66-blocks have effectively replaced binding posts. 66-blocks are easier and faster to use.

BinHex When sending files which aren't plain ASCII across a network — dial-up, leased line or the Internet — you basically have two options. First, you can attach them as a binary file (i.e. non-ASCII file). Or second, you can encode them into ASCII characters and send the file as part of your message. The first method is preferable. But you can typically only send binary files from one account to another on the same network, or between two networks that have agreed between themselves to a method of transferring files. But that's a rarity. And it certainly doesn't work in and around the Internet. Thus something called, MIME was created. MIME stands Multipurpose Internet Mail extensions. It is the name for encoding binary files into ASCII characters for transfer across the Internet and on-line services. Under MIME, there are a number of methods of encoding binary files, one of which is called BinHex, which is a popular encoding algorithm that uses Run-Length Encoding (RLE). To send a binary file, you encode it using BinHex encoding software. You then include the ASCII encoded file in your message. The recipient of your message then decodes the ASCII using BinHex decoding software, which may or may not be built into his electronic mail or browser package.

Biometric Device A device used in authenticating access to a system. A biometric device authenticates a user by measuring some hard-to-forge physical characteristic, such as a fingerprint or retinal scan. You see a lot of biometric devices in Hollywood thrillers, including Mission Impossible, a movie which was released in the summer of 1996, as I wrote this definition.

BIOS Basic Input/Output System of desktop computers. The BIOS contains the buffers for sending information from a program to the actual hardware device the information should go to.

BIOS Enumerator A Windows 95 term. Bios enumerator is responsible, in a Plug and Play system, for identifying all hardware devices on the motherboard of the computer. The BIOS supports an API that allows all Plug and Play computers to be queried in a common manner.

Biosensors Devices such as fingerprint readers and signature recognition systems.

BIP-N Bit Interleaved Parity N. A method of error monitoring. With even parity, an N bit code is generated by the transmitting equipment over a specified portion of the signal in such a manner that the first bit of the code provides even parity over the first bit of all N-bit sequences within the specified portion, etc. Even parity is generated by setting the BIP-N bits so that there are an even number of 1s in each of all N-bit sequences including the BIP-N.

Bipolar The predominant signaling method used for digital transmission services, such as DDS and T-1. The signal carrying the binary value alternates between positive and negative. Zero and one values are represented by the signal amplitude at either polarity, while no-value "spaces" are at zero amplitude.

Bipolar Coding The T carrier line coding system that inverts the polarity of alternate "one" bits.

Bipolar Signal A signal having two polarities, both of which are not zero. It must have two-state or three-state binary coding scheme. It is usually symmetrical with respect to zero amplitude.

Bipolar Violation The presence of two consecutive "one" bits of the same polarity on the T carrier line. See also BIPOLAR CODING.

Bird A satellite.

Birthdays

January 4, 1847	Thomas Edison born
February 16, 1982	Michael Allen Newton born
February 20, 1980	Claire Elizabeth Newton born
February 23, 1965	Michael Dell (Dell Computer) born
March 3, 1847	Alexander Graham Bell born
March 11, 1933	Ben Rosen (Compaq, SRX, etc.) born
March 15, 1949	Gerry Friesen born
April 6, 1939	John Sculley born
April 27, 1791	Samuel Morse born
May 11, 1948	International Communications Association born
June 10, 1942	Harry Newton born in Sydney Australia
June 13, 1961	TeleCommunications Association born
June 19, 1924	Ray Noorda (Novell) born
June 27, 1968	Carterfone decision handed down by FCC
July 4, 1943	Susan Newton born
August 31, 1962	President Kennedy signed Communications Satellite Act
September 2, 1936	Andy Grove (Intel)
September 12, 1948	Communications Managers Association
October 28, 1955	Bill Gates (Microsoft)
November 13, 1954	Scott McNealy (Sun Microsystems)
December 10,	William G. McGowan, founder of MCI

Bis 1. The French term for "second" or "encore." It is used by the CCITT/ITU to designate the second in a family of related standards. "ter" designates the third in a family. See V SERIES.
2. Border Intermediate System

BISDN Also spelled B-ISDN. Broadband ISDN. This is a vaguely defined term. It basically means any circuit capable of transmitting more than one Basic rate ISDN, i.e. 144 kbps. One definition I read recently suggested that BISDN is "a set of public network services that are delivered over ATM, including data, voice, and video. BISDN will provide services such as high-definition television (HDTV), multi-lingual TV, voice and image storage and retrieval, video conferencing. high-speed LANs, and multimedia." See B-ISDN for a longer explanation. See also ISDN.

Bistable Trigger Circuit A trigger circuit that has two stable states.

BISUP Broadband ISDN User's Part: A SS7 protocol which defines the signaling messages to control connections and services. See SIGNALING SYSTEM 7.

BISYNC (pronounced bye-sink). BISYNChronous Transmission. A half-duplex, character-oriented, synchronous data communications transmission method originated by IBM in 1964. See SYNCHRONOUS.

NEWTON'S TELECOM DICTIONARY

Bisynchronous Transmission Also called BISYNC. A data character-oriented communications protocol developed by IBM for synchronous transmission of binary-coded data between two devices. BISYNC uses a specific set of control characters to synchronize the transmission of that binary coded data. See also BINARY SYNCHRONOUS COMMUNICATION.

Bit Bit is a contraction of the term BInary digiT. It is the smallest unit of information (data) a computer can process, representing either high or low, yes or no, or 1 or 0. It is the basic unit in data communications. A bit can have a value of zero (a mark) or one (a space).

Bit Bucket Slang for throwing out bits — into a wastepaper bucket.

Bit Buffer A section of memory capable of temporarily storing a single BInary uniT (bit) of information. Used to make data transmission accurate or consistent.

Bit Check A bit added to a digital signal and used for error checking, i.e. a parity bit. See also PARITY.

Bit Depth The number of bits used to represent the color of each pixel in a graphic file or the sound in an audio file. The higher the number, the more information included in the file and the higher the quality of the data. Common graphic bit depths are 4-bit, 8-bit, 16-bit, 24-bit and 32-bit. Common sound bit depths are 8-bit and 16-bit.

Bit Duration The time it takes one bit to pass a point on the transmission medium. Used to measure delay times, especially in high speed communications.

Bit Error The value of an encoded bit can be changed due to a transmission problem (e.g. noise on the line) and then interpreted incorrectly by the receiver.

Bit Error Rate BER. The percentage of received bits in error compared to the total number of bits received. Usually expressed as a number to the power of 10. For example, 10 to the fifth power means that one in every 100,000 bits transmitted will be wrong. In transmitting data a high error rate on the transmission medium (i.e. some noise), doesn't mean there'll be lots of problems with the final transmission. It just means there'll have to be lots of re-transmissions — "until one gets it right." These re-transmissions reduce the amount of data transmitted in a unit of time and therefore, increase the time needed to send that information. If the BER gets too high, it might be worth while to go to a slower transmission rate. Otherwise, you would spend more time retransmitting bad packets than getting good ones through. The theory is that the faster the speed of data transmission the more likelihood of error. This is not always so. But if you are getting lots of errors, the first — and easiest — step is to drop the transmission speed.

Bit Interleaving A form of TDM for synchronous protocols, including HDLC, SDLC, BiSync and X.25. Bit interleaving retains the sequence and number of bits, so that correct synchronization is achieved between both ends. See BIT INTERLEAVING/MULTIPLEXING.

Bit Interleaving/Multiplexing In multiplexing, individual bits from different lower speed channel sources are combined one bit at a time/one channel at a time into one continuous higher speed bit stream. Compare with byte interleaving/ multiplexing.

Bit Oriented Used to describe communications protocols in which control information may be coded in fields as small as a single bit.

Bit Oriented Protocol BOP. A data link control protocol that uses specific bit patterns to transfer controlling information. Examples are IBM's Synchronous Data Link Control (SDLC) and the ITU-T High-Level Data Link Control (HDLC). Bit-oriented protocols are normally used for synchronous transmission only. Bit-oriented protocols are code transparent (meaning they work regardless of the character encoding method used), since no encoded characters are used in the control sequence.

Bit Oriented Transmission An efficient transmission protocol that encodes communications control information in fields of bits rather than characters or bytes.

Bit Parity A binary bit appended to an array of bits to make the sum of all the bits always odd or always even. See PARITY.

Bit Pattern A group of bits arranged in specified ways to represent numbers, letters or symbols, forming a unique binary number for each character. For example, the 7-bit ASCII code produces 128 different characters, i.e. $2 \times 2 \times 2 \times 2 \times 2 \times 2 \times 2 = 128$.

Bit Rate The number of bits of data transmitted over a phone line per second. You can usually figure how many characters per second you will be transmitting — in asynchronous communications — if you divide the bit rate by ten. For example, if you are transmitting at 1200 bits per second, you will be transmitting 120 characters per second. In real life, it's never this simple, however. The total bits transmitted will depend on re-transmissions, which depends on the noise of the line, etc. See BAUD RATE.

Bit Robbing A technique to signal in-band in digital facilities, which typically use out of band signaling, e.g. Signaling System 7. In bit robbing, we steal bits from the speech path a few line-signal bits. The remaining bits are adequate to recreate the original electrical analog signal (and ultimately, the original sound). Bit robbing typically uses the least significant bit per channel in every sixth frame for signaling. See BIT STUFFING and ROBBED BIT SIGNALING.

Bit Specifications Number of colors or levels of gray that can be displayed at one time. Controlled by the amount of memory in the computer's graphics controller card. An 8-bit controller can display 256 colors or levels of gray; a 16-bit controller, 64,000 colors: a 24-bit controller, 16.8 million colors.

Bit Stream A continuous stream of data bits transmitted over a communications line with no break or separators between the characters.

Bit Stuffing Process in some data communications protocols where a string of "one" bits is broken by an inserted "zero." This inserted zero is added by the sender and stripped by the receiver. The idea of inserting the zero is to avoid confusing the receiver into thinking the series of one bits mean something else, like a flag control character. See ZERO STUFFING.

Bit Transfer Rate The number of bits transferred per unit time. Usually expressed in Bits Per Second (BPS).

BITBLT BIT BLock Transfer. Microsoft Windows relies intensively on a type of operation called bit block transfer (BitBLT) to redraw rectangular areas of the image on the computer's screen. Generally, BitBLT operations are accomplished by software routines in the video driver, a cheap, but slow method that uses many of your CPU's clock cycles. If you add a separate video controller with a special processor to handle BitBLT, you will be able to offload video tasks from your main CPU and make your computer run faster.

Bitmap Representation of characters or graphics by individual pixels arranged in row (horizontal) and column (vertical) order.

Each pixel can be represented by one bit (simple black and white) or up to 32 bits (high-definition color). Bitmapped images can be displayed on screens or printed. See BIT SPECIFICATIONS.

Bitmapped Graphics Images which are created with matrices of pixels, or dots. Also called raster graphics. See BITMAP.

BITNET Because It's Time NETwork. An academic computer network based originally on IBM mainframes connected with leased 9600 bps lines. BITNET has recently merged with CSNET, The Computer+Science Network (another academic computer network) to form CREN: The Corporation for Research and Educational Networking. The network connects more than 200 institutions and has more than 900 computational nodes.

Bitnik A person who uses a coin-operated computer terminal installed in a coffee house to log into cyberspace.

Bitronix Hewlett Packard's term for its bidirectional parallel port communications "standard." It introduced this standard with its 600 dps LaserJet 4 plain paper printer in the fall of 1992. It is hoping other manufacturers will adopt the standard. The big plus of the standard is that it allows a printer to tell a connected computer that it (the printer) has run out of paper, or the paper has jammed, etc. Having that communications back and forth will allow the user to clear the problem and get the printer and up and running faster. It will also stop the computer locking up.

BITS Building Integrated Timing Supply. In North America, the clocks that provide and distribute timing to a wireline network's lower levels. Known in the rest of the world as an SSU (Synchronization Supply Unit).

Bits Per Second The number of bits passing a specific point per second.
A KILObit is one thousand bits per second.
A MEGAbit is one million bits per second (thousands of kilos).
A GIGAbit is one billion bits per second (thousands of millions).
A TERAbit is one trillion bits per second (thousands of billions).
A PETAbit is equal to 10 to the 15th or 1,000 terabits.

Bitslag All the useless rubble on the Net one has to plow through to get to the rich information core.

Bix A Northern Telecom trade name for an in-building termination and cross- connect system for unshielded twisted pair cables.

BL Business Line

Black Body A totally absorbing body that does not reflect radiation (i.e. light). In thermal equilibrium, a black body absorbs and radiates at the same rate; the radiation will just equal absorption when thermal equilibrium is maintained.

Black Box An electronic device that you don't want to take the time to understand. As in, "We'll put the data through a black box that will put it into X.25 format." The term has recently come also to mean PBX switches. While "Black Box" is a generic term, The Black Box Corporation of Pittsburgh, PA, has had the audacity (and brilliance) to register the term as a trademark. The Phone Phreak community has used the term black box to describe a device that's put on phone lines in electromechanical central office areas (they don't work under ESS offices). To the phone phreak community, a black box made up of a resistor, a capacitor and a toggle switch that would "fool" the central office into thinking the phone had not been picked up when receiving a long distance call. Since the call was not "answered," the call could not be billed. Clever, eh? (Illegal, too.) See BLUE BOX and RED BOX.

Black Facsimile Transmission 1. In facsimile systems using amplitude modulation, that form of transmission in which the maximum transmitted power corresponds to the maximum density of the subject copy.
2. In facsimile systems using frequency modulation, that form of transmission in which the lowest transmitted frequency corresponds to the maximum density of the subject copy.

Black Level The lowest luminance level that can occur in video or television transmission and which, when viewed on a monitor, appears as the color 'black.'

Black Matrix Picture tube in which the color phosphors are surrounded by black for increased contrast.

Black Recording 1. In facsimile systems using amplitude modulation, that form of recording in which the maximum received power corresponds to the maximum received power corresponds to the maximum density of the record medium.
2. In a facsimile system using frequency modulation, that form of recording in which the lowest received frequency corresponds to the maximum density of the record medium.

Black Signal 1. In facsimile, the signal resulting from the scanning of a maximum-density area of the subject copy.
2. In cryptographic systems, a signal containing only unclassified or encrypted information.

Blacksburg Electronic Village Blacksburg is a small town in the mountains of Western Virginia. It has a branch of Virginia Tech. University. They have wired Virginia Tech students and thousands their town's citizens with email, internet access and bulletin boards showing happening in town. The official explanation is: The Blacksburg Electronic Village (BEV) is a cooperative project of Virginia Tech, Bell Atlantic of Virginia and the Town of Blacksburg. It links the town's citizens, via computers lines to each other and to the Internet. Citizens gain access to the Internet from their home or office through a high-speed model pool or by using Ethernet LANS which are available in some offices and hundreds of apartment units in town. Blacksburg residents may take advantage of a full spectrum of services including the World Wide Web, Gopher, electronic mail, electronic mailing list and thousands of Usenet newsgroups. In addition to full Internet access, citizens enjoy the benefits of extensive online local resources. The Blacksburg Electronic Village is famous because it really is the first town to aggressively ensure that the bulk of citizenry could and would have access to electronic mail and to the various resources on the Internet. It has apparently made a major difference to how people live and communicate in Blacksburg. www.bev.net

Blank A character on teletype terminals that does not punch holes in paper tape (except for feed holes to push the paper through). Also the character between words, usually called a "Space" is referred to in IBM jargon as a Blank.

Blanking The suppression of the display of one or more display elements or display segments.

Blanking Interval Period during the television picture formation when the electron gun returns from right to left after each line (horizontal blanking) or from top to bottom after each field (vertical blanking) during which the picture is suppressed.

Blanking Pulses The process of transmitting pulses that extinguish or blank the reproducing spot during the horizontal and vertical retrace intervals.

Blast BLocked ASynchronous Transmission.

Blend To have outbound and inbound phone calls answered by the same agents. See the next two definitions.

Blended Agent A call center person who answers both incoming and makes outgoing calls. This idea of a blended agent is a new concept in a Call Center. In the past call centers have typically kept their inbound and outbound agents separate. The reason? Management felt that the necessary skills were very different and no one could master both.

Blended Call Center A telephone call center whose agents both receive and make calls. In other words, a call center whose phone system acts both as an automatic call distributor and a predictive dialer.

BLER Block Error Ratio. The ratio of the blocks in error received in a specified time period to the total number of blocks received in the same period.

BLERT BLock Error Rate Test.

BLF The Busy Lamp Field is a visual display of the status of all or some of your phones. Your BLF tells you if a phone is busy or on hold. Your Busy Lamp Field is typically attached to or part of your operator's phone. See BUSY LAMP FIELD.

Blind Transfer Someone transfers a call to someone else without telling the person who's calling. Also called Unsupervised Transfer. Contrast with SCREENED TRANSFER.

Blinking An intentional periodic change in the intensity of one or more display elements or display segments.

Blister Pack A pocketed polyvinyl chloride shipping container with a snap-on cover.

Bloatware Ever-fatter packages of "upgraded" software that, with each upgrade, come with dozens and dozens of new features. With each upgrade, the customer has less need to look elsewhere. At least that's the theory.

BLOB Binary Large OBjects. When a database includes not only the traditional character, numeric, and memo fields but also pictures or other stuff consuming of large space, a database is said to include BLOBs — binary large objects.

Block In data communications, a group of bits transmitted as a unit and treated as a unit of information. Usually consists of its own starting and ending control deliminators, a header, the text to be transmitted and check characters at the end used for error correction. Sometimes called a Packet.

Block Character Check BCC. The result of transmission verification algorithm accumulated over a transmission block, and normally appended at the end, e.g. CRC, LRC.

Block Diagram A graphic way to show different elements of a program or process by the use of squares, rectangles, diamonds and various shapes connected by lines to show what must be done, when it must be done and what happens if it's done this way or that. In short, it shows how all the small decision points add up to the whole process.

Block Error Ratio BLER. The ratio of the blocks in error received in a specified time period to the total number of blocks received in the same period.

Block Misdelivery Probability The ratio of the number of misdelivered blocks to the total number of block transfer attempts during a specified period.

Block Mode Terminal Interface BMTI. A device used to create (and break down) packets to be transmitted through a ITU-T X.25 network. This device is needed if block-mode terminals (such as IBM bisync devices) are to be connected to the network without an intermediate computer.

Block Multiplexer Channel An IBM mainframe input/output channel that allows interleaving of data blocks.

Block Parity The designation of one or more bits in a block as parity bits whose purpose is to ensure a designated parity, either odd or even. Used to assist in error detection or correction, or both.

Block Transfer The process of sending and receiving one or more blocks of data.

Block Transfer Attempt A coordinated sequence of user and telecommunication system activities undertaken to effect transfer of an individual block from a source user to a destination user. A block transfer attempt begins when the first bit of the block crosses the functional interface between the source user and the telecommunication system. A block transfer attempt ends either in successful block transfer or in block transfer failure.

Block Transfer Efficiency The average ratio of user information bits to total bits in successfully transferred blocks.

Block Transfer Failure Failure to deliver a block successfully. Normally the principal block transfer failure outcomes are: lost block, misdelivered block, and added block.

Block Transfer Rate The number of successful block transfers made during a period of time.

Block Transfer Time The average value of the duration of a successful block transfer attempt. A block transfer attempt is successful if 1. The transmitted block is delivered to the intended destination user within the maximum allowable performance period and 2. The contents of the delivered block are correct.

Blocked Calls Delayed A variable in queuing theory to describe what happens when the user is held in queue because his call is blocked and he can't complete it instantly.

Blocked Calls Held A variable in queuing theory to describe what happens when the user redials the moment he encounters blockage.

Blocked Calls Released A variable in queuing theory to describe what happens when the user, after being blocked, waits a little while before redialing.

Blocking When a telephone call cannot be completed it is said that the call is "blocked." Blocking is a fancy way to say that the caller is "receiving a busy." There are many places a call can be blocked: at the user's own telephone switch — PBX or key system, at the user's local central office or in the long distance network. Blocking happens because switching or transmission capacity is not available at that precise time.

The number of calls you try compared to the number of times you get blocked measures "the grade of service" on that network. Blocked calls are different from calls that are not completed because the called number is busy. This is because numbers that are busy are not the fault of the telephone switching and transmission network. One might think the fewer blocked calls, the better. From the user's point of view, the answer is obviously YES, it is better. Less blockage, fewer busies and less frustration. But as one designs a switching and transmission network for less and less blocking, the network becomes more and more expensive. Logarithmically so. We keep adding extra circuits and extra equipment. Thus, in any telecommunications network design there is always a trade-off: What are you prepared to pay, compared to what can you tolerate?

Everyone designs their network with different trade-offs depending on what they and their users or customers, can tolerate and/or are willing to pay. Most companies are willing to pay more for better service if someone explains the logic of telephone design to them. Many network salesmen, however, don't believe this. They practice the sales "theory" of selling better service for less money. This doesn't work in business, and especially not in telephony.

The "Grade of Service" is a measurement of blocking. It varies from almost zero (best, but most expensive case, no calls blocked) to one (worst case, all calls blocked). Grade of Service is written as P.05 (five percent blocking). "Blocking" used to be a technical term but has now become a sales tool especially among PBX manufacturers, who increasingly claim their switch to be "non-blocking." This means it will not, they claim, block a call in the switch.

There are several flaws in this logic: First, it's not logical or useful to buy a non-blocking PBX if the chances of being blocked elsewhere — the local lines, the local exchange or the long distance network — are very high. Second, a true non-blocking PBX can be very expensive, perhaps too much power and too much money for most peoples' needs. Third, most manufacturers define "non-blocking" differently. One defines it strictly in terms of switching capability and ignores the fact that his PBX might not have sufficient other "things," like devices which ring bells on phones (to indicate an incoming call) or devices which deliver dial tone to a phone (to indicate the PBX is ready to receive instructions).

Blocking Factor The number of records in a block; the number is computed by dividing the size of the block by the size of each record contained therein. Each record in the block must be the same size.

Blocking Formulas Specific probability distribution functions that closely approximate the call pattern of telephone users probable behavior in failing to find idle lines.

Blower A microphone.

Blowfish The name of the scrambling algorithm behind Philip Zimmerman's powerful encryption scheme called Pretty Good Privacy, or PGP, which lets you converse in total privacy over normal phone lines. See PGP for a fuller explanation.

Blowing Your Buffer Losing one's train of thought. This happens when the person you're speaking with won't let you get a word in edgewise or just said something so astonishing that your brain gets derailed. "Damn, I just blew my buffer!"

Blown Fiber Blown fiber is an installation method where a housing with lots of smaller tubes is installed without fiber. Once the housing is checked out, they take an air compressor and use air pressure to float the fiber down the tube. This greatly lowers the rate of failures during underground burying of fiber since it is empty during the installation. The air flow moves in the same direction as the fiber is installed and helps lower friction and any tugging on the fiber itself. One advantage to blown fiber is that the tubes can be installed with lots of joins (as needed to accommodate the installation problems, such as tight corners), but the fiber strands are later installed with no splices required. The idea of blown fiber is that you can upgrade your fiber when you needed it.

Blown Fuse A broken fuse.

BLSR See Bidirectional Line Switched Ring.

BLU Northern Telecom word for Basic Link Unit.

Blue Alarm Used in T-1 transmission. Also known as the AIS (Alarm Indication Signal). The blue alarm is turned on when two consecutive frames have fewer than three zeros in the data bit stream. A blue alarms sends 1's (ones) in all bits of all time slots on the span. See T-1 and AIS.

Blue Box A device used to steal long distance phone calls. The classic blue box was slightly larger than a cigarette container. It had a touchtone pad on the front and a single button on top. Typically, you went to a coin phone and dialed an 800 number. While the distant number was ringing, you punched the single button on the top of the blue box. That button caused the blue box's speaker to emit a 2600 Hz tone. This disconnected the ringing at the other end but left the user inside the long distance network. The user then punched in a series of digits on the touchtone pad. The phone network heard those tones and sent the call according to the instructions in the tones. The tones duplicated the tones which the touchtone pads of long distance operators emitted. They are different from those emitted by normal telephones. The first blue box was "discovered" at MIT in a small utility box that was painted blue, thus the term blue box. When they were young, Steve Jobs and Steve Wozniak, founders of Apple Computer, sold blue boxes, which Wozniak built. People who used blue boxes in their salad days included characters with adopted pseudonyms like Dr. No, The Snark and Captain Crunch, who got his name from the free 2600 Hz whistle included as a promotion in boxes of Captain Crunch breakfast cereal. With the advent of CCIS, Common Channel Interoffice Signaling (i.e. out-of-band signaling), blue boxes no longer work.

Blue Collar Computer A colloquial term for a handheld computer which is used by "blue collar" workers for tasks such taking inventory, tracking goods, etc. Such computer may have a pen, a large pen-sensitive or touch-sensitive screens, a bar code scanner and a modem. It may be able to capture signatures — useful for confirmation of the delivery of goods.

Blue Grommet The rubber collar over the joint between the handset and the armored cable on a pay phone. Blue identifies a "hearing aid compatible" handset.

Blue Pages Section of phone directory commonly used for government phone numbers, as distinct from white and yellow pages.

BM Burst Modem.

BMTI Block Mode Terminal Interface. A device used to create (and break down) packets to be transmitted through a ITU-T X.25 network. This device is needed if block-mode terminals (such as IBM bisync devices) are to be connected to the network without an intermediate computer.

BN 1. Bridge Number: A locally administered bridge ID used in Source Route Bridging to uniquely identify a route between two LANs. 2. An ATM term. BECN Cell: A Resource Management (RM) cell type indicator. A Backwards Explicit Congestion Notification (BECN) Rm-cell may be generated by the network or the destination. To do so, BN=1 is set, to indicate the cell is not source-generated, and DIR=1 to indicate the backward flow. Source generated RM-cells are initialized with BN=0.

BNA Burroughs Network Architecture. Communications architecture of Burroughs, now Unisys.
BNC A bayonet-locking connector for slim coaxial cables, like those used with Ethernet. BNC is an acronym for Bayonet-Neill-Concelman. Don't ask who Neill and Concelman are. Someone told me it also stands for British National Connector.
BNC Barrel Connectors These connectors join two lengths of thin Ethernet coaxial cable together. See also BNC.
BNC Connectors The connectors for thin Ethernet coaxial cable are BNC connectors. BNC connectors are used on thin Ethernet cable. The BNC connectors on each end of thin Ethernet cable connect to T-connectors, barrel connectors, and other network hardware. See BNC.
BNC Female To N-Series Female Adapter The BNC female to N-Series female adapter is a connector which enables you to connect thin coaxial cable to thick coaxial cable. The BNC female connector attaches to the thin cable and the N-Series male connector attaches to the thick Ethernet cable.
BNC T-connectors The top of the T in a BNC T-connector functions as a barrel connector and links two lengths of thin Ethernet coaxial cable; the third end connects to the SpeedLink/PC 16.
BNC Terminators 50-ohm terminators are used to block electrical interference on a Ethernet coaxial cable network and to terminate the network at certain spots. You attach a BNC terminator to one plug on a T-connector if you will not be attaching a length of cable to that plug. You may also need to use a BNC terminator with a grounding wire to ground the network. See BNC.
BNR Bell-Northern Research. Northern Telecom's research arm. Northern Telecom is now called NorTel.
BNS 1000/2000 AT&T's Broadband Networking Systems using fast-packet, cell relay switching providing dynamic bandwidth allocation on demand with interfaces to a variety of LAN routers, including AT&T's LCS200, Wellfleet, Cisco Systems and NCR StarWAN products.
BnZS Code A bipolar line code with n zero substitution.
Board 1. Short for printed circuit board. Phone systems have boards for all sorts of purposes — from boards to serve trunk lines, to boards to serve proprietary phone sets, to boards that serve T-1 lines, etc. Computers also have boards — ones for SCSI ports, for floppy and hard disks, for CD-ROMs, etc. In display/monitor terminology a board refers to the adapter (or controller) that serves as an interface between the computer and monitor.
2. An SCSA term. Any hardware module that controls its own physical interface to the SCbus or SCxbus. From a programming point of view, a board is an addressable system component that contains resources.
BOB Software from Microsoft, which describes it as a "superapplication" for Windows designed for consumers intimidated by computer technology. According to Microsoft, users will use Bob to write letters, send e-mail, manage their household finances, keep addresses and dates and launch full-blown Windows applications, all under the guidance of cartoon characters. Bob hasn't done well.
BOC Bell Operating Company. The local Bell operating telephone company. These days there are 22 Bell Operating Companies. They are organized into (i.e. owned by) seven Regional Bell holding companies, also called RBOCs, pronounced "R-bocks," or RHCs. See BELL OPERATING COMPANY.
Body The main informational part of a message. Body is the information, not the address nor the addressing information. There can be single or multiple parts to a body. For example, a single part could be text, or multiple parts could include text and graphics, or voice and graphics, etc. See BODY PART 14.
Body Part 14 BP 14. An X.400 messaging term. A non-specific, identifying body part. A binary attachment with identifying header to explain the nature of the content such as a particular spreadsheet, word processor, etc,
BOF Business Operations Framework. A wireless telecommunications term. A document compiled to describe the operations of a telecommunications business entity in a specific area.
Bogon Something that is stupid or nonfunctional.
BOIC Barring of Outgoing International Calls. A wireless telecommunications term. A supplementary service provided under GSM (Global System for Mobile Communications).
Bolt Pattern The pattern that the bolts on the back of the device make. The idea is that if you attach something to the device, you don't want to mess up its warranty by opening the box and putting your stuff inside. So you design your stuff to fit the bolt pattern. This does not mess up the warranty. His product is designed to fit the bolt pattern, but not to intrude into the box.
BOM An ATM term. Beginning of Message: An indicator contained in the first cell of an ATM segmented packet.
Bond The electrical connection between two metallic surfaces established to provide a low resistance path between them.
Bonding 1. In ISDN BRI transmissions, bonding refers to joining the two 64 Kbps B channels together to get one channel of 128 Kbps. Also known as dial-in channel aggregation.
2. Bonding is also the name of a group known as the Bandwidth ON Demand INteroperability Group (BONDING). The group's charter is to develop common control and synchronization standards needed to manage high speed data as it travels through the public network. This will allow equipment from vendors to interoperate over existing Switched 56 and ISDN services. Version 1.0 of the standard, approved on August 17, 1992, describes four modes of inverse multiplexer (I-Mux) interoperability. It allows inverse multiplexers from different manufacturers to subdivide a wideband signal into multiple 56- or 64- Kbps channels, pass these individual channels over a switched digital network, and recombine them into a single high-speed signal at the receiving end.
3. In electrical engineering, bonding is the process of connecting together metal parts so that they make low resistance electrical contact for direct current and lower frequency alternating currents.
Bong A tone that long distance carriers and value added carriers make in order to signal you that they now require additional action on your part — usually dialing more digits. See also VOICE MODEM.
BONTs Broadband Optical Network Terminations.
Booking Factor Booking factor is a percentage of the frame relay links used by frame relay paths, based on the sum of the Committed Information Rates (CIRs) of all the frame relay paths (FRPs) on the frame relay link (FRL).
Bookmark A gopher or Web file that lets you quickly connect to your favorite pre-selected page. Appropriately named. The

way it works: You connect to a home page. You decide you'd like to return at some other time. So you command your internet surfing software to mark this web site with a "bookmark." Next time you want to return to that web site, you simply go to your bookmarks, click on which one you want. And bingo, you're there. A bookmark is also known as a hot list. Most Web browsers have book marks or hot lists.

Boolean Expression An expression composed of one or more relational expressions; a string of symbols that specifies a condition that is either true or false.

Boolean Logic Boolean Logic is named after the 19th century mathematician, George Boole. Boolean logic is algebra reduced to either TRUE or FALSE, YES or NO, ON or OFF. Boolean logic is important for computer logic because computers work in binary — TRUE or FALSE, YES or NO, ON or OFF.

Boolean Operators See BOOLEAN LOGIC. Boolean operators are AND, OR, XOR, NOR, NOT. The result of an equation with one or more of the boolean operators is that the result will either be true or false.

Boolean Valued Expression An expression that will return a "true" or "false" evaluation.

Boot Abbreviation for the verb to bootstrap. A technique or device designed to bring itself into a desired state where it can operate on its own. For example, one type of boot is a routine whose first few instructions are sufficient to bring the rest of itself into memory from an input device. See BOOTSTRAP.

Boot Loader A Windows NT term. Defines the information needed for system startup, such as the location for the operating system's files. Windows NT automatically creates the correct configuration and checks this information whenever you start your system.

Boot Partition A Windows NT term. The volume, formatted for either an NTFS, FAT, or HPFS file system, that contains the Windows NT operating system and its support files. The boot partition can be (but does not have to be) the same as the system partition.

Boot Priority Which disk drive the computer looks to first for the files it needs to get started. Modern PCs start their boot cycle with the hard disk and then move to the floppy disk drive. Older PCs started their boot cycle with the floppy disk drive.

BOOTP Boot Protocol. A TCP/IP protocol, which allows an internet node to discover certain startup information such as its IP address.

Boot ROM A read-only memory chip allowing a workstation to communicate with the file server and to read a DOS boot program from the server. Workstations can thus operate on the network without having a disk drive. These are commonly called diskless PCs or diskless workstations.

Bootstrap The process of starting up a computer. Think about the following explanation in regard to your desktop MS-DOS machine. Usually, when the computer is turned on, it goes to a location of permanent Read Only Memory (See ROM) for instructions. These instructions, in turn, load the first instructions from the disk telling the computer what tasks to start performing. The name of this process comes from the expression "pulling oneself up by one's own bootstraps." The typical personal computer BOOT (startup) throws a message on the screen instructing the user to "insert a disk."

To confuse matters, there are WARM boots and COLD boots. Cold boots occur when the ac power switch on the computer is turned on. Warm boots occur when you hit the reset button (or CTL/ALT/DELETE) while the ac power switch stays on. A warm boot — reset — is done when you're changing disks or programs, or have done something dumb, like tried to access a drive that didn't access, or tried to print without connecting up to a printer.

You do a cold boot when the machine locks up rock hard and a warm boot doesn't work. To do a cold boot with your computer, turn the ac power off, count to ten and then turn it on. Remember: never leave disks in your computer when you're turning it on and off. The surge of electricity might destroy the disks. When modems give trouble, do a cold boot on them. In fact, when phone systems give trouble, do a cold boot on them also. See also BOOT RAM, DEVICE DRIVER and MS-DOS.

BOP Bit Oriented Protocol. See BIT ORIENTED TRANSMISSION.

Border Gateway Protocol BGP is a Transmission Control Protocol/Internet Protocol (TCP/IP) routing protocol for inter-domain routing in large networks. It is used in the Internet and is an alternative to EGP (Exterior Gateway Protocol).

Border Note An ATM term. A logical node that is in a specified peer group, and has at least one link that crosses the peer group boundary.

Borscht A group of functions provided in Line Circuits (LC). It stands for
B: Battery supply to subscriber line. O: Overvoltage protection. R: Ringing current supply. S: Supervision of subscriber terminal. C: Coder and decoder. H: Hybrid (2 wire to 4 wire conversion). T: Test.
Borscht is a group of functions provided to an analog line from a line circuit of a digital central office switch. An analog electronic switch can omit C and possibly H. A line circuit on a switch with a metallic matrix (SXS, Xbar, 1,2,3ESS) only detects call originations and disconnects itself.

Bottle A water-tight device shaped like a glass bottle which contains amplifiers, regenerators and other equipment and used at regular distances along an underwater cable.

Bottom Line A phrase that can mean net profit, the lowest possible price that someone will take or the basic meaning with all the frills and nonsense cut away.

Bounce What Your E-mail Does When It Cannot Get To Its Recipient. It Bounces Back To You, Unless It Goes Off Into The Ether, Never To Be Found Again.

Bounced Mail Mail that is returned to the originator due to an incorrect e-mail address.

Bound Mode In an optical fiber, a mode whose field decays monotonically in the transverse direction everywhere external to the core and which does not lose power to radiation. Except in a single-mode fiber, the power in bound modes is predominantly contained in the core of the fiber.

Boundary Conditions Boundary conditions are those that are found at the cusp of valid and invalid inputs and parameters. Many faults are found in a computer telephony system's ability to handle boundary conditions, especially when the computer telephony system is under load. For example, for a network that expects a switch to reset a trunk port within two seconds, the associated boundary conditions would be found at 1.9 to 2.1 seconds.

Boundary Function Capability in SNA sub-area node to handle some functions that nearby peripheral nodes are not capable of handling.

Boundary Node In IBM's SNA, a sub-area node that can provide certain protocol support for adjacent sub-area nodes, including transforming network addresses to local addresses, and vice versa, and performing session level sequencing and flow control and less intelligent peripheral nodes.

Bowser General Magic's mascot, a rabbit.

BP 14 Body Part 14, an X.400 electronic messaging term referring to a nonspecific body part, commonly used to transfer binary attachments.

BPAD Bisynchronous Packet Assembler/Disassembler.

BPDU An ATM term. Bridge Protocol Data Unit: A message type used by bridges to exchange management and control information.

BPI Bytes Per Inch. How many bytes are recorded per inch of recording surface. Typically used in conjunction with magnetic tape.

BPON Broadband Passive Optical Network.

BPP Bits Per Pixel. The number of bits used to represent the color value of each pixel in a digitized image.

BPS Bits Per Second. Not bytes per second. Bits per second is a measure of the speed of data communications. Virtually all digital transmission circuits — from WANs to LANs — are defined in bits per second. Virtually all hard disk drive transfer rates are in bytes per second. There are many ways to measure bits per second. So don't assume that just because one LAN or other data communications system has a faster bits per seconds, it will transmit your information faster. You have to factor in speed of writing and reading from the disk and the accuracy of transmission. All data communications schemes have error-checking systems, some better than others. Typically such systems force a re-transmission of data if a mistake is detected. You might have a fast, but "dirty" (i.e. lots of errors) transmission medium, which may need lots of re-transmissions. Thus, the "effective" bps of data communications network may actually be quite low. See also BAUD.

BPSK Binary Phase Shift Key.

BRA Basic Rate Access. A Canadian term for the ISDN 2B+D standard, which is called BRI in the U.S. — Basic Rate Interface. See ISDN.

Bragg Reflector A device designed to finely focus a semiconductor laser beam. Dennis Hall, a professor at the University of Rochester's Institute of Optics in New York, told the Economist Magazine in the Spring of 1993 that he and his colleague Gary Wicks have etched into the surface of his gallium-arsenide laser a grating of 600 concentric grooves, each a quarter of a millionth of a meter apart. The grating acts as what is known as a Bragg reflector. As the waves of laser light pass through each of its ridges, they are reflected by each of its ridges, a process which causes them to come together into an even, circular beam.

Braid A fibrous or metallic group of filaments interwoven cylindrically to form a covering over one or more insulated conductors.

Brain Fart A definition from Wired Magazine: A byproduct of a bloated mind producing information effortlessly. A burst of useful information. "I know you're busy on the Microsoft story, but could give us a brain fart on the Mitnik burst?" Variation of old hacker slang that had more negative connotations.

Brainerd, Paul S. Founder of Aldus Corporation of 1984, Mr. Brainerd is reputed to be "the father of desktop publishing." His program Aldus PageMaker allowed the average PC user to produce professional-looking documents.

Branch A path in the program which is selected from two or more paths by a program instruction. "To branch" means to choose one of the available paths.

Branch Feeder A cable between the distribution cable and the main feeder cable to connect phone users to the central office. An outside plant term.

Branching Filter A device placed in a waveguide to separate or combine different microwave frequency bands.

Branding A term for identifying the Operator Service Provider (OSP) to the caller. Picture calling from your hotel room. You dial long distance. You have no idea which carrier you're using. But a message comes on: "Thanks for using MCI." Now you know. That's called branding.

BRCS Business and Residence Customer Services. An approach that the AT&T 5ESS switch employs to provision revenue-generating services.

Breadboard A circuit board made by hand, usually in building a prototype. No one knows where the word breadboard came from. Maybe because it looks a little like a breadboard?

Break An interruption. As in "Make and Break." Make means contacts which are usually open, but which close during an operation. "Make and Break" accurately describes rotary dialing.

Break In The attendant can interrupt conversations and announce an emergency or an important call.

Break Key A Break Key is found on some PCs. It is usually used to interrupt the current task running on a remote host. Break is not an ASCII character, it is simply a period of start (space) polarity.

Break Optimization A call center term. The automatic adjustment of break start times for schedules in the Daily Workfile so as to more closely match staff to workload in each period of the day. The program can thus improve upon the originally scheduled break arrangement because it now has information about schedule exceptions, newly added schedules, and additional call volume in AHT (Average Handle Time) history. See BREAK PARAMETERS.

Break Out Box A testing device that permits a user to cross-connect and tie individual leads of an interface cable using jumper wires to monitor, switch, or patch the electrical output of the cable. The most common break out box in our industry is probably the RS-232 box. Some of these boxes have LEDs (Light Emitting Diodes), which allow you to see which lead is "live." See also BREAKOUT BOX.

Break Parameters A call center term. A group of scenario assumptions you set to govern the placement of breaks in employee scheduling. These are typically:

* Earliest allowable break start time * Latest allowable break start time * Duration of the break * Whether the break is paid or unpaid

NEWTON'S TELECOM DICTIONARY

Break Test Access Method of disconnecting a circuit, which has been electrically bridged, to allow testing on either side of the circuit. Devices that provide break test access include: bridge clips, plug-on protection modules, and plug-on patching devices. Break test access also provides a demarcation point.

Breakdown Voltage The voltage at which the insulation between two conductors breaks down.

Breaking Strength The amount of force needed to break a wire or fiber.

Breakout A wire or group of wires in a multi-conductor configuration which terminates somewhere other than at the end of the configuration.

Breakout Box A device that is plugged in between a computer terminal and its connecting cable to re-configure the way the cable is wired. When hooking up a terminal that is wired as if it were a computer itself (such as a VT-100), a break out box is used to break out, or fan out the 25 connections in the RS-232 cable. Each wire in the break out box goes through a switch that can be turned off, and a wire jumper is provided to connect each pin on one side to one or the other pin on the other side. This allows you, for example, to switch pins 2 & 3, thus fooling two computer devices into thinking one is talking to a terminal. (Now you have the essence of a null modem cable.) Break out boxes are necessary because there is no such thing as "standard" pinning on an RS-232 cable. To connect one computer to a printer one minute and to another computer the next minute, usually requires totally different wiring in the RS-232 cable, i.e. two sets of cables. This lack of standardization is why you'll always see dozens of RS-232 cables lying around where computers are used.

BRI Basic Rate Interface. There are two subscriber "interfaces" in ISDN. This one and PRI (Primary Rate Interface). In BRI, you get two bearer B-channels at 64 kilobits per second and a data D-channel at 16 kilobits per second. The bearer B-channels are designed for PCM voice, slow-scan video conferencing, group 4 facsimile machines, or whatever you can squeeze into 64,000 bits per second full duplex. The data (or D) channel is for bringing in information about incoming calls and taking out information about outgoing calls. It is also for access to slow-speed data networks, like videotex, packet switched networks, etc. See BASIC RATE INTERFACE and ISDN.

Brick A large hand-held cellular phone or handheld two-way radio. In more technical language, a "brick" is a station in the mobile service consisting of a hand-held radiotelephone unit licensed under a site authorization. Each unit can work while being hand-carried.

Bridge First, we'll explain bridge as a noun. A bridge is a data communications device that connects two or more networks and forwards packets between them. The bridge may use similar or dissimilar media and signaling systems such as Ethernet, Token-Ring, and X.25. It operates at the data-link layer of the OSI model. Bridges sometimes work within a PC and sometimes within a special purpose computer. Bridges read and filter data packets and frames. A bridge is also called a data link relay or level 2 relay." When used for connecting LANs, a bridge connects LANs at the Medium Access Control (MAC) sub-layer of the OSI model's Data Link Layer. Bridges forward packets destined for another LAN. Bridges can be classified as source routing bridges, which are used in IBM environments, or transparent bridges, which operate primarily in non-IBM environments. A third bridging method, Source Routing Transparent (SRT), enables a bridge to act as both a transparent and a source routing bridge. In voice and videoconferencing, a bridge connects three or more telecommunications channels so that they can all communicate together. In video conferencing, bridges are often called MCUs — Multipoint Conferencing Units. One feature of some video bridges is their ability to figure who's speaking and turn the camera onto that person and have that person be on everyone's screen.

Second, we'll explain bridge as a verb, as in "to bridge." Imagine a phone line. It winds from your central office through the streets and over the poles to your phone. Now imagine you want to connect another phone to that line. A phone works on two wires, tip and ring (positive and negative). You simply clamp each one of the phone's wires to the cable coming in. That's called bridging. Imagine bridging as connecting a phone at a right angle. When you do that, you've made what's known as a "bridged tap." The first thing to know about bridging is that bridging causes the electrical current coming down the line to lose power. How much? That typically depends on the distance from the bridged tap to the phone. A few feet, there's no significant loss. But that bridged tap can also be thousands of feet. For example, the phone company could have a bridged tap on your local loop, which joined to another long-defunct subscriber. If you want the cleanest, loudest phone line, the local loop to your phone should not be bridged — but be a direct run from your central office to your phone. Bridging can be a real problem with digital circuits. Circuits above 1 million bits per second — e.g. T-1 and Ethernet — should never, ever be bridged. Because of the power loss, they simply won't work or will work so poorly they won't be worth having. ISDN BRI channels are also digital. But they were specifically designed to work with the existing telephone cable plant, which has a huge number of bridged circuits. Telephone companies typically will install ISDN BRI circuits with up to six bridged taps and about 6,000 feet of bridged cabling. But that's a rule of thumb. Once they get an ISDN BRI circuit installed, the first thing a phone company will do is to check it out. Below I've excerpted some words from a Nynex Networking & Technology Planning document, called "Network Loops and Customer Premises Wiring."

Says Nynex, the design objective for the ISDN Basic Access DSL (Digital Subscriber Loop) is to allow transmission through as much as practical of the North American loop plant. The majority of this loop plant was built using the "Resistance Design" rules to control the total signaling resistance and transmission loss. Up to a working length of loop of 18 kilofeet, mixed gauges of wire are used to limit the total de resistance typically to 1300 ohms. This corresponds to a maximum loss at 1000 Hz of about 9 dB. Total length of bridged taps is limited to 6 kft. Note that not every network loop meets Resistance Design criteria and some bridged tap lengths, even for Resistance Design nonloaded loops, appear as quarter-wave length shorting stubs for critical DSL transmission frequencies. To limit resistance and loss for loops longer than 18 kft, coarser gauge wire used and 88 mH loading coils are placed at 6 kft intervals. The series inductive loading combines with the cable shunt capacitance acts as a low pass filter with a sharp cutoff about 3000 Hz, and effectively prevents high frequency DSL operation (what ISDN requires). About one quarter of the total loop plant was loaded in 1988. The customer-network interface for Basic Access is at the end of this loop plant, at the intersection with the customer on-premises wiring. Note that customers are free to construct their own premises wiring and to purchase wiring from any source. There is potential for reduced DSL transmission performance due to inappropriate cable types, gauge changes or excessive bridged taps existing at the customer end of the loop. This may create opportunity for significant exposure to induced noise and crosswalk plus increased transmission loss. In recognition of the relatively uncontrolled customer portion of the DSL cable pair, the members of the T-1 work group typically used 1000 feet of customer wiring as a target for analytical transmission studies. For problems dealing with direct current, a nominal de resistance of 100 ohms was typically used to represent customer wiring. Planners must therefore take into account the possible effects of customer premises wiring on DSL operation when the customer premises is located near the extremes of the DSL loop operating range or when unusual premises wiring conditions exist. See INTERNETWORKING, LOADING COIL, ROUTERS, SOURCE ROUTING, and TRANSPARENT ROUTING.

Bridge Amplifier An amplifier installed on a CATV trunk cable to feed branching cables.

Bridge Battery A small supplementary battery on a laptop which holds the contents of the memory and the system status for a few minutes while you replace a drained battery. NEC uses the term on its UltraLite Versa laptops.

Bridge Clip A small metal clip that used to electrically connect together two sides of a 50 pair block. Removing the bridging clips breaks the circuit. You might remove the clips when you want to insert a piece of test gear and check to see which side the trouble is on.

Bridge Equipment Equipment which connects different LANs, allowing communication between devices. As in "to bridge" several LANs. Bridges are protocol-independent but hardware-specific. They will connect LANs with different hardware and different protocols. An example would be a device that connects an Ethernet network to a StarLAN network. With this bridge it is possible to send signals between the two networks, and only these two networks.

These signals will be understood only if the protocols used on each LAN are the same, e.g. XNS or TCP/IP, but they don't have to be the same for the bridge to do its job for the signals to move on either LAN. They just won't be understood. This differs from gateways and routers. Routers connect LANs with the same protocols but different hardware. The best examples are the file servers that accommodate different hardware LANs. Gateways connect two LANs with different protocols by translating between them, enabling them to talk to each other. The bridge does no translation. Bridges are best used to keep networks small by connecting many of them rather than making a large one. This reduces the traffic faced by individual computers and improves network performance.

Bridge Lifter A device that removes, either electrically or physically, bridged telephone pairs. Relays, saturable inductors, and semiconductors are used as bridge lifters.

Bridge Tap An undetermined length of wire attached between the normal endpoints of a circuit that introduces unwanted impedance imbalances for data transmission. Also called bridging trap or bridged tap. See BRIDGED TAP.

Bridged Ringing A system where ringers on a phone line are connected across that line.

Bridged Tap A bridged tap is multiple appearances of the same cable pair at several distribution points. A bridged tap is any section of a cable pair not on the direct electrical path between the central office and the user's offices. A bridged tap increases the electrical loss on the pair — because a signal traveling down the pair will split its signal between the bridge and main pairs. You can't run high-speed digital circuits, e.g. T-1, over cable that has bridged taps in it. But you can run ISDN circuits over cable with a limited number of bridged taps. See BRIDGE and LOADING COIL.

Bridging Bridging across a circuit is done by placing one test lead from a test set or a conductor from another circuit and placing it on one conductor of another circuit. And then doing the same thing to the second conductor. You bridge across a circuit to test the circuit by listening in on it, by dialing on it, by running tests on the line, etc. You can bridge across a circuit by going across the pair in wire, by stripping it, etc. You can bridge across a pair (also called a circuit path) by installing external devices across quick clips on a connecting block.

Bridging Adapter A box containing several male and female electrical connectors that allows various phones and accessories to be connected to one cable. Bridging adapters work well with 1A2 key systems and single line phones, but usually not with electronic or digital key systems and electronic or digital telephones behind PBXs.

Bridging Clip A small piece of metal with a U-shape cross-section which is used to connect adjacent terminals on 66-type connecting blocks.

Bridging Connection A parallel connection by means of which some of the signal energy in a circuit may be extracted, usually with negligible effect on the normal operation of the circuit. Most modern phone systems don't encourage bridging connections, since the negligible is rarely negligible.

Bridging Loss The loss at a given frequency resulting from connecting an impedance across a transmission line. Expressed as the ratio (in decibels) of the signal power delivered to that part of the system following the bridging point before bridging, to the signal power delivered to that same part after the bridging.

Bridle Cards Proprietary Basic Rate ISDN Dual Loop Extension lets ISDN service be provided up to 28,000 feet away. See ISDN.

Brightness An attribute of visual reception in which a source appears to emit more or less light. Since the eye is not equally sensitive to all colors, brightness cannot be a quantitative term.

BRISC Bell-Northern Research Reduced Instruction Set Computing.

Brite Cards And Services Basic Rate Interface Transmission Extension lets telephone companies extend service from ISDN-equipped central offices to conventional central offices. See ISDN.

Brittle Easily broken without much stretching.

Broadband A transmission facility that has a bandwidth (capacity) greater than a voice grade line of 3 kHz. (Some say that to be "broadband" it should be 20 kHz.) Such a broadband facility — typically coaxial cable — may carry numerous voice, video and data channels simultaneously. Each "channel" will take up a different frequency on the cable. There'll be "guardbands" (empty spaces) between the channels to make sure each channel doesn't interfere with its neighbor. A coaxial CATV cable is the "classic" broadband channel. Simultaneously it carries many TV channels. Broadband cables are used in some office LANs. But more common are the baseband variety which have the capacity for one channel only. Everything on that cable to be transmitted or received must use that one channel. That one channel is very fast, so each device needs only to use that high speed channel for only a little of the time. The problem is getting onto the channel. See BASEBAND.

Broadband Bearer Capability A bearer class field that is part of the initial address message.

Broadcast To send information to two or more receiving devices simultaneously — over a data communications network, a voice mail, electronic mail system, a local TV or radio station or a satellite system.

Broadcast List A list of two or more system users to whom messages are sent simultaneously. Master Broadcast Lists are shared by all system users and are set up by the System Administrator. Personal Lists are set up by individual subscribers.

Broadcast Message A message from one user sent to all users. Just like a TV station signal. On LANs, all workstations and devices receive the message. Broadcast messages are used for many reasons, including acknowledging receipt of information and locating certain devices. On voice mail systems, broadcast messages are important announcement messages from the system administrator that provide information and instructions regarding the voice processing system. Broadcast messages play before standard Voice Mail or Automated Attendant messages.

Broadcast Net A British Telecom turret feature that allows each trader single key access to a group of outgoing lines. This is designed primarily for sending short messages to multiple destinations. The "net" function allows the user to set up and amend his broadcast group.

Broadcast Storm A pathological condition that may occur in a TCP/IP network that can cause large number of broadcast packets to be propagated unnecessarily across an enterprise-wide network, thereby causing network overload. Broadcast storms happen when users mix old TCP/IP routers with routers supporting the new releases of TCP/IP protocol. Routers use broadcast packets to resolve IP addressing requests from stations on LANs. If a station running an old version of TCP/IP sends such a request, TCP/IP routers in an enterprise-wide network misunderstand it and send multiple broadcasts to their brother and sister routers. In turn, these broadcasts cause each router to send more broadcasts, and so on. This chain reaction can produce so many broadcast messages that the network can shut down. It should be noted that this is extremely rare and it happens only in TCP/IP networks that use two specific TCP/IP protocol releases.

Broadcast Transmission A fax machine feature that allows automatic transmission of a document to several locations.

Brochureware A pejorative term for what companies can pull off with a Web site and $10,000.

Broker A company (or person) that buys and sells equipment often without taking ownership. A broker does not test or refurbish the equipment. Often, it never sees the equipment it buys and sells. Instead, it has the equipment shipped from the supplier to the customer, relying on the supplier to have tested and refurbished the equipment. Its specialty is knowing who has what equipment nationwide and selling it, possibly, at below-market price. See SECONDARY EQUIPMENT.

Brokernet A virtual private dedicated network offering from New York Telephone and provided within Manhattan aimed at brokerage, banking and message industries. It uses digital switching to provide virtual private lines, specifically "hot line" service.

Bronze Allow of copper and tin, widely used and known since ancient times. Copper content in bronze varies between 89% and 96%.

Brouter Concatenation of "bridge" and "router." Used to refer to devices which perform both bridging and routing functions. In local area networking, a brouter is a device that combines the dynamic routing capability of an internetwork router with the ability of a bridge connect dissimilar local area networks (LANs). It has the ability to route one or more protocols, such as TCP/IP and XNS, and bridge all other traffic.

Brownout When you lose all your electricity, it's called a blackout. When your voltage drops more than 10% below what it's meant to be, it's a brownout.

Browser 1. Software which moves documents on the World Wide Web to your computer. A Web Browser is software which allows a computer user (like you and me) to "surf" the Internet. It lets us move easily from one World Wide Web site to another. Every time we alight on a Web Page, our Web Browser moves a copy of documents on the Web to your computer. A Web Browser uses HTTP — the HyperText Transfer Protocol. Invisible to the user of a Web Browser, HTTP is the actual protocol used by the Web Server and the Client Browser to communicate over the Internet. The most famous Web Browsers are Netscape and Microsoft's Internet Explorer. See BROWSING, INTERNET and SURF. www.netscape and www.microsoft.
2. A developed tool used to inspect a class hierarchy in an object-oriented software system.

Browsing The act of searching through automated information system storage to locate or acquire information without necessarily knowing of the existence or the format of the information being sought.

Brush A computer imaging term. A paint package's most basic image-creation tool. Most packages let you select a variety of sizes and shapes. Many let you customize shapes.

Brute Force Attack A cracker term. Brute force attack means hurling passwords at a system until it cracks.

BSA See both BASIC SWITCHING ARRANGEMENT and OPEN NETWORK ARCHITECTURE.

BSC 1. Binary Synchronous Communication. A set of IBM operating procedures for synchronous transmission used in teleprocessing networks.
2. Base Station Controller. A wireless telecommunications term. The BSC is a device that manages radio resources in GSM (Global System for Mobile Communications), including the BTS, for specified cells within the PLMN (Public Land Mobile Network).

BSD Berkeley Software Distribution. Term used when describing different versions of the Berkeley UNIX software, as in "4.3BSD UNIX".

BSE 1. Basic Switching Element. See OPEN NETWORK ARCHITECTURE.
2. Basic Service Elements. A term used in voice processing to describe technical telephone system features such as ANI, DID trunks, call forwarding, stutter dial tone, suppressed ringing, and directory database access.

BSGL Branch Systems General Licence. A British term. A licence that must be obtained by any organization seeking to link a private network to the British PSTN (Public Switched Telephone Network). A separate licence must be held on each site.

BSI British Standards Institution. The UK standards body responsible for input to European and international standards setting bodies like ISO and the ITU-T.

NEWTON'S TELECOM DICTIONARY

BSIC Base Station Identity Code. An attribute of a GSM (Global System for Mobile Communications) cell which is a code allowing a distinction between local cells having the same radio frequency.

BSP Bell System Practice. A very defined way of writing and presenting instruction and installation manuals. BSPs also establish standards for splicing cable, for installing phones, answering phones, collecting debts, finding phone taps, climbing poles. They are (or once were) the instruction manuals that dictated how to do everything. Divestiture has changed the rules. BSPs are not as important as they were when AT&T handed down all the BSPs.

BSS 1. Base Station System. A wireless telecommunications term. A GSM (Global System for Mobile Communications) device charged with managing radio frequency resources and radio frequency transmission for a group of BTSs. See also GSM.
2. Business Support System. The system used by network operators to manage business operations such as billing, sales management, customer-service management and customer databases.

BT 1. British Telecom.
2. Burst Tolerance: BT applies to ATM connections supporting VBR services and is the limit parameter of the GCRA.

BTA 1. Basic Trading Area. A wireless telecommunications term. The United States is broken down into 493 major trading areas for economic purposes. These boundaries were used for licensing PCS wireless phone systems. Several BTAs make up Metropolitan Trading Area, an area defined by the FCC for the purpose of issuing licenses for PCS. Thus, each MTA consists of several Basic Trading Areas (BTAs).
2. Broadband Telecommunications Architecture, an architecture introduced by General Instrument's Broadband Communications Division at the Western Cable Television Show on December 1, 1993. General Instrument said the plant is built to 750 MHz and can support reduced node size and add services such as video-on-demand, telephony, interactivity, data services, etc.
3. Business Technology Association (was originally called NOMDA/LANDA).

BTAM Basic Telecommunications Access Method. One of IBM's early host-based software programs for controlling remote data communications interface to host applications, supporting pre-SNA protocols. See IBM.

BTag An ATM term. Beginning Tag: A one octet field of the CPCS_PDU used in conjunction with the Etag octet to form an association between the beginning of message and end of message. See ATM.

BTI British Telecom International.

BTL Bell Telephone Laboratories.

BTN Billed Telephone Number or Billing Telephone Number. The primary telephone number used for billing, regardless of the number of phone lines associated with that number. According to Bellcore, BTN is sometimes known as a screening telephone number, the BTN is a telephone number used by the AMA process as the calling-party number for recording purposes.

BTOS A UNIX program which translates binary files into ASCII.

BTRL British Telecom Research Laboratories.

Btrieve Btrieve is a key-indexed database record management system. You can retrieve, insert, update, or delete records by key value, using sequential or random access methods. First introduced in 1983, Btrieve was one of the first databases designed for LANs. Novell bought the company in the late 1980s. Now called NetWare Betrieve, it is included with most versions of NetWare 2.x and NetWare 3.x. There are Btrieve developer products for DOS, OS/2 and Windows.

BTS Base Transceiver Station. A wireless telecommunications term. A GSM (Global System for Mobile Communications) device used to transmit radio frequency over the air interface.

BTSM BTS (Base Transceiver Station) Management. A wireless telecommunications term. Devices configured to manage BTS functions and equipment.

BTU Basic Transmission Unit.

BTV See Business Television.

BTW By The Way. An acronym used in electronic mail on the Internet to save words or to be hip, or whatever.

Buffer 1. In data transmission, a buffer is a temporary storage location for information being sent or received. Usually located between two different devices that have different abilities or speeds for handling the data. The buffer acts like a dam, capturing the data and then trickling it out at speeds the lower river can handle without, hopefully, flooding or overflowing the banks. A buffer can be located in the sending or receiving device.
2. A coating material used to cover and protect the fiber. The buffer can be constructed using either a tight jacket or loose tube technique.

Buffer Coating Protective material applied to fibers. Increases apparent fiber size. May be more than one layer. Stated in microns. Usually thicker or multi-coated on tight-buffer cables.

Buffer Memory Electronic circuitry where data is kept during buffering. See BUFFER.

Buffer Storage Electronic circuitry where data is kept during buffering. See BUFFER.

Buffered Repeater A device that amplifies and regenerates signals so they can travel farther along a local area cable. This type of repeater also controls the flow of messages to prevent collisions.

Bug 1. A concealed microphone or listening device or other audio surveillance device.
2. To install the means for audio surveillance.
3. A semiautomatic telegraph key.
4. A problem in software or hardware. The original computer bug, a moth, is enshrined at the Washington Navy Yard. It was the cause of a hardware failure in an early computer in 1945. The story goes like this: a team of top Navy scientists was developing one of the world's first electronic computers. Suddenly, in the middle of a calculation, the computer ground to a halt, dead. Engineers poured over every wire and every inch of the massive machine. Finally, one of the technicians discovered the cause of the problem. Buried deep insides its electronic innards, crushed between two electric relays, lay the body of a moth. These days, "bugs" in telecom or computer systems are not insects. They're indescribable glitches that adversely affect smooth operations. Bugs usually originate in software. Some programmers call bugs "undocumented features." And they are, indeed. All the

above is the story the navy likes to put out. In actual fact, the word "bug" for problem in design has been around for eons. It's mentioned in a 1910 book called, "Edison, His Life and Inventions" by Frank Lewis Dyer. The book talks about Edison harassing his employees to "get all the bugs out."

Bug Mix A silly term for the precise collection of bugs in a particular piece of software.

Bug Rate The frequency with which new bugs are found during the testing cycle is referred to as the bug rate. As these bugs are fixed, and as time passes, bugs will become more and more difficult to find, and new bugs will be found less frequently, i.e., the bug rate decreases. Usually, when the bug rate drops to zero and all major bugs (and most of the minor bugs) have been fixed, a product is ready for the next stage in its lifecycle.

Building Distribution Frame Somewhere in the building where all telephony wiring for the building is. All the cables from the outside the building would first come here and be punched down on the building distribution frame. Additional cables would take the telephone circuits upstairs.

Building Entrance Area The area inside a building where cables enter the building and are connected to riser cables within the building and where electrical protection is provided. The network interface as well as the protectors and other distribution components for the campus subsystem may be located here. Typically this area is the end of the local telephone company's responsibility. From here on it's your responsibility. You should protect your equipment inside the building from spikes and surges and other electrical nonsenses which the phone company's cables might bring in. For the best disaster protection, it's wise to have two building entrances by which your telecommunications cables can enter. And they should enter from separate telephone central offices. Some telephone companies, e.g. New York Telephone, are now tariffing such services.

Building Footing The concrete base under the foundation of a building in which copper wire may be laid to form an electrical ground.

Building Integrated Timing Supply BITS. A clock, or a clock with an adjunct, in a building that supplies DS1 and/or composite clock timing reference to all other clocks in that building. See BITS.

Building Out The process of adding a combination of inductance, capacitance,and resistance to a cable pair so that its electrical length may be increased by a desired amount to control impedance characteristics.

Building Steel The structural steel beams that make up the frame of a building. If the steel frame is buried deep in the earth, it can be used as an electrical ground. But you'd better be careful: Unbalanced three-phase power is also probably using the frame as a ground, and you may pick up huge quantities of 60 hz hum.

Bulk Billing A method of billing for long distance telephone services where no detail of calls made is provided. WATS is a bulk billed service. Therein lies the problem for the cost conscious user. There's no verification of calls made. See CALL DETAIL RECORDING.

Bulk Encryption Simultaneous encryption of all channels of a multichannel telecommunications trunk.

Bulk Storage Lots of storage. Usually reels of magnetic tape or hard disks.

Bulletin Board System A fancy name for an electronic message system running on a microcomputer. Call up, leave messages, read messages. The system is like a physical bulletin board. That's where the name comes from. Some people call bulletin board systems electronic mail systems. See also BBS.

Bumper Beeper Radio beacon transmitter, hidden in or on a vehicle for use with radio tailing equipment.

Bunched Frame-Alignment Signal A frame-alignment signal in which the signal elements occupy consecutive digit positions.

Bundle 1. A group of fibers or wires within a cable sharing a common color-code.
2. In T-1, specifically M44 Multiplexing, a bundle consists of 12 nibbles (4 bits) and may represent 11 channels of 32 Kbps compressed information plus a delta channel. A bundle is typically a subset of a DSI and treated as an entity with its own signaling delta channel.

Bundle Fodder Junk software included on CD-ROMs or packaged with peripherals such as modems and designed to bulk up the presumed value of the total package to entice an unsuspecting consumer to buy. See SHOVELWARE.

Bundled Combining several services under one telephone tariff item at a single charge.

Bundled Rates Several service combined into one offering for one charge.

Burden Test A semi-legitimate test used in regulation to determine if the offering of a new or continued service will cause consumers of other services to pay prices no higher than if the service were not offered. In other words, the question is "Who carries the Burden?" It's sometimes called the "avoidable cost test."

Burford Courier An MCI definition. A communications software package developed for use with Wang VS mainframes enabling users to communicate directly with MCI International's Telex messaging services.

Burn In To run new devices and printed circuits cards, often at high temperatures, in order to pinpoint early failures. The theory is all semiconductor devices show their defects — if any — in the first few weeks of operation. If they pass this "burn-in" period, they will work for a long time, so the theory goes. "Burn-in" should probably be 30-days under full power and working load. Burn-in should also take place in a room with lots of heat and at least 50% humidity, since this will simulate the poorly-ventilated places most people install telephone systems.

Burn Out A condition, where stress causes agents to be apathetic and lethargic, caused by intensity of calling, lack of variety and poor working conditions. It is particularly associated with outbound cold calling and inbound complaint handling, both of which are stressful for agents if not carefully managed.

Burn Rate The speed at which a new company using up its cash en route to developing a product before it turns cash positive.

Burning A Pole Slang expression to describe when an installer accidentally slides down a telephone pole.

Burrus Diode A surface-emitting LED with a hole etched to accommodate a light-collecting fiber. Named after its inventor, Charles Burrus.

Burst 1. In data communication, a sequence of signals, noise, or interface counted as a unit in accordance with some specific criterion or measure.
2. To separate continuous-form or multipart paper into discrete sheets.

Burst Isochronous Isochronous burst transmission. See ISOCHRONOUS.

Burst Mode A way of doing data transmission in which a continuous block of data is transferred between main memory and an input/output device without interruption until the transfer has been completed.

Burst Switching In a packet-switched network, a switching capability in which each network switch extracts routing instructions from an incoming packet header to establish and maintain the appropriate switch connection for the duration of the packet, following which the connection is automatically released. In concept, burst switching is similar to connectionless mode transmission, but it differs from the latter in that burst switching implies an intent to establish the switch connection in near real time so that only minimum buffering is required at the node switch.

Burst Traffic Burst traffic is many phone calls (usually incoming telephone calls) that simultaneously arrive at a computer telephony system. Burst traffic tests are usually performed as part of a systems load and stress testing. Burst traffic tests are particularly important for computer telephony systems as usually they must perform a lot of processing to set up to handle a call. The arrival of many calls simultaneously, such as in response to a television sales offer (or when Oprah asks you to call 1-800-xxxxxxx to vote on whether OJ is guilty or innocent), can place significant strain on a computer telephony system.

Burst Transmission 1. A method of transmission that combines a very high data signaling rate with very short transmission times.
2. A method of operating a data network by interrupting, at intervals, the data being transmitted. The method enables communication between data terminal equipment and a data network operating at dissimilar data signaling rates.

Bursty Information Information that flows in short bursts with relatively long, silent intervals between.

Bursty Seconds Bursty seconds is a measure of the amount of time spent at maximum data transfer rate.

Bus 1. An electrical connection which allows two or more wires or lines to be connected together. Typically, all circuit cards receive the same information that is put on the Bus. Only the card the information is "addressed" to will use that data. This is convenient so that a circuit card may be plugged in "anywhere on the Bus." There are two common buses inside a PC — the older ISA bus, capable of only five megabytes per second and the newer PCI bus, capable of transmitting up to 132 megabytes per second. All computers and most telephone systems use buses of some type. Computer buses are typically open. Telephone system buses are typically closed. See also BACKPLANE and BUS NETWORK.
2. An ATM term. BUS. Broadcast and Unknown Server: This server handles data sent by an LE Client to the broadcast MAC address ('FFFFFFFFFFFF'), all multicast traffic, and initial unicast frames which are sent by a LAN Emulation Client.

Bus Enumerator A Windows 95 term. A new type of driver required for each specific bus type, responsible for building ("enumerating") the hardware tree on a Plug and Play system.

Bus Hog A device connected to a transmission bus which, after gaining access to the transmission medium, transmits a large number of messages regardless of whether other devices are waiting.

Bus Master A VME board (usually a CPU) that can contend for, seize and control the VME bus for the purpose of accessing bus resources such as voice boards or even other CPUs. See VME.

Bus Mastering A method of transferring data through a bus in which the device takes over the bus and directly controls the transfer of data to the computer's memory. Bus mastering is a method of Direct Memory Access (DMA) transfer.

Bus Mouse Mouse that takes up an expansion slot in a PC, rather that a serial port (those are called "serial mice"). There are generally more expansion slots available than serial ports.

Bus Network All communications devices share a common path. Typically in a bus network, a "conversation" from each device is sampled quickly and interleaved using time division multiplexing. Bus networks are very high-speed — millions of bits per second — forms of transmission (e.g. on a local area network) and switching. They often form the major switching and transmission backbone of a modern PBX. The printed circuit cards which connect to each trunk and each line are plugged into the PBX's high-speed "backbone" — i.e. the bus network. See also BUS.

Bus Slave A VME board (usually a subsystem or I/O board) which can only respond to VME bus accesses mapped to its address. Slaves can usually interrupt the VME bus on one of 7 levels. See VME.

Bus Speed The speed at which the computer's CPU (central processing unit) communicates with other elements of the computer. For example, the speed at which data moves between the CPU and your serial ports.

Bus Topology A network topology in which nodes are connected to a single cable with terminators at each end.

Business Audio At one stage business audio was thought to involve voice annotating spreadsheets. Now it means making phone calls through and by your PC.

Business ID An MCI definition. A five-digit numeric code identifying the business to which a customer is assigned. The first two digits indicate division number, the third indicate service type, and the fourth and fifth indicate billing cycle. These are assigned during order entry and passed to MCI A/R with the customer install/ transaction.

Business Support System See BSS.

Business Technology Association An association formerly known as Nomda/Landa.

Business Television BTV. Point-to-multipoint videoconferencing. Often refers to the corporate use of video for the transmission of company meetings, training and other one-to-many broadcasts. Typically uses satellite transmission methods and is migrating from analog to digital modulation techniques.

Businessview A Rolm term. Businessview for Call Centers is a centralized system for collecting, integrating, analyzing and reporting call center data.

Busy In use. "Off-hook". There are slow busies and fast busies. Slow busies are when the phone at the other end is busy or off-hook. They happen 60 times a minute. Fast busies (120 times a minute) occur when the network is congested with too many calls. Your distant party may or may not be busy, but you'll never know because you never got that far.

Busy Back A busy signal.

Busy Call Forwarding When you call a busy phone extension, your call is automatically sent to another predetermined telephone extension.

Busy Hour The hour of the day (or the week, or the month, or the year, — check) during which a telephone system carries the most traffic. For many offices, it is 10:30 A.M. to 11:30 A.M. The "busy hour" is perhaps the most important concept in traffic engineering — the science of figuring what telephone switching and transmission capacities one needs. Since the "busy hour" represents the most traffic carried in a hour, the idea is if you create enough capacity to carry that "busy hour" traffic, you will be able to carry all other traffic. In actuality, one never designs capacity sufficient to carry 100% of the busy hour traffic. That would be too expensive. So, the argument then comes down to, "What percentage of my peak busy or busy hour traffic am I prepared to block?" This percentage might be as low as half of one percent or as high as 10%. Typically, it's between 2% and 5%, depending on what business you're in and the cost to you — in lost sales, etc. — of blocking calls.

Busy Hour Usage Profile The busy hour usage profile identifies how a system will normally be used (i.e., who the users are and what type of transactions they are performing) during the busy hour. Different things that users of CT systems do stress the CT system in different manners. For example, sending a broadcast message in voice mail (where one message is automatically sent to many recipients) may force the system to perform 10 times as much disk I/O as sending message to single recipients. Or, updating an account balance in an IVR transaction forces strain on the IVR to mainframe link. Or, setting up conference calls across a switch or network stresses the use of the network database resources. Understanding and characterizing the type and mix of calls that will take place during the busy hour is key to designing and placing a real-world load on the CT system. Load testing a system should always incorporate tests using the busy hour call profile.

Busy Lamp A light on a telephone showing a certain line or phone is busy. See BUSY LAMP FIELD.

Busy Lamp Field A device with rows of tiny lights that shows which phones in a telephone system have conversations on them, which phones are ringing, which phones are on hold. Each light corresponds to a telephone extension on the system. The busy lamp field usually sits attached to the attendant's console, telling the attendant if an extension is busy, free, on hold, etc. The benefit of having a busy lamp field is that the operator doesn't have to dial the number to find out what's happening with the extension. This saves the attendant time in handling incoming calls and gives the caller better service. A busy lamp field is often combined with DSS (Direct Station Select.) Next to each light on the busy lamp field there is a button which the operator can push which will dial the corresponding extension (i.e. directly select it) and will typically transfer the call automatically. This button is like an autodial button. This saves the time of dialing the two, three or more numbers of the extension. These days, busy lamp fields are often built into phones on a key system, and everyone, not just the operator, can have one. This gives everyone information on what's happening in the system. It makes transferring calls, etc. easier. See also DIRECT STATION SELECT.

Busy Out To cause a line to return a busy signal to a caller. Busying out lines going into a computer is useful when the computer is not available, i.e. during maintenance periods. This way callers do not get connected to modems with no computers to talk to. This is also known as "taking the phone off the hook." In a voice phone system with trunks that rotary (or hunt) on, sometimes busying one or more broken trunks out helps calls rotary on to trunks that are still working. This way, someone doesn't end up on your third trunk with endless ringing, while your 4th, 5th, 6th etc. trunks are free, leaving them wondering where you are and wondering why you're not getting any calls.

Busy Override A feature of some PBXs which allows the attendant or other high priority user to barge in on a telephone conversation. A warning tone is usually thrown into the conversation to alert the parties of an override. The feature is also called "Barge-In." Sometimes when conversations are overridden, only the person within the organization can hear the barge-in.

Busy Season An annual recurring and reasonably predictable period of maximum busy hour requirements — normally three months of the year, and typically the three months preceding Christmas.

Busy Signal A signal indicating the line called is busy. The busy signal is generated by the central office. There are two types of busy signals. See BUSY.

Busy Test A method of figuring whether something which can carry traffic is actually doing so or whether it's broken or free and available for use.

Busy Verification Of Station Lines 1. An attendant can confirm that a line is actually in use by establishing a connection (dialing in and listening) to that apparently busy line.
2. In the public switched telephone network, a switching system service feature that permits an attendant to verify the busy or idle state of station lines and to break into the conversation. An alternating tone of 440 Hz is applied to the line for 2 seconds, followed by a 0.5-second burst every 10 seconds to alert both parties that the attendant is connect to the circuit.

Butt Set See BUTTINSKY.

Butt Splice Connecting the ends of two wires with a "butt splice connector," from such manufacturers as AT&T and 3M.

Buttinsky Or Butt Set. The one-piece telephone carried on the hips of telephone technicians. It's called buttinsky or butt set because it allows technicians to "butt in" on phone calls, not because the device is worn on their butts. Butt sets used to be essentially telephones without ringers. But now they are much more sophisticated. They will pulse out in rotary or dial out in touchtone and allow you to talk or to monitor a call. They will run computerized tests on the line. Some even have the equivalent of an asynchronous computer terminal built in, which can be used to talk to a distant computer over a phone line. This distant computer could assign them their next jobs, allow them to check and assign features (touchtone, rotary, hunt), report the time spent on this job, etc. In short, the terminal and the computer could replace a raft of clerks and a deluge of paperwork. The derivation of the term "buttinsky" has long been lost in history. Butt can also mean to attach the end of something to the end of something else. If the clips of a butt set are attached to a pair of wires, it is a "butt" connection. There are those people who also think that the term buttinsky came from some middle European language — Polish or Yiddish — and is slang for someone who butts in a lot, which is what you can do with a Butt Set. I shall keep researching this one. If you can help, let me know, please.

Button Caps Interchangeable plastic squares fit over the buttons of electronic telephones, and are used to label the features programmed onto each programmable button location. Button caps can be either pre-printed or have clear windows which allow features, lines, and Autodial numbers to be labelled on the button.

Buzz 1. To check the continuity of a cable pair by putting an audible buzzer on one end and then checking with a "buttinsky" to see if you can hear the buzz and thus identify the correct cable pair.
2. A feature of a Rolm CBX which lets the user signal one Rolm desktop product without picking up the handset. Only one buzz per extension is permitted.

Buzzer An electromechanical device that makes a buzzing noise when power is applied, often used to signal someone to answer an intercom call. Battery-powered buzzers were once used to help trace phone circuits. TONE GENERATORS are more common today, but old terminology is still used, as in "buzzing out a line."

Buzzer Leads The wires inside a telephone intended for the connection of a buzzer, usually as part of an intercom system.

BVA AT&T's exclusively-owned Billing Validation Application database. Today, BVA contains all the Regional Bell Operating Companies (RBOCs) calling cards, and other billing information such as billed number screening and payphone numbers. The RBOCs and AT&T access that database today. Prior to Divestiture three market players, the RBOCs, AT&T and most Independent Companies, dominated the "0" Operator Services business which provided alternate billing arrangements such as collect calls, bill to third number and charging calls to calling cards. The three market players still exclusively employ BVA which allows them to validate or authorize alternate billing arrangements. No other long distance carrier or a company needing access to the data for billing validation can use the system. The database is owned by AT&T and is updated daily by the RBOCs and Independent Companies with local exchange information, billing number screening and calling card information. The scenario is further complicated by the 1984 AT&T Plan of Reorganization's exclusive BVA access restrictions. In other words, the three original market players (AT&T, the RBOCs and Independent Companies) have exclusive access to BVA for a predetermined contract length. In most cases these arrangements run into the 1990's. See BVS.

BVS Business Validation Service. US West Service Link was the first in the nation to develop and make available a nationwide Billing Validation Service open to any company that needs to verify the legitimacy of their callers' requests to place charge calls to their local telephone calling cards. US West Service Link developed BVS in 1987 and turned up the system for "on-line" customers in early 1988. Today, the US West BVS system is a national validation source containing calling card data of customers served by the RBOCs, GTE, Southern New England Telephone, United Telecommunications, Cincinnati Bell, Rochester Telephone and Telecomm Canada, a consortium representing all of Canada's local telephone operating companies. In all, more than 60 million records are stored. US Sprint, MCI and ITI are among the carriers using BVS. BVS uses X.25 and SS7 protocol. See BVA.

BW Bandwidth: a numerical measurement of throughput of a system or network.

BX.25 AT&T's rules for establishing the sequences of events and the specific types and forms of signals required to transfer data between computers. BX.25 includes the international rules known as X.25 and more.

Bypass A term coming from the idea of using a method of local communications other than the local telephone company. The theory is you bypass the local phone company for typically two reasons: 1. Because the phone company is too expensive. 2. Because the phone company can't get you the bandwidth, the quality, the fast delivery or the service you want.
Bypass means you might be transmitting between two of your offices in the same city, perhaps between your office and your factory, etc. You might also be transmitting directly to your friendly long distance telephone company, which then carries your calls to distant cities. Bypass means you might be using one, several or many technologies, including your own private microwave, your own cable (metal or fiber), using the local CATV company, using the local bypass company, of which there are now dozens. Bypass is a word created by the local telephone industry to sound very threatening. The theory is major users will bypass their local phone company, depriving their phone company of needed revenues. This will drive the phone company close to imminent bankruptcy, or at the very least, to the state regulators for huge rate increases, hurting everyone. The reality of this threat has not been proven. Nevertheless, the rhetoric frightens sufficient regulators to look at the evils of bypass and to outlaw it, or at least severely restrict it — as several states have done. This will strengthen the phone company's local monopoly, the objective of the exercise in the beginning. The reality is that bypass in all its forms is expanding, under encouragement from large users and from the FCC. See also 700 SERVICE.

Bypass Cabling Bypass cabling or relays are wired connections in a local area ring network that permit traffic between two nodes that are not normally wired next to each other. Such bypass cabling might be used in an emergency or while other parts of the system are being serviced. Usually such bypass relays are arranged so that any node can be removed from the ring and the two nodes on either side of the removed node can then talk.

Bypass Trunk Group A trunk group circumvents one or more tandems in its routing ladder.

BYTE A popular microcomputer magazine, which derives its name from the true meaning of byte, which is eight bits of information composed of zeros or ones, one of which may include a parity bit. Most character sets, e.g. ASCII or EBCDIC, use one byte per character of information such as a letter, a number or digit, a punctuation mark or a symbol, such as $. A byte is to a bit what a word is to a character. In some circles a byte is called an octet. See BYTE COUNT.

Byte Count The number of 8-bit bytes in a message. Since ASCII characters typically have 8-bits, the byte count is also called the character count.

Byte Count Protocol A class of data link protocols in which each frame has a header containing a count of the total number of data characters in the body of the frame.

Byte Multiplexer Channel An IBM mainframe input/output channel that allows for the interleaving, or multiplexing, of data in bytes. Compare with block multiplexer channel.

Byte Multiplexing A byte (or character) from one channel is sent as a unit and bytes from different channels follow in successive time slots.

Byte Stuffing The process whereby dummy bytes are inserted into a transmission stream so that the net data transmission rate will be lower than the actual channel data rate. The dummy bytes are identified by a single controlling bit within the byte.

Byte Timing Circuit Optional X.21 circuit used to maintain byte or character synchronization.

BZT Bundesamt fur Zulassungen in der Telekommunikation. The name of the German telecom approval authority. It was established in 1982 under the name of the Central Approvals Office for Telecommunications. The name was changed to BZT on March 10, 1992. It is currently based in Saabrucken.

NEWTON'S TELECOM DICTIONARY

C The programming language AT&T uses for several of its central office switches. It is also used as the programming language of choice for interactive voice response (voice processing) systems. C operates under UNIX, MS-DOS, Windows and other operating systems. It is very powerful and is becoming somewhat of a standard for programming telecom switches.

C Band Portion of the electromagnetic spectrum used heavily for satellite transmission. The uplink frequency is at 6 GHz and the downlink is at 4 GHz. Contrast with KU BAND.

C Battery A source of low potential used in the grid circuit of a vacuum tube to cause operation to take place at the desired point on the characteristic curve.

C Lead The third of three wires which make up trunk lines between central office switches. There are three wires — positive, negative, and the "c lead." The purpose of the "c lead" is to control the grounding, holding and releasing of trunks.

C Message Weighting This definition from James Harry Green, author of the excellent Dow Jones Handbook of Telecommunications. C Message Weighting is a factor in noise measurements to approximate the lesser annoying effect on the human ear of high and low-frequency noise compared to mid-range noise.

C Wire C Wire is what the phone company calls the last piece of its wire that comes into your house or office. It is typically the piece of underground cable that comes in from its pedestal on the street to your network interface box on the side of your house or building.

C&C Computers and Communications. An NEC slogan which focused on the deployment of computer and telephony elements to create an integrated environment. Later on, NEC changed it to "Computing and Communicating" and expanded it into a "Fusion" strategy. See FUSION.

C-DTE Character mode Data Terminal Equipment. A term to describe most PCs (personal computers) and printer-terminals that use asynchronous signals for data communications.

C-Link A signaling link used to connect mated pairs of Signal Transfer Points (STPs). An Ericsson term.

C-Message Weighting A type of telephone weighting network that allows for equal attenuation of all frequencies within the voice band in the same manner as it appears to be attenuated by the media.

C/A Code The standard Clear/Acquisition GPS (Global Positioning Code) — a sequence of 1023 pseudo-random, binary biphase modulations on the GPS carrier at the chip rate of 1,023 MHz. Also known as the "civilian code." See GPS.

C++ A high-level programming language developed by Bjarne Stroustrup at AT&T's Bell laboratories. Combining all the advantages of the C language with those of object-oriented programming, C++ has been adopted as the standard house programming language by several major software vendors.

C.E.R.T. Computer Emergency Response Team. Founded in 1988 by the Pentagon's Advanced Research Projects Agency, after a program written by a graduate student jammed more than 6,000 Internet-connected computers nationwide. C.E.R.T. is based at the Software Engineering Institute of the Carnegie Mellon University in Pittsburgh. C.E.R.T. is Internet's "the fire department." It counsels computer operators on how to keep viruses and intruders out of their computers.

C2 Command and Control. The exercise of authority and direction by a properly designated commander over assigned forces in the accomplishment of the mission. Command and control functions are performed through an arrangement of personnel equipment, communications, facilities, and procedures employed by a commander in planning, directing, coordinating, and controlling forces and operations in the accomplishment of the mission. See C3.

C3 Command, Control and Communications. The capabilities required by military commanders to exercise command and control of their forces. See C2.

C7 European equivalent of the North American System Signaling 7. C7 is not 100% compatible with North American System Signaling 7 and that's where gateway switches come in. These switches convert the signaling between one and the other and do it in real time.

CA Call Appearance

CAAGR Compound Annual Average Growth.

Cable May refer to a number of different types of wires or groups of wires capable of carrying voice or data transmissions. The most common interior telephone cable has been two pair. It's typically called quad wiring. It consists of four separate wires each covered with plastic insulation and with all four wires wrapped in an outer plastic covering. Quad wiring is falling into disrepute as it is increasingly obvious that it does not have the capacity to carry data at high speeds. The wire and cable business is immense. The assortment of stuff it produces each year is mind-boggling. In telecommunications, there is one rule: The quality of a circuit is only as good as its weakest link. Often that "weak link" is the quality of the wiring or cabling (we used the words interchangeably) that the user himself puts in. Please put in decent quality wiring. Don't skimp. See CATEGORY OF PERFORMANCE.

Cable Act Of 1984 An Act passed by Congress that deregulated most of the CATV industry including subscriber rates, required programming and fees to municipalities. The FCC was left with virtually no jurisdiction over cable television except among the following areas: (1) registration of each community system prior to the commencement of operations; (2) ensuring subscribers had access to an A-B switch to permit the receipt of off-the-air broadcasts as well as cable programs; (3) carriage of television broadcast programs in full without alteration or deletion; (5) non-duplication of network programs; (5) fines or imprisonment for carrying obscene material; and (6) licensing for receive-only earth stations for satellite-delivered via pay cable. The FCC could impose fines on CATV systems violating the rules. This Act was superseded by the Cable Reregulation Act of 1992.

Cable Assembly A completed cable and its associated hardware ready to install.

Cable Bays Lots of cable arranged like bays in a harbor.

Cable Business A magazine on cabling run by Steve Paulov and family in Mesquite (Dallas), TX. A great magazine. 214-270-0860.

Cable Cutoff Wavelength For a cabled single-mode optical fiber, Cable Cutoff Wavelength specifies a complex inter-relation of specified length, bend, and deployment conditions. It is the wavelength at which the fiber's second order mode is attenuated a measurable amount when compared to a multimode reference fiber or to a tightly bent single-mode fiber.

Cable Diameter Expressed in millimeters or inches. Affects space occupied, allowable bend radius, reel size, length on a reel and reel weight. Also affects selection of pulling grips.

Cable Dog Slang expression. In the West, lifelong cable installer who seeks no upward mobility. In the East, worker who deals with underground cable.

Cable/Information Technology Convergence Forum This forum is a new organization of the cable television, telephone, computer and switching network industries. The Forum was created to promote greater communication between vendors in the information technology industry, cable television companies and CableLabs, the research and development consortium serving most of the cable operators in North America. The Convergence Forum, based in Louisville, CO, was conceived and sponsored by CableLabs. Companies that have agreed to join the Forum include Apple Computer, Bay Networks, Cisco Systems, Compaq, Digital Equipment Corp., Fore Systems and LANCity.

Cable Loss The amount of radio frequency (RF) signal attenuated (lost) while it travels on a cable. There are many reasons for cable loss, including the cable's shape, its type, its size, its length and what it's made of. For coaxial cable, higher frequencies have greater loss than lower frequencies and follow a logarithmic function. Cable losses are usually calculated for the highest frequency carried on the cable. See ATTENUATION.

Cable Management Companies have oodles of telephone cables in and around their buildings. Cable management is the science and art of managing those cables. Typically, cable management covers keeping track of where the cables are (maps are useful), what type and quality the cable is, and what is attached to either end of it. Cable management for corporations is critical, since stringing, laying and snaking of new cable can be inordinately expensive.

Cable Mapping Cable mapping is the task of trying to track every single pair of wire or circuit from beginning to end. You will need to know where all cables reside, not just the circuits that are in use. Cable mapping is critical for any organization — from company to university — which has a lot of cables floating around. Installing more of it — when there are plenty of spare pairs — is stupid and expensive. Thus, the need for cable mapping.

Cable Modem A cable modem is a modem designed for use on TV coaxial cable circuit. The idea is simple: Put the cable modem on a home cable TV line, provide the subscriber a circuit on which he can dial up Prodigy, CompuServe, America On Line, the Internet, etc. And charge extra for the privilege.

Cable Plant A term which refers to the physical connection media (optical fiber, copper wiring, connectors, splicers, etc.) in a local area network. It is a term also used less frequently by the telephone to mean all its outside cables — those going from the central office to the subscribers' offices.

Cable Protection There are three basic types of protection in addition to standard plastic cladding:
- ElectroMagnetic (EM) Shielding: Prevents passive coupling. EM shielding can be a metallic conduit or metal wrapping-with appropriate grounding-on the wires.
- Penetration-Resistant Conduit: Used to secure the cable from cutting or tapping. Note, however, not all penetration-resistant conduits provide EM shielding.
- Pressurized Conduit: Detects intrusion by monitoring for pressure loss. Fiber optic cable is extremely difficult to tap and if tapped, the intrusion can be detected through signal attenuation. But since fiber optic cable can be cut, penetration-resistant conduit is recommended to protect the cable.

Cable Racking Framework fastened to bays to support cabling between them.

Cable Reregulation Act Of 1992 Reregulation Bill 1515 passed Congress in October 1992, forcing the FCC to reregulate cable television and cable television rates (after the Cable Act of 1984 effectively de-regulated the cable TV industry). After the Act was passed, the FCC forced the industry to reduce its rates by 10% in 1993 and then again by 7% in 1994.

Cable Riser Cable running vertically in a multi-story building to serve the upper floors.

Cable Run Conduit used to run cables through a building. Also, path taken by a cable or group of cables.

Cable Shield A metallic component of the cable sheath which prevents outside electrical interference and drains off current induced by lightning.

Cable Television Relay Station CARS. A fixed or mobile station used for the transmission of television and related audio signals, signals of standard and FM broadcast stations, signals of instructional television fixed stations, and cablecasting from the point of reception to a terminal point from which the signals are distributed to the public.

Cable Type The type of cable used. Also called the media. Examples are coaxial, UTP (Unshielded Twisted Pair), STP (Shielded Twisted Pair) and fiber. Factors including cost, connectivity and bandwidth are important in determining cable type. Choosing cable is getting more and more complex. Our tip: Choose and buy well in advance of when you'll need it. The cable you want will not always be in stock.

Cable Vault Room under the main distribution frame in a central office building. Cables from the subscribers lines come into the building through the cable vault. From here they snake their way up to the main distribution frame. The cable vault looks like a bad B-movie portrayal of Hell, replete with thousands of dangerous black snakes. Cable vaults are prime targets for the spontaneous starting of fires. They should be protected with Halon gas, but usually aren't because some parts of the phone industry think Halon is too expensive.

Cable Weight Expressed in lbs. per 1000 (without reel weight included). Affects sag, span and size of messenger in aerial applications.

NEWTON'S TELECOM DICTIONARY

Cablegram Service An MCI definition. An MCI International service which provides cablegram communication to International destinations through the use of a computerized message switching center in New York City.

Cableport Intel Corporation's new technology, which brings high speed multimedia-rich interactive services to personal computers in the home via cable.

Cablespan A Tellabs Operations, Inc. product which deliver two-way voice and data services over coaxial cable used by cable TV operators. Tellabs is in Lisle, Illinois.

Cableway An opening in a work surface that allows access to cords or cables from below, or mounting of an electrical receptacle or telephone jack. Cableways typically come with removable plastic grommets.

CAC 1. Customer Administration Center. A type of terminal used by a PBX user to maintain and troubleshoot his PBX.
2. Connection Admission Control is defined as the set of actions taken by the network during the call setup phase (or during call re-negotiation phase) in order to determine whether a connection request can be accepted or should be rejected (or whether a request for re-allocation can be accomplished).

CACH Call Appearance Call Handling.

Cache A cache is high-speed memory designed to hold upcoming to-be-accessed and/or recently-accessed data. A cache speeds up a computer's operation because high-speed memory sends information to and from the computer's central microprocessor much faster than a hard disk could. Cache memory gives the computer user the impression that the computer and its programs are actually running a lot faster than they are theoretically capable of. Cache memory comes in RAM on a computer's motherboard and on hard disk controller cards. The value of a cache is directly related to how good the software is that runs it. There are wide variations in the quality of caching software. A cache is especially important in a LAN file server. It can be used to "speed up" access to a database stored on the file server, thereby saving workstation users time and aggravation. Caches work because of a phenomenon known as the locality principle which states that a von Neumann CPU (i.e. one that performs instructions one after another) tends to access the same memory locations over and over again. A cache works like this: when the CPU needs data from memory, the cache hardware and software checks to see if the information is already in the cache. If it is, it grabs that information. This is called a cache hit. If it isn't, it's called a cache miss and the computer has to access the disk, which is slower. Data retrieved during a cache miss is often written into the cache. See also CACHE MEMORY.

Cache Controller A chip, such as the Intel 82385, that manages the retrieval, storage and delivery of data to and from memory or the hard disk.

Cache Hit When the data you want is actually in cache. Thus you don't have to access your hard disk and your computing is fast. See CACHE and CACHE MEMORY.

Cache Memory Available Random Access Memory that you set up to allow your computer to "remember" stuff — so the next time your computer wants that information, it can find it fast from RAM, instead of searching through a hard disk which is slow. This high speed cache memory eliminates the CPU wait state. When the CPU reads data from main memory, a copy of this data is stored in the cache memory. The next time the CPU reads the same address, the data is transferred from the cache memory instead of from main memory. Novell's NetWare, for example, uses cache memory to improve file server access time. In NetWare, cache memory contains the directory and file caches, along with the FAT (File Allocation Table), the turbo FAT, the Hash table, and an open space for other functions. See also CACHE.

Cache Miss When the caching software guesses wrongly and you have to read your data off your hard disk rather than reading it from the cache. See also CACHE and CACHE MEMORY.

CAD 1. Computer Aided Dispatch.
2. Computer Aided Design. A computer and its related software and terminals used to design things. A CAD system might be as simple as computerized drafting tools or as complex as detailed layouts of integrated circuits. CAD systems often have terminals on peoples' desks and a central maxi-computer in the company's main computer room. CAD terminals are often run over LANs (local area networks) or through telephone systems. The terminals are often moved, thus having universal wiring and a universal switching system — a LAN or a phone system — is extremely useful.

CAD/CAM Computer Aided Design/Computer Aided Manufacturing. See CAD.

CADB Calling Area Data Base. An MCI definition. An MCI System that stores reference data for various MCI Systems and reconciles MCI Calling Areas with those of Bell.

Caddy The shell of an optical disc. Protects it from grubby fingerprints, and includes write protection devices. AKA case.

Cadence In voice processing, cadence is used to refer to the pattern of tones and silence intervals generated by a given audio signal. Examples are busy and ringing tones. A typical cadence pattern is the US ringing tone, which is one second of tone followed by three seconds of silence. Some other countries, such as the UK, use a double ring, which is two short tones within about a second, followed by a little over two seconds of silence.

CADS Code Abuse Detection System.

CAE Computer Aided Engineering.

Cage Antenna An antenna having conductors arranged cylindrically.

CAGR Compound Annual Growth Rate.

CAI 1. Computer Assisted Instruction. See also CAD for a discussion on telecom needs.
2. A British Term. Common Air Interface. The CT2/telepoint international standard which allows any compliant equipment, i.e. a wireless handset, to be used on any network of the same type. CAI-compliant telepoint handsets from different vendors may therefore be used on a telepoint network.

Cairo Microsoft's planned heavy duty successor to Windows NT. Cairo is due out the later part of the decade. It is rumored to include a radically new file structure that will let you easily find information among many thousands of files.

Calendar Routing A call center term for directing calls according to the day of the week and time of day. See also SOURCE/DESTINATION ROUTING, SKILLS BASED ROUTING and END-OF-SHIFT ROUTING.

Calibrate To test and reset a measuring or timing device against a standard to make sure it is functioning correctly.

Call My definition: Two people on the phone. One calls the other. People who speak on phones are called "parties." No one knows why. Bellcore's definition: An arrangement providing for a relation between two or more simultaneously present users for the purpose of exchanging information. The ATM Forum's definition: A call is an association between two or more users or between a user and a network entity that is established by the use of network capabilities. This association may have zero or more connections.

Call Abandons Also called ABANDONED CALLS. Call Abandons are calls that are dropped by the calling party before their intended transaction is completed. The call may be dropped at various points in the process. The point in the call at which the call is abandoned will have varying impacts on a computer telephony system. Many callers upon hearing an automated system will hang up. For systems that expended significant energy in setting up to answer a call, a large percentage of call abandons can negatively impact the call capacity of the system.

Call Accepted Signal A control signal transmitted by the called equipment to indicate that it accepts an incoming call.

Call Accounting System A computer, a magnetic storage device (floppy or hard disk), software and some mechanical method of attaching itself to a telephone system. A call accounting system is used to record information about telephone calls, organize that information and upon being asked, prepare reports — printed or to disk. The information which it records (or "captures") about telephone calls typically includes from which extension the call is coming, which number it is calling (local or long distance), which circuit is used for the call (WATS, MCI, etc.), when the call started, how long it lasted, for what purpose the call was made (which client? which project?). A call accounting system might also include information on incoming calls — which trunk was used, where the call came from (if ANI or interactive voice response was used), which extension took the call, if it was transferred and to where and how long it took.

There are eleven basic uses for call accounting systems:

1. Controlling Telephone Abuse. It's the 90-10 rule. 10% of your people sit on long distance calls all day to their friends and family. The others work. Some people still think WATS calls are free. Knowing who's calling where and how much they're spending is useful. Often they appreciate being told they're spending money. Big money...and they stop.

2. Controlling Telephone Misuse. I figured once you could call between two major cities for five cents a minute and $1 a minute. That's a 20-fold difference! Often you need different lines. Often a company has different lines. Sometimes the phone system makes the dialing decision. Sometimes the person makes the dialing decision. Whoever's doing it can be wrong. A call accounting system is a good check to see if you're spending money needlessly.

3. Allocating telephone calling costs among departments and divisions. Telephones — voice, data, video and imaging — are some of your biggest expenses. They're a cost that should be allocated to the products you're making, or the departments or divisions in your company. Telephone costs can determine which product is profitable. Which isn't. Item: A software company recently dropped one of its three "big" software packages because phone calls for support got too expensive.

4. Billing Clients and Projects back for telephone charges incurred on their behalf. Every lawyer, government contractor, etc. does it. Makes sense.

5. Sharing and Resale of long distance and local phone calls, as in a hotel/motel, hospital, shared condominium, etc. Someone's got to send out the bills. And it's not the phone company. In fact, with a call accounting system you can be your own phone company!

6. Motivation of Salespeople. The more phone calls they make the more they sell. This rule is as obvious as the nose on your face. You WANT salespeople to make more calls? Hang a list of all their calls on the wall. Give prizes to those make the most! Or those who make more than last week. Or those who set a new record.

7. Personnel Evaluation. Which employees are doing better at being productive on the phone (however you define "productive"). You want them to get on and off the phone fast? Or you want them to stay on and coddle your customers? You can now correlate phone calls with income — from service or just straightout sales.

8. Network Optimization. Two fancy words for figuring which is the best combination of MCI, AT&T, MCI, Sprint, Wiltel, etc. lines. And which is the best combination of all the various services each offer. A rule of thumb: There's a 20-fold difference in per minute telephone calling costs between any two major cities in the US. And — amazing — you won't hear any difference in quality, despite the huge difference in price. I think it's the biggest price difference in any product anywhere. It's amazing.

9. Phone System Diagnostics. Is the phone system working as well as it should? Are all the lines working? Are all the circuit packs (circuit cards) working? Call accounting systems can tell you which lines you're getting no traffic on. Or which line carried the 48 hour call to Germany (it's happened). Either way, you can figure quickly which lines are working and which aren't.

10. Long Distance Bill Verification. Was the bill accurate we received from our chosen long distance phone company? Mostly it isn't. In fact, there's no such thing as an accurate phone bill. That's an oxymoron. Using your call accounting systems to check your long distance gives you some peace of mind. It's cheap peace of mind. Everyone should have one.

11. Tracing Calls. True story: Every third or fourth Friday afternoon a large factory in the south received bomb threats. They'd clear the factory, search the factory and not find anything. By the time they'd checked, it was too late to start up production. One day they checked their call accounting records. The calls were coming from a phone on the factory floor. The whole thing was a ruse to get an afternoon off...And now that many phones give you the number of who's calling, call accounting systems are turning out to be great for checking the effectiveness of regional ad campaigns, figuring the profitability of direct mailings and even figuring the profitability of individual customers.

Call Announcement A telephone operator or person acting as a telephone operator can announce a call to the called party before putting the call through. All modern phone systems have this feature.

Call Attempt A try at making a telephone call to someone. Tally up call attempts and compare them to completions and you'll have some idea of corporate frustration and thus, the need for more lines or more phone equipment. The measures in this in call attempts, calls answered, calls overflowed, and calls abandoned.

Call Back A security procedure in which a user, dials into access a system and requests service. The system then disconnects and calls the user back at a preauthorized number to establish the access connection. Same as dial-back. See also INTERNATIONAL CALL BACK.

Call Barring The ability to prevent all or certain calls from reaching to or from a phone.

Call Before Dig A preventive maintenance measure in which signs are posted near buried cables advising people to phone before digging in the area.

Call Blending A phone system has a bunch of people answering and making calls. The calls are coming (say in response to an ad). The calls are going out, courtesy of a dialing machine (perhaps a predictive dialer). The idea is to keep the calls at a constant level. The idea is to blend incoming with outgoing calls. Some predictive dialers have call blending. Others don't. They need a dedicated workforce. Call blending automatically transfers staff members between outbound and inbound programs as call volumes change. Some predictive dialers let you choose which workstations will be used for call blending, to avoid training of every staff member.

Call Block A name for a Pacific Bell (and possibly other local telephone companies') service which helps you avoid unwanted calls by rejecting calls from numbers you specify. BellSouth describes its service called "Call Block" as "prevents the number received from getting through again. It also lets the subscriber pre-select up to six phone numbers to be blocked.

Call Blocking Check into a hotel. Dial a 0+ call. You're connected to an Alternate Operator Service company. But you know their rates may be high. You ask to be connected to AT&T or MCI, or whoever is the carrier of your choice. Sadly, the AOS cannot connect and neither can (nor will) your hotel's operator. This is called "Call Blocking." The FCC has disbarred the practice. But it continues. See also CALL SPLASHING.

Call Card A British term. A paper record, used in manual telebusiness systems, to record the results of a call.

Call Center A place where calls are answered and calls are made. A call center will typically have lots of people (also called agents), an automatic call distributor, a computer for order-entry and lookup on customers' orders. A Call Center could also have a predictive dialer for making lots of calls quickly. The term "call center" is broadening. It now includes help desks and service lines. For more information on Call Centers, please read CALL CENTER Magazine. For subscriptions call 215-355-2886. See CALL CENTRE.

Call Centre A British term. An area in an organization where business is conducted by phone in a methodical and organized manner. Call centres are typically based on the integration of a computerized database and an automatic call distribution system. (Note that this definition contains British spellings.) See CALL CENTER.

Call Clear Packet An information packet that ends an X.25 communications session, performing the equivalent of hanging up the phone.

Call Clearing The process by which a call connection is released from use.

Call Collision 1. Contention that occurs when a terminal and a DCE simultaneously transfer a call request and an incoming call specifying the same logical channel. The DCE will proceed with the call request and cancel the incoming call.
2. That condition arising when a trunk or channel is seized at both ends simultaneously, thereby blocking a call.

Call Completion Rate The ratio of successfully completed calls to the total number of attempted calls. This ratio is typically expressed as either a percentage or a decimal fraction.

Call Control Call control is the term used by the telephone industry to describe the setting up, monitoring, and tearing down of telephone calls. There are two ways of doing call control. A person or a computer can do it via the desktop telephone or a computer attached to that telephone, or the computer attached to the desktop phone line (i.e. without the actual phone being there). That's called First Party Call Control. Third-party call control controls the call through a connection directly to the switch (PBX). Generally third-party call control also refers to the control of other functions that relate to the switch at large, such as ACD queuing, etc.

Call Control Procedure Group of interactive signals required to establish, maintain and release a communication.

Call Control Signal Any one of the entire set of interactive signals necessary to establish, maintain, and release a call.

Call Data Call data refers to any data about a phone call that is passed by a switch to an attached computer telephony system. Call data is usually used by the computer telephony application to process the call more intelligently. Call data may be passed In-Band, over the same physical or logical link as the call — usually via tones, or Out-Of-Band, over a separate link — usually a serial link. Call data may also be passed as part of the data designed to control telephone networks, such as SS7 (Signaling System 7) links. In addition to information about the call, status about the call and even control over the call, can be available as part of the call data link services. Call data almost always includes what number dialed the call (ANI) and/or what number called (DNIS). More complex call data links used for "PBX integration" may also indicate why the call was presented (such as forwarded on busy), tell what trunk the call is coming in on, or to pass message waiting on or off indications, and other functions. Full blown computer telephony links, such as are now being offered by many switching vendors, enhance the call data path, providing additional status information about calls and can even provide a level of call control to the attached computer telephony system. The above definition courtesy of Steve Gladstone, author of a great book called Testing Computer Telephony Systems, available from 212-691-8315.

Call Delay The delay encountered when a call reaches busy switching equipment. In normal POTS telephone service, the delay is considered OK if no more than one and a half percent of the calls are delayed by three seconds during the busy hour.

Call Detail Recording CDR. A feature of a telephone system which allows the system to collect and record information on outgoing and incoming phone calls — who made/received them, where they went/where they came from, what time of day they happened, how long they took, etc. Sometimes the data is collected by the phone system; sometimes it is pumped out of the phone system as the calls are made. Which ever way, the information must be recorded elsewhere — dumped right into a printer or into a PC with call accounting software. See also CALL ACCOUNTING SYSTEM.

Call Diverter 1. A device which when connected to a called telephone number intercepts calls to that number and connects them to a telephone operator or prerecorded message.
2. An ancillary device which is connected to a telephone line. The device will, when the called telephone rings, initiate a telephone call on another line to a different telephone number. The calling party may or may not be aware that his call has been diverted to another telephone.

Call Duration The time from when the call is actually begun (i.e. answered) to the instant either party hangs up. Call Duration is an important concept for traffic engineering.

Call Establishment The process by which a call connection is created.

Call Forwarding A service available in many central offices, and a feature of many PBXs and some hybrid PBX/key systems, which allows an incoming call to be sent elsewhere. There are many variations on call forwarding: Call forwarding busy. Call forwarding don't answer. Call forwarding all calls, etc.
Call forward is a useful feature. For example, you're going to a meeting but you're expecting an important call. Pick up your phone, punch in some digits and all your calls will go to the new number — perhaps the phone outside the meeting room. The big disadvantage is that many people return to their offices but forget they forwarded their calls elsewhere. As a result, they usually miss a whole bunch of important calls. Some electronic phones now have a reminder light or message on them saying "all calls are being forwarded." Some people program their PBXs to cancel all call forwards at noon and at midnight every day. This makes sense.
Call forwarding is used to send calls to voice mail systems. For example, tell your PBX that if your phone isn't answered in four rings, send that call to your voice mail.
If you are getting call forwarding service from a central office in North America, the code to begin call forwarding is 72# and the number you want to be forwarded to. To cancel it, you punch in 73#.

Call Frame Harris' PBX to computer link. Harris' protocol for linking its PBX to an external computer and having that computer control the movement of calls within a Harris PBX. See also OPEN APPLICATION INTERFACE.

Call Handoff A cellular phone term. Call handoff happens when a wireless call is transferred to another cell site in mid-conversation.

Call Hold If you hang the phone up, you lose the caller. Call hold — a feature of most phone systems — allows you to "hold" the call, so the other person can't hear you. You can then return to the conversation by pushing a button on your phone, typically the button flashing which shows which line the person is sitting on hold. Call hold is useful when you have someone on another line calling you.

Call Identifier A network utility that is an identifying name assigned by the originating network for each established or partially established virtual call and, when used in conjunction with the calling DTE address, uniquely identifies the virtual call over a period of time.

Call In Absence Horn Alert A cellular car phone feature that sounds your car's horn when you are receiving a call.

Call In Absence Indicator A cellular car phone feature that ensures that power to the cellular phone is not lost if the car's ignition is turned off.

Call In Progress Override A cellular car phone feature that keeps power to the phone during a call even though you've turned off the car's ignition.

Call Letters Certain combinations of letters assigned to radio stations by the FCC. The group of letters assigned the U.S. by the International Radiotelegraph Convention are all three and four letter combinations beginning with N and/ or W and all combinations of KDA to KZZ inclusive.

Call Me Card A special AT&T Card number which permits others to call you, and only you, and have the call charged to your telephone. Give an AT&T CALL ME Card to your very best customers to encourage them to call you. A Call Me Card is like having a private unlisted 800 toll-free number.

Call Me Message A Rolm/IBM feature on their CBX that allows internal users to leave other users a message showing the time, date and the caller's extension number.

Call Mix Call mix is the pattern of call types (each call type defines what the caller will do for that call) that goes into creating a busy hour call profile or other call profile. A voice mail system's busy hour call profile call mix may consist of 10% call abandons, 20% login and send one message, 30% login and listen to one message, and so on. Varying the call mix can often be useful to stress particular parts of a system. For example, a call mix of 100% call abandons is frequently used to stress a computer telephony system's ability to handle high traffic call setup scenarios. This definition courtesy of Steve Gladstone, author of the book Testing Computer Telephony Systems, available from 212-691-8215.

Call Model An abstraction of the call processing functionality of the architecture and the relationship that exists between the functionality of the Service Switching FE in an ASC and the Service Logic and Control FE in a SLEE (Service Logic Execution Environment). The call model consists of two components: Connection View and Basic Call State Model. Definition from Bellcore in reference to Advanced Intelligent Network.

Call Not Accepted Signal A call control signal sent by the called terminal to indicate that it does not accept the incoming call.

Call Packet A block of data carrying addressing and other information that is needed to establish an X.25 switched virtual circuit (SVC).

Call Park The phone call is not for you. Or maybe it is, but you don't want to answer it on your phone. Put it into CALL PARK, then you or anyone else can answer it from any other phone. Call Park is similar to placing a call on hold, but you're not there. So he places the call in Call Park, pages you and tells you the call is in Call Park. You pick up the nearest phone, dial one or two digits (the code for grabbing the call out of Call Park) and you have the call. It's faster than looking for you, then telling you to hang up while she transfers the call.

Call Pickup A phone is ringing but it's not yours. With call pickup, you can punch in a button or two on your phone and answer that person's ringing phone. Saves time. See CALL PICKUP GROUP.

Call Pickup Group CPUG. All the phones in an area that can be answered by each other by simply punching in a couple of digits. See CALL PICKUP.

Call Processing The system and process that sets up the intended connection in a switching system. The system scans the trunk and/or station ports for any "requests" for service. Upon detecting a request, the system checks the stored instructions and look-up tables and sets the connection up accordingly.

Call Progress The status of the telephone line; ringing, busy ring/no answer, voice mail answering, telephone company intercept, etc. See CALL PROGRESS ANALYSIS and CALL PROGRESS TONE.

Call Progress Analysis As the call progresses several things happen. Someone dial or touchtones digits. The phone rings. There might be a busy or operator intercept. An answering machine may answer. A fax machine may answer. Call progress analysis is figuring out what is occurring as the call progresses. This analysis is critical if you're trying to build an automated system, like an interactive voice response system.

Call Progress Signaling All telephone switches use the same three general types of signals: + Event Signaling initiates an event, such as ringing. + Call Progress Signaling denotes the progress (or state) of a call, such as a busy tone, a ringback tone, or an error tone. + Data Packet Signaling communicates certain information about a call, for example, the identify of the calling extension, or the identity of the extension being called.

Call Progress Tone A tone sent from the telephone switch to tell the caller of the progress of the call. Examples of the common ones are dial tone, busy tone, ringback tone, error tone, re-order, etc. Some phone systems provide additional tones, such as confirmation, splash tone, or a reminder tone to indicate that a feature is in use, such as confirmation, hold reminder, hold, intercept tones.

Call Queuing Incoming or outgoing calls may be queued pending an answer. The idea of call queuing is to save money. See also CALLBACK QUEUING.

Call Record The data record of a call transaction. The record is made up of event details that typically include date, time, trunk(s) used, station(s) used and duration. In an ACD, these events may also include time in queue, call route used, system disposition flag, inbound or outdialed digits and wrap-up data entered.

Call Reference Information element that identifies to which call a Layer 3 message pertains.

Call Release Time The time it takes from sending equipment a signal to close down the call to the time a "free condition" appears and the system is ready for another call.

Call Reorigination Caller reorigination allows a caller with a telephone debit card account number or a telephone credit card account to make unlimited calls without hanging up and redialing their access and their account numbers. At the end the first call, the caller remains on the line. The caller then presses the pound key (the # key) for a prescribed number of seconds and receives a confirmation tone (which sounds like a high-pitched dial tone). After receiving the tone, the caller immediately dials their next phone number. And so on. Sometimes you can hit the # button after the person you were talking to has hung up. My experience has been that it's better to hit the # before the person has hung up. Just tell them what you're doing. All this allows the card account holder to make a series of calls without ever hanging up and redialing the often lengthy access card account numbers. This saves a lot of time for callers with a long list of calls to make. It also saves money for callers from hotels, which charge for each connection to the long distance provider, but don't charge based on the length of the call (especially if it's a local call or toll-free 800 call). The prescribed number of seconds that the user needs to hold down the pound key is configurable. Depending on the application the time may range from 1 to 5 seconds. With most carriers, it's two to three seconds. The above definition was kindly provided by Karen Shelton, Systems Engineer, IEX Corporation, Richardson, TX.

Call Request Packet In packet data switching, a call request packet carries information, such as sender and recipient identification, that is needed to establish an X.25 circuit. In more technical terms, a call request packet is sent by the originating data terminal equipment (DTE) showing requested network terminal number (NTN), network facilities and either X.29 control information or call user data.

Call Restrictor Equipment inserted in a telephone line or trunk which restricts outgoing calls in some way. Usually from making a toll call.

Call Return A name for a Pacific Bell (and possibly other local telephone companies') service which allows you to dial the last caller, even if you did not answer the telephone.

Call Screening There are several definitions. Here are two. 1. A PBX feature that looks at the digits dialed by the caller to figure whether the call should be completed.
2. A receptionist or secretary answers the "executive's" phone and checks out that the person calling is important enough be put through to the almighty executive whose calls are being screened.

Call Second A unit for measuring communications traffic. Defined as one user making one second of a phone call. One hundred call seconds are called "ccs," as in Centum call seconds. "ccs" is the U.S. standard of telephone traffic. 3600 call-seconds = 1 call hour. 3600 call-seconds per hour = 36 CCS per hour = 1 call-hour = 1 erlang = 1 traffic unit. See also ERLANG and TRAFFIC ENGINEERING.

Call Selector A local phone company service which alerts the subscriber with a distinctive ring that one of the six numbers your pre-selected is calling.

Call Sequencer A call sequencer, also called an Automatic Call Sequencer, is a piece of equipment which attaches to a key system or a PBX. The Call Sequencer's main function is to direct incoming calls to the next available person to answer that phone. It typically does this by causing lights on telephones to flash at different rates. The light with the fastest flashing is the one whose call has been waiting longest. This call is answered first. Call Sequencers also might answer the phone, deliver a message and put the person on hold. They might keep statistical tabs of incoming calls, how fast they were answered, how long the people waited, how many people abandoned (hung up while they were on hold waiting for their call to be answered by a human being), etc. Call Sequencers are usually simple and inexpensive. Better, but much more expensive devices for answering incoming phone calls are Automatic Call Distributors. These are the devices which typically answer when you call an airline. See AUTOMATIC CALL DISTRIBUTOR.

NEWTON'S TELECOM DICTIONARY

Call Setup The first six PICs (Point In Call) of the Originating BCSM (Basic Call State Model), or the first four PICs of the Terminating BCSM. Definition from Bellcore in reference to Advanced Intelligent Network.

Call Setup Time The amount of time it takes for a circuit-switched call to be established between two people or two data devices. Call set-up includes dialing, wait time and time to move through central offices and long distance services. You don't pay for call set-up, but you will need extra lines to take care of it. See also ANSWER SUPERVISION and TRAFFIC ENGINEERING.

Call Shedding In many states, laws require that a real-life breathing person be available for a phone call being outdialed with an automated device. The reason these laws were enacted is because a lot of times call-center managers have their predictive dialers going so crazy in search of human answerers that they don't have an agent ready when someone does actually pick up the phone. Most systems simply hang up the connection when this occurs. This is called "call shedding" and is illegal in many states.

Call Spill-over In common-channel signaling, the effect on a traffic circuit of the arrival at a switching center of an abnormally delayed call control signal relating to a previous call, while a subsequent call is being set up on the circuit.

Call Splashing A "splash" happens when an Alternate Operator Service (AOS) company, located in a city different to the one you're calling from, connects your call to the long distance carrier of your choice in the city the AOS operator is in. Let's say you're calling from a Hotel in Chicago. You ask AT&T handle your call. The AOS, located in Atlanta, "splashes" your call over to AT&T in Atlanta. But you're calling Los Angeles. Bingo. Your AT&T call to LA is now more expensive than it would be — if you had been connected to AT&T in Chicago.

Call Splitting A feature allowing a phone user to speak privately with either party of a conference call by alternating between the two. Call splitting by an attendant allows the attendant to speak to the called person privately while effectively putting the calling person on hold, or vice versa.

Call Stalker An AT&T PC-based product which gives the 911 attendant the phone number and address of the person calling.

Call Store The temporary memory used in a stored program control switch (SPC) to hold records of calls in progress. These records are then transferred to permanent memory.

Call Stream British Telecom's premium rate service.

Call Tag A term used in the secondary telecom equipment marketplace. A ticket directing a freight carrier (e.g., UPS) to pick up equipment at another site. The company issuing the ticket pays the freight charge. Normally used to return defective equipment, it ensures the dealer a quick return and an accurate tracking mechanism.

Call Trace A name for local telephone company service which permits the tracing of the last call received and holds the results for later use by an authorized law enforcement agency. (Results of the trace are not available to the customer.)

Call Transfer Allows you to transfer a call from your phone to someone else's. On some phones you do this by punching in a bunch of numbers. Some you do it by hitting the "transfer" button and then the number you want to send the call to. The fewer buttons and numbers you have to punch, the easier it will be for your people.
If you're choosing a telephone system, check how easy it is to transfer a call. It is the most commonly used (and misused) feature on a phone system. How many times have you been told, "I'll transfer you to Mr. Smith, but if we get disconnected, please call back on extension 234." If your people are saying this to your customers or prospects, you are giving the outside world the wrong impression of your business. And since 97% of your prospects' contact with your company is first through your phone system, you could be losing precious business.

Call Type A call center term used in Rockwell ACDs. A portion of your call center traffic corresponding to one or more ACD gates or splits. This division of the total ACD traffic is the level at which forecasting and scheduling are done. At setup time, each Call Type is defined in the ACD by a unique three-letter code and specific gate or split number(s) that identifies the corresponding ACD report data.

Call User Data In packet data networking technology, user information transmitted in a call request packet to the destination data terminal equipment (DTE).

Call Waiting Call Waiting is a feature of phone systems that lets you know someone is trying to call you. You're speaking on the phone. A call comes in for you. You might hear a beep in your ear or see a light on your phone turn on. Or you might hear a beep and see a message come across the screen of your phone. When you hear the beep, you can, if you wish, put the present call on hold and answer the new one. Or you can ignore the new one, hoping it will go away, and perhaps send it to your attendant/operator, or voice mail. Call Waiting can be done manually by your telephone system operator. Or it can be a service which you buy as a monthly from your local phone company.
A major problem with call waiting is if you're on a data call from your PC, the call waiting "beep" will often cause your modem to hang up, thus destroying your data call. There are two solutions to this, the obvious one being turn off call waiting. Some phone systems will allow you to turn it off. The less obvious one is modify your modem's initialization string. Here's how. In all Hayes and Hayes-compatible modems, there's a S10 register. It tells the modem how long before it hangs up after losing carrier. In Hayes modems, the S10 register is set for 1.4 seconds. The typical call waiting tone is 1.5 seconds. Solution, increase the S10 register to six seconds (to be sure). Use your communications software. Go into terminal mode, then type: ATS10=60. You must put this command in every time you power up, because the Hayes 1200 modem (and others) have volatile memory. But the Hayes 2400 and higher speed asynchronous modems have non-volatile memory. They remember the six seconds after they've been switched off. The command to write this to memory is ATS10=60&W. The "&W" means write it to memory.

Callback 1. A feature of some voice and data telephone systems. You dial someone. Their phone or computer is busy. You hit a button or code for "call-back." When their phone becomes free, the phone system will call you and them simultaneously. You can only use this call-back feature on things internally in your phone system — calling other people, calling long distance lines (which might be busy), calling the dictation pool, etc. See CALL WAITING, CALLBACK MODEM and CALLBACK QUEUING.
2. A quick way of referring to international callback, which works thus: Calling the United States from many countries abroad is far more expensive than calling those countries from the United States. A new business called International Callback has started. It works like this. You're overseas. You dial a number in the United States. You let it ring once. It won't answer. You

105

hang up. You wait a few seconds. The number you dialed in the U.S. knows it was you calling. There is a piece of equipment on that number that "hears" it ring and knows it's you since no one else has that number. That was your special signal that you want to make a call. A switch attached to that line then calls you instantly. When you answer (overseas, obviously) it conferences you with another phone line in the United States and gives you U.S. dial tone. You can then touchtone from overseas your American number, just as if you would, were you in the U.S. There are huge savings. U.S. international callback operators can offer as much as 50% savings on calls from South America, where international calling rates are very high. The process of international callback is being automated with software and dialing devices. International callback is also helping to bring down the high cost of calling the U.S. from overseas. A company called kallback in Seattle, WA has received a service mark from the U.S. Patent and Trademark Office for the words "callback" and "kallback" and sends letters to and threatens law suits against companies who use "their" words. See CALLBACK.

Callback Modem A modem that calls you back. Here's how it works. You dial into a network. A modem answers. You put your password in. It accepts the password. It says "Please hang up. I will now call you back." You hang up. It calls you back. There are two reasons for doing this instead of allowing you to just go straight into the network. 1. It's better security. You have to be at a pre-determined place — an authorized phone number. 2. It may save on phone calls. The modem uses the company's communications network, which is probably cheaper than what the person calling in can use.

Callback Queue The queue used to hold callers who have requested a busy pool or extension. See CALLBACK QUEUE.

Callback Queuing An option on a telephone system which allows outgoing calls to be put in line for one or several trunks. When a trunk becomes available, the phone system calls the user, his phone rings and then the phone system dials the distant party on the trunk it grabbed before calling the user. Phone systems typically have two types of queuing. The first is called Hold-On Queuing. With this, the user dials his long distance number, the phone system searches for the correct trunk, finds it's not available and tells the user with a beep or message. The user then elects to stay on the line and wait. The instant the trunk becomes free, the phone system connects the user to it. The second type of queuing is called Callback Queuing. The user hangs up and the phone system calls you back, as we explained above.

There are tradeoffs between the two types of queuing. Callback queuing obviously can tolerate longer queues. The longer you wait, the more chance you have of reaching a very low-cost trunk. But users don't like waiting so long for a trunk. And when the call does come, it may likely reach a phone, newly-deserted by a user who's gone to the bathroom.

In contrast, hold-on queuing is more efficient of the user's time, but less efficient of the user's trunks. The less time you wait, the less chance you have of reaching a low-cost trunk. Life is a trade-off. Queuing is no exception. See also QUEUING.

Callbridge Rolm to non-IBM computers open architecture interface. A method of connecting a Rolm CBX (telephone system) to a non-IBM computer, so that the computer may "talk" to the PBX and make certain things happen, e.g. moving a screen of client information around simultaneously with the phone call from the client. This feature is especially useful in customer service and customer order-entry environments — for example with direct mail order catalog companies, etc. See OPEN APPLICATION INTERFACE.

Calldisplay A member of the RolmCenter family of products. It is a software product that runs on an IBM PS/2 under the SCO Xenix operating system. CallDisplay enhances the Rolm Automatic Call Distribution (ACD) current status and system summary displays. It operates in the Rolm 9751 CBX, the Rolm CBX II 9000 and the Rolm CBX II 8000 environments.

Called DTE A DTE which receives a call from another DTE.

Called Line Identification Facility A service feature provided by a network (private or public), which enables a calling terminal to be notified by the network of the address to which the call has been connected. See CALLER ID.

Called Line Identification Signal A sequence of characters transmitted to the calling terminal to permit identification of the called line.

Called Party Subaddress Information element that is passed transparently by the SPCS (if certain conditions are met) and can be used to further identify the destination party.

Called Party Camp-on A communication system service feature that enables the system to complete an access attempt in spite of issuance of a user blocking signal. Systems that provide this feature monitor the busy user until the user blocking signal ends, and then proceed to complete the requested access. This feature permits holding an incoming call until the called party is free.

Caller ID Your phone rings. A name pops upon on your phone's screen. It's the name of the person calling you. Or it may be just the caller's phone number. It's called Caller ID and the information about name and/or calling phone number is passed to your phone by your telephone company's central switch. There are basically two forms of "caller ID" — one provided by your local phone company and one provided by your long distance company (chiefly on 800 calls). Caller ID, generic term, is a term most commonly applied to the service your local phone company provides, usually called CLASS. In CLASS, the information about who's calling and/or their phone number is passed to your phone between the first and second ring signaling an incoming call. See also CLASS and ANI.

Caller Independent Voice Recognition Having a voice response unit recognize the voice of a caller without having been trained on the caller's voice.

Calling A procedure which consists of transmitting address signals in order to establish a link between devices that want to talk to each other.

Calling Card A credit card issued by Bell operating companies, AT&T, MCI, Sprint and other phone companies (local and long distance) and used for charging local and long distance calls. Typically, the number on your calling card is the phone number at which you receive bills (home or business phone) plus a four digit Personal Identification Number (PIN number). Increasingly often it's not. I prefer to carry a calling card with digits completely different to my phone number since this provides me with greater security. It's harder for someone to figure out my calling card. Some phone companies — local and long distance — charge more for a call made with a Calling Card. Some don't. Bell Canada claims they trademarked the term "Calling Card" in Canada. If they did, good luck protecting it, since the term "calling card" is generic. See BREAKAGE, DEBIT CARD and PREPAID CALLING CARD.

Calling DTE A DTE (Data Terminal Equipment) which places a call to another DTE.
Calling Line ID You are called. As the call comes in, you receive the phone number of the person calling you. Calling Line ID is another term for AUTOMATIC NUMBER IDENTIFICATION. Calling Line ID is used increasingly for the phone numbers of local calls, while ANI tends to be used for the phone numbers of distant phone numbers. Essentially they're the same. See also ANI, CALLING NUMBER DISPLAY, ISDN and SIGNALING SYSTEM 7.
Calling Line Identification Facility A service feature, provided by a network, that enables a called terminal to be notified by the network of the address from which the call has originated. See CALLING LINE ID.
Calling Number Display Your phone has a LCD (Liquid Crystal Display) or LED (Light Emitting Diode) display. When your phone rings, it will show which telephone number (internal or external) is calling you. Some phone systems allow you to add the person's name to the calling number display. See also ANI and CALLER ID.
Calling Party The person who makes (originates) the phone call.
Calling Party Camp-on A feature that enables the system to complete an access attempt in spite of temporary unavailability of transmission or switching facilities. Systems that provide this feature monitor the system facilities until the necessary facilities become available, and then proceed to complete the requested access. Such systems may or may not issue a system blocking signal to let the caller know of the access delay.
Calling Party Identification A telephone company service which tells the person being called the number and sometimes the name of the person calling them. They can then decide to answer or not answer it. See ANI, which stands for Automatic Number Identification.
Calling Party Number Information element that identifies the number of the originating party. An AIN term.
Calling Party Subdividers Information element that is passed transparently by the SPCS (if certain conditions are met) and can be used to further identify the originating party. An AIN term.
Calling Pattern Telecommunications managers are great at looking at phone bills, seeing patterns and smelling out calls that don't fit into those calling patterns. A simple example: zillions of calls over a weekend to Pakistan. The company doesn't do business with Pakistan. Nor does anyone work on the weekend.
Calling Sequence A sequence of instructions together with any associated data necessary to perform a call.
Calling Tone See CNG.
CallPath IBM's telephone system link to IBM's computers. See CALLPATH CALLPATH COORDINATOR, CALLPATH SERVICES ARCHITECTURE, CALLPATH CICS, and CALLPATH HOST.
CallPath CICS Enabling software that connects your telephone systems with your IBM 370 or 390 (i.e. the mainframe version of CallPath/400, which works on the AS/400 platform).
CallPath CallCoordinator CallCoordinator is IBM's integrated call management application that uses CallPath Services APIs to integrate data processing applications with telephone systems. IBM has versions of CallCoordinator for MVS CICS, OS/2 and Windows workstations. CallCoordinator provides features such as Intelligent Answering (based on ANI, DNIS, or Calling Line ID), Coordinated Voice and Data Transfer, Consultation (both voice and data), Conferencing (both voice and data), Transfer Load balancing between a single or multiple telephone systems, Outbound dialing, Event logging for Management Information Reporting, Personal Dialing Directory (Windows Only), Personal telephony facilities (answer phone, disconnect, transfer, etc.), Integration with CallPath DirectTalk/2 and CallPath DirectTalk/6000, Customizable Application Programming Interfaces. CallCoordinator integrates with existing 3270 or 5250 applications, and on the workstation versions, has the ability to communicate with existing applications via Dynamic Data Exchange or standard LAN communications protocols (such as TCP/IP). See CALLPATH SERVICES ARCHITECTURE.
CallPath Services Architecture CSA is IBM's architecture that defines the protocols for communication between computers and telephone switches. CallPath Services Architecture, announced in 1991, provides an Application Programming Interface (API) that enables a call management application to interact with telephone systems, with little regard to the protocols or communications interface provided by the telephone system. The idea is that with CallPath a call will arrive at a computer terminal simultaneously with the database record of the caller. And such call and database record can be transferred simultaneously to an expert, a supervisor, etc. CallPath has especial value in telephone call centers. As of writing, IBM provided connectivity to PBXs (AT&T Definity Generic 3, Northern Telecom Meridian 1, ROLM 9751, Siemens Hicom, NEC), central office switches (AT&T 5ESS and Northern Telecom DMS-100), and has agreements with other PBX manufactures that provide CallPath connectivity for Aspect, Ericsson, SDX, Telenorma, and Cortelco. IBM's CallPath products provide support for locally attached applications and client/server applications. IBM has CallPath APIs available for mainframes, minicomputers and workstations, in particular IBM System 390 and ES9000, AS/400, RISC System/6000, OS/2 workstations, Windows workstations, Sun Solaris, HP UX, and SCO UNIX workstations. See OPEN APPLICATION INTERFACE and DIRECTTALK.
Callpower A Rockwell ACD term. An integrated voice and data workstation for use in combining ACD capabilities with host computer database management.
CallWare CallWare is a company in Salt Lake City, UT, which makes computer telephony software that runs on the Novell NetWare operating system. CallWare software includes voice mail, autoattendant, IVR database lookup, etc.
CAM 1. Call Applications Manager. The name of the Tandem software interface which provides the link between a call center switch telephone switch (either a PBX or an ACD) and all Tandem NonStop (fault tolerant) computers. CAM supports most major PBXs and automatic call distributors.
2. Computer-Aided Manufacture. The actual production of goods implemented and controlled by computers and robots. Often used in conjunction with CAD. Only a few factories are completely automated. Usually, there is some human intervention in the actual construction of the product, often to make sure a part is placed in the robot correctly.
3. Controlled Attachment Module. Intelligent Token-Ring hub.
CAMA Centralized Automatic Message Accounting. See CAMA/LAMA.

NEWTON'S TELECOM DICTIONARY

CAMA/LAMA Centralized Automatic Message Accounting/Local Automatic Message Accounting. Specific versions of AMA in which the ticketing of toll calls is done automatically at a central location for several COs (CAMA) or only at the local office for that office's subscribers.

Camcorder A camera and a video recording system packaged as a whole.

Cameo Macintosh-based personal videoconferencing system, announced by Compression Labs in January of 1992. Developed jointly with AT&T and designed to work over ISDN lines and, most recently, Ethernet LANs. The Cameo transmits 15 fps of video and needs an external handset for audio.

Camp-on You're calling a telephone an extension or you want to transfer a call to a phone but it's busy. This telephone system feature will allow you to lock the call you're trying to transfer onto the line that's busy. When it becomes free, the phone will ring and the "camped-on" call will be connected automatically.

Campus Backbone Wiring between buildings.

Campus Environment An environment in which users — voice, video and data — are spread out over a broad geographic area, as in a university, hospital, medical center, prison. There may be several telephone systems. There may be several LANs on a campus. They will be connected with bridges and/or routers communicating over telephone, microwave or fiber optic cable.

Campus Subsystem The part of a premises distribution system which connects buildings together. The cable, interbuilding distribution facilities, protectors, and connectors that enable communication among multiple buildings on a premises.

Cancel By touching the "cancel" button on a phone system you're telling the phone system to ignore the last command you gave it. That command might have been transfer, hold, park, etc. The "cancel" button is often mistakenly confused with the "release" button. The "release" button acts the same as hitting "Enter" on a computer system, i.e. it tells the system to go ahead and do what you just told it to do, no matter how stupid your command. In short, "Cancel" means kill the last command. You use when you make a mistake. "Release" means "Enter" — Do it and do it now.

Cancel Call Waiting On a touchtone phone in North America, you typically can cancel the feature, Call Waiting, by touchtoning *70.

Cannibalize To devour a phone system by stripping parts from it to repair another system. A common technique for maintaining equipment whose original manufacturer no longer supplies parts. Before you cannibalize, check out the monthly publication Telecom Gear. That publication lists sources of secondary telecom equipment. Good stuff, too.

CAP 1. Competitive Access Provider. An alternative, competitive local exchange carrier such as MCI Metro and MHS.
2. Customer Administration Panel. A simplified alternative to CAC, which is a Customer Administration Center. These are AT&T words.
3. Computer Aided Professional Publishing. The computerization of professional publishing (as opposed to desktop operations), including true color representation of the layout on the workstation screen.
4. Cellular Array Processor.

Cap'N Crunch see CAPTAIN CRUNCH.

Capacitance The capacity of a medium (wire, cable, resistor, bus) to store an electrical charge. Capacitance is measured in farads.

Capacitive Coupling The transfer of energy from one circuit to another by virtue of the mutual capacitance between the circuits. The coupling may be deliberate or inadvertent. Capacitive coupling favors transfer of higher frequency components, whereas inductive coupling favors transfer of lower frequency components.

Capacity 1. The information carrying ability of a telecommunications facility. What the "facility" is determines the measurement. You might measure a date line's capacity in bits per second. You might measure a switch's capacity in the maximum number of calls it can switch in one hour, or the maximum number of calls it can keep in conversation simultaneously. You might measure a coaxial cable's capacity in bandwidth.
2. The measure of the amount of electrical energy a condenser can store up. The unit of capacity is the farad.

Capacity Transfer Control A Northern Telecom term for a feature which permits single allocation of capacity to be shared among members in a digital switched broadcast connection. For teleconferencing, for instance, a conference leader can transfer transmission capacity among the digital ports in the circuits. 95% of such transfers will take place within 10 seconds.

Capcode A capcode is a four or seven digit number on either side or rear of the casing of a pager, the type you wear on your belt. This number is a paging system necessity to know how to generate the right sequence of tones to alert the pager.

CAPS 1. Code Abuse Prevention System.
2. Competitive Access Providers to the local telephone network i.e., Teleport or Metropolitan Fiber System.

Capsizing When downsizing, rightsizing and upsizing fail. Contributed by Fred Schindler of IBM.

Capstan 1. A flangeless pulley used to control speed and motion of magnetic tape through a recorder or playback unit.
2. A rotating drum or cylinder used for pulling cables by exerting traction upon a rope or pull line passing around the drum.

CAPTAIN Character And Pattern Telephone Access Information Network System. A form of videotext developed in Japan and operated through the public switched telephone network. Displays are on a TV set. It's interactive.

Captain Crunch At one point in the 1960s, a breakfast cereal had a promotion. It was a toy whistle. When you blew the whistle, it let out a precise 2,600 Hz tone. If you blew that whistle into the mouthpiece of a telephone after dialing any long distance number, it terminated the call as far as the AT&T long distance phone system knew, while still allowing the connection to remain open. If you dialed an 800 number, blew the whistle and then touchtoned in a series of tones (called MF tones) you could make long distance and international calls for free. The man who discovered the whistle was called John Draper and he picked up the handle of Cap'n Crunch in the nether world of the late 1960s phone phreaks. A marvelous account of the exploits of phone phreaks was published in the October 1971 issue of Esquire Magazine. That article described how the Cap'n would call himself (he needed two lines) — choosing to route the connection through Tokyo, India, Greece, South Africa, South America, London, New York and California — to make his second phone next to him ring. He'd have a wonderful time talking to himself, albeit with a round-the-world delay (despite the speed of light) of as long as 20 seconds. Later, AT&T closed the loophole Cap'n Crunch had discovered. AT&T turned from in-band signaling to out-of-band signaling. Cap'n Crunch's legacy (he got put in jail four times during the 1970s) is System Signaling 7, a system of immense benefit to us all. See 2600 TONE, MULTI-FREQUENCY SIGNALING and SIGNALING SYSTEM 7.

Captive Effect An effect associated with the reception of frequency-modulated signals in which, if two signals are received on or near the same frequency, only the stronger of the two will appear in the output. The complete suppression of the weaker carrier occurs at the receiver limiter, where it is treated as noise and rejected. Under conditions where both signals are fading randomly, the receiver may switch from one to the other.

Captive Screw Let's say you have a couple of screws at the front of an industrial grade computer. You unscrew the two screws and a panel pops down. You can then get access to something inside the computer. The screw, however, doesn't leave the computer. No matter how much you unscrew it, you cannot remove the screw. It's "captive."

Capture Effect An effect associated with the reception of frequency-modulated signals in which, if two signals are received on or near the same frequency, only the stronger of the two will appear in the output. The complete suppression of the weaker carrier occurs at the receiver limiter, where it is treated as noise and rejected. Under conditions where both signals are fading randomly, the receiver may switch from one to the other.

Capture Ratio The ability of a tuner or receiver to select the stronger of two signals at or near the same frequency. Expressed in decibels, the lower the figure, the better.

Car Phone The type of cellular phone that's installed in a vehicle. There are four types of cellular phones being sold today — mobile, transportable, portable and handheld. A car phone (also called a mobile unit) is attached to the vehicle, its power comes from the vehicle's alternator (or battery if the car if not running) and the car phone has an external antenna, which works best if it's mounted in the middle of the highest point of the car and wired directly with no breaks in the wire. Many window-mounted antennas have a break in their wiring. The wiring ends at the inside. There is no electrical connection between the inside of the window and the antenna glued onto the outside of the window. The "connection" is done through signal radiation. In North America, the car phone transmits with a standard three watts of power.

Carbon Block A device for protecting cable from lightning strikes. The carbon block consists of two electrodes spaced so that any voltage above the design level is arced from line to ground. Carbon block protectors are used commonly in both local customer offices and central offices. They are effective, but can be destroyed if high voltage is directly applied — as in a direct strike by lightning. A more expensive, but more effective method of protection is the gas tube. These are glass capsules that are connected between the circuit and the ground. When a voltage higher than the design voltage strikes the line, the gas ionizes and conducts the excess voltage to ground. When the voltage is gone, the protector restores itself to normal. Gas tubes, however, take a tiny time to ionize. This may not be fast enough for very sensitive things, like PBX circuit cards. So gas tube protectors are often equipped with diodes, which clamp the interfering voltage to a safe level until the gas tube ionizes.

Carbon Fiber A strong synthetic material that is low in mass with excellent damping characteristics, used in the manufacture of tonearms.

Carbon Rheostat A rheostat using carbon as the resistance material. See RHEOSTAT.

Carbon Transmitter The microphone of an telephone set from yesteryear which uses carbon granules and a diaphragm. The diaphragm responds to our voice and varies the pressure on the granules and hence, their resistance. If your carbon mike isn't working well, the humidity has got to it, tap it lightly on your desk. The carbon granules will line up and it will work much better. Carbon microphones are very reliable but are being increasingly replaced with more sensitive electret microphones.

Card Authorization Center CAC. A computer directly linked to MCI switches for authorization and determination of billing center ID for MCI card calls.

Card Cage A frame in a telephone system or computer for mounting circuit cards, power supply, backplane and other equipment.

Card Dialer A device attached to a telephone which accepts a special plastic card and then automatically dials the number on the card as indicated by the holes punched in it. A card dialer is now obsolete except for unusual applications, like systems whereby you carry your card with you and use it as a security device.

Card Services The software layer above Socket Services that coordinates access to PCMCIA cards, sockets and system resources. Card Services is a software management interface that allows the allocation of system resources (such as memory and interrupts) automatically once the Socket Services detects that a PC Card has been inserted. This is called "hot swapping." The idea is that you can slide PCMCIA cards in and out of PC at will and your Socket and Card services will recognize them and respond accordingly. It's a great theory. In practice, it doesn't work because certain cards, like network cards, simply can't be connected and disconnected at will. Socket Services is a BIOS level software interface that provides a method for accessing the PCMCIA slots of a computer. Card Services is a software management interface that allows the allocation of system resources (such as memory and interrupts) automatically once the Socket Services detects that a PC Card has been inserted. Both of these specifications are contained in the PCMCIA Standards document. You do not need either Socket or Card Services to successfully use PCMCIA cards in your desktop or laptop. You simply need the correct device drivers and the proper memory exclusions. See PCMCIA, SOCKET SERVICES and SLOT SIZES.

Card Slot A place inside a phone system or computer into which you slide a printed circuit board. See BOARD.

CardBus The new name for the computer bus into which you slide PCMCIA cards. These cards are increasingly known as CardBus. The new CardBus, which will combine the PCI bus, will have a 32-bit interface and support 132 Mbps. The CardBus specification is the successor to the previous PC Card standard. See PCMCIA.

Caret The symbol which is found above 6 on most keyboards. Also used to indicate the "Ctrl" key in some instruction manuals.

CAROT Centralized Automatic Reporting On Trunks. A test and maintenance facility associated primarily with electronic toll switching systems like the AT&T Communication's #4-ESS. CAROT is a computerized system that automatically accesses and tests trunks for a maximum of fourteen offices simultaneously. It enables rapid routine testing of all trunks to ensure quick identification of faults and potential failures.

NEWTON'S TELECOM DICTIONARY

Carpal Tunnel Syndrome Carpal tunnel syndrome is a serious disorder of the arm caused by fast, repetitive work, such as typing without support for your wrists or with insufficient time for rest. In carpal tunnel syndrome, the tendons passing through the wrist bones swell and press on the median nerve. Surgery to take pressure off the nerve can relieve numbness and pain, but it's not always effective and many victims remain permanently disabled. The best prevention is using a wrist rest and undertaking specific exercises. There is a good book on the subject — Conquering Carpal Tunnel Syndrome by Sharon J. Butler, New Harbinger Publications, Oakland, CA.

Carriage Return By hitting this key, the printing head or the cursor on your screen will return to the left hand margin. Usually hitting a Carriage Return or the "Enter" key includes a line feed, i.e. the paper will move up one line or the cursor will drop down one line. "Usually" does not mean always. So check. You can usually correct the problem of not having a line feed with a carriage return by moving a dip switch on the printer, changing one of the parameters of the telecommunications software program (the part where it says something about auto linefeed) or changing the computer's operating system (by doing a "Config" or the like). In most microcomputers, a Carriage Return is equivalent to a "Control M," or ASCII 13. A line feed is a "Control J".

Carried Load The traffic that occupies a group of servers on a LAN.

Carried Traffic The part of the traffic offered to a group of servers that successfully seizes a server on a LAN.

Carrier 1. A company which provides communications circuits. Carriers are split into "private" and "common." A private carrier can refuse you service. A "common" carrier can't. Most of the carriers in our industry — your local phone company, AT&T, MCI, US Sprint, etc. — are common carriers. Common carriers are regulated. Private carriers are not.
2. An electrical signal at a continuous frequency capable of being modified to carry information. It is the modifications or the changes from the carrier's basic frequency that become the information carried. Modifications are made via amplitude, frequency or phase. The process of modifying a carrier signal is called modulation. A carrier is modulated and demodulated (the signal extracted at the other end) according to fixed protocols. Some of the wideband (i.e. multi-frequency) circuits are also called "carriers." T1, which typically has 24-channel PCM voice circuits, is known as a carrier system.

Carrier Band The range of frequencies that can be modulated to carry information on a specific transmission system. See also CARRIERBAND.

Carrier Bypass A long distance phone company provides a direct link between its own switching office and a customer's office, thus bypassing the local phone company. Bypass is done to save the customer or the long distance company money. Bypass is also done to get service faster. Sometimes the local phone company simply can't deliver fast enough.

Carrier Common Line Charge Also called Access Charge. The charge which long distance phone companies pay to local phone companies to complete their long distance calls. These charges typically are much more than the local phone company gets from its normal business and residential users for completing local phone calls of the same distance. High access charges are designed to help the local phone company financially. That they do. They also encourage bypass and may, in the long term, be self-defeating.

Carrier Detect Circuitry Electronic components which detect the presence of a carrier signal and thus determine if a transmission is about to happen. Used in modems.

Carrier Frequency The frequency of a carrier wave. The frequency of an unmodulated wave capable of being modulated or impressed with a second (information-carrying) signal. In frequency modulation, the carrier frequency is also referred to as the "center frequency."

Carrier Identification Codes CIC. Three digit numbers used by end-user customers to reach the services of interexchange carriers through equal access arrangements. The primary carrier of choice is reached by dialing "1" plus the area code and called party number. Secondary IX carriers can still be reached by dialing 10 plus the CIC assigned to the carrier desired. CIC numbers are used to dial around the carrier presubscribed to the calling telephone (e.g. 10-XXX or 950-0XXX where XXX is the CIC).

Carrier Leak The unwanted carrier remaining after carrier suppression in a suppressed carrier transmission system.

Carrier Liaison Committee CLC. A committee formed to help industry participants work together to resolve the issues of implementing 800 Portability. CLC is sponsored by the Exchange Carriers Standards Association (ECSA) and is comprised of the LECs (local exchange carriers), long distance carriers and users of 800 service.

Carrier Loss In T-1, carrier loss means too many zeros. A carrier loss in T-1 is said to occur when 32 consecutive zeros appear on the network. Carrier is said to return when the next 1 is detected.

Carrier Noise Level The noise level resulting from undesired variations of a carrier in the absence of any intended modulation.

Carrier Power (of a Radio Transmitter) The average power supplied to the antenna transmission line by a transmitter during one radio frequency cycle taken under the condition of no modulation. Does not apply to pulse modulation or frequency-shift keying.

Carrier Provided Loop MCI is responsible for ordering coordinating, maintaining, and billing the local loop.

Carrier Select Keys Buttons at the bottom of a payphone used to choose a long distance carrier.

Carrier Selection As a result of Judge Greene's Modified Final Judgment which lead to the breakup of the Bell System, most local phone companies must offer their customers (business and home) the opportunity to select which long distance company they would like to be use on a "primary" basis. That means when you dial 1+ (one plus) you get that carrier. To use any other long distance company you have to dial more digits, e.g. 1-0288 (for AT&T). See NANP.

Carrier Sense In a local area network, a PC or workstation uses its network card to detect if another station is transmitting. See CSMA.

Carrier Sense Multiple Access CSMA. In local area networking, CSMA is a way of getting onto the LAN. Before starting to transmit, personal computers on the LAN "listen" to make sure no other PC is transmitting. Once the PC figures out that no other PC is transmitting, it sends a packet and then frees the line for other PCs to transmit. With CSMA, though stations do not transmit until the medium is clear, collisions still occur. Two alternative versions (CSMA/CA and CSMA/CD) attempt to reduce both the number of collisions and the severity of their impact. See CSMA/CA and CSMA/CD.

NEWTON'S TELECOM DICTIONARY

Carrier Sense Multiple Access/Collision Avoidance CSMA/CA. A protocol that requires the PC to sense if another PC is transmitting. If not, it begins transmitting. Under CSMA/CA, a data station that intends to transmit sends a jam signal; after waiting a sufficient time for all stations to pick up the jam signal, it sends a transmission frame; if while transmitting, it detects another station's jam signal, it stops transmitting for a designated time and then tries again.

Carrier Sense Multiple Access/Collision Detection A network control scheme. It is a contention access control scheme. It "listens" for conflicting traffic to avoid data collisions. The Ethernet LAN uses CSMA/CD, then waits a small amount of time and then tries again. See CSMA/CD and ETHERNET.

Carrier Shift 1. A method of keying a radio carrier for transmitting binary data or teletypewriter signals, which consists of shifting the carrier frequency in one direction for a marking signal and in the opposite direction for a spacing signal.
2. In amplitude modulation, a condition resulting from imperfect modulation whereby the positive and negative excursions of the envelope pattern are unequal, thus effecting a change in the power associated with the carrier. There can be positive or negative carrier shift.

Carrier Signal A continuous waveform (usually electrical) whose properties are capable of being modulated or impressed with a second information-carrying signal. The carrier itself conveys no information until altered in some fashion, such as having its amplitude changed (amplitude modulation), its frequency changed (frequency modulation) or its phase changed (phase modulation). These changes convey the information.

Carrier Synchronization In a radio receiver, the generation of a reference carrier with a phase closely matching that of a received signal.

Carrier System A system where several different signals can be combined onto one carrier by changing some feature of the signals transmitting them (modulation) and then converting the signals back to their original form (demodulation). Many information channels can be carried by one broadband carrier system. Common types of carrier systems are frequency division, in which each information channel occupies an assigned portion of the frequency spectrum; and time division, in which each information channel uses the transmission medium for periodic assigned time intervals.

Carrier Terminal The modulation, demodulation and multiplex equipment used to combine and separate individual channels at the ends of a transmission system.

Carrier To Noise Ratio CNR. In radio receivers, the ratio, expressed in decibels, of the level of the carrier to that of the noise in the receiver bandwidth before any nonlinear process such as amplitude limiting and detection takes place.

Carrier Wave The radio frequency wave generated at a transmitting station for the purpose of carrying the modulated or audio frequency wave.

Carrierband Same as single-channel broadband. See also CARRIER BAND.

CARS Cable Television Relay Service Station. A fixed or mobile station used for the transmission of television and related audio signals, signals of standard and FM broadcast stations, signals of instructional television fixed stations, and cablecasting from the point of reception to a terminal point from which the signals are distributed to the public.

Carterfone A device for connecting a two-way mobile radio system to the telephone network invented by Thomas Carter. It was electrically connected to the base station of the mobile radio system. Its electrical parts were encased in bakelite. When someone on the radio wanted to speak on a "landline" (the phone system), the base station operator would dial the number on a separate phone then place the telephone handset on the Carterfone device. The handset was acoustically, not electrically, connected to the phone system. No more than 4,000 Carterfones were ever installed, yet the Bell System thought they were the most dangerous device ever invented. Tom Carter died in Gun Barrel, TX where he lived, in the early part of 1991. He died not a rich man. See CARTERFONE DECISION.

Carterfone Decision In the summer of 1968 the FCC said that the Carterfone and other customer phone devices could be connected to the nation's phone network — if they were "privately beneficial, but not publicly harmful." The Carterfone decision was a landmark. It allowed the connection of non-telephone company equipment to the public telephone network. This decision marked the beginning of the telephone interconnect business as we know it today. The Carterfone decision made a lot of lawyers rich before all the rules on connection to the network got cleared up, and finally codified in something called Part 68 of the FCC's Rules. See CARTERFONE, NATA and NETWORK HARM.

Cartridge 1. A device which holds magnetic tape of some kind.
2. A device to translate (transduce) stylus motion to electrical energy in a phonograph. It comes in three basic types — moving magnetic coil, induced magnet and ceramic. A phono cartridge is also call a pickup. Most record players use ceramic cartridges because they have higher output than the three magnetic types and can work with a less powerful (i.e. cheaper) amplifier.

CAS 1. Centralized Attendant Service. One group of switchboard operators answers all the incoming calls for several telephone systems located throughout one city. CAS is used by customers with several locations in the same geographic area, i.e. retail stores, banks.
2. Communicating Applications Specification. A high-level API (Application Programming Interface) developed by Intel and DCA that was introduced in 1988 to define a standard software API for fax modems. CAS enables software developers to integrate fax capability and other communication functions into their applications.
3. Channel Associated Signaling: A form of circuit state signaling in which the circuit state is indicated by one or more bits of signaling status sent repetitively and associated with that specific circuit.

Cascade 1. To connect the output of a device into the input of another device, which then may in turn be connected to another device. Image the organization of a company. At the top is the president. Reporting to him are three vice presidents. Reporting to each of these vice presidents are five directors. Reporting to each of these directors are five managers. If each of these positions were a piece of network gear, you'd have a classic cascaded topology.
2. A Windows term. When windows cascade, they are arranged in an overlapping pattern so that the title bar of each window remains visible.

NEWTON'S TELECOM DICTIONARY

Cascade Amplification The method of successively using two or more vacuum tubes for amplification at radio, intermediate or audio frequencies.

Cascaded Stars Local area network topology in which a centralized multiport repeater serves as the focal point for many other multiport repeaters.

Cascaded Topology See CASCADE.

Cascading Faults Faults that cause other faults. Typically faults in a network causing other faults.

Cascading Menu A Windows term. A menu that is a submenu of a menu item. Also known as a hierarchical menu.

Cascading Notification A feature of some sophisticated voice mail systems. Let's say someone leaves a message for you in your voice mail box. Your voice mail system then automatically goes out to find you, i.e. to notify you. It may start by lighting your message light, calling your home phone number, calling your cellular phone, calling your beeper, etc. I like this feature because when I want you, I want you. And a little mechanized help is much appreciated. I first saw the feature in Macrotel's MVX voice mail series.

CASE Computer Aided Software Engineering. CASE is a new, faster, more efficient way of writing software for some applications. The idea with CASE is to sketch out relations between databases, events, and options and then have the computer write the code.

CASE Tools These tools provide automated methods for designing and documenting software programming. Computer-aided software engineering (CASE) sketches relations between databases, events, and options. It then provides a language in which the computer writes the code, letting programmers develop applications faster. It's new and seems to work in limited instances.

Case Method A traditional way of load testing computer telephony systems, the case method involves gathering many individuals together in a room full of telephones along with several cases of an appropriate libation (frequently beer), and using these individuals to simulate real users calling into (or being called by) the computer telephony system. Case method testing usually continues until all the cases have been consumed, the testing is completed, it becomes too late in the evening to continue, or the perspective of the gathered individuals becomes too subjective to be of use any longer to those conducting the test. This definition courtesy of Steve Gladstone, author of the book Testing Computer Telephony Systems, available from 212-691-8215.

Case Sensitive This means that uppercase letters must be typed in uppercase on your keyboard, and that lowercase letters must be typed in lowercase. It is important to key in your data in the exact combination of upper or lower case characters. Inputting in the wrong case could make your entry invalid for some fields (for example, password). DOS and Windows are much less case sensitive than Unix, for example.

Cassegrain Antenna An antenna in which the feed radiator is mounted at or near the surface of a concave main reflector and is aimed at a convex secondary reflector slightly inside the focus of the main reflector. Energy from the feed unit illuminates the secondary, reflects it back to the main reflector, which then forms the desired forward beam. This technique is adapted from optical telescope technology and allows the feed monitor radiator to be more easily supported.

Cassette Tape A slow, inefficient method of storing and retrieving data which uses the same technology as audio cassettes — like the Sony Walkman. Some PBXs use cassette tape to backup their user programming and database.

Castellation A series of ribs and metallized indentations that defines edge contact regions.

CAT 1. An AT&T Merlin term. Call Accounting Terminal. A stand-alone unit with a built-in microprocessor and data buffer that provides simple call accounting at a relatively low cost.
2. Shortened way of saying "Category," as in Cat 1 cabling. See CATEGORY 1 through 5 and CATEGORY OF PERFORMANCE.

Category 1 See CATEGORY OF PERFORMANCE.

Category 2 There's no such thing as Category 2. See CATEGORY OF PERFORMANCE for a fuller explanation.

Category 3 Cabling and cabling components designed and tested to transmit cleanly 16 megahertz of communications. Used for voice and data/LAN traffic to 10 megabits per second. Cat three cabling is the most common type of cabling used for voice telephone cabling. Category 3 technical specifications are defined by FCC Part 68, ANSI/EIA/TIA-568, TIA TSB-36 and TIA TSB-40. Category 3 safety requirements are defined by UL 1459 (Telephone), UL 1863 (Wire and Jacks) and NEC 1993, Article 800-4. See CATEGORY OF PERFORMANCE for a fuller explanation.

Category 4 Cabling and cabling components designed and tested to transmit cleanly 20 megahertz of communications. Few people today buy Category 4 cabling and cabling components. Category 4 technical specifications are defined by FCC Part 68, EIA/TIA-568, TIA TSB-36, and TIA TSB-40. Category 4 safety requirements are defined by UL 1459 (Telephone), UL 1863 (Wire and Jacks) and NEC 1993, Article 800-4. See CATEGORY OF PERFORMANCE for a fuller explanation.

Category 5 Cabling and cabling components designed and tested to transmit cleanly 100 megahertz of communications. Used for voice and data/LAN traffic to 100 megahertz. Cat 5 is now the most common cabling being installed for data / LAN use. Increasingly, Cat 5 cabling is being installed for data and voice use. Increasingly, Cat 5 is the cabling of choice for forward-looking offices. Category 5 technical specifications are defined by FCC Part 68, EIA/TIA-568, TIA TSB-36, and TIA TSB-40. Category 4 safety requirements are defined by UL 1459 (Telephone), UL 1863 (Wire and Jacks) and NEC 1993, Article 800-4. See CATEGORY OF PERFORMANCE for a fuller explanation.

Category Of Performance As we try and push more and data faster and faster down a pair or two of wires, so the quality of the wires and the components they connect to has become increasingly important. You can't shove 100 million bit per second and down junky phone lines. As a result, the telecommunications industry has defines cabling and cabling component standards. The idea is that if your stuff conforms to the standard, users will be able to achieve the data rates and reliability they want. Some standards specify physical characteristics, such as thickness of cable, plastic material used in the outer jacket, etc. These "Category of Performance" standards, which the telecommunications industry has adopted, in the main do not specify materials. They specify tests which the cabling and cabling components must pass. There were originally five categories of tests. But now there are effectively only two categories that anyone buys — Category 3 and Category 5. In simple terms, if you want voice and data to 10 megabits per second, i.e. standard Ethernet, then buy Category 3. If you are looking to the future when you may be transmitting 100 million bits per second on your local area network, then you should buy Category 5. Today the difference in price is about 20% to 30%. In other words, Cat 5 is 20% to 30% more expensive than Cat 3. In my unhumble opinion, all LAN cable going in today should be Cat 5, while all voice cable going in should be Cat 3, which will happily support ISDN BRI, when that happens in the not too distant future.

Let's talk about how they measure Cat 3 and Cat 5. First, you should understand that all the tests are self-certifying, which means that while there are standards, each manufacturer is himself responsible for conforming. No one will put a manufacturer in jail if his Cat 5 stuff doesn't perform to Cat 5 standards. The only thing likely to happen to the manufacturer is that the world will find out his stuff is garbage and he'll go broke.

The concept of the test is simple. The test for cable and components is a swept frequency test. Cat 3 must pass all signals from one through 16 megahertz. (Note megahertz, not megabits). In Cat 5, it must pass signals from one through 100 megahertz. Cat 3 and Cat 5 cables and components are designed to support any applications intended to operate over those frequencies. That includes all forms of modulation of any carrier waves within those frequencies and all forms of digital pulses. Handling digital pulses is more difficult since digital pulsing typically uses a much broader range of frequencies (than simply modulating one or two carrier waves as in AM or FM modulation). Digital pulsing needs a broader range of frequencies in order to get its edges more square. Interestingly, both Cat 3 and Cat 5 cabling are the same thickness, namely 24 gauge. The twist structure on Cat 5 is tighter. Insulation is better. Connecting hardware is definitely different. Below is Harry's quick rule of thumb for what you should buy based on what you want to transmit, as against the swept frequency test:

Category	Data Rata Supported
1,2	Voice (Obsoleted in early 1995)
3	10 Mbps
4	20 Mbps
5	100 - 155 Mbps
150 OHM STP-A	Full bandwidth video

If you make the following mistakes, your nice new Cat 5 cables probably won't work at 100 Mbps.

NEVER!:
- Use Staples
- Use Tie Wraps - other than VERY loosely to hold a bundle in place. Velcro is a better idea.
- Untwist the wire before you punch it down on the block or terminate it on a jack - the twists should be right up to the termination!
- Strip more jacket off the wire than is needed to terminate it.
- Pull too hard on the cable - especially around a corner.
- Pull on a cable to straighten out a kink or loop. Always go back and untwist it!
- Stuff too many cables in too small a conduit. See COMMERCIAL BUILDING TELECOMMUNICATIONS WIRING STANDARD.

Catenet A network in which hosts are connected to networks are interconnected by gateways (routers). The Internet is an example of a catenet.

Cathode The heated element which emits electrons in a vacuum tube. It may be a filament, or may be a separate element, heated by proximity to a filament. It is maintained at a negative potential in respect to the anode or plate. Cathodes have other applications, also.

Cathode Ray The beam of electrons emitted by a cathode. See CATHODE RAY TUBE.

Cathode Ray Tube CRT. A TV screen. A CRT is a tube of glass, used in television, oscilloscope and computer terminals, from which air has been removed (i.e. vacuum tube). At the back of the CRT is an electron gun which directs an electron beam to the front of the tube. The inside front of the tube has been coated with fluorescent material which reacts to and lights up once the electron beams hit. CRTs are very reliable if they are vented, since the electron gun gets hot. CRTs have a "memory." They will memorize what's been left on their screen for a while, i.e. the image is burned into the screen. And you'll see it even though the screen is turned off. In short, turn your screen off when you're not using it. Or run a "CRT-saving" program which varies the image on the screen.

CATI Computer Assisted Telephone Interviewing, a market research term for a call center based on the use of a computerized database.

CATLAS AT&T software standing for Centralized Automatic Trouble Locating and Analysis System. CATLAS is used as a maintenance tool for locating and diagnosing problems in AT&T electronic central offices.

CATS Consortium for Audiographics Teleconferencing Standards, San Ramon, CA. 510-831-4760. CATS describes itself as a non-profit corporation dedicated to promoting standards for this technology, which it describes as enhancing audioconferencing by allowing people at different sites to work together in real time to create, manipulate, edit, annotate and reference still images. Now called IMTC.

CATV Community Antenna TeleVision or CAble TeleVision. CATV is a broadband transmission facility. It generally uses a 75-ohm coaxial cable which simultaneously carries many frequency-divided TV channels. Each channel is separated by guard channels. See ADDRESSABLE PROGRAMMING and BROADBAND.

CAU 1. Northern Telecom term for Connection Arrangement Unit.
2. Controlled Access Unit. Intelligent Token-Ring hub.

Cavity A volume defined by conductor-dielectric or dielectric-dielectric reflective boundaries, or a combination of both, and having dimensions designed to produce specific interference effects (constructive or destructive) when excited by an electromagnetic wave.

CB Why 10-4, good buddy, that stands for Citizens Band. Also known as Children's Band, not because of Radio Shack's toy walkie talkies, but for the inane chatter that sometimes goes on in these channels. In short, CB is low-power (up to four WATTS permitted) public radio. You do not need permission from the FCC to transmit or receive at these frequencies. Thus CB's great popularity. CB went through a boom (perhaps a craze?), then it ran out of radio frequencies and public enthusiasm. Its original frequencies were 26.965 to 27.225 Mhz. Now the FCC's given it new frequencies — 462.55 to 469.95 MHz. These new frequencies are much better, clearer and less congested. If you buy a CB set, make sure you get one that operates in these higher frequencies. In some countries they use different frequencies. CB radio is not allowed in many countries, even some civilized countries, though it will obviously work there.

113

CBDS Connectionless Broadband Data Service: A connectionless service similar to Bellcore's SMDS defined by European Telecommunications Standards Institute (ETSI). In short, the European version of SMDS.
CBEMA Computer Business Equipment Manufacturers Association. A lobbying group created to protect the interests of its members.
CBF Computer Based Fax.
CBK Change BacK.
CBR Constant Bit Rate. CBR refers to processes such as voice and video that require a constant, repetitive or uniform transfer of information. The ATM Forum defines it as "an ATM service which supports a constant or guaranteed rate to transport services such as video or voice as well as circuit emulation which requires rigorous timing control and performance parameters."
CBTA Canadian Business Telecommunications Alliance. The largest organization of business telecom users in Canada.
CBX Computerized Branch eXchange. CBX is a registered trademark of the ROLM Corporation, Santa Clara, CA. Rolm is now owned by Siemens and IBM, but largely by Siemens. What CBX is to Rolm, PBX or PABX is to other companies. The term CBX has not received wide acceptance, except at Rolm.
CBX II A Rolm communications controller for larger systems. Two versions exist: the CBX II 8000 (16-bit CPU) and the CBX II 9000 (32-bit CPU).
CC 1. Call Control. A wireless telecommunications term. A term used to refer to circuit communications management.
2. Country Code. The portion of an international telephone number used to identify the country of the called party. That country code may be one, two or three digits. An international phone number consists of a Country Code (CC) and a NSN (National Significant Number). Until December 31, 1996, the CC can be up to three digits and the NSN up to 11 digits — for a total of no more than 12 digits. After December 31, 1996, the CC stays at up to 3 digits and the NSN goes to up to 14 digits, for a total of no more than 15 digits.
3. Carbon Copy. That's where it comes from. It refers to the person who received one of the copies of a memorandum or document. Now CC is used electronic mail to mean that someone will receive a copy of the electronic mail.
CCB Common Carrier Bureau. One of the largest divisions of the FCC, this Bureau regulates interstate telephone systems — licensing them, monitoring their charges, the conditions they offer service under, etc. The Common Carrier Bureau opened long distance communications to competition in America.
CCBM Came Clear By Magic. This term created (but not contributed) by Nynex. This is pseudo-technical lingo used by Nynex when they repair something that they broke (i.e. the problem was their fault), but they are unwilling to admit it was their fault. They say this "CCBM" to interconnect companies who are trying to get their customers' phone lines fixed.
CCC 1. Clear Coded Channel. A 64 kbps channel in which all 64 kbps is available for data.
2. Clear Channel Capability. The bandwidth of a data transmission path available to end users after control and signaling bits are accounted for.
3. Communications Competition Coalition. Lobbying organization established to encourage competition in telecommunications in Canada.
CCD 1. Charge Coupled Device. The "eyes" of a scanner or "digital camera." CCDs are small electronic devices with arrays of light-sensitive elements. The number of these elements and the width determine the scanner's or camera's resolution. Light is bounced off the image onto the CCD, which translates the varying intensities of the reflected light into digital data. CCD technology is used also in "digital still cameras" such as the Sony Mavica. The small size of the array itself — approximately microchip size — and the high resolution — around 1,000 by 1,018 elements — of these cameras have greatly improved "image acquisition" capabilities and opened up new applications in manufacturing quality control and in medicine.
2. Change Coupled Device.
CCDN Corporate Consolidated Data Network. It is the name for IBM's main internal data communications network. It used to be managed by IBM. It's now managed by Advantis, an IBM spin-off company, which is majority-owned by IBM and the rest by Sears.
CCFL Cold Cathode Fluorescent Lamp. A technology several laptop computer manufacturers use to light their LCD screens.
CCIA Computer and Communications Industry Association. A trade organization of computer, data communications and specialized common carrier services companies headquartered in Arlington VA. It runs seminars, does lobbying and generally tries to take care of the common interests of its members. See CBEMA.
CCIR Comite Consultatif International des Radiocommunications (Consultative Committee on International Radiocommunications France). Also used to describe 625-line television system used primarily in Western Europe. The US system of TV is 525. The high definition TV is scheduled to be 1,125.
CCIR 601 An internationally agreed-upon standard for the digital encoding of component color television that was derived from the SMPTE RP125 and the EBU 324E standards. It uses the 4:2:2 sampling scheme for Y, U and V with luminance sampled at 13.5 MHz and chrominance (U and V components) sampled at 6.75 MHz. After sampling, 8-bit digitizing is used for each channel. These frequencies are used because they work for both 525/60 (NTSC) and 625/50 (SECAM and PAL) television systems. The system specifies that 720 pixels be displayed on each line of video. The D1 digital videotape format conforms to CCIR 601. See CCIR 656.
CCIR 656 The international standard defining the electrical and mechanical interfaces for digital TV operating under the CCIR 601 standard. It defines the serial and parallel interfaces in terms of connector pinouts as well as synchronization, blanking and multiplexing schemes used in these interfaces.
CCIS Common Channel Interoffice Signaling. A way of carrying telephone signaling information along a path different from the path used to carry voice. CCIS occurs over a separate packet switched digital network. CCIS is separate from the talk path. A special version of CCIS is called Signaling System #7. SS#7 is integral to ISDN. CCIS offers basically two benefits: first, it dramatically speeds up the setting up and tearing down of phone calls. Second, it allows much more information to be carried about the phone call than what is carried on in-band (old-fashioned) signaling. That information can include the calling number, a message, etc.
Signaling for a group of voice telephone circuits is done on CCIS by encoding the information digitally on one of the voice circuits. In the previous method of signaling — the one replaced by CCIS — multi-frequency tones were sent down the same talkpath and

the conversation would eventually travel. By taking the signaling information out of the talk path, the "phone phreak" community could no longer get free calls by using so-called "blue boxes" which duplicated the multi-frequency tones used by switching machines. CCIS is a much more efficient method of signaling, since it doesn't require a full voice grade channel just to check if the called party in LA is free and whether the call coming in from New York should be put through. See also COMMON CHANNEL SIGNALING, SYSTEM SIGNALING 7, ISDN and COMMON CHANNEL INTEROFFICE SIGNALING.

CCITT Comite Consultatif Internationale de Telegraphique et Telephonique, which, in English, means the Consultative Committee on International Telegraphy and Telephony. The CCITT is one of the four permanent parts of the International Telecommunications Union, the ITU, based in Geneva Switzerland. The scope of its work is now much broader than just telegraphy and telephony. It now also includes telematics, data, new services, systems and networks (like ISDN). The ITU is a United Nations Agency and all UN members may also belong to the ITU (at present 182), represented by their governments. In most cases the governments give their rights on their national telecom standards to their telecommunications administrations (PITs). But other national bodies (in the US, for example, the State Department) may additionally authorize Recognized Private Operating Agencies (RPOAs) to participate in the work of the CCITT. After approval from their relevant national governmental body, manufacturers and scientific organizations may also be admitted, as well as other international organizations. This means, says the ITU, that participants are drawn from the broad arena. The activities of the CCITT divide into three areas:
Study Groups (at present 15) to set up standard ("recommendations") for telecommunications equipment, systems, networks and services.
Plan Committees (World Plan Committee and Regional Plan Committee) for developing general plans for a harmonized evolution of networks and services.
Specialized Autonomous Groups (GAS, at present three) to produce handbooks, strategies and case studies for support mainly of developing countries.
Each of the 15 Study Groups draws up standards for a certain area - for example, Study Group XVIII specializes in digital networks, including ISDN. Members of Study Groups are experts from administrations, RPOAs, manufacturing companies, scientific or other international organizations - at times there are as many as 500 to 600 delegates per Study Group. They develop standards which have to be agreed upon by consensus. This, says the ITU, can sometimes be rather time-consuming, yet it is a democratic process, permitting active participation from all CCITT member organizations.
The long-standing term for such standards is "CCITT recommendations." As the name implies, recommendations have a non-binding status and they are not treaty obligations. Therefore, everyone is free to use CCITT recommendations without being forced to do so. However, there is increasing awareness of the fact that using such recommendations facilitates interconnection and interoperability in the interest of network providers, manufacturers and customers. This is the reason why CCITT recommendations are now being increasingly applied — not by force, but because the advantages of standardized equipment are obvious. ISDN is a good example of this.
CCITT has no power of enforcement, except moral persuasion. Sometimes, manufacturers adopt the CCITT specs. Sometimes they don't. Mostly they do. The CCITT standardization process runs in a four-year cycle ending in a Plenary Session. Every four years a series of standards known as Recommendations are published in the form of books. These books are color-coded to represent different four cycles. In 1980 the CCITT published the Orange Books, in 1984 the Red Books and, in 1988, the Blue Books. The CCITT is now more commonly called the ITU, after its parent. See CCITT STUDY GROUPS and CCITT V.XX below.
The CCITT has now been incorporated into its parent organization, the International Telecommunication Union (ITU). Telecommunication standards are now covered under Telecommunications Standards Sector (TSS). ITU-T (ITU-Telecommunications) replaces CCITT. For example, the Bell 212A standard for 1200 bps communication in North America was referred to as CCITT V.22. It is now referred to as ITU-T V.22.

CCITT Study Groups The CCITT operates as a series of groups considering specialist areas. Key study groups applicable to networking and communications are: Study Group VII responsible for terminal equipment for telematic services, including fax and higher level OSI standards; Study Group X covering Languages and methods for telecommunications applications; Study Group XI covering ISDN, telephone network including V-series Recommendations; Study Group XVIII covering digital networks including ISDN. See CCITT above and ITU-T.

CCITT V.XX A set of evolving telecom standards. For more on those standards, see under the letter "V."

CCIU Northern Telecom word for Conference Control Interface Unit.

CCMS A Rolm term. It stands for Call Center Management System.

CCO System DEC VAX and IBM mini/mainframe telecommunications management software from the folks at Telco Research Nashville, TN. Includes cost allocation, directory, inventory management, problem recovery, traffic statistics, trouble tickets, work orders, student resale, tenant resale, bill verification, cable and wire, and network optimization applications. Also has a PC-based graphical interface to Telco Research's mini/mainframe telemanagement system.

CCR 1. Customer Controlled Reconfiguration. An AT&T service that lets users make changes in their digital access and cross connect (DACS) network configurations at a DS0 Level in either real time or according to a preplanned schedule.
2. An ATM term. Current Cell Rate: The Current Cell Rate is an RM-cell field set by the source to its current ACR when it generates a forward RM-cell. This field may be used to facilitate the calculation of ER, and may not be changed by network elements. CCR is formatted as a rate.

CCS 1. Centi Call Seconds. One hundred call seconds or one hundred seconds of telephone conversation. One hour of telephone traffic is equal to 36 CCS (60 x 60 = 3600 divided by 100 = 36) which is equal to one erlang. CCS are used in network optimization. Lee Goeller calls CCS an obsolete traffic unit. He says "When given traffic in CCS, always divide by 36 immediately. It is not obvious that 5 trunks cannot carry 185 CCS, but you don't have to be a rocket scientist to know that you can't average 5.14 calls on a five trunk group." See also TRAFFIC ENGINEERING.
2. Common Channel Signaling: A form of signaling in which a group of circuits share a signaling channel. Refer to SS7.

CCS/SS7 A Bellcore term for Common Channel Signal/System Signaling 7. See SYSTEM SIGNALING 7.

CCS7 COMMON CHANNEL SIGNALING 7. See ISDN.
CCSA Common Control Switching Arrangement. A private network set up by AT&T for very large users and using parts of the public switched network. One important feature of a CCSA is that any user anywhere in a CCSA network can reach any other user by dialing only seven digits. Only very large customers subscribe to this service. It's expensive. AT&T has fewer than 100 customers.
CCSA ACCESS A PBX feature which allows a PBX user to get into a CCSA network. See CCSA.
CCT CONTINUITY CHECK TONE.
CCTV Closed Circuit TV.
CCU 1. Communication Control Unit. A processor, often a minicomputer, associated with a host mainframe computer that performs a number of communications-related functions. Compare with cluster control unit.
2. Camera Control Unit.
CD 1. Carrier Detect. CD is a signal generated by dial-up modem. CD indicates its connection status. If your CD light is on, then your modem is speaking to another modem.
2. Compact Disc. A 4.5 inch diameter disk containing digital audio or digital computer information, which can be played back and (now recorded) on a laser-equipped player. It was introduced by Sony and Philips in 1982. A compact disc originally came in only one flavor — read only. And most music tapes can only be listened to, not recorded to. For music it was a major breakthrough. It recorded music digitally (that is, coded as the zeros and ones of computer-speak) instead of trying to make an electrical copy of the sound waves themselves as devices like audio cassettes and LP records had. The Economist described the CD well. It said, "Instead of using a needle, the sound was plucked from the CD's surface by a tiny beam of laser light and then processed by a microcomputer. To the ear, the leap in performance between a compact disc and a long-playing record was even greater than the difference between color and black and white TV was to this eye."
Each month, a division of Ford Motor Co. publishes a CD-ROM inventory of 300,000 parts for its 2,400 dealers. A CD can typically hold up to 650 megabytes of information. That is the equivalent of 1,500 floppies or 250,000 pages of print. Most computer CD-ROM drives can play audio CD disks — if they have the software and the speakers. Audio CD players, though, cannot play computer CD-ROM discs. But most computer CD-ROM players can play CD audio discs. As CD-ROM have become more popular and their makers have tried to do more and more with them, so CD-ROM formats have proliferated and some are not compatible with each other. See CD-R, CD-ROM, CD-I, CD-V and WORM.
CD I See CD-I below.
CD UDF This format defines specifications for a unified logical file format for CD-Recordable. The new format, known as CD UDF, defines a common scheme of "packet writing" to assure interchangeability of CD-R (CD Recordable) discs and to greatly increase performance and flexibility when storing large and small files on recordable disc media. As a result, CD-Recordable drives can be integrated into computer systems to behave much like other removable disk storage products. www.osta.org
CD-Audio Sometimes called "Redbook audio," is the digital sound representation used by CD-ROMs. CD audio is converted to analog sound output within the CD-ROM drive. The sampling frequency for CD audio is 44.1 KHz.
CD-I Compact Disc Interactive. Geared toward home entertainment, the drive connects to a television.
CD-Plus A format for CDs created by Sony and PHilips Electronics which makes the multimedia track on CDs invisible to CD players. The problem: there are few enhanced CDs — discs for both CD player and CD-ROM drives. The reason: the format prior to CD-Plus puts the multimedia data on track one, which listeners must skip over on their CD players. Thus the new format.
CD-R CD Recordable. A standard and technology allowing you to write to and read from a Compact Disc, but not erase or change what you record. This technology is compatible with existing CDs, i.e. you are able to read these discs in existing CD-players, and often in both PC and Macintosh machines. See CD-ROM and MULTI-SESSION.
CD-ROM Compact Disc Read Only Memory. Also called CD or CD-ROM. The familiar 4.7 inch Compact Disc which you see in the audio stores, but now made for computers. These discs hold huge amounts of data — as much as 660 megabytes per disk, or 330 pages of ASCII test. Put into a computer drive, time to retrieve information off the CD-ROM is much slower than from a hard disc. But CD-ROMs are catching up. Most of today's newer CD-ROMs can now be used in multimedia applications and virtually all CD-ROM drives available today support the Multimedia PC or MPC standard. As of writing, the first desktop device to record CD-ROM disks were beginning to appear. They cost several thousand dollars, but are, of course, coming down in price. 1993 was the first year where sales of encyclopedias on CD-ROM exceeded the number of sales of encyclopedias on paper. See CD, CD-ROM XA, CD-V, CD-WO and SHOVELWARE. See also DVD for an update on what CD-ROMs are becoming.
CD-ROM XA Stands for Compact Disc - Read Only Memory eXtended Architecture. Microsoft's extensions to CD-ROM that let you interleave audio with data. Though it is not a video specification, limited video can be included on disc. Demand for multimedia applications is increasing use of CD-ROM XA. To use it, you must have a drive that reads the audio portions of the disc and audio card in your computer that translates the digital into sound. Not all drives can recognize the extensions. See CD-WO.
CD-V Compact Disc Video. A format for putting 5 minutes of video on a 3-inch disc. This format has come and gone. Video is shifting towards CD-ROM XA.
CD-WO Compact Disc Write Once. A CD-ROM version of the WORM (Write Once Read Many) technology. For companies performing all CD-ROM publishing in-house, this format is useful for creating test discs before sending the master for duplication. CD-WO discs conform to ISO 9660 standards and can be played in CD-ROM drives.
CDAR Customer Dialed Account Recording.
CDC Customer Data Change.
CDCS Continuous Dynamic Channel Selection.
CDDI Copper Distributed Data Interface is a version of FDDI (Fiber Distributed Data Interface— a 100 million bit per second local area network) that runs on unshielded twisted-pair cabling rather than optical fiber.
CDEV Control panel DEVice. An Apple Macintosh term.
CDF An ATM term. Cutoff Decrease Factor: CDF controls the decrease in ACR (Allowed Cell Rate) associated with CRM.

NEWTON'S TELECOM DICTIONARY

CDFP Centrex Data Facility Pooling.
CDFS Compact Disc File System, which controls access to the contents of CD-ROM drives in PCs.
CDH Interface An interface once required by the Bell System to protect their phone lines from "foreign" (i.e. non-AT&T) phone equipment. CDH devices were eventually ruled a total waste of money and the phone companies refunded the money — at least to the subscribers who asked. If you still have the stuff installed, you may be due a huge refund. Watch out for the statute of limitations.
CDLC Cellular Data Link Control. A public domain data communications protocol used in cellular telephone systems. In other words, you can attach a data terminal to a cellular telephone and send and receive information. There are more 5,000 modems using CDLC on the Vodaphone Cellular System in the UK, where it is the de facto standard for cellular data communications. Features like improved synchronization field, forward error correction, bit interleaving, and selective retransmission make CDLC ideal for cellular transmissions, according to Millidyne who makes the CDLC modems in the US.
CDLRD Confirming Design Layout Report Date. The date a common carrier accepts the facility designed proposed by the Telco.
CDMA 1. Call Division Multiple Access.
2. Code Division Multiple Access, also called Spread Spectrum, is a name for a new form of digital cellular phone service. Motorola, a leading cellular manufacturer, says CDMA is a spread spectrum technology that assigns a code to all speech bits, sends a scrambled transmission of the encoded speech over the air and reassembles the speech to its original format. The major benefits of CDMA is increased capacity (up to 10 times analog) and more efficient use of spectrum. More importantly, CDMA technology provides three features that improve system quality: 1) The "soft hands-off" feature ensures that a call is connected before handoff is completed, reducing the probability of a dropped call. 2) Variable rate vocoding allows speech bits to be transmitted at only the rates necessary for high quality which conserves the battery power of the subscriber unit. 3) Multipath signal processing techniques combines power for increased signal integrity. Additional benefits to the subscriber include increased talk times for handportable units, more secure transmissions and special service options such as data, integrated voice and data, fax and tiered services.
CDMA works by combining each phone call with a code which only one cellular phone plucks from the air. Business Week said CDMA works "by spreading all signals across the same broad frequency spectrum and assigning a unique code to each. The dispersed signals are pulled out of the background noise by a receiver which knows the code. This method, developed by a San Diego company called Qualcomm Inc. is very new. According to the Wall Street Journal, CDMA systems are said to offer up to 20 times more call handling capacity than the conventional cellular systems by assigning a special electronic code to each call signal, allowing more calls to occupy the same space and be spread over an entire frequency band. Much of the equipment to support CMDA, like cellular switches, however, have not yet been developed."
CDMA is about to be used in inside-building wireless PBX conversations by companies including SpectraLink of Boulder, CO. SpectraLink's explanation: "One of several technologies used to separate multiple transmissions over a finite frequency allocation. CDMA operates in conjunction with spread spectrum transmission. Spread spectrum takes the original information signal and combines it with a correlating code, resulting in a signal which occupies a much greater bandwidth than the original. By assigning a unique correlating code to each transmitter, several simultaneous conversations can share the same frequency allocation. The process of using spread spectrum in conjunction with individual correlating codes is known as Code Division Multiple Access." See also CODE DIVISION MULTIPLE ACCESS.
CDO Community Dial Office. A small automatic central office switching system that is completely unattended. Routine maintenance is provided by a traveling technician once or twice each year, or as troubles develop. Such an office usually serves a small community with a few hundred lines in a rural area.
CDP Customized Dial Plan.
CDPD Cellular Digital Packet Service. A technology that folds packets of data into envelopes (packets) that are sent at very high speeds during pauses in cellular phone conversations. By adding CDPD to an existing analog cellular system (most systems worldwide are cellular), cellular operators can transmit data eight times faster without having to go to the trouble of building a whole new digital world. CPDP is new and just becoming available.
CDPD is a new, open standard for the use cellular as a means of wireless data transmission. The technology allows data files to be broken into packets and sent along idle channels of existing (but CDPD-upgraded) cellular voice networks. CPDP was originally developed by IBM and is backed by a group of companies including Ameritech, AT&T, McCaw Cellular, Motorola and Nynex Mobile. CDPD lets you send data at 19.2 Kbps over an enhanced cellular network. It uses a full voice channel, but it can move your connection from one channel to another to avoid congesting voice communications. The drawback to CDPD is that it requires you to have CDPD modem and your cellular supplier to upgrade its base stations to CDPD, which not all are willing to do. On July 21, 1993, the group of cellular carrier that supports the Cellular Digital Packet Data (CDPD) project released the complete version 1.0 of its open specification designed to enable customers to send computer data over an enhanced cellular network. The group said the packet data approach is ideally suited to those applications that require the transmission of short bursts of data, for example, authorizing a credit card number, exchanging e-mail messages or making databases queries. The cellular networks deploying CDPD will enable mobile workers to use a single device to handle all of their voice and packet data needs. This version of the specification includes input from parties that reviewed the earlier release (0.8 and 0.9). The new version provides details of the CDPD architecture, airlink, external network interfaces, encryption and authentication, network support services, network applications services, network management, radio resource management and radio media access control. To get a copy of specification, call CDPD Project Coordinator Tom Solazzo at Pittiglio Rabin Todd & McGrath 714-545-9400 extension 235. See CDPD FORUM.
CDPD Forum The CDPD Forum Inc. is an industry association that brings together companies with an interest in CDPD. It is based in Chicago on 312-644-6610 or 800-335-CDPD.
CDR Call Detail Recording (as in Call Accounting) or Call Detail Record, as a record generated by customer traffic later used to bill the customer for service. See CALL ACCOUNTING.

117

NEWTON'S TELECOM DICTIONARY

CDR Exclude Table A table listing local central office codes which are not monitored (i.e. ignored) by a call accounting system.

CDV 1. Compression Labs Compressed Digital Video, a compression technique used in satellite broadcast systems. CDV is the compression technique used in CLI's SpectrumSaver system to digitize and compress a full-motion NTSC or PAL analog TV signal so that it can be transmitted via satellite in as little as 2 MHz of bandwidth. (A normal NTSC signal takes 6 Mhz.)

2. An ATM term. Cell Delay Variation: CDV is a component of cell transfer delay, induced by buffering and cell scheduling. Peak-to-peak CDV is a QoS delay parameter associated with CBR and VBR services. The peak-to-peak CDV is the ((1a) quantile of the CTD) minus the fixed CTD that could be experienced by any delivered cell on a connection during the entire connection holding time. The parameter "a" is the probability of a cell arriving late. See CDVT.

CDVT Cell Delay Variation Tolerance-ATM layer functions may alter the traffic characteristics of ATM connections by introducing Cell Delay Variation. When cells from two or more ATM connections are multiplexed, cells of a given ATM connection may be delayed while cells of another ATM connection are being inserted at the output of the multiplexer. Similarly, some cells may be delayed while physical layer overhead or OAM cells are inserted. Consequently, some randomness may affect the inter-arrival time between consecutive cells of a connection as monitored at the UNI. The upper bound on the "clumping" measure is the CDVT.

CE An ATM term. Connection endpoint: A terminator at one end of a layer connection within a SAP.

CE Mark A type of pan-European equipment approval. Obtaining the CE Mark allows a product to be sold into 18 European countries without any further in-country testing. Several country regulatory bodies are set up to do the testing and thus the certification. These bodies are called "notified bodies." When one company, Larscom, reported one of its NDSUs had been granted a CE Mark, it said that the European "Notified Body" issuing the approval for its product was the British Approval Board for Telecommunications (BABT).

The CE Mark will become a requirement for all telecommunications terminal equipment (TTE) products sold into the European Union, effective January 1, 1996. Countries covered are Austria, Belgium, Demark, Finland, France, Germany, Greece, Ireland, Italy, Luxemburg, the Netherlands, Portugal, Spain, Sweden, and the United Kingdom. CE marketing confirms that a product has been tested and meets the essential requirements of the European Telecom Directive to market it throughout the EU. The European TTE Directive 91/263/EEC specifies approved products to meet appropriate telecommunications technical standards. These include personal safety, protection of public networks, interoperation with public network equipment, and electromagnetic compatibility.

CeBIT CeBIT is the world's largest computer and office automation show. It attracts 600,000 or so people to Hannover, Germany in March or so of each year. It is also called the Hannover Fair. It is about five times the size of Comdex, which is North America's largest computer show. Many of the "booths" at Hannover are really small buildings, which are used year-round. May of the "booths" are three stories high, with an open air restaurant on the top floor. Space at the show is sometimes rented for four years.

CED 1. CallEd station iDentification. A 2100 Hz tone with which a fax machine answers a call. See CNG.

2. Capacitance Electronic Disc. System of video recording a grooved disc, employing a groove-guided capacitance pickup.

CED Compression A method of compression used in faxing.

CEDAR A Rolm desktop computer phone that is part IBM PC, part Rolm phone.

CEI 1. Comparable Efficient Interface. The idea is that the telephone industry will let all its information providers have this interface — defined by technical specs and pricing — and, if it does, then the phone companies can themselves use this information to become information providers themselves. The concept has merit. Implementation has been agonizingly slow.

2. An ATM term. Connection endpoint Identifier: Identifier of a CE that can be used to identify the connection at a SAP.

Ceiling Distribution Systems Cable distribution system that use the space between a suspended or false ceiling and the structural floor for running cable. Methods used in ceiling distribution systems include zone, home-run, raceway, and poke-through.

CEKS Centrex Electronic Key Set.

Cell 1. The basic geographic unit of a cellular system. It derived its name "cell" from the honeycomb pattern of cell site installations. Cell is the basis for the generic industry term "cellular." A city or county is divided into smaller "cells," each of which is equipped with a low-powered radio transmitter/receiver. The calls can vary in size depending upon terrain, capacity demands, etc. By controlling the transmission power, the radio frequencies assigned to one cell can be limited to the boundaries of that cell. When a cellular phone moves from one cell toward another, a computer at the Mobile Telephone Switching Office (MTSO) monitors the movement and at the proper time, transfers or hands off the phone call to the new cell and another radio frequency. The handoff is performed so quickly that it's not noticeable to the callers. For a longer explanation, see CMTS.

2. The basic unit of a battery, consisting of plates, electrolyte and a container. A chemical device that produces electricity through electrolysis.

3. A unit of transmission in ATM. A fixed-size frame consisting of a 5-octet header and a 48-octet payload.

Cell Header ATM Layer protocol control information.

Cell Interarrival Variation CIV. An ATM term. "Jitter" in common parlance, CIV measures how consistently ATM cells arrive at the receiving end-station. Cell interarrival time is specified by the source application and should vary as little as possible. For constant bit rate (CBR) traffic, the interval between cells should be the same at the destination and the source. If it remains constant, the latency of the ATM switch or the network itself (also known as cell delay) will not affect the cell interarrival interval. But if latency varies, so will the interarrival interval. Any variation could affect the quality of voice or video applications.

Cell-loss Priority Field CLP. A priority bit in the ATM cell header; when set, it indicates that the cell may be discarded if necessary.

Cell Relay A form of packet switching using fixed length packets which results in lower processing and higher speeds. Cell relay is a generic term for a protocol based on small fixed packet sizes capable of supporting voice video and data at very high speeds. Information is handled in fixed length cells of 53 octets (bytes). A cell has 48 bytes of information and 5 bytes of address. The objective of cell relay is to develop a single high-speed network based on a switching and multiplexing scheme that works for all data types. Small cells (like 53 bytes) favor low-delay, a requirement of isochronous service. The downside to small cells is that the address information is almost 10 percent of the total packet. That's a high overhead. See ATM.

Cell Reversal The reversal of the polarity of the terminals of a battery cell as the result of discharging.

Cell Site A transmitter/receiver location, operated by the WSP (Wireless Service Provider), through which radio links are established between the wireless system and the wireless unit. The area served by a cell site is referred to as a "cell". A cell site consists of an antenna tower, transmission radios and radio controllers. The cell site of an analog cellular radio system handles up to 5,000 users (but not all at once).

Cell Site Controller The cellular radio unit which manages the radio channels within a cell.

Cell Splitting A means of increasing the capacity of a cellular system by subdividing or splitting cells into two or more smaller cells.

Cell Switching A term that refers to how cellular calls are switched. Cellular systems are built to accommodate moving phones — ones in cars, buses, etc. These phones are low-powered. The "moving-ness" and the low power of the phones poses major design constraints on the design of cellular switching offices, which are called Mobile Telephone Switching Office (or MTSO). First, you have to build many cellular switching sites. That way each phone is close to a cell site. Thus there's always a cell site which can pick up the transmission. Second, because of the closeness of the cell sites, any phone conversation may be simultaneously heard by several MTSOs. As a result, the MTSO constantly monitors signal strength of both the caller and the receiver. When signal strength begins to fade, the MTSO locates the next best cell site and re-routes the conversation to maintain the communications link. The switch from cell site to another takes about 300 milliseconds and is not noticeable to the user. All switching is handled by computer, with the control channels telling each cellular unit when and where to switch.

The Cellular Mobile Telephone System is a low-powered, duplex, radio/telephone which operates between 800 and 900 MHZ, using multiple transceiver sites linked to a central computer for coordination. The sites, or "cells", named for their honeycomb shape, cover a range of three to six, or more, miles in each direction. Their range is limited only by certain natural or man-made objects.

The cells overlap one another and operate at different transmitting and receiving frequencies in order to eliminate cross-talk when transmitting from cell to cell. Each cell can accommodate up to 45 different voice channel transceivers. When a cellular phone is activated, it searches available channels for the strongest signal and locks onto it. While in motion, if signal strength begins to fade, the telephone will automatically switch signal frequencies or cells as necessary without operator assistance If it fails to find an acceptable signal, it will display an "out of service" or "no service" message, indicating that it has reached the limit of its range and is unable to communicate.

Each mobile telephone has a unique identification number which allows the Mobile Telephone Switching Office (MTSO) to track and coordinate all mobile phones in its service area. This ID number is known as the Electronic Security Number (ESN). The ESN and Telephone Number are NOT the same. The ESN is a permanent number engraved into a memory chip called a PROM or EPROM, located in the telephone chassis. This number cannot be changed through programming as the telephone number can, although it can be replaced. Each time the telephone is used, it transmits its ESN to the MTSO by means of DTMF tones during the dialing sequence. The MTSO can determine which ESN's are good or bad, thus individual numbers can be banned from use within the system. See CELLULAR RADIO.

Cellphone A British term for a cellular telephone — whether car-based or handheld.

Cellular Data Link Control CDLC is a public domain data communications protocol used in cellular telephone systems. In other words, you can attach a data terminal to a cellular telephone and send and receive information. There are more 5,000 modems using CDLC on the Vodaphone Cellular System in the UK, where it is the de facto standard for cellular data communications. Features like improved synchronization field, forward error correction, bit interleaving, and selective retransmission make CDLC ideal for cellular transmissions, according to Millidyne who makes the CDLC modems in the US.

Cellular Digital Packet Data CDPD is an open standard developed by a group of cellular carriers led by McCaw. The specification provides a standard for using existing cellular networks for wireless data transmission. Packets of data are sent along channels of the cellular network. See CDPD for a much more detailed explanation.

Cellular Digital Packet Data Group In the summer of 1993, a group of cellular carriers that supports the Cellular Digital Packet Data (CDPD) project has released the complete - version 1.0 - of its open specification designed to enable customers to send computer data over an enhanced cellular network. According to the group, the packet data approach is ideally suited to those applications that require the transmission of short bursts of data, for example, authorizing a credit card number, exchanging e-mail messages or making databases queries. The cellular networks deploying CDPD will enable mobile workers to use a single device to handle all of their voice and packet data needs. The 1.0 specification provides network and customer equipment manufacturers the parameters for building to this nationwide approach that sends packets of data over existing cellular networks. This version of the specification includes input from parties that reviewed the earlier release (0.8 and 0.9) and provides details of the CDPD architecture, airlink, external network interfaces, encryption and authentication, network support services, network applications services, network management, radio resource management and radio media access control. Copies of the specification can had from CDPD Project Coordinator Tom Solazzo at Pittiglio Rabin Todd & McGrath, 714-545-9400 extension 235.

Cellular Floor Method A floor distribution method in which cables pass through floor cells, constructed of steel or concrete, that provide a ready-made raceway for distributing power and communication cables.

Cellular Geographic Service Area CGSA. The geographic area served by the wireless (cellular) system within which a WSP is authorized to provide service.

Cellular Mobile Telephone Service See CMTS.

Cellular Phone Service See CMTS.

Cellular Protocols Conventions and procedures which relate to the format and timing of device communications. In data transmission communications, there are currently three major protocols, which are converging into a de facto standard: MNP, SPCL, and PEP.

NEWTON'S TELECOM DICTIONARY

Cellular Radio A mobile radio system. In the old days, there was one central antenna and everything homed in on that and emanated from it. With cellular radio, a city is broken up into "cells," each maybe no more than several city blocks. Every cell is handled by one transceiver (receiver/transmitter). As a cellular mobile radio moves from one cell to another, it is "handed" off to the next cell by a master computer, which determines from which cell the strength is strongest. Cellular mobile radio has several advantages:
1. You can handle many simultaneous conversations on the same frequencies. One frequency is used in one cell and then re-used in another cell. You can't do this on a normal mobile radio system.
2. Because one cellular system can accommodate many more subscribers than a normal mobile radio system, and therefore because it can achieve certain economies of scale, it has the potential of achieving much lower transmission costs.
3. Because the transceiver is always closer to the user than in a normal mobile system, and the user's radio device thus needs less power, the device can be cheaper and smaller. Cellular radios started at over $5,000 and are now well under $500. From the first portable units, weight has already dropped to under one pound. There are several units that will fit in your breast pocket and not overly stretch your suit.
The following are specific cellular radio terms, or general telecom terms that mean something special in cellular radio:
A/B Switch Permits user to select either the wireline (B system) or the nonwireline (A system) carrier when roaming.
Alphanumeric memory Capability to store names with phone numbers.
Call-in-absence horn alert User-activated feature that sounds car horn upon receiving a call.
Call-in-absence indicator Feature that displays what calls came in while user was absent.
Call-in-progress override Insures that power to the phone is not lost if the car's ignition is turned off.
Call restriction Security feature that limits phone's use without completely locking it. Variations might include dial from memory only, dial last number only, seven-digit dial only, no memory access, etc.
Call timer Displays information on call duration and quantity. Variations might include present call, last call, total number of calls, or total accumulated time since last reset. Call-timer beep serves as a reminder to help keep calls brief. It might be set to go off once a minute, ten seconds before the minute, for example.
Continuous DTMF (touch-tones) Sends DTMF (dual-tone, multi-frequency) tones — also called touchtones — allowing access to voice mail and answering machines that require long-duration tones. "Continuous" means you get the tone so long as your finger is on the button. This may seem obvious to you and me, except that some "modern" phones just give a short tone no matter how long you keep your finger on the touchtone button.
Dual NAM Allows user to have two phone numbers with separate carriers (see multi-NAM).
Electronic lock Provides security by completely locking phone so it can't be used by unauthorized persons.
Expanded spectrum Full 832-channel analog cellular spectrum currently available to users.
Hands-free operation Allows user to receive calls and converse while leaving handset in cradle (similar to office speakerphone).
Hands-free answering Phone automatically answers incoming call after a fixed number of rings and goes to hands-free operation.
Memory linkage Allows programming specific memory locations to dial a sequence of other memory locations.
Multi-NAM A cellular telephone term to allow a phone to have more than two phone numbers, each of which can be on a different cellular system if desired. This lets the user register with both carriers in home city, expanding available geographic coverage.
Mute Silences the telephone's microphone to allow private conversations without discontinuing the phone call. Audio mute turns off the car stereo automatically when the phone is in use, and turns it back on when the call is completed.
NAM Numerical Assignment Module. Basically, your cellular phone number, although it refers specifically to the component or module in the phone where the number is stored.
On-hook dialing Allows dialing with the handset in the cradle.
Roaming Using any cellular system outside your home system. Roaming usually incurs extra charges.
Scratch pad Allows storage of phone numbers in temporary memory during a call. Silent scratch pads allows number entry into scratch pad without making beep tones.
Signal strength indicator Displays strength of cellular signal to let user know if a call is likely to be dropped.
Speed dialing Dialing phone number from memory by pressing a single button.
Standby time Maximum time cellular phone operating on battery power can be left on to receive incoming calls.
Talk time Maximum time cellular phone operating on battery power can transmit.
Voice-activated dialing Your cellular phone recognizes your words and dial accordingly. You say "Dial Mom" and it dials mom.
Cellular Radio Switching Office The electronic switching office which switches calls between cellular (mobile) phones and wireline (i.e. normal wired) phones. The switch controls the "handoff" between cells and monitors usage. Different manufacturers call their equipment different things, as usual.
Cellular Switching See CELL SWITCHING.
Celluplan II A proposed national standard to place packets of data between idle spaces on a cellular voice network.
CELP Code Excited Linear Prediction. An analog-to-digital voice coding scheme. Sun is proposing the use of CELP so that a user could send realtime voice communications over local area or wide are network — bypassing the phone system!
CEMH Controlled Environment ManHole. Environmental control of the CEMH is maintained by a heat pump (a fancy name for an airconditioner — cooler and heater).
CEN/CENELEC Comite European de Normalisation/Comite European e Normalisation ELECtrotechnique) Two committees that have combined to address the standardization of information technology. The main part of their work is aimed at the development of functional standards for OSI and related standards. Its membership comprises national standards bodies such as the UK's BSI as well as other European standards bodies.
CENLEC European Electrical Standards Institute (Comite European de Normalization Electrique).
Centel Corporation A telephone company serving over one million customer lines. It's based near the Chicago O'Hare Airport. During 1992 it was absorbed by Sprint.

Central Office Telephone company facility where subscribers' lines are joined to switching equipment for connecting other subscribers to each other, locally and long distance. Also called CO, as in See-Oh. Sometimes the term central office is the same as the overseas term "public exchange." Sometimes, it means a wire center in which there might be several switching exchanges.

Central Office Battery A group of wet cells joined in series to provide 48 volts DC. Central office batteries are typically charged off the main 120 volts AC. The batteries have two basic functions. 1. To provide a constant source of DC power for eight hours or so after AC powers drops, and 2. To isolate the central office from glitches on the AC line.

Central Office Code Part of the national numbering plan, the central office code is the second three digits of a subscriber's telephone number, which identifies the local switching office. Here is Bellcore's definition: A 3-digit identification under which up to 10,000 station numbers are subgrouped. Exchange area boundaries are associated with the central office code that generally have billing significance. Note that multiple central office codes may be served by a central office. Also called NXX code or end office code.

Central Office Connection COC. An MCI charge (monthly and installation) for each local access channel. If no access channels are provided, one COC is charged for each serving MCI terminal.

Central Office Override A third party may interrupt or join in your conversation.

Central Office Trunk 1. A trunk between central offices. It may be between major switches or between a major and a minor switch. 2. A trunk between public and private switches.

Central Processing Unit CPU. The part of a computer which performs the logic, computational and decision-making functions. It interprets and executes instructions as it receives them. Personal computers have one CPU, typically a single chip. It is the so-called "computer on a chip." That chip identifies them as an 8-bit, 16-bit or 32-bit machine.
Telephone systems, especially smaller ones, are not that different. Typically they have one main CPU — a chip — which controls the various functions in the telephone. Today's telephone systems are in reality nothing more than special purpose computers. As phone systems get bigger, the question of CPUs - - central processing units — becomes harder to figure. The design of phone systems has, of late, tended away from single processor-controlled telephone systems (as in single processor controlled PCs). There are several reasons for this move. First, it's more economical for growth. Make modules of "little" switches and join little ones together to make big ones. Second, it's more reliable. It's obviously better not to rely on one big CPU, but to have several. In short, the issue of Central Processing Units — CPUs — is blurring. But the concept is still important because by understanding how your telephone switch works (its architecture), you will understand its strengths and weaknesses.

Centralized Attendant Service Calls to remote (typically branch) locations are automatically directed to operators at a central location. Imagine four retail stores in a town. There are three branch stores and one main, downtown store, each having their own local phone numbers, which customers call. It's clearly inefficient to put operators at each of the stores — when one group is busy, the other will be free, etc. What this feature does is to direct all the calls coming into each of the stores into one bank of operators, who then send those calls back to the outlying stores.
Despite the extra schlepping of calls around town, having one large group of operators is cheaper than maintaining many small groups. Each store has its own local Listed Directory Number (LDN) Service. Special Release Link Trunk circuits connect each unattended location (each store) to the main attendant location. These trunks are only temporarily used during call processing. An incoming call to an unattended store seizes such a trunk circuit for completion of the call to the centralized attendant, who then uses the same trunk circuit to process the call to the remote location's internal extension. (After all if the caller was calling that store, they obviously want to talk to someone in that store.) The circuit is then released and is available for other calls. Since such special trunk circuits are only used during that part of a call that requires connection between locations, such trunks are more efficient than normal tie trunk circuits.

Centralized Automatic Message Accounting CAMA. The recording of toll calls at a centralized point.

Centralized Ordering Group COG. An organization provided by some communications service providers (like a local phone company) to coordinate services between the companies and vendors.

Centrex Centrex is a business telephone service offered by a local telephone company from a local central office. Centrex is basically single line telephone service delivered to individual desks (the same as you get at your house) with features, i.e. "bells and whistles," added. Those "bells and whistles" include intercom, call forwarding, call transfer, toll restrict, least cost routing and call hold (on single line phones).
Think about your home phone. You can often get "Custom Calling" features. These features are typically fourfold: Call forwarding, Call Waiting, Call Conferencing and Speed Calling. Centrex is basically Custom Calling, but instead of four features, it has 19 features. Like Custom Calling, Centrex features are provided by the local phone company's central office.
Phone companies peddle Centrex is leased to businesses as a substitute for that business buying or leasing its own on-premises telephone system — its own PBX, key system or ACD. Before Divestiture in 1984, Centrex was presumed dead. AT&T was, at that time, intent on becoming a major PBX and key system supplier. Then Divestiture came, and the operating phone companies recognized they were no longer part of AT&T, no longer had factories to support, but did have a huge number of Centrex installations providing large monthly revenues. As a result, the local operating companies have injected new life into Centrex, making the service more attractive in features, price, service and attitude. Here are the main reasons businesses go with Centrex as opposed to going with a stand-alone telephone system:
1. Money. Centrex is typically cheaper to get into (the central office already exists). Installation charges can be low. Commitment can also be low, since most Centrex service is leased on a month-to-month basis. So it's perfect for companies planning an early move. There may be some economies of scale, also. Some phone companies are now offering low cost, large size packages.
2. Multiple locations. Companies with multiple locations in the same city are often cheaper with Centrex than with multiple private phone systems and tie lines, or with one private phone system and OPX lines. (An OPX line is an Off Premise eXtension, a line going from a telephone system in one place to a phone in another. It might be used for an extension to the boss' home.)
3. Growth. It's theoretically easier to grow Centrex than a standalone PBX or key system, which usually has a finite limit. With

Centrex, because it's provided by a huge central office switch, it's hard, theoretically, to run out of paths, memory, intercom lines, phones, tie lines, CO lines, etc. The limit on the growth of a Centrex is your central office, which may be many thousands of lines.
4. Footprint Space Savings. You don't have to put any switching equipment in your office. All Centrex switching equipment is at the central office. All you need at your office are phones.
5. Fewer Operators because of Centrex's DID features. Fewer operator positions saves money on people and space.
6. Give better service to your customers. With Centrex, each person has their own direct inward dial number. Many people prefer to dial whomever they want directly rather than going through a central operator. Saves time.
7. Better Reliability. When was the last time a central office crashed? Here are some of the features built into modern central offices: redundancy, load-sharing circuitry, power back-up, on-line diagnostics, 24-hour on-site personnel, mirror image architecture, 100% power failure phones, complete DC battery backup and battery power. Engineered to suffer fewer than three hours down time in every 40 years.
8. Non-blocking. Trunking constraints are largely eliminated with Centrex, since a central office is so large.
9. Minimal Service Costs. Repair is cheap. Service time is immediate. People are right next to the machine 24-hours a day. Phones and wires are the only things that require repair on the customers' premises. You can easily plug new phones in, plug them out yourself. All other equipment is in the central office. You need not hold inventory or test equipment.
10. No technological obsolescence. Renting Centrex means a user has the ultimate flexibility — ability to jump quickly into new technology. Central offices are moving quickly into new technologies, such as ISDN.
11. Ability to manage it yourself. You can now get two important features previously available only on privately-owned self-contained phone systems (like PBXs): 1. The ability for you, the user, to make changes to the programming of your own Centrex installation without having to personally call a phone company representative. 2. The ability to get call detail accounting by extension and then have reports printed by a computer in your office. The phone company does this call accounting by installing a separate data line which carries Centrex call records back to the customer as those calls are made.
The above arguments are pro-Centrex. There are also anti-Centrex arguments. And there's plenty of evidence to argue exactly the opposite. For example, central offices often run out of capacity. The "big" key to Centrex traditionally comes down to price. And, in fact, in some cities the price of Centrex lines is lower than "normal" PBX lines. Of course, you can buy Centrex lines and attach your own PBX or key system to those Centrex lines. The big disadvantage of Centrex is that there are no specialized Centrex phones able to take better advantage of Centrex central office features than normal electronic phones can.
Centrex is known by many names among operating phone companies, including Centron and Cenpac. Centrex comes in two variations — CO and CU. CO means the Centrex service is provided by the Central Office. CU means the central office is on the customer's premises. See the following CENTREX definitions.

Centrex Call Management A Centrex feature that provides detailed cost and usage information on toll calls from each Centrex extension, so you can better manage your telephone expenses.

Centrex CCRS Centrex Customer Rearrangement System. Computer software from New York Telephone that allows their Centrex customers to make certain changes in their own line and features arrangements. Other phone companies have similar services under different names.

Centrex CO Indicates that all equipment except the attendant's position and station equipment is located in the central office. See CENTREX.

Centrex CU Indicates that all equipment including the dial switching equipment, is located on the customer's premises. See CENTREX.

Centrex Extend Service The name of a Bell Atlantic service. If you maintain offices in multiple locations — or have work-at-home employees — this service allows you to tie all your locations into one phone system. So everyone in your company can take advantage of Centrex features and services on a cost-efficient, call-by-call sharing basis. With Centrex Extended Service, you can even tie non-Centrex locations into your Centrex system.

Centrex LAN Service Put a modem on a dedicated central office line. Connect that line to a switch. Bingo you have a switched, relatively high-speed service that can connect synchronous and asynchronous terminals and other equipment at speeds up to 19.2 Kbps. Centrex LAN service works with existing wiring Centrex central office to customer premise wiring. Centrex LAN is a name given this service by Nynex. Other phone companies have similar services under different names. It's not a very successful service, since it's very slow. Compare 19,200 bits per second to Ethernet, which is 10 million bits per second!

Centrex SMDI Have you ever called someone and been forwarded to their voice mail, and then had to enter their extension again. This is because their voice mail does not know where the call originated. The voice mail system does not know who you just called, All it knows is that it just received another call. SMDI is simply a modem link back to the central office supplying your Centrex system. This modem link will feed a computer at your location the information about incoming calls as they are forwarded through your system. It feeds the originating number and why the call is transferred, so you know whether the user didn't answer their phone, or that it was busy.
To indicate the health of the SMDI link, it generates a heartbeat message every few seconds. If you don't receive a heartbeat within the time window, you know there is a problem with the SMDI link and know to restart the link. The one good thing about SMDI is the option that your voice mail system can control the status of message waiting indicators on the user's phone. They can either have a message waiting lamp on their phone, or use the "stutter dial tone" which causes a broken dial tone when the user picks up their phone.
Problems to watch out for... Since SMDI is a communications link, it can be broken. If the SMDI link goes down, make sure you build in the old two step method of finding out what extension the caller was attempting to call. Make sure you also offer a user directory in case the caller does not know the extension number. When the link goes down, make sure the system does not continue to spend the message waiting status commands down an inactive link. SMDI stands for Standard Message Desk Interface or Simplified Message Disk Interface. See also SMDI.

Centronics The name of the printer manufacturer whose method of data transmission between a computer and a parallel printer has become an industry standard. See CENTRONICS PRINTER STANDARD.

Centronics Printer Standard The Centronics standard was developed by the Centronics company which makes computer printers. The Centronics standard is a 36-pin single plug/connector with eight of the 36-pins carrying their respective bits in parallel (eight bits to one character), which means it's much faster than serial transmission which sends only one bit a time. There are several types of Centronics male and female plugs and receptacles. So know which you want before you buy. The pinning — the location of and function of each of the 36-individual wires, is standard from one Centronics cable to another. The Centronics printer standard has been adopted by many printer and PC companies, including IBM. It is a narrower standard than the RS-232-C standard. The Centronics works only between a computer and a printer. It won't work over phone lines, unless conversion is done at either end. However, it is standard and has none of the dumb interface problems the RS-232-C standard does.

Centum Call Second 1/36th of an erlang. The formula for a centum call second is the number of calls per hour multiplied by their average duration in seconds, all divided by 100.

CEO Chief Executive Officer.

CEPT Conference des administrations Europeenes des Postes et Telecommunications (European Conference of Postal and Telecommunications Administrations). Standards-setting body whose membership includes European Post, Telephone, and Telegraphy Authorities (PTTs). It in turn participates in relevant areas of the work of CEN/CENELEC. It was originally responsible for the NET standards, but these have subsequently been passed on to ETSI.

CEPT Format Defines how the bits of a PCM carrier system of the 32 channel European type T-1/E-1 will be used and in what sequence. To correctly receive the transmitted intelligence, the receiving end equipment must know exactly what each bit is used for. CEPT format uses 30 VF channels plus one channel for supervision/control (signaling) and one channel for framing (synchronizing). All 8 bits per channel are used to code the waveshape sample. For a much better explanation, see T-1.

CER An ATM term. Cell Error Ratio: The ratio of errored cells in a transmission in relation to total cells sent in a transmission. The measurement is taken over a time interval and is desired to be measured on an in-service circuit.

CERB Centralized Emergency Reporting Bureau. A Canadian term similar to PSAP — Public Safety Answering Position. See PSAP.

CERN European Laboratory for Particle Physics Research in Geneva, Switzerland.

Certificates A cryptography term.

Certified Several companies in the "secondary" industry test used equipment, parts and/or systems. They have various ways of testing them. Typically they test with working phones operating for extended periods at different temperatures. The idea is to check that this used equipment works the way it's meant to work — to the original manufacturer's design specification. Once these tests have been completed a secondary dealer will "certify" such equipment, usually in writing. Such certification carries the assurance that the used equipment works as it's meant to. Sometimes certified equipment is upgraded to the most current revision level of hardware and software. Sometimes it's not. You, the buyer, must check. Certified equipment typically carries a guarantee — that guarantee being as good, obviously, as the company that backs it.

Certified Equipment A term used in the secondary telecom equipment business. Equipment carrying the written assurance that it will perform up to the manufacturer's specifications. It qualifies for addition to existing maintenance contracts.

CES An ATM term. Circuit Emulation Service: The ATM Forum circuit emulation service interoperability specification specifies interoperability agreements for supporting Constant Bit Rate (CBR) traffic over ATM networks that comply with the other ATM Forum interoperability agreements. Specifically, this specification supports emulation of existing TDM circuits over ATM networks.

Cesium Clock A clock containing a cesium standard as a frequency-determining element. It's a very accurate clock. See CESIUM STANDARD.

Cesium Standard A primary frequency standard in which a specified hyperfine transition of cesium-133 atoms is used to control the output frequency. Its accuracy is intrinsic and achieved without calibration.

CEV Controlled Environmental Vault. A below ground room that houses electronic and/or optical equipment under controlled temperature and humidity.

CFA Carrier Failure Alarm. The alarm which results from an out-of-frame or loss of carrier condition and which is combined with trunk conditioning to create a CGA.

CFAC Call Forward All Calls.

CFAMN Call Forwarding Address Modified Notification.

CFB 1. Call Forward Busy.
2. Call Forwarding on mobile subscriber Busy. A wireless telecommunications term. A supplementary service provided under GSM (Global System for Mobile Communications).

CFDA Call Forward Don't Answer.

CFGDA Call Forward Group Don't Answer.

CFM Cubic Feet per Minute. A measure of how much air you move through the fan of a PC.

CFNRc Call Forwarding on mobile subscriber Not Reachable. A wireless telecommunications term. A supplementary service provided under GSM (Global System for Mobile Communications).

CFNRy Call Forwarding on No Reply. A wireless telecommunications term. A supplementary service provided under GSM (Global System for Mobile Communications).

CFO Chief Financial Officer.

CFP Channel Frame Processor.

CFRP Carbon Fiber Reinforced Plastic. A light and durable material, which has been used (for the wings of advanced fighter jets) in the defense business and which Toshiba introduced in 1991 as casing for a line of notebook sized computer laptops.

CFUC Call Forwarding UnConditional. A wireless telecommunications term. A supplementary service provided under GSM (Global System for Mobile Communications).

NEWTON'S TELECOM DICTIONARY

CFV Call For Votes. Begins the voting period for a Usenet newsgroup. At least one (occasionally two or more) email addresses is customarily included as a repository for the votes.
CFW Call Forward.
CGA 1. Carrier Group Alarm. A service alarm generated by a channel bank when an out-of-frame (OOF) condition exists for some predetermined length of time (generally 300 milliseconds to 2.5 seconds). The alarm causes the calls using a trunk to be dropped and trunk conditioning to be applied.
2. Color Graphics Adapter. An obsolete IBM standard for displaying material on personal computer screens. The simplest (and conventional) CGA displays 320 horizontal picture elements, known as pels or pixels, by 200 pels vertically. There is also an Enhanced CGA, which is 640 x 400, or 128,000 pixels per screen. Older portables may use CGA monochrome mode. CGA has essentially been obsoleted by VGA. See MONITOR and VGA.
CGI Common Gateway Interface. An Internet term. Programs or Scripts, usually executed on the Web server, that perform actions (like searching or running applications) when the user clicks on certain buttons or parts of the Web Screen. CGI actually refers to the pre-defined way in which these programs communicate with the Web Server but has lately come to refer to the programs themselves.
CGI Joe A Wired Magazine definition. A hardcore CGI script programmer with all the social skills and charisma of a plastic action figure.
CGM Computer Graphics Metafile. A standard format that allows for the interchanging of graphics images.
CGSA Cellular Geographic Service Area. The actual area in which a cellular company provides cellular service. CGSAs are usually made up of multiple counties and often cross state lines.
CGSA Restriction If you own a cellular phone, you are prevented from making calls outside your own local Cellular Geographic Service Area. This restriction is an option that is available to subscribers in most cellular cities.
Chad 1. The little solid round dots of paper made when paper tape is punched with information.
2. CHAnge Display.
Chad Tape Punched tape used in telegraphy/teletypewriter operation. The perforations, called "chad," are severed from the tape, making holes representing the characters.
Chadless Tape 1. Punched tape that has been punched in such a way that chad is not formed.
2. A punched tape wherein only partial perforation is completed and the chad remains attached to the tape. This is a deliberate process and should not be confused with imperfect chadding. See CHAD.
Chain Mailboxes Mailboxes that are connected together to provide a service or a number of messages (e.g. Directory, Product Information, etc.).
Chaining A programming technique linking one activity to another, as in a chain. Each link in the chain may contain a pointer to the next link, or there may be a master control or program instructing the programs to link together.
Challenge-Response A type of authentication procedure into a system in which a user must respond correctly to a challenge, usually a secret key code, to gain access.
Channel 1. Typically what you rent from the telephone company. A voice-grade transmission facility with defined frequency response, gain and bandwidth. Also, a path of communication, either electrical or electromagnetic, between two or more points. Also called a circuit, facility, line, link or path.
2. An SCSA term. A transmission path on the SCbus or SCxbus Data Bus that transmits data between two end points.
3. A channel of a GPS (Global Positioning System) receiver consists of the circuitry necessary to tune the signal from a single GPS satellite.
Channel Attached Describing the attachment of devices directly to the input/output channels of a (mainframe) computer. Devices attached to a controlling unit by cables rather than by telecommunications circuits. Same as locally attached (IBM).
Channel Bank A multiplexer. A device which puts many slow speed voice or data conversations onto one high-speed link and controls the flow of those "conversations." Typically the device that sits between a digital circuit — say a T-1 — and a couple of dozen voice grade lines coming out of a PBX. One side of the channel bank will be connections for terminating two pairs of wires or a coaxial cable — those bringing the T-1 carrier in. On the other side are connections for terminating multiple tip and ring single line analog phone lines or several digital data streams. Sometimes you need channel banks. Sometimes, you don't. For example, if you're shipping a bundle of voice conversations from one digital PBX to another across town in a T-1 format — and both PBXs recognize the signal — then you will probably not need a channel bank. You'll need a Channel Service Unit (CSU). If one, or both, of the PBXs is analog, then you will need a channel bank at the end of the transmission path whose PBX won't take a digital signal. See CHANNEL SERVICE UNIT and T-1.
Channel Capacity A measure of the maximum possible bit rate through a channel, subject to specified constraints.
Channel Gate A device for connecting a channel to a highway, or a highway to a channel, at specified times.
Channel Identification Information element that requests or identifies the channel to be used for a call. An AIN term.
Channel Loopback In network management systems, diagnostic test that forms a loop at the multiplexer's channel interface that returns transmitted signals to their source. See also LOOPBACK.
Channel Mode An AT&T term for a method of communications whereby a fixed bandwidth is established between two or more points on a network as a semi-permanent connection and is rearranged only occasionally.
Channel Modem That portion of multiplexing equipment required to derive a desired subscriber channel from the local facility.
Channel Packing A technique for maximizing the use of voice frequency channels used for data transmission by multiplexing a number of lower data rate signals into a single higher speed data stream for transmission on a single voice frequency channel.
Channel Queue Limit Limit on number of transmit buffers used by a station to guarantee that some receive buffers are always available.
Channel Seized The time when a connection is established between the cellular user's mobile equipment and the mobile telephone switching office (MTSO). Channel seizure occurs before the number dialed begins to ring.

NEWTON'S TELECOM DICTIONARY

Channel Service Unit CSU. A device used to connect a digital phone line (T-1 or Switched 56 line) coming in from the phone company to either a multiplexer, channel bank or directly to another device producing a digital signal, e.g. a digital PBX, a PC, or data communications device. A CSU performs certain line-conditioning, and equalization functions, and responds to loopback commands sent from the central office. A CSU regenerates digital signals. It monitors them for problems. And it provides a way of testing your digital circuit. You can buy your own CSU or rent one from your local or long distance phone company. See also CSU and DSU.

Channel Surfing Flipping channels on a TV set. A person who channel surfs is called a MOUSE POTATO.

Channel Tier An AT&T term for the tier within the Universal Information Services network that partitions transmission capacity into channels and offers the channels to the nodes' higher tiers.

Channel Time Slot A time slot starting at a particular instant in a frame and allocated to a channel for transmitting a character, in-slot signal, or other data. Where appropriate a modifier may be added.

Channel Translator Device used in broadband LANS to increase carrier frequency, converting upstream (toward the head-end) signals into downstream signals (away from the head-end).

Channelization Subdividing a wideband channel into many smaller channels so it may carry many different streams of information.

CHAP/PAP Challenge Handshake Authentication Protocol/Password Authentication Protocol; Standard authentication protocols for PPP connections. An Internet term.

Chapter 7 See Chapter 11.

Chapter 11 The Chapter 11 process is started when a company files a reorganization petition with the federal Bankruptcy Court. From that moment on, creditors are prevented from suing the company, and any creditor lawsuits in process are halted, pending the outcome of the Chapter 11 reorganization. Creditors of the company file claims with the Bankruptcy Court. A creditors committee, usually made up of the seven creditors who have filed the largest claims against the company, represents the interest of all creditors. Under Chapter 11 protection, the company's management usually continues to manage the company's business, subject to judicial review. In rare circumstances, such as fraud, a party may ask the court to appoint a trustee to manage the company during reorganization. The ultimate objective of a Chapter 11 reorganization is to restructure creditors' claims so that the company can move ahead with its business. Company management includes a negotiated partial payment to creditors. The plan also can include exchanges of debt for equity, a moratorium on repayment or a combination of these actions. In some cases, more than one plan may be proposed. For example, a creditor, or group of creditors, may develop its own plan. The complex process of reaching a consensual plan entails extensive negotiations among the company, its creditors and its shareholders.

Once developed, the company's reorganization plan — or one of the competing plans — must be accepted by specified margins of creditors and shareholders. Creditors representing two-thirds of the total dollar amount of bankruptcy claims against the company and 51 percent of the total number of those voting must accept the plan, and two-thirds of the amount of shares represented by shareholders voting on the plan must approve it for a plan to be accepted. Once accepted, the Bankruptcy Court reviews the plan to ensure that it conforms to certain additional statutory requirements before confirming it. With a restructured balance sheet, the company then emerges from Chapter 11 protection to implement the plan. Some companies emerge from Chapter 11 and become normal operating companies again. Some don't and move into Chapter 7 bankruptcy, which is complete and relatively immediate liquidation of the company (i.e. sale of all the company's assets).

Character A letter, a number or a symbol. A character is sometimes described by the digit represented by the bit pattern that makes up the Character. i.e., the letter A is ASCII code 65, a carriage return is ASCII code 13.

Character Cell In text mode on a PC, each pel is called a character cell. Character cells are arranged in rows and columns. A typical PC will support two text modes — 80 columns by 25 rows and 40 columns by 25 rows. The default text mode on virtually all PCs is 80 x 25.

Character Code One of several standard sets of binary representations for the alphabet, numerals and common symbols, such as ASCII, EBCDIC, BCD.

Character Distortion In telegraphy, the distortion caused by transients that, as a result of previous modulation, are present in the transmission channel. It effects are not consistent. Its influence upon a given transition is to some degree dependent upon the remnants of transients affecting previous signal elements.

Character Generator A functional unit that converts the coded representation of a character into the graphic representation of the character for display.

Character Impedance The impedance termination of an electrically uniform (approximately) transmission line that minimizes reflections from the end of the line.

Character Interleaving A form of TDM used for asynchronous protocols. A 20% saving can be obtained by omitting the start and stop bits. This can be used either with extra channels or by carrying RS232-C control signals.

Character Interval The total number of unit intervals (including synchronizing, information, error checking, or control bits) required to transmit any given character in any given communication system. Extra signals that are not associated with individual characters are not included. An example of an extra signal that is excluded in the above definition is any additional time added between the end of the stop element and the beginning of the next start element as a result of a speed change, buffering, etc. This additional time is defined as a part of the intercharacter interval.

Character Oriented Protocol A communications protocol in which the beginning of the message and the end of a block of data are flagged with special characters. A good example is IBM Corp's. Binary Synchronous Communications (BSC) protocol. Character-oriented protocols are used in both synchronous and asynchronous transmission.

Character Oriented Windows Interface COW. An SAA-compatible user interface for OS/2 applications.

Character Printer A device which prints a single character at a time. Contrast with a line printer, which prints blocks of characters and is much faster.

Character Set All the letters, numbers and characters which a computer can use. The symbols used to represent data. The ASCII standard has 256 characters, each represented by a binary number from 1 to 256. This set includes all the letters in the alphabet, numbers, most punctuation marks, some mathematical symbols and some other characters typically used by computers. See ASCII.

Character Stuffing A technique used to ensure that transmitted control information is not misinterpreted as data by the receiver during character-based transmission. Special characters are inserted by the transmitter and then removed by the receiver.

Character Terminal A computer terminal that cannot show graphics, only text.

Characteristic Frequency A frequency that can be easily identified and measured in a given emission. A carrier frequency may, for example, be designated as the characteristic frequency.

Characteristic Impedance The impedance of a circuit that, when connected to the output terminals of a uniform transmission line of arbitrary length, causes the line to appear infinitely long. A line terminated in its characteristic impedance will have no standing waves, no reflections from the end, and a constant ratio of voltage to current at a given frequency at every point on the line.

Characters Per Second CPS. A data transfer rate generally estimated from the bit rate and the character length. For example, at 2400 bps, 8-bit characters with Start and Stop bits (for a total of ten bits per character) will be transmitted at a rate of approximately 240 characters per second (cps). Some protocols, such as USR-HST and MNP, employ advanced techniques such as longer transmission frames and compression to increase characters per second.

Charged Coupled Device CCD. The full name of the term is Interline Transfer Charge-Coupled Device or IT CCD. CCD are used as image sensors in an array of elements in which charges are produced by light focused on a surface. They consist of a rectangular array of hundreds of thousands of light-sensitive photo diodes. Light from a lens is focused onto the photo diodes. This frees up electrons (charges) which accumulate in the photo diodes. The charges are periodically released into vertical shift registers which move them along by charge-transfer to be amplified.

Charlie-Foxtrot Slang. Seriously beyond all hope. Very badly broken.

Chat A common name for a type of messaging done over a network, involving short, messages sent from one node to another. Chatting sometimes happens in real-time, sometimes in just short messages, replied to quickly. Sometimes, chatting software is RAM-resident, meaning it can be "popped up" inside an application program. Users are usually notified of an incoming chat by a beep and a message at the bottom of their screens.

Cheapernet A slang name for the thin wire coaxial cable (0.2-inch, RG58A/U 50-ohm) that uses a smaller diameter coaxial cable than standard thick Ethernet. Thin Ethernet is also called "Cheapernet" due to the lower cabling cost. Thin Ethernet systems tend to have transceivers on the network interface card, rather than in external boxes. PCs connect to the Thin Ethernet bus via a coaxial "T" connector. Thin Ethernet is now the most common Ethernet coaxial cable, though twisted pair is gaining. Thin Ethernet is also referred to as ThinNet or ThinWire. See also 10BASE-T.

Check Bit A bit added to a unit of data, say a byte or a word, and used for performing an accuracy check. See also PARITY.

Check Characters Characters added to the end of a block of data which is determined by an algorithm using the data bits which are sent. The receiving device computes its own check characters. It compares them with those sent by the transmitter. If they do not match, the receiver requests the sender to send the block again. If the check characters match, then all the bits used to compute the check characters have been received properly.

Check-In Mailbox The Centigram VoiceMemo II mailbox used to assign names and passcodes for guests checking into a hotel.

Check-Out Mailbox The Centigram VoiceMemo II mailbox used to clear out guest mailboxes when the guest checks out of the hotel.

Checkpoint Cycle HDLC error recovery cycle formed by pairing an F bit with a previous P bit or vice versa.

Checkpointing HDLC error recovery based on pairing of P and F bits and giving the equivalent of a negative acknowledgment without using either REJ or SREJ.

Checksum The sum of a group of data items used for error checking. Checksum is computed by the sending computer based upon an algorithm that counts the bits going out in a packet. The check digit is then sent to the other end as the tail, or trailer of the packet. As the packet is being received, the receiving computer goes through the same algorithm, and if the check digit it comes up with is the same as the one received, all is well. Otherwise, it requests the packet be sent again.

Cheesing When a buffered fiber cable appears to stretch during stripping and then cheeses (creeps) back into the outer jacket of the cable, to resume its original place.

Chemical Rectifier A chemical device for changing alternating current to pulsating direct, usually used to storage battery charging.

Chemical Stripping Soaking an optical fiber in a chemical to remove its coating.

Chemical Vapor Deposition Technique CVD. In optical fiber manufacturing, a process in which deposits are produced by heterogeneous gas-solid and gas-liquid chemical reactions at the surface of a substrate. The CVD method is often used in fabricating optical fiber preforms by causing gaseous materials to react and deposit glass oxides. The preform may be processed further in preparation for pulling into an optical fiber.

Cherry Picking A call center term. Calls come in and are identified in some way — by ANI (automatic number identification), Caller ID, or caller touchtone input. The identity of the callers is known to the agents in the call center, who can now answer the callers they wish. They decide to answer those callers who they think will buy the most and presumably give them the highest commission or best reward. Thus the expression "cherry picking."

Chicago An early code word for Microsoft's Windows 95, due out originally in mid-1994. It finally appeared on August 24, 1995.

Chiclet 1. Another term for a B Connector. See B CONNECTOR.
2. IBM once came out with a PC that had small keys. The press said the PC had a chiclet keyboard, after the chewing gum.

Chief Information Officer The person responsible for planning, choosing, buying, installing — and ultimately taking the blame for — a company's computer and information processing operation. Originally, CIOs were called data processing managers. Then they became Management Information System (MIS) managers. Then, CIOs. The idea of calling them CIOs was to reflect a new idea that the information they controlled was a critical corporate advantage and one that could give the company a competitive edge over its competitors — if played correctly.

Child Group In some systems, a new group of users created under a parent group is called a child group. Child groups sometimes have more properties than their parent groups.

Child Node An ATM term. A node at the next lower level of the hierarchy which is contained in the peer group represented by the logical group node currently referenced. This could be a logical group node, or a physical node.

Child Peer Group An ATM term. A child peer group of a peer group is any one containing a child node of a logical group node in that peer group. A child peer group of a logical group node is the one containing the child node of that logical group node.

CHILL Ccitt HIgh Level Language. A computer language developed by the CCITT for the standardization of software in telecommunications switches. Not widely adopted. C is more widely adopted.

Chime An electromechanical or electronic substitute for the conventional telephone bell, that sounds like a musical chime being struck, typically in a "bing-bong" sequence.

Chimney Effect Picture a phone system. We have an upright, rectangular cabinet full of printed circuit cards and all getting hot. How to cool them? Simple, raise the machine a little off the ground, put holes in the bottom of the cabinet and holes in the top of the cabinet. Hot air rises. Bingo, air will rise through the top of the cabinet and cool air will get sucked in the bottom of the cabinet. And bingo, you don't need a fan. This natural cooling technique is called the Chimney Effect and many modern phone systems now use it.

Chip 1. An integrated circuit. The physical structure upon which integrated circuits are fabricated as components of telephone systems, computers, memory systems, etc.
2. The transition time for individual bits in the pseudo-random sequence transmitted by the GPS satellite.

Chip Head Anyone whose education, entertainment and employment is primarily derived from computer-based devices. Also called a BIT HEAD.

Chip Jewelry A Wired Magazine definition: Chip Jewelry is a euphemism for old computers destined to be scrapped or turned into decorative ornaments. "I paid three grand for that Mac SE, and now it's nothing but chip jewelry."

Chirping 1. A rapid change (as opposed to a long-term drift) of the wavelength of an electromagnetic wave. Chirping is most often observed in pulsed operation of a source.
2. A pulse compression technique that uses (usually linear) frequency modulation during the pulse.

Choice Chip Your new TV will come with electronics that will allow you to program it not to receive certain programs, e.g. violent ones, you choose not to receive. The idea is that shows will be rated. Before they start, the show will broadcast a digital signal containing its rating. The "choice" chip in your TV will recognize the rating, check it against your instructions and block it or allow it through. The "choice" chip is so named as to give parents a choice of programs they and their children will watch. The provision for a choice chip was contained in telecommunications reform legislation passed by the Senate in mid-1995.

Choke An obsolete term: An inductance with either an air or iron core, designed to retard certain frequencies; as a radio frequency choke or an audio frequency choke.

Choke Coil A coil so wound as to offer a retarding or self inductance effect to an alternating current.

Choke Packet Packet used for flow control. Node detecting congestion generates choke packet and sends it toward source of congestion, which is required to reduce input rate.

Chooser A desk accessory on the Apple Macintosh that allows a user to choose items such as a printer or file server by clicking on an icon of the device.

Chopper A device for rapidly opening and closing a circuit. An ancient radio term.

Chroma The level of saturation or intensity of a color. Name sometimes applied to color intensity control in a receiver.

Chroma Key Method of electronically inserting the image from one video source into the picture from another video source using color for discrimination. A selected "key color" is replaced by the background image.

Chromatic Dispersion Chromatic dispersion is one of the mechanisms that limits the bandwidth of optical fibers by producing pulse spreading because of the various colors of light traveling in the fiber. Different wavelengths of light travel at different speeds. Since most optical sources emit light containing a range of wavelengths, each of these wavelengths arrive at different times and thereby cause the transmitted pulse to spread as it travels down the fiber.
Chromatic dispersion is the sum of material and waveguide dispersion. Dispersion can be positive or negative because it measures the change in the refractive index with wavelength. Thus, the total chromatic dispersion can actually be zero (really close to zero). For example, step-index single-mode fibers have zero dispersion at 1300nm, almost exactly at the same wavelength where the optical loss of the fiber is at a minimum. This is what allows single-mode fibers to have low loss and high bandwidth.

Chrominance The color portion of the video signal. Chrominance includes hue and saturation information but not brightness. Low chroma means the color picture looks pale or washed out; high chroma means the color is too intense, with a tendency to bleed into surrounding areas. Black, gray and white have a chrominance value of 0. Brightness is referred to as luminance.

Chromium Dioxide Tape whose coating is of chromium dioxide particles. Noted for its superior frequency output.

CHS Cylinder-Head Sector. The method of identifying a given location on a hard drive used by the original PC-AT BIOS (INT 13) and original IDE specification. Differences between details of the two methods resulted in the 528 MB limit on IDE drives. Enhanced IDE-compliant BIOSes can translate between the two methods, allowing drive sizes up to 8.4 GB. See ENHANCED IDE and IDE.

Chuck Hole Also known as Pot Hole. Slang for when your system hangs up on-line.

NEWTON'S TELECOM DICTIONARY

Churn Cellular phone and beeper users drop their monthly subscriptions often. The industry calls This phenomenon "churn." Churn is defined as the level of disconnects from service relative to the total subscriber base of the system. Often referred to on a percentage basis monthly, quarterly or annually. Sometimes it's as high as 2% or 3% a month. It drives the cellular and beeper business mad. It's very expensive to sign up a new customer.

Chutzpah A New York Jewish word which means unmitigated gall (audacity). The word is best exemplified by the story of the 15-year old who goes into court having killed his father and mother and falls on the mercy of the court that he's now an orphan. Now that's Chutzpah! Telecom companies without competition show they understand the meaning of chutzpah very well in the way they typically treat their customers.

CI 1. Customer Interface.
2. Certified Integrator.
3. An ATM term. Congestion Indicator: This is a field in a RM-cell, and is used to cause the source to decrease its ACR. The source sets CI=0 when it sends an RM-cell. Setting CI=1 is typically how destinations indicate that EFCI has been received on a previous data cell.

CI Labs Component Integration Labs. www.cilabs.org

CIC See Carrier Identification Code.

CICS Customer Information Control System. An IBM program environment designed to allow transactions entered at remote computers to be processed concurrently by a mainframe host. Also, IBM's Customer Information Control System software.

CID A generic term in Britain to identify a customer identity, client identity or contract identity. It is a single record and all the fields of information associated with it; for example, name, address, phone number, contact history and so on.

CIDB Calling Line Identification Delivery Blocking. A "feature" of central offices which lets you block the sending of your phone number to the person you're calling.

CIDR An internetworking term. Classless Inter-Domain Routing. A method for using the existing 32-bit Internet Address Space more efficiently.

CIF 1. A videophone ISDN standard which is part of the ITU-T's H.261/Px64 standard. It produces a color image of 288 non-interlaced luminance lines, each containing 352 pixels to be sent at a rate of 30 frames per second. The format uses two B channels, with voice taking 32 kbps and the rest for video.
2. CIF stands for Cost, Insurance and Freight are included. That means the seller pays the freight. The opposite of CIF is FOB, which stands for Free On Board. What this means is that you buy something, F.O.B. The seller puts it on a truck or railroad, plane, i.e. some carrier. He's responsible for getting it on the carrier. You — the buyer — are responsible for paying for the cost of the freight of getting you the goods you ordered.
3. An ATM term. Cells In Flight: An ABR service parameter, CIF is the negotiated number of cells that the network would like to limit the source to sending during idle startup period, before the first RM-cell returns.

CIFS Common Internet File System. A remote collaborative file sharing technology that, according to Microsoft, dramatically reduces the time it takes to open and work with remote files. Data General, Digital Equipment, Intel, Intergraph Corp, Network Alliance and Microsoft announced support for CIFS n June 13, 1996. According to Microsoft, the Common Internet File System is an enhanced version of the native file-sharing technology used in the Microsoft MS-DOS, Windows and Windows NToperating systems and IBM OS/2, and widely available on leading UNIX systems. It enables millions of computer users to open and share remote files directly on the Internet, expanding the Internet's ability to support interactive computing. CIFS technology provides reliable direct read and write access to files stored on remote computers without first requiring users to download or copy the files to a local machine, as done previously on the Internet. According to Microsoft's June 13 release, CIFS can improve the performance of many types of file access. Because CIFS is based on existing standards, users will be able to use thousands of existing applications over the Internet as well as integrate them with browser applications designed for the World Wide Web.
The proposed Common Internet File System protocol runs over TCP/IP and is an enhanced version of the open, cross-platform protocol for distributed file sharing called Server Message Block (SMB). The SMB protocol is the standard way that millions of PC users already share files across corporate intranets and is the native file-sharing protocol in Windows 95, Windows NT and OS/2. The proposed Common Internet File System protocol has been enhanced over previous versions of the SMB protocol in ways that make it well suited for use on the Internet. CIFS, for example, supports the Internet's Domain Name Service (DNS) for address resolution. The protocol runs optimally over slow-speed dial-up lines, helping improve performance for the vast numbers of users today who access the Internet using a modem. In addition to remote file sharing, CIFS has mechanisms to support remote printer sharing.

CIGOS Canadian Interest Group on Open Systems. Canadian organization which promotes OSI.

CIIG Canadian ISDN Interest Group. Canadian organization which promotes ISDN.

CIM 1. Computer-Integrated Manufacturing.
2. An MCI definition. Customer Information Manager.

CIP An ATM term. Carrier Identification Parameter: A 3 or 4 digit code in the initial address message identifying the carrier to be used for the connection.

Ciphertext The result of processing plaintext (unencrypted information) through an encryption algorithm. Ciphertext is thus the content of an encrypted message. See CLIPPER CHIP.

CIR 1. Committed Information Rate. It is used in the Frame Relay arena. CIR refers to the amount of data guaranteed to be transmitted over the carrier's network. The customer is always free to "burst" up to the port's speed but any amount of data over the CIR will be marked as DE or Discard Eligible. All data marked as DE will be discarded in the event of network congestion. CIR comes in increments of 16 Kbps (16-32-48-64) with 16 being the lowest. An example would be: A 56k frame relay connection with a CIR of 16k (very common) would allow the customer to transmit or burst data up to the full 56 Kbps. If they bursted up to the full 56k they would only be guaranteed delivery of 16k. See also COMMITTED INFORMATION RATE.
2. An ATM term. Committed Information Range: CIR is the information transfer rate which a network offering frame relay services (FRS) is committed to transfer under normal conditions. The rate is averaged over a minimum increment of time.

NEWTON'S TELECOM DICTIONARY

Circuit The physical connection (or path) of channels, conductors and equipment between two given points through which an electric current may be established. Includes both sending and receiving capabilities. A circuit can also be a network of circuit elements, such as resistors, inductors, capacitors, semiconductors, etc., that performs a specific function. A circuit can also be a closed path through which current can flow.

Circuit Board Same as a Printed Circuit Board, namely a board with microprocessors, transistors and other small electronics components. Such a board slides into a the slot in a telephone system or personal computer. Also called a circuit card.

Circuit Breaker A special type of switch arranged to open a circuit when overloaded, without injury to itself. A circuit breaker is basically a re-usable fuse.

Circuit Card Same as a Circuit Board. See CIRCUIT BOARD.

Circuit Emulation A connection over a virtual channel-based network providing service to the end user that is indistinguishable from a real, point-to-point, fixed bandwidth circuit.

Circuit Identification Code CIC. The part of CCS/SS7 signaling message used to identify the circuit that is being established between two signaling points (14 bits in the ISDNUP).

Circuit Mode 1. An AT&T term for the method of communications in which a fixed bandwidth circuit is established from point to point through a network and held for the duration of a telephone call.
2. An AIN term for a type of switching that causes a one-to-one correspondence between a call and a circuit. That is, a circuit or path is assigned for a call between each switching node, and the circuit or path is not shared with other calls.

Circuit Noise Level At any point in a transmission system, the ratio of the circuit noise at that point to some arbitrary amount of circuit noise chosen as a reference.

Circuit Order Management System COMS. An automated processing system of MCI circuit- and service-related information. Processes hardwire service circuit orders from order entry through scheduling and completion. COMS also provides circuit order data, hardwire customer data, and circuit inventory data to other MCI systems in Finance, Engineering, and Operations.

Circuit Order Record COR. Report generated by the COR Tracking System within NOBIS, indicating circuit installations, changes, and disconnects.

Circuit Provisioning The telephone operating company process that somehow organizes to get you a trunk or other special service circuit.

Circuit Segregation Differentiating between services that are maintained by separate technicians or departments. Can be accomplished through visual and/or mechanical means.

Circuit Switched Digital Capability CSDC. A service implemented by some regional Bell Operating Companies that offers users a 56-Kbps digital service on a user-switchable basis.

Circuit Switching The process of setting up and keeping a circuit open between two or more users, such that the users have exclusive and full use of the circuit until the connection is released. There are basically three types of switching — CIRCUIT, PACKET and MESSAGE.
— PACKET SWITCHING is like circuit switching in that you can also switch information between people or devices, but in packet switching (as in circuit switching), no circuit is left open on a dedicated basis. Circuit switching is like having your own railroad track for your conversation to travel on that's yours as long as you keep the connection open. Once you hang up, the next caller gets to use that track.
— PACKET SWITCHING is a data switching technique only. In packet switching, the addresses on your packets are read by the switches as they approach, and are switched down the tracks. The next packet is read to throw the switches to send THAT packet where it needs to go. The data conversation is sent in packets. Each packet can be sent along different tracks as they are open. The packets are assembled at the other end — typically in the last switching office before the packets reach the distant computer or distant user.
— MESSAGE SWITCHING sends a message from one end to the other. But it's not interactive, as in packet or circuit switching. In message switching, the message is typically received in one block, stored in one central place, then retrieved or sent in one clump to the other end.

Circuit Tier An AT&T term for the tier within the Universal Information Services network that provides real-time circuit switching of channels.

Circuit, Four Wire A path in which four wires are presented to the terminal equipment (phone or data), thus allowing for simultaneous transmission and reception. Two wires are used for transmission in one direction and two in the other direction.

Circular Extension Network Permits two or more single-line phones connected to a PBX, each with its own extension, to operate like a "square" key telephone system. An incoming call directed to any non-busy phone in the group will ring at all of the non-busy phones. The first extension to answer will be connected to the incoming call. At any time, a non-busy extension can make or receive calls.

Circular Hunting When calling a phone, the switching system makes a complete search of all numbers within the hunting group, regardless of the location within that group of the called number. For example, the hunt groups is 231, 232, 233 and 234, the call is directed to 233. If it is busy, the equipment will search 234, 231, and 232 to find a non-busy phone or line. Essentially it goes around the ring, remembering where it last connected and then goes to the next line or phone in the circle. See also HUNT GROUP and TERMINATED HUNT GROUP.

Circular Mil The measure of sectional area of a wire.

Circular Polarization In electromagnetic wave propagation, polarization such that the tip of the electric field vector describes a circle in any fixed plane intersecting, and normal to, the direction of propagation. The magnitude of the electric field vector is constant. A circularly polarized wave may be resolved into two linearly polarized waves in phase quadrature with their planes of polarization at right angles to each other.

NEWTON'S TELECOM DICTIONARY

Circulator 1. In networking, a passive junction of three or more ports in which the ports can be accessed in such an order that when power is fed into any port it is transferred to the next port, the port counted as following the last in order.
2. In radar, a device that switches the antenna alternately between the transmitter and receiver.
Circumnaural A type of headphone that almost totally isolates the listener from room sounds.
CIS Contact Image Sensor. A type of scanner technology in which the photodetectors come in contact with the original document.
CISC Complex Instruction Set Computing. PC Magazine defines CISC as a microprocessor architecture that favors robustness of the instruction set over the speed with which individual instructions are executed. The Intel 486 and Pentium are both examples of CISC microprocessors. See also RISC — Reduced Instruction Set Computing. See RISC.
Ciscoworks A series of router management applications developed and marketed by Cisco Systems for managing Cisco-based router networks.
CIT Computer Integrated Telephone is Digital Equipment Company's program, announced in October 1987, that provides a framework for functionally integrating voice and data in an applications environment so that the telephone and terminal on the desktop can be synchronized, the call arriving as the terminal's screen on the caller arrives. CIT uses the DEC VAX line of computers. According to DEC, CIT supports both inbound and outbound telecommunications applications. In an inbound scenario, the application may recognize the caller's originating phone number through Automatic Number Identification (ANI) and/or the dialed number through Dialed Number Identification Service (DNIS), match the information to corresponding data base records and automatically deliver the call and the data to the call center agent. In an outbound application, dialing can be automated, increasing the number of connected calls. In either scenario, the telephone calls and associated data can be simultaneously transferred to alternate locations within an organization, adding a new level of customer service to call center applications. Digital made its first CIT announcements at Telecom '87 in Geneva, Switzerland. The CIT product set, consisting of client and server software implementing a variety of switch-to-computer link protocols, and providing a robust applications interface, was first shipped in 1989. The company announced its latest release, CIT Version 2.1, in January 1991. See also OPEN APPLICATION INTERFACE.
Citizens Band One of two bands used for low power radio transmissions in the United States — either 26.965 to 27.225 megahertz or 462.55 to 469.95 megahertz. Citizens band radio is not allowed in many countries, even some civilized countries. In some countries they use different frequencies. CB radios, in the United States, are limited by FCC rule to four WATTS of power, which gives each CB radio a range of several miles. Some naughty people boost their CBs with external power. The author of this dictionary has actually spoken to Australia while driving on the Santa Monica Freeway in Los Angeles. See also CB.
CIV Cell Interarrival Variation. An ATM term. See CELL INTERARRIVAL VARIATION
CIVDL Collaboration for Interactive Visual Distance Learning. A collaborative effort by 10 US universities that uses dial-up videoconferencing technology for the delivery of engineering programs.
CIX Commercial Internet Exchange; an agreement among network providers that allows them to exchange commercial traffic. A connection point between the commercial Internet service providers. Pronounced "kicks."
CL Connectionless service: A service which allows the transfer of information among service subscribers without the need for end-to-end establishment procedures.
Cladding 1. The transparent material, usually glass, that surrounds the core of an optical fiber. Cladding glass has a lower refractive index than core glass.
2. When referring to a metallic cable, a process of covering with a metal (usually achieved by pressure rolling, extruding, drawing, or swaging) until a bond is achieved.
Cladding Diameter The diameter of the circle that includes the cladding layer in an optical fiber.
Cladding Mode In an optical fiber, a transmission mode supported by the cladding; i.e., a mode in addition to the modes supported by the core material.
Cladding Mode Stripper A device for converting optical fiber cladding modes to radiation modes; as a result, the cladding modes are removed from the fiber. Often a material such as the fiber coating or jacket having a refractive index equal to or greater than that of the fiber cladding will perform this function.
Cladding Ray In an optical fiber, a ray that is confined to the core and cladding by virtue of reflection from the outer surface of the cladding. Cladding rays correspond to cladding modes in the terminology of mode descriptors.
Claim Process A technique used to determine which station will initialize an FDDI ring.
Claire Harry Newton's favorite daughter. Why "favorite?" Simple. She's his only daughter. Her full name is Claire Elizabeth Newton. In 1996, she was 16.
CLAMN Called Line Address Modification Notification
Clamper An electronic circuit which sets the level of a signal before the scanning of each line begins to insure that no spurious electronic noise is introduced into the picture signal from the electronics of the video equipment.
CLAS Centrex Line Assignment Service. A service from Nynex and other Bell operating companies, which allows Centrex subscribers to change their class of service by dialing in on a personal computer, reaching the phone company's computer and then changing things themselves — without phone company personnel assisting or hindering.
Load your PC with communications software. Dial your local central office. Change your Centrex phone numbers. Turn on, turn off features. Change pickup groups. Add numbers to speed dialing, etc. Your on-line changes are checked by the phone company's computers. If they make sense (i.e. one change doesn't conflict with another), they take effect by early the following day — at which time you can call up and get a report on which took, which didn't and who's got what. Saves calling in person. Is more accurate. And, best of all, saves money. Typically just one flat monthly charge. No charge for any of your changes.
CLASS 1. Custom Local Area Signaling Services. It is based on the availability of channel interoffice signaling. Class consists of number-translation services, such as call-forwarding and caller identification, available within a local exchange of Local Access and Transport Area (LATA). CLASS is a service mark of Bellcore. Some of the phone services which Bellcore promotes for CLASS are Automatic Callback, Automatic Recall, Calling Number Delivery, Customer Originated Trace, Distinctive

130

Ringing/Call Waiting, Selective Call Forwarding and Selective Call Rejection. See also CALLING LINE IDENTIFICATION.
2. In an object-oriented programming environment, a class defines the data content of a specific type of object, the code that manipulates it, and the public and private programming interfaces to that code. See ANI and ISDN.
3. See also CLASS 1 and 2, below.

Class 1 Also called Class 1/EIA-578. It's an American standard used between facsimile application programs and facsimile modems for sending and receiving Class 1 faxes. The Class 1 interface is an extension of the EIA/TIA's (Electronics Industry Association and the Telecommunications Industry Association) specification for fax communication, known as Group III. Class 1 is a series of Hayes AT commands that can be used by software to control fax boards. In Class 1, both the T.30 (the data packet creation and decision making necessary for call setup) and ECM/BFT (error-correction mode/binary file transfer) are done by the computer. A specification being developed (fall of 1991) Class 2, will allow the modem to handle these functions in hardware. Industry analysts believe Class 2 will be the standard for the long haul, but approval is slow. Even so, some modem makers will shortly deliver data/fax modems. See also CLASS 1 OFFICE.

Class 1 Office A regional toll telephone switching center. The highest level toll office in AT&T's long distance switching hierarchy. There are essentially five levels in the hierarchy, with the lowest level — Class 5 — being those central offices owned by the local telephone companies. Each of the classes can complete calls between themselves. But, if the routes are busy, then calls automatically climb the hierarchy. A Class 1 office is the office of "last resort."

Class 2 Also known as Class 2.0/EIA-592. An American standard used between facsimile application programs and facsimile modems for sending and receiving Class 2.0 faxes. This class places more of the task of establishing the fax connection onto the fax modem, while continuing to rely on the host's processor to send and receive the image data. The Class 2 standard (known as PN-2388) is still under study by the EIA's (Electronic Industries Association) TR.29 committee, with further revisions expected. See CLASS 1.

Class 2 Office The second level in AT&T's long distance toll switching hierarchy.

Class 3 Office The third level in AT&T's long distance toll switching hierarchy.

Class 4 Office The fourth level in AT&T's long distance toll switching hierarchy — the major switching center to which toll calls from Class 5 offices are sent. In U.S. common carrier telephony service, a toll center designated "Class 4C" is an office where assistance in completing incoming calls is provided in addition to other traffic. A toll center designated "Class 4P" is an office where operators handle only outbound calls, or where switching is performed without operator assistance.

Class 5 Office An end office. Your local central office. The lowest level in the hierarchy of local and long distance switching which AT&T set up when it was "The Bell System." A class 5 office is a local Central Office that serves as a network entry point for station loops and certain special-service lines. Also called an End Office. Classes 1, 2, 3, and 4 are toll offices in the telephone network.

Class A Certification A Federal Communications Commission (FCC) certification that a given make and model of computer meets the FCC's Class A limits for radio frequency emissions, which are designed for commercial and industrial environments. See CLASS B CERTIFICATION.

Class B Certification A Federal Communications Commission (FCC) certification that a given make and model of computer meets the FCC's Class b limits for radio frequency emissions, which are designed to protect radio and television reception to residential neighborhoods from excessive radio frequency interference (RFI) generated by computer usage. Class B computers also are shielded more efficiently from external interface. Computers used at home are more likely to be surrounded by radio and television equipment. If you plan to use your computer at home, avoid computers that have only Class A certification (that is, they failed Class B).

Class n Office The way a telephone company defines its switching facilities. Class 5 is an end office (local exchange), Class 4 is a toll center, Class 3 is a primary switching center, Class 2 is a sectional switching center, and Class 1 is a regional switching center. See CLASS 1, CLASS 2, CLASS 3, CLASS 4 and CLASS 5.

Class Of Emission The set of characteristics of an emission, designated by standard symbols, e.g., type of modulation of the main carrier, modulating signal, type of information to be transmitted, and also if appropriate, any additional signal characteristics.

Class Of Office A ranking assigned to switching points in the telephone network, determined by function, interfaces and transmission needs.

Class Of Service 1. Each phone in a system may have a different collection of privileges and features assigned to it, such as access to WATS lines. Class of Service assignments if properly organized, can become an important tool in controlling telephone abuse.
2. A subgrouping of telephone users for the sake of rate distinction. This may distinguish between individual and party lines, between Government lines and others, between those permitted to make unrestricted international dialed calls and others, between business or residence and coin, between flat rate and message rate, and between restricted and extended area service.
3. A category of data transmission provided in a public data network in which the data signaling rate, the terminal operating mode, and the code structure (if any) are standardized. This is defined within ITU-T Recommendations X.1.

Classical IP A set of specifications developed by the Internet Engineering Task Force (IETF) for the operation of LAN-to-LAN IP connectivity over an ATM network.

Classmark A designator used to describe the service feature privileges, restrictions, and circuit characteristics for lines or trunks accessing a switch; e.g., precedence level, conference privilege, security level, zone restriction. See CLASS OF SERVICE.

CLC Carrier Liaison Committee. A committee formed to help industry participants work together to resolve the issues of implementing 800 Portability. CLC is sponsored by the Exchange Carriers Standards Association (ECSA) and is comprised of the LECs (local exchange carriers), long distance carriers and users of 800 service.

CLD Coalition Inc. Competitive Long Distance Coalition, a Washington, DC-based lobbying group.

Clear To cause one or more storage locations to be in a prescribed state, usually that corresponding to a zero or to the space character.

Clear Channel 1. A digital circuit where no framing or control bits (i.e. for signaling) are required, thus making the full bandwidth available for communications. For example, a 56 Kbps circuit is typically a 64 Kbps digital circuit with 8 Kbps used for signaling. Sometimes called Switched 56, DDS or ADN. Each of the carriers have their own name for clear channel service. The phone companies are obsoleting the 56 Kbps service in favor of the more modern ISDN BRI, which has two 64 Kbps circuits and one 16 Kbps packet service, part of which is used for signaling on the 64 Kbps channels.
2. An SCSA term. A channel which is used exclusively for data transmission, with no bandwidth required for administrative messages such as signaling or synchronization. All SCbus data channels are clear.
Clear Collision Contention that occurs when a DTE and a DCE simultaneously transfer a clear request packet and a clear indication packet specifying the same logical channel. The DCE will consider that the clearing is completed and will not transfer a DCE clear confirmation packet.
Clear Confirmation Signal A call control signal to acknowledge reception of the DTE clear request by the DCE or the reception of the DCE clear indication by the DTE.
Clear To Send CTS. One of the standard attributes of a modem in which the receiving modem indicates to the calling modem that it is now ready to accept data. One of the standard pins used by the RS-232-C standard. In ITU-T V.24, the corresponding pin is called Ready For Sending.
Clearinghouse A service company that collects and processes roaming and billing information from a number of carriers. It then transfers the compiled data to the proper carriers for credits and billing.
Clearline 1.5 A Sprint name for T-1 service. It is an all digital 1.544 Mbps private line service that connects two customer sites via dedicated T-1 access lines.
Clearline 45 A Sprint name for DS-3 service. This high-capacity point-to-point private line service transmits voice, data, and video at 44.736 Mbps.
Clearline Fractional 1.5 A Sprint name for all digital private line service which transmits voice, data, and video at speeds from 112/123 Kbps up to 672/768 Kbps - a fraction of a T-1, also called a DS-1. The service may be ordered in 56/64 Kbps increments from two channels (112/128 Kbps) to 12 channels (672/768 Kbps). Point-to-point service connects customer sites via dedicated T-1 access lines.
Cleaving The controlled breaking of a fiber so that its end surface is smooth.
CLEOS Conference of the Lasers and Electro-Optics Society.
CLI Compression Labs, Inc. One of the foremost codec makers and developer of some of the first "low-bandwidth" codecs in the US. Their VTS 1.5 codec was one of the first two codes (the other was from NEC) to compress full-motion video to 1.5 Mbps (T-1 speed).
Click Tones Indicates acknowledgment by the cellular system that the cellular system's computer is processing the call.
Clickstreams The paths a user takes as he or she navigates cyberspace. Advertisers and online media providers are developing software that can accurately track user's clickstreams.
CLID Calling Number IDentification. Same as ANI (Automatic Number Identification), except CLID tends to refer to identification of local calling numbers. CLID tends to be referred to as a service of CLASS. See also CALLER ID, CLASS and ANI.
Client Clients are devices and software that request information. Clients are objects that uses the resources of another object. A client is a fancy name for a PC on a local area network. It used to be called a workstation. Now it is the "client" of the server. See also CLIENT SERVER, CLIENT SERVER MODEL, FAT CLIENT, MAINFRAME SERVER, MEDIA SERVER and THIN CLIENT.
Client Application Any computer program making use of the processing resources of another program.
Client Operating System Operating System running on the client platform. See CLIENT.
Client Server A computer on a local area network that you can request information or applications from. The idea is that you — the user — are the client and it — the slave — is the server. That was the original meaning of the term. Over time, client server began to refer to a computing system that splits the workload between desktop PCs (called "workstations") and one or more larger computers (called "servers") joined on a local area network (LAN). The splitting of tasks allow the use of desktop graphic user interfaces, like Microsoft's Windows or Apple Macintosh's operating system, which are easier to use (for most people) than the host/terminal world of mainframe computing, which placed a "dumb terminal" on a user's desk. That dumb terminal could only send and receive simple text-based material. And the less it sent, the faster it worked (lines were slow), so some of the "human interfaces" were very cryptic. You often were forced to spend weeks at school learning simple mainframe programs.
A good analogy of client-server computing, according to Peter Lewis of the New York Times is to think of client server as a restaurant where the waiter takes your order for a hamburger, goes to the kitchen and comes back with some raw meat and a bun. You get to cook the burger at your table and add your favorite condiments. In computerese, this is client/server, distributed computing, where some processing work is done by the customer at his or her table, instead of entirely in the kitchen (centralized computing in the old mainframe days). It sounds like more work, but it has many advantages. The service is faster. The food is cooked exactly to your liking, and the giant, expensive stove in the kitchen can be replaced by lots of cheap little grills. See CLIENT SERVER MODEL, DOWNSIZING, REENGINEERING and SERVER.
Client Server Model In most cases, the "client" is a desktop computing device or program "served" by another networked computing device. Computers are integrated over the network by an application, which provides a single system image. The server can be a minicomputer, workstation, or microcomputer with attached storage devices. A client can be served by multiple servers. See CLIENT SERVER.
Client Software An Internet access term. Multiprotocol PPP client software allowing dial-in access to the public Internet or corporate LANs via a dialup switch or a remote access server. The client software dialer is responsible for establishing/terminating the dial-in connection. The client software PPP driver manages the traffic sent/received across the network link.
Client Telescript A General Magic term. Telescript that is integrated into the various client platforms including Macintosh, DOS/Windows, UNIX, PenPoint, and Newton.

CLIP Calling Line Identification Presentation, A wireless telecommunications term. A supplementary service provided under GSM (Global System for Mobile Communications).

Clip On Toll Fraud Clip on toll fraud occurs when someone connects a phone between someone else's phone (typically a coin phone) and the central office and makes unlawful toll calls. The term "clip on" comes because the telephone service thief "clips on" to the line. Clip on toll fraud is often done on COCOT (Customer Owned Coin Operated Telephone) phone lines because these lines do not enjoy the same protection from toll fraud which is afforded to coin phone lines which local telcos provide to their own coin phones.

Clipboard A generic term for a place in software which holds text, pictures or images that you are copying or moving between applications. The clipboard is a temporary holding place only. When you cut or copy a new item it will replace the current clipboard contents. In other words, there's usually only one clipboard. Windows uses a clipboard and so does the Apple Macintosh.

Clipbook A Windows NT term. Permanent storage of information you want to save and share with others. This differs from the Clipboard which temporarily stores information. You can save the current contents of the clipboard, which temporarily stores information. You can then share that information, allowing others to connect to the Clipboard. See CLIPBOOK PAGE.

Clipbook Page A Windows NT term. A unit of information pasted into a local ClipBook. The ClipBook page is permanently saved. Information in a ClipBook page can be copied back onto the Clipboard and then pasted into documents. You can share ClipBook pages on the notework.

Clipbook Service A Windows NT term. Supports the ClipBook Viewer application, allowing pages to be seen by remote ClipBooks.

Clipper A circuit or device that limits the instantaneous output signal amplitude to a predetermined maximum value, regardless of the amplitude of the input signal. See CLIPPER CHIP.

Clipper Chip A microprocessor chip, officially known as the MYK-78, which the Federal Government wants to add to phones and data communications equipment. The chip would ensure that conversations in Clipper-equipped communicating equipment would be private — from everybody except the Government. With a court-approved wiretap, an agency like the FBI, could listen in, since the Government would have the key to Clipper. On February 4, 194, the Clinton White House announced its approval of the Clipper chip and the "Crypto War" broke out — with many companies and individuals urging a stop to Clipper. In late Spring, 1994 an AT&T Bell Labs researcher revealed that he had found a serious flaw in the Clipper technology. As of writing it wasn't clear what would happen to the Clipper Chip. See NSA.

Clipping Clipping has two basic meanings. The first refers to the effect caused by a simplex (one way at a time) speakerphone. Here the conversation goes one way. When the other person wants to talk, the voice path has to reverse (to "flip"). While the flipping takes place, a few sounds are "clipped" from that person's conversation. This phenomenon happens on some long distance and many overseas channels. These channels are so expensive, they are simultaneously shared by many conversations. Gaps in your conversation are filled by other people's conversation. But when you start talking, the equipment has to recognize you're now talking, find some capacity for your conversation, and send it. In the process of doing this, your first word or part of your first word might be "clipped" and the conversation will sound "broken."

The second way your voice is clipped is what happens every day on the telephone. You're squeezing your own voice which typically spans 10,000 Hertz into a voice channel which is only 3,000 Hertz. This clips the extremes of your conversation — the higher sounds. As a result, your voice sounds flatter over the phone. As you become more economical and try to squeeze your voice into smaller capacity channels, so it becomes increasingly clipped.

CLIR Calling Line Identification Restriction, A wireless telecommunications term. A supplementary service provided under GSM (Global System for Mobile Communications). See GSM.

CLLI Code Pronounced silly code. When a phone company institutes trapping and tracing on a phone line, one of the pieces of information it can receive on incoming calls is the CLLI Code, a 11 character code that might look like "nycmny18dso." That says the call is coming in from New York City, Manhattan from a central office called 18DSO (which I happen to know is an AT&T 5E located on West 18th Street. See ANNOYANCE CALL BUREAU, TRAP and TRACE, and WIRE TAP.

CLM Career Limiting Move. An ill-advised activity. Trashing your boss while he or she is within earshot is a serious CLM.

CLNP Connectionless Network Protocol. An OSI network layer protocol that does not require a circuit to be established before data is transmitted. The OSI protocol for providing the OSI Connectionless Network Service (data gram service). CLNP is the OSI equivalent to Internet IP, and is sometimes called ISO IP.

CLNS See Connectionless Network Service.

Clock Exactly as it sounds. An oscillator-generated signal that provides a timing reference for a transmission link. A clock provides providing signals used in a transmission system to control the duration of certain functions such as the duration of signal elements or the sampling rate. It also generates periodic, accurately spaced signals used for such purposes as timing, regulation of the operations of a processor, or generation of interrupts. In short, a clock has two functions: 1. To generate periodic signals for synchronization on a transmission facility. 2. To provide a time base for the sampling of signal elements. Used in computers, a clock synchronizes certain procedures, such as communication with other devices. It simply keeps track of time, which allows computers to do the same things at the same time so they don't "bump into each other."

Clock Bias The difference between the GPS clock's indicated time and true universal time. GPS is Global Positioning System. See GPS.

Clock Cycle The time that elapses from one read or write operation to another in the main memory of a computer's central processing unit (CPU). The more tasks that can be accomplished per cycle, the more efficient the chip. Some chips like the i860 chip can execute two instructions and three operations per clock cycle.

Clock Difference A measure of the separation between the respective time marks of two clocks. Clock differences must be reported as algebraic quantities measured on the same time scale. The date of the measurement should be given.

Clock Speed Each CPU contains a special clock circuitry which is connected to a quartz crystal (same as the one in your watch). The quartz crystal's vibrations, which are very fast, coordinate the CPU's operation, keeping everything in step. CPU clock speeds are measured in megahertz, or MHz, which stands for "million cycles per second." The clock speeds of today's computer range from a slow of 4.77 MHz (the original IBM PC) to 25 MHz with some Intel 80386 based machines. Clock speed is a misleading term. It is only one way of measuring the speed of a computer. One other critical way is how fast you can read and write information to the hard disk. How important that is depends on whether you're running a program with lots of access to your hard disk (like a database program) or running a program which uses a lot of calculations in RAM, i.e. a spreadsheet.

Clock Tolerance The maximum permissible departure of a clock indication from a designated time reference such as Coordinated Universal Time.

Clocking In synchronous communication, a periodic signal used to synchronize transmission and reception of data and control characters.

Clockwise Polarized Wave An elliptically or circularly polarized electromagnetic wave in which the direction of rotation of the electric vector is clockwise as seen by an observer looking in the direction of propagation of the wave.

Clone Fraud A way of using cellular phones to steal phone calls. In clone fraud, a legitimate serial number is programmed into an imposter's cellular telephone. This allows unauthorized calling to go on until a huge bill appears on the mailbox of the bewildered subscriber to whom the serial number actually belongs. Crooks get the numbers because the numbers are broadcast with every cellular call and can be picked up by ordinary radio scanners, which you can often buy at your local electronics store. See CLONED PHONE and TUMBLING.

Cloned Phone A cellular phone has two basic ways it identifies itself to the cellular phone company it wants to use — its own telephone number (which can be changed) and a special secret number that's embedded into silicon inside the phone. That number is called an Electronic Serial Number, or ESN. When the phone wants to make a call, it sends those numbers and the cellular carrier uses them to check if the call is authorized. But because the information is traveling through the air, anyone with a scanner can pick up the information and retransmit it later, thus creating a "cloned phone" and pretending that he's authorized to make the call. Of course, the owner of the cloned phone ultimately gets the bill and a nasty shock.

Close Coupling The condition in which two coils are placed in close magnetic relation to each other, thus establishing a high degree of mutual induction.

Close Talk A voice recognition term. An arrangement where a microphone is fewer than four inches from the speaker's mouth.

Closed Architecture Proprietary design that is compatible only with hardware and software from a single vendor of single product family. Contrast with OPEN ARCHITECTURE.

Closed End The end of a Foreign Exchange — FX — line which ends on a PBX, a key system or a telephone. The closed end is the end of the circuit beyond which a call cannot progress further. The other end of the FX circuit is called the "open end," because calls can progress further.

Closed Loop System A closed electrical circuit into which a standard signal is feed and received instantly. A measure of the difference between the input signal and the output signal is a measure of the error, and potentially what's causing it.

Closed User Group A group of specified users of a data network that is assigned a facility that permits them to communicate with each other but precludes communications with other users of the service or services.

Closure A cabinet, pedestal, or case used to enclose cable sheath openings necessary for splicing or terminating fibers.

Cloud Some of the newer high-speed data, phone company-offered services resemble a local area network. You connect to them directly. To make a call, you don't actually dial a number as you do on a circuit-switched service, you just transmit, putting an address at the front of your transmission. The service reads the address and sends it where you want. Like a LAN, everything is connected and on line. The concept is get stuff sent from one place to another much faster than would be possible if you had to wait to dial, for the circuit to be set up, for the machine at the other end to answer, etc. In these high-speed services, the circuit is "always set up." The provider (the phone company) refers to its network as a "cloud." And when you see diagrams of these newer high-speed services, like SMUTS and frame relay, you see the carrier portion drawn as a cloud (like the one you see in the sky). Services with "clouds" are also called "connectionless." See the various definitions below beginning with CONNECTIONLESS.

CLP Cell Loss Priority: This bit in the ATM cell header indicates two levels of priority for ATM cells. CLP=0 cells are higher priority than CLP=1 cells. CLP=1 cells may be discarded during periods of congestion to preserve the CLR of CLP=0 cells. See also Cell-loss Priority Field

CLR Cell Loss Ratio: CLR is a negotiated QoS (Quality of Service) parameter and acceptable values are network specific. The objective is to minimize CLr provided the end-system adapts the traffic to the changing ATM layer transfer characteristics. The Cell Loss Ratio is defined for a connection as: Lost Cells/Total Transmitted Cells. The CLR parameter is the value of CLR that the network agrees to offer as an objective over the lifetime of the connection. It is expressed as an order of magnitude, having a range of 10-1 to 10-15 and unspecified.

CLTS ConnectionLess Transport Service.

Cluster 1. Collection of terminals or other devices in a single location. A cluster control unit and a cluster controller in IBM 3270 systems are devices that control the input/output operations of a group (cluster) of display stations.
2. Unit of storage allocation used by MS-DOS usually consisting of four or more 512-byte sectors.
3. Physical grouping of workstations that share one or more panel runs.

Cluster Controller A device that can control input/output operations of more than one device connected to it (e.g. a terminal). An interface between several bisynchronous devices and a PAD, NC or communication facility. The cluster controller handles remote communications processing for its attached devices. Most common types are IBM 327X.

Cluster Size An operating function or term describing the number of sectors that the operating system allocates each time disk space is needed.

CLUT An imaging term. Color Look-Up Table. The palette used in an indexed color system. Usually consists of 256 colors.

CM 1. Computing Module.
2. Configuration Management. A wireless telecommunications term. The tracking, coordination, and administration of software and hardware related to telecommunication or information systems. Versions are controlled and tracked.
CMC Common Messaging Calls. A messaging standard defined by the X.400 API Association. CMC 1.0 defines a basic set of calls to inject and extract messages and files and access address information. CMC is intended to define a useful common denominator across a wide variety of messaging systems. The idea is that an electronic mail system, no matter how crude, should be able to support a CMC front end. CMC's major "competition" is MAPI — Messaging Application Programming Interface — from Microsoft, though simple MAPI is almost identical to CMC.
CMIP Common Management Information Protocol. An ITU-TSS standard for the message formats and procedures used to exchange management information in order to operate, administer maintain and provision a network. In short, CMIP is the protocol used for exchanging network management information. Typically, this information is exchanged between two management stations. CMIP can, however, be used to exchange information between an application and a management station. CMIP has been designed for OSI networks, but it is transport independent. Theoretically, it could run across a variety of transports, including, for example, IBM's Systems Network Architecture. See CMIP/CMIS.
CMIP/CMIS Common Management Information Protocol/Common Management Information Services. An OSI network management protocol/service interface created and standardized by ISO for managing heterogeneous networks.
CMISE Common Management Information Service Element. A wireless telecommunications term. The functionality provided by CMIP in transporting network management information.
CMOL Short for "CMIP Over Logical Link Control". An implementation of the CMIP protocol over the second layer of the OSI protocol stack, to be proposed as a standard by 3 Com Corp. and IBM. The goal of CMOL is to create agents that require significantly less memory than CMIP implemented over OSI, or SNMP implemented over UDP.
CMOS Complementary Metal Oxide Semiconductor. A technology for making integrated circuits known for requiring less electricity. See also COMPLEMENTARY METAL OXIDE SEMICONDUCTOR.
CMOS RAM Complementary Metal Oxide Semiconductor Random Access Memory. Memory which contains a personal computer's configuration information. CMOS RAM must have continuous power to preserve its memory. This power is typically supplied by a lithium battery.
CMOS Setup A program which prepares the system to work. CMOS setup records your PC's hardware configuration information into CMOS RAM. It must be modified when you add, change or remove hardware.
CMOT CMIP Over TCP/IP. An Internet standard defining the use of CMIP for managing TCP/IP local area networks.
CMP Communications Plenum Cable.
CMR An ATM term. Cell Misinsertion Rate: The ratio of cells received at an endpoint that were not originally transmitted by the source end in relation to the total number of cells properly transmitted.
CMS 1. Call Management System. This is the AT&T label for an inbound call distribution management reporting package. CMS is found on the Horizon, the Merlin, S/75 and S/85 PBX/ACDs. CMS is the successor product to AEMIS/PRO 150/500.
2. Call Management Services. Canadian term for local calling features based on CLID (Calling Line Identification).
CMS 8800 Cellular Mobile Telephone Service (North American version).
CMTS CMTS stands for the Cellular Mobile Telephone System. The original and still, most common CMTS is a low-powered, duplex, radio/telephone which operates between 800 and 900 MHZ, using multiple transceiver sites linked to a central computer for coordination. The sites, or "cells,", named for their honeycomb shape, cover a range of one to six, or more, miles in each direction. The cells overlap one another and operate at different transmitting and receiving frequencies in order to eliminate crosstalk when transmitting from cell to cell. Each cell can accommodate up to 45 different voice channel transceivers. When a cellular phone is activated, it searches available channels for the strongest signal and locks onto it. While in motion, if the signal strength begins to fade, the telephone will automatically switch signal frequencies or cells as necessary without operator assistance. If it fails to find an acceptable signal, it will display an "out of service" or "no service" message, indicating that it has reached the limit of its range and is unable to communicate. Each cellular telephone has a unique identification number which allows the Mobile Telephone Switching Office (MTSO) to track and coordinate all mobile phones in its service area. This ID number is known as the Electronic Security Number (ESN). The ESN and cellular phone's telephone Number are NOT the same. The ESN is a permanent number engraved into a memory chip called a PROM or EPROM, located in the telephone chassis. This number cannot be changed through programming as the telephone number can, although it can be replaced. Each time the telephone is used, it transmits its ESN to the MTSO by means of DTMF tones during the dialing sequence. The MTSO may be able to determine which ESNs are good or bad, thus individual numbers can be banned from use within the system. See also CELL and CELLULAR.
CMY A computer imaging term. A color model used by the printing industry that is based on mixing cyan, magenta, and yellow. It's also referred to as CMYK, with the K denoting black. The K was added after printers discovered they could obtain a darker black using special black colorants rather than by combining cyan, magenta, and yellow alone. See also CMYK.
CMYK A computer imaging term. A color model used by the printing industry that is based on mixing cyan, magenta, yellow and black (called "K.") It used to be called CMY. The K was added after printers discovered they could obtain a darker black using special black colorants rather than by combining cyan, magenta, and yellow. CMYK is the basis of what's known now as "four-color" printing. But there is also five, six, seven and eight color printing, etc. Each of these "extra" colors are basically "colors" which are better printed as their own color rather than by printing a combination the basic three. Silver, copper, gold, aluminum, etc. are all printed traditionally as extra colors, and extra passes through the printer.
CNA Cooperative Network Architecture.
CNC Complementary Network Service. See OPEN NETWORK ARCHITECTURE.
CND 1. Calling Number Delivery.
2. Calling Number Display.

CNE Certified (local area) Network Engineer. When you graduate from Novell's third level class, you become a certified network engineer. CNEs are an elite group in the LAN industry.

CNET Centre National d'Etudes de Telecommunication. The French organization that approves telecommunications products for sale in France.

CNG Also called Auto Fax Tone, or Calling Tone. This tone is the sound produced by virtually all fax machines when they dial another fax machine. CNG is a medium pitch tone (1100 Hz) that lasts 1/2 second and repeats every 3 1/2 seconds. A fax machine will produce CNG for about 45 seconds after it dials. The CNG tone is useful for owners of fax/phone/modem switches. Such switches answer an incoming call. If they hear a CNG tone, they will transfer the call to a fax machine. If they don't, they'll transfer the call to a phone, answering machine or perhaps a modem. Depends on how they're set up. Some fax machines do not transmit a CNG tone with manually-dialed transmissions — i.e. where the caller picked up the handset on the fax machine, dialed and waited for high-pitched squeal before pushing his fax machine's "start" button. A manual dialed fax transmission will "fool" fax/voice switches. See CED and FACSIMILE.

CNIS Calling Number Identification Services.

CNM An ATM an d SMDS (Switched Megabit Data Service) term. Customer Network Management. All activities that customers perform to manage their communications networks. SMDS CNM service enables customers to directly manage many aspects of the SMDS service provided by telecommunications carriers. See CUSTOMER NETWORK MANAGEMENT.

CNO Corporate Networking Officer, a term invented by William Y. O'Connor, CEO of Ascom Timeplex.

CNR 1. Telephone company term for re-scheduling a telephone installation appointment because the "Customer is Not Ready." 2. An ATM term. Complex Node Representation: A collection of nodal state parameters that provide detailed state information associated with a logical node.

CNRI Corporation for National Research Initiatives.

CNS Complementary Network Service. CNSs are basic services associated with end user's lines that make it easier for Enhanced Service Providers (ESPs) to offer them enhanced services. Some examples of CNSs include Call Forwarding Busy/Don't Answer, Three Way Calling, and Virtual Dial Tone. See OPEN NETWORK ARCHITECTURE.

CO Central Office. In North America, a CO is that location which houses a switch to serve local telephone subscribers. Sometimes the words "central office" are confused with the switch itself. In Europe and abroad, the words "central office" are not known. The more common words are "public exchange." But those words tend to refer more to the switch itself, rather than the site, as in North America. CO was the name of a magazine published by Telecom Library Inc., the publisher of this dictionary. See also CENTRAL OFFICE or PUBLIC EXCHANGE.

CO Lines These are the lines connecting your office to your local telephone company"s Central Office which in turn connects you to the nationwide telephone system.

CO Location See COLOCATION.

CO Simulator A desktop device which pretends to act like a mini-central office. The smallest version will consist of two lines and two REJ-11 jacks. Plug a phone into both jacks. Pick up one phone. You hear dial tone. Dial or touchtone two or three digits. Bingo, the second phone rings. You pick up the second phone. You can have a conversation with yourself or with a machine — like a voice processing system. Most central office simulators can simulate normal on-hook, off-hook, dialing, answering, speaking, etc. Some now can simulate caller ID features — including number of person calling.

Co-location The ability of a someone who is not the local phone company to put their equipment in the phone company's offices and join their equipment to the phone company's equipment. That "someone" who might co-locate their equipment on the phone company's premises might be an end-user or it might be another local or long distance telecommunications company. It might even be a competitor of the local phone company, i.e. another local phone company. The idea of co-location is to save money, give better service, ensure better interconnection, get technical problems solved faster, etc. Not all local phone companies offer their customers co-location. New York Telephone and New England Telephone (now Nynex) are two that now do. Also spelled collocation. See also VIRTUAL CO-LOCATION.

COAM Customer Owned And Maintained equipment.

Coasting Mode In timing-dependent systems, a free-running operational timing mode in which continuous or periodic measurement of timing error is not available. In some systems, operation in this mode can be enhanced for a period of time by using clock or timing error (or correction) information obtained during a prior tracking mode to estimate clock or timing corrections to be made in the free-running mode.

Coated Filament A vacuum tube filament coated with a metallic oxide to provide greater electron emission and longer life.

Coating A protective material (usually plastic) applied to the optical fiber immediately after drawing to preserve its mechanical strength and cushion it from external forces that can induce microbending losses.

Coaxial Cable A cable composed of an insulated central conducting wire wrapped in another cylindrical conducting wire. The whole thing is usually wrapped in another insulating layer and an outer protective layer. A coaxial cable has great capacity to carry great quantities of information. It is typically used to carry high-speed data (as in connections of 327X terminals to computer hosts) and in CATV (multiplexed TV stations).

COB Close Of Business.

COBOL Common Business Oriented Language. A very popular computer programming language for business applications.

COBRA Common Object BRoker Architecture.

Cobweb Site A World Wide Web site that hasn't been updated for a long time. A dead Web page.

COC Central Office Connection. Separately tariffed part of T-1 circuit.

COCOT Customer Owned Coin Operated Telephone. See also CLIP ON TOLL FRAUD.

COD An ATM term. Connection Oriented Data: Data requiring sequential delivery of its component PDUs to assure correct functioning of its supported application, (e.g., voice or video).

Code The system of dots and dashes used to represent the letters of the alphabet, numerals, punctuation and other symbols.
Code Bit The smallest signaling element used by the Physical Layer for transmission on the fiber cable.
Code Blocking A switch's ability to block calls to a specified area code, central office code or phone number.
Code Breaker See KEY and KEY HOLDER.
Code Call Access A very useful PBX feature. It allows attendants and extension users to activate, by dialing an access code followed by a two or three digit called code, customer-provided signaling devices throughout the premises. The signaling devices then issue a series of tones or visual coded signals corresponding to the called code. The called or paged party responds by dialing a meet-me answering code from any phone and is then connected to the paging party.
Code Coverage Modern computer telephony systems are composed largely of software, or "code." Invariably this code has many different logic paths and options. During normal system usage, many code paths are used only infrequently, if at all, meaning that normal usage will really only test a small portion of the total system. Code coverage refers to the amount of the system code that has been accessed during the testing of the system. The greater the code coverage, the more code that has been tested. 100% code coverage means that all the code has been tested. The amount of code coverage that has been achieved is usually determined through the use of a code coverage tool. Code coverage tools are available for most computer operating systems. Code coverage is especially important for computer telephony applications because many features are only infrequently used or are turned on or off based on the user's class of service. Also, many features interact with other features and the use of one feature often turns off another. Functional anomalies frequently exist in little-used paths and feature interactions. Tests should be designed to make sure that all code paths have been accessed and are adequately exercised. This definition courtesy of Steve Gladstone, author of the book Testing Computer Telephony Systems, available from 212-691-8215.
Code Conversion A process which converts the codes coming in from one network into codes that can be recognized on another network, such as converting from the Baudot code in a telex network to the ASCII code on the TWX network. Usually, the hardware will convert differences in transmission speed.
Code Division Multiple Access CDMA, also called Spread Spectrum, is a name for a new form of digital cellular phone service. The idea is that each phone call is combined with a code which only one cellular phone plucks from the air. Business Week said CDMA works "by spreading all signals across the same broad frequency spectrum and assigning a unique code — the company says one of 42 billion — to each. The dispersed signals are pulled out of the background noise by a receiver which knows the code. This method, developed by a San Diego company called Qualcomm Inc. is very new. Much of the equipment to support it — like the cellular switches have not yet been developed." CDMA is also being used by wireless PBXs. See CDMA for a longer and better explanation.
Code Excited Linear Prediction CELP. An analog-to-digital voice coding scheme.
Code Independent Data Communication Data communication mode using a link procedure associated with the character and not dependent on the set of characters or the code used by the data source.
Code Level Number of bits used to represent a character.
Code Of Federal Regulations CFR. CFR is a codification of the general and permanent rules published in the Federal Register. It is divided into 50 titles that represent broad areas subject to federal regulation. Title 47 of the CFR pertains to telecommunications and contains the rules covering Part 22 Common Carriers and Part 90 Private Carriers.
Code Violation Violation of a coding rule; for example, the AMI coding rule is corrupted by a bipolar violation.
CODEC Originally CODEC stood for CODer-DECoder, i.e. microprocessor chip. Now the PC industry thinks it stands for COmpression/DEcompression, i.e. an overall term for the technology used in digital video and stereo audio. The original CODEC (still in big use in today's telephony industry) converts voice signals from their analog form to digital signals acceptable to modern digital PBXs and digital transmission systems. It then converts those digital signals back to analog so that you may hear and understand what the other person is saying. In some phone systems, the CODEC is in the PBX and shared by many analog phone extensions. In other phone systems, the CODEC is actually in the phone. Thus the phone itself sends out a digital signal and can, as a result, be more easily designed to accept a digital RS-232-C signal.
CODEC Conversion The back-to-back transfer of an analog signal from one codec into another codec in order to convert from one proprietary coding scheme (for instance, that used by CLI) to one used by another codec manufacturer (PictureTel, VTEL, GPT, BT, NEC, etc). The analog signal, instead of being displayed to a monitor, is delivered to the dissimilar codec where it is redigitized, compressed and passed to the receiving end. This is obviously a bi-directional process. Conversion service is offered by carriers such as AT&T, MCI and Sprint.
Coded Character Set A set of unambiguous rules that establish a character set and the one-to-one relationships between the characters of the set and their coded representations.
Coded Image A representation of an image in a form suitable for storage and processing.
Coded Trunks You buy several trunks. They hunt on. The main number is 555-3000. If the main number is busy, the call goes to the next line. There are two types of "next lines." One type can have an actual number, like 555-3001, which you can call directly. The other can be a coded trunk with no actual number and which you can't call directly. It's better to have no coded trunks because it's hard to test coded trunks. You can't dial them directly. Actual dial-able numbers are better.
Coder An analog-to-digital converter that changes analog voice signals to their digital equivalents. See CODEC.
Codial Office CDO. A small central office designed for unattended operation in a distant community. Usually a community dial office is fairly small, rarely more than 10,000 lines.
Coding Theory Mathematical theory describing how to encode data into streams of digital symbols at transmitter and decode it at receiver to maximize accuracy of data presented to user.
Coherence Area In optical communications, the area in a plane perpendicular to the direction of propagation over which light may be considered highly coherent.
Coherence Length The propagation distance over which a light beam may be considered coherent. See COHERENT LIGHT.

Coherent Light Light emitted from laser and some light-emitting-diodes. It is made up of light of a single frequency in which all the light waves are in phase (i.e., their wave peaks and troughs are all in alignment so the waves reinforce or amplify each other.)
Coil A number of turns of wire, so wound as to afford inductance.
Coil Antenna One consisting of one or more complete turns of wire. See LOOP ANTENNA.
Coin Acceptor/Rejector A mechanical or electromechanical device that checks and validates the coins deposited in a coin pay phone. They measure the coin's size and weight and steel content. These coin acceptor/rejector units transmit the value of the coin deposits to the processing part of a smart payphone or they signal the information to the telephone company central office via coded tones.
Coin Supervisory Trunk Group A trunk group that lets a switchboard operator collect overtime monies due on coin phones and check for stuck coins.
Coin Telephone A pay telephone that takes coins. The coin telephone was invented by William Gray, an American whose previous inventions included the inflatable chest protector for baseball players. Mr. Gray's first phone lacked a dial. Its instructions read: "Call Central in the usual manner. When told by the operator, drop coin in proper channel and push plunger down."
In today's nomenclature, Mr. Gray's original phone is known as a post-pay coin phone. See also PAYPHONE and several entries following it.
Cold Docking Docking is to insert a portable computer into a base unit. Cold docking means the computer must begin from a power-off state and restart before docking. Hot docking means the computer can be docked while running at full power. See COLD START.
Cold Start Everything starts from scratch. The power to the computer or telephone system is turned off. Everything in the system's volatile memory is erased. A cold start may be needed on a microcomputer when something has happened to "lock up" the keyboard and the Reset button (if there is one) doesn't clear the problem completely. A Cold Start is also needed when you want to load a new operating system. When your phone system gives troubles you find hard to diagnose, turn it off, count to ten and turn it on. This cold boot to your phone system will often fix the problem, as it will typically do on a computer system.
Collaboration A multimedia term. Collaboration involves two or more people working together in real-time, or in a "store-and-forward" mode. Applications will enable a group of people to collaborate in real-time over the network using shared screens, shared whiteboards, and video conferencing. Collaboration can range from two people reviewing a slide set on line to a conference of doctors at different locations sharing patient files and discussing treatment options.
Collaboration Software Software that lets two or more people do a task together. See COLLABORATION.
Collaborating Two or more people working together on a project to share information and ideas, view suggestions, and make modifications. Computers can enable users to collaborate in real-time over a network or phone line using tools such as shared documents, shared whiteboards, and video conferencing, or time-efficient workflow such as document forwarding.
Collapsed Backbone The backbone network connecting all network segments is contained within a hub, allowing for the possibility of speeds far exceeding the standard backbone media.
Collect Call A telephone call in which the called person pays for the call. The person calling calls a number and asks that the call be made "collect." Sometimes collect calls are handled by live operators, sometimes by machines. In a collect call, the phone company has to get some authorization from the person receiving the call that he will pay for it. This may be done by saying "Yes" or hitting a button a touchtone phone.
Collector Ring Metallic ring generally on the armature of a generator in contact with brushes for completing the circuit to a rotating member.
Collimate The condition of parallel light rays.
Collimation The process by which a divergent or convergent beam of electromagnetic radiation is converted into a beam with the minimum divergence or convergence possible for that system (ideally, a parallel bundle of rays).
Collinear Antenna A cellular car antenna which looks like a pigtail, because it has a little curlicue in the middle. The curlicue is not a spring, but a clever bit of electro-mechanical magic known as a phasing network, which allows the antenna to boost the effective power of the transmitter's signal. Typically a collinear cellular car antenna is 13 inches high.
Collision The result of two workstations trying to use a shared transmission medium (cable) simultaneously. The electrical signals, which carry the information they are sending bump into each other. This ruins both signals and both will have to re-transmit their information. In most systems, a built in delay will make sure the collision does not occur again. The whole process takes fractions of a second. Collisions in LANs make no sound. Collisions do, however, slow a LAN down. See ALOHA, COLLISION DETECTION and CONTENTION.
Collision Detection The process of detecting that simultaneous (and therefore damaging) transmission has taken place. Typically, each transmitting workstation that detects the collision will wait some period of time and try again. Collision detection is an essential part of the CSMA/CD access method. Workstations can tell that a collision has taken place if they do not receive an acknowledgement from the receiving station within a certain amount of time (fractions of a second). See ALOHA, CONTENTION and ETHERNET.
Collision Window The time it takes for a data pulse to travel the length of the network. During this interval, the network is vulnerable to collision.
Colocation A competing local phone company can locate its switches within a local exchange company's (LEC) central office. See also CO LOCATION.
Colophon Did you ever notice a paragraph at the end of a book describing the typefaces used, the production methods, and so forth? That little paragraph is called a colophon.
Color Code A color system for circuit identification by use of solid colors, contrasting stripes, tracers, braids, surface marking, etc.
Color Difference Signal The first step in encoding the color television signal. The color difference signals are formed by subtracting the luminance information from each primary color: red, green or blue. Color difference conventions include the Betacam format, the SMPTE format, the EBU-N10 format and the MII format.

NEWTON'S TELECOM DICTIONARY

Color Model A technique for describing a color (see CMY, HSL, HSV, and RGB).
COLT Cellular On Light Trucks. A temporary, mobile installation. I first heard the expression when I heard that BellSouth was using them to help expand its communications capacity in Atlanta for the 1996 Summer Olympics.
Column A database definition: The logical equivalent of a field, a column contains an individual data item within a row or record.
COM Continuation of Message: An indicator used by the ATM Adaptation Layer to indicate that a particular ATM cell is a continuation of a higher layer information packet which has been segmented.
Combat Net Radio CNR. A radio operating in a network, providing a half-duplex circuit employing a single radio frequency or a discrete set of radio frequencies (frequency hopping). Combat net radios are primarily used for command and control of combat, combat support, and combat service support operations between and among ground, naval, and airborne forces.
Combination System An alternative to upgrading older telephone equipment, combination systems makes it possible to add network-based features to an equipment-based telephone system.
Combination Trunk A central office trunk circuit which is available as either an incoming or outgoing circuit to the attendant and also available through dial access to internal phone users for outgoing calls.
Combined Distribution Frame CDF. A distribution frame that combines the functions of main and intermediate distribution frames. The frame contains both vertical and horizontal terminating blocks. The vertical blocks are used to terminate the permanent outside lines entering the station. Horizontal blocks are used to terminate inside plant equipment. This arrangement permits the association of any outside line with any desired terminal equipment. These connections are made with equipment. These connections are made with twisted pair wire, normally referred to as jumper wire, or with optical fiber cables, normally referred to as jumper cables. In technical control facilities, the vertical side may be used to terminate equipment as well as outside lines. The horizontal side is then used for jackfields and battery terminations.
Combined Station MDLC station containing both a primary and a secondary and used in asynchronous balanced mode.
Comcode AT&T's numbering system for telecom equipment, replacing older KS-prefix numbers, and supplements standard industry part designations. Comcode No. 102092848 is touchtone Princess phone with a transparent plastic housing. How many would you like to order? See also KS NUMBER.
ComForum The Network Reliability ComForum is a gathering of senior telecommunication industry executives and government officials that serve on the Network Reliability Council. The ComForum meets to report on the Council's findings. See USITA.
COMINT COMunications INTelligence.
Comma-Free Code A code constructed such that any partial code word, beginning at the start of a code word but terminating prior to the end of that code word, is not a valid code word. The comma-free property permits the proper framing of transmitted code words, provided that: (a) external synchronization is provided to identify the start of the first code word in a sequence of code words, and (b) no uncorrected errors occur in the symbol stream. Huffman codes (variable length) are examples of comma-free codes.
Command See COMMAND SET.
Command And Control C2. The exercise of authority and direction by a properly designated commander over assigned forces in the accomplishment of the mission. Command and control functions are performed through an arrangement of personnel equipment, communications, facilities, and procedures employed by a commander in planning, directing, coordinating, and controlling forces and operations in the accomplishment of the mission.
Command And Control System The facilities, equipment, communications, procedures, and personnel essential to a commander for planning, directing and controlling operations of assigned forces pursuant to the missions assigned. See COMMAND, CONTROL and COMMUNICATIONS.
Command Buffer A segment of memory used to temporarily store commands. The command buffer only holds a copy of the last command issued.
Command Conference System A conference calling arrangement in a Northern Telecom PBX which allows a designated phone to originate a conference to and between a group of PBX extensions. Any phone that is busy when the conference begins is automatically connected to the conference as soon as that phone becomes free.
Command, Control And Communications C3. The capabilities required by military commanders to exercise command and control of their forces.
Command Line Interpreter CLI. A Rolm user interface to the CBX software and used for things like testing.
Command Net A communications network which connects an echelon of command with some or all of its subordinate echelons for the purpose of command control. See C2 and C3.
Command Path The list of path names that tells MS-DOS where to look for files that aren't in the current directory.
Command Port In network management systems an interface used to monitor and control the system.
Command Processor The MS-DOS program, COMMAND.COM, that contains all DOS's internal commands, like DIR, ERASE and REName. Once you have your hard disk set up, it's a good idea to made your COMMAND.COM "read only." This way it will be difficult for anyone to erase the file. If you don't have COMMAND.COM on your disk, or don't have it in a place where MS-DOS can find it, your disk will not boot (i.e. start) your computer. The command to make COMMAND.COM read only is
ATTRIB +R COMMAND.COM
A mistake many novices make is to "open" COMMAND.COM using their word processor, find that it's full of "junk," then save the file. Saving COMMAND.COM with a word processor destroys COMMAND.COM. The next time you start your computer it will "hang." The solution: Boot from a floppy. Erase COMMAND.COM and REName COMMAND.BAK to COMMAND.COM. Reboot your machine again. This time it should work. If it doesn't, copy COMMAND.COM from your original MS-DOS disk.

Command Prompt The MS-DOS command prompt appears on the screen as the default drive letter followed by a greater-than > sign. The command prompt lets you know MS-DOS is ready to receive a command. You can use the PROMPT command to change the command prompt. The PROMPT I prefer is
    prompt.$_   DATE $d   TIME $t $_$P$g
And it looks like this
    DATE Tues 10-11-1994 TIME 14:49:00.00
    C:\BIN\HARRY>
Don't forget to type the period after the word "prompt."

Command Save A Rockwell ACD term. The introduction of a new demand command defines up to 10 commands per terminal position to enhance the productivity of both IST and non-IST supervisors.

Command Set In computer telephony, a command set is a collection of special software instructions that do special jobs. These software instructions are often called function calls. For example, the command M_Make_Call (plus parameters) tells Northern's Norstar phone system to have telephone set number 21 dial a phone number. Northern's Norstar and other open phone systems (those that can be commanded by an external computer which you and I can program) all have their own command sets. Each command set is made up of function calls with funny words like M_Make_Call. Typically those function calls work in C, a common software language. A function call will reach into the specialized driver that controls the phone system (an exact analogy is the driver that drives a laser printer) and get the phone system to do something. A programmer must use these function calls if he/she wants to control the phone system from software. Exactly how M_Make_Call works is typically not revealed to the programmer. That keeps the manufacturer's technology proprietary and secret. It also saves the programmer the time and expense of writing the driver.

COMMDesk Banker A communications software package offered by MCI International that provides all the capabilities of COMMDesk, plus a security feature essential for financial transactions.

COMMDesk Manager An MCI definition. A communications software package designed to run on an IBM PC/XT/AT or compatible that gives the user full access to all MCI International and MCI communications services.

Commercenet A nonprofit group that works to accelerate the application of electronic commerce on the Internet. www.commerce.net/information/about.html

Commercial Building Telecommunications Wiring Standard In 1985, the Electronic Industries Association undertook the task of developing a standard for commercial and industrial building wiring. Approved and published on July 9, 1991, the EIA/TIA-568 "Commercial Building Telecommunications Wiring Standard" defines a generic wiring system which will support a multiproduct, multivendor environment and which will have a useful life of over 10 years. The EIA/TIA standard is based on star topology in which each workstation is connected to a telecommunications closet situated within 90 meters of the work area. Backbone wiring between the communications closets and the main cross-connect is also organized in a star topology. However, direct connections between closets are allowed to accommodate bus and ring configurations. Distances between closets and the main cross-connect are dependent on backbone cable types and applications. Each workstation is to provided with a minimum of two communications outlets (which may be on the same faceplate). One outlet is supported by a four-pair, 100-ohm ushielded twisted-pair (UTP) cable. The other may be supported by (a) an additional four-pair UTP cable, (b) a two-pair, 150 ohm shielded twisted pair (STP) cable or (c) a two-fiber 62.5 /125 micron fiber optic cable. For more on cabling and cabling components, see CATEGORY OF PERFORMANCE.

Commercial Internet The part of the Internet provided by commercial services. Allows business usage of the Internet without violating the appropriate usage clause of the National Science Foundation NETwork (NSFNET), who actually runs the Internet.

Commercial Internet Exchange Association This is a non-profit trade association for public data internetworking service providers. www.cix.org/cixhome.html

Committed Information Rate CIR. When a customer buys a shared service like frame relay, he commits to a certain minimum data transmission rate he will use. The monthly bill a customer will receive will include two elements — usage and a charge for the committed rate. In the early days of frame relay when few people were using the service, customers were opting for a low committed information, thus keeping their bills low, but knowing that they could always get their transmissions through — because there were few other people on the service. As the service got more popular, many customers found they had to hike their committed information rate if they wanted to get their information through. For a more technical explanation, see CIR.

Common Audible The same as Common Bell. Ringer wiring is such that ringing occurs on more than one CO or PBX line.

Common Battery A battery (or several batteries) that acts as a central source of energy for many pieces of equipment. In many telecommunications applications, a common battery provides 48 volts of power to a central office switch and to all its phones.

Common Battery Signaling A system in which the signaling power of a telephone is supplied by the battery at the servicing switchboard. Switchboards may be manual or automatic, and "talking power" may be supplied by common or local battery.

Common Bell A bell or ringer which sounds when any of the lines terminating on that phone rings. A term harking back to 1A2 key system days.

Common Business Line CBL. An option with 800 Service that has been replaced by 800 Business Line.

Common Carrier A company that furnishes communications services to the general public. It is typically licensed by a state or federal government agency. A common carrier cannot refuse to carry you, your information or your freight as long as you conform to the rules and regulations as filed with the state or federal authorities. See OTHER COMMON CARRIER.

Common Carrier Bureau A department of the Federal Communications Commission responsible for recommending and implementing regulatory policies on interstate and international common carrier (voice, video, data) activities.

Common Channel Interoffice Signaling CCIS. A way of transmitting all signaling information for a group of trunks by encoding that information and transmitting it over a separate channel using time-division digital methods. By transmitting that signaling information over a separate channel, CCIS saves huge long distance bandwidth, which in the past was used to switch calls across the country only to find a busy signal and then come all the way back again to signal the calling party a busy. For the biggest explanation of common channel signaling, see SIGNALING SYSTEM 7. See also MTP, SCCP, ISUP, ISDN and TCAP.

Common Channel Signaling This is a Bellcore definition: A network architecture which uses Signaling System 7 (SS7) protocol for the exchange of information between telecommunications nodes and networks on an out-of-band basis. It performs three major functions: 1. It allows the exchange of signaling information for interoffice circuit connections. 2. It allows the exchange of additional information services and features, e.g. CLASS, database query/response, etc. 3. It provides improved operations procedures for network management and administration of the telecommunications network. For the biggest explanation of common channel signaling, see SIGNALING SYSTEM 7. See also ISDN, ISUP, MTP, SCCP, STP and TCAP.

Common Channel Transit Exchange An intermediate exchange where networking of common channel signaling systems occurs.

Common Control An arrangement in which items of control equipment in a switching system are shared. They are associated with a given call only during the periods required to accomplish the control functions. All crossbar and electronic switching systems have common control.

Common Control Switching Arrangement CCSA. An AT&T offering for very big companies. Those big companies can create their own private networks and dial anywhere on them by dialing a standard seven digit number, similar to a local phone number. The corporate subscriber rents private, dedicated lines and then shares central office switches. CCSA uses special CCSA software at the central office.

Common Costs Costs of the provision of some group of services that cannot be directly attributed to any one of those services.

Common Equipment In telephone systems Common Equipment are items that are used by several or all phones for processing calls. On a key system, the device that permits a light on any instrument to flash on and off may be Common Equipment when used to control all lights on all instruments.

Common Intermediate Format A videophone ISDN standard which is part of the ITU-T's H.261. It produces a color image of 352 by 288 pixels. The format uses two B channels, with voice taking 32 kbps and the rest for video.

Common Mail Calls New APIs (Application Programming Interfaces) from Microsoft which allow you to move information around your various mail services — the ones on your LAN, on your wireless pager, etc.

Common Mode The potential or voltage that exists between neutral and ground. Electronic equipment requires this to be as close to 0 volts as possible or not to exceed 1/2 volt.

Common Mode Interference 1. Interference that appears between signal leads, or the terminals of a measuring circuit and ground. 2. A form of coherent interference that affects two or more elements of a network in a similar manner (i.e., highly coupled) as distinct from locally generated noise or interference that is statistically independent between pairs of network elements.

Common Mode Rejection Ratio CMRR. The ratio of the common mode interference voltage at the input of a circuit to the corresponding interference voltage at the output.

Common Mode Voltage 1. The voltage common to both input terminals of a device.
2. In a differential amplifier, the unwanted part of the voltage between each input connection point and ground that is added to the voltage of each original signal.

Common Peer Group An ATM term. The lowest level peer group in which a set of nodes is represented. A node is represented in a peer group either directly or through one of its ancestors.

Common Return A return path that is common to two or more circuits and that serves to return currents to their source or to ground.

Common Return Offset The dc common return potential difference of a line.

Common Trunk In telephone systems having a grading arrangement, a trunk accessible to all groups of the grading.

Common User Circuit A circuit designated to furnish a communication service to a number of users.

Common User Network A system of circuits or channels allocated to furnish communication paths between switching centers to provide communication service on a common basis to all connected stations or subscribers.

Commonality 1. A quality that applies to material or systems: (a) possessing like and interchangeable characteristics enabling each to be utilized, or operated and maintained by personnel trained on the others without additional specialized training; (b) having interchangeable repair parts and/or components; (c) applying to consumable items interchangeably equivalent without adjustment.
2. A term applied to equipment or systems that have the quality of one entity possessing like and interchangeable parts with another equipment or system entity.

Communicating Applications A General Magic term. An application whose design presupposes the user's desire to send and receive messages. For a Personal Intelligent Communicator to be effective, it needs to be equipped with a suite of communicating applications. All Magic Cap applications are built to communicate.

Communicating Applications Platform A General Magic term. The Cap in Magic Cap. Software on which Personal Intelligent Communicators are based. It is designed to make it easy for developers to create communicating applications and services. Magic Cap can run on dedicated devices as well as other computer operating systems.

Communicating Objects A term created in the fall of 1992 by Mitel's VP Tony Bawcutt for a new Mitel division which specializes in making PC printed cards for and software drivers and developer tools for those cards. Those cards are designed to be the building blocks of what Mitel calls multimedia applications — but what are more properly called PC-based voice and call processing telecom developer building blocks. One of the first cards Mitel introduced was an ISDN S-access card which converts PCs into ISDN telephones, also called voice and data workstations.

Communicating Word Processor A dedicated word processor that includes software for sending word processed files over phone lines. Communicating word processors have now largely been replaced by PCs (Personal Computers) running word processing programs and asynchronous communications software programs.

NEWTON'S TELECOM DICTIONARY

Communication Channel A two-way path for transmitting voice and/or data signals. See also CIRCUIT.

Communication Controller Another name for a Front End Processor, a specialized computer which was common in 3270 data communications networks. The FEP acted as a data communications "traffic cop," removing the communications traffic routing and controlling burden from the mainframe computer which lay behind the FEP. In short, the FEP designates a device placed between the network and an input/output channel of a processing system (i.e. the computer).

Communication Endpoint An ATM term. An object associated with a set of attributes which are specified at the communication creation time.

Communication Server A dedicated, standalone system that manages communications activities for other computers.

Communication Workers Of America CWA. A national union of telephone industry employees, currently very worried about its future membership growth given the phone industry's propensity to let its surplus workers go.

Communications Act Of 1934 Federal legislation which established national telecommunications goals and created the Federal Communications Commission to regulate all interstate and international communications.

Communications Adapter Device attached to an IBM System 3X computer or an IBM PC that allows communications over RS-232 lines.

Communications Control Character A character intended to control or help transmission over data networks. There are ten control characters specified in ASCII which form the basis for character-oriented communications control procedures.

Communications Protocol Specifications which define hardware and software requirements for that type of communications. In short, the format governing the communication and the transfer of data between two or more devices.

Communications Satellite A satellite circling the earth, usually at a distance of about 22,000 miles, with electronic equipment for relaying signals received from the earth back to other points on the earth. See GEOSTATIONARY SATELLITE.

Communications Server Also called an asynchronous server or asynchronous gateway. A communications server is a type of gateway that translates the packetized signals of a LAN to asynchronous signals, usually used on telephone lines or on direct connections to DEC and other minicomputers and mainframes. It handles different asynchronous protocols and allows nodes on a LAN to share modems or host connections. Usually one machine on a LAN will act as a gateway, sharing its serial ports or an RS-232 connection to a minicomputer. All devices on the LAN can use this machine to get to the modems and the minicomputer.

Communications Settings Settings that specify how information is transferred from your computer to a device (usually a printer or modem).

Communications System Engineering The translation of user requirements for the exchange of information into cost-effective technical solutions of equipment and subsystems.

Communications Toolbox An extension of he Apple Macintosh operating system that provides protocol conversion and the drivers needed for communications tasks.

Communications Zone A military term: Rear part of theater of operations (behind but contiguous to the combat zone), which contains the lines of communications, establishments for supply and evacuation, and other agencies required for the immediate support and maintenance of the field forces.

Communicator A British term. An alternative, and probably more meaningful, name for a telebusiness agent. A communicator is called a telemarketer in North America.

Community Antenna Television CATV. Signals from distant TV stations are picked up by a large antenna, typically located on a hill, then amplified and piped all over the community below on coaxial cable.

Community Of Interest A grouping of telephone users that call each other with a high degree of frequency. Often several Communities of Interest exist within an organization. This phenomenon can influence design for service when new switches are planned.

Commutator A device used on a dynamo to reverse the connection periodically in order to cause the current flow in one direction, i.e., to produce direct current.

Compact Disc A standard medium for storage of digital audio data, accessible with a laser-based reader. CDs are 12 centimeters (about 4 3/4") in diameter. CDs are faster and more accurate than magnetic tape for audio. Faster, because even though data is generally written on a CD contiguously within each track, the tracks themselves are directly accessible. This means the tracks can be accessed and played back in any order. More accurate, because data is recorded directly into binary code; mag tape requires data to be translated into analog form. Also, extraneous noise (tape hiss) associated with mag tape is absent from CDs. See CD-ROM and CVD.

Compact Disc Interactive A compact disc format, developed by Philips and Sony, which provides audio, digital data, still graphics and limited motion video. See CD-I.

CompactPCI CompactPCI is a variation of the PCI bus, which a bunch of industrial grade PC makers have designed for two reasons: First, to be able to put more PCI cards into one PC (8 versus 4). Second, to make somewhat more rugged, i.e. better able to withstand shaking, etc. The physical configuration of the hardware conforms to the Eurocard (VMR-style) standard. The cards, are identical to VME cards in size. They differ, however, in that they use a high density 2mm (contact spacing) pin-and-socket connector for interface to a passive backplane. The PC makers told me that "CompactPCI is an adaptation of the Peripheral Component Interconnect (PCI) Specification for industrial and/or embedded applications requiring a more robust mechanism form factor than desktop PCI. CompactPCI uses industry standard mechanical components and high-performance connector technologies to provide a system optimized for rugged applications. CompactPCI is electrically compatible with the PCI Specification, allowing low-cost PCI chipsets to be used in a mechanical form factor suited for rugged environments."

Compander See COMPANDING.

Companding The word is a contraction of the words "compressing" and "expanding." Companding is the process of compressing the amplitude range of a signal for economical transmission and then expanding them back to their original form at the receiving end.

Compartmentation A military/government term: A method employed to segregate information of different desired accessibilities from each other. It may be used for communications security purposes.

Compatible A widely misused word. In the computer world, two computers are said to be compatible when they will produce the identical result if they run identical programs. Another meaning is whether equipment — peripherals and components — can be used interchangeably with each other, from one computer to another. Compatibility used to be regarded as a useful trait in 1A2 key telephones. But then the electronics revolution came along and now there are no compatible electronic telephones from different manufacturers. This means you can't take an electronic Mitel PBX phone, example, and have it work behind a Northern Telecom PBX. And a Northern Telecom PBX phone won't work behind an AT&T PBX. The major compatibility in our industry is at the lowest common denominator — the tip and ring analog phone. See INTEROPERABILITY TESTING.

Compatible Sideband Transmission That method of independent sideband transmission wherein the carrier is deliberately reinserted at a lower level after its normal suppression to permit reception by conventional AM receivers. The normal method of transmitting compatible SSB (AME) is the emission of the carrier plus the upper sideband.

Compelled Signaling A signaling method in which the transmission of each signal in the forward direction is inhibited until an acknowledgment of the satisfactory receipt of the previous signal has been sent back from the receiver terminal.

Competitive Long Distance Coalition A Washington, D.C. lobbying group.

Compilation The translation of programs written in a language understandable to programmers into instructions understandable to the computer. Think of programmers writing in every language but Greek and computers understanding only Greek. In this case, Greek is called machine language. The other languages (the programmer languages) are called things like COBOL, FORTRAN, Pascal, dBASE. A compiler is a special program that translates from all these other languages into machine language.

Compiler A program that takes the source code a programmer has written and translates it into object code the computer can understand. A computer program used to convert symbols meaningful to a human into codes meaningful to a computer. For example, a compiler takes instructions written in a "higher" level language such as BASIC, COBOL or ALGOL and converts them into machine code which can be read and acted upon by a computer. Compilers converts large sections of code at one time, compared to an interpreter which translates commands one at a time.

Complementary Metal Oxide Semiconductor CMOS. A method of building chips which produces a logic circuit family which uses very little power.

Complementary Network Service CNS. CNSs are basic services associated with end user's lines that make it easier for ESPs to offer them enhanced services. Some examples of CNSs include Call Forwarding Busy/Don't Answer, Three Way Calling, and Virtual Dial Tone. See OPEN NETWORK ARCHITECTURE.

Complete Document Recognition The ability to perform recognition on documents, retaining as much information as possible about the features and formatting of the original, and including the ability to capture images as well as text.

Completed Call Careful with this one. In telephone dialect, a Completed Call is one that has been switched to its destination and conversation has begun but has not yet ended.

Completion Ratio The proportion of the number of attempted calls to the number of completed calls.

Complimentary Network Services CNS. The means for an enhanced-service provider's customer to connect to the network and to the enhanced service provider. Complimentary network services usually consist of the customer's local service (e.g.,business or residence line) and several associated service options, e.g., call-forwarding service.

Component Software Component software is software constructed from reusable components. It was popularized by Microsoft Visual Basic and its successful custom control architecture. This architecture allows third party software components to "plug" into and extend the Visual Basic development environment. Hundreds of third party components, or custom controls, exist — for everything from accessing a mainframe database to programming a computer telephony board. Component-based software development is a productive way to build software. System developers benefit from being able to tailor their development environment for a specific need. Consider the development of an IVR system that allows callers to access their account balance stored on an IBM mainframe. To build this system, Visual Basic developers extend their development environment with a custom control for telephony and another that provides access to an IBM mainframe. There's no need to learn a new and proprietary language for telephony development. Plus, every control is accessed through a common interface of actions, properties, and events.

Component Video Transmission and recording of color television that stores separate channels of red, green and blue.

Composite Output of a multiplexer that includes all data from the multiplexed channels. Contrast with AGGREGATE.

Composite Link The datastream composed of all the input channels and control and signaling information in a multiplexed circuit. The Composite Link Speed is the transmission speed of the circuit.

Composite Materials Composite materials consist of two or more components. They make it possible to combine the best properties of different materials; for example, the compression strength and low price of concrete with the tensile strength of reinforcing rods. Composite materials include: Reinforced concrete, fiber-reinforced plastic, fiber-reinforced metals, plywood, chipboard and ceramics. The composites mainly considered for antennas are fiber-reinforced plastics. They combine the low weight and protective properties of plastics with the stiffness and strength of fiber.

Composite Signaling A direct current signaling system that separates the signals from the voice band by filters. Two pairs (a quad) provide talking paths and full-duplex signaling for three channels. Also called CX Signaling.

Composite Video A television signal where the chrominance signal is a sine wave that is modulated onto the luminance signal which acts as a subcarrier. This is used in NTSC and PAL TV systems. Composite video is the visual wave form representation used in color television. Composite video is analog and must be converted to digital to be used in multimedia computing. See also DIGITAL VIDEO.

Composited Circuit A circuit that can be used simultaneously for telephony and dc telegraphy, or signaling; separation between the two being accomplished by frequency discrimination.

Compound A term used to designate an insulating or jacketing material made by mixing two or more ingredients.

Compound Document The simple explanation: A compound document contains information created by using more than one application. It is a document often composed of a variety of data types and formats. Each data type is linked to the application that created it. A compound document might include audio, video, images, text, and graphics. Compound documents first became possible to the world of PCs with the introduction of Windows 3.1, which included OLE (Object Linking and Embedding). OLE allows you to write a letter in your favorite Windows word processor, embed a small voice icon in your document, send your letter to someone else, have them open your letter, place their mouse on the voice icon and hear whatever comments you recorded. To make this possible, both you (the creator) and your recipient would need access to programs that could read both the text and the voice. Ideally, you would both be on a LAN (Local Area Network) and would both get access to the identical applications software, resident, presumably, on the LAN's file server. See COMPOUND MAILBOX.

Compound Document Mail See COMPOUND DOCUMENT.

Compound Mailbox A mailbox for mail from all sources — fax, voice mail, e-mail, pager, etc. See COMPOUND DOCUMENT.

Compressed Video Television signals transmitted with much less than the usual bit rate. Full standard coding of broadcast quality television typically requires 45 to 90 megabits per second. Compressed video includes signals from 3 mb/s down to 56 Kbps. The lower bit rates typically involve some compromise in picture quality, particularly when there's rapid motion on the screen. See MPEG.

Compression Reducing the representation of the information, but not the information itself. Reducing the bandwidth or number of bits needed to encode information or encode a signal, typically by eliminating long strings of identical bits or bits that do not change in successive sampling intervals (e.g., video frames). Compression saves transmission time or capacity. It also saves storage space on storage devices such as hard disks, tape drives and floppy disks.

Compression Algorithm The arithmetic formulae which convert a signal into smaller bandwidth or fewer bits.

Compressor See COMPANDING.

Compromise Equalizer Equalizer set for best overall operation for a given range of line conditions. This is often fixed but may be manually adjustable.

COMPSURF COMPrehensive SURFace Analysis. A Novell program that checks the surface of a hard disk, marks off sections that are lousy and therefore shouldn't be written to, and then low level formats the disk. The program is slow, but thorough and rigorous. No hard disk should ever be used on a file server on a Novell local area network without being subjected to this wonderful program. Don't believe Novell when it says that you don't need to subject new disks to COMPSURF. You should submit ALL disks.

CompuCALL Northern Telecom DMS central office link to computer interface. An open architecture specification. Northern Telecom spells it as CompuCALL. According to Northern Telecom's own words, CompuCALL employs the Switch Computer Application Interface (SCAI) open architecture standard to connect the central office with customers' general-purpose business computers. CompuCALL consists of:
• The CompuCALL base software (NTXJ59AA) in the Northern Telecom DMS-100 switch, or the Meridian Automatic Call Distribution (ACD) Server, that sends and receives SCAI messages; and
• The CompuCALL transport mechanism that physically links the switch to the computer or other external processor and carries the SCAI message.
Northern Telecom also supplies the Meridian ACD CompuCALL Options (NTXJ39AA) and other applications software which rely on the CompuCALL base. Computers and software vendors provide application programming interface (API) software as well as business application software. The API, which resides in the business computer, converts SCAI messages into information that can be used by computer-based business application software.
The first applications of CompuCALL integrate computer databases with voice telephony:
• CompuCALL Coordinated Voice and Data provides an agent a screen of information about a caller concurrently with receipt of a call.
• CompuCALL Voice Processing Integration uses Interactive Voice Response (IVR) systems and Voice Response Units (VRU) to obtain additional information about callers and direct them to the appropriate agent.
• CompuCALL Third-Party Call Control lets the customer's computer place outgoing calls.
Successful implementation of CompuCALL requires interaction with application software on a business computer. CompuCALL consists of:
• The CompuCALL base software residing in the Northern Telecom DMS-100 central office switch that sends and receives SCAI messages; and
• The CompuCALL transport mechanism that physically links the switch to the computer or other external processor and carries the SCAI message.
Computer and software vendors provide application programming interface (API) software as well as business application software. The API, which resides in the business computer, converts SCAI messages into information that can be used by computer-based business application software. CompuCALL is based on the seven-layer protocol defined by the International Telegraph and Telephone Consultative Committee's (CCITT) Open System Interconnection (OSI) reference model. The product is transport independent. With BCS33, CompuCALL is scheduled to be available for Verifications Office (VO) on X.25. See also OPEN APPLICATION INTERFACE.

Compunications A recent creation meaning the combination of telephones, computers, television and data systems.

CompuServe An on-line, dial-up service — one of the largest worldwide. CompuServe has everything from electronic mail to manufacturer-sponsored forums where you can download files for updated drivers, etc. CompuServe is one of the hardest on line services to find your way around. See the following definitions.

CompuServe Electronic Mail You can send electronic mail to CompuServe addresses. Here's the formula: All CompuServe addresses are either of the form 7xxxx,xxx or 1xxxxx,xxx. (where each "x" signifies a digit from 0 to 7). There can

be from 2 to 4 digits following the comma. To send mail to such an address from the Internet, change the comma to a period and attach "@CompuServe.com" as is shown in the following examples:
   74906.1610@compuserve.com or 100906.1610@compuserve.com

CompuServe B+ File Transfer This file transfer protocol is used by the CompuServe information service and no one else. Recovery of interrupted transfers is supported. In CompuServe B+, the host initiates the transfer. In contrast, in XMODEM, the receiver initiates the transfer, i.e. tells the distant computer to begin sending the file.

CompuServe Mail Hub A facility of CompuServe which enables users on a local area network operating Novell Message Handling Service (MHS) software to exchange electronic messages with other MHS users, CompuServe Mail subscribers and users of other E-mail services that can be reached via a CompuServe gateway.

Computer This is a definition straight from AT&T Bell Laboratories. "An electronic device that accepts and processes information mathematically according to previous instructions. It provides the result of this processing via visual displays, printed summaries or in an audible form."

Computer Aided Dialing A newer (and allegedly less offensive) term for predictive dialing. See also PREDICTIVE DIALING.

Computer Aided Professional Publishing CAP. The computerization of professional publishing (as opposed to desktop operations), including true color representation of the layout on the workstation screen.

Computer And Business Equipment Manufacturers Association CBEMA. Association active before Congress and the FCC promoting the interests of the competitive terminal, computer and peripheral equipment industries.

Computer And Communications Industry Association CCIA. Organization of data processing and communications companies which promotes their interests before Congress and the FCC.

Computer Fraud Deliberate misrepresentation, alteration or disclosure of computer-based data to obtain something of value.

Computer Inquiry A series of ongoing FCC proceedings examining the distinctions between communications and information processing to determine which services are subject to common carrier regulation. The FCC decision in 1980 resulting from the second inquiry was to limit common carrier regulation to basic services. Enhanced services and customer premises equipment are not to be regulated. This meant the Bell operating companies had to set up separate subsidiaries if they were to offer non-regulated services.
Computer Inquiry III, adopted by the FCC in May, 1986, removed the structural separation requirement between basic and enhanced services for the BOCs and for AT&T. CI III replaced that requirement with "nonstructural safeguards." This action resulted in the imposition of such concepts as "comparably efficient interconnection" (CEI) and Open Network Architecture (ONA). The FCC's jurisdiction regarding Computer Inquiry I, II and III has now been usurped by Judge Greene, who insists on fairly tight control over the non-basic telephone company activities of the Bell operating companies. Sometimes he gives dispensations (waivers). Sometimes he doesn't. His word these days is final law on what the Bell operating companies can and can't do.

Computer Support Telephony See CST.

Computer Telephony Computer telephony is the adding of computer intelligence to the making, receiving, and managing of telephone calls. Harry Newton coined the term in 1992. Computer telephony has two basic goals: to make making and receiving phone calls easier, i.e. to enhance one's personal productivity and second, to please corporate customers who call in or who are called for information, service, help, etc. Computer telephony encompasses six broad elements:
1. Messaging.
Voice, fax and electronic mail, fax blasters, fax servers and fax routers, paging and unified messaging (also called integrated messaging) and Internet Web-vectored phones, fax and video messaging.
2. Real-time Connectivity.
Inbound and outbound call handling, "predictive" and "preview" dialing, automated attendants, LAN / screen-based call routing, one number calling / "follow me" numbers, video, audio and text-based conferencing, "PBX in a PC," collaborative computing.
3. Transaction Processing and Information Access via the Phone.
Interactive voice response, audiotex, customer access to enterprise data, "giving data a voice," fax on demand and shopping on the World Wide Web.
4. Adding Intelligence (and thus value) to Phone Calls.
Screen pops of customer records coincident with inbound and outbound phone calls, mirrored Web page "pops," smart agents, skills-based call routing, virtual (geographically distributed) call centers, computer telephony groupware, intelligent help desks and "AIN" network-based computer telephony services.
5. Core Technologies.
Voice recognition, text-to-speech, digital signal processing, applications generators (of all varieties — GUI to forms-based to script-based), VoiceView, DSVD, computer-based fax routing, USB (Universal Serial Bus), GeoPort, video and audio compression, call progress, dial pulse recognition, caller ID and ANI, digital network interfaces (T-1, E-1, ISDN BRI and PRI, SS7, frame relay and ATM), voice modems, client-server telephony, logical modem interfaces, multi-PC telephony synchronization and coordination software, the communicating PC, the Internet, the Web and the "Intranet."
6. New Core Standards.
The ITU-T's T.120 (document conferencing) and H.320 (video conferencing), Microsoft's TAPI — an integral part of Windows 95 and NT, Novell's TSAPI — a phone switch control NLM running under NetWare. Intel's USB and InstantON. Natural MicroSystems / Mitel's MVIP and H-MVIP. Dialogic has SCSA. And the industry has ECTF.
That's today. But what really excites is the potential. It's huge. Despite the above, phone calls today are dumb, seriously bereft of common sense. Few phones have "backspace erase." 75% of business calls end in voice mail! Often in voice mail jail. Every call not completed is an irritated customer and a lost sale. Computer telephony addresses the waste. Computer telephony adds intelligence to the making and receiving of phone calls. Bingo, happier customers and more completed transactions.
The best news: We now have the technology, the resources, the computer power, the new standards and the muscle to back our

NEWTON'S TELECOM DICTIONARY

hype. We also have many new players who are, thankfully, not burdened by the assumptions of yesteryear's telecommunications industry. We also have legions of developers and systems and integrators who are grabbing these computer telephony tools and are cranking out hundreds of customer-pleasing, productivity-enhancing solutions for your business. Computer telephony delivers. And fortunately, industry now wants it.

Once a year, typically in March, industry leaders meet at a trade show called Computer Telephony Conference and Exposition. There is also a monthly magazine covering the industry called Computer Telephony Magazine. For more, call 1-800-677-3435 or 215-355-2886. See also TELEPHONY, TELEPHONY SERVICES and WINDOWS TELEPHONY.

Computer-Like Transport An AT&T term for the carrying of digital information with the potential for acting on that information at any network node as appropriate.

Computername The name by which the lan identifies a server or a workstation in lan Manager terminology. Each computername must be unique on the network.

COMSAT The COMmunications SATellite corporation was created by Congress as the exclusive provider to the U.S. of satellite channels for international communications. COMSAT is also the U.S. representative to Intelsat and Inmarsat, two international groups responsible for satellite and maritime communications. Comsat is now being merged with Continental Telephone, also known as Contel.

COMSPEC COMmand SPECification. A configuration command that tells MS-DOS where to find the program that interprets what is typed at the prompt. Except in unusual circumstances, that program is COMMAND.COM. COMSPEC is typically used when you have the program COMMAND.COM somewhere other than in the root directory of your C: drive (if you have a hard disk machine) or your A: drive if you have a floppy system. Let's say you want to tell your computer to find COMMAND.COM on your D: drive, the command in your AUTOEXEC.BAT program would be:
SET COMSPEC=D:\COMMAND.COM

COMSTAR A domestic communications satellite system from Comsat.

Concatenation Joining several fibers together end-to-end.

Concentration A fundamental concept to telephony. Communications from a number of phones are sent out on a smaller number of outgoing lines. The theory is that, since not all the phones are being used at any one time, fewer trunks than phones are needed. Some phone system designs assume that only 5% of the phones will be in use at any one time. Some phone systems design assume 10%. In some phone-intensive industries, you can't make any assumptions about concentration. You have to assume one line per phone. No concentration. TELECONNECT Magazine is around one phone per line. We are very phone intensive. See CONCENTRATION RATIO.

Concentration Ratio The ratio between lines and trunks in a concentrated carrier system or line concentrator.

Concentrator 1. This telecommunications device allows a number of circuits (typically slow speed ones) to be connected to a smaller number of lines for transmission under the assumption that not all of the larger group of lines will be used at the same time. The concentrator allows a shared transmission medium to accommodate more data sources than there are channels currently available within the transmission medium. A concentrator's job is to save money on transmission.

2. Some LANs use concentrators, or access units, that allow network devices to be interconnected through a central point. The wiring is star, also called home run. Attaching devices through a central concentrator typically simplifies the maintenance of the LAN. A concentrator is used when connecting a LAN over the "normal" twisted wires in a building which were put there originally to carry voice and which now have to be used because — for one reason or another — it's not possible to connect the LAN using its normal bus coaxial cable connection.

Concentricity In a wire or cable, the measurement of the location of the center of the conductor with respect to the geometric center of the circular insulation.

Concentricity Error The amount by which a fiber's core is not centered in its cladding. The distance between the center of the two concentric circles specifying the cladding diameter and the center of the two concentric circles specifying the core diameter.

Concert A basket of international telecommunications services offered by British Telecom and MCI.

Concert VNS Concert VNS is a worldwide, flexible virtual private voice and data network that offers customers a consistent feature set available in multiple countries. The network combines the geographic reach of both MCI's Vnet and BT's International Virtual Network (IVN) physical networks and is connected to public networks around the world. Concert VNS provides a comprehensive set of virtual network capabilities for international voice and data applications.

Concert Network Management System British Telecom operating environment.

Concurrency The shared use of resources by multiple interactive users or applications at the same time. Concurrency often means that a company need only buy as many licenses to a program as it has people using the program at one time — concurrent users, in other words. See SOFTWARE METERING.

Concurrency Control A feature that allows multiple users to execute database transactions simultaneously without interfering with each other.

Concurrent Site License Companies that buy software for multiple computers typically buy one copy of the program and a license to reproduce it up to a certain number of times. This is called a site license, though it may apply to its use throughout an organization. Site licenses vary. Some require that a copy be bought for each potential user — the only purpose being to indicate the volume discount and keep tabs. Others allow for a copy to be placed on a network server but limit the number of users who can gain simultaneous access. This is called a Concurrent Site License. And many network administrators prefer this concurrent license, since it gives them greater control. For example, if the software is customized, it need be customized only once, namely on the server.

Condenser A device for storing up electrical energy and consisting of two or more conducting surfaces or electrodes separated by an insulating medium called a dialectic.

Condenser Antenna An antenna consisting of two capacity areas.

NEWTON'S TELECOM DICTIONARY

Condenser Microphone Microphone which operates through changes in capacitance caused by vibrations of its conductive diaphragm.
Conditions Busy. Voice Mail. Out of service. All the situations that a phone line is likely to find itself in.
Conditioned Circuit A circuit that has conditioning equipment to obtain the desired characteristics for voice or data transmission. See CONDITIONING.
Conditioned Loop A loop that has conditioning equipment to obtain the desired line characteristics for voice or data transmission. See CONDITIONING.
Conditioning Electrical treatment of transmission lines to improve their performance for specific uses such as data transmission. The "tuning" or addition of equipment to improve the transmission characteristics of a leased voice-grade line so that it meets the specifications for higher-speed data transmission. Voice-grade lines often have too much "noise" on them. By altering the equipment at both ends of the line, this noise on the line can be overcome. This allows transmission of data, which is much more sensitive to noise than voice.
Here is another definition of conditioning: A procedure used to make circuit transmission impairments lie within certain acceptable limits which are specified in a tariff (typically used on telephone lines leased for data transmission to improve transmission speed and quality). Usually done with special equipment or routing.
Condofiber A shared tenancy cable or shared ownership facility such as a transatlantic fiber cable. Multiple vendors such as Sprint, MCI and AT&T may all own a group of fibers with responsibility for maintaining their own operation while at the same time paying an overall "association" fee for the common maintenance of the overall cable.
Conductance The opposite of resistance; a measure of the ability of a conductor to carry an electrical charge. Conductance is a ratio of the current flow to the potential difference causing the current flow. The unit of conductance is Mho (a reversed spelling of Ohm).
Conducting Materials Substances which offer relatively little resistance to the passage of an electric current.
Conductivity A term used to describe the ability of a material to carry an electrical charge. The opposite of specific resistance. Usually expressed as a percentage of copper conductivity - copper being one hundred percent. Conductivity is expressed for a standard configuration of conductor.
Conductor Any substance, usually a wire or cable, that can carry (i.e. offer a relatively small opposition to the passage of) an electrical current.
Conduit A pipe, usually metal but often plastic, that runs either from floor to floor or along a floor or ceiling to protect cables. A conduit protects the cable and prevents burning cable from spreading flames or smoke. Many fire codes in large cities thus require that cable be placed in metal conduit. In the riser subsystem when riser closets are not aligned, conduit is used to protect cable as well as to provide the means for pulling cable from floor to floor. In the horizontal wiring subsystem, conduit may be used between a riser or satellite closet and an information outlet in an office or other room. Conduit is also used for in-conduit campus distribution, where it is run underground between buildings and intermediate manholes and encased in concrete. Multiduct, clay tile conduit may also be used.
Conduit Run The path taken by a conduit or group of conduits.
Conferee Participant in a conference call who is not the call controller. This definition courtesy Hayes. According to Hayes, a "controller" is the person who sets up the conference call.
Conference Bridge A telecommunications facility or service which permits callers from several diverse locations to be connected together for a conference call. The conference bridge contains electronics for amplifying and balancing the conference call so everyone can hear each other and speak to each other. The conference call's progress is monitored through the bridge in order to produce a high quality voice conference and to maintain decent quality as people enter or leave the conference.
Conference Call Connecting three or more people into one phone conversation. You used to have to place conference calls through an AT&T operator (you still can). But now you can also organize conference calls with most modern phone systems or a conference bridge. If conferencing is important to you, make sure your conferencing device has amplification and balancing. If not, it will simply electrically join the various conversations together and people at either end won't be able to hear each other. There are different types of conference devices you can buy, including special teleconferencing devices that sit on conference tables and perform the function of a speakerphone, albeit a lot better. There are also dial-in devices called conference bridges. But, however, you use these devices, they will requires lines (and/or trunks). If you install one inside your phone system, be careful have the extra spare extensions. For a conference of 10 people, you'll typically need 10 extensions connected to your conference bridge. See CONFERENCE BRIDGE.
Conference, Meet-Me A conference call in which each of the people wishing to join the conference simply dials a special "Meet-Me" Conference phone number, which automatically connects them into the conference. It is a feature of some PBXs and also some special Conferencing Equipment. See CONFERENCE BRIDGE.
Conferencing Several parties can be added to a phone conversation through Conferencing.
Confidencer A noise-cancelling microphone for use on a telephone in noisy places. A confidencer is not an easy device to use.
Confidential Reception The ability to receive a facsimile transmission directly into memory which can be printed out or viewed at a later time.
Confidential Transmission A facsimile message that is sent confidentially into memory or a private mailbox, to be retrieved by the receiver at a later time. It's usually retrieved by using a confidential passcode or password.
Config.sys A file of commands used to set up a PC in MS-DOS. MS-DOS reads it each time it starts. Config.sys is where you put commands for device drivers for a mouse, a RAM drive, instructions on where to put DOS, how much memory to set for your cache, etc. Other configuration parameters like how many open files are allowed, how many buffers, how many drives, etc. are also put in this file. You'll find Config.sys under Windows 95 also. See also AUTOEXEC.BAT.
Configuration 1. The hardware and software arrangements that define a computer or telecommunications system and thus determine what the system will do and how well it will do it. This information can be entered in the CMOS and EEPROM setup programs.

2. An ATM term. The phase in which the LE Client discovers the LE Service.

Configuration Databases Rolm/IBM words for those databases which represent unique user specifications relating to system and phone features. These databases can be entered on-site and are not part of the generic software which runs the phone system.

Configuration File An unformatted ASCII file that stores initialization information for an application.

Configuration Management One of five categories of network management defined by the ISO (International Standards Organization). Configuration management is the process of adding, deleting and modifying connections, addresses and topologies within a network.

Configuration Manager 1. A SCSA system service which manages configuration information and controls system startup.
2. An Intel Plug'n Play term. A driver, such as the ISA Configuration Utility, that configures devices and informs other device drivers of the resource requirements of all devices installed in a computer system. The Windows 95 Resource Kit defined configuration manager as the central component of a Plug and Play system that drives the process of locating devices, setting up their nodes in the hardware tree, and running the resource allocation process. Each of the three phases of configuration management-boot time (BIOS), real mode, and protected mode-has its own configuration manager.

Configuration Registry A database repository for information about a computer's configuration.

Confirming Design Layout Report Date CDLRD. The date a common carrier accepts the facility designed proposed by the Telco.

Conformance Test A test performed by an independent body to determine if a particular piece of equipment or system satisfies the criteria of a particular standard, sometimes a contract to buy the equipment.

Conforming End Office Central office with the ability to provide originating and terminating feature group D local access and transport area access service.

Congestion Control The process whereby packets are discarded to clear buffer congestion in a packet-switched network.

Congestion Management The ability of a network to effectively deal with heavy traffic volumes; solutions include traffic scheduling and enabling output ports to control the traffic flow. See BECN.

Connect Time Measure of computer and telecommunications system usage. The interval during which the user was on-line for a session.

Connected 1. On line.
2. A voice recognition term for words spoken clearly in succession without pauses. For recognition to occur, words or utterances must be separated by at least 50 milliseconds (1/20th of a second). Generally refers to digit recognition and sometimes used to describe fast discrete recognition.

Connected Time The length of time a path between two objects is active.

Connected User A Windows NT term. A user accessing a computer or a resource across the network.

Connecting Arrangement The manner in which the facilities of a common carrier (phone company) and the customer are interconnected.

Connecting Block A plastic block containing metal wiring terminals to establish connections. from one group of wires to another. Usually each wire can be connected to several other wires in a bus or common arrangement. A 66-type block is the most common type of connecting block. It was invented by Western Electric. Northern Telecom has one called a Bix block. There are others. These two are probably the most common. A connecting block is also called a terminal block, a punch-down block, a quick-connect block, a cross-connect block. A connecting block will include insulation displacement connections (IDC). In other words, with a connecting block, you don't have to remove the plastic shielding from around your wire conductor before you "punch it down."

Connection 1. A path between telephones that allows the transmission of speech and other signals.
2. An electrical continuity of circuit between two wires or two units, in a piece of apparatus.
3. An SCSA term which means a TDM data path between two Resources or two Groups. It connects the inputs and outputs of the two Resources, and may be unidirectional (simplex) if either of the Resources has only an input or an output. Otherwise it is bi-directional (dual simplex). It usually has a bandwidth that is a multiple of a DS0 (64kbit) channel. Inter-group connections are made between the Primary Resource of each Resource Group.
4. An ATM connection consists of concatenation of ATM Layer links in order to provide an end-to-end information transfer capability to access points.

Connection Master Software from Mitel, which brings the Connection Control Standard to an even higher level for the MVIP developer. Connection Master interacts with circuit switches on multiple MVIP cards to make connections and resolve switching contention. It also interfaces between applications and makes connections in such a way that simple one-chassis applications become networked applications. Connection Master fully supports MC-MVIP, Multi-Chassis MVIP. See also MVIP.

Connection Oriented The model of interconnection in which communication proceeds through three well-defined phases: connection establishment, information transfer, connection release. Examples include X.25, Internet TCP and OSI TP4 and ordinary telephone calls. See CONNECTION SERVICE.

Connection Oriented Network Service CONS. An OSI protocol for packet-switched networks that exchange information over a virtual circuit (a logical circuit where connection methods and protocols are pre-established); address information is exchanged only once. CONS must detect a virtual circuit between the sending and receiving systems before it can send packets.

Connection Oriented Operation A communications protocol in which a logical connection is established between communicating devices. Connection-oriented service is also referred to as virtual-circuit service.

Connection Oriented Transmission Data transmission technique involving setting up connection before transmission and disconnecting it afterward. A type of service in which information always traverses the same pre-established path or link between two points. See CONNECTIONLESS SERVICE.

Connection Service A circuit-switching service whereby a connection is switched into place at the beginning of a session and held in place until the session is completed. Also referred to as circuit switching. The circuit switched in place may be real or virtual. See CIRCUIT SWITCHING.

NEWTON'S TELECOM DICTIONARY

Connectionless Mode Transmission In packet data transmission, a mode of operation in which each packet is encoded with a header containing destination address sufficient to permit the independent delivery of the packet without the aid of additional instruments. See also CLOUD, CONNECTIONLESS NETWORK and CONNECTIONLESS SERVICE.
Connectionless Network A type of communications network in which no logical connection (i.e. no leased line or dialed-up channel) is required between sending and receiving stations. Each data unit (datagram) is sent and addressed independently. IEEE 802 LAN standards specify connectionless networks. Connectionless networks are becoming more common in broadband city networks now increasingly offered by phone companies.
Connectionless Network Service CLNS. Packet-switched network where each packet of data is independent and contains complete address and control information; can minimize the effect of individual line failures and distribute the load more efficiently across the network.
Connectionless Service A networking mode in which individual data packets in a network (local or long distance) traveling from one point to another are directed from one intermediate node to the next until they reach their ultimate destination. Because packets may take different routes, they must be reassembled at their destination. The receipt of a transmission is typically acknowledged from the ultimate destination to the point of origin. A connectionless packet is frequently called a datagram. A connectionless service is inherently unreliable in the sense that the service provider usually cannot provide assurance against the loss, error insertion, misdelivery, duplication, or out-of-sequence delivery of a connectionless packet. Connectionless networks may be better than connection-oriented ones for obtaining management information over a failing network because a device doesn't have the additional burden of maintaining a network connection.
Connectionless Transmission Data transmission without prior establishment of a connection.
Connections Per Circuit Hour CCH. A unit of traffic measurement; the number of connections established at a switching point per hour.
Connectivity Property of a network that allows dissimilar devices to communicate with each other.
Connector A device that electrically connects wires or fibers in cable to equipment, or other wires or fibers. Wire and optical connectors most often join transmission media to equipment (host computers and terminal devices) or cross connects. A Connector at the end of a telephone cable or wire is used to join that cable to another cable with a matching Connector or to some other telecommunications device. Residential telephones use the REJ-11C connector. Computer terminals with an RS-232-C interface, use the DB-25 connector. The RS-232-C standard is actually the electrical method of using the pins on a DB-25. See RS-232-C.
CONS Connection Oriented Network Service. See also CONNECTION SERVICE.
Consent Decree 1982 The agreement which divested the Bell Operating Companies from AT&T. It took effect at midnight on December 31, 1983.
Conservation Of Radiance A basic principle stating that no passive optical system can increase the quantity L/n2, where L is the radiance of a beam and n is the local refractive index. Formerly called conservation of brightness, or the brightness theorem.
Console 1. A large telephone which a PBX attendant uses to answer incoming calls and transfer them around the organization. Before you buy a PBX for your company, make sure your operator has checked out its console. Some are very difficult to use. Some are easy. Some operators hate some consoles. Some consoles hate some operators. You can measure the efficiency of consoles by comparing keystrokes to do simple jobs and comparing them — e.g. answer an incoming line, dial an extension and transfer the call. How many keystrokes does your PBX take?
2. The device which allows communications between a computer operator and a computer.
3. The console is the Novell NetWare name for the monitor and keyboard of the file server. Here you can view and control the file server or router activity. At the console, you can enter commands to control disk drives, send messages, set the file server or router clock, shut down the file server, and view file server information. NetWare commands you can enter only from the console (for example, MONITOR) are called console commands. Keep your file server locked up and away from prying eyes. It's clearly not just a case of changing passwords and getting in and mucking around. There have been examples of thieves simply removing the file server's hard disk, putting it in their briefcase and walking off with it.
Consoleless Operation Some PBXs can work without a console. Some must have a console. It's good to check. Consoles are expensive. If you don't want one — because your company is small — you don't want to be forced to buy one, only to have it sit idly by.
Consolidated Carrier Carriers that provide connection both as interexchange carriers as well as international carriers.
Constant Bit Rate CBR. A data service where the bits are conveyed regularly in time (following a timing source or clock just as members of a marching band follow the beat of the drummer). Voice and video traffic has to be transported at constant bit rate because they are sensitive to variable delay and, as such, have to be transported without any interruptions in the flow of data.
Constant Carrier Physical line specification selection indicating full duplex line in bisync network.
Construction Zone The building of the new information infrastructure by telecommunications and cable companies.
Consult To seek another's endorsement of a decision you've already made.
Consultant See CONSULT and CONSULTANT LIAISON PROGRAMS.
Consultant Liaison Programs Large users often use communications consultants to help them choose systems and long distance phone lines. In recognition of the important role consultants play, many suppliers have consultant liaison programs. Such programs typically consist of a toll-free number and somebody on the other end to answer technical and pricing questions, a three-ring containing information on all the company's products and services, occasional seminars and, for those extra-privileged consultants, trips to all expense paid trips to exotic places and "something" else. With MCI that "something else" is a dial-up, toll-free, bulletin board. Dial it up with your PC, you can download MCI's latest prices and services. It's truly splendid as most of the paperwork others issue is obsolete the moment it's issued.
Consultation See CONSULTATION HOLD.
Consultation Hold PBX feature which allows an extension to place a call on hold while speaking with another call. The idea is "consulting with" someone while you have someone else on the phone.

Contact A strip or piece of metal which makes an electrical contact when some electromechanical device like a relay or a magnet operates. Contacts are often plated with precious metal to prevent them from oxidizing (i.e. rusting) and thus messing up the switch.

Contact Gateway A Rockwell ACD term. A hardware/software package that acts as a gateway to integrate various ACD, voice response, host computer and live agent processes.

Contact History A log of all the contacts, either by phone or letter, made with a prospect or customer. This is an important factor in building up a marketing database which can be used to accurately target prospects.

Contact Image Sensor Uses a flat bar of light-emitting diode that directly touches the original. It eliminates the step of having the diodes move through the lens, which causes poorer resolution. This method is a more sophisticated than the charged-coupled device scanning method.

Contact Management A business has customers and prospects. In computerese, they're called "contacts." Software to "manage" your customers and prospects is called contact management software. It has three elements: First, a screen or two of information about that contact (address, phone number, notes about your conversations, etc.) Second, the ability to print lists, and mailers, etc. And third, often a tie in with your phone system to let your computer dial your clients and fax them stuff. With many newer phone systems, you have one extra benefit — namely when your phone rings, your contact management software will receive the calling phone number and pop up the screen or two about your contact. This way you'll be a little prepared before you answer the phone. See also ANI and CLID.

Contact Region The section of the jack wire inside the plug opening as shown in Subpart F of FCC rule 6B, figures 6B.500 (a) (3) and 6B.500 (b) (3).

Contended Access In local area networking technology it's the shared access method that allows stations to use the medium on a first-come, first-served basis.

Contending Port A programmable port type which can initiate a connection only to a preprogrammed port or group of ports.

Content In today's information rich and hyped society, "carriage" is the new name for transmission. And "content" is the new name for what we carry. Content is a more than just phone calls, of course. It's movies, music, games, on-line books, information, etc. See also CONTENT SUPPLIER.

Content Processing Voice processing is the broad term made up of two narrower terms — call processing and content processing. Call processing consists of physically moving the call around. Think of call processing as switching. Content consists of actually doing something to the call's content, like digitizing it and storing it on a hard disk, or editing it, or recognizing it (voice recognition) for some purpose (e.g. using it as input into a computer program.

Content Provider A fancy name for a writer, also called a language therapist by William Safire in the Sunday New York Times Magazine of January 28, 1996.

Content Supplier Content is a new fancy name for what telecommunications facilities carry. It includes movies, music, games, on-line books, information, etc. Content suppliers are thus movie studios, publishers, and music companies.

Contention Contention occurs when several phones are vying for access to a line and only one of them can get it at one time. Some method is usually established for selecting the winner (first in, first out, campon, etc.) and accommodating the loser(s) (.e.g. giving them a busy tone). When you cannot get an outside line from your extension you have been in contention and lost.

Context Dependent Soft Keys Many telephones now have an LCD screen. Sometimes such screens have unmarked keys underneath them and/or at their side. What these keys do depends on the "labels" appearing on the screen. They are called "context dependent" because what those keys do depends on where the call is at that time. The first context dependent soft keys were on the Mitel SuperSet 4 phones. When the handset was resting on the phone, only three of the six context sensitive keys had meaning. One said "Program," one said "Msg" and one said "Redial." When you picked the phone up, three buttons would now be alive. One would say "Page," one would say "Redial" and one would say "Hangup." If the phone rang and you picked it up, one button would now say "transf/conf" (meaning transfer/conference. When another phone was ringing. one button would say "Pickup," letting you push that button and answer someone else's phone. And so on. The neatest implementation of context sensitive keys was probably on the Telenova (now no longer manufactured). At one point when you were in voice mail, this phone's six buttons looked exactly like a cassette recorder —- record, play, fast forward, fast reverse, etc. It was brilliant. No one has ever made using voice mail so easy.

Context Sensitive A term from the computer industry which means that "Help" is only a keystroke away. Hit F1 and Help information will flash on the screen. That information will be relevant to what you're doing now, i.e. that help is within the context of what's going on right this moment. See also CONTEXT DEPENDENT SOFT KEYS.

Context Switch The technique with which an Intel 80x86 microprocessor handles multitasking is called a context switch. The CPU performs a context switch when it transfers control from one task to another. In the process, it saves the processor state (including registers) of one task, then loads the values for the task that is taking control. Context switching is the kind of multitasking that is done in standard mode Windows, where the CPU switches from one task to another, rather than allocating time to each task in turn, as in timeslicing.

Contiguous Port Ports occurring in unbroken numeric sequence.

Contiguous Slotting This term refers to the process of selecting individual DS-0 circuits, within a DS-1 circuit or DS-3 circuit, which are adjacent to one another. Due to the timing difference which can result when non-adjacent channels are selected, contiguously slotted channels are preferable when the end equipment is designed to multiplex the individual low-speed channels into a single, higher speed connection.

Contiguous United States The area within the boundaries of the District of Columbia and the 48 contiguous states as well as the offshore areas outside the boundaries of the coastal states of the 48 contiguous states, (including artificial islands, anchored vessels and fixed structures erected in such offshore areas for the purpose of exploring for, developing, removing and

NEWTON'S TELECOM DICTIONARY

transporting resources therefrom) to the extent that such areas appertain to and are subject to the jurisdiction and control of the United States within the meaning of the Outer Continental Shelf Land Act, 43 U.S.C. Section 1331, et seq.

Continental Telecom Inc. CONTEL. A telephone company made up of more than 600 small phone companies. In 1990 it merged with GTE in a tax-free swap of shares. Contel was formed and grown by Charles Wohlstetter, an ex-stockbroker, who became comfortable in the process of growing Contel. In late 1990, Contel merged with GTE, which is a euphemism for GTE buying Contel.

Continuity An uninterrupted electrical path.

Continuity Check A check to determine whether electrical current flows continuously throughout the length of a single wire on individual wires in a cable.

Continuity Check Tone CCT. A single frequency of 2000 Hz which is transmitted by the sending exchange and looped back by the receiving exchange. Reception of the returned indicates availability of the channel. See ITU-T Recommendation.271.

Continuous A word used in voice recognition to mean a type of recognition that requires no pause between utterances.

Continuous DTMF This is a feature of some phones (especially cellular phones) that sends touchtone sounds for as long as the key is held down, allowing access to services such as voice mail and answering machines that need long-duration tones. Some phones automatically have continuous DTMF; some don't. It's worth checking. Continuous DTMF makes a lot more sense.

Continuous Information Environment A term for the world we live in — in which information (text, voice, video, images, etc.) is flowing at us continuously. And our job is, somehow, to manage the information. The idea is to use the new computer telephony terms to manage the information.

Continuous Waves A series of wave or cycle of current all of which have a constant or unvarying amplitude.

Continuously Variable Capable of having one of an infinite number of values, differing from each other by an arbitrarily small amount. Usually used to describe analog signals or analog transmission.

Contract For the purposes developing applications in the telecommunications industry, there are two types of contracts: Active and Passive. An active contract is one you must sign. A passive contract is the type of contract you find in a software package. By opening the shrink wrapped package, you are committing yourself to the terms of the contract inside the package — the terms of which mostly consist of not duplicating the software in an authorized way.

Control Cable A multiconductor cable made for operation in control or signal circuits.

Control Character A non-printing ASCII character which controls the flow of communications or a device. Control characters are entered from computer terminal keyboards by holding down the Control key (marked CTRL on most keyboards) while the letter is pressed. To ring a bell at remote telex terminal, an operator could hold down the CTRL key, and tap the "G" key, since Control-G is the BELL character. Most computers display Control as the " " character in front of the designated letter. For example,
M is the Carriage Return character.

Control Circuit X.21 interface circuit used to send control information from DTE to DCE.

Control Connections A Control VCC links the LEC to the LECS. Control VCCs also link the LEC to the LES and carry LE_ARP traffic and control frames. The control VCCs never carry data frames.

Control Equipment 1. The central "brains" of a telephone system. That part which controls the signaling and switching to the attached telephones. Known as the KSU (or key service unit) in a key system.
2. Equipment used to transmit orders from an alarm center to remote site to enable you to do things by remote control.

Control Field Field in frame containing control information.

Control Head Roam Lights Indicates that the cellular phone is outside the "home" system.

Control Of Electromagnetic Radiation 1. Measures taken to minimize electromagnetic radiation emanating from a system or component, or to minimize electromagnetic interference. Such measures are taken for purposes of security and/or the reduction of interference, especially on ships and aircraft.
2. A national operational plan to minimize the use of electromagnetic radiation in the United States and its possessions and the Panama Canal Zone in the event of attack of imminent threat thereof, as an aid to the navigation of hostile aircraft, guided missiles, or other devices.

Control Of Flow Language Programming-like constructs (IF, ELSE, WHILE, GOTO, and so on) provided by Transact-SQL so that the user can control the flow of execution of SQL Server queries, stored procedures, and triggers. This definition from Microsoft SQL server.

Control Package Network A Rolm/IBM term for a multinode network used by the processors in one CBX node to communicate with the processors in other nodes. CPN-1 is used with RolmBus 74. CPN-2 is used with Rolmbus 295. Formerly called HUB.

Control Panel The control panel on the Apple Macintosh is for general hardware and software settings. Icons allow a user to customize the system or application, or select a particular service, such as a specific printer, set the sound level, the date and time and choose an Ethernet connection through the network control panel.

Control Point A program that manages an APPN network node and its resources, enabling communications to other control points in the network.

Control Segment A worldwide network of Global Positioning System monitoring and control installations that ensure the accuracy of satellite positions and their clocks.

Control Signal 1. In the public network, control signals are used for auxiliary functions in both customer loop signaling and interoffice trunk signaling. Control signals are used in the customer loop for Coin Collect and Coin Return and Party Identification. Control signals used in interoffice trunk signaling include Start Dial (Wink or Delay Dial) signals, Keypulse (KP) signals or Start Pulse (ST) signals.
2. In modem communications, control signals are modem interface signals used to announce, start, stop or modify a function.

151

NEWTON'S TELECOM DICTIONARY

Here's a table showing common RS-232-C and ITU-T V.24 control signals

Pin	Control Signal	From	To
4	Request-To-Send (RTS)	DTE	DCE
5	Clear-To-Send (CTS)	DCE	DTE
6	Data Set Ready (DSR)	DCE	DTE
8	Carrier Detect (CD)	DCE	DTE
20	Data Terminal Ready (DTR)	DTE	DCE
22	Ring Indicator (RI)	DCE	DTE

Control Station On a multiaccess link, a station that is in charge of such functions as selection and polling.

Control Tier An AT&T term for the tier within the Universal Information Services network node that provides the transport network's connection control function.

Control Unit An architectural component of a processor chip which orchestrates processor activity and handles timing to make sure the processor doesn't overlap functions.

Controlled Access When access to a system is limited to authorized programs, processes or other systems (as in a network).

Controlled Environment Vault CEV. It is a low maintenance, water-tight concrete or fiberglass container typically buried in the ground which provides permanent housing for remote switches, remote line concentrators, pair gain and fiber transmission systems. Because it is buried, it can often be installed in utility easements or other places where local building laws may be a problem. This below ground room that houses electronic and/or optical equipment is under controlled temperature and humidity conditions.

Controller 1. In the truest sense, a device which controls the operation of another piece of equipment. In its more common data communications sense, a device between a host and terminals that relays information between them. It administers their communication. Controllers may be housed in the host, can be stand-alone, or can be located in a file server. Typically one controller will be connected to several terminals. The most common controller is the IBM Cluster Controller for their 370 family of mainframes. In an automated radio, a controller is a device that commands the radio transmitter and receiver, and that performs processes, such as automatic link establishment, channel scanning and selection, link quality analysis, polling, sounding, message store and forward, address protection, and anti-spoofing.
2. Participant in a conference call who sets up the conference call.

Controller Card Also called a hard disk/diskette drive controller. It's an add-in card which controls how data is written to and retrieved from your PC's various floppy and hard drives. Controller cards come in various flavors, including MFN and SCSI. Controller cards are the devices used to format hard drives. Controller cards are not hard drive specific (except within categories). Controller cards will format many drives. But once you have a hard drive that has been formatted by that one controller card, it tends to prefer talking to that controller card forever. If you switch your hard disk to another machine, switch the controller card along with it. If you switch your hard disk to another machine, but not the controller card, then format the hard disk. That's not a "100% Do It Or Else You'll Be Disappointed" rule. But just a "Play It Safe and Switch Them" rule.

Conturing In digital facsimile, density step lines in received copy resulting from analog-to-digital conversion when the original image has observable gray shadings between the smallest density steps of the digital system.

CONUS A military term for CONtiguous United States (lower 48 states). See CONTIGUOUS UNITED STATES.

Convection Cooling Design techniques used in switching system construction to permit safe heat dissipation from the equipment without the need for cooling fans.

Convector The device which covers the steam heating radiator in buildings and typically sits underneath a window. Also called a weathermaster.

Convector Area An area allocated for heat circulation and distribution. Convector areas, typically built into a wall, can be used as a satellite location only if a more suitable area is unavailable.

Convergence 1. A measure of the clarity of a color monitor. A measure of how closely the red, green and blue guns in a color monitor track each other when drawing a color image. The other measures are focus and dot pitch.
2. A LAN term. The point at which all the internetworking devices share a common understanding of the routing topology. The slower the convergence time, the slower the recovery from link failure. 3. The word to describe a trend, now that most media can be represented digitally, for the traditional distinctions between industries to blur and for companies from consumer electronics, computer and telecommunications industries to form alliances, partnerships and other relationships, as well as to raid each others markets.

Conversation Path The route from originating port to terminating port for a two-way call. A conversation thus typically requires two ports on most PBXs.

Conversation Time The time spent on a conversation from the time the person at the other end picks up to the time you or him hang up. Conversation time plus dialing, searching and ringing time equal the time your circuit will be used during a call.

Conversational Mode Telex An MCI International product providing real time exchange between Telex terminals or other compatible devices that allows instantaneous, two-way conversations in writing.

Converter 1. A vacuum tube which combines the functions of oscillator and mixed tube.
2. A device for changing AC to DC and vice versa. An ancient radio term.
3. An adapter, such as one that allows a modular phone to be plugged into a 4-hole jack.
4. A British term. A repeater that also converts from one media type to another, such as from fibre (British spelling) to copper. Often called a media adaptor.

Convolutional Code Error protection code encoding data bits in a continuous stream. An error-correction code in which each m-bit information symbol to be encoded is transformed into an n-bit symbol (n>m) where the transformation is a function of the last k information symbols, and k is referred to as the constraint length of the code. Convolutional codes are often used to improve the performance of radio and satellite links.

Cooperative Processing Mainframe and intelligent workstations dividing application code between them.

Copland Apple's code name for a flashy new operating system — the one designed to make people forget about Windows 95. I found mention of Copland about two weeks before Windows 95 was due to be released by Microsoft, August 24, 1995.
Copolymer Compound resulting from the polymerization of two different monomers.
Coprocessor An additional processor which takes care of specific tasks, the objective being to reduce the load on the main CPU. Many IBM PCs and IBM clones have the capacity to install a coprocessor chip which does only arithmetic functions. This significantly speeds up your computer if you do a lot of calculations. See MATH COPROCESSOR.
COPW Customer Owned Premises Wire. You own the telephone wiring in your office.
Copy A nice new telephone system programming feature. We found it on Northern Telecom's Norstar phone. With this button, certain programmed settings can be copied from one line to another, or from one telephone to another. Line programmable settings that can be copied on the Norstar are Line Data, Restrictions, Overrides, and Night Service. Telephone settings that can be copied are Line Access, Restrictions, Overrides, and Permissions.
Copyright A copyright protects the original author of a story, software program, song movie, piece of sculpture, or other original work from direct copying. Copying may be inferred where alleged copyist had access to the copyrighted work. The copyright notice (the © symbol, or the word "Copyright", the year of creation, and the name of the copyright owner) should be provided on each copy of the work. Copyrights may also be registered with the Library of Congress, but this is not necessary in all cases. Copyright protects the expression of an idea, not the idea itself. (In appropriate cases, patents can be used to protect the idea.) Where the idea is so simple that there is only one way to express it, the idea and its expression may merge, preventing copyrightability. This logic was used successfully in defense of several suits involving "clean room" reverse engineering of microcode: A first group hacked out the code, and prepared a complete functional specification defining the function of each instruction. A second group then wrote new code implementing these functions. Since there had been no copying, there could be no copyright infringement; the fact that both versions of the code for some instructions were identical merely showed that the idea and expression had merged. There are only so many ways to code an ADD instruction, after all. See INTELLECTUAL PROPERTY, PATENT and TRADE SECRET.
Cord A small, flexible insulated wire.
Cord Board The earliest manual PBX. Usually an elegant wooden device consisting of lots of cords with plugs on them. These cords sat horizontally sticking up, like missiles in a silo. Each cord corresponded to an extension. Whenever the phone rang, the cord board attendant would answer it. Each incoming line was a vertical hole. When the operator had figured for whom the call was, he/she would simply plug the cord corresponding to the desired extension into the hole corresponding to the incoming trunk. The operator would reverse the process if the internal user wanted to make an external call. Either the operator would dial the call first, or simply plug in the user's extension and thus allow the user to dial the call directly. The tip of the plug and the circular ring of the plug gave the term "tip and ring" to telephony. In electronics, it's known as positive and negative. See CORD CIRCUIT.
Cord Circuit A switchboard circuit, terminated in two plug-ended cords, used to establish connections manually between user lines or between trunks and user lines. A number of cord circuits are furnished as part of the manual switchboard position equipment. The cords may be referred to as front cord and rear cord or trunk cord and station cord. In modern cordless switchboards, the cord circuit is switch operated. See CORD BOARD.
Cord Lamp The lamp associated with a cord circuit that indicates supervisory conditions for the respective part of the connection. See CORD BOARD.
Cordboard See CORD BOARD.
Cordless Telephone A telephone with no cord between handset and base. Each piece contains a radio transmitter, receiver, and antenna. The handset contains a rechargeable battery; the base must be plugged into an AC outlet. Depending on product design, radio frequency, environmental conditions, and national law, range between handset and base can be 10 feet to several miles. Cordless phones were once all analog. Now a breed of digital ones is out. They work much better in electrically noisy environments — like the typical office.
Cordless Switchboard A telephone switchboard in which manually operated keys are used to make connections. See CORD BOARD.
Core The central glass element of a fiber optic cable through which the light is transmitted (typically 8-12 microns in diameter for single mode fiber and 50-100 microns in diameter for multimode fiber). This light conducting portion of the fiber is defined by the high refraction index region. The core is normally in the center of the fiber, bounded by the cladding material.
Core Gateway The primary routers in the Internet. Historically, one of a set of gateways (routers) operated by the Internet Network Operations Center at BBN. The core gateway system formed a central part of Internet routing in that all groups would advertise paths to their networks from a core gateway, using the Exterior Gateway Protocol (EGP).
Core Non-Circularity The percent that the shape of the core's cross section deviates from a circle. Sometimes referred to as core ovality.
Core Size Primary description of a fiber. Stated in microns. Does not include cladding. Determines end surface area which accepts and transmits light.
Corner Reflector 1. A device, normally consisting of three metallic surfaces or screens perpendicular to one another, designed to act as a radar target or marker.
2. In radar interpretation, an object that, by means of multiple reflections from smooth surfaces, produces a radar return of greater magnitude than might be expected from the physical size of the object.
3. A reflected electromagnetic wave to its point of origin. Such reflectors are often used as radar targets.
4. Passive optical mirror, that consists of three mutually perpendicular flat, intersecting reflecting surfaces, which returns an incident light beam in the opposite direction. 5. A reflector consisting of two mutually intersecting conducting flat surfaces.
Corporate Account Service An MCI specific service involving a single, unified reporting system for multiple business that the customer owns, franchises, manages, or directs.

153

Corporate ID Number The MCI term for the number which identifies a customer on a corporate level. (Not all MCI customers have a corporate ID number.)

Corporate Network Also called an internetwork or a wide area network. A network of networks (the mother of all networks) that connects most or all of a corporation's voice, data, and video resources using various methods, including the phone system, LANs, private data networks, leased telecommunications lines, and public data networks. Connections between networks are made with bridges and routers. Corporate networks come in many shapes and sizes. Often, they will consist of networks within the same building or facility. Here, networks are combined using bridges and routers. Corporate networks may also span great distances. Such internetworks require different types of connections than single-facility internetworks, though the fundamentals are similar. Internetworks that connect remote facilities usually rely on some type of public or leased data communications network provided by the phone company or a data network service company. Bridges and routers are still required to connect networks to the long-distance data service, whether it's an X.25 packet switched network, a T-1 line, or even a regular phone line. See also BRIDGE and ROUTER.

Corporation For Open Systems International COS. A not-for profit consortium based in Fairfax, VA that helps resolve open systems issues, both technically and businesswise. It helps organize ISDN Solutions '94, video teleconferencing, open set top boxes, cooperative routers and SONET issues.

Correlation The AMA (Automatic Message Accounting) function that permits the association of AMA data generated at the same network system or at physically separate network systems. There are three levels of correlation that affect Advanced Intelligent Network Release 1: record level, service level, and customer level. Definition from Bellcore in reference to Advanced Intelligent Network.

Corresponding Entities Peer entities with a lower layer connection among them.

Corridor Optional Calling Plan Nynex offers a discounted way for subscribers in the 212 and 718 area codes to call five northern New Jersey counties — Bergen, Essex, Hudson, Passaic and Union. See CORRIDOR SERVICE.

Corridor Service A term that Bell Atlantic and Nynex are using for calls to and from the New York City area to and from Northern New Jersey, or between Philadelphia and Southern New Jersey.

Corrosion The destruction of the surface of a metal by chemical reaction.

COS 1. See CLASS OF SERVICE.
2. Compatible for Open Systems.
3. Corporation for Open Systems international. A Federal Government blessed organization which aims towards standardizing OSI and ISDN. COS members includes everyone from end-users to manufacturers. COS deals with private and public networking issues.

COSINE Cooperation for Open Systems Interconnection Networking in Europe. A program sponsored by the European Commission aimed at using OSI to tie together European research networks.

COSName Identifies class of service SNA.

Cost Of Service Pricing A procedure, rationale or methodology for pricing services strictly on the basis of the cost to provide those services.

COT 1. Continuity Check Message. The second of the ISUP call set-up messages. Indicates success or failure of continuity check if one is needed. See ISUP and COMMON CHANNEL SIGNALING.
2. Central Office Terminal

COTS COnnection Transport Service.

Couch Commando A couch potato who insists on taking charge of what he and the rest of the couch potatoes are watching on the TV.

Couch Potato A person who spends their life sitting on a couch surfing TV channels with remote control TV device. See MOUSE POTATO.

Coulomb The quantity of electricity transferred by a current of one ampere in one second. One unit of quantity in measuring electricity.

Counter-Rotating Ring An arrangement whereby two signal paths, whose direction opposite to each other, exists in a ring topology.

Counterpoise A system of electrical conductors used to complete the antenna system in place of the usual ground connection.

Country Code 1. The one, two or three digit number that, in the world numbering plan, identifies each country or integrated numbering plan in the world. In short, the one, two or three digits that precede the national number in an international phone call. This code is assigned in and taken from Recommendation E.163 (Numbering Plan for International Service) adopted by the ITU-T. A list of the country codes can be found in Section 1.8 of the Local Exchange Routing Guide (LERG).
2. In international record carrier transmissions, the country code is a two or three alpha or numeric abbreviation of the country name following the geographical place name.

Coupled Modes 1. In fiber optics, a condition wherein energy is transferred among modes. The energy share of each mode does not differ after the equilibrium length has been reached
2. In microwave transmission, a condition where energy is transferred from the fundamental mode to higher order modes. Energy transferred to coupled modes is undesirable in usual microwave transmission in a waveguide. The frequency is kept low enough so that propagation in the waveguide is only in the fundamental mode.

Coupling Any means by which energy is transferred from one conductive or dielectric medium (e.g., optical waveguide) to another, including fortuitous occurrences. Types of electrical coupling include capacitive (electrostatic) coupling, inductive coupling, and conductive (hard wire) coupling. Coupling may occur between optical fibers unless specific action is taken to prevent it. Coupling between fibers is very effectively prevented by the polymer overcoat, which also prevents the propagation of cladding modes, and provides some degree of physical protection.

Coupling Loss The power loss suffered when coupling light from one optical device to another.

Coupon From Britain: A tear-off slip to encourage response to advertisements or to a promotion on packaging. The information is keyed into a telebusiness system which automatically handles the follow-up. This may be a phone call, acknowledgement letter, brochure, distribution of lead to a distributor and so on.

Courier Dispatch The Courier Dispatch service offered by MCI International allows customers to generate and send high-priority messages from their own Telex terminals to any destination in the Continental U.S. and Hawaii.
Coverage The percent of completeness with which a metal braid covers the underlying surface.
Coverage Area The geographic area served by a cellular system; that is, the area in which service is available to users of the system. Once the mobile telephone number has traveled outside the coverage area, the mobile telephone will show "NO SERVICE."
COW Cellsite On Wheels. A trailer with antenna and transmitting/receiving hardware used to provide temporary service in emergencies, special events, remote testing and repair, or until a normal tower can be erected. Comes with climate control, diesel generator, and self-supporting wind-resistant 84' antenna mast. Both full-size COWs and Mini-COWs are available.
COW Interface Character-Oriented Windows Interface. An SAA-compatible user interface for OS/2 applications.
CP Connection Point in Northern Telecom parlance.
CP/M Control Program for Microcomputers. An erstwhile popular operating system for primarily 8-bit microcomputer systems based on the family of Intel 8080 family of microprocessor chips. CP/M system was originally written by Gary Kidall a programmer and consultant who later formed a company called Intergalatic Digital Research (later just Digital Research). Sadly, that company never upgraded CP/M to 16-bit machines. Thus it left the way open for Bill Gates and the company he formed, Microsoft, to create MS-DOS, which, in its initial form, bore a remarkable resemblance to CP/M.
CPC 1. Calling Party Control.
2. Calling Party Connected.
CPCS An ATM term. Common Part Convergence Sublayer: The portion of the convergence sublayer of an AAL that remains the same regardless of the traffic type.
CPCS-SDU An ATM term. Common Part Convergence Sublayer-Service Data Unit: Protocol data unit to be delivered to the receiving AAL layer by the destination CP convergence sublayer.
CPE Customer Provided Equipment, or Customer Premises Equipment. Originally it referred to equipment on the customer's premises which had been bought from a vendor who was not the local phone company. Now it simply refers to telephone equipment — key systems, PBXs, answering machines, etc. — which reside on the customer's premises. "Premises" might be anything from an office to a factory to a home. GTE once used CPE to refer to "Company Provided Equipment." It doesn't any longer.
CPI Computer to PBX Interface. This proprietary hardware/software interface provides direct connectivity between a PBX's switching network and a host computer to allow switched access between the host computer and data terminal equipment connected with the PBX. The interface is based on the North American Standard T-Carrier specification (24 multiplexed 64 kbps channels operating at a combined speed of 1,544 Mbps). Developed by Northern Telecom, Inc. this interface uses in-band signaling and provides bidirectional data transmission at speeds up to 56 Kbps synchronous per channel. See OPEN APPLICATION INTERFACE.
CPI-C IBM SAA Common Programming Interface-Communication between SNA and OSI environments.
CPM 1. Customer Premise Management.
2. Critical Path Method. See also CP/M.
3. Cable Plant Management. See CABLE MANAGEMENT.
CPN Computer PBX Network.
CPNI Customer Proprietary Network Information. Information which is available to a telephone company by virtue of the telephone company's basic service customer relationship. This information may include the quantity, location, type and amount of use of local telephone service subscribed to, and information contained on telephone company bills. This is the definition of CPNI that the independent voice mail and live telephone answering industry uses.
CPODA Compression Priority Demand Assignment. Another protocol for converting voice into data bits. See also PCM.
CPS Characters per second, or cycles per second. In asynchronous communications, there are typically ten bits per characters — 8 bits for the character and one stop and one start bit.
CPU The Central Processing Unit. The computing part of a computer. The "brain" of the computer. It manipulates data and processes instructions coming from software or a human operator. See CENTRAL PROCESSING UNIT.
CPUG Call Pickup Group. All the phones in an area that can be answered by each other by simply punching in a couple of digits. See CALL PICKUP.
CR 1. Carriage Return. The key on a computer called Carriage Return or sometimes "ENTER." Touching this key usually signals the computer that the entry has been completed and is now ready for processing by the computer. See CARRIAGE RETURN.
2. Critical (alarm status). Indicates a failure affecting more than 96 customers. An AT&T definition.
3. Call Reference.
Cracker A person who "cracks" computer and telephone systems by gaining access to passwords, or by "cracking" the copy protection of computer software. A cracker usually does illegal acts. A Cracker is a "Hacker" whose hacks are beyond the bounds of propriety, and usually beyond the law. The term "cracker" is said to derive from the word "safecracker." See HACKER.
CRAFT Cooperative Research Action For Technology.
Craft Terminal A PCS wireless term. A craft terminal is a device built specifically to provide a man-machine interface that is otherwise not available. The interface is customized to provide a view into a particular device's operation such as a proprietary switch or BSS, which is a Base Station Sub-system charged with managing radio frequency resources and radio frequency transmission for a group of BTSs, which is a Base Transceiver Station, used to transmit radio frequency over the air.
Craftsperson In the phone industry, a craftsperson has two distinct meanings. First, it is the person who toils to instal phones, repair outside plant, fix problems inside central offices. This person typically carries tools and dresses in jeans. Second, craftspeople are at the bottom of the management hierarchy in most phone companies. They typically belong to a union. Craftspeople are not "in management." See LEVEL.

Crankback An ATM term. A mechanism for partially releasing a connection setup in progress which has encountered a failure. This mechanism allows PNNI to perform alternate routing.
Crapplet A poorly written or totally useless Java applet. "I just wasted 30 minutes downloading this awful crapplet!"
Crash The complete failure of a hardware device or a software operation. Usually used to mean a "fatal" crash in which the device or software must be started from a "power up" condition. See BOOT.
CRC Cyclic Redundancy Check. A process used to check the integrity of a block of data. A CRC character is generated at the transmission end. Its value depends on the hexadecimal value of the number of ones in the data block. The transmitting device calculates the value and appends it to the data block. The receiving end makes a similar calculation and compares its results with the added character. If there is a difference, the recipient requests retransmission. CRC is a common method of establishing that data was correctly received in data communications. See CRC CHARACTER and CYCLIC REDUNDANCY CHECK.
CRC Character A character used to check the integrity of a block of data. The character is generated at the transmission end. Its value depends on the hexadecimal value of the number of ones in the data block. And it is added to the data block. The receiving end makes a similar calculation and compares its results with the added character. If there's a difference, there's been a mistake in transmission. So, please, re-send the data.
CRD Contention Resolution Device.
Cream Skimming Selecting only the most profitable markets or services to sell into. Choosing the cream of the market. An erstwhile popular economic theory to deny new entrants into the telephone industry.
Credentials A way of establishing, via a trusted third party, that you are who you claim to be.
CREDFACS Conduit, Risers, Equipment space, Ducts and FACilitieS. Collective term for pathway elements used in communications cabling.
Credit Card Phone A pay telephone that accepts credit cards with magnetic strips on them instead of coins.
CRF 1. An ATM term. Cell Relay Function: This is the basic function that an ATM network performs in order to provide a cell relay service to ATM end-stations.
2. An ATM term. Connection Related Function: A term used by Traffic Management to reference a point in a network or a network element where per connection functions are occurring. This is the point where policing at the VCC or VPC level may occur.
Critical Angle The smallest angle at which a ray will be totally reflected within a fiber.
Critical Mess An unstable stage in a software project's life when any single change or bug fix can result in two or more new bugs. Continued development at this stage leads to an exponential increase in the number of bugs.
Critical Technical Load That part of the total technical power load required for synchronous communications and automatic switching equipment.
CRM An ATM term. Cell Rate Margin: This is a measure of the difference between the effective bandwidth allocation and the allocation for sustainable rate in cells per second.
Cross Assembler An assembler that can run symbolic-language on one type of computer and produce machine-language output for another type of computer.
Cross Border Digital Data Service CDDS. An MCI International digital, private-line service that provides customers with service that provides customers with 56 kbps dedicated terrestrial channels between the U.S. and Canada.
Cross Border Terrestrial Digital Data Service CTDDS. An MCI International point-to-point dedicated, leased channel service enabling customers to transmit traffic between the U.S. and Canada over digital terrestrial facilities at a transmission speed of 1.544 Mbps.
Cross Connect Distribution system equipment used to terminate and administer communication circuits. In a wire cross connect, jumper wires or patch cords are used to make circuit connections. In an optical cross connect, fiber patch cords are used. The cross connect is located in an equipment room, riser closet, or satellite closet.
Cross Connect Field Wire terminations grouped to provide cross connect capability. The groups are identified by color-coded sections of backboards mounted on the wall in equipment rooms, riser closets, or satellite closets, or by designation strips placed on the wiring block or unit. The color coding identifies the type of circuit that terminates at the field.
Cross Connection Cross-connection is the connection of one wire to another usually by anchoring each wire to a connecting block and then placing a third wire between them so that an electrical connection is made.
Cross Coupling The coupling of a signal from one channel, circuit, or conductor to another, where it becomes an undesired signal. See COUPLING.
Cross Extension Cable When you make an REJ-11 extension cable, the wiring crosses over. Conductor 1 becomes 4. Conductor 2 becomes 3. Conductor 3 becomes 2. And conductor 4 becomes one. Next time you have an REJ-11 extension cable in your hand, hold the REJ-11s next to each other and compare them. You'll notice the cross-over of the conductors.
Cross Plan Termination The conversion of ten-digit telephone numbers to seven digits, or vice versa.
Cross Polarization The relationship between two radio waves where one is polarized vertically and the other horizontally.
Cross Subsidization Supporting one area of a business from revenues generated by another area. Local phone companies in the U.S. have long argued that their ability to compete
Cross Wye A cable used at the host system, or network interface equipment which changes pin/signal assignment in order to conform to a given wiring standard (USOC, AT&T PDS, DEC MMJ, etc).
Crossbar A switching system that uses a centrally-controlled matrix switching network of electromechanical switches which worked with magnets and which connected horizontal and vertical paths to establish a path through the network. Crossbar is known for its reliability but is now largely obsolete because it takes up a lot of space and isn't programmable.
Crossbar Tandem A 2-wire common-control switching system with a space-division network used as local tandem, toll tandem, and CAMA switching. While originally designed to switch trunks, some systems have been locally modified to accept loop-start or ground-start lines.

Crosscomm A networking company specializing in token ring internetworking and enterprise-wide networks. Located in Marlboro, MA.

Crossed Pinning Configuration that allows two DTE devices or two DCE devices to communicate. See also CROSSOVER CABLE.

Crosslink An X.25 link connecting two XTX NCs on the same level.

Crossover Cable Another word for NULL MODEM CABLE. A cable used to join two computers together. Usually conductors 2 and 3 are reversed in an RS-232 serial connection.

Crosspoint A single element in an array of elements that comprise a switch. It is a set of physical or logical contacts that operate together to extend the speech and signal channels in a switching network.

Crosspolarized Operation The use of two transmitters operating on the same frequency, with one transmitter-receiver pair being vertically polarized and the other pair horizontally polarized (orthogonal polarization).

Crossposting Crossposting is putting one copy of an electronic file up on the Internet in such a way that it can be viewed from any of several newsgroups (discussion areas). Today's Internet software lets readers avoid seeing a widely crossposted article more than once. They see it in the first group they find it. Crossposting is frowned upon in the Internet when it becomes excessive and off-topic. Crossposting is a less serious offense than spamming, which is seriously frowned upon. See SPAMMING.

Crosstalk Crosstalk occurs when you can hear someone you did not call talking on your telephone line to another person you did not call. You may also only hear half the other conversation. Just one person speaking. There are several technical causes for crosstalk. They relate to wire placement, shielding and transmission techniques. CROSSTALK is also the name of once popular telecommunications software program for 8- and 16-bit microcomputers.

Crosstalk Attenuation The extent to which a communications system resists crosstalk.

CRS An ATM term. Cell Relay Service: A carrier service which supports the receipt and transmission of ATM cells between end users in compliance with ATM standards and implementation specifications.

CRT Cathode Ray Tube. The glass display device found in television sets and video computer terminals. See CATHODE RAY TUBE.

CRTC Canadian Radio Television and Telecommunications Commission. Canada's federal telecom regulator. It's based in Ottawa.

Cryptanalysis 1. The steps and operations performed in converting encrypted messages into plain text without initial knowledge of the key employed in the encryption.
2. The study of encrypted texts. The steps or processes involved in converting encrypted text into plain text without initial knowledge of the key employed in the encryption.

Crypto A term used to describe encrypted information. The use of encryption on data communications circuits lessens the chance that the information will be successfully copied by eavesdroppers.

Cryptochannel A complete system of crypto-communications between two or more holders. The basic unit for naval cryptographic communication. It includes: (a) the cryptographic aids prescribed; (b) the holders thereof; (c) the indicators or other means of identification; (d) the area or areas in which effective; (e) the special purpose, if any, for which provided; and (f) pertinent notes as to distribution, usage, etc. A cryptochannel is analogous to a radio circuit.

Cryptography The process of concealing the contents of a message from all except those who know the key. Cryptography is unregulated in the United States. See CLIPPER CHIP.

Crystal Microphone A microphone, the diaphragm of which is attached to a piezo-electric crystal, which generates electrical currents when torque is applied, due to the vibration of the diaphragm. The earliest form of microphone, now obsolete. See also CONDENSER and ELECTRET MICROPHONE.

CS Convergence Sublayer. The upper portion of BISDN Layer 3. As an ATM term, it covers the general procedures and functions that convert between ATM and non-ATM formats. It describes the functions of the upper half of the AAL layer. It is also used to describe the conversion functions between non-ATM protocols such as frame relay or SMDS and ATM protocols above the AAL layer.

CS-1 Capability Set 1. Term used by ITU-T to refer to their initial set of Advanced Intelligent Network (AIN) standards. Contains 18 trigger detection points. Bellcore (Bell Communications Research) plans to adopt the CS-1 terminology for its own AIN.

CSA 1. Callpath Services Architecture — IBM's computer host to PBX interface. It links computer and telephone systems. See also CALLBRIDGE and OPEN APPLICATION INTERFACE.
2. Canadian Standards Association — a non-profit, independent organization which operates a listing service for electrical and electronic materials and equipment. It is the body that establishes telephone equipment (and other) standards for use in Canada. The Canadian counterpart of the Underwriters Laboratories.

CSA T-528 Canadian equivalent of EIA-606 standard, harmonized.

CSA T-529 Canadian equivalent of EIA-568 standard, harmonized.

CSA T-530 Canadian equivalent of EIA-569 standard, harmonized. defined. Also has Rcv Clock, and both Xmit Clocks for synchronous systems. For use with modems, the lack of Ring Indicate is a drawback.

CSC 1. Customer Service Center
2. Customer Service Consultant
3. Customer Service Coordinator
4. Customer Support Center
5. Customer Support Consultant.

CSDC Circuit Switched Digital Capability. AT&T defines it as a technique for making end-to-end digital connections. Customers can place telephone calls normally, then use the same private connection to transmit high-speed data. CSDC is a circuit-switched, 56-kbps, full-duplex data service that provides high-speed data communications over regular telephone lines.

CSFI IBM Communications Subsystem For Interconnection: networking software.

NEWTON'S TELECOM DICTIONARY

CSI 1. Called Subscriber Identification. This is an identifier whose coding format contains a number, usually a phone number from the remote terminal used in fax.
2. Capability Set I. A set of service-independent building blocks for the creation of IN services developed by the European Telecommunications Standards Institute and the ITU-T.

CSMA Carrier Sense Multiple Access. In local area networking, CSMA is a way of getting onto the LAN. Before starting to transmit, personal computers on the LAN "listen" to make sure no other PC is transmitting. Once the PC figures out that no other PC is transmitting, it sends a packet and then frees the line for other PCs to transmit. With CSMA, though stations do not transmit until the medium is clear, collisions still occur. Two alternative versions (CSMA/CA and CSMA/CD) attempt to reduce both the number of collisions and the severity of their impact. See CSMA/CA and CSMA/CD.

CSMA/CA Carrier Sense Multiple Access (CSMA) with Collision Avoidance. In local area networking, CSMA technique that combines slotted time-division multiplexing (TDM) with carrier sense multiple access/collision detection (CSMA/C) to avoid having collisions occur a second time. CSMA/CA works best if the time allocated is short compared to packet length and if the number of stations (these days PCs) is small. See CARRIER SENSE MULTIPLE ACCESS/COLLISION AVOIDANCE and CSMA.

CSMA/CD Carrier Sense Multiple Access with Collision Detection. In local area networking technology, CSMA technique that also listens while transmitting to detect collisions. CSMA/CD is a leading control technique for getting onto and off a local area network. All devices attached to the network listen for transmissions in progress (i.e. carrier sense) before starting to transmit (multiple access). If two or more begin transmitting at the same time and their transmissions crash into each other, each backs off (collision detection) for a different amount of time (determined by an algorithm) before again attempting to transmit.
If you didn't understand the above definition, try this one: CSMA/CD: Abbreviation for Carrier Sense Multiple Access with Collision Detection, a method of having multiple workstations access a transmission medium (multiple access) by listening until no signals are detected (carrier sense), then transmitting and checking to see if more than one signal is present (collision detection). Each workstation attempts to transmit when they "believe" the network to be free. If there is a collision, each workstation attempts to retransmit after a preset delay, which is different for each workstation. It is one of the most popular access methods for PC-based LANs. Think of it as entering a highway from an access road, except that you can crash and still try again. Or think of it as two polite people who start to talk at the same time. Each politely backs off and waits a random amount of time before starting to speak again. Ethernet-based LANs use CSMA/CD. See Ethernet and IEEE 802.3.

CSMDR Centralized Station Message Detail Recording.

CSMI Call Screening, Monitoring and Intercept.

CSO Central Services Organization; an Internet service that makes it easy to find user names and addresses.

CSR 1. Customer Station Rearrangement (as in Centrex).
2. Customer Service Representative. A customer care agent that provides direct customer support.
3. Customer Service Record. Computer printout that details the fixed monthly charges billed by your local telephone company. The CSR is composed of computer codes called USOCs, which in turn correspond to a particular tariffed service. USOCs tell the telephone company's billing system what tariff rate should be billed for a particular service. In order to ensure your telephone bill is correct you must request and review this document. No telecom manager should be without this important document.

CST Computer Supported Telephony, a term coined by Siemens. Here is an explanation from Dr. Peter Pawlita of Siemens. "More people communicate by telephone than by any other means. The reason is simple: The telephone bridges any distance, saves travel time and can be used spontaneously and is universally available. Unfortunately telephone usage is often associated with annoying delays and frayed nerves resulting from such things as time wasted in finding a number, dialing errors, and the absence of the dialed party. Added to this the person to whom you are speaking does not have the knowledge you require, or has to spend a long time looking or documents. What could be more obvious, therefore, than to turn these problems over to the computer — to implement Computer Supported Telephony (CST). CST denotes the functional connection of a computer system to a PBX at the application level. CST applications can automatically initiate calls, receive incoming calls, and provide "just-in-time" business data, documents and notes on the screen. All this makes telephony more convenient, time-saving, efficient and largely error-free."

CSTA Computer Supported Telephony Application. A standard from the European Computer Manufacturers Association (ECMA) for linking computers to telephone systems. Basic CSTA is a set of API call agreed upon by the ECMA. See also CST and OPEN APPLICATION INTERFACE.

CSTP Customer Specific Term Plan. See CUSTOMER SPECIFIC TERM PLAN.

CSU 1. Channel Service Unit. Also called a Channel Service Unit/Data Service Unit or CSU/DSU because it contains a built-in DSU device. A device to terminate a digital channel on a customer's premises. It performs certain line coding, line-conditioning and equalization functions, and responds to loopback commands sent from the central office. A CSU sits between the digital line coming in from the central office and devices such as channel banks or data communications devices. A Channel Service Unit is found on every digital link and allows the transfer of data at a range greater than 56 Kbps. A 56 Kbps circuit would need a 56 Kbps DSU on both ends to transfer data from one end to the other. A CSU looks like your basic "modem," except it can pass data at rates much greater and does not permit dial-up functions (unless it has an asynch dial-backup feature).
2. Channel Sharing Unit. Line bridging device that allows several inputs to share one output. CSUs exist to handle any input/output combination of sync or asynch terminals, computer ports, or modems and thus these units are variously called modem sharing units, digital bridges, port sharing units, digital sharing devices, modem contention units, multiple access units, control signal activated electronic switches or data-activated electronic switches.

CSU/DSU See CSU.

CSUA Canadian Satellite Users Association. Trade association of satellite users.

CT 1. Call Type.
2. Cordless Telephone.

158

NEWTON'S TELECOM DICTIONARY

CT-2 Interim Standard for the Telepoint service favored by UK Memorandum of Understanding with other network operators.

CT1 Cordless Telephony Generation 1. A new type of low-cost public cordless telephone system getting popular in Europe. You carry a cheap handset. You go to within several hundred yards of a local antenna and you make your phone call. You can't receive calls as you can on a cellular radio. You can't make calls unless you're close to the antenna. The service helps overcome the serious lack of street-side coin and public phones in Europe. CT1 is the analog version of the interface specification. See CT2, CT2+, CT3 and DECT.

CT2 Cordless Telephony Generation 2, interface specification for digital technology, currently in use in the U.K. for telepoint (payphone) applications. Think of telepoint phones as cellular phones but using micro-cells. By having smaller cells than normal cellular cells, CT2 phones can be smaller, cheaper and lighter. The first generation of these phones didn't do well, since they weren't smaller and lighter; there weren't many micro-cells and you couldn't receive an incoming call. See CT1.

CT2+ An expansion of the CT2 interface specification that would extend network capabilities and allow backwards compatibility with CT2 handsets. See CT1 and CT2.

CT3 Ericsson's proprietary cordless phone system.

CTA Competitive Telecommunications Association. Trade association of alternate long distance carriers (resellers) in Canada.

CTCA Canadian Telecommunications Consultants Association. Professional organization of telecommunications consultants.

CTD 1. Continuity Tone Detector.
2. An ATM term. Cell Transfer Delay: This is defined as the elapsed time between a cell exit event at the measurement point 1 (e.g., at the source UNI) and the corresponding cell entry event at measurement point 2 (e.g., the destination UNI) for a particular connection. The cell transfer delay between two measurement points is the sum of the total inter-ATM node transmission delay and the total ATM node processing delay.

CTE 1. Channel Translation Equipment.
2. Coefficient of thermal expansion.

CTI Computer Telephone Integration. A term for connecting a computer (single workstation or file server on a local area network) to a telephone switch (e.g. a PBX or an ACD) and have the computer issue the switch commands to move calls around. The classic application for CTI is in call centers. Picture this: A call comes in. That call carries some form of caller ID — either ANI or Class Caller ID. The switch "hears" the calling number, strips it off, sends it to the computer. The computer does a lookup, sends back the switch instructions on what to do with the call. The switch follows orders. It might send the call to a specialized agent or maybe just to the agent the caller dealt with last time. There are emerging standards for CTI. See also TSAPI and WINDOWS TELEPHONY.

CTIA 1. Cellular Telecommunications Industry Association. Washington, D.C.-based industry association. 202-785-0081.
2. Computer Technology Industry Association, a Lombard, Illinois trade association.

CTIP Computer Telephony Interface Products. Adapters that allow telephones to work with computers. An example is the Konnexx connector, which fits between the handset and the phone, allows a connection to a PC modem or fax machine. This definition contributed by Larry Kettler of San Diego.

CTS 1. Clear To Send. Pin 5 on the 25-conductor RS-232-C interface or an RS-232-C signal used in the exchange of data between the computer and a serial device.
2. Communication Transport System. CTS is The Siemon Company's proprietary structured wiring system. It consists of the methodology and the connecting hardware products to plan, design, and implement the communications wiring infrastructure for commercial buildings (for more information see the company's CTS Design Workbook and CTS Training Videotape). The Siemon Company is based in Watertown, CT.
3. Conformance Testing Services.

CTTS Coax To The Curb. An approach that provisions a multiline remote terminal to deliver voice and data to concentrated residential applications.

CTTU Centralized Trunk Test Unit. An operational support system providing centralized trunk maintenance through a data link on a switch.

CTX Centrex.

CUA Common User Access.

Cuckoo-Clock Telecom slang for some 6- and 10-button models of AT&T 1A2 wall phones shaped vaguely like traditional cuckoo-clocks. These were probably the first multi-line phones to come with handsets that plugged into the base with Trimline-style 5-pin plugs, before the current modular connectors were adopted. Often seen in hospitals on TV shows.

CUG Closed User Group. Selected collection of terminal users that do not accept calls from sources not in their group and also often restricted from sending messages outside the group.

Curie Point The temperature at which certain elements (usually so-called "rare earth" elements) relax their resistance to magnetic changes. In a magneto-optic disk drive the surface to be marked is heated briefly by a laser light to its Curie point. Magnetism is then applied in the proper polarity to make the spot a "1" or a "0". It cools, and is locked in that position, until it re-heated and changed again. This is how magneto-optic drives can be erasable.

Current A measure of how much electricity passes a point on a wire in a given time frame. Current is measured in amperes, or amps. The abbreviation for current is I. See OHM'S LAW.

Current Carrying Capacity The maximum current an insulated conductor can safely carry without exceeding its insulation and jacket temperature limitations.

Current Loop Transmission technique that recognizes current flows, rather than voltage levels. It has traditionally been used in teletypewriter networks incorporating batteries as the transmission power source. In this serial transmission system, a pair of wires connecting the receiving and sending devices transmit binary 0 (zero) when no current flows and a binary 1 (one) when current is flowing.

NEWTON'S TELECOM DICTIONARY

Cursor A symbol on a screen indicating where the next character may be typed. Cursors may be solid, blinking, underlines, etc. Many programs, computers and phone systems allow you to reprogram the cursor to what you like. One author of this dictionary, Harry Newton, likes a non-blinking solid block, which came standard with his original CP/M version of WordStar, but doesn't any longer.

Cursor Submarining A liquid crystal display on a computer laptop screen doesn't write to screen very fast. When you move a cursor across your screen or move your mouse quickly across the screen, the cursor disappears. This phenomenon is known as cursor submarining. Cute.

Curves And Arcs A computer imaging term. Paint packages handle curves and arcs in a variety of ways. Examples include spline curves, where-in you specify a series of points and the package draws a curve that smoothly approaches those points, and "three point" curves, in which the first two points anchor the ends of the curve and third selects the apex.

CUSEEME An Internet videoconferencing system that enables up to eight users to see and hear each other on their computer screens. Pronounced "See You, See me."

Custom Calling A group of special services available from the central office switching system which the telco can offer its subscribers without the need for any special terminal equipment on their premises. Basic custom calling features now available include call waiting, 3-way calling, abbreviated dialing (speed calling), call forwarding, series completing (busy or no answer) and wake up or reminder service.

Custom ISDN A version of ISDN BRI (Basic Rate Interface) provided off an AT&T 5ESS central office. It actually offers more features and is easier to install than a National ISDN-1 BRI line. We are all awaiting the specifications on National ISDN-2, which is meant to be "standard." Meantime, Custom ISDN is the most popular, most versatile and most understood ISDN service in North America. See ISDN.

Custom Local Area Signaling Services CLASS. A generic term (like WATS) describing several enhanced local service offerings such as incoming-call identification, call trace, call blocking, automatic return of the most recent incoming call, call redial, and selective forwarding and programming to permit distinctive ringing for incoming calls. See CLASS.

Customer Care Center A term created by Alex Szlam, the president of Melita International, Norcross, GA to describe a telephone call center with three basic elements: First, the database technology and the marketing savvy to fill that database with individual customer preference information. Second, the ability to intelligently handle inbound phone calls. Third, the ability to intelligently make outbound calls. See also CUSTOMER SENSITIVITY KNOWLEDGE BASE.

Customer Contact Zone A term invented by Keith Dawson, editor of Call Center Magazine. It refers to all the information a customer requires which is delivered through multiple media, including manned call centers, interactive voice response machines, fax back devices, etc.

Customer Control An AT&T term for the ability for an end user to monitor, choose, modify, redesign and/or program the type of service received from a network.

Customer Information Manager CIM. An MCI definition. A component of the NCS which supports the creation and maintenance of customer databases for Vnet customers. Customers have remote access to and control over their portion of the NCS database via a terminal at the customer's location.

Customer Network Management CNM. An ATM term. CNM allows users of ATM public networks to monitor and manage their portion of the carrier's circuits. Thus far, the ATM forum has agreed that the CNM interface will give users the ability to monitor physical ports, virtual paths, usage parameters, and quality of service parameters.

Customer Premises Equipment CPE. Terminal equipment — telephones, key systems, PBXs, modems, video conferencing devices, etc. — connected to the telephone network and residing on the customer's premises.

Customer Proprietary Network Information CPNI. Information which is available to a telephone company by virtue of the telephone company's basic service customer relationship. This information may include the quantity, location, type and amount of use of local telephone service subscribed to, and information contained on telephone company bills. This is the definition of CPNI that the independent voice mail and live telephone answering industry uses.

Customer Provided Loop The customer assumes responsibility for ordering, coordinating, maintaining, and billing for the local loop.

Customer Provided Terminal Equipment Or just Customer Provided Equipment (CPE). Terminal equipment connected to the telephone network which is owned by the user or leased from a supplier other than the local telephone operating company.

Customer Sensitivity Knowledge Base A term created by Alex Szlam of Melita International, Norcross, GA to describe a complex database that would keep track of your customers' preferences. Such database would be updated almost automatically based on every contact you had with the customer. The database would probably be object-oriented since the idea is define customer preferences based on individual preferences, not on a statistical analysis of conglomerate preferences such as those typically gleaned from existing character databases.

Customer Service Center CSC. MCI organization responsible for installing, verifying, and maintaining MCI customers and customer service.

Customer Service Record CSR. Computer printout that details the fixed monthly charges billed by your local telephone company. The CSR is composed of computer codes called USOCs, which in turn correspond to a particular tariffed service. USOCs tell the telephone company's billing system what tariff rate should be billed for a particular service. In order to ensure your telephone bill is correct you must request and review this document. No telecom manager should be without this important document.

Customer Service Unit CSU. A device that provides an accessing arrangement at a user location to either switched or point-to-point, digital circuits. A CSU provides local loop equalization, transient protection, isolation, and central office loopback testing capability. See also CSU/DSU.

Customer Specific Term Plan A Customer Specific Term Plan is an option offered by AT&T on the purchase of its 800 services whereby customers can earn additional discounts by committing to a multi year contract. This also is one of two plans used by aggregators to resell 800 services. The other is the Revenue Volume Pricing Plan.

Cut To transfer a service from one facility to another.

Cut Back Technique A technique for measuring optical fiber attenuation or distortion by performing two transmission measurements. One is at the output end of the full length of the fiber. The other is within 1 to 3 meters of the input end. Without disturbing the source-to-fiber coupling, access to the short length output is accomplished by "cutting back" the test fiber.

Cut Down A method of securing a wire to a wiring terminal. The insulated wire is placed in the terminal groove and pushed down with a special tool. As the wire is seated, the terminal cuts through the insulation to make an electrical connection, and the tool's spring-loaded blade trims the wire flush with the terminal. Also called punch down.

Cut Through 1. Cut-through is a voice processing term. It's what stops voice prompt playback when a key is pressed. Some of the speech recognition solutions also add cut-through that will stop voice prompt playback as soon as you start talking. Only voice cards that support continuous speech recognition are able to provide cut-through. Cut-though can be a problem in some cases. Imagine yourself at the airport trying to make a call using a speech recognition system. At the start of a new prompt, the airport public address system blares out a last boarding call for a flight. If cut-through is active, it would stop playing the prompt and wait on your response. Now what do you do?

2. The act of connecting one circuit to another, or a phone to a circuit. This is when a user dials the access code for the circuit and is immediately "cut through" to the tie line. The user controls the call. It is a tie line operation.

Cut Through Resistance A measure of an insulation's ability to withstand penetration by sharp edges.

Cutoff Attenuator A waveguide of adjustable length that varies the attenuation of signals passing through the waveguide.

Cutoff Frequency 1. The frequency above which, or below which, the output current in a circuit, such as a line or a filter, is reduced to a specified level.

2. The frequency below which a radio wave fails to penetrate a layer of the ionosphere at the angle of incidence required for transmission between two specified points by reflection from the layer.

Cutoff Mode The highest order mode that will propagate in a given waveguide at a given frequency.

Cutover The physical changing of lines from one phone system to another, or the installation of a new system. It's usually done over the weekend, accompanied by heavy praying that everything will go right. There are two types of cutovers — flash cuts and parallel cuts. Parallel cuts occur when the old phone system is left functioning and the new one, central switching equipment and phones, is installed around it. This means that for some weeks there are two sets of phones, two sets of wires, two switches, two sets of phone lines, etc. The parallel cut is a far more reliable method of cutting over a new switch. But it's also more expensive.

A "flash cut" occurs in a flash. On Friday, everyone is using the old switch. When everyone comes to work on Monday, the old switch and its phones have disappeared. In its place, there's a brand new system. Sometimes it works. More often than not, there are remaining nagging problems. With any Cutover, it's a good idea to set up a Complaint or Cutover Number. Thus, if anyone's having trouble with their phone, they can call this number and get their problems taken care of. How well these problems are taken care of will determine how well the cutover went and how well the employees perceive the new switch is working. Perception, not reality, is what's at stake here.

CV Old Bell-Speak for single-line phone. It stands for Combined Voice. In old Bell-Speak it meant that the two parts of the phone that dealt with voices were combined into one unit (the handset). Before this, there were phones like the HH (Hand-Held) where there was a piece you spoke into and another piece you put to your ear. From CV, you get CVW (CV Wall phone) and later on, KV (Key Voice) and KVW (Key Wall phone). Later on all this crept into the USOC codes — the Universal Service Order Code numbering systems the local Bell operating phone companies used to identify products and services. See USOC.

CVD Chemical Vapor Deposition.

CVF Compressed Volume File. A Microsoft term which refers to a file on a compressed disk. The term was first introduced in MS-DOS 6.0, which first had double-your-disk-space technology. That technology was later removed when Stac Electronics, originator of Stacker disk doubling technology, took Microsoft to court and won.

CVP 1. Certified Vertical Partner.

2. A British term: Co-operative Voice Processing, gives the caller the ability to move seamlessly between an Interactive Voice Processing device and a live agent.

CVSD Continuously Variable Slope Delta modulation. A method for coding analog voice signals into digital signals that uses 16,000 to 64,000 bps bandwidth, depending on the sampling rate.

CW 1. Call Waiting (as in Custom Calling Service).

2. Continuous Wave.

CWA Communications Workers of America. A national union of telephone industry employees, currently very worried about its future membership growth given the phone industry's propensity to let its surplus workers go.

CX Signaling A direct current (DC) signaling system that separates the signal from the voice band by filters. Also called Composite Signaling.

CXR Carrier.

Cyber Five letters which can, seemingly, be attached to a word and made into a noun or a verb. The first Cyber word was Cyberspace, a term coined by science fiction writer William Gibson in his 1984 fantasy novel "Neuromancer" to describe the "world" of connected computers and the society that gathers around them. The idea of Cyberspace is that this world of computer networks can be explored with the proper addresses and codes. People who use the system for hours on end are said to be lost in cyberspace. Today, many people say that world has arrived in the form of the Internet. And, so with projections that there will be 100 million users of the Internet by the year 2000, the word Cyber has become popular. There's "The Cyberbrary of

Congress" (books Congress has on on-line). According to William Safire writing in the New York Times Magazine of December 11, 1994, "cyber is the hot combining form of our time. If you don't have cyberphobia, you are a cyberpiliac." The US News & World Report labels its election night on-line forum a cybercast. The Washington Post wrote that "battlefield valor belongs not to the brawny soldier but to the astrophysics major who invented smart bombs," somebody who's called a cyberwonk.

Cyberbusiness A company that does most of its business on the Internet is called a Cyberbusiness.

Cybercash A name for electronic money transmitted in and around the Internet.

Cyberia Electronic stuff for the cyberpeople. An advertisement in the November 12, 1995 issue of the New York Times (Sunday) Magazine showed Cyberia covering everything from modern chairs to laptop computers, to cellular phones to an Apple Newton PDA.

Cybernetics A term invented in 1948 by Norbert Wiener, the automation genius, who declared "We have decided to call the entire field of control and communications theory, whether in the machine or the animal, by the same term Cybernetics."

Cyberpunk A work coined by a book called "Cyberpunk: Outlaws and Hackers on the Computer Frontier" by Katie Hafner and John Markoff. The book defines Cyberpunk as what you and I know as a computer hacker — a person who manages to get into other people's computer systems. He does this usually through telephone lines. In most cases, hackers see themselves as harmless electronic joyriders. But they occasionally steal data, inject viruses and misleading information and disrupt legitimate business and research. Sometimes they get caught.

Cybersex Adult-oriented computer games, images and chat lines. A place where people can discuss their sex lives and wanton desires with total strangers in online (over phone line) forums, even falling in love without having ever met face to face.

Cyberskating Browsing the Internet. See CYBERSPACE.

Cyberspace A term coined by science fiction writer William Gibson in his fantasy novel Neuromancer to describe the "world" of connected computers and the society that gathers around them. The idea of Cyberspace is that this world of computer networks can be explored with the proper addresses and codes. People who use the system for hours on end are said to be lost in cyberspace. Today, many people say that world has arrived in the form of Internet. John Perry Barlow, a rock-'n'-roll lyricist turned computer activist, defined cyberspace in Time magazine as "that place you are in when you are talking on the phone." Thus by Barlow's definition, just about everybody has already been to cyberspace. I prefer Gibson's definition.

Cybrarian A person who makes a living doing online research and information retrieval (comes from cyberspace librarian). According to one definition I read, a cybrarian is a futurist librarian who swims in the electronic ocean or cyberspace. The term is alleged to have been coined by Michel Bauwens of BP Nutrition) A cybrarian is also known as a data surfer or a super searcher.

Cycle One complete sequence of an event or activity. Often refers to electrical phenomena. One electrical cycle is a complete sine wave. (A complete set of one positive and one negative alternation of current.) In the battery business, a cycle is the process of one complete battery discharge and recharge. See CYCLE LIFE.

Cycle Life In the battery business, cycle life is the useful life of a rechargeable battery, expressed as the total number of discharges and recharges.

Cycle Manager Extraction An MCI system which selects processable calls from Distribution and forwards them to the appropriate MCI Reference System for billing.

Cycle Pools Where dial-up call records are stored in MCI's Revenue System until extracted for billing.

Cycle Slip A discontinuity in the measured carrier beat phase resulting from a temporary loss-of-lock in the carrier tracking loop of a Global Positioning System receiver.

Cycle Time The time to complete a cycle. In microcomputers, it's the time between successive RAM read or write operations.

Cyclic Distortion In telegraphy, distortion that is neither characteristic, bias, nor fortuitous, and which in general has a periodic character. Its causes are, for example, irregularities in the duration of contact time of the brushes of a transmitter distributor or interference by distributing alternating currents.

Cyclic Redundancy Check CRC. A check performed on data to see if an error has occurred in the transmitting, reading or writing of the data. A CRC is performed by reading the data, calculating the CRC character and comparing its value to the CRC character already present in the data. If they are equal, the new data is presumed to be the same as the old data. Otherwise, it's wrong. Re-send the data. A CRC character is figured by treating a block of data as a string of bits that are equal to a binary number. Then we divide that number by another predetermined binary number and append the remainder from the division to the data. We call that appended number the CRC character.

Cylinder A hard disk drive contains a number of platters, which are divided into tracks. A cylinder is a collection of all corresponding tracks on all sides of the platters in a disk drive. Think of a hard disk consisting of dozens of concentric cylinders, each of slightly different diameters. These distinct concentric storage areas on the hard disk roughly correspond to the tracks on a floppy diskette. Generally, the more cylinders a hard disk has, the greater its storage capacity.

Cypress A Rolm/IBM desktop device which combines a data terminal, a digital phone and several time management software features.

D Channel In an ISDN interface, the "D" channel (the Data channel) is used to carry control signals and customer call data in a packet switched mode. In the BRI (Basic Rate Interface, i.e. the lowest ISDN service) the "D" channel operates at 16,000 bits per second, part of which will carry setup, teardown, ANI and other characteristics of the call. 9,600 bps will be free for a separate conversation by the user. In the PRI (Primary Rate Interface, i.e. ISDN equivalent of T-1), the "D" channel runs at 64,000 bits per second. The D channel provides the signaling information for each of the 23 voice channels (referred to as "B channels"). The actual data which travels on the D channel is much like that of a common serial port. Bytes are loaded from the network and shifted out to the customer site in a serial bit stream. The customer site of course responds with its serial bit stream, too. An example of a data packet sent from the network to indicate a new call has the following components:
- Customer Site ID
- Type of Channel Required (Usually a B channel)
- Call Handle (Not unlike a file handle)
- ANI and DNIS information
- Channel Number Requested
- A Request for a Response

This packet is responded to by the customer site with a format similar to:
- Network ID
- Channel Type is OK
- Call Handle

The packets change as the state of the call changes, and finally ends with one side or the other sending a disconnect notice. The important concept here is the fact the information on the D channel could actually be anything — any kind of serial data. It could just as well be sports scores! So with that in mind, consider the Channel Number Requested packet above. This is the networks' selected channel for the customer site to use. Normally, this number is between 1 and 23, but could be a higher number if needed. This is what NFAS is all about. NFAS (Non Facility Associated signaling, pronounced N-FAST without the T) allows a D channel to carry call information regarding channels which may not even exist in the same PBX or PC system. See also ISDN and DS-0.

D Conditioning A type of conditioning which controls harmonic distortion and signal-to-noise ratio so that they lie within specified limits.

D Link Diagonal Link. A SS7 signaling link used to connect STP pairs that perform work at different functional levels. These links are arranged in sets of four (called quads).

D Region That portion of the ionosphere existing approximately 50 to 90 kilometers above the surface of the Earth. Attenuation of radio waves, caused by ionospheric free-electron density generated by cosmic rays from the sun, is pronounced during daylight hours.

D Type The standard connector used for RS-232-C, RS-423 and RS-422 communications. D-type connectors are typically seen in nine, 15 and 25 pin configurations.

D-AMPS Digital Advanced Mobile Phone Service. A term for digital cellular radio in North America. In Europe, it's called GSM and in Japan PDC.

D-Bank Another name for channel bank. A device that multiplexes groups of 24 channels of digitized voice input at 64 Kbps into T-1 aggregate outputs of 1,544 Mbps.

D-Bit Also called DBIT. The delivery confirmation bit in an X.25 packet used to indicate whether or not the DTE wishes to receive an end-to-end acknowledgment of delivery. In short, a bit in the X.25 packet header that assures data integrity between the TPAD and the HPAD.

D-Inside Wire Direct-Inside Wire. Made of 24-gauge, annealed-copper conductors with color-added PVC, which allows it to be pulled in conduit without the aid of lubricants. Generally used in the horizontal subsystem.

D/A Digital to Analog conversion.

D/A Converter Digital to Analog converter. A device which converts digital pulses, i.e. data, into analog signals so that the signal can be used by analog device such as amplifier, speaker, phone, or meter.

D/I Drop and Insert. See BIT STUFFING and BIT ROBBING.

D1, D1D, D2, D3, D4 and D5 T-1 framing formats developed for channel banks. All formats contain a framing bit in every 193rd bit position. The Superframe (introduced in D2 channel banks) is made up of 12 193-bit frames, with the 193rd bit sequence being repeated every 12 frames. D2 framing also introduced robbed bit signaling, where the eighth bit in frames 6 and 12 were "robbed" for signaling information (like dial pulses). D1D was introduced after D2 to allow backwards compatibility of Superframe concepts to D1 banks.

D2-MAC One of two European formats for analog HDTV.

D3 Format 24 data channels on one standard (North American standard) T1/D3 span line. Each data channel is 8-bits wide and has a bandwidth of 8KHz. See also DS-1.

D3/D4 Refers to compliance with AT&T TR (Technical Reference) 62411 definitions for coding, supervision and alarm support. D3/D4 compatibility ensures support of digital PBXs, M24 services, Megacom services and Mode 3 D3/D4 channel banks at a DS-1 level.

D4 In T-1 digital transmission technology, D4 is the fourth-generation channel bank. A channel bank is the interface between the T-1 carrier system and a analog premises device such as an analog PBX (private branch exchange).

D4 Channelization Refers to compliance with AT&T TR (Technical Reference) 62411 in regards to the DSI frame layout (the sequential assignment of channels and time slot numbers within the DSI).

D4 Framing First read T-1 FRAMING. The most popular framing format in the T1 environment is D-4 framing. The name stems from the way framing is performed in the D-series of channel banks from AT&T. There are 12 separate 193-bit frames in a super-frame. The D-4 framing bit is used to identify both the channel and the signaling frame. In voice communications, signaling is an important function that is simulated and carried by all the equipment in the transmission path. In D-4 framing, signaling for voice channels is carried "in-band" by every channel, along with the encoded voice. "Robbed-bit-signaling" is a technique used in D-4 channel banks to convey signaling information. With this technique, the eighth bit (least significant bit) of each of the 24 8-bit time slots is "robbed" every sixth frame to convey voice related signaling information (on-hook, off-hook, etc.) for each voice channel. See also EXTENDED SUPER-FRAME FORMAT.

DA 1. Doesn't answer, as in "The phone rang DA."
2. Directory Assistance.
3. Demand Assignment.
4. Discontinued Availability. Meaning a circuit that was once available is now no longer.
5. Destination Address, a field in FDDI, Ethernet and Token Ring packets which identifies the unique MAC (Media Access Control, the lower part of ISO layer two) address of the recipient. A six octet value uniquely identifying an endpoint and which is sent in IEEE LAN frame headers to indicate frame destination.
6. Desk Accessory. Standard desk accessories on the Apple Macintosh include a calculator, alarm clock and the chooser. Desk accessories are available to the user regardless of the application currently in use, networked or non-networked. Desk accessories are installed in the Apple menu and accessed from there.

DAA Data Access Arrangement. A device required before the FCC registration program if a customer was going to hook up CPE (Customer Provided Equipment), usually modems and other data equipment, to the telephone network. Today, equipment is FCC registered (under Part 68) meaning that the device itself is approved for connection to the phone network. DAAs can still be found in old DP (data processing) installations.

DAB Dynamically Allocable Bandwidth.

DAC Digital to Analog Converter. 1. A device which converts digital pulses, i.e. data, into analog signals so that the signal can be used by analog device such as amplifier, speaker, phone, or meter. 2. A chip that converts the binary numbers that represent particular colors to analog red, green and blue signals that a color monitor displays.

DACC 1. Digital Access Cross-Connect.
2. Directory Assistance Call Completion.

DACD Digital Automatic Call Distributor. Nynex's name for its central office provided ACD.

DACOMNET A packet-switched network in South Korea.

DACS Digital Access and Cross-connect System. A digital switching device for routing and switching T-1 lines, and DS-0 portions of lines, among multiple T-1 ports. It performs all the functions of a normal "switch," except connections are typically set up in advance of the call, not together with the call, as in most, normal low bandwidth communications systems (e.g. voice-band voice and data). A DACS is in essence a manual T-1 switch.

DACS/CCR Digital Access Crosscontrol System/Customer Controlled Reconfiguration is a feature of AT&T Accunet T1.5 service. DACS/CCR allows Accunet subscribers to redirect T-1 trunk or DS-0 data traffic over the public network from a terminal on their own premises.

Daemon An agent program which continuously operates on a UNIX server and provides resources to client systems on the network. Daemon is a background process used for handling low-level operating system tasks.

DAF Destination Address Field.

Daisy Chain A method of connecting devices in a series. The computer's signals are passed through the chain from one device to the next. Jack 1 is connected to jack 2, which is connected to jack 3 and so on. The last jack in the chain is not connected to jack 1. A SCSI adapter is a daisy chain. It can support a daisy chain of up to seven devices. Intel's Universal Serial Bus also is a daisy chain.

Daisy Wheel Printer A printer that prints its characters on paper using a rotating wheel with "fingers." Mounted on the fingers are the various characters. The wheel rotates, bringing the character in line with the paper, then a hammer strikes the character against the ribbon, forming an image on the paper. Daisy wheel printers are typically used for "letter quality" correspondence and are popular for word processing. Their letters are crisp and attractive. But the speed of printing is slow because of the time it takes to rotate the wheel into place. These days, more and more of what used to be printed by daisy wheel printers is now printed by laser and inkjet printers which print faster, more quietly and do a nicer job.

DAL Dedicated Access Line. A private tie line from you to your long distance or local phone company. The line may be analog or digital, e.g. a T-1 circuit.

DAMA Demand Assigned Multiple Access. A way of sharing a channel's capacity by assigning capacity on demand to an idle channel or an unused time slot.

Damped Wave A wave consisting of a series of oscillations or cycles of current gradually decreasing amplitude.

Damping 1. The decreasing of the amplitude of oscillations caused by resistance in the circuit. 2 The progressive diminution with time of certain quantities characteristic of a phenomenon. 3. The progressive decay with time in the amplitude of the free oscillations in a circuit. 4. More generally, decreasing some dimension of a phenomenon, such as its power.

DAP Directory Access Protocol. The protocol used between a Directory User Agent (DUA) and Directory System Agent (DSA) in an X.500 directory system.

DAP Information Distributor DID. An MCI definition. Software which directs the distribution of network information from the Operational Data Integrator to the Data Access Points.

Dark Current The external current that, under specified biasing conditions, flows in a photosensitive detector when there is no incident radiation.

Dark Fiber Unused fiber through which no light is transmitted or installed fiber optic cable not carrying a signal. Sometimes dark fiber is sold by a carrier without the (usually) accompanying transmission service. It's "dark" because it's sold without light communications transmission. The customer is expected to put his own electronics and signals on the fiber and make it light.

DARPA Defense Advanced Research Projects Agency. Formerly called ARPA, it is a US government agency that funded research and experimentation with the ARPANET and later the Internet. The group within DARPA responsible for the ARPANET is ISTO (Information Systems Techniques Office), formerly IPTO (Information Processing Techniques Office). See also DARPA INTERNET.

DARPA Internet World's largest internetwork, linking together thousands of networks around world. Sponsored by U.S. Defense Advanced Research Projects Agency.

DASD Direct Access Storage Device. Any on-line data storage device. Usually refers to a magnetic disk drive, because optical drives and tape are considered too slow to be direct access devices. Pronounced DAZ-dee. The term is said to have been invented by IBM.

DASS Direct Access Secondary Storage. Same as near-line: storage on pretty-fast storage devices (e.g., rewritable optical) that are less expensive than hard drives but faster than off-line devices.

DASS1 Digital Access Signaling. A British term. The original British Telecom (BT) ISDN signalling developed for both single line and multi-line Integrated Digital Access but used in the BT ISDN pilot service for single line IDA only.

DASS2 A British Term. A message-based signalling system following the ISO-based model developed by British Telecom to provide multi-line IDA interconnection to the BT network.

DAT Digital Audio Tape used to identify a type of digital tape recorder and player as well as the tape cassette. DAT tape machines record music that is much crisper, and free of the hisses and pops that mar traditional analog recordings. The drawback with DAT tape machines is they require considerable tape to store music digitally. In a DAT machine, the music is recorded by sampling the music 48,000 each second. Each of those samples is represented by a number that is written as a 16-digit string of zeros and ones. There are two such signals, once for each stereo channel, meaning that storing a single second of music requires about 1.5 million bits. On top of that, extra bits are added to allow the system to mathematically correct errors and help the machine automatically find a particular song on the tape. All together, according to Andrew Pollack writing in the New York Times, a single second of music on a digital audio tape requires 2.8 million bits. But compression techniques are cutting down the amount of information required to be recorded.

Data This is AT&T Bell Labs' definition: "A representation of facts, concepts or instructions in a formalized manner, suitable for communication, interpretation or processing." Typically anything other than voice.

Data Abstraction A term in object-oriented programming. An object is sometimes referred to as an instance of an abstract data type or class. Abstract data types are constructed using the built-in data types supported by the underlying programming language, such as integer and date. The common characteristics (both attributes and methods) of a group of similar objects are collected to create a new data type or class. Not only is this a natural way to think about the problem domain, it is a very efficient way to write programs. Instead of individually describing several dozen instances, the programmer describes the class once. Once identified, each instance is complete with the exception of its instance variables. The instance variables are associated with each instance, i.e., each object; methods exist only with the classes. See OBJECT ORIENTED PROGRAMMING.

Data Access Arrangement DAA. Equipment that allows you to attach your data equipment to the nation's phone system. At one stage, DAAs were required by FCC "law." Now, their limited functions are built into directly attached devices, such as terminals, computers, etc.

Data Access Point DAP. MCI computer that holds the number translation and call-routing information for 800 and Vnet services. These computers respond to inquiries from MCI switches on how to handle these calls.

Data Arrangement In the public switched telephone networks, a single item or group of items present at the customer side of the purposes, including all equipment that may affect the characteristics of the interface.

Data Attribute A characteristic of a data element such as length, value, or method of representation.

Data Bank A collection of data in one place. The data is not necessarily logically related, nor is it necessarily consistently maintained. See DATABASE.

Data Base See DATABASE, which is our preferred spelling.

Data Base Administrator DBA. A computer at MCI that maintains the master file of Vnet translation information. The master file is created when a customer begins service and can be changed at anytime through CIM. The updated copies of the database are downloaded each night to the DAPs.

Data Base Lookup A software program which allows telephone users to find information on someone calling via the LCD window on their phone. This information comes to the user via CLID (Calling Line IDentification) or ANI (Automatic Number Identification). See also CLASS.

Data Bubble A new organization within BellSouth to provide high-speed digital services. No one seems to know why it's called "Data Bubble," except that someone inside BellSouth clearly thinks the term is cute.

Data Burst Burst transmission.

Data Bus A bus transmits and receives data signals throughout the computer or telephone system. See BUS.

Data Circuit Terminating Equipment See DATA COMMUNICATIONS EQUIPMENT and DCE.

Data Circuit Transparency The capability of a circuit to transmit all data without changing its content or structure.

Data Cleansing And Scrubbing A process of removing redundancies and inconsistencies in operational data.

Data Communications The transfer of data between points. This includes all manual and machine operations necessary for this transfer. In short, the movement of encoded information by means of electrical transmission systems. See DATA COMMUNICATIONS EQUIPMENT.

Data Communications Channel A three-byte, 192 kbps portion of the SONET signal that contains alarm, surveillance and performance information. It can be used for internally or externally generated messages, or for manufacturer specific messages.

Data Communications Equipment DCE. A definition of an interface standard between computers and printers. A device typically comes configured as a DCE or DTE. See DCE and DTE. Which way it comes determines how you connect it to another device.

Data Compression Reducing the size of a file of data by eliminating unnecessary information, such as blanks and redundant data. The idea of reducing the size is to save money on transmission or to save money on storing the data. The file or program which has been compressed is useless in its compressed form and must be "decompressed," i.e. brought back to normal before use. One method of data compression replaces a string of repeated characters by a character count. Another method uses fewer bits to represent the characters that occur more frequently. See also COMPRESSION.
Here's another definition, courtesy the US Department of Commerce: 1. The process of reducing (a) bandwidth, (b) cost, and (c) time for the generation, transmission, and storage of data by employing techniques designed to remove data redundancy. 2. The use of techniques such as null suppression, bit mapping, and pattern substitution for purposes of reducing the amount of space required for storage of textual files and data records. Some data compaction methods employ fixed tolerance bands, variable tolerance bands, slope-keypoints, sample changes, curve patterns, curve fitting, floating- point coding, variable precision coding, frequency analysis, and probability analysis. Simply squeezing noncompacted data into a smaller space, e.g., by transferring data on punched cards onto magnetic tape, is not considered data compression. See DATA COMPRESSION TABLE.

Data Compression Protocols All current high-speed dial-up modems also support data compression protocols. This means the sending modem will compress the data on-the-fly (as it transmits) and the receiving modem will decompress the data (as it receives it) to its original form. There are two standards for data compression protocols, MNP-5 and ITU-T V.42 bis. Some modems also use proprietary data compression protocols. A modem cannot support data compression without using an error control protocol, although it is possible to have a modem that only supports an error control protocol but not any data compression protocol. A MNP-5 modem requires MNP-4 error control protocol and a V.42 bis modem requires V.42 error control protocol. Note that although V.42 include MNP-4, V.42 bis does not include MNP-5. However, virtually all high-speed modems that support ITU-T V.42 bis also incorporate MNP-5. The maximum compression ratio that a MNP-5 modem can achieve is 2:1. That is to say, a 9,600 bps MNP-5 modem can transfer data up to 19,200 bps. The maximum compression ratio for a V.42 bis modem is 4:1. That is why all those V.32 (9,600 bps) modem manufacturers claim that their modems provide throughput up to 38,400 bps.
Are MNP-5 and V.42 bis useful? Don't be fooled by the claim. It is extremely rare, if ever, that you will be able to transfer files at 38,400 or 57,600 bps. In fact, V.42 bis and MNP-5 are not very useful when you are downloading files from online services. Why? How well the modem compression works depends on what kind of files are being transferred. In general, you will be able to achieve twice the speed for transferring a standard text file (like the one you are reading right now). V.42 bis and MNP-5 modem cannot compress a file which is already compressed by software. In the case of MNP-5, it will even try to compress a precompressed file and actually expand it, thus slow down the file transfer! The above information courtesy modem expert, Patrick Chen.

Data Compression Table A term from US Robotics, makers of fine modems. A data compression table is a table of values assigned for each character during a call under data compression. Default values in the table are continually altered and built during each call. The longer the table, the more efficient the throughput gained.

Data Compressors Also called compactors. These devices take over where high speed modems and statistical multiplexers leave off. They save phone lines by a doubling of data throughput by further compressing async or sync data streams.

Data Concentrator A device which permits the use of a transmission media by a number of data sources greater than the number of channels currently available for transmission.

Data Connections An ATM term. Data VCCs connect the LECs to each other and to the Broadcast and Unknown Server. These carry Ethernet/IEEE 802.3 or IEEE 802.5 data frames as well as flush messages.

Data Contamination Data corruption.

Data Control Block A data block usually at the beginning of a file containing descriptive information about the file.

Data Conversion Converting data from one format to another. Conversion typically falls into three basic categories. 1. To convert to a form usable by the equipment you have, e.g. you convert some data from tape to disk (because you don't have a tape drive). Or you may convert from one method of encoding data to another, say from EBCDIC to ASCII, because you don't have software which can understand IBM's EBCDIC method of coding. 3. Or you may convert from one format to another, e.g. from the dBASE method of encoding databases to the Paradox method, or from WordStar to WordPerfect. There are many service bureaus whose job is to convert computer data from one form to another and there are now many programs out to do the conversion. Our favorite programs are Word-for-Word for converting word processing formats and Data Junction for converting database formats. See also DATA COMPRESSION.

Data Country Code A 3-digit numerical country identifier that is part of the 14-digit network terminal number plan. This prescribed numerical designation further constitutes a segment of the overall 14-digit X.121 numbering plan for a ITU-T X.25 network.

Data Dialtone Networking as widespread as the telephone. The Internet is widely thought to contain the beginnings of Data Dialtone. See INTERNET, METCALFE's LAW, and WORLD WIDE WEB.

Data Dictionary 1. A part of a database management system that provides a centralized meaning, relationship to other data, origin, usage, and format.
2. An inventory that describes, defines, and lists all of the data elements that are stored in a database.

Data Diddling Unauthorized altering of data before, during or after it is input into a computer system.
Data Directory An inventory that specifies the source, location, ownership, usage, and destination of all of the data elements that are stored in a database.
Data Element A basic unit of information having a unique meaning and subcategories (data items) of distinct units or values. Examples of data elements are military personnel grade, sex, race, geographic location, and military unit.
Data Encryption Standard DES. A cryptographic algorithm for the protection of unclassified computer data issued as Federal Information Processing Standard Publication. The DES, which was promulgated by the National Institute of Standards and Technology (NIST)—formerly the National Bureau of Standards (NBS)—is intended for public and Government use.
Data Entry Using an I/O device (input/output device), such as a keyboard on a terminal, to enter data into a computer.
Data File A database typically contains multiple files of information. Each file contains multiple records. Each record is made up of one or more fields. Each field contains one or more bytes of data. The terms file, record and field find their roots in manual office filing systems.
Data Fill One name for the specifications your ISDN phone lines. Ask for your ISDN Data Fill. It will give you useful information, such as how your lines are set up — voice, data, data/voice, etc.
Data Frame An SCSA term. A set of time slots which are grouped together for synchronization purposes. The number of time slots in each frame depends on the SCbus or SCxbus Data Bus data rate. Each frame has a fixed period of 125us. Frames are delineated by the timing signal FSYNC.
Data Grade Circuit A circuit which is suitable for transmitting data. High speed data needs better quality phone lines than normal dial-up phone circuits. You can acquire such circuits from many telephone companies. To upgrade voice phone lines to high-speed data circuits, you must sometimes "condition" the phone line. See CONDITIONING.
Data Group A number of data lines providing access to the same resource.
Data Hunt Group An AT&T Merlin term. A group of analog or digital data stations that share a common extension number. Calls are connected in a round-robin fashion to the first available data station in the group.
Data Integrity The concept of data being protected so that it retains its original meaning and value and is not messed up. Data integrity is also a performance measure based on the rate of undetected errors. A measure of how consistent and accurate computer data is. In data transmission, error correcting protocols, such as LAP-M, MNP and X-modem — provide methods of ensuring that data arrives at the destination in its full integrity. This is done in many ways, including retransmitting messed-up blocks.
Data integrity can be threatened by hardware problems, power failures and disk crashes, but most often by application software. In a database system, data integrity can be threatened if two users are allowed to update the same item or record of data at the same time. Record locking, where only a single user at a time is allowed access to a given data record, is a method of insuring data integrity.
Data Line Interface The point at which a data line is connected to a telephone system.
Data Line Monitor A measuring device that bridges a data line and looks at how clean the data is, whether the addressing is accurate, the protocol, etc. Being only a monitor, it does not in any way affect the information traveling on the line. See DATA MONITOR.
Data Line Privacy Prohibits activities which would insert tones on a data station line used by a facsimile machine, a computer terminal or some other device sensitive to extraneous noise.
Data Link A term used to describe the communications link used for data transmission from a source to a destination. In short, a phone line for data transmission. Or, A fiber optic transmitter, cable, and receiver that transmits digital data between two points.
Data Link Control DLC. Characters used in data communications that control transmission by performing various error checking and housekeeping functions — connect, initiate, terminate, etc.
Data Link Escape The first control character of two-character sequence used exclusively to provide supplementary line-control signals.
Data Link Layer The second layer of the Open Systems Interconnection data communications model of the International Standards Organization. It puts messages together and coordinates their flow. Also used to refer to a connection between two computers over a phone line. See OSI MODEL.
Data Mart A small, single-subject warehouse used by individual departments or groups of users.
Data Message A message included in the GPS (Global Positioning System) signal which reports the satellite's location, clock corrections and health. Included is rough information on the other satellites in the constellation.
Data Mining Sophisticated data search capabilities that use statistical algorithms to discover patterns and correlations in data.
Data Monitor A device used to look at a bit stream as it travels on a circuit. It will show the user what is going down both sides of a data channel. It will show what the user at his terminal is typing, and what the computer is responding with. Extremely useful for troubleshooting data communications problems. Also called a Data Scope.
Data Multiplexer A device allowing several data sources to simultaneously use a common transmission medium while guaranteeing each source its own independent channel. See MULTIPLEXER.
Data Network Identification Code DNIC. In the ITU-T International X.121 format, the first four digits of the 14-digit international data number; the set of digits that may comprise the three digits of the data country code (DCC) and the 1-digit network code (which is called the "network digit"). See DNIC.
Data Numbering Plan Area DNPA. In the U.S. implementation of the ITU-T X.25 network, the first three digits of a network terminal number (NTN). The 10-digit NTN is the specified addressing information for an end-point terminal in an X.25 network.
Data Object An individually addressable unit of information, specified by a data template and its content, that can persist independently of the invocation of a service.
Data Over Voice A device that takes a voice grade line and multiplexes it so it can carry a voice and a data signal. Typically the data is carried in analog form. Thus, to put data on this type of circuit, you need a modem. It is called Data Over Voice, because the data streams (transmission and reception) travel at a higher frequency than the voice conversations using a tech-

nique called frequency division multiplexing (FDM).

Data Packet Signaling All telephone switches use the same three general types of signals: + Event Signaling initiates an event, such as ringing. + Call Progress Signaling denotes the progress (or state) of a call, such as a busy tone, a ringback tone, or an error tone. + Data Packet Signaling communicates certain information about a call, for example, the identify of the calling extension, or the identity of the extension being called.

Data Packet Switch System-common equipment that electronically distributes information among data terminal equipment connected to a data transmission network. The switch distributes information by means of information packets addressed to specific terminal devices.

Data PBX A PBX for switching lots of low-speed asynchronous data. A switch that allows a user on an attached circuit to select from among other circuits, usually one at a time and on a contention basis, for the purpose of establishing a through connection. Distinguished from a PBX in that only digital transmissions, and not analog voice, is supported. Like a telecommunications PBX that makes and breaks phone connections, a data PBX makes and breaks connections between computers and peripherals. In response to dynamic demand, it establishes communications paths between devices attached to its input/output ports by receiving, transmitting and processing electrical signals. Usually, data PBXs work off PCs' serial ports rather than through cable attached to a network interface card. For that reason, they are restricted to serial speeds, topping out at about 19.2K bits per second. For switching lots of low-speed asynchronous data, a data PBX (also called a line selector) can be better than a LAN. Total throughput can actually be higher. See also LINE SELECTOR.

Data Phase A phase of a data call during which data signals may be transferred between DTEs that are interconnected via the network.

Data Port Point of access to a computer that uses trunks or lines for transmitting or receiving data.

Data Protection A means of ensuring that data on the network is safe. Novell's NetWare protects data primarily by maintaining duplicate file directories and by redirecting data from bad blocks to reliable blocks on the NetWare server's hard disk. A hard disk's Directory Entry Table (DET) and File Allocation Table (FAT) contain address information that tells the operating system where data can be stored or retrieved from. If the blocks containing these tables are damaged, some or all of the data may be irretrievable. NetWare reduces the possibility of losing this information by maintaining duplicate copies of the DET and FAT on separate areas of the disk. If one of the blocks in the original tables is damaged, the operating system switches to the duplicate tables to get the location data it needs. Data protection within standard NetWare also involves such features as read-after-write verification, Hot Fix, and disk mirroring or duplexing. See DISK MIRRORING and HOT FIX.

Data Rate The measurement of how quickly data is transmitted. Expressed in bps (bits-per-second).

Data Record See DATA FILE and DATABASE MANAGEMENT SYSTEM.

Data Scope See DATA MONITOR.

Data Scrambler A device used in digital transmission systems to convert an input digital signal into a pseudo random sequence free from long runs of marks, spaces, or other simple repetitive patterns.

Data Secure Line A single tip and ring line off the PBX which is protected against any tones (like call waiting) or break-ins that would otherwise mess up any ongoing data transmission call.

Data Security The protection of data from unauthorized (accidental or intentional) modification, destruction, disclosure, or delay.

Data Segment A pre-defined set of data elements ordered sequentially within a set, beginning and ending with unique segment identifies and terminators. Data segments combine to form a message. Their relation to the message is specified by a Data Segment Requirement Designator and a Data Segment Sequence. Data Segment is an EDI (Electronic Data Interchange) term.

Data Segment Requirement Designator An EDI (Electronic Data Interchange) requirement designator determines that if and when the data segment will occur in a message: - MANDATORY. The segment must appear in the message. - CONDITIONAL. The segment will occur in the message depending on agreement conditions. The relevant conditions must be given as part of the message definition. - OPTIONAL. The segment may or may not occur.

Data Segment Sequence In Electronic Data Interchanges, each data segment has a specific place within a message. The data segment sequence determines exactly where a segment will occur in a message: - HEADING AREA. A segment occurring in this area refers to the entire message. - DETAIL AREA. A segment occurring here is detail information only will override any similar specification in the header area. - SUMMARY AREA. Only segments containing total or control information may occur in this area (e.g., invoice total, etc.)

Data Service Unit DSU. Device designed to connect a DTE (Data Terminal Equipment like a PC or a LAN) to a digital phone line to allow fully-digital communications. A DSU is sort of the digital equivalent of a modem. In more technical terms, a DSU is a type of short haul, synchronous data line driver, normally installed at a user location that connects a user's synchronous equipment over a 4-wire circuit to a serving dial-central-office. This service can be for a point-to-point or multipoint operation in a digital data network. DSUs are typically used for leased lines. For switched digital services, you need a CSU/DSU also called a DSU/CSU. See CSU/DSU and DSU/CSU.

Data Set In AT&T jargon, a data set is a modem, i.e. a device which performs the modulation/demodulation and control functions necessary to provide compatibility between business machines which work in digital (on-off) signals and voice telephone lines. In IBM jargon however, a data set is a collection of data, usually in a file on a disk. See also MODEM.

Data Set Ready One of the control signals on a standard RS-232-C connector. It indicates whether the data communications equipment is connected and ready to start handshaking control signals so that transmission can start. See RS-232-C and the Appendix.

Data Sheets What Business Communications Review calls its statistical and descriptive material comparing PBXs. BCR includes these data sheets in its excellent BCR Manual of PBXs. This manual is the most extensive write-up of larger (more than 200 line) PBXs in the world. It is available from Telecom Library on 1-800-LIBRARY.

Data Signaling Rate The total of the number of bits per second in the transmission path of a data transmission system. A measurement of how quickly data is transmitted, expressed in bps, bits-per-second.

Data Sink Part of a terminal in which data is received from a data link.

Data Source The originating device in a data communications link.

Data Steward A new role of data caretaker emerging in business units. Individual takes responsibilities for the data content and quality.

Data Stream 1. Collection of characters and data bits transmitted through a channel.
2. An SCSA term. A continuous flow of call processing data.

Data Surfer A person who makes a living doing online research and information retrieval. Also known as a Cybrarian (comes from cyberspace librarian) or a super searcher. See CYBRARIAN.

Data Switching Exchange DSE. The equipment installed at a single location to perform switching functions such as circuit switching, message switching, and packet switching.

Data Synchronization The process of keeping database data timely and relevant by sending and receiving information between laptops, between desktops in the field and between bigger computers at headquarters. See also SYNCHRONIZATION and REPLICATION.

Data Terminal Equipment DTE. A definition of hardware specifications that provides for data communications. There are two basic specs your hardware can conform to, DTE (Data Terminal Equipment) or DCE (Data Communications Equipment). See DCE and DTE.

Data Terminal Ready One of the control signals on a standard RS-232-C connector. It indicates if the data terminal equipment is present, connected and ready and has had handshaking signals verified. See RS-232-C and the Appendix.

Data Transfer Rate The average number of bits, characters, or blocks per unit time passing in a data transmission system.

Data Transfer Request Signal A call control signal transmitted by a DCE to a DTE to indicate that a distant DTE wants to exchange data.

Data Transfer Time The time that elapses between the initial offering of a unit of user data to a network by transmitting data terminal equipment and the complete delivery of that unit to receiving data terminal equipment.

Data Typing When converting a database from one format to another, several conversion programs will convert the data to a common format before converting it to the final version. During the conversion process a program may check through the data in the database to determine what it is and arbitrarily make one field numeric, one field character, one field memo, etc.

Data Warehouse A database warehouse consolidates information from many departments within a company. This data can either be accessed quickly by users or put on an OLAP server for more thorough analysis. Data warehouses often use OLAP servers. OLAP stands for On Line Analytical Processing, also called a multidimensional database. According to PC Week, these databases can slice and dice reams of data to produce meaningful results that go far beyond what can be produced using the traditional two-dimensional query and report tools that work with most relational databases. OLAP data servers are best suited to work with data warehouses.

Database A collection of data structured and organized in a disciplined fashion so that access is possible quickly to information of interest. There are many ways of organizing databases. Most corporate databases are not one single, huge file. They are multiple databases related to each other by some common thread, e.g. an employee identification number. Databases are made up of two elements, a record and a field. A record is one complete entry in a database, e.g. Gerry Friesen, 12 West 21 Street, New York, NY 10010, 212-691-8325. A field would be the street address field, namely 12 West 21 Street.
Databases are stored on computers in different ways. Some are comma delineated. They differentiate between their fields with commas — like Gerry's record above. A more common way of storing databases is with fixed length records. Here, all the fields and all the records are of the same length. The computer finds fields by index and by counting. For example, Gerry's first name might occupy the first 15 characters. Gerry's last name might be the next 20 characters, etc. Where Gerry's names are too short to fill the full 15 or 20 characters, their fields are "padded" with specially-chosen characters which the computer recognizes as padded characters to be ignored. The most important thing to remember about databases is that all the common database programs, like dBASE, Paradox, Rbase, etc. don't automatically make backups of their files like word processing programs do. Therefore, before you muck with a database file — sort it, index it, restructure it, etc. Please make sure you make a backup of the main database file.

Database Management System DBMS. Computer software used to create, store, retrieve, change, manipulate, sort, format and print the information in a database. Database management systems are probably the fastest growing part of the computer industry. Increasingly, databases are being organized so they can be accessible from places remote to the computer they're kept on. The "classic" database management system is probably an airline reservation system.

Database Object One of the components of a database: a table, view, index, procedure, trigger, column, default, or rule.

Database Server A specialized computer that doles out database data to PCs on a LAN the way a file server doles out files. Where a traditional DBMS runs both a database application and the DBMS program on each PC on the LAN, a database server splits up the two processes. The application you wrote with your DBMS runs on your local PC, while the DBMS program runs on the database server computer. With a regular file server setup, all the database data has to be downloaded over the LAN to your PC, so that the DBMS can pick out what information your application wants. With a database server, the server itself does the picking, sending only the data you need over the network to your PC. So a database server means vastly less network traffic in a multi-user database system. It also provides for better data integrity since one computer handles all the record and file locking. See SERVER.

Datagram A transmission method in which sections of a message are transmitted in scattered order and the correct order is re-established by the receiving workstation. Used on packet-switching networks. The Dow Jones Handbook of Telecommunications defines it as, "A single unacknowledged packet of information that is sent over a network as an individual packet without regard to previous or subsequent packets." Here's another definition I found. A finite-length packet with sufficient information to be independently routed from source to destination. In packet switching, a self-contained packet, independent of other packets, that carriers information sufficient for routing from the originating data terminal equipment to the destination data terminal equipment, without relying on earlier exchanges between the equipment and the network. Unlike virtual call service, there are no call establishment or clearing procedures, and the network does not generally provide protection against loss, duplication, or misdelivery. Datagram transmission typically does not involve end-to-end session establishment and may or may not entail delivery confirmation acknowledgment. A datagram is the basic unit of information passed across the Internet. It contains a source and destination address along with data. Large messages are broken down into a sequence of IP datagrams. See CONNECTIONLESS MODE TRANSMISSION.

Datagram Packet Network The type of packet-switched network in which each packet is individually routed. This may result in a loss of sequence within a message because of alternate routing, or a loss of portions of a message because of packet elimination for congestion control. See DATAGRAM.

Datakit VCS The Datakit VCS is an AT&T packet switch that can switch over 44,000 packets per second and support up to 3,500 simultaneous virtual circuits. According to AT&T, the Datakit VCS is a digital virtual circuit switch with an architecture that combines the advantages of LANs, PBXs, data circuits and X.25 packet switches.

DATAP The programmer and system house in Atlanta which provides several long distance carriers with value-added services — what Sprint calls its SCADA system manufacturer.

Datapak A packet-switched network run in Denmark, Sweden, Finland and Norway and operated by their respective governments.

Datapath A name that's becoming generic for a data service that provides digital, full-duplex data transmission at speeds of 300 bps through 19.2 Bps asynchronous and 1,200 through 64 Kbps synchronous. Datapath has built in autobaud and handshaking protocols.

Dataphone A service mark of AT&T for various data hardware products such as modems and printers and services. See also DDS.

DataSPAN DataSPAN is generally characterized as a "fast packet" service and is based on frame relay standards recommended by the International Consultative Committee for Telephone & Telegraph (CCITT)(now called the ITU-T) and the American National Standards Institute (ANSI). Northern Telecom has introduced DataSPAN as a new DMS SuperNode value-added, data communications service that is targeted toward connecting high-speed Local Area Networks. Northern Telecom asserts that DataSPAN's rapid and efficient data transport assures reliable delivery and substantial performance improvement over current LAN interconnect solutions. DataSpan switching and transmission delay is less than 3 ms per node; X.25 switching and transmission delay can be up to 50 ms per node. Using Frame Relay, wide-area packet switching can be accomplished with the same level of performance that is traditionally limited to complex, dedicated private-line networks. DataSPAN is accessed through standard DS-0 or DS-1 links.

Date And Time Stamp Many voice mail systems will append the date and time of receipt of voice messages for their users/subscribers.

Datel Services An MCI International service that carries data for medium-volume users. This service provides two-way data and voice transmission for simplex, half-duplex, and full duplex operations.

Datex P An Austrian packet-switched network in Austria and Germany and run by their respective governments.

Dating Format The format employed to express the time of an event. The time of an event on the UTC time scale is given in the following sequence; hour, day, month, year; e.g., 0917 UT, 30 August 1997. The hour is designated by the 24-hour system. UTC stands for Coordinated Universal Time, believe it or not.

Daughterboard First there is the motherboard. That's the main circuit board of a computer system. The motherboard contains edge connectors or sockets so other PC (printed circuit) boards can be plugged into it. Those PC boards are called Fatherboards. Some fatherboards have pins on them into which you can plug smaller boards. Those boards are called Daughterboards. In a voice processing system, you might have a Fatherboard to do faxing. And you might have a range of Daughterboards, which allow you to connect different types of phone connections. Different boards exist for standard analog tip and ring, digital switched 56, etc. See also MOTHERBOARD.

Daughtercard Same as DAUGHTERBOARD.

DAVIC Digital Audio/Video Council

Day-Of-Week Factors A call center term. A historical pattern consisting of seven factors, one for each day of the week, that defines the typical distribution of call arrival throughout the week. Each factor measures how far call volume on that day deviates from the average daily call volume. This is a historical pattern.

DB 1. Should really be written as dB. Decibel. A unit of measure of signal strength, usually the relation between a transmitted signal and a standard signal source. Therefore, 6 dB of loss would mean that there is 6 dB difference between what arrives down a communications circuit and what was transmitted by a standard signal generator. See DECIBEL for a better explanation.

2. Database

DB Loss Budget Should be written as dB. dB Loss Budget is the amount of light available to overcome the attenuation in the optical link and still maintain specifications.

DB-9 This is the standard nine-pin RS-232-C serial port on the IBM AT and most laptop computers. The term DB-9 is used to describe both the male and female plug. So be careful when you order. See also the Appendix in this Dictionary for more information on the pinning of RS-232-C plugs.

DB-15 A standardized connector with 15 pins. It can be used Ethernet transceivers. It can also be used for connecting VGA monitors.

DB-25 The standard 25-pin connector used for RS-232-C serial data communications. In a DB-25 there are 25-pins, with 13 pins in one row and 12 in the other row. DB-25 is used to describe both the male and female plug. So be careful when you order. See also the Appendix in this Dictionary for more information on the pinning of RS-232-C plugs.

NEWTON'S TELECOM DICTIONARY

DB2 IBM's relational database system that runs on System 370-compatible mainframes under the MVS operating system.

DBA 1. Dynamic Bandwidth Allocation.

2. Database administrator. The individual in the organization with the responsibility for the design and control of databases.

DBIT Also called D-BIT. The delivery confirmation bit in an X.25 packet that is used to indicate whether or not the DTE wishes to receive an end-to-end acknowledgment of delivery. In short, a bit in the X.25 packet header that assures data integrity between the TPAD and the HPAD.

DBM Decibels below 1mW. This should be written as dBm. Output power of a signal referenced to an input signal of 1mW (Milliwatt). Similarly, dBm0 refers to output power, expressed in dBm, with no input signal. (0 dBM = 1 milliwatt and -30 dBm = 0.001 milliwatt). See DECIBEL.

DBMS Database management system. A computer program that manages data by providing the services of centralized control, data independence, and complex physical structures. Advantages include efficient access, integrity, recovery, concurrency control, privacy, and security. A DBMS enables users to perform a variety of operations on data, including retrieving, appending, editing, updating, and generating reports.

DBRN A ratio expressed in DeciBels above Reference Noise.

DBS Direct Broadcast Satellite.

DBT Deutsche Bundespost Telecom.

DBU Decibels below 1uW.

DC Direct Current. The flow of free electrons in one direction within an electrical conductor, such as a wire. DC also stands for Delayed Call.

DC SIGNALING A collection of ways of transmitting communications signals using direct current — the type of current produced by a dry cell household "D" cell battery. DC signaling is only used on cable. It's an out-of-band signal.

DCA 1. Defense Communication Agency. US government agency responsible for installation of Defense Data Networks, including the ARPANET and MILNET, and PSNs. The DCA writes contracts for operation of the DDN (Defense Data Network) and pays for network services.

2. Document Content Architecture. The IBM approach to storing documents as two types of document group: draft documents and final form documents. For presentation, the draft document is transformed into a final document through an office system.

DCC 1. Data Communications Channel. Channels contained within section and line overhead used as embedded operations channels to communicate to each network element. An AT&T SONET term.

2. An ATM term. Data Country Code: This specifies the country in which an address is registered. The codes are given in ISO 3166. The length of this field is two octets. The digits of the data country code are encoded in Binary Coded Decimal (BCD) syntax. The codes will be left justified and padded on the right with the hexadecimal value "F" to fill the two octets. An ATM term.

3. Digital Compact Cassette. A digital version of the familiar analog audio cassette. A DCC recorder can play and record both analog and digital cassettes. But the digital ones will sound a lot better.

DCD 1. Data Carrier Detect. Signal from the DCE (modem or printer) to the DTE (typically your PC), indicating it (the modem) is receiving a carrier signal from the DCE (modem) at the other end of the telephone circuit.

2. Dynamically Configurable Device. A dynamically configurable device is a fancy name for a Plug and Play device, so-called because you don't have to reboot the system after installing one.

DCE Data Communications Equipment. In the RS-232-C "standard" developed by the Electronic Industries Association, there are DCE devices (typically modems or printers) and DTE (Data Terminal Equipment) devices, which are typically personal computers or data terminals. The main difference between a DCE and DTE is the wiring of pins two and three. But there is, of course, no standardization. When wiring one RS-232-C device to another, it's good to know which device is wired as a DCE and which as a DTE. But it's actually best to go straight to the wiring diagram in the appendix of the device's instruction manual. Then you compare the wiring diagram of the device you want to connect and build yourself a cable that takes into account the peculiar (i.e. strange) vagaries of the engineers who designed each product. In short, with an RS-232-C connection, the modem is usually regarded as DCE, while the user device (terminal or computer) is DTE. In a X.25 connection, the network access and packet switching node is viewed as the DCE. DCE devices typically transmit on pin 3 and receive on pin 2. DTE (Data Terminal Equipment) devices typically transmit on pin 2 and receive on pin 3. See also the Appendix.

DCH D-Channel Handler.

DCM Digital Circuit Multiplication. A means of increasing the effective capacity of primary rate and higher level PCM hierarchies, based upon speech coding at 64 Kbit/s. Also a Digital Carrier Module.

DCP Digital Communications Protocol.

DCP Telephone A digital voice telephone of the AT&T model 7400 series, operated with the Digital Communications Protocol (DCP) used in System 75/85 digital PBXs.

DCS 1. Distributed Communications System. See DISTRIBUTED SWITCHING.

2. Digital Crossconnect System. A device for switching and rearranging private line voice, private line analog data and T-1 lines. A DCS performs all the functions of a normal "switch," except that connections are typically set up in advance of when the circuits are to be switched — not together with the call. You make those "connections" by calling an attendant who makes them manually, or dialing in on a computer terminal — one similar to an airline agent's. See also NETWORK RECONFIGURATION SERVICE.

3. Digital Cellular System.

4. Digital Communications System.

5. DCS. A packet-switched network implemented in Belgium and operated by the Belgian government.

DCS 1800 Digital Cellular System at 1800 MHz. A GSM (Global System for Mobile Communications) standard for cellular mobile telephony established by ETSI (European Telecommunications Standards Institute) for operation at 1800 MHz.

DCT 1. Digital Carrier Termination.

2. Discrete Cosine Transform. A form of coding used in most of the current image compression systems for bit rate reduction.
DCTI Desktop Computer Telephone Integration. Basically, providing a way for your computer to control your telephone set.
DCV Digital Compressed Video.
DDA Domain Defined Attribute. A way of adding additional information to the address of your electronic mail in order to avoid confusion between people of the same or similar name.
DDB Digital Data Bank.
DDC Direct Department Calling.
DDCMP Digital Data Communications Message Protocol. A byte-oriented, link-layer protocol from Digital Equipment Corp., used to transmit messages between stations over a communications line. DDCMP supports half- or full-duplex modes, and either point-to-point or multipoint lines in a DNA (Digital Network Architecture) network.
DDD Direct Distance Dialing. The "brand name" the Bell System used to call its Message Toll Telephone Network. It used the words "Direct Distance Dialing" to convince the public to dial their own long distance calls directly without the help of an operator.
DDE See DYNAMIC DATA EXCHANGE
DDEML Dynamic Data Exchange Management Library. A feature of Microsoft Windows.
DDI Direct Dialing Inward. A British term. It is a service where a call made to a DDI number arrives direct, without the intervention of an organization's operator, at an extension or, if routed via an ACD, a group of extensions. A specific DDI number is assigned to each campaign. When an agent answers a call made to one of these DDI numbers, the relevant script and screen are instantly displayed, so the agent can give an appropriate response.
DDM Distributed Data Management Architecture. An IBM SNA LU 6.2 transaction providing users with facilities to locate and access data in the network. It involves two structures: DDM Source, and DDM Target. The DDM Source works with a transaction application to retrieve distributed data and transmits commands to the DDM Target program on another system where the data that has been requested is stored. The DDM Target interprets the DDM commands, retrieves the data and sends it back to the DDM Source that originated the request.
DDN Defense Data Network. A network that provides long haul and area data communications and interconnectivity for DoD systems, and supports the DoD suite of protocols (especially TCP and IP). All equipment attached to the DDN by military subscribers must incorporate, or be compatible with, the DoD Internet and transport protocols. DDN was split off from ARPANET to handle U.S. military needs. It is also called MILNET.
DDP Distributed Data Processing. See DISTRIBUTED SWITCHING.
DDS 1. Dataphone Digital Service, also called Digital Data System. DDS is private line digital service, typically with data rates at 2,400; 4,800; 9,600 and 56,000 bits per second. DDS is offered on an inter-LATA basis by AT&T and on an intra-LATA basis by the Bell operating companies. AT&T has now incorporated it into their Accunet family of offerings. But, by the time your read this, they may have a new name for the service, like Accunet.
2. Digital Data Storage. A DAT format for storing data. It is sequential — all data that is recorded to the tape falls after the previous block of data.
DDSD Delay Dial Start Dial. A start-stop protocol for dialing into a switch.
DE Discard Eligible. The frame relay standard specifies that data sent across a virtual connect in excess of that connection's Committed Information Rate (CIR) will be marked by the network as being eligible for discard in the event of network congestion. DE data is the first to be discarded by the network when congestion occurs, thus providing protection for data sent within the parameters of the CIR. It is the responsibility of the intelligent end equipment and/or protocol to recognize the discard and respond by resending the information.
De Facto Used to describe standards. Reflecting current or actual practice, but not having approval or sanction by any official standards-setting organization. Usually created by an individual manufacturer or developer — hence, often used as a synonym for proprietary. The Hayes AT auto-dial modem command language is an example of a de facto standard. Very often, de facto standards form the basis for de jure standards (Ethernet was the basis of Institute of Electrical and Electronic Engineers or IEEE standard 802.3). Contrast with DE JURE.
De Forest, Lee Lee de Forest invented the vacuum tube in 1907. Until the invention of the transistor after World War II, all radio, long distance telephony and complicated electronics, including electronic computers, were derived from de Forest's invention.
De Jure Used to describe standards approved or sanctioned by an official standards setting organization. ITU-T Recommendation X.25 for packet data networks is an example of a de jure standard. Contrast with DE FACTO.
Dead Sector In facsimile, the elapsed time between the end of scanning of one line and the start of scanning of the following line.
Dead Spot A cellular radio term. It denotes an area within a cell where service is not available. A dead spot is usually caused by hilly terrain which blocks the signal to or from the cell tower. It can also occur in tunnels and indoor parking garages. Excessive foliage or electronic interference can also cause dead spots. If you encounter a dead spot, tell your cellular radio supplier. They often will do something.
Dead Tree Edition The paper version of a publication available in both paper and electronic forms. If you are reading my dictionary on paper, you are reading the dead tree edition.
Deadline During the Civil War, captured troops were placed in a field. They were told they could move wherever they wanted in the field except beyond a line that was drawn around the field. If they did, they would be shot. Thus the origin of the word deadline.
Deadlock See DEADLY EMBRACE.
Deadly Embrace Stalemate that occurs when two elements in a process are each waiting for the other to respond. For example, in a network, if one user is working on file A and needs file B to continue, but another user is working on file B and needs file A to continue, each one waits for the other. Both are temporarily locked out. The software must be able to deal with this.
Dealer Dealer is simply a person who sells equipment — hardware or software — made by someone else. That dealer may be a distributor of Novell LAN software. Or that dealer may be in the secondary telecom equipment business, i.e. a company that

buys and sells secondary market telephone equipment. Generally in the secondary equipment business, a dealer takes ownership of the equipment (see BROKER). He tests and refurbishes it before remarketing. An authorized dealer has a contract with the manufacturer to buy equipment at a preset price. He also has the added support of a manufacturer's warranty. An independent dealer has no formal agreement with the manufacturer. He uses a variety of sources to obtain equipment and his warranty is backed only by the company's internal resources.

Dealer Board British term meaning "trading turret."

Dealing Room British term meaning "trading room."

Debbie Warren A remarkable woman of great stamina, great perseverance and great tolerance. How do we know this? Anyone married to Stuart desperately needs these character traits. Favorite pastimes: Rachael and Jennifer.

Debit Card The term telephone "debit card" covers three categories of a new type of telephone calling card, variously called "calling card" and "prepaid telephone card." The definitions are in flux. Here's the best shot at debit card: A telephone debit card is a piece of credit-card size plastic with some technology on it or embedded into which represents the value of the money remaining on the debit card. Such money can be used to make phone calls. The technology on the card is most typically an integrated circuit, a magnetic strip, bar codes which can be read by an optical reader. See BREAKAGE, CALLING CARD and PRE-PAID TELEPHONE CARD.

Debug A MS-DOS program to examine or alter memory, load and look at sectors of data from disk and create simple assembly-language programs. MS-DOS DEBUG.COM lets you write some small programs. You can use DEBUG to correct problems in some programs.

DEC Digital Equipment Corporation. A leading manufacturer of minicomputers. The Unix operating system, developed at Bell Labs, runs on DEC computers. DEC, with its headquarters in Massachusetts and having sold so many computers to Western Electric for inclusion in central office switches and toll switches, is sometimes referred to as "Eastern Electric." Kenneth H. Olsen founded DEC in 1957 with three employees in 8,500 square feet of leased space in a corner of an old woolen mill.

DEC LAT A proprietary data communications protocol developed by DEC.

DECags DEC ASAI Gateway Services. Two-directional link to AT&T's Definity. See also OPEN APPLICATION INTERFACE.

Decibel dB. A unit for measuring the power of a sound or the strength of a signal. It is expressed as the ratio of two values. Here is an explanation from James Harry Green's excellent book The Dow Jones-Irwin Handbook of Telecommunications. (Get a copy from Telecom Library 1-800-LIBRARY.) Mr. Green writes, "The power in telecommunications circuits is so low that it is normally measured in milliwatts. However, the milliwatt is not a convenient way to express differences in power level between circuits. Voice frequency circuits are designed around the human ear, which has a logarithmic response to changes in power. (Likewise, the human eye has a logarithmic response to changes in light.) Therefore in telephony, the decibel, which is a logarithmic rather than a linear measurement, is used as a measure of relative power between circuits or transmission level points. A change in level of 1 dB is barely perceptible under ideal conditions...Increases or reductions of 3 dB result in doubling or halving the power in a circuit. This ratio is handy to remember when evaluating power differences. The corresponding figure for doubling or halving voltage is 6 dB. See DB.

Decide A deadly computer virus.

Decimal Our normal numbering system. It is to the base 10.

Decision Circuit A circuit that measures the probable value of a signal element and makes an output signal decision based on the value of the input signal and a predetermined criterion or criteria.

Decision Instant In the reception of a digital signal, the instant at which a decision is made by a receiving device as to the probable value of a signal condition.

Decision Support Systems DSS. Computerized systems for transforming data into useful information, such as statistical models or predictions of trends, which is used by management in the decision-making process. There are several aspects of the best decision support systems: First, they are connected to mainframe databases. Second, they are accessible by executives from their desktops. Third, there are usually lots of programs for producing graphs, charts and writing simple reports, i.e. for the executives to extract and portray information in forms that are most useful to them.

DECNET DEC's proprietary Ethernet LAN that works across all of the company's machines, endowed with a peer-to-peer methodology. See DEC.

Decoder A device that converts information from one form to another — typically from analog to digital and vice versa. See DECODING.

Decoding Changing a digital signal into analog form or into another type of digital signal. The opposite of Encoding. See also MODEM. Decoding and coding should not be confused with deciphering and ciphering. See DES.

Decollimation In optics, that effect wherein a beam of parallel light rays is caused to diverge or converge from parallelism. Any of a large number of factors may cause this effect, e.g., refractive index inhomogeneities, occlusions, scattering, deflection, diffraction, reflection, refraction.

Decolumnization The process of reformatting multi-column documents into a signal column. Generally, when you are processing a document for use in a word processing program, a single column of text is preferable to multiple columns.

Decompression Decompression is the process of expanding a compressed image or file so it can be viewed, printed, faxed or otherwise processed.

Decrypt To convert encrypted text into its equivalent plain text by means of a cryptosystem. This does not include solution by cryptanalysis. The term decrypt covers the meanings of decipher and decode.

DECT Digital European Cordless Telecommunication. The pan-European standard based on time division multiple access used for limited-range wireless services. Based on advanced TDMA technology, and used primarily for wireless PBX systems, telepoint and residential cordless telephony today, potential uses for DECT include paging and cordless LANs. DECT frequency is 1800-1900 MHz.

Dedicated Access A connection between a phone or phone system (like a PBX) and an InterExchange Carrier (IXC) through a dedicated line. All calls over the line are automatically routed to a particular IXC.

Dedicated Array Processor A microprocessor on a hardware-based RAID array that controls the execution of RAID array-specific functions, such as rebuilding. See RAID.

Dedicated Attendant Link Assures that there will always be an intercom link available for your attendant or receptionist or operator to announce incoming calls.

Dedicated Bypass A connection between a phone or phone system (like a PBX) and an InterExchange Carrier (IXC) through a dedicated line that is not provided by the dominant local provider of local phone service. For example, I live in New York. I might order a leased T-1 to MCI. If that line is not provided by Nynex, it is "dedicated bypass." Such bypass circuits are often cheaper and better quality than what the dominant local carrier can provide.

Dedicated Channel Or Circuit A channel leased from a common carrier by an end user used exclusively by that end user. The channel is available for use 24 hours a day, seven days a week, 52 weeks of the year, assuming it works that efficiently.

Dedicated Circuit A circuit designated for exclusive use by specified users. See also DEDICATED CHANNEL.

Dedicated Feature Buttons The imprinted feature buttons on a telephone: Conf or Conference, Drop, HFAI (Hands Free Answerer on Intercom), Hold, Mute or Microphone, Speaker or Speakerphone, Transfer, Message, and Recall.

Dedicated Line Another name for a private leased line or dedicated channel. A dedicated line provides the ability to have constant transmission path from point A to point B. A dedicated line may be leased or owned. It may be assigned a single purpose, such as monitoring a distant building. It may be part of a network, with the ability for many to dial into it. It may be a tie-line between your offices or it may be a line to a long distance carrier. In this case, you do not have to dial a local connection number or put in an authorization code. A WATS line is in effect a Dedicated Line to AT&T or whomever you purchased WATS from.

Dedicated Machine A computer designed to run only one program or do one thing. This machine cannot easily be re-programmed to do another task, as, for example, a general-purpose machine can. A general-purpose machine, however, can be dedicated to running only one task if programmed to do so.

Dedicated Mode When a file server or router on a local area network is set up to work only as a file server or router, it runs in dedicated mode.

Dedicated Server A computer on a network that performs specialized network tasks, such as storing files. The word "dedicated" means that the computer is used exclusively as a server. It is not used as a workstation, which means no one is sitting in front of it, using it. A dedicated server sits all alone, attached to its network, working happily all by itself.

Dedicated Service A communication network devoted to a single purpose or group of users, e.g., AUTOVON, FTS. It may also be a subset of a larger network; e.g., AUTOVON, FTS.

Dedicated Trunk A trunk which bypasses the Attendant Console and rings through to a particular phone, hunt group or distribution group.

Deep Space Space at distances from the Earth approximately equal to or greater than the distance between the Earth and the Moon.

Deep UV Printed circuits are made by optical lithography, by shining light through a negative, or mask. In the beginning, it was visible light, but lately it's been made with more precise ultraviolet light, which can print lines as thin as 0.15 micron wide. They call this light "deep uv" and it is ultraviolet light emitted by an exciter or pulsing laser.

Defacto Standards Standards, widely accepted and used, but lacking formal approval by a recognized standards organization. The Hayes AT command set, MS-DOS, Windows, TCP/IP etc. are defacto standards.

Default The default is a factory-set hardware of software setting or configuration. It is the preset value that the program or equipment comes with. It will work with default values in the absence of any other command from the user. For example, communications software programs, such as Crosstalk, Blast, etc., have as their default settings 300 baud, 8 bit, one stop bit, no parity. If you want to run at 1,200 baud, you have to change that "default" setting.

Default Carrier Generic name given to the long distance carrier which will carry the traffic of customers who haven't pre subscribed to a long distance carrier.

Default Node Representation An ATM term. A single value for each nodal state parameter giving the presumed value between any entry or exit to the logical node and the nucleus.

Defense Data Network DDN. The Department of Defense integrated packet switching network capable of worldwide multilevel secure and non-secure data transmission.

Deferred Processing Performing operations as a group or batch, all at once. Using batch mode, you can quickly prescan your documents, capturing just the image of each page, then perform recognition on these images later, freeing your computer for interactive work.

Definition A figure of merit for image quality. For video-type displays, it is normally expressed in terms of the smallest resolvable element of the reproduced received image.

Degauss To demagnetize. To degauss a magnetic tape means to erase it. See DEGAUSSER.

Degausser Device to demagnetize color picture tube for color purity.

Degradation In communications, that condition in which one or more of the established performance parameters fall outside predetermined limits, resulting in a lower quality of service.

Degraded Service State The condition wherein degradation prevails in a communication link. For some applications e.g., automatic switching to a non degraded standby link, degradation must persist for a specified period of time before a degraded service state is considered to exist.

Degree Of Coherence A measure of the coherence of a light source.

Degree Of Isochronous Distortion In data transmission, the ratio of (a) the absolute value of the maximum measured difference between the actual and the theoretical intervals separating any two significant instants of modulation (or demodulation) to (b) the unit interval. These instants are not necessarily consecutive. The degree of isochronous distortion is usually expressed as a percentage. The result of the measurement should be completed by an indication of the period, usually limited, of the observation.

Degree Of Start-Stop Distortion 1. In asynchronous data transmission, the ratio of (a) the absolute value of the maximum measured difference between the actual and theoretical intervals separating any significant instant of modulation (or demodulation) from the significant instant of the start element immediately preceding it to (b) the unit interval.
2. The highest absolute value of individual distortion affecting the significant instants of a start-stop modulation. The degree of distortion of a start-stop modulation (or demodulation) is usually expressed as a percentage.

Deinstallation A term used in the secondary telecom equipment business. The shutoff and disconnect of machine power and the disassembly of a PBX switch to prepare for its removal from a switch room facility. A properly executed deinstallation will include all necessary parts for the reassembly, operation, maintenance and acceptance of the switch and any of its components at the next location.

Dejitterizer A device for reducing jitter in a digital signal, consisting essentially of an elastic buffer into which the signal is written and from which it is read at a rate determined by the average rate of the incoming signal. Such a device is largely ineffective in dealing with low-frequency impairments such as waiting-time jitter.

Delay The wait time between two events, such as the time from when a signal is sent to the time it is received. There are all sorts of reasons for delays, such as propagation delays, satellite delays, etc.

Delay, Absolute The time elapsed between transmission of a signal and reception of the same signal.

Delay Announcements These are pre-recorded announcements to incoming callers that they are being delayed and being placed in an ACD queue. Sample: "Please wait. All our agents are permanently busy. You are being placed on Eternity Hold. Don't go away or you'll never be allowed back." Some announcements are giving callers sales pitches and some idea of how long they'll have to stay on line until someone helps them.

Delay Distortion The difference, expressed in time, for signals of different frequencies to pass through a phone line.

Delay Encoding A method of encoding binary data to form a two-level signal. A binary zero caused no change of signal level unless it is followed by another zero, in which case a transition takes place at the end of the first bit period. A binary "1" causes a transition from one level to the other in the middle of the bit period. Used primarily for encoding of radio signals since the spectrum of the encoding of signal contains less low frequency energy than an NRZ signal and less high frequency energy than a biphase signal.

Delay Equalizer A corrective piece of electronic circuitry designed to make communications circuit delays constant over a desired frequency range. A delay equalizer is a device that adds a delay to analog signals, which will travel through a medium faster than other frequencies used to transmit portions of the same data. The objective is to create a medium that transfers the information on all the used frequencies in the same time over the same distance, thus eliminating transmission delay distortion.

Delay Line A transmission line, or equivalent device, designed to introduce delay.

Delay Modulation A modulation scheme that uses different forms of delay in a signal element. Frequently used in radio, microwave and fiberoptic systems.

Delayed Delivery Hold a message for delivery later. Just as the words say.

Delayed Delivery Facility A facility that employs storage within the data network whereby data destination for delivery to one or more addresses may be held for subsequent delivery at a later time.

Delayed Ring Transfer An optional KTU facility that provides for automatic transfer to the ringing signal from a principal telephone set to the attendant telephone station after an adjustable number of rings.

Delayed Sending A feature of fax machines which allows the machine to be programmed to send its transmissions at a later time — to take advantage of lower phone rates, for example.

Delayed Transmission A fax machine feature that allows a document to be transmitted automatically at a specific time.

Delimiter A special word or character that allows a computer to recognize the beginning or end of a portion of a program or segment of data.

Delivered Block A successfully transferred block.

Delivered Overhead Bit A bit transferred to a destination user, but having its primary functional effect within the telecommunication system.

Delivered Overhead Block A successfully transferred block that contains no user information bits.

Delivery Confirmation Information returned to the originator indicating that a given unit of information has been delivered to the intended addresses.

Delivery Envelope An X.400 term.

DELNI Digital Ethernet Local Network Interconnect

Delphi Forecasting One of the silliest methods of forecasting the future. Namely, to ask a bunch of alleged experts (often academic eggheads) what they think might happen and then averaging out their opinions, sort of.

Delta Channel In T-carrier/ISDN communications, a delta channel/"D channel" contains signaling and status information.

Delta Frame Also called Difference Frame. Contains only the pixels different from the preceding Key Frame. Delta frames reduce the overall size of the video clip to be stored on disk or transmitted on phone lines.

Delta Modulation A method for converting analog voice to digital form for transmission. It is the second most common method of digitizing voice after Pulse Code Modulation, PCM. Sampling is done in all conversion of analog voice to digital signals. The method of sampling is what distinguishes the various methods of digitization (Delta vs.' PCM, etc.). In delta modulation, the voice signal is scanned 32,000 times a second, and a reading is taken to see if the latest value is greater or less than it was at the previous scan. If it's greater, a "1" is sent. If it's smaller, a "0" is sent.

Delta modulation's sampling rate of 32,000 times a second is four times faster than PCM. But Delta records its samples as a zero (0) or a one (1), while PCM takes an 8-bit sample. Thus PCM encodes voice into 64,000 bits per second, while Delta codes it into 32,000. Because delta has fewer bits, it could theoretically produce a poorer representation of the voice. In actual fact, the human ear can't hear the difference between a PCM and a Delta encoded voice conversation.

Delta modulation has much to recommend it, especially its use of fewer bits. Unfortunately no two delta modulation schemes are compatible with each other. So to get one delta-mod digital PBX to speak to another, you have to convert the voice signals back to analog. With AT&T making T-1 a de facto digital encoding scheme, PCM has become the de facto standard for digitally encoding voice. And although there are three types of PCM in general use, they can be made compatible on direct digital basis (i.e. without having to go back to analog voice). One problem with PCM is that American manufacturers typically put twenty four 64,000 bit per second voice conversations on a channel and call it T-1. The Europeans put 30 conversations on their equivalent transmission path. Thus, you can't directly interface the American and the European systems. But there are "black boxes" available...(In this business, there are always black boxes available.)

Delta Sigma Modulation A variant of delta modulation in which the integral of the input signal is encoded rather than the signal itself. A normal delta modulation encoder by an integrating network. See DELTA MODULATION.

Delta Technology An Internet access tern. Delt technology is specialized remote adaptive routing protocols for optimizing bandwidth. It prevents unnecessary traffic from being sent over slow WAN connections by only sending the changes (deltas).

Deltas The changes.

Delurking Coming out of online "lurking mode," usually motivated by an irresistible need to flame about something. "I just had to delurk and add my two cents to that conversation about a woman's right to abortion."

Deluxe Queuing A feature that allows incoming calls from phone users, tie trunks and attendants to be placed in a queue when all routes for completing a particular call are busy. The queue can be either a Ringback Queue (RBQ)— the user hangs up and is called back when a trunk becomes available — or an Off-Hook Queue (OHQ) — the user waits off-hook and is connected to the next available trunk. Deluxe Queuing is a term used mainly by AT&T. Most modern PBXs have this feature. Most have simpler names, however.

Demand Assigned Multiple Access See DAMA.

Demand Assignment A technique where users share a communications channel. A user needing to communicate with another user on the network activates the required circuit. Upon completion of the call, the circuit is deactivated and the capacity is available for other users.

Demand Factor The ratio of the maximum demand on a power system to the total connected load of the system.

Demand Load In general, the total power required by a facility. The demand load is the sum of the operational load (including any tactical load) and non operational demand loads. It is determined by applying the proper demand factor to each of the connected loads and a diversity factor to the sum total.

Demand Paging The common implementation in a PC of virtual memory, where pages of data are read into memory from storage in response to page faults.

Demand Priority Access method providing support for time sensitive applications such as video and multimedia as part of the proposed 100BaseVG standard offering 100Mbit/s over voice grade UTP (Unshielded Twisted Pair) cable. By managing and allocating access to the network centrally, at a hub rather than from individual workstations, sufficient bandwidth for the particular application is guaranteed on demand. Users, say its proponents, can be assured of reliable, continuous transmission of information.

Demand Publishing The production of just the number of printed documents you need at the present time, as in "just in time." In short, the immediate production of printed documents which have been created and stored electronically.

Demand Service In ISDN applications, a telecommunications service that establishes an immediate communication path in response to a user request made through user-network signaling.

Demarc or Demark (Pronounced D-Mark.) The demarcation point between the wiring that comes in from your local telephone company and the wiring you install to hook up your telephone system — your CPE (Customer Provided Equipment) wiring. A Demarc might be anything as simple as an RJ-11C jack (one trunk) or an RJ-14C (two trunks) or an RJ-21X (up to 25 trunks) or a 66-block. A 66-block is a punchdown block on one side of which the telephone company may punch down its trunks and on the other, you may punch down your connections into your phone system. On a 66 block, there are little metal clips called "bridging clips" between you and the phone company. Lifting these clips off cuts your equipment from the phone company's trunks. This way you can quickly see whose fault it is. Often it's yours. But, if you think it's theirs and they come and find it's yours, they'll bill you. See DEMARCATION POINT or DEMARCATION STRIP.

Demarcation Point The point of a demarcation and/or interconnection between telephone company communications facilities and terminal equipment, protective apparatus, or wiring at a subscriber's premises. Carrier-installed facilities at or constituting the demarcation point consist of a wire or a jack conforming to Subpart F of Part 68 of the FCC Rules.

Demarcation Strip The terminal strip or block (typically a 66 block) which is the physical interface between the phone company's lines and the lines going directly to your own phone system. See also DEMARC.

DEMKO Denmark Elektriske MaterielKOntrol (Denmark Testing Laboratory).

Democratically Synchronized Network A mutually synchronized network in which all clocks in the network are of equal status and exert equal amounts of control on the others.

Demodulation The process of retrieving an electrical signal from a carrier signal or wave. The reverse of modulation. See MODEM.

Demultiplex DEMUX. To separate two or more signals previously combined by compatible multiplexing equipment. As an ATM term, it is a function performed by a layer entity that identifies and separates SDUs from a single connection to more than one connection.

Demultiplexer A device that pulls several streams of data out of a bigger, fatter or faster stream of data.

Demultiplexing A process applied to a multiplex signal for recovering signals combined within it and for restoring the distinct individual channels of the signals.

NEWTON'S TELECOM DICTIONARY

DEMUX Jargon for demultiplexer.
DENet The Danish Ethernet Network which consists of many Ethernet networks in universities connected together by bridges.
Denial Of Service You're no longer allowed to use a service. That service might be anything from normal phone service (you didn't pay your bills) or not being allowed into the company's email because you were just fired.
Denis The Little The 6th century monk who decided that history should be split between B.C. and A.D.
Density 1. The number of bits (or bytes) in a defined length on a magnetic medium. Density describes the amount of data that can be stored.
2. The number of circuits that can be packed into an integrated circuit.
3. In a facsimile system, a measure of the light transmission or reflection properties of an area, expressed by the logarithm of the ratio of incident to transmitted or reflected light flux.
Denwa Japanese for telephone. An aka denwa is a red telephone. Some coin phones in Japan are red and are often known as aka denwas.
Departmental Firewall NEC PrivateNet Systems Group issueed a White Paper called Connecting Safely to the Internet — A study in Proxy-Based Firewall Technology. In that White Paper, they defined an Departmental Firewall: A departmental firewall is identical to an Internet firewall except that it controls access to and from a single department in a larger organization. It is used to protect sensitive corporate data, such as financial information and personnel records, from access by unauthorized people. A departmental firewall tends to be more generous in the access it allows, but if insecure services, such as NFS (Network File System), are allowed through a departmental firewall, the purpose of installing the firewall in the first place might be defeated. See FIREWALL and NFS.
Departure Angle The angle between the axis of the main lobe of an antenna pattern and the horizontal plane at the transmitting antenna.
Depersonalization In 1879 a flu epidemic in Lowell, MA made it likely that all four of the telephone operators would get sick simultaneously. To help substitute operators, management numbered each of the exchange's two hundred plus customers. No problem. The customers accepted the change easily.
Deplaning Getting off a plane. See DETRAINING.
Depolarization 1. In electromagnetic wave propagation, that condition wherein a polarized transmission being transmitted through a nonhomogeneous medium has its polarization reduced or randomized by the effects of the medium being traversed.
2. Prevention of polarization in an electric cell or battery.
Depopulate A technique to reduce the traffic load on a switch by removing devices from the shelf or cabinet. Depopulating reduces the effective device capacity of a switch but can increase switching capacity. This is a ploy used to give older PBX systems traffic capacity nearer true ACD systems.
Depressed Cladding Fiber An optical fiber construction, usually single mode, that has double cladding, the outer cladding having an index of refraction intermediate between the core and the inner cladding.
Deregulation The removal of regulatory authority to control certain activities of telephone companies. An attempt by federal authorities to make the telephone industry more competitive. Deregulation is meant to benefit the consumer. Sometimes it does. Sometimes it doesn't. Often, it's a scapegoat for whatever subsequently goes wrong.
Derivation Equipment Produces narrow band facilities from a wider band facility. Such equipment can, for instance, derive telegraph grade lines from the unused portion of a voice circuit.
DES 1. Data Encryption Standard. An algorithm for encrypting (coding) data designed by the National Bureau of Standards so it is impossible for anyone without the decryption key to get the data back in unscrambled form. The DES standard enciphers and deciphers data using a 64-bit key specified in the Federal Information Processing Standard Publication 46, dated January 15, 1977. DES is not the most advanced system in computer security and there are possible problems with its use. Proprietary encryption schemes are also available. Some of these are more modern and more secure. The quality of your data security is typically a function of how much money you spend. See CLIPPER CHIP and NSA.
2. Destination End Station: An ATM termination point which is the destination for ATM messages of a connection and is used as a reference point for ABR services. See SES.
Descenders Those parts (or tails) of the letters p, y, j and g which descend below the base line. This type style is much easier to read than one in which the tails rest on the base line. Watch out for true (i.e. real) descenders when you're buying a system. Most have them these days, but many telephone screens don't. Careful.
Descrambler A device which corrects a signal (often video) that has been intentionally distorted to prevent unauthorized viewing. Used with satellite TV systems. See DES.
DESI Strip A slang term for Designation Strip, the small printed piece of paper or card that slides into or attaches onto a telephone and tells you which button answers which line or which button does what in the way of features or intercom.
Design Layout Report DLR. A record containing the technical information that describes the facilities and terminations provided by a local telephone company to a long distance telephone company. The technical information is needed by the long distance carrier to design the overall service and includes such items as cable makeup (gauge, loading, length, etc.), carrier channel bank type and system mileage, signaling termination compatibility, etc. The DLR is sent to the designated Carrier representative via the local Telephone Company's Engineering Department.
Designation Strip Also called Desi Strip. A designation Strip is the small printed piece of paper that slides into or attaches onto a telephone and tells you which button answers which line or which button does what in the way of features or intercom.
Deskewing An imaging term. Adjusting — straightening — an image in software to compensate for a crooked scan.
Desktop The computer's working environment. The screen layout, the menu bar, and the program icons associated with the machine's operating environment. Apple's Macintosh (introduced on January 24, 1984) really started the idea that the computer's screen was a desktop. With the introduction of Windows, IBM PCs and clones now also have desktops.

177

NEWTON'S TELECOM DICTIONARY

Desktop Collaboration Using ISDN lines or analog lines with high speed lines, you can link your desktop computers so teleworkers, suppliers and clients can share documents and work together no matter where they are.

Desktop Connection AT&T's code name for hardware which includes an AT&T serial port adapter, DSS cable and a 9 to 25 pin connector and some software to make it work. The new name for desktop connection is PassageWay. It attaches to the back of the AT&T phone. The cable connects the adapter to your PC's serial port. In AT&T's words, the AT&T PassageWay integrates telephone functions with a Microsoft Windows 3.1 or greater application, facilitating outdialing using the Hayes Command Set used by modems, linking caller identification to PC business applications, and paving the way to an open interface between the PC software and the telephone. As a development platform, AT&T PassageWay offers an application layer access to features of the telephone including receiving information on visual/audible alerts and telephone displays, activating button presses, turning the speakerphone on or off, adjusting the volume of the telephone and dialing, including some call state progress. In addition, operating with Windows provides a unique way of accessing and sharing information between applications that co-reside on a PC. Windows provides a Graphic User Interface (GUI) that allows software packages to be easily developed. Windows also offers the Dynamic Data Exchange (DDE) interface between co-resident applications that is a standard for sharing information on DOS PCs. AT&T PassageWay also passes information using Microsoft's DDE standard. See also AT&T APPLICATIONS PARTNER.

Desktop Metaphor A desktop metaphor is the conceptual way a workstation screen area is used to emulate a user's physical desktop through graphic icon images. The icon maps directly to its real life function. For example, a trash can icon will allow a user to "throw out" a document. Gives an application a "user friendly" feel. Desktop metaphors are consistent throughout all Windows and windows-like applications, like Sparc's OPEN LOOK.

Desktop Pattern A Windows 3.1 term. A design that appears across the desktop. You can use Control Panel to create your own pattern or choose a pattern provided by Windows.

Desktop Video Communications that rely either on video phones or personal computers offering a video window.

Desktop Videoconferencing By combining ISDN technology and individual PCs, people can meet "face-to-face" without leaving their offices. It's a way to reduce costly and time-consuming travel, maybe.

Despun Antenna Of a rotating communications satellite, an antenna, the direction of whose main beam with respect to the satellite is continually adjusted so that it illuminates a given area on the surface of the Earth, i.e., the footprint does not move.

Destination Address That part of a message which indicates for whom the message is intended. Usually a collection of characters or bits. Just like putting a destination address on an envelope. On a token ring network this is a 48-bit sequence that uniquely defines the physical name of the computer to which a LAN data packet is being sent. The IEEE assures that in the world of LANs no two devices have the same physical address. It does so by assigning certain numbers to vendors of token ring adapters, the devices that connect computers to a token ring network.

Destination Address Filtering A feature of bridges that allows only those messages intended for the extended LAN to be forwarded.

Destination Code See DESTINATION FIELD.

Destination Document The document into which an object is linked or embedded via OLE. The destination document is sometimes also called the container document.

Destination Field The field in a message header that contains the network address of the individual for whom the message is meant and who will (with luck and good management) receive the message.

Destination Node Those system nodes which receive messages over the control packet network from the source or transmitting node.

Destination Point Code DPC. The part of a routing label that identifies where the SS7 signaling message should be sent.

Destination Service Access Point DSAP. The logical address of the specific service entity to which data must be delivered when it arrives at its destination. This information may be built into the data field of an IEEE 802.3 transmission frame.

Destuffing The controlled deletion of stuffing bits from a stuffed digital signal, to recover the original signal.

Detector 1. In a radio receiver, a circuit or device which converts or rectifies high frequency oscillations into a pulsating direct current or which translates radio frequency power into a form suitable for the operation of an indicator. This is most frequently a vacuum tube, less commonly a crystal. Coherers and delicate chemical rectifiers were used in former years.

2. In an optical communications receiver, a device that converts the received optical signal to another form. Currently, this conversion is from optical to electrical power; however, optical-to-optical techniques are under development.

Detem An opto-electronic transducer that combines the function of an optical detector and emitter in a single device or module. Do not confuse with DTERM, a name for one of NEC's telephones that work on its NEAX 2000 and 2400 PBXs.

Detent Tuner "Click" type of TV tuner.

Detour Difficulty in gaining access to the information highway. Often involves a Highway Construction Supervisor solving your access problem.

Detraining Getting off a train. An absolutely ghastly word invented by the railroad industry to keep them on a par with new, awful language invented by the airline industry. See also DEPLANING.

Device Contention 1. Occurs when more than one application is trying to use the same device, such as a modem or printer. Some of the newer operating systems do a better job handling handling device contention.

2. The way Windows 95 allocates access to peripheral devices, such as a modem or a printer, when more than one application is trying to use the same device.

Device Control A multimedia definition. Device control enables you to control different media devices over the network through software. The media devices include VCRs, laser disc players, video cameras, CD players, and so on. Control capabilities are available on the workstation through a graphical user interface. They are similar to the controls on the device itself, such as play, record, reverse, eject, and fast forward. Device control is important because it enables you to control video and audio remotely — without requiring physical access.

178

Device Driver 1. A special type of software (which may or may not embedded in firmware) that controls devices attached to the computer, such as a printer, a scanner, a voice card, a diskette drive, a CD-ROM, a hard disk, monitor or mouse. Device drivers are typically loaded low into the memory of PCs in the MS-DOS CONFIG.SYS command, or they're loaded high using the Devicehigh command. A device driver is software that expands an operating system's ability to work with peripherals. A device driver controls the software routines that make the peripherals work.
2. A program that enables a specific piece of hardware (device) to communicate with Windows 95. Although a device may be installed on your computer, Windows 95 cannot recognize the device until you have installed and configured the appropriate driver.
Device ID A Plug and Play term. A code in a device's Plug and Play extension that indicates the type of device it is. The device ID and the vendor ID create a unique identifier for each PnP (Plug N Play) device.
DFA Doped Fiber Amplifier.
DFI Digital Facility Interface. An 5ESS switch circuitry in a DTLU responsible for terminating a single digital facility and generating one PIDB.
DFT Direct Facility Termination. A telephone company trunk that terminates directly on one or more telephones.
DG Directorate General (CEC).
DGPS Differential Global Positioning Service — a new venture of the US Coast Guard, which it hopes to have ready by 1996. It will use an existing network of radio beacons throughout the US to create a fixed grid of known reference points in order to improve the accuracy of the Defense Department's GPS signal. The Coast Guard hopes to achieve an accuracy of about 10 meters.
DGT Direccion General de Telecommunicaciones (Spanish General Directorate of Telecommunications).
DHCP Dynamic Host Configuration Protocol. A TCP/IP protocol that Its hots get temporary IP addresses out of a pool for centrally-administered servers. the host runs the DHCP server and the workstation runs the DHCP client.
Dhrystones Benchmark program for testing the speed of a computer. It tests a general mix of instructions. The results in Dhrystones per second are the number of times the program can be executed in one second. The Dhrystone benchmark program is used as a standard indicating aspects of a computer system's performance in areas other than its floating-point performance, for instance, integer processes per second, enumeration, record and pointer manipulation. Since the program does not use any floating-point operations, performs no I/O, and makes no operating system calls, it is most applicable to measuring the performance of systems programming applications. The program was developed in 1984 and was originally written in Ada, although the C and PASCAL versions became more popular by 1989. See WHETSTONES.
DIA/DCA Document Interchange Architecture/Document Content Architecture. IBM promulgated architectures, part of SNA, for transmission and storage of documents over networks, whether text, data, voice or video. Becoming industry standards by default.
Diagnostic Programs Programs run by the computer portion of a PBX to detect faults in the system. Such programs may run automatically at regular intervals or continuously. The goal of diagnostic programs is to detect faults before they become serious and to alert someone — typically the attendant — to do fix it. Some diagnostic programs stop running when the switch gets too busy. Some don't. You can dial into some diagnostic programs from afar. And you can't in some. Remote diagnostic programs are probably the greatest boons to improved reliability of telephone systems.
Diagnostics A term used in the secondary marketplace. Original Equipment's Manufacturer (O.E.M) prescribed test procedure whose successful completion is normally required for maintenance acceptance of a switch, cabinet, or peripheral piece of equipment. Comment: A new maintenance contract will not go into effect until the maintenance company accepts the results of the diagnostics.
Dial A Prayer A sarcastic name for the local 611 number run by the local telephone companies as their centralized number for repair.
Dial-Around A method used by callers to purposely bypass a payphone company's local or long distance carrier services. Such methods include calling cards and alternative carrier's collect services, such as 1-800-COLLECT or 1-800-CALL-ATT. Payphone operators receive little or no revenue from such calls.
Dial Backup 1. A network scheme using dial-up phone lines as a replacement for failed leased data lines. In one typical case, two dial-up lines can be used. One dial-up link is used to transmit data and the other to receive data, thus giving us full-duplex data transmission.
2. A security feature that ensures people do not log into modems that they shouldn't have access to. When a connection is requested, the system checks the username presented for validity, then "dials back" the number associated with that username.
Dial By Name You can dial someone by spelling their name out on the touchtone pad. Typically, the system plays a recorded announcement giving directions for using the Dial by Name feature: the caller then inputs the appropriate digits/letters. When the system recognizes a match, a recorded announcement states the name of the dialed party for confirmation by the caller before automatically completing the call. If the input digits are not uniquely associated with a particular station the system may ask the caller to pick a name from a menu of choices. Dial by name is getting cheaper. Automated attendants are being programmed to have the feature. And you shouldn't buy an auto attendant unless it has this feature.
Dial Call Pickup A phone user on a PBX or hybrid can dial a special code and answer calls ringing on any other phone within his own predefined pickup group.
Dial Dictation Access A service feature available with some switching systems that permits dialing a special number to access centralized dictation equipment.
Dial In Banner An Internet Access Term. Optional pop-up window for dial-in connections. Allows network managers to display information or warning messages when users dial into remote networks.
Dial In Channel Aggregation The ability to use more than one communications channel per connection. By aggregating both 64 kbit/s ISDN B channels, users ers can take advantage of 128 kbit/s dial-in connections. Fast 128 kbit/s data transfer rates reduce large file transfer times. The same as BONDING.
Dial In Tie Trunk A Dial In Tie trunk is a trunk that may be accessed by dialing an access code and then seizing a dedicated transmission path to a distant PBX (or another PBX a short distance away). Once the trunk is seized in the distant PBX, the

caller may then use the features of that PBX, depending on the class-of-service and restrictions assigned to the trunk.

Dial IT 900 Service A special one-way mass calling service that allows prospects, customers and others to reach you from anywhere across the country. In contrast to 800-service, the caller pays the 900 charge, generally one charge for the first minute, with a lesser charge for each additional minute. DIAL-IT 900 Service is a great way to involve your customers and prospects in a promotion! Premium Billing lets you select a rate above standard DIAL-IT 900 rates. The long distance carriers (through their deals with local phone companies) handle the billing. You, the information provider, split the revenues with the long distance provider. International DIAL-IT 900 service is currently available from a growing number of countries.

Dial Level The selection of stations or services associated with a PBX, based on the first digit(s) dialed.

Dial Pick-Up PBX feature. A phone on a PBX can answer another ringing phone by dialing a few digits. Also called an access code.

Dial Pulse Signaling A type of address signaling in which dial pulse is implemented to signal the distant equipment. See DIAL PULSING.

Dial Pulsing A means of signaling consisting of regular momentary interruptions of a direct or alternating current at the sending end in which the number of interruptions corresponds to the value of the digit or character. In short, the old style of rotary dialing. Dial the number "five" and you'll hear five "clicks." See DIAL SPEED, DIAL TRAIN and DTMF.

Dial Repeating Trunks PBX tie trunks used with terminating PBX equipment capable of handling telephone signaling without attendant involvement.

Dial Service Assistance DSA. A service feature associated with the switching center equipment to provide operator services, such as information, intercepting, random conferencing, and precedence calling assistance.

Dial Speed The number of pulses a rotary dial can send in a given period of time, typically 10 per second. A Hayes modem with a communications package, like Crosstalk, can send 20 pulses per second.

Dial String A Dial String is the sequence of characters sent to a device which can dial a phone number. Such a device might be a modem or a voice processing card. Here are some "digits" in a dial string: ! — flashhook (TAPI standard); & — flashhook (Dialogic); T — use tone dialing; , — pause (typically of half a second to two seconds); W — wait for dial tone.

Dial String/Command String A sequence of characters and digits used for dial-in access; ATDT5107861000,,,,,,,,123456, H<CR> for example.

Dial Through A technique, applicable to access circuits, that allows an outgoing routine call to be dialed by the PBX user after the PBX attendant has established the initial connection.

Dial Tone The sound you hear when you pick up a telephone. Dial tone is a signal (350 + 440 Hz) from your local telephone company that it is alive and ready to receive the number you dial. If you have a PBX, dial tone will typically be provided by the PBX. Dial tone does not come from God or the telephone instrument on your desk. It comes from the switch to which your phone is connected to.

Dial Tone Delay The specific time that transpires between a subscriber's going off-hook and the receipt of dial tone from a servicing telephone central office. It's a measure of the time needed to provide dial tone to customers. Many of the local public service commissions in the United States say that 90% of customers should receive dial tone in fewer than three seconds.

Dial Tone First Coin Service A type of pay phone service in which dial tone is received when the caller goes off-hook and coins must be inserted only after the call is connected.

Dial Train The series of pulses or tones sent from the phone that's calling and the switching system it's attached to in order to signify the call's destination.

Dial Up The use of a dial or push button telephone to create a telephone or data call. Dial-up calls are usually billed by time of day, duration of call and distance traveled. A connection to the Internet, or any network, where a modem and a standard telephone are used to make a connection between computers. See DIAL UP LINE.

Dial Up Line A telephone line which is part of the switched nationwide telephone system. Typically a "dial up line" is a standard analog POTS line. These days, ISDN lines are dial up, also. So this definition is changing also.

Dial Up Modem A modem that connects to a remote resource for the duration of an individual call.

Dial Up Networking A Windows 95 definition. It is a service that provides remote networking for telecommuters, mobile workers, and system administrators who monitor and manage servers at multiple branch offices. Users with Dial-Up Networking on a computer running Windows 95 can dial in for remote access to their networks for services such as file and printer sharing, electronic mail, scheduling, and SQL database access.

Dial Zero Phone A telephone on a Northern Telecom Norstar phone system which is assigned to ring when someone dials 0 (zero) from another Norstar telephone.

DIALAN DMS Integrated Access Local Area Network

Dialback Security Dialback security is a telecom security feature. If a person calls in wanting remote access, the system asks for a password. Once it receives a correct password, it hangs up on the caller and dials back a pre-defined remote number, only then giving the caller access. Unless the hacker has you tied up in your living room, it makes things very secure. It can be made even more secure with multiple passwords and features like voice recognition.

Dialed Number Identification Service See DNIS and 800 SERVICE.

Dialed Number Recorder Also called a Pen Register. An instrument that records telephone dial pulses as inked dashes on paper tape. A touchtone decoder performs the same thing for a touchtone telephone.

Dialing Parity This definition from the Telecommunications Act of 1996: The term `dialing parity' means that a person that is not an affiliate of a local exchange carrier is able to provide telecommunications services in such a manner that customers have the ability to route automatically, without the use of any access code, their telecommunications to the telecommunications services provider of the customer's designation from among 2 or more telecommunications services providers (including such local exchange carrier).

Dialing Pattern Dialing pattern refers to the digits you need to dial to place local, long distance, collect calls, or other phone calls. Dialing patterns will vary due to different types of telephone carrier switching equipment, computer software and the host carrier's credit policies (e.g. automatic roaming versus credit card roaming when using a cellular phone).

Dialing Plan A description of the dialing arrangements for customer use on a network.

Dialog Box A dialog box is a temporary window which prompts you to input information or make selections necessary for a task to continue.

Dialogic Dialogic Corporation, Parsippany, NJ, is a leading manufacturer of interactive voice processing equipment, most of it PC-based. They sell equipment through value added resellers, dealers and distributors. Many of their dealers "add value" to the Dialogic components by doing their own specialized software programming, tailoring Dialogic products to particular specialized (and useful) applications. See also AEB. PEB and SCSA.

Dialup Switch An Internet Access Term. Category of switching equipment designed to manages the dialup connections between the PSTN and either the Internet or a corporate LAN internetwork, providing security, accounting, and service management capabilities.

Diaphragm The thin flexible sheet which vibrates in response to sound waves (as in a microphone) or in response to electrical signals (as in a speaker or the receiver of telephone handset).

Dibit A group of two bits which can be represented by a single change of modulation of the carrier signal. On phase modulation, one of four phases in four-phase modulation is used to represent 00, 01, 10 or 11.

Dichroic Filter An optical filter designed to transmit light selectively according to wavelength (most often, a high-pass or low-pass filter).

Dichroic Mirror A mirror designed to reflect light selectively according to wavelength.

Dictation Access and Control A telephone system feature which allows a user to dial a dictation machine and use that machine (giving it instructions by push button) as if it were in his office. Typically, the material on that dictation machine is taken off by one or several typists of a centralized pool and word processed into letters, reports, legal briefs, etc. Telephone suppliers usually don't supply the dictation equipment. Newer telephone dictation machinery is, in reality, a specialized application of voice processing equipment. See VOICE PROCESSING.

Dictation Tank A recording gadget which receives messages dictated through the telephone system. This tank contains tape which can then be transcribed into letters or documents. See DICTATION ACCESS AND CONTROL.

DID Direct Inward Dialing. You can dial inside a company directly without going through the attendant. This feature used to be an exclusive feature of Centrex but it can now be provided by virtually all modern PBXs and some modern hybrids, but you must connect via specially configured DID lines from your local central office. A DID (Direct Inward Dial) trunk is a trunk from the Central office which passes the last two to four digits of the Listed Directory Number to the PBX or hybrid phone system, and the digits may then be used verbatim or modified by phone system programming to be the equivalent of an internal extension. Therefore, an external caller may reach an internal extension by dialing a 7-digit central office number. Notice: DID is different from a DIL (Direct-In-Line) where a standard, both-way central office trunk is programmed to always ring a specific extension or hunt group. DID lines cannot be used for outdial operation, since there is no dialtone offered.

DIEL Advisory committee on telecommunications for DIsabled and Elderly People (UK).

Dielectric A nonconducting or insulating substance which resists passage of electric current, allowing electrostatic induction to act across it, as in the insulating medium between the plates of a condenser. Also an insulating material otherwise used (e.g. a Bakelite panel, or the cambric covering of a wire is a dielectric material). See also SEMICONDUCTOR.

Dielectric Absorption The penetration of a dielectric by the electric strain during a period of time.

Dielectric Constant The ratio of the capacity of a condenser with a given dielectric to the capacity of the same condenser with air as the dielectric.

Dielectric Lens A lens made of dielectric material that refracts radio waves in the same material that an optical lens refracts light waves.

Dielectric Process A printing process that uses a specially treated, charge-sensitive paper. Paper is roller-fed past an electrode array where an electrical charge is applied on line-by-line to form a latent image, then passed through a toner. The toner adheres to the charged image and heat fuses the toner to the paper to create the printed document.

Dielectric Sheath Or Cable A sheath or cable that contains no electrically conducting materials such as metals. Dielectric cables are sometimes used in areas subject to high lightning or electro-magnetic interference. Synonym for nonmetallic cable.

Dielectric Strength The property of material which resists the passage of an electric current. It is measured in terms of voltage required to break down this resistance (such as volts per mil.).

Dielectric Test A test in which a voltage higher than the rated voltage is applied for a specified time to determine the adequacy of the insulation under normal conditions.

Differential Manchester Encoding A digital signaling technique in which there is a transition in the middle of each bit time to provide clocking. The encoding of a zero or one is represented by the presence (absence) of a transition at the beginning of the bit period.

Differential Modulation Modulation in which information is encoded into differences of successive signals. Delta modulation is an example.

Differential Phase-Shift Keying DPSK. A modulation technique in which the relative changes of the carrier signal's phase are coded according to the data to be transmitted. See PHASE SHIFT KEYING.

Differential Positioning Precise measurements of the relative positions of two receivers tracking the same GPS (Global Positioning System) signals.

Diffie-Hellman Key A technique of changing encryption techniques on the fly. In a landmark 1976 paper, called New Directions in Cryptograph, IEEE Transactions on Information Theory, W. Diffie and M. Hellman describe a method by which a secret key can be exchanged using messages that do not need to be kept secret. This type of "public" key management provides a significant cost advantage by eliminating the need for a courier service. In addition, security can be considerably enhanced by permitting more frequent key changes and eliminating the need for any individual to have access to the key's actual value.

Diffraction The deviation of a wavefront from the path predicted by geometric optics when a wavefront is restricted by an opening or an edge. Diffraction is usually most noticeable for openings of the order of a wavelength.

Diffraction Grating An array of fine, parallel, equally spaced reflecting or transmitting lines that mutually enhance the effects of diffraction at the edges of each so as to concentrate the diffracted light very close to a few directions depending on the spacing of the lines and the wavelength of the diffracted light.

Digicash A name for electronic money transmitted in and around the Internet.

Digiline Southwestern Bell's name for ISDN BRI service.

Digipath A Nynex service, Digipath Digital Service provides point-to-point and multi-point full duplex, two-way, private digital data communications at synchronous data rates of 2.4, 4.8, 9.6, 19.2 and 56 Kbps within a New York State Regional Calling Area (RCA).

Digipath Option: Secondary Channel Capability An optional feature for Nynex's 56 Kbps Digipath service. It allows customers a second channel at 2.6 Kbps, independent of the primary channel. Customers usually use this "extra" channel for diagnosing the line; for running network management over the same line; or to transmit a second, lower speed data stream through the same line.

Digirepeater Digital Repeater

Digit Any whole number from 0 to 9.

Digit Deletion It's nice to make it easy for people to dial their desired numbers. Part of making it "nice" is to keep their pattern of dialing consistent. The charm of our ten-digit numbering system in North America — the three digit area code and seven digit local number — is its consistency, making for easy use and easy remembering. Some corporate networks, however, don't use a common numbering scheme. They might use tie trunks to get to Chicago, and insist on the user dialing 69, instead of the more common 312 area code. They might insist on the user dialing 73 when he wants to go to Los Angeles. But if he wants to reach the LA office, he might dial 235. This can be awfully confusing. So some switches — central office and PBXs — have the ability to insert or delete digits. That is, they will recognize the number dialed and change it as it progresses through the network. The user, however, knows nothing of this. He simply dials a normal phone number and listens as his call progresses normally. Digit insertion and digit deletion are components of a PBX feature called common number dialing.

Digit Insertion See DIGIT DELETION.

Digital 1. In displays, the use of digits for direct readout.
2. In telecommunications, in recording or in computing, digital is the use of a binary code to represent information. See PCM (as in Pulse Code Modulation.) Analog signals — like voice or music — are encoded digitally by sampling the voice or music analog signal many times a second and assigning a number to each sample. Recording or transmitting information digitally has two major benefits. First, the signal can be reproduced precisely. In a long telecommunications transmission circuit, the signal will progressively lose its strength and progressively pick up distortions, static and other electrical interference "noises."
In analog transmission, the signal, along with all the garbage it picked up, is simply amplified. In digital transmission, the signal is first regenerated. It's put through a little "Yes-No" question. Is this signal a "one" or a "zero?" The signal is reconstructed (i.e. squared off) to what it was identically. Then it is amplified and sent along its way. So digital transmission is much "cleaner" than analog transmission. The second major benefit of digital is that the electronic circuitry to handle digital is getting cheaper and more powerful. It's the stuff of computers. Analog transmission equipment doesn't lend itself to the technical breakthroughs of recent years in digital. See also PCM, as in Pulse Code Modulation.

Digital Access And Cross-Connect System See DACS and DIGITAL CROSS-CONNECT SYSTEM.

Digital Audio The storage and processing of audio signals digitally. It usually requires at least 16 bits of linear coding to represent each digital sample.

Digital Audio Tape See DAT.

Digital Cash Once there were only stores. To buy something you needed money. Then they invented checks. And storekeepers took them. Then they invented credit cards, which were sort of checks that you paid later. Once we had credit cards, we could invent direct mail catalogs and 800 lines and call centers that took your orders via the 800 lines and you paid for what you bought by giving them your credit card number. Then came the Internet and vendors started to put catalogs on the Internet in the hope that somebody would buy from them. But they needed a way to get paid. Credit cards worked but many journalists wrote about how the Internet was "insecure" and anyone could steal your number and go on a spending binge. No one asked the journalists to cite instances of spending binges. Nor did they ask the journalists about the Federal Government legislation which limits credit card liability in the case of fraud to $50. But scaring people was a good story. Meantime, some entrepreneurs thought there was an opportunity to solve people's paranoia by creating "digital cash." No one exactly knows what digital cash is, yet. Lots of people are working on variations of it. But the idea is that it will be some form of encoded information transfer that contains instructions to take money from one person and pay it to another. We'll see how it evolves.

Digital Cellular The state of the art in cellular communications technology. Implementation will result in substantial increases in capacity (up to 15 times that of analog technology). In addition, digital will virtually eliminate three major problems encountered by users of analog cellular: static, loss/interruption of signal when passing between cells (during handoff), and failure to get a connection because of congested relays. Specifications for TDMA digital systems have been developed in North America (D-AMPS), in Europe (GSM) and in Japan (PDC).

Digital Circuit Multiplication DCM is a variation of analog TASI — Time Assigned Speech Interpolation. In DCM, speech is encoded digitally and advanced voice band coding algorithms are applied to TASI's old speech interpolation techniques. DCM delivers a four to fivefold increase in the effective capacity of normal pulse code modulation (PCM) T-1 links operating at 1.544 megabits per second. DCM equipment is used on the TAT-8 transatlantic optical fiber submarine cable. Most DCM equipment has three operating elements: a speech activity detector. An assignment mapping and message unit, and a speech reconstitution unit. See February, 1987 issue of Data Communications for more.

Digital Command Signal Signal sent by a fax machine or card when the caller is transmitting, which tells the answerer how to receive the fax. Modem speed, image width, image encoding and page length are all included in this frame.

Digital Communications Manager DCM. An MCI monitoring system that maintains communications through the network with the Site Controllers, the Extended Super-frame Monitoring Units, and the I/O DXCs. The DCM issues requests for data and collects alarm and performance information, which is processed and stored in real-time for further computation and display.

Digital Compact Cassette DCC. A digital version of the familiar analog audio cassette. A DCC recorder can play and record both analog and digital cassettes. But the digital ones will sound a lot better.

Digital Convergence A Microsoft term for getting all the digital devices of the office and the home together working in a seamless architecture. See AT WORK.

Digital Cross-Connect System DACS or DCS. A specialized type of high-speed data channel switch. It differs from a normal voice switch, which switches transmission paths in response to dialing instructions. In a digital cross-connect system, you give it separate and specific instructions to connect this line to that. These instructions are given independently of any calls that might flow over the system. This contrasts with a normal voice switching in which switching instructions and conversations go together. Commands to a digital cross-connect system can be given by an operator at a console or can be programmed to switch at certain times. For example, you might want to change the T-1 24-voice conversation circuit to Chicago at 11 A.M. each day to allow for the president's 30 minute video conference call.

Digital Echo Canceller A Digital Echo Canceller is an echo canceller as opposed to an echo suppressor. An echo canceller filters out unwanted echoes among incoming signals, while the echo suppressor shut-offs the entire signal, by using an analog voice switch. The Digital Echo Canceller is one application of a digital transversal filter.

Digital Enveloping Digital enveloping is an application in which someone "seals" a message m in such a way that no one other than the intended recipient, say "Bob," can "open" the sealed message. The typical implementation of digital enveloping involves a secret-key algorithm for encrypting the message (i.e., a content-encryption algorithm) and a public-key algorithm for encrypting the secret key (i.e., a key-encryption algorithm).

Digital Ethernet Local Network Interconnect DELNI. The product offered by Digital Equipment Corp. that allows up to eight active devices to be connected to a single Ethernet transceiver. A similar device is manufactured by many other suppliers under various names. The DELNI can be thought of as "Ethernet in a box."

Digital European Cordless Telecommunication See DECT.

Digital Facilities Management System A Northern Telecom software which integrates the maintenance of all types of digital facilities from T-1 to the high-bit fiber. Largely used by telephone companies.

Digital Facsimile Equipment Facsimile equipment that employs digital techniques to encode the image detected by the scanner. The output signal may be either digital or analog. Examples of digital facsimile equipment are ITU-T Group 3, ITU-T Group 4, STANAG 5000 Type I and STANAG 5000 Type II.

Digital Frequency Modulation The transmission of digital data by frequency modulation of a carrier, as in binary frequency-shift keying.

Digital LAT Protocol The LAT protocol, announced by Digital in the mid '80s, is today one of the industry's most widely used protocols for supporting character terminals over Ethernet networks. See LAT.

Digital Line Protection Many extensions behind PBXs deliver greater voltage to the desk than do normal tip and ring analog lines. This higher voltage can damage a PCMCIA modem inside a laptop. In fact, it can destroy the modem. Newer PCMCIA cards (now called PC cards) have Digital Line Protection, which protects against that higher voltage — what one manufacturer called "innovative isolation circuitry."

Digital Loopback A diagnostic feature on a modem, a short haul microwave or some other digital transmission equipment which allows the user to loop a signal back from one part of the system to another to test the circuit or the equipment. Digital loopbacks can be as long or as short as are necessary to isolate the problem. By looping a signal back and measuring it at both ends of the loop (at the beginning and at the end), you can see if the device carried the message cleanly and is thus, operating correctly.

Digital Microwave A microwave system in which the modulation of the radio frequency carrier is digital. The carrier is still a standard microwave radio wave. The digital modulation may be frequency or phase shift, but the control of that modulation is the digital bit stream.

Digital Modem A term used to describe a piece of hardware that can support digital communications lines (i.e. ISDN) — also called an ISDN terminal adapter.

Digital Monitor Receives discrete binary signals at two levels; one level corresponds to Logic 1 (true) while the other corresponds to Logic O (false). Monitors generally were of this type before VGA models appeared. Digital monitors do not have as wide a range of color choices as analog types; digital EGA monitors, for example, can display just 16 colors out of a palette of 64.

Digital Multiplex Hierarchy An ordered scheme for the combining of digital signals by the repeated application of digital multiplexing. Digital multiplexing schemes may be implemented in many different configurations depending upon the number of channels desired, the signaling system to be used, and the bit rate allowed by the communication medium. Some currently available multiplexers have been designated as D1-, DS-, or M-series, all of which operate at T-carrier rates. Extreme care must be exercised when selecting equipment for a specific system to ensure interoperability, because there are incompatibilities among manufacturers' designs (and various nations' standards).

Digital Multiplexer A device for combining digital signals. Usually implemented by interleaving bits, in rotation, from several digital bit streams either with or without the addition of extra framing, control, or error detection bits. In short, equipment that combines by time division multiplexing several signals into a single composite digital signal.

Digital Network A network in which the information is encoded as a series of ones and zeros rather than as a continuous-

NEWTON'S TELECOM DICTIONARY

ly varying wave — as in traditional analog networks. Digital networks have several major pluses over analog ones. First, they're "cleaner." They have far less noise, static, etc. Second, they're easier to monitor because you can measure them more easily. Third, you can typically pump more digital information down a communications line than you can analog information.

Digital Phase-Locked Loop A phase-locked loop in which the reference signal, the controlled signal, or the controlling signal, or any combination of these, is in digital form.

Digital Phase Modulation The process whereby the instantaneous phase of the modulated wave is shifted between a set of predetermined discrete values in accordance with the significant conditions of the modulating digital signal.

Digital Port Adapter DPA. A device which provides conversion from the RS449/422 interface to the more common interfaces of RS-232-C, v.35, WE-306 and others.

Digital Recording A system of recording by conversion of musical information into a series of pulses that are translated into a binary code intelligible to computer circuits and stored on magnetic tape or magnetic discs. Also called PCM - Pulse Code Modulation.

Digital Reference Signal DRS. A digital reference signal is a sequence of bits that represents a 1004-Hz to 1020-Hz signal.

Digital Selective Calling DSC. A synchronous system developed by the International Radio Consultative Committee (CCIR), used to establish contact with a station or group of stations automatically by radio. The operational and technical characteristics of this system are contained in CCIR Recommendation 493.

Digital Signal A discontinuous signal. One whose state consists of discrete elements, representing very specific information. When viewed on an oscilloscope, a digital signal is "squared." This compares with an analog signal which typically looks more like a sine wave, i.e. curvy. Usually amplitude is represented at discrete time intervals with a digital value.

Digital Signal Processor A specialized digital microprocessor that performs calculations on digitized signals that were originally analog (e.g. voice) and then sends the results on. There are two main advantages of DSPs — first, they have powerful mathematical computational abilities, much more than normal computer microprocessors. DSPs need to have heavy mathematical computation skills because manipulating analog signals requires it. For example, DSPs are often called upon to compress video signals. Each sample must be examined and processed. And all done in very little time. The second advantage of a DSP lies in the programmability of digital microprocessors. Just as digital microprocessors have operating systems, so DSPs are now acquiring their very own operating systems. DSPs are used extensively in telecommunications for tasks such as echo cancellation, call progress monitoring, voice processing and for the compression of voice and video signals. They are also used in devices from fetal monitors, to anti-skid brakes, seismic and vibration sensing gadgets, super-sensitive hearing aids, multimedia presentations and low cost desktop fax machines. DSPs are replacing the dedicated chipsets in modems and fax machines with programmable modules — which, from one minute to another, can become a fax machine, a modem, a teleconferencing device, an answering machine, a voice digitizer and device to store voice on a hard disk, to a proprietary electronic phone. DSPs will do (and are already doing) for the telecom industry what the general purpose microprocessor (e.g. Intel's 80286 or 80386) did for the personal computer industry. DSPs are made by Analog Devices, AT&T, Motorola, NEC and Texas Instruments, among others.

Digital Signature The network equivalent of signing a message so that you cannot deny that you sent it and that the recipient knows it must have come from you. In short, an electronic signature which cannot be forged. It verifies that the document originated from the individual whose signature is attached to it and that it has not been altered since it was signed.

Digital Speech Interpolation DSI. A type of multiplexing. A way of sharing bandwidth among a larger number of users than we really have circuits for. DSI allocates the silent periods in human speech to active users. At least 50% of a voice conversation is always quiet. The technique was originally called TASI (Time Assigned Speech Interpolation). TASI and DSI are lousy for data because they "clip" the first little bit of every new snippet of conversation — unless you hog the channel the whole time by talking incessantly or transmitting continuously. If you pause, you'll get clipping as the system drops you and then reconnects you. Clipping can ruin the meaning of the beginning of data conversations, unless the header knows that TASI or DSI is coming up or the data transmission is following some reasonable protocol and can resend the data. Unfortunately, this slows transmission. DSI and TASI are somewhat similar techniques to STATISTICAL MULTIPLEXING.

Digital Storage Media Command And Control DSMCC. Network protocols specified in Part 6 of MPEG-2 (ISO 13818) standards dealing with user-to-network and user-to-user signaling and communications.

Digital Subscriber Line A fancy name for an ISDN BRI channel. Here's AT&T's definition: "A three-channel digital line that links the ISDN customer's terminal to the telephone company switch with four ordinary copper telephone wires. Operated at the Basic Rate Interface (with two 64-kilobit per second circuit switched channels and one 16-kilobit packet switched channel), the DSL can carry both voice and data signals at the same time, in both directions, as well as the signaling date used for call information and customer data. With the introduction of the AT&T 5E5 generic, up to eight different users can be served by a single DSL." See also ISDN.

Digital Switching A connection in which binary encoded information is routed between an input and an output port by means of time division multiplexing rather than by a dedicated circuit.

Digital To Analog Conversion A circuit that accepts digital signals and converts them into analog signals. A modem typically has such a circuit. It also has other circuits, such as those doing with signaling. See MODEM.

Digital Versatile Video See DVD.

Digital Video Digital video is video recorded and played digitally, i.e. in on-on bits. Traditional analog video — the one we have seen in our homes for eons — is recorded and played back in analog format, i.e. in analog wave forms. Why Digital Video? Due to its versatility, digital video has several advantages over analog video. You can edit it, store it, and transmit it easily. Digital video may be taken from analog source — such as standard over-the-air National Television Systems Committee (NTSC) analog source. Or it me be taken from an analog video camera and a VCR. To convert analog video into digital video typically requires a board inside a PC. Analog video is typically recorded on tape, such as a VCR. Digital video is typically recorded on a hard disk (magnetic or optical) or on a CD-ROM. Two common digital video technologies are Intel's Digital Video Interactive

184

(DVI) and Microsoft's Audio/Video Interleaved (AVI). See ANALOG, DIGITAL VIDEO INTERACTIVE and INDEO VIDEO.

Digital Video Interactive DVI. A compression and playback technology originally developed by RCA's Sarnoff Research Institute and eventually acquired by Intel Corp. DVI is not a compression technique per se but a brand name for a set of processor chips that Intel is developing to compress video onto disk and to de-compress it for playback in real time at the U.S. standard motion video rate of 30 frames per second. The chip set includes both a pixel processor, which performs most of the decompression and also handles special video effects, and a display processor, which performs the rest of the decompression and produces the video output. DVI's greatest long-term advantage, according to Nick Arnett writing in PC Magazine, is that its microprocessors are programmable, so DVI can be adapted to a variety of compression and decompression schemes.

Digital Voice Coding Technology by which linear audio (voice) samples are collected and then compressed using an encoding algorithm. Typically used to store voice data for future decoding.

Digital Wireless Standards

Standards	Japan	US	Europe
Cordless	PHS	PACS	DECT
Cellular	PDS	IS54,IS136,IS95	GSM
PCS	PHS	IS136,IS95 PCS1900 (GSM) PACS Omnipoint BB CDMA	DCS1800 (GSM)
Data	PDC RAM PHS	CDPD (AMPS) RAM, ARDIS IS136,IS95 PACS Omnipoint 802.11 (Wireless Lan) BBCDMA	GSM RAM DCS11800 HyperLan DECT

Digitize Converting an analog or continuous signal into a series of ones and zeros, i.e. into a digital format. See DELTA MODULATION and PULSE CODE MODULATION. See also the Appendix.

Digitized Voice Analog voice signals represented in digital form. There are many ways of digitizing voice. See PULSE CODE MODULATION for the most common.

Digitizer 1. A device that converts an analog signal into a digital representation of that signal. Usually implemented by sampling the analog signal at a regular rate and encoding each sample into a numeric representation of the amplitude value of the sample. 2. A device that converts the position of a point on a surface into digital coordinate data.

Digroup Two groups of 12 digital channels combined to form one single 24-channel T-1 system.

Dijkstra's Algorithm An algorithm that is sometimes used to calculate routes given a link and nodal state topology database.

Dikes A wire-cutter.

DIL Direct-In-Line. A standard, both-way central office trunk is programmed to always ring a specific extension or hunt group within the PBX. This contrasts with Direct Inward Dialing, which allows an external caller to reach an internal extension by dialing a 7-digit central office number. A DID (Direct Inward Dial) trunk is a trunk from the Central office which passes the last two to four digits of the Listed Directory Number into the PBX, thus allowing the PBX to switch the call to and thus ring the correct extension.

Dilution Of Precision The multiplicative factor that modifies the ranging error. It is caused solely by the geometry between the user and his set of GPS (Global Positioning System) satellites. Known as DOP or GDOP.

DIM Document Image Management. The electronic access to and manipulation of documents stored in image format, accomplished through the use of automated methods such as high-powered graphical workstations, sophisticated database management techniques and networking.

Dimension An analog PBX that used PAM techniques, first introduced in the late 1970s by Western Electric (now AT&T Technologies) for AT&T. Now effectively discontinued except for the hotel/motel version. Some claim Archie McGill was responsible for Dimension. Others claim it was Bob Hawke.

Dimmed In Windows, dimmed means unavailable, disabled, or grayed. A dimmed button or command is displayed in light gray instead of black, and it cannot be chosen.

DIMS Document Image Management System.

DIN Deutsche Institut fur Normung (German Institute for Standardization). DIN specifications are issued under the control of the German government. Some are used on a worldwide basis to specify, for example, the dimensions of cable connectors, often called DIN connectors. See also VME.

DINA Distributed Intelligence Network Architecture.

DIOCES Distributed Interoperable and Operable Computing Environments and Systems.

Diode This wonderful explanation comes from George Gilder's book Microcosm: "Named from the Greek words meaning two roads, an ordinary diode is one of the simplest and most useful of tools. It is a tiny block of silicon made positive on one side and negative on the other. At each end it has a terminal or electrode (route for electrons). In the middle of the silicon block, the positive side meets the negative side in an electrically complex zone called a positive-negative, or p-n, junction. Because a diode is positive on one side and negative on the other, it normally conducts current only in one direction. Thus diodes play an indispensable role as rectifiers. That is, they can take alternating current (AC) from your wall and convert it into direct current (DC) to run your computer.

"In this role, diodes demonstrate a prime law of electrons. Negatively charged, electrons flow only toward a positive voltage. They cannot flow back against the grain. Like water pressure, which impels current only in the direction of the pressure, voltage impels electrical current only in the direction of the voltage. To attempt to run current against a voltage is a little like attaching a gushing hose to a running faucet.

"It had long been known, however, that if you apply a strong enough voltage against the grain of a diode, the p-n wall or junction will burst. Under this contrary pressure, or reverse bias, the diode will eventually suffer what is called avalanche breakdown. Negative electrons will overcome the p-n barrier by brute force of numbers and flood "uphill" from the positive side to the negative side. In erasable programmable read-only memories (EPROMs), this effect is used in programming computer chips used to store permanent software, such as the Microsoft operating system in your personal computer (MS-DOS). Avalanche breakdown is also used in Zener diodes to provide a stable source of voltage unaffected by changes in current."

Diode Matrix Ringing A method of connecting a common audible line to a system so that all stations do not ring on all lines. See also MATRIX RINGING.

Diodea A semiconductor device which allows electricity to pass through it in only one direction, restricting flow the other way.

DIP 1. See DIP SWITCH.

2. Document Image Processing. A term for converting paperwork into electronic images manipulable by a computer. Components of include input via scanner, storage on optical media and output via video display terminal, printer, fax, micrographics, etc.

DIP Switch Dual In-Line Package. A teeny tiny switch usually attached to a printed circuit board. It may peek through an opening in a piece of equipment. It may not. It usually requires a ball point pen or small screwdriver to change. There are only two settings — on or off. Or 1 or 0. But printed circuit boards often have many DIP switches. They're used to configure the board in a semipermanent way. The DIP switches are similar to integrated circuit chips which have two rows (dual) of pins in a row (in-line) that fit into holes on a printed circuit board. If something doesn't work when you first install it, check the dip switches first. Then check the cable connecting it to something else.

Diphones Speech segment beginning in the middle of one phoneme and concluding in the middle of another. See PHONEME.

Diplexer A device that permits parallel feeding of one antenna from two transmitters at the same or different frequencies without the transmitters interfering with each other. Diplexers couple transmitter and receiver to the same antenna for use in mobile communications.

Dipole Antenna fed from the center. Name often applied to "rabbit ear" antenna.

DIR An ATM term. This is a field in an RM-cell which indicates the direction of the RM-cell with respect to the data flow with which it is associated. The source sets DIR=0 and the destination sets DIR=1.

Direct Bond An electrical connection using continuous metal-to-metal contact between the things being joined.

Direct Broadcast Satellite DBS. A satellite transmitting TV programs which can be received by small and somewhat inexpensive dish antennas often installed in back yards or on the roofs of houses. Direct broadcast satellites have generated more interest in Europe, which suffers from a lack of TV programs (good and bad), than in the U.S. which has plenty. The back yard satellite antennas you see in America are not receiving signals from Direct Broadcast Satellites. They are receiving them from satellites designed to transmit and receive from large, commercial dishes.

Direct Connect A term describing a customer hooking directly into a long-distance telephone company's switching office, bypassing the local phone company. Such "direct connect" could be via a leased copper pair, a specially-run copper pair, a fiber optic or a private microwave system. See DIRECT ELECTRICAL CONNECTION, which is different. See DIRECT CONNECT MODEM.

Direct Connect Modem A modem connected to telephone lines using a modular plug or wired directly to the outside phone line. It thus transfers electrical signals directly to the phone network without any intermediary protective device. Direct connect modems must be certified by the FCC. Direct connect modems are more much reliable and more accurate than acoustically coupled modems. Virtually all modems these days are directly connected. One day pay phones will even come with RJ-11 jacks into which you can plug the modem of your portable laptop computer.

Direct Connection Connection of terminal equipment to the telephone network by means other than acoustic and/or inductive coupling.

Direct Control Switching The switching path is set up directly through the network by dial pulses without the use of central control. The telex network is an example of direct control switching. A step by step central office also uses direct control switching.

Direct Current DC. A flow of electricity always in the same direction. Contrast with alternate current (AC).

Direct Current Signaling DX. A method whereby the signaling circuit E & M leads use the same cable pair as the voice circuit and no filter is required to separate the control signals from the voice transmission.

Direct Department Calling DDC. A telephone service that routes incoming calls on a specific trunk or group of trunks to specific phones or groups of phones.

Direct Distance Dialing DDD. A telephone service which lets a user dial long distance calls directly to telephones outside the user's local service area without operator assistance.

Direct Electrical Connection A metallic connection between two things. The normal electrical way of connecting two things. This dumb definition is included in this dictionary because there was a time back in the early 1970s and before, when you couldn't (i.e. weren't allowed to) directly electrically connect your own phone or phone system to the nation's phone network. Those were the "good old days" when they (the Bell System) were trying to convince the world that electrically connecting anyone else's phones could harm the network. They never did prove this, and so today we have direct electrical connection of FCC-certified phone equipment. It's certified so it won't cause any harm to the network. See PART 68.

Direct In Line See DIL.

Direct In Termination Incoming calls on a PBX may be programmed to route directly to pre selected telephones without the attendant intervening. DIT features may be assigned to trunk circuits on a day, night or full time basis. Direct In Termination is slightly different from DIRECT INWARD DIALING, though how different depends on whose PBX you're using.

Direct InterLATA Connecting Trunk Groups Those trunk groups used for switched LATA access that interconnect Interexchange Carriers (IXCs) used to connect that Point of Presence (POPs) directly with the Bell Operating Company (BOC) end office switching system.

Direct Inward Dialing DID. The ability for a caller outside a company to call an internal extension without having to pass through an operator or attendant. In large PBX systems, the dialed digits are passed down the line from the CO (central office). The PBX then completes the call. Direct Inward Dialing is often proposed as Centrex's major feature. See also DIRECT INWARD SYSTEM ACCESS (DISA) for another approach to DID.

Direct Inward System Access DISA. This feature of a telephone system allows an outside caller to dial directly into the telephone system and to access all the system's features and facilities. DISA is typically used for making long distance calls from home using the company's less expensive long distance lines, like WATS or tie lines. It's also used for leaving dictation for the typing pool. With DISA, you can dial individual extensions without the aid (or hindrance) of an operator. To use DISA, one must punch in from your touchtone phone a short string of numbers as a password code.

The problem with DISA is that "phone phreakers" (i.e. unauthorized people) often acquire that number or figure it out and run up expensive long distance phone calls. It's best to restrict DISA to trusted people and check the numbers called and bills generated. Changing the password code from time to time can help prevent this. DISA is acquiring a whole new life. It's becoming something called AUTOMATED ATTENDANT. An additional piece of equipment, called an automated attendant, is placed next to the phone system. You dial a special phone number (as you do with DISA). You're answered by a recording that says "Dial the extension you want." In DISA, the response is typically just a tone. An automated attendant is designed to save on operators and speed up outside people getting to talk to your inside people. Automated attendant is being suggested as a lower cost alternative to Centrex.

The following is excerpted from a document Northern Telecom sent to its PBX users. Read it. It's well-done:

PBX features that are vulnerable to unauthorized access include call forwarding, call prompting and call processing features. But the most common ways hackers enter a company's PBX is through Direct Inward System Access (DISA) and voice mail systems. They often search a company's trash for directories or call detail reports that contain 800 numbers and codes. They have also posed as systems administrators and conned employees into telling them PBX authorization codes. More "sophisticated" hackers use personal computers and modems to break into databases containing customer records showing phone numbers and voice mail access codes, or simply dial 800 numbers with the help of sequential number generators and computers until they find one code that gives access to a phone system. Once these thieves have the numbers and codes, they can call into the PBX and place calls out to other locations. In many cases, the PBX is only the first point of entry for such criminals. They can also use the PBX to access the company's data system. Call-sell operators can even hide their activities from law enforcement officials by using "PBX-looping" - using one PBX to place calls out through another switch in another state.

To minimize the vulnerability of the Meridian 1 system to unauthorized access through DISA, the following safeguards are suggested:

1) Assign restricted Class of Service, TGAR and NCOS to the DISA DN. 2) Require users to enter a security code upon reaching the DISA DN. 3) In addition to a security code, require users to enter an authorization code. The calling privileges provided will be associated with the specific authorization code. 4) Use Call Detail Recording (CDR) to identify calling activity associated with individual authorization codes. As a further precaution, you may choose to limit printed copies of these records. 5) Change security codes frequently. 6) Limit access to administration of authorization codes to a few, carefully selected employees.

Direct Line Terminations The term refers to central office/PBX lines which terminate directly on telephones, and are generally common to all instruments within the system. In a square configuration on a Key Telephone System, these lines must appear at the same button location on each phone.

Direct Memory Access DMA is a technique in which on adapter bypasses a computer's CPU, and handles the transfer of data between itself and the system's memory directly.

Direct Outward Dialing DOD. The ability to dial directly from an extension without having to go through an operator or attendant. In PBX and hybrid phone systems, you dial 9, listen for a dial tone, and then dial the number you want to reach. In some phone systems, you don't have to listen for the second dial tone. You can dial straight through. All phone systems now have DOD. The older ones didn't, especially cordboard PBXs. Some Club Meds and lots of cheap hotels (especially the ones Harry — the editor — stays in) do not have DOD.

Direct Set An ATM term. A set of host interfaces which can establish direct layer two communications for unicast (not needed in MPOA).

Direct Station Select DSS. A piece of key system equipment usually attached to an operator's phone set. When the operator needs to call a particular extension he/she simply touches the corresponding button on the Direct Station Select equipment. Typically DSS equipment/feature is part of a Busy Lamp Field (BLF), which shows with lights what's happening at each extension. Is it busy? Is it on hold? Is it ringing? See BUSY LAMP FIELD.

Direct Station Select Intercom DSS. An interoffice caller can punch one button on his or her phone and dial his desired person, instead of dialing the full intercom number. Direct station select is like having an auto dial or speed dial button for everyone in the office. DSS saves time, but adds more buttons to the phone — one button for each extension the user wants to dial.

Direct Termination Overflow DTO. An optional MCI Vnet and 800 Service feature, which allows a call to "overflow" to shared lines for completion by the local telephone company if the dedicated line is busy.

Direct Trunk A trunk between two class 5 central offices.

Direct Trunk Access A PBX feature. By dialing some digits, the attendant can directly access any specific trunk. You'd do this if you want to check the trunk for problems, etc.

Direct Trunk Select Permits you, the user, the attendant, or an attached computer telephony system to access an individual outgoing trunk instead of one chosen by the PBX from a group of trunks. You may want to grab a special trunk to get access to a specially conditioned data line, for example. Direct trunk select is particularly important in testing computer telephony systems and facilities as it is usually the best method to use to address specific ports on a VRU (Voice Response Unit) or switching system.

Directed Call Pickup An extension user on a phone system user can answer calls — ringing or holding — on any other phone by dialing a unique answer code. If the call has already been answered by the called phone, the user who dials the answer code will join the connection in conference. Some tones will alert the conversing parties to the intrusion.

Directional Antenna An antenna which impels electrical waves with more energy in one direction than in another, or which receives electrical waves more readily from one direction than from another.

Directional Coupler 1. A device put in a microwave system's waveguide to couple a transmitter and receiver to the same antenna.
2. A transmission coupling device for separately sampling (through a known coupling loss) either the forward (incident) or the backward (reflected) wave in a transmission line. A directional coupler may be used to sample either a forward or backward wave in a transmission line. A unidirectional coupler has available terminals or connections for sampling only one direction of transmission; a bidirectional coupler has available terminals for sampling both directions. For optical fiber applications.

Directories Are places within a hard disk volume where you can store files or subdirectories. The term subdirectory is relative. A directory is a subdirectory only in relation to the directory above it. To a directory below it, the same directory is a parent directory.

Directory 1. A list of all the files on a floppy diskette or hard disk. A directory may also contain other information such as the size of the files and the amount of free space remaining.
2. Also a telephone directory.

Directory Assistance DA. Formerly known as "Information", but changed because they were getting too many stupid questions. DA is provided by the local telephone company. It allows the subscriber to call for phone numbers. In most states, the local phone company charges for this service, or will begin to very shortly. AT&T and MCI charge for long distance directory assistance. Most local phone companies will give you the person's address as well as his phone number if you ask for it.

Directory Caching A method of decreasing the time it takes to find a file's location on a PC's disk. The FAT (File Allocation Table) and directory entry table are written into the file server's memory. The area holding all directory entries is called the directory cache. The file server can find a file's address (from directory cache) and the file data (from the file cache) much faster than retrieving the information from disk.

Directory Information Base DIB. Made up of information about objects. The collective information held in the directory. An X.500 term.

Directory Information Tree DIT. Information that outlines the structure of an X.500 directory.

Directory Name As defined in the X.500 Recommendations, the Directory Name is an ITU term for the name of a directory entry. For example, used to retrieve an O/R (Originator/Recipient) Name for message submission.

Directory Number 1. The full complement of digits associated with the name of a subscriber in a telephone directory. This is a very long way of saying the obvious, namely your phone number.
2. A unique phone number which is automatically assigned to each telephone or Data Terminal during System Startup. The DN, also referred to as an intercom number, is often used to identify a telephone when settings are assigned during programming.

Directory Package The process of adding, deleting and moving people attached to PBX or Centrex phones is more than simply programming in new extensions. Or should be. First, there are the changes necessary for the call accounting system. Second, there are changes necessary for "The Corporate Phone Directory". In the past, the directory bore little relation to the telephone system and it was often months, and sometimes years behind the actual phone system. Now some phone systems are incorporating various Directory Software Packages into the PBXs.
Some features included in these systems are the ability to dial someone by name — both for the attendant and for the individual phone user, i.e. dial HARRY, or HAR, instead of 3245. Some also include the ability to interface to the call accounting system. So that who's in what corporate department corresponds to which department's bill. There's also that important thing called a Telephone Directory. It would be useful if you could hook a laser printer to the telephone system and tell it to print, in neat, photo-ready columns, an alphabetical (by last name), departmental or any other sorted telephone directory. Some of the newer PBXs have "directory package" features which include some or all of the above. Most users find today's necessity of at least three different systems to be a pain in the behind. Rightly so. The three different systems are CDR, phone directory and extension dialing in the PBX. The three are often out of synch.

Directory Replication A Windows NT term. The copying of a master set of directories from a server (called an export server) to specified servers or workstations (called an import computers) in the same or other domains. Replication simplifies the task of maintaining identical sets of directories and files on multiple computers, because only a single master copy of the data must be maintained. Files are replicated when they are added to an exported directory and every time a change is saved to the file.

Directory Replicator Service A Windows NT term. Replicates directories, and the files in those directories, between computers. See DIRECTORY REPLICATION.

Directory Service 1. A simple term for the information service which the telephone company runs on 411 or 555-1212.
2. A computer networking term. The facility within networking software that provides information on resource available on the network, including files, users, printers, data sources, applications, and so on. The directory service provides users with access to resources and information on extended networks.

Directory Services A service that provides information about network objects. DNS (Domain Name System) provides node address information. An X.500 Directory service provides any appropriate information an enterprise wishes to include in the

NEWTON'S TELECOM DICTIONARY

X.500 directory itself.

Directory Synchronization The reconciliation of user directories from two electronic mail post offices. Many gateways and messaging switches have software to automate reconciling these directories.

Directory Tree A list of directories. A directory tree looks like an organizational chart and shows how your directories and subdirectories are related. Our favorite program that shows the directory tree in your hard disk is called CTREE. You can pick it up from many electronic bulletin boards for free, including from TELECONNECT Magazine's on 212-989-4675.

DirectTalk DirectTalk is IBM's family of voice processing products, introduced in 1991. According to IBM, its new CallPath DirectTalk product line lets business automate routine operations and also provide callers with easy access to many kinds of information over the telephone. There are two versions of DirectTalk: DirectTalk/2 which requires OS/2 Version 2, and runs on any OS/2 certified Microchannel (MCA) or Industry Standard Architecture (ISA) bus personal computer; DirectTalk/6000 which requires AIX and runs on any IBM RISC System/6000. DirectTalk has features, including:
Application Developers tools and aids, 3270 and 5250 terminal emulation support, Voice recognition, Text to speech, Telecommunications Devices for the Deaf (TDD), Analog/T1/E1 support, Dynamic invocation of multiple applications, Voice messaging, Integrated Voice mail support (DirectTalk/6000 only), and integration with CallPath CallCoordinator. DirectTalk/2 ranges from 2 to 48 voice ports with voice storage capacity in excess of 100 hours. DirectTalk/6000 ranges from 12 to 96 voice port with voice storage capacity over 144,000 hours. See also CALLPATH SERVICES ARCHITECTURE.

DIRTBAGS Digitally Initiated Resale of Telecommunications Bypass Applications by Scumbags. A term created by Ron Adams of Marylhurst College in Washington. There is some confusion between another definition — Digitally Initiated Reorigination of Telecommunications Bypass Access Generated by Scumbags. It is ascribed to Karen Corcoran of MCI's Atlanta office to describe as a group the various hackers, phreakers, and others who invade our networks and/or steal long distance service.

Dirty Power Dirty power typically refers to alternating current that is not a perfect sine wave and not perfectly 120 volt. There are all sorts of ways electricity can be made "dirty." It can be affected by spikes. Spikes are transient impulses (sometimes called glitches) of relatively high amplitude but very short duration. Spikes so short that a very high-speed oscilloscope is needed to observe them can often cause problems. Many spikes can occur in a fraction of a cycle.
Power can also be affected by sags and surges. Sags and surges are rapid changes in the amplitude of an AC voltage. These are generally caused by abrupt changes in the load on a power source or circuit (such as when an air conditioner starts up), and can range from a fraction of a cycle to several complete cycles. Power can also be delivered consistently beyond its rating. In New York Con Edison guarantees 120 volts plus or minus 10%. Any level below 108 volts or above 132 volts Con Edison would consider dirty power. These are called low or high average variations. And they occur when the average voltage is above or below a desired level for significant periods of time, usually measured in seconds, minutes or longer.
Other kinds of dirty power include: Blackouts or brownouts. They occur when the power is switched off or lost completely (blackout), or when the voltage feeding a load is deliberately or inadvertently reduced significantly for a sustained period (brownout). Common mode noise is a small (+1V-2V) signal that appears between a neutral line and ground (earth) where there should be no signal. High/low frequency variations occur when the instantaneous frequency of an AC power source differs from its normal frequency, e.g., 60 Hz, by 0.5 Hz or more. Phase angle variations can be observed in three-phase systems whenever the phase relationships vary from their normal 120-degrees.

DIS Draft International Standard. As specified by ISO, a development step representing near final status on a specification. Once a specification has reached DIS status, companies are encouraged to develop actual products based on it.

DISA 1. Direct Inward System Access. DISA is a way of dialing into a phone system. It can be used to access one's voice mail, often over 800 lines. It can be used to make long distance calls, often on the theory that it's cheaper for employees to dial over the company's leased line phone network than to dial direct from their homes. DISA has been the major way crooks have dialed into PBXs and stolen toll calls. See DIRECT INWARD SYSTEM ACCESS.
2. Data Interchange Standards Association, Inc. DISA is the Secretariat and administrative arm of the Accredited Standards Committee (ASC) X12 which has responsibility for developing Electronic Data Interchange (EDI) standards. DISA OnLine is an electronic messaging and information system designed for use by DISA's member constituency. www.disa.org

Disable You figured this one. It means to prevent a hardware device from working. Unplugging it is the easiest way. It also refers to a tone or other signal which you send over a phone line to disable the equipment at the other end.

Disaster Recovery A generic term for all the tools and planning you need to bring your telecommunications facilities back to where they were before the disaster hit you.

Disc An older method of spelling DISK, as in Floppy DISK. Disc (spelled with a C) is now more commonly used to refer to optical storage devices, like CD (Compact Discs) and MO (Magneto Optical) discs. See also MS-DOS.

Discharge Block A protective device through which unwanted voltages discharge to ground.

Disco Tech Slang expression for a technician who only handles disconnections.

Disconnect 1. The breaking or release of a circuit connecting two telephones or data devices.
2. The occasional April Fools issue of TELECONNECT Magazine, New York City. Sometimes funny. Sometimes not.

Disconnect Frame Indicates in a fax call that the call is done. The sending fax machine sends the disconnect frame before hanging it. It does not wait for a response.

Disconnect Signal The signal sent from one end to indicate to the other to shut down the connection.

Disconnect Supervision The change in electrical state from off-hook to on-hook. This indicates that the transmission connection is no longer needed.

Disconnect Switch In a power system, a switch used for closing, opening, or changing the connections in a circuit or system or for purposes of isolation. It has no interrupting rating and is intended to be operated only after the circuit has been opened by some other means, such as by a circuit breaker or variable transformer.

Discontinuity An interruption or drop out of the optical signal.

Discounted Payback Period The number of years in which a stream of cash flows, discounted at an organization's cost of money, repays an initial investment.

Discounting The process of computing the present worth of a future cash flow by reducing it by a factor equivalent to the organization's cost of money (or some other measure of the value of money as measured by an interest rate) and the time until the cash flow occurs.

Discrete In voice recognition, refers to an isolated word. A discrete word is preceded and followed by silence, hence isolated in speech. Discrete words need to be separated by about half a second of silence when spoken to a discrete recognizer.

Discrete Cosine Transform DCT. A pixel-block based process of formatting video data where it is converted from a three dimensional form to a two dimensional form suitable for further compression. In the process the average luminance of each block or tile is evaluated using the DC coefficient. Used in the ITU-T's Px64 videoconferencing compression standard and in the ISO/ITU-T's MPEG and JPEG image compression recommendations.

Discrete Multi-Tone See DMT.

Discretionary Preview Dialing A single button dialing technique where the agent initiates a call with a single key stroke. Often used in association with a CRT tied to a database. Upon hitting a single button the system selects the phone number field from the screen and dials the number. Contrast this with Forced Preview Dialing. When the call ends, the computer brings up the next screen and starts dialing the call without the agent helping or hindering.

Discriminating Ringing See DISTINCTIVE RINGING.

Discussion A method of confirming others in their errors.

Disengagement Denial Disengagement failure due to excessive delay by the telecommunication system.

Disengagement Failure Failure of a disengagement attempt to return a communication system to the idle state, for a given user, within a specified maximum disengagement time.

Dish Typically a parabolic microwave antenna — used for receiving line-of-sight terrestrial signals or signals from satellites.

Disintermediation The decline of middlemen companies that today operate between the buyer and maker of goods. Pundits predict this will happen with the rise of commerce on the Internet.

Disk A piece of plastic or metal upon which a coating has been applied and which can thus, record computer information magnetically. The present convention is that a "disk" with a K refers to magnetic storage, while "disc" with a C refers to optical storage. See also MS-DOS.

Disk Array A disk subsystem combined with management software which controls the operation of the physical disks and presents them as one or more virtual disks to the host computer.

Disk Cache On a PC, a disk cache is the part of RAM that is set aside to temporarily hold data read from disk. A disk cache doesn't have to hold an entire file, as a RAM disk does, but can hold parts of running application software or parts of a data file. Disk-caching software manages the process of swapping data to and from the disk cache. See DISK CACHING.

Disk Caching A technique used to speed up processing. Each time your application retrieves data from the disk, a special program, called a disk caching program, stores data read from the disk in an area of RAM. When the application next requests more data, some of it may already be in RAM, thereby dramatically speeding the retrieval of data. See DISK CACHE and CACHING.

Disk Controller A hardware device that controls how data is written to and retrieved from the disk drive. The disk controller sends signals to the disk drive's logic board to regulate the movement of the head as it reads data from or writes data to the disk.

Disk Dancers Teenagers who uses the AOL disks given away in magazines and via direct mail to hop from one free account to another.

Disk Drive A device containing motors, electronics and other gadgetry for storing (writing) and retrieving (reading) data on a disk. See DISK DUPLEXING, DISK MIRRORING, DISK OPERATING SYSTEM and other definitions below starting with the work "Disk." See also IDE and ENHANCED IDE.

Disk Drive Performance Three basic things affect your perception of the speed of your hard disk drive: 1. The disk transfer rate — the speed at which data can be read from or written to the drive's media. This speed is governed mainly by drive mechanics — especially rotational speed, latency and seek time. 2. The controller transfer rate — the speed at which the drive's controller electronics can move data across the interface. This is governed by the design of the drive controller. 3. The host transfer rate — the speed at which the computer can transfer data across the interface. This is a matter of both CPU and bus speeds as well as the hardware that provides the bridge between the host bus and the disk interface.

Disk Duplexing A method of failsafe protection, used on file servers on local area networks. Disk duplexing involves copying data onto two duplicate hard disks, each using a separate controllers and a separate disk channel. Disk duplexing protects data against the failure of a hard disk or of the hard disk channel between the disk and the file server. The hard disk "channel" includes the disk controller and interface cable. If any component on one channel fails, the other channel continues to operate normally. (You hope.) The operating system sends a warning message to the workstations to indicate the failure. It's a good idea then to fix the problem fast. See also DISK MIRRORING.

Disk Mirroring A technique for protecting the information on your hard disk. Disk mirroring writes data simultaneously to two identical hard disks using the same hard disk controller. Here's how it works: You have a special hard disk controller card. That's the card which organizes getting information into and off your hard disk. When you come to write information to your hard disk, your hard disk controller writes first to the first hard disk, called the primary hard disk. The controller retains that information in its memory and then writes it to the second hard disk. This causes a 50% degradation in performance since it now takes twice as long to write to the disk. When the controller comes to read, it reads only from the primary disk. Thus there is no performance degradation in reading. Mirroring is designed to protect against mechanical problems with one of your hard disks. If one of the hard disks break, the other one will take over instantly. You will get a warning message. You will be told to repair the broken disk and you will be told to designate the other disk now as your primary disk (it may already be). That primary disk will now become your bootable disk, the one you boot your computer from. Mirroring does not protect against viruses, or corrupt data or losing data. Any idiocy you can perform on one hard disk you can now happily perform on two. Mirroring does not protect you against a lightning strike which could knock out both your hard disks. Mirroring does not protect against the loss of a controller card since you're only using one. Another protection technique called disk duplexing uses two separate controllers to drive two separate hard disks. See also DISK DUPLEXING.

Disk Operating System See DOS.

Disk Pack A series of disks mounted horizontally and arranged as a single unit. A disk pack contains more space for storing and retrieving information than one single disk.

Disk Sector Magnetic diskettes are typically divided into tracks, each of which contains a number of sectors. A sector typically contains a predetermined amount of data, such as 256 bytes.

Disk Server A device equipped with disks and a program that permits users to create and store files on the disks. Each user has access to their own section of the disk. It gives users disk space which they would not normally have at their own personal computers. The disk server is linked to the PCs via a LAN. The next level of sophistication would be a file server, which would allow users to share files.

Disk Striping Writing data in stripes across a volume that has been created form areas of free space on from 2 to 32 disks. A Windows NT definition.

Disk Subsystem A collection of physical disks and the hardware required to connect them to a host computer or network.

Disk/File Server A mass storage device that can be accessed by several computers, usually through a local area network (LAN).

Diskless PC Just what it says: a PC without a disk drive. Used on a LAN, a diskless PC runs by booting DOS from the file server. It does this via a read-only memory chip on its network interface card called a remote boot ROM. Diskless PCs are cheaper than PCs with disks, they're more compact and they offer better security since users can't make off with floppy disks of important and sensitive data or add their own virus-ridden programs to file servers. Diskless PCs appeal primarily to users interested in security. One system of diskless PCs allow the system operator to disable and physically lock the machine's various ports and the computer case itself. Should something go wrong with the machine, the ports can be restored to operating condition by letting the system allow a technician to attach a laptop computer and run diagnostic programs.

Diskless Workstation See DISKLESS PC.

DISN Defense Information Systems Network.

DISOSS IBM's Distributed Office Support System.

Dispatch A radio communications technique where one communicates to many through short bursts of communication. Users of dispatch services include taxis, trucking companies and service personnel.

Dispersion A term used to describe how an electromagnetic signal is distorted because the various frequency components of that signal have different propagation characteristics. Dispersion is the degree of scattering taking place in the light beam as it travels along the fiber optic. Or the overlapping of a light signal on one wavelength to different wavelengths because of reflected rays and the different refractive index of the core fiber material.

Dispersion-Shifted Single-Mode Fibers These types of fibers have a different internal configuration. This changes the zero total chromatic dispersion point to 1550nm. This is important because the attenuation at this wavelength is only about half as much as at 1300nm.

Display The visual presentation of information, usually on a TV-like screen or an array of illuminated digits.

Display Driver A piece of software which translate instructions from the software you are running into thousands of colored dots, or pixels, that appear on your video monitor. A display driver is also called a Video Driver. Symptoms of a video driver giving trouble can range from colors that don't look right, to horizontal flashing lines to simply a black screen. In the Macintosh world, Apple rigidly defined video drivers. Windows, in contrast, is a free-for-all. Windows 3.1 defined the lowest common denominator of displays — namely 16 colors at 640 x 480 pixels. But most multimedia programs and many games won't run with only 16 colors. They require at least 256 colors.

Distance Learning Video and audio technologies into education so students can attend classes in a location distant from where the course is being presented.

Distance Sensitive Pricing Product pricing based on the distance (airline mileage) between the originating and terminating locations of a call/data transmission.

Distance Vector An approach, or algorithm, used by network equipment in selecting the best available network path by calculating the total distance over which a packet would travel on each alternative route. The shortest distance is usually the most preferred, however, it is possible to consider other factors in the decision as well.

Distant Learning A Pacific Bell term for students sitting in front of TVs and phones and participating in classes that are being held and delivered elsewhere. In one of PacBell's trials, they used a T-1 signal, so the distant lecturer could see and hear his distant students using full-color video.

Distinctive Dial Tones In some phone systems, dial tones sound different. An internal dial tone sounds different to an external dial tone. The logical reason for this is simply to alert the user as to whether he or she is making an intercom or an outside local or long distance call.

Distinctive Ringing Enables telephone users to distinguish types of incoming calls such as outside calls and inside (intercom) calls by the tone of the ringing.

Distort To change some characteristic of a signal during its transmission. See DISTORTION.

Distortion 1. The difference in values between two measurements of a signal — for example, between the transmitted and received signal. "Distortion" typically refers to analog signals.

2. In imaging, distortion is any deformation of the on-screen image. Two common types of distortion are pincushion and barrel.

Distributed Capacity The capacity in a coil due to the proximity of the turns.

Distributed Common Control There are two elements of telephone switching: The switching itself and the control of that switching. The earliest step-by-step telephone switches had their "Control" built into them. The dialing information at the beginning of the call physically moved switches. You could say, as a result, that control was distributed throughout the switching system. Then came the 1940s and crossbar exchanges, and the economics pointed to centralizing control. Then came computerized or stored program control (SPC) switches in which large computers were used centrally to perform virtually all the functions of the erstwhile electromechanical senders, registers, markers, etc. — those things which affect the setting up and tearing down of the call. As computers got smaller and as microprocessors appeared (the so-called computer on a chip), it became economical and efficient to place inexpensive microprocessors in the telephone circuits themselves, in essence getting much of the processing done before it hits the central processing unit. Increasingly, as special microprocessors (so-called "computers on a chip") for telecommunications evolve, we will see more and more of the processing being distributed to further and further away from the central point and closer and closer to the originating telephone instrument. It will be rare in coming years for telephones to come without microprocessors. One day, each phone will have its own switch and the rest of the system will just be one gigantic loop of cable — not unlike today's local area networks.

Distributed Computing Environment DCE. A new idea from Digital Equipment Corporation. Distributed Computing Environment (DCE) is a comprehensive integrated set of services that supports the development, use, and maintenance of distributed applications. Digital's DCE is an implementation of the Open Software Foundation's DCE (OSF DCE). In response to OSF's request for distributed computing technology, Digital submitted for consideration four of Digital's established distributed computing technologies:
Remote Procedure Call (RPC), a joint effort with HP/Apollo; Threads Service, based on Digital DECthreads; Cell Directory Service (CDS), based on the Digital Distributed Name Service (DECdns); Distributed Time Service (DTS), based on the Digital Distributed Time Service (DECdts).

Distributed Data Processing DDP. A data processing arrangement in which the computers are decentralized — i.e. scattered in various places. Hence, processing occurs in a number of distributed locations and only semi-processed information is communicated on data communications lines from remote points to the central computers. The object of DDP is to save telecommunications charges and to improve network response time.

Distributed Environment Refers to a network environment, or topology, in which decision making, file storage and other network functions are not centralized but instead are found through the network. This type of environment is typical for client-server applications and peer-to-peer architectures.

Distributed File System A type of file system in which the file system itself manages and transparently locates pieces of information from remote files and distributes files across a network. It can recognize multiple servers and be accessed independently of where it physically resides on the network.

Distributed Management Environment A compilation of technologies now being selected by the Open Software Foundation to create a unified network and systems management framework, as well as applications. Those technologies will complement OSF's own Unix implementation, OSF/1, as well as other operating systems.

Distributed Microprocessor Common Control This means that the system employs many individual microprocessors to control system and telephone phone functions. The microprocessors may be located in central processing equipment or in the telephones themselves.

Distributed Name Service A technique for storing network node names so that the information is stored throughout the network (either one LAN or many joined together), and can be requested from, and supplied by, any node.

Distributed Network Service Introduced in March 1991, AT&T's Distributed Network Service was designed expressly for the switchless resale community unlike SDN. It allows resellers to purchase large volumes of services and receive progressive discounts on all direct dial domestic and international calls. Resellers may designate any number of locations to participate in the plan with the flexibility of adding locations.

Distributed Nodes PBX and its "slave" switches which are physically in separate buildings, in separate areas of the campus, in separate parts of the town.

Distributed Processing A network of computers such that the processing of information is initiated in local computers, and the resultant data is sent to a central computer for further processing with the data from other local systems. The term also covers computing jobs "farmed out" from a central site to remote processors where faster processing or specialized databases are available. Distributed Processing is often a more efficient use of computer processing power since each CPU can be devoted to a certain task. A LAN is the perfect example of distributed processing. See also DISTRIBUTED DATA PROCESSING.

Distributed Queue Dual Bus DQDB. A connectionless packet-switched protocol, normally residing in the Medium-Access Control sublayer of the data link layer. Definition from Bellcore in reference to Advanced Intelligent Network.

Distributed Switching When electronics and computers were expensive it made sense to centralize them and run individual lines out for miles to subscribers. Then the economies changed. Electronics and computers became cheaper and running phone lines for miles became very expensive. So switching companies started building small switches which they could put closer to subscribers. Thus, individual local loops would be shorter and the long lines going back to the larger central office would be more efficiently used — namely by more people. The remote, or distributed switches, are called everything from remote switches to slave switches (because they slave off the main one which is distant). Usually these remote switches are unattended.

Distribution Refers to the arrangement of premises wiring runs and their associated hardware required to implement the planned customer premises wiring system extending from the Network Interface Jack to each Communications Outlet.

Distribution Cable Part of the outside cable plant connecting feeder or subfeeder cables to drop wires or buried service wires that connect to the customer's premises. In simpler language, it's the cable from the serving area interface — a box on a pole, in the ground, etc. — to the lightning protection at the entrance to the customer's premises. See also FEEDER PLANT and DROP WIRE.

Distribution Cable, Inside Plant Cables usually running horizontally from a closet on a given floor within a building. Distribution cables may be under carpet, simplex, duplex, quad, or higher fiber count cables.

Distribution Cable, Outside Plant The cable running from a central office or remote terminal to the side of a subscriber's lot.

Distribution Frame Cables coming in from thousands of subscribers need to connect to the correct ports on a central office. Similarly, cables coming in from many PBX extensions need to connect to the PBX. The cables could be directly wired to the CO or to the PBX. This would be inflexible. It would make future moves and changes a nightmare. So the solution is something called a Distribution Frame. Basically it's a giant wire connecting devices made of metal. There are no electronics in it whatsoever. On one side we punch down the wires coming in from the outside world. On the other side, we punch down the wires coming in from the CO or PBX. Both sides are connected with wire that's called "jumper" wire. By pulling off one end of the jumper wire and moving it to another location we can quickly change phone numbers, add or subtract cabling (one, two or three pairs for normal or electronic phones, etc.). In big central offices, distribution frames can span whole city blocks and the "jumper" wires can be several hundred yards long. Designing distribution frames and their layout in advance is critical, otherwise it becomes a mess and tracing where jumper wires go becomes an enormously time consuming job.

Distribution Group 1. A group made of phone extensions on a PBX arranged to share the load. In the Rolm PBX, each group is assigned a dummy extension number called a pilot number.
2. A group of telephone extensions on an automatic call distributor (ACD). The ACD answers the incoming calls then checks to see if any agents' phones are free. If none are free, it delivers the caller a message and then puts the caller on hold. Which line the call has come in on may determine which group of agents should handle that call. They would be called a Distribution Group. Once the call is released from hold, it may be sent to a member of that Distribution Group following some pre-determined mathematical formula — for example, so that everyone's workload is kept constant, or a group of people are kept busy.

Distribution Service In ISDN applications, a telecommunications service that allows one-way of information from one point in the network to other points in the network with or without user individual presentation control.

Distribution Voltage Drop The voltage drop between any two defined points of interest in a power distribution system.

Distributor A company with a contractual relationship with a manufacturer to buy equipment at a preset price. The manufacturer provides training, advertising and warranty support. Often called an authorized dealer, although a dealer may be one step lower in the distribution chain. A distributor is often used as a generic term for any supplier. Therefore you should clarify whether a distributor is an authorized distributor.

DIT Directory Information Tree. The global tree of entries corresponding to information objects in the OSI X.500 Directory.

Dithering Dithering is an imaging term with at least two meanings. One meaning that it's the processing of an image containing more colors than a system can handle to an image containing exactly the right number of colors that the system can handle. For example, some of the color images on my laptop contain 16 million colors. But my laptop (the way I have it set up) will only handle 256 colors. If I ask my image display software to display that image, it will "dither" it to 256 colors. This means it will give its best shot guess at what the image should look like.
In another meaning, dithering is patterning black and white dots to approximate shades of grey on a scanned image.

Diurnal Phase Shift The phase shift of electromagnetic signals associated with daily changes in the ionosphere. The major changes usually occur during the period of time when sunrise or sunset is present at critical points along the path. Significant phase shifts may occur on paths wherein a reflection area of the path is subject to a large tidal range. In cable systems, significant phase shifts can be occasioned by diurnal temperature variance.

Diversity In microwave communications, the strength of a microwave signal can decrease for many reasons — heat, rain, fog, etc. This is not good if the objective is to get reliable communications. One solution is to simultaneously send and receive two microwave signals at slightly different frequencies. Since different frequencies respond differently to weather problems, the likelihood is that at least one will get through well. This is called diversity.

Diversity Combiner A circuit or device for combining two or more signals carrying the same information received via separate paths or channels with the objective or providing a single resultant signal that is superior in quality to any of the contributing signals.

Divestiture On January 8, 1982 AT&T signed a Consent Decree with the U.S. Department of Justice, stipulating that on midnight December 30, 1983, AT&T would divest itself of its 22 telephone operating companies. According to the terms of the Divestiture, those 22 operating Bell telephone companies would be formed into seven regional holding companies of roughly equal size. Terms of the Divestiture placed business restrictions on AT&T and the BOCs. Those restrictions were threefold: The BOCs weren't allowed into long distance, equipment manufacturing, or information services. AT&T wasn't allowed into local telecommunications (i.e. to compete with the BOCs). But it was allowed into computers. The federal Judge overseeing Divestiture, Judge Harold Greene, is slowing the lifting the restrictions against the BOCs being allowed into information services. He has stayed firm on the other two — equipment manufacturing and long distance.

DIW Type D Inside Wire. Originated as a specific AT&T cable. Now commonly used to describe any 22, 24, or 26 gauge PVC jacketed twisted-pair cable used primarily for inside telephony wiring.

DIX Connectors A local area network connector. DIX connectors on the transceiver local area network cable link it to the network; the male DIX connector plugs into the SpeedLink/PC16 and the female DIX connector attaches to an external transceiver.

DIX Ethernet The DEC, Intel, Xerox Ethernet standard, also known as Version 1 or Bluebook Ethernet. There are subtle differences between IEEE 802.3 and the DIX Ethernet.

DL 1. Distribution List.
2. Distance Learning.

DLC 1. Digital Loop Carrier. Network transmission equipment used to provide a pair gain function. Used in situations in which the cost of the DLC equipment is more than offset by the savings in copper distribution accomplished by eliminating need for as many individual copper pairs. Digital loop carrier systems consist of two parts, a Central Office Terminal and a Remote Terminal. The Central Office Terminal provides the multiplexing/demultiplexing function of individual voice signals to the composite multiplexed signal at the interface between the switching equipment and the DLC. The Remote Terminal provides the multiplexing/demultiplexing function at the interface between the individual subscriber pairs and the DLC.

2. See DATA LINK CONTROL.

3. Direct Line Console. An AT&T Merlin term. An answering position used by system operators to answer calls, transfer calls, make calls, set up conference calls, and monitor system operations. Calls can ring on any of the line buttons, and several calls can ring simultaneously (unlike the QCC where calls are sent to a common QCC queue and wait until a QCC is available to receive a call).

DLCI Data Link Connection Identifier. The frame relay virtual circuit number corresponding to a particular destination which is part of the frame relay header and is usually ten bits long.

DLE Data Link Escape. A control character used exclusively to provide supplementary line control signals, control character sequences or DLE sequences. In packet switching, Data Link Escape is a name applied to the Control P non-print character which is used to swap the PAD from the data mode to the command mode in packet switched networks.

DLL 1. Dynamic Link Library. A feature of OS/2 and Windows that allow executable code modules to be loaded on demand and linked at run time. This lets library code be field-updated — transparent to applications — and then unloaded when they are no longer needed. Unlike a standard programming library, whose functions are linked into an application when the application's code is compiled, an application that uses DLL links with those DLL functions at runtime. Hence, the term dynamic.

2. Data Link Layer driver. A driver specification developed by DEC primarily to work with DECnet PCSA for DOS. DLL is a shared driver specification, allowing multiple protocol stacks to share a single network interface card.

DLPI An ATM term. UNIX International, Data Link Provider Interface (DLPI) Specification: Revision 2.0.0, OSI Work Group, August 1991.

DLS Data Link Switching. IBM's method for carrying SNA and NetBIOS over TCP/IP operating at the Data Link layer. DLS, now an open Internet spec, can be used with OSPF or PPP.

DLSE Dial Line Service Evaluation.

DLSw DataLink Switching Workgroup. This workgroup has issued a new interoperability standard for integrating SNA and NetBIOS over the TCP/IP protocol. According to Cisco, the new DLSw standard provides interoperability and functionality not currently offered by Informational RFC 1434 or existing DLSw implementations.

DLTU Digital Link Trunk Unit. An AT&T term for a device which provides the interface to digital trunks and lines such as T1, EDSL, and remote line units.

DM Delta Modulation.

DMA 1. Direct Memory Access. A fast method of moving data from a storage device or LAN interface card directly to RAM which speeds processing. In essence, DMA is direct access to memory by a peripheral device that bypasses the CPU.

2. Direct Marketing Association.

3. Document Management Alliance.

DMA Channel A channel for direct memory access that does not involve the microprocessor, providing data transfer directly between memory and a disk drive. See DMA.

DMB Digital Multipoint Bridge.

DMI 1. Digital Multiplexed Interface. AT&T's Digital Multiplexed Interface. A PBX to computer interface that divides the T-1 trunk into 23 user channels and one signaling channel. Also used as a T-1 PBX to computer interface. See OPEN APPLICATION INTERFACE.

2. Desktop Management Interface, a protocol independent management interface developed by the DMTF, the Desktop Management Task Force, a group working on improving network printing. Making network printers easier to use — more standard and easier to use — requires three basic elements: standardized definitions of printer objects, a protocol for communicating with those objects and an application interface.

Question: What is DMI?

Answer: The Desktop Management Interface (DMI) is a result of a cooperative, industry-wide effort to make PC systems easier to manage, use and control. Specifically, DMI is a specification developed by the Desktop Management Task Force (DMTF), a consortium of hardware, software, and peripheral vendors. DMI describes hardware and software components in a format that can be easily accessed over a phone line by PC management applications and technical support personal. Thanks to DMI, technical support personnel and PC management applications can:

• Access an inventory of hardware and software components • Access and change parameter values and settings • View data generated by software agents and diagnostics routines

With DMI, technical support applications and PC management applications use a Management Interface (MI) to access data stored in Management Information Files (MIFs).

DMI-BOS Digital Multiplexed Interface-Bit Oriented Signaling. A form of signaling, which uses the 24th channel of each DSI to carry signaling information, allowing clear channel 64 Kbps functionality.

DMO Digital Modification Order.

DMS Digital Multiplex System. Also the name of a line of digital central office switches from Northern Telecom. There are DMS-10s, DMS-100s, DMS-100Fs and DMS-200s and, by the time you read this, probably more. See also DMS SUPERNODE.

DMS Integrated Access Local Area Network Shortened to DIALAN. A central office provided local area network offering completely digital, full duplex data transmission at speeds of 300 bps through 19.2 asynchronous and 1,200 bps through 64 kbps synchronous. DIALAN users use existing telephone sets and an Integrated Voice and Data Multiplexer (IVDM) that plugs into a telephone jack.

DMS Supernode The SuperNode is a very flexible central office switch from Northern Telecom which can be configured as a high-capacity local or tandem switch with Common Channel Signaling 7 (CCS7) Service Switching Point (SSP) function, a Signaling Transfer Point (STP), a Service Control Point (SCP), a Digital Network CrossConnect (DNX) system or a network service node with custom programming applications. In 1988, the DMS SuperNode offered twice the capacity of the DMS-100 switch (the older Northern Telecom central office) based on the NT40 processor.
DMS-INode DMS Integrated STP/SSP Node.
DMT Discrete Multi-Tone. A new technology using digital signal processors to pump more than 6 megabits per second of video, data, image and voice signals over today's existing one pair copper wiring. DMT technology, according to Northern Telecom, provides the following:
• Four "A" channels at 1.5 Mbps. Each "A" channel may carry a "VCR"- quality video signal, or two channels may be merged to carry a "sports"- quality real-time video signal. In the future, all four channels operating together will be able to transport an Extended Definition TV signal with significantly improved quality over anything available today. ("A" channels are asymmetric — carrying information only from the telephone company to the subscriber's residence. All other channels within ADSL are symmetric or bi-directional.)
• One ISDN "H zero" channel at 384 kbps (kilobits per second). This channel is compatible with Northern Telecom's multi rate ISDN Dialable Wideband Service or equivalent services. This channel could also be used for fast, efficient access to corporate LANs for work-at-home applications, using Northern Telecom's DataSPAN or other frame-relay services.
• One ISDN Basic Rate channel, containing two "B" channels (64 kbps) and one "D" channel (16 kbps). Basic Rate access allows the home user to access the wide range of emerging ISDN services without requiring a dedicated copper pair or the expense of a dedicated NT1 unit at the home. It also permits the extension of Northern Telecom's VISIT personal video conferencing to the home at fractional-T1 rates (Px64).
• One signaling/control channel, operating at 16 kbps giving the home user VCR-type controls over movies and other services provided on the "A" channel including fast-forward, reverse, search, and pause.
• Embedded operations channels for internal system maintenance, audits, and telephone company administration.
• Finally, the home user can place or receive telephone calls over the same copper pair without affecting the digital transmission channels listed above. And since ADSL is passively coupled to the POTS line, the subscriber's POTS capability is unimpaired in the event of a system failure.
DMTE Desktop Management Task Force
DMTF Desktop Management Task Force. See DMI.
DN Directory Number or subscriber number or telephone number entries. DN is typically the phone number associated with a telephone line. The international format for a phone number consists of four items:
1. A + plus, which indicates that the caller should dial whatever digits are necessary to get him internationally. Iin the U.S., the prefix is 011.
2. The country code. The country code for the U.S. and Canada is 1. For Australia, it's 61. For England, it's 44.
3. The routing code, which is also called an area code.
4. The local number. In North America it's 7-digits. But in other countries it varies. At one stage in Sydney, Australia, some local numbers were five digits; some were six, some were seven and some were eight.
See DIRECTORY NUMBER.
DNA 1. Digital Network Architecture. The framework within which Digital Equipment Corporation (DEC) designs and develops all of its communications products. DNA includes many standards of the OSI Model. Some of these standards will be adopted into ISDN. Acronym also by Network Development Corporation for their network offering.
2. Dynamic Node Access. A high-speed bus invented by Dialogic to join together multiple voice processing PEB-based systems. PEB stands for PCM Expansion Bus. See PEB.
DNAR Directory Number Analysis Reporting.
DNC Dynamic Network Controller.
DNIC Data Network Identification Code. An address to reach a host computer system residing on a different packet switched network than the one you're on. The data equivalent of a telephone number with country code and area code. Typically the DNIC is a four digit number. The first three digits of a DNIC specify a country. The fourth digit specifies a public data network within that country. See also DATA NETWORK IDENTIFICATION CODE.
DNIS Dialed Number Identification Service. DNIS is a feature of 800 and 900 lines that provides the number the caller dialed to reach the attached computer telephony system (manual or automatic). Using DNIS capabilities, one trunk group can be used to serve multiple applications. The DNIS number can be provided in a number of ways, inband or out-of-band, ISDN or via a separate data channel. Generally, a DNIS number will be used to identify to the answering computer telephony system the "application" the caller dialed. For example, a 401K status program may be offered by a service provider to a number of different companies. The employees of each company are provided their own 800 number to call to access their account status. When the computer telephony system sees the incoming DNIS number, it will know to which company the call was directed, and can so answer the phone correctly with a customized "you have reached the 401K line for xyz company. Please enter your personal account code and password..."
Here's another application: You use one 800 phone number for testing your advertisements on TV stations in Phoenix; another number for testing your ads on TV stations in Chicago; and yet another for Milwaukee. The DNIS information can be used in a multitude of ways — from playing different messages to different people, to routing those people to different operators, to routing those people to the same operators, but flashing different messages on their screens, so the operators answer the phone differently. In Ireland, incoming toll free phone calls from the rest of Europe arrive with DNIS. As a result a phone call arriving from Germany is routed to a computer telephony system playing messages in German. The advantage of DNIS is basically econom-

ic: You simply need fewer phone lines. Without DNIS you would need at least one phone line for every different 800 or 900 number you gave out to your callers. Make sure you understand the difference between DNIS and ANI and Caller ID. DNIS tells you the number your caller called. ANI or Caller ID is the number your caller called from. See 800 SERVICE and 900 SERVICE.

DNPA Data Numbering Plan Area. In the U.S. implementation of the ITU-T X.25 network, the first three digits of a network terminal number (NTN). See also DATA NUMBERING PLAN AREA.

DNPIC Directory Number Primary InterLATA Carrier.

DNR 1. Dialed Number Recorder. Also called a Pin Register. An instrument that records telephone dial pulses as inked dashes on paper tape. A touchtone decoder performs the same thing for a touchtone telephone.
2. Dynamic Network Reconfiguration. Allows IBM networks to change addresses without reloading and bringing the network down.

DNS 1. The Domain Naming System is a mechanism used in the Internet for translating names of host computers into addresses. The DNS also allows host computers not directly on the Internet to have registered names in the same style. The DNS is a distributed database system for translating computer names (like ruby.or-a.com) and vice-versa. DNS allows you to use the Internet without remembering long lists of numbers. On TCP/IP networks (like the Internet), the Domain Naming System provides IP address translation for a given computer's domain name. DNS would change a computer name such as harry.newton.com to the machine' actual numeric IP address, which is in the format xxx.xxx.xxx.xxx. The DNS makes it easier to remember where you want to go.
2. See DISTRIBUTED NETWORK SERVICE.

DO A word in a high-level language program which comes before a collection of things to be done, i.e. statements to be executed.

Do-Not-Disturb Makes a telephone appear busy to any incoming calls. May be used on intercom-only, by extension line-only or both.

Do-While A programming statement used to perform instructions in a loop while a certain condition exists — i.e. do something while the variable Y is less than 20.

DOA Dead On Arrival. A term several manufacturers use to refer to equipment which arrives at the customer's premises not working. A person who receives a DOA machine will ask the company for a NPR number — New Product Return number. This allows them to return the product and have the factory replace it with another new one.

DOC 1. Department of Communications. Canadian government department.
2. See DYNAMIC OVERLOAD CONTROL.

DOC-IT Okidata's name for a combination scanner, printer, copier, fax machine. A truly wonderful machine.

Dock To insert a portable computer into a base unit. Cold docking means the computer must begin from a power-off state and restart before docking. Hot docking means the computer can be docked while running at full power.

Docket Formal FCC/State regulatory commission proceeding, also referred to as a case.

Docking Station Base station for a laptop that includes a power supply, expansion slots, monitor, keyboard connectors, CD-ROM and extra hard disk connectors. A user slides his laptop into a base station and in effect, gets the equivalent of a desktop machine.

Docobjects See DOCUMENT OBJECTS.

Document Camera A specialized camera on a long neck that is used for taking pictures of still images — pictures, graphics, pages of text and objects which can then be sent stand alone or as part of a video conference.

Document Database An organized collection of related documents.

Document Image Management DIM. The electronic access to and manipulation of documents stored in image format, accomplished through the use of automated methods such as high-powered graphical workstations, sophisticated database management techniques and networking.

Document Objects DocObjects for short. Microsoft Office for Windows95 has an app called Office Binder which is a "document container that enables a user to manipulate text files, spreadsheets, graphics presentations, and other documents as a single entity." DocObjects is the "core technology that makes Office Binder work." This technology can support document containers other than Office Binder and document server as well. "The DocObjects technology consists of a set of extensions to OLE Documents. One application for this technology is in Internet browsers; a user would only need one navigation tool to browse and view all documents, whether local or network-based. A document provided from a DocObjects server is essentially a full-scale, conventional document that is embedded as an object within another DocObjects container (binder, browser, etc.)."

Document Recognition The ability to capture all the information on a page (text and images) and perform not only character recognition, but page structure analysis as well.

Documentation Written text describing the system, how it works and how to work it. In most cases of high technology products, documentation is awful. Better documentation helps sell equipment and software. Please write your instruction manuals better. Please.

DOD Department of Defense.

DOD Master Clock The U.S. Naval Observatory master clock, which has been designated as the DOD Master Clock to which DoD time and frequency measurements are referenced. This clock is one of two standard time references for the U.S. Government in accordance with Federal Standard 1002; the other standard time reference is the National Institute for Standards and Technology (NIST) master clock.

Dolby A system of noise/hiss reduction invented by Ray Dolby, widely used in consumer, professional and broadcast audio applications.

Domain The part of a computer network in which the data processing resources are under common control. That's the formal technical definition. But when the Internet came along, the definition got a little looser. On the Internet, most people think of a domain as a place you can visit (with a Web browser) and which you can send e-mail to. "A place you can visit" might be part of a computer — a subdirectory or two. It might be a single computer or it might be a group of computers masquerading as a single computer. On the Internet, the domain is thus whatever the address is that gets you there. An domain address is computertelephony.com or teleconnect.com. A domain's address on the Internet is a hierarchical sequence of names (labels) separated by periods (dots), e.g., computer.telephony.com. In fact, the central naming registry on the Internet — which is called InterNIC — will only register the word before the .com, the .edu, the .gov and the .mil. In this case, they will register the domain name "telephony.com" or "computertelephony.com." (They are different sub-domains.) What you put in front of your domain name is your concern. Everything that is "something.something.telephony.com" will come to the domain "telephony.com" for resolution (further routing, processing, etc.)

Some people think of a domain as a group of computers on a network that share something in common, such as the job they do, the organization type they belong to, or their location. Top level domains on the Internet (yet another definition of domain) include .COM for commercial organizations; .EDU for educational organizations; .GOV for government organizations, .ORG (miscellaneous organizations, usually non-profit), .NET (network service providers), .INT (international), .AUS for Australia, .NZ for New Zealand, and numerous others specifying foreign countries.

Domain is a term defined differently by various electronic mail vendors. PC Magazine's Frank Derfler defines a domain as referring to a set of hosts on a single LAN that needs only one intermediary post office to move mail from one host to another. A domain may also consist of only one host, depending on its design and implementation. In IBM's SNA, a domain is a host-based systems services control point (SSCP), the physical units (PUs), logical units (LUs), links, link stations and all the affiliated resources that the host (SSCP) can control.

In Microsoft networking, a domain is a collection of computers that share a common domain database and security policy that is stored on a Windows NT Server domain controller. Each domain has a unique name. See also DOMAIN CONTROLLER, DOMAIN NAME, DOMAIN NAMING SYSTEM and WORKGROUP.

Domain Controller For a Windows NT Server domain, the primary or backup domain controller that authenticates domain logons and maintains the security policy and the master database for a domain.

Domain Defined Attribute DDA. In X.400 addressing, the DDA is a special field that may be required to assist a receiving E-mail system in delivering a message to the intended recipient. Up to four DDAs are allowed per address, with each DDA address entry made up of two parts, a type and a value. For example, if I were a subscriber to MCI Mail and I wanted to send a message to Harry Newton, this is how I would address it:
TO: Harry Newton EMS: CompuServe / 592-7515 MBX: P=CSMAIL MBX: DDA=ID=70600,2451
If I were a subscriber to CompuServe, the only addressing information I would need would be the number 70600,2451.

Domain Name In networks using the TCP/IP (Transfer Control Protocol/Internet Protocol), the full domain name consists of a sequence of names (labels) separated by periods (dots), e.g., "pictures.computertelephony.com." See DOMAIN and TCP/IP.

Domain Name Servers Domain Name Servers are computers on the Internet which contain the the programs and files which make up a domain's name database. Using a name server is much like calling directory assistance; you provide the operator (name server) with a name and it is translated to a telephone number (IP address), which you then use to establish connection with the person (host computer). the telephone network (Internet) addresses by telephone number (IP address), not really by the person's (domain's) name.

Domain Naming System DNS. in Unix-based networks, of which the Internet is the largest, a domain naming system is the commonly accepted way of giving attached computers names. Domain naming system is sometimes referred to as the BIND (berkeley internet name domain) service in BSD Unix, a static, hierarchical naming service for tcp/ip hosts. A DNS server computer maintains a database for resolving host names and IP addresses, allowing users of computers configured to query the DNS to specify remote computers by host names (in words) rather than IP addresses (which are only numbers). DNS domains should not be confused with Windows NT networking domains, although Windows NT does support and can use the Internet's DNS scheme. See DOMAIN.

Domestic Arc The portion of the geostationary orbit allowing a satellite to return a footprint that almost covers the continental United States.

Dominant Carrier The long distance service provider which dominates a particular market and is subject to tougher regulation than its competitors. An FCC term. Essentially it means AT&T.

DOMSAT Domestic Communications Satellite.

Don't Answer Recall Allows an extension user on a PBX to automatically retry a call by dialing a special digit code.

Done Deal A term used in the secondary telecom equipment business. Term used between seller and buyer to signify that a sale has been agreed to and an oral contract is now in effect, binding both parties to the agreed-to-sale as if a written contract has been signed. A written contract submitted later that does not conform to the original oral agreement is not justification for dissolving the original agreement unless both parties agree to the new written contract.

Dongle A device to prevent copies made of software programs. A dongle is a small device supplied with software that plugs into a computer port. The software interrogates the device's serial number during execution to verify its presence. If it's not there, the software won't work. A dongle is also called a hardware key. See HARDWARE KEY.

Door A software program that allows access to files and programs not built into an electronic bulletin board system, thus letting users run them on-line.

Doppler Aiding A signal processing strategy that uses a measured doppler shift to help the GPS Global Positioning System) receiver smoothly track the GPS signal. It allows more precise velocity and position measurement.

Doppler Shift The apparent shift in the frequency of a signal caused by the relative motion of the transmitter and the receiver.

DOS Disk Operating System, as in MS-DOS, which stands for MicroSoft Disk Operating System. A disk operating system is software that organizes how a computer reads, writes and reacts with its disks — floppy or hard — and talks to its various input/output devices, including keyboards, screens, serial and parallel ports, printers, modems, etc. Until the introduction of Windows, the most popular operating system for PCs was MS-DOS from Microsoft, Bellevue, WA.

Dot When you want to impress the Internet veterans you meet at parties, say "dot" instead of "period," for example: "My address is john at site dot domain dot com."

Dot Addressable Graphics Refers to the mode of operation on a dot matrix printer which allows you to control each element in the dot matrix printhead. With this feature, you may produce complex graphics drawings.

Dot Pitch A measure of the clarity of a color monitor. Dot pitch measures the vertical distance between the centers of like-colored phosphors on your screen. The smaller the distance, the sharper the monitor. Dot pitch is the major determinant in the clarity of an image on screen. And you can't do anything about it. When you buy a monitor, you buy it with a certain dot pitch and you're stuck with that dot pitch. You may be able, however, to do something about improving convergence and focus — the other measures of the clarity of a color monitor.

Dot Zero When new software is issued, it often bears the number .0 (i.e. dot zero) as in MS-DOS 5.0. The theory among software gurus is that you should always avoid a "Dot Zero" revision, since it will likely contain bugs and that one should wait for 4.01 or 5.01 etc. This theory has some validity, although MS-DOS 5.0 came out very clean and was not revised until 6.0.

Dotted Quad A set of four numbers connected with periods that make up an Internet address; for example, 147.31.254.130.

Double Buffering The use of two buffers rather than one to temporarily hold data being moved to and from an I/O device. Double buffering increases data transfer speed because one buffer can be filled while the other is being emptied.

Double Camp-On Indication A PBX feature. A phone attempting to camp on to another phone which is already being "camped on" shall receive a distinctive audible signal and may be denied the ability to camp-on.

Double Click With a Mac and an IBM-compatible PC running Windows, double-clicking carries out an action, such as beginning a new program. Press and release the mouse button twice in rapid succession to double-click. If you don't double click fast enough, it won't work. Some mouse software allow you to assign the left hand button on the mouse to a single click and the right hand button to a double click. This is very useful.

Double Crucible Method A method of fabricating optical fiber by melting core and clad glasses into two suitably joined concentric crucibles and then drawing a fiber from the combined melted glass.

Double Density Refers to a diskette which can contain twice the amount of data in the same amount of space as a single-density diskette. For example, a double-density 360k diskette has a 720k storage capacity. These days double density is an obsolete term, since there are now disks that are "double double" density. Most 3 1/2 inch disks will now hold twice 720K — or 1,440,000 bytes. These are called high density disks. Toshiba has introduced a "double double double" floppy, which will hold 2,880,000 bytes. But it hasn't caught on, yet.

Double Ended Synchronization A synchronization control system between two exchanges, in which the phase error signals used to control the clock at one exchange are derived from comparison with the phase of the internal clock at both exchanges.

Double Interrupted Ring Two quick rings followed by a period of silence indicating the arrival of an outside call in some systems.

Double Modulation Modulation of a carrier wave of one frequency by a signal wave, this carrier then being used to modulate another carrier wave of different frequency.

Double Pole A double pole switch is one which opens and closes both sides of the same circuit simultaneously. Most electrical circuits open and close with only one side being broken.

Double Pull A method for pulling cable into conduit or duct liner that is similar to backfeed pulling except that it eliminates the need to lay the backfeed cable on the ground.

Double Sideband Carrier Transmission That method of transmission in which frequencies produced by the process of amplitude modulation are symmetrically spaced above and below the carrier. The carrier level is reduced for transmission at a fixed level below that which is provided to the modulator. Carrier is usually transmitted at a level suitable for use as a reference by the receiver except in those cases where it is reduced to the minimum practical level (suppressed carrier).

Doublewide Two trailer homes stapled together with a modest gabled roof.

Doubly Clad Fiber An optical fiber, usually single mode, that has a core surrounded by an inner cladding of lower refractive index, which is in turn surrounded by an outer cladding, which has a higher refractive index than the inner cladding. This type of construction is often employed in single-mode fibers to reduce bending losses.

DOV Data Over Voice. A technology used primarily with local Centrex services or special customer premises PBXs for transmitting data and voice simultaneously over twisted-pair copper wiring. Typical data rates for Centrex operation are 9.6 Kbps and 19.2 Kbps.

DOVPATH Transport Service A Nynex service which lets you transmit both voice and data simultaneously on one phone line. The data is carried on a higher frequency, i.e. "Over the Voice." Hence the name Data Over Voice, or DOV. DOVPATH lets you transmit up to 19.2 Kbps asynchronous data, which could be fast enough for talking to a distant mainframe computer.

Down-Converter A device for performing frequency translation in such a manner that the output frequencies are lower in the spectrum than the input frequencies.

Downline Loading A system in which programs are loaded into the memory of a computer system, such as a LAN bridge, router or server, via the same communication line(s) the system normally uses to communicate with the rest of a network. As opposed to systems in which all programs are loaded into the computer from a disk or tape associated with the computer. A PC connected to a LAN may use this type of loading when it is first turned on in the morning to get the information it needs from a file server. Diskless PCs always work this way.

Downlink 1. The part of a transmission link reaching from a satellite to the ground. Some satellite transmission circuits, especially international ones, are priced and billed separately for the uplink and the downlink. This is because their transmissions are provided by different carriers.

2. In packet data communications, a downlink is a link from an NC or PAD to another NC or PAD on a different level. The defining of downlinks and uplinks depends on the network configuration of PADs, their relationships to each other and the direction

of data transmission.

Download To receive data from another computer (often called a host computer or host system or just plain host) into your computer. It's also called to RECEIVE. The opposite is UPLOAD or TRANSMIT. You have to be very careful distinguishing between the two. Choosing the "Download" option in some communications programs automatically erases a file of the same name that was meant for transmission. If that happens, stop everything. Grab your file unerase program and use it. Don't wait. If you wait, you may write over your erased file and never get it back. See DOWNLOADED FONTS.

Downloaded Fonts Fonts that you send to a printer either before or during the printing of a document. When you send a font to a printer, it is stored in printer memory until it is needed. Downloaded fonts are one reason for loading lots of RAM memory into your printer.

Downsizing Downsizing is what happens when companies move from large computer systems to smaller systems. There are four major reasons companies downsize from mainframe-based computer to local area network-based computing: 1. They save money. There are several reasons: a. Mainframe computers cost lots each month in maintenance. They require costly maintenance agreements with the supplier, e.g. IBM. Servers are usually bought without maintenance agreements. When they break, the managers or the workers simply replace the broken parts themselves. b. Mainframe computers cost lots to program. There are comparatively few programs available for mainframes, compared to the plethora of off-the-shelf programs available for workstations. c. Servers require a far less costly home to live. You don't need air-conditioning, special buildings with raised floors, etc. 2. Servers today have the power of mainframes 10 years ago. In fact, servers are now beginning to acquire more power than mainframes of ten years ago. And as servers increasingly acquire several processors, they will leap in power beyond what mainframes have. 3. Servers are typically manufactured from off-the-shelf, standard components that are usually available from several manufacturers. As a result, there is constant competition and constant improvement in quality and features. 4. Servers are much more flexible tools to design networks. You can start with one baby network containing one server and several workstations (a.k.a. clients) and confined to one floor of one small building and grow to a huge, complex network containing thousands of workstations, dozens of servers and spanning the globe.

To most people, downsizing is not only swapping out the "big iron" (the mainframe) and bringing in servers and local area networks. It's also a new way of thinking about the way corporations are organized. Downsizing is often accompanied by re-engineering, which is basically re-organizing for a greater focus on the customer — a focus which means responding faster to customer needs. See SERVERS.

Downstream Refers to the relative position of two stations in a local area network ring topology. A station is downstream if it receives a token after the previous station. See also DOWNSTREAM CHANNEL.

Downstream Channel The frequency multiplexed band in a CATV channel which distributes signals from the headend to the users. Compare with UPSTREAM CHANNEL, the band of frequencies on a CATV channel reserved for transmission from the user to the CATV company's headend.

Downtime The total time a telephone system is not working due to some software or hardware failure. You know your vendor is about to lie when he begins to answer a question about "downtime."

DP 1. Dial Pulse (as in dialing a phone) or Data Processing. Also called EDP for Electronic Data Processing. Now more commonly called Management Information Systems — or MIS.

2. Demarcation Point. The point of a demarcation and/or interconnection between telephone company communications facilities and terminal equipment, protective apparatus, or wiring at a subscriber's premises. Carrier-installed facilities at or constituting the demarcation point consist of a wire or a jack conforming to Subpart F of Part 68 of the FCC Rules.

DPA 1. Digital Port Adapter. A Northern Telecom word.

2. Demand Protocol Architecture. A technique for loading protocol stacks dynamically as they are required. It is associated with adapter cards in workstations and servers. Only the protocol stacks that are required for a particular communications sessions are loaded. Examples of such stacks include TCP/IP, XNS, SPX/IPX and NetBios.

DPBX Digital PBX. Not a common term. Most PBXs these days are digital.

DPCM Differential Pulse Code Modulation

DPLB Digital Private Line Billing.

DPMI An acronym for DOS Protected Mode Interface. DPMI is an industry standard that allows MS-DOS applications to execute code in the protected operating mode of the 80286 or 80386 processor. The DPMI specification is available from Intel Corporation. It is a superset of the VCPI (Virtual Control Program Interface) specification for controlling multiple programs inside a PC, as well as programs that use protected mode.

DPMS Display Power Manager Signaling is a power reduction feature that places a computer monitor in reduced power when the monitor is still on but has been idle for some time.

DPN Northern Telecom term for Data Packet Network.

DPN Lanscope A Northern Telecom software package for remote management of local area networks over wide area networks based on Northern's DPN-100 digital data packet networking system. DPN Lanscope provides fault and performance monitoring, resource management, software distribution and usage tracking for geographically dispersed LANs. Network configuration, status, alarms and performance information are presented to a centralized LAN administrator through a graphics interface.

DPNSS Digital Private Network Signaling System. A standard in Britain which enables PBXs from different manufacturers to be tied together with E-1 lines and pass calls transparently between each — as easily as if the phones were extensions off the same PBX and were simply making intercom calls. The international version of DPNSS is called Q.SIG, but apparently it's not as well developed.

DPP Distributed Processing Peripheral.

DPU Dynamic Path Update. Allows IBM networks to add new network nodes or change backup routing paths while the front-end is still operating.

DPX DataPath loop eXtension.

DQDB Distributed Queue Dual Bus. The metropolitan area network technology that operates by maintaining a queue at each station to determine when the station may access its dual buses. The dual buses provide bidirectional transmission between each station. DQDB is IEEE's standard 802.6 for MANs (Metropolitan Area Networks). Formerly called QPSX (Queued Packet and Synchronous Exchange). DQDB was devised for metropolitan area networks (MAN). It consists of two uni-directional buses connected as an open ring (i.e., each bus is not directly interconnected). It provides for full duplex operation between any two nodes and the bandwidth available provides for signaling speeds in excess of 155 Mbit/s. Each transmission cell or packet consists of a 5-byte header plus 48 bytes of message. The header bytes include bits of control, priority and an error check sequence. See also ATM.

Draft Proposal An ISO standards document that has been registered and numbered but not yet approved.

Drag Dragging is a way of moving an item on the screen using your mouse. To drag a window in Windows 3.1, for example, move the mouse pointer onto a window's title bar, then hold down the mouse button while moving the mouse across your desktop. When you release the mouse button, the window will remain in its new location. Apply this technique to drag any data object, such as icons or list box items.

Drag And Drop The "drag and drop" definition defines how objects from one desktop application can be "dragged" out of that application, through clicking on the object with a mouse, across the desktop and "dropped" on another application. Most of the graphics operating systems, like Windows, Apple's Macintosh and Sun Sparc use Drag and Drop.

Drag Line A length of rope or string used to pull wire and cable through conduit or inaccessible spaces. Drag lines are often inserted in wall and ceilings during construction to ease future wire installation.

Drain Wire In a cable, an uninsulated wire laid over the component or components and used as a ground connection.

DRAM 1. Dynamic Random Access Memory chip. Pronounced "dee-ram." The readable/writable memory used to store data in PCs. DRAM stores each bit of information in a "cell" composed of a capacitor and a transistor. Because the capacitor in a DRAM cell can hold a charge for only a few milliseconds, DRAM must be continually refreshed to retain its data. In contrast, static RAM, or SRAM, requires no refresh and delivers better performance, but it is more expensive to manufacture. See also EDO RAM, SRAM, MICROPROCESSOR (for a full explanation). See also DYNAMIC RANDOM ACCESS MEMORY.
2. Digital Recorder Announcer Module.

Drawing In the manufacture of wire, pulling the metal through a die or series of dies in order to reduce the diameter to a specified size.

Drawing Tools A computer imaging term. The means of creating freehand lines or basic geometric shapes. Paint packages often provide an ellipse-drawing function as a variation of the circle (or vice versa) and a square drawing function as a variation of the rectangle. Virtually all packages offer filled geometric figures, the fill item being either a solid color or a pattern.

Dressing Cable You "dress" cable by taking multiple cables and joining them together neatly with cable ties. A nicely-dressed, clean installation is a sign of telephony professionalism. It still exists.

Drift A slow change in a normally constant signal. The term typically applies to movements in the signal's frequency.

Drill Down Microsoft jargon for to learn more about a subject.

Drive Icon A Windows NT term. An icon in a directory window in File Manager that represents a disk drive on your system. Different icons depict floppy disk drives, CD-ROM drives, network drives, etc.

Drive Mappings A Novell NetWare term. Drive mappings provide direct access to particular locations in the directory structure. They are a "shorthand" method for accessing directories on a disk. Instead of typing in the complete path name of a directory that you want to access, you can simply enter a drive letter that has been assigned to that directory. NetWare recognizes two types of drives (physical drives and logical drives) and three types of drive mappings (local, network, and search drive mappings).

Drive Type A number representing a standard configuration of physical parameters (cylinders, heads and sectors) of a particular type of hard disk drive. You need to know your drive's drive type otherwise the bios of your machine will not recognize your drive on boot up and your PC will not work. Normally this information of your drive's type resides in memory kept alive by a small lithium battery. However, should the lithium battery die, your PC will "forget" which drive it has and it will ask you. If you don't know, you're in big trouble. Your PC simply won't work. My recommendation: write the drive type on two labels — one to stick on the drive and the other to stick on the bottom of your machine. This way you'll be able to find it easily. You will usually find the drive type's number on paperwork sent originally with your machine.

Drivebar A Windows NT term. Allows you to change drives by selecting one of the drive icons.

Driver A driver (which is always software) provides instructions for reformatting or interpreting software commands for transfer to and from peripheral devices and the central processor unit (CPU). Many printed circuit boards which you drop into a PC require a software driver in order for the other parts of the computer and the software you're running to work correctly. In other words, the driver is a software module that "drives" the data out of a specific hardware port. The port in question will usually have another device connected, such as a printer or modem, and the driver will be organized in software (i.e. configured) to communicate with the device.

Drop 1. A wire or cable from a pole or cable terminal to a building.
2. That portion of a device that looks toward the internal station facilities, e.g., toward an AUTOVON 4-wire switch, toward a switchboard, or toward a switching center.
3. Single channel attachment to the horizontal wiring grid (wall plate, coupling, MOD-MOD adapter).
4. The central office side of test jacks.
5. To delete, intentionally or unintentionally, part of a signal for some reason, e.g., dropping bits.

Drop And Insert That process wherein a part of the information carried in a transmission system is demodulated (dropped) at an intermediate point and different information is entered (inserted) for subsequent transmission.

Drop Cable 1. The outside wire pair which connects your house or office to the transmission line coming from the phone

company's central office. See also DROP WIRE, which is different.
2. In local area networks, a cable that connects a network device such as a computer to a physical medium such as an Ethernet network. Drop cable is also called transceiver cable because it runs from a network node to a transceiver (a transmit / receiver) attached to the trunk cable.

Drop Channel Operation A type of operation where one or more channels of a multichannel system are terminated (dropped) at any intermediate point between the end terminals of the system.

Drop Loop The segment of wire from the nearest telephone pole to your home or business.

Drop Outs Drop outs are one major cause of errors in data communications circuits. The technical definition is that the signal level drops more than 12 dB (decibels) for more than 4 milliseconds. It means some of your data will not arrive. A four millisecond drop out in a transmission at 2,400 bits per second will lose about ten bits. A "drop out" is similar to a person's voice fades away in a telephone conversation. To correct the problem of drop out, we will ask the person (or computer) at the other end to repeat what they just said. "Huh?" This is called retransmission of data. In telephony, drop outs are defined as incidents when signal level unexpectedly drops at least 12 dB for more than 4 milliseconds. (Bell standard allows no more than two drop outs per 15 minute period.)

Drop Repeater A repeater that is provided with the necessary equipment for local termination (dropping) of one or more channels of a multichannel system.

Drop Set All parts needed to complete connection from the drop (wall plate, coupling, MOD-MOD) to the terminal equipment. This would typically include a modular line cord and interface adapter.

Drop Shipment Equipment shipped to a buyer from a location separate from the seller's premise. If this third location is a different company, then this third-party supplier bills the seller and the seller bills the buyer. This saves time, but the seller loses control over the equipment's condition.

Drop Side Defines all cabling and connectors from the terminal equipment to the patch panel or punch down block designated for terminal equipment at the distribution frame.

Drop Wire Wires going from your phone company to the 66 Block or protector in your building. See also DISTRIBUTION CABLE.

Dropout A short period of time during which a transmission service looses the ability to transmit data. Bell System specifications define a dropout as any such loss which lasts for more than four milliseconds. See DROP OUTS for a longer explanation.

Dropped Call A call in which the radio link between the cellular customer and the cell site is broken. Dropped calls can happen often, and for many reasons, including terrain, equipment problems, atmospheric interference, and traveling out of range. In short, a dropped call is a call terminated by other than the calling or called party.

DRU 1. DACS Remote Unit.
2. Digital Remote Unit. An NEC term. It's a multiplexer used to distribute NEC Dterm digital telephones and analog sets throughout the user's communications network, whether that network is local or geographically dispersed. The multiplexing technology used is North American Standard T1 and European Standard E1.

Drum Factor In facsimile systems, the ratio of drum length to drum diameter. Where drums are not used, it is the ratio of the equivalent dimensions.

Drum Speed The angular speed of the facsimile transmitter or recorder drum, measured in revolutions per minute.

Drunken Swede A way of describing the sound of a computer doing text-to-speech conversion. "Why, he sounds like a drunken Swede." This great definition from Stuart Segal of Phone Base Systems, Inc. in Vienna, VA. Says Stuart, "Our people think that a drunken Swede has recorded this message." It is possible to have a computer generate speech that doesn't sound like a drunken Swede if you throw sufficient horsepower (MIPS and memory) at it. Throwing sufficient horsepower, however, has been expensive, until recently. Drunken Swedes are going to get less and less common as horsepower gets cheaper and cheaper.

Dry Cell A type of primary cell in which the electrolyte is in the form of a paste rather than that of a liquid.

Dry Circuit A circuit over which voice signals are transmitted and which carries no direct current.

Dry Contact A dry contact refers to a circuit with an energy level such that a spark is not created in a mechanical relay or switch contracts when the circuit is opened. As a result, no cleaning of the contacts takes place (sparking vaporizes contact materials thereby continually exposing fresh contact material). Sometimes general purpose contacts are gold-flashed so the relay or switch can be used on dry circuits; when used on higher energy circuits the gold coating is destroyed, but no damage is done except that the contact should not be used to carry dry circuit signal levels again. A dry contact might operate a relay which might turn something of higher power on or off. For example, a low voltage signal in a key system might cause a dry contact to close, thus causing much higher voltage to flow to a bell, a klaxon, a strobe light. And, yes, there is a "wet" contact. The term "mercury-wetted relay" refers to a relay or switch in which the movable contact of the device makes contact with a pool of mercury. In fact, before solid-state devices, this was a common technology for switching dry circuits.

Dry Electrolytic Condenser An electrolytic condenser in which the electrolyte is in the form of a paste or jelly rather than that of a liquid.

Dry Loop Powering Refers to local (not span) powering, and a transmission medium other than copper wire (microwave/fiber optic).

Dry T-1 A T-1 line with an unpowered interface. A T-1 line with a power is called "Wet."

Dry Twisted Pair A normal telephone twisted pair without loading coils. In some towns you can rent dry twisted pairs from the phone company and run whatever you want on the wire — voice, video, data, etc. Most phone companies, however, don't like renting dry cable, since it deprives them of revenue, or so they believe.

DS 1. Danske Standardiseringsrad (Danish Standards Institution).
2. An ATM term. Distributed Single Layer Test Method: An abstract test method in which the upper tester is located within the system under test and the point of control and observation (PCO) is located at the upper service boundary of the Implementation Under Test (IUT) - for testing one protocol layer. Test events are specified in terms of the abstract service primitives (ASP) at the

upper tester above the IUT and ASPs and/or protocol data units (PDU) at the lower tester PCO.

DS- A hierarchy of digital signal speeds used to classify capacities of lines and trunks. The fundamental speed level is DS-0 (64-kilobits per second) and the highest is DS-4 (about 274 million bits per second). Here are the definitions: DS-1, DS-1C, DS-2, DS-3, DS-4. They correspond to 1.544, 3.152, 6.312, 44.736, and 274.176 Mbps. DS-1 is also called T-1.

DS-0 Digital Service, level 0. It is 64,000 bps, the worldwide standard speed for digitizing one voice conversation using pulse code modulation (PCM). There are 24 DS-0 channels in a DS-1.

DS-1 Digital Service, level 1. It is 1.544 Mbps in North America, 2.048 Mbps elsewhere. Why there's no consistency is one of those wonderful, unanswered, questions. The 1.544 standard is an old Bell System standard. The 2.048 standard is a ITU-T standard. Standard for 1.544 Mbps is 24 voice conversations each encoded at 64 Kbps. Standard for 2.048 megabits is 30 conversations.

DS-1C Digital Service, level 1C. It is 3.152 Mbps in North America and is carried on T-1.

DS-2 Digital Service, level 2. It is 6.312 Mbps in North America and is carried on T-2.

DS-3 Digital Service, level 3. Equivalent of 28 T-1 channels, and operating at 44.736 Mbps. Also called T-3.

DS-4 Digital Signal, level 4. 274,176,000 bits per second. 168 T1s. 168 DS1s. 4032 standard voice channels.

DS0 Digital Signal, level Zero. Pronounced "D-S Zero." DS0 is 64,000 bits per second. It is equal to one voice conversation digitized under PCM. Twenty-four DS0s (24x64 Kbps) equal one DS1, which is T-1 or 1.544 million bits per second. See DS-0.

DS0-A Refers to a process where a subrate signal (2.4, 4.8, or 9.6 Kbps) is repeated 20, 10 or 5 times, respectively to make a 64 Kbps DS-0 channel.

DS0-B Refers to a process performed by a subrate multiplexer where twenty 2.4 Kbps, ten 4.8 Kbps, or five 9.6 Kbps signals are bundled into one 64 Kbps DS-0 channel.

DS1 Digital Signal, level One. A 1.544 Mbps digital signal carried on a T-1 transmission facility. See DS-1.

DS2 Digital Signal, level Two. 6,312,000 bits per second. Four T1s. Four DS1s. 96 standard voice channels.

DS3 Digital Signal, level Three. 44,736,000 bits per second. 28 T1s. 28 DS1s. 672 standard voice channels.

DS3 PLCP An ATM term. Physical Layer Convergence Protocol: An alternate method used by older T carrier equipment to locate ATM cell boundaries. This method has recently been moved to an informative appendix of the ATM DS3 specification and has been replaced by the HEC method.

DS4 Digital Signal, level Four. 274,176,000 bits per second. 168 T1s. 168 DS1s. 4032 standard voice channels.

DSA 1. Distributed Systems Architecture, the network architecture developed by Honeywell.
2. Directory System Agent. The software that provides the X.500 Directory Service for a portion of the directory information base. Generally, each DSA is responsible for the directory information for a single organization or organizational unit.

DSAP Destination Service Access Point. The logical address of the specific service entity to which data must be delivered when it arrives at its destination. This information may be built into the data field of an IEEE 802.3 transmission frame.

DSAT Digital Supervisory Audio Tones. A supervisory signaling scheme used in NAMPS — a new form of digital cellular radial called Narrow-band Advanced Mobile Phone service. See also NAMP.

DSC Digital Selecting Calling. A synchronous system developed by the International Radio Consultative Committee (CCIR), used to establish contact with a station or group of stations automatically by radio. The operational and technical characteristics of this system are contained in CCIR Recommendation 493.

DSCWID Call Waiting Display with Disposition.

DSDC Direct Service Dialing Capability. Network services provided by local switches interacting with remote data bases via CCIS.

DSE 1. Data Switching Equipment.
2. An ATM term. Distributed Single-Layer Embedded (Test Method): An abstract test method in which the upper tester is located within the system under test and there is a point of control and observation at the upper service boundary of the Implementation Under Test (IUT) for testing a protocol layer, or sublayer, which is part of a multi-protocol IUT.

DSI Digital Speed Interpolation. A technique for squeezing more voice conversations onto a line. DSI digitizes speech so it can be cut into slices, such that no bits are transmitted when no one is speaking. As soon as speech begins, bits flow again. See DIGITAL SPEECH INTERPOLATION.

DSL See DIGITAL SUBSCRIBER LINE.

DSMCC Digital Storage Media Command And Control. Network protocols specified in Part 6 of MPEG-2 (ISO 13818) standards dealing with user-to-network and user-to-user signaling and communications.

DSN 1. Distributed Systems Network, the network architecture developed by Hewlett-Packard.
2. Double Shelf Network.

DSOB An AT&T Digital Data Service standard that specifies a means of multiplexing several subrate data channels within one DS0. Five 9.6, ten 4.8 or 20 2.4 kbps subrate channels may be multiplexed within one DS0.

DSP 1. Display System Protocol
2. Digital Signal Processor. A Digital Signal Processor is a specialized computer chip designed to perform speedy and complex operations on digitized waveforms. Useful in processing sound and video. See DIGITAL SIGNAL PROCESSOR for a much better explanation.

DSP Modem Digital Signaling Processor chip set used for analog modem emulation. The software is programmable for easy upgrades.

DSP Module An internal module of the National Sensor Conductor NS32FX16 processor that handles vector operations on complex variables, enhancing DSP performance of the device.

DSPRM Microsoft's DSP Resource Manager (DSPRM) is a new layer of code scheduled to be part of the new Chicago operating system. DSPRM is the layer below TAPI and should make the addition of DSP resources to desktop machines far easier and be a boon to computer telephony applications. Microsoft has even gone so far as to specify a recommended hardware platform. When Chicago (Windows 4.0) ships, computer telephony should be the next killer application.

DSR Data Set Ready. This signal is on pin 6 of the RS-232-C connector. It means the modem (which some telephone companies call a "data set") is ready to send data from the terminal. Some modems use Data Set Ready. Some don't. Modems that are snooty enough to give you the DSR signal, are obnoxious enough to not work until they receive the DTR (Data Terminal Ready) signal from the terminal on pin 20. By bridging pins 6 and 20 on the connector at the modem, you can usually get it to work. If it doesn't, bridge in pin 8 (carrier detect) as well.

DSRR Digital Short Range Radio.

DSS Direct Station Select. A piece of key telephone equipment usually attached to an operator's phone set. When the operator needs to call a particular extension he/she simply touches the corresponding button on the Direct Station Select equipment. Typically DSS equipment/feature is part of a Busy Lamp Field (BLF), which shows with lights what's happening at each extension. Is it busy? Is it on hold? Is it ringing?
DSS is also Decision Support Systems. Computerized systems for transforming data into useful information, such as statistical models or predictions of trends, usually in a graphical format, which is used by management in the decision-making process. See DECISION SUPPORT SYSTEMS.

DSS2 Setup An ATM term. Digital Subscriber Signaling #2: ATM Broadband signaling.

DSTN Double Super Twisted Nematic. A display technology which uses two layers of crystal to correct distortions caused by the first layer and so improve readability.

DSU Digital Service Unit, also called Data Service Unit. Converts RS-232-C or other terminal interface to DSX-1 interface. See DATA SERVICE UNIT.

DSU/CSU The devices used to access digital data channels channels are called DSU/CSUs (Data Service Unit/Channel Service Units). At the customer's end of the telephone connection, these devices perform much the same function for digital circuits that modems provide for analog connections. For example, DSU/CSUs take data from terminals and computers, encode it, and transmit it down the link. At the receive end, another DSU/CSU equalizes the received signal, filters it, and decodes it for interpretation by the end-user.

DSVD Digital Simultaneous Voice and Data. Technology in a modem that allows you to send and receive voice and data (fax, images, files, etc.) on the same "conversation" on one analog phone line. DSVD allows the simultaneous transmission of data and digitally-encoded voice signals over a single dial-up analog phone line. DSVD modems use for V.34 modulation (up to 33.6 kilobits per second), but may also use V.32 bis modulation (14,400 kilobits per second). DSVD modems reserve eight kilobits per second for voice transmissions. The remaining bandwidth is available for data transmission. The DSVD voice coder is a modified version of an existing specification and is defined as G.729 Annex A. The DSVD voice/data multiplexing scheme is an extension of the V.42 error correction protocol widely used in modems today. DSVD also specifies fallbacks that enable DSVD modems to communicated with standard data modems (i.e. V.34, V.32 bis, V.32 and V.22).
What Can You Do With DSVD? (Below is good manufacturer promotion on DSVD. The words came from Rockwell.)
Have you ever promised someone on the telephone that you'd fax a document to them, then forget to send it? Have you ever traveled cross-country or halfway around the world to make a presentation? Have you considered expensive video conferencing hardware and applications as a way to communicate? Ever been frustrated waiting online, hoping to get some technical support? When you need to work on a project, spreadsheet or presentation with someone from another location, a DSVD modem lets you do it without leaving your office or requiring costly video conferencing equipment. Both parties can simultaneously collaborate on a shared document, each viewing and discussing the additions and modifications made. This can slash time and money previously wasted exchanging faxes and relying on overnight express mail deliveries. Computer and software demonstrations can be easily conducted through a DSVD modem. This is a great way to highlight or test market new features and get immediate customer feedback. For the computer company, DSVD means fewer support follow-up calls and lower telephone bills. With a DSVD modem, you will be able to dial your favorite online catalog showrooms, browse the aisles, ask a salesperson a question or two, then place an order. Unlike other online "malls", with DSVD the human interface is still intact. While you're interactively shopping you can talk with salespeople or even other shoppers, just like the real mall. See V.70.

DSX Digital System Cross-connect frame. A bay or panel to which T-1 lines and DS1 circuit packs are wired and that permits cross-connections by patch cords and plugs. A DSX panel is used in small office applications where only a few digital trunks are installed. See also DACS.

DSX-1 Digital Signal Cross-connect Level 1. The set of parameters for cross connecting DS-1 lines.

DSX-3 The designation for the DS3 point of interface (cross-connect).

DTC 1. Digital Trunk Controller.
2. Digital Transmit Command.

DTE 1. Data Terminal Equipment. In the RS-232-C standard specification, the RS-232-C is connected between the DCE (Data Communications Equipment) and a DTE. The main difference between a DCE and a DTE is that pins two and three are reversed. See also DCE and the Appendix.
2. Defense Technology Enterprise.

DTE-DCE Rate Data terminal equipment/data communications equipment rate. A designation for the maximum rate at which a modem and a PC can exchange information, expressed in kilobits per second (kbps). For maximum performance, a modem must support a DTE-DCE rate in excess of its maximum theoretical throughput.

Dterm A line of proprietary electronic phones made by NEC for use with its PBXs. The Dterm terminal derives its intelligence from its own microprocessor, which detects events and accepts direction from the PBX.

DTL An ATM term. Designated Transit List: A list of nodes and optional link IDs that completely specify a path across a single PNNI peer group.

DTL Originator An ATM term. The first switching system within the entire PNNI routing domain to build the initial DTL stack for a given connection.

NEWTON'S TELECOM DICTIONARY

DTL Terminator An ATM term. The last switching system within the entire PNNI routing domain to process the connection and thus the connection's DTL.
DTLU Digital Trunk and Line Unit. Provides system access for T1-carrier lines used for inter office trunks or remote switching module umbilicals.
DTMF Dual Tone Multi-Frequency. A fancy term describing push button or Touchtone dialing. (Touchtone is a not registered trademark of AT&T, though until 1984 it was.) In DTMF, when you touch a button on a push button pad, it makes a tone, actually a combination of two tones, one high frequency and one low frequency. Thus the name Dual Tone Multi Frequency. In U.S. telephony, there are actually two types of "tone" signaling, one used on normal business or home push button/touchtone phones, and one used for signaling within the telephone network itself. When you go into a central office, look for the test board. There you'll see what looks like a standard touchtone pad. Next to the pad there'll be a small toggle switch that allows you to choose the sounds the touchtone pad will make — either normal touchtone dialing (DTMF) or the network version (MF).
The eight possible tones that comprise the DTMF signaling system were specially selected to easily pass through the telephone network without attenuation and with minimum interaction with each other. Since these tones fall within the frequency range of the human voice, additional considerations were added to prevent the human voice from inadvertently imitating or "falsing" DTMF signaling digits. One way this was done to break the tones into two groups, a high frequency group and a low frequency group. A valid DTMF tone has only one tone in each group. Here is a table of the DTMF digits with their respective frequencies. One Hertz (abbreviated Hz.) is one cycle per second of frequency.

Digit	Low frequency	High frequency
1	697 Hz.	1209 Hz.
2	697	1336
3	697	1477
4	770	1209
5	770	1336
6	770	1477
7	852	1209
8	852	1336
9	852	1477
0	941	1336
*	941	1209
#	941	1477

There are four other digits defined in the DTMF system and usable for specialized applications that cannot be generated by standard telephones They are:

A	697 Hz.	1633 Hz.
B	770	1633
C	852	1633
D	941	1633

Normal telephones (yours and mine) have 12 buttons, thus 12 combinations. Government Autovon (Automatic Voice Network) telephones have 16 combinations, the extra four (those above) being used for "precedence," which in Federal government parlance is a designation assigned to a phone call by the caller to indicate to communications personnel the relative urgency (therefore the order of handling) of the call and to the called person the order in which the message is to be noted. See also LONG TONES and the four following definitions.
DTMF Automatic Routing This is a term relating to a fax server operating on a Novell file server. In this system, the fax software assigns a four-digit number to each user. A fax sender dials the fax line, and after the fax server answers, it sends a special auto routing request signal. The sender dials the four-digit number for the correct user, and the fax is automatically sent to the user's workstation on the LAN.
DTMF Cut-Through The capability of a voice response system to receive DTMF tones while the voice synthesizer is delivering information, i.e. during speech playback. This capability of DTMF cut-through saves the user waiting until the machine has played the whole message (which typically is a menu with options). The user can simply touchtone his response anytime during the message — when he first hears his selection number, when the message first starts, etc. When the voice processor hears the touchtoned selection (i.e. the DTMF cut-through), it stops speaking and jumps to the chosen selection. For example, the machine starts to say, "If you know the person you're calling, touchtone his extension in now." But before you hear the "If you know" you push button in 230, which you know is Joe's extension. Bingo, the message stops and Joe's extension starts ringing. DTMF Cut-Through is also known as touchtone type-ahead.
DTMF Register A printed circuit card in a switch that converts the DTMF signals coming from the phone into signals which can be used by the switch's stored program control, central computer to do its switching, etc.
DTMF To Dial Pulse Conversion A PBX feature. DTMF (push button) phones are very popular. But sometimes you install a PBX with push button phones in an area which doesn't have a central office which will respond to push button tones. It's old. In this case, anyone dialing on a push button phone will find that the PBX converts that dialing to rotary pulsing when the PBX accesses a trunk which can't handle push button dialing. All this doesn't speed up the time the call takes to get through. It just speeds up the user's dialing and makes him or her feel she is dealing with a more modern phone system.
DTP DeskTop Publishing.
DTR Data Terminal Ready. A control signal sent from the DTE to the DCE that indicates that the DTE is powered on and ready to communicate. DTR can also be used for hardware flow control.
DTS Digital Termination Systems. Microwave based transmission technology designed for bypass functions for short-hop, line-

of-sight applications. It never converts to analog. Is useful in high-volume, pure-data applications in urban settings where line costs are high. It requires FCC license and is referred to formally by FCC as Digital Electronic Message Service, or DEMS.
DTSR Dial Tone Speed Recording.
DTT Digital Trunk Testing.
DTU Digital Test Unit.
DTX Battery-saving feature on a cellular phone that cuts back the output power when you stop speaking.
DU Fiber Optic Connector developed by Nippon Electric Group.
DUA Directory User Agent. The software that accesses the X.500 Directory Service on behalf of the directory user. The directory user may be a person or another software element.
DUAL Distributed Update Algorithm. A routing algorithm that provides fast rerouting (convergence) with minimal consumption of resources.
Dual Attachment Concentrator A concentrator that offers two connections to the FDDI network capable of accommodating the FDDI dual ring, and additional ports for connection of other FDDI devices.
Dual Attachment Station A device used in local area networks to allow access to two separate cable systems at the same time, providing protection against cable failure or damage.
Dual Cable A two-cable system in broadband LANs in which coaxial cable provides two physical paths for transmission, one for transmit and one for receive, as against dividing the capacity of a single cable.
Dual Coat An optical fiber coating structure consisting of a soft inner coating and a hard outer coating.
Dual Fiber Cable A type of optical fiber cable that has two single-fiber cables enclosed in an extruded overjacket of polyvinyl chloride with a rip cord for peeling back the overjacket to access the fibers.
Dual Headset Also known as an integrated headset. A special type of headset for the blind. One jack plugs into a telephone and another jack plugs into a telephone and another jack plugs into a specially configured PC. This PC provides voice synthesized output. The dual headset allows a visually impaired TSR (Telephone Sales Representative) hands-free capability. Example: The Social Security Administration has numerous blind TSRs handling incoming public calls. Dual headsets allow these blind TSRs to perform their duties with no deterioration in public service. This definition provided by Matt Gottlieb, telecommunications specialist for the Social Security Administration.
Dual Homing 1. The process of using two geographically diverse frame relay port connections, each with its own set of virtual circuits, to support a network location running critical business applications which cannot afford network down time.
2. A method of cabling FDDI concentrators and stations that permits a alternative path to the dual ring. Can be used in a tree or dual ring of trees topology.
Dual Line Registration The ability to have two cellular telephone numbers in a single cellular telephone. This allows the user to have service on two cellular systems without "roaming" in a second city (and paying higher toll charges) or to have one number on the wireline and one number on the non-wireline system. This assures the cellular user of backup. I know a salesman who lives in Los Angeles, but spends much of his week serving customers in Phoenix. He has a cellular phone with Los Angeles number and a Phoenix number.
Dual Line Service Telephone service where two pairs of wires are connected to the premises. One or both could be in service.
Dual Mode Dual mode is the cellular industry's term for a cellular phone which will transmit and receive, i.e. work for both analog and digital cellular phone systems. The cellular phone industry is going digital. Today's analog phones won't work on tomorrow's digital systems. But some phones — dual mode — will work on both. You need to buy them over today's analog phones. You also need to be careful that the digital mode which you get in your dual mode cellular phone will work with the digital technology of your local carrier. That's not as standard, as yet, as today's analog technology, which works universally in North America.
Dual NAM Allows a cellular phone user to have two phone numbers with the same or separate carriers. Very useful for someone who spends half his life in one place and half in another. For example, a friend of mine lives in LA, but works weekdays in Phoenix. His handheld cellular has phone numbers from LA and Phoenix carriers.
Dual Processing An SFT II configuration under Novell's NetWare that assigns parts of the operating system to separate processors. Because SFT II is split into two engines (the IOEngine and the MSEngine), it is possible to install each engine on a separate CPU, creating dual processing system. However, unless such a system is extremely busy, the extra CPU will not help network performance. Dual processing improves performance only when the servers are being used at near-maximum capacity.
Dual Ring Of Trees Topology An FDDI network topology of concentrators and nodes that cascades from concentrators on a dual ring.
Dual Tone Multi-Frequency DTMF. A way of signaling consisting of a push button or touchtone dial that sends out a sound which consists of two discrete tones, picked up and interpreted by telephone switches — either PBXs or central offices. See DTMF for a bigger explanation.
Duct A pipe, tube or conduit through which cables or wires are passed. Duct space is always at a premium. If you ever install a duct, make sure it's twice the diameter you think you need. If you're lucky, it will last a couple of years. The cost of putting in thicker or extra ducts is peanuts compared to the cost of having to install additional ones later. Digging up places is getting very expensive, despite Ditch Witch, a company that makes the greatest backhoe trenching equipment. And also has the greatest name.
Duct Cycle The relationship between the time a device or facility is used and the time it is idle.
Duct Liner A small diameter pipe or tubing placed inside conventional underground conduit so you can install fiber optic or cables. Its main purpose is to provide a clean, continuous path with known frictional characteristics.
Ductile Capable of being drawn out, hammered thin or being flexed or bent without failure.
Due Date The date an event is to occur, i.e. an installation, a change or a connection. Some vendors give accurate due dates. Penalty clauses work most effectively in ensuring due dates are met.
Dumb Switch A slang word for a telecommunications switch that contains only basic switching software and relies on instruc-

tions sent it by an outside computer. Those instructions are typically fed the "dumb" switch through a cable from the computer to one or more RS-232 serial ports which the dumb switch sports. The switch makes no demands on what type of computer it talks to, but simply insists that it be able to feed the computer questions and promptly receive responses in a form that it (the switch) can understand. Plain ASCII is OK. For example, the dumb switch might signal the computer, "A call is coming in on port 23, what do I do now?" The computer might reply "Answer it and transfer it to extension 23." Or it might say "answer it and put it on hold," or "answer it, put it on hold and play it recording number three." In essence, a dumb switch is anything but. It is in reality an empty cage containing whatever network interface cards the user has chosen. Each of these network interface cards is designed to "talk" to one type of telephone line. That line might be a T-1 line. It might be a normal tip and ring loop start line. It might be a tie trunk with E&M signaling. The card may handle one or many lines, but always of the same type. The card knows how to answer a call or pulse out a call on that particular type of line. It has all the telephony smarts. What it lacks is the intelligence of what to do with the calls. That is provided by the outside computer. Well, almost. Most "dumb" switches do contain rudimentary intelligence — a small computer and some memory. That computer is usually programmed to handle "default" calls — and to handle calls should the link to the outside computer fail, or the outside computer itself fail. Dumb switches come in flavors all the way from residing in their own cabinet to being printed circuit cards which reside in one or more of the personal computer's slots. Dumb switches are programmed to do "specialized" telecom applications, for example emergency 911, added value 800 services, cellular switching, automatic call distributors, predictive dialers, etc. They can, of course, be programmed to be "normal" PBXs. The question increasingly being asked is "If I want to program a specialized telecom application should I use a dumb switch or should I use an open PBX?" And the answer is "It depends." Depends on what you want to do. Depends on what software is available, etc. See also OAI.

Dumb Terminal A computer terminal with no processing or no programming capabilities. Hence, it derives all its power from the computer it is attached to — typically over a local hardwire or a phone line. A dumb terminal does not employ a data transmission protocol and only sends or receives data one character at a time, sequentially. There are many reasons for "dumb" terminals. They're cheap. They're foolproof. Operators don't have to mess with floppy disks, etc. The require minimal training. Dumb terminals are typically used for simple data entry and data retrieval tasks. Their disadvantage is that everything must come from the central computer — not only the information (data record) but also the form in which to put it. This has led to the creation of "intelligent" terminals, which have a modicum of capabilities — such as an inbuilt (with software) form, some smart function keys and perhaps, a modicum of processing power, etc.

Dumb Terminal Access An Internet Access Term. Telnet shell command that allows remote terminals to connect to a LAN host. Basic terminal server support over the same modems and phone lines used for remote access.

Dummy Load A dissipative impedance-matched network, used at the end of a transmission line to absorb all incident power, usually converted to heat.

Dump To copy the entire contents of something — memory, a file on a disk, a complete disk — to a printer or another magnetic storage medium. A dump is often called a "core dump," which is a bigger dump than dump.

Dumpster Diving Searching for access codes or other sensitive information in the trash. In North America, large trash receptacles — those found outside buildings — are often known as dumpsters.

Duobinary Signal A pseudobinary-coded signal in which a "0" ("zero") bit is represented by a zero-level electric current or voltage; a "1" ("one") bit is represented by a positive-level current or voltage if the quality of "0" bits since the last "1" bit is even. and by a negative-level current or voltage if the quantity of "0" bits since the last "1" bit is odd. Duobinary signals require less bandwidth than NRZ. Duobinary signaling also permits addition of error-checking bits.

Duopoly Similar to monopoly except there are two licensed competitors per market instead of one. The cellular business in the United States is a duopoly. Each major city in the United States has two licensed cellular providers — a feat accomplished by the FCC. As a result, prices in most cities as between the carriers are identical.

Duplex 1. Simultaneous two-way transmission in both directions. A data communications term.
2. Two-sided printing.

Duplex Circuit A telephone line or circuit used to transmit in both directions at the same time. Also referred to as full duplex as opposed to half duplex which allows transmission in only one direction at one time.

Duplex Operation The simultaneous transmission and reception of signals in both directions.

Duplex Signaling DX. A direct current signaling system that transmits signals directly on the cable pair. Duplex signaling is a facility signaling system and range extension technique that used bridge type detection of small dc changes. Duplex signaling is typically used on long metallic trunks.

Duplex Transmission The simultaneous transmission of two series of signals by a single operating communicating device. A data communications term.

Duplexer 1. A device which splits a higher speed source data stream into two separate streams for transmission over two data channels. Another duplexer at the other end puts the two slower speed streams back together into one higher-spread stream.
2. A waveguide device designed to allow an antenna to be used both transmission and reception simultaneously.

Dustbuster A phone call or e-mail message you send to someone after a long silence — just to "shake off the dust" and see if the connection still works.

Duty Cycle The ratio of operating time to total elapsed time for a device that operates intermittently. Usually expressed as a percentage.

DV-1 As in Northern Telecom's PBX/computer product called the Meridian DV-1.

DVB The Digital Video Broadcasting Group is a European organization which publishes its work through ETSI that has authored many specifications for satellite and cable broadcasting of digital signals. Part of the DVB work has been focused on conditional access and has produced draft recommendations for a common CA module and the DVB Superscrambling algorithm.

DVD Digital Versatile Disc. A specification announced in early December, 1995 by nine companies for a new type of digital

videodisk, similar to CD-ROMs but able to store far more music, video or data in a common format. DVDs will be 5 inches in diameter (same diameter as today's CDs) and will be able to store as much as 17 gigabytes, compared to the 650 megabytes on today's CD-ROMs. Current CD-ROMs are 1.2 mm thick. The DVD disc used two bonded 0.66 mm substrates, reducing the amount of distance between the surface of the disc and the physical pits on the disc that hold information. The laser doesn't have to penetrate as much plastic so it can be focused on a smaller area. Therefore, the pits can be made smaller and packed more tightly together, increasing capacity. At the same time, two different lens apertures can be included in the pick-ups, so information can be read by both DVDs and standard CDs. Not only can the disc hold more information, but having two bonded sides strengthens it and prevents warping. Here is a comparison of DVD storage:
Single-sided, single layer DVD5 = 4.7 gigabytes
Single-sided, dual layer DVD9 = 8.5 gigabytes
Double-sided, single layer DVD10 = 9.4 gigabytes (i.e. 2 x 4.7 gigabytes)
Double-sided, double layer DVD18 = 17 gigabytes (i.e. 2 x 8.5 gigabytes)
Write Once DVD-R (Recordable) = 3.8 gigabyte per side
Overwrite DVD RAM "More than 2.6 gigabyte per side
DVD Specifications:
Smallest Pit Size = 0.4 microns
Track Pitch: 0.74 microns
Sector Layout: Constant Linear Density
Modulation: 8 to 16
Laser Diode Wave Length: 650/635 nm
Lens Numericl Aperture: 0.6
Logical Format: UDF/Micro UDF/iSO-9660
Error Correction: Reed Solomon Product Code

DVD-R Digital Versatile Disc — Record. A write once 5" doubled-sided disc with a capacity of 3.8 gigabytes per side. See DVD.
DVI Digital Video Interactive. A name for including still and moving video pictures in material shown on a PC's screen. DVI is part of multimedia. DVI is also Intel's old name for its scheme for digitizing and compressing video and audio for storage, editing, playback and integration into PC applications. The name has been replaced on the software side with Indeo video technology, on the retail side with Smart Video Recorder and on the hardware side with i750 Processors. See DIGITAL VIDEO INTERACTIVE for a much bigger explanation.
Dvorak Keyboard A keyboard, invented mainly by August Dvorak, on which letters and characters are arranged for faster and easier typing than on the standard QWERTY keyboard. The QWERTY keyboard was actually designed to be difficult to use, to slow down typists so they wouldn't jam the old typewriters' mechanisms! In that respect the QWERTY keyboard resembles the present touchtone in that it also was designed to be slow and difficult to use so that it wouldn't confuse the early and slow central offices of the time.
DVTS Pronounced "Divitz." Stands for Desktop Video conferencing Telecommunications System.
DWS Dialable Wideband Service.
DX Signaling A form of DC (direct current) signaling in which the differences in voltage on two pairs of a four-wire trunk indicates the supervision information, i.e. the call's beginning, its end, etc. See DUPLEX SIGNALING.
DXI Data eXchange Interface. A specification developed by the SMDS (Switched Megabit (or Multi-megabit) Data Services) Interest Group to define the interaction between internetworking devices and CSUs/DSUs that are transmitting over an SMDS access line. SMDS is a way for a corporate network to dial up switched data services as fast as 45 megabits per second. The ATM Forum defines DXI as "a variable length frame-based ATM interface between a DTE and a special ATM CSU/DSU. The ATM CSU/DSU converts between the variable-length DXI frames and the fixed-length ATM cells."
Dye Sublimation a spectacular printing process where exactly measured temperatures control the amount of ink transferred from colored ribbons to paper. Under high temperature and pressures, the ink is not melted, but is transformed directly to gas, which hardens on the paper after passing through a porous coating. Dye sub printers create very nearly continuous tones, making them great for natural images. Because the gas makes "fuzzy" dots, dye sub is not recommended for sharp-edged "computer-y" graphics or type. But it does turn out gorgeous photo-like images.
Dynamic Events are constantly changing.
Dynamic Answer This a term typically used in Automatic Call Distributors. The ability to dynamically assign the number of ring cycles (interrupt, more or less) to the queue period when agents are unavailable. The implication of being able to assign this number allows return supervision to the calling in person to be delayed and thus not allow billing on 800 INWATS lines to begin. This is a money saving feature. But it can cost you some customers if they get bored waiting for your phones to pick up.
Dynamic Backup A backup made while the database is active.
Dynamic Bandwidth Allocation The capability of subdividing large, high-capacity network transmission resources among multiple applications almost instantaneously, and providing each application with only that share of the bandwidth that the application needs at that moment. Dynamic bandwidth allocation is a feature available on certain high-end T-1 multiplexers that allows the total bit rate of the multiplexer's circuits to exceed the bandwidth of the network trunk. This works because the multiplexer only assigns channels on the network trunk to circuits that are transmitting.
Dynamic Beam Focusing When you have a curved cathode ray tube, the distance between the gun which shoots the electrons and all the parts of the screen are equal. When you have a flat screen, the distance varies slightly. Some beams have to travel further. When some have to travel not so far, Dynamic beam focusing, a term I first heard used by NEC, focuses each electron to the precise distance it must travel, thus ensuring edge-to-edge clarity on the screen.
Dynamic Capacity Allocation The process of determining and changing the amount of shared communications capacity

assigned to nodes in the network based on current need.

Dynamic Configuration Registry A part of Chicago (Windows 4.0) which contains a list of all the various hardware bits and pieces that make up your computer. The dynamic configuration registry is a vital element of what Microsoft calls "Plug and Play," which is the ability to remove and add bits and pieces of hardware while the machine is running and have the machine automatically recognize those hardwares and alert applications accordingly.

Dynamic Data Exchange DDE. A form of InterProcess Communication (IPC) in Microsoft Windows and OS/2. When two or more programs that support DDE are running simultaneously, they can exchange information, data and commands. In Windows 3.xx this capability is enhanced with Object Linking and Embedding (OLE). See OLE.

Dynamic Host Configuration Protocol DHCP. A protocol for automatic TCP/IP configuration that provides static and dynamic address allocation and management.

Dynamic IP addressing An Internet Access Term. Allows dial-in client to update a variety of TCP/IP protocol stacks with a dynamically acquired address. Removes the burden of assigning per-user IP addresses and addresses the issue of dialing into different subnets. Virtually all the 30 million or so people who regularly surf the Internet do so through dynamic IP addressing.

Dynamic Link Library DLL. An executable code module for Microsoft Windows that can be loaded on demand and linked at run time, and then unloaded when the code is no longer needed.

Dynamic Load & Stress Testing An advanced and accurate form of load testing, dynamic load and stress testing more accurately presents the variety of stimulus of external callers, systems and networks to a computer telephony system. It will test your system using real-world user actions and a realistic call mix and busy hour usage profile. Most modern computer telephony network and systems provide a number of services to their users. Each of those services may stress the system in a different fashion. As the usage patterns of those users is varied and what each user will do can vary significantly, it is critical to load test systems with traffic patterns as dynamically and as close as possible to the way the system will be used in the real world. This definition courtesy of Steve Gladstone, author of the book Testing Computer Telephony Systems, available from 212-691-8215.

Dynamic Load Balancing A technique where a switching system, particularly multiple connected ACDs, apportion incoming calls (the load) to balance the workload. This is done dynamically in real time.

Dynamic Loud Speaker A loud speaker in which the diaphragm is driven by means of a small "voice coil" suspended in a powerful magnetic field.

Dynamic Memory The most common form of memory, used for RAM, with an access speed ranging from about 60 to 150 nanoseconds. (A nanosecond is thousandth of a second.) Dynamic memory is an inexpensive but relatively complicated form of semiconductor memory with two states: presence and absence of electrical charge. Dynamic memory requires a continuous electrical current. All data is lost when the power is cut. Frequent saving files to disk helps preserve your data.

Dynamic Microphone A microphone, the coil of which is moved in a strong magnetic field by vibrations striking the diaphragm to which it is attached. Electrical currents are thus generated in the moving coil.

Dynamic Node Access A high-speed bus invented by Dialogic to join together multiple voice processing PEB-based systems. PEB stands for PCM Expansion Bus. See PEB.

Dynamic Overload Control DOC. The feature of a switch which uses its translation tables and intelligence to allow the switch to adapt to changes in traffic loads by re-routing and blocking call attempts.

Dynamic Port Allocation In a voice processing system running multiple applications, dynamic port allocation is automatic allocation of ports based on the traffic being used by each application.

Dynamic RAM RAM memory that requires data to be refreshed periodically to prevent its loss in memory.

Dynamic Random Access Memory RAM which requires electronic refresh cycles every few milliseconds to preserve its data. See also RANDOM ACCESS MEMORY.

Dynamic Range In a transmission system, the ratio of the overload level to the noise level of the system, usually expressed in decibels. The ratio of the specified maximum level of a parameter (e.g., power, voltage, frequency, or floating point number representation) to its minimum detectable or positive value, usually expressed in decibels.

Dynamic Resource Allocation The assignment of network capacity to specific users and specific services as required on a moment-to-moment basis.

Dynamic Routing Routing that adjusts automatically to changes in network topology or traffic.

Dynamic Storage Allocation The allocation of memory space while a program is running. The memory is released when the program is complete.

Dynamic Variation A short time variation outside of steady-state conditions in the characteristics of power delivered to communication equipment.

Dynamically Adaptive Routing An algorithm, used for route determination in packet-switched networks that automatically routes traffic around congested, damaged, or destroyed switches and trunks and allows the system to continue to function over the remaining portions of the network.

Dynamo An electrical machine which generates a direct current.

Dynamotor A direct current machine having two windings on its armature: one acting as a motor, the other as a generator.

NEWTON'S TELECOM DICTIONARY

E (Echo) Channel 16 Kbps ISDN basic rate channel echoing contents of DCEs to DTEs. Used in bidding for access to multipoint link.

E & M In telephony, a trunking arrangement that is generally used for two way (either side may initiate actions) switch-to-switch or switch-to-network connections. It also frequently used for computer telephony system to switch connections. See E & M LEADS.

E & M Leads The pair of wires carrying signals between trunk equipment and a separate signaling equipment unit. The "M" lead transmits a ground or battery conditions to the signaling equipment. The "E" lead receives open or ground signals from the signaling equipment. These leads are also known as Ear and Mouth Leads. The Ear lead typically means to receive and the Mouth lead typically means to transmit. Changes of voltage on these leads convey such information as seizure of circuit, recognition of seizure, release of circuit, dialed digits, etc. In the old days it was the PBX operators who originated trunk calls by asking the long distance carrier for free trunks using their mouth or M lead. If the carrier had a free trunk, the PBX heard about it through its ear or E lead. See also E & M SIGNALING.

E & M Signaling In telephony, an arrangement that uses separate leads, called respectively the "E" lead and "M" lead, for signaling and supervisory purposes. The near end signals the far end by applying -48 volts dc (vdc) to the "M" lead, which results in a ground being applied to the far end's "E" lead. When -48 vdc is applied to the far end "M" lead, the near-end "E" lead is grounded. The "E" originally stood for "ear," i.e., when the near-end "E" lead was grounded, the far end was calling and "wanted your ear." The "M" originally stood for "mouth," because when the near-end wanted to call (i.e., speak to) the far end, -48 vdc was applied to that lead.
When a PBX wishes to connect to another PBX directly or to a remote PBX or extension telephone over a leased voice grade line, a channel on T-1, the PBX uses a special line interface which is quite different from that which it uses to interface to the phones it's attached directly to (i.e. with in-building wires). The basic reason for the difference between a normal extension interface and the long distance interface is that the signaling requirements differ — even if the voice signal parameters such as level and two-wire, 4-wire remain the same. When dealing with tie lines or trunks it is costly, inefficient and too slow for a PBX to do what an extension telephone would do, i.e. go off hook, wait for dial tone, dial, wait for ringing to stop, etc. The E&M tie trunk interface device is the closest thing there is to a standard that exists in the PBX, T-1 multiplexer, voice digitizer telco world. But even then it comes in at least five different flavors. E&M signaling is the most common interface signaling method used to interconnect switching signaling systems with transmission signaling systems. See E & M LEADS.

E Link Extended Link. A SS7 signaling link used to connect a Signaling End Point (SEP) to an STP pair not considered its home STP pair.

E Mail Electronic Mail.

E Mail Gateway A LAN application that fetches messages from one electronic mail system, translates them to the format of another electronic mail system, and then sends them to the "post office" of that other system. The post office is the public entry point — the place you put mail you want the other system to receive.

E Purse Electronic purse. An electronic monetary transaction card being proposed by several government agencies.

E-1 The European equivalent of the North American 1.544 Mbps T-1,except that E-1 carries information at the rate of 2.048 megabits per second. This is the rate used by European CEPT carriers to transmit 30 64 Kbps digital channels for voice or data calls, plus a 64 Kbps channel for signaling, and a 64 Kbps channel for framing (synchronization) and maintenance. CEPT stands for the Conference of European Postal and Telecommunication Administrations. Since robbed-bit signaling is not used (as it is for T-1 in North America) all 8 bits per channel are used to code the waveshape sample. See E1, E2, E3, and T-1.

E-2 Interim data signal that carries four multiplexed E-1 signals. Effective data rate is 8.448 Mbps.

E-3 CEPT signal which carries 16 CEPT E-1s and overhead. Effective data rate is 34.368 Mbps.

E-911 Service Enhanced 911 service. Dial 911 in most major cities and you'll be connected with an emergency service run typically by a combination of the local police and local fire departments. 911 service becomes enhanced 911 emergency reporting service when there is a minimum of two special features added to it. E-911 provides ANI (Automatic Number Identification) and ALI (Automatic Location Information) to the 911 operator. Picture: A call comes in. Someone is dying. The 911 operator's screen comes alive as his phone rings. The number calling is on the screen. The caller is dying and needs an ambulance. The operator punches a button or two and his screen immediately indicates the location of the ambulance dispatch center nearest the caller. The operator contacts the dispatch center, another button may dispatch a fax of a map of how to get there to the ambulance and an ambulance gets there in short order and saves a live. (Remember, this is a book, not the real world.)

E-Bend A smooth change in the direction of the axis of a waveguide, throughout which the axis remains in a plane parallel to the direction of electric E-field (transverse) polarization.

E-Commerce Electronic Commerce. Some experts think using electronic commerce to place orders could slash the cost of a typical company purchase order from $150 to $25.

E-IDE Enhanced IDE. An enhancement to the original IDE disk drive found on many PCs. E-IDE raises the storage capacity limit from 504 megabytes to 8,033 megabytes and the data transfer rate from up to 3 megabytes per second to up to 16.6 megabytes per second. See also ENHANCED IDE.

E-Mail Electronic Mail.

E-TDMA Extended Time Division Multiple Access. A proposed, new, standard for cellular. Other standards are TDMA (Time Division Multiple Access), CDMA (Code Division Multiple Access) and NAMPS (Narrow Advanced Mobile Phone Service). Refers to the extended (digital cellular) transmission technology developed by Hughes Network Systems. E-TDMA is alleged to have 15 times the capacity of today's analog cellular phone systems — in other words to allow the simultaneous use of 15 times as many cellular phones as today's analog cellular phone system.

E-Zine Magazines that are published (i.e. made public) on the World Wide Web. Typically an e-zine is available for anyone to read who wants to visit the site the electronic magazine is located at. An e-zine is also called a Webzine.

E.164 1. ITU-T recommendation for international telecommunication numbering, especially ISDN, BISDN, and SMDS. An evolution of normal telephone numbers. In short, a scheme to assign numbers to phone lines.
2. An ATM term. A public network addressing standard utilizing up to a maximum of 15 digits. ATM uses E.164 addressing for public network addressing.

E.169 An ITU-T recommendation which will allow International Freephone Service customers to be allocated a unique Universal International Freephone Number (UIFN) which will remain the same throughout the world, regardless of country or telecommunications carrier. "Freephone" is a service which permits the cost of a telephone call to be charged to the called party, rather than the calling party. In North America, 800 and 888 numbers are "freephone" numbers.

E1 Name given to the CEPT (Conference of European Postal and Telecommunication Administration) digital telephony format devised by the ITU-T that carries data at the rate of 2.048 million bits per second (DS-1 level). It's designed to carry 32 64 Kbps digital channels. E1 is the rate used by European CEPT carriers to transmit 30 64 Kbps digital channels for voice or data calls, plus a 64 Kbps channel for signaling, and a 64 Kbps channel for framing and maintenance. Plesiochronous means "almost synchronous." In the network sense, when two networks operate with clocks of sufficiently high quality such that the signals in the two networks are nearly synchronous, the networks are plesiochronous. In the synchronization hierarchy, stratum I clocks are required for plesiochronous operation. See E-1 and T-1, which is the North American equivalent.

E2 Data signal that carries four multiplexed E-1 signals. Effective data rate is 8.448 million bits per second.

E3 CEPT signal which carries 16 CEPT E-1 circuits and overhead. Effective data rate is 34.368 million bits per second. Also known as CEPT3.

E4 Carries four E3 channels — or 139,264 million bits per second, or 1920 simultanous conversations.

E5 Carries four E4 channels — or 565,148 million bits per second, or 7,680 simultaneous voice conversations.

EA SEE EQUAL ACCESS.

EACEM The European Association of Consumer Electronics Manufacturers.

EADAS Engineering and Administrative Data Acquisition System.

EADAS/NWM EADAS NetWork Management.

Early Packet Discard EPD. A congestion control technique that selectively drops all but the last ATM cell in a Classical IP over ATM packet. When congestion occurs, EPD discards cells at the beginning of an IP packet, leaving the rest intact. The last cell is preserved because it alerts the switch and the destination station of the beginning of a new packet. Because IP packets from cells have been discarded receive no acknowledgment from the source. Most vendors expect EPD to be used in conjunction with unspecified bit rate (UBR) service. Switches simply junk UBR cells when congestion occurs, without regard for application traffic. By discarding ceiling selectively, so that whole IP packets are resent, EPD makes UBR a safer option.

Early Token Release This is a method of token passing which allows for two tokens to exist on the network simultaneously. It is used primarily in 16-Mbps token ring LANs. On a regular 4-Mbps token ring LAN, the token is passed on only after the sending computer receives its message back from the destination computer. With early token release, the sending computer does not wait for its message to return before passing the token. This means there are two tokens on the network at the same time. This is done to take advantage of the idle time created on the faster token ring. While the message is moving to its destination and back the sending computer is idle. On the 4-Mbps token ring, this is not much of a problem since most of the time the message is on the ring. On the 16-Mbps token ring less of this idle time is transmission time and more is taken by copying the message to the token ring card. That is, on the faster ring, there is more of a window for a second token. Early token release is especially helpful when traffic is heavy.

EARN European Academic Research Network. A network using BITNET technology connecting universities and research labs in Europe.

EAROM An acronym for Electrically Alterable Read-Only Memory. A type of ROM chip which can be erased and reprogrammed without having to be removed from the circuit board. An EAROM is reprogrammed electrically faster and more conveniently than an EPROM (Erasable Programmable Read-Only Memory). An EAROM chip does not lose its memory when power is turned off.

Earth Ground The connection of an electrical system to earth. This connection is necessary to provide lightning and static protection as well as to establish the zero-voltage reference for the system. See EARTH GROUNDING.

Earth Ground Connection The conductor which connects directly to earth ground usually via a water pipe or possibly via a copper rod driven into the earth. This ground is different than the logic ground used in electronic circuits.

Earth Ground Electrode The conducting body in contact with the earth. The grounding electrode may be a metallic cold water pipe when used in conjunction with a driven rod, a mat, a grid, etc. Earth should never be used as the sole equipment grounding conductor.

Earth Grounding The purpose of earth grounding is essentially threefold: 1. Lightning protection; 2. Static protection; and 3. Establish a zero voltage reference. See GROUND and GROUNDING.

Earth Station The antenna and associated equipment used to receive and/or transmit telecommunications signals via satellite.

Earthing There are two distinct and unique categories in the broad area called grounding. One is earthing, which is designed to guard against the adverse effects of lightning, assist in the reduction of static and bring a zero-voltage reference to system components in order that logic circuits can communicate from a known reference. The other category of grounding is known as equipment grounding. This is the primary means of protecting personnel from electrocution. According to the Electric Power Research Institute, "electrical wiring and grounding defects are the source of 90% of all equipment failures." Many telephone system installer/contractors have found that checking for and repairing grounding problems can solve many telephone system problems, especially intermittent "no trouble found" problems. As electrical connections age, they loosen, corrode and become subject to thermal stress that can increase the impedance of the ground path or increase the resistance of the connection to earth. Equipment is available to test for proper grounding. One of our favorite devices is made by Ecos Electronics in Oak Park, IL. Before you attach any equipment (computer, telephone, hi-fi set, etc.) to an improperly grounded electrical outlet, you should have the problems corrected. See also GROUND and GROUNDING.

EAS Extended Area Service. A novel name for a larger than normal local telephone calling area. The local phone company extends its subscribers the option of paying less per month for a small calling area and paying extra per individual call outside that area (i.e. the extended area), or paying more per month flat rate but having a larger calling area (i.e. having extended area service).

EASE A voice processing applications generator from Expert Systems, Inc. in Atlanta GA.

EASY Sabre Consumer oriented version of American Airlines' Sabre reservation system, accessible as an option on most on line services (American On line, CompuServe, Delphi, GEnie, Prodigy, etc). It will give you airline flights and prices, and it will allow you to make reservations.

EASYNET The name of DEC's internal corporate network.

EAX Electronic Automatic eXchange. Term used throughout the non-Bell telephone industry to refer to an electronic central office. Similar to ESS (Electronic Switching System), the term used by AT&T and the Bell operating telephone companies.

EBCDIC (Pronounced Eb-si-dick.) Extended Binary Coded Decimal Interexchange Code. It is the way IBM codes characters, letters and numbers into a digital binary stream for use in its larger computers. EBCDIC codes characters into eight bits. This gives it 256 possible characters, $2 \times 2 \times 2 \times 2 \times 2 \times 2 \times 2 \times 2 = 256$. See also EXTENDED CHARACTER SET.

EBCDIC is mainly used in IBM mainframes and minicomputers, while ASCII is used in IBM and non-IBM desktop microcomputers. EBCDIC is not compatible with ASCII, meaning that a computer which understands EBCDIC will not understand ASCII. But there are many real-time and non-real time translation programs that will convert text files back and forth. A good program is Word-For-Word from MasterSoft. See ASCII.

EBDI Electronic Business Data Interchange. Term for EDI (Electronic Data Interchange). See EDI.

EBITDA Earnings Before Interest, Taxes, Depreciation and Amortization. Otherwise referred to as cash flow. Most commonly used for the valuation of wireless telecom stocks because of the heavy fixed-cost nature of the business.

EBONE European Backbone. A pan-European network backbone service.

EBS 1. Electronic Business Set.

2. Enhanced Business Service (also known as P-Phone) is an analog Centrex offering provided by Northern Telecom. It operates over a single-pair subscriber loop., providing normal full duplex audio conversations and a secondary 8 KHz half-duplex amplitude shift-keyed signal, which is used to transmit signaling information to and from the Northern Telecom-equipped central office.

3. Emergency Broadcast System. Some local radio stations have volunteered their services to be part of a group of radio stations which would broadcast information should there be a public emergency. They operate the EBS under the aegis of the Federal Communications Commission. Such emergency would be a natural disaster, a technological disaster or a war.

EBU European Broadcasting Union

EC European Community. Also known as the Common Market; an organization of 12 nations in Western Europe that has its own institutional structure and decision-making framework. The intent of the organization is to promote trade and reduce barriers. Member nations are Belgium, Denmark, France, Germany, Greece, Ireland, Italy, Luxembourg, the Netherlands, Portugal, Spain and the United Kingdom.

ECC RAM Error-correcting code memory, often used in local area network servers. ECC memory tests for and corrects errors on the fly, using circuitry that generates checksums to correct errors greater than one hit.

Eccentricity Like concentricity a measure of the center of a conductor's location with respect to the circular cross section of the insulation. Expressed as a percentage of center displacement of one circle within the other.

ECH Enhanced Call Handling. ECH systems are those in which a telephone call is handled "intelligently" b a variety of network, human, computer and telecommunications resources. ECH systems cover those from voice mail, to interactive voice response to computer telephone integration to fax-on-demand to complex telephone networks.

Echelon The name of a startup company in Palo Alto that is making a microprocessor chip destined for mundane household appliances like toasters, air conditioners, ovens, etc. The idea is that chip will be used by these devices to talk to other devices and thus coordinate their coming on and going off and doing things. The chips are destined to be networked together. Early uses for the chip includes smoke detectors which call you when they detect smoke and wall switches that detect when you come into a room and turn the lights on.

ECHO 1. European Commission Host Organization.

2. Exactly what you expect it to mean. You hear yourself speak. Echoes happen in both voice and data conversation. Echoes are good and bad. In voice, an echo happens when the equipment meant to amplify the voice of the party at one end, picks up the signals from the party at the other end, and amplifies them back to that party. Some echo is acceptable (in fact, almost a necessity) in voice conversations. It's called SIDETONE. When the speaker can hear himself speak through the receiver, sidetone gives the speaker some feeling his conversation is actually going through. But too much (i.e. too loud) echo is unacceptable. There are devices called echo suppressors, which do exactly that.

In low speed or on-line data conversations, an echo is positively vital. An echo in a data conversation is where I send my words to the distant computer which "echoes" them back to me and my screen displays them. This way I can visually check if the distant computer received my words/data accurately. In some software programs there's a command called ECHO or ECHOPLEX.

Switch it one way, the distant computer echoes the words to my screen. Switch it the other, the words I'm typing on my keyboard are put on my screen.

There are two basic transmission modes in data communications — full duplex and half duplex. In full duplex, I have simultaneous two-way data flowing. In full duplex, I therefore have the capacity to "echo" back the data I am sending and have it displayed on my screen. But this "echoing" depends on the capability and/or programming of the computer at the other end. All dial-up services — Tymnet, GTE Telenet, MCI Mail, etc. — will echo my data back to me so I can check it.

Some computers, such as the extremely dumb Compugraphic typesetter which typeset the first edition of this book, won't send an echo. In this case, if I want to see the data I am sending, I can change the parameters on my communications software to "half duplex." This way I will see what I am sending, but I will not see what the computer is receiving. Which may be very different. (And with our Compugraphic often is.) Of course, watching characters being echoed across your screen is a very poor method of data transmission and only useful in on-line transmissions. Some form of error-checking protocol is much better.

An echo is also a public discussion group that extends over more than one BBS (bulletin board system) via echomail.

Echo Attenuation In a communications circuit (4- or 2-wire) in which the two directions of transmission can be separated from each other, the attenuation of echo signals that return to the input of the circuit under consideration. Echo attenuation is expressed as the ratio of the transmitted power to the received echo power in decibels. See also ATTENUATION.

Echo Cancellation Technique that allows for the isolation and filtering of unwanted signals caused by echoes from the main transmitted signal. See ECHO CANCELLATION.

Echo Canceller Device that allows for the isolation and filtering of unwanted signal caused by echoes from the main transmitted signal. In data communications networks, echo cancelers are used in the same way as PADS are in the network, but some brands of echo cancelers have the ability of being disabled by a 2100 Hertz tone transmitted by the data device prior to the exchange of the data device's handshaking protocol. If the echo canceler cannot be disabled by the data device, it will block the data call from completing.

Echo Check A technique for verifying data sent to another location by returning the received data (echoing it back) to the sending end.

Echo Modeling A mathematical process where an echo is conceptually created from an audio waveform and subtracted from that form. The process involves sampling the acoustical properties of a room and guessing what form an echo might take, then removing that information from the audio signal.

Echo Return Loss ERL. The difference between a frequency signal and the echo on that signal as it reaches its destination.

Echo Suppressor Used to reduce the annoying effects of echoes in telephone connections. The worst echoes occur on satellite circuits. An echo suppressor works by turning off transmission in the reverse direction while a person is talking, thus effectively making the circuit one way. An echo suppressor obviously impedes full-duplex data — data flowing both ways simultaneously. Echo suppressors are turned off by the high-pitched tone (typically 2025 Hz) in the answering modem, which it uses to signal it's answered the phone and is ready for a data conversation.

Echo Suppressor Disabler An echo suppressor disabler is a device which causes an echo suppressor to be disabled (i.e. turned off). Echo suppressors are turned off by the high-pitched tone (typically 2025 Hz) in the answering modem, which it uses to signal it's answered the phone and is ready for a data conversation. A disabled echo suppressor stays disabled until the circuit is disconnected and restored to its "ready" connection. Because an echo suppressor hinders full duplex transmission in data communications, it is necessary to disable the echo suppressor.

Echomail A public message area or conference on a bulletin board system (BBS) that is "echoed" to other systems in a BBS network. EchoMail is organized into different groups, each with a different topic and the term normally references communications on a FidoNet network. Also a term referring to the electronic transfer of messages between bulletin board systems.

Echoplex A way of checking the accuracy of data transmitted whereby the data received are returned to the sender for comparison with the original data. Somewhat time consuming. Used typically in slow speed transmissions. See ECHO.

ECITC European Committee for Information Technology testing and Certification.

ECL Emitter Coupled Logic.

ECM Error Correction Mode. An enhancement to Group 3 fax machines. Encapsulated data within HDLC frames providing the received with an opportunity to check for, and request retransmission of garbled data. See FACSIMILE and V.17.

ECMA European Computer Manufacturers Association. See also CEPT.

ECOC European Conference on Optical Communication.

Economic Bandwidth An AT&T term for the maximum bandwidth that a physical medium can support without a significant increase in its cost.

Economist The standing joke is that an economist is someone who didn't have the personality to make accountant. There is a theory floating around in some academic circles that God invented economists to make weather forecasters and astrologists look good. See ACCOUNTANT.

Economy Of Scale As throughput gets bigger, so the per unit cost comes down. This is the argument used by economists to justify monopolies — namely that the per unit costs of one supplier are far lower than having two suppliers. The economy of scale argument is used to justify having only one water company in town. It makes more sense in that industry than in the telephone industry.

It was once used in the telephone industry to justify one combined local phone company/long distance phone company/one supplier of terminal equipment. This argument does not really apply to telephony as technological breakthroughs have brought down the cost of getting into the telephone industry and have allowed smaller, competitive companies to become cost effective. Some large telecommunications monopolies, in fact, are experiencing diseconomies of scale. In this case, their cost of per unit business starts to rise as they get very large. Diseconomies of scale are caused by bloated bureaucracies and inertia in management decision making.

Econopath Calling Plans Discount plans for New York City businesses to make calls in their regional calling area. For example, Econopath offers a Manhattan business a discount for calls to the East Suffolk region.
ECP 1. Enhanced Call Processing. An Octel term for an interactive customized menu in its voice mail system which provides levels of call routing. See ENHANCED CALL PROCESSING.
2. Extended Capabilities (Parallel) Port. An upgrade to the original parallel port on a PC, which gives you: Transfer rates of more than two million bytes per second; bidirectional 8-bit operation)a standard parallel port has only 4 input bits); support for CD-ROM and scanner connections; 16-byte FIFO buffer; support for run length coding data compression. See also USB.
ECPA Electronic Communications Privacy Act.
ECS Energy Communications Services.
ECSA Exchange Carriers Standards Association. See EXCHANGE CARRIERS STANDARDS ASSOCIATION.
ECTEL The European Telecommunications and Professional Electronics Industry.
ECTF Enterprise Computer Telephony Forum is a California non-profit mutual benefit corporation, formed to focus on the technical challenges of interoperability among Computer Telephony Integration (CTI) products. It was formed in the Spring of 1995. Information on the ECTF is available from The ECT Forum, Foster City, CA 415-578-6852. ectf@sbexpos.com. On July 31, 1995, the Versit companies cooperating in computer telephony integration (Apple, AT&T, IBM, Novell and Siemens) announced that they will bring the results of their CTI development efforts (Versit CTI Encyclopedia including Versit TSAPI) to the ECTF, enabling review and use by the companies represented in this forum. In support of these efforts, the Versit CTI companies are also joining ECTF.
ECTUA European Council of Telecommunications Users Association.
EDA Electronic Design Automation.
EDAC Error Detection And Correction.
EDCH Enhanced D-Channel Handler.
EDDA The European Digital Dealers Association. An European association of DEC resellers. EDDA members include VARs, systems integrators, leasing companies and service organizations.
Eddy Current Losses Losses in electrical devices using iron, due to the currents set up in it by magnetic action.
EDF Erbium-Doped Fiber.
EDFA Erbium-Doped Fiber Amplifier. A form of fiber optical amplification where the transmitted light signal passes through a section of erbium-doped fiber and is amplified by means of a laser pump diode. EDFA is used in transmitter booster amplifiers, in-line repeating amplifiers, and receiver preamplifiers.
Edge Connector A connector made of strips of brass or other conductive metal found at the edge of a printed circuit board. The connector plus into a socket of another circuit board to exchange electronic signals.
Edge Device A physical device which is capable of forwarding packets between legacy interworking interfaces (e.g., Ethernet, Token Ring, etc.) and ATM interfaces based on data-link and network layer information but which does not participate in the running of any network layer routing protocol. An Edge Device obtains forwarding descriptions using the route distribution protocol.
Edge Site A remote network site. A site at the edge of the network.
EDH Electronic Document Handling.
EDI Electronic Data Interchange. A series of standards which provide computer-to-computer exchange of business documents between different companies' computers over phone lines. These standards allow for the transmission of purchase orders, shipping documents, invoices, invoice payments, etc. between an enterprise and its "trading partners." A trading partner in EDI parlance is a supplier, customer, subsidiary, or any other organization with which an enterprise conducts business. EDI software translates fixed field or "flat" files that are extracted from applications into a standard format and hands off the translated data to communications software for transmission. EDI standards are supported (i.e. have been adopted) by virtually every computer company in the country and increasingly, by every packet switched data communications company. For example, you can use IBM VAN — IBM's Value Added Network for Electronic Data Interchange. The formats used to convert the documents into EDI data are defined by international standards bodies and by specific industry bodies. See also IES.
Edison Battery A type of storage battery in which the elements are nickel and iron and the electrolyte is potassium hydroxide. An old type of battery.
Edison Effect The phenomenon attributed to Edison, that when a filament is incandescent a current will flow between it and another electrode in the tube. In other words, the light bulb.
Editing Editing is a familiar process of changing the content of files to achieve more effective communication by cutting, pasting, cropping, resizing, or copying. Multimedia editing can be done on all types of media: voice annotations, music, still images, motion video, graphics and text. Tools for editing vary from simple tools for email voice annotations to more sophisticated tools for video manipulation. See also ELECTRONIC MAIL.
Editor A software program used to modify programs or files while they are being prepared or after they are (allegedly) complete. An editor is really a very rudimentary word processing program. When you buy MS-DOS, one of the programs you get is a rudimentary editing program called EDLIN. Most programmers, however, now prefer more advanced editors, like ZEdit. This dictionary was written in ZEdit.
EDLIN The MS-DOS line editor that comes with DOS and which you can use to create and edit batch files and other small text files. It's not very good. There are far better editors around, including QEdit.
EDM Electronic Document Management. EDM unites the disparate workflow, document management and imaging.
EDMS An imaging term. Engineering Document Management System.
EDO RAM Extended Data-Out Random Access Memory. A form of DRAM (Dynamic Random Access memory) that, according to PC Magazine, speeds accesses to consecutive locations in memory by (1.) assuming that the access to next memory will target an address in the same transistor row as the previous one and (2.) latching data at the output of the chip so it can be read

even as the inputs are being changed for the next memory location. EDO RAM reduces memory access times by an average of 10 percent over with standard DRAM chips and costs only a little more to manufacture. EDO RAM has already replaced DRAM in many computers, and the trend is expected to continue. See also DRAM and EDRAM.

EDP Electronic Data Processing. Also DP, as in Data Processing. Basically, a machine (also called a computer) that receives, stores, operates on, records and outputs data. The word "electronic" was added to Data Processing when the industry moved away from tab cards — the 80 column "do not spindle," etc. — and was able to accept data electronically, instead of electro-mechanically as with the tab cards. People in the industry used to be called EDPers. Now the term MIS — Management Information System — is more common.

EDRAM 1. Enhanced Digital Announcement Machine

2. Enhanced Dynamic Random Access Memory. A form of DRAM (Dynamic RAM) that boosts performance by placing a small complement of static RAM (SRAM) in each DRAM chip and using the SRAM as a cache. Also known as cached DRAM, or CDRAM.

EDS See 1962.

EDSL Extended Digital Subscriber Line. The ISDN EDSL combines 23 B-channels and one 64-Kbps D-channel on a single line. Also called the Primary Access Rate.

EDTV See EXTENDED DEFINITION TV.

Edutainment The answer to the question "What do you get when you cross educational material with interactive video?" A term coined by "someone who obviously knows nothing about either education or entertainment," says Laura Buddine, president of multimedia games maker Tiger Media. But it is becoming popular in residences and it's typically played on PCs with CD-ROM players.

EE End to End signaling. Punch DTMF are sent through the lines to signal the end of a conversation. A tone code is also used to access long distance carriers, to signal your answering machine, or to access your voice mail.

EEC European Economic Community.

EEHLLAPI Entry Emulator High Level Language Applications Programming Interface. An IBM API subset of HLLAPI.

EEHO Either End Hop Off. In private networks, a switch program that allows a call destined for an off-net location to be placed into the public network at either the closest switch to the origination or the closest switch to the destination. The choice is usually by time of day and is usually done to take advantage of cheaper rates.

EEMA European Electronic Messaging Association.

EEPROM Electronically Erasable Programmable Read Only Memory. A read only memory device which can be erased and reprogrammed. Typically, it is programmed electronically (not electromagnetically) with ultraviolet light. EEPROMs don't lose their memory when you lose power. EEPROM used to be often used in PBXs and were the way manufacturers of older style PBXs upgraded their software. In other words, every time they sent you a software upgrade, they'd send you a bunch of chips. You'd pull out a bunch of chips on one of the main boards in your PBX. And you'd replace them with the new chips. When you don't have a disk drive (and in the olden days disk drives were very expensive), EEPROMs were the only way to go. EEPROM Setup in a computer allows it to recognize certain system board configurations during initialization. You, the user, can then choose options such as the type of memory chips installed and base memory size without changing jumpers on the system board.

EES Escrow Encryption Standard.

EETDN End-to-End Transit Delay Negotiation.

EF&I Engineer, Furnish and Install.

EFCI Explicit Forward Congestion Indication: EFCI is an indication in the ATM cell header. A network element in an impending-congested state or a congested state may set EFCI so that this indication may be examined by the destination end-system. For example, the end-system may use this indication to implement a protocol that adaptively lowers the cell rate of the connection during congestion or impending congestion. A network element that is not in a congestion state or an impending congestion state will not modify the value of this indication. Impending congestion is the state when a network equipment is operating around its engineered capacity level.

EFD Event Forwarding Discriminator. A wireless telecommunications term. Software that contains a discriminator that determines if a notification should be forwarded on to a particular destination.

Efficiency Factor In data communications, the ratio of the time to transmit a text automatically and at a specified modulation rate, to the time actually required to receive the same text at a specified maximum error rate.

EFOC European Fiber Optics and Communications conference.

EFS Error Free Seconds: A unit used to specify the error performance of T carrier systems, usually expressed as EFS per hour, day, or week. This method gives a better indication of the distribution of bit errors than a simple bit error rate (BER). Also refer to SES.

EFT Electronic Funds Transfer. The moving of bits of data from one bank to another. Done in place of moving little green pieces of paper, called money.

EFTA European Free Trade Association.

EFTPOS Electronic Funds Transfer Point Of Sale.

EGA Enhanced Graphics Adapter. Second color video interface standard established for IBM PCs. Maximum resolution is 640 x 350 pixels. See MONITOR.

Ego Surfing Scanning the Net, databases, print media, or research papers looking for mentions of your own name.

EGP Exterior Gateway Protocol. An Internet protocol for exchanging routing information between autonomous systems.

Egress The exit point. This typically refers to information being sent out of, as opposed to being sent in to a frame relay port connection or other network element.

EHz Exahertz (10 to the 18th power hertz). See also SPECTRUM DESIGNATION OF FREQUENCY.

NEWTON'S TELECOM DICTIONARY

E1 Refers to Europe's 2.048 Mbps (T-1 type) digital carrier system.
EIA Electronic Industries Association. A Washington, D.C. trade organization of manufacturers which sets standards for use of its member companies, conducts educational programs and lobbies in Washington for its members' collective prosperity. In April 1988, a new association called the Telecommunications Industry Association was formed by a merger of the US Telecommunications Suppliers Association (USTSA) and the Electronics Industries Association's Information and Telecommunications Technologies Group (EIA/ITG). See EIA INTERFACE.
EIA 232-D New version of RS-232-C physical layer interface adopted in 1987.
EIA 530 Interface using DB-25 connector, but for higher speeds than EIA-232. Has balanced signals (like EIA-422) except for three maintenance signals which are EIA-423.
EIA 561 EIA-232E interface on DIN-8 connector (like Macs use).
EIA 568 An EIA standard for commercial building wiring. The standard covers four general areas; the medium, the topology of he medium, terminations and connections, and general administration. See EIA/TIA 568 below.
EIA 569 Commercial Building Standard for Telecommunications Pathways & Spaces - EIA/TIA, 1991. Lays out guidelines for sizing telecom closets, equipment rooms, conduit, etc . Every architect doing commercial buildings should have to memorize it. Few have even heard of it!
EIA 574 EIA 232E interface on DE-9 connector. (OK, DE is a bit pedantic. Most call it a DB-9, though you'll never find one in a parts book).
EIA 606 Telecommunications Administration Standard for Commercial Buildings. Guidelines covering design and identification of two- level backbone cabling for individual buildings and for campuses.
EIA Interface A set of signal characteristics (time, duration, voltage and current) set up by the Electronic Industries Association to standardize the transfer of information between different electronic devices, like computers, modems, printers, etc. The most famous EIA interface is the RS-232-C (now called the RS-232-E.) EIA-232 specifies three things: the functions of the interchange circuits, the electrical characteristics AND the connector (EIA-232-E includes two different connectors).
In contrast, the ITU-T's V.24 specifies ONLY the interchange circuit FUNCTIONS. V.28 specifies electrical characteristics compatible with EIA 232. ISO 2110 is the internal standard that defines the 25-pole D-shell connector compatible with EIA-232. Following the merger of EIA and the ITG part of EIA, all formed EIA telecommunications standards are now EIA/TIA publications and the standard referred to is now known as EIA/TIE-232-D, edition D, being the most recent. See EIA, EIA/TIA-RS-232-E and RS-232-E.
EIA/TIA-232-E The latest version of the familiar RS-232-C serial data transfer standard for communicating between Data Terminal Equipment (DTE) and Data Circuit Terminating equipment employing serial binary data interchange. (EIA/TIE's exact words.) This standard defines the serial ports on computers, which communicate with such things as external modems, serial printers, data PBXs, etc. You can get your very own copy of the 36-page standard from the EIA Standard Sales Office, 2001 Pennsylvania Avenue NW, Washington D.C. 20006 202-457-4966. See also the APPENDIX.
EIA/TIA 568 EIA/TIA 568 Commercial Building Wiring Standard. This telecommunications standard in early 1991 was out for industry review under draft specification SP-1907B. Its purpose is to define a generic telecommunications wiring system for commercial buildings that will support a multi-product, multi-vendor environment. It covers topics such as:
.Recognized Media .Topology .Cable Lengths/Performance .Interface Standards .Wiring Practices .Hardware Practices .Administration
EIDE Enhanced IDE. See IDE and ENHANCED IDE.
Eight Hundred Service 800-Service. A generic and common (and not trademarked) term for AT&T's, MCI's, US Sprint's and the Bell operating companies' IN-WATS service. All these IN-WATS services have "800" as their "area code." Dialing an 800-number is free to the person making the call. The call is billed to the person or company being called.
800 Service works like this: You're somewhere in North America. You dial 1-800 and seven digits. Your local central office sees the "1" and recognizes the call as long distance. It ships that call to a bigger central office (or perhaps processes the call itself). At that central office it's processed, a machine will recognize the 800 "area code" and examine the next three digits. Those three digits will tell which long distance carrier to ship the call to.
Until 800 portability, happened in May, 1993 each 800 provider (local and long distance company) was assigned specific 800 three digit "exchanges." For example, MCI had the exchange 999. AT&T had the exchange 542. If you wanted a phone number beginning with 800-999, then you had to subscribe to MCI 800 service. If you wanted a phone number beginning with 800-542, you had to subscribe to AT&T 800 service. With 800 Portability that is no longer the case.
Here is a history of what the phone industry calls 800 Data Base Access Service. It comes courtesy Bellcore:
"After divestiture (1984), the seven regional telecommunications companies began to provide limited 800 Service on their own as well as in conjunction with interexchange carriers. The regional companies transported 800 calls only within their own calling areas. The 800 number — containing 10 digits in accordance with the North American Numbering Plan (NANP) — was routed onto the long distance carrier's networks.
"The Common Carrier Bureau of the Federal Communications Commission (FCC) endorsed an incremental approach that would ultimately give the seven companies the right to create their own 800 Service architecture and eliminate reliance on the only existing signaling system (AT&T's). That approach involved assigning to 800 service providers one or more special numbers from the NANP. These numbers, known as "NXX codes," allowed carriers (MCI, Sprint, NY Telephone, etc.) to identify their own 800 numbers and offer their customers 800 numbers.
"Bellcore — Bell Communications Research — began to develop a new network architecture that would allow an 800 Service subscriber to change to another carrier without changing their existing 800 number (full number portability), in accordance with a September 1991 FCC order. That order declared that 800 data base service should be implemented by March 1993 (later extended to May, 1993) and that the old NXX plan be eliminated as long as access times met certain FCC standards.

NEWTON'S TELECOM DICTIONARY

"In September 1991, the FCC endorsed the plan initially set forth by the Bell operating companies, which provided that the administration of the Number Administration and Service Center (NASC) be transferred from Bellcore to an independent third party outside the telecommunications industry. Lockheed Information Management Systems Company (IMS) was selected by competitive bid to succeed Bellcore as NASC administrator.

"How 800 data base service works: The telecommunications network architecture that supports 800 Data Base Access Service is considered "intelligent" because data bases within the network supplement the call processing function performed by network switches. The Service uses a Common Channel Signaling (CCS) network and a collection of computers that accept message queries and provide responses. When a caller dials an 800 number, a Service Switching Point (SSP) recognizes from the digits "8-0-0" that the call requires special treatment and processes that call according to routing instructions it receives from a centralized database. This database, called the Service Control Point (SCP), can store millions of customer records.

"Although each regional company maintains whatever number of SCPs it needs to provide 800 Data Bases Access Service, information about how an 800 call should be handled is entered into the SCP through the off-line Bellcore support system called the Service Management System (SMS). SMS is a national computer system which administers assignment of 800 numbers to 800 service providers. It is located in Kansas City, maintained by Southwestern Bell Telephone Company, and administered by Bellcore with information received from 800 Number Administration and Service Center (NASC). The NASC provides user support and system administration for all 800 Service providers who access the SMS/800."

Because 800 service is essentially a data base lookup service, there are an endless "800 services" you can create. Here are the variables that can be used to influence how an 800 phone call is handled and where an 800 phone call ultimately gets sent:
- The number calling. Virtually all 800 calls in North America (excepting Mexico) now come with the information as to from which number the call came.
- The number being called.
- The time of day, week, month etc.
- The instructions given at that particular moment. A computer might say "Sorry, our phone system is busy. We can't take any more calls in New York. Please send this one and all subsequent ones — until informed otherwise — to our phone system in Kansas City.

Here are a few examples of the services 800 providers have created using the above variables:
- TIME OF DAY ROUTING: Allows you to route incoming calls to alternate, predetermined locations at specified days of the week and times of the day.
- PERCENTAGE ALLOCATION ROUTING: Allows you to route pre-selected percentages of calls from each Originating Routing Group (ORG) to two or more answering locations. Allocation percentages can be defined for each ORG (typically an area code), for each day type and for each time slot.
- SINGLE NUMBER: The same 800 number is used for intrastate and interstate calling.
- CALL BLOCKAGE: You can block calling areas by state or area code. The caller from a blocked area hears the message: "Your 800 call cannot be completed as dialed. Please check the number and dial again or call 1- 800-XXX-XXXX for assistance." (You may want to block callers from areas which didn't see your special commercial, for example.)
- POINT OF CALL ROUTING: Allows a customer to route calls made to a single 800 number to different terminating locations based on the call's point of origin (state or area code.) You establish Originating Routing Groups (ORGs) and designate a specific answering location for each ORG's call.
- CALL ATTEMPT PROFILE: A special service that allows subscribers to purchase a record of the number of attempts that are made to an 800 number. The attempts are captured at the Network Control Point, and from this data a report is produced for the subscriber.
- ALTERNATE ROUTING: Allows a customer to create alternate routing plans that can be activated by the 800 carrier upon command in the event of an emergency. Several alternate plans can be set up using any features previously subscribed to in the main 800 routing plan. Each alternate plan must specify termination in a location previously set up during the order entry process.
- DIALED NUMBER IDENTIFICATION SERVICE: DNIS. Allows a customer to terminate two or more 800 numbers to a single service group and to receive pulsed digits to identify the specific 800 number called. DNIS is only available on dedicated access lines with four-wire E & M type signaling or a digital interface. The customer's equipment must be configured to process the DNIS digits.
- ANI: The carrier will deliver to you the incoming 800 call plus the phone number of the calling party. See also ANI, COMMON CHANNEL INTEROFFICE SIGNALING and ISDN.
- COMMAND ROUTING: Allows the customer to route calls differently on command at any time his business requires it.
- FOLLOW ME 800: Allows the customer to change his call routing whenever he wants to.

Now to the question of how to complete an 800 call. There are essentially two ways to terminate an 800 call. You can end the call on your normal phone line — business or residence. This is the phone line you use for normal in and out calling. That's called not having a dedicated local loop. Or you can end the call on a dedicated phone line. By "dedicated," we mean there's a leased line between your office and the local office of your 800 provider, local or long distance carrier. There are several ways this dedicated "line" might be installed. It could be part of a T-1 circuit. It could be one circuit on one single copper pair. It could even be a phone number dedicated to your 800 number — a phone number you can't make an outgoing call on.

There is one major problem with 800 lines. They're hard to test. You may have bought an 800 number to cover the country, but you may be unreachable from certain parts of the country for weeks on end and not know it. That's part of the problem of a service which uses multiple databases lookup tables and relies on many exchanges to carry the calls. Many companies — like the airlines — recruit their distant employees to call their 800 number regularly. The only part of your 800 IN-WATS line you can test is your local loop (assuming you have one) from the local central office to your office. If you have a dedicated, leased line, you may have local Plant Test Numbers — standard seven digit numbers. You can call these numbers. If they work, you know that the end parts of your lines are working. One of our WATS lines is 1-800-LIBRARY. When it had a dedicated local phone number, it had a plant test number of 212-206-6870. So we could call this and all the subsequent hunt-on numbers every day first thing. Just to check. And when we go traveling, we call our own numbers. Just to check. Now we don't have any dedicated local phone numbers, we rely on prayer. The most common problem we have with our 800 numbers is at our local central office. Seems that it crashes every so often for very short amounts of time. When it starts up, it's meant to load all the tables to give us the features we're paying for — like hunting. Sadly, it doesn't always do this. We then report the trouble to our local phone company. It's usually fixed within an hour or so. Depends on how busy they are. See also 800 at the front of this dictionary.

EIGRP Enhanced Interior Gateway Routing Protocol. Cisco System's newest version of its proprietary routing algorithm, IGRP.

EIP Early Implementers Program. A term Novell coined to refer to those companies who had early on committed to adopt its Telephony Services architecture. See TELEPHONY SERVICES.

EIR Equipment Identity Register. A database repository used to verify the validity of equipment used in mobile telephone service. It can provide security features such as blocking calls from stolen mobile stations and preventing unauthorized access to the network.

EIRPAC A ITU-T X.25 packet-switched network operating in Ireland under the control of the Irish government.

EISA In a computer a "bus" is an electrical channel for getting information and commands in and around the computer. It is the way the central microprocessor running the computer gets its information and commands to the various peripheral devices or device controllers, such as video controllers, hard disk controllers, etc. The original IBM PC was "balanced" in that the microprocessor matched the speed of the bus that came in the machine. But the microprocessor got faster and more powerful and the bus lagged behind. So there has been much effort to speed the bus up, including EISA, which stands for Extended Industry Standard Architecture. EISA is the independent computer industry's alternate to IBM's Micro-Channel data bus architecture which IBM uses in some of its high end PS/2 line of desktop computers. EISA, like Micro-Channel (also called MCA), is a 32-bit channel. But, unlike IBM's Micro-Channel, plug-in boards which work inside the XT and AT-series of IBM and IBM clone desktop computers will work within EISA machines. They won't work in Micro-Channel machines. EISA expands the 16-bit ISA (Industry Standard Architecture) to 32-bit. EISA technology is useful in computing environments where multiple high performance peripherals are operating in parallel. The intelligent bus master can share the burden on the main CPU by performing direct data transfers into and out of memory. EISA capabilities are valuable when the system is being used as a server on a local area network or is running a multi-user operating system such as UNIX or OS/2. As of writing, over 200 manufacturers had endorsed EISA. Broader, wider buses than EISA are now available. 64-bit is not uncommon, especially among servers. See LOCAL BUS, PCI and VESA for examples of newer, faster buses.

Either End Hop Off EEHO. In private networks, a switch program that allows a call destined for an off-net location to be placed into the public network at either the closest switch to the origination or the closest switch to the destination. The choice is usually by time of day and is usually done to take advantage of cheaper rates.

Either Way Operation Same as half-duplex.

EIU Ethernet Interface Unit

EKTS Electronic Key Telephone System.

ELAN Emulated Local Area Network: A logical network initiated by using the mechanisms defined by LAN Emulation. This could include ATM and legacy attached end stations.

Elastic Buffer A variable storage device having adjustable capacity and/or delay, in which a signal can be temporarily stored.

Elasticity Of Demand The relationship between price and the quantity sold. The theory is the lower the price, the more you'll sell. In telecommunications, this has traditionally been true, though sometimes it has taken time for demand to catch up with dramatic price cuts.

ELDAP The beginnings of an Internet directory protocol.

Electric Banana Telecom installers' slang for TONE PROBE.

Electric Lock A cellular phone feature that provides security by locking a cellular phone so it can't be used by unauthorized persons.

Electrically Powered Telephone A telephone in which the operating power is obtained either from batteries located at the telephone (local battery) or from a telephone central office (common battery).

Electrodeposition The deposition of a conductive material from a plating solution by the application of electric current.

Electroluminescence The direct conversion of electrical energy into light.

Electrolysis The production of chemical changes by passage of current through an electrolyte.

Electrolyte A chemical solution used in batteries, chemical rectifiers, and certain types of fixed condensers.

Electrolytic Process A printing process where paper is treated with an electrolyte and a stylus passes the signal current through the paper to produce an image. Paper is roll-fed past the stylus and changes color depending on the intensity of current passing through the stylus.

Electromagnetic Compatibility EMC. The ability of equipment or systems to be used in their intended environment within designed efficiency levels without causing or receiving degradation due to unintentional EMI (Electromagnetic Interference). EMI is reduced by, amongst other things, copper shielding.

Electromagnetic Emission Control The control of electromagnetic emissions. e.g., radio, radar, and sonar transmissions, for the purpose of preventing or minimizing their use by unintended recipients. A military term. Electromagnetic emission is reduced by, amongst other things, copper shielding.

Electromagnetic Interface EMI. interference in signal transmission or reception caused by the radiation of electrical and magnetic fields. That's the easy explanation. Here's a more comprehensive explanation: Any electrical or electromagnetic phenomenon, manmade or natural, either radiated or conducted, that results in unintentional and undesirable responses from, or performance degradation or malfunction of, electronic equipment.

Electromagnetic Interference See ELECTROMAGNETIC INTERFACE.
Electromagnetic Lines Of Force The lines of force existing about an electromagnet or a current carrying conductor.
Electromagnetic Wave The electric wave propagated by an electrostatic and magnetic field of varying intensity. Its velocity is 186,300 miles per second.
Electromechanical Ringing The traditional bell or buzzer in a telephone which announces incoming calls.
Electron The smallest known particle of matter assumed to be a particle or charge of negative electricity.
Electron Gun Device in a television picture tube from which electrons are emitted toward screen.
Electron Tube Rectifier A device for rectifying an alternating current by utilizing the flow of electrons between a hot cathode and a relatively cold anode.
Electronic Blackboard This is a teleconferencing tool. At one end there's a large "whiteboard." Write on this board and electronics behind the board pick up your writing and transmit it over phone lines to a remote TV set. The idea is that remote viewers can hear your voice on the phone and see the presentation on the electronic blackboard. The product has not done well because it is expensive — typically several hundred dollars a month just for rent, plus extra hundreds for transmission costs. In Japan, there are similar boards called OABoards — Office Automation Boards. They do one thing differently — they will print a copy on normal letter-size paper of what's written on the board. This takes about 20 seconds. Some of these Japanese OABoards will also transmit their contents over phone lines. So far, neither the OABoards nor the electronic blackboards have found a sizable market in the United States.
Electronic Bulletin Board A computer, a modem, a phone line and a piece of software. Load communications software in your computer, dial the distant electronic bulletin board. The system will answer and present you with a menu of options. Typically those options will include leave messages, pick up messages, find out information, fill in a survey and upload and download a file.
Electronic Business Card Also known as vCard. See VERSIT and VERSITCARD.
Electronic Call Distribution Another term for Automatic Call Distribution. See AUTOMATIC CALL DISTRIBUTOR.
Electronic Commerce Using electronic information technologies to conduct business between trading partners, not necessarily using EDI (Electronic Data Interchange). See EDI.
Electronic Custom Telephone Service Provides deluxe key telephone features and simplified access to certain AT&T Dimension PBX phones.
Electronic Data Interchange EDI. The process whereby standardized forms of documents are transferred between systems often run by different companies. EDI is used for placing orders, etc. The form and format of such documents may be defined by vendor specifications, ITU-T standards, the ANSI X.12 standard, or the United Nations EDIFACT standard. See EDI for a fuller explanation. See the next definition.
Electronic Data Interchange Association EDIA. An organization which works to provide a common platform to communicate global EDI activity, bypassing language conventions and national boundaries.
Electronic Data Processing See EDP.
Electronic Frontier Foundation EFF. A foundation established by Mitch Kapor, founder of Lotus, to address social and legal issues arising from the impact on society of the increasingly pervasive use of computers as the means of communication and information distribution.
Electronic Funds Transfer EFT. A system which transfers money electronically between accounts or organizations without moving the actual money.
Electronic Image Mail The transmission of slow scan TV or facsimile via "Store and Forward." Not a common term.
Electronic Key System A key telephone system in which the electromechanical relays and switches have been replaced by electronic devices — often in the phone and in the central cabinet. The innards of the central cabinet of an electronic key system more resemble a computer than a conventional key system. These days, virtually all key systems are electronic. Production of electromechanical key systems (such as 1A2) has been severely curtailed and most manufacturers have ceased making it.
Electronic Leash Pagers or beepers are often called "electronic leashes" because they allow your boss to contact you, to control you, to keep you on a leash. At one stage, beepers were carried by doctors and technicians who could never be "out of touch." As a result, beepers got a bad reputation and "real" people, i.e. bosses, wouldn't carry them. But that's getting better now and real people are now carrying them.
Electronic Lock Lets you lock your cellular phone so no one can use it. If you use Electronic Lock, you'll have to punch in some extra digits — like a password — to unlock the lock.
Electronic Mail A term which usually means Electronic Text Mail, as opposed to Electronic Voice Mail or Electronic Image Mail. Sometimes electronic mail is written as E-Mail. Sometimes as email. These days electronic mail is everything from simple messages flowing over a local area network from one cubicle to another, to messages flowing across the globe on an X.400 network. Such messages may be simple text messages containing only ASCII or they may be complex messages containing embedded voice messages, spreadsheets and images. See ELECTRONIC TEXT MAIL, ELECTRONIC VOICE MAIL, ELECTRONIC IMAGE MAIL and WINDOWS, WINDOWS TELEPHONY.
Electronic Mail Gateways A collection of hardware and software that allows users on an E-mail system to communicate and exchange messages with other mail systems that use a different protocol.
Electronic Message Registration A system to detect and count a phone user's completed local calls and then tell the central office the number of message units used. Also used in hotels.
Electronic Messaging Association EMA. The trade association for electronic messaging and information exchange. EMA is based in Fairfax, VA. EMA was formerly called the Electronic mail Association.
Electronic Order Exchange EOE. Inter-company transactions between buyers and sellers handled electronically via standard data communications protocols. EOE can be employed to send purchase orders, price and product listings and order-relat-

ed information.

Electronic Phone General description for most phones designed after about 1980, where many mechanical and electrical parts are replaced by smaller, lighter, and cheaper electronic parts. Features such as mute, redial and memory became popular with these phones, which range in price from $5 to hundreds of dollars.

Electronic Receptionist A fancy name for a voice processing automated attendant, except that in addition to all the normal auto attendant features, it also sends messages to personal PCs on LANs telling the owner who's calling and giving the owner (the called party) the choice of doing something with the call — like answering it or putting it into voice mail.

Electronic Redlining A term for disenfranchising people and institutions because of their lack of telecommunications services and apparatus. In December, 1993, Vice President of the United States, Al Gore, told the National Press Club, When it comes to ensuring universal service, our schools are the most impoverished institution in society. Only 14% of our public schools used educational networks in even one classroom last year. Only 22% possess even one modem. Video-on-demand will be a great thing. It will be a far greater thing to demand that our efforts give every child access to the educational riches we have in such abundance. The recent article in the Washington Post on the proposed video communication network in the D.C. area is a wake-up call to all of us concerned about "electronic redlining."

Electronic Ringer A substitute for the conventional telephone bell, that uses music synthesizer circuitry to generate an attention-getting signal played through a speaker. Typical sounds include warbles, chirps, beeps, squawks, and chimes. The writer of this entry, Michael Marcus, once installed a phone with a chirp sound. A few days later, the customer complained that she had not been receiving any calls, and the birds in her yard were chirping much more than usual.

Electronic Serial Number ESN. A unique ID number of a cellular phone that has been embedded in its circuits by the manufacturer. See ESN.

Electronic Sweep Variation in the frequency of a signal over a whole band as a means of checking the response of equipment under the test.

Electronic Switching System A telephone switch which uses electronics or computers to control the switching of calls, their billing and other functions. The term is now vaguely defined, with each manufacturer defining it as something somewhat different. In fact, every telephone switch sold today is electronic. The term originally came about because early telephone switches were entirely electro-mechanical. The switch consisted entirely of a moving switch. Devices like relays physically moved in order to send the call through the exchange and on its way. These things moved in direct response to the digits dialed by the telephone subscriber. These switches contained no "intelligence" — i.e. no ability to deviate from a set number of very simple tasks which could be accomplished by electromechanical relays.

Then someone said: it would be more efficient if the "instruction part" of the process were divorced the switching mechanism. This lead to the creation of the "electronic" switch in which the "brains" of the switch are separated from the switching mechanism itself. Thus the "brains" can do simple things like collect the dialed digits as they are slowly dialed and pulse them out quickly to the switch — as fast as it can handle them. Now, the "brains" are typically a digital computer.

Electronic Tandem Network 1. Two or more switching systems operating in parallel as part of providing network services (usually voice) to large users.
2. A telephone company switching device used to connect telephone company toll offices located in the same geographic area.

Electronic Telephone Directory Service A PBX feature which stores and produces, on demand, a directory of all extension phone numbers. The directory may include all users in a network. A CRT with keyboard and/or printer is usually required for input and retrieval. In some systems, the CRT or another type of alphanumeric display is part of the Attendant Console. In some systems, the directory may also include names and telephone numbers of frequently called outside people, especially those in the speed calling system. The directory may also be enhanced to include SMDR data such as client codes, account codes and client telephone numbers.

Electronic Text Mail A "Store and Forward" service for the transmission of textual messages transmitted in machine readable form from a computer terminal or computer system. A message sent from one computer user to another is stored in the recipient's "mailbox" until that person next logs onto the system. The system then can deliver the message. Telex, in which a machine readable form of message transmission takes place, is also considered an Electronic Text Mail medium, albeit a very slow one. For an example of electronic mail, please dial our electronic mail system on 212-989-4675. It's free. Parameters are 300, 1200 or 2400 baud, 8 data bits, one stop bit, and no parity.

Electronic Voice Mail A system which stores messages usually spoken over a telephone. These messages can be retrieved by the intended recipient when that person next calls into the system. Also called Voice Mail, it operates just like a touch-tone controlled answering machine.

Electronic Warfare See EW.

Electrophotographic Printing A printing method that uses light to modify electrostatic charges on a photoconductive substrate.

Electrostatic Charge An electric charge at rest.

Electrostatic Discharge ESD. Discharge of a static charge on a surface or body through a conductive path to ground. Can be damaging to integrated circuits.

Electrostatic Printing A method of printing, very common in photocopying, in which charges are beamed onto the surface of paper. The charges attract particles of a very fine (typically black powder) which sticks to the charges. The black powder is fused permanently on the paper by great heat. "XEROXing" is electrostatic printing. In xeroxing, the black powder is called toner.

Elegant An elegant program is one that is efficiently written to use the smallest possible amount of main memory and the fewest instructions.

Elevator Eyes A term used in sexual harassment to mean viewing someone up and down.

Elevator Seeking Organizes the way data is read from hard disks and logically organizes disk operations as they arrive at the Novell NetWare local area network server for processing. A queue is maintained for each disk driver operating within the server. As disk read and write requests are queued for a specific drive, the operating system sorts incoming requests into a priority based on the drive's current head position. As the disk driver services the queue, subsequent requests are located either in the vicinity of the last request or in the opposite direction. Thus, the drive heads operate in a sweeping fashion, from the outside to the inside of the disk. Elevator seeking improves disk channel performance by significantly reducing disk head thrashing (rapid back-and-forth movements of the disk head) and by minimizing head seek times. Imagine how inefficient an elevator would be if the people using it had to get off the elevator in the order they got on.

ELIU Electrical Line Interface Unit.

Elongation The fractional increase in length of a material stressed in tension.

ELOT Hellenic Organization for Standardization (Greece).

Ethernet LAN Service Unit An ELSU provides 12 independent virtual Ethernet bridges for running over ATM networks. ELSUs are designed for flexible deployment, either local to an ATM switch or at a remote site. ELSUs are designed for LAN internetworking services over ATM networks.

Elvis Year The peak year of something's popularity.

EM Element Manager. Software and hardware used to manage and monitor components of a telecommunications network at their lowest level.

EMA Electronic Messaging Association, Arlington, VA. www.ema.org

EMACS A standard Unix text editor preferred by Unix types that beginners tend to hate.

EMAG ETSI MIS Advisory Group.

Email A colloquial term for electronic mail. See EMAIL address.

Email Address The UUCP or domain-based address by which a user is referred to. My email address is HARRYNEWTON@MCIMAIL.COM.

Email Gateway An email gateway is typically a PC on LAN. The PC has one or more modem and/or fax/modem cards. Its job is to send and receive e-mails and/or send and receive faxes for everyone on the LAN. To pick up emails, it might dial once an hour into various mail systems, like MCI Mail, CompuServe, and download all the messages for all the people on the LAN. Once it has those messages, it brings them onto its hard disk and then alerts the recipients that they now have an e-mail. See SERVER.

Email Reflector An Internet electronic mail address which automatically sends you back a reply (i.e. reflects mail to you) if you include certain key words in your message to it. Such key words might be "subscribe" or "lists help."

Email Server See EMAIL GATEWAY.

Embarc Motorola's company which does wireless electronic mail to people carrying laptops and palmtops. Embarc, according to Motorola, stands for Electronic Mail Broadcast to A Roaming Computer. Actually Embarc does more than mail. It also broadcasts snippets of news.

Embed To insert information (an object) that was created in one document into another document (most often the two documents were created with different applications). The embedded object can be edited directly from within the document. To embed under Windows 3.1, you must be using applications that support object linking and embedding.

Embedded Base Equipment All customer-premises equipment that has been provided by the Bell Operating Companies (BOCs) prior to January 1, 1984, that was ordered transferred from the BOCs to AT&T by court order.

Embedded Code Formatting ECF. A NetWare definition. This is something of a programming language, in which faxing commands or other program that automatically generates information, formats it, and faxes it without user intervention.

Embedded Customer-Premises Equipment Telephone-company-provided premises equipment in use or in inventory of a regulated telephone utility as a December 31, 1982.

Embedded Object A Windows term. An embedded object is information in a document that is a copy of information created in another application. By choosing an embedded object, you can start the application that was used to create it, while remaining in the document you're working in.

Embedded SCSI A hard disk that has a SCSI (Small Computer System Interface) and a hard disk controller built into the hard disk unit. See also SCSI.

Embedded SQL SQL statements embedded within a source program and prepared before the program is executed.

Embedded System Processors National Semiconductor's line of high-performance microprocessors used in dedicated systems, such as fax machines and laser printers.

Embossing A means of marker identification by thermal indentation leaving raised lettering on a cable's sheath material.

EMC ElectroMagnetic Compatibility.

Emergency Access An alarm system built into some PBXs. In an emergency it rings all phones.

Emergency Broadcast System EBS. The EBS is composed of AM, FM, and TV broadcast stations; low-power TV stations; and non-Government industry entities operating on a voluntary, organized basis during emergencies at national, state, or operational (local) area levels.

Emergency Dialing A variation on speed calling to call numbers for police, fire department, ambulance, etc. Typically found as special buttons on an electronic phone.

Emergency Hold "Emergi-hold" allows a 911 caller's line to be held open in the event that a caller attempts to hang up. This gives the PSAP (Public Service Answering Position) agent full control of the call. It will not be released until the agent finishes the call.

Emergency Ringback This feature enables the 911 PSAP (Public Service Answering Position) attendant to signal a caller who has either hung up or left the phone off hook. Emergency Ringback enables the PSAP agent to ring a phone which has been hung up or issue a loud "howling" sound from the customer's phone if it has been left off hook.

Emergency Telephone A single line telephone that becomes active when there is no commercial AC power to the Key

Service Unit.

EMI ElectroMagnetic Interference or ElectroMagnetic Interface. Leakage of radiation from a transmission medium due to high-frequency energy. Interference to other equipment or radio services may result if sufficient power from these signals escape the equipment enclosures or transmission media. EMI is reduced by copper shielding. National and international regulatory agencies (FCC, CISPR, etc.) set limits for these emissions. Class A is for industrial use and Class B is for residential use.

EMI/RFI Filter A circuit or device containing series inductive (load bearing) and parallel capacitive (non-load bearing) components, which provide a low impedance path for high-frequency noise around a protected circuit.

Emission 1. Electromagnetic energy propagated from a source by radiation or conduction. The energy thus propagated may be either desired or undesired and may occur anywhere in the electromagnetic spectrum.
2. Radiation produced, or the production of radiation, by a radio transmitting station. For example, the energy radiated by the local oscillator of a radio receiver would not be an emission but a radiation.

Emissivity Ratio of flux radiated by a substance to the flux radiated by black body at the same temperature. Emissivity is usually a function of wavelength.

Emitter The source of optical power.

EML Element Management Layer. A layer representing the management and monitoring of components, at their lowest level, in a telecommunications network. In short, an abstraction of the functions provided by systems that manage each network element on an individual basis.

EMM An acronym for Expanded Memory Manager, the software that controls expanded memory on an IBM PC or clone. Drivers written for the 80386 microprocessors usually allow you to use extended memory to simulate expanded memory. See EXPANDED MEMORY.

EMM386.EXE An expanded memory manager that comes with MS-DOS, Windows and Windows for Workgroups. It provides access to upper memory and simulates expanded memory using extended memory. This is useful for MS-DOS based applications that can use expanded memory. EMM386 also makes it possible to load programs and device drives into upper memory blocks (UMBs). Although EMM386 can provide more free conventional memory by loading programs and device drivers into UMBs, this only helps when running MS-DOS applications. Windows and Windows for Workgroups uses extended memory (XMS) to run the Windows operating systems and Windows-based applications. Accessing device drivers and application code from UMBs is slower than accessing the same code directly from conventional memory. To maximize the performance when accessing MS-DOS device drives, Microsoft recommends that EMM386 (or similar third party memory managers) not be used. According to Microsoft, "in addition to possible performance penalties, EMM386.EXE will use about 150K of XMS (extended memory) memory to provide a mappable memory range for the UMA in which to load device drivers." Recommended Microsoft, "If you are not running MS-DOS applications that require more conventional memory than is available after loading the necessary device drives and programs in conventional memory, and you are not using MS-DOS applications that require EMS, you may remove EMM386.EXE from your CONFIG.SYS without adversely affecting your system. See EMS and SMARTDRV.EXE.

Emotags Mock HTML tags (<smile>, <smirk>) used in WWW-related e-mail and newsgroups in place of ASCII emoticons, for example:
"<flames> Someone tell that jerk to shut up, I'm sick of his vapid whining!
</flame>." Definition from Wired Magazine. See EMOTICON.

Emoticon From Emotional Icon, one of a growing number of typographical cartoons used on BBSs (Bulletin Board Systems) to portray the mood of the sender, or indicate physical appearance. They are meant to be looked at sideways. Some examples:

:-D writer talks too much
:-# writer's lips are sealed
:-o writer is surprised
:-& writer is tongue-tied
ALL CAPS writer is shouting
:) is a smiley face
;) is a smile with a wink
;(is a frown with a wink
(:(is very sad
;? is a bad guy
[:0 is a wide-open mouth and a crewcut
(:{>X is bald with a handlebar mustache and bow tie
:-I is Wayne Newton

{8<)# is Michael Marcus, the writer of this entry: balding, glasses, mustache, smiling, beard.

Here's another collection of emoticons which I found on the Internet:
:-)	Smile	:-D	Laughing
:)	Smile	:-}	Grin
:-]	Smirk	:-(	Frown

;-)	Wink	:-X	Close-mouthed
8-)	Wide-eyed	:-0	Open-mouthed
:-l	I wear a moustache	:-Q	But I don't inhale
:-o	Oh, no!	<g>	Grin
<ggg>	Wide Grin	<g....g>	Very wide grin

EMP A large and fast-moving electromagnetic pulse caused by lightning.

Emphasis In FM transmission, the intentional alteration of the amplitude-versus-frequency characteristics of the signal to reduce adverse effects of noise in a communication system. The higher frequency signals are emphasized to produce a more equal modulation index for the transmitted frequency spectrum, and therefore a better signal-to-noise ratio for the entire frequency range.

EMS 1. Expanded Memory Specification. Several years ago, three computer companies — Lotus, Intel and Microsoft — jointly developed EMS. This standard defines how an MS-DOS program can access memory beyond 640KB while running under MS-DOS. Applications that conform to EMS (sometimes called LIM-EMS for Lotus/Intel/Microsoft Expanded Memory Specification) can take advantage of the computer's memory beyond 640KB of RAM. LIM-EMS uses a portion of the reserved memory area (between 640KB and 1MB) to access RAM beyond 1MB. Software that supports expanded memory uses this window to pass pages of data to and from expanded RAM as needed. See also EXPANDED MEMORY.
2. Enterprise Messaging Server. A new Microsoft concept which allows users to transparently access the messaging engine from within desktop applications to route messages, share files, or retrieve reference data. According to Microsoft, corporate developers will be able to add capabilities using Visual Basic and access EMS by writing either to the X.400 Application Program Interface Association's (XAPIA's) Common Mail Calls (CMC) or to Microsoft's Messaging API (MAPI). See MAPI.
3. Element Management System: A management system that provides functions at the element Management Layer.

EMT Electrical Metal Tubing. In many towns you must run your electrical AC wire inside metal tubing. In other towns you can run normal plastic insulated wiring. Theoretically, EMT is a safer fire hazard. What you are allowed to run depends on local laws and regulations. Tip: Dimmers for incandescent lights raise havoc with LAN data. Solution: Put the plastic electrical wires inside EMT (Electrical Metal Tubing) and ground the conduit.

Emulate To duplicate one system with another. To imitate a computer or computer system by a combination of hardware and software that allows programs written for one computer or terminal to run on another. The most common data terminal is a DEC VT-100. Our communications program, Crosstalk, allows us to "emulate" a DEC-VT100 on our IBM PCs and PC clones.

Emulation What happens one gadget emulates another. See EMULATE.

Emulator A device or computer program which can act as if it is a different device or program, that is Emulate (i.e. pretend to be) another device. Certain computer terminals are necessary in specific systems and a terminal that is not that type may be able to act as if it was. If it can, it is an Emulator. This is not a common term. See also the verb EMULATE.

Enable To make something happen. Or, in more complex language, to set various hardware and software parameters so that the central computer will recognize those parameters and start doing what you want.

Enabler An "enabler" is a strange name for a piece of software.

Enabling Signal A signal that permits the occurrence of an event.

Encapsulated Postscript File EPS. A file that prints at the highest possible resolution for your printer. An EPS file may print faster than other graphical representations. Some Windows NT and non-Windows NT graphical applications can import EPS files.

Encapsulating Bridge A LAN/WAN term. A special bridge type usually associated with backbone/subnetwork architectures. Encapsulating bridges place forwarded packets in a backbone-specific envelope — FDDI, for example - and send them out onto the backbone LAN as broadcast packets. The receiving bridges remove the envelope, check the destination address and, if it is local, send the packet to the destination device.

Encapsulation 1. Encasing a splice or closure in a protective material to make it watertight.
2. In object-oriented programming, the grouping of data and the code that manipulates it into a single entity or object. Encapsulation refers to the hiding of most of the details of the object. Both the attributes (data structure) and the methods (procedures) are hidden. Associated with the object is a set of operations that it can perform. These are not hidden. They constitute a well-defined interface — that aspect of the object that is externally visible. The point of encapsulation is to isolate the internal workings of the object so that, if they must be modified, those changes will also be isolated and not affect any part of the program. See OBJECT-ORIENTED PROGRAMMING.
3. An electronic messaging term. The technique used by layered protocols in which a layer adds header information to the PDU (Protocol Data Unit) form the layer above. As an example, in Internet terminology, a packet would contain a header from the physical layer, followed by a header from the network layer (IP), followed by a header from the transport layer (TCP), followed by the application protocol data.
4. A networking term. It means carrying frames of one protocol as the data in another. Often the encapsulating protocol will be TCP/IP.
5. See also ENCAPSULATION BRIDGING.

Encapsulation Bridging Method of bringing dissimilar networks where the entire frame from one network is simply enclosed in the header used by the link-layer protocol of the other network.

Encoding The process of converting data into code or analog voice into a digital signal. See also PCM and ADPCM.

Encryption The transformation of data into a form unreadable by anyone without a secret decryption key. Its purpose is to ensure privacy by keeping the information hidden from anyone for whom it is not intended. In security, encryption is the ciphering of data by applying an algorithm to plain text to convert it to ciphertext.

End Access End Office EAEO. An end office that provides Feature Group D.

End Delimiter ED. Sequence of bits used by IEEE 802 MAC to indicate the end of a frame. Used in token bus and ring networks, with nondata bits making ED easy to recognize.

End Distortion In start-stop teletypewriter operations, the shifting of the end of all marking pulses except the stop pulse from their proper positions in relation to the beginning of the next pulse. Shifting of the end of the stop pulse would constitute a deviation in character time and rate rather than being an end distortion. Spacing end distortion is the termination of marking pulses before the proper time. Marking end distortion is the continuation of marking pulses past the proper time. Magnitude of the distortion is expressed in percent of a perfect unit pulse length.

End Finish Surface condition at the optical fiber face.

End Instrument A communication device that is connected to the terminals of a circuit.

End Node A node such as a PC that can only send and receive information for its own use. It cannot route and forward information to another node.

End Of File EOF. A control character or byte used in data communications that indicates the last character of the last record of a file has been read.

End Of Medium EM. A control character used to denote the end of the used (or useful) portion of a storage medium.

End Of Message EOM. A control character used in data communications to indicate the end of a message.

End Of Shift Routing A call center term for a process that calls won't be left in limbo when a shift ends. See also SOURCE/DESTINATION ROUTING, SKILLS-BASED ROUTING and CALENDAR ROUTING.

End Of Text Message ETX. A control character used in data communications to indicate the end of a text message. See ETX.

End Of Transmission Block A communications control character indicating the end of a block of Bisync data for communication purposes.

End Of Transmission Block Character A control character used in data communications to indicate the end of a block where data are divided into blocks for transmission purposes.

End Office A central office to which a telephone subscriber is connected. Frequently referred to as a Class 5 office. The last central office before the subscriber's phone equipment. The central office which actually delivers dial tone to the subscriber.

End Station An ATM term. These devices (e.g., hosts or PCs) enable the communication between ATM end stations and end stations on "legacy" LAN or among ATM end stations.

End To End Loss The loss of an installed transmission path. The loss consists of the loss of the transmission cable or fiber, splices and connectors.

End To End Signaling A signaling system capable of generating and transmitting signals directly from the originating station to the terminating end after the connection is established, without disturbing the connection. Touchtone dialing is such a system, allowing the user to send tones to a remote computer for data or other access. See POINT TO POINT.

End User A highfalutin' term for a user. It's actually the occupant of the premises who uses and pays for the telephone service received and does not resell it to others. Bellcore's definition: A user who uses a loop-start, ground-start, or ISDN access signaling arrangement. Definition from Bellcore is part of its concept of the Advanced Intelligent Network.

Endurability The property of a system, subsystem, equipment, or process that enables it to continue to function within specified performance limits for an extended period of time, usually months, despite a potentially severe natural or man-made disturbance, e.g., nuclear attack, and a subsequent loss of external logistic or utility support.

Energy Communications EC. A PBX feature which communicates with energy consuming and monitoring devices and perform functions like dimming the lights or turning down the heat in a vacant hotel room. See also ENERGY CONTROL.

Energy Control Indicates that phone system has software and hardware necessary to control and regulate the energy consuming devices in a user's facility (heating, ventilating, air conditioning, electrical machinery etc.). The system's processor transmits control signals, over existing telephone wiring where possible, to control units at each power-consuming device. This feature always includes user reconfiguration of the system's control parameters in response to operational and/or environmental changes. At one stage, AT&T and some other telephone equipment manufacturers sold energy control as a integral feature of their phone systems. The idea didn't take off for a lot of reasons.

Energy Density A beam's energy per unit area, expressed in joules per square meter. Equivalent to the radiometric term "irradiance."

ENET Enhanced Network.

ENFIA Exchange Network Facilities for Interstate Access. A tariff providing a series of options for connecting long distance carriers with local exchange facilities of the local telephone company.

Engineering Orderwire EOW. A communication path for voice or data, or both, that is provided to facilitate the installation, maintenance, restoral, or deactivation of segments of a communication system by equipment operators, attendants, and controllers.

Enhanced 800 Services A name MCI uses for a family of 800 services with additional features added to them. It includes time of day and day of week routing.

Enhanced 911 Enhanced 911 is an advanced form of 911 service. With E-911, the telephone number of the caller is transmitted to the Public Safety Answering Point (PSAP) where it is cross-referenced with an address database to determine the caller's location. That information is then displayed on a video-monitor for the emergency dispatcher to direct public safety personnel responding to the emergency. This enables police, fire departments and ambulances to find callers who cannot orally provide their precise location. See also E-911.

Enhanced Call Processing An Octel term for the interactive voice response option in its voice mail system. Here's how

Octel defines the term: "Companies and departments that receive a heavy volume of calls can use ECP to create menus that are presented to callers. When the system answers a call, a recorded voice instructs the caller how to use a touch-tone telephone to send call routing instructions to the system. Depending on which option is chosen, ECP's customized call routing feature allows a caller to press a single key to reach a predetermined extension, a voice messaging mailbox where he can leave a message, an Information Center Mailbox where he can listen to a series of recordings giving frequently requested information or additional levels of ECP menus. ECP menus are easily custom-built by the customer to meet its specific needs. Each menu can offer as many as ten options."

Enhanced DNIS Enhanced DNIS is a combination of ANI and DNIS delivered before the first ring on a T1 span. The number of digits delivered is configurable on a per span basis.

Enhanced IDE An improved interface to the IDE hard disk interface. Enhanced IDE allows you to attach hard disks of larger than 528 megabytes (the largest normal IDE will handle) up to a maximum of 8.4 gigabytes. Enhanced IDE has a data transfer rate of between 11 and 13 megabytes per second, compared to the 2 to 3 megabytes per second, which normal IDE drives sport. See IDE.

Enhanced Mode The Intel 8088 and 8086 microprocessors, used in the earliest PCs, run DOS programs using real mode. Real mode causes problems when you try to run more than one program at a time because nothing prevents a poorly designed program from invading another program's memory space, resulting in a system crash. The Intel 80386 microprocessor introduced several technical improvements. For compatibility, an 80386 can run in real mode, but also offers protected mode. In protected mode, the 80386 can address up to four gigabytes of RAM, far more than you can install in any PC. The chip also can simulate more than one 8086 "machines, are protected from one another, preventing memory conflicts.
Running a DOS program in the protected mode of an 80386 computer requires software to manage the memory. Like a traffic cop, this software — called memory-management — puts DOS programs into their own 640K virtual machines, where they work away without interfering with other programs. The most popular memory-management available for 80386 and 80486 is Windows. In 386 Enhanced mode, Windows takes advantage of the 80386/80486/Pentium's virtual memory capabilities. Virtual memory is a way of extending RAM. Most DOS applications swap program instructions and data back and fourth from disk rather than keep them in memory. See EXPANDED MEMORY and EXTENDED MEMORY.

Enhanced Parallel Port EPP. A new hardware and software innovation (and now a standard) which allows computers so equipped to send data out their parallel port at twice the speed of present parallel ports. There's no difference in the shape of the plug or the number of conductor. See EPP for a fuller explanation.

Enhanced Private Switched Communications Service EPSCS (pronounced EP-SIS). A private line networking offering from AT&T which provides functions similar to CCSA. Big companies are its customers.

Enhanced Serial Interface ESI. A new, broader serial interface announced by Hayes Microcomputer Products, Norcross, GA, and placed in the public domain. The ESI is an extension of the familiar COM card used in personal computers. ESI includes the definition of I/O, control registers, buffer control, Direct Memory Access (DMA) to the system and interaction with attached modem devices. ESI specification is available from Hayes Customer Service at no charge. Combined with Hayes' announcement of ESI was their announcement of new Enhanced Serial Port hardware products for the IBM microchannel and IBM XT/AT or EISA bus personal computers. According to Hayes, the ESI spec and the supporting ESP hardware provide a "cost-effective" communication coprocessor to manager the flow of data between an external high speed modem and PC. This technology prevents loss of data resulting from buffer overflow errors and provides maximum data throughput for high speed modems. Hayes says that the combination of ESP and ESI will allow through-the-phone modem speeds of up to 38.4 Kbps.

Enhanced Serial Port See ENHANCED SERIAL INTERFACE.

Enhanced Service Provider ESP. An ESP is a company that provides enhanced or value-added services to end users. An ESP typically adds value to telephone lines using his own software and hardware. Also called an IP, or Information Provider. An example of an ESP is a public voice mail box provider or a database provider, for example, one giving the latest airline fares. An ESP is an American term, unknown in Europe, where they're most called VANs, or Value Added Networks. See also OPEN NETWORK ARCHITECTURE and INFORMATION PROVIDER.

Enhanced Services Services offered over transmission facilities which may be provided without filing a tariff. These services usually involve some computer related feature such as formatting data or restructuring the information. Most Bell operating companies (BOCs) are prohibited from offering enhanced services at present. But the restrictions are disappearing.
The FCC defines enhanced services as "services offered over common carrier transmission facilities used in interstate communications, which employ computer processing applications that act on the format, content, code, protocol or similar aspects of the subscriber's transmitted information; provide the subscriber additional, different or restructured information; or involve subscriber interaction with stored information." In other words, an enhanced service is a computer processing application that messes in some way with the information transmitted over the phone lines. Value-Added Networks, Transaction Services, Videotex, Alarm Monitoring and Telemetry, Voice Mail Services and E-Mail are all examples of enhanced services.

Enhanced Small Device Interface An interface which improves the rate of data transfer for hard disk drives and increases the drive's storage capacity.

Enhanced Unshielded Twisted Pair EUTP. UTP (Unshielded Twisted Pair) cables that have enhanced transmission characteristics. Cables that fall under this classification include Category 4 and above.

ENIAC Electronic Numerical Integrator and Computor (spelled with an O). Early computer, built in 1944.

ENQ ENQuiry character.

ENQ/ACK Protocol Hewlett-Packard communications protocol in which the HP3000 computer follows each transmission block with ENQ to determine if the destination terminal is ready to receive more data. The terminal indicates its readiness by responding with ACK.

Enriched Services Providers Those third-party service providers (other than Network Providers) who provide value-

added services that are accessed through telecommunications networks.
ENS Emergency Number Services.
ENSO ETSI National Standardization Organizations (ETSI).
ENTELEC ENergy TELECommunications and electrical association, the oldest nationwide user group in telecommunications. It is an association of communications managers and engineers in the oil, gas, pipeline and utility industries. ENTELEC played an important role in the early opening of competition in the telecommunications industry, including the famous "Above 890" decision, which allowed private companies to build their own long distance microwave system. The decision was called "Above 890" because electromagnetic waves in the radio frequency spectrum above 890 Megahertz (million cycles per second) and below 20 Gigahertz (billion cycles per second) are typically called microwave. Microwave used to be a common method of transmitting telephone conversations and was used by common carriers as well as by private networks. Now fiber is far more common. Microwave signals only travel in straight lines. In terrestrial microwave systems, a single transmission is typically good for 30 miles, at which point you need another repeater tower. Microwave is the frequency for communicating to and from satellites. ENTELEC was formerly known as the Petroleum Industry Electrical Association.
Enterprise Computing Enterprise means the whole corporation. Enterprise computing refers to the computing applications on which a company's life depends: order entry, accounts receivable, payroll, inventory, etc. It is also known by the phrase "mission critical." See also ENTERPRISE NETWORK.
Enterprise Network The word Enterprise was invented by IBM. It means the whole corporation. An enterprise-wide network is one covering the whole corporation. Local PBXs. Local area networks. Internetworking bridges. Wide area networks, etc, etc. See also CORPORATE NETWORK and ENTERPRISE COMPUTING.
Enterprise Number A service provided by AT&T and the Bell operating companies (a.k.a. the Bell System) years ago which allowed people to make collect calls and have their calls automatically accepted by the company at the other end. It was very expensive. It has largely been replaced with 800 IN-WATS service, which is much more successful.
Enterprise Solution Software that enables individuals and groups (either within an organization or part of a virtual organization beyond one company) to use computers in a networked environment to access information from a wide range of sources, collaborate on projects, and communicate easily with text, graphics, video, or sound.
Entity An active element within an OSI layer or sublayer.
Entity Coordination Management The portion of connection management which controls bypass relays and signals connection management that the medium is available.
Entrance And Exit Ramps The companies who control access to the internet and other networks of the information superhighway, whatever that is.
Entrance Facility Point of interconnection between the Network Demarcation Point and/or campus backbone and intra building wiring. The Entrance Facility includes overvoltage protection and connecting hardware for the transition between outdoor and indoor cable.
Entrapment The deliberate planting of apparent flaws in a system for the purpose of detecting attempted penetrations.
Entry Border Node An ATM term. The node which receives a call over an outside link. This is the first node within a peer group to see this call.
Enumerator A Windows 95 term. A Plug and Play device driver that detects devices below its own device node, creates unique device IDs, and reports to Configuration Manager during startup. For example, a SCSI adapter provides a SCSI enumerator that detects devices on the SCSI bus.
Envelope 1. The boundary of the family of curves obtained by varying a parameter of a wave.
2. The part of messaging that varies in composition from one transmittal step to another. It identifies the message originator and potential recipients, documents its past, directs its subsequent movement by the MTS (Message Transfer System) and characterizes its content.
Envelope Delay Circuit characteristics which result in some frequencies arriving ahead of others, even though they were transmitted at the same time.
Envelope Delay Distortion The distortion that results when the rate of change of phase shift with frequency over the bandwidth of interest is not constant. It is usually stated as one-half the difference between the delays of the two frequency extremes of the band of interest.
Environment The place your telephone system's main cabinet and main electronics live. While most PBX vendors will specify the room's characteristics, the ultimate responsibility for the room is yours, the user. Not designing your telephone system's environment correctly is tantamount to jinxing your telephone system from the start.
Here are some things to watch out for (your vendor has a more comprehensive list): 1. Sufficient air conditioning? Telephone systems give off heat. You need some way of getting rid of the heat. If you don't, you will blow some of your phone system's delicate electronic circuitry. 2. Sufficient space? Is there room for technicians to get in and around your telephone system so they can repair it? Will you have room for additional cabinets when you need to grow your phone system? 3. Sufficient and correct power? Will you have sufficient clean commercial AC power? Will you require isolation regulators? Or you will require extensive wet cell batteries? Will you have space? 4. Will you have a solid electrical ground? Can you find somewhere solid to ground your telephone system to — other than the third wire on the AC power, which is not suitable for most telephone systems? Beware of cold water pipes which end in PVC plastic pipes.
Environment Variable A Windows 95 and NT definition. A string consisting of environment information, such as a drive, path, or filename, associated with a symbolic name that can be used by Windows 95 or Windows NT. You use the System option in Control Panel or the set command from the Windows NT command prompt to define environment variables.
Envoy 1. A palmtop communicator introduced by Motorola in March of 1994. The device lets its users receive and transmit messages via Ardis, a network owned by Motorola and IBM. Envoy contains software from General Magic.

2. **Spectrum Envoy** is a DSP-based PC-board used for "telephone management" from a company called Spectrum Signal Processing, Burnaby, BC. Telephone management includes voice mail, contact manager, upgradable fax/modem, business audio, etc.

EO 1. End Office. Typically your own central office.

2. Erasable Optical drive. EO drives act like hard drives yet offer virtually unlimited storage because their cartridges are removable. Each cartridge sports at least 650 MB. Some sport 1 gigabyte.

3. EO was a startup in Mountain View, CA which did wireless data. It made a device called EO Personal Communicator 440 and 880. It uses GO's PenPoint operating system and the Hobbit microprocessor made by AT&T, which is "optimized" for telecommunications. In fall of 1994, AT&T closed EO down and stopped the sale of EO devices. It was too expensive and wasn't selling. An excellent book was written about EO. It is called "Startup; A Silicon Valley Adventure Story." It was written by Jerry Kaplan, one of EO's founders. The book is published by Houghton Mifflin.

EOA End Of Address. A header code.

EOB End Of Block. A control character or code that marks the end of a block of data.

EOC Embedded Operations Channel.

EOE See ELECTRONIC ORDER EXCHANGE.

EOF The abbreviation for End Of File. MS-DOS files and some programs often mark the end of their files with a Ctrl Z — or ASCII 26.

EOM End of Message (indicator). In ATM network, EOM is an indicator used in the AAL that identifies the last ATM cell containing information from a data packet that has been segmented.

EOT End of Transmission, End of Tape.

EOTC European Organization for Testing and Certification.

EOW Engineered OrderWire.

EPABX Electronic Private Automatic Branch eXchange. A fancy name for a modern PBX. Other fancy names include CBX, Computerized Branch Exchange.

EPD Early Packet Discard. See EARLY PACKET DISCARD.

Ephemeris The predictions of current satellite position that are transmitted to the user in the data message of a GPS (Global Positioning System) satellite message.

EPLANS Engineering, PLanning and ANalysis Systems. Software offered by Western Electric (now called AT&T Technologies) to help operating telephone company people run their business better.

Epoxy A liquid material that solidifies upon heat curing, ultraviolet light curing, or mixing with another material. Epoxy is sometimes used for fastening fibers to other fibers or for fastening fibers to joining hardware.

EPP Enhanced Parallel Port. A new hardware and software innovation (and now a standard) which allows computers so equipped to send data out their parallel port at twice the speed of older parallel ports, i.e. those that came on the original IBM PC. The EPP conforms to the EPP standard developed by the IEEE (Institute of Electrical and Electronics Engineers) 1284 standards committee. The EPP specification transforms a parallel port into an expansion bus that theoretically can handle up to 64 disk drives, tape drives, CD-ROM drives, and other mass-storage devices. EPPs are rapidly gaining acceptance as inexpensive means to connect portable drives to notebook computers. There's no difference in the shape of the ordinary, 25-pin D-connector plug/connector or the number of conductors. The Enhanced Parallel Port (EPP) was developed by Intel Corp., Xircom Inc., Zenith, and other companies that planned to exploit two-way communications to external devices. Many laptops built since mid-1991 have EPP ports. See also ECP.

EPROM Erasable Programmable Read Only Memory. A read only memory device which can be erased and reprogrammed. Typically, it is programmed electronically (not electromagnetically) with ultraviolet light. EPROMS are typically returned to the vendor or factory for reprogramming. An Eprom on a graphics card might contain the default or ROM character set. EPROM chips normally contain UV-permeable quartz windows exposing the chips' internals. See also ROM and EEPROM.

EPS Encapsulated PostScript. A desktop publishing and imaging term.

EPSCS (Pronounced Ep-Sis.) Enhanced Private Switched Communications Service. An AT&T offering for large businesses with offices scattered all over the country. This service allows such businesses to rent space on AT&T electronic switches and join that switching capacity to leased lines. EPSCS customers get a network control center in their offices which gives them information on the continuing operation of their network and allows them some limited options for changing their services.

EPSN Enhanced Private Switched Network.

EQ Abbreviation for EQUALIZATION.

Equal Access All long distance carriers must be accessible by dialing 1 — and not a string of long dialing codes. This is laid down in Judge Green's Modified Final Judgment (MFJ), which spelled out the terms of the Divestiture of the Bell Operating phone Companies (BOCs) from their parent, AT&T. Under the terms of this Divestiture, all long distance common carriers must have Equal Access for their long distance caller customers. City by city telephone subscribers are being asked to choose their primary carrier who they will reach by dialing 1 before their long distance number. All other carriers (including AT&T, if not chosen as primary) can be reached by dialing a five digit code (10XXX), thus providing Equal Access for all carriers. Not all long distance companies will opt for full equal access since this involves considerable expense to the local phone companies. See also FEATURE GROUP A, B, C and D.

Equal Access End Office A central office capable of providing equal access. See also EQUAL ACCESS.

Equal Gain Combiner A diversity combiner in which the signals on each channel are added together. The channel gains are all equal and can be made to vary equally so that the resultant signal is approximately constant.

Equalization The process of reducing distortion over transmission paths by putting in compensating devices. The telephone network is equalized by the spacing and operation of amplifiers along the way. In recording, equalization is frequency manipulation to meet the requirements of recording; also the inverse manipulation in playback to achieve uniform or "flat" response.

Also called Compensation. See EQUALIZATION CIRCUIT.

Equalization Circuit A compensation circuit designed into modems to counteract certain distortions introduced by the telephone channel. Two types are used: fixed (compromise) equalizers and those that adapt to channel conditions. U.S. Robotics high speed modems use adaptive equalization.

Equatorial Orbit An orbit with a zero degree inclination angle, i.e. the orbital plane and the Earths' equatorial plane are coincident.

Equipment Cabinet The metal box which houses relays, circuit boards or other phone apparatus. Usually also contains the power supply, which converts the 120 volt AC current into the low voltage direct current necessary to run the telephone system.

Equipment Compatibility One computer system will successfully do the same thing that another computer will do with the same data. There are many levels of "equipment compatibility." The only true compatibility, however, is identical machinery. And identical means "identical" down to the very last chip and very last integrated circuit. We have found that some computers — even those consecutively numbered — do not always perform the same. We have empirically proven this for both IBM and AT&T computers.

Equipment Identity Register See EIR.

Equipment Wiring Subsystem The cable and distribution components in an equipment room that interconnect system-common equipment, other associated equipment, and cross connects.

Equipped For Capacity The maximum number of lines and trunks that can be supported by the available hardware. It is not a totally effective measure of the size of a PBX. See WIRED-FOR-CAPACITY.

Equivalent Four-Wire System Transmission using frequency division to get full duplex transmission over only one pair of wires. Normally two pairs are needed for full duplex.

Equivalent Network 1. A network that may replace another network without altering the performance of that portion of the system external to the network.
2. A theoretical representation of an actual network.

Equivalent PCM Noise Through comparative tests, the amount of thermal noise power on an FDM or wire channel necessary to approximate the same judgment of speech quality created by quantizing noise in a PCM channel.

ER An ATM term. Explicit Rate: The Explicit Rate is an RM-cell field used to limit the source ACR to a specific value. It is initially set by the source to a requested rate (such as PCR). It may be subsequently reduced by any network element in the path to a value that the element can sustain. ER is formatted as a rate.

Erasable Programmable Read-Only Memory See EPROM.

Erasable Storage A storage device whose contents can be changed, i.e. random access memory, or RAM. Compare with read-only storage.

Erase Head On a magnetic tape recorder — voice or video — this is the "head" which erases the tape by demagnetizing it immediately before a new recording is placed on the tape by the adjacent record head.

Erbium A rare earth element that when added to fiber optic cabling could obviate the need for repeaters every 20 miles on undersea cables and expand fiber optic cabling to capacities of trillions of bits a second. See ERBIUM-DOPED FIBER AMPLIFIER.

Erbium-Doped Fiber Amplifier EDFA. Erbium-Doped Fiber Amplifiers have become the dominant method for signal amplification in long-haul lightwave transmission systems. EDFAs differ from the normal method of regenerative or electro-optic repeaters in that light does not have be converted to an electrical signal, amplified, and then converted back to light. Optical amplifiers contain a length of erbium-doped (a rare earth) that provides the gain medium, an energy source or "pump" from a laser source as the correct frequency, and a coupler to couple the pump laser to the doped fiber. Both the signal to be amplified and the pump energy are coupled into the doped fiber section of the transmission system. The pump laser puts the erbium-doped fiber into an excited state where it is able to provide optical gain through emission stimulated by a passing signal photon. One of the most important features, after the fact that EDFAs are amazingly simple, is that they are not frequency dependent, and therefore allow bandwidth upgrades (within limits) without replacing the entire transmission systems. Undersea transmission systems, such as Americas 1, TAT-12/13, and TCP-5 use EDFA technology.

Erector Set Telecom In North America, there's a children's game of building blocks called Lego. The game comes with hundreds of small plastic blocks, which can be assembled into all sorts of wonderful designs, from castles to gas stations. In England, Lego sets are also called Mechano sets. The generic term for Lego and Mechano sets is erector sets. The term "erector set telecom" is a concept created by Harry Newton as a way of explaining "the new open" telecommunications equipment, namely that you build your own computer telephony system from freely-available, non-proprietary hardware and software components. In short, a telecom industry along the same open hardware and software lines as the PC industry. See COMPUTER TELEPHONY.

Ergonomics The science of determining proper relations between mechanical and computerized devices and personal comfort and convenience; e.g., how a telephone handset should be shaped, how a keyboard should be laid out.

Erlang A measurement of telephone traffic. One Erlang is equal to one full hour of use (e.g. conversation), or 60 x 60 = 3,600 seconds of phone conversation. You convert CCS (hundred call seconds) into Erlangs by multiplying by 100 and then dividing by 3,600 (i.e. dividing by 36). Numerically, traffic on a trunk group, when measured in erlangs, is equal to the average number of trunks in use during the hour in question. Thus, if a group of trunks carries 12.35 erlangs during an hour, a little more than 12 trunks were busy, on the average.

Erlang gets its name from the father of queuing theory, A. K. Erlang, a Danish telephone engineer, who, in 1908, began to study congestion in the telephone service of the Copenhagen Telephone Company. A few years later he arrived at a mathematical approach to assist in designing the size of telephone switches. Central to queuing theory are basic facts of queuing life. First, traffic varies widely. Second, anyone who designs a telephone switch to completely handle all peak traffic will find the switch idle for most of the time. He will also find he's built a very expensive switch. Third, it is possible, with varying degrees of certainty to predict upcoming "busy" periods. See also ERLANG, A.K., ERLANG B, ERLANG C and POISSON.

Erlang, A. K. In 1918, A. K. Erlang, a Danish telephone engineer, published his work on blocking in "The Post Office Electrical Engineers' Journal," a British publication. Like E.C. Molina, an AT&T engineer, Erlang assumed a Poisson distribution of calls arriving in a given time. Molina had assumed a constant holding time for all calls, whereas Erlang assumed an exponential distribution for holding times. That means that longer calls occur less frequently than shorter calls. Erlang assumed that blocked calls are immediately cleared and lost and do not return. A formula that Erlang worked out based on these assumptions (Erlang B) is still in use in telephone engineering. See ERLANG, ERLANG B, ERLANG C and POISSON.

Erlang B A probability distribution developed by A.K. Erlang to estimate the number of telephone trunks needed to carry a given amount of traffic. Erlang B assumes that, when a call arriving at random finds all trunks busy, it vanishes (the blocked calls cleared condition). Erlang B is also known as "Lost Calls Cleared." Erlang B is used when traffic is random and there is no queuing. Calls which cannot get through, go away and do not return. This is the primary assumption behind Erlang B. Erlang B is easier to program than Poisson or Erlang C. This convenience is one of its main recommendations. Using Erlang B will produce a phone network with fewer trunks than one using Poisson formulae. See also ERLANG, ERLANG A. K., ERLANG C, and TRAFFIC ENGINEERING.

Erlang C A formula for designing telephone traffic handling for PBXs and networks. Used when traffic is random and there is queuing. It assumes that all callers will wait indefinitely to get through. Therefore offered traffic (see ERLANG) cannot be bigger than the number of trunks available (if it is, more traffic will come in than goes out, and queue delay will become infinite). Erlang C is not a perfect traffic engineering formula. There are none that are.

Erlang Formula A mathematical way of making predictions about randomly arriving work-load (such as telephone calls) based on known information (such as average call duration). Although traditionally used in telephone traffic engineering (to determine the required number of trunks), Erlang formulas have applications in call center staffing as well. See ERLANG.

ERMES 1. European Radio MEssaging System.
2. One of the communications protocols used between paging towers and the mobile pagers/receivers/beepers themselves. Other protocols are POCSAG, ERMES, FLEX, GOLAY and REFLEX. The same paging tower equipment can transmit messages one moment in GOLAY and the next moment in ERMES, or any of the other protocols.

ERP Effective Radiated Power.

Error Burst A sequence of transmitted signals containing one or more errors but regarded as a unit in error in accordance with a predefined measure. Enough consecutive transmitted bits in error to cause a loss of synchronization between sending and receiving stations and to necessitate resynchronization.

Error Checking And Correction Error checking is the process of checking a "packet" being transmitted over a network to determine if the package, or the data content within the package, has been damaged. If checked and found wanting, damaged packets are discarded. Error correction is the process of correcting the damage by resending a copy of the original packet. In public frame relay services, the network performs the function of error checking, but not error correction. That function is left to the intelligent end equipment (at the user's site).

Error Control Various techniques which check the reliability and accuracy of characters (parity) or blocks of data sent over telecommunications lines. V.42, MNP and HST error control protocols (three common dial-up phone line modem protocols) use error detection (CRC) and retransmission of errored frames (ARQ). See ERROR CONTROL PROTOCOLS.

Error Control Protocols Besides high-speed modulation protocols, all current models of high-speed dial-up modems also support error control and data compression protocols. There are two standards for error control protocols: MNP-4 and V.42. The Microcom Networking Protocol, MNP, was developed by Microcom. MNP 2 to 4 are error correction protocols. V.42 was established by ITU-T. V.42 actually incorporates two error control schemes. V.42 uses LAP-M (Link Access Procedure for Modems) as the primary scheme and includes MNP-4 as the alternate scheme. V.42 and MNP-4 can provide error-free connections. Modems without error control protocols, such as most 2400 bps Hayes-compatible modems, cannot provide error-free data communications. The noise and other phone line anomalies are beyond the capabilities of any standard modem to deliver error-free data. V.42 (and MNP 2-4) copes with phone line impairments by filtering out the line noise and automatically retransmitting corrupted data. The filtering process used by V.42 (and MNP 2-4) is similar to the error correction scheme used by file transfer protocols (such as XMODEM). The two modems use a sophisticated algorithm to make sure that the data received match with the data sent. If there is a discrepancy, the data is re-sent.

What is the difference between error control protocols (such as V.42) and file transfer protocols (such as XMODEM)? For one thing, file transfer protocols provide error detection and correction only during file transfers. File transfer protocols do not provide any error control when you are reading e-mail messages or chatting on line. Even though an error control protocol is "on" all the time, we still need file transfer protocols when two modems establish a reliable link. A modem works with bit streams, timing and tones. It does not understand what a file is. When you download or upload a file, your communications software needs to take care of the details related to the file: the filename, file size, etc. This is handled by the file transfer protocol which does more than error-checking.

The other benefit of V.42 (or MNP 4) is that it can improve throughput. Before sending the data to a remote system, a modem with V.42 (or MNP 4) assembles the data into packets and during that process it is able to reduce the size of the data by stripping out the start and stop bits. A character typically takes up 1 start bit, 8 data bits and 1 stop bit for a total of 10 bits. When two modems establish a reliable link using V.42 or MNP 4, the sending modem strips the start and stop bits (which subtracts 20% of the data) and sends the data to the other end. The receiving modem then reinserts the start and stop bits and passes the data to the computer.

Therefore, even without compressing the data you can expect to see as much as 1150 characters per second on a 9600 bps connection. Although the modem subtracts 20% of the data, the speed increase is less than 20% due to the overhead incurred by the error control protocol.

The above definition with great thanks to modem expert Patrick Chen.

Error Correcting Protocol A method of transmitting bit streams in a mathematical way such that the receiving computer verifies to the sending computer that all bits have been received properly. SNA and XMODEM protocols, in the mainframe and microcomputer environments respectively, are Error Correcting Protocols. See ERROR CONTROL PROTOCOL.

Error Correction Code In computers, rules of code construction that facilitate reconstruction of part or all of a message received with errors.

Error Correction Mode A method of transmitting and receiving data that eliminates errors.

Error Free Second A Bellcore definition. An error-free second is, surprise, surprise, a one second time interval of digital signal transmission during which no error occurs. That's it.

Error Level A numeric value set by some programs that you can test with the errorlevel option of the "If" batch command. It works as follows. Some programs set the DOS errorlevel to a certain number depending on a certain input or response to an event. Let's say when you type the letter "Y" in response to a question the errorlevel is set to 32. Once this is done, you may condition other events based upon this number using an If command in a batch file. You can say "IF ERRORLEVEL = 32 THEN GOTO END." That way, when you type "Y" you will get whatever is at END. This can be very helpful in batch files and other programs for providing "branching" from one event to another based on certain inputs.

Error Logical An error in the binary content of a signal, for example, bit error.

Error Rate In data transmission, the ratio of the number of incorrect elements transmitted to the total number of elements transmitted.

Error Suspense An MCI definition. An automated process which allows billable MCI calls on switch tapes to be processed for billing, while calls with errors are held in the Error Suspense File (a separate file for each switch).

Error Trapping In software programming, an exception is an interruption to the normal flow of a program. Common exceptions are division-by-zero, stack overflow, disk full errors and I/O (input/output) problems with a file that isn't open. The quality of a software program depends on how completely it checks for possible errors and deals with them. Code used for trapping errors can be excessive. Some programming languages have error trapping built in. Others don't and you have to program it in.

Erstwhile An English word meaning previous. I define this word because I use it in this dictionary and lots of readers have told me they don't know it.

ES 1. Errored-Second.
2. End System: A system where an ATM connection is terminated or initiated. An originating end system initiates the ATM connection, and terminating end system terminates the ATM connection. OAM cells may be generated and received.

ES/9000 IBM Enterprise System/9000: mainframe computer family.

ESA Emergency Stand Alone.

ESC The ESC key on the keyboard. Often used to leave (escape) a program. Appears on the upper left of some keyboards on the IBM or compatibles but moves around with IBM's latest keyboard redesign whim. See also ESCAPE.

Escalation A formal word for taking your trouble up through the levels of management at the vendor — until you get your problem resolved. Some users have formal Escalation Charts, which detail action to be taken depending on how many hours the problem persists, etc. Escalation sometimes works and sometimes doesn't, depending on the vendor. Usually it does. The rule in telecommunications (and we guess most other industries) is that "the squeaky wheel gets the most attention." Escalation works well with honey, flowers, plants and chocolates.

Escape 1. The button on many computer keyboards which allows you to "escape" the present program. ESCape is the ASCII control character — code 27. It is often used to mark the beginning of a series of characters that represent a command rather than data. So called "ESCAPE" because it escapes from the usual meaning of the ASCII code and allows commands to be interspersed in a file of data, especially for data transmission to peripheral devices such as printers and modems. See ESCAPE SEQUENCE.
2. A means of aborting the task currently in progress.
3. A code used to force a smart modem back to the command state from the on-line state.

Escape Guard Time An idle period of time before and after the escape code sent to a smart modem, which distinguishes between data and escapes that are intended as a command to the modem.

Escape Sequence A series of characters, usually beginning with the escape character, that is to be interpreted as a command, not as data. Escape sequences are used with ANSI.SYS to change the color of a screen. They are mostly used to send print commands to printers. The name Escape is due to the fact that it "escapes" from the usual meaning of the ASCII code, letting characters be commands instead of data, yet interspersed with data in a transmission.

ESCON Enterprise Systems Connection. A 10 to 17 megabytes per second fiber optic LAN for linking IBM mainframes to disk drives and other mainframes. Not very popular with smaller users.

Escrow Bucket A hopper at the outlet of a coin phone's acceptor/rejector that is tipped electrically to return money through the Coin Return or to send the money to the Cash Box as a collection for a completed call.

ESD Electrostatic Discharge.

ESDI 1. Enhanced Small Device Interface. An interface which improves the rate of data transfer for hard disk drives and increases the drive's storage capacity.
2. Northern Telecom term for Enhanced Serial Data Interface.

ESF Extended Super Frame or Extended Superframe Format. A T-1 format that uses the 193rd bit as a framing bit. ESF provides frame synchronization, cyclic redundancy checking and data link bits. Frames consist of 24 bits instead of the previous standard 12 bits as in the D4 format. The standard allows error information to be stored and retrieved easily, facilitating network performance monitoring and maintenance. Electronic Tandem Network

ESI 1. See ENHANCED SERIAL INTERFACE.
2. End System Identifier: This identifier distinguishes multiple nodes at the same level in case the lower level peer group is partitioned.

ESMA Expanded Subscriber Module-100A.
ESMR Enhanced Specialized Mobile Radio. Two-way dispatch service with the capability to provide wireless voice telephone service to compete against cellular. It uses TDMA technology to put six voice conversations into one 25 kilohertz UHF radio channel in the 806 to 821 megahertz band.
ESMTP Extended Simple Mail Transport Protocol (see SMTP), described in RFC 1651. Seldom used outside the Unix community.
ESMU Expanded Subscriber Carrier Module-100 URBAN.
ESN 1. Emergency Service Number. An ESN is a "list" that corresponds to a particular ESZ (Emergency Service Zone). This list has to do with 911 service. Usually this ESN list is unique and contains a listing of the corresponding police, figure and ambulance dispatch centers for the caller's area. This "list" is used for selective routing and one button transfer to secondary PSAPs — Public Safety Answering Positions. The ESN/ESZ concept is especially useful in fringe areas.
2. Electronic Serial Number. Each cellular phone is assigned a unique ESN, which is automatically transmitted to the base station every time a cellular call is placed. The Mobile Telephone Switching office checks the ESN to make sure it is valid, that the phone has not been reported stolen, that the user's monthly bill has been paid, etc., before permitting the call to go through. At least that's the theory. It doesn't always work this way on calls made from roaming cellular phones. And some cellular phones have been known to have their ESNs tampered with, which tends to mess up the billing mechanisms.
3. Electronic Switched Network.
ESOs European Standardization Organizations.
ESP 1. Enhanced Serial Port. The Hayes Enhanced Serial Port (ESP) adapter, introduced in late 1990, replaces and extends the traditional COM1/COM2 serial port adapter. The ESP combines dual 16550 UARTS with an on-board communications coprocessor. The ESP has two distinct modes of operation to provide both old and new standards in the same package: Compatibility Mode and Enhanced Mode. Each ESP port can be independently operated in either mode. Default modes are configured via DIP switches and can be modified by ESP commands. The MCA-bus version of the ESP uses Programmable Option Selection (POS) rather than DIP switches. See ENHANCED SERIAL INTERFACE.
2. Enhanced Service Provider — a vendor who adds value to telephone lines using his own software and hardware. Also called an IP, or Information Provider. An example of an ESP is a public voice mail box provider or a database provider, say one giving the latest airline fares. An ESP is an American term, unknown in Europe, where they're most called VANs, or Value Added Networks. See also OPEN NETWORK ARCHITECTURE and INFORMATION PROVIDER.
3. EncapSulated Postscript File.
ESPA European Selective Paging Association.
ESPRIT European Strategic Program for Research and development in Information Technology. A $1.7 billion research and development program funded by the European Community.
ESS 1. Electronic Switching System. ESS was originally a designation for the switching equipment in Bell System central offices but has slightly more general use now. In the independent telephone company industry, the abbreviation for the same thing is EAX.
2. European Standardization System.
ESS No 4 AT&T's large toll telephone switch. It will handle over 100,000 trunks and over 500,000 attempts at making a call each hour. It's large and sophisticated and can probably be configured to be the largest telephone switch in the world.
ESS No 5 AT&T's Class 5 digital central office. See also END OFFICE.
Essential Service 1. A service provided by a telecommunications provider, such as an operating telephone company or a carrier, for delivery of priority dial tone. Generally, only up to 10 percent of the customers may request this type of service.
2. A service that is recommended for use in conjunction with NS/EP (national emergency) telecommunications services.
ESSX ESSX (pronounced essex) is some local phone companies' name for Centrex. See CENTREX.
ESZ Emergency Service Zone. This term is used in conjunction with 911 emergency service. An ESZ is a geographic area that is served by a unique mix of emergency services. Each ESZ has a corresponding ESN (a list of Emergency Service Numbers) which enables 911 service to properly route incoming calls.
ET Exchange Termination. Refers to the central office link with the ISDN user.
ETACS Extended TACS. The cellular technology used in the United Kingdom and other countries. It is developed from the U.S. AMPS technology. See also AMPS, TACS, NTACS and NAMPS.
ETB End of Transmission Block.
Eternity Hold Our own creation for what happens when someone puts you on long-term hold. Governmental agencies, airlines and police departments (especially when you need them) tend to be firm believers in placing their callers on Eternity Hold. A new service adjunct to Eternity Hold is Conference Hold. Here everyone on Eternity Hold can speak to each other. We made this up. It doesn't exist, but we think it would be great if it did.
Ether The medium which, according to one theory, permeates all space and matter and which transmits all electromagnetic waves.
Ethernet A local area network used for connecting computers, printers, workstations, terminals, servers, etc., within the same building or campus. Ethernet operates over twisted wire and over coaxial cable at speeds up to 10 Mbps. For LAN interconnection, Ethernet is a physical link and data link protocol reflecting the two lowest layers of the DNA/OSI model. The theoretical limit of Ethernet, measured in 64 byte packets, is 14,800 packets per second (PPS). By comparison, Token Ring is 30,000 and FDDI is 170,000.
Ethernet specifies a CSMA/CD (Carrier Sense Multiple Access with Collision Detection). CSMA/CD is a technique of sharing a common medium (wire, coaxial cable) among several devices. As Byte Magazine explained in its January, 1991 issue, Ethernet is based on the same etiquette that makes for a polite conversation: "Listen before talking." Of course, even when people are trying not to interrupt each other, there are those embarrassing moment when two people accidentally start talking at the same time. This is essentially what happens in Ethernet networks, where such a situation is called a collision. If a node on the network detects a collision, it alerts the other nodes by jamming the network. Then, after a random pause, the sending nodes try again.

The messages are called frames (see the diagram).

AN ETHERNET FRAME					
Preamble	Destination address	Source address	Type	Data up to 1500 bytes	Frame check sequence
8 bytes	6 bytes	6 bytes	2 bytes		4 bytes (contains CRC check)

The first personal computer Ethernet LAN adapter was shipped by 3Com on September 29, 1982 using the first Ethernet silicon from SEEQ Technology. Bob Metcalfe created the original Ethernet specification at Xerox PARC and later went on to found 3Com. In the October 31, 1994 issue of the magazine InfoWorld, Bob Metcalfe explained that Ethernet got its name "when I was writing a memo at the Xerox Palo Alto Research Center on May 22, 1973. Until then I had been calling our proposed multimegabit LAN the Alto Aloha Network. The purpose of the Alto Aloha Network was to connect experimental personal computers called Altos. And it used randomized retransmission ideas from the University of Hawaii's Aloha System packet radio network, circa 1970. The word ether came from luminiferous ether — the omnipresent passive medium once theorized to carry electromagnetic waves through space, in particular light from the Sun to the Earth. Around the time of Einstein's Theory of Relativity, the light-bearing ether was proven not to exist. So, in naming our LAN's omnipresent passive medium, then a coaxial cable, which would propagate electromagnetic waves, namely data packets, I chose to recycle ether. Hence, Ethernet."
According to Metcalfe, "Ethernet has been renamed repeatedly since 1973. In 1976, when Xerox began turning Ethernet into a product at 20Mbps, we called it The Xerox Wire. When Digital, Intel, and Xerox decided in 1979 to make it a LAN standard at 10Mbps, they went back to Ethernet. IEEE tried calling its Ethernet standard 802.3 CSMA/CD — carrier sense multiple access with collision detection. And as the 802.3 standard evolved, it picked up such names as Thick Ethernet (IEEE 10Base-5), Thin Ethernet (10Base-2), Twisted Ethernet (10Base-T), and now Fast Ethernet (100Base-T)."
Ethernet PC cards now come in a couple of basic varieties — for connecting to an Ethernet LAN via coaxial cable or via two twisted pairs of phone wires, called 10Base-T. See also 10Base-T, ETHERNET CONTROLLER, ETHERNET IDENTIFICATION NUMBER, ETHERNET SWITCH, ETHERTALK, THINNET and TOKEN RING.
Ethernet Controller The unit that connects a device to the Ethernet cable. An Ethernet controller typically consists of part of the physical layer and much or all of the data link layer and the appropriate electronics.
Ethernet Identification Number This is a unique, hexadecimal Ethernet number that identifies a device, such as a PC/AT with a SpeedLink/PC16 network interface card installed, on an Ethernet network.
Ethernet Switch A new device that connects local area networks. Here's a definition of the capabilities which an Ethernet switch must have from a company called Kalpana Inc. in Santa Clara CA. "What capabilities must a device have to be an Ethernet Switch? Ethernet switching is being embraced as the next milestone solution for bandwidth-constrained Ethernets. To qualify as an Ethernet Switch, the device must: Be capable of switching packets from one Ethernet segment to another "on-the-fly;" Avoid using slower store-and-forward technologies to route packets from one segment to another; Exhibit very low port-to-port latency (the elapsed time between receiving and transmitting a LAN packet is measured in 10s of microseconds, not 100s); Offer a busless, scaleable architecture that increases network carrying capacity as switched connections are added; Support all higher-level products; Provide the technology for creating a massively parallel system with a tens-of-gigabits per second capacity."
EtherTalk AppleTalk protocol governing Ethernet local area network transmissions. Also the Apple Computer Ethernet adapter and drivers. Apple's implementation of Ethernet is compliant with IEEE specification 802.3.
Ethertype A two-byte code indicating protocol type in an Ethernet local area network packet.
ETI Electronic Telephone Interface.
ETISALAT Emirates Telecommunications Corporation is the sole provider of telecommunications services throughout the United Arab Emirates. The head office is in Abu Dhabi.
ETM Electronic Ticketing Machine. A machine that looks like a banking Automated Teller Machine (ATM), except that it will dispense airline tickets and possibly, hotel reservations, car rental agreements, etc.
ETN Electronic Tandem Network.
ETNO The European Public Telecommunications Network Operations Association (ETNO).
ETS 1. European Telecommunications Standard. A standard defined by the European Telecommunications Standards Institute (ETSI). 2. Electronic Tandem Switching. See ELECTRONIC TANDEM SWITCHING.
ETS 300 211 Metropolitan Area Network (MAN) Principles and Architecture.
ERS 300 212 Metropolitan Area Network (MAN) Media Access Control Layer and Physical Layer Specification.
ETS 300 217 Connectionless Broadband Data Service (CBDS).
ETS Set A Northern Telecom term for an electronic Telephone Set.
ETSI European Telecommunications Standards Institute, is the European counterpart to ANSI, the American National Standards Institute. ETSI is based in Sophia-Antipolis, near Nice, France. ETSI's task is to pave the way for telecommunications integration in the European community as part of the single European market program. ETSI was founded in 1988 as a result of an initiative of the European Commission. It was established to produce telecommunications standards by democratic means, for users, manufacturers, suppliers, administrations, and PTTs. ETSI's main aim is the unrestricted communication between all the

member states by the provision of essential European standards. It is now an independent, self-funding organization that has progressively merged since 1990, with both CEPT and EBU.

ETX End of Text. Indicates the end of a message. If multiple transmission blocks are contained in a message in Bisynch systems, ETX terminates the last block of the message. ETB is used to terminate preceding blocks. The block check character is sent immediately following ETX. ETX requires a reply indicating the receiving station's status.

EUCL End User Common Line charge. An amount added by your local phone company to your monthly phone charges for the privilege (it seems) of providing you phone service. No one seems to understand the reason for this charge. EUCL is (or was) an FCC idea. And the FCC seems to mandate what the local operating phone companies should charge each subscriber for EUCL each month.

Eudora An electronic mail sending and receiving mail that runs on Macs and under Windows and is probably the most common e-mail service used by people on the Internet.

EURESCOM EUropean institute for REsearch and Strategic studies in TeleCOMmunication.

EUROBIT European Association of Business Machines Manufacturers and Information Technology Industry.

Euro-ISDN The European implementation of ISDN. It differs from North American National ISDN-1 in that Euro-ISDN is very limited in the options it offers. In the United States, ISDN comes with many options, including two call appearances, conference calling, call forwarding variable, call forwarding — busy, call forwarding — no answer, voice mail with indicator, two secondary directory numbers, etc. That makes North American ISDN more full-featured, but much less easy to order. Users, can however, call from the United States to Europe and complete ISDN calls. They can not carry their end-user ISDN equipment from the United States and use it in most places in Europe.

European Space Agency The European equivalent of the U.S.'s NASA — National Aeronautics and Space Administration.

EUROSINET-EUROTOP International ISDN pilot project for travel agents.

EUROTELDEV EUROpean TELecommunications DEVelopment. An organization involved in telecommunications standardization.

EUTELSAT EUropean TELecommunications SATellite organization. Inter-governmental organization that aims to provide and operate a communications satellite for public intra-European international telecommunications services. The segment is also used to meet domestic needs by offering leased capacity, primarily for television. U.K. and France are the largest shareholders, with about 25 member countries in total.

EUTP Enhanced Unshielded Twisted Pair. UTP Cables that have enhanced transmission characteristics. Cables that fall under this classification include Category 4 and above.

EV European Videotelephony.

EVA Economic Value Added. A financial measure of whether you're making more money with your plant, factory, assets, etc. in your present business than you would be if you sold everything and stuck the proceeds in a investment. The common assumption is that you can get a 10% or 11% return. If you earn more than that, you're EVA positive. EVA is a term you hear a lot around AT&T. You can become EVA positive in any ways — writing down the value of your capital is one way. Sacking people works too. And, so does selling more (presumably, at a decent price).

Even Parity In data communications there's something called a PARITY BIT that's used for error checking. The transmitting device adds that parity bit to a data word to make the sum of all the "1" ("one" bits) either odd or even. If the sum is odd, the result is called ODD parity. If it's even, it's called EVEN PARITY. See also PARITY.

Event An unsolicited communication from a hardware device to a computer operating system, application, or driver. Events are generally attention-getting messages, allowing a process to know when a task is complete or when an external event occurs.

Event Code A code that an agent in a call center enters at the conclusion of a call. Event codes can trigger a variety of follow-up activities such as an acknowledgement letter, or inclusion in a list for a subsequent campaign.

Event Driven A style of programming under which programs wait for messages to be sent to them and react to those messages. See EVENT DRIVEN ALARMS/TRIGGERS/TICKLERS.

Event Driven Alarms/Triggers/Ticklers In a parallel process, an event trigger can be set to move the processing forward when a set of criteria is met (ex. the last piece of documentation is added to the file). Alarms can also be time-driven, as when a folder is automatically routed to exception processing if no action is taken within a specified time frame. This term is often found in workflow management.

Event History A history of the activities that have been carried out on a record, (customer or prospect). For example, phone calls and mailers, offers and so on.

Event Mask The set of events that the SLEE (Service Logic Execution Environment) designates the ASC (AIN Switch Capabilities) to report for a particular connection segment, and an indication for each event if the ASC should suspend processing events for that connection segment until the SLEE sends a message back. Definition from Bellcore in reference to its concept of the Advanced Intelligent Network.

Event Signaling All telephone switches use the same three general types of signals: + Event Signaling initiates an event, such as ringing. + Call Progress Signaling denotes the progress (or state) of a call, such as a busy tone, a ringback tone, or an error tone. + Data Packet Signaling communicates certain information about a call, for example, the identify of the calling extension, or the identity of the extension being called.

EW Electronic Warfare. The military use of radar, electronic counter measures and electronic counter-counter measures to keep an enemy from finding invading forces, on land or in the air. It covers such methods as sending planes equipped with equipment which transmit thousands of signals purporting to be signals that an enemy radar might see on locating an incoming plane. By sending thousands of such signals, the enemy's radar becomes a myriad of "radar" signals, of bright spots. Thus it's impossible for the enemy to read any intelligent information. There are also anti-radiation missiles which home in on and destroy air-defense radar facilities. the only defense against such anti-radiation missiles is to turn off the radar. Electronic warfare also covers such techniques as jamming radio frequencies, anti-jamming. It is not the state of Judge

Greene's courtroom or the boardroom at the FCC.
eWORLD Apple Computer's on-line, dial-up information service.
EWOS European Workshop for Open Systems.
EWP Electronic White Pages.
Exabyte An exabyte is equal to 10th the 18th power. An exabyte is equal to 1,000 petabytes. Here are the abbreviations: MHz = Megabyte (10 to the 6th power) GHz = Gigabyte (10 to the 9th power) THz = Terabyte (10 to the 12th power) PHz = Petabyte (10 to the 15th power) EHz = Exabyte (10 to the 18th power)
Exalted Carrier Reception A method of receiving either amplitude- or phase-modulated signals in which the carrier is separated from the sidebands, filtered and amplified, and then combined with the sidebands again at a higher level prior to demodulation.
EXCA Exchangeable Card Architecture. ExCA is a hardware and software architectural implementation of PCMCIA 2.0 from Intel that allows card interoperability and exchangeability from system to system, regardless of manufacturer. See PCMCIA.
Exception 1. In telecom, when something happens that's "unusual," it's an exception. The key is to define what's "unusual." For example, you might define that every phone call of longer than 15 minutes is an "exception." Now you have defined an "exception," the question is how to use that information. You might ask the phone system to print out each "exception" call on a printer next to your desk immediately after the call is over. Or you might ask the machine to print "Exceptions" reports at the end of the month listing all the calls over 15 minutes. These reports might be by perpetrator. Or in chronological order, or order of phone number called, etc. In short, any event you define by certain strict parameters can be an "exception." Management reports printed in full are almost useless because they contain so much information, so much paper. Management reports which list only previously-defined "exceptions" are more useful. They show you where to focus your attention so as to improve your or your company's performance.
2. In software programming, an exception is an interruption to the normal flow of a program. Common exceptions are division-by-zero, stack overflow, disk full errors and I/O (input/output) problems with a file that isn't open. The quality of a software program depends on how completely it checks for possible errors and deals with them. Code used for trapping errors can be excessive. Some programming languages have error trapping built in. Others don't and you have to program it in.
Exception Condition In data transmission, the condition assumed by a device when it receives a command that it cannot execute.
Exception Reports Reports generated by "exceptions," often detailing extra long calls or indications of bad circuits. See EXCEPTION.
Excess Insertion Loss In a optical fiber coupler, the optical loss associated with that portion of the light which does not emerge from the operational ports of the device.
Excessive Zeros More consecutive zeros received than are permitted for the selected coding scheme. For AMI-encoded T-1 signals, 16 or more zeros are excessive. For B8ZS encoded serial data, 8 or more zeros are excessive.
Exchange 1. Sometimes used to refer to a telephone switching center — a physical room or building. Outside North America, telephone central offices are called "Public Exchanges."
2. A geographic area in which there is a uniform set of charges for telephone service.
Exchange Access In the telephone networks, the provision of exchange services for the purpose of originating or terminating interexchange telecommunications. Such services are provided by facilities in an exchange area for the transmission, switching, or routing of interexchange telecommunications originating or terminating within the exchange area.
The Telecommunications Act of 1996 defined it as follows: The term `exchange access' means the offering of access to telephone exchange services or facilities for the purpose of the origination or termination of telephone toll services.
Exchange Area Geographic area in which telephone services and prices are the same. The concept of exchange is based on geography and regulation, not equipment. An exchange might have one or several central offices. Anyone in that exchange area could get service from any one of those central offices. It's good to ask which central offices could serve your home or office and take service from the most modern. There will be no difference in price between being served by a one-year old central office, or a 50-year old step-by-step central office.
Exchange Carrier A Bellcore definition. A company that provides telecommunication within a franchised territory.
Exchange Carriers Association An organization of long distance telephone companies with specific administrative duties relative to tariffs, access charges and payments. See EXCHANGE CARRIERS STANDARDS ASSOCIATION.
Exchange Carriers Standards Association ECSA. According to their literature ECSA is "the national problem-solving and standards-setting organization where local exchange carriers, interexchange carriers, manufacturers, vendors and users rationally resolve significant operating and technical issues such as network interconnection standards and 800 database trouble reporting guidelines. The Association was created in 1983. The major committees sponsored by ECSA are The Carrier Liaison Committee (to coordinate and resolve national issues related to provision of exchange access); the Telecommunications Industry Forum (TCIF) (to respond to the growing need for voluntary guidelines to facilitate the use of new technology that offers cost savings throughout the telecommunications industry — e.g. EDI, bar coding, automatic number identification); and the Information Industry Liaison Committee (IILC) (an inter industry forum for discussion and voluntary resolution of industry wide concerns about the provision of Open Network Architecture (ONA) services and related matters and Committee T1-Telecommunications (an accredited standards group under ANSI to develop technical standards and reports for US telecommunications networks. In October, 1993, The Exchange Carriers Standards Association changed its name to the Alliance for Telecommunications Industry Solutions (ATIS). It is based in Washington, D.C.
Exchange Facilities Those facilities included within a local access and transport area.
Exchange Network Facilities For Interstate Access See ENFIA.
Exchange, Private Automatic Branch (pabx) A private telephone exchange which transmits calls internally and to

and from the public telephone network.

Exchange Service A name that BellSouth gives to its local phone services, which it also calls Plain Old Telephone Service (POTS).

Exchange Termination ET. In Integrated Services Digital Network (ISDN) nomenclature, ET refers to the central office link with the end user.

Exclude A memory management command-line option that tells the memory manager in an MS-DOS machine not to use a certain segment of memory. For example, you may exclude upper memory locations D200 through D800 (hexadecimal) because your network adapter card uses that space. The reciprocal term — include — specifically directs the memory manager to use an area of memory.

Exclusion A PBX feature that prevents the attendant from silently monitoring a call once he/she has extended it.

Exclusive Hold Only the telephone putting the call on hold can take it off. This feature assures that the call on hold will not be picked up by someone at another telephone who can then listen to your call.

Exclusive Hold Recall When a call is placed on "Exclusive Hold" and is not picked up after a predetermined amount of time, you will hear a beeping at that phone, which indicates the call is still on hold.

Exclusive Or Private Unit A circuit card installed in each key telephone set sharing the same line or intercom path that causes the first caller on the line to lock out (exclude) all other stations from using or listening in, until the line is released (or privacy feature is defeated by the active caller).

Execunet An intercity switched telephone service introduced by MCI in 1975. Execunet was the first dial-up switched service introduced by a long distance phone company in competition with AT&T. At that time, all of AT&T's competitors, including MCI, were selling full-time private lines and shared private lines. The service was named by Carl Vorder-Bruegge, MCI's VP marketing at that time and introduced and made successful by Jerry Taylor, who was MCI's regional manager in Texas and is now one of the company's most senior executives. The service was the forerunner of what is today a $10 billion plus industry — the non-AT&T provided switched long distance business. MCI no longer uses the word Execunet to describe its switched long distance service. It's just plain long distance. Jerry Taylor started Execunet using a 104-port Action WATSBOX in Dallas, Texas. He deserves a place in the history books, not just a dictionary.

Executable File A computer program that is ready to run. Application programs, such as spreadsheets and word processors, are examples of executable files. Such files in PCs running MS-DOS and Windows usually end with the BAT, COM or EXE extension.

Execute To complete a task.

Execution Time The time needed to complete a task.

Executive Barge-In See EXECUTIVE OVERRIDE.

Executive Busy Override See EXECUTIVE OVERRIDE.

Executive Camp-On A feature for use by executives or other privileged people. When they call a someone lowly, that low person hears a special distinctive tone or sees a special light or sees a special signal that their phone has been camped on by someone significant. These days many PBXs let you know who's calling — even though you're on the phone. So executive camp-on is not that useful.

Executive Override A feature of some telephone systems which permits certain users to intrude on conversations on other extensions. In some systems, executive barging-in will not be heard by the person outside the office, only the one inside the office. In some systems, such as the Mitel SX series with the Mitel Superset 4 phones, this feature activates the hands-free speakerphone of the called party, who is using his other line to speak on a normal phone conversation.

Exit Border Node An ATM term. The node that will progress a call over an outside link. This is the last node within a peer group to see this call.

Exit Event An event occurring in an ASC (AIN Switch Capabilities) that causes call processing to leave a PIC (Point in Call). Definition from Bellcore in reference to its concept of the Advanced Intelligent Network.

EXM Exit Message. The seventh ISUP message. It's a message sent in the backward direction from the access tandem to the end office indicating that call setup information has successfully proceeded to the adjacent network. See ISUP and COMMON CHANNEL SIGNALING.

EXOS Abbreviation for EXtension OutSide; a phone connected to a key system based in another building. The wiring belongs to the telephone company, even though the phone equipment may not. Unlike an OPX, the circuit between the two locations does not pass through a central office.

Exosphere This region lies beyond an altitude of about 400km from the surface of the earth. The density is such that an air molecule moving directly outwards has an even chance of colliding with another molecule or escaping into space.

Expanded Memory MS-DOS running on the Intel 80286, 80386, 80486 and Pentium family of microprocessors can only address one megabyte at one time. Expanded memory is memory located between base memory (either 512K or 640K) and one megabyte. Expanded memory is reserved by MS-DOS for "housekeeping" tasks such as managing output to the screen. As programs got larger and more hungry for memory (640K was no longer enough), people started jealously eyeing the memory between 640K and 1024K (one megabyte). The first technique was a standard called LIM-EMS, named after the three companies which developed it — Lotus, Intel and Microsoft. Essentially LIM grabs 64K of the 640-1024 memory and uses it to swap pages of other memory in and out quickly. This fools DOS into thinking that it has actually more memory. LIM-EMS lets you work on bigger spreadsheets, and do other jobs faster.

There are many ways of using expanded memory, including special memory management application program or DOS 5.0 or higher. 80386, 80386SX, 80486 and Pentium computers can create expanded memory readily by using the EMS (expanded memory specification) driver provided with DOS, through Windows 3.xx or Windows 95, or through a memory manager such as Quarterdeck QEMM or Qualitas 386. To use expanded memory, a program must be EMS-aware or run under an environment

such as Microsoft Windows. See also EXTENDED MEMORY, which is memory above 1MB.
Expanded Spectrum A cellular telephone term for having the full 832-channel analog cellular spectrum currently available to you, the user of the cellular phone.
Expander That device in a transmission facility which expands the amplitude of received compressed signals to their approximate normal range. The receiving side of a compandor.
Expandor See EXPANDER.
Expansion The switching of a number of input channels, such as telephone lines onto a larger number of output channels.
Expansion Carrier An AT&T Merlin term. A carrier added to the control unit when the basic carrier cannot house all the modules needed. An expansion carrier houses a power supply module and up to six additional modules.
Expansion Slots In a computer there are card slots for adding accessories such as internal modems, extra drivers, hard disks, monitor adapters, hard disk drivers, etc. Most modern PBXs are actually cabinets with nothing but expansion slots inside. Into these slots we fit trunk cards, line cards, console cards, etc. Some phone systems have "universal" slots, meaning you can put any card in any slot. Some phone systems have dedicated expansion slots, meaning that they expect only a certain card in that slot. In the PC industry, many manufacturers make cards for IBM and IBM compatible slots. In the phone industry, nobody makes cards for expansion slots in anyone else's phone system. See also EISA and MCA.
Expert System A very sophisticated computer program consisting of three parts. 1. A stock of rules or general statements, e.g. Some long distance phone calls are free. These rules are generally based on the collective wisdom of human "experts" who are interviewed. 2. A set of particular facts, e.g. Three companies provide the bulk of long distance service in the United States. 3. Most importantly, a "logical engine" which can apply facts to rules to reach all the conclusions that can be drawn from them — one of which might be "Three companies give away long distance phone calls." (Which would be wrong.) The idea of expert systems is to help people solve problems. For example, Compaq is trying to improve its customer service by installing automated assistants that work on the principle that reasoning is often just a matter of remembering the best precedent. The simplest expert systems, according to the Economist Magazine, assume that their rules and facts tell them everything there is to know. Any statement that cannot be deduced from the system's rules and facts is assumed to be false. This can lead machines to answer "YES" or "NO," when they should say "I don't know." Slowly we are beginning to find ways of dealing with the inflexibility of machines. One such gadget is a "truth maintenance machine" invented by Dr. Jon Doyle of MIT. As each fact is fed into the system, Dr. Doyle's program checks to see if it (or the deductions derived from it) contradict any of the facts or deductions already in the system. If there is a contradiction, the machine works backward along its chain of reasoning to find the source and dispose of that troublesome fact or deduction. So the system maintains one consistent set of beliefs.
Exponential Back-Off Delay The back-off algorithm used in IEEE 802.3 systems by which the delay before retransmission is increased as an exponential function of the number of attempts to transmit a specific frame.
Export Imagine you have a software program, like a spreadsheet or a database. And you have information in that program. Let's say it's Microsoft Word or Lotus 123. And you want to get it into a different program, say to give it to a workmate who uses WordPerfect or Excel. You have to convert it from one format to another. From Word to WordPerfect or from Lotus to Excel. That process is typically called "exporting." And you'll typically see the word "EXPORT" as a choice on one of your menus. The opposite is called importing. See IMPORT.
Export Script First read my definition of EXPORT. An export script is a series of specifications which control the export process. It contains the fields to be sent, which records to be sent, the name of file to send as well as the name of the import script (if there is one) located at the receiver's end which will control the merge. See EXPORT.
Export Server A Windows NT term. In directory replication, a server from which a master set of directories is exported to specified servers or workstations (called import computers) in the same or other domains.
Express Call Completion Someone calls an information operator. "What is the name?" the operator answers. "Here is the number. Would you like me to get that number for you now? If so, please hit 1." Express Call Completion lets the operator complete the call for you while you're on line. Express Call Completion was begun in September of 1990 by Pacific Bell using a Northern Telecom central office. Express Call Completion is part of Northern Telecom's Automated Directory Assistance Call Completion (ADACC) software and Traffic Operator Position Systems Multipurpose (TOPS MP).
Express Client Installation An Internet Access Term. A client installation scripting utility that enables network managers to establish defined defaults that make client installation and deployment easier.
Express Orderwire A permanently connected voice circuit between selected stations for technical control purposes.
EXT See EXTENSION.
Extend A verb used by the phone industry to describe an operator transferring a call to a telephone extension. The word is used thus: The operator extended the call to Mr. Smith on extension 200. "Putting a call through" is a clearer way of saying "extending" a call. The word "extend" probably comes from the old days when the operator extended her arm to plug you in on her cordboard.
Extended Addressing In many bit-oriented protocols, extended addressing is a facility allowing larger addresses than normal to be used. In IBM's SNA, the addition of two high-order bits to the basic addressing scheme.
Extended Area Service An option offered by local telephone companies where a subscriber can pay a higher rate and reach more areas as local calls (no long distance charges).
Extended Binary Coded Decimal Interchange Code EBCDIC. (Pronounced Eb-Si-Dick.) An IBM standard of coding characters. It's an 8-bit code and can represent up to 256 characters. A ninth bit is used as a parity bit. See PARITY and EBCDIC.
Extended BIOS Data Area In PCs, extended BIOS data area is 1KB of RAM located at 639KB. It is used to support extended BIOS functions including support for PS/2.
Extended Call Management A Northern Telecom term for a collection of features being added to its DMS Meridian central office Automatic Call Distribution (ACD) service. Using Switch-to-Computer Applications Interface (SCAI), ECM will work

with user-provided computer equipment to integrate call processing, voice processing (recorded announcements, voice mail and voice response) and data processing. For example, ECM will allow an outboard computer device to coordinate the presentation of customer data on the ACD agent's computer screen with an incoming call. The D channel of an ISDN Basic Rate Interface (BRI) serves as the transport mechanism from the DMS-100 central office switch to an outboard computing device. Communication is peer-to-peer, meaning that neither the switch or the computer is in a "slave" relationship to the other. The application layer messaging — i.e. layer 7 messaging as defined by the Open Systems Interconnection (OSI) reference model — is in the Q.932 format and is designed to conform to the T1S1 SCAI message protocol.

Extended Character Set The characters assigned to ASCII codes 128 through 255 on IBM and IBM-compatible microcomputers. These characters are not defined by the ASCII standard and are therefore not "standard." See EXTENDED GRAPHICS CHARACTER SET.

Extended Definition Television EDTV. Television that includes improvements to the standard NTSC television system, which improvements are receiver-compatible with the NTSC standard, but modify the NTSC emission standards. Such improvements may include (a) a wider aspect ratio, (b) higher picture definition than distribution-quality definition but lower than HDTV, and/or (c)any of the improvements used in improved-definition television. When EDTV is transmitted in the 4:3 aspect ratio, it is referred to simply as "EDTV." When transmitted in a wider aspect ratio, it is referred to as "EDTV-Wide."

Extended Digital Subscriber Line The ISDN EDSL combines 24 B-channels and one 64-Kbps D-channel on a single line, ISDN primary rate interface.

Extended Graphics Character Set The characters assigned to ASCII codes 128 through 255 on IBM and IBM-compatible microcomputers. These characters are not defined by the ASCII standard and are therefore not "standard." The original ASCII code used a seven bit one-or-zero code. There are two to the seventh power, or 128 possible combinations. The IBM PC uses a 16-bit CPU with an eight bit data bus and thus transmits data internally in eight big bytes. Instead of using the seven bit ASCII code, the PC uses the equivalent eight bit code, by simply making the left most digit, a zero. In seven bit code, an R is 1010010. In 8-bit, it's 01010010. The only difference between the first 128 characters and the second 128 characters is that in the second, the first bit is a 1.

Extended Key Code The two digit code that represents pressing a key outside the typewriter portion of the keyboard, such as a function key, cursor-control key or combinations of CTRL (control) and ALT keys with another key. The first number is always 0 (zero) and is separated from the second number by a semicolon.

Extended LAN A collection of local area networks connected by protocol independent devices such as bridges or routers.

Extended Memory Memory beyond 1 megabyte in 80286, 80386, 80486 and Pentium computers. Windows uses extended memory to manage and run applications. Extended memory can be used for RAM disks, disk caches, or Microsoft Windows, but requires the processor to operate in a special mode (protected mode or virtual real mode). With a special driver, you can use extended memory to create expanded memory. Extended memory typically is not available to non-Windows applications or MS-DOS. See also EXPANDED MEMORY.

Extended Superframe Format ESF. A new T1 framing standard used in Wide Area Networks (WANs). With this format 24 frames — instead of 12 — are grouped together. In this grouping, the 8,000 bps frame is redefined as follows:
• 2,000 bps for framing and signaling to provide the functions generally defined in the D-4 format.
• 2,000 bps are CRC-6 (Cyclic Redundancy Check-code 6) to detect logic errors caused by line equipment, noise, lightning and other interference. Performance checking is done by both the carrier and the customer without causing any interference with the T-1 traffic.
• 4,000 bps are used as a data link. This link is to perform functions such as enhanced end-to-end diagnostics, networking reporting and control, channel or equipment switching, and/or optional functions or services. See also T-1 FRAMING and D-4 FRAMING.

Extended Superframe Monitoring Unit ESFMU. An MCI definition. Placed on customer data circuits to provide performance monitoring throughout MCI's Digital Data Network.

Extended Telephony Level The lowest level of service in Windows Telephony Services is called Basic Telephony and provides a guaranteed set of functions that corresponds to "Plain Old Telephone Service" (POTS - only make calls and receive calls). The next service level is Supplementary Telephone Service providing advanced switch features such as hold, transfer, etc. All supplementary services are optional. Finally, there is the Extended Telephony level. This API level provides numerous and well-defined API extension mechanisms that enable application developers to access service provider-specific functions not directly defined by the Telephony API. See WINDOWS TELEPHONY SERVICES.

Extended Text Mode Standard text mode is 80 columns wide. So-called extended text mode is 132 columns wide. This mode allows you to view more text on-screen when using such applications as Lotus 1-2-3.

Extensible In strictest terms, the word means "capable of being extended." When Microsoft introduced its At Work operating system on June 9, 1993 it said that one of the operating system's key features was that it was "extensible." Microsoft's explanation: The software is designed to allow both manufacturers and customers to add new features. For example, local area network connectivity will be able to be added easily by installing an optional LAN hardware module and a software driver. Additional memory will be able to be added to the system, and the system will automatically make use of this memory. New image-processing software and communications protocols will be able to be added on the premises, and it will even be able to be done over the phone line, allowing manufacturers to create basic models that can be enhanced in many different ways to fit the needs of different user groups.

Extension 1. An additional telephone connected to a line. Allows two or more locations to be served by the same telephone line or line group. May also refer to an intercom phone number in an office.
2. The optional second part of an PC computer filename. Extensions begin with a period and contain from one to three characters. Most application programs supply extensions for files they create. Checking a file's extension often tells you what the file

does or contains. For example, most BASIC files use a filename extension of .BAS. Most backup files have an extension of .BAK. MS-DOS programs have .EXE or .COM. dBASE database files have the extension .DBF and .DBT. Paradox files have the extension .DB. Files of sounds have their own extensions. Here are the typical extensions on sound files of various computers: Microsoft Windows — .wav Apple — .aif NeXT — .snd MIDI — .mid and .nni Sound Blaster — .voc
Here are the typical extensions on graphics formats:
.TIFF, .EPS, .CGM, .PCX, .DRW, .WMF, and .BMP

Extension Cord A multi-conductor, male/female modular line cord generally used to permit greater separation between the Communications Outlet and the telephone equipment. Available in various lengths up to 25 feet. May be of tinsel or stranded wire construction.

Exterior An ATM term. Denotes that an item (e.g., link, node, or reachable address) is outside of a PNNI routing domain.

Exterior Link An ATM term. A link which crosses the boundary of the PNNI routing domain. The PNNI protocol does not run over an exterior link.

Exterior Reachable Address An ATM term. An address that can be reached through a PNNI routing domain, but which is not located in that PNNI routing domain.

Exterior Route An ATM term. A route which traverses an exterior link.

External Memory Storage devices, such as magnetic disks, drums or tapes which are outside (externally attached) to the main telephone or computer system.

External Modem A modem external to the computer, it sits in its own little box connected to a computer through the computer's serial port. Compare with an internal modem, which typically comes on one printed circuit card and is placed into one of the computer's expansion slots and thus connects to the computer through the computer's "backplane." Internal modems cost less because they don't need any external housing and separate power supply. But because they're mounted inside the computer, it's harder to see what they're doing. You can't see the various status lights, like OH (for Off-Hook) and CD (for Carrier Detect). They also take up valuable slots instead of a serial port.

External Photoeffect In fiber optics, an external photoeffect consists of photon-excited electrons that are emitted after overcoming the energy barrier at the surface of a photo-emissive surface.

External Storage See also EXTERNAL MEMORY.

External Timing Reference A timing reference obtained from a source external to the communications system such as one of the navigation systems. Many of which are referenced to Coordinated Universal Time (UTC).

EXTN Extension.

Extranet The opposite of the Internet. Synonym for the Internet. A stupid word.

Extremely High Frequency EHF. Frequencies from 30 GHz to 300 GHz.

Extremely Low Frequency ELF. Frequencies from 30 Hz to 300 Hz.

Extrinsic Joint Loss For an optical fiber, that portion of a joint loss that is not intrinsic to the fibers, e.g., loss caused by end separation, angular misalignment, or lateral misalignment.

Extrusion Method of continuously forcing plastic, rubber or elastometer material through an orifice to apply insulation or jacketing over a conductor or cable core.

Eye Pattern An oscilloscope display used to visually determine the quality of an equalized transmission line signal being received. So called because portions of the pattern appearing on the scope resemble the elliptical shape of the human eye.

Eye Phone Several researchers are studying something they call "virtual reality." One version of it, a system developed by a company called VPL Research, Redwood City, CA, is based around three things: a three-dimensional glove worn on the head (called an Eye Phone), an electronic glove (the Data Glove) and a high-speed computer. The whole system cost $250,000 in the fall of 1990.

Eyeball A viewing audience. "There are plenty of new eyeballs available in this time slot."

EZTV A software created by Apple which will enable consumers to order movies, go shopping and play games on their TV sets. The product is mostly on the drawing boards.

F Link Fully Associated Link. A link used to connect two SS7 signaling points when there is a high community of interest between them and it is economical to link them. Also called associated signaling.
F Port In Fiber Channel, refers to the port residing on the fabric side of the link, or the fabric port.
F Type Connector A low cost connector used by the TV industry to connect coaxial cable to equipment.
F2F Face to Face. When you actually meet those people you been corresponding electronically with.
Fab Factory that makes ("fabricates") IC chips.
Fabric An Internet term. In Fabric Channel, refers to a non-blocking switch which interconnects all nodes to which it's attached.
FAC See FORCED AUTHORIZATION CODE.
Faceplate A cover that fits around the pushbuttons or rotary dial of a telephone. Hotels and motels put instructions on them. More businesses should also.
Facilities A stupid, imprecisely defined word that means anything and everything. To me it sounds like toilets. But it's not. It can mean the equipment and services which make up a telecom system. It can mean offices, factories, and/or building. It can be anywhere you choose to put telecom things. Oops, I nearly said telecom facilities. So "facilities" means practically anything you want it to mean so long as it covers a sufficiently broad variety of "things" which you haven't got a convenient name for. "Facilities" sounds better than things, especially if you want to sound pompous.
Facilities Administration And Control A PBX feature which allows you, the subscriber, to assign to your users features and privileges like authorization codes, restriction levels and calling privileges.
Facilities Assurance Reports This feature allows a subscriber to get an audit trail of the referrals produced by the automatic circuit assurance feature of some PBXs. The audit trail will identify the trunk circuit, the time of referral, the nature of the problem and if a test was performed, the outcome of the test.
Facilities Based Carrier Long distance carriers which own call switching equipment and transmission lines regionally or nationally are called facilities based carriers. The major national facilities based carriers are AT&T, MCI and Sprint. Facilities based carriers other than AT&T along with resellers also are known as Other Common Carriers. The national facilities based long distance carriers have switching offices in all service areas of the country and provide originating service nationwide. Major facilities based carriers sell their services to business and residential users and other long distance companies which resell those services.
Facilities Management Also called Outsourcing, facilities management is having someone else run your computers or your telecommunications system. The concept is that you're a great bank and you should concentrate on being in the banking business. Your outside facilities manager should concentrate on running your computers or telecom systems. He can do it cheaper, allegedly. Ross Perot's Electronic Data Systems (EDS) probably started facilities management. Mr. Perot incorporated EDS on June 27, 1962. At that time he was a leading IBM salesman. See also OUTSOURCING.
Facilities Restriction Level Which types of calls a PBX user is entitled to make.
Facility A telephone industry term for a phone or data line. Sometimes (but rarely) used to describe equipment. See FACILITIES.
Facility Grounding System The electrically interconnected system of conductors and conductive elements that provides multiple current paths to the earth electrode subsystem. The facility grounding system consists of the earth electrode subsystem, the lightning protection subsystem, the signal reference subsystem, and the fault protection subsystem. Faulty grounding causes more phone and computer problems than any other single factor.
Facility Work Order An order to a phone company to rearrange things.
FACS 1. How they abbreviate the word facsimile in Bermuda.
2. Facilities Access Control Systems. A collection of dozens of interrelated computer applications developed by the former AT&T Bell Operating Companies which manage the local loops connecting customers to the Public Switched Telephone Network.
Facsimile Equipment FAX. Equipment which allows hard copy (written, typed or drawn material) to be sent through the switched telephone system and printed out elsewhere. Think of a fax machine as essentially two machines — one for transmitting and one for receiving. The sending fax machines typically consists of a scanner for converting material to be faxed into digital bits, a digital signal processor (a single chip specialized microprocessor) for reducing those bits (encoding white space into a formula and not an endless series of bits representing white), and a modem for converting the bits into an analog signal for transmission over analog dial-up phone lines. The receiving fax consists of a modem and a printer which converts the incoming bits into black and white images on paper. More modern and more expensive machines also have memory — such that if the machine runs out of paper, it will still continue to receive incoming faxes, storing those faxes into memory until someone fills the machine with paper and it prints the faxes out.
There are five internationally accepted specifications for facsimile equipment. Group 1, Group 2, Group 3, Group 3 Enhanced and Group 4. Only 1, 2, 3 and 3 Enhanced will work on "normal" analog dial-up phone lines. Group 4 is designed for digital lines running at 56/64 Kbps, e.g. ISDN lines. Among the analog line fax machines, Group 2 is faster than Group 1. Group 3 is faster than Group 2, etc. Virtually all machines sold today are Group 3, though an increasing percentage are Group 3 enhanced,

which has speeded up Group 3's transmission speed from 9,600 bps to 14,400 bps and improved its error correction. Group 3 faxes send an 8-1/2 x 11 inch page over a normal phone line in about 20 seconds. How much time it actually takes depends on how much stuff is actually on the paper. Unlike older machines, Group 3 machines are "intelligent." They only transmit the information that's on the paper. They do not transmit white space, as earlier machines did. Group 3 fax machines are now available in "slimy" (i.e. chemically coated) paper and plain paper, the same paper your photocopy machines now uses.

When a fax machine calls a phone line, it emits a standard ITU-T-defined, "CNG tone" (calling tone) — 1100 Hz tone every three seconds. When the receiving fax machine hears this tone, it knows it's an incoming fax call and it can automatically connect. With this tone it is possible to insert a "fax switch," which would "listen" for an incoming fax call and switch it to a fax machine if it heard the CNG tones or to something else — like a phone or answering machine — if it didn't. It is not possible to do this with a modem. A calling modem does not issue any tones whatsoever. A modem works backwards — when the receiving modem answers the phone, it emits a tone.

Typically, a Group 3 machine can speak to a Group 2 and a Group 1 machine. A Group 2 can speak to a Group 1. Speaking down means slowing down. Fax machines are dropping in price. "Personal" fax machines are emerging. Most fax machines today at Group 3 or Group 3 enhanced.

A Group 4 standard has been promulgated by the ITU-T. Group 4 facsimile machines are 100% digital and directly attach to the B (bearer channel) of a digital ISDN line. They will transmit a sheet of 8 1/2 x 11 paper in under six seconds. The author of this dictionary has seen a working Group 4 fax machine. It's mighty impressive.

Some warnings on fax machines:

1. All analog Group 3 and enhanced Group 3 fax machines pose a security risk. Anyone can attach a normal audio cassette recorder to a phone line, record the incoming or outgoing fax "tones" of an analog fax machine. By playing back to another fax machine at a later time, you'll get a perfect reproduction of the fax. There are now fax encryption devices which make the fax transmission unintelligible to any machine other than the one it's intended for — i.e. the one that has a similar un-encryption device.

2. Some plain paper fax machines present a different security risk. Some (not all) use a carbon ribbon the width of their paper. As a result, if you want to read what came in, you simply read the carbon ribbon, which you open like a scroll, which the cleaning lady finds in the trash. These machines are increasingly less common, as plain paper fax machines acquire laser printing engines.

3. Most fax machines record all the digits dialed into them which were used to set up a fax call. If a fax machine is sitting behind a PBX (as many are these days) it will capture all the confidential authorization codes of all the company's employees. To get those codes all you need do is ask the machine to print out a report. There is no easy solution to this problem as at the time of writing this dictionary, except that some fax makers have told me they intend to obscure these numbers on their reports, at some stage. Some may, by the time you read this.

4. Slimy paper fades. How long it takes to fade depends on a bunch of factors — from what's sitting on top of the fax, to the temperature in the room, to whether it's exposed to sunlight, etc. Recommendation: If you want to retain a slimy fax, make a photo copy of it the moment you get it and throw out the original.

5. Poor quality slimy fax paper can abrade the fax machine's drum and cause a costly repair. Don't buy cheap slimy fax paper.

6. Plain paper fax machines cost more to buy, but less to run. You can buy a second tray for some plain paper fax machines which will hold 8 1/2" wide x 14" long paper, which is useful for receiving faxes from outside the US where they use longer paper. This way you save a sheet of paper.

7. It makes sense to have banks of fax machines attached to phones which roll over — also called "hunt." It makes absolutely no sense to have multiple fax machines on separate phone lines that don't hunt, i.e. one for everybody in the office. Two fax machines in rotary can receive and transmit more than twice the number of faxes that two machines on separate, non-hunting phone lines can send and receive. "Personal" fax machines should be out. Banks of fax machines should be in. Egos, though, usually prevail over logic.

8. The paper feed mechanism on plain fax machines has a tendency to jam. Slimy paper fax machines don't jam because their paper typically comes in rolls. And roll paper doesn't jam. The feed mechanism is much simpler.

9. Plain paper fax machines, like laser printers (which many are) use supplies, like toner, which run out. When the supplies run out, such machines usually accept incoming faxes into memory — until that runs out. Then they just ring and ring and ring. Which means that incoming faxes don't get through and don't roll over to the next machine. There is no simple solution since the FCC (Federal Communications Commission) has ruled that fax machines must not return a busy signal to the central office if it runs out of supplies or paper. We have a separate machine that automatically busies out a line if it failed to answer on the fifth ring. But so far, the device is not commercially available. I don't know the answer to this problem except to make sure your fax machine is always stuffed with supplies. Especially check every Friday night. A final note: If your plain paper fax machine is missing supplies, but stuffed with incoming messages in memory, don't turn it off, since you'll lose the messages. Simply replace the supplies and pray your messages will emerge.

10. Some slimy fax paper rolls are coated on the inside of the paper. Others are coated on the outside. When you put one in a fax machine and images don't appear on the paper, then turn the roll over and feed it from underneath. In short, ignore what the instruction book says.

11. Fax modem switches only work when they're called automatically by a fax machine — not by a person using a fax machine manually and is waiting the sound of the distant fax prior to pushing the "Send" button. Make sure you warn your senders. It's remarkable how many people manually dial their faxes and thus penetrate fax modem switches.

12. Think about putting your fax machine on "fine." You'll transmit better quality faxes and may only cost yourself 10% more in transmission time. But that savings depends on the quality of the fax machine at the other end. If it's an older machine, it may cost you as much as double the transmission time. Here are the numbers: Standard is 203 x 98 dpi. Fine is 203 x 196 dpi. "Fine" faxes obviously look much better.

13. Printed circuit cards which slide into slots of PCs and allow you to transmit and receive faxes work well — when transmit-

ting faxes. They work far less well when receiving faxes — largely because of the difficulty of reading faxes. Faxes conform to one type of digital encoding and PC screens conform to another. Moreover a PC screen is landscape (i.e. horizontal), while a fax message is portrait (i.e. vertical). Viewing vertical images on horizontal screens is difficult. Here is a comparison of how fax machines and how personal computer screens encode their images. Obviously, the more digits or pixels, the clearer the end picture. Notice that the encodings are completely dissimilar:

FAX ENCODING
Standard, Group III .... 203 x 98
Fine, Group III .... 203 x 196
Superfine, Group III .... 203 x 391
Standard, Group IV .... 400 x 400

PC SCREEN ENCODING
CGA .... 320 x 200
Enhanced CGA .... 640 x 400
EGA .... 640 x 350
Hercules .... 720 x 348
VGA .... 640 x 480
Super VGA .... 800 x 600 8514/A (also called XGA) 1,024 x 768

See also 1966, 1978, 1980, AT WORK, DEMODULATION, FACSIMILE CONVERTER, FACSIMILE RECORDER, FACSIMILE SIGNAL LEVEL, FACSIMILE SWITCH, FAX, FAX AT WORK, FAX BACK, FAX BOARD, FAX DATA MODEM, FAX DEMODULATION, FAX MAILBOX, FAX MODEM, FAX PUBLISHING, FAX SERVER, FAX SWITCH, FAXBIOS, GROUP 1, 2, 3 BIS AND 4, PHASE A thru E, and WINDOWS TELEPHONY.

Facsimile Converter A facsimile device that changes the type of modulation from frequency shift to amplitude and vice versa.

Facsimile Recorder That part of the facsimile receiver that performs the final conversion of the facsimile picture signal to an image of the original subject copy on the record medium.

Facsimile Signal Level The facsimile signal power or voltage measured at any point in a facsimile system. It is used to establish the operating levels in a facsimile system, and may be expressed in decibels with respect to some standard value such as 1 milliwatt.

Facsimile Switch A new breed of "black box." Its purpose is to avoid having to lease a separate phone line for your facsimile machine, for your phone and for your modem. You buy this box, connect it to an incoming line, connect it to your fax machine, your phone and, possibly, your modem. When a call comes in, the fax switch answers the call, listens if the call coming in is from a fax machine (it can hear the fax machine's CNG calling tone) and switches the call to the fax machine, or switches the call to your modem if a computer is calling. It knows if a computer is calling because the calling computer will, when it hears the fax switch answer, send out some ASCII characters — e.g. 22. (You must put those numbers in your modem dialing stream.) And it knows if a person is calling because it hears neither a CNG tone from a fax machine nor touchtones from the dialing stream of a modem.
The above are the basics of how fax switches work. There are variations on this theme. Some fax switches work automatically. Some work by the incoming caller punching in digits. Some allow you to switch from fax machine to modem to phone and back again. And some fax switches will answer and connect to three modems and one fax or other combinations. The major problem with fax switches is that they typically send a DC ringing tone to whatever device they're trying to connect you (the incoming caller with). Sometimes some devices — for example, high-speed 9600 baud and higher modems — have difficulties responding to low power, DC ringing signals. And they just sit there not answering. Better to buy one that sends standard telephone company AC ringing signals. In short, before you buy a fax/modem/phone switch, test it on your favorite 9,600 or 14,400 bps modem. The more expensive switches tend to work better.

Factoid Factoids are paragraph size pieces of "Gee Whiz" information. They were originally made famous in the newspaper, USA Today.

Factory Refurbished A term used in the secondary telecom equipment business. Equipment that has been returned to the factory and the factory has replaced plastic, repaired, upgraded boards, or otherwise reconditioned.

Fade A reduction in a received signal which is caused by reflecting, refraction or absorption. See also FADING.

Fade Margin The depth of fade, expressed in dB, that a microwave receiver can tolerate while still maintaining acceptable circuit quality.

Fading The reduction in signal intensity of one or all of the components of a radio signal.

FADS Force Administration Data System. A system which takes basic statistics on telephone traffic and gives hints as to how many operators should be employed to answer the incoming calls and when they should be present.

Fail Safe A specially designed system that continues working after a failure of some component or piece of the system. There are precious few, genuinely fail safe systems. To be genuinely fail safe, a system needs to be completed duplicated. It is prohibitively expensive for most commercial users to duplicate every part of their system. But you can duplicate selectively and bring yourself closer to "fail safe." The extent of the duplications you choose (and thus the cost of your telephone equipment and transmission system) depends on how important it is that your system function as close to 100% as possible. The idea is to identify those things most likely to break and to duplicate them. Power is clearly the first area to focus on. These days, the words "FAIL SAFE" are increasingly being replaced with "FAULT TOLERANT." Given the number of times your local, friendly airline has told you that its "computer is down," you can understand the reason for the wording change.

Fair Condition A term used in the secondary telecom equipment business. One step up from "as is" condition. Equipment may have been tested; i.e., product is in working order but looks semi-awful.

Fair Market Value See FMV.

Fairness An ATM term. As related to Generic Flow Control (GFC), fairness is defined as meeting all the agreed quality of service (QOS) requirements, by controlling the order of service for all active connections.

Fake Root A subdirectory on the file server of a local area network that functions as a root directory, where you can safely assign rights to users. Fake roots only work with NetWare shells included with NetWare v2.2 and above. If you use older versions of the workstation shell, you will not be able to create fake roots.

Fall.com An early virus which made the characters on a screen fall to the bottom.

Fallback Rate A modem speed that is lower than its normal (that is, maximum) speed of operation. May be used when communicating with a slower, compatible modem, or to help transmission over a line that is too noisy for full speed operation.

False Ringing False ringing is a recording of a telephone ringing signal (two seconds on, four seconds off, which is played while a call is transferred or while a switching device listens for modem for facsimile CNG (calling) tones.

Falsing In telecom signaling, DTMF tones are created using specific combinations of frequencies to prevent the possibility of "falsing." Falsing is the condition where a DTMF detector incorrectly believes a DTMF is present when in fact it is actually a combination of voice, noise and/or music.

Fan Antenna An aerial consisting of a number of wires radiating upwards from a common terminal to points on a supporting wire.

Fanatic Someone who's overly enthusiastic about something in which you have zero interest.

FAP Formats And Protocols. The set of rules that specifies the format, timing, sequencing and/or error checking for communication between clients and servers.

FAQ Either a Frequently Asked Question, or a list of frequently asked questions and their answers. Many Internet USENET news groups, and some non-USENET mailing lists, maintain FAQ lists (FAQs) so that participants won't spend lots of time answering the same set of questions.

Far End Crosstalk Crosstalk which travels along a circuit in the same direction as the signals in the circuit. The terminals of the disturbed channel at which the far-end crosstalk is present and the energized terminals of the disturbing channel are usually remote from each other.

Far Field Pattern Synonym for Far-Field Radiation Pattern.

Far Talk In voice recognition, far talk is an arrangement where a microphone is more than four inches from the speaker's mouth. The opposite is CLOSE TALK, where the microphone is closer than four inches.

Farad The practical unit of capacity. A capacity which retains a charge of one coulomb with a potential difference of one volt. See FARADAY and FARADAY CAGE.

Faraday As a Faraday shield: refers to the protection a material or container provides to electronic devices to keep them from exposure to electrostatic fields. Named after M. Faraday, the English physicist.

Faraday Cage A structure designed to isolate a sensitive electronic system or device from outside interference, usually constructed of metal screens. Named for 19th century inventor Michael Faraday, whose name also gave us the FARAD, the unit of measuring capacitance.

Farms of Mainframes Picture a hall full of largeframe computers. Picture a hall full of American Airline mainframe computers, lined one after another. Now you have the concept of a farm of mainframes.

Fast Busy A busy signal which sounds at twice the normal rate (120 interruptions/minute vs. 60/minute). A "fast busy" signal indicates all trunks are busy.

Fast Ethernet A way to run Ethernet at 100 million bits per second, up tenfold from today's Ethernet, which runs at 10 Mbps. Fast Ethernet, running on two pairs of standard unshielded telephone copper wire, is positioned as an alternative to CDDI and FDDI, which is also delivering 100 Mbps on unshielded twisted pairs. We'll see.

Fast Ethernet Alliance A group of vendors who are writing the 100Base-X technical and wiring specifications, which would allow fast Ethernet (100 megabits a second) to run over Category 3, voice-grade unshielded twisted pair wiring.

Fast Network An AT&T term for a network with low delay relative to the needs of the application.

Fast Packet Multiplexing Multiplexing is putting more than one "conversation" onto one circuit. You can do this in either of two ways — by splitting the channels sideways into subchannels of narrower frequency. This is called Frequency Division Multiplexing. Or you could split it by time. Like a railroad train. The first car carries "Conversation 1." The second carries "Conversation 2." And split them apart at the other end.

Fast packet multiplexing is a combination of three techniques — time division multiplexing, packetizing of voice and other analog signals and computer intelligence. Here are the main advantages fast packet multiplexing has over today's industry standard time division multiplexing:

1. Fast packet multiplexing doesn't blindly slot in "information" from devices if there's no information to send. Most other multiplexing techniques, including the most common — time division and frequency division — slot in capacity, whether the device is "talking" or not.

2. The fast packet multiplexer can start sending a packet before it has completely received the packet. This speed of movement is critical to voice, for example, which must move ultra-fast. Delays are devastating. (No one can afford to replace the phone instruments broken in anger.)

3. Fast packet multiplexing can interrupt the delivery of one packet in favor of sending another. It's OK to delay a packet of data by several milliseconds. It's not OK to delay a packet of voice or video.

Fast Packet Switching A recent wide area networking technology capable of transmitting data, digitized voice and digitized image information. It makes use of short, fixed length packets (or cells) that are all the same size. The underlying switching technology is based on the statistical multiplexing of data and voice in fixed length cells. Any of these packets could carry digital voice, data or digital image information. All the packets travel at Level Two of the OSI Model, and routing is performed on the basis of the Level Two addressing. Fast packet is claimed to be very effective way of make best use of available bandwidth. It is claimed to offer the benefits of conventional multiplexing techniques and circuit switching techniques because of the way it operates. It is one of the transmission technologies being developed for use with B-ISDN (Broadband ISDN). The switch used to route packets in a fast packet network is termed a fast packet switch. Also, fast packet technology can carry data transmissions that enter the network using frame relay. For particularly high speed networking, an implementation of fast packet switching known as ATM is being developed. See ATM and FAST PACKET MULTIPLEXING.

Fast Scan Receiver A cellular term. A piece of equipment that scans all 1,300 channels in an entire cellular network. It is a quick way to determine channel usage and signal strength.

Fast Select In packet switched networks, a calling method which allows the user to send a limited amount of information along with a "call req packet" rather than after the packet. A more technical explanation: An optional user facility in the virtual call service of ITU-T X.25 protocol that allows the inclusion of user data in the call request/connected and clear indication packets. An essential feature of the ITU-T X.25 (1984) protocol.

Fast Stat MUX MICOM's advanced statistical multiplexer that uses data compression, priority echoplex handling and fast packet technology to improve throughput.

Fast Switching Channel A single channel on a GPS (Global Positioning System) which rapidly samples a number of satellite ranges. "Fast" means that the switching time is sufficiently fast (2 to 5 milliseconds) to recover the data message.

FAT File Allocation Table. The FAT is an integral part of the MS-DOS operating system. It is like an index of a hard or floppy disk. It keeps track of where the various pieces of each file on a disk are stored. A hard disk's directory and file allocation tables are extremely important because they contain the address and mapping information the operating system needs to figure where to store and retrieve data. If any of the data storage blocks containing these tables is damaged, some or all of the data may be lost. For a much longer explanation, see MS-DOS. Novell's NetWare maintains duplicate copies of directory entries and file allocation tables on separate areas of the hard disk, thus reducing the chance of a catastrophic loss of data. See FAT FILE SYSTEM and VFAT.

FAT Client Clients are devices and software that request information. Client is a fancy name for a PC on a local area network. It used to be called a workstation. Now it is the "client" of the server. Clients come in two varieties — Fat and Thin. Here's a definition of Fat Client, courtesy of Oracle Corporation, writing in early 1994: "Since the early 1980s, users have loaded their personal computers will more and more software and data. PCs often are connected to file servers that store information. With each loaded PC costing thousands of dollars, the fat client model has a high cost per machine. Example, a PC or a Macintosh. See also CLIENT, CLIENT SERVER, CLIENT SERVER MODEL, MAINFRAME SERVER, MEDIA SERVER and THIN CLIENT.

FAT File System A file system based on a file allocation table, maintained by the operating system, to keep track of the status of various segments of disk space used for file storage. The 32-bit implementation in Windows 95 is called the Virtual File Allocation Table (VFAT). See FAT.

FATbits A computer imaging term. Extreme magnification of individual pixels to allow easy pixel-by-pixel editing of images.

Fatherboard First there is the motherboard. That's the main circuit board of a computer system. The motherboard contains edge connectors or sockets so other PC (printed circuit) boards can be plugged into it. Those PC boards are called Fatherboards. Some fatherboards have pins on them into which you can plug smaller boards. Those boards are called Daughterboards. In a voice processing system, you might have a Fatherboard to do faxing. And you might have a range of Daughterboards, which allow you to connect different types of phone connections. Different boards exist for standard analog tip and ring, digital switched 56, t-1, etc.

Fatwa An Islamic ruling issued by one of the many Muslim religious leaders in the middle east. Fatwas cover everything from recommended clothing to calls for the death of non-popular people. The most famous fatwa was the one issued in 1989 by the Ayatollah Khomeini of Iran for the murder of the novelist Salman Rushdie.

FAU Fixed Access Unit. A fixed, wireless telephone placed in a user's home or business using cellular or PCS (Personal Communications Service). A new, lower powered, higher-frequency technology. The device provides local telephony service circumventing existing LEC (Local Exchange Carrier) transmission equipment using wired connections.

Fault Isolation The process of determining where a network problem, or fault located.

Fault Management Detects, isolates and corrects network faults. It is also one of five categories of network management defined by the ISO (International Standards Organization).

Fault Tolerant A method of making a computer or network system resistant to software errors and hardware problems. A fault tolerant LAN system tries to ensure that even in the event of a power failure, a disk crash or a major user error, data isn't lost and the system can keep running. Cabling systems can also be fault tolerant, using redundant wiring so that even if a cable is cut, the system can keep running. True fault tolerance is very difficult to achieve.

Fax An abbreviation for FACSIMILE. See FACSIMILE and FACSIMILE SWITCHES.

Fax At Work Fax At Work is a subset of Microsoft's office equipment architecture called At Work which it announced on June 9, 1993. Microsoft's idea is to put a set of software building blocks into both office machines and PC products, including: • Desktop and network-connected printers. • Digital monochrome and color copiers. • Telephones and voice messaging systems. • Fax machines and PC fax products. • Handheld systems. • Hybrid combinations of the above.

According to Microsoft, the Microsoft At Work architecture focuses on creating digital connections between machines (i.e. the ones above) to allow information to flow freely throughout the workplace. The Microsoft At Work software architecture consists of several technology components that serve as building blocks to enable these connections, including

• Microsoft At Work operating system. A real-time, preemptive, multitasking operating system that is designed to specifically address the requirements of the office automation and communication industries. The new operating system supports Windows-compatible application programming interfaces (APIs) where appropriate for the device.

• Microsoft At Work communications. Will provide the connectivity between Microsoft At Work-based devices and PCs. It will support the secure transmission of original digital documents, and it is compatible with the Windows Messaging API and the

Windows Telephony API of the Windows Open Services Architecture (WOSA).
• Microsoft At Work rendering. Will make the transmission of digital documents, with formatting and fonts intact, very fast and, consequently, cost-effective; will ensure that a document sent to any of these devices will produce high-quality output, referred to as "What You Print Is What You Fax Is What You Copy Is What You See."
• Microsoft At Work graphical user interface. Will make all devices very easy to use and will make sophisticated features accessible; will provide useful feedback to users. Leveraging Microsoft's experience in the Windows user interface, Microsoft At Work-based products will use very simple graphical user interfaces designed for people who are not computer users.
• Microsoft At Work desktop software for Windows-based PCs. Will provide Windows-based PC applications the ability to control, access and exchange information with any product based on Microsoft At Work. Desktop software is the one piece of the Microsoft At Work architecture that will reside on PCs.
Despite the dramatic growth in fax usage in recent years, the fax machine, according to Microsoft (and it's right) is still a fairly primitive communications tool. Difficulty using machine features, low-quality printing, lack of integration with the work environment and lack of security are all commonly identified problems. Microsoft At Work-based fax capabilities will address these deficiencies with a broad spectrum of fax solutions that will transform the fax from an independent tool to a powerful, integrated part of the modern communications process. Microsoft At Work-based fax products include the following:
• Microsoft At Work-based departmental fax machines. A standalone multi functional machine (fax, print, scan, copy) will be able to be used by a single person or an entire depart of people. In addition to being great standalone devices, they will be able to be integrated completely with PCs. They will also be shareable via direct network connection or via direct PC connection.
• Microsoft At Work-based fax servers. High-volume, LAN-connected fax solutions will offer the ideal platform for automating wide area communication tasks, such as forms automation, billing and invoicing with suppliers, distribution of information to a field organization, etc. A host of new applications will be possible when users and developers can count on a widely deployed, secure, anywhere-to-anywhere messaging platform that will be provided by Microsoft At Work-based systems that are also integrated with Windows-based PCs.
• Microsoft At Work-based PC faxes. Any user of Windows with an industry-standard fax board will benefit from rich document transmission through Microsoft At Work desktop software that will become a standard part of the Windows operating system and other Microsoft At Work-based products and services.
• Microsoft At Work-based fax-enabled network. Public networks will be adding support for Microsoft At Work communications, allowing users to benefit from their high-volume broadcasting capabilities and the ability to access integrated, public mailboxes from any location.
Key benefits of Microsoft At Work-based faxes (as implemented in any of the above products or services) will include the following:
• Far Greater Ease of Use
While today's fax machines have dozens of advanced features, few are ever used. In fact, many users do not even know these features exist. The problem is that the small, cryptic display on today's fax machines makes these features inaccessible to users. To compound the problem, owners' manuals are rarely available when a problem occurs.
Microsoft At Work-based fax machines will use a graphical, touch-sensitive display to make every feature simple to use. Context-sensitive features will help guide users through tasks. For example, if a jam occurs, a picture showing users how to clear it will be displayed.
• Ability to Send Original-quality "Published" and Editable Documents
Today's fax machines send fuzzy pages that are often difficult to read. As a result, people either don't use a fax when the document has to look professional, or they send another "good copy" via courier. Today's machines also don't allow users to send editable documents, which would enable wide area joint authoring and the automation of many communication tasks.
Using Microsoft At Work rendering technology, a fax machine will become a remote publishing tool, allowing users to distribute final, laser-quality versions of documents directly from PC applications. Users will also be able to send editable versions of documents to reviewers and co-authors so that changes can be made directly without re-keying information, and then be returned to the author.
• Full Document Security
The most frequently faxed documents include contracts, internal correspondence and purchase orders. Despite the sensitive nature of these documents, anyone can walk by and read or pick up received faxes, and there's no guarantee that the document will even get to the proper recipient. Some fax machines advertise security features, but these machines are not really secure. Passwords are included in the message, but the encryption method is easy to break. Moreover, all of these methods require both sending and receiving machines to be from the same manufacturer, a rare occurrence in today's market.
Microsoft At Work-based faxes will have strong built-in security that will allow users to encrypt messages so that documents aren't read by others. It will also be able to ensure that documents are delivered to intended recipients and verify document contents as authentic. By implementing Microsoft At Work security in both Windows-based applications and in devices from a broad base of manufacturers, secure messaging could become as commonplace in the future as regular fax transmissions are today.
• Strong PC Connectivity
While the vast majority of all documents today are created on PCs, most users still print documents and manually feed them into fax machines. Users who choose to investigate PC fax alternatives find them unreliable and not well-integrated into their PC environment.
The Microsoft At Work-based fax is designed to fully integrate faxes with the rest of a PC's messaging environment by integrating this functionality into the operating system. Users will be able to send fax messages in the same way they send other messages, simply by selecting "Send" from their mail package or their favorite application. Received faxes will be automatically delivered into the user's mailbox - the same mailbox where e-mail and voice mail messages are received. They will be able to forward and reply to the message with a single button click.

NEWTON'S TELECOM DICTIONARY

- MIS Support

Today's fax machines are a nightmare for MIS professionals. They can't be centrally managed, so someone has to walk around to every machine to update fax numbers, change settings, collect activity reports and fix problems. Most machines don't support "default settings," so they can't set up machines to do things such as automatically send faxes when telephone rates are lowest. Users frequently can't enter accounting codes to track costs, and even when they can, the resulting reports can only be printed out, so the data must be manually entered into the accounting system. Finally, fax machines represent a network entirely distinct from the advanced data networks that they pay every month to maintain. MIS professionals should rightly wonder why they can't send all that "charged-per-minute" fax traffic over the data lines for which they pay a fixed monthly fee. Microsoft At Work-based fax machines are designed to let MIS manage fax services in the same way as other corporate communication resources. They will be centrally administrable so that any settings can be changed directly, and common resources such as address books will be able to be maintained and downloaded automatically. They are programmable so that faxes queued up after 4 p.m., for example, will be able to automatically be sent at discount rates, saving between 25 percent and 40 percent on toll charges. Activity reports will be able to automatically be sent in binary format to accounting, where they could be entered into accounting systems directly. They will automatically send trouble reports when problems occur. They will easily route traffic over existing corporate data networks, saving 95 percent of the transmission costs.

- Features to Reduce Fax Costs

Long-distance charges and the employee time required to send and receive faxes account for more than three-fourths of total fax costs, while the cost of the fax machine accounts for only 15 percent. Yet today's fax machines do little to address these costs. Moreover, the few features that manufacturers have added are virtually unused because they are so difficult to access. For example, while most mid- to higher-end machines offer the ability to delay fax transmissions until rates decrease, few people know how to use this feature today.

Microsoft At Work-based fax machines will have features to dramatically reduce fax costs. In addition to making cost-saving features easy to find and use, Microsoft At Work rendering will reduce file sizes and, as a result, transmission times. Digital cover sheets can decrease the cost of a typical four-page fax by up to 25 percent. Cost savings assumes that digitally transmitted header eliminates the majority of cover sheet data that is currently sent in bitmap form. . As noted above, simple access to discount transmissions can reduce toll charges by 25 percent to 40 percent and integration with corporate networks would reduce toll charges by as much as 95 percent.

Fax Back You dial a computer using the handset of your fax machine. The distant computer answers. "What documents would you like? Here's a menu." You touchtone in 123. It says "Touch your Start button." You do. Seconds later your fax machine disgorges the document you wanted. Fax-back is the generic term for the process of ordering fax documents from remote machines. Fax-back uses a combination of fax and voice processing technology. Fax-back is also called fax on demand.

Fax Board A specialized synchronous modem for designed to transmit and receive facsimile documents. Many fax boards also allow for binary synchronous file transfer and V.22 bis communication. See also FAX SERVER.

Fax Data Modem See FAX MODEM.

Fax Demodulation A technique for taking a Group III fax signal and converting it back to its original 9.6 Kbps. It works like this: When a sheet of paper is inserted into a fax machine, the fax machine scans that paper into digital bits — a stream of 9600 bps. Then, for transmission over phone lines, that 9.6 Kbps is converted into an analog signal. But if you wish to transmit the fax signal over a digital line, then it makes sense to convert it back to its original 9.6 Kbps. That means you can put several fax transmissions on one 56 Kbps or 64 Kbps line — the capacity you'd normally need if you transmitted one voice conversation, or one erstwhile analog fax transmission. See FAX/DATA modem.

Fax Mailbox Companies can send facsimiles of documents to be stored for later retrieval to a fax mailbox. Fax mailbox is like voice mail for faxes. Travelers can check their fax mailboxes and have the faxes sent to convenient locations, like a hotel front desk. You can do fax mailboxes with fax servers. You can also subscribe to a fax mailbox, as you can subscribe to a voice mailbox.

One company with a fax mailbox service is Bell Atlantic. Here's their explanation of their service, which they call FAX Mailbox. They say it gives you the freedom to leave the office and pick-up your faxes on the run. Wherever you need them, they're always as close as the nearest fax machine. With FAX Mailbox, you can control your faxes as though you were in the office. Every fax sent to you is kept strictly confidential in a private mailbox, instead of being left by the machine for anyone to see. You can access the faxes in your mailbox by calling from any touchtone phone and entering your personal code. Then you can review, store or delete faxes — or send them to any machine you choose. There's no need to be dependent on office support to receive and send faxes. Or have to delegate or postpone work because you're out of the office. You determine when and where you receive faxes, according to your schedule and itinerary.

Fax Mode The mode in which the fax modem is capable of sending and receiving files in a facsimile format. See FAX MODEM.

Fax Modem A combination facsimile machine/modem. A device which lets you send documents from a computer to a fax machine. It comes in many shapes and sizes. It may come as a card which you slip into a vacant slot in your desktop PC (called an internal fax/modem). It may come as a PCMCIA card which you slip into your laptop. It may come as a small box which you connect by a cable to your computer's serial port. It may also come as a small self-contained package about the size of a cigarette package. The technology of "fax modems" is changing radically. Originally they contained dedicated fax/modem chipsets, i.e. microprocessors designed as fax modems and good for nothing else. Increasingly, fax modems are now coming with powerful, general purpose digital signal processors (DSPs), instead of dedicated fax modem chipsets. These DSP devices become fax modems when you load the appropriate software. When you load other software they can also become the equivalent of sound blaster cards, or become a Microsoft Sound System, etc.

There are big advantages to sending faxes from a fax modem, as compared to sending it from a fax machine. First, faxes sent

244

are cleaner because they're not scanned but computer generated. Second, sending faxes directly from your computer is faster than printing the document, then sliding it in a fax machine, dialing and sending it. Third, a fax modem is typically cheaper than a fax machine. Fourth, because a fax modem uses computer software it may have some neat features, like the ability to send faxes when phone costs are low, like running the fax software in the background while you're doing something else.
There are two main disadvantages:
1. Viewing incoming faxes on your PC's screen is not easy. A PC screen is horizontal. Most faxes are vertical. It's also not easy to translate an incoming fax into the pixels on your computer screen. Here's why:

FAX ENCODING

Standard, Group III	203 x 98
Fine, Group III	203 x 196
Superfine, Group III	203 x 391
Standard, Group IV	400 x 400

PC SCREEN ENCODING

CGA	320 x 200
Enhanced CGA	640 x 400
EGA	640 x 350
Hercules	720 x 348
VGA	640 x 480
Super VGA	800 x 600 8514/A (also called XGA) 1,024 x 768

2. For these two reasons, you may still have to print your incoming faxes.
3. Keeping faxes on your hard disk is also pretty consuming of hard disk space. A typical one page fax can easily use between 40,000 and 50,000 bytes. Twenty pages and you've used up a megabyte.
See also FAX DEMODULATION, FAX SERVER AND FAX SWITCH.

Fax On Demand You dial a computer using the handset of your fax machine. The distant computer answers. "What documents would you like? Here's a menu." You touchtone in 123. It says "Touch your Start button." You do. Seconds later your fax machine disgorges the document you wanted. Fax on demand is one term for the process of ordering fax documents from remote machines. Fax on demand uses a combination of fax and voice processing technology. Fax on demand is also called fax-back. See FAX SERVER for a more complete explanation.

Fax Publishing Fax publishing allows a caller to have electronically stored information automatically faxed to them via a touchtone telephone. By pressing touchtone keys, callers can have timely information, including product brochures, business forms and benefits information, automatically faxed to them anytime, anywhere. See also FAX SERVER.

Fax Server 1. A relatively high-powered computer which sits on a LAN and has one or more PC fax boards in its slots. The fax server receives incoming faxes over phone lines, stores them on its hard disk and, if it knows for whom the faxes are meant, it will alert that person over the LAN. If it doesn't know for whom the faxes are meant, it may send the faxes to a printer or alert a supervisor to manually check the incoming faxes and distribute them — electronically or on paper. The fax server also accepts from workstations on the LAN, stores them and gets them ready for sending out over phone lines. It might send the faxes immediately or wait until later, when phone calls are cheaper.
2. A fax server is also a specialized interactive voice response system which you call. When you call it, it answers, reads you a menu of options — including various documents it can send you. You choose which documents you want by touchtoning in numbers. Then you designate to which fax machine you want the documents sent. The fax machine you designate might be the one you're calling from (i.e. you dialed using your fax machine's handset).
There are two types of interactive voice response fax servers. One is a one-call machine. The caller calls from his own fax machine. When he's chosen his faxes and he's ready to receive a fax, he simply hits the "Start" button on his fax machine and his machine receives the chosen faxes. There is also a two-call machine. The caller will call from a phone and touchtone in the phone number of a fax machine he wants the fax of his desired documents sent. One-call IVR fax servers are the newer breed, harder to build than the older two-call machines. There are obvious advantages to both. The one call machine — in which the user pays the phone bill — will, I suspect, become the more popular type. See also FAX PUBLISHING and other FAX definitions.

Fax Switch A device which allows you to share one phone line with a fax machine, a phone and a modem. Here's how it works. A call comes in. The device answers the call. The switch listens for the distinctive CNG (Calling) tone which a calling fax machine emits (the "cry" of the fax machine). When it hears this sound, it switches the call to the fax machine. If it doesn't hears nothing (or at least nothing it can recognize) it switches the call to the phone. If it hears some touchtones — e.g. 44, or *6 — it will switch the call to the modem (and therefore the attached computer) or whatever other device you've designated, including a modem-equipped cash register, etc. Some fax switches allow you to have a data conversation with one device (the cash register), then switch to another device (the second cash register) and another, etc. — all on the one conversation.
The advantage of a fax switch is that it saves having to buy several phone lines. Phone lines are expensive compared to fax switches. There are disadvantages to a fax switch — it typically must hear an incoming CNG tone to switch the call to the fax machine. This means if your friend wanting to send you a fax is dialing manually (i.e. not letting his fax machine do it), your fax switch may not ever send the call to your fax machine. Also you have to set up to dial those extra digits for your distant computer to "dial through" your fax switch. And finally, some fax switches don't send the "right" ringing signal to their attached devices. Some 9,600 baud and 14,400 baud modems, for example, are very sensitive and won't answer certain fax switches' ringing signals, especially if the fax switch's ringing signal is a DC square wave, not an AC sine wave. All this can be solved, however, with intelligence, checking and proper programming. I use a fax modem switch every day. It saves me money and is convenient. See CNG.

Fax Waiting Service The name of a Bell Atlantic service. FAX WAITING Service is like Call Waiting for your fax machine. If your fax line is busy, a second incoming fax is electronically stored. When your machine is clear, FAX Waiting service sends it through. So customers, prospects and suppliers can get their faxes through on the first call — without the frustration of busy signals. And employees can use your fax machine without interrupting your important fax communications. In fact, FAX Waiting is the next best thing to a second fax machine. A great idea.

FaxBios The FaxBios Association is an organization of fax printed circuit card manufacturers who have formed an association in order to promulgate a standard applications programming interface (API) which they are calling FaxBios. Phone 801-225-1850; 2625 Alcatraz Avenue, Berkeley CA 94705.

Faxed The past tense of the new verb "to fax," as in "I faxed the document to him."

FB Framing bit.

FBT Fused Biconic Tape.

FBU Functional Business Unit. A fancy name for a group of workers inside a company. An FBU might be your sales department, your accounting department, etc.

FBus Frame Transport Bus.

FC 1.Fiber optic Connector (developed by NTT).
2. Frame Control. On Token Ring networks, this data supplies the frame type.
3. Feedback Control: Feedback controls are defined as the set of actions taken by the network and by the end-systems to regulate the traffic submitted on ATM connections according to the state of network elements.

FC and PC Face Contact and Point Contact. Designations for fiber optic connectors designed by Nippon Telegraph and Telephone which feature a movable anti-rotation key allowing good repeatable performance despite numerous matings.

FCB The abbreviation for File Control Block. FCBs are used by older MS-DOS application programs to create, open, delete, read, and write files. One FCB is set up for each file you open.

FCC Federal Communications Commission. See FEDERAL COMMUNICATIONS COMMISSION.

FCC Registration Number A number assigned to specific telephone equipment registered with the FCC, as set forth in FCC docket 19528, part 68. The presence of this number affixed to a device indicates that the FCC has approved it as being a compatible device for direct connection to telephone line facilities.

FCC Tariff #9 The FCC tariff for private line services including Accunet T-1.5, DDS, Voice Grade circuits, and Accunet T45.

FCC Tariff #11 AT&T's tariff file at the FCC for local private line services.

FCC Tariff #12 AT&T's tariff filed at the FCC tariff for custom-designed integrated services. A special tariff that allows AT&T to develop custom network solutions, including allowing customers to install their networking multiplexers in AT&T central offices and letting AT&T manage the network.

FCC Tariff #15 AT&T's FCC tariff filed at the FCC that allows AT&T to lower rates after all bids are placed to be competitive with other carriers.

FCFS A silly abbreviation for First Come First Served. See FIFO (First In, First Out).

FCN Abbreviation for Function. This button enables your cellular phone or fax machine or other telecom device to access special features, like switching from one cellular phone company to another. See also DUAL NAM.

FCOS Fully programmable classes of service that control user (Feature Class of Service) access to mailbox features, operations and options. Feature Classes of Service (FCOS) are entirely independent of Limits Classes of Service (LCOS).

FCOT See Fiber Control Office Terminal.

FCS 1. Frame Check Sequence. Any mathematical formula which derives a numeric value based on the bit pattern of a transmitted block of information and uses that value at the receiving end to determine the existence of any transmission errors. In bit-oriented protocols, a frame check sequence is typically a 16-bit field that contains transmission error checking information, usually appended to the end of the frame. See FRAME CHECK SEQUENCE.
2. Federation of Communications Services.
3. An MCI term for Fraud Control System.

FDD Floppy Disk Drive.

FDDI Fiber Distributed Data Interface. FDDI is a 100 megabits per second fiber optic LAN. It is an ANSI standard. It uses a "counter-rotating" token ring topology. FDDI is ANSI X3T12 standard for a dual-ring LAN operating at 100 Mbps and using token passing; FDDI rings may use up to 200 km of optical fiber, or may employ twisted copper pairs for short hops. FDDI is compatible with the standards for the physical layer of the OSI model. An FDDI LAN is often known as a "backbone" LAN. It is used to join file servers together and to join LANs together. The theoretical limit of Ethernet, measured in 64 byte packets, is 14,800 packets per second (PPS). By comparison, Token Ring is 30,000 and FDDI is 170,000 pps. See FDDI TERMS and FDDI-II.

FDDI Follow-On LAN A faster FDDI. Said to operate at up to 2.4 gigabits per second.

FDDI Terms DAC Dual Attachment Concentrator DAS Dual Attachment Station ECF Echo Frames ESF Extended Service Frames LER Link Error Rate LLC Logical Link Control MAC Media Access Control MIC Media Interface Connector NIF Neighborhood Information Frame NSA Next Station Addressing PDU Protocol Data Unit PHY Physical Protocol PMD Physical Media Department PMF Parameter Management Frames RAF Resource Allocation Frames RDF Request Denied Frames SAC Single Attachment Concentrator SAS Single Attachment Station SDU Service Data Unit SIF Station Information Frames SMT Station Management SRF Status Report Frame THT Token Holding Timer TRT Token Rotation Timer TTRT Target Token Rotation Timer TVX Valid Transmission Timer UNA Upstream Neighbor Address

FDDI-II Fiber Distributed Data Interface-II is a recently standardized enhancement to FDDI. It still runs at 100 megabits per second on fiber or on twisted copper pairs, but in addition to transporting conventional packet data like other LANs, FDDI-II allows portions of the 100 Mbps bandwidth to carry low delay, constant bit rate, isochronous data like 64 Kbps telephone channels. This means the same LAN that carries computer packet data can carry live voice or live video calls. Some additional terms used with FDDI-II are: I-MAC which stands for Isochronous Media Access Control; P-MAC which stands for Packet Media Access Control; and WBC which stands for Wide Band Channel. See FDDI, FDDI TERMS, ISOCHRONOUS and ISOETHERNET.

FDM Frequency Division Multiplexing. A technique in which the available transmission bandwidth of a circuit is divided by frequency into narrower bands, each used for a separate voice or data transmission channel. This means you can carry many conversations on one circuit. The conversations are separated by "guard channels." At one point, FDM was the most used method of multiplexing long haul conversations when they were transmitted in analog microwave signals. No more. Fiber optic transmission (today's preferred method) uses TDM — Time Division Multiplexing.

FDMA Frequency Division Multiple Access. One of several technologies used to separate multiple transmissions over a finite frequency allocation. FDMA refers to the method of allocating a discrete amount of frequency bandwidth to each user to permit many simultaneous conversations. In cellular telephony, for example, each caller occupies approximately 25 kHz of frequency spectrum. The cellular telephone frequency band, allocated from 824 MHz to 849 MHz and 869 MHz, consists of 416 total channels, or frequency slots, available for conversations. Within each cell, approximately 48 channels are available for mobile users. Different channels are allocated for neighboring cell sites, allowing for re-use of frequencies with a minimum of interference. This technique of assigning individual frequency slots, and re-using these frequency slots throughout the system, is known as FDMA. See CDMA, TDMA.

FDS Frequency Division Switching. Seldom used for voice switching. Primarily used for radio and TV broadcasting.

FDX See FULL DUPLEX.

FE Extended Framing ("F sub E"). An old name for ESF, also known as Extended SuperFrame, a T-1 carrier framing format that provides a 64 Kbps clear channel, error checking, 16 state signaling and some other nice data transmission features.

FE D4 Superframe Extended Another designation for AT&T's ESF (Extended Super Frame).

Feather An imaging term. An effect in which the edges of a pasted selection or paint tool fade progressively at the edges for a seamless blend with the background.

Feature Buttons Think of a feature button on a telephone as a collection of numbers stored in a bin. When you hit the button, the bin quickly disgorges all the numbers one after another. Feature buttons are fast ways of doing things. You have a feature button labelled "Conference." Hit the button, set up a conference call. Without a feature button, you'd probably have to hit the switch hook and some numbers on your touchtone pad. In computer terms, a feature button on a phone is the same as a macro — an easy way of doing something. On most phones with feature buttons, the feature buttons are "programmable." This means you can assign different features to different buttons, i.e. the ones you want. For example, I always assign "Last Number Redial," "Saved Number Redial" and "Conference Call" to the buttons of any phone I'm programming. Some phones have many feature buttons. Some don't.

Feature Cartridge A replaceable software cartridge containing software features. The Feature Cartridge is inserted into the central cabinet, or Key Service Unit (if it's a key system). Several small phone systems (under 100 lines) use cartridges to upgrade their software. The manufacturers find cartridges are cheaper than equipping their phone systems with a floppy drive and the associated electronics.

Feature Code This is a number that is used to activate a particular feature on a phone system.

Feature Creep A term to show how features tend to get added to telecom equipment as time passes and new models appear. The term "feature creep" makes no judgments about whether the new features are actually useful. In book called "Startup; A Silicon Valley Adventure Story," Jerry Kaplan, the author, describes "Feature creep as "the irresistible temptation for engineers to load a product down with their favorite special features."

Feature Group A, B, C, D FGA, FGB, FGC, FGD, are four separate switching arrangements available from local exchange carrier (LEC) end central offices to interexchange (long distance) carriers. These switching arrangements allow the LEC's end-users to make toll calls via their favorite LEC. Feature groups are described in a tariff filed by the National Exchange Carrier Association with the FCC. The feature group used by each IX (InterExchange) carrier together with any special access surcharge determines the service they can provide their customers and the carrier common line access fee they will pay to the local exchange carrier involved. The most common Feature Group now is D. See the next four definitions. See FEATURE GROUP A, FEATURE GROUP B, FEATURE GROUP C, FEATURE GROUP D

Feature Group A Offers access to the local exchange carrier's network through a subscriber-type line connection rather than a trunk. It is a continuation of the ENFIA arrangement used in the early days of OCCs, until equal access using an access tandem central office is available. Remember, without equal access the IX carrier had to require its customers to dial a local number to reach their long distance facilities, then dial an identification number, then dial long distance numbers of the called party desired. This service handicap, compared to AT&T's superior connections, qualifies the OCC for a discount off the FGA rate until access is equal. The IX carrier is billed by the LEC based upon actual monthly use rather than the ENFIA method of projected "minutes of use" rate.

Feature Group B Is similar to FGA, but provides a higher quality trunk line connection from end CO to the IX carrier's facilities, instead of the subscriber-type line. The IX customer can originate a call from anywhere within the LATA, while FGA requires customers to initiate the call from within the local exchange of the exchange carrier connecting to the IXC. FGB billings to the IX are on a flat usage basis, and a discount is applicable.

Feature Group C Is the traditional toll service arrangement offered by LECs to AT&T prior to breakup of the Bell System. Quality is superior, and the service includes automatic number identification of the calling party, answerback, and disconnection supervision, and the subscribers can use either a dial or touchtone pad. This FGC service is offered only to AT&T without a discount.

Feature Group D Is the class of service associated with equal access arrangements. All IX carriers (i.e. long distance phone companies) enjoy identical connections to the local exchange carrier. All customers dial the same number of digits, and can reach the predetermined IX of their choice by dialing 1 plus the telephone number being called. Eventually, all other feature groups convert to FGD and the IX is billed for actual measured use, without discount. In some cases an IX carrier may desire to maintain FGA or FGB arrangements, but the FGD equal access rates will apply.

Feature Keys Same as FEATURE BUTTONS. A key is to a telephone man what a switch is to an electrical man.

Feature Phone A generic name for a telephone that has extra features (often speed dial buttons) designed to simplify and speed making and receiving phone calls.

FEBE Far End Block Error: A maintenance signal transmitted in the PHY overhead that a bit error(s) has been detected at the PHY layer at the far end of the link. This is used to monitor bit error performance of the link.

FEC Forward Error Correction. A technique used by a receiver for correcting errors incurred in transmission over a communications channel without requiring retransmission of any information by the transmitter. Typically involves a convolution of the transmitter using a common algorithm. See FORWARD ERROR CORRECTION.

FECN Forward Explicit Congestion Notification. This bit notifies the user that congestion-avoidance procedures should be initiated where applicable for traffic in the same direction as the received frame. It indicates that this frame, on this logical connection, has encountered congested resources.

FED Field Emission Display. A new way of making TV and computer screen displays. FED screens are flat and potentially cheap. Like conventional glass screens, they emit light. LCDs, by comparison, don't. A typical FED screen packs millions of tiny individual emitters between two ultra-thin glass layers. Each emitter fires electrons simultaneously across a minuscule vacuum gap onto a phosphor coating very much like a CRT's. See also FIELD EMISSION DISPLAYS.

FED-STD A system of standards numbered FED-STD-1001 to 1008 which set modulation specifications for data transmission.

Federal Communications Commission FCC. The federal organization in Washington D.C. set up by the Communications Act of 1934. It has the authority to regulate all interstate (but not intrastate) communications originating in the United States. The FCC is the U.S. federal regulatory agency responsible for the regulation of interstate and international communications by radio, television, wire, satellite and cable. Established by the Communications Act of 1934, it is responsible directly to Congress and is directed by five Commissioners appointed by the President and confirmed by the Senate for 5-year terms. The President designates one of the Commissioners to serve as Chairman. The Chairman's tenure is at the pleasure of the President. No more than three Commissioners may be members of the same political party. None can have a financial interest in any Commission-related business.

Stripped of all the extensive regulatory and legal mumbo jumbo, the FCC does three things: 1. It sets the prices for interstate phone, data and video service. 2. It determines who can or cannot get into the business of providing telecommunications service or equipment in the United States. 3. It determines the electrical and physical standards for telecommunications equipment and services. The FCC's powers, although strong, are tempered (limited) by the Federal Courts. Anyone who disagrees with FCC rulings can appeal them to a Federal Court. The FCC's power and rulings are also affected by the Justice Department (The Justice Department changed the industry with Divestiture), Congress and The 50 state public service commissions. The FCC changed with the passage of the Telecommunications Act of 1996 — the first telecom act passed by Congress since the Communications Act of 1934.

Federal Information Processing Standards FIPS. The identifier attached to standards developed to support the U.S. government computer standardization program. The FIPS effort is carried out by the U.S. Department of Commerce, Springfield, VA. See the next definition.

Federal Telecommunications Standards Committee FTSC. A U.S. government agency established in 1973 to promote standardization of communications and network interfaces. FTSC standards are identified by the designator FED-STD. The FTSC's address is General Services Administration, Specification Service Administration, Bldg 197, Washington Navy Yard, Washington DC 20407.

Federal Telecommunications System FTS. The private network used primarily by the civilian agencies of the federal government to call other government locations and to place calls to phones connected to the public network.

Federal-State Joint Board An organization with representatives from the FCC and the state public service commissions which tries to resolve Federal and State conflicts on telecommunications regulatory issues. Sometimes successfully and sometimes not successfully.

Feedback The return of part of an output signal back to the input side of the device. Think of the high-pitched squeal you hear when someone brings a microphone too close to the loudspeaker. Not all feedback is as obvious or as irritating. Some feedback is good. See SIDETONE, which is what happens when you hear a little in the receiver of you're saying in the transmitter of a phone.

Feeder Cable A group of wires, usually 25-pair or multiples of 25-pair, that supports multiple phones in a single cable sheath. These cables may or may not be terminated with a connector on one or both ends. Feeder cable typically connects an intermediate distribution frame (IDF) to a main distribution frame (MDF). But the term "feeder cable" is also used in backbone wiring. And Bellcore defines the term slightly differently: A large pair-size loop cable emanating from a central office and usually placed in an underground conduit system with access available at periodically place manholes.

Feedholes Holes punched in paper or papertape which allow the paper or paper tape to be driven by sprocket wheels.

Feedware Software designed to get demand for a product or a new market segment started. Feedware is typically a less-full featured piece of software than the software you're really trying to sell. Feedware typically costs very little. It may even be free. See also SEEDWARE.

FEFO First Ended, First Out. A rule for dealing with things in a queue. For example, higher priority messages will be sent before lower priority messages.

Femtosecond One-millionth of a billionth of a second. Femtoseconds are used in laser transmission and in other measures of very small happenings. It's 10 to the minus 15. There are as many femtoseconds in one second as there are seconds in thirty million years. There are 1,000,000,000,000,000 femtoseconds in one second. How small is a femtosecond? In a little more than a second, light can travel from the moon to the earth, but in a femtosecond it only travels one hundredth the width of a human hair.

FEP Front End Processor. The "traffic cop" of the mainframe data communications world. Typically sits in front of a mainframe computer and is designed to handle the telecommunications burden, so the mainframe computer can concentrate on handling the processing burden. Here's a more technical definition: A dedicated communications system that intercepts and handles activity for the host. Can perform line control, message handling, code conversion, error control, and such applications functions as control and operation of special-purpose terminals. Designed to offload from the host computer all or most of its data communications functions. Front end processors are not used in the client/server world.

FER Frame Error Rate. A computation determined based on the number of frames received with errors compared to frames received without errors.

Fernsprechvermittlungsstelle A central office in German. In Europe, they call a central office a "public exchange," or just plain "exchange." They look at you kinda strange when you say the North American word, namely "central office."

Ferreed Assembly A glass enclosed reed relay switch in which the reeds are made of some metal which can be opened or closed by an external magnetic field.

Ferri Chrome A tape comprising a layer of ferric oxide particles and a layer of chromium dioxide particles and combining the attributes of both.

Ferric Oxide A tape whose coating is of red iron oxide, the original material used for magnetic recording tapes.

Ferrite A type of ceramic material having magnetic properties and consisting of a crystalline structure of ferric oxide and one or more metallic oxides, such as those of nickel or zinc. See BARIUM FERRITE, HARD FERRITE and SOFT FERRITE.

Ferrule A component of a fiber optic connection that holds a fiber in place and aids in its alignment.

FET Field Effect Transistor. Very thin and small transistors are used to control pixels in a TFT (Thin Film Transistor) display.

Fewer A smaller number. The word "fewer" is always confused with the word "less." According to the Oxford American Dictionary, the word "less" is used of things that are measured by amount (for example, eat less butter, use less fuel). Its use with things measured by numbers is regarded as incorrect (for example in "we need less workers"; correct usage is "fewer workers").

FF Form Feed. A printer function used to skip to the top of the next page or form.

FFDI Fast Fiber Data Interface. A proprietary 100 megabit per second local area network that uses fiber optic, coax, shielded twisted pair or unshielded twisted pair. It is manufactured by PlusNet, Phoenix, Arizona.

FFOL Fiber Follow On LAN. Emerging LAN technology.

FG An ATM term. Functional Group: A collection of functions related in such a way that they will be provided by a single logical component. Examples include the Route Server Functional Group (RSFG), the IASG (Internetwork Address Sub-Group), Coordination Functional Group (CFG), the Edge Device Functional Group (EDFG) and the ATM attached host Behavior Functional Group (AHFG).

FGB See FEATURE GROUP B.

FGC See FEATURE GROUP C.

FGC-EA See FEATURE GROUP C and EQUAL ACCESS.

FGD See FEATURE GROUP D.

FGD-EA Feature Group D - Equal Access. See FEATURE GROUP D, also See EQUAL ACCESS.

FIAT Fix It Again Tony.

Fiber A shortened way of saying "fiber optic." Fiber is made of very pure glass. In Bill Gates' book called "the Road Ahead," he says that optical fiber is so smooth and pure that if you looked through a wall of it 70 miles thick, you'd be able to see a candle burning on the other side. Digital signals. in the form of modulated light, travel on strands of fiber for long distances. As Gates points out, a signal doesn't move any faster through fiber optic cable than it does through copper wire. Both go at the speed of light. But the big advantage that fiber has over copper is that it can carry far, far more information. The short history of fiber optic for communications is that scientists keep discovering more and more ways of putting more and more information down one single strand of fiber. And based on my own personal researches, no one has any idea what the eventual capacity limit of a strand of fiber optic might be. I have personally asked many scientists and all seem to think there must be a theoretical limit. But they don't know what it is. And they believe we have many many years of breakthroughs in fiber still to go. As I wrote this entry in the early summer of 1996, MCI had just announced new technology allowing it to add electronics onto the end of its existing fiber optic network and increase each strand's capacity from 2.5 Gbps (gigabits per second) to 10 Gbps, a fourfold increase. Fiber is the American spelling. English, European, Canadian, Australian and Zealand spelling is Fibre. See also the following definitions beginning with FIBER. See OPTICAL FIBER for an essay on the advantages of fiber as a communications medium.

Fiber Axis In an optical fiber, the line connecting the centers of line circles that circumscribe the core, as defined under "tolerance field."

Fiber Buffer The material surrounding and immediately adjacent to an optical fiber that provides mechanical isolation and protection. Buffers are generally softer than jackets.

Fiber Bundle An assembly of parallel unbuffered optical fibers, in intimate contact with one another and secured, usually with an epoxy or other adhesive. Each endface of the bundle is typically finished to a flat or other optical surface, usually at right angles to the axis of the bundle. Such bundles are used to transmit optical power or images. Bundles used to transmit images must maintain spatial coherence amongst the relative positions of the respective fibers at each end (aligned bundles). There is no requirement for this if the bundle is used to transmit optical power only. Fiber bundles were employed in early, short-distance communication applications, but have become obsolete in modern telecommunications.

Fiber Channel Fiber Channel is an ANSI X3T9.3 standard high speed serial interface. It offers a scaleable data rate beginning at 133 Mbits/sec, but ranging to over 1.06 Gbits/sec. It is an ANSI standard developed as a high-speed interface for linking mainframes and peripherals. Fibre Channel can run on single or multimode fiber as well as coax cable and twisted pair. It is best suited as a backbone for a private network.

Fiber Control Office Terminal FCOT is a generic term for a fiber terminal that can be configured for full digital, full analog or mixed communications.

Fiber Distributed Data Interface FDDI. A set of ANSI/ISO standards that, when taken together, define a 100 Mbps, timed-token protocol, Local Area Network that uses fiber optic cable as the transmission medium. The standards define Physical Layer Medium Dependent, Physical Layer, Media Access Control, and Station Management entities. The standard specifies: multi-mode fiber, 50/125, 62.5/125, or 85/125 core-cladding specification; and LED or laser light source; and 2 kilometers for unrepeatered data transmission at 40 Mbps.

Fiber Identifier A test instrument that can differentiate between live and dead fibers in a working cable and can identify a preselected fiber to which a special transmitter has been attached.

Fiber Loss The attenuation (deterioration) of the light signal in optical fiber transmission.

Fiber Optic See FIBER.

Fiber Optic Amplifier As light, like electricity, travels, it loses intensity. At some point in a communications transmission system, you must take your increasingly weak signal and bring it back to its original form. In analog systems, you simply amplify the signal. This amplifies the noise it picked up along the way. In digital signals, you first regenerate the signal, then amplify it, then send it on its way. In the early says of fiber transmission systems, it was necessary to convert the original light signal on the fiber to electrical impulses, regenerate the signal, then amplify it, then convert it back to light pulses, then send it on its way. This took significant energy and equipment — not altogether convenient for an underwater cable of several thousand miles. With new fiber optic amplifiers you no longer need to convert the light signal to electrical impulses. A fiber optic amplifier uses special fiber doped with erbium to act as the amplifier. Light comes into this special fiber, is pumped with the correct frequency laser and is amplified with extremely high gain and very low noise. It's truly amazing technology. See ERBIUM-DOPED FIBER AMPLIFIER for a more technical explanation.

Fiber Optic Cable See FIBER.

Fiber Optic Transmission System See FOTS.

Fiber Optic Waveguide A relatively long thin strand of transparent substance, usually glass, capable of conducting an electromagnetic wave of optical wavelength (visible region of the frequency spectrum) with some ability to confine longitudinally directed, or near-longitudinally directed, lightwaves to its interior by means of internal reflection.

Fiber Optics A technology in which light is used to transport information from one point to another. More specifically, fiber optics are thin filaments of glass through which light beams are transmitted over long distances carrying enormous amounts of data. Modulating light on thin strands of glass produces major benefits in high bandwidth, relatively low cost, low power consumption, small space needs, total insensitivity to electromagnetic interference and great insensitivity to being bugged. All these benefits have great attraction to anyone who needs vast, clean transmission capacity, to the military and to anyone who runs a factory with lots of electronic machinery.
The first field trial of an AT&T lightwave system took place in Chicago in 1977. There has been a rapid improvement in cost effectiveness of fiber systems, expressed as cost per bit per kilometer. A one hundredfold decrease in cost in one five year period — from about 1980 to 1985! Some versions of fiber optics new carry two billion pulses of light a second to carry more than 30,000 times the information of electronic signals over copper wire. See LEDs.

Fiber Pigtail A short length of optical fiber, permanently fixed to a component, used to couple power between the component and the transmission fiber.

Fiber Remote Fiber Remote extends what Northern Telecom calls Intelligent Peripheral Equipment using dark single or multimode fiber cable. Fiber Remote operates over a range of typical campus distances — from thousands of feet to several miles. Distance is site specific, determined by variables such as the type and quality of the fiber installed and the number of connectors and splices. The signal attenuation between the local PBX system and the remote IPE shelves should not exceed a 13 dB loss. Fiber Remote is used when users prefer a single switch; when users have a campus with right of way for running fiber where the distance between the local and remote site is within 6 miles; where there's limited riser space in a large high rise building; where a user needs to alleviate switch room congestion and where security is of utmost importance (e.g. in military bases).

Fiberoptichead Slang expression to describe a customer who thinks he knows everything about cable. Usage: "That fiberoptichead wouldn't know a drop from a fish job."

Fiberphone A battery-powered device that connects to both ends of a fiber optic cable allowing people (typically craftspeople) to talk over the cable. The complete device (two ends) costs $1,000 to $2,000 and often comes with a headset.

Fiberworld In simultaneous media events in Washington and Montreal on October 12, 1989, Northern Telecom and Bell Northern Research (BNR) unveiled "FiberWorld," which they referred to "as a vision and commitment to deliver the world's first completely family of fiber-optic access, transport, and switching products."

Fibre The European, Australian, Canadian, British and New Zealand spelling. The American spelling is fiber. See FIBER.

Fibre Channel A high speed serial bus. For more contact The Fibre Channel Association, c/ Ed Frymoyer at H-P, Cupertino CA, 408-447-2839.

FID Field Identifier, an ISDN SPID term.

Fidonet An electronic bulletin board technology for transfer and receipt of messages. According to PC Magazine, the origins of FidoNet date back to the early 1980s, when the two authors of the BBS software Fido, who lived on opposite coasts, needed an easy way to exchange modifications they made to the source code. They designed a system where, as a nightly event, the board would shut down and run utilities that automatically transferred the changed files between the author's BBSs. The logical next step was to permit the exchange of private mail messages called NetMail, between the sysops. The author found these capabilities so useful that they include them as part of the Fido BBS (Bulletin Board Software) package. It didn't take long for an informal network of Fido nodes to come into existence, all running the Fido software and exchanging various utility and program files and NetMail among sysops. Like other BBSs, the FidoNet BBSs had their own SIGs, or Special Interest Groups, where users with similar interests could exchange messages in a way similar to what on-line services call conferences or forums. By 1986 a Fido sysop had extended the NetMail concept to allow SIGs to share public messages among the BBSs, and EchoMail was born. In the years since, BBS authors and FidoNet users and sysops extended these capabilities to other BBS packages, and FidoNet grew. It currently has over 11,000 nodes covering most of the world. Many of the existing public and private networks go through FidoNet gateways into the Internet Mail system, which carries e-mail over a group of interconnected networks to universities, government agencies, military branches, and corporations. FidoNet technology uses store-and-forward messaging and is based on point-to-point communications between nodes.

Field 1. One half (every other line) of a complete television picture "frame", consisting of every other analog scan line. There are 60 fields per second in American television.
2. A place with no phones or other communications capability where an important person inevitably is when you need some vital information, service or device that only he or she can provide. "I'm sorry, the chief technician is in the field today, and can't be reached." Few "fields" are actually fields. They're usually downtown office buildings.
3. The specific location of data within a record. In the jargon of database management systems, many fields make up one record. Many records make up one file. A field is one of the basic subdivisions of a data record. The record on you in your company's database might include your name, your address, your salary, etc. A field is simply one of these — e.g. your salary, your last name, or your street address. All the records of all the employees in your company make up a file, also called a database.
4. The name given to that part of an electrical system in which electromagnetic lines of force are established.
5. In Windows, the field is the empty line in a dialog box where you enter data.
6. In call center jargon. A field is a single piece of data, such as an employee ID, stored in a record. The fields are organized under column headings.

Field Emission Displays FED. Another way of making thin, flat, lightweight computer displays for laptops, planes, etc. The other way is called "active matrix liquid crystal display." In field emission displays, a tiny color cathode ray tube sits behind each of the many pixels in the screen. This results in a brighter picture that uses less energy than the active matrix LCD displays. See also FED.

Field Intensity The irradiance of an electromagnetic beam under specified conditions. Usually specified in terms of power per unit area, e.g., watts per square meter, milliwatts per square centimeter.

Field Interlacing In television, field interlacing is the process of creating a complete video frame by dividing the picture into two halves with one containing the odd lines and the other containing the even lines. This is done to eliminate flicker.

Field Measurement Refers to both signal strength and qualitative field tests of wireless networks.

Field Programmable The ability of a system to have changes made in its program while it is installed — without having to be returned to the factory.

Field Repairable A characteristic of an unfortunately-decreasing number of electronic devices, that allows users or technicians to fix them where they are used ("in the field"), instead of having to send them to a centralized repair facility where esoteric parts and tools are available.

Field Rheostat A variable resistance used in the field circuit of a generator or motor to control the field current and consequently the strength of the electromagnetic field, thereby regulating the speed or power of the motor, or the output of the generator.

Field Sequential System Field sequential system was the first broadcast color television system, approved by the FCC in 1950. It was later changed to the NTSC standard for color broadcasting.

Field Strength The intensity of an electric, magnetic, or electromagnetic field at a given point. Normally used to refer to the rms value of the electric field, expressed in volts per meter, or of the magnetic field, expressed in amperes per meter.

Field Strength Meter Electronic instrument that measures the intensity of the magnetic field.

Field Upgradable A desirable characteristic of telecom equipment, computers, etc., that allows new features to be added and other improvements to be made, where the device is used, rather than having to return it to the manufacturer or a repair facility.

Field Wire A flexible insulated wire used in field telephone and telegraph systems. WD-1 and WF-16 are types of field wire. Usually contains a strength member.

FIFO First In, First Out. All telephone networks are a trade-off. It's simply too expensive to build a phone network which will be ready to give everyone dial tone and a circuit — if everyone picked up the phone simultaneously and tried to make a call. There are basically two ways of handling calls which cannot be sent on their way — i.e. for which there's no present available capacity. First, you can "block" the call. This means giving the caller a busy or a "nothing" (also called "high and dry"). Second, you can put the call into a queue. Now you have people waiting in queue, how do you handle them? The most equitable — the way most queues work — is to handle the calls on the basis of First In, First Out. (First call to come in is handled first.) There are other ways of handling calls in a queue — including First In, Last Out, by priority (e.g. which line you came in on and how much it cost, or how high you are in the corporation, etc.)

FIFO also is a term used in data communications. It is a buffering scheme in which the first byte of data that enters the buffer is also the first byte retrieved by the CPU. This scheme is used in the 16550 (the UART chip which controls the serial port on most PCs and most other serial-buffering designs, because it closely mimics the way serial data is actually transmitted; that is, one bit at a time.

Fifth Generation Fifth generation computers and telephone systems will be based on artificial intelligence. A fifth generation phone system may make far more sophisticated decisions about routing calls across networks. Those decisions may be made on how many calls have already happened so far that month, the choice of carrier by the likely quality of his connection, etc.

FIGS FIGure Shift. A physical shift in a terminal using Baudot Code that enables the printing of numbers, symbols and upper-case letters.

Filament An electrically heated wire in an evacuated glass bulb, forming one element (the cathode) of a vacuum tube.

File 1. A set of similarly structured data records (such as personnel records using a standardized form). See FIELD.
2. A call center term. A logical division of the data stored on a disk or diskette; for example, employee information vs. supervisor information vs. call volume history. Files generally consist of one or more records of a certain structure.

File Allocation Table MS-DOS' method of storing and keeping track of files on a disk (hard or floppy) is referred to as the File Allocation Table (FAT) file system. It's a map that DOS uses to store and retrieve files on disk. The FAT's map also shows available and used disk space. See FAT.

File Caching A Novell local area network NetWare file server can service requests from workstations up to 100 times faster when it reads from and writes to the file server's cache memory (in RAM) rather than executing direct reads from and writes to the file server's hard disks.

File Extensions MS-DOS files can have an 8-character filename followed by a period and a 3-character file extension. While most extensions are arbitrarily assigned by users or companies, some extensions are reserved for special purposes or circumstances. Many local area networks follow the MS-DOS naming conventions.
- EXE DOS executable file
- BAT DOS executable batch file
- DAT ASCII text file (usually)
- COM DOS executable command file
- ERR Error log file
- OVL Overlay file used often to contain
- HLP Help screens which appear by pressing F1
- SYS Operating system file
- BAS Basic language file
- GIF Graphic Interchange Format
- PS Postscript file
- TAR UNIX tape archive format Z
- UNIX compressed file
- ZIP DOS compressed file

You'll find a definition of the extensions you're using in your Windows WIN.INI file. Here are some of mine
- cal=calendar.exe^.cal

trm=terminal.exe^.trm
txt=notepad.exe ^.txt
ini=notepad.exe ^.ini
pcx=pbrush.exe ^.pcx
bmp=pbrush.exe ^.bmp
wri=write.exe ^.wri
rec=recorder.exe ^.rec
doc=C:\WINWORD\winword.exe ^.doc
dot=C:\WINWORD\winword.exe ^.dot
rtf=C:\WINWORD\winword.exe ^.rtf
pub=C:\MSPUB\mspub.exe ^.pub
ppt=C:\powerpnt\powerpnt.exe ^.ppt

In Windows 3.xx, there's a program called File Manager, which lets you associate a three character extension with an application. This tells File Manager that, when you click twice on the file, File Manager will know which application to launch. For example, you may tell Windows that the .QXD extension is associated with QuarkXpress. When you click on a QXD file, File Manager will launch Quark and load that particular file.

File Gap A short length of blank tape used to separate files stored on linear magnetic tape.

File Locking Picture a cabinet of file folders. Now I remove a folder to work on it. I make a photocopy of the folder in the cabinet and leave the original. You come along and remove the original because you want to work on it. You make changes and replace the changed copy in the cabinet. Ten minutes later I pull your file out and replace it with mine. Bingo, all your changes are lost. But let's say when I remove the file to work on it, I staple the remaining folder shut. That's a message to anyone else — including you that you shouldn't mess with the file. When I return, I unstaple the file, and add my changes. Now it's ready for you to do your thing. File locking ensures that a file will be updated correctly before another user, applications, or process will be allowed to write to the file. When a file is locked, no one else can write to it. Without file locking, one user could overwrite the file update of another user. In contrast to file locking, record locking allows many users to access the same file at once, but have only one access the record. See also ATTRIB and RECORD LOCKING.

File Maintenance The job of keeping your data base files up to date by adding, changing or deleting data.

File Management The system of rules and policies for maintaining a set of files — including how files can be created, accessed, retrieved and deleted.

File Server A file server is a device on a local area which "serves" files to everyone on that local area network (LAN). It allows everyone on the network to get to files in a single place, on one computer. It typically is a combination computer, data management software, and large hard disk drive. A file server directs all movement of files and data on a multi-user communications network, namely the LAN. It allows the user to store information, leave electronic mail messages for other users on the system and access application software on the file server — e.g. word processing, spreadsheet. In computer telephony applications, potentially many users or voice channels need to access data on a file server. The file server may therefore present a significant bottleneck for computer telephony especially if it is used to store large files, such as voice prompts. The ability for the file server to handle the transaction load planned for your computer telephony application is therefore a key design consideration and issue to test.

File Server Console Operator A user or a member of a group to whom a Novell NetWare SUPERVISOR delegates certain rights in managing the file server. A file server console operator has rights to use FCONSOLE to broadcast messages to users, to change file servers, to access connection information, to monitor file/lock activity, to check LAN driver configurations and to purge all salvageable files.

File Sharing A topology-independent feature of Apple Macintosh's System 7 operating system which allows users to share files and folders on their disks with other users across the LAN. File sharing is slow but acceptable for sharing small numbers of files among small groups. For larger networking, the user must consider AppleShare, Netware, Vines, etc.

File Transfer Protocol FTP. A service that supports file transfer between local and remote computers, including the Internet. FTP supports several commands that allow bidirectional transfer of binary and ASCII files between computers. The FTP client is installed with the TCP/IP connectivity utilities. See FTP and FILE TRANSFER PROTOCOLS.

File Transfer Protocols One problem with transmitting information over phone lines is the noise on the phone line. One way to overcome the problem of noise is a file transfer protocol. The idea is simple: send your information in bundles (called packets). Accompany those packets with a special number derived in some way from the information in the packet. Send it all to the other end. Have the computer at the other check the number and see if corresponds to the packet. If not, send a signal back, saying "Something went wrong. Please send the packet of information back again."
Most asynchronous file transfer protocols use some form of error detection, typically checksum or cyclic redundancy check (CRC). Both the checksum and the CRC are values derived from the data being sent (or received) according to mathematical algorithms. The protocol sends the value long with the information (the bits) in the packet. The receiving program compares with the check values with the values it calculates. If the check values do not match, the receiver asks the sending computer to retransmit the packet. Older protocols required a positive acknowledgement (an ACK) before they sent another packet. But newer protocols allow transmission of several packets before they receive an acknowledgement. This is particularly useful for circuits with long delays, especially satellites. See also XMODEM.

Fill Bit Stuffing

Fills A computer imaging term. Designated areas that are flooded with a particular color. Most paint packages let you create geometric shapes in filled form. All packages also let you fill irregular closed regions. Two types of such fills exist: A seed fill floods all connected regions with the color specified by the mouse or stylus pointer; a boundary fill floods a color until the algorithm encounters a specified boundary color.

Filter 1. A device which transmits a selected range of energy. An electrical filter transmits a selected range of frequencies, while stopping (attenuating) all others. It is used to suppress unwanted frequencies or noise, or to separate channels in communications circuits. Such a filter might be called a BANDPASS filter. You can also use a filter to remove certain characters you might be receiving over a data communications channel, for example control characters or higher-order nonstandard ASCII bits.
2. An operating parameter used in LAN bridges and routers that when set will cause these devices to block the transfer of packets from one LAN to another. Filters can be set to prevent the internetworking of several types of messages. They may be set to block all packets originating from a specific destination, called source address filtering, or all packets heading for a particular destination, called destination address filtering. Filters may also be set to exclude packet of a particular protocol or any particular filed in a LAN packet.

Filtering A process used in both analog and digital image processing to reduce bandwidth. Filters can be designed to remove information content such as high or low frequencies, or example, or to average adjacent pixels, creating a new value from two or more pixels.

Filtering Agent A new form of smart agent whose basic job is to filter out the stuff you don't want and find the stuff you do want — such as information gleaned from the Internet.

Filtering Traffic This is the process of selecting which traffic will be allowed into a certain portion of a network, such as the wide area network. It is also the process of determining which traffic is transmitted first, then next, and so on. The traffic is compared to a filter, or a set of specifications, to determine if it can pass through or not.

Final Trunk Group A last-choice trunk group that receives overflow traffic and which may receive first-route traffic for which there is no alternate route.

Finder The user interface portion of the Apple Macintosh operating system. Unlike running Windows on top of DOS, tight integration of the finder and system requires both to be running.

Finger A standard protocol specified in RFC-742. A program implementing this protocol lists who is currently logged in on another host. In short, finger is a computer command that displays information about people using a particular computer, such as their names and their identification numbers.

Finite State Machine A computer system with a defined set of possible states and defined transitions form state to state. Given the same inputs, two identical state machines will change states identically.

FIPS Federal Information Processing Standard. See also FIPS PUBS nn.

FIPS PUBS nn Various standards for data communications.

Fire To discharge someone. The story goes that in the early part of the 20th century, if an NCR salesman lost an order, when he returned to his office, they put his desk out on the front lawn and burned the desk. Then they "fired" the salesman.

Firewall A combination of hardware and software which limits the exposure of a computer or group of computers to an attack from outside. A firewall is a system or combination of systems that enforce a boundary between two or more networks. A network-level firewall, or packet filter, examines traffic at the network protocol packet level. An application-level firewall examines traffic at the application level — for example, FTP, E-mail, or Telenet. An application-level firewall also often readdresses outgoing traffic so it appears to have originated from the firewall rather than the internal host.

NEC PrivateNet Systems Group issueed a White Paper called Connecting Safely to the Internet — A study in Proxy-Based Firewall Technology. In that White Paper, they defined an Internet firewall:

The primary purpose of an Internet firewall is to provide a single point of entry where a defense can be implemented, allowing access to resources on the Internet from within the organization, and providing controlled access from the Internet to hosts inside the organization's internal networks. The firewall must provide a method for a security or system administrator to configure access control lists to establish the rules for access according to local security policies. All access should be logged to ensure adequate information for detailed security audit.

A traditional firewall is implemented through a combination of hosts and routers. A router can control traffic at the packet level, allowing or denying packets based on the source/destination address of the port number. This technique is called packet filtering. A host, on the other hand, can control traffic at the application level, allowing access control based on a more detailed and protocol-dependent examination of the traffic. The process that examines and forwards packet traffic is known as a proxy.

A firewall based on packet filtering must permit at least some level of direct packet traffic between the Internet and the hosts on the protected networks. A firewall based on proxy technology does not have this characteristic and can therefore provide a higher level of security, albeit at the cost of somewhat lower performance and the need for a dedicated proxy for each type of connectivity.

Each organization needs to choose one of these basic types of technologies. The right choice depends on the organization's access and protection requirements.

FireWire It's a 100 Kbps serial bus, also known as IEEE 1394. It is geared to become a digital interface for consumer video electronics and hard-disk drives. It's designed for up to 4.5 meters per segment and features six pins per connector. See USB, UNIVERSAL SERIAL BUS.

Firm Order Confirmation FOC. The form the local Telco submits to MCI indicating the date when the ordered circuits will be installed.

Firmware Software kept in semipermanent memory. Firmware is used in conjunction with hardware and software. It also shares the characteristics of both. Firmware is usually stored on PROMS (Programmable Read Only Memory) or EPROMs (Electrical PROMS). Firmware contains software which is so constantly called upon by a computer or phone system that it is "burned" into a chip, thereby becoming firmware. The computer program is written into the PROM electrically at higher than usual voltage, causing the bits to "retain" the pattern as it is "burned in." Firmware is nonvolatile. It will not be "forgotten" when the power is shut off. Handheld calculators contain firmware with the instructions for doing their various mathematical operations. Firmware programs can be altered. An EPROM is typically erased using intense ultraviolet light.

First In, First Out See FIFO.

First Office Application The first Ericsson AXE central office to put a new feature into service.

First Party Call Control A call comes into your desktop phone. You can transfer that call. When the phone call has left your desk, you can no longer control it. That is called First Party Call Control. If you were still able to control the call (and let's say, switch it elsewhere) that would be called Third Party Call Control. First party call control is mostly done at your desk with your telephone or with a card in your PC, which emulates a telephone. Third party call control is usually done via a computer (often a server on a LAN) attached to a special link directly into your PBX. There are some evolving standards in call control — chiefly Microsoft's Windows Telephony and Novell's TSAPI (Telephony Services API). There is no such animal as Second Party Call Control. See CALL CONTROL, TELEPHONY SERVICES and WINDOWS TELEPHONY.

First Ring Suppression Caller ID in North America comes in just after the first ring. You don't want to answer the call before the second ring otherwise you will mess up your receiving of Caller ID information. You can simply not answer until the second ring. Or you can get a trunk-based gadget which will turn off the first ring so you or your voice processing equipment won't hear it.

Fish To push a stiff steel wire or tape through a conduit or interior wall. Pull through wires, cable or a heavier pulling-in is then attached to one end of the steel wire. The other end is then pulled until the wire or cable appears.

Fish Job Running cables inside walls. Usage: "That fish job is too tough for a rookie."

Fish Tape Non-conductive tape with a reinforced fiberglass core and slippery outer nylon coating which slides easily through conduit without jamming. The idea is to push the tape through, attach it to cable and pull the cable back. You might use something called wire pulling lubricants, which come in various formulations — for use in different temperatures, for pulling different cable, etc.

FISK Fax a dISK. Method of sending information on 3 1/2" disks painlessly across phone lines. Plug your disk into a Fisk machine, choose which files you want to send, dial the number you want to send your files to and walk away. First product is from a company called Fisk Communications in San Diego.

FITC Fiber To The Curb.

FITH Flber To The Home. (I kid you not. That's what it stands for.)

FITL Fiber In The Loop.

Five By Five Slang expression meaning satisfactory transmission in both directions.

FIX Federal Internet Exchange. A connection point between the North American governmental internets and the Internet. The FIXs are named after their geographic region, as in "FIX West" East" (College Park, Maryland).

Fixed Condenser A condenser, the plates of which are stationary and the capacity of which cannot be changed.

Fixed Format A way of communicating in which everything to be sent follows a predetermined sequence, i.e. it fits into a specific length and format. The idea is to allow you to predict message length, the location of the message, where the control characters are, etc.

Fixed Length Records A set of data records all having the same number of characters in them. Think of a database of name, address, city, state, zip. Clearly, not everyone's record will be the same length. In order to make a fixed length record, the computer will pad the record with "padding characters" which the computer will ignore when it reads the record. But by including the padding characters it has effectively given everyone the same fixed length record.

Fixed Loop A services feature available in some switching systems that permits an attendant on an assisted call to retain connection through the attendant position for the duration of the call. The attendant will normally receive a disconnect signal when the call has been completed.

Fixed Priority-Oriented-Demand Assignment FPODA. Medium access technique in which one station acts as master and controls channel based on requests from stations.

Fixed Rate A fixed monthly price. See also FLAT RATE.

Fixed Satellite Service A radiocommunication service between Earth stations as specified fixed points when one or more satellites are used; in some cases this service includes satellite-to-satellite links, which may also be effected in the inter-satellite service, the fixed-satellite service may also include feeder linker for other space radiocommunication services.

FL Fault Locating.

Flag 1. A variable in a program to inform the program later on that a condition has been met.
2. In synchronous transmission, a flag is a pattern of six consecutive 1 bits used to mark the beginning and end of a "Frame" (IBM jargon for a packet). The character representation of the six consecutive "1" bits is 01111110. See ZERO STUFFING.
3. Fiberoptic Link Around The Globe. A consortium of phone companies owning an underwater submarine cable made of multiple strands of fiber, each strand carrying information at five gigabits per seconds. Each strand of fiber is unidirectional, i.e. you need two to make a conversation, one for going and one for coming.

Flag Sequence (HDLC, SDLC, ADCCP). The unique sequence of eight bits (01111110) employed to delimit the opening and closing of a frame.

Flame An outpouring of verbal abuse that network users write about other users who break the rules. A wonderful term for getting mad via electronic mail. People who frequently write flames are known as "flamers." You can flame by simply sending messages ALL IN CAPS!!!!!!!!! See FLAME FEST, FLAME WAR and MAIL.

Flame Fest Massive flaming. See FLAME.

Flame Mail Slang term for rude electronic mail. Bill Gates, Microsoft chairman, is said to be famous for the flame mail he sends to employees who don't perform according to his likings. Mr. Gates is famous for flame mail sent by him between midnight and 2:00 AM.

Flame Resistant Insulated wire which has been chemically treated so it will not aid the spread of flames.

Flame Retardant Constructed or treated so as not to be able to convey flame.

Flame War What happens when people send too much flame mail at each other.

Flaming To send an insulting message, usually in the form of a tirade, sent via online postings but also as personal. Flaming is the verb. Flame is the noun. And too much of flaming can cause a nasty flamewar.

Flammability Measure of a material's ability to support combustion.

Flash Quickly depressing and releasing the plunger in or the actual handset-cradle to create a signal to a PBX or Centrex that special instructions will follow such as transferring the call to another extension.

Flash Button A button on a phone which performs the same thing as quickly pressing the switch hook on a phone. See FLASH, FLASH HOOK and FLASHPHONE.

Flash Cut The conversion from an old to a new phone system occurs instantly as one is removed from the circuit and the other is brought in. There are advantages and disadvantages to Flash Cuts. For one, they're likely to be much more dangerous than the opposite view, known as a Parallel Cut, in which the two phone systems run side by side for a month or so. Also known as Hot Cut.

Flash EPROM A type of EPROM that can be electronically erased. It differs from EEPROM in that generally the entire memory must be erased at once.

Flash Hook Another name for Switch Hook. The little button on the telephone that you place your receiver into. It obviously hangs the phone up, releasing that line to receive another call. If you push the flash hook quickly, you can signal the switch at the other end (central office or PBX) to do something, such as place a call on hold and switch to the incoming one (call waiting), or transfer the call to another phone. See FLASH and FLASH BUTTON above.

Flash Memory A technology developed by Intel and licensed to other semiconductor companies, Flash memory is non volatile storage that can be electrically erased in the circuit and reprogrammed. Flash memory occupies little space and doesn't need continuous power to retain its memory. Some laptop companies, like Toshiba, are using flash memory as nonvolatile storage for the BIOS (Basic Input/Output System) and the instructions that start the company (the bootstrap loader). See also MEMORY CARDS.

Flash ROM Flash Read Only Memory. Read Only Memory that can be erased and reprogrammed, but stays on when power to your computer is turned off. Flash Rom is used in modems, for example, to hold software known as firmware. When a later software release comes out, you dial a distant computer which downloads new software into your Flash Rom, updating it. See FLASH MEMORY.

Flashphone A Rolm single line analog phone that has a "flash" button. If you pushed this button, it gave the PBX the same signal as if you had flicked the touch hook up and down. The PBX would then give you a dial tone and you could pushbutton in codes to get access to the PBX's features, like transfer, etc. See FLASH.

FlashROM See FLASH ROM.

Flat Rate Service An erstwhile common method of pricing local phone calls in the United States. The concept was that for a fixed amount of money — say $10 a month — you received a plain old desk telephone and an unlimited number of

NEWTON'S TELECOM DICTIONARY

local calls. For years, most residential and most business phones were on a flat rate service. The first thing to go was the phone instrument. You had to pay a dollar or so a month to continue renting it, or you could send it back and buy your own. Second to go was the size of the local calling area you could call. It got smaller.
Third to go were the phone calls themselves. This happened first with businesses and now increasingly with residential service. Under this new "pay-per-call" you get charged a "message unit" for each local calls. A message unit is typically eight to ten cents. But psychologically, "message units" sound better than dimes. Fourth to go was the definition of local calls. What was now a "local" call got smaller, i.e. you could call less far for the price of a local call. And what was now a "local" long distance call changed. Calls which, years ago, were free (i.e. on flat rate service) have now become long distance calls. You can witness this phenomenon of changing local pricing in California, New York and Jersey. In other states, it's taking a little longer. There are cities where flat rate service still exists. Treasure them. They're disappearing, too.

Flat Top Antenna An aerial consisting of one or more parallel horizontal wires supported between masts. The "T" type and the inverted "L" type belong in this class.
Flattery The art of telling someone exactly what he thinks of himself.
Flex One of the communications protocols used between paging towers and the mobile pagers/receivers/beepers themselves. Other protocols are POCSAG, ERMES, GOLAY and REFLEX. The same paging tower equipment can transmit messages one moment in GOLAY and the next moment in ERMES, or any of the other protocols.
Flex Life The measurement of the ability of a conductor or cable to withstand repeated bending.
Flexible Dialing Pattern A PBX dialing pattern that allows you to set your PBX so it can have one, two, three or four digit numbers for its extensions. See also FLEXIBLE NUMBERING OF STATIONS.
Flexible Drill Bit A long drill bit that bends and is used for pulling cable and wire through walls in one operation. This means you drill the hole and then reverse the drill and it pulls the cable through the hole, while the drill bit is still inserted. Diversified Manufacturing of Graham, NC, makes such a marvelous product.
Flexible Intercept Allows you to assign "operator intercept" service to those extensions you wish for whatever reason, unassigned number, temporary disconnect, etc.
Flexible Line Ringing A PBX feature which allows different phones to have different ringing for incoming calls from inside the building and from outside. Different ringing for intercom calls, different for inter-net calls, different ringing for outside calls, etc.
Flexible Numbering of Stations A PBX feature which allows you some flexibility in the way you number the extensions off your PBX. How much flexibility depends on the particular PBX and the number of extensions you have. Hotels like giving their hotel phones the same number as the rooms. Makes sense. See FLEXIBLE STATION NUMBERING.
Flexible Pricing Tariffs A regulatory procedure which permits rates for certain services to be changed quickly to meet market conditions, i.e. competition.
Flexible Release The ability of the switching system to release a connection when either party hangs up.
Flexible Routing The ability to choose different physical paths through a network for different calls as circumstances warrant.
Flexible Station Numbering A feature that allows telephone extensions to be numbered according to their physical location or departmental location, etc. No rewiring is required for in-place telephones. It's all done in software. See FLEXIBLE NUMBERING OF STATIONS.
Flexibility The quality of a cable or cable component which allows for bending under the influence of outside force, as opposed to limpness which is bending due to the cable's own weight..
FLEXPATH FLEXPATH service provides a 1.544 Mbps interface between a digital PBX and a Nynex central office and supports Direct Inward Dialing (DID) and Direct Outward Dialing (DOD). In short, it allows an incoming call to be dialed directly to a phone associated with a digital PBX on the user's premises. It also allows outgoing phone calls to be dialed directly (i.e. dial 9), without using a PBX attendant.
Flicker The wavering or unsteady image sometimes seen on monitors. A major cause is a refresh rate that's too low. Above 60 Hz, flicker disappears completely. See MONITOR.
FLINK A FLash and a wINK makes a flink signal.
Flip Flop A device or circuit which can assume either of two stable states. Flip flop devices are used to store one bit of information.
Floating Point Arithmetic Calculations performed on floating point, or exponential numbers. These numbers have two parts, a mantissa and an exponent. The mantissa designates the digits in the number, and exponent designates the position of the decimal point. Essentially floating point arithmetic allows the representation of very large numbers using a small number of bits. The speed of scientific computers is often rated in the Millions of FLoating Operations Per Second (MFLOPS) they can perform.
Floating Selection An imaging term. A selected area that is conceptually floating above the image, allowing it to be manipulated without affecting the background (for example, the contents of the Clipboard).
Floating Virtual Connection FVC. The ability to resume an on-demand connection on a port other than the port on which the original on-demand connection was established.
Flood Projection In facsimile, the optical method of scanning in which the original is floodlighted and the scanning spot is defined by a masked portion of the illuminated area.
Flood Search Routing A routing method that employs an algorithm that determines the optimum route for traffic within a network, avoiding failed and congested links.
Flooding A packet-switched network routing method whereby identical packets are sent in all directions to ensure that they reach their intended destination.

Floppy Disk A thin, flexible plastic disk resembling a phonograph record upon which computer data is stored magnetically. Called a floppy disk because it is flexible and can (and will) flop inside a drive as it is being turned. And it may sound as though it is flopping. Floppy disks were never designed as the permanent storage many people are using them for at present. Floppy disks were designed by IBM as a way of having its sellers and engineers carry programs and program updates to its customers. Floppy disks were lighter and less cumbersome than carrying heavy spools of magnetic tape. IBM designed its floppy disks to be thrown away once their information was loaded into the mainframe computer. The moral of this story is that floppy disks are NOT permanent reliable storage. Anything stored on floppy disks should be backed up at least once and, if possible, twice. Floppy disks come in three diameter sizes — 3 1/2, 5 1/4 and 8 inches. Floppy disks can be now safely put through X-ray machines at US airports.

Floppy Mini A floppy disk smaller than the traditional 5 1/4 inch diameter floppy disk. Now most commonly the 3 1/2 inch size invented by Sony, and used by the Apple Macintosh, among others. All MS-DOS laptop computers have 3 1/2 inch disks.

Floptical Technology The combination of optical servo track positioning and magnetic read-and-write technologies used in 3 1/2-inch Very High Density floppy disk drives. Floptical is a registered trademark of Insite Peripherals.

Flow Control The hardware, software and procedure for controlling the transfer of messages or characters between two points in a data network — such as between a protocol converter and a printer — to prevent loss of data when the receiving device's buffer begins to reach its capacity. In flow control, you can also deny access to additional traffic that would further add to congestion. (Think about flow control and the airlines.) See FLOW CONTROL PROCEDURE.

Flow Control Parameter Facility X.25 facility that allows the negotiation of packet and window sizes in both directions of transmission.

Flow Control Procedure The procedure for controlling the rate of transfer of data among elements of a network, e.g., between a DTE and a data switching exchange network, to prevent overload.

Flowchart A graphic or diagram which shows how a complex operation, such as programming, takes place. The flowchart breaks that operation down into its smallest, and easiest-to-understand events.

Flush Outlet A Communication Outlet recessed into the wall. This type of outlet may be positioned approximately 12" above the floor level for the connection of portable communications devices or at shoulder height to accept permanent wall telephone installations.

Flushofone A cordless headset with a noise cancelling microphone so you can't hear a toilet flush when on the phone in the bathroom.

Flush Protocol An ATM term. The flush protocol is provided to ensure the correct order of delivery of unicast data frames.

Flutter A rapid change in an electrical signal. The change may be in strength, frequency or phase. Distortion due to variation in loss resulting from the simultaneous transmission of a signal to another frequency.

Flux In soldering, a substance used to remove oxides from metal so the metal can be wet with molten solder for soldering.

Fly-By-Wire In traditional airplanes, the controls pilots moved were attached to heavy cables and hydraulic systems which themselves physically moved the rudder or the flaps, etc. Fly-by-wire replaced these wires and the hydraulic systems with computers and thin electrical wires. There are two main advantages to fly-by-wire. The computers can continuously adjust the aircraft's controls without the input of the pilot, trimming control surfaces so that the plane slides through the air with a minimum of air drag. Second, by eliminating heavy control cables and cutting down on hydraulic lines you can cut several hundred pounds off the weight of the plane, thus saving huge amounts of fuel over the life of the plane.

Flywheel A flywheel is a large heavy wheel used in electrical power generation. It's connected to an electrical power generator and will keep the generator spinning after the power source (a waterfall, or whatever) is unavailable.

FM 1. Fault Management. A network management function designed to receive fault information into a centralized management function. The faults are monitored, tracked, and resolved.
2. Frequency Modulation. See FREQUENCY MODULATION.

FM Blanketing That form of interference to the reception of other broadcast stations, which is caused by the presence of an FM broadcast signal of 115 dBu (562 mV/m) or greater signal strength in the area adjacent to the antenna of the transmitting station. The 115-dBu contour is referred to as the "blanking area."

FM Stereo Separation A measure of a radio tuner's ability to separate the left and right hand channels of a stereo broadcast. The higher the number, the greater the separation. The unit of measure is the Decibel (dB), a logarithmic unit which expresses the ratio between two voltage, current or power levels, usually relating to a standard reference level, or a background noise level.

FM Subcarrier One-way data transmission using the modulation of an unwanted portion of an FM broadcast station's frequency band.

FMAS Facility Maintenance and Administration System.

FMIC Flexible MVIP Interface Circuit. The FMIC provides a complete MVIP compliant interface between the MVIP bus and a variety of processors, telephony interfaces and other circuits. A built-in digital time slot switch provides Enhanced-Compliant MVIP switching between the full MVIP bus and any combination of up to 128 full duplex local channels of 64 kbps each. An 8-bit microprocessor port allows real-time control of switching and programmable device configuration. On board clock circuitry, including both analog and digital phase-locked loops, supports all MVIP clock modes. The local interface supports ST_BUS (Mitel), PCM Highway (Siemens), CHI (AT&T) signal formats at programmable rates of 2.048 MHz, 4.096 MHz, and 8.192 MHz as well as parallel DMA through the microprocessor port. See MVIP.

FMS Fraud Management System. An MCI term.

FMV Fair Market Value. A special lease for IRS purposes. Be careful. With Fair Market Value (FMV) leases, there is a catch to having the lessor guarantee the dollar amount or the percentage of your buyout. In order to be a FMV lease there must be a risk. That is why you are paying a lower rate of interest. If you agree on an amount up front, make sure it is not in writing, otherwise it does not meet the IRS test for a FMV lease. With A FMV lease, the lessor owned the asset and depreciates it; lessee expenses monthly payments and deducts them for tax purposes. If the buyout is determined in writing and the IRS can prove it, then it is a financing lease and lessee owns asset and depreciates it. Beware of this. This advice from Jane A Blank, telecom consultant, Westerville OH.

FNA A Brussels-based strategic alliance, which exists to facilitate global communications connections for companies in the financial services sector. The 12 founding FNA companies are Stentor of Canada, AOTC of Australia, RTT-Belgacom of Belgium, France Telecom, Deutsche Bundespost Telekom of Germany, Hong Kong Telecom, Italcable of Italy, KDD of Japan, Singapore Telecom, Telefonica of Spain, Mercury Communications of the United Kingdom, and MCI of the United States.

FNC Federal Networking Council. The body responsible for coordinating networking needs among U.S. Federal Agencies. A US group of representatives from those federal agencies involved in the development and use of federal networking, especially those networks using TCP/IP, and the connected Internet. The FNC coordinates research and engineering. Current members include representatives from the DoD, DOE, DARPA, NSF, NASA and HHS.

FNF Future of Numbering Forum. This definition, according to Lee Goeller: A series of meetings which Bellcore hopes will allow the industry to decide numbering issues without resorting to violence or the FCC.

FNR Fixed Network Reconfiguration

FNS Fiber Network Systems

FO Fiber Optics.

FOA 1. First Office Application. A telephone company term for what you and I know as beta testing.
2. Fiber Optic Amplifier. See FIBER OPTIC AMPLIFIER for a full definition.

FOB Free On Board. Term indicating where the seller's responsibility ends and the buyer's begins. You buy something, F.O.B. The seller puts it on a truck or railroad, plane, i.e. some carrier. He's responsible for getting it on the carrier. It's FOB, the truck. You — the buyer — are responsible for paying for the cost of the freight of getting you the goods you ordered. The opposite of F.O.B. is C.I.F. That stands for Cost, Insurance and Freight are included. That means the seller pays the freight. See FOB, FOB DESTINATION, FOB PLACE OF DELIVERY and FOB SHIPPING POINT.

FOB Destination Seller retains ownership until delivered to buyer. See FOB.

FOB Place of Delivery Seller retains ownership until delivered to buyer. See FOB.

FOB Shipping Point Seller responsibility ends when item is turned over to carrier. The buyer is responsible for payment if goods are damaged in transit. The buyer also handles any insurance claim. See FOB.

Focal Point An IBM Network management term, it consolidates the functions needed to manage centrally all parts of a network. It provides an end-to-end network view and receives information from entry points and service points. NetView is IBM's key implementation of the focal point.

Focus A measure of the clarity of a color monitor. Focus relates to the sharpness of a monitor's electron beam as it paints the face of a Cathode Ray Tube (CRT). The other measures are convergence and dot pitch.

FOD Fax On Demand. See FAX BACK.

Foil A slang term for an overhead transparency. The express "he gives good foil" reflects an executive's ability to make great presentations using overhead transparencies. the 1970s and early 1980s, so many managers at IBM made presentations that some senior executives actually got overhead projectors built into their desks.

FOIRL Fiber Optic Inter Repeater Link. Defined in IEEE 802.3 and implemented over two fiber links, transmit and receive, this medium may be up to one kilometer in length, depending on the number of repeaters in the network. A FOIRL is the perfect transmission medium to join a local area network on the eleventh floor to the fourth floor of the same building.

Folder The equivalent on an Apple Macintosh of an MS-DOS subdirectory.

Follow Me 800 Service Basically, Follow Me 800 Service is call forwarding of your personal 800 line. MCI announced this service in the Spring of 1991. It differs from local call forwarding in that you can dial into MCI from anywhere in the world and change the number your 800 line will send its calls to. Your 800 number always stays the same. What changes is the number it calls. A simple explanation: We buy a personal 800 line from MCI. The number is 800-555-6534. When someone calls that number, MCI looks up a database, checks where to send the number and sends it to my office at 212-691-8215. However, one day I go traveling. So I call another MCI 800 number, punch in my identification number and then give it the new number I will be at — namely 212-206-6660. From then on, MCI will send all my calls to that number — until I call and change the number again.

Follow Me Call Forwarding ROLM's way of saying Progressive Call Forwarding. Allows a previously forwarded call to be forwarded from that to another extension.

Follow Me Services Also called One Number Services. Follow me systems and services are based on the premise that people are mobile (e.g., they move around a lot in and out of the office), and have many phone numbers or places they might be. A person could have an office number, a cellular number, a voice mail number, a home number, and a pager number. Which phone number will a caller be at? Follow me systems will "track-down" the user being called no matter where they are and connect the caller to the user. The caller need only dial a single phone number. Usually, network or local switch provided call data (or data gathered by a voice response unit) is used to identify each call as being intended for a specific user. Based on options the user has selected, the caller will hear an answering prompt customized to that user, and the one number system will then automatically attempt to locate the user at one of several locations. The tracking-down process varies considerably between follow me systems. Some systems try multiple locations at once, others will try the possible destination locations sequentially. Almost all follow me services provide the caller an exit to voice mail at various points of the call.

Follow The Sun Dialing A technique used in call centers whereby the agents call those parts of the country where it's convenient to call and move the calling across the country as the sun moves. Our agents might call New York households between 6 P.M. and 9 P.M. When the time hits 9 P.M., the agents stop calling New York households and start focusing calls on households in the central time zone. To accomplish Follow The Sun Dialing, a call center needs software which knows in which phone numbers are in which time zones.

Fonline 800 Sprint's inbound service for small to medium-sized customers with applications up to 500 hours per month.

Font Alphanumeric and other characters in a distinctively shaped type style or type face. The words you are now reading are in the Helvetica normal type font, also called Helios by some people. The words are in Fat Futura. In Windows 3.1 terminology, font includes the type size.

Font Family A group designation that describes the general look of a font. For example, in Windows 3.1 the Times New Roman font family contains proportionally spaced fonts with serifs.

Font Size See point size.

Fonview A Sprint term for a combination of analytical software and monthly usage data diskettes which allow small and medium sized customers to perform traffic analysis, summarize usage, analyze historical usage, allocate cost to departments or clients and perform accounting functions on an IBM compatible personal computer.

Foo An Internet term. A place-holder for nearly anything — a variable, function, procedure, or even person. "A given user foo has the address 'foo@bar.com'."

Footprint 1. The area on the earth's surface where the signals from a specific satellite can be received. A footprint is shown as a series of concentric contour lines that show the area covered and the decreasing power of the signal as it spreads out from the center.
2. The area on a desk a device occupies, i.e. the computer's footprint.

Force Administration Data System See FADS.

Force Feed An arrangement in an outbound telebusiness unit where agents are force fed with a new call which is automatically dialed, a pre-determined time after finishing the previous call.

Forced Account Code Billing A telephone feature which prevents call from being completed if the user does not pushbutton in a billing code. That billing code may correspond to the department within the company. Or it may conform to the client and to the client's matter number the call must be billed to.

Forced Authorization Code FAC. A PBX feature which requires all or certain users to enter a code before dialing an outside number.

Forced Perfect Terminator A type of terminator containing a sophisticated circuit that can compensate for variations in the power supplied by the host adapter, as well as variations in bus impedance of complex SCSI systems.

Forced Release/Disconnect The switching center's automatic hang-up if the calling party fails to do so at the end of a conversation.

Forced Route Override Allows a PBX user to automatically redirect an outgoing call to a different trunk if the first trunk is busy or the connection is poor.

FORD Fixed Or Repaired Daily. Found On the Road Dead.

Forecasting Taking historical data (what happened in the past) from your ACD and using that information to predict what might happen in the future. Has your call volume always doubled on Tuesday? It will probably double next Tuesday too. A very important function of call center management software.

Foreground Processing Automatic execution of computer programs designed to preempt the use of the computing facilities. Usually a real time, urgent program. Contrast this with Background Processing, which might be something less urgent, for example, diagnostics of the system.

Foreign Address An ATM term. An address that does not match any of a given node's summary addresses.

Foreign Area Translation Translating the office codes of a distant (foreign) area to codes that make sense to a PBX which has more than one way of completing the call to that area.

Foreign Central Office Service Getting telephone service in a multi-office exchange from a central office other than the one you are normally served by. Not a common term any longer. Foreign central office service is the same price as normal local telephone service. It typically just involves asking for service off another central office. For example, our main number in New York City is 212-691-8215. Our 691- central office is in the 18th Street Exchange, a tall building on 18th Street. There is another central office in the same building. It is 206- and it is a more modern central office. When we ordered additional lines, we ordered them from this central office. You can now also call us on 212-206-6660. Don't trust my definition, however. Ask you local telephone company. See also FOREIGN EXCHANGE SERVICE.

Foreign EMF Any unwanted voltage on a telecommunications circuit.

Foreign Exchange Service FX. Provides local telephone service from a central office which is outside (foreign to) the subscriber's exchange area. In its simplest form, a user picks up the phone in one city and receives a dial tone in the foreign city. He will also receive calls dialed to the phone in the foreign city. This means that people located in the foreign city can place a local call to get the user. The airlines use a lot of foreign exchange service. Many times, the seven digit local phone number for the airline you just called will be answered in another city, hundreds of miles away. See also FOREIGN CENTRAL OFFICE SERVICE and FOREIGN EXCHANGE TRUNK.

Foreign Exchange Trunk A Foreign EXchange (FEX) trunk provides a direct connection between a PBX switch and a remote central office other than the central office that serves the location of the PBX.

Foreign Numbering Plan Area FNPA. Any other NPA (Numbering Plan Area) outside the geographic NPA where the customer's number is located.

Foreign Prefix Service Getting dial tone in a multi wire center exchange from a foreign wire center other than the one you are normally served by. Similar to Foreign Central Office Service, except that you may get charged extra for Foreign

Prefix Service. Don't trust my definition, however. Ask you local telephone company. See also FOREIGN CENTRAL OFFICE.

Forklift Upgrades There is some telecom equipment that can be upgraded and expanded by simply inserting a card or two and perhaps swapping out a power supply. And there's some equipment that needs to be almost completely replaced — cabinet and all. These sorts of upgrades are called "Forklift Upgrades." And they typically cost an arm and a leg. The term "forklift upgrade" has come to be pejorative. And most switch manufacturers are trying to figure ways to avoid forcing their customers into forklift upgrades, when they grow.

Form 230 Form 730 Application Guide is a collection of literature you'll need to register your telephone/telecom equipment under Part 68 of Title 47 at the Federal Communications Commissions. To get this material (it's free) drop a line or call the Federal Communications Commission, Washington DC 20554. As I write this edition, the person at the FCC in charge is William H. Von Alven, who also puts out a newsletter for Part 68 applicants. See PART 68 for a much larger explanation.

Form Effectors FEs. Control characters intended for the layout and format of data on an output device such as a printer or CRT. Examples are CR (carriage return) and LF (line feed).

Form Factor Fancy way of saying shape and size (width, depth, height).

Formal Call Centre A British term. A telebusiness unit in which all of the staff are dedicated to telephone based work. See also Informal Call Centre.

Formal Standards Specifications which are approved by vendor-independent standards bodies, such as ANSI (American National Standards Institute), ISO (International Standards Organization), IEEE (Institute of Electrical and Electronic Engineers) and NIST (National Institute of Standards and Technology).

Format 1. Arrangement of bits or characters within a group, such as a word, message, or language.
2. Shape, size and general makeup of a document. As a verb, its most common usage is in "to format this disk."

FORTRAN FORmula TRANslating system. A computer programming language.

Fortuitous Conductor Any conductor that may provide an unintended path for intelligible signals, e.g., water pipes, wire or cable, metal structural members.

Fortune Cookie An inane/witty/profound comment that can be found around the Internet.

Forum A section within an online service (such as CompuServe, America Online, etc.) where you can find out information on a specific subject — computers made by Toshiba or printers made by Hewlett Packard. Forums may include a library from which you can download various files (programs, bug fixes, printer drivers, text, press releases of new products and so on). Many forums also include one or more "conference rooms" which users may "enter" for conversations (on-line or off-line) with representatives of companies or the person running the forum, who is typically called the "sysop," as in system operator. Most manufacturers run forums as a relatively cheap and painless way of getting help information to their customers.

Forward Busying That feature of a telecommunications system wherein supervisory signals are forwarded in advance of address signals to seize assets of the system before attempting to establish a call.

Forward Channel The communications path carrying data or voice from the person who made the call. The Forward Channel is the opposite of the Reverse Channel.

Forward Echo An echo propagating in the same direction as the original wave in a transmission line, and formed by energy reflected back from one irregularity and then onward again by a second. Forward echoes can occur at all irregularities in a length of cable, and, when they add systematically, can impair its performance as a transmission medium.

Forward Error Correction FEC. A system of data transmission in which redundant bits generated at the transmitted end are used at the receiving terminal to detect, locate and correct any transmission errors before delivery to the local data communications link. The idea of forward error correction is to avoid having to retransmit information sent incorrectly. This technique is consuming of bandwidth and can make the transmission take longer.

Forward Prediction A technique used in video compression, specifically compression techniques based on motion compensation, where a compressed frame of video is reconstructed by working with the differences between successive video frames.

Forwarding Description An ATM term. The resolved mapping of an MPOA Target to a set of parameters used to set up an ATM connection on which to forward packets.

FOSSIL Fido/Opus/Seadog Standard Interface Layer. This is the interface used as an add-on to mailer software packages to connect them to PCs that are not 100% IBM-compatible.

FOTS Fiber Optic Transmission System. Not the same as POTS. But a neat acronym, nevertheless.

Foundation Graphics A set of graphics libraries or imaging models that form the lowest level graphics programmer's interface in Sun's OpenWindows. Examples: a graphics sub-routine library that a program could call to draw graphics primitives like arcs, circles, rectangles, etc.

Four Horsemen of the Apocalypse War, Plague, Famine and Death.

Four Wavelength Wave Division Multiplexing 4WL-WDM, also called Quad-WDM. MCI announced this technology in the Spring of 1996 as a method of allowing a single fiber to accommodate four light signals instead of one, by routing them at different wavelengths through the use of narrow-band wave division multiplexing equipment. The technology allowed MCI to transmit four times the amount of traffic along existing fiber. At that time MCI's backbone network operated at 2.5 gigabits per second (2.5 billion bits) over a single strand of fiber optic glass. Using Quad-WDM the same fiber's capacity will rise to 10 gigabits — enough capacity to carry 64,500 simultaneous transmissions over one single strand of fiber.

Four-wire A transmission path that allows physical separation of transmit and receive channels. At one time, 4-wire was the only method for implementing full-duplex transmission.

Four-wire Adapter A device which allows the connection of two-wire telephone equipment to a four-wire line. See FOUR-WIRE CIRCUITS.

Four-wire Circuit A path in which four wires are presented to the terminal equipment (phone or data), thus allowing for simultaneous transmission and reception. Two wires are used for transmission in one direction and two in the other direction.

Four-wire Circuits Telephone lines using two wires for transmitting and two wires for receiving, i.e. four altogether. All long distance circuits are four-wire. Almost all local phone lines are two-wire. All analog phones are two-wire. Four-wire circuits offer much higher quality.

Four-wire Repeater A repeater with two amplifiers, one used to amplify the signal going. The other used to amplify the signal coming. See FOUR-WIRE CIRCUITS.

Four-wire Terminating Set An electrical device which takes a four-wire circuit — one pair coming and one pair going — and turns it into the "normal" tip and ring circuit you need for a typical telephone, key system or PBX. See FOUR-WIRE CIRCUITS.

Fourier's Theorem In the early 1800s, the French mathematician Emile Fourier proved that a repeating, time-varying function may be expressed as the sum of a (possibly infinite) series of sine and cosine waves. Digital data is a bit stream, which can be sent as a sequence of square waves. Fourier's Theorem shows that to send a square wave (digital signal), a series of sine waves (analog signals) are actually summed together. If 1,000 square waves are to be sent every second, for example, the frequency components of the sine waves that are summed together are 1 kHz, 3 kHz, 5 kHz, 7 kHz, etc. The point of this analysis is to show that high frequency signals are required to form a stable, recognizable square wave.
As the bit rate increases, the square wave frequency increases and the width of the square waves decrease. Thus, narrower square waves require sine waves of even higher frequencies to form the digital signal. Note, then, that there is insufficient bandwidth in the 3 kHz voiceband to send square waves due to the absence of frequency components above 3,300 Hz. Even low frequency square waves cannot be sent because sine waves below 300 Hz are also absent. Thus, the local loop, according to Fourier's Theorem, cannot be used for the transmission of digital signals! The last paragraph is, in fact, no longer totally correct, as the increasingly successful ISDN trials are proving.

Fourth Utility The non-vendor specific communications premise wiring system which you use for integrated information distribution (voice, data, video, etc.) Leviton in Bothell, Washington has trademarked the term Fourth Utility. They make a broad range of premise wiring products.

Fox Message A standard sentence for testing teletypewriter circuits because it uses all the letters on the keyboard. That sentence is The quick brown fox jumped over the lazy sleeping dog.

FP 1. Feature Package. A software release for a telephone system. Originated with AT&T's Dimension PBX, now manufacturer discontinued. 2. File Processor.

FPDL Foreign Processor Data Link. A link from a Rockwell ACD to an external computer.

FPG Feature Planning Guide.

FPGA Field Programmable Gate Array. An FPGA is a specialized microprocessor that has no physical connections between its logic gates when it leaves the factory. But it has a huge number of potential connections, which can be firmed up in the field by a programmer with the right tools. FPGAs are a competitor to the cheaper ASICs — Application Specific Integrated Circuits. See VIRTUAL COMPUTING.

FPI Formal Public Identifier. A string expression that represents a public identifier for an object. FPI syntax is defined by ISO 9070.

FPLMTS Future Public Land Mobile Telecommunication Systems. A subject under discussion among the world's standards bodies. FPLMTS's objective is global terminal mobility.

FPS 1. Fast Packet Switching.
2. Frames Per Second. A measure of the quality of a video signal. NTSC TV — the standard in North America — uses 30 fps. Film is 24 FPS. PAL/SECAM (European) is 25 FPS.

FPU Floating Point Unit. A formal term for the math coprocessors (also called numeric data processors, or NDPs) found in many PCs. The Intel 80387 is an example of an FPU. FPUs perform certain calculations faster than CPUs because they specialize in floating-point math, whereas CPUs are geared for integer math. Today, most FPUs are integrated with the CPU rather than sold separately. See also CPU and DSP.

FQDN Fully Qualified Domain Name. An Internet term. The FQDN is the full site name of an Internet computer system, rather than just its hostname. For example, the system lisa at Widener University has a FQDN of lisa.cs.widener.edu.

FR-1 A flammability rating established by Underwriters Laboratories for wires and cables that pass a specially designed vertical flame test. This designation has been replaced by VW-1.

Fractal Compression An asymmetrical compression technique that shrinks an image into extremely small resolution-independent files by storing it as a mathematical equation as opposed to storing it as pixels. The process starts with the identification of patterns within an image and results in collection of shapes that resemble each other but that have different sizes and locations within an image. Each shape-pattern is summarized and reproduced by a formula that starts with the largest shape and repeatedly displaces and shrinks it. These patterns are stored as equations and the image is reconstructed by iterating the mathematical model. Fractal compression can store as many as 60,000 images on one CD-ROM. One disadvantage of fractal compression is that it is time consuming, taking as long as four minutes to convert a 1.3 MB TIFF file to a 228 KB file. See FRACTALS.

Fractal Geometry The underlying mathematics behind fractal image compression, discovered by two Georgia Tech mathematicians, Michael Barneley and Alan Sloan.

Fractal Image Format FIF. A compression technique that uses on-board ASIC chips to look for patterns. Exact matches are rare and the process works on finding close matches using a function known as an affine map.

Fractals Along with raster and vector graphics, fractals are a way of defining graphics in a computer. Fractal graphics translate the natural curves of an object into mathematical formulas, from which the image can later be constructed. See FRACTAL COMPRESSION.

Fractional Services A British term. Bandwidth available from carriers in increments of 64Kbit/s such as Mercury's Switchband. See FRACTIONAL T-1 for the North American definition.

Fractional T-1 Fractional T-1 refers to any data transmission rate between 56 kbps (DS0 rate) and 1.544 megabits per second (Mbps), which is a full T-1. Fractional T-1 is simply a digital line that's not as fast as a T-1. Fractional T-1 is popular because it's typically provided by a phone company (local or long distance) at less money than a full T-1. FT-1 is typically used for LAN interconnection, video conferencing, high-speed mainframe connection and computer imaging. Fractional T-1 is typically provided on four-wire (two-pair) copper circuits. See also ASDS.

Fractional T-3 A telephone company service in which portions of a T-3 (44.7364-megabits per second) transmission service are leased to provide a service similar to a T-1 (1.544-Mbps) or T-2 (3.152-Mbps) channel, but normally at a lower cost.

FRAD Frame Relay Access Devices.

Fragment The pieces of a frame left on an FDDI ring, caused by a station stripping a frame from the ring.

Fragmentation 1. In messaging it is the process in which an IP (Internet Protocol) datagram is broken into smaller pieces to fit the requirements of a given physical network. The reverse process is termed "reassembly."
2. A condition that affects data stored on a disk. Adding and deleting records in a file, creates what is sometimes called the Swiss cheese effect. The operating system stores the data for an individual file in many different physical locations on the disk, leaving large holes between records. Fragmented files slow system performance because it takes time to locate all parts of a file.

Frame 1. Generally, a group of data bits in a specific format, with a flag at each end to indicate the beginning and end of the frame. The defined format enables network equipment to recognize the meaning and purpose of specific bits. The group of bits are sent serially (one after another). Generally a frame is a logical transmission unit. A frame usually contains its own control information for addressing and error checking. A frame is the basic data transmission unit employed in bit-oriented protocols. In this way, a frame is similar to a block.
2. In TV video, a frame is a single, complete picture in video or film recording. A video frame consists of two interlaced fields of either 525 lines (NTSC) or 625 lines (PAL/SEACAM), running at 30 frames per second (NTSC) or 25 frames per second (PAL/SEACAM). 24 frames are sent in moving picture films and a variable number, typically between 8 and 30, sent in video-conferencing systems, depending on the transmission bandwidth available. Up to about 12 frames a second looks "jerky."
3. One complete cycle of events in time division multiplexing. The frame usually includes a sequence of time slots for the various sub channels as well as extra bits for control, calibration, etc.
4. A distribution frame. A rectangular steel bar framework having "verticals and horizontals" which is used to place semipermanent wire cross connections to permanent equipment. Found in telephone rooms and central offices. See DISTRIBUTION FRAME.

Frame Alignment The extent to which the frame of the receiving equipment is correctly phased (synchronized) with respect to that of the received signal.

Frame Alignment Signal The distinctive signal inserted in every frame or once in n frames that always occupies the same relative position within the frame and is used to establish and maintain frame alignment, i.e. synchronization.

Frame Buffer A section of memory used to store an image to be displayed on screen as well as parts of the image that lie outside the limits of the display. Some systems have frame buffers that will hold several frames, in which case they should be called "frames buffers." But they're not.

Frame Check Sequence Bits added to the end of a frame for error detection. Similar to a block check character (BCC). In bit-oriented protocols, a frame check sequence is a 16-bit field added to the end of a frame that contains transmission error-checking information. In a token ring LAN, the FCS is a 32-bit field which follows the data field in every token ring packet. This field contains a value which is calculated by the source computer. The receiving computer performs the same calculation. If the receiving computer's calculation does not match the result sent by the source computer, the packet is judged corrupt and discarded. An FCS calculation is made for each packet. This calculation is done by plugging the numbers (1's and 0's) from three fields in the packet (destination address, source address, and data) into a polynomial equation. The result is a 32-bit number (again 1's and 0's) that can be checked at the destination computer. This corruption detection method is accurate to one packet in 4 billion.

Frame Dropping The process of dropping video frames to accommodate the transmission speed available.

Frame Duration The sum of all the unit time intervals of a frame. The time from the start of one frame until the start of the next frame.

Frame-DS1 The DS1 frame comprises 193 bit positions. The first bit is the frame overhead bit, while the remaining 192 bits are available for data (payload) and are divided into 24 blocks (channels) of 8 bits each.

Frame Error An invalid frame identified by the Frame Check Sum (FCS). See also FRAME ERRORS.

Frame Errors In the 12-bit, D4 frame word, an error is counted when the 12-bit frame word received does not conform to the standard 12-bit frame word pattern.

Frame Grab To capture a video frame and temporarily store it for later manipulation by a graphics input device.

Frame Grabber A PC board used to capture and digitize a single frame of NTSC video and store it on a hard disk. Also known as Frame Storer. See VIDEO CAPTURE BOARD.

Frame Ground FGD. Frame Ground is connected to the equipment chassis and thus provides a protective ground. Frame Ground is usually connected to an external ground such as the ground pin of an AC power plug.

Frame Multiplexing The process of handling traffic from multiple simultaneous inputs by sending the frames out one at a time in accordance with a specific set of rules. Instead of multiplexing traffic from a lower-speed connection into a higher speed connection based on a specific time duration for each low-speed channel, frame multiplexing using the length of a given frame as the measurement.

Frame Rate The number of images displayed per second in a video or animation file.

Frame Relay Frame relay switching is a form of packet switching that relies on today's higher quality phone lines to minimize errors. It does error checking but no correcting in the network. The end devices, like LAN servers and clients, make corrections when necessary. When a traditional protocol like SDLC is adapted to frame relay by a Frame Relay Access Device

(FRAD), the FRAD will correct errors. Frame Relay is very good at efficiently handling high-speed, bursty data over wide area networks. It offers lower costs and higher performance for those applications in contrast to the traditional point-to-point services (leased lines).

With frame relay, a pool of bandwidth is made instantly available to any of the concurrent data sessions sharing the access circuit whenever a burst of data occurs. An addressed frame is sent into the network, which in turn interprets the address and sends the information to its destination at up to 45 Mbps. Like traditional X.25 packet networks, frame relay networks use bandwidth only when there is traffic to send. Frame relay is often provided to the end user at three speeds — 56/64 kilobits per second, 256/383 kbps and 1.5/2 Mbps. Frame relay supports voice that has been compressed and packetized, even though voice traffic in its traditional uncompressed form on a TDM circuit is highly sensitive to variations in the transmission delay introduced by packet networks. Small variations in delay are not critical to data traffic. For voice to be supported satisfactorily in a packet network, the receiving end compensates for delay variation.

The American National Standards Institute (ANSI) describes frame relay service in the following documents:
ANSI T1.602 — Telecommunications — ISDN — Data Link Layer Signaling Specification for Application at the User Network Interface.
ANSI T1.606 — Frame Relaying Bearer Service — Architectural Framework and Service Description.
ANSI T1S1/90 - 175 - Addendum to T1.606 - Frame Relaying Bearer Service — Architectural Framework and Service Description.
T1.607-1990 ISDN Layer 3 Signaling Specification for Circuit-Switched Bearer Service for DSS-1
T1.618 DSS-1 Core aspects of Frame Protocol for use with frame relay bearer service, ANSI, 1991
ANSI T1.617a, Signaling specification for Frame Relay bearer service for DSS-1, 1994
See the next two definitions.

Frame Relay Forum Organization of frame-relay equipment vendors and service providers, working to speed the development and deployment of frame relay products and interfaces with other broadband technologies, such as ATM. The Frame Relay Forum is based in Foster City, CA. 415-578-6980. It was formed in May 1991 as a non-profit mutual corporation. It has over 300 members. See also FRAME RELAY IMPLEMENTORS FORUM and ATM.

Frame Relay Implementors Forum A group of companies which have announced their support for a common specification for frame relay connections to link customers premises equipment to networking equipment. The common specification was originally announced on September 4, 1990. The common specification is based on the standard frame relay interface proposed by the American National Standards Institute (ANSI). The common specification supports the proposed ANSI standard and defines the extensions to that standard, including a local management interface that allows the exchange of control information between the user device and the frame relay network equipment. The specification is available for review from Cisco Systems, Digital Equipment Corporation, Northern Telecom and StrataCom. See FRAME RELAY and FRAME RELAY FORUM.

Frame Slip That condition in a TDM network under which a receiver of a digital signal experiences starvation or overflow in its receive buffer due to a small difference in the speeds of clocks and the clock (transmission rate) at the transmitter. The receiver will drop or repeat of a full TDM frame (193 bits on a T-1 line) in order to maintain synchronization.

Frame Store A system capable of storing complete frames of video information in digital form. This system is used for television standards conversion, computer applications incorporating graphics, video walls and video production and editing systems.

Frame Switch A device similar to a bridge that forwards frames based on the frames' layer 2 address. Frame switches are generally of two basic forms, cut-through switch (on-the-fly-switching) or store and forward switch. LAN switches such as Ethernet, Token Ring, and FDDI switches are all examples of frame switches.

Frame Synchronization The process whereby a given digital channel (time slot) at the receiving end is aligned with the corresponding channel (time slot) of the transmitting end as it occurs in the received signal. Usually extra bits (frame synchronization bits) are inserted at regular intervals to indicate the beginning of a frame and for use in frame synchronization.

Framework A Taligent definition. A set of prefabricated software building blocks that programmers can use, extend, or customize for specific computing solutions. With frameworks, software developers don't have to start from scratch each time they write an application. Frameworks are built from a collection of objects, so both the design and code of a framework may be reused.

Framing An error control procedure with multiplexed digital channels, such as T-1, where bits are inserted so that the receiver can identify the time slots that are allocated to each subchannel. Framing bits may also carry alarm signals indicating specific alarms. In TDM reception, framing is the process of adjusting the timing of the receiver to coincide with that of the received framing signals. In video reception, the process of adjusting the timing of the receiving to coincide with the received video sync pulse. In facsimile the adjustment of the facsimile picture to a desired position in the direction of line progression

Framing Bit 1. A bit used for frame synchronization purposes. A bit at a specific interval in a bit stream used in determining the beginning or end of a frame. Framing bits are non-information-carrying bits used to make possible the separation of characters in a bit stream into lines, paragraphs, pages, channels etc. Framing in a digital signal is usually repetitive.

Framing Error An error occurring when a receiver improperly interprets the set of bits within a frame.

Franchise The exclusive right to operate telephone service in a community. This right — also called the franchise — is granted by some government agency. Some phone companies existed before the appropriate regulatory authority, so they're "grandfathered" in their exclusivity. Some phone companies have an exclusive area to serve more because of their presence than because of the legal right conferred on them. The question of who has a franchise to serve what community with what service is becoming increasingly unclear as competition penetrates all aspects of the phone industry.

FRD Fire RetarDant. A rating used for cable within duPont's Teflon or equivalent fluorpolymer material. FRD cable is used when local fire codes call for low flame and low smoke cable. FRD cable is typically run in forced air plenums as an alternative to metal conduits.

FRED 1. A system for searching the international X.500 user directory.

2. Friggin' Ridiculous Electronic Device (often applied to minicomputers, phone switches, and other "high tech" gadgets that work very hard to accomplish seemingly "simple" tasks.

Free Address Office Office arrangement in which all personal spaces are eliminated in favor of employees picking up their supplies at a front desk upon arriving and then choosing a temporary work area each day.

Free On Board FOB. Board. Term indicating where the seller's responsibility ends and the buyer's begins. You buy something, F.O.B. The seller puts it on a truck or railroad, plane, i.e. some carrier. He's responsible for getting it on the carrier. It's FOB, the truck. You — the buyer — are responsible for paying for the cost of the freight of getting you the goods you ordered. The term FOB is typically enhanced (or made clear), thus:

FOB Destination: Seller retains ownership until delivered to buyer.

FOB Place of Delivery. Seller retains ownership until delivered to buyer.

FOB Shipping Point. Seller responsibility ends when item is turned over to carrier. The buyer is responsible for payment if goods are damaged in transit. The buyer also handles any insurance claim.

The opposite of F.O.B. is C.I.F. That stands for Cost, Insurance and Freight are included. That means the seller pays the freight.

Free Space Communications Any form of telecommunications that doesn't use cable or fiber optic. In other words, free space communications is radio, microwave, satellite, wireless LANs, cellular, etc. Free space communications is in great demand today because of its use as a method of disaster prevention — when cables get sliced, etc.

FREENET An organization to provide free Internet access to people in a certain area, usually through public libraries.

Freephone A service which permits the cost of the call to be charged to the called party, rather than the calling party. Pioneered in the US in 1966, the freephone service now carries around 100 millions calls per day in the US alone. US companies currently hold around 90% of the world's 9 million freephone numbers.

Freephone has proves particularly popular with business subscribers, who are often willing to bear the cost of a telephone call in order to promote their services or to encourage customers to order their products by phone. Recent estimates by AT&T indicated some $100 billion is currently traded over the freephone service every year. Until the recent approval of the new ITU-T Recommendation E-169, however, companies have been restricted by only being able to use their freephone number in one country. Those organizations wishing to offer products or services to customers on an international basis have had no choice but to register a separate number in each country, which has proved unwieldy and often inefficient. The new standard for international freephone will greatly free up companies' abilities to operate across international markets, and will benefit consumers by allowing them to obtain information or to shop around for goods and services at no personal expense. It is hoped the new standard might also stimulate the market for freephone services in Europe and Asia-Pacific, regions that until now have been slow to take up the service.

The potential market for the new international freephone service is expected to be considerable. The globalization of markets via new technologies such as the Internet means that many companies are now able to offer their products and services to users in different countries, and will benefit from being able to advertise a single toll-free number to potential customers all over the world. Calls to the new global market number can also be routed to different destinations, allowing companies to direct their incoming calls to the most appropriate location for efficient handling.

Freeware Software that doesn't cost anything.

Freeze Frame The transmission of discrete video picture frames at a data rate which is too slow to provide the perception of natural motion, referred to as "full-motion." An uncompressed, digitized full-motion video signal is typically transmitted at many millions of bits per second. Freeze frame can be carried on anything from a simple voice grade phone line running at 9.6 Kbps (the same speed as a Group 3 facsimile machine).

Freq A term or abbreviation for a "file request" for a file from another node in a network. In FidoNet a node user usually freq's a file through mailer software which sends an appropriate request to a distant node that has the desired file. Freqing is the ability in FidoNet to transfer files back and forth between BBSs (bulletin board systems) automatically. Equivalent to file transfer in PCRelay.

Freqing See FREQ.

Frequency The rate at which an electrical current alternates, usually measured in Hertz. Hertz is a unit of measure which means "cycles per second." So, frequency equals the number of complete cycles of current occurring in one second.

Frequency Agile Modem A modem used on some broadband systems that can shift frequencies to communicate with stations in different dedicated bands.

Frequency Agility The ability of a cellular mobile telephone system to shift automatically between frequencies.

Frequency Band The portion of the electromagnetic spectrum within a specified upper- and lower-frequency limit. Also known as Frequency Range. See also BAND, FREQUENCY for a complete list of all the frequencies.

Frequency Diversity A way of protecting a radio signal by providing a second, continuously operating radio signal on a different frequency, which will assume the load when the regular channel fails. Here's another way of saying the same thing: Frequency diversity is a any method of diversity transmission and reception wherein the same information signal is transmitted and received simultaneously on two or more independently fading carrier frequencies.

Frequency Division Multiple Access A technique for sharing a single transmission channel (such as a satellite transponder) among two or more users by assigning each to an exclusive frequency band within the channel.

Frequency Division Multiplexing FDM. An older technique in which the available transmission bandwidth of a circuit is divided by frequency into narrower bands, each used for a separate voice or data transmission channel. This means you can carry many conversations on one circuit.

Frequency Frogging The interchanging of the frequency allocations of carrier channels to prevent singing, reduce crosstalk, and to correct for a transmission line frequency-response slope. It is accomplished by having the modulators in a repeater translate a low-frequency group to a high-frequency group, and vice versa. Because of this frequency inversion process, a channel will appear in the low group for one repeater section and will then be translated to the high group for the next section. This results in nearly constant attenuation with frequency over two successive repeater sections, and eliminates the need for large slope equalization and adjustment. Also, singing and crosstalk are minimized because the high-level output of a repeater is at a different frequency from the low-level input to other repeaters.

Frequency Hopping Another name for spread spectrum transmission. A technique developed by Hedy Lamarr, the actress, in the early part of the second world war to prevent the enemy from jamming or eavesdropping on conversations and on commands to steer torpedoes, etc. The idea is to hop from one frequency to another in split-second intervals as you transmit information. Attempts to jam the signal succeed only in knocking out a few small bits of it. So effective is the concept that it is now the principal antijamming device in the US military. Ms. Lamarr never got paid for the invention. But it was definitely hers. And she invented it because of her patriotism for the United States. She had fled Austria in 1937. And she received a U.S. patent in 1940. See also SPREAD SPECTRUM.

Frequency Modulation A modulation technique in which the carrier frequency is shifted by an amount proportional to the value of the modulating signal. The amplitude of the carrier signals remains constant. The deviation of the carrier frequency determines the signal content of the message. Commercial TV and FM radio use this technique, which is much less sensitive to noise and interference.

Frequency Offset Non-linear distortion that causes a shift in the frequency of a received signal.

Frequency Response The variation (dB) in relative strength between frequencies in a given frequency band, usually the voice frequency band of an analog telephone line.

Frequency Reuse The ability to use the same frequencies repeatedly within a single system, made possible by the basic design approach used in cellular. Since each cell is designed to use radio frequencies only within its boundaries, the same frequencies can be reused in other cells not far away with little potential for interference. The reuse of frequencies is what allows a cellular system to handle a huge number of calls with a limited number of channels.

Frequency Shift Keying FSK. A modulation technique for data transmission. It shifts the frequency above the carrier for a 1 and below the carrier for a 0 (zero). See also FSK.

Frequency Tolerance The maximum permissible departure by the center frequency of the band occupied by an emission from the assigned frequency or by the characteristic frequency of an emission from the reference frequency. By international agreement, frequency tolerance is expressed in parts per 10 (6) or in hertz. This includes both the initial setting tolerance and excursions related to short- and long-term instability and aging. In the United States, frequency tolerance is expressed in parts per 10(n), in hertz, or in percentages.

Frequency Translator In a split broadband cable system, a frequency translator is an analog device at the headend that converts a block of inbound frequencies to a block of outbound frequencies.

Fresnel Loss The loss at a joint that is caused by a portion of the light being reflected.

Fresnel Reflection In optical physics, fresnel reflection is the reflection of a portion of incident light at a planar interface between two homogeneous media having different reflective indices. Fresnel reflection occurs at the air-glass interfaces at entrances at entrance and exit ends of an optical fiber. Resultant transmission losses (on the order of 4 percent per interface) can be virtually eliminated by using antireflection coatings or index-matching materials. Fresnel reflection depends upon the index difference and the angle of incidence. In optical elements, a thin transparent film is sometimes used to give an additional Fresnel reflection that cancels the original one by interference. This is called an antireflection coating.

Fresnel Reflective Losses For optical fiber communication, the losses incurred at the terminus interface that are due to refractive index differences.

Fresnel Region In radio communications, the region between the near field of an antenna and the Fraunhofer region. The boundary between the two is generally considered to be at a radius equal to twice the square of antenna length divided by wavelength.

Fresnel Zone Fresnel zone is the line-of-sight path between two microwave antennas. It is an elliptical zone between the two antennas where the total path distance varies by more than half of the operating wavelength. The concept is extended to describe the distance by which the direct wave clears any intervening obstacle such as a mountain peak. If the total path distance between transmitter, peak and receiver, is 1 wavelength greater than the direct distance, then the clearance is said to be two Fresnel zones.

FRF92.02 Multiprotocol Interconnect over Frame Relay.
FRF92.07 Frame Relay Multicast Draft Service Description.
FRF92.08 Frame Relay Network-to-Network Interface Implementation Agreement Draft.
Frictional Electricity Static electricity produced by friction (e.g., by rubbing a hard rubber rod with a silk cloth).
Friendly Name A name, typically identifying a network user or a device, intended to be familiar, meaningful, and easily identifiable. A friendly name for a printer might indicate the printer's physical location (e.g. "Sales Department Printer").
Friesen, Gerry Mr. Friesen is half the operation which published this dictionary. Call him if you find any mistakes in this dictionary. He didn't write it. But he should take the blame. He's good at handling blame. His phone number is 215-355-2886. Don't call him collect, since I'll have to pay half the cost of the phone call.
FRL Facility Restriction Level. A term created by AT&T for its Dimension PBX. These levels define the calling privileges associated with a line; for example, intragroup calling only in the warehouse, but unrestricted calling from the boardroom.
FRND Frame Relay Network Device. A frame relay access device or frame relay network switch.
Frogging Frogging is the process of inverting line frequencies of a carrier system so that incoming high-frequency channels leave at low frequencies and vice versa. Frogging equalizes the transmission loss between high and low frequency channels.
Front End The client part of a client/server application that requests services across a network from a server, which is

known sometimes as the back end. It typically provides an interactive interface to the user, for example, a data entry front end, allowing database to be entered into a database server. The term "front end" is, of course, a contradiction in terms. But, who said language had to be consistent, or logical? See FRONT END CONTROLLER, FRONT END DEVELOPMENT TOOLS and FRONT END PROCESSOR.

Front End Controller See FRONT END PROCESSOR.

Front End Development Tools These tools let a programmer control the design and manipulation of applications using visual techniques. Front-end development creates a graphical user interface, providing more flexibility and making it easier to link users with data accessed from database servers.

Front End Equipment The equipment positioned between a computer and the communications line(s). Its purpose is to organize data being sent and received.

Front End Mailer A program that operates on a bulletin board system and determines if a caller is another computer that wants to exchange mail or a human that wants to exchange mail or a human that wants to access the BBS resources. Usually the mailer transmits the prompt "Press ESC" and upon receiving an ESC character, or the passing of a timeout period, considers the caller to be human and gives it the resources of the BBS. Also known as a mailer.

Front End Processor FEP. An FEP is a computer under the control of another, larger computer (typically a mainframe) in a network. The FEP does simple, basic "housekeeping" operations on the data streams as they arrive to be processed by the bigger computer. The FEP acts as a sort of intelligent traffic cop. It relieve the bigger, host computer of some of its telecommunications Input/Output burden, so that the host computer can concentrate on handling the processing burden. Depending on its sophistication, the front end processor might also perform serial to parallel conversion, protocol conversion, block or message assembly, etc. Here's a more technical definition: A dedicated communications system that intercepts and handles activity for the host. Can perform line control, message handling, code conversion, error control, and such applications functions as control and operation of special-purpose terminals. Designed to offload from the host computer all or most of its data communications functions. IBM 3705, 3725 and 3745 are Front End Processors. Front end processors are not used in client/server networks.

FRS Frame-Relay Service: A connection oriented service that is capable of carrying up to 4096 bytes per frame.

FRTT Fixed Round-Trip Time: This is the sum of the fixed and propagation delays from the source to the furthest destination and back.

FRU Field Replaceable Unit.

FSA Foreign Serving Arrangement.

FSK Frequency Shift Keying. A modulation technique for translating 1's and 0's into something that can be carried over telephone lines, like sounds. A "1" will be assigned a certain frequency of tone, and a "0" will be assigned to another tone. The transmission of the bits keys the sounds to shift from one frequency to the other. See also FREQUENCY SHIFT KEYING.

FSN Full Service Network. A term introduced by Time Warner Cable, a large cable television operator, in late 1994. According to Time Warner, FSN is a prototype for future interactive games being played over cable TV lines.

FSP File Service Protocol. It's a file-transfer protocol, similar to FTP except that it doesn't create much load on a server.

FT-1 Fractional T-1. Any part of a T-1 circuit that's smaller than a full T-1 circuit. Fractional T-1 circuits are cheaper than full T-1 circuits. That's their reason for existing. See FRACTIONAL T-1 for a bigger explanation.

FT1 Fractional T1.

FTAM File Transfer and Access Management. The OSI (Open Systems Interconnection) standard for file transfer (i.e., the communication of an entire file between systems), file access (i.e, the ability to remotely access one or more records in a file) and management (e.g., the ability to create/delete, name/rename a file). FTAM is also an international standard.

FTE Full Time Equivalent. A call center term. A scenario assumption used in budget forecasting and scheduling that defines the number of hours per week full-time employees are normally suppose to work. As a measure of staffing level, an FTE is equivalent to a full-time position, even though the hours may actually be filled by part-time schedules.

FTIP Fiber Transport Inside Plant.

FTP File Transfer Protocol. There are two allied meanings:
1. In local area networking technology, FTP is file-sharing protocol that operates at layers 5 through 7 of the Open Systems Interconnection (OSI) model. As specified in RFC-959, FTP provides full authentication of the user by requiring login on the remote host. It lets users transfer text and binary files to and from a PC, list directories on the foreign host, delete and rename files on the foreign host, and perform wildcard transfers between hosts.
2. On the Internet, the FTP, or File Transfer Protocol, is an Internet tool for accessing file archives around the world that are linked to the Internet. If you have a modem with a terminal emulation program, you have all the software you need to visit ftp sites. All you might be lacking is an on-line provider with Internet access. Once you have access, here's what it's like to visit the a ftp site. The site address may be ftp.pht.com. Most ftp sites accept anonymous login. That is, when the site asks who you are, you answer "anonymous." Then when they ask for your password, you give them your Internet email address. See ANONYMOUS FTP.

FTP Mail Server A server which permits the retrieval of files via e-mail. See FTP.

FTS Federal Telecommunications System is a private telephone network sometimes shared enthusiastically by all federal government agencies. And sometimes not. See FTS2000.

FTS2000 The U.S. General Services Administration in Washington, D.C. describes the FTS2000 as "the state of the art, digital, long distance telecommunications program that provides voice, data and video transmission services to federal government agencies." FTS2000 contracts, expiring in December, 1998 have an estimate total value of $25 billion. For more on FTS2000, call 703-285-1020.

FTTB Fiber To The Building.

FTTC Fiber To The Curb.

FUBAR F...d Up Beyond All Recognition. In short, a mess. A term often used in electronic mail messages.

FUD Fear, Uncertainty, Doubt. A marketing tactic which a dominant player (once IBM) has used to discourage its customers from buying from its competitors.

Fugitive Glue Glue used by printers to affix stuff into magazines. The glue is designed to stick until the magazine is delivered. At that point, the stuck-in thing becomes easier to remove and/or falls into your lap. This definition contributed by Rich Kubik.

Fugitive Odor A smell that leaks out of a composting plant or landfill.

Full Availability Idealized condition which exists when your phone system can provide connections for every telephone connected to it. Also called NON-BLOCKING.

Full Duplex Transmission in two directions simultaneously, or, more technically, bidirectional, simultaneous two-way communications. The best two-way phone conversations take place on four-wire circuits, two for transmission in one direction and two for transmission in the other. All long distance circuits are four wire. Most local lines are two wire, which means they're a compromise. Most speakerphones are half-duplex, meaning they only transmit in one direction at one time. The speakerphone flips its direction based on who's talking, or, more precisely, who's talking the loudest. Full duplex speakerphones are the best, but they're expensive. See SPEAKERPHONES and FOUR-WIRE CIRCUITS.

Full Echo Suppressor An echo suppressor in which the speech signals on each path are used to control the suppression loss in the other path of a 4-wire circuit. Used for long-distance communications. Compare with split echo compressor.

Full Motion Video Television transmission where images are sent and displayed in real-time and motion is continuous. Video reproduction at 30 frames per second (NTSC-original signals) or 25 frames per second. (PAL-original signals). Compare with freeze frame. See FREEZE FRAME VIDEO.

Full System Battery Backup This means there's sufficient battery power backing the phone system so that during a power outage, the telephone system will continue to work, i.e. you won't even know the commercial power has gone out. All programming will be intact. Calls will get through, etc. Full System Battery Backup is critical to many businesses, especially those in the "life or death" business, such as hospitals, police, fire departments, etc. Other businesses who depend heavily on the phone for their revenues — airlines, brokerage companies, hotel/motels, etc. — often also use full system battery backup.

Full Time Equivalent FTE. A call center term. A scenario assumption used in budget forecasting and scheduling that defines the number of hours per week full-time employees are normally suppose to work. As a measure of staffing level, an FTE is equivalent to a full-time position, even though the hours may actually be filled by part-time schedules.

Fully Connected Network A network topology in which each node is directly connected by branches to all other nodes. This architecture becomes impractical as the number of nodes in the network increases in complexity. Such networks normally go to distributed nodes.

Fully Perforated Paper tape on which information is represented by the holes punched through the paper.

Fully Qualified Domain Name FQDN. In Internet terms, the full name of a system, rather than just a host name. For example, if the host name is Harry, the full name is harry.company.com).

Fully Restricted Stations In a PBX, fully restricted stations (also called phones) can't place any outside calls. They can make intercom calls as well as receive incoming calls.

Function Key 1. One of up to 12 keys on a PC keyboard labeled with the letter F followed by a number. The effect (if any) of pressing a particular function key depends on what program you are running at the time.
2. An undefined key on a computer or telephone that can be defined to perform one function, which would normally require the user hitting one or several keys in succession.

Functional Entity FE. A set of functions that provides one or more specified capabilities. Seven FEs have been identified for the Advanced Intelligent Network Release 1 architecture: Network Access, Service Switching, Service Logic and Control, Information Management, Service Assistance, Automatic Message Accounting and Operations. Definition from Bellcore in reference to its concept of the Advanced Intelligent Network. See AIN.

Functional Group A collection of FEs (Functional Entities) that reside together in a system.

Functional Management Layer A communications layer in SNA that formats presentations.

Functional Profile A defined stack of ISO OSI-Layer elements, such as GOSIP, MAP or TOP. Functional profiles were developed in order to ensure that, when defined, ISO OSL stacks could interoperate. Due to the number of different protocol elements at each OSI layer, it was possible to define stacks that were syntactically correct, but would not be able to exchange information due to differences at particular layers. A functional profile that has been defined as a standard is termed a standardized profile. Likewise, an International Standard Profile is an ISO OSI functional profile.

Functional Resource An abstraction of physical entities (e.g., voice synthesizers) that the Service Assistance FE (Functional Entities) can manipulate.

Functional Signaling In an ISDN circuit, function signaling provides messages with unambiguous, defined meanings known to both the sender and receiver of the messages. Signaling is generated by the terminal.

Functional Specification A description of a system from a working point of view. It differs from a precise technical description which includes each piece of equipment precisely spelled out. A system can often work the same using different hardware and software configurations. By functionally describing a system, a user allows sellers to use their imagination to solve the problem in the most creative, cost-effective way. Most sellers prefer functional descriptions.

Functional Split A division within an automatic call distributor (ACD) which allows incoming calls to be directed from a specific group of trunks to a specific group of agents.

Functional Test A test carried out under normal working conditions to verify that a circuit or particular part of the equipment works properly.

Functional Transparency The ability of a network to carry any user information regardless of its form, so that user applications can operate through the network.

NEWTON'S TELECOM DICTIONARY

Functional User An entity external to the functional architecture that uses the functional architecture capabilities to exchange information with other functional users. Definition from Bellcore in reference to its concept of the Advanced Intelligent Network.

Fuse Verb: To blend together through melting. Noun: An electrical device typically consisting of a wire or strip of fusible metal that melts to interrupt an electrical circuit when current exceeds the rated level of the fuse. The idea is that in any electrical circuit, the fuse should be the weakest point — thus the point that heats up when things go wrong and melts. Better the fuse melts than your expensive PBX. See also CIRCUIT BREAKER.

Fused Quartz The precise term for glass made by melting natural quartz crystals.

Fusible Links Short lengths (about 25 feet) of fine-gauge wire pairs inside metallic sheath cable that melt to interrupt an electrical circuit and to prevent overheating in building wiring and equipment.

Fusion In 1995 NEC America invented the term "Fusion" for its program to bring a combination of computers and communications into the office. The Fusion strategy focuses on delivery of "multimedia information (text, diagrams, graphics, voice, images and video) to the information worker's desktop." Fusion has three elements — the NEC PBX, various application softwares, and integration links, which consist of Fusion Netlinks (trunk level interfaces), Desklinks (station level interfaces), CPULinks (Processor level interfaces such as OAI and other Ethernet interfaces) and Commandlinks (data stream level interfaces, such as property management system (PMS), Message Center Interface (MCI) and other RS-232 interfaces. See also C&C.

Fusion Splicing In optical transmission systems using solid transmission media, the joining together of two media by butting them, forming an interface between them, and then removing the common surfaces so that there be no interface between them. Thus, no reflection or refraction at the former interface occurs.

Fuzzy Logic Fuzzy logic is the newest wrinkle in the ancient science of controlling processes that involve constantly changing variables. Contrary to its name, fuzzy logic is a very precise sub discipline in mathematics. It was invented in the 1960s by University of Berkeley's Russian-born Iranian computer science professor Lotfi Zadeh. It enables mathematicians and engineers to simulate human thinking by quantifying concepts such as hot, cold, very far, pretty close, quite true, most usually, almost impossible, etc. It does this by recognizing that measurements are much more useful when they are characterized in linguistic terms that when taken to the fourth decimal point. Fuzzy logic reduces a spectrum of numbers into a few categories called membership groups. Within five years virtually all consumer goods will come with fuzzy logic. Already fuzzy logic is inside video camcorders (to reduce the motion of the camera), in washing machines (to figure the optimum mix of washing conditions for that weight and filth). Fuzzy logic chips are made by companies such as InfraLogic Inc. of Irvine, CA.

FVR Flexible Vocabulary Recognition.

FWIW Abbreviation for "For What It's Worth;" commonly used on E-mail and BBSs (Bulletin Board Systems).

FYI An Internet term. An abbreviation for the phrase "for your information." There is also a series of RFCs put only by the Network Information Center called FYIs. They address common questions of new users and many other useful things.

FYI Service An MCI International service that provides a summary of the latest news on business, current events, and sports in the USA via Telex.

FX 1. Foreign Exchange. A Central Office trunk which has access to a distant central office. Dial Tone is returned from that distant Central Office, and a location can be reached in the area of the foreign Central Office by dialing a local number. This will provide easier access for customers in that area and calls may be made anywhere in the foreign exchange area for a flat rate. See also FOREIGN EXCHANGE and FXO.

FXO Foreign Exchange Office. Foreign exchange (FX) service is a service that can be ordered from the telephone company that provides local telephone service from a central office which is outside (foreign to) the subscriber's exchange area. In its simplest form, a user can pick up the phone in one city and receive a dial tone in the foreign city. This type of connection is provided by a type of trunk called foreign exchange (FX) trunks. FX trunk signaling can be provided over analog or T-1 links. Connecting POTS telephones to a computer telephony system via T-1 links requires a channel bank configured with FX type connections. To generate a call from the POTS set to the computer telephony system, you will need a FXO (foreign exchange office) connection configured. To generate a call from the computer telephony system to the POTS set, you will need a FXS connection configured. See FX.

FXS Foreign Exchange Station. See FXO.

FZA Fernmeldetechnisches Zentralamt. Telecom approval authority Austria, literally translated "long distance communications technical central office." All that in two words. Not bad!

G 1. Giga, meaning a billion or one thousand million. In computers it is 1024 times mega and is actually 1,073,741,824. One thousand gigas are a tera. One thousand teras one peta, which is equal to 10 to the 15th.
2. Abbreviation of "Grin," commonly typed within pointy brackets as <G>, at the end of an item uploaded to a BBS (Bulletin Board System), where the sender wants to make sure that readers realize that the message was meant to be humorous or sarcastic, and not to be taken literally. Example: "If my wife makes meat loaf one more time, I'm going to cut her fingers off <G>." Usage is similar to appending Wayne's-World usage of "Not" to reverse the meaning of a sentence.

G Recommendations A series of standards defined by the ITU-T covering transmission facilities. Namely: G.703 transmission facilities running at 2.048 megabits/second (E1) and 64 kilobits per second. G.703 is the ITU-T standard 1984 current version for the physical and logical traits of transmission over digital circuits. G.703 now includes specifications for the US 1.544 megabits per second as well as the European 2.048 megabits/second, and circuits with larger bandwidths on both continents. G.703 is still generally used to refer to the standard for 2.048 megabits per second; G.821 is the ITU-T Recommendation that specifies performance criteria for digital circuits for ISDN.

G-Style Handset A G-style handset is a standard round screw-in, screw-out handset, as compared to the K-style handset, which is the newer square handset with the two screws in the middle.

G,D&R Abbreviation for "Grinning, Ducking and Running;" commonly used on E-mail and BBSs (Bulletin Board Systems).

G.703 ITU-T Recommendation G.703, "Physical/Electrical Characteristics of Hierarchical Digital Interfaces". See G RECOMMENDATIONS.

G.704 ITU-T Recommendation G.704, "Synchronous Frame Structures Used at Primary and Secondary Hierarchy Levels".

G.711 An algorithm used for digital telephone sets on digital PBX and ISDN channels, which produces digital audio at 64 Kbps. Support for this algorithm is required for ITU-T compliant videoconferencing (the H.320 standard).

G.721 This algorithm is used for digital audio at 64 Kbps.

G.722 This algorithm produces digital audio at a rate of 32 Kbps. G.722 is optional for H.320 compliance. See H.320.

G.723 This algorithm is for compressed digital audio over POTS line — Plain Old Telephone Lines. It is the voice part of H.324. See H.324.

G.728 This algorithm produces digital audio at a rate of 16 Kbps and, like G.722 is optional for H.320 compliance.

G.729 The International Telecommunications Union's standard voice algorithm — CS-ACELP (Conjugate Structure Algebraic Code Excited Linear Predictive) voice algorithm for the coding of speech signals in telecommunications networks. G.729 is supported by, inter alia, AT&T, France Telecom and Japan's NTT. See V.70, the specification for DSVD, which uses G.729.

G.804 ITU-T Recommendation G.804, "ATM Cell Mapping into Plesiochronous Digital Hierarchy (PDH)".

G.DSVD Voice digitizer for V.DSVD.

G.O.O.D. Job A "Get-Out-Of-Debt" job. A well-paying position you take to pay off your debts, and one you'll quit as soon as you're solvent.

GA 1. Generally Available or General Availability. A vague term manufacturers use to refer to when their new product will be generally available.
2. Abbreviation for "Go Ahead," used in real-time computer communications to indicate that you have finished a sentence and are awaiting a reply.

GAB Group Access Bridging. A service for bridging of multiple calls to create a conference call.

Gaff The spur on a telephone lineman's climbing iron.

Gain The increase in signaling power that occurs as the signal is boosted by an electronic device. It's measured in decibels (dB).

Gain Hits A cause of errors in data transmission over phone lines. Usually the signal surges more than 3dB and lasts for more than four milliseconds. AT&T's standard calls for eight or fewer gain hits in a 15-minute period.

Galaxy The brand name for an early Rockwell automatic call distributor.

Gallium Arsenide A substance from which microprocessor and memory chips are made. Compared with silicon, GaAs is three to ten times faster or, depending on its speed, uses as little as one-tenth the power; GaAs can detect, emit and convert light into electrical signals, opening the possibility of providing optoelectronic properties on a single chip; GaAs can resist up to 10,000 times the radiation; GaAs can withstand operating temperatures of 200 degrees Centigrade, and GaAs have a higher electron mobility.

Galvanometer A delicate instrument used for measuring minute currents.

Galvo Man Telco-talk for Galvanometer Man, a technician who uses a galvanometer to find and repair circuit faults. It's common for the phone company rep to tell you that your new lines can't be installed when promised because the galvo man hasn't finished his work.

Gammic Ferric Oxide The type of magnetic particle used in conventional floppy disks.

GAN Global Area Network.

Gap An open space in a circuit through which a condenser discharges for producing electric oscillations.

Gap Loss That optical power loss caused by a space between axially aligned fibers. For waveguide-to-waveguide coupling, it is commonly called "longitudinal offset loss."

Garage Silicon Valley, according to contemporary lore, started in a garage in Palo Alto in 1939. In that year Bill Hewlett and Dave Packard started Hewlett Packard with $538. Hewlett was the inventor and Packard the manager. Their first product, an audio oscillator, was an immediate success. Walt Disney used it in making Fantasia.

Garbage Collection Routine that searches memory for program segments or data that are no longer active in order to reclaim that space.

Garbage In, Garbage Out GIGO. If the input data is wrong or inaccurate, the output data will be inaccurate or wrong. GIGO is problem with data entered by hand into computer systems. Ask yourself how many times you've received "junk" mail with the wrong spelling of your name? That's called Garbage In, Garbage Out.

Garbitrage Sending garbage from one city to another, usually organized by garbitrageurs on the phone.

Gas Pressurization A method for preventing water from entering openings in splice closures or cable sheaths by keeping the cables under pressure with dry gas.

Gas Tube A method of protecting phone lines and phone equipment from high voltage caused by lightning strikes. See CARBON BLOCK (another protection technology) for a more detailed explanation.

Gaseous Conductors The gases which, when ionized by an electric field, permit the passage of an electric current.

Gate This term is typically used in Automatic Call Distributors, devices used for handling many incoming telephone calls. Gate refers to a telephone trunk or business transaction grouping that may be handled by one group of telephone answerers (called attendants, operators, agents or telemarketers). That one group of telephone answerers is called "the gate." All calls coming into that gate can, theoretically, be handled by any of the telephone answerers. A telephone call is homogeneous throughout the gate. An automatic call distributor may have one gate — all calls coming in can be handled by everyone. Or it may have many gates, each one consisting of the line (or lines) bringing the call in — e.g. Band 5 WATS, New York City foreign exchange line. Or it may have two gates — one for orders and one for service. ACDs with multiple gates will establish rules for moving the calls between the gates, should one gate become overloaded.

Gate Array A circuit consisting of an array of logic gates aligned on a substrate (a piece of silicon) in a regular pattern.

Gate Assignments Used in context of ACD (Automatic Call Distribution) equipment. Gates are made up of trunks that require similar agent processing. Individual agents can be reassigned from one gate to another gate by the customer via the supervisory control and display station. Also called splits.

Gate D Gateway Daemon. A popular routing software package which supports multiple routing protocols. Developed and maintained by the GateDaemon Consortium at Cornell University.

Gateway A gateway is what it sounds like. It's an entrance and exit into a communications network. That "communications network" may be huge, for example, at the point where AT&T Communications ends and Comsat begins — for taking my satellite call overseas. Gateways may be small — between one LAN and another LAN. Technically, a gateway is an electronic repeater device that intercepts and steers electrical signals from one network to another. Generally, the gateway includes a signal conditioner which filters out unwanted noise and controls characters. In data networks, gateways are typically a node on both two networks that connects two otherwise incompatible networks. For example, PC users on a local area network may need a gateway to gain access to a mainframe computer since the mainframe does not speak the same language (protocols) as the PCs on the LAN. Thus, gateways on data networks often perform code and protocol conversion processes. Gateways also eliminate duplicate wiring by giving all users on the network access to the mainframe without each having a direct, hard-wired connection. Gateways also connect compatible networks owned by different entities, such as X.25 networks linked by X.75 gateways. Gateways are commonly used to connect people on one network, say a token ring network, with those on a long distance network. According to the OSI model, a gateway is a device that provides mapping at all seven layers of the model. A gateway may be used to interface between two incompatible electronic mail systems or for transferring data files from one system to another. Electronic mail systems that sit on local area networks often have gateways into bigger e-mail systems, like Internet or MCI Mail. For example, I might use MCI Mail to send a e-mail to someone's internal LAN e-mail. It might travel from MCI Mail to Internet via a gateway and then from Internet via another gateway to the company's e-mail on its own LAN.

Gateway City A city where international calls must be routed. New York, Washington, DC, Miami, New Orleans, and San Francisco are the five gateway cities in the United States.

Gating 1. Enabling or disabling a signal through applied logic. If it's turned on, the signal gets through. If not, the signal doesn't get through.
2. The process of selecting only those portions of a wave between specified time intervals or between specified amplitude limits.

Gauge A term for specifying the thickness (diameter) of cables. Thicker cables have a lower number in the American Wire Gauge (AWG) scale. (It doesn't make sense. But that's the way it is.) Thicker gauge cables can carry phone conversations further and more cleanly than thinner gauge cable. But thicker cables cost more and take up more room, especially when you bundle them together and put them in a duct. When buying a phone system it is good to specify the thickness of the cables that will be installed — especially if some of your extensions will be a great distance from the central telephone switch, if you intend to carry high-speed data on them or you intend to live with your cabling scheme for more than a few months. You should, of course, not only specify the cable's thickness, but also whether it's stranded or solid core, coax, etc. Gauge is but one part of a cable description. See AWG.

Gauge, Wire The method of specifying the size of wire. The two important American gauges are the American Wire Gauge (Brown and Sharpe) and the Steel Wire Gauge.

Gauss The unit of magnetic field intensity in terms of the lines of force per square centimeter.

Gaussian Noise Gaussian noise is white noise uniform across the whole range of frequencies. "Gaussian" refers to the measurement of the noise. It is essentially the assumption you make when working on the problem of noise mathematically.

Gazillion An extremely large, indetermined amount.

NEWTON'S TELECOM DICTIONARY

GBCS AT&T Global Business Communications Systems. The new name for the merger of two divisions into one division now responsible for all the business telephone system sales of AT&T worldwide. The two divisions that were merged into GBCS were called General Business Systems (office systems 80 lines and smaller) and Business Communications Systems (systems bigger than 80 lines). The merger happened during the summer of 1992.
GBH Group Busy Hour.
GCAC An ATM term. Generic Connection Admission Control: This is a process to determine if a link has potentially enough resources to support a connection.
GCRA An ATM term. Generic Cell Rate Algorithm: The GCRA is used to define conformance with respect to the traffic contract of the connection. For each cell arrival the GCRA determines whether the cell conforms to the traffic contract. The UPC function may implement the GCRA, or one or more equivalent algorithms to enforce conformance. The GCRA is defined with two parameters: the Increment (I) and the Limit (L).
GCT Greenwich Civil Time.
GDDM An SNA definition: Graphical Data Display Manager (GDDM) system software used for graphics display and printer devices and performs the same functions as QuickDraw in Macintosh computers.
GDF Group Distribution Frame.
GDI Graphics Device Interface. The part of Windows that allows applications to draw on screens, printers, and other output devices. The GDI provides hundreds of convenient functions for drawing lines, circles, and polygons; rendering fonts; querying devices for their output capabilities; and more.
GDMO Guidelines for the Definition of Managed Objects.
Gender Connectors, plugs and receptacles are assigned a gender to describe their physical type. Ones with pins are male, and those with holes into which the male pins slide are female. See GENDER BENDER.
Gender Bender A device which changes the gender of a connector, plug or receptacle. A gender bender is typically a small plug with all male pins on one side and all male pins on the other. By plugging a female connector into one side of a gender bender, you've effectively changed the female gender of the cable to male. Alternatively, a gender bender could be female on either side. But a gender bender must be the same on both sides.
Gender Changer Another name for a gender bender. See GENDER BENDER.
General Availability How a product gets to market varies from one company to another. But typically, along the way, there's something called an alpha — the first version of hardware or software. It typically has so many bugs you only let your employees play with it. A beta is the next version. It's a pre-release version and selected customers (and the press) become your guinea pigs. They give you feedback. After beta, and when the bugs are removed and the features have been fine-honed, comes "general availability." That's when the product is finally available for buying by the general public.
General Call The letters CQ in the international code and used as a general inquiry call.
General Magic An alliance of Apple, AT&T, Fujitsu, Matsushita, Motorola, Philips, Sony and Toshiba based in Mountain View, CA to establish a new class of personal communication products and services. Magic Cap and Telescript technologies are at the core of the company's offerings. The company's first public announcement was February 8, 1993. General Magic started in January 1989. Apple Computer established a project led by Marc Porat to develop business opportunities beyond traditional personal computing. The team examined how people will need to deal with the complexities of everyday life in the coming century. Code-named Paradigm and Pocket Crystal, the project evolved into a set of comprehensive plans, including products and services that would require expertise in computing, communications and consumer electronics. It became clear that no one company could deliver all the components of personal intelligent communications. In February 1991, Sony and Motorola joined Apple, becoming investors in General Magic and licensees of its technology. AT&T joined the alliance in January of 1992, Philips in November of 1992 and Matsushita in January of 1993. Sony, Motorola and Philips are building devices using Magic Cap to support personal intelligent communications. AT&T is providing a new kind of public messaging service based on Telescript. Apple will incorporate General Magic technologies into its products. See TELESCRIPT.
General Protection Fault GPF. A General Protection Fault is an indication that Windows 3.xx has tried to assign two or more programs to the same area in memory. Obviously that's not possible, since two things can't occupy the same area in memory. As a result, your screen stops and says "General Protection Fault." If you can save what you're doing, do it. If you have other programs open, try and save the material in them. Close Windows and then do a cold reboot. Do not continue to work after you have received a General Protection Fault. You must reboot. Better do a cold reboot, too.
General Purpose Network An AT&T term for a network suitable for carrying many forms of communication — voice and data, circuit and packet, image, sensor or signaling, for example.
General Release When software is finally finished and ready to be sold to the general public, it said to be in "General Release." Before it is in general release, it is still in beta. Beta software is not alleged to be bug-free. General release software is meant to be bug-free. Sometimes it is.
Generations, Computer As computers have improved, so the industry's pundits have assigned "generations" to those improvements. The concept of generations is not perfect nor finite. Here's our best shot on generations in computers:
• First generation: 1951-1958, core memory 8 Kbytes to 32 Kbytes.
• Second generation: 1958-1964, transistor technology, memory 32 Kbytes to 64 Kbytes.
• Third generation: 1964-1975, integrated circuitry.
• Fourth generation: 1975-date, non procedural languages, software driven.
• Fifth generation: into the 1990s, natural language programming, parallel processing and super computing.
• Sixth generation: in the 2000, will process knowledge rather than data.
Generations, PBX As PBXs have improved, so the industry's pundits have assigned "generations" to those improvements, as they did in computers. The concept of generations is not perfect nor finite. Here's our best shot on generations in PBXs:

NEWTON'S TELECOM DICTIONARY

First generation 1920s to the late 1960s. Step-by-step mechanical equipment. The first and the last of the step-by-step Bell PBXs switches was called a 701. Lee Goeller says the 701 "was the best PBX ever built. It was infinitely flexible. It was just too BIG. In fact, it was usually bigger than the office it served. This era of the stepper will be remembered as the era the Bell System was intact and had the gaul to rent operator chairs."

Second Generation Late 1960s: Bell 801 reed relay switch. Stromberg Carlson 800 series reed relay switch. GTE had a series, also. Reed relay switches were not very popular.

Third Generation 1974 and 1975: Rolm introduces its first CBX, an electronic, solid-state PBX. AT&T introduces Dimension. Digital Telephone Systems introduced the D1200. Northern introduced its first stored-program controlled SL-1. Some of these PBXs switched voice digitally, though they used different techniques, including PCM, PAM, and Delta Modulation. The codecs were in the switch, not in the instruments.

Fourth generation early 1980s. Distributed processing. Northern, Rolm and NEC and others introduced remote modules — slaves to the master switch at headquarters. These switches also added the capability of handling data without using modems. Switches like Lexar and InteCom were designed from the beginning to handle data without modems, thus requiring digital capability out to the set.

Fifth generation CXC, Anderson Jacobson and Ztel and others called themselves "fourth generation." When they started to fail, some people called them the "fifth generation." It wasn't clear exactly what that generation was. But they all got lots of publicity and the PBXs from CXC, Anderson Jacobson and Ztel ultimately failed.

Sixth generation Networked PBXs. Sit in New York. Operate your national network as if it were in the same building. Bingo, you can transfer calls across the country. All your messaging is the same wherever you sit on the network. Lee Goeller, however, says you can network stepper PBXs. In fact, in 1971 he says he managed one of the biggest integrated voice and data networks in the US using step-by-step electromechanical PBXs (701s made by AT&T). It was the world's largest dial tandem network. He had 63 different locations and three hubs.

Seventh generation Open Architecture. You can now program your own PBX. For more, see the NORSTAR COMMAND SET.

Eighth generation Dumb switches. You can now buy completely dumb phone systems which are just basically switches. To get them to do anything they require an external computer (and software programming that computer) to drive the dumb switch. Often the "driving" is done through one or more serial ports.

In reality, the concept of generations amongst PBXs is very flimsy. But it's the stuff dictionaries are made of.

Generator A machine which converts mechanical energy, such as the power from a piston engine into electrical energy.

Generic Flow Control Field GFC. Four priority bits in an ATM header. Setting any of the bits in the GFC field tells the target end-station that the switch may implement some form of congestion control.

Generic Program A set of instructions for an ESS or electronic PBX that is the same for all installations of that equipment. Detailed differences for each individual installation are listed in a separate parameter table. Here's a more formal definition, from Bellcore. A generic program is a set of instructions for an electronic switching system or operations system that is the same for all central offices using that exact type of system. Detailed differences for each individual office are usually listed in a separated parameter table.

Generic Services Framework GSF. Generic Services Framework is a set of software designs being implemented by Bell Northern Research (a subsidiary of Northern Telecom) to accelerate the development and testing of new features and provide a platform for the later stages of DMS SuperNode interworking with Advanced Intelligent Networking (AIN). GSF applies the principles of object-oriented programming to DMS SuperNode software, and delivers enhancements that simplify feature development and reduce testing needs. The GSF has three distinct elements:

1. Call Separation - The new architecture uses separate software data and processes to handle the two "halves" of a call (originating and terminating).
2. GSF Agent Interworking Protocol (AIP) - The AIP uses a standardized set of instructions to allows the two call halves to communicate.
3. Event-Driven Call Processing (EDCP) - Just as the AIP mediates communications between call halves, EDCP handles communications within each call half. EDCP also uses a standardized set of instructions to simplify messaging.

Genlock A video devices which determines the exact moment at which a video frame begins. Genlock allows multiple devices (video recorders cameras, etc.,) to be used together with precise timing so that they capture a scene in unison.

Geographic Information Systems GIS. Computer applications involving the storage and manipulation of electronic maps and related data. Applications include resource planning, commercial development, military mapping, etc.

Geographic Interface First there was the ASCII interface - a screen containing nothing but ASCII letters. You saw them on airline reservation terminals. Just plain boring green type against a dark, unlit, black background. Then you got the GUI - the Graphical User Interface. Windows is the most famous GUI. It used icons to represent actions. An eraser to indicate you could remove something. A calendar to indicate that you could enter your day's schedules, etc. Now there is the "Geographic Interface." It attempts to depict objects from the real world — or at least some circumscribed part of it, like your office. The first of these products was Magic Cap from General Magic. Another was the opening screen for Apple's eWorld on-line service. Another is Novell's Corsair technology. Click on a filing cabinet to get to its database contents. Click on a bloodhound find something. One idea behind these geographic interfaces is that there should be a libraries of objects. A set useful for an auto mechanic. Another set useful for a bond trader. The idea is that everyone gets to choose the bits and pieces of the geographic interface that he or she is most comfortable with. Some of these new geographic interfaces are endearing. Some are tiresome. We'll see if they endure.

Geometric Dilution of Precision The multiplicative factor that modifies the ranging error. It is caused solely by the geometry between the user and his set of GPS (Global Positioning System) satellites. Known as DOP (Dilution Of Precision) or GDOP. See GPS.

GEOPort GEOPort, developed by Apple Computer (and this is Apple's definition), is a plug-and-play serial interface which is backward compatible with the serial ports used in most personal computers, but offers over 200 times the bandwidth. The serial communications architecture of Geoport is optimized for computer-telephony integration:
It allows any telephone (or telephone line, up to T1/E1 rates) to be connected to any computer, in any country in the world.
It supports any set of Telephony APIs (application programmatic interfaces) such as AT&T/Novell's TSAPI, IBM's CallPath, Microsoft's TAPI, or Apple's Telephone Manager.
It is inexpensive to implement,and uses existing technology.
It supports any type of information: computer data, voice, fax, modem, voice, video.
It allows multiple simultaneous streams of information — including real time information like voice and video — to pass through a telephone connection in order to be processed by the computer.
These services will allow vendors and developers the opportunity to offer such features as:
An integrated mail-box for voice mail, e-mail and facsimiles.
Fax and modem capability over a digital telephone connection, like that found in an ISDN line or in many PBX environments.
Application sharing or other simultaneous voice and data applications over conventional phones lines.
Video conferencing over a PBX or ISDN connection.
Telephone assistant services, like automated call screening, call forwarding, and call tracking. See also GEOPORT TECHNOLOGY MODEL and USB.

GEOPort Technology Model The definition has been provided by Apple: GEOPort is a cross platform, plug-and-play media communications interface for corporate desktop users. It delivers voice, data, audio and video communications over any analog (POTS) or digital (PBX) telephone line, including ISDN, to any computing device — whether a personal computer, Personal Digital Assistant (PDA) or workstation — running any operating system. Additionally, GEOPort is a high performance serial interface for devices like high-speed modems, hand-held scanners, and videophones whose current capabilities are limited by the serial interface commonly used in most of today's computing devices. GEOPort supports speeds of over 2Mb/sec, which is over 200 times the performance of the serial interface found in most personal computers. And unlike higher speed interfaces such as Small Computer System Interface (SCSI), which requires bulky cables, terminators, and configuration IDs, GEOPort provides a flexible means of simultaneously attaching multiple peripherals through a single compact mini-DIN connector.
There are three main parts to the GEOPort architecture:
First, the GEOPort Adapter provides the electrical, physical and logical connection to a given peripheral, telephone line or telephone set. Some adapters will support just a single function, such as providing a localized telephone interconnect mechanism, while others will provide several services, such as a combination of several telephone or peripheral connections.
Second, the GEOPort communication protocols provide the basic initialization and set-up routines which allow the various adapters to configure themselves for operation in the various host environments. This configuration occurs after the adapter is connected to the machine and the machine itself is turned on.
Finally, on the host, GEOPort provides a highly scalable hardware and software architecture. The basic hardware requirements are minimal, and provide vendors with a great deal of flexibility in designing their Input/Output (I/O) architecture as well as their memory and real-time architectures. Also significant is the GEOPort serial driver multiplexes and demultiplexes the various data streams going to and coming from the adapter. It also connects the individual data streams with the appropriate host hardware or software interfaces. Since different vendors will choose to include different real-time services based on cost, and other considerations, this driver software, while similar for all machines, will be different for different hosts. See also USB.

Geostationary Orbit An orbit, any point on which has a period equal to the average rotational period of the Earth, is called a synchronous orbit. If the orbit is also circular and equatorial, it is called a stationary or geostationary orbit.

Geostationary Satellite A satellite in geostationary orbit, also called geosynchronous orbit. A satellite placed in an geosynchronous orbit — 22,300 miles (or 42,164 kilometers) directly over the earth's equator — will appear to be stationary in the sky, turning synchronously with the earth. This means you can plant a satellite receiving/transmitting antenna on the ground, and point it at that one place in the sky to receive signals from and transmit signals to "the bird", as satellites are sometimes called. Most communications satellites are in geostationary or geosynchronous orbit. The Russians have some satellites that orbit the earth and require antennae which move. These satellites are used to transmit to far northern communities which are difficult to reach with normal geosynchronous satellites.

Geosynchronous Orbit Synchronous with the Earth. An orbit above the earth's equator where satellites circle at the same rate as the earth's rotation, thereby appearing stationary to an earth-bound observer. See also GEOSTATIONARY ORBIT.

Get A Life What your kids say to you when you start talking too much about the information superhighway, or the Internet, of fiber optic, or something that doesn't interest them.

GFC Generic Flow Control. GFC is a field in the ATM header which can be used to provide local functions (e.g., flow control). It has local significance only and the value encoded in the field is not carried end-to-end.

GFCI Ground Fault Circuit Interrupter. A device intended to interrupt the electrical circuit when the fault current to ground exceeds a predetermined value (usually 4 to 6 milliamps) that is less than required to operate the overcurrent protection (fuse or breaker) for the circuit. This device is intended to protect people against electrocution. It does not protect against fire from circuit overload. GFCI outlets are typically installed in bathrooms, kitchens and garages because the presence of water in these area increases the possibility of electric shock. Sometime GFCI circuits are incorrectly wired. The way to find out if your GFCI is wired correctly is to press the test button on its face. This should shut off power to the GFCI outlet and to those outlets connected to it.

GFI Group Format Identifier. In packet switching, refers to the first four bits in a packet header. Contains the Q bit, D bit and modulus value.

GFLOPS One billion FLoating point Operations Per Second. (G stands for GIGA, meaning billion). Today's fastest supercomputers are able to maintain a sustained throughput of over one billion floating point operations per second (GFLOPS) while performing real-world applications. By contrast, a 25-MHz 486 personal computer can sustain about one million floating operations per second (one MFLOP), or about one-thousandth the throughput of a supercomputer. See also G.

GGP Gateway to Gateway Protocol. The protocol that core gateways use to exchange routing information, GGP implements a distributed shortest path routing computation.

GHz One billion, or one thousand million, hertz, or cycles per second. See also BANDWIDTH.

GIF pronounced JIF, as in peanut butter. GIF stands for Graphics Interface Format, which is pronounced "Jiff." It is a format for encoding images (pictures, drawings, etc.) into bits so that a computer can "read" the GIF file and throw the picture up on a computer screen. GIF can only handle 256 colors. The major advantage of GIF is that it compresses the image (photograph, drawing, etc.), thus making it faster to transmit across phone lines. CompuServe, the on-line service, invented GIF in 1987. There are thousands of GIF images on CompuServe, which its subscribers can download. GIF is based on a mathematical algorithm called LZW, which is a set of mathematical formulae used to compress images into GIF files. Unisys has a patent on LZW. The GIF format has become a standard in the electronic bulletin board world, and on the Internet. Most of the graphics you see on the Internet are stored in GIF format. GIF is a popular format also for photographs of scantily-clad women. According to Jack Rickard, publisher of the excellent Boardwatch Magazine, "some of the photographs are reasonably good, but most feature strikingly plain women rather artlessly photographed by those whose higher calling is probably more aptly found in the building trades or automotive repair." Other graphics formats include BMP, PCX, TIFF, etc. See also INTERNET.

Giga Prefix meaning one billion, which is one thousand million. 1,000,000,000. See also GFLOPS.

Gigabit Ethernet Alliance A multi-vendor forum committed to driving the industry's adoption of networking standards at up to 1,000 megabits per second. The forum supports the CSMA/CD (Carrier Sense Multiple Access with Collision Detection) protocol of the original Ethernet standard. According to the Alliance, gigabit Ethernet will initially be deployed in backbone environments as the preferred interconnection for switches which support multiple transmission speeds between Ethernet segments (10 Mbps and 100 Mbps). Technical proposals developed by Alliance members will be submitted to the IEEE 802.3z standards committee, furthering efforts to standardize 1,000 Mbps Ethernet technology.

Gigabits One thousand million bits. One billion bits. Or more precisely 1,073,741,824 bits. See GHz.

Gigabyte GB. One thousand million bytes. One billion bytes. More precisely, 1,073,741,824 bytes. That's enough space to hold 200 copies of this dictionary. Here is a summary of sizes:

MH = Megabyte (10 to the 6th power) GH = Gigabyte (10 to the 9th power) TH = Terabyte (10 to the 12th power) PH = Petabyte (10 to the 15th power) EB = Exabyte (10 to the 18th power)

Gigahertz GHz. A measurement of the frequency of a signal equivalent to one billion cycles per second, or one thousand million cycles per second.

GII Global Information Infrastructure, a term first advocated in March, 1994 and since used as a concept around which to form many international committees. For more, see the IEEE Communications Magazine, June 1996.

GIM Group Identification Mark. The Group ID mark is a two digit number used by cellular sites other than your home system to determine if your cellular phone should be allowed to make phone calls, i.e. access on "roam" status. This feature is not yet fully implemented. As cellular systems are upgraded, the GIM will be on line real time, requiring all NAM information, including the Mobile Identification Number (MIN), to be validated before a subscriber is allowed to call outside of their home area.

GIS Geographic Information Services. Computer applications involving the storage and manipulation of electronic maps and related data. Applications include resource planning, commercial development, military mapping, etc. Raw input comes often from satellite photographs.

GIX Global Internet eXchange. A common routing exchange point which allows pairs of networks to implement agreed upon routing policies. The GIX is intended to allow maximum connectivity to the Internet for networks all over the world.

GL Graphics Library.

Glare Glare occurs when both ends of a telephone line or trunk are seized at the same time for different purposes or by different users. Most embarrassing. See GLARE RESOLUTION.

Glare Resolution Ability of a system to ensure that if a trunk is seized by both ends simultaneously, one caller is given priority and the other is switched to another trunk.

Glasgal Communications Glasgal publishes one of the best networking and data communications catalogs around. Their annual catalog contains the best technical explanations. Their president and founder, Ralph Glasgal, is an absolutely wonderful person. When he first started, he used to peddle datacom equipment out of catalogs he carried in a used shopping bag, peddling from door to door on Wall Street. He was, thus, one of the original shopping bag men. He's authored a number of books on data communications. And once he authored a column called Baudy Bits. These days he likes music, professionalism and success. He's good at all three. 201-768-8082.

Glass House A colloquial word for a mainframe computer. It derives from the fact that all mainframe computers were once housed in a separate, locked room, with glass windows. You typically needed to pass through heavy security to gain admittance to the glass room.

Glass Insulators Glass insulators were widely used in the 1800s to fasten open wire to telephone poles and to protect insulator pins from moisture so they couldn't conduct electricity. This was a technique developed by the telegraph industry over a 40-year period of experimentation and was one of the few basic telegraph practices carried over into telephone line construction. Insulators are found in a variety of different shapes and colors depending on the time period they were developed and on their application. Most have a greenish color from traces of iron oxide from the sand used to make the glass. When insulator design was in its heyday in the mid-to-late 1800s, hundreds of patents to improve the product were issued. For example, the double petticoat, a second lip on the bottom of the insulator, was added to reduce the amount of moisture that could travel up the inside of the cap. The above explanation courtesy Tellabs of Lisle, IL.

Glass Mount Antenna A type of car phone antenna used in cellular service. A glass mount antenna is glued to a car's rear window. Many window-mounted glass antennas have a break in their wiring. The wiring ends at the inside. There is no electrical connection between the inside of the window and the antenna glued onto the outside of the window. The "connection" is done through signal radiation. This type of antenna is not as efficient as one in which the wire goes unbroken from the radio to the antenna.

Glass Terminal A keyboard and screen that conveys data generated by the user directly to a computer or network without buffering or otherwise acting upon the data, and also returns data unchanged from the computer to the user. This terminal type does not provide for cursor addressing or escape sequences.

GLines GSF Lines.

Glitch A jargon term used in data communications to describe an extraneous bit that has been introduced into a bit stream usually by a noise source. It can also be a problem or a delay. Can be a noun. "What's the glitch?" Or a verb: "Who glitched this thing up?" Glitch is also a momentary interruption in electrical power.

Global Universal. An adjective meaning the whole world. See GLOBAL SEARCH.

Global Access A new service of MCI Mail. It allows you, an MCI Mail user, to use your computer and its modem to dial a local number in a foreign (i.e. non-North American city) and reach a port of a packet switched operation called InfoNet. When you reach InfoNet you will then punch in a few letters and reach MCI Mail in the U.S. You can then leave MCI Mail messages, send telexes, send faxes and send paper mail, i.e. do all the normal services MCI Mail allows you to do. The advantage of Global Access is that you don't have to dial back to the U.S. (which usually doesn't work because of all the garbage on the line) or subscribe to a foreign packet switched operation (they have them in all industrialized countries). Sadly, it usually takes weeks to subscribe to a foreign packet switched operator.

Global Directory Imagine a bunch of local area networks all connected together. Today you log onto one server on one LAN, tell it who you are and what your password is. You want to connect to another server? You have to tell it who you are and what your password is, which may be different from the first time. Global Directory, a feature of NetWare 4.x, gives you a central directory. You establish your user name once and associated with your name is your authorized service on all the connected LANs. This way you don't have to sign on again for another server. If telephone systems can attach to NetWare (see Computer Telephony), then they should also be able to benefit from this central directory — for phone bill allocation, people location, phone moves and changes, etc. The reason NetWare never had a central directory is historical. Novell introduced NetWare in 1983 as software to allow a handful of personal computers to share a single hard disk, which at that stage was a costly and scarce resource. As hard disks became more available, the product evolved to allow the sharing of printers and file servers. It was always designed as a departmental computing solution. It's only recently, with more powerful desktop machines, that the Client-Server LAN concept has become more corporate-wide in concept.

Global Directory Service Allows desktop clients to transparently access data on servers across a network.

Global Network Navigator GNN. An application developed at CERN in Switzerland which provides information about new services available on the Internet, articles about existing services, and an online version of Internet related books. The GNN is a World Wide Web (WWW) based information service.

Global Positioning System See GPS.

Global Satellite System A new breed of high-power satellite systems that propose to offer global telecom services. According to the Economist, they will offer three basic services:

• Voice: Satellite telephones will be able to make calls from anywhere on earth to anywhere else. That could make them especially useful to remote, third-world villages (some of which already use stationary satellite telephones), explorers and disaster-relief teams. Today's mobile telephones depend on earth-bound transmitters, whose technical standards vary from country to country. So business travelers cannot use their mobile phones on international trips. Satellite telephones would make that possible.

• Messaging: Satellite messages have the same global coverage as satellite telephones, but carry text alone, which could be useful for those with laptop computers. Equipped with a small screen like today's pagers, satellite messagers will also receive short messages.

• Tracking: Voice and messaging systems will also tell their users where they are to within a few hundred meters. Combined with the messaging service, the location service could help rescue teams to find stranded adventurers, the police to find stolen cars, exporters to follow the progress of cargoes, and haulage companies to check that drivers are not detouring to the pub. America's military Global Positioning System (GPS) satellites will provide better positioning information to anyone who has a receiver for their signals, but GPS does not carry messages, so such a receiver cannot be used on its own for tracking or rescue. By the mid-1990s GPS receivers will be able to tell people where they are to within 70 meters anywhere in the world, and to within a meter or so in areas where the service is supplemented by a ground-based transmitter.

Global Search A word processing term meaning to automatically find a character or group of characters wherever they appear in a document.

Global Search and Replace A word processing term meaning to automatically find a character or group of characters wherever they appear in a document and replace them with something else.

Global Security Service Provides networkwide security functions, including single log-in to multiple systems.

Global Software Defined Network AT&T's international virtual private network. It provides business customers with point-to-point, two-way voice and voiceband data communications between the US and various overseas countries. Features include:
• International two-way on-net calling. • Abbreviated user defined 7-digit dialing. • Usage sensitivity. • SDN call screening.

Global Title An address such as customer-dialed digits that does not explicitly contain information that would allow routing in the SS7 signaling network, that is, the SCCP translation function is required.

Global Transaction Network See GTN.

GlobalNet A free, electronic amateur bulletin board system network which operates based upon FidoNet technical standards. GlobalNet nodes are located in North America and Europe.

Globalstar Imagine a hand-held, light, low-cost telephone that looks like a cell phone, but works by talking to a satellite. Several companies have proposed a collection of low-orbiting satellites. The idea being the closer the satellite the stronger the signal on the ground and thus the smaller the needed telephone. Companies proposing such systems include Iridium, Teledesic and Globalstar. Here is information on Globalstar, kindly supplied by their public relations department:

Globalstar is a low-earth-orbiting (LEO) satellite-based digital telecommunications system that will offer wireless telephone and other telecommunications services worldwide beginning in 1998. Globalstar will provide low-cost, high quality telephony and other digital telecommunications services such as data transmission, paging, facsimile and position location to areas currently underserved or not served by existing wireline and cellular telecommunications systems.

Globalstar service will be delivered through a 48-satellite LEO constellation that will provide wireless telephone service in virtually every populated area of the world where Globalstar service is authorized by the local telecommunications regulatory authorities. (Globalstar service providers will be required to obtain such approvals before beginning to offer Globalstar service in their territories.) Globalstar will begin launching satellites in the second half of 1997 and will commence initial commercial operations via a 24-satellite constellation in 1998. Full 48-satellite coverage will occur in the first half of 1999. Globalstar expects to require capital of about $2 billion from inception through the full coverage date and expects to begin generating positive cash flow in 1999. Users of Globalstar will make or receive calls using hand-held or vehicle-mounted terminals similar to today's cellular telephones. Because Globalstar will be fully integrated with existing fixed and cellular telephone networks, Globalstar's dual-mode handsets units will be able to switch from conventional cellular telephony to satellite telephony as required. In remote areas with little or no existing wireline telephony, users will make or receive calls through fixed-site telephones, similar either to phone booths or ordinary wireline telephones. Each subscriber terminal will communicate through a satellite to a local Globalstar service provider's interconnection point (a gateway) which will in turn connect into existing telecommunications networks. Based in San Jose, California, Globalstar is a limited partnership founded by Loral Corporation of New York City, and QUALCOMM Inc., of San Diego, California, and includes ten strategic partners representing the world's leading telecommunications service providers and equipment manufacturers. See LEO.

GlobeSpan Trademark for AT&T Paradyne's DSL technology.

Glueware The trend of joining software applications to physical networks through the deal AT&T and Novell have struck to adapt Novell local area networking software to communicate over AT&T's long-distance network. Intel and Microsoft are considering similar arrangements, according to The Wall Street Journal.

GMD Gesellschaft fur Mathematik und Datenverarbeitung: a German government computer science research institute.

GMDSS Global Maritime Distress and Safety System.

GMSC Gateway Mobile services Switching Center. A wireless telecommunications term. A means to route a mobile station call to the MSC (Mobile Switching Center) containing the called party's HLR (Home Location Register).

GMSK Gaussian Minimum Shift Keying. A wireless telecommunications term. A means of radio wave modulation used specifically in the GSM (Global System for Mobile Communications) air interface.

GNE Gateway Network Element. A SONET Network Element (NE) that provides a direct OS/NE interface. The GNE provides an indirect OS/NE interface. The GNE provides an indirect OS/NE interface for other NEs in its own management subnetwork.

GNN See GLOBAL NETWORK NAVIGATOR.

GNU A non-USENET news group hierarchy devoted to the discussion of things related to the GNU Project of the Free Software Foundation.

GNX Development Tools A set of fax software development tools offered by National Semiconductor.

Goats People in our population whose voices cannot — under any circumstances — be recognized by voice recognition machines. No one seems to know where this term came from. I first heard it from Ed Tagg, VP Engineering of Voice Control Systems, Dallas, TX.

Go Local A command typically given in a microcomputer asynchronous data communications program to tell the computer that it will connect to something without a modem over a null modem cable. The command "Go Local" also refers to modem connections and can tell one modem to overlook some of the handshaking and assume it's already taken place.

GO-MVIP Global Organization for MVIP. GO-MVIP is a non-profit trade association established in 1993 to take over coordination of the Multi-Vendor Integration Protocol. Information is available by calling 1-800-NOW-MVIP (800-669-6847). See MVIP.

Going Cyrillic Going cyrillic is when a graphical display (LED panel, bit-mapped text and graphics) starts to display garbage. "The thing just went cyrillic on me."

GOLAY One of the communications protocols used between paging towers and the mobile pagers/receivers/beepers themselves. Other protocols are POCSAG, ERMES, FLEX and REFLEX. The same paging tower equipment can transmit messages one moment in GOLAY and the next moment in ERMES, or any of the other protocols.

Gold Codes Named after Robert Gold, Gold codes are used in direct-sequence spread spectrum transmission. Each transmitted signal is assigned to unique Gold code, which correlates the original information signal into a pseudo random sequence. This sequence is then modulated and transmitted as a spread spectrum signal. The receiver, which uses the same Gold code, is able to de-correlate the spread spectrum signal and recover the original information. Gold codes possess two very desirable qualities which are important in a high quality communications system. The first quality is called "auto-correlation." When a receiver is subjected to several spread spectrum signals, it must extract the desired information and reject the remainder. Auto-correlation allows for an excellent signal recovery when the transmitted code matches the reference code in the receiver.

The second quality is called "cross-correlation." Cross-correlation simply means that an undesired transmitted code cannot produce a false match at the receiver. The advantage of Gold codes is that they consistently exhibit superior cross-correlation performance, which is critical in an environment with multiple transmitters, each representing a potential interfering source. See SPREAD SPECTRUM.

Gold Disk You have finally finished your new software. And you're now ready to go into production, to have your software reproduced onto disks you can sell. That final, completed version of software, from which you reproduce commercial production disks, is called your "Gold Disk." From what I can see, it's a "Gold Disk," though there may be more than one disk in your package.

Gold Number Also called vanity number. A service of Nynex. It's a phone number that's easy for your customers to remember, e.g. 555-LIMO. But occasionally hard for them to dial. Tip: If you buy a vanity number, make sure it doesn't have numbers in it. 555-LIMO is harder to remember than CAR-RENT.

Golden Rolodex, The The small handful of experts who are always quoted in news stories and asked to be quests on discussion shows. Example: Henry Kissinger appears to be in The Golden Rolodex under foreign policy.

Good Condition A term used in the secondary telecom equipment business. One step up from fair condition. Product is in working condition and looks good.

Gopher Programmers at the University of Minnesota — home of the Golden Gophers —- developed a kind of menu to "go for" items on Internet, bypassing complicated addresses and commands. If you want to connect to the State Library in Albany you select that option off the menu. Time Magazine once described Gopher as a tool used for "tunneling quickly from one place on the Internet to another." Hence the term Gopher. See also GOPHERSPACE.

Gopherspace The vast number of servers and areas of interest accessible through the Internet gopher. See GOPHER.

GORIZONT The Russian geostationary telecommunications satellite.

GOS Grade of Service. Telecom traffic term. The probability that a random call will be delayed, or receive a busy signal, under a given traffic load.

GOSIP Government Open Systems Interconnection Profile. The U.S. government's version of the OSI protocols. GOSIP compliance is typically a requirement in government networking purchases. GOSIP addresses communication and inter operation among end systems and intermediate systems. It provides specific peer-level, process-to-process and terminal access functionality between computer system users within and across government agencies.

Gotcha Law, The This law comes from "Got You." It's my favorite law. Think about life. You buy a beautiful laptop. You fall in love with it. You want to buy one for your wife. When you go back to the store all hot to do your wife a wonderful favor, they tell you, "Sorry, it's been discontinued." Gotcha! Compare with MURPHY's LAW.

Government Radio Publications Publications on radio subjects by the Bureau of Standards and Signal Corps and sold by the superintendent of Documents, Government Printing Office, Washington D.C.

GPA General Purpose Adapter. An AT&T Merlin device that connects an analog multi line telephone to optional equipment such as an answering machine or a FAX machine.

GPF General Protection Fault. A problem that happens too often under Windows 3.xx. A General Protection Fault is an indication that Windows 3.xx has tried to assign two or more programs to the same area in memory. Obviously that's not possible, since two things can't occupy the same area in memory. As a result, your screen stops and says "General Protection Fault." If you can save what you're doing, do it. If you have other programs open, try and save the material in them. Close Windows and then do a cold reboot. Do not continue to work after you have received a General Protection Fault. You must reboot. Better do a cold reboot, too.

GPI GammaFax Programmers Interface. C-level programming language. Real-time applications for fax switched and gateways.

GPIB An interconnection bus and protocol that allows connection of multiple instruments in a network under the direction of a controller. Also known as the IEEE 488 bus, it allows test engineers to configure complete systems from off-the-shelf instruments and control those systems with a single, proven interface.

GPS Global Positioning System. A system to allow us all to figure out precisely where we are anywhere on earth. The GPS will eventually consist of a constellation of 21 satellites orbiting the earth at 10,900 miles — they circle the earth twice a day. In a way, you can think of them as "man-made stars" to replace the stars that we've traditionally used for navigation. The US Government is investing over $10 billion to build and maintain the system. Applications are almost limitless: Delivery vehicles will be able to pinpoint destinations. Emergency vehicles will be more prompt. Cars will have electronic maps that will instantly show us the way to any destination. Planes will be able to land in zero visibility. GPS is based on satellite ranging. That means we figure our position on earth by measuring our distance from a group of satellites in space. The satellites act as precise reference points. To get your precise position, latitude, longitude and altitude (and in some applications velocity), you need to get a "fix" off at least four satellites. That's the main reason for the 21 satellites.

Each GPS satellite transmits on two frequencies: 1575.42 MHz referred to as L1, and 1227.60 MHz referred to as L2. It transmits a host of somewhat complicated data occupying about 20 MHz of the spectrum on each channel. But basically, it boils down to three items. The satellite transmits its own position, its time, and a long pseudo random noise code (PRN). The noise code is used by the receiver to calculate range. If we know precisely where a satellite is located and our precise range from it, and if we can obtain similar readings from other satellites, we can calculate precisely our own location and even altitude by triangulation. Satellite position and time are derived from on-board celestial navigation equipment and atomic clocks accurate to one second in 300,000 years. But the ranging is the heart of GPS. Both in the receiver, and in the satellite, a very long sequence of apparently random bits are generated. By comparing internal stream of bits in the receiver to the precisely duplicate received bits from the satellite, and "aligning" the two streams, a shift error or displacement can be calculated representing the precise travel time from satellite to receiver. Since the receiver also knows the precise position of the satellite, and its range from the receiver, a simple triangulation calculation can give two dimensional position (lat/long) from three satellites and additional elevation information from a fourth.

There are actually two PRN strings transmitted: a course acquisition code (C/A code) and a precision code (P code). The coarse code sequence consists of 1023 bits repeated every 266 days. But each satellite transmits a seven day segment re-initialized at midnight Saturday/Sunday of each week. By using both codes a very accurate position can be calculated. By transmitting them at different frequencies, even the signal attenuating effects of the ionosphere can to some degree be factored out. At the present time, civilian users are only authorized to use the coarse acquisition code and this is referred to as the GPS Standard Positioning Service (SPS), a best accuracy of about 5 meters. Military users use both the coarse acquisition code and the precision code in what is referred to as the

Precise Positioning Service or PPS accuracy to centimeters. The Department of Defense (DOD) can at any time encrypt the precision code with another secret code on demand. This is referred to as "anti-spoofing" and ensures that no hostile military forces can also use the GPS service. The DOD can also purposely degrade the accuracy of the coarse acquisition code referred to as "selective availability" to about a 100 meter accuracy. But other than during brief test periods and national emergencies, the service is generally available to all. New techniques now make the civilian use of GPS almost as accurate as the military use.

GPS receivers are now accurate to within 70 meters virtually anywhere in the world. How accurate GPS receivers are depends on how many GPS satellite signals they pick up. Some low-cost receivers can pick up five satellites and are thus accurate to within ten meters.

GR Generic Requirement.

Graceful Close Method terminating a connection at the transport layer with no loss of data.

Graceful Degradation A condition in which a system continues to operate, providing services in a degraded mode rather than failing completely. See ABEND.

Grade 1 Cable Twisted pair cables specifically designed for analog voice circuits and data transmissions up to 1 Mbps. Applications — Key systems, analog and digital PBX, low speed data, RS-232, etc.

Grade 2 Cable Twisted pair cables designed to meet the IBM Type 3 specification. These cables are capable of data transmissions at 4 Mbps, IBM 3270, STAR-LAN I, IBM PC Network, ISDN, etc.

Grade 3 Cable Twisted pair grade 3 LAN cables have performance characteristics that permit data transmissions at 10 Mbps. Each have been tested to insure they meet the EIA/TIA 568 emerging standard. Applications — 802.3 10BASE-T at 10 Mbps, STARLAN 10 and 802.5 token ring at 4 Mbps.

Grade 4 Cable The highest quality twisted pair cables available. Super grade cables have been tested up to speeds of 20 Mbps, 802.5 token ring at 4 Mbps and 802.3 10Base-T at 10 Mbps.

Grade 5 Cable These are the IBM-type individually shielded 2 pair twisted data cables. They're currently being tested for data rates at 100 Mbps. Applications — IBM Cabling System, 802.5 token ring at 16 Mbps and FDDI at 100 Mbps.

Grade Of Service GOS. A term associated with telephone service indicating the probability that a call attempted will receive a busy signal, expressed as a decimal fraction. Grade of service may be applied to the busy hour or to some other specified period. A P.01 Grade of Service means the user has a 1% chance of reaching a busy signal. See TRAFFIC ENGINEERING.

Graded Index A characteristic of fiber optic cable in which the core refractive index is varied so that it is high at the center and matches the refractive index of the cladding at the core-cladding boundary. This reduces dispersion, which is fiber's equivalent of fading.

Graded Index Fiber Graded index fiber is a multimode fiber optic cable that is made with progressively lower refractive index fiber toward the outer core. This reduces dispersion, which is fiber's equivalent of fading.

Gradient In graphics, having an area smoothly blend from one color to another, or from black to white, or vice versa. See GRADIENT FILL.

Gradient Fill A computer imaging term. A fill composed of a smooth blend from a starting color to an ending color. There are many variations on this theme. Most programs let you apply textures, and others have "smart" gradient fill routines that lend a three-dimensional appearance.

Grandfather Clause See GRANDFATHERED.

Grandfather Tape The first backup of a program or a data record, saved so that you can always go back to step one if something goes wrong.

Grandfathered Something that has a right to be a thing or own a thing by reason of it being or owning that thing before laws or rules were introduced to formalize the process. The derivation of term goes back to the Civil War. Grandfathering was a provision in several southern state constitutions designed to enfranchise poor whites and disfranchise blacks by waiving voting requirements for descendants of men who voted before 1867. The word derives from a "grandfather clause." Grandfather clauses stated that the right to vote was only available to those Americans whose grandfathers had been eligible to vote. These clauses were used, primarily in the South, to discriminate against blacks and immigrants shortly after Lincoln's issuance of the Emancipation Proclamation and congressional ratification of the Fourteenth Amendment. As a result, "grandfather" has come to mean something allowable because it was allowable before prohibitive legislation. See also GRANDFATHERED EQUIPMENT.

Grandfathered Equipment Non-FCC registered telephone equipment that was directly connected to the telecommunications network without a phone company-provided protective connecting arrangement (PCA) prior to the formalized FCC registration program. See GRANDFATHERED.

Granularity Microsoft jargon for complexity. For example, when you "achieve granularity," you grasp the complexity of the issue or problem.

Grapevine The name of the Xerox Corporation internal company network.

Graphic Character A character, other than a character representing a control function (like Ctrl G being, in WordStar and dBASE nomenclature, to delete the character on the right) that has a visual representation normally handwritten, printed, or displayed, and that has a coded representation consisting of one or more bit combinations.

Graphic Equalizer A device that permits altering the tonal balance of an audio component/system at several frequency levels.

Graphical User Interface GUI. A fancy name probably originated by Microsoft which lets users get into and out of programs and manipulate the commands in those programs by using a pointing device (often a mouse). Microsoft's own definition is more elaborate. Namely that GUI puts visual metaphor that uses icons representing actual desktop objects that the user can access and manipulate with a pointing device.

Graphics Coprocessor A programmable chip that speeds video performance by carrying out graphics processing independently of the computer's CPU. Among the coprocessor's common abilities are drawing graphics primitives and converting vectors to bitmaps.

Graphics Engine The print component that provides WYSIWYG (What You See Is What You Get) support across devices.

Graphics Interface Format Graphics Interface Format (GIF) is pronounced "Jiff." GIF is a format for encoding images (pictures, drawings, etc.) into bits so that a computer can "read" the GIF file and throw the picture up on a computer screen. CompuServe pioneered the bit-graphics format titled Graphics Interface Format. The format has become a standard in the electronic bulletin board world and it is used primarily to carry photographs of women in semi-clad and naked poses. According to Jack Rickard, publisher of Boardwatch Magazine, "some of the photographs are reasonably good, but most feature strikingly plain women rather artlessly photographed by those whose higher calling is probably more aptly found in the building trades or automotive repair."

Graphics Mode PCs work in two modes — text and graphics. In graphics mode, the pixels are individually addressable. In text mode, the graphics card inside your PC throws a type font on your screen. And that's it. You can't change it. In graphics mode, you can. Graphics and text modes are mutually exclusive.

Grasshopper Fuse A fuse that indicates that it has been blown by the movement of a piece of springy metal.

Grating Lobes Secondary main lobes.

Gravity Cell A closed circuit cell used where a continuous flow of current is desired. This type consists of copper and zinc electrodes with copper sulfate and zinc sulfate electrolyte. These are separated because of difference in their specific gravity.

Gray Elisha Gray was an inventor who filed for a patent on his own telephone design a few hours after Alexander Graham You-know-who. Gray was involved in a number of lawsuits with the young AT&T, and ultimately co-founded (with Enos BARTON) electrical equipment maker Western Electric, which was later sold to AT&T.

Gray Code A code used for translating certain analog representations into binary representations, such as the representation of angle. Gray code may also be called cyclic binary code or reflective code.

Gray Scale The spectrum, or range, of shades of black that an image has. An optical pattern consisting of discrete steps or shades of gray between black and white. Early facsimile machines could only receive black and white images and print them in black or white. Now they can print 16 shades of gray. This way if they receive a photo, it will look like a photo.

Gray Whale One of the first non-Bell key systems sold in North America. It was a 1A2 electromechanical key system from TIE/communications. TIE was a shortening of the words Telephone Interconnect Equipment. It was a wonderfully reliable phone system. A gray whale is now a prized possession.

Graybar Probably the oldest distributor of phone equipment, as well as various electrical products. Named for Elisha GRAY and Enos BARTON, who formed Western Electric. The company is headquartered in St. Louis, though there is a famous Graybar building in Manhattan, attached to Grand Central Terminal. A plaque in the terminal near stairs leading into the Graybar building shows Gray and Barton (if it hasn't been defaced or stolen).

Grayscale Monitor Any monitor capable of showing levels of gray and not just black or white.

Great Circle A circle defined by the intersection of the surface of the Earth and any plane that passes through the center of the Earth. The shortest distance, over the idealized surface of the Earth, between two points, lies along a great circle.

Greek Prefixes Remember the word "chronous?" It's used to mean the process of adjusting intervals or events of two signals to get the desired relationship between them. Here are the Greek prefixes that describe different timing conditions: asyn - not with hetero - different homo - the same iso - equal meso - middle piesio - near syn - together

Green Green is a term being applied by manufacturers to mean their equipment uses less electricity than other equipment. Classic PC equipment include screens that shut themselves almost down if they haven't been accessed for a minute or two. Much of the technology of "green" has already been used in laptops. When it moves to the desktop, there are occasionally compromises — such as moving from a 3/4rds off to full-on might take a moment or two. Are you willing to live with it?

Greenwasher A business that uses the fact that it recycles to promote itself.

Greenwich Mean Time GMT. Mean solar time at the meridian of Greenwich, England; formerly used as a basis for standard time throughout the world. Also called Zulu time and World Time. Normally expressed in 24-hour time. GMT is an obsolete term. Because the second is no longer defined in terms of astronomical phenomena, "Universal Time" (which is defined in terms of cesium-133 transitions) is the most accurate measure of time, and the primary name for time is "Coordinated Universal Time."

Greetings Only Mailboxes Mailboxes that deliver a message to incoming callers but do not allow a message to be left. The Greeting Only Mailbox may transfer a caller to a designated telephone number.

Grid 1. That element in a vacuum tube having the appearance of a grid and which controls the flow of electrons from the filament to the plate. "Grid" generally refers to the control grid.
2. Global Resource Information Database: part of the United Nations environment program.

Gritch A computer complaint.

Groom/Fill In telephony, terms associated with more efficient use of T-1 trunks by combining partially filled input T-1 trunks into fully filled outgoing T-1 trunks.

Grooming Managing bandwith on a wide area, public or private network to use the long haul transmission facilities as effectively as possible.

Gross Additions A cellular industry term. The amount of new subscribers signing up for the service before adjusting for disconnects (churn).

Ground 1. A problem that exists when a circuit is accidentally crossed with a grounded conductor.
2. Connecting equipment by some conductor (wire) to a route that winds up in the earth (Ground) for electrical purposes. One purpose of a "ground" wire is to carry spurious voltage (e.g. lightning strikes) away from the electrical and electronic circuits it can cause harm to. Incorrect grounding is probably the major cause of telephone systems problems. See GROUNDING (the major explanation), GROUND RETURN and GROUND START.

Ground Absorption The loss of energy in transmission of radio waves due to dissipation in the ground.

NEWTON'S TELECOM DICTIONARY

Ground Button A button needed on phones used for power failure transfer behind a PBX. You need the button because many trunks behind a PBX are ground start (as compared to loop start).

Ground Constants The electrical constants of the earth, such as conductivity and dielectric constant. The values vary with frequency, and also with local moisture content and chemical composition of the earth.

Ground Fault In AC electricity, a ground fault is any unintended connection between a supply conductor and ground (i.e.: hot conductor in contact with the metal case of a piece of equipment). A ground fault will cause a high current flow and should operate the overcurrent protection (fuse or breaker provided such devices are functionally adequate) only if the ground path impedance is sufficiently low — but under no circumstances greater than two ohms.

Ground Fault Circuit Interrupter GFCI or GFI. A device intended to interrupt the electrical circuit when the fault current to ground exceeds a predetermined value (usually 4 to 6 milliamps) that is less than required to operate the overcurrent protection (fuse or breaker) for the circuit. This device is intended to protect people against electrocution. It does not protect against fire from circuit overload.

Ground Fault Protector GFP. A device designed to protect electrical service equipment from arcing ground faults. A GFP does not provide protection for people.

Ground Lead The conductor leading to the ground. Connection.

Ground Loop This occurs when a circuit is grounded at one or more points. It can cause telephone system problems.

Ground Plane The surface existing or provided, that serves as the near-field reflection point for an antenna.

Ground Potential The electrical potential of the earth with respect to another body or region. The ground potential of the earth will vary with locality and also as a function of certain phenomena such as meteorological disturbances.

Ground Return If a battery is connected to a closed electrical circuit, an electric current will flow in the circuit. In the early days of the telegraph, the circuit consisted of a long wire, the telegraph key, the electromagnet of a telegraph sounder and a return path through the ground, which served as a conductor. Thus the current flowed from one terminal of the battery through the wire, through the electromagnet, to a metal stake driven into the ground (a "ground" electrode), back through hundreds of miles of earth to the distant stake at the distant telegraph office and then to the other terminal of the battery. In later telecommunications, the ground return path was replaced by a second wire.

Ground Return Circuit A circuit in which the earth serves as one conductor. A circuit in which there is a common return path, whether or not connected to earth ground.

Ground Start A way of signaling on subscriber trunks in which one side of the two wire trunk (typically the "Ring" conductor of the Tip and Ring) is momentarily grounded (often to a cold water pipe) to get dialtone. There are two types of switched trunks typically for lease by a local phone company — ground start and loop start. PBXs work best on ground start trunks, though many will work — albeit intermittently — on both types. Normal single line phones and key systems typically work on loop start lines. You must be careful to order the correct type of trunk from your local phone company and correctly install your telephone system at your end — so that they both match. In technical language, a ground start trunk initiates an outgoing trunk seizure by applying a maximum local resistance of 550 ohms to the tip conductor. See LOOP START.

Ground Start Supervision Telephone circuitry developed to prevent Glare.

Ground Station A cluster of communications equipment, usually including signal generator, transmitter, receiver and antenna that receives and/or transmits to and from a communications satellite. Also called a satellite earth station.

Ground Wave In radio transmission, a surface wave that propagates close to the surface of the Earth. The Earth has one refractive index and the atmosphere has another, thus constituting an interface. These refractive indices are subject to spatial and temporal changes. Ground waves do not include ionospheric and tropospheric waves.

Grounding There are two distinct and unique categories in this broad area called grounding. One is earth grounding. The purpose of an earth grounding system is essentially threefold: 1. To guard against the adverse effects of lightning, 2. To assist in the reduction of static and 3. To bring a zero-voltage reference to system components in order that logic circuits can communicate from a known reference. The other category of grounding is known as equipment grounding. The purpose of equipment grounding is threefold: 1. To maintain "zero volts" on all metal enclosures under normal operating conditions. This provides protection from shock or electrocution to personnel in contact with the enclosure. This is the safety aspect. This is the primary means of protecting personnel from electrocution. 2. To provide an intentional path of high current carrying capacity and low impedance to carry fault current under ground fault conditions; and 3. To establish a zero voltage reference for the reliable operation of sensitive electronic equipment. Effective equipment grounding is defined in the National Electrical Code, Article 250-51 and the Canadian Electrical Code Article 10-500. These Codes read almost identically. They say, "The path to ground from circuits, equipment and metallic enclosures for conductors shall;

1. Be permanent and continuous.
2. Have the capacity to conduct safely any fault current likely to be imposed on it, and
3. Have sufficiently low impedance to limit the voltage to ground and to facilitate the operation of the circuit protective devices in the circuit.

The Earth shall not be used as the sole equipment grounding conductor."

According to the Electric Power Research Institute, "electrical wiring and grounding defects are the source of 90% of all equipment failures." Many telephone system installer/contractors have found that checking for and repairing grounding problems can solve many telephone system problems, especially intermittent "no trouble found" problems. As electrical connections age, they loosen, corrode and become subject to thermal stress that can increase the impedance of the ground path or increase the resistance of the connection to earth. Equipment is available to test for proper grounding. (Ecos Electronics of Oak Park, IL makes some.) Before you install power conditioning equipment such as voltage regulators, surge arresters, etc. you should test for and correct any problems you have with grounding and wiring. See also AC, AC POWER and BATTERY.

Grounding Strap A device worn on the wrist or on the shoe when handling a static-sensitive component to prevent static shocks (sparks) which could damage the component. Don't even think of touching a printed circuit card without wearing a grounding strap. See GROUNDING.

Group 1. In call centers or in automatic call distributors, a group is the same as GATE or SPLIT. A group is an ACD routing division that allows calls arriving on certain telephone trunks or calls of certain transaction types to be answered by specific groups of employees.
2. A group is a collection of voice channels, typically 12. In AT&T jargon, a group is 12 channels. A supergroup is 60 channels. A mastergroup is 10 supergroups or 600 voice channels.
3. An SCSA definition. A group is an associated set of one or more Resource Objects. Groups encapsulate the functionality of the Resource Objects that are associated with them. Resource Objects within a Group have defined connectivity. The Group provides three services to the application: implicit management of connectivity between group members; representation of a single entity to the applications (group ID); and reservation of all physical resources (CPU, memory, time slots) required to provide the application with exclusive use of configured resources.

Group 1, 2, 3, 3 bis & 4 These relate to the facsimile machine business. They are essentially standards of speed and sophistication. They were created by the ITU-T in Geneva, Switzerland to make sure facsimile machines from one maker could speak to facsimile machines of another maker.
Group 1 transmits an 8 1/2 by 11-inch page in around six minutes. It conforms to ITU-T Recommendation T.2.
Group 2 transmits an 8 1/2 by 11-inch page in around three minutes. It conforms to ITU-T Recommendation T.3.
Group 3 — the most common fax in the world today — transmits an 8 1/2 by 11-inch page (also called A4)in as little as 20 seconds. It is a digital machine and includes a 9,600 baud modem. It transmits over dial up phone lines. Group 3 standards for facsimile devices were developed by ITU-T adopted in 1980 and modified in 1984 and 1988. Group 3 defines a resolution of 203 x 98 dots per inch and 203 x 196 for "fine." Group 3 uses modified Huffman code to compress fax data for transmission. For example, a white line with no text, called a run, extending across an 8.5" page equals 1728 bits. Modified Huffman Code compresses the 1728 bits into a 17-bit code word. The lengths for all possible white runs are grouped together into 92 binary codes that will handle any white run length from 0 to 1728.
Group 3 bis. This is an update to Group 3. It includes an image resolution of 406 x 196 dpi and a transfer rate of 14,400 bits per second. Fax machines that are Group 3 bis can transmit 50% faster to fax machines that are also Group 3 bis. A big speed improvement. Group 3 bis can drop to Group 3 if there's a Group 3 on the other end. Most of the modern plain paper fax machines and most of the fax modems are Group 3 bis.
Group 4 Fax. The latest and fastest international standard for facsimile machines. It specifies a machine which operates at 64 Kbps, which can only work on a digital channel and which takes six seconds to transmit a 8 1/2 x 11 inch page. The Group 4 standard was promulgated in January, 1987. Group 4 fax machines are designed to use one of the 64,000 bit per second B (Bearer) channels on ISDN.
Most Group 3 machines will transmit and receive from Group 1, Group 2 and Group 3 machines (but at their slower speeds). Group 1, 2, 3 & 4 are international standards. Group 3 is now by far and away the most common. All Group 3 machines will transmit and receive from each other. Some manufacturers have improved on the standards by offering Group 3 "fine," for example. These "fines" can talk to the same machines. But often can't talk to other "fines." If you're buying a facsimile machine and want super-quality transmission, check its compatibility with other machines. Or, easier, buy all identical machines. See FACSIMILE.

Group Address A single address that refers to multiple network devices. Synonymous with multicast address.

Group Addressing In transmission, the use of an address that is common to two or more stations. On a multipoint line, where all stations recognize addressing characters but only one station responds.

Group Busy Hour GBH. The busy hour offered to a given trunk group.

Group Call A special type of station (i.e. extension) hunting that requires a special access number to permit a call to the special access number and ring the first available phone in that group.

Group Distribution Frame GDF. In frequency-division multiplexing, a distribution frame that provides terminating and interconnecting facilities for the modulator output and demodulator input circuits of the channel transmitting equipment and modular input and demodulator output circuits for the group translating equipment operating in the basic spectrum of 60 kHz to 108 kHz.

Group Hunting Automatically finds free telephones in a designated group. See HUNT

Group Scheduling Software Software designed to coordinate and manage both worker schedules and office resources such as equipment and conference rooms.

Group Velocity 1. The velocity of propagation of an envelope produced when an electromagnetic wave is modulated by, or mixed with, other waves of different frequencies. The group velocity is the velocity of information propagation. In optical fiber transmission, for a particular mode, the reciprocal of the rate of change of the phase constant with respect to angular frequency.

Grouping A facsimile term for periodic error in the spacing of recorded lines.

Groupware A term for software which runs on a local area network and which allows people on the network (typically a team) to participate in a joint (often complex) project. According to Fortune Magazine, March 23, 1992, using groupware, "Boeing has cut the time needed to complete a wide range of team projects by an average of 91%, or to one-tenth of what similar work took in the past." Groupware can be used in a meeting, with everyone sitting around a conference table and typing their ideas into the PC in front of them. Groupware can also be used off-line, with members of the "team" in different cities adding their comments. The "bellwether" of groupware software is Lotus' program called notes.

GRSU Generic Remote Switch Unit.

GS Trunk Ground-Start Trunk. A trunk on which the communications system, after verifying that the trunk is idle (no ground on tip), transmits a request for service (puts ground on ring) to a telephone company. The other and more common type of trunk is called a Loop-Start Trunk.

GSA General Services Administration.
GSDN See GLOBAL SOFTWARE DEFINED NETWORK.
GSF See GENERIC SERVICES FRAMEWORK.
GSM GSM stands for Groupe Speciale Mobile, now known as Global System for Mobile Communications, is the standard digital cellular phone service you will find in Europe and Japan. GSM actually is a set of ETSI standards specifying the infrastructure for a digital cellular service. To ensure interoperability between countries, these standards address much of the network wireless infrastructure, including the radio interface (900 MHz), switching, signaling, and intelligent network. An 1,800 MHz version, DCS1800, has been defined to facilitate implementation in some countries, particularly the UK. Since GSM is limited to technical standards, an association of GSM operators called the Memorandum of Understanding (MoU) ensures service interoperability, allowing subscribers to roam across networks. GSM has gained widespread acceptance in several parts of the world, most notably Europe, with deployment in 52 countries by mid year '94. GSM subscriber data is carried on a Subscriber Identity Module (SIM) or "smartcard" which is inserted into the phone to get it going. As a result, the subscriber potentially has the option of either SIM card mobility or terminal mobility across multiple networks.
One of GSM's major problem is that people wearing hearing aids can't use GSM cell phones, because GSM phones produce unpleasant high-pitched squealing the closer they get to the cell phone.
GSM technical Characteristics: Receiver frequency: 935.2 — 959.8 MHz. Transmitter frequency: 890.2 — 914.8 MHz. Access method: mixed TDMA & FDMA with optional frequency hopping. Security: Optional radio interface encryption. Carrier frequency division: 200 KHz. Users per carrier frequency: 8. Speech bit rate (transfer rate): full rate (13 kbps) or half rate. Total bit rate: 21 Kbps. Bandwidth per channel: 25 KHz." A good book on GSM is "The GSM System for Mobile Communications" by Michael Mouly and Marie-Bernadette Pautet, both of France. The authors, in fact, contributed to the development of GSM. It's really the definitive document on GSM. See also BSS, which stands for BASE STATION SYSTEM. See also SIM CARD.
GSM-900 Global System for Mobile Communications at 900 MHz. A wireless telecommunications term. A GSM (Global System for Mobile Communications) standard for cellular phone systems operating at 900 Megahertz. See GSM.
GSTN General Switched Telephone Network. Same as public telephone network.
GTA 1. Government Telecommunications Association. An association of local, state and federal telecommunications professionals in Washington, D.C.
2. Government Telecommunications Agency. Specialized agency which provides telecom service to Canadian federal government departments.
GTE General Telephone and Electronics. A major telecommunications company, whose main business is owning and operating "independent" (i.e. non Bell) telephone companies. In 1990 it bought Continental Telephone. GTE also has a small part of a long distance company called US Sprint, a packet switched operation called Telenet and various equity interests in various manufacturing operations making most things in telecommunications — from central offices to PBXs. Its partners in these joint ventures include Fujitsu and Siemens. See also GTE SPRINT and GTE TELENET.
GTE Sprint A long distance service once provided by GTE, then a 50-50 joint venture of GTE and United Telecom. Now majority owned (80%) by United Telecom and 20% or so by GTE. Now, GTE Sprint is just called Sprint. And Sprint is no longer owned at all by GTE.
GTE TELENET A public data network which operates by using the ITU-T approved X.25 packet switching protocol. The new name for GTE TELENET is SPRINTNET. It's owned and run by Sprint, which is 80% owned by United Telecom and 20% by GTE.
GTN Global Transaction Network. An AT&T service which adds smarts to the routing of inbound 800 calls. It offers six call processing services: Next available agent routing, call recognition routing, transfer connect service (allows agents to transfer calls to distant ACDs), network queuing, 800 select again service and multiple number database (allows multiple 800 numbers to be assigned to a single routing plan in the network, rather than each 800 number having its own unique routing plan).
GTP General Telemetry Processor.
GTT Global Title Translations.
Guard Band A narrow bandwidth between adjacent channels which serves to reduce interference between those adjacent channels. That interference might be crosstalk. Guard bands are typically used In frequency division multiplexing. They are not used in time division multiplexing, because the technology is completely different.
Guarding The process of holding a circuit busy for a certain interval after its release to assure that a necessary minimum disconnect interval will occur between calls.
Guest Mailbox A mailbox used by a hotel or motel to set up temporary mailboxes for their guests. At least that was the original definition. Now it seems every voice mail system comes with guest mailboxes that could be used for visitors, employees from out of town, etc. Same application as a hotel — temporary use.
GUI Graphical User Interface. A generic name for any computer interface that substitutes graphics for characters. GUIs usually work with a mouse or trackball. Windows 3.0 and Windows 3.1 are the most famous GUIs. Second most famous is the Apple Macintosh. GUI is pronounced "GOO-ey." See GRAPHICAL USER INTERFACE.
GUID A Versit term. Globally Unique IDentifier.
Guided Ray In an optical waveguide, a ray that is completely confined to the core.
Guided Wave A wave whose energy is concentrated near a boundary or between substantially parallel boundaries separating materials of different properties and whose direction of propagation is effectively parallel to these boundaries.
Gutta-Percha a latex substance first discovered in 1847 and derived from the sap of Malayan evergreen trees. It first found use circa 1851 as the insulation in the first international telegraph cable which ran between England and France. Gutta-Percha was also the first insulator to survive in under seas applications (particularly for submarine cables) and was still the insulator material of choice for golf balls and telephone receivers until about 1947, when polyethylene finally began to gain acceptance.
Guy Wire A wire used to support a radio mast.

H-Channel The packet-switched channel on an ISDN BRI (Basic Rate Interface) which is designed to carry user information streams at varying rates, depending on type: H0 — 384 Kbps; H11 — 1,536 Kbps; and H12 — 1,920 Kbps. In short, H-channels are ISDN bearer services that have pre-defined speeds, starting and stopping at locations on an ISDN PRI circuit. They are contiguously transported from one PRI site through networks to another ISDN PRI site. See ISDN and PRI.

H-MVIP The original MVIP standard, now called MVIP-90, was developed in 1989-90 and first deployed in 1990. MVIP-90 supports up to 512 telephony channels of 64 Kbps each between circuit boards within a single computer chassis. MVIP-90 has been widely adopted for voice, FAX, data and video services as well as for telephony switching applications. Other MVIP standards have since been developed to address multi-chassis MVIP systems and higher level software APIs. Beginning in 1993, there was interest in higher capacities within individual single-chassis MVIP nodes. Specific applications include large audio conferences, multi-media servers, and the termination and switching of all traffic on a dual FDDI-II fiber ring (as used in MC2 standard Multi-Chassis MVIP) as well as the termination of and switching of traffic from T-3/E-3 telephone trunks and SONET/SDH links at OC-3 (155 Mbps) rates (as used in MC3 standard Multi-Chassis MVIP). H-MVIP addresses this need for higher telephony traffic capacity in individual computer chassis. H-MVIP defines three major items that together make a useful digital telephony transport and switching environment. These are the H-MVIP digital telephony bus with up to 3072 "time-slots" of 64 Kbps each; a bus interface with digital switching that allows a group of H-MVIP interfaced circuit boards to provide distributed telephony switching and a logical device driver model and standard software interface to that logical model. The bus definition includes the mechanical, electrical and timing requirements for a high capacity telephony bus that is a super-set of the existing MVIP-90 standard for computer-based telephony. Several levels of capacity expansion are defined to support a range of system implementations. Among them, H-MVIP "24/2" is a wider version of MVIP-90, while the full H-MVIP bus is both wider and faster. See ECTF, SCSA and TAO.

H.221 A framing recommendation which is part of the ITU-T's H.320 family of video interoperability Recommendations. The Recommendation specifies synchronous operation where the coder and decoder handshake and agree upon timing. Synchronization is arranged for individual B channels or bonded 384 Kbps (H0) connections.

H.222.0 Defines the general form of elementary stream multiplexing as the Moving Picture Experts Group 2 (MPEG-2) system part. See H.222.1.

H.222.1 Specifies the parameters of H.222.0 for communication use.

H.230 A multiplexing recommendation which is part of the ITU-T's H.320 family of video interoperability recommendations. The recommendation specifies how individual frames of audiovisual information are to be multiplexed onto a digital channel.

H.231 A recommendation, formally added to the ITU-T's H.320 family of recommendations in March, 1993, which specifies the multipoint control unit used to bridge three or more H.320-compliant codecs together in a multipoint conference.

H.233 An Recommendation, part of the ITU-T's H.320 family, which specifies the encryption method to be used for protecting the confidentiality of video data in H.320-compliant exchanges. Also called H.KEY.

H.242 Part of the ITU-T's H.320 family of video interoperability Recommendations. This Recommendation specifies the protocol for establishing an audio session and taking it down after the communication has terminated.

H.244 Recommendation on a channel aggregation method for audiovisual communications. This enables several ISDN B-channels to behave as a single higher-rate channel.

H.245 Multimedia system control, defining negotiation procedures and signals. This recommendation is for B-ISDN and GSTN (dial-up phone lines). H.246 was incorporated in H.245.

H.246 See H.245.

H.261 The ITU-T's H.261 is the standards watershed in videoconferencing. Also known as p x 64, H.261 specifies the video coding algorithms, the picture format, and forward error correction techniques to make it possible for video codecs from different manufacturers to successfully communicate. Announced in November 1990, it relates to the decoding process used when decompressing video conferencing pictures, providing a uniform process for codecs to read the incoming signals. Other important standards are H.221: communications framing; H.230 control and indication signals and H.242d: call set-up and disconnect. Encryption, still-frame graphics coding and data transmission standards are still being developed. See H.320.

H.310 Recommendations for a videoconferencing terminal in an ATM environment.

H.320 The most common family of ITU-T videoconferencing standards. These standards allow dissimilar videoconferencing systems and videophones to communicate with each other. "H.320 compatible videoconferencing systems" are now the most common videoconferencing systems. They work on ISDN BRI circuits. I've personally had several H.320 compatible videoconferencing systems and videophones on my desk and have received from and made videoconferencing calls to many different H320 compatible video phones. The quality is not brilliant. But you can recognize the person at the other end. And they can recognize you. Most H.320 systems allow you to bond together the two B channels of a 2B+D ISDN BRI channel and thus get better video. See all the H.2NN explanations above. See also G.711.

H.321 The adaptation of the ATM environment of H.320 videoconferencing standards.

H.322 The adaption to a guaranteed quality of service LAN of H.320 terminals.

H.323 The H.323 spec defines packet standards for terminal, equipment and services for multimedia communications over LANs communicating to systems connected to telephony networks such as ISDN.

283

H.324 Standard for analog POTS telephone line based teleconferencing.

H.R.nnnn A proposed law introduced into the House of Representatives by a Congressman. Typically, four digits follow the H.R., signifying the proposed bill's number. The reason for including this definition in this dictionary is that every few months since divestiture some Congressman has attempted to introduce a bill into the House of Representatives changing the Communications Act of 1934. Such a bill is generally supported by a bevy of Bell telephone companies trying to use the proposed bill to remove those restrictions placed on them by Divestiture — manufacturing, creating information content and getting into long distance.

H0 Channel An H zero channel is a 384 kbps channel that consists of six contiguous DS0s (64 kbps) of a T-1 line.

H10 Channel The H ten channel is the North American 1,472 Kbps channel from a T-1 or primary rate carrier. It is equivalent to twenty-three (23) 64 kbps channels.

H11 Channel The North American primary rate used as a single 1,536 Kbps channel. This channel uses 24 contiguous DS0s (DS zeros) or the entire T-1 line except for the 8 Kbps framing pattern.

H12 The European primary rate used as a single 1,920 kbps channel (30 64 kbps channels) or the entire E-1 line except for the 64 kbps framing and maintenance channel.

HA Horn Alert. A cellular car phone feature that automatically blows the car's horn if a call is coming in.

Hack The output of a hacker. Usually good programs, but sometimes just something clever of no discernible use. Just a "good hack", or something done for the "hack value."

Hacker A person who "hacks" away at a computer until his program works. The term has been wrongly used by the press to mean people who break into computer systems. The word hacker has gone through many meanings. At one stage it was a badge of honor conferred on an elite programmer or computer hardware designer. But in 1983 the movie "War Games" presented another view of the hacker mentality — someone who tries to break into computer systems for fun and sport. These days to call someone a hacker is not so prestigious. See CRACKER.

HAL 1. The computer from the movie 2000: A Space Odyssey. HAL is an acronym for "Heuristically programmed ALgorithmic computer." The one-letter-shift transposition to "IBM" (The I became H. The B became A. The M became L) was noted shortly after the film's release and was widely accepted as a subtle Kubrick/Clarke joke directed at the computer giant. Clarke himself has pointed out that when he named HAL, he didn't catch the one-letter-shift bit, and if he had, he would have changed the name. IBM had been a huge supporter of the film project, and he wouldn't have dreamed of poking fun at them, however subtly. (Incidentally, according to Clarke, the one intentional joke in the film was the scene where a knuckle-chewing Heywood Floyd read the instructions for the Zero-Gravity Toilet.)

Half-Bridge Apple Computer term for a device linking LANS over a low-speed link such as a telephone line or X.25 link. It is termed a half-bridge as one is required at each end of the link.

Half-Duplex A circuit designed for data transmission in both directions, but not at the same time. Telex is an example of a half duplex system, as is speaking on with most speakerphones. (The best speakerphones are full duplex. They're rare and expensive.)

Half-Life In science, the time it takes for half the radioactivity of a substance to disappear. Among techies, it is a gauge of an individual's usefulness. "He may have a short half-life here."

Half-Repeater A device which extends the distance a LAN can cover by joining two lengths of cable over another communication medium.

Half-Tone Any photomechanical printing surface or the impression therefrom in which detail and tone values are represented by a series of evenly spaced dots in varying size and shape, varying in direct proportion to the intensity of tones they represent.

Half-Wave Antenna An antenna which is half as long as the wave being received.

Halo Effect, The Websters defines a halo as a conventional, geometric shape, usually in the form of a disk, circle, ring or rayed structure representing a radiant light around or above the head of a divine or sacred personage. A company or person acquiring "the Halo Effect" suggests that the company is doing that most outside observers really view as "right." In short, the company or person is on a major roll.

Hamming Code An error correcting code named after R. W. Hamming of Bell Labs. The code has four information bits and three check bits per character.

Han Characters Han characters are Chinese language symbols which are used to represent whole words or concepts in Chinese, Japanese and Korean. See UNICODE.

Hand-Off The process of transferring cellular-based calls from one cell site to another as the mobile or portable moves through the service area.

Handhole A buried box whose lid is even with the surface of the ground. It provides a space for splicing and terminating cables.

Handle 1. In the Windows 95 user interface, an interface added to the object that facilitates moving, sizing, reshaping, or other functions pertaining to an object.

2. In programming, a pointer to a pointer- that is, a token that lets a program access a resource identified in another variable. See also object handle.

3. The name you use in an online computer service. It's typically not your own name. You adopt this name to give yourself anonymity for whatever reasons you find convenient. A handle is called a "Nom de ligne" in some circles.

Handled Call A call center term. A call that is answered by an employee, as opposed to being blocked or abandoned.

Handoff 1. The process by which the mobile telephone switching office passes a cellular phone conversation from one radio frequency in one cell to another radio frequency in another cell. It is performed so quickly (allegedly 250 milliseconds) that callers don't notice. Or we hope they don't notice.

2. An SCSA definition. The change of ownership of a Group (and therefore, typically, a call) from one session to another. For example, if a call center application discovers that a caller wishes to access a technical support audiotex database, it hands off the call to an application servicing that database.

Handover Word The word in the GPS (Global Positioning System) message that contains synchronization information for the transfer of tracking from C/A to P code.

Handset The part of a phone held in the hand to speak and listen, it contains a transmitter and receiver. In the old days, the transmitter was a carbon mike. Now it's mostly electronic. Some electronic mikes are awful. Some phone makers are going back to carbon mikes. There are two basic types of telephone handsets in North America: the G-style handset, which has round, screw-in ear and mouthpiece, and the new K-style handset, which has square ear and mouthpieces and has the two screws in the middle. I prefer the older G-style one. I think it's sturdier.

Handsfree A feature that's included with most of today's car phones. It permits the driver to use his cellular car phone without lifting or holding the handset to his ear. This is an important safety feature. See also HANDSFREE DIALING.

Handsfree Answerback This feature, when activated, automatically turns ON the microphone at a telephone receiving a call so that the person receiving the call can respond without lifting the receiver. Handsfree answerback is typically used on intercom calls.

Handsfree Dialing A telephone feature which allows the user to place outside calls and listen to the progress of those calls without lifting the handset of his telephone. This feature is unbelievably useful when calling airlines which inevitably put you on "eternity hold." (We made that term up.)

Handsfree Monitoring You can dial an outside call and hear the call's progress without having to lift your handset. Similar to hands-free dialing. With hands-free monitoring, you can only listen. To speak, you must pick up the handset. To be able to speak, you need a full speakerphone. Be careful of the distinction. Many people have been caught.

Handsfree Telephone Could be another word for a speakerphone or for a phone that does hands-free dialing.

Handshake Two modems trying to connect. Two modems trying to agree on how to transfer data. The series of signals between a computer and another peripheral device (for example, a modem) that establishes the parameters required for passing data.

Handshake In HIN. A general purpose control signal sent from the DTE to DCE in a Newbridge Networks RS-232-C connection. HIN can be used in place of Request to Send (RTS), Carrier Detect (CD) or Ring Indicator (RI).

Handshake Out HOUT. A general purpose control signal sent from the DCE to the DTE. For example, in the case of a Newbridge Networks Mainstreet Data Controller with ports configured as DCE, HOUT is sent from the Data Controller to an attached device. HOUT can be used in place of Clear to Send (CTS), Carrier Detect (CD) or Ring Indicator (RI).

Handshaking The initial exchange between two data communications systems prior to and during data transmission to ensure proper data transmission. A handshake method is part of the complete transmission protocol. A serial (asynchronous) transmission protocol might include the handshake method (XON/XOFF), baud rate, parity setting, number of data bits and number of stop bits. Just as people shake hands, and go through a perfunctory "Hi, how are ya?", computers must go through a procedure of greeting the opposite party, verifying the identity of the other party, and other functions that can be described by this "humanizing term." As with human contacts, once the Handshaking is complete, the business of communications begins.

Handwriting Recognition A system for taking handwritten generated with a stylus on a computer pad or directly onto the computer screen, and converting then into machine-readable text.

Hang Up Hang up lets you disconnect from an ISDN call. To hang up from the phone set you must depress and hold the receiver button for a specified amount of time. By default, the time is set for 0.8 of a second.

Hard Copy Anything on paper. It is all well and good to have information flash by on your CRT or video display terminal, but there are times when you want to take a Hard Copy with you. This dictionary was written on a computer screen. Now you have a hard copy in your hands. In this case, that's a lot more useful than having a disk.

Hard Disk Assembly A sealed mass storage unit used for storing large amounts of data. Now available on personal computers.

Hard Drawn Copper Wire Copper wire that has been drawn to size and not annealed.

Hard Ferrite Ferrite that remains permanently magnetized. Used to make magnets.

Hard RAM Carve some memory out of a computer's RAM; power it continuously and bingo you have Hard RAM, also called a Virtual disk. Setting up a RAM disk lets you use your computer's conventional, extended or expanded memory to simulate a disk drive (or drives). The primary advantages of a RAM disk are its very fast access speed and its battery power-saving properties. It has not mechanical element to slow it down or use additional power. Hard RAM exists in memory beyond 640KB.

Hard Rubber A hard insulating material made of rubber, and having a dielectric constant of from two to four.

Hard Sectoring Physically marking the sector boundaries of a magnetic disk by punching holes in the disk where there's space available to store data. Hard sectored disks are not very common these days. Most disks — like those used on the IBM PC — are soft-sectored.

Hard Tubes Vacuum tubes having a high vacuum.

Hard Wired 1. Describes a circuit designed to do one task (e.g. a leased line).
2. A person with a very narrow and rigid view of his or her job. "That security guard is really hard-wired."

Hardened Resistant to disaster. Facilities with protective features that have been designed to withstand an explosion, a natural disaster, or ionizing radiation.

Hardware The actual physical computing machinery, as opposed to Software which is the list of instructions to operate the hardware, or the Firmware which is combination hardware/software that is "burned into" a Programmable Read Only Memory chip or chips. See FIRMWARE and SOFTWARE.

Hardware Address Also called physical address or MAC-layer address, a data-link layer address associated with a particular network device. Contrasts with network or protocol address, which is a network layer address.

Hardware Flow Control Hardware flow control is the method used by the UART chip (that chip controls the serial port) to modulate the flow of data. It does this by controlling the Clear to Send/Ready to Send (CTS/RTS) lines of the serial port's interface. For example, it can turn off or re-enable the flow of data. Most high-speed sessions require hardware flow control due to their need for precise, instantaneous control over the flow of incoming and outgoing data.

Hardware Interrupt See INTERRUPT.

Hardware Tree A Windows 95 term. A record in RAM of the current system configuration information for all devices in the hardware branch of the Registry. The hardware tree is created each time the computer is started or whenever a dynamic change occurs to the system configuration.

Hardwire To permanently connect by wire two or more devices rather than to connect them temporarily through connectors or switches. Hardwire is a term also used to represent a leased line.

Hardwire Services An MCI definition. Services providing intercity communications facilities dedicated to the use of a specific customer, and provided through a dedicated access line from the customer to the MCI switch.

Hardwire Terminating City City of circuit termination for hardwire services.

Harm See NETWORK HARM.

Harmonic A frequency which is an exact multiple of a fundamental frequency.

Harmonic Distortion A problem caused when the nonlinearities in communication channels cause the harmonics of the input frequencies to appear in the output channel.

Harmonic Ringing A way of stopping users on a party line from hearing other than their own ring. We do this by tuning the ringer in their phone to a given ringing frequency, so it only rings when their frequency comes down the line.

Harmonica A device attached to the end of a connectorized feeder cable that converts the 25 pair into individual 4, 6 or 8 wire modular channels.

Hash Total Adding up one or more information fields in order to provide a check number for error control. The addition is not intended to have any meaning other than for checking.

HASP Houston Automatic Spooling Program. A control protocol adopted by IBM for transmitting data processing files and jobs to IBM 360 and 370 computers. An early job control language still in limited use.

Hayes At Command Set Before 1981, the modem was a dumb device. It had no memory or ability to recognize commands. It simply modulated and demodulated signals between the telephone line and the computer or terminal. In 1981, Hayes Microcomputer Products, Inc. in Norcross, GA produced the first "smart" modem, appropriately named the Smartmodem 300. It was "smart" because it understood commands, such as "ATD" which means "ATtention, Dial the phone." The Hayes Standard AT Command Set (its full name) — a language for modems — has been accepted as a standard by the modem industry. And now many modems claim to be 100% Hayes compatible, which may mean they are and may mean they aren't. As in all cases of claimed compatibility, one should check. You'll find the complete Hayes AT Command Set spelled out in virtually every manual of every modem which purports to be "100% Hayes Compatible." See also CLASS 1.

Hayes Microcomputer Products Inc. The company that made modems for PCs affordable and intelligent. Continues to make great modems. Still the standard against which most PC modems are measured. Based in Norcross, GA. See also HAYES COMMAND SET.

Hayes ULTRA 96 New Hayes 9600 bps modem with ITU-T V.32, V.42 error correction and V.42 bis data compression. 9600 bps communications does not require data compression, although most modem manufacturers do have compression at 9600 bps. What happens is two 9600 bps modems will first negotiate modulation, then they'll negotiate error-control, and finally, they'll negotiate data compression. Hayes' ULTRA 96 has "Automatic Feature Negotiation" which means this whole process occurs automatically without the user's help. All this means is that two Hayes Ultras talking to each other can "talk" as fast as 38,400 bps.

HBA Host Bus Adapter. A printed circuit board that acts as an interface between the host microprocessor and the disk controller. The HBA relieves the host microprocessor of data storage and retrieval tasks, usually increasing the computer's performance time. A host bus adapter (or host adapter) and its disk subsystems make up a disk channel.

HBFG Host Behavior Functional Group: The group of functions performed by an ATM-attached host that is participating in the MPOA service.

HCI Host Command Interface. Mitel SX-2000 PBX to computer link. HCI is designed to work with Digital Equipment Corporation computers. See OPEN APPLICATION INTERFACE.

HCL Hardware Compatibility List. See HCL.

HCS 1. Hundred Call Seconds. One hundred seconds of telephone conversation. See CCS.
2. Hard Clad Silica.

HCT Hardware Compatibility Test. Microsoft came up with this definition and concept when it found several computers didn't run its software as well as they should. Basically the Microsoft Hardware Compatibility Test (HCT) is a series of tests for verifying the compatibility of hardware systems with Windows NT. The TCT Test Manager is an application that provides a way to launch the tests, keep track of test results and return the results to Microsoft. Microsoft maintains a Windows NT Hardware Compatibility List (HCL). If you send back the test results and your stuff passes, your hardware will be included on Microsoft's list of hardware that works with NT.

HD Half Duplex circuit.

HDB3 High Density Bipolar 3. A bipolar coding method that does not allow more than 3 consecutive zeros.

HDD Hard Disk Drive.

HDLC High level Data Link Control. An ITU-TSS link layer protocol standard for point-to-point and multi-point communications. In HDLC, control information is always placed in the same position. And specific bit patterns used for control differ dramatically from those used in representing data, so that errors are less likely to occur. SDLC and ADCCP are similar protocols. See also HIGH LEVEL DATA LINK CONTROL.

HDMAC Another potential high definition TV standard. HDMAC was spawned by Britain's Independent Broadcasting Authority. Unlike Japan's Hi-Vision, HDMAC has the attraction of being compatible with existing TV sets, i.e. those in Europe.

HDSL High bit rate Digital Subscriber Line. A technology to put an up to two-way T-1 on a normal unshielded, bridged (but not loaded) twisted pair (the stuff common in local loops) without using repeaters over twisted pair up to 20,000 feet. In Europe,

where T-1 is E-1 (2 megabits per second), one method for getting HDSL on copper wire is to use two pairs — one pair with 784 Kbps and one pair with 1,168 Kbps. See also ADSL.
HDT Host Digital Terminal.
HDTP Hoofddirectie Telecommunicatie en Post (Directorate for Telecommunications and Posts, The Netherlands).
HDTV High Definition TeleVision. Today's typical TV set in North America contains 336,000 pixels. A high definition TV set — one giving at least the definition of a movie theater, or 35 mm slide — will require at least two million pixels. Researchers are pursuing at least two dozen technologies to achieve this level of quality. The ideal HDTV would be flat screen, cheap, reliable and require very little electrical power. One standard, recommending the doubling of the current 525 lines per picture to 1050 lines and increasing the screen aspect ratio (width:height) from the current 12:9 to 16:9, which would create a television screen shaped more like a movie screen. See HIGH DEFINITION TELEVISION.
HDX Half DupleX.
HE See HEAD END.
Head A device that reads, writes, or erases data on a storage medium. The device which comes in contact with or comes very close to the magnetic storage device (disk, diskette, drum, tape) and reads and/or writes to the medium. In computer devices, it performs the same function as the head on a home cassette tape recorder.
Head End 1. The originating point of a signal in cable TV systems. At the head-end, you'll often find large tall TV and dish satellite receiving antennae.
2. A central control device required within some LAN/MAN systems to provide such centralized functions as remodulation, retiming, message accountability, contention control, diagnostic control, and access.
Head Landing Zone An area of the hard disk set aside for take off and landing of the heads when the drive is turned on and off.
Head Slap Similar to head crash but occurs while the drive is turned off. It usually occurs during mishandling or shipping. Head slap can cause permanent damage to a hard disk drive.
Head Thrashing A term for rapid back and forth movements of the disk head of a hard drive.
Header 1. Protocol control information located at the beginning of a protocol data unit.
2. The portion of a message that contains information that will guide the message to the correct destination. This information contains such things as the sender's and receiver's addresses, precedence level, routing instructions, and synchronization pulses.
Header Area The area containing preliminary information for the entire document, such as the data, company name, address, purchase order, terms, etc. An EDI (Electronic Data Interchange) term.
Header Error Control A CRC code located in a data cell which is used for checking the integrity of the transmission.
Headset A telephone transmitter and receiver assembly worn on the head. Headsets are becoming very light and very comfortable and are no longer worn only by switchboard attendants and airline clerks. Northern Telecom says the following are traditional applications for headset use:
Receptionists. Console attendants. Telemarketers. Customer service reps. Order entry reps. Financial Service professionals. Stockbrokers. Sales reps and reservation agents.
Headset Jack A place on a phone or console into which you can plug a headset.
Hearing Aid Compatible A hearing aid compatible phone may be used with inductively coupled hearing aid devices. You can find hearing aid compatible coin phones by looking for the blue grommet between the handset and the cord.
Heartbeat Ethernet-defined SQE signal quality test function, defined in IEEE 802.3. Heartbeat is created by a circuit (normally part of the transceiver) that generates a collision signal at the end of a transmission. This signal is used by the controller interface for self-testing.
Heat Electromagnetic waves of a frequency between that of light waves and radio waves. A form of energy.
Heat Coil An electrical protection device used to prevent equipment from overheating as a result of foreign voltages that do not trigger voltage limiting devices. It typically consists of a coil of fine wire around a brass tube that encloses a pin soldered with a low-melting alloy. When abnormal currents occur, the coil heats the brass to soften the solder, allowing the spring-loaded pin to move against a ground plate directing currents to ground.
HEC Header Error Control - a CRC code located in the last byte of an ATM (Asynchronous Transfer Mode) cell header used for checking integrity only. Using the fifth octet in the ATM cell header, ATM equipment may check for an error and corrects the contents of the header. The check character is calculated using a CRC algorithm allowing a single bit error in the header to be corrected or multiple errors to be detected.
HEHO Head-End Hop Off. You have a private network. You overflow a long distance call to WATS or DDD at the originating end (the end the call is coming from). This HEHO (Head-End Hop Off) is done because it's usually cheaper than carrying the call part way through the network, then jumping off the network at that point (because the network is busy or it won't reach the end point). The opposite of HEHO is TEHO — Tail-End Hop Off. In TEHO, you carry the call as far as possible through the network, then pass it off to WATS or DDD as close to its destination as possible. The decision to go HEHO or TEHO has to do with economics, primarily which is cheaper.
Held Call A held call is a call to which you are connected but which is on hold.
Held Orders A telephone company term for requests for telephone lines which the phone company cannot fill. Thus it is "holding" the orders. The reasons for holding customer orders might range from lack of capacity at the serving central office to a lack of local cable plant.
Helical Antenna An antenna that has the form of a helix. When the helix circumference is much smaller than one wavelength, the antenna radiates at right angles to the axis of the helix. When the helix circumference is one wavelength, maximum radiation is along the helix axis.
Helical Scan Storage method that increases media capacity by laying data out in diagonal strips. Used in video tape recorders, etc.

Helical Stripe A continuous, colored, spiral stripe applied over the outer perimeter of an insulated conductor for circuit identification purposes.

Helix A spiral. The shape of screw.

Hello When the phone was first invented, no one was sure how to begin the conversation. Thomas Edison saw the telephone as being used by businesses with permanently open lines. How would anyone know that the other party wanted to speak? A letter was found from Thomas Edison, dated August 15, 1877 to TBA, David, president of the Central District and Printing Telegraph Co. in Pittsburgh. "Friend David, I don't think we need a call bell as Hello! can be heard 10 to 20 feet away. What do you think? " Edison. At that time Alexander Graham Bell insisted on answering the telephone with "Ahoy." Hello! became the standard as the first telephone exchanges were set up across the country. Hello first appeared in the Oxford English Dictionary in 1883. In September of 1880, the first National Convention of Telephone Companies was held in Niagara Falls. "Hello" was used on everyone's name tag for the first time. Besides electricity, the phonograph and hundreds of other inventions, we can thank Edison for the "Hello" greeting. The above from New Pueblo Communications in Tucson, AZ.

Hello Packet A type of PNNI Routing packet that is exchanged between neighboring logical nodes.

Help Desk A centralized location where queries about product usage, installation, problems or services are answered. Sometimes help-desks are provided by the manufacturer of the product. Sometimes help desks are provided by outside companies — systems integrators and integrators, independent software developers and third party companies.

Henry The inductance in a circuit in which the electromotive force induced is one volt when the inducing current varies at the rate of one ampere per second. It is 1,000,000.000 electromagnetic units, and is the unit of inductance.

HEPNET An Internet term. A non-USENET set of newsgroups devoted to discussing the topic of high-energy nuclear physics.

HER High Energy Radio Frequency Gun. Shoots a high-powered radio signal at an electronic target (such as a computer) and puts it out of commission.

Hercules Graphics Hercules graphics adheres to the Hercules standard of monochrome graphics on a monochrome PC monitor. That standard is 720 x 348 pixel resolution and 64K screen memory. This encoding was never adopted as a color standard and is now pretty well obsolete. See also MONITOR and FACSIMILE.

Hermaticity Test A fine and gross leak test of a hermetically sealed IC to see if there are any leaks in the seal. The gross leak test uses a fluorocarbon fluid, and the fine leak uses a light gas such as helium.

Hermetic Coating A coating applied over the cladding of a fiber that retards the permeation of moisture and hydrogen into the fiber.

Heroinware Doom played on corporate networks.

Hertz Abbreviated Hz. A measurement of frequency in cycles per second. A hertz is one cycle per second.

Hertzian Wave A name sometimes given to electromagnetic waves.

Hetero The Greek prefix meaning different.

Heterodyne To generate new frequencies by mixing two or more signals in a nonlinear device such as a vacuum tube, transistor, or diode mixer. A superheterodyne receiver converts any selected incoming frequency by heterodyne action to a common intermediate frequency where amplification and selectivity (filtering) are provided. A frequency produced by mixing two or more signals in a nonlinear device. See HETERODYNING.

Heterodyne Repeater A repeater for a radio system in which the received signals are converted to an intermediate frequency, amplified, and reconverted to a new frequency band for transmission over the next repeater section.

Heterodyning Here is an explanation from James Harry Green's book, the Dow Jones Handbook of Telecommunications: Analog microwave repeaters use either of two techniques to amplify the received signal for retransmission: HETERODYNING or BASEBAND. In a baseband repeater, the signal is demodulated to the multiplex (or video) signal at every repeater point. In heterodyne repeaters the signal is demodulated to an intermediate frequency, typically 70 MHz, and modulated or heterodyned to the transmitter output frequency. Heterodyne radio is reduced to baseband only at main repeater stations where the baseband signal is required to drop off voice channels. The primary advantage of baseband radio is that some carrier channel groups can be dropped off at repeater stations. Heterodyne radio has the advantage of avoiding the distortions caused by multiple modulation/demodulation and amplification of a baseband signal. Therefore, heterodyne radio is employed for transcontinental use with drop-off points only at major junctions.

Heterogeneous Networks Networks composed of hardware and software from multiple vendors usually implementing multiple protocols.

Heterojunction A junction between semiconductors that differ in their doping level conductivities, and also in their atomic or alloy compositions.

Heuristic Using much trial and error to arrive at a solution to a problem.

Hexadecimal A numbering system of 16 characters, ten digits and six letters. It is used to condense the long strings of zeros and ones in large binary numbers. This base-16 numeric notation system is frequently used to specify addresses in computer memory. It makes life simpler for programmers. In hexadecimal notation, the decimal number numbers 0 through 15 are represented by the decimal digits 0 through 9 and the alphabet "digits" A through F (A=decimal 10, B=decimal 11, and so forth).

HF Hands Free.

HFAI Hands Free Answer on Intercom. A desirable feature of several phone systems.

HFC Hybrid Fiber Coaxial cable. Many buildings and campuses are being wired for HFC networks. The concept is that such a network will be able to provide extensive video needs and accommodate the organization's communications needs for many years.

HFU Hands Free Unit

HGC Hercules Graphic Card; long the standard monochrome graphics adapter for PCs and compatibles. Maximum resolution is 720 x 348 pixels.

Hi-Lo Tariff A long distance private line tariff filed by AT&T whereby private lines between major cities were priced lower than private lines between smaller cities. In effect, those "larger" cities were those MCI operated in and those "smaller" cities were those MCI didn't operate it. Eventually the tariff was thrown out by the FCC and it figured in anti-trust suits by MCI and the Federal Government against AT&T.

Hibernation You turn your computer off in the middle of a program. It grows to sleep. When you turn it on, it returns to exactly where you left it, without the need for rebooting. Compaq invented the term after copying the feature from Toshiba, who called it Resume. These days most laptops have some form of hibernation or resume.

Hidden Markov Method HMM. A common algorithm in voice recognition which uses probabilistic techniques for recognizing discrete and continuous speech.

Hierarchical Network A network that includes two or more different classes of switching systems in a defined homing arrangement, meaning to home in on the telephone you wish to be connected to. This is a fancy Bell System (Oops, I mean AT&T - Old habits die hard) term meaning that when direct circuits between two switches are busy or too far apart to be directly connected, the machinery will seek a higher level of switches to route the call through.

Hierarchical Routing The process of establishing a network data path to a destination based on addresses with some kind of ranking criteria.

Hierarchically Complete Source Route An ATM term. A stack of DTLs representing a route across a PNNI routing domain such that a DTL is included for each hierarchical level between and including the current level and the lowest visible level in which the source and destination are reachable.

Hierarchy A hierarchy is a group of things arranged in order of rank. It is a set of transmission speeds arranged to multiplex successively higher numbers of circuits. See also HIERARCHICAL NETWORK. It is also an Internet USENET newsgroup hierarchy which refers to the set of all news-groups contained within a specific broad subject category.

High and Dry What happens when you dial into a long distance network and nothing happens. You don't hear anything. Your call doesn't go anywhere. You're simply left High and Dry!

High ASCII ASCII characters whose values exceed 127. In most bulletin board networks, The use of high-ASCII in messages is prohibited since some types of personal computers cannot correctly interpret those characters.

High Bandwidth A person who is super intelligent is said to have "high bandwidth." The term is believed to have originated at Microsoft in Seattle, Washington. See also BANDWIDTH.

High Capacity Service Generally refers to tariffed, digital-data transmission service equal to, or in excess of T1 data rates (1,544 Mbits.)

High Definition TV HDTV. A system for transmitting a TV signal with far greater resolution than the standard NTSC standards. NTSC is the National Television System Committee and the standard for TV broadcast and reception in North America. HDTV has approximately twice the horizontal and twice the vertical emitted resolution specified by the NTSC standard. HDTV employs a wide aspect ratio. The total number of pixels is therefore more than four times that of NTSC. HDTV may include any or all IDTV and EDTV improvements. Several standards have been proposed in North America and, as of writing, the FCC had not agreed on one. Here are some of the proposed HDTV systems for North America:

Firms:	GI-MIT	GI	ATT-Zenith	ARTC	NHK
Systems:	CC-DigiCipher	DigiCipher HDTV	DSC HDTV	DSC MUSE	Narrow
Lines:	787.5	1,050	787.5	1,050	1,125
Screens:	16:9	16:9	16:9	16:9	16:9
Transmission:	Digital	Digital	Digital	Digital	Analog

High Dome Synonym for "egghead." A scientist.

High Energy Radio Frequency Gun HER. Shoots a high-powered radio signal at an electronic target (such as a computer) and puts it out of commission.

High Fidelity Systems of radio transmission and reception which permit a wide band of audio frequencies to be transmitted and/or reproduced.

High Frequency Noise A signal frequency more than 1,000 times the normal AC power line frequency of 60 cycles. The frequency will lie between 3 and 30MHz.

High Level Data Link Control HDLC. A communications protocol that is bit oriented in which control codes differ according to their bit positions and patterns.

High Level Languages Essentially any of the computer languages whose code is not unique to the hardware or architecture of a particular computer. High level languages are more like human language than the machine language which computers talk. High level languages translate human instructions into the machine language computers can understand, but which humans don't have to (in order to tell the computer what to do). Computer languages such as BASIC, FORTRAN, COBOL and Pascal are high level languages. They are a number of levels (at a High Level) away from the actual bit manipulation (machine language, also called "bit twiddling" by the Hackers). Compare with LOW LEVEL.

High Low Tariff A tariff in which two prices are given for something — a high price and a low price. The first high/low tariff from AT&T was for leased voice lines where a lower charge was made per mile for connections between routes that have much traffic (High Density) and greater charges per mile are made for all other (Low Density) routes. The High/Low tariff was significant because it was AT&T's response to competition from long distance carriers like MCI and it was one of the first moves away from nationwide rate averaging, which was the way things were done under monopoly.

High Memory Area HMA. High Memory Area is the first 64KB of extended memory. If you're using MS-DOS 5.0 or 6.0, you can save some conventional memory (i.e. below 640K memory) by loading the operating system into HMA. Add the line DOS=HIGH to your CONFIG.SYS to use HMA for the operating system.

High Pass Filter A filter which passes frequencies above a certain frequency and stops (attenuates) those below.

High Performance Computing Act An Act passed by Congress in 1991 to foster the creation of computer "superhighways" linking computers at universities, national laboratories and industrial organizations. One objective of the High Performance Computing Program is the establishment of a gigabit/second National Research and Education Network (NREN) that will link the government, industrial and higher education communities involved in general research activities. Such a gigabit network would provide a significant increase in bandwidth compared with the existing National Science Foundation network, which is evolving from a 1.5 megabit per second (T-1) backbone to 45 megabit per second (T-3).

High Performance Routing HPR. A local area networking term. HPR is the next-generation APPN — referred to in the past as APPN+ — that adds IP-like dynamic networking — e.g., dynamic alternate routing in the event of path failure — features to APPN, and uses a routing mechanism that works at Layer 2 using a RIF concept similar to that found in SRB.

High Power Amplifier HPA. A device which provides the high power needed to shoot signals 22,000 miles plus from an earth station to a satellite.

High Rejection The ability of a voice recognition system active vocabulary words and reject those sounds that do not match closely with those words.

High Resolution TV Television with over 1000 lines per screen, about double the resolution of present systems. Sometimes called HDTV, for high-definition television. We're still awaiting standards for high resolution TV.

High Sierra Format A standard format for placing files and directories on CD-ROM, revised and adopted by the International Standards Organization as ISO 9660.

High Speed Digital Subscriber Loop See HDSL.

High Speed Local Network HSLN. A local network designed to provide high throughput between expensive, high-speed devices, such as mainframes and mass storage devices.

High Speed Printer Any printer which can print at over 100 lines a minute. Like many definitions, this one is arbitrary. Some people claim a dot matrix is "high speed" and a letter quality, daisy wheel is a "low speed" printer. Laser printers could be classed as high speed printers, maybe.

High Speed Register Set Registers are storage locations within the CPU that are used to hold both the data to be operated on and the instructions to accomplish the operations.

High Speed Signal An AT&T definition for a signal traveling at the DS3 rate of 44.736 mbps (million bits per second) or at either 90 mbps or at 180 mbps (Optical mode).

High Split A broadband cable system in which the bandwidth used to send toward the head-end (reverse direction) is about 5 MHz to 174 MHz, and the bandwidth used to send away from the head-end (forward direction) is about 232 MHz to 400 MHz. The guardband between the forward and reverse directions (180 MHz to 220 MHz) provides isolation from interference. High split requires a frequency translator.

High Usage Groups Trunk groups established between two central office switching machines to serve as the first choice path between the machines and thus, handle the bulk of the traffic. See HIGH USAGE TRUNK GROUP.

High Usage Trunk Group A Bellcore definition. A trunk group that is designed to overflow a portion of its offered traffic to an alternate route.

Highway 1. Another word for BUS. A common path or set of paths over which many channels of information are transmitted. The channels of the highway are separated by some electrical technique.
2. The Information Superhighway. In 1995, a consulting firm called Ovum define the superhighway as a mechanism for providing access to electronic information and content held on network servers. It has four key features, according to Ovum: A. it supports two way communications. B. It offers more than just simple voice telephony. C. It is interactive and provides real-time, co-operative communications. D. It supports electronic screen-based applications.
According to Ovum, the superhighway will allow services to be interconnected using the existing telephony network, as well as infrastructure which has yet to be constructed. The three waves of development will be:
Wave One (now). Wave one will be driven by the dramatic increase in computing power and the associated reduction in cost. The Internet has developed in Wave One, providing cheap access to basic communications. The Internet demonstrates the potential of the superhighway, but in its present form, its future is limited. For example, over the next two years the development of video and voice-processing applications on PC will overload its limited capacity.
Wave Two. In the second wave suppliers will build on Wave One and provide added functionality, using ISDN technology. ISDN will give users a wider choice of still images, low quality moving images. Set up times will speed up and costs will fall.
Wave Three (by 2000). By the third wave suppliers will be offering broadband capacity at the 1.5Mbit/s and above needed to provide broadcast-quality moving pictures and high-speed LAN interconnect. The new bandwidth capacities will require additional equipment, both in the local loop and in terminal equipment. Broadband capacity for business customers will be in both directions, but residential customers only need broadband to the home.

Highway Construction Supervisor A consultant to provide assistance in specification, installation and/ or operation of systems and software for accessing the information highway.

Highway Patrol The U.S. Congress.

HIPPI High Performance Parallel Interface. It's ANSI X3T9.3. HIPPI was originally developed as the interface to move data between supercomputers and peripherals such as disk arrays and frame buffers. More recently HIPPI has been extended and is used to network supercomputers, high-end workstations and peripherals using crossbar-type circuit switches. HIPPI provides for transfer rates of 800 megabits a second over 32 twisted pair copper wires (single HIPPI) and 1600 megabits a second over 64 pairs (double HIPPI). HIPPI connections are limited to 25 meters, although a serial HIPPI standard is being developed to extend this range for more than 10 kilometers. HIPPI is currently the most common interface in supercomputing environments.

Histogram An imaging term. A display plotting the density of the various colors and/or values in an image.

HIT 1. Electrical interference that causes the loss or introduction of spurious bits into a data stream.
2. The unit of measure most commonly cited by companies that have set up shop on the Internet's World Wide Web. Hits is a measure of file openings done on a Web site. It is not a measure of how many people have visited your Web site. Since one file is needed for every chunk of text and every graphic element on a Web page, one mouse click may count as a dozen or more hits, depending on the complexity of the page. Also, because most Web sites contain more than one page, hits can rapidly multiply. So, using "hits" as a measure of the popularity of a Web site is not accurate. The industry is working on a measure of people visits that's a little more accurate.
HITS Matches found in a search; e.g., a Veronica search for "NASA" will return a long list of hits for your query.
HIVR Host Interactive Voice Response. tying a voice response unit into a mainframe computer which has lots of data. Applications which can be produced include bank-by-phone, reservations-by-phone, etc. See INTERACTIVE VOICE RESPONSE.
HKSW Abbreviation for HOOK SWITCH, the actual electrical switch inside a phone that is controlled by the motion of the SWITCH HOOK.
HLC High Level Committee of ITU (International Telecommunication Union).
HLF High Level Function.
HLLAPI High Level Language Applications Programming Interface. An IBM API.
HLR Home Location Register. A wireless telecommunications term. A permanent database used in GSM (Global System for Mobile Communications) to identify a subscriber and to contain subscriber data related to features and services.
HMA See HIGH MEMORY AREA.
HMI 1. Novell's Hub Management Interface.
2. Human-to-Machine Interface.
HMM Hidden Markov Method. A common algorithm in voice recognition which uses probabilistic techniques for recognizing discrete and continuous speech.
Hobbit A microprocessor chip developed by AT&T's Bell Labs and used in the EO handheld devices. As of this writing, EO was 50% owned by AT&T.
HOBIC HOtel Billing Information Center used by hotels for getting immediate charges for long distance calls placed by their guests. A service of AT&T and the local telephone operating company delivered through a distinct and separate trunk and usually terminating on a telex machine, which prints guest long distance charges.
HOBIS HOtel Billing Information System.
Hold To temporarily leave a phone call without disconnecting it. You can return to the call at any time, sometimes from other extensions. There are several types of "HOLD" on a telephone system. How they work and what lamping they put on instruments varies from phone system to phone system.
Exclusive Hold: Prevents every other telephone from picking up the call. Only the telephone instrument that put the call on hold can retrieve it.
I-Hold: Effectively the same as Exclusive Hold.
Line Hold: The call is on hold. Anyone with a phone with the held line appearing on it can pick up the phone.
Hold Recall A telephone system feature which reminds you periodically that you've put someone on hold.
Holding Tank A queue in which a call is held until it can either use its assigned route or overflow into the next available route.
Holding Time The total time from the instant you pick up the handset, to dialing a call, to waiting for it to answer, to speaking on the phone, to hanging up and replacing the handset in its cradle. You are never billed for holding time. You are always billed for conversation time which is shorter than holding time. But holding time is an important figure to know when you're trying to determine how many circuits you need. For you will need sufficient circuits to take care of dialing, etc. — even though you're not being billed for that time.
Holiday Factor A call center term. A historical factor associated with a specific date and multiplied by the forecast call volume for that data in order to take into account an expected increase or decrease in the call volume. For example, if on a given day only half the usual number of calls occur for that day of the week and that time of year, the holiday factor for that date would be .5.
Hollerith Card A punched-hole 80 column card used for storing information for input into a computer. Remember the cards you got telling you "not to fold, bend, punch, spindle, etc."? They were Hollerith Cards. They're now falling into disfavor as other, less tamper-proof methods appear.
Hollerith Code Twelve level punched card code.
Hollow Pipeline Jargon for a broad bandwidth circuit that has no framing. See T-1.
Holographic Data Storage A technology still in the labs. It uses lasers and crystals rather than magnetic medium to store bits and bytes in holograms. Holographic data storage portends major reduction in physical storage space and much shorter seek time.
Holy War Arguments that involve certain basic tenets of faith, about which one cannot disagree without setting one of these off. For example: PCs are superior to Macintoshes.
Home The beginning place of a cursor on a CRT screen. Usually it's the top left hand corner. The function key on an IBM PC or clone marked "Home" will take the cursor to the home position, namely the top left hand corner.
Home Carrier The cellular operating company which a subscriber is registered with and pays the monthly service charge and usage charges to.
Home Location Register See HLR.
Home Page The Internet is a gigantic network of connected computers. The World Wide Web is on the Internet. The Web is the universe of accessible information available on many computers attached to the Internet. The Web has a body of software, a set of protocols and a set of defined conventions for getting at the information on the Web. The Web uses hypertext and multimedia techniques to

make the web easy for anyone to roam, browse and contribute to. The Web makes publishing information (i.e. making your information public) very easy. You'll need a computer, a telecommunications connection to the Internet and software to make your information accessible to anyone browsing the Internet (also known as net surfing). The first page that browsers see of the information you have posted on your computer attached to the World Wide Web is your "home page." It's a "welcome" page. It says "Welcome to my site, my home." It typically contains a table of contents to more information which a visitor (browser, surfer, etc.) will find at your site by clicking onto hypertext links you've created. See HTML, INTERNET, STREAMING and WORLD WIDE WEB.

Home Run Phone system wiring where the individual cables run from each phone directly back to the central switching equipment. Home run cabling can be thought of as "star" cabling. Every cable radiates out from the central equipment. All PBXs and virtually all key systems work on home run cabling. Some local area networks work on home run wiring. See LOOP THROUGH.

Home Screen An AT&T Merlin term. The "home base" of the display screen on digital telephones, which shows time, date, and call information, and when a feature is in use.

Homeostasis The state of a system in which the input and output are exactly balanced, so there is no change.

Homes Passed An expressed of the number of dwellings that a CATV provider's distribution facilities pass by in a given cable service area and an expression of the market potential of the area.

Homing When you dial a long distance number, your central office will choose a special set of trunks to send your call onto the next switching center for movement through the nationwide toll system. Those trunks are said to be the homing trunks for your central office. In other words, your central office is said to home on these trunks. If you're consistently encountering lousy long distance lines (and so are others on your central office), then ask your telephone company to check these trunks out.

Homo 1. The Greek prefix meaning the same.
2. Home Office Mobile Office. See also SOHO, which stands for Small Office Home Office.

Homogeneous Networks Composed of similar hardware from the same manufacturer.

Homologation Conformity of a product or specification to international telephony connection standards.

Honeycomb Coil A type of inductance in which the turns do not lie adjacent to each other.

Hook-up Wire A wire used for low current, low voltage (under 1000 volts) applications within enclosed electronic equipment.

Hookemware Free software that contains a limited number of features designed to entice the user into purchasing the more comprehensive version. See also HYPERWARE, MEATWARE, SHOVELWARE, SLIDEWARE and VAPORWARE.

Hookflash Momentarily depressing the hookswitch (up to 0.8 of a second) can signal various services such as calling the attendant, conferencing or transferring calls. In ISDN a hookflash signals the System Adapter to perform an operation, such as placing such as placing a call on hold. To hookflash, simply depress and release the receiver button. By default, the Hayes ISDN System Adapter recognizes a hookflash when the receiver button is depressed less than 0.8 of a second. You can change the default.

Hooking Signal An on-hook signal of 0.1 to 0.2 seconds duration used to indicate that a subscriber intends to initiate a new process such as "add-on."

Hookswitch Also called SWITCHHOOK or switch hook. The place on your telephone instrument where you lay your handset. A hookswitch was originally an electrical "switch" connected to the "hook" on which the handset (or receiver) was placed when the telephone was not in use. The hookswitch is now the little plunger at the top of most telephones which is pushed down when the handset is resting in its cradle (on-hook). When the handset is raised, the plunger pops up and the phone goes off-hook. Momentarily depressing the hookswitch (up to 0.8 of a second) can signal various services such as calling the attendant, conferencing or transferring calls. See HOOKING SIGNAL.

Hookswitch Dialing You can make phone calls by depressing the hookswitch carefully. If you push it five times, you dial five. Push it ten times you dial 0. Some coin phones discourage hookswitch dialing. Some don't.

Hoot'n'Holler Hoot'n'Holler are special 24-hour a day phone circuits which stay open 24-hours a day, seven days a week. Anyone picking up a phone on the circuit can listen and talk to whoever's on the line, or whoever might be within earshot. The idea is that to get someone to speak to you, you "hoot and holler," i.e. make a noise. A hoot'n'holler circuit is also called a Junkyard Circuit, Holler Down, Shout Down, Open Speech Circuit, Squawk Box System, FP or Full Period (as in FP 123456 circuit #). Hoot'n'Holler is a circuit consisting of 4 wires (2 pair — a transmit pair and a receive pair).
Technically, this is how it works: Audio energy present on the transmit pair at any location will appear on the receive of all the other locations, usually a multipoint circuit. The transmitted audio will not return on the receive pair of the originating location. Hoot'n'Holler circuits are voice conferencing oriented party lines and are non private by nature. There is no signaling on a hoot and holler circuit except when one "shouts down" to open speakers and "listening" at the distant out points. A Hoot'n'Holler circuit is a dedicated full time voice network. Individual four wire "drops" are connected via various bridging mechanisms. These "bridges" can be analog or digital and act as mix minus devices (mixes everyone else minus yourself). They provide all drops with connectivity with each other.

Hop-by-Hop Route A route that is created by having each switch along the path use its own routing knowledge to determine the next hop of the route, with the expectation that all switches will choose consistent hops such that the call will reach the desired destination. PNNI does not use hop-by-hop routing.

Hop Count A LAN definition. The number of nodes (routers or other devices) between a source and a destination. In TCP/IP networks, hop count is recorded in a special field in the IP packet header and packets are discarded when the hop count reaches a specified maximum value.

Hop Off When you make a phone call on the Internet you can call from one phone attached to the Internet to another phone attached to the Internet or you can call from one phone attached to the Internet and, at the other end, go into a PC stuffed with voice and switching cards and which is attached to local phone lines. The process of leaving the Internet is called "Hop Off."

Hops Term describing the number of times a message traverses different nodes.

Horizontal H. In television signals, the horizontal line of video information which is controlled by a horizontal synch pulse.

Horizontal Blanking Interval The period of time during which an electron gun shuts off to return from the right side of a monitor or TV screen to the left side in order to start painting a new line of video.

Horizontal Cross-Connect A cross-connect in the telecommunications closet or equipment room to the horizontal distribution cabling.

Horizontal Distribution Frame Located on the floor of a building. Consists of the active, passive, and support components that provide the connection between inter-building cabling (i.e. cabling coming from outside the building) and the intra-building cabling for a building.

Horizontal Link An ATM term. A link between two logical nodes that belong to the same peer group.

Horizontal Resolution Detail expressed in pixels that provide chrominance and luminance information across a line of video information.

Horizontal Scan Rate The frequency in Hz (hertz) at which the monitor is scanned in a horizontal direction; high horizontal scan rates produce higher resolution and less flicker. Thus, the EGA horizontal scan rate is 21.5Hz, while the VGA standard scan rate is 31.4Hz. Some displays now offer even higher scan rates, as much as 70 Khz. See MONITOR.

Horizontal Wiring The portion of the wiring system extending from the workstation's outlet to the BHC (Backbone to Horizontal Cross-Connect) in the telecommunications closet. The outlet and cross-connect facilities in the telecommunications closet are considered part of the horizontal wiring. See HORIZONTAL WIRING SUBSYSTEM.

Horizontal Wiring Subsystem The part of a premises distribution system installed on one floor that includes the cabling and distribution components connecting the riser subsystem and equipment wiring subsystem to the information outlet via cross connects, components of the administration subsystem.

Horn In radio transmitting, a waveguide section of increasing cross-sectional area used to radiate directly in the desired direction or to feed into a reflector that forms the desired beam.

Horn Alert HA. A cellular car phone feature that automatically blows the car's horn if a call is coming in.

Horsepower A unit of power equivalent to 550 foot pounds per second or to 746 watts.

Horton A software program which provides an automatic method for creating a directory of e-mail addresses. Users can look up electronic addresses via a search key which can be a fragment of a person's name.

Hose And Close A pattern of behavior exhibited by phone tech-support people who spout a bunch of jargon you don't understand, ask you to perform a bunch of procedures you don't follow, and then abruptly hang up. This definition from Wired Magazine.

Host 1. An intelligent device attached to a network.
2. A mainframe computer.
3. A computer with full two-way access to other computers on the Internet. A host can use virtually any Internet tool, such as WAIS, Mosaic and Netscape.

Host Apparent Address A set of internetwork layer addresses which a host will directly resolve to lower layer addresses.

Host Bus Adapter HBA. A printed circuit board that acts as an interface between the host microprocessor and the disk controller. The HBA relieves the host microprocessor of data storage and retrieval tasks, usually increasing the computer's performance time. A host bus adapter (or host adapter) and its disk subsystems make up a disk channel.

Host Carrier The cellular operating company a subscriber from another cellular system would be billed roamer charges.

Host Computer A computer attached to a network providing primarily services such as computation, database access or specific programs of special programming languages.

Host Interactive Voice Response A voice response system that can communicate with a host computer, typically a mainframe. Applications which can be produced include bank-by-phone, reservations-by-phone, etc.

Host Name The name given to a mainframe computer.

Host Name Resolution A mechanism that provides static and dynamic mechanisms for resolving host names into numeric addresses. The Internet Name Server Protocol accesses an Internet name server that provides dynamic name-to-number translation (this process is specified in IEN 116). The Domain Name Protocol accesses a Domain Name Server that provides dynamic name-to-number translation (this process is specified in RFC-1034 and RFC-1035). A static local host table can also be accessed for name-to-number translation.

Host Number The part of an internet address that designates which node on the (sub)network is being addressed. See DOMAIN.

Host Processor Same as HOST COMPUTER.

Host Server A device which connects to a LAN and then allows a computer, which cannot directly support the LAN protocols, to connect to it, providing all necessary LAN support.

Host Site In the transfer of files, the host site is the location receiving a file. When two individuals are exchanging files, the one who receives the file first would be the host, the other would be considered the remote.

Host Switch A central office switching system which provides certain functions to a smaller switch located remotely.

Host Table 1. A list of TCP/IP hosts along with their IP addresses.
2. In Windows 95, the host table is HOSTS or LMHOST file that contains lists of known IP addresses mapped to host names or NetBIOS computer names.
3. An ASCII text file where each line is an entry consisting of one numeric address and one or more names associated with that address. Host tables are used to resolve hostnames into numeric addresses.

Hot Live wire. A conductor carrying a signal is said to be a hot conductor, i.e. the wire carrying the signal or the ground as opposed to the neutral or ground wire.

Hot And Ground Reversed In AC electrical power, the correct connection of the Hot and Ground wires is reversed. This is an extremely dangerous condition because the GROUND path will rise to 129 Volts and can present a lethal shock hazard to anyone in contact with equipment powered from this outlet or any outlet using the same ground path.

Hot And Neutral Reversed Also called reversed Polarity. A symptom of poor AC electrical wiring. In this case the correct connection of the Hot and Neutral conductors is reversed. Dangers include increased leakage current, and damage to electronic equipment or motors and appliances requiring correct polarity.

Hot Chat An Internet term. Sex talk, in real time, online, usually between two or more consenting people (through not necessarily adults).

Hot Cut The conversion from an old to a new phone system which occurs instantly as one is removed from the circuit and the other is brought in. There are advantages and disadvantages to Hot Cuts. For one, they're likely to be much more dangerous than a Parallel Cut, in which the two phone systems run side by side for a month or so. Also known as Flash Cut.

Hot Desk An employee of a company no longer has a permanent office. He works out of their home, visits customers and communicates with the office through fax and electronic mail. Occasionally that person finds it necessary to visit an office of the company. He is allocated a desk and perhaps an office for his stay. That stay might be as short as an hour or as long as several weeks. Once he checks out, someone else gets the desk. This arrangement is called a "Hot Desk."

Hot Docking Docking is to insert a portable computer into a base unit. Cold docking means the computer must begin from a power-off state and restart before docking. Hot docking means the computer can be docked while running at full power.

Hot Fix A feature of Novell's NetWare LAN (local area network) operating system in which a small portion of the hard disk's storage area is set aside as a "Hot Fix Redirection Area." This area is set up as a table to hold data that are "redirected" there from faulty blocks in the main storage area of the disk. It's a safety feature.

Hot Key Combination A combination of keys on the keyboard that are pressed down simultaneously to make the computer perform a function. For example, the Ctrl, Alt, Del hot-key combination will warm boot an MS-DOS computer.

Hot Line A private line dedicated between two phones. When you pick up either phone or do some act of signaling (like push a button), the other phone rings instantly. Hot lines are useful in emergencies and other areas where time is of the essence — e.g. trading currencies.

Hot Line Service When you pick up the phone, you're automatically connected with a phone number. Such Hot-Line Service on a PBX typically gets you emergency service, etc. See also HOT-LINE.

Hot Links A methodology that references and can connect information from one document to another, regardless of the type of application used.

Hot List A gopher or Web file that lets you quickly connect to your favorite pre-selected page. Appropriately named. The way it works: You connect to a home page. You decide you'd like to return at some other time. So you command your internet surfing software to mark this web site on your hot list (also called marking it with a "bookmark." Next time you want to return to that web site, you simply go to your hot list or your bookmarks, click on which one you want. And bingo, you're there. A hot list is also known as a bookmark. Most Web browsers have book marks or hot lists.

Hot On Neutral, Hot Unwired In AC electrical power, the HOT wire is connected to the NEUTRAL terminal of the outlet and the HOT terminal in UNWIRED. Dangers include shock hazard from excessive leakage current and fire hazard. Depending on other conditions, equipment may or may not operate.

Hot Potato Routing A packet-switching routing technique that retransmits a packet as soon as possible after receiving a packet, even if it means making a poor routing choice.

Hot Racking A Navy term referring to the practice in submarines of having sailors sleep in the same bunk at different times. This occurs because of the shortage of bunks on submarines. Hot racking is the reason the US Navy gave in May of 1995 for vetoing the idea of having women serve on crowded submarines.

Hot Redundancy A term used in conjunction with very critical telecom and computing systems, such as 911 service. With Hot Redundancy, the component or the system runs in parallel with an identical "twin." Should one twin fail, the other is already running and provides full service without interruption.

Hot Restart Imagine a corporate telephone system, perhaps a PBX or an ACD. It's handling phone calls to and from customers every second of the day. It's mission critical. You can't allow it to crash for even a second a day. But phone systems are nothing more than specialized computers with specialized software. They will crash or lock up just like your PC, though perhaps not as often. But they will crash and lock up. What happens when your PC crashes and locks up? You will lose data. You will probably then reboot your PC, probably from its hard disk. This will cost you anywhere from 30 to 60 seconds. You can tolerate this on your PC. You can't tolerate this on your phone system. An integral part of most phone systems' software is a feature called "Hot Restart." When the phone system's software crashes or locks up, there is "Hot Restart" software that enables the phone system to restart itself (another word for reboot) without losing the phone calls in progress or without taking 30 to 60 seconds to load a new operating system from hard disk or tape. How this exactly is done seems to vary from one telephone system maker to another. Some may keep an operating system in RAM. Some may keep two identical processors chugging away simultaneously and switch from one to another, when things go awry. How manufacturers do "Hot Restarts" is something of a trade secret.

Hot Standby Backup equipment kept turned on and running in case some equipment fails.

Hot Swap The process of replacing a failed component — e.g. a RAID drive — while the rest of the system (in this case, the disks) continue to provide function normally, i.e. providing data to the network users and providing a place for them to store their data.

Hotel/Motel Console A specialized PBX console or a normal console programmed to work specifically in hotels and motels. The console will often show room status information.

Hotfix A Novell program that dynamically marks defective blocks on the hard disk so they will not be used. See HOT FIX.

Hotjava Java is a programming language from Sun Microsystems designed primarily for writing software to leave on World Wide Web sites and downloadable over the Internet to a PC owned by you or me. HotJava, its brother, is another piece of software installed on a Web browser at your desktop. HotJava enables Java programs delivered over the Web to run on your desktop PC. In short, Java is the programming language the programs on the Web are written in. HotJava is the software that will sit on your PC. Java is basically a new virtual machine and interpretive dynamic language and environment. It abstracts the data on bytecodes so that when you develop an app, the same code runs on whatever operating system you choose to port the Java compiler/interpreter to. What's a Java application? According to Wired Magazine, point to Ford Motor Company's website today, for instance, and all you'll get are words and pictures of the latest cars and trucks. Using Java, however, Ford could relay a small application (called an applet) to a customer's computer (the one on your desk which are using the surf the Internet).

NEWTON'S TELECOM DICTIONARY

The customer could then customize options on an F-series pickup while calculating the monthly tab on various loan rates offered by a finance company or local bank. Add animation to these applications and you could get to "drive" the truck.

Hotline There are several definitions. In the cellular business, a hotline is a system restriction that allows a cellular customer to call only one prearranged number. In the landline phone business, a hotline is often a dedicated line. But it may also be a phone which only dials one number. In this case it's often call a virtual private line. See HOTLINE VIRTUAL PRIVATE LINE SERVICE.

Hotline Virtual Private Line Service A Nynex offering that simulates private line service. HotLine lets you automatically dial a predetermined line within the HotLine network, just by picking up the handset. Hotline uses facilities, switches and programmed intelligence of the public network to create a closed network of simulated private lines.

House Cable Communication cable within a building or a complex of buildings. House cable owned before divestiture by the Bell System and after divestiture by the Regional Bell Operating Companies will eventually be fully depreciated and will then belong to the customer.

Howler A device which produces a loud sound to a subscriber's phone or private branch exchange (PBX) extension to indicate that the handset is off-hook and it ought to be put back on hook.

Howler Tone A tone which gets increasingly louder over a short period of time. It is used to notify a user that his phone handset is off its hook.

Howling Howling is typically heard in a speakerphone or conferencing unit when there is "Acoustic coupling" between the microphone and the speaker. This is due to putting the microphone too near the speaker. New circuits called acoustic echo cancelers allow you to operate the microphone and the speaker simultaneously and much closer to each other.

HP Hewlett-Packard Company. HP was formed in 1939 by Bill Hewlett and David Packard. David lost the toss to Bill, which is why the company is called Hewlett-Packard, not Packard-Hewlett. They started the company in a small garage in Addison Avenue in Palo Alto, CA. Their first sale was an audio oscillator used in a Disney film, "Fantasia." See also MBWA.

HP Openview Hewlett-Packard's Openview network management products allow network administrators to monitor and control network devices from an MS-DOS PC or UNIX workstation.

HPA See HIGH POWER AMPLIFIER.

HPAD Host Packet Assembler/Disassembler. The HPAD can link to a host or FEP with native protocol data, or if the host can accept it, with X.25 input. The 4400 PAD functions as either an HPAD or a TPAD See TPAD.

HPFS High-performance file system (HPFS); primarily used with the OS/2 operating system version 1.2 or later. It supports long filenames but does not provide security. OS/2 can use any file system it wants, thanks to its installable file system (IFS) architecture. Two choices available are the FAT file system, used by MS-DOS, and the High Performance File System (HPFS). You can mix and match each and select one at boot time, thanks to OS/2's Dual Boot option. IBM, which created OS/2, claims HPFS is much more efficient than FAT. It tries to store all files on disk contiguously and uses its own built-in cache. However, HPFS' most notable attribute is the long 254-character file names and case preservation. OS/2 remembers file names as upper and lower case (though it's not case-sensitive to commands).

HPO High Performance Option. A way of improving equipment transmission characteristics. For instance, the upgrading of a voice-grade line to meet standards for data transmission.

HSCS High-Speed Circuit Switched

HSDA High Speed Data Access.

HSDL High-speed Subscriber Data Line. A Bellcore idea for a two pair phone line coming into a house or business that is a full-duplex T-1 line. See also ADSL.

HSDU High Speed Data Unit.

HSL A computer imaging term. A color model based on hue, saturation, and luminance. Hue is the attribute that gives a color its name (e.g., red, blue, yellow, or green). In this model, saturation refers to the strength, or purity, of the color. If you mix watercolors, saturation would specify how much pigment you added to a given amount of water. Luminance identifies the brightness of a color. For example, full luminance yields white, while no luminance yields black. See also HSV.

HSLN High-Speed Local Network. A local network designed to provide high throughput between expensive, high-speed devices, such as mainframes and mass storage devices.

HSSI High Speed Serial Interface. A serial data communication interface optimized for high speeds up to 52 Mbps. Used for connecting an ATM switch to a T-3 DSU/CSU, for example.

HST High Speed Technology, a U.S. Robotics proprietary signaling scheme, design and error control protocol for high-speed modems. HST incorporates trellis-coded modulation, for greater immunity from variable phone line conditions, and asymmetrical modulation for more efficient use of the phone channel at speeds of 4,800 bps and above. HST also incorporates MNP-compatible error control procedures adapted to asymmetrical modulation.

HSV A computer imaging term. A color model based on hue, saturation, and value. Hue specifies the color, as in the HSL model. In this model, saturation specifies the amount of black pigment added to or subtracted from the hue. Value identifies the addition or subtraction of white pigment from the hue.

HTML HyperText Markup Language. This is the authoring software language used on the Internet's World Wide Web. HTML is used for creating World Wide Web pages. HTML is basically ASCII text surrounded by HTML commands in angle brackets, which your browser interprets whichever way it feels. That means (and I reiterate this) different browsers will display the exact same HTML code differently.) Here is an example of a simple line of HTML code:
<H2>Call Center Magazine Editorial Calendar</H2>
This line says to the browser: Display those words as a type 2 headline. You can also include an image in an HTML page.
<img src="photos/harry.gif">
Your browser would go find the picture "harry.gif" in the subdirectory called "photos" on the computer which had the URL and the Web page you were visiting. An HTML document has three types of content: tags (which define type styles, like the one

above), comments (words to tell the HTML author what he's doing, but which aren't displayed) and text (i.e. the words Call Center Magazine Editorial Calendar. Tags also let you do hyperlinking, which lets a person browsing click your HTML document, click on that hyperlink and go elsewhere — either to another page which you wrote, or to another Web page (i.e. URL) across the world.

Here's an example of text and tags you might put in your HTML document:
<h2>For more information on computer telephony, please visit</h2> <a href="http:/www.computertelephony.com"> </a>
When someone browsing your page paces their mouse on the words "www.computertelephony.com," their cursor changes to a hand. They click on it. A few seconds later, they see the computer telephony home page begin to download.

HTR Hard-To-Reach.

HTTP HyperText Transfer Protocol. Invisible to the user, HTTP is the actual protocol used by the Web Server and the Client Browser to communicate over the "wire". In short, the protocol used for moving documents around the Internet. See DOMAIN, INTERNET, SURF and WEB BROWSER.

Hub The point on a network where a bunch of circuits are connected. Also, a switching node. In Local Area Networks, a hub is the core of a star as in ARCNET, StarLAN, Ethernet, and Token Ring. Hub hardware can be either active or passive. Wiring hubs are useful for their centralized management capabilities and for their ability to isolate nodes from disruption.

Hub Junction Box A box used to connect a Hub Interface when a node is placed at a remote location.

Hub Management Interface A network management protocol developed by Novell to allow network managers to manage hubs anywhere on a NetWare LAN.

Hub Polling A polling system in which a polled station sends its traffic and passes the polling message (after it's sent its message) to the next station.

Hub Site The location(s) on a network where many circuits are brought together to be multiplexed into a single higher speed connection.

Hue The attribute by which a color may be identified within the visible spectrum. Hue refers to the spectral colors of red, orange, yellow, green blue and violet.

Huffman Encoding A popular loss-less data compression algorithm that replaces frequently occurring data strings with shorter codes. Some implementations include tables that predetermine what codes will be generated for a particular string. Other versions of the algorithm build the code table from the data stream during processing. Huffman encoding is often used in image compression.

Huge Pipes A high-bandwidth Internet connection, typically T-1.

Hundred Call Seconds Known by the initials CCS where C is the roman numeral for Hundred. One CCS is 36 times the traffic expressed in Erlangs. See CCS.

Hunt Refers to the progress of a call reaching a group of lines. The call will try the first line of the group. If that line is busy, it will try the second line, then it will hunt to the third, etc. See also HUNT GROUP.

Hunt Group A series of telephone lines organized in such a way that if the first line is busy the next line is hunted and so on until a free line is found. Often this arrangement is used on a group of incoming lines. Hunt groups may start with one trunk and hunt downwards. They may start randomly and hunt in clockwise circles. They may start randomly and hunt in counter-clockwise circles. Inter-Tel uses the terms "Linear, Distributed and Terminal" to refer to different types of hunt groups. In data communications, a hunt group is a set of links which provides a common resource and which is assigned a single hunt group designation. A user requesting that designation may then be connected to any member of the hunt group. Hunt group members may also receive calls by station address. See also TERMINAL HUNT GROUP.

Hunting See ROLLOVER LINES.

HVAC Heating, Ventilating, and Air-Conditioning systems. High voltage stuff. Keep your telecommunications cables away from the motors in HVAC systems, please.

HVQ Hierarchical Vector Quantization — a method of video compression introduced by PictureTel in 1988 which reduced the bandwidth necessary to transmit acceptable color video picture quality to 112 Kbps.

Hybrid A device used for converting a conversation coming in on two pairs (one pair for each direction of the conversation) onto one pair and vice versa. This is necessary because all long distance circuits are two pairs, while most local circuits are one pair. Here is a longer explanation from "Signals, The Science of Telecommunications," by John Pierce and Mike Noll:

The telephone instrument in your home is connected to a single pair of wires called the subscriber loop or local loop, which carries both the outgoing voice signal and the incoming one. This pair of wires creates an electrical circuit for each of the two signals. A device in your phone called a hybrid or hybrid coil keeps the two signals separate, more or less, so that what you say into your phone's transmitter doesn't blast into your ear from the receiver.

In contrast, all multiplex systems provide separate talking paths in two directions. Separate paths are necessary because the amplifiers placed along the lines between terminals amplify signals traveling in one direction only. When two people talk between New York and San Francisco the call goes from one phone through a local two-wire voice circuit to a multiplex terminal. There the call is transferred to a four-wire long distance circuit that consists of two separate one-way circuits. At the end of the system, a hybrid reconverts each four-wire circuit into a two-wire circuit.

Hybrid Cable A communication cable that contains two or more types of conductors that bear electrical signals, a mixture of signal-bearing electrical conductors and optical fibers, and/or two or more different types of optical fibers. A communication cable containing signal-bearing media and electric power conductors.

Hybrid Coil Transformer used in a balancing network to connect a 4-wire line to a 2-wire line.

Hybrid Communication Network A communication system that uses a combination of trunks, loops, or links, some of which are capable of transmitting (and receiving) only analog or quasi-analog signals and some of which are capable of transmitting (and receiving) only digital signals.

Hybrid Connector A connector containing both optical fiber and electrical conductors.

Hybrid Coupler In antenna work, a hybrid junction forming a directional coupler. The coupling factor is normally 3 dB.

Hybrid Disk 1. A CD-ROM term. Under the Orange Book standard for recordable CD, a hybrid disc is a recordable disc on which one or more sessions are already recorded, but the disc is not closed, leaving space open for future recording. However, in popular use the term "hybrid" often refers to a disc containing both DOS/Windows platform is seen as a ISO 9660 disc, while on a Mac it appears as an HFS disc.
2. A CD-ROM disc which works in both a Macintosh and an MS-DOS/Windows. Most these days do.
Hybrid Junction A waveguide or transmission line arrangement having four ports that, when terminated in their characteristic impedance, have the property that energy entering any one port is transferred (usually equally) to two of the remaining three ports. Widely used as a mixing or dividing device.
Hybrid Key System Term used to describe a system which has attributes of both Key Telephone Systems and PBXs. The one distinguishing feature these days is that a hybrid key system can use normal single line phones in addition to the normal electronic key phones. A single line phone behind a hybrid works very much like a single line phone behind a PBX. The second distinguishing feature of a hybrid is that it's "non-squared." This means that not every trunk appears as a button on every phone in the system — as occurs on virtually every electronic key system manufactured today.
Hybrid Local Network An integrated local network consisting of more than one type of local network (e.g. LAN, HSLN, digital PBX).
Hybrid Mode A mode possessing components of both electrical and magnetic field vectors in the direction of propagation.
Hybrid Network 1. A communications network which has some links capable of sending and receiving only analog signals and other links capable of handling only digital signals. The current public switched telephone network is Hybrid. A Hybrid Network is also a network with a combination of dissimilar network services, such as frame relay, private lines and/or X.25.
2. An amalgam of public and private network transmission facilities.
Hybrid Set Two or more transformers interconnected to form a network having four pairs of accessible terminals to which may be connected four impedances so that the branches containing them may be made interchangeable.
Hydra A 25-pair cable that at one end has an Amphenol connector (typical of what 1A2 phone systems were connected with) and at the other has many individual 2, 4, 6 and 8 wire connectors, typically male RJ-11s. A hydra cable is named for a mythological multi-headed monster. It's more commonly called an octopus cable. The reason it's called an octopus is that it looks a bit like an octopus — one body and many arms.
Hydrogen Loss Increases in optical fiber attenuation that occur when hydrogen diffuses into the glass matrix and absorbs some light.
Hydrometer Instrument for determining the density of liquids. Formerly in wide use for testing radio storage "A" batteries.
Hygroscopic Capable of absorbing moisture from the air.
Hyperchannel An SCSA term. A data path on the SCbus or SCxbus Data Bus made up of more than one time slot. By bundling time slots into a hyperchannel, data paths with a bandwidth greater than 64 Kbps can be created.
Hyperfiction According to the New York Times, hyperfiction is a new narrative art form, readable only on the computer, and made possible by the developing technology of hypertext and hypermedia. Not all adults have familiarized themselves with hypertext, but most children have, for it is the basis of many of their computer games and is fast becoming the dominant pedagogical tool of our digitalized times.
Hyperlink A way to connect two Internet resources via a simple word or phrase on which a user can click to start the connection.
Hypermedia A way of delivering information that provides multiple connected pathways through a body of information. Hypermedia allows the user to jump easily from one topic to related or supplementary material found in various forms, such as text, graphics, audio or video.
Another definition we found is: Non-linear media, of which multimedia can be a form. Just as hypertext is a non-sequential, random-access arrangement of text, hypermedia is a non-sequential, random-access arrangement of multiple media such as video, sound and computer data.
A third definition: Hypermedia is a type of authoring and playback software through which you can access multiple layers of multimedia information related to a specific topic. The information can be in the form of text, graphics, images, audio, or video. For example, suppose you received a hypermedia document about the Sun file system. You could click on a hotspot (such as the words file system) and then read a description. You could then click an icon to see an illustration of a file structure, and then click the file icon to see and hear information in a video explaining the file system.
Hyperstream A name for frame relay service from Stentor and MCI.
Hypertext Also called hypermedia; software that allows users to explore and create their own paths through written, visual, and audio information. Capabilities include being able to jump from topic to topic at any time and follow cross-references easily. Hypertext is often used for Help files. See HYPERTEXT TRANSFER PROTOCOL.
Hypertext Transfer Protocol HTTP is the transport protocol in transmitting hypertext documents around the Internet.
Hyperware New hardware that has been announced and perhaps even publicly demonstrated, but is not being shipped to commercial customers. Vaporware is software which has been announced, but is not yet shipping to commercial customers. Years can pass between public announcement and actual commercial shipment. Be wary.
Hysterises An uninterruptible power supply (UPS) definition. The voltage output from the wall will continually shift within a certain range, causing some UPSes to constantly switch back and forth from AC to battery power.
Hytelnet A hypertext system which contains information about the Internet, such as accessible library catalogs, Freenets, Gophers, bulletin boards, etc.
HZ See Hertz.

NEWTON'S TELECOM DICTIONARY

I Used on switches to mean "ON." The "OFF" setting is "O," (the letter coming after N.)

I&R Installation and Repair. The telephone company department responsible for these jobs. I&R refers to a person's job area, a department, or tools and test gear made for I&R

I-CF ISDN Call Forwarding.

I-CFDA ISDN Call Forwarding Don't Answer.

I-CFDAIO ISDN Call Forwarding Don't Answer Incoming Only.

I-CFIB ISDN Call Forwarding Interface Busy.

I-CFIBIO ISDN Call Forwarding Interface Busy Incoming Only.

I-CFIG ISDN Call Forwarding IntraGroup only.

I-CFIO ISDN Call Forwarding Incoming Only.

I-CFPF ISDN Call Forwarding over Private Facilities.

I-CFV ISDN Call Forwarding Variable.

I-CFVCG ISDN Call Forwarding Variable facilities for Customer Groups.

I-CNIS ISDN Calling Number Information Services.

I-ETS Interim European Telecommunications Standard.

I-HC ISDN Hold Capability.

I-Hold Indication A telephone system feature. If I put someone on hold at my phone, all the other phones which have the same line appearing on them will start flashing — indicating that the call is on hold.

I-MAC Isochronous Media Access Control. An FDDI-II term. See FDDI-II.

I-MUX Inverse Multiplexer. See INVERSE MULTIPLEXER.

I-Series Recommendations ITU-T recommendations on standards for ISDN services, ISDN networks, user-network interfaces, and internetwork and maintenance principals.

I-TV Interactive TV.

I-Use Shows a user which line the phone is connected to when the receiver is off-hook. It does this by illuminating a small light below that line button. Most key sets and PBX sets have I-Use buttons. Most of the newer two-line phones do also.

I-Way An acronym for the Information SuperHighway, that nebulous concept which refers to interconnected telecommunications channels snaking their way into every household, every company, every college, every university in the world. Essentially, the Information SuperHighway is a fancy term for the nation's phone network overlaid with heavy data communications ability.

I.122 ITU-T description of the general "bearer" services offered by ISDN networks, including both packet-switched and frame relay data services.

I.356 ITU-T Specifications for Traffic Measurement.

I.361 B-ISDN ATM Layer Specification.

I.362 B-ISDN ATM Layer (AAL) Functional Description.

I.363 B-ISDN ATM Layer (AAL) Specification.

I.430 Basic rate physical layer interface defined for ISDN. The ITU-T Layer 1 specification for the ISDN BRI S/T-interface, which consists of four wires. I.430 specifies ASI line coding. The beginning and end of each frame is marked with deliberate bi-polar violations. Each BRI frame is forty-eight bits in length including the bi-polar violations; and repeated 4,000 times per second for a total line rate of 192 Kb/s in each direction. The point-to-point limit is one kilometer. the passive bus is limited to about 10 meters.

I.431 Primary rate physical layer interface defined for ISDN. The ITU-T Recommendation for Layer 1 of the ISDN PRI. Specifies operation on North American T-1 at 1.544 megabits per second (23B+D) or European E-1 at 2.048 megabits per second (30B+2D).

I.432 ITU-T Recommendation for B-ISDN User-network Interface.

I.440 The ITU-T specification, commonly known as Q.920, which describes the general network aspects of the LAPD protocol (also known as DSS1).

I.450 (Q.930) The ITU-T specification describing the general network aspects of the ISDN D channel Layer 3 protocol.

I.451 See Q.931

I.452 See Q.932

I.R.D.A. Infrared Data Association. See INFRARED.

I/G Bit Bit in IEEE 802 MAC address field distinguishing between individual and group addresses.

I/O Input/Output. See the following definitions.

I/O Bound When a computer systems spend much of its time waiting for peripherals like the hard disk or video display, it is said to be I/O bound. If your computer is I/O Bound, going to a faster CPU (like a 386 or 486) might make little perceived difference. What you need is a faster hard disk or faster video card, etc.

I/O Channel Equipment forming part of the input/output system of a computer.

I/O Controller Provides communications between the central processor and the I/O devices.

NEWTON'S TELECOM DICTIONARY

I/O Device An input/output device, which is a piece of hardware used for providing information to and receiving information from the computer, for example, a disk drive, which transfers information in one of two directions, depending on the situation. Some input devices, such as keyboards, can be used only for input; some output devices (such as a printer or a monitor) can be used for output. Most of these devices require installation of device drivers.
I/O Request Packet IRP. Data structures that drivers use to communicate with each other.
IX An InterExchange carrier, i.e. a long distance phone company.
iXXX The little i stands for Intel and the numbers that follow refer to the particular microprocessor chip. For example, i386 refers to the 80386 chip.
i386SL One version of Intel's '386 family of microprocessors. The i386SL's special feature is that it can be slowed to 0 megahertz and still maintain register integrity (memory) practically indefinitely. This results in significant power savings for computers (especially laptops) that advantage of this feature.
i750 Name of the programmable video processor family from Intel.
IA5 International Alphabet No. 5
IAB Internet Architecture Board. A policy setting and decision-review board for the TCP/IP-based Internet.
IAL Intel Architecture Labs, home of the ISA Bus, Plug and Play, Universal Serial Bus and other PC advances designed to sell more Intel products.
IAM Initial Address Message in the ISUP call set up messages. A mandatory message sent in the forward direction to initiate seizure of an outgoing circuit and to transmit address and other information relating to the routing and handling of a call. See also ISUP and COMMON CHANNEL SIGNALING.
IANA Internet Assigned Numbers Authority. The entity responsible for assigning numbers in the Internet Suite of Protocols.
IAO IntraOffice SONET Signal. Standard SONET signal used within an Operating Company central office, remote site, or similar location.
IAP Internet Access Provider. Same as Internet Service Provider. See ISP.
IASG Internetwork Address Sub-Group: A range of internetwork layer addresses summarized in an internetwork layer routing protocol.
IBDN The Integrated Building Distribution Network (IBDN) is an unshielded twisted pair/fiber optic based structured wiring system based on the EIA\TIA 568 wiring standard. IBDN is a creation of Northern Telecom. IBDN is an open wiring system meaning that it can support any standards based data or voice application available today on unshielded twisted pair horizontal wiring.
IBERPAK A ITU-T X.25 packet-switched network operated in Spain by the Spanish government.
IBM International Business Machines. Also known affectionately as I've Been Moved, International Big Mother, Itty Bitty Machines, It's Better Manually, along with others not suitable for printing (but used in the back of better computer rooms all across the North American continent). IBM is powerful in some parts of the computer industry. This is why many of the IBM terms in this book are described as how IBM defines them.
IBM 8514/A Graphics standard introduced by IBM with 1,024 x 768 resolution. Many current monitors are 8514/A-compatible. 8514/A is also called XGA, or eXtended Graphics Array, which is IBM's high-resolution extension to its VGA adapter. It provides a resolution of 1,024 horizontally x 768 vertically, yielding 786,432 possible bits of information on one screen, more than two and a half times what is possible with VGA.
IBM PC Introduced in the summer of 1981 using the 16-bit Intel 8088 processor.
IBM Token Ring A local area network using star wiring architecture of two pair cabling to each drop location — one pair from the hub to the workstation and one pair from the workstation back to the hub to continue the ring. The IBM 8228 Multiple Access Unit (MAU) will support communications for eight PCs (workstations). Up to 33 IBM 8228 MAUs may be connected together into a single ring, supporting up to 260 data devices. MAU to MAU connection is accomplished with data connectors equipped with Type 1 cables from a MAU's RO (Ring Out) connection to the next MAU's RI (Ring In). The final MAU's RO connects back to the initial MAU's RI to complete the ring. See also TOKEN PASSING and TOKEN RING.
IBND Interim Billed Number Database
IBR Bellcore spec 54019, which covers specs on delivering fractional T-1.
IBS Intelligent Battery System. A conventional battery system interfaces to its host product and charger through a power and perhaps a simple sensor port. An intelligent battery system has state sequential intelligence, memory, and a data communications protocol to the conventional battery sensor package. Thee IBS additions allow sensor data, events and memory access to take place between the battery pack and the host device and charger.
IBX Integrated Business eXchange. Another name for a PBX. This is the name InteCom uses for their PBX family. InteCom is now owned by Matra, a French company.
IC 1. InterCom.
2. Integrated Circuit.
3. Intermediate Cross-connect. An interconnect point within backbone wiring. for example, the interconnection between the main cross-connect and telecommunications closet or between the building entrance facility and the main cross-connect.
IC DRAM Integrated Circuit Dynamic Random Access Memory.
ICA International Communications Association. ICA is the biggest trade association of the largest corporate telecommunications users — the people whose companies spend the most. In 1992, ICA has more than 700 corporate members, representing $23.7 billion in combined annual purchasing power. It was founded in 1949 as the National Committee of Communications Supervisors. The ICA holds an annual telecommunications trade show. It happens every May. A schedule is published in the back of TELECONNECT Magazine. The ICA is based in Dallas, TX. The other large user association is called the TCA — TeleCommunications Association. They're West Coast oriented.
ICAPI International Call Control API.

ICD International Code Designator: This identifies an international organization. The registration authority for the International Code Designator is maintained by the British Standards Institute. The length of this field is two octets.
ICEA Insulated Cable Engineers Association.
ICFA International Computer Facsimile Association. The mission of this new organization is to create awareness of the benefits and uses of computer fax to increase worldwide market size.
ICI 1. Interexchange Carrier Interface. The interface between carrier networks that support SMDS.
ICIT International Center for Information Technologies. A part of MCI.
ICM See INTEGRATED CALL MANAGEMENT.
ICMP Internet Control Message Protocol. As specified in RFC-792, ICMP provides a number of diagnostic functions and can send error packets to hosts. ICMP uses the basic support of IP and is an integral part of IP.
iCOMP Intel Comparative Microprocessor Performance Index. A test for measuring how fast microprocessors are.
Icon An icon is a picture or symbol representing an object, task, command or choice you can select from a piece of software (e.g. a trash can for a deletion command).
Iconography The science of icons. A fancy name for talking about icons — those little visual representations of objects, tasks or commands you find in Windows and other GUI (Graphical User Interface) programs and operating systems.
ICP 1. Intelligent Call Processing. The ability of the latest ACDs to intelligently route calls based on information provided by the caller, a database on callers and system parameters within the ACD such as call volumes within agent groups and number of agents available.
2. Instituto das Communicacoes de Portugal (The Portuguese Institute of Communications).
3. Intelligent Call Processor. The name of an AT&T service. ICP allows users to directly link their customer premise equipment to its network for individual call processing based on customer-specific information.
ICR Initial Cell Rate: An ABR service parameter, in cells/sec, that is the rate at which a source should send initially and after an idle period.
ICS 1. See INTERACTIVE CALL SETUP.
2. Integrated Communications System. A Northern Telecom definition: A telecommunications based platform with advanced processing power and capacity that enables integration and orchestration of typical business equipment (telephones, fax, etc.) through open architecture interfaces, as in the Norstar-PLUS Modular ICS.
ICTA The International Computer-Telephony Association (ICTA), based at Campus Box 350, University of Colorado, Boulder, CO 80302.
ICV Integrity Check Value is a digest of a message which provides a high level of assurance that the message has not been tampered with. Also referred to as Message Authentication Code.
ID Identifier.
IDA Integrated Data Access or Integrated Digital Access.
IDC Insulation Displacement Connection. A type of wire connection device in which the wire is "punched down" into a double metal holder and as it is the metal holders strip the insulation away from the wire, thus causing the electrical connection to be made. The alternate method of connecting wires is with a screw-down post. There are advantages and disadvantages to both systems. The IDC system, obviously, is faster and uses less space. But it requires a special tool. The screw system takes more time, but may produce a longer-lasting and stronger, more thorough (more of the wire exposed) electrical connection. The most common IDC wiring scheme is the 66-block, originally invented by Western Electric.
IDCMA Independent Data Communications Manufacturers Association, a lobbying and education group based near Washington, DC.
IDDD International Direct Distance Dialing. The capability to directly dial telephones in foreign countries from your own home or office telephone.
IDDS Installable Device Driver Server. A Dialogic term.
IDE 1. Integrated Drive Electronics. IDE is a hard disk drive standard interface for PCs. It appeared in 1989 as a low-cost answer to two other standard hard disk interfaces, ESDI and SCSI. The distinguishing feature of the IDE interface is that it incorporates the drive controller functions right on the drive. Instead of connecting to a controller card, an IDE drive attaches directly to the motherboard with a 40-pin connector. IDE drives offer a data transfer rate of three megabytes per second, which is not very fast. Several methods of data encryption can be used with the IDE interface, including MFM and RLL. Many laptops use IDE drives. IDE has a limit of 528 megabytes. Enhanced IDE drives, which appeared around 1994 to solve the problem that computers had gotten much faster and IDE wasn't keeping up, have a data transfer rate of between 11 and 13 megabytes per second and can handle drives of up 8.4 gigabyte. See ENHANCED IDE.
2. Integrated Development Environment. A term for products such as Microsoft's Visual C++ and Borland's Delphi that combine a program editor, a compiler, a debugger, and other software development tools into one integrated software package. The first of the IDEs, Borland's Turbo Pascal changed the way programmers write code by allowing programs to be edited and compiled within the same application.
iDEN The technology being developed by Motorola for use by ESMR operators.
Identification Failure Automatic Number Identification (ANI) equipment in the originating office failed to identify the calling number. See ANI.
Identified Outward Dialing Same as AIOD. It's a PBX feature which provides identification of the PBX extension making outward toll calls. This identification may be provided by automatic equipment or by attendant identification of the extension.
Identifier The name of a database object (table, view, index, procedure, trigger, column, default, or rule). An identifier can be from 1 to 30 characters long.
IDF Intermediate Distribution Frame.

IDLC See Integrated Digital Loop Carrier.
Idle 1. Not being used but ready.
2. An SCSA term. A state of the SCbus or SCxbus Message Bus where no information is being transmitted and the bus line is pulled high.
Idle Channel Code A repetitive pattern (code) that identifies an idle channel.
Idle Channel Noise Noise which exists in a communications channel when no signals are present.
Idle Line Termination An electronic network which is switch controlled to maintain a desired impedance at a trunk or line terminal when that terminal is in an idle state.
Idling Signal Any signal that indicates no data is being sent.
IDN Integrated Digital Network.
IDSU Intelligent Data Service Unit from ADC Kentrox.
IDT Inter-DXC Trunk
IDTV Improved Definition TeleVision. See IMPROVED DEFINITION TELEVISION.
IDU Interface Data Unit: The unit of information transferred to/from the upper layer in a single interaction across the SAP. Each IDU contains interface control information and may also contain the whole or part of the SDU.
IEC 1. InterExchange Carrier. Also called an IXC (as in IntereXchange Carrier). In practice, an IEC or IXC is any common carrier authorized by the FCC to carry customer transmissions between LATAs. In practice this means anyone and his brother who print up stationery, rent a few lines and proclaim themselves to be in the long distance phone business. Except for AT&T, regulation of long distance carriers by the FCC is perfunctory. It is less perfunctory by the local state authorities, some of whom still think competition in telecommunications is a mild form of insanity.
2. International Electrotechnical Committee.
IEEE Institute of Electrical and Electronic Engineers, a publishing and standards-making body responsible for many telecom and computing standards, including those standards used in LANs — e.g. the 802 series.
IEEE 1394 An IEEE data transport bus that supports up to 63 nodes per bus, and up to 1023 buses. The bus can be tree, daisy chained or any combination. It supports both asynchronous and isochronous data. 1394 is a complimentary technology with higher bandwidth (and associated cost) than Universal Serial Bus. Intel told me in summer of 1996 that it was supporting USB for most devices that attach to PC up through audio and video conferencing. Intel told that they are "supporting IEEE 1394 as the preferred interface for higher bandwidth applications such as high quality digital video editing, and connection to new digital consumer electronics equipment. We expect that 1394 will show up in low volumes in 97, and ramp into high volumes in 99." See USB.
IEEE 488 IEEE 488 is the most widely-used international standard for computer-to-electronic instrument communication. It is also known as GPIB and HPIB.
IEEE 802.1 This IEEE committee defines the LAN Management and bridging standards.
IEEE 802.1d An algorithm, the original version of which was invented by Digital Equipment Corporation, that is used to prevent bridging loops by creating a spanning tree. The algorithm is now documented in the IEEE 802.1d specification, although the Digital algorithm and the IEEE 802.2 handles errors, framing, flow control, and the Layer 3 service interface.
IEEE 802.2 A data link layer standard used with the IEEE 802.3, 802.4 and 802.5 standards. 802.2 is the more modern form of 802.3. Novell recommends that it be used on its NetWare networks in preference to the 802.3 (which will still work). Most Ethernet networks support both 802.2 and 802.3. For more on the 802 series, see the numbers definitions at the front of this dictionary.
IEEE 802.3 A Local Area Network protocol suite commonly known as Ethernet. Ethernet has either a 10 Mbps or 100 Mbps throughput and uses Carrier Sense Multiple Access bus with Collision Detection CSMA/CD. This method allows users to share the network cable. However, only one station can use the cable at a time. A variety of physical medium dependent protocols are supported. This is the most common local area network specification. See 802.2.
IEEE 802.3 10 BASE-T This is the standard for Ethernet over twisted pair cabling using home runs. See 802.3
IEEE 802.3 10Broad36 This IEEE standard describes a long-distance type of Ethernet cabling with a 10-megabit-per-second signaling rate, a broadband signaling technique, and a maximum cable-segment distance of 3,600 meters.
IEEE 802.4 A physical layer standard specifying a LAN with a token-passing access method on a bus topology. Used with Manufacturing Automation Protocol (MAP) LANs. Arcnet can work this way. Typical transmission speed is 10 megabits per second.
IEEE 802.5 A physical layer standard specifying a LAN with a token-passing access method on a ring topology. Used by IBM's Token Ring hardware. Typical transmission speed is 4 or 16 megabits per second. Typical topology is star.
IEEE 802.6 This IEEE standard for metropolitan area networks (MANs) describes what is called a Distributed Queue Dual Bus (DQDB). The DQDB topology includes two parallel runs of cable — typically fiber-optic cable — linking each node (typically a router for a LAN segment) using signaling rates in the range of 100 megabits per second.
IEEE 802.9 This IEEE committee deals with integrated voice and data LANs, i.e. Isochronous Ethernet.
IEEE 802.10 This committee deals with LAN security.
IEN Internet Experimental Note. A standards document similar to an RFC, and is available from the Network Information Center (NIC). IENs contain suggestions and proposals for Internet implementations or specifications.
IES 1. Inter-Enterprise Systems is EDI (Electronic Data Interchange) and inter-company electronic mail, fax, electronic funds transfer, videotex/online databases and the exchange of CAD/CAM graphics. See also EDI. 2. Information Exchange Services.
IETF Internet Engineering Task Force. One of two technical working bodies of the Internet Activities Board. The little-known all volunteer group, the IETF, meets three times a year to set the technical standards that run the Internet. The IETF is the primary working body developing new TCP/IP (Transmission Control Protocol/Internet Protocol) standards for the Internet. The IETF was formed in 1986 when the Internet was evolving from a Defense Department experiment into an academic network. Of late, the IETF has been the forum where engineers and programmers have cooperated to solve the succession of crises caused by the

NEWTON'S TELECOM DICTIONARY

Internet's phenomenal growth. The IETF's home page is: www.letf.cnrl.reston.va.us/home.html

IF Intermediate frequency.

IFG The minimum idle time between the end of one frame transmission and the beginning of another. On Ethernet 802.3 LANs the minimum interframe gap is 9.6 micro-seconds.

IFIP International Federation for Information Processing. A research organization that performs substantive pre-standardization work for OSI. IFIP is noted for having formalized the original Message Handling System (MHS) model. www.dit.upm.es/cdk/Iflp.html

IFRB International Frequency Registration Board.

IG 1. AT&T's ISDN Gateway. A set of specs for hooking up an outside computer to an AT&T switch. Under IG, information travels in one direction — from the switch to the host. See also ITG (which is two-directional), ASAI and OPEN APPLICATION INTERFACE. 2. Isolated Ground. In AC electricity, an isolated ground is a type of outlet characterized by the following features and uses:
- It may be orange and must have a Greek "delta" on the front of the outlet. (A delta looks like a triangle.)
- It must be grounded by an insulated green wire.
- It must have an insulator between the ground terminal and the mounting bracket.
- It is used primarily to power electronic equipment because it reduces the incidence of electrical "noise" on the ground path.

IGMP Internet Group Management Protocol. A protocol used by IP hosts to report their multicast group memberships to an adjacent multicast router.

Ignition Key A rod arranged to strike the arc in an arc generator of high frequency currents.

IGP Interior Gateway Protocol. The protocol used to exchange routing information between collaborating routers in the Internet. RIP and OSPF are examples of IGPs.

IGRP Interior Gateway Routing Protocol. A distance-vector routing protocol developed by Cisco Systems for use in large, heterogeneous networks.

IGT Ispettorato Generale delle Telcomunicazioni (General Inspectorate of Telecommunications, Italy).

IILC Information Industry Liaison Committee (part of ATIS). See EXCHANGE CARRIERS STANDARDS ASSOCIATION.

IIR Interactive Information Response.

IISP The Information Infrastructure Standards Panel formed by ANSI in July 1994 to accelerate the development and acceptance of standards critical to the establishment and deployment of the information superhighway. See ANSI.

IIW The ISDN Implementors Workshop, a group within the North American ISDN Users Forum.

ILAN A protocol independent router for token ring and Ethernet networks from CrossComm Corporation, Marlboro, MA.

ILLP Inter Link-to-Link Protocol.

ILMI Interim Link Management Interface: An ATM Forum-defined interim specification for network management functions between an end user and a public or private network and between a public network and a private network. ILMI is based on a limited subset of SNMP capabilities.

ILS Input buffer Limiting Scheme. A flow control scheme used in data communications that blocks overload by limiting the number of blocks arriving at a buffer.

ILSR IPX Link State Router. Novell's improvement on its RIP distance vector-based routing protocol.

IMA Interactive Multimedia Association. Formed in 1991 (rooted in IVIA, Interactive Video Industry Association), an industry association chartered with creating and maintaining standard specifications for multimedia systems. www.ima.org

Image Antenna A hypothetical, mirror-image antenna considered to be located as far below ground as the actual antenna is above ground.

Image Resolution The fineness or coarseness of an image as it was digitized, measured in Dots Per Inch (DPI), typically from 200 to 400 DPI.

IMAP Internet Unix server protocol which allows users to scan message headers and then download only selected messages. Also makes administration of Unix e-mail easier. See also MIME and POP.

IMAS Intelligent Maintenance Administration System. Northern Telecom software which is a menu-driven PC-based program that provides enhanced maintenance and administrative capabilities for DMS-10 central offices.

IMEI International Mobile station Equipment Identity. A wireless telecommunications term. An equipment identification number, similar to a serial number, used to identify a mobile station.

IMHO Abbreviation for "In My Humble Opinion;" commonly used on E-mail, on the Internet and BBSs (Bulletin Board Systems).

Immediate Ringing A PBX feature which makes the called telephone begin ringing the instant the phone has been dialed. Normally there's a small wait between dialing the number and having the phone ring.

IMNSHO In My Not So Humble Opinion. An acronym used in electronic mail on the Internet to save words or to be hip, or whatever. See IMHO.

IMO Abbreviation for "In My Opinion;" commonly used on E-mail and BBSs (Bulletin Board Systems). See IMHO.

IMPACS An MCI International packets switching service that is useful to firms with overseas remote computing needs, and to scientific, educational or commercial organizations that need periodic access to U.S. database facilities. IMPACS also provides overseas users with communications links to their own computers in the USA for applications such as order entry, inventory control, billing, payroll, and sales statistics.

Impact Strength A test designed to ascertain the abuse a cable configuration can absorb, without physical or electrical breakdown. Done by impacting with a given weight, dropped from a given height, in a controlled environment.

Impact Tool Also called a "punch down" tool. See PUNCH DOWN TOOL.

Impaired When an individual circuit exceeds the transmission limits or its signaling functions (e.g., seizure, disconnect, ANI) are experiencing failures.

Impedance The total opposition (i.e. resistance and reactance) a circuit offers to the flow of alternating current. It is measured in ohms and the lower the ohmic value, the better the quality of the conductor. Low impedance will help provide safety and fire protection and a reduction in the severity of common and normal mode electrical noise and transient voltages. For telecommunications, impedance varies at different frequencies. Ohm's law says that voltage equals the product of current and impedance at any single frequency. The unit of impedance is the ohm.

Impedance Matching The connection of additional impedance to existing impedance one in order to improve the performance of an electrical circuit. Impedance Matching is done to minimize distortion, especially to data circuits.

Implied Acknowledgment Implied acknowledgment is a process whereby negative acknowledgment of a specific packet of information implies that all previously transmitted packets have been received correctly. See also PIPELINING.

Import Imagine you have a software program, like a spreadsheet or a database. And you have information in that program. Let's say it's Microsoft Word or Lotus 123. And you want to get it into a different program, say to give it to a workmate who uses WordPerfect or Excel. You have to convert it from one format to another. From Word to WordPerfect or from Lotus to Excel. That process is typically called "exporting" and the process of your workmate getting it into his computer is called "importing." And you'll typically see the words "EXPORT" and "IMPORT" as choices on one of your menus.

Import Computers A Windows NT. In directory replication, the servers or workstations that receive copies of the master set of directories from an export server.

Import Script First read my definition of IMPORT. An import script is a series of specifications which control the merging processes. It contains a series of merge rules which specify how the fields are to be merged and a record precedence rule which governs which records to merge of the ones received.

Improved Definition Television IDTV. Television that includes improvements to the standard NTSC television system, which improvements remain within the general parameters of NTSC television emission standards. These improvements may be made at the transmitter and/or receiver and may include enhancements in parameters such as encoding, digital filtering, scan interpolation, interlaced scan lines, and ghost cancellation. Such improvements must permit the signal to be transmitted and received in the historical 4:3 aspect ratio.

Improved Mobile Telephone Service IMTS. In the beginning, there was dispatch mobile service. The base operator broadcast a message to you. Everyone could hear it. You responded. Then they had mobile telephone service. You picked up the phone in your car, the operator responded. You asked for the number you wanted and she/he dialed it and connected you. You had the channel to yourself but others could still tune in. Then came Improved Mobile Telephone Service (IMTS). Now you could dial from your car without using an operator with some assurance of privacy. The latest development is cellular mobile telephone service. See CELLULAR.

Impulse A surge of electrical energy usually of short duration, of a non repetitive nature.

Impulse Hits Errors in telephone line data transmission that are caused by voltage surges lasting from 1/3 to 4 milliseconds and at a level within 6 dB of the normal signal level (Bell standard allows no more than 15 impulse hits per 15 minute period).

Impulse Noise High level, short duration noise that comes on a circuit. You can get impulse noise from electromechanical relays. These noise "spikes" have little effect on voice transmission but can be devastating to data. You can get a piece of test equipment called an impulse noise measuring set. Such a machine establishes a threshold and counts the number of impulses (hits) above that threshold. IMS/VS Information Management System/Virtual Storage. An IBM host operating environment.

Impurity Level An energy level outside the normal energy band of the material, caused by he presence of impurity atoms. Such levels are capable of making an insulator semiconductor.

IMSI International Mobile Subscriber Identity. A CCITT specification used to uniquely identify a subscriber to mobile telephone service. It is used internally to a GSM (Global System for Mobile Communications) network.

IMT InterMachine Trunk. A circuit which connects two automatic switching centers, both owned by the same company.

IMTC International Multimedia Teleconferencing Consortium. At last count, the IMTC consisted of 90 Companies worldwide focused on delivering standards based conferencing solutions. It has six technical activity groups developing APIs, doing interoperability testing and assisting the ITU-T standards development. It had a major role in the development of the family of T.120 standards. The IMTC is open to all organizations that support open standards. http://www.imtc.org/imtc

IMTS See IMPROVED MOBILE TELEPHONE SERVICE.

IMUX Inverse Multiplexer

IN Intelligent Network. Ericsson has done focus groups on Mobile Intelligent Network Services. Among the new IN (Intelligent Network) services, Ericsson identified: • Enhanced number translation services functions • Enhanced screening services, i.e. selective call diversion • Selective forwarding of calls • Location-dependent call forwarding • Improvements to voice announcements • Services to support fixed and mobile integration, i.e. personal communications services, PCS and universal personal telecommunications, UPT, and • Enhanced billing. See AIN.

In-band Signaling Signaling made up of tones which pass within the voice frequency band and are carried along the same circuit as the talk path that is being established by the signals. Virtually all signaling — request for service, dialing, disconnect, etc. — is in-band signaling. Most of that signaling is MF — multi-frequency dialing. The more modern form of signaling is out-of-band. Several local and long distance companies provide ANI (Automatic Number Identification) via in-band signaling. Some long distance companies provide it out-of-band, using the D-channel in a PRI ISDN loop. See ISDN and SS7 (CCITT Signaling System Number 7).

In-line Device Hardware that is physically attached between two communications lines.

In-Safe An inbound store-and-forward MCI International Telex service that automatically answers a subscriber's incoming calls, provides an answerback, and accepts the messages.

INA Information Networking Architecture.

Inactivity Time-outs Dial-in users can be disconnected after specific periods of inactivity. By eliminating idle connections, you reduce the number of ports required for remote access to a network.
Inbound Path On a broadband LAN, the transmission path used by stations to transmit packets toward the headend.
incAlliance incAlliance stands for the Isochronous Network Communication Alliance. It was announced publicly on June 13, 1995. It was formed by high-technology businesses, including several Fortune 500 companies, to promote the use of isochronous Ethernet (also called isoEthernet) to provide interactive multimedia applications to the enterprise desktop over existing cable infrastructures. isoEthernet combines ISDN channels with standard ethernet packet traffic to facilitate real-time interactive voice/video/data transmission over 10Base-T local area networks. Isochronous Ethernet is a draft standard which is expected to be ratified by the IEEE 802.9a multimedia committee by June 1995. The incAlliance will act as a plenary liaison and resource to industry groups such as the IEEE, ITU, Electronic Computer Telephony Forum (ECTF), International Multimedia Teleconferencing Consortium, ATM Forum and the Multimedia Communications Forum. The incAlliance's mission is to promote the value of high quality, real-time, interactive multimedia including new isochronous LAN technologies such as isoEthernet; educate the industry on differences and requirements between voice/video and data for LAN/WAN implementation; foster industry growth through joint applications development and interoperability testing; provide a vision and roadmap for upgrading existing data networks. Two key players in the formation of this group are National Semiconductor's Interactive Multimedia Group and Dialogic Corporation.
INCC Internal Network Control Center
Incentive Regulation Prices of services provided by the local regulated phone company are fixed or capped but incentives are provided to improve earnings through cost savings. Earning levels are flexible, within a range of rates, allowing opportunity for earnings improvement.
Incident Angle The angle between an incident ray and a line perpendicular to an optical surface.
Inclined Orbit Any nonequatorial orbit of a satellite. Inclined orbits may be circular or elliptical and may be synchronous or nonsynchronous. Inclined orbits are used for many reasons — for photographing, for reaching places in the extreme north and south which normal geosynchronous satellites can't reach.
Incoherent Light A random form of light whereby the phase of the light is unpredictable. LEDs emit incoherent light.
Incoming Call Identification ICI. Some way of telling the user who's calling. It might be the caller's extension number on an LCD screen or even the caller's name spelled out, e.g. "KATE BRODIE-DAVID CALLING." Today, most "incoming call identification" is done totally within one PBX. However, the days of ISDN and CCITT Signaling System Number 7 are arriving. They promise to deliver to us the phone number of everyone calling us — from within the PBX or key system and from the outside world.
Incoming Calls Barred An interface configuration option that blocks call delivery attempts. Only outgoing calls are allowed.
Incoming WATS An incoming WATS (INWATS) trunk is used exclusively for received incoming calls from a defined geographical area to a customer's PBX. An incoming WATS trunk can only be used to receive calls via the dial-up telephone network. Originally WATS lines came in only lines that could receive calls or only lines that could make calls. Now, you can buy a WATS line that handles both incoming and outgoing lines. See WATS.
Increment A small change in the value of a quantity.
Incremental Cursor Control The user-controlled function that moves the focus in increments dictated by the application. In character-based text editing, the increment is typically one character in the horizontal direction and one line in the vertical direction.
Indeo Video Developed by Intel Architecture Labs in 1992, Indeo video technology is Intel's video recording and playback software technology. Indeo Video provides personal computer users with a standard format for recording digital video, just as VHS provides a standard format for recording motion video on analog tapes. More specifically, Indeo video is a family of software compression techniques which together form the software equivalent of COmpressor/DECompressor (codec) hardware used for recording and compressing video data.
Indeo video is what Intel calls an ingredient technology licensed free of charge by Intel to software application developers, CD-ROM publishers, operating system vendors, computer OEMs, as well as graphics hardware manufacturers who, in turn, include its technology within their products. PC users find Indeo technology already built into the multimedia extensions of operating systems such as Microsoft's Video for Windows, IBM's OS/2 Multimedia Presentation Manager, Apple's QuickTime for the Macintosh and QuickTime for Windows.
Indeo video offers personal computer users more than a family of techniques for compressing and decompressing video data. Indeo video also incorporates some features to help make recording, playing, or distributing digital video applications on a PC easy and affordable: Real-Time, One-Step Capture and Compression, Scalable Playback, Software-Only Playback and Cross Platform Compatibility.
Indeo Video is used in personal conferencing products like the Intel ProShare Video System, which comes with two boards, a camera and a microphone/speaker. With ProShare and its ISDN implementation you can share applications on your computer with a distant person, while simultaneously seeing them and speaking with them. Another way to use Indeo Video is to use your video camera (e.g. camcorder) to record a "movie." Maybe it's a shot of a building you want your boss to buy? You pop your "movie" into a VCR, which you've attached to a compression board inside your PC. That board might be called the Intel Smart Video Recorder. You play your movie through the board inside your PC, which contains software. The combination card/PC software turns your incoming NTSC (National Television Standards Committee) analog video into digital signals, which it then compresses and stores on the computer's hard disk. Now your analog "movie" has become a digital computer file. This means you can send it by modem to your boss who's thousands of miles of away. You can edit it. You can add titles to it. You can embed it in other documents, like word-processed documents, or spreadsheets. To play the "movie" back (i.e. to see it) you'll need only software in your Indeo Video equipped PC. To hear the video, you'll need a sound card or a sound system built into your com-

puter. Some — like some laptops — come with them already built in. Indeo video supports playback of 320 x 240 pixels at 30 frames per second (fps). Indeo technology records at 24 bit, but "scales" to 8- 16- or 24-bit during playback. The Intel Smart Video Recorder Pro can capture 320x240, 30 fps; but with a "higher-end" capture card that captures MJPEG at 640x480, 30 fps, captured video clips can be compressed with Indeo video. Indeo supports the Intel/Microsoft Display Control Interface (DCI) standard, which allows software video playback to be enhanced using video hardware on Windows graphics accelerator cards. At the time of writing, we were anticipating the release of Indeo 4, which features new performance scalability, quality, and inter-activity features that are dramatically different to Indeo 3.

Independent Clocks A communication network timing subsystem using precise free running clocks at the nodes for synchronization purposes. Variable storage buffers installed to accommodate variations in transmission delay between nodes are made large enough to accommodate small time (phase) departures among the nodal clocks that control transmission. Traffic is occasionally interrupted to reset the buffers.

Independent Sideband Transmission ISB. That method of double sideband transmission in which the information carried by each sideband is different. The carrier may be suppressed.

Independent Software Vendor ISV. Typically a company which writes and sells software, but not hardware. Manufacturers of hardware and operating systems, i.e. IBM or Northern Telecom, often contract with ISVs to produce specialized software to make their hardware and operating system more attractive.

Independent Telephone Company A telephone company not affiliated with one of the "Bell" telephone companies. There are about 1,400 independent phone companies. They serve more than half the geographic area of the United States, but only around 15% of its telephones. The independent phone companies used to be represented by the United States Independent Telephone Association (USITA). But once Divestiture happened, the association dropped the word "Independent" from its name, accepted membership of the Bell operating companies (but not AT&T) and became USTA.

Index Think of a filing cabinet. It contains oodles of information. Think of a computer hard disk. Same thing as a filing cabinet. Oodles of information. Now think of putting everything in the filing cabinet into filing folders and putting them in alphabetical order. Makes finding things a lot easier. Now think of a computer. You ask it to find you the name of a file folder. Nothing sophisticated here. Except it's dumber than you. It starts at the top and searches down. Of course, it searches fast. But it still searches from the top down. The fastest way for it to search is to give it less stuff to search through. Thus you make an index. Just as you do in a book. Only, compared to you and me, a computer is willing to do more stupid work. It will index in alphabetical order. It will index in date order. It will index in order of how much you sold the guy recently. It will index in any order you ask it to. And many database software programs will let you keep several indexes concurrently, thus allowing you to find things quickly. The rules of database are simple: The more indexes you keep concurrently, the more time your computer will take to update its indexes every time you enter a new record or update an old record.

Index Dip In an optical fiber, a decrease in the refractive index at the center of the core, caused by certain manufacturing techniques.

Index Matching Material In fiber optics, a material (liquid, gel, or cement) whose refractive index is nearly equal to the fiber core index. It is used to reduce Fresnel reflections from a fiber end face.

Index Of Cooperation In facsimile, the product of the total line length in millimeters times the lines per millimeter divided by r. For rotating devices, the index of cooperation is the product of the drum diameter times the number of lines per unit length.

Index Of Refraction The ratio of the speed of light in vacuum to the speed of light in a material.

Index Profile In an optical fiber, the refractive index as a function of radial distance from the optical axis.

Indication Circuit X.21 circuit used to send control information from DCE to DTE.

Indication Of Lights, bells and buzzers indicating that something has or is about to happen. For example, indication of camp-on to a station: short bursts of tone are periodically transmitted to the busy phone to indicate that another call is camped on and waiting.

Indirect Control In digital data transmission, the use of a clock at a higher standard modulation rate, e.g., 4, 8, 128 times the modulation rate, rather than twice the data modulation rate, as is done in direct control.

Indium Gallium Arsenide InGaAs, a semiconductor material used in lasers, LEDs, and detectors.

Individual Load Cycling Feature This is one feature of AT&T's Dimension Energy Communications Service Adjunct. Individual Load Cycling reduces energy consumption by turning devices on and off (e.g. Air-conditioning) on an hourly basis.

Individual Speed Calling A key system or PBX feature by which a user can dial a longer number by punching one or two buttons on his phone. Sometimes this speed dial ability is programmed into the phone. Sometimes it's programmed into the system. Whichever it is, each user has a bunch of numbers he/she can speed dial. These are his/her own. No one else can speed dial them.

Induced Uplink An ATM term. An uplink "A" that is created due to the existence of an uplink "B" in the child peer group represented by the node that created uplink "A". Both "A" and "B" share the same upnode, which is higher in the PNNI hierarchy than the peer group in which uplink "A" is seen.

Inductance Inductance is the property of an electric force field built up around a conductor. Inductance allows a circuit to store up electrical energy in electromagnetic form. When current flows through a wire, lines of force are built up around the wire. The field created by DC current is steady. When AC flows through a wire, the lines of force are constantly building and collapsing. An inductor is formed by winding a conductor into a coil. In long local loops, conversation gets difficult because the long wires encounter capacitive resistance. To counter this, inductors known as load coils are connected in series, increasing the inductance. When load coils, or inductors are connected in parallel, they reduce the inductance. See also INDUCTIVE CONNECTION.

Induction Electromagnetic transfer of energy from one coil to another.

Induction Coil A coil having a high turn ratio used for raising the voltage. A step-up transformer.

Inductive Connection A connection between a telephone instrument and another device by means of the electromagnetic field generated by the telephone instrument. No direct electrical connection is established between the two. See INDUCTANCE.

Inductive Coupling The transfer of energy from one circuit to another by means of the mutual inductance between the circuits, i.e. energy jumping from one circuit to another without actually touching it copper wire to copper wire. The coupling may be deliberate and desired as in an antenna coupler or may be undesired as in powerline inductive coupling into telephone lines. See also INDUCTANCE.

Inductive Pickup A coil used to tap phone lines without direct connection.

Inductive Tap Wiretap that is not physically connected to the telephone wires. A voltage proportional to the varying line current is induced into a coil.

Inductively Coupled Receiver A radio receiver in which the energy in the antenna circuit is transferred to the secondary circuit by induction.

Inductivity A term sometimes used to denote the dielectric constant or the specific inductive capacity.

Inductor See INDUCTANCE and INDUCTIVE CONNECTION.

INE Intelligent Network Element. Network equipment which contains autonomously intelligent computing capabilities.

INET Institutional Network. Generally dedicated to linking government and other public buildings for such uses as training, meetings, data and voice. Such INETs are provided by cable TV operators pursuant to 47 U.S.C. 531 for public, educational and governmental use. It is not available for consumers.

INF File A file that provides Windows 95 Setup with the information required to set up a device, such as a list of valid logical configurations for the device, the names of driver files associated with the device, and so on. An INF file is typically provided by the device manufacturer on a disk.

Inference Engine The AI (Artificial Intelligence) heart of a knowledge base system. The inference engine is the technology which directs the reasoning process. The inference engine contains the general problem-solving knowledge such as how to interact with the user and how to make the best use of the domain information.

Infinite Loop A state in which specific steps of a program are executed repeatedly, not allowing the program execution to advance further.

Info Look Gateway Service A Nynex service. Use your PC, modem and communications software to dial Info-Look and check on news, weather, health, finance, sports. Do some shopping. Info-Look connects you to many computer services and databases, none of which are provided by Nynex or New England Telephone. That's why Info-Look is called a "gateway." It's a gateway to a myriad of services. As of writing this there was talk Nynex was closing Info-Look down. And it eventually did.

Infobahn A new term for the Information Superhighway.

Infomercial A short segment shown on the video system of a plane purporting to be informational/newsy and educational. In fact, the segment is a commercial paid for by the company whose products and/or services are featured. According to research, infomercials create three types of phone calls — order calls, inquiry/incomplete calls and customer service call.

InfoNet A consortium of about 10 European telephone companies and MCI. InfoNet has built a worldwide packet switched network that is great for data communications. It offers one major advantage over packet switched networks owned by the individual phone companies — You don't have to separately sign up to use InfoNet in all the countries it operates. You just dial Infonet up, punch in and communicate. Not having to sign up is a big savings in time, since it can take weeks to sign up for European packet switched networks and cost minimum monthly fees. For MCI Mail subscribers, using Infonet means you only get one bill — directly from MCI. See also MCI Mail and MCI FAX. Infonet is based in El Segundo, CA 90245.

Infopath Packet Switching Service A New York and New England Telephone service using a switched data network, where users share all network facilities. Offered for only the NY Metropolitan RCA. Its Low Throughput (private dial access) is dedicated to one customer, Medium Throughput uses a Data Over Voice channel and High Throughput uses a Digipath Digital Service facility.

Information At Your Fingertips At Fall Comdex 1990 Bill Gates, Microsoft chairman, suggested the idea. With Information at your fingertips, he said, PC users can easily access company wide information "anywhere at anytime" through an icon-based graphical user interface. In the speech, Gates demonstrated applications that used Object Linking and Embedding (OLE), Dynamic Data Exchange (DDE), handwriting recognition, cellular communications and multimedia. See OLE.

Information Bearer Channel 1. A channel provided for data transmission that is capable of carrying all the necessary information to permit communication, including user's data, synchronizing sequences, control signals, etc. It may therefore, operate at a greater signaling rate than that required solely for the user data.
2. A basic communication channel made available by the circuit provider with no enhanced or value-added services included other than the bandwidth transmission capability.

Information Center Mailboxes An Octel term for a voice bulletin board on a voice mail system. Here's their explanation: Multiple callers can access, directly or indirectly, recorded announcements containing information that would otherwise have been given live by employees. Callers are frequently "outside" users of the system. One type of "listen only" mailbox simply plays the messages to the callers. This technology, sometimes known as audiotex, makes it possible to create a verbal database so callers can select which information they want to hear. Another type of Information Center Mailbox prompts callers to reply to announcements. Callers wanting further information can be given the opportunity to leave their names and phone numbers after listening to a product description. They can also be transferred to a designated employee who can immediately take an order. If desired, a password can be required before confidential or controlled access information can be heard.

Information Element The name for the data fields within an ISDN Layer 3 message.

Information Engineering Coined by James Martin, the most prolific writer in data processing, the term refers to systems within data processing and their impact on giving the corporation a greater competitive edge. In short, a fancy term for Management Information Systems (MIS), which itself was a fancy term for DP, namely Data Processing.

Information Frame Frame in HDLC, DDCMP, or related protocols containing user data.

Information Highway A term coined by Al Gore. This fact affirmed by Dan Lynch, the man who started the trade show, InterOp and who was very heavily involved with Internet from the very beginning. As the term got developed and more people got turned on by the idea, it became known as The Information Superhighway. See also INFORMATION SUPERHIGHWAY.

Information Outlet IO. Sort of like an AC power outlet, but a little more cerebral. A connecting device designed for a fixed location (usually a wall in the office) on which horizontal wiring subsystem cable pairs terminate and which receives an inserted plug; it is an administration point located between the horizontal wiring subsystem and work location wiring subsystem. Although such devices are also referred to as jacks, the term information outlets encompasses the integration of voice, data, and other communication services that can be supported via a premises distribution system.

Information Page Mapping See ADSI.

Information Provider A business or person providing information to the public for money. The information is typically selected by the caller through touchtones, delivered using voice processing equipment and transmitted over tariffed phone lines, e.g., 900, 976, 970. Typically, billing for information providers' services is done by a local or long distance phone company. Sometimes the revenues for the service are split by the information provider and the phone company. Sometimes the phone company simply bills a per minute or flat charge. A typical "information provider" is American Express, which provides a service — 1-900-WEATHER. By dialing that number you can touchtone in city names and find out temperatures, weather forecasts, etc. Calling 1-900-WEATHER costs several dollars a minute.

Information Service The Telecommunications Act of 1996 defined Information Service as: The term `information service' means the offering of a capability for generating, acquiring, storing, transforming, processing, retrieving, utilizing, or making available information via telecommunications, and includes electronic publishing, but does not include any use of any such capability for the management, control, or operation of a telecommunications system or the management of a telecommunications service.

Information Signals A Bellcore definition. Information signals inform the customer or operator about the progress of the call. They are generally in the form of universally understood audible tone (for example, dial tone, busy, ringing) or recorded announcements (for example, intercept, all circuits busy).

Information Superhighway A very vague concept which Senator Al Gore created in the early 1990s and which gained great popularity when he became vice president and the Clinton/Gore administration started pushing the concept. The Information Superhighway is a term sufficiently vague that it can mean anything to anyone. It can mean a gigantic Internet reaching everybody in North America, or the planet (if you're that expansive). It could just as easily mean a combination 500-channel interactive cable TV system with full video on demand to every household in North America. Somewhere in all this is the idea that easy access to large amounts of information will enrich our lives immeasurably. Who's going to get first access to it all, what the precise technical details will be, and who's going to pay for it are, naturally, minor details to be worked out. We can be assured that the details will be worked out, since the idea originated in Washington, DC., home of so many practical ideas.

Information Technology IT. A fancy name for data processing, which became management information systems (MIS), which became information technology. All the same thing, essentially. See also IT.

Information To Go A term coined by Digital Equipment Corporation to refer to the transmission of data over airwaves instead of fixed wires.

Infrared The band of electromagnetic wavelengths between the extreme of the visible part of the spectrum (about 0.75 um) and the shortest microwaves (about 100 um). This portion of the electromagnetic spectrum is used for fiber-optical transmission and also for short haul (up to 20 feet) through-the-air data transmission. With the adoption of new infrared standards at a meeting of over 50 manufacturers in June 1994, many PC devices will begin sporting something called the "Infrared Serial Data Link," with speeds up to 1.5 million bytes per second. This standard is designed to insure that products sporting this link will work together and interchangeably.

Infrared Data Association More than 90 PC equipment manufacturers are attempting to set a standard for infrared data transmission. Imagine that you're carrying around a small portable laptop, PDA or other device and you want to exchange data with your desktop. You'll simply aim the device at your desktop PC and transmit information back and forth. Maximum transmission distance is slated to be about a yard.

Infrared Fiber Optical fibers with best transmission at wavelengths of 2 um or longer, made of materials other than silica glass.

Infrared Serial Data Link As a result of a meeting at Microsoft of over 50 manufacturers in June 1994, many PC devices will begin sporting something called the "Infrared Serial Data Link," an infrared through-the-air (up to 20 feet) link with speeds up to 1.5 million bytes per second. This standard is designed to insure that products sporting this link will work together and interchangeably. There is now an organization called I.R.D.A., the Infrared Data Association, representing over 80 manufacturers.

Ingredient Technology See INDEO VIDEO.

Ingress The incoming traffic, or incoming direction into a network device.

Inheritance A term from object oriented programming. Data abstraction can be carried up several levels. Classes can have super-classes and subclasses. In moving to a level of greater specificity, the application developer has the option to retain some attributes and methods of the super-class, while dropping or adding new attributes or methods. This allows greater flexibility in class definition. It is even possible in some languages to inherit from more than one parent. This is referred to as multiple inheritance. See OBJECT ORIENTED PROGRAMMING.

INIC ISDN Network Identification Code.

INIT An INIT is the Macintosh System 7 equivalent of a terminate and stay resident (TSR) program . An init might load to initialize a fax modem, screen saver, etc. Similar to the DOS environment, some inits conflict. When troubleshooting operating system problems, remove inits first.

Initial Address Message. IAM. A SS7 signaling message that contains the address and routing information required to establish a point-to-point telephone connection.

NEWTON'S TELECOM DICTIONARY

Initial Answer Initial answer refers to the point in time at which a computer telephony system answers an incoming call. Many computer telephony systems require significant processing to set up to answer incoming calls. For example, the system may examine the incoming ANI, DNIS, or PBX integration data to determine how to answer (which prompt to use), or where to switch the call. This can involve significant database access and processing time. Therefore, the ability to handle large number of incoming calls (especially in burst mode) may delay the initial answer. The delay from when a call reaches a computer telephony system until the computer telephony system answers the call (the initial answer) is usually a key response time to understand when testing a computer telephony system.

Initial Period The minimum billing period on a call. For interstate or inter-LATA AT&T calls, the initial period is one-minute. Some non-AT&T long distance companies have initial periods under one-minute. This also applies to local calls in Measured areas.

Initial Program Load The initial loading of generic and/or configuration software into a PBX or other phone system. The Initial Program Load is a pain in the rear end. But an even bigger pain is what happens when you lose your programming and you've forgotten to back it all up.

Initial Sequence Number ISN. Generated at each end of TCP connection to help to uniquely identify that connection.

Initialization String A group of commands sent to the modem by a communications program at start-up — before the number has been dialed. Such a string tells the modem to set itself up in a way that will make it easy to correctly communicate with a distant modem.

Initializing Terminals An ISDN term. These devices, sometimes called self-initializing terminals, are basically ISDN terminals that can generate their own terminal identification number. This makes it easier for the network and the terminal to agree on the number to use.

Initiating Event An event that causes the ASC (AIN Switch Capabilities) to assign an ID to a connection segment for a certain user ID and to communicate with a SLEE (Service Logic Execution Environment) for the first time with respect to the combination of the specific user ID and connection segment ID.

Initiator A SCSI device, usually a host system, that requests an 1/0 process to be performed by another SCSI device.

Injection Laser Another name for a semiconductor or diode laser. See INJECTION LASER DIODE.

Injection Laser Diode ILD. A solid-state device that works on the laser principle to produce a light source for optical fiber.

INL See INTERNODE LINK.

INMAC International Network Management Center

INMARSAT The INternational MARitime SATellite service that has eight satellites and provides mobile communications to ships at sea, aircraft in flight, vehicles on the road and small stationary satellite antennas. Inmarsat provides dial-up telephone, telex, fax, electronic mail and data connections to more than 65,000 customers. Typically you buy a small earth station and install it on your ship, or truck and pay Comsat a per minute charge for the use of the single phone and single telex service most customers subscribe to. As of writing, a ship earth station cost $50,000 and a minute of calling was $10 anywhere in the world. Not cheap! Inmarsat describes itself as the international mobile satellite organization. It is based in London and has 79 member countries. Established in 1979, Inmarsat says its members provide worldwide mobile satellite telephone, telex, fax and data communications on land, at sea and in the air. In London, Inmarsat's phone number is 011-171-728-1000.

Inmate Call Management The P2000 ESS inmate call management application package serves the special requirements of correctional institutions. By assigning Personal identification Numbers (PINs), inmates may be allowed debit calling only, collect calling only, or a combination of both. The P2000 ESS can also access external commissary databases, allowing immediate debit of an inmate's account in real time.

Other outstanding features of the inmate call management application package are:
- billed number screening, reducing billing fraud
- blocking of calls after commissary funds are exhausted, minimizing lost revenue on commissary funds
- identifying announcement to called party
- time-of-day restrictions to specified destination numbers
- temporary call prohibition for specific PINs.
- restricted destination numbers for specific PINs
- system-wide deny numbers

Inmate Calling Dennis Squires from Bell Atlantic says, "Prison inmates can get very creative when they have a lot of time on their hands, like 20 years." So, to avoid stealing of phone calls by prisoners, some phone companies have a service called "Inmate Calling." The inmate enters his authorization code, which allows him to make phone calls — but only to authorized phone numbers, i.e. mothers, lawyers, bail bondsmen, etc.

INN See InterNode Network.

INode Integrated Node (SSP and STP).

INOS Intelligent Network Operations System. An operations system designated to manage, monitor, and control elements of an intelligent telecommunications network.

INP Board Intelligent Network Processor board for synchronous transmission on an Hewlett Packard 3000 system.

INPA Interchangeable Numbering Plan Area. An area code that looks like an office code. There is no particular name for an office code that looks like an area code, according to Lee Goeller.

INPUT A signal fed into a circuit.

Input Buffer Limiting Buffering strategy that divides buffer at a mode into two classes, both available to transit packets but only one available to packets input at the node.

Input Circuit The grid circuit of an electron tube.

Inquiry A request for specific information.

Insertion Gain The gain resulting from the insertion of a device in a transmission line, expressed as the ratio of the power delivered to that part of the line following the device to the power delivered to that same part before insertion. If more than one component is involved in the input or output, the particular component used must be specified. If the resulting number is negative, an "insertion loss" is indicated. This ratio is usually expressed in decibels. See INSERTION LOSS.

Insertion Loss The difference in the amount of power received before and after something is inserted into the circuit (viz. another telephone instrument) or a call is connected. In an optical fiber, insertion loss is the optical power loss due to all causes, usually expressed as decibel/kilometer. Causes of insertion loss may be absorption, scattering, diffusion, dispersion, microbending, or methods of coupling power outside the fiber. In lightwave transmission systems, the power lost at the entrance to a waveguide due to causes, such as fresnel reflection, packing fraction, limited numerical aperture, axial misalignment, lateral displacement, initial scattering, or diffusion. See INSERTION GAIN.

Inside Dial Tone A tone users hear when they are connected to an intercom line.

Inside Link An ATM term. Synonymous with horizontal link.

Inside Plant 1. The portion of a cable network that resides inside buildings.
2. Everything inside a telephone company central office. See also INSIDE WIRING.

Inside Wiring That telephone wiring located inside your premises or building. Inside Wiring starts at the telephone company's Demarcation Point and extends to the individual phone extensions. Traditionally, Inside Wiring was installed and owned by the telephone company. But now you can install your own wiring. And most companies installing new phone systems are installing their own new wiring because of potential problems with reusing the old telephone company cable. See also INSIDE PLANT.

Inspect Screen An AT&T Merlin term. A display screen on digital telephones that allows users to preview incoming calls and see a list of the features programmed on line buttons.

Instabus1080 and Instabus1480 Trademarks for MICOM's direct host attachment products.

Instalink 1. An MCI International service that allows access to a host computer in the U.S.A. from a Telex machine anywhere overseas. This allows easier retrieval of information from a U.S. database.
2. A trademark for MICOM's data-over-voice products.

Installation The physical hook-up and diagnostic testing of a PBX switch, cabinet, or peripheral item prior to a cutover and maintenance acceptance by the maintaining vendor.

Installed Base How many of whatever are in and working. Installed base is often confused with annual shipments. They're very different. Shipments is what goes out the factory. Installed base is what's out there. The equation is: Installed base at beginning of year plus annual shipments less equipment taken out of service during the year is equal to the installed base at the end of the year.

Instance ID An ATM term. A subset of an object's attributes which serve to uniquely identify a MIB instance.

Instanet Trademark for MICOM's family of local data distribution and data private automatic branch exchange (PABX) products.

Instant On Buy a PC (Personal Computer). Turn it on. Bingo, it's already loaded with Windows or OS/2. Instant On is a new term for preloading software onto hard disks of new computers and shipping those computers already pre-loaded with that software.

Instantaneous Override Energy Function IOEF. A feature of the AT&T PBX Dimension Energy Communications Service Adjunct (ECSA), which allows the user to turn all the ECSA energy functions ON or OFF. IOEF is most often used for periodic maintenance, or to adjust to sudden changes in weather.

Institute for Telecommunications Sciences ITS is the research and engineering branch of the National Telecommunications and Information Administration (NTIA), which is part of the U.S. Department of Commerce (DoC). The Institute is located at 325 Broadway, Boulder, CO 80303. www.its.bldrdoc.gov/its.html

Instruction Register The register which contains the instruction to be executed and functions as the source for the subsequent operations of the arithmetic unit.

Insulated Wire Wire which has a nonconducting covering.

Insulating Materials Those substances which oppose the passage of an electric current through them.

Insulation A material which does not conduct electricity but is suitable for surrounding conductors to prevent the loss of current.

Insulation Displacement Connection IDC. A type of wire connection device in which the wire is "punched down" into a double metal holder and as it is, the metal holders strip the insulation away from the wire, thus causing the electrical connection to be made. The alternate method of connecting wires is with a screw-down post. There are advantages and disadvantages to both systems. The IDC system, obviously, is faster and uses less space. But it requires a special tool. The screw system takes more time, but may produce a longer-lasting and stronger, more thorough (more of the wire exposed) electrical connection. The most common IDC wiring scheme is the 66-block, invented by Western Electric. See IDC.

Insulation Resistance That property of an insulating material which resists electrical current flow through the insulating material when a potential difference is applied.

Insulator A material which does not conduct electricity and is traditionally used to separate or protect wiring used for carrying signals.

INT14 A software interrupt designed to communicate with the com (serial) port in a PC. Communications programs use interrupt 14h to talk to a modem physically attached to another computer on the network.

Integer A computing procedure for solving or finding the optimum solution for complex problems in which the variables are based on integers. Integers include all the natural numbers, the negatives of these numbers, or zeros.

Integrated Access An AT&T term for the provision of access for multiple services such as voice and data through a single system built on common principles and providing similar service features for the different classes of services.

Integrated Call Management ICM. A family of Rolm networking products and services for single-site and multi-site CBX installations.

Integrated Circuit IC. A single electronic device that contains hundreds or thousands of previously separate (discrete) components. The so-called "computer on a chip" is a special purpose integrated circuit.

Integrated Development Environment IDE. A Windows program within which a developer may perform all the essential tasks of development including editing, compiling and debugging.

Integrated Digital Loop Carrier IDLC. Access equipment which extends Central Office services; it connects to a SONET ring on the network side while providing telephony services on the subscriber side (POTS, ISDN, leased lines, etc.).

Integrated IS-IS Formerly Dual IS-IS. Routing protocol based on the OSI routing protocol IS-IS, but with support for IP or other networks. Integrated IS-IS implementations send only one set of routing updates, regardless of protocol type, making it more efficient than two separate implementations.

Integrated Messaging Also called Unified Messaging. Integrated messaging is one of many benefits of running your telephony via a local area network. Here's the scenario: Voice, fax, electronic mail, image and video. All on the one screen. You arrive in the morning. Turn on your PC. It logs onto your LAN and its various servers. In seconds, it gives you a screen listing all your messages — voice mail, electronic mail, fax mail, reports, compound documents .... Anything and everything that came in for you. Each is one line. Each line tells you whom it's from. What it is. How big it is. How urgent. Skip down. Click. Your PC loads up the application. Your LAN hunts down the message. Bingo, it's on screen. If it contains voice — maybe it's a voice mail or compound document with voice in it — it rings your phone and plays the voice to you. Or, if you have a sound card, it can play the voice through your own PC. If it's an image it may hunt down (also called launch) an imaging application which can open the image you have received, letting you see it. Ditto, if it's a video message.

Messages are deluging us. To stop them is to stop progress. Run your eye down the list, one line per entry. Pick the key ones. Junk the junk ones. Postpone the others.

It gets better. You're out. Dial in on a gateway with your laptop. Skim your messages. Dial in on a phone. Punch in some buttons. Hear your voice mail messages. Or if you're not on laptop, have your e-mail read to you. Better, have your fax server OCR your faxes and image mail and have it read them to you. A LAN server is the perfect repository for messages. It can search for them, assemble them, process them, store them, convert them, compress them, shape them, shuffle them, interpret them. Integrated messaging essentially applies intelligence and order to the messages deluging you each day. See TELEPHONY SERVICES.

Integrated Personal Computer Interface IPCI. A ROLM-designed communications printed circuit card designed to provide an IBM PC with asynchronous data transmission over two-strand wiring to and from a Rolm CBX PBX.

Integrated Photonics Integrated photonics are devices that include optical waveguides embedded in a semiconductor or ferroelectric substrate, and which perform some type of signal processing function under electrical control. These functions include: routing of light signals in different directions, filtering out one or several wavelengths, emitting light or modulating the intensity and/or phase of an incoming light signals. An optical waveguide consists of a region in which the refractive index is higher than in the surrounding material so that a light signal can propagate without spreading (diffraction). In any applications, only single mode waveguides are useful. This definition courtesy Ericsson.

Integrated Services Digital Network See ISDN and SIGNALING SYSTEM 7.

Integrated Services Digital Network User Part ISDN-UP The part of SS7 (Signaling System Number 7) that encompasses the signaling functions required to provide voice and non-voice services in ISDN and pre-ISDN architectures. The basic service offered by the ISDN-UP is the control of circuit switched connections between subscriber line exchange terminations. Definition from Bellcore in reference to its concept of the Advanced Intelligent Network.

Integrated Solution-II/III Integrated Solution II is an AT&T Product that combines Audix Voice Power (Voice Mail), Call Accounting and System Administration Software. Integrated Solution Systems with Audix includes a tape drive that backs up all of the information on the computer's hard disk, including voice messages, auto attendant greetings, system programming, and call accounting records.

Integrated Voice Data There are many different meanings to this concept. The most common (we'll get arguments on this) is that a workstation or a combination telephone/personal computer on a desk can combine voice and data signals over a single communications channel. That channel might be carried digitally on one pair of wires. That is "the most integrated" voice/data. Less integrated is when you carry voice and data digitally on two pairs — one pair for transmitting and one pair for receiving. Even less integrated are some systems which use three pairs of cabling set up as one voice analog pair, one digital data pair and one power/signaling pair. In short, "integrated voice/data" means different things to different people and depends on the technology. See also ISDN.

Integrated Voice Data Workstation See ISDN, IVDT and INTEGRATED VOICE/DATA.

Intel i860 Microprocessors Family Intel Corporation's RISC-based microprocessor line, introduced in 1989, incorporates many firsts, including more than 1 million transistors on a single chip. The line currently comprises the i860 XR, with 1.2 million transistors, and the i860 XP, with 2.5 million transistors.

Intel Blue When ordering your ISDN phone line and you want to use it for data or video, tell them it's "Intel Blue." That should tell your local phone company the correct technical specifications for your line. And when you come to plug in your ISDN equipment (assuming your chosen manufacturer has made it compatible with Intel Blue), it should work. This is not a guarantee, but a probability. See ISDN.

Intel Proshare A popular collaborating computing tool that allows two computer users to share a computer application or whiteboard across analog or digital phone lines. Some models of ProShare also include a video camera (so you can see the person at the other end) and a microphone (so you can hear the person at the other end). See INDEO VIDEO.

Intellectual Property Property produced by effort of the mind, as distinct from real or personal property. Intellectual property may or may not enjoy the benefit of legal protection: See COPYRIGHT, PATENT, TRADEMARK and TRADE SECRET.

Intellidial Intellidial is basically a Nynex Centrex service for small offices of big businesses or small businesses. You can get from two to nine lines. And you get to buy "packages" of features. Features in the packages include touchtone, three-Way calling, call transfer, call hold, call pick-up, intercom, call forward-variable, call forward busy, call forward-don't answer, speed calling and call waiting.

Intelligence The part of a computer which performs the arithmetic and logic functions. Also, the information impressed or modulated on a transmission carrier — either voice or data.

Intelligent Agent Software that has been taught something of your desires or preferences and acts on your behalf to do things for you. It might, for example, search through incoming material on networks (e-mail and news) and find what you're interested in or looking for. It might, for example, monitor your TV viewing habits, accept general instructions about your preferences and then, on its own, browse through huge databases of available videos and make recommendations about programs you might be interested in viewing.

Intelligent Answering A Rolm term, explained thus: "When your customers call — or you call them — the Rolm 9751 CBX system can use automatic number identification (ANI) or dialed number identification service (DNIS) to identify the caller and the reason for that call."

Intelligent Assistance A concept Apple is pushing for its Newton PDA. Newton can anticipate what you want to do and provide a bit of help. This is how Fortune Magazine explained it: For example, scrawl "lunch with John Thursday." My Newton would assume that Thursday means next Thursday and that John is the John I've been meeting with a lot lately, John Sculley, and that I want to eat at 12:30, my usual lunch hour. Newton updates my calendar, and presto, displays the entry for my approval. I can okay it or change it.

Intelligent Battery System See IBS.

Intelligent Concentrator A concentrator which receives signals from a device on one port and retransmits them to devices on other ports. An intelligent concentrator is one that has software and therefore has programming capabilities.

Intelligent Hub A hub that functions both as a bridge and multiprotocol router.

Intelligent Network IN. A network that allows functionality to be distributed flexibly at a variety of nodes on and off the network and allows the architecture to be modified to control the services; (In North America) an advanced network concept that is envisioned to offer such things as (a) distributed call-processing capabilities across multiple network modules, (b) real-time authorization code verification, (c) one-number services, and (d) flexible private network services (including (1) reconfiguration by subscriber, (2) traffic analyses, (3) service restrictions, (4) routing control, and (5) data on call histories). Levels of IN development are:

• IN/1. A protocol intelligent network targeted toward services that allow increased customer control and that can be provided by centralized switching vehicles serving a large customer base.

• IN1+. A protocol intelligent network targeted toward services that can be provided by centralized switching vehicles, e.g., access tandems, serving a large customer base.

• IN/2. A proposed, advanced intelligent-network concept that extends the distributed IN/1 architecture to accommodate the concept called the "service independence." Traditionally, service logic has been localized at individual switching systems. The IN/2 architecture provides flexibility in the placement of service logic, requiring the use of advanced techniques to manage the distribution of both network data and service logic across multiple IN/2 modules. See AIN, which stands for Advanced Intelligent Network.

Intelligent Peripheral IP. A network system in the Advanced Intelligent Network Release 1 architecture containing an Resource containing an Resource Control Execution Environment (RCEE) functional group that enables flexible information interactions between a user and the network.

Intelligent Phone When the Bell operating companies get bored they occasionally fantasize about applications for the networks they provide. Here are some of their ideas for what intelligent phones could, if motivated, do:
Select entertainment on demand (movies, music, video); Order groceries or other services or products; Record customized news and sports programming; Enroll and participate in education programs from the convenience of subscribers' living rooms; Find up-to-minute medical, legal and encyclopedic information; Pay bills and manage finances; Make airline, rental car and hotel reservations and buy sports and entertainment tickets.

Intelligent Premises Equipment This refers to modern equipment, such as routers and intelligent switches. These devices are often capable of taking on roles traditionally performed by the network service, such as error correction.

Intelligent Routing A voice call comes in. Your voice mail machine recognizes it as being urgent, so it gives the caller a message, "Please hold. Harry is away from his desk. I'll find Harry for you." Meantime, it dials several numbers looking for me. It also beeps me. Eventually I call in. It tells me, "John Smith is calling for you. You want him?" Yes, I say and we're connected. This is a simple form of a broad concept that many are beginning to call intelligent routing. See also AT WORK and WINDOWS TELEPHONY.

Intelligent Terminal A terminal is an input/output device to a distant computer. The terminal may communicate with the computer over a dedicated collection of wires or over phone lines. In the early days, terminals contained no processing power. They simply reflected what the user typed in and what the distant computer responded. As computers became cheaper and with the advent of the "computer on a chip," so it was economically possible to put computing power into a terminal. This reduced the load on the main computer and cut down on communications costs. There are levels of "intelligence" in terminals. An intelligent terminal might perform simple arithmetic functions or it might check the accuracy of input data (does the zip code match the state?). It may perform far more comprehensive processing — as doing virtually all the local processing, and only transmitting summary results to corporate headquarters once a day. A personal computer can be used and act as an Intelligent Terminal. Many personal computer communications software can emulate terminals, the most common being the DEC VT-100.

Intelligible Crosstalk Crosstalk from which information can be derived.

INTELLIHUB A Nynex service which offers its larger customers large, local bandwidth. The service was developed primarily as a response to the threat of bypass. INTELLIHUB integrates and modifies FLEXPATH Digital PBX Service, INTELLIPATH, INTEL-

LIPATH II Digital Centrex Service and INFOPATH Packet Switching Service for large multi-location business customers within an regional calling area.

INTELLIMUX A new service from Bell Atlantic. This service allows Bell Atlantic's customers to do real-time reconfigurations of their network. Changes may be made to specific individual voice grade and digital grade services within a T-1 and/or individual channels of a T-1 service connected at digital cross connect systems.

Intellipath Digital Centrex Service Intellipath Digital Centrex Service is a Nynex and New England Telephone digital Centrex service for smaller businesses with fewer than 99 lines. It has the basic features — such as call transfer, hold and three-way calling. One nice feature: each user can have his own 30-number speed calling list.

Intellipath II Digital Centrex Service INTELLIPATH II Digital Centrex Service is a Nynex and New England Telephone service for businesses needing over 100 lines. It has voice and data features. It has over 100 voice features — from automatic call back, to call pickup, to intercept and three-way calling. Its data features let you plug computers and terminals directly into a digital circuit.

INTELSAT INternational TELecommunications SATellite organization. A worldwide consortium of national satellite communications organizations. As of writing Intelsat is owned by 138 governments and Intelsat itself owns 24 satellites worldwide. INTELSAT owns and operates the world's most extensive global communications satellite system. INTELSAT provides international, regional and domestic telephone and television services. It also offers, business services such as international video, teleconferencing, facsimile, data and telex. Comsat acts as the exclusive manager for Intelsat. Comsat is also the exclusive U.S. representative. Comsat stands for Communications Satellite Corporation.

Intensity Modulation IM. In optical communication, a form of modulation in which the optical power output of a source is varied in accordance with some characteristic of the modulating signal. In intensity modulation, there are no discrete upper and lower sidebands in the usually understood sense of these terms, because present optical sources lack sufficient coherence to produce them. The envelope of the modulated optical signal is an analog of the modulating signal in the sense that the instantaneous power of the envelope is an analog of the characteristic of interest in the modulating signal. Recovery of the modulating signal is by direct detection, not heterodyning.

Inter- Means between two things, as opposed to INTRA-, which means inside one thing. INTERSTATE means communications between states. INTRASTATE means communications inside one state.

Interactive The ability to interact with a computer or to be in a conversational mode with a computer. Interactive processing is very time dependent since a user is waiting for the computer to ask him/her questions. The opposite of Interactive processing is batch processing. See BATCH.

Interactive Call Setup A Rolm definition. An interface between a station user and the Rolm CBX that provides prompts and error messages to process the setting up of data calls, and provides for the displaying of data line parameters.

Interactive CATV Interactive CATV is a two-way cable system from which subscribers can receive and send signals. They will probably do this by punching buttons on their cable TV's remote control, which may look more like a computer keyboard than a traditional cable TV handheld remote signaling device.

Interactive Data Transaction A single (one-way) message, transmitted via a data channel to which a reply is required for work to proceed logically.

Interactive Kiosk Interactive kiosks represent a powerful new product delivery vehicle for "non-store" marketers, to increase sales by offering their products and services in high traffic areas, such as airports. According to research analyst Warren Hersch, financial services kiosks are not confined to full-service bank locations; they are found in supermarkets, shopping malls and auto dealerships. Self-service terminals offering government services reside in discount stores, libraries, outdoor pavilions and subway stations. Kiosks purveying travel-related services occupy office complexes, colleges and universities, pharmacies and other retail outlets.

Interactive Video The fusion of video an computer technology. A video program and a computer program running in tandem under the control of the user. In interactive video, the user's actions, choices, and decisions affect the way in which the program unfolds. See INDEO VIDEO.

Interactive Voice Response IVR. Think of Interactive Voice Response as a voice computer. Where a computer has a keyboard for entering information, an IVR uses remote touchtone telephones. Where a computer has a screen for showing the results, an IVR uses a prerecorded human voice that is stored (digitized) on a hard drive. In addition it can use a synthesized voice (computerized voice) for read back information that is constantly changing. (The synthesized voice is commonly referred to as Text-to-Speech.) Whatever a computer can do, an IVR can too, from looking up train timetables to moving calls around an automatic call distributor (ACD). The only limitation on an IVR is that you can't present as many alternatives on a phone as you can on a screen. The caller's brain simply won't remember more than a few. With IVR, you have to present the menus in smaller chunks.

Interaxial Spacing Center to center conductor spacing between any two wires.

Interbreak Interval A call center term. A Scenario scheduling assumption specifying the minimum amount of time that must elapse between the end of one break and the beginning of another.

Interbuilding Cable The communications cable that is part of the campus subsystem and runs between buildings. There are four methods of installing interbuilding cable: in-conduit (in underground conduit), direct-buried (in trenches), aerial (on poles), and in-tunnel (in steam tunnels).

Interbuilding Cable Entrance The point at which campus subsystem cables enter a building.

Interbuilding Wiring Consists of underground or aerial telephone wire/cables used on the premises to connect structures remote from the primary building to the premises telephone system.

Intercast New plug-in cards from Intel which will allow your PC to simultaneously receive TV pictures, and in the blank spaces of TV signals, Internet Web pages and text.

Intercept Calls which cannot reach their destination may be intercepted and diverted to a station attendant, a recording or some other place. See INTERCEPT RECORDING and INTERCEPT SERVICE.

Intercept Recording You make a phone call. It doesn't go through. The phone company intercepts that call and sends it somewhere. Intercept Recording is a recording telling you your call cannot be completed and has been intercepted on its way to the destination number for some reason that will be explained by the recording. The most common voice you hear on intercept announcements is Jane Barbie's. See BARBIE, JANE and INTERCEPT SERVICE.

Intercept Service A service of the local phone in which a phone call is redirected by an operator or a recording to another phone number or a message.

Interchange Carrier IC. A common carrier that provides services to the public local exchanges on an intra or interLATA basis in compliance with local or Federal regulatory requirements and that is not an end user of the services provided.

Interchangeable NPA Code Code in the NXX format used as a central office code (NNX format), but that can also be used as an NPA code. Interchangeable NPA codes will be introduced on or after January 1, 1995.

Intercom Intercommunication. An internal communication system which allows you to dial another phone in your building, office complex, factory or home. There are three types of intercom: 1. Dial: It allows you to dial or pushbutton another extension; 2. Automatic: One phone goes off hook and automatically dials another; and 3. Manual: The user can manually signal another phone by pushing a button for that phone. An example is a buzzer between a boss and a secretary.

Intercom Blocking A PBX feature by which phones with a particular Class Of Service (COS) are blocked from calling certain phones. A rare feature.

Interconnect A circuit administration point, other than a cross connect or an information outlet, that provides capability for routing and re-routing circuits. It does not use patch cords or jumper wires, and typically is a jack-and-plug device used in smaller distribution arrangements or that connects circuits in large cables to those in smaller cables. See INTERCONNECT COMPANIES.

Interconnect Agreement An agreement between an established local phone company and a new local phone company for both companies to allow their subscribers to dial each other. Such agreement covers issues such as sharing of revenues and if a subscriber, who changes local phone companies, can keep his phone number.

Interconnect Companies Companies which sell, install and maintain telephone systems for end users, typically businesses. AT&T coined the word "interconnect" as a pejorative word — to indicate that these companies "interconnected" to AT&T's telephone network — but didn't really belong there. These "interconnect" companies contrasted with true-blue companies belonging to AT&T which did a sterling job — such as AT&T itself or its local operating companies. Anyway, the term stuck but the nasty associations went away. Now the irony is that the CPE (Customer Premise Equipment) equipment subsidiaries and divisions of both AT&T and the local Bell operating telephone companies are, in fact, interconnect companies, though they don't call themselves that. The industry is looking for a better word.

We at TELECONNECT Magazine started a campaign to make "TELECONNECT" a better replacement for interconnect. But our lawyers and the lawyers for a manufacturing/interconnect company called Teleconnect told us to lay off and stop trying to make the word generic. We rather like the terms "Telecommunications Systems Integrator," "Telecommunications VAR" or "Telecom Developer." They seem to be catching on.

Interdrive The name of the FTP Software client implementation of the Sun NFS protocol.

Interenterprise Communications Communications exchanged between multiple organizations, e.g., between business trading partners, collaborators, affiliates or a business and its customers.

Interexchange Carrier IXC. A telephone company that is allowed to provide long-distance telephone service between LATAs but not within any one LATA. See IXC.

Interexchange Channel IXC. A communications channel or path between two or more telephone exchanges.

Interexchange Customer Service Center ICSC. The Telephone Company's primary point of contact for handling the service needs of all long distance carriers.

Interface 1. A mechanical or electrical link connecting two or more pieces of equipment together.
2. A shared boundary. A physical point of demarcation between two devices where the electrical signals, connectors, timing and handshaking are defined. The procedures, codes and protocols that enable two entities to interact for a meaningful exchange of information.
3. To bring two things or people together to allow them to talk.
4. A poorly-defined word often used when the speaker is incapable of figuring precisely what he means. See also INTERFACE DEVICE.

Interface Device A device which meets a standard electrical interface on one side and meets some other nonstandard interface on the other. The purpose of the device is to allow a device with a nonstandard interface to connect to a device with a standard interface. See also INTERFACE.

Interface Functionality The characteristic of interfaces that allows them to support transmission, switching, and signaling functions identical to those used in the enhanced services provided by the carrier. As part of its comparably efficient interconnection (CEI) offering, the carrier must make available standardized hardware and software interfaces that are able to support transmission, switching, and signaling functions identical to those used in the enhanced services provided by the carrier.

Interface Message Processor IMP. A processor-controlled switch used in packet-switched networks to route packets to their proper destination.

Interface Overhead the interface overhead is the remaining portion of the bit stream after deducting the information payload. The interface overhead may be essential (e.g. framing for an interface shared by users) or ancillary (e.g. performance monitoring).

Interface Payload The portion of the bit stream which can be used for telecommunications services. Any signaling is included in the interface payload. See also INTERFACE OVERHEAD.

NEWTON'S TELECOM DICTIONARY

Interface Shelves Shelves in a Rolm PBX cabinet containing the printed circuit card groups that connect telephones, terminals, lines and trunks to CBX interface channels. These shelves also contain shared electronics cards.

Interference Energy you receive with a signal. You don't want the energy. You want the signal. Getting rid of the interference may be a pain. The interference may be manmade (e.g. elevator motors) or it may be GOD-made, e.g. lightning, thunderstorms. Some media (fancy word for cabling) may be more immune to interference than other. Media immune to interference, in order 1. Optical fiber 2. Coax 3. Shielded twisted pair 4. Unshielded twisted pair 5. Unshielded untwisted pair

Interference Emission Emission that results in an electrical signal being propagated into and interfering with the proper operation of electrical or electronic equipment. The frequency range of such interference may be taken to include the entire electromagnetic spectrum.

Interferometer An instrument that employs the interference of light waves for measurement.

Interflow The ability to establish a connection to a second ACD and overflow a call from one ACD to the other. This provides a greater level of service to the caller.

Interframe See INTERFRAME CODING.

Interframe Coding Compression techniques which track the differences between frames of video and eliminates redundant information between frames. Stores only once those pixels that don't change. Then multiple frames access those pixels during decompression. This results in more compression over a range of frames than intraframe coding, which compresses information within a single frame.

Interframe Encoding A way of video compression that transmits only changed information between successive frames. This saves bandwidth.

Interframe Gap IFG. The minimum idle time between the end of one frame transmission and the beginning of another. On Ethernet 802.3 LANs the minimum interframe gap is 9.6 micro-seconds.

Interior An ATM term. Denotes that an item (e.g., link, node, or reachable address) is inside of a PNNI routing domain.

Interlace In TV, each video frame is divided into two fields with one field composed of the odd- numbered horizontal scan lines and the other composed of the even-numbered horizontal scan lines. Each field is displayed on an alternating basis. This is called interlacing. It is done to avoid flicker. See also INTERLACING.

Interlacing Regular TV signals are interlaced. In the US there are 525 scanning lines on the regular TV screen. This is the NTSC standard. Interlaced means the signal refreshes every second line 60 times a second and then jumps to the top and refreshes the other set of lines also 60 times a second. Non-interlaced signals, which are used in the computer industry, means each line on the entire screen is refreshed X times. X times depends on what the video card is outputting to the color monitor. The more expensive the card and the monitor, the more often the monitor will be refreshed. The more it's refreshed, the better and more stable it looks — the less perceived flicker. For example, text on an NTSC United States TV set tends to "flicker." It doesn't on a non-interlaced monitor. Typical non-interlaced computer monitors refresh at 60 to 72 times a second. But good ones refresh at higher rates. Generally, anything over 70 Hz (i.e. 70 times a second) is considered to be flicker-free and therefore preferred, if you can afford it. In short, buy an non-interlaced monitor. You'll like it better.

InterLATA Telecommunications services that originate in one and terminate in another Local Access and Transport Area (LATA). Under provisions of Divestiture, the Bell operating companies cannot provide Inter-LATA service, but can provide Intra-LATA service. Some LATAs are very large. So some "local" phone companies provide the equivalent of long distance service. And some of these phone companies have different pricing packages. Some of these packages are cheap, but not highly-publicized. See also LATA.

InterLATA Call A call that is placed within one LATA (Local Access Transport Area) and received in a different LATA. These calls are currently carried by a long distance company.

InterLATA Competition Currently being considered by the Public Utility Commission (PUC) in ARF (Alternative Regulatory Framework) Phase III. Would allow long distance phone companies to compete with local monopolies to carry intraLATA toll calls.

Interleave 1. The transmission of pulses from two or more digital sources in time-division sequence over a single path.
2. A data communication technique, used in conjunction with error-correcting codes, to reduce the number of undetected error bursts. In the interleaving process, code symbols are reordered before transmission in such a manner that any two successive code symbols are separated by I-1 symbols in the transmitted sequence, where I is called the degree of interleaving. Upon reception, the interleaved code symbols are reordered into their original sequence, thus effectively spreading or randomizing the errors (in time) to enable more complete correction by a random error-correcting code.
3. Interleaving also refers to the way a computer writes to and reads from a hard disk. Understanding interleaving is critical if you want to get your hard disk to work at its maximum speed (without in any way damaging the disk). Let's look at the way MS-DOS reads information from a hard disk. All hard disks are controlled by a special card called a hard disk controller card. Let's say your computer wants a file. It tells the hard disk controller card it wants the file. The controller searches the disk for the first sector of the requested file, reads that sector (usually 512 bytes) to your computer's RAM and then transfers the information to

Interleave	Time to copy file Setting
3	1 min 15 seconds
4	1 min 17 seconds
5	1 min 1 second
6	35 seconds
7	41 seconds
8	1 min 10 seconds

the CPU to be processed. When this is complete, the controller goes back to the hard disk and searches for the file's second sector. The process continues in this way until the file is completely read.
The problem is that while the controller and the CPU are doing their things, the hard disk itself is spinning 3,600 times a minute. By the time the controller reads one sector and it is ready to return to the disk, the next consecutive file sector has spun past the read/write head. If the file is stored in contiguous sectors, the controller must wait for the disk to complete its revolution before it can read the next file sector. To solve this problem, hard drive makers developed a concept called interleave setting, which tells the hard drive controller to skip a certain number of sectors when it writes a file to disk. Thus, when the file is later read back, the appropriate file sectors should fall under the read/write heads at the appropriate time.
If the controller reads or writes one sector and then skips a sector, the interleave is 2 (every other sector is used to store logically consecutive blocks). The interleave is sometimes written as 2:1. If the controller writes to one sector and then skips two, the interleave is 3 or 3:1. The interleave factor is usually established by the manufacturer or reseller of the hard disk/controller combination. If someone else assembles the hard disk/controller combination, that person may need to experiment to determine the correct interleave factor — i.e. the one that works fastest without messing up.
Setting the "correct" interleave settings on your hard disk is critical to getting maximum performance out of your hard disk. Here's a test that a writer for PCResource Magazine did. He copied the same files from one part of his hard disk to another part using different interleave settings:
Clearly his best interleave setting is six. There are two ways of choosing the correct interleave setting. You can do it by trial and error as the writer did. His test took three hours. Or buy a program and do it in seconds. The best program is called Disk Technician. It's from a company called Prime Solutions in San Diego. Sadly, the program doesn't work on certain laptops and on certain controller card/hard disk combinations.

Intermediate Assist A method for pulling cables into conduits or duct liners in which manual labor or machines are used to assist the pulling at intermediate manholes.

Intermediate Cross-Connect an interconnect point within backbone wiring. for example, the interconnection between the main cross-connect and telecommunications closet or between the building entrance facility and the main cross-connect.

Intermediate Distribution Frame IDF. A metal rack designed to connect cables and located in an equipment room or closet. Consists of bits and pieces that provide the connection between inter-building cabling and the intra-building cabling, i.e. between the Main Distribution Frame (MDF) and individual phone wiring. There's usually a permanent big, fat cable running between the MDF and IDF. The changes in wiring are done at the IDF. This saves confusion in wiring. See also FEEDER CABLE and CONNECTING BLOCK.

Intermediate Frequency Transformer A transformer designed to amplify the intermediate frequencies generated in a superheterodyne radio receiver. These are normally sharply tuned to a single frequency band.

Intermediate High-Usage Trunk Group A Bellcore definition. A high-usage trunk group that receives route-advanced overflow traffic and may receive first-route traffic and/or switched-overflow traffic.

Intermediate Reach Intermediate reach refers to optical sections from a few km to approximately 15 km. An AT&T SONET term.

Intermediate System An OSI term which refers to a system that originates and terminates traffic, as well as forwarding traffic to other systems.

Intermittent Problems Intermittent problems are issues or bugs that come to light only after systems have been running for some time, or certain infrequently performed sequences of events are performed. Often many thousands of calls need to be put through before they are discovered. And bugs may only be seen occasionally, perhaps one of every 100 times something is done. Intermittent problems are among the hardest to find and duplicate.

Intermodulation IM. The production, in a nonlinear element of a system, of frequencies corresponding to the sum and difference frequencies of the fundamentals and integral multiples (harmonics) of the component frequencies that are transmitted through the element.

Intermodulation Distortion IMD. Nonlinear distortion characterized by the appearance of frequencies in the output, equal to the sum and difference frequencies of integral multiples (harmonics) of the component frequencies present in the input. Harmonic components also present in the output are usually not included as part of the intermodulation distortion.

Internal Bus See LOCAL BUS.

Internal Modem A modem on a printed circuit card which is inserted into one of the slots on a PC (personal computer). The other type of modem for a PC is an external modem — essentially a modem with the same circuitry as an internal modem but with a metal or plastic case. An internal modem costs slightly less than an external one. Internal modems are good if you're short of desk space and afraid your external modem will be stolen. External modems have lights so it's easier to tell what's going on. Everybody has their theories on which type of modem is best. We prefer the external ones — largely for their lights and ease of moving around.

Internal Reachable Address An ATM term. An address of a destination that is directly attached to the logical node advertising the address.

International 800 Service You can now have your customers overseas call you for free on an 800 line, just as your domestic customers do. The service is available from countries including Australia, Brazil, France, Hong Kong, Israel, Italy, Japan, Sweden, Switzerland and the United Kingdom. Overseas 800 service is often known as "Freephone."

International Alphabet No. 5 IA5. Internationally standardized alphanumeric code with national options. ASCII is United States version.

International Ampere The current which will in one second deposit 0.001118 gram of silver from a neutral solution of silver nitrate.

International Callback Calling the United States from many countries abroad is far more expensive than calling those countries from the United States. A new business called International Callback has started. It works like this. You're overseas. You dial a number in the United States. You let it ring once. It won't answer. You hang up. You wait a few seconds. The number you dialed in the U.S. knows it was you calling. There is a piece of equipment on that number that "hears" it ring and knows it's you since no one else has that number. That was your special signal that you want to make a call. A switch attached to that line then calls you instantly. When you answer (overseas, obviously) it conferences you with another phone line in the United States and gives you U.S. dial tone. You can then touchtone from overseas your American number, just as if you would, were you in the U.S. There are huge savings. U.S. international callback operators can offer as much as 50% savings on calls from South America, where international calling rates are very high. The process of international callback is being automated with software and dialing devices. International callback is also helping to bring down the high cost of calling the U.S. from overseas. A company called Kallback in Seattle, WA, has received a service mark from the U.S. Patent and Trademark Office for the words "callback" and "kallback" and sends letters to and threatens law suits against companies who use "their" words. See CALLBACK.

International Carrier A carrier that generally provides connections between a customer located in World Zone 1 and a customer located outside of World Zone 1, but with option of providing service to World Zone 1 points in North American Numbering Plan area codes outside the U.S.

International Center For Information Technologies A Washington "think tank" whose mission is to bring together discussion on new telecommunications and computer technologies. Targeted at senior executives who are looking for ways to apply new technologies to gaining competitive edges for their company.

International Denial An optional restriction on your cellular phone that prevents the cellular number from marketing international calls. Some carriers place this restriction on all subscribers using their service.

International Direct Distance Dialing IDDD. Being able to automatically dial international long distance calls from your own phone. The direct calling by the originating customer to the distant (international) called customer via automatic switching. IDDD is synonymous with the phrases international direct dialing and international subscriber dialing.

International Gateways The switches in the various domestic long distance networks (e.g. MCI, AT&T and Sprint) which interface their networks with International telecommunications networks. All US International calls are routed through an international gateway.

International Organization For Standardization ISO. An organization established to create standards. See ISO.

International Prefix The combination of customer-dialed digits prior to dialing of the country code required to access the automatic outgoing international equipment in the originating country.

International Record Carrier IRC. One of a group of carriers that, until recently, was part of a monopoly of U.S. common carriers certified to carry data and text to locations outside the U.S. In recent years, regulation of this type of service has been markedly relaxed. Most of the IRCs got bought by MCI.

International Standards Organization ISO. An international standards-setting organization.

International Switching Carrier ISC. An exchange whose function is to switch telecommunications traffic between national network and the networks of other countries. Also known as an international gateway office.

International Telecommunication Union-Radiocommunication Sector ITU-R. Formerly the International Radio Consultative Committee (CCIR). The technical study branch of the International Telecommunication Union responsible for the study of technical and operating questions relating specifically to radio communications. See also ITU-T.

International Telecommunications Union ITU. The specialized agency of the United Nations which tries to establish standardized communications procedures and practices. Its most successful work is done in the allocation of radio frequencies worldwide — including satellites, etc. See CCITT and ITU-T.

International Telephone Address A four-part code specifying a unique address for any telephone company in the world.

Internet It is very hard to define the Internet in a way that is either meaningful or easy to grasp. To say the Internet is the world's largest computer network is to trivialize it. Once a cooperative research effort of the Federal Government (called ARPA), which tied universities to other universities to their military customers, the Internet has essentially become a new publishing and commerce medium. Several people (including myself) believe that its invention is as important to the dissemination of knowledge and to peoples' life styles as the invention of the Gutenberg Press was in 1455. At its heart, the Internet is several large computer networks joined together over high-speed data links ranging from ISDN to T1, T3, FDDI, SONET, SMDS, OT1, etc. The most prominent of these national nets are MILNET (Military Network), NSFNET (National Science Foundation NETwork), and CREN (Corporation for Research and Educational Networking). Commercial networks from AT&T, MCI, SPRINT, WILTEL, and many others are now beginning to carry the bulk of the traffic. Note: There "Internet" is now private. NSNET no longer does the funding it once did. But there are a myriad of regional and local campus networks joined in all over the world. And increasingly, businesses are joining their computers to the Internet. In 1995, the Government Accounting Office (GAO) said that the Internet linked 59,000 networks, 2.2 million computers and 15 million users in 92 countries. The Internet traces its origins to a network called Arpanet which was set up in 1969 by the Defense Department. The Internet's backbones and university-government hookups are financed by federal government monies. But the Internet's networking technology is very smart. Every time, someone hooks a new computer to the Internet, the Internet adopts that hookup as its own and begins to route Internet traffic over that hookup and through that new computer. Thus as more computers are hooked to the Internet, its network grows exponentially. The invention of the Internet will go into the history books as one of the twentieth century's most important inventions. See various INTERNET definitions following. See also DOMAIN, DOMAIN NAMING SYSTEM, INTERNET APPLIANCE, INTERNET PROTOCOL, SURF, TCP/IP and WEB BROWSER.

Internet Appliance A sub-$500 machine specially designed for Internet browsing and first proposed in the late Fall of 1995 by Larry Ellison, head of database software company Oracle. Part of its appeal to people outside Microsoft and Intel is that the Internet Appliance would not have to be based on standard PC technology. It need have an Intel chip and need not run Windows.

NEWTON'S TELECOM DICTIONARY

This device is also called an Internet Terminal, a Network Computer or an IPC, an Interpersonal computer. The original description of the Internet Appliance was that it would come with 4mb of RAM, 4mb of flash memory, processor, monitor, keyboard and mouse — all for under $500.

Internet Architectures Board The Internet Architectures Board oversees Internet protocols and procedures and the creation of Internet standards. According to Computerworld Magazine, it has 16 members. www.iab.org/lab/

Internet Assigned Numbers Authority IANA. This group coordinates assignment of Internet parameters, domain names and Internet addresses. www.isi.edu/div7/iana/

Internet Control Message Protocol ICMP. The protocol used to handle errors and control messages at the IP player. ICMP is actually part of the IP protocol.

Internet Engineering Steering Group IESG. The executive committee of the IETF (Internet Engineering Task Force).

Internet Engineering Task Force IETF. One of two technical working bodies of the Internet Activities Board. The IETF is the primary working body developing new TCP/IP (Transmission Control Protocol/Internet Protocol) standards for the Internet. It has more than one thousand active participants. Home page: www.ietf.cnri.reston.va.us/home/html

Internet Gateway Internet gateways are devices which typically sit on a local area network and handle all the translations between IPX traffic on your LAN (IPX is the NetWare protocol) and the TCP/IP traffic on the Internet. TCP/IP is the protocol used on the Internet. See also INTERNET SERVERS and other definitions beginning with INTERNET.

Internet Group Name In Microsoft networking, a name registered by the domain controller that contains a list of the specific addresses of computers that have registered the name. The name has a 16th character ending in 0x1C.

Internet MIB Subtree A tree-shaped data structure in which network devices on a local area network and their attributes can be identified within the confines of a network management scheme. The name of an object or attribute is derived from its location on this tree.
For example, an object in MIB-I might be named 1.2.1.1.1.0. the first 1 indicates the object is on the Internet. The 2 denotes that it falls within the Management category. The second 1 shows the object is part of the first fully defined MIB, known as MIB-I. The third 1 indicates which of the eight object groups is being referenced. And the fourth 1 is a textual description of the network component. The 0 indicates there is only one object instance. An object instance links a particular object to a specific node on the network. The numbering system is infinitely extendible to accommodate additions to this base identification scheme. This common naming structure permits equipment from a variety of vendors to be managed by a single management station that uses SNMP. The four main categories of the tree are Directory, Management, Experimental and Private/Enterprises.

Internet Number The dotted-quad address used to specify a certain system. The Internet number for cs.widener.edu is 147.31.130. A resolver is used to translate between hostnames and Internet addresses.

Internet Packet Exchange IPX. Novell NetWare's native LAN communications protocol, used to move data between server and/or workstation programs running on different network nodes.

Internet Protocol IP. Part of the TCP/IP family of protocols describing software that tracks the Internet address of nodes, routes outgoing message, and recognizes incoming messages. Used in gateways to connect networks at OSI network Level 3 and above. See INTERNET PROTOCOL DATAGRAM.

Internet Protocol (IP) Address A unique, 32-bit number for a specific TCP/IP host on the Internet. IP addresses are normally printed in dotted decimal form, such as 128.127.50.224. Once your domain is assigned a group of numbers by the Internet's central registry, it can house one or several domains and/or hosts, i.e. computertelephony.com and teleconnect.com. People looking for those domains will be pointed to that server where they will find all information in the domain — perhaps a home page, or a place to leave e-mail, etc. There are three classes of IP address A, B, and C — the most common of which is a class "C" address block. A class "C" address block can address about 256 hosts (e.g., 128.10.70.*). a class "B" address block can contain about 256*256 (e.g., 128.10.*.*) hosts. Some ip addresses are reserved for broadcasts in respective domains. See DOMAIN and INTERNET.

Internet Protocol Datagram The fundamental unit of information passed across the Internet. Contains source and destination addresses along with data and a number of fields which define such things as the length of the datagram, the header checksum, and flags to say whether the datagram can be (or has been) fragmented, This is a self-contained packet, independent of other packets.

Internet Protocol Suite A suite of network protocols which have been adopted as the main de facto protocols for LANs.

Internet Relay Chat IRC. Sort of like CB radio, but run on the Internet, and far more confusing than CB radio.

Internet Server An Internet server is a device which users on the Internet access to get services. Such services might be electronic mail, news, a Web page, etc. A company will have one or more Internet servers attached to the Internet when it wants to deliver services to people on the Internet. Such Internet servers could be called e-mail servers, FTP servers, News servers and World Wide Web servers. Internet servers most commonly run on Unix. But Microsoft Windows NT is increasingly gaining popularity.

Internet Server API See ISAPI.

Internet Service Provider ISP. A vendor who provides direct access to the Internet. The ISP also usually provides a core group of internet utilities and services like E-mail, News Group Readers and sometimes weather reports and local restaurant reviews. The user reaches his ISP by either dialing-up with their own computer, modem and phone line, or over a dedicated line installed by the telephone company. The ISP could also

Internet Society ISOC. A non-profit organization that fosters the voluntary interconnection of computer networks into a global communications and information infrastructure. According to Computerworld, the Internet Society is concerned with the evolution of the Internet and its social, political and technical issues. The ISOC is the umbrella organization for the IAB, IETF and IRTF. The ISOC has 5,000 members in 120 countries. Its Web site is www.isoc.org

Internet Suite Of Protocols The combination of TCP (Transmission Control Protocol) and IP (Internet Protocol). See TCP/IP.

NEWTON'S TELECOM DICTIONARY

Internet Telephony Technology that lets you make voice phone calls over the Internet using your PC. You'll need appropriate software on both
The basic There are several ways of making calls. Requires appropriate software on sender and receiver PCs.

Internet Terminal A sub-$500 machine specially designed for Internet browsing and first proposed in the late Fall of 1995 by Larry Ellison, head of database software company Oracle. Part of its appeal to people outside Microsoft and Intel is that the Internet Appliance would not have to be based on standard PC technology. It need have an Intel chip and need not run Windows. This device is also called an Internet Terminal, a Network Computer or an IPC, an Interpersonal computer. The original description of the Internet Appliance was that it would come with 4mb of RAM, 4mb of flash memory, processor, monitor, keyboard and mouse — all for under $500.

Internet Worm This software program caused a major part of the Internet network to crash by replicating and generating spurious data.

Internetwork See INTERNETWORKING.

Internetwork Management A generic term used to describe the actions that help maintain, a complex network.

Internetwork Packet Exchange IPX. A network layer protocol developed by Novell Inc. and used in NetWare implementations. See IPX.

Internetwork Router In local area networking technology, an internetwork router is a device used for communications between networks. Messages for the connected network are addressed to the internetwork router, which chooses the best path to the selected destination via dynamic routing. Internetwork routers function at the network layer of the Open Systems Interconnection (OSI) model. Also known as a network router or simply as a router.

Internetworking 1. Communication between two networks or two types of networks or end equipment. This may or may not involve a difference in signaling or protocol elements supported. And, in the narrower sense — to join local area networks together. This way users can get access to other files, databases and applications. Bridges and routers are the devices which typically accomplish the task of joining LANs. Internetworking may be done with cables — joining LANs together in the same building, for example. Or it may be done with telecommunications circuits — joining LANs together across the globe.

InterNIC The Internet's central naming registry. InterNIC is the place you apply to register your domain name. The InterNIC Registration Services Host computer contains information on Internet Networks, ASNs, Domains, and POCs). Its address is http://internic.net. For MILNET information (i.e. military servers), use the whois server at nic.ddn.mil. See DOMAIN and DOMAIN NAME.

Internode Communication paths which originate in one node and terminate in another.

Internode Link A data line for high-bandwidth connections between PBXs.

Interoffice Between two telephone company switching offices.

Interoffice Channel A transmission path between two AT&T serving offices.

Interoffice Trunk A trunk circuit connecting two local telephone company central offices.

Interoperate The ability of multi vendor computers to work together using a common set of protocols. With interoperability, PCs, Macs, Suns, DEC VAXes, IBM mainframes, etc. all work together allowing one host computer to communicate with and take advantage of the resources of another.

Interoperability The ability to operate software and exchange information in a heterogeneous network, i.e. one large network made up of several different local area networks.

Interoperability Testing In our industry there are several levels of testing of new products and services. Clearly, the manufacturer does one level of testing. His job is to ensure that his product meets the claims that he promotes for it. Then there is a interoperability testing. That is testing to ensure that his product works with other products that allegedly conform to the same standard. The most public example of interoperability testing in the telecom industry is that which has occurred so successfully in the ATM Forum. That organization was established by a bunch of manufacturers who make ATM products. Their objective was not to establish new standards, but to make sure that their products worked successfully with each other without hardware modification. Plug the equipment together. Configure the software. Bingo it would work.

InterPBX Calls coming into one PBX can be transferred to extensions on another PBX using direct tie lines between the two PBXs.

Interpersonal-Message IPM. The term used in the 1984 X.400 recommendations to refer to a message in the Interpersonal Messaging System. The 1988 X.400 recommendations use the term "interpersonal message" (IPM).

Interpersonal-Message (IPM) User Agent A class of cooperating user Agents capable of processing Interpersonal (IP) messages. An x.400 term.

Interpersonal Messaging IPM. Electronic exchange of information between two or more persons.

Interpolation A video technique used in motion compensation where a current frame of video is reconstructed by using the differences between it and past and future frames. This technique is also known as forward and backward prediction.
Intel (originator of Indeo Video) defines interpolation slightly differently, namely: The process of averaging pixel information when scaling an image. When the size of an image is reduced, pixels are averaged to create a single new pixel; when an image is scaled up in size, additional pixels are created by averaging pixels of the smaller image.

Interposition Calling One operator in a multi-position system calling another.

Interposition Transfer Transfer of a call from one operator to another.

Interposition Trunk 1. A connection between two positions of a large switchboard so that a line on one position can be connected to a line on another position.
2. Connections terminated at test positions for testing and patching between testboards and patch bays within a technical control facility.

Interprocess Communications this is the capability of programs to share information. At the most basic level, it consists of cutting and pasting information between two programs. Above that ranks the "live" paste, in which information shared between two documents is updated whenever one of the documents is modified. This is referred to as Dynamic Data Exchange (DDE). In advanced DDE, programs can send messages as well as data to other programs running locally or remotely. Beyond DDE is Object Linking and Embedding (OLE), which lets one program borrow the specialized capabilities of another program loaded on the machine (say, advanced chart creation) rather than having to implement that capability redundantly.

Interrogate To determine the state of a device or unit.

Interrupt A temporary suspension of a process caused by an event outside of that process. More specifically, an interrupt is a signal or call to a specific routine. An interrupt setting allows the hardware in a file server, router, workstation or PC to send an interrupt signal to the processor. The interrupt signal temporarily suspends the other station tasks while the processor performs the task requested by the interrupting device. After the routine is completed, the processor then continues with the original tasks. Each piece of hardware (serial and parallel ports and network boards) installed in the same computer needs a unique interrupt. Interrupts are divided into two general types, hardware and software. A hardware interrupt is caused by a signal from a hardware device, such as a printer. A software interrupt is created by instructions from within a software program.

TIP: When you slide a new card into one of the empty slots on your PC and things go awry, check that the new card's Interrupt is not the same as one of the other cards in your bus. An interrupt is also called a hardware interrupt or an InterRupt reQuest (IRQ). For a listing of normal IRQs see IRQs. See also INTERRUPT REQUEST and POLLING.

Interrupt Driven Someone who moves through a workday responding to interruptions rather than the work goals as originally set.

Interrupt Flag In a PC, there is a configuration control (addressed as bit IF of the processor flag register). This control process manages the CPU's ability to receive and process interrupt requests. The flag is often set to zero (which means interrupts disabled) by device drivers or other I/O privilege-level code that needs exclusive access to the CPU during critical operations. See also INTERRUPT, INTERRUPT HANDLING ROUTINE and UART.

Interrupt Handling Routine This program, which is often part of a device driver, handles all requests from a particular interrupt line. Interrupt-handling routines are defined in the CPU's Interrupt Descriptor Table (IDT). When the CPU (the Central Processing Unit of your PC) receives an interrupt request, it looks up the matching interrupt-handling routine in the IDT, then transfers control to the routine until it (the CPU) gives an interrupt return call (IRET), indicating the task is complete.

Interrupt Latency The delay in servicing an interrupt request is known as interrupt latency. It is not a problem with devices that are not sensitive to timing inconsistencies (such as hard-disk controllers or video boards). But it is a problem with high-speed, asynchronous communications (9,600 bps and above), which are highly time-sensitive operations.

Interrupt Overhead The cumulative demand on your computer's central microprocessor by peripheral devices that generate interrupt requests is referred to as interrupt overhead. Such devices include hard-disk controllers, network interface cards, parallel and serial ports.

Interrupt Request IRQ. This is the communications channel through which devices issue interrupts to the interrupt handler of an IBM PC or IBM compatible PC's microprocessor. It's the channel through which these devices get the microprocessor's attention. Different IRQs are assigned to different devices. This assignment pattern differs from PC to PC. Many LAN interface cards use an IRQ to get to the microprocessor. You must be sure that your LAN interface card is not trying to use the IRQ assigned to another peripheral, like the hard disk controller or EGA card. See also IRQ for a different and longer explanation.

Interrupt Request Lines Hardware lines over which devices can send signals to get the attention of the processor when the device is ready to accept or send information. Typically, each device connected to the computer uses a separate IRQ.

Interrupter An automatically operated electromechanical device used to turn lights, bells or other signals on and off in timed sequences. An interrupter makes lights wink on and off on a key system. Or did, when everything was electromechanical. It was used on 1A key telephone systems.

Interrupting Equipment Motor-driven mechanical devices used to break the ringing generator's output into ringing and silent periods, creating the busy and ringback tone pulses.

InterSpan The full name is InterSpan Frame Relay and it's AT&T's new frame relay data communications service, announced in the late fall of 1991.

Intersputnik A Russian satellite system similar in concept to the West's Intelsat, except that it's set up by Russia and the Eastern bloc countries. Two US carriers, AT&T and IDB Communications, once used Intersputnik to alleviate their shortage of US-Russia circuits. See INTELSTAT.

Interstate Between states (crossing a state line).

Interstices In cable construction, the spaces, valleys or voids between or around the cable's components.

Interswitch Trunk A circuit between two switching machines.

Intersymbol Interference Distortion of signals due to preceding or following pulses affecting desired pulse amplitude at time of sampling.

Intertandem Trunk Groups A category of trunk groups that interconnects tandems.

Intertoll Trunks Trunks connecting Class 4 and higher switching machines in the AT&T long distance network.

Interval Time. Pulse interval, for example, means the time from the start of one pulse to the start of the next.

Interworking The ability to seamless communicate between devices supporting dissimilar protocols, such as frame relay and ATM, by translating between the protocols, not through encapsulation. Many carriers are planning to implement the necessary equipment and conversion algorithms to allow the network itself to transparently convert from frame relay to ATM, and vice versa.

INTFC Interface.

Intra- Intra- means inside. Intrastate means inside the state. Interstate means between states.

Intracalling This is an outside plant term. Intracalling refers to the ability of a remote line concentrator to interconnect users served by the same concentrator without providing two trunks directly back to the central office.

Intraday Distribution A call center term. A historical pattern consisting of factors for each intra-day period of the week that define the typical distribution of call arrival or average handle time throughout each day. Each factor measures how far call volume or average handle time in that half hour or quarter hour deviates from the average half-hourly or quarter-hourly figure for that day. This information enables the program to forecast intra-day call volumes and staffing requirements.

Intraenterprise Communications Communications that are exchanged within a single organization (including multiple sites of the organization).

Intraflow This is an automatic call distribution term. It refers to the ability to select a second or subsequent group of agents to backup the primary agent group. This is designed to allow the caller to be serviced more efficiently and less expensively.

Intraframe Coding A way of video compression that compresses information within a single frame. Compare to INTER-FRAME CODING.

IntraLATA Telecommunications services that originate and end in the same Local Access and Transport Area.

Intramodal Distortion In an optical fiber, the distortion resulting from dispersion of group velocity of a propagating mode. It is the only form of multi mode distortion occurring in single-mode fibers.

Intranet A private network that uses Internet software and Internet standards. In essence, an Intranet is a private Internet reserved for use by people who have been given the authority and passwords necessary to use that network. An Intranet might use circuits also used by the Internet or it might not. Companies are increasingly using Intranets — internal Web servers — to give their employees easy access to corporate information.

Intranode Communications path which originates and terminates in the same node.

Intraoffice Call A call involving only one switching system.

Intraoffice Trunk A telephone channel between two pieces of equipment within the same central office.

Intrapreneur An entrepreneur who works inside a big company. Hence, intra, as in inside. It's hard to imagine it actually happening. But the word has became popular as a way for large companies to motivate their employees to take personal career risks and introduce new products.

Intrastate Remaining entirely within the boundaries of a single state and, therefore, if related to telephone, falling under the jurisdiction of that state's telephone regulatory procedures.

Intrinsic Joint Loss That loss in optical power transmission, intrinsic to the optical fiber, caused by fiber parameters, e.g., dimensions, profile parameter, mode field diameter, mismatches when two non identical fibers are joined.

Intrinsics Intrinsics are a component of many windows toolkits. The windows toolkit intrinsics definition has been developed by the MIT X Consortium. The intrinsics define the function of specific graphical user interface and window objects. They do not define any particular look or feel, just the function. Example: A pull down menu intrinsic would define the function of a pull down menu within a toolkit but not the appearance of it.

Inverse Fourier Transform Inversion of Fourier transform to convert frequency representation of signal to time representation.

Inverse Multiplexer I-Mux. Inverse multiplexers do two things, according to which manufacturer is making the device. So I have two definitions: First, the older definition: An inverse multiplexer is the opposite of a "normal" multiplexer, which takes many small circuits and makes one big one out of them. An inverse multiplexer takes one big circuit — e.g. a T-1 — and pulls smaller ones out of it. A "classic" inverse multiplexer is a device which takes a T-1 circuit and lets you pull one, two or three DS-0s (56 Kbps or 64 Kbps) circuits out of it. The reason you use an inverse multiplexer is that it is sometimes cheaper to buy one T-1 circuit than to buy several digital channels.

Second — and this is the newer meaning — an inverse multiplexer will dial up and combine several lower speed circuits into a virtual broadband circuit. It will take in any multiple up to an aggregate bandwidth (what the device is capable of, but typically up to 2.048 megabits per second, which is European E-1). An inverse multiplexer, under this application, gives users bandwidth on demand for high-speed applications such as LAN-to-LAN connectivity and video conferencing. The I-Mux also provides the synchronization necessary to recombine the multiple channels into a single integrated transmission at the receiving end. This is necessary because once the transmission has been subdivided into multiple channels some parts of the transmission may take a different and longer path to the destination. This results in a synchronization problem. The job of the I-Mux is to make sure that the channels are recombined into a cohesive single transmission.

Inverse Multiplexing The process of splitting a single high-speed channel into multiple signals, transmitting the multiple signals over multiple facilities operating at a lower rate than the original signal, and then recombining the separately-transmitted portions into the original signal at the original rate.

Inverter A device which converts direct current electricity to alternating current electricity, often used to power AC devices in a car.

Invitation To Send A character or sequence of characters which calls for a station to begin transmission. Usually this is part of a polling arrangement.

Inward Restriction A Centrex service feature which stops Centrex lines from receiving certain incoming calls.

Inward Trunk Used only for incoming calls, these trunks cannot dial out. "800" lines, for example, can only be used to receive calls.

INWATS INward Wide Area Telephone Service. A service of AT&T, MCI, Sprint, the Bell operating companies and the independent phone companies in North America which allows subscribers to receive calls from specified areas (depending on the rate band chosen) with no charge to the person who's calling. See 800 SERVICE for a much bigger explanation.

IOC 1. Inter-Office Channel
2. Independent Operating Carrier

IOD Identified Outward Dialing. See also AIOD and CALL ACCOUNTING SYSTEM.

IOEngine Input/Output Engine. The part of the Novell SFT III operating system that handles physical processes, such as net-

work and disk I/O, hardware interrupts, device drivers, timing, and routing. SFT III is split into two parts: the IOEngine and the MSEngine (Mirrored Server Engine). The IOEngine routes packets between the network and the MSEngine. To network workstations (i.e. PCs on the LAN), the IOEngine appears as a standard NetWare router or bridge. The primary server and the secondary server each have an IOEngine, but they share the same MSEngine. Because the IOEngines are not mirrored, NetWare Loadable Modules (NLMs) and applications that directly interface with hardware, such as backup NLMs may be installed in the IOEngines on both the primary and the secondary server. See MSENGINE.
IOF Inter Office Facility
Ion Exchange Technique A method of fabricating a graded-index optical fiber by an ion exchange process.
Ionization The process of breaking up molecules into positively and negatively charged carriers of electricity called ions.
IONL Internal Organization of the Network Layer. The OSI standard for the detailed architecture of the Network layer. Basically, it partitions the Network layer into subnetworks interconnected by convergence protocols (equivalent to internet working protocols), creating what the Internet community calls a catenet or internet.
Ionosphere That part of the atmosphere in which reflection and/or refraction of electromagnetic waves occurs. It extends from about 70 to 500 kilometers. At that point, ions and free electrons exist in sufficient quantities to reflect electromagnetic waves.
Ionospheric Absorption Attenuation of the energy in a radio wave due to the interaction between it and gas molecules. Deviative absorption describes the appreciable bending that occurs in an ionospheric layer at close to critical frequency. Non-deviative absorption describes the condition where little or no bending occurs as the wave passes through an ionized layer. See IONOSPHERE.
Ionospheric Cross Modulation Nonlinearities within the medium can produce nonlinear absorption. This can lead to the modulation on a strong signal being transferred to a weaker carrier. Sometimes described as the Luxembourg effect.
Ionospheric Disturbance An increase in the ionization of the D region of the ionosphere, caused by solar activity, which results in greatly increased radio wave absorption. See IONOSPHERE.
Ionospheric Focusing A variation in the curvature of the ionospheric layers can give rise to a focusing/defocusing effect at a receiving antenna. This may produce either an enhancement or attenuation in the received field strength due to signal phase variations.
Ionospheric Refraction The change in the propagation speed of a signal as it passes through the ionosphere.
IOP 1. Input/Output Processor.
2. Interoperability: The ability of equipment from different manufacturers (or different implementations) to operate together.
IOS The International Organization for Standardization based in Geneva, Switzerland. The IOS develops and publishes hundreds of international ISO standards such as ISO-9000 for quality assurance or ISO 14000 for environmental performance. The IOS is comprised of more than 90 member standards bodies worldwide plus other international associations, government and non-government bodies. The U.S. member of the IOS is ANSI. See ANSI and ISO.
IP 1. The Internet Protocol. Originally developed by the Department of Defense to support interworking of dissimilar computers across a network. This protocol works in conjunction with TCP and is usually identified as TCP/IP. A connectionless protocol that operates at the network layer (layer 3) of the OSI model. IP is the most important of the protocols on which the Internet is based. The IP Protocol is a standard describing software that keeps track of the Internetwork addresses for different nodes, routes outgoing messages, and recognizes incoming messages. It allows a packet to traverse multiple networks on the way to its final destination. See IP ADDRESS and INTERNET.
2. Internal Protocol.
3. Intelligent Peripheral. A device in an IN (Intelligent telecom Network) that provides capabilities such as voice announcements and help guidance.
IP Address An internetworking term. The IP (Internet Protocol) address is a 32-bit address, used in IP routing, which includes a Network address identifier assigned by a central authority and a Host ID (an end station identifier assigned by the LAN administrator). In other words, the network and the computer itself are both represented in a sequence of numbers.
IP Spoofing An attack whereby a system attempts to illicitly impersonate another system by using its IP network address. See IP, IP ADDRESS and IP ROUTER.
IP Router A computer connected to a multiple physical TCP/IP networks that can route or deliver IP packets between networks. See also GATEWAY.
IPARS The International Passenger Airline Reservation System. An IBM-originated term.
IPC Interprocess Communications. A system that lets threads and processes transfer data and messages among themselves; used to offer services to and receive services from other programs. Supported IPC mechanisms under MS OS/2 are semaphores, signals, pipes, queues, shared memory, and dynamic data exchange.
IPCH Initial Paging CHannel is the channel number used by your cellular provider to "page" the phones on the system. The term "paging" refers to notifying a particular phone that it has an incoming call. All idle, turned-on phones on a system monitor the data stream on the IPCM. Non-wireline cellular carriers use channel 0333 as the IPCH, while wireline providers (those operated by a telephone company use channel 0334.
IPCI See INTEGRATED PERSONAL COMPUTER INTERFACE.
IPCP IP Control Protocol; protocol for transporting IP traffic over a PPP connection.
IPDS Intelligent Printer Data Stream. It's IBM's host-to-printer page description protocol for printing. You can now buy kits which let you use your present printer to emulate an IBM printer.
IPE Intelligent Peripheral Equipment. Northern Telecom's term for being able to extend all the features of its PBX over distances longer than a normal extension in a building. See FIBER REMOTE.
IPEM If the Product Ever Materalizes.
IPL Initial Program Load.

IPM Interruptions Per Minute or Impulses Per Minute.

IPNG IP Next Generation. Collective term used to describe the efforts of the Internet Engineering Task force to define a new version of the Internet Protocol (IP) which can handle larger IP addresses to cope with the explosive growth of the Internet. At the time of this writing there were three candidate protocols for IPng: CATNIP, TUBA and SIPP.

IPNS International Private Network Service. It actually international private line service and it's typically a circuit from 9.6 Kbit/sec up to T-1 or E-1. Domestically you would simply call it "Private line data service."

IPO Initial Public Offering. Start a company. Some years later, take it public. Come out at $12. A year later, your stock is at $24. You're a success, and rich. IPOs are critical in saying "Thank You" to all your hardworking employees.

IPS Internet Protocol Suite.

IPv6 The new proposed Internet Protocol designed to replace and enhance the present protocol which is called TCP/IP. IPv6 has 128-bit addressing, auto configuration, new security features and supports real-time communications and multicasting.

IPX Internet Packet eXchange. Novell NetWare's native LAN communications protocol, used to move data between server and/or workstation programs running on different network nodes. IPX packets are encapsulated and carried by the packets used in Ethernet and the similar frames used in Token-Ring networks. IPX supports packet sizes up to 64 bytes. Novell's NCP and SPX both use IPX. See also IPX.COM.

IPX.COM The Novell IPX/SPX (Internetwork Packet eXchange/Sequenced Packet eXchange) communication protocol that creates, maintains, and terminates connections between network devices (workstations, file servers, routers, etc.). IPX.COM uses a LAN driver routine to control the station's network board and address and to route outgoing data packets for delivery on the network. IPX/SPX reads the assigned addresses of returning data and directs the data to the proper area within a workstation's shell or the file server's operating system. See also NETWARE.

IPX/SPX Internetwork Packet Exchange/Sequenced Packet Exchange. Two network protocols. IPX is NetWare protocol for moving information across the network; SPX works on top of IPX and adds extra commands. In the OSI model, IPX conforms to the network layer and SPX is the transport layer.

IPXCP IPX Control Protocol; protocol for transporting IPX traffic over a PPP connection.

IPXWAN A Novell specification describing the protocol to be used for exchanging router-to-router information to enable the transmission of Novell IPX data traffic across WAN (Wide Area Network) links.

IR Infrared. The band of electromagnetic wavelengths between the extreme of the visible part of the spectrum (about 0.75 um) and the shortest microwaves (about 100 um).

IRC 1. International Record (i.e. non-voice) Carrier. One of a group of common carriers that, until a few years ago, exclusively carried data and text traffic from gateway cities in the U.S. to other countries. The distinction between international companies providing "record" and data has eroded and now both types of companies provide voice and data services internationally.

2. Internet Relay Chat; an Internet-based application that allows realtime conversations among many users. Sort of like CB radio, but much more confusing. See INTERNET.

IRD Integrated Receiver/Descrambler. A receiver for satellite signals that also decodes encrypted or scrambled signals. Especially used in the cable TV business.

IRDA The InfraRed Data Association. More than 120 PC equipment manufacturers are working together to set a standard for infrared data transmission. Imagine that you're carrying around a small portable laptop, PDA or other device and you want to exchange data with your desktop. You'll simply aim the device at your desktop PC and transmit information back and forth. Maximum transmission distance is slated to be about a yard. IRDA defines a link protocol for serial infrared links, called Infrared Link Access Protocol (IrLAP). IrLAP explains how links initialization, device address discovery, connection start-up (including link data rate negotiation), information exchange, disconnection, link shutdown, and device address conflict resolution occur on an IR connection. IrLAP implements the high-level data-link control (HDLC) communications protocol for infrared environments and adds procedures for infrared-based link initialization and shutdown plus connection start-up, disconnection, and information transfer. The rules for discovery and address-conflict resolution are IrLAP's most significant departure from HDLC. See IrLAP and IrLMP.

IREQ The Interrupt Request signal between a PCMCIA Card and a socket when the I/O interface is active.

Iridium The name for Motorola's ambitious satellite project "to bring personal communications to every square inch of the earth." According to Motorola, "for the first time, anyone, anywhere, at any time can communicate via voice, fax, or data." Motorola has targeted the band 1610 to 1626.5 MHz. Motorola estimates the service costing $3 a minute. The idea is that we all carry an Iridium handset — a device larger than today's cellular phone — and that we talk directly from the phone to the satellite (one of 66) and then down to the satellite closest to the called person, then down to an Iridium phone on the ground or to a satellite dish, through landlines to the phone of the called person. The big benefit is that the system knows who you are and where you are the moment you turn on your phone. This way it can always complete calls from somewhere calling you who doesn't know — or doesn't need to know — where you precisely are. It's called Iridium after the element called Iridium which has 77 electrons, which used to be the number of satellites needed. In 1994, the number got cut to 66. But the name stuck. In November of 1992, Business Week estimated that putting up the full Iridium system would cost $3.4 billion. See also TELEDISC.

IrLAP InfraRed Link Access Protocol from IRDA, the InfraRed Data Association. IrLAP defines a link protocol for serial infrared links. IrLAP explains how link initialization, device address discovery, connection start-up (including link data rate negotiation), information exchange, disconnection, link shutdown, and device address conflict resolution occur on an IR (Infrared) connection. IRLAP implements the high-level data-link control (HDLC) communications protocol for infrared environments and adds procedures for infrared-based link initialization and shutdown plus connection start-up, disconnection, and information transfer. The rules for discovery and address-conflict resolution are IrLAP's most significant departure from HDLC. Until recently, IRDA's standards characterized infrared ports as serial links operating at speeds up to 115 Kbps. IRDA's latest standards allow transmission rates as high as 4 Mbps and provide for LAN access via a new IRLAN protocol. The 4 Mbps mode uses pulse-position-modulation data encoding with four possible chip or time-slice positions per data symbol. The system can recognize and prevent interference with UART-based systems by including a Serial Infrared physical-layer link Interaction Pulse (SIP) at least every 500 milliseconds. IRDA has developed APIs for accessing the infrared port. The first, IRCOMM, emulates existing communications device drivers to handle legacy serial and parallel-port connectivity. There's also a native API that infrared-aware programs can use to locate and communicate with each other.

IrLMP InfraRed Link Management Protocol. See IRDA.

IROB 1. In Range of Building. An Underwriters Laboratories term to define where the protection of UL 1459 will apply. See UL 1459.
2. An AT&T Merlin term. In-Range Out-of-Building Protector. A surge protection device for off-premises telephones at a location within 1000 ft. (305 m) of cable distance from the communications system control unit.

IRP I/O Request Packet. Data structures that drivers use to communicate with each other. See INTERRUPT and IRQs.

IRQs Interrupt ReQuests. IRQs are found in PCs. IRQs are also called hardware interrupts. They are the way a device signals the data bus and the CPU that it needs attention. In more technical terms, an IRQ is a signal sent to the central processing unit (CPU) to temporarily suspend normal processing and transfer control to an interrupt handling routine. Interrupts may be generated by conditions such as completion of an I/O process, detection of hardware failure, power failures, etc. Devices that use hardware interrupts include the serial and parallel ports, mouse interface cards, modems, game ports, and even the hard disk on XTs. The original IBM PC and PC-XT had only seven hardware interrupts. The bigger AT bus extended that to 15. Until the advent of the 32-bit PS/2 micro-channel and the 32-bit EISA buses, hardware interrupts could not be shared by two or more devices within the PC. Thus if one device had a specific hardware interrupt, even though you weren't using it that time, nothing else could use it. When you start filling your PC with devices (and remember most PCs still use the old AT bus) — like serial ports, modems and mice, you may suddenly find your modem no longer works. There are two solutions — change the interrupts (either in software or using jumpers), making sure no two devices are trying to share the same interrupt — or simply remove one of the printed circuit devices you're not using from the bus. (That's typically my solution.) These are the "normal" IRQs used by current hardware devices in PCs. See list on next page of 16-bit IRQs as they have become used.

Irradiation In insulations, the exposure of the material to high energy emissions for the purpose of favorably altering the molecular structure by crosslinking.

Irritainment Annoying entertainment and media spectacles you're unable to stop watching. The O.J. trial is a prime example.

IRSG Internet Research Steering Group. See IRTF.

IRTF Internet Research Task Force. The IRTF is a community of network researchers, generally with an Internet focus. The work of the IRTF is governed by its Internet Research Steering Group (IRSG).

IRTU Integrated Remote Test Unit.

IRU Indefeasible Right of User. A measure of currency in the underseas cable business. Someone owning An IRU means he has the right to use the cable for the time and bandwidth the IRU applies to. An IRU is to a submarine cable what a lease is to a building.

IS 1. Information Separator. A type of control character used to separate and qualify data logically. Its specific meaning has to be defined for each application.
2. Interim Standard.
3. Intermediate System: A system that provides forwarding functions or relaying functions or both for a specific ATM connection. OAM cells may be generated and received.

IS-III An AT&T Merlin term. Integrated Solution. One or more UNIX-based applications for improving voice and data communications and automating office operations.

IS-IS Intermediate System to Intermediate System. OSI link-state hierarchical routing protocol, based on DECnet Phase V routing, whereby intermediate systems (routers) exchange routing information based on a single metric to determine network topology.

IS41+ A signaling protocol used in the North American standard cellular system.

IS54 Interim Standard number 54. It is the dual mode (analog and digital) standard for cellular phone service in North America. In its analog form, it conforms to the AMPS standard. IS-54 is a TIA and CTIA standard.

IS55 Standard for TDMA digital cellular service, which is three times the capacity of today's analog cellular service. IS-55 is a fully digital cellular system.

IS95 Digital CDMA standard for U.S. cellular radio systems.

ISA 1. Interactive Services Association. http:/policy.net/isa
2. Industry Standard Architecture. The most common bus architecture on the motherboard of MS-DOS computers. The ISA bus was originally pioneered by IBM on its PC, then its XT and then its AT. ISA is also called classic bus. It comes in an 8-bit and 16-bit version. Most references to ISA mean the 16-bit version (which carries data at up to 5 megabytes per second). Many machines claiming ISA compatibility will have both 8- and 16-bit connectors on the motherboard. In 1987 IBM introduced a 32-bit bus which it called MCA for Micro Channel Architecture, which is the internal bus inside some of IBM's line of PS/2 MCA machines. But MCA isn't popular because it is incompatible with ISA, so the industry (excluding IBM) invented a 32-bit bus called EISA which stands for Extended Industry Standard Architecture, which is compatible with ISA. EISA, however, suffered from some of the same problems as the MCA bus, namely it was complicated to program to get the card's full benefit. As a result, other buses have been invented, including the VL bus from VESA (the Video Electronics Standards Association) and Intel's PCI bus. PCI stands for Peripheral Component Interconnect. Both the VL and PCI buses claim to be more than 20 times faster than the ISA bus. The PCI bus, according to Intel, can transfer data at up to 132 megabytes per second. Until the advent of complex Windows graphics and imaging programs, the speed of the ISA was not a gating factor in a computer. As of writing, a VL bus could only handle three drop-in printed cards, while a PCI bus could handle 10. See also EISA and MICROCHANNEL.

8 Bit XT Bus		16 Bit AT Bus
IRQ0	TIMER SERVICES	IRQ0 TIMER SERVICES
IRQ1	KEYBOARD	IRQ1 KEYBOARD
IRQ2	UNUSED	IRQ2 SLAVE INTERRUPT
IRQ3	COM2 & COM4	IRQ3 COM2 & COM4 *
IRQ4	COM1 & COM3	IRQ4 COM1 & COM3
IRQ5	LPT2	IRQ5 LPT2 *
IRQ6	FLOPPY DISK	IRQ6 FLOPPY DISK
IRQ7	LPT1	IRQ7 LPT1
		IRQ8 REAL TIME CLOCK
		IRQ9 IRQ2 VECTOR
		IRQ10 Available *
		IRQ11 Available *
		IRQ12 Available *
		IRQ13 MATH COPROCESSOR
*Available for assigning to new devices, such as network and video cards:		IRQ14 HARD DISK CONTROLLER
		IRQ15 Available *

IRQs in alphabetical order:

Device	IRQ	
ARCnet card	2	These assignments
Bus Mouse	2	are not hard and fast.
Cascade	2	Experimentation
CD-ROM drive	5	works best.
COM1	4	
COM2	3	
COM3	4	
COM4	3	
Diskette Controller	6	
Ethernet card (old ones)	5	
Ethernet card (new ones)	10 or 15	
Floppy Drive	6	
Hard disk drive	14	
Keyboard	1	
LPT1 (PARALLEL)	7	
LPT2 (PARALLEL)	5	
Math Coprocessor	13	
PC Timer	0	
Printer 1	7	
Printer 2	5	
PS/2 Mouse	12	
Real time	8	
Reserved	9	
Reserved	10	
Reserved	11	
Reserved	15	
Scanner	7	
Sound Card	10	
Tape Backup	5	

ISA Configuration Utility The Intel Architecture Lab has been co-developing the Plug and Play specifications with industry partners to ensure long-term compatibility across cards, systems and software. IAL has openly licensed the necessary BIOS software to PC manufacturers so they can add Plug and Play capabilities to their systems. Intel Architecture Labs also designed the ISA Configuration Utility for system and add-in card manufacturers to include with their products. This software utility makes it easier for users to install existing ISA cards in their PCs. The software tells the user which resources are available, but configuration is still done manually. The utility also allows the user to optimize the way resources are assigned, which is particularly important for memory addresses.

ISAM Indexed Sequential Access Method. It is a procedure for storing and retrieving data from a disk file. When the programmer designs the format of the file, a set of indexes is created which describes where the records of the file are located on the disk. This provides a quick method of retrieving the data, and eliminates the need to read all the data from the beginning to find the desired information. The indexes can be stored as part of the data file or in a separate index file.

ISAPI Internet Server API, a new API from Microsoft and Process Software Corp. which will allow Internet browsers supporting it to access remote server applications from Microsoft and others. ISAPI was first talked about by Microsoft at the Fall, 1995 InterOp show in Atlanta.

Isarithmic Flow Control Approach to flow control in which transmission permits circulate throughout network. Node wishing to transmit must first capture permit and destroy it, then recreate permit after transmission finished.

ISB InterShelf Bus.

ISD Incremental Service Delivery.

ISDN Integrated Services Digital Network. ISDN comes today in two basic flavors — BRI, which is 144,000 bits per second and designed for the desktop, and PRI which is 1,544,000 bits per second and designed for telephone switches, computer telephony and voice processing systems. Neither ISDN BRI or ISDN PRI is a standard service, though there are several "standard" configurations. ISDN BRI is a wonderful service in your home or office because it can give you videoconferencing, and ultrafaster data communications. But it is not an easy service to get up and running. The best advice I can give you is: 1. Figure out what you want to do with your ISDN. 2. Find which equipment you're going to need that will do the best job for you. 3. Call the manufacturer of that equipment, tell him where you're located and ask him which ISDN service to order. 4. After he tells you, order your ISDN service from your local phone company. 5. Then buy the equipment. 6. Allow yourself at least a month to get up and running. 7. Any ISDN equipment you install in a PC will cause major interrupt problems. Make sure you know which interrupts your PC is using for what. See IRQs.

ISDN is essentially a totally new concept of what the world's telephone system should be. According to AT&T, today's public switched phone network has the following limitations: 1. Each voice line is only 4 KHz, which is very narrow, which limits also the speed you can send data across. 2. Most signaling is in-band signaling, which is very consuming of bandwidth (i.e. it's expensive and inefficient). 3. The little out-of-band signaling that exists today runs on lines separate to the network. This includes signaling for PBX attendants, hotel/motel, Centrex and PBX calling information. 4. Most users have separate voice and data networks, which is inefficient, expensive and limiting. 5. Premises telephone and data equipment must be separately administered from the network it runs on. 6. There is a wide and growing variety of voice, data and digital interface standards, many of which are incompatible.

ISDN's "vision" is to overcome these deficiencies in four ways: 1. By providing an internationally accepted standard for voice, data and signaling. That standard has pretty well achieved, though don't try and take North American ISDN equipment to Europe. 2. By making all transmission circuits end-to-end digital. 3. By adopting a standard out-of-band signaling system. 4. By bringing significantly more bandwidth to the desktop.

One of the best features of ISDN is the speed of dialing. Instead of 20 seconds for a call to go through on today's old analog network, with ISDN it takes less than a second. It's beautiful. Here are some sample ISDN services:

Call waiting A line is busy. A call comes in. The user knows who is calling. He can then accept, reject, ignore, transfer the call.

Citywide Centrex A myriad of services: Specialized numbering and dialing plans. Central management of all ISDN terminals, including PBXs, key systems, etc.

Credit card calling Automatic billing of certain or all calls into accounts independent of the calling line/s.

Calling line identification presentation Provides the calling party the ISDN "phone" number, possibly with additional address information, of the called party. Such information may flash across the screen of an ISDN phone or be announced by a synthesized voice. The called party can then accept, reject or transfer the call. If the called party is not there, then his/her phone will automatically record the incoming call's phone number and allow automatic callbacks when he/she returns or calls back in from elsewhere.

Calling line identification restriction Restricts presentation of the calling party's ISDN "phone" number, possibly with additional address information, to the called party.

Closed user group Restricts conversations to or among a select group of phone numbers, local, long distance or international.

Collaborative Computing Work on the same document or drawing or design with someone 10,000 miles away. With ISDN, it doesn't really matter where members of the design team live.

Desktop videoconferencing I have an ISDN desktop videoconferencing device on my desk. It's wonderful to see the person at the other end. It makes for a far more meaningful conversation.

E-Mail (a.k.a. Personal mailbox) ISDN can carry information to and from unattended phones as long as they're equipped with proper hardware and software.

Internet Access It's much nicer to browse the Internet at 128 Kbps than at 28.8 Kbps which is the fastest I can do it today with the fastest analog modem I can buy.

Shared Screen — Switched data services provided via ISDN lets two people in remote locations, both equipped with a computer terminal, view the same information on their screens and discuss its contents while making changes — all over one telephone line.

Simultaneous Data Calls Two users can talk and exchange information over the D packet and/or the B circuit or packet switched channel.

There are two major problems to the widespread acceptance of ISDN: First, the cost of ISDN terminal equipment is too high. Second, the cost of upgrading central office hardware and software to ISDN is too high. Both costs are coming down. In early, 1995 Pacific Bell announced that it would install one million lines of ISDN BRI by the end of 1998. And, to make this happen, it dropped its ISDN monthly prices to an affordable $24.95 a month for residences and $26.50 a month for businesses.

There are three basic configurations you can get ISDN:

1. The 2B+D "S" interface (also called the "T" interface). The 2B+D is called the Basic Rate Interface (BRI). The "S" interface uses four unshielded normal telephone wires (two twisted wire pairs) to deliver two "Bearer" 64,000 bits per second channels and one "data" signaling channel of 16,000 bits per second. An S-interfaced phone can be located up to one kilometer from the central office switch driving it. Each of the two 64 kpbs "bearer" or B channels can be used to carry a voice conversation, or one high speed data or several data channels, which are multiplexed into zone 64 kbps high speed data line. The "D" channel of 16 kbps will carry control and signaling information to set up and break down the voice and data calls. The "D" channel can also carry data up to 9600 bits per second in addition to the control and signaling information. Signaling and control on the D channel conforms to a protocol (LAPD) and a messaging structure (Q.931). These two allow intelligent endpoints and switching nodes from different vendors to talk a common language and thus be able to transfer features across a network, from one switch to another, e.g. to transfer a Centrex call across town through several switches and to have it arrive at the end phone with the calling party's name.

2. The 2B+D "U" interface. This "U" interface delivers the same two 64 kbps bearer channels and one 16 kbps data channel, except that it uses 2-wires (one pair) and can work at 5-10 kilometers from the central office switch driving it. The "U" interface is the most common ISDN interface. It carries 160,000 bits per second from the central office to your home or office. Of those 160,000 bits, two are used for 64,000 bps Bearer (B) channels and one is used by the subscriber for 16,000 bps of data (the D channel). The other 16,000 bps is used by the network for signaling between the black box on the subscriber premises and the central office. The idea is to get the ISDN "U" interface working to 18,000 feet — the average length of a North American subscriber local loop. You connect the two "U" wires (local loop pair) coming in from your local ISDN CO into a black box about the size of desk printing calculator, called an NT-1. Out the side of the black box comes four wires, which are called the "S Bus." Onto these four wires you can attach, in a loop configuration (also called single bus), as many as eight ISDN terminals — telephones, fax machines, etc. See ISDN.

3. The 23B+D or 30B+D. This is called the Primary Rate Interface (PRI). At 23B+D, it is 1.544 megabits per second. At 30B+D, it is 2.048 megabits per second. The first, 23B+D is the standard T-1 line in the U.S. which operates on two pairs. The second 30B+D is the standard T-1 line in Europe, which also operates on two pairs.

Integral to ISDN's ability to produce new customer services is CCITT Signaling System 7. This is a ITU-T recommendation which does two basic things: First, it removes all phone signaling from the present network onto a separate packet switched data network, thus providing enormous economies of bandwidth. Second, it broadens the information that is generated by a call, or call attempt. This information — like the phone number of the person who's calling — will significantly broaden the number of useful new services the ISDN telephone network of tomorrow will be able to deliver.

ISDN has "enjoyed" many "meanings," including I Still Don't Know to, its most recent, I Smell Dollars Now.

For more on ISDN, see EURO-ISDN, INTEL BLUE, ISDN STANDARDS, ISDN TELEPHONE, ISUP, ISDN NETWORK TERMINATION 1, PERSONAL COMPUTER TERMINAL ADAPTER, PROSHARE, Q.931, ROBBED BIT SIGNALING, SIGNALING SYSTEM 7, SPID and TCAP.

ISDN 30 The name of an ISDN service which delivers 30 ISDN bri lines over a single line. ISDN-30 is a fancy name for ISDN on an E-1 line. You find it in countries outside North America.

ISDN Basic Link Facility Here is a Nynex definition: The ISDN Basic Link Facility consists of a local transmission facility terminated in the local central office and in a suitable network interface device that is capable of supporting ANSI standard ISDN Basic Rate 2 Binary 1 Quaternary U interface line coding scheme. The standard ISDN Basic Link Facility is 18,000 cable feet or less from the central office termination or served via appropriate electronic equipment, to the Network Termination One device located on the customer's premises.

ISDN BRI Service 2B+D - Two bearer channels and one D channel to your desktop. There are many varieties of ISDN BRI service. The three most common are National ISDN-1 compliant, AT&T 5ESS Custom (an older form of ISDN BRI) and Northern Telecom DMS 100 ISDN. With ISDN BRI, you choose your ISDN equipment first, figure out what it needs, check whether what it needs is available from your local telephone company, then get your line installed, then buy your equipment. Changing your line specs later is expensive and slow. ISDN is still a very first generation product, though when you get it working, it usually works thereafter relatively flawlessly. See ISDN.

ISDN Modem A special kind of ISDN Terminal Adapter (see ISDN Terminal Adapter) with enhanced functionality (software, hardware, etc.) to allow it to connect and exchange data with conventional analog modems. Ordinary ISDN terminal adapters can only support data communications with other ISDN devices. They lack the ability to communicate with conventional modems. Because of this, the many on-line services and other analog modem-based applications are not available to users of normal ISDN terminal adapters. The extra functionality in ISDN modems allow their users to make use of these and other existing analog modem-based services. This definition supplied by Paul D. Cook or Ameritech, who once wrote a White Paper describing the obvious need for such a device.

ISDN Network Termination Device You can't plug your ISDN phone directly into an ISDN line like you can with today's analog lines. You needs a black box, called a Network Termination device, called an NT1, as in Network Termination 1. In North America you can pick one of these devices up for $250. The NT1 provides an interface between the ISDN loop and an S or T interface terminal, such as an ISDN phone, or the PCTA (Personal Computer Terminal Adapter). The PCTA is the device which turns a PC into an ISDN terminal/phone. The NT1 is the classic ISDN "black box." It sits on the subscribers premises at the end of the subscriber loop coming in from the phone company. It talks to the ISDN central office. And, in turn, all ISDN terminals, phones and other devices on the subscriber premises are plugged into this black box. The basic NT1 functions are:
• Line transmission termination.
• Layer 1 line maintenance functions and performance monitoring.
• Layer 1 multiplexing, and
• Interface termination, including multi drop termination employing layer 1 contention resolution. See ISDN.

ISDN Overflow/Diversion A feature of Rockwell Galaxy ACDs. ISDN Overflow/Diversion allows Galaxy ACD users to overflow calls between multiple switches using PRI D-channels and B-channels through the public network. This gives the user a virtual private network without the cost of dedicated trunks. By using ISDN messages to overflow a call, specific information associated with the call can be passed to the destination switch, such as ANI, DNIS, and delay time in queue at the originating switch.

ISDN PRI PRI stands for Primary Rate Interface. In North America, ISDN PRI can be thought of as "enhanced T-1." And some long distance carriers are only delivering T-1 in this format. In this ISDN-PRI format, it has major benefits — chiefly the extra bandwidth and benefits derived from the much richer and much faster out-of-band signaling. For example, ANI (Automatic Number Identifcation, DNIS (Dialed Number Identification Service), etc. are delivered much better this way. ISDN PRI is 24 B (bearer) channels, each of which is a full 64,000 bits per second. One of these channels is typically used to carry signaling information for the 23 other channels. If you're running voice on a single ISDN-PRI, you get 23 voice channels, compared to 24 if you run voice on a T-1 line. However, if you get multiple ISDN-PRI lines, you can often carry the signaling for those voice lines on that one B channel and thus get 24 on each of the others. For example, some voice processing cards will let you support up to eight ISDN PRI channels on one B channel — i.e. the first PRI gives you 23 voice channels. All seven others will give you 24 channels. In Europe, ISDN PRI is 30 bearer channels of 64 Kbps and two signaling channels, each of 64 KBps.

ISDN Repeaters ISDN repeaters let telephone companies extend ISDN lines up to 48,000 feet from their central office. This means they can send ISDN to people further away.

ISDN Standards The path of a call in an ISDN network is based on standards:
1. ISDN User A signals the public network over a standard interface. The 2B1Q protocol is used by the terminal as well as the line card in the telephone company central office. The 2B1Q arranges the bits of a digital ISDN signal in a standard manner over the twisted pair connecting the user and the central office. Bellcore TR268 establishes protocols for signaling between the caller and the network.
2. TR444 and TR448 define the standard protocols that allow ISDN services to be carried by the SS7 network.
3. TR317 defines the protocols for standard SS7 networking between the LEC's intraLATA switches that, when combined with TR444 and TR448, can deliver ISDN services.
4. TR394 defines the protocols for standard SS7 networking of the interLATA switches that, when combined with TR444 and TR448, provide the interface for ISDN services and the interexchange network.

5. ISDN UserB, equipped with the standard network interface and using the 2B1Q and TR268 protocols, is prepared to receive the ISDN call from User A. The network delivers the ISDN call information carried over SS7 from User A to User B in the call set-up message.

ISDN Teladapter A Northern Telecom term for a device which connects a national ISDN 1 telephone and a Macintosh to a Northern Telecom switch via an ISDN line card.

ISDN Telephone An ISDN phone can attach to an ISDN basic rate interface. It typically has one digital voice (at 64 kbps) channel and two data options — one for packet switched services (up to 9600 bps) and another for circuit switched data (up to 64 kbps). It will also have an RS-232-C connector on its back and a two line, 48-character LCD adjustable display. It will also have a bunch of dedicated buttons for standard stuff — last number redial, speed dial, on-hook dialing, listen-on-hold, etc. Some ISDN phones work behind most ISDN central offices — e.g. the Telrad phones. Most don't. They have to work behind the central office they were designed for. Or did have to. In February, 1991, Bellcore issued a technical specification for a standard ISDN phone line. The idea of National ISDN-1 is that it be a set of standards which every manufacturer can conform to. A consumer can buy an ISDN phone (one conforming to National ISDN-1) at his local Radio Shack (or other store) take it home, plug it in and know it will work, irrespective of whose central office he's connected to. At time of writing most ISDN phones cost over $600. Some cost nearly $1,000. Some cost more than a personal computer. In late fall of 1992 there was increasing talk that PCs will soon come with telecom ports — able to accept the ISDN signal directly from the central office, without the need for a separate (and expensive ISDN) phone instrument. The PC and the software within it, will then become the phone (presumably with a handset, headset or earset attached to the back of the PC). See ISDN.

ISDN Terminal Adapters ISDN terminal adapters are devices that typically allow analog devices to speak on digital ISDN lines. Terminal adapters are essentially similar to modems, with the following difference. Modems connect terminals to the traditional analog network, and terminal adapters connect those analog-network (non-ISDN) terminals to the digital ISDN network.

ISDN.ISO 8877 ITU-T is the standards group responsible for ISDN-Integrated Services Digital Network. ISDN is an international standard that provides end-to-end digital connectivity to support a wide range of voice, data, and video services. It uses a single communications channel for all forms of information transfer. The ISDN interface connector is specified as an 8-position connector (plug and jack). The four center contacts are assigned to transmit and receive pairs. The four outer contacts are for power.

ISDN/AP Northern Telecom's host to SL1 PBX protocol, which supports NT's Meridian Link. See MERIDIAN LINK and ISDN.

ISE Integrated Switching Element.

ISG Incoming Service Grouping. A fancy name for hunting or rollover. You receive many incoming calls. You don't want to miss a call, so you set your phone lines up to roll over, also called hunt, also called ISG in telephonese. You order five lines in hunt. The calls come into the first. If the first one is busy, the second rings. If it's busy, the third rings. If they're all busy, then the caller receives a busy.

There are two types of hunting — Sequential and Circular hunting. Sequential hunting starts at the number dialed and ends at the last number in the group. Circular hunting hunts all the lines in the hunting group, regardless of the starting point. Circular hunting, according to our understanding, circles only once (though your phone company may be able to program it circle a couple of times). The differences between sequential and circular are subtle. Circular seems to work better for large groups of numbers.

You don't need consecutive phone numbers to do rollovers. Nowadays you can roll lines forwards, backwards and jump around. Rollovers are now done in software. This also has its downside, since software fails. Our recommendation: Test your rollovers at least twice a day. In particular, test that your callers ultimately get a busy if all your lines are busy. Nothing worse than your customer should receive a ring-no-answer.

ISL Northern Telecom term for ISDN Signaling Link.

ISM Industrial Scientific Medical. A term the FCC uses to name several of the frequency bands in the spectrum. The 902-928 Mhz is in the ISM band, for example. It is also referred to as Part 15.247 of FCC regulations.

ISN AT&T's Information Systems Network.

ISO 1. The Greek prefix which means equal or symmetrical, as in isometric. See also ISOCHRONOUS.

2. Most people believe that ISO stands for The International Standards Organization in Paris, which it doesn't. Actually the organization is strictly called the International Organization for Standardization (IOS) and is based in Geneva. It publishes ISO standards for international and national data communications. The U.S. representative to the ISO is ANSI. IOS or ISO (it's known as both) is a voluntary, non-treaty organization chartered by the United Nations, established in 1949, to define international standards. It is best known for the 7-layer OSI (Open Systems Interconnection) Reference Model. See ANSI and IOS. www.iso.ch/welcome.html

3. ISO also stands for Independent Service Organizations or Independent Sales Organizations in the computer sales community.

ISO 8877 Information Processing Systems Q Interface Connector and Contact Assignment for ISDN Basic access interface located at reference points S and T - International Organization for Standardization. Part of this standard describes pin/pair assignments for 8-line modular connectors. The assignments are the same as EIA's T-568A.

ISO 9000 Series The ISO 9000 series, published in 1987, outlines the requirements for the quality system of an organization. It is a set of generic standards that provide quality assurance requirements and quality management guidance. It is now evolving into a mandatory requirement, especially for manufacturers of regulated products such as medical and telecommunications equipment. ISO 9001, the most comprehensive of three compliance standards — 9001, 9002, 9003 — is a model for quality assurance for companies involved with design, test, manufacture, delivery and service of products. ISO 9002 covers manufacturing and installation only. ISO 9003 covers product testing and final inspection of standards. ISO is the International Standards Organization based in Paris. See ISO 9001.

ISO 9001 ISO 9001 is a rigorous international quality standard covering a company's research and development, design, production, installation and service procedures. Compliance with the standard is of increasing significance for vendors trading in international markets, in particular in Europe where ISO 9001 registration is widely recognized as an indication of the integrity of a supplier's quality processes. ISO is the International Standards Organization. ISO 9001 is the most rigorous of the three standards. See ISO 9000 SERIES.

ISO 9002 ISO 9002 covers manufacturing and installation only. See ISO 9000 Series.

ISO 9003 ISO 9003 covers testing and final inspections of manufactured products. See ISO 9000 Series.

ISO 9660 The CD-ROM logical file format standard adopted by ISO in 1987. Describes a table of contents but not the format of the actual data. This has led to incompatibilities between different computers. Based on a specification developed by the High Sierra Group (HSG) which included Apple, Microsoft, 3M, Philips, Hitachi, DEC. Also know as Yellow Book and High Sierra.

ISO/IEC International Standards Organization international and the International ElectroTechnical Commission. www.iso.ch

ISOC The Internet Society: a membership organization whose members support a worldwide information network. It is also the governing body to which the IAB reports. See INTERNET.

Isochronous Isochronous transmission means "two-way without delay." Normal everyday voice conversations are isochronous. They have always been isochronous. We could not tolerate delays. We just never called them isochronous. The word isochronous appeared when we started digitizing voice, then joining it with data on a single channel. The data guys suddenly woke up to the fact that users would only tolerate joining voice and data — if the voice went through without delays. So they came up with this fancy new term "isochronous." By accepting this realization, they then could design buses — e.g. Universal Serial Bus — where other flows of data (e.g. printing, keyboard entry, data communications from the Internet) could be delayed minute amounts of time.

Isochronous comes from the Greek "iso" (equal) and "chronous" (time). Isochronous transmission is used to move stuff which must get to its destination with absolutely no delays. Voice and video need isochronous transmission. Let me explain. In the beginning, the phone network switched a call from A to B. It kept the circuit open. Whatever you said at one end went to the other end at the speed of light, effectively instantly. Then they invented other methods of transmission, where the circuit isn't open 100% from end to end during the "conversation." One example is packet switching, used widely for sending data. If you're sending an electronic mail, it clearly doesn't matter if your electronic letter arrives half a second faster or slower. It does matter with voice and video.

In isochronous data transmission, timing is derived from the signal carrying the data. No timing or clock lead is provided at the customer interface. In isochronous data transmission, data has no embedded timing - send it slower and it is still valid, just late. Voice and video are intimately tied to timing. Send voice slower and it sounds very different. With TDM, there is a direct relationship between the signal rate used to digitize the voice samples and the bearer channel rate, allowing accurate reconstruction of the voice (or other signals) at the far end. With packet technologies no such relationship exists. See ISOETHERNET and UNIVERSAL SERIAL BUS.

Isochronous Ethernet A 10 Mbps LAN topology that sets aside 96 ISDN channel to carry voice, data and video. See ISOCHRONOUS and ISOETHERNET.

ISODE ISO Development Environment. An implementation of OSI's upper layers on a TPC/IP protocol stack. Pronounced "eye-so-dee-eee".

ISODE Consortium X.500 directories (www.isode.com/overview.htm)

IsoENET Also called ISOETHERNET. Developed by National Semiconductor in cooperation with IBM, which helped establish the call set-up procedure, IsoENET adds isochronous services to established Ethernet LANs without affecting, i.e. degrading, normal Ethernet's data traffic. Isoenet allocates a six megabit isochronous channel which can be divided into 97 standard 64 kbps ISDN-like B-channels, 96 of which can be used for real-time interactive multimedia connections, such as voice and video, and one, the D channel, is used for signaling. Isoenet may also be used in combination with Ethernet's standard 10 Mbps to provide faster access to local area network servers. See ISOCHRONOUS.

IsoEthernet An extension to the Ethernet LAN standard proposed by IBM and National Semiconductor and first demonstrated at Fall Comdex 1992. IsoEthernet adds 6 Mbps of capacity to regular Ethernet, specifically to carry low delay, constant bit rate, isochronous data, especially voice and video. This isochronous capacity appears as up to 97 telephony channels of 64 Kbps each — 96 for transmission of information (voice, video, data, etc.) and one (called the D channel) for signaling. Like FDDI-II, IsoEthernet has the potential to carry both live voice or video calls together with LAN packet data on the same cable. See ETHERNET, ISOCHRONOUS, FDDI-II.

Isolated Ground IG. In AC electricity, an isolated ground is a type of outlet characterized by the following features and uses:
- It may be orange and must have a Greek "delta" on the front of the outlet. (A delta looks like a triangle.)
- It must be grounded by an insulated green wire.
- It must have an insulator between the ground terminal and the mounting bracket.
- It is used primarily to power electronic equipment because it reduces the incidence of electrical "noise" on the ground path.

Isolation See POWER CONDITIONING.

ISP 1. Internet Service Provider. A vendor who provides direct access to the Internet. The ISP also usually provides a core group of internet utilities and services like E-mail, News Group Readers and sometimes weather reports and local restaurant reviews. The user reaches his ISP by either dialing-up over normal phone lines with their own computer and modem, or over a dedicated line installed by the telephone company.
2. ISDN Signal Processor.
3. Information Services Platform.

ISPBX Integrated Services Private Branch eXchange.

ISPT Instituto Superiore delle Poste e delle Telecomunicazioni. (Superior Institute for Posts and Telecommunications, Italy).

ISSI InterSwitching Interface. An interface between two SMDS switching systems within a LATA.

ISSN Integrated Special Services Network.

ISTF Integrated Services Test Facility.

ISUP Integrated Services Digital Network User Part. The call control part of the SS7 protocol. ISUP determines the procedures for setting up, coordinating, and taking down trunk calls on the SS7 network. ISUP is defined by ITU-T recommendations Q.761 and Q.764. ISUP also provides:
- Calling party number information (including privacy indicator).
- Call status checking, to keep trunks in consistent states at both ends.
- Trunk management, and
- Relates of trunks and the application of tones and/or announcements in the originating switch upon encountering error, blockage or busy conditions.

There are seven ISUP Messages: Initial Address Message (IAM), Continuity Check Message (COT), Address Complete Message (ACM), Answer Message (ANM), Release Message (REL), Release Complete Message (RLC) and Exit Message (EXM). For you to benefit from these capabilities, your phone equipment must first be able to access the CCS7 network. One suggested way (but not the only way) is through the ISDN primary rate access (PRA) standard, which supports Q.931 protocol. See IAM, ISDN and COMMON CHANNEL SIGNALING.

ISV Independent Software Vendor. Typically a company which writes and sells software, but not hardware. Manufacturers of hardware and operating systems, i.e. IBM or Northern Telecom, often contract with ISVs to produce specialized software to make their hardware and operating system more attractive.

IT Information Technology. A fancy name for data processing, which became management information systems (MIS), which became information technology. All the same thing, essentially. IT may have come from Europe. I heard it first from Siemens and Nixdorf who merged in 1989.

IT&T Information Technology and Telecommunications.

It's A shortened form of "It is." Not to be confused with ITS, which is the possessive of it. Its house. It's a house.

ITAA Information Technology Association of America. ITAA was founded in 1961 as ADAPSO. ITAA says its members are companies who create and market products and services associated with computers, communications and data. ITAA is based in Arlington, VA.

ITB Intermediate Block Character. A transmission control character that terminates an intermediate block. A Block Check Character (BCC) usually follows. Using ITBs allows for error checking of smaller blocks in data communications.

ITC 1. International Teletraffic Congress.
2. Japan's Telecommunications Technology Committee.

ITCA International Teleconferencing Association, a professional association organized to promote the use of teleconferencing (audio and videoconferencing). Located in Washington DC.

Iterative Process The process of repeatedly processing a bunch of instructions. Each repetition, theoretically, comes progressively closer to the desired result, the "correct" answer, etc.

ITG AT&T's Integrated Telemarketing Gateway. This is a set of specs for hooking up an outside computer to an AT&T switch. Under ITG, information travels in both directions — from the switch to the host computer and from the host computer to the switch. See also IG (one-directional link), ASAI and OPEN APPLICATION INTERFACE.

ITI 1. Idle Trunk Indicator.
2. Information Technology Industry Council. ITI, according to ITI, represents the leading U.S. providers of information technology products and services. Its members had worldwide revenues of $323 billion in 1994. They employ more than one million people in the United States. 202-626-7375.

ITM See Information Technology Management.

ITSEC The European Information Technology Security classification and evaluation initiative.

ITT International Telephone and Telegraph. A company that once was the largest manufacturer of telecommunications equipment outside the U.S.

ITU International Telecommunications Union. See below.

ITU-T International Telecommunications Union Telecommunications. ITU-T is an organization established by the United Nations with membership from virtually every government in the world. ITU-T's objective is to set telecommunications standards, allocate frequencies to various uses and hold trade shows every four years. As satellites have become more important as a method of long distance communications, so the ITU-T's allocation of scarce satellite frequencies among countries has become a hot bed of controversy. There are many who believe the ITU-T to be the most important telecommunications organization in the world. The organization owes its origins to Union Telegraphique which was formed in 1865, with the specific aim of developing standards for the telegraph industry. In 1947, under a United Nations charter, it was reformed as the ITU-T. This body has three main aims:
a. To maintain and extend, international cooperation for the improvement and interconnectivity of equipment and systems, through the establishment of technical standards.
b. To promote the development of the technical and natural facilities (the spectrum) for most efficient applications.
c. To harmonize the actions of national standards bodies to attain these common aims. In particular, to encourage the growth of communications facilities in developing countries.

Due to the rapid growth of the telecommunications industry, it was necessary to set up the International Consultative Committees (CCITT, CCIR and IFRB) within the ITU-T's jurisdiction in order to adequately manage this expansion. The aims are achieved by organizing international conferences and meetings, by sponsoring technical cooperation, and by publishing information and promoting world exhibitions. Currently, the ITU-T has about 170 member nations. I.T.U., Place des Nations, CH-1211 Geneve 20, Switzerland. Tel +41 22 99 51 11. Fax +41 22 33 72 56. International Telecommunication Union (www.itu.ch). See the following definitions.

ITU E.169 An ITU-T Recommendation allowing International Freephone customers to be allocated a unique Universal Freephone

NEWTON'S TELECOM DICTIONARY

Number (UIFN) which will remain the same throughout the world, regardless of country or telecommunications carrier.

ITU H.222 An ITU-T Study Group 15 standard that addresses the multiplexing of multimedia data on an ATM network.

ITU Q.2100 B-ISDN Signaling ATM Adaptation Layer Overview.

ITU Q.2110 B-ISDN Adaptation Layer - Service Specific Connection Oriented Protocol.

ITU Q.2130 B-ISDN Adaptation Layer - Service Specific Connection Oriented Function for Support of Signaling at the UNI.

ITU Q.2931 The signaling standard for ATM to support Switched Virtual Connections. This is based on the signaling standard for ISDN.

ITU Q.931 The signaling standard for ISDN to support SVCs. The basis for the signaling standard developed for Frame Relay and ATM. See Q.931 for a much fuller explanation.

ITU Q.933 The signaling standard for Frame Relay to support SVCs. This is based on the signaling standard for ISDN.

ITUSA Information Technology Users' Standards Association.

ITV Interactive TV. Also abbreviated to I-TV.

IU Interface Unit

IUT Implementation Under Test: The particular portion of equipment which is to be studied for testing. The implementation may include one or more protocols.

IUW The ISDN Users Workshop.

IVDM Integrated Voice and Data Multiplexer. A device that Northern Telecom uses to provide DIALAN, a central office provided local area network offering completely digital, full duplex data transmission at speeds of 300 bps through 19.2 asynchronous and 1,200 bps through 64 kbps synchronous. DIALAN users use existing telephone sets and an Integrated Voice and Data Multiplexer (IVDM) that plugs into a telephone jack.

IVDS Interactive Video and Data Service. A service that the FCC figures will be used for new narrow band PCS licenses.

IVDT Integrated Voice/Data Terminal. A device with a terminal keyboard/display and a voice telephone with or without its own processing power. See INTEGRATED VOICE/DATA TERMINAL.

IVR Interactive Voice Response. Think of IVR as a voice computer. Where a computer has a keyboard for entering information, an IVR uses remote touchtone telephones. Where a computer has a screen for showing the results, an IVR uses a prerecorded human voice that is stored (digitized) on a hard drive, in addition it can use a synthesized voice (computerized voice) for read back information that is constantly changing. (The synthesized voice is commonly referred to as Text-to-Speech.)

Whatever a computer can do, an IVR can too — from looking up train timetables to moving calls around an automatic call distributor (ACD). The only limitation on an IVR is that you can't present as many alternatives on a phone as you can on a screen. The caller's brain simply won't remember more than a few. With IVR, you have to present the menus in smaller chunks. See also AUDIOTEX and COMPUTER TELEPHONY.

The benefits of Interactive Voice Response are obvious. By automating the retrieval and processing of information by phone, you can "give data a voice" and "add intelligence to the phone call." By doing that, you can:

Put information to work. The classic IVR "killer app" takes an existing database (e.g., a magazine's article archives, a freight company's package-tracking system) and makes it available by phone (or other media, such as fax, e-mail, or DSVD — Digital Simultaneous Voice and Data). You can automate telephone-based tasks. From "bank by phone" to "find my package," to "sell me an airline ticket," to "validate my new credit card," IVR gives access to and takes in information; performs record-keeping, and makes sales, 24 hours a day — supplementing or standing in for human personnel. IVR can add value to communications. Any call-handling operation can profit from IVR. Used as a front-end for an ACD, an IVR system can ask questions (e.g., "what's your product serial code?") that help routing and enable more intelligent and informed call processing (by people or automatic systems). IVR far supersedes more rudimentary technologies (such as Caller ID) in such applications. Used in place of traditional on-hold programming, IVR can add interactive value to what would otherwise be wait-time. The IVR can be used to distribute info, make callers aware of specials — even provide entertainment. The result: fewer callers drop off queue; you make more sales.

IVS Interactive Voice Service.

IW Interworking.

IWS Intelligent Workstation.

IX InterExchange. Any service which crosses exchange boundaries.

IXC InterExchange Channel, or InterExchange Carrier — as contrasted to the LEC — the Local Exchange Carrier, a new word for a local phone company. InterExchange Carriers used to be called "Other Common Carriers," except that didn't include AT&T. Now, AT&T, MCI, U.S. Sprint and all the long distance carriers are called InterExchange Carriers.

J-Hook 1. In a microwave or satellite antenna, a j-hook is the name for a length of waveguide with one end turned through 180 degrees. This passes through the reflector vertex to illuminate the reflector's surface from an electronics unit mounted behind the microwave or satellite structure. It looks like a J-Hook.
2. J-hooks are also pieces of J-shaped pieces of bent metal used to hold cables in an equipment rack.

Jabber To jabber. In local area networking technology, continuously sending random data (garbage). Normally used to describe the action of a station (whose circuitry or logic has failed) that locks up the network with its incessant transmission.

Jack Contacts Metallic elements of telephone jacks that carry the central office currents/voltages to the CPE plus contacts.

Jack Pins See JACK CONTACTS.

Jack Positions A numbering scheme to permit consistent identification of the Jack Contact(s) position. Position identification helps assure compatibility between the wiring system and the associated terminal equipment.

Jacket The protective and insulating housing of a cable. Not part of the fiber or the fiber buffer.

Jacket Material The material used as the outer insulator of a cable.

Jacks A receptacle used in conjunction with a plug to make electrical contact between communication circuits. Jacks and their associated plugs are used in a variety of connecting hardware applications including cross connects, interconnects, information outlets, and equipment connections. Jacks are used to connect cords or lines to telephone systems. A jack can be female or male.

Jam In an IEEE 802.3 network, the jam signal, which is normally produced by fixing the minimum number of data bytes that must be transmitted, is used to ensure that if a collision is produced, all devices on the network will detect it. See JAM SIGNAL.

Jam Signal A signal generated by a printed circuit card to ensure that other cards know that a packet collision on a local area network has taken place.

Jamming The interference with through-the-air radio transmission, the object being to hinder the receiver's ability to pick up and understand the signal. An example is the Russians' jamming Radio Free Europe.

Jane Barbie The electronic "Voice With A Smile" on most telephone company intercept recordings. Ms. Barbie does her work for the Electronic Telecommunications Inc., Atlanta, GA. See BARBIE, JANE.

JANET Joint Academic Network. A university network in the U.K. In recent years, with the advent of higher speed links, renamed to "Super-JANET." I have a wonderful lady who works for me called Janet Lindsley. I think of her also as Super-Janet.

JATE The Japanese equivalent of the U.S. FCC part 68 certification for equipment to be attached to the Japanese telephone network. It stands for Japan Approvals Institute for Telecommunications Equipment. Getting JATE approval is expensive, complex and immensely time consuming. At least that's what I wrote in the ninth edition of my edition. In the autumn of 1995, JATE wrote me and suggested that I correct my definition as follows: 1. Getting JATE approval is not expensive. It is not expensive compared with the same approval in U.S.A. We don't require to give detailed descriptions on a testing machine nor environmental tests, while FCC requires those. Therefore it is easier to get approval in JATE than in FCC. In some cases, FCC registration costs more than JATE approval, with fee for application and for a test laboratory. 2. Getting JATE approval is not complex. Japanese technical conditions only require not to harm network, like condition in U.S.A. does so. It's just simple. 3. Getting JATE approval is not time consuming. The average period is 23.4 days for documentation examination process, from the date an application is received by JATE to the date it is completed. And 90% were completed within 38.4 days.

Java Java is a programming language from Sun Microsystems designed primarily for writing software to leave on World Wide Web sites and downloadable over the Internet to a PC owned by you or me. HotJava, its brother, is another piece of software installed on a Web browser at your desktop. HotJava enables Java programs delivered over the Web to run on your desktop PC. In short, Java is the programming language the programs on the Web are written in. HotJava is the software that will sit on your PC. Java is basically a new virtual machine and interpretive dynamic language and environment. It abstracts the data on bytecodes so that when you develop an app, the same code runs on whatever operating system you choose to port the Java compiler/interpreter to. What's a Java application? According to Wired Magazine, point to Ford Motor Company's website today, for instance, and all you'll get are words and pictures of the latest cars and trucks. Using Java, however, Ford could relay a small application (called an applet) to a customer's computer (the one on your desk which are using the surf the Internet). The customer could then customize options on an F-series pickup while calculating the monthly tab on various loan rates offered by a finance company or local bank. Add animation to these applications and you could get to "drive" the truck. http://java.sun.com/starter.html or E-mail java@java.sun.com.

JB7 Jam Bit 7. It is the same zero suppression format found on T-1s as AMI, which is Alternate Mark Inversion. See AMI.

JBIG Joint Bitonal Image Group. Standard for black and white, and grayscale image representation.

JCL Job Control Language.

JEDEC Joint Electronic Devices Engineering Council. An organization of the U.S. Semiconductor manufacturers and users that sets package outline dimension standards for packages made in the U.S.

JEMA Japan Electronic Messaging Association.

Jeopardy A wonderful AT&T word meaning anything occurring during the course of accomplishing scheduled work which might cause the scheduled completion date to slip.

JES Job Entry Subsystem. Control protocol and procedure for directing host processing of a task in an IBM host environment.

Also the specific IBM software release, host-based, that performs job control functions.

Jitter If a network provides various latency (i.e. different waiting times) for different packets or cells, it introduces jitter, which is particularly disruptive to audio communications because it can cause audible pops and clicks. Jitter is also the tendency towards lack of synchronization caused by mechanical or electrical changes. Technically, jitter is the phase shift of digital pulses over a transmission medium. Three forms of jitter exist: Data Dependent Jitter (DDJ), Duty Cycle Distortion (DCD), and Random Jitter (RJ). Data Dependent Jitter is caused by limited bandwidth characteristics and imperfections in the optical channel components as it relates to the transmitted symbol sequence, according to Information Gatekeepers. This jitter results from less than ideal individual pulse responses and from variation in the average value of the encoded pulse sequence which may cause baseline wander and may change the sampling threshold level in the receiver. DCD Jitter is caused by propagation delay differences between low-to-high and high-to-low transitions. DCD is manifested as a pulse width distortion of the nominal baud time. RJ is the result of thermal noise.

Jitterati What the digital generation becomes after tanking up on too much coffee. This definition courtesy Wired Magazine.

Job A file, typically sent in batch mode. Specifically a set of data, including programs, files and instructions to a computer, that together amount to a unit of work to be done by a computer.

Join A basic operation in a relational system that links the rows in two or more tables by comparing the values in specified columns.

Joining An ATM term. The phase in which the LE Client establishes its control connections to the LE Server.

Joint Costs A regulatory concept. Joint costs are essentially overhead costs. They cover the costs of providing more than one service. Most costs in the telecommunications industry are joint. And being "joint" they give regulators enormous pleasure trying to allocate those costs to various services and therefore trying to figure what prices for those services should be.

Joint User Service An arrangement whereby a corporation, association, partnership or individual whose telecommunications needs do not warrant the provision of separate leased service, is permitted to use the service of another customer by mutual agreement. The primary objective of joint user service is to save money by buying circuits in bulk.

Joule The unit of work or energy. The energy expended when a current of one ampere flows through a resistance of one ohm for one second. Joule's Law says the heat produced in a circuit in joules is proportional to the resistance, to the square of the current and to the time.

Journal Printers These are special purpose printers which provide hard copy output for audit trail and demand printing functions associated with hotel/motel management features.

Joy Clicker One who nervously fiddles with a mouse.

JPEG Worldwide standard for still image compression devised by the Joint Photographic Experts Group, experts sanctioned by the International Standards Organization (ISO) and the ITU-T. The JPEG compression standard works this way, according to the New York Times: A color image is converted into rows of dots, called pixels, each with a numerical value that represents brightness and color. The picture is then broken down into blocks, each 16 pixels x 16 pixels and then reduced to 8 pixels by 8 pixels by subtracting every other pixel. The software uses a formula that computes an average value for each block, permitting it to be represented with less data. Further steps subtract even more information from the image. To retrieve the data, the process is simply reversed to decompress the image. A specialized chip decompresses the images hundreds of times faster than is possible on a standard desktop computer. PC Magazine defines JPEG as "a lossy image-compression algorithm that often reduces the size of bitmapped images by a factor of 10 or more with little or no discernible image degradation. JPEG compression works by filtering out an image's high-frequency information to reduce the volume of data and then compressing the resulting data with a lossless compression algorithm. Low-frequency information does more to define the characteristics of an image, so losing some high-frequency information doesn't necessarily affect the image quality." See also JPEG ++ and MPEG — a standard for compression of full motion digital video.

JPEG++ Storm Technology's proprietary extension of the JPEG algorithm. It lets users determine the degree of compression that the foreground and background of an image receive; for example, in a portrait, you could compress the face in the foreground only slightly, while you could compress it in the background to a much higher degree. See JPEG.

JT-2 6.312 Mb/s data rate. Same as T-2. Signal compatible with ITU-T document G.704 signal specification.

Judge Harold Greene Judge Greene presided over the 1982 AT&T Antitrust settlement, enforcing its provisions and making decisions about requests from the participants to modify or reinterpret the provisions of the settlement. As long as he doesn't allow AT&T to be completely free of regulation, Judge Greene will probably always be involved in figuring the future of the telecommunications industry.

Jughead A database of Gopher links which accepts word searches and allows search results to be used on many remote Gophers.

Jukebox A jukebox is a piece of hardware that holds storage media, such as optical disks or cartridge tapes. Jukeboxes are typically designed to hold as few as five and as many as 120 devices. Like old-fashioned record playing jukeboxes, media is moved by a robot-like device from the storage slot to the drive reading it. This lets the user share one drive among several cartridges or disk. Jukeboxes are typically used for secondary and archival storage. Access to information is not fast. See JUKEBOX MANAGEMENT.

Jukebox Management In a network, tasks like retrieval and writes to a jukebox come randomly from all the users. These tasks vary in urgency — retrievals are higher priority than writes, for example. Jukebox management software sorts out requests from the network by priority. Management also enhances the performance of a jukebox, by intelligently reordering requests. For example, if there are three requests for images on platter 1 and two from platter 2 and the another from platter 1, jukebox management means the requests from platter 1 will get handled together, then go to platter two. Sometimes it's called "elevator sorting" — responding to requests in logical order, not in the order in which they were made.

Jumbo Group A 3,600 channel band of frequencies formed from the inputs of six master groups. See MASTERGROUP and SUPERGROUP.

Jump Hunting See NONCONSECUTIVE HUNTING.
Jumper 1. A wire used to connect equipment and cable on a distributing frame.
2. A patch cable or wire used to establish a circuit, often temporarily, for testing or diagnostics.
3. Jumpers are pairs or sets of small prongs on adapters and motherboards. Jumpers allow the user to instruct the computer to select one of its available operation options. When two pins are covered with a plug, an electrical circuit is completed, When the jumper is uncovered the connection is not made. The computer interprets these electrical connections as configuration information. 4. When errors are found on printed circuit boards, a Jumper cable is sometimes soldered in to correct the problem.
Jumper Wire A short length of wire used to route a circuit by linking two cross connect points. See also JUMPER.
Junction Box Metal box inside which cables are connected.
Junctor A connection or circuit between inlets and outlets of the same or different switching networks.
JUNET Japan UNIX Network.
Juniper A Rolm/IBM product that consists of a digital phone and a printed circuit for dropping into an IBM PC or clone.

NEWTON'S TELECOM DICTIONARY

K 1. See K-STYLE HANDSET below.
2. In metric terms it means one thousand (1,000), taken from the Greek word kilo. It is often appended to a measurement such as KHertz or kHz, which means 1,000 Hertz. In data communications, a kilobit means a thousand bits per second. In computer memory terms, it means 1,024, which is the figure for two raised to the 10th power, i.e. $2 \times 2 \times 2 \times 2 \times 2 \times 2 \times 2 \times 2 \times 2 \times 2$.
K Band That portion of the electromagnetic spectrum in the high microwave/millimeter range — from 10.9 GHz to 36 GHz. See also Ka BAND.
K Plans Also call keysheets. When designing a phone system, you need to assign features and line assignments to each extension. A keysheet or a K Plan is an organized way of figuring and keeping track of those features and assignments. Typically it's one page per extension. These days, keysheets are often done on computer.
K Plant Old Bell System lingo for equipment used in key systems.
K Style Handset A K-style handset is the newer, square telephone handset. The older, round handset is called the G-style handset.
K10 Old Bell-Speak for 10-button (9-lines) key telephone.
K20 Old Bell-Speak for 20-button (19 lines) key telephone.
K30 Old Bell-Speak for 30-button (29 lines) key telephone.
KA9Q A popular implementation of TCP/IP (Transmission Control Protocol/Internet Protocol) and associated protocols for amateur packet radio systems.
Ka Band That portion of the electromagnetic spectrum in the high microwave/millimeter range — approximately 33 to 36 GHz.
Kaleida The much-heralded joint venture between Apple and IBM to make multimedia products. Multimedia is the bringing together of textual, graphical, video, sound, telephone and audio material. Kaleida was closed down in mid-November, 1995.
KBIT/S KiloBITs per Second. Standard measure of data rate and transmission capacity. See K.
KBPS Should be written kbps. KiloBits Per Second. One thousand bits per second.
KDD Kokusai Denshin Denwa Company, Ltd. The Japanese supplier of international telecom service.
Kearney System An AT&T numbering scheme for telecom parts. See KS NUMBER.
Keep-Alive Signal A generic term for a T-1 signal transmitted when a DTE detects a loss of input from the customer's equipment for 150 msec. (sometimes called a blue signal or AIS).
Kerberos A security system for client/server computing. Kerberos is a scheme, developed at MIT in the 1980s, to enable secure multiple system access to a client/server computing environment. Named for the three-headed dog, Cerberus, who guarded the gates of Hades in Greek mythology. Kerberos is a UNIX-based distributed database used for user authentication.
Kermit An asynchronous file transfer protocol designed for PC communications at Columbia University. The protocol has become popular because of its flexibility. Kermit is found most frequently on DEC VAX computers, IBM mainframes and other minicomputers. One of the clearest advantages of Kermit is its ability to be tailored for virtually any equipment. Protocols break a file into equal parts called blocks or packets. These packets are sent and the receiving computer checks the arriving packet and sends an acknowledgement (ACK) back to the sending computer. Because modems use phone lines to transfer data, noise or interference on the line will often mess up the block. When a block is damaged in transit, an error occurs. The purpose of a protocol is to set up a mathematical way of measuring if the block came through accurately. And if it didn't, ask the distant end to re-transmit the block until it gets it right. Because of Kermit's emphasis on flexibility, Kermit is very slow and should not be used when faster options (i.e. Zmodem) are available. www.columbia.edu/kermit and www.cc.columbia.edu/kermit
Kernel The level of an operating system or networking system that contains the system-level commands or all of the functions hidden from the user. In Unix, the kernel is a program that contains the device drivers, the memory management routines, the scheduler, and system calls. This program is always running while the system is operating. See KERNEL-BASED WINDOW SYSTEM.
Kernel Based Window System Kernel-based window systems are those in which the software application executes and displays in the same physical machine. Examples include personal computers and Macintoshes. The advantage is speed. The disadvantage is that applications are closely tied to the system environment and are therefore not portable. Kernel-based window systems also do not allow users/developers to use the network as a means of sharing computer resources.
Kernel Driver A Windows NT term. A driver that accesses hardware.
Kerr Effect When polarized light is shined onto a magnetized surface, the light is reflected back at an angle and in a different direction, depending on the polarity of the magnetism. This quirk of nature is called the Kerr Effect and it is the basis of magneto-optical (erasable) discs.
Kevlar A strong synthetic material used in cable strength members. The name is a trademark of the Dupont Company. Kevlar is also used in bulletproof vests worn by police.
Key 1. One or more characters or perhaps a field within a data record used to identify the data and perhaps control its use.
2. The physical button on a telephone set. What normal people call a "Switch," telephone people call a "Key."
3. In encryption, a key is a data string which, when combined with the source data according to an algorithm, produces output that is unreadable until decrypted. A key can also be used to decrypt a data string. See KEY HOLDER.

4. The device which unlocks your front door or perhaps your terminal or computer.

Key Holder In encryption, a key is a data string which, when combined with the source data according to an algorithm, produces output that is unreadable until decrypted. A key can also be used to decrypt a data string. In the mid-1990s in the United States, there was great controversy about software that could encrypt electronic communications. Under U.S. laws, such software could not be exported. There was a movement to change the law and create organizations called "key holders." These would be organizations that would be given copies of an individual's decryption key or codebreaker. Such organizations, the theory went, would, under court order, give an individual's decryption key to a law enforcement agency. This might happen with or without the individual's consent.

Key Illumination A lamp under a button (called a "key" in telephony) which flashes at different rates to signal an incoming call, a steady busy and "wink" (fast) hold.

Key Map A MIDI patch-map entry that translates key values for certain MIDI messages, for example, the keys used to play the appropriate percussion instrument or a melodic instrument in the appropriate octave.

Key Pad The touchtone dial pad on a pushbutton phone. Contrary to popular belief, touchtone is not a registered trademark of AT&T. See TOUCHTONE.

Key Pad State An AT&T enhanced fax term. The KEYPAD state can be: NULL: The keypad is in use by a feature and is not available for use on a call. NON-NULL. The keypad is available for use in originating a call or for sending DTMF tones on an existing call.

Key Pulse In multi-frequency (MF) tone signaling, a signal used to prepare the distant equipment to receive digits.

Key Pulsing A pulsing system in which digits are transmitted using pushbuttons. Each button corresponds to a digit and generates a unique set of tones.

Key Service Panel An old 1A2 key telephone term. Wired or unwired connector panel for modular expansion of key system service by allowing the installation of additional Line Cards and/or other KTUs. Typically, Key Service Panels are available in different jack configurations to accommodate 18-, 20-, 36- and 40-pin KTUs. Commonly abbreviated as KSP. Most KSPs are supplied as rack-mount equipment.

Key Service Unit KSU. This is a small metal cabinet which contains all the switching electronics of a business key telephone system. The KSU fits between the lines coming in from the central office and the lines going to the individual phones. Be careful where you place the Key Service Unit. That place should be well-ventilated as the KSU gets hot. It should be near a power outlet (it needs one). Unless it's a very small phone system, it should be plugged into a power outlet dedicated to it (other devices, such as typewriters, plugged into the same electrical circuit could affect it). And the power outlet should be above the reach of the mops and brooms of the local cleaning lady. Otherwise the plug will get knocked out and the phone system won't work the next day. See KEY TELEPHONE SYSTEM.

Key Set Also called Key Telephone Set. A telephone set having several buttons which can be used for call holding, line pickup, auto-dialing, intercom and other features. Ericsson calls a keyset a touchtone telephone.

Key Strip The row(s) of additional buttons on key telephone sets used for call hold, intercom and other KTU function control.

Key System Power Supply The local source for all DC voltages required for talking and lamp signaling within the Key Telephone System. The power supply may or may not also provide an AC voltage output for ringing. If it does, a separate Ringing Generator will normally not be required.

Key Telephone System A system in which the telephones have multiple buttons permitting the user to select outgoing or incoming central office phone lines directly. With a key system you don't have to dial 9 to get obtain an outside line. With a PBX, you have to dial "9" to make a call outside the building. Dialing 9 is the major difference these days between a key telephone systems and PBXs. PBXs are usually, but not always, larger.

Key Telephone Unit A modular 1A2 Key Telephone System building block that plugs into a KSU or KSP. Commonly abbreviated as KTU. Typical KTU examples include 4000 Series Line Cards, 4448 Delayed Ring Transfer Card, 6606 Interrupter, etc.

Key To Disk A method of entering data whereby it's sent directly from the keyboard to a disk, usually a hard disk.

Keyboard A series of switches, arranged somewhat like a standard "QWERTY" typewriter that allows you to send information to a computer. There is no such animal as a "standard" computer keyboard. For speed typists, this is a terrible pity. If you are buying multiple computers for your office, check out your peoples' preference for keyboards. Getting the right one can make a big difference. You can often buy PCs without keyboards and buy third party differently designed keyboards.

Keyboard Buffer A temporary storage area in memory that keeps track of keys that you typed, even if the computer did not immediately respond to the keys when you typed them.

Keyboard Call Setup Allows you to set up a data call using the buttons of a telephone, or it allows you to set up a voice call using the keyboard of your PC. Which definition you choose — setting up a voice or a data call — depends on which manufacturer you're working with.

Keyboard Plaque The disgusting buildup of dirt and crud found on computer keyboards. "Are there any other terminals I can use? This one has a bad case of keyboard plaque."

Keyboarding A really stupid word for typing.

Keyed A term used in data communications whereby the RJ-45 male plug has a small, square bump on its end and the female RJ-45 plug is shaped to accommodate the plug. A keyed RJ-45 plug will not fit into a female, non-keyed (i.e. normal) RJ-45. The purpose of keying a plug is to differentiate it from a "normal" non-keyed plug. Keyed RJ-45 plugs are typically used for data communications. See also RJ-11, RJ-22 and RJ-45.

Keying Modulation of a carrier signal, usually by frequency or phase, to encode binary (digital) information, (as in FSK or Frequency Shift Keying).

Keypad See KEY PAD.

Keystone A monitor distortion where one end of the screen — either side to side, or top to bottom — is larger than the other end.

KHz KiloHertz. One thousand hertz. See K.

Kill Message A recorded message played at the beginning of a call to a 900 (or other pay-per-call) number that warns the caller of the charges and gives him the option to hang up before it starts.

Killer App Killer Application. The high-tech industry's lifelong dream. That dream is to discover a new application that is so useful and so persuasive that millions of customers will rush in, and throw money at you to buy your killer app. The term derives from the PC industry where a killer app was so powerful that it alone justifies the purchase of a computer. The first PBX killer app was probably being able to dial out without an operator (i.e. dial 9). The second was probably least cost routing. The third was probably call accounting. The next may be some form of hookup to PCs, with desktop and LAN connectivity. In the PC industry, killer apps have been spreadsheets, word processing and databases. Finding that one killer app that will make them wealthy beyond their wildest dreams is what drives many software programmers and entrepreneurs. See WINDOWS TELEPHONY.

Kilo One thousand.

Kilobyte 1,024 bytes. Roughly the amount of information in half a typewritten page.

Kilocharacter One thousand characters. Used as a measure of billing for data communications by some overseas phone companies. See also KILOSEGMENT.

Kilosegment 64,000 characters. Used as a measure of billing for data communications by some overseas phone companies. See also KILOCHARACTER.

Kingsbury Commitment December 13, 1913: A letter from Nathan C. Kingsbury, VP AT&T, to the Attorney General of the United States committed AT&T to dispose of its stock in Western Union Telegraph Company. It also promised to provide long distance connection of Bell System lines to independent phone companies (where there was no local competition) and further agreed not to purchase any more independent telephone companies, except as approved by the Interstate Commerce Commission which regulated the phone industry at that time. See also DIVESTITURE.

Kittyhawk A trade name for a line of very small hard disks manufactured by Hewlett Packard.

Kludge A hardware solution that has been improvised from various mismatched parts. A slang word meaning makeshift. A kludge can also be in software. It may not be elegant and is probably only a temporary fix. As in, "That patch to the software is a real kludge."

Kluge Another way of spelling kludge. We think spelling it kludge is correct. See KLUDGE.

KNET Kangaroo Network Hardware/software product (Spartacus/Fibronics) that lets IBM mainframes communicate over networks using the TCP/IP protocol.

Knowbots Intelligent computer programs that automate the search and gathering of data from distributed databases. The creation of knowbots is part of a research project headed up by the Corporation for National Research Initiatives, Reston, VA. Two knowbot-based databases for the medical field are expected to be available in 1991. Knowbots could become more widespread for general use, according to networking experts.

Knowledge Base System In its most simple term, it means knowledge that is known by the system. Software in which application specific information is programmed into something called the "knowledge base" in the form of rules. The system uses artificial intelligence (AI) procedures to mimic human problem solving. It applies the rules stored in the knowledge base and the facts supplied to the system to solve a particular business problem.

Knowledge Worker A person who uses a computer. John Perry Barlow thinks the expression was created by the "droids" who run Microsoft and Apple. Mr. Barlow is a cattle rancher, computer hacker, poet, and a lyricist for the rock band The Grateful Dead.

Konexx A modem to telephone connection product made by Unlimited Systems in San Diego, CA. This is a device which lets you connect your laptop modem to a screwy, (i.e. non-analog) phone system you encounter in travels. That screwy phone system might be a digital one, or a proprietary electronic one, or anything that isn't plain old tip and ring. With the Konexx, you simply unplug your handset, plug in the Konexx and plug in your modem. You can dial, connect and transfer e-mails and faxes. A useful device, but not one that works on all phones. Check.

KS Number Abbreviation for Kearney System number. AT&T's Western Electric division had a major manufacturing and distribution facility in Kearney, New Jersey. Thousands of items were assigned part numbers with KS prefixes. KS numbers still appear on certain basic telecom hardware items made by various manufacturers. AT&T now uses both a Code number which reflects standard industry numbering and a different "Comcode" number on most products. The Code No. 259C modular-to-Amphenol adapter is Comcode No. 103339396, and KS No. 21997L15. Very confusing. Kearney, by the way, is pronounced "carny."

KSR Keyboard Send Receive. A combination teleprinter/transmitter/receiver with transmission capability from the keyboard only, i.e. there is no punch paper tape device and no magnetic memory device, such as a floppy disk.

KSU Key Service Unit. The main cabinet containing all the switching equipment and electronics necessary to run a key telephone system. See also KEY SERVICE UNIT.

KSU-Less Phone System KSU stands for Key Service Unit. It's a funny term for the main cabinet which contains all the switching equipment and electronics necessary to run a key telephone system. When you pick up your key telephone's handset and punch a button for an outside side, the KSU connects you to an outside line. When you pick up the handset and punch a button to make an intercom call, the KSU gives you intercom dial tone and receives your dialing instructions, rings the correct extension and gives you a talk path. When you have a KSU key system, you plug outside phone lines into the KSU and run lines in a star configuration to each phone. A KSU-less phone system, on the other hand, has all its electronics and switching in the phone sets themselves. You run the outside phone lines into each KSU-less phone and typically one pair of wires looping to each one of the KSU-less phones. A KSU-less phone system is typically a very small system, consisting of usually no more than six phones. A KSU-less phone can be very easy to install. And that's its primary charm. It may also be cheaper than having a small KSU phone system. I always think of a KSU-less phones as multiple single line phones joined together in one plastic multi-line phone with one intercom path to other KSU-less phones.

KTA Key Telephone Adapter. A Rolm multiplexing unit which connects a standard 1A2 key telephone to a three-pair cable coming in from the Rolm CBX.
KTI Key Telephone Interface.
KTILA Development Centre for Telecommunications (Greece).
KTS Key Telephone System.
KTU Key Telephone Unit. The circuit cards found in a KSU that control telephone sets and their features in a key system.
KU Band Portion of the electromagnetic spectrum in the 12 to 14 GHz range. Used for satellites.
KV Old Bell-Speak for key telephone. K stands for Key; V stands for Voice. CV stood for Combined Voice, a single line, simple telephone. All this had to do with letters included in USOC codes — the numbering systems the local Bell operating phone companies used to identify products and services. See USOC.
KVW Old Bell-Speak for a key telephone that's designed specifically for wall mounting. Most modern key telephones can be used on a desk or wall.
KWH KiloWatt Hour. One thousand WATTS of electricity used for one hour.

L Band Portion of the electromagnetic spectrum commonly used in satellite and microwave applications, with frequencies in the 390 MHz to 1550 Mhz. The GPS (Global Positioning System) frequencies are in the L-Band. GPS uses 1227.6 MHz and 1575.42 MHz.

L Carrier A long haul frequency division multiplexed coax-cabled long haul carrier system. It was first introduced just before the second World War. Eventually it grew to a capacity of 13,300 voice channels over a pair of coaxial tubes. L Carrier systems are still used today. They are the most widely-used analog long distance transmission system.

L Multiplex A system of analog multiplexers built up through groups, supergroups, master groups and jumbo groups of circuits. See L CARRIER.

L-to-T Connector A device that mates two FDM (Frequency Division Multiplexed) groups with one TDM (Time Division Multiplexed) digigroup to allow 24 voice conversations in analog form to talk to (tie into) a DS-1 line — a T-1 line.

Label A set of symbols used to identify or describe an item, record, message or file. It can also be the same as the address in storage.

Labeling Algorithm Algorithm for shortest path routing or similar problems which labels individual nodes, updating labels as appropriate to reach a solution.

LAD LATA Architecture Database

LADT Local Area Data Transport. A method by which customers will send and receive digital data over existing customer loop wiring.

LAI Location Area Identity. A wireless term. A LAI is part of GSM (Global System for Mobile Communications), used in the radio interface. An LAI is part of the GSM Temporary Mobile Subscriber Identity (TMSI). The TMSI is allocated by the GSM network on a location basis. More simply put, it identifies the cell that a mobile telephone user is in.

LAMA Local Automatic Message Accounting. A process using equipment in the central office which records the information necessary to bill your local phone calls by your local phone company.

Lambda The Greek letter used as a symbol for wave length.

Laminate The whole structure in the mold, consisting of several piles. A layer in a composite is called a "ply."

Laminations Thin sheets of steel used as the magnetic core in electrical apparatus, (e.g., the core of an audio frequency transformer is normally composed of laminations).

Lamp The technically correct term for light bulb, which nontechnical folks put into their lamps.

Lamp Battery A steady (unpulsing) 10 volt AC source of power to operate the lamps in key telephone sets; usually one of the outputs of the local Key System Power Supply.

Lamp Flash A pulsed 10 VAC source of lamp power sent to a key telephone set to indicate a CO or PBX line is ringing in. Pulse repetition rate is normally 60 Hertz with a duty cycle of .5 sec on and .5 sec off. This signal is usually provided by the local Interrupter KTU.

Lamp Leads Lamp and Lamp Ground (L&LG) wires connected to all lamps in the key telephone set over which steady and pulsed 10 VAC signals from the Line Card KTU are sent.

Lamp Steady A steady (unpulsed) 10 VAC source of lamp power sent to a key telephone set to indicate that the line is in use. See also Lamp Battery.

Lamp Wink A pulsed 10 VAC source of lamp power sent to a key telephone set to indicate that the line is on Hold status; pulse repetition rate is normally 120 Hertz with a duty cycle of .4 sec on and .1 sec off. This signal is usually provided by the local Interrupter KTU.

LAMs Line Adapter Modules.

LAN Local Area Network. A short distance data communications network (typically within a building or campus) used to link together computers and peripheral devices (such as printers) under some form of standard control. For a longer, more detailed explanation, See LOCAL AREA NETWORK.

LAN Adapter A PC-compatible circuit card that provides the PC-to-LAN hardware connection. In addition, LAN software drivers and LAN operating systems need to be run on the PC for it to function as a LAN station.

LAN Aware Applications that have file and record locking for use on a network.

LAN Emulation Also known as LAN-E, it is a set of specifications developed by the ATM Forum for the operation of LAN-to-LAN bridged connectivity over an ATM network, allowing ATM to be deployed on a legacy LAN or with legacy LAN applications.

LAN Ignorant Applications written for single users only. These are not recommended for use on LANs (local area networks).

LAN Intrinsic Applications written for client-server networks.

LAN Manager 1. A person who manages a LAN. Duties can includes adding new users, installing new hardware and software, diagnosing network problems, helping users, performing backup and setting up a security system. Unlike MIS managers, LAN managers are rarely formally trained in LAN management. Sometimes they're called LAN Network Managers.
2. The multi-user network operating system co-developed by Microsoft and 3Com. LAN Manager offers a wide range of network-management and control capabilities. It has been superseded by Windows NT Advanced Server.

LAN Network Manager An IBM-developed network management tool. It is a software program that runs under OS/2 and which provides management and diagnostics tools needed to manage a Token Ring LAN. A PS/2 running LAN Network Manager collects vital statistics and special management data packets on the ring to which it is connected. When multiple rings are involved, the LAN Network Manager relies on the token ring bridges and routers to help in managing those token ring LANs that are not directly connected to the LAN Network Manager station. IBM has installed software in its bridges called the LAN Network Manager Agent. The agent software acts as the eyes and ears for the LAN Network Manager station so that the station can manage the remote rings as if it were connected directly to them. If there were no such agents, managers of networks would be blind to what's going on these LANs. Remote management with LAN Network Manager includes the ability to perform ring testing, analyze traffic and error statistics, and force adapters off the network.

LAN Server IBM's implementation of LAN manager, now largely superseded by OS/2 2.1.

Land Line A telephone circuit that travels over terrestrial circuits, be they wire, fiber or microwave. A call may originate from a source not connected to the "terrestrial" network, as from a car telephone or a ship to shore radio, and the call be completed via a Landline.

LANDA The Local Area Network Dealers Association. It runs a number of excellent trade shows each year. Its members are LAN resellers, distributors, manufacturers, and consultants. It's based at Suite 260, 360 West Butterfield Road, Elmhurst, IL. 708-279-2255. In 1993 it merged with NOMDA, the National Office Machine Dealers Association.

Landscape Most computer screens are horizontal, i.e. they are wider than they are high. In the new language of computer screens, such screens are called "landscape." When a computer screen is higher than it is wide, it's called "portrait." Some computer screens can actually work both ways. Some even have a small mercury switch in them that determines which way the screen is standing (portrait or landscape) and will adjust their image accordingly.

LANE An ATM term. LAN Emulation: The set of services, functional groups and protocols which provide for the emulation of LANS using ATM as a backbone to allow connectivity among LAN and ATM attached end stations.

Lane A single interactive multimedia network.

Language Computer software that allows you to write programs.

Language Interpreter Any processor, assembler or software that accepts statements in one software language and then produces equivalent statements in another language.

LAP Link Access Procedure.

LAP-B LAPB. Link Access Procedure Balanced, the most common data-link control protocol used to interface X.25 DTEs with X.25 also specifies a LAP or link access procedure (not balanced). Both LAP and LAP-B are full-duplex, point-to-point bit-synchronous protocols. The unit of data transmission is called a frame. Frames may contain one or more X.25 packets. LAP-B is the data link level of X.25 in a packet switched network. Same as a subset of the asynchronous balanced mode of HDLC. It is the link initialization procedure that establishes and maintains communications between the data terminal equipment (DTE) and data communications equipment (DCE). All public packet data networks (PDNs) support LAPB.

LAP-D Or LAPD. Link Access Procedure-D. Also called Link Access Protocol for the D channel. Link-level protocol devised for ISDN connections, differing from LAPB (LAP-Balanced) in its framing sequence. Likely to be used as basis for LAPM, the proposed ITU-T modem error-control standard.

Laplink The word is originally the name of a DOS program that transferred files between laptop and desktop computers. It came with a special cable that attached from one computer's serial port to the other. You plugged the cable in, loaded the LapLink software and transferred files back and forth. You could do the transferring from either machine. It was a nifty program, the brainchild of Mark Eppley and he formed a company to sell it, called Traveling Software in Bothell, WA. Eventually the program became so successful that the word "to laplink" became a common verb to connote the transferring of files between computers, as in "I'll go laplink these files over to Mary's machine." Eppley's program is now in its sixth revision. It's in Windows format. You can transfer across a parallel port cable (which is faster than the serial port). You can also transfer files across phone lines. You don't even have to be in the same city to transfer files. And Eppley has a new feature called SpeedSync that really cuts transfer time down by sending only the changed parts of the file. We use LapLink every day. It's one of those "must have" pieces of software.

Laptop Computer A portable computer you can use on your lap. Usually weighing less than 15 pounds. Probably the most useful gadget to come along in years. I wrote much of this dictionary in planes, trains, airports, etc. on a laptop computer — initially on a Toshiba T1200HB, sporting one megabyte of RAM, one 21 megabyte drive, one 3 1/2 inch disk drive with a capacity of 730K and a 1200 baud modem, then later on a more modern laptop called the Toshiba T1200XE, sporting five megabytes of RAM, one 42 megabyte hard drive, one 3 1/2 inch disk drive with a capacity of 1.4 megs, and one 2400 baud modem, then later on a Toshiba T2000SX with four megs of RAM and a 60 meg hard disk, then later on a Toshiba T3300SL sporting 8 megs of RAM, a 386SL 25 MHz processor and a 130 meg hard disk. I finished it on a Toshiba T4800CT sporting a 486 75 MHz processor, 24 megs of RAM and a 520 meg hard disk. There's a new term for a laptop computer, namely a notebook computer. A notebook laptop weighs less than eight pounds and is roughly the shape of a 8 1/2" x 11" sheet of paper. Sub notebooks are under 4 lbs.

Large Squaring Capability A feature on some key systems which permits all lines to appear on all telephone sets.

Laser An acronym for Light Amplification by Stimulated Emission of Radiation. It is a device which produces a single frequency light. By turning on and off the laser light signal quickly, you can transmit the ones and zeros of a digital communications channel. Lasers carried through glass fiber are ideal for telecommunications transmission for two major reasons. 1. Glass fiber of such purity has now been developed that only a very minute portion of the laser light traveling through it is lost. In telecom terms, this means very little of the laser signal is attenuated. The signal maintains its strength and thus reduces the need for frequent and expensive repeaters (the digital word for "amplifiers.") Laser transmission systems can now carry thousands of voice conversations for hundreds of miles without repeaters on two fibers no thicker than a human hair. (You need two fibers — one for the conversations coming and one for them going.) 2. Laser fiber optic telecom systems are totally immune to electromagnetic interference of any kind. There's no humming from electrical motors. You can't pick up the local TV station in the background. You can't pick up any interference from adjacent cables.

In recent years, laser fiber optic transmission has been getting cheaper, more reliable and more powerful at roughly the same rate as computers, and like computers, nobody believes there is an end in sight. As we were writing this dictionary, Russell Dewitt of Contel delivered a paper entitled Evolution of fiber optics in rural telephone networks. In it, he talked about "Fiber Optics Progress" and said:

"In 1860, the Pony Express could deliver a letter from St. Louis to San Francisco in ten days. For three typed pages the data transmission rate was about three bits per minute. In comparison, today in Contel we are transmitting at the rate of 565 Mbps (million bits per second) over single mode fiber. This is a capacity of 8064 voice channels. Gigabit per second systems will be available for use this year (a gigabit is a million million) and a 20 Gbps system has been demonstrated in the laboratory. For the future, the ultimate potential of a single mode fiber has been estimated. It is about 25,000 Gbps (25,000,000,000,000,000 bits per second). At that rate you could transmit all the knowledge recorded since the beginning of time in 20 seconds. See also LASER FAX, LCD and FIBER OPTICS.

Laser Fax A conventional laser printer that is also capable of being used as a FAX machine when combined with an optional plug-in cartridge and used with a personal computer.

Laser Optical System of recording on grooveless discs using a laser-optical-tracking pickup. Originally, the technology was WORM — Write Once (i.e. not erasable) Read Many. It's now erasable.

Laser Printer A high speed non impact dot matrix printer which uses a laser beam to electrostatically form characters on paper. The printer then heats the paper which melts a metallic dust attracted to the electrostatic areas which form the inked images on the paper. Laser printers are fast and the quality of their printing beautiful, rivaling that produced by conventional photo typeset (the way this book was produced).

Lashing Attachment of a cable to a support strand by wrapping steel wire or dielectric filament around the cable.

LASS See LOCAL AREA SIGNALING SERVICES

LAST Local Area Systems Technology. A Digital (DEC) protocol

Last Digit Dialed Signal Allows the use of the # sign on Touchtone telephones to indicate that the last digit has been dialed on outgoing calls. This signal enables the PBX to process calls more rapidly, since some PBXs count the time after a digit was dialed. If nothing else is dialed within a certain time, it assumes that the dialing is complete and then pulses out the call.

Last Extension Called Same as LAST NUMBER REDIAL.

Last In First Out LIFO. The last phone call (or data) arriving is the first call (or data) to leave — to be processed, to be saved, whatever. The term LIFO comes from accounting. It's one of several ways to value an inventory.

Last Number Redial Most modern electronic phones have a button on them called "Last Number Redial." When you touch this button, your phone will automatically dial the last number you dialed. If you also have speed dial numbers on your phone, any number you dialed with a Speed Dial button will not appear in Last Number Redial. Most Last Number Redial buttons on electronic phones attached to a PBX will only recognize completely-dialed numbers. Last Number Redial buttons are useful. Also useful — but less common — is a stored number dial button. Dial a number, punch in "save," then that number will be saved to that button, ready to be dialed later, even though you might dial some other numbers in the meantime.

LAT Local Area Transport. A proprietary communications protocol developed by DEC for terminal-to-host communications. LAT allows terminal emulators to access VAX and VMS systems over Ethernet. See LAT PROTOCOL.

LAT Protocol The LAT protocol, announced by Digital Equipment Corporation in the mid-80s, is today one of the industry's most widely used protocols for supporting character terminals over Ethernet networks. LAT is currently licensed by more than 40 third party hardware and software developers, and is compatible with the products of more than 30 major system vendors, from Apollo, and Apple to IBM, Tandem and Wang. The basic function of the LAT protocol is to permit a terminal server to connect multiple asynchronous devices — video display terminals, printers or plotters— to a host timeshare computer. To do this, LAT (or any other terminal server protocol) puts data into packets that can be understood by both the asynchronous device and the host. Essentially, a terminal server protocol is responsible for establishing lower level communications connections, and for routing appropriate transmissions to their destinations.

LATA Local Access and Transport Area. One of 161 local geographical areas in the US within which a local telephone company may offer telecommunications services — local or long distance. AT&T is expressly prohibited from offering intraLATA calls by the terms of the Divestiture. Other competitors, such as MCI and Sprint, are not, though rules vary by state, according to state regulation.

Latency A fancy term for waiting time or time delay. The time it takes to get information through a network. Real-time, interactive applications such as voice and desktop conferencing are sensitive to accumulated delay, which is referred to as latency. For example, telephone networks are engineered to provide less than 400 milliseconds (ms) round-trip latency. You can get latency in several ways:

1. From propagation delay — the length of time it takes information to travel the distant of the line. This period is mostly determined by the speed of light; therefore, the propagation delay factor is not affected by the networking technology in use.
2. From transmission delay — the length of time it takes to send the packet across the given media. Transmission delay is determined by the speed of the media and the size of the packet.
3. From processing delay - the time required by a networking device for route lookup, changing the header, and other switching tasks. In some cases, the packet also must be manipulated; for instance, changing the encapsulation type, changing the hop count, and so on. Each of these steps can contribute to the processing delay.
4. Rotation delay. The delay in accessing data which comes from waiting for a disk to rotate to the currant location.

In a bridge or a router, latency is the amount of time elapsed between receiving and retransmitting the LAN packet. The length of time the packet is stuck in a bridge or router. See INTERRUPT LATENCY.

Latitude Communications A wonderful new company that makes an audio conferencing server that fits on a LAN and allows users on the LAN to schedule their own conference calls, managing them through touchtone or through the PC on their desk.

Launch A new term for starting a program from within another. Typically what might happen is you're working in a messaging program, which has individual lines showing you've just received several faxes, voice mails, electronic mail documents. You click on one of the lines. Your program recognizes that it's an electronic mail message and says "quickly open the electronic messaging software and get it to read the message." So it "launches" the messaging software.

Laurus Unit A rack for T1 interface units.

Lay A term used in cable manufacturing to denote the distance of advance of one member of a group of spirally twisted members, in one turn, measured axially.

Layer See LAYERING.

Layer Entity An active element within a layer.

Layer Function A part of the activity of the layer entities.

Layer Service A capability of a layer and the layers beneath it that is provided to the upper layer entities at the boundary between that layer and the next higher layer.

Layer User Data Data transferred between corresponding entities on behalf of the upper layer or layer management entities for which they are providing services.

Layered Network Architectures Currently the basis of all telecommunication network architecture standards, with functions allocated to different layers and standardized interfaces between layers. The OSI Model is a layered network architecture. See LAYERING.

Layered Network Sales Strategy An AT&T sales strategy which involves segmenting customer network needs into four usage levels or layers, then pursuing sales opportunities at these various entry points. The layers are Premises Networks, Office Networks, "metropolitan" or " campus" networks and Wide Area Networks.

Layering Layering is a technique to write complex software faster and more easily. Layering is often used with public, open software. The idea is to have layers of software on top of other layers. Each performs a specific task. The idea is that if your software works at one layer — i.e. conforms to the rules of that layer — it should be compatible (i.e. work with) the layers of software above and below it. The most famous layered software is the seven-layer OSI (Open Systems Interconnection) model. It breaks each step of a transmission between two devices into a discrete set of functions. These functions are grouped within a layer according to what they are meant to accomplish. The data link layer, for example, is concerned with the transmission of frames of data between devices and covers protocols that are aimed at packaging raw data characters into frames, detecting and correcting errors when frames get lost or mutilated, arranging for retransmission and adding flags and headers so that DTE can recognize the beginning and end of a frame. Other layers serve other purposes. Each layer communicates with its counterpart through header records. The flexibility offered through the layering approach allows products and services to evolve. Accommodating changes are made at the layer level rather than having to rework the entire OSI model. Another layered software architecture is Microsoft/Intel's Windows Telephony. It has three layers. At the lowest is SAPI, which is the Service providers' API. In the center is the actual Windows Telephony code. At the top is the TAPI — Telephony applications API. See OSI and Windows Telephony.

Lays The twists in twisted pair cable. Two single wires are twisted together to form a pair; by varying the length of the twists, or lays, the potential for signal interference between pairs is reduced.

LB An ATM term. Leaky Bucket: Leaky Bucket is the term used as an analogous description of the algorithm used for conformance checking of cell flows from a user or network. See GCRA, UPC and NPC. The "leaking hole in the bucket" applies to the sustained rate at which cells can be accommodated, while the "bucket depth" applies to the tolerance to cell bursting over a given time period.

LBA An abbreviation for Logical Block Address. See Logical Block Address.

LBO Electrical Line Build Out.

LBRV Low Bit Rate Voice. Digitized voice that requires a bandwidth of fewer than 32 Kbps. LBRV digitizing techniques include packetized voice, APV, DSI, and LPC.

LCD Liquid Crystal Display. An alphanumeric display using liquid crystal sealed between two pieces of glass. The display is divided into hundreds or thousands of individual dots, which are charged or not charged, reflecting or not reflecting external light to form characters, letters and numbers. LCD displays have certain advantages. They use little electricity and react reasonably quickly — though not nearly as quickly as a glass cathode ray tube or a gas plasma screen. They are reasonably legible. They require external light to reflect their information to the user. The newer so-called "supertwist" LCDs are much more readable. You see LCDs on computer laptops and telephone screens. The reason computer laptop LCD screens are brighter than phone screens is that laptops use fluorescent or other light sources to illuminate the LCD (typically from the back or the side). Phones typically don't have the power. The only way for a phone to get the power is to plug the phone and its screen into an AC outlet. Most users, however, don't want to have to plug their phone into both AC and phone outlets. It's cumbersome. In newer LCDs, called active matrix displays, the circuit board contains individual transistors for each pixel, or dot on the screen. The enables the crystals to shift quickly, resulting in a higher quality image and the ability to display full-motion video. Active matrix displays in color are hard to manufacture. Low production yields are common, though improving.

LCF-PMD The ANSI X3T9.5 standard which defines the requirements for the transmission of data over low cost fiber in an FDDI topology. Also refers to the ANSI working group responsible for the development and perpetuation of the standard.

LCM 1. Line Control Module.

2. Line Concentrating Module is a cabinetized peripheral which contains two duplicated Enhanced Line Concentrating Modules (LCME) to interface analog circuits.

LCR Least Cost Routing. A telephone system feature that automatically chooses the lowest cost phone line to the destination. What actually is the "lowest cost" is determined by algorithms, equations and decision trees programmed into the PBX. Least Cost Routing typically works with "look-up" tables in the PBX's memory. These tables are put into the PBX by the user. The PBX does not automatically know how to route each call. It must be told so by the user. That "telling" might be as simple as saying

"all 312 area codes will go via the AT&T FX line."
Or it might be as complex as actually listing which exchanges in the 312 area code go by which method. LEAST COST ROUTING tells the calls to go over the lines which are perceived by the user to be the least cost way of getting the call from point A to point B. There are typically two types of "least cost routing" translation — that which examines the first three digits of the phone number (i.e. just the area code) and the first six digits of the phone number (i.e. the area code and the three digits of the local central office). Six digit translation is preferred because it allows you more flexibility in routing — particularly to big area codes, like 213 in LA, where there are long distance calls within the area code. See also AUTOMATIC ROUTE SELECTION, ALTERNATE ROUTING and SIX DIGIT TRANSLATION.

LCS Live Call Screening. This is a wonderful phone/voice mail feature, at this stage confined to Panasonic phone systems. Here's how it works. Someone calls you and ends up in your voice mail. You can hit a LCS button on your phone (or it will happen automatically) and you will hear the message being left in your voice mail. If you choose to speak to this person (they're important), you press a button or simply pick up your handset. There are various options with this feature. You can listen through your speakerphone or through your handset. You can have it happen on every voice mail call or just the ones you choose. You'll know if someone is leaving you a voice mail because a special light lights on your phone. There's an allied feature, called Remote Live Screening, which lets you plug a single line cordless phone into your Panasonic phone and monitor incoming voice mails — even though you're in the bathroom. This way you can always be assured you will always receive the important call you're waiting for. These features are presently only available on the Panasonic Super Hybrid Digital PBX with its integrated digital Panasonic voice mail system. But they should be on every phone system.

LD 1. Long Distance.
2. Loop Disconnect.
LDA Long Distance Alerting.
LDBS Local Data Base Services.
LDM See LIMITED DISTANCE MODEM.
LDN Listed Directory Number. Your main phone number. The one you list in the telephone directory and Directory Assistance.
LDT Line appearance on a Digital Trunk.
LDTV Low Definition TeleVision (e.g. VHS).
LE LAN Emulation. Refer to LANE.
LE_ARP An ATM term. LAN Emulation Address Resolution Protocol: A message issued by a LE client to solicit the ATM address of another function.
Lead Agent The first agent in an ACD group. See also AUTOMATIC CALL DISTRIBUTOR.
Lead In 1. Wire or cable from antenna to TV set.
2. The conductor from the antenna to the radio receiver.
Leader The section at the beginning of a roll of magnetic tape which holds no data and often is not even magnetic tape. A leader is used to feed the magnetic tape through the tape mechanism and secure it onto the roll.
Leadership Priority An ATM term. The priority with which a logical node wishes to be elected peer group leader of its peer group. Generally, of all nodes in a peer group, the one with the highest leadership priority will be elected as peer group leader.
Leading Current The phase difference in a capacitive alternating current where the current leads the E. M. F.
LEAF Law Enforcement Access Field. See CLIPPER CHIP and the NSA.
Leaky Bridge A type of LAN bridge that forwards packets from one LAN to another even though the packet should not be forwarded. Usually due to poor engineering.
Leaky Bucket An ATM term. An informal term for the Generic Cell Rate Algorithm.
Leaky PBX One of those really silly terms the phone industry is famous for. Picture this. You dial a distant PBX. If you can dial on — either locally or make a long distance call, that distant PBX is referred to as "leaky." All PBXs are, of course, "leaky," or at least capable of being made leaky.
LEAS LATA Equal Access System.
Leased Circuit Same as LEASED LINE or PRIVATE LINE.
Leased Line Same as a LEASED or DEDICATED CIRCUIT, PRIVATE LINE, LEASED CHANNEL. A telephone line rented for exclusive use 24-hour a day, seven days a week from a telephone company — a local phone company (like Nynex) or a long distance company like AT&T or MCI.
Least Cost Routing LCR. A telephone system feature which automatically chooses the "least cost" long distance line to send out a long distance call. The user typically dials "9" and then his 10-digit long distance number which is routed over the least costly service. See also LCR.
Leave Word Calling For AT&T System 75 and 85 users for internal messages. The caller's name, time of call and extension number are taken and can be retrieved on a Digital Display Monitor or BCT. The service is accompanied by an Integrated Directory — a simple electronic listing of employee names and extension numbers, access by a terminal equipped with a DDM or via the attendant console.
LEC 1. Local Exchange Carrier. The local phone companies, which can be either a Bell Operating Company (BOC) or an independent (e.g. GTE) which provide local transmission services. Prior to divestiture, the LECs were called telephone companies or telcos.
2. An ATM term. LAN Emulation Client: The entity in end systems which performs data forwarding, address resolution, and other control functions.
LECID An ATM term. LAN Emulation Client Identifier: This identifier, contained in the LAN Emulation header, indicates the ID of the ATM host or ATM-LAN bridge. It is unique for every ATM Client.
LECS An ATM term. LAN Emulation Configuration Server: This implements the policy controlled assignment of individual LE

clients to different emulated LANs by providing the LES ATM addresses.

LED Light Emitting Diode. A semiconductor diode which emits light when a current is passed through it. In lightwave transmission systems, light emitting diodes or lasers are used as sources of light. These devices are fabricated from multi layered structures of compound semiconductors epitaxially grown on a single-crystal substrate. LEDs are used as sources for optical data link applications in which the data rates are less than about 500 megabits per second and the transmission distances do not exceed a few kilometers. LEDs are also used in alphanumeric displays on calculators and computer devices. LEDs use less power than normal incandescent light bulbs, but more power than LCDs (Liquid Crystal Displays).

Leg My definition: A segment of a multipoint circuit which lies between any two of the points. Bellcore's definition: An object within a connection view that represents a communication path toward some addressable entity.

Legacy 1. All the stuff you have on hand — equipment, software, files and paperwork. In short, everything of a data processing/telecommunications nature in your business today. The use of the word "legacy" suggests that you've inherited all this stuff from previous generations of obsolete management. Most English dictionaries define the word legacy as "Anything handed down from the past. The idea is that you're forced to update it, without junking it altogether — which is expensive and potentially problematical. All you have is the legacy of previous generations. Preserve it, because no one can afford to junk it. Or so the theory goes."
2. Microsoft defines legacy in its Windows 95 Resource Kit as hardware and devices cards that don't conform to its Plug and Play standard.

Legal Holiday A call center term. Any holiday for which special wages are paid to employees who work on that day.

Lego Lego is a Danish-created child's game of many standard plastic pieces which fit together to make wonderful objects. I include this term in this dictionary because many people see computer telephony as "lego telephony." They see computer telephony comprised of standard building blocks — hardware and software — which fit together to make wonderful objects, just like LEGO. See LEGOWARE TELECOM.

Legoware Telecom The British term for what North Americans know as Lego Set or the British term Erector Set Telecom.

LEN Line Equipment Number.

Length The number of bits or bytes in a computer word, a field, a record, etc.

LEO Low Earth Orbit satellites. Also called LEOS, as in Low Earth Orbiting Satellites. Low Earth Orbit satellites move a few hundred miles in orbit around the earth. To maintain communications, you must have many of these satellites above you at any times. LEOs satellites are being promoted for functions as diverse as worldwide paging with acknowledgement, worldwide handheld telephone service and tracking cargo (with the truck sending up a continuous stream of info about its whereabouts). The primary advantage of LEOs is that the transmitting terminal — the one on earth — doesn't have to be very powerful, because the LEO satellite is so much closer than traditional geostationary satellites, which are satellites placed in an geosynchronous orbit — 22,300 miles (or 42,164 kilometers) directly over the earth's equator. See IRIDIUM and GLOBALSTAR.

LEOS Low Earth Orbiting Satellites. Motorola's proposed Iridium is one of these. See LEO and IRIDIUM.

LEP Large Electron Positron Collider.

LERG The Local Exchange Routing Guide.

LES An ATM term. LAN Emulation Server: This implements the control coordination function for the Emulated LAN, examples are enabling a LEC to join an ELAN, resolving MAC to ATM addresses.

Less Than Of smaller quantity. Of less importance. The word "less" is always confused with the word "fewer." According to the Oxford American Dictionary, the word "less" is used of things that are measured by amount (for example, eat less butter, use less fuel). Its use with things measured by numbers is regarded as incorrect (for example in "we need less workers;" correct usage is fewer workers.)

Letter Of Agency A letter sent by an end user to a telephone company — local or long distance — authorizing the end user's equipment vendor to deal on the end user's behalf with the phone company. A letter of agency is actually a specialized Power of Attorney.

Letters Shift A physical shift in a teletypewriter, specifically Telex, which enables the printing of alphabetic characters.

Level 1. The power of a signal measured at a certain point in the circuit.
2. Your management position (i.e. "level" in the management structure) in a telephone company. In AT&T and members of the operating Bell telephone companies, employees are identified by their "Levels." At the bottom of the totem pole are crafts people, the installers, the repair people, the trench diggers, etc. They do not have a level. They are often unionized. Management begins one level above the union. They are called first level. They are often called supervisors. Above them are second level managers. Above them are third level managers. They are called district managers. Above them are fourth level managers. They are division managers. Fifth level managers are assistant vice presidents. Sixth level managers are vice presidents. Above vice presidents, levels get fairly vague. Salary is contingent upon level. There are several levels with different salary levels within each level. It is not uncommon for AT&T or for a Bell operating company to have as many as 16 different management levels. At one stage, there was talk about eliminating the fourth level altogether.
3. As wiring got to carry faster and faster data flows, so the quality of wiring has become increasingly important. Thus more and more companies have started specifying cabling standards. Here is a series of standards, which Anixter has promoted:
• Level 1 VOICE
Level 1 cables are MADE to meet minimum telecommunication cable requirements. Typical uses include analog and digital voice plus low speed data (20 Kbps). Plenum constructions are available in shielded and unshielded designs while PVC constructions are available in shielded designs only.
• Level 2 ISDN & LOW SPEED DATA
Cables support the IBM Type 3 Media requirement. Most uses are defined through the IBM Cabling System guidelines. This specification defines electrical requirements through 1 MHz. These products are available in both plenum and PVC UTP (unshielded twisted pair) constructions. There are no shielded options in Level 2.

- Level 3 LAN & MEDIUM SPEED DATA
These products support the ANSI/EIA/TIA-568 Commercial Building Telecommunications Wiring Standard specification horizontal cable (also known as Category 3). This standard defines cable performance through 16 MHz and thus supports high speed LAN applications. Shielded constructions are available.
- Level 4 EXTENDED DISTANCE LAN
Level 4 identifies the first 100 ohm premises cables specifically designed for LAN applications. Most UTP LANs require a higher degree of performance than the standard telecommunications design offers. Level 4 cables require performance testing through 20 MHz and provide outstanding crosstalk isolation and attenuation. They are ideal for extended distance 10Base-T and 16 Mbps Token Ring. The specification for Level 4 is referenced from TIA TR41.8.1 Category 4 and NEMA "Low Loss."
- Level 5   HIGH SPEED LAN 100 OHM
This level requires the ultimate design for 100 ohm UTP cable. TIA TR4 and the NEMA Premises Wiring Task Force have recently defined this new specification for 100 ohm cable tested through 100 MHz. These cables are intended to be used up to and including 100 Mbps CDDI applications.
- HIGH SPEED LAN 150 OHM - DGM
The 150 ohm shielded twisted pair (STP) data grade media is the cornerstone of the IBM Cabling System. In addition to the many IBM applications, this cable is now supported by a consortium of five system vendors for 100 Mbps twisted pair transmission until the ANSI X3T9.5 standard is complete.

Level 1 Relay Another name for a repeater. Level 1 indicates that the device operates at the lowest layer (physical layer), as defined by the Open Systems Interconnect (OSI) architecture.

Level 2 Relay Another name for a bridge. Level 2 indicates that the device operates at the second layer (data link layer), as defined by the Open Systems Interconnect (OSI) architecture.

Level 3 Relay Another name for a router. Level 3 indicates that the device operates at the third layer (network layer), as defined by the Open Systems Interconnect (OSI) architecture.

Level 7 Relay Another name for a gateway. Level 7 indicates that the device operates at the seventh layer (application layer) of the Open Systems Interconnect (OSI) architecture.

There are other standards. See also CATEGORY OF PERFORMANCE.

Level Mode Interrupt A method of transmitting an Interrupt Request from a PCMCIA Card to a socket using the IREQ signal. In this mode, the IREQ signal is asserted when the Card initiates an interrupt and is negated when the Host acknowledges to the PC Card that the interrupt has been serviced. The method of acknowledgment is specific to devices on the PCMCIA Card.

Level Sensitive Interrupt A host system interrupt which causes repeated interrupts as long as the interrupt request signal is in the asserted state and the interrupt request is not disabled. Used in Micro Channel Architecture bus hosts and available in EISA hosts.

Level Sensitive Interrupt Trigger These adjustable triggers are the key to the operation of the new UART chip, called the 16550. They determine both the amount of data (in bytes) that the UART can receive before generating an interrupt request and the remaining buffer space available to store additional, incoming data. See 16550 and UART.

LF Line Feed. ASCII character 10. This character is now identified in the ASCII code set as New Line. See NEW LINE.

LGC Line Group Controller.

LGCI ISDN Line Group Controller.

LGN An ATM term. Logical Group Node: LGN is a single node that represents the lowest level peer groups in the respective higher level peer group.

Liberation A Northern Telecom line of headsets designed for use with Meridian telephones and attendant consoles. The Liberation product line includes monaural and binaural styles. Northern Telecom says the following are traditional applications for headset use:
Receptionists. Console attendants. Telemarketers. Customer service reps. Order entry reps. Financial Service professionals. Stockbrokers. Sales reps and reservation agents.

Library 1. A file that stores related modules of complied code.
2. As in TELECOM LIBRARY, the source of all good books on telecommunications, voice processing and local area networks. Ask for your FREE catalog today. Call 212-691-8215 or 1-800-LIBRARY.

LIDB Line Information Data Bases which are being developed by the Regional Bell Operating Companies and all the local phone companies will include such services as Originating Line number Screening, Calling Card Validation, Billing Number Screening, Calling Card Fraud and Public Telephone Check. The LIDB systems contain all valid telephone and calling card numbers in their regions, and have the necessary information to perform billing validation. The LIDB systems operational. A national system connecting them all together started working at the beginning of 1992.

Life Cycle A test performed on a material or configuration to determine the length of time before failure in a controlled, usually accelerated, environment.

Lifeline Service A minimal telephone service designed for the poor and elderly to assure they can be reached by phone and have a "Lifeline" to the world in case of emergency. The basic rate for this limited service is less than standard rates and varies by state. Typically, Lifeline Service entitles you to a phone line, a listing in the directory and a minimal number of outgoing local calls, e.g. 10. Some people who are neither poor nor elderly, subscribe to Lifeline Service and use it for incoming calls — for an answering machine or a computer electronic mail or bulletin boards. There's no difference in the quality of service provided by Lifeline Rates and normal phone lines.

LIFO Last In First Out. A method of organizing queues. See LAST IN FIRST OUT.

Light Technically, light is electromagnetic radiation visible to the human eye. The term is also applied to electromagnetic radiation with properties similar to visible light, including the invisible near-infrared "light" (or more technically correct, radiation)

that carries signals in most fiber optic communication systems. Light consists of electromagnetic waves ordinarily applied to those having a wave length of from .000075 cm. (the red ray) to .000038 cm. (the violet ray).
Light Emitting Diode See LED.
Light Pen A video terminal input device which is a light sensitive stylus connected by a cable to the video terminal. The user brings it to the desired point on the screen surface and presses a button. A light pen is used to select options from a menu on the screen or to draw images by dragging the cursor around the screen on a graphics terminal.
Light Piping Use of optical fibers to illuminate.
Lightgate Service A BellSouth INtraLATA optical fiber-based private line services that allows high-volume customers integrated voice, data and video transmission. Lightgate is the equivalent of 672 voice or data, private line or dial up circuits.
Lightwave Communications Fiber Optic communications using light to carry information.
Lightwave Transmission This term now means laser communications systems shot through the air (as opposed to glass fiber). Also called "free space lightwave communications." Typically, a signal is radiated directly from a light transmitter to a receiver less than a mile away. Advantages to lightwave transmission: easy to install, no digging of cables, wide bandwidth, reliable, cheap, no FCC frequency clearance approvals required and the receiving and transmitting equipment occupy little space. Disadvantages: only works for a mile or so and is subject to attenuation (fading) from fog and dust. It's perfect for between downtown buildings, where installing cables is too expensive, too cumbersome, too slow, etc. See LASER.
LIJP An ATM term. Leaf Initiated Joint Parameter: Root screening options and Information Element (IE) instructions carried in SETUP message.
Like New A term used in the secondary telecom equipment business. It means in excellent condition. Under normal conditions, the like new equipment could pass as new (i.e., not used, but not necessarily in the O.E.M. packaging). See LIKE NEW REPAIR AND UPDATE.
Like New Repair And Update LNRU. A term in the industry which repairs telecom equipment. It means all equipment is repaired and updated to the current manufacturer's specifications. New plastic is used to refurbish to a "like new" status. Also added are a new coil cord, line cord and address tray. Included is a full diagnostic test with a burn-in (if required) and an operational system test. Definition courtesy Nitsuko America. See also REPAIR AND QUICK CLEAN and REPAIR, UPDATE AND REFURBISH.
LIM Link Interface Module.
LIM-EMS The abbreviation for Lotus Intel Microsoft-Expanded Memory Specification. A software technique that allows MS-DOS to access memory beyond one megabyte by mapping the memory into a window in an area that MS-DOS can access. LIM-EMS is one of the greatest techniques for speeding up getting in and out of programs. For example, when my calendar program called Maxi-Calendar is not running, it occupies only 7K of normal RAM and 350K of expanded RAM. When I need it, it swaps itself quickly out of expanded RAM into normal RAM, taking less than half a second. If I didn't have expanded memory, it would take as long as 15 seconds to swap the program onto and off my hard disk, which is the other alternative. LIM stands for Lotus/Intel/Microsoft, the founding organizations that developed the Expanded Memory Specification. AST Research was also part of the driving force behind EMS, though its name doesn't appear in the acronym.
Limited Distance Modem LDM. A special purpose conversion device designed to connect two DTEs (data communications devices) over a relatively short distance, typically up to several miles. An LDM is not really a modem since it does not perform a digital-to-analog conversion, but transmits a special type of digital signal to the other LDM on the circuit. Also called a line driver, local dataset or short-haul modem.
Limiter A circuit which shapes a signal sent through it to conform to certain preset tolerances, used in both audio and video to regulate signal flow and prevent overloading, which would lead to distortion and the introduction of spurious noise.
Line The word line is confusing. In traditional telecom, a line is an electrical path (two wires) between a phone company central office and a subscriber, usually with an individual phone number that can be used for incoming and outgoing calls. A line, in this definition, is the most common type of loop. In carrier systems, a line is the portion of a transmission system that extends between two terminal locations. The line includes the transmission media and associated line repeaters. A line is also used to indicate the side of a piece of central office equipment that connects to or toward the outside plant. The other side of the equipment is called the drop side. And finally, a line is a family of equipment or apparatus designed to provide a variety of styles, a range of sizes, or a choice of service features. The confusion over the word line starts with an office phone system. Some people believe a line to be the same animal as a trunk — i.e. the line coming in from the central office to the PBX. Other people think a line is an extension, i.e. the line from the PBX to the phone on the user's desk.
Line Build Out Because T-1 circuits require the last span to lose 15-22.5 dB, a selectable out put attenuation is generally required of DTE equipment (typical selections include O.O, 7.5, and 15 dB of loss at 772 KHz).
Line Card A plug-in electronic Printed Circuit (PC) card that operates lamps, ringing, holding and other features associated with one or several telephone lines or telephones in a telephone system.
Line Circuit The sensor in the CO which detects and advises the switching system that one of its subscribers has gone off-hook and wishes to make a call. One line circuit is dedicated to each line of each subscriber.
Line Conditioning A service offered by telephone companies to reduce envelope delay, noise and amplitude distortion. By doing this, you allow for transmission of higher speed data than over a traditional dial-up phone line.
Line Cord The connecting cord between the phone and the jack in the wall.
Line Current Electrical current measured on an idle telephone line. Typical range is 20 - 100 ma DC, with 40 - 50 ma considered optimum for proper operation of phone equipment.
Line Discipline Archaic term for communications protocol.
Line Driver A short haul communications device used when cable lengths between RS-232 devices begin to the alleged 50-foot RS-232 limit. A line driver is a signal converter that conditions the digital signal transmitted by an RS-232 interface to

346

ensure reliable transmission beyond the 50-foot RS-232 limit and often up to several miles; it is a baseband transmission device. Also called baseband modem, limited distance modem, or short-haul modem. See also LIMITED DISTANCE MODEM.

Line Equipment Equipment in a central office which is there to serve one phone line. That line equipment includes a line relay or equivalent which starts to work when the customer's telephone goes off-hook.

Line Feed The act of moving a cursor or the head or a printer or telex machine down one line. These days, on the keyboards of most equipment — Personal Computers, etc. — there's no single key that says "Line Feed." There's usually a key that's labelled "Enter" or "Return." This key does two functions — a line feed and a carriage return (i.e. sending the print head or cursor to the left hand side of the carriage or the screen). In many (but not all) programs, a line feed is control J or ASCII character 10. The name for this character has been changed to New Line. See NEW LINE.

Line Finder The first switching element of a step-by-step phone system which recognizes a calling party is waiting for dial tone to make a call, identifies the party, and connects that party to the switching system so that the processing of the call may begin. Normally line finders serve 100 or 200 subscriber lines.

Line Hit Electrical interference that causes a hit, that is, a loss or introduction of spurious bits into a data stream.

Line Hold Provides a winking, blinking flash on the line lamp at every telephone which has the line appearing on it.

Line Insulation Test LIT. A test performed from the central office, which measures resistance and voltages on local lines to find faults.

Line Link Frame LLF. An arrangement that permits a crossbar office to transmit dial pulse information over a line to a PBX for switching Direct Inward Dial (DID) calls to the indicated phone. See CROSSBAR SWITCHING

Line Lockout When a phone stays off-hook for longer than a predetermined time, line lockout provides some loud noise and then puts the phone line out of service — until someone puts the phone back on hook again.

Line Noise Spurious signals introduced into a line by static, or other imbalances in the circuit. Line Noise is the most common cause of "Hits" or problems in data calls.

Line Of Sight Some through the air transmission media — such as microwave, infrared, and laser — operate at a frequency which transmit in a perfectly straight line. Or in "line of sight." In other words, the area between a transmitter and a receiver must be clear of obstructions.

Line Pool A Line Pool is a specific group of lines in certain key systems used for making outside calls. In Northern Telecom's Norstar, three Line Pools give phone access to outside lines without taking up too many Line buttons on each phone instrument.

Line Powered Telephone equipment that is powered solely by the CO talk battery supplied in a standard phone line.

Line Preference User selects the line to be used simply by pressing the button associated with that line.

Line Printer A type of printer which prints an entire line of text at one time. This printer is obviously a high speed printer. It is used, for example, to print TELECONNECT Magazine's monthly mailing list

Line Protocol Rules for controlling transmission on a synchronous data transmission line. Includes rules for bidding for the line, for positive and negative acknowledgments, requests for retransmission, and transmitter time-outs.

Line Queuing Dial an outside line (typically a long distance line). It's busy. Your phone system will put you in queue for that line. The queue might involve your waiting a few seconds on hold; or it might involve your hanging up and having the phone system call you back. There are thus two types of Line Queuing — hold-on and callback queuing.

Line Ringing Provides the user with an audible indication of a call on a specific line that appears on his/her telephone.

Line Side Connection A connection of a transmission path to the customer side of a local exchange switching system. Line side connections generally provide support for standard POTS telephones and a variety of vendor specific proprietary feature phones.

Line Speed The maximum number of bits you can transmit over a line in a certain defined time, say one second.

Line Status Indication Provides a visual indication on an ECTS (electronic telephone set) telephone of the idle, busy, ringing or held state for each line appearing on the telephone.

Line Switching Another term for circuit switching. See CIRCUIT SWITCHING and SWITCHING.

Line Termination LT. An ISDN term. Electronics at the ISDN network side of the user-network interface, that complements the electronics equipment. The classic line termination devices is the NTI at the user side of the interface.

Line Turnaround Time The delay in a circuit as the direction of communications changes, usually in half duplex communications. When one side of the communications stops sending, there is a delay before the other party stars sending in return. This is the Line Turnaround delay.

Line Voltage Voltage measured on a telephone circuit; typically 48 volts DC when phone is idle. Voltage may be lower at great distance from the central office, or when carrier equipment is used to multiplex several phone lines on one pair of wires. Line voltage on PBX systems is typically 24 volts.

Linear Distortion Amplitude distortion wherein the output signal envelope is not proportional to the input signal envelope. This distortion is often caused by part of the signal being bounced off something, while part arrives free and clear. Thus the receiver hears the same signal but bits of it arrive earlier and later than other bits, causing distortion. Anyone coming across a more understandable definition, please send it to the editors for the next edition.

Linear Predictive Coding A speech coding method that analyzes a speech waveform to produce a time-varying filter as a model of the human vocal tract. See also Digital Signal Processing.

Linear Programming Techniques in Operations Research (OR) to find an optimum solution to a linear function, given certain restrictions and typically expressed in many equations. A typical linear programming problem might be to find the least expensive, most efficient route between various pick-up and drop-off points in a transportation route.

Linearity The straightness of a frequency response curve as an indication of true or accurate sound reproduction. In an A/D (Analog to Digital) or D/A (Digital to Analog), linearity measures the precision with which the digital output/input tracks the analog input/output.

Linearly Polarized A mode of operation of fiber optics for which the field components in the direction of propagation are small compared to components perpendicular to that direction.

Lineman A person who fixes the telephone company's outside aerial plant — typically the wires hanging from poles dotted across the country-side. See also LINEMAN'S CLIMBERS.

Lineman's Climbers Telephone pole climbing irons which are strapped to the telephone lineman's legs, allowing him or her to climb a wooden telephone line. You can tell when a pole has been climbed by the holes left in it by the lineman's climbers.

Lines A computer imaging term. The line tool draws straight lines, typically from point to point. Most paint packages let you continue lines in a fashion that permits rapid creation of polygons.

Lines Of Force The directional lines of magnetic or static field which represent the stresses.

Lines Per Minute The way of measuring the speed of a line printer. Like any measure of speed, the speed you will get from your printer may be different from what the manufacturer says. Your speed will depend on how fast you feed the printer from your computer — a function of how fast you're transmitting, what software you're running, how fast that software can get the information to be printed off your disk, etc.

Link 1. Another name for a communications channel or circuit. The ATM Forum defines link as an entity that defines a topological relationship (including available transport capacity) between two nodes in different subnetworks. Multiple links may exist between a pair of subnetworks. Synonymous with logical link.
2. A Windows command that takes several programs and subprograms that were meant to be used together, but were written separately, and combines them into one. Usually used to create an executable program out of modules that were not themselves directly executable.
3. Microsoft Windows 95 Resource Kit as (v.) to form a connection between two objects. (n.) for OLE, a reference to an object that is linked to another object, and for networking, a connection at the LLC layer that is uniquely defined by the adapter's address and the Destination Service Access Point (DSAP). Also refers to shortcuts.

Link Aggregation Token See AGGREGATION TOKEN.

Link Attached Describing devices that are connected to a network, a communications data link, or telecommunications circuit; compare with channel-attached.

Link Attribute A link state parameter that is considered individually to determine whether a given link is acceptable and/or desirable for carrying a given connection.

Link Connection An ATM term. A link connection (e.g., at the VP-level) is a connection capable of transferring information transparently across a link without adding any overhead, such as cells for purposes for monitoring. It is delineated by connection points at the boundary of the subnetwork.

Link Constraint A restriction on the use of links for path selection for a specific connection.

Link Control Facility A fiber channel hardware facility which attaches to the end of a link and manages the transmission and reception of data.

Link Converter A device for an InteCom S/80 which connects distributed switching modules to the centralized switching equipment through a coaxial cable or a fiber optic cable.

Link Layer Access Method The algorithm that determines when any given network interface in a PC/TCP local area network is allowed to transmit. It is also known as the access method. CSMA/CD is the access method for the Ethernet.

Link Metric An ATM term. A link parameter that requires the values of the parameter for all links along a given path to be combined to determine whether the path is acceptable and/or desirable for carrying a given connection.

Link Optimization ISDN feature that prevents administration packet from opening the communications link and allows only user data to open the line. Link optimization ensures that remote connections are not kept open unnecessarily, which saves usage costs.

Link Protocol The set of rules by which a logical data link is set up and by which data transfers across the link. It includes formatting of the data.

Link Set A group of signaling links directly connecting two signaling points.

Link State A type of routing protocol in which updates to routing tables are exchanged when modifications are made to the table, but are not exchanged on a highly regular and predetermined basis. Link State protocols consume less networking resources and reduce network congestion by only providing updates when needed, and only sending the changes. Periodically, the entire route table will be sent as a precautionary procedure.

Link State Parameter Information that captures an aspect or property of a link.

Link Status Signal Unit LSSU. A packet sent between MTPs (Message Transfer Part of the SS7 Protocol) to provide SS7 information about the sending node and its links. This information is sent during the initial alignment of the links, when there is an associated processor outrage, or when link congestion is detected.

Link Time This is a specific time delay that allows access to PBX or Centrex features through a telephone system. Link Time is also referred to as a Hookswitch Flash or Recall.

Linked Object A representation or place holder for an object that is inserted into a destination document. The object still exists in the source file and, when it is changed, the linked object is updated to reflect these changes.

Links 1. The transmission portion of the local loop.
2. An affectionate name for Apple's electronic mail system, called AppleLink.

Linux Linux (Officially pronounced Lih-nucks) is the creation of Linus Torvalds, who started the Linux kernel as a research project at the University of Helsinki, Finland. Since then, Linux has evolved into a full-featured, powerful and robust Unix operating system. Linux is among the most powerful and feature-packed Unix operating systems available for the PC, offering a large base of hardware peripheral compatibility. Linux hosts an impressive array of compilers and development environments, including C/C++, Perl, Pascal, SmallTalk, and complete X-windows system that rivals many commercial offerings. For most of its life, Linux has been shareware. Linux has many script languages and parsers such as Awk, Sed, Yacc as well as all popular shells (Borne, Korn, C, BASH, etc.). The Linux kernel was written for the Intel 386/486/Pentium processor. In many ways, Linux is a better performing UNIX than many of the commercial versions of UNIX. Ray Noorda, one of the genusies behind Novell, has recently become involved with Linux and is commercializing it. For more information, contact Linux Journal, Seattle, Washington.

Liquid Crystal Display LCD. A low power display that aligns material suspended in a liquid under the influence of a low voltage so it reflects ambient light and displays alphanumeric characters. LCD displays are finding great use as methods of displaying information on new electronic telephones, especially those positioned behind PBXs. The advantage of putting such displays on telephones is that the power to drive the display is very small. The display can be line powered — i.e. powered by the one or two pairs coming from the PBX. This avoids the necessity and cost of a transformer/rectifier — the little black box you plug into the wall to run your answering machine or to power up your rechargeable calculator/laptop computer. Such LCD displays on electronic phones can perform many functions. The most useful is that of "walking" the user through the phone call — showing him/her how to transfer a call, to make a conference call, to split a conference, etc. An LCD can also alert you as to who's calling you.

LIS Link Interface Shelf.
List Box A Windows term. In a dialog box, a box that lists available choices, for example, a list of all files in a directory. If all the choices do not fit in the list box, there is a scroll bar.
List Server An automated mailing list distribution system.
Listed Directory Number LDN. Incoming exchange network calls to the PBX via assigned listed local telephone directory number are directed to the attendant.
Listen Before Talk LBT. Same as carrier sense multiple access (CSMA). Compare with LISTEN WHILE TALK.
Listen While Talk LWT. Same as carrier sense multiple access with collision detection (CSMA /CD). Compare with LISTEN BEFORE TALK.
LIT See LINE INSULATION TEST.
Lithium Ion Type of highly efficient rechargeable battery, often used in computer laptops and cellular portables.
Little Endian A format for storage or transmission of binary data in which the least significant byte (bit) comes first.
LIU Line Interface Unit.
LIU7 Line Interface Unit for CCS7.
Live Bug Colloquialism used to refer to a leaded integrated circuit package when the leads are down, like a bug that is alive and standing upright.
Live Call Screening See LCS.
Livelock A request for an exclusive lock that is repeatedly denied because a series of overlapping shared locks in a shared database keeps interfering. A SQL server will detect the situation after several denials, and refuse further shared locks.
Liveware People.
LLB Line LoopBack. A maintenance and/or diagnostic mode of operation whereby a CSU regenerates a signal received from a span line and retransmits that signal back onto the span towards its point of origin.
LLC Logical Link Control. A protocol developed by the IEEE 802.2 committee for data-link-level transmission control. It is the upper sublayer of the IEEE Layer 2 (OSI) protocol that complements the MAC protocol. IEEE standard 802.2 includes end-system addressing and error checking. It also provides a common access control standard and governs the assembly of data packets and their exchange between data stations independent of how the packets are transmitted on the LAN. See 802 STANDARDS.
LLC2 Logical Link Control 2. The frame format used to carry 3270 traffic on Token Ring LANs.
LLF 1. Line Link Frame. See CROSSBAR SWITCHING.
2. Low Layer Functions.
LM Long distance Marketer.
LMDS LEC Message Distribution System.
LMI 1. Local Management Interface. A specification for the use of frame-relay products that define a method of exchanging status information between devices such as routers.
2. Logical Modem Interface. The core of the Microsoft Fax interface. LMI lets third-party licensed vendors write plug-in modules to provide instant and transparent access to diverse underlying systems. An easy analogy for the LMI is to consider the Windows print manager. To the user, simply installing the printer driver suited to their printer is all that is required. According to Microsoft, "The LMI interface provides a similar layer between the internal fax components of Windows 95 and the fax hardware or, in our case, the fax server."
LMOS Loop Maintenance Operations System.
LMS Local Message Switch.
LMSS Land Mobile Satellite Service.
LMU Line Monitor Unit.
LNA Low Noise Amplifier.
LNNI An ATM term. LANE NNI: The standardized interface between two LAN servers (LES-LES, BUS-BUS, LECS-LECS and LECS-LES).
LNP Local Number Portability.
LNRU Like New Repair and Update. A term in the industry which repairs telecom equipment. It means all equipment is repaired and updated to the current manufacturer's specifications. New plastic is used to refurbish to a "like new" status. Also added are a new coil cord, line cord and address tray. Included is a full diagnostic test with a burn-in (if required) and an operational system test. Definition courtesy Nitsuko America. See also REPAIR AND QUICK CLEAN and REPAIR, UPDATE AND REFURBISH.
LO Local Operator. In the PCS sense, a functional entity providing local wireless service to customers in a geographical region.

The LO (Local Operator) is serviced by the National Services Organization in the PCN (Personal Communications Network) for long-distance communications and for marketing/sales.

LOA Letter Of Agency. A letter that you give to someone whom you allow to represent you and act on your behalf. For example, a letter of agency is used when your interconnect company orders lines from your local phone company on your behalf. Letters of Agency are also used when companies switch their long distance service from one carrier to another.

Load 1. The act of taking a program or data from external storage — a cassette, a floppy or hard disk, etc. and storing it in the computer's main RAM memory.
2. The load is any electric or electronic appliance or gadget plugged into an AC electrical outlet. It completes the circuit from the transformer through the hot conductor, to the load, through the neutral conductor and back to the utility transformer. See AC, AC POWER, GROUND and GROUNDING.

Load Balancing The practice of splitting communication into two (or more) routes. By balancing the traffic on each route, communication is made faster and more reliable. In telephone systems, you can change phone and trunk terminations in order to even out traffic on the network. An example: You have a PBX of three separate cabinets, each of which are joined by tie lines. Instead of having each cabinet serve anyone in the building, you might figure which groups talk to each other the most and concentrate them into specific cabinets. The objective is to maximize the number of calls that can be handled inside each cabinet and reduce the number of calls that need to travel between the cabinets. This makes the calls go faster and reduces the need for inter- cabinet lines.

In data internetworking, bridges and routers perform load balancing by splitting LAN-to-LAN traffic among two or more WAN links. This allows for the combination of several lower speed lines to transmit higher speed LAN data simultaneously. In local area networking, load balancing is a function performed by token ring routers. In data networking, load balancing can also be a form of inverse multiplexing where data packets are alternated over all available circuits. At the receiving end, the packets are reassembled in their proper order.

In disk arrays, load balancing means using multiple power supplies within a disk array so that power usage is spread equally across all the power supplies. The failure of one supply will not cause the entire array to fail.

Load Coil A bridged tap is multiple appearances of the same cable pair at several distribution points. A bridged tap is any section of a cable pair not on the direct electrical path between the central office and the user's offices. A bridged tap increases the electrical loss on the pair — because a signal traveling down the pair will split its signal between the bridge and main pairs. Since most existing telephone cable pair is bridged, the phone company puts loading coils in the circuit. The effect of load coils is to modify the loss versus frequency response of the pair so it is nearly constant across the voice band. This works for voice. However the loss above the voice band due to load coils increases rapidly. ISDN and other digital circuits operates above the voice band. So, when the phone company installs digital circuits, it must remove the load coils. Bridged taps, on the other hand, are acceptable to some degree, especially with slower speed digital circuits, such as ISDN. But you can't run high-speed digital circuits, e.g. T-1, over cable that has bridged taps in it. See BRIDGE and LOADING COIL.

Load Leveling Distributing traffic over more than one route.

Load Number Load number is the Canadian equivalent of the U.S. concept of Ringer Equivalence. The idea is that each phone or "phone thing" you buy (e.g. answering machine) comes with a number. You add the numbers together and if you get above a certain number, you are drawing too much current and none of the bells on the phones will ring. In Canada, single line phones are typically rated at 10 for the newer ones with electronic "bells" or 20 for the older electro-mechanical ones with real metal bells. In Canada, the rule is not more than 100 points on a line. In the U.S., phones are typically one and the rule is not more than five points on a line.

Load Service Curves The output from load and stress testing on a computer telephony system is a set of load service curves. Load service curves identify how individual areas of the system respond under various load amounts. Traffic is provided to the computer telephony system at defined steps (perhaps at 1,000 call per hour increments) until the system design threshold is reached. Measurements are taken at each step, and usually shown graphically, in a "curve". Most computer telephony systems are designed to handle up to a specified number of busy hour calls with specific response times. For example, the time that passes between the point in time a caller enters a DTMF digit and the point the computer telephony system speaks a response should usually be no more than 1 second 97% of the time, nor more than 3 seconds 99% of the time. A load service curve would be used to illustrate the response time at each step of increasing load. When the load curve shows the response time is slower than the above parameters, the system has reached its capacity. Of course, the load placed on the system must accurately mimic the real world load the system will experience or it is largely meaningless. This definition from Steve Gladstone, president, Hammer Technologies, makers of fine computer telephony testing systems, 508-694-9759.

Load Sharing In data processing, load sharing is the technique of using two computers to balance the processing normally assigned to one of them. In local area networking, load sharing is performed by token ring routers when connecting remote LANs. It allows combining Ethernet and Token Ring traffic over a common WAN (Wide Area Network) link such as T-1 or 56 Kbps circuit. Loads sharing eliminates the need for duplicate WAN links (and bridges or routers) each serving a different type of LAN.

Load Testing Also known as stress testing, the goal of load testing is to make sure the system will meet or exceed its busy hour load capacity objectives under all operating conditions. This requires stressing the system in incremental steps until it breaks and understanding what happens when the system is operating under its full rated transaction load? Beyond its load? Does it slow down? How? Does it fail? Where? How is service restored after an outage? Is service restoration graceful or must the system reboot? Is restart manual or will the system reset itself? Individual load tests may be performed to understand the impact of load on specific system bottlenecks. Most significant architectural problems will come to light under load testing. It is critical that any load placed on a computer telephony system be dynamic, and mimic the load characteristics the system will experience under real-world usage and conditions. See also DYNAMIC LOAD TESTING and LOAD SERVICE CURVES.

Loaded Line A telephone line equipped with loading coils to add inductance in order to minimize amplitude distortion. See LOADING and LOADING COIL.

Loading A method of improving the voice quality of a phone line. Telephone companies put load coils on local lines. What this loading does is to insert inductance in a local loop circuit to offset the effect of capacitance in the cable. Loading "tunes" the circuit to the voice frequency band (500 to 2500 Hz) and thus improves the quality at the expense of overall bandwidth. You usually have to ask that the loading coils be removed if you're planning to transmit high-speed data exclusively on that circuit. See LOADING COIL.

Loading Coil An induction device employed in local loops exceeding 18,000 feet in length, that compensates for wire capacitance and boosts voice grade frequencies. Loading coils are often removed for higher speed data services, as distortion will occur at frequencies higher than those used for voice. See LOADING.

Loading High A memory management verb for loading a device driver or TSR (Terminate and Stay Resident) program into upper memory, out of conventional memory. Under DOS, the loading high commands are DEVICEHIGH for device drivers and LOADHIGH (or LH) for TSRs. Third party memory managers use their own routines to load high, though they can sometimes borrow DOS commands.

Loadware A Rolm term for software that can be downloaded directly to the Rolm 9751 CBX high-density circuit cards.

LOC An ATM term. Loss of Cell Delineation: A condition at the receiver or a maintenance signal transmitted in the PHY overhead indicating that the receiving equipment has lost cell delineation. Used to monitor the performance of the PHY layer.

Local Access Transport Area LATA. The geographical area within which a Local Exchange Company (a local phone company) can provide communications services. California is divided into 10 LATAs. Currently, local phone companies, such as Pac Bell, have a monopoly on all calls, including toll calls, made within a LATA.

Local Airtime Detail This cellular telephone carrier option (which means it costs money) provides a line-itemized, detailed billing of all calls, including call attempts and incoming calls to the mobile. What you get for free is generally a non-detailed, total summary of all calls.

Local Area And Transport Area See LATA.

Local Area Data Transport LADT. A service of your local phone company which provides you, the user, with synchronous data communications.

Local Area Network LAN. A short distance data communications network (typically within a building or campus) used to link computers and peripheral devices (such as printers, CD-ROMs, modems) under some form of standard control. Older data communications networks used dumb terminals (devices with no computing power) to talk to distant computers. But the economics of computing changed with the invention of the personal computer which had "intelligence" and which was cheap. LANs were invented as an afterthought — after PCs — and were originally designed to let cheap PCs share peripherals — like laser printers — which were too expensive to dedicate to individual PCs. And as time went on, what LANs were used for got broader and broader. Today, LANs have four main advantages: 1. Anyone on the LAN can use any of the peripheral devices connected to the LAN. 2. Anyone on the LAN can access databases and programs running on client servers (super powerful PCs) attached to the LAN; and 3. Anyone on the LAN can send messages to and work jointly with others on the LAN. 4. While a LAN does not use common carrier circuits, it may have gateways and/or bridges to public telecommunications networks. See LAN MANAGER, TOKEN RING and ETHERNET.

Local Area Signaling Services LASS is a group of central office features provided now by virtually all central office switch makers that uses existing customer lines to provide some extra features to the end user (typically a business user). They are based on delivery of calling party number via the local signaling network. LASS can be implemented on a standalone single central office basis for intra office calls or on a multiple central office grouping in a LATA (what the local phone companies are allowed to serve) for interoffice calls. Local CCS7 (Common Channel Signaling Seven) is required for all configurations. The following features typically make up LASS:

Automatic Callback: Lets the customer automatically call the last incoming call directory number associated with the customer's phone when both phones become idle. This feature gives the customer the ability to camp-on to a line.

Automatic Recall: Lets the customer automatically call the last outgoing call currently associated with the customer's station when both stations become idle. This feature gives the customer the ability to camp-on to a line.

Customer-Originated Trace: Lets the terminating party request an automatic trace of the last call received. The trace includes the calling line directory number and time and date of the call. This information is transmitted via an AM IOP channel to a designated agency, such as the telephone company or law enforcement agency.

Individual Calling Line Identification: Consists of two distinct features: 1. Calling Number Delivery which transmits data on an incoming call to the terminating phone. 1. Directory Number Privacy which prevents delivery of the directory number to the terminating phone.

Also, LASS has some selective features:

Selective Call Acceptance: Allows users to restrict which incoming voice calls can terminate, based on the identity attribute of the calling party. Only calls from parties identified on a screening lists are allowed to terminate. Calls from parties not specified on a screening list are rerouted to an appropriate announcement or forwarded to an alternate directory number.

Selective Call Forwarding: Allows a customer to pre-select which calls are forwarded based on the identity attribute of the calling party.

Selective Call Rejection: Allows a customer to reject incoming voice calls from identity attributes which are on the customer's rejection list. Call attempts from parties specified on the rejection list are prevented from terminating to the customer and are routed to an announcement which informs the caller that his/her call is not presently being accepted by the called party.

Selective Distinctive Alert: Allows a customer to pre-select which voice calls are to be provided distinctive alerting treatment based on the identify attributes of the calling party.

Users can, at their convenience, activate or modify any of these features by sending commands to the central switch from their existing touchtone telephones.

Local Automatic Message Accounting LAMA. A combination of automatic message accounting equipment and automatic number identification equipment in your telephone company's central office and used by them to bill your local phone calls.

Local Battery Having "local battery" means the telecom equipment — the telephone, the PBX, the key system, etc. — has its own source of power and does not draw from the power coming down the phone line. The term came from telegraphy and was used to distinguish the battery which provided power to the telegraphic station as against the power that went to drive the line and the signal traveling down it. See BATTERY.

Local Bridge A bridge between two or more similar networks on a local site (within same building).

Local Bus A microprocessor inside a PC must communicate with certain integral devices, including memory, video controllers, hard disks. This is typically called an internal bus. That is to distinguish it from the "external" bus, such as the AT, ISA, EISA, MCA buses, which define the communications between the motherboard and the various peripheral devices, such as the I/O cards like those handling modems and LAN connections. As microprocessors have gotten faster, so they have begun to outpace the speed of their computer's internal bus, which has tended to narrow the stream of data in and out of the CPU, slowing the computer. A Local Bus is a new type of internal bus. It is a faster bus. The idea is to get a broader path between your critical components — memory, video and disk controller — and your microprocessor. The idea is to get the data in and out of the microprocessor at the same speed as the microprocessor's system clock. Local Bus is an emerging standard. See also EISA, PCI and VESA.

Local Call Any call within the local service area of the calling phone. Individual local calls may or may not cost money. In many parts of the US, the phone company bills its local service as a "flat" monthly fee. This means you can make as many local calls per month as you wish and not pay extra. Increasingly this luxury is dying and local calls are costing money.

Local Call Accounting Computes the dollar amount for local calls based on the total message units stored for each phone.

Local Call Billing Computes the dollar amount for local calls placed by guests based on total message units.

Local Central Office Switching office in which a subscriber's lines terminate.

Local Channel Controller An AT&T name for its family of 3270 compatible cluster controllers.

Local Composite Loopback In network management systems. Composite loopback test that forms the loop at the output of the local multiplexer that returns transmitted signals to their source. See loopback.

Local Dataset Signal converter that conditions the digital signal transmitted by an RS-232 interface to ensure reliable transmission over a dc continuous metallic circuit without interfering with adjacent pairs in the same telephone cable. Normally conforms with Bell 43401. Also called baseband modem, limited distance modem, local modem, or short-haul modem. See line driver.

Local Distribution Frame LDF. Another word for an Intermediate Distribution Frame. It's a device for cross connecting cables — from one thing to another. On one side of the LDF are the pairs from individual phones in that part of the building or area. On the other side are trunks coming in from a central office or cables coming in from the central, larger PBX. LDFs typically help with the organization of cables in a building or area. See also INTERMEDIATE DISTRIBUTION FRAME.

Local Echo A modem feature that enables the modem to send copies of keyboard commands and transmitted data to the screen. When the modem is in Command mode (not online to another system) the local echo is invoked through the ATE1 command. The command causes the modem to display your typed commands. When the modem is online to another system, the local echo is invoked through the ATFO command. This command causes the modem to display the data it transmits to the remote system.

Local Exchange The telephone company exchange where subscribers lines are terminated. Also called an "End Office."

Local Exchange Carrier A local phone company. See also LEC.

Local Heap A memory storage area limited to 64K in size.

Local Loop The physical wires that run from the subscriber's telephone set, or PBX or key telephone system, to the telephone company central office. Increasingly, the local loop now goes from the main distribution frame in the basement to the phone company. And the subscriber is responsible for getting his/her wires from the box in the basement to his phone, PBX or key system.

Local Management Interface LMI. A specification for the use of frame-relay products that define a method of exchanging status information between devices such as routers.

Local Measured Service LMS. Years ago virtually all phone lines in the United States were FLAT RATE. That meant that for a fixed amount of money each month, you, the customer (a.k.a. subscriber) were allowed to make as many local calls as you wanted. For many reasons, the U.S. phone industry has progressively moved to LOCAL MEASURED SERVICE for local calls. Typically this means that for a fixed amount of money each month, you, the customer, can receive as many calls as you want and can make a finite number of outgoing local calls — typically 50. Each additional call beyond the 50 (or whatever the number is) costs extra. How much that call costs depends on the distant the call travels, the time of day, the day of the week, and the local company's tariffs.

Local Net The broadband architecture used in Sytek's work. Also the product name of their network. Sytek is in Sunnyvale, CA.

Local Order Wire A communications circuit between a technical control center and selected terminal or repeater locations.

Local Phone Service When I dial the pizza store on the corner, I'm making a local phone call. But when I'm calling the pizza store 50 blocks away, is that a local phone call? Answer, it could be. It depends. What's local phone service? What do you charge for it? Once upon a time, most Americans didn't pay for local phone service. They paid a flat monthly fee. Then the phone companies needed money, so they started charging for local service. A few cents per call. Then the phone companies timed the call and charged more the longer you talked. Then they started charging for longer local calls — maybe for calls of ten miles and further. In short, the definition and pricing of local phone calls is changing. Now local calls are looking increasingly like long distance calls — charged by time, distance, and day of the week.

Local Printer A printer that is directly connected to one of the ports on your computer.

Local Service Area The geographic area that telephones may call without incurring toll charges. A flat rate calling area. Increasingly rare.

Local Switch The term local switch refers to the switch (PBX, ACD, dumb) to which the computer telephony system is directly connected. Usually, the local switch will provide better integration with the computer telephony system (more comprehensive call data) than connections that take place over the network. Additionally, the local switch will have both line and trunk side connections and will also support connection to whatever agents or desktop users that may use the computer telephony application.

Local Tandem A central office, usually in large metropolitan areas, serving as a transit switch between noncontiguous class 5 exchanges. It connects end office trunks.

Local Test Desk A testing system that is used to test local loops and central office subscriber line equipment from a central point, typically a central office.

Local Trunk Trunks between Class 5 offices (local central offices).

Locality A measure of how close commonly-accessed files are to one another on a hard disk. "High locality" means the files reside on sectors or tracks which are close to each other. When this is the case, seek times during operation are shorter than average.

LocalTalk Apple Computer's proprietary local area network for linking Macintosh computers and peripherals, especially LaserWriter printers is called Appletalk. AppleTalk's LAN hardware is called LocalTalk. Appletalk is a CSMA/CD network that runs at 230.4 kilo bits per second and is therefore, incompatible with any other local area network. It is also a lot slower than the present top speeds of Ethernet (10 Mbps) and Token Ring (16 Mbps). Outside manufacturers, however, make gateways which will connect an Appletalk LAN to other local area and telecommunications networks — LANs, WANs and MANs. See also APPLETALK.

Location One definition is the place where a telephone jack is located. This location is given a number. The wire going to that location is given a number. All this in hope of being organized for installation, moves, changes and maintenance.

Locator A term used in the secondary telecom equipment business. A locator is a company that assists both a buyer and seller to quickly find each other. A locator contracts with dealers to provide them with daily lists of potential customers. The list develops from phone calls to an 800-number asking for a specific component.

Lock And Load Originally a military term. Then it became software speak for freezing code on a program in development. Then it became "Let's make a decision and get on with it."

Lock Code The lock code locks a cellular telephone to prevent unauthorized use. The lock code is programmed into the NAM (Numerical Assignment Module) and is frequently factory set to either 1234 or 00004.

Lock On The process by which an earth station initially acquires the signal from a satellite.

Lock Out In a satellite telephone circuit controlled by an echo suppressor, one or both subscribers can't get through because of excessive noise at one end. You get this also with speakerphones. The person with the speakerphone can simply hog the conversation because his speakerphone keeps transmitting his voice. There are weird variations on this. Sometimes you might call someone on your speakerphone and wait for them to pick up. They do. They shout into the phone "I'm here." All they can hear is you at the other end talking or typing. The sound at your end is hogging the channel and thus locking out the person at the other end. The solution? Turn the "mute" button on your speakerphone. This will stop your end transmitting and allow the other end to say "Hello, I'm here." See also LOCKOUT.

Locked Resources An Intel Plug and Play term. Resources that must be used by the same card each time the system is booted. The configuration manager cannot assign these resources to any other card.

Locker Telco-speak for a storage area, often an urban storefront, where phone company installers and repairman can pick up and drop off tools and installation material such as phones, wire and hardware. These places are prime targets for burglary and robbery, and often have no identification to show their valuable contents.

Locking Preventing several people getting to and changing the same data in a shared database simultaneously. Locks may be permanent and prevent access completely, or they may be "advisory." A user is warned the data is being used by someone else and that the data is not presently available. Locks prevent the destruction of data that can occur if two people access a file at the same time. In any data base or other computer system, there are typically two types of "locks" — record and file locks. A record lock occurs when an airline agent pulls up your travel plans. No other travel agent can access those records at that time. A file lock occurs when the whole file is locked up. This might occur in a centralized word processing program. The whole document will be locked when it is being used by someone.

Lockout A PBX feature. Denies the attendant the ability to re-enter an incoming central office connection directly terminated or held on her position, unless specifically recalled by the phone user.

LOF An ATM term. Loss of Frame: A condition at the receiver or a maintenance signal transmitted in the PHY overhead indicating that the receiving equipment has lost frame delineation. This is used to monitor the performance of the PHY layer.

Log In The process of identifying and authenticating oneself to a computer system. Used to control access to computer systems. See LOGIN SCRIPT.

Log Off 1. To type in the needed keystrokes for ending a session that's on-line with a computer. Often those keystrokes are "Logoff." Usually it's very easy to Log Off. It's more difficult to Log On.
2. Employees wear their photo-ID badges in little plastic holders attached by a clip to their clothing. It's a good idea to "log off" when you're in public. This means to turn your badge around so nobody can see whom you work for.

Log On To enter the needed keystrokes to start an on-line session with a computer. "Logging On" may be done with a computer that's local or one that's long distance and your work is done over communications lines.

Logic Logic is the application of mathematical analysis and deductive reasoning to propositions that may or may not be true or false. The logic we're interested in owes much to the work of George Boole in the mid-19th century. He formulated a system that could be applied to the relationships between propositions to which only a binary choice of truth existed, i.e. yes or no. The first application of the Boolean Algebra that derived from this was Shannon's research into the analysis of relay switching circuits in 1938.

Logical Block Address A logical block address is a sequential address for accessing blocks on storage media. The first block of the media is addressed as block 0 and succeeding blocks are numbered sequentially until the last block is encountered. This is the traditional method for accessing peripherals on a SCSI bus.

Logic Bomb Program routine that destroys data. For example, a logic bomb may reformat the hard disk or insert random bits into data files. It may be brought into a personal computer by downloading a corrupt public domain program. Once executed, it does its damage right away and then stops, whereas a virus keeps on destroying. Another definition of a logic bomb is that it is a A resident computer program that lies dormant for a period, and then triggers an unauthorized act when a certain event, such as a date, occurs.

Logical Bus A LAN topology, such as Ethernet, which shares a common communications channel.

Logical Channel X.25 concept referring to a virtual connection operated over a physical connection that can support one or more virtual connections simultaneously.

Logical Channel Number Virtual circuit identified at the packet level of X.25. See LOGICAL CHANNEL.

Logical Drive A disk drive recognized by the operating system. A computer's logical drives may differ from its physical drives. For example, a single hard disk drive may be partitioned into two or more logical drives. Before MS-DOS 5.0, the biggest logical "drive" that DOS could address was 32 megabytes. If you had a larger hard disk, you partitioned the bigger driver into logical drives of 32 megabytes.

Logical Formatting The third step in structuring a data medium so that data may be written to it. Logical formatting must follow physical formatting (also called low-level formatting) and partitioning (figuring into how many drives you wanted to slice the one drive into).

Logical Group Node An ATM term. A logical node that represents a lower level peer group as a single point for purposes of operating at one level of the PNNI routing hierarchy.

Logical ID An AT&T Merlin term. A numbering sequence used to identify station and trunk locations on the communications system control unit.

Logical Link An abstract representation of the connectivity between two logical nodes. This includes individual physical links, individual virtual path connections, and parallel physical links and/or virtual path connections.

Logical Link Control LLC; A protocol developed by the IEEE 802 committee, common to all of its LAN standards, for data link-level transmission control; the upper sublayer of the IEEE Layer 2 (OSI) protocol that complements the MAC protocol; IEEE standard 802.2; includes end-system addressing and error checking.

Logical Modem Interfaces LMIs are to Microsoft's MAPI what SPIs are to TAPI. LMIs serve network-based fax servers and multi-port fax boards. MAPI sits on top of Microsoft's Exchange and other MAPI-compliant messaging systems.

Logical Node An abstract representation of a peer group or a switching system as a single point.

Logical Node ID A string of bits that unambiguously identifies a logical node within a routing domain.

Logical Provisioning An AT&T term for the establishment of network services by changing software controls, rather than by physically installing or rearranging hardware.

Logical Ring A network which is treated logically as a ring even though it maybe cabled as a physical star topology.

Logical Unit Interface See LU 6.2.

Logiciel French for software.

Login Script When users log into a local area network, they may wish to do many things or the network supervisor may wish them to do several things. These commands are part of something called a "login script." In computerize (Novell's words), a login script contains commands that initialize environmental variables, map network drives, and/or control the user's program execution. Login scripts are similar to batch files. The familiar AUTOEXEC.BAT can be thought of as an MS-DOS login script.

LOI Letter of Intent

Long Distance Any telephone call to a location outside the local service area. Also called toll call or trunk call.

Long Haul Communications That type of phone call which reaches outside a local exchange or serving area.

Long Key A long key is a character held down for a prescribed period of time. The time period is generally longer than the time for other keys. In call reorigination, the # sound is defined as a "long key." This definition was kindly provided by Karen Shelton, Systems Engineer, IEX Corporation, Richardson, TX.

Long Lines AT&T Long Lines. The department of AT&T which operates long distance toll service. It is no longer called Long Lines. It is called AT&T Communications.

Long Reach Long reach refers to optical sections of approximately 25 km or more in length and is applicable to all SONET rates.

Long Tones First, we invented touchtone, also called DTMF, Dual Tone Multi Frequency tones. You'd punch your number with tones, instead of dialing them. Then someone thought you could control telephone response gadgets, like voice mail, interactive voice response, etc. with touch tones. For these gadgets to work, they had to "hear" the tones you sent. No one really set standards as to the minimum length tone they would hear. But it was generally conceded that they were to be 120 milliseconds. So some manufacturers of telephone equipment started to make phone equipment that, if you pushed a touchtone button, the machine would only sent a touchtone of 120 millisecond duration. That was called a short tone. It wasn't very useful because the manufacturers quickly discovered that many pieces of equipment couldn't respond that quickly. And the manufacturers got complains that their customers couldn't call their voice mail, their bank, etc. As a result, some manufacturers of equipment brought out new hardware (replacing the old) to allow you to send "long tones," which are now defined as touchtones that last for as long as you hold down the button — just as it is (and has always been) on a normal single line, non-electronic, non-digital telephone. Isn't progress wonderful? See also DTMF for a much longer explanation of tone dialing.

Long Wavelength Light whose wavelength is greater than about 1 um.

Longest Available This is a method of distributing incoming calls to a bunch of people. This method selects an agent based on the amount of time that each agent has been on the phone. This allows for an equitable distribution of calls to each agent. See also TOP DOWN and ROUND ROBIN.

Longevity Testing When you're building a computer telephony system, many problems do not come to light until it's been running under high traffic for a long term. According to Steve Gladstone, author of the book "Testing Computer Telephony Systems," (available from 212-691-8215), there can be slow memory leaks, counter overflows, or disk fragmentation which slows down data access, system resets that cause calls to be dropped inadvertently, and so on. Frequently these issues are discovered only after thousands of calls are placed or after some random call pattern has occurred. Longevity tests, according to Gladstone, provide a high load to a system over an extended period of time. The most effective way to do this is to run a load test based on your busy hour usage profile over several days and track failures over the entire testing period. If failure rates increase, or specific failures occur, the events leading up to the fault can be relatively easily duplicated. If the system generating the load provides good error tracking and event logging, it may be possible to immediately identify the sequence that caused the fault, the corresponding reactions of the computer telephony system, and even immediately duplicate the fault causing scenario.

Longitudinal Balance A measure of the electrical balance between the two conductors (tip and ring) of a telephone circuit; specifically, the difference between the tip-to-ground and ring-to-ground AC signal voltages, expressed in decibels.

Longitudinal Redundancy Check LRC. An error checking technique based on an accumulated collection of transmitted characters. An LRC character is accumulated at both the sending and receiving stations during the transmission of a block. This accumulation is called the Block Check Character (BCC) and is transmitted in the last character in the block. The transmitted BCC is compared with the accumulated BCC character at the receiving station for an equal condition. When they're equal, you know your transmission of that block has been fine.

Longitudinal Transmission Check LTC. An even or odd parity check at fixed intervals during data transmission.

Longitudinal Wrap A tape applied longitudinally with the axis of the core being covered, a opposed to a helical, or spiral, tape wrapped core.

Look-Up Table 1. A translation table. You dial a certain number. But the number is meaningless. For the phone system to complete the call, it needs routing instructions. It gets that by "looking up" that number in a table, which translates that number to another number that is now meaningful to the switching network. There are lots of applications for Look-up Tables. Least Cost Routing tables are essentially look up tables. IN-WATS dialing works by looking up the 800 number and finding its real ten digit normal number. Most private networks use look up tables to translate the number dialed by the internal user into a number that the network can recognize.

2. A set of addresses (source and destination) used by a bridge or router to determine what should be done with a packet. As the packet comes in, its address information is read and compared with the information in the look-up table. Depending on the information, the bridge may forward the packet, or discard it, leaving it for the local LAN. Many bridges and routers can build their look-up tables as they operate. See also 800 SERVICE, EIGHTHUNDRED SERVICE, BRIDGE, PERSONAL 800 NUMBER and ROUTER.

Loop 1. Typically a complete electrical circuit.

2. The loop is also the pair of wires that winds its way from the central office to the telephone set or system at the customer's office, home or factory, i.e. "premises" in telephones.

3. In computer software. A loop repeats a series of instructions many times until some prestated event has happened or until some test has been passed.

Loop Antenna An antenna consisting of one or more complete turns of wire, both ends of which are to be connected to the input circuit of the radio receiver.

Loop Back A diagnostic test in which a signal is transmitted across a medium while the sending device waits for its return. See LOOKBACK.

Loop Checking A method of checking the accuracy of transmission of data in which the received data are returned to the sending end for comparison with the original data.

Loop Circuit Generally refers to the circuit connecting the subscriber's set with the local switching equipment.

Loop Current Detection When a modem, telephone or fax card (etc.) seizes the line (i.e. completes the connection between tip and ring terminals of the telephone cable) current flows from the positive battery supply in the telephone central office, through the twisted pair in the loop, through the card (or phone) and back to the central office negative terminal where it is detected, showing that this telephone or telephone device is off hook. The fax card or modem can detect problems such as disconnects, shutting down the connection or a busy signal.

Loop Extender Device in the central office that supplies augmented voltage out to subscribers who are at considerable distances. It provides satisfactory signaling and speech for such subscribers.

Loop Plant Telco-talk for all the wires and hardware and poles and manholes used to connect their central offices to their customers.

Loop Reverse-battery A method of signaling over interoffice trunks in which changes associated with battery reversal are used for supervisory states. This technique provides 2-way signaling on 2-wire trunks; however, a trunk can be seized at only one end. It cannot be seized at the office at which battery is applied. It is also called reverse-battery signaling.

Loop Signaling A method of signaling over circuit paths that uses the metallic loop formed by the line or trunk conductors and terminating circuits.

Loop Signaling Systems Any of three types of signaling which transmit signaling information over the metallic loop formed by the trunk conductors and the terminating equipment bridges.

Loop Start LS. You "start" (seize) a phone line or trunk by giving it a supervisory signal. That signal is typically taking your phone off hook. There are two ways you can do that — ground start or loop start. With loop start, you seize a line by bridging through a resistance the tip and ring (both wires) of your telephone line. The Loop Start trunk is the most common type of trunk

found in residential installations. The ring lead is connected to -48V and the tip lead is connected to 0V (ground). To initiate a call, you form a "loop" ring through the telephone to the tip. Your central office rings a telephone by sending an AC voltage to the ringer within the telephone. When the telephone goes off-hook, the DC loop is formed. The central office detects the loop and the fact that it is drawing DC current and stops sending the ringing voltage. In ground start trunks, ground Starting is a handshaking routine that is performed by the central office and the PBX prior to making a phone call. The central office and the PBX agree to dedicate a path so incoming and outgoing calls cannot conflict, so "glare" cannot occur. See GLARE. Here are two questions that help in understanding:
How does a PBX check to see if a CO Ground Start trunk has been dedicated?
To see if the trunk has been dedicated, the PBX checks to see if the TIP lead is grounded. An undedicated Ground Start Trunk has an open relay between 0V (ground) and the TIP lead connected to the PBX. If the trunk has been dedicated the CO will close the relay and ground the TIP lead.
How does a PBX indicate to the CO that it requires the trunk?
A CO ground start trunk is called by the PBX CO Caller circuit. This circuit briefly grounds the ring lead causing DC current to flow. The CO detects the current flow and interprets it as a request for service from the PBX.
See also POTS.

Loop Test A way of testing a circuit to find a fault in it by completing a loop and sending a signal around that loop. See LOOPBACK.

Loop Through A type of phone system wiring that allows phones to connect to one cable in parallel going to the common central switching equipment. The most common type of Loop Through wiring is that which you have in your home. You have one cable with two conductors — a red and a green — winding through your home. Whenever you want to connect a phone, you simply attach it to the red and green conductors. The other way of connecting phones is called HOME RUN. In that system, every phone has its own one, two or three pairs of conductors which wind their lonely way back to the central PBX or key system cabinet. In Loop Through wiring, many phones share one set of cables. In Home Run Cabling, only one phone sits on that line.

Loop Timing A way of synchronizing a circuit that works by taking a synchronizing clock signal from incoming digital pulses.

Loop Up/Loop Down In T-1, there are generally two loopback types, LLB (line loopback) and TLB or DLB (terminal or DTE loopback). Loop Up refers to activating one of these loop backs, where as Loop Down refers to deactivating one of these loopbacks.

Loopback Type of diagnostic test in which the transmitted signal is returned to the sending device after passing through a data communications link or network. This allows a technician (or built-in diagnostic circuit) to compare the returned signal with the transmitted signal and get some sense of what's wrong. Loopbacks are often done by excluding one piece of equipment after another. This allows you to figure out logically what's wrong. (It's called Sherlock Holmes deductive reasoning.) See LOOPBACK TEST.

Loopback Test A test typically run on a four-wire circuit. You take the two transmit leads and join them to the two receive leads. Then you put a signal around the loop and see what happens. Measuring differences between the sent and the received signal is the essence of a loopback test. See LOOPBACK.

Looping Problem encountered in distributed datagram routing in which packets return to a previously visited node.

Loopstart Circuit The standard world-wide telephone circuit. For the phone to signal the phone system that it wants to make a call, it applies a DC termination across the phone line. See LOOP START for a longer explanation.

Loose Tube Buffer A cable construction in which the optical fiber is placed in a plastic tube having an inner diameter much larger than the fiber itself. The loose tube isolates the fiber from the exterior mechanical forces acting on the cable. The space between the tube and the fiber is often filled with a gel to cushion the fiber.

Loosely Coupled A computer system architecture consisting of multiple computer systems, each with its own dedicated memory and its own copy of the operating system, connected over a communications link. See also TIGHTLY COUPLED.

LOP An ATM term. Loss of Pointer: A condition at the receiver or a maintenance signal transmitted in the PHY overhead indicating that the receiving equipment has lost the pointer to the start of cell in the payload. This is used to monitor the performance of the PHY layer.

LORAN A radio-navigation system which helps you find where you are. It works by timing times the difference in reception of pulses from one or more fixed transmitters, usually on land. It's a radio based systems that's pretty good for coastal waters where there are LORAN transmitters. But it won't work further than 1,000 miles offshore. And it doesn't cover much of the rest of the earth and its accuracy varies depending on electronic interference and geographic variations. For a better system, see GPS.

LOS Loss of Signal: A condition at the receiver or a maintenance signal transmitted in the PHY overhead indicating that the receiving equipment has lost the received signal. This is used to monitor the performance of the PHY layer.

Loss The drop in signal level between two points on a network. It is important to distinguish between LOSS and LEVEL. Level is measured at a finite point. Loss is the difference between levels. Loss occurs constantly throughout telephony — from long distance circuits to switches. Loss is usually measured in dB — decibels. Loss is cumulative. Add two circuits each with a loss of 10 dB. You will have 20 dB loss in the total circuit. The human ear can detect a 3 dB loss.

Loss Budget A loss budget is the maximum amount of signal degradation a data communications network can withstand before it becomes susceptible to errors and/or loss of data. The idea is a establish a "loss budget" by consulting your equipment vendors for recommended wire types and maximum allocable lengths of cable before you build the network. We first heard the idea of a loss budget from The Siemon Company, Watertown CT.

Lossless Compression Image- and data-compression applications and algorithms, such as Huffman Encoding, that reduce the number of bits a picture would normally take up without losing any data. In this way, no information is lost or altered in the compression and/or transmission process. See LOSSY.

Lossy Compression Methods of image compression, such as JPEG, that reduce the size of an image by disregarding some pictorial information.

Lost Call Attempt A call attempt that cannot be further advanced to its destination due to an equipment shortage or failure in the network.
Lost Calls Cleared Traffic engineering assumption used in Erlang C that calls not satisfied on the first attempt are held (delayed) in the system until satisfied.
Lost Calls Held Traffic engineering assumption used in Poisson that calls not satisfied on the first attempt are held in the phone system for a period not exceeding the average holding time of all calls.
Lotus Express An e-mail communications software program for PC users of MCI Mail that allows the user to automatically send and receive messages, as well as binary files, such as spreadsheets or documents.
Loudspeaker Paging Access Interface to customer-provided paging equipment.
Lovelace, Ada Augusta Ada Augusta Lovelace was the daughter of Lord Byron, the English poet. Miss Lovelace is regarded as the first computer programmer because she worked for the computer pioneer Charles Babbage. A computer language was named after her. That language is called ADA.
Low Battery Cutoff An uninterruptible power supply (UPS) definition. This UPS feature automatically switches off battery power before the batteries discharge beyond safe limits. Without this feature, batteries can be taken into deep discharge, making them useless.
Low Entry Networking LEN. A peer-oriented extension to SNA, first implemented on IBM's System/36, that allows networks to be more easily built and managed by such techniques as topology database exchange and dynamic route selection.
Low Frequency The band of frequencies between 30 and 300 kilohertz.
Low Level Formatting The first step in preparing a drive to store information after physical installation in complete. The process sets up the "handshake" between the drive and the controller. Most drivers are now low level formatted at the factory. See PHYSICAL FORMATTING.
Low Level Language A programming language that uses symbols — one step away from the machine language of a computer. Low level computer languages, such as Assembler and C, actually manipulate the bits in computer registers. Higher level languages such as Basic and Fortran will take care of the piddling details of doing specific functions when you give it a broad command like "PRINT". In a lower level language, you must provide all the details of instruction necessary in the code (program) to perform the operation. It is possible to do this by calling standard routines, but still takes up the programmers' time in deciding which routines, and keeping the registers straight as he designs the program.
Low Noise Amplifier LNA. Typically a parametric amplifier in a satellite earth station.
Low Pass Filter A device that cuts frequencies off above a certain point and allow all other frequencies to pass. Opposite of high pass.
Low Speed Loopback A closed circuit feature useful for maintenance or testing.
Low Speed Signal Signal traveling at the DS1 rate of 1,544 Mb/s or at the DS1C rate of 3,152 Mb/s.
Low Voltage A low voltage condition exist when fewer than 105 volts AC is present at a 120 VAC outlet or HOT conductor. This figure was chosen by many manufacturers of electronic and telephone equipment. It is also the test of "low voltage" tested by a wonderful AC electric outlet-testing product called the Accu-Test II made by Ecos Electronics Corporations of Oak Park, Illinois. Below 105 volts, motors deteriorate and electronic circuits overheat. Long-term damage can occur to most gadgets plugged into an electrical outlet which consistently delivers below 105 volts. Also 105 volts is below the stated tolerance levels of all North American power utilities who state that their acceptable power is 120 volts plus or minus 10 percent. If your power is consistently below 105 volts, you should contact your local power utility.
LPC Linear Predictive Coding. Low bit rate voice (LBRV) digitizing technique that requires a bandwidth of only 2.4 or 4.8 Kbps. This technique may result in poor quality voice signals.
LPD Line Printer Daemon, a process on Berkeley spooler implementations that provides LPR systems.
LPDA-2 IBM's protocol under NetView for monitoring of dial-up modems for error correction.
LPF Low Pass Filter: In an MPEG-2 clock recovery circuit, it is a technique for smoothing or averaging changes to the system clock.
LPI Lines Per Inch. The number of lines both horizontal and vertical, that a facsimile machine will print in a square inch.
LPP Link Peripheral Processor or Link Peripheral Processing.
LPR The LPR command is used to queue print jobs on Berkeley queuing systems.
LPS Line Profile System.
LPT Port A logical designation for a series of I/O (Input/Output) addresses that allows the computer to communicate with a parallel printer.
LPT1 The first or primary parallel printer port on the IBM PC or clone. LPT2 is the second parallel port. COM1 is the first serial port. LPT1 is usually the default printer port, i.e. the one your computer will print to, if you don't tell it something else.
LQA Link Quality Analysis. See AUTOMATIC LINK ESTABLISHMENT.
LRC Longitudinal Redundancy Check.
LRN Location Routing Number.
LS Trunk Loop-Start Trunk. A trunk on which a closure between the tip and ring leads is used to originate or answer a call. High-voltage 20-Hz AC ringing from the telephone company signals an incoming call. See LOOP START.
LS1A-A Single Mode, Single-Fiber Interconnection Cable. An AT&T definition.
LSAP An ATM term. Link Service Access Point: Logical address of boundary between layer 3 and LLC sublayer 2.
LSB Least Significant Bit and Least Significant Byte. That portion of a number, address or field which occurs right most when its value is written as a single number in conventional hexadecimal or binary notation. The portion of the number having the least weight in a mathematical calculation using the value.
LSCIE Lightguide Stranded-Cable Interconnect Equipment.
LSCIM Lightguide Stranded-Cable Interconnect Module.

LSCIT Lightguide Stranded-Cable Interconnect Terminal.

LSI Large Scale Integration. Refers to micro electronic components which combine many hundreds of transistors on an integrated circuit. See CHIP.

LSL Link Support Layer. A layer within the Novell Open Data-Link Driver specification. This layer lets multiple protocol stacks access a network card simultaneously.

LSR An ATM term. Leaf Setup Request: A setup message type used when a leaf node requests connection to existing point-to-multipoint connection or requests creation of a new multipoint connection.

LSSGR LATA Switching System General Requirements.

LT 1. Logical Terminal

2. An ATM term. Lower Tester: The representation in ISO/IEC 9646 of the means of providing, during test execution, indirect control and observation of the lower service boundary of the IUT using the underlying service provider.

LTB Last Trunk Busy.

LTC Line Trunk Controller.

LTCI ISDN Line Trunk Controller.

LTE SONET Lite Terminating Equipment: ATM equipment terminating a communications facility using a SONET Lite Transmission Convergence (TC) layer. This is usually reserved for end user or LAN equipment. The SONET Lite TC does not implement some of the maintenance functions used in long haul networks such as termination of path, line and section overhead. In short, line terminating equipment includes network elements which originate and/or terminate line (OC-N) signals. LTEs originate, access, modify, and/or terminate the transport overhead.

LTRS Letters Shift. 1. Physical shift in a terminal using Baudot Code that enables the printing of alphabetic characters. 2. Character that causes the shift.

LTS Loop Testing System.

LU 1. Line Unit.

2. Logical Unit, access port for users in SNA. In a bisync network, a port through which the user gains access to the network services. A LU can support sessions with the host-based System Services Control Point (SSCP) and other LUs.

3. Local Use flag. Occasionally used to initialized approval for local cellular calls. The Cellular carrier insures that local users are registered with a local system.

LU 6.2 Logical Unit Interface. Version 6.2. An IBM SNA protocol that allows for peer-to-peer or program-to-program communications. The LU 6.2 protocol standard frees application programs from network specific details. On an IBM PC, a LU 6.2 program accepts commands and passes them on to an SDLC card to communicate directly with the mainframe or a token ring handler. LU 6.2 enables users to develop applications programs for peer-to-peer communications between PC's and IBM host systems. It increases the processing power of the PC user without the constraints of mainframe-based slave devices, i.e. 3274/3276 controllers. It creates a transparent environment for application-to-application communications, regardless of the types of systems used or their relative locations. Also referred to as Advanced Program-to-Program Communication (APPC).

LU Type 1 LU 1 is the SNA protocol that describes generic input/output devices (e.g. line printer).

LU Type 3 LU 3 is the SNA protocol that describes a print output device that uses 3270 data streams.

Lucent Technologies In early 1996, AT&T split its manufacturing operations and its Bell Labs off from its company and floated those operations as a separate company. It called that company Lucent Technologies, and Lucent's shares are now trading publicly. Business Communications Systems (BCS) is the unit of Lucent that develops, manufactures, markets and services advanced communications products for business customers of all types and sizes around the world. www.att.com/gbcs/services

Lug Something which sticks out onto which a wire may be connected by wrapping or soldering.

Luma The brightness signal in a video transmission.

Lumina A single line telephone adjunct device produced by Northern Telecom. It is called the Lumina 200. It will display the name and number of the calling person, if that information is provided by the local central office. Lumina sits between the incoming central office line and the telephone and records the incoming number and name of the caller delivered via Calling Line IDentification (CLID) services on Custom Local Area Signaling Services (CLASS). Lumina receives the call identification information between the first and second rings. To capture this information, the telephone must be allowed to ring twice before lifting the receiver to answer a call. See LUMINA CALL DISPLAY.

Lumina Call Display A small Northern Telecom device which sits between the incoming central office line and the telephone and records the incoming number and name of the caller delivered via Calling Line IDentification (CLID) services on Custom Local area Signaling Services (CLASS). The Lumina will also store the date and time of the call, whether the call is new or a repeated call, whether the calls comes from an area not equipped for calling number delivery, whether the call has previously been displayed during a review of the caller's list. The Lumina receives the call identification information between the first and second rings. To capture this information, the telephone must be allowed to ring twice before lifting the receiver to answer a call. See also LUMINA and MAESTRO.

Luminance That part of the video signal which carries the information on how bright the TV signal is to be.

LUNI An ATM term. LANE UNI: The standardized interface between a LE client and a LE Server (LES, LECS and BUS).

LUNS Logical Unit Numbers. An identification number given to devices connected to a SCSI adapter. Each SCSI ID can have eight LUNs. Normally, there is only one device with LUN 0. See DAISY CHAIN and SCSI.

Lurker A person who "hangs around" online bulletin boards and forums, browsing through the messages and, if moved, replying to some of them. That's one definition. Here's another: A visitor to a newsgroup or online service who only reads other people's posts but never posts his or her own messages, thus remaining anonymous.

Lux A contraction of luminance and flux and a basic unit for measuring light intensity. A Lux is approximately 10 foot candles.

LUXPAC A ITU-T X.25 packet switched network operated in Luxembourg by the Luxembourg government.

LZS Lempel-Ziv-Stac. A data compression algorithm developed by Stac Electronics and sometimes used by routers.

LZW The Lempel-Ziv-Welsh compression algorithm is another way of reducing the number of bits to transfer. Northern Telecom uses this compression algorithm for its Distributed Processing Peripheral (DPP) — the Automatic Message Accounting Transmitter (AMAT) for the DMS-100 family of central office switches. Northern Telecom selected this non proprietary protocol and helped promote it as an industry standard. The nominal compression ration is 2.8:1, without considering field suppression. Transmitting data compressed at a ratio of 2.8:1 at 9600 bps is equivalent to transmitting non-compressed data at 27 Kbps. Compatible compression collectors can poll the DPP in compressed or non-compressed mode. DPPs equipped with the Data Compression feature can transmit in either compressed or non-compressed mode, based on the collector's polling request for a specific polling session. To preserve data integrity, AMA data are still stored in non-compressed form on the DPP disks. LZW is a lossless data-compression algorithm. See LOSSLESS and LOSSY.

m (small letter) Milli. One-thousandth. M (big letter) Mega. One million, e.g. Mbps or Mbit/s, one million bits per second.

M Bit The More Data mark in an X.25 packet that allows the DTE or DCE to indicate a sequence of more than one packet.

M Hop The transmission of satellite signals through an uplink to a satellite that downlinks to a receiving station half-way or between the final receiving station. The intermediate receiving station uplinks the signal again to a satellite for delivery to the final, targeted receiving station. As an example, data from a company in the UK for a business in Japan may be sent to a satellite covering North America and Europe. The signal is downlinked to an earth station in Colorado. The signal is then uplinked to a satellite covering Asia and North America for final delivery to an earth station in Japan. The signal flow forms an 'M', thus the name.

M Patch Bay A patching facility designed for patching and monitoring of digital data circuits at rates from 1 Mbps to 3 Mbps.

M Port The port in an FDDI topology which connects a concentrator to a single attachment station, dual attachment station, or implemented in a concentrator. This port is only implemented in a concentrator.

M VTS Marconi Video Telephone Standard sends color pictures over regular phone lines at up to 10 frames per second at 14.4 Kbps. The MCI Video Phone, which conforms to this standard, has a resolution of 128 by 96 pixels.

M1 1. Multiplexer in the U.S. digital signal hierarchy. See M1, which is listed as if it were M-ONE in this dictionary.
2. Management Interface 1: The management of ATM end devices.

M2 Management Interface 2: The management of Private ATM networks or switches.

M3 Management Interface 3: The management of links between public and private networks.

M4 Management Interface 4: The management of public ATM networks.

M5 Management Interface 5: The management of links between two public networks.

M12 A designation for a multiplex which interfaces between four DSIs and one DS2 circuit.

M13 The multiplexer equivalent of T-1. In the U.S. digital hierarchy, multiplexers are called by the digital signal levels they interface with. For example, a multiplexer, which joins DS-1 channels to DS-3 is called a M1-3. A M1-3 takes 28 DS-1 inputs and combines them into a single 45 megabit per second stream. (The bit stream is actually 44.736 megabits.)

M24 A T-1 service that allows a user to multiplex up to 24 voice or data channels into a single T-1 link, compatible with AT&T central office based channel banks (M24 compatibility generally refers to compliance with the channelization and coding techniques specified by AT&T TR62411).

M28 The telephone company multiplexing scheme that multiplexes 28 T-1 data streams onto a single carrier system, the T-3.

M34 A designation for a multiplex which interfaces between six DS3s and one DS4 circuit.

M44 A T-1 service that allows up to 44 voice channels (48 without signaling) to operate over a single T-1 link by using ADPCM, and is compatible with AT&T central office based equipment. MJ44 compatibility generally refers to the ability to accept the 44 channel T-1 aggregate and break out one of the channels individually for routing purposes.

MA Abbreviation for MILLIAMP or MILLIAMPERES, unit of electric current.

Ma Bell A term used to refer affectionately to the old AT&T and the old Bell System. Several Women's Lib organizations objected to it some years ago on the basis that there were no women in the higher corporate structure of AT&T, and that women, as over supervised operators, were the downtrodden majority within the Bell System. There was a movement afoot to change it Pa Bell. But then came Divestiture and the breakup of the Bell System. The term, MA BELL, now largely belongs in the history books. And there are now a handful of women in senior management in the Bell operating companies.

MAC 1. Moves, Adds and Changes. When you first install a phone system it will cost money to run wires and install phones all over the building. Very quickly you will notice that you'll need to move people and their phones, add phones for new people and change phones around. This will cost money, often lots of it. How much it will cost you depends on the arrangement you have negotiated with the vendor of your phone system. It's a good idea to get a good deal on Moves, Adds and Changes later on BEFORE you sign your original deal to buy the phone system.
2. IEEE specifications for the lower half of the data link layer (layer 2) that defines topology dependent access control protocols for IEEE LAN specifications. MAC is a media-specific access control protocol within IEEE 802 specifications. It currently includes variations for the token ring, token bus and CSMA/CD. The lower sublayer of the IEEE's link layer (OSI) which complements the Logical Link Control (LLC).

MAC Address The address for a device as it is identified at the Media Access Control layer in the network architecture.

MAC Name A MAC name (also called a MAC address) is a 48-bit number, unique to each local area network card, that is programmed into the card, usually at the time of manufacture. Unlike Network Layer Addresses, MAC names are location-independent. Destination and source MAC names are contained in the LAN packet and are used by bridges to filter and forward packets. See also NETWORK LAYER ADDRESS.

MACs Moves, Adds and Changes.

Machine Code Same as Machine Language. See MACHINE LANGUAGE.

Machine Dependent Software which will only run (i.e. is dependent) on a certain computer.

Machine Directory A Windows 95 definition. For shared installations, the directory that contains the required configuration files for a particular computer. The machine directory contains WIN.COM, the Registry, and startup configuration files.

Machine Language A computer language composed of machine instructions that can be executed directly by a computer without further compilation. Instructions and data coded in binary code. Machine language is the native language of computer hardware. Machine language is the only language recognized by the microprocessor that controls all the operations in your PC. All programs and all data to be processed by your computer (PC, mini or mainframe) have to be translated into machine language at some stage.

Macintosh A family of Apple microcomputers that represents the first wide scale PC (personal computer) deployment of icons, windows, "mice" and a consistent user interface, now called a Graphical User Interface (GUI). Apple introduced the Macintosh in January, 1984.

Macmail The name of a piece of communications software that makes using MCI Mail very easy. The software was written by Pete Maclean who can be reached on MCI Mail 263-7215.

Macro Language A collection of instructions by which any kind of information in the system can be located and manipulated and by which new information types can be added to the system.

Macrobending In an optical fiber, all macroscopic deviations of the fiber's axis from a straight line; distinguished from microbending.

Macrocell macrocell is a new word for what we used to call a "cell," as in cellular radio. This is what some researchers at Bell Northern Research (BNR) wrote: "Today's cellular networks employ macrocells and are optimized to serve users in automobiles, moving at relatively high speeds. Yet, a growing proportion of cellular traffic is originating from users who are not driving in vehicles, but are on foot... If a portion of the radio frequencies in a geographic area were transferred from macrocell (optimized for cars) to microcell technology (optimized for pedestrians), cellular traffic would increase up to a hundredfold. Microcell networks will entail the deployment of many more transceivers than today's macrocell systems. However, microcell equipment will be less costly because it is low power, simpler and smaller, say the researchers at Bell Northern Research.

Macros A common name for a microcomputer software program which lets you alter the definitions of what the keys on your computer keyboard are. With a "macro" software program, you could change the letter "M" on your keyboard to type "Michael" every time you hit it. But this would be stupid. Better to hit a combination of letters to get "Michael." Most personal computers have extra non-alphabetic keys, like Control and Alternate. And some macro software programs typically allow you to define a "Supershift" key. This means you can define one of your keys as a second Shift key. When you hit it (or Control or Alt) and then hit another key on the keyboard that key will type something different.
For example, when we hit ALT T on the macro we've programmed for this dictionary, we get T103 in square brackets, which means to change the typeface to this, which is called Helvetica Bold. When we first started writing this dictionary, we were using Radio Shack Model II machines running an operating system called CP/M. On that machine, we had no ALT key. It had only a control key. So we defined our Macro key to be "Control 6." When we hit the Control Key and the "6" key simultaneously, nothing happened on the screen, but the next character, number or letter was different, depending on how we defined it. For example, if we hit Control 6 and then T, we'd get T103 in square brackets. Our biggest use for macros is typesetting commands.
We have about 40 macro translations loaded on every MS-DOS machine in our office. Macros are among the most useful and, curiously least expensive, of all PC software programs. Increasingly, macro capability is being built into most major word processors. ZEdit, the word processor we used to write this dictionary, has extensive macros. You can pretty well define any key or combination of keys to type out words or executive commands.

Macstar An AT&T 3B2 computer-based software system that interfaces with the Remote Memory Administration System (RMAS) to effect customer moves and rearrangements on the 5ESS switch. The user needs only a terminal and printer.

MADN Multiple Appearance Directory Number.

MAE-East MAE stands for Metro Area Ethernet. MAE is a huge interconnection point for Internet Service Providers (ISPs). They use MAEs for routing their internet traffic from customers on their network to Internet sites on other peoples' Internet networks. There are two MAEs — MAE-East in Vienna, Virginia and MAE-West in San Jose, California. Physically a MAE is a building with zillions of wires, gigantic switches and computers with routing tables containing the location of Internet sites and how to get to them. MAEs are run by networking companies (MAE-East is run by MFS) and ISPs pay a fee to these networking companies to locate equipment there and for the transmission, switching and interconnect services. interconnection services.

MAE-East++ An expansion of MAE-East, located in a nearby building. i++ (plus plus) gets its term from the C programming language where it means "use the current variable i and add 1." Plus plus has come to mean an expansion, an improvement, an upgrading, etc. See MAE-East.

MAE-West See MAE-East.

MAE-West++ An expansion of MAE-West, located in a nearby building. i++ (plus plus) gets its term from the C programming language where it means "use the current variable i and add whatever one to it." Plus plus has come to mean an expansion, an improvement, an upgrading, etc. See MAE-East.

Magalog A mail order catalog disguised as a magazine in the hopes of confusing its recipients.

Magazine Hardware unit, mounted in a frame, containing printed board assemblies.

Magellan A term used by Northern Telecom to describe its family of broadband multimedia products.

Magic On Hold A term peculiar to AT&T's Merlin phone system which refers to customized Music-on-Hold that promotes the customer's products and services.

Maggots Telco slang for B-connectors, which are about the size and shape of fly larvae.

MAESTRO A Northern Telecom telephone designed for the home for CLASS service, in particular the CLID — Calling Line IDentification service available with SS7. Maestro has a 16-character display and logging capability that retains and displays on command the last 15 unanswered incoming directory numbers, and allows return dialing from the unanswered call log. Maestro's ten programmable one-touch memory keys can be used either for storing phone numbers or for one-button access to CLASS and other custom calling features that are otherwise accessed by dialing a code. MAESTRO also has a Call Forwarding Lamp which makes it easy to tell when call forwarding has been activated from the Maestro. List price for the Maestro was set originally at $136. See also CLASS and LUMINA.

Magic Cap Magic Cap is a "Communicating Applications Platform," which, according to General Magic of Mountain View, CA, allows developers to build a new class of software and services, centered on communications and easy for users to personalize.

Magic Wand One of the devices used by con men when pretending to do TSCM (Technical Surveillance CounterMeasures). A magic wand is typically a field strength meter or box with many fancy lights.

Magnetic Bubble A device in which information is stored in a magnetic film as a pattern of oppositely directed magnetic fields. Magnetic bubble devices hold their memory even if you lose power.

Magnetic Disk A computer storage device that records data bits as tiny spots on magnetic-coated disk platters. Hard disks and floppy disks are variations on magnetic disks. See also MAGNETO OPTICAL and HOLOGRAPHIC DATA STORAGE.

Magnetic Ink An ink that contains particles of a magnetic substance whose presence can be detected by magnetic sensors. Typical is the ink on your checks, which carry your name, your account number and the check's number.

Magnetic Medium Any data-storage medium and related technology including diskettes and tapes, in which different patterns of magnetization are used to represent the values of stored bits and bytes.

Magnetic Storage Any medium (generally tape or disk) upon which information is encoded as variations in magnetic polarity. The hard disk on your computer is magnetic storage.

Magnetic Stripe A strip of magnetic material, usually tape, attached to a credit card containing data relating to the card holder. You have a magnetic stripe now on the back of most of your credit cards. That stripe tells a computer who you are, what your account number is, etc.

Magnetic Tape A tape made of magnetic material upon which data may be stored for later retrieval by a computer.

Magneto A small hand-cranked AC generator which uses permanent field magnets and can make electricity to ring telephone bells.

Magneto Optic Relating to the change in a material's refractive index under the influence of a magnetic field. Magneto-optic materials generally are used to rotate the plane of the polarization. This phenomenon is how magneto optical disk drives work.

Magneto Optical Drive A computer data storage device that writes data using magnetism (in the form of a magnetic field called the bias field) and light (a laser beam) to write to a disk that resembles a CD-ROM disk. A magneto optical drive holds huge amounts of information — as much as 500 megabytes on a single disk. Introduced in 1988, the magneto optical drive (also spelled with a dash between the magneto and optical), the drive provides the convenience of the removability of floppies and the Bernoulli Box, the random access convenience of hard disk, the reliability of CD-ROMs and the promise of DAT-like capacity. But, according to PC Magazine, before you rush out and buy this ultimate storage solution, note that they're expensive, can't provide as much storage on one side of a disk as the largest hard disks and they're slower than today's hard disks. An explanation of how magneto optical drives work, courtesy PC Magazine: The recording layer on the disk stores the equivalent of binary 1s and 0s on the magnetic domains. The disk is designed so that the bias field by itself is to too weak to change the polarity of the magnetic domains. But when a spot on the disk is heated by a high-powered laser beam, its resistance to changing polarity drops. The bias field can now change the disk area's polarity. To read the disk, the drive uses a laser beam that is not hot enough to allow the bias field to change the disk area's polarity.

Magneto Phone A magneto phone has a crank handle on the side and dial pad. It is a very old phone — typically manufactured around the turn of the century. It is totally manual. You crank the handle which turns a bunch of permanent magnets on the inside which in turns generates electricity which in turn is used to ring a bell at the central office. When that bell rings, an operator answers, asks where the person wants to dial, dials and connects the call.

MAHR Abbreviation for MILLIAMPERES HOUR, 1/1000th of an ampere hour. Term is commonly used with small rechargeable battery packs, such as those used by portable phones and laptop PCs.

Mail Enable Applications Applications that use mail as a way of addressing and transporting information to and from users on a network.

Mail Exploder Part of an electronic mail delivery system which allows a message to be delivered to a list of addresses. Mail exploders are used to implement mailing lists. Users send messages to a single address (e.g., smith@somehost.edu) and the mail exploder at somehost.edu takes care of delivery to the individual mailboxes on the lists, including Smith.

Mail Gateway A machine that connects two or more electronic mail systems (especially dissimilar mail systems on two different networks) and transfers messages between them.

Mail Path A series of machine names used to direct electronic mail from one user to another.

Mail Reader Software which enables a user to select unread electronic mail and unread conferences messages and have them downloaded for reading off-line. Most mail readers also permit users to create responses off-line and upload them at their convenience.

Mail Reflector An Internet term. A special mail address; electronic mail sent to this address is automatically forwarded to a set of other addresses. Typically, used to implement a mail discussion group.

Mail Server Mail Server is the "post office" of a messaging network. Mail server is a computer host and its associated software that offer electronic mail reception and (optionally) forwarding service. Users may send messages to, and receive messages from, any other user in the system.

Mailer Also known as the front end; a program that allows BBSs (bulletin board systems) using different software to "talk" to each other. The mailer acts as a transfer layer for the messages passed. Some BBSs require mailers to talk to some netmail systems (for example, Spitfire for FidoNet). Seadog is generally acknowledged to be the market leader, according to PC Magazine.

Mailer-Daemon The Mailer-Daemon is not a person, but rather a computer program on Internet which runs automatically to perform the service of telling you why it could not deliver your message.

Mailgram An overnight electronic mail service of Western Union. The letter is phoned in or sent by computer to a central Western Union computer, from where it is sent to teleprinter machines located in post offices in major cities. When it's printed at the post office, the Mailgram is placed in an envelope and hand delivered by your friendly, local mailman in next day's mail.

Main PBX or Centrex switch into which other PBXs or remote concentration of switching modules are homed. A PBX or Centrex connected directly to an electronic tandem switch (ETS). Also, a power source.

Main Cross-Connect The interconnect point where wiring from the Entrance Facility and from the Workstation is connected to telecom equipment.

Main Distribution Frame MDF. A wiring arrangement which connects the telephone lines coming from outside on one side and the internal lines on the other. A main distribution frame may also carry protective devices as well as function as a central testing point. See MAIN DISTRIBUTION FRAME FILL, DISTRIBUTION FRAME and FRAME.

Main Distribution Frame Fill The central office mainframe is the termination point for outside plant cables. The "fill" is the percentage of pairs used by customers of the total number of pairs on the frame. Optimum fills vary based on the size of the central office and the amount of growth in the area. A low fill means idle lines and wasted investment in outside plant. A high fill, plus unexpected growth, forces budget busting and crisis construction projects.

Main Feeder Feeder cable that transports pairs from the central office to branching or taper points.

Main Lobe The main lobe is the area with the maximum intensity in the pattern of radiation produced by an antenna. One presumes it's called "lobe" because the pattern in a microwave signal of the main lobe typically looks like a ear lobe.

Main Memory The principal random storage area inside the computer. Used for storing data and programs and under the direct control of the CPU — the main processor. Also called RAM memory.

Main Network Address In IBM's SNA, the logical unit (LU) network address within ACF/VTAM used for SSCP-to-LU sessions for certain LU-to-LU sessions. Compare with auxiliary network address.

Main PBX A main PBX is one which has a Directory Number (DN) and can connect PBX stations to the public network for both incoming and outgoing calls. A main PBX can have an associated satellite PBX, and can be part of a tandem tie trunk network (TTTN). If the main PBX provides tandem switching for tie trunks, it is called a tandem PBX. In the context of ESN (Electronic Switched Network), a main PBX has tie trunks to only one node. See PBX.

Main Satellite Service A PBX feature that allows multi-location customers to concentrate their attendant positions at one location referred to as the Main. Other unattended locations are referred to as Satellites.

Main Service Entrance In AC electricity, the main service entrance is the necessary equipment, usually consisting of main circuit breakers or fuses, a switch and branch circuit breakers or fuses, in a grounded enclosure (panel) connected directly to earth. Located in the building at the point of entrance of the supply conductors from the power utility. Other panels in the building are referred to as branch, service or supply panels.

Main Station A subscriber's telephone instrument, terminal or workstation used to originate and receive calls. Very often if two instruments have the same extension number (are bridged), one becomes the Main Station and the other is a bridged station for inventory purposes.

Mainframe A powerful computer, almost always linked to a large set of peripheral devices (disk storage, printers, and so forth), and used in a multipurpose environment at the corporate or major divisional level. A mainframe is a a large-scale computer typically containing hundreds of megabytes of main memory and hundreds of gigabytes of disk storage. It is capable of "serving" thousands of "on-line" terminals. The term — main frame — derives from the racks that typically hold a large computer and its memory.

Mainframe Gateway A hardware/software system that allows PCs on a LAN (Local Area Network) to communicate with a mainframe. A single, usually dedicated, PC acts as the gateway. PCs on the LAN share its hardware and its communication link, communicating with it over the LAN cable. The most common mainframe gateway is an SNA gateway, which hooks a LAN into an IBM mainframe.

Mainframe Server Clients are devices and software that request information. Client is a fancy name for a PC on a local area network. It used to be called a workstation. Now it is the "client" of the server. A mainframe server is a large computer that stores lots of information and manages libraries of information. Here's a definition of Thin Client, courtesy of Oracle Corporation, writing in early 1994: "Mainframe systems store lots of data, but they're expensive, slow and difficult to use. Because all the processing happens on one large computer, they can't move large amounts of multimedia information to large numbers of users. Example, the IBM ES/9000, Amdahl's 5995-1400 or any plug compatible mainframe." See also CLIENT, CLIENT SERVER, CLIENT SERVER MODEL, FAT CLIENT and MEDIA SERVER.

Mains Some countries call their normal commercial power outlets — "mains." In Europe the frequency of commercial power is 50 Hz. In the United States, its frequency is 60 Hz. It's hard to convert the frequency of commercial power. It's easier to convert voltage. In Europe and Australia, normal voltage is 240 volts. In the U.S., it's 120 volts.

Mains Modem A modem which is part of a system called remote metering which monitors electricity usage and allows electric companies to offer such services as electronic mail, burglar alarms and energy management. The idea of energy management is that if the electric companies could turn off unnecessary appliances for a few hours during peak times, they might not have to build expensive new power stations. In exchange for that favor, they undoubtedly would be prepared to offer their customers price reductions.

Maintenance 1. All work needed to keep the telephone system operating properly, including periodic testing, repairs, etc. See PREVENTIVE MAINTENANCE.
2. All work needed to keep a software program operating properly, operating on new machinery and operating with new management needs. Often, software maintenance means substantially rewriting the original software program. Most of the work done by data processing departments in large companies involves maintaining old programs. This is not a put-down.

Maintenance Acceptance A term used in the secondary telecom equipment business. The point at which a maintenance company has tested a system, component, or peripheral device and determined that it meets manufacturer's specifications. The product can now be added to a maintenance contract. Once under contract, the maintenance company is responsible for repairing or replacing any defective components.

Maintenance Contract Contract guaranteeing the repair of a PBX switch to support it at operational levels for a predetermined fixed term and fixed price.

Maintenance Control Center MCC. A central place in a stored program control central office from which system configuration and trouble testing are controlled.

Maintenance Control Circuit MCC. A voice circuit used by maintenance personnel over microwave links for coordination. This is not available to operations or technical control personnel.

Maintenance Release An euphemism for a new piece of software that fixes a buggy piece of old software. When software is released, it's usually buggy. A maintenance releases attempts to fix the bugs. A more correct term for the new software would be a "bug fix." A maintenance release often carries the number one as the second digit after the decimal point, e.g. 3.51 or 4.01.

Maintenance Services In IBM's SNA, network services performed between a host SSCP and remote physical units (PUs) that test links and collect and record error information. Related facilities include configuration services, management services and session services.

Maintenance Termination Unit MTU. A MTU is an electronic circuit that is owned and deployed by a telephone company to aid in fault sectionalization and is installed at the network interface. The MTU should meet the requirements of Bellcore Technical Advisory TSY-000324 and be testable with the Mechanized Loop Test System (MLT). The MTU is designed to work on single line residence or business service.

Maintenance Update A euphemism for a piece of software which fixes bugs in a previously-released version of the software. A maintenance update rarely has any new features and rarely costs anything. Software companies send them because they find bad bugs in their software and want to fix those bugs asap, or because they can't stand the heat from complaining customers.

Majordomo A Majordomo is a mail-processing software program. It maintains multiple mailing lists. It automatically interprets (i.e. understands) commands in emails sent to it by individuals who wish to subscribe to, unsubscribe from, receive periodic summaries of, or otherwise become associated with a mailing list. It also handles the mass mailing of messages to members of the list. Most MajorDomo applications are unix based. You find Majordomo all over on the Internet. For example of how Majordomo mailing lists work, here's a note on Subscription Information for an electronic newsletter called WinNews published by Microsoft. "If you know someone who might be interested in WinNews, please instruct them to:
1. Send Internet e-mail to: ENEWS99@MICROSOFT.NWNET.COM 2. Send the message from the account that you wish to subscribe (some people use more than one e-mail account). 3. Subject line should be blank. 4. Body of message should ONLY have in the text: SUBSCRIBE WINNEWS If you wish to stop receiving WinNews, send mail to enews@microsoft.nwnet.com with a blank subject line and the body of the message should only save in the text: UNSUBSCRIBE WINNEWS."

Make Busy To make a communication circuit unavailable for connection. The technical term for taking the phone off the hook and leaving it off hook.

Malicious Call Tracing An ISDN service which enables to User to Network message to be sent while the call is in progress, ensuring that origination details are captured at the local exchange.

MAN Metropolitan Area Network. A high-speed data intra-city network that links multiple locations within a campus, city, or LATA. Typically extends as far as 50-kilometers, operates at speeds from 1 Mbit/s to 200 Mbps and provides an integrated set of services for real-time data, voice and image transmission. The IEEE 802.6 standard defines MAN standards and SMDS is the MAN service offered by local phone companies. Private and public MANs may use the ANSI FDDI standard.

Man Machine Interface A term coined by James Martin to designate the ease (or lack of ease) of a person working with a computer.

Man Page Manual page. On-line documentation that commonly comes bundled with computers running Unix.

Managed System An entity that is managed by one or more management systems, which can be either Element Management Systems, Subnetwork or Network Management Systems, or any other management systems.

Management Domain MD. An X.400 term describing a set of messaging systems. At least one system contains, or realizes, an MTA (Messaging Transfer Agent) managed by a single organization. It is a primary building block in the organizational construction of an MHS (Messaging Handling System), referring to an organizational area for the provision of messaging services. MD is used in a similar manner in X.500, consisting of at least one DSA (Directory System Agent). A management domain may or may not be identical with a geographical area. See DOMAIN.

Management Information Base See MIB.

Management Information System MIS. Management information provided by computer data processing. Once upon a time called data processing.

Management Services In IBM's SNA, network services performed between a host SSCP and remote physical units (PUs) that include the request and retrieval of network statistics.

Management System An entity that manages a set of managed systems, which can be either NEs, subnetworks or other management systems.

Manchester Encoding A digital encoding technique in which each bit period is divided into two complementary halves. A negative-to-positive (voltage) transition in the middle of the bit period designates a binary "1" while a positive-to-negative transition represents a "0". This encoding technique is self clocking (the receiving device can recover transmitted clock from the data stream).

Mandrel Wrapping A technique in which excess higher order modes launched into an optical fiber filtered out by wrapping several turns of fiber around a mandrel approximately 1 cm in diameter.

Manual Exclusion A PBX extension user, by entering a certain code, can block all other phones on that line from entering the call. Assures privacy on the line.

Manual Gain Control MGC. There are two electronic ways you can control the recording of something — Manual or Automatic Gain Control (AGC). AGC is an electronic circuit in tape recorders, speakerphones and other voice devices which is used to maintain volume. AGC is not always a brilliant idea since it attempts to produce a constant volume level. This means it will try to equalize all sounds — the volume of your voice and, when you stop talking, the circuit static and/or general room noise which you undoubtedly do not want amplified. Sometimes it's better to have quiet, when you want quiet. Manual Gain

Control is preferred in professional applications. Manual Gain Control is simply an elegant way of saying there's a record volume control. Never record a seminar or speech using AGC. The end result will be decidedly amateurish.

Manual Hold The method of placing a line circuit on "hold' by activating a non-locking "hold" button on the phone, usually one colored red.

Manual Intercom A crude, single-path communications link between telephones without the ability to signal the receiving party.

Manual Modem Adapter An external device for the Merlin key system from AT&T. It allows connection of single line accessories to any Merlin telephone. The device, in effect, draws a standard tip and ring line out of the Merlin proprietary cabling/signaling scheme. Some other key systems have similar devices. Comdial calls theirs a "data port" and their phones contain extra RJ-11 jacks.

Manual Originating Line Service The attendant must complete all outgoing calls. All other calls are blocked. This "feature" is used to cut down on long distance phone abuse. There's a wonderful story. When many of the PBXs in Europe went from manual originating line service to automatic dial "9" long distance, the number of long distance calls doubled within two months. Some of these calls were legitimate. Some were not. How much abuse there was varied from company to company. Typically, those companies with employees who were more bored suffered (or enjoyed?) more abuse.

Manual PBXs Refers to PBXs which are not automatic and which require that all calls, including intercom calls, be placed through the attendant. Such PBXs are still used today, though in limited applications. You can still find manual PBXs in vacation hotels, nursing homes and in the data communications departments of some firms, who use manual PBXs as manual dataPBXs. These are especially useful in places where long data calls and sold metal-to-metal connections are an advantage.

Manual Ringdown Tie Trunk A direct talk path between two distant phones. Signaling must be done manually from either phone. Contrast this with Automatic Ringdown Tie Trunk, in which the signaling occurs the moment one of the phones is lifted off hook.

Manual Signaling Pushing a button on a telephone sends an audible signal to a predetermined phone. Manual signaling can be used for secretary/boss communications.

Manual Telephone A telephone without a dial. Taking the receiver off hook automatically rings a predetermined number. A courtesy phone.

Manual Terminating Line Service Provides extension lines that require all calls be completed by the attendant. For a better explanation see MANUAL ORIGINATING LINE SERVICE.

Manufacturing And Automation Protocol See MAP.

Manufacturing Message Specification MMS. An International Standards Organization (ISO) application layer protocol that defines the framework for distributing manufacturing messages within a network. This specification is used in MAP 3.n.

MAP 1. A new term for multiplexing, implying more visibility inside the resultant multiplexed bit stream than available with conventional asynchronous techniques.
2. Mobile Application. Part Application protocols used in SS7 for the specific needs of mobile, wireless networks and allowing roaming by mobile stations.
3. Maintenance and Administration Position. See MAP/MAAP below.
4. Manufacturing Automation Protocol. A protocol initially developed as an internal specification for its own factory floor equipment and now championed by General Motors as the industry standard to facilitate communications among the diverse automation devices found in production environments. AT&T, IBM and DEC have endorsed this standard and have already or will introduce MAP-compatible products. TOP (Technical and Office Protocol) was initiated by Boeing Computer Services (one of the nine companies that helped form the MAP Users Group in 1984) and is designed for use in the engineering and office environment and to move information from the factory floor to other parts of the company. Implementation of these protocols would lead to GM's factory of the future concept.

MAP/MAAP Maintenance and Administration Panel. A device attached to a PBX to allow you to maintain and administer the system — to change phone features, etc.

MAP/TOP Manufacturing Automation Protocol/Technical Office Protocol.

MAPI Microsoft's Windows Messaging Application Programming Interface, which is part of WOSA (Windows Open Services Architecture). MAPI is a set of API functions and as OLE interface that lets messaging clients, such as Microsoft Exchange, interact with various message service providers, such as Microsoft Mail, Microsoft Exchange Server, Microsoft Fax and various computer telephony servers running under Windows NT server. Overall, MAPI helps Exchange manage stored messages and defines the purpose and content of messages — with the objective that most end users will never know or care about it. A friend of mine, who's a great programmer, Pete MacLean, explained MAPI as: MAPI is Microsoft's new foundation for a modular mail system. You can pick and choose among various email clients, address books, message stores (foldering systems), and transports (the message-service specific pieces) and build your own custom mail system. See also AT WORK, MICROSOFT EXCHANGE, WINDOWS 95, WINDOWS TELEPHONY and WOSA. The biggest explanation of MAPI is in the definition for WINDOWS 95.

Mapping 1. In network operations, the logical association of one set of values, such as addresses on one network, with quantities or values of another set, such as devices on another network (e.g. name-address mapping, internetwork-route mapping).
2. A Novell NetWare term. To assign a drive letter to a chosen directory path on a particular volume of a particular file server. For example, if you map drive F to the directory SYS:ACCTS\RECEIVE, you will access that directory every time you enter "F:" at the DOS prompt. See also DRIVE MAPPINGS.
3. In EDI (Electronic Data Interchange), mapping defines the translation between a company's unique data layout and an EDI formal structure.

Marathon A family of products that are combination fast packet multiplexer, data compression, voice compression and fax de-modulation devices that fit many, voice, data, fax and LAN "conversations" onto one leased circuit — analog or digital. The idea of Marathon is to save money on long distance telecommunications charges. The Marathon family of products is made by Micom Communications Corporation, Simi Valley CA.

NEWTON'S TELECOM DICTIONARY

Marconi, Guglielmo Guglielmo Marconi, born in Bologna, Italy in 1874, was on a holiday when he read of the electromagnetic wave experiments of Hertz. This article established the thought in Guglielmo's mind that electromagnetic waves could free telegraphy from the wires and submarine cables, which at that time constrained its use. Finding out if electromagnetic waves could be used to communicate at a distance became an obsession for Marconi. His mother allowed him to use two large rooms on the top floor of their house as a laboratory. She also helped persuade Guglielmo's father to provide (albeit grudgingly) the money necessary for the batteries, wire and other equipment Guglielmo needed. Marconi started by repeating Hertz's experiments. His oscillator was an induction coil equipped with four spheres for the spark discharge. The frequency of the oscillations was in what we, today, call the VHF range. The detector he used with his receiving coil was a Branly coherer, similar to that used by Oliver Lodge. The coherer provided much greater sensitivity than the spark-gap equipped loop of wire Hertz had used. Marconi placed a curved metal detector behind his oscillator to direct the waves toward the detecting circuit. Soon, Marconi was able to cause a bell, located thirty feet away, to ring when the oscillator was keyed. Through trial-and-error experimentation, he was able to increase the sensitivity of the coherer significantly over what others had achieved. The following spring, Marconi took his experiments outdoors. Connecting metal plates to the oscillator's spark gap lowered the frequency and strengthened the intensity of the oscillations produced. Similar plates were connected to each side of the coherer. By chance, Marconi found that if one of the metal plates was elevated high in the air and the other was laid on the ground, the range at which oscillations could be detected increased to over one-half mile. Soon, the elevated plates at the oscillator and detector were replaced by long vertical wires. The plates which had lain on top of the ground now were buried. This arrangement increased the distance at which signals could be received to one and one-quarter miles. An intervening hill was found to be no barrier to the reception of the signals. The combination of using lower-frequency oscillations and using the Earth as an element in his antenna system were crucially important achievements. Another demonstration was held in March of 1897. This time longer wavelengths were used in conjunction with wire antennas raised some 120 feet above the ground by means of kites and balloons. This arrangement resulted in signals being received over a distance of four and one-half miles. In May of 1887, Marconi demonstrated that wireless signals could span significant lengths across water by sending signals between the shore and an island in the Bristol Channel, a distance of 8.7 miles. This was a crucial test because the submarine cable that normally provided communications to the island had failed several times in recent months. Repairing the cable was costly both in time and in money so Marconi's system must have appeared as an excellent alternative. Marconi established the Wireless Telegraph and Signa. Ltd. in July of 1897. In 1899, he changed the name of his company to The Marconi Wireless Telegraph Co. Ltd. A major goal Marconi had in mind was to show the value of wireless for communicating with ships. In 1897, he returned home to Italy to convincingly demonstrate that wireless could communicate between naval warships. The Italian Navy soon adopted the Marconi wireless system. In 1896, in England, Marconi obtained the first patent on the wireless. In 1901 he succeeded in transmitting signals across the Atlantic. In 1909 he received jointly with C. F. Braun the Nobel Prize in Physics. Marconi was made a Marchese and a member of the Italian senate. He died in Rome on July 20, 1937. See also TESLA, Nikola.

Marine Telephone Marine telephones operate on assigned radiotelephone frequencies much as a radio broadcast does. Marine telephones can be used to contact other marine telephones or to reach land-based telephones through an operator.

MARISAT A satellite for marine use. Conversations on MARISAT are crystal clear. Call Comsat and ask them for a demo call to a ship somewhere in the world. It's very exciting.

Maritime Mobile Service A mobile service between coast stations and ship stations, or between ship stations, or between associated on-board communication stations. Survival craft stations and emergency position-indicating radiobeacon stations may also participate.

Maritime Mobile Satellite Service A mobile satellite service in which mobile earth stations are located on board ships. Survival craft stations and emergency position-indicating radiobeacon stations may also participate in this service.

Maritime Radio Navigation Service A radio navigation service intended for the benefit and for the safe operation of ships.

Mark 1. A term that originated with the telegraph. It currently indicates the binary digit "1" (one) in most coding schemes.
2. A call center term. To flag a record in a browse listing for some special purpose. Typically you mark a record that you want to copy information from. As long as the record is marked, you can continue copying the information to other records.

Mark-Hold The normal no-traffic line condition where a steady mark is transmitted.

Marker The logic circuitry in a crossbar central office that controls call processing functions.

Marker Beacon A transmitter in the aeronautical radio navigation service which radiates vertically a distinctive pattern for providing position information to aircraft.

Marker Tape A tape laid parallel to the conductors under the sheath in a cable imprinted with the manufacturer's name and the specification to which the cable is made. Other information such as date of manufacture may also be included.

Marker Thread A colored thread lain parallel and adjacent to the strands of an insulated conductor which identifies the cable manufacturer. It may also denote a temperature rating or the specification to which the cable is made.

Market Price Prices set at market rates but, in most cases, are not permitted to be less than cost.

Marking Bias The uniform lengthening of all marking signal pulses at the expense of all spacing pulses.

Martian Packets that turn up unexpectedly on the wrong network because of bogus routing entries. Also used as a name for a packet which has an altogether bogus (non-registered or ill-formed) Internet address.

MASER Microwave Amplification by Stimulated Emission of Radiation. A device that generates electromagnetic signals in the microwave range, known for relatively low noise.

Mask 1. A field made up of letters or numbers and wildcard characters, used to filter data. For example, a mask 800xxxxxxx may be applied to the dialed digits field of a call record to identify toll-free calls.
2. A computer imaging term. The electronic equivalent of placing transparent tape over selected regions of an image, a mask marks pixels that remain unchanged by subsequent painting operations. For example, you might mask out a mountain range and add background clouds to the sky. In the final image, the clouds will appear between the peaks.

Masquerade To pretend to be someone else by using another person's password or Token.

MASS860 An organization of computer system vendors formed to promote open system standards and the writing of applications software for the Intel i860 microprocessor. Members include Intel, Oki Electric Industry Co., Ltd., IBM, Stratus, Olivetti, Alliant, Samsung and Stardent Computer.

Massively Parallel Systems Tightly coupled multi processing computers that house 100 or more CPUs, each with its own memory.

Master Term applied to the data communications equipment at one end of a synchronous digital transmission network that supplies the clock timing signal that determines the rate of transmission in both directions.

Master Clock An electronic timing circuit which synchronizes the entire data communications network. The source of timing signals, or the signals themselves, that all network stations use for synchronization.

Master Control Unit MCU. An InteCom word for the device which controls the main operating functions of the system.

Master Frequency Generator In FDM, equipment used to provide system end-to-end carrier frequency synchronization and frequency accuracy of tones over the system.

Master Group MG. In frequency division multiplexing (the old way of putting many voice conversations onto on communications line) a master group consists of 300 voice-grade (4 kHz) channels.

Master Number Hunting When a call is directed to the pilot number of a hunt group, it will hunt to the first non-busy station in that group. If a call is directed to a specific station in that hunt group it will go directly to that station and not hunt to another station in the group.

Master Slave Switching System A configuration consisting of a central switch and one or more remote switches. The master switch typically controls all I/O (input/output) information. The slave system performs tasks as directed by the master, including switching calls between phones attached to that remote module — without sending those calls back to the central switch. There are enormous savings in wiring since not every remote phone has to have a pair back to the central switch.

Master Slave Timing In a communication system, a timing subsystem wherein one station or node supplies the timing reference for all other interconnected stations or nodes.

Master Station 1. The main phone or station in a group. The one controlling the transmission of the others.

2. The unit which controls all the workstations on a LAN, usually through some type of polling. The master station on a token-passing ring allows recovery from error conditions, such as lost, busy or duplicate tokens, usually by generating a new token. Sometimes servers are referred to as master stations.

3. In navigation systems employing precise time dissemination, a station whose clock is used to synchronize the clocks of subordinate stations.

4. In basic mode link control, a data station that has accepted an invitation to ensure a data transfer to one or more slave stations. At a given instant, there can be only one master station on a data link.

Mastergroup A mastergroup consists of 600 voice channels, or 10 supergroups. Six mastergroups are equal to one jumbogroup.

Masters MCI's Administrative & Student TElecommunications Services program. MCI's Masters program reduces students' long distance calling costs while raising college and university revenues. Students are issued calling cards with built-in credit limits, no monthly fees, toll-free customer service, a parent billing option, an itemized monthly statement, and low calling rates. The MASTERS program easily ties in with the growing use of debit cards (modeled after Florida State University's FSUCard) which are being implemented by other colleges and universities. The integration of the calling and debit card gives students a single plastic card from which to make banking transactions, gain access to student services such meal programs and copy machines, and provide access to telecom features such as voicemail and fax machines.

Matched Junction A waveguide component having four or more ports, and so arranged that if all ports except one are terminated in the correct impedance, there will be no reflection of energy from the junction.

MATEL A Multiplex Automatic TELephone system. Picture one long, up to two miles wire (any decent quality works). You roll it out, then you clip phones anywhere into the wire. Then you have, in effect, a seven channel bus PBX. You have two digit extension dialing between the phones (up to 60). You also have conferencing, broadcast, call back, DID and connection to one central office line and one radio channel. The uses? String it around a rioting prison, a siege, an airport hijack, an emergency in New York City subway, etc. Three advantages: instant communications, communications where radio is bad and radio silence — you keep the press and the bad guys in the dark. This definition from John McCann, general manager of Racal Acoustics Limited, Frederick, MD, which makes the MATEL.

Material Dispersion Material dispersion occurs because of pulse of light in a fiber includes more than one wavelength. Because of the refractive index of a material varies with wavelength (check out a prism in the sun!) different wavelengths travel down the fiber at different paths. See also CHROMATIC DISPERSION.

Material Scattering In an optical fiber, that part of the total scattering attributable to the properties of the materials used for fiber fabrication.

Math Coprocessor A coprocessor is a special purpose microprocessor which assists the computer's main microprocessor in doing special tasks. A math coprocessor performs mathematical calculations, especially floating point operations. Math coprocessors are also called numeric and floating point coprocessors. If you do a lot of mathematical tasks on your PC, like recalculating large spreadsheets, then installing a math coprocessor makes huge sense. Intel included a math coprocessor with its 486DX chip, but removed it for the 486SX. No other Intel chip has a math coprocessor built in. When you buy a math coprocessor make sure it's the same speed as your existing processor.

MATR Minimum Average Time Requirement.

Matrix 1. A switch. A device for moving calls from one input to the desired output. There are many types of switching matrices — from simple step-by-step matrices to complex digital pulse code modulated matrices. Most switching matrices are "blocking." They do not have sufficient capacity to switch every call. There are some switches that are "non-blocking." These have the ability to switch every call simultaneously. By definition, non-blocking matrices are more expensive. They are only needed in special situations of high traffic.

2. The encompassing material in the composite (i.e. the plastic).

Matrix Ringing Two key system phones picking up the same extensions with different lines ringing on different phones. These days, with electronic phones, matrix ringing is easy. In the old days, with 1A2 phones, you needed to do considerable wiring.

Matrix Switch Device that allows multiple channels connected via serial interfaces (typically RS-232C) to connect under operator control to designated remote or local analog circuits or other serial interfaces.

MATV Master Antenna System, such as used in apartment buildings and motels.

MAU 1. Math Acceleration Unit.

2. Multistation Access Unit. A MAU is a wiring concentrator used in Local Area Networks. In token ring networks it's called a Multi-station Access Unit. In Ethernet networks it's called a Medium Attachment Unit. Basically a MAU is a device that allows terminals, PCs, printers, and other devices to be connected in a star-based configuration to Token Ring or Ethernet LANs. MAU hardware can be either active or passive. Each computer is wired directly to the MAU which then provides the connection to all other computers. MAUs themselves can be connected to expand the network. The MAU is a small box with eight or sixteen connectors and an arrangement of relays that function as bypass switches. When only one MAU is used, its relays and internal wiring arrange themselves so that the MAU and the connected computers form a complete electrical ring. MAUs can be cascaded to create bigger rings. The MAU listens for the "I'm here" signal sent by a computer when the computer is attached to the MAU. If the token ring adapter card in a computer is not working properly or the computer is turned off, the MAU no longer hears the "I'm here" signal and automatically disconnects the computer from the ring using the bypass relay.

A managed MAU contains on-board intelligence which enables it to communicate with network management software. This software can then be used to control the MAU. This type of MAU provides the network administrator with an "out-of-band" (a separate line) method of port control, monitoring, and diagnostics. For example, in the event that a token ring network is down, conventional network management may not be possible as it relies on the network itself to communicate with the various components on the token ring LAN. The out-of-band management provided by the managed MAU gives the administrator the ability to diagnose and reconfigure the network without physically inspecting and reconfiguring each MAU in the network.

MaxCR Maximum Cell Rate: This is the maximum capacity usable by connections belonging to the specified service category.

Maximize Button The maximize button in Windows is the up-arrow button at the far right of a title bar in the Windows operating system. Click on the maximize button to enlarge the IMARA Lite window to full size. See also, Minimize button and Restore button.

Maximum Access Time Maximum allowable waiting time between initiation of an access attempt and successful access.

Maximum Block Transfer Time Maximum allowable waiting time initiation of a block transfer attempt and completion of a successful block transfer.

Maximum Calling Area Geographic calling limits permitted to a particular access line based on requirements for a particular line.

Maximum Keying Frequency In facsimile systems, the frequency in hertz numerically equal to the spot speed divided by twice the X-dimension of the scanning spot.

Maximum Modulating Frequency The highest picture frequency required for a given facsimile transmission system. The maximum modulating frequency and the maximum keying frequency are not necessarily equal.

Maximum Stuffing Rate The maximum rate at which bits can be inserted or deleted.

Maximum Transmission Unit MTU. The largest possible unit of data that can be sent on a given physical medium. Example: The MTU of Ethernet is 1500 bytes.

Maximum Usable Frequency MUF. The upper limit of the frequencies that can be used at a specified time for radio transmission between two points and involving propagation by reflection from the regular ionized layers of the ionosphere. MUF is a median frequency applicable to 50 percent of the days of the month, as opposed to 90 percent cited for the lowest usable high frequency (LUF) and the optimum traffic frequency (OTF).

Maximum User Signaling Rate The maximum rate, in bits per second at which binary information can be transferred (in a given direction) between users over the telecommunication system facilities dedicated to a particular information transfer transaction, under conditions of continuous transmission and no overhead information.

MAYPAC A X.25 packet-switched network operated in Malaysia by the Malaysian government.

MBE Micom Business Exchange. Trademark for MICOM's series of communications processors.

MBG Multilocation Business Group.

MBONE Multicast Backbone. A collection of Internet routers that support IP multicasting. The MBONE is used as a "broadcast (actually multicast) channel" on which various public and private audio and video programs are sent.

MBPS Mbps. M/bps. MegaBits Per Second. Million bits per second.

MBS 1. Maximum Burst Size: In the signaling message, the Burst Tolerance (BT) is conveyed through the MBS which is coded as a number of cells. The BT together with the SCR and the GCRA determine the MBS that may be transmitted at the peak rate and still be in conformance with the GCRA.

2. Meridian Business Set.

MBus A Sun Microsystems definition: An open specification for connecting multiple CPUs (such as those in SPARC modules) with a 64-bit, 320-MB/second data path. Designed by Sun Microsystems; available from SPARC International.

MBWA Management By Walking Around. A technique pioneered by David Packard, one of the two founders of Hewlett-Packard which he and Bill Hewlett started in 1939 in a small garage in Addison Avenue in Palo Alto, CA.

MC 1. Main Cross-connect. The interconnect point where wiring from the entrance facility and from the workstation is connected to telecom equipment.

2. Multi-Carrier.

MC-MVIP Multi-Chassis MVIP. See MVIP.

MC1 Cable The inter-PC chassis MVIP bus cable, which can support up to 20 PCs. It allows a developer to distribute MVIP's resources across all the connected computers. MC-MVIP type MC1 media provides 1536 x 64 Kbps of inter-chassis connectivity using twisted-pair copper cables. See MC2 and MVIP.

MC2 Cable The advanced inter-PC chassis MVIP bus cable, which leverages FDDI-II to provide up to 3072 x 64 Kbps of inter-chassis connectivity on fiber or copper. MC3 leverages SONET/SDH fiber technology at 155 Mbps to provide 2400/4800 x 64 Kbps of inter-chassis telephony. See MC1 and MVIP.

MCA Micro Channel Architecture. The internal 32-bit bus inside some of IBM's PS/2 machines. It was originally introduced by IBM as a proprietary bus which manufacturers of IBM clone PCs would have to pay IBM large royalties if they wanted to include the bus in their machines. Sadly, this strategy backfired and few people wanted IBM PCs with the MCA bus. ISA remains the most popular PC bus. See ISA and EISA.

MCC Mobile Country Code. A portion of the LAI and the IMSI (International Mobile Subscriber Identity).

MCCS Mechanized Calling Card Service was formerly known as ABC Service. MCCS is a CO switch facility that automatically bills credit card calls made on DDD without the involvement of an operator.

MCDV An ATM term. Maximum Cell Delay Variance: This is the maximum two-point CDV objective across a link or node for the specified service category.

MCF Message Confirmation Frame. Confirmation by the receiver in a fax transmission that the receiver is ready to receive the next page.

MCI 1. Once upon a time it was called Microwave Communications Inc. Thus the MCI. But now it's just MCI and it stands for nothing. It is the largest long distance phone company in the US after AT&T. In MCI's early days, the initials were said to stand for "Money Coming In." Now MCI is a $10 billion plus corporation and its stock is selling for more than AT&T's (or was at the time of writing). MCI is a full-service long distance company offering every service from switched single channel voice to leased T-1. See also MCI FAX and MCI MAIL.
2. Media Control Interface. A standard control interface for multimedia devices and files. Using MCI, a multimedia application can control a variety of multimedia devices and files. Windows provides two MCI drivers; one controls the MIDI sequencer, and one controls sound for .WAV files.
3. Message Center Interface. An interface in some PBXs which allows you to connect an external PC and do voice mail/IVR (interactive voice response).

MCI Fax A way of sending facsimile messages. You can send them directly from fax machine to fax machine. Or you can send them from your personal computer (equipped with a modem) to any fax machine anywhere in the world. If you send faxes from your PC, you don't need a fax machine and you can send the same fax to a lot of people very easily and very quickly. MCI FAX is cheaper than sending faxes by dialing directly. A good service. See also MCI Mail and INFONET.

MCI Fiberline An MCI International service that uses TAT-8 (the new trans-Atlantic digital fiber optic cable) to provide private-line channels to major business locations throughout Europe, in a wide range of transmission speeds from 56/64 to 1.544/2.048 Mbps.

MCI Information Manager MIM. A program on the NCS host which allows MCI employees access to the Data Base Administrator (DBA), primarily for updating, maintaining, and trouble-shooting purposes for Vnet services.

MCI International This MCI subsidiary handles MCI's International business. MCII, which includes Western Union International and RCA Globecom, provides private-line, Telex, cablegram, MCI Mail and other services to more than 100 overseas countries, and voice service to direct-dialable countries.

MCI Mail A way of sending electronic ASCII messages, binary files and faxes over phone lines and paper mail (through the US Postal Service) from your PC instantly to other people who have MCI Mail Mailboxes, to anyone on an electronic mail system connected to MCI Mail (like Internet, AT&T Mail, Sprint Mail, AppleLink, CompuServe, etc.) to anyone who has a fax machine anywhere in the world or to anyone who receives US Mail (i.e. all of us). MCI Mail is an easy, fast, not-costly way to keep in touch in North America and in most industrialized countries around the globe. The charm of electronic mail is that increasingly executives respond to electronic mail while they don't always respond to the voice mail messages. Another charm of electronic mail is that you get to organize your thoughts, instead of blabbing them out as we often do with voice mail messages. Also electronic mail is more accurate. Numbers are clear; information comes cleanly across. In case you hadn't got the message, the author's prefers and enjoys electronic mail. If you want to send the author a message, send a message to HARRY NEWTON, who is an MCI Mail subscriber. Harry's MCI Mail number is 101-5032. Some of the definitions in this dictionary came to him via MCI Mail. Here's a quick reference to MCI Mail. Ever subscriber has five "folders." They are INBOX (incoming unread mail), DESK (incoming mail already read), OUTDESK (mail you have sent) and PENDING (unsent mail you have prepared). Here are MCI Mail's most important commands: READ (read incoming messages), CREATE (prepare a message to send), ANSWER (respond to a message), PRINT (dump all my incoming messages to me nonstop), FORWARD (send a message I received to someone else). Typically, you get things done on MCI Mail by combining messages, e.g. PRINT INBOX, which dumps all your incoming, unread messages. See also AT&T Mail, MCI FAX and INFONET.

MCL Mercury Communications Limited (UK). The second long distance company in England. It is competitor of British Telecom, the erstwhile monopoly local and long distance company in the U.K.

MCLR An ATM term. Maximum Cell Loss Ratio: This is the maximum ratio of the number of cells that do not make it across the link or node to the total number of cells arriving at the link or node.

MCNC Microelectronic Center of North Carolina.

MCR An ATM term. Minimum Cell Rate: An ABR service traffic descriptor, in cells/sec, that is the rate at which the source is always allowed to send.

MCT Algorithm A compression algorithm introduced in 1986 by PictureTel. MCT reduced the bandwidth necessary to transmit acceptable picture quality from 768 kbps to 224 kbps making two-way videoconferencing convenient and economical at relatively low data rates (for those times).

MCTD An ATM term. Maximum Cell Transfer Delay: This is the sum of the fixed delay component across the link or node and MCDV.
MCU Multipoint Control Unit. A PBX-like device for switching and conferencing video calls, announced by AT&T on March 22, 1993.
MCVF Multi-Channel Voice Frequency.
MD Mediation Device. A SONET device that performs mediation functions between network elements and OSs. Potential mediation functions include protocol conversion, concentration of NE to OS links, conversion of languages, and message processing.
MDA Monochrome Display Adapter.
MDC Meridian Digital Centrex. A Northern Telecom abbreviation.
MDF See MAIN DISTRIBUTION FRAME.
MDQ Market Driven Quality. An IBM term of the mid-1980s.
MDT Mean Down Time.
MDUs Multiple Dwelling Units. A telephony jargon word for high-rise apartment buildings.
Meaconing A system for receiving radio beacon signals and retransmitting them on the same frequency to confuse navigation and cause inaccurate bearings to be obtained by aircraft or ground stations.
Mean Busy Hour For a telephone line or group of lines or a switch the Mean Busy hour is the 60 minute period where traffic is the greatest.
Mean Power Of A Radio Transmitter The average power supplied to the antenna transmission line by a transmitter during an interval of time sufficiently long compared with the lowest frequency encountered in the modulation taken under normal operating conditions. Normally, a time of 0.1 second, during which the mean power is greatest, will be selected.
Mean Time Between Failure MTBF. The average time a manufacturer estimates before a failure occurs in a component, a printed circuit board or a complete telephone system. One must check, since MTBFs are cumulative.
Mean Time Between Outages MTBO. The mean time between equipment failures or significant outages which essentially render transmission useless. See MEAN TIME BETWEEN FAILURE.
Mean Time To Repair MTTR. The vendor's estimated average time required to do repairs on equipment.
Mean Time To Service Restoral MTSR. The mean time to restore service following system failures that result in a service outage. The time to restore includes all time from the occurrence of the failure until the restoral of service.
Measured Rate A message rate structure in which the monthly phone line rental includes a specified number of calls within a defined area, plus a charge for additional calls. See LOCAL MEASURED SERVICE.
Measured Service Also known as USAGE SENSITIVE PRICING (USP). A local phone company method of pricing used to bill local phone calls. Measured service is often charged on the number of calls, the time of day, the distance traveled and the length of the call. See LOCAL MEASURED SERVICE.
Meatware People.
MECCA Multiplex Engineering Control Center Activity
Mechanic A programmer.
Mechanical Hold A very basic line-holding mechanism used on simple two- and three-line phones that operated by placing a short circuit or a resistor across one phone line while talking on another. Chief disadvantage was that a call put on hold at one phone could not be taken off hold at another phone. Inexpensive multi-line phones with electronic holds largely replaced mechanical holds in the 1980s.
Mechanical Splice A splice in which optical glass fibers are joined mechanically (e.g., glued or crimped in place) but not fused (i.e. melted) together.
Mechanical Stripping Removing the coating from a fiber using a tool similar to those used for removing insulation from wires.
Mechanized Calling Card Service MCCS was formerly known as ABC Service. MCCS is a central office switch feature that automatically bills credit card calls made on DDD (direct distance dial) rates without the involvement of an operator.
Mechanized Line Testing MLT - The system provides computer control of accurate and extensive loop testing functions in the customer contact, screening, testing, dispatch and closeout phases of trouble report handling. It also provides full diagnostic outputs instead of just pass/fail indications.
Media In the context of telecommunications, media is most often the conduit or link that carries transmissions. Transport media include coaxial cable, copper wire, radio waves, waveguide and fiber.
Media Access Control MAC. The lower half of the data link layer as defined in 802.3. The MAC layer contains specifications for LAN frame format as well as the rules for accessing the physical portions of the network.
Media Compatible Usually used to refer to floppy disk media. Even though two different computers (e.g. an AT&T PC 6300 and an Apple IIe) both use 5 1/4 inch floppy disks, the information recorded on them is recorded in a different format and thus, they are not media compatible. You can put one disk in another's machine. But it won't work. You'll get a dumb error message.
Media Interface Connector An optical fiber connector which links the fiber media to the FDDI node or another cable.
Media Path Same as wire run. The means by which telephone signals are conveyed from the Network Interface Jack to the Communications Outlet.
Media Server A new term for a file server on a local area network which contains files containing voice, images, pictures, video, etc. In short, a media server is a repository for media of all types. Media servers are also called file servers. Here's a definition of Media Server, courtesy of Oracle Corporation, writing in early 1994: "The media server harnesses the power of the fastest computers ever built — massively parallel computers. Media servers support diverse clients so that users don't have to discard existing systems. They provide storage, network interfaces and memory — plus support for all forms of multimedia information. Because they use thousands of low-cost microprocessors, media servers offer astonishing performance at low cost."
Mediation System A wireless telecommunications term. A mediation system provides for three functions for transporting data from one device to another. These include protocol conversion, message routing, and store-and-forward processing.

Medium Any material substance that can be used for the telecommunications transmission of signals. "Mediums"' (or media) include optical fiber, cable, wire, dielectric slab, water, air or free space.

Medium Access Control MAC. Sublayer in a local area network, that part of the OSI Data Link Layer that supports topology-dependent functions and uses the services of the physical layer to provide services to the logical link control sublayer.

Medium Frequency MF. Frequencies from 300 KHz to 3000 KHz.

Medium Interface Connector MIC. In LAN/MAN systems, the connector at the interface point between the bus interface unit and the terminal, termed the medium interface point.

Meet-Me Conference A conference arrangement by which any telephone can dial a specific access code to join a conference with other users who have dialed the same code. Conference participants may call in at a preset time or may be directed to do so by an attendant or secretary. Some phone systems restrict this to intercom circuits only. In almost all phone systems there is a maximum number of parties that can be connected in such conference at one time.

Meet-Me Intercom Conference Dial a special number ("access code") and any telephone can join an intercom conference call.

Meet-Me Page A feature which allows a person to answer an intercom page from any phone in the system.

Meet-Point Billing This is a billing arrangement that applies when two Local Exchange Carriers in the same LATA are used to complete a private line circuit.

Mega A prefix meaning one million, also represented as an M. MEGABIT = one million bits. MEGABYTE = one million bytes. MEGAHERTZ = one million cycles per second. See also MEGABYTE.

Megabyte Megabyte. A unit of measurement for data storage equal to 1,048,576 bytes. Here is a summary of sizes: MH = Megabyte (10 to the 6th power) GH = Gigabyte (10 to the 9th power) TH = Terabyte (10 to the 12th power) PH = Petabyte (10 to the 15th power) EB = Exabyte (10 to the 18th power)

Megacenter An MCI definition. An MCI facility providing concentrated telemarketing. A megacenter does not handle incoming calls or customer service.

Megacom Megacom 800 and Megacom. Two AT&T services which are variations on INWATS and OUTWATS. Both are normal WATS services except that the local lines to and from the customer's office to and from AT&T's serving class 4 central office are the responsibility of the customer. Such lines tend to be T-1. And they are usually direct wiring through the local phone company, or short-haul microwave, or leased fiber optic.

Megaflops Million Floating point Operations Per Second. A measure of computing power usually associated with large computers. Mega means million. Also known as MFLOPS.

Megahertz MHz. A unit of frequency denoting one million Hz or one million cycles per second. See BANDWIDTH and HERTZ.

Megalink Name for BellSouth's leased T-1 service.

Megastream British Telecom's brand name for a service of 30 64-Kbps channels (i.e. E-1).

Megohm A resistance of 1,000,000 ohms.

Memo 1. A telephone feature that enables the user to store a phone number for calling in the future. For example, while speaking to a Directory Assistance operator, you can put the number she gives you into memory, and then call that number by pushing one or two buttons.

2. A call center term. A free form field used to store descriptive text or comments. The information in a memo field can be of any length and type.

Memory The part of a computer or sophisticated phone system which stores information or instructions for use. Memory comes in many variations. There is memory which is lost when the power is switched off. There is memory which is retained when power is turned off.

Memory Administration MA. A set of functions that provide network system database updates, network system database integrity, network system database security and network system database backup and restoration. Definition from Bellcore in reference to its concept of the Advanced Intelligent Network.

Memory Board An add-on board designed to increase a computer's amount of RAM.

Memory Caching A technology for increasing hardware performance by storing frequently used sequences of instructions in a memory cache separate from the computer's main memory where they can be more quickly accessed by the CPU.

Memory Call Service A family of central office based voice messaging services from BellSouth.

Memory Cards The memory card is a bunch of memory chips crammed into a small plastic cartridge about the size of a credit card and about three times the thickness. It is used in several palmtop computers. As this dictionary was being written, we were awaiting the release of a 16 megabyte memory card. In contrast to flash memory, a memory card requires small batteries, typically the same ones as used in watches.

Memory Interface A PCMCIA definition. The memory interface is the default interface after power up, PCMCIA Hard Reset and PCMCIA Soft Reset for both PCMCIA cards and sockets. This interface supports memory operations as defined in PCMCIA Release 1.0 and later and is used by both Memory Cards and I/O Cards.

Memory Map An indication of what type of data is stored where in a computer's RAM memory.

Memory Protection The structuring of memory resources in Novell's NetWare 4.0 that guards the NetWare server memory from corruption by NLMs. Memory protection allows you to run NLMs in a separate memory domain called the OS PROTECTED domain. Once you determine the NLM to be safe, you can load it into the OS domain, where it can run most efficiently.

Memory Reserve Power The operating voltage, generally provided by a battery, which supplies power to the memory modules when your commercial power fails. You should check your memory reserve power before it's too late. You should test it even when you don't need it.

Memory Technology Driver A PCMCIA card definition. A memory technology driver is a memory device specific software that interfaces to Card Services to mask the details of accessing different memory technologies.

Menu Options displayed on a computer terminal screen or spoken by a voice processing system. The user can choose what he wants done by simply choosing a menu option — either typing it on the computer keyboard, hitting a touchtone on his phone or speaking a word or two. There are basically two ways of organizing computer or voice processing software — menu-driven and non-menu driven. Menu-driven programs are easier to use but they can only present as many options as can be reasonably crammed on a screen or spoken in a few seconds. Non-menu driven screens allow more alternatives but are much more complex and frightening. It's the difference between receiving a bland "A" or "C" prompt on the screen — as in MS-DOS and receiving a menu of "Press A if you want Word Processing," "Press B if you want Spread Sheet," etc. It's very easy to write menus in MS-DOS using BATch files. See also AUDIO MENUS and PROMPTS.

Menu Bar call center term. The part of the menu system visible as a single row across the top of the display.

Mercury-Wetted Relay A relay or switch in which the movable contact of the device makes contact with a pool of mercury. Before solid-state devices, this was common technology for switching dry circuits. See also DRY CONTACT.

Merging Traffic The telecommunications, cable, consumer electronic and media conglomerates all vying for access to the same markets. See SILIWOOD.

Meridian A Northern Telecom name for its family of PBXs. In recent years, Northern's PBXs have evolved from the SL-1 family to the Meridian SL-1 family to the Meridian 1 family, which Northern says will support from 30 to 60,000 lines.

Meridian Link Meridian Link from Northern Telecom enables Meridian 1 Communication Systems (i.e. PBXs and ACDs) to exchange information with host computers and the application software that resides on those computers. This means that the application software can use information such as Automatic Number Identification (ANI) and Dialed Number Identification System (DNIS) to automatically perform a series of routines, such as record look-up. According to Northern, its customer installations have proven that these capabilities increase employee productivity, reduce call lengths and result in higher revenues. Meridian Link offers these capabilities to application software:
• Intelligent Answering. Intelligent Answering provides agents with the capability of knowing who is calling (ANI) and why that person is calling (DNIS). Intelligent answering improves agent productivity by saving the record look-up time.
• Computer Assisted Dialing. This provides a productive means of passing dial requests from a computer to Meridian 1. Computer assisted dialing eliminates errors and improves productivity.
• Intelligent Dialing. With predictive dialing, the computer makes calls and monitors call progress, transferring active (or answered) calls to available agents.
• Coordinated Transfer of Voice and Data. A telephone call and a screen of information are simultaneously transferred. A customer does not have to repeat information.
• Telephone Operations from a Terminal. Capabilities include making a call, disconnecting, transferring, or conferencing by a keyboard command.

Meridian Link Interface A Northern Telecom definition. The Meridian Link interface is the link between the host computer and the Meridian, through the Meridian Link Module. It uses LAB protocol between the MLM and Meridian 1, and either X.25 or LAB protocol between the MLM and host computer. See MERIDIAN LINK.

Meridian Link Module MLM. A Northern Telecom definition. The MLM supports the Meridian Link interface to host processors. It is packaged within the Application Equipment Module. See MERIDIAN LINK.

Meridian Mail Northern Telecom's voice messaging system. Meridian Mail provides voice processing capability to Meridian Link applications.

Meridian Teladapter TCM Connects the digital telephone and Macintosh to the switch. The TelAdapter connects to a QPC578 or NT8D02AB digital line card.

Meridional Ray In fiber optics, a ray that passes through the optical axis of an optical fiber. This contrasts with a skew ray, which does not.

Merlin AT&T's first electronic key system, distinguished by futuristic styling, horrible membrane line keys, nonstandard 8-conductor wiring scheme, and expensive accessories. Later versions were much better. See below.

Merlin Classic Term applied to original series of AT&T's MERLIN 206, 410 and 820 electronic key telephone systems.

Merlin II Third generation of AT&T's Merlin electronic key telephone system. Uses digital technology, and either older analog Merlin phones, dedicated single-line phones or newer digital multi-line phones. Programming is menu-driven. Features include hospitality functions and automatic route selection. Accepts Legend circuit boards and phones.

Merlin Legend Fourth Generation of AT&T's Merlin phone system, actually considered an enhancement of previous Merlin II. Maximum size is 90 lines and 144 phones (but not both maximums in same system). Operates in either key system, PBX or "behind switch" mode. T-1 compatible. Works with either older Merlin phones or new line of MLX digital phones, with more conventional styling than older Merlin phones.

Merlin Plus Second Generation of AT&T's Merlin electronic key telephone system. It introduced such features as call forwarding, automated attendant, remote system access, auto-busy-redial, and direct station access.

MERS Most Economical Route Selection. A term used by GTE and some other PBX manufacturers to mean Least Cost Routing. See LEAST COST ROUTING.

MESA Architecture Centigram's trademarked brand name for its architecture that stands for Modularly Expandable System Architecture: 1. The capability to expand and/or enhance a VoiceMemo II system hardware and software in a modular fashion. 2. The capability to expand a single module VoiceMemo II into a multi-module system with a single database and centralized control.

Mesa-Flex Centigram's service design utility that allows a VoiceMemo II system to be fully customized through individually designed Feature Class of Service, independent Limits Classes of Service, Group Classes of Service and Network Classes of Service.

Mesalink Centigram's registered trademark for its inter-processor high speed bus that carries control information to the distributed processors.

Mesh Connectivity A data network that offers multiple paths between points. See MESH NETWORK.
Mesh Network A data network that provides multiple paths between points. Internetworking devices choose the most efficient paths in moving data from one point to another. A mesh network might be constructed for greater reliability (one path goes down, another can take over) or it might be constructed because all the points on the network need to be connected (that's the more typical case).
Meso The Greek prefix meaning the middle.
Mesochronous The relationship between two signals such that their corresponding significant instants occur at the same average rate.
Message 1. A sequence of characters used to convey information or data. In data communications, messages are usually in an agreed format with a heading which establishes the address to which the message will be sent and the text which is the actual message and maybe some information to signify the end of the message. A Northern Telecom Norstar definition: A message, which appears on the telephone display that informs the recipient to call the person who sent the message. Messages can only be sent within the Norstar system.
2. The Layer 3 information in the OSI model that is passed between the CPE and SPCS for signaling.
3. A SCSA definition. The transport container for SCSA requests, replies and events. Assumes a set of conventions for directing the delivery of the message to the proper entity, either a client or service provider. See also SCSA Message Protocol.
Message Alert A cellular phone term, also called "call-in-absence" indicator. A light or other indicator announcing that a phone call came in, an especially important feature if the cellular subscriber has VOICE MAIL.
Message Alignment Indicator In a signal message, data transmitted between the user part and the message transfer part to identify the boundaries of the signal message.
Message Backbone A single format message transport system designed for the electronic mail and messaging needs of an entire corporation or enterprise.
Message Center A centralized place within the corporation where messages are taken and (occasionally) delivered. Message centers are good if they are staffed with competent, motivated people and the various phones in the place have message waiting lights (like they do in hotels). If staffed by talented people, message centers work a lot better than the amateur message takers called secretaries and their part-time short-term replacements.
Message Detail Recording See CALL DETAIL RECORDING and CALL ACCOUNTING SYSTEMS.
Messaging Enabled Applications MEA. This term defined by the Electronic Messaging Association, Arlington, VA. Applications that directly access the messaging service as a way of addressing and transporting information to and from objects on a network. Messaging Enabled Applications differ from Mail Enabled Application because they do not require electronic mail to successfully navigate the network. They may or may not have their own directory and also have the ability to directly access Directory Services. For example: A student requests a class via electronic mail to the education system. The education system automatically schedules the student (this an electronic mail enabled application).
Message Format The rules for placing information necessary for an electronic message. The format includes where the heading is and how long it will be as well as other control information.
Message Frame A SCSA term. A data link layer frame the encapsulates control and signaling data transmitted on the SCbus or SCxbus Message Bus. The form of a Message Bus frame is fully compliant with ISO HDLC UI (Unnumbered Information) Frame specifications. See SCSA.
Message Handling Services MHS. Software whose primary function is the movement of messages between application programs. MHS was developed by Action Technologies and Novell acquired full marketing and development rights to MHS for Netware based Lans. Under MHS each application sends messages to the server's \ mhs \ mail \ snd directory. MHS delivers messages to an application by placing them in the application's assigned directory. MHS is most commonly used for e-mail (electronic mail). Various e-mail programs use the MHS format for exchanging e-mail messages.
MHS also means the standard defined by ITU-T as X.400 and by ISO as Message-Oriented Text Interchange Standard (MOTIS). MHS is the X.400 family of services and protocols that provides the functions for global electronic-mail transfer among local mail systems.
Message Header The header before a string of data containing information regarding the destination of the data, usually in a packet in X.25 format.
Message Management A new term for managing all your voice mail, fax mail and electronic mail. The major concept is to join the control of the devices that produce voice, fax and e-mail through your desktop PC connected over your LAN. For example, you come into work in the morning, turn your PC on, and immediately see a screenful of messages — one line per message. By clicking on that line with your mouse, your PC would pull up the application that will then let you read or hear your message. The benefit is that you can handle your messages faster and be more discriminating about which ones you pay most and least attention to.
Message Packet A unit of information used in network communication. Messages sent between network devices (workstations, file servers, etc.) are formed into packets at the source device. The packets are reassembled, if necessary, into complete messages when they reach their destination. A message packet might contain a request for service, information on how to handle the request, and the data that will be serviced. An individual packet consists of headers and a data portion. Additional headers are appended to the data portion as the packet travels through the layers of the communication protocol. Any message that exceeds the maximum size is partitioned and carried as several packets. When the packet arrives at its destination, the headers are stripped off, the message delivered and the request serviced.
Message Rate A method of billing local phone calls that varies from one place to another. Phone calls are billed as "message units." Message units are a combination of length of call and distance of the call. In a city like New York you might buy "basic" phone service and be entitled to 50 "message units." That may mean you can call the local pizza house for under 5 minutes 50 times. Or it may mean that you can call from Manhattan to the Bronx 25 times — assuming each call is two message units.

Message Register Leads Terminal equipment leads at the interface used solely for receiving dc message register pulses from a central office at a PBX so that message unit information normally recorded at the central office only is also recorded at the PBX.
Message Registration A phone system feature that records the number of message units incurred by each phone. Useful in hotels which bill local calls by message units.
Message Retrieval The ability of a fax machine to store material already transmitted so it can be retransmitted.
Message Signal Unit Signal unit of CCS that carries a message corresponding to the information part or packet of the HDLC frame plus a message transfer part corresponding to the HDLC frame header.
Message Store An X.400 electronic mail term: A staging point, similar to a post office, in which messages are temporarily held for later transmission to one or more recipients.
Message Switch A message switch is another term for an electronic mail gateway, which is a LAN application that fetches messages from one electronic mail system, translates them to the format of another electronic mail system, and then sends them to the "post office" of that other system. The post office is the public entry point — the place you put mail you want the other system to receive.
Message Switching A technique for receiving a message, storing it for a while and then sending it on. Message switching is normally used when the desired recipient is not there. The message switch will keep attempting delivery, freeing the calling party to handle other work. Unlike voice phone calls no direct connection is made in message switching between the incoming and outgoing messages. Each message, like a Western Union telegram, contains a destination address and is recipient through intermediate nodes. Each node along the way receives the message, stores it briefly, and then passes it on to the next node.
Message Switching Network A public data communications network over which subscribers send primarily text messages to one another.
Message Telecommunications Service MTS. The regulatory term for long distance or message toll voice service. Misnamed. Actually there's nothing "message-y" about this service. This is a 100% switched telephone service (which can, obviously, be used for voice, data, video or fax).
Message Telephone Service MTS. Official designation for tariffed long-distance, or toll, telephone service. See MESSAGE TELECOMMUNICATIONS SERVICE.
Message Toll Voice Service See Message Telecommunications Service.
Message Transfer MT. The carriage of information between parties using computers as intermediaries. It is one aspect of message handling.
Message Transfer Agent An X.400 electronic mail term: Software usually residing in a LAN server or host computer that moves messages between senders and recipients.
Message Transfer Part MTP. The part of SS7 signaling node that is used to place formatted signaling messages into packets, strip formatted signaling messages from packets, and send or receive packets.
Message Unit 1. The charge for one unit of local telephone service. How many message units you will find on your bill is a function of how many calls you made, how far the calls traveled, what time of the day or night you called and for how long you talked.
2. In IBM Corp's Systems Network Architecture, the portion of the data within a message that is passed to, and processed by, a specific software layer.
Message Unit Detail MUD. A service offered by local telephone companies in which they give you a report listing the phone number of all local calls made from each of your billing numbers. The billing number may be the main number for a PBX or the individual extensions if Centrex service is used. MUD reports are usually available to a telephone company customer at additional cost. MUD reports generally have to be requested in advance. Some telephone company central offices cannot generate MUD reports. When available, MUD reports may not be in machine processable form.
Message Waiting A light on the phone or some letters or characters on the phone's display indicating there's a message waiting somewhere for the owner of the phone. That message might be with a special message center (as in a hotel or a larger company), with the operator or with a computer attached to the phone system or with someone else in the company. Message waiting lights are incredibly useful at hotels. It's amazing more companies don't use them also.
Message-To-Slave Directory Propagation In electronic mail on a local area network, message-to-slave directory propagation is a way of updating user addresses where changes in the master post office are sent to the slaves, but changes in the slave post offices are not sent to the master.
Messages Roses French term for pay telephone pornographic messages. What the English call phone sex lines.
Messaging Application Programming Interface MAPI. A set of calls used to add mail-enabled features to other Windows-based applications. See MAPI.
Messaging Middleware Lets applications communicate and exchange data via asynchronous messages and queues.
Messenger A piece of heavy metal cabling attached to a pole line to support aerial phone cable.
MET Multibutton Electronic Telephone for an old phone system called AT&T Horizon.
Metal Oxide Varistor Metal Oxide Varistor. A voltage dependent resistor which absorbs voltage and current surges and spikes. This low-cost, effective device can sustain large surges and switch in 1 to 5 nanoseconds. It is used as a surge protector and suppressor. It often the first electronic component that electrons coming in on an incoming phone line hit. Many trunk boards inside PBX are protected by MOVs. If the voltage or current is high, it will blow the MOV, thus protecting the remaining the far more valuable devices on the board.
Metal Tape Recording tape coated with iron particles and noted for its wide dynamic range and wide frequency response.
Metallic Circuit A circuit completely provided by metallic wire conductors, and not containing any carrier, radio or fiber and in which the ground or earth forms no part.
Metallic Voltage A potential difference between metallic conductors, as opposed to a potential difference between metallic conductor and ground.

Metallization Metallization is necessary if composites are to be used in antenna applications. Antennas are mainly metallized with copper or gold. Composite systems that lend themselves to metallization are those epoxy and thermoplastic composites with Kevlar, glass or carbon fiber reinforcement.

Metaphor See DESKTOP METAPHOR.

Metasignaling ATM Layer Management (LM) process that manages different types of signaling and possibly semipermanent virtual channels (VCs), including the assignment, removal and checking of VCs.

Metasignaling VCs An ATM term. The standardized VCs that convey metasignaling information across a User-Network Interface (UNI).

Metcalfe's Law Named after Robert N. Metcalfe, co-inventor of Ethernet. Metcalfe's Law states that the value of a network — defined as its utility to a population — is V=A x N x N+B x N+C where V is the value and N the number of users of a network. The value of a network grows with the square of the number of its users. Often, for small N, the COST of a network exceeds its VALUE. This means there is a critical mass phenomenon in networks. Therefore, small pilots of some networks might fail where a larger operational network might succeed wildly. It's tough getting them started, but then BAM! off they go. A simple example: One telephone is useless. Two telephones are better, but not much. Only when a good proportion of the population has a telephone does the network gain the power to change the society. Some people estimate that a totalitarian government cannot survive even one telephone per hundred people. According to many observers, the Internet owes its extraordinary growth and impact to its ability to harness both Metcalfe's Law and Moore's Law (which says that computing power and capacity double every 18 months). See also DATA DIALTONE.

Meteor Burst Communications Communications by radio signals reflected by ionized meteor trails.

Meter The metric unit of length, equivalent to 39.37 inches. An instrument for measuring quantities of length.

Metering Pulses In virtually all foreign countries, periodic pulses are returned from the distant exchange to the exchange (central office) of the calling number. These pulses determine the cost of the call, local or long distance. Typically, all pulses cost the same. However, the farther you call, the quicker the pulses come. This system contrasts to the North American long distance billing scheme which typically charges a certain amount for the first minute — no matter how much of the minute is actually used in conversation. After the first minute in the U.S., conversations are billed in one minute increments. Overseas pulses can be as short as three or four seconds (especially for international calls).

Method An SCSA definition. The specific implementation of an operation for a class; code that can be executed in response to a request.

Metric A benchmark that has nothing to do with a 100.

Metropolitan Area Network MAN. A loosely defined term generally understood to describe a broadband network covering an area larger than a local area network. It typically interconnects two or more local area networks, may operate at a higher speed, may cross administrative boundaries, and may use multiple access methods. It may carry data, voice, video and image.

Metropolitan Dial The common rotary dial or touchtone pad that contains both numbers and letters. Dials and pads are also available without the letters. Presumably metropolitan areas required the letters because of multiple central office exchanges, but rural areas with few subscribers and only one CO, required just a few digits and no letters.

Metropolitan Fiber Ring A metropolitan fiber ring is an advanced, high-speed local network that can also be used to connect businesses and residences directly to a long distance carrier's network, and provide alternatives to the local telecommunications services they have today. This definition, courtesy MCI.

Metropolitan Statistical Area MSA. Sometimes known as SMSA, MSAs are areas based on countries as defined by the U.S. Census Bureau that contain cities of 50,000 or more population and the surrounding countries. Using data from the 1980 census, the FCC allocated two cellular licenses to each of the 305 MSAs in the United States.

MF Multi-Frequency.

MFD Abbreviation for Microfarad; one thousandth of a farad, the unit of measuring capacitance. The capacitor is a common electrical device that can store electric charges, and pass AC but not DC. Most phones use capacitors to disconnect the bell during conversations.

MFJ The Modified Final Judgment is the federal court ruling that set up the rules and regulations concerning deregulation and divestiture of AT&T and the Bell system.

MFLOPS Million Floating point Operations Per Second. A measure of computing power usually associated with large computers. Also known as MEGAFLOPS.

MFM 1. Modified Frequency Modulation. An encoding scheme used to record data on the magnetic surfaces of hard disks. It is the oldest and slowest of the Winchester hard disk interface standards. RLL (Run Length Limited encoding) is a newer standard, for example.

2. Multi Function Module. A term in the AT&T Merlin phone system. MLM is an adapter that has a tip/ring mode for answering machines, modems, FAX machines, and tip/ring alerts, and a Supplemental Alert Adapter mode for 48VDC alerts. It supplies the connection of optional equipment such as answering machines, external alerts, and FAX machines to a Merlin MLX telephone. The MFM is installed inside the MLX telephone.

MFOS MultiFunction Operations System. An AT&T term.

MFP See MULTI-FUNCTION PERIPHERAL.

MFS Multifunction Peripherals. A gadget you connect to your computer that can print, photocopy, fax and scan.

MFSK Multiple Frequency Shift Keying.

MH Modified Huffman data compression method.

MHO The unit of conductivity.

MHS Message Handling Service. A program developed by Action Technologies (and others) and marketed by those firms and Novell to exchange files with other programs and send files out through gateways to other computers and mail networks. It is used particularly to link dissimilar electronic-mail systems. A company running e-mail on their internal LAN will dedicate one computer on the network to be a MHS machine. Every hour or so it will call MCI Mail, CompuServe, etc. and download e-mail messages for people and upload messages from people on the network. Once it has the messages downloaded it will distribute them to the people on the LAN the messages are destined for. See MHS MESSAGE HANDLING SYSTEM.

MHS Enterprise A messaging installation either on a local or corporate-wide level that uses MHS as its backbone between several messaging applications such as E-mail, scheduling, fax, workflow and more. Gateways are used to connect to X.400 systems, public carriers and mainframe systems.

MHS Message Handling System An ISO standard Application Layer protocol that defines a framework for distributing data from one network to several others. It transfers relatively small messages in a store-and-forward manner (defined by ITU-T as X.400 and by ISO as MOTIS/Message-Oriented Text Interchange Standard). See MHS.

MHz An abbreviation for Megahertz. One million Hertz. One million cycles per second. Used to measure band and bandwidth. See BAND and BANDWIDTH. Megahertz is also used by the computer industry to mean millions of clock cycles per second, a measure usually applied to the computer's main microprocessor. Everything that happens in a computer is timed according to a clock which ticks millions of times every second. Higher MHz computers work faster than lower MHz computers. But megahertz is not an accurate measure of a microprocessor's speed. Other factors, such as wider data paths and the ability to execute more than one instruction per clock cycle, affect the actual speed of a microprocessor. Which is why a 100 MHz Pentium chip outpaces a 100 MHz Intel DX4 chip. When comparing the speed of one PC to another, there are other factors also, such as the amount and speed of the system's random access memory (RAM).

MI/MIC Mode Indicate/Mode Indicate Common, also called Forced or Manual Originate. Provided for installations where other equipment, rather than the modem, does the dialing. In such installations, the modem operates in Dumb mode (no Auto Dial capability), yet must go off hook in Originate mode to connect with answering modems.

MIB Management Information Base. A directory listing the logical names of all information resources residing in a network and pertinent to the network's management. Within the Internet MIB employed for SNMP (Simple Network Management Protocol)-based management, ASN.1 (Abstract Syntax Notation One) is used to describe network management variables. These variables, which include such information as error counts or on/off status of a device, are assigned a place on a tree data structure. MIB is used in X.400 electronic mail. See INTERNET MIB TREE.

MIB Attribute An ATM term. A single piece of configuration, management, or statistical information which pertains to a specific part of the PNNI protocol operation.

MIB-I The initial collection of objects and attributes defined by the TCP/IP (Transmission Control Protocol/Internet Protocol) standards community. MIB-I was elevated to Internet standard status in May 1990.

MIB Instance An ATM term. An incarnation of a MIS object that applies to a specific part, piece, or aspect of the PNNI protocol's operation.

MIB Object An ATM term. A collection of attributes that can be used to configure, manage, or analyze an aspect of the PNNI protocol's operation.

MIC 1. Microphone.
2. Medium Interface Connector. FDDI de facto standard connector.
3. A designation for a multiplex which interfaces between two DSIs and one DSIC circuit.

Michael Harry's favorite son. His only one. Surprise. Surprise. Once Michael couldn't read what was in this dictionary. Now he's disagreeing with the definitions, especially this particular definition. How time flies.

Mickey Unit of space that a mouse moves, measured at 1/200th of an inch.

Micom A manufacturer of data and voice equipment located in Simi Valley, CA.

MICR Magnetic Ink Character Recognition. A process of character recognition where printed characters containing particles of magnetic material, are read by a scanner and converted into a computer-readable digital format.

Micro One-millionth.

Micro Cassette Miniaturized version of the standard audio cassette.

Micro Channel A proprietary bus developed by IBM for its PS/2 family of computers' internal expansion cards. Also offered by NCR, Tandy and other vendors. See ISA and EISA.

Micro Components Miniaturized audio components that provide the benefits of traditional sized components in far less space.

Micro Farad One millionth of a farad. This is the common unit for designation capacitance in electronics and communications.

Micro To Mainframe Link The telecommunications path over which data between a microcomputer and a mainframe computer travels.

Micro To Mainframe Software Software which provides the logic by which data can be transferred back and forth between a microcomputer and a mainframe computer.

Microbend Loss In an optical fiber, that loss attributable to macrobending.

Microcell 1. A PacTel technology to improve cellular calling coverage in areas where high quality cellular service has never been available. PacTel MicroCell, according to PacTel, will reach into canyons, freeway underpasses and high-rise areas where it has sometimes been impossible to provide good cellular coverage. MicroCell uses very small antennas that can be mounted on utility poles or billboards or in other inconspicuous places. The microcell technology can direct the cellular signal right into an isolated trouble spot, leaving broader coverage to conventional cellular sites.
2. A cellular radio cell aimed at serving pedestrians, not automobile drivers. Microcells are smaller, lower-powered, simpler and smaller than macrocells. See MACROCELL.

Microchannel A proprietary bus developed by IBM for its PS/2 family of computers, which it introduced in 1987. The bus was one of IBM's "weapons" against the disturbing increase in clone PCs (disturbing to IBM). IBM figured that introducing a proprietary bus like the microchannel (called MCA for the MicroChannel Architecture), it could slay the clones or control them by charging large amounts of money to license the microchannel architecture. The microchannel, however, engendered a big yawn from consumers, who didn't at the time need its speed, nor its greater complexity, nor its higher cost. As the need to process large amounts of Windows' video information became more urgent in the late 1980s and early 1990s, so faster buses have become more critical. And two new buses, VL and PCI, have been created. For a fuller explanation, see ISA, EISA, PCI and VESA.

Microcode Programmed instructions that typically are unalterable. Usually synonymous with firmware and programmable read-only-memory (PROM).

Microcom Network Protocol. The Microcom Networking Protocol, MNP, is a defacto standard protocol that provides error correction and data compression in dial-up modems. The protocol's design allows for a broad range of services to be implemented, while maintaining compatibility among modems with different levels of MNP capabilities. For example, a modem capable of MNP Class 5 and Class 7 data compression can talk to a modem that lacks MNP data compression.

According to Microcom, "MNP is an error correction protocol accepted by international standards authorities (ITU-T Rec. V.42). MNP offers a reliable and widely accepted method of correcting errors in transmissions over dial-up communications lines. MNP incorporates three different data compression methods, including the ITU-T recommendation, V.42bis.

Since its original definition, MNP has evolved through nine classes of enhancements. Of those nine classes, the first four provide error control and are in the public domain. Classes 5 through 7 may be licensed from Microcom. Currently MNP error control (Classes 2,3 and 4) has been adopted, along with the LAPM protocol, as mandatory elements of the Consultative Committee on International Telegraphy and Telephony (ITU-T) V.42 recommendation for modem error control.

An Overview Of MNP Service Classes

Class 1
This is the first level of MNP performance. MNP Class 1 uses an asynchronous byte-oriented half-duplex method of exchanging data. MNP Class 1 implementations make minimal demands of processor speeds and memory storage. MNP Class 1 makes it possible for devices with few hardware resources to communicate error-free. Class 1 implementations are no longer included in modems.

Class 2
MNP Class 2 uses asynchronous byte-oriented full-duplex data exchange (i.e., data goes in both directions at once). All microprocessor-based modems are capable of supporting MNP Class 2 performance.

Class 3
This class uses synchronous bit-oriented full-duplex data exchange, eliminating the overhead of start and stop bits used in byte-oriented asynchronous communications. The user still sends data asynchronously to the modem while communications between modems is synchronous.

Class 4
This class introduces two new concepts. Adaptive Packet Assembly and Data Phase Optimization, both of which further enhance performance. Adaptive Packet Assembly means that the size of the packets in which data is sent between modems is altered according to the quality of the physical link. The higher the line quality, the larger the packets. Larger packets, while more efficient (the ratio of user data to control data is higher), are also more susceptible to errors. Data Phase Optimization means that repetitive control information is removed from the data stream to make packets more efficient. Both techniques, when combined with Class 3, yield a protocol efficiency of about 120 percent (A V.22bis 2400 bps modem will realize approximately a 2900 bps throughput).

Class 5
This class implements MNP basic data compression to realize a net throughput efficiency of 200 percent on average. (A 2400 bps modem will realize 4800 bps). Class 5 uses a real-time adaptive algorithm to compress data. The real-time aspects of the algorithm allow the data compression to operate on interactive terminal data as well as file transfer data. The adaptive nature of the algorithm means data compression is always optimized for the user's data. The compression algorithm continuously analyzes the user data and adjusts the compression parameters to maximize data throughput.

Class 6
This class implements Universal Link Negotiation. and Statistical Duplexing. The first feature allows a single modem to operate at a full range of speeds between 300 and 9600 bps, depending on the maximum speed of the modem on the other end of the link. Modems begin operation at a common slower speed and negotiate the use of an alternative high speed modulation technique. The Microcom AX/9624c modem is an example of a modem that uses Universal Link Negotiation, starting with 2400 bps. V.22bis technology and shifting to 9600 bps V.29 fast train technology, if the other modem has that technology too. Statistical Duplexing allows the modem to simulate full-duplex service on the half-duplex V.29 modem connection.

Class 7
This class implements a more efficient data compression method than the one used in Class 5. The difference between the two classes is that Class 5 realizes an average 200 percent speed improvement over a non-MNP modem, versus an average 300 percent improvement for Class 7. Class 7 data compression uses Huffman encoding with a predictive algorithm to represent user data in the shortest possible Huffman codes.

In addition to Class 5 and Class 7 data compression, MNP also supports V.42bis data compression. Based on the Lempel Ziv data compression model, V.42bis supports an average 400 percent efficiency improvement.

Class 8
Not defined.

Class 9
This class reduces the amount of time required for the modem to perform two frequently occurring administrative activities: to acknowledge that a message was received- and to retransmit information following an error. Message acknowledgment is streamlined by "piggy-backing" the acknowledgment in its own dedicated packet. Retransmission is streamlined by indicating in the error or Negative Acknowledgment Packet (NAK) the order sequence number of each of the failed messages. Rather then sending all the messages over again (even the good ones) from the point of the error, as is usually done with error correcting protocols, only the failed messages are resent.

Class 10
MNP Class 10 consists of Adverse Channel Enhancements that optimize performance in environments with poor or varying line quality, such as cellular telephones, international telephone calls, and rural telephone service. These enhancements fall into four categories:
1. Multiple aggressive attempts at link setup 2. Adapting packet size to accommodate varying levels of interference 3. Negotiating transmission speed shifts to achieve the maximum acceptable line speed 4. Dynamically shifting to the modem speed most suitable to transmission line conditions.
See ERROR CONTROL PROTOCOLS and LZW.

Microcomputer The combination of CPU (Central Processing Unit) and other peripherals (I/O, memory, etc.) that form a basic computer system. See MICROPROCESSOR.

Microfloppies The latest generation of floppy disks at 3 1/2 inches diameter, invented by Sony. The microfloppy is used in the Apple Macintosh and most MS-DOS laptop computers. Used in an MS-DOS machine, a 3 1/2 inch microfloppy diskette will currently format to carry 1.44 million bytes of data — equivalent to about 500 pages of double spaced text.

Microform Microform means Microfiche and Microfilm.

Micrographics Conversion of information into or from microfilm or microfiche.

Micron One thousandth of a millimeter. Or one millionth of a meter. A unit of measurement corresponding to 1/25,000 of an inch. A micron can be used to specify the core diameter of fiber-optic network cabling. This diameter should match your hardware vendor's requirements; but if you install fiber before you buy the equipment, specify the 62.5-micron size.

Microphone A transducer that changes the air pressure of sound waves into an electrical signal that can be recorded, amplified and/or transmitted to another location.

Microprocessor An electronic circuit, usually on a single chip, which performs arithmetic, logic and control operations, with the assistance of internal memory. The microprocessor is the fabled "computer on a chip," the "brains" behind all desktop personal computers. Typically, the microprocessor contains read only memory — ROM — (permanently stored instructions), read and write memory — RAM, and a control decoder for breaking down the instructions stored in ROM into detailed steps for action by the arithmetic logic unit — ALU — which actually carries out the numerical calculations. There's also a clock circuitry which connects the chip to an exterior quartz crystal whose vibrations coordinate the chip's operations, keeping everything in step. And finally, the input/output section directs communications with devices on the outside of the chip, such as the keyboard, the screen and the various disk drives.

The Fortune Magazine issue of May 6, 1991 contained a very good explanation of chips and microprocessors (usually used interchangeably). Here is the article, slightly condensed:

Chips today can store and retrieve data, perform a simple mathematical calculation, or compare two numbers or words in a few billionths of a second. And they can carry out tens of thousands of such tasks in the blink of an eye. Today's chips contain millions of transistors, capacitors, diodes, and other electronic components, all connected by metallic threads a fraction of the diameter of a human hair. A single chip the size of a fingernail can store dozens of pages of text or combine circuits that can perform scores of tasks simultaneously.

Most chips fall into one of two categories - memory chips and logic chips. Memory chips have the easier job: They merely store information that will be manipulated by the logic chips, the ones with the smarts. Today's biggest-selling memory chip (mid-1991) is the one-megabit dynamic random access memory, or DRAM. Each DRAM is a slice of silicon embedded with a lattice of 1,000 vertical and 1,000 horizontal aluminum wires that circumscribe one million data cells. The densest DRAM designed so far has 64 million cells.

Think of those wires as streets and those cells as blocks. Each block contains a transistor that can be turned on or off — to signify 1 or 0 — and that can be identified by it's unique "address" in the wire grid, much like a house in a suburban subdivision. Each digit, letter, or punctuation mark is represented by 1's or 0's stored in eight-cell strings. (See ASCII.) The word "chip" takes up 32 cells in a memory chip. Most PCs sold today have at least eight one-megabit DRAMs.

It's the job of the logic chips to turn those transistors in the DRAMs on or off, and to retrieve and manipulate that information once it's stored. The most important and complex logic chips are microprocessors like Intel's 80386DX, the brains of the more powerful IBM-compatible PCs sold today. If the structure of a memory chip is a suburban subdivision, the layout of a microprocessor is more like an entire metropolitan area, with distinct neighborhoods devoted to different activities. A typical microprocessor contains among other things:
- A timing system that synchronizes the flow of information to and from memory and throughout the rest of the chip.
- An address directory that keeps track of where data and program instructions are stored in the DRAMs.
- An arithmetic logic unit with all the circuits needed to crunch numbers.
- On-board instructions that control the sequence of microprocessor operations.

Other logic chips in a computer take their cues from the microprocessor millions of times each second to draw images on the screen, to feed instructions from a spreadsheet program, say, out of the disk drives into DRAMs, or to dispatch data to a modem or a printer. Perhaps most amazing of all, memory and logic chips can accomplish all this with just a trickle of electricity - far less than it takes to light a flashlight bulb.

Ted Hoff at Intel invented the microprocessor in 1971. See also 1971 in the beginning of this dictionary.

Microsecond One millionth of a second.

Microsegmenting The process of configuring Ethernet and other LANs with a single workstation per segment. The objective is to remove contention from Ethernet segments. With each segment having access to a full 10 Mbps of Ethernet bandwidth, users can do things involving significant bandwidth, such as imaging, video and multimedia.

Microsegmentation Division of a network into smaller segments, usually with the intention of increasing aggregate bandwidth to devices.

Microserf A person who works at Microsoft.
Microsoft Founded in 1975 by Bill Gates and Paul Allen, Microsoft is (or was at the time of writing this edition of this dictionary) one of the largest software companies in the world. It is the originator of At Work, MICROSOFT AT WORK, MS-DOS, Windows, Windows NT and Windows Telephony. See AT WORK, MS-DOS, WINDOWS, WINDOWS NT and WINDOWS TELEPHONY.
Microsoft At Work A new architecture announced by Microsoft on June 9, 1993 and then put into retirement a couple of years later. Many of its features and ideas surfaced in Windows 95. It consisted of a set of software building blocks that will sit in both office machines and PC products, including:
• Desktop and network-connected printers. • Digital monochrome and color copiers. • Telephones and voice messaging systems. • Fax machines and PC fax products. • Handheld systems. • Hybrid combinations of the above.
According to Microsoft, the Microsoft At Work architecture focuses on creating digital connections between machines (i.e. the ones above) to allow information to flow freely throughout the workplace. The Microsoft At Work software architecture consists of several technology components that serve as building blocks to enable these connections. Only one of the components, desktop software, will reside on PCs. The rest will be incorporated into other types of office devices (the ones above), making these products easier to use, compatible with one another and compatible with Microsoft Windows-based PCs. The components, according to Microsoft, are:
• Microsoft At Work operating system. A real-time, preemptive, multi tasking operating system that is designed to specifically address the requirements of the office automation and communication industries. The new operating system supports Windows compatible application programming interfaces (APIs) where appropriate for the device.
• Microsoft At Work communications. Will provide the connectivity between Microsoft At Work-based devices and PCs. It will support the secure transmission of original digital documents, and it is compatible with the Windows Messaging API and the Windows Telephony API of the Windows Open Services Architecture (WOSA).
• Microsoft At Work rendering. Will make the transmission of digital documents, with formatting and fonts intact, very fast and, consequently, cost-effective; will ensure that a document sent to any of these devices will produce high-quality output, referred to as "What You Print Is What You Fax Is What You Copy Is What You See."
• Microsoft At Work graphical user interface. Will make all devices very easy to use and will make sophisticated features accessible; will provide useful feedback to users. Leveraging Microsoft's experience in the Windows user interface, Microsoft At Work-based products will use very simple graphical user interfaces designed for people who are not computer users.
• Microsoft At Work desktop software for Windows-based PCs. Will provide Windows-based PC applications the ability to control, access and exchange information with any product based on Microsoft At Work. Desktop software is the one piece of the Microsoft At Work architecture that will reside on PCs.
See also FAX AT WORK, VOICE SERVER, WINDOWS, WINDOWS 95 and TELEPHONY, WOSA.
Microsoft Exchange A family of products that offers enterprise computing and information sharing. According to the Windows 95 Resource Kit, Windows 95 includes the Microsoft Exchange client, an advanced messaging application that retrieves messages into one inbox from many kinds of messaging service providers, including Microsoft Mail, The Microsoft Network and Microsoft Fax. Its integration with Microsoft Fax software allows you to send rich-text documents as faxes or mail messages. With Microsoft Exchange client, you can do the following:
• Send or receive electronic mail in a Win 95 workgroup
• Include files and objects created in other applications as part of messages
• Use multiple fonts, font sizes and colors, and text alignments in messages
• Create a Personal Address Book or use address books from multiple service providers
• Create folders for storing related messages, files, and other items
• Organize and sort messages in a variety of ways
• Send and receive messages to and from the following service providers Microsoft Mail, the Microsoft Network (online service), Microsoft Fax and other messaging services that use MAPI service providers.
Microsoft Fax Microsoft Fax is an integral part of Windows 95. It is the first Microsoft operating system to include built-in faxing capabilities. I plucked the following explanation from the Windows 95 Resource Development Kit: With Microsoft Fax, users with modems can exchange faxes and editable files as easily as printing a document or sending an electronic mail message. Microsoft Fax is compatible with the millions of traditional Group 3 fax machines worldwide, yet it provides advanced security and binary file transfer (BFT) features that make sharing information by means of a fax easier and more powerful. To use Microsoft Fax, you must install Microsoft Exchange. Microsoft Fax has been integrated into Microsoft Exchange as a Messaging Application Programming Interface (MAPI) service provider. All faxes sent to Microsoft Fax are received in the Microsoft Exchange universal inbox. You can send a fax by composing a Microsoft Exchange message, or by using the Send option on the File menu of a MAPI-compatible application (such as Microsoft Excel or Microsoft Word). In addition, Microsoft Fax includes a fax printer driver so that users can "print to fax" from within any Windows-based application. Microsoft Fax provides the following key features.
Fax at your fingertips. With Microsoft Fax, sending traditional faxes to Group 3 fax machines is as easy as printing a document. Additionally, Microsoft Fax uses the highest transmission speed and image compression supported by recipient fax machine. Faxes sent in this way cannot be edited by the recipient.
Delivery by address type. The MAPI service provider architecture allows you to mix different types of recipients in the same message. For example, it is possible to send a message simultaneously to Microsoft Mail, CompuServe, Internet, and Microsoft Fax users as long as profiles for these destinations have been defined within Microsoft Exchange. A recipient's fax address can be selected from the Microsoft Exchange Personal Address Book, or the fax can be addressed by using an address that you use just once such as (fax:555-1212).
Binary file transfer (BFT). Microsoft fax supports Microsoft At Work BFT, which makes it possible to attach an editable doc-

ument to a Microsoft Exchange mail message. These editable documents can be sent to users of Windows 95, Windows for Workgroups 3.11, and other Microsoft At Work enabled platforms.
Security. Microsoft Fax lets you securely exchange confidential documents by using public key encryption or digital signatures. Any security specified by user is applied before the message is passed to the modem or connected fax device.
Network fax service. You can install a fax device in one computer and share it with other users within a workgroup. Individual computers can have their own fax devices installed and still use the shared fax device.
Microsoft Fax Viewer. The Microsoft Fax Viewer displays outgoing fax messages that have been queued to a local fax modem or to a Microsoft Fax network fax that are queued for transmission. You can also browse multipage faxes in thumbnail or full-page views.
"Best available" fax format. When you make a fax connection in Windows 95, Microsoft Fax queries and exchanges its fax capabilities with the recipient. This exchange of capabilities determines whether the recipient is a traditional Group 3 fax machine, which can only receive rendered faxes, or if the recipient has Microsoft Fax capabilities, and can receive editable files. Windows 95, Windows for Workgroups 3.11, and Microsoft At Work fax platforms are all capable of receiving binary files and traditional faxes.
If the receiving fax device supports Microsoft Fax capabilities and an editable document is attached to a Microsoft Exchange message, then the file is transferred in its native format, in the same way as electronic mail.
If the receiving fax device is a traditional Group 3 fax machine, then Microsoft fax converts the document to the most compressed type of fax supported by the machine (MH, MR or MMR compression type) and transmits the image by using the best available communications protocol supported by the mutual connection (that is, V.17, V.29 or V.27).
If Microsoft Fax sends a noneditable fax to another Microsoft Fax user, then the fax is transmitted by using the Microsoft At Work rendered fax format. This special format is much more compressed, on average, than Group 3 MMR. Therefore, the exchange of noneditable faxes between Microsoft Fax users is always faster than between Group 3 fax machines.
Compatibility with popular fax modems. Microsoft fax is compatible with Class 1 and Class 2 fax modems, and provides support for high-speed fax communications (V.17, V.29 and V.27).
Custom fax cover pages. With Microsoft fax, you can create new fax cover pages with over page designer that lets you incorporate graphics and text, or you can customize one of the predefined cover pages included with Microsoft Fax.
Connecting to fax information service. Microsoft Fax easily connects to fax-on-demand systems by using a built-in, poll-retrieve feature that allows you to retrieve rendered faxes or editable documents from a fax information service.
Before you install and configure Microsoft Fax on the network, you will need to decide the following:
- Which users need Microsoft Fax. For every user who needs Microsoft fax, you must also install Microsoft Exchange. All faxes sent to that user are received in the Microsoft Exchange inbox.
- Whether to install fax modems on individual computers or to designate a computer running Windows 95 to host a Microsoft fax network fax service for other members of a workgroup.
- Which computer within a workgroup will host the Microsoft Fax network fax service. If the computer will also be used as a workstation, then a 80486-based computer with at least 12 MB of RAM is recommended. Depending on fax volumes, a shared fax service with this configuration could support up to 25 network fax users.
When the computer hosting the shared fax modem receives faxes, Microsoft Fax does not automatically route them to individual inboxes. The workgroup administrator must use Microsoft Exchange to send a received fax to the recipient's Microsoft Exchange inbox.
- Whether you want to control or restrict access to the shared fax service. You can control access by defining a shared fax password.
- Whether your workgroup's fax needs might be better served by a LAN fax server or commercial fax service. This depends on whether your organization has high fax volumes and inbound routing requests and as a result needs more detailed cost tracking and management.
You can send faxes either by using the mail client or the Microsoft fax printer driver. In each case, the message is sent to the Microsoft Fax service provider by using MAPI. If you sent the message from a mail client, it might contain text, embedded OLE formats, and attachment to the mail message.
MAPI allows messages to be preprocessed based on the transport protocol used to send them. The transport protocol chooses the correct modem connection, uses TAPI to create a dial string, and sends the message into a fax form to be printed by a fax machine. The rendered format is attached to the original message as a message property and is deleted either when the message is sent or when the transport protocol tries to send the message but determines it cannot. If the message does not have to be rendered, the message is converted form its original binary format to a line image (also called a linearized form), and then it is compressed.
After the message is submitted, the transport protocol determines what type of recipient the message is intended for as follows:
For each local modem that has been installed by Windows 95, information is stored in the key named TAPInnnnnnnn where nnnnnnnn is an arbitrary number assigned by TAPI. The TAPI key contains subkey values. Values that are important to the system administrator are described in the following tables.
Microsoft Network An interactive online service announced by Microsoft in November 1994. Software to access this new service will be part of Windows 95. It will include information on these subjects: arts and entertainment; news and weather; business and finance; sports, health and fitness; science and technology; computers and software; community and public affairs; home and family. Other services will include e-mail, "chat", BBSes, Internet access and download libraries.
Microsoft Solution Provider This is Microsoft's definition: Microsoft Solution Providers are independent organizations that have teamed with Microsoft to use technology to solve business problems for companies of all sizes and industries. SPs use the Microsoft Solutions Platform of products as building blocks and offer various value-added services, such as integration, consulting, software customization, developing turnkey applications and technical training and support. All Solution Providers have at least one Microsoft Certified Professional on staff who has demonstrated expertise in developing, implementing, and supporting Microsoft solutions.

Microsoft Speech API The Microsoft Speech API is a set of applications programming interfaces (APIs) which allow applications to incorporate speech recognition and text-to-speech in their Win32 applications (for Windows 95 or Windows NT). All of the APIs are accessed through the OLE Component Object Model and derive benefits from this object-oriented approach. Like other Windows Open Services Architecture (WOSA) services (e.g. TAPI), the Microsoft Speech API provides a set of standard APIs which allow Windows developers to add speech capabilities to their applications without being tied directly to a specific speech engine technology.

Here are some specific areas where speech recognition and text-to-speech can improve a user-interface in any application, and a few uses to avoid. This list of pros and cons is courtesy Mike Rozak, Microsoft software design engineer.

What speech recognition is good for...

- Fast access to complex features. Some applications have features that are frequently used but which are difficult to pre
- Magic keystrokes. Many applications have overloaded the keyboard with not only text entry features, but also commands. Some applications have so many keyboard accelerators that they distinguish between ctrl-f2, shift-f2, alt-f2, etc. These are difficult for the end-user to memorize. Using voice commands to replace these makes life easier for the user because voice commands are easier to memorize.
- Macros. Macros are related to magic keystrokes. Many applications allow users to create macros that speed up frequently-done or difficult tasks. To activate these macros, users often have to invent and memorize a magic keystroke. The keystroke can be replaced with speech, whose commands are easier to memorize and are less likely to cause an accelerator conflict.
- Global commands. At times, an application wants to have an input hook that is always active, even when the application doesn't have keyboard focus. If the application uses a keyboard hook then it's likely that the chosen key will already be used by another application, causing conflicts. Because there are so many different possible sentences, an application can provide a global command that is always active and not worry about conflict. Example: A PC-based phone can use global commands for "Call <name>", so that a user can make a phone call even when the phone isn't the active application.
- Anthropomorphize the computer. If you want the user to perceive the computer as a person, then the application should use speech (synthesized or recorded) to talk to the user information, and speech recognition to get information from the user. The computer can listen for commands, or answers to questions, using speech recognition.
- Interaction with characters. Applications that have characters, especially adventure games, have a lot to gain from speech. Simulating characters is a specialized anthropomorphizing of the system.
- Form entry - Applications can use speech recognition for form entry, both for selecting elements from a list, and entering numbers. Because speech can have so many commands active at once, it can be listening for all of the possible values for all of the fields at once, and infer which field the user wanted filled in. Example: If a user spoke, "Male, sixteen, one hundred and fifty pounds," the application could correctly identify which field was being referred to by "male", "sixteen", and "one hundred and fifty pounds", set the focus to the proper field for each command, and place in the proper data. The user doesn't have to worry about tabbing around.
- Sit back and relax. If you watch users sitting in front of a computer, you'll notice that whenever they type or use the mouse they hunch forward. If you want the user to relax and sit back while using your application, you should provide a speech recognition interface. Example: A consumer title might want to use speech recognition in order to make the computer seem less work-oriented because the user can sit back in his chair without touching the keyboard and mouse.
- Dictation for poor typists. Current dictation technology (which requires that users leave pauses between words) is good for people who cannot type or who are poor typists. The only reason not to include dictation is that dictation systems require about 8 megabytes of extra RAM.
- Access over the telephone. With the arrival of voice-modems in more and more PCs, applications will start providing "phone-based" UIs. Speech recognition is a much better interface than maneuvering through menus with touch-tones, especially since most Europeans don't have touch-tone phones. Example: A user will call up his computer and ask to speak to his PIM application, which then allows him to look up his schedule, address book, etc., all over the phone. The user can merely say, "Give me the phone number for John Smith
- Accessibility. Some people cannot use the keyboard or mouse effectively. The most common disability is carpel tunnel syndrome or equivalent. Adding speech recognition enables them to use the application.
- Mouse overloaded. Some applications (especially CAD systems) drag, drop, and select objects with the mouse. If the user wants to do any more complex actions, he either has to memorize a magic keystroke (which may not have the information bandwidth), or move the mouse to a menu/toolbar. Instead, the user could just speak, "Rotate this sixty two degrees."
- No keyboard/mouse available. Sometimes a keyboard or mouse is not available. This happens a lot in kiosk situations, where the kiosk has the keyboard and mouse hidden so that users don't feel intimidated, won't break the devices, and won't steal the devices. In order to get input from the users, current systems use a touch-screen. Why not use speech recognition?
- Hands busy. If a user's hands are busy, then speech is useful. Example: Some dentists' offices use speech recognition to enter charts about the patient's teeth because their hands are busy probing around the patient's mouth.
- Multiple people in front of machine. Speech recognition is good for applications that have several people sitting in front of the same computer, each participating in the application. Normally, an application which is to be used by several users would require that they pass the mouse and/or keyboard around. Since speech recognition is effective up to several feet, users would just have to sit around the computer and talk to it.

Where not to use speech recognition...

- Selecting from a large list of words. Most recognition systems break down when more than 100 words/commands are active, so keep lists below this number. For example, allowing the user to address electronic mail to any of 10,000 employees will not work well; instead, it would be better to allow any name from the 100 (or 75, or 50) employees to whom the user frequently sends mail.

- Spelling. Asking a user to spell a word does not work well because many letters such as "m" and "n" sound the same. Instead, it would be better to ask the user to type the word or offer a list of possibilities. Spelling with communications code words_"alpha," "bravo," "charlie," and the like_may be appropriate for certain vertical markets but not for general-purpose applications.
- Entering long sequences of numbers. Most engines will have a very high error rate for long series of digits that are spoken continuously. For phone numbers or other long series, either break the number into groups of four or fewer digits or have the user speak each digit as an isolated word.
- Pointing device. Do not use speech as a pointing device. Speaking "up" five times in a row is very annoying to the user.
- Action games. Because of the background noise (music and sound effects) that is typically present, speech recognition does not work well for action games. The speech-recognition engine spends time processing audio from the computer's speakers and may even recognize it as commands. Speech-recognition works for action games only if the user wears a close-talk microphone (for example, a headset). Additionally, by the time that the user finishes saying, "Fire," it's probably too late.

Microsoft SQL Server A Microsoft retail product that provides distributed database management. Multiple workstations manipulate data stored on a server, where the server coordinates operations and performs resource intensive calculations.

Microsoft TAPI See TAPI and Windows Telephony.

Microtac A Motorola name for a family of small hand held cellular phones.

Microvax II DEC's minicomputer operating system which runs on Berkeley UNIX (also known as 4.2).

Microwave Electromagnetic waves in the radio frequency spectrum above 890 Megahertz (million cycles per second) and below 20 Gigahertz (billion cycles per second). (Some people say microwave refers to frequencies between 1 GHz and 30 GHz.) Microwave is a common form of transmitting telephone, facsimile, video and data conversations used by common carriers as well as by private networks. Microwave signals only travel in straight lines. In terrestrial microwave systems, they're typically good for 30 miles, at which point you need another repeater tower. Microwave is the frequency for communicating to and from satellites.

Microwave Multi-Point Distribution System MMDS. Microwave Multi-point Distribution System. A means of distributing cable television signals, through microwave, from a single transmission point to multiple receiving points. Often used as an alternative to cable-based cable TV. According to an April, 1995 press release from Pacific Telesis, which was starting an MMDS service, "in digital form, it will provide more than 100 channels to a radius of approximately 40 miles from the transmitter. The MMDS transmitter delivers video to homes that are in its 'line of sight.' MMDS transmissions are limited by the terrain and foliage of a given market. The microwave signal is received by an antenna on the subscriber's home, then sent down coaxial cable to a box atop the customer's TV set. The box decodes and decompresses the digital signal."

Microwave Pulse Generator MPG. A device that generates pulses at microwave frequencies.

MID An ATM term. Message Identifier: The message identifier is used to associate ATM cells that carry segments from the same higher layer packet.

Mid-Air Passenger Exchange Grim air-traffic-controller speak for a head-on collision. Midair passenger exchanges are quickly followed by "aluminum rain." Definition from Wired Magazine.

Mid-Span Meet Sonet's ability to mix the terminal, multiplexing and cross-connect equipment from different vendors. A major accomplishment for standardization. Wish more telecom systems could meet mid-span. Sadly, most can't.

Midband Microsoft's word for telecom speeds that are faster than phone lines but slower than broadband networks. ISDN BRI is midband.

Middleware Middleware is software which sits between layers of software to make the layers below and on the side work with each other. On that broad definition, middleware could be almost any software in a layered software stack. And, to be sure, middleware is not a 100% cast-in-concrete term. It is an evolving term, with many people giving the software the name "middleware" — if they see it fashionable that week. In computer telephony, middleware tends to be software that sits right above that part of the operating system dealing with telephony — TSAPI in NetWare or TAPI in Windows — but below the computer telephony application above it which the user sees on their desktop. In short, middleware is software invisible to the user which takes two or more different applications and makes them work seamlessly together.

MIDI Musical Instrument Digital Interface, a standard for connecting musical instruments, synthesizers and computers. The MIDI standard provides a way to translate music into computer data, and vice versa. A file with a MIDI extension means that it's a file containing music. In contrast, a file with the extension WAV typically means it's a sound file, which means it may contain a voice recording.

Midrange System Medium-scale computer that functions as a workstation or as a multiuser system handling several hundred terminals.

Midsplit A broadband cable system in which the cable bandwidth is divided between transmit and receive frequencies. The bandwidth used to send toward the head-end (reverse direction) is about 5 MHz to 116 Mhz, and the bandwidth used to send away from the head-end (forward direction) is about 168 MHz to 400 Mhz. The guard band between the forward and reverse directions (100 MHz to 160 MHz) provides isolation from interference. Requires a frequency translator.

MIF Minimum Internetworking Functionality. A general principle within the ISO that calls for minimum local area network station complexity when interconnecting with resources outside the local area network.

Migration An AT&T marketing strategy designed to encourage all phone equipment month-to-month renters into long-term contracts for AT&T's "flagship products", i.e., System 75, 85 and Definity PBXs.

Mike Microphone.

MIL 1/1000 of an inch.

Mileage See AIRLINE MILEAGE.

Millennium Bug Apparently, lots of COBOL-based business systems are going to break, on or around the turn of the century, because their underlying mechanisms can't reconcile two-digit 20th century dates (e.g., '00' - '99') with four-digit (or whatever) 21st century dates. It's caused because many mainframe programs used 95, instead of 1995 in the date field. that means that when it hits 00 or 01, the computer won't be able to figure that the difference between 01 and 95 = 6. There's an Expo on the

subject ("Millennium World") scheduled for April 16-18, 1996. They're calling it the "Coming IS Apocalypse." They estimate the total cost for repair will be between $100 and $300 billion. That sounds like a cute way to sell seminar registrations.
Milli One thousandth. Millisecond equals one thousand of a second.
Milliwatt One thousandth of a WATT. Used as a reference point for signal levels at a given point in a circuit.
MILNET One of the DDN networks that make up the Internet; devoted to non-classified U.S. military communications. It was built using the same technology as the ARPAnet.
MIM Metal-Insulator-Metal. A display technology which uses active matrix technology that uses diodes behind each pixel to produce images. It is an improvement on passive displays but a step behind TFT (Thin Film Transistor) technology. See also LCD.
MIMD Multiple Instruction Multiple Data is a type of parallel processing computer, which includes dozens of processors. Each processor can run different parts of the same program and execute those instructions on different data. This makes it more flexible, though more expensive than a computer running SIMD — single instruction multiple data.
MIME Multipurpose Internet Mail Extension. Developed and adopted by the Internet Engineering Task Force, this protocol is designed for transmitting mixed-media files across TCP/IP networks. The MIME protocol, which is actually an extension to SMTP, covers binary, audio and video data. Here's an explanation: When sending files which aren't plain ASCII across a network — dial-up, leased or the Internet — you basically have two options. First, you can attach them as a binary file (i.e. non-ASCII file). Or second, you can encode them into ASCII characters and send the file as part of your message. The first method is preferable. But you can typically only send binary files from one account to another on the same network, or between two networks that have agreed between themselves to a method of transferring files. But that's a rarity. And it certainly doesn't work in and around the Internet. Thus something called, MIME was created. MIME stands for Multipurpose Internet Mail extensions. It is the name for encoding binary files into ASCII characters for transfer across the Internet and on-line services. Under MIME, there are a number of methods of encoding binary files, including Uuencoding, BinHex. To send a binary file, for example, you encode it using BinHex encoding software. You then include the ASCII encoded file in your message. The recipient of your message then decodes the ASCII using BinHex decoding software, which may or may not be built into his electronic mail or browser package. See BinHex and UUENCODE.
MIMJ Modified Modular Jack. These are the 6-pin connectors used to connect serial terminal lines to terminal devices. MIMJ jacks can be distinguished from the familiar RJ11 jacks by having a side-looking tab, rather than a center-mounted one.
MIN Mark Can be 0 or 1. Your home cellular station sends extended address data upon origination and page response. See MIN1.
MIN1 Mobile Identification Number. The MIN is the actual seven digit cellular telephone number assigned by the cellular carrier exclusively to your phone, used for both billing and for receiving calls. 7 digits in length. See MIN2.
MIN2 The area code of your cellular phone number.
Mini-COW See COW.
Mini-floppy A floppy disk that is 3 1/2 inches in diameter. Also called a microfloppy. See MICROFLOPPY.
Mini-Link A generic name for an Ericsson family of compact radio links of one or more 2 megabit per second channels over distances of up to 30 km per hop.
Mini-MAP Mini-Manufacturing Automation Protocol. A version of MAP consisting of only physical, link and application layers intended for lower-cost process-control networks. With mini-MAP, a device with a token can request a response from an addressed device. Unlike a standard MAP protocol, the addressed Mini-MAP device need not wait for the token to respond.
MiniCLAS A service from Nynex which allows a small (up to 199 lines) Centrex III user to make certain changes to his own Centrex features and lines with an easy-to-use computer program or a touchtone phone. See also CLAS.
Minimal Regulation Regulated local telephone company under limited state regulation has the ability to file price lists for services and those price lists are usually effective on 10-days' notice to local state commission and customers. This means there are no extensive hearings.
Minimize Button The minimize button is the down-arrow button at the immediate right of a title bar in Windows software. Clicking the minimize button will shrink a window to its icon.
Minimum Internetworking Functionality MIF. A general principle within the ISO that calls for minimum local network station complexity when interconnecting with resources outside the local network.
Minitel French name for videotex. Very popular. See VIDEOTEX and packet switched networks.
MIPS Millions of Instructions Per Second. A measure of computer speed that refers to the average number of machine language instructions performed by the CPU in one second. A typical Intel 80386-based PC is a 3 to 5 MIPS machine, whereas an IBM System 370 mainframe typically delivers between 5 and 40 MIPS. MIPS measures raw CPU performance, but not overall system performance.
MIR Maximum Information Rate: See also PCR.
Mirror A term used to reference Internet FTP sites that copy files from other archives every day or so. By accessing a mirror site close to your location you reduce transmission over the Internet.
Mirrored Server Link MSL. A dedicated, high-speed, point-to-point connection between the primary server and the secondary server. The MSL can either be a coaxial or fiber-optic cable (with maximum distances of 100 feet and 2.5 miles, respectively). See MSL.
Mirroring A fault tolerance method in which a backup data storage device maintains data identical to that on the primary device and can replace the primary if it fails. Mirroring will typically cost you a 50% performance degradation when your write to disk and 0% performance degradation when you are reading. For a full explanation, see DISK MIRRORING.
MIS 1. Management Information System. A fancy name for Data Processing. MIS are also the first three letters of the word MISanthrope and MISguided and lots of other MIS words. MIS departments are taking over corporate telecommunications departments which is why there's little love lost between the two.
2. IBM-speak for Management Initiated Separation. Translation: You're fired.
Miscellaneous Common Carrier A communications common carrier (typically one using microwave) which is not offering switched service to the public or to companies. A miscellaneous common carrier usually provides video and radio leased line transmission services to TV and radio networks.

Miscellaneous Trunk Restrictions Denies preselected lines access to preselected trunk groups (e.g., FX or WATS trunks). A call attempt over a restricted group routes to an intercept tone.
Mission Critical Systems Systems on which the future success of an organization depends.
Mitel Mitel is an eminent PBX and computer telephony component maker. The word Mitel actually stands for MIke and TErry's electric Lawn mower. Here's how it happened. In the early 1970s, Mike Cowpland and Terry Matthews were engineers working in the semi-conductor factory of Northern Telecom in Kanata, Ontario, Canada. They decided they wanted to go out on their own. Their first idea was cordless, electric lawn mowers. Apparently they thought there was a demand such devices. They formed Mitel and ordered lawn mowers from a manufacturer in England. The mowers arrived late, after the onset of the Canadian winter. No one wanted lawn mowers. So they decided to produce telephone systems. They went more successfully. Mitel went public. Most of the company was later sold to British Telecom, who lost their shirt on it and eventually sold it on a fire sale to a venture capital company.
Mixed Cable Cables have characteristic impedances which vary according to the cables' physical parameters. When cables of different characteristic impedance are mixed, an impedance imbalance will occur. It is thus bad practice to mix different wire gauges, and twisted or non-twisted pair cables. However, it is accepted practice to combine long runs of twisted pair cable with short lengths of modular patch cords and line cords. Baluns are an exception to this rule.
Mixed Mode An imprecise term which suggests that one digital bit stream can carry voice, data, facsimile and video signals.
Mixed Mode Night Service After-hours answering of incoming calls in which Assigned Night Answer (ANA) is specified for some trunks, and Universal Night Answer (UNA) specified for others.
Mixed Station Dialing A telephone system feature which allows you to install both rotary and pushbutton phones on your phone system. Most modern PBXs have this feature.
MJ Modular Jack. A jack used for connecting voice cables to a faceplate, as for a telephone. See RJ (Registered Jack).
MK2 MK2 is the code name for the first product resulting from the best-of-breed combination of SCO's SCO OpenServer Release 5 operating system with Chorus Systems' open microkernel technology. MK2, a binary product initially targeted at the Intel ix86 platform, is specifically designed for telecommunications manufacturers. According to its maker, it provides the modularity and scalability needed to meet a wide range of telecommunications products ranging from switches to telephony servers. Scheduled to ship in early 1997, MK2, says its maker, enables telecommunications vendors to benefit from the open systems software world while protecting their current investment in legacy telecommunications-specific software.
MLHG MultiLine Hunt Group
MLID Multiple Link Interface Driver. A layer of the Novell Open DataLink Interface specification. The MLID layer controls a specific network interface, and works below the Link Support Layer.
MLL Monthly Leased Lines.
MLM Meridian Link Module. A Northern Telecom term for an Application Module that provides a link to a host processor through the Meridian Link interface.
MLPP See MULTILEVEL PRECEDENCE PREEMPTION.
MLT Mechanized Loop Testing.
MLT-3 Multi Level Transmit - 3 Levels. The ANSI approved modulation scheme used for the transmission of data on an FDDI network over shielded and unshielded copper twisted pair media (see TP-PMD).
MMCF Multimedia Communications Forum. Formed in June 1993, the MMCF is an non-profit research and development organization of telecommunications service providers, multimedia application and equipment developers, and end users who realize the revolutionary potential of multimedia communications. Forum members are dedicated to accelerating market acceptance and multivendor interoperability of multimedia communications worldwide. MMCF acts as a central clearinghouse for all multimedia communications-related standards, specifications and recommendations.
MMDS Microwave Multi-point Distribution System. A means of distributing cable television signals, through microwave, from a single transmission point to multiple receiving points. Often used as an an alternative to cable-based cable TV. According to an April, 1995 press release from Pacific Telesis, which was starting an MMDS service, "in digital form, it will provide more than 100 channels to a radius of approximately 40 miles from the transmitter. The MMDS transmitter delivers video to homes that are in its 'line of sight.' MMDS transmissions are limited by the terrain and foliage of a given market. The microwave signal is received by an antenna on the subscriber's home, then sent down coaxial cable to a box atop the customer's TV set. The box decodes and decompresses the digital signal." MMDS is increasingly being called "Wireless Cable." See WIRELESS CABLE.
MMF Multimode Fiber optic Cable: Fiberoptic cable in which the signal or light propagates in multiple modes or paths. Since these paths may have varying lengths, a transmitted pulse of light may be received at different times and smeared to the point that pulses may interfere with surrounding pulses. This may cause the signal to be difficult or impossible to receive. This pulse dispersion sometimes limits the distance over which a MMF link can operate.
MMFD Abbreviation for micromicrofarad; one millionth of a farad, the unit of measuring capacitance.
MMI Machine-to-Machine Interface.
MMIC Microwave Monolithic Integrated Circuit.
MMJ A six wire modular jack with the locking tab shifted off to the right hand side. Used in the DEC wiring system.
MML Man Machine Language.
MMR Modified Modified Read data compression method used in newer Group 3 facsimile machines.
MMSU Modular Metallic Service Unit.
MMTA MultiMedia Telecommunications Association. The successor organization to NATA. MMTA is organized around five divisions — computer telephony integration, conferencing/collaboration and messaging, LAN/WAN internetworking, Voice/Multimedia and Wireless Communications. MMTA is on 202-296-9800. See ACTAS and NATA.
MNA Multimedia Network Applications.

MNA7 Multiple CCS7 Network Addresses

MNC Mobile Network Code. A part of the IMSI (International Mobile Subscriber Identity) or LAI.

Mnemonic A shorthand label or term that is easy to remember, usually made up from the initials of the words in a term or process. LASER is a mnemonic. See LASER.

Mnemonic Dial Plan Pronounced "nemonic." A way of dialing using characters typed on the keyboard of a terminal. The word Mnemonic comes from the same roots as memory. It's a memory jogging way of remembering something, like a way to dial. See MNEMONIC PROMPTS.

Mnemonic Prompts System commands represented by the appropriate alphabet letter rather than by a number, (for example, "P" to "Play", "A" to "Answer"). See MNEMONIC DIAL PLAN.

MNP Microcom Networking Protocol. A proprietary error-correcting and data compression protocol for dial-up modems from 2.4 to 14.4 Kbps. See MICROCOM NETWORKING PROTOCOL and ERROR CONTROL PROTOCOLS for full explanations.

Mobile Cellular Phone The cellular handset unit permanently mounted in a vehicle. See CAR PHONE.

Mobile Cellular Service A fancy name for cellular phone service or cell phone service. See CELL, CELL SWITCHING, CELLULAR RADIO and IN.

Mobile Data Mobile data is a generic term used to describe data communications through the air from and to field workers — from package deliverers, to car rental companies (to track cars), to field service personnel, to law enforcement officials checking license plates.

Mobile Mounting Kit An optional cellular phone accessory that allows a transportable or portable to be connected to a vehicle's power supply and antenna lead, thereby boosting power and improving reception. Sometimes referred to as a car kit or car mounting kit. Some of these kits are very expensive. Check the price of the kit before you buy your phone.

Mobile Phone One term for a cellular phone. There are four main types of cellular phones — mobile (also called car phone), transportable, portable and personal. A mobile phone is attached to the vehicle, the vehicle's battery and has an external antenna. The mobile phone (the car phone) transmits with a standard three watts of power. Mobile telephone service is provided from a broadcast point located within range of the moving vehicle. That range is called a "cell." The broadcast point in turn is connected to the public network so that calls can be completed to or from any stationary telephone, i.e. one connected to a land line. See CELLULAR and CAR PHONE.

Mobile Terminated The term used to describe a call where the destination of the call is a mobile (i.e. cellular) telephone.

Mobile Unit The cellular telephone equipment installed in a vehicle. It consists of a transceiver, control head, handset and antenna.

Mobilink A "unified" cellular phone service covering 83% of North America's population. It is a consortium of six Bell cellular operators and some Canadian cellular operators. The idea is simple. Anyone calling a subscriber of one of these companies would have his call automatically routed to the subscriber, no matter where in Mobilink that subscriber was. Before Mobilink, you had to know where the person was you wanted and then dial a bunch of complex codes to get to him.

Mobitex Ericsson's cellular land radio based packet switched data communications system which is used by RAM Mobile Data, a leading provider of messaging to radio pagers.

Modal Dispersion Modal dispersion can be thought of as the blurring of the input signal in a fiber by the several modes that may propagate down a fiber. Each mode may take a separate path down the fiber, and thus that signal may arrive at slightly different times. The dispersion depends on the fiber's internal characteristics and its length. In short, modal dispersion is pulse rounding in lightwave communications that takes place because of the slightly different paths followed by the laser light rays as they arrive at the detector slightly out of phase.

Modal Distribution In an optical fiber operating at a single wavelength, the number of modes supported by the fiber, and their propagation time differences. In an optical fiber operating at multiple wavelengths simultaneously, the separation in wavelengths among the modes being supported by the fiber.

Modal Loss In an open waveguide, such as an optical fiber, a loss of energy on the part of an electromagnetic wave due to obstacles outside the waveguide, abrupt changes in direction of the waveguide, or other anomalies, that cause changes in the propagation mode of the wave in the waveguide.

Mode 1. A mode is a stable propagation of state of an electromagnetic wave in an optical fiber. In rough terms a mode can be thought of bundles of light rays, of the same wavelength, that enter a fiber at the same angle. See MODAL DISPERSION.
2. Mode is essentially a switch inside a computer that makes it run like another computer, usually an older one.

Modem Acronym for MOdulator/DEModulator. Equipment which converts digital signals to analog signals and vice-versa. Modems are used to send data signals (digital) over the telephone network, which usually is analog. The Modem modulates the "1's" and "0's" into tones which can be carried by the phone network. At the other end, the demodulator part of the modem converts the tones back into digital 1's and 0's. See MODEM ELIMINATOR, MODEM POOL, MODEM STANDARDS, MODULATION PROTOCOLS and SERIAL PORT.

Modem Cowboy A slang term for someone who typically lives in the mountain states of Western America and does most of his work via modem.

Modem Eliminator A wiring device designed to replace two modems; it connects equipment over a distance of up to several hundred feet. In asynchronous systems, this is a simple cable. Here is a specific application using a Modem Eliminator: You can connect a PC to a printer, or a PC to another printer using a cable. But you can only go a certain distance — maybe 100 feet. After that, the traditional solution has been to use a modem and go over traditional phone lines. Instead, you can connect the two devices directly by wire using a Modem Eliminator. There are two advantages of a Modem Eliminator over a normal modem. The eliminator is cheaper and it can often transmit faster. According to Glasgal Communications, there are many cases where it is either unnecessary, cumbersome or too expensive to interconnect terminals using modems or line drivers in an experimental or a very short-haul environment. A modem eliminator functionally resembles two modems back-to-back on a leased line and therefore saves the cost of two modems and a line in many situations.

Modem Fallback When the telephone line quality is not good enough to accommodate the top rated speed of a modem — for example 14,400 bps — the modem drops down to lower speeds — initially to 9,600 bps, then if necessary, to 4,800 bps, or even down to 2,400 bps.

Modem Pool A collection of modems which a user can dial up from his terminal, access one and use that one to make a data call over the switched telephone network. Modem pools are obviously designed to allow many users to share few modems, thus saving on modems. Now that modems have become less expensive, the advantages of modem pooling are no longer as great. There are also some advantages in having a modem right next to your terminal or computer — namely you can see how it is functioning. And modems have lights to indicate what they're doing. One of the most useful lights on most modems indicates whether the line is "off hook" or not. It is possible for your computer to instruct your modem to hang up the line and for your modem to forget to do it, leaving you with a huge phone bill. One problem with giving people their own modems is they (the modems) have a tendency to get pinched. It's hard to screw down modems. Harder, anyway, than computers and disk drives.

Modem Server A networked computer with a modem or group of modems attached to it that allows network users to share the modems for outbound calls.

Modem Standards Definitions of electrical and telecommunications characteristics which enable modems from dissimilar manufacturers to speak to each other. Bell 103...US standard for 300 bps; ITU-T V.21...International standard for 300 bps; Bell 212A...US standard for 1200 bps; ITU-T V.22...International standard for 1200 bps; ITU-T V.22 bis...US and international standard for 2400 bps; ITU-T V.23...International videotex standard (1200/75 bps or 75/1200 bps). See also HAYES COMMAND SET and V.SERIES recommendations, i.e. V.34.

Modem Turnaround Time The time needed for a half-duplex modem (an old-fashioned one) to reverse its transmission direction.

Moderator A moderator is a person who controls what gets posted to a particular Internet newsgroup. A moderator is used to ensure that a newsgroup's article stick to the agreed upon subject matter. A newsgroup may or may not have a moderator.

Modified Frequency Modulation MFM. An encoding scheme used to record data on the magnetic surfaces of hard disks. It is the oldest and slowest of the Winchester hard disk interface standards. RLL (Run Length Limited encoding) is a newer standard, for example.

Modified Huffman A one-dimensional data-compression scheme that compresses data in a horizontal direction only. Allows no transmission of redundant data.

Modified Read A two-dimensional coding scheme for facsimile machines that handles the data compression of the vertical line and that concentrates on space between the lines and within given characters.

Modular Equipment is said to be modular when it is made of "plug-in units" which can be added together to make the system larger, improve its capabilities or expand its size. There are very few phone systems that are truly modular.

Modular Jack A device that conforms to the Code of Federations, Title 47, part 68, which defines size and configuration of all units that are permitted for connection to the public telephone network.

Modular Plug A series of connecting devices adopted by the FCC as the standard interface for telephone and data equipment to the public network. The most common is the RJ-11, used to connect a single line phone.

Modular Windows Microsoft's name for a family of stripped down versions of Windows that are intended to be built into a wide variety of devices other than PCs, including telephone systems and office machines, such as fax machines. See also AT WORK and WINDOWS TELEPHONY.

Modulated Waves Alternating current waves which have their amplitude varied periodically. The signals transmitted by a radio station are examples of a modulated wave.

Modulation The process of varying some characteristic of the electrical carrier wave as the information to be transmitted on that carrier wave varies. Three types of modulation are commonly used for communications, Amplitude Modulation, Frequency Modulation and Phase Modulation. And there are variations on these themes called Phase Shift Keying (PSK) and Quadrature Amplitude Modulation (QAM).

Modulation Index In angle modulation, the ratio of the frequency deviation of the modulated signal to the frequency of a sinusoidal modulating signal. The modulation index is numerically equal to the phase deviation in radians.

Modulation Protocols Modem stands for MOdulator/DEModulator. A modem converts digital signals generated by the computer into analog signals which can be transmitted over an analog telephone line. It also transforms incoming analog signals into their digital signals for inputting into a computer. The specific techniques used to encode the digital bits into analog signals are called modulation protocols. The various modulation protocols define the exact methods of encoding and the data transfer speed. In fact, you cannot have a modem without modulation protocols. A modem typically supports more than one modulation protocol. The raw speed (the speed without data compression) of a modem is determined by the modulation protocols. Here are the main internationally accepted modulation protocols:

INTERNATIONALLY ACCEPTED MODEM MODULATION PROTOCOLS					
Standard	Speed	Modulation	Duplex	Symbol Rate	Bits per symbol
Bell 103	300	FSK	Full	300	1
v.21	300	FSK	Full	300	1
v.22	1200	DPSK	Full	600	2
v.23	1200/75	FSK	Half	1200	1
Bell 202	1200/75	FSK	Half	1200	1
Bell 212A	1200	DPSK	Full	600	2
v.22bis	2400	QAM	Full	600	4
v.32	9600	QAM	Full (EC)	2400	4
v.32bis	14400	TCM	Full (EC)	2400	6
v.32ter	19200	TCM	Full (EC)	2400	8

Two modems can establish a connection only when they support the same modulation protocol. A modem with a proprietary modulation protocol can only establish a connection with another modem which also supports that modulation protocol. That protocol is typically from the same manufacturer, or from one of several manufacturers that say they are supporting it. For example, there once was a modulation protocol called V.FAST, which delivered 28.8 Kbps over normal analog phone lines. Several manufacturers supported it. Later, the ITU-T came out with V.34. Every modem maker adopted it and V.FAST went away, leaving some modem owners with modems that only worked at 28.8 Kbps with other proprietary modems.

Modulation Rate The reciprocal of the measure of the shortest nominal time interval between successive significant instants of the modulated signal.

Modulation Suppression In the reception of an amplitude-modulated signal, an apparent reduction in the depth of modulation of a wanted signal, caused by presence, at the detector, of a stronger unwanted signal.

Modulator A device which converts a voice or data signal into a form that can be transmitted.

Modulo Term used to express the maximum number of states for a counter. Used to describe several packet-switched network parameters, such as packet number (usually set to modulo 8 — counted from 0 to 7). When the maximum count is exceeded, the counter is reset to 0.

Modulo N In communications, refers to a quantity, such as the number of frames or packets to be counted before the counter resets to zero. Relates to the number of frames or packets that can be outstanding from a transmitter before an acknowledgement is required from the receiver. Also indicates the maximum number of frames or packets stored, in case a retransmission is required (i.e., Modulo 8 or Modulo 128).

MOH Music On Hold.

Moire In a video image, a wavy pattern caused by the combination of excessively high frequency signals. Mixing of these signals causes a visible low frequency that looks a bit like French watered silk, after which it is named.

Moisture Barrier Bag MBB. A three-ply bag with characteristics that allow minimal moisture transmission, thereby preserving plastic surface-mount packages, which are packed into the bag, in a dry state.

Moisture Resistance The ability of a material to resist absorbing moisture from the air or when immersed in water.

Molding Raceway Method A cable-distribution method in which hollow metal or wood moldings support cables. Small sleeves of pipe are placed in the wall behind the molding to allow cable to pass through the wall.

Monitor 1. To listen in on a conversation for the purpose of determining the quality of the attendant's or agent's response and politeness to customers.
2. Video monitor. TV screen and surrounding electronics. IBM personal computer monitors come in monochrome and color. The more you pay the better quality the monitor is. The quality of monitors is typically measured by how monitor's screen resolution, which is measured by pixels, or picture elements. In monochrome monitors, the "standard" is Hercules Graphics, which offers a resolution of 720 pixels horizontally by 348 vertically. The worst color monitor is called CGA, which stands for Color Graphics Adapter. It offers 320 x 200. Most low-end laptops have this resolution screen. Next better is EGA, Enhanced Graphics Adapter. It offers 640 x 350 pixels. Next better is VGA, or Video Graphics Array, which offers 640 x 480. Next up is 8514/A also called XGA, or eXtended Graphics Array, which is IBM's high-resolution extension to its VGA adapter. It provides a resolution of 1,024 horizontally x 768 vertically, yielding 786,432 possible bits of information on one screen, more than two and a half times what is possible with VGA. It will only work on IBM machines with the Micro Channel Architecture (MCA). Here is a chart showing the pixel coding of various PC standard screens.

PC SCREEN ENCODING

CGA	320 x 200
Enhanced CGA	640 x 400
EGA	640 x 350
Hercules	720 x 348
VGA	640 x 480
Super VGA	800 x 600
8514/A (also called XGA)	1,024 x 768

Monitor On Hold A telephone feature. If the person you're speaking with puts you "on hold," you can turn your speaker on your phone and hang up your handset but keep listening until the other person comes back to the phone.

Monochromatic Consisting of one color or wavelength. Although light in practice is never perfectly monochromatic, it can have a narrow wavelength.

Monochrome Monitor A monitor with 720 x 348 pixel resolution in a single color. Most monochrome monitors display in paper white, green or amber.

Monospaced Font A font in which all characters have uniform widths. See also PROPORTIONAL FONT.

Month To Month The standard way of paying for telephone service. Some services now come in "rate stability" packages, which means if you commit to keeping the service for a while — typically three or five years — it's cheaper each month.

Monthly Factors A call center term. A historical pattern consisting of 12 factors, one of each month, that tells the program how much that much that month can be expected to deviate from the average monthly traffic year after year. For example, a monthly factor of .75 means that the month will be 25% slower than average, while a factor of 1.15 means that the month will be 15% busier than average.

MOON Magneto Optical On Network.

Moore's Law Gordon Moore, one of the cofounders of Intel Corporation, in 1965 in an Article in Electronics Magazine forecasted that computer chip complexity would double every twelve months for the next ten years. Ten years later his forecast proved true. He then forecasted that the doubling would occur every two years for the next ten years. Again history demonstrated his accuracy. The average of the two estimates is often stated as doubling every 18 months. He has again estimated this doubling will continue for another decade. His estimating is often referred to as "Moore's Law." There are many interpretations of what Moore actually said. For example, the July 4, 1994 Business Week said that in 1964, Moore "observed that, by shrinking the size of the tiny lines that form transistor circuits in silicon by roughly 10% a year, chip makers unleash a new generation of chips every three years — with four times as many transistors. In memory chips, according to Business Week, this has quadrupled the capacity of dynamic random memories (DRAMs) every three years. In microprocessors, the addition of new circuits — and the speed boost that comes from reducing the distance between them — has improved performance by four or five times every three years since Intel launched its X86 family in 1979 (the family that has powered IBM-compatible PCs).

MOP Maintenance Operations Protocol, a DEC protocol used for remote communications between hosts and servers.

MOPS Millions of operations per second. Refers to a processor's performance. In the case of DVI (Digital Video Interactive) technology, more MOPS translates to better video quality. Intel's video processor can perform multiple video operations per instruction, thus the MOPS rating is usually greater than the MIPS rating.

Morph Computer animation technique that allows figures to change from one shape to another in increments you choose.

Morphology A cellular radio term. Morphology describes population density. Higher population densities cover more POPs per cell, leading to economies of scale. Lower densities imply improved propagation characteristics and a greater coverage area.

Morse Code The International Morse code, named for Samuel F. B. Morse who invented telegraph, is referred to as International Telegraph Alphabet 1. Mr. Morse first demonstrated the telegraph in 1844. It is still used in radio telegraphy to ships at sea, though many ships now carry satellite antennae and subscribe to MARISAT. Morse code represents letters by combinations of long and short signals. Morse code can be written in dots and dashes or signaled with flashlights and radio bleeps. In 1912 the easily-memorized letters SOS were chosen as the international distress signal. "Save Our Souls" was the catch phrase devised later.

```
A .-    B -...   C -.-.
D -..   E .     F ..—.
G —.    H ....  I ..
J .——   K -.-   L .-..
M ——    N -.    O ———
P .——.  Q ——.-  R .-.
S ...   T -     U ..-
V ...-  W .——   X -..-
Y -.——  Z ——..
```

See also MORSE, SAMUEL

Morse, Samuel Samuel Finley Breese Morse was born in Charlestown (now a part of Boston) on April 27, 1791. He entered Yale University at the age of 14 and graduated in 1810. Although he attended lectures on electricity while at Yale, after graduation he went to England to study art. He returned to the United States in 1815 and became a well-known painter. In 1832, while returning from Europe, he and a fellow passenger discussed the electromagnet and Morse conceived the idea of his telegraph. He made a working model about 1835 and filed for a patent in 1838 and in 1844 inaugurated public service between Washington and Baltimore with his famous message, "What hath God wrought?" He died in 1872. See MORSE CODE.

MORT AT&T's database of its dead employees.

MOS Metal Oxide Semiconductor. Technology describing a transistor composed of a semiconductor layer including "source" and "drain" regions separated by a channel. Above the channel is a thin layer of oxide and over that a metal electrode called a gate. A voltage applied to this gate controls the current between the source and drain regions, or in another format, stops a flow between the two areas. (Definition courtesy of Bell Labs)

Mosaic A graphical piece of software that enables you to surf the Internet's Worldwide Web. Mosaic lets you see hypertext documents with embedded graphics and occasionally sound, movie clips and animation. Mosaic was the first popular software that allowed people to browse around the Web by pointing and clicking. In short, Mosaic is an interface to the WWW (World Wide Web) distributed-information system. Like Gopher, the WWW is functionally split into two parts, the server and the client. Using a GUI (Graphical User Interface) interface (like Mosaic), it has the ability to display:
- Hypertext and hypermedia (sounds, movies, extended character sets, and interactive graphics) documents.
- Electronic text in an enormous variety of fonts.
- Text in bold and italic
- Layout elements such as paragraphs, bulleted lists, and quoted paragraphs. It is mostly distinguishable by its support of multiple hardware platforms, and the WWW HTML (Hypertext Markup Language) document format. Mosaic used to be the most popular Web browser. At the time I wrote this, the most popular one was Netscape and browsers based on Netscape. See NETSCAPE.

Most Economical Route Selection MERS. Used by several phone companies and several manufacturers to mean Least Cost Routing — the feature of a telephone system which automatically chooses the least cost route for a long distance call. See LEAST COST ROUTING.

Most Favored Nation Clause A clause added to a purchase contract with a vendor saying that for a certain period after signing the contract, if the buyer finds out that the product has been bought for less, then the seller will refund the difference. The idea is to give the purchaser the assurance of the least expensive price.

MOTD Message Of The Day.

Mother Of The largest, greatest, grandest, of something. An expression coined in 1990 by Saddam Hussein, Iraqi dictator. The Mother Of all telephone switches would be the largest, most powerful, most elaborate etc. Such a device doesn't exist, as yet — as, in fact, Saddam Hussein's Mother of all Battles (the one against the Allies in 1990-1991) didn't exist. He lost in the Mother Of all defeats.

Motherboard The main circuit board of a computer system. The motherboard contains edge connectors or sockets so other PC (printed circuit) boards can be plugged into it. (Those boards are typically called Fatherboards, because they plug into the Motherboard.) On IBM's new OS series of personal computers, the motherboard is called the PLANAR BOARD. Motherboards are common in key systems and hybrid key/PBXs. They are not common in PBXs, where all the electronics are typically on printed circuit cards which slide into the PBX's cage and which attach to a backplane, which is typically a wiring scheme connecting the PBX's printed circuit cards. See also DAUGHTERBOARD.

NEWTON'S TELECOM DICTIONARY

Motif Motif is the name given to the Open Software Foundation's (OSF) toolkit (Application Programming Interface) and look and feel. The Motif toolkit is based on intrinsics.

MOTIS Message Oriented Text Interchange System. Original name for the ISO (International Organization for Standardization) standard now being changed to MHS (Messaging Handling System).

MOTO Mail Order Telephone Order. A credit/debit card classification by the banking and finance industry reflecting what the banking industry thinks are its highest risk transaction type.

Motorola According to the family which founded Motorola in the 1920s, the name Motorola was chosen to mean "Music in motion" to signify one of the company's first products — a car radio.

MOU Memorandum Of Understanding.

Mount The method in NFS and other networks by which modes access network resources. The word "mount" is often used as a verb, as in my workstation "mounts" the file server, called DALLAS2.

Mouse A computer input device that generates the coordinates of a position indicator on your computer screen (e.g. a hand, an arrow) and is worked by being moved on a flat surface. In contrast, a trackball is a stationary device

Mouse Blur Move your mouse quickly across your screen and if you're running an LCD (for example on a laptop), the mouse's pointer will blur — due to the screen's inability to change as fast as you can move the mouse. Another term for mouse blur is Cursor Submarining.

Mouse Potato A person who uses his mouse to view educational or entertainment on his computer. Museums are afraid, for example, that if they sell the electronic rights of the art hanging on their walls, every one will stay at home, become mouse potatoes and never visit the museums. The concept of a mouse potato derives from a couch potato — namely someone who sits on his couch and changes channels on his TV set using a remote device.

MOV Metal Oxide Varistor. A voltage dependent resistor which absorbs voltage and current surges and spikes. This low-cost, effective device can sustain large surges and switch in 1 to 5 nanoseconds. It is used as a surge protector and suppressor. It often the first electronic component that electrons coming in on an incoming phone line hit. Many trunk boards inside PBX are protected by MOVs. If the voltage or current is high, it will blow the MOV, thus protecting the remaining the far more valuable devices on the board.

Moves, Adds and Changes MACs. Any of the above ancillary work performed on a PBX switch, cabinet, or peripheral item after installation. See MAC for a fuller explanation.

Mozilla What the guys at Netscape called their Mosaic Killer, their new Web browser software (later to be known as Netscape Navigator).

MPC1 Published in 1991, the original Multimedia PC (MPC) specification was adopted worldwide as the basic multimedia extension of the PC standard. IN 1993 it was was followed by MPC2. MPC3, the latest, does not replace MPC2, but takes it one step further. See MPC3.

MPC2 See MPC3.

MPC3 MPC3 is the latest specification for multimedia PCs defined by the Multimedia PC Working Group, an independent special interest group of the Software Publishers Association (SPA). Minimum requirements for MPC3 machines include: 1. Support for MPEG-1 and other software-implemented video codecs; 2. 75 MHz Pentium or similar processor; 3. Quadruple speed CD-ROM drive; 4. Wavetable sound card.

MPEG MPEG is commonly known as a series of hardware and software standards designed to reduce the storage requirements of digital video, i.e. video recorded digitally or converted into digital bits. MPEG is most commonly known as an compression scheme for full motion video. The word MPEG is actually the acronym for the Motion Pictures Experts Group, a joint committee of the International Standards Organization (ISO) and the International Electrotechnical Commission (EG). The first MPEG specification, known as MPEG-1, was introduced by this committee in 1991. The common goal of all MPEG compression is to convert the equivalent of about 7.7 meg down to under 150 Kb, which represents a compression ratio of about 52 to one. The two requirements of MPEG-1 are 30 frames per second of Standard Image Format (SIG) of 352 pixels x 240 pixels and CD-quality sound at 44.1 Khz, 16 bit stereo. MPEG image scheme offers more compression than the other poplar JPEG image compression scheme, which is largely for still images. MPEG takes advantage of the fact that full motion video is made up of many successive frames consisting of large areas that are not changed — like blue sky background. While JPEG compresses each still frame in a video sequence as much as possible, MPEG performs "differencing," noting differences between consecutive frames. If two consecutive frames are identical, the second can be stored in remarkably few bits. MPEG condenses moving images about three times more tightly than JPEG. See also JPEG.

There are two types of MPEG Playback: Software and Hardware. Software MPEG playback is the decompression of MPEG video and audio files using the processing power of the CPU. Hardware MPEG Playback uses an add-in card to deliver full-screen, full-motion, full-color video and CD-quality audio at the full NTSC video frame rate of 30 frames per second, with no dropped frames. The card plays the video from a computer file that has been compressed using the MPEG video standard. Hardware playback is typically much better quality than software playback.

There are actually two MPEG standards: MPEG-1 and MPEG-2. A third, MPEG-4, is currently under development. MPEG -1 is a small-picture mode of MPEG geared to a resolution of 352 by 240 pixels at 30 frames per second (U.S.), with full CD-quality audio. MPEG-1 was originally designed to handle much larger picture sizes than 352 by 240 through interpolation or scaling, but MPEG-2 is more efficient. MPEG-2 offers a "main profile at main level" resolution of 720 by 480 pixels at 30 frames per second (U.S.), with full CD-quality audio. This picture size enables full-screen playback on PCs or TVs. MPEG-2 can incorporate a range of compression ratios, which trade off economies of storage and transmission bandwidth against picture quality. At compression ratios of 30:1 and smaller, MPEG-2 offers the perception of broadcast-quality TV. For greater economy, MPEG-2 supports up to 200:1 compression. MPEG-2 decodes such as the IBM decoder chip can also recognize and decode MPEG-1 bitstreams, enabling the IBM chip to support both compression standards.

MPEG-3 has been dropped. It was focused on HDTV with sampling dimensions up to 1,920 by 1,080 at 30 frames per second. The standard was to address bit rates between 20 and 40 Mbit/sec. Nevertheless, it was discovered that with a little tweaking, MPEG-2 and MPEG-1 work extremely well at the HDTV rate. HDTV is now part of the MPEG-2 High-1440 Level specification. MPEG-4 is currently in the application identification phase, with a target of November 1998 for the official sanction of the proposed standard. Intended for very narrow bandwidths, MPEG-4 is exploring ideas in frame reconstruction. Much like MIDI music creates realistic sound from a narrow bandwidth command string, using pre-existing sound components, MPEG-4 is considering speech and video synthesis, fractal geometry, computer visualization and artificial intelligence to build accurate pictures from minimal data. If you want to find out even more gory detail about MPEG, hyperlink over to the Moving Pictures Experts Group Web site in Italy at http://www.crs4.it/luigi/MPEG/mpegfaq.html.

MPEG-1 See MPEG.

MPEG-2 MPEG-2 is one of the most important standards developed by the Moving Pictures Expert Group, an International Standards Organization (ISO) group responsible for the standardization of coded representations of video and audio signals. MPEG-2 has been chosen as a leading digital video compression for a broad range of future video and broadcast applications. See MPEG.

MPEG-4 See MPEG.

MPG Microwave Pulse Generator. A device that generates electrical pulses at microwave frequencies.

MPI Multi-Path Interface. Between a transmitter and receiver, the radio wave can take a direct path and one or more reflected paths. The direct radio wave always arrives prior to the reflected waves. If the reflected waves are of sufficient amplitude, they will interfere with the direct wave. The relationship of the amplitude and time delay between the direct and reflected waves create peaks and nulls at the receiver, causing momentary signal fading or loss. In a digital system, this can result in very significant degradation, as the receiver loses signal acquisition and frame synchronization during each fade. The net effect is an increase in the residual bit error rate.

MPOA Multiprotocol Over ATM. This protocol, which is part of a working group of the ATM Forum, involves standards to allow seamless Network Layer routing over ATM. MPOA will allow companies to build scalable enterprise-wide LAN internetworks based on ATM.

MPOA Client An ATM term. A device which implements the client side of one or more of the MPOA protocols, (i.e., is a SCP client and/or an RDP client. An MPOA Client is either an Edge Device Functional Group (EDFG) or a Host Behavior Functional Group (HBFG).

MPOA Server An ATM term. An MPOA Server is any one of an ICFG or RSFG.

MPOA Service Area An ATM term. The collection of server functions and their clients. A collection of physical devices consisting of an MPOA server plus the set of clients served by that server.

MPOA Target An ATM term. A set of protocol address, path attributes, (e.g., internetwork layer QoS, other information derivable from received packet) describing the intended destination and its path attributes that MPOA devices may use as lookup keys.

MPS Multi Page Signal. A frame sent in fax transmission if the sender has more pages to transmit.

MPTN Multi Protocol Transport Networking. IBM scheme addressing multi protocol network support including TCP over SNA and SNA over IP.

MPTy Multiparty. A wireless telecommunications term. A supplementary service provided under GSM (Global System for Mobile Communications).

MPX Multiplex.

MR Modified Read. Relative element address differentiation code. A two-dimensional compression technique for fax machines that handles the data compression of the vertical line and that concentrates on space between the lines and within given characters. See MMR.

Mrm An ATM term. An ABR service parameter that controls allocation of bandwidth between forward RM-cells, backward RM-cells, and data cells.

MS 1. Mobile Station. A wireless telephone allowing mobility so that calls may be placed locally or in another geographic region. 2. Message Switch.

MS-DOS MicroSoft Disk Operating System. Being dumb, a computer needs to be told from the very basic steps what to do. MS-DOS is the basic command system for IBM and IBM clone personal computers. It explains to the computer how to talk to its keyboard, to its screen, to its various "ports" — serial and parallel — and how to talk to its internal bus — the broadband communications system that sends messages between PC (printed circuit) cards that are slid into the empty slots most personal computers come with. One critical MS-DOS function is to "talk" to its disk/s — hard and/or soft. Let me explain: In MS-DOS when you format a disk, you actually put temporary dummy data all throughout the disk. The smallest unit of storage on a disk is 512 bytes. When you create a file (say a letter), MS-DOS stores the file in clusters, which are typically groups of sectors. MS-DOS doesn't store these clusters contiguously (i.e. next to each other). Rather, it stores them all over the disk. Each cluster thus contains only a portion of the file. This is efficient, since MS-DOS then doesn't require long uninterrupted areas of space on the disk for each file.

If your files are stored all over the disk, clearly you need an index or roadmap to where the files are. When you format a disk, in addition to laying down dummy data, MS-DOS makes two other things — first, an area of the disk containing the boot-up logic MS-DOS needs to start the machine. And second, two copies of something called the File Allocation Table. The FAT is like an index to what's on the disk. When you make a new file, MS-DOS automatically checks the FAT to figure where the chunks of your new file should be stored. When you do a DIR, you can't see the FAT. It's hidden away. But it's there. And it's duplicated (just in case).

Sometimes the FAT becomes corrupted. This is a nice term for "messed up." When the FAT and the directory entries fall out of synch with each other, you could be in trouble. The most common problem is a lost cluster. MS-DOS wrote a portion of a file to the disk, but didn't connect it with a file. In other words, MS-DOS has simply misplaced the cluster. To solve the problem you must use SCANDISK or the older CHKDSK command. You do this by typing at C: prompt
CHKDSK /F

NEWTON'S TELECOM DICTIONARY

That means check the disk and fix it. One more point: How to bring back a file you've erased. MS-DOS doesn't actually "erase" a file. It marks the directory entry of the file as "erased" and frees the sectors (i.e. space) listed in the FAT. When you create a new file or add to an existing one, MS-DOS may then use these freed-up sectors. The key to recovering a file you've erased is not to have written anything to disk after you erased the file. If you erase a file, say "Drat" and then instantly use a program like Norton Utilities' UNERASE program. You'll be able to bring your erased file back to life with absolutely no problem and totally intact. The key is NOT to write anything over it. According to the Economist Magazine of May 22, 1993, some jealous Microsoft rivals claim that MS-DOS now stands for Microsoft Seeks Domination Over Society. See also DEVICE DRIVER, QDOS and WINDOWS.

MS-Net Microsoft Network, the MS-DOS networking systems software product shipped in 1985. Largely superseded by Microsoft LAN Manager. Now largely superseded by Windows NT Advanced Server.

MSA 1. Metropolitan Statistical Area. When it issued cellular radio licenses, the Federal Communications Commission divided the United States into MSA markets and Rural Service Area (RSA) markets. There are 306 MSAs in the United States, all of which now have cellular service.
2. Message Service Application.

MSAT Mobile SATellite. Technology for transmitting and receiving satellite transmissions in moving vehicles.

MSAU Multi-Station Access Unit.

MSB Abbreviation for Most Significant Bit and Most Significant Byte. That portion of a number, address or field which occurs leftmost when its value is written as a single number in conventional hexadecimal or binary notation. The portion of the number having the most weight in a mathematical calculation using the value.

MSC Mobile services Switching Center. A switch providing services and coordination between mobile users in a network and external networks.

MSCDEX Microsoft DOS extensions for CD-ROM. Allows the DOS operating system to recognize a CD-ROM as a DOS volume.

MSEngine Mirrored Server Engine. The part of the Novell SFT III operating system that handles nonphysical processes, such as the NetWare file system, queue management, and the bindery. SFT III is split into two parts: the IOEngine (Input/Output Engine) and the MSEngine) and the MSEngine. The primary server and the secondary server each have a separate IOEngine, but they share the same MSEngine. The file system, receive buffers, and queue management system all reside in the MSEngine. Applications and NetWare Loadable Modules (NLMs) that do not address hardware directly can be mirrored by loading them in the MSEngine. If one server fails, applications and NLMs in the MSEngine continue to run. The MSEngine keeps track of active network processes; it provides uninterrupted network service when the primary server fails and the secondary server takes over. See IOENGINE.

MSISDN Mobile Station ISDN number. An ISDN number provisioned to a mobile station subscriber and used to place a call to the subscriber.

MSL Mirror Server Link. A dedicated, high-speed connection between Novell NetWare SFT III primary and secondary servers. The mirrored server link is essentially a bus extension from the primary to the secondary server. It requires similar boards in each server, directly connected by fiber-optic or other cables.

MSNF Multisystem Networking Facility. An optional feature of certain IBM telecommunications access methods that allows more than one host running ACF/TCAM or VTAM to jointly control an ACF/NCP program.

MSO Multiple System Operator. A company that operates more than one cable TV system.

MSP The name of a general purpose programmable switch made by Redcom Laboratories.

MT An ATM term. Message Type: Message type is the field containing the bit flags of a RM-cell. These flags are as follows: DIR = 0 for forward RM-cells = 1 for backward; RM-cells BN = 1 for Non-Source Generated (BECN), RM-cells = 0 for Source Generated RM-cells CI = 1 to indicate congestion = 0 otherwise NI = 1 to indicate no additive increase allowed = 0 otherwise RA - Not used for ATM Forum ABR.

MTA 1. Message Transfer Agent. An OSI application process used to store and forward messages in the X.400 Message Handling System. Equivalent to Internet mail agent.
2. Apple Computer's Macintosh Telephony Architecture.
3. Major Trading Area. An area defined by the FCC for the purpose of issuing licenses for PCS. Each MTA consists of several Basic Trading Areas (BTAs). The United States is broken down into 51 major trading areas for economic purposes. These boundaries were used for licensing PCS.

MTBF Mean Time Between Failure. The length of time a user may reasonably expect a device or system to work before an incapacitating fault occurs.

MTD Memory Technology Drive.

MTM Maintenance Trunk Monitor.

MTP Message Transfer Part of the SS7 Protocol. It provides functions for basic routing of signaling messages between signaling points. It is Level 1 through 3 protocols of the SS7 protocol stack. MTP 3 (Level 3) is used to support BISUP.

MTS 1. Message Telecommunications Service. AT&T's name for standard switched telephone service. Also called DDD, for Direct Distance Dial.
2. Member of the Technical Staff. A common term at AT&T Bell Labs, Bellcore and other R&D labs.

MTSO Mobile Telephone Switching Office. This central office houses the field monitoring and relay stations for switching calls between the cellular and wire-based (land-line) central office. The MTSO controls the entire operation of a cellular system. It is a sophisticated computer that monitors all cellular calls, keeps track of the location of all cellular-equipped vehicles traveling in the system, arranges handoffs, keeps track of billing information, etc.

MTTR Mean Time to Repair. The average time required to return a failed device or system to service.

MTU Maximum Transmission Unit. The largest possible unit of data that can be sent on a given physical medium. Example: The MTU of Ethernet is 1500 bytes.

MU Monitoring Unit. A wireless telecommunications term. Devices added to circuit configurations that use sophisticated trending rules with fault and topology information to determine potential outages.

MU-LAW The PCM voice coding and companding standard used in Japan and North America. A PCM encoding algorithm where each digitized sample is represented by eight bits, thus a 64k transmission rate — the standard rate for encoding voice — is actually 8K samples/second. A sample consists of a sign bit, a three bit segment specifying a National logarithmic range, and a four bit step offset into the range. All bits of the sample are inverted before transmission. See PCM.

MUA An acronym for Mail User Agent, is the end user's mail program, like Eudora.

MUD Multi-User Dungeons. A term that Time Magazine in its 9/13/1993 issue called "the latest twist in the already somewhat twisted world of computer communications." Time called it "a sort of poor man's virtual reality" — created by using words, not expensive head-mounted displays. The first MUD apparently was invented in 1979 as a way for British university students to play the fantasy game Dungeons & Dragons by networked computers.

Mudbox An unsheltered item of equipment that is sufficiently rugged to withstand adverse environments. It is expected to work perfectly though it sits outdoors in good and bad weather.

MUDS Multi-User Dungeons. A cyberspace term. MUDS are elaborate fictional gathering places that users create one room at a time. All these "spaces" have one thing in common, according to cyberspace wisdom, they are egalitarian. Anybody can enter the rooms (provided he has the correct equipment) and everybody is afforded the same level of respect. See also USENET.

MULDEM A contraction for Multiplexer Demultiplexer, referring to a piece of equipment which performs both functions and generally operates between two of the AT&T digital hierarchy rates (i.e., DSI to DS3).

Multi Vendor Integration Protocol See MVIP.

Multi- In this dictionary, I include a dash between words beginning with "multi" and ending with something else. See below for examples.

Multi-Access The ability of several users to communicate with a computer at the same time with each working independently on their own job.

Multi-Address Calling Facility A system service feature that permits a user to nominate more than one addressee for the same data. The network may accomplish this sequentially or simultaneously.

Multi-Alternating Routing Alternate routing with provision for advancing a call to more than one alternate route, each of which is tested in sequence in the process of seeking an idle path. A Bellcore definition.

Multi-Carrier Modulation MCM. A technique of transmitting data by dividing the data into several interleaved bit streams and using these to modulate several carriers. MCM is a form of frequency division multiplexing.

Multi-Cast The broadcast of messages to a selected group of workstations on a LAN, WAN or internet. See MULTI-CAST PACKETS.

Multi-Cast Packets Multi-cast packets are addressed to a group of devices on a LAN. LAN stations use multi-cast packets to deliver information to a specific set of devices such as routers, file servers, and hosts.

Multi-Cast User Message A user message generated at the source node and distributed to two or more destination nodes.

Multi-Casting The ability of one network node to send identical data to a number of end points - known as broadcast in other circles; one example is if new software or addressing updates need to be distributed to all users; also, a point-to-multipoint video transmission is a multi-cast operation.

Multi-Channel The use of a common channel to make two or more channels either by splitting the frequency band of the common channel into several narrower bands (called frequency division multiplexing) or by allocating time slots in the entire channel (time division multiplexing).

Multi-Conductor More than one conductor within a single cable complex.

Multi-Domain Network In IBM Systems Network Architecture technology, a network that contains more than one host based System Services Control Point (SSCP).

Multi-Drop A communications circuit with multiple terminals and peripherals. Only one may transmit at any time.

Multi-Drop Line A communications channel that services many data terminals at different geographical locations and in which a computer (node) controls utilization of the channel by polling one distant terminal after another and asking it, in effect, "Do you have anything for me?"

Multi-Fiber A fiber that supports propagation of more than one of a given wavelength. See MULTI-MODE.

Multi-Frame In PCM systems, a set of consecutive frames in which the position of each frame can be identified by reference to a multi-frame alignment signal. The multi-frame alignment signal does not necessarily occur, in whole or in part, in each multi-frame.

Multi-Frequency Monitors Also known as multisync or multiscan monitors. They can show images in several resolution standards. Such versatility makes them more expensive than single-resolution monitors (e.g. a standard VGA) but also less prone to instant obsolescence. A multisync monitor showing a VGA may or may not look better than VGA monitor showing a VGA image. That depends on the screen's other attributes.

Multi-Frequency Pulsing An in-band address signaling method in which ten decimal digits (the numbers on the touchtone pad) and five auxiliary signals are each represented by selecting two frequencies and combining them into one "musical" sound.. The frequencies are selected from six separate frequencies — 700, 900, 1100, 1300, 1500 and 1700 Hz. See also CAPTAIN CRUNCH.

Multi-Function Peripherals MFS. These are devices which take on two or more functions generally associated with individual peripherals and combine these into one product or in a linker series of modules. A multi-function peripheral might combine a fax machine with a photocopier with a computer printer with a scanner. The term is not very precise, but it tends increasingly to mean a computer device that will print, photocopy, fax and/or scan.

Multi-Homed Host A computer connected to more than one physical datalink. The data links may or may not be attached to the same network.

Multi-Hop An example of a single hop system is a microwave system between one building (let's say downtown San Francisco) and another across town (let's say uptown San Francisco). Each with one microwave antenna on its roof. Let's say we wanted to extend that system to Oakland. We'd put a second antenna on the uptown San Francisco building and shoot across to an antenna in Oakland. That building would now have a multi-hop transmission system.

Multi-Leaving In communications, the transmission (usually via bisync facilities and using bisync protocols) of a variable number of data streams between user devices and a computer.

Multi-Level Precedence Preemption MLPP. A system in which selected customers may exercise preemption capabilities to seize facilities being used for calls with lower precedence levels.

Multi-Line Hunt The ability of switching equipment to connect calls to another phone in the group when other numbers in the group are busy.

Multi-Line Telephone Any telephone set with buttons which can answer or originate calls on one or more central office lines or trunks. Originally all multi-line telephones were 1A2 and they came in sizes of 2, 3, 5, 9, 11, 17, 19, 29, and 60 lines. Now, skinny wire electronic key systems come in all sizes. See KEY TELEPHONE SYSTEM.

Multi-Line Terminating System Premises switching equipment and key telephone type systems which are capable of terminating more than one local exchange service line, WATS access line, FX circuit, etc.

Multi-Link Procedure A procedure (defined in ITU-T Recommendations X.25 and X.75) that permits multiple links connecting a single pair of nodes to provide service to the network layer on a shared basis. Such sharing provides greater effective throughput capacity and availability than a single link.

Multi-Location Billing Multilocation Billing is an option whereby a long distance carrier bills separate locations and applies volume discounts pro-rated to each site based on usage, or fixed percentage with pro-rated discounts to sites based on usage.

Multi-Master An inherent mode of the VME bus which allows the controlled sharing of the bus by multiple CPUs under a flexible priority structure.

Multi-Media Multi-media is the combination of multiple forms of media in the communication of information. Multimedia enables people to communicate using integrated media: audio, video, text, graphics, fax, and telephony. The benefit is more powerful communication. The combination of several media often provides richer, more effective communication of information or ideas than a single media such as traditional text-based communication can accomplish. Multi-media communication formats vary, but they usually include voice communications (vocoding, speech recognition, speaker verification and text-to-speech), audio processing (music synthesis, CD-ROMs), data communications, image processing and telecommunications using LANs, MANs and WANs in ISDN and POTS networks. Multimedia technology will ultimately take the disparate technologies of the computer, the telephone, the fax machine, the CD player, and the video camera and combine them into one powerful communication center. Technologies that were once analog — video, audio, telephony — are now digital. The power of multimedia is the integration of these digital technologies. To many people, "multimedia" (as defined above) is a disparate collection of technologies in search of a purpose. And it's true: most of the merger of media (as above) is taking place in business communications in the moving around of compound documents. Meantime, multimedia has moved into training and in the home for education and entertainment. See AUTHORING, COMPOUND DOCUMENTS, HYPERMEDIA, OLE, SHARED SCREENS, SHARED WHITEBOARDS, and SYNCHRONIZATION.

Multi-Media Capabilities The ability to run simultaneous voice, image, data and video applications on a computer. A technology that requires enormous bandwidth and processing power. See MULTI-MEDIA.

Multi-Media Network A network capable of carrying multiple forms of user information such as voice text, sounds, etc.

Multi-Media PC The Multimedia PC Council now defines a multimedia PC as a PC having a minimum of two megabytes of memory, a 30 megabyte hard drive, a CD ROM drive, digital sound support and Microsoft's Multimedia Extensions for Windows. See MULTI-MEDIA.

Multi-Media Protocol A protocol suitable for handling multiple forms of information such as voice, text, pictures, numbers, etc.

Multi-Mode In fiber-optics, an optical fiber designed to allow light to carry multiple carrier signals distinguished by frequency or phase at the same time (contrasts with single-mode). (Also spelled multimode.)

Multi-Mode Distortion In an optical fiber, a result of different values of the group delay for each individual mode at a single wavelength. It isn't the same as "multi-mode dispersion."

Multi-Mode Dispersion Dispersion fattens or smears the transmitted signal, making it difficult to identify the original information sent. Dispersion limits the amount of information that can be transmitted. Dispersion is much worse on the information carried on multi-mode fibers, because there's simple more going on. As a result, dispersion limits the rate and how you can send the information. Single mode is the preferred mode for long distance.

Multi-Mode Fiber Multi-mode fiber allows many modes of light to propagate down the fiber-optic path. Multimode fibers are generally used for short-distance data links, as they provide limited bandwidth due to modal dispersion. The relatively large core of a multimode fiber allows good coupling from inexpensive LEDs, and the use of inexpensive couplers and connectors. Multi-mode fiber typically has a core diameter of 25 to 200 microns. The core is much larger than single-mode fiber and allows several modes of light to be passed through it. Multimode fiber was the original medium specified for FDDI. See also SINGLE MODE FIBER.

Multi-Mode Optical Fiber An optical fiber that will allow many modes to propagate. The fiber may be either graded-index or step index fiber. Multimode optical fibers have a much larger core diameter than a single-mode fiber. See MULTI-MODE FIBER for a better explanation. See also SINGLE MODE FIBER.

Multi-NAM A cellular telephone term to allow a cellular phone to have two (or more) phone numbers, each of which can be on a different cellular system. This lets you get service from many cellular phone companies. For example, you could subscribe to one carrier in your home city and another in a distant city — perhaps one you travel to often. This saves you paying high roaming charges.

Multi-Path Multiple routes taken by RF energy between the transmitter and the receiver. Signal can cancel or reinforce. Varying multipath (at sunset or sunrise) causes varying signal strength that sounds like a train's steam engine starting up.

Multi-Path Error Errors caused by the interference of a signal that has reached the receiver antenna by two or more different paths. Usually caused by one path being bounced or reflected.

Multi-Point A configuration or topology, designed to transmit data between a central site and a number of remote terminals on the same circuit. Individual terminals will generally be able to transmit to the central site but not to each other.

Multi-Point Circuit A circuit connecting three or more locations. It is often called a multidrop circuit. See also MULTI-POINT and MULTI-DROP LINE.

Multi-Point Distribution Service A one-way domestic public radio service rendered on microwave frequencies from a fixed station transmitting (usually in an omnidirectional pattern) to multiple receiving facilities located at fixed points.

Multi-Point Grounding System A system of equipment bonded together and also bonded to the facility ground at the nearest location of the facility ground.

Multi-Point Line A single communications link for two or more devices shared by one computer and or more computers or terminals. Use of this line requires a polling mechanism. It is also called a multidrop line.

Multi-Processing A type of computing characterized by systems that use more than one CPU to execute applications. Multi-processing is not multi-tasking, which is the ability to have more one application running on a system at the same time. The technique is not associated with multi-processing, nor does it require multi processing to take place. Multi-tasking typically uses a computer with one CPU (e.g. your desktop or laptop). Multi-processing uses a computer with several CPUs, often a server. See MULTI-TASKING.

Multi-Processor Kernel Software that enables a computer operating system to use more than one CPU chip simultaneously. Such software is typically used on servers. It is used to speed up a server.

Multi-Programming Computer system operation whereby a number of independent jobs are processed together. Multi-Protocol Message Routers A device which converts different electronic mail formats. Such a router would be used to move electronic mail from a cc mail equipped-LAN to a Davinci e-mail LAN to a Wang mini-computer based system.

Multi-Rate ISDN "Nx64" or switched-fractional T1 lets users combine multiple ISDN PRI "B" channels on a call-by-call basis. Since each "B" channel operates at 64 kbps, users could get from 128 kbps to 1.544 Mbps of bandwidth in 64 kbps increments.

Multi-Server Network A single local area network (one cabling system) that has two or more file servers attached. Network addresses assigned to the LAN drivers are the same in each file server because the network boards are attached to the same cabling system. On a multi-server network, users can access files from any file server to which they are attached (if they have access rights). A multi-server network should not be confused with an inter network (two or more networks linked together through an internal or external router).

Multi-Session When pictures are placed on Kodak Photo CDs the first time, that's called the first session. And the result is a single session CD. The next time more photos are put on the disk, the disk is now called a multi-session Photo CD. To read such a disk, you need a CD-ROM player, which is specifically called a multi-session CD-ROM player.

Multi-Stage Queuing This a term typically used in Automatic Call Distributors. It is the ability to array a number of agent groups in a routing table. The notion of multiple agent groups being addressed may mean that the system may be able to "lookback" and "look forward" as it searches for a free agent in the right group to take the call presently holding.

Multi-Station Any network of stations capable of communicating with each other on one circuit or through a switching center.

Multi-Tasking The concurrent management of two or more distinct tasks by a computer. Although a computer with a single processing unit (as virtually all PCs are) can only execute one application's code at a given moment, a multi tasking operating system can load and manage the execution of multiple applications, allocating processing cycles to each in sequence. Because of the processing speed of computers, the apparent result is the simultaneous processing of multiple tasks. Standard mode Windows performs multi tasking only in the form of context switching. 386 enhanced mode allows multi tasking in the form of time slicing. See MULTI-PROCESSING.

Multi-Tenant Sharing The capability of a PBX to serve more than one tenant in a building. This process is a new option for building owners. They now become the Telephone Company for tenants in the building. There is money in this business, but chiefly on the resale of long distance phone calls.

Multi-Tester Usually an alternate name for VOLT-OHM-MILLIMETER, but may also apply to other MULTI-FUNCTION testing devices.

Multi-Threading Concurrent processing of more than one message by an application program. One of OS/2's advantages over Windows is that IBM designed it as a multi-threaded operating system. Each program in OS/2 can start two or more threads, which carry out various interrelated tasks with less overhead than two separate programs would require. For example, a communications program could have three threads running: one that waits for characters to be received, another that monitors the keyboard, and a third that displays information. This is more efficient than running multiple tasks because it doesn't require the overhead of an operating system context switch.

In short, a new and complex method of programming. Here is a definition from Sun Microsystems. A traditional UNIX process has a single thread of control. A thread of control, or more simply a thread, is a sequence of instructions being executed in a program. A thread has a program counter (PC) and a stack to keep track of local variables and return addresses. A multi-threaded UNIX process is no longer a thread of control of itself; instead, it is associated with one or more threads. Threads execute independently. There is in general no way to predict how the instructions of different threads are interleaved, though they may have execution priorities that can influence the relative speed of execution. In general, the number or identities of threads that an application process chooses to apply to a problem are invisible from outside the process. Threads can be viewed as execution resources that may be applied to solving the problem at hand.

Threads share the process instructions and most of its data. A change in shared data by one thread can be seen by the other threads in the process. Threads also share the operating system. Each sees the same open files. For example, if one thread opens a file, another thread can read it. Because threads share so much of the process state, threads can affect each other in sometimes surprising ways. Programming with threads requires more care and discipline than ordinary programming because there is no system enforced protection between threads. See WINDOWS NT.

NEWTON'S TELECOM DICTIONARY

Multi-Tier Tariffs A way of paying for something (i.e. equipment) from your local phone company. The idea is that one tier of your monthly payments is to pay off the equipment, and after a finite period, this tier payment drops to zero. The next tier is to pay for your monthly service and it is ongoing. Other tiers are for other reasons. As this technique was practiced by the Bell System, it was called "two tier." You will no longer find two tier tariffs in common use.

Multi-User PC A microcomputer that has several terminals attached to it, so that multiple users can simultaneously use its resources. Multi-user PCs can either slice up the time of a single microprocessor or can give each terminal-based user his own microprocessor. Multi-user PCs are an alternative to LANs and are typically used in specialized, one-application solutions, such as a doctor's office billing system.

Multi-User Software An application designed for simultaneous access by two or more network nodes, i.e. two or more users on a network. It typically employs file and/or record locking. It is not associated with multi processing, nor does it require multi processing to implement.

Multi-Vendor Integration Protocol See MVIP.

Multi-Way Communication A multimedia definition. Multi-way communication goes between two people, or between groups of people in all directions. Multi-way communication can be in real-time, or in store-and-forward mode. Examples of multi-way communication include a video conference, where one individual is giving a presentation to a group of people who listen and ask questions from their workstations; and group conferencing, where several people collaborate, supported by audio, video, and graphics on their workstation screens.

Multifunction Peripheral MFS. A single computer device that prints, photocopies, faxes and scans.

Multimedia See MULTI-MEDIA.

Multimedia Communications Forum See MMCF.

Multimode See MULTI-MODE above.

Multiple Access The ability of several personal computers connected to a Local Area Network to access one another through a common addressing scheme and protocol.

Multiple Address Message A message to be delivered to more than one destination.

Multiple Console Operation A phone system with this feature can use more than one attendant console. It's good to know the maximum number of consoles your chosen PBX can use.

Multiple Customer Group Operation A PBX shared by several different companies, each having separate consoles and trunks.

Multiple Domains A set of domains on a single LAN, each of which has its own domain-wide post office. Hosts within each domain can exchange mail by going through one domain post office. Hosts in different domains must generally send mail through two intermediary post offices: the sender's domain post office and the receiver's domain post office.

Multiple Frequency-Shift Keying MFSK. A form of frequency-shift keying in which multiple codes are used in the transmission of digital signals. The coding systems may use multiple frequencies transmitted concurrently or sequentially.

Multiple Homing Connecting your phone so it can be served by one or several switching centers. This service may use a single directory number. It may also use several directory numbers (another term for phone numbers). It all depends on how you set the service up with your local phone company. The idea is to give you more ways of reaching the switched network — in case one or more of your local loops breaks.

Multiple Listed Directory Number Service Permits more than one listed directory number to be associated with a single PBX.

Multiple Name Spaces The association of several names or other pieces of information with the same file. This allows renaming files and designating them for dissimilar computer systems such as the PC and the Mac.

Multiple Parallel Processing A method of fault tolerance used with host computers. Several CPUs cooperate to process data. It one CPU fails, its processing tasks are automatically assigned to other processors.

Multiple Protocol Router A communications device designed to make decisions about which path a packet of information will take. The packets are routed according to address information contained within, and can route across different protocols.

Multiple Routing The process of sending a message to more than one recipient, usually when all destinations are specified in the header of the message.

Multiple Spot Scanning In facsimile systems, the method in which scanning is carried on simultaneously by two or more scanning spots, each one analyzing its fraction of the total scanned area of the subject copy.

Multiple Token Operation Variant of token passing for rings in which a free token on a LAN is transmitted immediately after the last bit of the data packet, allowing multiple tokens on ring (but only one free token) simultaneously.

Multiple Tuned Antenna An antenna with connections through inductances to ground at more than one point and so determined that the total reactances in parallel are equal to those necessary to give the antenna the desired natural frequency.

Multiplex To transmit two or more signals over a single channel.

Multiplex Aggregate Bit Rate The bit rate in a time division multiplexer that is equal to the sum of the input channel data signaling rates available to the user plus the rate of the overhead bits required.

Multiplex Baseband In frequency division multiplexing, the frequency band occupied by the aggregate of the signals in the line interconnecting the multiplexing and radio or line equipment.

Multiplex Hierarchy In the U.S. frequency division multiplex hierarchy,
    12 channels = 1 group
    5 groups (60 channels) = 1 supergroup
    10 supergroups (600 channels) = 1 mastergroup
    6 mastergroups = 1 jumbo group
In contrast, the ITU-T standard says 5 supergroups (i.e. 300 channels) = 1 mastergroup.

Multiplexed Channel A communications channel capable of carrying the telecommunications transmissions of a number of devices or users at one time.

Multiplexer Electronic equipment which allows two or more signals to pass over one communications circuit. That "circuit" may be a phone line, a microwave circuit, a through-the-air TV signal. That circuit may be analog or digital. There are many multiplexing techniques to accommodate both.

Multiplexing Efficiency Figure of merit for multiplexers. The ratio of the aggregate channel input data rate to the composite output data rate. Many statistical multiplexers achieve a multiplexing efficiency of 8 or more.

Multipoint Access User access in which more than one terminal equipment (TE) is supported by a single network termination.

Multipoint Grounding System A system of equipment bonded together and also bonded to the facility ground.

Multipoint-to-Multipoint Connection A Multipoint-to-Multipoint Connection is a collection of associated ATM VC or VP links, and their associated nodes, with the following properties:
1. All Nodes in the connection, called endpoints, serve as a Root Node in a Point-to-Multipoint connection to all of the (N-1) remaining endpoints.
2. Each of the endpoints on the connection can send information directly to any other endpoint, but the receiving endpoint cannot distinguish which of the endpoints is sending information without additional (e.g., higher layer) information.

Multipoint-to-Point Connection A Point-to-Multipoint Connection may have zero bandwidth from the Root node to the Leaf Nodes, and non-zero return bandwidth from the Leaf Nodes to the Root Node. Such a connection is also known as a Multipoint-to-Point Connection. Note that UNI 4.0 does not support this connection type.

Multiport Card A circuit board with two or more ports for modems or other devices. Useful for enabling one PC to handle multiple incoming or outgoing calls at one time.

Multiport Repeater A repeater, either standalone or connected to standard Ethernet cable, for interconnecting up to eight ThinWire Ethernet segments.

Multiport Switch A local area network term. A device which allows packets to switch from one cable to another.

Multiprocessing A type of computing characterized by systems that use more than one CPU to execute applications. Multi processing is not multi tasking, which is the ability to have more one application running on a system at the same time. The technique is not associated with multi processing, nor does it require multi processing to take place. Multi tasking typically uses a computer with one CPU (e.g. your desktop or laptop). Multi processing uses a computer with several CPUs, often a server. See MULTI-PROCESSING, MULTI-TASKING and MULTI-THREADED.

Multiprotocol Over ATM MPOA. A proposed ATM Forum spec that defines how ATM traffic is routed from one virtual LAN to another. MPOA is key to making LAN emulation, Classical IP over ATM, and proprietary virtual LAN schemes interoperate in a multiprotocol environment. At this point, its unclear how MPOA will deal with conventional routers, distributed ATM edge routers (which shunt LAN traffic across an ATM cloud, while also performing conventional routing functions between non-ATM networks), and route servers (which centralize lookup tables on a dedicated network server in a switched LAN).

Multipurpose Internet Mail Extension MIME. An extension to electronic mail (Internet and other e-mail systems) which provides the ability to transfer non-ASCII data, such as graphics, software, audio, binary and fax. MIME was developed and adopted by the Internet Engineering Task Force. It was designed for transmitting mixed-media files across TCP/IP networks. The MIME protocol, which is actually an extension to SMTP, covers binary, audio and video data. Essentially what happens with Mime is that you pick a binary file (any file that isn't ASCII) to send along with your e-mail. Your e-mail software then converts your binary file to ASCII text which can easily be transmitted across one or more e-mail systems. All e-mail networks will transmit ASCII (i.e. ASCII 128 and below). The technique that makes MIME encoding possible is called UUencoding. Here is an example. Notice that every line begins with a M. Every line is the same length. At one end you UUencode the file (and it looks as below). When you receive the file, you must UUdecode it. Sometimes the coding and encoding is done automatically by your e-mail program. Sometimes you must decode it manually. Don't try and UUdecode the following file. It won't work. I chopped the middle of the file out. It's 64K bytes long. I'm running it below purely as an example.
See UUENCODE.

Uuencoded File Attachment: ANTARESN.DOC

---

begin 660 ANTARESN.DOC
MVZ4M`````"00``````````````````@`$``$``$%```!!``P````````````*
M``!`````!```\\````\`````\`*`````\`*````.(T```
M_____
M_____
9_____
\`_____T16
end

Multisession An incrementally updated Kodak Photo CD. See MULTI-SESSION for a fuller explanation.

Multitasking See MULTI-TASKING.

Multithreaded See MULTI-THREADED.

Muriel Muriel Fullam worked for me for ten years before she worked with me on this dictionary. That had to be the longest apprenticeship ever. It shows. She did great work. Thanks. She is married to Jerry Fullam who is absolutely the best husband and will love it when he sees his name here.

Murphy's Law When something can go wrong, it will. Compare with the GOTCHA LAW.

MUSE Multiple Sub-Nyquist Sampling Encoding. The Japanese bandwidth compression system to accommodate HDTV transmission within an existing satellite channel. The Japanese called one of their HDTV proposals MUSE. They demonstrated to Congress in Washington in 1987.

Music Source An external music source such as a radio, can be connected to the Key Service Unit for Music on Hold, Background Music, or both.

Music-On-Camp Audio source input for use with attendant camp-on. See also MUSIC-ON-HOLD.

Music-On-Hold Background music heard when someone is put on hold, letting them know they are still connected. Some modern phone systems generate their own electronic synthesized music. Most phone systems have the ability to connect any sound-producing device, e.g. a radio or a cassette player. Most companies, unfortunately, devote little attention to the sound source they select. Sometimes competitors will deliberately advertise on the radio station that callers will hear on hold. Thus, Macys is now selling Gimbels. It's better to use pre-recorded music. Better yet are tapes of "specials" and other happenings around your firm. Use the "Music-on-Hold" feature as another method of selling. "Ask the operator about our special on ladies' underwear." "Ask our operator to send you a copy of our latest annual report."

Musical Instrument Digital Interface (MIDI) A standardized communications protocol. MIDI files contain musical note information and other performance data that can "play" a MIDI instrument or sound module to produce music.

Mute A feature which disconnects the handset microphone or speakerphone microphone so that side conversations won't be heard.

Mutual Capacitance The capacitance between two conductors when all other conductors, including the shield, are short circuited to ground.

Mutual Synchronization A timing subsystem not employing directed control, by which the frequency of the clock at a particular node is controlled by some weighted average of the timing on all signals received from neighboring nodes.

Mutually Synchronized Network A network-synchronized arrangement in which each clock in the network exerts a degree of control on all others.

MUX See MULTIPLEXER.

MVIP Multi-Vendor Integration Protocol. MVIP is a family of standards to let telephony products from different vendors interoperate within a computer or group of computers. MVIP started in 1990 with a telephony bus for use inside a single computer. Picture a printed circuit card that fits into an empty slot in a personal computer. The slot carries information to and from the computer. This is called the data bus. Printed circuit cards that do voice processing typically have a second "bus" — the voice bus. That "bus" is actually a ribbon cable which connects one voice processing card to another. The ribbon cable is typically connected to the top of the printed circuit card, while the data bus is at the bottom. As of this writing (late 1993), there are at least four such buses defined. Several (AEB, PEB, SCSA) have been defined by Dialogic Corporation, Parsippany, NJ for connecting devices with Dialogic products and Dialogic compatible devices. The MVIP Bus was defined by Natural MicroSystems, Natick, MA with assistance from Mitel, Promptus Communications and Rhetorex as a vendor-independent means of connecting telephony devices within a computer chassis. MVIP was introduced by a seven company group in 1990 and has been distributed by Natural MicroSystems, Mitel and NTT International (part of the Japanese telephone company). By 1993, MVIP had attracted several hundred participating companies including companies manufacturing a variety of board-level MVIP products including telephone line interfaces, voice boards, FAX boards, video codecs, data multiplexers and LAN/WAN interfaces. MVIP now has its own trade association — the Global Organization for MVIP (see GO-MVIP), which develops extensions such as higher-level APIs and multi-chassis switching. Here is a write-up on the original MVIP Bus from Mitel's Communicating Objects Division:

"The MVIP Bus consists of communications hardware and software that allows printed circuit cards from multiple vendors to exchange information in a standardized digital format. The MVIP bus consists of eight 2 megabyte serial highways and clock signals that are routed from one card to another over a ribbon cable. Each of these highways is partitioned into 32 channels for a total capacity of 256 full-duplex voice channels on the MVIP bus. These serial link from one card to another. They are electronically compatible with Mitel's ST_BUS specification for inter-chip communications. By letting expansion cards exchange data directly, the MVIP bus opens the PC architecture to voice/data applications that would otherwise overburden the PC processor with data transfers. The MVIP bus is equivalent to an extra backplane that is capable of routing circuit switched data.

"MVIP systems generally have two types of cards ; network cards and resource cards. They differ by the switching they provide and in the way they are wired to the bus. Network cards almost always provide more flexible switching and can drive either the input or the output side of the bus, although they usually drive the output side of the bus. Resource cards usually provide very little switching and are only able to drive the input side of the bus. Resource cards usually rely on the network cards to do most or all of the switching on the MVIP bus."

According to Brough Turner, chairman of the MVIP Technical Committee, MVIP Evangelist and vice president of Natural MicroSystems, "The original MVIP bus distributes switching elements but only requires switches on telephone network interface cards. This has simplified access to MVIP for many communications service providers and has permitted easy inter-connection of pre-existing and proprietary technology to the common bus. More recently, the Flexible MVIP Interface Circuit (see FMIC) has reduced the complete MVIP interface, including clocks and switching, to one chip, further simplifying MVIP connections.

"MVIP's distributed switching architecture provides software controlled digital switching within the PC chassis. MVIP software driver standards simplify access for application developers and support the integration of components from multiple vendors. As a result of work by more than a dozen companies active in the MVIP Technical Committee, MVIP now addresses multi-chassis connections and higher-level APIs. Multi-Chassis MVIP (MC-MVIP) provides telephony connections and distributed digital switching between MVIP-based PCs and supports interconnection of MVIP PCs with proprietary telephony equipment including PBXs, ACDs, and voice processing systems. MC-MVIP includes redundant clocks and specifically permits plugging and unplugging connections while a multi-chassis system is operating. This allows application developers to construct fault-tolerant systems." See FMIC, GO-MVIP, NMS, NTT, PEB and SCSA.

MVS Multiple Virtual Storage.
MW Abbreviation for MILLIWATT.
MX3 A designation for a multiplex which interfaces between any of the following circuit combinations: 28 DSIs to one DS3 (M13), 14 DSICs to one DS3 (MC3), or 7 DS2s to one DS3 (M23).
MXR Multiplexer
MZI Mach Zehnder Interferometer.

NEWTON'S TELECOM DICTIONARY

N Series Terminators Most local area networks are bus configurations. This means one long piece of cable (coaxial or fiber) with workstations connected along the way, typically with "T" connectors. For a network to work properly, you need to place resistance at the end of the bus, to terminate it. A thin wire Ethernet typically requires a 50 ohm resistance at either end of the bus. Thin wire Ethernet terminators are commonly called N series terminators. They may also be used with grounding wires to ground the network. Networks don't work well without resistance at the end of their buses.

N-ary Code A code used to encode real-world characters that use n different code elements. Examples are binary (two states) and tertiary (three states).

N-ISDN Narrowband ISDN (Integrated Services Digital Network). Narrowband ISDN is an unkind name for the present form of ISDN presently implemented. In short, anything under an ISDN PRI. According to Professor Michael L. Dertouzos of MIT, narrowband ISDN suffers from the same constraints as classical voice telephony. For example, N-ISDN can carry reasonably be altered to accommodate the great information only in fixed chunks of 64 kilobits per second. This would be like a road system closed to everything except motorcycle. An 18-wheel truck carrying produce or a heating fuel tanker would be "welcome" to use this road as long as it could repackage its cargo into (and out of) motorcycle size chunks. See also B-ISDN.

N1 The first short-haul multiplex carrier in the U.S. by the Bell System was Western Electric's N1. The N1 transmitted 12 voice frequency channels over separate transmit and receive pairs in a single cable. N carriers have been progressively improving.

N2 In packet data networking technology, parameter used to specify the allowable number of retransmissions before disconnection.

NA See NIGHT ANSWER.

NACN The North American Cellular Network. Various operators nationwide have linked up to offer seamless roaming services so that roaming subscribers can make calls anywhere nationally without speaking to an operator.

NADF North American Directory Plan. An association of electronic mail providers who are figuring standards and ways of sending mail between their subscribers. They plan to use X.500 Recommendations.

Nailed Connection A permanent circuit of a previously switched circuit/s. Defined by Fujitsu, their Omni SI PBX provides up to eight dedicated (constant) trunk-to-trunk connections for special applications. See also NAILED DATA CONNECTION and NAILED-UP CIRCUIT.

Nailed Data Connection A permanently established connection between two InteCom IBM S/80 data interfaces. This connection is defined in the IBM database and the switch paths are established when the call processing software is first brought on line.

Nailed Up Circuit A circuit semi permanently established through a circuit switching facility for point-to-point connectivity.

NAK Negative AcKnowledgement. NAK is a control character in ASCII that means a packet arrived with the check digits in error. It is sent from the computer receiving the packets to the sender, implying that the packet should be retransmitted so that all bits will arrive intact in the next go-round. See CHECK DIGITS and ACK.

NAM Numerical Assignment Module. Basically, your cellular telephone number, although it refers specifically to the component or module in the phone where the number is stored. Phones with dual or Multi-NAM features offer the user the option of registering the phone so that it will have two or more phone numbers.

NAMAS NAtional Measurement Accreditation Service.

Name A name, as opposed to an address, is a location. Independent description of an end-station or node on a network (LAN or WAN) that contains no information about where the name entity is located. Certain protocols, such as IBM NetBIOS, make extensive use of a naming scheme.

Name Registration A Windows 95 definition. The method by which a computer registers its unique name with a name server on the network. In a Microsoft network, a WINS server can provide name registration services.

Name Resolution The process of mapping a name to the corresponding address. The process used on the network for resolving a computer address as a computer name, to support the process of finding and connecting to other computers on the network.

Name Server 1. An AIN (Advanced Intelligent Network) term. A directory service located in the SLEE (Service Logic Execution Environment) that provides a mapping between a resource's global name and its physical location in the network.
2. An electronic messaging term. A program which provides information about network objects, such as domains and hosts within a domain, by answering queries.

Named Pipe A connection used to transfer data between separate processes, usually on separate computers. Named pipes are the foundation of interprocess communications (IPC). An administrator can set permissions on named pipes, but only LAN Manager and network applications can create them. See also NAMED PIPES.

Named Pipes A technique used for communications between applications operating on the same computer or across the local area network. It includes an applications programming interface, providing application programmers with a way to create interprogram communications using routines similar to disk-file opening, reading, and writing. In Microsoft's words, named pipes allow two or more processes to communicate with each other. Any process that knows the name of a named pipe can access it (subject to security checks).

Naming Authority An authority responsible for the allocation of names.

NAMPS Narrowband Analog Mobile Phone Service. A proposed new standard for cellular radio. NAMPS combines current voice processing with digital signaling. According to Motorola, NAMPS triples the capacity of today's cellular AMPS system, reduces the number of dropped calls and offers a range of new performance enhancements and digital messaging services. The other cellular standards include E-TDMA (Extended Time Division Multiple Access), TDMA (Time Division Multiple Access) and CDMA (Code Division Multiple Access).

Nanometer One billionth of a meter. Written nm.

Nanosecond One billionth of a second. Written nsec.

NANP The North American Numbering Plan. It assigns area codes and sets rules for calls to be routed across North America (i.e. the US and Canada). It is administered by Bellcore. The new one, put into effect in January, 1995 has one major change: The middle number in a North American Area code will no longer be a 1 or a 0 (one or zero). It will range between 0 and 9. See NORTH AMERICAN NUMBERING PLAN.

NAP Network Action Point. An AT&T term describing the switching point through which a call is processed. The NAP switches the call based on routing instructions received from the NCP.

NAPI Numbering/Addressing Plan Identifier.

NAPLPS North American Presentation-Level Protocol Syntax. A protocol for videotex text graphics and screen formats, developed by AT&T and since standardized within ANSI, based on Canada's Telidon videographics protocol.

NAR 1. Nothing Added Reseller. In contrast to a VAR, which is a Value Added Reseller.
2. Network Access Register. Centrex / ESSX term required for a voice conversation. Each conversation requires a LINE and a NAR to complete a connection.

Narrative Traffic Messages normally prepared in accordance with standardized procedures for transmission via optical character recognition equipment or teletypewriter. In contrast to data pattern traffic, narrative messages must contain additional message format lines.

Narrowband 1. An imprecise term. Some people think it's sub-voice grade channels capable of only carrying 100 to 200 bits per second. Others think it means lines or circuits able to carry data up to 2400 bits per second. So as lines get broader, narrowband gets broader. The latest definition of narrowband is up to and including T-1 — or 1.544 megabits per second. See also N-ISDN and B-ISDN.
2. In cellular radio terminology, narrowband refers to the methodology of gaining more channels (and hence more capacity) by splitting FM channels into channels that are narrower in bandwidth. See NAMPS and NTACS.

Narrowband ISDN Any ISDN speed up to 1.544 Mbps, which is called PRI or PRA. But this definition is imprecise. And as speeds get faster, so the definition of narrowband ISDN means faster and faster. See N-ISDN and B-ISDN.

Narrowband Signal Any analog signal or analog representation of a digital signal whose essential spectral content is limited to that which can be contained within a voice channel of nominal 4-kHz bandwidth.

NARTE National Association of Radio and Telecommunications Engineers, based in Salem, OR. 800-933-1470.

NARUC National Association of Regulatory and Utility Commissioners. Members are representatives of state regulatory agencies concerned with telephone regulatory and other utility matters.

NAS NAS is Digital Equipment Corporation's implementation of standards-based software for open computing and is the foundation for delivering the Open Advantage through a combination of open technology, open services, and open business practices.

NASA National Aeronautics and Space Administration.

NASC Number Administration and Service Center. Provides centralized administration of the Service Management System (SMS) database of 800 numbers. The NASC keeps track of the 800 numbers that are in use, or available for use, by new 800 users.

NATA North American Telecommunications Association. A trade association of manufacturers and distributors of telephone equipment that was formed in 1970, two years after the FCC's Carterfone decision, which said that the Carterfone and other customer phone devices could be connected to the nation's phone network — if they were "privately beneficial, but not publicly harmful." The Carterfone decision was a landmark. It allowed the connection of non-telephone company equipment to the public telephone network. This decision marked the beginning of the telephone interconnect business as we know it today. And NATA's mission was to fight all the restrictive rules and regulations which the telephone companies subsequently threw up to make connection of customer-owned equipment difficult and expensive. At that time the phone companies (especially AT&T and GTE) were among the largest manufacturers of phone equipment. Eventually NATA won the legal fight and lost its mission, its reason for being. In the early 1990s, NATA collapsed. The remaining bits of it were re-assembled into a new trade association called the MMTA, which stood for MultiMedia Telecommunications Association. MMTA is organized around five divisions — computer telephony integration, conferencing/collaboration and messaging, LAN/WAN internetworking, Voice/Multimedia and Wireless Communications. MMTA is on 202-296-9800, NATA's old phone number. See CARTERFONE and NETWORK HARM.

National Association Of Regulatory Utility Commissioners NARUC. An organization supporting the needs of the commissioners of U.S. federal and state regulatory agencies.

National Bureau Of Standards NBS. The U.S. government organization that helps prepare non-Department of Defense communications standards and operates a testing service to indicate conformity to existing standards. Address: Institute for Computer Sciences and Technology, National Bureau of Standards, Gaithersburg, MD 20899.

National Cable Television Association NCTA. An organization representing the major U.S. cable television operators in the United States.

National Communications System NCS. The organization established by Section 1(a) of Executive Order No. 12472 to assist the President, the National Security Council, the Director of the Office of Science and Technology Policy, and the Director of the Office of Management and Budget, in the discharge of their national security emergency preparedness telecommunications functions.

National Coordinating Center NCC. The joint telecommunications industry-Federal Government operation established by the National Communications System to assist in the initiation, coordination, restoration, and reconstitution of NS/EP telecommunication services or facilities.

National Electrical Code NEC. A nationally recognized safety standard for the design, construction, and maintenance of electrical circuits. The NEC also gives rules for the installation of electrical and telephone cabling. The NEC is developed by the NEC Committee of the American National Standards Institute (ANSI), sponsored by the National Fire Protection Association (NFPA, Boston) and identified by the description ANSI/NFPA 70-1990. This code has been adopted and enforced by many states and municipalities as law.

National Exchange Carrier Association NECA. An association of local exchange carriers mandated by the U.S. Federal Communications Commission upon the divestiture of AT&T.

National Information Infrastructure What the Clinton Administration prefers to call the Information Superhighway — basically a switched, broadband network that could, in theory, deliver everything from switched video to high-speed access to the Internet.

National Institute For Standards And Technology NIST. The U.S. government agency that oversees the operation of the U.S. National Bureau of Standards. The NIST is based in Gaithersburg, MD.

National ISDN-1 National ISDN-1 is a specification for a "standard" ISDN phone line. It is based on technical references (TRs) specified by Bellcore that lay the groundwork for a national ISDN infrastructure. Bellcore issued the National ISDN-1 document, SR-NWT-001937, Issue 1, in February 1991. The idea of National ISDN-1 is that it be a set of standards which every manufacturer can conform to. A consumer can buy an ISDN phone (one conforming to National ISDN-1) at his local Radio Shack (or other store) take it home, plug it in and know it will work, irrespective of whose central office he's connected to. Sadly, National ISDN-1 has been a total bomb. We are currently awaiting National ISDN-2 which is supposed to do what National ISDN-1 was meant to do — provide "Plug 'N Play" ISDN. See CUSTOM ISDN and ISDN.

National ISDN-2 See NATIONAL ISDN-1.

National Park A government-provided repository of on-line information and entertainment, such as is provided by the Smithsonian.

National Security Agency See NSA.

National Television Standards Committee Standard NTSC. The North American standard for the generation, transmission, and reception of television communication wherein the 525-line picture is the standard. The picture information is transmitted in AM and the sound information is transmitted in FM. Compatible with CCIR Standard M. This standard is used also in Central America, a number of South American countries, and some Asian countries including Japan. See NTSC.

Native Address An address that matches one of a given node's summary addresses.

Native Applications 1. Software that runs directly on the computer's operating system — without requiring an emulation or other program to sit between the software and the operating system. The term "native" acquired new meaning when Apple moved from its Motorola 680x0 microprocessor line of computers to the Power Macintosh line. See NATIVE MODE.
2. An ATM term. A term given to any application written to use any communications protocol prior to ATM.

Native LAN Services Carrier-provided services which interconnect LANs at their full speed, i.e., the native speed of the LANs.

Native Mode 1. Uncompressed.
2. Able to run software directly in the computer's indigenous operating system without intervening emulation software. The term became important when Apple moved from its Motorola 680x0 microprocessor line of computers (called Quadra, Macintosh IIX, Performa, etc.) to the Power Macintosh line of computers powered by the RISC microprocessors called PowerPC 601. This new line of computers runs an operating system different to what Apple had been running on its Macintoshes. The new Power Macintoshes can run software for the Macintosh 680x0 machines — but only in emulation. This makes the software run slowly. So, for the software to run at speeds the new faster machines are capable at running, application software (word processing, desktop publishing, etc.) has to be rewritten to run in native mode.

Native Signal Processing Analog signals, such as voice and video, often are converted into digital signals — for transmission, for computer storage, for compression, etc. Once an analog signal becomes digital, it often has to be manipulated — compressed (to save storage space or transmission time), to be edited, to be conferenced, etc. These speciality tasks of working on digital signals that originally were analog have largely been done by a specialized microprocessor, called a digital signal processor (DSP). DSPs are used extensively in telecommunications for tasks such as echo cancellation, call progress monitoring, voice processing and for the compression of voice and video signals. Lately, as the PC's general purpose microprocessor has become more powerful, the makers of these general purpose microprocessors (especially Intel) have started talking about using some of their spare MIPS (spare processing power) to do some of the tasks previously done by DSPs. They call this new idea — Native Signal Processing.

Native X.400 A term describing a messaging system or service developed using the protocol specifications and service definitions in the X.400 recommendations. Typically used to describe full X.400 services native to a user's home mail environment.

NATOA The National Association of Telecommunications Officers and Advisors, an affiliate of the National League of Cities. An association based in Washington, DC.

Natural A voice recognition term for a language as in normal spoken conversational sentences. The vocabulary would include words like fifty, sixty, hundred etc. and be used in digit recognition.

Natural Computing As computers get more and more powerful, they get easier to use as they are endowed with human-like senses — fluid speech, a good ear (speech recognition) and keen vision. IBM coined the term "natural computing" to describe this.

Natural Frequency The natural frequency of an antenna, the lowest frequency at which the antenna resonates without the addition of any inductance or capacitance.

Natural Microsystems NMS. A manufacturer of voice and call processing components and software development tools located in Natick, MA. They sell equipment through OEMs, system integrators, value-added resellers, dealers and distributors. Most of their dealers "add value" to the NMS components by doing their own specialized software programming, tailoring NMS products to their own specialized (and useful) applications. NMS was the prime mover behind MVIP. See MVIP.

Natural Monopoly A term used by economists to justify regulation. The idea is that one company can provide certain services (such as gas, water, or telecommunications) considerably cheaper than two or three. Therefore, let one company have the monopoly on the service. But substitute government regulation for free competition and this way keep prices down. How well the theory and the practice of government regulation has worked is the subject of acres of learned prose. Suffice, the theory of "natural monopoly" has evaporated in most areas it was practiced — from airlines to telephones, local and long distance. When regulation is removed, prices have usually fallen.

NAU Network Addressable Unit. SNA term for LU, PU and SSCP. Each unit in SNA has a unique address.

NAUCS Network Access Usage and Cost System.

Nautical Mile 6,076 feet. 15% longer than a normal mile, which is 5,280 feet. A measure of distance equal of one minute of arc on the Earth. An international nautical mile is equal to 1,852 meters or 6,076.11549 feet.

NAV Network Applications Vehicle.

Navigation The ability to route telecommunications traffic over diverse circuit options to achieve communications continuity between the desired end customer stations.

NBFCP NetBios Framing Control Protocol; protocol for transporting NetBIOS traffic over a PPP connection.

NBI Some people think this once famous company's name stood for "Nothing But Initials." Others thought it stood for Nectum Bilinium Inc. It actually stood for "Nothing But Initials."

NBP Name Binding Protocol. AppleTalk protocol for translating device names to addresses.

NBS National Bureau of Standards. A US government agency that produces Federal Information Processing Standards (FIPS) for all other agencies except the Department of Defense (DoD).

NBS/ICST National Bureau of Standards/Institute for Computer Sciences and Technology. The NBS directorate, based in Gaithersburg, MD, is concerned with developing computer and data communications.

NC Network Computer. Larry Ellison of Oracle's idea of a $500 (or so) PC that lacks a hard disk and may lack a monitor but can be used to browse the Internet. Ellison, who is Oracle's chairman, sees the NC as a "universal digital appliance."

NCA Number of Calls Abandoned. The number of calls accepted into the ACD on the trunks only but lost before being connected to a person.

NCC 1. Network Control Center. A central location on a network where remote diagnostics and network management are controlled.
2. National Computer Conference, once upon a time the largest annual conference of the computer industry.

NCCF Network Communication Control Facility. An IBM term. See NETVIEW.

NCCS Network Control Center System.

NCD Network Call Distribution. A Sprint service that enables customer premise equipment to interact with Sprint's intelligent network to determine how to route incoming 800 calls. The main service is called SiteRP. But, with NCD, Sprint is monitoring the status of user ACDs to provide load balancing across all agents. With NCD, a processor in the Sprint network polls each ACD over a dedicated X.25 link approximately every 30 seconds. With the resulting information, NCD knows how to balance traffic among agents, making sure one agent is not burdened with more calls than any other agent. As of writing, NCD worked only with Rockwell ACDs. See also AIN and SiteRP. See NET CARD DETECTION.

NCEO NonCompliant End Office.

NCL 1. Network Control Language. A command line interface language used by Digital Equipment Corp.'s Digital Network Architecture (DNA).
2. Non-Computing Module Load.

NCO Network Control Office.

NCOP Network Code Of Practice.

NCP 1. In AT&T language, Network Control Point. A routing, billing, and call control data base system for DSDC which uses the AT&T 3B2OD computer as the feature processor.
2. In IBM language, Network Control Program, which is a program that controls the operations of the communication controllers, 3704 and 3705 in an IBM SNA network.
3. In Northern Telecom language, it means Network Configuration Process.
4. In Novell language, it means NetWare Core Protocol. It is NetWare's format for requesting and replying to requests for file and print services.

NCR National Cash Register, now called AT&T Global Information Solutions. The name change was announced in January 1994. AT&T acquired NCR in 1991.

NCRA National Cellular Resellers Association. A Washington lobbying organization.

NCSANET National Center for Supercomputer Applications Network. A regional TCP/IP network which connects users in Illinois, Wisconsin and Indiana. NCSAnet is also a mid-level network in NSFNET.

NCTA National Cable Television Association. A trade organization representing U.S. cable television carriers.

NCTE Network Channel Terminating Equipment. The general name for equipment that provides line transmission termination and layer-1 maintenance and multiplexing, terminating a 2-wire U-interface. Another name for a CSU, a Channel Service Unit. Also called a Data Service Unit. A device to terminate a digital channel on a customer's premises. It performs certain line conditioning and equalization functions, and responds to loop back commands sent from the central office. A CSU sits between the digital line coming in from the central office and devices such as channel banks or data communications devices. Recent FCC decisions have established that most NCTE is customer premises equipment (CPE) and may therefore, by supplied by third party vendors as well as the telephone company.

NCUG National Centrex Users Group

NDA Non Disclosure Agreement. Check carefully before you sign one. There are precious few new ideas in the history of the world.

NDC National Destination Code. A wireless telecommunications term. The second part of an MSISDN (Mobile Station ISDN) number. It is used to identify a PLMN (Public Land Mobile Network) within a country.

NDD Network Descriptive Database.

NDIS Network Driver Interface Specification. A device driver specification co-developed by Microsoft and 3Com and used by LAN Manager and Vines and supported by many network card vendors. Besides providing hardware and protocol independence for network drivers, NDIS supports both MS-DOS and OS/2. It offers protocol multiplexing so that multiple protocol stacks can coexist in the same host. NDIS is conceptually similar to ODI.

NDS NetWare Directory Services. A new feature of Novell's NetWare 4.0 and higher. Log onto NetWare, you're part of a group. That group gives you various "directories," i.e. access to files. Those directories may be on one or many servers.

NDT No Dial Tone.

NDX NetWare Directory Services. A feature of Novell's NetWare operating system which allows users to sign on to multiple NetWare servers through one simple sign-on. The ultimate idea is that NDS will become somewhat of a front-end sign-on to other networked devices, e.g. telephone systems, voice mail, electronic mail, Lotus Notes, etc. Sign on in one place, get into many. A big user time-saving benefit. A competitor to NDS is Microsoft's ODSI Open Directory Services Interface.

NE Network Element. A single piece of telecommunications equipment used to perform a function or service integral to the underlying bearer network. In ATM networks, NE is a system that supports at least NEFs and may also support Operation System Functions/Mediation Functions. An ATM NE may be realized as either a standalone device or a geographically distributed system. It cannot be further decomposed into managed elements in the context of a given management function.

NE&I Network Engineering and Implementation. An area of responsibility in a telecommunications organization that identifies business processes and conducts activities that identify, plan, design, and construct network and system resources.

Near Instantaneous Companding NIC. The very fast quantizing of an analog signal into digital representation.

Near-end Crosstalk In wires packed together within a cable, the signals generated at one end of the link can drown out the weaker signals coming back from the recipient.

Near-line A storage term typically used as "near-line storage." A digital audio tape (DAT) might be considered near-line storage. A DAT would typically store a database of information sequentially. To find a record might take anywhere from three to 23 seconds. That's an eternity in most computer applications, but it may be adequate for finding something you rarely need to find — maybe several times a year. Thus the concept "near-line" storage.

NEARNET A commercial Internet access service run by Bolt Beranek and Newman, Inc., Cambridge, MA.

NEBS 1. New Equipment Building Systems.
2. Network Equipment Building Standards. NEBS defines a rigid and extensive set of performance, quality, environmental and safety requirements developed by Bellcore, the R&D and standards organization owned by the seven regional Bell operating companies (RBOCs). NEBS compliance is often required by telecommunications service providers such BOCs (Bell Operating Companies) and Interexchange Carriers (IEC) for equipment installed in their switching offices. NEBS defines everything from fire spread and extinguish ability test to Zone-4 earthquake tests to thermal shock, cyclic temperature, mechanical shock, and electro-static discharge. Conforming to NEBS is not inexpensive.
Here is a more detailed explanation of NEBS: NEBS, Generic Equipment Requirement, is a subset of the Family of Requirements as published by Bellcore known as LSSGR, LATA Switching System Generic Requirement. Bellcore has published a technical reference (TR), TR-NWT-000063, Issue 5, 1994, which outlines the NEBS "standard." Bellcore states that the intent of this TR is to "inform the industry of Bellcore's view of the proposed minimum generic requirements that appear to be appropriate for all new telecommunications equipment systems used in central offices (CO) and other telephone buildings of a typical BCC (Bellcore Client Company)."
The requirements defined in this TR are broken into two major categories: 1) spacial requirements and 2) environmental requirements. Spacial requirements apply to the equipment systems' cable distributing and interconnecting frames, power equipment, operations support systems and cable entrance facilities. Compliance with these requirements is intended to improve the use of space in the CO, simplify building-equipment interfaces and help make the planning and engineering of central offices simpler and more economical. In general, the spacial requirements relate to the size and location of the equipment going into a CO, how to power it and how to cable it all together.
The second category relates to environmental requirements that define the conditions under which the equipment could potentially be exposed to and still operate reliably or not cause any catastrophic situation such as fire spread. These environmental requirements include temperature and humidity, fire resistance (which seems to be at the top of many BCCs' lists), shock and vibration (which covers transportation, office and earthquake), electrostatic discharge (susceptibility and immunity) and electromagnetic compatibility (emission and immunity). Bellcore has also defined test methods in this TR which provide procedures on how to test the equipment to ensure that it meets the requirement.

NEC See NATIONAL ELECTRIC CODE.

NEC Requirements The National Electrical Code (NEC) is written and administered by the National Fire Protection Agency (NFPA). The latest 1990 version states that any equipment connected to the telecommunications networks must be listed for that purpose. Listing is acquired through Underwriters Laboratories (UL) or a similar approved lab. Listing requirements for premises wiring between the Network Interface Device and the modular jack at the work area took effect on October 1, 1990.

NECA National Exchange Carrier Association. Nonprofit organization established by the FCC in 1983 to implement the access charge objectives being introduced replacing interstate division of revenue procedures. Membership includes all U.S. local exchange carriers. NECA files interstate access charges with the FCC, and after approval, pools such collections from the LECs and distributes these revenues equitably to the members based on each one's contribution in terms of expenses and capital investment for interstate toll.

NECTAR Nectar is a collaboration among Bell Atlantic/Bell of PA, Carnegie-Mellon University, the Pittsburgh Supercomputing Center and Bellcore.

NEF Network Element Function: A function within an ATM entity that supports the ATM based network transport services, (e.g., multiplexing, cross-connection).

Negated A signal is negated when it is in the state opposed to that which is indicated by the name of the signal. Opposite of Asserted.

Negative Absorption Amplification. The positive difference between stimulated and absorbed radiation.

Negative Acknowledgement NAK. A communications control character sent by a receiving station to indicate that the last message or block received was not received correctly.

Neighbor Node An ATM term. A node that is directly connected to a particular node via a logical link.

NEL Network Element Layer: An abstraction of functions related specifically to the technology, vendor, and the network resources or network elements that provide basic communications services.

Nelson, Lorraine Routh The lady who records the messages on AT&T's Audix voice mail system. She records her timeless messages at the AT&T Bell lab in Denver. She lives in Oregon.

NEMA National Electrical Manufacturers Association.

NEMKO Norwegian Board for Testing and Approval of Electrical Equipment.

NEP See NOISE EQUIVALENT POWER.

Neper Np. A method of expressing the ratio between two quantities (Q1, Q2). Np=log(to the base e)xQ1/Q2. One Neper equals 8.686 decibels.

Nest 1. To embed a set of instructions or a block data within another.
2. Novell Embedded Systems Technology. Software that Novell wants to put into cars, office machines, telephones and other noncomputer products. It is a variant of NetWare.

Net 1. The Internet. See INTERNET.
2. NETwork.
3. Normes Europeenes de Telecommunication (European Telecommunications Standards).

Net 1000 A data communications network and processing service of AT&T that never got off the ground and was closed down at a cost of about $1 billion. At one stage it was called Advanced Communications Service.

Net Additions A cellular term. The amount of new subscribers signing up for service after adjusting for disconnects (churn).

Net Address The location on a network where an addressee's mail is held, usually in storage until the user logs into the system. The system delivers the message at that time. The Net Address in the header of a message gives the information required by an automated message processing or message switch, to deliver a message to an intended addressee. A net address of RA3Y#SAIL would designate that the message was addressed to the user with the "Net Name" of "RA3Y" (pronounced "Ray". The 3 is silent) located at (signified by the symbol "#") the system in the network called "SAIL", in this case, the Stanford (University) Artificial Intelligence Laboratory.

Net Card Detection Library NCD. Part of Microsoft's Windows for Workgroups. Its purpose is to determine the network adapted installed in the workstation and minimize mistakes made by the user.

Net Data Throughout NDT. The rate at which data is transferred on a communications channel, normally specified in bits per second.

Net Logon Service A Windows NT term. For Window NT Server, performs authentication of domain logons, and keeps the domain's database synchronized between the primary domain controller and the other backup domain controllers of the Windows NT Server domain.

Net Loss The signal loss encountered in a transmission facility or network, or the sum of all the losses the signal encounters on its way to its final destination.

Net Staffing A call center term. The actual number of staff minus the required number of staff in a given period. Net staffing that is positive indicates overstaffing; net staffing that is negative, understaffing. See NET STAFFING MATRIX.

Net Staffing Matrix A call center term. A report that shows the actual number of staff, required number of staff, and net staffing for each period of a given day. See NET STAFFING.

Net.citizen An inhabitant of Cyberspace. One usually tries to be a good net.citizen, lest one be flamed, i.e. insulted through e-mail.

Net.personality Somebody sufficiently opinionated/flaky/with plenty of time on his hands to regularly post in dozens of different Usenet newsgroups, whose presence is known to thousands of people. See INTERNET.

Net.police Derogatory term for those who would impose their standards on other users of the Internet. Often used in vigorous flame wars (in which it occasionally mutates to net.nazis).

Net/One The family of local area network products, bridges, gateways, network interfaces and software from Ungermann-Bass, Santa Clara, CA.

NETBEUI NetBIOS Extended User Interface. The transport layer driver frequently used by Microsoft's LAN Manager, Windows for Workgroups and Windows NT. NetBEUI implements the OSI LLC2 protocol.

NETBIOS Network Basic Input/Output System. A layer of software originally developed by IBM and Sytek to link a network operating system with specific hardware. Originally designed as the network controller for IBM's PC Network LAN, NetBIOS has now been extended to allow programs written using the NetBIOS interface to operate on the IBM Token ring. NetBIOS has been adopted as something of an industry standard and now it's common to refer to Netbios-compatible LANs. It offers LAN applications, a variety of "hooks" to carry out inter-application communications and data transfer. Essentially, NetBIOS is a way for application programs to talk to the network. Other applications interfaces are also being used these days, such as IBM's APPC. To run an application that works with NetBIOS, a non-IBM network operating system or network interface card must offer a NetBIOS emulator. More and more hardware and software vendors offer these emulators. They aren't always perfectly compatible, though. Today, many vendors either provide a version of NetBIOS to interface with their hardware or emulate its transport layer communications services in their network products.

NETBIOS Emulator An emulator program provided with NetWare that allows workstations to run applications that support IBM's NetBIOS calls.

NETDDE Network Dynamic Data Exchange, a feature of Microsoft Windows.

Netfind A white pages service which enables a person to query one service and have that service search other databases for addresses matching the originally entered query.

NetFX Consortium NetFX consortium is a group of companies that have come together to provide solutions for delivering real-time multimedia over the Internet's World Wide Web. The members are Diamond Multimedia Systems (http://www.diamondmm.com), NETCOM (http://www.netcom.com/), SAIC (http://merkury.saic.com/demo), Template Graphics Software (http://www.sd.tgs.com/template), and Xing Technology Corp (http://www.xingtech.com).

Netgod Also spelled NET.GOD. A person very visible on a network and who played an important role in the development of the network.

Nethopper A wonderfully cute name for a device that organizes data to hop from one local area network to another. Nethoppers are also called bridges and routers.

NetHub The home of the RIME network located in Bethesda, MD. All Super-Regional Hubs in the RIME network call the NetHub to exchange mail packets.

NETI Network Information Table.

Netiquette A pun on "etiquette". Contraction for "Network Etiquette". It means proper behavior on Internet. Because the recipients of text-only messages cannot see your face or hear the inflections of your voice, special care must be taken to avoid misunderstanding and to convey the "flavor" that you intended your messages to have. Some common techniques are the use of smileys (also called emoticons): punctuation that suggests sideways faces, bracketed abbreviations like <g> for grin and <ggg> for Big Grin, and careful capitalization as in I AM NOW SHOUTING AT YOU!

Netmail A private message transmission capability on FidoNet in which nodes directly communicate with one another on a point-to point (in FidoNet) or hub-routed (in PCRelay). NetMail was originally developed for use by SYSOPS to communicate with one another and is available on some BBs for regular users. See ECHO.

Netscape The most famous Web Browser for now is Netscape. It owes its genesis to a software program called Mosaic, which Marc Andreessen and a small team of student programmers working for $6.85 an hour in 1993 wrote at the National Center for Supercomputing Applications at the University of Illinois at Urbana-Champaign. A year later, after becoming annoyed at the way the University had taken over his Mosaic creation, Mr. Andreessen proposed a "Mosaic Killer" — a new and improved version of his own creation. The team was back at work by April 1994 in a company called Netscape. And by October, they had created a new version of Mosaic, called Netscape Navigator. Netscape went public in August of 1995 in one of the most successful IPOs (Initial Public Offerings) ever.

Netserv A file server used for distributing files directly related to the BITNET network.

Netsploitation Flick Any one of the Hollywood films about the Internet.

Netview An IBM product for management of heterogeneous networks that integrates the functions of three formerly separate Communications Network Management (CNM) software programs: 1. NCCF. Network Communication Control Facility. 2. NLDM. Network Logical Data Manager, which uses functions from NCCF and helps pinpoint problems along the logical connection/path of an SNA session. 3. NPDA. Network Problem Determination Application, which displays various alerts using IBM equipment located at strategic points in the network and allows diagnostic information to be displayed. Also, NetView incorporates some of the functions from two other programs: VNCA (Virtual Telecommunications Access Method/Node Control Application) which monitors the status and current activity of all resources in a domain, and NMPF (Network Management Productivity Facility) which helps the network operator to install, learn and use many network management products. See also NETWORK and NETWORK MANAGEMENT.

NetWare NetWare is an extremely popular and extremely good operating system for a local area network from Novell, Orem, UT. NetWare is actually its own operating system. This means it is the link between machine hardware (file servers, printers, modems, etc.) and people who want it use that hardware. NetWare is neither DOS nor OS/2 though it can be made to look and act like them. That's part (a small part) of its popularity. Because of NetWare's power, it is not easy to learn. Most of its complexity can easily be insulated from a normal user. See NETWARE MHS, NETWARE WORKSTATION FILES, NETx.COM and NOVELL.

NetWare Bindery Centralized authentication database for NetWare 3.xx LANs.

NetWare Directory Services. See NDS.

NetWare Global MHS Novell's implementation of MHS as a NetWare Loadable Module (NLM), providing powerful integration with NetWare services. This supports additional modules to connect to X.400, SNA and SMTP systems.

NetWare Loadable Module NLM. An driver that runs in a server on a local area network under Novell's NetWare operating system and can be loaded or unloaded on the fly as it's needed. In other networks, such applications could require dedicated PCs. A telephony NLM might allow a workstation on a LAN to control a PBX attached to a NetWare file server. It might also allow the workstation to control one or more voice processing cards sitting on in a NetWare server. In early 1993, AT&T became the first PBX maker to ink a deal with Novell, the creator of NetWare, to put telephony onto Novell LANs. AT&T created a PC-card resident in a Novell File server. The card connects to the ASAI (Adjunct Switch Applications Interface) BRI port on the AT&T Definity PBX. Anyone with a PC on the Novell network and an AT&T phone on their desk can use telephone features, such as auto-dialing, conference calling and message management (a new term for integrating voice, fax and e-mail on your desktop PC via your LAN). The Novell/AT&T deal intends to create open Application Programming Interfaces (APIs) that third party developers can work with. A Novell/AT&T example of what could be developed: A user could select names from a directory on his PC. He could tell the Definity PBX through the PC over the LAN to place a conference call to those names. At the same time, a program running under NetWare would automatically send an e-mail to the people, alerting them to the conference call and giving them the agenda. All participants would have access to both the document and the conference call simultaneously. See TELEPHONY SERVICES.

NetWare MHS Netware MHS, which is software that provides store-and-forward capability. Fax and E-mail systems that support MHS format their message transmissions according to MHS specifications. MHS reads compatible transmissions, determines the intended recipient and his location, and then sends the message to that location, regardless of the type of fax or E-mail system at the different ends. See MHS.

NetWare Telephony Services See TELEPHONY SERVICES.

NetWare Workstation Files Novell NetWare files that, along with a network board, convert a standalone computer into a network workstation. These files are memory-resident programs — IPX.COM (a protocol/LAN driver file) and NETx.COM (the NetWare shell). When a DOS workstation is booted, NETx.COM loads the NetWare shell into the workstation's RAM as a TSR (Terminate-and-Stay Resident) program that provides an interface between the application and DOS. The shell intercepts application requests and determines whether to route them locally to DOS or to the network file server for service by the NetWare operating system using NCP (NetWare Core Protocol). If the request needs services from the network, the shell makes any necessary protocol conversions.

Network 1. Networks are common in our lives. Think about trains and phones. A networks ties things together. Computer networks connect all types of computers and computer related things — terminals, printers, modems, door entry sensors, temperature monitors, etc. The networks we're most familiar with are long distance ones, like phones and trains. But there are also Local Area Networks (LANs) which exist within a limited geographic area — like the few hundred feet of a small office, an entire building or even a "campus," such as a university or industrial park. There are also Metropolitan Area Networks (MANs). See also LAN and MAN.

Network Access Control Electronic circuitry that determines which workstation may transmit next or when a particular workstation may transmit.

Network ACD Network ACD allows ACD agent groups, at different locations (nodes), to service calls over the network independent of where the call first entered the network. NACD uses ISDN D-channel messaging to exchange information between nodes.

Network Address Every node on an Ethernet network has one or more addresses associated with it, including at least one fixed hardware address such as "ae-34-2c-1d-69-f1" assigned by the device's manufacturer. Most nodes also have protocol specific addresses assigned by a network manager.

Network Addressable Unit NAU. In IBM's SNA, a logical unit (LU), physical unit (PU) or system services control point (SSCP), which is host-based, that is the origin or destination of information transmitted by the path control portion of an SNA network.

Network Application Architecture A generalized architecture allowing interoperability at the application level. Examples are Digital Equipment Corp.'s Network Application Support (NAS) and IBM Corp.'s Systems Application Architecture (SAA).

Network Application Support Digital Equipment Corporation's set of open software which allegedly allows its customers to integrate, port and distribute applications across different computer systems, including VMS, UNIX, MS-DOS, OS/2 and Apple MacIntosh.

Network Architecture The structure and protocols of a computer network. See ARCHITECTURE.

Network Balancing 1. Lumped circuit elements (inductances, capacitances and resistances) connected so as to simulate the impedance of a uniform cable or open-wire circuit over a band of frequencies.
2. Moving circuits around in a multi-node switching network such the switching loads on each of similar switching modules are roughly equal.

Network Basic Input/Output System NETBIOS. Within the context of the MS-DOS operating system, the software or software and firmware services that implement the interface between applications and a network adaptor, such as a CSMA\CD or token-ring adaptor.

Network Board 1. A circuit board installed in each network station to allow stations to communicate with each other and with the file server.
2. An SCSA term. A board device designed to act as an interface between a computer-based signal processing system and a telephone network.

Network Byte Order The Internet standard way of ordering of bytes corresponding to numeric values.

Network Channel Terminating Equipment NCTE. A device or devices at the user's premises used to amplify, match impedance or match network signaling to the customer's equipment connected to the network. Basically, network channel terminating equipment is a general name for equipment linking the network to a customer's premises. When NCTE connects to digital circuits, it typically consists of DSUs and CSUs. They are used for balancing of signals and providing for loop-back testing.

Network Control Center A physical point within a network where various management and control functions are implemented.

Network Control Language NCL. A command line interface language used by Digital Equipment Corp.'s Digital Network Architecture (DNA).

Network Control Point In IBM SNA networks, a host-generated program that controls the operation of a communications controller (such as an IBM 3705 or 3725).

Network Control Program NCP. An IBM Systems Network Architecture (SNA) term. This is the program that switches the virtual circuit connections into place, implements path control, and operates the Synchronous Data Link Control (SDLC) link. The Network Control Program is normally resident in the communications controller or the host processor.

Network Control Signaling The transmission of signals used in the telecommunications system which perform functions such as supervision, address signaling and audible tone signals to control the operation of switching machines in the telecommunication system.

Network Control Signaling Unit A telephone set that controls the transmission of signals into the telephone system which will perform supervision, number identification and control of the switching machines.

Network Controller A powerful microprocessor device designed to perform communications protocol translations between various terminals and computers and an X.25 packet switching network.

Network DDE Service A Windows NT definition. The Network DDE (dynamic data exchange) service manages shared DDE conversations. It is used by the Network DDE service.

Network Demarcation Point The network demarcation point is the point of interconnection between the local exchange carrier's facilities and the wiring and equipment at the end user's facilities. The demarcation point is located on the subscriber's side of the telephone company's protector.

Network Device Driver Software that coordinates communication between the network adapter card and the computer's hardware and other software, controlling the physical function of the local area network adapter cards.

Network, Distributed See DISTRIBUTED NETWORK.

Network Diversity A simply concept that says if you have a network which is important you need to have multiple ways of moving information around that network. Some of those ways should be provided by circuits from one vendor and some should be provided by circuits from several vendors. In other words, there should be diversity in both circuits and vendors. As an example of the necessity of taking network diversity very seriously, consider the year 1991. In 1991, nearly 30 million residential and business customers, more than one hundred thousand airline passengers and an alphabet soup of state and federal agencies were temporarily crippled by nine major telephone network outages. Also in a 12-month period, the FAA (Federal Aviation Commission) reported a total of 114 phone service failures that disrupted air traffic facilities around the nation. Between June 10 and July 1, 1991, six major outages in three states comprised the largest series of network outages in history.

Network Drive A disk drive that is available to multiple users and computers on a network. Network drives often store data files for many people in a work group.

Network Elements NE. Entities defined by the functionality of the functional groups they contain and by their interfaces. Definition from Bellcore in reference to its concept of the Advanced Intelligent Network.
In SONET, five basic network elements are: add/drop multiplexer; broadband digital cross-connect; wideband digital cross-connect; digital loop carrier; and switch interface.

Network Equalizing A device connected to a transmission path in order to alter the characteristics of that path in a specified way. Often used to equalize the frequency response characteristic of a circuit for data transmission.

Network Extension Unit NEU. An AT&T thing which sits in a telephone satellite closet (their words) which links up to 11 Starlan daisy chained clusters of PCs, etc. in a star configuration through standard telephone wiring, modular cords and plugs.

Network Fax Server An network fax server is typically a PC on LAN. The PC has one or more fax/modem cards. Its job is to send and receive faxes for everyone on the LAN. As a result it is connected to dial out phone lines. See SERVER.

Network File System NFS. A method developed by Sun Microsystems Inc. for distributing files within a heterogeneous network. NFS has become a de facto standard. It requires special software drivers to suit specific vendor hardware and operating system software. NFS hardware and software gateway products are also available.

Network Harm The reasoning behind the Bell System's insistence that the public could not hook its own equipment to the telephone line for fear it would produce "Irreparable Harm To The Network". A Bell System publication quoted the National Academy of Sciences as identifying four areas of potential harm. They were: Excessive signal power, hazardous voltage, improper network control signaling and line imbalance. Since the Carterfone Decision in the summer of 1968, the FCC has seen fit to allow devices, which pass an FCC Registration program, to be connected to the network, and while the quality of the stuff is occasionally poor, the fears of Network Harm have proven groundless. The fears were, of course, raised to preserve the Bell System's erstwhile almost monopoly on the manufacture of telephone equipment in the U.S.

Network Information Center NIC. Centers providing user assistance, document service, training on the Internet.

Network Interface The point of interconnection between Telephone Company communications facilities and terminal equipment, protective apparatus or wiring at a subscriber's premises. The network interface or demarcation point shall be located on the subscriber's side of the Telephone Company's protector, or the equivalent thereof in cases where a protector is not employed, as provided under the local telephone company's reasonable and nondiscriminatory standard operating practices.

Network Interface Card Electronic circuitry connecting a workstation to a network. Usually a card that fits into one of the expansion slots inside a personal computer. It works with the network software and computer operating system to transmit and receive messages on the network.

Network Interface Controller Same as a Network Interface Card. See above definition.

Network Interface Device 1. NID. A device wired between a telephone protector and the inside wiring to isolate the customer's equipment from the network.
2. A device that performs, functions such as code and protocol conversion, and buffering required for communications to and from a network.
3. A device used primarily within a local area network to allow a number of independent devices, with varying protocols, to communicate with each other. This communication is accomplished by converting each device protocol into a common transmission protocol.

Network Interface Module Electronic circuitry connecting a system (typically a PC) to the telephone network. Network interface modules come in as many versions as there are ways of connecting to the telephone network. They range from a simple loop start telephone line to complex ISDN PRI circuits. Usually the network interface modules slides into one of the expansion slots inside a PC. The card transmits and receives messages from the resource modules and provides access to the telephone network.

Network Interface Unit A communications controller that attaches to a local network. It implements the local network protocols and provides an interface for device attachment.

Network Inward Dialing NID. A service feature of an automatically switched telephone network that allows a calling user to dial directly to an extension number at the called user facility without operator intervention.

Network Layer Third layer of the OSI model of data communications, sometimes called the packet layer. Involves routing data messages through the network on alternate routes. See OSI STANDARDS.

Network Layer Address NLA. An address appended to the LAN packet that, unlike a MAC name, indicates exactly where a computing device is located within an internetwork. TCP/IP, DECNet and IPX support network layer addressing and each has its own unique NLA format. Protocol dependent routers use the NLA to make routing decisions.

Network Management A set of procedures, software, equipment and operations designed to keep a network operating near maximum efficiency. Network management generally falls into five areas:
1. Configuration Management deals with installing, initializing, "boot" loading, modifying and tracking the configuration parameters of network hardware and software.
2. Fault Location and Repair Management Tools let you find out what's going wrong with what equipment or lines and give you the ability to fix those resources — by re-routing traffic on different lines or reporting problems to the carrier, or suggesting to whoever that certain equipments should be replaced. Fault location and management tools have strong error and alarm characteristics.
3. Security Management Tools allow the network manager to restrict access to various resources in the network. There are devices such as password protection schemes, giving users different levels of access to different network resources.
4. Performance Management Tools provide real-time and historical statistical information about the network's operation. Such tools show, for example, how many packets are being transmitted at any given moment, the number of users logged into a specific server and use of network lines.
5. Accounting Management applications help users allocate the costs of the various network resources — from lines to PBXs, from access to a mainframe to time used on printers.
See NETVIEW.

Network Management Control Center NMCC. A central place from which the network is maintained and changed and from where statistical information is collected.

Network Management Forum See NM Forum.

Network Management Software The software that manages and controls all network functions within a network.

Network Management System 1. A comprehensive system of equipment used in monitoring, controlling and managing a data communications network. Usually consists of testing devices, CRT displays and printers, patch panels and circuitry for diagnostics and reconfiguration of channels, generally housed together in an operator console unit.
2. An MCI definition. An MCI system that is used for the automated transmission of data billing information is passed from the switches to an AOM sub-system within each division and then passed on to the billing centers to be edited and processed. This automated system of transfer of traffic information is much faster than the "switch tape" method, in which a physical tape is made and then mailed to the Billing Centers.

Network Redundancy In a network, the state of having more connecting links than the minimum required to provide a connecting path between all nodes.

Network Synchronization Unit See NSU.

Network Virtual Terminal A process whereby a network accommodates terminal devices with different characteristics by converting the varying characteristics into a single format so all terminals appear the same.

Networked Multimedia This definition courtesy Intel. Just as standalone PCs bring you rich audio, video, animation and three dimensional graphics, so networked multimedia brings these same capabilities to PCs connected to networks and the Internet. You could create an electronic letter with an embedded video clip of yourself in the salutation. Include hotlinks to your company's web site for more information on new products. Allow the receiver to have the entire letter read out aloud by a real human voice. Hold a virtual conference in shared interactive spaces, with people in different locations virtually meeting and interacting. Videoconferencing and broadcasting over the Internet will become commonplace.

NetworkMCI MCI's six-year development plan, called networkMCI, defines the company's core strategy for the coming era, as telephony, computers and television converge into interactive multimedia and wireless communications. A term coined by MCI during the first few days of 1994. Here are some MCI words:

"Our first strategic initiative under this vision is our nationwide information superhighway, which uses powerful SONET fiber-optic technology at 2.5 gigabits per second — 15 times faster than any SONET network commercially available. today. We have opened our information superhighway to its first user, the National Science Foundation, known as NSFNET, one of the fastest and most powerful of the tens of thousands of networks known collectively as the Internet.

"The second strategic component of networkMCI is our entry into the local telecommunications services marketplace through a new subsidiary, MCI Metro. MCI believes the time has come to enter the local market for three important reasons.

"First are the exorbitant access charges we pay the local exchange companies (LECs) to reach our customers. MCI today pays the LECs more than 45 percent of every revenue dollar we take in - as do our major competitors. Our entry into the local market will not only offset these charges, it also will stimulate local competition.

"Second is to assure the superior local access facilities we will need to offer our customers the wide range of new products and services that will result from the convergence of telephones, computers and entertainment.

"Third is the attractiveness of the local market as a business opportunity. This market today is approaching $90 billion and can be expected to grow rapidly as more and more households have personal computers, the era of intelligent computer networks gets underway and wireless communications and interactive multimedia take hold. This market is a tremendous opportunity for growth for MCI."

Network Name Automatic Routing A system for getting network-delivered faxes delivered to individual users. The fax network assigns each user a name (usually the user's network login name), and the sender uses this name. The receiving fax system detects this name and routes the fax to the intended recipient.

Network News Transfer Protocol NNTP. A common method by which articles over Internet's Usenet are transferred.

Network Number The part of an Internet address that specifies the network to which the host belongs.

Network Operating System NOS. The software side of a LAN. The program that controls the operation of a network. It allows users to communicate and share files and peripherals. It provides the user interface to the LAN, and it communicates with the LAN hardware or network interface card. It is important to note that a network operating system is different from a network interface card: IBM's Token Ring, for example, is an interface card, not a NOS. IBM's NOS is called the PC LAN Support Program. A variety of other vendors also offer NOSs, including Novell and 3Com. Most have drivers to support a variety of network interface cards.

Network Operations Functions that provide, maintain and administer services supported by the network systems. These functions reside in network systems or network operations applications that interface directly to network systems and include memory administration, surveillance, testing, network traffic management and network data collection. Definition from Bellcore in reference to its concept of the Advanced Intelligent Network.

Network Operations Business Information System NOBIS. An MCI definition. An MCI system supporting the construction and operation of MCI's telecommunications network.

Network Operations Center A name for the central place which monitors the status of a corporate network and sends out instructions to repair bits and pieces of the network when they break. In more formal terms, monitoring of network status, supervision and coordination of network maintenance, accumulation of accounting and usage data and user support.

Network Outward Dialing NOD. A service feature of an automatically switched telephone network that allows a calling user to dial directly all network user numbers without operator intervention.

Network Printer A printer shared by multiple computers over a network. See also local printer.

Network Problem Determination Application NPDA. A host resident IBM program product that aids a network operator in interactively identifying network problems from a central point.

Network Professional Association An organization that runs a program called CNP, which stands for Certified Network Professional. That program is designed for individuals whose career is to design, integrate, manage and maintain networked computing environments. The Association can be reached at 801-373-7888.

Network Protection A term used to describe an array of strategies to protect your network from crashing around your ears should a disaster happen. The ultimate network protection means duplicating every item in the network — including the people who operate it. Obviously anything else (which is what we are all forced to do) is a compromise. The typical network protection these days tends to focus on alternate routing and duplication of network lines, including local loops and long distance lines.

Network Protection Device NPD. A device which provides isolation between PBX circuits and CO trunks or tie lines.

Network Reconfiguration Service NRS. A Nynex service that lets you control, rearrange and switch your private line voice, private line analog data and Superpath 1.544 Mbps lines. You might wish to do this because you want to temporarily join one office to another for a high-speed data dump, or for a videoconference. Or in case something happens and you may want to route yourself around the trouble. You can reconfigure you own network in one of two ways — using a computer terminal or PC to dial in your instructions or call an 800 number and have an Nynex attendant do it for you. Rearrangements can be made in near-real time or scheduled for an appointed time.

Network Redundancy Network redundancy means that the network topology has been constructed such that a failure of a network component can be automatically and/or rapidly recovered by using identical components engineered into the network for recovery purposes. For example, IXC network redundancy may take the form of spare network capacity on geographically diverse IXC circuits so that applications can be recovered to an alternative network path in the event of a failure on the primary path.

Network Relay A device which allows interconnection of dissimilar networks.

Network Reliability ComForum The Network Reliability ComForum is a gathering of senior telecommunication industry executives and government officials that serve on the Network Reliability Council. The ComForum meets to report on the Council's findings. See USITA.

Network Server A powerful computer on a LAN designed to serve the needs of all the people on the LAN with everything from database, email, fax, images, voice messages, communications, etc. See also EMAIL SERVER and FAX SERVER.

Network Services In IBM's SNA, the services within network addressable units (NAUs) that control network operations via sessions to and from the SSCP.

Network Synchronization Within a data transmission network, especially one using multiplexers, the need for network synchronization is two-fold: 1. The network must transmit data to the receiving DTE device at the same rate it is being received from the transmitting DTE; and 2. The data originating from the low-speed channels must be capable of being inserted into the composite link (and vice versa) without the loss of any information (channel signaling rate too low) or the creation of unwanted information (channel signaling too high). The need for network synchronization is necessary with networks with time division multiplexing and synchronous data. It does not arise with time division multiplexing in which only asynchronous data is transmitted, because asynchronous data contains "start" and "stop" bits and therefore doesn't need the synchronization of a single master clock. When time division multiplexers are to multiplex synchronous channels, it is essential that the composite links and the low-speed channels are strictly synchronized, i.e. that their clocks operate at precise multiples of the basic clock rate. A single clock, therefore, must be in overall control of the whole network. See SYNCHRONOUS.

Network Terminal Number NTN. The number assigned to a data terminal under the Data Network Identification Code system. In the ITU-T International X.121 format, the sets of digits that comprise the complete address of the data terminal end point. For an NTN that is not part of a national integrated numbering format, the NTN is the 10 digits of the ITU-T X.25 14-digit address that follow the Data Network Identification Code (DNIC). When part of a national integrated numbering format, the NTN is the 11 digits of the ITU-T X.25 14-digit address that follow the DNIC.

Network Terminal Option NTO. An IBM program product that enables an SNA network to accommodate a select group of non-SNA asynchronous and bisynch devices via the NCP-driven communications controller.

Network Terminating Interface 1. NTI. The point where the network service provider's responsibilities for service begin or end.
2. NTI. The interface between DCE and its connected DTE.

Network Terminating Unit NTU. The part of the network equipment which connects directly to the data terminal equipment.

Network Termination Type 1 NT-1 or NT1. The NT-1 is the first customer premise device on a two-wire ISDN circuit coming in from the ISDN central office. It does several things. It converts the two-wire ISDN circuit (called "U" interface) to four-wire so you can hook up several terminals — like a voice phone and a videophone. The NT-1 typically has several lights on it which indicate if it's working. An ISDN central office can usually "talk" to the NT-1 and do testing and maintenance by instructing the NT-1 to loop signals back to the central office. An NT-1 will support up to eight terminal devices, though I've never seen it work with that many. Sometimes, NT1s are built into ISDN terminal devices, like phones or PC cards. The basic NT1 functions are:
• Line transmission termination.
• Layer 1 line maintenance functions and performance monitoring.
• Layer 1 multiplexing, and
• Interface termination, including multi drop termination employing layer 1 contention resolution.
See ISDN.

Network Time Protocol NTP. A protocol built on top of TCP that assures accurate local time keeping with reference to radio and atomic clocks located on the Internet. This protocol is capable of synchronizing distributed clocks within milliseconds over long time periods.

Network-to-Node Interface A set of ATM Forum-developed specifications for the interface between two ATM nodes in the same network. Two variations are being developed: an interface between nodes in a public network and an interface between nodes in a private network.

Network Topology The geometric arrangement of links and nodes of a network. The geography of a network. Networks are typically of either a star, ring, tree or bus topology, or some hybrid combination.

Network Traffic Management NTM. Functionality that maximizes the traffic throughput of the network during times of overload or failure and minimizes the impact of one service on the performance of others. This is done through centralized surveillance of maintenance conditions and traffic, and centralized control of traffic volumes being originated in network systems. Definition from Bellcore in reference to its concept of the Advanced Intelligent Network.

Network Virtual Terminal A communications concept wherein a variety of DTEs, with different data rates, protocols, codes and formats, are accommodated in the same network. This is done as a result of network processing, where each device's data is converted into a network standard format and then converted into the format of the receiving device at the destination end.

Networking Blueprint IBM's latest and most ambitious network initiative. The Blueprint is structured to allow applications to work over a wide range of networking schemes, e.g., SNA applications over TCP/IP or OSI, TCP/IP over APPN or OSI, DECnet applications over SNA.

Networking Systems IBM Networking Systems is new name (summer, 1991) for its Communication Systems group, formed in January, 1988.

Netx.com This Novell NetWare workstation shell file provides an interface between the application and DOS, monitoring all data transmissions that move in or out of either one. When a function request needs network service, such as a call to read a file from the file server, the shell begins the process of protocol conversion and network transmission. Netx intercepts requests by taking over software interrupts 21h (used to call standard DOS functions), 24h (DOS's critical error handler vector), and 17h (used to send data to local printer ports). In general, the shell intercepts all interrupt 21h DOS requests and inspects each one. The "x" typically stands for which version of DOS the workstation is running. NET3.COM would run on a workstation running

MS-DOS 3.X, i.e. 3.1, 3.2 and 3.3. (There is no 3.4).

Neural Network A neural network is designed to take a pattern of data and generalize from it. Thus, if the data are daily temperatures in New York city over two years, the neural network should emerge with a simple undulating curve that describes the way temperature rises in summer and falls in winter. It does this in effect by a sophisticated form of trial and error, or, in the jargon, varying the strengths of connections between individual processors until the input yields the right output. The two essential features of neural network technology are that it improves its performance on a particular task by trial and error (neural networkers prefer to say that "it learns") and that it can be a "black box." That is, you do not need to know what mathematical equation describes its output, although you could find out if you wanted to. (Thanks for the Economist Magazine for help on this definition.)

Neural Network Computer A very different kind of computer. Neural network computers are built from webs of randomly connected electronic neurons. These machines are designed to be trained, not programmed. In design and in function, they are meant to closely resemble the human brain. As in the brain, the neurons send signals to one another through thousands of adjustable connections, or synapses. As the machine learns, the settings of these controls are automatically turned up or down. And the chaos of connections evolves into a finely tuned machine, one that can read handwritten letters or recognize the spoken words. With a neural network computer, a trainer simply speaks words to the machine, rewarding it if it acts correctly and punishing it when it acts incorrectly. There is hope that neural network computers may perform better than traditional computers in running artificial intelligence software. See NEURAL NETWORKS.

Neutral The ac power system conductor that is intentionally grounded on the supply side of the service disconnect. It is the low potential (white) side of a single phase ac circuit or the low potential fourth wire of a three-phase Wye distribution system. The neutral provides a current return path for ac power currents whereas the safety ground (green) conductor should not, except during fault conditions.

Neutral Ground An intentional ground applied to the neutral conductor or neutral point of a circuit, transformer, machine, apparatus, or system.

Neutral Transmission Unipolar transmission. A form of signaling which employs two distinct states, one of which represents the existence of a space as well as the absence of current.

New A term used in the secondary telecom equipment business. Generally defined as being sold by authorized vendor of the O.E.M. and carrying the O.E.M.'s standard warranty. See UNUSED.

New Line ASCII character 10, abbreviated "NL". This character replaced the abbreviation "LF" which meant Line Feed. Newer terminals accept the NL character to mean both CR (Carriage Return) and LF functions for "end of line" sequences. The function of transmitting an "end of line" sequence to a printer or remote computer system is to designate to a computer to print what follows on the next line of a print out. This usually consists of transmitting the characters CR, LF (Carriage Return, followed by Line Feed). On older printing terminals, these two characters would be followed by a number of null characters whose job would be to waste time until the print head got back to the left margin before printing the first letter of the next line on the page. Otherwise, that character would be in the middle of the page, and the second character would start the line. Telex machines usually send the sequence CR, LF, CR as a New Line sequence to give the head enough time to get back.

New Wave New Wave is a user environment available for personal computers from Hewlett-Packard. It includes a look and feel specification, a toolkit and a desktop metaphor set of tools and definition.

New York Telephone A phone company that once was part of the Bell System. When AT&T was broken up in 1984, it was lumped with New England Telephone to become Nynex. Then in January of 1994, Nynex decided to eliminate the names New York Telephone and New England Telephone and call them both Nynex. A public relations spokesman told me the reason for this name change had something to do with consistency of marketing image. Frankly, I never understood the necessity for the name change, nor its cost — which included painting the 16,700 trucks the two companies owned.

Newbie A newcomer to cyberspace; usually applied condescendingly, the way sophomores talk about freshmen. Somebody new to the Net. Sometimes used derogatorily by net.veterans who have forgotten that, they, too, were once newbies who did not know the answer to everything. "Clueless newbie" is always derogatory.

Newbridge Networks A manufacturer of good T-1 muxes, channel banks and other digital communications devices founded by Terry Matthews, co-founder of Mitel and Peter Madsen, who also made his fortune with Mitel. Newbridge is one of our favorite companies. Terry and Peter will be pleased when they see their name in this dictionary.

Newsgroup A USENET newsgroup is a place on the Internet where people can have conversations about a well-defined topic. Physically it is made of the computer files that contain the conversation elements to the discussions currently in progress about the agreed-upon topic.

Newsgroup Distribution Software A set of computer programs used to manage the distribution of USENET news groups and articles.

Newsgroup Feed One remote computer receiving USENET newsgroups from another remote computer.

Newsgroup Management Software A set of computer programs used to manage USENET newsgroups and articles.

Newsreader A newsreader is a program that organizes the conversation elements in a sensible and presentable manner. The news reader allows the person using it to read and/or participate in those conversations.

Newton 1. The name of Apple's personal communicator.
2. The author of this dictionary, Harry Newton, has reached a pinnacle of achievement. According to Susan, his wife of over 20 years, he has become a sex symbol for women who no longer care. The photo on the back cover (in case you hadn't already figured) is years old. Maybe decades old.

NEXT Equipment that must concurrently receive on one wire pair and transmit on another wire pair in the same cable bundle must accommodate NEXT interference. NEXT is the portion of the transmitted signal that leaks into the receive pair. Since at this point on the link the transmitted signal is at maximum and the receive signal has been attenuated, it may be difficult to maintain an acceptable ACR with the received signal if the cable media allows large amounts of crosstalk leakage to occur. Foiled or shielded cables generally have less crosstalk than unshielded varieties.

2. Steve Jobs' "new" computing company — the one he started after leaving Apple. It used to be in hardware. It's now only in software.

Nextstation The name of Steve Jobs' new computer. It is made by his company which is called Next Inc. Nextstation comes with a user interface control panel (also called a user interface) called NextStep, which makes the computer especially easy to use. Next's color computer is called Nextdimension.

NextStep NextStep is the software development environment shipped by Next Computer. IBM's recently announced RISC System 6000 workstation family will support the NextStep environment for non-networked users. IBM has licensed the technology from Next.

NF Noise Figure.

NFAS Network Facility Associated Signaling. See also D CHANNEL and ISDN.

NFPA National Fire Protection Agency. See NEC REQUIREMENTS.

NFS Network File System. One of many distributed-file-system protocols that allow a computer on a network to use the files and peripherals of another networked computer as if they were local. This protocol was developed by Sun Micro-Systems and adopted by other vendors.

NGDLC Next Generation Digital Loop Carrier. DLC can receive and aggregate large amounts of Bandwidth (higher than T-1).

NI 1. National ISDN.

2. Network Interface. Demarcation point between PSTN and CPE.

NI-1 National ISDN 1. It's a BRI circuit. See ISDN.
NI-2 National ISDN 2. It's a BRI circuit. See ISDN.
NI-3 National ISDN 3. It's a BRI circuit. See ISDN.

Ni-MH Nickel Hydride is a new technology used in batteries for portable devices, such as cell phones and laptops, etc. Allegedly, these batteries do not have a "memory effect." Today, Ni-MH batteries cost about 30% more for a 20% or so gain in capacity. The average life of Ni-MH batteries is about 500 cycles. See NICKEL METAL HYDRIDE.

NIA Network Interface Adapter. An IBM hardware device that with certain software, will allow an SNA device to communicate over a packet switching network.

Nibble Four bits or half a byte. Usually described by one hexadecimal digit.

Nibble Interleaving/Multiplexing A technique where 4 bit nibbles (one at a time) from each lower speed input a channel are used to build the higher speed frame output of a multiplexer.

NIC 1. Near Instantaneous Companding. This describes the very fast, essentially real-time, process of quantizing an analog signal into digital symbols.

2. Network Interface Card. The attachment that connects a device to a network. The NIC, usually a PC expansion board, executes the code needed by the connected device to share a cable or some other media with other stations.

3. Network Information Center; any organization that's responsible for supplying information about any network.

4. The InterNIC, which plays an important role in overall Internet coordination.

NIC Network Information Center As close as an Internet-style network gets to a hub; it's usually where you'll find information about that particular network.

Nickel Metal Hydride Ni-MH or NiMH. A new type of rechargeable battery, now being used in laptop computers. Most of today's rechargeable batteries are nickel cadmium. These suffer from several problems: 1. They have a "memory." Which means they must be fully discharged once a month or they won't deliver their full potential. 2. They are hard to dispose of. Nickel hydride is a new technology with some benefits: It's easier to dispose of. It can hold at least 1.25 times as much power per unit as a standard nickel cadmium battery. It doesn't have a "memory." (Neither do car or telephone batteries, which are typically lead acid.) But there are some downsides: Nickel hydride batteries lose their charge faster than nickel cadmium. Nickel hydride batteries are ABOUT 30% more expensive than nickel cadmium. The Nickel hydride battery also can not put out power as fast as a nickel cadmium one can, so it probably won't be suitable for power tools and appliances that drain batteries quickly. But it's great for "slow-burning" items, like laptops and cellular phones.

Nicname A LAN protocol specified in RFC-812. It requests information about a specific user or hostname from the Network Information Center (or NIC) name database service.

NID Network Inward Dialing. A service feature of an automatically switched telephone network that allows a calling user to dial directly to an extension number without operator intervention.

NIF Network Interface Function.

Night Answer Incoming calls to a switchboard during evening and weekend hours are automatically rerouted to ring only at designated night answering phones such as the security desk. See the next few definitions.

Night Answer — Assigned Night answer going to specific, assigned telephones.

Night Answer — Offsite Phones being answered after hours by people and machinery located offsite — someplace else.

Night Answer — Universal Anybody and everybody can answer the incoming calls from any phone. In TELECONNECT, if the phone rings after hours, we simply touch "6" on any phone and we can answer that incoming call — whether it's a local, normal DDD or incoming WATS.

Night Audit This feature provides automatic printout of message registration data for all quest rooms at the front desk console.

Night Chime An auxiliary ringer, usually wall mounted, used to indicate a ringing trunk during night operations or used as a "phantom" extension for overflow applications.

Night Console Position Provides an alternate attendant position which can be used at night in lieu of the regular console. Usually cheaper than buying a second normal console.

Night Patch Assigned Night Answer, Fixed Night Service, Programmable Night Connections. Provides arrangements (which are prewired into the system) to route incoming central office calls normally answered at the attendant position, to pre-elected stations within the PBX system when the attendant is not on duty.

Night Service Your operator goes home, puts your phone system into "night service." A call now comes in. The night bell rings. You hear it. You go to a phone and hit a code (such as #6) or hit the flashing "Nigh Answer" button. And you answer the call. Not all phone systems need to be placed in Night Service for you to answer an incoming call from any phone. And you can program the night bell to ring all day.

Night Service Automatic Switching Should the attendant neglect to place the console in the night answering mode, after a certain period of timed ringing from an incoming central office call, the entire system will automatically jump to the night service.

Night Service — Expanded Service Routes calls normally directed to the attendant to preselected station lines within the system when it is arranged for night service. Calls to specific exchange trunks can be arranged to route to specific station lines and can be assigned on a flexible basis. Trunk Answer From Any Station capability is provided for calls which are handled by assigned night stations.

Night Service — Fixed Service Provides arrangements to route calls normally directed to the attendant to preselected station lines within the PBX system when regular attendant positions are not in use. In addition, calls to specific trunks can be arranged to ring on specific phones. The receiving phone can then transfer the call.

Night Station The phone assigned to automatically handle incoming calls after the main switchboard has shut down for the day.

NIH 1. Not Invented Here. The tendency of organizations to reject ideas and inventions which they didn't think of. A major and continuing problem in the telephone industry.
2. National Information Highway. A term coined by Al Gore when he was Vice President of the United States under President Clinton. See also NII.

NII National Information Infrastructure. A clumsy name for the Information SuperHighway. NII is a term that came from a paper which the Gore-Clinton Administration released, called "National Information Infrastructure; Agenda for Action." According to one engineer, NII is "multiple, interconnected interoperable networks."

NiMH Nickel Metal Hydride. See NICKEL METAL HYDRIDE.

NIMQ (pronounced "nihm-kyoo") Acronym for "Not in My Queue." Said in response to suggestions to take on more work when you're already overwhelmed. Similar to the more common "It's not my job."

NIOD Network Inward/Outward Dialing.

NIOSH National Institute of Occupational Safety and Health.

NIS 1. Network Information Service. A way of centralizing user configuration files in a big distributed computing environment. NIS was created by Sun Microsystems.
2. Network Imaging Server. A local area network based server largely devoted to storing, retrieving and possibly manipulating images. See also SERVER.

NISDN-1 See National ISDN-1.

NIST The National Institute of Standards and Technology, which is part of the U.S. Department of Commerce. It used to be called the National Bureau of Standards. (www.nist.gov) See also the NSA and the NORTH AMERICAN ISDN USERS' FORUM.

NIU Network Interface Unit.

NIUF North American ISDN Users Forum.

NKT Nederlands Keuringsinstituut voor Telecommunicatieapparatuur (Private laboratory for regulatory testing, The Netherlands).

NL See NEW LINE.

NLA Network Layer Address. An address appended to the LAN packet that, unlike a MAC name, indicates exactly where a computing device is located within an internetwork. TCP/IP, DECNet and IPX support network layer addressing and each has its own unique NLA format. Protocol dependent routers use the NLA to make routing decisions.

NLDM Network Logical Data Manager. An IBM term. See also NETVIEW.

NLM NetWare Loadable Modules. NLMs are applications and drivers that run in a server on a local area network under Novell's NetWare operating system and can be loaded or unloaded on the fly. The category of applications ranges from complex voice mail/auto attendant programs and electronic mail gateways to simple programs like drivers for PCI cards. In some other networks, such applications could require dedicated PCs. See NETWARE LOADABLE MODULE for a longer explanation.

NLS Name Lookup Service. An electronic directory service which is designed to respond to external queries for general information about a large group of users. Once installed, NLS can be accessed via finger or whois.

NLSP NetWare Link Services Protocol. A link state protocol from Novell to improve performance of IPX traffic in large internetworks.

NM Network Management. A business process in a telecommunications organization that constantly monitors the state of the network and invokes changes to the network when outages or problems occur. In an ATM network, NM is the body of software in a switching system that provides the ability to manage the PNNI protocol. NM interacts with the PNNI protocol through the MIB.

NMAA National Multimedia Association of American (www.nmaa.org)

NMC Network Management Center. A centralized location at the network management layer used to consolidate input form various network elements to monitor, control, and manage the state of a network in a telecommunications organization.

NM Forum Network Management Forum. International forum for network management; part of ITU.

NML Network Management Layer. The layer in the Telecommunications Management Network (TMN) standard addressing network functions including monitoring, management, and control. The network management layer is an abstraction of the functions provided by systems which manage network elements on a collective basis, so as to monitor and control the network end-to-end.

NMOS N-channel Metal Oxide Semiconductor.

NMP Network Management Protocol. An AT&T-developed set of protocols designed to exchange information with and control the devices that govern various components of a network, including modems and T1 multiplexers.

NMS Network Management System. The system responsible for managing a portion of a network. The NMS talks to network management agents, which reside in the managed nodes, via a network management protocol. The NMS is the entity that implements functions at the Network Management Layer. It may also include Element Network Management Layer functions. A Network Management System may manage one or more other Network Management Systems.

NMT Another form of cellular service, common in some overseas countries.

NMVT Network Management Vector Transport. A network management protocol which provides alert problem determination statistics and other network management data within SNA management services.

NN Network Node. In frame relay, a network node is typically the frame relay service port connection and its associated virtual circuits.

NNI 1. Network Node Interface. An Asynchronous Transfer Mode (ATM) term. The interface between two public network pieces of equipment (contrast that to UNI, which stands for User Network Interface).
2. Network to Network Interface. A protocol defined by the Frame Relay Forum and the ATM forum to govern how ATM switches establish connection and how ATM signaling requests are routed through an ATM network. Equivalent to a routing protocol in a router environment.
3. Network to Network Interface. There are two types of network interfaces specified by frame relay standards. The first is a user-to-network (UNI) interface and the second is a network-to-network interface (NNI). The NNI describes the connection between two public frame relay services, and includes elements such as bidirectional polling, to assist the network services providers with gaining information on the status of the public networks being interconnected.

NNTP Network News Transport Protocol. An extension of the TCP/IP local area network protocol that provides a network news transport service. NNTP is the standard for Internet exchange of Usenet messages, published in RFC-977, Network News Transfer Protocol: A Proposed Standard for the Stream-Based Transmission of News.

NNX A three-digit code to identify the local central office in which N represents any digit from 2 to 9 and X is any digit. NNX means the first three digits of a North American local telephone number. Originally, only NNX codes were used to identify and number local central offices. Now all subscribers dial 1+ when making a direct distance dialed long distance call. And central offices are being changed to be NXX — which allows local central offices to have numbers which look like area codes, e.g. 206, 210, etc. This gives us more central office numbers.

No Answer Transfer A service provided by a cellular carrier that automatically transfers an incoming cellular call to another phone number if the cellular subscriber is unable to answer. Most no-answer transfer systems will automatically transfer an incoming cellular call to another phone number if the cellular subscriber is unable to answer (it's not turned on) or if it's not answered after the third or fourth ring if the phone is turned on.

No Attendant Option CBX systems with Direct Inward Dialing may be designed (configured) without an attendant console.

No Bill Phone A name for a cellular phone from which you can make free phone calls — local or long distance. The name "no bill" was given by crooks, largely in Southern California in the Spring of 1992, to phones they had modified to emulate other legitimate cellular phones. Thus all calls made by "no bill" phones are billed to legitimate users, who happen to have other cellular phones with identical codes and now have huge cellular bills. Sorry about that.

No Busy Test A circuit used to connect to a busy subscriber's line number.

No Hold Conference/transfer In the event that a call to 911 is being transferred to a secondary or what they call a downstream PSAP (Public Service Answering Position), it is important that the caller never be left in disconcerting silence. After all it is an emergency call. "No hold" features allow the conference or transfer to be done while the PSAP 911 agent is in full uninterrupted communication with the caller.

No Incoming Calls A carrier restriction that prevents incoming calls to the assigned cellular number. Only outgoing calls are permitted. This is an optional feature that some cellular phone uses subscribe to because it saves them

No Line Preference Requires the user to manually select (i.e. punch down) a line for each call.

No Op Instruction that does nothing. It is used to hold the place for future insertion of a machine instruction.

NOAA National Oceanic and Atmospheric Administration. A US government agency which runs satellites used, inter alia, to track wildlife movements. Two satellites, called NOAA-11 and NOAA-12, orbit the earth via the poles every 100 minutes. As the Earth revolves beneath them, they are able to scan every point on its surface. The satellites work by listening for the "Doppler Shift" in the signal being transmitted by a collar round an animal's neck. The shift is a change in the perceived frequency of the radio signal — similar to what a pedestrian hears happen to the pitch of a police siren as the police car speeds by. Using this technique, you can track animals to with an accuracy of a little more than half a mile, or one kilometer.

NOC Networks Operations Center, a group which is responsible for the day-to-day care and feeding of a network. Each service provider usually has a separate NOC, so you need to know which one to call when you have problems.

NOD Network Outward Dialing. A service feature of an automatically switched telephone network that allows a calling user to dial directly all user numbers on the network without operator intervention. See also DIRECT OUTWARD DIALING.

Nodal Architectures Also called Hub architectures. Nodal network architectures means that network traffic from many locations are connected into a single network site for consolidation and aggregation to higher speed circuits. The traffic is then transmitted from this hub site to a central network site, or to other hub sites. Nodal architectures save money, but also increase the risk of a single network failure affecting multiple network locations.

Nodal Attribute A nodal state parameter that is considered individually to determine whether a given node is acceptable and/or desirable for carrying a given connection.

Nodal Clock The principal clock or alternate clock located at a particular node that provides the timing reference for all major functions at that node.

Nodal Constraint A restriction on the use of nodes for path selection for a specific connection.

Nodal Metric A nodal parameter that requires the values of the parameter for all nodes along a given path to be combined to determine whether the path is acceptable and/or desirable for carrying a given connection.

Nodal State Parameter Information that captures an aspect or property of a node.

Node 1. A point of connection into a network. In multipoint networks, it means it's a unit that's polled. In LANs, it's a device on the ring. In packet switched networks, it's one of the many packet switches which form the network's backbone.
2. An SCSA term. An independent SCSA unit in a distributed processing SCSA network, consisting of one or more resource and/or network boards, and one or more SCxbus adapter boards. Communication between nodes take place via the SCxbus. From a device programming point of view, a node is simply an addressable system unit which contains boards connected by an SCbus.

Node Address The unique identifier used to describe a specific node. See NODE NUMBER.

Node Number A node number identifies a network board on a local area network. Every station on a network must contain at least one network board. Each network board must have a unique node number to distinguish it from all the other network boards in that network. In a file server with more than one network board, the node number in LAN A is designated for all traffic addressed to that server. Node numbers can be set in a variety of ways, depending on which network board you use: (1) with jumpers or switches on boards such as Arcnet, (2) at the factory for Token-Ring and Ethernet boards, or (3) with software.

Node Type In IBM's SNA, the classification of a network device based on the protocols it supports and the network addressable units (NAUs) it can contain. Type 1 and Type 2 nodes are peripheral nodes. Type 4 and 5 nodes are subarea nodes.

NOI Notice Of Inquiry. The first public notification that the FCC is about to hold a public inquiry into a particular subject.

Noise Unwanted electrical signals introduced into telephone lines by circuit components or natural disturbances which tend to degrade the performance of the line. Also known as Line Noise.

Noise Cancelling Headset manufactures have long sought to reduce the background noise transmitted via headsets. One approach is the use of noise cancelling microphones. These microphones consist of two separate microphones, one directed at the headset user's mouth, the other in the opposite direction. The room side element will pick up ambient room noise along with some ambient user sound. The microphone directed at the user will receive the same amount of ambient room noise as the other microphone, but a much greater amplitude of the user's voice. Both signals are then transmitted to the amplifier. At this point, signals common to both microphones are cancelled out. What remains is the extra voice signals received by the user side microphone. This signal is then amplified and transmitted to the party on the receiving end of the call. This approach has one drawback. It demands perfect microphone positioning, because without it, the headset user's voice is cancelled. The technology works well with highly-trained people such as pilots, astronauts, and military personnel, but can be difficult to implement in the office environment where less skilled personnel struggle to properly position sensitive microphones. Headset manufacturers compromised by using noise cancelling microphones with more limited capabilities but that were easier to use.
A second approach to noise reduction is the use of voice switching technology. This technique only allows the microphone to transmit when volume reaches a predetermined level. When the headset user is not talking, or is pausing during the conversation, no sound is transmitted. When the headset user speaks at a normal level, the microphone is "live" and will transmit in a normal fashion. This approach also has it drawbacks. When the microphone is "live" it picks up not only the voice of the person using the headset, but any and all background noise. Voice switching helps the headset user hear what is being said more clearly, but does little to help the person to whom they are talking.
As a solution, some headset manufacturers have merged the two technologies. By using a noise cancelling microphone and voice switching, they achieve near perfect noise reduction. Each manufacturer offers noise cancelling technology on some of their headsets.
Noise cancelling is important in a telephone call center. In a large center, as room noise rises, agents speak louder. For those employees, noise is more than just an inconvenience, or a black spot on a professional image, it directly affects productivity. When conversations must be repeated, call durations increase. Multiply this by enough calls, and staffing and equipment must also be increased. The above information from headset distributor, CommuniTech.

Noise Equivalent Power At a given data-signaling rate or modulation frequency, operating wavelength, and effective noise bandwidth, the radiant power that produces a signal-to-noise ratio of unity at the output of a given optical detector. Information Gatekeepers defines NEP as a measurement in fiber optics that at a given modulation frequency, wavelength, and for a given effective noise bandwidth, the radiant power that produces a signal-to-noise ratio of 1 at the output of a given detector. Some manufacturers and authors, according to Information Gatekeepers, define NEP as the minimum detectable power per root unit bandwidth; when defined in this way, NEP has the units of watts/(hertz) 1/2. Therefore, the term is a misnomer, because the units of power per watts. Some manufacturers define NEP as the radiant power that produces a signal-to-dark current noise ratio of unity. This is misleading when dark-current noise does not dominate, as is often true in fiber systems.

Noise Measurement Units A series of terms used to express circuit noise. These units include: -dB RN — Decibel rated noise -dBrnC — C message weighting refers to the noise measured at 1000 Hz.

Noise Suppressor Filtering or digital signal-processing circuitry in a receiver or transmitter that automatically eliminates or reduces noise.

Noise Voltage In optical communication, an rms component of the optical detector electrical output voltage which is incoherent with the signal radiant power.

Noise Weighting A method of assigning a specific value to the transmission impairment due to the noise encountered by an average user operating a particular telephone. Noise weightings generally in use have been established by regulatory agencies concerned with public telephone service. They are tightened as technology improves.

Nom De Ligne The name you use in an online computer service. It's typically not your own name. You adopt this name to give yourself anonymity for whatever reasons you find convenient. Nom de ligne is called a "handle" in some circles.

Nomadic Computing An enterprise application describing laptop users dialing over the public telephone network, from hotel rooms, from client offices, from home or from airports to access information on the corporate LAN.

NOMDA National Office Machine Dealers Association, now merged with LANDA and now called NOMDA/LANDA.

Nominal Bit Stuffing Rate The rate at which stuffing bits are inserted (or deleted) when both the input and output bit rates are at their nominal values.

Nominal Linewidth In facsimile systems, the average separation between centers of adjacent scanning or recording lines.

Non-adaptive Routing A routing method that cannot adapt to or accommodate changes in a network.

Non-blocking A PBX is Non-Blocking if all callers can be switched through the PBX even during the busiest time of day. However, the caller may find no trunks are available and thus may not be able to complete the call. The trunk availability is a different problem than the ability of the switch to handle all the traffic it gets. But if your call does not go through do you care? A more technical explanation from InteCom of non-blocking: "Indicates that the internal network of the switching system is such that the total number of available transmission paths is equal to the number of ports. Therefore, all ports can have simultaneous access through the network."

Non-blocking Switch A switching system where a connection path always exists for attached device. See NON-BLOCKING.

Non-concur IBM-speak for to withhold approval, as in I non-concur with this proposal.

Non-consecutive Hunting Often referred to as "jump hunting." NON-CONSECUTIVE lines, trunks or extensions can be accessed or "searched" by the switching equipment upon dialing the initial number in the hunting group to find a connection to the first non-busy phone. NON-CONSECUTIVE hunting can be used on incoming and outgoing lines. For example, you could order four trunks from your local phone company which are to "hunt" on. Should the first be busy, the call will go to the next. If the next is busy, then it will hunt to the next. TELECONNECT Magazine has "consecutive" numbering in its first four phone numbers — 212-691-8215, 8216, 8217 and 8218. But let's say we only started with one number and then, because of growth, we needed three more. It's possible consecutive numbers might be taken. Therefore to get the hunting feature, the phone company might have assigned us NON-CONSECUTIVE hunting numbers like 691-8220, 8256, 8678. Sometimes the phone company will also assign us "coded" trunks as part of our trunk group. These are trunks which have no dialable number associated with them. It's best to get real numbers so you can test them by calling them individually. You cannot test coded trunks.

Non-critical Technical Load That part of the technical load not required for synchronous operation.

Non-data Bit Bit with encoding violating normal format: used for special control purposes.

Non-dedicated Server A node on which user applications are available while network resource maintenance applications execute in the background.

Non-deterministic A term which refers to the inability of being able to predict the performance or delay on a network where collision is possible.

Non-deterministic Network A network where access to the transmission medium within some specified period, cannot be guaranteed.

Non-erasable A switch where a through traffic path always exists for each attached phone. Generically, a switch or switching environment designed never to experience a busy condition due to call volume. See NON-BLOCKING, which is a better term.

Non-impact Printer Refers to printers that do not strike a hammer to a platen as typewriters do. Usually a heat sensitive paper or laser printing technology is involved.

Non-intelligible Crosstalk Crosstalk which is not of sufficient level to be understood by a listener but which is more annoying than other crosstalk because you think it's intelligible. Or think it should be intelligible.

Non-interlaced There's a good definition on interlacing and non-interlacing under INTERLACING.

Non-ISDN Line Any connection from a CPE to a central office switch that is not served by D-channel signaling.

Non-ISDN Trunk In ISDN language, a non-ISDN trunk is any trunk not served by either SS7 or D-channel signaling.

Non-linear Distortion Amplitude distortion of a signal in which the output signal does not have a linear relationship to the input signal.

Non-loaded Lines Cable pairs or transmission lines with no added inductive loading coils. In short, straight, raw copper pairs. See LOAD COILS.

Non-persistent In local area networking technology, describes a carrier sense multiple access (CSMA) local area network (LAN) in which the stations involved in a collision do not try to retransmit immediately — even if the network is quiet. Compare with persistent and p-persistent.

Non-printing Character A character in a transmission code which performs a control function but is not reproduced when the transmission is printed.

Non-proprietary LAN A Local Area Network that can connect the equipment of many vendors. See PROPRIETARY LAN.

Non-published There are various interpretations of what constitutes an "non-published," "unpublished" and "unlisted" phone number in North America. Some phone companies use these words interchangeably. Some don't. In California, Pacific Bell offers unpublished phone service. Your phone number is not listed in the paper phone directories, but is listed with dial up "Directory Assistance." Pacific Bell also has a more expensive service called "Unlisted Service." Here, your phone number is not included in the paper phone directories or given out to callers to Directory Assistance. Nynex (and I presume other phone companies) has a service whereby you can leave a message for the owner of an unlisted number. "Please call me. You've won the lottery." The owner of the unlisted number then has the choice to return the call or not. He doesn't pay to receive this message. Some telephone companies confuse the definitions and some invent new ones. For example, some phone companies use the term "non-published" number. You won't find the number in a phone book or by calling Directory Assistance. Over 25% of many private phone numbers in major metropolitan areas are unlisted, unpublished or non-published — a "service" their subscribers pay extra for. To my simple brain, it's a lot easier to simply publish your name as "Apple Plumpudding." See UNPUBLISHED.

Non-receipt Notification In X.400, a non-receipt notification is a report prepared by a recipient UA (User Agent) or Access Unit (upon request) and sent to the originating UA or Access Unit when a message is deemed unreceivable by a recipient.

Non-routable Protocols LAN protocols, such as IBM's NetBIOS, LAN Server and SNA, that use names and not Network Layer Addresses to identify devices and therefore supply no routing information. Internetworking devices must find other ways to route traffic in networks that use nonroutable protocols.

Non-sent Paid Utility industry term for calls made as third party billings, reversed charges or with a Calling Card.

Non-session A type of synchronous access mode that can exist between an end user and a host computer in an Amdahl packet network. In non-session access, the calling line (PU) is mapped to a specific offering line (LU). The offering line is reserved for that calling line at all times.

Non-simultaneous Transmission Half duplex transmission. Transmission in one direction at a time. This mode of transmission may be the result of limitations of the transmission channel or of the transmitting/receiving equipment.

Non-synchronous Communications See ASYNCHRONOUS COMMUNICATIONS.

Non-traffic Sensitive Plant Telephone company facilities which are unaffected by changes in volume of telephone activity.

Non-transparent Mode A transmission environment, mainly of bisynch transmission, in which control characters and control-character sequences are recognized through the examination of all transmitted data. Compare with transparent mode. Also called normal mode.

Non-trivial Microsoft word for "hard."

Non-volatile Memory which is not lost (i.e. that does not "forget") when the power is shut off.

Non-wireline Also called the block "A" carrier. The "A" originally stood for "alternate." The FCC, in setting up the licensing and systems in each market. It reserved one for the local wireline telephone company, and opened the second system — the Block A system — to other interested applicants. The distinction between Block A and Block B is meaningful only during the licensing phase at the FCC. Once a system is constructed, it can be sold to anyone. Thus in some markets today, both the A and B systems are owned by a telephone company. One happens to be the local phone company, and the other is a phone company that decided to buy a cellular system outside its home territory. Non-Wireline, or Block A systems, operate on the radio frequencies from 824 to 849 Megahertz.

NORC Network Operators Research Committee.

Normal Mode The AC voltage that exists between the normal current-carrying wires, that is, between neutral and hot or live.

Normal Response Mode NRM. HDLC mode for use on links with one primary and one or more secondaries. Under NRM, a secondary can transmit only after receiving a poll addressed to it by a primary. It may then send a series of responses. But after it sets the F bit in a response, it cannot transmit any more until it receives another poll.

Normalize A call center term. To change an unusual call statistic reported by the ACD so as to reflect what would have been usual for that period of the day or that day of the week. Normalizing is something you do before updating historical patterns so that the historical patterns will not be distorted by the unusual data.

Normalized Average Transfer Delay Average transfer delay divided by packet transmission time at the clock rate of the medium.

Normalized Network Throughput Network throughput in packets per second divided by maximum throughput possible at clock rate of medium. Less than one.

Normalized Offered Traffic The average number of attempted packet transmissions per second divided by the average number of packet transmissions/second possible at the clock rate of the medium. May exceed one.

Norstar A programmable phone system from Northern Telecom which can accommodate up to 100 people. See ACCESS TOOLKIT.

Norstar Access Toolkit See ACCESS TOOLKIT

Norstar Manager A feature which permits remote programming and administration of a Norstar telephone system.

Nortel Nortel is a shorthand version of Northern Telecom's corporate name (which remains Northern Telecom Limited). According to the company's annual report, it is designed to reflect the corporation's heritage, while signaling a new direction, presenting a single face that reinforces Northern Telecom's global presence. The new logo, with its Globemark "O", creates a dynamic visual symbol for a corporation whose business knows no boundaries and whose spirit is one of leadership, innovation, dedication and excellence. In the U.S., Northern Telecom's principal subsidiary is called Northern Telecom Inc. www.nortel.com

North American Area Codes See NPA.

201 Northern New Jersey	281 Texas (Houston, pagers and cell phones)
202 Washington DC	301 Maryland
203 Connecticut	302 Delaware
204 Manitoba, Canada	303 Colorado (Denver)
205 Alabama (Birmingham)	304 West Virginia
206 Washington (Seattle)	305 Florida (Miami)
207 Maine	306 Saskatchewan, Canada
208 Idaho	307 Wyoming
209 California	308 Nebraska (North Platte)
210 San Antonio, Texas	309 Illinois (Peoria)
212 New York (Manhattan, Bronx)	310 Southern California
213 California (part of Los Angeles)	312 Illinois (Chicago)
214 Texas (Dallas)	313 Michigan (Detroit)
215 Pennsylvania (Philadelphia)	314 Missouri (St Louis)
216 Ohio (Cleveland)	315 New York (Syracuse)
217 Illinois (Springfield)	316 Kansas (Wichita)
218 Minnesota (Duluth)	317 Indiana (Indianapolis)
219 Indiana (South Bend)	318 Louisiana (Shreveport)
250 British Columbia	319 Iowa (Dubuque)

330 Ohio (Akron, Canton)	419 Ohio (Toledo)
334 Alabama	423 Tennessee
352 Florida (Gainesville)	441 Bermuda
360 Washington State	501 Arkansas
401 Rhode Island	502 Kentucky (Louisville)
402 Nebraska	503 Oregon
403 Alberta, Yukon, Northwest Territories, Canada	504 Louisiana (New Orleans)
404 Georgia (Atlanta)	505 New Mexico
405 Oklahoma (Oklahoma City)	506 New Brunswick
406 Montana	507 Minnesota
407 Florida (Boca Raton)	508 Massachusetts (Worcester)
408 California (Silicon Valley, Cupertino)	509 Washington (Spokane)
409 Texas (Beaumont)	510 Northern California
410 Maryland	512 Texas (San Antonio)
412 Pennsylvania (Pittsburgh)	513 Ohio (Cincinnati)
413 Massachusetts (Springfield)	514 Quebec (Montreal) Canada
414 Wisconsin (Milwaukee)	515 Iowa (Des Moines)
415 California (San Francisco)	516 New York (Long Island)
416 Ontario (Toronto) Canada	517 Michigan (Lansing)
417 Missouri (Springfield)	518 New York (Semi-Upstate)
418 Quebec (Quebec City) Canada	519 Ontario, Canada

520 Arizona	617 Massachusetts (Boston)
540 Virginia	618 Illinois (Centralia)
541 Oregon	619 California (San Diego)
573 Missouri (Jefferson City)	630 Illinois
601 Mississippi	700 Information and other services by local/long distance phone companies.
602 Arizona 603 New Hampshire	701 North Dakota
604 British Columbia, Canada	702 Nevada
605 South Dakota	703 Virginia (Arlington)
606 Kentucky (Covington)	704 North Carolina (Charlotte)
607 New York (Binghamton)	705 Ontario (North Bay) Canada
608 Wisconsin (Madison)	706 Georgia (Atlanta)
609 New Jersey (Trenton)	707 California (Santa Rosa)
610 Pennsylvania (Allentown)	708 Illinois (Naperville, Addison)
612 Minnesota (Minneapolis)	709 Newfoundland and Labrador, Canada
613 Ontario (Ottawa) Canada	712 Iowa (Council Bluffs)
614 Ohio (Columbus)	713 Texas (Houston)
615 Tennessee (Nashville)	714 California (Anaheim)
616 Michigan (Grand Rapids)	

715 Wisconsin (Eau Claire)
716 New York (Buffalo)
717 Pennsylvania (Harrisburg)
718 New York (Brooklyn, Queens, Staten Island)
719 Colorado
770 Georgia, outside Atlanta
800 IN-WATS Service
801 Utah 802 Vermont
803 South Carolina
804 Virginia (Richmond)
805 California (Bakersfield)
806 Texas (Amarillo)
807 Ontario (Thunder Bay) Canada
808 Hawaii
809 Puerto Rico, Virgin Islands & some (not all) Caribbean Islands
810 Michigan (Southfield)
812 Indiana (Evansville)
813 Florida (Tampa, St Pete)
814 Pennsylvania (Altoona)
815 Illinois (Rockford)
816 Missouri (Kansas City)
817 Texas (Fort Worth)
818 California (Burbank)
819 Quebec (Sherbrooke) Canada
847 Illinois
860 Connecticut
864 South Carolina Service (Equivalent to 800)
900 DIAL-IT Services
901 Tennessee (Memphis)
902 Nova Scotia, Prince Edward Island, Canada
903 Texas (Dallas suburbs)
904 Florida (Jacksonville)
905 Ontario (Hamilton) Canada
906 Michigan (Escanaba)
907 Alaska 908 New Jersey
909 California (Riverside)
910 North Carolina (Greensboro)
912 Georgia (Savannah)
913 Kansas (Topeka)
914 New York (Westchester County)
915 Texas (Sweetwater)
916 California (Sacramento)
917 New York (Manhattan, The Bronx) (Cellular and Beepers)
918 Oklahoma (Tulsa)
919 North Carolina (Raleigh)
941 Florida (Fort Meyers)
954 Florida (Miami)
970 Colorado

North American Cellular Network A Craig McCaw idea to join together a bunch of cellular phone companies providers who would provide a painless, simple way for someone making or receiving a cellular call in their territory. Up until now, traveling cellular users — what the industry calls "roamers" — have been forced to pay heavy charges for calling from the territory of a cellular company not theirs. And roamers have been in effect incommunicado from incoming calls. It's impossible to call someone on a cellular phone if they're not in the own territory and you don't know where they are. Craig McCaw is the chairman and founder of McCaw Communications, a large cellular phone company and now partly-owned by AT&T.

North American Directory Forum NADF. An association of electronic mail providers who are figuring standards and ways of sending mail between their subscribers.

North American ISDN Users Forum Here is the complete explanation of this important group, as excerpted from their brochure: The barriers to the widespread use of ISDN nationally and internationally is difficult because the technology is complex and developing rapidly. One underlying problem is the lack of standard implementations of ISDN applications. ISDN standards are currently developed by the International Telephone and Telegraph Consultative Committee (ITU-T) and by accredited standards committee T-1 under the umbrella of the American National Standards Institute (ANSI). But standards designed to meet many requirements offer multiple options which are open to diverse interpretations. As a result, services and products produced by different manufacturers are incompatible.
To solve this problem, NIST collaborated with industry in 1988 to establish the North American ISDN User's Forum (NIU-Forum). NIST's (the National Institute of Standards and Technology's) Computer Systems Laboratory (CSL) serves as chair of the forum and hosts the NIU-Forum Secretariat. Through support of the forum, CSL advances new uses of computer and telecommunications technology in government and industry. The purpose of the NIU-Forum is to create a strong user voice in the implementation of ISDN applications. The forum provides users of ISDN technology with the opportunity to work with implementors to assure that user's needs are met in the ISDN design process. Through the forum, users and manufacturers concur on ISDN applications, the selection of options from standards, and conformance tests, enhancing the strength of the U.S. telecommunications industry in the world marketplace.

North American Numbering Plan NANP. The method of identifying telephone trunks in the public network of North America, called World Numbering Zone 1 by the ITU-T. The Plan has three ways of identifying phone numbers in North America — a three digit area code, a three digit exchange or central office code and four digit subscriber code. Other countries have much more complicated numbering schemes. There are some countries, for example, where the length of the area code actually exceeds the subscriber code. There are many countries where there is no consistency in the length of phone numbers. Some are nine-digit. Some are 12-digit, etc. All these varying number lengths may be within 100 miles of each other.
Under the NAMP format prior to January 1, 1995, the second digit of the area code was always a one or a zero. With this system, approximately one billion telephone numbers, 152 area code combinations and 640 prefixes were available. Numbers started to run out as a result of increased use of fax, modem and cellular lines. With the new Numbering Plan introduced on January 1, 1995, the area code can be any combination of three digits. This will allow more than 6 billion telephone numbers and 792 area code combinations.

North American Telecommunications Association NATA. The national trade association for companies providing customer premise telephone equipment. NATA represents the interests of the industry before the Congress, the FCC and in court actions. See also NATA.

Norstar A family of telephone systems from Northern Telecom. See ACCESS TOOLKIT and NORSTAR COMMAND SET.

Norstar Command Set A series of standard commands for programming a telephone system. These commands were created by Northern Telecom to program their Norstar telephone system. The Norstar Command Set includes commands such as mDisconnectCall, mDialDigits, mJoinCalls, mOnEvent, mOnErrorEvent and mTermWrite (which means to write to the phone's screen).

In creating this command set, Northern Telecom saw two types of applications for its Norstar telephone system: 1. Server applications; and 2. Personal telephony applications.

Server applications would include automatic call distribution, dial by name, predictive dialing, dial from a computer and transfer a database record and the call to an agent. Personal telephony applications include on-screen dialing for telemarketing. placing a call in an electronic rolodex by hitting "ENTER."

Under the Norstar Command set, a programmer can take complete control of a phone system away from its central factory-programming. Using the Norstar command set, one moment a phone could be an ACD agent terminal, the next moment a predictive dialer, the next a data input terminal and the third an operator position. For more information on the Norstar Command Set, see page 14 of the December, 1990 issue of TELECONNECT Magazine.

NOS Network Operating System, the software for a network that runs in a file server and controls access to files and other resources from multiple users. It can run on top of DOS, and also provides security and administrative tools. Novell's NetWare, Banyan's VINES, Microsoft's Windows NT Server and IBM's OS/2 LAN Server are NOS examples.

Notch An out-of-phase impulse causing spontaneous dip in voltage. This is an under-voltage impulse, similar to a spike, but of reverse polarity to the instantaneous value of the AC sine wave. A notch normally is associated with the power company removing a generator from the power grid. See NOTCHED NOISE.

Notched Noise Noise in which a narrow band of frequencies has been removed. Normally used for testing devices or circuits.

NotePhone One of the first products of a collaboration of Apple Computer and Siemens Private Communications Systems Group. The NotePhone (which I haven't seen) will, according to Apple, Rolm and Siemens, combine Siemens and Rolm Telephony and Apple's Newton technology to provide access to telephone and fax features. Siemens and Rolm are planning to provide NotePhones for standard (analog) telephone lines commonly found in homes and hotels as well as for digital business communications systems (i.e. PBXs) from Siemens and Rolm.

Noteworks One of my favorite electronic mail software programs and the one we use every day and run our business on. It's made by a company called ON Technology, Cambridge, MA. Call them on 617-374-1400 and get yourself a free 30-day trial subscription.

NOTHS Network Operations Trouble Handling System.

Novell Novell is the reincarnation of NDSI, a computer firm that almost went under in the early 1980s. The then nearly 60-year-old Ray Noorda, who had had 20 years of experience in systems automation with General Electric before striking out for Silicon Valley, was called in to help NDSI prepare for a trade show. He spotted that it had some potentially interesting technology, and bought the ailing company in 1983. He still owns a sufficiently large stake in the company to make him a billionaire and Novell sold more networking software than anyone else in the world. Their software is called NetWare. See NETWARE.

Novell IPX See IPX.

Novell NLSP Novell NetWare Link Services Protocol. A link state protocol under development by Novell to improve performance of IPX traffic in large internetworks.

Novell SAP Novell Service Advertisement Protocol. See SAP.

Novell Telephony Services See TELEPHONY SERVICES.

NPA 1. In IBM language, it means Network Performance Analyzer, a product for network tuning, determining performance, degradation and determining the affect of network growth.

2. AT&T's National Pricing Agreement. A multi-year agreement an AT&T customer signs with AT&T which specifies that the customer will buy $xxx of equipment and services from AT&T at a "deep" discount. (This definition from AT&T.)

3. Numbering Plan Area. A fancy way of saying Area Codes. There are over 200 area codes in the United States, Canada, Bermuda, the Caribbean, Northwestern Mexico, Alaska and Hawaii. Within any of these area codes, no two telephone lines may have the same seven digit phone number. The middle number has been either "1" or "0" creating "N 1/0 x" codes. The number of codes available on this basis were nearing depletion. Bellcore has now modified the plan to obtain more area codes. Future area codes will be NXX like the central office numbering scheme. Switching systems in the national network will differentiate between the central office and area codes by recognizing the subscriber always dials 1+ or 0+ preceding an area code when direct dialing such long distance calls.

Here are the special, unassigned and reserved NPAs:
200 Reserved for special services
211 Assigned to local operators
300 Assigned to special services
311 Reserved for special local services
400 Reserved for special services
500 Reserved for special services
511 Reserved for special local services
600 Reserved for special services
700 Assigned special access code for interLATA carriers/resellers
711 Reserved for special local services
See NNX and NORTH AMERICAN AREA CODES.

NPC 1. Network Processing Card.
2. Network Parameter Control: Network Parameter Control is defined as the set of actions taken by the network to monitor and control traffic from the NNI. Its main purpose is to protect network resources from malicious as well as unintentional misbehavior which can affect the QoS of other already established connections by detecting violations of negotiated parameters and taking appropriate actions. Refer to UPC.
NPD See NETWORK PROTECTION DEVICE.
NPDA Network Problem Determination Application. A program which allows for the monitoring of an entire network from a single location, collection of statistics, and isolation of communication faults. An IBM term. See NETVIEW.
NPM Network Process Monitor.
NPR DOA is Dead On Arrival. And it's a term several manufacturers use to refer to equipment which arrives at the customer's premises not working. A person who receives a DOA machine will ask the company for a NPR number — New Product Return number. This allows them to return the product and have the factory replace it with another new one.
NPRM Notice of Proposed Rule Making. A term used in regulatory agencies. The agency runs an idea up the flagpole, then typically hold hearings to find out how people react.
NPSI IBM X.25 NCP Packet Switching Interface. Networking software package that allows Systems Network Architecture (SNA) 3270 traffic to be transmitted over an X.25 packet data network (PDN). See also QLLC. Contrast with DSP.
NRAM Nonvolatile memory. NRAM does not lose its memory when you turn off the computer or phone system. Many modems, for example, use nonvolatile memory to store configuration information (in place of the switches used on other modems). The command &W writes instructions to the NRAM in many PC modems — chiefly those 2400 baud and faster.
NRC 1. Non-Recurring Charge.
2. Network Reliability Council.
NREN The National Research and Education Network, an ultragigabit network established by legislation in 1991 by the U.S. House and Senate. Preliminary steps toward deploying this information superhighway (for talking between University computers) include some gigabit network testbeds and the cutover to a 45 megabit per second backbone for the National Science Foundation Network (NSFnet) used by the scientific research and university community nationwide.
NRM 1. NoRMal response of HDLC. See NORMAL RESPONSE MODE.
2. An ATM term. An ASP service parameter, Nrm is the maximum number of cells a source may send for each forward RM-cell.
NRN No Response Necessary. A Proposed e-mail convention to prevent endless back-and-forth acknowledgments: "Thanks for the info." " You're welcome ...hope it helps." "I hope so too. Thanks." By putting NRN at the bottom of your mail, you absolve the receiver from having to reply.
NRUG National Rolm Users Group.
NRZ Non-Return to Zero. A binary encoding scheme in which ones and zeroes are represented by opposite and alternating high and low voltages and where there is no return to a zero (reference) voltage between encoded bits. NRZ is now used as an encryption scheme for getting data onto and off hard disk fast. It eliminates the need for clock pulses and yields up to 18.5 kilobytes per track and high read/write speeds.
NRZI Non-Return to Zero Inverted. A binary encoding scheme that inverts the signal on a "one" and leaves the signal unchanged for a "zero". Where a change in the voltage signals a "one" bit, and the absence of a change denotes a "zero" bit value. Also called transition coding.
NS/EP Telecommunications Federal government definition. Telecommunications services that are used to maintain a state of readiness or to respond to and manage any event or crisis (local, national, or international) that causes or could cause injury or harm to the population, damage to or loss of property, or degrade or threaten the national security or emergency preparedness of the United States.
NSA 1. Network Service Address. NSAs are unique addresses that define physical or logical locations in equipment, such as a residential telephone number (TN).
2. National Security Agency. A super-secret agency of the Federal Government. NSA is the Federal agency responsible for the design and use of nonmilitary encryption technology, developing sophisticated codes to scramble data, voice or video information. In short, it is charged with signals intelligence and is widely assumed to monitor all communications traffic (phone, fax, data, video, etc.) into and out of the United States with foreign countries. It is barred from intercepting domestic communications. NSA grabbed the headlines in 1993 and 1994 when it adopted its most visible attempt to outgun cybervillains with something called the Clipper Chip. The idea is that the Clipper Chip (a microprocessor) would be installed in every phone, computer, and personal digital assistant in America would carry a device identification number or electronic "key" — a family key and unit key unique to each Clipper chip. The device key is split into two numbers that, when combined into what's called a Law Enforcement Access Field number, can unscramble the encrypted messages. The device keys and the corresponding device numbers, according to NSA proposals, would be kept by the US government through key escrow agents. Under a plan proposed, the attorney general would deposit the two device keys in huge, separate electronic database vaults. One key would be held by the National Institute for Standards and Technology (NIST) and the other by the Automated Systems Division of the U.S. Treasury. Access to these keys would be limited to government officials with legal authorization to conduct a digital wiretap. When a law enforcement agency wants to tap into information encrypted by the Clipper ship, they must obtain a court order and then apply to each of the escrow agents. The agents electronically send their key into to an electronic black box operated by the law enforcement agency. When these keys are electronically inserted, encrypted conversations stream into the black box and come as standard voice transmissions or as ASCII characters in the case of electronic mail. At least that's the theory. See NSA LINE EATER.
NSA Line Eater The more paranoid Internet users believe that the National Security Agency has a super-powerful computer assigned to reading everything posted on the Net. They will jokingly refer to this line eater in their postings. Goes back to the early days of the Internet when the bottom lines of messages would sometimes disappear for no apparent reason.

NSAP Network Service Access Point. The point at which the OSI Network Service is made available to a Transport entity. The NSAPs are identified by OSI Network Addresses. The NSAP is a generic standard for a network address consisting of 20 octets. ATM has specified E.164 for public network addressing and the NSAP address structure for private network addresses.
NSAPI Netscape Server API. A programming specification for Netscape's Web servers to access back-end applications.
NSC 1. Network Service Center. See NETWORK CONTROL POINT and SDN.
2. Non-Standard facilities Command. A response to the called fax DIS response.
NSEP Telecommunications See NS/EP TELECOMMUNICATIONS.
NSFNET National Science Foundation's TCP/IP-based NETwork, funded by the U.S. Government, that links supercomputing centers and over 2500 academic and scientific institutions across the world. See NREN.
NSG National Systems Group. Division of Canadian telcos responsible for multiple accounts — large customers with service in multiple telephone company territories.
NSO National Services Organization. An NSO provides the interconnecting network infrastructure, intelligent network services, network management, and roaming in a national Personal Communications Network (PCN). An NSO has ownership of resources that allow it to provide services to local operators. Operations such as marketing are centralized.
NSP Native Signal Processing. Intel's idea to use the "spare MIPS" on its Pentium, Pentium Pro and later versions to take over the processing that once dedicated chips, like DSPs, used to do. The idea, clearly, is to sell more Intel chips and fewer chips from other makers.
NSR Non-Source Routed: Frame forwarding through a mechanism other than Source Route Bridging.
NSS 1. Non Standard Facilities Setup command, a response to an NSF frame.
2. Network and Switching Sub-system. A wireless telecommunications term. The part of the GSM (Global System for Mobile Communications) system in charge of the management of calls and the interface with other networks.
NSU Network Synchronization Unit. A timing distribution system that provides synchronization signals to all electronic equipment within a wireline network node or office.
NT 1. Network Termination. Network Termination represents the termination point of a Virtual Channel, Virtual Path, or Virtual Path/Virtual Channel at the UNI. See NT-1.
2. New Technology, usually known as Windows NT. It's a new operating system from Microsoft which will let Windows run on high-end machines, such as file servers and workstations. NT has two sets of goals: to provide true multi-tasking, security, network connectivity and 32-bit power. Secondly, it's to provide a smooth upgrade path from Windows and MS-DOS. For a much larger explanation, see WINDOWS NT and WINDOWS NT ADVANCED SERVER.
NT-1 NT-1 or NT1. The NT-1 is the first customer premise device on a two-wire ISDN circuit coming in from the ISDN central office. It does several things. It converts the two-wire ISDN circuit (called "U" interface) to four-wire so you can hook up several terminals — like a voice phone and a videophone. The NT-1 typically has several lights on it which indicate if it's working. An ISDN central office can usually "talk" to the NT-1 and do testing and maintenance by instructing the NT-1 to loop signals back to the central office. An NT-1 will support up to eight terminal devices, though I've never seen it work with that many. The basic NT1 functions are:
• Line transmission termination.
• Layer 1 line maintenance functions and performance monitoring.
• Layer 1 multiplexing, and
• Interface termination, including multi drop termination employing layer 1 contention resolution.
Increasingly, NT-1s are being incorporated into customer premises equipment, like phones, PC ISDN data cards, etc. See ISDN.
NT-2 In ISDN, the Network Termination type 2 is an intelligent customer premise device terminates the 4-wire T interface. The NT-2 can perform switching and concentration, such as a digital PBX. It typically terminates primary rate access lines from the local ISDN CO switch. It may include the Layer 1 functions of a NT-1 and the protocol handling functions of Layers 2 and 3. See NETWORK TERMINATION 1.
NT1 Network Termination Type 1. A small box that attaches to an ISDN BRI line. It is the first piece of ISDN equipment at your office. It terminates the ISDN loop from the central office and provides the interface into which you, the user, can plug a telephone, a data, video device, a PC, etc. It contains a power supply and changes the signaling on the ISDN line from the two wires that come in from the central office to the four wires that most ISDN terminal equipment needs. These days some ISDN terminal devices contain their own built-in NT1. See NT-1 above for a fuller explanation, see NETWORK TERMINATION TYPE 1.
NT2 Network Termination 2. A unit that provides switching concentration of subscriber lines at the S-interface. This unit performs the functions of a customer-premises switch or multiplexer to multiplex B channel(s) and D channel onto one physical path and to route calls to the appropriate B or D channel. See NT-2 above for a fuller explanation.
NTACS Narrow TACS. Cellular radio system deployed in Japan using narrow band technology to increase capacity by splitting TACs channels into two narrow channels.
NTCA National Telephone Cooperative Association. A trade association representing primarily rural telephone cooperatives and other small telephone companies.
NTFS Windows NT file system; an advanced file within the Windows NT operating system. It supports file system recovery, extremely large storage media, and various features for the POSIX subsystem. It also supports object-oriented applications by treating all files as objects with user-defined and system-defined attributes.
NTI 1. Northern Telecom Inc. Now called NorTel.
2. Network Terminating Interface. A. The point where the network service provider's responsibilities for service begin or end, or B. The interface between DCE and its connected DTE.

NTIA National Telecommunications and Information Administration. An agency of the U.S. Department of Commerce concerned with spectrum management, public safety, and the communications (primarily telephony) standards. (www.ntia.doc.gov)
NTN See NETWORK TERMINAL NUMBER.
NTO Network Terminal Option. An IBM program product that enables an SNA network to accommodate a select group non-SNA asynchronous and bisynchronous devices via the NCP-driven communications controller.
NTP 1. Network Termination Point.
2. The Network Time Protocol was developed to maintain a common sense of "time" among Internet hosts around the world. Many systems on the Internet run NTP, and have the same time (relative to Greenwich Mean Time), with a maximum difference of about one second.
NTS 1. Network Test System.
2. Non-Traffic-Sensitive commercial line costs levied on the user.
3. Number Translation Services.
NTSC National Television Standards Committee of Electronic Industries Association (EIA) that prepared the standard of specifications approved by the Federal Communications Commissions in 1953 for commercial broadcasting. The initials are used to describe the standard method of television transmission in U.S., Canada, Japan, Central America and half of South America. The North American system uses interlaced scans and 525 horizontal lines per frame at a rate of 30 frames per second. The picture information is transmitted in AM (amplitude modulation) and the sound information is transmitted in FM (frequency modulation). NTSC is compatible with CCIR Standard M. PAL (which stands for Phase Alteration Line) is the name of the format for color TV signals used in West Germany, England, Holland, Australia and several other countries. It uses an interlaced format with 25 frames per second and 625 lines per frame. The two systems are not compatible. You cannot view an Australian videotape on a U.S. TV.
When TV engineers get together to hoist some brews, however, the initials for the various standards for TV broadcast take on other meanings, such as: NTSC, Never Twice the Same Color. PAL, Peace At Last. PAL-M, Peace At Last — Maybe. SECAM (French system) becomes System Essentially Contrary to the American Method. We bother mentioning it because if you try to take computer terminals outside the country and connect the video outputs to monitors, you will be made aware of the many TV and video monitor "standards" in use.
NTSC Signal National Television System Committee specified signal. De-facto standard governing the format of television transmission signals in the United States.
NTSI Format A color television format having 525 scan lines; a field frequency of 60 Hz; a broadcast bandwidth of 4 MHz; line frequency of 15.75 KHz; frame frequency of 1/30 of a second; and a color subcarrier frequency of 3.58 MHz.
NTT Nippon Telephone and Telegraph, the major Japanese local and long distance telephone company.
NTT FAN Inc. On July 31, NTT, AT&T and Sony formed a joint venture, NTT FAN Inc. to test an on-line multimedia communications service in Japan using General Magic's Telescript and Magic Cap platform technologies.
NTU Network Terminating Unit. A device which is placed at the final interconnect point between the PSTN and the customer owned equipment. NTUs allow the Carrier to isolate their facilities from those of their customers for testing, fault detection, and some service feature functionality.
NuBus (Pronounced "New Bus.") The name of the bus design for most Apple Macintosh computers. See NUBUS CARD.
NuBus Card An add-on card that fits inside the NuBus slots of a Modular Macintosh. Often used for video cards and modems.
Nucleus An ATM term. The interior reference point of a logical node in the PNNI complex node representation.
NUI Network User Identifier. A unique alphanumeric number provided to dial-up users to identify them to packet switched networks around the world. The number is used to get onto the network and for billing.
Null Having no value. A dummy letter, letter symbol, or code group inserted in an encrypted message to delay or prevent its solution, or to complete encrypted groups for transmission or transmission security purposes. See also NULL CHARACTERS.
Null Call_id AN ISDN term. A null call_id is the call_id used to convey information that does not pertain to a specific call between the ISDN System Adapter and the central office switch. Null call_ids are primarily used for accessing feature buttons that do not relate to a specific call.
Null Characters Characters transmitted to fill space, time or to "pad" something. They add nothing to the meaning of a transmitted message, but the null characters are expected by the system. On older teletype machines, for example, when the type head reaches the end of a line and the New Line sequence is transmitted, it usually includes a number of Null Characters in order to give the mechanical type head enough time to reach the left margin of the page before transmitting the next line to the terminal. In this manner, no characters are lost. MCI Mail uses five null characters at the beginning of every line — unless you tell it otherwise.
Null Modem A device that connects two or computers by emulating the characteristics of a modem. Similar to a modem eliminator. See NULL MODEM CABLE.
Null Modem Cable Special wiring of an RS-232-C cable such that a computer can talk to another computer without a modem (thus "null" modem) or a computer can talk to a printer. Essentially, a null modem cable reverses pins 2 and 3 on an RS-232-C cable. But there are no standard null modem cables. And other pins need changing and jumpering together. Control Cable in Baltimore, MD sells one that works for us.
Null Suppression A data compression technique whereby streams of null characters are identified at a transmission source and replaced by two or more control characters. The first character indicates the null suppression, and more characters indicate the number of null characters removed. The receiver uses this information to replace the removed data.
Null Value A parameter or field position for which no value is specified.
Number Administration And Service Center NASC. Provides centralized administration of the Service

Management System (SMS) database of 800 numbers. The NASC keeps track of the 800 numbers that are in use, or available for use, by new 800 customers.

Number Crunching A repetitive series of mathematical calculations.

Number Portability Once upon a time, 800 numbers belonged to the phone companies who supplied 800 service. So if you got 800 service from AT&T and you wanted to keep your number, you were stuck with AT&T. For example, we had (and still have) 800-LIBRARY (800-542-7279). The reason? The database of 800 numbers was maintained by Bellcore, who allocated certain "exchanges," like 800-542, to certain carriers. In the case of 542, it was AT&T. If we wanted to go to a different phone company, we had to give up your phone number. Since many companies had invested vast monies in promoting their 800 numbers, this was inconvenient and, in essence, forced companies to stay with the same long distance company. In 1993, things changed. The FCC mandated Number Portability, which allowed you to change your long distance carrier but keep your valued 800 number. The way the whole thing works now is simple. You pick up the phone. You dial an 800-number. Your local central office holds the call for a moment, while it checks an external central database of 800 numbers for the routing of that call. It typically checks via the packet-switched SS7 (Signaling System 7) network. When it receives a reply from that external database, it simply then sends the call to the appropriate long distance carrier which is providing service on that 800 number. To change the carrier of your 800 number, simply tell the database who the new carrier is. And that's number portability.

Numbering To have any sort of network work, everyone on the network has to have a unique number which we can all dial. I only included this definition because I found the following definition in a glossary of LAN terms and I thought, "Maybe someone doesn't know, though if they didn't know, this definition wouldn't help them much." Here's the LAN definition: "The assignment of unique identities to a user-network interface."

Numbering Plan 1. In Wide Area Networks, the method for assigning NNX codes to provide a unique telephone address for each subscriber, special line or trunk destination.
2. In PBXs, the method of assigning extension numbers and trunk designations at the local premises.

Numbering Plan Area NPA. A fancy term the Bell System came up with years ago to mean Area Codes. See NPA.

Numbers Shift A character in the Baudot code which establishes that the characters following in the transmission are to be interpreted as numeric characters. See LETTERS SHIFT.

Numeric Key Pad A separate section of a computer keyboard which contains all the numerals 0 through 9. Sometimes, some special keys are included — a plus sign, a minus sign, a multiplication sign and a division sign. The numeric key pad on a computer is the same as that found on calculators and adding machines. The top row is 789. The second top row is 456. The third top row is 123. The lowest row is typically 0, "." and "+". The numeric key pad is exactly opposite that of the touchtone telephone keypad, which was designed deliberately to be unfamiliar to users, so they may not input digits into the nation's telephone system faster than it could take them. Early touchtone central offices were very slow.

Numeric User Identifier According to the 1988 X.400 recommendations, a numeric user identifier is a standard attribute of an O/R (Originator/Recipient) address that consists of a unique sequence of numbers for identifying a user. (Numeric User Identifier was referred to as Unique Identifier in the 1984 recommendations.)

Numeris The French name for ISDN.

NVOD Near Video On Demand. Providing a consumer a multimedia item — movie, TV program, etc. — on a rotating schedule, thus giving the appearance of an on-demand system, i.e. VOD (Video On Demand).

NVS NonVolatile Storage is a storage device, like a disk or EPROM, that retains data when you turn off.

NWL Non-Wireline. Cellular radio licenses received from the FCC with no initial association to telephone company. Also referred to as A-Block.

Nx64K An ATM term. This refers to a circuit bandwidth or speed provided by the aggregation of nx64 kbps channels (where n= integer> 1). The 64K or DS0 channel is the basic rate provided by the T Carrier systems.

Nx384 N-by 384. The ITU-T's approach to creating a standard algorithm for video codec interoperability It is based on the ITU-T's H0 switched digital network standard, which was expanded into the Px64 or H.261 standard, approved in 1990.

NXX In a seven digit local phone number, the first three digits identify the specific telephone company central office which serves that number. These digits are referred to as the NXX where N can be any number from 2 to 9 and X can be any number. At one stage, many moons ago, it was not permissible to have a 1 or a 0 as the second digit in an NXX and it was called an NNX. But that was before everyone had to dial a "1" before making a direct distance dialed long distance call, whether within their own area code or outside it. This little trick of forcing everyone to dial "1" for long distance allowed us to introduce telephone exchanges with the same three digits as area codes. For example, one of our company's numbers is 212-206-6660. The "206" elsewhere is an area code for Seattle and other parts of Washington state.

NYNEX Corporation One of the seven Regional Holding Companies formed at Divestiture. It includes New England Telephone and New York Telephone Company and sundry service and cellular radio companies. The company says its name is spelled in all capitals, thus NYNEX. The New York Times spells it Nynex.

Nyquist Theorem In communications theory, a formula stating that two samples per cycle is sufficient to characterize an analog signal. In other words, the sampling rate must be twice the highest frequency component of the signal (i.e., sample 4 KHz analog voice channels 8000 times per second).

O Used on switches to mean "OFF." The "ON" setting is "I."
O&M Operations and Maintenance
O+ Calls Called "OH PLUS." O+ plus calls are calls made by dialing zero plus the desired telephone number. Calls made this way may be interrupted by a live operator requesting billing information, or a recorded announcement requesting the caller to enter the billing information.
O- Calls Called "OH MINUS." O- calls are operator-assisted calls. The caller dials zero and waits for the operator to pick up the line and talk to the caller.
O.E.M. Original Equipment Manufacturer. See OEM.
O/E Optic to Electric conversion.
O/R Short for Originator/Recipient in the X.400 MHS (Message Handling System).
O/R Name A CCITT (International Telegraph and Telephone Consultative Committee) term for the set of user attributes that identifies a specific MHS (Message Handling System) user. Example components are: given name, surname, ADMD (ADministrative Management Domain), country and PRMD (PRivate Management Domain), country and PRMD (PRivate Management Domain). An example of a complete O/R Address is: G=HARRY; S=NEWTON; I=HN; A=MARK400; O=PIPELINE; OUI=COMPUTER TELEPHONY; C=US.
OA Office Automation. Nobody knows what it means. But there are many consultants out there who will tell you for the right amount of money. Actually the early "office automation" seemed to translate into word processing for the masses. When someone discovered that word processing didn't enhance office productivity, office automation fell into disrepute.
OA&M Operations, Administration and Maintenance. This usually term refers to PBX management.
OACSU Off Air Call Set Up. A wireless telecommunications term. A method of establishing a call to a mobile telephone where the radio channel is set up at the last possible moment.
OAI Open Application Interface. Basically one or many openings in a telephone system that lets you link a computer to that phone system and lets the computer command the phone system to answer, delay, switch, hold etc. calls. The term is also called PHI — as in PBX-Host-Interface. The term OAI was first used by PBX makers, NEC and InteCom. And now the term has become somewhat generic, like all good things. Essentially every manufacturer of phone systems is evolving towards open application interfaces of their own. According to Probe Research, there are really two separate "markets" for OAI or PHI:
First, there's Horizontal/Office Automation applications. These are applications that support business functions across organizational groups or industry verticals in inter- or intra-department business settings. Examples include voice mail, electronic mail, message centers, corporate telephone directories, automated screen-based dialing, personal productivity tools, conferencing, PBX feature enhancements, ANI interfaces, time clocks, 911 emergency service and compound image (data, text, image and voice) processing. I see these applications as all those useful, productivity-enhancing things I always wished my telephone systems would do if only they would let me program the thing. (Until OAI, all phone systems were totally closed architecture.)
Second, there's transaction applications. These are applications that support an actual business transaction — customer service, inbound telephone order taking, outbound telemarketing, market research, data gathering, inventory inquiry, account time billing, credit collections, locator services with or without transfer to the local dealer. These always require access to a computer database. These applications are generally complicated, time-sensitive and customized for each installation.
A sample OAI arrangement: In early 1993, Novell and AT&T inked a deal to put telephony onto Novell LANs. The Telephony Server NetWare Loadable Module (NLM) will be the first product. It is an AT&T PC-card sitting in the Novell File server. The card connects to the ASAI (Adjunct Switch Applications Interface) port on the AT&T Definity PBX. Anyone with a PC on the network and an AT&T phone on their desk can use telephone features, such as auto-dialing, conference calling and message management (a new term for integrating voice, fax and e-mail on your desktop PC via your LAN). The Novell/AT&T deal intends to create open Application Programming Interfaces (APIs) that third party developers can work with. A Novell/AT&T example of what could be developed: A user could select names from a directory on his PC. He could tell the Definity PBX through the PC over the LAN to place a conference call to those names. At the same time, a program running under NetWare would automatically send an e-mail to the people, alerting them to the conference call and giving them the agenda. All participants would have access to both the document and the conference call simultaneously.
Probe figured that the OAI market will grow from $46 million in 1989 to $2.4 billion by 1993 in computers and applications software and $400 million in direct PBX/ACD enhancements. As a result of this bullishness several things have happened: Virtually every computer manufacturer has jumped in. Virtually every telephone manufacturer is hurrying to open his architecture or introduce a phone system that is open. And the author of this dictionary has started a trade show to concentrate on this area. It's called TELECOM DEVELOPERS. It happens in the Spring of every year. (If you'd like more information on the show, call 1-800-999-0345 or 1-800-LIBRARY). (If you'd like more on OAI, read TELECONNECT Magazine.) Meantime, here are some of the names which manufacturers of PBXs and computers have coined for their open application interfaces, also called PHIs:
ACL — Applications Connectivity Link — Siemens' protocol
ACT — Applied Computer Telephony — Hewlett Packard's generic application interface to PBXs
Application Bridge — Aspect Telecommunications' ACD to host computer link

ASAI — AT&T's Adjunct Switch Application Interface
CAM — Tandem's Call Applications Manager — the name of the Tandem software interface which provides the link between a call center switch telephone switch (either a PBX or an ACD) and all Tandem NonStop (fault tolerant) computers.
CIT — Digital Equipment Corporation's Computer Integrated Telephony (works with major PBXs)
CSA — Callpath Services Architecture — IBM's Computer to PBX link
Call Frame — Harris' PBX to computer link
Callbridge — Rolm's CBX and Siemens to IBM host or non-IBM host computer link
Callpath — IBM's announced, CICS application link to IBM's CSA, available on the AS400 in April of 1991
Callpath Host — IBM and ROLM's CICS-based integrated voice and data applications platform which links to ROLM's 9751 PBX
CompuCall — Northern Telecom's DMS central office link to computer interface
CPI — Computer to PBX Interface developed by Northern Telecom and DEC
CSTA — Computer Supported Telephony Application, RSL standard from ECMA
DECags — DEC ASAI Gateway Services. Two-directional link to AT&T's Definity
DMI — AT&T's Digital Multiplexed Interface, a T-1 PBX to computer interface
HCI — Host Command Interface. Mitel's digital PBX link to DEC computer
IG — AT&T's ISDN Gateway (one direction from the switch to the host)
ITG — AT&T's Integrated Telemarketing Gateway (two directional)
ISDN/AP — Northern Telecom's host to SL1 PBX protocol, which supports NT's Meridian Link
Meridian Link — Northern Telecom's host to PBX link available on the Meridian PBX
ONA — Open Network Architecture (for telephone central offices)
PACT — Siemens' PBX and Computer Teaming, protocols between Siemens PBXs and computers
PDI — Telenova/Lexar's Predictive Dialing Interface
PHI — PBX Host Interface (a generic term coined by Probe Research)
SAI — Stratus Computer Switch Application Interface
SCAI — Switch to Computer Application Interface, one name given by Northern Telecom to PHI
SCIL — Aristacom's Switch Computer Interface Link Transaction Link
Solid State Applications Interface Bridge — Solid's State's PBX to external computer link.
STEP — Speech and Telephony Environment for Programmers; WANG's link
Transaction Link — Rockwell's link from its Galaxy ACD to an external computer
Teleos IRX-9000 — Teleos' Intelligent Call Distribution platform
VoiceFrame — Harris Digital Telephone Systems Division Platform

OAM Operations, Administration and Maintenance. A group of network management functions that provide network fault indication, performance information, and data and diagnosis functions. Some switches have computers devoted to OAM. Northern Telecom, for example, uses a Sun Microsystems 3/80 B-4 processor in one of its central office switches.

OAM&P Operation, Administration, Maintenance and Provisioning.

Object 1. In its simplest form, it is a unit of information. It can be used much more broadly, depending on the application. In X.400, an object contains both attributes and method describing how the content is to be interpreted and/or operated on.
2. An entity or component, identifiable by the user, that may be distinguished by its properties, operations, and relationships.
3. Any piece of information, created by using a Windows-based application with OLE capabilities, that can be linked or embedded into another document.
See OBJECT ORIENTED PROGRAMMING.

Object Code Instruction code in machine language produced as the output of a compiler or a assembler. The original program, or code, is called the source Code.

Object Definition Alliance A group formed to establish software standards for interactive multimedia applications that will operate uniformly over a variety of hardware, operating systems and networks.

Object Encapsulation Data and procedures may be encapsulated to produce a single object, thereby hiding complexity.

Object Handle Code that includes access control information and a pointer to the object itself.

Object Inheritance The transfer of characteristics down a hierarchy from one to another.

Object Linking And Embedding OLE. An enhancement to DDE protocol that allows you to embed or link data created in one Windows application in a document created in another application, and subsequently edit that data in the original application without leaving the compound document. See DDE and OLE.

Object Oriented File System A file system, based on object-oriented programming, that allows permanent storage of objects and associated links.

Object Oriented Programming OOP. Object oriented programming is a form of software development that models the real world through representation of "objects" or modules that contain data as well as instructions that work upon that data. These objects are the encapsulation of the attributes, relationships, and methods of software-identifiable program components. Object-oriented methodology differs from conventional software programming where functions contained in code are found within an application. Although hype about object-oriented technology is fairly recent, the approach was first introduced in the programming language Simula developed in Norway in the late 1960s. The methodology was based on the way children learn (i.e., object + action = result). The idea of object oriented programming is make the writing of complex computer software much easier. The idea is to simply combine objects together to produce a fully-written software application. The work, goes into producing all the objects. They are the building blocks. Theoretically, libraries of objects will be worth a fortune to those companies who develop them.
Object oriented programming is not easy to understand. Here's a definition of object-oriented programming from Business Week,

September 30, 1991. "Software objects are chunks of programming and data that can behave like things in the real world. An object can be a business form, an insurance policy or even an automobile axle. The axle object would include data describing its physical dimensions and programming that describes how it interacts with other parts, such as wheels and struts. A system for a human resources department would have objects called employees, which would have data about each worker and the programming needed to calculate salary raises and vacation pay, sign up dependents for benefits, and make payroll deductions. Because objects have 'intelligence,' they know what they are and what they can and can't do. Thus objects can automatically carry out tasks such as calling into another computer, perhaps to update a file when an employee is promoted. The biggest advantage of objects is they can be reused in different programs. The object in an electronic-mail program that places messages in alphabetical order can also be used to alphabetize invoices. Thus, programs can be built from prefabricated, pretested building blocks in a fraction of the time it would take to build them from scratch. Programs can be upgraded by simply adding new objects."

The key concepts of object-oriented programming, according to ComputerWorld are

1. OBJECTS. The basic building block of a program is an object. Objects are software entities. They may model something physical like a person, or they may model something virtual like a checking account. Normally an object has one or more attributes (fields) that collectively define the state of the object; behavior defined by a set of methods (procedures) that can modify those attributes; and an identity that distinguishes it from all other objects. Some objects may be transient, existing temporarily during the execution of a program, i.e., only during run time. Others may be persistent, existing on some form of permanent storage (file, database, programming library) after the program finishes.

2. ENCAPSULATION. This concept refers to the hiding of most of the details of the object. Both the attributes (data structure) and the methods (procedures) are hidden. Associated with the object is a set of operations it can perform. These are not hidden. They constitute a well-defined interface — that aspect of the object that is externally visible. The point of encapsulation is to isolate the internal workings of the object so that, if they must be modified, those changes will also be isolated and not affect any part of the program.

3. MESSAGING. One object requests another object to perform its operation through messaging. The client object sends a message to the server object consisting of the identity of the server object, the name of the operation and, in some cases, optional parameters. The names of the operations are limited to those defined for that object. For example, the operations for a checking account object may be defined to be OPEN, DEBIT, CREDIT, COMPUTE INTEREST, ISSUE STATEMENT, SCHEDULE AUDIT, AND CLOSE.

4. DATA ABSTRACTION. An object is sometimes referred to as an instance of an abstract data type or class. Abstract data types are constructed using the built-in data types supported by the underlying programming language, such as integer and date. The common characteristics (both attributes and methods) of a group of similar objects are collected to create a new data type or class. Not only is this a natural way to think about the problem domain, it is a very efficient way to write programs. Instead of individually describing several dozen instances, the programmer describes the class once. Once identified, each instance is complete with the exception of its instance variables. The instance variables are associated with each instance, i.e., each object; methods exist only with the classes.

5. INHERITANCE. Data abstraction can be carried up several levels. Classes can have super-classes and subclasses. In moving to a level of greater specificity, the application developer has the option to retain some attributes and methods of the super-class, while dropping or adding new attributes or methods. This allows greater flexibility in class definition. It is even possible in some languages to inherit from more than one parent. This is referred to as multiple inheritance.

Object Request Broker Software which manages and routes communications between objects on a network.

ObjectVision For Windows A computer program from Borland, with the following explanation: There have been many attempts to create end user-programming tools based on cards and stacks with Basic-like scripting languages such as HyperCard and Toolbook. Yet these have not achieved widespread success because end users do not want to write code in any language. A more appropriate metaphor for business users is based on forms and databases. For many common applications, ObjectVision provides the built-in forms and database management capabilities as well as visual programming tools to handle business decisions. ObjectVision supports the creation of interactive business applications without the need for procedure programming (i.e. code). ObjectVision uses visual programming techniques to allow users to define business rules and directly access Paradox, dBASE and BTrieve databases across a network.

OC 1. Operator Centralization.
2. Optical Carrier. A SONET optical signal. See OC-1.

OC-1 Optical Carrier level-1. The optical counterpart of STS-1 (the basic rate, 51.840 Mbps, on which Sonet is based). Direct electrical-to-optical mapping of the STS-1 signal with frame synchronous scrambling. All higher levels are direct multiples of OC-1 (i.e., OC-3 equals three times OC-1, etc.). See OC-N.

OC-12 Optical Carrier level-12. Sonet channel of 622.08 Mbps.

OC-48 Optical Carrier level 48. Sonet channel of 2.4 Gbps.

OC-N Optical Carrier — n. The optical interface designed to work with the STS-n signaling rate in a Synchronous Optical Network (SONET). Optical Carrier level N. The optical signal that results from an optical conversion of an STS-N signal. N= 1, 3, 9, 12, 18, 24, 36, or 48. See OC-1.

OCC Other Common Carriers. Carriers providing long distance communications service other than AT&T. Nowadays, these companies (including AT&T) are called IXCs — IntereXchange Carriers.

Occupancy 1. The time a circuit or a switch is in use, i.e. occupied. Occupancy is normally expressed as a percentage, occupancy represents the actual usage versus the maximum amount of time available during a 1-hour period.
2. A call center term. The percentage of the scheduled work time that employees are actually handling calls or after-call wrap-up work, as opposed to waiting for calls.

OCDD On-line Call Detail Data. An AT&T method of accessing ANI (Automatic Number Identification) information from their computer within 48 hours after receiving the telephone call. You access the OCDD via a DEC VT100 and retrieve the phone numbers of the people who called you.

OCE See OPEN COLLABORATION ENVIRONMENT.

OCIS On-Line Customer Information System. An MCI definition: OCIS. An MCI customer service and sales system used for on-line access of customer information stored in the MCI Customer Database and for processing information to update the Customer Database.

OCR See OPTICAL CHARACTER RECOGNITION or OUTGOING CALL RESTRICTION.

OCR Automatic Routing Implemented by Optus and others, this technique also assigns a number to each user up to four digits long. The sender types this number within double parentheses anywhere on the cover sheet of a fax transmission, and the LAN-networked fax system uses optical scanning technology to read the number and route the fax to the intended recipient's workstation.

Octal A numbering system with the base eight.

Octathorp The character on the bottom right of your touchtone keyboard, which is also typically above the 3 on your computer keyboard. It's commonly called the pound sign, but it's also called the number sign, the crosshatch sign, the tic-tack-toe sign, the enter key, the octothorpe (also spelled octathorp) and the hash. On some phones it represents "NO." And on others it represents "YES." MCI and AT&T and some other long distance companies use it as the key for making another long distance credit card call without having to redial.

Octet An eight-bit byte.

Octopus A 25-pair cable that at one end has an amphenol connector (typical of what 1A2 phone systems were connected with) and at the other has many individual 2, 4, 6 and 8 wire connectors, typically male RJ-11s. The reason it's called an octopus is that it looks a bit like an octopus — one body and many arms. It's also called a Hydra.

Octothorpe The character on the bottom right of your touchtone keyboard, which is also typically above the 3 on your computer keyboard. It's commonly called the pound sign, but it's also called the number sign, the crosshatch sign, the tic-tack-toe sign, the enter key, the octothorpe (also spelled octathorp) and the hash. On some phones it represents "NO." And on others it represents "YES." MCI and AT&T and some other long distance companies use it as the key for making another long distance credit card call without having to redial.

OCUDP Office Channel Unit Data Port. A channel bank unit used to interface between the channel bank and a customer's DDS CSU or DSU.

OCX Ole Custom Control. According to InfoWorld, OCXes are the core of Microsoft's plans for packaging prewritten code that can be downloaded from Web servers.

ODA Office Document Architecture. ISO's standard 8613-1/8 for document architecture and interchange format adopted by MAP/TOP 3.0, GOSIP, and standardized by ECMA as ECMA-101.

ODBC Open Database Connectivity is a standard put out by Microsoft that allows databases created by various relational and non-relational database programs — such as dBASE, Microsoft Access, Microsoft FoxPro, and Oracle — to be accessed by a common interface. That interface is independent of the database file format. By relying on ODBC, one can write an application that uses the same code to read records from a dBASE file or a FoxPro file. ODBC drivers use a form of SQL to carry out database operations.

Odd Parity One of many methods for detecting errors in transmitted data. An extra bit is added to each character sent and that bit is given a value of 0 ("zero") or 1 ("one") such that the total number of ones in the character (including the parity bit) will be odd.

ODI 1. Operational Data Integrator. An MCI term: Combines data from the Customer Information Manager, the MCI Information Manager, and the Management Information Systems to build databases containing network information.
2. Open Data-link Interface. A device driver standard from Novell. ODI allows you to run multiple protocols on the same network adapter card. Interconnectivity strategy that adds functionality to Novell's NetWare and network computing environments by supporting multiple protocols and drivers. ODI's benefits allow you to:
• Expand your network by using multiple protocols without adding network boards to the workstation. ODI creates a "logical network board" to send different packet types over one board and wire.
• Communicate with a variety of workstations, file servers, and mainframe computers via different protocols without rebooting your workstation.
• Configure the LAP driver for any possible hardware configuration with NET.CFG, instead of using limited SHELL.CFG choices.

ODP Open Distributed Processing.

ODSI Open Directory Services Interface. Announced by Microsoft in the summer of 1995, two years after Novell's NetWare Directory Services (NDS), it is designed to do essentially what NDS does, namely give users a way to sign on to multiple servers through one simple sign-on. The ultimate idea is that NDS or ODSI will become a single place for a networked user (i.e. one on a local area network — a LAN) to sign on to multiple networked devices, e.g. telephone systems, voice mail, electronic mail, Lotus Notes and, of course, multiple file servers (wherever they're located). Sign on in one place, get into many. A big user time-saving benefit.

OEM 1. Original Equipment Manufacturer. The maker of equipment marketed by another vendor, usually under the name of the reseller. The OEM may only manufacture certain components, or complete computers, which are then often configured with software and/or other hardware, by the reseller.
2. Operations Enterprise Model. MCI's name for the mass of its internal productivity tools. Covers everything from group calendars to presentation tools.

OES Operations Evaluation System. An MCI internal system, which generates daily, weekly, and monthly switch data reports; used to scan high-level switch degradation problems and to analyze specific switch problems.

Off-Hook When the handset is lifted from its cradle it's Off-Hook. When you lift the handset, the hookswitch is moved by a spring (except in Rolm phones) and alerts the central office that the user wants the phone to do something like receive an incoming call or dial an outgoing call. A dial tone is a sign saying "Give me an order." The term "off-hook" originated when the early handsets were actually suspended from a metal hook on the phone. When the handset is removed from its hook or its cradle (in modern phones), it completes the electrical loop, thus signaling the central office that it wishes dial tone. Some leased line channels work by lifting the handset, signaling the central office at the other end which rings the phone at the other end. Some phones have autodialers in them. Lifting the phone signals the phone to dial that one number. An example is a phone without a dial at an airport, which automatically dials the local taxi company. All this by simply lifting the handset at one end — going "off-hook."

Off-Hook Call Announce A telephone system feature. A telephone has a speaker. If the person is speaking on the phone, another person (inside the building, on the same phone system) can "off-hook voice announce" you and can give you a message or speak with you. You will hear their voice coming through the speaker on your phone. (So may the person on the other end.) Depending on your phone, you may be able to put your hand over your telephone handset and whisper something back to the person who's "off-hook voice announcing" you. Otherwise you'll have to hang up or put the person on hold, and speak to the person on another line.

Off-Hook Queue There are two types of queuing: ON-HOOK and OFF-HOOK. In On-Hook Queuing, the user dials his number, the switch tells him the outgoing trunks are busy. The user then hangs up. The switch calls him back when a trunk becomes available. In Off-Hook dialing, the user waits with his receiver screwed into his ear until a trunk comes free and the PBX connects him to the next available trunk. Off-Hook queues are usually shorter than on-hook queues. If a trunk doesn't come free quickly in off-hook queuing, the call will often flow over onto the more expensive DDD trunks. Off-hook queuing costs more but keeps the user waiting less. On-hook queuing costs less by waiting for a cheaper trunk but can be tiring and frustrating for the workers waiting for their calls to go through.

Off-Hook Routing See OFFHOOK and RINGDOWN.
Off-Hook Voice Announce See OFF-HOOK CALL ANNOUNCE.
Off-Line Any equipment not actively connected to a phone line but which can be activated to work with that system is Off-Line. This concept also applies to computer systems. For example, a modem attached to or built into a microcomputer can be plugged permanently into a phone line. But the microcomputer can be used for word processing most of the time. While it's doing word processing, it is "off-line." When the user loads the communications software and turns on the modem, the microcomputer is now said to be "on line." Off-line computer storage is a place to put stuff which a computer cannot access "on-line," like a hard-disk. Off-line storage might be microfilm or microfiche.

Off-Line Storage Storage that is not under the control of a processing unit.
Off-Net Calling Phone calls which are carried in part on a network but are destined for a phone not on the network, i.e. some part of the conversation's journey will be over the public switched network or over someone else's network. MCI defines off-net calls as "Billable calls to non-tariffed cities. Can be MCI Off-Net or WATS Off-Net. Classified as Tier 2 for tariff purposes."

Off-Network Access Line ONAL. A circuit in a private network which allows the user to go off the private network and complete calls on the public dial network.

Off/On Hook A modem term. Modem operations which are the equivalent of manually lifting a phone receiver (taking it off hook) and replacing it (going on hook). See OFF-HOOK.

Off-Peak The periods of time after the business day has ended during which carriers offer discounted airtime charges. Usually, OFF-PEAK rates are available for cellular calls between 7:00 p.m. and 7:00 a.m. and on weekends and holidays, but times vary among carriers. Among landline carriers, the business day usually ends at 5 p.m., after which time residential calling builds, and that ends at 11 p.m., after which little happens, except rates drop once again until they rise at the beginning of the next business day at 8 a.m. the next morning.

Off-Premises Extension OPX. Now also called OPS for Off-Premises Station. A telephone located in a different office or building from the main phone system. The OPX is connected by a phone line dedicated to it. It acts as if it were in the same place as the main phone system and can use its full capabilities. Here's another explanation a reader sent in. OPX is the appearance of an actual telephone line (such as 212-691-8215) in two physically separate locations. For example, this line (212-691-8215) could appear and ring in my office and at my home without my home phone being a part of the office telephone system. An OPX is commonly used for answering services or for small businesses. Bell operating phone companies are sharply increasing the charge for dedicated OPS lines — often by several hundred percent per year.

Off-the-shelf When something has already been produced and is available for immediate delivery, it is said to be available "off-the-shelf." It is presumably sitting on the shelf in a warehouse waiting for your order. Sometimes called "shrinkwrapped," referring to the plastic wrap around the box. It means you don't are not meant to need intervention by a programmer or integrator to make the software work.

Offered Call 1. A call that is presented to a trunk or group of trunks. See TRAFFIC ENGINEERING.
2. A call center term. A call that is received by the ACD. Offered calls are then either answered by an employee (handled) or abandoned.

Offered Load The total telephone traffic load, including load that results from retries, submitted to a group of servers. See also OFFERED TRAFFIC.

Offered Traffic The total attempts to seize a group of servers.
Office Automation Nobody knows what office automation means, though there are many consultants out there who will tell you what it means for an grand sum of money. In reality, it's a benign, imprecise term for data processing when it applies to self-focused (as opposed to customer-focused) white collar-type activities — accounting, word processing, communications, document management.

Office Class Functional ranking of a telephone central office switch depending on transmission requirements and hierarchical relationship to other switching centers. (Awful mouthful!) There used to be five classes of switches in the U.S. telephone network hierarchy, with the one closest to the end-subscriber being a class 5 central office. But technology and marketing is changing things and, by distributing intelligence closer to the end user, it is diffusing our traditional definitions of network hierarchies and the class of switches. See CLASS 5 central office. See OFFICE CLASSIFICATION.

Office Classification Prior to divestiture, those numbers that were assigned to offices according to their hierarchical function in the U.S. public switched telephone network. The following class numbers are used:
Class 1: Regional Center (RC) Class 2: Sectional Center (SC) Class 3: Primary Center (PC) Class 4: Toll Center (TC) if operators are present, or else Toll Point (TP) Class 5: End Office (EO) (local central office)
Any one center handles traffic from one to two or more centers lower in the hierarchy. Since divestiture and with more intelligent software going into telephone switching offices, these designations have become less firm.

Office Code The first three digits of your seven-digit local telephone number. Also called NNX code.

Office Network A network within an office. An older term for a Local Area Network. User concern is with application sharing, file/database sharing, electronic mail, word processing and circuit switching.

Office Telesystem OTS. An AT&T office automation solution that runs on 3B computers and includes a family of allegedly low-cost, easy-to-use workstations.

Office User Interface A special shell program which sets up windows with menus of available utilities and applications.

Office Window Interface See OFFICE USER INTERFACE.

OfficeVision OfficeVision is a series of IBM applications programs that run 'cooperatively' on PCs and IBM mainframes and performs such tasks as manage electronic mail and extract data from a mainframe for use in a PC spreadsheet. OfficeVision is meant to be a concrete example of one of the benefits of IBM's Systems Application Architecture. OfficeVision was announced in May, 1989. Delivery was scheduled for the following year. See SYSTEMS APPLICATIONS ARCHITECTURE.

Offset The offset of a port or a memory location is the difference between the address of the specific port or memory address and the address of the first port or memory address within a contiguous group of ports or a memory window. This term is used when identifying the locations of registers located with respect to the base address of the 16 contiguous I/O ports in a PCMCIA card. It is also used when identifying the location of memory mapped registers with respect to the base address of the memory window. See also OFFSET PARABOLIC ANTENNA and OFFSET GEOMETRY.

Offset Geometry Shadow-free geometry. The feeder in the primary focus is mounted so that the effect of its shadow on the secondary radiation is negligible.

Offset Parabolic Antenna An offset antenna is a new form of satellite antenna that is taller than it is wide. According to the manufacturers, the antenna design makes for more efficient use of the antenna surface. What that means is that it captures more of the satellite signal hitting the antenna. Offset antennas are more expensive than the "normal" parabolic satellite antennas, which are called "prime focus parabolic antennas." They are more expensive because they cost more to make since they typically must be made out of one sheet of metal. Offset antennas are harder to carry around, since you can't make them out of several foldover sheets of metal.

Offsite Night Answer This mode allows incoming after-hours calls to be forwarded automatically to an off site location.

OFTEL OFfice of TELecommunications.

OGM OutGoing Message. The message an answering machine delivers to someone who calls. Sample, "I'm not here. Leave a message after the beep."

OGT OutGoing trunk.

OHD Optical Hard Drive. A term pioneered by Pinnacle Micro, Irvine, CA. OHD technology, according to Pinnacle, combines the advantages of magneto-optical technology with speeds faster than most hard drives.

OHM The practical unit of resistance. The resistance that will allow one ampere of current to pass at the electrical potential of one volt. Ohm's Law dictates the relation between the current, electromotive force and resistance in a circuit:
Amperes = Volts divided by Ohms Volts = product of Amperes and Ohms Ohms = Volts divided by Amperes

Ohms Measures of resistance. A resistance of one Ohm allows one Ampere to flow when a potential difference of one volt is applied to the resistance. See OHM'S LAW.

Ohm's Law The law which relates current measured as Amps (I), voltage (V) and resistance measured as Ohms (R). Ohm's Law is $E = I \times R$. It can also be expressed as $I = V/R$, or $R = V/I$. See OHMS.

OHQ Off-Hook Queue. See OFF-HOOK QUEUING.

OHR Optical Handwriting Recognition. Exactly what it says. Machine reading of handwriting.

OHSA Occupational Health and Safety Act. Specifically the Williams-Steiger law passed in 1970 covering all factors relating to safety in places of employment.

OHVA Off-Hook Voice Announce. A phone system feature that permits an intercom announcement to be heard through a speaker at a phone where the handset is in use on an outside call.

OLAP On-Line Analytical Processing, also called a multidimensional database. According to PC Week, these databases can slice and dice reams of data to produce meaningful results that go far beyond what can be produced using the traditional two-dimensional query and report tools that work with most relational databases. OLAP data servers are best suited to work with data warehouses. A database warehouse consolidates information from many departments within a company. This data can either be accessed quickly by users or put on an OLAP server for more thorough analysis.

OLCE On-Line Circuit Engineering. An MCI definition: A NOBIS subsystem that allows On-Line Circuit Engineering and Circuit Record Sheet (CRS) file storage.

OLE Object Linking and Embedding. It means tying one piece of information in one form into a document in another form, such that a change in one piece of information will be automatically reflected in the other document. Here's an explanation from the New York Times: Business reports may contain information in a variety of formats, including text and numbers, charts, tables, images, graphics, sound and video. Typically, these are created in separate applications programs (e.g. spreadsheet, word processing, charting, database, etc.) and are merged into a single document (i.e. the report). But when the numbers used to create a chart are changed the chart must be updated as well. The executive then has to track down all the various components of the report, call up their respective applications, make the changes and stitch everything back together. OLE promises to keep track of those links and update the various components as they change. Here's an explanation from PC Magazine: Ole is a complex specification that describes the interfaces used for such tasks as embedding objects created by one application within documents created by another, performing drag-and-drop data transfers within or between applications, creating automation servers that expose their inner functionality to other programs, extending the Windows 95 shell with custom DLLs, and much more. Version 1.0 of the specification was originally created for placing objects such as Excel spreadsheets inside documents created by other applications such as Microsoft Word for Windows. OLE 2.0 greatly expanded the scope of OLE and made the original name obsolete, but the name had achieved widespread recognition and was retained.

OLIU Optical Line Interface Unit

OLNS Originating Line Number Screening

OLTP OnLine Transaction Processing. A generic concept in the computer industry to cover everything from issuing airline tickets to dispensing money out of street-corner, automated teller machines.

OM Operational Measurement

OMAT Operational Measurement and Analysis Tool.

OMC Operation and Maintenance Center. Computer hardware and software assigned specifically to monitor and manage one part of a telecommunications network, usually employed in GSM (Global System for Mobile Communications).

OMC-Env Operations and Maintenance Center — Environment. An OMC dedicated to monitoring and managing the physical environment where telecommunications equipment resides.

OMC-IN Operations and Maintenance Center — Intelligent Network. A wireless telecommunications term. An OMC dedicated to monitoring and managing components of the Intelligent Network in GSM (Global System for Mobile Communications). This includes the HLR, AUC, EIR, SMSC, and VMS.

OMC-Misc Operations and Maintenance Center — Miscellaneous. An OMC that manages and monitors non-intelligent devices in a mobile telecommunications network under GSM (Global System for Mobile Communications).

OMC-R Operations and Maintenance Center — Radio. An OMC that manages and monitors the radio interface under GSM (Global System for Mobile Communications) including the BSS, BSC, and BTS.

OMC-S Operations and Maintenance Center — Switching. An OMC dedicated to monitoring and managing switches in a GSM (Global System for Mobile Communications) telecommunications network.

OMC-SS7 Operations and Maintenance Center — SS7. A wireless telecommunications term. An OMC dedicated to monitoring and managing the SS7 signaling network in a GSM (Global System for Mobile Communications) telecommunications network.

OMC-T Operations and Maintenance Center — Transmission. A wireless telecommunications term. An OMC dedicated to managing and monitoring transmission activities under a GSM (Global System for Mobile Communications) telecommunications network.

OMC-WAN Operations and Maintenance Center — Wide Area Network. A wireless telecommunications term. An OMC dedicated to managing and monitoring components of the Wide Area Network under GSM. The WAN links all or most devices in a GSM (Global System for Mobile Communications) telecommunications network together.

OMEGA A global radio navigation system that provides position information by measuring phase difference between signals radiated by a network of eight transmitting stations deployed worldwide. The transmitted signals time-share transmission on frequencies of 10.2, 11.05, 11.33, and 13.6 Khz. Since the transmissions are coordinated with UTC (USNO), they also provide time reference.

OMG Object Management Group. A group of major systems vendors involved in the definition of standards for object management.

Omnidirectional A microphone with a pickup pattern essentially uniform in all directions.

Omnidirectional Antenna An antenna whose pattern is nondirectional in azimuth. The vertical pattern may be of any shape.

OMNIPoint A program established by the ITU-T as a way to achieve a unified approach to making decisions with the help and guidance of experts from all over the telecommunications industry. OMNIPoint aims to minimize the investment outlays and risks of all concerned — commercial users, service providers, vendors and standards organizations.

OMR Optical Mark Recognition. Refers to machine recognition of filled-in "bubbles" on reader service bingo cards.

On Hook See ON-HOOK.

On-Demand Connection An ISDN BRI term. The ability to automatically suspend and resume a physical connection while "spoofing" network protocols, routing and applications. The physical connection is only brought up on-demand. This ensures that users' ISDN holding time charges are proportional to their useful holding time not the total holding time.

On-Demand Dialing ISDN cost-savings feature that sets up, transfers, and closes a call only if the ISDN device detects a data packet that is addressed to a remote network.

On-Going Maintenance A term used in the secondary telecom equipment business. A manufacturer's guarantee of maintainability from one owner to the next. There are all types of conditions which must be met, including: The machine has been under the manufacturer's maintenance contract until its time of deinstallation. There hasn't been any damage in storage or transit. And there have not been any modifications made.

On-Hook When the phone handset is resting in its cradle. The phone is not connected to any particular line. Only the bell is active, i.e. it will ring if a call comes in. On-Hook is thus the normal, inactive condition of a telephone system terminal device. See ON-HOOK DIALING and OFF-HOOK.

On-Hook Dialing Allows a caller to dial a call without lifting his handset. After dialing, the caller can listen to the progress of the call over the phone's built-in speaker. When you hear the called person answer, you can pick up the handset and speak or you can talk hands-free in the direction of your phone, if it's a speakerphone. Critical: Many phones have speakers for hands-free listening. Not all phones with speakers are speakerphones — i.e. have microphones, which allow you to speak, also.

On-Line When a device is actively connected to a PBX or a computer, it is On-Line. Terminals, PCs, modems and phones are often On-Line.

On-Line Circuit Engineering An MCI definition: OLCE. A NOBIS subsystem that allows On-Line Circuit Engineering and Circuit Record Sheet (CRS) file storage.

Online Fallback A modem feature. It allows high speed error-control modems to monitor line quality and fall back to the next lower speed if line quality degrades. Some modems fall forward as line quality improves.

On-Net Telephone calls which stay on a customer's private network, traveling by private line from beginning to end are said to be on-net. Here's MCI's definition: Billable calls to MCI-tariffed cities, including those cities reached via leased lines. Can be MCI On-Net or WATS On-Net. Classified as Tier 1 for tariff purposes.

On-Ramps A way of getting onto The Information Superhighway. Such an on-ramp could be anything from a phone line to a two-way cable TV channel. Many companies are trying to create on-ramps, however they define them.

ONA See OPEN NETWORK ARCHITECTURE.

ONAC Operations Network Administration Center.

ONAL Off Network Access Line.

ONC Open Network Computing. A distributed applications architecture developed by Sun Microsystems. Includes NFS, NIS and RPC. Now part of Solaris OS.

One Call 1. Fax-back system that requires you to call from your fax machine to send documents on the same line after picking from menu of verbal prompts. the other fax-back system is called a two-call system. You dial from one line and tell the fax-back machine on the other end that you want it to send your requested fax to second number, i.e. one where your fax machine will receive your requested message.
2. A service provided by Centigram which lets its users dial their Centigram voice mail machines and hear their voice mail messages and store and receive their faxes.

One Dimensional Coding A data compression scheme for fax machines that considers each scan line as being unique without referencing it to a previous scan line. One dimensional coding operates horizontally only.

One Hop Set A set of hosts which are one hop apart in terms of internetwork protocols.

One Number Presence This is the consistent use of one telephone number (particularly an 800 number) across all advertising media. The long distance vendors can arrange this for both in-state and national 800 services.

One Number Systems Also called Follow Me Systems. Follow me systems and services are based on the premise that people are mobile (e.g., they move around a lot in and out of the office), and have many phone numbers or places they might be. A person could have an office number, a cellular number, a voice mail number, a home number, and a pager number. Which phone number will a caller be at? Follow me systems will "track-down" the user being called no matter where they are and connect the caller to the user. The caller need only dial a single phone number. Usually, network or local switch provided call data (or data gathered by a voice response unit) is used to identify each call as being intended for a specific user. Based on options the user has selected, the caller will hear an answering prompt customized to that user, and the one number system will then automatically attempt to locate the user at one of several locations. The tracking-down process varies considerably between follow me systems. Some systems try multiple locations at once, others will try the possible destination locations sequentially. Almost all follow me services provide the caller an exit to voice mail at various points of the call.

One Plus Bulk Restriction This is the name for a local service provided by a Northern Telecom DMS central office switch. One-Plus Bulk Restriction allows subscribers to deny or permit all one-plus (i.e. long distance) calls from their phones by dialing a special PIN (Personal Identification Number).

One Plus Per-Call Restriction This is the name for a local service provided by a Northern Telecom DMS central office switch. Subscribers can restrict one-plus toll calls from their phones by requiring that a PIN (Personal Identification Number) be dialed prior to a one-plus call. If the PIN is valid, the caller hears a second dial tone, and the one-plus number can then be dialed. When a one-plus call is attempted without the PIN or with a wrong PIN, the caller is routed to a tone or announcement and is not able to place the one-plus call.

One Plus Restriction A central office service. Telephone subscribers can now limit one-plus toll calls by selecting an authorization code that must be dialed before any one-plus call will be connected. If the code is valid, the caller hears a second dial tone, permitting the number to be dialed and connected.

One Time Programmable OTP. A term describing memory that can be programmed to a specific value once, and thereafter cannot be changed (or can only be revised in a limited way.) OTP EPROMs are typically ordinary EPROMs that have been packaged in such as way that ultra-violet light cannot be used to erase the contents of the EPROM. Such packaging is usually less expensive.

One Way Operation See SIMPLEX.

One Way Splitting When the attendant is connected to an outside trunk and an internal phone, pushing a button on the console allows her to speak privately with the internal extension, thus "splitting" her off from the external trunk.

One Way Trade A call center term. A schedule trade in which only one employee is working the other's schedule.

One Way Trunk A trunk between a switch (PBX) and a central office, or between central offices, where traffic originates from only one end. You can, of course, still speak and listen on the trunk. It's just like a normal two-way trunk except that a one-way trunk can only be used for dialing out or only to receive calls.

Ones Density The requirement for digital transmission lines in the public switched telephone network that eight consecutive

zeros cannot exist in a digital data transmission. On a T-1 line, 0 means no voltage, no pulse. Too many zeroes and the repeaters lost count because they had no signal pulses to count. Ones density exists because repeaters and clocking devices within the network will lose timing after receiving eight zeros in a row. There are many techniques or algorithms used to insert a one after every seventh-consecutive zero. The question of how many consecutive zeros you can have on a digital line is changing. In the old days, the FCC actually said you could have 15 zeroes in a row. Now the FCC says you can have up to about 40 something (just like you) 0s without 'harming the network." For all practical purposes, seven consecutive zeros is the maximum today. See BIT STUFFING.
ONI 1. Operator Number Identification.
2. Optical Network Interface. A device which converts photons to electrons and vice versa. It's a device which converts an optical signal into an electrical signal that non-optical telecommunications transmission and switching devices can understand and vice versa.
Online Transaction Processing See OLTP.
ONMS Open Network Management System. Digital Communications Associates architecture for products confirming to ISO's CMIP.
ONP 1. One Night Process.
2. Open Network Provision.
OOF Out Of Frame. A designation for a condition defined as either the network or the DTE equipment sensing an error in framing bits. It's declared when 2 of 4 or 2 of 5 framing bits are missed (the OOF condition existing for 2.5 seconds generally creates a local Red Alarm). See LOF.
OOP Object-oriented programming (OOP) is simply combining objects to produce a finished software application. The real work is developing the objects that carry out tasks within the program. Programs can be produced by recycling objects from other programs and upgraded by replacing old objects.
OPAC Outside Plant Access Cabinet.
Open Means the circuit is not complete or that the fiber is broken. There is a break in it. A break does not necessarily mean it's malfunctioning, only that it's been turned off.
Open Air Transmission Referring to a transmission type or associated equipment, that uses no physical communications medium other than air. Most radio communications systems, including microwave, shortwave and FM radio and infrared are open-air (also called "through-the-air") transmission systems.
Open Application Interface See OAI.
Open Architecture See OAI.
Open Circuit A circuit is not complete. There is no complete path for current flow. Electrical current cannot flow in the circuit. In electrical engineering, a loop or path that contains an infinite impedance. In communications, a circuit available for use.
Open Collaboration Environment O.C.E. Apple's Open Collaboration Environment extends the Macintosh operating system to provide a platform for the integration of fax, voicemail, electronic mail, directories, telephony and agents. From a user's perspective, according to Apple, O.C.E.'s functionality will be seen through:
• System-wide directory services, including a desktop directory browser and electronic business cards.
• A compound mailbox for mail from all sources — fax, voice mail, e-mail, pager, etc.
• Application integration, with all applications having the ability to send documents.
Open Ended Access Term used to describe the ability to terminate a call to any public network destination. For example, on the open end of a foreign exchange circuit, the customer may call any number in the local calling area without being charged for long distance service.
Open Ground, Neutral Or Hot In AC electrical power, an "open" is a break, an extremely loose or an unconnected wire in any electrical path. Dangers of an "open" GROUND include serious shock and fire hazard and are life-threatening. Caution: an "open" GROUND will not stop equipment from operating. However it will stop a fuse or circuit breaker from operating should a ground fault occur.
Open Line Dealing This is a term used by British Telecom in its turrets. The system electronically recreates the original "pit" share trading environment in the stock exchange in which everybody could talk to everybody else. In open line dealing, the trader has the ability to program a number of parties onto speakers. Full duplex speech is achieved using the associated microphones and all lines receive a simultaneous broadcast.
Open Loop System A control system which does not use feedback to determine its output.
Open Network Architecture ONA. The "network" refers to the public switched network. The FCC wants to encourage companies to get into the value-added telecom business — voice mail, electronic mail, shopping by phone, etc. These companies may be called "value added providers" or "enhanced service providers." The FCC's idea of encouraging companies to add value to phone lines is a nice idea, except that all these companies will rely on phone lines provided by local phone companies who also want to be in value-added telecommunications business. And the local phone companies would prefer that the business be a monopoly (easier to manager, higher prices, etc.) If the FCC is to allow the Bell operating companies and other local phone companies into the value-added business and encourage others in, then it must figure a way the Bell operating companies don't organize things so they have an unfair advantage. The FCC's latest idea is called ONA — Open Network Architecture. Under this concept, the telephone companies are obliged to provide a certain class of service to their own internal value-added divisions and the SAME class of service to nonaffiliated (i.e. outside) valued-added companies. The concept is that the phone company's architecture is to be "open" and that everyone and anyone can gain access to it on equal footing. ONA is only a concept at present and still needs some rigorous defining. There is not much pressure from outside entrepreneurial companies for ONA access. Thus ONA at the FCC and elsewhere drags its feet.
The March 18, 1991 issue of Telephony Magazine said that as conceptualized by the FCC, ONA "is the overall design of a car-

rier's basic network facilities and services to permit all users of the basic network, including the enhanced services operations of a carrier and its competitors, to interconnect to specific basic network functions on an unbundled and "equal access" "basis." Selected regional Bell holding companies ONA services would include

1. Basic Serving arrangements (BSAs):
A BSA is the basic interconnection access arrangement which offers a customer access to the public network and provides for the selection of available Basic Service Elements (BSEs, see below). Basic serving arrangements are:
• Switched, line side connection • Switched, trunk side connection • Dedicated, metallic dedicated

2. Basic service elements (BSEs):
Basic Service Elements are optional basic network functions that are not required for an ESP to have a BSA, but when combined with BSEs can offer additional features and services. Most BSEs allow an ESP to offer enhanced services to their customers in a more flexible manner. BSEs fall into four general categories: Switching, where call routing, call management and processing are required; Signaling, for applications like remote alarm monitoring and meter reading; Transmission, where dedicated bandwidth or bit rate is allocated to a customer application; and Network Management, where a customer is given the ability to monitor network performance and reallocate certain capabilities. The selection of available BSEs is an ongoing process, with new arrangements being developed many times in response to customer demands. ANI, Audiotext "Dial-It" Services, and Message Waiting Notification are all examples of BSEs, which also include:
• Multiline hunt group • Uniform call distribution • Central office announcements • Three-way call transfer

3. Complementary network services (CNSs):
CNSs are basic services associated with end user's lines that make it easier for ESPs (Enhanced Service Providers) to offer enhanced services. Some examples of CNSs include
• Call forwarding Busy/Don't Answer • Three way calling • Call waiting • Virtual dial tone • Message waiting/indicator • Speed calling • Warm line

4. Ancillary services.
Ancillary Services. These are options available to an ESP which support and complement the provision of enhanced services. Examples of ancillary services are protocol conversion, and DID with third number billing inhibited.
In June, 1991, according to Communications Week, the FCC established a tariff structure that will determine how much the telcos can charge enhanced-services providers — and ultimately how much end users will have to pay for those services. FCC Chairman Alfred Sikes called the agency's action "one of our most pivotal steps" in the implementation of ONA. The FCC's idea is that ONA tariffs will be filed with the FCC in November, 1991 and will take effect February 1, 1992. They didn't. At the time I wrote this, the FCC had a new chairman, with a different agenda.
See also COMPUCALL, ENHANCED SERVICES, OAI, and OPEN APPLICATION INTERFACE.

Open Office Typically, this layout places the manager's desk in the foreground, within view of all other desks and enclosed spaces. The first open office appeared around 1960 when it was introduced in Germany as the "office landscape."

Open Skies When a government or government agency allows virtually anyone to sell satellite telecommunications service, you have "OPEN SKIES." The United States has an Open Skies satellite policy. Virtually anyone can apply to launch and operate a telecommunications satellite and, with a high degree of certainty, you'll be granted your wish. The European community is just now (summer, 1991) beginning to think of opening its skies. They have gone one small step — namely allowing anyone to buy and operate a receive-only satellite earth station not connected to the public network.

Open Solutions Developers A Dialogic term for companies which develop and sell end-user voice processing applications. We'd probably call these people value added resellers (VARs).

Open System Interconnect OSI. An ISO publication that defines seven independent layers of communication protocols. Each layer enhances the communication services of the layer just below it and shields the layer above it from the implementation details of the lower layer. In theory, this allows communication systems to be built from independently developed layers. See OSI.

Open Systems Open systems refers to that best of all possible worlds, where everyone would comply with a set of hardware and software standards. You could buy a server from company A, a client from company B, a networking system from companies C, D or E, and applications software from companies F-Z, and everything would work harmoniously. In real life, some "open" systems are more open than others. Some companies talk, without embarrassment, of their "proprietary open systems." Some things are open and closed. PBXs have many "open" ports on which you can attach things. But there are no PBXs I know of which have open backlanes, meaning you (or someone else) can design a board and plug it into the PBX's backplane. That's closed. The concept of open systems has been more popular in the computer industry.

Open Systems Interconnection See OSI.

Open Toolkit Developers A Dialogic term for outside developers (outside of Dialogic) who provide applications generators that simply application development and work in a variety of operating systems — including MS-DOS, UNIX, OS/2, Windows, etc.

Open Wire A transmission facility typically consisting of pairs of bare (uninsulated) conductors supported on insulators which are mounted on poles to form an aerial (above ground) pole line. Most basic of all practical types of transmission media. Open wire may be used in both communication and power.

OpenView Hewlett-Packard's suite of a network-management application, a server platform, and support services. OpenView is based on HP-UX, which complies with AT&T's Unix system.

Operand That which is being operated on. An operand is usually identified by the address part of an instruction.

Operating Environment Referring to the combination of (usually IBM) host software that includes operating system, telecommunications access method, database software and user applications. Some common operating environments include MVS/CICS and MVS/TSO.

Operating System A software program which manages the basic operations of a computer system. It figures how the computer main memory will be apportioned, how and in what order it will handle tasks assigned to it, how it will manage the flow of information into and out of the main processor, how it will get material to the printer for printing, to the screen for viewing, how it will receive information from the keyboard, etc. In short, the operating system handles the computer's basic housekeeping. MS-DOS, UNIX, PICK, etc. are operating systems.

Operating Time The time required for seizing the line, dialing the call and waiting for the connection to be established.

Operation Code The command part of a machine instruction.

Operational Data Integrator ODI. An MCI term: Combines data from the Customer Information Manager, the MCI Information Manager, and the Management Information Systems to build databases containing network information.

Operational Grammar A voice recognition term. A vocabulary structure where certain word sets activate other word sets.

Operational Load The total power requirements for communication facilities.

Operational Security OPSEC. A thorough on-site examination of an operation or activity to determine if there are vulnerabilities that would permit adversaries and exploitation of critical information during the planning, preparation, execution, and post-execution phases of any operation or activity. A Federal Government definition.

Operational Service Period A performance measurement period, or succession of performance measurement periods, during which a telecommunication service remains in an operational service state. An operational service period begins at the beginning of the performance measurement period in which the telecommunications service enters the operational service state, and ends at the beginning of the performance measurement period in which the telecommunications service leaves the operational service state.

Operations Evaluation System OES. An MCI internal system, which generates daily, weekly, and monthly switch data reports; used to scan high-level switch degradation problems and to analyze specific switch problems.

Operations Support System See OSS.

Operator Employee of telephone company, or an individual business or institution, who aids in the completion of phone calls. Traditionally a woman's occupation, now increasingly the role of men and machines. In some countries, like Germany, the phone company doesn't have operators.

Operator Assistant A software program from TELECONNECT Magazine which allows you to maintain a directory of personnel and their messages on a PC. The program runs on a network and people can pick up their messages from their computers attached to the network — without calling the company's telephone operator. The program is available for free. It is a premium for subscribing to TELECONNECT Magazine.

Operator Assisted A phone call placed with the assistance of the carrier's operator. You pay more when you use an operator.

Operator Console Same as attendant console. See ATTENDANT CONSOLE.

Operator Services OS. Any of a variety of telephone services which need the assistance of an operator or an automated "operator" (i.e. using interactive voice response technology and speech recognition). Such services include collect calls, third party billed calls and person-to-person calls.

Operator Workstation OWS. The OWS is an advanced voice and data workstation (typically a PC running a flavor or Windows) that streamlines and automates many of the routine tasks of an operator, thus reducing the amount of time needed for call handling. Color screens, pop-up windows, one-touch commands, and database look-up are some of the features that simplify the operator's tasks and speed call processing.

OPRE Operations Order Review

OPS 1. Operator Services

2. Off-Premises Station. See OFF PREMISES EXTENSIONS.

3. Other Peoples' Software. A term created by Rick Luhmann, editor of Computer Telephony Magazine.

Optical Amplifier A device to amplify an optical signal without converting the signal from optical to electrical back again to optical energy. The two most common optical amplifiers are erbium-doped fiber amplifiers (EDFAs), which amplify with a laser pump diode and a section of erbium-doped fiber, and semiconductor laser amplifiers.

Optical Attenuator In optical communications, a device used to reduce the intensity of the optical signal. In some optical attenuators used in optical fiber systems, the amount of attenuation depends on the modal distribution of the optical signal.

Optical Blank A casting consisting of an optical material molded into the desired geometry for grinding; polishing; or, in the case of some optical fiber manufacturing processes, drawing to the final optical/mechanical specifications.

Optical Cavity A region bounded by two or more cavity surfaces, referred to as mirrors or cavity mirrors, whose elements are aligned to provide multiple reflections of lightwaves. The resonator in a laser is an optical cavity.

Optical Character Recognition OCR. Reading data using a machine that visually scans the characters in a document and converts that data into standard form which can be stored on conventional magnetic medium, e.g. floppy or hard disk. OCR is not 100% accurate and so requires manual cleanup or fuzzy logic to make your OCRed relatively clean and thus retrievable through a search engine. See OPTICAL SCANNER.

Optical Combiner A passive device in which power from several input fibers is distributed among the smaller number (one or more) of output fibers.

Optical Computer A computer that uses photons, not electrons as in today's old-fashioned computers. Scientists think photon computers and photon switches could be a thousand times faster than present computers and switches, and, of course, totally impervious to electromagnetic interference.

Optical Connectors Connectors designed to terminate and connect either single or multiple optical fibers. Optical connectors are used to connect fiber cable to equipment and interconnect cables.

Optical Cross-Connect Panel A cross-connect unit used for circuit administration and built from modular cabinets. It provides for the connection of individual optical fibers with optical fiber patch cords.

Optical Disk Peripheral storage disk for programs and information. Optical disks are emerging as computer storage devices because of their tremendous storage capacities in comparison to magnetic disk. Optical disks are WORM — which stands for Write Once, Read Many Times, meaning you write once to the disk, and read that information many times. But with today's technology, you can't erase the information on the optical disk. It is not uncommon to be able to store 600 megabytes of information on an optical disk the size and shape of one you currently buy recorded music on. See OPTICAL STORAGE DEVICE.

Optical Fiber Any filament made of dielectric materials, that guides light, whether or not it is used to transmit signals. Optical fiber is an almost ideal transmission medium. It has these advantages:
- Transmission losses are very small.
- Bandwidth is greater than any other transmission medium we know of today. And we have no idea what the theoretical bandwidth of a strand of fiber might be.
- Fiber is immune to electromagnetic interference. This means it can operate in hostile or hazardous environments, like on the factory floor, in elevators shafts, on battleships, etc.
- Fiber does not radiate. You can't place a receiver next to it and figure out what's going on the fiber, as you can with cable.
- You can put many strands of fiber carrying much information in the same bundle and they won't interfere with each other, i.e. there won't be any significant cross-talk between the adjacent fibers.
- Its raw material, silica (sand) is the second most abundant element on earth.

The insulating properties of the glass fiber produces the only major disadvantage. Metallic conductors need to be included to power repeater amplifiers needed over long fiber runs. However, these can form the strength members that are necessary to aid the laying of fiber cables. Typical fiber applications range from use in local area network sections, where there is a high degree of electrical interference, to trans-oceanic telecommunications cables.

An optical fiber not carrying signals is typically called a "dark fiber." There are two types of optical fiber: Single Mode and Multimode. In single mode fiber, light can only take a single path through a core that measures about 10 microns in diameter. A micron is one millionth of a meter. Multimode fibers have thicker cores — typically 50 to 200 microns. Single mode fiber is more efficient. It offers low dispersion, travels great distances without repeaters and has enormous information-carrying capacity. The relatively large core of multimode fiber lightguide allows light pulses to zig-zag along many different paths. It's also ideal for light sources larger than lasers, such as LEDs (Light Emitting Diodes). Multimode fiber is not the preferred method of optical telecommunications any longer. See also FIBER, OPTICAL FIBER CABLE and OPTICAL SPECTRUM.

Optical Fiber Cable A transmission medium consisting of a core of glass or plastic surrounded by a protective cladding, strengthening material, and outer jacket. Signals are transmitted as light pulses, introduced into the fiber by light transmitter (either a laser or light emitting diode). Low data loss, high-speed transmission, large bandwidth, small physical size, light weight, and freedom from electromagnetic interference and grounding problems are some of the advantages offered by optical fiber cable. There are five common types: single, dual, quad, stranded, and ribbon. See OPTICAL FIBER.

Optical Fiber Facility Transmission system which uses glass fibers as the transmission medium. See OPTICAL FIBER.

Optical Fiber Preform Optical fiber material from which an optical fiber is made, usually by drawing or rolling.

Optical Fiber Ribbon A cable of optical fibers laminated in a flat plastic strip.

Optical Fiber Splice A permanent joint whose purpose is to couple optical power between two fibers.

Optical Fibre British spelling. In an optical fibre transmission system, the data is carried by pulses of light along glass fibres. This method of transmission has a much higher bandwidth than copper cables and is less subject to distortion and interference. It is safer than copper because it provides electrical isolation and as it carries no current can be used in flammable areas. See OPTICAL FIBER for a longer explanation.

Optical Interconnection Panel An interconnection unit used for circuit administration and built from modular cabinets. It provides interconnection for individual optical fibers. Unlike the optical cross-connect panel, the interconnection panel does not use patch cords.

Optical Receiver An optoelectric circuit that converts an incoming optical signal to an electrical signal.

Optical Repeater In an optical fiber communication system, an optoelectronic device or module that receives a signal, amplifies it (or, in the case of a digital signal, reshapes, retimes, or otherwise reconstructs it) and retransmits it. See also OPTICAL AMPLIFIER.

Optical Scanner A hardware device that recognizes images on paper, film and other media and converts them into digital form which can be stored in a conventional computer readable magnetic medium, such as floppy or hard disk. Optical scanners are getting better and better but are still not perfect. See OCR.

Optical Spectrum Generally, the electromagnetic spectrum within the wavelength region extending from the vacuum ultraviolet at 1 nm to the far infrared at 0.1 mm. The term was originally applied to that region of the electromagnetic spectrum visible to the normal human eye, but is now considered to include all wavelengths between the shortest wavelengths of radio and the longest of X-rays. See OPTICAL FIBER.

Optical Storage Device Optical storage devices use a source of coherent light — usually a semiconductor laser — to read and write the data. There are three big advantages to using a laser — size, safety and portability. Because you can focus a laser into approximately one micron in size — a far smaller area for encoding a bit of data than conventional drives — you can fit more data in.

Optical media are also more stable than metal-oxide disks. They aren't affected by light, normal temperatures or electromagnetic fields. (You can put through as many airport x-ray machines as you wish.) And best, the read/write head doesn't get as close to the recording medium as it does in conventional disk drives. Optical drives are interchangeable, also. You can remove them and store them. That makes them great for archiving.

Optical Time Domain Reflectometer OTDR. A device that measures distance to a reflection surface by measuring the time it takes for a lightwave pulse to reflect from the surface. Reflection surfaces include the ends of cables and breaks in fiber.

The reflectometer measures the ratio of incident and reflected light power. By using this device you can figure precisely where a fiber optic link is broken. This device operates like Radar. It sends a light pulse down the cable and waits for it to return. It measures the time taken and calculates the distance based on the speed of light through the fiber optic cable. The reflectometer usually also displays the reflected waves on a time axis for precise reading of, e.g., the leading edges of the transmitted and reflected waves. The reflectometer is also capable of launching a light pulse into the fiber optic transmission medium and measuring the time required for its reflection to return by backscattering or end reflection, thus indicating the continuity, crack, fracture, break, or other anisotropic features of the medium.

Optical Transmitter An optoelectric circuit that converts an electrical signal to an optical signal.

Optical Virtual Tributary Group OVTG contains four DS-1 signals packaged into an optical SONET virtual tributary, or VT.

Optical Waveguide Technically, any structure that can guide light. Sometimes used as a synonym for optical fiber, it also can apply to planar light waveguides.

Optical Waveguide Connector A device whose purpose is to transfer optical power between two optical waveguides or bundles, and that is designed to be connected and disconnected repeatedly.

Optically Active Material A material that can rotate the polarization of light that passes through it. An optically active material exhibits different refractive indices for left and right circular polarizations (circular birefringence).

Opto-Electrical Transducer A device which converts electrical energy to optical energy and vice versa. Used as transmitters and receivers in fiber optic communications systems.

Opto-Electronics The range of materials and devices that generate and transmit light (lasers and light-emitting devices), amplify light (optical amplifiers), detect light (photodiodes), and control light (electro-optic circuits). Each of these functions requires electrical energy to operate and depends on electronic devices to sense and control this energy.

OPX Off Premise Extension. An extension or phone terminating in a location other than the location of the PBX. The station uses a line circuit out of the PBX. OPX is commonly used to provide a company executive with an extension off the PBX in his home. This way he can pretend to be at the office when he's really at home. He can also make toll calls and have them easily charged to the office.

Orange Book The common name for the U.S. Department of Defense's Trusted Computer System Evaluation Criteria (TSEC).

ORB Office Repeater Bay.

Order Entry A voice processing application which allows someone with a touchtone phone to buy something, i.e. enter their order.

Order Wire 1. A circuit used by telephone personnel for fixing, installing and removing phone lines.
2. Equipment and the circuit providing a telephone company with the means to establish voice contact between central office and carrier repeater locations.

Order Wire Circuit A voice or data circuit used by telephone company technical control and maintenance personnel for the coordination and control action relating to activation, deactivation, change, rerouting, reporting and maintenance of communication systems and services.

Originate Mode The "originate mode" sets the modem to begin a data phone call — i.e. dial the phone, listen for a carrier tone from a remote modem and connect to that modem. The modem at the receiving end must be set to "Answer" mode. In any asynchronous data conversation, one side must be set to "Originate" and the other to "Answer." Such settings are usually made in software.

Originate/Answer The two modes of operation for a modem. Originate and answer states define the frequencies used to transmit and receive. In a two-way communication system, one modem must be set to originate and the other to answer.

Originating Restriction A phone line with this restriction cannot place calls at any time. Calls directed to the phone, however, will be completed normally.

Originator The user that is the ultimate source of a message or probe.

ORM Optically Remote Module. A type of switching module made by AT&T which connects directly to the 5ESS switch communications module via optical fibers.

Orphan A Windows NT term. A member of a mirror set or a stripe set with parity that has failed in a severe manner, such as a loss of power or a complete head crash. When this happens, the fault-tolerance driver determines that it can no longer use the orphaned member and directs all new reads and writes to the remaining members of the fault-tolerant volume.

Orphaned Member A Windows definition: A member of a mirror set or a stripe set with parity that has failed in a severe manner, such as a loss of power or a complete head crash. When this happens, the fault-tolerance driver determines that it can no longer use the orphaned member and directs all new reads and writes to the remaining members of fault-tolerance volume.

OS 1. Outage Seconds.
2. Operating System, as in MS-DOS (Microsoft Disk Operating System), Windows NT or OS/2. See OPERATING SYSTEM.
3. Operator Services. See OPERATOR SERVICES.
4. Operations System. Includes SCOTS, FMAS, etc.

OS/2 Operating System/2. An operating system originally developed by IBM and Microsoft for use with Intel's microprocessors and for use with IBM personal system/2 personal computers. Now OS/2 is the prime responsibility of IBM and it will run on many PCs, including those using the Intel family of PC microprocessors. OS/2 is a multitasking operating system. This means many programs can run at the same time. See OS/2 EXTENDED EDITION.

OS/2 2.0 A 32-bit version of the IBM's OS/2 operating system. Apple Macintosh's operating system is 32-bit.

Osborne Effect Once there was a personal computer company called Osborne Computer Company. One day, the president announced a revolutionary new computer. It was so good not one of his dealers wanted to (or could) sell the existing product and they sent all their inventory back. Meantime, it was six months before the company could deliver the new product. But without any sales in the meantime, it had no money and Osborne went broke. There is a lesson here for companies who are attempting to manage transition between old and new product lines. Be careful, or suffer the horrible consequences of The Osborne Effect.

Oscillator 1. A device for generating an analog test signal.
2. Electronic circuit that creates a single frequency signal.

Oscilloscope Electronic testing device that can display wave forms and other information on a TV-screen-like cathode ray tube. A basic fixture in sci-fi movies.

OSF Open Software Foundation (www.osf.org)

OSF/1 Version 1 of the Open Software Foundation's Unix-based operating system

OSI Open Systems Interconnection. The only internationally accepted framework of standards for communication between different systems made by different vendors. ISO's major goal is to create an open systems networking environment where any vendor's computer system, connected to any network, can freely share data with any other computer system on that network or a linked network. OSI was developed by the International Standards Organization. Most of the dominant communications protocols used today have a structure based on the OSI model. Although OSI is a model and not an actively used protocol, and there are still very few pure OSI-based products on the market today, it is still important to understand its structure. The OSI model organizes the communications process into seven different categories and places these categories in a layered sequence based on their relation to the user. Layers 7 through 4 deal with end to end communications between the message source and the message destination, while layers 3 through 1 deal with network access.

Layer 1 — The Physical layer deals with the physical means of sending data over lines (i.e. the electrical, mechanical and functional control of data circuits).

Layer 2 — The Data Link layer is concerned with procedures and protocols for operating the communications lines. It also has a way of detecting and correcting message errors.

Layer 3 — The Network layer determines how data is transferred between computers. It also addresses routing within and between individual networks.

Layer 4 — The Transport layer defines the rules for information exchange and manages end-to-end delivery of information within and between networks, including error recovery and flow control.

Layer 5 — The Session layer is concerned with dialog management. It controls the use of the basic communications facility provided by the Transport layer.

Layer 6 — The Presentation layer provides transparent communications services by masking the differences of varying data formats (character codes, for example) between dissimilar systems.

Layer 7 — The Applications layer contains functions for particular applications services, such as file transfer, remote file access and virtual terminals.

See also OSI STANDARDS.

OSI Model Open Systems Interconnection Model. See OSI.

OSI Network Address The address, consisting of up to 20 octets, used to locate an OSI Transport entity. The address is formatted into an Initial Domain Part which is the responsibility of the addressing authority for that domain and a domain-specific part which is the responsibility of the addressing authority for that domain.

OSI Presentation Address The address used to locate an OSI Application entity. It consists of an OSI Network Address and up to three selectors, one each for use by the Transport, Session, and Presentation entities.

OSI Standards The International Standards Organization (ISO) has established the Open Systems Interconnection (OSI). The idea of OSI is to provide a network design framework to allow equipment from different vendors to be able to communicate. Codex of Mansfield, MA. has published an excellent booklet called "The Basics Booklet of Local Area Networking". Here is a shortened excerpt of what Codex says about standards. "Standards allow us to buy items such as batteries and bulbs. Many of us have learned "the hard way" that the lack of computer standards can make it impossible for computers from different vendors to talk to each other. Because a major goal of a LAN (Local Area Network) is to connect varied systems, standards are being developed to specify the set of rules networks will follow. The OSI Model is a design in which groups of protocols, or rules for communicating, are arranged in layers. Each layer performs a specific data communications function. The concept of layered protocols is analogous to the steps we follow in making a phone call:

Step 1 — Listen for dial tone.
Step 2 — Dial a phone number.
Step 3 — Wait for a ring.
Step 4 — Exchange greetings.
Step 5 — Communicate message.
Step 6 — Say Good-bye.
Step 7 — Hang up.

Each of these steps, or OSI "layers," builds upon the one below it. Although each step must be performed in preset order, within each layer there are several options. Within the OSI model, there are seven layers. The first three are the PHYSICAL, DATA LINK, and NETWORK layers, all of which are concerned with data transmission and routing. The last three — SESSION, PRESENTATION and APPLICATIONS — focus on user applications. The fourth layer TRANSMISSION provides an interface between the first and last three layers. The X.25 PROTOCOL which created a standard for data transmission and routing is equivalent to the last three layers of the OSI Model. The OSI model is quickly becoming the standard for how LAN products should be built." See also OSI and X.25.

OSINet A test network sponsored by the National Bureau of Standards (NBS) designed to provide vendors of products based on the OSI model a forum for doing interoperability testing.

OSN Operations System Network.

OSP 1. Operator Service Provider. A new breed of long distance phone company. It handles operator-assisted calls, in particular Credit Card, Collect, Third Party Billed and Person-to-Person. Phone calls provided by OSP companies are often more expensive than phone calls provided by "normal" long distance companies, i.e. those which have their own long distance networks and which you see advertised on TV. You normally encounter an OSP only when you're making a phone call from a hotel or hospital phone, or privately-owned payphone. It's a good idea to ask the operator what the cost of your call will be before you make it.

2. On-line Service Provider, i.e. CompuServe, America On Line, Prodigy, MSN, and others.

OSPF Open Shortest Path First. A link-state routing algorithm that is used to calculate routes based on the number of routers, transmission speed, delays and route cost.

OSPFIGP Open Shortest-Path First Internet Gateway Protocol. An experimental replacement for RIP. It addresses some problems of RIP and is based upon principles that have been well-tested in non-internet protocols. Often referred to simply as OSPF. See OSPF.

OSPS An AT&T word for Operator Services Position System.

OSS Operations Support Systems. Methods and procedures (mechanized or not) which directly support the daily operation of the telecommunications infrastructure. The average LEC (Local Exchange Carrier) has hundreds of OSSs.

OSS7 Operator Services Signaling System Number 7.

OSTA The Optical Storage Technology Association. On September 25, 1995, the OSTA introduced their Universal Disk Format, which is an implementation of the ISO interchange standard for non-sequential recording of data — ISO 13346. www.osta.org

OTDR Optical Time Domain Reflectometer, a test and measurement device often used to check the accuracy of fusion splices and the location of fiber optic breakers. See OPTICAL TIME DOMAIN REFLECTOMETER.

OTGR Operations Technology Generic Requirements.

Other Common Carriers Providers of long distance telephone service in competition with AT&T. OCCs often (but not always) have lower rates than AT&T. All long distance carriers — including AT&T — are now called interexchange carriers.

OTOH Abbreviation for "On The Other Hand;" commonly used on E-mail and BBSs (Bulletin Board Systems).

OTS Operations Technical Support. See also OFFICE TELESYSTEM.

OUI Organizationally Unique Identifier: The OUI is a three-octet field in the IEEE 802.1a defined SubNetwork Attachment Point (SNAP) header, identifying an organization which administers the meaning of the following two octet Protocol Identifier (PID) field in the SNAP header. Together they identify a distinct routed or bridged protocol.

Out-Of-Band Network Management A method of managing LAN bridges and routers that uses telephone lines for communications between the network management station and the managed devices. This type of management is normally in addition to the conventional method which uses the LANs and WANs that are being connected by these devices. The principal advantage is that in the event of a system failure (which may take a LAN or a WAN down), a network supervisor can bypass the failed system and use a telephone link to reach a bridge/router to diagnose a network problem. Bridges and routers must have built-in telephone modems for this to work.

Out-Of-Band Signaling Signaling that is separated from the channel carrying the information — the voice, data, video, etc. Typically the separation is accomplished by a filter. The signaling includes dialing and other supervisory signals.

Out-Of-Frame In T-1 transmission, an OOF (Out Of Frame) error occurs when two or more of four consecutive framing bits are in error. When this condition exists for more than 2.5 seconds a RED alarms is sent by OOF detecting unit. Equipment receiving this RED alarm responds with a YELLOW alarm.

Out-Of-Order Tone A tone which indicates the phone line is broken.

Out-Of-Paper Reception The ability to receive a facsimile transmission into memory when the facsimile machine is out of paper. The facsimile paper will be printed when you put in new paper.

Out-Of-Service Or Used. A term used in the secondary telecom equipment business. Equipment taken from service. Can be in any condition. Expected to work and be complete. May not be.

Outage Service interrupted.

Outage Ratio The sum of all the outage durations divided by the time period of measurement.

Outgoing Access A CCITT description of the ability of a device in one network to communicate with a device in another network.

Outgoing Calls Barred A switch configuration option that blocks call origination attempts. Only incoming calls are allowed.

Outgoing Line Restriction The ability of the system to selectively restrict any outgoing line to "incoming only."

Outgoing Station Restriction The ability of the system to restrict any given phone from making outside calls.

Outgoing Trunk A line or trunk used to make calls.

Outgoing Trunk Queuing OTQ. Extensions can dial a busy outgoing trunk group, be automatically placed in a queue and then called back when a trunk in the group is available. This feature allows more efficient use of expensive special lines such as WATS or FX. Instead of having to redial the trunk access code until a line is free, the caller can activate OTQ. See also OFF-HOOK QUEUING.

Outgoing WATS An outgoing WATS (OUTWATS) trunk can only be used for outgoing bulk-rate calls from a customer's phone system to a defined geographical area via the dial-up telephone network. Originally WATS lines came in only lines that could receive calls or lines that could make calls. Now, you can buy a WATS line that handles both incoming and outgoing lines. See WATS.

Outlet A set of openings containing electrical contacts into which an electrical device can be plugged. See OUTLET TELECOMMUNICATIONS.

Outlet Telecommunications A single-piece cable termination assembly (typically on the floor or in the wall) and containing one or more modular telecom jacks. Such jacks might be RJ-11, RJ-45, coaxial terminators, etc.

Outlier An ATM term. A node whose exclusion from its containing peer group would significantly improve the accuracy and simplicity of the aggregation of the remainder of the peer group topology.

Outline Font Font is the design of printed letters, like the ones you see on this page. The first type was produced with raised metal or wooden blocks. Put ink on the blocks. Put paper on the inked blocks. Lift paper off. Bingo you have type on paper. Blocks came in fonts — styles of type, which has neat names like Times Roman, Helvetica, Souvenir, etc. Blocks also came in various sizes — 10 point, 12 point, 14 point, 36 point, etc. "Point" is simply the name for a way of measuring the size of type, like miles measure distance. When computers came along, they simply copied this technique. You picked type and you picked the size. Printers with print cartridges still work this way. They have to. They couldn't simply take one size font and enlarge or contract it because type enlarged or contracted doesn't look "right." Then two men, John Warnock and Martin Newell, said there had to be a better way and they came up with the idea of an outline font, originally called JaM, then Interpress and now PostScript. In PostScript letters and numbers become mathematical formulas for lines, curves and which parts of the character are to be filled with ink and which parts are not. Because they are mathematical, outline fonts are resolution independent. They can be scaled up or down in size in as fine detail as the printer or typesetter is capable of producing. PostScript outline fonts contain "hints" which control how much detail is given up as the type becomes smaller. This makes smaller type faces much more readable than they otherwise would be. Before outline fonts can be printed, they have to be rasterized. This means that a description of which bits to print where on the page has to be generated. And this is one reason printing outline fonts is so consuming of computer power (whether the power is in the computer or in the printer — usually it's in both). But it's also the reason why outline fonts, of which PostScript is the most successful and the most common, look so great.

Outpulse Dial A pushbutton dial which allows rotary dial users the convenience of "Touch-Tone" dialing. Pushing the buttons makes the phone pretend to be a rotary dial phone. This is necessary because touch-tones are not recognized everywhere.

Outpulsing The process of transmitting address information over a trunk from one switching center to another.

Output Data that flows out of a computer to any device.

Output Device A device by which a computer transfers its information to the outside world. For example, a printer.

Outside Link An ATM term. A link to an outside node.

Outside Node An ATM term. A node which is participating in PNNI routing, but which is not a member of a particular peer group.

Outside Plant The part of the telephone system that is physically located outside of telephone company buildings. This includes cables, supporting structures, and certain equipment items such as load coils. Microwave towers, antennas, and cable-system repeaters are not considered outside plant.

Outsourcing An allegedly more modern word for the term "facilities management." What outsourcing means is that a company contracts one of its internal functions out to an outside company. Those functions might include running the company's phone systems and telecom networks and/or running the company's computer system. A company might be motivated to do this because they lack the internal resources (typically people) or feel they can bring their phone costs into line and those phone costs (or at least certain of them) might now become able to be budgeted with some precision. This has appeal to senior management, who are trying to reduce their uncertainties. This usually has no appeal to lower level management who might be fired, especially if the new corporate outsourcing manager felt they were useless.
Here are some questions and answers about outsourcing taken from an information brochure from a company called Cyclix Communications Corporations, which does a lot of outsourcing:
Q: What is outsourcing?
A: Outsourcing is the process by which a company arranges for a third party to implement and manage a specific department or function of the company.
Q: Why would a company outsource?
A: A company would outsource for a number of reasons. In a recent Network World "Critical Issues Survey," two-thirds of the respondents noted lack of in-house resources, staff or expertise, while one-third noted cost effectiveness as reasons for outsourcing. Other reasons for outsourcing may include to cope with new technologies, to minimize capital expenses by having a third party provide, manage and maintain hardware and software, to acquire skilled personnel without having to train them or put them on the payroll and to remain focused on what the company can do best.
Q: What is an advantage of outsourcing?
A: The advantages of outsourcing will largely be dependent upon the objectives which were predetermined before the arrangement begins. For example, if a relatively small MIS department is responsible for both data processing and data networking for a national company, outsourcing one of those functions to an experienced vendor will allow the internal department to concentrate more fully on the other function.
Q: What is a disadvantage of outsourcing?
A: Once you decide to outsource, the transition back to managing those functions in-house can be costly. Companies must realize that outsourcing is a long term financial and competitive strategy. To simply try outsourcing creates false expectations and limits the long-term benefits which can be derived from outsourcing.

Outward Restriction Phone lines within the PBX can be denied the ability to access the exchange network without the assistance of the attendant. Restricted calls are routed to intercept tone.

Outward Trunk Queuing A process of holding outgoing long distance calls in queue until the appropriate long distance facility is available. See OFF-HOOK QUEUING.

OutWATS Outward Wide Area Telephone Service. See WATS.

OV Ground. See GROUND START.

Overbuild Adding radio capacity to a telecommunications network.

Overdrive Processor Intel's name for its line of single-chip performance upgrade chips. Based on Intel486 DX2 "speed doubling" technology, the Intel overdrive processors allow users of Intel486SX systems to double the internal speed of their computer's CPU by adding a single chip, without upgrading or modifying any other system components.

Overfloor Duct Method A distribution method that uses metal or rubber ducts to protect and conceal exposed wiring across floor surfaces.

Overflow Additional traffic beyond the capacity of a specific trunking group which is then offered to another group or line. For example, overflowing calls from WATS lines to (DDD) direct distance dial lines.

Overflow Load The part of an offered load that is not carried. Overflow load equals offered load minus carried load.

Overflow Tie-Line Enhancement A call center term. Using Overflow Tie-Line Enhancement, non-ISDN calls diverted to an overflow call center now convey the city-of-origin announcement prior to being connected to an agent.

Overflow Traffic The part of the offered traffic that is not carried, for example, overflow traffic equals offered traffic minus carried traffic.

Overhead In communications, all information, such as control, routing and error-checking characters, that is in addition to user-transmitted data. Includes information that carries network status or operational instructions, network routing information, as well as retransmissions of user-data messages that are received in error.

Overhead Bit A bit other than one containing information. It may be an error checking bit or a framing bit. See OVERHEAD BITS.

Overhead Bits Overhead bits are bits assigned at the source. They are transmitted with the information payload and are for functions associated with transporting that payload.

Overlay 1. Typically a piece of cut-out cardboard, which you place over several keys on a phone or console. When you punch in a certain code, the buttons become what's written on the programming overlay. Also called a Programming Overlay.
2. The ability to superimpose computer graphics over a live or recorded video signal and store the resulting video image on videotape. It is often used to add titles to videotape.
3. A Northern Telecom definition: Generally used to describe some software that is not always memory resident. It is loaded on request. In the Meridian 1 most configuration, administration and maintenance functions are done from a tty terminal using various overlays. Each overlay is designed for a specific task: For example, Overlay 10 is used to configure PBX (500/2500) sets, Overlay 11 is used to configure proprietary sets. Overlay 17 is used to configure I/O ports, Overlay 15 is used to configure customer data, Overlay 48 is used to configure link maintenance, and so on. The Meridian 1 has some 100 overlays. An overlay is loaded from the tty by typing LD nn where is the overlay number. Overlays are exited by typing ●●●●./

Overlay Network A separate network for a particular service covering most of the same geographical locations as the basic telephone network, but operating independently.

Overload Control How a system responds to being overstressed is called Overload Control. When a system is overloaded, frequently there are so many extra events being processed that the system's actual capacity or throughput goes down. Even though it may be rated at, say 10,000 busy hour calls, when overloaded, for example, with 11,000 calls, the computer telephony system may be only able to process only 8,000 calls.

Overload Management An AT&T term for handling peak demands by selectively delaying, degrading, or dropping only those portions of traffic flow that are tolerant of those particular types of impairments.

Overload Protection An uninterruptible power supply (UPS) definition. This feature automatically shuts the unit off when the battery is overloaded in order to protect it against overload damage.

Override When a circuit already in use is seized. For example, when your boss can break into your telephone conversation.

Override Prompts The ability of callers and users to key over system prompts.

Overrun Loss of data because the receiving equipment could not accept the data at the speed at which it was being transmitted.

Oversampling Time division multiplexing (TDM) technique where each bit from each channel is sampled more than once.

Overscan The image fills the screen from bezel to bezel. The bezel is the metal or plastic part — in short, the frame — that surrounds a cathode ray tube — a "boob" tube. NEC's term for overscan is "full scan."

Overspeed Condition in which the transmitting device runs slightly faster than the data presented for transmission. Overspeeds of 0.1% for data PABXs are typical.

Oversubscription In frame relay, this refers to a port connection state where the sum total of the committed information rates for the permanent virtual circuits supported by that port is greater than the port connection speed. While the port connection cannot simultaneously support transmissions at greater than its connection speed, its ability to dynamically allocate traffic makes oversubscription a cost-effective way to gain a high level of direct network interconnectivity.

Overtime Period Those minutes of use of a telephone service beyond the initially defined period for which a basic charge is quoted. The initial period on many calls is one or three minutes. After that, the next minute is overtime.

Overview Proteon's architecture for products conforming to SNMP.

Overwrite A call center term. To replace the contents of something — a file or record, for example — with new data.

OVS Open Video Systems, the successor to Video Dialtone.

OVTG See Optical Virtual Tributary Group.

Owner Microsoft jargon for person in charge.

OWS Operator Workstation. See OPERATOR WORKSTATION.

OWT Operator Work Time.

P-Code The Precise or Protected Code. A very long sequence of pseudo random binary biphase modulations on the GPS (Global Positioning System) carrier at a chip rate of 10.23 MHz which repeats every 267 days. Each one week segment of this code is unique to one GPS satellite and is reset each week.

P-Frame Predictive framing, specified by the MPEG Recommendation. Pictures are coded through a process of predicting the current frame using a past frame. The picture is broken up into 16x16 pixel blocks and each block is compared to a previous frame's block that occupies the same vertical and horizontal position.

P-MAC Packet Media Access Control. An FDDI-II term. See FDDI-II.

P-Phone Enhanced Business Service (also known as P-Phone) is an analog Centrex offering provided by Northern Telecom. It operates over a single-pair subscriber loop., providing normal full duplex audio conversations and a secondary 8 KHz half-duplex amplitude shift-keyed signal, which is used to transmit signaling information to and from the Northern Telecom-equipped central office.

P/AR Peak to Average Ratio. A standard analog transmission-line test signal of varying frequencies and amplitudes, which is then compared with the received signal. Composite results are a weighted number from 1 to 100 being the maximum. The P/AR is used increasingly as a standard quick test of a telecommunications channel comparative quality. Per Bell standard, the minimal acceptable P/AR rating for medium-speed data transmission is 48.

P.139 Serial Bus The P.139 Serial Bus is promulgated by the IEEE (202-371-0101). Initially it will operate at 100 million bits per second. But it has extensions for 200 and 400 megabits per second. The cable topology allows for branching and daisy-chaining of the peripherals attache, with a maximum of 32 4.5 meter hops using the baseline copper cable. That cable is a specially-designed combination consisting of two shielded twisted pairs for clock and data, two wires for power and ground and an overall shield that surrounds the twisted pairs and power and ground wires. See also USB, which stands for Universal Serial Bus.

P0 Protocol The protocol for messaging headers used for undefined messaging in an X.400 MHS (Message Handling System). Not officially part of the X.400 standards but became the U.S. intercept solution to P35 and X.435.

P01, P.01, Pnn or P.nn The Grade of Service for a telephone system. The digits following the P, i.e. nn, indicate the number of calls per hundred that are or can be blocked by the system. It is a goal or a measure of an event. In this example, P01 (also spelled P.01) means one call in a hundred (i.e. one divided by 100) can be blocked, so the system is designed to meet this criterion. See GRADE OF SERVICE and TRAFFIC ENGINEERING.

P1 Protocol P1 is the protocol defined in the X.400 standard used between MTAs (Message Transfer Agents) for relaying messages. Defined as the envelope, P1 is a portion of an X.400 message the identifies the message originator and potential recipients, records information about the message's path through the MTS (Message Transfer Service), directs the message's subsequent movement through the MTS, and includes characteristics of the message's contents. The composition of an envelope changes as the message is submitted, relayed and delivered.

P2 Format The 1984 defined message format, or protocol, used between cooperating User Agents in the IPM (Interpersonal Messaging System). The P2 heading is a component of an IP-message that indicates such items as the originator, recipients(s) and subject of the message.

P3 Protocol X.400 standard protocol used between a RUA (Remote User Agent) and an MTA (Message Transfer Agent) and an MTA (Message Transfer Agent) or between a remote MS (Message Store) and an MTA. P3 may be used optically between a co-resident UA (User Agent) and MTA.

P5 Protocol The 1984 X.400 protocol used between a MTA (Message Transfer Agent) and a Teletex Unit. This protocol is not popular.

P6 Intel's successor to the Pentium processor. The P6 is now called the Pentium Pro.

P7 Protocol The protocol used between a UA (User Agent) and a MS (Message Store). Defined in the 1988 X.400 standard.

P22 Format An enchanted version of the P2 format that appeared in the 1988 X.400 standard.

P35 Format The EDI (Electronic Data Interchange) message header enhancement to X.400 that enables EDI-specific addressing, routing and handling of EDI messaging. Defined in 1990 in the X.435 standard.

P6 Intel's name for the next generation chip to follow the Pentium. Among its specifications: four integer execution units, 32 kilobyte level one cache, 133 MHz clock, "dynamic execution" logic allowing chip to execute three instructions simultaneously. The first P6 should be about twice as fast as a 90MHz Pentium.

P802.11 P802.11 is a series of evolving standards in the wireless Local Area Network arena. The IEEE Program P802.11 Standards Working Group aims to define universal protocols for wireless LANs in the 900 MHz, 2.4 GHz and infrared frequency bands. Main components in the 802.11 standard are twofold: 1. The physical specifications for medium-dependent protocols. There are different physical specifications for each frequency band supported in 802.11. 2. The Medium Access Control (MAC) specifications for ad-hoc wireless networks and wireless network infrastructures. A single medium-independent MAC protocol provides a unified network interface between different wireless PHYs (Physical Specifications) and wired networks.

PA Public Address. Loud speaker system, sometimes used for paging.

PABX Private Automatic Branch eXchange. Originally, PBX was the word for a switch inside a private business (as against one serving the public). PBX means a Private Branch Exchange. Such a "PBX" was typically a manual device, requiring operator assistance to complete a call. Then the PBX went "modern" (i.e. automatic) and no operator was needed any longer to complete outgoing calls. You could dial "9." Thus it became a "PABX." Now all PABXs are modern. And a PABX is now commonly referred to as a

"PBX." Some manufacturers have tried to make their PBX appear different by calling it something else. Rolm/IBM calls theirs the "CBX" (Computerized Branch Exchange). Some others call theirs the "EPABX" (the Electronic Private Automatic Branch Exchange. Then there are the special ones, like Wang's WBX, SRX's SRX, NEC's IMS (Information Management System), etc. See PBX.

PAC Personal Activity Center. A combination IBM PC clone, alarm clock, answering machine, speakerphone, fax machine, modem, compact-disk player and AM/FM radio all rolled into one unit sitting on your desk.

Pac Bell Pacific Bell, one of the subsidiaries of Pacific Telesis.

Pacific Telesis One of the seven, now-independent Regional Holding Companies formed at the Divestiture of AT&T of 1983. It holds Pacific Bell, Nevada Bell and several non-regulated subsidiaries. In 1996, Southwestern Bell announced that it was buying Pacific Telesis.

Pacing Controlled rate of flow dictated by the receiving component, to prevent congestion. A method of flow control in IBM's SNA. See PACING GROUP.

Pacing Algorithm The mathematical rules established to control the rate at which calls are placed by an automatic dialing machine, also called a predictive dialer. See PREDICTIVE DIALER.

Pacing Control SNA term for flow control. See PACING GROUP.

Pacing Group In IBM's SNA, the number of data units (Path Information Units, or PIUs) that can be sent before a response is received. An IBM term for window.

Pack To compress data items so they take up less space. A process used by many database programs to remove records marked for deletion.

Packet A bundle of data, usually in binary form, organized in a specific way for transmission. Three principal elements are included in the packet: 1. Control information — destination, origin, length of packet, etc., 2. the data to be transmitted and 3. error detection and correction bits.

Packet Assembler/Disassembler PAD. A hardware/software combination that forms the interface between an X.25 network such as PDN and an asynchronous device such as a PC. The PAD generates call request, call clear, and other information packets in addition to the ones that contain user data.

Packet Buffer Memory set aside for storing a packet awaiting transmission or for storing a received packet. The memory may be located in the network interface controller or in the computer to which the controller is connected. See BUFFER.

Packet Burst Protocol A protocol built on top of IPX that speeds the transfer of NCP data between a workstation and a NetWare server by eliminating the need to sequence and acknowledge each packet. With packet burst, the server sends a whole set (or burst) of packets before it requires an acknowledgement.

Packet Controller The hub of the AT&T ISDN system. It acts as a fast packet switch providing virtual circuit services to the devices hooked to the system.

Packet Driver The specification developed by John Romkey at FTP Software to allow TCP/IP and other transport protocols to share a common network interface card. Packet Drivers have been written for a variety of network interface cards, and in many cases provide NetWare compatibility.

Packet Filter The ability to search a packet to determine its destination and to then route it or block it accordingly. This ability helps to control network traffic.

Packet Filtering The recognition and selective transmission or blocking of individual packets based on destination addresses or other packet contents.

Packet Forwarding Copying the packet to another node without looking at the destination address.

Packet Handler Function The packet switching function within an ISDN switch, for the packet mode bearer service.

Packet Interleaving Refers to the process of multiplexing multiple incoming packets from multiple channels on to a single outgoing channel by sampling one or more packets from the first channel, then the next, and so on.

Packet Level In packet data networking technology, level 3 of X.25. Defines how user messages are broken into packets, how calls are established and cleared over the packet data network (PDN) and how data flows across the entire PDN. The packet level also handles missing and duplicate packets.

Packet Level Procedure PLP. A full-duplex protocol that defines the means of packet transfer between a X.25 DTE and a X.25 DCE. It supports packet sequencing, flow control (including maintenance of transmission speed), and error detection and recovery.

Packet Overhead A measure of the ratio of the total packet bits occupied by control information to the number of bits of data, usually expressed as a percent.

Packet Radio Packet Radio is the transmission of data over radio using a version of the international standard X.25 data communications protocol adapted to radio (AX.25). It takes your information, and breaks it up into "packets" which are each sent and acknowledged separately. This assures error-free delivery from sender to receiver. A packet is a stream of characters consisting of a header, the information the user is sending, and a check sequence. The header gives the destination call sign, the call sign of the sender, and any digipeaters (digital repeater) call signs that will be used for relaying the packet. The check sequence makes certain that the data received is what was sent.

Packet Size The length of a packet, expressed in bytes.

Packet Switching Sending data in packets through a network to some remote location. The data to be sent is subdivided into individual packets of data, each packet having a unique identification and each packet carrying its own destination address. This way each packet can go by a different route. The packets may also arrive in a different order than how they were shipped. The packet ID lets the data be reassembled in proper sequence. Packet switching is a very efficient method of moving digital data around. It is not useful for voice, yet, though experiments are underway.

Packet Switching Network A network designed to carry data in the form of packets. See PACKET SWITCHING.

Packet Tracing The monitoring and reporting a particular packet addresses or types for diagnostic purposes.

Packet Type Identifier In packet data networking technology, the third octet in the packet header that identifies the packet's function and, if applicable, its sequence number.

Packetized Voice First read the definition of "Packet Switching" just above. The idea is to digitize voice and then slice it up into packets and sent those packets from the sender by various routes and assemble them as they get to the receiver. Packet switching for data makes sense. Packet switching for voice has not made sense because the voice is too sensitive to small delays. Yet the logic and the economics are there. And experiments continue.

PacketNet Sprint's internal X.25 Packet Network.

PACT Siemens' PBX And Computer Teaming. It defines protocols between Siemens PBXs and external computers.

Pad 1. A device inserted into a circuit to introduce loss.
2. Packet Assembler/Disassembler. A device that accepts characters from a terminal or host computer and puts the characters into packets that can be handled by a packet switching network. It also accepts packets from the network, and disassembles them into character streams that can be handled by the terminal or host.

Pad Characters In (primarily) synchronous transmission, characters that are inserted to ensure that the first and last characters of a packet or block are received correctly. Inserted characters that aid in clock synchronization at the receiving end of a synchronous transmission link. Also called fill characters.

Pad Switching A technique of automatically cutting a transmission loss pad into and out of a transmission circuit for different operating conditions.

PAF File A British term. Post Office Address file, a publicly available data file that, when integrated with an application, links postcodes to full addresses. When using a PAF file, an agent can save time by entering only the postcode. The PAF file automatically inserts post town, street and country.

Page A chunk of information, like a document or file, on the Web. A hypermedia document as viewed through a World Wide Web browser. Pages are the way you make information available on the Web. They can contain text, black and white and color photographs, audio and video.

Page Zone A local area in the office that can receive directed Page announcements independently of the remainder of the office.

Pager A small one-way receiver you carry with you. When someone wants you, they make your pager receiver sound an alarm. They can activate your pager in a number of ways, including dialing your pager digits directly into a computer, or giving your name and pager number to an operator who then punches out your numbers. Pagers have become small, cheap and very reliable. Monthly service costs have dropped and the area which you can be paged in most areas has widened dramatically. With some pagers you can be reached in most major metropolitan areas of the US. Some pagers also display small alphanumeric messages — like phone numbers to call and names of babies born. All this has made the pager far more useful. See also PAGING.

Paging To give a message to someone who is somewhere, but where we don't know. Paging can be done with a little "beeper" carried in her purse or on his belt. Paging can also be done through speakers in phones or from speakers in the ceiling. Most phone systems offer a paging channel access. You dial that number and page your party. The system comes "live" and your voice is heard everywhere. Similar to the stuff they have at airline terminals. Paging systems as an accessory to phone systems always cost extra. They're one of the most valuable features on a phone system. Don't skimp, however, on the quality of the speakers or the power of the paging system. If you do, your system will sound awful and people will not use it. See PAGER.

Paging Access, Rapid Your attendant and you or anybody else with an extension off your PBX can make a page (access the paging equipment) by pushbuttoning one or several digits. Sometimes you can dial one several numbers for different paging alternatives. One number gives you the tenth floor. One number gives you the fourth floor, etc. And one number to page everyone — also called an "ALL CALL" page. See also RAPID PAGING ACCESS.

Paging By Zone By dialing the appropriate access code, any phone is able to selectively page "groups" of pre designated phones or speakers.

Paging Code Call Access A feature of the ROLM Attendant Console which offers direct, one-touch access to the paging or code call features.

Paging Speakers Speakers in the telephone. Also external units located in ceilings, on walls, etc.

Paging Total System Upon dialing the appropriate special code, any station may make a paging announcement through all the loudspeakers.

Paid Call The usual type of a toll telephone call automatically billed to the calling telephone number.

Paid Hours A call center term. The time that an employee is either on duty-handling calls, doing other work, in meetings, etc., or on a paid schedule exception, such as an excused absence. In TCS's TeleCenter System this is calculated as scheduled hours minus any unpaid schedule exceptions that occur within those scheduled hours. TCS makes manpower scheduling packages. It is based in Nashville, TN.

Paintmonkey Someone with a less-than-glamorous, entry-level computer graphics job. A paintmonkey may spend months on a nanosecond of digitized film footage, painting mattes, or doing monotonous touch-ups.

Pair The two wires of a circuit. Those which make up the subscriber's loop from his office to the central office.

Pair Gain The multiplexing of x phone conversations over a lesser number of physical facilities. Pair gain usually refers to electronic systems used in outside plant — from the central office to the subscriber's premises. In "pair gain" you might do something as simple as take one pair of wires and carry two conversations on it. You might also take two pairs and carry 128 conversations. "Pair gain" is actually the number of conversations you get minus the number of wire pairs used by the system. AT&T Technologies has various subscriber pair gain devices called "SLC" (pronounced "slick" and standing for Subscriber Loop Carrier Systems). Other companies, like Rockwell, have comparable systems. The more circuits these devices produce, i.e. the more cable pairs they save you, the more they cost. The cost of subscriber pair gain equipment — like all electronics — has been dropping in recent years, reducing the phone company's need to install outside cable and thus making better use of the presently installed cable. T-1 is a type of subscriber pair gain equipment.

Paired Cable A cable in which all conductors are arranged in twisted pairs. This form of cable is the most common for communications.

PAL 1. Programmable Array Logic.
2. Proprietary ALgorithm. A designation for a privately designed and owned intelligence-based electronic method for performing a task (such as voice compression).
3. Phase Alternation Line. PAL is the format for color TV signals used in the United Kingdom, West Germany, Holland, much of the rest of western Europe, several South American countries, some Middle East and Asian countries, several African countries, Australia, New Zealand, and other Pacific island countries. It stands for Phase Alternating Line system. It was invented in 1961 and is used in England and many other European countries. With its 625-line scan picture delivered at 25 frames/second (primary power 220 volts), it provides a better image and an improved color transmission over the US system called NTSC, which uses interlaced scans and 525 horizontal lines per frames at a rate of 30 frames per second. SECAM, Sequential and Memory, is used in France and in a modified form in Russia. SECAM uses an 819-line scan picture which provides better resolution than PAL's 625-line and NTSC's 525. All three systems are not compatible. You cannot view an Australian or English videotape on a US TV. See PAL-M.

PAL-M A modified version of the phase-alternation-by-line (PAL) television signal standard (525 lines, 50 hertz, 220 volts primary power), used in Brazil. See also NTSC, PAL, SECAM.

Palette In some programs, a palette is a collection of drawing tools, brush widths, line widths and colors. In other programs it is the part of the color lookup table that determines the number and type of colors that will be displayed on the screen.

PALS A standard library database interface. An Internet term.

PAM Pulse Amplitude Modulation. Process of representing a continuous analog signal (a voice conversation) with a series of discrete analog samples. This concept is based on the information theory which suggests that the signal can be accurately recreated from a sufficient sample. Why bother? Sampling allows several signals to then be combined on a channel that otherwise would only carry one telephone conversation. PAM was used as part of a method of switching phones calls in several PBXs. It is not a truly "digital" switching system. PAM is the basis of PCM, Pulse Code Modulation. See PCM and T-1.

PAMA Pulse Address Multiple Access. Where carriers are distinguished by their time and space characteristics simultaneously.

Pancake Coil A type of inductance having flat spiral windings. An old radio definition.

Panel, Patch See PATCH PANEL.

Panel Office A very early type of central office switch.

Panel System Workstation defined by thin panels that provide privacy and insulation from noise.

Panne Fatale An equipment crash in Italian.

PANS Pretty Amazing New Services/Stuff. PANS is a term coined to describe ISDN Capabilities which should eventually replace POTS. Contrast with POTS.

PAP Packet-Level Procedure. A protocol for the transfer of packets between an X.25 DTE and an X.25 DCE.X.25 PAP is a full-duplex protocol that supports data sequencing, flow control, accountability, and error detection and recovery.

Paper Sizes

US	Europe and Japan
A = 8 1/2" x 11"	A3 = 11.7" x 16.5"
B = 11" by 17"	A4 = 8.3" x 11.7"
C = 18" by 24"	A5 = 5.8" x 8.3"
D = 24" by 36"	B4 = 10.1" x 14.3"
E = 34" by 44"	B5 = 7.2" x 10.1" B6 = 5.1" x 7.2"

Paper Tape A long thin paper roll on which data is stored in the form of punched holes. Usually used as input to other systems. Many old-fashioned telex machines still use paper tape as their storage medium. Punch up the message on the paper tape, rewind the paper tape, call the distant telex machine, then start the paper tape containing the message. The primary benefit of paper tape is that you save on transmission line cost. The paper tape will run through at the maximum speed of the line, while a human operator typing manually would be slower. The disadvantage of paper tape is that you can't change the message once you've typed it. Magnetic medium — floppy disks, hard disks, bubble memory — are much more flexible. They are rapidly replacing paper tape, even on telex machines, or on personal computers, which are replacing telex machines as telex data entry devices.

Paper Tape Punch A device to physically punch holes in a roll of paper tape in order to store information.

Paper Tape Reader A device which translates the holes in coded perforated tape into electrical signals suitable for further handling. The reader may be attached to a keyboard-printer or it may be a free standing device.

PAR Positive Acknowledgement Retransmit.

Parabola A shape which can focus a microwave signal into one narrow beam. All satellite and microwave antennae are parabolic, not spherical.

Parabolic Reflector The technical name for a dish antenna shaped like a perfect parabola.

Paradigm An assumption about the ways things work. The word paradigm (pronounced par-a-dime) is typically used by people who want to sound a little more pompous and intellectual than you and I. A yuppie word. If you want to talk about how things are changing you can talk about a "paradigm shift." According to the Economist Magazine, Thomas Kuhn invented the notion of the paradigm shift to explain what happens in scientific revolutions. A revolution happens, his theory goes, not because of startling new facts, but because of a change in the overall way the universe is seen. After this shift, old knowledge suddenly takes on new meaning. A classic paradigm shift is the way we have changed our concepts of computing from mainframe centralized mainframe computing to distributed LAN-based computing.

Parallel Circuit A circuit in which electricity can take more than one path. Household wiring is the most common type of parallel circuitry. Every outlet is parallel with every other outlet on the same circuit-breaker. So, if a bulb blows in a light fixture on that circuit, all the other devices on that circuit will still function. Another explanation. Imagine three 1.5 volt batteries. With the batteries connected in parallel, the circuit will deliver 1.5 volts. Connected in series it will deliver 3 x 1.5 volts, or 4.5 volts. By contrast, a string of Christmas tree lights is strung in a series circuit, one long continuous circuit. Should one bulb go out, usually the rest will also. See also PARALLEL DATA.

Parallel Connection A connection in which the current divides, only a part of the total current passing through each device.

Parallel Data The transmission of bits over multiple wires at one time. This is usually accomplished by having one wire for each bit of an eight-bit byte going from a device, usually a computer, to another device, usually a printer. Thus the word "Parallel." Data transmission in parallel is very fast, but usually happens only over short distances (typically under 500 feet) because of the need for huge amounts of cable. In contrast, the other common method of data transmission, serial transmission, takes place over one pair of wires and is usually slower than parallel transmission, but can happen over much longer distances, especially using phone lines. Parallel data transmission does not happen on phone lines. See SERIAL DATA and see the APPENDIX.

Parallel Interfacing A method of interfacing peripherals to computers, usually printers. Not as common as RS-232-C serial interfacing.

Parallel Networks Parallel networks, or segregated networks, exist when a single network location supports more than one physical wide area network connection for the purpose of supporting one or more applications.

Parallel Port An output receptacle often located on the rear of a computer. Unlike serial, there is no EIA standard for parallel transmission, but most equipment adheres to a quasi-standard called the Centronics Parallel Standard. Almost every PC since the original IBM PC has come with an ordinary, 25-pin D-connector parallel port. These low-speed ports are fine for sending output to a printer (which is usually the slowest device in a computer system). But when transferring data between two PC parallel ports or using the parallel port as a method of getting to and from external hard disks, the speed is too slow. As a result, there have been a number of attempts to speed up and add intelligence to the lowly parallel port. How can you tell what type of parallel port you have? A free utility called PARA14.ZIP is available on the Internet at netlab2.usu.edu/misc, or on CompuServe in the IBMHW forum, Library2. See EPP (Enhanced Parallel Port), ECP (Extended Capabilities Port) and USB (Universal Serial Bus).

Parallel Processing 1. A computer technology in which several or even hundreds of low-cost microprocessors are linked and able to work on different parts of a problem simultaneously.
2. A computer performs two or more tasks simultaneously. This contrasts with multi-tasking in which the computer works fast and gives the impression of performing several tasks at once.

Parallel Sessions In IBM's SNA, two or more concurrently active sessions between the same two logical units (LUs), using different network addresses. Each session can have different transmission parameters.

Parallel Tasking Technology which allows LAN adapters to transmit data to the network before an entire frame has been loaded from the computer to the adapter's buffer and to transmit the data to the computer's main memory before an entire frame has been received from the network. In effect, a frame can reside on the network, the adapter and in the computer memory simultaneously, thus boosting throughput.

Parallel Transmission 1. Method of information transfer in which all bits of a character are sent simultaneously as opposed to serial transmission where the bits are sent one after another.
2. Method of achieving higher system reliability through use of completely redundant transmission facilities.

Parameters The record in a stored program control central office's data base that specifies equipment and software and options and addresses of peripheral equipment for use in call processing.

Parametric Equalizer A device for manipulating sound by boosting and cutting selected frequencies by specific amounts. Basically, a much more elaborate and precise version of the bass and treble controls found on stereo systems. See EQUALIZATION.

PARC Palo Alto Research Center. A laboratory owned by Xerox and populated by it in the 1970s with some of the most creative scientists of the day. It is legendary for having pioneered technologies ranging from the laser printer to the Ethernet local area network and the graphical user interface for PCs. Sadly, Xerox senior management didn't recognize what it had and didn't recognize the value of its inventions, and didn't exploit most of them. Fortunately, other companies did.

Parasite A radio tap that takes its power from the phone line.

Parent Node The logical group node that represents the containing peer group of a specific node at the next higher level of the hierarchy.

Parent Peer Group The parent peer group of a peer group is the one containing the logical group node representing that peer group. The parent peer group of a node is the one containing the parent node of that node.

Parity A process for detecting whether bits of data (parts of characters) have been altered during transmission of that data. Since data is transmitted as a stream of bits with values of one or zero, each character of data composed of, say seven bits has another bit added to it. The value of that bit is chosen so that either the total number of one bits is always even if Even Parity error correction is to be obeyed or always Odd if odd Parity error correction is chosen.
Here's an explanation (better, but longer) from The Black Box Corporation in Pittsburgh: Many asynchronous systems append a parity bit following the data bits for error detection. Parity bits trap errors in the following way. When the transmitting device frames a character, it counts either the number of 0s or 1s in the data bits and appends a parity bit that corresponds to whether or not the count in the data bits was even or odd. The receiving end also counts the data bit 0s or 1s as it receives them and then compares the computation to the parity bit. If an error is detected, a flag can be set and retransmission may be requested. When even parity is chosen, the parity bit is set at 0 if the number of 1's in the data bits is even and it is set at 1 if the number of 1's is odd. Conversely, odd parity sets the parity bit at 1 if the number of 1's in the data bits is even, and it is set at 0 if the number of 1's is odd. Other parity selections include mark, space or off. Mark parity always sets parity at 1. Space parity always sets parity at 0, and "off" tells the system to ignore the parity bit.

Parity Bit A binary bit appended to an array of bits to make the sum of all the bits always odd or always even. See PARITY and ASCII. See PARITY.

Parity Check A method of error-detection in binary data transmission whereby an extra bit is added to each group of bits (usually a character of data). If parity is to be odd, then the extra or parity bit is assigned either a one or zero so the total number of ones in the character will be odd. If the parity is even, the parity bit is assigned a value so that the total number of ones in the character is even. This way errors can be detected. See PARITY.

Park 1. A telephone system feature that (like many features) may mean different things depending on who created it. One definition of "park" is that I dial another extension and park the call at that extension. It doesn't ring. Then I go over to that extension and pick up the phone and I'll be speaking with whoever I parked over there. This feature is useful if I have to go to another phone to find some information the caller wants. There's another definition of the telephone system meaning of "park." You have a single line phone. You put that call on a variation of hold. Then you or anyone else can pick up any phone in that pickup group and you will have your parked call.
2. In the language of hard disks, "parking" means moving the read/write head to a safe area of the hard disk when you're ready to turn the disk off. "Parking" places the heads of a hard disk in a locked position so that the storage medium (i.e. the hard disk) will not be damaged during transit. This is useful because it keeps the head from bouncing on data areas of the disk and damaging the disk. Some hard disks have a program called "park" which you run before you turn off the machine. Others do it (self-park) automatically. All hard disks on laptops are self-parking. Most modern disks are. You ought to check. It's very important.

Park Timeout This is the period of time before an unanswered Call Park call is redirected to the Prime Phone for the line the call is on.

Part 68 Requirements Specifications established by the FCC as the minimum acceptable protection communications equipment must provide the telephone network. Meeting these requirements does not certify that equipment performs any task. Part 68 is the section of Title 47 of the Code of Federal Regulations governing the direct connection of telecommunications equipment and premises wiring with the public switched telephone network and certain private line services, e.g., foreign exchange lines (customer premises end), the station end of off-premises stations associated with PBX and Centrex services, trunk-to-station tie lines (trunk end only), and switched service network station lines (common control switching arrangements); and the direct connection of all PBX (or similar) systems to private line services for tie trunk type interfaces, off-premises station lines, automatic identified outward dialing and message registration. These rules provide the technical, procedural and labeling standards under which direct electrical connection of customer-provided telephone equipment, systems, and protective apparatus may be made to the nationwide network without causing harm and without a requirement for protective circuit arrangements in the service provider's network. Form 730 Application Guide is a collection of literature you'll need to register your telephone/telecom equipment under Part 68 of Title 47 at the Federal Communications Commissions. To get this material (it's free) drop a line or call the Federal Communications Commission, Washington DC 20554. FCC's general # is 202-418-0200. The Bureau Chief's # is 202-418-1500. As I write this edition, the person at the FCC in charge is Reed Hunt who also puts out a very useful newsletter for Part 68 applicants. The newsletter is called "The Billboard." You can file Form 730 yourself, but the Form 730 Application Guide also contains a list of Part 68 Certification Laboratories, a list of technical references and a list of reference sources.

Partial Meshed Network A type of wide area network topology in which every remote location is not connected directly to every other remote location, but instead is connected directly to a small subset of locations. It is a topology with more direct connectivity than a star configuration, but less direct connectivity than a fully meshed configuration.

Partially Perforated Tape Same as chadless tape. See CHADLESS TAPE.

Partition A division of memory or hard disk. For example, the MS-DOS and MS OS/2 operating systems can allocate hard disk space for one or more partitions, each of which behaves as a physically distinct hard disk.

Partitioned Emulation Programming Extension PEP. An IBM special software package that, with the Network Control Program (NCP), allows the same communications controller to operate in split mode, controlling an SNA network while at the same time managing a number of non-SNA communications lines. It was developed by IBM to facilitate migration of users to SNA.

Partitions Sections on a hard disk. You can divide your hard disk into as many as four partitions to run four operating systems. See also 32-MEGABYTE BARRIER.

Party A stupid word for the person making or receiving a phone call. Also now used in the airline business, as in "How many people will there be in your Party?"

Party Identification Identifying the person who is placing a call on a party line, i.e. phone line with several people sharing it. Often found in rural locations.

Party Line 1. Saying what your company or boss wants you to say.
2. A telephone line with several subscribers sharing its use.

Party Line Service Telephone service which provides for two or more phones to share the same loop circuit. Party line service, which is becoming less common, is offered in two-party, four-party and eight-party versions. Interestingly, there is a version of ISDN in which several subscribers do share the same ISDN line — but they would rarely be affected by it because of ISDN's specialized signaling and the two phone lines in its 2B+D bandwidth.

Party Line Stations Two party phone service can be expanded to support to multi-party service.

PAS A Northern Telecom word for Port Adapter System.

Pascal A programming language designed for general information processing and noted for its structured design.

Pass Through The process of accessing one device via another device. The intermediate device that sends backs the transmitted messages for testing.

Pass Window The range of frequencies used in a transmission system to transmit voice or data signals. More often referred to as bandwidth. See BANDWIDTH.

PassageWay The name of the AT&T device which connects to your PC's serial port on the one end and your AT&T proprietary phone on the other and allows you to control your phone and all its incoming and outgoing calls with Windows software on your PC. AT&T's idea behind the name PassageWay is that it unlocks the hidden PassageWay between your computer and your telephone on your Desk. Stephen Smith of AT&T is credited with creating the term PassageWay.

Passband The range of frequencies that can pass through a filter without being attenuated (i.e. stopped).

Passive Bus ISDN feature which allows up to six terminal devices and two voice devices (also called telephones) to simultaneously share the same twisted pair, each being uniquely identifiable to the switched ISDN telephone network. See ISDN.

Passive Contract In the software business, there are two types of contracts. One you sign and one you don't. A passive contract is the one you don't sign. A passive contract typically comes with an over-the-counter, shrink-wrapped software package and you execute it by breaking the seal on the package. The passive contract spells out terms and conditions you agree to — like not copying the software, not selling, etc.

Passive Coupler A coupler that divides entering light among output ports without generating new light.

Passive Headend A device that connects the two broadband cables of a dual-cable system. It does not provide frequency translation.

Passive Hub A device used in certain network topologies to split a transmission signal, allowing additional workstations to be added. A passive hub cannot amplify the signal, so it must be connected directly to a workstation or an active hub.

Passive Leg A leg of a call representing the communication path toward another user. Definition from Bellcore in reference to its concept of the Advanced Intelligent Network.

Passive Optical Network See PON.

Passive Reflector A simple reflector used to change the direction of radiation from a microwave beam. For example, a reflecting surface mounted on a hill top and so positioned as to direct the energy down onto a valley receiving site. See PASSIVE REPEATER.

Passive Repeater A passive reflector system constructed from two reflectors that are simply coupled together with a short length of waveguide. The first reflector acts as a receiver while the second transmits but in a different direction. There's no electronics in the system.

Passive Side When describing a loopback test, the passive side is used to identify the device that sends back the transmitted messages for testing.

Passive Splicing Aligning the two ends of a fiber without monitoring its splice loss.

Passive Star A star-topology local network configuration in which the central switch or node is a passive device. Each station is connected to the central node by two links, one for transit and one for receive. A signal input on one of the transmit links passes through the central node where it is split equally among and output to all of the receive links. Also called a star coupler or a retransmissive star.

Passthrough Gaining access to one network through another element. Also spelled PASS THROUGH.

Password A word or string of characters recognized by automatic means permitting a user access to a place or to protected storage, files or input or output devices.

Password Control Of Changes A feature that makes it impossible to alter the performance of a piece of equipment without first entering a password.

Patch 1. Instructions added to a software program, usually at the last moment, to get the program to run. The "patch" is usually not part of the original planning or logic of the program. Too many patches make it difficult to effectively manage the program over a long term. Patches, like the main programming, should be extensively documented.
2. To connect circuits temporarily with a jack and a cable. Patching is typically done on devices called PATCH BAYS, PATCHBOARDS or PATCH PANELS. See also PATCHING.

Patch Bay A collection of hardware put together in such a way that circuits appear on jacks and can be connected together for transmission, monitoring and testing. See PATCH PANEL.

Patch Cord A short length of wire or fiber cable with connectors on each end used to join communication circuits at a cross connect.

Patch Panel A device in which temporary connections can be made between incoming lines and outgoing lines. It is used for modifying or reconfiguring a communications system or for connecting devices such as test instruments to specific lines. A patch panel differs from a distribution frame in that the interconnections on a distribution frame are intended to be permanent.

Patchboard Same as a patch bay. See PATCH BAY.

Patching Means of connecting circuits via cords and connectors that can be easily disconnected and reconnected at another point. May be accomplished by using modular cords connected between jack fields or by patch cord assemblies that plug onto connecting blocks.

Patchmaster A patch panel in which multiple pair lines can be interconnected as a group.

Patent A patent is a grant, limited in time and technological extent, of the right to exclude others from making, using or selling the invention. The temporal extent of a U.S. patent (other countries' laws are similar) is usually at least 17 years. The technological extent of protection is defined by the claims of the patent. Claims are allowed only after examination by a technically degreed examiner of the Patent and Trademark Office to ensure that an opinion is defined that is nonobvious with respect to the predecessor technology, or "prior art." Having a patent does not confer the right to manufacture the thing claimed; others may have dominating patents. Independent invention is no defense to a charge of infringement. Rights in an invention made by an employee in performance of his normal duties belong to the employer in most states, even without a written agreement.

The claims made for the invention in applying for the patent may define a "process, machine, manufacture, or composition of matter". The great software patent controversy has largely been settled with the understanding that software is normally best claimed as a process, although sometimes claims usefully intermix hardware and software. The only software that remains per se unpatentable is that defining a process operating on pure numbers, that is, not tied to any particular end use, control process, or the like. A Fast Fourier Transform algorithm would thus not be patentable although an unobvious method of using it to (say) remove harmonic noise from a signal might be.

Patent serve several functions. A patent is a technical disclosure, forming part of the scientific literature, as the invention must be described with sufficient particularity that others can use it "without undue experimentation." The idea is that the inventor is given a limited monopoly in exchange for the benefit to the public of having inventions made and disclosed. A patent is also a legal document, in that the claims define the exact extent of protection. The third function of a patent is to market the invention. A well-written patent explains the underlying technical problem, the deficiencies of the prior art, and the way in which the invention solves these problems, in clear and non-technical language. Only then can a federal judge or jury, or the CEO of a competitor being asked to pay damages, be expected to understand the relation of the claimed invention to the prior art and the allegedly infringing product.

Path 1. The route a telecommunications signal follows through a circuit or through the air.

2. A description of the location of an MS-DOS directory or file within your computer. The idea is that the Path command shows DOS a "path" to find a file or command you wish to execute. The path name of a file consists of a drive letter, followed by a directory name, one or more subdirectory names and a filename. You use a backslash to separate each name from the previous one. The MS-DOS command may look like this

path c:\bin;c:\wp;c:\wfw;c:\glite;c:\ccm;c:\bin\compare

In this way, you can find an executable file which might be located in any of the above directories. Writing about Paths, Van Wolverton says "as its name implies, the PATH is essentially a trail that DOS follows to get somewhere. But, unlike trails you find on a map, the DOS Path isn't a trail that someone at Microsoft or IBM to lead DOS to a specific place. Instead, the DOS Path is a trail you define so that DOS can find external commands, program files and batch files.

DOS will always search the current subdirectory for any command you want. Then, if it finds nothing, it will follow the trail set in the path.

Path Clearance In through-the-air microwave transmission, you must find a line-of-sight path, free of obstruction of buildings, trees, other microwave towers, etc. In microwave line-of-sight communications, the perpendicular distance from the radio-beam axis to obstructions such as trees, buildings, or terrain. The required path clearance is usually expressed, for a particular k-factor, as some fraction of the first Fresnel zone radius. That's the technical definition of path clearance.

Path Constraint A bound on the combined value of a topology metric along a path for a specific connection.

Path Control IBM Corp.'s implementation of what is normally referred to as the network layer in the International Standards Organization Open Systems Interconnect (OSI) layered network architecture.

Path Control Layer In IBM's SNA, the network processing layer that handles primarily the routing of data units as they travel through the network and manages shared link resources.

PATI Comdial's new parallel port, TAPI card-in-a-box / software computer telephony device. PATI stands for PC and Telephone Interface.

Pattern Recognition A small element of human intelligence. The ability to recognize and match visual patterns. (Auditory pattern recognition is the ability to recognize spoken words.) Pattern recognition basically works by having the computer seek out particular attributes of the character (assuming it's pattern recognition for reading words) and then having the computer compare what it finds to what's in its database of patterns. By a process of breaking down letters into curves and lines, and by a process of elimination, the computer can figure out what it's seeing. As Forbes said, "think of pattern recognition as a kind of super detective, a tireless if unimaginative collector of clues, distinguished not by brilliance, but by ceaseless legwork.

Pause This feature on some phone systems — especially the Northern Telecom Norstar — inserts a 1.5 second delay in a dialing sequence on an outside line.

PAX Private Automatic eXchange. Typically an intercom system not joined to the public telephone system. PAXs are more common in Europe, where is it common for business people to have two phones on their desk — one for internal intercom calls and one for external calls.

Pay Phone See PAYPHONE

Payload From the perspective of a network service provider: of a data field, block or stream being processed or transported, the part that represents information useful to the user, as opposed to system overhead information. Payload includes user information and may include such additional information as user-requested network management and accounting information. In Sonet, the STS-1 signal is divided into a transport overhead section and an information payload section (similar to signaling and data). See SPE (Synchronous Payload Envelope) for a description of what would be found in the payload.

Payload Type Indicator Field PTI. A three-bit field in the ATM cell header that indicates the type of information being carried in the payload.

Payphone Used to be just a public phone that accepted only coins. Now pay phones can be coinless and can read credit cards. Soon they will be acquiring keyboards, computer screens and dataports for plugging in fax machines and portable computers. The payphone was invented by William Gray, an American whose previous inventions included the inflatable chest protector for baseball players. Mr. Gray's first phone lacked a dial. Its instructions read:

"Call Central in the usual manner. When told by the operator, drop coin in proper channel and push plunger down."

In today's nomenclature, Mr. Gray's original phone is known as a post-pay coin phone. See other entries below.

Payphone-Postpay Calls are paid for after they are completed, typically with a credit card or calling card, etc.

Payphone-Prepay At a coin phone, calls must be paid for before they can be dialed. Virtually all local calls are prepay.

Payphone-Private Referred to as Customer Owned Coin Operated Telephone Companies (COCOTs). Installed and maintained by companies other than local exchange carriers who are rapidly entering this industry. COCOTs may have access to more than one IXC.

Payphone-Public A coin phone installed in a "public" place. The local operating Company is totally responsible for its installation. The phone company will typically pay someone — the city, the bus station owner — a commission on the calls made from this phone. Also see PAYPHONE-SEMI-PUBLIC.

Payphone-Semi-Public A coin phone installed for public use but installed in a "semi-public" place, such as a restaurant or bar. The proprietor of the establishment is obliged to guarantee that the phone company will receive a minimum amount of money out of the phone. The phone company will typically not pay a commission on this type of phone and takes all the money in the coin box for itself. What is a "public" and what is a "semi-public" phone is a decision made by the local telephone company for whatever reason it chooses. The pay phone business is rapidly deregulating. It is now legal to own your own payphone.

Paystation, Postpay Calls are paid for after they are completed, typically with a credit or calling card, etc.

Paystation, Prepay Calls must be paid for before they can be dialed. Virtually all local calls are prepay.

Paystation, Public A coin phone installed in a "public" place. The phone company is totally responsible for its installation. The phone company will typically pay someone — the city, the bus station owner — a commission on the calls made from this phone. See PAYSTATION, SEMI-PUBLIC.

Paystation, Semi-Public A coin phone installed for public use but installed in a "semi-private" place, such as a restaurant or bar. The proprietor of the establishment is obliged to guarantee that the phone company will receive a minimum amount of money out of the phone. The phone company will typically not pay a commission on this type of phone and takes all the money in the coin box for itself. What is a "Public" and what is a "Semi- Public" phone is a decision made by the local telephone company for whatever reasons it chooses. The pay phone business is rapidly deregulating. So the rules are changing. And it is now legal to own your own payphone.

PBX Private Branch eXchange. A private (i.e. you, as against the phone company owns it), branch (meaning it is a small phone company central office), exchange (a central office was originally called a public exchange, or simply an exchange). In other words, a PBX is a small version of the phone company's larger central switching office. A PBX is also called a Private Automatic Branch Exchange, though that has now become an obsolete term. In the very old days, you called the operator to make an external call, except in Europe. Then later someone made a phone system that you simply dialed nine (or another digit — in Europe it's often zero), got a second dial tone and dialed some more digits to dial out, locally or long distance. So, the early name of Private Branch Exchange (which needed an operator) became Private AUTOMATIC Branch Exchange (which didn't need an operator). Now, all PBXs are automatic. And now they're all called PBXs, except overseas where they still have PBXs that are not automatic.

At the time of the Carterfone decision in the summer of 1968, PBXs were electro-mechanical step-by-step monsters. They were 100% the monopoly of the local phone company. AT&T was the major manufacturer with over 90% of all the PBXs in the U.S. GTE was next. But the Carterfone decision allowed anyone to make and sell a PBX. And the resulting inflow of manufacturers and outflow of innovation caused PBXs to go through five, six or seven generations — depending on which guru you listen to. (See my definition for GENERATIONS in this dictionary). Anyway, by the fall of 1991 when I wrote this fourth edition of this dictionary, PBXs were thoroughly digital, very reliable, and very full featured. There wasn't much you couldn't do with them. They had oodles of features. You could combine them and make your company a mini-network. And you could buy electronic phones that made getting to all the features that much easier. Sadly, by the late 1980s the manufacturers seemed to have finished innovating and were into price cutting. As a result, the secondary market in telephone systems was booming. Fortunately, that isn't the end of the story. For some of the manufacturers in the late 1980s figured that if they opened their PBXs' architecture to outside computers, their customers could realize some significant benefits. (You must remember that up until this time, PBXs were one of the last remaining special purpose computers that had totally closed architecture. No one else could program them other than their makers.) Some of the benefits customers could realize from open architecture included:

• Simultaneous voice call and data screen transfer.
• Automated dial-outs from computer databases of phone numbers and automatic transfers to idle operators.
• Transfers to experts based on responses to questions, not on phone numbers.

And a million more benefits. We discuss them at our annual trade show called TELECOM DEVELOPERS held in May each year. Call 212-691-8215 for more information. For more on open architecture, see OAI.

An alternative to getting a PBX is to subscribe to your local telephone company's Centrex service. For a long explanation on Centrex and its benefits, see CENTREX. Here are some of the benefits of a PBX versus Centrex:

1. Ownership. Once you've paid for it, you own it. There are obvious financial and tax benefits.
2. Flexibility. A PBX is a far more flexible than a central office based Centrex. A PBX has more features. You can change them faster. You can expand faster. Drop another card in, plug some phones in, do your programming and bingo you're live.
3. Centrex benefits. You can always put Centrex lines behind a PBX and get the advantages of both. In some towns, Centrex lines are cheaper than PBX lines. So buy Centrex lines and put them behind your PBX. Make sure you don't pay for Centrex features your PBX already has. (It has most.)
4. PBX phones. There are really no Centrex phones — other than a few Centrex consoles. If you want to take advantage of Centrex features, you have to punch in cumbersome, difficult-to-remember codes on typically single line phones. PBXs have electronic phones, often with screens and dedicated buttons. They're usually a lot easier to work. A lot easier to transfer a call. Conference another, etc. A lot more productive.
5. Footprint savings. Modern PBXs take up room, more than Centrex. But the space they take up is far less than it used to be. PBXs are getting smaller.
6. Voice Processing/Automated Attendants. Centrex's DID (Direct Inward Dialing) feature was always pushed as a big "plus." You saved operators. However, you can now do operator-saving things with PC-based voice processing and automated attendants you couldn't do five years ago. These things work better with on-site standalone PBXs than with distant, central office based Centrex. Moreover, virtually every PBX in existence today supports DID. You can dial directly into PBXs and reach someone at their desk just as easily as you can dial directly using Centrex.
7. Open Architecture. Most PBXs have open architecture. See OAI for the benefits. Central offices don't.
8. Good Reliability. There have been sufficient central office crashes and sufficient improvement in the reliability of PBXs that you could happily argue that the two are on a par with each other today. Both are equally reliable, or unreliable. The only caveat, of course, is that you back your PBX up with sufficient batteries that it will last a decent power outage. Of course, that assumes that your people will be prepared to hang around and answer the phones during a blackout.
9. Expansion. Central offices are big. Allegedly you can grow your lines to whatever size you want. In contrast, PBXs have finite growth. It's true about PBXs. But it's equally true about central offices. I've personally heard too many stories about central office line shortages to believe in the nonsense about "infinite Centrex" growth. Fact is central offices grow out, just like PBXs. Given the tight economy of recent years, local phone companies have not been buying the central offices they should have. And they

have been filling central offices up a little too tight for my taste.

10. Technological obsolescence. Allegedly central offices are upgraded faster than PBXs and therefore are always up to date technologically. It's nonsense. The life cycle of a typical central office was 40 years until recently. It's now around 20 years. Think of what's happened to PCs in the past 10 years — the IBM PC debuted only in 1981 — and you can imagine how obsolete many of the nation's central offices are.

PBX Central Office Trunk
PBX central office (CO) trunks connect the PBX switch to the central office serving the PBX location. The trunks appear as station lines at the central office equipment.

PBX Driver Profiles Telephony Services (also called TSAPI) is software which AT&T invented to run on NetWare servers and allow those servers to communicate with PBXs. According to a presentation made by Oliver Tavakoli of Novell in late August, 1994, "The purpose of TSAPI is to make it easier for developers to create telephony applications. Given the broad scope of TSAPI and the number of ways in which PBX vendors can package and deliver solutions, applications being developed against TSAPI are unlikely to work with a large cross-section of PBXs. This is particularly problematic for application developers wishing to add simple telephony features to a product that is not telephony centric. In order to spur the growth of CTI application development, Novell has created 'profiles' that are placed on top of TSAPI. The profiles listed in this document (i.e. the speech) are based on documents created by other organizations:

"- A NOTA document entitled PABX Driver NLM Conference Meeting and Draft Recommendations dated July 20th, 1994

"- An ECMA document entitled 'Technical report on CSTA Scenarios/Second Draft' dated July 1994.

"Each profile is made up of the following components: Functions (and matching confirmation events) included in the profile; unsolicited events that may be encountered in the event flow scenarios in the profile; event flow scenarios that attempt to describe what should occur in the common scenarios encountered when using the functions in the profile.

"Group A Profile: Using the functions and event flows in the Group A profile, an application should be able to: provide "screen pops" to the end user; make outbound calls originating at an end user's device; provide hands-free operation for answering the phone and making calls (assuming the PBX and end user's device support hands-free operation)

"Group B Profile: Using the functions and event flows in the Group B profile, an application should be able to: perform operations involving two calls on a single device (the connection to one of the calls is in the held state); perform operations involving more than two parties on the same call; obtain information and unsolicited events from a call-centric view (as opposed to a device-centric view).

"Group C Profile: Using the functions and event flows in the Group C profile, an application should be able to: perform the rest of the telephony functions usually done from a phone — pickup, group pickup, call completion and perform more sophisticated monitoring functions.

(There is apparently no Group D profile.)

"Group E Profile: Using the functions in the Group E ('E' for Environmental) profile, an application should be able to: activate and deactivate features of a device, query the state of features on a device, query the generic information about a device."

Once the profiles are finalized, they will be incorporated into Novell's PBX Driver certification process; groups supported by the driver will be included in literature describing all certified drivers. See also NETWARE TELEPHONY SERVICES.

PBX Extension A telephone line connected to a PBX.

PBX Fraud Same as TOLL FRAUD.

PBX Generations See GENERATIONS, PBX.

PBX Integration A loose term to mean joining the PBX to any number of outside computer based gadgets and services, from voice mail to call accounting. To make voice mail integrate into PBXs, you minimally need the ability to provide a message waiting indicator (light or stutter dial-tone) at the user's phone when a message is received, and to forward a call to the user's mailbox when a call is sent to the recipient and they are on the phone or do not answer (forward on busy or ring no answer). This requires PBX "integration." Most PBX integrations provide the attached voice mail system call data that includes the calling phone number, the number of the caller, why the call was presented (such as forwarded on busy, or ring no answer), to pass message waiting on or off indications. PBX integration data may be implemented in-band or out-of-band on a separate link, most often a serial link. Some PBXs "integrate" with outside equipment better than others.

PBX Line Access An AT&T term. Provides for a PBX line to appear in a line access button on 16 ECTS telephones. A single-line telephone may be used for an appearance of a PBX.

PBX Profiles PBXs do things differently. To make a conference call, one PBX's phone may put the caller on hold automatically, while another may insist that you put that person on hold manually and then dial the next person to join the conference call. As Novell in the fall of 1994 attempted to get as many PBXs to conform to TSAPI, it discovered that PBX features often work very differently. So it decided to categorize PBXs and their features. This, it called, PBX profiles. The idea being that Profile A would contain the most common, easy-to-integrate-to-TSAPI features. B would contain the second most common, etc. Novell also calls PBX Profiles PBX Driver Profiles. See PBX DRIVER PROFILES for a much bigger explanation.

PBX Station Line A transmission path extending from the station (phone instrument) location to the switching equipment.

PBX Tie Line A tie line between two PBX's, permitting extensions in one PBX to be connected to extensions in the other without having to dial through the public switched network. See also OPX and OPS, which are different and are lines between PBXs and distant extensions, not tie lines between PBXs.

PBX Trunk A circuit which connects the PBX to the local telephone company's central office switching center or other switching system center.

PC 1. Personal Computer.
2. Peg Count.
3. Point Code.

4. Printed Circuit.
5. Product Committee.
6. Protocol Control.
7. Politically Correct.

PC Card A memory or I/O card compatible with the PCMCIA PC Card Standard. In short, PC Cards are a new name for PCMCIA cards. For a much fuller definition, see PCMCIA, which stands for the Personal Computer Memory Card International Association.

PC Centric There are two ways you can organize a computer to control telephone calls on an office telephone system. One way is to join a file server on a local area network to a phone system. Commands to move calls around are passed from the desktop PC over the LAN to the server and then to the phone system via the cable connection between the server and the system. A second way to get a computer to control phone calls is through a connection at the desktop. This is called PC Centric. There are two ways you can do this. The first is to join the desktop phone to the computer with a cable. This is often done via the PC's serial port connecting via cable to the phone's data communications port (if it has one — if it doesn't, you get one). The second way to be PC Centric is by simply replacing the standalone phone with a board that emulates in a phone and drop it into the PC's bus.

PC DOS What IBM calls a version of the operating system on its older personal computers. If PC-DOS runs on an IBM compatible, it is called MS-DOS, which stands for MicroSoft Disk Operating System. Microsoft of Bellevue, Washington, wrote MS-DOS and PC-DOS.

PC Network IBM's first LAN (Local Area Network).

PC Telephony Another term for Computer Telephony. See COMPUTER TELEPHONY.

PC's The plural of the word PC, according to the New York Times. However, every other computer and general magazine spells them PCs. And that's the spelling which this dictionary writer prefers also.

PCA 1. Premises Cabling Association. A association in Great Britain.
2. Protective Connecting Arrangement. A device that AT&T and members of the Bell System insisted be connected between a telecommunications device (like a phone) that wasn't made and sold by AT&T and a phone line provided by an AT&T Bell operating company. Many years later, the PCAs were found by the FCC to be totally unnecessary and AT&T and members of the Bell System were ordered to refund all payments received for rental of PCAs. The Bell System insisted on the PCAs as a way of protecting AT&T's effective monopoly of telecommunications equipment. See also PROTECTIVE CONNECTIVE ARRANGEMENT.

PCC Personal Companion Computer. What other companies call a PDA (Personal Digital Assistant), Intel calls a PCC. A PCC or PDA is meant to have significant telecommunications abilities — including wired and wireless. See PDA.

PCB Printed Circuit Board.

PCF Physical Control Fields. The AC (Access Control) and FC (Frame Control) bytes in a Token Ring header.

PCI 1. Protocol Control Information. The protocol information added by an OSI entity to the service data unit passed down from the layer above, all together forming a Protocol Data Unit (PDU).
2. Peripheral Component Interconnect, a 32/64 bit local bus inside a PC or a Mac designed by Intel for the PC. According to Intel, it can transfer data between the PC's main microprocessor (its CPU) and peripherals (hard disks, video adapters, etc.) at up to 132 megabytes per second, compared to only five megabytes per second which the ISA bus is capable of. PCI is one of two widely adopted local-bus standards. The other, the VL-Bus, is primarily used in 486 PCs. See also VLB.

PCL 1. Hewlett-Packard's Printer Control Language, developed by HP in 1984 as a way for the then-new PC to communicate with a new breed of laser printers — the HP LaserJet printer. HP's PCL language is now the de facto industry standard for PC printing. Most of the printers in the world today are equipped with PCL or a PCL-compatible language. PCL allows the type of sophisticated page creation generally referred to as "laser quality output." PCL supports such advanced features as fully scalable typefaces and rotation of text. PCL defines a standard set of commands enabling applications to communicate with HP or HP-compatible printers. PCL has become a de facto standard for laser and ink jet printers and is supported by virtually all printer manufacturers. On April 8, 1996 HP announced PCL 6 which it billed as "the next generation" of HP Printer Control Language). HP said that PCL 6 includes font synthesis technology for true what-you-see-is-what-you-get (WYSIWYG) printing and better document fidelity. PCL 6 commands were designed by HP to closely match Microsoft Windows GDI (Graphical Direct Interface) commands.
2. Product Compute-Module Load.

PCM Pulse Code Modulation. The most common method of encoding an analog voice signal into a digital bit stream. First, the amplitude of the voice conversation is sampled. This is called PAM, Pulse Amplitude Modulation. This PAM sample is then coded (quantized) into a binary (digital) number. This digital number consists of zeros and ones. The voice signal can then be switched, transmitted and stored digitally. There are three basic advantages to PCM voice. They are the three basic advantages of digital switching and transmission. First, it is less expensive to switch and transmit a digital signal. Second, by making an analog voice signal into a digital signal, you can interleave it with other digital signals — such as those from computers or facsimile machines. Third, a voice signal which is switched and transmitted end-to-end in a digital format will usually come through "cleaner," i.e. have less noise, than one transmitted and switched in analog. The reason is simple: An electrical signal loses strength over a distance. It must then be amplified. In analog transmission, everything is amplified, including the noise and static the signal has collected along the way. In digital transmission, the signal is "regenerated," i.e. put back together again, by comparing the incoming signal to a logical question: Is it a one or a zero? Then, the signal is regenerated, amplified and sent along its way.

PCM refers to a technique of digitization. It does not refer to a universally accepted standard of digitizing voice. The most common PCM method is to sample a voice conversation at 8000 times a seconds. The theory is that if the sampling is at least twice the highest frequency on the channel, then the result sounds OK. (See NYQUIST THEOREM.) Thus, the highest frequency on a voice phone line is 4,000 Hertz. So one must sample it at 8,000 times a second. Many PCM digital voice conversations are typically put on one communications channel. In North America, the most typical channel is called the T-1 (also spelled T1). It places 24 voice conversations on two pairs of copper wires (one for receiving and one for transmitting). It contains 8000 frames each of 8 bits of 24 voice channels plus one framing (synchronizing bit) bit which equals 1.544 Mbps, i.e. 8000 x (8 x 24 + 1) equals 1.544 megabits.

Countries outside of the United States and North America use a different scheme for multiplexing voice conversations. It is based not on 24 voice channels, but on 32. This scheme keeps two of the 32 channels for control, actually transmitting 30 voice conversations at a data rate of 2.048 Mbps. The European system is calculated as 8 bits x 32 channels x 8000 frames per second. European PCM multiplexing is not compatible with North American multiplexing. The two systems cannot be directly connected. Some PBXs in the U.S. conform to the U.S. standard only. Some (very few) conform to both. Both the European and North American T-1 "standards" have now been accepted as ISDN "standards." In addition to PCM, there are many other ways of digitally encoding voice. PCM remains the most common. See T-1 and VOICE COMPRESSION.

PCM Voice Transmission Synchronization There are three levels in PCM voice transmission synchronization:
1. Bit Level Synch — operating the transmitter and receiver at the same bit rate so the bits are not lost.
2. Frame Level Synch — phase alignment between the transmitter and receiver so the beginning of the frame can be identified.
3. Time Slot Synch — phase alignment between the transmitter and receiver so the time slots are lined up for information retrieval.

PCM-30 Short name of international 2.048 Mbps T-1 (also known as E1) service derived from the fact that 30 channels are available for 64 Kbps digitized voice each using pulse code modulation (PCM).

PCMCIA The Personal Computer Memory Card International Association (an awful mouthful) standardizes credit-card size packages for memory and input/output (modems, LAN cards, etc.) for computers, laptops, palmtops, etc. There are three physical standards for PCMCIA cards — Type I, II, and III and undefined standard called Type IV, which only Toshiba has at this moment. The cards are 69.2 millimeters (3.37 inches) long x 51.46 millimeters (2.126 inches) wide. All three types use the same 68 female pin edge connector for attachment to the computer, and differ only in thickness. The thickness for Type I, Type II and Type III are 3.3, 5.0, 10.5 millimeters respectively. Toshiba's Type IV is 16 mm. A Type I PC Card is typically used for various types of memory enhancements, including RAM, FLASH memory, one-time programmable (OTP) memory, and electronically erasable programmable read only memory (EEPROM). A Type II PC Card (the most common) is typically used for input/output such as modem, LANs, host communications and SCSI device connection. A Type III PC Card is twice the thickness of the Type II and is typically used for for I/O features that require a larger size, such as rotating mass storage devices (removable hard disk drives) and radio communication devices. Since Type I, Type II and Type III Cards all use the same interface, the size of the card chosen for the application is dependent on the miniaturization of the technology to be implemented.

PCMCIA's first standards were issued in September, 1991. The idea is that small computers will use these cards for modems, fax cards, hard disks, LAN connections, Ethernet connections, SCSI device connections, etc. A PCMCIA card is, in most cases, the only way to get to a laptop's bus without attaching a docking station. In terms of performance, the bus is comparable, but not equivalent to the ISA bus on desktops. The PCMCIA bus is 8 bits wide and does not allow for direct memory access (DMA) transfers or bus mastering. This means, for example, that you can't equip a laptop with a sound blaster compatible PCMCIA card, since it requires DMA transfers. A new spec called PCMCIA v3 (now called CardBus) calls for a number of improvements, including 32-bit data paths, DMA, bus mastering and significant improvements in speed with the bus as fast as the PCI bus.

PCMCIA standards exist so others can make these cards. Some computers — like some pre-summer 1991 notebooks and laptops from Compaq and Toshiba — don't comply because these computers were released before the standards were released. The PCMCIA v2.0 standard contains a software specification for XIP, the "eXecute In Place" mechanism that maps application software stored on the PCMCIA card into the system address space. This means application software will run directly from the card, start faster and not require precious RAM from the host computer.

One key element of the PCMCIA software architecture are Socket Services and Card Services. Socket Services is a BIOS level software interface that provides a way to access the PCMCIA sockets (slots) of a computer. Socket Services identifies how many sockets are in your computer system and detects the insertion or removal of a PC Card while the system is powered on. Socket Services is part of the PCMCIA Specification and interfaces with Card Services. Card Services is a software management interface that allows the allocation of system resources (such as memory and interrupts) automatically, once the Socket Services detects that a PC Card has been added. Card Services also releases these resources when the PC Card has been removed. Card Services also provides you with an interface to higher level software to load any needed hardware drivers.

The combination of PC Card hardware, Card Services software and Socket Services software provides an almost "plug-and-play" capability in the portable computing environment. Once the software has been installed, it is theoretically possible to add and remove PC Cards without powering off the system. But this is theory. And in practice, it has worked only intermittently for this writer. It doesn't work for network cards. It's meant to be possible, for example, to insert a modem PCMCIA Card to access another computer system, download information into the portable computer's memory, remove the modem PCMCIA Card, replace it with a flash PCMCIA Card, and store the downloaded information — all while your portable computer is still powered on. Great theory.

The PCMCIA has around 300 members, including manufacturers of semiconductors, connectors, peripherals and systems, as well as BIOS and software developers and related industries. Members include Intel, IBM, Toshiba, Lotus, Epson and Fujitsu. The association is based in Sunnyvale, CA. 408-720-0107. It has an electronic bulletin board — 408-720-9388. Its standards are also recognized by the Japanese Electronic Industry Development Association (JEIDA). The Association publishes a free book listing all the manufacturers making cards which comply to their standards.

Advice in buying PCMCIA cards: PCMCIA standards are still being formed and are thus not truly standards. Check that cards you buy work in your machine. The manufacturer should have tested it in your machine. Don't assume compatibility. Check your socket and card services supports that card. It may not. Use "enabler" software instead. Avoid PCMCIA cards with "pigtails." They break. You lose them. Use PCMCIA cards with an "X-JACK." You can plug directly into a PCMCIA card with an X-JACK, or equivalent. If you have to buy a card with a pigtail, buy a second pigtail, just in case.

In 1994, the PCMCIA started calling its specs, PC Card. Some people said it was because many people claimed that PCMCIA stood for "People Can't Memorize Computer Industry Acronyms."

In the latter part of 1995, the PCMCIA came out with a new specification called CardBus which is a 32-bit bus, as against the present 16-bit cards. CardBus is an extension of the PCI bus. This means that new CardBus will support 132 Mbps, much faster than the present 8 Mbps. See CARD SERVICES, PCMCIA STANDARDS, SOCKET SERVICES and SLOT SIZES.

PCMCIA Standards The complete set of all of the PCMCIA PC Card Standards. It includes the PC Card Standard Release v2.01, Socket Services Specification Release v2.0, Card Services Specification Release v2.0, ATA Specification Release v1.01, AIMS Specification Release v1.0, and the Recommended Extensions Release v1.0. Standard v3.0 has been proposed. See PCMCIA.

PCN Personal Communications Network. A new type of wireless telephone system that would use light, inexpensive handheld handsets and communicate via low-power antennas. When it was originally conceived, PCN was primarily seen as an a city communications system, with far less range than cellular. Subscribers would be able to make and receive calls while they are traveling, as they can do today with cellular radio systems, but at a low price. Now PCN is seen as what Dr. Sorin Cohn of Northern Telecom calls an "enabler of unplanned growth." One idea for PCN is to locate a PCN cell site (transmitter/receiver) in a residential community. When someone wanted a new phone line, they'd simply drop down to their local phone store, pick up a PCN portable phone and, by the time, they got back home, their frequency would be "switched on" and they'd be "live." See also PERSONAL COMMUNICATIONS NETWORK.

PCO Point of Control and Observation: A place (point) within a testing environment where the occurrence of test events is to be controlled and observed as defined by the particular abstract test method used.

PCP Program Clock Reference: A timestamp that is inserted by the MPEG-2 encoder into the Transport Stream to aid the decoder in the recovering and tracking the encoder clock.

PCR An ATM term. Peak Cell Rate: The Peak Cell Rate, in cells/sec, is the cell rate which the source may never exceed.

PCS 1. Personal Communications Service. A new, lower powered, higher-frequency competitive technology to cellular. Whereas cellular typically operates in the 800-900 MHz range, PCS would operate in the 1.5 to 1.8 Mhz range. The idea with PCS is that the phones would be cheaper, have less range, would probably be digital; the cells would be smaller and closer together and the airtime would be cheaper also. The concept of PCS is evolving. It is not clear exactly where it will end up. For a bigger explanation, see PERSONAL COMMUNICATIONS NETWORKS and PERSONAL COMMUNICATION SERVICES.
2. Personal Communications Software — available in ROLM Cedar, Cypress and Juniper — makes for easy access to external data bases, rapid transfer of data files and one-touch access to most telephone features. Other PCS features include built-in calculator, small phone list, reminder function and internal clock.

PCSA Personal Computing System Architecture. A PC implementation of DECnet, that lets PCs work in a DECnet environment. PCSA is a network architecture defined and supported by Digital Equipment Corporation for the incorporation of personal computers into server-based networks.

PCTA Personal Computer Terminal Adapter. A printed circuit card that slips into an IBM PC or PC compatible and allows that PC to be connected to the ISDN T-interface. See PERSONAL COMPUTER TERMINAL ADAPTER.

PCTE Portable Common Tool Environment.

PCTS Public Cordless Telephone Service. A Canadian digital cordless telephone service for residential, business and public use. For other variations of digital cordless telephone service, see CT1, CT2, CT2Plus, CT3, and DECT.

PCWG Personal Conferencing Work Group (www.gogcwg.org/pcwg/)

PCX Server Software PCX server software turns your PC into a graphics terminal front-end for Unix and X applications. Thus, your PC can display application output generated by remote X-based client applications.

PDA Personal Digital Assistant. A consumer electronics gadget that looks like a palmtop computer. Unlike personal computers, PDAs will perform specific tasks — acting like an electronic diary, carry-along personal database, multimedia player, personal communicator, memo taker, calculator, alarm clock. The communications will take place through the phone or through wireless. Apple has announced a PDA, which it has named Newton. When I added this definition in the late fall of 1992, sales of PDAs weren't doing well and some wag in Silicon Valley called them Probably Disappointed Again. IBM prefers to call them Personal Communicators. General Magic prefers to called them PICs, Personal Intelligent Communicators.

PDAU Physical Delivery Access Unit. A gateway device that facilitates the delivery of messages (excluding probes and reports) in physical form. This is an X.400 term.

PDH Plesiochronous Digital Hierarchy: PDH (plesiochronous means nearly synchronous), was developed to carry digitized voice over twisted pair cabling more efficiently. This evolved into the North American, European, and Japanese Digital Hierarchies where only a discrete set of fixed rates is available, namely, nxDS0 (DS0 is a 64 kbps rate) and then the next levels in the respective multiplex hierarchies.

PDI 1. A Versit term. Personal Data Interchange, a collaborative application area which involves the communication of data between people who have a business or personal relationship, but do not necessarily share a common computing infrastructure.
2. Telenova/Lexar's Predictive Dialing Interface which links the Telenova/Lexar PBX and an outboard predictive dialing computer processor. Telenova went Chapter 11. So much for that good idea.

PDL A page description language (PDL) is a is a clever short-cut for transmitting bit-mapped images from a PC application to a printer. They save processing time, by sending only "instructions" to a printer, rather than the entire bitmapped image. They also allow the printer to print any font, any size. PDL is the generic term. Hewlett-Packard has been the major proponent of the concept of PDLs. And they include something called PCL with all their printers. PCL stands for Printer Command Language. Postscript is also a PDL, but different to PCL. HP includes PCL with all its laser printers, but has only included
What we really meant was that HP had never built PostScript into their printers. They still haven't, really. You still have to buy PostScript as either a plug-in cartridge or an add-on "SIM" chip from HP, Adobe or third parties for most of the HP line. The exception is the top-of-the-line LaserJet IIIsi, for which built-in PostScript is an option that you pay extra for.
That may change. For now, if you're using Windows applications, Windows' built-in "True Type" scalable fonts — which do work on HP and other non-PostScript printers — will satisfy most of your printing needs.

PDM 1. Pulse Duration Modulation.
2. See PERSONAL DATA MODULE.

PDN 1. An ISDN Term. Primary Directory Number (7 digits for 5ESS switch; 10 digits fr DMS-100 switch)
2. Public Data Network. A public network for the transmission of data, particularly a network compatible with X.25 protocol. A public data network is to data what the Public Switched Voice Network is to voice. To access a public data network, you typically dial a local number, receive a carrier tone and then follow very specific instructions. Public data networks send their digital data in packets over high speed channels. The major reason to use a PDN is that it may be cheaper than dialing directly on a switched voice line. Also, they get you into some databases and services which are hard to get into by dialing direct. There are many public data networks in the United States. The two best-known public data networks are Tymnet and Telenet. But there are probably 30 more. Every industrialized foreign country has at least one, usually owned by the local government phone company.
PDP 1. See POWER DISTRIBUTION PANEL.
2. Plasma Display Panel.
PDP-11 On the last day of March, 1970 Digital Equipment shipped its first PDP-11/20 minicomputer to a customer in Tennessee. The PDP-11 family was among the first minicomputer that incorporated open standards into its operations. And the PDP-11 software platform has continued through subsequent generations of hardware.
PDS 1. Premise Distribution System. AT&T's proprietary building-wide telecommunications cabling system.
2. Processor Direct Slot. Some non-NuBus Apple Macintosh computers have one PDS that allows for expansion cards.
PDU Protocol Data Unit. A PDU is a message of a given protocol comprising payload and protocol-specific control information, typically contained in a header. PDUs pass over the protocol interfaces which exist between the layers of protocols (per OSI model). Basically, PDU is OSI (Open Systems Interconnection) terminology for "packet". A PDU is a data object exchanged by protocol machines (entities) within a given layer. PSUs consist of both data and control (protocol) information that allows the two to coordinate their interactions.
PE Processing Element.
Peak That part of the business day in which customers expect to pay full service rates. For cellular customers, peak hours are generally 7:00 a.m. to 7:00 p.m. For business landline customers, peak hours are generally 8:00 a.m. to 5:00 p.m.
Peak Emission Wavelength Of an optical emitter, the spectral line having the greatest power.
Peak Hour When used with an automatic call distributor, the peak hour is when the number of calls coming into your center are at their highest level. ACDs allow you to track and report on calls by hour. Some allow you to also track peak half-hours, or peak delays of the week or months of the year.
Peak Load A higher than average quantity of traffic. Peak Load is usually expressed for a one-hour period, often the busiest hour of the busiest day of the year. See BUSY HOUR.
Peak Position Requirements A call center term. The maximum number of base staff required in any half hour within a given date range.
Peak Rate The per-minute price for using a communications device in the "peak" time period. For cellular, "Peak" time generally includes hours such as evenings Monday through Friday, and all day Saturday, Sunday, and certain holidays. For normal landline phone service, peak rates will include workday days.
Peak To Average Ratio P/AR. An analog test that provides a measure of data circuit quality by sending a pulse into one end of a circuit and measuring its envelope at the distant end of the circuit. See P/AR.
Peaked Load The load that results from peaked traffic.
Peaked Traffic Random traffic that has a variance-to-mean ratio grater than one.
Peanut Tubes A name given to the smaller sizes of vacuum tubes.
PEB PCM Expansion Bus. A digital voice bus for sending voice across different voice processing cards and components in the same PC. PEB is from Dialogic, Parsippany, NJ. It is an open platform. Many companies make voice processing products connecting via PEB. See also DYNAMIC NODE ACCESS.
Pedestal A mounting device used in pay telephone installations where the instrument is not attached to a wall. Pedestals are usually too small and rarely have a shelf big enough to take notes on.
PEEK Taligent's Partners Early Experience Kit. I guess it's their cute name for their developer's kit.
Peek/Poke Instructions that view and alter a byte of memory by referencing a specific memory address. Peek displays the contents; poke changes it.
Peer Entities Entities within the same layer.
Peer Group A set of logical nodes which are grouped for purposes of creating a routing hierarchy. PTSEs are exchanged among all members of the group.
Peer Group Identifier A string of bits that is used to unambiguously identify a peer group.
Peer Group Leader A node which has been elected to perform some of the functions associated with a logical group node.
Peer Group Level The number of significant bits in the peer group identifier of a particular peer group.
Peer Node A node that is a member of the same peer group as a given node.
Peer Protocol The set of rules defining the procedures for communication between like entities. The identical entities may be devices or software modules at specific layers in a layered network architecture implementation.
Peer-Peer Directory Propagation A way of updating user addresses in which changes in any post office on a LAN (local area network) are sent to all other post offices.
Peer-To-Peer Network 1. A network (typically a local area network) in which every node has equal access to the network and can send and receive data at any time without having to wait for permission from a control node. While peer-to-peer resource sharing is effective in small networks, security and reliability issues prevent its widespread use in larger networks.
2. A new telephony term describing the relationship between a telephone system and the external computer working with it. Picture a telephone switch acting as an automatic call distributor and an outboard computer processor. The idea is to coordinate the call and the screen at the agent. Communication must take place between the switch and the computer. If that communication is peer-to-peer, as it is, for example, in the DMS Meridian ACD, then neither the switch nor the computer is in a "slave" rela-

tionship to the other. See EXTENDED CALL MANAGEMENT.
Peer-To-Peer Resource Sharing An architecture that lets any station contribute resources to the network while still running local application programs. 10-Net Local Area Network allows for peer-to-peer resource sharing.
PEG Public, Educational or Government access. A cable TV term to denote the local public access channel/s.
Peg Count A raw count of some event. Because this was originally maintained by moving pegs on a board with units of 1,10s,100, 1000s it became a peg count. A count of the number of calls placed or received at a certain point or over certain lines during a period such as an hour or day or week. A peg count simply tells you how many calls you made or received. It does not tell you how long they were or where they went or anything else. In the old days before we had accurate and relatively inexpensive call accounting equipment, we relied on Peg Counts to figure out how many circuits we needed. No more. The peg count method is too inaccurate.
PEL Picture ELement. A pel is the smallest area on a video screen that can be controlled by software. Pels are arranged on screens in a grid-like fashion. Depending on the screen mode selected by an application, a pel may be a single pixel or several pixels. (A pixel is the smallest visual element on a screen, which can be turned on or off or varied in intensity.) The larger the number of pixels in a pel, the greater the size of the palette of distinct colors. When a pel consists of multiple pixels, the number of colors can be quite large. The pel size determines the clarity of the image — called screen resolution. Larger individual pels reduce the total number of available pels, resulting in lower resolution. Smaller pels increase the number of pels that can fit on the screen, resulting in higher resolution and a clearer picture. See also PIXEL, FAX and VGA.
PEM Privacy Enhanced Mail. An Internet electronic mail capability which provides confidentially and message integrity using various encryption methods.
Pen Register Also called a Dialed Number Recorder (DNR). An instrument that records telephone dial pulses as inked dashes on paper tape. A touchtone decoder performs the same thing for a touchtone telephone.
Pen Windows A new Microsoft operating system for notebook size computers that uses a stylus instead of a keyboard.
Penetration Bypassing the security mechanisms of a system.
Penetration Tap A connection method used in installations that allows devices to be connected to cable without interrupting network operation. Penetration taps are most commonly found used with coaxial cable. A sharp, pointed probe is used to penetrate the outer insulation and grounding shield of the coaxial cable and to make direct contact with the inner conductor.
Penetration Testing Attempting to circumvent a system's security features to identify weaknesses.
PenPoint An operating system from Go Corp. for pen computers. Such computers are light, small, portable and have a stylus and a screen which recognizes the words and numbers you write on the screen with the stylus. PenPoint is to pen computers what MS-DOS (or more recently, Windows) is to desktop computers. PenPoint won't run existing MS-DOS software. But Microsoft's new Windows For Pens — what it calls an "operating environment" — promises to. But it also promises to be more difficult to use than PenPoint. As of writing this, (early 1992), I hadn't seen final versions of either operating system. So we'll see.
Pentium In the fall of 1992, Intel adopted the name Pentium for its 80586 chip, its successor to the 80486. It introduced the Pentium formally in early April, 1993. The chip is capable of 112 million instructions per second and is 80% faster than the fastest 80486. It contains more than three million transistors and is said to be a superscalar chip, which means it can execute two instructions at a time.
Pentium Pro Intel's name for the 80686 microprocessor chip, its successor to the Pentium.
PEP Packet Ensemble Protocol. A high-speed modulation method from Telebit Corporation for dial-up modems. Now obsolete.
Per Call Calling Identity Delivery Blocking Feature Allows a caller to toggle or override the value of a calling identity item's Permanent Presentation Status (PPS) for a particular call.
Percentage ATB Percentage of All Trunks Busy. Percentage of time during a reporting period that all trunks in a group or split were busy. This may be measured in two ways, actual simultaneous busies and call length per event, or backed into statistically. Neither technique is absolutely accurate as each depends on "snap shots" in a environment of random interleaved call events.
Percentage CA Percentage of Calls Abandoned. Indicates the percentage of calls abandoned by callers after being accepted by the ACD.
Percentage HLD Percentage of total calls HeLD in queue within a reporting group.
Percentage NCO Percentage of total of Number of Calls Offered to a particular reporting group.
Percentage TUT Percentage of Trunk Utilization Time. The percentage of a time during a reporting period that a trunk is in use and not idle.
Perforator An instrument for the manual preparation of a perforated tape, on which telegraph signals are represented by holes punched in accordance with a predetermined code.
Perforator, Paper Tape An electro-mechanical device which converts electrical signals into coded holes in a paper tape. See PAPER TAPE PUNCH.
Performance Management Measures and records resource utilization. It is one of the categories of network management defined by the ISO (international Standards Organization).
Periapsis In satellite systems, the point on a satellite's orbit at which it is closest to the center of the primary body about which it is orbiting. Where Earth-based satellite systems are concerned, the term is synonymous with perigee. See also GEOSTATIONARY ORBIT.
Perigee The point at which a satellite orbit is the least distance from the center of the gravitational field of the Earth. See also PERIAPSIS.
Perimeter Protection System A field disturbance sensor which uses buried leaky cables installed around a facility to detect any unauthorized entry or exit.
Period Of A Satellite The time elapsing between two consecutive passages of a satellite through a characteristic point on its orbit.
Periodic Postings Articles that are posted periodically to a newsgroup for the benefit of people who are new to the newsgroup. An Internet term.
Peripheral Device See PERIPHERAL EQUIPMENT or APPLICATIONS PROCESSOR.
Peripheral Equipment Equipment not integral to but working with a phone system. An example might be a printer or television screen on which calling traffic statistics are displayed. It might also be a voice mail system. AT&T once called PBX peripheral equipment "applications processors," because they process specific applications. Some people now call them Adjunct Processor or Outboard Processors.

PERL An interpreted scripting language, typically used in writing CGI (Common Gateway Interface) scripts.

Permalancer A permanent freelancer. A person hired on a per-project basis who lives a benefits-free existence.

Permanent Presentation Status PPS. A PPS has a public value of a "public" or "anonymous" and is used as the presentation status of a call if no per-call CIDB (Calling Line Identification Delivery Blocking) feature is active. A PPS should exist for each calling identity item. A Bellcore definition.

Permanent Shift Type A call center term. A shift definition that the program gives priority to in creating schedules but uses only as long as no overstaffing results in any intra-day period (at which point flexible shift types are used). When scheduling is done for more than a week at a time, the permanent schedules are always identical from one week to the next.

Permanent Signal A sustained off-hook supervisory signal originating outside a switching system and not related to a call in progress. Permanent signals can occupy a substantial part of the capacity of a switching system.

Permanent Virtual Circuit A virtual circuit that provides the equivalent of a dedicated private line service over a packet switching network between two DTEs. The path between users is fixed. A PVC uses a fixed logical channel to maintain a permanent association between the FTEs. Once a PVC is defined, it requires no setup operation before data is sent and no disconnect operation after data is sent. A new connection between the same users may be routed along a different path. See also VIRTUAL CIRCUIT and SDN for another type of virtual network.

Permissible Interference Observed or predicted interference which complies with quantitative interference and sharing criteria contained in these [Radio] Regulations or in CCIR Recommendations or in special agreements as provided for in these Regulations. (RR) See also accepted interference, interference.

Permission A Windows NT term. A rule tied to an object (usually a directory, file or printer) to regulate which users can have access to the object and in what manner.

Permissions 1. A call center term. Privileges granted to each user with respect to what data that user is allowed to access and what menu options or commands he or she is allowed to use. Permissions are under control of the System Administrator.
2. A Northern Telecom Norstar definition to define specific characteristics that can be assigned to an individual telephone. Permissions includes Full Handsfree, Handsfree Answerback, Pickup Group, Page Zone, Auxiliary Ringer, Receive tones, and Priority Call.

Permuter Table Routing table in TYMNET.

Perspective For VNet A new service from MCI which allows large companies to analyze long distance billing information electronically or via CD-ROM.

Perpetrator Someone who carries out an illegal or malicious act affecting information security.

Persistence 1. In a CRT, the time a phosphor dot remains illuminated after being energized. Long-persistent phosphors reduce flicker, but generate ghost-like images that linger on screen for a fraction of a second.
2. The probability that when a device on a local area network which has the data to transmit senses a free transmission line it will attempt the transmission. 1.0 persistence indicates that there is 100 percent probability that the device will always attempt to transmit. IEEE 802.3 and Ethernet both use 1.0 persistence. A 0.5 persistence indicates that when a device senses a free line it will only attempt to transmit 50 percent of the time.

Persistency A persistent configured site will attempt to re-establish the connection in the event of an unexpected line drop. A non-persistent configured site will not.

Persistent Communication A dialogue between an ASC (AIN Switch Capabilities) and a SLEE (Service Logic Execution Environment) that may involve a sequence of messages. Definition from Bellcore in reference to its concept of the Advanced Intelligent Network.

Persistent Information Information for which a permanent data object exists. Definition from Bellcore in reference to its concept of the Advanced Intelligent Network.

Person Call There will be two types of calls: person calls and place calls. I make a place call when I call a phone number which ends in one designated, fixed RJ-11 jack attached to the wall or floor. I make a person call when I call a phone number which doesn't necessarily end on a fixed place. A typical person call might be a cellular or wireless phone call. It might also be a service like MCI's Personal 800 Service. The major characteristic of a person call is that I am calling a person and I don't know where that person is. As a result, the person I called might answer my call anywhere — from his car, from vacation home, from his wireless phone, etc.

Person To Person Call The most expensive way to make a long distance call. Call the operator. Say "I want to speak to Andy Moore on 212-691-8215." The operator dials Mr. Moore's phone number, gets Mr. Moore on the phone. "Are you Mr. Moore?" Mr. Moore replies: "Yes, I am." The operator bows out of the conversation and sends you, the caller, a hefty bill for that personalized service. Until recently, person-to-person service was only offered by AT&T. Now it's offered by many companies. But prices are not regulated. And some companies charge an arm and a leg. Please be careful.

Personal 800 Number Several long distance companies are now offering Personal 800 numbers, which are basically party line 800 numbers with call routing. The way they work is as follows: You dial a number, e.g. 800-484-1000. A machine answers with a double beep. You punch in four or five digits on your touchtone pad. A voice response unit at the other end hears the digits, says "Thank you for using MCI" and dials out your long distance number which might be 212-691-8215 (mine). The long distance carriers are charging under $5 a month and 15 to 25 cents a minute for the service. The per minute charges are more expensive than normal 800 lines. One company, MCI, is also offering FOLLOW ME 800 which allows you to change the routing of your personal 800 number instantly with one phone line.

Personal Authentication Device P.A.D. A portable or fixed device which allows for precise identification and validation of each user. Used in fraud prevention and unauthorized access. Also called a "token".

Personal Central Office Trunk Line Allows a user to access a central office trunk line dedicated to him. A trunk line can appear in a line access button on his PBX phone or his 16 ECTS telephones on a AT&T PBX.

Personal Communications Networks PCN. A new type of wireless telephone system that would use light, inexpensive handheld handsets and communicate via low-power antennas. PCN is primarily seen as a city communications system, with far less range than cellular. Subscribers would be able to make and perhaps receive calls while they are traveling, as they can do today with cellular radio systems, but at a low price. There's talk that they'll put PCN antennas in communities and issue everyone in the community with a PCN phone that would hub off the PCN antenna. This would save the phone company wiring up each house. In this way, the PCN phone would resemble the common household cordless phone, but with a larger range. Dr. Sorin Cohn of Northern Telecom calls the new low-power, wireless, personal communications systems an "enabler of unplanned growth." Nice definition. See PCN and PERSONAL COMMUNICATIONS SERVICES.

In the fall of 1992 MCI broadened the definition of PCN. It proposed a national consortium of local PCNs joined together by a long distance carrier (namely it). In its release, MCI said that "PCNs are the next generation of digital wireless communications technology. PCNs use less power and are less expensive than the current cellular technology and permit the use of inexpensive pocket telephones with much longer battery life than cellular portables. PCN phones will have many more features than today's cellular or conventional telephones, and unlike cellular phones, will be usable in most areas of the world. PCNs will operate in the same frequency band in most countries, (1850-1990 MHz) while cellular is operating in several different frequency bands in various countries, and thus is not portable from country to country." See also PERSONAL COMMUNICATIONS SERVICES.

Personal Communications Services PCS are a broad range of individualized telecommunications services that let people or devices to communicate irrespective of where they are. Some of the services include:
- Personal numbers assigned to individuals rather than telephones
- Call completion regardless of location ("find me")
- Calls to the PCS customer can be paid for by the caller, or by the PCS customer
- Call management services giving the called party much greater control over incoming calls.

PCS can both find and complete a call to a person regardless of location, but give that person the choice of accepting or rejecting the call or sending it somewhere else. PCS will possibly use a new category of wireless, voice and data communications — using low power, lightweight pocket telephones and hand-held computers.

Personal Communicator AT&T and IBM's preferred name for what Apple calls a PDA, Personal Digital Assistant. Tandy calls theirs a Personal Information Processor. See also PDA. Here's what AT&T thought of them early 1993: "A personal communicator represents a new product category that provides a new level of mobility by marrying communications and computing. It is, AT&T believes, a communications-oriented device built around a high-performance computing engine that manages all of the communications and system functions so transparently, in the background, that we are never aware of the computer inside. A personal communicator will be an amalgam of today's mobile technologies and tools - cellular phone, portable fax and modem, pager, personal organizer and daytimer, video, and pen-based personal device - integrated into one totally new, breakthrough package.

"These devices will come in a variety of sizes shapes, and capabilities built for a broad cross-section of users — from on-the-go managers to in-the-field mobile workers, and mass-market consumers. The general or industry-specific application will drive the device's feature set and, thus, price — from a couple of hundred to a couple of thousand dollars. Personal communicators must be truly portable and rich in communications. One day they will seamlessly integrate voice, data, handwriting, fax electronic mail, still images, and full-motion video — wirelessly. Over time these products will become smaller and lighter weight, deliver increased performance yet consume less power, provide extended battery life, and access multiple network services to send and receive messages on demand.

"If these devices are to be adopted by a whole new class of user who may be uncomfortable with technology and who would not otherwise consider buying a computer, then using a personal communicator must be painless. This means no instruction manuals. This means simple, intuitive interfaces. A personal communicator must be as comfortable and as easy to use as a touchstone telephone. Above all, these products are not an extension or miniaturized version of the desktop personal computer. That is, the personal communicator is not the device for crunching numbers, manipulating relational databases, or creating lengthy documents. There will, however, be "bridges" enabling us to access those resources that reside on our desktop or corporate file server back at the office.

"In short, the personal communicator will serve as our intimate communications link to our personal and professional worlds. These devices will merge telecommunications with computing to deliver immediate access to voice, data, handwriting, fax and video communications. They will be "intensely personal" and inextricably linked to our daily routine (we'll "... never leave home without them"). Because of their "immediacy," they will enable us to react spontaneously to the information we receive. These machines will afford us a new dimension in mobility and portability yet be extremely easy to use. They will appeal to a much broader set of customers than the devices we use today. Personal communicators will be as revolutionary as the first telephone."

Personal Computer PC. A computer for personal single-user use, as opposed to other types of computers — mainframes and minis — typically shared by many users. A PC is also called a workstation. See also MICROPROCESSOR.

Personal Computer Memory Card International Association See PCMCIA.

Personal Computing Systems Architecture PCSA. A network architecture defined and supported by Digital Equipment Corp. for the incorporation of personal computers into server-based networks.

Personal Computer Terminal Adapter PCTA. A printed circuit card that slips into an IBM PC or PC compatible and connects the PC to the ISDN T-interface. As of writing, such devices cost between $1250 and $1750 and are available in almost-commercial form from about six manufacturers. The PCTA basically turns a normal PC into an ISDN phone and ISDN terminal, ready for voice and data communications. According to Northern Telecom, one of the manufacturers of such a device, the PCTA does:
- Functional signaling call setup.
- B channel for circuit switched data.
- X.25 packet data services on the B or D channel.
- Simultaneous operation of the B and D channels.
- M5317T digital telephone to PC messaging for integrated voice and data operation.
- NetBIOS interface to applications software, such as Microsoft Networks. For more, see Northern Telecom's brochure ISDN PC Terminal Adapter NetBIOS Interface Description (D307-1).

Personal Data Module A removable module, unique to the ROLM Cypress and Cedar, which stores all information entered by the user. Such items as phone numbers, terminal profiles and log-on sequences are stored in the PDM, which also features battery backup to protect the memory in the event of a power failure. The user can take his personal data module with him, plug it into another Cypress or Cedar and get all his speed dial and other personalized programming on his phone he just moved from.

Personal Digital Assistant PDA. A consumer electronics gadget that looks like a palmtop computer. Unlike personal computers, PDAs will perform specific tasks — acting like an electronic diary, carry-along personal database, multimedia player, personal communicator, memo taker, calculator, alarm clock. The communications will take place through the phone or through wireless. Apple has announced a PDA, which it has code-named, Newton.

Personal Identification Number PIN number. 1. An AT&T term meaning the last four digits of your AT&T, MCI Bell operating company Credit Card — the card you use for making long distance numbers. 2. Some banks and financial institutions issue credit cards for machine, teller-less banking. These machines, called Automated Teller Machines, ask you for a password consisting of several numbers or characters. These are not on your credit card. These numbers or characters, called PIN numbers, are designed to make sure the right person is using your card. It's not a good idea to use your birthday as your PIN number.

Personal Intelligent Communications What General Magic calls the products and services its alliance members will create using General Magic technologies that will help people remember, communicate and know things in new and powerful ways. According to General Magic, "the alliance's shared long-term vision is to bring personal communications to people who may not use a computer today, to people whose personal technology is a car, a television set and a telephone."

Personal Information Assistant Tandy's name for what Apple calls a Personal Digital Assistant. See PDA.

Personal Information Manager PIM. Software application which allow the user to organize personal information. similar to an appointment book but personalized and programmed into the PC. PIMs now support Caller ID. When the phone rings, the phone system passes the calling number to the PC, which does a lookup in the PIM on the phone number and throws information about the caller up on screen.

Personal IVR An Interactive Voice Response system running on your own personal PC and designed to serve the needs of only one person. See IVR.

Personal Line A feature which allows specific key telephones to have their own private Central Office line. Sometimes called an AUXILIARY LINE. You can typically receive and make calls on this line. No one else can answer it, since it does not appear on any other phone instrument in the office. You can give this number to your wife or girlfriend. It is not a good idea to give it to both.

Personal Name An X.400 term for a standard attribute of an O/R (Originator/Recipient) Address form that identifies a person relative to another attribute (e.g., an organization name). The personal name may include surname, given name, initials, and generation qualifier. Initials consists of the first letter of all the user's names except the user's surname.

Personal Productivity Tool Another term for a computer. John Perry Barlow thinks the expression was created by the "druids" who run Microsoft and Apple. Mr. Barlow is a cattle rancher, computer hacker, poet, and a lyricist for the rock band The Grateful Dead.

Personal Speed Dial Simplified ways of dialing. You do them by dialing a couple of digits. Or you punch in a button at your phone. Personal Speed Dial codes are programmed for each telephone, and can only be used at the telephone on which they are programmed. System Speed Dials, in contrast, can be used from every phone in the system.

Personal System/2 PS/2. IBM's current family of microcomputers some of whom sport one major difference from its predecessors, namely the existence of a 32-bit micro-channel bus. This "bus" serves the same purpose as a PBX's backplane — namely to move information from the printed circuit cards and to other printed circuit cards, which may contain their own individual microprocessors (computers on a chip) and which may communicate with the outside world through their own communications ports.

Personal Telephone The category of cellular telephones pioneered by Motorola's Pan American Cellular Subscriber Group with the introduction of the MicroTAC Digital Personal Communicator Telephone. Weighing less than one pound, they are so compact and lightweight, they fit comfortably into a shirt pocket or purse making them "body friendly."

Personal Video System A PC-based video system for Windows introduced by AT&T in conjunction with its NCR subsidiary in the spring of 1993.

Personality Module A small motherboard added to a voice board to give it the "personality" of a proprietary electronic PBX telephone. "Personality" means electrical characteristics and the same button configuration and responsiveness, all of which can be recreated on the screen of a PC which has the voice board installed.

Personalized Ringing A telephone feature which allows you to select different ringing sounds for your telephone. This feature is useful if you work in a big room with lots of other people and it's hard to tell whose phone is ringing.

PERT Project Evaluation and Review Technique. A variation on the Critical Path Method of organizing the completion of projects. Projects are examined for the their worst, best, average completion times. A critical path is determined and overall standards for completion times are created. The PERT technique was created by the military. It is used for organizing complex tasks.

PES Packetized Elementary Stream: In MPEG-2, after the media stream has been digitized and compressed, it is formatted into packets before it is multiplexed into either a Program Stream or Transport Stream.

Petabyte A petabyte is equal to 10 to the 15th. A petabyte is equal to 1,000 terabytes. Here is a summary:
  MH = Megabyte (10 to the 6th power)
  GH = Gigabyte (10 to the 9th power)
  TH = Terabyte (10 to the 12th power)
  PH = Petabyte (10 to the 15th power)
  EB = Exabyte (10 to the 18th power)

PF Xfer Power Failure Transfer.

PG Peer Group: A set of logical nodes which are grouped for purposes of creating a routing hierarchy. PTSEs are exchanged among all members of the group.

PGA Programmable Gain Amplifier, a part of the Analog Front End.

PGL Peer Group Leader: A single real physical system which has been elected to perform some of the functions associated with a logical group node.

PGP PGP stands for Pretty Good Privacy. It is a cryptography program for computer data, electronic mail and voice conversations. PGP was written in 1991 by Philip R. Zimmermann, who gave the software away. It was posted on a public computer on Internet and thousands have downloaded copies of it. So far, according to the New York Times Magazine, June 12, 1994, no has been able to crack information encoded with PGP. Originally, PGP was just used for the transmission of computer data and electronic mail. Then it got extended to voice conversations on the phone, called PGPfone. The idea is that you use modems to dial. Then the PCs "shake hands" and jointly agree on a complex number that is plugged in a scrambling algorithm equation. This notoriously complex scrambling algorithm, called Blowfish, recalculates the digital ones and zeroes of the sampled voices into a stream of numbers unintelligible even to highly sophisticated eavesdroppers. Finally, PGPfone unscrambles the stream to provide intelligible — though not great quality — sound. http://web.mit.edu/network/pgpfone/

PGPfone The name of the phone is used for Pretty Good Privacy. See PGP for a full explanation.

PH Packet Handler, or Packet Handling function.

Phantom Circuit A circuit derived from two suitably arranged pairs of wires, called side circuits, with each pair of wires being a circuit in itself and at the same time acting as one conductor of the phantom.

Phase The relationship between a signal and its horizontal axis, also called zero-crossing point. A full cycle describes a 360 degree arc. A sine wave that crosses the zero-point when another has attained its highest point is 90 degrees out of phase with the other. See PHASE SHIFT KEYING.

Phase A Phase A is the first part of a fax machine's call process. It is the call establishment. It occurs when transmitting and receiving units connect over the phone line, recognizing one another as fax machines. This is the start of the handshaking procedure. See PHASE B.

Phase B Phase B is the second part of a fax machine's call process. It is the premessage procedure, where the answering machine identifies itself, describing its capabilities in a burst of digital information packed in frames conforming to the HDLC standard. See PHASE C.

Phase C Phase C is the third part of a fax machine's call process. It is the fax transmission portion of the operation. This step consists of two parts C1 and C2 which take place simultaneously. Phase C1 deals with synchronization, line monitoring and problem detection. Phase C2 includes data transmission. See PHASE D.

Phase D Phase D is the fourth part of a fax machine's call process. This phase begins once a page has been transmitted. Both the sender and receiver revert to using HDLC packets as during Phase B. If the sender has further pages to transmit, it sends an MPS and Phase C recommences for the following page. See PHASE E.

Phase Displacement Antenna An antenna constructed from a driven element and a group of reflectors, the secondary radiation from which produces an antenna with directivity. The Yagi-Uda array is a member of this family.

Phase Delay See ENVELOPE DELAY.

Phase Distortion An unwanted modification of a transmitted signal caused by the non-uniform transmission of the different frequency components of the signal. Same as Delay Distortion.

Phase E Phase E is the fifth part of a fax machine's call process. This phase is the call release portion. The side that transmitted last sends a DCN frame and hangs up without awaiting a response.

Phase Hit In telephony, the unwanted and significant shifting in phase of an analog signal. As defined by AT&T: any case where the phase of a 1004 HZ test signal shifts more than 20 degrees. Also, error-causing events more severe than phase jitter, especially for data transmission equipment using PSK modulation.

Phase Jitter In telephony, the measurement, in degrees out of phase, that an analog signal deviates from the referenced phase of the main data-carrying signal. Phase jitter is often caused by alternating current components in a network.

Phase Lock Loop In electronics, a circuit that acts as a phase detector by comparing the frequency of a known oscillator with an incoming signal and then feeds back the output of the detector to keep the oscillator in phase with the incoming frequency.

Phase Modulation One of three ways to change a sine wave (or L signal) to let it carry information. In this case, the phase of the sine wave is changed as the information to be carried is changed. See PHASE SHIFT KEYING and MODULATION.

Phase Roll Variations in the phase of a transmitted signal and its echoed back modem verification. Phase roll is encountered most often in international systems.

Phase Shift A change in the time or amplitude that a signal is delayed with respect to a reference signal.

Phase Shift Keying Used for transmitting digital signals over analog phone lines. Some modems use phase shift keying. Picture a sine wave. The sine wave is the carrier. We want to place a digital signal on it. We want the signal to change in some way to reflect the presence of a "1" or the absence of a one, i.e. a zero. In phase shift keying, we simply change the phase of the signal.

Phase Lock Loop. PLL. Phase Lock Loop is a mechanism whereby timing information is transferred within a data stream and the receiver derives the signal element timing by locking its local clock source to the received timing information.

Phasing The process of ensuring that both sending and receiving facsimile machines start at the same position on a page.

Phasor Temporary buffer storage that compensates for slight differences in data rate between TDM I/O ports and devices.

Phenolic Insulating Materials A type of insulating materials, one of which is bakelite. Now no longer used.

PHF Packet Handling Function. The switching capability that processes and routes X.25 virtual calls.

PHI 1. PBX-to-Host Interface. The same as CPI (Computer to PBX Interface) but it puts the PBX first, which is the way many telephone manufacturers prefer to see it. It refers to a connection between a telephone system and a computer, such that the computer can signal the telephone system to switch calls and the telephone system can signal the computer when it has switched them. There are major advantages in joining a telephone system to a computer. For a much greater explanation, see OAI, which stands for OPEN APPLICATION INTERFACE.

2. PHI is also a Northern Telecom term for Packet Handler Interface.

Phoenix MCI's ill-fated order entry system that never rose from the dead, i.e. never worked.

Phone A simpler way of saying the word "telephone." An invention of the devil. See TELEPHONE.

Phone Bomb A phone bomb is a bobby-trapped cellular phone. I saw the term first used on the front page of the January 6, 1996 New York Daily News. The story read: "'The engineer,' an extremist Palestinian bomb maker which was No. 1 on Israel's most-wanted list was killed on the Gaza Strip yesterday when his head was blown up by a booby-trapped cellular phone. The phone was rigged with about two ounces of explosives." No one admitted to killing the man.

Phone Centric Users These are knowledge workers for whom the telephone plays a key role in the success of their business —, e.g. salesmen, stock brokers.

Phone Phreaks Communication hobbyists. People, usually kids, who like to figure out how the telephone network works and sometimes make free calls on the network by figuring a way to bypass billing mechanisms. Phone Phreaks have become Computer Phreaks with the boom of microcomputers and the advent of out-of-band signaling, making it a lot more difficult to make long distance calls for free.

PhoneMail A ROLM term for Voice Mail. Rolm's PhoneMail is a voice messaging system that provides telephone answering (with the user's own greeting), the capability to store and forward voice messages and the capability to turn on a message waiting light or message on the recipient's phone. PhoneMail can be used positively to speed the flow of information. It can also be used negatively to allow the user to "hide behind" the system and avoid the outside world and anyone in the outside world who might actually want to buy something. See also VOICE MAIL.

Phoneme A voice recognition term. The minimal significant structural unit in the sound system of any language that can be used to distinguish one word from another. For example, the p of pit and the b of bit are considered two separate phonemes, while the p of spin is not. These minimal sound units comprise words.

PhoneNet Farallon's twist on Apple's local area network called LocalTalk. PhoneNet uses standard one pair UTP (unshielded twisted pair) wiring for networking. PhoneNet is compatible with LocalTalk.

PhoneSmart A name for a bundle of features NY Telephone provides from selected digital central office. They include: Call Return — automatically dials the number of the last person who called. Repeat Dialing — redials the last number called. Call Trace — forwards the number, the date and time of the last call received to Nynex's Annoyance Call Bureau.

Phonetic Alphabet A list of standard words used to identify letters in a message transmitted by radio or telephone. The following are the authorized words, listed in order, for each letter in the alphabet: Alpha, Bravo, Charlie, Delta, Echo, Foxtrot, Golf, Hotel, India, Juliet, Kilo Lima, Mike, November, Oscar, Papa, Quebec, Romeo, Sierra, Tango, Uniform, Victor, Whiskey, X-ray, Yankee, Zulu.

Phosphor Substance which glows when struck by electrons. The back of a picture tube face is coated with phosphor.

Photo Diode The basic element that responds to light energy in a solid state imaging system. It generates an electric current that is proportional to the intensity of the light falling on it.

Photoconductive Effect Some non-metallic materials exhibit a marked increase in electrical conductivity when they absorb photon or light energy. This is called the photoconductive effect. The conductivity increase is due to the additional free carriers generated when photon energies are absorbed in electronic transitions. The rate at which free carriers are generated and the length of time they persist in conducting states (their lifetime) determines the amount of conductivity change.

Photoconductivity The conductivity increase exhibited by some nonmetallic materials, resulting from the free carriers generated when photon (i.e. light) energy is absorbed in electronic transitions. The rate at which free carriers are generated, the mobility of the carriers, the length of time they persist in conducting states (their lifetime) are some of the factors that determine the extent of conductivity charge. See PHOTOCONDUCTIVE EFFECT.

Photoconductor 1. Any transducer that produces a current which varies in accordance with the incident light energy. A fiber optic communications term.

2. Photoconductor is also material, available in many forms (sheets, belts, and drums), which changes in electrical conductivity when acted upon by light. Electrophotography (a form of facsimile machine printing) relies on the action of light to selectively change the potential of a charged photoconductive surface, creating areas receptive to an oppositely charged toner, thus making the latent charged-image visible.

Photocurrent The current that flows through a photosensitive device (such as a photodiode) as the result of exposure to radiant power. Internal gain, such as that in an avalanche photodiode, may enhance or increase the current flow but is a distinct mechanism.

Photodetector In a lightwave system, a device which turns pulses of light into bursts of electricity.

Photodiode See PHOTO DIODE.

Photoelectric Effect The emission of electrons by a material when it is exposed to light. Albert Einstein received a Nobel Prize for explaining this phenomenon. Amazingly, he never received one for his brilliant theories of relativity.

Photon The fundamental unit of light and other forms of electromagnetic energy. Photons are to optical fibers what electrons are to copper wires. Like electrons they have a wave motion.

Photonic Layer The lowest of four layers of Sonet capability, which specifies the kind of fiber to be used including sensitivity and laser type. See SONET.

Photonics The technology that uses light particles (photons) to carry information over hair-thin fibers of very pure glass.

Phototransistor A transistor that detects light and amplifies the resulting electrical signal. Light falling on the base-collector junction generates a current, which is amplified internally.

Photovoltaic Adjective to describe material which develops voltage and electrical current when light shines on it.

Photovoltaic Effect Using light to produce electricity. Shine light on a device, typically a "cell." If the device produces electricity, that's called the photovoltaic effect. It's not very efficient at present. Less than 10% of the light energy emerges as electricity. But it's getting better.

PHP Personal Handy Phone. Japan's standard for digital cordless phones.

Phreaking Employing technology to attack the public telephone system and get free long distance service.

PHY PHYsical, as in physical specifications. OSI Physical Layer: The physical layer provides for transmission of cells over a physical medium connecting two ATM devices. This physical layer is comprised of two sublayers: the PMD Physical Medium Dependent sublayer, and the TC Transmission Convergence sublayer. See PMD and TC.

Physical Address A six digit number giving the physical location of an extension within the Rolm CBX. The number is in the form of xxyyzz, where xx=shelf, yy=slot and zz=channel.

Physical Colocation A local exchange carrier (LEC) provides space within the building housing its central office to other phone companies, to interconnect companies and/or to users to place their equipment. Typically it's done to connect circuits — transmission or switching — to the phone company's central office equipment. The interconnector (i.e. the company placing the equipment) installs, maintains, and repairs its own equipment, while the LEC provides power, environmental conditioning, and conduit and riser space for the interconnector's cable.

Physical Connection The full-duplex physical layer association between adjacent PHYs in an FDDI ring.

Physical Delivery Delivery of a message in physical form through a Physical Delivery System; for example, delivery of a letter through the U.S. Postal Service. This term is used in X.400.

Physical Delivery Address Component An X.400 address component that describes how to physically deliver a message. For example, the name and mail stop to hand deliver a message after it is printed. The concept is that the X.400 address would cause a message to be printed on a printer and an individual would complete the hand delivery.

Physical Delivery Office Name Standard attribute of a Postal O/R (Original/Recipient) Address, in the context of physical delivery, specifying the name of the city, town, etc., where the physical delivery is to be accomplished. An X.400 term.

Physical Delivery Office Number Standard attribute in a Postal O/R (Originator/Recipient) Address that distinguishes between more than one physical delivery office within a city, etc. An X.400 term.

Physical Delivery Organization Name A free form name of the addressed entity in the postal address, taking into account the specified limitations in length. An X.400 term.

Physical Delivery Personal Name In a postal address a free form name of the addressed individual containing the family name and optionally the given name(s), the initial(s), title(s) and generation qualifier, taking into account the specified limitations in length. An X.400 term.

Physical Delivery Service The service provided by a Physical Delivery System. An X.400 term.

Physical Delivery Service Name Standard attribute of a Postal O/R (Original/Recipient) Address in the form of the name of the service in the country electronically receiving the message on behalf of the physical delivery service. An X.400 term.

Physical Formatting The second step in structuring a hard drive so that you may write to it. Physical formatting follows partitioning.

Physical Layer The wires, cables, and interface hardware that connect devices on the network. Within the OSI model, the physical layer is the lowest level of network processing, below the link layer, that is concerned with the electrical, mechanical and handshaking procedures over the interface that connects a device to a transmission medium.

Physical Layer Connection An association established by the PHY (OSI Physical Layer) between two or more ATM entities. A PHY connection consists of the concatenation of PHY links in order to provide an end-to-end transfer capability to PHY SAPs.

Physical Layer Medium Dependent The Physical Layer sublayer that defines the media independent portion of the Physical Layer in FDDI. Items defined by PMD include transmit and receive power levels, connector requirements, and fiber optic cable requirements.

Physical Layer Protocol The Physical Layer sublayer that defines the media independent portion of the Physical Layer in FDDI. Items defined by the PMD include transmit and receive power levels, connector requirements, and fiber optic cable requirements.

Physical Link A real link which attaches two switching systems.

Physical Media Any means in the world for transferring signals between OSI systems. Considered to be outside the OSI Model, and therefore sometimes referred to as "Layer 0." The physical connector to the media can be considered as defining the bottom interface of the Physical Layer, i.e. the bottom of the OSI Reference Model.

Physical Rendition The transformation of an MHS (Message Handling System) message to a physical message (e.g., by printing the message on paper and enclosing it in a paper envelope). An X.400 term.

Physical Signaling Sublayer PLS. In a LAN or MAN system, that portion of the OSI Physical Layer that interfaces with the medium access control sublayer and performs bit symbol encoding and transmission, bit symbol reception and decoding, and optional isolation functions.

Physical Slots Slots that are available to cards in a card shelf.
Physical Topology The actual arrangement of cables and hardware that comprise a network.
Physical Unit PU. In IBM's SNA, the component that manages and monitors the resources of a node, such as attached links and adjacent link stations. PU types follow the same classification as node types.
Physical Unit Control Point PUCP. In SNA, the component that provides a subset of the system-services control point (SSCP) within a node. Types 1, 2 and 4 nodes contain a PUCP, while a Type 5 (host) contains a SSCP.
PHz Petahertz (10 to the 15th power hertz). See also SPECTRUM DESIGNATION OF FREQUENCY.
PIA Personal Information Appliance. A name for a product most people call a Personal Digital Assistant (PDA).
PIC 1. See PRIMARY INTEREXCHANGE CARRIER.
2. Plastic Insulated Conductor. Conductors covered with an extruded coating of plastic.
3. Also an imaging term. Picture Image Compression. Intel-DVI Technology's on-line still image compression algorithm. See DVI.
4. Personal Intelligent Communicator. A General Magic term for a product most other people called a Personal Digital Assistant. See PDA.
5. Point In Call.
6. Programmable Interrupt Controller. A chip or device that prioritizes interrupt requests generated by keyboards, serial ports, and other devices and passes them on to the CPU in PC in order of highest priority. See also IRQ.
Pick The computer operating system of VMark Computer Inc. Pick is a neat operating system that unfortunately never really caught on.
Pick-And-Place The manufacturing operation in which components are selected and placed in the correct position on a substrate for the purpose of interconnection to the substrate. This is most commonly done with a programmable machine equipped with a robot arm.
PICMG PICMG stands for the PCI Industrial Computer Manufacturers Group. It is a consortium formed to push an open standard. The electrical and mechanical specifications of PICMG enable CPU boards and backplanes from different manufacturers of industrial-grade computers to be interchangeable. PICMG is the organization specifying the new CompactPCI standards. See CompctPCI.
Pickup Means you can answer a call from your phone. There all sorts of "pickups." The most common is GROUP PICKUP. Here you are part of a group and you can answer — from your phone — the call of anybody in that group, usually by punching a digit or a button or two. There's also NIGHT PICKUP, which typically allows anyone to answer an incoming call after hours, again by punching down a digit or a button or two. In TELECONNECT Magazine, we have one GROUP PICKUP. Everybody in the company belongs to that Group and everyone can answer everyone else's phone. We believe this simplifies things. It also allows anyone, anywhere, from any phone, to play telephone attendant, answering any and all incoming calls and transferring them through the system.
Pickup Group A telephone can be placed into one of four Call Pickup Groups. A call ringing at a telephone within a Pickup Group can be picked up by any other telephone within the same Pickup Group.
Pickup Pattern A determination of the directions from which a microphone is sensitive to sound waves. It varies with the mike element and mike design. The two most common pickup patterns are omni- and uni-directional.
Pico Prefix meaning one-trillionth, one-millionth of a millionth of a second. (or, for those in England, Australia or New Zealand, one thousand millionth.)
Picofarad One-trillionth of a farad. A unit of capacitance usually used to designate capacitance unbalance between pairs and capacitance unbalance of the two wires of a pair to ground.
Picosecond One-millionth of a millionth of a second. NANO is a billionth of a second, (or, for those in England, Australia or New Zealand, one thousand millionth of a second).
PICS 1. Protocol Implementation Conformance Statement: A statement made by the supplier of an implementation or system stating which capabilities have been implemented for a given protocol.
2. Product Inventory Control System.
3. USWest defines the term as Plug-In Inventory Control System. Plug-Ins, according to USWest, are circuit cards which fit into central office equipment and which control various aspects of transmission and carrier circuit functionality and usage.
Picture Element See PIXEL.
Picturephone AT&T's trademark for a video telephone that permitted the user to see as well as talk with the person at the distant end. AT&T introduced it at the 1964 World's Fair in Flushing Meadow, Queens, New York City. The device had a camera mounted on the top and a 5.25" x 4.75" screen. Audio signals were transmitted separately from video signals and the system could not use the public switched telephone network. It needed a transmission bandwidth of 6.3 Mbps and no one wanted to pay the price for the service. It never got off the ground. AT&T picked up on the name Picturephone and came out with an offering called Picturephone Meeting Service which provided full video teleconferencing. It was available through rented rooms or through equipment sold or rented to corporations. It also didn't do too well since the service was expensive and no one wanted to spend the time traveling to the few and far between conferencing room. We believe, as of writing, AT&T has abandoned Picturephone, the product, and has closed down the Picturephone Meeting Service rooms it rented to corporations. In January, 1992 AT&T introduced a product called VIDEOPHONE 2500, which transmitted moving (albeit slowly-moving) color pictures over normal analog phone lines. See VIDEOPHONE 2500.
Picturephone Meeting Service An AT&T service once provided under experimental tariff. It combined TV techniques with voice transmission. PMS is usually only available between telephone company-located picturephone centers. Most (we think all) have now been closed down. The venture was losing too much money.
Picturetel One of the two largest manufacturers of videoconferencing equipment (both desktop systems and room-based systems) in the world.

463

NEWTON'S TELECOM DICTIONARY

PID Protocol Identification. See OUI.
PIECE Productivity, Information, Education, Creativity, Entertainment. Microsoft's trick for remembering the big five multimedia computing applications.
Piepser The German word for beeper. Also the name of a small beeper made by Swatch and sold by BellSouth.
Piesio The Greek prefix meaning near.
Piezo-Electric Crystal A type of crystal which, when subjected to mechanical stress, generates current; or which, when subjected to varying electrical stresses, generates mechanical movement. Most familiar type is Rochelle Salts crystal. An old radio term.
PIF 1. Personal Communications Services Industry Forum.
2. Program Information File, a binary file, which contains information about how Windows should run an MS-DOS application, such as how much memory it needs, the path to the executable file, and whether the window in which the program is run closes automatically when the program terminates.
Piggy Back Data Slurp Imagine a data communications connection between two distant computers. Somewhere along the line, someone has attached a terminal or communicating computer and begun to capture the data as it flows across the line. In short, someone had slurped out data on a piggyback terminal or communicating computer. A piggyback data slurp could continue forever, without either party being aware their data and their conversations were being stolen — so long as the piggyback terminal didn't let its presence be known. The piggyback terminal, however, may insert itself into the conversation, pretending to send back authentic communications, thus misleading one or both of the parties. You could call this active deception or proactive espionage.
Piggybacking A technique used at the data link or transport layer in a layered network architecture that allows for transmission acknowledgments to be carried in transmission frames received from the destination.
Pigtail 1. Multiple pieces of short cable with single circuit connectors connected to a multi-conductor cable. See OCTOPUS.
2. A short, permanently attached piece of optical fiber used to link the transmitter and receiver to the transmission fiber.
Pigtail Antenna The standard cellular antenna for a car. The term "pigtail" refers to the spring-like section in the lower third of the antenna, the phasing coil.
Pilot Number Identifies a Hunt Group or Distribution Group. See also DISTRIBUTION GROUP.
Pilot-Make-Busy Circuit A circuit arrangement by which trunks provided over a carrier system are made busy to the switching equipment in the event of carrier system failure, or during a fade of the radio system.
PIM 1. Protocol-Independent Multicast. Multicast routing architecture that enables IP multicast routing on existing IP networks.
2. Personal Information Manager. A specialized form of software used by individuals and groups for keeping track of their contacts (and their addresses and phone numbers), their appointments, their lists of Things to Do, their reminder notes, their anniversaries, etc. PIMs are also called contact managers. Examples are Maximizer, TeleMagic, Borland's SideKick and Lotus' Organizer.
PIN 1. Procedure Interrupt Negative. A fax term.
2. Positive Intrinsic Negative. Type of photodetector used to sense lightwave energy and then to convert it into electrical signals. See PIN DIODE.
3. Personal Identification Number. See PERSONAL IDENTIFICATION NUMBER.
PIN Code Personal Identity Number code. A code used by a mobile telephone subscriber in conjunction with a SIM card to complete a call.
Pin Diode A photodiode made with an intrinsic layer of undoped material between doped P and N layers and used as a lightwave detector.
PIN Number Personal Identification Number. A group of characters entered as a secret code to gain access to a computer system, such as the one that completes long distance calls. See PERSONAL IDENTIFICATION NUMBER.
Pin Photodiode An optical detector that converts light into electricity. This type is the typical diode used in a fiber optic receiver.
Pincushion Distortion When a video screen is distorted — with the top, bottom and sides pushing in — the screen is said to be suffering pincushion distortion.
PING Packet InterNet Groper. A program used to test reachability of destinations by sending them an ICMP echo request and waiting for a reply. The term is used as verb: "Ping host X to see if it is up!" Ping is useful for testing and debugging networks. It sends an ICMP Echo packet to a specified host, and waits for a response. It reports success or failure and statistics about its operation.
Ping Pong A method of getting full duplex data transmission over a two wire circuit by rapidly alternating the transmission direction. See PING PONGING.
Ping Ponging Routing that causes a packet to bounce back and forth between two modes.
Pink Adult audiotext, i.e. dirty talking over the phone for money.
Pink Noise Noise in which power distribution is logarithmic through the spectrum, with an equal amount of power in each octave.
Pink Pages In Australia, at one stage, the list of businesses organized by industry, were printed on pink paper and called the Pink Pages. They are the equivalent of the North American "Yellow Pages."
Pinouts Pin configurations for cabling. In other words, which pin connects to which cable. Not all pins are always connected. Not all cables always connected.
Pinosecond One-trillionth of a second. One-millionth of a microsecond.
Pipe A communications process within the operating system that acts as an interface between a computer's devices (keyboard, disk drives, memory, and so on) and an applications program. A pipe simplifies the development of application programs by "buffering" a program from the intricacies of the hardware or the software that controls the hardware; the application developer writes code to a single pipe, not to several individual devices. A pipe is also used for program-to-program communications.
Pipelining 1. Executing instructions by breaking them into component parts and processing them in parallel on separate processors. This reduces reduce cycle time and increases the computer's performance.
2. In imaging, pipelining also lets an imaging card start compressing and writing the image to disk while it is still being scanned.
3. In networking, pipelining is a technique used at the transport layer or data link layer in a layered network architecture that

allows for the transmission of multiple frames without waiting to see if they are acknowledged on an individual basis. Each frame may have to be acknowledged later and in sequence, or a process of implied acknowledgment may be employed. Implied acknowledgment is a process whereby negative acknowledgment of a specific frame implies that all previously transmitted frames have been received correctly.

PITA Abbreviation for "Pain In The Ass;" commonly used on E-mail and BBSs (Bulletin Board Systems).

Pitch Control Variable control for increasing or decreasing the speed of a tape deck or turntable.

PIU Path Information Unit in SNA.

Pixilation A technique used by cinematographers and stage managers to make human performers appear to move as it artificially animated. Using a stop-frame camera, the pixilator can distort and speed up the motion of actors.

Pixel PIcture ELement. The smallest unit of area of a video screen image that can be turned on or off, or varied in intensity. The single point on a CRT display. The single point in a facsimile transmission. The image you see on your screen is the result of some pixels being on and others off. A pixel is the smallest part of the video screen that can be turned on or off or varied in intensity. It is one of the phosphor elements that coat the inside of a CRT tube. Pixels glow when struck by an electron beam. The number of pixels in the most common computer screen, a VGA monitor, is 640 x 480, with the first number (640) being the number of pixels in each horizontal row and the second number (480) the number of rows displayed. VGA stands for Video Graphics Array. Resolution (crispness and clarity of text and images) improves as the number of pixels displayed increases. If it's a color screen, a pixel is really three dots together, or clusters of red, green and blue — the triad of colors that, when energized, add up to white, or when the set is turned off, show as black. The phrase "picture element" was first used in 1927 in the magazine "Wireless World" writing about the mosaic of dots, or picture elements. See PEL.

PIXIT Protocol Implementation eXtra Information for Testing: A statement made by a supplier or implementor of an IUT (Implementation Under Testing) which contains information about the IUT and its testing environment which will enable a test laboratory to run an appropriate test suite against the IUT.

Pixrects Pixrects is the primary graphics programming interface in the SunView Window System from Sun Microsystems. It is replaced in the OpenWindows XView toolkit by the Pixwin interface, which is a thin layer on top of Xlib.

Pixwin Pixwin is the primary graphics programming interface in the XView toolkit from Sun Microsystems. Pixwin is a thin layer on top of Xlib.

PJ-327 A double RCA dipole plug for connecting a headset into a PBX console. When you buy a headset you need to specify — 4-pin modular jack (for connecting to a telephone) or two prong plug (for connecting to a PBX console).

PL Private Line.

PL/1 Programming Language One (IBM).

Place Call There will be two types of calls: person calls and place calls. I make a place call when I call a phone number which ends in one designated, fixed RJ-11 jack attached to the wall or floor. I make a person call when I call a phone number which doesn't necessarily end on a fixed place. A typical person call might be a cellular or wireless phone call. It might also be a service like MCI's Personal 800 Service. The major characteristic of a person call is that I am calling a person and I don't know where that person is. As a result, the person I called might answer my call anywhere — from his car, from vacation home, from his wireless phone, etc.

Plaintext A message which is not encrypted. A message which is encrypted is called a ciphertext. See CLIPPER CHIP.

Plan A route to an end or objective usually achieved accidentally and often written in hindsight. Derived from the expression: "Most people spend more time planning their annual vacation than they do planning their careers."

Plan File A file that lists anything you want others on the Internet to know about you. You place it in your home directory so your public-access site. Then, anybody who fingers (sees) you, will get to see this file.

Planar Array Antenna A planar array antenna is designed for use at microwave frequencies. It resembles a double-sided printed circuit board. One side of the substrate carries an etched pattern of microstrip whilst the other is left completely metallic to act as a ground plane.

Planar Board IBM's new name for a motherboard in their new series of System/2 Personal Computers. A motherboard is the main board in a PC on which the main CPU, the main memory, the clock and sundry other things like serial and parallel ports are mounted. Other boards, i.e. graphics boards, are plugged into the motherboard. Thus the expression "motherboard." No one knows why IBM dropped the word. Maybe it was too risque? Maybe they included more on their motherboards in the System/2 series that they would no longer function as motherboards? Maybe a feminist group of mothers objected?

Plant A general term for all equipment used by a telephone company to provide telecommunications services. Usually divided into outside and inside plant.

Plant Hump A very friendly term for a craftsperson in a phone company — installer, splicer, underground cable guy — who works hard with outside phone equipment, often under adverse conditions and whose labors tend to be undervalued, particularly by the white collar, pencil pushers back at headquarters. In British slang, "hump" means to exert oneself. There is a badge of honor to being a "plant hump." These are the people who bring phone service to your door. This definition contributed by Steve Marcus of Nynex, who readily acknowledges he is one of the pencil pushers back at headquarters. The word plant hump is occasionally written as one word, i.e. PLANTHUMP.

Plant Test Numbers Virtually every 800 IN-WATS number has a plant test number. This is its equivalent seven digit local number. That number looks like a normal local seven digit number, with a standard three-digit central office exchange code and a four-digit extension. The purpose of plant test numbers is to allow the telephone company to test the local part of the incoming 800 number by simply dialing that number. For example, TELECOM LIBRARY and TELECONNECT Magazine, which published this dictionary have an 800 number — 800-LIBRARY (or 800-542-7279). The plant test number of the first line of that 800-LIBRARY group is 212-206-6870. The second line is 212-206-6871 and so on. It is valuable to know the plant test numbers of your incoming WATS lines so you can test the local loop part of those lines. The local loop part is the part which typically gives the most problem. It is, unfortunately, the only part of your 800 lines you can test yourself — unless you ask someone (or several people) to call you regularly on your 800 lines, just to test them. You can get plant test numbers out of your local and/or your long distance carrier. When they tell you those numbers are "not available," beg a little. They are available and you are entitled to them. Calling plant test numbers costs exactly what a normal long distance IN-WATS call on that line costs. So keep your test calls short. You should call your plant test numbers once a day.

Planthump A colloquial word for a telephone company craftsperson. It derives from the term "plant," a telephone company word used to describe their "factory" — i.e. everything from their inside plant, their central office switch, to their outside plant, which includes wire strung on telephone poles. In British slang, "hump" means to exert oneself. Planthump is a term of endearment in the telephone industry. Definition courtesy, Steve Marcus, New York Telephone. The term is often spelled as two words, i.e. PLANT HUMP.

PLAR Private Line, Automatic Ringdown. In telecommunications, leased voice circuit that connects two single instruments together. When either handset is lifted, the other instrument automatically rings.

Plasma An ionized gas with a mixture of positive and negative electrons. See PLASMA DISPLAY.

Plasma Display Type of flat visual display device in which selected electrodes, part of a grid of crisscross electrodes in a gas-filled panel, are energized, causing the gas to be ionized and light to be emitted. Some computers use plasma displays. They're fabulous, and quite expensive. See PLASMA DISPLAY PANELS.

Plasma Display Panels PDPs. Plasma gases composed of helium, neon and xenon are sandwiched into cells between two vertical glass plates. Bursts of electricity are applied between transparent electrodes attached to one pane of glass. These bursts causes the plasma gases to emit ultraviolet rays. This activates red, blue and green phosphor dots, which emit visible light and form pictures on the screen. The more common cathode ray tube technology (e.g. computer monitors) uses an electron gun to direct a beam that lights up phosphors on a screen. Directing that beam requires CRT sets to be deep, heavy and unwieldy.

Plasmatron The name Sony chose for a flat screen display it demoed at Comex in the fall of 1995. The screen measured 20 inches diagonally. It was as bright as normal CRT screen, but was less than four inches wide. It is the beginning of flat screen entertainment screens for the home. It was spectacular.

Plastic Fiber Optics An optical fiber where the core transmission media is plastic in contrast to glass or silica cores. Proposed plastic fibers generally have larger attenuation and dispersion than glass fiber but may have applications where the distance is limited. Plastic systems may also offer lower cost connectors that may be installed with simple tools and a limited amount of training.

Plastic Optic Fiber POF. A fiber transmission medium made from a plastic material. Such a medium uses low-quality light sources and carry data at speeds greater than 10 Mbps over distances up to 100 meters. POF is evolving as a replacement for twisted-pair copper wire. see PLASTIC FIBER OPTICS.

Plasticizer A chemical agent added in compounding plastics to make them softer and more flexible.

Plat An imaging term. When a CAD/CAM plotter prints a large drawing, it's called a plat.

Plate The anode in a vacuum tube, which collects the electrons emitted by the filament.

Plate Battery The source of E.M.F. connected in the plate circuit to give the plate element its positive charge.

Plate Voltage The potential applied to the plate of the vacuum tube by the plate voltage supply.

Platen A cylinder in a printer or typewriter around which the paper goes and which the printing mechanism strikes to produce an impression.

Platform A loosely-defined word for a software operating system and/or open hardware, which an outsider could write software for. If every phone system were a platform, then every owner of that phone system could buy outside-produced software and have his phone system work more to his liking. That's the objective of creating a "platform." See also OAI and PLATFORM INDEPENDENCE.

Platform Independence A term from IBM and Metaphor Computer Systems. The idea, they say, is to produce a layer of software that would rest atop any operating system on any piece of hardware. The applications developers would write their software just once, rather than start from scratch each time they wanted get their software working on a different computer. If the whole idea sounds rather daunting, you're right.

Platter The round magnetic disk surfaces used for read/write operations in a hard disk system.

Playback Retrieval, decoding and transmission of encoded data. It is also a multimedia term. Playback is the process of viewing multimedia materials created by an author. Playback can include a range of activities, from viewing a single video clip to participating in a series of interactive multimedia training modules. Some playback applications (for example many training and presentation applications) are sold separately from their authoring applications. However, many developers are selling authoring and playback capabilities in a single product.

Playback Head The part which converts the magnetic information on the tape or disk into an electrical signal. Moving the magnetic fields on the medium (tape or disk) past the playback head generates a tiny voltage, which is picked up in a conductor (a coil) in the payback head and sent onto the electronic equipment where it is amplified or transmitted.

Player An SCSA definition. A resource object that plays TVM data. The audio data can come from a voice or audio encoded file, or from text that has passed through a text-to-speech service. The output of a player can be analog audio, TDD, ADSI, etc.

PLCP Physical Layer Convergence Protocol. The part of the physical layer that adapts the transmission facility to handle DQDB functions as defined in IEEE 802.6-1990. It is used for DS-3 transmission of ATM. ATM cells are encapsulated in a l25microsecond frame defined by the PLCP which is defined inside the DS3 M-frame.

Plenum In some modern buildings, the ducts carrying the air conditioning are not metal ducts but actually are part of the ceiling. This is called a plenum ceiling. Most cities now have rules and regulations which say that if you run cabling through these plenum ceilings, you must not use cabling sheathed in PVC (polyvinyl chloride), the standard jacketing of most electrical cable. The reason is that PVC burns and emits toxic smoke ferociously. Plenum cable is low smoking so that if it catches fire it won't circulate toxic smoke through the vent system and suffocate everyone. Plenum cabling is often made of teflon. It's much more expensive than normal cabling.

Plenum Area The space between the drop ceiling and the floor above. Continuous throughout the length and width of each commercial building floor.

Plenum Cable Cable specifically designed for use in a plenum (the space above a suspended ceiling used to circulate air back to the heating or cooling system in a building). Plenum cable has insulated conductors often jacketed with polyvinylidene diflouride (PVDF) material to give them low flame spread and low smoke-producing properties. Plenum cable has fully color coded insulated copper conductors and is available in various pair sizes. It can be either 22 or 24 AWG.

Plesiochronous Plesiochronous means "almost synchronous." In the network sense, when two networks operate with clocks of sufficiently high quality such that the signals in the two networks are nearly synchronous, the networks are plesiochronous. In the synchronization hierarchy, stratum I clocks are required for plesiochronous operation. In other words, two signals that are not sourced from the same clock but are nominally at the same frequency to a defined degree of precision. For example, because they are not sourced to the same atomic clock, DS1s (also called T-1s) from MCI and AT&T are plesiochronous. They are arbitrarily close but over the long term will cause slips (frame repeats or deletions due to buffer under or overflow) in digital signals crossing between the two carriers.

Plesiochronous Networks Network elements that derive timing from more than one primary reference source. Network elements accommodate minor frequency differences between nodes.

PLL Phase Lock Loop: Phase Lock Loop is a mechanism whereby timing information is transferred within a data stream and the receiver derives the signal element timing by locking its local clock source to the received timing information.

PLMN Public Land Mobile Network. A mobile telephone communications network established by a provider to facilitate mobile telecommunications services. This includes equipment, operations, and staff. A single provider may have more than one PLMN.

Plotter A type of computer peripheral printer that displays data in two-dimensional graphics form.

PLS Premises Lightwave System.

PLSC Private Line Service Center.

Plug A male element of a plug/jack connector system. In the Premises Wiring System it provides the means for the user to connect his communications devices to the Communications Outlet as well as the means to disconnect his service at the Network Interface Jack when trouble analysis is required.

Plug "N" Go A new product IBM introduced in September, 1991. The Plug "N" Go System Solution includes an AS/400 9402 Model D02, terminals, printers and pre-loaded software, including Operating System/400*, the IBM Application Program Driver and selected Plug "N" Go Applications. The Plug "N" Go System Solution carries a customer satisfaction guarantee. The entire system — hardware, software and application — can be returned within 90 days for a full refund. The guarantee is designed for small businesses and is limited to one per customer. With a pre-loaded Plug "N" Go System Solution, according to IBM, customers can install a Model D02 and start using the application approximately ten minutes after turning on the system.

Plug 'N Play 1. Manufacturers' concept of how easy it is to install their equipment. "Why it's just plug 'n play," says the manufacturer. In reality, nothing, absolutely nothing, is plug 'n play. It's a fantasy concept. See PLUG AND PLAY.

2. Also defined as a new hire who doesn't need any training. "The new guy, Harry, is great. He's 100% plug-and-play."

Plug And Play This explanation comes from an Intel Technology Primer: Since add-in cards first appeared over a decade ago, they've given users a lot of different ways to improve their PCs and given them a lot of installation headaches. In this brief, we'll tell you how Intel, together with industry leaders, has spent years developing Plug and Play technology to make add-in cards both easier to use and install. Never before has the PC had as many capabilities as it does today. That's due in part to the large number of add-in cards available, like those for multimedia and faxmodems. Yet, as more cards are added to a PC, their installation can become quite complex. Installing a card can be a time-consuming and technical process, and there's no guarantee it will even work the first time. Sometimes the user must configure the card manually, which means selecting a variety of system resources for each card. These include Interrupt Requests (IRQ), I/O and memory addresses, and Direct Memory Access (DMA) channels. Every PC has a limited number of these resources available. Each card is designed to use a small group of them. Assigning these resources means opening the computer and physically setting the jumpers and DIP switches. And since no standard has been set to determine which cards can use which resources, numerous conflicts can arise between cards. Often, it's a process of trial and error to determine which resources aren't already being used by other cards. Since the ISA bus was introduced, several new bus architectures have followed to solve the resource allocation problem. For example, the MCA and the EISA bus standards both defined a mechanism where add-in cards were configured somewhat automatically. These bus architectures allocated the resources, but the process wasn't always flexible and still required some manual intervention. And they still left the current ISA cards without a solution. Plug and Play technology, co-developed by Intel and other industry partners, consists of hardware and software components that card, PC, and operating system manufacturers incorporate into their products. With this technology, the user is responsible for simply inserting the card. Plug and Play makes the card capable of identifying itself and the resources it requires. The system's software automatically sets up a suitable configuration for the card. Newly developed PCI and Plug and Play ISA cards are all built to eliminate user intervention during the installation process. See PLUG AND PLAY BIOS EXTENSIONS.

Plug And Play BIOS Extensions Software code added to a PC's bios which purports to automatically recognize which peripherals are in the PC and automatically configure the PC for those peripherals — without the need for fiddling with dip switches or setting interrupts, etc. Plug and Play comes from Intel. And more and more PC cards are coming Plug and Play compatible.

Plug Compatible Devices made by different manufacturers that are totally interchangeable. The word derives from the fact that the devices are so completely interchangeable that you can simply unplug one device and plug in another device made by different manufacturer and it will work the same, or better.

Plug-In Stations The telephone need stays constant for all types of phones. For ease in phone "moves" and rearrangements, all phones are provided (and installed) as plug-in instruments.

Plugboard A telephone switchboard on which connections are made by a jack and an attached cord representing a trunk (the male jack and the cord) and a female plug (the telephone extension). Early plugboards needed an operator to place outside calls and connect incoming calls. All calls were completed by the operator. Plugboards were common in the days of "PBXs." Then came PABXs (Private Automated Branch Exchanges) and you could dial out without the help of an operator. Then electronic PABXs came in and you could dial directly in to many internal extensions, using a feature called DID (Direct Inward Dial). Now PABXs are called PBXs because they're all automatic. Plugboards are rapidly disappearing. They do have two great uses, however. First, operators who grew up with them, still like them. Operators who now live in nursing homes like them. Second, because the jack and plug make a pure metallic connection, they're great for data transmission and occasional data switching.

Plugs Circuit cards which control various aspects of transmission and carrier circuit functionality and usage.

PLV An imaging term. Production Level Video. DVI Technology's highest quality motion video compression algorithm. It's about 120-1 compression. Compression is done "off-line". i.e. non-real time, and playback (decompression) is real time. Independent of the technology in use, off-line compression will produce a better image quality than real time since more time and processing power is used per frame.

PLY One layer in a composite.

PM 1. Physical Medium: Physical Medium refers to the actual physical interfaces. Several interfaces are defined including STS-1, STS-3c, STS-12c, STM-1, STM-4, DS1, E1, DS2, E3, DS3, E4, FDDI-based, Fiber Channel-based, and STP These range in speeds from 1.544Mbps through 622.08 Mbps.
2. Performance Monitoring. Gives a measure of the quality of service and identifies degrading or marginally operating systems (before an alarm would be generated). Digital signal parameters, including errored seconds and out of frame, measure the integrity of a communication channel as defined in AT&T Compatibility Bulletin 149 (CB 149).
3. Peripheral Module.

PMD Physical Medium Dependent. This sublayer defines the parameters at the lowest level, such as speed of the bits on the media. The bottom half of BISDN Layer 1.

PMMU Paged Memory Management Unit. Macintosh computers equipped with a PMMU may use virtual memory with the System 7 operating system.

PMN Indicates loss of ac power at the far-end terminal.

PMR Poor Man's Routing. A technique used in any packet-switched networks to allow a source node to predefine the routing to the destination, bypassing the normal routing algorithm implemented at the network layer.

PMS 1. Picturephone Meeting Service. An AT&T service once provided under experimental tariff. It combined TV techniques with voice transmission. PMS is usually only available between telephone company-located picturephone centers. Most have now been closed down. The venture was losing too much money.
2. Property Management System, a software program and computer that controls all guest billing and guest services functions in a hotel. In short, the guts of a hotel's computer system. Some telephone systems have a PMS Interface, which allows various degrees of integration between the telephone system and the hotel's computer systems. For example, voice mail could be administered through the hotel's Property Management System.
3. The Pantone Matching System, a universal language for solid-color specification and reproduction. Colors defined by PMS receive a unique number and mixing formula. Consequently, when artists specify a PMS number they can be sure that the final printed product will match the chosen color. But, be careful, PMS colors look different when printed on different papers. The biggest perceived difference is when you print on glossy or matte paper.

PMS Interface An interface that allows telephone system functions (like voice mail) to be administered through a hotel's Property Management System.

PNI An ATM term. Permit Next Increase: An ABR service parameter, PNI is a flag controlling the increase of ACR upon reception of the next backward RM-cell. PNI=0 inhibits increase. The range is 0 or 1.

PNNI Private Network-Network Interface: A routing information protocol that enables extremely scalable, full function, dynamic multi-vendor ATM switches to be integrated in the same network.

PNNI Protocol Entity An ATM term. The body of software in a switching system that executes the PNNI protocol and provides the routing service.

PNNI Routing Control Channel An AIM term. VCCs used for the exchange of PNNI routing protocol messages.

PNNI Routing Domain An ATM term. A group of topologically contiguous systems which are running one instance of PNNI routing.

PNNI Routing Hierarchy An ATM term. The hierarchy of peer groups used for PNNI routing.

PNNI Topology State Element An ATM term. A collection of PNNI information that is flooded among all logical nodes within a peer group.

PNNI Topology State Packet An ATM term. A type of PNNI Routing packet that is used for flooding PTSEs among logical nodes within a peer group.

PNM Public Network Management.

PnP Plug and Play. The technology that lets Windows 95 and soon other operating systems automatically detect and configure most of the adapters and peripherals connected to or sitting inside a PC. A fully Plug and Play-enabled PC requires three PnP pieces: a PnP BIOS, PnP adapters and peripherals, and a PnP operating system. Adding a PnP-compliant CD-ROM drive, hard disk, monitor, printer, scanner, or other device to a PnP PC means little more than making the physical connection. The operating system, together with PnP logic present in the BIOS and in the device itself, handles the IRQ settings, I/O addresses, and other technical aspects of the installation to make sure that the thing will work. The idea of PnP is to make installation of complex gadgets — such as sound cards and modems — easy, taking care of the major bane of everyone's life: That your new device now conflicts with an old device, effectively killing both devices and maybe crashing your PC at the same time. PnP is a great idea. Its success has been slow in coming, because so many devices are not PnP compatible.

PNS Personal Number Service is a new concept in telecommunications that assigns a telephone number to a person, not a location, effectively allowing a subscriber to use one number for all calls and helping them manage their incoming communications. The service does not require the user to change any existing phone numbers. The subscriber simply provides the various numbers — office, cellular, pager, fax and home — and instructions on where and when the calls should be routed, and the PNS directs the calls in the order requested by the subscriber.

POCSAG One of the communications protocols used between paging towers and the mobile pagers/receivers/beepers themselves. Other protocols are GOLAY, ERMES, FLEX and REFLEX. The same paging tower equipment can transmit messages one moment in POCSAG and the next moment in ERMES, or any of the other protocols.

PODP Public Office Dialing Plan.

POF Plastic Optic Fiber. A fiber transmission medium made from a plastic material. Such a medium uses low-quality light sources and carry data at speeds greater than 10 Mbps over distances up to 100 meters. POF is evolving as a replacement for twisted-pair copper wire.

POFS Private Operation Fixed Systems. Microwave incumbents in the 2.0 Ghz band. Must be relocated with comparable alternative facilities funded.

POGO 1. Post Office Goes Obsolete. When MCI Mail was originally being planned, its code name was POGO. The idea was obvious. In September of 1994, I asked MCI what "POGO" meant and they answered: "Pogo" is an internal message format used by MCI for coding purposes.

POH Path OverHead. Overhead assigned to and transported with the payload until the payload is demultiplexed. It is used for functions that are necessary to transport the payload. These functions include parity check and trace capability. An AT&T SONET term. It is not implemented in SONET Lite.

POI Point Of Interface. The physical telecommunications interface between the LATA access and the interLATA functions. A POI is a demarcation point between LEC and a Wireless Services Provider (WSP). This point establishes the technical interface, the test point(s) and the point(s) for operational division of responsibility. See also POINT OF PRESENCE.

Point In Call PIC. A representation of a sequence of activities that the ASC (AIN Switch Capabilities) performs in setting up and maintaining a basic two-party call. PICs occur in Originating and Terminating BCSMs (Basic Call State Model).

Point Of Interface POI. The physical telecommunications interface between the LATA access and the interLATA functions. A POI is a demarcation point between LEC and a Wireless Services Provider (WSP). This point establishes the technical interface, the test point(s) and the point(s) for operational division of responsibility.

Point Of Presence POP. Physical place within a LATA where a long distance carrier or a cellular provider interfaces with the network of the local exchange carrier (LEC), also called the local telephone company. The point at which the local telephone company terminates subscribers circuits for cellular phone call or long-distance dial-up phone call or or leased-line communications.

Point Of Purchase Politics Politically correct shopping or cause-related marketing, such as that advocated by Benetton or Ben and Jerry's.

Point Of Sale Terminal A special type of computer terminal which is used to collect and store retail sales data. This terminal may be connected to a bar code reader and it may query a central computer for the current price of that item. It may also contain a device for getting authorizations on credit cards.

Point Of Termination POT. The point of demarcation within a customer-designated premises at which the telephone company's responsibility for the provision of access service ends.

Point Size The height of a printed character specified in units called points. A point equals 1/72 inch. Also known as font size.

Point To Multipoint A circuit by which a single signal goes from one origination point to many destination points. The classic example is a TV signal (say a Home Box Office program) being broadcast from one satellite to many CATV subscribers all around the country. Not to be confused with a multi-drop circuit. See POINT TO MULTIPOINT CONNECTION.

Point To Multipoint Connection A Point-to-Multipoint Connection is a collection of associated ATM VC (Virtual Channel) or VP (Virtual Path) links, with associated endpoint nodes, with the following properties:
1. One ATM link, called the Root Link, serves as the root in a simple tree topology. When the Root Node sends information, all of the remaining nodes on the connection, called Leaf Nodes, receive copies of the information.
2. Each of the Leaf Nodes on the connection can send information directly to the Root Node. The Root Node cannot distinguish which Leaf is sending information without additional (higher layer) information. (Note: UNI 4.0 does not support traffic sent from a Leaf to the Root.)
3. The Leaf Nodes cannot communicate directly to each other with this connection type. See ATM.

Point To Point A private circuit, conversation or teleconference in which there is one person at each end, usually connected by some dedicated transmission line. In short, a connection with only two endpoints. See also POINT TO MULTIPOINT.

Point To Point Connection An uninterrupted connection between one piece of equipment and another.

Point-To-Point Protocol PPP. An implementation of TCP/IP which is intended for transmission using telephone lines. PPP is the successor to SLIP (Serial Line Interpersonal Protocol). PPP provides router-to-router and host-to-network connections over both synchronous and asynchronous circuits.

Point To Point Signaling A signaling method where signals must be completely received by an intermediate station before that station can set up a call connection. See END TO END SIGNALING.

Point To Point Topology A network topology where one node connects directly to another node.

Pointer Processing Pointer processing accommodates frequency differences by adjusting the starting position of the payload within the frame. A pointer keeps track of the starting position of the payload.

Points Of Failure A simple term to indicate that in a complex network there are many places things can go wrong. Those places need to be identified so that you can anticipate and plan for things to go wrong.

Poisson See POISSON DISTRIBUTION.

Poisson Distribution A mathematical formula named after the French mathematician S. D. Poisson, which indicates the probability of certain events occurring. It is used in traffic engineering to design telephone networks. It is one method of figuring how many trunks you will need in the future based on measurements of past calls. Poisson distribution describes how calls react when they encounter blockage (see QUEUING THEORY for a detailed explanation of blockage). There are two main formulas used today in traffic engineering: Erlang B and Poisson. The Erlang B formula assumes all blocked calls are cleared. This means they disappear, never to reappear. The Poisson formula assumes no blocked calls disappear. The user simply redials and redials. If you use the Poisson method of prediction, you will buy more trunks than if you use Erlang B. Poisson typically overestimates the number of trunks you will need, while Erlang B typically underestimates the number of trunks you will need. There are other more complex but more accurate ways of figuring trunks — Erlang C (blocked calls delayed or queued) and computer simulation. Poisson has been used extensively by AT&T to recommend to its customers the number of trunks they needed. Since AT&T was selling the circuits and preferred its customers to have excellent service, it made sense to use the Poisson formula. As competition in long distance has heated up, as circuits have become more costly and as companies have become more economically-minded (more aware of their rising phone bills), Poisson has become widely ignored.

After I wrote the above definition, Lee Goeller, a noted traffic engineering expert contributed the following definition of Poisson Distribution: A probability distribution developed by E.C. Molina of AT&T in the early 1900s for use in solving problems in telephone traffic (see TRAFFIC ENGINEERING), although it has many other uses and is widely applied in many fields. When made aware of Poisson's prior effort (circa 1820), Molina gave him full credit and even taught himself French so he could read Poisson in the original. The Poisson distribution assumes a call is in the system for one holding time, whether it is served or not (blocked calls held); the first form of the distribution estimates the probability that exactly X calls will be in the system, while the second estimates the probability that X or more calls will be present. If there are only X trunks to serve the calls, the second form gives the probability of blocking. Although limited tabulations of the Poisson distribution had been made earlier, Molina published an extensive set of tables in 1942. The Poisson distribution slightly overstates the number of trunks needed when compared to the Erlang B distribution (see ERLANG B).

Poisson Process A kind of random process based on simplified mathematical assumptions which makes the development of complex probability functions easier. In traffic theory, the arrival of telephone calls for service is considered a Poisson process. Calls arrive "individually and collectively at random," and the probability of a new call arriving in any time interval is independent of the number of calls already present. A Poisson process should not be confused with the Poisson Distribution, which gives the probability that a certain number of calls will be present if certain additional assumptions are made. See POISSON DISTRIBUTION.

Poke-Through Method A distribution method that involves drilling a hole through the floor and poking cables through to terminal equipment from the ceiling space of the floor below. See also CEILING DISTRIBUTION SYSTEMS and NEWTON.

Polar Keying A transmission technique for digital signals in which the current flows in opposite directions for 1s and 0s or marks and spaces. It is used in telegraph signaling. It is also known as polar transmission.

Polar Relay A relay containing a permanent magnet that centers the armature. The direction of movement of the armature is governed by the direction of current flow.

Polarity Which side of an electrical circuit is the positive? Which is the negative? Polarity is the term describing which is which. Knowing polarity is not critical with rotary phones. They will work irrespective of which way the telephone circuit's polarity is. Touchtone phones, however, need correct polarity for their touchtone pads to work. How to tell? If you can receive an incoming call, can speak on the phone clearly, but can't "break" dial tone by touching a digit on your touchtone pad, then the polarity of your line is reversed. Simply reverse the red and green wires. Some electronic phones behind PBXs and key systems are also sensitive to polarity. If in doubt, simply reverse the wires. In video, reversed polarity results in a negative picture.

Polarization Characteristic of electromagnetic radiation (e.g. lightwave, radio or microwave) where the electric-field vector of the wave energy is perpendicular to the main direction, or vector, of the electromagnetic beam.

Pole Attachment Cost to cable TV and other telecom operators (including end users) to rent space to attach cables to telephone company and power company poles. There are charges and often significant restrictions on the attachment of your cable to their pole.

Polishing Preparing a fiber end by moving the end over an abrasive material.

Politeness The most acceptable hypocrisy. Mostly seen before the sale.

Politician John Maynard Keynes said that politicians are apt to be slaves to the ideas of long-deceased economists.

Politics A clash of self-interests masquerading as a clash of principles. Also, the technique by which most telephone systems are bought in large corporations.

Poll In data communications, an individual control message from a central controller to an individual station on a multipoint network inviting that station to send if it has any traffic to send. See POLLING.

Poll Cycle The complete sequence in which stations are polled on a polled network.

Poll/Final Bit Bit in HDLC frame control field. If frame is a command, bit is a poll bit asking station to reply. If frame is a response, bit is a final bit identifying last frame in message.

Polling Connecting to another system to check for things like mail or news. A form of data or fax network arrangement whereby a central computer or fax machine asks each remote location in turn (and very quickly) whether they want to send some information. The purpose is to give each user or each remote data terminal an opportunity to transmit and receive information on a circuit or using facilities which are being shared. Polling is typically used on a multipoint or multidrop line. Polling is done to save money on telephone lines.

Polling Delay Communications control procedure where a master station systematically invites tributary stations on a multipoint circuit to transmit data. Polling delay is a measure of the time to transmit and receive on a polled network versus a direct point-to-point circuit.

POLSK POLarization Shift Keying.

Polyethylene A family of insulating (thermoplastic) materials derived from polymerization of ethylene gas. They are basically pure hydrocarbon resins with excellent dielectric properties. Used extensively in cables.
Polymer A material having molecules of high molecular weight formed by polymerization of lower molecular weight molecules.
Polymerization A chemical reaction in which low molecular weight molecules unite with each other to form molecules with higher molecular weights.
Polyolefin Any of the polymers and copolymers of the ethylene family of hydrocarbons.
Polypropylene A thermoplastic similar to polyethylene but stiffer and having a higher softening point (temperature) and excellent electric properties.
Polyvinylchloride PVC. A thermoplastic material composed of polymers of vinyl chloride. A tough, water and flame-retardant thermoplastic insulation material that is commonly used in the jackets of building cables when fire retardant, but not smoke retardant properties are required. Unfortunately, it burns and gives out noxious gases which kill. PVC can't be run in air return ducts, also called plenum ducts and most towns, therefore, don't allow PVC to be run in their plenum ceilings. See PLENUM.
Polyvinylidene Difluoride PVDF. A fluoropolymer material that is resistant to heat and used in the jackets of plenum cable.
PON Passive Optical Network. A fiber-based network which uses passive splitters to deliver signals to multiple users.
Pony Express Out of the summer haze bursts a horse and rider, swiftly approaching a lonely sod building on the prairie. Arriving in a cloud of dust, the rider leaps from his horse and heads for a water barrel to quench his thirst. Meanwhile, a leather sack filled with mail is whisked off the tired horse and thrown over the saddle of a fresh mount. Within two minutes, the rider is gone, galloping toward the far horizon. This young man in a hurry was one of some 200 Pony Express riders who carried the mail in a giant relay between St. Joseph, Missouri, and Sacramento, California, a distance of 1,966 miles, in ten days or less. Changing horses every ten to fifteen miles at swing stations, and switching riders at home stations after a run of 75 miles or more, the riders averaged 250 miles a day. During the short time the Pony Express was in operation — from April 1860, through October, 1861 — its rider defied hostile Indians, blazing desert heat, and bone-chilling blizzards to travel a total of 650,000 miles with 34,753 pieces of mail. To save weight the letters they carried were written on tissue-thin paper as postage cost $10 an ounce, later cut to $2. The best time ever achieved was in March 1861, when Lincoln's inaugural address was carried from Missouri to California in seven days, 17 hours.
The Pony Express was organized by stagecoach operator William Hepburn Russell, who had been convinced by a group of prominent Californians that an overland mail route to their state was feasible. Russell's business partners opposed the venture because it was not protected by a U.S. mail contract. (They had competition and de-regulation even in those days.) But Russell went ahead, building stations and purchasing 500 top quality Indian horses. In advertising for riders, he hinted at the hazardous nature of the job by asking for "small, daring young men, preferably orphans." The riders received board and keep and were paid $100 to $150 a month. Their average age was 19, but one rider, David Jay, was 13, and William F. Cody, who became famous as "Buffalo Bill," was 15. In a further effort to save weight, a rider usually carried only a pistol and a knife. He was expected to out-run the Indians, not out-fight them.
The Pony Express days of glory ended abruptly in 1861 following completion of the transcontinental telegraph. Russell's firm lost more than $200,000 in the venture, but the daring of the Pony Express riders caught the imagination of every American, and their exploits became an important part of the legend and lore of the nation.
The above, copyright 1979 by Panarizon Publishing Corp.
Pool A collection of things available to all for the asking or the dialing. A modem pool is a collection of modems typically attached to a PBX. Dial a special extension and you can use the modem, which answers that extension (or one of the extensions in the hunt group) to make a data call. Pooling is sharing. The purpose of having a "pool" is to avoid buying everybody one of whatever it is you're pooling. Actually, "pooling" is a fancy word for something we've been doing in the telephone business for the past 100 years — sharing. We started sharing lines, then sharing switches, then sharing voice mail devices, now we're sharing equipment, like modems.
POP 1. Point Of Presence. The IXC equivalent of a local phone company's central office. The POP is a long distance carrier's office in your local community (defined as your LATA). A POP is the place your long distance carrier, called an InterExchange Carrier (IXC), terminates your long distance lines just before those lines are connected to your local phone company's lines or to your own direct hookup. Each IXC can have multiple POPs within one LATA. All long distance phone connections go through the POPs.
2. Short for "population." One "pop" equals one person. In the cellular industry, systems are valued financially based on the population of the market served.
3. Post Office Protocol. A system on the Internet you can use to get your mail. You employ a mail server to get the mail and then download it to your computer. POP is an Application protocol offering mailbox retrieval service. Used by many Internet personal computer users. POP is increasingly being replaced by IMAP.
Pop-Up A call center term. A button that displays, on demand, several items from which you can choose (by clicking on it, for example). In effect, this is a list box that you don't see until you push button. Pop-ups always have double lines on their right and bottom sides.
Pop-Up Electronic Mail An electronic mail system that runs as a terminate-and-stay-resident program (typically within DOS) and can be popped up inside any application to send or read mail. Our our office, we have a TSR electronic mail program that pops up. It is called Noteworks and we really love it.
Pop-Up Program A memory-resident program that is loaded into memory but isn't visible until you press a certain key combination or until a certain event occurs, such as receiving a message. See also TSR.
POPS A cellular industry term for its customers or its potential customers. (It varies with usage.) POPS, short for "population" (well, sort of) refers to members of the population. According to Ron Schneiderman's book on Wireless, "if the coverage area of a cellular carrier include a popular base of one million people, it is said to have one million POPS. The financial community uses the number of potential users as measuring stick to value cellular carriers."
Populated Occupied by chips on printed circuit boards.

Port 1. An entrance to or an exit from a network.
2. The physical or electrical interface through which one gains access. A point in the computer or telephone system where data may be accessed. Peripherals — like call accounting devices — are connected to ports. The two most common ports are the parallel and serial ports.
3. The interface between a process or program and a communications or transmission facility.
4. To move a process, program or subroutine from one processor to controller to another ("port it over").
5. Network access point for data entry or exit. In Internet terms, it is the identifier (16-bit unsigned integer) used by Internet transport protocols to distinguish among multiple simultaneous connections to a single destination host.

Port Connection The point of entry into a public frame relay network service.

Port Identifier The identifier assigned by a logical node to represent the point of attachment of a link to that node.

Port Multiplier A local area network interconnect, a concentrator providing connection to a network for multiple devices.

Port Replicators Low-cost docking station substitutes that provide one-step connection to multiple desktop devices.

Port Selector Another name for a dataPBX. Since the advent of LANs (local area networks) these devices have been getting a bad rap. Not fair. These gadgets are really great at transmitting and switching huge number of low-speed asynchronous lines. If you put this sort of traffic on a LAN, you could severely mess up its performance. Some port selectors have data throughput in excess of 20 million bits per second.

Port Sharing 1. A system which connects multiple lines to a single port by means of a manual or automatic line selection method.
2. In frame relay, where multiple virtual connections share the same port connection.

Port Sharing Device A system which connects multiple lines to a single port by means of a manual or automatic line selection method.

Port Switching According to 3Com, port switching is merely an electronic patch panel function, not the genuine switching capability that provides a performance boost. Port switching lets administrators configure their networks, allocating any port to any backplane on their hub. Unlike true switching, it doesn't increase the bandwidth available to the network manager.

Portability 1. The ability of a customer to take his telephone number from place to place and, for 800 numbers, from one long-distance company to another.
2. The ability of software designed for one computer system to be used on other systems. Little software outside MS-DOS software for IBM and IBM clone computers is portable. UNIX software is portable to an extent.

Portable A one-piece, self-contained cellular telephone — easily carried in a brief case or purse. Portables normally have a built-in antenna and rechargeable battery and operate with six-tenths of one watt (0.6 watt) of power. Car cellular phones operate with three watts.

Portable Cellular Phone Also known as a "hand-held phone". Refers to a lightweight, compact cellular handset that incorporates a battery power supply, and can be used without any peripheral power or antenna. See PORTABLE.

Portable Teletransaction Computers PTC. These are typically handheld devices used for retail (inventory), healthcare (tracking supplies), mobile field repair (reporting fixes), insurance (visiting car wrecks and other disasters), etc. The devices typically have telecommunications capabilities, sometimes wireless, sometimes landlines. And they typically include microprocessors, memories, displays, keyboards, touchscreens, character recognition software, barcode readers, printers, modems and local and/or wide area data radios.

Portrait Most computer screens are horizontal, i.e. they are wider than they are high. In the new language of computer screens, this is called "landscape." When a computer screen is higher than it is wide, it's called "portrait." Some computer screens can actually work both ways. Some even have a small mercury switch in them that determines which way the screen is standing (portrait or landscape) and will adjust their image accordingly. See also PORTRAIT MODE.

Portrait Mode 1. In facsimile, the mode of scanning lines across the shorter dimension of a rectangular original. ITU-T Group 1, 2 and 3 facsimile machines use portrait mode.
2. In computer graphics, the orientation of a page in which the shorter dimension is horizontal. The opposite is called landscape mode. See also PORTRAIT.

POS 1. Point Of Service. Also called Point of Presence. See POINT OF PRESENCE.
2. POS Device. A point of sale device such as a credit card scanner used for authorization when a purchase is made.

POSI Promoting Conference for OSI. Consists of executives from the six major Japanese computer manufacturers and Nippon Telephone and Telegraph. They set policies and commit resources to promote OSI.

Position A telephone console at a switchboard manned, er, staffed by an attendant, or operator, or agent, or whatever the latest fashionable word is.

Positive Action Digit A digit that must be dialed before a PBX will advance a call to a higher-cost route. The WATS lines are busy. Time on the queue is over. It's time to move the call to the more expensive direct distance dial. Before it can go that route, the caller must punch in a positive action digit. This affirms that the user knows he is now making a more expensive call. It causes him to think twice, allegedly.

POSIX Portable Operating System Interface uniX. A proposed universal UNIX interface to user-created application programs that would run on all vendor equipment, thereby improving system interoperability.

Post 1. To compose a message for an Internet Usenet newsgroup and then send it out for others to see.
2. Power-On Self-Test

Post Dial Delay PDD. The time from when the last digit is dialed to the moment the phone rings at the receiving location.

Post Office Any part of an e-mail system that directors or delivers mail. But, says PC Magazine's David Stone, to refer to a particular level of mail handling, the term post office needs a modifier. A local post office or host post office is the module on a LAN that users directly interact with to send and receive mail. A domain post office is the module that controls the mail delivery within a domain of multiple hosts on a single network.
In Windows 95, the postoffice is a temporary message store, holding the message until the recipient's workstation retrieves it. The postoffice exists as a directory structure on a server and has no programmatic components.

Post Pay A method of coin phone operation characterized by the operation of a lever or button that causes the collection of deposits after the called party answers. This method of "A" and "Buttons" is still used on coin phones overseas, especially in Great Britain.

Post Restante Address A standard attribute in a postal address indicating that physical delivery at the counter is requested. It may also carry a code.

Postalize To structure rates or prices so that they are not distance sensitive, but depend on other factors (such as duration of a call, etc.) See also POSTALIZED.

Postalized Long distance phone calls are traditionally billed by the distance they travel. The longer they travel, the more they cost. However, to be competitive (and different) some long distance companies started charging a flat rate per minute — irrespective of how far you called. They called this "postalized" charging — from the fact that Post Office also charges a flat rate irrespective of how far it carries the mail (within the country).

Posting An individual article sent to a USENET news group on Internet; or the act of sending an article to a USENET news group.

Postlink A program used by a RIME bulletin board system node in place of mailer software.

Postmaster The person responsible for taking care of mail problems, answering queries about users, and performing similar work for a given site.

Postoffice See POST OFFICE.

Postpay See POST PAY.

Postscript 1. PostScript is a popular format used to create World Wide Web screens and send documents over the Internet.
2. PostScript by Adobe Systems Inc. is the standard page description language for desktop computer systems. It describes type, graphics and halftones as well as the placement of each on the page. The big advantage of PostScript is that it is device independent. Thus if you create a postscript image (text and/or photo and/or drawing), you can print it to a relatively cheap, low quality printer like a laser printer or a magazine quality printer like a Linotronics. PostScript is a printer language, much the same that BASIC is a computer language. By sending your PostScript printer a series of commands, you can make it do almost anything from printing text in a circle to printing foot-high letters to printing halftones. If you need PostScript, buy a printer that has built-in PostScript. If your printer doesn't have built-in PostScript, you may be able to get an external software interpreter but that interpreter will slow down your printer and tie up your computer while printing. See OUTLINE FONT.

POT 1. Techie-slang for POTENTIOMETER.
2. Point Of Termination. The point of demarcation within a customer designated premises at which the telephone company's responsibility for the provision of access service ends.

POT Bay The POT Bay, or Point-Of-Termination bay, is a device placed between a competitor's network and the natural point of connection to the local exchange carrier (LEC) network. It is located between a LEC's main distribution frame (MDF), which directs traffic to proper channels for distribution throughout the LEC network and the interconnector's colocated equipment. The POT Bay, according to most new local phone companies (i.e. ones that are not a Bell operating company or GTE), is an unnecessary obstacle that adds to the costs of interconnection, serves no necessary engineering function, can degrade quality, and is nothing more than a latter-day "protective coupling arrangement." As of writing, Nynex BellSouth, U.S. West, Southwestern Bell, and Pacific Bell insist that a LEC-provided POT Bay be installed as a point of demarcation between the LEC and the competitive interconnector. Bell Atlantic and GTE do not require competitive interconnectors to use POT Bays.

Potential The difference in voltage between one point and another. One point is usually ground.

Potential Revenue A call center term. The revenue value per call times the number of calls forecast for a given period.

Potentiometer A variable RESISTOR, such as the ubiquitous volume control.

POTS Plain Old Telephone Service. Pronounced POTS, like in pots and pans. The basic service supplying standard single line telephones, telephone lines and access to the public switched network. Nothing fancy. No added features. Just receive and place calls. Nothing like Call Waiting or Call Forwarding. They are not POTS services. All POTS lines work on loop start signaling. See also LOOP START.

POTV Plain Old TV. A Microsoft definition. I kid you not.

Pound The # on a pushbutton touchtone key pad is called the pound key. It's also called the number sign, the crosshatch sign, the tic-tack-toe sign, the enter key, the octothorpe (also spelled octathorp) and the hash. Musicians call the # sign a "sharp."

Power The term which describes the amount of work an electric current can do in a unit of time. We measure power in WATTS (note we spell it with two "T"s.) A WATT measures the amount of work done in lifting a quarter-pound weight a distance of one yard in one second. Metric WATTS are a little more powerful. They go the distance of one meter. Power is the product of the current in amperes times the voltage, i.e. P = IV. See OHM'S LAW.

Power Conditioner A combination voltage regulating transformer and isolation transformer, providing smooth, regulated, noise-free, AC voltage. See POWER CONDITIONING.

Power Conditioning Power conditioning is a generic concept to encompass all the methods of protecting sensitive hardware against power fluctuations. When electricity leaves a commercial power generating plant, it is very clean. In fact, most power companies make sure the power they put out is a pure sine wave. Unfortunately, nearly all devices connected to power lines — and the worst are things with motors, like elevators, air conditioners, etc. — create disturbances that pollute the sine wave. As power travels through a wire away from the power plant, it picks up more of these interferences. A pure AC power sine wave appears as a smooth wave. The height of the wave is measured in volts. The wave starts at zero volts and moves to the highest point of 120 volts. The wave then cycles through a low point of -120 volts and back to zero. The speed at which it travels through this cycle is the frequency. Normal frequency in North America is 60 cycles per second (Hz). (In other places it's often 50 cycles per second.) Anything that disrupts this wave can cause hardware or data problems and needs to be regulated. Power disturbances can be categorized in several ways. A transient, sometimes called a spike or surge, is a very short, but extreme, burst of voltage. Noise or static is a smaller change in voltage. Brownouts and blackouts are the temporary drop in or loss of electrical power. Three types of protection against these three events are available: suppression, isolation, and regulation.

Suppression protects against transients. The most common suppression devices are surge protectors that include circuitry to prevent excess voltage. Although manufacturers originally designed surge protectors to prevent large voltage changes, most have also added circuitry to reduce noise on the line. Isolation protects against noise. Ferro-resonant isolation transformers use a transformer within the circuitry to envelop the sine wave at a slightly higher and lower voltage. Any voltage irregularity that extends beyond this envelope is clamped. Isolation transformers are usually expensive.

Regulation protects against brownouts and blackouts. Regulation modifies the power wave to conform to a nearly pure wave form. The Uninterruptible Power Supply (UPS) is the most commonly used form of regulation. A UPS comes in two varieties, on-line and off-line. An on-line UPS actively modifies the power as it moves through the unit. This is closer to true regulation than the off-line variety. If a power outage occurs, the unit is already active and continues to provide power. The on-line UPS is usually more expensive but provides a nearly constant source of energy during power outages. The off-line UPS monitors the AC line. When power drops, the UPS is activated. The drawback to this method is the slight lag before the off-line UPS jumps into action. That lag is getting shorter as electronics improves. So it's rarely a problem any longer.

Because UPS systems are expensive, most companies attach them only to the most critical devices, such as phone systems, network file servers, routers, and hard disk subsystems. Attaching a UPS to a local area network file server enables the server to properly close files and rewrite the system directory to disk. Sadly, most programs run on the workstation and data stored in their RAM is not saved during a power outage unless each workstation has its own UPS. If the UPS doesn't have its own form of surge protection, it is a good idea to install a surge protector to protect the UPS from transients. Proper use of power conditioning devices greatly reduces telephone system and network maintenance costs. Make sure that proper amperage is available for each system and that all outlets are grounded. Power conditioning devices connected to poorly-grounded outlets offer very little protection. Studies have shown that total local area network maintenance costs are higher with line-surge suppressors and ferro-resonant isolation transformers alone, than with uninterruptible power supplies.

Power Cross A situation in which AC current flows into a telephone circuit, as a result of contact with a power line.

Power Dialer A piece of hardware to which you feed a list of phone numbers you want called. It calls them one after another. It detects busies, no answers, fax machines, answering machines, voice mail machines, etc. When it hears these, it disconnects and dials the next number in the list. If it reaches a live human being, it will pass the call to the owner of the machine, who will presumably try and sell the fellow at the other end something, or try and collect money from him — the two main uses which power dialers are typically put to. The term "power dialer" is fairly new. I first saw it used on a single line device. It seemed to need a prompt from its owner to dial the next call, very much like a preview dialer. See also PREDICTIVE DIALING and PREVIEW DIALER.

Power Distribution Panel A part of the Rolm CBX power distribution system that receives voltages from the main power supply and distributes them to the cabinet shelves.

Power Down The sequence of things you to have to do to turn off a computer or telephone system. Not following the correct power down procedures can cause a loss of data.

Power Fail Bypass A feature that allows analog trunks to be answered if your commercial AC power or your telephone system crashes.

Power Failure Backup If your AC power fails, your telephone system can still operate by switching to a backup battery power supply, often called a UPS — Uninterruptible Power Supply.

Power Failure Transfer When the commercial AC power fails and there is no backup power source — such as a battery or a generator — this feature switches some of the trunks connected to the phone system to several single line phones, which don't need external power and can draw their power from the phone lines.

Power Level The measure of signal power at some point. The measure can be referenced to some power level in which case the measurement is expressed in dB (decibels). It may also be referenced to 1 milliwatt in which case the measurement is expressed in dBm.

Power Line Carrier An AC power line can be made to carry high frequency radio waves, which can carry "information," which could be a voice or data call. Over the years many companies have tried power line carrier — a seductively easy method of not having to install conventional phone lines. The results have been poor-to-awful. One day someone will figure a reliable, cheap method.

Power Mac A line of Macintoshes which use a RISC chip. If you buy a Power Mac, make sure you get software which runs in Power Mac native mode, which means that the software is written for that chip. When the first Power Macs were introduced, much software had to run in "emulation." And that software ran very slowly. The Power Mac runs very quickly if you run software in native mode.

Power On See POWER UP.

Power Open A new operating system which is planned to run on a new super-powerful PC manufactured by a joint IBM-Apple alliance. The idea of the IBM-Apple alliance is make a super-powerful PC that runs virtually every PC operating system imaginable, including MS-DOS, UNIX, Windows, OS/2, Macintosh. The new, all powerful operating system, would be called "Power Open."

Power PC The RISC-based microprocessor created by Apple, IBM and Motorola. Apple and IBM have created computers using this very versatile chip, which allegedly runs a bunch of operating systems, including Apple's MAC OS, IBM's AIX (its version of Unix) and OS/2, Microsoft Windows 95 and Windows NT.

Power Regulator Equipment that regulates the power delivered to a system. Designed to mitigate transients in the commercial electric power source.

Power Supply Most single line phones are powered by the electricity that comes in over the phone line. That's why they'll work when there's a power outage. Single line phones that have gadgetry associated with them, all ISDN phones and all multi-line phones (like key systems and PBXs) require a power supply. A power supply which is a device which converts the normal 120 or 240 volts AC power to AC and DC at the various voltages and frequencies needed by the components and circuits of the phone or computer system. Power supplies are usually the least reliable part of modern electronic gadgetry. This is because they take the hits form the lousy power the local utility sends in and also because many manufacturers skimp on the quality of their power supplies. A cheap power supply is not evident immediately. It may take time to break down. Whenever you're having intermittent problems with your phone system or computer, suspect the power supply. And, given a choice, buy the best quality power supply you can.

Power Systems A system that provides a conversion of a primary alternating current power to direct current voltages required by telecom equipment, and may generate emergency power when the primary alternating current source is interrupted.

Power Up The sequence of things you have to do in order to turn a computer or telephone system on. You can't cut corners starting up electronic equipment. It must be done carefully and in the correct order. Always count to ten after turning something off before turning it back on again. See also POWER DOWN.

Powerbook Apple's name for a line of laptop Macintosh computers it introduced in late 1991.

Power, Peak In a pulsed laser, the maximum power emitted.

PPDN Public Packet Data Network.

PPI Pixels Per Inch. See RESOLUTION.

PPM 1. Pulse Position Modulation.
2. Northern Telecom term for Periodic Pulse Metering.

PPP Point-to-Point Protocol; a protocol that allows a computer to connect to the Internet with a standard dial-up telephone line and a high-speed modem and enjoy most of the benefits of a direct connection, including the ability to use graphical front ends such as Mosaic and Netscape. PPP is considered to be better than SLIP, because it features error detection, data compression, and other elements of modern communications protocols which SLIP, the older Internet protocol, lacks. PPP is thus replacing SLIP. See also SLIP.

PPS 1. Pulses Per Second.
2. Precise Positioning Service. The most accurate dynamic positioning possible with GPS (Global Positioning System), based on the dual frequency P-code. 3. Packets Per Second. The theoretical limit of 10 Mbps Ethernet, measured in 64 byte packets, is 14,800 packets per second (PPS). By comparison, Token Ring is 30,000 and FDDI is 170,000 pps. 4. The Path Protection Switched ring defined by Bellcore TA-496. PPS is really a fancy name for a duplicated SONET signal traveling over diverse (i.e. different) routes. When one route of network crashes, the other will take over. This enables it to survive service outages caused by cable cuts, earthquakes, lightning strikes and equipment failures. The PPS ring gives a SONET route a greater degree of survivability than other Sonet transmission paths that don't have route diversity.

PPSN Public Packet Switched Network

PPSS Public Packet Switched Service. A connection-oriented, packet-switched data communication service that permits users to communicate with data terminals of other customers and on other packet networks.

PPTP Point-to-Point Tunneling Protocol, a new protocol that enables virtual private networking — enabling secure remote access to corporate networks over the Internet. The protocol was first demonstrated at InterOp, Spring 1996. According to the press release from U.S. Robotics and Microsoft which demonstrated the protocol, PPTP will help companies deploy remote access to employees more quickly, using fewer resources, by allowing them to take advantage of existing enterprise network infrastructures such as the Internet for remote access. U.S. Robotics developed the Windows NT PPTP driver, which will be integrated into Microsoft's Windows NT Server 4.0. Under an agreement with Microsoft, U.S. Robotics will license a variety of software components for PPTP to Microsoft. PPTP will be added as a standard feature to Microsoft's Windows NT Server and U.S. Robotics' Total Control NETServer remote access server platform. PPTP streamlines access in NT networking environments, and allows NT network clients to take full advantage of the services provided by Microsoft's RAS (Remote Access Service). For remote access, over analog or ISDN lines, PPTP creates a "tunnel" directly to the appropriate departmental NT Server on a network — even if there are hundreds of NT Servers. The PPTP specification builds on standards such as PPP and TCP/IP. PPTP 'tunnels' a remote user's PPP packets from the NETServer to a Windows NT server. By terminating the remote user's PPP connection at the NT server, rather than at the remote access hardware, PPTP allows network administrators to standardize security using the existing services and capabilities built into the Windows NT security domain. Using PPTP, network administrators can extend a virtual private network from their Windows NT server throughout the Internet and still retain control of their user passwords and accounts. NT provides its own knowledge of enterprise users, databases, allowed access and network addressing integrated into its RAS capabilities. With PPTP, users accessing their NT-based network will use these services, including DHCP and WINS, for access.

PQFP Plastic Quad Flat Pack, a format used in the design of PCMCIA devices. Another format is called TQFP, which stands for Thin Quad Flat Pack.

PRA PRIMARY RATE ACCESS. A Canadian term for ISDN's PRI, which is Primary Rate Interface or 23 64 Kbps channels and one 64 Kbps D (data) channel. See also PRIMARY RATE INTERFACE.

Practice The technical and installation manuals often used by Bell Operating Companies. A poor use of the word. A better word would be procedure.

PRBS Pseudo Random Bit Sequence/pattern. A test pattern having the properties of random data (generally 511 or 2047 bits), but generated in such a manner that another circuit, operating independently, can synchronize on the pattern and detect individual transmission bit errors.

Pre-Fetching. Imagine a call center with extensive customer records. Perhaps the company is a utility, a phone company or an insurance company. Now imagine that the phone rings. CallerID picks up the customer's name and pops a screen of information on the customer. Meantime, the database system has sent a signal to what's known as "near line" file storage. That near line file storage might be a magneto optical jukebox. It might be a tape file library. The "normal" time to retrieve a record from such systems might range from 30 seconds to two minutes. But if the signal is sent early, the records could be retrieved and loaded into a cache memory, ready if the customer asks a question, for example, about his old bills. This called pre-fetching. See also PRE-LOADING.

Pre-Loading Or pre-fetching. Documents are manually (by the administrator) retrieved from a jukebox or optical drive on a local area network in anticipation of some special need. Pre-loading may take advantage of the regular cycles of business, such as payroll processing on the 10th and 20th, personnel reviews the first week of each quarter, year-end processing of accounting reports. The network's administrator can take advantage of regular cycles to improve the system's response to users. See also PRE-FETCHING.

Preamplifier An electronic circuit which maintains or establishes an audio or video signal at a predetermined signal strength, prior to that signal being amplified for reproduction through a monitor or speaker.

Precedence Precedence is Federal government parlance to mean a designation assigned to a phone call by the caller to indicate to communications personnel the relative urgency (therefore the order of handling) of the call and to the called person the order in which the message is to be noted. Autovon phones which have "precedence" have an additional four touchtone buttons. You can find frequencies for those buttons under the definition for DTMF.

Precedence Prosign An introductory character or set of characters which indicate how a message is to be handled by the receiving unit.

Precise Positioning Service The most accurate dynamic positioning possible with GPS (Global Positioning System), based on the dual frequency P-code.

Predictive Dialer I love like predictive dialers because they deliver the benefits they promise. And they do it neatly and painlessly. I've never met a predictive dialer user who didn't realize the productivity improvements he was promised and figured on to justify the system. I have only one problem with predictive dialers. Most people use them for collecting money. Dial up the deadbeats. Dun them for money. B-O-R-I-N-G! I'm a mail order junkie. Everyone is. Mail order catalogs have become our escapist reading. I'm dressed in clothes from L.L. Bean, Eddie Bauer, Nordstrom, etc. I'm typing on a computer from PC Connection. My computer is tied into a server. We bought the wire and the server and the network cards by mail order. But I'm an unloved mail order junkie. No one ever calls me. Bean and Bauer know I like "large tall." They know I'm 11N in shoes. They know what I like and what I don't like. I've been buying from them for years. How come they never call me and tell me of their fantastic tall and narrow new products? I'm Platinum on American Airlines. Their most frequent flyer. So why don't they call me and offer me a special "weekend" surprise? Call me on Wednesday. Fill their planes on Friday and Sunday. Don't call me if their planes are full. To my naive brain, predictive dialers are the best sales tool ever invented. They let you sell when you've got something to sell. And not sell when you got nothing to sell. They let you sell "talls" to tall people and narrow shoes to narrowed-footed people. This is not rocket science. One day, some companies will figure there are millions of people (like me) waiting at the end of their phone line waiting for someone to call them and sell them something. That someone, if they're smart, will be using a predictive dialer. See PREDICTIVE DIALING.

Predictive Dialing An automated method of making many outbound calls without people and then passing answered calls to a person as the calls are answered. Here's the story: Imagine a bunch of operators having to call a bunch of people. Those calls may be for collections. They may be for employee callups. They may be for alumnae fund raising. When it's done manual, here's how it works: Before each call operators spend time reviewing paper records or computer terminal screens, selecting the person to be called, finding the phone number, dialing the numbers, listening to rings, listening to phone company intercepts, busy signals and answering machines. Operators also spend time updating the records after each call. Predictive dialing automates this process, with the computer choosing the person to be called and dialing the number and only passing it to an operator when a real live human being answers. There are enormous productivity gains made by screening out answering machines, busy signals, network busy signals, non-completed calls, operator intercepts etc. The result is productivity increases of 200% to 300%. According to generally accepted industry lore, a well-run manual dialing center can get its people talking on the phone for 25 minutes an hour. With a predictive dialer you can get them on the phone making sales, collecting money, etc. for 55 minutes an hour. It's a major productivity gain.

True predictive dialing should not be confused with automated dialing. True predictive dialing has complex mathematical algorithms that consider, in real time, the number of available telephone lines, the number of available operators, the probability of getting no answer, a busy signal, a disconnected number, operator intercept or an answering machine, the time between calls required for maximum operator efficiency, the length of an average conversation and the average length of time the operators need to enter the relevant data. Some predictive dialing systems constantly adjust the dialing rate by monitoring changes in all these factors.

Some people don't like the term "predictive dialing," since they think it's getting "a bad rap" in Washington, DC by being associated with junk phone calls. As a result some people would prefer to call it Computer Aided Dialing. See also PREVIEW DIALING.

Preemptive Multi-Tasking Operating System A multi-tasking operating system allows more than one task to be active at the same time. Under Windows 3.x, a task is defined as a single program. For instance, if you have ever had a word processor and a spreadsheet program open at the same time, then you have used Win 3.x multi-tasking. Win 3.x is a cooperative multi-tasking environment. In other words, applications must cooperate for multi-tasking to work. The system cannot pre-empt the program that has control of the CPU. It is the responsibility of each program to share the CPU. Windows 95 and Windows NT is a preemptive multi-tasking environment. This means the CPU is in charge and can seize control from applications when necessary. This environment reduces the risk of your system freezing up.

Preemptive Operating System An operating system scheduling technique that allows the operating system to take control of the processor at any instant, regardless of the state of the currently running application. Preemption guarantees better response to the user and higher data throughput. Most operating systems are not preemptive multitasking, meaning that task-switching occurs asynchronously and only when an executing task relinquishes control of the processor. See PREEMPTIVE, PREEMPTIVE MULTITASKING, REAL TIME SUPPORT.

Preemptive, Real Time Support When Microsoft released its At Work operating system, it said it had a number of key features, one of which was "pre-emptive, real-time support." Here's Microsoft's description:
* Pre-emptive, real-time support. Communication devices such as fax machines and phones are distinct from personal computers in that they have critical real-time needs. Consequently, the software in these devices must attend to communication hardware such as modems very frequently, so that pieces of the communication are not lost. To support this need, the operating system was designed to be able to put other processes "on hold" temporarily in order to service the communication hardware before continuing other functions. See AT WORK and WINDOWS TELEPHONY.

Preferred Call A local phone company services which lets you forward calls from a bunch of numbers you have pre-selected. The service uses the calling number ID as the basis for choosing which calls to forward.

Prefix One or several digits dialed or touchtoned in front of a phone number, usually to indicate something to the phone system. For example, dialing a zero in front of a long distance number in the United States would indicate to the phone company you wanted operator assistance on the call.

Preform Optical fiber source material. Preform is glass rod formed and used as source material for drawing an optical fiber. The glass structure is a magnified version of the fiber to be drawn from it.

Premise A thesis. a proposition supposed or proved as a basis of argument or inference. Often misused to mean the space occupied by a customer or authorized or joint user in a building or buildings on continuous or contiguous property (except railroad rights of way, etc.) not separated by a public road or highway. See PREMISES.

Premises the space occupied by a customer or authorized or joint user in a building or buildings on continuous or contiguous property (except railroad rights of way, etc.) not separated by a public road or highway.

Premises Distribution System PDS. There are two meanings for premises distribution system — a general one and a specific one, specific to AT&T. Here, first, is the general definition. A PDS is the transmission network inside a building or group of buildings that connects various types of voice and data communications devices, switching equipment, and other information management systems to each other, as well as to outside communications networks. It includes cabling and distribution hardware components and facilities between the point where building wiring connects to the outside network lines and back to the voice and data terminals in your office or other user work location. The system consists of all the transmission media and electronics, administration points, connectors, adapters, plugs, and support hardware between the building's side of the network interface and the terminal equipment required to make the system operational.
Here is the specific definition — A multi-functional distribution system from AT&T to support voice, data, graphics and video communications on premise. PDS includes cables, adapters, electronics, eight pin universal wall jacks and protective devices, all arranged in a logically coherent and economic fashion. It uses fiber optic cable and twisted pair copper wire and is suitable for single building, multi-tenant high rise or campus environment.

Premises Lightwave System The fiber optic part of the Premises Distribution System (PDS) from AT&T. PDS, which can replace the coaxial cables linking IBM terminals and printers, consists of two fiber optic interface units, one at a controller-end and the optic interface units and one at a terminal-end linked by a fiber optic pair. The fiber optic interface units connect to the terminals through four-pair building wiring and balun adaptors. Balun adaptors also enable direct connections of the terminals to the cluster controller through building wiring.

Premises Wire The twisted-pair, quad or other wire installed at the user's location to provide telephone service. Includes both intra-building and inter-building wiring.

Premises Wiring System The entire wiring system on the user's premises, especially the supporting wiring that connects the communications outlets to the network interface jack.

PReP PowerPC Reference Platform specification. It details the hardware, operating system and software elements necessary to build PowerPC-based systems that meet certain compatibility goals. According to P Magazine, these machines will run 32-bit operating systems and will resemble today's high-end desktop system with lots of memory, CD-ROM drives, stereo audio support, and PCI/PCMCIA expansion buses.

Prepaid Telephone Card A prepaid telephone card is a piece of credit-card size plastic entitling the owner to make phone calls. Prepaid telephone cards are sold in many denominations — typically $1, $5, $10, $20 and $50. Once the owner of the card makes phone calls, the value of the card decreases. The accounting is typically done in a remote switch which the user dials to make his calls. He often dials via an 800 number. Companies sell prepaid telephone calls as a business, often owning the switch and renting the long distance lines. Sometimes companies buy cards from a supplier, print up their names on the cards and give the cards away as promotion. It allows them to reward clients and customers with what one seller of cards calls "a universally valuable commodity — long distance service." Some people have started collecting prepaid telephone cards the way people collect baseball cards. Surprise, surprise, the value of some telephone cards is increasing. I know someone who has a collection of 30,000 telephone prepaid cards. He tells me he is going to retire on this collection.
A prepaid telephone card doesn't typically have technology on it or embedded into which represents the value of the money remaining on the debit card. Such technology might be an integrated circuit, a magnetic strip, bar codes which can be read by an optical reader. That's a debit card. See BREAKAGE, CALLING CARD and PREPAID TELEPHONE CARD.

Prepay The industry standard for coin phone operation which requires that the full cost of a call be deposited before a connection is attempted through the Central Office.

Preprocessor A device or information handling system which converts raw data into a form more easily processed with standard equipment.

Prereq Prerequisite

Presentation Layer The sixth layer of the OSI model of data communications. It controls the formats of screens and files. Control codes, special graphics and character sets work in this layer. See OSI STANDARDS.

Presentation Manager Presentation Manager is a look and feel specification and kernel-based toolkit development environment. It was developed for IBM by Microsoft with input from IBM. Presentation Manager is the standard graphical user interface and toolkit for the OS/2 operating system, which is a multi tasking operating system for personal computers. The screens are similar to those of Microsoft Windows.

Presentation Status For a particular call, an item that indicates if a calling identity item may be presented to the called party. If the presentation status is "public" presentation is allowed. If it is anonymous", presentation is restricted. Presentation status has to do with the presentation or not or calling line identification numbers.

Preset The "programming" of radio station frequencies on a tuner or receiver or musical selections on a tape, for instant recall at the push of a button.

Preset Call Forwarding Incoming calls will be re-routed to a pre-determined secondary number.

Press-To-Talk Telephone circuits are two way. Some circuits, such as mobile dispatch services for taxis, etc., are one-way. They use a microphone or handset with a button you must press-to-talk and release to listen. You can also buy a normal telephone handset with a press-to-talk button. Such a handset is useful in noisy places.

Pressurization Pumping inert gas into a heavy casing in which a couple of thick cables will be joined. The pressure is usually maintained at a few pounds above the surrounding atmospheric pressure. The idea is that the higher pressure inside keeps moisture out of the splice and thus improves the quality of phone service.

Prestel A videotex system used only in Britain.

Prestel Terminal Emulation Prestel is a character-based graphics emulation for communicating with the Viewdata service, particularly popular in the United Kingdom.

Presubscription A local Bell or local independent operating telephone company service that encourages each subscriber to select one long distance carrier he may use without having to dial a multiple digit access code. If you pre-subscribe to MCI, you will simply reach MCI by dialing "1" plus the 10-digit long distance number. AT&T's code is 1-0-288 (as in 1-0-ATT). Sprint's is 1-0-333. See PRE-SUBSCRIPTION.

Pretrip A central office malfunction that causes a phone to ring only once, and then stop, as if it had been answered. IT can be very confusing and difficult to get repaired, because most telco repair people have never heard of the problem and will insist your phone is at fault. Often fixed by replacing a faulty heat coil.

Pretty Good Privacy A powerful encryption scheme. See PGP for a full description.

Prevail An office automation UNIX software package which combines functions usually available only in individual programs, such as spreadsheet, a word processor, a database management system, communication capabilities and more.

Preventive Maintenance The periodic inspection, cleaning, adjusting and repair to eliminate problems before they affect service. Usually ignored.

Preview Dialing Preview dialing is a term used to describe an automatic dialer. Preview dialing is also called "screen dialing" or "cursor dialing." Typically the prospect's account information and/or phone number appears on the screen BEFORE the call is made. Thus the agent can "preview" the number, the screen, the customer. If the agent wants to make the call, the agent hits a key, such as "Enter" and the computer dials the number. In some preview dialing equipments, the agent must hit a key if he/she DOESN'T want the number dialed. Contrast preview dialing with Predictive Dialing where the computer makes all the dialing decisions and presents the calls to the agent only after they are connected. Predictive dialing is a lot faster than preview dialing. See PREDICTIVE DIALING.

Prewiring The practice of concealing telephone wiring or cable in the wall of a building while the building is being constructed.

PRI See PRIMARY RATE INTERFACE and ISDN.

PRI-EOP A fax signal. PRocedure Interrupt-End of Page.

PRI-MPS A fax signal. PRocedure Interrupt-MultiPage Signal.

Price Cap The phone industry has always been regulated on the basis of the profits it earns compared to the investment it had. That was called Rate of Return Regulation. How you figure profits — since you also have to figure what are allowable expenses — has been the subject of on-going debate for over 100 years. The latest idea in regulation is to replace the rate of return regulation with something called "Price caps" which allow the price of phone company services to rise by x% at maximum — the so-called price cap.

Primary Agent Group An automatic call distributor term: Primary agent group for which the inbound calls are intended. Intraflow goes to secondary and tertiary groups if the primary group does not have an agent available after the time or after overflow parameters are exceeded. Different ACD systems label this process differently.

Primary Alias A term used in the Sun Solaris Teleservices platform. See PROVIDER.

Primary Buffer A part of a computer's memory where fast incoming or outgoing data is kept until the computer has a chance to process it.

Primary Center A control center connecting toll centers — a Class 3 Central Office. It can also serve as a toll center for its local end offices.

Primary Domain Controller For a Windows NT Server domain, the computer that authenticates domain logons and maintains the security policy and the master database for a domain. See DOMAIN.

Primary Group A group of basic signals which are combined by multiplexing. The lowest level of the multiplexing hierarchy.

Primary High-Usage Trunk Group A high-usage trunk group that is offered first-route traffic only.

Primary Insulation The first layer of non-conductive material applied over a conductor to act as electrical insulation.

Primary Interexchange Carrier A primary Interexchange Carrier is the long distance company to which traffic from a given location is automatically routed when dialing 1+ in equal access areas. The PIC is identified by a code number which is assigned by the local telephone company to the telephone numbers of all the subscribers to that carrier to ensure the calls are routed over the correct network. When a subscriber switches long distance carriers, it often is referred to as a PIC change.

Primary Link The active LAN connection. When it fails the LAN is switched to the Backup link.

Primary Partition A portion of a physical disk that can be marked for use by an operating system. Under MS-DOS, there can be up to four primary partitions (or up to three, if there is an extended partition) per physical disk. A primary partition cannot be subpartitioned.

Primary Rate Interface The ISDN equivalent of a T-1 circuit. The Primary Rate Interface (that which is delivered to the customer's premises) provides 23B+D (in North America) or 30B+D (in Europe) running at 1.544 megabits per second and 2.048 megabits per second, respectively. There is another ISDN interface. It's called the Basic Rate Interface. It delivers 2B+D over either one or two pairs. In ISDN, the "B" stands for Bearer, which is 64,000 bits per second, which can carry PCM-digitized voice or data. See ISDN for a much better explanation.

NEWTON'S TELECOM DICTIONARY

Primary Resource An SCSA definition. The main resource around which a Group is constructed. Typically, the primary resource will be an interface to the telephone network, but it may also be a switch port.

Primary Server The SFT III Novell NetWare server that has been operating longer than its partner and is currently servicing the attached workstations. The primary server is the SFT III server that network workstations "see," and the one to which they send requests for network services. Routers on the internetwork see only the primary server and send routing packets to it. The primary server's IOEngine determines the order and type of events that are sent to the MSEngine. Only the primary server sends reply packets to network workstations. The secondary server is the SFT III NetWare server that is activated after the primary server. Either server may function as primary or secondary, depending on the state of the system. You cannot permanently designate which server is primary or secondary. System failure determines each server's role, that is, when the primary server fails, the secondary server becomes the new primary server. When the failed server is restored, it becomes the new secondary server.

Primary Station A network node that controls the flow of information on a communications link. Also, the station that, for some period of time, has control of information flow on a communications link (in this case primary status is temporary).

Primary Storage The main internal storage.

Primary Wire Center A switching center in the AT&T/Bell system hierarchy of exchange classes. The primary center is a Class 3 exchange. It is used to connect toll offices and less frequently to connect a toll center with a local end office. Primary centers are capable of connecting toll centers through sectional centers and then to local end offices to establish communication connections when simple routing possibilities are busy.

Prime Focus Parabolic Antenna There are two types of satellite antennas — a prime focus parabolic antenna and an offset parabolic antenna. A "normal" satellite antenna is a pure parabola. It's been around forever. A newer antenna, called the offset antenna is taller than it is wide. According to the manufacturers, the offset antenna design makes for more efficient use of the antenna surface than a traditional prime focus parabolic antenna. What that means is that it captures more of the satellite signal hitting the antenna. Offset antennas are more expensive than the "normal" parabolic satellite antennas, which are called "prime focus parabolic antennas." They are more expensive because they cost more to make since they typically must be made out of one sheet of metal. Offset antennas are harder to carry around, since you can't make them out of several foldover sheets of metal.

Prime Line A telephone can be programmed to automatically select a certain line type whenever a call is made by lifting the receiver.

Prime Line Preference When you pick up the handset on your key system or hybrid key, you are automatically connected to your preferred line (central office or intercom), rather than having to punch down an extra line button. Some phone systems tout this as a feature. Some have it set up where you simply leave one of the line buttons depressed and it doesn't pop up when you put your handset back into its cradle. Most 1A2 phone systems have this feature. Not all electronic phone systems do. Walker breaks the features into Prime Line Outgoing and Prime Line Incoming, and allows you to program the phones separately.

Primitive An abstract, implementation independent, interaction between a layer service user and a layer service provider. See PRIMITIVES.

Primitives Abstract representations of interactions across the service access points indicating information is passed between the service user and service provider. There are four types of primitives in the OSI Reference Model — request, indication, response and confirm.

Principal First or highest in importance. An owner or part-owner of a business. A person who authorizes another, as an agent, to represent him. Often confused with principle. See PRINCIPLE.

Principle A general or fundamental truth on which others are based. A rule of conduct. Often confused with principal. See PRINCIPAL.

Print Control Character A coded control character used to instruct the receiving unit on how a message is to be formatted in hard copy. Print control characters include carriage returns, back spaces, line feeds, tabs, etc.

Print Server A networked computer, usually consisting of fixed-disk storage and a CPU, that controls one or more printers that can be shared by users.

Print Spooler An application that manages print requests or jobs so that one job can be processed while other jobs are placed in a queue until the printer has finished with previous jobs. See PRINT SPOOLING.

Print Spooling A technique used to schedule printing tasks to one printer and to free up computer time from the slow task of feeding a slow printer (Any printer is slow compared to the speed of a computer). A small program or program/machine called the spooler does the scheduling. A user loads the print task to the spooler and when the print task's turn comes, the job is printed. Print spooling is handled several ways: You can allocate part of the computer's main memory to become a print spooler. You can allocate part of the company's disk memory to become a print spooler. You can get an external device called a print spooler. It will have all the storage space and software necessary. There are two primary advantages to Print Spooling: 1. You can use the spooler to save your and your computer's time. Dump the report to a print spooler at thousands of bits per second. Get on with something else on the computer. 2. You can use a print spooler to schedule several users' printing requests. This is particularly good in multi-user environments — for example, where the printer is a laser printer (and therefore expensive) and is attached to a LAN (Local Area Network).

Printed Circuit Board PCB. Flat material (fiberglass/epoxy) on which electronic components are mounted. A PCB also provides electrical pathways called traces, that connect components. Printed circuit boards are what PBXs and computers are made of these days. Be careful when you're replacing PCBs. They're usually very sensitive to static electricity. Handle them only when you're attached to a static electricity strap that is properly grounded. Lay them down only on a surface you're sure is static electricity free. And don't touch the components on PCBs whatever you do.

Printer A device which takes computer information and prints it on paper.

Printer Control Language PCL. See PCL.

Printer Driver A program that controls how your computer and printer interact. A printer driver file supplies information such as the printing interface, description of fonts, and features of the installed printer.

Printer Emulation A fax term for mimicking a printer-generated document. This way, the outgoing fax will look as if it has come from the printer attached to the computer. This can include full formatting, as well as letterhead, signature and graphic images.

Printer Font A font stored in your printer's memory, or soft fonts that are sent to your printer before a document is printed.

Printer Server A computer and/or program providing LAN (Local Area Network) users with access to a centralized printer. A person using the LAN will send a message to the printer server computer. This computer will then assign it a piece of memory or disk space to store its file while it waits to be printed. With a printer server, users can send to the printer any time. Their print jobs are usually handled in the order they are received. But "big bosses" can be given priority and can be bumped to the top of the queue. Print servers allow fewer printers to satisfy more users. Print servers are also especially useful for expensive, laser or high speed printers because they (the print servers) spread the cost of these expensive machines over many users, making them more affordable. See PRINT SPOOLING.

Printer, Wire A matrix printer which prints using a set of wire hammers which strike the page through a carbon ribbon to generate the matrix characters.

Prioritization The process of assigning different values to network users, such that a user with higher priority will be offered access or service before a user with lower priority. Increasingly available as an added option with network operation. Any procedure where different levels of precedence exist.

Prioritization Parameters The hierarchical rules which a network device, such as a router, applies to incoming traffic to determine which traffic should be handled first, next, and so on.

Priority A ranking given to a task which determines when it will be processed.

Priority Bumping The process during a link, trunk or facility failure where lower priority user access to network services is interrupted in order to offer those services or bandwidth to a pre designated higher priority user.

Priority Call 1. Emergency calls to the attendant bypass the normal queue and alert the attendant with some special signal.
2. The name of a Bell Atlantic service. While all callers are important, some are more critical to your business. So give your high-priority callers a ring of their own with Priority Call. Simply use your phone to program up to six callers' numbers. When any of these people call, you'll know right away because you'll hear a different ring. If you have Call Waiting, you'll hear a Priority "beep" when you're on the phone. This way, you'll only have to interrupt your work to take priority calls.

Priority Indicator A character or group of characters which determine the position in queue of the message in relation to the urgency of other messages. Priority indicators control the order in which messages are to be delivered.

Priority Ringing A name for a Pacific Bell (and possibly other local telephone companies') service which alerts you to have calls from selected numbers ring at another number.

Priority Transport The capability of a network for certain classes of traffic to have priority over others and thus have lower delay or otherwise better performance.

Priority Trunk Queuing Through user-chosen trunk access level, this PBX feature places any caller with this or higher level in the class of service assignment ahead of callers waiting for the same trunk group (or Agent Group in the case of incoming ACD calls).

Privacy Privacy usually means that once a caller "seizes" a line, no other user can access that same line even though it appears on his/her key set. Privacy can be automatic or selected for each call.

Privacy And Privacy Release All other extensions of a line are unable to enter a conversation in progress unless the initiating telephone releases the feature.

Privacy Enhanced Mail. PEM. An Internet electronic mail capability which provides confidentially and message integrity using various encryption methods.

Privacy Lockout Privacy automatically splits the connection whenever an attendant would otherwise be included on the call, i.e. the attendant can't listen in to a call she's just extended to someone. A tone warning is generated when the attendant bridges into a conversation in progress.

Privacy Override Activation of a special pushbutton allows the phone user to access a given busy line, even though the automatic exclusion facility is being used by the station on that line. This privilege of Privacy Override is usually only given to Big Bosses.

Private ATM Address A twenty-byte address used to identify an ATM connection termination point.

Private Automatic Branch Exchange PABX. A private telephone switch for a business or an organization in which people have to dial "9" to access a local line. In the old days, private branch exchanges were manual, meaning that operators/attendants were needed to manually place calls. Then the systems improved and you were able to dial the outside world from your extension without the help (or hindrance?) of an operator. Thus they became known as private automatic branch exchanges. But then all PBXs became Automatic. So these days, PABXs are all called PBXs, except in some countries outside North America, where they're still called PABXs. See also the next definition and PBX.

Private Branch Exchange PBX. Term used now interchangeably with PABX. PBX is a private telephone switching system, usually located on a customer's premises with an attendant console. It is connected to a common group of lines from one or more central offices to provide service to a number of individual phones, such as in a hotel, business or government office. For the biggest definition, see PBX. See also PABX.

Private Carrier An entity licensed in private services and authorized to provide communications service to others for money.

Private Dial-In Ports A packet network term. For customers who have many calls, the packet network operator provides dedicated, unpublished phone numbers. The idea is to give the preferred user better service.

Private Exchange PX. A telephone switch serving a particular organization and having no means of connection with a public exchange. In other words, a phone system just for intercom calls.

Private Line A direct channel specifically dedicated to a customer's use between specified points. A line leased from a carrier, local or long distance. A non-switched circuit. One end of the line is directly connected to the other end. Here's the AT&T definition of a private line. "A dedicated, nonswitchable link from one or more customer-specified locations to one or more customer-specified locations..."

Private Line Service An outside telephone number separate from the PABX, can be set up to appear on one of the buttons of a key telephone. Also called an Auxiliary Line.

Private Management Domain PRMD. An X.400 electronic mail term: A private domain to which MTAs (Message Transfer Agents) send mail. PRMDs are connected to ADMDs (Administrative Management Domains) for message routing over wide area links. Under X.400 addressing, the PRMD represents a private electronic messaging system that may be connected to a Administrative Management Domain. The PRMD is usually a corporate or government agency E-Mail system connected to an ADMD.

Private Message A message designation which prevents that message from being given to another mailbox.

Private Network A series of offices connected together by leased and non-leased phone lines with switching facilities and transmission equipment owned and operated by the user or by the carrier and leased to the user. This is a broad definition. But these days with SDN and other virtual private networks, it's hard to see the difference between using a private network and using the public network except that you hope your private network will be cheaper, better quality and perhaps offer a little more flexibility in the types of traffic — voice, video, data and fax — you carry on it. See PRIVATE VOICEBAND NETWORK.

Private Networks Marketing A Northern Telecom term which defines their organization for making and selling all telecom switches, except central offices. These products include the Meridian 1 PBX family, residential and business telephone sets, including Norstar and data communications.

Private Subscriber Network A virtual private network service supported by Public Packet Switched Service (PPSS) and incorporating interLATA transmission facilities owned or leased by the customer for private traffic. A Bellcore definition.

Private Voiceband Network A network that is made up of voice band circuits, and sometimes switching arrangements, for the exclusive use of one customer. These networks can be nationwide in scope and typically serve large corporations or government agencies.

Private Wire A private line. Derives its name from the old telegraph days when messages were carried on wires that strung across the nation.

Privileges The access rights to a directory, file or program over a local area network. Typically read, write, delete, create and execute.

PRMD PRivate Management Domain. An X.400 Message Handling System private organization mail system. Example: NASAmail.

PRO WATS PRO WATS I is AT&T's discount long distance service for companies which spend over $120 a month on long distance. AT&T PRO WATS II and III are designed for companies with higher calling volumes. Charges and discounts increase with plan size.

Pro750 An imaging term. A DVI product family introduced in 1989 consisting of Application Development Platform, boards and software.

Proactive Taking the initiative. Doing it before someone (most likely your competition) forces you to do it. The word is currently in vogue among those people who believe the telephone companies should do all the positive, forward-looking actions before the competition does them and gets the public kudos. The word has no real meaning, but serves a purpose as a cry to action. The word actually is grammatically incorrect. The real word is "active." It is the opposite of "reactive." The person who told us this is Norm Brust, a wonderful man and one of the more active (not proactive) people in our industry.

Probe 1. A sensing device, typically about the size and shape of a pencil, that is used to sense various physical conditions such as temperature, humidity, current flow, speed. Usually connected to a meter or oscilloscope which displays the condition being monitored.
2. An empty message that is sent to reach a particular address to determine if an address can be reached.

Probe Envelope In X.400, the envelope that encloses a probe in the MTS (Message transfer System). See PROBE.

Process A software application. Any activity or systematic sequence of operations that produces a specified result. Typically, a computer function that consists of, or involves, procedure code, data storage and an interface for communicating with other processes.

Process Manufacturing The making of things. This contrasts with flow manufacturing which is working on something — like oil — that flows through a production process.

Processing Gain In a spread spectrum transmission system, the original information signal is combined with a pseudo random correlating, or spreading code. The more random and the greater the length of the code, the more robust the resulting spread spectrum signal is against interference and interception. A measure of this robustness is referred to as processing gain. The FCC requires a minimum of 10 dB processing gain for non-licensed equipment operating in the Part 15 902-928 MHz, 2400-2483 MHz, and 5725-5850 MHz frequency bands. See also CDMA.

Processing, Batch A method of computer operation in which a number of similar input items are accumulated and sorted for processing. Compare with ON-LINE or INTERACTIVE PROCESSING.

Processor The intelligent central element of a computer or other information handling system. Also called the Central Processing Unit (CPU).

Processor Data Module AT&T's PDM performs conversion between the EIA signal protocol and the System 85 Digital Communications Protocol (DCP).

Processor Occupancy The time the telephone system processor is in use. There are two typical demands on the central processor in a telephone system, moving calls around and running self-diagnostics. Be sure you factor in the second when you're trying to figure out how many calls your telephone system processor will handle before it dies.

Procurement Lead Time The interval in months between the initiation of procurement action and receipt into the supply system of the production model (excludes prototypes) purchased as the result of such actions, and is composed of two elements: production lead time and administrative lead time.

Procr Processor.

Prod A device that resembles a pencil, but containing a metal tip in an insulated handle with a wire to connect it to a piece of test equipment, such as a VOM (volt-ohm-millimeter). the metal tip is touched to various points in an electrical circuit for measurements and trouble-shooting.

Prodigy A joint venture of IBM and Sears Roebuck & Company, Prodigy offers on-line computer services, at last count, roughly 800. These services are offered for a largely flat monthly fee — probably the only on-line service offering unlimited usage for one monthly price. The fees are kept low by allowing companies to run commercials on the bottom of subscriber screens. children get a kick out of Prodigy.

Productize This is a stupid word. But it means to make an idea into a product. What this means is to complete the R&D on it, to finish the customer documentation, to finish the packaging design, to assign a name, model number and stocking number, to pass the information onto product support, etc. Everything necessary to make it a product that can be sold. See also BETA.

Profile A set of parameters defining the way a device acts. In the LAN world, a profile is often used by one or more workstations to determine the connections they will have with other devices and those devices they will offer for use by other devices. Often called a login file. Profiles and login files usually work like batch files, automatically executing a number of commands when you turn on the machine.

PROFS PRofessional OFfice System. Interactive productivity software developed by IBM that is part of the Virtual Machine (VM) Productivity System and runs under the VM/CMS mainframe system. PROFS is frequently used for electronic mail and is said to give a user an edge in productivity in three areas: business communications (including electronic mail), time management and document handling.

Program Instructions given to a computer or automated phone system to perform certain tasks. Most vendors improve (update) their software programs continuously. It's a good idea to ask what the deal is with getting updates.

Program Circuit A voice circuit used for the transmission of radio program materials. It is a telephone circuit which has been equalized to handle a wider range of frequencies than are required for ordinary speech signals.

Program Counter A device inside a computer which keeps track of which instruction in the program is next, etc.

Program Evaluation Reviews An activity many of us are consigned to spend our aging years doing. In the early days of our careers, we used to do things — actually do tasks hands-on. Then many of us, sadly, got "successful." Our jobs then became telling younger people what to do and checking that they do what we told them, or doing something better (hopefully). To do all this, we sit in meetings. We call these meetings "Program Evaluation Reviews."

Program Evaluation Review Technique PERT. A management tool for graphically displaying projected tasks and milestones, schedules and discrepancies between tasks.

Program File A file that starts an application or program. A program file has an .EXE, .PIF, .COM, or .BAT filename extension. AKA executable.

Program Information File PIF. A file that provides information about how Windows should run a non-Windows NT application. PIFs contain such items as the name of the file, a start-up directory, and multitasking options for applications running in 836 enhanced mode. See PIF.

Program Logic The particular sequence of instructions in a program.

Programmable In telephony, the ability to change a feature or a function or the extension assigned to a telephone without rewiring.

Programmable Call Forwarding This feature of a telephone system allows a user to instruct his phone to send all his calls to another phone. That phone might be another extension in the same phone system or it might be another phone number altogether in a different part of the country. This feature is great. You're going to a meeting but don't want to miss that one special call. You can send your calls to a person close to the meeting and ask them to interrupt you if that "special" call comes in. The problem with this feature is that often people forget their phone is on "forwarding" and when they return to their office, they sit around all afternoon waiting for that special call, which got call forwarded elsewhere. There are two ways to overcome this. Some phones have lights or messages on their screens which indicate all calls are being forwarded. Also, some telecom managers program their total phone systems so that twice a day, all call forwarding shall cease and all calls shall return to their original phone.

Programmable Configuration Select Refers to the EEPROM setup routine which allows jumperless configuration of the system board.

Programmable Memory Memory that can be both read from and be written into by the processor. Synonym for RANDOM ACCESS MEMORY — RAM.

Programmable Terminal A user terminal that has some limited processing power. Also, intelligent terminal.

Programming Language A language used by a programmer to develop instructions for the computer. It is translated into machine language by language software called assemblers, compilers and interpreters. Each programming language has its own grammar and syntax.

Programming Overlay Typically a piece of cutout cardboard, which you place over certain of the keys on a phone or console. When you punch in a certain code, the buttons become what's written on the programming overlay.

Program Sharing The ability of several users or computers to use a program simultaneously.

Program Store Permanent memory in a stored program control central office that contains the machine's generic software program, parameters and translations.

Progressive Conference A PBX feature. Allows the extension user to create conferences of more than three people using the consultation hold and add-on conference features. To create a conference, an extension user typically uses the consultation hold, dials the desired internal or external number and effects an add-on conference. The conference may then be progressively expanded, in this same fashion, to the maximum capability of the phone system offering this feature. A good question to ask before you get sold this feature is "does the conferencing have amplification and balancing?" Without these features, the conferencing conversation will simply get more and more difficult to hear on.

Progressive Dialing A form of predictive dialing, progressive dialing is slightly more automated than preview dialing. The customer data is not displayed until the number is dialed, giving the agent less time to review it and a shorter time between calls. See also PREDICTIVE DIALING and PREVIEW DIALING.

Progressive Tuning A method of painting pictures on computer monitors or TV screens in which the picture is painted line by line. It is today's most common way of painting a picture or an image on a computer screen.

Project Evaluation Review Technique PERT. A technique for managing a project — say the installation of a PBX — which produces a guess at the project's critical path (longest task to complete) and of project milestone completion dates. See PERT.

PROM Programmable Read Only Memory. A PROM is a programmable semiconductor device in which the contents are not intended to be altered during normal operation. PROM acts like nonvolatile memory. When you install an autoboot PROM on a LAN network board, the workstation can boot up from the network server. This is particularly useful for diskless workstations.

Promiscuous Mode Most Ethernet cards ignore all the packets on the network that aren't destined for them. But in a Remote server — one serving multiple remote users all calling in over modems — the Ethernet LAN card has to get access to all the packets and grab those that are meant for it — so it can pass them over to the remote callers. I assume it's called "promiscuous mode" because it means that the Ethernet card has to have a relationship with all the packets traveling on the local area network. Another application for promiscuous mode is if you want to attach software or hardware to your computer, monitor and analyze all the packets flying around your network. You can set some (but not all) Ethernet cards to promiscuous mode.

Prompt An audible or visible signal to the system user that some process is complete or some user action is required. Also used to signify a need for further input and/or location of needed input. See also PROMPTS.

Prompt Tagging And Encoding According to Steve Gladstone, author of the book, "Testing Computer Telephony Systems," (available from 212-691-8215) many test applications require the test system "know" which prompt is being played by the computer telephony system. This is often accomplished by prompt tagging, also known as prompt encoding. Prompt tagging has the computer telephony system play-out an audio tone that can be heard by the test system with each prompt that is played. Prompt tagging is relatively inexpensive to perform, and inexpensive to automate. Prompt tagging can be accomplished in two ways, either by inserting tones into actual user prompts ("insert" mode), or by having a special programmatic switchable test mode that may be toggled on/off by the computer telephony system that will play out a tone sequence before or after each prompt is played ("append" mode).

Prompts 1. Recorded instructions delivered by voice processing units. Prompts may include MENUS or other information that is played each time you get into the system.
2. Messages from the computer instructing the user on how to use the system. See MENU and AUDIO MENU.

PRONET A family of token-passing local area networks from Proteon, Natick, MA. The fastest of Proteon's networks is an extremely fast 80 Mbps. See the next two definitions also.

PRONET-4/16 An IEEE 802.5 Token-Ring network sold by Proteon Inc.

PRONET-10 A proprietary star-shaped Token-Ring network operating at 10 Mbit/sec, and sold by Proteon Inc.

Proof Of Concept You're at a trade show. You go to a booth. You see some great new technology. You can't buy it. It's simply a demonstration of new technology. What's called "proof of concept." It proves that the idea works. It doesn't mean there's a market for it. The idea of "proof of concept" is to excite people — customers or security analysts. Maybe someone will place a big order or buy the company's stock?

Propagation Delay The delay caused by the finite speed at which electronic signals can travel through a transmission medium. Propagation delay is estimated at 160,000,000 meters per second in copper wire. Note: No signal can travel through any medium faster than the speed of light, which is 300,000,000 meters per second, or 186,000 miles per second.

Propagation Time Time required for an electrical wave to travel between two points on a transmission line.

Propagation Velocity The speed at which electrons or photons travel through a transmission medium.

Property Management Interface PMI. A telephone system's ability to talk to a hotel's computer system.

Proportional Font A font in which different characters have varying widths. All magazines and newspapers are printed in proportional characters, which make reading easier. By contrast, in a monospaced font, such as one on an old typewriter, all characters have the same widths.

Proprietary If something is proprietary it means it will only work with one vendor's telephone system. There are many telephones that are proprietary to one telephone system or one manufacturer. These Proprietary phones are usually the electronic and multi-line instruments. For example, Rolm's Cedar, Juniper and Cypress phones will only work connected to a Rolm PBX. Similarly with the CXC Rose TeleTerminal which will only worked connected to a CXC Rose PBX. However, the Mitel SuperSet 4s will work with Solid State PBXs and ACDs.

Proprietary LAN A LAN that runs the equipment of only one vendor. A proprietary LAN, for example, cannot join IBM PC's to DEC minicomputers. DEC and Wang both make proprietary LANs. Some people say such LANs are more "bug-free" because they have only one vendors' wares to deal with. They also tend to be more expensive. They also tend to tie you to one vendor, although some makers are now coming out with bridges which connect proprietary LANs to non-proprietary LANs.

Proprietary Network A network developed by a vendor that is not based on protocols approved by standards body or on standards that are "open." Typically, you won't be able to connect to a network with any equipment other than the manufacturer who created it.

Proprietary Telephone Sets Proprietary telephones are feature phones that are specific to a particular make of PBX, ACD or other switching system. They may be digital or analog. As they are custom designed for that system, they have non-standard interfaces and have non-standard protocols to communicate between the telephone and the switch. This has several implications: 1. You can't take a proprietary phone from one switch and expect it to run on another switch. It won't. 2. Proprietary phones are expensive, and are highly profitable to the their manufacturers. Hence the manufacturers' insistence on keeping them proprietary. 3. Signaling between proprietary phones and their switches is richer than signaling between switches and single line analog phones. As a result, it's preferable to integrate voice mail and automated attendants through proprietary phones. Sometimes the manufacturer of the switch will divulge his secret signaling scheme. Other times he won't. And you, as a voice mail or auto attendant manufacturer have to reverse engineer it, which is sometimes successful. ISDN is actually the first attempt to make "proprietary" sets standards. So far, only a few manufacturers have used ISDN-like phones as their proprietary phones.

ProShare An Intel hardware/software combination that uses phone lines (analog and ISDN digital) to do collaborative computing. One version of ProShare uses ISDN lines and allows you to see and be seen, to speak and be heard, and to share a document, like an Excel spreadsheet, and have both of you make changes to the spreadsheet during the phone call.

Prosody Intonation. In text to speech, prosody refers to how natural it sounds — the ups and downs of the sentence.

Prospective The opposite of Retrospective or Retroactive. Most regulatory commission rate cases are prospective, which means they relate to prices and things in the future. Some rate cases, however, are retroactive or retrospective, which means they apply to prices and things in the past. Most of these decisions involve forcing the company to return money to its subscribers in the form of a refund. Interestingly — try this one — most retroactive commission decisions are prospectively retroactive. In other words, they only take effect some time in the future, when the decision is voted upon by all the commission members.

Prospero UNIX software which helps you search archives connected to the Internet. Prospero uses a virtual file system which enables users to transparently view directions and retrieve files. In short, Prospero is a distributed directory service and file system that allows users to construct customized views of available resources while taking advantage of the structure imposed by others.

Protected Distribution System PDS. This is a US Federal Government definition: A wireline or fiber-optics telecommunication system which includes adequate acoustic, electrical, electromagnetic, and physical safeguards to permit its use for the unencrypted transmission of classified information. A complete protected distribution system includes the subscriber and terminal equipment and the interconnecting lines.

Protected Mode A computer's operating mode that is capable of addressing extended memory directly. The operating mode for the Intel 80286 and higher processors (the 80386, 80486 and Pentium) that supports multi tasking, data security, and virtual memory. The 80286 processor can run in either of two modes: real or protected. In real mode, it emulates an 8086 (it accesses a maximum of 640KB of RAM and runs only one software application at a time). Protected mode allows the 80286 processor to access up to 16MB of memory. It uses a 24-bit address bus. Since a bit can have one of two values, raising the base number of 2 to the power of 24 is equal to 16,777,316 unique memory addresses. Each memory address can store one byte of information (16,777,216 bytes equals 16MB). Protected mode operation also makes it possible to run more than one application at once and to handle more processes because more memory is available. Processes can be requests from an operating system or an application to perform disk I/O, memory management, printing, or other functions. Processes are assigned priority numbers in protected mode. The processor gives priority to those with higher numbers. Operating system processes always have higher priority than application processes. See also REAL MODE and VIRTUAL 8086 MODE.

Protective Connecting Arrangement PCA. A device leased from the telephone company and placed between your own (customer-provided) telephone equipment and the lines of the telephone company. The idea was to protect their lines from your junky equipment. No instance/case was ever proven of harm occurring to the network from faulty customer-provided equipment and the PCAs were thrown out and replaced by the FCC's Part 68 Registration Program. Under this program, customer-owned equipment which passes FCC tests can be registered and connected directly to the phone network without these devices. The phone industry eventually refunded most of the fees it charged on the PCAs. NATA and many manufacturers claimed the PCAs were designed to prevent the growth of the interconnect or customer-owned phone industry. They were probably right. The question is now moot, since the charges and the devices no longer exist, except in a museum or attached to very old equipment. See also PROTECTIVE COUPLING ARRANGEMENT and PCA.

Protective Coupling Arrangement PCA. A device placed between the phone company's trunks and your particular telephone gadget. The objective of the PCA is to isolate the telephone company's lines from your equipment and thus protect their lines from your equipment. The device is not needed if your equipment has passed FCC approval — under Part 68 of the FCC's rules. See also PROTECTIVE CONNECTIVE ARRANGEMENT, which is another term for the same thing.

Protector Block A device interconnected to an exchange access line to protect the connected equipment from over-voltage and/or over-current of 600 volts or greater. Hazardous voltages and currents are shunted to ground.

PROTEL PRocedure Oriented Type Enforcing Language. Protel is a block-structured, type-enforcing, high level, software language that enables extensive type checking on the source code at compile time. It was developed at Bell Northern Research, a subsidiary of Northern Telecom. Protel is used in the DMS-100, a family of Northern Telecom central office telephone switches. Both the central control CPU and the DMS SuperNode CPU are programmed in Protel.

Protn Protection.

Protocol A procedure for adding order to the exchange of data. A protocol is a specific set of rules, procedures or conventions relating to format and timing of data transmission between two devices. A standard procedure that two data devices must accept and use to be able to understand each other. Sort of like us both speaking English so we can communicate. The protocols for data communications cover such things as framing, error handling, transparency and line control. There are three basic types of protocol: character-oriented, byte-oriented and bit-oriented.

Protocols break a file into equal parts called blocks or packets. These packets are sent and the receiving computer checks the arriving packet and sends an acknowledgement (ACK) back to the sending computer. Because modems use phone lines to transfer data, noise or interference on the line will often mess up the block. When a block is damaged in transit, an error occurs. The purpose of a protocol is to set up a mathematical way of measuring if the block came through accurately. And if it didn't, ask the distant end to re-transmit the block until it gets it right. See PROTOCOLS for a list of the more common protocols.

Protocol Analyzer A specialized computer and/or program that hooks into a LAN and analyzes its traffic. Good protocol analyzers can record and display data on all levels of traffic on a LAN cable, from the lowest media access control packets to NetBIOS commands and application data. They are excellent for diagnosing network problems, but they require some expertise, as their data output can be obscure.

Protocol Control Protocol Control is a mechanism which a given application protocol may employ to determine or control the performance and health of the application. Example, protocol liveness may require that protocol control information be sent at some minimum rate; some applications may become intolerable to users if they are unable to send at least at some minimum rate. See MCR.

Protocol Control Information PCI. the protocol information added by an OSI (Open Systems Interconnection) entity to the service data unit passed down from the layer above, all together forming a PDU (Protocol Data Unit).

Protocol Conversion A data communications procedure which permits computers operating with different protocols to communicate with each other. See PROTOCOL and PROTOCOL CONVERTER.

Protocol Converter A device which does protocol conversion. It's your classic "black box." Glasgal Communications defines a protocol converter as any device which translates a binary data stream from one format to another according to a fixed algorithm. Compare with bridge and gateway, which are different animals and may contain protocol converters...and more.

Protocol Dependent Routing Any routing method in which routing decisions are made on the basis of information provided by the specific LAN protocol used by the communicating devices. TCP/IP and DECnet routers are protocol dependent routers. So are so-called multi protocol routers, because they must support each protocol running in the network. See also PROTOCOL INDEPENDENT ROUTING.

Protocol Filtering A feature available in some network bridges which allows it to be programmed to always forward or reject transmissions associated which specified protocols.

Protocol Independent Router A routing device that provides the functionality of protocol specific routers such as TCP/IP or DECnet routers but is independent of protocols. In addition to routing "routable" protocols like TCP/IP, DECnet or XNS, it routes IBM protocols which are not routable. The protocol independent router combines the latest in computer hardware with the new advanced routing technologies such as SPF (Shortest Path First) and IS-IS (OSI routing standard). It represents an alternative to conventional routers that use old routing technologies and are protocol dependent. Protocol independent routers provide easy-to-install-and-use enterprise-wide networks in a token ring or Ethernet environment.

Protocol Independent Routing A routing method in which routing decisions are made without reference to the protocol being used by the communicating devices. Protocol independent routers provide the functionality of protocol specific routers such as TCP/IP or DECnet routers, but can also route non-routable protocols. See also PROTOCOL INDEPENDENT ROUTER.

Protocol Stacks A group of drivers that work together to span the layers in the network protocol hierarchy. Also called protocol suites.

Protocol Suite A hierarchical set of related protocols.

Protocols For an explanation of protocols, see PROTOCOL. Here are the more common PC protocol types:

MODULATION PROTOCOLS
Bell 103: Low Speed (300 baud)
Bell 212: Low Speed (1200bps)
ITU-T V.22bis: Medium (2400bps)
ITU-T V.32: Medium Speed (9600bps)
ITU-T V.32bis: High Speed (14,400bps)
ITU-T V.34: High Speed (28,800bps)

ERROR CONTROL PROTOCOLS
Microcom Network Protocol (MNP)
ITU-T V.42 (Includes LAP-M & MNP)

DATA COMPRESSION PROTOCOLS
MNP/5
ITU-T V.42bis

All data compression requires an underlying error control protocol.

FILE TRANSFER PROTOCOLS
Kermit: 7-bit data path, quotes control characters.
XMODEM: 8 bit data path, ACK/NAK protocol.
YMODEM: 8-bit data path, batch capability.
ZMODEM: 8-bit data path, quotes some control characters.

Provider A process that represents an interface between the Sun Solaris Teleservices platform and an installed telephone device, such as a telephone line or a fax machine. Multiple instances of a provider can be configured, each based on the same information. Each configuration is identified by a unique primary alias. A primary alias is the primary label used to prefer to a provider configuration. The primary alias is a provider's default and primary name. This definition courtesy Sun.

Provisioning The act of supplying telecommunications service to a user, including all associated transmission, wiring, and equipment. In NS/EP telecommunication services, "provisioning" and "initiation" are synonymous and include altering the state of an existing priority service or capability.

Proxy A software agent that acts on behalf of a user. Also, the mechanism whereby one system "fronts for" another system in responding to protocol requests. Proxy systems are used in network management to avoid having to implement full protocol stacks in simple devices, such as modems. In SNMP, a proxy is a device which performs SNMP functionality for a separate managed device. The amount of responsibility may vary. Proxy ARPing refers to address recognition for another unit with SNMP capability, while a proxy agent provides an external SNMP agent for a managed device which does not have SNMP capability.

Proxy ARP The technique in which one machine, usually a router, answers ARP requests intended for another machine. By "faking" its identity, the router accepts responsibility for routing packets to the "real" destination. Proxy ARP allows a site to use a single IP address with two physical networks. Subnetting would normally be a better solution.

Proxy Server A type of fire wall, proxy servers provide extra security by replacing calls to insecure systems' subroutines. Proxy servers allow companies to provide World Wide Web access to selected people, by doing it through a network firewall.

PRSM Post Release Software Manager.

PRX Program.

PS/2 IBM Personal System/2 personal computer.
PSAI AT&T's Processor-to-Switch Applications Interface. See also ASAI and SCAI.
PSAP Public Safety Answering Position. PSAPs are customarily segmented as "primary," "secondary" and so on. The primary PSAP is the first contact a 911 caller will get. Here, the PSAP operator verifies or obtains the caller's whereabouts (called locational information), determines the nature of the emergency and decides which emergency response teams should be notified. In some instances, the primary PSAP may dispatch aid. In most cases, the caller is then conferenced or transferred to a secondary PSAP from which help will be dispatched. Secondary PSAPs might be located at fire dispatch areas, municipal police force headquarters or ambulance dispatch centers. Often the primary PSAP will answer for an entire region.
PSC/PUC Public Service Commission. Also known as Public Utility Commission. It's the state agency charged with regulating the local phone company utility. In reality, there are only two things the PSC can do: 1. Allow the phone company to increase its prices, and 2. Restrict competition to the phone company by creating all sorts of restrictive rules and regulations. As competition in the telecommunications industry grows — chiefly because of Federal rulings — the state PSCs are losing their power. This bothers them.
PSDS Public Switched Digital Service. A BOC service. AT&T Circuit Switched Digital Capability (CSDC), also known commercially as AT&T's Accunet Switched 56 service. It allows a full-duplex, dial up, 56-Kbit/s digital circuits on an end-to-end basis.
Pseudo Code P-CODE. A code which must be translated before it can be executed or acted upon. Here's a more technical explanation. Pseudo Code is a compiled program written for a hypothetical processor and interpreted at runtime by a P-code interpreter written for a native environment. P-code has many objectives in its different implementations, most often portability and space savings.
Pseudo Flaw A loophole planted in an operating system as a trap for intruders.
Pseudo Lite A ground based differential GPS (Global Positioning System) which transmits a signal like that of an actual GPS satellite, and can be used for ranging.
Pseudo Random Bit Pattern A test pattern consisting of 511 or 2,047 bits ensuring that all possible bit combinations can pass through a network without error.
Pseudo Random Test Signal A pseudo random test signal is a signal consisting of a bit sequence that approximates a random signal.
Pseudo Range A distance measurement based on the correlation of a GPS (Global Positioning System) satellite transmitted code and the local receiver's reference code, that has not been corrected for errors in synchronization between the transmitter's clock and the receiver's clock.
Pseudo Ternary A term used in ISDN Basic rate interface data coding. Refers to three encoded signal levels representing two-level binary data (binary "1"s are represented by no line signal, and binary "0"s by alternating positive and negative pulses).
PSI Packet Switching Interface.
PSK Phase Shift Keying. A method of modulating the phase of a signal to carry information. See PHASE MODULATION.
PSN Packet Switch Node. The modern term used for nodes in the ARPANET and MILNET. These used to be called IMPs (Interface Message Processors). PSNs are currently implemented with BBN C30 or C300 minicomputers.
PSP PCS Service Provider.
PSPDN Packet Switched Public Data Network.
Psophometer An instrument arranged to give visual indication corresponding to the aural effect of disturbing voltages of various frequencies. A psophometer usually incorporates a weighting network, the characteristics of which differ according to the type of circuit under consideration; e.g., high-quality music or commercial speech circuits.
PSTN Public Switched Telephone Network. An abbreviation used by the ITU-T, PSTN simply refers to the local phone company. See PUBLIC SWITCHED TELEPHONE NETWORK and PUBLIC DATA NETWORK.
PSU Packet Switch Unit.
Psychic ANI A term created by Howard Bubb from Dialogic to designate what happens when you call someone on one line while they're calling you on the other.
PT Payload Type: Payload Type is a 3-bit field in the ATM cell header that discriminates between a cell carrying management information or one which is carrying user information.
PTC 1. Portable Teletransaction Computers. These are typically handheld devices used for retail (inventory), healthcare (tracking supplies), mobile field repair (reporting fixes), insurance (visiting car wrecks and other disasters), etc. The devices typically have telecommunications capabilities, sometimes wireless, sometimes landlines. And they typically include microprocessors, memories, displays, keyboards, touchscreens, character recognition software, barcode readers, printers, modems and local and/or wide area data radios.
2. Personal Telecommunications Center. Infocorp's name for a product most people call a PDA, Personal Digital Assistant.
PTE Path Terminating Equipment. Network elements that multiplex and demultiplex the payload and that process the path overhead necessary to transport the payload. An AT&T SONET term.
PTI An ATM term. Payload Type Indicator: Payload Type Indicator is the Payload Type field value distinguishing the various management cells and user cells. Example: Resource Management cell has PTI=110, end-to-end OAM F5 Flow cell has PTI=101.
PTMPT Point-To-Multipoint: A main source to many destination connections.
PTN Public Telecommunications Network.
PTS Presentation Time Stamp: A timestamp that is inserted by the MPEG-2 encoder into the packetized elementary stream to allow the decoder to synchronize different elementary streams (i.e. lip sync).
PTS Public Telecommunications Systems.
PTSE An ATM term. PNNI Topology State Element: A collection of PNNI information that is flooded among all logical nodes within a peer group.

PTSP An ATM term. PNNI Topology State Packet: A type of PNNI Routing packet that is used for flooding PTSEs among logical nodes within a peer group.
PTT Post Telephone & Telegraph administration. The PTTs, usually controlled by their governments, provide telephone and telecommunications services in most foreign countries. In ITU-T documents, these are the Administrations referred to as Operating Administrations. The term Operating Administrations also refers to "Private Recognized Operating Agencies" which are the private companies that provide communications services in those very few countries that allow private ownership of telecommunications equipment.
PTY Party.
PU Physical Unit. In IBM's SNA, the component that manages and monitors the resources of a node, such as attached links and adjacent link stations. PU types follow the same classification as node types.
PU 2.0 & 2.1 IBM protocols which allow applications written to APCC and interpreted by LU 6.2 to access the mainframe (2.0) and token ring LAN (2.1).
PU Type 2 A physical unit (PU) refers to the management services in SNA node always contains one physical unit (PU), which represents the device and its resources to the network. PU Type 2 is often referred to as a cluster controller.
PU 4 An IBM SNA front end processor.
PU 5 An IBM SNA mainframe, such as a System/370 or System/390. It runs VTAM to handle data communications.
Public Access Terminal A kiosk (SK series), enclosure (TK series) or a system for special enclosures and custom applications (XE Series) which provides the public access to service. Contains a color monitor and keypad for customer interaction.
Public Announcement Trunk Group A trunk group used to provide multiple types of announcements, such as the weather, time and sports results.
Public Asynchronous Dial-In Port A term used in packet switching networks referring to the local phone number of a port into the packet switched network. Some networks provide different numbers for different speeds. Some provide different speeds on the same numbers. See also PRIVATE DIAL-IN PORTS.
Public Data Network A network available to the public for the transmission of data, usually using packet switching under the ITU-T X.25 packet switching protocol. See PACKET SWITCHED NETWORK.
Public Dial Up Port A port on a computer system or on a communications network which is accessible to devices operating over the public switched telephone network.
Public Exchange A British word for Central Office. Outside North America, central offices are all called "public exchanges." In the US, a public exchange is typically a local telephone switch. TELECONNECT's phone number in North America is 212-691-8215. The 212 is our area code. The 691 designates the central office or public exchange which serves us. That public exchange belongs to Nynex Company. See also CENTRAL OFFICE and CO.
Public Key A key given to a recipient of an encrypted message to decrypt the message.
Public Network A network operated by common carriers or telecommunications administrations for the provision of circuit switched, packet switched and leased-line circuits to the public. Compare with private network.
Public Room A video conferencing center that is arranged with transmission services. Public rooms can be rented by business customers who wish to have a video conference but do not have facility available at their own offices.
Public Safety Answering Point PSAP. A generic term for the person or group of people who answer 911 emergency phone calls.
Public Service Commission PSC. The state regulatory authority responsible for communications regulation. Also known as Public Utility Commission, Corporate Commission and in some states, the Railway Commission.
Public Switched Digital Service PSDS. A generic name for Bell telephone companies' service offerings that provide customer switches the capability of sending data at 56 Kbps over the public circuit switched network.
Public Switched Network Any common carrier network that provides circuit switching between public users. The term is usually applied to the public telephone network but it could be applied more generally to other switched networks such as Telex, MCI's Execunet, etc.
Public Switched Telephone Network Usually refers to the worldwide voice telephone network accessible to all those with telephones and access privileges (i.e. In the U.S., it was formerly called the Bell System network or the AT&T long distance network).
Public Telephone Station Coin phone. Pay phone.
Public Telephony A new term to describe what might be possible with a new type of public "phone" which could do more than make and receive analog phone calls. Perhaps it could do videoconferencing? Perhaps it could send faxes? Perhaps it could access remote databases? In short, it could do a bunch of multimedia things. Exactly what hasn't been defined, as yet. Having an RJ-11 plug in it so I could plug my laptop into would be a good beginning, however.
Public Utility Commission PUC. State body charged with regulating phone companies. Also called Public Service Commissions. See PUBLIC SERVICE COMMISSION.
Publications, Computer & Telecom
- Alcatel's Electrical Communication, a quarterly technical journal from Alcatel NV, 33 rue Emeriau, 75725 Paris Cedex 15, France
- America's Network, 312 West Randolph, Ste. 600, Chicago, IL 60606
- Boardwatch, 8500 West Bowlds Ave, Littleton, CO 80127
- Business Communications Review, 950 York Rd., Hinsdale, IL 60521
- BYTE, One Phoenix Mill Lane, Peterborough, NH 03458
- Cabling Business Magazine, 12035 Shiloh Road, Ste. 350, Dallas, TX 75228
- CALL CENTER Magazine, 12 West 21 Street, New York, NY 10010
- Communications News, 2504 N. Tamiami Trail, Nokomis, FL 34275
- Communications Week, 600 Community Drive., Manhasset, NY 11030
- Computer Telephony Magazine, 12 West 21 Street, New York, NY 10010

- Computerworld, 500 Old Connecticut Road, Framingham, MA 01701
- Data Communications, 1221 Avenue of the Americas, New York, NY 10020
- Dr. Dobb's Journal c/o Miller Freeman Inc., 600 Harrison Street, San Francisco, CA 94107
- Electronic Business, Cahners Bldg, 275 Washington St., Newton, MA 02158
- Federal Computer Week, 3110 Fairview Park Dr, Ste. 1040, Falls Church, VA 22042
- Flatiron Publishing, 12 West 21 Street, New York, NY 10010
- IEEE Spectrum, 345 East 47 Street., New York, NY 10017
- IMAGING Magazine, 12 West 21 Street, New York, NY 10010
- InfoWorld, 155 Bovet Rd., San Mateo, CA 94402
- Inform Business Systems, 10516 Summit Ave., Suite 200, Kingston, MD 20895
- Information Week, 600 Community Dr., Manhasset, NY 11030
- LAN Magazine, 500 Howard St., San Francisco, CA 94105
- MacUser, One Park Avenue, New York, NY 10016
- MacWEEK, P.O. Box 1763, Riverton, NJ 08077-9763
- MacWorld, Fifth Floor, 501 Second Street, San Francisco, CA 94107
- MSJ Microsoft Systems Journal, 411 Borel Avenue, Suite 100, San Mateo, CA 94402
- Mobile Office, 21600 Oxnard St., Suite 480, Woodland Hills, CA 91367
- Network Computing, 600 Community Drive, Manhasset, NY 11030
- Network World, 161 Worcester Road, Framingham, MA 01701-9172
- NeXTWORLD, P.O. Box 56430, Boulder, CO 80322-6430
- Payphone Magazine, P.O. Box 42371, Houston, TX 77242
- PC Magazine, One Park Avenue, New York, NY 10016
- PC WEEK, One Park Avenue, New York, NY 10016
- PC World, 501 Second Street, #600, San Francisco, CA 94107
- PC/Computing, 950 Tower Lane, Foster City, CA 94404
- Phone+, 4141 N. Scottsdale Road, Suite 316, Scottsdale, AZ 85251
- Public Communications Magazine, P.O. Box 42371, Houston, TX 77242
- Satellite Comm, 214 Massachusetts Ave. NE, Washington DC 20002
- Telecom Gear, 15400 Knoll Trail, Dallas, TX 75248
- TELECOM LIBRARY Inc, Central Source of All Books on Telecommunications, 12 West 21 Street, New York, NY 10010 212-691-8215. Free catalog.
- TeleManagement, 1400 Bayly Street, Office Mall 2, Suite 3, Pickering, Ontario, Canada L1W 3R2
- Telemarketing. 1 Technology Plaza, Norwalk, CT 06854
- Telecommunications, 685 Canton Street, Norwood, MA, 02062
- TELECONNECT Magazine, 12 West 21 Street, New York, NY 10010
- TeleProfessional, 209 West Fifth Street, Suite N, Waterloo, IA 50701
- UnixWorld, P.O. Box 571, Hightstown, NJ 08520-9331
- VARBusiness, 1 Jericho Plaza, Jericho, NY 11753
- Voice Processing Magazine, P.O. Box 42382, Houston, TX 77242
- Windows Magazine, 1 Jericho Plaza, Jericho, NY 11753
- Windows/DOS Developer's Journal, 1601 W. 23rd Street, Lawrence, KS 66046
- Windows Sources, One Park Avenue, New York, NY 10016
- Wired, 520 Third Street, Fourth Floor, San Francisco, CA 94107

Publish To make information public. These days you can "publish" in many ways — from paper to CD-ROM to the World Wide Web (i.e. via the Internet).
Publishing Making resources available to network users.
PUC See PUBLIC UTILITY COMMISSION.
Pull Box A box with a cover inserted in a long conduit run, particularly at a corner. It makes it easier to pull wire or cable into the conduits.
Pulling Eye A device on the end of a cable to which a pulling line is attached for pulling cable into conduit or duct liner.
Pulling Glass Laying fiber-optic cable.
Pulling Strength Expressed in lbs. The maximum force which may be applied to strength members of a cable. Pulling strength limits are specified for all Belden Fiber Optic cables in the General Line and Fiber Optic catalogs. Affects pulling methods, pulling tension and operation tension.
Pulp A type of older telephone twisted-pair cable whose wood-pulp "paper" insulation is formed on the cable during manufacture.
Pulse A quick change in the current or voltage produced in a circuit used to operate an electrical switch or relay or which can be detected by a logic circuit.
Pulse Address Multiple Access PAMA. The ability of a communication satellite to receive signals from several Earth terminals simultaneously and to amplify, translate, and relay the signals back to Earth, based on the addressing of each station by an assignment of a unique combination of time and frequency slots. This ability may be restricted by allowing only some of the terminals access to the satellite at any given time.
Pulse Amplitude Modulation PAM. A technique for placing binary information on a carrier to transmit that information. PAM is a technique for analog multiplexing. The amplitude of the information being modulated controls the amplitude of the modulated pulses. Samples of each input voltage are placed between voltage samples from other channels. The cycle is repeated fast enough so the sampling rate of any one channel is more than twice the highest frequency transmitted. See also PAM and PCM.

Pulse Cable A type of coaxial cable constructed to transmit repeated high voltage pulses without degradation.

Pulse Code Modulation PCM. The most common and most important method a telephone system in North America can use to sample a voice signal and convert that sample into an equivalent digital code. PCM is a digital modulation method that encodes a Pulse Amplitude Modulated (PAM) signal into a PCM signal. See PCM and T-1.

Pulse Density In T-1, since "0"s are represented by no pulse and "1"s by alternating pulses, pulse density refers to the number of no pulse ("0") periods allowed before a pulse ("1") must occur. Typically, no more than 15 no pulse periods ("0"s) are allowed before a pulse ("1") must occur.

Pulse Density Violation A pulse density violation occurs if a signal contains more than a specified number of zeros, or the percentage of ones in the signal is less than specified.

Pulse Dialing One or two types of dialing that uses rotary pulses to generate the telephone number. See ROTARY DIAL.

Pulse Dispersion The spreading out of pulses as they travel along an optical fiber.

Pulse Duration Modulation PDM. That form of modulation in which the duration of the pulse is varied in accordance with some characteristic of the modulating signal.

Pulse Link Repeater A signaling set that interconnects the E and M leads of two circuits. In E & M signaling, a device that interfaces the signal paths of concatenated trunk circuits. Such a device responds to a ground on the "E" lead of one trunk by applying -48Vdc to the "M" lead of the connecting trunk, and vice versa. This function is a built-in, switch-selectable option in some commercially available carrier channel units.

Pulse Modulation A general method of carrying digital information on any system that uses fixed-frequency pulses of transmit information from the source to the destination. Examples are pulse amplitude modulation and pulse duration modulation.

Pulse Overshoot In T-1, the amount of signal voltage that can remain at the trailing end of a pulse. It can be no more than 10-30% of the pulse amplitude. Also called afterkick.

Pulse Position Modulation PPM. That form of modulation in which the positions in time of the pulses are varied in accordance with some characteristic of the modulating signals, without modifying the pulse width. The time duration of the pulses is not varied.

Pulse Repetition Frequency PRF. In radar, the number of pulses that occur each second. Not to be confused with transmission frequency which is determined by the rate at which cycles are repeated within the transmitted pulse.

Pulse Stuffing When timing signals on digital circuits get out of whack, some method of allowing mismatches must be provided. In time division multiplexing, this is called pulse stuffing. One stream of data has bits added to it so its final rate is the same as the master clock.

Pulse To Tone Conversion Most of the world doesn't have touchtone service. They have rotary, make-and-break phone service and phones. Most computer telephony systems — from airline timetable audiotext machines to bank balance dispensers — require the user to punch in touch tone sounds. For people with rotary phones to get access to computer telephony systems, those systems must have a device called a pulse to tone converter — electronic circuitry which counts the clicks made by a rotary phone and converts them into touch tones. This technology is not 100% accurate and should be accompanied by programming which confirms the input. "You just entered 1034. If that is correct, please say YES or dial 1."

Pulse Train The resulting electronic impulses that transmit encoded information.

Pulse Width In T-1, refers to the width (at half amplitude) of the bipolar pulse (typically 324 + or -45 nsec).

Pulse Width Modulation Another but not very common method of modulating a signal, in which an analog input signal's DC level controls the pulse width of the digital output pulses. See PULSE CODE MODULATION and PULSE AMPLITUDE MODULATION.

Pulsenet Alert Transport Service A service from Nynex, PULSENET Alert Transport Service lets monitoring agencies use a business or residence customer's existing POTS telephone line to transport information (burglar, fire, medical, alert, etc.) without interfering with their basic voice telephone service.

Pulsing The method used for transmitting the phone number dialed to a telephone company switching office.

PUMA Product Upgrade Manager.

Punch 1. The process of perforating a paper tape or card in order to code information into machine readable form.
2. The process of connecting jumper interconnection wires on a distribution frame. It is called punching because of the tool which places the wire on the metal post of the frame. It is called a PUNCH and requires a heavy "punch" to make it strip its wires, then connect into the PUNCH-DOWN BLOCK.

Punch Down A term used to describe the connection of twisted pair wires to an insulation displacement block. (e.g. a 66 block, or a patch panel). See PUNCH DOWN TOOL.

Punch Down Tool A device used to connect twisted pair wires to an insulation displacement block (e.g. a 66 block or a patch panel). the punch down tool consists of a slotted blade attached to a heavy-duty plastic handle. Loop the wire between the prongs of an insulation displacement block, slide the slotted blade over the terminal and give the tool a downward push. A spring loaded mechanism in the handle complete the job automatically. The blade spreads the prongs just enough, strips the wire, drives it between the prongs, and neatly cuts the wire with a satisfying "ka chunk." Thus you've produced a perfect "gad tight" termination, which is important since the connection might corrode and the corrosion will insulate the connection. Result: Noise (buzz, hum) on the telephone line and bad voice or data transmission. The above explanation from Harris Dracon, a Camarillo, CA maker of communications craft tools.

Punchdown Block A device used to connect one group of wires to another. Usually each wire can be connected to several other wires in a bus or common arrangement. A 66-type block is the most common type of punchdown block. It was invented by Western Electric. Northern Telecom has one called a Bix block. There are others. These two are probably the most common. A punchdown block is also called a terminating block, a connecting block, a punch-down block, a quick connect block, a cross connect block. A punchdown block will include insulation displacement connections (IDC). In other words, with a connecting block, you don't have to remove the plastic shielding from around your wire conductor before you "punch it down."

Pure Aloha A random access technique developed by the University of Hawaii in the early 1970's. In this scheme, a user wishing to transmit does so at will. Collisions are resolved by retransmitting after a random period of time.

Purge Verb. To remove records from a database. To get rid of old voice mail messages.

Purged Customer Account An MCI definition. MCI customer account information removed from both A/R and OCIS.

Push Mechanism in TCP for ensuring immediate transmission of data.

Push-Down File Last-in, first-out

Push-To-Talk 1. In telephone or two-way radio systems, you have to push a button to talk and stop pushing to listen. Typically you say "over" to indicate it's the other person's turn to talk. In radio, it is used where the same frequency is employed by both transmitters.

2. A method of payphone operation in which a push button switch is touched by the caller when the called party answers. Once pushed, money is collected and handset microphone turned on. This system deprives the phone company of revenues for calls to 976 numbers, answering machines or answering services.

Pushbutton Dialing Instead of rotary dialing, buttons are pushed to generate the tones needed to place a phone call. Also called Touchtone and Touch-call. Some pushbutton phones do not produce tones, but generate the dial pulses of rotary dials. Some phones and phone systems will generate both rotary dial pulses and tone signaling. See TOUCHTONE.

Pushbutton Dialing To Stations A special attendant console feature in which the switching system is served by rotary dial central office trunk circuits. A ten-button keyset is provided on the console which allows fast dialing of extension numbers to complete incoming calls.

Pushbutton Originating Register A register used to store information about originating calls with pushbutton signals.

Put-Up Refers to the packaging of wire and cable. The term itself refers to the packaged product that is ready to be stored or shipped.

PVC 1. Premises Visit Charge.

2. PolyVinyl Chloride, a common type of plastic used for cladding telephone cable (except that to be run in plenum ceilings). Or,

3. Permanent Virtual Circuit, a permanent association between two DTEs established by configuration. A PVC uses a fixed logical channel to maintain a permanent association between the DTEs. Once a PVC is defined, it requires no setup operation before data is sent and no disconnect operation after. A new connection, i.e. new information to be sent between the same users may be routed along a different path.

PVCC Permanent Virtual Channel Connection: A Virtual Channel Connection (VCC) is an ATM connection where switching is performed on the VPI/VCI fields of each cell. A Permanent VCC is one which is provisioned through some network management function and left up indefinitely.

PVDF See POLYVINYLIDENE DIFLUORIDE.

PVN Private Virtual Network.

PVPC Permanent Virtual Path Connection: A Virtual Path Connection (VPC) is an ATM connection where switching is performed on the VPI field only of each cell. A Permanent VPC is one which is provisioned through some network management function and left up indefinitely.

PWB Printed Wire Board.

PWM Pulse Width Modulation. In communications, encoding information based on variations of the duration of carrier pulses. Also called Pulse Duration Modulation or PDM.

PWR Power.

Px64 Pronounced P-times-sixty four. Informal name for the ITU-T family of codec interoperability standards.

Pyramid Configuration A communications network in which the data link(s) of one or more multiplexers are connected to I/O ports of another multiplexer.

Pyramidal Horn A wave guide feed horn in which both opposite faces are tapered.

NEWTON'S TELECOM DICTIONARY

Q Queue.
Q Bit The qualifier bit in an X.25 packet that allows the DTE to indicate that it wishes to transmit data on more than one level.
Q.2100 B-ISDN Signaling ATM Adaptation Layer Overview.
Q.2110 B-ISDN Adaptation Layer - Service Specific Connection Oriented Protocol.
Q.2130 B-ISDN Adaptation Layer - Service Specific Connection Oriented Function for Support of Signaling at the UNI.
Q.2931 The signaling standard for ATM to support Switched Virtual Connections. This is based on the signaling standard for ISDN.
Q.921 Q.921 defines the ISDN frame format at the data link layer of the OSI/ISDN Model. It contains address information. The ITU-T/OSI Layer 2 protocol used in the D channel. It is synonymous with LAPD.
Q.922 ISDN Data Link Layer Specification for Frame Mode Bearer Services, ITU, Geneva, 1991.
Q.931 Q.931 is the powerful message-oriented signaling protocol in the PRI ISDN D-channel. It is also referred as ITU-T Recommendation I.451. This protocol describes what goes into a signaling packet and defines the message type and content. Specifically, Q.931 provides:
• call setup and take down.
• called party number, with type of number indication (private or public).
• calling party number information (including privacy and authenticity indicators).
• bearer capability (to distinguish, for example, voice versus data for compatibility check between terminals.
• status checking (for recovery from abnormal events, such as protocol failures or the manual busying of trunks), and
• release of B-channels and the application of tones and/or announcements in the originating switch upon encountering errors.
Q.931 makes it possible to interwork PBX features with features in the public network. In addition to offering users more access to a wider range of services, this interaction, according to Northern Telecom, will improve the revenue potential of service providers. Service provided over PRA, using Q.931, include:
• access to the public network, such as equal access, WATS, DDD, international DDD, dial-800 and other special number services and operator assisted calls.
• access to and from such private networks as Northern Telecom's Meridian Switched Network (previously call Electronic Switched Network — ESN), tandem tie networks, and extension dialing network, and
• integration of voice and circuit-switched data traffic (up to 64 Kbps).
The Q.931 protocol also enables corporations to use B-channels — that is voice and data channels — in ways currently not possible. Today, for example, a separate trunk from the PBX to the central office is often required for each different service, such as voice, data, foreign exchange, 800-service. With PRA, one common trunk between the PBX and the central office can carry multiple call types. Moreover each B-channel within the PRA trunk can be assigned dynamically to carry whatever service is needed at the moment.
Q.933 The signaling standard for Frame Relay to support SVCs. This is based on the signaling standard for ISDN.
Q-signal In the NTSC color system the Q signal represents the chrominance on the green-magenta axis.
QA Quality Assurance.
QAM, QSAM Quadrature Amplitude Modulation, Quadrature Sideband Amplitude Modulation. A sophisticated modulation technique, using variations in signal amplitude, that allows data-encoded symbols to be represented as any of 16 or 32 different states. Some QAM modems allow dial-up data rates of up to 9600 bits per second.
QBE Query By Example. A database front-end that requests the user to supply an example of the type of data he wants to retrieve. Typically, the user forms a query by filling in a table with examples of the requested information. IBM created QBE in the 1970s to simplify the process of retrieving information from mainframe databases; it was later implemented on the PC platform in such products as dBASE and Paradox. See also SQL.
QBF A test message containing the "Quick Brown Fox" text. Used to test data terminals. The text is "The Quick Brown Fox jumped over the lazy dog." It contains every letter of the alphabet. Check it out.
QC Laser Quantum Cascade laser, a new type of semiconductor laser that works like an electronic waterfall. According to AT&T Bell Labs who invented the QC laser, it is the world's first laser that can be tailored to emit light at a specific wavelength at nearly any point over a very wide range from the mid- to far- infrared spectrum. This can be done by simply varying the layer thickness of the laser, using the same combination of materials. Conventional semiconductor lasers, widely used in other applications such as lightwave communications and compact disk players, operate at wavelengths from the near infrared to the visible. When an electric current flows through the QC laser, electrons cascade down an energy staircase. Every time they hit a step they emit an infrared photon, or light pulse. At each step, the electrons make a quantum jump between well defined energy levels. The emitted photons are reflected back and forth between built-in mirrors, stimulating other quantum jumps and the emission of other photons until the amplified pulse escapes the laser cavity. The QC laser was invented by Federico Capasso and Jerome Faist in collaboration with Debbie Sivco, Carlo Sirtori, Al Hutchinson and Al Cho, according to AT&T Bell Labs.
QCIF Quarter Common Intermediate Format, a mandatory part of the ITU-T's H.261 standard which requires that non-interlaced video frames be sent with 144 luminance lines and 176 pixels.
QD Queuing Delay: Queuing delay refers to the delay imposed on a cell by its having to be buffered because of unavailability of

resources to pass the cell onto the next network function or element. This buffering could be a result of oversubscription of a physical link, or due to a connection of higher priority or tighter service constraints getting the resource of the physical link.

QDOS In 1980 IBM showed up on Bill Gates' doorstep seeking an operating system for its upcoming personal computer. Mr. Gates did not have one. But he knew someone that had one. A little firm down the road (in Seattle) had developed QDOS — the Quick and Dirty Operating System. It looked just right for IBM's PC. Mr. Gates bought QDOS for $100,000 and renamed it MS-DOS — Microsoft Disk Operating System. According to the Economist Magazine of May 22, 1993, some jealous Microsoft rivals claim that MS-DOS now stands for Microsoft Seeks Domination Over Society.

QDU Quantizing Distortion Units. ITU-T Recommendation G.113 defines one QDU as the amount of degradation introduced into a voice channel by a single conversion from analog to PCM and back to analog (analog-PCM-analog). Where several voice channels are connected in tandem, the end-to-end QDU rating for the whole circuit is calculated by adding the number of conversions from analog to PCM and back. For example: analog - PCM - analog - PCM - analog introduces 2QDUs.

QLLC Qualified Logical Link Control. Software package that allows Systems Network Architecture (SNA) commands to be transmitted over an X.25 packet data network (PDN). See also NPSI. Contrast with DSP.

QMS Queue Management System.

QNX A UNIX-like operating system that really works well for computer telephony applications.

QoR Query on Release.

QoS Quality of Service. See QUALITY OF SERVICE.

QPSK Quaternary Phase Shift Keying. Another way of sending data through a modem over an analog phone line.

QPSX Queued Packed and Synchronous Exchange. Technology used in Institute of Electrical and Electronic Engineers (IEEE) standard 802.6 for metropolitan area networks (MANs).

QSAM Quadrature Sideband Amplitude Modulation. A sophisticated modulation technique, using variations in signal amplitude, that allows data-encoded symbols to be represented as any of 16 or 32 different states.

QTC Quick Time Conference. Apple Computer's cross-platform, video-conferencing, collaborative computing and multimedia communications technology.

Quad A slang term for cable conductor with four single, plastic coated wires not twisted together and contained in a single plastic covering. Quad wiring has been traditionally used inside houses and small offices. Since it will not handle data well, it is no longer being recommended for installation anywhere, except in single-line analog (never data) applications. In the old days, a quad wire would support two analog phone lines. Color coding in quad wire in North America is red-green, yellow-black. When I showed this definition to a professional installer, he told me that quad wire was generally not used anymore except by ignorant do-it-yourselfers, cheap telcos (telephone companies), irresponsible contractors, etc. See QUAD WIRE.

Quad Fiber Cable A cable consisting of four single optical fiber cables placed inside a polyvinyl chloride jacket with a rip cord to peel back the jacket and gain access to each single cable.

Quad Wire A type of wire which contains four untwisted copper conductors in a plastic sheath. These four conductors are not two separate twisted pairs, although the four may have a very "slow" twist to them. Quad wiring is no longer recommended by the telephone industry for installation in other than analog single line applications. In short, quad is dead. See QUAD.

Quadded Cable A cable in which at least some of the conductors are arranged in the form of a quad.

Quality Of Service A measure of the telephone service quality provided to a subscriber. It's not easy to define "quality" of telephone service. It's very subjective. Is the call easy to hear? Is it "clear?" Is it loud enough, etc.? The state public service commissions have various measures which they insist phone companies conform to. They tend to be more measurable. They include the longest time someone should wait after picking up the handset before they receive dial tone (three seconds in most states). Quality of Service is more easy to define in digital circuits, since you can assign specific error conditions and compare them. For example if you were defining quality of service with respect to ATM, it would be defined on an end-to-end basis in terms of the following attributes of the end-to-end ATM connection:

Cell	Loss	Ratio
Cell	Transfer	Delay
Cell Delay Variation		

Quantization Take an analog voice signal, sample it and put numbers on those samples. That's called quantization. Here's an explanation from Understanding Telephone Electronics: "A circuit called a quantizer takes in the analog signal and produces an equivalent number. Threshold levels are established and numbers are assigned to the analog samples as their amplitudes fall within the bands formed by the threshold numbers. (For example, everything within the range 1000.05 and 1000.06 Hz is given the number 14568.) The assigned number in most costs is an approximation rather than a true value because the true value would require many more bits (i.e. more threshold limits) in the binary code...This approximation causes an error which is the difference between the approximate number and the true sample. This quantization error adds noise to the signal, called quantization noise, which is heard in the telephone as hissing. Quantization noise can be reduced by making the threshold bands narrower...However, providing more intervals requires more bits in the binary code. Therefore, more bandwidth is needed. There is a trade off between small quantizing intervals (higher bandwidth, lower noise), and fewer intervals (lower bandwidth, higher noise)."

Quantization Noise Signal errors which result from the process of digitizing (and therefore ascribing finite quantities to) a continuously variable signal. See QUANTIZATION.

Quantize The process of encoding a PAM signal (Pulse Amplitude Signal) into a PCM signal (Pulse Code Modulation). See QUANTIZATION.

Quantizing The second stage of pulse code modulation (PCM). The waveform samples obtained from each communication channel are measured to obtain a discrete value of amplitude. These quantized values are converted to a binary code and transmitted to a distant location to reconstruct the original waveform. See QUANTIZE and QUANTIZATION.

Quantizing Noise Noise caused by the inability of an analog signal to be exactly replicated in digital form.

Quantum In physics, quantum means a very small indivisible piece of energy. This word is widely misused by people who refer to "a quantum leap," meaning a big leap.

Quarter Speed An international leased teletype line capable of transmitting one quarter of Telex speed of 16 2/3 words per minute.

Quarter Wave Antenna An antenna, the length of which is 1/4 that of the wave length received.

Quartz Crystal A small piece of quartz which is cut to a precise size. When electricity is applied to the crystal, it vibrates at a specific and precise frequency. Quartz crystals are often used in watches. They vibrate quickly and make the watch far more accurate than a timing device which vibrates far more slowly, like a pendulum, for example, or a tick tock watch.

Quasi-Random Signal QR. A pseudo random test signal that has artificial constraints to limit the maximum number of zeros in the bit sequence.

Quenched Gap A spark gap so arranged that the spark is quenched quickly by a cooling effect. A method used to give impulse excitation. An old radio term.

Query 1. In data communications, it's the process by which a master station (or mainframe or boss computer) asks a slave station to identify itself and tell its status, i.e. is it busy, alive, OK, waiting, etc.?
2. In database, a query is a request for the retrieval of data.

Query Language A programming language designed to make it easier to specify what information a user wants to retrieve from a database.

Queue A stream of tasks waiting to be executed. A series of calls or messages waiting for connection to a line. See QUEUING.

Queue Management In a network, tasks like retrieval and writes to a jukebox come randomly from all the users. These tasks vary in urgency — retrievals are higher priority than writes, for example. Queue management sorts out requests from the network by priority. Queue management also enhances the performance of a jukebox, by intelligently re-ordering requests. For example, if there are three requests for images on platter 1 and two from platter 2 and the another from platter 1, queue management means the requests from platter 1 will get handled together, then go to platter two. Sometimes it's called "elevator sorting" — responding to requests in logical order, not in the order in which they were made.

Queue Service Interval The maximum length of time a queue will go unsampled.

Queued Mode Calls entering an Automatic Call Distributing system wait in a queue are presented, one at a time, to the first idle trunk in the chosen group.

Queued Telecommunications Access Method QTAM. A program component in a computer which handles some of the communications processing tasks for an application program. QTAM is employed in data collection, message switching and many other teleprocessing applications.

Queuing The act of "stacking" or holding calls to be handled by a specific person, trunk or trunk group. There are two reasons to queue telephone calls:
1. Because you simply don't have enough trunks.
2. Because you want to save money.
You can queue calls mechanically using your telephone switch or manually using a human operator or attendant. There are two ways you can queue calls — hold-on or call-back. In "hold-on" queuing, you dial, you get some queuing tone (or the operator tells you you're being queued), then you wait on-line until a line becomes free and you're connected. In "call-back" queuing, you tell the operator or the machine you want to dial a call. And you hang up. When the line becomes free you are called back and connected. There are advantages and disadvantages to both systems. In "hold-on" queuing, you waste your time but save on phone time. In call-back queuing, you waste less of your time, but more phone line time. In call-back queuing, the operator or the phone system has to grab the line you want and simultaneously call you. By then, you may have left your desk. The call may be wasted, etc. The line given to you could have been used by someone else, etc. Queuing calls as a method to save money on long distance calling makes sense ONLY:
1. IF you are out of trunks because of a temporary surge in telephone traffic — perhaps at your peak, peak busy time and it's very expensive to buy sufficient "cheap long distance" trunks to handle every conceivable peak, and
2. IF you never plan on having a queue longer than 20 seconds for a hold-on queue and 60 seconds for a call-back queue and
3. IF you are queuing calls into an expensive fixed-cost line. For example a tie line between New York and London. If you queue calls into a variable cost line, like an interstate WATS line, you will save money over throwing the call onto DDD, but the pennies you save usually won't be worth it — considering the aggravation you're going to cause your people. Queuing is a very sensitive subject in corporate telecommunications departments. People don't like to wait for telephone lines. They consider that insulting to them personally, damaging to their "productivity" and to heck with the cost. Queues do, however, make enormous sense. Even a queue as short as ten seconds can save big amounts of money. Queues of a maximum length of ten seconds are rarely noticed. These days some of the more modern PBXs will allow you to offer "selective" queuing, or levels of queuing. Upper management doesn't have to queue for the cheap long distance lines before it's bounced to the expensive ones. While lower management has to wait up to 30 seconds. And the worker bees (non-management) have to wait even longer. Queuing is also used on incoming trunks. See ACD and QUEUING THEORY.

Queuing Theory The study of the behavior of a system that uses queuing, such as a telephone system. Much of queuing theory derives from the science of Operations Research (OR). Dr. Leonard Kleinrock has written the authoritative books on the subject. He is probably a genius. His books are very difficult to understand for laymen. Here is an explanation of Queuing Theory from James Henry Green's Dow Jones-Irwin Handbook of Telecommunications (you can get a copy from Telecom Library at 1-800- LIBRARY or 1-212-691-8215.):
"The most common (telephone) network design method involves modeling the (phone) network according to principles of queu-

ing theory, which describes how customers or users behave in a queue. Three variables are considered in network design. The first is the arrival or input process that describes the way users array themselves as they arrive to request service...The second variable is the service process, which describes the way users are handled when they are taken from queue and admitted into the service providing mechanism. The third method is the queue discipline, which describes the way users behave when they encounter blockage in the network...Three reactions to blockage are possible:

- Blocked calls held (BCH). When users encounter blockage, they immediately redial and reenter the queue.
- Blocked calls cleared (BCC). When users encounter blockage, they wait for some time before redialing.
- Blocked calls delayed (BCD). When users encounter blockage, they are placed in a holding circuit until capacity to serve them is available. See QUEUE.

"Traffic engineers have different formulas or tables to apply, corresponding to the assumption about how users behave when they encounter blockage." See POISSON.

After I wrote the above definition, Lee Goeller, a noted traffic engineering expert contributed the following definition:

The study of systems in which customers wait in line for servers to become available, the "blocked calls delayed" condition in telephony (see TRAFFIC ENGINEERING). Although seldom used in designing voice networks (other techniques are usually more cost-effective), queuing is very important in the design of packet networks where speed of transmission more than offsets the delay of waiting for a transmission facility to become available, and in staffing Automatic Call Distributors.

QUICC Quad Integrated Communications Controller. A Motorola term.

Quick Connect Block Also called a 66-block or punch-down block. It's a 18" piece of metal and plastic which allows you to connect telephone wiring coming from two remote points. The quick-connect block has multiple metal "jaws" ranging horizontally and vertically. To connect up, you "punch" (or push) a wire between the two metal teeth of the "jaws." This both holds it firm and strips the wire's insulation, thus allowing for a good electrical connection. (There are special "punch-down" tools for punching wires into 66-blocks.) On a 66-block, one horizontal row of "jaws" is always the same conductor. To connect other wires to it, you simply punch those wires down along the row. Some 66-blocks have a gap between one side of the 66-block and the other. To connect one wire on one side to the wire on the other side, you have to use a BRIDGING CLIP. This is a small metal clip about one inch long. The bridging clip has one purpose: you can slip it off easily and thus cut one side of the circuit from the other. For example, if you connected central office trunks on one side of the 66-block and a PBX on the other, by removing the bridging clips, you can tell instantly if the trouble is in the PBX or in the central office. Two conductors on a 66-block makes a circuit — a trunk or a line. Therefore, the trunk 212-691-8215 (our main number) takes up the first two horizontal rows on our 66-block. The second two horizontal rows are taken up with 212-691-8216, and so on.

It is good to learn where your main 66-block is — the one that connects you to the telephone company's central office lines. The 66-block is what the telephone company calls the "demarcation point." And they (the phone company) usually install the 66-block. On one side (the trunk side) of their block, they're responsible. On the other (the PBX, key system or phone side), you're responsible. By knowing how to test your lines at this point, you can know whose fault it is — the phone company's or your equipment's. This can avoid having to wait until the phone company arrives, discovers it's not their problem and then sends you a hefty bill. Or the interconnect company arrives, finds out it's not their problem, and sends you a hefty bill, etc.

Quick Connect Blocks or 66-blocks are found in the Main Distribution Frame — where lines coming out of the PBX are connected to the individual wires going to the phones, or to big cables going to clumps of phones in other parts of the building. They're also found in Satellite Distribution Frames where they take big cable coming in from the main distribution frame and connect it to the individual cable pairs going to the individual phones. See CONNECTING BLOCK.

Quick Format A DOS program which deletes the file allocation table and root directory of a disk but does not scan the disk for bad areas.

Quick Plug A device which adapts a standard four wire telephone cord into a modular connector.

Quicktime Apple Computer's video environment (like Microsoft's Video For Windows). Quicktime video files must be converted to •.AVI format to run under Microsoft's Video For Windows. Indeo video technology is supported under MacOS.

Quiescent A fancy word for quiet. No noise. No activity. Quiescent time is the best time to write this dictionary. Sadly, it wasn't always to be.

Quorum A family of teleconferencing products linked in a system designed to meet a customer's teleconferencing needs.

QWERTY The name for a computer or typewriter keyboard. It got its name from the left side, top row of letter keys which spell QWERTY. One theory for the strange design of the QWERTY keyboard has to do with typewriter keyboards which had long metal arms that physically hit the paper. To keep the arms from jamming, they designed the QWERTY keyboard which split commonly used letters — i.e. a, i, o, e — to opposite sides of the keyboard. For years, people have argued that a keyboard called the Dvorak would be a much faster and more efficient. However, the U.S. General Service Administration in the 1950s contradicted the claims made by advocates of the Dvorak keyboard. The chief advocate was the patent owner, August Dvorak. According to September, 1995 Upside Magazine, his "book on the relative merits of QWERTY versus his own keyboard has about as much objectivity as a modern infomercial found on late night TV." With computers, there is no such thing as a "standard" QWERTY keyboard. Computer keyboards typically have 20 to 30 more keys than "standard" typewriter keyboards. Many of the keys are unique — on some keyboards, not on others. Many of the keys on computer keyboards are called "function" keys. If you hit one of them, they might perform a complete function on the computer, e.g. save a file, move to the end of the file, etc. There is absolutely no such thing as a standard computer keyboard.

QZ Special billing arrangement provided by your local telephone company. Before there was automatic call accounting and before there was Centrex, the phone company would give you "time and charges" on every outgoing call. This service was called "QZ" billing. It was used by engineers, lawyers, accountants, consultants and other service people who had to bill their calls back to their clients.

R Interface The 2-wire physical interface which is used for a single customer termination between the TE2 and TA.

R Reference Point An ISDN reference point between non-IDSN terminal equipment (TE2) and a terminal adapter (TA). Non-ISDN (TE2) terminal equipment connects to ISDN at the R-reference point through a terminal adapter.

R1 The ITU-T's name for a particular North American digital trunk protocol that happens to use multi-frequency (MF) pulsing. Some Europeans refer to any North American MF signaling protocol as R1 when distinguishing it from their own R2. See R2, MULTI-FREQUENCY PULSING.

R2 A whole series of ITU-T specs which refers to European analog and digital trunk signaling. It refers to a type of trunk found in Europe which uses compelled handshaking on every MF (multi-frequency) signaling digit.

RA Number Same a Return Material Authorization Number, or RMA. A code number provided by the seller as a prerequisite to returning product for either repair or refund. An indispensable tracking procedure, it operates like a purchase order system. If you return computer or telephone equipment without an RMA, chances are your equipment will be lost.

RACE An association in the European Economic Community. RACE stands for Research and development for Advanced Communications in Europe.

Raceway Metal or plastic channel used for loosely holding electrical and telephone wires in buildings. A raceway is usually located in the floor and is encased on three or four sides by concrete. A raceway is used for interior wiring and performs the same job as a conduit but is typically larger.

Raceway Method A ceiling distribution method in which open or closed metal trays are suspended in false ceilings from the structural floor above. The raceway method is generally used in large buildings or for complex distribution systems that demand extra support. When closed metal trays are embedded in the floor, this distribution method is often called underfloor raceways. See also CEILING DISTRIBUTION SYSTEMS and UNDERFLOOR DUCT METHOD.

Rack An open structure typically made of aluminum or steel, onto which equipment is mounted. A rack is typically attached to a building ceiling or wall. Cables are laid in and fastened to the rack. Sometimes a rack is called a tray. What a rack is to equipment, so a frame is to wiring. See also DISTRIBUTION FRAME.

Racon A radio navigation system transmitting, automatically or in response to a predetermined received signal, a pulsed radio signal with specific characteristics.

Rad 1. The unit used to measure the absorption of ionizing radiation.
2. A British Term. Recorded Announcement Device, a device which automatically answers a line and delivers a pre-recorded message. Often used to tell a caller to a telebusiness unit that the call is in a queue and will be dealt with soon. More sophisticated RADs gather information, take messages or work in conjunction with interactive fax machines.

Radar RAdio Detection And Ranging. See RADAR DETECTOR.

Radar Detector Picture a trooper sitting in his car aiming his radar gun down the highway. The gun emits a beam of electrons at microwave frequency. Those beams bounce off approaching vehicles and reflect back to the trooper's radar at an altered frequency (the Doppler Effect). By measuring the change in frequency, the trooper calculates the speed of the oncoming vehicle. The trouble is the radar beam fans out like a searchlight. At a distance of 1,000 feet, the beam is about as wide as the highway itself. That makes it difficult for the trooper to know which vehicle he's tracking.

Also, his reading can be thrown off by any number of operating errors or by interference from power lines, neon lights or even the fan motor in the trooper's car. According to some estimates, Esquire Magazine reported, as many as 30% of all radar-generated speeding tickets were given in error. In 1979 a Miami TV station showed a police radar clocking a house going 28 miles per hour and a banyan tree doing 86! Radar detectors are very much like FM receivers. They can pick up radar signals more than a mile from the source. At that distance the beam is too weak to bounce all the way back to the trooper's car but strong enough to make the detector beep.

Radar Screen 1. A CRT (cathode ray tube) showing

Radial Acceleration The rate at which a track on an optical disc accelerates toward and away from the center, because it is not perfectly aligned or perfectly round.

Radiant Energy Energy as measured in joules which is transferred via electromagnetic waves. There is no associated transfer of matter. And typically the giver or energy and the receiver of energy are not touching.

Radio The science of communicating over a distance by converting sounds or signals to electromagnetic waves and radiating them through the air or through space. Also called Wireless by the British and the Australians.

Radio Broadcast Data System RBDS. A new system designed to let radio stations broadcasters send text messages, such as emergency warnings and traffic alerts to radios equipped with special LCD screens. The system is designed ultimately to replace the Emergency Broadcast System

Radio Button A call center term. A button used for selecting from a group of options that are mutually exclusive. As with a car radio, selecting a particular button de-selects the previously selected button.

Radio Common Carrier RCC. A common carrier engaged in Public Mobile Service, which also is not the business of providing land line local exchange telephone service. These carriers were once known as Miscellaneous Common Carriers.

Radio Communication Any telecommunication by means of radio waves.

NEWTON'S TELECOM DICTIONARY

Radio Frequency That group of electromagnetic energy whose wavelengths are between the audio and the light range. Electromagnetic waves transmitted usually are between 500 KHz and 300 GHz.

Radio Frequency Filter Fit A Northern Telecom Norstar device designed to alleviate problems associated with radio frequency interference that may be experienced when a headset or external Auxiliary Ringer is used with a telephone.

Radio Frequency Interference The disruption of radio signal reception caused by any source which generates radio waves at the same frequency and along the same path as the desired wave.

Radio Frequency Interference Shield RFI Shield. A metal shield enclosing the printed circuit boards of the printer or computer to prevent radio and TV interference.

Radio Paging Access Provides attendant and phone user dial access to customer-owned radio paging equipment to selectively tone-alert, or voice-page individuals carrying pocket radio receivers. The paged party can answer by dialing an answering code from a phone within the PBX.

Radio Paging Access With Answer Back Allows access to customer-provided paging systems and provides the capability in the PBX to connect the paged party when the former answers the radio page by dialing a special code from any PBX.

Radio Wave Electromagnetic waves of frequencies between 10 KHz and 3MHz, propagated without guide in free space (air).

Radiogram A telegram sent by radio. Totally obsolete term, but cute.

Radiophone Apparatus for transmitting and/or receiving speech or music by radio. Totally obsolete term, but cute.

Radiosonde An automatic radio transmitter in the meteorological aids service usually carried on an aircraft, free balloon, kite, or parachute, and which transmits meteorological data.

Radiotelephony The science, art, and act of transmitting speech by means of radio. Now probably called telecommunications.

Radome A plastic cover for a microwave antenna. It protects the antenna from awful weather, but has little effect on the radiation pattern of the antenna.

RADSL Rate Adaptive Digital Subscriber Line. A new technology that will allegedly deliver seven megabits per second from the central office to the subscriber and one megabit from the subscriber back to the central office.

RAID Redundant Array of Inexpensive Disks. The idea is simple: Put several disk drives into a single housing. Then write your data over the disk drives in such a way that if you lose one or more of the drives, you won't have lost any of your data. Thus the term "redundant." At its simplest, RAID mirrors data to an equal number of disk drives, e.g. two sets of two. At its most complex, RAID writes data across a bunch of drives, so that if one goes the data can be retrieved from the remaining drives. See also REDUNDANT ARRAY OF INEXPENSIVE DISKS for a much more detailed explanation.

Rain Attenuation Signal losses due to absorption are common when radio signals encounter a heavily moisture laden atmosphere. Generally, the higher the radio frequency, the more attenuation (i.e. the more losses). Since microwave signals for satellite and for land line are essentially line-of-sight, microwave radio is very susceptible to signal attenuation in heavy rain. Modern microwave paths are engineered with weather patterns in mind. In areas where heavy rainfalls occur, microwave links may be closer together or more attention is paid to diverse routing.

Rain Barrel Effect Signal distortion of a voice telephone line caused by the under-attenuated echoes on the return path.

Raised Floor Method A floor distribution method in which square, steel and wood-laminated plates resting on aluminum locking pedestals are attached to the building floor. The plates are usually covered with cork, carpet, or vinyl tiles, and each plate can be removed for easy access to the cables below. Also referred to as access floor.

RAM Random Access Memory. The primary memory in a computer. Memory that can be overwritten with new information. The "random access" part of its name comes from the fact that the next "bit" of information in RAM can be located — no matter where it is — in an equal amount of time. This means that access to and from RAM memory is extraordinarily fast. By contrast, other storage media — like magnetic tape — have their information stored serially, one bit after another. Therefore you have to search for them. And your search time will depend on how far from the bit you're searching for you are. Floppy disks are faster than magnetic tape because their information is readily at hand, though the read/write head will have to search for it. Hard disks are even faster because there are multiple heads and because the disks spin faster and everything moves faster. RAM memory is the fastest of all. The problem with RAM memory is that it's volatile. This means when power is turned off (or power glitches occur) RAM memory is erased. RAM memory can be protected with rechargeable batteries — just remember to charge the batteries. See also DRAM, EDO RAM, SRAM, and VRAM.

RAM Base Address Random Access Memory Base Address. Starting address for memory dedicated to a specific task.

RAM BIOS BIOS transferred to RAM so things go faster.

RAM Disk A logical device made from semiconductor (i.e. chip) memory which emulates the functioning of a disk drive as closely as possible. Since most semiconductor memory (RAM) is volatile, most RAM disks are also volatile, i.e. they lose their memory when you turn off power.

RAM Mobile Data A public data communications wireless network (owned by RAM Broadcasting, Ericsson and BellSouth) that allows people carrying handheld devices to send and receive data messages. Such messages might be from sheriff standing in the street searching his department's database for unpaid parking tickets. A competitor to RAM Mobile Data is Ardis. RAM Mobile Data uses a hierarchical architecture to route messages. At the lowest level, base stations exchange wireless messages with nearby mobile computers. Base stations can route messages to other users who are within its service area, or the local switch to reach recipients who are in other areas, on LANs, or on public e-mail services. The local switch can either route the message to a different base station or to a regional switch.

Rambutan A symmetric cryptographic algorithm developed by Marconi.

RAMDAC Random access memory digital-to-analog converter. The chip on a VGA board that translates the digital representation of a pixel into the analog information needed for display on the monitor.

RAN 1. Return Authorization Number. A number you need for returning busted equipment to the factory. You call the factory, tell them what you want to return and its serial number, and the factory rep gives you an RMA, which

2. Recorded trunk ANnouncements. RAN devices are devices connected on 4-wires to older central offices (public exchanges). They are used to give recorded messages to callers, e.g. "The number you have called has been changed. Please make note of the new number..."

Random Scattered, unfocused, a non sequitur. A favorite expression of Bill Gates (chairman and founder of Microsoft) to dismiss ideas or strategies that lack logic (or he thinks lack logic).

Random Access Usually refers to computer memory or storage. Random Access is the ability to reach any piece of data in the memory directly without having to pass by other pieces of data. In telephony, this means the ability to reach any other subscriber through the telco switching network. See SEQUENTIAL ACCESS.

Random Access Memory See RAM.

Random Noise Interference to telephone communications occurring at irregular intervals.

Range Extender There are two definitions. I don't know which one is correct. I thought the first one was correct. That is that a range extender is a device that increases the length of a local loop by boosting battery voltage being sent out from the telephone company central office. Bellcore, however, says a range extender is a device that permits a central office to serve a line that has resistance that exceeds the normal limit for signaling. A range extender does not extend transmission range, according to Bellcore. See RANGE EXTENDER WITH GAIN.

Range Extender With Gain REG. A unit that provides range extension in a loop for both signaling and transmission.

RARE Reseaux Associes pour la Recherche Europeenne. European association of research networks.

Rare Earth Doping Here is an article on rare earth doping from The Economist Magazine of July 6, 1991: Optical fiber is the darling of the telecommunications world. Because light waves can be superimposed on one another, fiber can carry thousands of laser generated messages at the same time, over longer and longer distances. The longest fibers were once those which doctors use to explore their patients innards. Now they can stretch 70 km (40 miles). But even that does not get you across a sea much bigger than the English Channel without the messages fading. So today's transatlantic and transpacific optical cables are interrupted about every 70 km so that the messages can be sorted out, passed through an electronic amplifier, and then, turned into light again. These amplifiers are costly. Soon, though, they may be replaced.

The key technique is called rare earth doping, which was developed not by crooked bookmakers but by scientists at Southampton University in England, and AT&T Bell laboratories in New Jersey. The rare earths are a group of chemical elements with particularly restless electrons in their atoms. If these electrons are stirred up by a laser, they rise to higher energy levels inside their atoms. When they fall back again, they emit light. The frequency of the light emitted depends on the element. The trick is to pick one which emits at frequency used for telecommunications. By adding the right rare earth to a stretch of fiber, you can make it amplify signals. You can also make a laser out of the fiber itself.

In the optical amplifier developed at Southampton University, a laser is used to lift electrons in the rare earth atoms in a stretch of fiber up to higher energy levels. When a light signal comes along, it may knock one of these electrons off its perch. The falling electron gives off light, which boosts the signal. The enhanced signal then knocks down more electrons, gathering strength as it goes. Rare earths in the cable can be used for other things, as the team at Bell Labs has found. Normal light waves, even those in laser beams, spread out and dissipate as they travel.

Solitons, a special kind of wave, do not. Tidal bores, the best known form of soliton, can move up rivers for miles without losing their shapes. Light that traveled in solitons could travel much farther along an optical fiber between boosts. Solitons are created either by pumping the initial signal through an optical amplifier, or by using a laser made from doped fiber. The soliton holds its shape because the passage of light through the fiber temporarily increases the speed of light in that part of the fiber, so the back of the wave is always trying to travel faster than the front. The stronger the light, the stronger the effect.

Rare earth doping, with metals called erbium and praseodymium, has resulted in fibers which can handle billions of bits of data per second, and carry them thousands of kilometers. AT&T hopes to use erbium amplifiers in its new transoceanic cables in the 1990s. Other companies — such as British Telecom and NTT — also like praseodymium, which is harder to handle, but emits light at a more commonly used frequency.

RARP Reverse Address Resolution Protocol. A TCP/IP protocol governing the translation of a DLC (Data-Link Control) address to an IP (Internet Protocol) address.

Raster A pattern of horizontal scanning lines on a TV screen. Input data causes the beam of the TV tube to illuminate the correct dots to produce the required characters. See RASTER SCANNING and RASTERING.

Raster Scanning The method of scanning in which the scanning spot moves along a network of parallel lines, either from side to side or top to bottom.

Rastering The process by which a document image is converted to a stream of bits representing either black or white, or one of sixteen levels of gray, for each element of the image. For Group 3 faxes, there are either 98 or 196 raster lines per vertical inch, with a horizontal resolution of 203 lines per inch (yielding 1.86 or 3.72 million elements per 8 1/2 by 11 inch page). Sixteen levels of grey ("halftone " setting — requiring four bits per element, rather than the one bit required for black and white) can be specified, but are not typically used for documents containing only text and/or line drawings. The bit stream is compressed for transmission, and decompressed when received.

Rasterizing See RASTERING.

Rate The price of a particular service or piece of equipment from a telephone company. Telephone companies don't use the word "price." They use the word "rate." No one knows why, except that if they didn't cultivate their own jargon, there'd be no job for telecommunications dictionary writers. God forbid!

Rate Adaption The process of converting a digital stream of data into a different format and rate. For example, rate adaption allows 64 Kbps data to interoperate with 56 Kbps.

Rate Arrangements Telephone customer prices charged by tariffs for specified telephone services.

Rate Averaging Telephone companies' method for establishing uniform pricing by distance rather than on the relative cost (to them) of the particular route. The theory is that some routes are more heavily trafficked, have huge transmission equipment and achieve great economies of scale. Some routes, on the other hand, have little traffic, small transmission equipment and achieve no economies of scale. Therefore, it costs more to provide calls on these less-trafficked routes. But the phone industry doesn't charge more to call small towns than big cities. The phone industry simply charges by distance, averaging its costs. This is called rate averaging.

Rate Base A regulated telephone company's plant and equipment which forms the dollar base upon which a specified rate of return can be earned. The total invested capital on which a regulated company is entitled to earn a reasonable rate of return.

Rate Center Telephone company-designated geographic locations assigned vertical and horizontal coordinates between which airline mileages are determined for the charging of private lines. See AIRLINE MILEAGE.

Rate Chip A standard, nonvolatile memory device used to retain data base information on call pricing by Area Code and Central Office. Typically used in call accounting equipment.

Rate Design Utilities have a specific rate for every service provided. The rates must be approved by the PUC. In a major rate case, rates for many services will be changed in tandem. In a rate design hearing, different proposals as to rate levels are considered. The level of one rate can have an impact on what the level of another rate should be. The inter-relationship between rates and the impact of demand must all be considered in "designing" a rate structure.

Rate Elements The pricing structure of various telecommunications service offerings usually described in tariffs.

Rate Of Return The percentage of net profit which a telephone company is authorized (by a regulatory commission) to earn on its rate base. See RATE BASE.

Rate Period Dividing a day into various slices of time for the purpose of charging differently for long distance and local calls. There are three rate periods in force today in North America for intra-North America calls. One rate period is from 11:00 P.M. to 8:00 A.M.; one is from 8:00 A.M. to 5:00 P.M. and one is from 5:00 P.M. to 11:00 P.M. If you call outside the United States, there are different rate periods.

Rate Realignment In California's Alternative Regulatory Framework Phase III, rate realignment refers to redesigning telephone rates to reduce intraLATA toll rates and increase rates for other services to make up for the phone companies lost revenues. The Public Utility Commission (PUC) must approve all rate realignment proposals in the rate design stage of the proceeding.

Rate Stability Plan Commit yourself to keeping a Nynex service for several years and you'll pay less than if you keep it only from month to month. Other phone companies have similar schemes, typically by other names.

Rate Table A data base that contains the cost of calls referenced to the Area Code and/or number dialed plus time of day considerations. See RATE PERIOD.

Rate Zone A defined geographic division of an exchange area used as the primary basis for figuring toll rates.

Rated Temperature The maximum temperature at which an electric component can operate for extended periods without loss of its basic properties.

Rated Voltage The maximum voltage at which an electric component can operate for extended periods without undue degradation or safety hazard.

Rayleigh Scattering Scattering due to tiny impurities in the optical fiber which are fractions of the wavelength of the infrared rays.

RB Reverse Battery.

RBDS Radio Broadcast Data System. A new system designed to let radio stations broadcasters send text messages, such as emergency warnings and traffic alerts to radios equipped with special LCD screens. The system is designed ultimately to replace the Emergency Broadcast System

RBOC Regional Bell Operating Company. There are seven RBOCs each of which own two or more BOCs (Bell Operating Companies). The RBOCs were carved out of the old AT&T/Bell System by Judge Harold Greene when he signed off on the divestiture of the Bell operating companies from AT&T at the end of 1983. There is nothing magical about seven — nor the grouping of BOCs into RBOCs — except the Judge wanted to keep them all roughly the same size. The seven RBOCs are Ameritech, Bell Atlantic, BellSouth, Nynex, Pacific Telesis, Southwestern Bell and US West. In early October, 1994, Southwestern Bell changed its name to SBC Communications Inc. Its telephone companies, it said, would still operate under the Southwestern name. See BELL OPERATING COMPANY.

RBS Robbed-Bit Signaling. See ROBBED-BIT SIGNALING.

RCA 1. Regional Calling Area. The geographical area covered by a telephone company.
2. Once it stood for Radio Corporation of America.

RCA Globecom An International Telex and high-speed data communications company acquired by MCI from RCA in 1987. The acquisition of RCA Globecom gave MCI approximately 40% of the International Telex market and helped strengthen MCI's role in International data.

RCC Radio Common Carrier.

RCDD Registered Communication Distribution Designer, a title conferred on people who have acquired certain requisite education and experience by BICSI, the Building Industry Consulting Service International. BICSI is at 813-397-1991.

RCEE Resource Control Execution Environment. A term from Bellcore Advanced Intelligent Network model.

RCL ReCaLl.

RCP The Berkeley UNIX remote copy program.

RD An ATM term. Routing Domain: A group of topologically contiguous systems which are running one instance of routing.

RDF 1. Radio Direction Finding.
2. An ATM term. Rate Decrease Factor: An ABR service parameter, RDF controls the decrease in the cell transmission rate. RDF is a power of 2 from 1/32,768 to 1.

RDS Radio Data System. A way of sending data along with a standard FM radio broadcast
RDT 1. Recall Dial Tone.
2. Remote Digital Terminal.
RDY ReaDY.
RE-422 A high-speed electrical interface defined by the ITU-T, supporting data rates of up to 768 Kbps over up to 300 feet of cable.
Re-Engineering A term probably invented by Michael Hammer in the July-August, 1990 issue of Harvard Business Review. In that issue, he wrote "It is time to stop paving the cowpaths. Instead of embedding outdated processes in silicon and software, we should obliterate them and start over. We should 're-engineer' our business: use the power of modern information technology to radically redesign our business processes in order to achieve dramatic improvements in their performance." The term re-engineering now seems to me mean taking tasks presently running on mainframes and making them run on file servers running on LANs — Local Area Networks. The idea is to save money on hardware and make the information more freely available to more people. More intelligent companies also redesign their organization to use the now, more-freely available information. Also called VALUE DRIVEN RE-ENGINEERING.
Re-Initiation Time The time required for a device or system to restart (usually after a power outage).
Re-Installed Customer An MCI customer who is installed again with the same customer account number after having been previously canceled either at their, MCI's, or a third party's request.
REA Rural Electrification Administration. A federal agency which makes loans to extend telephone service into rural areas. The REA has been one of the most successful federal government programs ever.
Reactance The opposition offered to the flow of an alternating current which is due to the presence of inductance or capacity or both, in the circuit.
Read To glean information from a storage device, like a floppy disk. The opposite of READ is to WRITE. That's when you put information onto that storage device. Some storage devices can only be READ, but not written to. On a floppy disk that's called being "WRITE PROTECTED." See also WORM, which stands for Write Once, Read Many.
Read After Write Verification A means of assuring that data written to the hard disk matches the original data still in memory. If the data from the disk matches the data in memory, the data in memory is released. If the data doesn't match, the block location is recognized as "bad," and something happens. The data is transferred again. Or in Novell's NetWare, Hot Fix redirects the data to a good block location within the Hot Fix Redirection Area.
Read Only File A PC computer term. A read only file is a file that you can read but cannot make changes to. The read-only attribute specifies whether a file is read-only. To remove the read-only attribute, you would type the following command
ATTRIB -R FILENAME
Read Only Memory ROM. A memory device which is programmed at the factory and whose contents thereafter cannot be altered, even by a power breakdown, or being written to, or anything else. ROM memory is also random-access, which means accessing its information is very fast. See also MICROPROCESSOR and RAM.
Read Write Cycle Time of reading and writing data onto a memory device. See READ.
Reader 1. A device which converts information into a format recognized by a machine as input.
2. A device which interprets coded data in the process of transferring that data from one coded state of storage to another.
Readyline 800 A toll-free service designed for the small business. Receive "800" dialed calls over your existing telephone lines and equipment — no new lines to install, no new equipment needed. You can still use those same lines to make and receive local and long distance calls. Choose the geographic areas you want to cover — from a single area code to an entire state or the whole country. Even decide when you want your toll-free number to be available. You pay a one-time start-up charge and a low monthly fee. Calling prices are based on the market coverage you choose. There are time-of-day and day-of-week discounts, and a volume usage discount. Calls are priced on a mileage/distance-sensitive basis.
Real Mode Originally there was the first IBM PC and it was powered by an Intel 8086 chip which addressed a maximum of 1MB (megabyte of RAM). Real mode is the term that later generations of Intel chips came to call their ability to run programs written for the 8086. Real mode allows 80286, 80386, or 80486 processors to emulate an 8086 processor but perform better than the 8086 because they operate at a faster clock rate. Real mode is limited to a maximum of 1MB of addressable memory because the 8086 processor uses a memory address bus of 20 bits. This is calculated thus: Since a bit can have one of two values, raising the base number of 2 to the power of 20 is equal to 1,048,576 unique memory addresses. Each memory address can store 1 byte of information (1,048,576 bytes equal 1MB). See also PROTECTED MODE.
Real Time A voice telephone conversation is conducted in Real Time. That is, there is no perceived delay in the transmission of the voice message or in the response to it. This concept often applies to interaction between a computer and a terminal. See also REAL TIME CAPACITY. In data processing or data communications, real time means the data is processed the moment it enters a computer, as opposed to BATCH processing where the information enters the system, is stored and is operated on a later time.
Real Time Adherence Adherence is a term used in telephone call centers to connote whether the people working in the center are doing what they're meant to be doing. Are they at work? Are they on break? Are they answering the phone? Are they at lunch? All these activities are scheduled by work force management software. If they're in line, the workers is "in adherence." If not, they're "out of adherence." Some automatic call distributors have a real time adherence data link which connects the ACD to an external computer which then tracks and displays current service rep activity measured against a pre-defined schedule. The idea is to give call center supervisors tools to manage the call center's work force more efficiently. Supervisors are able to define the task, the start time of each task, and the task duration. In addition, thresholds and ranges of acceptable deviations for the call center can be set for each task or service rep work state. Once the schedules have been defined and thresholds set, real-time displays inform the supervisor of discrepancies between the work schedule and actual activity. Service rep information will auto-

matically appear should their status exceed the threshold, such as being on someone being break for too long.

Real Time Capacity The capacity of the central computer processor of a stored program control telephone system to process the instructions coming at it. Real Time Capacity is probably the most important measure of the size of a telephone system relying on a single main processor.

Real Time D-Channel Status Display This maintenance enhancement allows you to assess active or failed status of ISDN D-Channels in real-time. This saves time since ports no longer need to be evaluated.

Real Time Transport Protocol RTP. Developed by the IETF (Internet Engineering Task Force) it adds a layer to the Internet protocol. It is designed to address problems caused when real-time interactive exchanges such as video are transported over LANs were designed for data. Running video on LAN means you can encounter significant end-to-end latency. RTP's approach is to give video higher priority than connectionless data. RTP resides above the IP, Datagram Protocol and ST-II protocols.

Rearrangement A fancy word for moving phone extensions around.

Reasonableness Checks Tests made on information reaching a real-time system or being transmitted from it to ensure that the data lie within a given range.

Reassembly The process by which an IP datagram is "put back together" at the receiving host after having been fragmentation and MTU.

Reasserting Status An ISDN term. When the ISDN phone is being directly controlled by the application program, the set's physical status may be different from the status that has been received from the network. When direct control ends, the ISDN set reasserts the status received from the network to bring its physical condition back into conformity with the network status.

Reassignment Here is an explanation by Bill Etling, a senior planner for GTE. "Under the assigned plant concept, a pair is dedicated from the central office to the subscriber home and maintained at that address, even when idle. The likelihood of such a pair being reused, thus eliminating a field visit and extra assignment work, more than makes up for lost revenue while the pair is vacant. In areas of high cable fills, such a pair, when vacant, is often used to fill an order at a different address. Reassignment quickly snowballs, generating many installation field visits and assignment changes, increasing paperwork and the chance of errors."

Rebiller A rebiller, also called a switchless reseller, buys long distance service in bulk from a long distance company, such as AT&T, and resells that service to smaller users. It typically gets its monthly bill on magnetic tape, then rebills the bulk service to its customers. A rebiller owns no communications facilities — switches or transmission. It has two "assets" — a computer program to rebill the tape and sales skills to sell its services to end users. The profit it makes comes from the difference between what it pays the long distance company and what it is able to sell its services at. It's not an easy business to be in, since you are selling a long distance company's services to compete against itself.

Rebooting Repeating a Boot. Turning on or resetting the telephone system or the computer. The word derives from "bootstrapping." Starting from scratch. Pulling oneself up by one's own bootstraps. Booting a telephone system or a computer means starting it from scratch, usually by turning its AC power on. Rebooting a telephone system is done by simply turning it off, counting to ten and turning it back on again.

Rebooting is done to clear the volatile part of the telephone system's or computer's memory and its various processing and clock chips. You reboot typically when your PC "locks" inexplicably or when your telephone system does something you can't explain logically — like ring phones randomly or give strange error messages on the console. On a computer, "Lock" means that no matter which key or combination of keys you touch on your keyboard, you can't get your computer to do anything. In addition to "unlocking" your computer, you also reboot to clear RAM or RAM-resident programs. On an IBM or an IBM clone, rebooting is done by pressing the CONTROL, ALT and DELETE keys simultaneously. You can also reboot by pressing the reset button if your computer has one. (Not all do.)

You can reboot any computer by turning its power off, then turning it back on. This is usually not a good idea, since the surge of power that accompanies a computer being turned on and off will reduce the life of many of its electronic components. Some experts recommend leaving computers running full-time, though turning their hard disks off. They also recommend turning your screen off, or at least running a public domain program such as SCRNSAVE.COM or SCRN.COM which turn off your screen after several minutes of doing nothing (inactivity).

Rebuilding Imagine you have five hard disks in an array. Imagine that they are organized that data is being written to all five drives in such a way that if one drive fails, no data will be lost. That failed drive is now removed and replaced with a good drive. Immediately, the remaining four drives start writing data to the new, good, but empty drive. That process of rebuilding might take a few minutes, or an hour or two. It depends on how much data is in the system and how much activity is taking place. Typically, this rebuilding process happens in a system called RAID (which stands for Redundant Array of Inexpensive Disks). And typically RAID (which is not cheap) is found on servers on LANs. The process of rebuilding is also called reconstruction.

Recall The recall button on many phones provides a fresh dial tone without physically putting down and picking up the handset. Don't confuse it with REDIAL, which is a feature of a phone or phone system that allows a user to call the previously-dialed number by pressing one or a few buttons.

Recall Dial Tone A stutter or interrupted dial tone indicating to the extension user that the hookswitch flash has been properly used to gain access to system features.

Recall Key Used to get dial-tone or to transfer calls on a key system installed within a PBX. See also RECALL.

RECAPSS REmote CAble Pair Switching System is used to remotely handle cable transfers and related cable switching tasks by connecting a distribution cable pair to either an old cable pair or a new cable pair without interrupting service. The system accommodates both POTS and special services and the computer console operator can select one pair at a time or select thousands for sequential transfer.

Receipt Notification A report prepared by a recipient UA (User Agent) or Access Unit (upon request) and sent to the originating UA or Access Unit when a message is received by a recipient.

NEWTON'S TELECOM DICTIONARY

Receive Interruption The interruption of a transmission to a terminal to receive or send a higher priority message from the terminal.
Receive Only RO. Describing operation of a device, usually a page printer, that can receive transmissions but cannot transmit.
Received Line Signal Detector Modem interface signal defined in RS-232-C EIA interface which indicates to the attached data terminal equipment that it is receiving a signal from the distant modem.
Received Signal Level RSL. The strength of a radio signal received at the input to a radio receiver.
Receiver 1. Any device which receives a transmission signal.
2. Any portion of a telecommunications device which decodes an encoded signal into its desired form.
3. The earpiece portion of a telephone handset, which converts an alternating electric current into sound waves, usually through an electromagnet moving a diaphragm.
4. An electronic component capable of collecting radio frequency broadcasts and reproducing them in their original audio and/or video form, e.g. a TV or radio receiver.
Receiver Off-Hook Tone The loud tone sent by the central office to tell the telephone user that his/her phone is off the hook.
Receiver Sensitivity The magnitude of the received signal necessary to produce objective BER or channel noise performance.
Receiving Perforator REPERFORATOR. A telegraph instrument in which the received signals cause the code of the corresponding characters or functions to be punched in a tape.
Recent Change Changes to line and trunk translations in a stored program control switching machine that have not been merged with the permanent data base.
Reciprocal Agreements Also called Intercarrier Roaming Agreements. An agreement between two cellular carriers that allows the respective customers of the two carriers to use each others' systems automatically, without the necessity of registering as roamers.
Recognized Private Operation Agency RPOA. The ITU-T term for a packet interexchange carrier.
Recognizer A voice recognition term. A system that attempts to classify speech (input utterances) as words from an active vocabulary.
Reconfiguration A fancy word for rearranging equipment, features and options.
Record In a database, a record is a group of related data items treated as one unit of information — for example, your name, address and phone number. Each Record is made up of several fields. A field is simply your last name.
Record Communications Any form of communication which produces a "written" record of the transmission. Teletypewriter and facsimile are examples or record communications. Companies such as RCA Globecom, ITT Worldcom, TRT and MCI, which provide international telex, are known as international record carriers. Before deregulation, that business was exceptionally profitable.
Record Head The electromagnetic device which magnetizes the surface of a magnetic recording — tape, disk, etc. — in proportion to an electrical signal.
Record Length The number of bytes in a record. See RECORD.
Record Locking Think about an airline reservation. You call up. You want to change your reservation. While the airline has your record open, your travel agent calls up to change it. You change your reservation. Your travel agent changes it. Which one ends up in the "permanent" record? Confusion reigns. Clearly it makes sense to only allow one person to access one record at once and lock everyone else out. Record locking is the most common and most sophisticated means for multi-user LAN applications to maintain data integrity. In a record locking system, users are prevented from working on the same data record at the same time. That way, users don't overwrite other users' changes and data integrity is maintained. But though it doesn't allow users into the same record at the same time, record locking does allows multiple users to work on the same file simultaneously. So multi-user access is maximized. Contrast with file locking, which only allows a single user to work on a file at a time.
Recorded Announcement Intercept Provides a recorded message to an intercepted call indicating why the call cannot be completed, as an alternative to attendant intercept or intercept one for DID and CCSA calls to restricted or unassigned numbers.
Recorded Announcement Service A special type of central office trunk which when dialed, will connect the caller to a prerecorded message.
Recorded Answering Device See RAD.
Recorded Telephone Dictation Phone users can dial into centralized telephone dictation equipment. The dictation equipment is usually handled as a trunk connection or it can be wired on an extension level.
Recorder A device many large phone users use to record conversations with their callers. Recording truck dispatches can help a company gain the upper hand in customer service. Purchasing departments may use the recorder to remind vendors of their promises. The financial department can document money transfer orders and investments. Recorders come in several sizes. There are cassette recorders with standard speed and slow extended play speed. Open or reel-to-reel recorders have features similar to cassette recorders. Cassette recorders may be voice-operated (VOX) or started by a recorder coupler. Channel capacities available today include 7, 10, 14, 20, 28, 30, 40, 56 and 60 channels, depending on the manufacturer. Some recorders can search for and recall conversations recorded with an option called "autosearch."
Recorder Warning Tone A one-half second burst of 1400 Hz applied to a telephone line every 15 seconds to indicate to the called party that the calling party is recording the conversation. This tone is required by law to be generated as an integral part of any recording device used for the purpose and is required to be not under the control of the calling party. The tone is recorded together with the conversation.
Recoverability The way a computer or telephone system resumes operation after overcoming a problem with the hardware (say a power failure) or a program error. Some phone systems recover quickly by themselves. Some recover slowly by themselves. Some loose data. Some need human intervention. What causes a system to fail and how and how fast it recovers is key

501

to understand and verify during the test process. This definition from Steve Gladstone, author of the book "Testing Computer Telephony Systems."

Recovery The way a computer or telephone system resumes operation after overcoming a problem with the hardware (say a power failure) or a program error. Some phone systems recover quickly by themselves. Some recover slowly by themselves. Some need human intervention. These are the slowest. Check yours out. If your recovery is slow, and if you local power company is unreliable, you might consider backing your computer up with an uninterruptible power supply.

Rectifier A device for converting Alternating Current (AC) into Direct Current (DC). AC current comes out of the commercial power supply — 120 volts, 60 Hz. DC power is what drives telephone systems and the circuits that move the transmission around. Typically that DC power ranges from 5 to 48 volts.

Recursion The ability of a programming language to be able to call functions from within themselves.

Red Alarm In T-1, a red alarm is generated for a locally detected failure such as when a condition like loss of synchronization exists for 2.5 seconds, causing a CGA, (Carrier Group Alarm). See T-1.

Red Black Concept The separation of electrical and electronic circuits, components, equipment, and systems that handle classified plain text (RED) information in electrical signal form from those that handle encrypted or unclassified (BLACK) information.

Red Box A device that produces tones similar to those produced by dropping coins into a pay phone to inform the operator or automatic machinery that money has been deposited. The red box is used to defraud telephone companies. It is so named because they are usually built small enough to be placed in the "crush proof box" of a packet of Marlboro cigarettes. Red boxes are illegal.

Red Light District On-line pornography.

Redirector Networking software that accepts input/output requests for remote files, named pipes, or mailslots and then sends (redirects) them to a network service on another computer. Redirectors (also called network clients) are implemented as file system drivers in Windows 95. A redirector is a LAN software module loaded into every network workstation. It captures application programs requests for file- and equipment-sharing services and routes them through the network for action.

Reduce A Windows term. To minimize a window to an icon at the bottom of the desktop by using the Minimize button or the Minimize command. A minimized application continues running, and you can select the icon to make it the active application.

Redundancy 1. That part of any message which can be eliminated without losing the important information.
2. Having one or more "backup" systems available in case of failure of the main system.

Redundancy Check A technique of error detection involving the transmission of additional data related to the basic data in such a way that the receiving terminal, by comparing the two sets of data, can determine to a high degree of probability whether there has been an error in transmission.

Redundant Array Of Inexpensive Disks RAID. The idea is simple: Put several disk drives into a single housing. Then write your data over the disk drives in such a way that if you lose one or more of the drives, you won't have lost any of your data. Thus the term "redundant." At its simplest, RAID mirrors data to an equal number of disk drives, e.g. two sets of two. At its most complex, RAID writes data across a bunch of drives, so that if one goes the data can be retrieved from the remaining drives. RAID as a concept was first defined in 1987 by Patterson, Gibson and Katz of the University of California, Berkeley. As defined, RAID has three attributes:

1. It is a set of physical disk drives viewed by the user as a single logical device. 2. The user's data is distributed across the physical set of disk drives in a defined manner. 3. Redundant disk capacity is added so that the user's data can be recovered if one (but not more than one) drive fails.

The Berkeley engineers described five levels of RAID configurations called RAID-1 through RAID-5. RAID-0 and RAID-6 have since been added by industry usage. The distinguishing features among the various RAID levels are the way data is distributed and the way redundant capacity is implemented. Each RAID level represents very different trade-offs in terms of cost, availability and performance. Here's a simple explanation of the various levels of RAID:

Level 0 Disk striping across multiple disks. No error correction or redundancy provided.
Level 1 Disk mirroring or shadowing. One disk drive and an exact backup on a second disk, i.e. All data is redundantly recorded ("mirrored") on a second disk.
Level 2 Data is striped across multiple disks, and error checking and correcting (ECC) codes are written onto additional disks for use in fault recovery.
Level 3 Data is striped byte-by-byte across multiple disks and a single additional disk is dedicated to recording parity data.
Level 4 Similar to RAID-3, but stripes data in large chunks. Data is striped block-by-block across multiple disks and a single additional disk is dedicated to recording parity data.
Level 5 The most popular RAID. Data is striped block-by-block across multiple disk, and parity data is also spread out over multiple disks.
Level 6 RAID-5, plus redundant disk controllers, fans buses,etc.

A caveat: The above "levels" are overly simplistic. As Raid has appeared, most manufacturers have implemented different variations on the RAID theme. When I show them the above list, I usually get "Well, that's a beginning." And, of course, some levels are combined. The most popular RAID levels are 0/1 (Zero/One), which is an integral part of NetWare and Level 5, which is not but uses proprietary software techniques. The big difference is that level 0/1 maps the information on one drive to a second. You can always take one drive out, and read it. In Level 5, the data is spread across several drives. You can remove one drive and you won't lose any data. But you can't reconstruct your data from that removed drive. You need the others. When you replace a drive in Raid Level 5 (let's say because it is broken), the others will reconstruct the failed drive fairly quickly — often in less than an hour. Level 0/1 doesn't give you as much total storage space as Level 5.

Redundant Bits The extra bits included in a transmission for purposes of detecting and/or correcting errors. See REDUNDANCY CHECK.

Re-Engineer To redesign a business process. Re-engineering aims to use the power of information technology to radically redesign business processes to improve speed, service and quality. See DOWNSIZING.

Reed Relay Two tiny pieces of metal encapsulated in a tiny nitrogen-filled glass tube. When a current is passed through a magnet around the nitrogen-filled glass capsule, one arm of the metal reed relay moves and makes contact with the other. In this way it acts as a "switch." Reed relay switches are reliable. Because they are metal, they can carry great amounts of data. They are rapidly becoming obsolete.

Reed Solomon Forward error correcting code to mitigate effects of bit error bursts in receive bit stream.

Reengineer See RE-ENGINEER.

Reference Clock A clock of high stability and accuracy that is used to govern the frequency of a network and mutually synchronize clocks of lower stability.

Reference Level The measure of a value used as a starting point for further measurements. In communications applications this term usually refers to a power level of a signal or a noise. A common reference level is 0 dBm, that is, 1 milliwatt.

Reference Line In faxing, the reference line is the first scanning line in memory. The location of each black pixel of this line is kept in memory for the next scanned line. Depending on the compression technique used, more or fewer scan lines are necessary.

Reference Number Prompting An AT&T Enhanced Fax Mail term. Reference number prompting is an option that allows you to prompt anyone sending a fax message to your mailbox for a reference number of up to 16 digits.

Reference Track A special magnetic track placed on Floptical diskettes used by the drive to calibrate the optical tracking system with respect to the magnetic recording tracks.

Reference Noise RN. A reference level of noise power.

Referential Integrity Refers to a database's ability to link data in two or more files, so that adding data to a record in one file automatically updates data in another file.

Reflectance The ratio of reflected light power to incident light power. Synonym for "return loss."

REFLEX One of the communications protocols used between paging towers and the mobile pagers/receivers/beepers themselves. Other protocols are POCSAG, ERMES, GOLAY and FLEX. The same paging tower equipment can transmit messages one moment in GOLAY and the next moment in ERMES, or any of the other protocols.

Reflow Soldering A surface-mounting process for electronic components in which a solder paste is applied to the solder lands on the PCB and the components are properly aligned and placed on them. Upon heating the solder, it melts and forms a solder bond with the component terminals, electronically and mechanically bonding the component to the board.

Refractive Index A ratio of the velocity of light in a vacuum to the velocity of light in another medium, like glass.

Refresh Rate Also called Vertical Scan Frequency or Vertical Scan Rate. The phosphor coating on a monitor tube must be repainted or "refreshed" periodically. Typically, color displays use a low persistence phosphor that must be refreshed 60 times per second, or a rate of 60Hz to 70 Hz or more for VGA and higher resolution monitors. Generally, the faster the refresh rate, the less the flicker. Monochrome displays use a phosphor coating with longer persistence and typically are refreshed at a rate of 50 hertz; this difference accounts for the flicker sometimes seen on color monitors operating in a monochrome mode. Above 70 Hz, color monitors are considered flicker-free.

Refurbished A term used in the secondary telecom equipment business. Refurbishing means that telephone equipment has been cleaned, polished, resurfaced and whatever else it takes to return the equipment to a "like-new" appearance. Refurbishing usually means it has been completely tested and is ready for installation. But don't take my word for it. Get a written guarantee. See USED, CERTIFIED and REMANUFACTURED.

Regenerate To restore a signal to its original shape. Signals need to be restored because they become distorted and acquire noise during transmission. Analog signals cannot be regenerated because it is very hard for telecommunications equipment to distinguish between unwanted noise and wanted noise (i.e. your voice) in an analog signal. Digital signals can be more easily regenerated since they consist of "ones" and "zeros." If digital signals are flattened or distorted, a simple logic circuit — "Is it a zero or a one?" — can restore the signal to its original clean squared shape.

Regeneration The process of receiving and reconstructing a digital signal so that the amplitudes, waveforms, and timing of its signal elements are constrained within specified limits.

Regenerative Repeater A device which regenerates incoming signals and retransmits these signals on an outgoing circuit. See REGENERATE.

Regenerator A receiver and transmitter combination used to reconstruct signals for digital transmission. In an optical regenerator, the receiver converts incoming optical pulses to electrical pulses, decides whether the pulses are "1s" or "0s," generates "cleaned up" electrical pulses, and then converts them to squared off pulses for transmission.

Regional Bell Operating Company RBOC. One of the seven Bell operating companies set up after Divestiture. each of which own two or more BOCs (Bell Operating Companies). The RBOCs were carved out of the old AT&T/Bell System by Judge Harold Greene when he signed off on the divestiture of the Bell operating companies from AT&T at the end of 1984. There is nothing magical about seven — nor the grouping of Bell Operating Companies (BOCs) into RBOCs — except the Judge wanted to keep them all roughly the same size. The seven RBOCs are Ameritech, Bell Atlantic, Bell South, Nynex, Pacific Telesis, Southwestern Bell and US West. In early October, 1994, Southwestern Bell changed its name to SBC Communications Inc. But its telephone companies, it said, would still operate under the Southwestern name.

Regional Center A control center (Class 1 office) connecting sectional centers of the telephone system together.

Register A temporary-memory device used to receive, hold, and transfer data (usually a computer word) to be operated upon by a processing unit. The register holds the information for manipulation by the telephone system or a computer. In an automatic telephone system, a register receives dialed pulses or pushbutton tones and then uses that information to control the switch.

Computers typically contain a variety of registers. General-purpose registers perform such functions as accumulating arithmetic results. Other registers hold the instruction being executed, the address of a storage location, or data being retrieved from or sent to storage. Other words associated with "register" include buffer, fetch protection, M-sequence, read-only storage, permanent storage, random-access memory and shift register.

Registered Access In the context of message handling services, access to the service performed by subscribers who have been registered by the service provider to use the service.

Registered Jack RJ. Any of the RJ series of jacks, described in the Code of Federal Regulations, Title 47, part 68 used to provide interface to the public telephone network. See also RJ-11, RJ-45.

Registered Terminal Equipment Terminal equipment which is registered for connection to the telecommunications network in accordance with Subpart C of Part 68 of the FCC's Rules. If a terminal device has been properly registered it will have an identification number permanently affixed to it.

Registers An ISDN term. Registers are named storage areas for numbers or strings of characters that control the operation of the ISDN set.

Registration The address registration function is the mechanism by which Clients provide address information to the LAN Emulation Server.

Registration Number (FCC Part 68) Approval number given to telephone equipment to certify that a particular device passes the tests defined in Part 68 of the FCC Rules. These tests certify the phone won't cause any harm to the public network. They do not attest to the commercial value of the product, nor whether it will (or won't) sell. See REGISTRATION PROGRAM.

Registration Program The Federal Communications Commission program and associated directives intended to assure that all connected terminal equipment and protective circuitry will not harm the public switched telephone network or certain private line services. The program requires the registering of terminal equipment and protective circuitry in accordance with Subpart C of part 68, Title 47 of the Code of Federal Regulations. This includes the assignment of identification numbers to the equipment and the testing of the equipment. The registration program contains no requirement that accepted terminal equipment be compatible with, or function with, the network. In other words, a product registered under Part 68 doesn't mean that the product will actually work — i.e. make and receive phone calls (or whatever). Part 68 simply says it won't cause any harm to the network. See REGISTRATION NUMBER and PART 68.

Regression Analysis A method of forecasting the future by plotting events in the past and assuming there'll be some similarity in the future. About as accurate as any other pseudo scientific method.

Regulated 1. Controlled for uniformity. Many aspects of telecommunications are regulated — from the input voltage powering a telecom system to the output signal of a microwave system.
2. Adhering to the rules, regulations and sundry whims of a government agency. Most aspects of the telephone business are under the control of a government agency to some degree. Their rules cover everything from certifying of expenses which may be capitalized to specifying how many seconds the subscriber can be forced to wait for dial tone (three seconds). Stripped to bare essentials, a regulatory agency can only do two things. First, it can allow the regulated entity to raise its prices to a point where nobody wants to buy anymore. Second, it can stop competitors coming into the business. The first (high price) is the reason no one (or few people, anyway) send telegrams. The second (keep out the competition) reason gets stymied because new technology — e.g. cheap local microwave — comes along to force the regulatory agency's hand. In the long run, no regulated entity survives because it has a regulated monopoly. It survives because it provides good service at a fair price.

Regulated Charger An uninterruptible power supply (UPS) definition. Without a regulated charger, batteries can be insufficiently charged or blistered with too much charge voltage. Either case can cause permanent damage.

Regulation See POWER CONDITIONING.

Regulatory Groups Refers to local, State or Federal entities that issue orders, findings, etc. that are binding upon providers and users of telecommunications and services.

Rehoming An MCI definition. A major MCI network change which involves moving the appropriate customer service from a working switching center to a different switching center and establishing the necessary facilities. Rehomes can be any of the following: switch upgrade, switch move/decommission, off-to-on-net conversion.

REJ Abbreviation for REJect.

Rejection A word used in voice recognition to mean a type of recognition classification where the input utterance did not meet the criteria necessary to be classified as a word in the active vocabulary. Usually the speaker is asked to repeat the utterance.

REL RELease message. The fifth of the ISUP call set-up messages. A message sent in either direction indicating that the circuit identified in the message is being released due to the reason (cause) supplied and is ready to be put into the idle state on receipt of the Release Complete Message. See ISUP and COMMON CHANNEL SIGNALING.

Relation Synonym for table.

Relational Database A database that is organized and accessed according to relationships between data items. A relational database consists of tables, rows and columns. In its simplest conception, a relational database is actually a collection of data files that "relate" to each other through at least one common field. For example, one's employee number can be the common thread through several data files — payroll, telephone directory, etc. One's employee number might thus be a good way of relating all the files together in one gigantic data base management system (DBMS).

Relationship Routing A concept introduced by automatic call distributor manufacturer, Aspect Telecommunications, to have callers' calls routed to agents they had previously developed

Relative Transmission Level The ratio of the test tone power at one point to the test tone power at some other point in the system chosen as a reference point.

RELAX A release tool for an Ericsson AXE.

Relay An electrically activated switch used to operate a circuit. It connects one set of wires to another. Usually, the relay is operated by low voltage electric current and is used to open or close another circuit, which is of much higher voltage. Older telephone switches used many relays to switch (i.e. complete) their calls. Relays come in many forms. There are hermetically-sealed relays, in which thin metal contacts are sealed in an airtight glass or metal enclosure. There are also mercury relays in which a small tube of mercury tilts and completes or breaks a circuit. See also REED RELAY.

Relay Rack Open iron work designed to mount and support electronic equipment. A relay rack is to electronic equipment what a distribution frame is to wire. See DISTRIBUTION FRAME.

Relayer Allows a user to open or close a solenoid via the phone system.

Relaying A function of a layer by means of which a layer entity receives data from a corresponding entity and transmits it to another corresponding entity.

Release 1. A call comes into a switchboard. The operator calls you to tell you it's for you. Then he/she "releases" the call to you. On most switchboards there's a button labelled "RLS." That's the release button. On some phones (not consoles) the release button is the "hang-up" button. Hitting this button means disconnecting the call. Be careful.
2. The ending of an inbound ACD call by hanging up.
3. The feature key on most ACD instruments labelled Release.
4. A term used in the secondary telecom equipment business. The relinquishing of a piece of equipment to a purchaser or user upon fulfillment or anticipated fulfillment of contractual obligations, whether written or oral.

Release 1.0 A monthly newsletter from PC guru, Esther Dyson and her knowledgeable staff at EDventure Holdings on Park Avenue, New York City.

Release Button The release button — found always on operator consoles and occasionally on some phones — ends a call in the same way that hanging up the receiver does.

Release Link Capability The ability for an originating switching system, on receipt of a new destination address from the current terminating switching system, to release the transmission link to that terminating switching system and continue call processing using the new destination address. Definition from Bellcore in reference to its concept of the Advanced Intelligent Network.

Release Link Trunk RLT. Telecommunications channel used with Centralized Attendant Service to connect attendant-seeking calls from a branch location to a main location.

Release With Howler If a phone stays off-hook without originating a call (or the receiver is accidentally knocked off), the system transmits a loud tone over the line and then disconnects the line and the phone. The central office effectively then ignores them (the line and the phone) until someone puts the receiver back on-hook again.

Reliable Service Area RSA. The area specified by the field strength contour within which the reliability of communication service is 90 percent for a mobile unit.

Reliability A measure of how dependable a system is once you actually use it. Very different from MTBF (Mean Time Between Failures). And very different from availability. See MTBF.

Relocatable Code Machine language programs that can reside in any portion of memory.

Remailer Remailers are anonymous mail drops that computer hackers have set up on the Internet, untraceable electronic mail addresses where one can send or receive encrypted data. An article in the October, 1994 issue of High Times, a drug related magazine, offered plans for a similar security system as a remailer, adding one interesting twist. By incorporating a computer virus like Viper or Decide in the system, the computer could be programmed essentially to self-destruct as soon as it detected a security breach, thus rendering it worthless as evidence.

Remanufactured Equipment, parts and/or systems that have been repaired and upgraded to the latest higher revision level. The remanufacturing process makes the telecom equipment (used or new) into a finished product that is the latest release and ready for resale. Remanufactured is the term for the highest level of refurbishing equipment.

Remapping The practice of redefining the meaning of keys on the keyboard.

Remission IBM-speak to change the mission of a product or a facility.

Remodulator In a split broadband cable system, a digital device at the headend that recovers the digital data from the inbound ananlog signal and then retransmits the data on the outbound frequency.

Remote Access Sending and receiving data to and from a computer or controlling a computer with terminals or PCs connected through communications (i.e. phone) links.

Remote Access To PBX Services Allows a user outside the PBX to access the PBX by dialing it over a normal phone line. You dial the number. It answers. It may or may not say anything. It may just give you dial tone. You now punch in an authorization code. If your code is acceptable, the PBX gives you another dial tone. That dial tone is effectively the one all users within the PBX get. Once you have this dial tone, you can dial another extension, jump on the company's WATS network, get into the dictation unit, access its voice mail, or whatever. Suffice, you are inside the PBX. You can do whatever anyone else inside the PBX can do.

Remote Batch Processing Processing in a computer system in which batch programs and batch data are entered from a remote terminal or a remote PC (personal computer) over phone lines.

Remote Bridge A bridge between two or more similar networks on remote sites. Dial up or leased lines typically require a local bridge or gateway and a remote bridge or gateway an each end, in order to network.

Remote Call Forwarding RCF. This is a neat service. It allows a customer to have a local telephone number in a distant city. Every time someone calls that number, that call is forwarded to you in your city. Remote call forwarding is very much like call forwarding on a local residential line, except that you have no phone, no office and no physical presence in that distant city. Remote Call Forwarding exists purely in the central office. You can also think of it as measured Foreign Exchange. Companies buy Remote Call Forwarding for three reasons: 1. To encourage distant customers to call them by giving them a local number in their own city to call. (This the most obvious reason for an IN-WATS line, a FX or a RCF line); 2. They buy RCF over IN-WATS or FX lines because they don't have the volume to justify these potentially more expensive lines. 3. Companies buy RCF lines as overflow lines from IN-WATS and FX lines. They use their RCF lines when the other lines (FX and IN-WATS) get busy during peak busy periods. Remote Call Forwarding calls are typically charged at the same price as normal DDD calls (i.e. the most expensive to call). And you can't, as yet, reprogram RCF calls easily. You have to place an order with your friendly telco and wait for them to do the reprogramming. In 1994 Northern Telecom announced a service whereby remote call forwarded calls carried from its central office would carry the calling line ID number. This way when the call reached the distant number, it could be handled in an intelligent fashion.

RLCS Remote Live Call Screening. A Panasonic phone feature. RLCS is even better than live call screening feature. It lets you plug a single cordless phone into the External Data Port (XDP) on your Panasonic phone and monitor incoming voice mail remotely. When a call ends up in voice mail you use your cordless phone to screen your call. To take the call simply hit the same key again. This way you are always assured you will receive the important call you were waiting for — even if you're in a meeting. This feature is only available on the Panasonic Super Hybrid PBX with its integrated digital Panasonic voice mail system. But I liked the feature enough to include it in my dictionary.

Remote Concentrator A remote multiplexer. A device which places more than one distant user on two cable pairs. The idea of a remote concentrator is to substitute electronics for cable. It's simply cheaper to put electronics on either end of two cable pairs and drive many conversations through those wires, than running extra cables (digging streets, erecting poles, etc.)

Remote Control Remote control software allows a remote PC to connect to the network via a PC that is on the LAN. You must use such software for working from home, for sending in your work, checking on your email, etc. See REMOTE NODE.

Remote Diagnostics You own a phone system. You have a service company. There's some problem with it. Instead of sending a technician out, your service company dials your PBX from a data terminal or PC and "asks" your PBX in computerese what's wrong with it. If it isn't too broken, it will come back and give you some indication. This is called remote diagnostics. Some service companies call all their customers' phone systems every morning and run routine remote diagnostics on their switch. It's like going to the doctor for a daily physical. Sometimes this test may find a problem before the user is even aware. Sometimes the problem can be repaired on-line. If not, the service company will have to dispatch a technician. Remote diagnostics is a good idea. More phone systems should have it. To do remote diagnostics on a telephone system, you will typically need a phone line dedicated to the PBX and a modem on either end.

Remote Digital Loopback A test that checks the phone link and a remote modem's transmitter and receiver. Data entered from the keyboard is transmitted from the initiating modem, received by the remote modem's receiver, looped through its transmitter, and returned to the local screen for verification.

Remote Job Entry RJE. Remote Job Entry occurs in computer operations where work or input is sent in remotely over phone lines. That "work" might include the day's sales of a distant store.

Remote LAN Interconnection The connection of two or more LANs which are remotely located from each other so that LAN users can communicate with users and servers on any of the interconnected LANs.

Remote Line Concentrator A multiplexer. A device that "concentrates" several users' lines on a fewer number of trunks. Typically, a remote line concentrator is used because it's cheaper, easier or more flexible to substitute electronics for cable. See REMOTE CONCENTRATOR.

Remote Line Switch A line unit mounted near a cluster of users and equipped with intracalling capability.

Remote Line Unit A remote line concentrator without intracalling capability. See REMOTE CONCENTRATOR and REMOTE LINE SWITCH.

Remote Live Screening See LCS.

Remote Maintenance Facility See REMOTE DIAGNOSTICS.

Remote Node Remote Node software allows remote users to dial in to the corporate LAN and work with the applications and data on the LAN as if they were "actually in the office." By dialing in, they become a node on the LAN. Using a PC, Mac, or UNIX workstation; a modem; and a remote access server, employees can connect from any location in the world that has an analog, a switched digital, or a wireless connection.

Remote Office Test Line ROTL. A testing device that acts in conjunction with a central controller and a responder to make two-way transmission and supervision measurements.

Remote Operations Service Element ROSE. An application layer protocol that provides the capability to perform remote operations at a remote process. Definition from Bellcore in reference to its concept of the Advanced Intelligent Network.

Remote Order Wire An order wire is a line on which maintenance and monitoring is done. A remote order wire is an order wire that has been extended to a distant point that may be more convenient.

Remote Procedures Call RPC. A message-passing facility that allows a distributed program to call services available on various computers in a network. Used during remote administration of computers, RPC provides a procedural view, rather than a transport-centered view, of networked operations.

Remote Programming Dial your phone system with your friendly personal computer, modem and a communications software package and you can change the telephone system's programming remotely. This feature is great for companies with telephone systems in many locations. They can all be run from one central point. This feature is also great if you want some changes made on your system. It's obviously a lot cheaper for your vendor to make those changes from his office rather than have to visit yours. It's also a lot faster. See REMOTE DIAGNOSTICS.

Remote Site The remote site is the person or location doing the sending in a file transfer operation. An example: Sales reps in the field typically update the central database on a periodic basis. The central database location is known as the host and the sales reps in the field are doing so from remote locations.

Remote Site Location A location for a DCE device which is not at the central or control site. A typical application would have a terminal at the remote site and the host computer at the central or control site.

Remote Station Any piece of equipment attached to a LAN by a telephone company supplied link. Technically, that includes all devices that aren't servers. Usually it refers to a workstation at a distant location, linked to the main LAN by a modem and connected through a serial port "gateway." See MODEM.

Remote Station Lamp Field For use at multi-line phones, usually manned by secretaries who answer many phone lines.

Remote Switching System An switch that is away from its host or control office. All or most of the central control equipment for the RSU is located in the host or control office. See also REMOTE CONCENTRATOR.

Remote Terminal A terminal connected to a computer over a phone line.

Remote Traffic Measurement Traffic and feature usage data can be transmitted by the system to a distant service technician.

Remote Workstation A terminal or personal computer connected to the LAN (local area network) by a modem. A remote workstation can be either a standalone computer or a workstation on another network.

Remotely-Hosted An SCSA definition. The Client and the Server are different (i.e., the application is on a different physical box than the service provider).

Removable Cartridge System A high-capacity storage system that can be removed from the PC. A removable cartridge systems consists of a drive mechanism and the cartridges used to store data. The most well-known removable cartridge system is the Bernoulli Box by Iomega Corp.

Removable Media Diskettes or cartridges that can be removed from a computer drive. For example, a Bernoulli box uses removable cartridges.

REMS Remote Electronic Mail System. A REMS mailbox is MCI Mail's way of making the exchange of messages between networks (both public and private) and the people on them easier. A typical application for a REMS mailbox would be a company with an internal e-mail system (e.g. Microsoft Mail, cc Mail, Beyond Mail, Notework, DaVinci e-mail, Lotus Notes, BeyondMail, Notework, Futurus Team, etc.) that wishes to link that system to the outside world via MCI Mail, receiving messages from MCI Mail and passing them to MCI Mail for onward delivery. What happens typically is one of the workstations on the company's local area network (LAN) and running the company's e-mail package is designated as "The MCI Mail Gateway Machine." That workstation has a modem. At intervals set by the REMS system administrator, it will dial MCI, uploading all the messages for MCI and downloading all the messages which MCI has delivered into the company's REMS mailbox. When it has hung up from MCI, it will then distribute the messages to the individuals on the LAN. The way it does this is that when I send a message, I will address it thus:
TO: Chris Risley EMS: ON Technology Corporation / MCI ID: 413-9429 MBX: CRISLEY@NOTEWORK
The first line tells MCI Mail whom it's for. The second line tells it which REMS account (namely Notework Corporation, MCI Mail account 413-9429), and the third line (and optional additional lines) tells Notework Corporation's LAN e-mail gateway machine that recipient's local e-mail address.

REN Ringer Equivalency Number. Part of the FCC certification number approving a telephone terminal product for direct sale to the end user as not doing harm to the network. The REN consists of a number and a letter which indicates the frequency response of that telephone's ringer. "A" = 20 Hz or 30 Hz "B" = a range from 15.3 Hz to 68 Hz. The remaining letters represent ringers that will work on very narrow ranges such as "C" = 15.3 Hz to 17.4 Hz, etc. The number indicates the quantity of ringers which may be connected to a single telephone line and still all ring. The total of all RENs of the telephones connected to the one line must not exceed the value 5 or some or all of the ringers may not operate.

Reorder An announcement, or 120 interruptions per minute tone, returned to the caller when his call is blocked in the network. See REORDER TONE.

Reorder Tone The Reorder tone sounds like a busy signal but is twice as fast, i.e. a reorder tone is a tone applied 120 times per minute. The tone means that all switching paths are busy, all toll trunks are busy, there are equipment blockages, the caller dialed an unassigned code, or the digits he dialed got messed up along the way. Also called Channel Busy or Fast Busy Tone.

Reorg A shortened form of the word reorganization. It is used by people in companies which go through management reorganizations so often they don't have to figure what the latest organization means before the next one happens. And they certainly don't have the time to say the word "reorganization" in full. I first heard the word "reorg" from someone at Pacific Bell. He used the word as an excuse for not following up on something he had promised me.

Rep Repertory dialing. Speed dialing. Some cellular phones are capable of storing 100 numbers.

REPACCS REmote cable PAir Cross-Connect System is a PC controlled, metallic, automated cross-connect system that may be applied to Automated Distribution Frames, Building Terminals, Service Area Interfaces (SAIs) or cross-connect boxes and closures/terminals. It dramatically reduces dispatches, provides 100% record accuracy, facilitates multiple line testing (MLT), and operates without local or battery power as well as keeps people out of restricted or hazardous areas.

Repair And Quick Clean RQC. A term in the industry which repairs telecom equipment. It means all equipment is repaired and fully tested with a burn-in (if required) and an operational systems test. It also includes minor cosmetic cleaning of the unit. Definition courtesy Nitsuko America. See also LIKE NEW REPAIR AND UPDATE and REPAIR, UPDATE AND REFURBISH.

Repair, Update And Refurbish RUR. A term in the industry which repairs telecom equipment. It means equipment is repaired and updated to current manufacturer's specifications. Also includes minor cosmetic cleaning of metal cabinets, a full diagnostic test with burn-in (if required) and an operational test. Definition courtesy Nitsuko America. See also LIKE NEW REPAIR AND UPDATE and QUICK CLEAN.

Repair Only A term used in the secondary telecom equipment business. Equipment is repaired to original working condition, but does not include refurbishment or recycling except where required to bring equipment to working condition. See also REFURBISHED and REMANUFACTURED.

Repair Service Answering RSA. Functions that support the initial handling and entry of subscriber reported troubles. They enable subscribers to request trouble verification tests, to initiate a trouble report and to obtain information on the status of an open trouble report. Definition from Bellcore in reference to its concept of the Advanced Intelligent Network.

Repartee Active Voice's voice processing system that combines automated attendant, voice mail, audiotext and facsimile features. Active Voice is in Seattle, WA.

Repeat The act of a station receiving a code-bit stream (frame or token) from an upstream station and placing it onto the ring to its downstream neighbor. The repeating station may examine, copy to a buffer, or modify control bits in the code-bit stream as appropriate.

Repeat Call The name of a Bell Atlantic service. Dialing a busy number over and over is as time-consuming as it is frustrating. With Repeat Call, your phone will continuously monitor a busy number every 45 seconds for up to 30 minutes, without interrupting your incoming or outgoing calls. So you and your employees can do other things until your phone alerts you with a special ring when the call got through.

Repeat Dialing A name for a Pacific Bell (and possibly other local telephone companies') service which automatically checks a busy number and when the line is free, it rings you back and completes the call.

Repeater 1. An opto-electronic device inserted at intervals along a circuit to boost, and amplify an analog signal being transmitted. A repeater is needed because the quality and strength of a signal decays over distance. You will find repeaters in cables and in microwave systems. Repeaters may also regenerate a digital signal — "squaring it" and "cleaning" it up — but not changing it. Regenerating the signal removes noise and thus reduces the likelihood of errors. You can only regenerate digital signals. You cannot regenerate analog signals. You can regenerate digital signals because a machine can tell what's a signal and what's noise in a digital signal. But no machine exists to do that with an analog signal.
2. The simplest type of LAN interconnection device. A repeater moves all received packets or frames between LAN segments. The primary function of a repeater is to extend the length of the network media, i.e. the cable.

Repeating Coil A transformer which connects one telephone circuit with another without any DC connection between the circuits. Here's a more technical explanation: A voice-frequency transformer characterized by a closed core, a pair of identical balanced primary (line) windings, a pair of identical but not necessarily balanced secondary (drop) windings, and a low transmission loss at voice frequencies. It permits transfer of voice currents from one winding to another by magnetic induction, matches line and drop impedances, and prevents direct conduction between the line and the drop.

Reperforator In teletypewriter systems, a device used to punch a tape in accordance with arriving signals, permitting reproduction of the signals for retransmission. See also CHAD.

Reperforator/Transmitter RT. A teletypewriter unit consisting of a reperforator and a tape transmitter, each independent of the other.

Repertory Dialing Sometimes known as "memory dialing" or "speed-calling." A feature that allows you to recall from nine to 99 (or more) phone numbers from a phone's memory with the touch of just one, two or three buttons.

Replication Also known as data replication. It refers to the process of duplicating and updating data in multiple computers on a network, some of which are permanently connected to the networks. Others, such as laptops, may only be connected at intermittent times. The idea is twofold: Everyone should have access to the same information in the database/s. Second, many people can make changes to the same record and somehow, all those changes will meld themselves into the database/s. In the old days (i.e. pre-Lotus Notes), networked databases were stored in one place, e.g. an airline database of reservations. Everyone who wanted to access information in the database needed to be physically connected to the network through some form of phone line. That's still the case in most databases. Along came Lotus Notes whose major claim was everyone could create their own database/s and carry it with them on their laptops and everyone could put their own information in and Lotus Notes would update the central database and update everyone's database every time they logged into the network. If Lotus is confused, it sends messages out asking for clarification as to what the right database entry was. In short, replication is a far more complex process than what the traditional English language definition of replication is, namely making copies of itself. In data replication, it's the coordination, updating and reconciling of constantly-being-changed databases. That's the hard part. As I wrote this, Lotus Notes had big lead in this process of database replication. But others, like Oracle and Microsoft, were trying to catch up. The easiest replication strategy is one-way transfer. A simple case of one-way data replication is a mobile user who needs to update the information on his laptop, but not to update any information at the corporate site. See also SYNCHRONIZATION.

Reply 1. A transmitted message which serves as a response to an original message. (What else?)
2. An SCSA definition. An event which is a service provider's response to a synchronous or asynchronous request.

Report Mining Coined by Gartner Group, Inc. for migrating legacy report data to a server so it can be accessed by desktop query tools and regenerated into a new report.

Report Program Generator A computer language for processing large data files.

Report-Only Event An event that the ASC (AIN Switch Capabilities) reports to a SLEE (Service Logic Execution Environment) but the ASC does not suspend processing events for the connection segment. Definition from Bellcore in reference to its concept of the Advanced Intelligent Network.

Request For Comments A standard or a standard-defining document for the Internet. Individual RFCs define specific aspects of Internet operation. See RFC.

Request To Send RTS. One of the control signals on a standard RS-232-C connector. It places the modem in the originate mode so it can begin to send.

Rerouting A short-term change in the routing of telephone traffic. Rerouting may be planned and recurring or a reaction to a nonrecurring situation.

RES Residential Enhanced Service.

Resale Buying local and/or long distance phone lines in quantity at wholesale rates and then selling them to someone else, hopefully at a profit.

Resale Carrier A long distance company that does not own its own transmission lines. It buys lines from other carriers

NEWTON'S TELECOM DICTIONARY

and then resells them to its subscribers. Some resale carriers have their own switches. Some don't. Some have a mix of their own lines and leased lines. Most long distance carriers — including MCI and US Sprint — have a mix of their own lines and leased lines.

Resampling Reducing or increasing the number of pixels in an image to conform to a new size or resolution.

Reseller A company which purchases a block of cellular numbers from a cellular carrier for resale to its customers. Or a company which purchases a big block of long distance calling minutes or resale in smaller blocks to its customers. See AGGREGATOR.

Reserve Power A telephone system may be equipped with storage batteries to provide primary power during a commercial power failure. No loss of service will occur during transition to battery power. All this is a long way of saying your phone system is backed by batteries, typically lead acid (the same ones used in your car).

Reset Packet A packet that identifies error conditions on an X.25 communications circuit. The reset packet does not clear the session but rather notifies the communicating DTEs of error conditions at a known point in the data-packet transfer sequence.

Resident Command A command located in the personal computer's operating system itself, contained in the file COMMAND.COM.

Resident Program See RAM-RESIDENT PROGRAM.

Residential And Light Commercial Wiring Refers to the wiring system and all of its appurtenances required to provide convenient and useful telephone services to residences and light commercial buildings.

Residual Error Rate The ratio of the number of bits, unit elements, characters or blocks incorrectly received but undetected or uncorrected by the error-control equipment to the total number of bits, unit elements, characters or blocks sent.

Resipiscence The noun that comes from the verb to see the error of one's ways.

Resistance Any electrical conductor will resist the flow of electrical current. As it resists the flow of current, so the current becomes weaker. Resistance generates heat and occasionally light. It is technically defined as a property or a characteristic of a conductor, i.e. the metal through which the electricity flows. It is measured in Ohms.

Resistor A unit designed to oppose but not completely obstruct the passage of electrical current.

Resort An electronic destination you return to, time and again, to escape normal business.

Resource 1. Any facility of a computing system or operating system required by a job or task, including memory, input/output devices, processing unit, data files, and control or processing programs.
2. A network component such as a file, printer, or serial device that is shared by other components of the network.
3. An SCSA term. A voice processing technology, such as voice store and forward, fax processing, voice recognition, or text to speech. Here's the official definition: The abstraction of a standardized vendor-independent interface of a physical device used for call processing as seen by the Application. All Resources have common methods across all implementations. Examples of Resources are voice store and forward, fax send and receive, text to speech conversion, voice recognition, etc. Resources are assumed to have at a minimum one input or output of circuit switched TDM on the internal switch fabric of the system. Resources are shared among multiple applications. Once a Resource has been allocated to an application, it is locked from the use of any other application until freed. It is assumed that applications specify at some level their resource requirements to the server prior to accessing them. This may either be through explicitly attaching them to a Group, or having the server implicitly allocate them based on usage.
4. A Windows term. A resource is a program object, such as a button, menu or dialog box, that Windows treats differently than normal programs. Resources are developed either in a special resource language or using interactive tools. They can be loaded from separate files or bound directly to the executable file.

Resource Characteristics An SCSA definition. Resources have a set of characteristics that define their behavior beyond that defined by the methods on their interfaces. These may vary or be optional for this Resource Type. Resource may be selected by the application on the basis of their characteristics and a client may query for the characteristics of a resource which it has claimed. See RESOURCE.

Resource Class An SCSA definition. A set of methods (in object-oriented terms, a class) for controlling resource instances (or a resource). May be abbreviated as just "class." See RESOURCE.

Resource Group An SCSA term. A resource group is a dynamically formed group of resource units that can be made to work together as if they were a single device. See RESOURCE.

Resource Module A resource module is a card that slides into a PC and does everything from text-to-speech, to fax, to voice recognition, etc. Everything except interfacing to the network, which is done by another card or another part of the resource module card called the Network Interface Module. See NETWORK INTERFACE MODULE.

Resource Object An SCSA definition. An instance of a Resource Class.

Resource Unit An SCSA definition. An entity that communicates directly with a resource device. From the client application's point of view, it is the Resource Unit that provides services such as voice store-and-forward, fax send and receive, etc. See RESOURCE.

Resolution 1. The minimum difference between two discrete values that can be distinguished by a measuring device. High resolution does not necessarily imply high accuracy.
2. The degree of precision to which a quantity can be measured or determined.
3. A measurement of the smallest detail that can be distinguished by a sensor system under specific conditions.
4. A measure of the quality of a transmitted image. Beginning with the scan processing in the transmitter and ending with the display and/or printing process in the receiver, resolution is a basic parameter of any image transmission system. It affects the design of all its subsystems. In the scanner, the resolution is a function of the spot size which the scanner optics and associated electronics "look" at the scene and through which the system can uniquely identify the smallest distance along the scan line. Resolution is measured in picture elements (pixels) and is the total number of pixels (horizontal x vertical) used to display alphanumeric characters of graphic images on the screen. High resolution images are composed of more dots per inch and

appear smoother than low-resolution images. The higher the resolution, the better the display of details. See also MONITOR.

Resolver A program which connects user programs to domain name servers.

Resonance The condition that exists when inductive reactance equals capacitive reactance. In a series circuit it results in maximum current at the resonant frequency. In a parallel circuit it results in maximum voltage at the resonant frequency.

Resonant Cavity Closed metal container which has the characteristics of a parallel resonant circuit.

Resp Org Responsible Organization — the long distance company responsible for managing and administering the 800 subscriber's records in the 800 Service Management System (SMS/800). The SMS/800 only recognizes one RESP ORG for each 800 number. Management and record administration consists of data entry, changing records, accepting trouble reports and referring and/or clearing associated documents.

Responder A test line that can make transmission and supervision measurements through its host switch under control of a remote computer.

Response An answer to an inquiry. In IBM's SNA, the control information sent from a secondary station to the primary station under SDLC.

Response Time The time it takes a system to react to a given input. In voice recognition, response time typically refers to the amount of time required for a word (or utterance) to be recognized once the end of the word is detected. True response time is longer because silence often must occur before the end of the word can be declared. When operating a terminal connected to a computer, response time would be the time between the operator pressing the last key of a series of keys and the appearance of a response on the operator's display. In a data communications system, response time includes the transmission time, the processing time, the searching for records time and the transmission time back to the originator. Response time is very critical in applications like airline reservation systems. Here the customer is on the phone awaiting a reply. That time is critical in whether the customer perceives he's getting good or bad service. Response times of more than three seconds are not acceptable in situations where the customer is waiting on the phone to buy something. Response time is a function, inter alia, of the number of phone lines you lease or use. You can save a lot of phone line costs by cutting back on lines. But you'll extend response time. Life, as always, is a trade-off.

Responsible Organizations RespOrgs. Telecommunications providers that have responsibility for obtaining 800 Service numbers from the Service Management System and building and maintaining customer records. See EIGHTHUNDRED SERVICE.

Rest Stops Electronic malls and other online diversions provided for a small fee.

Restart A central office word or Apple Macintosh word for the rebooting a personal computer. In telephony, a system initiated action designed to restore overall service capacity.

Restart Packet A block of data that notifies X.25 DTEs that an irrecoverable error exists within X.25 network. Restart packets clear all existing SVCs and resynchronize all existing PVCs between X.25 DET and X.25 DCE.

Restocking Fee A fee for returning non-defective equipment. The fee is generally a percentage of the sales price, from 10 percent to 25 percent. Many buyers object to paying restocking fees. Check before you buy something — especially in the used telephone equipment business, where restocking fees are not uncommon.

Restore 1. Typically, to put a telephone system back into full operation.

Restore Button A Windows term. The small button containing both an up and down arrow at the right of the title bar. The Restore button appears only after you have enlarged a window to its maximum size. Mouse users can click the Restore button to return the window to its previous size. Keyboard users can use the Restore command on the Control menu.

Restriction Phone systems can disallow people or extensions from making certain calls. If they're not allowed to make long distance calls, this is called toll restriction. See TOLL RESTRICTION. There are other forms of restriction, like being able to only use the company's internal network.

Restriction From Outgoing Calls Phone users may be restricted from placing outgoing calls. See CLASS OF SERVICE.

Restriction Services These features allow the attendant to control the restriction of phones or groups of phones. It can be very useful in hotels and motels to turn off service to room phones during the time between check out and check in of guests. Here are some examples of restriction services:

• CONTROLLED OUTWARD RESTRICTION: Phones can be restricted from making dialed outgoing calls while inward calls are completed normally.

• CONTROLLED STATION-TO-STATION RESTRICTION: Originating phone calls to other extensions in the system are blocked, however, normal incoming and outgoing calls can be completed.

• CONTROLLED TERMINATION RESTRICTION: Phones can complete outgoing calls normally, but incoming calls are directed to either the attendant or an intercept tone or recording.

• CONTROLLED TOTAL RESTRICTION: Restricted phone lines cannot make or receive any calls.

Resynchronization The process of returning Novell NetWare SFT III (Novell's System Fault Tolerance) servers to a mirrored state after a failure. SFT III checks changes to all active servers and ensures that those changes are copied to the other servers. The time it takes the servers to complete resynchronization depends on the amount of memory and the disk storage in each server. Server memory synchronization is much faster than disk mirroring because disk mirroring speed is limited by the disk channel.

Retard Coil A coil having a large inductance which retards sudden changes of the current flowing through its winding.

Retractile Cord A coiled cord that springs back to its original length when you let it go. Telephone handset cords are the most common retractile cords. There are wide quality variations among retractile cords. Western Electric (oops AT&T Technologies) has set a very good standard for retractile cords. But not everyone conforms to it. If you want to quickly see the quality difference among various coil cords, take six from different manufacturers and hang them over your office door and come back in a week. You'll see a bunch touching the floor. Others will still be taut. Another way is simply to connect them to your phone, one by one, and listen to the differences. Some simply sound weaker. Cheaper ones tend to sound worse. (So what's new?)

Retraining Training is a feature of some modems which adjust to the conditions including amplitude response, delay distortions, timing recovery, and echo characteristic, of a particular telecommunications connection by a receiving modem. Retraining occurs after modems have been successfully transmitting and receiving data. Usually due to a change in line conditions.

Retransmission A method of error control in which hosts receiving messages acknowledge the receipt of correct messages and either do not acknowledge, or acknowledge in the negative, the receipt of incorrect messages. The lack of acknowledgment, or receipt of negative acknowledgment, is an indication to the sending host that it should transmit the failed message again.

Retransmissive Star In optical fiber transmission, a passive component that permits the light signal on an input fiber to be retransmitted on multiple output fibers. The signal comes in on one fiber, hits a star-type connector which splays the transmission out. A retransmissive star is formed by heating together a bundle of fibers to near their melting point. Such a device is used mainly in fiber-based local networks. It's also called star coupler. When you see one you'll be surprised how crude this device looks, despite its fancy name.

Retrial After failing to complete a call, a person tries again. This is called a "retrial." The term is used in traffic engineering. It's critical in figuring needed trunking capacity. See QUEUING THEORY, POISSON and TRAFFIC ENGINEERING.

Retrofit Kit A conversion kit which makes a standard pay phone into one which will accept credit cards.

Retry In the bisynchronous protocol, the process of resending the current block of data a prescribed number of times until it is accepted.

Return A carriage return. This key on some keyboards is also called "Enter." Touching the CR (Carriage Return) gives you two functions: a "line terminating function" and a "new line function", abbreviated "NL". Simply put, a Return at the end of a line, terminates that line and begins a new one.

Return Authorization Number See RMA.

Return Call The name of a Bell Atlantic service. Return Call automatically redials the number of the last person who called your business — whether you were able to answer the phone or not. If that number is busy, Return Call continues trying to get through for up to 30 minutes — without interrupting your incoming or outgoing calls. It signals you with a special ring when a connection is made. Besides making a return follow-up call quick and simple, this service also lets you call back anyone who hung up when you couldn't pick up the phone in time.

Return Loss A measure of the similarity of the impedance of a transmission line and the impedance at its termination. It is a ratio, expressed in decibels, of the power of outgoing signal to the power of the signal reflected back from an impedance discontinuity.

Return Material Authorization Number RMA. A code number provided by the seller as a prerequisite to returning product for either repair or refund. An indispensable tracking procedure, it operates like a purchase order system. If you return computer or telephone equipment without an RMA, chances are your equipment will be lost. RMA is also called a Return Authorization (RA) number.

Return To Zero RZ. Method of transmitting binary information such that, after each encoded bit, voltage returns to the zero level.

Return-To-Zero Code A code form having two information states called "zero" and "one" in which the signal returns to a rest state during a portion of the bit period.

Revenue Accounting Office RAO. A telephone company center using mainframe computers for billing other data processing. Functions performed include receipt and processing of AMA (Automatic Message Accounting) data and preparation of the subscriber's bill. Definition from Bellcore in reference to its concept of the Advanced Intelligent Network.

Revenue Volume Pricing Plan AT&T's Revenue Volume Pricing Plan gives discounts based on total monthly 800 and 900 billing after all other term discounts have been taken. One of two plans used by aggregators to resell 800 services, the other is Customer Specific Term Plan.

Revenue Requirement How much money a regulated phone company is allowed to earn is typically determined by its rate base (depreciated value of its assets). It is allowed to earn a percentage on its rate — just as you earn interest on your rate base (what you have deposited in the bank). So how the regulation works is (in principle) simple: Figure what the phone company's assets are. Figure what percentage you want the phone company to earn on its assets. Figure the calculation. Bingo you have the revenue requirement. Except you have to allow it to pay its expenses. So that gets added onto the revenue requirement. The formula is amount of return (rate base times rate of return — ROR) plus operations expenses. See also ROE.

Reverse Battery Signaling A type of loop signaling in which battery and ground are reversed on the tip and ring of the loop to give an "off-hook" signal when the called party answers. Some systems employ reverse battery, either for a short period or until the call is finished, to indicate that it is a toll call. In some PBXs this is used to provide toll diversion.

Reverse Battery Supervision A way of telling the originating central office that the called telephone has been answered (i.e. it has gone off-hook). The voltage of the line at the originating end is reversed. Reverse battery supervision, which puts a signal at the user's premises, is very useful for devices like call accounting systems (knowing precisely when to begin the billing cycle) and telemarketing systems (knowing precisely when to transfer the machined-dialed call over to the operator). See also ANSWER SUPERVISION.

Reverse Channel 1. A (typically) small-bandwidth channel used for supervisory or error-control signaling. Signals are transmitted in the opposite direction to the data that is sent.
2. The channel in a dial up telephone circuit from the called party to the calling party.

Reverse Interrupt In Bisync, a control character sequence (DLE) sequence sent by a receiving station instead of ACK1 or ACK0 to request premature termination of the transmission in progress.

Reverse Matching Attaching the name and address to a phone number. A It's a job usually done by a specialized service bureau. Called "reverse" matching because the service bureaus started in business by attaching phone numbers to lists containing names and their addresses. With ANI (Automatic Number Identification), we get the phone numbers of people calling us. But we don't get their names and addresses. We need to get this information for many reasons. The obvious being that getting this information on-line and fast saves asking the caller for it and typing all the stuff in. That saves time on the phone — as much as 20 seconds. And fewer questions about boring stuff like phone number, address, city, state, zip means less typing time (also called data entry time, less clerking time) and more time to explain the specials we're selling today. In short, the fewer questions we ask, the less we type and the more stuff we can sell. Reverse matching can be done instantly on-line via a direct data hookup to a distant specialized service bureau or it can be done at the end of the month when we receive our 800 phone bill containing the phone numbers of the people who called us that month.

Reverse Operation Briefly running a shredder in reverse to clear jams.

Reverse Transfer An Inter-Tel term for a phone feature in which a call on common hold at any phone may be retrieved from any phone anywhere in the phone system.

Reverse Video A video display with all the characters reversed. Characters which are normally white on the screen appear black. And blacks appear white. Reverse video is used to emphasize or enhance things — like those characters to be printed in italics or bold.

Revertive Pulse Ground pulses sent back to the sender in the originating panel office from the various selector frames to control the selection process.

Revertive Pulsing In telephone networks, a means of controlling distant switching selections by pulsing, in which the near end receives signals from the far end.

Revisable-Form Document An electronic document with its formatting information intact, readable and modifiable.

Rewritable Optical Disks They look like CD-ROMS but they're not. On one side you can store 284 megabytes or 335 megabytes depending on how large you make the sectors — either 512-byte sectors or 1,024-byte, respectively. All the optical disks conform to standards set up by ISO. They fit between Winchester hard drives and tape drives. Compared with hard drives, they are very very slow. And the optical drives are not cheap — at least $4,000 at this writing. However, the optical disks are cheap and make reasonably good secondary storage devices.

REX Routine Exercise.

RF Radio Frequency. Electromagnetic waves operating between 10 kHz and 3 MHz propagated without guide (wire or cable) in free space. If you have a home computer that lets you use your home TV set as a video display device, then the computer has an RF Generator. This means that this device is generating an RF carrier to carry the video signal information.

RFAC Restricted Forced Authorization Code. A forced authorization code that is valid only if it is entered from a specific telephone extension.

RFC Request For Comment. The development of TCP/IP standards, procedures and specifications is done via this mechanism. RFCs are documents that progress through several development stages, under the control of IETF, until they are finalized or discarded. RFC#### documents Internet "Request For Comment" documents (i.e., RFC822, RFC1521, etc.). The contents of an RFC may range from an official standardized protocol specification to research results or proposals. A set of papers in which the Internet's standards, proposed standards and generally agreed-upon ideas are documented and published. See the RFCs below.

RFC 1144 This RFC (Request For Comment) will provide overhead compression for the TCP/IP protocol down to 5 octets. It does this by anticipating that the next packet in a file transfer sequence will have the same address as the previous and will have the same sequence number plus one. This compression technique will be useful where SDLC encapsulation, or other bridging protocol encapsulation, is being used with low-speed PVCs (Private Virtual Circuits.) In these cases, the slight increase in processing power to perform the compression is more than balanced by the increase in application performance and throughput.

RFC 1294 This Request For Comment is Inverse ARP, which allows the automatic discovery of the addresses on the router at each end of another router's DLCIs. Right now, the RFC only applies to IP, but some equipment vendors have already expanded the protocol support to include Novell, AppleTalk, Vines, and DECnet. The benefit of the RFC is to simplify network configuration.

RFC 1315. This Request For Comment is the frame relay MIB (management information database), which standardizes what management information is made available on frame relay devices and where/how that information is accessed. This simplifies the process of integrating frame relay devices into your network monitoring and management process and programs.

RFC 1490 This Request For Comment, RFC 1294, now renumbered RFC 1490, is for multiprotocol encapsulation. The bottom line benefits are to increase interoperability between frame relay devices from different vendors. This means that you can use one vendor's routers (or other equipment type) at some locations, and a different vendor's equipment at other locations. This ability to mix and match allows you to pick the best and most cost-effective tool for the job.

RFC 1577 Under control of the IETF, Request For Comments (RFCs) are documents used to develop standards, procedures and specifications for TCP/IP. RFC 1577 is the document for classical IP.

RFC 1695 Definitions of Managed Objects for ATM Management or AToM MIB.

RFD Request For Discussion. A period of time during which comments on a particular subject are solicited. An Internet term.

RFI 1. Request For Information. General notification of an intended purchase of equipment or equipment and lines sent to potential suppliers to determine interest and solicit general descriptive product materials, but not prices or a formal request. See RFQ for a detailed explanation.

2. Radio Frequency Interference. All computer equipment generates radio frequency signals. The FCC regulates the amount of RFI a computing device can leak past its shielding. A Class A device is sufficient for office use. A Class B is a more stringent classification for home equipment use. See EMI and RADIO FREQUENCY INTERFERENCE.

RFP Request For Proposal. A detailed document prepared by a buyer defining his requirements for service and equipment sent to one or several vendors. A vendor's response to an RFP will typically be binding on the vendor, i.e. he will be obliged to deliver what he says in his RFP at the prices and following the conditions explained in that RFP. See RFQ for a detailed explanation.

RFQ Request For Quotation. A document prepared by a buyer defining his needs for service and equipment in fairly broad terms and sent to one or several vendors. The RFQ is much less detailed than the RFP. Let's start at the beginning of the buying process.

We have a buyer who wants a phone system. His first step may be to issue a formal or informal RFI — Request For Information. In effect, the RFI says "Please tell me what you have. I have a vague idea of what I want but I don't know exactly what is available to suit my needs. Please send me some information."

After a buyer gets his responses to his RFIs, he may issue an RFQ — Request For Quotation. An RFQ may include a tentative configuration of the type of phone system the user wants, plus some listing of features the buyer is interested in. In the RFQ, the buyer asks for a "ballpark" (approximation) of the possible price for such a system. Usually the "price" is within plus or minus 10% of where it will eventually be in the final configuration. In short, an RFQ's purpose is not to buy, but to find out what's out there and what it might cost. The purpose may be to allocate a budget or to put aside some money for the forthcoming purchase.

An RFP — Request For Proposal — is much more formal and definitive. Its purpose is simple. The buyer wants to buy something. The RFP contains a list of what the buyer wants, when he wants it, how it should be installed, how it should delivered, what financing may be necessary. It is now up to the vendor/s to respond with their configuration, their precise prices and their terms and conditions of sale. Whatever the vendor responds with — called a Response to an RFP — constitutes a definite offer. At this point, the buyer can negotiate the terms of the vendor's Response to his RFP. This will lead to the writing, and eventual signing of a contract. Or the buyer may simply decide to accept the Vendor's Response. Often that Response may have a line at the back of it — "I accept the terms and conditions of this response." If the buyer signs this, then the Response to the RFP becomes a valid contract.

RFS 1. Ready for Service.
2. Remote File System. The ability to mount a disk drive somewhere on a network — but it's not on your competitor.
RG-58 Coaxial cable with 50-ohm impedance used by Thinnet.
RG-62 Coaxial cable with 93-ohm impedance used by ARCnet.
RG/U RG is the military designation for coaxial cable. U stands for general utility.
RGB A color model based on the mixing of red, green, and blue — the primary additive colors used by color monitor displays and TVs. The combination and intensities of these three colors can represent the whole spectrum. Color television signals are oriented as three separate pictures: red, green and blue. Typically, they are merged together as a composite signal but for maximum quality and for computer applications the signals are segregated.
RGC 1604 An acronym used in a BellSouth proposal to the State of GA. The salesman who wrote the proposal didn't know what it was. It relates to Frame Relay Service.
RH Request Header or Response Header.
Rheostat A variable resistor.
Rhetorex A manufacturer of voice processing componentry based in Campbell, CA. Inspiration for the company's name came from the word rhetoric, which is the art of effectively using speech and language.
Rhombic Antenna An antenna composed of wire radiators describing the sides of a rhombus. It is usually terminated and unidirectional; when unterminated, it is bidirectional.
Ribbon Cable Multi-wire cable that is flat instead of round. In ribbon cable, the conductors are laid side by side. Ribbon cable can be more easily laid under carpeting because it is flat and thus, can extend phone and computer services to places otherwise hard to reach. There are disadvantages to ribbon cable. Because ribbon cable is flat, it's hard to twist its individual wire conductors around each other (thus humming can be a problem). It is hard to put a metal shielding around the twisted wire pairs. It is hard to put coax cable into ribbon cable. It is hard to make ribbon cable sufficiently strong to withstand thousands of high heels trampling it. It is hard to make ribbon cable which turns a corner... But there has been enormous progress in ribbon cable. And ribbon cable is finding greater use in buildings. These days it even carries commercial A.C. power.
Ribbon Fiber Cable A cable that accommodates one to 12 ribbons, each ribbon having 12 fibers for a cable size range of 12 to 144 fibers. Ribbon cables are designed for use in larger distribution systems where small cable size and high pulling strength are important.
Ribbon Of Highway Fiber optic cable.
RIF Rate Increase Factor: This controls the amount by which the cell transmission rate may increase upon receipt of an RM-cell.
RIFF Resource Interchange File Format. Platform-independent multimedia specification (published by Microsoft and others in 1990) that allows audio, image, animation, and other multimedia elements to be stored in a common format. See also Media Control Interface (MCI).
Right Hand Rule A rule for indicating the direction of magnetic effect. Grasp the wire with the right hand and with the thumb extended along the wire in the direction of current. The curved fingertips will indicate the direction of magnetic flow. Not totally relevant for including in this dictionary. But cute.
Right To Use See RTU.
Rightsizing Another term for re-engineering. See RE-ENGINEERING.
RILD Remote ISDN Line Drawer.
RIME RelayNet International Message Exchange, a multi-tier communications network which exchanges messages among member bulletin board systems.
Rimm Job A bogus academic study masquerading as legitimate science. Named after Marty Rimm, author of the dubious "cyberporn" study from Carnegie Mellon University that Time magazine gullibly took as gospel.
Ring 1. As in Tip and Ring. One of the two wires (the two are Tip and Ring) needed to set up a telephone connection.
2. Also a reference to the ringing of the telephone set.
3. The design of a Local Area Network (LAN) in which the wiring loops from one workstation to another, forming a circle (thus, the term "ring"). In a ring LAN, data is sent from workstation to workstation around the loop in the same direction. Each workstation (which is usually a PC) acts as a repeater by re-sending messages to the next PC in the ring. The more PC's, the slower the LAN. Network control is distributed in a ring network. Since the message passes through each PC, loss of one PC may

disable the entire network. However, most ring LANs recover very quickly should one PC die or be turned off. If it dies, you can remove it physically from the network. If it's off, the network senses that and the token ignores that machine. In some token LANs, the LAN will close around a dead workstation and join the two workstations on either side together. If you lose the PC doing the control functions, another PC will jump in and take over. This is how the IBM Token-Passing Ring works. See TOPOLOGY, BYPASS CABLE and TOKEN RING.

Ring Again The PBX remembers the last number called by a phone and will redial it when the feature is activated.

Ring Back Tone The sound you hear when you're calling someone else's phone. The tone you hear is generated by a device at your central office and may bear no relationship to the sound the phone at the other end is emitting — or not emitting. If your call didn't go through the first time, always call back at least once. See RINGING TONE.

Ring Banding A method of color coding insulated conductors by means of a small band of colored ink applied circumferentially at regular intervals along the axis of the insulated conductor.

Ring Battery Commonly unfiltered - 24 VDC source that supplies operating power to all local KSU components. Also called the B Battery.

Ring Conductor One conductor of a phone line.

Ring Cycle A ring cycle in North America is typically six seconds long, two of ringing, four of silence, then repeated.

Ring Down Box Ring down boxes, also known as CO simulators or telephone line simulators, are simple devices used for generating POTS calls — without a central office. You connect a phone to both sides of the device. When one side goes off-hook, the ring down box will "ring" the other side. When both sides are off-hook, both sides are coupled together, the line is powered and the sides can talk. Ring down boxes are available with various options and configurations. These include the ability to provide dial-tone to the caller side (required to test applications with modems, faxes, or other automated out dialing devices), caller ID, and disconnect supervision. They are generally available in one to four line sizes, although special configurations may support more. Ring-down boxes are used for giving demonstrations and testing. We use them in our test labs to test drive new computer telephony systems.

Ring Down Circuit A tie line connecting phones in which picking up one phone automatically rings the other phone. In a ringdown circuit, a ringing current (AC) is sent down the line. That current may light a lamp, set off a bell, buzz a buzzer. The idea is to alert the person at the other end to the incoming call. A ringdown circuit is often used in an elevator or other emergency situation.

Ring Down Interface A private line two-wire interface also called Loop Start Trunk.

Ring Generator A component of virtually all phone systems, ranging from large central offices to small key systems, that supplies the power to ring the bells inside phones, typically 90 volts AC at 20 Hz.

Ring Indicator Modem interface signal defined in RS-232-C which indicates to the data terminal equipment that a call is coming in.

Ring Isolator A device placed on a telephone line to disconnect the ringer when it is an idle state. It is used for noise prevention.

Ring Latency In a token-ring network, the time measured in bits at the data transmission rate, required for a signal to propagate once around the ring. Ring latency includes the signal propagation delay through the ring medium, including drop cables, plus the sum of propagation delays through each data station connected to the token-ring network. See also TOKEN-RING NETWORK.

Ring Network A network that links PBXs, computers, terminals, printers and other devices in a circular communications link. See RING.

Ring Signal The pulse ringing voltage output of the local Interrupter KSU. Typically, this signal is 105 VAC with a duty cycle of 2 seconds on and 4 seconds off.

Ring Topology A network topology in which nodes are connected to a closed loop, no terminators are required because there are no unconnected ends.

Ring Trip The process of stopping the AC ringing signal at the central office when the telephone being rung is answered.

Ring Wiring Concentrator A site through which pass the links between repeaters, for all or a portion of a ring.

Ring Wiring Configuration The same as daisy chain wiring, except the last jack is connected to jack 1, thus completing a ring.

Ringback The tone heard by a calling device when, at the called-device's end, the telephone is ringing or the system is otherwise being alerted of the incoming call.

Ringer A bell in a telephone which indicates if a phone call is coming in. These days "ringers" are electromechanical and clunky (old-style) or small and electronic (new style). The new electronic ones are cheaper, but less interesting to listen to. Most sound like bleating sheep in heat.

Ringer Equivalence Number REN. A number required in the U.S. for registering your telephone equipment with the phone company. Add together the RENs of all the telephones on a single line. The sum of those numbers should never exceed five otherwise none of your bells will work and you won't hear an incoming call. (Your central office simply doesn't send sufficient current down the line.) The alphabetic character after the number refers to the ringing frequency of the alternating current sent down the line to ring the bell. If the letter is "A", the ringer frequency is about 20 Hertz. Most single line phones have a Ringer Equivalence of 1.0A. If the letter is "B", the ringer will respond to any current coming down the line. Any other letter, and you are probably on a party line where the ringer frequency is used for party selection. In Canada, they use the term "Load Number" instead of Ringer Equivalence. The numbers are different, but the concept is the same. See LOAD NUMBER.

Ringer Isolator A device in the phone which disconnects the ringer when ringing voltage is not present.

Ringing Alternating Current (AC) sent out from the central office along the local loop to the subscriber. It's typically 70 to 90 volts at 17 Hz to 20 Hz. You can get a mild shock if you have your hands on a telephone circuit when ringing current comes along. The rest of the time, the lines are harmless.

Ringing Generator An optional KTU facility that generates the AC ringer voltage from the 24 VDC local Key System Power Supply source. Typically, this voltage is 90 to 115 (nominally 105) VAC at 30 Hertz.

Ringing Key A key that sends a ringing current.

Ringing Signal Any AC or DC signal transmitted over a line or trunk for the purpose of alerting someone or some thing at the distant end of an incoming call.

Ringing Tone A low tone which is one second ON and three seconds OFF. It indicates that ringing current is being sent by the central office to the person receiving the call. Ringing tone is not produced by the calling party's central office — but by the called party's central office. Thus, it is possible for you to hear ringing tone but for the person you are calling not to hear anything. As a general rule, if the person doesn't answer, call them a second time. Often, they'll say "The phone never rang." This will not be a lie, but simply a temporary glitch in their central office.

Ringing Transfer A PBX feature which allows you to choose which bells in a group of phones will ring when a call is coming in for that group.

Ringing Voltage In addition to talk battery, a Central Office provides ringing signaling. Ring Voltage is generally 70 to 90 volts at 17 Hz to 20 Hz. See also RINGING.

Ringmate A Nynex service which allows you to add one or two additional numbers to your existing phone number. You can tell which phone is ringing by the sound of the ringing. You can, of course, only speak on one line at once. The purpose of the service is to help you identify who's calling you and decide whether or not to answer.

Riogin Riogin is an application that provides a terminal interface between UNIX hosts using the TCP/IP network protocol. Unlike Telnet, Riogin assumes the remote host is (or behaves like) a UNIX machine.

RIP Routing Information Protocol. RIP is based on distance-vector algorithms that measure the shortest path between two points on a network in terms of the number of "hop" between those points.

Rip-Cord A cord placed directly under the jacket of a cable in order to facilitate stripping (removal) of the jacket.

RIPE Reseaux IP Europeens, a group formed to coordinate and promote TCP/IP based networks in Europe. RIPE holds periodic conferences to coordinate technical issues (similar to the IEFT) as well as running a Network Control Center (NCC) to handle operational issues such as the administration of the European domain name and routing tables.

RIPscrip Graphics Emulation Supports RIPscrip Graphics Emulation. Popular on many bulletin board systems, RIPscrip Graphics Emulation allows users to view screens mixing text and graphics. On-screen buttons can be clicked to send commands to a remote system.

RISC Reduced Instruction Set Computer. A microprocessor architecture that favors the speed at which individual instructions execute over the robustness of the instruction set. Computers based on RISC use an unusual high speed processing technology that uses a far simpler set of operating commands. These commands greatly speed a computer's performance, especially for calculation-intensive operations such as those performed by scientists and computer-aided design (CAD) and computer-aided manufacture (CAM) engineers. RISC is a design that achieves high performance by doing the most common computer operations very quickly. In contrast, the microprocessors used in most PCs are based on a design called CISC (Complex Instruction Set Computing). CISC does not execute instructions as quickly as RISC but it has more commands and accomplishes more with each command. Programs written for RISC are typically not compatible with those written for CISC processors. RISC is the prevailing technology for workstations today. The RISC semiconductor was an IBM baby, born in its Yorktown Heights, NY lab in 1974. But internal arguments over, and even whether the chip should be used kept IBM fiddling while Sun and other companies decisively powered ahead. IBM got its first good RISC product, the RS/6000, to market in 1990. The PowerPC architecture is one example of a RISC microprocessor design. See also CISC.

RISC System/6000 An IBM family of workstations and servers designed to run applications developed for the UNIX operating system.

Riser The conduit or path between floors of a building into which telephone and other utility cables are placed to bring service from one floor to another. Your risers should be twice the size you ever think you'll need in the next 30 years. It's expensive to build risers after the building is built. Very expensive.

Riser Cable High strength cables intended for use in vertical shafts between floors in multi-story buildings. Such shafts are called riser shafts. (What else would they be called?)

Riser Closet The closet where riser cable is terminated and cross connected to either horizontal distribution cable or other riser cable. The riser closet houses cross connect facilities, and may contain auxiliary power supplies for terminal equipment located at the user work location. See also SATELLITE CLOSET.

Riser Subsystem The part of a premises distribution system that includes a main cable route and facilities for supporting the cable from an equipment room (often in the building) to the upper floors, or along the same floor, where it is terminated on a cross connect in a riser closet, at the network interface, or other distribution components of the campus subsystem. The subsystem can also extend out on a floor to connect a satellite closet or other satellite location.

Risk A potential liability, caused by a threat.

Risk Assessment The process of quantifying the potential impact on an organization from various security threats.

RISLU Remote Integrated Services Line Unit. One of the remoting arrangements that the AT&T 5ESS switch architecture permits. The RISLU terminates DSLs and connects to the switch DLTU via T-1.

RIT Rate of Information Transfer. The amount of information that can be communicated from a sender to a receiver in a given length of time.

RJ-ICX	Single Tie Trunk, Type I or II E&M interface, 8 position.
RJ-IDC	Single-line, 4-wire, T/R, T1/R1, 6-position.
RJ-11C/W	Single-line, 2-wire, T/R, 6 position.
RJ-14C/W	Two-line, 2-wire, T/R, T(MR)/R(MR), T(OPS)/R(OPS) 6-position.
RJ-14X	Two-line, T1/R1, T2/R2, with sliding cover, 6-position.
RJ-15C	Single-line, T/R, weatherproof, 3-position.
RJ-17C	Single-line, T/R, used in hospital critical care areas, 6 positions.
RJ-18C/W	Single-line, T/R, with Make Busy leads, 6 positions.
RJ-2DX	12 lines, 4 wire, T/R, T1/R1, 50 positions.
RJ-2EX	12 Tie trunks, 2-wire, T/R, E&M Type I, 50 position.
RJ-2FX	8 Tie trunks, 2-wire, T/R, E&M SG/SB Type II 50 position.
RJ-2GX	8 Tie trunks, 4-wire, T/R, T1/R1, E&M, Type I 50 position.
RJ-2HX	6 Tie trunks, 4-wire, T/R, T1/R1, E&M, SG/SB, Type II, 50 positions.
RJ-2MB	12 lines, 2-wire, T/R, Make Busy leads, 50 position.
RJ-21X	25 lines, 2-wire, T/R, 50 position.
RJ-25C	3 lines, 2-wire, T/R, T (MR)/R(MR), T(OPS)/R(OPS), 6 position.
RJ-26X	8 lines, 2-wire, T/R, FLL, or Programmed data, 50 position.
RJ-27X	8 lines, 2-wire, T/R, Programmed Data, 50 position.
RJ-4MB	Single-line, 2-wire, T/R, MB/MB1, PR/PC, with Make Busy. 8 position, keyed and programmed.
RJ-41M	Up to 8 multiple installations of FLL or Programmed Data. 8 position, keyed.
RJ-41S	Single-line, 2-wire, T/R, FLL or Programmed Data, 8 position, keyed.
RJ-45M	Up to 8 multiple installations of Programmed Data. 8-position, keyed.
RJ-45	Single-line, 2-wire, T/R, PR/PC, programmed data, 8 position, keyed.
RJ-48C	Single-line, 4-wire, T/R, T1/R1, 1.544 Mbps, 8 position.
RJ-48H	Up to 12 lines. 4-wire, T/R, T1/R1, 1.544 Mbps, 50 position.
RJ-48M	Up to 8 lines, 4-wire, T/R, T1/R1, 1.544 Mbps, 50 position.
RJ-48S	One or two lines. T/R or T/R, T1/R1, LADC or subrate. 8-position, keyed.
RJ-48T	Up to 25 (2-wire) or 12 (4-wire), T/R OR T/R, TI/R1; LADC or Subrate,50-position.
RJ-48X	Single-line, 4-wire, T/R, T1/R1, 1.544 Mbps, 8-position with shorting bar.
RJ-61X	Up to 4 lines, T/R, 8-position.
RJ-M8	Single private line, 2/4 wire, T/R or T/R, T1/R1, Non-registered service, 8-position, keyed, w/wo loopback.

RJ Registered Jacks. They're telephone and data plugs registered with the FCC. RJ-XX (where X is a number) are probably the most common plugs in the world. Here is a table of the most common registered jacks, courtesy the FCC. Following the table are descriptions of the most common RJ jacks.

RJ-11 RJ-11 is a six conductor modular jack that is typically wired for four conductors (i.e. four wires). The RJ-11 jack (also called plug) is the most common telephone jack in the world. The RJ-11 is typically used for connecting telephone instruments, modems and fax machines to a female RJ-11 jack on the wall or in the floor. That jack in turn is connected to twisted wire coming in from "the network" — which might be a PBX or the local telephone company central office. In a home installation, the red and green pair would be used for carrying the phone conversation and the black and white might be used for carrying low voltage from a plugged-in power transformer to light buttons on the phone. In many offices, the tip and ring were used for the voice conversation and the black and white were used for signaling. Increasingly, these days more and more office phone systems use only one pair, i.e. the red and green conductors. See also RJ-22 and RJ-45.

RJ-14 A jack that looks and is exactly like the standard RJ-11 that you see on every single line telephone. Whereas the RJ-11 defines one line — with the two center, red and green, conductors being tip and ring, the RJ-14 defines two phone lines. One of the lines is the "normal" RJ-11 line — the red and green conductors in the center. The second line is the second set of conductors — black and yellow — on the outside.

RJ-21 Same as an RJ-21X. See RJ-21X.

RJ-21X An Amphenol connector under a different name. Here's the explanation: Amphenol is a manufacturer of electrical and electronic connectors. They make many different models, many of which are compatible with products made by other companies. Their most famous connector is probably the 25-pair connector used on 1A2 key telephones and for connecting cables to many electronic key systems and PBXs. The telephone companies call the 25-pair Amphenol connector used as a demarcation point the RJ-21X. The RJ-21X connector is made by other companies including 3M, AMP and TRW. People in the phone business often call non-amphenol 25-pair connectors, amphenol connectors.

RJ-22 RJ-22 is a four position modular jack that is typically used for connecting telephone handsets to telephone instruments. It is always wired with four conductors (also called wires). It is different and slightly smaller than the more common RJ-11 which is typically used for connecting telephone instruments, modems and fax machines to a female RJ-11 jack on the wall or in the floor. That jack in turn is connected to twisted wire coming in from "the network" — which might be a PBX or the local telephone company central office. See RJ-11 and RJ-45.

RJ-45 The RJ-45 is the 8-pin connector used for data transmission over standard telephone wire. That wire could be flat or twisted. And it's very important that you know what you're working with. You can easily use flat wire for serial data communications up to 19.2 Kbps. Up to that speed you're connecting with your wire to a data PBX, a modem, a printer or a printer buffer. If you wish to connect to a 10BaseT local area network, which you also do with a RJ-45, you must use twisted wire. You can typically tell the difference by looking at the cable. If it's flat grey satin (like a typical phone wire, only bigger) than it's probably untwisted. If it's circular, then it's probably twisted and therefore good for LANs. RJ-45 connectors come into two varieties — keyed and non-keyed. Keyed means that the male RJ-45 plug has a small, square bump on its end and the female RJ-45 plug is shaped to accommodate the plug. A keyed RJ-45 plug will not fit into a female, non-keyed (i.e. normal) RJ-45. See RJ-11 and RJ-22.

RJ-48C An 8-position keyed plug most commonly used for connecting T-1 circuits. The RJ-48C is an 8-position plug with four-wires (two for transmit, two for receive) commonly connected. When the phone company delivers T-1 to your offices, it usually terminates its T-1 circuit on a RJ-48C. And it expects you to connect that RJ-48C to your phone system or T-1 channel bank and then to your phone system.

RJE Remote Job Entry. A Remote Job Entry terminal is used for the transmission of "batch" data to a remote computer system. Processed information is then returned to the printer in the terminal. This type of processing from a remote site is a standard method of data transmission. See IBM.

RJXXX Registered Jack.

RLC ReLease Complete message. The sixth ISUP call set-up message. A message sent in either direction in response to the receipt of a Release Message, or if appropriate to, a Reset Circuit Message when the circuit concerned has been brought into the idle condition. See ISUP and COMMON CHANNEL SIGNALING.

RLCM Remote Line Concentrating Module.

RLE RLE, an acronym for Run Length Encoding, is a loseless compression algorithm, used in the BinHex version of MIME. See BinHex and MIME.

Run Length Encoding is also a compression algorithm used in Group III (fax-standard) compression. A scanner reads a horizontal line of pixels — the "scan line" — and counts the black pixels. This is the "run length" or "run count." The same is then done for white pixels. A scan line might have a run of 230 white pixels, followed by a run of 23 black, followed by 45 white, etc. The result is a set of run lengths for the line. That's RLE. The different run lengths are then given codes — shorter ones for the most frequently occurring run lengths, longer ones for the unique run lengths — to represent them. That's called "Huffman Coding." The codes are set in Group III compression; some compression schemes can assign new codes for each image.

RLL Run Length Limited. A type of data coding used for disk drives. The term Run Length Limited derives from the fact that the techniques limit the distance (run length) between magnetic flux reversals on the disk platter. An RLL certified hard drive can use an MFM controller card but the storage capacity and the data transfer rate will be reduced.

RLR Receive Loudness Rating.

RLT Release Link Trunks.

RM Resource Management: Resource Management is the management of critical resources in an ATM network. Two critical resources are buffer space and trunk bandwidth. Provisioning may be used to allocate network resources in order to separate traffic flows according to service characteristics. VPCs play a key role in resource management. By reserving capacity on VPCs, the processing required to establish individual VCCs is reduced. Refer to RM-cell.

RM-Cell An ATM term. Resource Management Cell: Information about the state of the network-like bandwidth availability, state of congestion, and impending congestion, is conveyed to the source through special control cells called Resource Management Cells (RM-cells).

RMA Returned Merchandise Authorization. A code number provided by the seller as a prerequisite to returning product for either repair or refund. An indispensable tracking procedure, it operates like a purchase order system. If you return computer or telephone equipment without an RMA number, chances are that it will be lost. RMA is also called a Return Authorization (RA) number.

RMAS Remote Memory Administration System.

RMATS Remote Maintenance and Test System. That equipment and programming used to run, maintain and test a telephone system remotely — usually by dialing in on a special phone line.

RMON Remote Network Monitoring Specification. A simple network management protocol used to manage networks remotely. It provides multi vendor interoperability between monitoring devices and management stations.

RMS Root Mean Square. Method of measuring amplifier power.

RMS-D1 Remote Measurement System Digital 1.

RMU Remote Mask Unit.

RNR Abbreviation for not ready to receive.

RO 1. Receive Only.

2. An ATM term. Read-Only: Attributes which are read-only can not be written by Network Management. Only the PNNI Protocol

entity may change the value of a read-only attribute. Network Management entities are restricted to only reading such read-only attributes. Read-only attributes are typically for statistical information, including reporting result of actions taken by auto-configuration.

Road Kill Companies who go out of business because they couldn't figure out how to compete in the all digital, mega-channel, interactive marketplace. See also ROAD PIZZA

Road Map Software that enables easier navigation to information desired. Especially helpful for Internet users.

Road Pizza Companies who get run over by their competitors. Also a code name for a previous new product from Apple Computer.

Roadblocks Slang for legislation, or lack thereof, which inhibits rather than promotes the growth of the markets using interactive multimedia.

Roadhog Any company trying to dominate the information highway through control of pieces of the infrastructure.

Roaming Using your cellular phone in a city besides the one in which you live. Roaming usually incurs extra charges. Roaming prices are usually at a premium to local home area prices.

Robbed-Bit Signaling This explanation from Gary Maier of Dianatel: ISDN is the key to future sophisticated telephone network services with its dynamic, highly configurable T-1 connection (also called PRI connection). Since T-1 is a common method of carrying 24 telephone circuits, many wonder about the uses for ISDN, especially when they learn ISDN signaling requires an entire voice channel, reducing today's T-1 from 24 voice channels to 23. But the popular signaling mechanism of "robbed bit" signaling in T-1 has serious limitations. Robbed bit signaling typically uses bits known as the A and B bits. These bits are sent by each side of a T-1 termination and are buried in the voice data of each voice channel in the T-1 circuit. Hence the term "robbed bit" as the bits are stolen from the voice data. Since the bits are stolen so infrequently, the voice quality is not compromised by much. But the available signaling combinations are limited to ringing, hang up, wink, and pulse digit dialing. In fact, the limitations are obvious when one recognizes DNIS and ANI information are sent as DTMF tones.

This introduces a problem: time. Each DTMF tone requires at least 100 milliseconds to send, which in a DNIS and ANI situation with 20 DTMFs will take at least two full seconds. There is also a margin for error in transmission or detection, resulting in DNIS or ANI failures. With the explosion of telephone related services, the telephone companies are turning to ISDN PRI to provide the more complicated and exact signaling required for new services. ISDN employs a more robust method of signaling. ISDN uses a T-1 circuit as 23 voice channels and one signaling channel. The term 23B plus D refers to 23 bearer (voice) channels and 1 Data (signaling) channel. The data channel carries the signaling information at a rate of 64 kilobits per second. This speed is many times greater than some of the most powerful modems available. Because of this high speed, telephone calls can be placed more quickly, and because of the protocol used, DNIS or ANI transmission failures are impossible.

Additionally, since no bits are "robbed" from the voice channels, the voice quality is better than that of Robbed Bit signaling on today's T-1 circuits. Also, computer modems and high speed faxes can use the voice channel for sending digital data instead of the traditional analog bit "noise." Therefore, ISDN PRI offers the end user countless new service capabilities. One channel could be used for faxing, another for modem data, several for video, another for a LAN and the remainder for voice. Suddenly, the average T-1 circuit becomes a pipeline for all communications! Increasingly long distance carriers are using ISDN PRI to provide inbound 800 calls with ANI and DNIS and re-routing skills. See BIT ROBBING.

Robust A term used by telecommunications switch manufacturers to describe the alleged hefty quality of their network connections — especially their switch-to-host links. The price of their link is often directly related to the number of times the manufacturer uses the word "robust" in a customer presentation.

Rodent Rubber Another term for a B connector. See B CONNECTOR.

ROE Regulatory commission authorized allowed rate of return on equity. See REVENUE REQUIREMENT.

ROFL Abbreviation for "Rolling On the Floor, Laughing;" commonly used on E-mail and BBSs (Bulletin Board Systems).

ROH Receiver Off Hook.

ROI Return On Investment.

Roll-Call Polling A technique in which every station is interrogated sequentially by a central computer system.

Roll About A totally self-contained videoconferencing system consisting of the codec, video monitor, audio system, network interfaces and other components. These roll about systems can, in theory, be moved from room-to-room but in fact are not because they are electronic equipment that does not benefit from jostling. It's also heavy.

Rollover Lines You receive many incoming calls. You don't want to miss a call, so you set your phone lines up to roll over, also called hunt, also called ISG in telephonese. You order five lines in hunt. The calls come into the first. If the first one is busy, the second rings. If it's busy, the third rings. If they're all busy, then the caller receives a busy.

There are two types of hunting — Sequential and Circular hunting. Sequential hunting starts at the number dialed and ends at the last number in the group. Circular hunting hunts all the lines in the hunting group, regardless of the starting point. Circular hunting, according to our understanding, circles only once (though your phone company may be able to program it circle a couple of times). The differences between sequential and circular are subtle. Circular seems to work better for large groups of numbers.

You don't need consecutive phone numbers to do rollovers. Nowadays you can roll lines forwards, backwards and jump around. Rollovers are now done in software. This also has its downside, since software fails. Our recommendation: Test your rollovers at least twice a day. In particular, test that your callers ultimately get a busy if all your lines are busy. Nothing worse than your customer should receive a ring-no-answer.

Rolm A telephone equipment manufacturer based in Santa Clara, CA and owned by Siemens.

RolmNet II Analysis And Support A service provided by the Rolm Analysis Center (RAC) that analyzes current call patterns and trunking facilities, chooses alternative facilities, and recommends the most cost effective long distance network based on the customer's needs. It also compares the present facilities to the proposed network and calculates cost savings.

ROLMphone Data RPD Interface A group of printed circuit cards forming the interface between Rolm digital desktop products and the CBX (Rolm's name for its PBX) for transmission of digital data. It accommodates ROLMphone telephones equipped with DataCom Modules, as well as Cedar, Cypress and Juniper desktop products. Compare with ROLMPHONE INTERFACE.

ROLMphone Interface RPI. A group of printed circuit cards forming the interface between Rolm digital desktop products and the CBX for transmission of digitized voice signals. In addition to the ROLMphones, the RPI also accommodates connections to Cypress, Cedar and Juniper. Compare with ROLMPHONE DATA INTERFACE.

ROLMphones A family of electronic telephones which only work on the Rolm line of PBXs (which Rolm calls CBXs). As of writing, these phones include the ROLMphone 400, ROLMphone 240 and ROLMphone 120. In addition, Rolm has telephone/workstation devices called Cypress, Cedar and Juniper and will probably have more by the time you read this.

Rolodex A Trademarked product which started life as paper card based device to keep names and address on. Now it has become more of a generic name to connote software to let you look up peoples' phone numbers and addresses. Software to do this is also called PIM — for Personal Information Manager.

ROM Read Only Memory. Computer memory which can only be read from. New data cannot be entered and the existing data is non-volatile. This means it stays there even when power is turned off. A ROM is a memory device which is programmed at the factory and whose contents thereafter cannot be altered. In contrast is the device called RAM, whose contents can be altered. See READ ONLY MEMORY and MICROPROCESSOR.

ROM Font The ROM Font is your PC's type font. It consists of a set of 256 characters which cannot be edited — unless you are running in video mode, in which case you can design your own type font.

Roofing Filter A low-pass filter used to reduce unwanted higher frequencies.

Room Cut-Off Hotel/motel guest telephones restricted from outgoing calls when the guest room is unoccupied.

Room Status And Selection Provides the capability to store and display the occupancy and cleaning status and the type number of each guest room. This helps housekeeping management, maid locating and room selection. Also, communications between the front desk and the housekeeper are speeded up via real-time maid activity and checkout audit printouts to indicate which rooms need cleaning next. The occupancy status is normally changed by the maid or inspector dialing from the room telephone.

Root Directory The top-level directory of a PC disk, hard or floppy. The root directory is created when you format the disk. From the root directory, you can create files and other directories.

ROSE 1. Remote Operations Service Element. An application layer protocol that provides the capability to perform remote operations at a remote process. Definition from Bellcore in reference to its concept of the Advanced Intelligent Network.
2. As in Bodin. The very thoughtful lady who helps us, our friends and customers. We couldn't survive without Rose. Call her and tell her that you also can't survive without Rose. 212-691-8215.

Rotary Dial The circular telephone dial. As it returns to its normal position (after being turned) it opens and closes the electrical loop sent by the central office. Rotary dial telephones momentarily break the DC circuit (stop current flow) to represent the digits dialed. The circuit is broken three times for the digit 3. The CO counts these evenly-spaced breaks and determines which digit has been dialed. You can hear the "clicks". The number "seven," for example consists of seven "opens and closes," or seven clicks. You can dial on a rotary phone without using the rotary dial. Simply depress the switch hook quickly, allowing pauses in between to signify that you're about to send a new digit. It's a good party trick.

Rotary Dial Calling The telephone system will accept dialing from conventional rotary dial sets.

Rotary Hunt You buy several phone lines. Let's say 212-691-8215, 212-691-8216, 212-691-8217, 212-691-8218. Someone dials you on your main number — 212-691-8215. It's busy. (That's our number.) The central office slides the call over to 212-691-8216. If that number is busy, it slides it over to 212-691-8217, and so on. This is called rotary hunt. It hunts to the next line in the rotary group. In the old days, the phone lines you could rotary hunt to had to be in numerical sequence. But now with modern stored program control central offices, your lines in rotary hunt can be very different as long as they're all on the same exchange.

Rotary Output To Central Office Most central offices are equipped to provide tone dial service. In cases where the telephone company central office trunks are not designed to accept tone signaling, your on premise phone system (PBX, key system or single line phone) will translate the number entered by a phone in tones into rotary dial pulses which can be processed by the central office.

Rotary Register A Rolm CBX printed circuit card that converts off-hook bits corresponding to dial pulses into digital words suitable for computer processing.

Rotary Sender A Rolm CBX printed circuit card that receives digital code words from the computer and converts them into digital pulses. These are routed to an interface card which performs the proper out pulsing conversion.

Rotating Cylinder (Drum) Scanner A scanning technique using a drum and a photocell scan head. The original is attached to the drum, enabling the scan head to travel along the length of the document. Reflected light from the document is concentrated on the scanner photocell, which causes an analog signal.

Rotating Helical Aperture Scanner Original is illuminated by a lamp when fed onto the platen, via a mirror and lens system, the document's image is focused first through a fixed horizontal slot, then through a rotating spiral slit disk series, and finally onto a photocell to generate an analogous electrical current.

Rotational Mailboxes Information only mailboxes whose information is automatically changed on a time sensitive or usage sensitive basis.

ROTFL I'm "Rolling on the Floor, Laughing." Used in e-mail.

ROTL 1. Remote Office Test Line. Provides the capability to originate automatic inter office trunk transmission test calls under the automatic control of CAROT from a remote location.

2. A popular online abbreviation, shorthand for "Rolling On The Floor Laughing"; an appropriate typed response to a particularly amusing online remark. Other common Net acronyms include IMHO ("In My Humble Opinion") and IMNSHO ("In My Not-So-Humble pinion").

Rotor The rotating part of a motor or other electrical machines.

ROTS Rotary Out-Trunk Switches.

Round Robin This is a method of distributing incoming calls to a bunch of people. This method selects the next agent on the list following the agent that received the last call. See also TOP DOWN and LONGEST AVAILABLE.

Roundtrip Propagation Delay Roundtrip propagation delay from a burst modem to a burst modem will be about 470 milliseconds to 570 milliseconds (About half a second). See SATELLITE TRANSMISSION DELAY.

Routable Protocols Protocols, such as TCP/IP, DECnet, and XNS, that support Network Layer addressing. Packets constructed using these protocols contain information about how data should move through a network. This information, carried in the NLA (Network Layer Address) field of the packet, is used by internetworking devices to make routing decisions.

Route The path that a message takes.

Route 66 A colloquial term for the Internet, with something interesting everywhere along the way.

Route Advance This feature routes outgoing calls over alternate long distance lines when the first choice trunk group is busy. The phone user selects the first choice route by dialing the corresponding access code. The phone equipment automatically advances to alternate trunks and trunk groups, based on the user's class of service. Route advance is a more primitive form of least cost routing. See LEAST COST ROUTING.

Route Daemon A program that runs under 4.2 or 4.3BSD UNIX systems (and derived operating systems) to propagate routes among machines n a local area network. Pronounced "route-dee."

Route Discovery Process through which a brouter can learn LAN topology by passing information about its address and the LANs it connects and receiving the same information from others.

Router Droppings The inclusions added to e-mail messages when a server or recipient cannot be found. Cryptic and foul-looking, their meaning is usually impossible to fathom. Also called "daemon droppings."

Route List A sequence of trunk groups that can be searched for a particular route. This list is comprised of trunk groups and configuration attributes (e.g. Class of Service) governing the use of a particular trunk group.

Route Indicator An address or group of characters in the heading of a message defining the final circuit or terminal to which the message is to be delivered.

Route Optimization The Rolm way of saying Least Cost Routing. See LEAST COST ROUTING and ROUTE OPTIMIZATION II.

Route Optimization II A Rolm product which offers automatic long distance cost control through Least Cost Routing, Queuing, Toll Restriction and the use of alternate long distance carriers. See also LEAST COST ROUTING, QUEUING and TOLL RESTRICTION.

Route Server An ATM term. A physical device that runs one or more network layer routing protocols, and which uses a route query protocol in order to provide network layer routing forwarding descriptions to clients.

Route Xpander Card A board manufactured by IBM for insertion into a PC which provides the PC with a wide area interface to a frame relay network, including handling all of the necessary protocol encapsulation.

Routed Protocol A protocol that can be routed by a router. To do so a router must understand the logical internetwork as perceived by that routed protocol. Examples of routed protocols include DECnet, AppleTalk, and IP.

Router 1. A system level function that directs a call to an application.
2. An interface between two networks. While routers are like bridges, they work differently. Routers provide more functionality than bridges. For example, they can find the best route between any two networks, even if there are several different networks in between. Routers provide network management capabilities such as load balancing, partitioning of the network, use statistics, communication priority, and trouble shooting tools that allow network managers to detect and correct problems even in a complex network of networks. Given these capabilities, routers are often used in building wide area or enterprise wide networks.
There are two types of routers, protocol dependent and protocol independent. Protocol dependent routers rely on the end stations (computers on a LAN) for routing information. Computers tell the routers where (on which network) the destination computer is located and the routers find the best way to get there. This means that routers must understand the language (LAN protocol) the computers are talking, this makes them protocol dependent. Protocol independent routers discover the location of the destination device on their own and without any assistance from the communicating computers. They therefore, do not need to understand the language the computers are using which makes them protocol independent.
Protocol independent routers are easier to install and use. Most importantly, they can route non routable protocols, like those used by IBM networks. See also INTERNETWORKING and BRIDGES.

Routine A program, or a sequence of instructions called by a program, that has some general or frequent use.

Routing The process of selecting the correct circuit path for a message.

Routing Code 1. Another name for area code. See DN (as in directory number).
2. The combination of characters or digits required by the switching system to route a transmission to its desired destination.

Routing Computation The process of applying a mathematical algorithm to a topology database to compute routes. There are many types of routing computations that may be used. The Djikstra algorithm is one particular example of a possible routing computation.

Routing Constraint A generic term that refers to either a topology constraint or a path constraint.

Routing Data Base Distance table in DNA.

Routing Flexibility The ability to send information over various network paths to avoid congestion and use portions of a total network that would otherwise be idle.

Routing Label The part of a signaling message identifying its destination.

Routing Metric The method by which a routing algorithm determines that one route is better than another. This information is stored in routing tables. Such tables include reliability, delay bandwidth, load, MTUs, communication costs, and hop count.

Routing Protocol A general term indicating a protocol run between routers and/or route servers in order to exchange information used to allow figuring of routes. The result of the routing computation will be one or more forwarding descriptions. In short, a protocol that accomplishes routing through the implementation of a specific routing algorithm. Examples of routing protocols include IGRP, RIP, and OSPF.

Routing Table 1. Incoming Phone Calls: A routing table is a user definable list of steps which are treatment instructions for an incoming call. Ideally these steps should be addressed and the call treatment begun before the call is answered. A routing table should consist of a minimum of steps that include agent groups, voice response devices, announcements (delay and informational) music on hold, intraflow and interflow steps, route dialing (machine based call forwarding). A significant issue in the structure of routing tables is "look-back" capability, where no one previously interrogated resource is abandoned by the system (i.e. an agent group is now ignored, even though an agent is now available, because the ACD does not consider previous steps in the routing table).
2. Outgoing Phone Calls: For a specific calling site, this table lists the long distance routing choices for each location to be dialed. There may be only one choice (route) listed for some or all destinations or there may be several choices for some destinations. (It depends how many outgoing lines and how many outgoing trunk groups you have.) If there are several choices then they will be ranked by some criteria (least cost, best quality, etc.).
3. In data communications, a routing table is a table in a router or some other internetworking device that keeps track of routes (and, in some cases, metrics associated with those routes) to particular network destinations. See ROUTING METRIC.

Routing Update A message sent from a router to indicate network and associated cost information. Routing updates are typically sent at regular intervals and after a change in network topology.

RPC Remote Procedure Call. 1. A protocol governing the method with which an application activates processes on other nodes and retrieves the results. A popular paradigm for implementing the client-server model of distributed computing. A request is sent to a remote system to execute a designated procedure, using arguments supplied, and the result returned to the caller. There are many variations and subtleties, resulting in a variety of different RPC protocols.
2. A mechanism defined by Sun Microsystems and described in RFC-1057 that provides a standard for initiating and controlling processes on remote or distributed computer systems.

RPD See ROLMPHONE DATA INTERFACE.

RPI See ROLMPHONE INTERFACE.

RPG Report Program Generator. A computer language for processing large data files.

RPL ROLM Processing Language. A high-level programming language that gives the programmer access to the underlying CBX II 9000 processor architecture so that real-time communications capabilities of the processor microcode can be used.

RPM Remote Packet Module.

RPOA Recognized Private Operating Agency. A term used by the ITU-T to describe those companies designated as operating telephone companies — if the country's phone networks are not run by government-owned administrations, such as the PTTs in Europe. A recognized Private Operating Agency is an organization that handles internetwork communications (e.g., long distance carriers). To identify some RPOAs, you must dial a prefix before your outgoing directory number. An RPOA can also refer to one or more DNICs that will connect two X.25 endpoints. For ISDN X.25, an RPOA is usually the DNIC for the ISDN's long distance carrier.

RPQ Request for Price Quotation. Solicitation for pricing for a specific component, software product, service or system. See also RFQ.

RQC Repair and Quick Clean. A term in the industry which repairs telecom equipment. It means all equipment is repaired and fully tested with a burn-in (if required) and an operational systems test. It also includes minor cosmetic cleaning of the unit. Definition courtesy Nitsuko America. See also LIKE NEW REPAIR AND UPDATE and REPAIR, UPDATE AND REFURBISH.

RR Abbreviation for Ready to Receive.

RS 1. Recommended Standard, as in RS-232.
2. Record Separator, in data processing terms.
3. An ATM term. Remote single-layer (Test Method): An abstract test method in which the upper tester is within the system under test and there is a point of control and observation at the upper service boundary of the Implementation Under Test (IUT) for testing one protocol layer. Test events are specified in terms of the abstract service primitives (ASP) and/or protocol data units at the lower tester PCO.

RS-170 The EIA (Electronics Industries Association) standard for the combination of signals required to form NTSC monochrome (black and white) video.

RS-170A The EIA standard for the combination of signals required to form NTSC color video. It has the same base as RS-170, with the addition of color information.

RS-232-C Also known as RS-232 and in its latest version EIA/TIA-232-E. A set of standards specifying various electrical and mechanical characteristics for interfaces between computers, terminals and modems. The RS-232-C standard, which was developed by the EIA (Electrical Industries Association), defines the mechanical and electrical characteristics for connecting DTE and DCE data communications devices. It defines what the interface does, circuit functions and their corresponding connector pin assignments. The standard applies to both synchronous and asynchronous binary data transmission.
Most personal computers use the RS-232-C interface to attach modems. Some printers also use RS-232-C. You should be aware that despite the fact that RS-232-C is an EIA "standard," you cannot necessarily connect one RS-232-C equipped device to another one (like a printer to a computer) and expect them to work intelligently together. That's because different RS-232-C devices are often wired or pinned differently and may also use different wires for different functions. The "traditional" RS-232-C

plug has 25 pins. The new IBM PC AT, most AT compatibles and the Toshiba T1100 Plus have a "new" RS-232-C plug with only nine pins. This smaller plug does essentially the same thing as its bigger cousin, but you need an adapter cable to connect one to another. They're widely available. See also interface and the RS-232-C diagram. See EIA/TIA-232-E and the APPENDIX for description of the pins and what they do.

RS-232 Fax Server A RS-232 fax server is software which connects a network server to a fax machine via an RS-232 port attached to the fax machine. There are not many fax machines with R2-232 so you need to chose carefully. The idea of this arrangement is to let users send faxes directly from their own PC via the fax server via the attached fax machine, or directly from the fax machine. Users can also use the fax machine as a scanner.

RS-328 October, 1966 the Electronic Industries Association issues its first fax standard: the EIA Standard RS-328, Message Facsimile Equipment for Operation on Switched Voice Facilities Using Data Communications Equipment. The Group 1 standard, as it later became known, made possible the more generalized business use of fax. Transmission was analog and it took four to six minutes to send a page.

RS-366 An EIA interface standard for auto dialing.

RS-422 A standard operating in conjunction with RS-449 that specifies electrical characteristics for balanced circuits (circuits with their own ground leads). RS-422 (now known as EIA/TIA-422) is a balanced electrical implementation of RS-449 for high-speed data transmission. RS stands for recommended standard.

RS-422-A Electrical characteristics of balanced-voltage digital interface circuits.

RS-423 A standard operating in conjunction with RS-449 that specifies electrical characteristics for unbalanced circuits (circuits using common or shared grounding techniques). Another EIA standard for DTE/DCE connection which specifies interface requirements for expanded transmission speeds (up to 2 Mbps), longer cable lengths, and 10 additional functions. RS-449 applies to binary, serial, synchronous or asynchronous communications. Half- and full-duplex modes are accommodated and transmission can be over 2- or 4-wire facilities such as point-to-point or multipoint lines. The physical connection between DTE and DCE is made through a 37-contact connector; a separate 9-connector is specified to service secondary channel interchange circuits, when used.

RS-423-A Electrical characteristics of unbalanced-voltage digital interface circuits. RS-423 (now known as EIA/TIA-423) is an unbalanced electrical implementation of RS-449 for RS-232-C compatibility.

RS-449 RS-449 (now known as EIA/TIA-449) is a popular physical layer interface. Essentially a faster (up to 2 Mbps) version of RS-232-C capable of longer cable runs. Another "standard" data communications connector. This one uses 37-pins and is designed for higher speed transmission. Each signal pin has its own return line, instead of a common ground return and the signal pairs (signal, return) are balanced lines rather than a signal referenced to ground. This cable typically uses twisted pairs, while a RS-232-C cable usually doesn't.

RS-485 Describes electrical characteristics of a balanced interface used as a bus for master/slave operation, for up to 32 users. Used in industry for the Process Field Bus, and in telco management networks.

RS-499-1 Addendum 1 to RS-449. (What else?)

RSA Rural Service (or Statistical) Area. The FCC designated 428 rural markets across the United States and licensed two service providers per RSA. See also MSA.

RSC Remote Switching Center.

RSC-S Remote Switching Center-S.

RSE An ATM term. Remote Single-layer Embedded (Test Method): An abstract test method in which the upper tester is within the system under test and there is a point of control and observation at the upper service boundary of the Implementation Under Test (IUT) for testing a protocol layer or sublayer which is part of a multi-protocol IUT.

RSFG An ATM term. Route Server Functional Group: The group of functions performed to provide internetworking level functions in an MPOA System. This includes running conventional interworking Routing Protocols and providing inter-IASG destination resolution.

RSL Request and Status Links. A generic term for linking computers and PBXs. Every manufacturer of phone systems is evolving towards open architecture and their own "RSL." The term RSL, which is too passive, is being replaced with PHI (PBX Host Interface), a term coined by Probe Research. Manufacturer PHI names include:

ACL — Applications Connectivity Link — Siemens' PHI link protocol
ACT — Applied Computer Telephony — Hewlett Packard's generic application interface to PBXs
Application Bridge — Aspect Telecommunications' ACD to host computer link
ASAI — AT&T's Adjunct Switch Application Interface
CIT — Digital Equipment Corporation's Computer Integrated Telephony (works with major PBXs)
CSA — Callpath Services Architecture — IBM's Computer to PBX link
Call Frame — Harris' PBX to computer link
Callpath Host — IBM and ROLM's CICS-based integrated voice and data applications platform which links to ROLM's 9751
Callpath — IBM's announced, CICS application link to IBM's CSA, available on the AS400 in April of 1991
Callbridge — Rolm's CBX and Siemens to IBM host or non-IBM host computer link
CompuCall — Northern Telecom's DMS central office link to computer interface
CPI — Computer to PBX Interface developed by Northern Telecom and DEC
CSP — Nabnasset's Communications Services Platform
CSTA — Computer Supported Telephony Application, PHI standard from ECMA
DECags — DEC ASAI Gateway Services. Two-directional link to AT&T's Definity
DMI — AT&T's Digital Multiplexed Interface, a T-1 PBX to computer interface
HCI — Host Command Interface. Mitel's digital PBX link to DEC computer
IG — AT&T's ISDN Gateway (one direction from the switch to the host)

ITG — AT&T's Integrated Telemarketing Gateway (two directional)
ISDN/AP — NT's PHI SL1 protocol supports NT's Meridian Link PHI
Meridian Link — NT's PHI product available on the Meridian PBX
OAI — Open Application Interface. InteCom's and NEC's PHI
ONA — Open Network Architecture (for telephone central offices)
PACT — Siemens' PBX and Computer Teaming, protocols between PBXs and computers
PDI — Telenova/Lexar's Predictive Dialing Interface
SAI — Stratus Computer Switch Application Interface
SCAI — Switch to Computer Application Interface, the name given by T1S1 to PHI
SCIL — Aristacom's Switch Computer Interface Link Transaction Link
STEP — Speech and Telephony Environment for Programmers; WANG's link
Transaction Link — Rockwell's link from its Galaxy ACD to an external computer
Solid State Applications Interface Bridge — Solid's State Systems' PHI
Teleos IRX-9000 — Teleos' Intelligent Call Distribution platform
For more information, see OAI (Open Architecture Interface.)

RSM 1. Remote Switching Module. An AT&T 5ESS switch standalone switching module that supports all line features and routes intro-RSM calls. It is either a single module or a multi module and can be situated up to 150 miles from the 5ESS switch host.
2. Radio Sub-system Management. A wireless telecommunications term. Management of radio channels including timing and frequency as well as all machines between the mobile station and the MSC.

RSRB Remote Source Route Bridging. Source-route bridging over wide-area links.

RSU Remote Service Unit. I saw the word in a Tellabs press release in April of 1996. In that release, they talked about how Cablespan RSUs "will be used to provide integrated delivery of CATV and telephony services over a standard coaxial drop at the subscriber's home. The RSUs, which will be installed on the outside of each subscriber's home, will function as a standard network interface device for termination of the customer's CATV and telephone service. The RSU provides the RF transceiver for the CATV network, modulation/demodulation of the broadband RF signals, multiplexing/demultiplexing of the digital signals, analog-to-digital conversion (as required), signaling conversion for subscriber loop operation, and diagnostics for problem isolation. The unit provides a 'virtual twisted pair' back to a local switching interface for transparent telephony operation of CLASS services and custom calling features."

RSVP New protocol being developed at MIT for voice communications over the Internet. Scheduled to be released in 1997.

RT 1. Reorder Tone.
2. Remote Terminal.

RTAN Real Time ANI.

RTC RunTime Control.

RTF Rich Text Format. A way of encoding documents such that the messages include boldface, italics and other limited text stylings. RTF is meant to be a cross word processing platform such that if you send one RTF document from one word processor to another word processor, that second word processor will be able to recreate the document's original format.

RTFM Read The Fantastic Manual. This acronym is often used when someone asks a simple or common question. The word "Fantastic" is usually replaced with one much more vulgar. Used on e-mail or Internet.

RTM 1. Ready To Manufacture.
2. Read The Manual.

RTNR A British term. Ring Tone No Reply, a telephone call which has not been answered. Typically a telebusiness system will automatically re-dial the number after a pre-determined period.

RTO-IS Ready To Order - In Service.

RTP Routing Table Protocol. Used in Banyan VINES routing with delay as a routing metric.

RTS Request To Send. One of the control signals on a standard RS-232-C connector. It places the modem in the originate mode so it can begin to send. See the APPENDIX.

RTSE Reliable Transfer Service Element. The OSI application service element responsible for transfer of bulk-mode objects.

RTTU Remote Trunk Test Unit.

RTU Right To Use. A term manufacturers have invented to stifle the used/secondary market in their equipment. Basically, the manufacturer says "Fine, you can sell your no-longer-needed product to some used equipment dealer. But if someone buys it from the dealer and wants to use it, they have to pay me a Right To Use fee." Without payment of this fee, the manufacturer won't contract to maintain the customer's equipment and certainly won't sell the customer software updates, etc. The right to use fee is exorbitant — typically considerably more than what the product actually sells on the used market for. A better approach for a manufacturer would be to innovate a little more and make the customer wants his new product more than his old price (despite the old product's lower price).

RTV Real-Time Video. DVI software that implements quick-and-dirty, realtime video compression. Once called "edit-level video," it stores video as only 10 frames per second. Meant for use while developing DVI applications.

RU 1. Request Unit or Response Unit. A basic unit of data in SNA.
2. Receive Unit

Rubber Bandwidth A term coined by Ascend, an inverse multiplexer manufacturer, to refer to the ability to support applications needing varying speeds. It breaks the original signal up into 56- or 64-Kbps chunks, and places these separate transmissions on the public switched digital network. See also INVERSE MULTIPLEXER.

Rule-Based System The most popular way to represent knowledge in an expert system. In general, a rule-based system's knowledge base contains both facts and IF..THEN production rules.

Run To start a software program.

Run/Stop On a Northern Telecom Norstar phone, this feature inserts a delay in a dialing sequence. The delay can be any length of time.

Run Length Encoding A facsimile encoding process that converts an expanse of white or black (i.e. identical bits) to a code corresponding to the length of the run. A string of identical bits is indicated by sending only one example, followed by a shorthand description of the number of times it repeats.

Run Time The time it takes to execute a software program. See RUNTIME.

Runt An Ethernet frame shorter than the minimum size. Often runts are collision fragments and therefore a normal occurrence.

Runtime A runtime environment is the software that plays back multimedia materials. The runtime material is created by the author. Examples of runtime applications are presentations are training, where the material cannot be edited but only viewed. The runtime software could be a slide show viewer, a software-only video playback application, or a hypermedia runtime document.

Runtime Control RTC. An SCSA definition. The mechanism by which one Resource Object can influence the behavior of another. Typically used for things such as terminating conditions and speed/volume control.

RUR Repair, Update and Refurbish. A term in the industry which repairs telecom equipment. It means equipment is repaired and updated to current manufacturer's specifications. Also includes minor cosmetic cleaning of metal cabinets, a full diagnostic test with burn-in (if required) and an operational test. Definition courtesy Nitsuko America. See also LIKE NEW REPAIR AND UPDATE and QUICK CLEAN.

Rural Service Area RSA. An area not included in either an MSA or a New England Country Metropolitan Area for which a common carrier may have a license to provide cellular service.

RVVP See REVENUE VOLUME PRICING PLAN.

RW An ATM term. Read-Write : Attributes which are read-write can not be written by the PNNI protocol entity. Only the Network Management Entity may change the value of a read-write attribute. The PNNI Protocol Entity is restricted to only reading such read-write attributes. Read-write attributes are typically used to provide the ability for Network Management to configure, control, and manage a PNNI Protocol Entity's behavior.

RZ Return to Zero. A method of transmitting binary information where voltage returns to a zero (reference) level after each encoded bit.

RZ Code Return to zero code. A code form having two information states called "zero" and "one" in which the signal returns to a rest state during a portion of the bit period.

NEWTON'S TELECOM DICTIONARY

S Designation the sleeve or control leads in electromechanical Central Offices which are used to make busy circuits, trunks and subscriber lines, as well as to test for busy conditions. It also designates the sleeve wire on a switchboard cord.
S Band 2 MHz to 2.4 MHz.
S Interface For basic rate access in ISDN, the S interface is the standard four-wire, 144-Kbps (2B+D) interface between ISDN terminals or terminal adapters and the network channel termination, which is two wires. The S interface allows a variety of terminal types and subscriber networks (e.g., PBXs, LANs, and controllers) to be connected to this type of network. At the S interface, there are 4,000 frames of 48 bits each, per second, for 192 Kbps. The user's portion is 36 bits per frame, or 144 Kbps. Out of that 144 Kbps, the user gets two B channels, each of 64 Kbps, and one D channel of 16 Kbps. The local telephone company usually needs a portion of the D channel for signaling. And often it will sell you a 9.6 Kbps packet switched service carved out of the D channel. See also T INTERFACE, U INTERFACE and ISDN.
S Mail Address Snail Mail address. Your post office address.
S Port Refers to the port in an FDDI topology which connects a single attachment station or single attachment concentrator to a concentrator.
S SEED Symmetric Self Electro-optic Effect Device. A switching device in which signals enter and exit as beams of light, not through electrical contacts. In 1990 AT&T Bell Labs built a general purpose digital optical processor/computer. The device contained 2,048 S-SEED chips which could be accessed simultaneously with separate beams of light. That means, that ultimately, such a computer could process huge amounts of information in parallel.
S Video Type of video signal used in Hi8, S-VHS and some laserdisc formats, It transmits luminance and color portions separately, using multiple wires. S-Video avoids composite video encoding, such as NTSC and the resulting loss of picture quality. Also known as Y-C Video.
S-HTTP Secure Secure Hypertext Transfer Protocol. An extension of HTTP for authentication and data encryption between a Web server and a Web browser.
S/DMS SONET Digital Multiplex System. A family of products for Northern Telecom's DMS family of central office switches. There were three initial members of the family as we went to press:
S/DMS TransportNode — a transport vehicle which provides SONET connectivity and bandwidth management for long haul, interoffice and local applications. It includes elements for the transport of SONET over fiber or radio, including integrated bandwidth management functions.
S/DMS AccessNode — a business service access vehicle which delivers switched and special services to the customer's premises, with bandwidth ranging from narrowband to broadband.
S/DMS SuperNode — an evolution of the DMS SuperNode system, which also supports the narrowband capabilities of a SONET network, while adding future broadband switching and network management capabilities.
S/W Software.
S10 Register Hayes, the modem people, invented their "Command Set." This command set lets you control your Hayes compatible modem. In the Command Set there are "S" registers which set how the modem responds to events like answering. Should it answer on the first, second, third, etc. ring. There are 27 registers. The most important S register is S10. This register sets the time between loss of carrier and internal modem disconnect. The factory setting is 1.3 seconds. Drop carrier for 1.3 seconds and your modem will turn itself off. This is long enough for all conditions, except the awful "call waiting" signal you get at hotels and at home. There is a solution: Get your communications software to "go local." Then type ATS10=20. That will increase your S10 register to two seconds. If you have a 300 or 1200 baud you'll have to do this every time you turn on your modem. If you have a 2400 baud modem (the only one to get), you type ATS10=20&W only once. The "&W" writes it into your 2400 baud's non-volatile memory. If you want to check to see if you did it right, type ATS10? That will reply by saying 020. That means 20 tenths of a second, or two seconds. If that still doesn't work for you, increase S10 to three seconds. Other S registers control how long your modem waits for the other end to answer, how long its dialing "pause" is, how quickly it outpulses tones for dialing, etc.
SA Source Address. The address from which the message or data originated. A six octet value uniquely identifying an end point and which is sent in an IEEE LAN frame header to indicate source of frame.
SAA 1. Systems Application Architecture. A set of specifications written by IBM describing how users should interface with applications and communications programs. The idea is to give all software "a common feel" so that training will be less burdensome. According to IBM advertising, "SAA will make it possible for everyone in an organization to access information regardless of its location. What's more, all software written to SAA specifications will provide similar screen layouts, menus and terminology." For a fuller explanation, see SYSTEMS APPLICATIONS ARCHITECTURE. 2. An AT&T Merlin term. Supplemental Alert Adapter. A device that permits 48VDC alerting equipment to be connect to an analog multiline telephone jack so that people working in noisy or remote areas of a building can be alerted to incoming calls.
SAAL Signaling ATM Adaptation Layer: This resides between the ATM layer and the Q.2931 function. The SAAL provides reliable transport of Q.2931 messages between Q.2931 entities (e.g., ATM switch and host) over the ATM layer; two sublayers: common part and service specific part.

525

SAC Single-Attached Concentrator.

SADL Synchronous Auto Dial Language. Created by Racal Vadic, SADL is a public domain auto-dialing protocol which defines procedures in BSC, SDLC (SNA) and HDLC for PCs and larger computers that wish to control synchronous modems directly under program control. SADL does for synchronous dialing systems what the Hayes "AT" command set has done for the async PC dialing world.

SAF-TE SCSI Accessed Fault Tolerant Enclosures specification is a non-proprietary, standardized alert detection and status reporting system for storage subsystems which can send and receive information via a standard SCSI interface. A SAF-TE compliant enclosure is designed to monitor and provide notification to the LAN administrator on the condition of disk drives, power and cooling systems and allow for communication to server-based software agents for network notification. Under the SAF-TE specification, the enclosure is typically implemented as an assignable SCSI target using a low-cost SCSI chip and 8-bit microcontroller. The microcontroller is attached to various alarm sensors and status lights / displays on the enclosure. The enclosure target ID is periodically polled (e.g. every 10-20 seconds) by the host to detect / send changes in status. Disconnect / reconnect and asynchronous event notifications area are not used.

SAFE A store-and-forward MCI International message switching system that provides customers with control of their Telex messages by enabling them to create messages, then specify the message handling parameters in a unique customer reference file. Customers can send and receive messages from the MCI Safe computer at speeds up to 9600 bps.

SAFENET Survivable Adaptable Fiber Network. A U.S. Navy experimental fiber-based local area network designed to survive conventional and limited nuclear battle conditions.

Sag The downward curvature of a wire or cable due to its weight.

Sagan A large quantity. Its derivation is thought to come from Carl Sagan, the astronomer, who used the term "billions and billions" on his TV series.

SAI Stratus Computer's PBX Switch to Stratus Computer Application Interface.

SAID SOD Speech-Activated Intelligent Dialing Stringing of Digits.

Sales Agent See AGGREGATOR.

Sales Automation See SALES FORCE AUTOMATION.

Sales Force Automation The use of computers and computer software by salespeople to boost their sales. There are two types of sales force automation — those totally self-contained on the computers of salespeople (mostly laptops) or those which communicate with headquarters computer over phone lines. There are many purposes of the phone communication — sending orders in, finding out about back orders, getting updates on "specials," dropping letters and memos in, getting new prices, new products, new technical specs, etc. Salespeople routinely show 10% to 20% sales gains armed with a laptop PC and sales automation software (also called "personal contact") software.

Sampling Converting continuous signals, like voice or video, into discrete values, e.g. digital signals. See also PCM and DIGITAL SIGNAL PROCESSING.

Sampling Rate 1. The number of times per second that an analog signal is measured and converted to a binary number — the purpose being to convert the analog signal to a digital analog. The most common digital signal — PCM — samples voice 8,000 times a minute.
2. The number of times per second that a digital audio sample is taken during recording or read during playback — expressed in kilohertz (kHz). An audio CD sampled at a rate of 44kHz has 44,100 bits of information per second.

Sanity Check A check to confirm the service capability of a switching system. This test has not been applied to the author of this dictionary.

SAP In local area networks, SAP is Service Access Point. The point at which the services of an OSI layer are made available to the next higher layer. The SAP is named according to the layer providing the services, e.g., Transport services are provided at a Transport SAP (TSAP) at the top of the Transport Layer. IBM defines SAP as a logical point made available by an interface card where information can be received and transmitted.
In an ATM network, a SAP is used for the following purposes:
1. When the application initiates an outgoing call to a remote ATM device, a destination_SAP specifies the ATM address of the remote device, plus further addressing that identifies the target software entity within the remote device.
2. When the application prepares to respond to incoming calls from remote ATM devices, a local_SAP specifies the ATM address of the device housing the application, plus further addressing that identifies the application within the local device.
There are several groups of SAPs that are specified as valid for Native ATM Services.

SAPI Service Access Point Identifier. The SAPI identifies a logical point at which data link layer services are provided by a data link layer entity to a Layer 3 entity. ISDN jargon. See also WINDOWS TELEPHONY.

SAR Segmentation And Reassembly. Method of breaking up arbitrarily sized packets. A layer of the ATM protocol stack responsible for the segmentation of data into standard ATM fixed-length cells for transmission of the data over an ATM network and for reassembly at the receiving end to reconstruct the original data. Also, the lower half of B-ISDN Layer 3.

SAS Simple Attachment Scheme.

SASG Special Autonomous Study Group. These ITU-T study groups are chartered to produce handbooks on basic telecommunications technical or administrative subjects for developing countries.

SASI Shugart Associates System Interface. The first SCSI interface specification defined by Shugart, a disk drive manufacturer. Later it was modified and renamed as the Small Computer System Interface (SCSI), pronounced Scuzzy. See also SCSI.

SASS Shared Access for Switched Services. An AT&T offering combining MEGACOM WATS and MEGACOM 800 to enable both outgoing and incoming MEGACOM service on the same T-1 circuit.

SAT Subscriber Access Termination. An SMDS term.

Satellite 1. A microwave receiver, repeater, regenerator in a geosynchronous orbit 22,300 miles above the earth. 2. Something distant to the main something. See MAIN DISTRIBUTION FRAME and SATELLITE TRANSMISSION.

Satellite Business Systems SBS. A satellite long distance carrier originally owned jointly by IBM, Aetna Insurance and Comsat, but now owned by MCI (which acquired it in 1986). SBS started out to serve the data communications transmission marketplace but found that marketplace too small to be profitable. It then started to serve the voice transmission marketplace and did somewhat better. Satellite Business Systems no longer exists as a separate entity. It has been merged into MCI. See SBS.

Satellite Cabinet Surface-mounted or flush-type wall cabinets for housing circuit administration hardware. Satellite cabinets, like satellite closets, supplement riser closets by providing additional facilities for connecting horizontal wiring subsystem cables from information outlets in user locations. Sometimes referred to as satellite location.

Satellite Closet A walk-in or shallow wall closet that supplements a backbone or riser closet by providing additional facilities for connecting riser subsystem cables to horizontal wiring subsystem cables from information outlets. Also referred to as satellite location. See also RISER CLOSET and BACKBONE CLOSET.

Satellite Communications The use of geostationary orbiting satellites to relay information.

Satellite Communications Control SCC. The earth station equipment that controls such communications functions as access, echo suppression, forward error correction and signaling.

Satellite Constellation The arrangement in space of a set of satellites.

Satellite Delay Compensator A device that compensates for the absolute delay in a satellite circuit communicating with data terminal equipment (DTE) with the DTE's own protocol.

Satellite Digital Audio Radio Service Special satellites transmitting digitally to tiny (fewer than two inches wide) antennas in the S band. The concept is that with digital technology, satellite broadcasters can stuff dozens of channels of CD-quality, interference-resistant programming into a narrow-band of frequencies.

Satellite Distribution Frame An intermediate point for connecting wires running between a group of phones and the Main Distribution Frame located elsewhere in the building. A fat multi-conductor cable comes from the main distribution frame to the satellite distribution frame, where it splits into individual cables to individual phones or workstations. The satellite distribution frame is usually located in a satellite wiring closet or cabinet. These wires are ultimately connected to the telephone system. See DISTRIBUTION FRAME.

Satellite Facility A transmission system using a satellite in a geostationary orbit above the earth and a number of earth stations.

Satellite Link Microwave link using a satellite to receive, amplify and retransmit signals. Typically that satellite is in a geosynchronous orbit.

Satellite Operation A configuration of multiple PBXs or one big PBX and several smaller PBXs. The configuration gives a company with several nearby locations a unified system of centralized trunks, centralized attendants, overall call detail recording and many of the advantages of a private network. The key advantage of satellite operation is that one big centralized telephone system can contain most of the intelligence and computer smarts for the total system. This advantage is heavily economic. A variation on satellite operation is called CENTRALIZED ATTENDANT SERVICE (CAS).

Satellite PBX A satellite PBX has no direct incoming connection from the public network. All incoming calls are routed from an associated main PBX over tie trunks. This definition places no restrictions on the handling of outgoing calls from the satellite PBX. A satellite PBX can have one-way outgoing trunks to the central office, in addition to outgoing service on trunks to the main PBX. A satellite PBX has no direct trunks to a node; however, calls to the node can be made through the main PBX.

Satellite Premises Channel This is the cable connecting arrangement between a dedicated earth station and the Customer Provided Equipment.

Satellite Processor A computer with little computing power used for operations that do not require the full processing power of the main machine.

Satellite Transmission A form of transmission which sends signals to an orbiting satellite which receives them, amplifies them and returns those signals back to earth. Satellite transmission provides great clarity but suffers from delay. See SATELLITE TRANSMISSION DELAY.

Satellite Transmission Delay Referring to the time it takes a signal to travel from one satellite earth station to the satellite in the sky then to the satellite earth station at the other end. Since most communications satellites orbit the earth at a distance of approximately 22,300 miles, the total distance the signal travels is 44,600 miles. Since radio waves travel at the speed of light (186,000 miles per second), simple arithmetic will show a delay of approximately one-quarter of one second thus, 44,600 divided by 186,000 = 0.239 second. If you are waiting for a reply, double this time. (You double the distance.)

SATPhone A telephone that works directly off a satellite. It comes with a small parabolic antenna which you aim at the satellite. You turn it on and talk. It's easy, though expensive — typically as much as $10 a minute. But for that you can talk from practically anywhere in the world, so long as you got lots of battery or easy close access to commercial power.

Saturn A family of excellent digital PBXs made by Siemens and sold by Siemens Information Systems of Boca Raton, FL.

Save A telephone feature that allows the user to put a phone number into memory for future calls, by pressing one or two buttons after dialing it the first time. See also SNR.

Save And Repeat Another way of saying "Autodial." Electronic phones may be able to save a number so you can dial it later by simply hitting one button on the phone. This feature is similar to a "Last Number Redial" button, except that button just dials the last number called. "Save and Repeat" puts a number into temporary storage for dialing at another time. Phones should have both auto-redial and save-and-repeat buttons.

SBC Communications Inc. In early October, 1994, Southwestern Bell Corporation, one of the seven regional Bell operating companies, changed its name to SBC Communications, Inc. The company's subsidiaries continue to be Southwestern Bell Telephone Co., Southwestern Bell Mobile Systems, Southwestern Bell Yellow Pages, Southwestern Bell Telecom and SBC International.

SBM DMS SuperNode Billing Manager.

SBS 1. Satellite Business Systems. A long distance satellite company that started out as a joint venture between Lockheed and MCI, was sold to IBM Aetna and Comsat and then eventually was given to MCI in exchange for shares issued to IBM. SBS never made any money. But that was irrelevant. Its job was to help IBM sell computer networks. See SATELLITE BUSINESS SYSTEMS.
2. Sick Building Syndrome. Phenomenon of employee discomfort and illness, or perceived illness, due mainly to a polluted indoor air supply. An office is diagnosed "sick" when more than 20% of its occupants exhibit typical symptoms, complaints persists for two weeks or more and disappear when sufferers are away from the building.

Sbus The type of system bus in Sun Microsystems, Inc., SPARC-based workstations and servers. The Sbus specification has been adopted by the IEEE (Institute of Electrical and Electronic Engineers) as a new bus standard.

SC2000 An SCSA definition. The SC2000 is a custom integrated circuit manufactured by VLSI Technology, Inc. that implements the switching handler interface defined in the SCSA Hardware Model specification. The SC2000 connects 32 internal timeslots on a device to the 2048 timeslots on the SCbus. The SC2000 also has compatibility modes for the PCM Expansion Bus (PEB) and Natural Microsystems' MVIP so that hardware developers can design a single hardware platform that with different software could be used as a PEB, and SCSA, or an MVIP device.

SCA 1. Selective Call Acceptance.
2. Supplemental Communications Authority. The authority granted by the Federal Communications Commission to transmit on a subcarrier.

SCAI Switch to Computer Applications Interface. A protocol that defines how switches talk to outboard computers, i.e. computers which are external to the switch and contain such a database of customer buying information. Using SCAI, calls and data screens about a calling customer can be presented to the agent simultaneously. See OPEN APPLICATION INTERFACE.

Scalable Something that can be made larger or smaller relatively easily and painlessly. And the cost to grow is straight line, not stair step, as in the old days of mainframes. At least that was the earlier, accepted definition. Then Microsoft started referring to Windows NT as "scalable," namely that it runs on everything from Intel to RISC processors and single to multi processor systems.

Scalable Video An Intel Indeo Video term. Scalable video is a playback format that can determine the playback capabilities of the computer on which it is playing. Using this information, it allows video playback to take advantage of high performance computer capabilities while retaining the ability to play on a lower performance computer.

Scalable Typeface A set of letters, numbers, punctuation marks, and symbols that are a given design (i.e. of one font) but can be scaled to any size.

Scaled Point Size A point size that approximates a specified point size for use on the screen. For example, text that prints at 10 point on the printer may be represented by a slightly larger font on the screen to make up for the screen's lower resolution.

SCAN 1. Switched Circuit Automatic Network. 2. To examine sequentially, part by part. 3. To examine every reference or every entry in a file routinely as part of a retrieval scheme. 4. In electromagnetic or acoustic search, one complete rotation of an antenna. 5. The motion of an electronic beam through space searching for a target. Scanning is produced by the motion of the antenna or by lobe switching.

Scan Time The time between two successive polls to a workstation on a data communications network.

Scanner 1. A radio receiver which automatically skips across selected frequencies, allowing you to listen in to any of the frequencies. You can buy scanners, for example, that let you scan all police, fire, cellular frequencies and let you listen in on any conversation that is presently occurring.
2. A program on a bulletin board system which scans the message base for previously entered e-mail and pulls a copy of each message and makes them available to the BBS (Bulletin Board System) mailer program.
3. A device used to input graphic images into the computer. Scanners look at or "scan" a piece of paper and put the image's information into digital form. The information can then be recognized by the computer. A scanner is also in a fax machine that "looks" at the original document and determines the brightness level of each pixel to be transmitted. See also Optical Character Recognition.

SCANNS Multiplexers which perform the vital functions of monitoring and control within System 75s and 85s Automated Building Management feature. The SCANNS continuously scan sensors and send the resulting data to local control units.

SCAPI SCSA Application Programming Interface. A high-level, object oriented hardware independent, technology independent programming model that permits the design and implementation of call processing applications.

Scattering A cause of lightwave signal loss in optical fiber transmission. The diffusion of a light beam caused by microscopic variations in the material density of the transmission medium. Scattering is a physical mechanism in fibers that attenuates light by changing its direction.

SCbus An SCSA definition. The standard bus for communication within an SCSA node. The SCbus features a hybrid bus architecture consisting of a serial Message Bus for control and signaling, and a 16-wire TDM data bus. The SCbus is a serial time division multiplexed bus for carrying information between hardware devices in a signal processing node. The SCbus can support up to 1024 bidirectional timeslots in a PC implementation or up to 2048 timeslots in a backplane implementation. The SCbus uses 16 synchronous data lines for carrying data and a dedicated messaging channel (SCbus Message Channel) for carrying signaling information and messages between devices. See SCBUS MESSAGE CHANNEL.

SCbus Message Channel An SCSA definition. The SCbus message channel is a 2.048 Mbps serial line for carrying signaling information and messages between hardware devices in an SCSA Hardware compliant server. The SCbus message channel uses an HDLC (high level data link controller) protocol. The SCbus message channel is an optional element of the SCSA Hardware Model and provides faster system performance by allowing for direct communication of messaging information between devices at the firmware level and without consuming any data timeslots.

SCC 1. Specialized Common Carrier. Another term for a long distance carrier in competition with AT&T. The word "Specialized" came about because these long distance carriers purported to provide "specialized" circuits for business customers. At one stage

they were also known as OCCs, or Other Common Carriers (i.e. other than AT&T). These days, both terms have fallen into disrepute. All long distance carriers — including AT&T — are called InterExchange Carriers (IXCs).
2. Standards Council of Canada.

SCCP Signaling Connection Control Part. Part of the ITU-T #7 signaling protocol. and of the SS7 protocol. It provides additional routing and management functions for transfer of messages other than call set-up between signaling points. A SS7 protocol that provides additional functions to the Message Transfer Part (MTP). It typically supports Transaction Capabilities Application Part (TCAP). See also SIGNALING SYSTEM 7 and COMMON CHANNEL SIGNALING.

SCCS Switching Center Control System.

SCDPI SCSA Device Programming Interface: A set of callable functions that allow SCSA application software to control SCSA hardware. The SCdpi consists of both common call processing services and technology specific modules for the application of particular resources to call processing tasks. See SCSA.

SCE Service Creation Environment. A term used in the jargon of intelligent networks (IN) to allow outside developers to define and create new value-added (i.e. intelligent) services by connecting pre-existing blocks of code into a flow chart that describes the logical processes the service will use to handle calls.

Sceptre A consumer videotex interactive terminal which, when used with a TV set and a phone line, allows the consumer to interact with a videotex data base service.

Schedule A call center term. A record that specifies when an employee is supposed to be on duty to handle calls. The complete definition of a schedule is the days of week worked, start time, break times and durations (as well as paid/unpaid status), and stop time. See the following six terms.

Schedule Exception A call center term. A specific date and period when an employee cannot handle calls or is engaged in some kind of special activity. An absence, meeting, or other work assignment creates an "exception" to the employee's daily work file schedule.

Schedule Inflexibility A call center term. A phenomenon that tends to create overstaffing in some periods when full coverage is the objective in creating a set of schedules. This is caused by the fact that it is impractical to have extremely short schedules for covering momentary peaks in call volume. To achieve a near perfect match of staff and workload at all times would require shifts of virtually every length; for example, 2-hour shifts, 45-minute shifts, even 15-minute shifts.

Schedule Preference A call center term. A description of the days and hours that an employee would like to work, used by the automatic assignment process to match the employee to a suitable schedule. In some call centers, each employee can have as many as 10 schedules preferences ordered by priority.

Schedule Test A call center term. A variation of the scheduling process that allows you to forecast the service quality that will result from using an existing set of schedules.

Schedule Trade A call center term. A situation in which two employees have agreed to work each other's schedules, or an employee has agreed to work the other's schedule, on a specific date or dates.

Scheduling Making the timetable of agent hours and shifts for your call center. Takes into account vacation days, breaks, training time, lengths of shifts and forecasting information. A call center software management package helps you do this.

Schematic Diagram which details the electrical elements of a circuit or system.

Schlepp Jewish word meaning to carry around, to drag around, as in "This machine is heavy. Schlepping it is a pain."

SCIL Aristacom's Switch Computer Interface Link. Aristacom is an external software company that has developed software for linking programs on an external computer to a PBX and having that computer control the movement of calls within the PBX. See also OPEN APPLICATION INTERFACE.

Scintillation In electromagnetic wave propagation, a random fluctuation of the received field strength about its mean value, the deviations usually being relatively small. The effects of this phenomenon become more significant as the frequency of the propagating wave increases.

SCM 1. Station Class Mark. A two digit number that identifies certain capabilities of your cellular phone. How the cellular network handles your call is based on these digits. The SCM tells the system if your phone transmits at standard power levels or low power levels, if it can use the full 832 channels or only the original 666 frequencies. The last attribute identified is whether your phone uses voice activated transmission (VOX).
2. Subscriber Carrier Mode.

SCM-100A Subscriber Carrier Module-100 Access; same as SMA.

SCO An AT&T specialized network maintenance organization providing a single point of contact for resolving customer network faults.

SCO Unix The Santa Cruz Operation's adaptation of Unix.

Scope 1. A slang term for cathode ray oscilloscope.
2. An ATM term. A scope defines the level of advertisement for an address. The level is a level of a peer group in the PNNI routing hierarchy.

SCOTS 1. Surveillance and COntrol of Transmission Systems. 2. Switched Circuit Ordering and Tracking System. MCI's automated tracking and order processing system for Dial up products, IMTs, and the MCI switched network.

SCP 1. Service Control Point. Also called Signal Control Point. A remote database within the System Signaling 7 network. The SCP supplies the translation and routing data needed to deliver advanced network services. The SCP translates an 800-IN-WATS number to the required routing number. It is separated from the actual switch, making it easier to introduce new services on the network. See also SCP and TCAP. For a full explanation of the Advanced Intelligent Network, see AIN.
2. Northern Telecom term for a Satellite Communications Processor.

SCR 1. Abbreviation for Silicon Controlled Rectifier, a semiconductor device that allows one electric circuit to control another; often replaces electromechanical relays.

2. SCSA Call Router.

3. Sustainable Cell Rate. Parameter defined by the ATM forum for ATM traffic management. The SCR is an upper bound on the conforming average rate of an ATM connection over time scales which are long relative to those for which the PCR is defined. Enforcement of this bound by the UPC could allow the network to allocate sufficient resources, but less than those based on the PCR, and still ensure that the performance objectives (e.g., for Cell Loss Patio) can be achieved.

Scrambler A device which deliberately distorts a voice or data conversation so that only another like device can figure out the content of the message. Analog scramblers invert the frequencies of speech. Digital scramblers first convert speech to digital form and then encrypt. Both types also perform the reverse process. The sophistication (i.e. complexity) of a scrambler determines its price.

Scratchpad A part of the random access memory of a computer or telephone system which can be used to temporarily store data. In a cellular phone system, scratch pad allows storage of phone numbers in temporary memory during a call. Silent scratch pads allows number entry into scratch pad without making beep tones. See also REGISTER.

Screen Buffer In Windows 95, the screen buffer is the size reserved in memory for the MS-DOS Prompt display.

Screen Font The font that is displayed on your screen. It is, hopefully, designed to match the printer font so that documents look the same on the screen as they do when printed. You typically need a graphics interface on your company — like Windows or X-Windows — to make the font you see on your screen the same as what you see when you print it out.

Screen Pop Screen Pop presents customer data and product and service information simultaneously with the incoming telephone call. Imagine a call center. An agent's phone is ringing. As it rings, the agent's computer screen pops up with information about the caller, what he ordered last, how much he owes, etc. This is called screen pop. The technology to make it happen typically comes from caller ID or ANI — information carried on the phone call just before the voice. It can also come from an IVR (Interactive Voice Response) system which answers the phone and asks the caller to punch in his phone or account number and then passes the phone call to the agent. In Screen Pop, the phone system typically listens for incoming digits and passes them across an attached local area network. See CALLER ID.

Screen Refresh Rate The rate at which your computer screen is re-drawn every second by a horizontal beam that scans from the top left hand corner to the bottom right hand corner. Screen refresh rates differ by the graphics standard you're running.

Screen Response Time The time it takes to refresh a terminal screen.

Screen Synch A colloquial term for sending an telephone service agent a phone call together with a screen of information about the incoming call, e.g. the customer's purchasing record or experiences with your product (if you're a help desk, for example.)

Screened Transfer You are transferring a call from your phone to your boss. You dial a code for transfer, then dial your boss. The caller you're transferring is automatically put on hold. You speak to your boss, tell her who you're putting through. She okays the transfer, then you hit another digit and the call goes right through. This is called screened transfer. An unscreened transfer occurs when you simply dial your boss's office and send the call through without announcing it. Most PBXs have the ability to do both screened and unscreened calls.

Screening Telephone Number The telephone number used by the phone company to bill, regardless of the number of phone lines associated with that number.

Script Files Some communications programs had script files that automate logging onto communications services, such as MCI Mail. The files are saved on your disk and read by your communications software when connecting to a remote service. Newer communications programs will "write" their own scripts by recording what you do in response to what questions from the remote service. This typically happens using a program feature called "Learn."

Script Language A software language that contains English-statements for commands. A statement might be as simple as WrapPara()
for wrap paragraph. Typically a script language contains commands that are specific to the type of task it's doing. For example, VOS from Parity Software in San Francisco is a script language for voice processing using Dialogic voice processing cards. A script language is more flexible than an Applications Generator, but is more difficult to program.

Scroll Bar A bar that appears at the right and/or bottom edge of a window or list box whose contents are not completely visible. Each scroll bar contains two scroll arrows and a scroll box, which enable you to scroll through the contents of the window or list box.

Scrolling Browsing through information at a video terminal. Scrolling is the continuous movement of information either vertically or horizontally on a video screen as if the information were on a paper being rolled under it.

SCS Structured Cabling System. What it sounds like

SCSA Signal Computing System Architecture. SCSA is a comprehensive architecture that describes how both hardware and software building blocks work together. It focuses on "Signal Computing" devices, which refer to any devices that are required to transmit information over the telephone network. Information can be transmitted via data modems, fax, voice or even video. SCSA defines how all these devices work together. Signal computing systems combine three major elements for call processing. Network interfaces provide for the input and output of signals transmitted and switched in telecommunications networks. Digital signal processors and software algorithms transform the signals through low-level manipulation. Application programs provide computer control of the processed signals to bring value to the end user.

SCSA is the common set of standards that telecommunication system manufacturers and computing system manufacturers can use to create computer telephony systems. The theory is no single company today can create the total solution for all customers. SCSA represents the common ground between the two fields so that manufacturers from each area can safely develop products that will work with other manufacturers. SCSA's coverage extends from low-level bus and hardware interfaces, like the inter-board switching bus that enables boards from different suppliers to work together, to high-level application programming and software interfaces, so that software designed to work with one set of hardware products, will work with different hardware. Dialogic Corporation of Parsippany, NJ announced SCSA in the Spring of 1993. Dialogic said that SCSA was defined and created with input from a number of leading computer and switch manufacturers, call processing suppliers, and technology developers. In many cases, SCSA has drawn on existing standards, like the T.611 fax standard endorsed by the European Computer Manufacturers Association, and in other cases SCSA has extended standards to make them more useful for call processing suppliers and users.

SCSA describes all elements of the system architecture from the electrical characteristics of the SCbus and SCxbus to the high level application programming interfaces (APIs). According to TELECONNECT Magazine, this SCSA standard is remarkable for several things:

1. On the day of its announcement over 60 telecom and voice processing companies publicly endorsed SCSA. In early 1994, over 150 companies public endorsed it.
2. With SCSA — a standard for PC/LANs and VME-backplaned computers — you can build much larger telecom switches and much larger call and voice processing boxes. Previous standards, like AEB, PEB and MVIP, were basically limited to what you could do with one PC. Now PCs can be joined together. With SCSA, you can put 16 T-1 lines, or 512 voice lines in one PC and join together 16 PCs, for a total of 16 x 16 x 24 = 6,144 lines! That's a central office built out of networked PCs. A mainframe built out of a LAN. The SCSA joining is not via LAN or LAN-emulation. That would be too slow and the transmission too bursty (great for data, lousy for voice). It's via an SCbus — something that looks and works like a PBX backplane.
3. SCSA incorporates virtually every other standard in PC-based switching — including the most popular ones, Mitel's ST-Bus, MVIP, Siemens PCM Highway, AEB and PEB.
4. It's a lot faster and more reliable. All signaling is out of band. There's clock fall back and time slot bundling. It's more modular, meaning you can start with one PC and grow one at a time. That makes it more "modular" (scaleable is the new word). It's also hot pluggable. You don't have to turn off to upgrade.
5. It has applications portability. Tandem, the highly-successful fault tolerant minicomputer maker, has an SCSA application in a call center. They call it the Tandem Non-Step Call Center. It uses the Tandem 2400 VRU and the 4800 VRU.

SCSA is open, truly open. All its specs and all levels of its specs are available. To that extent, SCSA represents a remarkable gamble by its creator, Dialogic, a telecom/voice processing hardware company. It is encouraging competing manufacturers to build hardware to its specs and gambling that it won't be left in the dust, as IBM was with its PC. (Compaq, not IBM, built the first '386 PC.)

SCSA, as an idea, is revolutionary (for telecom). No one in telecom has ever promulgated an open standard everyone can adopt — hardware and software vendors. Dialogic has done it to create opportunities by providing great economies of scale for the developers. Write one application, create one applications generator, design one piece of hardware. Erector set telecom/voice processing! Build small. Build large. Just join the bits and pieces together. See also AEB, ECTF, PEB, MVIP, SIGNAL COMPUTING (for a differently-worded definition) and TAO.

SCSA Call Router An SCSA definition. A system service of SCSA which provides the basic necessities of inbound and outbound call processing and call sharing to client applications, without those applications needing to be aware of the underlying telephony interface operations. See ECTF, SCSA and TAO.

SCSA Compatible An SCSA term. Able to function in an SCSA environment in its native mode.

SCSA Hardware Model An SCSA definition. The hardware layers of the SCSA specification. The SCSA Hardware Model defines an open architectural specification for a digital intra-node communication bus (SCbus), a switching model (SCSA Switching Model), and an multimode expansion capability (multimode Network Architecture, or MNA). The SCSA Hardware Model may be implemented independently of the SCSA Telephony Application Objects Framework. See TAO.

SCSA Message Protocol The open communications protocol by which entities communicate with one another in an SCSA system. The SCSA Message Protocol (SMP) is independent of the transport layers it is built upon, computer hardware, operating system, network topology (or lack thereof), and technology vendor. All SCSA-compliant AIAs will translate the functions called by client applications (via the API) into SMP messages; these are transmitted to service providers regardless of their location. Therefore, applications written to the API will be portable from one call processing environment to another. See TAO.

SCSA Message Protocol Interface The message presentation format required by, and used by, the service provider in delivering SPM information. Contrast with Service Provider Messages. See TAO.

SCSA Server A collection of service providers (objects) which in the aggregate implement the minimum set of services required for SCSA system conformance. The assumption is that these services are at a minimum provided to remote hosted client applications via common transports such as LANs, but may also be provided to client applications which are hosted on the SCSA server itself. Note that this is a logical image which may be implemented through multiple nodes (machines). See TAO.

SCSA Telephony Application Objects Framework The SCSA Telephony Application Objects (TAO) Framework originally defined the software layers of the SCSA open computer telephony specification. The SCSA TAO Framework defined a hardware-independent, open software architecture that simplifies design of distributed computer telephony systems. The SCSA TAO Framework includes a suite of interoperable, vendor-independent application programming interfaces (SCSA APIs), a set of System Services for handling various server management functions, and a set of messages and a standard transport for communication among various technology resources and system service providers (Service Provider Messages and SCSA Message Protocol). In 1995, TAO's development was taken over by a new organization ECTF — the Enterprise Computer Telephony Forum. The ECTF expanded the idea of TAO to make it the open software framework for the whole computer telephony world — to encompass hardware conforming to all major specifications, including SCSA and MVIP. Towards the end of January, 1996, ECTF promulgated TAO (now under a different name) as the software standard for the new computer telephony industry.

SCSI Small Computer System Interface. (Pronounced Scuzzie.) The brains of a computer is its microprocessor. That microprocessor (computer on a chip) does the computer's primary work (i.e. calculations). There must be a way for information to get into and out of the microprocessor. The history of computers could be written as a continuing race to figure new, faster and more efficient ways of getting information into and out of the microprocessor. The obsession with input/output stems from the fact that the microprocessor can work much faster than you can get information in and out and out of it. SCSI is a way for a devices such as magnetic hard disks, optical disk drives, tape drives, CD-ROM drives, printers and scanners to communicate with the computer's main processor. SCSI is a bus and an interface standard. The theory is that if you buy a SCSI device you can plug it into your computer's SCSI port and it will work — just as a parallel or serial port device will work. There are two good points about the SCSI interface — especially the newer SCSI-2 interface. First, is that it's fast. It's as fast as 10 megabytes per second data transfer rate. The second good point is that on one SCSI bus you can daisy chain up to seven different devices (so long as you remember to terminate the end of the chain.) (In reality, you rarely have more than four devices hooked up on one SCSI link, since protocol overhead and other factors begin to degrade system performance.) Each device will work quickly and each won't siphon excessive power from the computer's main processor. That's because the SCSI bus typically has its own controller/microprocessor which takes care of the SCSI input/output workload. SCSI disk drives work faster than a "normal" IDE hard drive, which is why many new computers are coming with SCSI drives, not IDS drives. ANSI (American National Standards Institute) has set several guidelines for SCSI connection. There is SCSI-1 and SCSI-2. The SCSI specifications are available from ANSI, 1430 Broadway, New York NY 10018. Phone 212-642-4900.

All Apple Macintosh computers come with built-in SCSI ports to which you can daisy chain one SCSI peripheral after another, until you have a total of seven. This is a fairly easy job, since Macintosh SCSI ports are standard and manufacturers of Macintosh SCSI peripherals will certify that their product works with the Macintosh SCSI standard. They wouldn't sell it if it didn't. One point: If you've removed the hard drive in your Macintosh (and replaced it with one or more SCSI-attached drives) your Macintosh may require a hard disk terminator. Some (not all) Macintoshes require a hard disk terminator (a $5 device) if their hard disk has been removed.

To add SCSI devices to a MS-DOS machine, you must first place a SCSI adapter card in your PC's bus or your MS-DOS laptop's PCMCIA slot and connect the SCSI devices to that card. Sadly, for MS-DOS machines, SCSI is not a universal plug-n-play standard. According to Keith Comer of Toshiba, when asked why Toshiba's computers didn't come with SCSI ports as they came with parallel and serial ports, said, "I an unconvinced of SCSI's universal compatibility. It's a nontrivial task to connect SCSI devices. All devices need their own drivers. And each need to be configured for the particular SCSI card you have. Further, many of the SCSI drivers are incompatible with memory managers. In short, for us as manufacturers it would be a support nightmare." The problem is lessening slowly. Corel (Ottawa, Canada) and others are creating "standard" SCSI Interface kits (software and/or hardware). These make connecting things less of a pain. But your desired-to-connect SCSI device (e.g. CD-ROM or magnetic optical drive) must be on the Corel list of approved devices. And — this is the Catch 22 of SCSI: If your SCSI device is new, you can be sure it will not be on Corel's list and probably will not work. In short, do not even bother trying to connect your "standard" SCSI device — unless someone has assured you that they have seen it work and it is on someone's list of approved SCSI devices. Yours truly has failed to connect many new SCSI devices using "standard" SCSI software. And when I asked the manufacturers (Corel, etc.) why they didn't work, I was told that my devices were too new and they hadn't released the necessary device driver software. But it's worse. Fingerpointing prevails. Manufacturers of SCSI deny responsibility for making SCSI device drivers to make their hardware work. And manufacturers of SCSI software and SCSCI adapter cards say they haven't been able to obtain/acquire/buy one of the new devices and figure out how to connect it. Further, there's no assurances that they will ever both to figure out how to connect that particular device. As I write this entry, I have two SCSI devices on my desk which I cannot connect through several "standard" SCSI adapter cards I am testing. I am able to connect them through a SCSI cable connected to my computer's parallel port. The throughout is very slow, . An example: Using a parallel cable I was able to transfer a one meg file in 21 seconds. Using the only PCMCIA SCSI adapter card I could coax to work, I was able to transfer the same one meg file in 12.3 seconds. See GEOPORT, SCSI-2, SCSI TRANSFER RATE.

SCSI-2 SCSI-2 (pronounced Scuzzie-Two) is a 16-bit implementation of the 8-bit SCSI bus. Using a superset of the SCSI commands, the SCSI-2 maintains downward compatibility with other standard SCSI devices while improving upon reliability and data throughput. SCSI-2 is capable of transferring data at rates up to 10 megabytes per second, twice as fast as SCSI-1. SCSI-2 defines more than a speed. It defines a command set and electrical characteristics.

SCSI Transfer Rate SCSI transfer rate is the speed of moving data between the SCSI adapter board and the SCSI device. Host transfer rate is the speed of moving data between the adapter board and the host PC. See SCSI and SCSI-II.

ScTP Screened Twisted Pair. Twisted pair cable with better specifications than Category 5 cable. ScTP has four twisted pairs and a foil shield (which is unlike UTP).

SCVF Single Channel Voice Frequency.

SCWID Spontaneous Call Waiting Display.

SCxbus An SCSA term. The standard SCSA bus for communication between nodes. The SCxbus features the same architecture as the SCbus. See SCxbus Adapter.

SCxbus Adapter Inter-box expansion adapter for the SCbus.

SDE 1. Synchronization Distribution Expander. 2. Secure Data Exchange as defined by the IEEE 802.10 security committee.

SDEC Northern Telecom term for Satellite Data Exchange Controller.

SDH Synchronous Digital Hierarchy. A set of standard fiber-optic-based serial standards planned for use with Sonet and ATM in Europe. Some of the SDH and SONET standards are identical. Standardized by the ITU.

SDI Northern Telecom term for Serial Data Interface.

SDF Sub Distribution Frame. Intermediate cross connect points, usually located in wiring or utility closets. A trunk cable or LAN backbone is run from each SDF to the MDF (Main Distribution Frame).

SDH Synchronous Digital Hierarchy. SDH is a ITU-T defined standard technically consistent with SONET. In short, another name for SONET.

SDK Software Development Kit.

SDL Signaling Data Link.

SDLC Synchronous Data Link Control. A bit-oriented synchronous communications protocol developed by IBM where the message may contain any collection or sequence of bits without being mistaken for a control character. SDLC is used in IBM's SNA — System Network Architecture. See HDLC and IBM.
SDLC-To-Token-Ring LLC Transformation A technique to integrate SDLC link-attached SNA devices into a LAN/WAN internet. A modified remote polling process is used to make the link-attached devices appear to be LAN-attached.
SDM 1. Subrate Data Multiplexing. A European term. In North America, it's called SRDM.
2. DMS SuperNode Data Manager.
SDMA Station Detail Message Accounting. See CALL ACCOUNTING.
SDN Software Defined Network. See SOFTWARE DEFINED NETWORK, SDN SERVING OFFICE, VIRTUAL NETWORK and the APPENDIX.
SDN Serving Office One of many AT&T-supplied switching nodes in an SDN network. See also SOFTWARE DEFINED NETWORK and the APPENDIX.
SDRM Sub-rate Data Multiplexing. Refers to a service where a DS0 (64 Kbps) channel may contain one 56 Kbps signal, five 9.6 Kbps signals, ten 4.8 Kbps signals or twenty 2.4 Kbps signals. Although speeds may be mixed, the highest speed determines the number of signals supported.
SDS IS Switched Digital Service Integrated Systems. A new (Spring, 1993) service of Pacific Bell which provides ISDN over one single business line, rather than the minimum two lines required previously.
SDSAF Switched Digital Services Applications Forum, a group of manufacturers and carriers whose objective to standardize the interconnection of switched 56 kilobit and n x switched 56 channel local and long distance services. The group is based in Reston, VA. Today a switched 56 Kbps "phone" call between multiple carriers probably wouldn't get through. In short, this group is trying to bring the simplicity of the voice dial up phone system into the switched data world.
SDT Structured Data Transfer: An AAL1 data transfer mode in which data is structured into blocks which are then segmented into cells for transfer.
SDU-SMDS Data Unit The user payload in an SMDS L3PDU packet. The SDU can contain up to 9,188 bytes.
SDU An ATM term. Service Data Unit: A unit of interface information whose identity is preserved from one end of a layer connection to the other.
SDV Switched Digital Video.
SDVN Switched Digital Video Network.
SE 1. Systems Engineering
2. An ATM term. Switching Element: Switching Element refers to the device or network node which performs ATM switching functions based on the VPI or VPI/VCI pair.
Sealed Case PC A fancy name for a PC that never needs to be opened up.
Sealing Current A designation for a powering situation that consists of a wet loop without span power.
Sealink Sealink is an error-correcting file transfer, data transmission protocol for transferring files between PCs. It is a variant of Xmodem. It was developed to overcome the transmission delays caused by satellite relays or packet-switching networks.
SEAL Simple and Efficient Adaptation Layer: An earlier name for AAL5.
Seals A way of telling if a device has been tampered with.
Seamless The word seamless means "perfectly smooth, without awkward transitions." In software, it means that what takes place between the user and the application accessed by the user is perfectly smooth to the user and the software being used by the used will work easily with other software the user is also using. On a LAN accessing "seamless" applications, the user doesn't perceive he's on a network because his programs run as though they were on his personal computer. In fact, the word "seamless" is a very vague term, meaning now that the "seamless" software you're about to use will work easily with other software you're using. But no one knows exactly what easily is, not how easy seamless is. And no one has a technical definition of seamless. Originally, the word meant "without the stitches showing." The 15-century word got boost from the phrase seamless stockings which filled a brief period between silk stockings — which had seams that always needed straightening — and panty hose that don't, sadly.
Search Drive A drive that is automatically searched by the operating system when a requested file is not found in the current (default) directory. A search drive allows a user working in one directory to transparently access an application or data file that is located in another directory.
Search Engine (On The Web) An Internet World Wide Web term. A search engine is a program that returns a list of Web Sites (URLs) that match some user-selected criteria such as "contains the words cotton and blouse." Basically, the procedure is simple. You surf to the search engine's site. You click a couple of times and type in what you're looking for. A few seconds later you get choices. You finally make your selection and you get instantly hotlinked over to that site. Here are the main Internet search engines and their addresses.
ALTA VISTA http://altavista.digital.com EXCITE http://www.excite.com LYCOS http://www.lycos.com INFOSEEK http://guide.infoseek.com OPENTEXT http://www.opentext.com:8080 INKTOMI http://www.inktomi.berkeley.edu/query.html WWW.WORM http://www.cs.colorado.edu/wwww/ WEBCRAWLER http:///www.webcrawler.com YAHOO http://akebono.stanford.edu/yahoo/
SEAS Signaling Engineering and Administration System.
Seatfone GTE's name for air-to-ground telephones which it installs in some airplanes. GTE is investing millions to make its service useful. It needs to.
SECAM Acronym for Systeme Electronique Couleur Avec Memoire. A television signal standard used in France, eastern European countries, Russia, the former Soviet Union and some African countries. SECAM uses an 819-line scan picture which provides a better resolution than PAL's 625-lines and NTSC's 525-lines (the US standard). All three two systems are not com-

patible. You cannot view an Australian, English or French videotape or through-the-air broadcast on a US TV.

SECAM Format 625 lines of resolution at 25 frames per second.

Second Dialtone 1. Dialtone given to the caller, on a phone system (e.g. PBX, Centrex or hybrid), after dialing an access code (e.g. 8 or 9) to make a call out of the system (e.g. a local or long distance). Sometimes referred to as outside dialtone. 2. Dialtone returned to a caller after they've dialed a local or long distance number and reached some type of switching device. That switching device might allow you to dial into a fax machine, a modem, a phone or an answering machine. It might allow you to dial into one of several cash registers or soda machines (to check if they need being refilled). It might even allow you to dial long distance through a voice mail system or through a long distance phone company. As you dial through networks you might encounter not only second, but also third and fourth dialtones.

Secondary Equipment Used telecommunications equipment. See also USED, CERTIFIED, REFURBISHED and REMANUFACTURED.

Secondary Market The market for used business telecommunications and computer equipment. The magazine serving this market is called TELECOM GEAR. It's a monthly. You can get your free subscription by calling 1-800-322-5156.

Secondary Protection Primary protection is a device that sits at your building entrance between your phone line coming in from outside and your line going into and up your building. The phone company is responsible for installing primary protection. Secondary protection sits on your floor just next to your phone system. Secondary protection is designed to protect your phone equipment from spikes, surges and high electricity that might affect your phone lines between the primary protection downstairs and the secondary protection upstairs. Secondary protection typically costs $20 to $30 a line. It's worth every penny.

Secondary Radar A radiodetermination system based on the comparison of reference signals with radio signals retransmitted from the position to be determined.

Secondary Radiation Particles (such as photons, Compton recoil electrons, delta rays, secondary cosmic rays, and secondary electrons) that are produced by the action of primary radiation on matter.

Secondary Resource An SCSA definition. Any resource that is attached to a Group after the Group has been created around a primary resource. See SCSA.

Secondary Ring One of the two rings making up an FDDI ring. The secondary ring is usually used in the event of failure of the primary ring.

Secondary Server Under NetWare, the secondary server is the SFT III NetWare server that is activated after the primary server that receives the mirrored copy of the memory and disk from the first server activated. The secondary server mirrors the disk and memory image of the primary server. Though it cannot be used to do additional work (because it uses all of its CPU cycles keeping up with the primary server), the secondary server can act as a router for the local network segments to which it is directly attached. In addition to mirroring the primary server, the secondary server provides split seeks. Either SFT III server may function as primary or secondary, depending on the state of the system. You cannot permanently designate which server is primary or secondary. System failure determines each server's role, that is, when the primary server fails, the secondary server becomes the new primary server. When the failed server is restored, it becomes the new secondary server. See also SERVER MIRRORING.

Secondary Service Area The service area of a broadcast station served by the skywave and not subject to objectionable interference and in which the signal is subject to intermittent variations in strength.

Secondary Station In a data communication network, the secondary station responsible for performing unbalanced link-level operations, as instructed by the primary station. A secondary station interprets received commands and generates responses.

Secondary Winding The minor winding on a relay having two windings. The winding on a transformer that is not connected to a AC source.

SECORD SEcure voice CORD board.

Secretarial Hunting The secretary's station number is programmed as the last number in one or more hunt groups. If all phones within a hunt group are busy the call will hunt to the secretary.

Secretarial Intercept A PBX feature. Causes calls for an executive to ring his/her secretary — even if the executive's direct extension number was dialed. The executive's phone will ring only if the secretary's phone is placed on "Do Not Disturb" or the secretary transfers the call in.

SECTEL Acronym for SECure TELephone.

Sectional Center A control center connecting primary telco switching centers. A Class 2 office. The next to the highest rank (Class 2) Toll Switching Center which homes on a Regional Center (Class 1).

Sector A pie-shaped portion of a hard disk. A disk is divided into tracks and sectors. Tracks are complete circuits and are divided into sectors. Under MS-DOS a sector is 512 bytes.

Sectoring The process of dividing a mobile cellular radio cell into 180-, 120- or 60-degree assigned to the cell.

Secure Kernel The core of a secure operating system.

Secure Public Dial A security term. Secure Public Dial is dialup switching functionality that allows the service provider to offer customers the security of a private connection with the economics of public dial. Also referred to as Virtual Private Networking (VPN) allows the service provider to support multiple enterprises' dial-in requirements, securely from the same dialup switch. Each separate customer has access to their own virtual network although they may share physical ports and access paths. The service provider manages the multicustomer net with the same ease as if it were one internetwork.

Secure Telephone Unit STU. A U.S. Government-approved telecommunication terminal designed to protect the transmission of sensitive or classified information — voice, data and fax.

Secure Voice Voice signals that are encoded or encrypted to prevent unauthorized listening.

Secure Voice Cord Board SECORD. A desk-mounted patch panel that provides the capability for controlling 16 wideband (50 Kbps) or narrowband (2400 bps) user lines and five narrowband trunks to AUTOVON or other DCS narrowband facilities.

NEWTON'S TELECOM DICTIONARY

SecureNet A service from Southwestern Bell in which customers' dedicated circuits are routed through a serving central office to a "primary" hub. In addition, a secondary path is routed via a SWBT-designated alternate serving central office and an alternate hub. The service is designed to give Southwestern Bell customers better reliability.

Security A way of insuring data on a LAN is protected from unauthorized use. Network security measures can be software-based, where passwords restrict users' access to certain data files or directories. This kind of security is usually implemented by the network operating system. Audit trails are another software-based security measure, where an ongoing journal of what users did what with what files is maintained. Security can also be hardware-based, using the more traditional lock and key.

Security Access Manager SAM. An AT&T computer-based building access security system.

Security Blanking The ability of a switch to blank out the called digits for certain extensions so no called number detail is printed. Senior executives in senior takeover negotiations find this feature useful. There have been instances of people figuring out which company another company is about to buy based on telephone calling records. If you have this information, you can buy the company's stock before the bid is announced and make a lot of money. This feature — security blanking — is designed to avoid such occurrences.

Security Cabinet A cabinet, usually on casters, used to store confidential materials under lock and key prior to shredding.

Security Code 1. A user identification code required by computer systems to protect information or information resources from unauthorized use. 2. A six-digit number used to prevent unauthorized or accidental alteration of data programmed into cellular phones. The factory default is 000000.

Security Equivalence A security equivalence allows one user to have the same rights as another. Use security equivalence when you need to give a user temporary access to the same information or rights as another user. By using a security equivalence, you avoid having to review the whole directory structure and determine which rights need to be assigned in which directories.

Security Management Protects a network from invalid accesses. It is one of the management categories defined by the ISO (International Standards Organization).

Security Modem A class of modem providing secure access.

Security Stud A cylindrically shaped metal finger that holds open the door to a Cash Box until the box is removed for collection.

Seedware Software designed to get demand for a product or a new market segment started. Software designed to "seed" a market. Seedware is typically a less-full featured piece of software than the software you're really trying to sell. Seedware typically costs very little. It may even be free. It's also called FEEDWARE.

Seek Time The time it takes to find some information on a disk (hard or soft) in a computer. Average seek time is a critical measure of the speed of a computer disk drive.

SEF Source Explicit Forwarding. Security feature that allows transmissions only from specified stations to be forwarded by bridges.

Segment 1. 64 characters. Use as a method of data communications billing by some overseas phone companies.
2. An electronically continuous portion of a network, usually consisting of the same wire.
3. A single ATM link or group of interconnected ATM links of an ATM connection.

Seize To access a circuit and use it, or make it busy so that others cannot use it.

Seizure Signal A signal used by the calling end of a trunk or line to indicate a request for service.

SEL Selector: A subfield carried in SETUP message part of ATM endpoint address Domain specific Part (DSP) defined by ISO 10589, not used for ATM network routing, used by ATM end systems only.

Select Call Forwarding A name for a Pacific Bell (and possibly other local telephone companies') service which allows you to have calls from selected numbers ring at another number.

Selectvideo Plus A service from Southwestern Bell which it describes as multirate ISDN, Nx64, wideband ISDN, dialable wideband ISDN or switched fractional DS1. According to Southwestern Bell, "it enables users to select a communications bandwidth on a per-call basis between 64 kilobits per second an 1.536 megabits per second." I called Southwestern Bell for more information. But the company's PR person, Chris Horner, couldn't explain how the service technically worked, nor what it cost.

Selection What most people call a "block" in a word processed document, Microsoft calls a "selection." If you want to print test you've blocked, tell Windows to print a selection.

Selective Calling The ability of the transmitting phone to specify which of several phones on the a line is to receive a message.

Selective Fading Fading in which the components of the received radio signal fluctuate independently.

Selective Paging To Station A phone can page to individual phone instruments.

Selective Ringing A method of ringing only the desired party on a party line.

Selective Signaling A method of inband signaling used on private networks to tell switches to switch the call.

Selectivity Ability of a tuner or receiver to get only a desired station while rejecting other adjacent stations. The higher the figure expressed in decibels (dB), the better the selectivity.

Selector The identifier (octet string) used by an OSI entity to distinguish among multiple SAPs at which it provides services to the layer above.

Self Diagnostics Your phone system tells you when something is wrong with it by sending you a "message" via the operator console or through one of the data ports on the phone system.

Self Electro-Optic Effect Devices SEEDs. Switches guided by light.

Self Extinguishing The characteristic of a material whose flame is extinguished after the igniting flame is removed.

Self Healing Brian Livingston, author of books on Windows and InfoWorld columnist, created this definition. He says that self healing software upgrades itself by periodically calling a central phone number (via bulletin board) or a Web site and automatically downloading and installing any new components that may have become available since the last product upgrade.

Self Hosted An SCSA definition. The Client and the Server are on the same computer platform. A server with one or more self-hosted applications may be a standalone unit which is not connected to any other system.

535

Self Test The capability of a PBX to run programs at regular intervals to test its own operation and signal when failures have occurred or are about to occur without human intervention.

Semaphore A message sent when a file is opened to prevent other users from opening the same file at that time. Its purpose is to preserve the integrity of data (i.e. stop it from being messed with) while you're using it.

Semi-Automatic Message Switching A network control technique whereby an operator manually switches a message to a destination according to the address information contained in the message header.

Semi-Rigid A cable containing a flexible inner core and a relatively inflexible sheathing material, such as a metallic tube but, which can be bent for coiling or spooling and placing in duct or cable run.

Semiconductor Material which has a resistance to electricity somewhere between a conductor (e.g. a copper wire) and an insulator (e.g. plastic). Hence the word "semi" conductor. Silicon and germanium are the two most commonly-used semiconductor materials. The flow of current in a semiconductor can be changed by light or the presence or absence of an electric or magnetic field.

Semipermanent Connection A connection established via a service order or via network management.

Sender Equipment in the originating telephone system which outpulses (sends out) the routing digits and the called person's number. Senders are necessary in computerized (stored program control) switches because the switch needs to know all the digits of the numbers you are calling before it chooses and seizes a trunk.

Sendmail Sendmail is the UNIX software that handles electronic mail. It is the most common form of electronic mail software on the Internet. Sendmail provides back-end message routing and handling for Simple Mail Transfer Protocol-based electronic mail communications.

Seniority A call center term. A field in each employee record establishing that employee's seniority for the purpose of automatic schedule assignment. This is typically the employee's hire date (year, month, and day) plus a three-digit "tie breaker."

Sensitivity 1. The input signal level required for a tuner, amplifier, etc., to produce a stated output. The lower the required input, the higher the sensitivity. Measured in mV (microvolts). 2. The degree to which a radio receiver responds to the wave to which it is tuned.

Sensitivity Analysis The process of rerunning a financial study to figure the degrees to which changing the assumptions changes the result of the analysis.

Sensor Applications Services involving the remote monitoring of instruments including burglar alarms, fire alarms, meter reading, energy management and load shedding.

Sensor Glove An interface device for experiencing virtual reality with the hand. Wired with sensors, it detects changes in finger, hand and arm movements and relays them to the computer, allowing users to manipulate and move things in a virtual environment. See VIRTUAL REALITY.

Sent Paid A utility industry term that describes all calls charged to the originating number or collected as coins in a pay telephone.

Separation Extent to which two stereo channels are kept apart. Expressed in decibels, the larger the number, the better the separation and stereo effect.

Separations And Settlements A complex set of accounting procedures developed by the traditional telephone industry. The procedures classify telephone plant as intrastate or interstate, and return revenues from long distance phone calls to local telephone companies. As part of Deregulation and Divestiture, this procedure is being replaced by per minute access charges.

Separator Pertaining to wire and cable, a layer of insulating material such as textile, paper, etc. placed between a conductor and its insulation, between a cable jacket and the components it covers or between various components of a multi-conductor cable. It can be utilized to improve stripping qualities and/or flexibility, or can offer additional mechanical or electrical protection to the components it separates.

SEPT Signaling End Point Translator, part of Signaling System 7. See SIGNALING SYSTEM 7.

Sequence A call center term. A pattern of days on and days off as defined in either a schedule preference or shift definition.

Sequencing Sequencing is the process of dividing a data message into smaller pieces for transmission, where each piece has its own sequence number for reassembly of the complete message at the destination end. Sequencing is thus also the process of properly ordering the receipt of packet data at their destination, regardless of the time they have taken to travel the X.25 network. It's similar to packetizing. See PACKET.

Sequencing Receivers All GPS (Global Positioning System) receivers must receive information from at least four satellites to calculate accurately where they (and thus you) are. Sequencing receivers use a single channel and move it from one satellite to the next to gather this data. They usually have less circuitry so they're cheaper and they consume less power than receivers which work on four satellites simultaneously. Unfortunately the sequencing can interrupt positioning and can limit their overall accuracy.

Sequential Pertaining to events occurring in a specific time or code order.

Sequential Access The need to read data — one record after another in sequence — before getting to the information you want. Magnetic tape, for example, requires you to read the entire tape up to where your information is. This is because the computer cannot tell where on the tape your information is because records on tape files are often of variable length. Most Random Access files, usually kept on a disk drive, require records to be of a fixed length, such as 80 characters per record. Then when you seek record 23, the computer seeks character 1840 in the file (23 x 80), and takes the next 80 characters as the record you want. Using the analogy of music recorded on records and magnetic tape, a phonograph record needle has the capability of random access because the needle can be set down in the spaces between cuts on the record. With mag tape, you must fast forward past all the music you don't want before you get to the music you want to hear. Also, if the computer tape drive is fast forwarding, it cannot count characters to find the record, and must read in the data you don't want (and throw it away), before it gets to the data you need. Random access is much faster than sequential access.

Sequential Logic Element A device that has at least one output channel and one or more input channels, all characterized by discrete states, such that the state of each output channel is determined by the previous states of the input channels.
Sequential Packet Exchange SPX. Novell's implementation of SPP for its NetWare local area network operating system.
Serial Bus Serial Bus was the original name for Intel's standard for a type of very local local area network that would be used for connecting peripherals to the motherboard of a PC. There'd be one plug on the back of the PC into which you'd daisy chain various peripherals, including a mouse, a keyboard, speakers, a microphone and a telephone. The idea of serial bus is to clear away all the clutter on the back of the PC. In March of 1995 when the first technical specs were released, serial bus' name was changed to Universal Serial Bus (I don't know why). See UNIVERSAL SERIAL BUS.
Serial Call Telephone system feature set up by the attendant when an incoming calling party wishes to speak with more than one person internally. When the first party hangs up, the call automatically moves to the second person the outside party wants to speak with. When that person hangs up, then the call automatically goes to the third person, etc.
Serial Communication Networks (local and long distance) use the RS-232 serial communications standard to send information to serial printers, remote workstations, remote routers, and asynchronous communication servers. The RS-232 standard uses several parameters that must match on both systems for information to be transferred. These parameters include baud rate, character length, parity, stop bit, and XON/XOFF.
Baud rate is the signal modulation rate, or the speed at which a signal changes. Since most modems or serial printers attached to personal computers send only one bit per signaling event, baud can be thought of as bits per second. However, higher-speed modems may transfer several bits per signal change. Typical baud rates are 300, 1200, 2400, 4800, 9600 and 19,200. The higher the number, the greater the number of signal changes and, therefore, the faster the transmission.
Character length specifies the number of bits used to form a character. The standard ASCII character set (including letters, numbers, and punctuation) consists of 128 characters and requires a character length of 7 bits for transmissions. Extended character sets (containing line drawings or the foreign characters used in IBM's extended character set) contain an additional 128 characters and require a character lengths of 8 bits. Parity error checking can only be used with character lengths of 7 bits.
Parity is a method of checking for errors in transmitted data. You can set parity to odd or even, or not use parity at all. When the character length is set to 8, parity checking cannot be done because there are no "spare" bits in the byte. When the character length is 7, the eighth bit in each byte is set to 0 or 1 so that the sum of bits (Os and 1s) in the byte is odd or even (according to the parity setting). When each character is received, its parity is checked again. If it is incorrect (because a bit was changed during transmission), the communications software determines that a transmission error has occurred and can request that the data be retransmitted.
Stop bit is a special signal that indicates the end of that character. Today's modems are fast enough that the stop bit is always set to one Slower modems used to require two stop bits.
XON/XOFF is one of many methods used to prevent the sending system from transmitting data faster than the receiving system can accept the information. See also EIA/TIA-232-E, RS-232-C and the APPENDIX.
Serial Interface The "lowest common denominator" of data communications. A mechanism for changing the parallel arrangement of data within computers to the serial (one bit after the other) form used on data transmission lines and vice versa. At least one serial interface is usually provided on all computers for the connection of a terminal, a modem or a printer. Sometimes also called a serial port. See EIA/TIA-232-E, RS-232-C, SERIAL INTERFACE CARD and the APPENDIX.
Serial Interface Card A printed circuit card which drops into one of the expansion slots of your computer and changes the parallel internal communications of your computer into the one-bit-at-time serial transmission for sending information to your modem or to a serial printer.
Serial Memory Memory medium to which access is in a set sequence and not at random.
Serial Port An input/output port (plug) that transmits data out one bit at a time, as opposed to the parallel port which transmits data out eight bits, or one byte at a time. Most personal computers (PCs) have at least one serial and one parallel port. In a typical configuration, the serial port is used for a modem while the parallel port is used for a printer. For a diagram of a typical 25-pin RS-232-C serial port, see the APPENDIX at the back of this book.
Serial Processing Method of data processing in which only one bit is handled at a time.
Serial Transmission Sending pulses one after another rather than several at the same time (parallel). When transmitting data over a telephone line there is only one set of wires. Therefore, the only logical way to transmit it is to send the data in serial mode. It is possible to use eight different frequencies to transmit a character all at once (parallel), but these modems are ridiculously expensive. See SERIAL PORT and PARALLEL.
Serialize To change from parallel-by-byte to serial-by-bit.
Series A connection of electrical apparatus or circuits in which all of the current passes through each of the devices in succession or on after another. See also PARALLEL.
Series 11000 An AT&T private line long distance tariff created in the 1970s and designed expressly to reduce MCI's chances of selling any private lines and thus of surviving. It was thrown out by the FCC and the tariff figured in MCI's and the Federal Government's antitrust against AT&T.
Series Connection A connection of electrical apparatus or circuits in which all of the current passes through each of the devices in succession or on after another. See also PARALLEL.
Series RF Tap A bugging device. It is a radio transmitter which is installed in series with one wire of the telephone circuit. Normally a parasite (i.e. takes power from the phone line). Transmits both sides of the conversation. It transmits only when the phone is off-hook.
Server A server is a shared computer on the local area network that can be as simple as a regular PC set aside to handle print requests to a single printer. Or, more usually, it is the fastest and brawniest PC around. It may be used as a repository and distributor of oodles of data. It may also be the gatekeeper controlling access to voice mail, electronic-mail, facsimile services. At

one stage, a local area network had only one server. These days networks have multiple servers. Servers these days have multiple brains, large arrays of big disk drives (often in redundant arrays) and other powerful features. New powerful servers are called superservers. A $35,000 superserver today can match the performance of a $2 million mainframe of ten years ago. Then again, according to Peter Lewis of the New York Times, the lowliest client today has more computing power than was available to the entire Allied Army in World War II. See DOWNSIZING for some of the benefits of running servers as against mainframes.

Server API A SCSA term. A communications protocol that allows a call processing application running on one computer to control SCSA hardware residing in another computer.

Server Application A Windows NT application that can create objects for linking or embedding into other documents.

Server Farm Imagine a room stuffed with PCs, ranged in racks along walls, ranged in racks in lines like a library's back room. The PCs are really servers — powerful PCs containing databases and other information they are dispensing to the thousands of PCs dialing into them from afar. A server farm may be owned by one company and used by one company, or it may be owned by one company and each of the machines leased to other companies. I first heard the term when MCI described a room it had in a place called Pentagon City. There it had hundreds of servers each of which it leased to other companies who used those servers as their Web sites.

Server Message Block SMB. The protocol developed by Microsoft, Intel, and IBM that defines a series of commands used to pass information between network computers. The redirector packages SMB requests into a network control block (NBC) structure that can be sent over the network to a remote device. The network provider listens for SMB messages destined for it and removes the data portion of the SMB request so that it can be processed by a local device. In short, SMB is basically a protocol to provide access to server-based files and print queues. SMB operates above the session layer, and usually works over a network using a NetBIOS application program interface. SMB is similar in nature to a remote procedure call (RPC) that is specialized for file systems.

Server Mirroring Server mirroring means you have two servers on your networks and each exactly what the other is doing simultaneously. It's a backup method. In Novell's NetWare, server mirroring requires two similarly configured NetWare servers. They should be evenly matched in terms of CPU speed, memory, and storage capacity. The servers are not required to be identical in terms of microprocessors type (386/486), microprocessor revision level, or clock speed. However, identical servers are recommended for NetWare SFT III 3.11. If the two servers are unequal in terms of performance, then SFT III performs at the speed of the slower server. The NetWare servers must be directly connected by a mirrored server link. SFT III servers can reside on different network segments, as long as they share a dedicated mirrored server link.

Server Operating System An SCSA definition. Operating System running on the SCSA Server.

Server Push Server push is a Internet term. With server push, the Web server sends data to display on the browser display, but leaves the connection open. At some point, the server sends additional data for the browser to display. Server push is used for displaying multimedia information on the browser.

Server Telescript A General Magic term. Telescript deployed on a server, enabling Telescript clients to participate in smart messaging.

Severely Errored Second SES. A second during which the bit error ratio is greater than a specific limit. During a severely errored second, transmission performance is significantly degraded. A Bellcore term.

Service Access Code SACs are 3-digit codes in the NPA (N00) format that are used as the first three digits of a 10-digit address, and that are assigned for special network uses. Whereas NPA codes are normally used for identifying specific geographical areas, certain SACs have been allocated in the NANP (North American Numbering Plan) to identify generic services or provide access capability. Currently only four SACs have been assigned and are in use: 600, 700, 800, and 900.

Service Affecting This definition courtesy Steve Gladstone, author of the book, "Testing Computer Telephony Systems" (available from 212-691-8255). These are major bugs that significantly impact the reliability or the functionality of CT systems. Comprehensive testing must uncover all service affecting problems with the goal that no computer telephony system should be installed at a live customer installation without an acceptable workaround for the service affecting problem. The goal for transition of a product from one phase to the next is that no service affecting bugs remain, and that the bug rate for new bugs be at or approaching zero.

Service Area The geographic area served by a supplier. The area in which the supplier, theoretically, stands ready to provide his service. The service area of New York Telephone is most (not all) of New York State.

Service Bureau A data processing center that does work for others. There are many ways of bringing work to a service bureau, including mailing it and transmitting it over phone lines. If it comes over phone lines, the service is likely to be called "time sharing."

Service Charge The amount you pay each month to receive cellular service. This amount is fixed, and you pay the same fee each month regardless of how much or how little you use your cellular phone. It usually ranges from about $10 to $65 per month, depending on the carrier's tariffs and the particular plan of service you select. In addition you page air time. Service Charge doesn't usually include any air time.

Service Charge Detail A listing of all the telephone equipment installed as part of a specific telephone system. Usually provided by the vendor or maintenance organization.

Service Code A 3-digit code in general use by customers to reach telephone company service, for example, 411 (Directory Assistance), 611 (Repair Service), 811 (Business Office) and 911 (Emergency Service).

Service Control Point SCP. The local versions of the national SMS/800 number database. SCPs contain the intelligence to screen the full ten digits of an 800 number and route calls to the appropriate, customer-designated long distance carrier. Bellcore defines SCP as the network system in the Advanced Intelligent Network Release 1 architecture that contains SLEE (Service Logic Execution Environment) functionality and communicates with AIN Release 1 Switching Systems in processing AIN Release 1 calls.

Service Creation The set of activities that must be performed to create a new service to be offered to subscribers and the associated service-specific operations capabilities to support the new service. Definition from Bellcore in reference to its concept of the Advanced Intelligent Network.
Service Creation Environment See SCE.
Service Display Displays a specific service being presently in effect.
Service Entrance The point at which network communications lines (telephone company lines) enter a building.
Service Function A primary or secondary service function.
Service Identification The information uniquely identifying an NS/EP telecommunications service to the service vendor and/or service user. NS/EP is a federal government definition. It refers to telecommunications services that are used to maintain a state of readiness or to respond to and manage any event or crisis (local, national, or international) that causes or could cause injury or harm to the population, damage to or loss of property, or degrade or threaten the national security or emergency preparedness of the United States.
Service Independent BAF The finite set of BAF structures and modules needed to record usage of Advanced Intelligent Network Release 1 services. The set is said to be robust because it will be designed to record future AIN Release 1 services that have not yet been identified. Definition from Bellcore in reference to its concept of the Advanced Intelligent Network.
Service Level Usually expressed as a percentage of a statistical goal. For example, if your goal is an average speed of answer of 100 seconds or less, and 80% of your calls are answered in 100 seconds or less, then your service level is 80%.
Service Line An exchange line associated with multiple data station installations to provide monitoring and testing of both customer and Telco data equipment.
Service Logic Execution Environment SLEE. A functional group residing in an SCP (Service Control Point) or Adjunct that contains the Service Logic and Control, Information Management, AMA (Automatic Message Accounting) and Operations FEs (Functional Entity). This composite set of capabilities, which includes FC routines, provides a functionally consistent interface to SLPs (Service Logic Program) independent of the underlying operating system. Definition from Bellcore in reference to its concept of the Advanced Intelligent Network.
Service Management System See SMS.
Service Negotiation The functionality needed to gather subscriber Advanced Intelligent Network Release 1 business needs; provide answers, AIN Release 1 service/feature descriptions, prerequisite non AIN services, availability and costs; reserve AIN Release 1 resources (e.g. 800 number and 900 number); identify required network resources; and identify available AIN Release 1 services/features by wire center. Definition from Bellcore in reference to its concept of the Advanced Intelligent Network.
Service Net-2000 A AT&T umbrella name for family of digital products and services.
Service Objectives If you're the phone company, service objectives are a statement of the quality of service that is to be provided to the customer; for example, no more than 1.5 percent of customers should have to wait more than three seconds for dial tone during the average busy hour or, the busy-hour blocking on a last choice trunk group not exceed 1 percent.
If you're designing phone systems, service objectives define the functional and performance goals for how your system will work and what will be experienced by the system's users and other systems. For example, a switch might have a service objective of providing dialtone to all users in less than two seconds. Or, of answering all incoming calls within ten seconds. Or, a VRU may have a service objective of responding to all user DTMF inputs in less than one second. When we buy or design systems, we often have many service objectives, rarely just one.
Service Observation 1. A generic word used by telephone companies to check the quality of the service they're providing. Some of it is done automatically with machinery. Some of it is done by senior operators who listen in on the conversations of other operators dealing with their subscribers. In short, the senior operators "observe" the service the junior operators are providing. 2. As a feature of some telephone systems, the Service Observation (SOB) command provides the capability to automatically record data about completed calls, incomplete calls and abnormal calls for the purpose of qualitative supervision of call traffic conditions.
Service Order Document or system generated group of information requesting installation or disconnection of, moving, or changing circuits.
Service Period The time during which the telephone company furnishes a circuit.
Service Points The points on the customer's premises where such channels or facilities are terminated in switching equipment used for communications with phones or customer-provided equipment located on the premises.
Service Profile Identifier See SPID.
Service Provider 1. A Windows Telephony Applications standard which lies between Windows Telephony and the network. It defines how the network — anything from POTS to T-1, from a Northern Telecom to an AT&T PBX — shall interface to Windows Telephony, which in turn talks to the Applications Programming Interface, which talks to the Windows telephony applications software. See WINDOWS TELEPHONY.
2. An organization that provides connections to a part of the Internet. If you want to connect your company's network, or even your personal computer, to the Internet, you have to talk to a "service provider."
3. A service provider is also a company which provides information to people who call up on a phone or on a modem.
4. An SCSA definition. An addressable entity providing application and administrative support to the client environment by responding to client requests and maintaining the operational integrity of the server.
Service Provider Messages An SCSA definition. The message information required by, and provided by, the service provider to perform its functions in the environment in which it is installed. Contrast with SCSA Message Protocol Interface. See SERVICE PROVIDER.
Service Quality A call center term. A measure of how well staffing matches workload, expressed often as average delay (in answering a call).

Service Terminal The equipment needed to terminate the channel and connect to the phone apparatus or customer terminal.

Service Traffic Management STM. The SLEE (Service Logic Execution Environment) functionality for detecting overloads associated with a specific service and for sending Automatic Code Gap messages to the appropriate entities. The SN&M (Service Negotiation and Management) OA (Operations Application) also provides STM(Service Traffic Management)-related capabilities.

Services Management System SMS. Administers 800 Data Base Service numbers on a national basis. Customer records for 800 Service are entered into the SCP through this system. See EIGHTHUNDRED SERVICE.

Services On Demand An AT&T term for the immediate provision of almost any network service through universal ports, whenever required by a user; as opposed to provision via an expensive, time consuming, inflexible service order process.

Serving Closet The general term used to refer to either a riser or a satellite closet; Satellite Cabinet; Satellite Closet.

Serving Office An office of AT&T or its Connecting or Concurring Carriers, from which interstate communications services are furnished.

Serving Wire Center The wire center from which service is provided to the customer.

Servo Short for servomechanism. Devices which constantly detect a variable, and adjust a mechanism to respond to changes. A servo might monitor optical signal strength bouncing back from a disc's surface, and adjust the position of the head to compensate.

SERVORD Service Order.

SES 1. Satellite Earth Stations.
2. Severely Errored Seconds. Errored seconds during which the error rate exceeded 10.
3. Source End Station: An ATM termination point, which is the source of ATM messages of a connection, and is used as a reference point for ABR services. See DES.

Session 1. An active communications connection, measured from beginning to end, between computers or applications over a network. Often used in reference to terminal-to-mainframe connections. Also a data conversation between two devices, say, a dumb terminal and a mainframe. It is possible to have more than one session going between two devices simultaneously.
2. As defined under the Orange Book, a recorded segment of a compact disc which may contain one or more tracks of any type (data or audio). The session is a purely logical concept; when a multisession disc is mounted in a multisession CD-ROM player, what the user will see is one large session encompassing all the data on the disc.

Session Layer The fifth layer — the network processing layer — in the OSI Reference Model, which sets up the conditions whereby individual nodes on the network can communicate or send data to each other. The session layer is responsible for binding and unbinding logical links between users. It manages, maintains and controls the dialogue between the users of the service. The session layer's many functions include network gateway communications.

Session Lead-In The data area at the beginning of a recordable compact disc that is left blank for the disc's table of contents. The session lead-in uses up 6750 blocks of space. See TRACK.

Session Lead-Out The data area at the end of a session which indicates that the end of the data has been reached. When a session is closed, information about its content is written into the disc's Table of Contents, and the lead-out and the pre-gap are written to prepare the disc for a subsequent session. The lead-out and the pre-gap together take up 4650 two-kilobyte blocks (nine megabytes). See TRACK.

Set 1. Set is another name for a telephone.
2. Switching System Exchange Termination. An SMDS term.

Set-Associative Mapping A caching technique where each block of main computer memory is assigned to a location in each cache set where the cache is divided into multiple sets.

Set-Top Box The electronics box which sits on top of your TV, connecting it to your incoming CATV signal and your TV's incoming coaxial cable. Set-tops vary greatly in their complexity with older models merely translating the frequency received off the cable into a frequency suitable for the television receiver while newer models can be addressable with a unique identity much like a telephone. That identity can be addressed from the cable headend. This allows the CATV operator to turn individual channels on and off, such as pay channels.

SETA SouthEastern Telecommunications Association, a user group.

SF 1. Single Frequency. A method of inband signaling. Single frequency signaling typically uses the presence or absence of a single specified frequency (usually 2,600 Hz). See SIGNALING.
2. SuperFrame: A DS1 framing format in which 24 DSO timeslots plus a coded framing bit are organized into a frame which is repeated 12 times to form the superframe.

SFBI Shared Frame Buffer Interconnect, a specification that makes it possible for hardware manufactures to produce a single-board video-graphics adapter for the PC.

SFC Switch Fabric Controller.

SFG Simulated Facility Group.

SFQL Structured Full-Text Query Language. A proposed standard for full-text databases. The primary focus of the proposed standard is interoperability of CD-ROMs. SFQL is based on the SQL (Structured Query Language) standard for relational databases.

SFT System Fault Tolerance. The capability to recover from or avoid a system crash. Novell uses a Transaction Tracking System (TSS), disk mirroring, and disk duplexing as its system recovery methods. System Fault Tolerance as a Novell NetWare term means data duplication on multiple storage devices. If one storage device fails, the data is available from another device. There are several levels of hardware and software system fault tolerance. Each level of redundancy (duplication) decreases the possibility of data loss.

SFTA Scalable Fault Tolerant Architecture.

SG Study Group. The ITU-T has formalized committees studying future telecommunications standards. These groups are called Study Groups.

SGML Standard Generalized Markup Language. A text-based language for describing the content and structure of digital documents. HTML, which has gained fame as the language to create World-Wide Web pages on the Internet, is a descendant of SGML. SGML documents are viewed with transformers, which render SGML data the way Web browsers render HTML data. SGML was adopted by the International Standards Organization in 1986. SGML allows organizations to structure and manage information in a cross-platform, application-independent way. It tags documents as a series of data objects rather than storing them as huge files. Theoretically, SGML can reduce errors, slice costs and speed work. SCML attempts to separate the informational content of a document from the information needed to present it, either on paper or on screen.

Shadow BIOS ROM Shadow BIOS ROM is a concept I first found in Toshiba laptops which use Flash ROM to hold the machine's BIOS. When you start the machine, the BIOS copies itself from the flash ROM to the Shadow BIOS area. Accessing the BIOS from the Shadow BIOS is much faster than from flash ROM. I learned later that Compaq actually started what they called shadowing the BIOS. According to InfoWorld, Compaq did it because PC-compatible systems available at the time could have no more than 16 megabytes of RAM. Compaq decided to use the memory address at the top of the 15-megabyte physical address space for the shadow RAM.

Shadow Mask The most common type of color picture tube in which the electron beam is directed through a perforated metal mask to the desired phosphor color element.

Shadow ROM A process used in many 386 machines to map ROM BIOS activities into faster 32-bit RAM memory. Shadow memory must be loaded with BIOS routines each time the computer boots. See also SHADOW BIOS ROM.

Shannon A measurement of the quality of information in a message represented by one or the other of two equally probable, exclusive and exhaustive states. See SHANNON'S LAW.

Shannon's Law A statement defining the theoretical maximum rate at which error-free digits can be transmitted over a bandwidth-limited channel in the presence of noise.

Shaping Descriptor An ATM term. N ordered pairs of GCRA parameters (I,L) used to define the negotiated traffic shape of a connection.

Shaping Network A network inserted in a circuit for improving or modifying the wave shape of the signals.

Share To make resources, such as directories, files, printers, and ClipBook pages, available to network users.

Shared Lock In a database a shared lock is created by non update (read) operations. Other users can read the data concurrently, but no transaction can acquire an exclusive lock on the data until all the shared locks have been released.

Shared Logic Simultaneous use of a single computer by multiple users.

Shared Memory Portion of memory accessible to multiple processes.

Shared Modem Pools Dial-out users share resources. Any authorized user attached to the network can dial out a port on the dial-up switch, reach a modem and go for it. Benefits: Reduced costs, improved management and security; It eliminates the need for separate modem and separate modem phone lines.

Shared Resource Any device, data, file or program that is used by more than one other device, program or person. For Windows NT, Windows 95 and Windows for Workgroups, shared resources refer to any resource that is made available to network users, such as directories, files, printers, and named pipes.

Shared Screens A multimedia concept. Shared screen applications enable two or more workstations to display the same screen simultaneously. For example, two users sharing a screen can work on the same spreadsheet. Changes made by one user can be seen by the other as they are made. Shared screens can be implemented in two ways. One way enables people to view each other's screen while one person makes changes. The other way enables people to run the same application on both screens so that both users can make changes simultaneously.

Shared Services Providing PBX-based communications and processing services to the unaffiliated tenants and/or the building manager/owner of a commercial building in a standalone or campus environment.

Shared Tenant Services Providing centralized telecommunications services to tenants in a building or complex.

Shared Whiteboards A multimedia concept. Shared whiteboards enable you to "mark-up" a screen using a mouse or stylus input device and have the results show on other screens, often communicating over long distance telephone lines. The concept is similar to a traditional whiteboard mark-up process where everyone has a different color marking pen to circle, write, or cross out items. The background board can be a window from the workstation such as a spreadsheet, image, or blank canvas, or it can be the entire workstation screen. The shared whiteboard can be used for either real-time or store-and-forward collaboration. In the store-and-forward scenario, the mark-ups can be implemented in a time-delayed fashion so everyone can follow the entire step-by-step process.

SHARP Self Healing Alternate Route Protection. A system typically employing redundant cables (often fiber) that carry traffic between two separate local exchange carrier offices along divergent paths.

Shear A computer imaging term. A tool for distorting a selected area vertically or horizontally.

Sheath The outer jacket (usually metal or plastic) surrounding copper and fiber cables that prevents water damage to the cables inside.

Sheath Miles The actual length of cable in route miles.

Shelf Life The useful life of components when not in use — such as being stored on a shelf as spare parts or in a warehouse awaiting shipment. Batteries tend to have the shortest shelf life of most telecommunications components. Today, the shelf life is less a problem of shelf decay and more a problem of technological obsolescence.

Shelfware Software that is bought, then placed on the shelf, but never used. See also FEEDWARE, HOOKEMWARE, HYPERWARE, MEATWARE, SEEDWARE, SHOVELWARE, SMOKEWARE SLIDEWARE and VAPORWARE.

Shell An outer layer of a program that provides the user interface, or the user's way of commanding the computer. Instead of presenting the user with a bland C prompt, i.e. C:> the shell presents a list of programs that the user can choose from, making it easier, allegedly, to figure out which program to run. The problem with shells is that they often take up precious memory. That memory might better be used in actually running a program faster, or more efficiently.

Shell Account An Internet term. A type of interface on a dial up connection in which you log in to the host computer and use a command shell to get to the Internet. Shell accounts are typically text-based-only interfaces controlled by host servers which normally don't allow for use of graphic Web browsers.

Shield A metallic layer consisting of type, braid, wire or sheath that surrounds insulated conductors in shielded cable. The shield may be the metallic sheath of the cable or the metallic layer inside a nonmetallic sheath. Shields reduce stray electrical fields and provide for safety of personnel. See SHIELD EFFECTIVENESS.

Shield Effectiveness The relative ability of a cable shield to screen our undesirable radiation. Frequently confused with the term shield percentage, which it is not.

Shielded Pair Two insulated wires in a cable wrapped with metallic braid or foil to prevent the wires acting as antennas and picking up external interference (e.g. a local TV station).

Shielded Twisted Pair A pair of insulated wires which are twisted together in a spiral manner. In addition, the pair is wrapped with metallic foil or braid, designed to insulate (i.e. shield) the pair from electromagnetic interference.

Shielding 1. The metal-backed mylar, plastic, teflon or PVC that protects a data-communications medium such as coaxial cable from Electromagnetic Interface (EMI) and Radio Frequency Interference (RFI).
2. The process by which electrical conductors are wrapped with metallic foil or braid to insulate them from interference and thus provide high quality transmission. Many devices can cause interference to cables (i.e. multiple conductors) carrying telecommunications conversations. Such things include high voltage AC power lines, machinery with motors, machines which make rays of some type (X-Ray systems, TV sets etc.). By wrapping conductors around the cable cores, these cables are less likely to be affected by these outside forces and the noise they create on telephone lines. Shielding will also lessen the chance that the information movement along the cable will interfere with signals on other, adjacent cables. The need for shielding stems from this phenomenon: If you send an electrical signal along one pair of cables, those cables will give off a small amount of electrical energy — called magnetic radiation. That radiation will cause electromagnetic interference with a cable close by. If you "shield" the pair carrying the electrical signal, you will cut down the susceptibility of those cables to interference from other cables. LANs should always be installed with the best quality shielded cable. They will run better with shielded cable. Never skimp on the quality of the cable you're installing for LANs. Most telephones don't require shielded cable unless the cable serving them is passing through some area of high electromechanical interference.

Shift 1. The movement of data to either the right or the left of an existing position in a data field. 2. The code control function of converting the characters from upper to lower case, or vice versa.

Shift Button This button acts exactly like a Shift button on a typewriter or computer. It gives the key you're touching a second meaning — either a capital letter or a second set of speed dial buttons, etc.

Shift Character The control character which defines the shift function.

Shift Definitions A call center term. A template from which the program can create schedules during a scheduling run. Each shift definition is a record in a Scenario giving more or less precise instructions on shift length, time of day, breaks, and how extensively such schedules can be used in meeting staffing requirements.

Shift Register A register in which a clock pulse causes the stored data to move to the right or left one bit position. See ZERO STUFFING.

Ship To Shore Telephone See MARINE TELEPHONES.

Shock A sudden stimulation of the nerve and convulsive contraction of the muscles caused by a discharge of electricity through the body. The severity depends on the amount and duration of the current and whether the path of the current is through a vital organ.

Shoebox A shoebox is a housing device with a power supply to support external peripherals. When the user's main computer case is filled to capacity, an external device is needed to handle the overflow. Hence a shoebox.

Short A circuit impairment that exists when two conductors of the same pair, which normally make up an operating electrical circuit, touch or are connected.

Short Bus A high-speed common channel in the AT&T ISDN packet controller over which all messages between sending and receiving devices pass.

Short Circuit A near zero resistance connection between any two wires that disrupts transmission where two pairs are involved usually called a "cross." It disrupts transmission and may cause an excessive current flow. In AC electricity, a short circuit is an unintended connection between two supply conductors (i.e.: HOT and Neutral conductors.) A short circuit will usually cause high current flow and will operate the over current protection (fuse or breakers) to interrupt the circuit.

Short Haul Between a few hundred yards and 20 miles. Many people would argue with this definition.

Short Haul Modem A data set designed for use in communicating data up to distances of 25 miles over a dedicated unloaded copper pair. Many people would argue with this definition.

Short Message Service Center See SMSC.

Short Reach Short Reach refers to optical sections of approximately 2 km or less in length. The sections may be interoffice or intra office sections. A SONET term.

Short Tones First, we invented touchtone, also called DTMF, Dual Tone Multi Frequency tones. You'd punch your number with tones, instead of dialing them. Then someone thought you could control telephone response gadgets, like voice mail, interactive voice response, etc. with touch tones. For these gadgets to work, they had to "hear" the tones you sent. No one really set standards as to the minimum length tone they would hear. But it was generally conceded that they were to be 120 milliseconds. So some manufacturers of telephone equipment started to make phone equipment that, if you pushed a touchtone button, the machine would only sent a touchtone of 120 millisecond duration. That was called a short tone. It wasn't very useful because the manufacturers quickly discovered that many pieces of equipment couldn't respond that quickly. And the manufacturers got complains that their customers couldn't call their voice mail, their bank, etc. As a result, some manufacturers of equipment brought out new hardware (replacing the old) to allow you to send "long tones," which are now defined as touchtones that last for as long as you hold down the button — just as it is (and has always been) on a normal single line, non-electronic, non-digital telephone. Isn't progress wonderful? See also DTMF for a much longer explanation of tone dialing.

Shoulder Surfing You're standing at a pay phone. You punch in your credit card numbers to make your long distance call. There's a fellow standing behind who. He's carefully watching what you're doing. He is memorizing the digits you have punched in. When you are through, he will write them down and sell them to someone else, who will use them to make fraudulent long distance phone calls. Our friend is indulging in a new "occupation." It's called "shoulder surfing."

Shovelware A term used to refer to the tendency of early CD-ROM disc producers to shovel anything they could to fill up their voluminous CD-ROM discs. They did this so they could tout all the great value they were offering in their CD-ROM discs. See also HOOKEMWARE, HYPERWARE, MEATWARE, SLIDEWARE, VAPORWARE and BUNDLE FODDER.

SHP Signaling Handoff Point, a type of equipment which acts as a gateway between two dissimilar Signaling System 7 networks, allowing information exchange between the two networks.

Shrink Wrap Software that requires no customization. So called because it comes in a package that is plastic shrink-wrapped. Of course, so does non-shrink wrap software, or software that has to be customized for your use. But this is a dictionary that explains what words mean. It doesn't explain the lack of logic behind the naming of those words.

Shrinkage Percentages A call center term. A group of Scenario budget assumptions that define the percentage of the time employees are scheduled to work but are not available to handle calls because of absence, breaks, vacation, non-productivity, training, and other activities.

SHT Short Hold Time.

SHTTP Secure Hypertext Transfer Protocol. An extension of HTTP for authentication and data encryption between a Web server and a Web browser.

Shunt Circuit An arrangement of apparatus or circuits in which the total current is subdivided. Same as Parallel Circuit.

SI Shift In.

SIA Securities Industries Association.

SIBB Service Independent Building Blocks, a term coined by Bellcore and the ITU-T for the Intelligent Network. Creation of SIBBs will in theory make it easier for non-software specialists to create new services by mixing and matching SIBBs. SIBBs are like software objects — capsules of resuable software that can be combined together to create wondrous and complex new computer telephony services.

SID System IDentification number. A five digit number that has been assigned to identify the particular cellular carrier from whom you are obtaining service. This number identifies your "home" system.

Side Circuit A metallic, single pair circuit arranged to derive a phantom circuit. The phantom circuit is derived by center tapping a repeating coil in each of two side circuits.

Side Hour Any hour that is not the Busy Hour.

Sideband The frequencies on either side of the main frequency in a telecommunications signal. In the early days these sideband frequencies were not used because they were too "noisy," unreliable and were not needed. Now, technology has improved and frequencies are in short supply, more and more transmission vendors are making use of their sidebands, and thus substantially broadening the throughput of their existing transmission paths. Sideband technology has made great strides especially in through-the-air microwave transmission.

Sidetone A part of the design of a telephone handset which allows you to hear your own voice while speaking. The idea is to let you know that the telephone you're speaking on is working. Too much sidetone becomes an echo and is bad. Too little sidetone makes the channel unerring.

SIDN Security Industry Digital Network.

SIG 1. Special Interest Group. A SIG is an ongoing discussion group held electronically via PCs. A SIG focuses on one area of interest. Members phone in with their PCs, read messages posted, contribute their wisdom, ask questions, etc. SIGs are ways people get up-to-date accurate information on a subject. SIGs are run on most BBS (Bulletin Board Systems). SIGs are to bulletin boards what on-line services call conferences or forums.

2. Standard Image Format. See MPEG.

SIG-SMDS Interest Group A consortium of vendors and consultants who are committed to advancing worldwide SMDS as an open, interoperable solution for high-performance data connectivity.

Sign-On To go through the process of beginning a working session between you, your data terminal or PC and a computer system.

Sign On/Sign Off The process of identifying oneself to a machine so as to gain access. In the case of an ACD system this process allows statistics to be kept for this person individually. It also allows for the movement of the person around the system while statistics are accumulated in one logical file.

Signal 1. An electrical wave used to convey information.

2. An alert.

3. An acoustic device (e.g. a bell) or a visual device (e.g. a lamp) which calls attention. To transmit an information signal or alerting signal.

4. Bellcore defines signal as a state that is applied to operate and control the component groups of a telecommunications circuit to cause it to perform its intended function. Generally speaking, says Bellcore, there are five basic categories of "signals" commonly used in the telecommunications network. Included are supervisory signals, information signals, address signals, control signals, and alerting signals.

Signal Computing Signal computing, as it has come to be, refers to the processing of analog signals for transmission over the worldwide telephone system. According to Analog Devices, the core of the Signal Computing concept is the ability to program an open, multi function chipset with easily upgraded software. The Signal Computing model gives non-traditional audio, video, speech, and communications technology providers opportunities to embrace PC channels. The term, signal computing, picked up steam in mid-1993 when Dialogic announced a new standard called SCSA (Signal Computing System Architecture). In Dialogic's words, SCSA is a comprehensive architecture that describes how both hardware and software building blocks work together. It focuses on "Signal Computing" devices, which refer to any devices that are required to transmit information over the telephone network. Information can be transmitted via data modems, fax, voice or even video. SCSA defines how all these devices work together.

Signal computing systems combine three major elements for call processing. Network interfaces provide for the input and output of signals transmitted and switched in telecommunications networks. Digital signal processors and software algorithms transform the signals through low-level manipulation. Application programs provide computer control of the processed signals to bring value to the end user.

SCSA is the common set of standards that telecommunication system manufacturers and computing system manufacturers can use to create computer telephony systems. No single company today can create the total solution for all customers. SCSA represents the common ground between the two fields so that manufacturers from each area can safely develop products that will work with other manufacturers.

SCSA's coverage extends from low-level bus and hardware interfaces, like the inter-board switching bus that enables boards from different suppliers to work together, to high-level application programming and software interfaces, so that software designed to work with one set of hardware products, will work with different hardware as well. SCSA has been defined with input from leading computer and switch manufacturers, call processing suppliers, and technology developers. In many cases, SCSA has drawn on existing standards, like the T.611 fax standard endorsed by the European Computer Manufacturers Association, and in other cases SCSA has extended standards to make them more useful for call processing suppliers and users.

Signal Computing System Architecture SCSA is an open hardware and software architecture supported by over 240 companies for developing multi-technology computer telephony systems using standard interfaces. SCSA consists of two independent layers: The SCSA Hardware Model and the SCSA Telephony Application Objects (TAO) Framework. While designed for interoperability, these two layers may be implemented independently of one another. See SIGNAL COMPUTING.

Signal Conditioning The amplification and/or modification of electrical signals to make them more appropriate for transmission over a certain medium — cable, microwave, etc.

Signal Control Point SCP. Computers that hold databases in which customer-specific information used by the "advanced intelligent network" (AIN) to route calls is stored. The AIN refers to a specific architecture — typically promulgated and created by a local or long distance phone company — that provides core capabilities in which customer-specific information held in databases within the network is used to intelligently process calls. An example of an AIN service is an 800 service that routes calls based on where the calls are coming from. The routing information is stored in the SCP, which is typically tied into the Signaling System 7 network. See SCP. For a full explanation of the Advanced Intelligent Network, see AIN.

Signal Converter The equipment which changes the data signal into a form suitable for the transmission medium, or the reverse. The converter can also work with DC/AC current. The signal converter comprises a modulator and/or demodulator.

Signal Ground SGD. In RS-232-C signaling, Signal Ground establishes a common reference level for the voltages of all other signals (such as RXD/TXD), except Frame Ground.

Signal Level The strength of a signal, generally expressed in either absolute units of voltage or power, or in units relative to the strength of the signal at its source.

Signal Processing Signal processing is a combination of computer telephony call control and media processing. Call control means moving telephone calls around — answering them, hanging up on them, transferring them, conferencing them, etc. — all the stuff you do on your office phone every day. Media processing means bringing computer power to bear on the media stream, the actual voice, video or data inside the phone call. You might save it to your hard disk. You might want to have your PC software listen for key words in what you saved, for example, the caller's phone number. You might even want your software to try transcribing the conversation into text you could use in your word processor.
See CALL CONTROL, DIGITAL SIGNAL PROCESSING, MEDIA PROCESSING and NATIVE SIGNAL PROCESSING.

Signal Processing Component An SCSA definition. An atomic bundle of signal processing functionality which can be allocated to a single group. It can be capable of supporting the functionality of one or more resources or a simple coder.

Signal Processing Element An SCSA definition. That part of a Signal Processing Component which is associated with a single Resource.

Signal Processing Platform A SCSA definition. This is a software component that supports a specific hardware package. This is typically an executable program and may control one or more instances of the vendor-specific hardware package. Each SPP may support several different types of signal processing functionality clustered into SPCs.

Signal Strength Indicator A display on a cellular radio that lets you know before you call about the relative strength of the cellular transmitter in your immediate area. On most cellular radios, the signal strength indicator has five bars, with five the strongest. It's best to call when you have four or five. Three is marginal. Below three, forget it. Go elsewhere and try again.

Signal Switching Point SSP. See SSP. For a full explanation of the Advanced Intelligent Network, see AIN.

Signal To Noise Ratio 1. The ratio of the usable signal being transmitted to the noise or undesired signal. Usually expressed in decibels. This ratio is a measure of the quality of a transmission.
2. The amount of useful information to be found in a given ratio Internet Usenet newsgroup. Often used derogatorily, for example: "the signal-to-noise ratio in this newsgroup is pretty low."

Signal Transfer Point The packet switch in the Common Channel Interoffice Signaling (CCIS) system. The CCIS is a packet switched network operating at 4800 bits per second. CCIS replaces both SF (Single Frequency) and MF (Multi-frequency) by converting dialed digits to data messages. See SCP, STP and SIGNALING SYSTEM 7. For a full explanation of the Advanced Intelligent Network, see AIN.

Signals Intelligence SIGINT. A federal government term. A category of intelligence information comprising, either individually or in combination, all communications intelligence, electronics intelligence, and foreign instrumentation signals intelligence, however transmitted.

Signaling Pertains to the transmission of electrical signals to and from the user's premises and the telephone company central office. Examples of central office signals to the user's premises are ringing (audible alerting) signals, dial tone, speech signals, etc. Signals from the user's telephone include off-hook (request for service), dialing (network control signaling), speech to the distant party, on-hook (disconnect signal), etc. See SIGNALING SYSTEM 7.

In new connection oriented services like ATM, signaling takes on a slightly different meaning. For example, in ATM, signaling is used to dynamically establish, maintain and clear ATM connections at the ATM User Network Interface. In simple words, signaling can be described as follows: a user requesting a connection transmits a setup request which passes across the network to the destination. Only if the destination agrees to form a connection, is a VCC (Virtual Channel Connection) set up between the two users.

Signaling Connection Control Part SCCP. Part of the SS7 protocol that provides communication between signaling nodes by adding circuit and routing information to the signaling message. The ISDN-UP (Integrated Services Digital Network User Part) and TCAP (Transaction Capabilities Application Part) use the SCCP (Signaling Connection Control Part) and the MTP (Message Transfer Part) to transport information. Definition from Bellcore in reference to its concept of the Advanced Intelligent Network.

Signaling Link Selection Code SLS. The part of a routing label that identifies the SS7 signaling link on which the message should be sent.

Signaling Point A node in a SS7 signaling network that either originates and receives signaling messages, or transfers signaling messages from one signaling link to another, or both. SPs are located at each switch in a Signaling System 7 network. They interface the switch with the Signal Transfer Points (STPs). See SIGNAL TRANSFER POINTS and SIGNALING SYSTEM 7.

Signaling Point Code A binary code uniquely identifying a SS7 signaling point in a signaling network. This code is used, according to its position in the label, either as destination point code or as originating point code.

Signaling Point Interface SPOI. The demarcation point on the SS7 signaling link between a LEC network and a Wireless Services Provider (WSP) network. The point established the technical interface and can designate the test point and operational division of responsibility for the signaling.

Signaling System 7 SS7. All phone systems need signaling. According to James Henry Green, author of the Dow Jones-Irwin Handbook of Telecommunications, signals have three basic functions:

1. SUPERVISING. Monitoring the status of a line or circuit to see if it is busy, idle or requesting service. Supervision is a term derived from the job telephone operators perform in manually monitoring circuits on a switchboard. On switchboards, supervisory signals are shown by a lit lamp indicating a request for service on an incoming line or an on-hook condition of a switchboard cord circuit. In the network (i.e. the automated part of the network), supervisory signals are indicated by the voltage level on signaling leads, or the on-hook/off-hook status of signaling tones or bits.

2. ALERTING. Indicates the arrival of an incoming call. Alerting signals are bells, buzzers, whoofers, tones, strobes and lights.

3. ADDRESSING. Transmitting routing and destination signals over the network. Addressing signals are in the form of dial pulses, tone pulses or data pulses over loops, trunks and signaling networks.

Most signaling today is MF (multi-frequency) and SF (single frequency) and is inband. This means that it goes along and occupies the same circuits as those which carry voice conversations. There are two problems with this. First, about 35% of all toll calls are not completed because the phone doesn't answer or is busy, or there are equipment problems along the way. The circuit time used in signaling is substantial, expensive and wasteful. Second, inband signaling is vulnerable to fraud. So the idea of out-of-band signaling came about. It got the name of Common Channel Interoffice Signaling (CCIS) because it used a communications network totally separate from the switched voice network. In North America, CCIS started out as an AT&T packet switched network operating at 4800 bits per second. Each of the packet switches in this network (they are no longer exclusively AT&T's) are called Signal Transfer Points — STPs. CCIS has the following advantages over SF/MF signaling:

Fraud is reduced. "Talk-off" is reduced. (Talk-off occurs when your voice contains enough 2600 Hz energy to activate the tone-detecting circuits in the central office.) Signaling is faster allowing circuits and conversations to be set up and torn down (i.e. disconnected) faster. Signals can be sent in both directions simultaneously and during voice conversation if necessary. Network management information is routed over the CCIS network. For example, when trunks fail, switching systems can be told with CCIS data messages to reroute traffic around problem areas.

The older CCIS signaling is being replaced with a newer out-of-band signaling system called CCITT Signaling System 7. According to an AT&T technical paper delivered at the International Switching Symposium in Spring, 1987, CCITT Signaling System 7 is being required by telecommunications administrations worldwide (i.e. all the local country-owned telephone companies) for their networks. AT&T continued with the introduction of digital switches and transmission equipment with 56 Kbps and 64 Kbps transmission rates, the International Telegraph and Telephone Consultative Committee (CCITT in French) in 1980 approved the CCITT 7 recommendations optimized for digital networks. This new protocol uses destination routing, octet oriented fields, variable length messages and a maximum message length allowing for 256 bytes of data. Addition of flow control, connectionless services and Integrated Services Digital Network (ISDN) capabilities were approved by CCITT in 1984. A major characteristic of CCITT #7 is its layered functional structure. Its transport functions are divided into four levels, three of which constitute the Message Transfer Part (MTP). The fourth consists of a common Signaling Connection Control Part (SCCP). The SS7 protocol consists of four basic sub-protocols:

• Message Transfer Part (MTP), which provides functions for basic routing of signaling messages between signaling points.

- Signaling Connection Control Part (SCCP), which provides additional routing and management functions for transfer of messages other than call setup between signaling points.
- Integrated Services Digital Network User Part (ISUP), which provides for transfer of call setup signaling information between signaling points.
- Transaction Capabilities Application Part (TCAP), which provides for transfer of non-circuit related information between signaling points.

Signaling System 7 provides two major capabilities:
1. Fast call setup, via high-speed circuit-switched connections.
2. Transaction capabilities which deal with remote data base interactions. What this means in its simplest terms and in one simple application is that Signaling System 7 information can tell the called party who's calling and, more important, tell the called party's computer. A scenario: when you call a direct mail order business, Signaling System 7 will send a signal as to which phone is calling. The agent's CRT screen will pop the caller's name and perhaps the caller's most recent buying information. The agent may answer the phone "Good morning, Mr. Newton. Did you enjoy the three khaki pants we sent you last week?..." Signaling System 7 will be an integral part of ISDN. It will enable us to extend full PBX and Centrex-based services like call forwarding, call waiting, call screening, call transfer, etc. outside the switch to the full international network. In effect, with Signaling System 7, the entire network will acquire the "smarts" of today's smartest electronic digital PBX. See also CAPTAIN CRUNCH, ISUP, MTP, SCCP, and TCAP.

Signaling Transfer Point STP. A signaling point with the function of transferring signaling messages from one signaling link to another and considered exclusively from the viewpoint of the transferrer.

Significant Hour Any hour that influences the sizing of a trunk group.

Silence Management See SILENCE SUPPRESSION.

Silence Suppression A term used in voice compression for transmission whereby silence in the voice conversation is filled with other transmissions — e.g. data, video, imaging, etc. According to AT&T, the average voice conversation is 62% quiet and 38% not quiet (i.e. actual conversation). You can figure that for yourself: One person is speaking at a time. That's 50% of the circuit silent. The person who's speaking doesn't speak continuously. He pauses, takes a breath, thinks, etc. That's another 12%. Thus 62% in silence. A company like Micom in Simi Valley, CA makes a product called Marathon which uses the 62% silence between syllables, words and sentences to transmit data, fax and video. Micom tells me that the typical talk spurt sequence is 300 milliseconds. And it can use very small time (as short as 300 milliseconds) between talk spurts to stuff data, fax or video into and transmit.

Silicon A dark gray, hard, crystalline solid. Next to oxygen, the second most abundant element in the earth's surface. Transistor chips are made from silicon, and it is the basic material for most integrated circuits and semiconductor devices. Silicon, a neutral element, is found primarily in raw form as sand. See SEMICONDUCTOR.

Silicon Avalanche Diode A fast-acting surge protector which features a narrow voltage clamping range.

Silicon Forest The area around Microsoft in the Redmond/Seattle area of Washington State is known as Silicon Forest because of all the high-tech companies that have sprouted.

Silicon Mudflats The part of the San Francisco area that is on the Oakland side of the bay and sits between San Francisco and Oakland. That area is home to a bunch of high-tech companies.

Silicon Valley Silicon Valley in Santa Clara County, California, south of San Francisco Bay, is known for its microelectronics innovation. It is one of the two places (the other being Dallas, Texas) where the microchip was invented and produced. There are over 3,000 microelectronic hardware and software firms within a radius of about 30 miles. Silicon Valley owes its good fortune to:
- The fact that W. Shockley, one of the three inventors of the transistor, being a native of Palo Alto, went there in 1955 after he left Bell Laboratories, to start his own industrial research center;
- The proximity of technology universities, especially Stanford University;
- A very active financial market in California and the enterprising attitude of capitalists willing to enter into "joint ventures," often very profitable;
- The West Coast orientation to Japan.

The above information courtesy of Electronics, Computers and Telephone Switching by Robert J. Chapuis and Amos E Joel, Jr.

Silicorn Valley A region in the state of Iowa that is fostering high-tech development.

Sillywood The convergence of Silicon Valley and Hollywood.

SILS A Standard being formulated for Interoperating LAN Security.

Silver Satin Once upon a time, phones came with cords that matched the color of the phone. This proved expensive and confusing to workers, who were color blind, who had to match the cord with the phone. So a manufacturer (we think it was AT&T) decided that the time was ripe for all phones to have a cord that matched every decor and every phone. In actuality, the color they settled upon — silver satin — matches no decor man or woman has ever created and certainly no phone has ever been produced in the silver satin color. However, the world is now stuck with every phone coming with one standard, ugly line cord, called silver satin. See also TOUCHTONE.

SIM 1. Subscriber Identity Module. A 'smart' card installed or inserted into a mobile telephone containing all subscriber-related data. This facilitates a telephone call from any valid mobile telephone since the subscriber data is used to complete the call rather than the telephone internal serial number. See GSM.
2. Single Interface Module (SIM). An NEC term. - The minimum equipment configuration for the NEAX2400 IMS is the Single Interface Module. The SIM, says NEC, is a fully featured, totally non-blocking digital switch capable of supporting all NEAX2400 IMS feature package application programs. Typical stations installed range from only a few to 300 lines.

SIM CARD Subscriber Identity Module, also called Smart Cards. Every mobile phone that conforms to the GSM (global system for mobile communications) and including many PCS (personal communications services) handsets has something called a SIM card. GSM phones won't work without these cards. Each SIM card contains a microchip that houses a microprocessor with eight kilobytes of memory. The card stores a mathematical algorithm that encrypts voice and data transmissions and makes it nearly impossible to listen in on calls. The SIM card also identifies the caller to the mobile network as being a legitimate caller. PCS and GSM cards come in two basic varieties. Some handsets work with slot-in SIM cards that are the size of conventional credit cards. Others come with smaller cards already built into the handset. GSM is the standard mobile service for Europe and Australia and most countries outside the U.S. Some PCS operators in the U.S. have adopted the GSM standard. See GSM.

SIMD Single Instruction Multiple Data is a type of parallel processing computer, which includes dozens of processors. Each processor runs the same instructions but on different data and one chip provides central coordination. See also MIMD.

SIMM Single In line Memory Module. Used on Macs and PCs. A form of chip packaging in which leads (pins) are arranged in a single row protruding from the chip. Connectors are attached to a stiff contact strip that permits a SIMM to be inserted into a slot like an expansion adapter. On PCs, SIMM-style RAM chips have virtually replaced the dual in-line package (DIP) chips, identifiable by two rows of protruding legs, that were popular in the 1980s. The most common SIMM is the 30-pin, 9-bit wide "1 by 9", which is the standard memory upgrade for PCs. See SIMM socket.

SIMM Socket The connector inside the Macintosh that holds the SIMM and connects it to the rest of the computer electronically.

Simple Mail Transfer Protocol SMTP. The TCP/IP protocol governing electronic mail transmissions and receptions. See SMTP.

Simple MAPI Simple MAPI is a subset of MAPI that lets developers easily create "mail-aware" applications capable of exchanging messages and data files with other network clients.

Simple Network Management Protocol SMNP. The protocol governing network management and monitoring of network devices and their functions. SNMP came out of the TCP/IP environment.

Simplegram A Versit term. A syntax for encoding the vCard in a clear-text encoding. Simplegrams are nominally, based on the ASCII, 7-bit character set.

Simplex 1. Operating a channel in one direction only with no ability to operate in the other direction. For example one side of a telephone conversation is all that could be carried by a Simplex line. 2. One-sided printing.

Simplex Loop Powering In T-1, refers to the powering of the digital signal pairs that are simplex in nature (Tip or Ring) and that may have voltage applied to maintain the required 60 mA dc current to control repeater signal regeneration, loopbacks, keep alive signals and alarms.

SIMTEL20 The White Sands Missile Range used to maintain a giant collection of free and low-cost software of all kinds, which was "mirrored" to numerous other ftp sites on the Internet. In the fall of 1993, the Air Force decided it had better things to do than maintain a free software library and shut it down.

Simultaneous Peripheral Operations On Line SPOOL. Temporarily storing programs, data or output on magnetic tape or in RAM for later output or execution. Many PCs use a small software spooling program which accepts material to be printed very quickly, stores it in a portion of RAM, then feeds that material to the printer at a speed the printer can handle. See SPOOLING.

Simulator A program in which a mathematical model represents an external system or process. For example, an engineer can simulate the forces that act on a building during an earthquake to find out how much damage is likely to be incurred.

Simulation A technique, often involving a computer, to guess the outcome of various events in the future. Where multitudes of complex events interact, simulation may well be the only way to deal with a given problem. Simulation is often used in traffic engineering instead of or in addition to the proven formula. Many people believe that simulation should NEVER be used when standard, proven formulas (such as Poisson, Erlang B and Erlang C) are appropriate. The difficulty with simulation is actually finding out what rules to use and then programming them correctly. Once this is done, it takes time, even on a fast computer, to run thousands of simulations to get a stable statistical estimate; a single run of simulation, like a single roll of the dice, is worse than useless. Simulators were built into hardware to predict the overall behavior of AT&T's No. 5 Crossbar switch when it was developed in the late 1940s; 1ESS behavior was simulated with software in the 1960s. These were major efforts involving many people and several years, but they dealt with problems far beyond the capabilities of standard traffic equations.

Singing An undesirable whistle or howl on a transmission circuit. Singing is usually caused by feedback, excessive gain, or unbalance of hybrid coils, or by some combination. Singing is the same effect observed when you increase the volume on a public address system until the system squeals or "sings."

Singing Return Loss The loss at which a circuit oscillates or sings at the extreme low and high ends of the voice band.

Single Address Message A message which is to be transmitted to only one specific terminal, as opposed to a broadcast or group message.

Single Attached Concentrator SAC. An FDDI (or CDDI) concentrator that connects to the network by being cascaded from an M (master) port of another FDDI(or CDDI) concentrator.

Single Attachment Concentrator A concentrator that offers one S port for attachment to an FDDI network and M ports for the attachment of stations or other concentrators.

Single Attachment Station An FDDI station that offers one S port for attachment to an FDDI ring.

Single Channel Broadband A local network scheme in which the entire spectrum of the cable is devoted to a single transmission path; frequency-division multiplexing is not used. Also known as carrierband.

Single Day Assignment A call center term. A method of automatic assignment (of employees to Master File schedules) that operates only on schedules of exactly one workday in length. Unlike multi-day assignment (which operates on schedules of any length), this makes it possible for each assigned employee to be scheduled for different hours each day.

Single Digit Dialing Provides for single-digit dialing to reach a preselected phone or group of phones.

Single Ended Terminal Device A device which terminates only one line at a given time.

Single Fiber Cable A plastic-coated fiber surrounded by an extruded layer of polyvinyl chloride, encased in a synthetic strengthening material and enclosed in an outer polyvinyl chloride sheath.

Single Frequency Signaling The use of one tone — typically 2600 Hz — to indicate if the phone line is busy or idle (supervision) and to convey dial pulse signals from one end of a trunk or line to the other, using the presence or absence of a single specified frequency.

Single-Frequency (SF) Signaling In telephone communications, a method of conveying dialing or supervisory signals, or both, with one or more specified single frequencies. Note: The DCS transmits dc signaling pulses or supervisory signals, or both, over carrier channels or cable pairs on a 4-wire basis using a 2600-Hz signal tone. The conversion into tones, or vice versa, is done by SF signal units. See also FREQUENCY, INBAND SIGNALING, SIGNAL.

Single In Line Memory Modules Basically memory packaged so it can be slipped into a PC or laptop PC much easier than present methods of installing memory — which typically consist of pushing memory chips with legs into printed circuit boards. The problem with memory with legs is that you're likely to bend one of the legs and thus have the installation go awry.

Single Line Instrument A telephone set normally used to access only one line. However when used with advanced telephone systems, additional lines can be accessed by dialing specific codes rather than by depressing keys.

Single Mode Fiber Single-mode fiber is a fiber that allows only a single-mode of light to propagate. This eliminates the main limitation to bandwidth, modal dispersion. However, the small core of a single-mode fiber makes coupling light into the fiber more difficult, and thus lasers must be used. The main limitation to the bandwidth of a single-mode fiber is chromatic dispersion. Laser sources must also be used to attain high bandwidth, because LEDs emit a large range of frequencies, and thus chromatic dispersion becomes significant. The current bandwidth of the fastest commercial systems (as of 1995) is approximately five gigabits/second, which can be transmitted for approximately 100-200 kilometers before a repeater or fiber-optic amplifier is required.

Single Number Service An optional feature for 800 IN-WATS Services which allows a subscriber who has or wants to have both intrastate and interstate 800 service to use the same 1-800 number for both services. If you'd like to buy another copy of this dictionary, call 1-800-LIBRARY. That phone number will be answered at our office in New York City. It will work for calls from inside and from outside New York State.

Single Party Revertive Ringing A feature from a Northern Telecom central office. Single-Party Revertive Ringing provides enhanced flexibility to residential subscribers' existing telephone service by turning extension phones into intercoms. This feature creates a type of home intercom system with which subscribers can reach people in other rooms in their home and/or on a remote part of their property (i.e., barn, garage, greenhouse, etc.). To initiate a home intercom call, the subscriber dials his own number and hangs up the phone. All extension phone rings. The intended party picks up the extension and is connected.

Single Protocol Router A communications device using the same mix of protocols which is designed to make decisions about which of several paths a packet of information will take. The packets are routed according to address information contained within the packet.

Single Sideband Transmission A system of transmission which suppresses one side-band of the carrier frequency at the source. Also applied to receiving systems designed to reproduce such transmissions.

Single Slot Current standard for coin phone construction that uses one slot for the deposit of all acceptable coins.

Single Use Batteries Batteries that aren't rechargeable — in other words, the Duracell and Eveready batteries at your local supermarket.

Single Wire Line A transmission path which uses a single conductor and a ground return to complete a circuit. Used a lot in rural areas.

Sink That part of a communications system which receives information.

SIO Serial Input/Output. The electronic methodology used in serial data transmission.

SIP 1. Single Inline Module with pins. 2. An SMDS term meaning SMDS Interface Protocol. 3. Simple Internet Protocol.

SIP L2PDU-SMDS Standard Interface Protocol Layer 2 protocol data unit. The 53-octet unit of information processed by the second layer of the SIP.

SIP L3PDU-SMDS An SMDS term. Standard Interface Protocol layer 3 protocol data unit. A variable-length (up to 9,220 octets long) unit of information processed by the third layer of the SIP.

SIP SMDS Interface Protocol An SMDS term. The protocol defined at the interface between the SMDS network and the end user.

SIPP 1. Simple Internet Protocol Plus. One of the 3-IPng candidates.

2. SMDS Interface Protocol: Protocol where layer 2 is based on ATM, AAL and DQDB. Layer 1 is DS1 and DS3.

SIR 1. Speaker Independent Recognition. See SPEAKER INDEPENDENT RECOGNITION.

2. An SMDS term meaning Sustained Information Rate. Determined at time of subscription, the SIR defines the long term average throughout an SMDS access line can carry. SIR is enforced through a Credit Manager resident of the SMDS network switches, and the Access Class-4 Mbps, 10 Mbps, 16 Mbps, 25 Mbps or 34 Mbps- the subscriber selects.

SIT Tones 1. Standard Information Tones. These are tones sent out by a central office to a pay phone to indicate that the dialed call has been answered by the distant phone, etc. 2. Special Information Tones. These are tones for identifying network provided announcements. Here's Bellcore's explanation: Automated detection devices cannot distinguish recorded voice from live voice answer unless a machine-detectable signal is included with the recorded announcement. The CCITT, which specifies signals that may be applied to international circuits, has defined Special Information Tones for identifying network provided announcement. The SIT used to precede machine-generated announcements also alerts the calling customer that a machine-generated announcement follows. Since SIT consists of a sequence of three precisely defined tones, SIT can be machine-detected, and therefore machine-generated announcements preceded by a SIT can be classified. At least four SIT encodings have been defined: Vacant Code (VC), Intercept (IC), Reorder (RO) and No Circuit (NC). With the exception of some small stored Program Control

Systems (SPCSs) and some customer negotiated announcements, Bell operating companies in North America now precede appropriate announcements with encoded SITs to detect and classify announcements.

SITA Society of International Aeronautical Telecommunications. The international data communications network used by many airlines.

Site Controller An industrial grade PC located at MCI terminals and junctions. It provides the operator local visibility into alarm and performance information from the Extended Superframe Monitoring Unit and 1/0 DXC, as well as enabling the operator to interact with both devices.

Site License Companies that buy software for multiple computers typically buy one copy of the program and a license to reproduce it up to a certain number of times. This is called a site license, though it may apply to its use throughout an organization. Site licenses vary. Some require that a copy be bought for each potential user — the only purpose being to indicate the volume discount and keep tabs. Others allow for a copy to be placed on a network server but limit the number of users who can gain simultaneous access. This is called a concurrent site license. And many network administrators prefer this concurrent license, since it gives them greater control. For example, if the software is customized, it need be customized only once.

SiteRP SiteRP stands for Sprint Interface to External Routing Processor. It is a service which Sprint offers to enable customer premise equipment (e.g. automatic call distributors) to interact with its long distance network to give that network information on how to route incoming 800 calls. With SiteRP, a call processing query is sent to the Sprint SCP (Signal Control Point). The SCP is instructed to check a user database kept on the customer premises for instructions on how to handle the call. Once that call handling instructions have been extracted from the user database, the customer premises equipment sends a message back to the Sprint SCP, which, in turn, uses that information to decide which call processing instructions it will pass back to the network switch. SiteRP allows such services as real-time switching of inbound calls, to allow for load balancing among agents or for specific calls to go to specific agents, etc. There are a million reasons why inbound calls from different places and different numbers should go to different agents. To use SiteRP, customers must directly interconnect their SiteRP processors, which can be a mainframe or a PC (or in between) to each of Sprint's five SCPs via 56 Kbps dedicated links. See also ACD, AIN, SCP, SSP and STP.

SIVR Speaker Independent Voice Recognition. See SPEAKER INDEPENDENT VOICE RECOGNITION.

Six Digit Translation We have a long distance number 212-691-8215. The ability of a switching system to do six digit translation means that it can "look at" 212-691 and figure how to route the phone call. The criterion of choosing which way to send the call is, most often, the least expensive way. Six digit translation is often an integral part of Least Cost Routing programs within the phone system which tell the calls to go over the lines perceived by the user to be the least cost way of getting the call from point A to point B. There are typically two types of "least cost routing" translation — that which examines the first three digits of the phone number (i.e. just the area code) and the first six digits of the phone number (i.e. the area code and the three digits of the local central office). Six digit translation is preferred because it allows you more flexibility in routing, particularly to big area codes, like 213 in LA, where there are long distance calls within the area code. See also LEAST COST ROUTING and ALTERNATE ROUTING.

Skew 1. The deviation from synchronization of two or more signals. 2. A computer imaging term. A tool that slants a selected area in any direction. 3. In parallel transmission, the difference in arrival time of bits transmitted at the same time. 4. For data recorded on multichannel magnetic tape, the difference in time of reading bits recorded in a single line. 5. In facsimile systems, the angular deviation for the received frame from rectangularity due to asynchronism between scanner and recorder. Skew is expressed numerically as the tangent of the angle of deviation.

Skew Ray In a multimode optical fiber, any bound ray that in propagating does not intersect the fiber axis (in contrast with a meridional ray). In a straight, ideal fiber, a skew ray traverses a helical path along the fiber, not crossing the fiber axis. A skew ray is not confined to the meridian plane.

Skills-Based Routing A call center term for routing incoming calls based on the type of service requested, assuring that calls go to agents with the skills to provide the highest quality of service to the calling customer. In other words, someone calling about a broken refrigerator should be directed to a refrigerator expert, not a vacuum cleaner expert. See also SOURCE/DESTINATION ROUTING, CALENDAR ROUTING, and END-OF-SHIFT ROUTING.

Skin Surface layer in a sandwich structure.

Skin Effect The tendency of a current to pass through the outer portion rather than through the center of a conductor.

Skip Key Use "skip keys" wherever possible. Skip keys are single key-presses that take the caller to the most commonly sought places, including information message boxes. For example, a skip key is Dial 1 for Books. Dial 3 for subscriptions. Dial 4 for our fax number. Ron Acher invented this term. Or thinks he did.

Skip Zone a ring-shaped region within the transmission range wherein signals from a transmitter are not received. It is the area between the farthest points reached by the ground wave and nearest points at which reflected sky waves come back to earth.

Skunkworks Term for usually-secret high-pressure/high-tech research group in a company or government, often populated by people who don't see much sunlight or soap. Hence the name, skunkworks.

SkyTel The company which provides SkyPager, SkyTalk and SkyWord, nationwide paging voice messaging services and alphanumeric text messaging. The company's primary services run at 931 MHz, which is the frequency of the SkyTel pager.

Slamming The practice of switching a telephone customer's long distance supplier without obtaining permission from the customer. A long distance company might do this to get itself some easy revenues. After a bitter court battle, AT&T and MCI settled on some proposed standards to allow a long distance to switch a customer's service under the following circumstances:
1. The customer initiates the switch by calling either their local telephone company, their current long distance company or the long distance company they want to switch to;
2. Someone outside the long distance company's sales department verifies the request to switch; or
3. The customer submits written authorization to the company.

Slave A device which operates under the control of a master — another device or system. Slave switching systems are common in rural areas. The master central office might be in town A. Twenty miles away there's a smaller town. It makes no economic sense to serve those subscribers each on single local dedicated loops from Town A. Best solution: place a "slave" central office in that distant town and drive its software, diagnostics and changes from the main central office in Town A. See also REMOTE CONCENTRATOR.

SLC-96 Pronounced "Slick 96." A product of AT&T Technologies. It's a short haul multiplexing device which enables up to 96 telephone customers to be served on three pairs of wires. SLC stands for Subscriber Loop Carrier. See SLCC.

SLCC Abbreviation for Subscriber Line (or Loop) Carrier Circuit, and pronounced "slick." It's a system that allows one pair of wires, that would normally provide one phone line, to carry multiple conversations. Various models are available, with capacity ranging from 2 to 96 lines. A SLCC is used between phone company central offices and areas where there are too many customers for the cable that is in place. It's much less expensive to install SLCCs than new cable, but the SLCC provides lower-than-normal line voltage, which may cause some phones to malfunction.

SLEE Service Logic Execution Environment.

Sleeves Short lengths of conduit, usually made from rigid metal pipes, used to protect cables entering a premises through a building wall or running through concrete floors between vertically aligned riser closets. Sleeves also provide for easy pulling of cable.

SLIC An AT&T term for Subscriber's Line Interface Circuit, a device that interfaces the SLC series of AT&T products of local subscriber pair gain or multiplexing devices. See SLC-96.

Slideware Slideware is hardware or software whose reason for existing (eventually) has been explained in 35-mm slides, foils, charts and/or PC presentation programs. Slide is best described as "virtual vaporware." Vaporware is software which has been announced, perhaps even demonstrated, but not delivered to commercial customers. Hyperware is hardware which has been announced but has not yet been delivered. Slideware is less real than vaporware or hyperware. Classic slideware is the ISDN deployment strategies of many of the telephone companies in the United States. See also HOOKEMWARE, HYPERWARE, MEATWARE, SHOVELWARE, and VAPORWARE.

SLIP Serial Line Internet Protocol. Used to turn home computers into Internet sites over a phone line. SLIP is a protocol that allows a computer to use the Internet protocols (and become a full-fledged Internet member) with a standard telephone line and a high-speed modem. SLIP is being superseded by PPP, but it's still used. "You got a slip connection?" means that your home computer can sport a home page. You won't get many "hits," because your connection to the Internet is slow speed. But a slip connection will get you up and running. Trumpet Winsock is a communications program, one of many used for accessing the Internet. In its instruction manual, it explains" "SLIP is a simple protocol which allows an Async serial connection to send Internet Protocol (IP). You usually need to have access to a server which can understand SLIP. SLIP is usually accessed via a phone line, and with the advent of high speed modems, TCP/IP is a reality over a dial-up connection." See PPP.

Slip/Slip Rate The loss (or rate of loss) of a data bit on a T-1 link due to a frame misalignment between the timing at a transmit node and timing at a receive node.

Slip-Controlled The occurrence at the receiving terminal of a replication or deletion of the information bits in a frame.

Slip-Uncontrolled The loss or gain of a digit position of a set of consecutive digit positions in a digital signal resulting from an aberration of the timing processes associated with the transmission or switching of the digital signal. The magnitude or the instant of the loss or gain is not controlled.

Slist Connected to a NetWare network? Type Slist in DOS on your PC and you will get a list of all the NetWare servers you are connected to.

SLM System Load Module.

Slot Sizes PCMCIA cards come now in three standard physical slot and card sizes: Type I: 3.3 mm thick, mostly for memory cards. Type II: 5 mm thick, fits most of the current cards on the market, including communications and networking cards. Type III: 10.5 mm, used for cards that have rotating hard disks in them. Toshiba, in its T4600, came out with a fourth size that is 16 mm thick. There is more to compatibility than just fit. The size of the slot is independent of the version of firmware that supports it. Some cards may only work with certain operating systems, BIOS, and drivers. Apple's Newton has a Type II slot but only supports cards that have been designed specifically for it. See CARD SERVICES, PCMCIA STANDARDS, SOCKET SERVICES and SLOT SIZES.

Slots Openings, typically rectangular, in the floor of vertically aligned riser closets that enable cable to pass through from floor to floor. A slot accommodates more cables than an individual slot.

Slotted Aloha An access control technique for multiple-access transmission media. The technique is the same as ALOHA, except that packets must be transmitted in well-defined time slots.

Slotted Ring A LAN architecture in which a constant number of fixed-length slots (packets) circulate continuously around the ring. A full/empty indicator within the slot header indicates when a workstation or PC attached to the LAN may place information into the slot. Think of a slotted ring LAN as an empty train that constantly travels in a circle, being filled and emptied at different terminals (workstations).

Slow Switching Channel A sequencing GPS (Global Positioning System) receiver channel that switches too slowly to allow the continuous recovery of the data message.

SLIP Serial Line Internet Protocol. An Internet protocol which is used to run IP over serial lines such as telephone circuits. IP is the Internet Protocol; the most important of the protocols on which the Internet is based. It allows a packet to traverse multiple networks on the way to its final destination. See INTERNET.

SLR Send Loudness Rating.

SLSA Single Line Switching Apparatus.

SLT Single-Line Telephone, as opposed to a phone that has buttons to select from several lines.

SMA Subscriber Carrier Module-100A; same as SCM-100A.

NEWTON'S TELECOM DICTIONARY

SMA 905/906 (Subminiature type A). A former microwave connector modified by Amphenol to become the "standard" fiber optic connector. The 905 version is a straight ferrule design, where as the 906 is a stepped ferrule design and uses a plastic sleeve for alignment.

Small Computer System Interface SCSI. Pronounced scuzzy, SCSI is a bus that allows computers to communicate with any peripheral device that carries embedded intelligence. The standard is covered by the American National Standards Institute (ANSI) and has developed into from SCSI-1 into SCSI-2 and, now SCSI-3. Different types of device can be connected in a daisy chain via a 50-pin cable (68-pin for SCSI-3), both ends of which must be terminated. All signals on the cable are common to all devices. To avoid bus contention, each device (up to a maximum of seven) connected to the bus is given a unique "SCSI" address with each address being allocated a degree of priority. The SCSI-1 bus carries 8-bit data, 1-bit parity and 9-bit control lines, to provide a maximum synchronous data transfer rate of 5 Mbyte/s.
The maximum bus length is dependent upon the type of bus driver/receiver used. For single-ended devices, the length is restricted to 6 meters, but you wouldn't want to go that far. A desktop is about the maximum distance for an un-amplified SCSI-1 bus. The 6 meters can be extended to 25 meters through the use of differential drivers/receivers. SCSI-2 has been extended to increase the data rate to 10 Mbyte/switch via either 16 or 32-bit processors. The maximum data rate is thus nearer to 40 Mbyte/s. The protocol has also been expanded to include tagged commands. This allows the execution of queued control commands according to a prescribed sequence. For a greater explanation, see SCSI, SCSI-2 and SCSI TRANSFER RATE.

Small Vocabulary A voice recognition term. Vocabularies containing fewer than 50 words.

Smart Agent A software tool that acts like you. It is your agent and it acts intelligent on your behalf. For example, you might ask assign it a series of tasks to do on your behalf every day, or even throughout the day. For example, "Which orders that I sold today have been shipped today?" See INTELLIGENT AGENT and SMART MESSAGING.

Smart Building Also referred to as the "intelligent" building. A centrally managed structure that offers advanced technology and perhaps shared tenant services. Central to the early ideas of a smart building was the concept of one gigantic phone system shared by all the tenants. The idea was that you needed a big switch to get lots of features and all the benefits of cheap long distance. That's no longer true. Small switches offer much the same benefits. Now buildings are often smart, but frequently they're smart, using those smarts to save energy and insure safety and less to share telephone service among the tenants, who have shown a tendency to fend for themselves.

Smart Envelopes A term coming from General Magic, Mountain View, CA. and AT&T Personal Link Services. Smart envelopes means you send the mail to a person and the system figures out how to get the mail to that person. The system figures which electronic mail system the person is on and it figures the way through the maze.

Smart Hunt Groups Picture a call center with people answering the phone from customers. The people answering the phone have different skills, know different things about the company's products, and are thus better able to answer some questions better than others. Now picture that the phone system the people is using is "smart" and somewhere in it and in the databases attached to it, it knows who of the people (also called agents) is good at what. What skills they have. What they know about which products, etc. Then, by asking questions of incoming callers through a combination of an attached interactive voice response system, the incoming caller's phone number and the number the caller dialed, it is able to route the call to the exact correct person. If that correct person is not available, it routes the caller to another person. And to another, etc. There may be many smart hunt groups, each consisting of the same people. If you think of a call center with several hundred people you can easily see it's possible to have potentially thousands of "smart hunt groups." Just depends on how complex your product line is and how sophisticated you want to get. The term was coined by Rick Luhmann, editor of Computer Telephony Magazine, New York, NY.

Smart Jack A common name for RJ48X.

Smart Mailboxes A term coming from General Magic, Mountain View, CA. and AT&T Personal Link Services. Smart mailboxes mean you can customize your mailbox to reflect your preferences. For example, You can have your electronic mailbox send specific incoming mails to specific people you designate. Sort of like automated task delegation. You can automatically file all messages about a particular subject, from a certain sender, in a specified folder. You might choose not to accept any mail from certain people or certain groups.

Smart Messaging A term coming from General Magic, Mountain View, CA. and AT&T Personal Link Services. Smart messages mean, for example, you can send an invitation containing a smart button that recipients can tap to schedule an appointment in their calendars and send a confirmation back to you. Or smart messaging can mean that you can specify is a message you send isn't read within a specified period of time, it should be sent to someone else, perhaps an assistant or other team member.

Smart Web Browsers A Web Browser is a piece of software which allows you to search the World Wide Web for the information you're seeking for. A smart Web Browser will contain a modicum of intelligence, allowing it to find what you're looking for faster, easier and less consuming of your time.

SMB Server Message Block. The protocol developed by Microsoft, Intel, and IBM that defines a series of commands used to pass information between network computers. The redirector packages SMB requests into a network control block (NBC) structure that can be sent over the network to a remote device. The network provider listens for SMB messages destined for it and removes the data portion of the SMB request so that it can be processed by a local device. In short, SMB is basically a protocol to provide access to server-based files and print queues. SMB operates above the session layer, and usually works over a network using a NetBIOS application program interface. SMB is similar in nature to a remote procedure call (RPC) that is specialized for file systems.
On June 13, 1996, Microsoft said, "The SMB protocol is an open technology widely available on UNIX, VMS and other platforms. It has been an Open Group (formerly X/Open) standard for PC and UNIX interoperability since 1992 (X/Open CAE Specification C209), and it is supported in products such as AT&T Advanced Server for UNIX, Digital's PATHWORKS, HP Advanced Server 9000, IBM Warp Connect, IBM LAN Server, Novell Enterprise Toolkit, and 3Com 3+Share, among others. SMB

551

is also the featured file and print sharing protocol of Samba, a popular freeware network file system available for LINUX and many UNIX platforms."

The above Microsoft words came in a press release announcing something called CIFS, which is essentially an enhanced version of CIFS. See CIFS.

SMBus System Management Bus. A two-wire bus for more intelligent handling of rechargeable batteries in portable stuff, like laptops.

Smart Card A credit card-sized card which contains electronics, including a microprocessor, memory and a battery. The card can be used to store the entire repair and maintenance history on the family automobile or any member of the family. There is no direct contact between the "smart card" and the device which reads it. This avoids the problem of wear that afflicts traditional credit cards — both their embossed numbers on the front and their magnetic strip on the back.

Smart Phone 1. A phone which can send, receive, sort, prioritize and notify users of communication transmissions, combining the capabilities of a PC with the familiarity of a telephone. They differentiate themselves from traditional telephones because they feature the following components: a microprocessor, an 8 to 32 bit CPU, memory, function keys, context-sensitive keys, and a built-in modem.

2. A phone created by AT&T Network Systems and due out some time in 1992 or 1993. The phone never appeared. But it was a neat idea. It was to have no buttons but simply a touch sensitive screen. From one moment to another the image on the screen would change, requesting information from the user — the number or person to call, the letters you'd like to transmit to an electronic mail system, the amount of money you'd like to transfer to your checking account, etc.

Smart Video Recorder Intel's PC Video single-step compression, real-time video capture card that uses Indeo technology and an Intel i750 processor.

Smartcall Smartcall is a name registered by GTE for services more advanced than conventional Custom Calling call enhancement services, such as speed calling or call waiting. Calling Number Identification is a GTE Smartcall service. Other services include Automatic Call Return, Automatic Busy Redial, Call Tracing Service, Protected Number Service, VIP Alert, Call Block, etc.

SmartHouse The National Society of Home Builders has registered SmartHouse. BellSouth has trademarked SuperHouse. And GTE has registered SmartPark. One day, fiber optic will snake to everyone's house, bringing the potential of immense information services. Until that day comes they'll be lots of interesting demonstrations at distant trade shows.

SmartPark GTE's registered trademark for a fiber-optic wired industrial park offering a wide variety of voice, data and video transmission and switching services. "The GTE SmartPark services help protect tenants from communications obsolescence and eliminates the need for relocation when their telecommunications requirements expand," says Harrison Coleman, national SmartPark product manager.

SMARTRING Self-Healing Multi-Nodal Alternate Route Topology which BellSouth describes as a dedicated ring-type special access DS3 service.

SMATV Satellite Master Antenna Television. A distribution system that feeds satellite signals to hotels, apartments, etc. Often associated with pay-per-view.

SMAS Switched Maintenance Access System.

SMC Standard Management Committee. Directs specialized working groups of the Architecture and Standards Steering Council.

SMDA Station Message Detail Accounting. Another name for telephone call accounting. See CALL ACCOUNTING SYSTEM.

SMDI Station Message Desk Interface or Simplified Message Desk Interface. The SMIS is the data link from the central office if you have ESSX, Centrex or Centron (etc.) that gives you your stutter dial tone or message waiting light. In essence, SMDI is a data line from the central office containing information and instructions to your on-premises voice mail box. With SMDI, the calling person is not required to re-enter the called phone number (or in any other way identify the called party) once the call terminates to the messaging system.

SMDR Station Message Detail Reporting. Another name for telephone call accounting. See SMDR PORT and CALL ACCOUNTING SYSTEM.

SMDR Port Modern PBXs and some larger key systems have an Station Message Detail Recording (SMDR) electrical plug, usually an RS-232-C receptacle, into which one plugs a printer or a call accounting system. The telephone system sends information on each call made from the system to the outside world through the SMDR port. That information — who made the call, where it went, what time of day, etc. — will be printed by the printer or will be "captured" by the call accounting system on a floppy or a hard magnetic disk and later processed into meaningful management reports. See CALL ACCOUNTING SYSTEM.

Smear The undesirable blurring of edges in a compressed image, often caused by the DCT which tends to eliminate the high-frequency portions of an image which represent sharp edges.

SMF-PDM The ANSI X3T9.5 PDM standard which defines the requirements for the transmission of data over single mode fiber in an FDDI topology. Also refers to the ANSI working group responsible for the development and perpetuation of the standard.

SMDS Switched Megabit (or Multi-megabit) Data Service. A connectionless high-speed data transmission service that provides wide area connectivity through the public telephone network. It is used to connect LANs, MANs and WANs to exchange data. SMDS is defined by Bellcore as a MAN service for telephone companies. It has several pluses. You can access it at five classes of speed — Class 1 — 4 megabits per second (same as low speed Token Ring); Class 2 — 10 megabits per second (same as Ethernet); Class 3 — 16 megabits per second (same as higher-speed Token Ring); Class 4 — 25 megabits per second (ATM speed) and Class 5 — 44.736 megabits per second (same as T-3). SMDS's major problem is that it does not have an isochronous channel for voice and video. See THE SMDS INTEREST GROUP.

SMDS Interest Group A non-profit industry consortium chartered to promote interoperability among SMDS products, applications and services. The SMDS Interest Group is based in Foster City, CA. www.sbexpos.com/sbexpos/associations/smds/home.html

SMDSU Switched Multimegabit Data Service Unit from ADC Kentrox.

NEWTON'S TELECOM DICTIONARY

SMF Single Mode Fiber: Fiber optic cable in which the signal or light propagates in a single mode or path. Since all light follows the same path or travels the same distance, a transmitted pulse is not dispersed and does not interfere with adjacent pulses. SMF fibers can support longer distances and are limited mainly by the amount of attenuation. Refer to MMF.

SMI Structure of Management Information. The set of rules and formats for defining, accessing and adding objects to the Internet MIB. SMI was elevated to full standard status in May 1990.

Smoke Test Test of new or repaired equipment by turning it on. If there's smoke, it doesn't work!

Smokeware When manufacturers announce new products, sometimes the products are not ready. Humorists have various words for these products — smokeware, slideware, foilware, vaporware and hyperware.

SMOP Small Matter of Programming. A little software and it will all work. Yeah!

SMP 1. Symmetric MultiProcessing. The use of several CPUs (i.e. several Intel Pentium chips) inside a PC or (more typically) a server to achieve a performance boost. This is achieved by an operating system that lets CPUs run operations in parallel. This is not a "normal" PC operating system. Normal PC operating systems — MS-DOS, Windows, etc. — expect that their work will be done on one processor, not on more than one. See also SYMMETRIC MULTIPPROCESSING.
2. SCSA Message Protocol.
3. Simple Management Protocol.

SMPI SCSA Message Protocol Interface.

SMPTE Time Code A video term. SMPTE time code is an 80-bit standardized edit time code adopted by SMPTE, the Society of Motion Picture and Television Engineers.

SMR Specialized Mobile Radio. A two-way radio dispatch service widely used for years by truck and taxi fleets. Many SMR operators are now converting their SMR networks to digital so they can deliver both voice and data to a single device and thus compete with cellular radio providers. SMR systems have far less radio spectrum than cellular has, but the signal can reach 25 times farther, which means it's cheaper to build a national network.

SMRP Simple Multicast Routing Protocol. Apple Computer's specialized network protocol for routing multimedia data streams on AppleTalk networks.

SMRT Single Message-unit Rate Timing. USA telephone company tariff under which local calls are timed in 5-minute increments — with a single message unit charge applied to each complete or partial increment.

SMS 1. Service Management System, a term coined by Bellcore for the Intelligent Network. The SMS allows provision and updating of information on subscribers and services in near-real time for billing and administrative purposes.
2. Short Message Service. A means to send or receive, short alphanumeric messages to or from mobile telephones.

SMS-R Subscriber Carrier Module-100S REmote.

SMS/800 The national database Service Management System that retains all 800 records. This database provides long distance carriers a single interface for 800 number reservations and record maintenance. Developed by Bellcore, the database has been in use by various Regional Bell Operating Companies (RBOCs) since 1988. The FCC has mandated that a neutral third party administer the database after 800 portability, which occurred in May, 1993.

SMSA Standard Metropolitan Statistical Area. A metropolitan area consisting of one or more cities as defined by the Office of Management and Budget and used by the FCC to allocate the cellular radio market.

SMSC Short Messaging Service Center. On a wireless network, allows short text messages to be exchanged between mobile telephones and other networks.

SMT 1. Surface Mounting Technology. 2. Station Management. The part of FDDI that manages stations on a ring.

SMTA Single-line Multi-extension Telephone Apparatus.

SMTP Simple Mail Transfer Protocol. In local area networking technology, protocol that operates at layers 5 through 7 of the Open Systems Interconnection (OSI) model. This TCP/IP protocol governs electronic mail transmissions and receptions.

SMU Subscriber Carrier Module-100 URBAN.

SN An ATM term. Sequence Number: SN is a 4 octet field in a Resource Management cell defined by the ITU-T in recommendation 1.371 to sequence such cells. It is not used for ATM Forum ABR. An ATM switch will either preserve this field or set it in accordance with 1.371.

SN Cell An ATM term. Sequence Number Cell: A cell sent periodically on each link of an AIMUX to indicate how many cells have been transmitted since the previous SN cell. These cells are used to verify the sequence of payload cells reassembled at the receiver.

SNA Systems Network Architecture. An IBM product. The most successful computer network architecture in the world. See IBM and SYSTEMS NETWORK ARCHITECTURE.

SNADS SNA Distribution Services. An IBM protocol that allows the distribution of electronic mail and attached documents through an SNA network.

SNAFU Situation Normal All Fouled Up. World War II military slang, that often describes frustrating facets of the telecom industry.

Snail Mail A term used to reference delivery of messages by your local postal service. In short, mail that comes through a slot in your front door or a box mounted outside your house.

Snake A flexible strip of metal, typically 1/4 to 1/2" wide and 10 to 100' long, used to pull or push wire and cable through conduit, ceilings, walls or crawl spaces where it is difficult or impossible for a human to fit.

SNAP Subnetwork Access Protocol. A version of the IEEE local area network logical link control frame similar to the more traditional data link level transmission frame that lets you use nonstandard higher-level protocols. The Subnet Access Protocol is an Internet protocol that operates between a network entity in the subnet and a network entity in the end system and specifies a standard method of encapsulating IP datagrams and ARP messages on IEEE networks. The SNAP entity in the end system makes use of the services of the subnet and performs three key functions: data transfer, connection management, and quality of services selection.

SNC Subnetwork Connection: In the context of ATM, an entity that passes ATM cells transparently, (i.e., without adding any overhead). A SNC may be either a stand-alone SNC, or a concatenation of SNCs and link connections.

SND Cellular language for SEND. You punch your digits into your phone. Check them the screen. If they're fine, hit the SND button. The calls goes through.

snd Format Sound resource format. A digital audio sound resource utilized by many Macintosh applications and by the Mac OS. Double-clicking on a snd file enables playback of the sound. Also called a System 7 sound.

Sneak Currents Unwanted but steady currents which seep into a communication circuit. These low-level currents are insufficient to trigger electrical surge protectors and therefore are able to pass them undetected. They are usually too weak to cause immediate damage, but if unchecked could potentially create harmful heating effects. Sneak currents may result from contact between communications lines and AC power circuits or from power induction, and may cause equipment damage due to overheating.

Sneak Fuse A fuse operated by a low-level current and capable of preventing sneak currents on communication lines. See also SNEAK CURRENTS.

Sneakernet A "network" for moving files between computers. It is the oldest "network." It consists of copying a file to a floppy disk, walking to another machine, loading it on that machine. The term "sneakernet" refers to the fact that the main method of moving the disks is by feet, presumably clad in sneakers.

Sniffer A program that monitors all traffic on a network and reports on problems on the network.

Sniglet Any word that does not appear in the dictionary, but should. A term invented by Rich Hall of the HBO Television program "Not Necessarily The News". An example of a sniglet is the definition of "Hozone." It's obviously where socks go when they don't come back from the laundry.

SNMP 1. Signaling Network Management Protocol.
2. Simple Network Management Protocol. In 1988, the Department of Defense and commercial TCP/IP implementors designed a network management architecture for the needs of the average Internet (a collection of disparate networks joined together with bridges or routers). Although SNMP was designed as the TCP's stack network management protocol, it can now manage virtually any network type and has been extended to include non-TCP devices such as 802.1 Ethernet bridges. SNMP is widely deployed in TCP/IP (Transmission Control Protocol/Internet Protocol) networks, but actual transport independence means it is not limited to TCP/IP. SNMP has been implemented over Ethernet as well as OSI transports. SNMP became a TCP/IP standard protocol in May 1990. SNMP operates on top of the Internet Protocol, and is similar in concept to IBM's NetView and ISO's CMIP. In 1991, Microsoft started referring to SNMP as SubNetwork Access Protocol. In November of 1993 Cisco Systems announced that its internetwork routers will support version 2 of the Simple Network Management Protocol (SMNP)and it has licensed SNMP v2 developed by SNMP Research, Inc. of Knoxville, TN.

Snow Random noise or interference appearing in a video picture as white specs.

SNR Abbreviation for SAVED NUMBER REDIAL, a phone system memory feature that allows the user to store a number for as long as it is useful, as opposed to other numbers that are stored more permanently. Some phones have two buttons — one for REDIAL and one for SNR. Redial will dial the last phone number you called. Saved Number Redial will dial one you dialed earlier and chose to save because you're going to call it back. Both buttons are very useful.

SO Serving Office. Central office where IXC (IntereXchange Carrier) has POP (Point Of Presence).

Soak A means of uncovering problems in software and hardware by running them under operating conditions while they are closely supervised by their developers.

SOC Software Optionality Control.

Social Computing A term that emerged in the summer of 1993. Defined by Peter Lewis in the New York Times of September 19, 1993, social computing is a "communications-rich brew," which is "expected to create new ways for businesses and their customers to communicate, over new types of wireless as well as wired pathways, using new types of computers called personal communicators." According to Peter Lewis, "The rise of social computing is expected to shift the emphasis of computing devices away from simple number crunching and data base management to wider-ranging forms of business communications...Where client server broke away from mainframe-based systems and distributed computing power to everyone in the organization, social computing goes the next step and extends the distribution of computing power to a company's customers."

Social Contract An arrangement between the local telephone company and its local regulatory authority whereby the telephone company's services are detariffed, but cannot be priced at less than cost. Quality of service standards apply.

Social Engineering Gaining privileged information about a computer system (such as a password) by skillful lying — usually via a phone call. Often done by impersonating an authorized user.

Socket 1. A synonym for a port.
2. The socket in a PC which is responsible for accepting a PCMCIA Card and mapping the host's internal bus signals to the PCMCIA interface signals.
3. A paring of an IP address and a port number.

Socket Number In TCP/IP, the socket number is the joining of the sender's (or receiver's) IP address and the port numbers for the service being used. These two together uniquely identifies the connection in the Internet.

Socket Services The software layer directly above the hardware that provides a standardized interface to manipulate PCMCIA Cards, sockets and adapters. Socket Services is a BIOS level software interface that provides a method for accessing the PCMCIA slots of a computer, desktop or laptop (but most typically a laptop). Ideally, socket services software should be integrated into the notebook's BIOS, but few manufacturers have done so to date. For PCMCIA cards to operate correctly you also need Card Services, which is (not are) a software management interface that allows the allocation of system resources (such as memory and interrupts) automatically once the Socket Services detects that a PC Card has been inserted. You can, however, happily operate PCMCIA cards in your laptop without using socket and card services. You simply load the correct device drivers for those cards. Such drivers always come with PCMCIA cards when you buy the cards. You will, however, have to load new drivers every time you change cards and allocate the correct memory exclusions. You will have to reboot if you disconnect your network card. Theoretically, with socket and card services loaded, you do not have to reboot every time you change cards. My experience is that this works, except with network cards, which cannot be hotswapped. See PCMCIA.

Sockets An application program interface (API) for communications between a user application program and TCP/IP. See SOCKET and SOCKET NUMBER.

Soft Copy 1. A copy of a file or program which resides on magnetic medium, such as a floppy disk, or any form that is not a hard copy — which is paper.
2. Old legacy systems term reapplied to distributed computing in which reports are created on-screen from data residing within different applications.

Soft Ferrite Ferrite that is magnetized only while exposed to a magnetic field. Used to make cores for inductors, transformers, and other electronic components. See BARIUM FERRITE, FERRITE and HARD FERRITE.

Soft Font A font, usually provided by a font vendor, that must be installed on your computer and sent to the printer before text formatted in that font can be printed. Also known as downloadable font.

Soft Key There are three types of keys on a telephone: hard, programmable and soft. HARD keys are those which do one thing and one thing only, e.g. the touchtone buttons 1, 2, 3, • and # etc. PROGRAMMABLE keys are those which you can program to do produce a bunch of tones. Those tones might be "dial mother." They might be "transfer this call to my home for the evening." They might be "go into data mode, dial my distant computer, log in and put in my password." SOFT keys are the most interesting. They are unmarked buttons which sit below or above on the side of a screen. They derive their meaning from what's presently on the screen. And what's on the screen will change based on where the call is at that moment — in a conference call, about to set up a conference call, about to go into voice mail, into voice mail, programming a speed dial number, etc.

Soft Sectored A floppy disk whose sector boundaries are marked with records instead of holes. Soft-sectored disks have typically one hole. Hard sectored disks have many holes. A soft-sectored disk won't work on disk drives which use hard-sectored drives and vice versa — even though the disks might be the same size. Soft-sectored disks are now much more common.

Soft Selectable/Soft Strappable Refers to an option that is controllable through software rather than hardware.

Softkey See SOFT KEY.

Softkey Mapping See ADSI.

Softlifting This definition came from Wired Magazine. Microsoft denies it. It was published just before Windows 95 came out. This is Wired's definition: Microsoft's technique of using a small worm program to interrogate computers on a network. The program gathers intelligence on what software is being run on the machines, and the reports back to Microsoft when the customer signs up for The Microsoft Network. This is reportedly a "feature" of a beta version of Windows 95.

Software The detailed instructions to operate a computer. The term was created to differentiate instructions (i.e. the program) from the hardware. See PROGRAM, HARDWARE and FIRMWARE.

Software Defined Network SDN. Generically, a software defined network refers to a virtual private network. Specifically, it refers to AT&T's Software Defined Network Service, which was introduced in 1985 for AT&T's largest customers and provided only dedicated access services. In 1989, AT&T extended its SDN Network to switched access. Currently, SDN is the most commonly resold of all long distance services. The AT&T Software Defined Network Service Description of July 1986 describes SDN as a service developed for multi-location businesses which allows network managers to tailor their network to their own specific communications needs. Call processing information is stored in a database that is accessed during a call. Calls are transferred over AT&T facilities to either a location that is a dedicated part of the network (for "on-net calling") or to non-dedicated facilities that are part of the network ("off-net") calling). Any company location can become part of a network through SDN dedicated access lines to an AT&T SDN serving office. Here is further explanation of SDN by Siemens Information Systems, whose Saturn PBX has the capability to interface to and mesh neatly with an AT&T SDN network. Here is Siemens' explanation (it's good): When a company establishes an SDN, each phone on the network has a unique seven digit number. This number may or may not be the same as the Listed Directory Number (LDN). When a call is placed at one PBX, it is sent over a dedicated access line to the long distance network. The call is received by the SDN Serving Office and digits are sent via a CCIS (Common Channel Interoffice Signaling) link to the Network Control Point (NCP) for analysis and routing. There is one NCP per SDN network. The NCP contains the unique database for the company using the SDN. The NCP analyzes the digits received against the database, determines whether it is an on-net or off-net call, and sets up the path over which the call will be rerouted on the long distance network. If it is an on-net call, the NCP translates the unique seven digit locator code to one that will be recognized by the AT&T network, sends the call over the network to another SDN Serving Office, and completes the call over dedicated lines to the PBX being called. There is a discount for any call which remains on-net throughout. If the call is off-net, the digits dialed are sent over the long distance network to a Central Office that is not part of the SDN. The call is then completed to the corporate PBX over DID (Direct Inward Dial) lines. Since the introduction of route selection features in PBXs, the caller now has the ability to dial the 10-digit LDN (Long Distance Number) and have the PBX make the decisions on the route and make the decisions on any digit translation or deletion that is necessary to route the call. The CCIS network which carries SDN call signaling is a packet switching network operating at 4,800 bits per second. It will eventually be replaced by CCITT 7 Signaling, a more powerful internationally-accepted signaling system. See SIGNALING SYSTEM 7.

Software Engineering A broadly defined discipline that integrates the many aspects of programming, from writing code to meeting budgets, to produce affordable software that works.

Software Interfaces Language between programs which allows one program to call upon another for assistance in processing.

Software Metering Software that monitors the use of applications — such word processing, spreadsheets, databases, etc. All metering programs tend to take advantage of a software feature called "concurrency." Concurrency means that a company need only buy as many licenses to a program as it has people using the program at one time — concurrent users, in other words.

Software Only Video Playback A multimedia term. Video software playback displays a stream of video without any specialized chips or boards. The playback is done through a software application. The video is usually compressed to minimize the storage space required.

Software MPEG Playback See MPEG.

Software Supervision "Answer Supervision" is knowing when the person at the other end answers the phone. The main reason for wanting to know this is so that a phone company can start billing the call. There are two ways of doing answer supervision. You can get it from the nation's phone system, i.e. the distant office signals back across the country when the called person picks up the phone. Or you can fake it with "software supervision." Essentially this means there's electronics which "listens" to the call. If it "hears" voice or something like voice, it assumes the conversation has started and it's time to start billing the call. Software supervision is not accurate. But when you haven't got access to real answer supervision (for whatever reason) it's better than the previous alternative, which was "timeout." In timeout answer supervision, the carrier simply assumed the call had begun after a certain number of seconds — like 30 — had elapsed with the calling person hanging up. This meant, for example, if you called Grandma and she wasn't there, but you left it ringing, 'cause you knew she took time to answer the phone, then you'd be charged for the call — even though she didn't answer phone!

SOH Start Of Header. A transmission control character used as the first character of the heading of an information message.

SOHO 1. Small Office Home Office. An acronym for a new telecom market opening up, which is part work at home, part commute from home. It's getting larger as companies downsize and their workers become "consultants" or small businesspeople. It's getting larger as companies close their distant sales offices ask their salespeople to work out of home. By the year 2000, some estimates show that more than half of the American workforce will work out of their home, or in some non-conventional job, but certainly not for what was typical in the 1950s through the end of the 1980s — a large, multi-office, multi-tier American corporation. 2. In New York City, there's an area called SOHO. It stands for South Of HOuston Street.

SOL Sure Out of Luck.

Solaris Teleservices See TELESERVICES.

Solder 1. An alloy of lead and tin having a low melting point. 2. To unite or join by solder.

Solenoid A coil consisting of a number of turns in cylindrical form.

Solid State Any semiconductor device that controls electrons, electric fields and magnetic fields in a solid material — and typically has no moving parts.

Solid State Applications Interface Bridge Solid's State Systems' PBX to external computer link. See OPEN APPLICATION INTERFACE.

Solid State Transfer An uninterruptible power supply (UPS) definition. Many UPSes use "mechanical relays" to switch from AC power to battery power. This technology requires 12 milliseconds switching time — slow enough to cause data loss. Solid state switching is faster and eliminates this type of problem.

Solitons Solitons are light pulses that maintain their shape over long distances. In early 1993, scientists at AT&T Bell Labs announced that they transmitted error-free solitons at 10 gigabits per second over one channel and at 20 gigabits per second more than 13,000 kilometers using two channels. To accomplish this feat, they used sliding-frequency guiding filters. See also RARE EARTH DOPING.

Solution IBM-speak to solve, as in "We've got to solution this problem if we're going to make the sale."

SONET Synchronous Optical NETwork. A family of fiber-optic transmission rates from 51.84 Mbps to 13.22 Gbps, created to provide the flexibility needed to transport many digital signals with different capacities, and to provide a standard for manufacturers to design from. Sonet is an optical interface standard that allows interworking of transmission products from multiple vendors (i.e. mid-span meets). It defines a physical interface, optical line rates known as Optical Carrier (OC) signals, frame format and an OAM&P protocol (Operations, Administration, Maintenance, And Provisioning). The OC signals have electrical called Synchronous Transport Signals (STSs). The base rate is 51.84 Mbps (OC-1/STS-1) and higher rates are direct multiples of the base rate.

SONET was created by the Exchange Carriers Standards Association for the American National Standards Institute (ANSI). The standard is being defined in phases — Phase 1, adopted in March of 1988 by ANSI, and Phase 2, planned or completion by February of 1990. SONET has also been adopted by the Consultive Committee for International Telephone and Telegraph (CCITT).

SONET was initiated by Bellcore on behalf of the Regional Bell Operating Companies to attain these goals: Multi-vendor interworking, to be cost effective for existing services on an end-to-end basis, to create an infrastructure to support new broadband services and for enhanced operations, administration, maintenance and provisioning (OAM&P).

SONET, according to Northern Telecom, offers many advantages over asynchronous transport including: Opportunity for back-to-back multiplexing, digital cross-connect panels; Easy evolution to broadband transport; Compatibility with evolving operations standards; Enhanced performance monitoring and Extension of OAM&P capabilities to end users.

SONET transmission equipment interleaves STSs in simple integer multiples to form a synchronous high speed signal. This permits easy access to low speed signals (e.g. DS-O, DS-1, etc.) without multi-stage multiplexing and demultiplexing. The low speed signals are mapped into sub-STS-1 signals called Virtual Tributaries (VTs).

SONET uses a 51.84 Mb/s STS-1 signal as the basic building block. Higher rate signals are multiples of STS-1 (e.g. the STS-12/OC-12 signal has a rate of 12 x 51.84 Mb/s or 622.080 Mb/s). The frame format consists of 90 x 9 bytes. The SONET frame format is divided into two main areas: Synchronous Payload Envelope (SPE) and Transport Overhead (TOH). The SPE contains the information being transported by the frame. The TOH supports the OAM&P functions of SONET, and includes a data communication channel that provides an OAM&P communication path between multiple interconnected SONET network elements. The SPE can handle payloads in any of three ways:

1. As a continuous 50.11 Mb/s envelope for carrying asynchronous DS-3, and other payloads requiring up to 50.11 Mb/s capacity in asynchronous (byte invisible) or byte visible format;
2. In a VT structured envelope to accommodate DS-1, DS-1C, DS-2, European CEPT1, or future VT based services (see chart below). These signals can have either an asynchronous or byte visible format; and
3. As concatenated payloads to accommodate services requiring more than 50.11 Mb/s capacity. For example, three STS-1 SPEs may be concatenated (combined) to transport a broadband ISDN signal of 135 Mb/s.
According to AT&T, the main Sonet characteristics are:
A family of rates at N x 51.84 Mb/s; Optical interconnect allowing mid-span meet; Intra office mixed vendor interconnects; Overhead channels for OAM & P functions and Synchronous networking.

SONET RATES ARE

OC Level	Line Rates	Capacity
OC-1	52 Mb/s	28 DS1s or 1 DS3
OC-3	155 Mb/s	84 DS1s or 3 DS3s
OC-9	466 Mb/s	252 DS1s or 9 DS3s
OC-12	622 Mb/s	336 DS1s or 12 DS3s
OC-18	933 Mb/s	504 DS1s or 18 DS3s
OC-24	1.2 Gb/s	672 DS1s or 24 DS3s
OC-36	1.9 Gb/s	1008 DS1s or 36 DS3s
OC-48	2.5 Gb/s	1344 DS1s or 48 DS3s

See also SONET INTERFACE LAYERS.
SONET Interface Layers The SONET standards define four optical interface layers. While conceptually similar to layering within the Open System Interconnection (OSI) reference model, SONET itself corresponds only to the OSI Physical Layer. The SONET interface layers are:
1. Photonic Layer: Handles bit transport across the physical medium; primarily responsible for converting STS (electrical) signals to OC (optical) signals. Electro-optical devices communicate at this layer;
2. Section Layer: Transports STS-N frames and Section Overhead (SOH) across the medium; functions include framing, scrambling, and error monitoring. Section Terminating Equipment (STE) communicate at this layer;
3. Line Layer: Responsible for the reliable transport of the Synchronous Payload Envelope (SPE) (i.e., user data) and Line Overhead (LOH) across the medium; responsibilities include synchronization and multiplexing for the Path Layer and mapping the SPE and LOH into an STS-N frame. An OC-N-to-OC-M multiplexer is an example of Line Terminating Equipment (LTE); and
4. Path Layer: Handles transport of services (e.g., DS-1, DS-3, E-1, or video) between Path Terminal Equipment (PTE); the main function is to map the services and Path Overhead (POH) information into the PTE includes SONET-capable switches with an interface to a non-SONET network, such as a T1-to-SONET multiplexer.
SONET Ring Fiber loop with Sonet protocols. Popular in delivering business services to metropolitan areas.
Sony Mini Disc A 2 1/2 inch silvery CD (compact Disc) that can record and play 74 minutes of sounds, almost as much as its five-inch forebear. To record on this disc, a laser momentarily heats a tiny spot on the disk to 400 degrees Fahrenheit, while a magnetic head writes the signal into the heated part of the magnetic layer. To play the disk an optical pickup analyzes the polarity of the light reflected from each spot.
SOP Standard Operating Procedure.
Sort To order a collection of records — for example, a telephone directory — in some specified way, say, in alphabetical order. Computers can sort in virtually any way you ask them to. Most companies don't produce sufficient "sorts" on their telephone directories.
Sort Scheme A call center term. A list of fields that tells the program how to sort a report or a list of records. This can be simple scheme that sorts by only one field or a complex scheme consisting of sorts within sorts.
Sound Files Files on PCs have their own extensions — the three letters which follow the name of the file. For example, a sound file of jungle noises might be called jungle.wav. Here are the typical extensions on sound files of various computers: Microsoft Windows — .wav Apple — .aif NeXT — .snd MIDI — .mid and .nni Sound Blaster — .voc Intel Indeo Video Movie clips — .avi
Sound Powered Telephone A telephone in which the operating power is derived from the speech input only.
Sound Waves The waves given off by a vibrating body, which are transmitted by an elastic material medium (such as the air) and which can be detected by the ear.
Source That part of a communications system which transmits information.
Source Address The part of a message which indicates who sent the message. Just like the top left-hand address on the envelope.
Source Code A set of instructions, written in a programming language, that must be translated to machine instructions before the program can be run on a computer. The program which finally runs on that computer is known as the object code.
Source Explicit Forwarding A feature that allows MAC-layer bridges on local area networks to forward packets from only source address specified by the network administrator.
Source Route A hierarchically complete source route. See SOURCE ROUTING.
Source Route Bridging SRB. Token-ring technique for establishing communications between devices on different LANs.

See SOURCE ROUTING.

Source Routing A method used by a bridge for moving data between two networks. Originally developed by IBM's token ring network, it relies on information contained within the token to route the packet between the two networks. Since the information in the token is supplied by the computer that sent the data packet, that computer must know on which network the destination computer is located. IBM developed a special protocol that lets computers discover that information. For source routing to work, every computer and every bridge on all networks must support this protocol. If some computers do not use this protocol, they will not receive packets from bridges that use source routing. See BRIDGE. Compare to TRANSPARENT ROUTING.

In IBM's method of routing local area network data across bridges, IBM's bridges can be configured as either single-route broadcast or all-routes broadcast. The default is single-route broadcast. Single-route broadcasting means that only one designated single-route bridge will pass the packet and only one copy of the packet will arrive at its destination. Single-route broadcast bridges can transmit both single-route and all-routes packets. All-routes broadcasting sends the packet across every possible route in the network, resulting in as many copies of the frame at the destination as there are all-routes broadcasting bridges in the network. All-routes broadcast bridges only pass all-routes broadcast packets.

Source Traffic Descriptor An ATM term. A set of traffic parameters belonging to the ATM Traffic Descriptor used during the connection set-up to capture the intrinsic traffic characteristics of the connection requested by the source.

Source/Destination Routing A term used in call centers for routing calls based on where they originate or terminate. See also SKILLS-BASED ROUTING, CALENDAR ROUTING and END-OF-SHIFT ROUTING.

Source/Sink Device A source/sink device is byte-synchronous with a byte orientation. Source devices originate; sink devices terminate.

Southwestern Bell Corporation One of the seven Regional Holding Companies formed at Divestiture. It includes Southwestern Bell Telephone Co., Southwestern Bell Mobile Systems, Southwestern Bell Yellow Pages, Southwestern Bell Telecom and SBC International. In early October, 1994, Southwestern Bell Corporation changed its name to SBC Communications Inc.

SP 1. Support Processor. 2. Sending Program. 3. Signal Present. 4. Signal Processor. 5. Signaling Point.

SPA Software Publishers Association. See MPC3.

Space 1. In digital transmission, the space is equated to the zero (0) and the mark is equated to the one (1). In telecommunications, space is the absence of a signal. Equivalent to a binary "0".

2. Space also stands for Service Creation and Customization.

Space Brokers Companies that provide all the facilities needed to start a 976 or 900 service. Those facilities include offices, computing equipment, voice processing software, telephone lines and numbers.

Space Diversity Protection of a radio signal by providing a separate antenna located a few feet below the regular antenna on the same tower to assume the load when the regular transmission path on the same tower fades because of rain, a bird flying through it, etc.

Space Division Multiplexing Each distinct signal or message travels over a separate physical path such as its own wire or wire pair within a cable.

Space Division Switching Method for switching circuits in which each connection through the switch takes a physically separate path.

Space Hold A no traffic line condition where a steady space is transmitted.

Space Segment 1. The part of a satellite system that is in space. 2. This is also the imprecise term used to describe the band of frequency purchased by the satellite customer. The customer can purchase a portion of the bandwidth of a single transponder or the customer can purchase one or more entire transponder bandwidths.

Spacecraft Switched Time Division Multiple Access SSTDMA. A method of sharing the capacity of a communications satellite by on-board switching of signals aimed at earth stations.

Spade Lug A metal connector attached to the end of a piece of wire, typically by soldering or by pressure. The metal spade lug is shaped like a "U." The idea is to slide the flat "U" shaped metal piece under a screw and then tighten the screw, thus making a connection. In the old days, all phones came with spade lug connectors. These days, there are other faster, more efficient ways of connecting phones — including modular jacks and punchdown tools.

SPAG 1. Europe's Standards Promotion and Application Group. 2. Standards Promotion and Application Group. A group of European OSI manufacturers which chooses option subsets and publishes these in a "Guide to the Use of Standards" (GUS).

Spaghetti Code A program written without thought, logic or structure. And whose "logic" is therefore very difficult to follow. Some would say this definition covers most software written today. That's unfair.

Spam Message posted to numerous Usenet newsgroups to which it has absolutely no relevance (also a verb). See SPAMMING.

Spamming Random indiscriminate posting of items (often advertisements) on computer bulletin boards. The term is derived from a brand of pink, canned meat that splatters messily when hurled. A milder form of spamming is called crossposting.

Span 1. Refers to that portion of a high speed digital system than connects a C.O. (Central Office) to C.O. or terminal office to terminal office.

2. Also called a T-Span Line. A repeatered outside plant four-wire, two twisted-pair transmission line.

3. A call center term. The total duration of a schedule from start time to stop time, including all breaks.

Span Line A T-1 link.

Span Powered In T-1, refers to the application of a varying voltage (+130V to -130V) to the digital cable pairs to maintain a 60mA DC current at each repeater and at the customer premises (this power is generally used for regeneration, loop backs, keep alive signals and alarms).

Spanning Tree Algorithm STA. An algorithm, the original version of which was invented by Digital Equipment Corporation, used to prevent logic loops in a bridged network by creating a spanning tree. The algorithm is now documented in the IEEE 802.1d specifications, although the Digital algorithm and the IEEE 802.1d algorithm are not the same, nor are they compatible. When multiple paths exist, says PC Magazine's Frank Derfler, STA lets a bridge use only the most efficient one. If that path fails, STA automatically reconfigures the network to make another path become active, sustaining network operations. This algorithm is used mostly by local bridges; it is not economical for use over leased telephone circuits connecting remote bridges.

Spanning Tree Protocol STP. Inactivation of links between networks so that information packets are channeled along one route and will not search endlessly for a destination.

SPARC A proprietary RISC microprocessor designed and marketed by Sun Microsystems and included in all Sun SPARCstations.

Spare Pairs In existing distribution systems, twisted pairs that are not being used and can be used to serve new communications devices. Spare pairs are exactly what they sound like — spare pairs of cables. Best to install as many spares as you can when you initially wire up a building or office. Remember Newton's Rule: You'll always need twice as much cabling as you ever dreamed in your wildest dreams you'd need.

Spark An arc of very short duration.

Spark Gap Terminals or electrodes designed to permit spark discharges to take place across a gap.

Spark Test A test designed to locate pin-holes in a wire's insulation by application of an electrical potential across the material for a very short period of time while the wire is drawn through an electrode field with one end of the wire grounded.

SPATA SPeech And daTA. Watch for this expression to pick up steam once true integration of voice and data occurs. The expression does not come from the sentence: "Spata to integrate today than tomorrow."

Spatial Data Management A technique which allows users access to information by pointing at picture symbols on the screen.

SPC 1. Stored Program Control. All phone systems these days are SPCs. There's stored software, which is the program, which controls the computer or microprocessor which in turn controls the operation of the switch. Thus switches are stored program control.
2. Signal Processing Component.

SPC Allocation Service An SCSA definition. A service which allocates SPCs (Signal Processing Components) to Groups.

SPCAS SPC Allocation Service.

SPCS Stored Program Controlled Switch. A digital switch that supports call control, routing, and supplementary services provision under software control. Pretty well switches made after 1970 in North America are SPCSs.

SPCL SPectrum CeLlular error-correction protocol.

SPE 1. Switch Processing Element or Signal Processing Element.
2. Synchronous Payload Envelope. In Sonet, three different payload structures are defined to address different input requirements: 1. Direct-to-STS-1 line rate multiplexing takes 28 DS-Is, 14 DS-ICs or 7 DS-2s directly into the 51.84 Mbps rate. Each is uniquely transported within the SPE; 2. Asynchronous DS-3 Multiplexing takes a complete asynchronous DS-3 bundle (the output of an M13 for example) into the SPE; 3. Synchronous DS-3 Multiplexing maps a Syntran DS-3 signal to the SPE.

Speaker Dependent Voice Recognition Technology capable of recognizing speech from a given user or others who sound like this user after completion of an enrollment procedure. It is not voice verification although it is sometimes confused with this technology.

Speaker Identification Speaker identification is used to determine the identity of a known speaker. It is accomplished by taking spoken input and searching a database of all known system users for a match. Due to its speaker dependent recognition characteristics, you must first be enrolled as a user prior to using the system. To enroll as a user, an individual is required to speak one or more password phrases which are recorded. These phrases create a reference templates which are stored in the system user database for later use during identification sessions. When in operation, the individual using the system is prompted for a specific password or password phrase. When speaking the prompted password as input it creates a new template. This template is then compared to all reference templates in the system for that particular password. The reference template with the closest match is selected. The uniqueness of each user's voice and the finite number of users of the system makes the identification accuracy quite high. With speaker identification the speaker does not claim to be a particular individual. He or she is identified from a group of common users. For the most part, this technology is used for hands free operation of a system where messages and other information specific to that identified individual are pulled-up for use at that time.

Speaker Independent Voice Recognition SIR or SIVR. Technology capable of recognizing any user without prior training or knowledge of the user. SIR converts speech to accurate and meaningful textual information (typically ASCII). SIR is used to accept input from callers to voice processors where the callers are using rotary dial phones instead of touchtone phones. SIR can substitute for the numbers on the DTMF keypad and can add the benefit of a few basic voice commands, e.g., Yes, No, Help, etc. Because computer processing demands are formidable with speaker independent recognition, accurate speaker independent products are created with limited vocabularies. In contrast, trainable or speaker dependent recognizers can feature larger vocabularies at lower prices. SIR has been slowly gaining acceptance in telephone applications. SIR is increasingly used in automated operator assistance applications. SIR will see increased use as system builders respond to pressures to provide voice processing functions to the enormous rotary phone installed base domestically and abroad.

Speaker Recognition Having a machine recognize human voice. This is an imprecise term.

Speakerphone A telephone which has a speaker and microphone for hands free, two-way conversation.

Special Access A dedicated line from a customer to a long distance company provided by a local phone company.

Special Billing Number A phone number assigned to certain customers for billing purposes. It cannot be called. It may be given to an operator as the calling number on an outgoing paid call, or it may be used as a "third number billed" number. It's designed as a measure of security and accounting convenience.

Special Characters 1. Microsoft calls special characters that ones not found on your computer's keyboard. In Windows 3.1

these characters are accessible through Character Map, an application in the Accessories group. 2. An AT&T Merlin term. A pause, stop, or end-of-dialing signal in a programmed dialing sequence such as an Auto Dial or Personal Speed Dial number.

Special Grade Access Line An AUTOVON access line specially conditioned, usually by providing amplitude and delay equalization, to give it characteristics suitable for handling special services; e.g., lower signaling rates of 600 to 2400 bits per second.

Special Night Answering Position Provides either a console or a pre-assigned single extension phone to answer all incoming night calls.

Special Routing Code A 3-digit code in the form 0XX and 1XX available for use within a network and used to modify routing or call-handling logic. End users are prevented from using system codes by the arrangement of the switching equipment to block all customer-dialed calls with a 0 or a 1 in the fourth digit of a 10-digit number, as well as 7-digit calls with a 0 or 1 in the first digit.

Special Services A variety of services that are separate from the public switched network.

Specialized Common Carrier A company providing domestic long distance telecommunications services other than AT&T. See OTHER COMMON CARRIERS.

Specific Gravity The ratio of the weight of any volume of substance to a weight of an equal volume of some substance taken as a standard, usually water for liquids and hydrogen for gases.

Specific Inductive Capacity The direct measure of the ability of a substance to store up electrical energy when used as a dielectric material in a condenser.

Speckle The bright and dark spots on the end face of a fiber caused by the interference of modes.

Spectral Efficiency The efficiency of a microwave system in its use of the radio spectrum, usually expressed in bits per Hz for digital radios and KHz per voice channel in analog radios.

Spectrogram A basic research tool for the speech scientist which provides a three-dimensional visual representation of speech.

Spectrum A continuous range of frequencies, usually wide in extent within which waves have some specific common characteristics. See SPECTRUM DESIGNATION OF FREQUENCY.

Spectrum Analyzer Tunable RF instrument which displays a portion of the RF spectrum with amplitude of signals on the vertical axis and frequency on the horizontal axis on a screen. Used in TSCM to analyze transmissions for the characteristics of an illegitimate transmitter (radio bug).

Spectrum Designation Of Frequency A method of referring to a range or band of communication frequencies. In American practice the designator is a two- or three-letter abbreviation for the name. In ITU practice, the designator is numeric.

FREQUENCY RANGE	TYPICAL AMERICAN DESIGNATOR	ITU FREQUENCY BAND DESIGNATOR
30 - 300 Hz	ELF (Extremely Low Frequency)	2
300 - 3000 Hz	ULF (Ultra Low Frequency)	3
3 - 30 kHz	VLF (Very Low Frequency)	4
30 - 300 kHz	LF (Low Frequency)	5
300 - 3000 kHz	MF (Medium Frequency)	6
3 - 30 MHz	HF (High Frequency)	7
30 - 300 MHz	VHF (Very High Frequency)	8
300 - 3000 MHz	UHF (Ultra High Frequency)	9
3 - 30 GHz	SHF (Super High Frequency)	10
30 - 300 GHz	EHF (Extremely High Frequency)	11
00 - 3000 GHz $\geq$	THF (Tremendously High Frequency)	12

FREQUENCY RANGE	ITU FREQUENCY BAND DESIGNATOR
3 - 30 THz	13
30 -300 THz	14
300- 3000 THz	15
3 - 30 PHz	16
30 - 300 PHz	17
300 - 3000 PHz	18
3 - 30 EHz	19
30 - 300 EHz	20
300 - 3000 EHz	21

Spectrus Name for BellSouth's general tariff service specifically for video, such as broadcast and commercial quality.
Speech API See MIcrosoft Speech API.
Speech Concatenation A term used in voice processing for economical digitized speech playback that uses independently recorded files of phrases or file segments linked together under application program control to produce a customized response in natural sounding language. For example, order status, bank balances, bus schedules or lottery results, etc. Concatenation is done for speed and economy. It lends itself to limited and structured vocabularies that are best stored in RAM (Random Access Memory) or speedily accessible from disk. Concatenation does not replace Text-To-Speech (TTS) as a method of getting the voice processor to deliver its responses. Concatenation, however, can be an excellent complement to TTS when a voice application demands broad, real time vocabulary production. See TEXT-TO-SPEECH.
Speech Digit Signaling Signaling in which digit time slots used primarily for encoded speech are periodically used for signaling (as, optionally, in ISDN). See also ISDN.
Speech Recognition Voice recognition is the ability of a machine to recognize your particular voice. This contrasts with speech recognition, which is different. It is the ability of a machine to understand human speech — yours and everyone else's. Voice recognition needs training. Speech recognition doesn't. For a good explanation of when to and when not to use speech recognition, see MICROSOFT SPEECH API.
Speed Bumps When your host system processes information faster than your network can handle and forces you to slow down.
Speed Dial A feature that enables a PBX or PBX phone to store certain telephone numbers and dial them automatically when a code is entered. See SPEED DIALING.
Speed Dialing Permits fast dialing of frequently used numbers. A repertory of numbers may be stored in the instrument and/or in the telephone switch. Usually a button or one, two or three digits are dialed to activate speed dialing.
Speed Of Light In A Vacuum 299×10^6 meters per second. Used in computing index of refraction.
Speedsync See LAPLINK.
SPF Shortest Path First.
SPI Service Provider Interface. See SERVICE PROVIDE INTERFACE and WINDOWS TELEPHONY.
SPID Service Profile IDentifier. When you order ISDN BRI service, you will order it with certain features, also called parameters. And your phone company will give you a SPID for every phone number you have. Each ISDN BRI line typically has two phone numbers. The SPID is an eight to 12 digit number that uniquely identifies the services you ordered. The SPID is actually a label identifier that points to a particular location in your telephone company's central office memory where the relevant details of your ISDN service are stored. You will need your SPID number because your ISDN phone, fax or PC software asks for it and often cannot be installed without it and certainly won't run without it. Also, when you make an ISDN call, your ISDN device — phone, fax, PC software, etc. — sends that number to your central office, which uses it to validate that you can use the services and features you are requesting. When you order a single BRI service, you will usually get two phone numbers — one for each of the BRI lines. This means that you should then have one SPID number — a different one — for each of those two lines. It's easy to lose your SPID number. Your PC may blow up. Your fax may break. Please write your SPID numbers on your ISDN wall jack. Write them in a file. Tattoo them on your tushy. Make sure you can find them when you need them. You can call your phone company and beg for them. But it's not worth the aggravation. Tip: If you find that the SPIDs which your telephone company gave you don't work, add a 1 or 0 or 000 at the end of the number and try that. By the way, Northern Telecom says that SPIDs for its DMS-100 switch are normally a 12 digit numerical code consisting of the user's three digit code, seven digit ISDN telephone number, followed by another zero.
SPIE Society of Photometric Industry Engineers or Society of PhotoOptical Instrumentation Engineering.
Spike An in-phase impulse causing spontaneous increases in voltage. Spikes are very fast impulses, less than 100 microseconds, of high-voltage electricity ranging from 400 volts to 5,600 volts superimposed on the normal 120V AC electrical sine wave. See also METAL OXIDE VARISTOR.
Spike Mike Contact microphone for listening through walls.
Spikes Electrical anomalies represented as short duration, instantaneous, very high voltage fluctuations on an electrical service.
Spill-Forward Feature A service feature, in the operation of an intermediate office, that, acting on incoming trunk service treatment indications, assumes routing control of the call from the originating office. This increases the chances of completion by offering the call to more trunk groups than are available in the originating office.
Spin Stabilization A method of preventing a satellite from tumbling by spinning it about its axis.
Spindle The rotating hub structure to which the disks in a hard disk system are attached.
Spindle Synchronization A process that coordinates all hard disks in a RAID array to use a single drive's spindle synchronization pulse.
Spiral Wrap A term given to describe the helical wrap of a tape or thread over a core.
SPIRIT European consortium focused on standardizing telecom operators' procurement processes.
Spirit Communications System A small electronic key system from AT&T. A SPIRIT telephone system supports as few as three outside lines and eight telephones (or, with an expansion unit, up to 24 outside lines with 48 phones).
Splanch SPlit Level rANCH.
Splash Tone Distinctive sound used on some phone systems to indicate that a command has been received, or that something has to be done. Vaguely resembles water being splashed.
Splashing A "splash" happens when an Alternate Operator Service (AOS) company, located in a city different to the one you're calling from, connects your call to the long distance carrier of your choice in the city the AOS operator is in. Splashing does not imply backhauling, but it often happens. For example, let's say you're calling from Hotel Magnificent in Chicago. You ask AT&T to handle your call. The AOS, located in Atlanta, "splashes" your call over to AT&T in Atlanta. But you're calling Los Angeles. Bingo. Your AT&T call to LA is now more expensive than it would be if you had been connected directly to AT&T from Hotel

Magnificent in Chicago.

Splice Verb. The joining of two or more cables together by splicing the conductors pair-to-pair.

Splint A pejorative word for US Sprint, a long distance company now owned largely (80%) by United Telecom, Kansas City and 20% by GTE and now called just Sprint.

Split A call center term. Split is an ACD routing division that allows calls arriving on specific trunks or calls of certain transaction types to be answered by specific groups of employees. Also referred to as gate or group. Same as Group. See ACD or AUTOMATIC CALL DISTRIBUTOR.

Split Access To Outgoing Trunks Two separate trunk groups provided for direct outward dialing which can be accessed by dialing the same trunk access code. Controlled on class of service basis.

Split Channel Modem A modem which divides a communications channel into separate send and receive channels. Most modems which use the dial-up phone network are split channel — meaning they can transmit and receive simultaneously over a two wire circuit. See also SPLIT STREAM MODEM, which is another term for the same thing.

Split Horizon The view a router has of a wide area network interface in a partial mesh environment where an incoming packet may need to be sent out the same interface over which it is received to reach its ultimate destination. Split horizon is normally disabled to ensure that this cannot occur and that routing loops are not created. This posed a problem with frame relay because packets should be sent back out the same physical interface over which they have been received, but not the same logical interface. Router vendors have now solved this problem, enabling routers to support partially meshed frame relay networks.

Split Link When one multiplexer uses two links to communicate to two separate multiplexers.

Split Pair Something that happens in cable splicing when one wire of a pair gets spliced to the wire of an adjacent pair. You could call it a split pair. It's more accurate to call it a mistake.

Split Seeks A process by which Novell NetWare SFT III splits multiple read requests between the two servers' disks for simultaneous processing and faster disk rads. Disk reads are split between both servers, with only one server doing a particular read and sending the data read over the MSL (if necessary) to the other server. MSL is the Mirror Server Link.

Split Stream Modem A modem which can handle multiple, independent channels over a single transmission path.

Split System A switching system which implements the functions of more than one logical node.

Splitting Permits operator to consult privately with one party on a call without the other party's hearing. Or permits a three-party telephone conference user to consult privately with one side of the conference while the other is effectively put on hold. Jumping from one party to the other is called "Swapping."

SPM 1. Subscriber Private Meter.

2. An AT&T Merlin term. System Programming and Maintenance. A DOS- or UNIX-based application for programming and maintaining the Merlin communications system.

3. Service Provider Messages.

SPN Subscriber Premises Networks.

SPOI Signaling Point Of Interface. The demarcation point on the SS7 signaling link between a LEC network and a Wireless Services Provider (WSP) network. The point established the technical interface and can designate the test point and operational division of responsibility for the signaling.

Spoofing 1. In COMSEC applications, the interception, alteration, and retransmission of a cipher signal or data in such a way as to mislead the receiver.

2. In automated-information-systems applications, an attempt to gain access to an automated information system by posing as an authorized user. In electronic mail applications, spoofing is where one person impersonates another person in order to gain access to that person's electronic mail.

3. Spoofing is a networking term. It is a method by which the client and/or router filters network traffic to keep unnecessary traffic from going over a WAN link. THe ability for a device to determine what is not "meaningful" traffic and rather than forwarding the traffic over the connection, the device responds to the source of the traffic with the response that would have been generated by the intended destination device.

SPOOL Simultaneous Peripheral Operation On Line. A program or piece of hardware that controls a buffer of data going to some output device, including a printer or a screen. A spool allows several users to send data to a device such as a printer at the same time, even when the printer is busy. The spool controls the transmission of data to the device by using a buffer or creating a temporary file in which to store the data going to the busy device. See SPOOLING.

SPOOLING Simultaneous Peripheral Operations On Line. Spooling means temporarily storing programs or program outputs on magnetic tape, RAM or disks for output or processing. The word "Spooling" is mostly associated with printers. Here's an example: Pretend that a lot of people on your Local Area Network all want to send their reports to the printer today. Instead of each person having control of the printer and relinquishing it only when they're through, each user tells the print Spooler what file they want printed. The program, called the spooler, places the print request in the print queue. When your request reaches the top of the queue, your report is printed out. Using a PC as print spooler slows it down. Best not to use it for much else.

Spot Beam Antenna A satellite antenna capable of illuminating or focusing on a narrow portion of the earth's surface.

SPOX An operating system for digital signal processors from Spectron Microsystems, Goleta, CA, now owned by Dialogic in Parsippany, NJ. Spox is a real-time, multitasking operating system that is optimized for use with fixed and floating point digital signal processors in both single- and multiprocessor systems. The SPOX environment is implemented as a library of relocatable, C-callable modules.

SPP 1. Sequenced Packet Protocol. XNS (Xerox Network Systems) protocol governing sequenced data.

2. Signal Processing Platform.

Spread Spectrum Also called frequency hopping, spread spectrum is a modulation technique in which the information content is spread over a wider bandwidth than the frequency content of the original information. Spread spectrum takes an input signal, mixes it with FM noise and "spreads" the signal over a broad frequency range, hopping from frequency to frequency at split-second intervals. The spread signal has greater bandwidth than the original message. Spread spectrum receivers recognize a spread signal, acquire and "de-spread" it and thus return it to its initial form (the original message). Spread spectrum is highly secure. Would-be eavesdroppers hear only unintelligible blips. Attempts to jam the signal succeed only at knocking out a few small bits of it. So effective is the concept that it is now the principal antijamming device in the U.S. Government's Milstar defense communications satellite system. Hedy Lamarr, the actress, created the concept of spread spectrum in 1940 and, two years later, received a U.S. patent for a "secret communication system." The patent was issued to her and George Antheil, a film-score composer, to whom Ms. Lamarr had turned for help in perfecting her idea. Spread spectrum was never of use during the World War II and did not come into use until 1962 when Sylvania installed it on ships sent to blockade Cuba. Ms. Lamarr never received one penny for her invention.

Spreadsheet An outsize ledger sheet used by accountants

Springboard Dialogic's term for a general purpose computing engine designed for a wide range of powerful voice processing applications. It contains two Motorola 56001 DSPs (Digital Signal Processors) and one Intel 80286 12.5 MHz microprocessor. It allows Dialogic's Technology Partners to extend D/12x functions.

Sprint The new name for United Telecom, Kansas City. Originally Sprint was a long distance telephone company owned 80% by United Telecom, Kansas City and 20% by GTE, now in Dallas, TX. Sprint used to be called GTE Sprint and once was called US Sprint. But now it's just called Sprint. In 1992, United Telecom bought GTE's 20%, thus making it the 100% owner of Sprint and changed the name of its complete company to Sprint Corporation. Here are some of the services Sprint provides (MCI and AT&T provide similar services, though under different names):
SPRINT DATACALL PLUS: offers global dial-up access to host computers, using the SprintNet data network (once called Telenet), for asynchronous and X.25 applications at 300 to 9600 bps, with end-to-end error correction.
SPRINT DAY PLUS: Sprint switched service for small business customers.
SPRINT EXPRESS: Using Sprint Express, you may call the United States from a number of overseas locations simply by dialing a toll-free number direct to a Sprint operator in the U.S. You do not have to contact a foreign operator or pay excessive overseas surcharges. You can bill the call to your FONCARD, local exchange carrier card or call collect. You will receive one consolidated bill, in U.S. currency.
SPRINT FAX: Send faxes. Broadcast Distribution allows a fax to be sent to multiple locations simply by making one phone call. Document on Demand automates information retrieval via fax from an 800 or 900 number.
SPRINT FONCARD: Sprint's name for its credit card.
SPRINTMAIL MESSAGING SERVICES: Allows you to send, receive and file messages 24 hours a day from a terminal, PC or communicating word processor via fax, telex or hard copy, anywhere in the world. SprintMail Fax allows messages to be sent directly from a desktop PC to virtually any fax machine in the world. SprintMail Post delivers electronically input messages from a desktop PC as laser-printed mail, in two days or less, anywhere in the continental U.S.
SPRINT MEETING CHANNEL: Allegedly the world's largest two-way digital videoconferencing network with over 950 locations in 30 countries world-wide. It's organized and run by Sprint.
SPRINT PLUS: Sprint PLUS is the name for Sprint's basic switched telephone service.
SPRINT QUICKCONFERENCE: A three-way conferencing capability, using Sprint's FONCARD when you're away from the office.
SPRINT VISA: Sprint VISA is a multi-functional credit card that combines a traditional VISA card with the Sprint FONCARD. It allows you to buy goods and services, access the VISA ATM network for cash and place a FONCARD call all with a single card. All FONCARD calls are billed to the VISA account.
SPRINT WORLD: Sprint WORLD gives residential customers flat rates to a designated number of countries during discount and economy time periods for a monthly fee.

SPS 1. Signaling Protocols and Switching.
2. Standard Positioning Service. The normal civilian positioning accuracy obtained by using the single frequency C/A code in the GPS (Global Positioning System) system.
3. Solution Provider, also called Microsoft Solution Provider. See MICROSOFT SOLUTION PROVIDER.

SPTS Single Program Transport Stream: An MPEG-2 Transport Stream that consists of only one program.

SPUD A special long-handled shovel used to loosen soil in a hole into which you're going to put a telephone pole.

Spurious A term used in voice recognition. A spurious error is said to occur when a sound that is not a valid spoken input is incorrectly accepted as an input speech utterance.

Spurious Emission Emission on a frequency or frequencies which are outside the necessary bandwidth and the level of which may be reduced without affecting the corresponding transmission of information. Spurious emissions include harmonic emissions, parasitic emissions, intermodulation products and frequency conversion products, but exclude out-of-band emissions.

Spurs 1. The sharp metal devices on the climbers used by telephone line-persons (people who climb telephone poles). Such climbing spurs make a mess of telephone poles. 2. The cowboy devices awarded by US WEST to privileged persons who have done US WEST some nice favor or are otherwise deserving of honor.

Sputnik Sputnik was the world's first artificial satellite. It was launched by the Russians on October 4, 1957.

SPX Sequenced Packet eXchange. 1. An enhanced set of commands implemented on top of IPX to create a true transport layer interface. SPX provides more functions than IPX, including guaranteed packet delivery. 2. Novell's implementation of SPP for its NetWare local area network operating system.

SQE Signal Quality Error. The 802.3 specification defines this for signals from the MAU to the NIC. "SQE test" is a signal indicating the MAU's collision detection circuitry is functioning properly. SQE test should be disabled on repeaters and on NICs which do not support SQE.

NEWTON'S TELECOM DICTIONARY

SQL Structured Query Language. Invented by IBM and first commercialized by Oracle in the early 1990s, SQL is a powerful database language used for creating, maintaining and viewing database data. It is becoming somewhat of a standard in the mainframe and minicomputer world, and it is on its way to becoming a PC standard. When it is a fully-accepted standard, different computer systems running different DBMSs will easily be able to communicate and exchange data with each other by simply trading SQL commands. SQL is commonly used with database servers, i.e. those running on a local area network. There is now an ANSI standard SQL definition for all computer systems. The largest purveyors of SQL databases are Gupta, Informix, Microsoft, Powersoft, Oracle and Sybase. See ODBC, QBE and SQL SERVER.

SQL Server Microsoft SQL Server. A Microsoft retail product that provides distributed database management. Multiple workstations manipulate data stored on a server, where the server coordinates operations and performs resource-intensive calculations.

Square Operation If there are fewer than eight lines in a Merlin system, all users can access all lines. See SQUARED KEY SYSTEM.

Square Wave A term used to refer to a digital signal where there are only two values, on and off. There's one major advantage to square "waves" (they're not really waves). They're much cleaner. When they reach a repeater site, they're not just amplified and sent on their way. They're regenerated. All the noise they picked up on their travels is removed. Then they're amplified and sent on their way. Square waves are also cheaper to produce. That's good. But if you send a square ringing "wave" to a device like a high-speed modem that's expecting an analog sine wave, such as one produced by a central office (i.e. public exchange), that high-speed modem often will not respond.

Squared Key System What is a squared key system? No one knows where the word "squared" came from. So if our explanation bears no relation to the word "squared," sorry. But it goes like this (we think): In the old days there were 1A2 phone systems. These 1A2 phones had buttons on them. These buttons could correspond to trunks — any trunk. These were called non-squared systems. Then came electronic key systems. Each trunk had to "appear" (i.e. be) the same button on each phone. These electronic key systems were called squared systems. There are advantages and disadvantages. Squared systems are portrayed as having one advantage: You can go to any phone anywhere in the system and punch any button for any trunk and know it to be the same button for the same trunk. Thus less confusion. But this means you can only have as many trunks on your key system as you have trunk buttons on your key telephones. In a non-squared system — a 1A2, the newer hybrids or some of the newer programmable key systems — you can have more trunks than you have buttons on each phone. Some phones will have trunks that others don't have and vice versa. Thus you can have more trunks on your phone system than you have buttons on your phones. This means, for example, that four executives can have each have private lines and access to four trunks on a six button phone. (The other button is for Hold.)

Squelch A circuit function that acts to suppress the audio output of a receiver.

SR Source Routing: A bridged method whereby the source at a data exchange determines the route that subsequent frames will use.

SRAM Static Random Access Memory. A form of RAM that retains its data without the constant refreshing that DRAM requires. SRAM is generally preferable to DRAM because it offers faster memory access times. but it also costs more to make because it has more electrical components. The most common use for SRAM is to cache data traveling between the CPU and a RAM subsystem populated with DRAM. This improves your PC's performance by reducing the number of DRAM accesses needed. See also DRAM and RAM.

SRC 1. Strategic Review Committee (ETSI).
2. Stupid Rich Customer. One who will buy anything.

SRDC SubRate Digital Cross-connect.

SRDM SubRate Data Multiplexer. The Europeans call it SDM. An SRDM typically subdivides DS-0 of 64 Kbps, into a number of circuits, each less than 64 Kbps.

SRF Specifically Routed Frame: A Source Routing Bridging Frame which uses a specific route between the source and destination.

SRP Suggested Retail Price.

SRS Statistics Repository System.

SRT 1. Station Ringing Transfer.
2. Source Routing Transparent, a token ring bridging standard that is jointly sponsored by the IEEE and IBM. It combines IBM Source Routing and Transparent Bridging (IEEE 802.1) in the same unit. This provides a way for universal bridging of token ring LANs supporting IBM and all non-IBM LAN protocols. An SRT bridge examines each data packet on the ring to discover whether the packet is using a source routing or non-source routing protocol. It then applies the appropriate bridging method. See also BRIDGE, SOURCE ROUTING and TRANSPARENT ROUTING.

SRTS Synchronous Residual Time Stamp: A clock recovery technique in which difference signals between source timing and a network reference timing signal are transmitted to allow reconstruction of the source timing at the destination.

SS7 Signaling System 7. See SIGNALING SYSTEM 7.

SSA In 1993, IBM, working with a committee of other major manufacturers (Conner, Western Digital, Micropolis, etc), announced an architecture, named Serial Systems Architecture (SSA). This SSA removed some of the constraints of SCSI, particularly the limitation on the attachment of storage devices. Where SCSI allows seven devices to be attached on a string, SSA attaches 127 devices on a loop. SCSI uses bulky and expensive straps where SSA employs low-cost thin cabling. Data rates are also increased from 10MB/s on SCSI to 80MB/s on SSA which is expected to increase to 160 MB/s in 1995. The use of SSA will enable faster data transfer, increase the maximum storage capacity and procedure smaller devices, all at lower costs than today's equivalents. In May of 1994, 17 companies issued a joint press release announcing their commitment to SSA. The number of companies working quietly on the development of SSA products and devices exceeds that figure. See also FIBRE CHANNEL, FIREWIRE and SCSI.

SSB Single SideBand. See SINGLE SIDEBAND.

SSB-SC Single-SideBand Suppressed Carrier.
SSCF Service Specific Coordination Function: SSCF is a function defined in Q.2130, B-ISDN Signaling ATM Adaptation Layer-Service Specific Coordination Function for Support of Signaling at the User-to-Network Interface.
SSCOP Service Specific Connection Oriented Protocol: An adaptation layer protocol defined in ITU-T Specification: Q.2110.
SSCP 1. System Services Control Point. A host based network entity in SNA that manages the network configuration, coordinates network operator and problem determination requests, maintains network address and mapping tables and provides directory support and session services.
2. Service Specific Convergence Sublayer: The portion of the convergence sublayer that is dependent upon the type of traffic that is being converted.
SSL Secure Sockets Layer, a transport level technology for authentication and data encryption between a Web server and a Web browser.
SSP Service Switching Point. Also called Signal Switching Point. A switch that can recognize IN (Intelligent Network) calls and route and connect them under the direction of an SCP (Service Control Point. A computer database that holds information on IN (Intelligent Network) services and subscribers. The SCP is separated from the actual SCP switch, making it easier to introduce new services on the network. See SCP. For a full explanation of the Advanced Intelligent Network, see AIN.
SST Spread Spectrum Technology. See SPREAD SPECTRUM.
SSU Session Support Utility. a DEC-proprietary protocol that allows multiple sessions to run simultaneously over a single serial cable. SSU is used to allow terminals to provide two session "windows" that can display session output simultaneously.
ST 1. STart signal to indicate end of outpulsing.
2. Straight Tip. A fiber-optic connector designed by AT&T which uses the bayonet style coupling rather than screw on as the SMA uses. The ST is generally considered the eventual replacement for the SMA type connector.
ST Connector See STRAIGHT-TIP CONNECTOR.
ST-506/412 Interface One of several industry standard interfaces between a hard disk and hard disk controller. The "intelligence" is on the controller rather than the drive.
ST-Bus Serial Telecom Bus. Mitel Semiconductor, which makes telecom componentry, has structured its digital component product line around the ST-BUS. The ST-BUS is a high speed, synchronous serial bus for transporting information in a digital format. Whether the digital information is voice, data, or video — or a mixture of each — the ST-BUS is designed to accommodate it. The ST-BUS consists of one or several serial data streams with a framing signal and clock signals. The framing signal always has a period of 125 us, resulting in 8,000 frames per second, with the original ST-BUS clock rate of 2.048 Mbit/s (thirty two 64 kbit/s channels). The ST-BUS standard now includes higher speed modes of 4.096 or 8.192 Mbit/s ST-BUS, resulting in 64 or 128 channels of 64kbit/s, respectively. This provides the bandwidth necessary for newer multimedia applications. According to Mitel, the advantages of using the ST-BUS are:
1. Printed circuit board area devoted to information transfer between functional modules is minimized.
2. Fewer tracks, backplane connections, and intra-shelf cables are needed compared to systems that use parallel paths.
3. The ST-BUS is designed to be divided down into individual channels of 64 kbit/s, resulting in improved efficiency and lower cost when several information paths are able to share the same ST-BUS.
4. Additional glue logic is not required when using ST-BUS compatible components.
5. From an IC perspective, the ST-BUS results in lower pin counts, improved reliability, and less power consumption.
STA Spanning Tree Algorithm. A technique based on an IEEE 802.1 standard that detects and eliminates logical loops in a bridged network. When multiple paths exist, STA lets a bridge use only the most efficient one. If that path fails, STA automatically reconfigures the network so that another path becomes active, sustaining network operations.
Stack A set of data storage locations that are accessed in a fixed sequence.
Staffing Requirements Forecast A call center term. A calculation of the number of employees required in each period of the day to handle the forecast call volume for that period.
Stage & Test A term used in the secondary telecom equipment business. The installation (stage) and diagnostic testing of a PBX switch, cabinet, part, or peripheral in a reconfiguration center facility — where a dealer tests the complete system as one entity before shipment.
Stagger In facsimile systems, periodic error in the position of the recorded spot along the recorded line.
Stair Stepping Using a low-level account to gain ever-higher levels of unauthorized access in a network.
Stako An Ericsson committee for component standardization.
Stand Alone Any device that can perform independently of something else.
Standard Industry Practice Terminology used to indicate normal rules used within the secondary telecom equipment business. These rules have developed over time and usage, but lack formal support by industry groups or dealers. In short, there is no clear definition as to what "standard industry practice" means in the secondary business.
Standard Jack The means of connecting Customer premises equipment to a circuit as specified in the FCC Registration Program.
Standard Metropolitan Statistical Area SMSA. A metropolitan area consisting of one or more cities as defined by the Office of Management and Budget and used by the FCC to allocate the cellular radio market.
Standard Test Zone A single-frequency signal with a standardization level generally used for level alignment of single links in tandem.
Standardized Test Tone A single frequency signal at a standardized power level.
Standards Agreed principles of protocol. Standards are set by committees working under various trade and international organizations. RS standards, such as RS-232-C are set by the EIA, the Electronics Industries Association. ANSI standards for data communications are from the X committee. Standards from ANSI would look like X3.4-1967 which is the standard for the ASCII

NEWTON'S TELECOM DICTIONARY

code. The CCITT (now called the ITU-T) does not put out standards, but rather, publishes "recommendations", owing to the international egos involved. "V" series recommendations refer to data transmission over the telephone network, while "X" series recommendations, such as X.25 (properly pronounced "Eks dot twenty five"), refer to data transmission over public data networks. Notice that the ANSI standards have the year they were approved as part of the name of the standard, while CCITT recommendations do not. The placement of the "dot" is another clue as to whose confusing standard belongs to whom.

When you're buying a phone system, at minimum it should conform to four standards:
- Emissions compliance according to the FCC Part 15.
- Telephone compliance according to the FCC Part 68.
- Safety standards set by the National Electric Code, OSHA and the Underwriters Laboratories 1459.
- Bellcore compliance (from the Network Equipment Building System publication and their Generic Physical Design Requirements for Telecommunications Products and Equipment publication. See STANDARDS BODIES.

Standards Bodies Here is a list of the major standards bodies in telecom and computing:
ANSI, American National Standards Institute, 11 West 42nd Street, New York, NY 10036. 212-642-4800
Bellcore, 445 South Street, Morristown NJ 07962, 201-829-4256
CSA, Canadian Standards Association, 178 Rexdale Blvd, Rexdale (Toronto), Ontario Canada M9W 1R3, 416-747-4044
ECTF, 303 Vintage Park Drive, Foster City, CA 94404 415-578-6852 Fax 415-378-6692
EIA, Electronics Industry Association, 2500 Wilson Boulevard, Arlington, VA 22201 703-907-7500
FCC, Federal Communications Commission, 1919 M Street, NW. Washington DC 202-418-0200
GO-MVIP, 3220 N Street NW, Suite 360, Washington DC 20007 508-650-1334 Fax 508-650-1375
ICEA, Insulated Cable Engineers Association, P.O. Box 440, South Yarmouth, MA 02664, 508-394-4424
NFPA, National Fire Protection Association, Batterymarch Park, Quincey, MA 02169, 617-770-3000 Fax 617-770-0700
OFR, Office of the Federal Register, National Archives & Records Administration, 800 North Capitol Street NW, Suite 700, Washington, D.C. 20408 202-523-3117 Fax 202-523-6866
TIA, Telecommunications Industry Association, 1722 I Street NW Suite 300, Washington, DC 20006, 202-347-7016
UL, Underwriters Laboratory, 333 Pfingsten Road, Northbrook, IL 60062 847-272-8800 Fax 847-272-8129

Standby Processor A spare computer exists which can direct PBX operations if the primary one fails. Some standbys are just sitting there, installed but not turned on. They require someone to turn them on. Some standbys are actually running all the time, as the main one is. If the main one crashes, the standby processor is ready to take over.

Standby Time The amount of time you can leave your fully charged cellular portable or transportable phone turned on to receive incoming calls before the phone will completely discharge the batteries. See TALK TIME.

Standing Wave When you look at it on an oscilloscope the pattern of the wave is perfectly flat, i.e. horizontal. It's caused by two sine waves of the same frequency moving in opposite directions. In transmission line theory the accepted definition is simply the superposition of two waves traveling in opposite directions.

Standing Wave Ratio SWR. The ratio of the amplitude of a standing wave at an anti-node to the amplitude at a node.

Star 1. A topology in which all phones or workstations are wired directly to a central service unit or workstation that establishes, maintains and breaks connections between the workstations. Virtually all phone systems are stars configurations. ISDN BRI bus will be the first phone system to operate on a bus. In datacom language, the center of a star is called the hub. The advantage of a star is that it is easy to isolate a problem node. However, if the central node fails, the entire network fails. The star network we're all most familiar with is our local telephone exchange. At the center (the hub) rests the central office. Spanning out in a star are the lines going to the individual workstations (telephones) in peoples' houses and offices. 2. Advanced telecommunications for the industrially less advanced regions of the European Community.

Star Coupler A device that couples multiple fibers at a central point and distributes the signal from one fiber into all others simultaneously.

Star Network A computer network with peripheral nodes all connected to one or more computers at a centrally located facility.

Star Quadded Cable Spiral-four cable. See STAR NETWORK.

Star Topology A LAN topology in which end points on a network are connected to a common central switch by point-to-point links. See STAR.

Starcontroller Farallon's controller allowing users to build Apple Macintosh LocalTalk networks in star configurations.

StarLAN A local area network developed by AT&T which uses twisted pair telephone wires in a star configuration. StarLan is AT&T's proprietary CSMA/CD product line, and includes a network operating system and Network Interface Cards (NICs). The original StarLan transmitted data at 1Mbps using the CSMA/CD access method, over unshielded twisted pair; it now offers standard 10Mbps CSMA/CD (also known as Ethernet) capabilities. It is compatible with AT&T's Premises Distribution System, a large cabling system. StarLan connects to the Information Systems Network, a larger data communications switching system for high-speed terminal-to-host and host-to-host connectivity. StarLan software provides transparency and ease of use to the end user through the AT&T menu program. AT&T is based in Bridgewater, NJ.

Start Bit In asynchronous data communications, characters are sent at arbitrary intervals, i.e. when the operator hits a key. In order for the computer to make heads or tails of what's coming in, each character starts its transmission with a Start Bit. This way, if the first bit of the character to be transmitted is a 1, the fact of receiving a Start Bit (always a 0) tips off the computer that the next bit is part of a transmitted character and not just part of the inter-character gap. See STOP BIT.

Start Element 1. The start pulse of a transmission character. It is used for synchronization of the following bits in a serial transmission process. 2. One of the input or output points in a communications system. This would include a telephone set, a data terminal, a computer communications port.

Start Of Heading SOH. A control character used in data communications that designates the beginning of the message header.

566

NEWTON'S TELECOM DICTIONARY

Start Of Heading Character SOH. A transmission control character used as the first character of a message heading.
Start Of Message SOM. A control character used in data communications that designates the beginning of the message.
Start Of Text STX. A control character used in data communications that designates the beginning of the information being transmitted.
Start Stop Transmission The technique of asynchronous data transmission wherein each character is comprised of a start element at its beginning and a stop element at its end. Start-stop elements allow the receiving device to determine where the transmitted bits for one character ends and the next begins.
Start Time Interval A call center term. A scheduling rule that governs the times at which schedules can start; for example, at 15-minute intervals as opposed to 30-minute intervals.
STAT MUX Informal for STATistical MUltipleXor.
Stat V Trademark for MICOM's series of combined statistical multiplexor and APV products.
State The instantaneous properties of an object that characterize that object's current condition. See STATE MACHINE PROGRAMMING.
State Machine Programming To control multiple telephone lines in a single voice processing program, a new program structure is required. Dialogic calls this technique state machine programming. Computer Science called state machines "Deterministic Finite State Automata."
State Public Service Commission PSC. The State legislative body responsible for among other things, regulating the operation of telephone companies and other persons involved in the furnishing of telephone service. Some states' PSCs are called Public Utilities Commissions. See the next definition.
State Utility Commissions Each state has a utility commission responsible for the regulation of telephone service provided wholly within that state. Regulation extends to introduction of new services, their prices, who will provide them, as well as discontinuance of existing services.
Statement 1. In computer programming languages, a language construct that represents a set of declarations or a step in a sequence of actions. 2. In computer programming, a meaningful expression or generalized instruction represented in a source language.
Static Interference caused by natural electric disturbances in the atmosphere, in your office, in your home. Static electricity can play havoc with telephone systems. Properly grounding your phone system to a true cold water pipe (not one that connects to a PVC plastic piper) is the most minimal protection.
Static Object Information that has been pasted into a document. Unlike embedded or "linked" objects, static objects cannot be changed from within the document. The only way you can change a static object is to delete it from the document, change it in the application used to create it, and paste it into the document again.
Static Positioning Location determination when the GPS (Global Positioning System) receiver's antenna is presumed to be stationary in the earth. This allows the use of various averaging techniques that improve the accuracy of figuring where you are by factors by over 1000.
Static RAM Static Random Access Memory chips do not require a refresh cycle like Dynamic RAM chips and thus can be accessed well over twice as quickly. Static RAM chips must have power to maintain the data they are holding. Static RAM chips cost more than Dynamic RAM chips, which are also called D-RAM.
Static Route A route that is manually entered into a routing table. Static routes take precedence over routes chosen by all dynamic routing protocols.
Station A dumb word for a telephone. Also called an instrument, or a telephone instrument. An extension station is one connected "behind" a PBX or key system. In other words, the PBX or key system is between the station and the telephone central office. We tried to remove the word "station" from this dictionary, but failed. We suspect the word comes from the very old days when the telephone industry was regulated by the Interstate Commerce Commission, (the ICC) which also regulated the railroad industry.
Station Adapters Cables and interface assemblies for connecting Dialogic network interface and switching products to telephones or analog telephone lines.
Station Apparatus The equipment which is installed on the customer's premises, including phones, ancillary electronics and small hardware.
Station Auxiliary Power Supply This device is used to provide power to an electronic phone that is connected more than 300 meters (or 1,000 feet) or so away from the Key Service Unit.
Station Battery A separate power source which provides the necessary DC power to drive a telephone system. Individual telephones are usually powered by a central source, i.e. their PBX or central office. The batteries may also power radio and telephone equipment as well as provide emergency lighting and controls for equipment. See BATTERY.
Station Busy Lamps Lamps located on a station instrument, providing visual indication of each busy phone in the system. Busy Lamp Fields (BLFs) often come on key systems and sometimes on smaller PBXs. They're very handy.
Station Busy Override Pre selected phones have the privilege and ability to preempt busy circuits and override a private conversation.
Station Call Transfer A phone user can transfer incoming and outgoing calls to another phone without attendant assistance.
Station Camp-On Phones can camp-on to a busy extension. The camped-on phone will be notified of the camp-on by a special beep signal. The person at the other end may or may not hear the signal.
Station Clock The principal clock or alternative clock located at a particular station providing the timing reference for all major telecommunications functions at that station. A station clock may also be used to provide timing or frequency signals to other equipment.
Station Code The final four digits of a standard seven or 10 digit telephone number.

Station Direct Station Selection The phone user places a call to an extension within the PBX by pushing a single push-button on his phone.

Station Equipment Telephone instruments and associated equipment furnished to subscribers. We suspect that the word "station" came from early telephone industry which was regulated by the same government agency which regulated the railroad business.

Station Hunting This feature allows a calling phone which places a call to a busy phone to proceed to the next idle phone in the hunt group. This jump is done automatically. See also ROTARY HUNT, which is the same thing for trunks.

Station Keeping The process on board a satellite for keeping it at its assigned longitude and inclination.

Station Line Cards Station line cards sit inside telephone systems and drive a bunch of connected phones. These cards translate the software inside the phone system into electrical impulses which tell the phones at the other what they're capable of and let them do things, like dial, transfer, conference, etc.

Station Line Protector Circuitry that protects the telephone system from high voltage hits and lightning strikes. Such circuitry is usually on every station line card.

Station Load The total power requirements of the integrated station facilities.

Station Message Detail Recording Now refers to the RS-232-C "port" or plug found on the back of most modern PBXs and some larger key systems. See CALL ACCOUNTING SYSTEM and CALL DETAIL RECORDING.

Station Message Registers Message unit information centrally recorded on a per-station basis for each completed outgoing call.

Station Message Waiting Special light on a phone to alert hotel/motel guests of messages waiting at the front desk.

Station Monitoring Selected phones can monitor (i.e. listen in on) any other phones in the system.

Station Override Security Designated phones can be shielded against executive busy override (presumably other executives).

Station Protector A station protector protects phones and other phone-like devices ("stations") from lightning. A station protector is typically a gas discharge, carbon block or other device that short circuits harmful voltages to ground in the event of lightning strikes on the phone line. Sometimes it works. Sometimes it doesn't. If a bad thunderstorm is about to erupt over your house or office, it's a good idea to unplug your phones from AC and from phone lines.

Station Rearrangement And Change Allows a user to move phones, change the features and/or restrictions assigned to phones and administer features associated with telephones.

Station Review A study of how people in an organization use the telephones and what communications needs are not being satisfied.

Station Ringer Cutoff Allows the ringer on the telephone to be turned off. Not always a good idea, since calls may still come in for that phone, but no one may pick them up because they don't hear it ring.

Station To Station Call A directly dialed call. No operator is used. Most calls are now directly dialed. Some long distance companies don't even have operators to help complete calls. AT&T still does.

Station Tone Ringing Electronic tone ringer that replaces the bell.

Station Transfer Security If trunk call is transferred from one phone to another, and the second phone does not answer within a predetermined time, the trunk call will automatically go to the attendant.

Station Visual Signaling Lamp on a phone which indicates flashing incoming, steady busy, and "wink" hold visual conditions associated with that phone.

Stationary Orbit An orbit, any point on which has a period equal to the average rotational period of the Earth, is called a synchronous orbit. If the orbit is also circular and equatorial, it is called a stationary or geostationary orbit.

Statistical Multiplexing A multiplexing technique that differs from simple multiplexing in that the share of the available transmission bandwidth allocated to a given user varies dynamically. In other words, in statistical multiplexing, a channel is only assigned to communicating devices (voice, modem, etc.) when they actually have data to send or receive. This makes statistical multiplexing more powerful than normal FDM or TDM multiplexing. In other words, statistical multiplexing typically allows you to cram more information down a pipe than you can with FDM or TDM multiplexing. See also FAST PACKET MULTIPLEXING.

Statistics Numbers looking for an argument. The best statistics are those for which you are the sole source or those which are repeated sufficiently often for them to become an integral part of society's conventional wisdom.

Statistics Port In network management systems, interface for reporting events and status.

STATMUX A statistical multiplexer. See STATISTICAL MULTIPLEXING.

Status Information Information about the logical state of a piece of equipment.

Status Signal Unit Signal unit of CCS used to initiate transmission on a link or to recover from loss of transmission.

Statute Mile A unit of distance equal to 1.609 km, 0.869 nmi, or 5,280 ft.

STC 1. Society of Telecommunications Consultants. A professional society for telecommunications consultants. They endeavor to set standards of behavior for the consulting community.

2. System Time Clock: The master clock in an MPEG-2 encoder or decoder system.

STD Subscriber Trunk Dialing. An non-North American term for direct distance dialing, i.e. dialing long distance calls directly without an operator's assistance. Pricing for long distance calls is typically done by billing a standard amount of money, e.g. a German Mark, for a length of speaking time, which shortens the further you call. For example. you might get one minute for a Mark if you're calling 50 miles. If you call 200 miles you might only get 20 seconds.

STDM Statistical Time Division Multiplexer. STDMs are TDMs (Time Division Multiplexers) with an added microprocessor that provides more intelligent data flow control and enhanced functionality, such as error control and more sophisticated user diagnostics. The major difference between TDMs and STDMs is that stat muxes dynamically allocate time slots on the link to inputting devices on an as-needed basis (rather than in round-robin fashion where all devices are polled in preordained order). Therefore, there is no idle time on the link because a device does not have information to send. Unlike TDMs, STDMs have

buffers for holding data from attached devices. They can handle a combined input speed (aggregate speed) that exceeds the speed of the communications link.
STE 1. Station Terminal Equipment.
2. Section Terminating Equipment. SONET equipment that terminates a section of a link between a transmitter and repeater, repeater and repeater, or repeater and receiver. This is usually implemented in wide area facilities and not implemented by SONET Lite. STE Network elements perform section functions such as facility performance monitoring. The section is the portion of a transmission facility between a lightwave terminal and a line repeater or between two line repeaters.
3. Spanning Tree Explorer: A Source Route Bridging frame which uses the Spanning Tree algorithm in determining a route through a network. Often used in ATM networks.
Steady-state Condition 1. In a communication circuit, a condition in which some specified characteristic of a condition, such as value, rate, periodicity, or amplitude, exhibits only negligible change over an arbitrarily long period of time. 2. In an electrical circuit, a condition, occurring after all initial transients or fluctuating conditions have damped out, in which currents, voltages, or fields remain essentially constant or oscillate uniformly without changes in characteristics such as amplitude, frequency, or wave shape. 3. In fiber optics, synonym for equilibrium mode power distribution.
Steerable Beam Antenna An antenna whose main beam can be directed in various directions either by an electrical or mechanical drive system.
Stentor A new name for the long distance network of nine regional Canadian phone companies. Inspiration for the name came from the Greek poet Homer. He had immortalized Stentor, a warrior in the Trojan war, "whose voice was as powerful as the voices of 50 other men." And so the adjective "stentorian" survived to be applied to someone with a powerful voice, often a politician. The nine members of Stentor are Bell Canada, British Columbia Telephone (BC Tel), AGT Limited, Manitoba Telephone Systems, SaskTel, Maritime Telephone, New Brunswick Telephone, Island Telephone and Newfound4land Telephone.
Step 1. One movement of an electromechanical switch which typically corresponds to one impulse from a rotary dial or one impulse from a touch tone phone which has been converted to a rotary dial. 2. Wang's name for its telephony link is STEP, which stands for Speech and Telephony Environment for Programmers.
Step By Step SXS. An automatic dial-telephone system in which calls go through the switching equipment by a succession of switches that move a step at a time, from stage to stage, each step being made in response to the dialing of a number. SXS is electromechanical switching. It was invented in the 1920s.
Step Call The phone user can, upon finding that the called phone is busy, call an idle nearby phone by merely dialing an additional digit.
Step Down 1. Reduce the voltage. 2. A feature of fax machines that makes them drop their transmission speed when the quality of the phone lines they are transmitting over begins to deteriorate. Dropping the transmission speed is the major way of getting the faxes through on "dirty" lines. All Group III fax machines have "step-down" as a built in feature, or should have.
Step Down Transformer A transformer wound to give a lower voltage on the secondary side than that impressed on the primary. The current, however, will be stepped up. A step down (often spelled step-down) transformer has more primary than secondary turns.
Step Index Fiber An optical fiber with a core having a uniform refractive index.
Step Index Optical Fiber A fiber that has a constant refractive index at its core but a different refractive index as the outer cladding is approached. This design minimizes loses at the core-cladding interface and is preferred for single-mode, long-distance transmission.
Step Index Profile For an optical fiber, a refractive index profile characterized by a uniform refractive index within the core and a sharp decrease in refractive index at the core-cladding interface.
Step Up Transformer A transformer wound to give a higher voltage on the secondary side than that impressed on the primary. The current, however, will be stepped down. It has fewer primary than secondary turns.
Stepped Index Referring to a type of optical fiber which exhibits a uniform refractive index at the core and a sharp decrease in the refractive index at the core-cladding interface.
Stereophonic Crosstalk An undesired signal occurring in the main channel from modulation of the stereophonic channel or that occurring in the stereophonic channel from modulation of the main channel.
Stereophonic Sound Subcarrier A subcarrier within the FM broadcast baseband used for transmitting signals for stereophonic sound reception of the main broadcast program service.
Stereophonic Sound Subchannel The band of frequencies from 23 kHz to 99 kHz containing sound subcarriers and their associated sidebands.
STG An imaging term. Scale To Gray. STG uses gray pixels to fill in jagged edges of document images. STG improves readability. According to a study commissioner by Cornerstone and done by Dr. Jim Sheedy, the ability to read STG images was improved between 4% and 19%, depending on the resolution, and symptoms such as headaches, tired back, blurred vision were cut way down.
STFS Standard Time and Frequency Signal.
STICI Pronounced sticky. It stands for Self Teaching Interpretive Communicating Interface. It's being touted as the "next wave of user interface" (the next wave after GUI). Apparently STICIs are found on gadgets like the Apple Newton MessagePad. See also GUI.
According to BIS Strategic Decisions, features of STICI include:
Self-Teaching: The operating system (OS) and interface use agents (special background processes) to study how the user makes use of the device. For example, agents will track exactly how a user use the OS and applications.
Interpretive: The system is able to make inferences based on the information it has collected about the user. The system does some interpretation, moving from the traditional interface approach of "Do what I say" to "Do what I mean." Unlike the static

menus and dialog boxes of GUI systems, STICI systems will dynamically adapt their operation, thereby better anticipating user needs. Agents will automate common tasks, based on observed usage patterns.
Communicating: The system will be able to manage all the different communications functions offered, such as store and forward, cellular, logging on and off wireless LANs and linking to the user's desktop PC. These communications will be transparent to the user, leaving him or her free to concentrate on the task at hand.
Interface: The interface will be oriented around documents. Traditionally, interfaces have centered on applications, such as a word processor or a spreadsheet. With the STICI, the interface is centered on documents. The user creates a document, writing or drawing freely, and the various applications needed are simply tools accessed to create a chart, to write or to show numbers within that document. Users will be able to seamlessly link applications from multiple vendors, as descendent technologies evolve from OLE and Publish & Subscribe. Both Microsoft and Apple are developing document-oriented interfaces for the next generations of their respective desktop operating systems.

Sticky A sticky shift key lets you access the shifted functions (such as capital A) by pressing the shift key first and then pressing the second key. Sticky keys may stay down for a second or two. Or you may have to hit them again to unstick them — somewhat like the CapsLock key.

STID Service Termination Identifier. An ISDN Service Profile term.

Stimulated Emission Radiation emitted when the internal energy of a quantum mechanical system drops from an excited level to a lower level when induced by the presence of radiant energy at the same frequency. An example is the radiation from an injection laser diode above lasing threshold.

STL 1. Standard Telegraph Level. 2. Studio-To-Transmitter link — typically through the air microwave.

STM Synchronous Transfer Mode. A transport and switching method that depends on information occurring in regular and fixed patterns with respect to a reference such as a frame pattern. A time division multiplex-and-switching technique to be used across the user's network interface for a broadband ISDN. It gives each user up to 50 million bits per second simultaneously — regardless of the number of users.

STM-1 Synchronous Transport Module 1: SDH standard for transmission over OC-3 optical fiber at 155.52 Mbps.

STM-n Synchronous Transport Module "n" : (where n is an integer) SDH standards for transmission over optical fiber (OC-'n x 3) by multiplexing "n" STM-1 frames, (e.g., STM-4 at 622.08 Mbps and STM-16 at 2.488 Gbps).

STM-nc Synchronous Transport Module "n" concatenated: (where n is an integer) SDH standards for transmission over optical fiber (OC-'n x 3) by multiplexing "n" STM-1 frames, (e.g., STM-4 at 622.08 Mbps and STM-16 at 2.488 Gbps, but treating the information fields as a single concatenated payload).

STN 1. Statens Telenamd (Swedish National Telecommunications Council).
2. Super Twist Nematic is a passive matrix technology now used in screens in some laptop computers. In a passive matrix color screen the current travels along transparent electrodes printed on the glass screen. These electrodes are driven by transistors placed around the edges of the display. Horizontal and vertical electrodes form a grid-like matrix, with a pixel at every intersection. A major problem with passive technology arises when current is lost in crosstalk as the electrodes criss-cross each other. This crossing over effect greatly diminishes overall display quality. See ACTIVE MATRIX, LCD and TFT.

Stop Bit The Stop Bit is an interval at the end of each Asynchronous Character that allows the receiving computer to pause before the start of the next character. The Stop Bit is always a 1. See START BIT.

Stop Element The last element of a character in asynchronous serial transmission, used to ensure recognition of the next start element.

Stop-Record Signal In facsimile systems, a signal used for stopping the process of converting the electrical signal to an image on the record sheet.

Stop/Start Transmission A method of transmission in which a group of bits are preceded by a start bit and followed by a stop bit. Also called asynchronous transmission. See ASYNCHRONOUS.

Storage Unit A device in which information can be recorded and retained for later retrieval and use.

Store And Forward S/F. In communications systems, when a message is transmitted to some intermediate relay point and stored temporarily. Later the message is sent the rest of the way. Not very convenient for voice conversations, but useful for telex type, and other one-way transmission of messages. Telephone answering machines, as well as voice mailboxes are considered forms of Store and Forward message switching.

Stored Procedures Compiled code on a database server that reduces the processing burden on clients.

Stored Program Computer A computer controlled by internally stored instructions, that can synthesize and store instructions, and that can subsequently execute those instructions. See also STORED PROGRAM CONTROL.

Stored Program Control SPC. The routing of a phone call through a switching matrix is handled by a program stored in a computer-like device, which may well be a special-purpose computer. Before SPC switches came along, the rotary dialing of the phone caused the elements of the switch to directly "step" through their dialing path. This was slow and cumbersome, since dialing can be slow. Also subscribers can abort half way (they made a mistake) and this can mess up the switch's efficiency. Thus the move to stored program control switches was very significant. These days virtually all switches as stored program control. Nothing happens in the switching matrix until the stored program control receives all the dialing digits and decides what to do with them.

Stovepiping In a call center, agents typically need access to many databases. In the past they've used dumb terminals. They log into one computer, get into one database, go further into it. When they need information out of another database, they've typically had to climb out of the previous database, the previous computer, log into another and climb down into it. This is called stovepiping, because it follows the contours of a stovepipe. These days, agents have intelligent computers as terminals. They can access several databases at once, by simply having different windows open on their screen or having a front end program that populates a screen with information from several databases, most likely using a GUI interface.

NEWTON'S TELECOM DICTIONARY

STP 1. Shielded Twisted Pair. Twisted pair (TP) wiring with a metal foil sheath around it to limit interference.
2. Signal Transfer Point. The packet switch in the nation's emerging Common Channel Interoffice Signaling (CCIS) system. The CCIS is a packet switched network operating at 4800 bits per second. CCIS replaces both SF (Single Frequency) and MF (Multifrequency) by converting dialed digits to data messages. It will run at 56,000 bps with the introduction of Signaling System 7. See SIGNALING SYSTEM 7. For a full explanation of the Advanced Intelligent Network, see AIN.

Straight-Through When wiring up phone and some data extensions, there are basically two ways of doing it — straight-through and crossover. Straight-through occurs when you wire both ends identically so the signals pass straight through. This is typically done with patch panels and modular EIA adapters. Crossover wiring has a reverse order of wiring. As an example, let's take a four conductor, RJ-11. In a crossover wiring (e.g. an RJ-11 phone extension cord), conductor 1 would be connected to hole 1 on one plug and one 4 on the other end. Conductor 2 would be connected to 4. And 3 would be connected to hole 2.

Straight-Tip Connector ST Connector. An optical fiber connector used to join single fibers together at interconnects or to connect them to optical cross connects.

Straightforward Outward Completion Operator can place an outgoing call for phone user. Also called "Through Supervision."

Strand A single uninsulated wire.

Strand Lay The distance of advance of one strand of a spirally stranded conductor, in one turn, measured axially.

Stranded Conductor A conductor composed of groups of wires twisted together.

Stranded Copper A type of electrical wire conductor comprised of multiple copper wires twisted together forming a single conductor and then covered with an insulating jacket.

Stranded Fiber Cable A fiber cable in which individual optical fibers are twisted around strength members. Both campus and building versions are used. The campus version, unlike the one for buildings, is environmentally protected for outside use.

Strap A permanent, wired connection between two more points.

Stray Current Current through a path other than the intended path. See also SPURIOUS EMISSION.

Strawman This concept is widely used in selling. The simple idea is to set up a Buyer's Checklist and tell your prospective customer that this Checklist is objective. Any product that meets all the criteria is worth buying. Of course, there's only one product that meets all the criteria. It's yours.

Stream An SCSA term. One of 16 physical data lines making up the SCbus or SCxbus Data Bus.

Streamer Streaming tape drive.

Streaming An Internet term. A Web page typically consists of text and graphics images. To receive the text typically takes much less time than receiving the graphics images. One of the features which Netscape's improved browser introduced was the streaming of graphical image. This allowed the user to look at a new page of text on screen as the graphics streamed in (came in over the phone lines). See INTERNET and HOME PAGE.

Streaming Media After Netscape defined the concept more narrowly — see Streaming — Bill Gates of Microsoft defined more broadly to video coming to you in packets over the Internet. The stream is so fast that the packets become full-blown 30 frames per second video, similar to commercial, over-the-air TV.

Streaming Tape Backup A device to back up files and programs. A streaming tape backup looks very much like a large audio cassette. It records data sequentially.

Streaming Tape Drive A magnetic tape unit especially designed to make a nonstop dump or restore magnetic disks without stopping at interblock gaps.

Streams An architecture introduced with Unix System V, Release 3.2 that provides for flexible and layered communication path between processes (programs) and device drivers. Many companies market applications and devices that can integrate through Streams protocols.

Street Price The real or typical selling price of computers, hardware, and software. Most laptop and desktop computers sell for about 25 percent below list price. Software may be discounted even more.

Street Talk The Banyan-developed protocol for discovering and maintaining resource information distributed among the servers connected to Banyan's VINES network operating system. Also known as a global naming service.

Streetsweeper A heavy duty shotgun with a revolving round magazine typically holding 18 12-gauge or 20-gauge shotgun shells. This word crept into a story the Wall Street Journal ran on cellular fraud. When the Feds rang a cellular phone store as a sting operation, one customer offered to trade his streetsweeper in on a phone. That's how dependent Detroit's drug-traffickers had become on cellular phones and beepers.

String A sequence of elements of the same type, such as characters, considered as a unit (a whole) by a computer. A data structure composed of a sequence of characters, usually in human-readable text.

Strobe An electrical pulse used to call for the transfer of information.

Stroke A straight line or arc that is used as a segment of a graphic character.

Stroke Edge An imaging and OCR term. In character recognition, the line of discontinuity between a side of a stroke and the background, obtained by averaging, over the length of the stroke, the irregularities resulting from the printing and detecting processes.

Stroke Speed In facsimile systems, the number of times per minute that a fixed line perpendicular to the direction of scanning is crossed in one direction by a scanning or recording spot. In most conventional mechanical systems, this is equivalent to drum speed. In systems in which the picture signal is used while scanning in both directions, the stroke speed is twice the above figure.

Stroke Width In character recognition, the distance measured perpendicularly to the stroke centerline between the two stroke edges.

Strowger, Armond The man who invented the telephone dial and the earliest automatic telephone switch as a method of allowing the user to complete calls without using the Operator. In Kansas City in the late 1800's, Ol' Armond was an undertaker who wasn't getting much business. That's because the girlfriend of a rival undertaker was a telephone operator, and when she got a call asking for the local undertaker, she forwarded the calls to her boyfriend. This story may or may not be apocryphal. But it's a great story.

Structured Query Language SQL. A relational database language (ANSI Standard) that consists of a set of facilities for defining, manipulating and controlling data.

Structured Programming A technique for organizing and coding (computer) programs in which a hierarchy of modules is used, each having a single entry and a single exit point, and in which control is passed downward through the structure without unconditional branches to higher levels of the structure. Three types of control flow are used: sequential, test, and iteration.

Structured Wiring As data flows have sped up in recent years and as moves, adds and changes have proliferated, so the erstwhile idea of wiring up a building with plain old analog voice telephone wire has become increasingly not intelligent. The idea then came up of defining wiring standards and schemes so that a user could feel comfortable about choosing a complete solution for wiring phones, workstations, PCs, LANS and other communicating devices throughout the building, the campus, the network and the company. Consistency of design, layout and logic are the keys to structured wiring systems. Typically a structured wiring system consists of three elements:
1. Manufacturer-originated standard components that link wires together in a systematic, intelligent way.
2. A set of rules for building smart wiring. The ANSI/TIA/EIA-568-A and ISO/IEC 11801 standards specify the minimum requirements for telecommunications cabling within a commercial building. The Commercial Building Telecommunications Wiring Standard is available from Global Engineering Documents, Englewood CO 314-726-0444 / 800-854-7179.

A structured cabling system will improve performance in five ways, according to Anixter, a leading supplier of structured wiring systems:
1. It eases network segmentation, the job of dividing the network into pieces to isolate and minimize traffic, and thus congestion.
2. It ensures that proper physical requirements, such as distance, capacitance, and attenuation are met.
3. It means adds, moves, and changes are easy to make without expensive and cumbersome rewiring.
4. It radically eases problem detection and isolation.
5. It allows for intelligent, easy and computerized tracking and documentation.

"Structure" brings order to what has often been an afterthought — wiring. The main pieces of a structured wiring system are:
1. Drop cable. The cable that runs from the computer to a network outlet.
2. Cable run. The cable that runs from the outlet to the wiring closet.
3. Patch panel. A board that collects all the cable runs in one place and "patches" them to different parts of the wiring concentrator. Network managers (users or their secretaries, it's that simple) change the LAN layout by plugging and unplugging "patch cables" between the patch panel and the wiring concentrators. No rewiring is necessary to move one user from one network segment to another.
4. Wiring concentrator. It makes the network connections. Some wiring concentrators are dumb, making only physical connections between network segments. Others are intelligent, making networking decisions and providing network diagnostics. A wiring concentrator can have bridges and routers that divide the network into segments. It can have the hardware necessary to change from one media, say twisted pair, to another, say fiber optic. And it can contain the hardware to change from one network type to another, say from Ethernet to Token Ring.

Here is a glossary of structured wiring words, with thanks to Anixter.

Access Method The method of "communicating" on the wire. Examples include Ethernet, Token Ring, AppleTalk, AS400 and 3270.

Cable Type (Media) The type of cable used in the system. Examples are coaxial, UTP, STP and fiber. Factors including cost, connectivity and bandwidth are important in determining cable type.

Data Speeds Different interconnect products (cables and connectors) are capable of supporting different data rates. For instance, Level 3 cable supports data rates up to 10 Mbps. (See LEVEL).

Environment Where the structured wiring system is found. The large majority of systems are located in office environments as opposed to factory or industrial environments.

Life Cycle How long the cable is physically anticipated to be in place. For example, if a customer intends to be in a large office for 10 years, fiber installation may be considered.

Methodology The physical means of getting the wiring system to the user (its distribution path). Examples include modular furniture, surface mounts, fixed wall, recessed wall, raised floor and undercarpet wiring.

Topology The way the cable is physically laid out or configured. Examples include star, ring, daisy chain and backbone.

Structured Wiring System See STRUCTURED WIRING.

STS Synchronous Transport Signal. A SONET electrical signal rate. See STS-1.

STS-1 Synchronous Transport Signal level 1. An electrical signal that is converted to or from Sonet's optically based signal; equivalent to the OC-1 signal of 51.84 Mbps.

STS-3 Synchronous Transport Signal level 2. (yes, that's right. It's STS-3, though it's level 2.) ATM Physical Layer implementation supporting 155 Mbps.

STS-n Synchronous Transport Signal "n" : (where n is an integer) SONET standards for transmission over OC-n optical fiber by multiplexing "n" STS-1 frames, (e.g., STS-3 at 155.52 Mbps STS-12 at 622.08 Mbps and STS-48 at 2.488 Gbps).

STS-nc Synchronous Transport Signal "n" concatenated: (where n is an integer) SONET standards for transmission over OC-n optical fiber by multiplexing "n" STS-1 frames, (e.g., STS-3 at 155.52 Mbps STS-12 at 622.08 Mbps and STS-48 at 2.488 Gbps but treating the information fields as a single concatenated payload).

STU Secure Telephone Unit.

STU-III

The third generation of secure telephone units used by the military and its suppliers.
Studio-To-Transmitter Link STL. Any communication link used for transmission of broadcast material from a studio to the transmitter. It's typically microwave radio but it may also be a conventional landline link.
Stunt Box A device to 1. control the nonprinting functions of a teletypewriter terminal, such as a carriage return and line feed and 2. a device to recognize line control characters.
Stutter Dial Tone Stutter Dial Tone is the broken-up tone a user hears on a phone when they pick up their phone to make a call and they have a message waiting in voice mail. This is used to notify users that they have a voice mail message when the phones don't or can't have a message-waiting light. They are given stutter dial tone instead of regular dial tone.
Stylus A pen-shaped instrument (usually made out of plastic) that is used to enter text, draw images, or point to choices on a computer — desktop or PDA. The pen is designed to make writing on the screen feel just like writing on paper. Yes!
Sub Substitute Character. A control character used in the place of a character that has been found to be invalid or in error.
Subaddressing A name for an ISDN service which enables many different types of terminals — phones, fax machines, PCs, etc. — to be connected to the ISDN user interface and uniquely identified during a call request. See ISDN.
Subcarrier A carrier which modulates a main carrier so that two different modulating signals can be transmitted simultaneously, one on the main carrier and one on the subcarrier.
Sublayer A logical sub-division of a layer.
Submarining Same as CURSOR SUBMARINING. When you drag your cursor across a screen and the cursor disappears as you move it. That's called Cursor Submarining. It happens most on monochrome LCD screens because they change slowly — much slower than active matrix screens or CRTs or VDTs (glass screens).
Subnet a portion of a network, which may be a physically independent network, which shares a network address with other portions of the network and is distinguished by a subnet number. A subnet is to a network what a network is to an internet.
Subnet Number A part of the internet address which designates a subnet. It is ignored for the purposes internet routing.
Subnetwork 1. A collection of OSI end systems and intermediate systems under the control of a single administrative domain and utilizing a single network access protocol. Examples: private X.25 networks, collection of bridged LANs.
2. A token ring LAN that is used to serve the communication needs of a department. Subnetworks are normally connected to token ring backbones via token ring bridges or routers so that they can communicate with other subnetworks via the backbone or with computers directly connected to the backbone.
3. An ATM term. A collection of managed entities grouped together from a connectivity perspective, according to their ability to transport ATM cells.
Subnetwork Access Protocol SNAP. A version of the IEEE local area network logical link control frame similar to the more traditional data link level transmission frame that lets you use nonstandard higher-level protocols.
subNMS An ATM term. Subnetwork Management System: A Network Management System that is managing one or more subnetworks and that is managed by one or more Network Management Systems.
Subroutine A functionally isolated program or sequence of instructions for a specific function that is often called by a program. A piece of software that performs a useful function that will be needed often. The code for the subroutine is stored on disk (like a letter, etc.) and dropped into a larger program as needed. A nicely-written subroutine saves you "re-inventing the wheel" and allows you to re-use your code in many programs.
Subscriber A person or company who has telephone service provided by a phone company. In other industries, subscribers are called customers. Some telephone companies are beginning to call their subscribers customers. Thank goodness.
Subscriber Line The telephone line connecting the local telco central office to the subscriber's telephone instrument or telephone system.
Subscriber Line Charge A monthly charge on subscribers created by the Federal Communications Commission and paid to the local telephone company. The logic for this charge has something to do with reimbursing the local phone companies for some costs which it is not recovering elsewhere. In reality, it's just another rate increase.
Subscriber Loop The circuit that connects the telephone company's central office to the demarcation point on the customer's premises. The circuit is most likely a pair of wires. But it could be three wires if some external signaling is being used. It could also be four wires if the circuit was a four-wire full duplex leased line.
Subscriber Loop Carrier See SLC-96.
Subscriber Number The number that permits a user to reach a subscriber in the same local network or numbering area (same as Directory Number or DN).
Subscriber Plant Factor A planning factor used by common carriers to allocate investment in phone equipment, subscriber lines and the non traffic sensitive portion of the central office equipment.
Subscriber Trunk Dialing STD. The European version of direct-distance dialing. Pricing for long distance calls is typically done by billing a standard amount of money, e.g. a German Mark, for a length of speaking time, which shortens the further you call. For example. you might get one minute for a Mark if you're calling 50 miles. If you call 200 miles you might only get 20 seconds.
Subscriber's Drop Wire which runs from a cable terminal or distribution point to the subscriber's premises.
Subset A contraction for Subscriber Set, or telephone set.
Subsplit A method of allocating frequencies in a broadband transmission system. Transmit frequencies are in the range of 5 to 32 megahertz, and receive frequencies are in the range of 54 to 300 megahertz.
Substation An additional phone which has been established as an extension to the main phone or primary line.
Substitute Character A transmission control character used in place of a character found to be in error.
Substitution A word used in voice recognition to mean a type of error that occurs when a word within the active vocabulary is spoken correctly but classified as another word within the vocabulary. This error is usually dealt with during a verification

stage in an application, i.e. " you said, 1,2,3...correct?"

Subvoice-Grade Channel A communications channel of bandwidth narrower than a standard 3Hz voice line. A subvoice-grade channel is usually used for slow data transmission such as teletype or telemetry.

Suite A collection. A suite of software tools is a collection of software tools.

Summary Address An ATM term. An address prefix that tells a node how to summarize reachability information.

Summary Billing Nynex will give you one monthly consolidated phone bill — no matter how many Nynex number accounts you have in your billing area.

Summation Check A check based on the formation of the sum of the digits of a numeral. The sum of the individual digits is usually compared with a previously computed value.

Sun Microsystems Sun Microsystems is a Californian microcomputer manufacturer. SUN stands for Stanford University Network.

Sundown Rule A rule in voice mail which says that all messages should be returned that day, before the sun goes down.

Sun Transit Outrage Satellite circuit outage caused by direct radiation of the sun's rays on an earth station receiving antenna.

SUNOS SunOS is Sun Microsystems' implementation of UNIX.

SunView SunView is Sum Microsystems' kernel-based window system.

Super Server A file server with more than one CPU (Central Processing Unit). At time of writing this dictionary, a high-end super server might contain four Intel Pentium chips. To take advantage of these super servers, you need an operating system capable of asymmetrical multi-processing, such as Unix and Windows NT Advanced Server.

Super Speed Calling A feature of Northern Telecom's DMS line of central offices. Subscribers can use a four-letter dialable name, preceded by an octothorpe (#), to speed-call up to 14 digits. For example, "#WORK" or "#HOME" could be used as a code for a longer number. Dialing is faster, and subscribers have the added convenience of using easier-to-remember names instead of forgettable numbers on their speed calling list. The list size can be set by the telephone operating company at initial setup on a per-system basis. Northern's default is 12 names per list.

Supercomputing A term applied to a class of high-speed computers employing advanced technologies such as simplified instruction sets, wide data paths and pipelining.

Superconductors Superconductors are materials which have no resistance to the flow of electricity. They are widely believed to have great potential for dramatically faster telecommunications switches and computers. In the past, the superconducting state — zero resistance to the flow of electricity — could be achieved only by cooling certain metal alloys to temperatures of near absolute zero, or about 460 degrees below zero. Starting in 1986, researchers discovered that ceramic materials could reach superconductivity at temperatures as high as 235 degrees below zero.

Superdrive The name for Apple's 1.44 Mb floppy that can read and write MS-DOS formatted floppies and Mac formatted disks. DOS floppies require Apple File Exchange or a third party product to read the DOS format.

Superframe Format The superframe transmission structure consists of 12 DS1 frames (2316 bits). The DS1 frame comprises 193 bit positions, the first of which is the frame overhead-bit position. Frame overhead bit positions are used for the frame and signaling phase alignment only.

Supergroup Sixty voice channels. In more technical terms: the assembly of five 12-channel groups occupying adjacent bands in the spectrum for the purpose of simultaneous modulation or demodulation.

Superheterodyne A type of radio receiver operating on the heterodyne or beat principle. See HETERODYNE.

Superhouse The National Society of Home Builders has registered SmartHouse. BellSouth has trademarked SuperHouse. And GTE has registered SmartPark. One day, fiber optic will snake to everyone's house, bringing the potential of immense information services. Until that day comes they'll be lots of interesting demonstrations at distant trade shows.

Superimposed Ringing A way of stopping party line phone users from hearing each other's ring by superimposing a DC (direct current) voltage over the ringing signal and using it to alert a vacuum tube or semiconductor device in only the phone instrument that we want to ring. See also SUPERPOSED RINGING.

Superpath New York and New England Telephone's Superpath services offer dedicated end-to-end digital transmissions at 1.544 Mbps (T-1) or 44.736 Mbps (T-3).

Superposed Circuit An additional channel obtained from one or more circuits, normally provided for other channels, in such a manner that all the channels can be used simultaneously without mutual interference.

Superposed Ringing Party-line telephone ringing in which a combination of alternating and direct currents is used, the objective being to only ring the bell in the phone of the one whose call is coming in. See also SUPERIMPOSED RINGING for a better explanation.

Supertrunk A cable that carries several video channels between facilities of a cable television company. A trunk between the master and the hub headends in a hub CATV system.

Supervised Transfer A call transfer made by an automatic device such as voice response unit which attempts to determine the result of the transfer — answered, busy, ring no answer — by analyzing call progress tones on the time.

Supervision Supervision of a phone call is detecting when a called party has picked up his phone and when that party has hung up. Supervision is used primarily for billing purposes. Not all long distance carriers have supervision capability. It depends on how "equal accessed" they have chosen to be. See ANSWER SUPERVISION, SOFTWARE SUPERVISION and SIGNALING SYSTEM 7.

Supervisor The person responsible for day to day maintenance and operation of a phone system. Typically used in conjunction with an ACD — automatic call distributor.

Supervisory Call This service feature allows the attendant, after connecting an incoming CO line or tie line call to the wanted ed phone, to continuously supervise the call in progress.

Supervisory Control Characters or signals which automatically actuate equipment or indicators at a remote terminal.

NEWTON'S TELECOM DICTIONARY

Supervisory Lamp A lamp which shows the operator whether the person is speaking (off-hook) or is not speaking (on-hook). These days such lamps are called BUSY LAMP FIELDS. In some smaller key systems, all phones have them. Busy lamp fields are an operator's best friend.

Supervisory Program 1. A program, usually part of an operating system, that controls the execution of other computer programs and regulates the flow of work in a data processing system. 2. A computer program that allocates computer component space and schedules computer events by task queuing and system interrupts. Control of the system is returned to the supervisory program frequently enough to ensure that demands on the system are met.

Supervisory Relay A relay which, during a call, is controlled by the transmitter current supplied to a subscriber line to receive from the associated phone signals that control the actions of operators or switching mechanisms.

Supervisory Routine A routine that allocates computer component space and schedules computer events by task queuing and system interrupts. Control of the system is returned to the supervisory program frequently enough to ensure that demands on the system are met.

Supervisory Signal 1. Supervisory signals are the means by which a telephone user initiates a request for service; or holds or releases a connection; or flashes to recall an operator or to initiate additional features, for example, 3-way calling. Supervisory signals are also used to initiate and terminate charging on a call. A signal also indicates whether a circuit is in use, or not in use.
2. A signal used to indicate the various operating states of circuit combinations.

Supplementary Services Telephone company talk for services above basic ability to make a phone call. Supplementary services include fast dialing, calling line ID, call waiting, call forwarding, and videoconferencing features. Here's a definitions of supplementary services as applied to ISDN service: Additional services, such as hold, conference, and call forwarding, offered to an ISDN customer. Supplementary services are always present if activated at the switch. Although the specific features and call appearances may differ between service providers, supplementary service generally provides users with the ability to connect and disconnect new calls when one (or more) call exists.

Supplementary Telephone Service The lowest level of service in Windows Telephony Services is called Basic Telephony and provides a guaranteed set of functions that corresponds to "Plain Old Telephone Service" (POTS - only make calls and receive calls). The next service level is Supplementary Telephone Service providing advanced switch features such as hold, transfer, etc. All supplementary services are optional. Finally, there is the Extended Telephony level. This API level provides numerous and well-defined API extension mechanisms that enable application developers to access service provider-specific functions not directly defined by the Telephony API. See WINDOWS TELEPHONY SERVICES.

Support A verb that typically means "This hardware or software has the following feature." An example: "This software supports 32-bit file sharing."

Support Hardware The racks, clamps, cabinets, brackets, trays, and other equipment that provide the physical means to hold the transmission media and connecting hardware. An AT&T definition.

Suppressed Carrier Single-Sideband Emission A single-sideband emission in which the carrier is virtually suppressed and not intended to be used for demodulation.

Suppressed Carrier Transmission A transmission technique in which only the sidebands (one or both) are transmitted and the main carrier is not transmitted and thus not used.

Suppressors, Echo Echo is controlled in long distance circuits with devices called echo suppressors. These devices automatically insert loss in the return path of a four-wire circuit. All long distance circuits are four-wire — two wires for each of the two paths (receiving and transmitting). The echo suppressor jumps back and forth between the two transmission paths. Properly adjusted, an echo suppressor puts only sufficient loss in a circuit so a listener can interrupt the talker. With very long circuits — 22,300 miles — in satellites, a better way is needed. They're called echo cancelers.

Surcharge A charge imposed in accordance with the Commission's Access Reconsideration decision in CC Docket 78-72, Phase 1, FCC 83-356. released August 22, 1983 and updated too many times since. The monthly charge is about $2.00 and is going up to $3.50. This charge is said to compensate the local phone company for long distance commissions (called settlements and separations) lost and now replaced with per minute access charges.

Surf The word "surf" in its classic sense means to ride the crest of a wave, skimming quickly across the water underneath. The electronic world have adopted that definition of skimming quickly and included it in several definitions. For example, shoulder surfing is gazing quickly over someone's shoulder while they're making a call at a payphone and writing down the user's credit card numbers. Channel surfing is moving from one TV channel to another quickly. Surfing the Web means moving from one Web site to another, jumping around in search of knowledge or amusement. See also INTERNET and WEB BROWSER.

Surface Mount With surface mount technology, Components sit on the surface of printed circuit boards and are soldered to conductive pads. In the "thru-the-hole" process, component leads are placed through holes in the boards and are sent through wave soldering for attachment. Surface mount technology is more cost-effective, as it allows for denser packaging on the board and components can be mounted on both sides of the surface.

Surface Outlet A Communications Outlet (modular jack) that is installed on the surface of the mounting location. The premises wire serving such an outlet may or may not be concealed behind the mounting surface.

Surface Wave A wave that is guided along the interface between two different media or by a refractive index gradient. The field components of the wave diminish with distance from the interface. Optical energy is not converted from the surface wave field to another form of energy and the wave does not have a component directed normal to the interface surface. In optical fiber transmission, evanescent waves are surface waves. In radio transmission, ground waves are surface waves that propagate close to the surface of the Earth, the Earth having one refractive index and the atmosphere another, thus constituting an interface surface.

Surfing See SHOULDER SURFING and CHANNEL SURFING.

Surge An increase in line voltage that lasts longer than one cycle of the line frequency of 60Hz, the North American frequency

NEWTON'S TELECOM DICTIONARY

or 50Hz in many other countries, especially those running at 240 volts.

Surge Protector A device which plugs between the phone system and the commercial AC power outlet. It is designed to protect the phone system from high voltage spikes (also called surges) which might damage the phone system. When a surge occurs on the power line, the surge protector sends the overload to ground. How fast it sends it to ground is a subject that could fill a book. The type of surge protector that you buy will be determined mostly by the speed you need to protect your equipment.

Surges The increased flow of current through an electrical device brought about by an instantaneous change in its resistance or impedance.

Survivability A property of a system, subsystem, equipment, process, or procedure that provides a defined degree of assurance that the device or system will continue to work during and after a natural or man-made disturbance; e.g. nuclear attack. This term must be qualified by specifying the range of conditions over which the entity will survive, the minimum acceptable level of post-disturbance functionality, and the maximum acceptable outage duration.

Survivable Adaptable Fiber Network SAFENET. A U.S. Navy experimental fiber-based local area network designed to survive conventional and limited nuclear battle conditions.

Susan A nice name for a wife. Everyone should have such a good one. She is the mother of my two children, Claire and Michael.

Susceptiveness In telephone systems, the tendency of circuits to pick up noise and low frequency induction from power systems. It depends on telephone circuit balance, transpositions, wiring spacing, and isolation from ground.

Suspended Customer An MCI definition. An MCI customer who has requested service but has not yet been installed due to insufficient network capacity or some other operational/administrative constraint.

Suspended On-Demand Connection The mode of an on-demand connection where the communications line is dropped and the connection sites are actively spoofing.

SUT An ATM term. System Under Test: The real open system in which the Implementation Under Test (IUT) resides.

SVC Switched Virtual Circuit. A connection across a network. It is established on an as-needed basis and can provide connection to any other user in the network. The connection lasts only for the duration of the transfer. It is the datacom equivalent of a dialed phone call.

SVCC Switched Virtual Channel Connection: A Switched VCC is one which is established and taken down dynamically through control signaling. A Virtual Channel Connection (VCC) is an ATM connection where switching is performed on the VPI/VCI fields of each cell.

SVD Simultaneous Voice Data. In the fall of 1994, the term began to apply to several techniques for putting voice conversations and data transfers on the same analog phone line. Some of these techniques involve interrupting the voice conversation while data is transferred. Others involve transferring the data and voice simultaneously on different bandwidths.

SVGA Super Video Graphics Array. An extension of the VGA video standard. SVGA enables video adapters to support resolutions of up to 1,024 by 768 pixels with up to 16.7 million simultaneous colors, which is known as true color because it's the number of colors someone once figured is in a great Kodachrome slide. See also VGA.

SVPC Switched Virtual Path Connection: A Switched Virtual Path Connection is one which is established and taken down dynamically through control signaling. A Virtual Path Connection (VPC) is an ATM connection where switching is performed on the VPI field only of each cell.

SWACT Switch of Activity.

Swap File Some operating systems and applications let you use more memory than what you have in RAM. They do this by pretending that part of your hard disk is RAM memory. They do this by creating a swap file on your hard disk and swapping memory back and forth. Some computer systems call this virtual memory. You need to be careful with swap files. Never turn your machine off when you have applications running. If you do you're likely to leave a huge swap on your hard disk, which you may not find (it's hidden) and which your system may not dispose of. To get back the space on your hard disk, you'll need to erase it separately.

SWATS Standard Wireless AT Command Set. An extension to the Hayes AT command set to support wireless modems, such as those used in standard AMPS analog cellular phones.

SWEDAC Swedish Board for Technical Accreditation. They have established two standards, which effectively limit radiation emissions, MPR1 and MPR2. These standards specify maximum values for both alternating electric fields and magnetic fields and provide monitor manufacturers with guidelines in creating low emission monitors. There is, as yet, no definite proof of harm from normal computers monitors. But the argument goes that they weren't so sure about nicotine 30 years ago.

Sweep Acquisition A technique whereby the frequency of the local oscillator is slowly swept past the reference to assure that the pull-in range is reached.

SWHK Abbreviation for SWITCH HOOK. Originally referred to an actual hook on older phones that held the receiver, and sprang upward to close a switch and activate the phone when the receiver was picked up. Today the term refers to any of various buttons and plungers that are pressed down and released when the handset is put down (physically "hung up" in the old days) and picked up.

SWIFT Society for Worldwide Interbank Financial Telecommunications.

Swim Slow, graceful, undesired movements of display elements, groups, or images about their mean position on a display surface, such as that of a monitor. Swim can be followed by the human eye, whereas jitters usually appears as a blur.

Swiss Army Knife The one tool to have when you can't have more than one. Be sure it has a normal screwdriver, a phillips head screwdriver and a pair of scissors. A corkscrew also is useful.

Switch A mechanical, electrical or electronic device which opens or closes circuits, completes or breaks an electrical path, or selects paths or circuits.

Switch Based Resellers Switch-based resellers lease facilities from national carriers or large private line networks. They resell services provided over those facilities under their own name and provide sales, customer service, billing and technical

576

support. Switch-based resellers own or lease their own switching equipment and, in some cases, own their transmission facilities. they typically provide originating service on a regional basis.

Switch Busy Hour The busy hour for a single switch.

Switch Domain An SCSA definition. A single instance of a particular technology-specific connection type.

Switch Fabric An SCSA definition. The facility for connecting any two (or more) transmitting or receiving Service Providers.

Switch Fabric Controller An SCSA definition. A technology-specific, replaceable ASP within the SCSA server. The SFC is designed to support both the internal connectivity within the group and the complex, multiparty call processing applications not directly addressed by the functionality of the Group.

Switch Hook It is also called the Hook Switch. A switch hook or hook switch was originally an electrical "switch" connected to the "hook" on which the handset (or receiver) was placed when the telephone was not in use. The switch hook is now the little plunger at the top of most telephones which is pushed down when the handset is resting in its cradle (on-hook). When the handset is raised, the plunger pops up (the phone goes off-hook). Momentarily depressing the switch hook (under 0.8 of a second) can signal various services such as calling the attendant, conferencing or transferring calls.
In ISDN, the AT&T ISDN sets have several switch hooks; one for the handset, one for the speakerphone, a "virtual" switch hook, and if an adjunct is attached, an adjunct switch hook. If all switch hooks are "on-hook" or hung up, the ISDN set is on-hook. If any switch hook is "off-hook," then the ISDN set is off-hook. If more than one switch hook is off-hook, the ISDN set uses a complex algorithm to determine whether the handset, the speakerphone, or the adjunct has precedence (only one can be used at a time).

Switch Hook Flash A signaling technique whereby the signal is originated by momentarily depressing the switch hook. See SWITCH HOOK.

Switch Over When a failure occurs in the equipment, a switch may occur to an alternative piece of equipment.

Switch Port An SCSA definition. A resource that allows a Group to communicate with another Group. All Groups implicitly possess a Switch Port as a secondary resource, but in order to use it, the application must explicitly connect the Switch Ports of two Groups.

Switch Redirect A central office service which instantly, on command, redirects thousands of phone numbers to different phone numbers. Such a service has great use in a disaster.

Switch Room The room in which you put phone equipment. Also called the Phone Room. (What else?) The Phone Room should be large, clean and should stay at roughly seventy degrees and 50% humidity. You, the customer, are responsible for the quality and condition of your phone room. The messier it is, the hotter it is, the dirtier it is, the poorer your phone system (and its technicians) will function.

Switch Tender In the old, old days, the switch tender was the person who took care of the switch that moved trains from one track to another. That person often stayed for many hours a day in a small hut next to the track. Based on timetables and telegraph and phone communications with central dispatch, he would change the switch and thus move incoming trains to the right track. The job of being a switch tender is now obsolete as switches are now changed remotely by signals over phone lines. The expression "sleeping at the switch" came from the switch tender profession. Sleeping at the switch could train derailments and dead passengers. It was not a good idea.

Switch Train In a telecom circuit (typically a step-by-step central office), the series of switching devices which a call moves through in sequence.

Switchboard The attendant position of a PBX. Most of them don't actually have "boards" (they were big), they have consoles (they're much smaller and they fit on desks). Switchboards are desks.

Switchboard Cable A cable used within and between the central office main frames and the switchboard.

Switched 56 A switched data service which lets you dial someone else and transmit at 56 kilobits per second. It is a circuit switched service, letting the user transmit data at full duplex, digital synchronous 56 Kbps for usually the price of a phone call. Switched 56 can be delivered on one pair or two pairs. Switched 56 is used for videoconferencing, high speed data transfer, digital audio broadcasting, Group IV fax and remote LAN access for telecommuters. Switched 56 data service is probably the most widely used switched digital service available in North America. It has been deployed by telephone companies in strange, expensive ways — like, overlay networks, etc. — that many phone companies are not pleased with. (I'm not quite sure why, except for a general telephone company aversion to overlay networks.) Most phone companies are phasing Switched 56 service out in favor of ISDN, which has the major advantage of potentially delivering more bandwidth, i.e. 144 Kbps versus 56 Kbps. The phone companies are doing this by raising the price of their Switched 56 service and dropping the price of their ISDN service. Over time, Switched 56 will disappear for good in North America and ISDN will become the pre-eminent service. Over time.

Switched Access A method of obtaining test access to telecommunications circuits by using electromechanical circuitry to switch test apparatus to the circuit.

Switched Carrier In data terms, physical line specification selection indicating a half duplex line in a bisync network.

Switched Circuit Automatic Network SCAN. A service arrangement at certain Telco premises to interconnect private line telephone service channels of a switched service network provided to certain agencies of the federal Government.

Switched Circuit Ordering And Tracking System SCOTS. MCI's automated tracking and order processing system for Dial up products, IMTs, and the MCI switched network.

Switched Connection A connection established via signaling.

Switched Digital Services Applications Forum See SDSAF.

Switched Line A circuit which is routed through a circuit switched network, such as the telephone or telex network.

Switched Local Service You pick up the phone. You dial a local number. Bingo, you have switched local phone service. The reason this trivial definition is even in this dictionary is because many states in the U.S. now — finally — allow companies to offer local switched telephone service in competition with the established company, e.g. Nynex or Southern Bell.

Previously, they had only allowed competition in leased lines. And then previous to that they had not allowed any competition in any area of local phone service. So things are changing, albeit very very slowly.

Switched Loop In telephony, a circuit that automatically releases connection from a console or switchboard, once connection has been made, to the appropriate terminal. Loop buttons or jacks are used to answer incoming listed directory number calls, dial "0" internal calls, transfer requests, and intercepted calls. The attendant can handle only one call at a time.

Switched Loop Operation Each call requiring attendant assistance is automatically switched to one of several switched loops on an attendant position.

Switched Multi-Megabit Digital Service SMDS. A 1.544 Mbps public data service with an IEEE 802.6 standard user interface. It can support Ethernet, Token Ring and FDDI (OC-3c) LAN-to-LAN connections. See SMDS and SMDS Interest Group.

Switched Network See PSTN.

Switched Private Line Network A network which results from combining point-to-point circuits with switches.

Switched Service Network A private line network that uses scan and/or CCSA type common control switching.

Switched Transport A name for telephone traffic between the local exchange carriers' Central Offices and an interexchange carrier's point of presence (POP). Switched transport is generally provided on a monopoly basis as part of a LEC's network.

Switched Virtual Circuit A call which is only established for the duration of a session and is then disconnected.

Switchhook A synonym for hookswitch or hook switch. Also spelled switch hook. See SWITCH HOOK.

Switching Connecting the calling party to the called party. This may involve one or many physical switches.

Switching Arrangement A circuit component which enables a Customer to establish a communications path between two phones on a network.

Switching Centers There are four levels in the North American switching hierarchy run at AT&T. They are: Class 1 — Regional Center, Class 2 — Sectional Center, Class 3 — Primary Center, Class 4c — Toll Center and Class 4P — Toll Point. In addition, the local Bell operating companies run a fifth level in the hierarchy, called the Class 5 — End Office.

Switching Equipment Premises equipment which performs the functions of establishing and releasing connections on a per call basis between two or more circuits, services or communications systems.

Switching Hub A multiport hub that delivers the full, uncontested bandwidth between any pair of ports. An intelligent switching hub also provides bridging and mutliprotocol routing capabilities.

Switching Point Same as end office and intermediate office.

Switching System An ATM term. A set of one or more systems that act together and appear as a single switch for the purposes of PNNI routing.

Switchless Resellers A switchless reseller buys long distance service in bulk from a long distance company, such as AT&T, and resells that service to smaller users. It typically gets its monthly bill on magnetic tape, then rebills the bulk service to its customers. A switchless reseller owns no communications facilities — switches or transmission. It has two "assets" — a computer program to rebill the tape and some sales skills to sell its services to end users. The profit it makes comes from the difference between what it pays the long distance company and what it is able to sell its services at. Switchless resellers are also called rebillers. It's not an easy business to be in, since you are selling a long distance company's services to compete against itself. See also AGGREGATOR.

Switchway Nynex and New England Telephone's Switchway is switched 56 Kbps service. Switchway lets you dial up and transmit data at up to 56,000 bits per second. It is a totally digital service — from one end to the other. Think of it as a dial-up service for computers. It's billed like a voice line — a monthly charge plus so much per minute. Switchway lines are assigned normal telephone numbers. Switchway lines require either a 2-wire or 4-wire non-loaded metallic connection. You can dial in and around New York City using Switchway. You can call further across the country by using Switchway to connect to long distance companies, like AT&T, MCI and Sprint. Switchway was constructed as a separate data network overlaid on top of the normal voice network. It is thus far more expensive to use than a voice call, though it is largely carried digitally.

SX A family of excellent PBXs made by Mitel of Kanata, Ontario Canada. There's the analog SX-10, SX-20, SX-50, SX-100 and SX-200, and the digital SX-2000.

SXS Step by Step switching system. An automatic dial-telephone system in which calls go through the switching equipment by a succession of switches that move a step at a time, from stage to stage, each step being made in response to the dialing of a number.

SYFA Virtual Network. The network architecture used by Computer Automation.

Symbol An abbreviated, predetermined representation of any relationship, association or convention.

Symbolic Language A computer programming language used to express addresses and instructions with symbols convenient to humans rather than machines.

Symbolic Logic The discipline in which valid arguments and operations are dealt with using an artificial language designed to avoid the ambiguities and logical inadequacies of natural languages.

Symmetric Connection A connection with the same bandwidth value specified for both directions.

Symmetric Multiprocessing SMP. A type of multiprocessing in which more than one processor can execute kernel-level code at the same time. The degree of symmetry can vary from limited, where there is very little concurrency of execution, to the theoretically ideal fully-symmetric system where any function can be executed on any processor at any time. Processors within the same system share all processes, including disk I/O, network I/O and memory. Compare to ASYMMETRIC MULTIPROCESSING, wherein processors in the same or different systems are dedicated to specific tasks, such as disk I/O, network I/O or memory management. They off-load these tasks from the main system CPU, which generally is responsible for running the operating system. Each processor usually has its own dedicated memory. See SMP.

Symmetrical Channel A channel in which the send and receive directions of transmission have the same data signaling rate.

Symmetrical Compression A compression system which requires equal processing capability for compression and decompression of an image. This form of compression is used in applications where both compression and decompression will be utilized frequently. Examples include: still-image databasing, still-image transmission (color fax), video production, video mail, videophones, and videoconferencing.

Symmetrical Pair A balanced transmission line in a multipair cable having equal conductor resistances per unit length, equal impedances from each conductor to earth, and equal impedances to other lines.

Syn, Syn Character, Synchronous Idle In synchronous transmission. Control character in character-oriented protocols used to maintain synchronization and as a time-fill in the absence of data. The sequence of two SYN characters in succession is used to maintain synchronization following each line turnaround. Contrast with flag.

Syn(c) Synchronization character.

Sync Bits Synchronizing bits (more properly bytes or characters) used in synchronous transmission to maintain synchronization between transmitter and receiver.

Synchronet Service Dedicated point to point and multipoint digital data transmission service offered by BellSouth at speeds of 2.4, 4.8, 9.6, 19.2, 56 and 64 Kbps.

Synchronization 1. A networking term which means that the entire network is controlled by one master clock and transmissions arrive and depart at precise times so that information is neither lost nor jumbled. For a bigger explanation, see NETWORK SYNCHRONIZATION and SYNCHRONOUS.
2. An uninterruptible power supply (UPS) definition. Specially designed circuitry is "synchronized" to your AC power outlet to ensure continuity of power. Without this feature, power reversal can occur on the input.
3. A multimedia term. Synchronization is very precise real-time processing, down to the millisecond. Some forms of multimedia, such as audio and video, are time critical. Time delays that might not be noticeable in text or graphics delivery, but are unacceptable for audio and video. Workstations and networks must be capable of transmitting this kind of data in a synchronized manner. Where audio and video are combined, they must be time stamped so that they can both play back at the same time.
4. Start with a database on your server. Now, take a copy of part of it on your laptop — for example, your very own sales leads. Go traveling. Come back in a week. You want to update the database with your changes. But you don't want to destroy other peoples' changes. Some people are calling this "file synchronization." Synchronization is a critical part of what is increasingly being called "Groupware." See also REPLICATION.

Synchronization Bit A binary bit used to synchronize the transmission and receipt of characters in data communications.

Synchronization Bits Bits transmitted from source to destination for the purpose of synchronizing the clocks of the transmitting and receiving devices. The term "synchronization bit" is usually applied to digital data streams, whereas the term "synchronization pulse" is usually applied to analog signals.

Synchronization Code In digital systems, a sequence of digital symbols introduced into a transmission signal to achieve or maintain synchronism.

Synchronization Pulses Bits transmitted from source to destination for the purpose of synchronizing the clocks of the transmitting and receiving devices. The term "synchronization pulse" is usually applied to analog signals, whereas the term "synchronization bit" is usually applied to digital data streams.

Synchronize The word synchronize means "to cause to match exactly." When you're synchronizing, you're causing one file on one computer to precisely match another one on another computer. Why would you want to do this? Let's say you have a database of sales contacts on a file server. One of your salesman takes a copy of his sales contacts with him on his laptop. He travels and makes changes to his contacts. Now he dials into the office via modem and wants to "synchronize" his changed database with the now-changed main database, and make them both the same, i.e. into synch. This process is far more difficult than it sounds because it means allowing for the changes made at the server and by the salesman. You have to set up elaborate rules. In operating systems, such as Windows NT, the word "synchronize" has a narrower meaning. Windows NT instruction manual defines "synchronize" as "to replicate the domain controller to one server of the domain, or to all the servers of a domain. This is usually performed automatically by the system, but can also be invoked manually by an administrator." See also REPLICATE.

Synchronizing Achieving and maintaining synchronism. In facsimile, achieving and maintaining predetermined speed relations between the scanning spot and the recording spot within each scanning line.

Synchronizing Pilot In FDM, a reference frequency used for achieving and maintaining synchronization of the oscillators of a carrier system or for comparing the frequencies or phases of the currents generated by those oscillators.

Synchronous The condition that occurs when two events happen in a specific time relationship with each other and both are under control of a master clock. Synchronous transmission means there is a constant time between successive bits, characters or events. The timing is achieved by the sharing of a single clock. Each end of the transmission synchronizes itself with the use of clocks and information sent along with the transmitted data. Synchronous is the most popular communications method to and from mainframes. In synchronous transmission, characters are spaced by time, not by start and stop bits. Because you don't have to add these bits, synchronous transmission of a message will take fewer bits (and therefore less time) than asynchronous transmission. But because precise clocks and careful timing are needed in synchronous transmission, it's usually more expensive to set up synchronous transmission. Most networks are synchronous these days. See ASYNCHRONOUS and NETWORK SYNCHRONIZATION.

Synchronous Control Character SYN. A transmission control character used in synchronous transmission to provide a signal in the absence of any other character. SYN is defined by IBM Corp's Binary Synchronous Communications (BSC) protocol.

Synchronous Data Link Control SDLC. A data communications line protocol associated with the IBM Systems Network Architecture. See SYSTEMS NETWORK ARCHITECTURE.

Synchronous Data Network A data network in which synchronism is achieved and maintained between data circuit-terminating equipment (DCE) and the data switching exchange (DSE), and between DSEs. The data signaling rates are controlled by timing equipment within the network. See NETWORK SYNCHRONIZATION.

Synchronous Data Transfer A physical transfer of data to or from a device that has a predictable time relationship with the execution of an I/O (Input/Output) request. See SYNCHRONOUS.

Synchronous Digital Hierarchy SDH. Term used by the International Telegraph and Telephone Consultative Committee to refer to Sonet.

Synchronous Idle Character A transmission control character used in synchronous transmission systems to provide a signal from which synchronism or synchronous correction may be achieved between data terminal equipment, particularly when no other character is being transmitted.

Synchronous Network A network in which all the communication links are synchronized to a common clock.

Synchronous Orbit An orbit, any point on which has a period equal to the average rotational period of the Earth. If the orbit is also circular and equatorial, it is called a stationary (geostationary) orbit.

Synchronous Payload Envelope The major portion of the SONET frame format used to carry the STS-1 signal divided into an information payload section and a transport overhead system. SPE is used to address three payload structures: direct to STS-1 line rate multiplexing; asynchronous DS-3 multiplexing; and synchronous DS-3 multiplexing.

Synchronous Request An SCSA definition. A request where the client blocks until the completion of the request. Contrast with asynchronous request.

Synchronous Satellite A satellite in a synchronous orbit. See SYNCHRONOUS ORBIT.

Synchronous TDM A multiplexing scheme in which timing is obtained from a clock that in turn controls both the multiplexer and the channel source.

Synchronous Time-Division Multiplexing STDM. A time-division multiplexing method whereby devices have access to a high-speed transmission medium at fixed time periods independent of likely load.

Synchronous Transfer Mode A proposed transport level, a time division multiplex-and-switching technique to be used across the user's network interface for a broadband ISDN. See STM.

Synchronous Transmission Transmission in which the data characters and bits are transmitted at a fixed rate with the transmitter and receiver synchronized. Synchronous transmission eliminates the need for start and stop bits. See SYNCHRONOUS and ASYNCHRONOUS.

Synchronous Transmission Mode STM. The synchronous transmission capability of a system that is capable of both synchronous and asynchronous capabilities of Broadband Integrated Services Digital Network (B-ISDN) service.

Synchronous Transport Module 1 STM-1. SDH standard for transmission over OC-3 optical fiber at 155.52 Mbps.

Synchronous Transport Signal Level 1 STS-1. The basic signaling rate for a Synchronous Optical Network (SONET) transmission medium. The STS-1 rate is 51.84 Mbps.

Synchronous Transport Signaling Level n STS-n. A definition of the transmission speed a Synchronous Optical Network (SONET) transmission medium where n is an integer between 1 and 48 and relates to the multiplier to be applied to the basic STS-1 51.8-Mbps transmission speed. STS-48 is 48 times faster than STS-1, with a speed of 2.5 gigabits per second.

Syncordia A subsidiary of British Telecommunications which specializes in telecom outsourcing.

Syndrome Basic element of decoding procedure. Identifies the bits in error.

Syntax The rules of grammar in any language, including computer language. Specifically, it is the set of rules for using a programming language. It is the grammar used in programming statements.

Syntax Error An error caused by incorrect programming statements according to the rules of the language being used. Sometimes the computer will throw up "SN" to indicate a syntax error.

Synthesized Voice Human speech approximated by a computer device that concatenates basic speech parts (or phonemes) together. Usually has a metallic, Germanic sound.

Syntonization The process of setting the frequency of one oscillator equal to that of another.

Syntran Synchronous Transmission A restructured DS-3 signal format for synchronous transmission at the 4.736 Mbps DS-3 level of the North American Hierarchy.

SYSGEN Acronym for SYStem GENeration.

SYSOP The SYStem OPerator of a PC-based electronic bulletin board/mail service or on-line computer service, such as CompuServe or America On Line. SYSOPs (pronounced sis-ops) typically put computers and modems on phone lines, then published the phone number, then invited people with computers to call them and leave them messages and interesting software programs which they had written. These programs then became "public domain," or freeware. And other callers were invited to download these programs for their own use. Lead Sysops are called Wizops.

SYSREQ System request; the seldom used key used to get attention from another computer.

System An organized assembly of equipment, personnel, procedures and other facilities designed to perform a specific function or set of functions.

System 25 The System 25 PBX from AT&T is a digital phone system for businesses with 20 to 150 telephones.

System Administrator The person or persons responsible for the administrative and operational functions of a computer and a telecom system that are independent of any particular application. The system Administrator is likely to be a person with the best overview of all the applications. The System Administrator advises application designers about the data that already exists on the various services, makes recommendations about standardizing data definitions across applications, and so on.

System Build This is the original manufacturer system building that occurs when the order is placed by the buyer with the vendor. The basic configuration is set up to reflect the users needs at that point in time. Thereafter, if any changes occur to reflect changes in the operating environment, the manufacturer must reconfigure the system to reflect this change. There is usually a reprogramming charge and a delay associated with the change.

System Clock The clock designated as the reference for all clocking in a network of electronic devices such as a multiplexer or transmission facilities management system.

System Common Equipment The equipment on a premises that provides functions common to terminal devices such as telephones, data terminals, integrated work station terminals, and personal computers. Typically, the system common equipment is the PBX switch, data packet switch, or central host computer. Often called common equipment.
System Connect The method by which connection is physically made to the cost computer or local area network.
System Coordinator This is the title assigned to the person responsible for administration programming and the training of workers on your phone system.
System Disk A disk that has been formatted as a system disk. MS-DOS system disks have two hidden files and the COMMAND.COM file. You can start the computer using a system disk.
System Fault Tolerance SFT. The ability of computer to work fully regardless of component failures.
System Feature A telephone switch feature that is typically available all the users.
System Gain The amount of free space path loss that a radio can overcome by a combination of enhancing transmitted power and improving receiver sensitivity.
System Message Messages that are not associated with a mailbox.
System Redundancy The duplication of system components to protect against failure. For protection against failure, install redundant cabling, power supplies, disk storage, gateways, routers, network boards, printers, switches and other mission-critical network components.
System Reload A process allowing stored data to be written from a tape into the system memory. Picture: your telephone system goes dead. For whatever reason it loses all memory of its generic programming and your specific programming (whose extension gets what, etc.). You have to quickly grab the backup (hopefully you have it on tape or magnetic disk) and load it back into your telephone system's memory. This is called system reload. Sometimes it's done automatically. Sometimes you have to do it manually.
System Renumbering An AT&T Merlin term. A process used to change the extension numbers assigned to telephones, adjuncts, Calling Groups, Paging Groups, Call Park Zones, Remote Access, and lines/trunks.
System Segment A conceptual subset of a system, usually referring to one which can be functionally replaced without damaging the capability of the system.
System Service Provider An SCSA definition. An entity that provides system wide services, such as session management and security, and the allocation and tracking of resources and groups.
System Side Defines all cabling and connectors from the host computer or local area network to the cross connect field at the distribution frame.
System Speed Dial Simplified ways of dialing. You do them by dialing several digits. System speed dial numbers can be used by everyone on the phone system — whether they are on an electronic phone or just a simply single line phone.
System Test This definition courtesy Steve Gladstone, author, "Testing Computer Telephony Systems": System test is the phase of the product life cycle that examines the entire system as a "whole" to assure it is ready to go to a true alpha or beta test. System testing is also more oriented to inter system functions as opposed to earlier phases. To pass a system test, all features and functions are expected to work correctly (function to specification) in all areas of the system — features, administration, maintenance, billing, etc. Additionally, the system must function as an "architectural whole," including all hardware and software components. Representative databases must be loaded to simulate site applications. Full load and stress testing is performed. It is in this phase that the bulk of system level testing will take place. System testing has a major focus on external load and other stimuli.
System V Interface Definition SVID. A UNIX application-to-system software interface developed and supported by AT&T. The interface is similar to POSIX.
System/36 The System/36 is a small computer that IBM calls a "departmental computer". It is designed to satisfy computing needs of a department whereas the mainframe is designed to fill the computing needs of the entire company. System/36 computers can be directly connected to a Token Ring LAN using TIC cards.
System/370 IBM's flagship mainframe computer is very popular in large corporations. It is used to store and manipulate very large amounts of information such as airline reservation applications or banking transaction processing. System/370 computers can be attached to token ring LANs via intermediate Communications Controllers such as 3720 or 3725.
Systems Analysis Analyzing an organization's activities to figure the best way of applying computer systems to its organization.
Systems Analyst A person who performs systems analysis and who follows through with methods, techniques and programs to meet the need.
Systems Application Architecture SAA. An ambitious IBM plan to create software to weave together its incompatible computers. According to IBM advertising, "SAA will make it possible for everyone in an organization to access information regardless of its location. What's more, all software written to SAA specifications will provide similar screen layouts, menus and terminology." IBM announced SAA in March 1987 and told its customers, according to Business Week, that they would be able to share programs and shuttle information easily among mainframes, minicomputers and desktop systems. Business Week said "The initial (SAA) concept was admittedly sketchy. But IBM committed massive monies to SAA and eventually it assigned an estimated 30,000 programmers to SAA. In May 1989 IBM showed a concrete example of how SAA could work. It introduced OfficeVision, which is series of IBM applications programs that run 'cooperatively' on PCs and IBM mainframes and performs such tasks as manage electronic mail and extract data from a mainframe for use in a PC spreadsheet. OfficeVision is meant to be a concrete example of one of the benefits of IBM's Systems Application Architecture. Delivery of OfficeVision was scheduled for 1990. But didn't make that year. See SYSTEMS APPLICATIONS ARCHITECTURE.
Systems Integrator A systems integrator is a company that specializes in planning, coordinating, scheduling, testing, improving and sometimes maintaining a computing operation (sometimes companywide, sometimes just locally). In the old days, this was done almost exclusively by the International Business Machines Corporation. Somewhere along, companies discovered they could get more flexibility and computing power at a lower cost by shopping around. Today, hundreds of companies contribute

various components — hardware, software, wiring, communications and so on — to a company's computer operation. But the added flexibility can bring stunning complexity. Systems integrators try to bring order to the disparate suppliers.

Systems Integration Interface SII. As used in the definition of the proposed multivendor integration architecture sponsored by Nippon Telegraph and Telephone (NTT) of Japan, SII specifies any set of standardized services used to connect computer based-systems.

Systems Network Architecture SNA. IBM's successful computer network architecture. At one stage the most successful computer network architecture in the world. In the days of mainframe computers, it was as successful in the computer networking world as AT&T's telephone network design was in telecommunications. The best explanation we've ever read of SNA is in James Henry Green's Dow Jones-Irwin Handbook of Telecommunications. (You can get a copy of this great book by calling the Telecom Library at 1-800-LIBRARY.) Here is an excerpt:

"SNA is a tree-structured architecture, with a mainframe host computer acting as the network control center. The boundaries described by the host computer, front-end processors, cluster controllers and terminals are referred to as the network's domain. Unlike the switched telephone network that establishes physical paths between terminals for the duration of a session, SNA establishes a logical path between network nodes, and it routes each message with addressing information contained in the protocol. The network is therefore incompatible with any but approved protocols. SNA uses the SDLC data link protocol exclusively. Devices using asynchronous or binary synchronous can access SNA only through protocol converters...SNA works in seven layers roughly analogous to ISO's seven level OSI model. Unlike OSI, however, SNA is fully defined at each level. SNA was first announced in 1974 and is the basis for much of the OSI model, but it differs from OSI in several significant respects." For more on these differences, see page 96 in Green's Handbook.

The following is a description we received from IBM's PR department: "What is SNA?" In general, SNA is the description of the rules that enable IBM's customers to transmit and receive information through their computer networks. SNA may also be viewed as three distinct but related entities: a specification, a plan for structuring a network and a set of products. First, SNA is a specification governing the design of IBM products that are to communicate with one another in a network. It is called an architecture because it specifies the operating relationships of those products as part of system. Second, SNA provides a coherent structure that enables users to establish and manage their networks and, in response to new requirements and technologies, to change or expand them. Third, SNA may be viewed as a set of products: combinations of hardware and programming designed in accordance with the specification of SNA. In addition to a large number of computer terminals for both specific industries and general applications, IBM's SNA product line includes host processors, communication controllers, and adapters, modems and data encryption units. The SNA product line also includes a variety of programs and programming subsystems. Telecommunications access methods, network management programs, distributed applications programming and the network control program are examples.

Systems Network Interconnection SNI. A service defined by IBM that allows for the interconnection of separately defined and controlled Systems Network Architecture (SNA) networks.

Systems Services Control Point SSCP. An IBM Corp. Systems Network Architecture (SNA) term for the software that manages the available connection services to be used by the Network Control Program (NCP). There is only one SSCP in an SNA network domain, and the software normally resides in the host processor, which is a member of the IBM System/370 mainframe family.

Systems Software A type of program used to enhance the operating systems and the computer systems they support.

Systemview The network management program that purports to let UNIX-based computers be managed along with other IBM systems.

Systimax A structured cabling system from AT&T. Described by AT&T, Systimax is a modular system comprised of unshielded twisted-pair copper and fiber optic cables, connectors and components that provides integrated, efficient signal transport.

Systimax SCS Systimax Structured Cabling System. See SYSTIMAX.

T 1. Tip.
2. Tera, which is 10 raised to the 12th power, or 1,000,000,000,000.
T 1 See T-1
T Carrier Generic name for any of several digitally multiplexed carrier systems. The designators for T carrier in the North American digital hierarchy correspond to the designators for the digital signal (DS) level hierarchy. T carrier systems were originally designed to transmit digitized voice signals. Current applications also include digital data transmission. The table below lists the designators and rates for current T carrier systems. If an "F" precedes the "T", it's an optical fiber cable system, but the same rates.

North American Designator (DS LEVEL)

T1 (DS 1)	1.544 Mbps	24 voice channels
T1C	3.152 Mbps	48 voice channels
T2 (DS 2)	6.312 Mbps	96 voice channels
T3 (DS 3)	44.736 Mbps	672 voice channels
T4 (DS 4)	274.176 Mbps	4032 voice channels

JAPANESE HIERARCHY

DS 1	1.544 Mbps	24 voice channels
DS 2	6.312 Mbps	96 voice channels
DS 3	32.064 Mbps	480 voice channels
DS 4	97.728 Mbps	1440 voice channels
DS 5	400.352 Mbps	5760 voice channels

EUROPEAN HIERARCHY (CEPT)

DS 1	2.048 Mbps	30 voice channels
DS 2	8.448 Mbps	120 voice channels
DS 3	34.368 Mbps	480 voice channels
DS 4	139.268 Mbps	1920 voice channels
DS 5	565.148 Mbps	7680 voice channels

T Connection T-shaped three-way conductor for distributing an incoming signal in two outgoing ways. Same shape as a T-connection in the road.
T Connector A T-shaped device with two female connectors and one male BNC connector used with Ethernet coaxial cable and used on local area networks.
T Interface 4-wire ISDN BRI circuit. Picture this: You order an ISDN circuit from your local phone company. They deliver it on a normal phone line — one copper pair. At your offices, you plug in a small device called a network termination device. That device converts the two-wire circuit called a U interface, into a four-wire S or T interface which you'll use to plug in your ISDN terminal equipment, which might be a phone, a computer, a PBX, a videoconferencing device, etc. It may be all the above. The S or T interface is designed to allow you hook up to eight terminal devices on one ISDN line. Definitions are changing, especially in ISDN. At one stage, the T Interface (more properly the T-Reference Point) needed an NT1 rather than an NT2. These days, some people believe the T interface refers to an ISDN electrical connection to a PBX. While, the S bus refers to a connection to other devices, like phones and videoconferencing devices. There's no electrical difference between an S and a T interface. But I may be proven wrong by some phone company that changes the specs. One of ISDN's most charming features is its ability to acquire different specs and features from one provider to the next. See S INTERFACE and U INTERFACE and ISDN.
T Reference Point See T INTERFACE.
T Span A telephone term for a transmission medium through which a T-carrier system is operated. Also called a Span line. See T CARRIER and T-1.
T Span Line Also called a Span line. A repeatered outside plant four-wire, two twisted-pair transmission line. See T CARRIER and T-1.
T Tap A passive line interface used for extracting data from a circuit. Also, for extracting optical signals from a fiber cable or electrical signals from a coaxial cable.
T-1 Also spelled T1. A digital transmission link with a capacity of 1.544 Mbps (1,544,000 bits per second). T-1 uses two pairs of normal twisted wires, the same as you'd find in your house. T-1 normally can handle 24 voice conversations, each one digitized at 64 Kbps. But, with more advanced digital voice encoding techniques, it can handle more voice channels. T-1 is a standard for digital transmission in the United States, Canada, Hong Kong and Japan. (For the complete T Carrier hierarchy see the definition for T Carrier above.)
T-1 lines are used for connecting networks across remote distances. Bridges and routers are used to connect LANs over T-1 networks. There are faster services available. T-1 links can often be connected directly to new PBXs and many new forms of short haul transmission, such as short haul microwave systems. It is not compatible with T-1 outside the United States and Canada. In Europe T-1 is called E-1 or E1.
Outside of the United States and Canada, the "T-1" line bit rate is usually 2,048,000 bits per second. France and West Germany impose slight variations that make their formats unique. Only one element remains constant — the DS-0. The 64 kilobit per channel is universal. Most often it represents a PCM voice signal sampled at 8,000 times per second. However, the form of PCM encoding differs between T-1 (mu-law) and E-1 (A-law companding). According to Bill Flanagan's book, the differences are not so great that a multiplexer cannot convert between them.

Conversion of E-1 to T-1 involves both the compression law and the signaling format.

At the higher rate of 2,048,000, 32 time slots are defined at the CEPT interface, but two are used for signaling and other housekeeping chores. Typically 30 channels are left for user information — voice, video, data, etc. CEPT is the Conference of European Postal and Telecommunications administrations. Standards-setting body whose membership includes European Post, Telephone, and Telegraphy Authorities (PTTs).

For a full explanation of T-1 see Bill Flanagan's book The Guide to T-1 Networking. (Call 1-800-LIBRARY or 212-691-8215 for your copy.) See also PRI, PRA and T CARRIER and the following five definitions.

T-1 Framing Digitization and coding of analog voice signals requires 8,000 samples per second (two times the highest voice frequency of 4,000 Hz) and its coding in 8-bit words yields the fundamental T-1 building block of 64 Kbps for voice. This is termed a Level 0 Signal and is represented by DS-0 (Digital Signal at Level 0). Combining 24 such voice channels into a serial bit stream using Time Division Multiplexing (TDM) is performed on a frame-by-frame basis. A frame is a sample of all 24 channels (24 x 8 = 192) plus a synchronization bit called a framing bit, which yields a block of 193 bits. Frames are transmitted at a rate of 8,000 per second (corresponding to the required sampling rate), thus creating a 1.544 Mbps (8,000 x 193 = 1,544 Mbps) transmission rate, the standard North American T-1 rate. This rate is termed DS-1. See also D-4 FRAMING and EXTENDED SUPER-FRAME FORMAT. See also T CARRIER, T2, T3 and T4.

T-1 Span Line See SPAN.

T-BERD See T-BIRD.

T-BIRD also T-BERD. Colloquial term for a T-1 carrier analyzer, used by T-1 circuit technicians. Taken from the brand name of a leading device, T Berd 90A.

T-I Channel Either of the two external ports of a TDI or RDI which provides for transmitting or receiving eight TDM channels. Do not confuse with T-1 (as in T-one).

T.120 The most important transmission protocol standard for document conferencing over transmission media ranging from analog phone lines to the Internet. T.120 is the International Telecommunications Union (ITU-T) standards suite for document conferencing. Virtually all major players in the document conferencing industry have announced support for this standard. Document conferencing adds a visual dimension to voice-only conference calls by allowing groups of people to share computer documents in real-time while participating in a standard voice conference call. Whatever materials would normally be distributed in a face-to-face meeting — graphs, spreadsheets, diagrams, or documents — can be shared on-line, in real-time. Participants can easily connect to a conference anywhere in the world, with the only requirements being a Windows PC, a modem and a document conferencing software program. T.120 series standards provide a framework to enable multi-point data conferencing across LANs and WANs. The T.120 architecture relies on a multilayered approach with defined protocols and service definitions between layers. Each layer presumes the existence of all layers below. The lower level layers (T.122, T.123, T.124 and T.125) specify an application-independent mechanism for providing multi-point data communications services to any application that can use these facilities. The upper level layers (T.126 and T.127) define protocols for specific conferencing applications, such as shared whiteboarding and binary file transfer. See also H.320, standards which extend data conferencing into video conferencing.

T.122 Multipoint Communications Service for Audiographics and Audiovisual Conferencing - Service Definition. (ITU approved 1993.) See T.120.

T.123 Protocol Stacks for Audiographics and Audiovisual Teleconference Applications. (ITU approved 1993/1994.) See T.120.

T.124 Generic Conference Control for Audiovisual Services (GCC) (ITU voted approval in 3/95.) T.125: Multipoint Communication Service - Protocol Specification (MCS). (ITU approved 1994.) See T.120.

T.125 Multipoint Communication Service - Protocol Specification (MCS). (ITU approved 1994.) See T.120.

T.126 Provides shared whiteboard and document conferencing protocols. See T.120.

T.127 Provides multipoint binary file transfer. See T.120.

T.30 ITU-T standard. Handshake protocol. This standard describes the overall procedure for establishing and managing communication between the two fax machines. There are five phases of operation covered: call set up, pre message procedure (selecting the communication mode), message transmission (including phasing and synchronization), post message procedure (end-of-message and confirmation) and call release (disconnection). See CCITT.

T.35 ITU-T recommendation proposing a procedure for the allocation of ITU-T members' country or area codes for non-standard facilities in telematic services.

T.4 ITU-T standard for Group 3 fax machines, using T.30 and various V series standards. It also describes the data compression methods MH and MR.

T.434 The concept is simple: use a fax machine equipped with a disk drive to send and receive binary files as easily as you send a fax. The system would benefit from the fax's technical sophistication and ease of use, such as calling tone and the called unit's identification of capabilities. T.434 is an evolving ITU-T recommendation which defines a format used to encode a binary file and its attributes into a set of octets. This encoded binary file can then by sent over phone lines using error-corrected T.30 fax pages. The union of these two elements (file format and T.30 ECM) is known as "Binary File Transfer." The T.434 file attribute encoding is independent of ECM's block and page segmentation. T.434 defines 27 attributes which are used to describe a file. These attributes include protocol version, filename, permitted actions, contents type, storage account, date and time of creation, date and time of last modification, date and time of last read access, identity of creator, identity of last modifier, identity of last readers, filesize, future filesize, access control, legal qualifications, private use, structure, application reference, machine, operating system, recipient, character set, compression, environment, pathname, user visible string and data file content. Fisk Communications of San Diego, CA has done extensive work in the area of T.434 and has granted an irrevocable royalty-free and compensation free license to the Telecommunications Industry Association for basically all Fisk's work on T.434. Fisk hopes to make T.434 an extensive standard so that its equipment, which sends binary file transfers, can communicate with and receive from other machines. Microsoft also has defined a binary file transfer in the fax portion of its Microsoft At Work architecture. Some observers believe the At Work binary file transfer architecture is richer and more robust than T.434, supporting password and public/private-key encryption as well as digital signature verification.

As of writing, it is not clear which standard will win.

T.6 ITU-T recommendation for Group 3 fax machines using T.30 and various V series standards. It also describes compression methods (Modified Huffman and Modified READ).

T1.601 The ANSI specification for ISDN BRI outside wire, known as the U interface. T1.601 uses the 2B1Q line code operating at 160 Kb/s (144 Kb/s of 2B+D plus Layer 1 overhead bits). The electrical signal can tolerate a maximum loss of 42 dB at 40 KHz, which usually limits the local loop length to 18 Kft or less.
T.611 Also known as Appli/Com. A messaging standard proposed by France and Germany defining a Programmable Communication Interface (PCI) for Group 3 fax, Group 4 fax, teletex and telex service.
T1 1. A committee accredited by ANSI and sponsored by the Alliance For Telecommunication Industry Solutions (ATIS). The committee's role is to establish U.S. standards for digital telephony, particularly T1. ATIS provides all the administrative and logistical support. See ALLIANCE FOR TELECOMMUNICATIONS INDUSTRY SOLUTIONS.
2. See T-1 above (as in T-ONE).
T1-606/T1-6ac/T1-gfr ANSI's frame relay service specifications.
T-120 See T.120
T1C 3.152 million bits per second. Capable of handling 48 voice conversations. T1C is further up the North American digital carrier hierarchy. See T CARRIER and T-1.
T1E1 An ANSI standards sub-committee dealing with Network Interfaces.
T1M1 An ANSI standards sub-committee dealing with T-1 Inter-Network Operations, Administration and Maintenance.
T1S1 An ANSI standards sub-committee dealing with services, architecture and signaling.
T1Q1 An ANSI standards sub-committee dealing with performance.
T1S1 T1S1 is a technical subcommittee to T1 responsible for standards related to services,architectures, and signaling.
T1X1 T1X1 is a technical subcommittee to T-1 responsible for standards pertaining to synchronous interfaces and hierarchical structures relevant to interconnection of network transport signals.
T2 6.312 million bits per second. Capable of handling at least 96 voice conversations depending on the encoding scheme chosen. T2 is four times the capacity of T1. T2 is further up the North American digital carrier hierarchy. See T-1.
T3 28 T1 lines or 44.736 million bits per second. Commonly referred to as 45 megabits per second. Capable of handling 672 voice conversations. T3 runs on fiber optic and is typically called FT3. T3 is further up the North American digital carrier hierarchy. See T1.
T3POS Transaction Processing Protocol for Point of Sale. T3POS is a transaction switching and transport protocol designed to provide existing Point-Of-Sale (POS) equipment and future POS terminals with efficient and economical switching and transport service over an X.25 based packet network.
T4 274.176 million bits per second. Capable of handling 4032 voice conversations. T4 has 168 times the capacity of T1. T4 can run on coaxial cable, waveguide, millimeter radio or fiber optic. T4 is further up the North American digital carrier hierarchy. See also T1.
TA 1. A Terminal Adapter allows existing non-ISDN terminals to operate on ISDN lines. It provides conversion between a non-ISDN terminal device and the ISDN user/network interface.
2. Technical Advisory. These publications are documents describing Bellcore's preliminary view of proposed generic requirements for products, new technologies, services, or interfaces.
3. Termination Attempt.
Table Driven Describing a logical computer process, widespread in the operation of communications devices and networks, in which a user-entered variable is matched against an array of pre defined values. Frequently used in network routing, access security and modem operation. It involves a table look up that is a reference to a collection of pre defined values.
Table Hook-Up Method An information retrieval system in which the input information and the related output information are stored as a pair. When a particular input is given, the table is accessed and the output data which coincides with the input is taken out.
Tables A collection of data in which each item is arranged in relation to the other items. Many telephony functions use "look up tables" to determine the routing of calls. These tables solve the problem, "If the call is going to this exchange in this area code, then use this trunk and this routing pattern." See TABLE DRIVEN.
TABS AT&T's Telemetry Asynchronous Block Serial protocol. A polled point-to-point or multi point "master-slave" (remote-monitored equipment) communication protocol that supports moderate data transfer rates over intra office wire pairs. The remotes send "requests" or "polls" to monitored equipment. The monitored equipment answers the request with "responses." Defines two physical interfaces for direct connection between the telemetry remote and the monitored equipment:
. RS422 Point-to-Point
. RS485 Point-to-Point or Multi-Point.
Four wire, two Tx (remote to monitored) and two to Tx (monitored to remote), 22 or 21 gauge twisted pair, max 4 kft remote-to-monitored.
TAC Total Access Communications systems.
TACACS Centralization mechanism allowing access to UNIX user-databases such as NIS or Kerberos. See KERBEROS.
TACT TransActive Cordless Telephony, a new idea being developed by company in the personal communications field called American Telezone, based in Houston, TX.
Tactical Automatic Digital Switching System TADSS. A transportable store-and-forward, message-switching system designed for rapid deployment in support of tactical forces. A military definition.
Tactical Command And Control (C2) Systems The equipment, communication, procedures, and personnel essential to a commander for planning, directing, coordinating, and controlling tactical operations of assigned forces pursuant to the missions assigned. A military definition.
Tactical Communication A military term. A method or means of conveying information of any kind, especially orders and decisions from one command, person, or place to another within the tactical forces, normally by means of electronic equipment (including communications security equipment). Excluded from this definition are communications provided to tactical forces by DCS, to non tactical forces by DCS, to tactical forces by non tactical military commands, and to tactical forces by civil organizations.
Tactical Communication System A system configured by various types of fixed-size, self-contained assemblages, such as radio terminals and repeaters; switching, transmission, and terminal equipment; and interconnect and control facilities, that are used within or in sup-

port of tactical military forces. The system provides securable voice and data communications and among mobile users to facilitate command and control within, and in support of, tactical forces.

Tactical Data Information Link TADIL. A military term. A Joint-Chiefs-of-Staff-approved standardized communication link suitable for transmission of digital information. A TADIL is characterized by its standardized message formats and transmission characteristics.

Tactical Data Information Link — A TADIL—A. A military term. A netted link in which one unit acts as a net control station and interrogates each unit by roll call. Once interrogated, that unit transmits its data to the net. This means that each unit receives all the information transmitted. This is a direct transfer of data and no relaying is involved.

Tactical Data Information Link — B TADIL—B. A military term. A point-to-point data link between two units which provides for simultaneous transmission and reception of data (duplex).

Tactical Load A military term. That part of the operational load required by the host service consisting of weapons, detection, command control systems, and related functions.

TAD Telephone Answering Device.

TADIL Tactical Data Information Link.

TADSS Tactical Automatic Digital Switching System.

TAFAS Trunk Answer From Any Station. The ability to answer an incoming phone call from any telephone attached to the system.

Tag Image File Format TIFF provides a way of storing and exchanging digital image data. Aldus Corp., Microsoft Corp., and major scanner vendors developed TIFF to help link scanned images with the popular desktop publishing applications. It is now used for many different types of software applications ranging from medical imagery to fax modem data transfers, CAD programs, and 3D graphic packages. The current TIFF specification supports three main types of image data: Black and white data, halftones or dithered data, and grayscale data. Some wags think TIFF stands for "Took It From a Fotograf." It doesn't.

TAI inTernational Atomic tIme.

Tail An echo cancellation term. The tail, measured in milliseconds, is the amount of your conversation which returns to you in the echo, as measured in milliseconds. A tail of zero milliseconds clearly means there's no echo.

Tail Circuit 1. A feeder circuit or an access line to a network. Typically, a connection from a satellite, microwave receiver to a user's equipment location.
2. A point-to-point circuit connecting a remote terminal to a local terminal via two modems at an intermediate site. A crossover cable connects the two modems at the intermediate site.

Tail-End-Hop-Off TEHO. In a private network with several nodes (locations), TEHO occurs when a call placed from one location on the network to a location not on the network leaves the network at the node closest to its destination.

Taligent A joint venture of IBM and Apple to create a new operating system for personal computers.

Talk A service on Internet whereby you hold conversations with others by typing into your computer. And they reply by typing into their computer.

Talk Battery The DC voltage supplied by the central office to the subscriber's loop so as to allow you to have a voice conversation. Also known as A BATTERY. Typically somewhere between 5 and 25 volts. See B BATTERY.

Talk Off Talk off (also called Talk-Off) is one hazard of inband signaling. The classic definition of talk-off is that it occurs when your voice has enough 2600 Hz energy to activate the 2600 Hz tone-detecting circuits in the central office. The 2600 Hz tone is used for inband signaling. In voice processing, talk-off happens when a person is recording an audio file (say, leaving a voice mail message) and the frequencies in his voice happen to match those of a touch-tone digit. The voice board reacts as if a touchtone digit were press. Which means it may terminate the recording immediately and move to another menu selection.

Talk Path The tip and ring conductors of a telephone circuit.

Talk Set, Optical An instrument for talking over fibers — used when installing and testing the cable.

Talk Time 1. The amount of time agents spend on the phone, as opposed to the time between calls spent updating records, sending out literature or going to the bathroom.
2. The length of time you can talk on your portable or transportable cellular phone from a fully charged battery without standby time. The battery capacity of a cellular portable or transportable is usually expressed in terms of so many minutes of talk time or so many hours of standby time. When you are talking, the phone draws more power from the battery.

Talk-Off See TALK OFF.

Talkdown Missed signals in the presence of speech. Commonly used to describe the performance of a DTMF receiver when it fails to recognize a valid DTMF tone due to cancellation of that tone by speech.

Talker Provides interaction with a Dialogic board, allowing for Voice Mail functionality.

Talking Head That part of the person seen in the typical business videoconference; the head and shoulders. This type of image is fairly easy to capture with compressed video because there is very little motion in a talking head image.

Talkoff See TALK OFF.

TAM 1. Telephone Answering Machine.
2. Telecommunications Access Method.
3. Novell's Telephony Access Module. A piece of free software Novell announced to telephony enable every Windows application you already own. According to Novell, Telephony Access Module will provide NetWare Telephony Services users with instant telephony enablement as well as access to MAPI-based address books, Novell GroupWise address books and NetWare Directory Services for name and phone number lookup service. TAM also provides a DDE interface for integration into application supporting DDE as well as directory engines that provide a DDE interface.

Tandem The connection of networks or circuits in series. That is, the connection of the output of one circuit to the input of another. See TANDEM SWITCH.

Tandem Architecture A physical network topology where connectivity between locations is achieved by linking several locations together in a chain using private line circuits. In a tandem architecture, a packet may have to pass through several intermediate locations before reaching its final destination. A single network failure can affect connectivity between several locations, a primary weakness of the topology.

NEWTON'S TELECOM DICTIONARY

Tandem Call A call processed by two or more switches. Also used to designate this type of call at a switch where a connection is established from one trunk to another (tandem trunking). See TANDEM SWITCH.

Tandem Center In a communication system, an installation in which switching equipment connects trunks to trunks, but not any customer loops.

Tandem Data Circuit A data channel passing through more than two data circuit-terminating equipment (DCE) devices in series.

Tandem Office A major phone company switching center for the switched telephone network. It serves to connect central offices when direct interoffice trunks are not available.

Tandem PBX A main PBX is one which has a Directory Number (DN) and can connect PBX stations to the public network for both incoming and outgoing calls. A main PBX can have an associated satellite PBX, and can be part of a Tandem Tie Trunk Network (TTTN). If the main PBX provides tandem switching for tie trunks, it is called a tandem PBX.

Tandem Point An intermediate location in a tandem architecture.

Tandem Switch Tandem is a telephony term meaning to "connect in series." Thus a tandem switch connects one trunk to another. A tandem switch is an intermediate switch or connection between an originating telephone call location and the final destination of the call. The tandem point passes the call along. A PBX can often handle tandem calls from other/to other locations as well as process calls to, from and within its own location.

Tandem Tie Trunk Switching The PBX permits tie lines to "tandem" through the switch. This means an incoming tie line call from a distant PBX receives a dial tone instead of automatically connecting with the operator. The caller can then dial a connection with either a phone on the PBX or an outgoing line. The outgoing line can be a local trunk in which case the distant PBX has access to a form of foreign exchange service, or another tie line which links a third system. This system of tie lines is widely used to form a corporate communications system, allowing economical connections between distant offices. To provide tie line tandeming ability, the PBX must be able to detect when either tie line goes on-hook at the distant end so that it can break its tandem connection and allow the tie lines to be used for other calls.

Tandem Trunks Trunks between an end office and a tandem switching machine or between tandem switching machines. Tandem trunks can provide direct routing or alternate routing capability when direct trunks are occupied.

Tank Test A voltage dielectric test in which the wire or cable test sample is submerged in water and voltage is applied between the conductor and water as ground.

TANSTAAFL There Ain't No Such Thing As A Free Lunch.

TAO Telephony Application Object. Part of the SCSA programming framework. See SCSA TELEPHONY APPLICATION OBJECTS FRAMEWORK.

Tap 1. An electrical connection permitting signals to be transmitted onto or off a bus. The link between the bus and the drop cable that connects the workstation to the bus. Also a device used on CATV cables for matching impedance or connecting subscriber drops.
2. Telocator Alphanumeric Paging Protocol. A 7-bit messaging protocol which allows someone sitting at a terminal or computer to send a message to a pager (also known as a beeper). For a more modern 8-bit protocol, see TDP, which stands for Telocator Data Protocol.
2. Telocator Alphanumeric Paging
Protocol. A 7-bit messaging protocol which allows someone sitting at a terminal or computer to send a message to a pager (also known as a beeper). For a more modern 8-bit protocol, see TDP, which stands for Telocator Data Protocol.
It also provides an error detecting link from the sender to the paging service provider. Invalid data, bad PINs, and other common errors are reported back to the sender.
Information about TAP can be gotten from PCI in Washington DC @ 202-467-4770

Tap Button A button found on single line phones behind a PBX or Centrex. The tap button gives a precisely measured Hookswitch flash. The purpose of this button is to signal the PBX that it is about to receive a command — typically a transfer. To transfer a call on a single line phone, you typically depress the hookswitch, then punch out the extension you want to transfer the call to, announce the call when someone answers, then hang up and the PBX or Centrex transfers the call. The problem with using a hookswitch to make this transfer is that if you depress the hookswitch for too long you will cut the call off. As a result, some manufacturers put a tap button on their single line phones. This button gives the precise hookswitch signal for the precise length of time necessary — no more, no less. The Tap Button is also called a Flash button or a Tap Key.

Tap Key Also called Tap Button or Flash Key. A button on a phone that accomplishes the same function as a switch hook but is not a switch hook. See TAP BUTTON.

TAPAC Terminal Attachment Program Advisory Committee. Body which recommends telecom standards to the Canadian Federal Government.

Tape Drive The physical unit that holds, reads and writes magnetic tape.

Tape Reader A device which reads information recorded on punched paper tape or magnetic tape.

Tape Relay A method of retransmitting TTY traffic from one channel to another, in which messages arriving on an incoming channel are recorded in the form of perforated tape, this tape then being either fed directly and automatically into an outgoing channel, or manually transferred to an automatic transmitter for transmission on an outgoing channel.

Tapered Rate 1. A proposal by Nynex to give discounts to large customers who might be considering bypass.
2. Prices drop as usage increases.

TAPI Telephone Application Programming Interface. Also called Microsoft/Intel Telephony API. A term that refers to the Windows Telephony API. TAPI is a changing (i.e. improving) set of functions supported by Windows that allow Windows applications (Windows 3.xx, 95 and NT) to program telephone-line-based devices such as single and multi-line phones (both digital and analog), modems and fax machines in a device-independent manner. TAPI essentially does to telephony devices what Windows printer system did to printers — make them easy to install and allow many application programs to work with many telephony devices, irrespective of who made the devices. TAPI is one of numerous high-level device interfaces that Windows offers as part of the Windows Open Services Architecture (WOSA). TAPI simplifies the process of writing a telephony application that works with a wide variety of modems and other devices supported by TAPI drivers. See also DIAL STRING, MICROSOFT FAX, WINDOWS 95 and WINDOWS TELEPHONY for fuller explanations.

TARGA Truevision Advanced Raster Graphics Adapter.

587

NEWTON'S TELECOM DICTIONARY

Target A SCSI device that performs an operation requested by an initiator.

Tariff Documents filed by a regulated telephone company with a state public utility commission or the Federal Communications Commission. The tariff, a public document, details services, equipment and pricing offered by the telephone company (a common carrier) to all potential customers. Being a "common carrier" means it (the phone company) must offer its services to everybody at the prices and at the conditions outlined in its public tariffs. Tariffs do not carry the weight of law behind them. If you or the telephone company violate them, no one will go to jail. The worst that can happen to you, as a subscriber, is that your service will be cut off, or threatened to be cut off. Regulatory authorities do not normally approve tariffs. They accept them — until they are successfully challenged before a hearing of the regulatory body or in court (usually Federal Court). Many tariffs were accepted by regulatory commissions only to be struck down in court as unlawful, discriminatory, not cost-justified, etc. Monies collected under the tariff have been refunded and unnecessary equipment removed. In these new, competitive days, many telephone companies are violating their own tariffs by charging less money than their tariffs say they should, or bundling services together at a discount. They are also providing service and equipment on terms less onerous than outlined in their tariffs. Many users now regard tariffs as starting bargaining points, rather than ending bargaining points.

Tariff 12 A user-specific long distance tariff of AT&T. Tariff 12 gives AT&T the ability to price its long distance services for one company practically any which way it feels — giving them a mix of services at stable prices over the long term with significant volume discounts. As this dictionary was going to the printer, a federal appeals court overturned the Federal Communications Commission's April 1989 decision allowing AT&T to offer custom networks and ordered the FCC to reopen its investigation into the legality of the Tariff 12 deals. There are still some users. They are "grandfathered" until we get a final say on the tariff. And, as we go to press, AT&T can offer Tariff 12 customized services to any company — but cannot include 800 services in its Tariff 12 pricing.

Tariff 15 A user-specific long distance tariff of AT&T. Tariff 15 gives AT&T the ability to price its long distance services for one company practically any way it feels. Tariff 15 is single-customer discounting. Some of AT&T competitors claim the tariff is "illegal."

TARM Telephone Answering and Recording Machines.

TAS Telephone Answering Service.

TASC Telecommunications Alarm, Surveillance, and Control system. Expands the scope of maintenance from the traditional alarm monitoring and control functions to include performance monitoring and fault locating.

TASI Time Assignment Speech Interpolation. A money saving analog multiplexing procedure which keeps the connection to the circuit as long as someone is speaking and lets other conversations use the circuit during the intervals (measured in microseconds) when there's no speaking. Since a long distance circuit is usually only half used — one person speaking, one person listening — at least 50% of the circuit can be used by someone else. TASI is typically used by long distance companies on submarine cable across the Atlantic and the Pacific. Unfortunately, the flip-flopping around of circuit allocation by TASI means that the first tiny bit of a conversation is often lost. This can be disastrous for data. The key in data is to keep transmitting. The key in voice is to say something like "Ah" to seize the channel and then say what you want. TASI is very much like a very fast version of mobile dispatch radio. A modern version of TASI is called DIGITAL SPEECH INTERPOLATION. TASI is somewhat comparable to statistical multiplexing of data.

Task Management Allocating resources and overseeing the sequence of tasks completed by the computer.

Task Switching You have a computer and you want to have it do several tasks at once. There are two ways. One is multi tasking. The computer will keep working on several tasks at once, though you may not see them on your screen. For example, you pay start a spreadsheet recalculating. And then you may call your electronic mail system. While you're receiving your mail, your spreadsheet is still recalculating. When you're through, you can switch back to your spreadsheet and see the final results. That's called multi tasking and MS-DOS doesn't have it. MS-DOS 5.0 through its Shell has something called task switching, whose idea is that you can load several programs into your computer and switch quickly between them. But the programs you put in background won't run. They stop the moment you put them in background. When you cycle back to them, they will start running again. Sadly, MS-DOS 5.0 Shell's task switching capabilities are very weak. Your programs will lock up and you will lose your data. I do not recommend using Task Switching in MS-DOS 5.0. Windows 3.1 is alleged to have a form of multi tasking. I don't trust it either yet. Windows has locked up on me on several occasions.

Task-To-Task Communication The process whereby one computer program exchanges data with another. May also be called program-to-program communications.

TAXI Transparent Asynchronous Transmitter Receiver Interface. 100-Mbps ATM physical interface specification based on the FDDI PHY.

TAT TransAtlantic Telephone cable.

TB An ATM term. Transparent Bridging: An IETF bridging standard where bridge behavior is transparent to the data traffic. To avoid ambiguous routes or loops, a Spanning Tree algorithm is used.

TB/S TeraBits per Second.

TBA To Be Announced. Pricing of a product that may exist and that may, one day, be priced. You often see TBA after products which are hyperware, vaporware, mirrorware or smokeware.

TBB Transnational Broadband Backbone.

TBBS The Bread Board System. A proprietary name for one of the earliest and still one of the best electronic bulletin boards. From a company called eSoft in Aurora CO. The president is Phil Becker. He's great. Tell him, Harry says Hello.

TBD To Be Determined.

TBE Transient Buffer Exposure: This is a negotiated number of cells that the network would like to limit the source to sending during start-up periods, before the first RM-cell returns.

TBM Northern Telecom's Transport Bandwidth Manager.

TBOS Telemetry Byte Oriented Serial protocol. TBOS is a protocol for transmitting alarm, status, and control points between NE and OS. TBOS defines one physical interface for direct connection between telemetry remote and the monitored equipment. This is a point-to-point communication, RS 422A modified four wire, two to Tx (remote to monitored) and two to Tx (monitored to remote), 26 gauge, max 4 kft remote-to-monitored. Remote sees a 100 to 180 ohms resistor at monitored terminal.

TBR Timed BReak.

TC 1. Transmission Control.

2. **Transmission Convergence.** The TC sublayer transforms the flow of cells into a steady flow of bits and bytes for transmission over the physical medium. On transmit, the TC sublayer maps the cells to the frame format, generates the Header Error Check (HEC), sends idle cells when the ATM layer has none to send. On reception, the TC sublayer delineates individual cells in the received bit stream, and uses the HEC to detect and correct received errors.

3. **Telecommunications Closet.** A closet which houses telecommunications wiring and telecom wiring equipment. It contains the BHC (Backbone to Horizontal Cross-connect). It may also contain the network demarcation, or MC (Main Cross-connect). The telecommunications closet is used to connect up telecom wiring. The closet typically has a door. It's a good idea to lock the door and not put anything else in the closet, like mops, buckets and brooms.

TCA TeleCommunications Association. A major association of West-of-the-Rockies telecommunications management professionals. Every September, TCA runs the TCA Show in September each year in San Diego.

TCAM TeleCommunications Access Method. A popular telecommunications software package to run on IBM 370 computers. See IBM.

TCAP Transactional Capabilities Application Part. Provides the signaling function for network data bases. TCAP is an ISDN application protocol. In addition to PRA and ISUP, the third major ISDN protocol in the delivery of advanced network services is TCAP, a CCS7 application protocol that provides the platform to support non-circuit related, transaction-based information exchange between network entities. This capability is required by transaction-based services that must exchange information between a pair of signaling nodes in a CCS7 network. Examples of these services include enhanced dial-800 service, automated credit card calling and virtual private networking. The TCAP protocol enables these services to access remote databases called service control points (SCPs) to process part of the call. The SCP supplies the translation and routing data needed to deliver advanced network services — like translating a dial calls into the required routing number. TCAP is useful also in coordinating some enhanced call-related services. For example, network ring again requires the connection of two users when both stations become idle. In this case, TCAP is used to coordinate between the users' switches while waiting for each line to become idle. And it can do this without tying up network trunks.

One of the major advantages of TCAP is that it provides a set of protocol building blocks for use in a variety of service definitions. The TCAP building blocks are subdivided into the transaction sublayer and the component sublayer. For more on TCAP, see the 1988-3 issue of Northern Telecom's Telesis publication.

TCAS T-Carrier Administration System. Provides mechanized support for the facility maintenance and administration center to achieve centralized administration and control of the digital network.

TCF Training Check Frame. Last step in a series of signals in a fax transmission called a training sequence, designed to let the receiver adjust to telephone line conditions.

Tchotchke A New York Jewish word meaning trinkets, best exemplified by the giveaway junk we often pick up at telecommunications trade shows.

TCIF The TeleCommunications Industry Forum, a major committee sponsored by the ECSA, the Exchange Carriers Standards Association. See EXCHANGE CARRIERS STANDARDS ASSOCIATION.

TCM 1. Traveling Class Mark.
2. Trellis Coding Modulation.
3. Time Compression Multiplexing. A digital transmission technique that permits full duplex data transmission by sending compressed bursts of data in a "ping-pong" fashion.

TCNS Thomas Conrad Networking System is a 100 million bit per second proprietary networking system (LAN) based on ARCnet that can use most standard ARCnet drivers on any network operating systems.

TCP 1. Transmission Control Protocol. ARPAnet-developed transport layer protocol. Corresponds to OSI layers 4 and 5, transport and session. TCP is a transport layer, connection-oriented, end-to-end protocol. It provides reliable, sequenced, and unduplicated delivery of bytes to a remote or local user. TCP provides reliable byte stream communication between pairs of processes in hosts attached to interconnected networks. It is the portion of the TCP/IP protocol suite that governs the exchange of sequential data. See TCP/IP for a much longer explanation.
2. An ATM term. Test Coordination Procedure: A set of rules to coordinate the test process between the lower tester and the upper tester. The purpose is to enable the lower tester to control the operation of the upper tester. These procedures may, or may not, be specified in an abstract test suite.

TCP/IP I wrote and re-wrote my definition of TCP/IP several times. Then I found a great definition in Microsoft's Windows for Workgroups Resource Kit: Transmission Control Protocol/Internet Protocol (TCP/IP) is a networking protocol that provides communication across interconnected networks, between computers with diverse hardware architectures and various operating systems. TCP (Transmission Control Protocol) and IP (Internet Protocol) are only two protocols in the family of Internet protocols. Over time, however, "TCP/IP" has been used in industry to denote the family of common Internet protocols. The Internet protocols are a result of a Defense Advanced Research Projects Agency (DARPA) research project on network interconnection in the late 1970s. It was mandated on all United States defense long-haul networks in 1983 but was not widely accepted until the integration with 4.2 BSD (Berkeley Software Distribution) UNIX. The popularity of TCP/IP (Harry's note: it's the Internet's networking protocol) is based on:

• Robust client-server framework. TCP/IP is an excellent client-server application platform, especially in wide-area network (WAN) environments.

• Information sharing. Thousands of academic, defense, scientific, and commercial organizations share data, electronic mail and services on the connected Internet using TCP/IP.

• General availability. Implementations of TCP/IP are available on nearly every popular computer operating system. Source code is widely available for many implementations. Additionally, bridge, router, and network analyzer vendors all offer support for the TCP/IP protocol family within their products.

TCP/IP is the most complete and accepted networking protocol available. Virtually all modern operating systems offer TCP/IP support, and most large networks rely on TCP/IP for all their network traffic. Microsoft TCP/IP provides cross-platform connectivity and a client-server development framework that many software vendors and corporate developers are using to develop distributed and client-server applications in heterogeneous enterprise networks over TCP/IP.

How TCP Works: TCP is a reliable, connection-oriented protocol. Connection-oriented implies that TCP first establishes a connection between the two systems that intend to exchange data. Since most networks are built on shared media (for example, several systems sharing the same cabling), it is necessary to break chunks of data into manageable pieces so that no two communicating Systems monopolize the network. These pieces are called packets. When an application sends a message to TCP for transmission, TCP breaks the message into packets, sized appropriately for the network, and sends them over the network.

Because a single message is often broken into many packets, TCP marks these packets with sequence numbers before sending them. The sequence numbers allow the receiving system to properly reassemble the packets into the original message. Being able to reassemble the original message is not enough, the accuracy of the data must also be verified. TCP does this by computing a checksum. A checksum is a simple mathematical computation applied, by the sender, to the data contained in the TCP packet. The recipient then does the same calculation on the received data and compares the result with the checksum that the sender computed. If the results match, the recipient sends an acknowledgment (ACK). If the results do not match, the recipient asks the sender to resend the packet. Finally, TCP uses port IDs to specify which application running on the system is sending or receiving the data.

The port ID, checksum, and sequence number are inserted into the TCP packet in a special section called the header. The header is at the beginning of the packet containing this and other "control" information for TCP.

How IP Works: IP is the messenger protocol of TCP/IP. The IP protocol, much simpler than TCP, basically addresses and sends packets. IP relies on three pieces of information, which you provide, to receive and deliver packets successfully: IP address, subnet mask, and default gateway.

The IP address identifies your system on the TCP/IP network. IP addresses are 32-bit addresses that are globally unique on a network. They are generally represented in dotted decimal notation, which separates the four bytes of the address with periods. An IP address looks like this: 102.54.94.97

Although an IP address is a single value, it really contains two pieces of information: (a.) Your system's network ID, and (b.) Your system's host (or system) ID.

The subnet mask, also represented in dotted decimal notation, is used to extract these two values from your IP address. The value of the subnet mask is determined by setting the network ID bits of the IP address to ones and the host ID bits to zeros. The result allows TCP/IP to determine the host and network IDs of the local workstation. Here's how to understand an IP address. For example:

When the IP address is 102.54.94.97 (specified by the user) And the subnet mask is 255.255.0.0 (specified by the user)
The network ID is 102.54 (IP address and subnet mask)
And the host ID is 94.97 (IP address and subnet mask)

OK. the above was Microsoft's definition. Here's my definition, which covers some areas Microsoft doesn't. TCP/IP is a set of protocols developed by the Department of Defense to link dissimilar computers across many kinds of networks, including unreliable ones and ones connected to dissimilar LANs. TCP/IP is the protocol used on the Internet. It is, in essence, the glue that binds the Internet. Developed in the 1970s by the U.S. Department of Defense's Advanced Research Projects Agency (DARPA) as a military standard protocol, its assurance of multi vendor connectivity has made it popular among commercial users as well, who have adopted TCP/IP. Consequently, TCP/IP now is supported by many manufacturers of minicomputers, personal computers, mainframes, technical workstations and data communications equipment. It is also the protocol commonly used over many Ethernet LANs (as well as X.25) networks. It has been implemented on everything from PC LANs to minis and mainframes.

TCP/IP currently divides networking functionality into only four layers:

A Network Interface Layer that corresponds to the OSI Physical and Data Link Layers. This layer manages the exchange of data between a device and the network to which it is attached and routes data between devices on the same network.

An Internet Layer which corresponds to the OSI network layer. The Internet Protocol (IP) subset of the TCP/IP suite runs at this layer. IP provides the addressing needed to allow routers to forward packets across a multiple LAN inter network. In IEEE terms, it provides connectionless datagram service, which means it attempts to deliver every packet, but has no provision for retransmitting lost or damaged packets. IP leaves such error correction, if required, to higher level protocols, such as TCP.

IP addresses are 32 bits in length and have two parts: the Network Identifier (Net ID) and the Host Identifier (Host ID). Assigned by a central authority, the Net ID specifies the address, unique across the Internet, for each network or related group of networks. Assigned by the local network administrator, the Host ID specifies a particular host, station or node within a given network and need only be unique within that network.

A Transport Layer, which corresponds to the OSI Transport Layer. The Transmission Control Protocol (TCP) subset runs at this layer. TCP provides end-to-end connectivity between data source and destination with detection of, and recovery from, lost, duplicated, or corrupted packets — thus offering the error control lacking in lower level IP routing. In TCP, message blocks from applications are divided into smaller segments, each with a sequence number that indicates the order of the segment within the block. The destination device examines the message segments and, when a complete sequence of segments is received, sends an acknowledgement (ACK) to the source, containing the number of the next byte expected at the destination.

An Application Layer, which corresponds to the session, presentation and application layers of the OSI model. This layer manages the function required by the user programs and includes protocols for remote log-in (Telnet), file transfer (FTP), and electronic mail (SMTP). See OSI.

TCR 1. Transaction Confirmation Report. A report from a fax machine listing the faxes received and transmitted. It provides details about each fax, including date, time, the remote fax's number, results, total pages.

2. An ATM term. Tagged Cell Rate: An ABR service parameter, TCR limits the rate at which a source may send out-of-rate forward RM-cells. TCR is a constant fixed at 10 cells/second.

TCS 1. Transmission Convergence Sublayer: This is part of the ATM physical layer that defines how cells will be transmitted by the actual physical layer.

2. TCS is a manufacturer of excellent call center manpower scheduling software packages. It is based in Nashville, TN.

TDC Time Division Controller.

TDD 1. Time Division Duplex.

2. Telecommunications Device for the Deaf. Under the Communications Act of 1934, a TDD is defined as a machine "that employs graphic communication in the transmission of coded signals through a wire or radio." TDD devices (which typically look like simple computer ter-

minals) use the Baudot method of communications. Most TDD devices are acoustically coupled and are slow, running at 300 baud. There is a special TDD/TTY Operator Services number. It's 800-855-1155. Users of TDDs often abbreviate commonly used words or expressions to save time. Here are some of the most frequently used:

ANS	Answer	R	Are
CUD	Could	REC	Receive
GA	Go Ahead	SK	Stop Keying
LTR	Letter	THRU	Through
MSG	Message	THX	Thank You
MIN	Minute	U	You
NITE	Night	UR	Your
PLS	Please	WUD	Would
QQ	Question	XOX	Hugs & Kisses

TDF Trunk Distributing Frame.
TDHS Time Domain Harmonic Scaling
TDI Transmit Division Intertie.
TDM See TIME DIVISION MULTIPLEXING.
TDM Network A Rolm term: The assembly of interface cards and buses for connecting phones to each other, to outside trunks, etc., and to certain shared electronics within the CBX (Rolm term for PBX).
TDMA Time Division Multiple Access. One of several technologies used to separate multiple conversation transmissions over a finite frequency allocation of through-the-air bandwidth. As with FDMA (Frequency Division Multiple Access), TDMA is used to allocate a discrete amount of frequency bandwidth to each user, in order to permit many simultaneous conversations. However, each caller is assigned a specific timeslot for transmission. A digital cellular telephone system using TDMA assigns 10 timeslots for each frequency channel, and cellular telephones send bursts, or packets, of information during each timeslot. The packets of information are reassembled by the receiving equipment into the original voice components. TDMA promises to significantly increase the efficiency of cellular telephone systems, allowing a greater number of simultaneous conversations. See CDMA, FDMA.
TDMS 1. Technical Document Management Systems.
2. Time Division Multiplex System.
3. Transmission Distortion Measuring Set.
TDP Telocator Data Protocol. A new 8-bit protocol for sending messages and binary files (images, spreadsheets, word processing files, executables, etc.) to pagers (also known as beepers). The older (and more common) 7-bit messaging protocol now widely in use is called TAP, which stands for Telocator Alphanumeric Paging Protocol. This protocol can only send simple alphanumeric messages, like "Your shares in XYZ are now $23, up 98%."
TDR Time Domain Reflectometer.
TDS MCI's name for Terrestrial Data Services, i.e. services that run through on-the-ground fiber, rather than through-the-air satellite services.
TDSAI Transit Delay Selection And Indication
TE 1. ISDN Terminal Equipment. See the next two definitions.
2. Terminal Equipment: As an ATM term, terminal equipment represents the endpoint of ATM connection(s) and termination of the various protocols within the connection(s).
TE1 Terminal Equipment Type 1. ISDN-compatible terminals. A TE1 could be an ISDN telephone, a personal computer capable of working with ISDN, a videophone, etc. In short, any device that can attach to and work with ISDN.
TE2 Terminal Equipment Type 2. Non-ISDN terminal equipment (e.g. analog telephone) linked at the RS-232, RS-449, or V.35 interfaces.
TEC NIS Telecommunications and Electronics Consortium in the Newly Independent States. An organization based in Moscow and administered by TIA (Telecommunications Industry Association) to assist US telecommunications and telecommunications-related electronics companies with doing business in the region.
Techinfo A common campuswide information system developed at MIT. An Internet term.
Technical Control Center A testing center for telecommunications circuits. The center provides test access and computer-assisted support functions to aid in circuit maintenance.
Technical Control Facility A federal government term. A term plant, or a designated and specially configured part thereof, containing the equipment necessary for ensuring fast, reliable, and secure exchange of information. This facility typically includes distribution frames and associated panels, jacks, and switches; and monitoring, test, conditioning, and order wire equipment.
Technical Load A military term. The portion of the operational load required for communications, tactical operations, and ancillary equipment including necessary lighting, air conditioning, or ventilation required for full continuity of communications.
Technical Means A term in the spy business which means spy satellites and electronic eavesdropping stations, typically costing tens of billions of dollars.
Technical Office Protocol TOP. A seven-layer network architecture designed for office automation that uses International Standards Organization (ISO) or CCITT specifications at each level. TOP was defined by Boeing Vertol Corp. and is now controlled by the MAP/TOP (Manufacturing Automation Protocol/Technical Office Protocol) Users Group.
Technology Partners Dialogic term for companies which make specialized voice processing devices which can be used with Dialogic products to build end user products. Technology partners include those with skills in facsimile, voice recognition and text-to-speech. Their products will work with Dialogic products.
TED Trunk Encryption Device.
Teddy Bear A stuffed animal named after President "Teddy" Roosevelt, a keen hunter who once took pity on a baby bear.
TEDIS Trade Electronic Data Interchange Systems.
Teen Service A feature of some central offices which allows two telephone numbers to be assigned to a single party phone line. Each num-

ber has a distinctive ringing pattern so that the called parties can recognize which line is ringing. The inventor of this service named it after the fact that his teenage children were always receiving phone calls. And he wanted a way for them to recognize when the calls were for them and when they were for the parents. Sadly, this phenomenon now begins earlier in life, with children as young as six receiving their own calls. We speak from experience. Teen service is now used by home businesses, roommates, boarders, college dorm suite-mates, and live-in relatives.

TEHO See TAIL END HOP-OFF (traffic engineering).

TEI Terminal Endpoint Identifier. Up to eight devices can be connected to one ISDN BRI line. The TEI defines for a given message which of the eight devices is communicating with the Central Office switch. In general, more than one of the eight may be communicating.

TEK Traffic Encryption Key.

Telabuse A term coined by John Haugh of Telecommunications Advisors in Portland, OR to include "insider" toll fraud, waste and abuse.

Telanophy A product of Active Voice of Seattle, WA, which includes modules to allow you to view all your incoming messages — fax and email — and to listen to your voice mail. And to run the whole process by clicking on one screen of single line messages. Neat program. See also UNIFIED MESSAGING.

Telbanking Banking transactions conducted through telecommunications.

Telco The local telephone company. Often a term of endearment. Americanism for telephone company.

Telco Splice Block What some parts of the data communications industry call a 66-block, i.e. a terminating block for twisted pair voice and data cable.

Tele Tele comes from the Greek work meaning "far." Telecommunications is therefore communicating over a distance.

Teleaction Service In ISDN applications, a telecommunications service using very short messages with very low data transmission rates between the user and the network.

Teleadmin A means in GSM (non-North America cellular digital standard) to update a Subscriber Identity Module (SIM) card via a short message sent by the network operator using Short Message Services (SMS). In GSM, the SIM card is located in the cell phone's handset and is customized to a specific subscriber's service options. TeleAdmin is also known as remote SIM card updating.

Telebusiness A British term for telemarketing. Here's a definition I found in England: Activities conducted by telephone in a planned and controlled manner. The term encompasses telesales, telemarketing, customer service and information broadcast. Telebusiness can be conducted between an organization and its customers and prospects, or conducted as in internal service.

Telecom A shortened and perfectly acceptable way of saying the word "telecommunications."

Telecom Developers A trade show started by the author of this dictionary and devoted to the notion that users ought to have the ability to program, control and build their own telephone systems. The show is now called COMPUTER TELEPHONY Conference and Exposition. It's held every Spring in March. For more information, call 212-691-8215 or 800-999-0345 or 800-LIBRARY. See ERECTOR SET TELECOM.

Telecommunication Service Any service provided by a telecommunication provider. A specified set of user-information transfer capabilities provided to a group of users. The telecommunication service provider has the responsibility for the acceptance, transmission, and delivery of the message.

Telecommunication Architecture The governing plan showing the capabilities of functional elements and their interaction, including configuration, integration, standardization, life-cycle management, and definition of protocol specifications, among these elements.

Telecommunicationally Challenged A politically correct term for being under-phoned, i.e. having too few phones. This definition contributed by John Warrington of Ashland University, Ashland, OH.

Telecommunications 1. The art and science of "communicating" over a distance by telephone, telegraph and radio. The transmission, reception and the switching of signals, such as electrical or optical, by wire, fiber, or electromagnetic (i.e. through-the-air) means.
2. A fancy word for "telephony," which it replaced and which many thought meant only analog voice, but didn't.

Telecommunications Broker A person or an organization which buys telecommunications services at bulk rates and resells these services at below "normal" i.e. retail prices.

Telecommunications Closet A closet which houses telecommunications wiring and telecom wiring equipment. Contains the BHC (Backbone to Horizontal Cross-connect). May also contain the Network Demarcation, or MC (Main Cross-connect). The telecommunications closet is used to connect up telecom wiring. The closet typically has a door. It's a good idea to lock the door and not put anything else in the closet, like mops, buckets and brooms.

Telecommunication Facilities The aggregate of equipment, such as telephones, teletypewriters, facsimile equipment, cables, and switches, used for various modes of transmission, such as digital data, audio signals, image and video signals.

Telecommunications Information Networking Architecture Consortium See TINA-C.

Telecommunications Lines Telephone and other communications lines used to transmit messages from one location to another.

Telecommunications Network The public switched telephone exchange network.

Telecommunications Reform Act of 1996 A bill passed on January 3, 1996 "to promote competition and reduce regulation in order to secure lower prices and higher quaity services for American telecommunications consumers and encourage rapid deployment of new telecommunications technologies." The Act, amongst other things, allowed the local regional Bell operating phone companies into long distance once they had met certain conditions about allowing competition in their local monopoly areas. See Dialing Parity below. You can download a copy of this Act (all 391,861 bytes) from http://thomas.loc.gov/cgi-bin/query/1?c104:./temp/~c104iLoA:: The following are definitions contained in the Act.

Affiliate — The term `affiliate' means a person that (directly or indirectly) owns or controls, is owned or controlled by, or is under common ownership or control with, another person. For purposes of this paragraph, the term `own' means to own an equity interest (or the equivalent thereof) of more than 10 percent.

AT&T Consent Decree — The term `AT&T Consent Decree' means the order entered August 24, 1982, in the antitrust action styled United States v. Western Electric, Civil Action No. 82-0192, in the United States District Court for the District of Columbia, and includes any judgment or order with respect to such action entered on or after August 24, 1982.

Bell Operating Company — The term `Bell operating company'
(A) means any of the following companies: Bell Telephone Company of Nevada, Illinois Bell Telephone Company, Indiana Bell Telephone Company, Incorporated, Michigan Bell Telephone Company, New England Telephone and Telegraph Company, New Jersey Bell Telephone

NEWTON'S TELECOM DICTIONARY

Company, New York Telephone Company, U S West Communications Company, South Central Bell Telephone Company, Southern Bell Telephone and Telegraph Company, Southwestern Bell Telephone Company, The Bell Telephone Company of Pennsylvania, The Chesapeake and Potomac Telephone Company, The Chesapeake and Potomac Telephone Company of Maryland, The Chesapeake and Potomac Telephone Company of Virginia, The Chesapeake and Potomac Telephone Company of West Virginia, The Diamond State Telephone Company, The Ohio Bell Telephone Company, The Pacific Telephone and Telegraph Company, or Wisconsin Telephone Company; and (B) includes any successor or assign of any such company that provides wireline telephone exchange service; but (C) does not include an affiliate of any such company, other than an affiliate described in subparagraph (A) or (B).

Customer Premises Equipment — The term `customer premises equipment' means equipment employed on the premises of a person (other than a carrier) to originate, route, or terminate telecommunications.

Dialing Parity — The term `dialing parity' means that a person that is not an affiliate of a local exchange carrier is able to provide telecommunications services in such a manner that customers have the ability to route automatically, without the use of any access code, their telecommunications to the telecommunications services provider of the customer's designation from among 2 or more telecommunications services providers (including such local exchange carrier).

Exchange Access — The term `exchange access' means the offering of access to telephone exchange services or facilities for the purpose of the origination or termination of telephone toll services.

Information Service — The term `information service' means the offering of a capability for generating, acquiring, storing, transforming, processing, retrieving, utilizing, or making available information via telecommunications, and includes electronic publishing, but does not include any use of any such capability for the management, control, or operation of a telecommunications system or the management of a telecommunications service.

Telecommunications Relay Service Telecommunication Relay Service (TRS), formerly called Dual Party Relay Service, is available to hearing impaired customers. Defined by the Communications Act of 1934, telecommunications relay services mean "transmission services that provide the ability for an individual who has a hearing impairment or speech impairment to engage in communication by wire or radio with a hearing individual in a manner that is functionally equivalent to the ability of an individual who does not have a hearing impairment or speech impairment to communicate using voice communication service by wire or radio."

Telecommunications Resellers Association TRA. An association of long distance companies in North America. The association is based in Washington, D.C. The TRA acts as the resale industry's collective trade group. TRA holds two exhibitions each year, and acts as the resale industry's lobbying group and consumer watchdog. TRA says it has a strong code of ethics, and most member resellers are viewed as having reputable business standards.

Telecommuting Working off-site. The process of commuting to the office through a communications link rather than transferring one's physical presence. In short, working at home on a computer, a modem and maybe a facsimile machine, rather than going into the office. There are benefits: you can live somewhere charming. There are disadvantages, especially accentuated if you work with others: "When you're getting data from afar, you're not in touch with the soul of the business, anymore," according to one telecommuter interviewed by the New York Times. He went on to say, "All the electronic communications are simply backup, I just hadn't factored the importance of personal loyalty and contact into the equation. And I was very wrong."

Telecomputer Telecomputer appears to be the couch potato's ultimate toy. Peter Coy writing in Business Week of November 1, 1993 called telecomputer a "computerized television." He said "the idea is that you can watch anything in the (on-line) video library anytime. Your telecomputer lets you scroll through a menu of programs, click on your choice, and send an order up the line." James H. Clark, chairman of Silicon Graphics Inc., a manufacturer of computers with heavy video skills calls telecomputer a term for a combination computer/CATV controller that is being popularized by the new media industry. The idea is to use the telecomputer to do interactive games, choose a movie to play out of thousands of choices, buy things, send electronic mail, etc.

Teleconference A telephone conversation with three or more people. They may be distant from each other. They may all be in the same office.

Teleconferencing A conference of more than two people linked by telecommunications. Can be audio only; can be video one-way and audio the other; can be video both ways. For years, teleconferencing has been heralded as a great coming event, and a significant replacement for travel. The death of travel has not happened because teleconferencing remains expensive, cumbersome, lacking in the human touch and difficult to initiate. Few companies have teleconferencing facilities in their offices. So, to hold a teleconference you often have to travel somewhere — a local hotel ballroom, the local AT&T office, etc. Teleconferencing will not displace travel. But it is growing and growing rapidly. It is developing its own special niche in the telecommunications travel-continuum. That niche is increasingly for events like new product announcements, price changes, new Vision statements — happenings which need to be broadcast quickly to many people, and which often generate questions from the dispersed audience.

TELECONNECT Magazine TELECONNECT is a monthly magazine covering developments in telecommunications equipment. It's the largest monthly telecommunications magazine. Its job is to help its readers choose, install and maintain telecom equipment. TELECONNECT Magazine contains critical reviews of new products and product comparisons. Subscriptions are available from 1-800-677-3435 and 215-355-2886 for $15 a year.

Telecopier A fancy word for a facsimile machine.

Teledensity A measure of the number of phone lines per 100 of population. Between 40 and 50 lines per 100 of population indicates pretty good density. Under 10 indicates pretty bad density. Teledensity is a measure of a country's economic development.

Teledesic Teledesic is undertaking to deploy a global constellation of hundreds of low-Earth-orbit satellites to provide affordable access to a broad array of advanced information services — ranging from high-quality voice channels to broadband channels supporting videoconferencing, interactive multimedia and other real-time, two-way digital data applications. Teledesic, which plans to begin service in the year 2001, does not intend to market services directly to users. Rather, it will provide an open network for the delivery of such services by others. The Teledesic network will enable local telephone companies and government authorities in host countries to cost-effectively modernize existing communications systems and bring fiber-like services to areas that would never get such capabilities through wireline means. Teledesic was founded in June of 1990 and is headquartered in Kirkland, Washington near Seattle. Its principal shareholders are Craig O. McCaw, founder of McCaw Cellular Communications, and William H. Gates III, co-founder of Microsoft.

NEWTON'S TELECOM DICTIONARY

Telefax 1. European term for fax.
2. A high-speed, 64 kilobit per second facsimile service that uses Group 4 fax machines and one Bearer channel of an ISDN circuit, or any other 64 Kbps circuit. Group 4 fax machines take about six seconds to transmit a page. They're fast and impressive.

Telegaming Using communications lines — from dial up through local area networks through WANs — to play games interactively with someone at the other end. The upcoming expected explosion of telegaming is what's driving the growth of DSVD (Digital Simultaneous Voice Data) modems.

Telegram Hard-copy information, in written, printed or pictorial form, routed to the general telegraph service for transmission and delivery to the addressee. Telegrams are dying due to the high cost of delivery.

Telegraph A system employing the interruption of, or change in, the polarity of DC current signaling to convey coded information.

Telegraph Key A type of switch for making and breaking a circuit at will for the purpose of transmitting dots and dashes.

Telegraphy Aging data transmission technique characterized by maximum data rates of 75 bits per second and signaling where the direction, or polarity, of DC current flow is reversed to indicate bit states.

Teleguerilla A term coined by Telecom Australia for "the first wave of informal and unofficially sanctioned telecommuters — those who occasionally work from home with the informal approval of their immediate boss." Says Telecom Australia, "They're the ones whose bosses say, 'I don't care where you work as long as you get the project finished.'"

Telemagic Our favorite contact management software program. It's from Remote Control in Carlsbad, CA.

Telemanagement A new term which involves using computer technology and techniques to better manage the telephone and telecommunications expenses of a corporation. Telemanagement includes every function the corporate telecommunications manager does today — from call accounting to ordering of new circuits, from managing the corporate inventory of phones, lines and other equipment, to choosing the right number of operators to man the phones at the right time, etc. Increasingly, you now see personal computers sitting on top of telephone systems, collecting and processing information. That's part of telemanagement, too.

Telemarketing Marketing and sales conducted via the telephone. There are two sides to telemarketing — incoming and outgoing. Incoming telemarketing is largely run through 800 toll-free IN-WATS numbers and local FX (foreign exchange) lines. Outgoing telemarketing is organized over OUT-WATS lines. An expanding range of telecom gadgetry is being developed to automate telemarketing — including automated outbound dialers, voice processing technology and automatic call distributors. The tone recognition, voice detection and transaction audiotex and transaction processing capabilities of voice processing gear can be used to enhance all telemarketing applications.

Telematic Agent TLMA. An X.400 AU (Access Unit) serving Teletex users of other telematic services (using Teletex, Fax, etc.).

Telemedicine The provision of health-care services from a distance using networks supporting audio, video, and computer data transmissions. Telemedicine traditionally uses videoconferencing to diagnose illness and provide medical treatment over a distance. Often used to view or teach surgical procedures. Used also in rural areas where health care is not readily available and to provide medical services to prisoners. At the ICA Show in 1994, Southwestern Bell demonstrated telemedicine applications including a dermatology microscope, a video scope, an electronic stethoscope and a telepathology system that allows a pathologist to exercise computer control over a remote microscope.

Telemetry A communications system for the transmission of digital or analog data which represents status information on a remote process, function or device.

Telenet A private, commercially available network providing both packet-switched and circuit-switched service to subscribers in North America, Europe and some parts of Asia.

Telenet Remote Login Protocol A virtual terminal service specified by the U.S. Department of Defense and implemented by most versions of UNIX.

Telenet International Quotations TIQ. A market data information subscription service operated by Telrate International Co. over a network that uses proprietary protocols to enhance security and other functions.

Teleos IRX-9000 Teleos' Intelligent Call Distribution platform. Teleos makes a programmable switch which can be controlled by an external computer. It calls the switch the IRX-9000 and the ability to write software for it, its intelligent call distribution platform.

Teleparents Parents who equip their children with pagers before allowing them to go out.

Telephone 1. An invention of the devil.
2. The most intrusive device ever invented.
3. The biggest time waster of all time, as in: "What did you do all day?" "Nothing. Just spent the day on the phone."
4. Also a truly remarkable invention. Here's a list of the eight things a telephone does, according to Understanding Telephone Electronics:
a. When you lift the handset, it signals you wish to use the worldwide phone system.
b. It indicates the phone system is ready for your wish by receiving a tone, called a dial tone.
c. It sends the number of the telephone to be called.
d. It indicates the progress of your call by receiving tones — ringing, busy, etc.
e. It alerts you to an incoming call.
f. It changes your speech into electrical signals for transmission to someone distant. It also changes the electrical signals it receives from the distant person to speech so you can understand them.
g. It automatically adjusts for changes in the power supplied to it.
h. When you hang up, it signals the phone system your call is finished.
And, most remarkably, most simple telephones cost under $50.

Telephone Amplifier A device to amplify the sound of the receiver. Something no phone should be without. Some devices work strictly on line power. They can only increase volume by 10 dB, which is often not enough (especially if you're over 40). The best telephone amplifiers are powered by AC/DC adapters. Newer ones are powered by nicad batteries. They will amplify to 20 dB.

Telephone Answering A feature of some voice mail systems in which incoming callers are immediately directed to the called party's voice mailbox where they hear a personalized greeting in the called party's voice and are prompted to leave a detailed message.

Telephone Channel A transmission path suitable for carrying voice signals. Defined by its ability to transmit signals in a frequency range of about 300 to 3000 Hz.

NEWTON'S TELECOM DICTIONARY

Telephone Circuit Electrical connection permitting the establishment of telephone communication in both directions between two telephone exchanges.

Telephone Exchange A switching center for connecting and switching phone lines. A European term for what North Americans call central office.

Telephone Frequency Any frequency within that part of the audio frequency range essential for the transmitting speech, i.e. 300 to 3000 Hz.

Telephone Line Simulator Also called ring-down box. Ring down boxes, also known as CO simulators, are simple devices used for generating calls from a POTS line to a computer telephony system (or vice versa). When one side goes offhook, the ring down box will "ring" the other side. When both sides are offhook, both sides are coupled together and the line is powered. Ring down boxes are available with various options and configurations. These include the ability to provide dialtone to the caller side (required to test applications with modems, faxes, or other automated outdialing devices), caller ID, and disconnect supervision. They are generally available in one to four line sizes, although special configurations may support more. Ring-down boxes are used for giving demonstrations and testing. We use them in our test labs to testdrive new computer telephony systems.

Telephone Management System The term originally meant a system for controlling telephone costs by:
1. Automatically selecting lower-cost long distance routes for placed calls; 2. Automatically restricting certain people's abilities to make some or all long distance calls; and 3. Automatically keeping track of telephone usage by extension, time of day, number called, trunk used and sometimes by person calling and client or account to be billed for call.
These days the terms means those three functions plus a whole lot more, typically those associated with professionally managing the corporate or government telecommunications expenses, including (but not limited to):
- Computerized inventory monitoring,
- Computerized traffic engineering and network design,
- Departmental telephone bill allocation and invoicing,
- Automated telephone directory, etc.
- Project tracking,
- Automated equipment and service ordering.
In short, all the functions of professional telecommunications management that can be automated or organized in some way on a computer. The telecommunications management system thus refers to the computer hardware and the software. For more on this subject see the latest June issue of TELECONNECT Magazine. See also CALL ACCOUNTING SYSTEM.

Telephone Manager Apple's telephony API for the Macintosh world.

Telephone Message Management System TMMS. This IBM software product permits detailed personalized call answering in the CBX II 9000 environment. Which sounds like voice mail to us. We stole this definition from a glossary Rolm supplied us.

Telephone Receiver Telephone earpiece. Device that converts electrical energy into sound energy, designed to be held to the ear.

Telephone Relay Service TRS. A voice/data system that enables communications with the hearing impaired.

Telephone Service Representative TSR. Another word for agent — the person who answers the phone on an automatic call distributor. See AGENT.

Telephone Set A fancy name for a telephone.

Telephone Set Emulation The concept is simple: Emulate the proprietary electronic phone on a printed circuit card inside a PC. Let the PC do everything a human using the phone could do. Only the PC will do it more efficiently and the human will find it easier to use all his phone's features because the PC's screen is bigger and the PC's keyboard easier to use than the phone's keyboard. Attach the phone emulation card to voice and call processing cards, like voice synthesis, voice recognition, voice mail, touchtone generation and recognition, etc. And bingo, phone systems acquire all the benefits of integrated voice and call processing. It's powerful concept. As I wrote this, a handful of telephone phone emulation cards had appeared. Within a little while, there won't be a phone worldwide that you won't be able to emulate on a printed circuit card you can drop into a vacant slot inside your PC.

Telephone Signaling Device A gadget which indicates that the phone is ringing. May also be hooked up to lamps or overhead lighting to cause those lights to flash when the phone is ringing.

Telephone Tag I call you. But you're not there. I leave a message. You call me back. But I'm not there. You leave a message. And so on. We're now playing telephone tag.

Telephony The science of transmitting voice, data, video or image signals over a distance greater than what you can transmit by shouting. The word derives from the Greek for "far sound." For the first hundred years of the telephone industry's existence, the word telephony described the business the nation's phone companies were in. It was a generic term. In the early 1980s, the term lost fashion and many phone companies decided they were no longer in telephony, but in telecommunications — a more pompous sounding term that was meant to encompass more than just voice. The pomposity of the word may have added some value to the stock of telecommunications companies. In the early 1990s, as computer companies started entering the telecommunications industry, the word telephony was resurrected. And in a white paper on Multimedia from Sun Microsystems, the company said that telephony refers to the integration of the telephone into the workstation. For instance, making or forwarding a call will be as easy as pointing to an address book entry. Caller identification (if available from the telephone company) could be used to automatically start an application or bring up a database file. Voicemail and incoming faxes can be integrated with e-mail (electronic mail). Users can have all the features of today's telephones accessible through their workstations, plus the added benefits provided by integrating the telephone with other desktop functions. See also COMPUTER TELEPHONY.

Telephony Access Module See TAM.

Telephony Interface Control A Telephony Interface Control resource is any resource that interfaces with the telephone network (public or private). This is usually claimed as the primary member of a group.

Telephony Server A telephony server is a computer whose major function is to control, add intelligence, store, forward and manipulate the various voice, data, fax and e-mail calls flowing into and out of a computer telephony system. The traditional function of a telephony server is to move call control commands from client workstations on a LAN to an attached PBX or ACD. (This is what it does under the paradigm called "Telephony Services.") A telephony server can also be a voice response system. It can also be a fax on demand system. It can also be a conferencing device. It can also be switch. And it can be all these capabilities, which traditionally run on physically separate servers, all rolled into one machine, called generically a "telephony server." See TELEPHONY SERVICE APPLICATION PROGRAMMING INTERFACE and TELEPHONY SERVICES.

Telephony Server Application Programming Interface. TSAPI. Described by AT&T, its inventor, as "standards-based API for call control, call/device monitoring and query, call routing, device/system maintenance capabilities, and basic directory services." For a better explanation, see TELEPHONY SERVICES.

Telephony Server NLM Telephony Server NetWare Loadable Module. A product announced in early 1993 by AT&T. The Telephony Server NLM is an AT&T-made card sitting in a file server on a Novell LAN. The card connects directly to the ASAI (Adjunct Switch Applications Interface) port on the AT&T Definity PBX. Anyone with a PC on the network and an AT&T phone on their desk will be able to use telephone features, such as auto-dialing, conference calling and message management (a new term for integrating voice, fax and e-mail on your desktop PC). Novell/AT&T have announced that they intend to create open Application Programming Interfaces (APIs) third-party developers can work with. A Novell/AT&T example of what could be developed: A user could select names from a directory on his PC. He could tell the Definity PBX through the PC over the LAN to place calls to them. At the same time, a program running under NetWare would automatically send an e-mail to the people in the conference call. All participants would have access to both the document and the conference call simultaneously.

A Novell White Paper in Spring of 1993 said "Telephony Services represents a significant piece of NetWare's overall strategy of providing a comprehensive set of integrated network services. Through the combining of Telephony Services for NetWare with the functionality of other network services such as fax, video, messaging and others, further gains are easily demonstrated. Today, organizations view the telephone and computer as essential parts of conducting daily business. In fact, they rely so heavily on the technology of these two instruments that productivity can often be jeopardized when access is interrupted. With such heavy reliance, the next stop in the technology evolution is to merge and integrate the capabilities of these two technologies to improve the return and benefits provided to customers.

"Telephony Services for NetWare provides benefits to three main customer segments. First, applications are being developed to provide increased productivity to everyday computer desktop users. Second, call-centers take advantage of this technology as it provides a right-sizing cost-effective solution. Finally, benefits will be available to telecommunications/IS administrators by providing the ability to reduce administrative costs through easier management of user databases.

"Computer-Telephone Integration (CTI) combines telephone and computer technology to provide access and control telephone functionality from a computer terminal. It combines the easy access and usable graphical interface of the computer desktop with the features of the telephone. CTI is not a new concept. Traditionally, however, CTI has only been available in a mini and mainframe computer environments. These solutions are expensive and can be cost-justified only in large call-center applications. Consequently, the penetration of CTI solutions has been very small.

"However, providing CTI in NetWare environment brings this technology mainstream. Not only does this solution provide a more cost-effective implementation, it also allows integration with the rich set of NetWare services. In the simplest example, a Telephony Services for NetWare application allows users to make a phone call by clicking on a name from a calling list displayed on a desktop computer and having the desktop computer dial the number. Possibilities exist for applications that will allow similar functionality with the addition of conference calling capability. Instead of clicking on a single name, the user can highlight a number of names, click on a conference-call icon and have the system place the calls to all parties. The benefits which are derived from the integration of telephony with other NetWare services is far reaching. As part of continued development efforts, applications are becoming available which allow desktop video phone calls. Callers can see each other and talk on the phone, while simultaneously viewing and editing image documents.

"Other capabilities include integrating voice-mail, fax and e-mail into a single message-management application. Possibilities also exist utilizing number recognition technology to integrate computer database records with caller-id. Administrators can manage a single user database utilized by the computer network, the PBX and the voice-mail system.

"Telephony Services for NetWare takes advantages of client/server technology to provide a broad framework for creating first-party and third-party call-control applications. These applications answer the customer demand for integrated business tools and solutions. This technology provides a logical connection between the desktop computer and the telephone. The only physical connection is established between the PBX and a NetWare server. This architecture is cost-effective and efficient by utilizing a company or organization's existing equipment. The initial product deliverables include the following components:
- Client/Server API
- Telephony Server NLM
- PBX Driver
- PBX Link Hardware
- Passageway Application

"The Telephony Server NLM is the mechanism for passing information between the PBX and the NetWare server. As part of the NLM, an open PBX Driver Interface allows PBX manufacturers to write drivers which communicate with their respective PBX models. The client/server API provides support across multiple desktop operating systems. It also allows call control at either the client or the server. The Passageway application provides the user with basic autodialing and notes capability.

"Telephony Services for NetWare provides a key opportunity for developers. Open APIs which support multiple desktop operating systems provide a development platform for both traditional telecommunications developers and new or experienced NetWare developers." See TELEPHONY SERVICES.

Telephony Services Telephony Services' real name is Telephony Services for NetWare. NetWare is the name of the best-selling local area network software supplied and sold by Novell, Orem, Utah. Telephony Services for NetWare basically consists of an addition to the NetWare operating system, called Telephony Server NLM. (See TELEPHONY SERVER NLM.) That addition handles communications between a NetWare file server (a PC loaded with NetWare) and an attached telephone switch, e.g. a PBX or ACD (automatic call distributor). The concept is very simple. Picture your department's LAN. You're sitting in front of your PC which is on your department's LAN. You have a telephone which is an extension off your company's or department's PBX. You click on an icon that says "Phone." Bingo, a screen comes up with icons and pull down menus. You can now look up Joe, click on his name. Your PC sends a command to your NetWare file server, which in turn sends a command to your PBX which tells it to dial Joe from your phone. Once it's completed dialing, it might turn on your phone's speakerphone or your telephone headset. You'll then hear Joe say "Hello." Telephony Services for NetWare is basically the software in the NetWare file server which takes care of interpreting your PC commands into commands your switch can understand and respond to.

Telephony Services requires a link to your switch. Each telecom switch manufacturer has been implementing that link in a different way technically. That's fine, because Telephony Services for NetWare insulates the user and the developer. This means that computer telephony applications written for Telephony Services for NetWare will work on any switch conforming to the Telephony Services standard. As of writing, virtually every switch manufactured in North America and many made in Japan and Europe is conforming to Novell's Telephony Services.

According to a White Paper issued by Novell in March 1994 and called NetWare Telephony Services, "The three main components of Telephony Services are call control, voice processing, and speech synthesis. Call control provides the core service for PBX-to-NetWare communication and an Application Programming Interface (API) for developing client server applications. With call control, users can enjoy features, says Novell, such as making calls, transfers, or conference calling. Voice processing functions include voice mail and interactive voice response. Speech synthesis will be a key area for integrating multiple media types. Through speech/text conversion, users can access voice mail, e-mail, and fax documents through audio or text media types. The initial products Novell is delivering include:

"Client/Server API
"Telephony Server NetWare Loadable Module (NLM)
"PBX Driver

"The Telephony Server NLM is the mechanism for passing information between the PBX and the NetWare server. As part of the NLM, an open PBX Driver Interface allows PBX manufacturers to write drivers that communicate with their respective PBXs. The client/server API provides support across multiple desktop operating systems and allows call control at either the client or the server."

Client server computer telephony (which is what all this is called) delivers ten benefits:

1. Synchronized data screen and phone call pop. Your phone rings. The call comes with the calling number attached (via Caller ID or ANI). Your PBX or ACD passes that number (via Telephony Services) to your server, which does a quick database look up to see if it can find a name and database entry. Bingo, it finds an entry. It passes the call and the database entry simultaneously to whoever is going to answer the phone: The attendant. The boss. The sales agent. The customer service desk. The help desk. All this saves asking a lot of questions. Makes customers happier.

2. Integrated messaging. Also called Unified Messaging. Voice, fax, electronic mail, image and video. All on the one screen. Here's the scenario. You arrive in the morning. Turn on your PC. Your PC logs onto your LAN and its various servers. In seconds, it gives you a screen listing all your messages — voice mail, electronic mail, fax mail, reports, compound documents .... Anything and everything that came in for you. Each is one line. Each line tells you whom it's from. What it is. How big it is. How urgent. Skip down. Click. Your PC loads up the application. Your LAN hunts down the message. Bingo, it's on screen. If it contains voice — maybe it's a voice mail or compound document with voice in it — it rings your phone (or your headset) and plays the voice to you. Or, if you have a sound card in your PC, it can play the voice through your own PC. If it's an image, it will hunt down (also called launch) imaging software which can open the image you have received, letting you see it. Ditto, if it's a video message.

Messages are deluging us. To stop them is to stop progress. But to run your eye down the list, one line per entry. Pick the key ones. Junk the junk ones. Postpone the others. That's what integrated messaging is all about. Putting some order back into your life.

3. Database transactions. Customer look ups. There are bank account balances, ticket buys, airline reservations, catalog requests, movie times, etc. Doing business over the phone is exploding. Today, the caller inputs his request by touchtone or by recognized speech. The system responds with speech and/or fax. Today's systems are limited in size and flexibility. The voice processing application and the database typically share the same processor, often a PC. Split them. Spread the processing and database access burden. Join them on a LAN (for the data) and on new, broader voice processing "LANs," like SCSA or MVIP. You've suddenly got a computer telephony system that knows no growth constraints. You could also get the system to front-end an operator or an agent. Once the caller has punched in all his information, then the call and the screen can be simultaneously passed to the agent.

4. Telephony work groups. Sales groups. Collections groups. Help desks. R&D. We work in groups. But traditional telephony doesn't. Telephony today is BIG. Telephony today is one giant phone system for the building, for the campus. Everyone shares the same automated attendant, the same voice mail, the same ubiquitous, universal, generic telephone features. But they shouldn't. The sellers need phones that grab the caller's phone number, do a look-up on what the customer bought last and quickly route the call to the appropriate (or available) salesperson. The one who sold the customer last time. The company's help desk needs a front end voice response system that asks for the customer's serial number, some indication of the problem and tries to solve the problem by instantly sending a fax or encouraging the caller to punch his way to one of many canned solutions. "The 10 biggest problems our customers have." When all else fails, the caller can be transferred to a live human, expert at diagnosing and solving his pressing problem. A development group might need e-mails and faxes of meeting agendas sent, meeting reminder notices phoned and scheduled video conferences set up. All automatically. The accounts receivable department needs a predictive dialer to dial all our deadbeats. The telemarketing department also needs a predictive dialer, but different programming.

5. Desktop telephony. There are two important aspects. Call control and media processing services. Call control (also called call processing) is a fancy name for using your PC to get to all your phone system's features — especially those you have difficulty getting to with the forgettable commands phone makers foist on us. *39 to transfer? Or it is *79. With attractive PC screens, you point and click to easy conferencing, transferring, listening to voice mail messages, forwarding, etc. There are enormous personal productivity benefits to running your office phone from your PC: You can dial by name, not by number you can't remember. You can set up conference calls by clicking on names and have your PC call the participants and call you only when they're all on the phone. You can transfer easily. You can work your voice mail more easily on screen, instead of having to remember "Dial 3 for rewind," "Dial 2 to save," and other obscure commands. Here's a wonderful quote from Marshall R. Goldberg, Developer Relations Group at Microsoft. He says "Voice mail systems that could benefit through integration with the personal computer largely remain isolated, difficult to use, and inflexible. Browsing, storing messages in hierarchical folders, and integration of address books — functions just about everyone could use — are either unavailable or unusable."

The second benefit is media control. Media control is a fancy name for affecting the content of the call. You may wish to record the phone call you're on. You may wish to have all or part of your phone call clipped and sent to someone else — as you often today with voice mail messages. You may wish to simply file your conversations away in appropriate folders. You may wish to be able to call your PC and get it to read you back any e-mails or faxes you received in the last day or so.

6. Applying intelligence. A PC is programmable. The typical office phone isn't. A PC can be programmed to act as your personal secretary, handling different calls differently. It can be programmed to include commands, such as "If Joe calls, break into my conversation and tell me." "If Robert calls, send him to voice mail." etc.

7. The Compound Document. The typed document lacks life. But add voice, image and video clips to it and it gets life. The LAN makes the compound document easier to achieve. The Compound Document gets attention.

8. Management of phone networks. Today, phone networks are very difficult to manage. Often the PBX is managed separately from the voice mail, which is managed separately from the call accounting, etc. It's a rare day in any corporate life when the whole system is up to date, with extensions, bills and voice mail mailboxes reflecting the reality of what's actually happening. The latest generations of LAN software — NetWare 4.1 and Windows NT — have solid enterprise-wide directories and far easier management tools. Integrate these LAN management tools with telecommunications management, and potentially all you need is to make one entry (for a new employee, a change, etc.) and the whole system — telecom and computing — could update itself automatically, including even issue change orders to the MIS and telecom departments and vendors.

9. No dedicated hardware in the PC. With only one link — from the switch to the LAN — there's no need to open the desktop PC and place specialized telephony hardware in each PC that wants to take advantage of the new LAN-based telephony features.

10. Switch elimination. The ultimate potential advantage of LAN-based telephony is to eliminate the connection to the switch (PBX or ACD) by simply populating the LAN server (now called a telephony server) with specialized computer telephony cards and run the company's or department's phones off the telephony server directly.

See TELEPHONY SERVER NLM and TELEPHONY SERVICES DEVELOPMENT TOOLS

Telephony Services Development Tools These are the main Telephony Services software development tools: TSCALL.EXE — a client rolodex program, CSTASpy.EXE — troubleshooting tool to track TSAPI messages; SIM.NLM — simulates a physical PBX outside of a live environment and CLSim.EXE — works with SIM.NLM to simulate a client session working.

Telephony Workgroup A concept that says work is done in groups and those groups need special telephony features and services. This is in contrast to most telephone installations today, where one giant phone system serves the company. Everyone shares the same automated attendant, the same voice mail, the same ubiquitous, universal, generic telephone features. But the concept of a telephony is that they shouldn't. Each group has different telephony needs: The sellers need phones that grab the caller's phone number, do a look-up on what the customer last bought and quickly route the call to the appropriate (or available) salesperson. The one who sold the customer last time.

The company's help desk needs a front end voice response system that asks for the customer's equipment serial number, some indication of the problem and tries to solve the problem by instantly sending a fax or encouraging the caller to punch his way to one of many canned solutions. "The 10 biggest problems our customers have." When all else fails, the caller can be transferred to a live human, expert at diagnosing and solving his pressing problem.

A development group might need e-mails and faxes of meeting agendas sent, meeting reminder notices phoned and scheduled video conferences set up. All automatically.

Computer telephony on a LAN can do to the workgroup what products like e-mail and Lotus Notes are doing — substantially improve productivity (or at least, the pleasure of work). Except that the telephone is still less intimidating.

There are probably as many specialized telephony workgroups features needed as there are computer workgroup features needed. And since computer workgroup features are often provided on a local area network, it makes sense to provide many telephony features for that workgroup on the same local area network.

Telepoint The British name for the new generation of cheap, digital mobile phones. They're also called CT2. Think of telepoint phones as cellular phones but using micro-cells. By having smaller cells than normal cellular cells, CT2 phones can be smaller, cheaper and lighter. The first generation of these phones didn't do well, since they weren't smaller and lighter; there weren't many micro-cells and you couldn't receive an incoming call. See CT1, CT2 and CT2+.

Teleports The definition written by Gary Stix in the August 12, 1986 issue of Computer Decisions reads, "High bandwidth telecommunications distribution systems that allow major local users to obtain local, private services and long distance services. The most notable example is the New York Teleport," which is located on Staten Island. Teleports traditionally consist of two things — a fiber optic/coaxial cable network around a city and a collection of nearby satellite antennas. The cable network collects transmissions from larger customers and takes them to the antennas for shipping to and from distant offices. Teleport companies are now more successful as local communications companies than they are as long distance gateways. Which is understandable, since the cost of local calls has gone up, while the cost of long distance calls has gone down.

Telepresence In 1985 a team of researchers at NASA invented the notion of telepresence — projecting yourself into someone else's virtual reality. In one version of telepresence, according to Discover Magazine, a computer prompts a robot to mimic your movements. As you manipulate objects in your virtual world a robot somewhere else does the same thing to real objects. Telepresence will be especially useful for hazardous jobs like repairing a nuclear reactor or satellite. Or going on a blind date? See VIRTUAL REALITY.

Teleprinter A teletypewriter. Also called a telex machine.

Teleprocessing An IBM term for data communications.

Telesales British. Sales activities conducted by telephone in a planned and controlled manner.

Telescript Trademark name of General Magic's communications language technology. An interpreted, object oriented language that creates and manages intelligent agents as they move through a network of computers and personal communicators. Telescript is implemented as a portable interpreter, called Telescript engine. Programs written in Telescript are called scripts. General Magic says that it expects Telescript to become an industry standard, licensed by major suppliers of personal intelligent communicators, computers, communication networks, and information services. For more information, ask for a brochure called "Telescript Technology; The Foundation for the Electronic Marketplace," a General Magic White Paper, Mountain View, CA. See GENERAL MAGIC.

Telescripts A call center term. Rockwell's software vectors for configuring the many options of routing a call in progress, a call in queue, a call in the system, etc.

Teleservices 1. A generic term for services offered on phone links. Includes e-mail and facsimile features.
2. A product of SunSoft, a division of Sun MicroSystems, Mountain View, CA. According to SunSoft, "Sun's vision of the impact of widespread use of teleservices suggests that the computer workstation will become the new communications center, combining many existing communication media with new ones, while creating new paradigms for the expression and sharing of ideas. Information in the form of charts and pictures, schedules and plans, and audio and video will merge through application programs that provide a collaborative vehicle for decisions in the 1990s and beyond. The desktop will become the platform for a new set of productivity tools, seamlessly integrated into the critical business activities and methodology of today's companies, and providing a competitive edge for facing the global challenges of tomorrow. Individuals will gain new freedom in where they work and how they access information. And ideas will be communicated in more expedient and creative ways." SunSoft has developed Solaris Teleservices to provide a platform for next-generation workstation applications which leverage the benefits and capabilities of the telephone network and the commonplace use of it. Teleservices applications, according to SunSoft, include:
Desktop Teleservices. Workstation based telephone and answering machine applications allow users to efficiently plant, receive and manage telephone calls.
Remote Access. Users can place calls for their workstation from any telephone and access applications and data through DTMF signaling, or perhaps through speech, using a workstation's speech recognition capabilities.
Wide Area Networking. The ubiquity of telephone networks allow for the complete connectivity of all computers. Network links can be brought up or taken down on demand merely by placing or tearing down a telephone call.
The Solaris Teleservices Platform is called XTEL, which is a multilayered software architecture based on client server computing model. XTEL consists of four key components:
A client side library. Providing a high-level, object oriented application programming interface (API) to application programmers. Using the XTEL API, an application can place and retrieve telephone calls. The API library consists of a collection of C++ objects which is linked to applications that wish to use the systems teleservices resources.
A server. Providing multi-client and multi-device support, the server is the central point of contact for all teleservices, resource management and security are provided by the server. Communication between the XTEL API and the server occurs through the XTEL Server Protocol (XTELS).
One or more providers. Manages each telecommunication device connected to the system. The Teleservices server communicates with an XTEL provider using the XTEL provider protocol (XTELP).
A data stream multiplexor (Sun's spelling). The universal multiplexor (Umux) provides a uniform means for applications to access and share data channels associated with a telephone call. Umux is a streams pseudo-device driver used to connect data channels to applications. An XTEL application is linked with the XTEL API library, through which communicates with the server using the XTELS protocol. XTELS is a synchronous, symmetric messaging protocol using the Solaris loopback transport mechanism. The XTELS protocol is essentially the XTEL provider protocol with extensions to support multiple clients and multiple XTEL providers. For more information, see a document called Solaris Teleservices Architectural Overview, available from SunSoft, Mountain View, CA. See TELESERVICES API.

Teleservices API The Teleservices Application Program Interface (API) library defines the model and control mechanism to develop SunSoft Solaris Teleservices applications. The Teleservices API includes functionality to place a call, to register with the services to answer a call, to enable call control features, to send and detect DTMF tones, and to enable security and the sharing of calls between processes. See TELESERVICES.

Teletel Terminal Emulation Teletel is a popular character-based graphics emulation for communicating with the Minitel service, found primarily in France.

Teletex A new ITU-T standard for text and message transmission which is replacing Telex. Teletex operates at 2400 baud, about 50 times faster than telex. Teletex uses ASCII to encode its characters for transmission.

Teletext A data communications information service used to transmit information from remote data banks to viewers. It was transmitted over the air in the vertical blanking interval of the TV signal of the BBS (British Broadcasting Service). Teletext was originally designed for public consumption. It gave out weather information, sports results, headlines, etc. Teletext is proving somewhat more successful among corporations for the internal dissemination of information.

Teletraffic Optimizer Program Derives data by processing actual calls instead of using an analytical model based on estimates or summaries.

Teletraining Education and training through telecommunications.

Teletype A specific type of teletypewriter.

Teletyper Input Method Teletyper Input Method (TIM) is specially designed for using pushbutton phone or mobile phone to input text, command and instruct your PC through the public telephone network. Teletyper Telephone-Input Method (TIM) and Teletyper Plus use three base keys on the phone pad with the combination of other keys to form alphabets, symbols, utility and control functions in order to command your PC remotely. Information could then be sent to fax, pager or voice output.

Teletyper Input Method (TIM)

*1 = A	01 = B	#1 = C	#9 = space
*2 = D	02 = E	#2 = F	** = Enter
*3 = G	03 = H	#3 = I	00 = Zero
*4 = J	04 = K	#4 = L	## = Backspace
*5 = M	05 = N	#5 = O	1 to 9 press 1 to 9
*6 = P	06 = Q	#6 = R	
*7 = S	07 = T	#7 = U	
*8 = V	08 = W	#8 = X	
*9 = Y	09 = Z		

Teletypewriter TTY. A telegraph device capable of transmitting and receiving alphanumeric information over communications channels. It may also contain a keyboard similar to that of a typewriter or computer but usually with fewer keys. See TELETYPE.

NEWTON'S TELECOM DICTIONARY

Teletypewriter Control Unit TCU. A device that serves as the control and coordination unit between teletypewriter devices and a message switching center when controlling teletypewriter operations.

Teletypewriter Exchange Service TWX. A switched teletypewriter service in which suitably arranged teletypewriter stations are provided with lines to a central office for access to other such stations. TWX and Telex are commercial teletypewriter exchange services. They are currently both owned by AT&T. These days their revenues are in decline. A computer with a modem is a lot faster than TWX or telex.

Teletypewriter Signal Distortion The shifting of signal-pulse transitions from their proper positions relative to the beginning of the start pulse. The magnitude of the distortion is expressed in percent of a perfect unit pulse length.

Television A standard TV channel today fits into a frequency bandwidth of six megahertz.

Telewatch Nynex's Telewatch is a maintenance service which periodically tests Nynex analog DID trunks for transmission loss, noise and trouble.

Telework The combination of computer and telecommunications technology which enables office workers to work at home or away from the main office on a part-time or full-time basis.

Teleworker A person who works from his home or some place distant from his company's office. A teleworker may send his completed work in and pick his new work up via a modem in his PC. A teleworker may also be on the phone at home answering calls on behalf of his company and entering the results of those calls (i.e. reservations on airlines, orders for catalogs) in on a PC connected by phone lines to his company. He may use one phone line, like an ISDN BRI line or he may simply use two analog phone lines — one for talking on and one for PC's data. Or he may simply use one analog phone line and use a protocol such as VoiceView.

Telex A worldwide switched message service. Telex service is offered in the United States by the Western Union Telegraph Company, MCI, ITT, RCA, FTCC and TRT. Telex has one gigantic advantage: Overseas it's very popular and widely used. Contacting overseas businesses by telex is often far more reliable and faster than contacting them by telephone. Telex is good for overseas time zone differences because you can send a message to an unattended telex machine. It also delivers a printed record. Telex is relatively inexpensive usually costing a little less than a phone call. Telex has one disadvantage: It's very slow and not very accurate with virtually no data communications error checking procedures. Telex is being rapidly displaced by faster, more accurate forms of data communications, including the public packet switched networks and the various electronic mail services, and most recently by massive competition from low-cost facsimile machines. See TELETEX.

Telex Access Unit TLXAU. An X.400 AU (Access Unit) serving Telex users.

Telnet A program that lets you connect to other computers on the Internet. The process by which a person using one computer can sign on to a computer in another city, state or country. Telnet is the terminal-remote host protocol developed for ARPAnet. Using Telnet, you can work from your PC as if it were a terminal attached to another machine by a hard-wired line. The format of the the telnet command is: telnet address.domain or telnet address.domain port#

These days, most users are insulated from TELNET by GUI browsers, such as Mosaic.

Telnet Port The port address on a computer which supports remote telnet access. Normally port 23 is the default telnet port.

Telpak A discontinued AT&T service that gave large customers discounts on purchases of multiple analog private lines. Telpak was to discourage users from building their own private microwave systems. Users who bought Telpak, however, were not allowed to resell any of the circuits, though they were allowed to share them. The FCC ruled Telpak as being discriminatory against competition. (It was too cheap.) It may still exist on an intrastate basis in some states. Telpak typically came in bundles of 12, 24, 60 and 240 voice lines. The bigger the bundle, the cheaper the per circuit cost.

Telstar 1 Telstar 1 was the world's first active communications satellite. It was launched on July 10, 1962. (Sputnik was launched on October 4, 1957.) There is some argument about Telstar's claim to fame as the first. The engineers at the RCA Astro-Electronics Division, Princeton NJ (now Martin-Marietta Aerospace) claimed that they launched and successfully used a satellite to broadcast the coronation of the pope a little earlier than Telstar. But it was Telstar that got all the fame and glory, and that RCA engineers don't deny that.

Telstra OTC Telstra OTC is Australia's largest local and overseas phone company. It was formed in February of 1992 by the merger of national (Telecom) and international (OTC) carriers. According to a press release of June 1993, the company has 76,000 employees, 8 million customers and 120 products and services. It claims to be the sixth largest user of Intelsat and the third largest owner of submarine cable in the world.

TEMA Telecommunication Engineering and Manufacturing Association limited.

Temperature Rating The maximum temperature at which the insulating material may be used in continuous operation without loss of its basic properties.

Tempest Unclassified name referring to the investigation, study, and control of compromising emanations from electrical and electronic equipment. Devices which are tempest-secure mean they do not send emanate electromagnetic signals which can, potentially, be received by others, i.e. enemies.

Template 1. A voice processing term. A pattern of information as a function of time, which is intended to represent an entire word.
2. A Northern Telecom Norstar definition: A system wide setting assigned during System Startup. The most important effects of a template are the number of lines assigned to the telephones, and the assignment of Line Pool Access. Templates will also assign other system wide defaults, such as Prime Line and Ringing Line assignment. It is important to understand that a template is only provided as a convenience, and that any settings effected by the template can be changed.

Temporal Coding Compression that is achieved by comparing frames of video over time to eliminate redundancies between frames.

Temporary Signaling Connections On August 14, 1995, AT&T announced Temporary Signaling Connections, which it billed as the first service that lets banks, retail outlets and other data-intensive businesses link their Software Defined Network locations together on demand using virtual connections created in AT&T's national signaling networks. Businesses can use Temporary Signaling Connections to verify credit card transactions, update inventory databases, exchange data with automatic cash machines. The service uses a portion of the D channel capacity of an ISDN PRI channel and passes information to other ISDN PRI locations.

Temporary Station Disconnection Allows the attendant to completely remove selected phones from service at any time on a temporary basis.

Tenant Service Some businesses acquire a telephone system too large for their needs so they sell parts of the service to smaller offices in their own building or in the surrounding community. There are two ways to make money on tenant service — renting phone equipment

600

or re-selling long distance lines. There's more money on re-selling long distance lines.
Tensile Strength A term denoting the greatest longitudinal tensile stress a substance can bear without tearing apart or rupturing.
Tera- A million million, or a thousand giga-. See TERABYTE.
Terabyte A million million bytes. One terabyte is equal to 10 to 12th. See TERA-. A megabyte is one million bytes. A gigabyte is one thousand million bytes. A terabyte is one thousand gigabytes, or one million megabytes (ten to the 12th power). A petabyte is equal to 1,000 terabytes. An exabyte is equal to 1,000 petabytes. Here is a summary:
  MH = Megabyte (10 to the 6th power)
  GH = Gigabyte (10 to the 9th power)
  TH = Terabyte (10 to the 12th power)
  PH = Petabyte (10 to the 15th power)
  EB = Exabyte (10 to the 18th power)
Teraflop A trillion floating point instructions per second. A measure of a computer's speed.
Terahertz THz. A unit denoting one trillion (10 to the 12th) hertz. See TERA-.
Terminal 1. The point at which a telephone line ends, or is connected to other circuits of a network.
2. An input/output device for communicating with computers. Typically has a keyboard and a CRT (TV screen) display. See TERMINALS.
Terminal Adapter The terminal adapter is a protocol converter (little black box) that adapts PCs, workstations and other equipment to the peculiar world of ISDN. Some ISDN devices include built-in terminal adapters. Some don't. In technical terms, a terminal adapter is an interfacing device employed at the "R" reference point in an ISDN environment that allows connection of a non-ISDN terminal at the physical layer to communicate with an ISDN network. Typically, this adapter will support standard RJ-11 telephone connection plugs for voice and RS-232C, V.35 and RS-449 interfaces for data. See also ISDN.
Terminal Block A device used to connect one group of wires to another. Usually each wire can be connected to several other wires in a bus or common arrangement. A 66-type block is the most common type of connecting block. It was invented by Western Electric. Northern Telecom has a terminal block called a Bix block. There are others. A terminating block is also called a connecting block, a punchdown block, a quick-connect block, a cross-connect block. A connecting block will include insulation displacement connections (IDC). In other words, with a connecting block, you don't have to remove the plastic shielding from around your wire conductor before you "punch it down."
Terminal Configuration The functional interconnection of the components of a terminal. For example, a keyboard-printer may be configured to transmit keystrokes without printing them. Printing is only performed on data retrieved from the communication line. Terminals with multiple components can be configured in a variety of ways.
Terminal Emulation An application that allows an intelligent computing device such as a PC to mimic or emulate the operation of a dumb terminal for communications with a mainframe or minicomputer. It does this with special printed circuit boards inserted into its motherboard and/or special software. For example, TELECONNECT uses the communications software program called Crosstalk to emulate a DEC VT-100, a Digital Equipment Corporation VT-100 terminal. We do this because emulating a DEC VT-100 works better with certain software programs we call up remotely.
Terminal Equipment Terminal Equipment usually refers to the telephones and other equipment at the end of telephone lines. See also CPE.
Terminal Equipment Type 1 TE1. In Integrated Services Digital Network (ISDN) technology, TE1 is a a type of terminal compatible with ISDN.
Terminal Equipment Type TE2. In Integrated Services Digital Network (ISDN) technology, a type of terminal that must be connected to ISDN via a specially designated point, normally an RS 232 or RS 449 interface.
Terminal Hunt Group Also called Terminated Hunt Group. It's another name for a "top down" hunt group. You have a bunch of phone lines. They are in a "hunt" group which means that, if one is busy, the switch sends the call to the next available line in the hunt group. In a terminal hunt group, the switch always starts at the top of the hunt group and goes down, searching for the first available line from the top. This contrasts with a circular hunt group, where the switch remembers the last line it connected and, starting there, hunts down to the next available line, searching basically in a circle. A terminated or "top down" hunt group puts more calls on the first lines in the group. A circular hunt group tends to distribute the calls evenly. There are reasons why you might choose one type of hunt group over another. You might choose to evenly distribute your calls over a bunch of humans answering incoming calls or you might choose to send the calls top down to a voice mail system and watch the usage statistics carefully to tell if you need more or fewer lines and more or fewer voice cards.
Terminal Impedance The impedance as measured at the unloaded output terminals of transmission equipment or a line that is otherwise in normal operating condition. The ratio of voltage to current at the output terminals of a device, including the connected load.
Terminal Node In IBM Corp.'s Systems Network Architecture (SNA), a network device that cannot be programmed by the user.
Terminal Number TN. A Northern Telecom definition. The physical address of a device (such as, telephone set, a truck, and attendant) on the Meridian 1 PBX. The TN is composed of the loop, shelf, card and unit IDs.
Terminal Repeater A repeater for use at the end of a trunk line.
Terminal Server A small, specialized, networked computer that connects many terminals to a LAN through one network connection. Any user on the network can then connect to various network hosts. A terminal server has a single network interface and several ports for terminal connections. The advantage of a terminal server is it allows many terminals to be connected to a host via a single existing LAN cable, rather than a variety of point-to-point cables. Digital Equipment's DECnet makes heavy use of terminal servers.
Terminal Table An ordered collection of information that identifies each line, phone, component or application program from which a message can be sent.
Terminals The screws or soldering lugs to which an external circuit can be connected.
Terminals To Long Distance Operator Commonly known as "toll terminals", they provide special trunks directly to the long distance telephone company operators. Upon completion of long distance calls, the toll operator will ring the attendant (or hotel operator) and give them "time and charges" for the phone call just ended.
Terminate 1. To connect a wire conductor to something, typically a piece of equipment.
2. To end one's telecommunications service or equipment rental.
Terminated 1. The condition of a wire or cable pair which is connected to (terminated on) binding posts or a terminal block.

2. The condition of a circuit connected to a network which has the same impedance the circuit would have if it were infinitely long.

Terminated Line A telephone circuit with a resistance at the far end equal to the characteristic impedance of the line, so no reflections or standing waves are present when a signal is entered at the near end. Compare with bridge tap.

Terminating Channel The name for the circuit in a private line channel that connects a local central office with the CBX/PBX or telephone instrument at the customer's premises.

Terminating Office The switching center (i.e. the central office) of the person you're calling (the "called party").

Terminating Resistor A grounding resistor placed at the end of a bus, line, or cable to prevent signals from being reflected or echoed. Sometimes shortened to terminator.

Termination Termination involves the placement of impedance matching circuits on a bus to prevent signals from being reflected or echoed.

Termination Restriction Prevents a user from receiving any calls on the phone line. A DID call to the restricted termination routes to an attendant, an announcement or intercept tone at customer option. All other calls route to intercept tone.

Termination Of Service The end of service of a line or equipment. All pursuant to the regulations set forth in the tariff.

Terminator Some communications facilities — e.g. local area networks — are bus configurations. This means one long piece of cable with workstations connected along the way, typically with "T" connectors. For a network to work properly, you need to place resistance at the end of the cable that serves to absorb the signal on the line. A thin wire Ethernet typically requires a 50 ohm resistance at either end of the bus. You can buy these Ethernet terminators already included in a connector.

Terminus A device used to terminate an optical fiber that provides a means to locate and contain an optical fiber within a connector.

Terrestrial Long distance facilities which are entirely on land and do not use satellites. This includes microwave, coaxial cable, optical fiber, normal cable, etc. There are reasons to prefer terrestrial facilities over satellite facilities:

1. No echo or delay in voice conversation. Some people find satellite conversations disturbing because of the delay; and 2. No significant reduction in data throughput. Most data communications protocols send their data in "chunks" and require an acknowledgement from the other end when one chunk has been received before the next chunk can be sent. When it takes a long time for an acknowledgement to be received (as in a satellite circuit), the effective throughput of data becomes very slow.

When you're trying to send one one-way signal to many locations, satellites often do a better and cheaper job.

Tesla, Nikola Nikola Tesla is regarded as one of the most mysterious and least recognized scientific pioneers in modern history. He is commonly associated with high-frequency electrical devices, radio transmission, and the invention of the multi-phase alternating current system in use today. Born on July 9, 1856 in Smiljan, Lika, Croatia, Yugoslavia, Tesla was the fourth of five children and son of a reverend of the Serbian Orthodox Church. Tesla was educated at the polytechnical school at Graz, Austria (1875), where he acquired an interest in the study of electrical engineering and mathematics. Tesla was dismayed at the inefficient design of DC motors. While working as a telegraph operator in Budapest (1881) and later as a telephone engineer for Edison in Paris (1882), he developed a multi-phase alternating current power system, named the Tesla Polyphase System. Other engineers had attempted AC designs modeled after direct current systems with a single circuit, but none were ever successful. Tesla's multi-circuit system included new polyphase induction motors, dynamos, and transformers, the patents to which were purchased in 1888 by the Westinghouse Electric Company from the Tesla Electric Company.

Tesla's 60 cycle AC Polyphase System was the primary competitor to the Continental Edison Company's DC Current System, and caused Thomas Edison to personally wage a massive propaganda campaign again Tesla and the "dangers of alternating current". However, AC could travel over hundreds of miles and at much higher voltages, while DC traveled only much shorter distances and required a generator every two miles. Despite Edison's efforts, the war of the currents was won by Tesla. The Chicago World's Fair in 1893 featured the first electrically powered pavilion, designed entirely by Tesla using his polyphase system and financed by Westinghouse. This pavilion featured Tesla demonstrating many high-energy devices, including the wireless transmission of energy using a tuned circuit, a device that would come to be known as radio. Guglielmo Marconi used several of Tesla's patents in his 1901 "invention" of the wireless radio transmitter. Marconi's initial patent claims for this device were rejected based on Tesla's prior patents (645,576 and 649,621 granted in 1900) for the wireless transmission of energy. Marconi's later patent claims for a signal communications device were granted. Despite a lawsuit filed by Tesla in 1915 and other efforts, history has erroneously painted Marconi as the inventor of the radio. Some redemption to this injustice occurred when the Supreme Court ruled on June 21, 1943 in the case of Marconi Wireless Telegraph Company of America vs. the United States that Tesla's radio patented predated those owned by Marconi. In effect, this ruling declared Tesla the inventor of radio; it came five months after his death.

Tesla continued to experiment with high-frequency energy. He believed that energy in phase with the natural vibrations of thunderstorms and the Earth (7.68 Hz) could be broadcast to anyplace. Using this theory, he received backing from J.P. Morgan to build a world-wide telegraphy system. Tesla also believed that electrical power was present everywhere, in unlimited quantities, and free for the taking. Tapping into such a natural source of energy could replace all other fuels. The global wireless transmission of power, however, did not interest Morgan, who did not see the benefit in providing free electricity to humanity. Tesla created hundreds of inventions and improvements during the course of his life, including the telephone repeater, wireless communications, radio, antennas, ground connections, aerial ground circuits with inductance and capacitance, tuned circuits, emitters and receivers tuned to resonance, the electronic tube, fluorescent lighting, the electromechanical audio speaker, AND and OR logic gates, radio remote-control, robotics, radar, and diathermy.

Tesla's patented inventions includes a robotic submarine (613,809), vertical take-off and landing aircraft (1,655,114), disk (bladeless) turbine engine (1,329,559), ozone generator (568,177), electro-dynamic induction lamp (514,170), and superconduction (685,012). His more fantastic concepts include anti-gravity propulsion and a thought photography machine. The more than 700 patents Tesla was awarded during his lifetime represent only a fraction of his total number of inventions and discoveries. Tesla often didn't bother to patent many of his inventions (he received his last patent in 1928) and often failed to document his work. Tesla died on January 7, 1943 secluded in his New York City apartment at the age of 86. The world-renown Tesla was highly intelligent, very highly strung, neurotic, charismatic, germ-phobic, and always very well-dressed. Over his lifetime his financial backers were a diverse group which included George Westinghouse, J.P. Morgan, and even Thomas Edison. His friends and supporters, an equally diverse group, included Albert Einstein, Samuel Clemens, and Eleanor Roosevelt. His manor and mystery as depicted by the press inspired the creation of evil comic book geniuses that did battle with Superman and Captain Marvel using a myriad of strange inventions and energy devices. Tesla's secret work on "death ray" energy devices fueled much of this, as did the U.S. government's confiscation of many of his inventions after his death.

Test And Validation Physical measurements taken to verify conclusions obtained from mathematical modeling and analysis.

Test Antenna An antenna of known performance characteristics used in determining transmission characteristics of equipment and associated propagation paths.

Test Board A switchboard equipped with testing apparatus.

Test Center Equipment for detecting and diagnosing faults and problems with communications lines and the equipment attached to them. If centralized, a facility where a network manager or technician can gain access to (almost) any circuit in a network for the purpose of running diagnostic testing. Also called a network control center.

Test Desk A desk equipped with equipment to test and repair subscriber lines. See also TEST CENTER.

Test Repeatability This definition courtesy Steve Gladstone, author, "Testing Computer Telephony Systems," (available from 212-691-8215) says that a key component in any successful test program is the ability to repeat tests simply and quickly. If problems are found, tests must be rerun, both to help recreate and document the problem, as well as to verify the bug fix. Repeating tests can be very time consuming as many tests will change the state of either the test system or the computer telephony system under test. For example, checking that messages are correctly deleted means starting the test from a known state where messages are in a mailbox. When the test is completed usually there are fewer messages left than when the test started. Repeating the test may therefore necessitates reinitializing the computer telephony system to the state before the test started.

Test Set A telephone handset with extra electronics designed to test telephone circuits. Also called a butt set, since it typically hangs on the technician's tool belt near the wearer's butt... Well, that's one explanation. Another is that it's called a "butt" set because it allows the use to "butt in" to a conversation and listen to its quality, etc.

Test Shoe A device that is applied to a circuit at a distributing frame to gain test access to circuit conductors.

Test Tone A tone used to find trouble on phone lines. A more technical explanation: A tone sent at a predetermined level and frequency through a transmission system to facilitate measurement and/or alignment of the gains and/or losses of devices in the transmission circuit.

Test, Friendly Busy A test to see if a line is busy. The subscriber does not know the line is being tested. Such a test is usually performed if someone calling that number requests it.

Tetrode A four-element vacuum tube, consisting of filament (or cathode), grid, screen grid and plate.

Text Transmitted characters which make up the body of a message.

Text Enriched
The successor to MS-DOS text/richtext, is a simple text markup language for MIME that lets you mark up the document (using commands enclosed in angle brackets) without making the text unreadable to someone without the software to interpret it. See MIME.

Text File A file containing only letters, numbers, and symbols. A text file contains no formatting information (like bolding and underlining and type fonts and sizes), except possibly line feeds and carriage returns. A text file is an ASCII file. A text file can be read by every word processor and editor. A text file the lowest common denominator in the word processing world. I wrote this file with an editor called ZEdit, which produces only text files. I did this because this dictionary has to be sent to a Macintosh for "typesetting" and to a DEC for distribution on CD-ROM. And a text file is the form both most easily recognize.

Text-To-Speech Synthesis TTS. Technologies for converting textual (ASCII) information into synthetic speech output. Used in voice processing applications requiring production of broad, unrelated and unpredictable vocabularies, e.g., products in a catalog, names and addresses, etc. This technology is appropriate when system design constraints prevent the more efficient use of speech concatenation alone. See SPEECH CONCATENATION.

TFT Thin Film Transistor. A display technology which uses modern active matrix technology that operates by assigning a tiny transistor to each pixel, making it possible to control pixels independently of each other. TFT screens are very fast, have a high contrast ratio and a wide viewing area.

TFTP Trivial File Transfer Protocol. A simplified version of FTP that transfers files but does not provide password protection or user-directory capability. It is associated with the TCP/IP family of protocols. TFTP depends on the connectionless datagram delivery service, UDP.

TFTS Terrestrial Flight Telephone System.

TGB Trunk Group Busy.

TGC Transmission Group Control in IBM's SNA.

TGW Trunk Group Warning.

TH Transmission Header, an SNA term.

THD+N Total Harmonic Distortion Plus Noise. A measure of the audio clarity of a voice system. The best measure is 0%., or close. Pretty bad is 25%. The average PC voice card is around 10%.

The Grand Alliance A consortium of seven organizations which produced a workable version of HDTV — High Definition TV. The companies are AT&T, General Instrument Corporation, The Massachusetts Institute of Technology, the David Sarnoff Research Center, Philips Consumer Electronics, Thompson Consumer Electronics and the Zenith Corporation.

The Phone Company Also known by the initials TPC. In the 1966 James Coburn movie "The President's Analyst", the evil worldwide conspiracy was run by TPC, which turned out to be The Phone Company. Some people believe the movie was not fiction.

Thermal Noise Noise created in an electronic circuit by movement and collision of electrons.

Thermionic Emission The emission of electrons or ions under the influence of heat, as in a vacuum tube cathode.

Thermistor A device made from mixtures of metal oxides that exhibits large negative coefficient of resistance changes as the temperature increases.

Thermocouple Two dissimilar wires joined together that generate a voltage proportional to temperature when their junction is heated.

Thermoplastic Material that will soften and distort from its formed shape when heated above a critical temperature peculiar to the material.

Thermoset A plastic material which is crosslinked by a heating process known as curing. Once cured, thermosets cannot be reshaped.

Thermostat Device responding to temperature changes, used to control heating and cooling equipment.

THF Tremendously High Frequency.

THI Telephone Headset Integrator. A new form of headset manufacturer who will make headsets that do new tasks, like take a phone off hook without physically having to lift the receiver off the phone.

NEWTON'S TELECOM DICTIONARY

Thick Ethernet Cable Thick Ethernet cable is 0.4-inch diameter, 50-ohm coaxial cable. Thick Ethernet cable can be bought in pre-cut lengths, with standard N-Series male connectors installed on each end. It also is available in bulk cable without connectors. Any of the following types of connectors will work:
Belden 9880 or Belden 89889
Montrose CBL5688 or Montrose CBL5713
Malco 250-4315-0004 or Malco 250-4314-0003
Inmac 1784 or Inmac 1785
Thicknet Jargon used to describe thick Ethernet coaxial cable. See THICKWIRE.
Thickwire 0.4 inch diameter, 50-ohm, Ethernet IEEE 802.3 coaxial cable. See also THICK ETHERNET CABLE and THINWIRE.
Thin Client Clients are devices and software that request information. Client is a fancy name for a PC on a local area network. It used to be called a workstation. Now it is the "client" of the server. Clients come in two varieties — Fat and Thin. Here's a definition of Thin Client, courtesy of Oracle Corporation, writing in early 1994: "The thin client modem stores and processes more data on the server, but keeps the user interface and application functions on the client device. Example: a television with a set-top box, Apple's Newton Personal Digital Assistant, or a low-end PC. See also CLIENT, CLIENT SERVER, CLIENT SERVER MODEL, FAT CLIENT, MAINFRAME SERVER and MEDIA SERVER.
Thin Ethernet A coaxial (0.2-inch, RG58A/U 50-ohm) that uses a smaller diameter coaxial cable than standard thick Ethernet. Thin Ethernet is also called "Cheapernet" due to the lower cabling cost. Thin Ethernet systems tend to have transceivers on the network interface card, rather than in external boxes. PCs connect to the Thin Ethernet bus via a coaxial "T" connector. Thin Ethernet is now the most common Ethernet coaxial cable, though twisted pair is gaining. Thin Ethernet is also referred to as ThinNet, ThinWire or Cheapernet. See also 10BASE-T.
Thinnet Jargon used to describe thin Ethernet coaxial cable. Referred to ThinNet, ThinWire or Cheapernet.
Thinwire The 75-ohm coaxial cable listed in IEEE 802.3 specifications and used in some Ethernet local area network installations.
Third Party Call Any call charged to a number other than that of the origination or destination party. It's not a good idea to let your employees make third party calls to one or more of your phone numbers. Best to ask them to place the calls on their personal phone credit cards. This way, they will spend a modicum of time justifying their exorbitant phone calls.
Third Party Call Control A call comes into your desktop phone. You can transfer that call. When the phone call has left your desk, you can no longer control it. That is called First Party Call Control. If you were still able to control the call (and let's say, switch it elsewhere) that would be called Third Party Call Control. Some Computer Telephony links allow only first party call control. Some allow third party as well. If you control the switch — the PBX or the ACD — you will typically have Third Party Call Control. If you just control the desktop, you'll typically have only First Party Call Control. There is no such animal as Second Party Call Control.
Third Wire Tap The activating of a telephone handset microphone by using a third wire, thus bypassing the hook switch.
Thread 1. A sequence of computing instructions that make up a program. A multi-threaded process can have multiple threads, each executing independently and each executing on separate processors. A multi-threaded program, if running on a computer with multiple processors, will run much faster than a single-threaded program running on a single processor machine. Windows NT is scheduled to be the first generally available multi-threaded PC operating system.
2. A topic on a Usenet group; e.g., "Baby Bop's voice" as a subcategory, or topical thread, that readers may follow in the alt.barney.die.die newsgroup.
In its Window 95 resource manual, Microsoft defined thread as
1. An executable entity that belongs to a single process, comprising a program counter, a user-mode stack, a kernel-mode stack, and a set of register values. All threads in a process have equal access to the processor's address space, object handles, and other resources. In Windows 95, threads are implemented as objects.
2. A collection of electronic mail messages organized chronologically and hierarchically to reflect the flow of the discussion.
Threaded A method of presenting articles within a newsgroup in a way that shows which articles refer to which other ones.
Threaded Code Threaded code is also known as Threaded Pseudo Code, or threaded p-code. It was first popularized in a software language called Forth. Eventually Microsoft fixed Forth and changed it into Basic. See THREAD, THREADS and MULTITHREADING.
Threads Individual processes within a single application.
Threat Analysis Examination of all actions and events that might adversely affect a system, a network or an operation.
Three Finger Salute Ctrl Alt Delete.
Three Slot An obsolete pay phone that is identified by three separate coin slots.
Three-Way Calling A local phone company feature that allows a phone user to add another user to an existing conversation and have a three party conference call.
Three-Watt Booster Optional equipment for use with a cellular phone car-mounting kit that raises a portable phone's maximum transmission power from 0.6 watts to 3.0 watts.
Three-Way Conference Transfer A PBX feature. By depressing the switch hook, a user can dial another extension and either hang up and transfer the call, get information from the called party and then resume the first call or bridge all parties together for a three-way conference call.
Three-Way-Handshake The process whereby two protocol entities synchronize during connection establishment.
Threshold The minimum value of a signal that can be detected by the system or sensor under consideration.
Threshold Of Pain 1. The present price of local telephone service.
2. Unbearable noise.
Through Dialing Allows the attendant on a phone system to select a trunk and pass dial tone to a restricted phone user so that user may directly dial an outside call.
Throughput The actual amount of useful and non-redundant information which is transmitted or processed. Throughput is the end result of a data call. It may only be a small part of what was pumped in at the other end. The relationship of what went in one end and what came out the other is a measure of the efficiency of that communications link — a function of cleanliness, speed, etc.
Thunking A Microsoft Windows 95 term for the transformation between 16-bit and 32-bit formats, which is carried out by a separate layer

in the VM. the fundamental idea is to make older 16-bit programs work better in newer 32-bit operating systems.
THZ Terahertz (10 to the 12th power hertz). See also SPECTRUM DESIGNATION OF FREQUENCY.
TIA 1. Telecommunications Industry Association. A new Washington lobbying and trade association, the result of the merger of the USTA (United States Telephone Association) and the EIA (Electronic Industries Association. TIA is half owner of an annual trade show called Supercomm held in the U.S. The other half owner is the U.S. Telephone Association (USTA). According to the TIA's own literature, it "is a full-service trade organization with membership of 600 large and small companies which provide communications and information technology products, materials, systems, distribution services and professional services in the United States and countries around the world. TIA represents the telecommunications industry in association with the Electronic Industries Association." www.tiaonline.org The TIE is at 703-907-7700. See also EIA/TIA-232-E.
2. Thanks In Advance.
3. The Internet Adapter, available from SoftAware, Inc. in Marina Del Ray, CA is software used to adapt a dial-up connection on Unix shell accounts.
TIA/EIA The US Telecommunications Industries Association and Electronics Industries Association which have merged. Now just called the Telecommunications Industry Association.
TIA/EIA IS-95 See TIA/EIA IS-96.
TIA/EIA IS-96 IS-96 is the Speech Service Option Standard for Wideband Spread Spectrum Digital Cellular System, a new vocoder standard. The standard supports IS-95, the North American spread spectrum digital standard based on Code Division Multiple Access (CDMA), which was published in July, 1993. The engineering effort to produce IS-96 was done in TIA Technical Subcommittee TR-45,5, Wideband Spread Spectrum Digital Technologies Standards. The specific vocoder described in IS-96 is a variable rate implementation, chosen because of its combination of high voice quality and low average transmission rate. The IS-96 vocoder provides variable vocoder rates depending on voice activity. This variability typically results in an average transmission rate of under 4 Kbps and yet provides for reasonably quality voice transmission. The IS-96 also provides a variable noise threshold which tracks and eliminates much of the background noise from the speaker's environment. You can get copies of both standards from Global Engineering Documents, 15 Inverness Way East, Englewood, CO 80112.
TIC 1. An AT&T term for a digital carrier facility used to transmit a DS-1 formatted digital signal at 3.152 Mbps.
2. Token-Ring Interface Coupler. An IBM device that allows a controller or processor to attach directly to a Token-Ring network. This is an optional part of several IBM terminal cluster controllers and front-end processors. See TIC CARD.
3. See TELEPHONY INTERFACE CONTROL.
TIC Card Token Ring Interface Coupler is the IBM name for a variety of token ring adapter cards used to connect IBM controllers to token ring LANs. See TIC.
Tick Tone Clicking noise heard on some PBX lines indicating that the digits dialed will shortly be repeated to the central office.
Ticker A one-way telex machine used to typically report stock or commodity prices. The machine prints on ticker tape, which is about one inch wide and perfect for throwing out windows at passing celebrities. Thus the term "Ticker Tape Parade."
Ticket A telephone industry term for a filled-out form, usually a form for billing someone for a call. There are all sorts of tickets, including ones on paper, ones on computer and ones automatically generated without human intervention.
TID Terminal Identification. Used for all National-1 ISDN services, a two digit number between 00 and 62 entered after the SPID.
TIE 1. Joining cables and/or wires together.
2. Time Interval Error.
3. Trusted Information Environment, an encryption scheme.
TIE/Communications A PBX and key system distributor to end users based in Seymour, CT. TIE was one of the original manufacturers of interconnect equipment in the US. TIE stood for Telephone Interconnect Equipment. The company fell on hard times in the late 1980s. It ceased manufacturing equipment and ceased acting as a wholesaler of others' equipment. Now it's simply a distributor and Nitsuko, its primary Japanese supplier, has taken over the sale of its own equipment directly.
Tie Down Verb meaning to terminate a wire on a main, intermediate or satellite distribution frame.
Tie Line A dedicated circuit linking two points without having to dial the normal phone number. A tie line may be accessed by lifting a telephone handset or by pushing one, two or three buttons.
Tie Trunk A dedicated circuit linking two PBXs.
Tie Trunk Access Allows a phone system to handle tie lines which can be accessed either by dialing a trunk group access code or through the attendant.
TIFF Tag Image File Format. TIFF provides a way of storing and exchanging digital image data. Aldus Corp., Microsoft Corp., and major scanner vendors developed TIFF to help link scanned images with the popular desktop publishing applications. It is now used for many different types of software applications ranging from medical imagery to fax modem data transfers, CAD programs, and 3D graphic packages. The current TIFF specification supports three main types of image data: Black and white data, halftones or dithered data, and grayscale data. Some wags think TIFF stands for "Took It From a FotograF." It doesn't.
TIES Time Independent Escape Sequence, a feature of modems.
Tiger Team A group hired by an organization to defeat its own security system to learn its weaknesses.
Tight Jacket Buffer A buffer construction which uses a direct extrusion of plastic over the basic fiber coating. This construction serves to protect the fiber from crushing and impact loads and to some extent from the microbending induced during cabling operations. See also LOOSE TUBE BUFFER.
Tightly Coupled Describing the interrelationship of processing units that share real storage, that are controlled by the same control program and that communicate directly with each other. Compare with loosely coupled.
Tightly Coupled CPUs Term used to describe multiple-processor computers in which several processors share the same memory and bus.
TIM Teletyper Input Method. See TELETYPER INPUT METHOD.
Time-based Authoring Tool A multimedia creation tool that uses time as a metaphor for building a project. Generally, objects are set up to happen at a certain time in a project, rather than in a certain place.
Time Assignment Speech Interpolation TASI. A voice telephone technique whereby the actual presence of a speech signal acti-

vates circuit use. The result is clipping of the first bit of the speech, but more efficient use of the transmission facility. TASI is used on expensive circuits, such as long submarine cables. See TASI.

Time Congestion The time resources (outgoing trunks) are busy.

Time Divert To Attendant A system feature which automatically transfers a phone to the attendant if the phone has been left off-hook too long.

Time Diversity A method of transmission wherein a signal representing the same information is sent over the same channel at different times. Often used over systems subject to burst error conditions and with the spacing adjusted to be longer than an error burst.

Time Division Controller TDC. A device which commands functions, monitors status and connects channels of TDM cards.

Time Division Multiple Access TDMA. A technique originated in satellite communications to interweave multiple conversations into one transponder so as to appear to get simultaneous conversations. A variation on TASI. A technique now used in cellular and other wireless communications. See TDMA.

Time Division Multiplex TDM. A technique for transmitting a number of separate data, voice and/or video signals simultaneously over one communications medium by quickly interleaving a piece of each signal one after another. Here's our problem. We have to transport the freight of five manufacturers from Chicago to New York. Each manufacturer's freight will fit into 20 rail boxcars. We have three basic solutions. First, build five separate railway lines from Chicago to New York. Second, rent five engines and schlepp five complete trains to New York on one railway track. Or, third, join all the boxcars together into one train of 100 boxcars and run them on one track. The train might look like this: Engine, Boxcar from Producer A, Box Car from Producer B, Producer C, Producer D, Producer E, and then the order begins again...Boxcar from Producer A, Producer B...Moving one large train of 100 boxcars is likely to be cheaper and more efficient than moving five smaller trains each of 20 boxcars on five separate railway tracks. Time Division Multiplexing, thus, represents substantial savings over have five separate networks (five separate tracks) and sending five separate transmissions (five separate trains).

This is what Time Division Multiplexing is all about. And the analogy is perfect. Take one large train (fast communications channel) and interleave pieces (boxcars) from each conversation one after another. If you do this fast enough, you'll never notice you've broken the conversations apart, moved them separately, and then put them back together at the distant end. In TDM, you "sample" each voice conversation, interleave the samples, send them on their way, then reconstruct the several conversations at the other end. There are several ways to do the sampling. You can sample eight bits (one byte) of each conversation, or you can sample one bit. The former is called word interleaving; the latter bit interleaving. The basic goal of multiplexing — whether it be time division multiplexing, or any other form

— is to save money, to cram more conversations (voice, data, video or facsimile) onto fewer phone lines. To substitute electronics for copper. See also the following three definitions.

Time Division Multiplexer TDM. A device which derives multiple channels on a single transmission facility by connecting bit streams one at a time at regular intervals. It interleaves bits or characters from each terminal or device using the time. See TIME DIVISION MULTIPLEX.

Time Division Signaling Signaling over a time division multiplex system in which all voice channels share a common signaling channel, with time division providing the separation between signaling channels. See SIGNALING SYSTEM 7.

Time Division Switching The connection of two circuits in a network by assigning them to the same time slot on a common time division switched bus.

Time Domain Reflectometer TDR. A testing device that acts on radar-like principles to determine the location of metallic circuit faults.

Time Guard Band A time interval left vacant on a channel to provide a margin of safety against interference in the time domain between sequential operations, such as detection, integration, differentiation, transmission, encoding, decoding, or switching.

Time Jitters Short-term variation or instability in the duration of a specified interval.

Time Marker A reference signal, often repeated periodically, enabling the correlation of specific events with a time scale. markers are used in some systems for establishing synchronization.

Time Multiplexed Switch The space switch of which the cross point settings are changed in each time slot.

Time Of Day Display The time and date displays on phones. Actually, it's very useful information. Sometimes it's not displayed on the operator's console. As a result, the operator may never know that every phone in the office is showing the wrong time and date.

Time Of Day Routing 1. This feature automatically changes access to certain types of lines at times when the lines change from being expensive to cheap, or vice versa. For example, it's cheaper to use WATS lines before 8:00 AM in the morning. A company has offices in New York and Los Angeles. It might be cheaper to route calls to Chicago in the morning over the tie lines to LA and then out the LA WATS lines to Chicago, than to go directly out the New York WATS lines. This is a way to allocate bandwidth for LAN traffic over corporate T-1 Networks. By programming T-1 multiplexers, customers can allocate the amount of T1 bandwidth that can be used by voice, data, and LAN traffic on a time of day basis. For example, during the day, most of the T1 bandwidth can be allocated for voice. At night, after employees go home, more bandwidth can be allocated to LAN and other computer data traffic so that file transfers can be done faster. This is particularly useful in IBM mainframe environments where large amounts of data needs to be transferred form remote offices/divisions to the headquarters.

Time Out In telecommunications and computer networks, an event which occurs at the end of a predetermined interval of time is called Time Out. For example if you lift the phone off the cradle and do not proceed to dial, after a certain number of seconds you will hear either a voice telling you to get on with it or a howling sound of some sort. Data networks have the same thing. Don't do anything for x minutes and the system will knock you off the air, i.e. hang up on you. In more technical terms, time out is the amount of time that hardware or software waits for an expected event before taking corrective action. In its most common form, time out is the amount of time an OCC or telephone system waits after your call goes through before it begins billing or timing the call. Also see ANSWER SUPERVISION.

Time Sharing A mode of operation that provides for the interleaving of two or more independent processes on one functional unit. Its most common use is the interleaved use of time on a computing system enabling two or more users to execute computer programs concurrently. Time sharing of computer resources is now relatively obsolete. See also TIMESHARING below.

Time Slice In a multi tasking environment, each task is allotted a portion of the CPU's overall processing power. This portion is called a time-slice. And it's usually measured in milliseconds. The CPU switches between tasks, and those with higher priority receive more time-slices than lower-priority tasks. See TIME SLICING.

Time Slicing The term used to describe the dividing of a computer resource so multiple applications or tasks requesting the resource are allocated some amount of the resource's time. See TIME SLICE.

Time Slot 1. In time division multiplexing or switching, the slot belonging to a voice, data or video conversation. It can be occupied with conversation or left blank. But the slot is always present. You can tell the capacity of the switch or the transmission channel by figuring how many slots are present.
2. An SCSA term. The smallest switchable data unit on the SCbus or SCxbus Data Bus. A time slot consists of eight consecutive bits of data. One time slot is equivalent to a data path with a bandwidth of 64 Kbps.

Time Space Time System TST. The most common form of switching matrix for small digital telephone exchanges in which a space switch is sandwiched between two time switches.

Time Switch A device incorporating a clock which arranges to switch equipment on or off at predetermined times.

Time Varying Media An SCSA definition. Time-varying media, such as audio data (as opposed to space-varying media, such as image data).

Timed Detection As a substitute for answer supervision, some long distance phone companies use call timing and estimate that a call is completed if the caller remains off-hook for 30 seconds or more. This is not necessarily accurate, of course. The caller might be holding, thinking the person is in the shower, out in the garden, etc. Little does the caller know he is now being charged to listen to ringing signals. A long distance phone company that is "equal accessed" doesn't have this problem. A long distance company that isn't equal accessed — one that you have to dial directly with a local call — might well have this problem. Rule: When in doubt, don't wait too long on the phone listening to endless ringing. Hang up. Count to ten. Then redial.

Timed Purge A feature of interactive voice response systems, especially fax-back systems. If the document isn't requested for x number of days or weeks or if the document ages to a certain point, the system automatically deletes the document.

Timed Recall Your PBX can be instructed to place a call at a designated time. When the time comes, your PBX rings your phone. When you answer your phone, the PBX places the call.

Timed Reminders At 20-second intervals, timed reminders will alert an attendant that a call is still waiting, a called line has not yet been answered or a call is still on hold. Timed reminders can be made longer or shorter. They can alert attendants to all sorts of events and non-events.

Timeout Two computers are "talking." One (for any reason) to respond. The other computer will keep on trying for a certain amount of time, but will eventually "give up." This is called time-out. A time also happens in a single computer. If a device (e.g. a printer) is not performing a task or responding, timeout is the amount of time the computer should wait before detecting it as an error. That period is called time-out.

Timesharing The use of one computer by many users at one time. Each user is typically sitting in front of a data terminal and connected to the master computer through communications lines — local or long distance. The user asks the computer to work on his task, whether it be a simple as looking up some stock prices, checking an airline reservation or doing some accounting calculations. It appears to each user as if he/she has a computer dedicated to his own task, but the computer is large and powerful, and is moving rapidly from one user's task to the next. Timesharing's advantages are twofold:
1. The user may find it cheaper to time share a computer than to buy his own. 2. The computer may have valuable and extensive information in it, which would be virtually impossible to duplicate or handle in many stand-alone computers. Timesharing was more popular when computers were more expensive.

Timing Jitter Deviation of clock recovery that can occur when a receiver attempts to recover clocking as well as data from the received signal. The clock recovery will deviate in a random fashion from the transitions of the received signal.

Timing Recovery The derivation of a timing signal from a received signal.

Timing Signal The output of a clock. A signal used to synchronize connected equipment.

Timing Slip A sudden timing delay change during high-speed digital transmission often caused by using T-1 carriers from different suppliers.

TINA-C Telecommunications Information Networking Architecture Consortium. An international standards body.

Tinned Wire Copper wire coated with tin to make soldering easier.

Tinsel A component of some phone line cord conductors. Tinsel is made by rolling copper into very thin, narrow rolls and then winding several strands of tinsel around a non-metallic core (a string) and then placing an insulating cover over the resulting conductor. A cord is then built up of two or more conductors encased in a plastic jacket. The essential reason for this type of construction is to obtain good cord flexibility and long life.

Tint Another name for hue.

Tip The first wire in a pair of phone wires. The second wire is called the "ring" wire. The tip is the conductor in a telephone cable pair which is usually connected to positive side of a battery at the telephone company's central office. It is the phone industry's equivalent of Ground in a normal electrical circuit. See TIP & RING.

Tip & Ring An old fashioned way of saying "plus" and "minus," or ground and positive in electrical circuits. Tip and Ring are telephony terms. They derive their names from the operator's cordboard plug. The tip wire was connected to the tip of the plug, and the ring wire was connected to the slip ring around the jack. A third conductor on some jacks was called the sleeve. That's it. Nothing more sinister. Nothing more interesting. See TIP, RING & GROUND.

Tip Cable A small cable connecting terminals on a distributing frame to cable pairs in the cable vault.

Tip Conductor The first conductor of a customer line.

Tip Side That conductor of a circuit which is associated with the tip of a plug, or of a telephone circuit.

Tip, Ring, Ground The conductive paths between a central office and a phone. The tip and ring leads constitute the circuit that carries a balanced speech or a data signal. The ground path in combination with the conductor is used occasionally for signaling.

TIPI Telephone Industry Price Index.

TIQ Telrate International Quotations.

TIS Technical Information Sheets.

Titanic On December 21, 1993 Vice President, Al Gore, told the National Press Club in Washington, "There is a lot of romance surrounding the sinking of the Titanic 91 years ago. But when you strip the romance away, a tragic story emerges that tells us a lot about human beings — and telecommunications. Why did the ship that couldn't be sunk steam full speed into an ice field? For in the last few hours before the Titanic collided, other ships were sending messages like this one from the Mesaba: "Lat42N to 41.25 Long 49W to Long 50.30W. Saw much heavy pack ice and great number large icebergs also field ice." And why, when the Titanic operators sent distress signal after distress signal did so few ships respond?

The answer is that — as the investigations proved — the wireless business then was just that, a business. Operators had no obligation to remain on duty. They were to do what was profitable. When the day's work was done — often the lucrative transmissions from wealthy passengers — operators shut off their sets and went to sleep. In fact, when the last ice warnings were sent, the Titanic operators were too involved sending those private messages from wealthy passengers to take them. And when they sent the distress signals, operators on the other ships were in bed."

Titles In the language of multimedia, when an author sells what he or she has created, it is called a title. The encyclopedias, dictionaries, musical works, and games available on CD are all "titles." Someone authors the material, and sells it to users who can play it back but not change the content.

TJF Test Jack Frame.

TL Tie Line.

TL1 Transaction Language 1. A machine to machine communications language which is a subset of ITU-T's man machine language.

TLA Three Letter Acronym. A form and usage common to our acronym-happy industry.

TLB Test LoopBack. A CSU (Channel Service Unit) operating mode that loops the telco's T-1 transmission facility back towards itself at itself at the same time it loops the CPE back toward itself.

TLF Trunk Link Frame.

TLP Transmission Level Point.

Token Ring Lan Service Unit The ATM TLSU provides a powerful tool for offering internetworking services over ATM networks. Emulated token rings consist of up to 64 TLSU token ring ports located anywhere in the ATM network, interconnected with PVCs. These emulated token ring networks can be completely isolated form one another to ensure security and fairness among the attached LANs. The TLSUs are designed for flexible deployment, either local to an ATM switch or at a remote site. See ATM ETHERNET LAN SERVICE UNIT.

TM 1. Trouble Management. The responsibilities associated with receiving any network events that impact customer service whether they are generated via customer contact or from internal network elements. Trouble Management tracks all problems, groups them together (if possible), and relays trouble tickets for problem resolution. A mobile phone term.

2. Traffic Management. As an ATM term, traffic Management is the aspect of the traffic control and congestion control procedures for ATM. ATM layer traffic control refers to the set of actions taken by the network to avoid congestion conditions. ATM layer congestion control refers to the set of actions taken by the network to minimize the intensity, spread and duration of congestion. The following functions form a framework for managing and controlling traffic and congestion in ATM networks and may be used in appropriate combinations.

    Connection Admission Control
    Feedback Control
    Usage Parameter Control
    Priority Control
    Traffic Shaping
    Network Resource Management
    Frame Discard
    ABR Flow Control

TMA Telecommunication Managers Association.

TMAP A piece of software announced by Northern Telecom in the summer of 1994 and designed to map Windows Telephony commands to Novell Telephony Services (TSAPI) commands. Tmap runs on the local workstation (also called client) and translates TAPI commands into TSAPI commands that are sent through the local area network to the telephony file server and thence to the PBX, causing the PBX to dial, or conference or ring a phone, etc. The idea is that computer telephony software running on the user's desktop PC could be implemented as easily through a card in the PC or through the company PBX attached to the LAN — and the whole process would be seamless to the user.

TMN Telecommunications Management Network. A concept where all OMCs (Operation and Maintenance Centers) are linked together to form a network. A centralization occurs to facilitate control, monitoring, and management of all devices in the communications network. This is an idea conceived as part of GSM (Global System for Mobile Communications). This is a mobile telecommunications term.

TMP Test Management Protocol: As an ATM term, it is a protocol which is used in the test coordination procedures for a particular test suite.

TMS 1. Time Multiplexed Switch. In the AT&T 5ESS switch CM, the TM provides switch paths between switching modules and passes control messages to and from the message switch, and functions as the hub for clock distribution to the switching modules.

2. TOPS Message Switch.

TN 1. Telephone Number.

2. Twisted Nematic. Most used display technology for calculators, watches and measuring equipment. TN uses liquid crystals sandwiched between two plates of glass with integrated transparent electrodes which can be made transparent and non-transparent by applying an electric current to them. See LCD.

TN3270 Delivery of a 3270 data stream via Telnet, provided as part of the TCP/IP protocol suite.

TNC A small connector used on coaxial cable, commonly used for cellular antennas, and some data and test equipment.

TNL Terminal Net Loss.

TNPP A protocol used to send paging messages from terminal to terminal on LANs and WANs over a wire circuit.

TNS Transit Network Selection: As an ATM term, it is a signaling element that identifies a public carrier to which a connection setup should be routed.

TNSS Non-Synchronous test line provides for rapid testing of ringing, tripping and supervisory functions of toll completing trunks. This test line provides an operation test which is not as complete as the Synchronous test but which can be made more rapidly.

TOA/NPI Type Of Address/Numbering Plan Identifier.

TOF Time Out Factor: As an ATM term, it is an ABR service parameter, TOF controls the maximum time permitted between sending forward

RM-cells before a rate decrease is required. It is signaled as TOFF where TOF=TOFF+1. TOFF is a power of 2 in the range: 1/8 to 4,096.

TOFF Time Out Factor: See TOF.

Toggle 1. A flip-flop switch that changes for every input pulse.
2. Any simple two-position switch.

Token 1. In networking, a unique combination of bits used to confer transmit privileges to a computer on a local area network. It also carries important information for routing messages over the network, such as source and destination addresses, access control information, route control information, and date checking information. When a LAN-attached computer receives a token, it has been given permission to transmit. On a token ring network, the token is 24 bits long. See TOKEN PASSING and TOKEN RING.
2. Here is a Rolm definition: The floating master message which coordinates use of the CBX control packet network among the nodes connected to it.

Token Bus A local network access mechanism and topology in which all phones or workstations attached to the bus listen for a broadcast token or supervisory frame. That token confers on them the right to communicate over the share channel, the token bus. An example of a Token-Bus is IEEE 802.4. See TOKEN PASSING.

Token Latency The time it takes for a token to be passed around the local area network ring.

Token Passing A method whereby each device on a local area network receives and passes the right to use the single channel on the LAN. The key to remember is that a token passing, or token ring LAN has only one channel. It's a high-speed channel. It can move a lot of data. But it can only move one "conversation" at a time. The Token acts like a traffic cop. It confers the privilege to send a transmission. Tokens are special bit patterns or packets, usually several bits in length, which circulate from node to node when there is no message traffic. Possession of the token gives exclusive access to the network for transmission of a message. The token is generated by one device on the network. If that device is turned off or fails, another device will assume the token creation task. When the package of token and message reaches its destination, the computer copies the message. The package is then put back on the network where it continues to circulate until it returns to the source computer. The source computer then releases the token for the next computer in the sequence.

With token passing it is possible to give some computers more access to the token than others. Usually one device on the network is designated the token manager. It generates the token. If that device is turned off or fails, another device will assume management of the token. There is a complicated sequence of events that result in the generation of a token and that deal with the eventuality of token loss or destruction. The logic for this process is built into token ring cards that fit inside computers. In some manufacturers' products, the logic is slightly different and can cause incompatibilities. See TOKEN, TOKEN RING and TOKEN RING PACKET

Token Ring A ring type of local area network (LAN) in which a supervisory frame, or token, must be received by an attached terminal or workstation before that terminal or workstation can start transmitting. The workstation with the token then transmits and uses the entire bandwidth of whatever communications media the token ring network is using. A token ring is a baseband network. Token ring is the technique used by IBM, Arcnet, and others. A token ring LAN can be wired as a circle or a star, with all workstations wired to a central wiring center, or to multiple wiring centers. The most common wiring scheme is called a star-wired ring. In this configuration, each computer is wired directly to a device called a Multi-station Access Unit (MAU). These are usually grouped together in a wiring closet for convenience. The MAU is wired in such a way as to create a ring between the computers. If one of the computers is turned off or breaks or its cable to the MAU is broken, the MAU automatically recreates the ring without that computer. This gives token ring networks great flexibility, reliability, and ease of configuration and maintenance. Despite the wiring, a token ring LAN always works logically as a circle, with the token passing around the circle from one workstation to another. The advantage of token ring LANs is that media faults (broken cable) can be fixed easily. It's easy to isolate them. Token rings are typically installed in centralized closets, with loops snaking to served workstations. Some other LANs require your going up in the ceiling or into walls and finding coax taps. All the work on a token ring can be done on one or several panels. These panels allow you to isolate workstations, and thus isolate faults.

Token Ring LANs can operate at transmission rates of either 4M bits per second or 16M bits per second. The number of computers that can be connected to a single Token Ring LAN is limited to 256. The typical installation is usually less than 100. Large installations connect multiple token ring LANS with bridges. The theoretical limit of Ethernet, measured in 64 byte packets, is 14,800 packets per second (PPS). By comparison, Token Ring is 30,000 and FDDI is 170,000. See FDDI-II and FDDI TERMS. Help on this definition courtesy Tad Witkowicz of Crosscomm, Marlboro, MA, Tim Becker, Lanquest Group, Santa Clara, CA and Elaine Jones, VP Marketing, Coral Network Corporation, Marlborough, MA. See also BRIDGE. IBM TOKEN RING, MAU, TOKEN PASSING, TOKEN RING, TOKEN RING CARD and TOKEN RING PACKET.

Token Ring Card Name given to the circuit board inserted into a computer device for connection to a token ring LAN. This board provides the physical connection to the LAN. It also participates in the collective management of the token by sending various messages to other token ring cards. Usually, one token ring card on the network is designated the token manager. It automatically generates a token as soon as it discovers one is missing, often with the help of other token ring cards. The sending of messages between token ring cards can be used to gather information about what is taking place on the network. Statistics may be collected. These may indicated that the network should be altered in some way to improve performance. This management capability is a distinct advantage of token ring LANs. One possible drawback is that various manufacturers' token ring cards may differ slightly in how they implement token management, thereby making them incompatible in certain management features. Virtually all token ring cards will work together in basic token passing.

Token Ring Packet Packets on a token ring network are made up of nine fields: starting delimiter, access control, frame control, destination address, source address, routing information, the data, frame check sequence, and ending delimiter.

Starting Delimiter (SD): This is an 8-bit binary (1s and 0s) sequence which marks the beginning of a data packet.

Access Control (AC) and Frame Control (AC): These are two 8-bit sequences that are used by the computers for maintenance purposes.

Destination Address (DA): This is a 48-bit sequence that uniquely identifies the physical name of the computer to which the data packet is being transmitted. Each computer on a ring examines this field to determine if the packet is for it.

Source Address (SA): This is a 48-bit sequence that uniquely identifies the physical name of the computer that send the data packet. This is used by the receiving computer to formulate its acknowledgement.

Routing Information (RI): This is a variable-length sequence used if the data packet is being sent to a computer located on another token ring LAN. (This information can make it impossible for some bridges to route some packets. See BRIDGE.)

NEWTON'S TELECOM DICTIONARY

Data: This is a variable-length sequence (up to 17,800 bytes) that is the actual data being sent from source to destination.
Frame Check Sequence (FCS): This 32-bit sequence is used to protect the contents of the packet from being corrupted during transmission. See FRAME CHECK SEQUENCE.
Ending Delimiter (ED): This is an 8-bit sequence that signals the end of a packet.

Token Tree LAN A type of local area network with a topology in the form of branches interconnected via active hubs. Using a token-passing scheme, the active hubs grant nodes access to the medium. See TOKEN RING.

Tokentalk The Apple Macintosh implementation of the Token Ring local area network.

Toll Call A call to any location outside the local service area. A long distance call.

Toll Center A Class Four central office.

Toll Connecting Trunk A trunk used to connect a Class 5 office (local central office) to the direct distance dialing network.

Toll Denial Permits phone user to make local calls but denies completion of toll calls or calls to the toll operator without the assistance of the attendant. See TOLL RESTRICTION.

Toll Diversion A system service feature by which users are denied the ability to place toll calls without the assistance of a human attendant. Toll diversion affects the entire switching system instead of discriminating between individual extensions.

Toll Fraud Theft of long distance service. Today's most common forms of toll fraud are DISA, voice mail and shoulder surfing. According to John Haugh of Telecommunications Advisors in Portland, OR, there are three distinct varieties of toll fraud:
"First Party" Toll Fraud, which is helped along by a member of the management or staff of a user. An example would be the telecommunications manager at the Human Resources department of New York City (an "insider") who sold his agency's internal code to the thieves, who in turn ran up unauthorized long distance charges exceeding $500,000.
"Second Party" Toll Fraud, which is facilitated by a staff member or subcontractor of a long distance carrier IXCs, vendor or local exchange telephone company selling the information to the actual thieves, or their "middlemen." An example would be a "back office clerk" working for one of these concerns who sells the codes to others.
"Third Party" Toll Fraud is facilitated by unrelated "strangers" who, though various artifices, either "hack" into a user's equipment and learn the codes and procedures, or obtain the needed information through some other source, to commit Toll Fraud.

Toll Office A central office used primarily for supervising and switching toll traffic.

Toll Restriction To curb a telephone user's ability to make long distance calls. Toll restriction capability on modern PBXs and key telephone systems has been increasing in sophistication. Some PBXs now allow selective restriction based on specific extensions, users or geography. In other words, Joe Smith, the president, could call everywhere. John Doe in accounting might only be allowed to call Chicago and Houston, where our two factories are located. Mary Johnson, the seller for the western U.S., might only be allowed to call Denver and points west. There's considerable debate as to how useful toll restriction really is.

Toll Station A Telco phone from which established long distance message rates are charged for all messages sent over company lines.

Toll Switching Trunk A trunk connecting one or more end offices to a toll center as the first stage of concentration for intertoll traffic. Operator assistance or participation may be an optional function. In U.S. common carrier telephony service, a toll center designated "Class 4C" is an office where assistance in completing incoming calls is provided in addition to other traffic; a toll center designated "Class 4P" is an office where operators handle only outbound calls, or where switching is performed without operator assistance.

Toll Testboard Manual test position at which toll circuits are tested and repaired.

Toll Terminal A phone only furnished with long distance service.

Toll Terminal Access Allows hotel/motel guest phones to access toll calling trunks.

Toll Ticket Ticket is the telephone company term for a bill. A toll ticket is a bill containing the calling number, called number, time of day, date and call duration. Some phone systems generate their own bills automatically. Some still need an operator. It depends on the equipment and the type of call.

Toll Trunk A communications channel between a toll office and a local central office.

Tone An audio signal consisting of one or more superimposed amplitude modulated frequencies with a distinct cadence and duration. See TONE SET.

Tone Alternator A motor-driven AC generator that produces audio-frequency tones.

Tone Dial A pushbutton telephone dial that makes a different sound (in fact, a combination of two tones) for each number pushed. The correct name for tone dial is "Dual Tone MultiFrequency" (DTMF). This is because each button generates two tones, one from a "high" group of frequencies
— 1209, 1136, 1477 and 1633 Hz — and one from a "low" group of frequencies — 697, 770, 852 and 841 Hz. The frequencies and the keyboard, or tone dial, layout have been internationally standardized, but the tolerances on individual frequencies vary between countries. This makes it more difficult to take a touchtone phone overseas than a rotary phone.
You can "dial" a number faster on a tone dial than on a rotary dial, but you make more mistakes on a tone dial and have to redial more often. Some people actually find rotary dials to be, on average, faster for them. The design of all tone dials is stupid. Deliberately so. They were deliberately designed to be the exact opposite (i.e. upside down) of the standard calculator pad, now incorporated into virtually all computer keyboards. The reason for the dumb phone design was to slow the user's dialing down to the speed Bell central offices of early touch tone vintage could take. Today, central offices can accept tone dialing at high speed. But sadly, no one in North America makes a phone with a sensible, calculator pad or computer keyboard dial. On some telephone/computer workstations you can dial using the calculator pad on the keyboard. This is a breakthrough. It a lot faster to use this pad. The keys are larger, more sensibly laid out and can actually be touch-typed (like touch-typing on a keyboard.) Nobody, but nobody can "touch-type" a conventional telephone tone pad. A tone dial on a telephone can provide access to various special services and features — from ordering your groceries over the phone to inquiring into the prices of your (hopefully) rising stocks.

Tone Disabling A method of controlling the operation of communications equipment by transmitting a certain tone over the phone line.

Tone Diversity A method of Voice Frequency Telegraph (VFTG) Transmission wherein two channels of a 16-channel VFTG carry the same information. This is commonly achieved by twinning the channels of a 16-channel VFTG to provide eight channels with dual diversity.

Tone Probe A testing device used to detect signals from a TONE GENERATOR to identify phone circuits, often the size of a fat pencil or

skinny banana. Some models contain speakers; others must be used with a HEADSET or BUTT SET.
Tone Ringing Either a steady or oscillating electronic tone at the phone to tell you someone is calling.
Tone Sender 1. A printed circuit card in Rolm CBX which supplies the data bus with the digital representations of the following tones: dial, ring, busy, error, howler (off-hook timeout) and pulse (after flashing).
2. A printed circuit card which generates the following tones: dial, ring, busy, error, howler (off-hook timeout) and pulse (after flashing).
Tone Set A collection of tones which are customarily used as a set for the purposes of call setup and teardown (e.g., DTMF, R1 MF, R2 MF). In the case of DTMF, the tone set can also be used by the client application during the conversation portion of a call.
Tone Signaling The transmission of supervisory, address and alerting signals over a telephone circuit by means of tones. Typically inband. See also SIGNALING SYSTEM #7.
Tone To Dial Pulse Conversion Converts DTMF (Dual Tone Multiple Frequency) signals to dial pulse signals when trunks going to carry outgoing calls are not equipped to receive tone signals. A lot of electronic phones with touchtone dials have a sliding switch that allows you to choose whether the phone will outpulse in rotary, or whether it will touchtone out. You choose whichever your trunk line will accept.
Tone/Pulse Switchable Most phones in North America come with a pushbutton dial. Many of these phones have a switch that says "Tone/Pulse." By sliding the switch one way, the pushbutton pad will dial by sending out touchtones. By sliding the switch the other way, the pushbutton pad will dial by rotary pulses. See ROTARY DIAL.
Tool In some computer languages, a small program executed as a shell command. In other computer languages, such as BASIC, it is called a "utility."
Toolbar A series of shortcut buttons providing quick access to commands. Usually located directly below the menu bar. Not all windows have a toolbar.
Toolkit A Dialogic word for an Applications Generator.
Toolkit Developer Program A strategic alignment by Dialogic with suppliers of voice processing applications development software to provide high-level application development tools.
Tone Dialing Same as touchtone dialing. See TOUCHTONE.
TOP Technical Office Protocol. A version of the Open Systems Interconnection (OSI) model for use in the non-shop floor environment, i.e. in the office, developed by Boeing, the people who make the planes. TOP was designed from the outset to be compliant with the ISO OSI seven-layer model. Development has been merged with MAP, and the two functional profiles share a common integration strategy, and have a single (MAP/TOP) user group.
Top Down This is a method of distributing incoming calls to a bunch of people. It always starts at the top of a list of agents and proceeds down the list looking for an available agent. See also ROUND ROBIN and LONGEST AVAILABLE.
Top Level Domain A certain segment of a network in the Transmission Control Protocol/Internet (TCP/IP) UNIX environment. A network is segmented into a hierarchy of domains or groupings. In the Internet in the United States, there are six top-level domains: com (commercial organizations), edu (education organizations), gov (government agencies), mil (Military milnet hosts), net (networking organizations), and org (nonprofit organizations). The next lower level relates to specific companies, and the level below to devices within a company.
Topology Network Topology. The geometric physical or electrical configuration describing a local communication network — the shape or arrangement of the system. The most common distribution system topologies are the bus, ring, and star.
Topology Aggregation The process of summarizing and compressing topology information at a hierarchical level to be advertised at the level above.
Topology Attribute A generic term that refers to either a link attribute or a nodal attribute.
Topology Constraint An ATM term. A topology constraint is a generic term that refers to either a link constraint or a nodal constraint.
Topology Database As an ATM term, it is the database that describes the topology of the entire PNNI routing domain as seen by a node.
Topology Metric A generic term that refers to either a link metric or a nodal metric.
Topology State Parameter A generic term that refers to either a link parameter or a nodal parameter.
TOPS 1. Traffic Operator Position System. A specialized console designed for telephone company operators to help them complete toll calls.
2. A computer operating system, which originally stood for the transcendental operating system.
3. The operating system used by Digital Equipment Corp.'s DECSYSTEM-10 and DECSYSTEM-20 computers. These computers have been discontinued, but many are still in use.
TOPS MPX Northern Telecom's Traffic Operator Position System designed on a token ring for interface between operator positions and the IBM Directory Assistance system database.
Torn Tape Relay An antiquated tape relay system in which the perforated tape is manually transferred by an operator to the appropriate outgoing transmitter. In short, it's a torn tape relay is a store and forward message switching system which uses punched paper as the storage medium.
Total Harmonic Distortion The ratio of the sum of the powers of all harmonic frequency signals (other than the fundamental) to the power of the fundamental frequency signal. This ratio is measured at the output of a device under specified conditions and is expressed in decibels.
Total Internal Reflection The reflection that occurs when light strikes an interface at an angle of incidence (with respect to the normal) greater than the critical angle.
Total Transaction Call Processing A Rockwell term. Rockwell's philosophy. It guides their approach to call centers. It involves managing the success of a call center, not merely supplying the ACD (Automatic Call Distributor). It could include software development, CTI integration, network management, consulting services, IVR and voice processing systems. Rockwell says it will act as the prime contractor or as a single provider for a call center solution.
Tower A name for a PC in a vertical or upright case. Tower PCs (if they're correctly designed) have a big benefit. Heat rises and escapes more easily than in traditional horizontal machines. Heat and power surges are the most damaging threats to PC.
Touchtone Touchtone is not a trademark of AT&T, despite what editions one through six of Newton's Telecom Dictionary said. It is a generic term for pushbutton telephones and pushbutton telecommunications services and the term "touchtone" may be used by anyone. At one stage it was a trademark of AT&T. At divestiture in 1984, AT&T gave it to the public. And that's who owns it now. The public. If you don't believe this, call Frank L. Politano, AT&T Trademark and Copyright Counsel, in Basking Ridge, New Jersey. For a full explanation of touchtone, see DTMF

which stands for Dual Tone Multi Frequency signaling, i.e. touchtone.

Touchtone Type Ahead Also known as DTMF Cut-Through. Touchtone Typed Ahead is the ability of a voice response system to receive DTMF tones while the voice synthesizer is delivering information, i.e. during speech playback. This capability of DTMF cut-through saves the user waiting until the machine has played the whole message (which typically is a menu with options). The user can simply touchtone his response anytime during the message — when he first hears his selection number, when the message first starts, etc. When the voice processor hears the touchtoned selection (i.e. the DTMF cut-through), it stops speaking and jumps to the chosen selection. For example, the machine starts to say, "If you know the person you're calling, touchtone his extension in now." But before you hear the "If you know" you push button in 230, which you know is Joe's extension. Bingo, the message stops and Joe's extension starts ringing.

Tourists People who take training classes just to get a vacation from their jobs. "We have about three serious students in the class. The rest are tourists."

TP Abbreviation for Transport Protocol or Twisted Pair.

TP1, TP2, TP3, TP4, TP5 The various service levels of the ISO IS 8073 Transport Protocol. TP4 is the most popular service level for information system networks and is specified in the U.S. government GOSIP architecture. TP4 stands for OSI Transport Protocol Class 4 (Error Detection and Recovery Class). This is the most powerful OSI Transport Protocol, useful on top of any type of network. TP4 is the OSI equivalent to TCP.

TP-4 Transport Protocol 4. An OSI layer-4 protocol developed by the National Bureau of Standards. See TP1.

TP-4/IP A term given to the ISO protocol suite that closely resembles TCP/IP.

TP-MIC Twisted-Pair Media Interface Connector: This refers to the connector jack at the end user or network equipment that receives the twisted pair plug.

TP-PMD Twisted-Pair Physical Media Dependent, Technology under review by the ANSI X3T9.5 working group that allow 100 Mbps transmission over twisted-pair cable. Also referred to as CDDI or TPDDI.

TPAD Terminal Packet Assembler/Disassembler linked to a cluster controller or terminal device, taking native protocol input and converting it to X.25 for transmission over a packet network.

TPC TOPS Position Controller.

TPCC Third Party Call Control: As an ATM term, it is a connection setup and management function that is executed from a third party that is not involved in the data flow.

TPDDI Twisted Pair Distributed Data Interface. Also known as ANSI X3T9.5.-TPDDI. TPDDI is a new technology that allows users to run the FDDI standard 100 Mbps transmission speed over twisted-pair wiring. Unshielded twisted-pair has been tested for distances over 50 meters (164 ft.). TPDDI is designed to help users make an earlier transition to 100 Mbps at the workstation. Also known as CDDI, Copper Distributed Data Interface.

TPDU Abbreviation for Transport Protocol Data Unit.

TPI Tracks Per Inch. A measurement of how much data can be stored on a disk.

TPON 1. Telephone Passive Optical Network.

2. OSI Transport Protocol Class 0 (Simple Class). This is the simplest OSI Transport Protocol, useful only on top of an X.25 network (or other network that does not lose or damage data).

TQFP Thin Quad Flat Pack, a format used in the design of PCMCIA devices. Another format is called PQFP, which stands for Plastic Quad Flat Pack.

TQM Total Quality Management. Doing what management should have been doing all along.

TR 1. Trouble Report or Technical Reference (BELLCORE).

2. Technical Reference.

3. Technical Requirement. These publications are the standard form of Bellcore-created technical documents representing Bellcore's view of proposed generic requirements and standards for products, new technologies, services, or interfaces. See ISDN.

TR-303 A defacto standard published by Bellcore. It amounts to an industry standard high level control interface to dumb switches. It also applies to Fiber In The Loop (FITL).

TR-444 A defacto standard published by Bellcore which spells out how the Bell regionals want long distance companies to connect to the Bell regionals' local networks. Several observers compare the TR-444 specs to simple direct dial long distance voice phone service.

TRA The Telecommunications Resellers Association is an organization representing the interests of more than 100 companies involved in the switchless resale of long distance telecommunications services. TRA can be reached at P.O. Box 8361, McLean, Va, 22106-8361, 703-734-1225, Fax 703-734-8572.

TRAC Northern Telecom term for Technical Recommendations Approval Committee.

Trace Agent This is a command used in the Infoswitch product line to report all the events and transactions an agent has been involved in over a defined period of time.

Trace Block See trailer.

Trace Packet A special kind of packet in a packet-switching network which functions as a normal packet but causes a report of each stage of its progress to be sent to the network control center.

Trace Program A computer program that performs a check on another computer program by showing the sequence in which the instructions are executed and usually the results of executing the instructions.

Traceroute A more useful superset of PING used to evaluate the hops taken from one end of a link to the other.

Tracer Stripe When more than one color coding stripe is required, the first or widest stripe is the base stripe. The other, usually narrower stripes are the tracer stripes.

Track 1. A storage channel on a disk or tape which can be magnetically encoded.

2. On a data medium, a path associated with a single read/write head as data move past the head.

3. Every time you write to a CD, you will create at least one track, which is preceded by a pre-gap and followed by a post-gap. Any session may contain one or more tracks, and the tracks within a session may be of the same of or different types (for example, a mixed-mode disc contains data and audio tracks).

NEWTON'S TELECOM DICTIONARY

Track Access Time The time it takes to move the pickup head on a disk drive from one track to another.
Track Density The number of tracks per unit length, measured in a direction perpendicular to the tracks.
Track Speed The maximum speed which a train can travel over a section of railway tracks.
Trackball An upside-down MOUSE; a rotatable ball in a housing used to position the cursor and move images on a computer screen. A mouse needs desktop room to work, a trackball stays in one place, and can even be part of a keyboard or built into a laptop computer. It's hard to see why anyone uses a mouse instead of a trackball. This dictionary was typeset by a fine lady called Jennifer Cooper-Farrow, who used a trackball and a Macintosh computer.
Tracking 1. Figuring where a satellite is and keeping track of it. This is not an easy job, given the vastness of space.
2. The effect created in compressed video when the speed of the transmission is not great enough to keep up with the speed of the action. Tracking creates a tearing effect on the video picture.
3. A call center term. A software feature that models actual events and activities in your call center to aid you in short-term planning and evaluation of employee and call center performance. The tracking functions include employee information scheduling assignment, daily activity, and intra-day performance.
Tractor Feeder A device which attaches to a computer printer and allows the printer to use continuous, sprocket-fed, paper. Such paper has a row of evenly spaced holes on both sides. Those holes coincide with the pins on the tractor feeder. In all tractor-fed printers, the tractor moves the paper, not the printer's platen.
Trade Secret A trade secret can be any information, knowledge, data, or the like which is useful in business and not commonly known. A trade secret is anything from a customer list to the formula for Coke syrup. Enforcing a trade secret, for example, in order to enjoin a former employee from working for a competitor normally requires proof that the secret allegedly taken was suitably identified as such, that the employee was subject to a written contract including an obligation of confidentiality, and that physical access to the secret was suitably restricted.
Trademark A trademark can be any word, symbol, slogan, design, musical jingle, or the like capable of differentiating one party's goods or services from another's. The question is whether a member of the relevant segment of the public would be misled as to the source of the goods. Thus, a descriptive mark ("Frigidaire") is less powerful than a coined mark ("Xerox"), and the same mark can be used by different parties, if on differing goods ("Cadillac" for dog food versus "Cadillac" for automobiles.) The r symbol indicates that a mark has been registered by the Federal government, while the tm or sm symbols merely indicate that the user does not intend to waive his rights in the mark. It is not legally necessary to use the statements commonly seen that certain trademarks are the property of their owners, or to use the r symbol in text, but it does prevent any accusation of misappropriation.
Trader Turret A very large key telephone used by traders of commodities, securities, etc. Turrets typically have many line buttons. Each one corresponds to a trunk, an autodial or tie-line circuit to another trader or a financial institution. The objective of turrets is to allow the trader to be in instant communication with others who might want to buy or sell that which he is trying to sell or buy. See the June issue of TELECONNECT for an annual roundup of turrets.
Traffic Bellcore's definition: A flow of attempts, calls, and messages. My definition: The amount of activity during a given period of time over a circuit, line or group of lines, or the number of messages handled by a communications switch. There are many measures of "traffic." Typically it's so many minutes of voice conversation, or so many bits of data conversation. Note that Bellcore includes attempts in its definition of traffic. I don't. The decision is yours. But you should be aware of what you include in your calculations. See also TRAFFIC ENGINEERING and QUEUING THEORY.
Traffic Capacity The number of CCS (hundred call seconds) of conversation a switching system is designed to handle in one hour. This is the simple definition. See TRAFFIC ENGINEERING.
Traffic Carried See TRAFFIC OFFERED AND CARRIED.
Traffic Concentration The average ratio of the traffic during the busy hour to the total traffic during the day.
Traffic Data To Customer The owner of a call accounting system can poll his PBXs daily or hourly and get traffic measurements, including peg counts, usage and overflow data. Summary reports, exception reports and complete traffic register outputs can be obtained.
Traffic Engineering The science of figuring how many trunks, how much switching equipment, how many phones, how much communications equipment you'll need to handle the telephone, voice, data, image and video traffic you're estimating. Traffic engineering suffers from several problems:
1. You are basing your future needs on past traffic.
2. Most traffic engineering is based on one or more mathematical formulas, all of which approach but never quite match the real world situation of an actual operating phone system. Computer simulation is the best method of predicting one's needs, but it's expensive in both computer and people time.
3. Many people in the telecommunications industry do not understand traffic engineering, have not worked with it sufficiently and make dumb and costly mistakes.
4. Since there are now several hundred long distance companies in the United States, and several thousand differently-priced ways of dialing between major cities, traffic engineering has become very complex.
After I wrote the above definition, Lee Goeller, disagreed with me and contributed this definition.
Traffic Engineering: The application of probability theory to estimating the number of servers required to meet the needs of an anticipated number of customers. In telephone work, the servers are often trunks, and the customers are telephone calls, assumed to arrive at random (see POISSON PROCESS). When arriving calls, upon finding all trunks busy, vanish, a "blocked call cleared" situation obtains (see ERLANG B). When a call stays in the system for a given length of time, whether it gets a trunk or not, "blocked calls held" applies (see POISSON DISTRIBUTION). If a call simply waits around until a trunk becomes available and then uses the trunk for a full holding time, the correct term is "blocked calls delayed" (See ERLANG C and QUEUING THEORY). Like any form of predicting the future on the basis of past behavior, traffic engineering has its limitations; however, when used by those who have taken the trouble to learn how it works, its track record is surprisingly good, and vastly better than most forms of simulation (see SIMULATION).
Traffic Intensity A measure of the average occupancy of a facility during a period of time, normally a busy hour, measured in traffic units (erlangs) and defined as the ratio of the time during which a facility is occupied continuously or cumulatively) to the time this facility is avail-

able. A traffic intensity of one traffic unit (one erlang) means continuous occupancy of a facility during the time period under consideration, regardless of whether or not information is transmitted. See also TRAFFIC ENGINEERING.

Traffic Load Total traffic carried by a trunk during a certain time interval.

Traffic Measurement Memory and other software in a telephone system which collect telephone traffic data such as number of attempted calls, number of completed calls and number of calls encountering a busy. The objective of traffic measurement is to enter the results into traffic engineering and so arrange one's incoming and outgoing trunks to get the best possible service. See TRAFFIC ENGINEERING.

Traffic Monitor PBX feature that provides basic statistics on the amount of traffic handled by the system.

Traffic Offered And Carried People pick up the phone and try to place their calls. This is "Traffic Offered" to the switch. The calls that get through the switch and onto lines is called "Traffic Carried." The difference between traffic offered and carried is the traffic that was lost or delayed because of congestion. There are two basic ways of measuring traffic — erlangs and CCS (or hundred call seconds).

Traffic Overflow Occurs when traffic flow exceeds the capacity of a particular trunk group and flows over to another trunk group.

Traffic Path A path over which individual communications pass in sequence.

Traffic Recorder A device which measures traffic activity on a transmission channel. It's a recorder, not a processor. It's dumb.

Traffic Table A computer database into which a PBX enters a count of feature activity. Certain detected operating errors are also entered in the traffic table.

Traffic Usage Recorder A device for measuring and recording the amount of telephone traffic carried by a group, or several groups, of switches or trunks.

Trail As an ATM term, it is an entity that transfers information provided by a client layer network between access points in a server layer network. The transported information is monitored at the termination points.

Trailer 1. A nonstandard way of standard way of sending data. Trailers are used on some networks by 4BSD UNIX and some of its derivatives.
2. A block of controlling information transmitted at the end of a message to trace error impacts and missing blocks. Also referred to as a trace block.

Train The creation of word reference data by presenting words to a recognizer. A voice recognition term.

Training A feature of some modems which adjust to the conditions including amplitude response, delay distortions, timing recovery, and echo characteristic, of a particular telecommunications connection by a receiving modem. See TRAINING UP.

Training up A technique that adjusts modems to current telephone line conditions. The transmitting modem sends a special training sequence to the receiving modem, which makes necessary adjustments for line conditions.

Transaction 1. It is a completed event that can be assembled in chronological sequence for an audit trail.
2. An entry or an update in a database.

Transaction Capabilities Function that controls non-circuit-related information transfer between two or more nodes via a SS7 signaling network.

Transaction Capabilities Application Part TCAP. The application layer protocol of SS7. Transaction capabilities in the SS7 protocol are functions that control non-circuit related information transferred between two or more signaling nodes. Definition from Bellcore in reference to its concept of the Advanced Intelligent Network.

Transaction Detail The detail of a transaction record.

Transaction File A collection of transaction records. A transaction data entry program allows for the creation of new transaction files used to update the data base.

Transaction Link Rockwell's link from its Galaxy ACD to an external computer. See OPEN APPLICATION INTERFACE.

Transaction Tracking Your software keeps track of each transaction as it happens. And if a component of your network fails, your transaction tracking software backs out of the incomplete transaction. This allows you to maintain your database's integrity. You may, however, lose the single transaction you were working on when your network got sick.

Transactional Integrity A term that describes how your computing/telecom system handles making sure that the transaction you just made is solid and clean and that the next time you want to get to the results of the transaction you can. "Transactional integrity" becomes critical when you're storing bits and pieces of your transactions on different media, in different places. For example, you might want to store your data on a magnetic hard drive and your associated images on a separate optical drive.

Transceiver 1. Any device that transmits and receives. In sending and receiving information, it often provides data packet collision detection as well.
2. In IEEE 802.3 networks, the attachment hardware connecting the controller interface to the transmission cable. The transceiver contains the carrier-sense logic, the transmit/receive logic, and the collision-detect logic.
3. A device to connect workstations to standard thick Ethernet-style (IEEE 802.3).

Transceiver Cable In local area networks, a cable that connects a network device such as a computer to a physical medium such as an Ethernet network. A transceiver cable is also called drop cable because it runs from a network node to a transceiver (a transmit / receiver) attached to the trunk cable. See TRANSCEIVER.

Transcoder A device that combines two 1.544 megabit per second bit streams into a single 1.544 megabit per second bit stream to enable transmission of 44 or 48 voice conversations over a DS-1 medium.

Transcriptionist A person who listens to a tape recording and types the words he hears. The word, transcriptionist, derives from the verb to transcribe. The most common employment of transcribers is in the medical industry, where busy doctors talk into tape recorders telling good and bad news of their patients. And even busier transcriptionists type those words into the patient's medical records, or whatever.

Transducer A device which converts one form of energy into another. The diaphragm in the telephone receiver and the carbon microphone in the transmitter are transducers. They change variations in sound pressure (your voice) to variations in electricity, and vice versa. Another transducer is the interface between a computer, which produces electron-based signals, and a fiber-optic transmission medium, which handles photon-based signals.

Transfer A telephone system feature which provides the ability to move a call from one extension to another. It is probably the most commonly used and misused feature on a PBX. Before you buy a PBX, check out how easy it is to transfer a call. If you have a single line phone, you should simply hit the touch hook, hear a dial tone and then dial the chosen extension number and hang up. This sounds easy in principle, but many people find it difficult since they associate the touch hook with hanging up the phone. Some companies have gotten around this by putting a "hook flash" button on the phone itself. Such a button is like having an autodial button which just makes the exact short tone you make when you quickly hit the hook flash button. An even better solution is an electronic phone with a button specially marked "transfer," or a button next to a screen which lights up "transfer." Failing to efficiently transfer a call is the easiest way to give your customers the wrong impression of your firm. Think of how many times have you called a company only to be told it wasn't the fellow's job and he will transfer the call, but "If we get cut off, please call Joe back on extension 2358." There are typically four types of Transfer: Transfer using Hold, Transfer using Conference, and Transfer with and without Announcement.

Transfer Callback After a specified number of rings, an unanswered transferred call will return to the telephone which originally made the transfer.

Transfer Delay A characteristic of system performance that expresses the time delay in processing information through a data transmission system.

Transfer Protocols Protocols are all of the packaging" that surround actual user data to tell the network devices where to send the data, who it comes from, and how to tell if it arrived. Transfer protocols are designed for the efficient moving of larger chunks of user data.

Transfer Rate The speed of data transfer — in bits, bytes or characters per second — between devices.

Transformer An electrical device used to reduce or increase the voltage in a circuit or to transfer energy from one circuit to another. One reason for using a transformer is that commercial power is typically 120 or 240 volts while many phone systems (and other computer-type "things") work best on 48, 24 or lower voltage.

Transhybrid Loss The transmission loss between opposite ports of a hybrid network, that is between the two ports of the four-wire connection.

Transient Any high-speed, short duration increase or decrease impairment that is superimposed on a circuit. Transients can interrupt or halt data exchange on a network. See HIT.

Transient Mobile Unit A mobile unit communicating through a foreign base station.

Transit Delay 1. In ISDN, the elapsed time between the moment that the first bit of a unit of data (such as a frame) passes a given point and the moment that bit passes another given point plus the transmission time of that data unit.
2. As an ATM term, it is the time difference between the instant at which the first bit of a PDU crosses one designated boundary and the instant at which the last bit of the same PDU crosses a second designated boundary.

Transit Exchange The European equivalent of a tandem exchange.

Transit Timing A method of eliminating looping between nodes used in the network layer of some packet-switched systems. This method is used in the Internet Protocol (IP) portion of Transmission Control Protocol/Internet Protocol (TCP/IP).

Transition Probabilities Probabilities of moving from one state to another.

Transition Zone The zone between the far end of the near-field region and the near end of the far-field region. The transition is gradual.

Transistor The transistor was invented in 1947 by John Bardeen, Walter H. Brattain and William Shockley of Bell Laboratories. Here is an explanation of how a transistor works, taken from "Signals, The Science of Telecommunications" by John Pierce and Michael Noll:
"To understand how a transistor works, we must look at the laws of quantum mechanics. We commonly picture an atom as a positive nucleus surrounded by orbiting electrons ... Vacuum tubes rely on the ability of electrons to travel freely with any energy through a vacuum. Transistors rely on the free travel of electrons through crystalline solids called semiconductors ... Semiconductors (such as silicon or gallium arsenide) differ from pure conductors, such as metals, in how full of electrons are the energy bands that allow free travel."
And depending on their design, transistors can act as amplifiers or switches. See TRANSISTOR RADIO.

Transistor Milestones Point-contact transistor 1948

Single-crystal Germanium	1950
Grown junction transistor	1951
Alloy junction transistor	1952
Zone melting and refining	1952
Single-crystal Silicon	1952
Diffused-base transistor	1955
Oxide masking	1957
Planar transistor	1960
MOS transistor	1960
Epitaxial transistor	1960
Integrated circuits	1961

Transistor Radio Sony unveiled the first transistor radio in 1955.

Translate To change the digits dialed on your phone into digits necessary for routing the call across the country. See TRANSLATIONS.

Translating Bridge A special bridge that interconnects different LAN types using different protocols at the physical and data link layers, such as Ethernet and Token Ring. A translating bridge supports the physical and data link protocols of both LAN types. When they forward packets from one LAN to another, they manipulate the packet envelope to conform to the physical and data link protocols of the destination LAN.

Translation The interpretation by a switching system of all or part of a destination code to determine the routing of a call. See TRANSLATIONS.

Translations Translations are changes made by the network to dialed telephone numbers to allow the call to progress through the network. Sometimes the translations are made automatically. Take one series of dialed numbers; convert them to another. Sometimes, translations are done with the help of "look up" tables, also called databases. Here's an example of translations done with the help of a database. TELECONNECT Magazine has a WATS line, 1-800-LIBRARY. If you dial it on the phone, you'll see it is really 1-800-542-7279. But this is not its real number. When someone in California dials 1-800-LIBRARY, AT&T's long distance network recognizes the "1-800" portion of the call and sends it to a special AT&T central office in Utah. When the call arrives, a computer looks up the number 542-7279 in its database and translates that to 1-212-206-6871 and puts the call back into the network. Within seconds, that number in New York, 212-206-6871, which we think is our IN-WATS line, rings.

Transliterate To convert the characters of one alphabet to the corresponding characters of another alphabet.

Translator 1. A communications device that receives signals in one form, normally in analog form at a specific frequency, and retransmits them in a different form.
2. A device that converts information from one system into equivalent information in another system.
3. In telephone equipment, it is the device that converts dialed digits into call-routing information.
4. In computers, it is a program that translates from one language into another language and in particular from one programming language into another programming language.
5. In FM and TV broadcasting, it's a repeater station that receives a primary station's signal, amplifies it, shifts it in frequency, and rebroadcasts it.

Transliterate To convert the characters of one alphabet to the corresponding characters of another alphabet.

Transmission Sending electrical signals carrying information over a line to a destination. Bellcore says that transmission has the following definitions: (a) Designates a field work, such as equipment development, system design, planning, or engineering, in which electrical communication technology is used to create systems to carry information over a distance. (b) Refers to the process of sending information from one point to another. (c) Used with a modifier to describe the quality of a telephone connection: good, fair, or poor transmission. (d) refers to the transfer characteristic of a channel or network in general or, more specifically, to the amplitude transfer characteristic. You may sometimes hear the phrase, "transmission as a function of frequency."

Transmission Block A group of bits or characters transmitted as a unit, with an encoding procedure for error control purposes.

Transmission Bus The circuits mounted on the backplane of the AT&T ISDN packet controller that carry data packets from sending device interface modules into the switch and back to the receiving device interface modules. See also BACKPLANE.

Transmission Channel All of the transmission facilities between the input (to the channel) from an initiating node and the output (from the channel) to a terminating node. In telephony, transmission channels may be of various bandwidths: e.g. nominal 3-kHz, nominal 4-kHz, or nominal 48-kHz (group). "Transmission channel" should not be confused with the more general term "channel."

Transmission Code A code by which information is sent and received on a transmission system.

Transmission Coefficient The ratio of the transmitted field strength to the incident field strength when an electromagnetic wave is incident upon an interface surface between media with two different refractive indices. In a transmission line, the ratio of the complex amplitude of the transmitted wave to that of the incident wave at a discontinuity in the line. A number indicating the probable performance of a portion of a transmission circuit. The value of a transmission coefficient is inversely related to the quality of the link or circuit.

Transmission Control Category of control characters intended to control or help transmission of information over telecommunication networks.

Transmission Control Characters A group of characters used to facilitate or control data transmission. Examples are NAK (Not acknowledge) and EOT (end of transmission).

Transmission Control Protocol TCP. A specification for software that bundles outgoing data into packets (and bundles incoming data), manages the transmission of packets on a network, and checks for errors. TCP is the portion of the TCP/IP protocol suite that governs the exchange of sequential data.

Transmission Electronics Any of the various devices used in conjunction with different transmission media to convert from one transmission method to another. Transmission electronics devices typically include multiplexing equipment and Asynchronous Data Units.

Transmission Facility A piece of a telecommunications system through which information is transmitted, for example, a multi pair cable, a fiber optic cable, a coaxial cable, or a microwave radio.

Transmission Frame A data structure, beginning and ending with delimiters, that consists of fields predetermined by a protocol for the transmission of user and control data.

Transmission Level The power of a transmission signal at a specific point on a transmission facility. See DECIBEL.

Transmission Level Point TLP. A designated point on a circuit where the transmission level has been specified by the designer. Referencing this point in relation to others in the network can determine the performance of the network.

Transmission Limit The wavelengths above and below which the fiber ceases to be transparent and therefore, can no longer transmit information.

Transmission Loss Total loss encountered in transmission through a system.

Transmission Media Anything, such as wire, coaxial cable, fiber optics, air or vacuum, that is used to carry an electrical signal which has information. Transmission media usually refers to the various types of wire and optical fiber cable used for transmitting voice or data signals. Typically, wire cable includes twisted pair, coaxial, and twinaxial. Optical fiber cable includes single, dual, quad, stranded, and ribbon.

Transmission Objectives A stated set of desired performance characteristics for a transmission system. Characteristics for which objectives are stated include loss, noise, echo, crosstalk, frequency shift, attenuation distortion, envelope delay distortion, etc.

Transmission Payload The interface bit rate minus the overhead bits.

Transmission Protocol 0 (TP0) OSI (Open Systems Interconnection) Transmission Protocol Class 0 (Simple Class). This is the simplest OSI Transmission Protocol, useful only on top of an X.25 network (or other network that does not lose or damage data).

Transmission Protocol 4 (tp4) OSI (Open Systems Interconnections) Transmission Protocol Class 4 (error detection and recover class). This is the most powerful OSI Transmission Protocol, useful on top any type of network. TP4 is the OSI equivalent to TCP (Transmission Control Protocol).

Transmission Security Key TSK. A key that is used in the control of transmission security processes such as frequency hopping and spread spectrum.

Transmissive The way many LCD (liquid crystal display) screens on laptops reflect light.

Transmit Bus In AT&T's Information Systems Network (ISN), the circuit on the backplane of the packet controller that transports message packets from sending device interface modules to the switch module.

Transmit Digital Intertie TDI. A 16-channel serial converter which converts the TDM Data Bus from parallel format to serial format for transmission between nodes.

NEWTON'S TELECOM DICTIONARY

Transmittance The ratio of transmitted power to incident power. In optics, frequently expressed as optical density or percent; in communications applications, generally expressed in decibels.

Transmitter The device in the telephone handset which converts speech into electrical impulses for transmission.

Transmitter Distributor A device in a teletypewriter system which converts the information from the parallel form in which it is used in the keyboard-printer to and from the serial form which it is transmitted on the transmission line.

Transmitter Start Code A coded control character or code sequence transmitted to a remote terminal instructing that terminal to begin sending information.

Transmobile The transmobile (not to be confused with a TRANSPORTABLE) is another type of cellular phone. It is essentially a standard 3-watt mobile unit — without an external battery pack — that can be quickly and easily moved from one vehicle to another. It draws its power from the vehicle's battery via a cigarette lighter plug.

Transmultiplexer A device that takes a bunch of voice analog phone conversations and converts them directly into a T-1 1.544 megabit per second bit stream — without the need for de-multiplexing the bunches down to individual conversations, then digitizing them, then bundling them up into a T-1 digital bit stream. A transmultiplexer does it all in one go.

Transparency 1. A data communications mode that allows equipment to send and receive bit patterns of virtually any form. The user is unaware that he is transmitting to a machine that receives faster or slower, or transmits to him faster or slower, or in a different bit pattern. All the translations are done somewhere in the network. He is unaware of the changes occurring — they are transparent. ISDN is planned to be transparent.

Transparency/Opacity An imaging term. A setting available in many image-processing functions that allows part of the underlying image to show through. 80 percent opacity is equivalent to 20 percent transparency.

Transparent When applied to telephone communications, the provision of a feature or service such as Automatic Route Selection in a such a way that the user is unaware of it and it has no affect on the way he uses the telephone. It's "transparent" to him. Translations are transparent to the telephone user. See TRANSLATIONS and TRANSPARENCY.

Transparent Bridging Transparent bridging is so called because the intelligence necessary to make relaying decisions exists in the bridge itself and is thus "transparent" to the communicating workstations. It involves frame forwarding, learning workstation addresses and ensuring no topology loops exist.

Transparent Mode 1. The operation of a digital transmission facility during which the user has complete and free use of the available bandwidth and is unaware of any inter- mediate processing. Generally implies out-of-brand signaling (also called Clear Channel).

2. In BSC data transmission, the suppression of recognition of control characters, to allow transmission of raw binary data without fear of misinterpretation.

3. An operational mode supported by the T3POS PAD which enables the use of existing credit authorization and data capture link level protocols. This mode requires minimal modifications to the POS (Point Of Sale) terminal, and no modification to the ISP/Credit Card Association (CCA) host system software.

Transparent Networking Transport TNT. A service for transporting of LAN data across WANs in which all responsibility for the WAN transport is assumed by the WAN and is therefore invisible to the LAN.

Transparent Routing A method used by a bridge for moving data between two networks. With this type of routing, the bridge learns which computers are operating on which network. It then uses this information to route packets between networks. It does not rely on the sending computers for its decision-making routine. A special kind of bridge combines the practice of transparent routing with source routing. It is called a source routing transparent (SRT) bridge. It examines each packet that comes by to see if it is using IBM's special source routing protocol. If so, this protocol is used to forward the packet. If not, the transparent method is used. Thus, the SRT bridge will support both IBM and non-IBM network protocols. See also BRIDGE and SRT. Compare with SOURCE ROUTING.

Transponder There are two meanings: 1. A transponder is a fancy name for radio relay equipment on board a communications satellite. Just like its domestic microwave counterpart (which you see along highways), a transponder will receive a signal, amplify it, change its frequency and then send it back to earth. Transponders typically have 36 MHz bandwidth. Full motion, full color TV video requires a 6 MHz analog channel. 2. A transponder on an airline is a slightly different kettle of fish. When a radar signal strikes a airline, it activates an electronic transmitter called a transponder. The transponder sends out a coded signal to the ground radar. The code appears next to the radar image of the plane, allowing the controller to identify each plane under his control.

Transport Driver A network device driver that implements a protocol for communicating between Lan Manager and one or more media access control drivers. The transport driver transfers Lan Manager events between computers on the local area network.

Transport Efficiency An AT&T term for the ability to carry information through a network using no more resources than necessary. Transport efficiency is achieved, for example, by statistical transport, which removes silent intervals from voice, data or other traffic and carries only the bursts of meaningful user information.

Transport Layer Layer 4 in the Open Systems Interconnection (OSI) data communications reference model that, along with the underlying network, data link and physical layers, is responsible for the end-to-end control of transmitted information and the optimized use of network resources. Layer 4 defines the protocols governing message structure and portions of the network's error-checking capabilities. Also serves the session layer. Software in the transport layer checks the integrity of and formats the data carried by the physical layer (layer 1, the network wiring and interface hardware), managed by the data link layer (layer 2) and possibly routed by the network layer (layer 1, which has the rules determining the path to be taken by data flowing through a network). See OSI.

Transport Medium The actual medium over which transmission takes place. Transport media include copper wire, fiber optics, microwave and satellites.

Transport Protocol A protocol that provides end-to-end data integrity and service quality on a network. Windows 95 Resource Kit defines transport protocol as how data should be presented to the next receiving layer in the networking model and packages the data accordingly. It passes data to the network adapter driver through the NDIS interface. See also TRANSPORT PROTOCOL CLASS FOUR.

Transport Protocol Class Four TP4. An International Standards Organization (ISO) transport layer protocol designated as ISO IS 8073 Class Four Service. TP4 has been adopted by the U.S. Department of Defense and specified in the U.S. Government OSI Profile (GOSIP).

Transportable Cellular Phone The transportable cellular phone is a standard 3-watt mobile phone that can be removed from the car

617

and used by itself with an attached battery pack. The entire unit is generally mounted or built into a custom carrying case to make it easy to carry on your shoulder. Although technically "portable," the transportable should not be confused with the true portable one-piece cellular phone. Also known as a "bag phone" or "briefcase phone"; refers to a cellular handset that is packaged with a larger carrying case containing a full-scale power supply.

Transposition Interchanging the relative position of conductors at regular intervals to reduce crosstalk. In data transmission, a transmission defect in which, during one character period, one or more signal elements are changed from one significant condition to the other, and an equal number of elements are changed in the opposite sense.

Transverse Interferometry The method used to measure the index profile of an optical fiber by placing it in an interferometer and illuminating the fiber transversely to its axis. Generally, a computer is required to interpret the interference pattern.

Transverse Parity Check Type of parity error checking performed on a group of bits in a transverse direction for each frame. See PARITY CHECK.

Transverse Scattering The method for measuring the index profile of an optical fiber or preform by illuminating the fiber or preform coherently and transversely to its axis, and examining the farfield irradiance pattern. A computer is required to interpret the pattern of the scattered light.

Trap 1. See TRAP and TRACE.
2. A programming term. A programmer sets a trap for something to happen when something else happens. You might say "Wait for the mouse to come by, when it does, close the trap." A trap might be sprung when a phone rings or when someone hangs up. In network management, a trap is a mechanism permitting a device to automatically send an alarm for certain network events to a management station. Typically, network management information is gained by polling network nodes on a regular basis. This strategy can be modified when a trap is set from a network node. With traps, a node alerts the management station of a catastrophic problem. The management station can then immediately initiate a polling sequence to the node to determine the cause of the problem. This strategy is often called trap-directed polling.

Trap And Trace A telephone company term. Trap and Trace is the term for equipment and procedures for determining the source of an incoming call (typically a harassing call). The phone company uses traps to trace the source of the incoming call. There are two types of traps — the Terminating Trap and the Originating Trap. A terminating trap sits on the receiving phone line. In the old days, a terminating trap was a physical piece of equipment. These days, with electronic central offices, it's basically a command to the computer running the central office to keep track of all information about the source of all incoming calls. That information might be the originating telephone number. It might be the trunk number on which the call came in on. Such trunk number might look like TGN701. Or it might be the CLLI code — which stands for the customer locator identifier. The CLLI code (pronounced "silly") consists of 11 characters. A sample CLLI code is "nycmny18dso." That says the call is coming in from New York City, Manhattan from a central office called 18DSO (which I happen to know is an AT&T 5E located on West 18th Street. Once the terminating trap identifies the possible direction /source / incoming trunk of the offending phone calls, the phone company will work its back towards the originating line. It will attach an Originating Trap to the offending trunk, then to the offending tandem office, then to the central office. This can be a tedious and time consuming business. With the advent of Caller ID — both local and nationwide — trapping and tracing is getting a whole lot faster and easier. Now if you receive an harassing call, you simply hit *57 the moment the call is over. This "tags" the incoming call's number and other information in the your central office's records. You, as subscriber, can't get access to that information. But a law enforcement agency (i.e. one investigating your annoying calls) can. See ANNOYANCE CALL BUREAU and WIRE TAP.

Trap Door Hidden software or hardware mechanism that, when triggered, allows system-protection mechanisms to be circumvented.

Trashware Software that is so poorly designed that it winds up in the garbage can.

TRAU Transcoder and Rate Adapter Unit. A transmission function of the BSS that converts speech from the user of a mobile station into digital representation needed for an ISDN, wireless network.

Travel Card Another name for a telephone credit calling card. Travel card calls that are placed against a travel card number issued by the service provider, typically a phone company. As each call is completed, the long distance switch increases that card's account balance by the amount of each call. During the processing of a call, if the travel card is invalid or if the caller does not respond to a system prompt, the serving switch will typically ask the caller to hold the line for a live operator, and transfers the call to an Operator Workstation. When the operator answers, the OWS screen shows call information, including card number (if already entered), destination number (if already entered), trunk identification, and a failure code.

Traveling Class Mark TCM. A code that accompanies a long distance call. When Automatic Route Selection (ARS) or Uniform Numbering/Automatic Alternate Routing (UN/AAR) selects a tie trunk to a distant tandem PBX, the traveling class mark (TCM) is sent over the tie trunk. It is then used by the distant system to determine the best available long distance line consistent with the user's calling privileges. The TCM indicates the restriction level to be used based on the phone, trunk or attendant originating the call or the authorization code, if dialed.

Tray See RACK.

Tree 1. A network topology shaped like a branching tree. (What else?) It is characterized by the existence of only one route between any two network nodes. Most CATV distribution networks are tree networks.
2. In MS-DOS, a tree describes the organization of directories, subdirectories, and files on a disk.

Tree Hugger IBM-speak for an employee who resists a move or any other change.

Tree Mailbox A special function mailbox that provides the caller with a menu and allows selections from the menu using single digit commands.

Tree Search In a tree structure, a search in which it is possible to decide, at each step, which part of the tree may be rejected without further search.

Tree Structure Describes the organization of directories, subdirectories, and files on a disk.

Tree Topology A network cabling architecture in which nodes are connected by cables to a central, or trunk, cable with a central retransmission capability.

Trellis Coding A method of forward error correction used in certain high-speed modems where each signal element is assigned a coded binary value representing that element's phase and amplitude. It allows the receiving modem to determine, based on the value of the preceding signal, whether or not a given signal element is received in error. See V.32 and V.32 bis.

Trellis Coding Modulation TCM. A modem modulation technique in which sophisticated mathematics are used to predict the best fit between the incoming signal and a large set of possible combinations of amplitude and phase changes. TCM provides for transmission speeds of 14,400 bps and above on single voice grade phone lines. See V.32 and V.32 bis.

Tremendously High Frequency Frequencies from 300 GHz to 3000 GHz.

TRFR TRansFeR.

TRI-CWDM On August 30, 1995, MCI Communications announced the deployment of a technology that will enable it to increase the capacity of its network by 50 percent without any additional fiber optic lines. The technology, known as Tri-Color Wave Division Multiplexing (Tri-CWDM), allows existing fiber to accommodate three light signals instead of two, by routing them at different light wavelengths, through the combined use of narrow and wide band wave division multiplexing.
With this method, lightwaves are transmitted at 1557 nanometers (nm) and 1553 nm to a wide band WDM device, where a 1310 nm signal is added. Once combined, the three signals are routed through a single fiber to the next site where they are separated and sent to the receivers. Transmitting three signals in each direction allows for three different transmit pairs on just two fibers, effectively increasing the total network capacity from 5 gigabits to 7.5 gigabits. MCI officials say the technology will be particularly valuable in major metropolitan areas, where the company is enjoying outstanding growth in voice and data traffic.

Triaxial Cable A cable construction having three coincident axes, such as a conductor, first shield and second shield all insulated from one another.

Tribit Transmission A transmission technique used by some modems in which three bits are transmitted simultaneously.

Tributary The lower rate signal input to a multiplexer for combination (multiplexing) with other low rate signals to form an aggregate higher rate signal.

Tributary Circuit A circuit connecting an individual phone to a switching center.

Tributary Office A local office, located outside the exchange in which a toll center is located, that has a different rate center from its toll center.

Tributary PBX An exchange within the main PBX configuration but with its own listed number. The only difference between a satellite and a tributary PBX is that the tributary PBX has a direct incoming connection from the public network. See SATELLITE PBX.

Tributary Station In a data network, a station other than the control station. On a multi point connection or a point-to-point connection using basic mode link control, any data station other than the control station.

Trichromatic The technical name for RGB representation of color to create all the colors in the spectrum.

Trickle The name of a BITNET mail server package which provides access to anonymous FTP archive sites via e-mail.

Trigger The combination of the occurrence of an event and the satisfaction of trigger criteria. Definition from Bellcore in reference to its concept of the Advanced Intelligent Network.

Triggering The process of detecting a word (or utterance) and capturing the speech data associated with that word (or utterance) for subsequent processing.

Triggers Uncompiled code residing on an intelligent database server.

Triode A combination of a heated cathode, a relatively cold anode, and a third electrode for controlling the current flowing between the other two; the whole enclosed in an evacuated bulb. Variously called, audion, pliotron, radiotron, oscillion, audiotron, aerotron, electron tube, vacuum tube, etc.

Trivial File Transfer Protocol TFTP. A UNIX-based file protocol. TFTP is a simplification of the earlier Simple File Transfer Protocol (SFTP).

TRL Transistor Resistor Logic.

Trojan Horse Software that appears to do something normal but which contains a trap door or attack program. A Trojan Horse program can be used to break into a network through a World Wide Web site. A Trojan Horse is dangerous software.

Troposphere The lower layers of the earth's atmosphere. You can bounce certain frequency radio signals off it and use it as an elementary transmission reflector. The troposphere is the region where clouds form, convection is active, and mixing is continuous and more or less complete. The lower layers of the Earth's atmosphere, between the surface and the stratosphere, in which about 80 percent of the total mass of air is located and in which temperature normally decreases with altitude. The thickness of the troposphere varies with season and latitude; it is usually 16 km to 18km over topical regions and 10 km or less over the poles. See TROPOSPHERIC SCATTER and TROPOSPHERIC WAVE.

Tropospheric Scatter The propagation of radio waves by scattering as a result of irregularities or discontinuities in the physical properties of the troposphere. The propagation of electromagnetic waves by scattering as a result of irregularities or discontinuities in the physical properties of the troposphere. A method of transhorizon communications using frequencies from approximately 350 MHz to approximately 8400 MHz. The propagation mechanism is still not fully understood, though it includes several distinguishable but changeable mechanisms such as propagation by means of random reflections and scattering from irregularities in the dielectric gradient density of the troposphere, smooth-Earth diffraction, and diffraction over isolated obstacles (knife-edge diffraction).

Tropospheric Wave A radio wave that is propagated by reflection from a place of abrupt change in the dielectric constant or its gradient in the troposphere. In some cases, the ground wave may be so altered that new components appear to arise from reflection in regions of rapidly changing dielectric constant. When these components are distinguishable from the other components, they are called "tropospheric waves."

Trouble Number Display The operator will know what the trouble is with the phone system by seeing a number pop up on her/his console. That number may pop up automatically or the operator may have to hit the ALM (for ALARM) or similar button.

Trouble Ticket Form used to report problems. Often incorrectly filled-in. Check.

Trouble Unit A weighting figure applied to telephone circuit or circuits to indicate expected performance in a given period.

Troubles Per Hundred Troubles per hundred is a criterion for acceptable customer service which telephone companies and public utility commissions have agreed upon. It's measured in terms of the number of complaints received per hundred telephones in one month. Six complaints per hundred is considered the maximum for acceptable service. See QUALITY OF SERVICE.

TRS-80 One of the early PCs. It was introduced by Radio Shack through its Tandy stores. It was based on the Zilog Z80 chip and began selling in the late 1970s. Along with with Apple and Commodore it proved the viability of personal computers. The TRS-80 helped insure the success of later generations of PCs by introducing a spreadsheet called Visicalc, word processing such as Electric Pencil, WordStar and databases such as dBASE. It also was one of Microsoft's earliest customers for their Basic language package. There were several TRS-80 models including the original Model 1, Model 2, Xenix (Unix), Model 3 with integrated drives and Model 4. Some of the models could run CP/M or TRSDOS. Tandy later also introduced the Model 100 which was, arguably the first Laptop/Notebook, and for which Bill Gates is alleged to have written much of the software code.

TRU Inside a telephone system, the TRU (the Tone Receiver Unit) is used to retrieve and interpret touch-tone data received. Those tones might be sent by a TSU — Tone Sender Unit.

Tru Server Telco Research's client/server-based telemanagement and facilities management software system that automates and integrates both voice and data communications management operations. See TRU SYSTEM.

Tru System PC-based call accounting software from the fine people at Telco Research Nashville, TN. Includes call recording and polling, directory, ISDN recording, Centrex call detail capture, cost allocation, traffic statistics, network optimization, bill verification, client billback, inventory management, hospitality and toll fraud detection applications.

TrueSpeech TrueSpeech is a low-bandwidth method of digitizing speech, which was created by a company called DSP Group, Inc. Santa Clara, CA. TrueSpeech uses compression to drop one minute of voice down to 62 kilobytes with remarkably little degradation. It is used in many digital telephone answering devices for storing and reproducing voice. TrueSpeech's compression is not meant for high fidelity music. But it is more than acceptable for such business applications as voice mail, voice annotation, dictation, and education and training. The small file size means that it can be transferred more easily to other users by using either a corporate network. The DSP Group describe TrueSpeech an enabling technology for speech compression in personal computers and future personal communications devices. Speech compression is key technology to the effective convergence of personal computers and telephony. TrueSpeech compression is a technology based on complex mathematical algorithms which are derived from the way airflow from our lungs is shaped by the throat, mouth, and tongue when we speak. This shaping is what our ear finally hears. TrueSpeech is 5 to 15 times more efficient than other methods of digital speech compression. For example, a one minute long speech file which uses other PC audio technology would consume as much as 960 kilobytes. With TrueSpeech, the same file would be just over 60 kilobytes. TrueSpeech is used in the Microsoft Sound System, which also lets you choose the voice sampling you wish when you're recording material. Here is Microsoft Sound System's recording options:

	Sampling Rate	Technology	
TrueSpeech	62K per minute	8 KHz	Proprietary
Voice	234K per minute	8 KHz	ADPCM
Radio	322K per minute	11 KHz	ADPCM
Tape	1291K per minute	22 KHz	PCM
CD	5176K per minute	44 Khz	PCM

The above is for mono recordings. For stereo, double the amount of space.

Truetype A Windows 3.1 feature. Fonts that are scalable and sometimes generated as bitmaps or soft fonts, depending on the capabilities of your printer. TrueType fonts can be sized to any height, and they print exactly as they appear on the screen. Using TrueType, you'll be able to create documents that retain their format and fonts on any Windows 3.1 machine — even if the fonts aren't installed on that computer. This makes Windows 3.1 documents portable.

Truevoice 1. In the fall of 1993, AT&T announced that it was introducing new voice quality throughout its long distance network. And that it was calling that quality "true voice." AT&T set up a demo line. Some people thought they could notice an improvement. Some thought they couldn't. I personally thought true voice sounded pretty good.
2. The trademark name of Centigram's text-to-speech product, which they acquired from SpeechPlus.

Truncated Binary Exponential Back Off Another name for exponential back off used in IEEE 802.3 local area networks. In an exponential back-off process, the time delay between successive attempts to transmit a specific frame is increased exponentially.

Truncation In data processing, the deletion or omission of a leading or a trailing portion of a string in accordance with specified criteria.

Trunk A communication line between two switching systems. The term switching systems typically includes equipment in a central office (the telephone company) and PBXs. A tie trunk connects PBXs. Central office trunks connect a PBX to the switching system at the central office.

Trunk Access Number 1. The number of the trunk over which a call is to be routed.
2. The number that needs to be dialed in order to gain access to an outbound trunk. This applies to both local and local distance trunks, as the access number can be different.

Trunk Answer From Any Phone When a call comes in, something rings. You can now answer the incoming call from any phone. To do so, you must dial a special code or hit a special feature button on your phone. When TELECONNECT'S PBX bells ring, all we have to do is to touch "6" on any phone and we can answer the incoming call.

Trunk Data Module TDM. Provides the interface between the DCP signal and a modem or Digital Service Unit (DSU).

Trunk Direct Termination An option on switchboards which terminates a trunk group on one key (or button) on the console.

Trunk Encryption Device TED. A bulk encryption device used to provide secure communication over a wideband digital transmission link. It is usually located between the output of a trunk group multiplexer and a wideband radio or cable facility.

Trunk Exchange A telephone exchange dedicated primarily to interconnecting trunks.

Trunk Group A group of essentially like trunks that go between the same two geographical points. They have similar electrical characteristics. A trunk group performs the same function as a single trunk, except that on a trunk group you can carry multiple conversations. You use a trunk group when your traffic demands it. Typically, the trunks in a trunk group are accessed the same way. You dial your Band 5 WATS trunk group by dialing 62, for example. If the first trunk of that group is busy, you choose the second, then the third, etc. See TRUNK HUNTING.

Trunk Group Alternate Route The alternate route for a high-usage trunk group. A trunk group alternate route consists of all the trunk groups in tandem that lead to the distant terminal of the high-usage trunk group.

Trunk Group Multiplexer TGM. A time division multiplexer whose function is to combine individual digital trunk groups into a higher rate bit stream for transmission over wideband digital communication links.

Trunk Group Warning Alerts the attendant when a preset number of trunks in a group are busy. See TRUNK GROUP.

Trunk Holding Time The length of time a caller is connected with a voice processing system. Defined from the time when the system goes off-hook to the time the port (i.e. the trunk) is placed back on hook.

Trunk Hunting Switching incoming calls to the next consecutive number if the first called number is busy.

Trunk Make Busy A fancy name for saying that, by punching a few buttons on the console, you can make any trunks in your PBX or key system busy, effectively putting the trunk out of service. You may want to do this if your trunk is acting up. By busying it out at the console, you are effectively denying its use to anyone in the company. Thus you are protecting yourself from further complaints. Hopefully, it will be repaired promptly.

Trunk Monitoring Feature which allows individual trunk testing to verify supervision and transmission. You dial an access code and then the specific trunk number from the attendant console. You want the ability to test a specific trunk because normally you might be only accessing a trunk group when you dial an access code. Thus, each time you dial into the trunk group, you might end up on another individual trunk. Some PBXs have a variation of trunk monitoring, whereby if a user encounters a bad trunk, he can dial a specific code, then hang up. The PBX recognizes these digits and makes a trouble report on that specific trunk, possibly reporting it to the operator, keeping it in memory for later analysis or dialing a remote diagnostic center and reporting its agony.

Trunk Number Display The specific trunk number of an incoming call can be displayed on the attendant console, enabling your attendant to instantly identify the origin of certain calls. For example, if you have several tie lines to branch offices, your attendant knows immediately which office is calling. Many newer PBXs have displays on individual telephones, which show the actual trunk being used for outgoing and incoming calls. This provides an additional measure of control. You might, for example, speak faster if you knew the call was coming in on your IN-WATS line. You might also answer the call differently if you know what trunk it's coming in on. For example, you might be running several, totally-separate businesses from the same console. Each business has a different number. The only way you know what to answer — Joe's Bakery or Mary's Real Estate — is by the trunk.

Trunk Occupancy The percentage of time (normally an hour) that trunks are in use. Trunk occupancy may also be expressed as the carried CCS per trunk.

Trunk Order A document (or data system equivalent) used in an operating telephone company to request a change to a trunk group.

Trunk Queuing A feature whereby your phone system automatically stacks requests for outgoing circuits and processes those requests on, typically, a first-in/first-out basis. See QUEUING THEORY.

Trunk Reservation The attendant can hold a single trunk in a group and then extend it to a specific phone. This means, for example, that a WATS line can be held for someone special — a heavy caller, the president of the firm, etc.

Trunk Restriction Some people may not be allowed to use certain trunks at certain times. The sophistication of trunk restriction depends on the switch and the way it's programmed.

Trunk Segment The main segment of cable in an Ethernet network is called the trunk segment.

Trunk Side Trunk side connections are those that attach to other switching systems.

Trunk Type TT. Trunks that use the same type of equipment going to the same terminating location.

Trunk Type Master File TTMF. An MCI definition. A comprehensive listing of all trunk assignments on the MCI network for shared and dedicated services, necessary for processing and billing MCI customer calls.

Trunk To Tie Trunk Connections The ability of the switching system to provide the attendant with the capability of extending an incoming trunk call to a tie trunk terminating some place else.

Trunk To Trunk By Station A PBX feature which permits the user who established a three-way conference involving himself and two trunks to drop from the call without disconnecting the trunk-to-trunk connection.

Trunk To Trunk Connections The attendant can establish connections between two outside parties on separate trunks. Call your office on your IN-WATS. Ask the operator to extend that call to the VP who happens to be at his home. The operator must place an outside call to the VP on an outside trunk and join that call to the incoming call. Sometimes it works.

Trunk To Trunk Consultations Allows a phone connected to an outside trunk circuit to gain access to a second outside trunk for "outside" consultation. No conference capability is available with this feature.

Trunk Transfer By Station Permits the user who established a three-way conference involving two lines to drop from the call without disconnecting the trunk-to-trunk connection.

Trunk Verification By Customer Provides the attendant or phone user access to individual lines in a trunk group to check their condition. See also TRUNK MONITORING.

Trunk Verification By Station Provides a warning tone if a phone user enters a busy trunk.

Trunks In Service The number of trunks in a group in use or available to carry calls. Trunk in service equals total trunks minus the trunks broken or made busy for any reason.

Truth Table An operation table for a logic operation. A table that describes a logic function by listing all possible combinations of input values and indicating, for each combination, the output value.

TRX Transmitter and Receiver (Transceiver). A wireless telecommunications term. A function of the radio channel device for receiving and transmitting signals or information on the radio channel.

TRXXX Various Bellcore standards. See ISDN.

TS Transport Stream. As an ATM term, it is one of two types of streams produced by the MPEG-2 Systems layer. The Transport Stream consists of 188 byte packets and can contain multiple programs.

TS Traffic Shaping: Traffic shaping in an ATM network is a mechanism that alters the traffic characteristics of a stream of cells on a connection to achieve better network efficiency, while meeting the QoS (Quality of Service) objectives, or to ensure conformance at a subsequent interface. Traffic shaping must maintain cell sequence integrity on a connection. Shaping modifies traffic characteristics of a cell flow with the consequence of increasing the mean Cell Transfer Delay.

TS Time Stamp: As an ATM term, Time Stamping is used on OAM cells to compare time of entry of cell to time of exit of cell to be used to determine the cell transfer delay of the connection.

TSAC Time Slot Assigner Circuit; a circuit that determines when a CODEC will put its eight bits of data on a RCM bit stream.

TSAP Abbreviation for Transport Service Access Point in the OSI transport protocol layer.

TSAPI Telephony Server Application Programming Interface. Described by AT&T, its inventor, as "standards-based API for call control, call/device monitoring and query, call routing, device/system maintenance capabilities, and basic directory services." For a better explanation, see TELEPHONY SERVICES.

TSCM Technical Surveillance CounterMeasures. Commonly called debugging, sweeps or electronic sweeping.

TServer Telephony Server. Name of NetWare Telephony Services LAN server which is joined physically (by wire) to an adjacent PBX. See TELEPHONY SERVICES.

TSI 1. Time Slot Interchange or Interchanger. A way of temporarily storing data bytes so they can be sent in a different order than they were received. Time Slot Interchange is a way to switch calls.
2. Transmitting Subscriber Information. A frame that may be sent by the caller, with the caller's phone number, which may be used to screen calls, etc.

TSIC Time Slot Interchange Circuit; a device that switches digital highways in PCM based switching systems. In short, a digital crosspoint switch.

TSIU Time Slot Interchange Unit. Switching module hardware unit that provides the digital time switching function.

TSK Transmission Security Key.

TSM Terminal Server Manager (TSM), a program that allows terminal servers on a network to be remotely managed from another node. It is supported on VMS systems running the LAT protocol (and is incompatible with TCP/IP-only networks).

TSO Time Share Operation.

TSP 1. Telecommunication Service Priority.
2. Terminal Service Profile.

TSPS Traffic Service Position System permits operator positions serving public phones and HOBIC operations to be located remotely from the CO which services the pay phone or the hotel, or the hospital, etc.

TSR 1. Telephone Service Representative. See also AGENT.
2. Terminate and Stay Resident. A term for loading a software program in an MS-DOS computer in which the program loads into memory and is always ready for running at the touch of a combination of keys, e.g. Alt M, or Ctrl ESC. Here's some information from Jackie Fox writing in PC Today: You can't load TSRs willy-nilly and expect them to work with each other. Some will get along with each other. Others won't. When you install a TSR, it goes to a location in RAM (Random Access Memory) called the Interrupt Vector Table.

The interrupt vector table is like a hotel lobby, and TSRs are like guests waiting for messages. The TSR watches every incoming keystroke to see if it's the special hot key combination (message), the TSR is waiting for. If it isn't, the TSR passes it back to the regular program. What if you have four or five TSRs loaded? The one you loaded last has seniority. It checks the incoming keystrokes first. If the TSR recognizes the keystroke combination as its own hot-key combination, it takes over. If not, it passes it along to the next TSR. This process is called interrupt handler chaining.

If none of the TSRs recognize that particular combination, they pass it along to DOS so it can process it as a regular keystroke combination. Not all TSRs pass instructions along the way they should. Some TSRs intercept keystrokes and never pass them on. Some TSRs never restore their original addresses. Sometimes two TSRs fight over the same hot key combination. Then you end up with a frozen keyboard. The basic problem is there are no rules for loading and running TSRs.

TSRM Telecommunication Standards Reference Manual.

TST Time-Space-Time system.

TSTN Triple Super Twisted Nematic. A display technology often used on laptop computers which uses three layers of crystal to give better contrast and more grey scales.

TSTS Transaction Switching and Transport Services. In 1992, the Regional Bell Operating Companies (RBOCs) agreed to provide uniform transaction processing capabilities under a banner called Transaction Switching and Transport Services.

TSU Tone Sender Unit. A device inside a telephone system. The TSU passes along touch-tone digits to telephone extension cards within the phone system. See also TRU (the Tone Receiver Unit) which is used to retrieve and interpret touch-tone data received.

TSYN The SYNCHronous test line provides for testing of ringing, tripping and supervisory functions of toll completing trunks. See TNSS.

TTC The Telecommunications Technology Committee, a Japanese standards committee.

TTCN Tree and Tabular Combined Notation: The internationally standardized test script notation for specifying abstract test suites. TTCN provides a notation which is independent of test methods, layers and protocol.

TTD Temporary Text Delay. The TTD control sequence (STX ENQ) is sent by a sending station in message transfer state when it wants to retain the line but is not ready to transmit.

TTI Transmit Terminal Identification. A fax machine's stupid term for its telephone number and the name of its owner. When you receive a fax from someone, the top line of the fax typically will have a phone number and a name on it. That phone number and name does NOT come from the phone or the phone company. It comes from what the person who owns the machine programmed into his machine. He typically did that by punching buttons on his fax machine. He'll do that if he can understand the instruction booklet which came with his fax (which he probably won't). The point of all this is twofold: First, don't forget to put your name and phone number into your fax machine. Second, don't assume that what you read at the top of any fax you receive is accurate.

TTL Transistor Transistor Logic.

TTR Touch Tone Receiver. A device used to decode touchtones dialed from single-line telephones or Remote Access telephones.

TTS 1. Text-To-Speech. A term used in voice processing. See TEXT-TO-SPEECH.
2. Transaction Tracking System. A Novell NetWare feature that protects database applications from corruption by backing out incomplete transactions that result from a failure in a network component. When a transaction is backed out, data and index information in the database are returned to the state they were in before the transaction began.

TTTN Tandem Tie Trunk Network.

TTY A teletypewriter. Typewriter-style device for communicating alphanumeric information over telecom networks. TTY is the most widely used type of emulation for PC computer communications.
TTY/TDD A unique telecommunication device for the deaf, using TTY principles.
TU Transmit Unit. Term used in a DS-3 channel bank.
TUA Telecommunications Users Association (UK).
TUANZ Telecommunications Users Association of New Zealand.
TUBA TCP and UDP with Bigger Address. One of the 3 IPng candidates.
TUG Telecommunication User Group.
TUI Telephony User Interface. Like GUI, except that it means that someone is using the telephone to get to information in a computer. See GUI.
Tumbling A form of cellular fraud. The crook uses a cellular telephone with a bogus electronic serial number that disguises it as a telephone from outside the local calling area. By the time the local cellular phone network operator has checked with the network where the bogus telephone supposedly is registered and discovered the fraud, the crook has tumbled the telephone, or changed its electronic serial number (typically by one digit) and is ready to make more free calls. See also CLONE FRAUD.
Tunable Operating System Parameters Tuning an operating system is the same as optimizing it, in that you rewrite commands and programs so they operate faster and more efficiently. Any new operating system needs to be tuned to the specific machine on which it is running.
Tungsten A metallic element used in ceramic IC packaging to provide the traces within the package that connect the device circuitry to the external terminals pads or leads.
Tunneling A local area network term. Tunneling means to temporarily change the destination of a packet in order to traverse one or more routers that are incapable of routing to the real destination. For example, to route through a backbone whose internal routers don't contain entries for external destinations, the entry border router must "tunnel" to the exit border router.
Tunneling Ray Leaky ray.
TUR Traffic Usage Recorder.
Turing Machine A mathematical model of a device that changes its internal state and reads from, writes on, and moves a potentially infinite tape, all in accordance with its present state, thereby constituting a model for computer-like behavior. This is the same Alan Turing who coined the Turing test which I mention under artificial intelligence.
Turnaround Time On a half-duplex channel, the time during which the transmission flow is stopped each time the direction of transmission is reversed.
Turbo FAT Turbo FAT is an index NetWare v2.2 creates to group all the FAT (File Allocation Table) entries corresponding to a file larger than 262,144KB. The first entry in the turbo FAT index table consists of the first FAT number of the file. The second entry consists of the second FAT number of the file, etc. The turbo FAT enables a large file to be accessed quickly.
Turbocharging A phone fraud term. Turbocharging is the practice of increasing phone charges by adding on extra time onto each call in the hope that the customer won't notice.
Turn Toward Utility Rate Normalization. A non-profit education organization that gets it funding from the California Public Utilities Commission. Turn's goal is "to help consumer organizations better educate their clients or members about the changes taking place in the telecommunications industry and in the way the industry is regulated. It's based in San Francisco.
Turnaround Time The actual time required to reverse the direction of transmission from sender to receiver or vice versa when using a half-duplex circuit. The turnaround time is needed for line propagation effects, modem timing and computer reaction.
Turnkey System An entire phone system with hardware and software assembled and installed by a vendor and sold as a total package. The term "turnkey" means the buyer is presented with the key to the thing he has just bought. He turns the key and the system will do everything it is supposed to do, including work. Most telephone systems are purchased Turnkey. An integral part of a contract to buy a turnkey phone system is the terms and conditions for the acceptance of the system. Someone has to define what it means for the thing to work, what you expect from it — so you, the buyer, can formally accept the system and thus incur an obligation to pay for it. Defining Acceptance Conditions is no small task on bigger phone systems.
Turret A very large key system for financial traders, emergency teams at nuclear power stations and others who need single phone button access to hundreds of people. By simply pushing one button, the user can dial one of hundreds of people. These buttons may be connected to tie lines, foreign exchange lines. They may even be DDD lines with autodial capability. Like all good key systems, the buttons have a lamping display which shows if the particular line is idle, busy, ringing, on hold, etc.
TVC See TRUNK VERIFICATION BY CUSTOMER.
TV/PC Bill Gates' new name for a TV with PC smarts.
TVM See Time-Varying Media.
TVM Object An SCSA definition. An encapsulation of an atomic piece of time-varying media. This encapsulation may be the data itself or a reference to the data.
TVS See TRUNK VERIFICATION BY STATION.
TWAIN TWAIN is a protocol intended to make easy the communication between software applications and image-acquisition devices, such as cameras and scanners. It is, in essence, a cross-platform application interface standard for image capturing. It allows you to bring images into imaging programs (like HiJaak Pro, Hotshot Graphics, DocuWare, Documagic) from your graphics hardware — for example, desktop scanners, hand-held scanners, slide scanners, frame grabbers and digital cameras. If your hardware is TWAIN compliant and if you have installed the correct driver, you should be able to use that imaging hardware with any TWAIN complaint application. The TWAIN protocol is the most popular protocol for imaging sources and has become an industry standard. Any application that supports TWAIN can communicate with any TWAIN-compliant imaging device. TWAIN is spearheaded by Hewlett-Packard, Logitech, Eastman Kodak, Aldus, Caere and other imaging hardware and software vendors. It was previously known as CLASP and "Direct Connect" during its development stage. Apparently the term TWAIN comes from "Toolkit Without An Interesting Name." Despite its silly name, it is a very serious standard. There are several key elements to TWAIN including:
Application Layer — This is the application that controls and uses the TWAIN resource.
Protocol Layer — This contains the TWAIN Source Manager (the code that communicates between

the application and the Source).

Acquisition Layer — This is software that controls the image acquisition device. The application layer is developed by the device manufacturer. It can be thought of as a hardware device driver.

Device Layer — This is the physical device, such as a scanner.

TWAIN Working Group An industry organization dedicated to developing and advancing software standards for the imaging world. See TWAIN.

Tweak Freak A computer techie obsessed with finding the root of all tech problems, regardless of the relevance. A tweak freak might spend hours trying to track down something that could instantly be fixed by reinstalling the software.

Twin Cable A cable composed of two insulated conductors laid parallel and either attached to each other by the insulation or bound together with a common covering.

Twinax Twinaxial Cable made up of two central conducting leads of coaxial cable. See TWINAXIAL CABLE.

Twinaxial Cable Two insulated conductors inside a common insulator, covered by a metallic shield, and enclosed in a cable sheath. Because it carries high frequencies, twinaxial cable is often used for data transmission and video applications, especially for cable television.

Twinning (pronounced twin-ning.) The act of paralleling systems to work together. An example is connecting a wireless system to the same CO line as a key system, so the user can use either instrument to access the trunk.

Twinplex A frequency-shift-keyed, carrier telegraphy system in which four unique tones (two pairs of tones) are transmitted over a single transmission channel (such as one twisted pair). One tone of each tone pair represents a "mark," and the other, a "space."

Twist 1. A change, as a function of temperature, in the response characteristic of a transmission line.
2. The amplitude ratio of a pair of DTMF tones. Because of transmission and equipment variations, a pair of tones that originated equal in amplitude may arrive with a considerable difference in amplitude. In short, signals at different frequencies are transmitted with differing response by the transmission system. Twist usually refers to distortion of DTMF signals.

Twisted Pair Two insulated copper wires twisted around each other to reduce induction (thus interference) from one wire to the other. The twists, or lays, are varied in length to reduce the potential for signal interference between pairs. Several sets of twisted pair wires may be enclosed in a single cable. In cables greater than 25 pairs, the twisted pairs are grouped and bound together in a common cable sheath. Twisted pair cable is the most common type of transmission media. It is the normal cabling from a central office to your home or office, or from your PBX to your office phone. Twisted pair wiring comes in various thicknesses. As a general rule, the thicker the cable is, the better the quality of the conversation and the longer cable can be and still get acceptable conversation quality. However, the thicker it is, the more it costs. Gives new meaning

Twitch Game A computer or arcade game that's all hand-eye coordination and little brain. Similar to "thumb candy."

Two Dimensional Coding A data compression scheme in facsimile transmission that uses the previous scan line as a reference when scanning a subsequent line. Because an image has a high degree of correlation vertically as well as horizontally, two-dimensional coding schemes work only with variable increments between one line and the next, permitting higher data compression. See ONE DIMENSIONAL CODING.

Two Electrode Vacuum Tube A vacuum tube having a hot cathode and a relatively cold anode, i.e., one with filament and plate only.

Two Hots In Outlets In AC electrical power, more than one HOT conductor has been incorrectly connected to the terminals in the outlet being tested. Dangers include extreme fire hazard and/or major damage to equipment plugged into the outlet.

Two Out Of Five Code A decimal code system in which each decimal digit is represented by five binary bits, two of which are ones and three are zeroes.

Two Party Hold On Console Allows an attendant to hold a call with both a calling and a called phone (or trunk) connected. Such a feature is required for activation of Attendant Lockout, Serial Call and Trunk-to-Trunk connections features.

Two Party Station Service PBX system with two internal phones, each with selective ringing. Resembles rural two party service of old.

Two Pilot Regulation In FDM systems, the use of two pilot frequencies within a band so that the change in attenuation due to twist can be detected and compensated for by a regulator.

Two Tier Pricing A complex and now largely obsolete AT&T pricing plan which imposed two monthly "rate elements" on every hardware piece of an AT&T telephone system. Tier A was a fixed rate, not subject to rate increases. It was fixed for a certain number of months, say 60. It was, allegedly, to pay for the system. At the end of the 60 months, Tier A disappeared, as though it were a full-payout lease and you now owned the equipment (which you didn't.) Tier B is the second element in this pricing scheme. It covers maintenance, and it is subject to rate increases. AT&T doesn't offer two-tier pricing any longer. Many two-tier contracts are now finding their Tier A payments ceasing. Remember, the equipment still belongs to AT&T.

Two Tone Key Same as frequency shift keying.

Two Tone Keying In telegraphy systems, a system employing a transmission path composed of two channels in the same direction, one for transmitting the "space" binary modulation, the other for transmitting the "mark" of the same modulation; or that form of keying in which the modulating wave causes the carrier to be modulated with a single tone for the "marking" condition and modulated with a different single tone for the "spacing" condition.

Two Way Alternate Operation Transmission in one direction or the other but not in both simultaneously. Most often referred to as half-duplex transmission.

Two Way Simultaneous Operation Transmission and reception at the same time. More often referred to as full-duplex transmission.

Two Way Splitting PBX feature. Allows a telephone user to jump back and forth between two calls. Try this: Someone calls you. You both decide you want to speak to a third person. You call that person and conference the three of you together. Then you decide you want to consult with one of the people confidentially. So you "split" one from the other and you speak to one. Then you swap back and forth between the two, speaking to one and then the other in complete privacy. It's easier to do this sort of complicated phone transaction on a phone with a LCD screen. Fortunately, these are becoming more common these days.

Two Way Trade A call center term. A schedule trade in which both employees are working each other's schedules.

Two Way Trunk A trunk which can be seized from either end. Can be used to carry conversations into or out of a telephone system, i.e. most trunks. Some trunks are set up as one-way only. A classic one-way trunk is a IN-WATS line. It is designed to only receive calls.

NEWTON'S TELECOM DICTIONARY

Two Wire Circuit A transmission circuit composed of two wires — signal and ground — used to both send and receive information. In contrast, a four wire circuit consists of two pairs. One pair is used to send. One pair is used to receive. All trunk circuits — long distance circuits — are four wire. A four wire circuit costs more but delivers better reception. All local loop circuits — those coming from a Class 5 central office to the subscriber's phone system — are two wire, unless you ask for a four-wire circuit and pay a little more.

TWT Traveling Wave Tube.

TWX (Pronounced TWIX.) Teletype Writer eXchange. An automatic teletypewriter (i.e. telex-like) switching service where subscribers may dial any other subscriber and send and receive a message. Formerly owned by AT&T and sold to Western Union in 1972. It differed from Telex in that TWX used AT&T's normal long distance phone network, was thus more ubiquitous, was faster than Telex and was incompatible with Telex, which Western Union owned. However, Western Union, in a major accomplishment, got them to talk to each other.

TYMNET A packet switched company, offering its services to the public.

Type A Intelligent Network term describing IN (Intelligent Network) services invoked by, and affecting, a single user. Most of them can only be invoked during call setup or teardown.

Type Ahead Imagine a voice processing service. It says "punch in your zip code at the beep." If you are able to punch in your zip code before you hear the beep or before the talking stops, you have "type ahead." If you are unable to punch in your zip code before you hear the beep, you don't have "type ahead." Better interactive voice response systems have "type ahead."

Type B Intelligent Network term describing IN (Intelligent Network) services invoked at any point by, and affecting directly, several users.

Type Approval A concept in which a design is approved by an agency and all devices subsequently manufactured according to that design are automatically approved.

Type 3 Cable An unshielded twisted-pair wire that meets IBM specifications for use in 4-megabit-per-second Token-Ring networks.

Type 66 Punchdown Block A standard, solderless terminal wiring block used today. Invented by Western Electronic.

Type I PC Card The thinnest PCMCIA Card from factor at 3.3 mm thick. The Type I format is typically used for various memory enhancements, including RAM, Flash, OTP, SRAM, and EEPROM.

Type II PC Card A PCMCIA Card which is 5 mm thick. This card is typically used for I/O such as modem, LAN, and host communications.

Type III PC Card The thickest PCMCIA Card type at 10.5 mm thick, the Type III Format is primarily used for memory enhancements or I/O capabilities that require more space, such as rotating media and wireless communication devices.

Typebar Linear type element in a printer containing the printable symbols.

Typing Reperforator Same as receive only typing reperforator.

NEWTON'S TELECOM DICTIONARY

U Interface Two wire ISDN circuit — essentially today's standard one pair telephone company local loop made of twisted-wire. The "U" interface is the most common ISDN interface. It carries 160,000 bits per second from the central office to your home or office. Of those 160,000 bits, two are used for 64,000 bps Bearer (B) channels and one is used by the subscriber for 16,000 bps of data (the D channel). The other 16,000 bps is used by the network for signaling between the black box on the subscriber premises and the central office. The idea is to get the ISDN "U" interface working to 18,000 feet — the average length of a North American subscriber local loop. You screw the two "U" wires (local loop pair) coming in from your local ISDN CO into a black box about the size of desk printing calculator, called an NT-1. Out the side of the black box comes four wires, which are called the "S Bus." Onto these four wires you can attach, in a loop configuration (also called single bus), as many as eight ISDN terminals — telephones, fax machines, etc. See ISDN.

U-Law A voice amplitude compression/expansion quasi-logarithmic curve, based on the approximation with 15 linear segments. Used for PCM encoding/decoding in North America. See PCM.

U Reference Point In the U.S., the point that defines the line of demarcation between user-owned and supplier-owned Integrated Services Digital Network (ISDN) facilities.

UA User Agent. An OSI application process that represents a human user or organization in the X.400 Message Handling System. A US creates, submits, and takes delivery of messages on the user's behalf.

UART Universal Asynchronous Receiver/Transmitter. PCs have a serial port, which is used for bringing data into and out of the computer. The serial port is used for data movement on a channel which requires that one bit be sent (or received) after another, i.e. serially. The UART is a device, usually an integrated circuit chip that performs the parallel-to-serial conversion of digital data to be transmitted and the serial-to-parallel conversion of digital data that has been transmitted. The UART converts the incoming serial data from a modem (or whatever else is connected to the serial port) into the parallel form which your computer handles. UART also does the opposite. It converts the computer's parallel data into serial data suitable for asynchronous transmission on phone lines. UART chips control the serial port/s on personal computers. Now read the next definition. See also INTERRUPT, INTERRUPT LATENCY, INTERRUPT OVERHEAD, INTERRUPT REQUEST and 16550.

UART Overrun UART overruns are errors received from a universal asynchronous receiver/transmitter when receiving equipment cannot match transmission speed. UART overrun occurs when the UART's receive buffer is not serviced quickly enough by the CPU, and the next incoming byte of data crashes into the previous byte. The previous byte is then lost, forcing the communications driver to report and error. Your communications software must then ask for a retransmission of the lost data. High interrupt overhead is the most common cause of a UART overrun. The easiest way of solving UART overrun is to get yourself a UART with a 16 byte buffer (like the 16550), not today's typical one byte buffer UART. See 16550 and UART.

UBR Undefined Bit Rate or Unspecified Bit Rate. Traffic class defined by the ATM Forum. UBR is an ATM service category which does not specify traffic related service guarantees. Specifically, UBR does not include the notion of a per-connection negotiated bandwidth. No numerical commitments are made with respect to the cell loss ratio experienced by a UBR connection, or as to the cell transfer delay experienced by cells on the connection.

UCA Utility Communications Architecture. An architecture for networks used to monitor and control electric power distribution systems.

UCC Uniform Commercial Code.

UCD Uniform Call Distributor. A device for allocating incoming calls to a bunch of people. Less full-featured than an Automatic Call Distributor. For a bigger explanation see UNIFORM CALL DISTRIBUTOR.

UDF Universal Disk Format. See OSTA.

UDI Unrestricted Digital Information.

UDK A dumb GTE abbreviation, for Universal Dialing Keyset, a key pad that is switchable for either TONE or PULSE dialing. Outside GTE's private world, a keyset would mean a KEY TELEPHONE, not part of a phone.

UDOP The ultimate dumb, open programmable (UDOP) switch built from multi-vendor SC-based products. A term coined by Dialogic.

UDP User Datagram Protocol. A TCP/IP protocol describing how messages reach application programs within a destination computer. This protocol is normally bundled with IP-layer software. UDP is a transport layer, connectionless mode protocol, providing a (potentially unreliable, unsequenced, and/or duplicated) datagram mode of communication for delivery of packets to a remote or local user.

UDP/IP User Datagram Protocol/Internet Protocol. See UDP.

UEM Universal Equipment Module. A Northern Telecom acronym. The basic unit of Meridian 1 PBX modular packaging. A UEM is a self-contained hardware cabinet housing a card cage, with a power supply, backplane, and circuit cards. If the UEM has the card cage for an AEM installed, it functions as an AEM.

UFGATE A program which enables a FIDO compatible bulletin board system to exchange UUCP mail with UUCP sites.

UG UnderGround.

UHF The Ultra High Frequency part of the radio frequency spectrum ranging between 300 Megahertz and 3 Gigahertz.

UI 1. User Interface, as in GUI, or Graphical User Interface.

626

2. UNIX International is a consortium of computer hardware and software vendors which is interested in the development of open software standards for the UNIX industry. Prominent members include AT&T, Sun, UNISYS and Fujitsu.

UIFN Universal International Freephone Number. In early June 1996, the ITU-T released its approval of a new standard, called E.169. This standard will allow International Freephone Service customers to be allocated a unique Universal International Freephone Number (UIFN) which will remain the same throughout the world, regardless of country or telecommunications carrier. "Freephone" is a service which permits the cost of a telephone call to be charged to the called party, rather than the calling party. In North America, 800 and 888 numbers are "freephone" numbers. According to the ITU-T, a UIFN is composed of a three digit country code for global service application, 800, and an 8-digit Global Subscriber Number (GSN), resulting in an eleven digit fixed format, which allows companies to choose the digits they wish and embed existing freephone numbers into the available number space.

UIS Universal Information Services. AT&T's vision of a single fully-integrated, user-defined digital network with a universal port of entry. Very similar to ISDN, now aggressively adopted by AT&T.

Uiterbi Decoder Algorithm for decoding Trellis encoded signals.

UL Underwriters Laboratories, a privately-owned company that charges manufacturers a stiff fee to make sure their products meet the safety standards which UL itself develops. A UL label on a product has a very specific message. It says the product confirms to the safety standards UL has developed — nothing more. It does not affirm that the product will work. UL is now beginning to concern itself with adopting and promulgating standards (which have nothing to do with safety standards) including those relating to cabling. See UL APPROVED, UL CABLE CERTIFICATION PROGRAM and UL NNNN.

UL 1449 A method of rating and approving surge suppressors. This Underwriters Laboratories measurement is important as it tells if you're buying a true surge suppressor or just an extension cord. This listing measures how much voltage actually reaches the attached equipment after going through the surge suppressor. It's on a scale from about 330 volts to 6,000 volts. The lower the rating, the greater the protection. Decent surge suppressors tend to be rated around 400 volts for the basic units and 340 for the advanced and superior models. In short, check for UL 1449 rating on your surge arrestor before you buy it.

UL 1459 Effective 7/1/91, telephone equipment manufacturers will be required to provide protection from current overloads and power line crosses on equipment systems. Equipment systems covered under this listing requirement include single- and multi-line telephones, PBXs, key systems and central office switches. In general, the UL 1459 requirements apply to any location where wires enter a building from the public network, as well as in most IROB (In Range Out of Building) situations. See also NEC REQUIREMENTS and UNDERWRITERS LABORATORIES.

UL 1863 This requirement covers miscellaneous accessories intended to be electrically connected to the telecommunications network. The listing requirement applies to components that comprise the premises communications wiring system from the point of demarcation up to and including the final outlet providing modular plug and jack connection (or equivalent). Requirements are listed under Communication Circuit Accessories, UL 1863. Listing equipment for all other equipment will be covered under UL 1459, effective July 1, 1991. See also NEC REQUIREMENTS and UNDERWRITERS LABORATORIES.

UL 497 & 497A According to the National Electrical Code, primary and secondary protection systems that will be used on a telephone circuit must be listed for that purpose. The listing requirements are UL 497 for primary protection systems and UL 497A for secondary protection systems. See also NEC REQUIREMENTS and UNDERWRITERS LABORATORIES.

UL Approved Tested and approved by the Underwriters Laboratories. The Underwriters Laboratories, Inc. was established by the National Board of Fire Underwriters to test equipment affecting insurance risks of fire and safety. Most phone systems are tested and approved. Most of the testing focuses on the power supply feeding the phone system. The power supply is that little black box that plugs into the AC wall outlet at one end, takes 120 volt AC and converts it to low voltage DC power that the phone system typically runs on. If the power supply tests OK, then that's usually sufficient UL testing. For it is the power supply — and what happens to the commercial AC power that feeds into the power supply — that determines the potential fire hazard of your phone system. After many fire deaths in recent years, most local communities are a lot more concerned about UL Approval of installed telephone equipment. Fire departments have been known to zealously enforce these rules. In addition to the UL approval, the other major fire concern is the use of proper wire in new building construction, with especial emphasis on teflon-covered cable in plenum ceilings. See also UL, an entry which talks about UL's expanding certification business. See also UL CABLE CERTIFICATION PROGRAM.

UL Cable Certification Program United Laboratories, in conjunction with companies such as Anixter, has developed a Data-Transmission Performance-Level Marking Program that covers UL Listed communications cable or power-limited circuit cable. The UL program identifies five levels of performance. UL evaluates cable samples to all of the tests required for each level. Only Levels II through V require testing.

LEVEL I: Level I cable performance is intended for basic communications and power-limited circuit cable. There are no performance criteria for cable at this level.

LEVEL II: Level II cable performance requirements are similar to those for Type 3 cable (multi-pair communications cable) of the IBM Cabling System Technical Interface Specification (GA27-3773-1). These requirements apply to both shielded and unshielded cable constructions. Level II covers cable with two to 25 pair twisted pairs of conductors.

LEVEL III: Level III data cable complies with the transmission requirements in the Electrical Industries Association/Telecommunications Wiring Standard for Horizontal Unshielded Twisted-Pair (UTP) Cable and with the requirements for Category 3 in the proposed EIA/TIA Technical Systems Bulletin PN-2841. These requirements apply to both shielded and unshielded cables.

LEVEL IV: Level IV cable complies with the requirements in the proposed National Manufacturer Association (NEMA) Standard for Low-Loss Premises Telecommunications Cable. Level IV requirements are similar to Category 4 requirements of the proposed Electronic Industries Association / Telecommunication Industry Association (EIA/TIA) Technical Systems Bulletin PN-2841. These requirements apply to both shielded and unshielded cable constructions.

LEVEL V: Level V cable complies with the requirements in the proposed National Electrical Manufacturers Association (NEMA) Standard for Low-Loss Extended-Frequency Premises Telecommunications Cable. Level V requirements are similar to Category 5 requirements of the proposed Electronic Industries Association/Telecommunication Industry Association (EIA/TIA) Technical Systems Bulletin PN-2841. These requirements apply to both shielded and unshielded cable constructions.

UL evaluates communications and data transmission cable to one of two UL Safety Standards: UL 444, the Standard for Safety for Communications Cable; and UL 13, the Standard for Safety for Power-Limited Circuit Cable.

ULANA An ongoing U.S. Air Force project aimed at creating a series of interconnected local area networks using Transmission Control Protocol/Internet Protocol (TCP/IP) as the unifying transport layer.

ULP Abbreviation for Upper Layer Protocol. Layer above TCP.

ULS User Location Service (ULS) provides a mechanism for users of Microsoft's NetMeeting to locate other people on the Internet, even if their Internet addresses change. A sample of the ULS can be found at http://uls.microsoft.com/.

ULSI Ultra Large Scale Integration, the technique of putting millions of transistors on a single integrated circuit. Compare with LSI (Large Scale Integration) and VLSI (Very Large Scale Integration).

Ultimedia IBM's product in multimedia — combining sound, motion video, photographic imagery, graphics, text and touch into a unified, natural interface representing, in IBM's words, the ultimate in multimedia solutions. Ultimedia supports both Ultimotion and Indeo video. Ultimedia was coined in the Spring of 1992.

Ultimotion IBM's video compression algorithm. Although IBM supports Indeo video technology in OS/2 and Windows systems, IBM feels several OS/2 vertical applications are adequately served by the Ultimotion algorithm. Ultimotion does not offer software scalable playback or single step compression. See ULTIMEDIA.

Ultra 800 Sprint service for large users of 800 service. Access is provided over a T-1.

Ultra Forward Service The name of a Bell Atlantic service. Ultra Forward service allows you to automatically forward your calls to whatever number you choose — from any location. That's because you can activate Ultra Forward service remotely from any touchtone phone (even your mobile phone) and reroute your calls while on the go.

Ultra Hi-Res Ultra high resolution. Properly speaking, the term should be for monitors with resolutions of 1,200 x 800, 1,024 x 1024 or better, but it is sometimes used to describe monitors with 800 x 600 resolution and above.

Ultra High Frequency Frequencies from 300 MHz to 3000 MHz.

Ultra WATS Plus Sprint's Ultra WATS PLUS is a packaging of Sprint Dial 1 WATS, Ultra WATS, and FONCARD services.

Ultrasonic Bonding The use of ultrasonic energy and pressure to join two materials.

Ultraviolet That portion of the electromagnetic spectrum in which the wavelength is just below the visible spectrum, extending from approximately 4 nanometers to approximately 400 nanometers. Some scientists place the lower limit at values between 1 and 40 nanometers, 1 nm being the upper wavelength limit of X-rays. The 400-nm limit is the lowest visible frequency, namely violet. "Light" in the ultraviolet spectrum is used for erasing EPROMS.

Ultraviolet Fiber Special fiber which extends the usable range into the UV region of the spectrum.

Ultrix Ultrix is Digital Equipment Corporation's proprietary implementation of UNIX. It runs on DEC's RISC-based workstations.

UM Micron (10-6 meters).

UMA An acronym for Upper Memory Area. See UPPER MEMORY AREA.

UMB An acronym for Upper Memory Block, an area of upper memory (the area between 640KB and 1MB of RAM) in an MS-DOS PC that has been remapped with usable RAM. This allows device drivers and TSRs to be loaded high, into the UMB and out of conventional memory. See UPPER MEMORY AREA.

UME UNI Management Entity: The software residing in the ATM devices at each end of the UNI circuit that implements the management interface to the ATM network.

UMIG Universal Messaging Interoperability Group. This group is an offshoot of AMIS (Audio Messaging Interchange Specification). The UMIG is being technically facilitated (their words) by the Information Industry Association in Washington, DC. The UMIG has two top priorities: First, to foster the development of "universal messaging," which entails integrating platforms supporting different types of messaging, such as voice processing, electronic mail and facsimile messaging to allow users to move easily among the three media. The second priority entails working towards standardized addressing and directory schemes that make it easy and intuitive for users to message one another.

UMTS Universal Mobile Telecommunications Systems.

UN Industry jargon for UNreachable, an unsuccessful call where the agent is unable to speak to the contact or decision maker.

Unassigned Cells A cell identified by a standardized Virtual Path Identifier (VPI) and Virtual Channel Identifier (VCI) value, which has been generated and does not carry information from an application using the ATM Layer service.

Unattended Equipment working without a human attendant or operator. There are pros and cons to operators. On the pro side, they offer a personalized service that's absolutely critical to customer goodwill. On the con side, they can be slow and cumbersome. They can be very irksome when you know you could do that task yourself, but have to wait for the operator. Some companies have only one main number. Some companies use a main number and DID — on their Centrex and their PBX. Some companies use an automated attendant and an operator. There's more flexibility with DID and a main number, or Centrex DID and a main number. Customers without knowledge dial the main number. Customers with knowledge can dial direct DID numbers. See also AUTOMATED ATTENDANTS.

Unattended Call Calls placed by a computerized dialing system in anticipation of an agent being available to answer the call. A called party is detected answering the phone and no agent is available to serve the call. The system hangs up on the party so as not to create any greater nuisance than has already occurred. The telemarketing industry does not believe that an unattended call can be queued for the next available agent.

Unbalanced Line A telephone circuit in which the voltages on the two conductors are not equal with respect to ground. Unbalanced lines give poor phone service. Lines can become unbalanced when they come from the central office or when they

are in the PBX or the on site phone system. Problems can and should be repaired for decent quality results.
Unbundled Services, programs, software and training sold separately from the hardware.
UNC Universal naming convention. See also UNC NAMES.
UNC Names Filenames or other resources names that begin with the string \\, indicating that they exist on a remote computer.
Uncontrolled Terminal A user terminal that is on line at all times and that does not contain the logic that would allow it to be polled, called, or otherwise controlled by the device to which it is connected.
Underfill A condition for launching light into a fiber in which not all the modes that the fiber can support are excited (i.e. turned on).
Underfloor Duct Method A floor distribution method using a series of metal distribution channels, often embedded in concrete, for placing cables. This method uses one or two levels depending on the complexity of the system. Sometimes referred to as underfloor raceways. See also RACEWAYS METHOD.
Underflow In computing, a condition occurring when a machine calculation produces a non-zero result that is smaller than the smallest non-zero quantity that the machine's storage unit is capable of storing or representing.
Underground Cable installed in buried conduit. Does not typically include cables buried directly in the ground.
Underlap In facsimile, a defect that occurs when the width of the scanning line is less than the scanning pitch.
Underlying Carrier A common carrier providing facilities to another common carrier which then provides services to end users.
Understaffing Limit A call center term. The percentage by which you'll allow the scheduling process to fall short of the required staffing level in any period. This typically provides more economical coverage during the least-busy periods of the day.
Underwriters Laboratories, Inc. A non-profit laboratory which examines and tests devices, materials and systems for safety, not for satisfactory operation. It also has begun to establish safety standards. See also UL APPROVED.
Undesired Signal Any signal that tends to produce degradation in the operation of equipment or systems.
Undetected Error Ratio The ratio of the number of bits, unit elements, characters, or blocks incorrectly received and undetected, to the total number of bits, unit elements, characters, or blocks sent.
Undisturbed Day A day in which the sunspot activity or ionospheric disturbance does not interfere with radio communications.
Unequal Access Refers to long distance phone companies who do not take advantage of Judge Harold Greene's Equal Access divestiture provisions. Rather than a carrier selection code, unequal access carriers require you to dial a local seven digit number and punch in an authorization code. If the carrier elected to pay for Equal Access, you would just dial directly the same 10 digits you do today, and your local telephone company would give your billing number to your long distance company.
Unerase A command for getting back files you've accidentally erased. See MS-DOS.
Ungrounded Not connected to ground. PBXs, key systems and other phone systems will not work well when not connected to a solid ground because they have no place to send high voltage spikes (static electricity, lightning strikes, etc.) Improper grounding is probably the most common cause of phone system faults. Our feeling: the better the ground, the better the phone system performance. One way of grounding is the third wire of an electrical outlet. This may be OK if you check where that wire is ultimately connected to. You can ground to the metal cold water pipe. But that may connect to a plastic PVC pipe one floor below. Best to check. A ground ultimately ending firmly routed a dozen feet below the ground is best. Books have been written on the subject of grounding for telephony. See Lees abc Teletraining books, available from the Telecom Library at 1-800-LIBRARY or 212-691-8215.
UNH IOL University of New Hampshire Interoperability Lab. A testing organization affiliated with the Research Computing Center of the University of New Hampshire which tests FDDI products for vendor interoperability.
UNI The ATM Forum has developed User-Network Interface (UNI) specifications for the procedures and protocols between a user and the ATM network so a user can take advantage of ATM services and capabilities. The UNI Interface is the physical and electrical demarcation point between the user and the public network service provider. An Asynchronous Transfer Mode (ATM) term.
UNI Interface See UNI.
UNIBOL A UNIX version of COBOL.
Unicast Sending information to one person on a network. See MULTICASTING.
Unicasting As an ATM term, it is the transmit operation of a single PDU by a source interface where the PDU reaches a single destination.
Unicode Unicode is a 16 bit system for encoding letters and characters of all the world's languages. At 16 bits it can encode 65,536 characters. That's two raised to the 16th power. Work it out: Multiply 2 x 2 x 2 x 2 x 2 x 2 x 2 x 2 x 2 x 2 x 2 x 2 x 2 x 2 x 2 x 2. Sixteen-bit characters (like Unicode) are also called Wide Characters. The first 128 codes of Unicode are identical to ASCII. Just add another zero byte to each ASCII character to convert to Unicode. Unicode contains over 20,000 Han characters, which are used to represent whole words or concepts in Chinese, Japanese and Korean.
Unidirectional The transmission of information in one direction only.
Unidirectional Bus A distribution conductor or set of conductors that can transfer information in one direction only.
Unidirectional Path Switched Ring UPSR. A SONET transport method in which working traffic is transmitted in one direction. UPSR is preferred for interconnected rings with numerous signals crossing the rings.
Unified Messaging Another name for integrated messaging. See INTEGRATED MESSAGING and TELEPHONY SERVICES.
Unified Network Management Architecture AT&T's architecture for providing end-to-end network management. There are three levels within the architecture — network elements, element management systems, and network management integration.
Uniform Access Number See UNISERV.
Uniform Call Distribution See UNIFORM CALL DISTRIBUTOR.
Uniform Call Distributor UCD. A device for distributing many incoming calls uniformly among a group of people (typi-

cally called "agents" because of the early use of these machines by the airline, hotel and car reservation industry). These days the term Uniform Call Distributor is falling into disrepute as the newer term, Automatic Call Distributor comes in. According to incoming call experts, a Uniform Call Distributor is generally less "intelligent," and therefore less costly than an ACD. A UCD will distribute calls following a predetermined logic, for example "top down" or "round robin." It will not typically pay any heed to real-time traffic load, or which agent has been busiest or idle the longest. Also, a UCD's management reports tend to be rudimentary, consisting of simple pegs counts, as opposed to an ACD, which can produce reports on the productivity of agents.

Uniform Encoding An analog-to-digital conversion process in which, except for the highest and lowest quantization steps, all of the quantization subrange values are equal.

Uniform Linear Array An antenna composed of a relatively large number of usually identical elements arranged in a single line or in a plane with uniform spacing and usually with a uniform feed system.

Uniform Numbering Plan A uniform seven-digit number assignment made to each phone in a private corporate network. Such a plan allows routing of calls to distant phones from any on-net telephone without any differences in the dialed number. Without a uniform numbering plan, you would dial your boss in New York differently if you were in the company's Chicago office and differently again if you were in your company's San Francisco office. With a uniform numbering plan, it would be the same from all locations. The nation's long distance network has, obviously, a uniform numbering plan.

Uniform Resource Locator URL. A standardized way of representing different documents, media and network services on the World Wide Web. An Internet term.

Uniform Service Order Code See USOC.

Uniform Spectrum Random Noise Noise distributed over the spectrum in such a way that the power per unit bandwidth is constant. Also known as "white" noise.

Unimodem Name used to refer to the universal modem driver in TAPI (Telephony Applications Programming Interface). It is a driver-level component that uses modem description files to control its interaction with the communications driver, VCOMM. UniModem is part of Windows 95. (Yes, it's spelled with a capital M in the middle.)

Unintelligent Crosstalk Crosstalk giving rise to unintelligent signals.

Uninterruptible Power Supply UPS. A device providing a steady source of electric energy to a piece of equipment. A continuous on-line UPS is one in which the load is continually drawing power through the batteries, battery charger and invertor and not directly from the AC supply. A steady off-line UPS normally has the load connected to the AC supply. When the line is weak or down, it transfers the load without any user intervention. UPS are typically used to provide continuous power in case you lose commercial power. An UPS is typically a bank of wet cell batteries (similar to automobile batteries, but often much, much larger) engineered to power a phone system up to eight hours without any re-charging. A UPS system can also include a gasoline-powered generator. And if the generator works (make sure it has gas), you can power your phone system for much longer. According to Bell Labs, however, over 90% of all power outages last less than five minutes. The cost of Uninterruptible Power Supplies is typically a direct function of how large the battery/batteries are. The larger the batteries, the higher the cost. Many file servers on local area networks are also backed by UPSes. Many NetWare file servers, which are protected by a UPS, often are attached a printed circuit card inside the server. This card acts as an early warning system. When AC power drops, and the UPS takes over, it signals the file server through the card what has happened. The file server then will send a message to all the workstations on the network that the file server has lost AC power, is running on battery power, is running out and would everyone kindly log off the server. This protects the network.

Unipolar Signal A two-state signal where one of the states is represented by voltage or current and the other state is represented by no voltage or no current. The current flow can be in either direction.

Unique Addressing The addressing of a node by using the software-programmable address assigned to each one upon system initialization. For example, TELECONNECT's LAN has a "unique" addressing scheme. Each workstation is known by the operator's first name.

Uniserv Service A BellSouth service which provides a seven-digit local phone number that a subscriber may use in multiple LATAs or local calling area. The subscriber's clients will be able to dial one number from all locations within the calling area, and the call will then be routed to one specified location within the calling area. The service was formerly called Uniform Access Number.

Unit Interval In a system using isochronous transmission, that interval of time such that the theoretical durations of the significant intervals of a signal are all whole multiples of this interval. The unit interval is the shortest time interval between two consecutive significant instants.

United States Telephone Association USTA. The largest trade association of telephone companies. It represents more than 1,200 telephone companies. Based in Washington, its primary role is to ensure no regulation or legislation is passed that is unfavorable to the telephone industry. Its secondary aim is to hold trade shows/conventions each year. Its third aim is education. It has a sister organization, the United States Telephone Suppliers Association, which is now merged with the EIA, forming the TIA. The USTA's phone number is 202-326-7282. See the next definition.

United States Telephone Suppliers Association USTSA. This is the association of suppliers — manufacturers and wholesalers — affiliated with the United States Telephone Association (USTA). It is based in Chicago. Its aims are to promote sales of its members' products to international and domestic telephone companies through conventions. It has merged with the Electronic Industries Association to become the TIA — Telecommunications Industry Association. See TIA.

United Telecom Local operating telephone holding company which changed its name to Sprint Corporation in February, 1992 after it became the sole owner of Sprint, the long distance company.

Unity Gain Refers to the balance between signal loss on a broadband network and signal gain through amplifiers.

Univercell ATM switching technology from AT&T. Trademarked name.

Universal Access Number A single number dialed from anywhere in the country which will route a customer to one or several locations for service, advice, etc. The definition varies depending on whose networking scheme you're dealing with.

Universal Addressing The addressing of a node by the use of the universal addresses which all nodes recognize.

Universal Agent A telephone agent who answers incoming calls and also makes outgoing calls. This duality feature may not seem worth of its inclusion in this dictionary. But the fact is that agents have largely been just "inbound" or just "outbound" — because managers felt that most agents were not capable of doing both. The skills were, allegedly, too different. Now the idea is to "empower" the agent with more flexibility and make them "universal," i.e. capable of being used for both inbound and outbound.

Universal Asynchronous Receiver-Transmitter UART. A device that converts outgoing parallel data from your computer to serial transmission and converts incoming serial data to parallel for reception. See UART for a bigger explanation.

Universal Circuit Card See UNIVERSAL TRUNK CARD.

Universal Device A SCSA device. A call processing device which has every conceivable resource for the handling of calls. The SCSA programming applies resources from many different physical devices to a call processing task. These then act as if they were a single universal device.

Universal Serial Bus See USB.

Universal Service Fund There is money somewhere in Washington in something called the Universal Service Fund, whose goal is to provide telephone service to every household in the U.S. with at least one access line for basic telephone service. This fund gets money from a surcharge on phone lines. The fund's goal is to offset operating costs of small telcos.

Universal Information Services UIS. AT&T's vision for the future of the telecommunications industry, where network providers everywhere can give any customer any kind of voice, data or image service in any place, at any time, in any combination and with maximum convenience and economy. UIS is very much like ISDN. The word UIS seems to be disappearing as AT&T embraces ISDN.

Universal Integrated Services See UNIVERSAL INFORMATION SERVICES.

Universal Mailbox Allows a user of unified messaging services to have single access to all messages from internal and external electronic mail systems, fax systems and voice mail systems. A really neat idea given today's lack of standardization among electronic mail services.

Universal Night Answer A feature of telephone systems that permits any phone to pick up any incoming trunk call when the Attendant's console is unmanned (unpersonned?) and the phone system is set up (typically at the console) for "Night Answer."

Universal Pay Phone Description for a coin-and-credit-card phone.

Universal Ports A modern telephone system is typically an empty cabinet into which you slide printed circuit cards. Those cards have an edge connector and they slide into a connector at the rear of the cabinet. That connector connects via wires to other connectors in what is typically called the phone system's bus. In the old days, phone systems had dedicated slots — meaning you could only slide one type of printed circuit card into that particular slot. As phone systems got more advanced, they acquired "universal ports." Our definition of a universal port is that all the slots are totally flexible — namely that you can slide any trunk or phone card (either electronic or single line phone) into any slot in the phone system. The advantage of this is obviously a far more flexible phone system, able to accommodate lots of phones and few trunks or vice versa.

Universal Power Supply A power supply which you can plug into electricity ranging from 100 volts to 240 volts AC. With a universal power supply (now standard with many laptops) you can travel the world, plugging yourself into virtually any power outlet and have your device work perfectly, without the need for a transformer. What you'll need, however, is a plug that converts the plug you have into the necessary plug for that country. Such a converter plug shouldn't cost you more than $2.

Universal Sender Allows the dialed number to be sent out by the user.

Universal Service Originally conceived by the first chairman of the Bell System, Theodore Vail, universal service has been the goal of the entire telephone industry, including its Federal and state regulators, since the 1920s. The idea of universal service is to have residential telephone service priced sufficiently low so anyone in the United States can afford it. Keeping residential service low has been the reason why local business service is usually priced much higher — though the two services are usually identical. Universal Service is now effectively accomplished. It has become a political rallying cry, used to justify the strange and wonderful pricing schemes the phone industry and its regulators create with little regard for the actual costs of providing those services.

Universal Trunk Cards Most PBXs have different circuit boards (or circuit cards - same thing) for combination trunks and for DID (direct inward dial) trunks. A Universal Circuit Board enables you to use the ports on the board for either combination trunks OR DID trunks. This capability makes the PBX more flexible. If you have spare ports on a trunk circuit board you may use them for either a combination or a DID type of trunk.

Universal Turret A very large key system for financial traders, emergency teams at nuclear power stations and others who need single phone button access to hundreds of people. By simply pushing one of the button in front of them, the user can dial one of hundreds of people. These buttons may be connected to tie lines, foreign exchange lines. They may even be DDD lines with autodial capability. Like all good key systems, the buttons have a lamping display which shows if the particular line is idle, busy, ringing, on hold, etc.

Universal Wall Jack There's really no such animal. Every manufacturer of installation gadgetry is trying to propagate the idea that their jack is universal, when it really isn't. The "universal" wall jack we installed in our new offices is actually four jacks — 1. Four pairs for two PBX voice lines (one electronic two-pair phone and one tip and ring phone) and one spare. 2. One RS-232-C 12-conductor shielded cable for connecting to centralized printers, for connecting to a dataPBX and for permanent null-modem connection of computers. 3. One for connecting to our high-speed, one megabit per second LAN, and 4. One spare twisted, shielded, stranded pair for a second LAN, or whatever comes along.

UNIX An immensely powerful and complex operating system for computers for running data processing and for running telephone systems. UNIX provides multi-tasking, multi-user capabilities that allow both multiple programs to be run simultaneously and multiple users to use a single computer. On a single-user system, such as MS-DOS, only one person at a time, on an individual task basis, can use a computer's files. programs, and other resources. UNIX works on many different computers. This means you can often take applications software which runs on UNIX and move it — with little changing — to a bigger, different computer, or to a smaller, computer. This process of moving programs to other computers is known as "porting." Today, the UNIX operating system is available on a wide range of hardware, from small personal computers to the most powerful mainframes, from a multitude of hardware and software vendors. UNIX was developed in 1969 by Ken Thompson of AT&T Bell Laboratories. UNIX is also a trademark of UNIX Systems Laboratories, Inc. which is now owned by Novell Inc.

UNIX-To-UNIX Copy Program UUCP. A standard UNIX utility for exchanging information between two UNIX-based machines in a network. UUCP may also be referred to as the UNIX-TO-UNIX Communications Protocol and is widely used for electronic mail transfer.

Unlisted Number There are various interpretations of what constitutes an "unlisted, an "unpublished" or a "non-published" phone number in North America. Some phone companies use these words interchangeably. Some don't. In California, Pacific Bell offers unpublished phone service. Your phone number is not listed in the paper phone directories, but is listed with dial up "Directory Assistance." Pacific Bell also has a more expensive service called "Unlisted Service." Here, your phone number is not included in the paper phone directories or given out to callers to Directory Assistance. Nynex (and I presume other phone companies) has a service whereby you can leave a message for the owner of an unlisted number. "Please call me. You've won the lottery." The owner of the unlisted number then has the choice to return the call or not. He doesn't pay to receive this message. Some telephone companies confuse the definitions and some invent new ones. For example, some phone companies use the term "non-published" number. You won't find the number in a phone book or by calling Directory Assistance. Over 25% of many private phone numbers in major metropolitan areas are unlisted, unpublished or non-published — a "service" their subscribers pay extra for. To my simple brain, it's a lot easier to simply publish your name as "Apple Plumpudding." See UNPUBLISHED.

Unloaded Line A telephone line with its loading coils removed to increase the distance and speed with which data may be transmitted over the line. A fee is usually charged for removing the coils.

Unlock Code This is a three-digit number required to unlock a cellular phone when you have electronically locked it to prevent unauthorized use. You might lock it when you park your car in a hotel. The factory default is 123.

UNMA AT&T's Unified Network Management Architecture.

UNMR Universal Network Management Record.

Unnumbered Command In a data transmission, a command that does not contain sequence numbers in the control field.

Unnumbered Information Frame U1 frame. A transmission frame generated by the High-Level Data Link Control (HDLC) data link protocol, where no flow control and no error control are implemented.

Unpublished Phone Number There are various interpretations of what constitutes an "unpublished" phone number in North America. Some phone companies use these words interchangeably. Some don't. In California, Pacific Bell offers unpublished phone service. Your phone number is not listed in the paper phone directories, but is listed with dial up "Directory Assistance." Pacific Bell also has a more expensive service called "Unlisted Service." Here, your phone number is not included in the paper phone directories or given out to callers to Directory Assistance. Nynex (and I presume other phone companies) has a service whereby you can leave a message for the owner of an unlisted number. "Please call me. You've won the lottery." The owner of the unlisted number then has the choice to return the call or not. He doesn't pay to receive this message. Some telephone companies confuse the definitions and some invent new ones. For example, some phone companies use the term "non-published" number. You won't find the number in a phone book or by calling Directory Assistance. Over 25% of many private phone numbers in major metropolitan areas are unlisted, unpublished or non-published — a "service" their subscribers pay extra for. To my simple brain, it's a lot easier to simply publish your name as "Apple Plumpudding." See UNPUBLISHED.

Unrestricted Digital Information An ISDN term. An information sequence of bits is transferred at its specified bit rate without alteration.

Unshielded Wiring not protected by a metal sheathing from electromagnetic and radio frequency interference, but covered with plastic and/or PVC.

Unshielded Twisted Pair UTP. A cable medium with one or more pairs of twisted insulated copper conductors bound in a single plastic sheath. Now becoming the most common method of bringing telephone and data to the desktop. There is now a strong belief that twisted pair cables carrying data over 16 megabits per second should definitely be carried on unshielded twisted wire — since at such speeds, such cables emanate radiation which should be allowed to escape (otherwise it would interfere with itself). Hence the unshielded cables.

Unsolicited Event Events in switching that happen without control of a program that allegedly is controlling your phone and phone system. Such an unsolicited event might be a user picking up his phone and hanging it up or simply pushing a random button on the phone. Or it may be that the switch actually does something you or your controlling computer don't expect it to do.

Unsuccessful Call A call attempt that does not result in the establishment of a connection.

Unsupervised Transfer Someone transfers a call to someone else without telling the person who's calling. Also called Blind Transfer.

Unused A term used in the secondary telecom equipment business. Equipment never used and still in O.E.M. original packaging with all appropriate documents and user guides. Such equipment may or may not carry the O.E.M. standard warranty. In other words, manufacturers don't want to see independent (unauthorized) dealers advertising their equipment as new, but "unused" is acceptable.

Up Sell To sell a higher value product to an existing customer. For example to lease a more sophisticated photocopier to an existing customer.

Up-Converter A device for performing frequency translation in such a manner that the output frequencies are higher than the input frequencies.

UPC Usage Parameter Control: As an ATM term, Usage Parameter Control is defined as the set of actions taken by the network to monitor and control traffic, in terms of traffic offered and validity of the ATM connection, at the end-system access. Its main purpose is to protect network resources from malicious as well as unintentional misbehavior, which can affect the QoS of other already established connections, by detecting violations of negotiated parameters and taking appropriate actions.

UPCS Unlicensed Personal Communications Services. The FCC designated 1890-1930 MHz to UPCS. 1920-1930 MHz is presently being assigned by UTAM, primarily for short range, wireless PBX applications. Here are the specific designations: 1890-1910 MHz isochronous, 1910-1920 MHz asynchronous, 1920-1930 MHz isochronous. Isochronous communication is good for voice and asynchronous communication is for bursty data.

Uplink 1. In satellites, it's the link from the earth station up to the satellite. The link from the satellite down to the earth station is called the downlink. The two are differentiated because customers often buy uplinks and downlinks from different suppliers — especially in international telecommunications.
2. In data transmission, an uplink is from a data station to the headend or mainframe.
3. As an ATM term, it represents the connectivity from a border node to an upnode.

Upload To transmit a data file from your computer to another computer. The opposite of download, which is receiving a file on your computer from another computer. Upload means the same as TRANSMIT, while DOWNLOAD means the same as receive. Before you upload or download, check at least three times you're going the direction you want. It's very easy to erase files (weeks of work) if you make a mistake and confuse uploading and downloading. (Don't laugh. We've done it several times. Dumb!)

Upnode As an ATM term, it is the node that represents a border node's outside neighbor in the common peer group. The upnode must be a neighboring peer of one of the border node's ancestors.

Upper Memory Area In an IBM compatible PC, upper memory is the area between 640KB and 1MB of RAM. This area is made up of Upper Memory Blocks (UMBs) of various sizes. Access to this area is possible only with a special memory drive such as MS-DOS's EMM386.EXE.

Upper Memory Blocks See UPPER MEMORY AREA.

UPS 1. See UNINTERRUPTIBLE POWER SUPPLY.
2. United Parcel Service, a package deliverer that regularly lives up to its self-imposed, relatively speedy delivery schedule.

UPS Monitoring UPS monitoring allows a local area network file server to monitor an attached Uninterruptible Power Supply (UPS). When a power failure occurs, NetWare notifies users. After a time out specified with SERVER.CFG and ROUTER.CFG, the server logs out any remaining users, closes any open files, and shuts itself down. If you install a Novell-approved UPS, you must also install a printed circuit board in the file server to monitor the UPS. If you have a file server with a microchannel bus (as compared to the more common AT bus), the UPS is monitored through the mouse port and does not require a board.

Upspeak According to Newsweek, upspeak is the annoying way teenagers speak.

Upstream In a communications circuit, there are two circuits — coming to you and going away from you. Upstream is another term for the name of the channel going away from you. In a broadband TV network, the definition of the upstream channel or signal is different. It is the channel from the transmitting stations to the CATV headend. See UPSTREAM CHANNEL.

Upstream Channel In a communications circuit, there are two circuits — coming to you and going away from you. Upstream is another term for the name of the channel going away from you. In a broadband TV network, the definition of the upstream channel or signal is different. It is the channel from the transmitting stations to the CATV headend. In yet another definition, in the cable TV industry, the upstream channel is a collection of frequencies on a CATV channel reserved for transmission from the terminal next to the user's TV set to (upstream to) the CATV company's computer. Such signals might be requests for pay movies. See UPSTREAM.

Upstream Operations Functions that provide a BCC (Bellcore Client Company) control of features and service configurations and subject to BCC control, some service management capabilities for subscribers. These functions include Service Negotiation and Management, Service Provisioning and Repair Service Answering/Work Force Administration. Definition from Bellcore in reference to its concept of the Advanced Intelligent Network.

UPT Universal Personal Telecommunications. According to L.M. Ericsson, Swedish telecom manufacturer, UPT is a "new service concept in the field of telecommunications which aims at making telecommunications both universal and personal. instead of calling a telephone line or a mobile terminal, you call the person you wish to get in touch with and leave it to the network to locate the line or terminal where he/she can be reached." There was an article on UTP in the 1993 No.4 issue of the Ericsson Review. An article in the June, 1996 issue of IEEE Communications Magazine described UPT as a service that enables users to access various services through personal mobility. It enables each UPT user to participate in a user-defined set of subscribed services, and to initiate and receive calls on the basis of a personal, network transparent UPT number across multiple networks on any fixed or mobile terminal, irrespective of geographical location. This service is limited only by terminal and network capabilities and restrictions imposed by the network operator. In short, UPT is still not totally defined and is under discussion by the world's major standards bodies. For more information on UPT, see ITU-T Recommendation F.850, Principles for Universal Personal Telecommunications, Geneva, 1993.

Uptime Colloquial expression for the uninterrupted amount of time that network or computer resources are working and available to a user. In short, time between failures or periods of nonavailability (as for maintenance).

Upward Compatible Any device that can be easily organized, fixed or configured to work in either a different, expanded operating environment or some enhanced mode. Software is said to be upward compatible if a computer larger than the one for which it was written can run the program.

Urban Service Any of the grades of service regularly furnished inside base or locality rate areas, or outside base or locality rate areas at base or locality rates plus zone connection charges or incremental rates. Another way of saying expanded metropolitan phone service.

URL Uniform Resource Locator. An Internet term. A URL is a fancy name for an address on the World-Wide Web. In more technical terms, a URL is a string expression that can represent any resource on the Internet or local TC/IP system. RFC 1738 defines the syntax for an URL. An address for a Website such as http://www.flatironpublishing.com brings you to the opening screen — or home page — of the publisher of this dictionary's web site. In general http:\\ can be safely omitted with most browsers and you'll still get to the site. See UNIFORM RESOURCE LOCATOR.

US Sprint An old name for Sprint, the long distance carrier.

US West One of the seven Regional Holding Companies formed at divestiture. It includes Mountain Telephone, Northwestern Bell and Pacific Northwest Bell among other service entities and entrepreneurial adventures.

USA Direct An AT&T service designed for those calling the U.S. from overseas. Callers in foreign countries dial a toll-free number in that country. That call gets them to an AT&T operator in the U.S. who lets them make calls to the U.S. with their AT&T calling card or by making a collect call. AT&T calls the service USADirect — one word. It would look like a mistake if I spelled it that way in this dictionary. MCI has a similar service called "Call USA."

USACII See ASCII. The name change was a result of the name change of the standards organization. When the name changed again to ANSI, most people simply reverted to ASCII.

Usage A measurement of the load carried by a server or group of servers, usually expressed in CCS. Usage may also be expressed in erlangs.

Usage Based Usage-Based refers to a rate or price for telephone service based on usage rather than a flat, fixed monthly fee. Until a few years ago, most local phone service in the United States was charged on a flat rate basis. Increasingly, phone companies are switching their local charging over to usage-based. Flat-rate calling will probably disappear within a few years. Allegedly, usage based phone service pricing is fairer on those phone subscribers who don't use their phone much. Usage based pricing is not consistent throughout the U.S. Typically, you get charged for each call. And the charging is very much like that for long distance — by length of call, by time of day and by distance called. See also FLAT RATE.

Usage Sensitive A form of Measured Rate Service. See USAGE BASED.

USART Universal Synchronous/Asynchronous Receiver/Transmitter. An integrated circuit chip that handles the I/O (input/output) functions of a computer port. It converts data coming in parallel form from the CPU into serial form suitable for transmission, and vice versa.

USB Universal Serial Bus. In March of 1995, Compaq, Digital, IBM, Intel, Microsoft, NEC and Northern Telecom announced a new "open and freely licensed" serial bus called Universal Serial Bus — USB for short. The bus (which could also be called a special purpose local area network) is 12 megabits per seconds and supports up to 63 devices. The idea of Universal Serial Bus is to replace the PC cable clutter. USB's proponents showed a diagram of a future PC with only three ports out the back — a USB, a graphics port (for your monitor) and a LAN port. Gone were the parallel, serial, graphics, modem, sound/game and mouse ports. USB is designed to handle a broad range of devices — telephones (analog, digital and proprietary), modems, printers, mice, joysticks, scanners, keyboards, tablets. USB is designed to be "completely Plug and Play," meaning that devices will be correctly detected and configured automatically as soon they are attached. USB also has "Hot attach/detach," which allows adding and removing devices at any time, without powering down or rebooting." USB uses a connector, currently in design. The mockup I saw was the size of your pinky finger. Topology is tiered star, with up to five meters per segment. At each star is a hub or pod with connections to other devices and pods. Hubs function as repeaters, providing power for devices, routing signals in each direction and providing terminations for each line. Some devices, e.g. proprietary PBX phones, are expected to come with their own built in pods. A phone pod will allow a PBX manufacturer to pass call and media control to the desktop PC. PBX makers are likely to go for this choice, since it enables them to continue selling proprietary PBX phones, which, today, are a very profitable part of their business. USB is sophisticated in that it will handle certain "important" data streams — e.g. voice and video — with preference. Serial Bus has three basic types of data transfer:

• Isochronous or streaming real time data which occupies a prenegotiated amount of Serial Bus bandwidth with a prenegotiated latency. (This would be for voice and video.)

• Asynchronous interactive data such as characters or coordinates with few human perceptible echo or feedback responsible characteristics (e.g. tele-gaming).

• Asynchronous block transfer data which is generated or consumed in relatively large and bursty amounts and has wide dynamic latitude in transmission constraints.

Does USB mean T-1 or E-1 To every desktop? Potentially, yes. It's certainly powerful enough. USB is 7.8 times the speed of T-1. In the meantime, ISDN may be the major beneficiary, according to its proponents. Timetable: 0.9 specification: call 1-800-433-3652 and for $35, they'll send you the 250-page document. Or pick it up for free through the World Wide Web. The address is http://WWW.teleport/.com/~USB. 1.0 spec. available in June 1995. Developers' kits in Q3, 1995. USB is expected to be a standard feature in PC chipsets (from Intel and others) in Q4 1995) and PCs in Q1 1996. USB connectors have four pins.

USDLA United States Distance Learning Association. Their mission: The delivery of education or training through electronically mediated instruction including satellite, video, audiographic, computer, multimedia technology and other forms of learning at a distance. 800-275-5162 or 510-606-5160.

Used Equipment which was previously in service (i.e. used someplace else) and may not have been tested, refurbished or remanufactured before you bought it. Used simply means it's no longer new. No more, no less. See also CERTIFIED and REFURBISHED.

Used Or Out Of Service A term used in the secondary telecom equipment business. Equipment taken from service. Can be in any condition. Generally expected to work and be complete.

USENET The USENET is an informal, rather anarchic, group of computer systems that exchange "news." News is essentially similar to "bulletin boards" on other networks. USENET actually predates the Internet network. These days, the Internet is used to transfer much of the USENET's traffic. You can find newsgroups covering every conceivable subject from nude sunbathing to molecular physics. See INTERNET, MUDS and USENET NEWSGROUP ORGANIZATION.

USENET Newsgroup Organization USENET conversations are organized in hierarchical newsgroup trees. There are seven core newsgroup hierarchies or trees: comp (computers), misc (miscellaneous topics), news (newsgroup information), rec (recreation), sci (science), soc (society), talk (conversation). Each tree branches into different levels of newsgroup sub-topics.
UserID A compression of "user identification"; the unique account signature of an Internet user; that which precedes the @ sign in an E-mail address.
User Accessible Tables There are many tables (databases) inside a phone system. They include the extensions with privileges and long-distance dialing selections (see LEAST COST ROUTING). In the old days, most PBX and phone system tables were not accessible to the user, on the assumptions that 1. The user would screw the tables up, and/or 2. Really didn't care about getting access. Things have changed. Users now want faster and greater control over their own destiny. So, many manufacturers are making their tables user accessible.
User Account Each user has a user account that is part of local area network security and controls the user environment. Some account features are assigned to each user automatically, some must be assigned, some are optional.
User Agent 1. Generally refers to the windows and menus used to make interfacing to UNIX easier.
2. UA. An OSI (Open Systems Interconnection) process that represents a human user, or organization, or application in the X.400 MHS (Message Handling System). The user creates, submits, and takes delivery of messages on the user's behalf, and in some cases, can even create the message. UA is thus an X.400 electronic mail term: It is software that prepares the message for transmission to the Message Transfer Agent. The user can be an individual or a distribution list. Users are known by their originator/recipient (O/R) addresses.
User Context A user session created by an operating system in response to a logon request, and typically characterized by privilege sets that strictly define the user's authority to access system resources and information on a LAN. Contexts restrict unauthorized access to facilities and data and protect the system itself from user and applications interference, accidental or otherwise. Contexts are a feature of most multi user operating systems, usually integrated with the security system.
User Datagram Protocol UDP. A packet format included in the Transmission Control Protocol/Internet Protocol (TCP/IP) suite and used for short user messages and control messages. The transmission of UDPs is unacknowledged.
User Data X.25 call control field used to transfer information concerning layers above X.25 between the originating and terminating DTEs.
User Event Input from the user (for example, clicking the mouse, pressing a key, etc.) that causes a multimedia project to perform a specific function (for example, play a movie, change pages, etc).
User ID 1. Persistent information in an ASC (AIN Switch Capabilities) that the ASC communicates to the SLEE (Service Logic Execution Environment) as a parameter in a message to the SLEE. The SLEE uses this parameter to identify the set of information related to the user (e.g., customer record) that service logic needs to perform its task. If a user does not subscribe to any Advanced Intelligent Network Release 1 feature but invokes an AIN Release 1 feature, the user ID in the ASC may correlate in the SLEE to a set of default information to be used by service logic.
2. A compression of "user identification"; the unique account signature of an Internet user; that which precedes the @(at) sign in an E-mail address.
User Loop A 2- or 4-wire circuit connecting a user to a PBX or other phone system.
User Message Part of a CPN message directing a destination node to accomplish some task.
User Segment The part of a satellite system that includes the receivers on the users' premises.
User To User Messaging An ISDN service enabling voice and computer data to be transmitted simultaneously — for example, enabling one person to transmit a spreadsheet file to another so both can examine it on their individual screens, and for each to have the ability to change the spreadsheet and have the changes appear instantly on the other person's screen, and then discuss the changes.
User's Set Apparatus located on the premises of a communications user. Designed to work with other parts of his system.
UserID A compression of "user identification"; the unique account signature of an Internet user; that which precedes the @ sign in the email address.
USITA United States Independent Telephone Association, the old name for the United States Telephone Association. See USTA.
USOC Universal Service Order Code. (Pronounced "U-Sock.") An old Bell System term identifying a particular service or equipment offered under tariff. There was nothing "Universal" or consistent about USOC codes. Many services and equipment were called different things by different Bell operating companies. Since Divestiture, there is absolutely no consistency and little relevance left in USOC codes. Each Bell operating company is busily creating its own billing codes and terms.
According to some oldtimers there was an order to USOCs designation of types of telephones. USOCs were five characters, 12345; shown 123++ meant two or more characters were to follow; 123+x meant one or more character to follow; 123xx meant this code was complete. Let's focus on the last two characters, "++". The First "+" indicated the color, "B" for black, "W" for white. "E" for beige because "B" was used for black, and so on.
The second "+" indicated the type of instrument: "K" for key (remember those clear lighted pick-up buttons and the red hold button?) "C" for combined meaning all was contained in one telephone, and sat on a desk, etc. So, a "BK" was a black key set and, a "BC" was a black desk set.
At one time, just about all Bell operating companies used the standard USOC. All but New York Telephone that is. In the late seventies/early eighties, before the big "D" (divestiture), a new order was put to USOC. What had been New York Tel USOCs, now pretty much became everyone else's USOCs. For consistency of explanation, the definitions of the "K" and the "C" remain the same (K for key and C for combined). Color never did matter to Nynex. So, CV was A Combined Voice (simple telephone) and a KV was a Key Voice (multi line telephone). Obviously the "V" was for voice.
USOP User Service Order Profile.
USOS Universal Operations Services. A software application that supports UIS by providing traditional network operations functions.

USP Usage Sensitive Pricing. A tariff for local service under which the subscriber only pays for the telephone service he uses. This is done for gas and electric service.
USPID User Service Profile Identifier. An ISDN term.
USRT Universal Synchronous Receiver/Transmitter. Integrated circuit that performs conversion of parallel data to serial for transmission over a synchronous data channel.
USTA See UNITED STATES TELEPHONE ASSOCIATION.
USTSA See UNITED STATES TELEPHONE SUPPLIERS ASSOCIATION.
UT 1. Universal Time.
2. Upper Tester. An ATM term. The representation in ISO/IEC 9646 of the means of providing, during test execution, control and observation of the upper service boundary of the IUT, as defined by the chosen Abstract Test Method.
UTAM Unlicensed Transition and Management for Microwave. Relocation in the 2.0 Ghz Band. UTAM Inc. is an open industry organization pledged to relocate incumbents (POFS) presently operating in the 1890 to 1930 MHz band. Designated by the FCC as the frequency coordinator for the UPCS spectrum. "Memorandum and Opinion and Order - June 1994," FCC Docket 90-314.
UTC Universal Time Coordinated; also known as UCT, for Universal Coordinated Time, which is time kept by the "i" laboratory, where i is any laboratory cooperating in the determination of UTC. In the United States, the official UTC is kept by the U.S. Naval Observatory and is referred to as UTC.
UTDR Universal Trunk Data Record.
Utility Communications Architecture UCA. An architecture for networks used to monitor and control electric power distribution systems.
Utility Program A computer program in general support of the processes of a computer; for example, a diagnostic program.
Utility Routine A routine in general support of the processes of a computer; for example, an input routine.
Utilize An absolutely awful word created the people who believe that speaking or writing in big words is a demonstration of their superior intelligence.
Utility In some computer languages, a small program executed as a shell command is called a "tool." In other computer languages, such as BASIC, it is called a "utility."
Utility Program A program which performs generalized functions.
Utility Routine A routine in general support of the operation of a computer, including input/output, diagnostic, tracing or monitoring.
UTOPIA Universal Test and Operations Interface for ATM: Refers to an electrical interface between the TC and PMD sublayers of the PHY layer.
UTP Unshielded Twisted Pair. A cable medium with one or more pairs of twisted insulated copper conductors bound in a single plastic sheath. Now becoming the most common method of bringing telephone and data to the desktop. There is now a strong belief that twisted pair cables carrying data over 16 megabits per second should definitely be carried on unshielded twisted wire — since at such speeds, such cables emanate radiation which should be allowed to escape (otherwise it would interfere with itself). Hence the unshielded cables.
UTR Universal Tone Receiver.
UTS Universal Telephone Service.
Utterance A word used in voice recognition to mean a vocalized sound that is typically a word.
UUCP UNIX-to-UNIX Copy Program. A standard UNIX utility for exchanging information between two UNIX-based machines in a network. UUCP may also be referred to as the UNIX-TO-UNIX Communications Protocol and is widely used for electronic mail transfer. In short, UUCP is the name of a Unix command, but it is now also used to refer to the protocols used by it to transfer files between Unix machines. There are a number of such protocols, and the two machines choose between the ones supported by each. Free implementations also exist for VMS and MS-DOS. The Internet newsgroup comp.mail.uucp has more information.
UUD Short for UUDecode, which is a software utility used to decode UUEncoded files. See UUENCODE.
UUE Short for UUEncode, which is a de facto encoding protocol used to transfer binary files across the Internet and on-line services. See UUENCODE.
UUEncode Software the allows you to take a binary computer file, e.g. a Word for Windows document or a PowerPoint presentation, convert it to ASCII for sending the binary file across the Internet or some other e-mail service and then, once received, converting it back to binary. In its Uuencoded form, the file is visible as ASCII, but it makes absolutely no sense. Direct e-mail services, like MCI Mail or CompuServe, allow you to "attach" a binary file to a message — without ever affecting the message. The person at the other end receives it as a binary file. However, once the message passes across an X.400 gateway, from one e-mail service to another (including or not including the Internet), you need Uuencoding. Also called MIME encapsulation. See MULTIPURPOSE INTERNET MAIL EXTENSION for an example of UUencoding.
UV 1. UltraViolet.
2. A microvolt.

NEWTON'S TELECOM DICTIONARY

V Abbreviation for VOLT.
V & H Vertical and Horizontal grid coordinates. AT&T Bell Labs assigned pair of numbers that locates each telephone company's central office or rate center on a grid of the North American Continent. The V & H numbers are used to determine the "airline distance" between any two rate centers. Long distance phone calls are priced on the basis of airline distance between two rate centers or telephone central offices. You calculate airline distance based on Pythagoras' theorem. The formula is thus:
Step One. Calculate the sum of V1 minus V2 all squared plus H1 minus H2 all squared.
Step Two. Divide the answer by 10.
Step Three: Take the square root of this number.
Example: San Diego to Boston.
(V1 = 9,468) - (V2 = 4422) = 5046
(5046 x 5046) = 25,462,116
(H1 = 7629) - (H2 = 1249) = 6380
6038 x 6038 = 40,704,400
25,462,116 + 40,704,400 = 66,166,516
Divide 66,166,516 by 10 = 6,616,652
The square root of 6,616,652 = 2,572 airline lines

V & H COORDINATES OF MAJOR CONTINENTAL US CITIES

	V	H		V	H
	coordinates			coordinates	
ALABAMA			Sunnyvale	8576	8643
Birmingham	7518	2304	Van Nuys	9197	7919
Huntsville	7267	2535	**COLORADO**		
Mobile	8167	2367	Denver	7501	5899
Montgomery	7692	2247	Fort Collins	7331	5965
ARIZONA			Grand Junction	7804	6438
Flagstaff	8746	6760	Greeley	7345	5895
Phoenix	9135	6748	Pueblo	7787	5742
Tucson	9345	6485	**CONNECTICUT**		
Yuma	9385	7171	Bridgeport	4841	1360
ARKANSAS			Hartford	4687	1373
Fayetteville	7600	3872	New Haven	4792	1342
Hot Springs	7827	3554	New London	4700	1242
Pine Bluff	7803	3358	Stamford	4897	1388
CALIFORNIA			**DELAWARE**		
Anaheim	9250	7810	Wilmington	5326	1485
Bakersfield	8947	8060	**DISTRICT OF COLUMBIA (D.C.)**		
Fresno	8669	8239	Washington	5622	1583
Long Beach	9217	7856	**FLORIDA**		
Los Angeles	9213	7878	Clearwater	8203	1206
Oakland	8486	8695	Daytona Beach	7791	1052
Redwood City	8556	8682	Fort Lauderdale	8282	0557
Sacramento	8304	8580	Jacksonville	7649	1276
San Bernardino	9172	7710	Miami	8351	0527
San Diego	9468	7629	Orlando	7954	1031
San Francisco	8492	8719	Tallahassee	7877	1716
San Jose	8583	8619	Tampa	8173	1147
Santa Monica	9227	7920			
Santa Rosa	8354	8787			

	V	H
	\multicolumn{2}{c}{coordinates}	

	V	H
GEORGIA		
Atlanta	7260	2083
Augusta	7089	1674
Macon	7364	1865
Savannah	7266	1379
IDAHO		
Boise	7096	7869
Pocatello	7146	7250
ILLINOIS		
Chicago	5986	3426
Joliet	6088	3454
Peoria	6362	3592
Rock Island	6276	3816
Springfield	6539	3518
INDIANA		
Bloomington	6417	2984
Fort Wayne	5942	2982
Indianapolis	6272	2992
Muncie	6130	2925
South Bend	5918	3206
Terre Haute	6428	3145
IOWA		
Burlington	6449	3829
Cedar Rapids	6261	4021
Des Moines	6471	4275
Dubuque	6088	3925
Iowa City	6313	3972
Sioux City	6468	4768
KANSAS		
Dodge City	7640	4958
Topeka	7110	4369
Wichita	7489	4520
KENTUCKY		
Danville	6558	2561
Frankfort	6462	2634
Madisonville	6845	2942
Paducah	6982	3088
Winchester	6441	2509
LOUISIANA		
Baton Rouge	8476	2874
New Orleans	8483	2638
Shreveport	8272	3495
MAINE		
Augusta	3961	1870
Lewiston	4042	1391
Portland	4121	1384
MARYLAND		
Baltimore	5510	1575
MASSACHUSETTS		
Boston	4422	1249
Framingham	4472	1284
Springfield	4620	1408
Worcester	4513	1330
MICHIGAN		
Detroit	5536	2828
Flint	5461	2993
Grand Rapids	5628	3261
Kalamazoo	5749	3177
Lansing	5584	3081
MINNESOTA		
Duluth	5352	4530
Minneapolis	5777	4513
St. Paul	5776	4498
MISSISSIPPI		
Biloxi	8296	2481
Jackson	8035	2880
Meridian	7899	2639
MISSOURI		
Joplin	7421	4015
Kansas City	7027	4203
St. Joseph	6913	4301
Springfield	7310	3836
MONTANA		
Billings	6391	6790
Helena	6336	7348
Missoula	6336	7650
NEBRASKA		
Grand Island	6901	4936
Omaha	6687	4595
NEVADA		
Carson City	8139	8306
Las Vegas	8665	7411
Reno	8064	8323
NEW HAMPSHIRE		
Concord	4326	1426
Manchester	4354	1388
Nashua	4394	1356
NEW JERSEY		
Atlantic City	5284	1284
Camden	5249	1453
Hackensack	4976	1432
Morristown	5035	1478
Newark	5015	1430
New Brunswick	5085	1434
Trenton	5164	1440
NEW MEXICO		
Albuquerque	8549	5887
Las Cruces	9132	5742
Santa Fe	8389	5804
NEW YORK		
Albany	4639	1629
Binghamton	4943	1837
Buffalo	5076	2326
Nassau	4961	1355
New York City	4977	1406

	V	H		V	H
	coordinates			coordinates	
Poughkeepsie	4821	1526	Knoxville	6801	2251
Rochester	4913	2195	Memphis	7471	3125
Syracuse	4798	1990	Nashville	7010	2710
Troy	4616	1633	**TEXAS**		
Westchester	4912	1330	Amarillo	8266	5076
NORTH CAROLINA			Austin	9005	3996
Asheville	6749	2001	Corpus Christi	9475	3739
Charlotte	6657	1698	Dallas	8436	4034
Fayetteville	6501	1385	El Paso	9231	5655
Raleigh	6344	1436	Fort Worth	8479	4122
Winston-Salem	6440	1710	Houston	8938	3563
NORTH DAKOTA			Laredo	9681	4099
Bismarck	5840	5736	Lubbock	8596	4962
Fargo	5615	5182	San Antonio	9225	4062
Grand Forks	5420	5300	**UTAH**		
OHIO			Logan	7367	7102
Akron	5637	2472	Ogden	7480	7100
Canton	5676	2419	Provo	7680	7006
Cincinnati	6263	2679	Salt Lake City	7576	7065
Cleveland	5574	2543	**VERMONT**		
Columbus	5872	2555	Burlington	4270	1808
Dayton	6113	2705	**VIRGINIA**		
Toledo	5704	2820	Blacksburg	6247	1867
OKLAHOMA			Leesburg	5634	1685
Lawton	8178	4451	Lynchburg	6093	1703
Oklahoma City	7947	4373	Norfolk	5918	1223
Tulsa	7707	4173	Richmond	5906	1472
OREGON			Roanoke	6196	1801
Medford	7503	8892	**WASHINGTON**		
Pendleton	6707	8326	Bellingham	6087	8933
Portland	6799	8914	Kennewick	6595	8391
PENNSYLVANIA			North Bend	6354	8815
Allentown	5166	1585	Seattle	6336	8896
Altoona	5460	1972	Spokane	6247	8180
Harrisburg	5363	1733	Yakima	6533	8607
Philadelphia	5257	1501	**WEST VIRGINIA**		
Pittsburgh	5621	2185	Clarksburg	5865	2095
Reading	5258	1612	Morgantown	5764	2083
Scranton	5042	1715	Wheeling	5755	2241
RHODE ISLAND			**WISCONSIN**		
Providence	4550	1219	Appleton	5589	3776
SOUTH CAROLINA			Eau Claire	5698	4261
Charleston	7021	1281	Green Bay	5512	3747
Columbia	6901	1589	La Crosse	5874	4133
Spartanburg	6811	1833	Madison	5887	3796
SOUTH DAKOTA			Milwaukee	5788	3589
Aberdeen	5992	5308	Racine	5837	3535
Huron	6201	5183	**WYOMING**		
Sioux Falls	6279	4900	Casper	6918	6297
TENNESSEE			Cheyenne	7203	5958
Chattanooga	7098	2366			
Johnson City	6595	2050			

V Fast A new higher speed over-normal-phone-line modem called V.Fast Class (V.FC) for 28,800 bits per second speed. See V.34.

V Interface The 2-wire ISDN physical interface used for single-customer termination from a remote terminal. See ISDN and U INTERFACE.

V Path Custom Network Service V PATH is a network service from Nynex for voice and data transmissions specifically designed for large, multi-location business customers within a regional calling area. V PATH is designed to provide the features of a private network using public switched lines.

V Reference Point The proposed interface point in an ISDN environment between the line termination and the exchange termination.

V Series Recommendations ITU-T standards dealing with data communications operation over the telephone network. The idea of standards is simple. If you have them and if every manufacturer conforms, then every modem can talk to every other one. That's the idea. But it's not always that simple. Sometimes you have to conform to several standards. For example, in the higher speed modems, for example those at 9,600 bps, you have to conform to speed. That's one standard. You have to conform to error control. That's another standard. And you also have to conform to data compression — if you are using data compression. ISDN terminal adapters are V series recommendations, too. ITU-T uses the term "bis" to designate the second in a family of related standards and "ter" designates the third in a family.

V.110 Terminal rate adaptation protocols for the ISDN B channel with a V-type interface. Includes V.120.

V.120 Terminal rate adaptation protocols for the ISDN B channel with a V-type interface. Includes V.110.

V.13 ITU-T standard for simulated carrier control. Allows a full-duplex modem to be used to emulate a half-duplex modem with interchange circuits changing at appropriate times.

V.14 ITU-T standard for asynchronous-to-synchronous conversion without error control. Allows a modem that is actually synchronous to be used to carry start/stop (async) characters. If a V.42 modem connects with another modem that doesn't have error-control, it falls back to V.14 operation to work without error-control.

V.17 ITU-T standard for simplex (one-way transmission) modulation technique for use in extended Group 3 Facsimile applications only. Provides 7200, 9600, 12000, and 14400 bps trellis-coded modulation (the modulation scheme is similar to V.33), MMR (Modified Modified Read) compression and error-correction mode (ECM).

V.21 ITU-T standard for 300 bit per second duplex modems for use on the switched telephone network. V.21 modulation is used in a half-duplex mode for Group 3 fax negotiation and control procedures (ITU-T T.30). Modems made in the U.S. or Canada follow the Bell 103 standard. However, the modem can be set to answer V.21 calls from overseas.

V.21 CH 2 ITU-T standard for 300 bps modem, describing the operation of modems at 300 bps, and used for critical control and handshaking functions. This low speed is highly tolerant of noise and impairments on the phone line. Fax machines use only Channel 2 of the V.21 recommendations (half duplex channel).

V.21 Fax An ITU-T standard for facsimile operations at 300 bps.

V.22 ITU-T standard for 1,200 bit per second duplex modems for use on the switched telephone network and on leased circuits. V.22 is compatible with the Bell 212A standard observed in the U.S. and Canada.

V.22 bis ITU-T standard for 2,400 bit per second duplex modems for use on the switched telephone network. V.22 bis also provides for 1200bps operation for V.22 compatibility. Bis is used by the ITU-T to designate the second in a family of related standards. "ter" designates the third in a family. The standard includes an automatic link negotiation fallback to 1200 bps and compatibility with Bell 212A/V.22 modems.

V.23 V.23 is the standard for a modem with a 600 bps or 1200 bps "forward channel" and a 75 bps "reverse" channel for use on the switched telephone network.

V.24 ITU-T definitions for interchange circuits between data terminal equipment (DTE) and data communications equipment (DCE) equipment. In data communications, V.24 is a set of standards specifying the characteristics for interfaces. Those standards include descriptions of the various functions provided by each of the pins. This standard is similar (but not identical) to the RS-232-C as established by the American TIA/EIA — Telecommunications Industry Association / Electronics Industries Association.

V.25 Automatic calling and/or answering equipment on the general switched telephone network, including disabling of echo suppressors on manually established calls. Among other things, V.25 specifies an answer tone different from the Bell answer tone. Many modems, including U.S. Robotics modems, can be set with the BO command so that they use the V.25 2100 Hz tone when answering overseas calls.

V.25 bis An ITU-T standard for synchronous communications between the mainframe or host and the modem using the HDLC or character-oriented protocol. Modulation depends on the serial port rate and setting of the transmitting clock source.

V.26 V.26 is the ITU-T standard for 2400 bps modem for use on 4-wire leased lines.

V.26 bis ITU-T standard for 1.2/2.4 Kbps modem. It is important to note that V.26 bis is a half-duplex modem (1200 or 2400 bps in only one direction at a time); it provides an optional 75 bps reverse channel as well.

V.26 ter V.26 ter is a FULL DUPLEX 2400 bps modem, like V.22 bis. The difference is that V.26 ter uses echo cancellation (like V.32) instead of frequency division (like V.22 bis), making it more expensive than V.22 bis. It was intended to serve as a fallback mode from V.32, but most manufacturers ignored it and provide V.22 bis as a fallback instead (V.26 ter is used only in a few installations in France, as far as we know).

V.27 ITU-T standard for 4,800 bits per second modem with manual equalizer for use on leased telephone-type circuits. May be full-duplex on four wire leased lines, or half-duplex on two wire lines.

V.27 bis ITU-T standard for 2,400 / 4,800 bits per second modem with automatic equalizer for use on leased telephone-type circuits. 2.4 Kbps modem for 4-wire leased circuits. Either speed (2,400 is a fallback) can be used on either 4-wire leased lines (full duplex) or 2-wire leased lines (half-duplex). It also provides an optional 75 bps reverse channel.

V.27 ter ITU-T standard for 2,400 / 4,800 bits per second modem for use on the switched telephone network. Half-Duplex only.

V.27 ter is the modulation scheme used in Group 3 Facsimile for image transfer at 2400 and 4800 bps. 4800 bps is a common "fallback" speed.

V.28 V.28, entitled "Electrical Characteristics for Unbalanced Double-Current Interchange Circuits" provides the ITU-T equivalent of the electrical characteristics defined in EIA-232.

V.29 ITU-T standard for 9,600 bits per second modem for use on point-to-point leased circuits. Virtually all 9,600 bps leased line modems adhere to this standard. V.29 uses a carrier frequency of 1700 Hz which is varied in both phase and amplitude. V.29 also provides fallback rates of 4800 and 7200 bps. V.29 can be full-duplex on 4-wire leased circuits, or half-duplex on two wire and dial up circuits. V.29 is the modulation technique used in Group 3 fax for image transfer at 7200 and 9600bps.

V.3 ITU-T specification that describes communications control procedures implemented in 7-bit ASCII code.

V.32 ITU-T standard for 9,600 bit per second two wire full duplex modem operating on regular dial up lines or 2-wire leased lines. If you're buying a 9,600 bps modem for use on the normal dial up switched phone lines, make sure it conforms to V.32. If your modem also conforms to V.42 bis, you should be able to transmit and receive at up to 38,400 bps with other modems that conform to these two specifications. I personally use a number of V.32/V.42 bis modern and they work wonderfully fast. V.32 also provides fallback operation at 4,800 bps. See also V.32 bis, V.42 bis ERROR CORRECTION and V.42 bis DATA COMPRESSION and MODULATION PROTOCOLS.

V.32 bis New higher speed ITU-T standard for full-duplex transmission on two wire leased and dial up lines at 4,800, 7,200, 9,600, 12,000, and 14,400 bps. Provides backward compatibility with V.32. Modems running at V.32 bis at its highest speed of 14,400 bps are actually transmitting that many bits per seconds. They do not rely on compression to achieve that high speed. However, with data compression — such as V.42 and V.42 bis — they can achieve higher speeds. The V.32 bis standard also includes "rapid rate renegotiation" feature to allow quick and smooth rate changes when line conditions change. See MODULATION PROTOCOLS, V.42 and V.42 bis.

V.32 terbo Modulation scheme that extends the V.32 connection range: 4800, 7200, 9600, 12K and 14.4K bps. V.32 bis terbo modems fall back to the next lower speed when line quality is impaired, and fall back further as necessary. They fall forward to the next higher speed when line quality improves.

V.33 ITU-T standard for 14,400 and 12,000 bps modem for use on four wire leased lines.

V.34 V.34 is the international standard for dial up modems of up to 28,800 bits per second. Since the standard suggests speeds twice as fast as the top standard they replace, they carry the nickname "V.Fast." New V.34 modems have a feature called line probing that will allows them to identify the capacities and quality of the specific phone line and adjust themselves to allow, for each individual connection, for maximum throughput. The standard also supports a half-duplex mode of operation for fax applications. The new V.34 technology includes an optional auxiliary channel with a synchronous data signaling rate of 200 bits/second. Data conveyed on this channel consists of modem control data. V.34 modems contain multidimensional trellis coding, which is used to gain higher immunity to noise and other phone line impairments. V.34 modems are the first modems to identify themselves to telephone network equipment (handshaking). V.34 technology has been long in coming and has had to overcome many obstacles. At one point, members of the modem manufacturing industry became so impatient, that some of them began shipping their own proprietary versions of what they thought V.34/V.Fast/28,800 bps modems would be. Many of these modems are only compatible, at higher than 14,400 bps speeds, with themselves. See V.34bis.

V.34 bis A faster version of the data communications standard, V.34, which supports up to 28,800 bps. V.34 bis adds two optional higher data rates to V.34. These speeds are 31,200 bits per second and 33,600 bits per second.

V.35 ITU-T standard for trunk interface between a network access device and a packet network that defines signaling for data rates greater than 19.2 Kbps. V.35 was a definition of a GROUP band modem (meaning, one that used the bandwidth of several telephone circuits). V.36 and V.37 are other group band modems. V.35 just happened to describe, in an appendix, electrical characteristics for a high speed interface. IBM decided to use this for other things, but it was never standardized for those purposes. The ITU-T, in the 1988 Blue Book, says "It is the opinion of the ITU-T that the information contained in Recommendation V.35 is out of date. Therefore it is not recommended to use the techniques described in this Recommendation for new designs. Alternative techniques are described in Recommendations V.36 and V.37. It should be noted that other Recommendations make reference to the electrical characteristics described in Appendix II to this Recommendation [V.35]. As these characteristics are expected to allow interworking with V.11 characteristics, use of V.11 circuits is recommended in those cases." V.35 is no longer published by the ITU-T.

V.35 Gender Changers and Adaptors See photo this page.

V.42 Error Correction ITU-T error-correction standard specifying both MNP4 and LAP-M. The ITU-T title says "Error-correcting procedures for DCEs using Asynchronous-to-Synchronous Conversion". It also notes in the text that it applies only to full-duplex devices. The ITU-T modulation schemes with which V.42 may be used are V.22, V.22 bis, V.26 ter, and V.32, and V.32 bis. LAPM, based on HDLC, is the "primary" protocol, on which all future extensions will be based. The Alternative Protocol specified in Annex A of the Recommendation is for backward compatibility with the "installed base" of error-correcting modems. See V.42 bis.

V.42 bis Data Compression ITU-T data compression standard. It compresses files "on the fly" at an average rate of 3.5:1 and can yield file transfer speeds of up to 9,600 bps on a 2,400 bps modem, 38,400 bits per second with a 9,600 bps modem, 57,600 bps with a 14,400 bps V.32 bis modem, or 115,600 bit/s on a 28,800 bps modem. On-the-fly data compression only has value if you use it to transfer and receive material that is not already compressed. Compressing stuff a second time yields no significant improvement in speed (assuming your compression technique worked the first time around). So the decision to buy a V.42 bis modem depends on the material you're working with and your pocketbook. V.42 bis modems are more expensive. V.42 bis was approved by the ITU-T because of its technical merits. Existing data compression methods (MNP 5 for example) only provided up to two-to-one compression. Also, V.42 bis provides for built-in "feedback" mechanisms, so that the modem can monitor its own compression performance. If the DTE starts send pre-compressed or otherwise uncompressible data, V.42 bis can automatically suspend its operation to avoid expansion of the data. It continues to monitor performance even when sending data "in the clear," and when a performance improvement can be gained by reactivating compression, it will do so automatically.

V.42 bis was selected because it would work with a wide variety of different implementations — different amounts of memory, different processor speeds, etc. Because of this, there WILL be differences between various manufacturer's products in terms of THROUGHPUT performance (although they will all properly compress and decompress, some will do it faster than others). If maximum throughput is important, you should check published benchmark tests to find the modem that provides the best performance.

This chart, courtesy Hayes, shows the speedup that's possible. It includes information on a modem called the Hayes Optima 288, which includes a proprietary (i.e. not compatible with anyone else) Hayes enhanced implementation of V.42 bis.

V.32	9,600 + data compression	=	38,400 bit/s
V.32 bis	14,400 + data compression	=	57,600 bit/s
V.34	28,800 + data compression	=	115,600 bit/s
Hayes Optima 288	28,800 + Hayes enhanced V.42 bis	=	230,400 bit/s

V.54 ITU-T standard for loop test devices in modems, DCEs (Data Communications Equipments) and DTEs (Data Terminal Equipment). Defines local and remote loopbacks. There are four basic tests — a local digital loopback test that is used to test the DTE's send and receive circuits; a local analog loopback test that is used to test the local modem's operation; a remote analog loopback test that is used to test the communication link to the remote modem; and a remote digital loopback test that is used to test the remote modem's operation. If a modem has V.54 capability (most V.32 and V.32 bis modems do), its manual should include documentation on performing the various tests. Version 7 of the Norton Utilities (from Symantec) also includes a local digital loopback test for your PC's COM ports, for which you will need the optional jumper plug offered with the software. Where a modem supports local digital loopback testing, it simulates the jumper plug and does not, therefore, need to be disconnected.

V.70 ITU-T standard for Digital Simultaneous Voice and Data (DSVD) modems. DSVD allows the simultaneous transmission of data and digitally-encoded voice signals over a single dial-up analog phone line. DSVD modems use for V.34 modulation (up to 33.6 kilobits per second), but may also use V.32 bis modulation (14,400 kilobits per second). The DSVD voice coder is a modified version of an existing specification and is defined as G.729 Annex A. The DSVD voice/data multiplexing scheme is an extension of the V.42 error correction protocol widely used in modems today. DSVD also specifies fallbacks that enable DSVD modems to communicated with standard data modems (i.e. V.34, V.32 bis, V.32 and V.22). See DSVD for a bigger explanation.

V.75 ITU-T recommendations which specify DSVD control procedures. See V.70.

V.76 ITU-T recommendations which define V.70 multiplexing procedures.

V.8 A way V.34 modems negotiate connection features and options.

V.8 bis New start-up sequence for multimedia modems.

V.ASVD Analog Simultaneous Voice and Data modem.

V.AVD Alternating Voice and Data. This is the same function as provided by VoiceView products.

V.DSVD Digital Simultaneous Voice and Data.

V.GMUX The multiplexer for V.DSVD.

V.Fast V.FC. An interim modem standard to support speeds to 28,800 bits per second for uncompressed data transmission rates over regular dial up, voice-grade lines. V.FAST stands for Very Fast. V.Fast was a "standard" that only a few manufacturers of modems adopted. These manufacturers adopted V.Fast because they were impatient with the ITU's slowness. Eventually, however the ITU did adopt a new standard, called V.34. See V.34 and V.34bis.

V.FC Version Fast Class. It is an interim standard that was developed for use until the ITU-T ratified V.Fast, i.e. V.34, which is the speed that a V.34 modem communicates at — namely at 28,800 bits per second. V.FC was eventually obsoleted by V.34, which the ITU-T eventually adopted. See V.34.

VAB Value Added Business partner. A term which Hewlett-Packard uses for developers which write software for its computers. HP helps its VABs sell software. Clearly, by doing so, it helps sell more HP computers.

VAC Voice Activity Compression.

Vacant Code An unassigned area code, central office or station code.

Vacant Code Intercept Routes all calls made to an unassigned "level" (first digit dialed) to the attendant, a busy signal, a "reorder" signal or to a recorded announcement.

Vacant Number Intercept Routes all calls of unassigned numbers to the attendant, a busy signal or a prerecorded announcement.

VACC Value Added Common Carrier. A common carrier that provides some network service other than simple end-to-end data transmission. Services include least-cost routing, accounting data, and delivery clarification.

Vacuum Tubes The first electronic computers, Eniac and Univac, built in the wartime secrecy of the 1940s, employed vacuum tubes. They have an average life span of about 20 hours, but with thousands of hot glowing tubes in a single machine, some computers shut down every seven to twelve minutes. Vacuum tubes imposed a limit on the size and power of planned next generations of computers. The second generation of computers, never used vacuum tubes. It used transistors, which were invented in 1947.
VAD 1. Value Added Dealer. Another term for Value Added Reseller (VAR). Essentially, VARs or VADs are companies who buy equipment from computer or telephone manufacturers, add some of their own software and possibly some peripheral hardware to it, then resell the whole computer or telephone system to end users, typically corporations.
2. Voice Activated Dialing.
Vail, Theodore N. Theodore N. Vail began his career with the Bell System as general manager of the Bell Telephone Company in 1878. He later became the first president of the American Telephone & Telegraph Company in 1885. He left AT&T two years later. After pursuing other interests for 20 years, he returned as president of AT&T in 1907, retiring in 1919 as chairman of the board. Vail believed in "One policy, one system, universal service." He regarded telephony as a natural monopoly. He saw the necessity for regulation and welcomed it.
VAIVR Voice Activated Interactive Voice Response.
Validation 1. Generally, all long distance carriers, operator service providers and private pay phone companies will not put a call through unless they can "validate," the caller's telephone company calling card, home/business phone number or credit card. Until the advent of US West's Billing Validation Service and other similar databases in 1987, the companies who needed to validate their callers' billing requests had to turn back the caller or accept the call on faith. Validating a user's calling card is, simply, a Yes-No. If the card number is validated, it is Yes. Getting the validation involves a data call from the provider to the owner of the database. There are many ways of doing this, including a dedicated trunk and an port through an X.25 network. Here's an explanation from material put out by Harris, maker long distance switches, including the P2000V: "Validation processing starts with a check of the P2000V's own internal database of invalid 'billed to' numbers. This database contains numbers that the system administrator wishes to temporarily block. If a call's 'billed to' number does not appear in the database, the P2000V then queries the external validation service. The P2000V directly accesses external validation services via an X.25 modem connected to a leased line. The P2000V can also access Line Information Database (LIDB) through LIDB service bureau providers."
2. Tests to determine whether an implemented system fulfills its requirements. The checking of data for correctness or for compliance with applicable standards, rules, and conventions. The portion of the development of specialized security test and evaluation, procedures, tools, and equipment needed to establish acceptance for joint usage of an automated information system by one or more departments or agencies and their contractors.
Validity Check Any check designed to insure the quality of transmission.
Value Added Refers to a voice or data network service that uses available transmission facilities and then adds some other service or services to increase the value of the transmission.
Value Added Carrier VAC. A voice or data common carrier that adds special service features, usually computer related, to services purchased from other carriers and then sells the package of service and features.
Value Added Common Carrier VACC. A common carrier that provides some network service other than simple end-to-end data transmission. Services include least-cost routing, accounting data, and delivery clarification.
Value Added Network VAN. A data communications network in which some form of processing of a signal takes place, or information is added by the network. No one knows, however, exactly what a VAN is. The general idea is that a VAN buys "basic" transmission and sometimes switching services from local and long distance phone companies and adds something else — typically an interactive computer with a database, a computer and massive storage. In this way, the VAN adds value to basic communications services. Dial up stock market quoting services are VANs. Electronic mail providers are VANs. But VANs can also simply be basic X.25 packet switching networks which are open to the public. Such a network will use X.25 packet switching to provide error correction, redundancy, and other forms of network reliability. Private organizations (companies, universities, etc.) may set up their own value-added networks, or — as in the case of PDNs (Public Data Networks) — another fancy name for a VAN that offers its services to the public. The classic VAN is a packet-switched operation like Tymnet, GTE Telenet, MCI Mail or AT&T Mail.
Value Added Reseller See VAR.
Value Driven Re-Engineering A fancy term for Re-Engineering, which is a term probably invented by Michael Hammer in the July-August, 1990 issue of Harvard Business Review. In that issue, he wrote "It is time to stop paving the cowpaths. Instead of embedding outdated processes in silicon and software, we should obliterate them and start over. We should 're-engineer' our business: use the power of modern information technology to radically redesign our business processes to achieve dramatic improvements in their performance." The term re-engineering now seems to me mean taking tasks presently running on mainframes and making them run on file servers running on LANs — Local Area Networks. The idea is to save money on hardware and make the information more freely available to more people. More intelligent companies also redesign their organization to use the now, more-freely available information. See RE-ENGINEERING.
Value Added Service A communications facility using common carrier networks for transmission and providing extra data features with separate equipment. Store and forward message switching, terminal interfacing and host interfacing features are common extras. See also VALUE ADDED NETWORK.
Valuflex A Nynex service which lets you make and receive regular phone calls and 800-number calls from local areas on your normal business phone lines. There are big advantages here. You don't have to rent additional phone lines. You don't have to expand your existing phone system — or buy a new one (if getting extra lines means you'd grow out). And you can combine your incoming 800 lines with features you can get on business lines — like call forwarding, conferencing, etc. Have your 800 calls come into your office during the day. Have them call forwarded at night to your home. 800 Valuflex is aimed at smaller business.
Valve The original British word for an electron tube.

Vampire Tap In local area networking technology, a cable tap that penetrates through the outer shield to make connection to the inner conductor of a coax cable.

VAN See VALUE ADDED NETWORK.

Van Allen Belts Two layers of charged particles emitted from the sun that are trapped within the earth's magnetic influence. These are named after the discoverer, J. Van Allen. The inner layer exists from about 2,400 to 5,600 km altitude above the earth's surface and consists of secondary charged particles. The outer layer lies between about 13,000 and 19,000 kilometers and is thought to consist of the original particles released from the sun's surface.

Van Eck Detection Kit A receiver that monitors the electromagnetic radiation given off by a computer screen, allowing an eavesdropper to monitor the contents of a victim's screen from a distance (say in the bushes outside a company).

VAPD Voice Activated Premier Dialing.

VAPN Voice Access to Private Network.

Vaporware A semi-affectionate slang term for software which has been announced, perhaps even demonstrated, but not delivered to commercial customers. Hyperware is hardware which has been announced but has not yet been delivered. Slideware is hardware or software whose reason for existing (eventually) has been explained in 35-mm slides, foils, charts and/or PC presentation programs. Slideware is usually less real than vaporware or hyperware, though some people would argue with this. Allegedly the term vaporware cam as a result of the many delays in releasing Windows after Bill Gates of Microsoft announced it at the Fall, 1983 Comdex show in Las Vegas. See also HOOKEMWARE, HYPERWARE, MEATWARE, SLIDEWARE and SHOVELWARE.

VAR Value Added Reseller. Typically VARs are organizations that package standard products with software solutions for a specific industry. VARs include business partners ranging in size from providers of specialty turn-key solutions to larger system integrators.

Vari-A-Bill A 900 service of AT&T whereby the call's price varies depending on certain events — the caller punching out some tones on his phone, or a service technician coming on line, etc.

Variable Call Forwarding An optional feature of AT&T's 800 IN-WATS service. It allows the subscriber to route calls to certain locations based on time of day or day of week.

Variable Format Message A message in which the page format of the output is controlled by format characters embedded in the message itself. The alternative is to have the format determined by prior agreement between the origin and the destination.

Variable Length Buffer A buffer into which data may be entered at one rate and removed at another, without changing the data sequence. Most first-in, first-out (FIFO) storage devices serve this purpose in that the input rate may be variable while the output rate is constant or the output rate may be variable while the input rate is fixed. Various clocking and control systems are used to allow control of underflow or overflow conditions.

Variable Length Record A file in a database containing records not of uniform length and in which the distinctions between fields are made with commas, tabs or spaces. Records become uniform in length either because they are uniform to start with or they are "padded" with special characters.

Variable Quantizing Level VQL. A speech-encoding technique that quantizes and encodes an analog voice conversation for transmission at 32,000 bits per second.

Variable Resistor A resistance element which may be varied to afford various values.

Variable Term Pricing Plan VTPP. A rate plan developed by AT&T to replace two-tier pricing. VTPP generally provides for two, four, five or six year contracts, over which period the customer is promised stable prices for some — not all — of the equipment and/or tariffed services he uses. Generally, under VTPP, the customer does not end up owning any of the equipment. VTPP has now been replaced by more normal ways of doing commercial business — outright sale, leasing, etc.

Variable Timing Parameter Timing durations for features such as hold recall, camp-on recall, off-hook duration, and many other programmable telephone system services.

Variolosser A device with a variable level of attenuation which is controlled by an external signal. Often this signal is the level of the signal being attenuated, that is the higher the level of the signal the more it is attenuated.

VARTI Value Added Reseller Telephone Integrator. A term coined at Telecom Developers '92. It refers to the VARs and interconnects of the 90s that are combining telephony and personal computers to offer products that tie the telephone network to personal computer applications.

VAX A line of minicomputers made by Digital Equipment Corporation (DEC).

VAX Mailgate For MCI Mail An MCI product that enables users of DEC's ALL-IN-1 Integrated Office and Information System to communicate with others outside their internal network via MCI Mail.

VBI Vertical Blanking Interval. The vertical blanking interval is the portion of the television signal which carries no visual information and appears as a horizontal black bar between the pictures when a TV set needs vertical tuning. The VBI is used for carrying close-captioned signals for the hearing impaired. Digitized data can also be inserted into the VBI for transmission at rates greater than 100,000 bps. Information services such as stock market quotations and news offerings are now available via the VBI of a CATV signal. The data embedded in the VBI signal is retrieved from a standard cable or satellite receiver wall outlet by a receiver set, which connects to a RS-232 port on a microcomputer. Software packages then allow subscribers instant access to the information, which may be displayed in a number of formats.

VBNS Very high-speed Backbone Network Service. It's a service run by MCI for the National Science Foundation (NSF). It runs at speeds up to 155 megabits per second.

VBR Variable Bit Rate. A voice service over a an ATM switch. Voice conversations receive only as much bandwidth as they need, the remaining bandwidth is dynamically allocated to other services that may need it more at any given moment. Northern Telecom refers to this approach as "making bandwidth elastic." VBR also refers to networking processes such as LANs which generate messages in a random, bursty manner rather than continuously.

VC 1. Virtual Channel or Virtual Circuit — a communications path between two nodes identified by label rather than fixed physical path.
2. Virtual Circuit. In packet switching, network facilities that give the appearance to the user of an actual end-to-end circuit. A dynamically variable network connection where sequential user data packets may be routed differently during the course of a "virtual connection." Virtual circuits allow many users to share transmission facilities simultaneously.
3. An ATM term. Virtual channels, defined in the UNI 3.0 specification, are physical connections between end points that are not assigned dedicated bandwidth. Instead, bandwidth is allocated on demand by the network at a user's request. In short, a communications channel that provides for the sequential unidirectional transport of ATM cells. See UNI.
4. Virtual Container. SDH defines a number of "containers". The container and the path overhead from a "Virtual Container" (VC) in Europe or "Virtual Tributary" (VT) in North America (ref: CCITT G.709).
VCA See VOICE CONNECTING ARRANGEMENT.
vCard A Versit term. An electronic, virtual information card that can be transferred between computers, PDAs, or other electronic devices through telephone lines, or e-mail networks, or infrared links. vCard defines a common electronic business card format which moves standard business cards to the electronic age. With vCard, individuals can consistently identify themselves without restating or rekeying their information. vCards include data such as name, address, phone number, e-mail user ID, with multimedia support for photographs, sound clips, company logos, and more. vCard is the result of a collaborative industry effort between Versit and multiple vendors. The vCard SDK is available in Windows 95, OS/2, and MacOS versions. The SDK includes the vCard Specification, a demonstration application, vCard file examples, and free source code. A printed copy of the V2.0 Specification may be obtained by calling 1-800-803-6240 (outside the U.S., call 201-327-2803).
VCC Virtual Channel Connection. As an ATM term, it is a concatenation of VCLs that extends between the points where the ATM service users access the ATM layer. The points at which the ATM cell payload is passed to, or received from, the users of the ATM Layer (i.e., a higher layer or ATM-entity) for processing signify the endpoints of a VCC. VCCs are unidirectional.
VCEP Video Compression/Expansion Processor chip.
VCI Virtual Channel Identifier. The address or label of a VC (a virtual circuit). As an ATM term, it is a unique numerical tag as defined by a 16 bit field in the ATM cell header that identifies a virtual channel, over which the cell is to travel.
VCL Virtual Channel Link: An ATM term. A means of unidirectional transport of ATM cells between the point where a VCI value is assigned and the point where that value is translated or removed.
VCN Virtual Corporate Network. Stentor's name for a service it later changed to Advantage VNet. It's similar to MCI's VNet.
VCO Voltage Controlled Oscillator: An oscillator whose clock frequency is determined by the magnitude of the voltage presented at its input. The frequency changes when the voltage changes.
VCOS Visible Caching Operating System. VCOS is a realtime multitasking DSP operating system for the AT&T DSP3210 Digital Signal Processor. Visible Caching means the programmer caches the program and the data onchip, in contrast to logic caching where state machines (implemented in silicon) perform all caching.
VCPI An acronym for the Virtual Control Program Interface, a standard developed by Quarterdeck and Phar Lap Software for running multiple programs and controlling the Virtual-86 mode of 386 microprocessors. A program that's VCPI-compatible and can run in the protected mode under DOS without conflicting with other programs in the system.
VCR VideoCassette Recorder (or Player).
VD Virtual Destination. See VS/VD.
VDE Verband Deutscher Elektrotechniker. Federation of German Electrical Engineers similar in form to the IEEE.
VDI Video Device Interface. A software driver interface that improves video quality by increasing playback frame rates and enhancing motion smoothness and picture sharpness. VDI was developed by Intel and will be broadly licensed to the industry.
VDISK Virtual DISK. Part of the computer's Random Access Memory assigned to simulate a disk. VDISK is a feature of the MS-DOS operating system.
VDM Voice Data Multiplexer.
VDRV Variable Data Rate Video. In digital systems, the ability to vary the amount of data processed per frame to match image quality and transmission bandwidth requirements. DVI symmetrical and asymmetrical and asymmetrical systems can compress video at variable data rates.
VDS Vocabulary Development System.
VDSL An updated version of Asymmetrical Digital Subscriber Line, scheduled to run work at two flavors — 26 million bits per second to the subscriber (downstream) and 2 million from the subscriber back (upstream) or 52 million bits per second to the subscriber and 3 million from the subscriber back (upstream). See ADSL.
VDT 1. Video Display Terminal. A data terminal with a TV screen. Another name for computer monitor. VDT is the term you hear in Europe.
2. Video Dial Tone. The new concept of getting home entertainment, information and interactive services to residences over some form of new broadband network stretching into the nation's homes. Video Dial Tone is a term used by traditional telephone companies. They're the ones allegedly building this broadband network to provide "Video Dial Tone."
VDU Visual Display Unit.
Vector A quantity in the visual (video) telecommunications industry that describes the magnitude and direction of an object's movement — for example, a head moving to the right. See VECTOR IMAGES.
Vector Graphics Images defined by sets of straight lines, defined by the locations of the end points.
Vector Images Images based on lines drawn between specific coordinates. A vector image is based on the specific mathematics of lines. In contrast, a raster image is a bit-mapped (i.e. bit-drawn) image. A vector engineering image is more useful for engineering, since it can be changed easier than a bit-mapped image. A vector image can easily be converted to a raster image. But it's much more difficult to go from a raster image to a vector image. Some storage systems now store images as combination raster/vector.

Vector Processor Array Processor.

Velocity Of Light The speed of light in a vacuum is 186,280 miles per second, or 299,792 kilometers per second. The speed of light is very important because today we can measure time more accurately than length. In effect, we define the meter as the time traveled by light in 0.000000003335640952 of a second as measured by the cesium clock.

Velocity Of Propagation The speed at which a signal travels from a sender, through a transmission line and finally arrives at the receiver.

Velocity Of Sound The velocity of sound varies with the medium carrying it. In air at 0 degrees centigrade, it's 331 meters per second. In glass at 20 degrees centigrade, its 5485 meters per second.

Velveeta An Internet Usenet posting, often commercial in nature, excessively cross-posted to a large number of newsgroups. Similar to Spam, although that term is often used to describe an identical post that's been loaded onto lots of inappropriate newsgroups, one group at a time (rather than cross-posted). This definition courtesy Wired Magazine.

Vendor Code Software written by the same company that manufactured the computer system on which it is running (or not running...).

Vendor ID A Plug and Play term. Vendor ID is the 32-bit vendor ID that indicates the manufacturer, specific model, and version of a device. It is this number which helps Plug and Play configure the PC to run the device.

Vendor Independent Hardware or software that will work with hardware and software manufactured by other vendors. The opposite of proprietary.

Vendor Independent Messaging Group A group of software and software companies who are trying to create non-proprietary, standard programming interfaces to help software and corporate developers write messaging and mail-enabled applications. Ultimately, end users should be able to work together more effectively and be able to exchange information from within desktop applications in a work group environment regardless of vendor platform. Members of the group include Apple, Borland, IBM, Lotus, Novell and WordPerfect.

VENUS-P A CCITT X.25 packet-switched network operated in Japan by Nippon Telephone and Telegraph (NTT) Co.

Verification A service of a phone company operator who dials into a busy or otherwise impossible-to-reach line and checks that line and reports on that check to the caller. Phone companies are beginning to charge for this service. As of writing, AT&T, for example, was charging 40 cents to verify the line was busy and 70 cents additional for the operator to interrupt the conversation and say another call was coming in.

Verification Trunk A trunk to which an operator has access and which will switch through to a called line even if the line is busy.

Verified Off-Hook In telephone systems, a service provided by a unit that is inserted on each of a transmission circuit for the purpose of verifying supervisory signals on the circuit. Off-hook service is a priority telephone service for key personnel, affording a connection from caller to receiver by the simple expedient of removing the phone from its cradle or hook.

Verifier A device that checks the correctness of transcribed data, usually by comparing with a second transcription of the same data or by comparing a retranscription with the original data.

VERONICA Very Easy Rodent Oriented Netwide Index to Computerized Archives. An Internet service that allows users to search Gopher systems for documents.

Versit Versit (201-327-2803) is a loose association between Apple, AT&T, IBM and Siemens Rolm. Its "vision"? To "enable diverse communication and computing devices, applications and services from competing vendors to interoperate in all environments. Communicate and collaborate with anyone, any time, anywhere.." The products include PDAs, notebooks, phones, servers and "collaboration products." One early thrust: standardize on call control within Novell/AT&T's Telephony Services. Background: Call control among PBXs "conforming" to Telephony Services is not standard. PBXs often do the same things differently. Example: Conference a call on one PBX, the PBX may put one call automatically on hold as the other is dialed. Another PBX may expect it to be done manually. The good news: IBM has agreed to pass all its CallPath call control standards over to Versit. Versit's members (including Novell coopted for this task) are now working on making Telephony Services call control more standard. Upshot: Developers won't have to test their "standard" telephony services software on each and every PBX. What works on one will work on the others. That's the goal. In late July 1995, Versit effectively merged all its activities into another association, called ECTF. (www.versit.com) Versit@cup.portal.com. See ECTF and PBX DRIVER PROFILES.

VersitCard A vCard.

Vertical That part of a wiring grid which connects the host computer of Main Distribution Frame (MDF) to equipment located on other fields.

Vertical And Horizontal Coordinates V & H Coordinates. For purposes of determining airline mileage between locations, vertical and horizontal coordinates have been established across the United States. These V&H coordinates are derived from geographic latitude and longitude coordinates. See V & H.

Vertical Blanking Interval The interval between television frames in which the picture is blanked to enable the trace (which "paints" the screen) to return to the upper left hand corner of the screen, from where the trace starts, once again, to paint a new screen. Several companies are eyeing the vertical blanking interval as a place to send digital data, including news and weather information. The vertical blanking interval was the basis of teletext, a 1970s technology that, with the help of a decoder, displays printed information on the TV screen. Teletext has never caught on in the U.S. in part because the amount of data that could be transmitted comfortably was small.

Vertical Redundancy Check VRC. A check or parity bit added to each character in a message such that the number of bits in each character, including the parity bit, is odd (odd parity), or even (even parity). See PARITY.

Vertical Service Options that the customer can add to his basic service such as touchtone, conference calling, speed dialing, etc. No one can explain why it's called "vertical" service.

Very High Frequency VHF. Frequencies from 30 MHz to 300 MHz.

Very Large Scale Integration VLSI. Semiconductor chip with several thousand active elements or logic gates — the

equivalent of several thousand transistors on a single chip. VLSI is the technique for making the micro chip, the so-called "computer on a chip."
Very Low Frequency VLF. Frequencies from 3 KHz to 300 KHz.
VESA Video Electronics Standards Association, San Jose, CA. Along with eight leading video board manufacturers, NEC Home Electronics founded VESA in the late 1980s. The association's main goal is to standardize the electrical, timing, and programming issues surrounding 800 x 600 pixel resolution video displays, commonly known as Super VGA. VESA has also issued a standard called "local bus," a new high-speed bus for the PC designed to move video between the CPU and the screen a lot faster than the conventional AT bus.
Vestigial Sideband Transmission VSB. A modified double-sideband transmission in which one sideband, the carrier, and only a portion of the other sideband are transmitted.
VF 1. Voice Frequency.
2. Variance Factor. An ATM term. VF is a relative measure of cell rate margin normalized by the variance of the aggregate cell rate on the link.
VFast More commonly spelled V.FAST. A future modem standard in the early stages of development, expected to be released in the early part of 1994. If approved by the ITU-T, it would raise modem speeds to about 19,200 bps.
VFAT Virtual File Allocation Table. A fat file system is a file system based on a file allocation table, maintained by the operating system, to keep track of the status of various segments of disk space used for file storage. The 32-bit implementation in Windows 95 is called the Virtual File Allocation Table (VFAT). An extension of the FAT file system in DOS and Windows 3.xx, VFAT supports long filenames while retaining some compatibility with FAT volumes. See also FAT.
VFDN Voice Frequency Directory Number. A Northern Tom term.
VFG Virtual Facility Group.
VFTG Voice Frequency TeleGraph.
VG Voice Grade.
VGA Variable Graphics Array. A graphics standard developed by IBM for the IBM PC. VGA allows the PC's screen to generate any of four levels of resolution — with one of the sharpest being 640 horizontal picture elements, known as pels or pixels, by 480 pels vertically with 16 colors. VGA is superior to earlier graphics standards, such as CGA and EGA. VGA is barely adequate for CAD-CAE. See MONITOR for all the numbers on pixels in various screens. VGA was the graphics standard introduced for IBM PS/2 line and quickly adopted by PC compatibles; supports analog monitors with a 31.5 Hz horizontal scan rate.
VGPL Voice Grade Private Line.
VHD Very High Density. Techniques of recording 20 megabytes and more on a 3 1/2" magnetic disk.
VHF Very High Frequency. The portion of the electromagnetic spectrum with frequencies between 30 and 300 MHz.
VHI Virtual Host Interface.
VHS Video Home System using half-inch tape introduced by Matsushita/JVC in 1975 and now the most popular form of video tape. There is also a VHS at 3/4". It's often used inside ad agencies for previewing work in progress. Industrial video tape — the stuff the TV stations use — is one inch. And it shows a much better quality picture than half-inch VHS.
Via Net Loss VNL. A planning factor used in allocating the attenuation losses of trunks in a transmission network. A specified value for this loss is selected to obtain a satisfactory balance between loss and talker echo performance. The lowest loss in dB at which it is desirable to operate a trunk facility considering limitations of echo, crosstalk, noise and signing.
Vibratory Plow A plow that rips open the ground by vibrating a plow share.
VIDCAP Microsoft's Video For Windows program to capture video input to RAM or hard disk memory.
Video Capture Video Capture means converting a video signal into a format that can be saved onto a hard disk or optical storage device and manipulated with graphics software. This is accomplished with a device internal in a computer called a "frame grabber" or video capture board. Images thus captured are digitized, and can be dropped into a document or database record and may be transmitted locally on a LAN or long distance over a WAN. See VIDEO CAPTURE BOARD.
Video Capture Board To capture a single frame of motion video successfully, you need a board inside your PC that can capture the two fields comprising a single video frame. The best source of single frame video images is a laser disk player which can pause and display a perfect frame of video without noise or jitter. Video cameras or camcorders aimed at a static, non-moving image also work well. VCR, which produces a jittery image when the tape is paused, are the poorest source. See also FRAME GRABBER.
Video Codec The device that converts an analog video signal into digital code.
Video Compression A method of transmitting analog television signals over a narrow digital channel by processing the signal digitally. You can compress an analog TV signal into one T-1 signal of 1.544 megabits per second. More advanced compression techniques will enable video signals to be compressed into fewer bits per second. One increasingly common method allows a full-color reasonably full-motion video to be compressed into two 56 Kbps channels.
Video Conference See VIDEOCONFERENCE.
Video Dial Tone Video dial tone in telco-speak means the phone company, in competition with the cable TV business, provides video to houses and offices. It does not affect the content of that video signal in any way. Thus the term video dial tone, which is like voice dial tone, whose content the phone company also does not affect or change in any way, shape or form. The definition contributed by Stephen Butera, staff director, technical training, New England Telephone.
This is Northern Telecom's explanation of video dial tone: "Recent advances in communications and computer technology, such as fiber optic cables, digital switching and hyperspeed computing, make it possible to transmit extraordinary volumes of interactive electronic information in digital form through telephone networks. Consumers may soon be able to access an intriguing array of multimedia electronic entertainment and information services from the comfort of their homes via a gateway service called Video Dial Tone, which is part of what multimedia is all about. Multimedia means interactive full-motion video, sound, text and graphics all available on your TV, computer terminal or advanced intelligent telecomputer. There may soon be a prolif-

eration of so-called 'intelligent phones' that will transform the touchtone telephone into a versatile home computer. By means of simple pushbutton commands, customers will be able to:
"Select entertainment on demand (movies, music, video).
"Order groceries or other services or products.
"Record customized news and sports programming.
"Enroll and participate in education programs from the convenience of their living rooms.
"Find up-to-minute medical, legal and encyclopedic information.
"Pay bills and manage finances.
"Make airline, rental car and hotel reservations.
"Buy sports and entertainment tickets."
The successor to Video Dialtone is now called Open Video Systems (OVS), according to a press release of the TIA dated April 1, 1996. See also VIDEO ON DEMAND and VIDEO SERVER.

Video Driver A piece of software which translate instructions from the software you are running into thousands of colored dots, or pixels, that appear on your video monitor. A video driver is also called a display Driver. Symptoms of a video driver giving trouble can range from colors that don't look right, to horizontal flashing lines to simply a black screen. In the Macintosh world, Apple rigidly defined video drivers. Windows, in contrast, is a free-for-all. Windows 3.1 defined the lowest common denominator of displays — namely 16 colors at 640 x 480 pixels. But most multimedia programs and many games won't run with only 16 colors. They require at least 256 colors.

Video Electronics Standards Association See VESA.

Video For Windows An extension to Microsoft's Windows operating environment. It includes a set of low-end software tools for the playback, capture and editing of video. It includes three different codecs: Intel's Indeo, Microsoft Video 1 an a RLE codec. It also includes a CD-ROM with hundreds of sample video clips. The product offers "scalable performance." It automatically adapts the size and frame rate of a video image to the computer hardware available. See INDEO VIDEO.

Video Mail Electronic mail that includes moving or still images.

Video On Demand VOD. Punch some buttons. Order up Gone With the Wind to start playing at your house at 8:26 P.M. on Channel 35. Bingo, you have video on demand. It's a great concept with two major problems: The equipment to provide the service is complex and expensive. Second, there's little consumer research on whether consumers are prepared to pay the high price that will be necessary. Still, it's a neat idea. In short, video on demand is a service that allows many users to request the same videos at the same time or anytime. Video on demand requires a high-end video server with hundreds of gigabytes of storage. See NVOD, VIDEO DIAL TONE and VIDEO SERVER.

Video Server A device that could store hundreds, if not thousands of movies, ready for watching by subscribers at their individual whim. A video server could be jukebox like device that would stack several hundred movies. Or it could be a powerful, large computer with several large hard disks and/or optical disk drives. The device would be used in conjunction with the local telephone companies' service called video dial tone — providing movies over normal phone lines to their subscribers or it could be used with the CATV industry's Video On Demand service.

Video Signal Transmission of moving frames or pictures of information requiring frequencies of 1 to 6 Megahertz. A commercial quality full-color, full-motion TV signal requires 6 MHz.

Video Teleconferencing Also called Videoconferencing. The real-time, and usually two-way, transmission of digitized video images between two or more locations. Transmitted images may be freeze-frame (where television screen is repainted every few seconds to every 20 seconds) or full motion. Bandwidth requirements for two-way videoconferencing range from 6 MHz for analog, full-motion, full-color, commercial grade TV to two 56 Kbps lines for digitally-encoded reasonably full motion, full color, to 384 Kbps for even better video transmission to 1,544 Mbit/s for very good quality, full-color, full motion TV. See also VIDEOCONFERENCING.

Video Telephony Real time video call similar to a voice call.

Video Wall Multi-screen video system where a large number of video monitors (typically 16 monitors arrayed in a 4 x 4 matrix) or back projection modules together produce one very large image or combinations of images. Video walls come with their own software, which lets you program the video effects you want. Typically, you can feed a video wall everything from VGA computer output to moving TV (NTSC) signals. Video walls are used for exhibitions and trade shows. They're not cheap. But, when programmed properly, they ARE spectacular.

Video Windows A Bellcore invention which is basically a large, high capacity video conferencing device. Bellcore's Video Windows are connected by two optical links, each carrying 45 million bits of information per second. Though impressive, Bellcore's Video Windows is not considered "high definition" TV. For that to happen, you'd probably need 100 to 150 million bits being transmitted in both directions each second.

Videoconference Videoconference is to communicate with others using video and audio software and hardware to see and hear each other. Audio can be provided through specialized videoconferencing equipment, through the telephone, or through the computer. Videoconferencing has traditionally been done with dedicated video equipment. But, increasingly personal computers communicating over switched digital lines are being used for videoconferencing. See also VIDEOCONFERENCING.

Videoconferencing Video and audio communication between two or more people via a videocodec (coder/decoder) at either end and linked by digital circuits. Formerly needing at least T-1 speeds (1.54 megabits per second), systems are now available offering acceptable quality for general use at 128 Kbit/s and reasonable 7 KHz audio. Factors influencing the growth of videoconferencing are improved compression technology, reduced cost through VLSI chip technology, lower-cost switched digital networks — particularly T-1, fractional T-1, and ISDN — and the emergence of standards. See VIDEOCONFERENCING STANDARDS.

Videoconferencing Standards ITU-T H.261 was the standards watershed. Announced in November 1990, it relates to the decoding process used when decompressing videoconferencing pictures, providing a uniform process for codecs to read the incoming signals. Originally defined by Compression Labs Inc. Other important standards are H.221: communications framing; H.230 control and indication signals and H.242d: call setup and disconnect. Encryption, still-frame graphics coding and data transmission standards are still being developed.

Videophone 2500 In January, 1992 AT&T introduced a product called VIDEOPHONE 2500, which transmitted moving (albeit slowly-moving) color pictures over normal analog phone lines. The phone carried a price tag $1,500 a piece. It was not compatible with one MCI later introduced, made for it by GEC-Marconi of England and costing only $750 retail. Videophone 2500 relies on video compression from Compression Labs, Inc. of San Jose, CA. According to the New York Times, the phone took two years, about $10 million and 30 full-time people at AT&T to develop. The January 3, 1993 New York Times carried a quote from John F. Hanley, group VP for AT&T consumer products division, "We could make an AT&T phone talk to an MCI phone. It would be in both of our interests."

Videotex Two-way interactive electronic data transmission or home information retrieval system using the telephone network. Videotex has not been successful because of its (erstwhile) need for expensive, proprietary (i.e. dedicated) equipment and lack of variety in information offered. There are various forms of videotex. The "classic" European version of interactive videotext typically works at 75 baud going out from the terminal and 1200 baud coming in from the central office. Some American versions ape the European system. Some have 1200 baud both ways. In interactive videotext, you can do everything from sending serious electronic mail to your business suppliers to holding raunchy conversations with perverts in distant cities. As long as you pay your bills, no European PTT seems to care about what you transmit or receive. In France, videotex is called Minitel. And it's a success because the French phone company funds it.

Vidicon Camera An image sensing device that uses an electron gun to scan a photosensitive target on which a scene is imaged.

View 1. In satellite communications, the ability of a satellite to "see" a satellite earthstation, aimed sufficiently above the horizon and clear of other obstructions so that it is within a free line of sight. A pair of satellite earthstations has a satellite in "mutual" view when both enjoy unobstructed line-of-sight contact with the satellite simultaneously. 2. An alternative way of looking at the data in one or more database tables. A view is usually created as a subset of columns from one or more tables.

ViewCall A module of Active Voice's TeLANophy that allows users to manage incoming calls on screen via Microsoft Windows. For instance, they can ask callers to hold, route callers to another extension, or play customized pre-recorded greeting that ask callers to leave a message. Active Voice is in Seattle, WA. See also ViewFax and ViewMail.

Viewdata An information retrieval system that uses a remote database accessible through the public telephone network. Video display of the data is on a monitor or television receiver. Another name for Videotex, the original English (UK) name for it. See VIDEOTEX.

ViewFax A module of Active Voice's TeLANophy that allows users to receive and preview their faxes on their personal desktop computer. Each fax document is stored electronically until the receiver wants to see it on screen or send it to a printer. Active Voice is in Seattle, WA. See also ViewFax and ViewMail.

ViewMail A module of Active Voice's TeLANophy which allows users to see all voice, fax and e-mail message on the desktop PC, allowing users to retrieve them in practically any desired order, including LIFO (Last In, First Out) as against LILO (Last In, Last Out), which is what most voice mail systems do. Active Voice is in Seattle, WA. See also ViewCall and ViewFax.

VIM Vendor Independent Messaging. A new E-mail protocol developed by Lotus, Apple, Novell and Borland to provide a common layer where dissimilar messaging programs can share data and back-end services. A group called the Vendor Independent Messaging Group will is intent on developing an open, industry-standard interface that will allow e-mail features to be built into a variety of software products. See also MAPI, which is the E-mail protocol developed by Microsoft.

VINES VIrtual NEtwork System. A network operating system developed and marketed by Banyan Systems, Westboro, MA. Vines is based on Unix system V. This network operating system provides transparent communication across heterogeneous networks and is more expansive, although attracting far fewer users, than Novell's NetWare. VINES is based on the UNIX operating system.

Viper A deadly computer virus.

Virtual In the telephone industry, "Virtual" is something that pretends to be something it isn't, but can be made to appear to be that thing. A virtual private line is effectively a dial up phone line with an auto-dialer on it. To the user, it appears to be a private line. (But the phone company can re-sell that capacity when it's not in use.) The concept of "virtual" is to give the telephone company an excuse to lower the price to the end user. See VIRTUAL NETWORK.

Virtual 8086 Mode Virtual 8086 mode allows the Intel 80386 and beyond microprocessors to emulate multiple real mode processors and still switch to and from protected modem. The processor can load and execute real mode applications (in virtual 8086 mode), then switch to protected mode and load and execute another application that requires access to the full extended memory available. The microprocessor, together with a control program like Microsoft Windows 3.x or OS/2 assumes the responsibility of protecting applications from one another. See REAL MODE and PROTECTED MODE.

Virtual Banding 1. In WATS services, virtual banding is the ability of trunks to carry traffic to all WATS bands, with billing based on the end points of the call instead of the band over which the traffic went. 2. MCI's definition: Allows customers of MCI's, PRISM, Hotel WATS, and University WATS to call nationwide while only paying for the distance to the actual area. For example, if a customer calls to a Band 1 area, Band 1 pricing is used. Similarly, if a call is placed to a Band 4 area, Band 4 pricing is used.

Virtual Bypass Virtual bypass is a way smaller users can fill the unused portion of local T-1 dedicated loops going from a user site to a local office of a long distance company, called a POP (Point of Presence).

Virtual Call Capability 1. Provides setup and clearing on a per call basis. Each call placed appears to have a dedicated connection for the duration of the call.
2. A data communications packet network service feature in which a call setup procedure and a call-clearing procedure will determine a period of communication between two DTEs. This service requires end-to-end transfer control of packets within a network. Data may be delivered to the network before the call setup has been completed but it will not be delivered to the destina-

tion address if the call setup is not successful. The user's data are delivered from the network in the same order in which they are received by the network. See also VIRTUAL CIRCUIT.

Virtual Cell A call, established over a network, that uses the capabilities of either a real or virtual circuit by sharing all or any part of the resources of the circuit for the duration of the call.

Virtual Channel Switch An ATM term. A network element that connects VCLs. It terminates VPCs and translates VCI values. It is directed by Control Plane functions and relays the cells of a VC.

Virtual Circuit A communications link — voice or data — that appears to the user to be a dedicated point-to-point circuit. Virtual circuits are generally set up on a per-call basis and disconnected when the call is ended. The concept of a virtual circuit was first used in data communications with packet switching. A packetized data call may send packets over different physical paths through a network to its destination, but is considered to have a single virtual circuit. Virtual circuits have become more common in ultra-high speed applications, like frame relay or SMDS. There the connection might be permanently connected like a LAN. When the user wants to transmit he simply transmits. There's no dialing in the conventional sense, just the addition of an address field on the information being transmitted. A virtual circuit is referred to as a logical, rather than physical path for a call. A virtual voice circuit is anything from as simple as a phone with an auto dialer in it to a high-speed link in which voice calls are digitized and send on the equivalent of a ultra high-speed, wide-area equivalent of a local area network. There are two basic reasons people buy virtual circuits. They're cheaper and faster. See PERMANENT VIRTUAL CIRCUIT.

Virtual Circuit Capability A network service feature providing a user with a virtual circuit. This feature is not necessarily limited to packet mode transmission. e.g., an analog signal may be converted at its network node to a digital form, which may then be routed over the network via any available route. See VIRTUAL CIRCUIT.

Virtual Co-Location There are several definitions of this evolving. First: Someone other than the local phone company (called an interconnector) can designate his choice of transmission equipment to be located within a local exchange carrier's central office and dedicated to its use. The interconnector would have the right to remotely monitor and control the equipment, but the local exchange carrier would install, maintain, and repair it. Second; And this is a more restricted definition. Connection is done from sites near telephone central offices, but not within them.

Virtual Computing A new term for software that shapes computing hardware into hardware that never was. Virtual computing uses FPGAs — Field Programmable Gate Arrays. See FPGAs.

Virtual Connection A logical connection that is made to a virtual circuit.

Virtual Container See VC.

Virtual Device A device that software can refer to but that doesn't physically exist.

Virtual Disk A portion of RAM (Random Access Memory) assigned to simulate a disk drive. Also called a ram disk. See RAM DISK.

Virtual Fax A device consisting of a personal computer and an image scanner that can duplicate the functions of a facsimile machine.

Virtual File Allocation Table VFAT. A fat file systems is a file system based on a file allocation table, maintained by the operating system, to keep track of the status of various segments of disk space used for file storage. The 32-bit implementation in Windows 95 is called the Virtual File Allocation Table (VFAT). See FAT.

Virtual Hard Drive Memory Factor The available space on a hard drive partition that Windows can address as physical memory.

Virtual LAN A logical grouping of users regardless of their physical locations on the network. Racal-Datacom defines a virtual LAN as "a LAN extended beyond its geographical limit and flexibly configured to add or remove locations." LANs are typically extended beyond their geographical limits (i.e. several thousand feet within a building or campus) by using telephone company facilities, like T-1, T-3, Sonet, etc.

Virtual Machine Facility VM/370. An IBM system control program, essentially an operating system that controls the concurrent execution of multiple virtual machines on a single System/370 mainframe.

Virtual Machine VM. Software that mimics the performance of a hardware device. For Intel 80386 and higher processors, a virtual machine is protected memory space that is created through the processor's hardware capabilities.

Virtual Memory 1. In computer systems, the memory as it appears to the operating programs running in the CPU. Virtual memory is typically the addition of RAM memory and swapfile memory — portion of a hard disk devoted solely to swapfile memory.
2. The term used with Apple Macintoshes to connote the ability to use disk swap files as RAM. This requires the Macintosh to be running System 7 and PMMU.
3. The space on your hard disk that various versions of Windows (including Windows for Workgroups and NT) use as if it were actually memory. Windows NT does this through the use of swap files. The benefit of using virtual memory is that you can run more applications at one time than your system's physical memory would otherwise allow. The drawbacks are the disk space required for the virtual-memory swap file and the decreased execution speed when swapping to the hard disk is required.

Virtual Memory Manager Virtual Memory Manager is a software-only approach to Expanded Memory. These work almost identically to the EMS emulators, except that they use your hard disk rather than extended memory as the storage medium for blocks of memory copied out of your program. As you can imagine, this is painfully s-l-o-w. Use this approach only as a last resort.

Virtual Network A network that is programmed, not hard-wired, to meet a customer's specifications. Created on as-needed basis. Also called Software Defined Network by AT&T. See SOFTWARE DEFINED NETWORK and VIRTUAL PRIVATE NETWORK.

Virtual Path Switch An ATM term. A network element that connects VPLs. It translates VPI (not VCI) values and is directed by Control Plane functions. It relays the cell of the VP.

Virtual Printer Memory In a PostScript printer, virtual printer memory is a part of memory that stores font information. The memory in PostScript printers is divided into banded memory and virtual memory. Banded memory contains graphics and page-layout information needed to print your documents. Virtual memory contains any font information that is sent to your printer either when you print a document or when you download fonts.

Virtual Printer Technology VPT. Virtual Printer Technology is the enterprise network printer architecture developed by Dataproducts Corporation that enables a printer to become an intelligent node in a networked computing environment and provide printing services to other network nodes through a Client/Server type relationship.

Virtual Private Network VPN. A carrier-provided service in which the public switched network provides capabilities similar to those of private lines, such as conditioning, error testing, and higher speed, full-duplex, four wire transmission with a line quality adequate for data. A virtual private network eliminates or partly eliminates the need for fixed point-to-point private line because it provides on demand dial up circuits or bandwidths that can be dynamically allocated. AT&T, a major provider of virtually private networks, defines them as the equivalent of a private network designed logically within a public network, thus achieving the economy of scale of a public network while offering the user control of the simulated private network. Virtual private network resources are occupied only while information is transiting the network.

Virtual Private Network Service An MCI software-defined network service with multiple locations that customers can access through the use of customized dialing plans.

Virtual Reality VR. The publisher of Virtual Reality Report says, "Virtual reality is a way of enabling people to participate directly in real-time, 3-D environments generated by computers." Virtual reality involves the user's immersion in and interaction with a graphic screen/s. Using 3-D goggles and sensor-laden gloves, people "enter" computer-generated environments and interact with the images displayed there. Says Business Week, "Imagine the difference between viewing fish swimming in an aquarium and donning scuba gear to swim around them. That's the sensory leap between regular computer graphics and virtual reality. There are three kinds of VR (Virtual Reality) immersion. First, the toe in the water experience of beginners who stand outside the imaginary world and communicate by computer with characters inside it. next, wading up to the hips, are the "through the window" users, who use a "flying mouse" to project themselves into the virtual, or artificial, world. Then there are the hold-the-nose plungers: "first persona interaction within the computer-generated world via the use of head-mounted stereoscopic display, gloves, bodysuits and audio systems providing binaural sound. The trick with virtual reality is not only to simulate another world but to interact with it — pouring in data affecting its plots, changing its characters and introducing real-world unpredictability into this "mirror world." Once virtual reality was called artificial reality. But artificial means "fake," while virtual means "almost." The father of virtual reality is Joran Lanier. A term close to virtual reality is telepresence. See TELEPRESENCE.

Virtual Route Virtual circuit in IBM's SNA. See SYSTEMS NETWORK ARCHITECTURE.

Virtual Route Pacing Control SNA congestion control at the path control level. See SYSTEMS NETWORK ARCHITECTURE.

Virtual Storage Storage space that may be viewed as addressable main storage to a computer user, but is actually auxiliary storage (usually peripheral mass storage) mapped into real addresses. The amount of virtual storage is limited by the addressing scheme of the computer.

Virtual Telecommunication Access Method VTAM (Pronounced "Vee-Tam.") A program component in an IBM computer which handles some of the communications processing tasks for an application program. VTAM also provides resource sharing, a technique for efficiently using a network to reduce transmission costs.

Virtual Terminal VT. A universal terminal. The ISO virtual terminal (VT) protocol is designed to describe the operation of a so-called universal terminal so any terminal can talk with any host computer.

Virtual Terminal Protocol VTP. An International Standards Organization (ISO) standard for virtual terminal service.

Virtual Tributary VT. A structure designed for transport and switching of sub-DS3 payloads. A unit of sub-Sonet bandwidth that can be combined, or concatenated, for transmission through the network; VT1.5 equals 1.544 Mbps; VT2 equals 2.048 Mbps; VT3 equals 3 Mbps; VT6 equals 6 Mbps.

Virtualization The process of implementing a network based on virtual Network segments.

Virus A software program capable of replicating itself and usually capable of wreaking great harm on the system.

Visible Light Electromagnetic radiation visible to the human eye at wavelengths of 400-700 nm.

Vision One Vision One "is a strategy for evolution of telecommunications developed by Siemens AG, Siemens companies worldwide and GPT. Vision One comprises all the innovative, homogeneous and compatible network elements required for the high performance universal network of the future." This from an advertisement Siemens ran in conjunction with Telecom '91 in Geneva Switzerland.

Visit Northern Telecom's name for its desktop multimedia conferencing system. There's Visit Voice, Visit FastCall and Visit Messenger. See below.

Visit FastCall A low-cost PC-based call center from Northern Telecom. Allegedly, Visit Fastcall will go as low as one agent.

Visit Video A desktop video conferencing system from Northern Telecom.

Vista A videotext service offered in Canada.

Vistium A desktop videoconferencing device which AT&T introduced and then withdrew from the market towards the end of 1995.

Visual Basic A version of the programming language BASIC written by Microsoft Corporation for Windows. The new program promises to make it much easier for businesses to develop customized Windows applications. Some programmers are calling the software a major breakthrough in ease of programming. When I wrote this, Microsoft had sold over one million copies of Visual Basic.

Visual Display Unit VDU. Another term for a computer monitor. VDU is preferred in Europe.

Visual Solutions A family of AT&T products which do videoconferencing, first announced on March 23, 1993.

Visual Voice Messaging A term created by Microsoft as part of its At Work announcement in June of 1993. There'll be At Work-based visual voice messaging servers sitting on a LAN. Messages for PC users on the LAN will be able to be displayed in a list, much like electronic mail, including the caller's name or number, the time he or she called and the length of the call. This information would let the user browse all messages and select the order for listening to the messages. Administrative options, such as creating a new greeting, will be accessed with a single button. Operations that are difficult today, such as forwarding a voice message to multiple people, will be dramatically simplified, according to Microsoft. One will simply select the recipients from the phone book and broadcast the message. Using visual voice messaging, users will be able to bypass today's inconsistent, time consuming and confusing audio menus and access their voice messages with the push of a button or the click of a mouse on a Windows type icon. Messages will be able to be retrieved in any order and even delivered to a single mailbox along with other messages such as e-mail and faxes. These visual voice messaging servers will, according to Microsoft, provide applications beyond basic voice messaging, such as supporting voice annotation of PC documents or reading electronic mail over the phone to a traveler.

Visualization A combination of computerized graphics and imaging technology that provides high-resolution, video-like results on the workstation or personal computer's screen.

Visually Impaired Attendant Service Visually impaired attendant service capability is achieved by augmenting the normal visual signals provided on a standard attendant position with special tactile devices and/or audible signals which enable a visually impaired person to operate the position.

VITA VFEA International Trade Association. A widely supported industry trade group in Scottsdale, AZ. VITA is chartered to promote the growth and technical excellence of the VME bus and Futurebus based microcomputer board market. VITA is chartered to submit standards for ANSI registration. See VME.

Vitreous Silica Glass consisting of almost pure silicon dioxide.

Vivid Newbridge Networks' brand name for some ATM equipment it sells.

VL-Bus A new PC bus from VESA — the Video Electronics Standards Association. The VL bus is up to 20 times as fast as an ISA bus, the most common PC bus and the one common to the original PC, the PC XT and the PC AT and clones. VL-Bus was popular on 486 PCs. Pentium-based machines now largely use the newer, PCI bus. As a result, VL-Bus is pretty obsolete. See ISA, MICROCHANNEL and PCI.

VLAN Virtual Local Area Network: Work stations connected to an intelligent device which provides the capabilities to define LAN membership.

VLC Variable length coding.

VLF Very Low Frequency. That portion of the electromagnetic spectrum having continuous frequencies ranging from about 3 Hz to 30 kHz.

VLM Novell Virtual Loadable Module network client architecture uses packet burst technology, so ample packets are sent without waiting for packet acknowledgement. VLM support (compared with the old IPX) improves transfer times, especially for compressible files, since there is no waiting for acknowledgement.

VLR Visitors' Location Register. A wireless telecommunications term. A local database to an MSC (Mobile services Switching Center) for registering visiting mobile station users. This information is retained as long as the user resides in the geographical area of the visited MSC. The VLR obtains information from the HLR of the user. See MSC.

VLSI Very Large Scale Integration. The art of putting hundreds of thousands of transistors onto a single quarter-inch square integrated circuit. Compare with LSI and ULSI.

VM 1. Voice Mail, Voice Messaging or Virtual memory. See VIRTUAL STORAGE and VOICE MAIL.
2. Virtual Machine. IBM's mainframe operating system.

VME Acronym for "VersaModule-Europe". A one through 21 slot, mechanical and electrical bus standard originally developed by the Munich, Germany division of Motorola in the late 70s. VME uses most of the bus structure from then current Motorola's VersaBus board standard along with the newly developed DIN 41612 standard pin-in-socket connector for enhanced reliability. After years of work, VME was finally adopted by the ANSI/IEEE in 1987 (as ANSI/IEEE-1014). VME is known in Europe as the IEC 821 bus. This makes it an open standard. The VME backplane runs at 80 Mbytes per second. It is the most common bus on big open computers (i.e. ones larger than the PC). As of writing, there were over 300 vendors offering more than 3,000 off-the-shelf VME products. The IEEE standard is soon to lapse and be replaced by an extended VME64 specification, now in ANSI ballot being conducted by VITA.

VME64 An enhanced VME bus standard which includes multiplexed address and data cycles with 40 and 64 bit address modes and 64 bit data transfer modes allowing up to 80 MB/s transfer speed. This standard is under the ANSI ballot process conducted by VITA. See VME.

VME64 Extensions A VITA draft standard that provides extra functionality to VME64 including 5 row J1/P1 and J2/P2 connectors that support live insertion on both 3U and 6U VME boards. Other features: 3.3V power, more grounds, ETL (slew rate) drivers, geographic addressing (slot ID) as well as support for parity, a serial diagnostic bus, JTAG test support and lots of user I/O. Some mechanical features: locking extractors, RFI gasketing, and ESD chassis discharge strips. See VME.

VMEC Voice Messaging Educational Committee. An organization formed by voice messaging manufacturers and service providers to promote a better understanding of voice mail and its business benefits, and to help business implement voice mail systems in ways that meet the needs of callers and mailbox owners alike. See VME.

VMI Voice Messaging Interface.

VMF Validation Message Fraud

VMR Violation Monitoring and Removal. The process of removing a violations which are detected, so that violations do not propagate beyond the maintenance span.

VMS Virtual Memory System

VMS OSI Transport Services VOTS. A Digital Equipment Corp. software product that modifies Digital's DECnet transport layer to conform to the International Standards Organization (ISO) Transport Protocol Class Four (TP4).

VMUF Voice Messaging User Interface Forum. A standards body formed by voice messaging end users, service providers and manufacturers to define a minimum set of common human interface specifications for voice messaging systems.

VMX Voice Message Exchange. One day in 1979, Gordon Matthews came back from lunch and noticed that he had received the usual half dozen messages that had been randomly taken down semi-correctly by a harried receptionist. Already a noted inventor, Matthews saw an opportunity to build an adjunct device to the company PBX which would these messages to be recorded by the caller without an intermediary and would allow the recipient to store these messages, forward them to others or to directly reply to them if they were generated from another internal extension. He called this device the Voice Message Exchange, and the company later became VMX, which later got bought by Octel.
VNET Virtual private NETwork. An MCI term for a service it offers to customers who want to join geographically dispersed switches (typically PBXs). Instead of private lines joining the PBX, Vnet uses fast switched lines.
VNL Via Net Loss. A loss objective for trunks, the value of which has been selected to obtain a satisfactory balance between two data terminals for the duration of the call.
VO Verification Office.
VOC In February, 1996 a newspaper called, Investor's Business Daily, ran a story entitled "Is Your Office at Home Making You Sick?" It said that more than 20 million American workers are telecommuting, with another 20 million owning home based businesses. "Experts are finding that home based offices outfitted with the latest fax machines, photocopiers, laser printers and personal computers often foster unhealthy environments. Office machinery emits air pollutants called volatile organic compounds, or VOCs, which can make your head hurt and irritate your eyes, nose and throat. VOCs are also known to cause more serious health problems, including kidney and liver damage, experts say."
Vocabulary Development Development of specific word sets to be used for speaker independent recognition applications.
VOD Video On Demand. At one stage it was considered a gigantic potential money maker for the phone companies who were real interested in getting into this market. Now it doesn't seem so hot. See NVOD and VIDEO ON DEMAND.
VODAS Voice Operated Device Anti-Sing. A device used to prevent the overall voice frequency singing of a two-way telephone circuit by ensuring that transmission can occur in only one direction at any given instant.
Vocoder An early type of voice coder, consisting of a speech analyzer and a speech synthesizer. The analyzer circuitry converts analog speech waveforms into digital signals. The synthesizer converts the digital signals into artificial speech sounds. For COMSEC purposes, a vocoder may be used in conjunction with a key generator and a modulator-demodulator device to transmit digitally encrypted speech signals over normal narrowband voice communication channels. These devices are used to reduce the bandwidth requirements for transmitting digitized speech signals. There are analog vocoders that move incoming signals from one portion of the spectrum to another portion.
VOFR Voice Over Frame Relay.
VOGAD Voice-Operated Gain Adjusting Device. A voice-operated compressor circuit that is designed to provide a near-constant level of output signal from a range of input amplitudes. Such a circuit has a fast attack time with a relatively slow release time to avoid excess volume compression at the system output.
Voice Activated Dialing A feature that permits you to dial a number by calling that number out to your cellular phone, instead of punching it in yourself. See VOICE ACTIVATED VIDEO.
Voice Activated Video A microphone/camera that is activated in response to voice. Imagine you're watching a videoconference going on in four locations. You can hear what everyone is saying. What you need is to be able to see the person who is speaking the loudest, and therefore, presumably the principal speaker — the person whose attention everyone should be focused on. In voice activated video, the videoconferencing system senses who's speaking the loudest and throws that person's face up on everyone's screen.
Voice Activity Compression VAC. A method of conserving transmission capacity by not transmitting pauses in speech.
Voice Applications Program System software providing the necessary logic to carry out the functions requested by telephone system users. It is responsible for actual call processing, making the various voice connections and providing user features, such as Call Forwarding, Speed Dialing, Conference, etc.
Voice Board Also called a voice card or speech card. A Voice Board is an IBM PC- or AT-compatible expansion card which can perform voice processing functions. A voice board has several important characteristics: It has a computer bus connection. It has a telephone line interface. It typically has a voice bus connection. And it supports one of several operating systems, e.g. MS-DOS, UNIX. At a minimum, a voice board will usually include support for going on and off-hook (answering, initiating and terminating a call); notification of call termination (hang-up detection); sending flash hook; and dialing digits (touchtone and rotary). See VOICE BUS and VRU.
Voice Body Part An X.400 term. A body part sent or forwarded from an originator to a recipient which conveys voice encoded data and related information. The related information consists of parameters which are used to assist in the processing of voice data. These parameters include information detailing the duration of the voice data, the voice encoding algorithm used to encode the voice data, and supplementary information.
Voice Bulletin Boards These are voice mailboxes which contain pre-recorded information that can be updated as frequently as the provider of the mailboxes desires and can be accessed by the public 24 hours a day. Voice bulletin boards can be used by city or county departments which receive a large number of calls asking for routine information, e.g., summer programs for kids as listed by a parks and recreation department; jobs currently open in the city as listed by the personnel departments; etc.
Voice Bus Picture an open PC. Peer down into it. At the bottom of the PC, you'll see a printed circuit board containing chips and empty connectors. That board is called the motherboard. Fatherboards are inserted into the connectors on the motherboard. These fatherboards do things on the PC — like pump out video to your screen or material to your printer or your local area network. The motherboard controls which device does what WHEN by sending signals along the motherboard's data bus — basically a circuit that connects all the various fatherboards through their connectors. That data bus was not designed for voice. For voice you need another bus. Several voice processing manufacturers have addressed that need by creating a voice bus at the top of their PC-based voice processing cards. They have tiny pins sticking out of their cards. You attach a ribbon cable from one set

653

of pins on one voice processing card to the next set on the adjacent card and then the next. There are several voice bus "standards." Two come from Dialogic. One is called AEB, Analog Expansion Bus. And one is called PEB, PC Expansion Bus (a digital version). One comes from a consortium of companies and is called MVIP. There are many advantages to having a voice bus. It gives you enormous flexibility to mix and match voice processing boards, like voice recognition, voice synthesis, switching, voice storage, etc. You can build really powerful voice processing systems inside today's fast '386 and '486 PCs with the great variety of voice processing now available. For more information on this exciting field, read TELECONNECT Magazine. 212-691-8215. See MVIP.

Voice Call A telephone call established for the purpose of transmitting voice, rather than data.

Voice Calling One manufacturer describes this as allowing a phone user to have calls automatically answered and connected to his phone's loudspeaker. Not a common definition. A Northern Telecom Norstar definition: This feature allows a voice announcement to be made, or a conversation to begin, through the speaker of another telephone in the system.

Voice Circuit A circuit able to carry one telephone conversation or its equivalent, i.e. the typical analog telephone channel coming into your house or office. It's the standard subunit in which telecommunication capacity is counted. It has a bandwidth between 300 Hz and 3000 Hz. The U.S. analog equivalent is 3 KHz. The digital equivalent is 56 Kbps in North American and 64 Kbps in Europe. This is not sufficient for high fidelity voice transmission. You'd probably need at least 10,000 Hz. But it's sufficient to recognize and understand the person on the other end.

Voice Coil The element in a dynamic microphone which vibrates when sound waves strike it. The coil of wire in a loudspeaker through which audio frequency current is sent to produce vibrations of the cone and reproduction of sound.

Voice Compression Refers to the process of electronically modifying a 64 Kbps PCM voice channel to obtain a channel of 32 Kbps or less for the purpose of increased efficiency in transmission.

Voice Connecting Arrangement VCA. A device that, once upon a time, was necessary for connecting your own phone system to the nation's switched telephone network. Most phones now meet FCC (and other) safety standards, so VCAs are no longer necessary. Most phone systems (as opposed to phones) do have internal protection circuitry, as shown by the "F" (for fully protected) in their FCC registration number.

Voice Coupler An interface arrangement once provided by the telephone company to permit direct electrical connection of customer-provided voice terminal equipment to the national telephone network. No longer needed because of the FCC's Registration Program.

Voice Data An SCSA definition. Encoded audio data.

Voice Dialing The ability to tell your phone to dial by talking to it. Say, "Call Police" and it will automatically dial the police. This feature has enormous benefits for handicapped people. It will have greater benefits for normal people when the technology of voice recognition improves.

Voice Digitization The conversion of an analog voice signal into binary (digital) bits for storage or transmission.

Voice DTMF Forms Applications This Voice DTMF (DUAL TONE MULTIPLE FREQUENCY) application allows a use of a voice mail system to take specific information from its customers 24 hours a day. By prompting callers to respond by speaking or pressing the keys of their touchtone phones, a city department, for example, could plan service calls, building inspections or send out appropriate forms.

Voice Frame See VOICE FRAME.

Voice Frequency VF. An audio frequency in the range essential for transmission of speech. Typically from about 300 Hz to 3000 Hz. See VOICE FREQUENCIES.

Voice Frequencies VF. Those frequencies lying within that part of the audio range that is employed for the transmission of speech. In telephony, the usable voice frequency band ranges from a nominal 300 Hz to 3400 Hz. In telephony, the bandwidth allocated for a single voice frequency transmission channel is usually 4 KHz, including guard bands.

Voice Frequency Telegraph System A telegraph system permitting use of up to 20 channels on a single voice circuit by frequency division multiplexing.

Voice Grade A communications channel which can transmit and receive voice conversation in the range of 300 Hertz to 3000 Hertz.

Voice Hogging See VOICE SWITCHED.

Voice Mail Voice Mail allows you to receive, edit and forward messages to one or more voice mailboxes in your company or in your universe of friends. With voice mail, employees can have their own private mailboxes. Here's an explanation of how it works: You call a number. A machine answers. "Sorry. I'm not in. Leave me a message and I'll call you back." It could be a $50 answering machine. Or it could be a $200,000 voice mail "system." The primary purpose is the same — to leave someone a message. After that, the differences become profound. A voice mail system lets you handle a voice message as you would a paper message. You can copy it, store it, send it to one or many people, with or without your own comments. When voice mail helps business, it has enormous benefits. When it's abused — such as when people "hide" behind it and never return their messages — it's useless. Some people hate voice mail. Some people love it. It's clearly here to stay.

In the fall of 1991, the Wall Street Journal carried a story negative on voice mail. Les Lesniak, Rolm's Senior VP Marketing disagreed. His reply published in the Journal is one of the finest explanations of voice mail's virtues:

"The writer's observations ignore the way today's voice communication technology is making communication between people easier and more convenient, and is elevating the level of service savvy companies provide their customers. Manufacturers use it to take orders after hours and on weekends. Financial services companies use it to provide account information to customers on a 24-hour basis. Colleges use it to register students. A retail executive uses it to broadcast messages to her staff. And a lawyer uses it to respond to calls when traveling.

"Voice messaging keeps calls confidential, simplifies decision making, saves time and money, eliminates inaccurate messages and "telephone tag," allows people to use their time more productively. In short, it keeps communication crisp, clear and constant. The writer's line of thinking would demand that people remain at their desk 24 hours a day. If they don't, the phone goes unanswered, a receptionist answers the phone and takes a message, or an answering machine records the message and cuts off the caller at will. None of these scenarios is ideal.

"To be successful, voice mail technology must be understood by users and supported by top management, And it must meet the needs of the customer. Training for all employees must be mandatory and the system must be administered and managed properly. 'Must answer' lines and greetings that are changed daily are only two ideas that make voice mail not just helpful, but essential to customer service and an enhanced company image.

"Contrary to the writer's view, voice mail contributes to effective business communication and is far superior to an unanswered phone call, a misplaced message or an answering machine."

here are some statistics which add weight to voice mail's logic:
- 75% of all business calls are not completed on the first attempt.
- This can easily waste $50 to $150 per employee per month in toll charges.
- Half of the calls are for one-way transfers of information.
- Two thirds of all-phone calls are less important than the work they interrupt.
- The average length of a voice mail message is 43 seconds. The average long distance call is 3.4 minutes. Voice mail is 80% faster.

Here are the standard benefits of voice mail:
1. No more "telephone tag." Voice mail improves communications. It lets people communicate in non-real time.
2. Shorter calls. When you leave messages on voice mail, your calls are invariably shorter. You get right to the point. Live communications encourage "chit chat" - wasting time and money.
3. No more time zone/business hour dilemma. No more waiting till noon (or rising at 6 A.M.) to call bi-coastally or across continents.
4. Reduce labor costs, Instead of answering phones and taking messages, employees are free to do more vital tasks.
5. Fewer callbacks. In some cases, as many as 50%.
6. Improved message content. Voice mail is much more accurate and private than pink slips. Messages are in your own voice, with all the original intonations and inflections.
7. Less paging and shorter holding times.
8. Less peakload traffic.
9. 24-hour availability.
10. Better customer service.
11. Voice mail allows work groups to stay in contact - morning, noon and night.
12. Voice mail reduces unwanted interruptions.
See Also VOICE MAIL JAIL and VOICE MAIL SYSTEM.

Voice Mail Jail What happens when you reach a voice mail message and you try and reach a human by punching "0" (zero) and you get transferred to another voice mail box and you try again by punching "0" or some other number you're told to punch...and you never reach a human. You're stuck forever inside the bowels of a voice mail machine, being instructed to go from one box to another, never reaching a real human. You're in voice mail jail.

Voice Mail System A device to record, store and retrieve voice messages. There are two types of voice mail devices — those which are "stand alone" and those which profess some integration with the user's phone system. A stand alone voice mail is not dissimilar to a collection of single person answering machines, with several added features. You can instruct the machines (voice mail boxes) to forward messages among themselves. You can organize to allocate your friends and business acquaintances their own mail boxes so they can dial, leave messages, pick up messages from you, pass messages to you, etc. You can also edit messages, add comments and deliver messages to a mailbox at a pre-arranged time. Messages can be tagged "urgent" or "non-urgent" or stored for future listening. The range of voice mail options varies among manufacturers.

An integrated voice mail system includes two additional features. First, it will tell you if you have any messages. It does this by lighting a light on your phone and/or putting a message on your phone's alphanumeric display. Second, if your phone rings for a certain number of rings (you set the number), the phone will transfer your caller automatically to your voice mail box, which will answer the phone, deliver a little "I am away" message and then receive and record the caller's message.

There are other levels of integration. You might have a phone which has "soft" buttons and an alphanumeric display. That display might label your phone's soft buttons like those on a cassette recorder — forward, reverse, slow, fast, stop, etc. so you can go through your messages any way you like. Telenova has such a phone. It's very impressive.

There are pros and cons to voice mail systems. Some employees will hide behind them, forwarding calls from their customers into voice mail boxes and never returning them. Some employees will make good use of them. They dial in for their messages, research what the customer wants and return the voice mail calls quickly. Many voice mail systems are being combined with automated attendants. Many are being combined with interactive voice processing systems, including sophisticated tie-ins to mainframe databases. Some people hate voice mail systems. Others love them. It all depends on how the system is used, managed and sold. See also VOICE MAIL, AUDIOTEX, AUTOMATED ATTENDANTS, INFORMATION CENTER MAILBOXES, ENHANCED CALL PROCESSING and VOICE PROCESSING.

Voice Message Service A leased service typically over dial up phone lines which provides the ability for a phone user to access a voice mail system and leave a message for a particular phone user. See VOICE MAIL SYSTEM.

Voice Message Exchange See VMX.

Voice Messaging Recording, storing, playing and distributing phone messages. Essentially voice messaging takes the benefits of voice mail (such as bulk messaging) beyond the immediate office to almost any phone destination you select. Voice messaging is often done through service bureaus. Nynex has an interesting way of looking at voice messaging. Nynex sees it as four distinct areas: 1. Voice Mail, where messages can be retrieved and played back at any time from a user's "voice mailbox"; 2. Call Answering, which routes calls made to a busy/no answer extension into a voice mailbox; 3. Call Processing, which lets callers route themselves among voice mailboxes via their touchtone phones; and 4. Information Mailbox, which stores general recorded information for callers to hear.

Voice Modem A new type of modem which handles both voice and data over standard analog phone lines. A voice modem is the classic computer telephony device, since it applies intelligence to the making and receiving of normal analog phone calls. Such voice modem might be a full-duplex speakerphone and an answering machine / voice mail device. Such modem might be able to detect incoming and outgoing touchtone and other signals, such as Caller ID. Such modem might also include music on hold, pager dialing, bong and SIT tone detect, line break detect, local phone on / off detect, extension off hook detect, remote ring back detection and VoiceView. The thrust towards voice modems is coming from chip manufacturers, including Sierra Semiconductor, Rockwell and Cirrus Logic. Some standards bodies are working on voice modems. Two standards are emerging — IS-101 and PN-3131.

Voice On The Net Coalition An organization formed to stop regulatory attempts to stifle the growth of voice on the Internet. See VON Coalition. 802-878-9884 and www.von.org

Voice Operated Relay (VOX) Circuit A voice-operated relay circuit that permits the equivalent of push-to-talk operation of a transmitter by the operator.

Voice Over TIE/communications name for a totally wonderful feature on a phone system — namely that while you are speaking to someone on the phone, your operator can talk to you "over" the conversation you're having. What happens is that you hear your operator in your telephone's handset receiver, but the person you're speaking with can't. You can reply to the operator (telling him/her you'll be one minute, please call back, etc.) by hitting a DND/MIC (Do Not Disturb/Microphone) button on your phone. Voice Over has major benefits. It saves on long distance calls you don't have to return. It closes deals that can't wait. And it gives customers immediate answers. In short, it improves corporate efficiency and customer satisfaction.

Voice Paging Access Gives attendants and phone users the ability to dial loudspeaker paging equipment throughout the building. An unbelievably useful feature, if your people are prone to wander.

Voice Print A voice recognition term. A voice print is a speech template used to "train" systems, in particular voice patterns. When a system is operating, the user's speech is compared to the stored voice prints. If they match, the system recognizes the word and executes the command.

Voice Processing Think of voice processing as a voice computer. Where a computer has a keyboard for entering information, a voice processing system recognizes touchtones from remote telephones. It may also recognize spoken words. Where a computer has a screen for showing results, a voice processing system uses a digitized synthesized voice to "read" the screen to the distant caller.

Whatever a computer can do, a voice processing system can too, from looking up train timetables to moving calls around a business (auto attendant) to taking messages (voice mail). The only limitation on a voice processing system is that you can't present as many alternatives on a phone as you can on a screen. The caller's brain simply can't remember more than a few. With voice processing, you have to present the menus in smaller chunks.

Voice processing is the broad term made up of two narrower terms — call processing and content processing. Call processing consists of physically moving the call around. Think of call processing as switching. Content consists of actually doing something to the call's content, like digitizing it and storing it on a hard disk, or editing it, or recognizing it (voice recognition) or some purpose (e.g. using it as input into a computer program.) See VOICE BOARD, VOICE RESPONSE UNIT and VOICE SERVER.

Voice Recognition The ability of a machine to recognize your particular voice. This contrasts with speech recognition, which is different. Speech recognition is the ability of a machine to understand human speech — yours and most everyone else's. Voice recognition needs training. Speech recognition doesn't. See SPEAKER DEPENDENT and SPEAKER INDEPENDENT VOICE RECOGNITION.

Voice Response Unit VRU. Think of a Voice Response Unit (also called Interactive Voice Response Unit) as a voice computer. Where a computer has a keyboard for entering information, an IVR uses remote touchtone telephones. Where a computer has a screen for showing the results, an IVR uses a digitized synthesized voice to "read" the screen to the distant caller. An IVR can do whatever a computer can, from looking up train timetables to moving calls around an automatic call distributor (ACD). The only limitation on an IVR is that you can't present as many alternatives on a phone as you can on a screen. The caller's brain simply won't remember more than a few. With IVR, you have to present the menus in smaller chunks. See IVR and VOICE BOARD.

Voice Ring Multiple Digital Intertie Buses connected in series to all nodes. Provides extra channels for voice data transmission when direct link (DI) channels are busy.

Voice Server A PC sitting on a LAN (Local Area Network) and containing voice files which are accessible by the PCs on the LAN. Such voice files may be transmitted on the LAN or over phone lines under the control of the PCs on the LAN. A voice server might contain voice mail. It might contain voice annotated electronic mail. Its primary function is to store voice in such a way that it's accessible easily. Voice servers are typically faster, have more disk capacity and more backup provisions than normal PCs. According to a letter I received in early May, 1993 from the lawyers for a company called Digital Sound Corporation, that company owns federal trademark registration number 1,324,258 for the mark Voiceserver, spelled as one word, not two.

VoiceSpan VoiceSpan is a new class of modem technology that combines existing modem technology with the ability to simultaneous transfer voice (without digitizing the voice). This general technology has been termed analog simultaneous voice and data (ASVD) to contrast it from digital simultaneous voice and data (DSVD) where the voice is first digitized and then multiplexed with the data. Currently V.asvd is based on technology similar to V.32bis (maximum data rate 14,400 bit/s) but the technology allows the possibility of expansion to include V.34 like capabilities (maximum data rate 33,600 bit/s).

The advantages of ASVD verses DSVD are likely to include better sound quality, rapid ability to detect pauses in voice and send the highest data rate, simpler technology and reduced voice delay between speakers. The disadvantages are likely to be a lower data rate when voice is present and the possibility of less flexibility compared to the all digital approach (DSVD).

Currently two different Recommendations: V.8 and V.25 are used to define what happens when a telephone call is started in fax or data mode. These Recommendations are responsible for the initial sounds (`beeps and squawks'), you hear when your modem connects to a remote modem. However these Recommendations do not easily allow a modem to connect after a voice telephone conversation has started. Most conferencing calls would likely begin with a voice conversation and then switch to data and/or video conferencing. So there is considerable interest in defining a new way to accomplish this. Although some ITU members believe that just making some changes to V.8 would accomplish the same as V.8bis, more simply.

In the market Radish Communications Inc. has established a protocol to accomplish the start-up of data communications during a call as part of an alternating voice and data system called VoiceView(TM). Radish proposed to SG 14 the use of the VoiceView start-up sequence for V.8bis. Unfortunately, while no insurmountable technical problems were identified with the VoiceView approach, the standards committees could not agreed to make V.8bis compatible with the VoiceView start up. The proposed draft V.8bis created at this meeting of SG 14 is independent (the two start-up mechanisms will not confuse each other) of the VoiceView proposal. Therefore manufacturers can make equipment that supports both the VoiceView start-up and the proposed draft V.8bis if they wish.

The biggest remaining problem to SG 14 members is compatibility. Whenever a new sequence of `beeps and squawks' is created all the different types of fax machines, modems, voice response systems, DTMF detectors, etc. currently installed on and in the telephone networks worldwide need to be tested to make sure that the new signals do not cause any serious compatibility problems. The current draft of V.8bis just hasn't been tested yet. This makes the members of SG 14 uncomfortable about its compatibility.

As a result of this concern the US, United Kingdom and France all expressed reservations about Study Group 14 supporting the current draft of V.8bis. This will result in a delay in the approval of V.8bis until at least March 1996. Such a delay is in favor of Radish and the VoiceView partners. The delay allows them to develop an installed base of alternative voice and data VoiceView users. However the likely approval of V.8bis in March 1996 means that the simultaneous voice and data applications will use V.8bis as their start-up sequence. And over the long term markets usually move away from the proprietary solutions to the ITU standard, e.g. MNP4 to V.42, MNP5 to V.42bis, V.FC to V.34.

Two different start-up sequences, three if you count V.8, for different applications is not the best solution. But it appears that it will work. And the output of standards work is often the best that can be agreed, not the best that can be conceived.

VP An ATM term. Virtual Path is a unidirectional logical association or bundle of VCs, which are communications channels that provide for the sequential unidirectional transport of ATM cells.

VPC An ATM term. Virtual Path Connection: A concatenation of VPLs (Virtual Path Links) between Virtual Path Terminators (VPTs). VPCs are unidirectional.

VPI An ATM term. Virtual Path Identifier is an eight bit field in the ATM cell header which indicates the virtual path over which the cell should be routed.

VPL An ATM term. Virtual Path Link is a means of unidirectional transport of ATM cells between the point where a VPI value is assigned and the point where that value is translated or removed.

VPT Virtual Path Terminator. As an ATM term, it is a system that unbundles the VCs of a VP for independent processing of each VC.

VS Virtual Scheduling. As an ATM term, it is a method to determine the conformance of an arriving cell. The virtual scheduling algorithm updates a Theoretical Arrival Time (TAT), which is the "nominal" arrival time of the cell assuming that the active source sends equally spaced cells. If the actual arrival time of a cell is not "too" early relative to the TAT, then the cell is conforming. Otherwise the cell is non-conforming.

VS Virtual Source. Refer to VS/VD.

VS/VD Virtual Source/Virtual Destination. An ATM term, a VS/VD is an ABR connection may be divided into two or more separately controlled ABR segments. Each ABR control segment, except the first, is sourced by a virtual source. A virtual source implements the behavior of an ABR source endpoint. Backwards RM-cells received by a virtual source are removed from the connection. Each ABR control segment, except the last, is terminated by a virtual destination. A virtual destination assumes the behavior of an ABR destination endpoint. Forward RM-cells received by a virtual destination are turned around and not forwarded to the next segment of the connection.

VSE A British Term. Voice Services Equipment, a generic term for voice response unit, interactive voice response, voice processing unit and so on.

VSELP Vector Sum Exited Linear Prediction. A speech coding technique used in U.S. and proposed Japanese DMR standards. Second generation European DMR will probably use some version of VSELP.

Voice Store And Forward Voice mail. A PBX service that allows voice messages to be stored digitally in secondary storage and retrieved remotely by dialing access and identification codes. See VOICE MAIL SYSTEM.

Voice Switched A device which responds to voice. When the device hears a voice, it turns on and transmits it, muting the receive side. The most common voice-switched device is the desk speakerphone. With voice switching, it's easy to hog a circuit. Just keep making a noise. Watch out for voice hogging. If you're calling someone and waiting for them by listening in on your speakerphone, mute your speakerphone. This way you'll hear them when they answer.

Voice Switching Equipment used in voice and video conferences. The equipment is activated by sounds of sufficient amplitude; hopefully speech, but also loud noises. Fast switching activates microphones so that only one conference participant can speak at a time. See also VOICE ACTIVATED VIDEO.

Voice Terminal A pretentious AT&T term for a TELEPHONE.

Voice Verification The process of verifying one's claimed identity through analyzing voice patterns.

Voiceband A transmission service with a bandwidth considered suitable for transmission of audio signals. The frequency range generally is 300 or 500 hertz to 3,000 or 3,400 hertz — the frequency range the common analog home phone service is made at.

Voiceframe VoiceFrame is Harris Digital Telephone Systems' name for their open application platform for voice and call processing. According to Harris, VoiceFrame allows businesses to create a comprehensive set of applications that link computers, telephone networks and the telephone. Examples of applications that use the VoiceFrame platform include
- touchtone driven transaction processing such as telebanking
- operator services
- advanced paging systems
- intelligent call routing via host computer database inquiries
- other call center applications.

VoiceFrame serves as a communications controller whose primary responsibility is the disposition of inbound and outbound call traffic under computer control. VoiceFrame has these abilities
- It interprets call signaling information and translates it into protocols for host computer use.
- It accepts commands from the host and interprets them to switch, route and complete calls.
- It provides host computer access to private and public network services, such as DNIS, DID, SMSI, ANI, 950, 900 and 800 services.
- It allows the host computer to perform those tasks for which it is best suited — real-time call routing decision making, database look up, complex calculations and detailed billing.

VOICEMUX 100 Trademark for MICOM's combined TDM and APV products.

VoiceView The family name for the concept and the product line from Radish Communications, Boulder, CO. VoiceView describes the protocol, the transaction, and the platform. Essentially, VoiceView is technology for switching between voice, data, fax and binary image transmission on the same conversation on a standard 3 Khz analog phone line — the one you have in your home. Three types of information transfer modes are available: Modem Data Mode, Fax Data Mode and VoiceView Data Mode. Imagine that you're talking to your travel agent and want to fly from Newark to Chicago. Your travel agent starts reading off the eight different flights. You begin to scribble down the information as she reads through them as quickly as they come up on her screen. As you struggle to get the information down, she says, "Do you have VoiceView?" You purchased a modem with VoiceView capability a month ago and loaded on the software just in case, so you say, "Yes!" She says, "Great! I'll just transfer the information on my screen to your PC screen and it will save both of us a lot of time and hassles — it will only take about 3 to 10 seconds (depending on the amount of information on the screen), and then we can begin talking again. Here it comes!" At your end you experience being put on hold briefly. You hear a faint chirp at the beginning of the transmit time and another a few seconds later. You can then begin talking again with the information you need right in front of you. In the same way, you can also send and receive faxes and binary files.

VoiceView is the protocol that allows all of this to happen. The VoiceView technology must be embedded in a separate box and/or modem through which the conversation and your various communications devices — PCs, fax machine, modem — etc. are connected. To use VoiceView, you must either own a ViewBridge, made by Radish, or a VoiceView certified modem. VoiceView software will be included with the hardware. Microsoft has released a dynamic link library on the Microsoft Developers Network that will help application developers more easily take advantage of VoiceView from Microsoft Windows, with which it is compatible, since VoiceView is simply implemented as a set of AT-command set extensions. The VoiceView software in Windows95 and other Windows products will allow for the transfer and receipt of the VoiceView file. It will not allow for the viewing of the file. For that, you will need extra software. Here are some typical VoiceView applications:
- Collaborative computing. Spreadsheets and documents are easily exchanged, discussed and modified.
- Call Centers can be VoiceView enabled, for example, letting agents send complex flight times, availabilities and itineraries to customers — while the agent is on the phone, or while the customer is in the queue.
- Mixed messaging systems with VoiceView combine the retrieval of voice messages, faxes and electronic mail with one standard phone call.
- Transactions, such as sales orders, processed or confirmed using both visual and audible interchange.
- Hardware and software support is simplified with VoiceView. During a single phone call, the support person can capture diagnostic files, can send diagrams showing how to fix things, etc. troubleshoot and install fixes.
- Other apps include business card exchange and remote presentations where both audio and visual materials are sent via VoiceView. See also VOICEVIEW AGENT, VOICEVIEW BRIDGE, VOICEVIEW PEER and VOICEVIEW SET.

VoiceView Agent VoiceView Agent, also known as Agent Software, is the original software developed by Radish to launch the VoiceView Protocol. It enables the ViewBridge or VoiceView enabled modems to send PC screens, files, graphics, or bit-maps to other VoiceView compatible users. This includes the VoiceView Formatting Toolkit and is available from Radish. See VOICEVIEW.

VoiceView Bridge VoiceView Bridge, also known as a Viewbridge, is a device originally developed to allow a provider to send to or exchange information with another VoiceView-equipped PC user, VRU, IVR database or computer. Used in conjunction with VoiceView Peer software to receive or send data and is required at both endpoints. The Viewbridge attaches to the PC via an RS-232 interface. It connects directly to an analog telephone line and has a modular jack for connecting to the telephone. The ViewBridge can also emulate a ViewSet. See VOICEVIEW.

VoiceView Peer Also known as Peer software, the software that allows a caller equipped with a PC and a Viewbridge to send and receive VoiceView messages. This software has now largely been superseded by Windows DLLs from Microsoft. See VOICEVIEW.

VoiceView Set Also known as a Viewset. A display unit used by callers to receive visual information during VoiceView phone conversation. The Viewset connects directly to an analog telephone line and has a modular jack for connecting to your telephone. See VOICEVIEW.

Volatile Storage Computer storage that is erased when power is turned off. RAM is volatile storage.
Volser An MCI term used to denote a volume of calls. Based on the words "Volume Serial." The term "Volser" can be applied to the manual collection of calls from a switch on a switch tape or through call data transmitted via NEMAS.
Volt The unit of measurement of electromotive force. Voltage is always expressed as the potential difference in available energy between two points. One volt is the force required to produce a current of one ampere through a resistance or impedance of one ohm.
Volt Meter An instrument for measuring voltages, resistance and current.
Voltage Electricity is a essentially a flow of electrons. They're pushed into a gadget — toaster, computer, phone — on one wire and they sucked out on the other wire. For this movement of electrons to occur there must be "pressure," just as there must be pressure in the flow of water. The pressure under which a flow of electrons moves through a gadget is called the electric voltage. Voltage doesn't indicate anything about quantity, just the pressure. The amount of electricity moving through a wire is called its current and is measured in amps. You figure the power in an electron flow (i.e. in electricity) by multiplying the flow's current by the voltage under which it flows.
Voltage Drop The voltage differential across a component or conductor due to current flow through the resistance or impedance of the component or conductor.
Voltage Rating The highest voltage that may be continuously applied to a conductor in conformance with standards or specifications.
Voltage Spike An extremely high voltage increase on an electrical circuit that lasts only a fraction of a second, but can damage sensitive electronic equipment like telephone systems or can cause it to act "funny." If your phone system starts acting "funny," one "cure" is to shut it off, count to ten, and then turn it on again. This sometimes clears the problem.
Voltage Standing Wave Ratio VSWR. The ratio of the maximum effective voltage to the minimum effective voltage measured along the length of mis-matched radio frequency transmission line.
Voltmeter A device for measuring the difference of potential in volts.
Volume 1. A volume is a partition or collection of partitions that have been formatted for use by a computer system. A Windows NT volume can be assigned a drive letter and used to organize directories and files. In NetWare a volume is a physical amount of hard disk storage space. Its size is specified during installation. NetWare v2.2 volumes, for example, are limited to 255MB and one hard disk, but one hard disk can contain several volumes. A NetWare volume is the highest level in the NetWare directory structure (on the same level as a DOS root directory). A NetWare file server supports up to 32 volumes. NetWare volumes can be subdivided into directories by network supervisors or by users who have been assigned the appropriate rights.
2. Under ISO 9660, a single CD-Rom disc.
Volume Label A name you can assign to a floppy or hard disk in MS-DOS. The name can be up to 11 characters in length. You can assign a label when you format a disk or, at a later time, using the LABEL command.
Volume Serial Number A number assigned to a disk by MS-DOS. The FORMAT command creates the serial number on a disk.
Volume Unit VU. The unit of measurement for electrical speech power in communications work. VUs are measured in decibels above 1 milliwatt. The measuring device is called a VU meter.
VOM Abbreviation for VOLT-OHM-MILLIAMETER, probably the most common form of electronic test equipment. It measures voltage, resistance and current, and may have either a digital or analog meter readout. Some VOMs have other test functions such as audible continuity signals and special tests for semiconductors.
Vomit Comet A plane used to simulate zero-G for astronaut flight training. Trainers often get motion sickness inside.
VON Voice On the Internet. See the next definition.
VON Coalition The "Voice on the Net" (VON) Coalition is an Internet organization devoted to creating global awareness of audio, video, multimedia, and Internet Telephony products and software. It was formed, inter alia, to provide a forum against those telephone companies who wanted to ban telephone conversations over the Internet. VON Coalition, 802-878-9884 or www.von.org
Vote ACK Also known as Mass ACK; in Usenet, the posting of the e-mail address of each person that voted for or against a newsgroup proposal.
Voting Receivers A group of mobile base phone receivers operating on the same frequency as a control unit to pick the best signal from among them.
VOTS VMS OSI Transport Services. A Digital Equipment Corp. software product that modifies Digital's DECnet transport layer to conform to the International Standards Organization (ISO) Transport Protocol Class Four (TP4).
VOX Voice Operated eXchange. Your voice starts it. When you stop speaking, it stops. Tape recorders use it to figure when to start recording and when to stop. There are pros and cons to VOX. With VOX you often miss the beginning of the conversation. And the tape goes on for 3 or 4 seconds after you've stopped talking. Also if ambient noise is high, VOX might mistake it for speaking and turn the recorder on and keep it running. Cellular phones also use VOX to save battery. A cellular phone without VOX is continuously transmitting a carrier back to the cell cite the entire time your call is in progress. The VOX operation used in smaller phones allows the phone to transmit only when you're actually talking. This reduces battery drain and enables handheld phones to operate longer on a smaller battery.
Voycall An early key system manufacturer, which made a combination 1A2 handsfree intercom telephone system. It was wood grained, inlaid into black plastic. An impressive phone system. Sadly, no more.
VP Virtual Path. A collection of VCs (Virtual Channels) all traveling between common points.
VPI Virtual Path Identifier. The address of a Virtual Path. Combined with the VPI, defines a virtual circuit through an ATM network.
VPI/VCI Virtual Path Identifier/Virtual Channel Identifier. Combined, these fields identify a connection on an ATM network.
VPDN Virtual Private Data Network. A private data communications network built on public switching and transport facilities rather than dedicated leased facilities such as T1s.

VPDS Virtual Private Data Services. MCI's equivalent of Vnet for data.

VPIM Voice Profile for Internet Messaging, a proposed Internet messaging protocol for transferring messages between disparate voice messaging servers. With VPIM, users will be able to make an Internet voice message from any telephone and deliver it directly into the recipient's voice mailbox.

VPN Virtual Private Network. Virtual Private Network is a software-defined network offering the appearance, functionality and usefulness of a dedicated private network, at a price savings. Here's how it works: Your company buys a bunch of leased lines from your offices to the nearest local offices of your chosen long distance carrier. You're in your New York offices. You want to dial your offices in Chicago. You pick up the phone, dial perhaps seven digits. The phone rings in Chicago. What's happened is that your local PBX has recognized that call as belonging to your VPN. So it shunts the calls over the dedicated local loop to your long distance carrier. Your carrier then checks your dialed number, perhaps changing it with the aid of a database look up table, and completes the call over the carrier's own switched telephone facilities (fiber optic, microwave, copper, etc.). These are the same facilities which you and I use when we dial 1 and the long distance number (assuming we're equal accessed to that carrier).
There are several differences between a VPN and normal dial service:
1. VPN's price per minute is cheaper, often a lot cheaper.
2. You dial fewer digits with VPN. Sometimes you can get right to the distant desk, without going through the operator at the distant end.
3. You have to pay for the dedicated phone lines at the various ends of the VPN which have those dedicated phone lines. But they're often T-1, and thus not expensive on a per voice circuit basis.
4. You have to commit to use VPN for much longer than you do with normal dial up service — which is typically month-to-month.

VPN 56 Sprint's Switched 56 Kbps service, supports advanced voice, data and image network communication tools including Group IV Fax, high resolution image transfer, file transfer, videoconferencing and switched data service via access to SprintNet, a large public data network.

VPOTS Very Plain Old Telephone Service. No automated switching.

VPU 1. Virtual Physical Unit.
2. Voice Processing Unit.

VQL Variable Quantizing Level. Speech-encoding technique that quantizes and encodes an analog voice conversation for transmission, nominally at 32 Kbps.

VR 1. Voice Recognition. See VOICE RECOGNITION. 2. Virtual Reality. See VIRTUAL REALITY.

VRAM Video RAM. Memory used to buffer an image and transfer it onto the display. It is a form of DRAM specially suited for video. VRAM differs from common DRAM in that it has two data paths — a technique known as dual porting — rather than the single path of traditional RAM; thus, it can move data in and out simultaneously. Two devices can access it at once. The CRT controller, which converts bits and bytes in video memory to pixels on the screen, and the CPU, which manipulates the contents of video memory, can access VRAM simultaneously. Conventional DRAM chips allow one read or write operation at a time. Video RAM supports simultaneous read/write, read/read and write/write operations. It's most often used in graphic accelerators. In video boards fitted with the less expensive DRAM, performance suffers somewhat because the CRT controller and the CPU must takes turns getting to the video buffer held in VRAM. See also DRAM and WRAM.

VRC Vertical Redundancy Check.

VREPAIR A Novell NetWare program somewhat analogous to MS-DOS's CHKDSK program. VREPAIR fixes FAT (File Allocation Table) and DIR (Directory) Tables. It's a most useful program. Highly recommended.

VRML Virtual Reality Modeling Language. A language for writing 3D HTML applications. VRML, according to PC Magazine, is an open standard for 3-D imaging on the World-Wide Web that paves the way for virtual reality on the Internet. The way VRML code describes a 3-D scene is analogous to four points describing a square, or a center point and radius describing a sphere. VRML viewers, similar to HTML Web browsers, interpret VRML data downloaded from the Web and render it on your computer. This allows the bulk of the processing to be performed locally and drastically reduces the volume of information that must be transmitted from the Web — a key consideration if rendering is to be performed in real time.

VROOMM Virtual Runtime Object Oriented Memory Manager. VROOMM is Borland's proprietary technology designed to save memory by reducing commands to compact objects. Because of VROOMM Quattro Pro has had three fully-featured releases without compromising spreadsheet capacity or performance on current machines — even if it's an 8088-based PC with 512K or RAM. How does VROOMM make more memory available for your spreadsheet data? According to Borland, most spreadsheet programs work with only a small number of large overlays, loading one or another into memory depending on the operations being performed. In contrast, VROOMM moves hundreds of small program modules, which are in effect code objects in and out of memory as necessary. Individual areas of Quattro Pro use different code objects, so it's possible to have very large spreadsheets in memory while significantly increasing functionality. Quattro Pro's graphics subsystem also take advantage of VROOMM by storing the graphics you, the user, create in available EMS.

VRPRS Virtual Route Pacing Response in SNA.

VRU See VOICE BOARD and VOICE RESPONSE UNIT.

VRU Connect Signaling A Rockwell ACD definition. When the incoming call is answered by a VRU, a caller typically inputs information via a DTMF keypad. This user-defined information can now be "whispered" as an audio announcement by the VRU to a service representative. After the VRU has conveyed the caller's information to the agent, the VRU drops out and the call is transferred to the agent. The most notable benefits are 1) enabling agents to better anticipate and consequently address the needs of callers, 2) eliminating the need for agents to request information already gathered by the VRU, 3) presenting a better organized and professional appearance to callers, and 4) expediting the transaction process.

V-Chip The Communications Reform Act of 1996, signed into law by President Clinton in early 1996, requires manufacturers of television sets to install a special V-chip in every TV set to allow parents to automatically block any program with a special code. For the V-chip, which is some sort of special microprocessor (not defined in the Act), to be effective, broadcasters must develop a system for deciding which shows are violent and then transit a signal which the V-chip can recognize and act upon, i.e. shut the TV set off, if required to by the parents.

V-Series Recommendations Sets of telecommunications protocols and interfaces defined by ITU-T Recommendations.

VS See VIRTUAL STORAGE.

VS&F Voice Store and Forward. Voice is digitally encoded, sent to large storage devices and later forwarded to the recipient. See VOICE MAIL.

VSAT Very Small Aperture Terminal. A relatively small satellite antenna, typically 1.5 to 3.0 meters in diameter, used for transmitting and receiving one channel of data communications. You see VSATs on top of retail stores which use them for transmitting the day's receipts and receiving instructions for sales, etc.

VSB Vestigial SideBand. A form of AM modulation that compresses required bandwidth.

VSCBX A Rolm PBX called a Very Small Computerized Branch EXchange.

VSE Virtual Storage Extended.

VSS Voice Server System.

VSWR Voltage Standing Wave Ratio. The ratio of the maximum effective voltage to the minimum effective voltage measured along the length of mis-matched radio frequency transmission line.

VT Virtual Tributary. A structure designed for transport and switching of sub-DS3 payloads. VT1.5 equals 1.544 Mbps; VT2 equals 2.048 Mbps; VT3 equals 3 Mbps; VT6 equals 6 Mbps. These are measures of speed in Sonet. See SONET.

VT1.5 Virtual Tributary equals 1.544 megabits per second.

VT2 Virtual Tributary equals 2.048 megabits per second.

VT3 Virtual Tributary equals 3 megabits per second.

VT6 Virtual Tributary equals 6 megabits per second.

VT100 A terminal-emulation system. Supported by many communications program, it is the most common one in use on the Internet. VT102 is a newer version.

VTA Virtual Trunk Agent.

VTAM Virtual Telecommunications Access Method. A program component in an IBM computer which handles some of the communications processing tasks for an application program. In an IBM 370 or compatible, VTAM is a method to give users at remote terminals access to applications in the main computer. VTAM resides in the host. It performs addressing and path control functions in an SNA network that allows a terminal or an application to communicate with and transfer data to another application along some sort of transmission medium. VTAM also provides resource sharing, a technique for efficiently using a network to reduce transmission costs. See SYSTEMS NETWORK ARCHITECTURE.

VTP Virtual Terminal Protocol. An International Standards Organization (ISO) standard for virtual terminal service.

VTTH Video To The Home. The general ability to provide interactive multimedia services to people in their homes.

VU Meter VU is the unit of measurement for electrical speech power in communications work. VUs are measured in decibels above 1 milliwatt. The measuring device is called a VU meter.

VUI First came the CLI (Command-Line Interface). Then came the GUI (Graphical User Interface). Get ready for the VUI: the Video User Interface. Actually, you don't need to get ready for it any time soon, but you might start wondering how to use it.

VW-1 A test used by Underwriters Laboratories to classify wires and cables by their resistance to burning. (Formerly designated as FR-1.)

NEWTON'S TELECOM DICTIONARY

W 1. Abbreviation for WATT.
2 The Hayes AT Command Set describes a standard language for sending commands to asynchronous modems. One of the commands is "W." If you embed a W in your dialing string, i.e. 212-691-8215-W-10045, the modem will dial 212-691-8215 and wait until it hears dial tone. When it hears dial tone, it will dial out 10045. That is the standard Hayes command set interpretation of W. There is another. When using some of the communications software products from Crosstalk (now a subsidiary of DCA) you can place a [W] in your dialing string. If you do, your modem will dial the number until it encounters a [W]. It will then wait until you hit any button on your keyboard. The purpose of W commands is to allow you to dial through private networks (your own), through public networks (MCI, Sprint, etc.), through fax/modem/telephone switches and through any other device or network.
W-DCS Wideband Digital Cross-connect System. W-DCS is an electronic digital cross-connect system capable of cross-connecting signals below the DS3 rate.
W3 An abbreviation for the Internet's World Wide Web. See WORLD WIDE WEB.
W3C World Wide Web Consortium based in Cambridge, MA. An organization of commercial firms who are trying to drive standards into the Internet. In some ways, its ambitions are not that different to the IETF, the Internet Engineering Task Force, except that its members are commercial firms. The IETF's members are volunteers and tend to come from academia. Its Web site is www.w3.org/hypertext/WWW/Consortium.
WACK Wait before transmitting positive ACKnowledgement. In Bisynch, this DLE sequence is sent by a receiving station to indicate it is temporarily not ready to receive.
Wafer A thin disk of a purified crystalline semiconductor, typically silicon, that is cut into chips after processing. Typically, a wafer is about one fiftieth of an inch thick and four or five inches in diameter.
Wafer Fabs Wafer fabs are a slang term for ultraclean factories that fabricate chips on silicon wafers.
WAIS Wide-Area Information Servers. A very powerful system for looking up information in databases (or libraries) across the Internet. WAIS allows you to perform a keyword search. WAIS is like an index, whereas Gopher, which is sometimes used as a complement to WAIS, is like a table of contents.
Wait On Busy An English term for the American term "Camp On" or "Call Waiting." A service allowing the subscriber to make a call to a busy phone line, wait until the call is over, then be connected automatically.
Wait State A period of time when the processor does nothing; it simply waits. A wait state is used to synchronize circuitry or devices operating at different speeds. Wait states are introduced into computers to compensate for the fact that the central microprocessor might be faster than the memory chips next to it. For example, wait states used in memory access slow down the CPU so that all components seem to be running at the same speed. A wait state is a "missed beat" in the cycle of information to and from the CPU that is necessary for a memory transaction to be completed.
WAL A direct dial WATS line, as compared to a WATS line connected via a T-1 line.
Walkaways People who walk away from coin phones though they owe extra money. You can tell a phone that has just been visited by a walkaway: It's typically ringing. And when you answer it, the operator will ask you to deposit some additional coins.
Walkie-Talkie Hand-held radio transmitter and receiver. Like the police carry. Probably the best named device in telecom. You walkie, you talkie.
Walk Time The time required to transfer permission to poll from one station to another.
Wall Outlet A phone outlet positioned at shoulder height to accept a wall telephone set. The typical installation includes a special modular jack containing two mounting bosses that insert into key-hole slots in the base of the telephone set. Electrical connection is made by a short cord or a lug element that is integral to the telephone set base.
Wall Phone A phone that is mounted on the wall. Where else would a wall phone be mounted? Some new phones — especially some key systems — come so you can use them on the desk or mount them on a wall, without extra hardware. Some desk phones cannot be mounted on a wall. This is a disadvantage when you run out of space on your desk, as you will with all the computers and workstations you'll be putting there.
Wall Thickness A term expressing the thickness of a layer of applied insulation or jacket.
Wallpaper The area of your Windows desktop on your PC behind and around your windows and icons. The color and pattern you put on it through the desktop manager is called wallpaper.
WAN Wide Area Network. Uses common carrier-provided lines that cover an extended geographical area. Contrast with LAN. This network uses links provided by local telephone companies and usually connects disperse sites. See WIDE AREA NETWORK.
Wander Long-term random variations of the significant instants of a digital signal from their ideal position in time. Wander variations are usually considered to be those that occur over a period greater than 1 second. See also WANGNET. As I wrote this, WANG had just filed for Chapter 11 protection under US Bankruptcy laws. For more, see CHAPTER 11.
Wang Wang's name for its telephony link is STEP, which stands for Speech and Telephony Environment for Programmers.
WangNet Wang Laboratories' proprietary broadband LANs. Used as the brand name for Wang's LAN products.
War Dialer A program that tries a set of sequentially changing numbers (i.e., telephone numbers or passwords) to determine which ones respond positively.

War Room Also called a "solutions room." This is an enclosed area with a large table used for decision- or strategy-making.

Warble Tone A tone changing in frequency at a slow enough rate to give the effect of warbling. A warble tone is the sound of an electronic ringer, according to many people.

WARC World Administrative Radio Conference. Sets international frequencies. Just before Telecom '87, WARC allocated important new frequencies for satellite-based land mobile (satellite to truck, etc.) and radio determination navigation services (electronic maps for your car). WARC is part of the 154-member International Telecommunication Union. ITU-T is part of the ITU. See ITU-T.

Warm Start Restarting or resetting a computer without turning it off (also called "soft boot"); press Ctrl + Alt + Del on an IBM or IBM compatible.

Warranty Span of time that equipment will be repaired or replaced due to failure. Usually does not include reimbursement of engineer's fees required for replacement. May not include equipment failure due to abuse or destruction by either intentional or unintentional means. Lightning, floods, and other Acts of God are not covered under warranty.

Watch Commands Watch Commands are found in programming. They allow you to "watch" the value of selected application variables while the application is executing (e.g., see the last-entered touchtone digits from a caller).

Watching Timer A circuit used in ETHERNET transceivers to ensure that transmission frames are never longer than the specified maximum length.

Water Pipe Ground A water pipe to which connection is made for the ground.

WATS Wide Area Telecommunications Service. Basically, a discounted toll service provided by all long distance and local phone companies. AT&T started WATS but forgot to trademark the name, so now every supplier uses it as a generic name. There are two types of WATS services — in and out WATS, i.e. those WATS lines that allow you to dial out and those on which you receive incoming calls (the typical 800 line service). You subscribe to in- and out-WATS services separately. In the old days you needed separate in and out lines to handle the in and out WATS services. But these days you can choose to have in- and out-WATS on the same line. This is not particularly brilliant traffic engineering, since you can't receive an incoming 800 call if you're making an outgoing call. But I do know someone who has an 800 line on his cellular phone!
Many users inside companies think their company's WATS lines (and thus their WATS calls) are free, so they speak longer. This can kill the idea of buying WATS lines to save money. In the old days, interstate WATS was charged at effectively a flat rate and thus, there was some reason to believe that marginal WATS calls were 'free.' These days EVERY WATS call costs money. EVERY one! Without exception. See 800 SERVICE and PLANT TEST NUMBER.

Watt The unit of electricity consumption and representing the product of amperage and voltage. the power requirement of a device is listed in watts, you can convert to amps by dividing the wattage by the voltage (e.g., 1,200 watts divided by 120 volts, equal 10 amps). See OHM's LAW. Don't confuse WATTS (the measure of electricity) with WATS, which stands for Wide Area Telecommunications Service. See WATS.

Wave Audio Also called "waveform audio," is a digital representation of actual sound waves. Wave audio "samples" the sound waveforms at regular intervals. The three standard sampling frequencies are 11.025 KHz, 22.05 Khz, and 44.1 KHz. Higher sampling frequencies yield higher fidelity sound.

Wave Length The distance between peaks of an electromagnetic (or other) wave. The distance traveled by a wave during one complete cycle. See also WAVELENGTH.

Waveform The characteristic shape of a signal usually shown as a plot of amplitude over a period of time.

Waveform Editor A word processor for sound. You record something. Then you "play" it back on your PC's screen. Your PC screen now looks like an oscilloscope. Then you use this wave form editor to edit (i.e. change, replace. amplify, echo, fade in or out, cut out noise, cut/paste from other files, or generally muck with) the sound. A wave form editor is used in voice processing.

Waveguide A conducting or dielectric structure able to support and propagate one or more modes. More specifically, a waveguide is a hollow, finely-engineered metallic tube used to transmit microwave radio signals from the microwave antenna to the radio and vice versa. Waveguides comes in various shapes — rectangular, elliptical or circular. They are very sensitive and should be handled very gently. Waveguides may contain a solid or gaseous dielectric material. In optical, a waveguide used as a long transmission line consists of a solid dielectric filament (optical fiber), usually circular. In integrated optical circuits an optical waveguide may consist of a thin dielectric film.

Waveguide Scattering Scattering (other than material scattering) that is attributable to variations of geometry and refractive index profile of an optical fiber.

Wavelength The length of a wave measured from any point on one wave, to the corresponding point on the next wave, such as from crest to crest. In other words, a wavelength is the distance an electromagnetic wave travels in the time it takes to oscillate through a complete cycle. There is a direct proportion between the wavelength of a radio signal and its frequency.

Wavelength Division Multiplexing WDM. A way of increasing the capacity of an optical fiber by simultaneously operating at more than one wavelength. With WDM you can multiplex signals by transmitting them at different wavelengths through the same fiber. Wavelength division multiplexing works similar to frequency division multiplexing. In optical fiber communications, WDW is any technique by which two or more optical signals having different wavelengths may be simultaneously transmitted in the same direction over one strand of fiber, and then be separated by wavelength at the distant end. See also FOUR WAVELENGTH WAVE DIVISION MULTIPLEXING.

WAW Waiter-Actor-Webmaster. Used to describe fly-by-night graphic designers and Web consultants trying to cash in on the Web boom. "Can you believe they hired that clueless WAW for $60K a year?!" This definition courtesy Wired Magazine.

Way Operated Circuit A circuit shared by three or more phones on a party line basis. One of the phones usually operates as the control point.

Way Station One of the phones, other than the central controller, on a way operated circuit. See WAY OPERATED CIRCUIT.

WBC Wide Band Channel. An FDDI-II term. See FDDI II.

WCV See Weighted Call Value.

WDM Wavelength Division Multiplexing. A technique in fiber-optic transmission for using different light wavelengths to send data parallel-by-bit (one discrete wavelength per bit), serial-by-character. One multimode fiber can act as an 8-bit parallel bus.

WDMA Wavelength-Division Multiple Access: a technique which is used to provide multiple channels on different wavelengths on the same fiber-optic cable.

Weather Trunk Group A trunk group used to provide customers with weather information.

Weathermaster Method A distribution method where the unused wall space inside heating and cooling units beneath windows is used for satellite location. Cables are fed from a riser or other serving closet to the location through baseboards, conduit, or underfloor system.

Web An abbreviation for the Internet's World Wide Web. See WORLD WIDE WEB and INTERNET.

Web Art A definition for the artwork that you are beginning to see proliferating the Internet, especially on home pages.

Web Browser Clients software which navigates a web of interconnected documents on the World Wide Web. A Web Browser is software which allows a computer user (like you and me) to "surf" the Internet. It lets us move easily from one World Wide Web site to another. Every time we alight on a Web Page, our Web Browser moves a copy of documents on the Web to your computer. A Web Browser uses HTTP — the HyperText Transfer Protocol. Invisible to the user of a Web Browser, HTTP is the actual protocol used by the Web Server and the Client Browser to communicate over the Internet. The most famous Web Browser is currently Netscape. Many online commercial services such as AOL, CompuServe, Microsoft, Prodigy, etc. distribute Web Browsers to surf their "intranet" as well as the "internet." See INTERNET, MOSAIC, NETSCAPE and SURF.

Web Master See WEBMASTER.

Web Mistress See WEBMISTRESS.

WEB Search Engines An Internet World Wide Web term. A search engine is a program that returns a list of Web Sites (URLs) that match some user-selected criteria such as "contains the words cotton and blouse." Basically, the procedure is simple. You surf to the search engine's site. You click a couple of times and type in what you're looking for. A few seconds later you get choices. You finally make your selection and you get instantly hotlinked over to that site. Here are the main Internet search engines and their addresses. ALTA VISTA http://altavista.digital.com EXCITE http://www.excite.com LYCOS http://www.lycos.com INFOSEEK http://guide.infoseek.com OPENTEXT http://www.opentext.com:8080 INKTOMI http://www.inktomi.berkeley.edu/query.html WWW.WORM http://www.cs.colorado.edu/wwww/ WEBCRAWLER http://www.webcrawler.com YAHOO http://akebono.stanford.edu/yahoo/

Web Server A Web Server is a powerful computer which is connected to the Internet or an Intranet. It stores documents and files — audio, video, graphics or text — and can display them to people accessing the server via hypertext transfer protocol (http). A Web server derives its name because it is part of the World Wide Web. See WORLD WIDE WEB.

Web Service Provider A vendor who provides customers with Web Pages on the vendor's computer/s. Frequently, a Web Service Provider will provide additional services such as design help and usage statistics. Often they will just provide the computer space and leave the rest to you. A Web Service Provider may or may not also be an Internet Service Provider.

Web Site Any machine on the Internet that is running a Web Server to respond to requests from remote Web Browsers is a Web Site. In more common usage it refers to individual sets of Web Pages that can be visited with Web Browsers. It is also spelled as one word, namely WEBSITE. See also INTERNET.

Weblock The Internet data traffic version of gridlock.

Webmaster An Internet term. The Webmaster is the administrator responsible for the management and often design of a company's World Wide Web site. See also WEBMISTRESS.

Webmistress An Internet term. The Webmistress is the female administrator responsible for the management and often design of a company's World Wide Web site. See also WEBMASTER.

Website See WEB SITE.

Webzine Magazines that are published (i.e. made public) on the World Wide Web. Typically a Webzine is available for anyone to read who wants to visit the site the electronic magazine is located at. A Webzine is also called an e-zine.

WECO Western Electric COmpany. The company is now called Lucent Technologies. It used to be the equipment manufacturing arm of AT&T, but in early 1996, AT&T spun it off to the public as a separate, publicly-traded company, called Lucent Technologies. Many oldtimers are sad that about the name change and new ownership. Western Electric had a wonderful reputation and is remembered with great fondness. It has an excellent reputation for high quality products and is still used as a brand name on some products.

Weighted Average A call center term. A method of averaging several numbers in which some numbers are increased before averaging because they have more significance relative to the other numbers.

Weighted Call Value WCV. The average handling time of a call transaction. ACD vendors count this differently. Typically, a combination of the talk time and the after-call work or wrap-up time.

WESTAR Family of communications satellites owned and operated by Western Union.

Western Union International WUI. Acquired by MCI in 1982 to establish MCI in the International Telex and communications market. WUI is now part of MCI International.

WestNet One of the National Science Foundation funded regional TCP/IP networks that covers the states of Arizona, Colorado, New Mexico, Utah, and Wyoming.

Wet Circuit A circuit carrying direct current.

Wet Loop Powering Defined as local power (non-Span provided) with use of copper pairs (power is looped at the last repeater).

Wet T-1 A T-1 line with a telephony company powered interface.

Wetting Agent A chemical which reduces surface tension in a liquid, motivating the liquid to spread more evenly on a surface.

WFWG Windows For Workgroups. See WINDOWS.

WG An abbreviation for workstation, i.e. a computer on a desktop that isn't a server.

Wheatstone Bridge An instrument for measuring resistances.

Whetstones How well does a computer work? Let's test it. The Whetstone benchmark program, developed in 1976, was designed to simulate arithmetic intensive programs used in scientific computing. It is applicable in CAD and other engineering areas where floating-point and trigonometric calculations are heavily used. The Whetstone program is completely CPU-bound and performs no I/O or system calls. The speed at which a system performs floating point operations is measured in units of Whetstones per second or floating point operations per second (flops). Whetstone I tests 32-bit, and Whetstone II tests 64-bit operations. See also DHRYSTONES.

Whisper Technology A call comes into a call center. The voice response unit prompts the caller to the enter their account number. When the call is transferred to the agent, the VRU "whispers" the account number to the agent, who then manually types it into his computer. This technology is now obsolete, since VRUs can now transfer their account number directly into the agent's database and have the look up done automatically. And the call is transferred simultaneously.

White Board See Whiteboard below.

White Facsimile Transmission In an amplitude-modulated facsimile system, that form of transmission in which the maximum transmitted power corresponds to the minimum density of the subject copy. In a frequency-modulated system, that form of transmission in which the lowest transmitted frequency corresponds to the minimum density of the subject copy.

White Line Skip A facsimile transmission technique used to speed up the transmission time by bypassing redundant areas such as white space. (Also known as skip scan.)

White Noise A signal whose energy is uniformly distributed among all frequencies within a band of interest. Seldom occurring in nature, white noise is a useful tool for theoretical research. White noise is also used less scientifically to simply mean background noise. When the first digital PBXs came out, their intercom circuits were so "clean," they spooked users out who were used to some noise on the line. And some PBX manufacturers added a little "white noise" to their PBXs.

White Pages 1. In many countries, including the U.S., Canada and Australia, the phone company publishes two types of telephone directories. One called the "White Pages" lists all the subscribers in alphabetical order. The other, called the Yellow Pages, lists businesses by industry. On the Internet, the White Pages are the lists of Internet users that are accessible through the Internet.

White Signal In facsimile, the signal resulting from the scanning of a minimum-density area of the subject copy.

Whiteboard a device which lets you share images, text and data simultaneously as you speak on the phone with someone else. That someone might be in the next office. Or that someone might be 3,000 miles away. The transport mechanism might be a local area network or an analog phone line running a special modem designed for whiteboarding or it might be an ISDN digital line running special PC software and hardware. The concept of whiteboarding is new; there are no standards. As a result to do whiteboarding successfully, you typically need the same equipment (hardware and software) on either end. Whiteboarding has the potential to be one of the most successful "multimedia" applications around.

Who-Are-You Code WRU. A control character which operates the answerback unit in a terminal (typically a telex terminal) for identification of sending and receiving stations in a network.

Whois 1. A command on some systems that reveals the user's name, based on that person's network username.
2. An Internet program which allows users to query a database of people and other Internet entities, such as domains, network, and hosts. The information for people shows a person's company name, address, phone number and electronic mail address.

Whole Person Paradigm This is one of the more fascinating telecom concepts in a while. General Magic created it as some sort of psychological basis for the product/s it is producing. Here's General Magic's definition: A psychological or behavior model of needs that all people experience. This paradigm is the design center for General Magic's personal intelligent communication products and services. It consists of three elements. 1. Remember - managing your internal agenda, such as things to do and people to see. 2. Communicate - maintaining relationships with your friends, family, and associates. 3. Know - getting information about the world.

Wide Area Network WAN. An data network typically extending a LAN (local area network) outside the building, over telephone common carrier lines to link to other LANs in remote buildings in possibly remote cities. A WAN typically uses common-carrier lines. A LAN doesn't. WANs typically run over leased phone lines — from one analog phone line to T1 (1.544 Mbps). The jump between a local area network and a WAN is made through a device called a bridge or a router. Bridges operate independently of the protocol employed. They will work, according to Jeff Weiss, of Cryptall Communications, with all present and expected future communications packages. Routers are specific to the protocol being employed. New routing software is needed for each new protocol or protocol deviation. See BRIDGE, ROUTER, CORPORATE NETWORK and DIGITAL HIERARCHY.

Wide Area Telephone Service See WATS and 800 SERVICE.

Wide Characters 16-bit characters. See UNICODE.

Wide Frequency Tolerant Power Plant PBX power facilities are provided that will operate from AC energy sources which are not as closely regulated as commercial AC power. The wide tolerant plant will tolerate average frequency deviations of up to plus or minus 3 Hz or voltage variations of -15% to +10% as long as both of the conditions do not occur simultaneously. This feature permits operation with customer provided emergency power generating equipment.

Wideband Refers to a channel wider in bandwidth than a voice-grade channel.

Wideband Channel Any channel wider in bandwidth that a single voice-grade channel. That's the more modern definition. An older definition is a communication channel of a bandwidth equivalent to twelve or more analog voice-grade channels.

Wideband Modem A modem whose modulated output signal can have an essential frequency spectrum that is broader than that which can be wholly contained within a voice channel with a nominal 4-kHz bandwidth. A modem whose bandwidth capability is greater than that of a narrow band modem.

Wideband Packet Transport Transmission of addressed, digitized message fragments (packets) interleaved among the addressed fragments of other messages at a rate high enough to support general purpose telecommunications services.

Wideband Switch Switch capable of handling channels wider in bandwidth than voice-grade lines. Radio and TV switches are examples of wideband switches.

NEWTON'S TELECOM DICTIONARY

Wildcards Special characters you use to represent one or more characters in an MS-DOS filename. An asterisk (*) represents several characters and a question mark (?) represents a single character. For example, the command ERASE *.BAK would erase all the files with the suffix "BAK." The command ERASE *.?A? would erase all the files with "A" as the middle letter in a three-letter suffix.

Wildfire The all-hearing, all-doing computer telephony slave from a company called Wildfire Communications, Lexington MA. The product uses very sophisticated voice recognition software so that its "master" (i.e. the user) can get Wildfire to take messages, find him, connect his calls, transfer his calls and act as a super intelligent on-line, computerized, 24-hour a day, never resting, all obedient secretary. A real breakthrough product, first introduced in the fall of 1994. And one deserving of its own definition in this illustrious dictionary.

WiLL A name Motorola uses for its Wireless Local Loop (WiLL) product, which was developed to serve the basic telephony needs of people in urban and difficult to reach rural areas. Cellular based, WiLL technology is intended to provide fixed telephony services in areas with little or no existing wireline telephone service or as a supplement to the existing wireline service. It uses very few cellular transmit/receivers — often just one at the end of the landline.
The WiLL system provides three major benefits to the telecom operator looking to expand their service area: more rapid deployment of telephone service; lower cost alternative to copper wire installation, and increased flexibility in system implementation and design. A WiLL system can be operational in weeks, compared to the huge amounts of time it would take to lay and install copper wire from an end office to each of the subscriber points in a typical local loop. Although WiLL is cellular-based, the system does not require a cellular switch. This makes the WiLL system a lower cost alternative to using "typical" cellular systems for fixed telephony applications because the total system outlay costs as well as associated backhaul and maintenance costs are reduced. WiLL has three elements: the WiLL System Controller (WiSC), a Digital Loop Concentrator (DLC), and a Motorola cellular base station. It interfaces directly to the central office switch via 2-wire analog subscriber loops.

Willful Intercept The act of intercepting messages intended for a station experiencing a line or equipment malfunction.

WIMP Interface Stands for Windows, Icons, Menus and a Pointing device. A derogatory reference to GUI. Some people think WIMP is on the way out. See also GRAPHICAL USER INTERFACE.

WIN Wireless In-building Network. WIN is a technology from Motorola which uses microwaves to replace local area network cabling.

WIN32 API A 32-bit application programming interface for both Windows 95 and Windows NT. It updates earlier versions of the Windows API with sophisticated operating system capabilities, security, and API routines for displaying text-based applications in a window.

WIN95 See WINDOWS 95.

Winch A machine for pulling cable into conduit (in the street or in the building) or duct liner. A winch has a rotating drum that winds up the pulling line.

Winchester Disk A sealed hard disk. The Winchester magnetic storage device was pioneered by IBM for use in its 3030 disk system. It was called Winchester because "Winchester" was IBM's code name for the secret research project that led to its invention. A Winchester hard disk drive consists of several "platters" of metal stacked on top of each other. Each of the platter surfaces is coated with magnetic material and is "read" and "written" to by "heads" which float across (but don't touch) the surface. The whole system works roughly like the old-style Wurlitzer jukebox. There are several advantages to a Winchester disk system: 1. It can store, read and write enormous quantities of information. Some Winchesters have a capacity of over 100 megabits; 2. You can access information on a Winchester faster than on most computer storage medium (RAM and ROM are obviously faster); and 3. Winchesters are reliable and relatively inexpensive. There are also disadvantages: 1. They are very sensitive to rough handling (they hate being moved); 2. They are very sensitive to the organization of their directory track (lose that and you're in big trouble); and 3. When Winchesters "crash" (i.e. the heads touch the surface of the rotating platters), you can lose an enormous amount of precious data — possibly millions of bytes of data.

Winding Coils of wire usually found in transformers and used to boost inductance.

Window 1. A band of wavelengths at which an optical fiber is sufficiently transparent for practical use in communications applications.
2. A flow-control mechanism in data communications, the size of which is equal to the number of frames, packets or messages that can be sent from a transmitter to a receiver before any reverse acknowledgment is required. It's called a pacing group in IBM's SNA.
3. A box on the CRT (cathode ray tube) of your personal computer or terminal. A software program is running inside the box. It's possible with new "windows" software to run several programs simultaneously, each accessible and visible through the "window" on your CRT.
4. A technique of displaying information on a screen in which the viewer sees what appears to be several sheets of paper much as they would appear on a desktop. The viewer can shift and shuffle the sheets on the screen. Windowing can show two files simultaneously. For example, in one window you might have a letter you're writing to someone and in another window, you might have a boilerplate letter from which you can take a paragraph or two and drop it in your present letter. Being able to see the two letters on the screen makes writing the new letter easier.

Window Control A credit or token scheme in which a limited number of messages or calls are allowed into the system.

Window Size The minimum number of data packets that can be transmitted without additional authorization from the receiver.

Window Treatment You take the world's most beautiful window and you screw it up with expensive stuff you affix around it. Paula Friesen invented the term.

Windowing A technique of running several programs simultaneously — each in running a separate window. For example, in one window you might run a word processing program. In another, you might be calculating a spreadsheet. In a third, you might be picking up your electronic mail.

WinISDN WinISDN is ISDN*Tek's (San Gregorio, CA — 415-712-3000) API for talking to internal ISDN modems. It supports all of the high level functions for call setup and answering on an ISDN modem. Most of the more popular internal ISDN modems support WinISDN. One of the ways WinISDN helps increase throughput is by handling data transfers in large blocks rather than one byte at a time. The overhead on single byte transfers is much higher than handling a single block.

Windows A Microsoft operating system that hides the cryptic DOS system of typed commands behind a graphical facade (also called a Graphical User Interface, GUI). Windows let you issue commands (i.e. run programs and complete tasks within programs) by pointing (with or without a mouse) at symbols or menu items and clicking, or hitting "Enter." Most Windows programs have the same "look and feel" to them. So issuing commands becomes almost intuitive. The idea is that "use one Windows program, you can use them all." Sort of. The latest versions of Windows were 3.1 and 3.11 (also called Windows for Workgroups). These versions contained two big improvements over 3.0 — namely OLE (Object Linking and Embedding) and DLL (Dynamic Link Library). Windows 3.1 and 3.11 is now about to be obsoleted by a new version, called Windows 95. See DLL, OLE, WINDOWS 95, WINDOWS FOR WORKGROUPS, WINDOWS NT and WINDOWS TELEPHONY.

Windows 95 Windows 95 is a new operating system from Microsoft which first shipped on August 24, 1995. The August, 1995 issue of our Computer Telephony Magazine said the following about Windows 95:

Win95 is the first Windows operating system designed for communications. It does for modems and phones (of all sorts — from single line analog to proprietary ISDN phones) what Windows did for printing — insulate the suffering user from the idiocies of device drivers. Win95 does wonders for fax, for sending color pictures of the kids to Grandma and for making the world one gigantic personal local area network. And, for the first time ever, an operating system is treating voice as it should be treated — another media stream no different, no more complex than printing a pretty document.

Computer Telephony (voice, fax, e-mail) is a major focus of Windows 95. It will have a revolutionary impact on computer telephony's desktop interface and CT-enabling hardware, from simple off-the-shelf SOHO apps built on inexpensive multimedia modems to full-blown unified-messaging systems humming on the LAN. One of the key improvements of Win95 over Windows 3.xx is the replacement of the latter's monolithic communications driver (COMM.DRV) with a far more flexible communications architecture that splits communication tasks into three primary areas: Win32 communications APIs and TAPI; the universal modem driver; and comm port drivers.

VCOMM is the new communications device driver. It protect-modes services and lets Windows apps and drivers use ports and modems. To conserve system resources, comm drivers are loaded into memory only when in use by an app. VCOMM also uses new Plug and Play services in Windows 95 to help configure and install comm devices. The Win32 communications APIs provide an interface for using modems and comm devices in a device-independent fashion. Applications call the Win32 APIs to configure modems and perform data I/O through them. Through TAPI, meantime, apps can control modems or other telephony devices.

The universal modem driver (Unimodem) is a layer that provides services for data and fax modems and voice so that users and app developers don't have to learn or maintain difficult modem AT commands to dial, answer and configure modems. Rather, Unimodem does these tasks automatically by using mini-drivers written by modem vendors. Unimodem is both a VCOMM device driver and a TAPI service provider. Other service providers (like those supporting things such as an ISDN adapter, a proprietary PBX phone or an AT-command modem) can also be used with TAPI. Port drivers are specifically responsible for communicating with I/O ports, which are accessed through the VCOMM driver. Port drivers provide a layered approach to device communications. For example, Win95 provides a port driver to communicate with serial communications and parallel ports, and other vendors can provide port drivers to communicate with their own hardware adapters, such as multiport voice and fax cards. With the port driver model in Win95, it's not necessary for vendors to replace the communications subsystem as they did in Windows 3.xx, whose COMM.DRV forced people to completely replace the comm driver if something new was needed by a hardware device. The Win95 driver means we no longer have to be "hard wired" to a 16550 UART. Previous versions of Windows assumed this type of port hardware. This means new ports like USB (Universal Serial Bus) can be slipped in with full apps compatibility.

Besides this strong attempt at "virtualizing" many of the communications hardware interface problems that plagued Windows developers in the past, Win95 also strengthens itself considerably by acknowledging voice as a data type, filtered into its Plug and Play world of communication device compatibility via the Windows Telephony API (TAPI). TAPI-aware apps, for example, no longer need to provide their own modem support list because interaction with a modem is now centralized by the OS. All comm services provided with Win95 use these services. (The analogy is printing under Windows 3.xx.)

TAPI provides a standard way for communications apps to control telephony functions for data, fax and voice calls. It manages all signaling between a computer and phone network, including basic functions such as dialing, answering and hanging up a call. It also includes supplementary call-handling things such as hold, transfer, conference and call park that are often found in PBXs, Centrex, ISDN and other phone systems. In general, TAPI services arbitrate requests from apps to share comm ports and devices. Win32-based apps can use TAPI to make outgoing calls while others are waiting for incoming calls. Of course, only one call can be performed at a time, but users no longer have to close apps that are using the comm port.

TAPI does not need local hardware. It can also use drivers that work on a LAN, which support multiple systems today (e.g. Genesys, Dialogic's CT Connect, Northern Telecom's Tmap, etc.) Microsoft is making the client side ubiquitous and is planning on dropping the server side in (i.e. Windows NT Server) shortly.

MODEMING: Win95 lets you install and configure a modem once to work for all communications apps, just as you do for a printer. Benefits: Centralized modem and COMM port configuration through the "modems" option in the control panel for all comm apps created for Windows 95; Support for hundreds of modems, including automating the detection of them (again, just like Windows did with printers); Modem connections and configuration using point-and-click options rather than annoying AT commands.

The Windows OS includes three tools here: 1. HyperTerminal. This lets you connect two computers through a modem and TAPI for transferring files. It also automatically detects data bits, stop bits and parity. 2. Phone Dialer. This lets you use your computer to dial numbers for voice calls. It includes a phone dialpad, user-programmable speed dials and a call log. 3. Microsoft File Transfer. This lets you send and receive files while talking on the phone. This works with VoiceView-enabled modems.

E-MAILING: Also in Win95 is its included Microsoft Exchange client capability. Exchange is a messaging app that retrieves messages into one inbox from many kinds of messaging service providers, including, Microsoft Mail, The Microsoft Network and Microsoft Fax. With Microsoft Exchange client, you can: send or receive e-mail in a Win95 workgroup; include files and objects created in other apps as part of messages; use multiple fonts, font sizes, colors and text alignments in messages (via an included OLE-compatible text editor); create a personal address book or use books from multiple service providers; and create folders for storing related messages, files and other items.

According to Microsoft, Exchange client will work with any electronic mail system or unified messaging app that has a MAPI service provider, which architects very similar to the TAPI schematic. The MAPI service provider specifies all the connections and addressing settings needed to talk with a mail server on one end and with the Exchange client on the other.

MAPI is a set of API functions and OLE interface that lets messaging clients, such as Microsoft Exchange, interact with various message service providers, such as Microsoft Mail, Microsoft Exchange Server, Microsoft Fax and various computer telephony servers running under Windows NT server. Overall, MAPI helps Exchange manage stored messages and defines the purpose and content of messages — with the objective that most end users will never know or care about it.

FAXING: As part of Windows 95, Microsoft included a new technology called Microsoft Fax. With this embedded Group 3-compatible software, users with modems and running Microsoft Fax can exchange faxes and editable files (pictures, binary files, *.EXE software, etc.) as easily as printing a document or sending an electronic mail message. To use it, you must install Microsoft Exchange. Microsoft integrated the two as a messaging application programming interface (MAPI) service provider. All faxes sent to Microsoft Fax are received in the Exchange universal inbox. You can send a fax by composing a Microsoft Exchange message or by using the Send option on the File menu of a MAPI-compatible application (such as Microsoft Excel or Microsoft Word). In addition, Microsoft Fax includes a fax printer driver so that users can "print to fax" from within any Windows-based application.

Key features: 1. Delivery By Address Type. The MAPI service provider architecture lets you mix different types of recipients in the same message. For example, it's possible to send a message simultaneously to destination addresses in Microsoft Mail, CompuServe, Internet, normal fax and Microsoft Fax as long as profiles for these destinations have been defined within Microsoft Exchange. A recipient's fax address can be selected from the Microsoft Exchange Personal Address Book or the fax can be addressed by using an address that you use just once, such as [fax:555-1212].

1. Binary File Transfer (BFT). Microsoft Fax supports Microsoft At Work BFT, which makes it possible to attach an editable document to a Microsoft Exchange mail message. These editable documents can be sent to users of Windows 95, Windows for Workgroups 3.11, and other Microsoft FAX BFT enabled platforms.

2. Security. Microsoft Fax lets you securely exchange confidential documents by using public key encryption or digital signatures. Any security specified by user is applied before the message is passed to the modem or connected fax device.

3. Network Fax Service. You can install a fax device in one computer and share it with other users within a workgroup. Individual computers can have their own fax devices installed and still use the shared fax device.

4. Microsoft Fax Viewer. The Microsoft Fax Viewer displays outgoing fax messages that have been queued to a local fax modem or to a Microsoft Fax network fax that are queued for transmission. You can also browse multipage faxes in thumbnail or full-page views.

5. Connecting to Fax Information Services. Microsoft Fax can connect to fax-on-demand systems by using a built-in, poll-retrieve feature that allows you to retrieve rendered faxes or editable documents from a fax information service.

6. "Best Available" Fax Format. When you make a fax connection in Windows 95, Microsoft Fax queries and exchanges its fax capabilities with the recipient. This exchange of capabilities determines whether the recipient is a traditional Group 3 fax machine, which can only receive nonedited faxes, or if the recipient has Microsoft Fax capabilities, and can receive editable files. Windows 95, Windows for Workgroups 3.11 and Microsoft At Work fax platforms are all capable of receiving binary files and traditional faxes. Perks:

If the receiving fax device supports Microsoft Fax capabilities and an editable document is attached to a Microsoft Exchange message, then the file is transferred in its native format. If the receiving fax device is a traditional Group 3 fax machine, then Microsoft fax converts the document to the most compressed type of fax supported by the machine (MH, MR or MMR compression type) and transmits the image by using the best available communications protocol supported by the mutual connection (that is, V.17, V.29 or V.27). If Microsoft Fax sends a noneditable fax to another Microsoft Fax user, then the fax is transmitted by using the Microsoft At Work rendered fax format. This special format is much more compressed, on average, than Group 3 MMR. Therefore, the exchange of noneditable faxes between Microsoft Fax users is always faster than between Group 3 fax machines.

Overall, you can send faxes either by using the mail client or the Microsoft Fax printer driver. In each case, the message is sent to the Microsoft Fax service provider by using MAPI. If you sent the message from a mail client, it might contain text, embedded OLE formats and attachment to the mail message. MAPI allows messages to be preprocessed based on the transport protocol used to send them. The transport protocol chooses the correct modem connection, uses TAPI to create a dial string and sends the message into a fax form to be printed by a fax machine. The rendered format is attached to the original message as a message property and is deleted either when the message is sent or when the transport protocol tries to send the message but determines it cannot.

If the message does not have to be rendered, the message is converted from its original binary format to a line image (also called a linearized form), and then it is compressed. After the message is submitted, the transport protocol determines what type of recipient the message is intended for through TAPI subkey values.

See also DIAL STRING, MICROSOFT FAX and WINDOWS TELEPHONY for fuller explanations.

Windows Application A term used in this document as a shorthand term to refer to an application that is designed to run with Windows and does not run without Windows. All Windows applications follow similar conventions for arrangement of menus, style of dialog boxes, and keyboard and mouse use.

Windows Character Set The character set used in Windows and Windows applications. Most TrueType fonts have a set of about 220 characters.

Windows For Workgroups Windows for Workgroup is a local area networked version of Microsoft Windows operating system version 3.1 that offers integrated file sharing, electronic mail (Microsoft Mail) and workgroup scheduling (Schedule+), thus bringing the graphical user interface to the workgroup. Windows For Workgroups also has Network DDE, which allows users to create compound documents that share data across network. Most importantly, Windows for Workgroups 3.11 lets you do 32-bit disk access and 32-bit file access.

Windows Internet Name Service WINS. A name resolution service that resolves Windows networking computer names to IP addresses in a routed environment. A WINS server, which is a Windows NT Server computer, handles name registrations, queries, and releases.

Windows MetaFile WMF. A method of encoding files. Other methods include EPS, PCX and TIFF.

Windows NT Windows New Technology is a 32-bit operating system from Microsoft, designed to replace Windows and MS-DOS. As an operating system, Windows NT is targeted at the top 10% "power" users who need the power of a big, powerful operating system. Other less demanding souls will continue to run Windows and Windows for Workgroups and MS-DOS and possibly access Windows NT through a local area network to a server running Windows NT AS. Windows NT require a minimum of 12 megabytes of RAM and 100 megs of free hard disk space. Here are the main advantages of Windows NT, as explained by Microsoft:
• Interoperability. Windows NT delivers support for open computing benefits through its protected subsystem architecture. Windows NT was also designed to be protocol independent. As such it will interoperate with all leading network systems, regardless of the native protocol of the system.
• Portability. Windows NT was designed to be portable across a variety of hardware systems. The Hardware Abstraction Layer (HAL) limits and isolates the amount of code necessary to port Windows NT to a new platform. Windows NT will run on processors other than those made by Intel. MS-DOS, for example, doesn't.
• Scalability. Windows NT scales to work on both single and multi processor computer systems. This scalability gives users the flexibility to implement their own solutions, today or over time, on machines that meet the performance needs of sophisticated client server solutions.
• System Management. Windows NT supports SubNetwork Access Protocol (SNMP) and NetView network management standards.
• Published Interfaces. The interfaces to the Windows NT operating system are fully documented and published. Software developers are free to add functionality to the system based on their interface definitions.
• Support of Industry Standards. These include POSIX.1, OSF DCE, TCP/IP and WOSA, which is Microsoft's Windows Open Services Architecture. WOSA is a standard set of interfaces to connect a variety of applications with a range of back-end devices and services, such as messaging, telephony, databases, etc. Windows Telephony is part of WOSA. See WINDOWS NT ADVANCED SERVER.

Windows NT File System NTFS. An advanced file system designed for use specifically with the Windows NT operating system. NTFS supports file system recovery and extremely large storage media. It also supports object-oriented applications by treating all files as object with user-defined and system-defined attributes.

Windows NT Server Windows NT Server is a specialized (and larger) version of the Windows NT operating system, which is designed to run a server on a local area network. To run it effectively, you'll need a server with at least 150 megabytes of free space on a hard disk, 20 megs of RAM and a high speed processor. Here are words from Microsoft which give a good idea of what it's meant to do. (Ignore the obvious epithets.)
"The Windows NT Server is designed to be a dedicated server in a client server environment. With its power, scalability, enhanced fault tolerance and standards-based openness, the Windows NT Server will make an excellent applications server on Novell NetWare, Banyan VINES and Microsoft Networks, providing a platform for sophisticated business solutions such as financial, accounting and vertical applications. As an application server, it provides a powerful platform for the following:
"Database servers, such as Microsoft SQL Server and ORACLE Server, on top of which customers can build retail, banking, insurance and similar solutions
"Communications servers such as the Microsoft SNA Server
"Mail servers such as Microsoft Mail
"For network management, the Windows NT Server also provides customers with centralized security and server management, along with graphical tools to manage multiple systems, and gives users a single log on for the enterprise.
"A superset of the Windows NT operating system, Windows NT Server provides additional reliability by supporting advanced fault tolerance including RAID 5. It also delivers Macintosh connectivity and the Remote Access Service — providing network and client server connections to telecommuters, traveling executives and remote system managers over phone lines, X.25 and ISDN networks."

Windows Open Services Architecture See WOSA.

Windows Telephony Introduced in the spring of 1993 jointly by Microsoft and Intel, Windows Telephony is a piece of software called a Windows Telephony DLL AND two standards. The first standard is the Service Provider Interface (SPI). If a hardware manufacturer's product honors that SPI, that product can happily talk to the Windows Telephony DLL. The second standard is called the Application Programming Interface and it is directed at software developers who write applications programs. If those developers' programs adhere to the API, they can take advantage of the Windows Telephony DLL to drive whatever telephony devices or services adhere to the SPI. The Windows Telephony API is affectionately called TAPI. DLL stands for Dynamic Link Library. It is a Windows feature that allows executable code modules to be loaded on demand and linked at run time.
Windows Telephony should bring about an explosion of shrink-wrapped Windows based telephone software applications — from simple personal rolodexes to power dialers, to customized phone systems for banks and for bakers. It should also bring about an explosion of new telephony hardware devices — from telephones that look more like PCs than phones, to PCs that are phones, to blackbox telephony devices that hook to laptops and transform hotel phones.
Windows Telephony effectively removes earlier overwhelming barriers to creating PC-driven telephony applications, namely the wide enormity of telephony "network" services — from the many telephone company interfaces (POTS to T-1), to the many more proprietary interfaces behind dozens of proprietary PBXs, key systems and hybrid phone systems.
The goal is to bundle the Windows Telephony DLL in the next major release of Windows, sometime in 1994. It will also be included in Windows NT and Windows NT Server. Although the Windows NT code may be different, the API and SPI interfaces will be the same, thus causing no re-write of software code or necessitating redesign of telephony hardware.
The original work on the Windows Telephony DLL was done by Herman D'Hooge, a senior software architect with Intel's

Architecture Development Lab in Hillsboro, Oregon. The final effort is a result of joint development effort with Microsoft, where the team was headed by Charles Fitzgerald. It also includes input from 40-odd companies — including virtually all major telecom switch vendors and several major telephony developers.

The goal of joint Microsoft/Intel Windows Telephony is to get rid of the bottleneck to bringing the power of the PC to telephony. Intel and Microsoft believe that the bottleneck exists because of two factors:

First, it has been incredibly difficult to interface to the variety of telecom switches in existence today. For example, no manufacturer's switch will talk to another's manufacturer's proprietary phone.

Second, it has been incredibly redundant and time consuming for software to talk to the various switches. The big analogy is word processing in the old days. In those days, each word processing software company could easily spend 99% of his R&D budget writing drivers to get his program to work with yet another new printer. That is no longer necessary under Windows. Windows takes care of interfacing the printers. All you have to do, as a developer is to make sure you conform to Windows specs.

According to Microsoft, Windows Telephony products will include ones, such as those including:
* Visual interface to telephone features. * Personal communication management. With a graphical user interface, people can have their PCs handle incoming telephone calls, automatically controlling which calls reach them. For example, people will be able to ensure that they receive important calls by requesting that certain calls be forwarded automatically to locations at which they expect to be working. * Telephone network access. A personal information management can be used not only to look up phone numbers, but also to actually place calls to those numbers. * Integrated messaging. People will be able to check their messages — electronic mail, fax mail and voice mail — from a single place, namely their telephony empowered PC. Also voice mail messages can be accessed randomly, which is far more efficient than the serial access provided on most telephone based voice mail-mail systems. * Integrated meetings. Here's Microsoft's explanation: "One of the most attractive capabilities of the computer is that it can store, communicate and present information that spans the entire spectrum of media — text, data, graphics, voice and video in any combination. By itself, the telephone can communicate and present voice information only. By combining the functions of the computer and the telephone, people in geographically separate locations can participate in interactive meetings and share visual as well as audio information. That means they can hold meetings over the telephone network that are nearly as rich in information content as in-person meetings."

The Windows Telephony Software Development Toolkit is available from several sources:
* CompuServe. Either GO MSL (Microsoft Software Library) and search for the keyword "S14400," or GO WINNEXT (the Windows Extensions Forum) and download the file TAP110.ZAP from Library 3.
* Internet. Use anonymous FTP to ftp.microsoft.com(198.105.323.1) and look in the directory "/devtools/tapi" for the file TAP110.ZIP.
* Microsoft Download Service BBS. Dial 1-206-936-6735, log in, and look for the file TAPISDK.EXE.
* Microsoft Windows Development Platform Level II CD-ROM. This set of four CD-ROMs contains all Microsoft SDKs, DDKs and operating systems. This quarterly offering for developers accompanies the MSDN Level 1 Development Library, which contains all documentation for the development platform (but no code). Call 1-800-759-5474 or 1-402-691-0173.

Private questions or feedback about Windows Telephony can be sent via Internet to "telephon@microsoft.com or via fax to Windows Telephony Coordinator at 1-206-936-7329.

See AT WORK, FAX AT WORK, TELEPHONY SERVICES, WINDOWS TELEPHONY SERVICES, WINDOWS TOOLKITS and WOSA.

Windows Telephony Services

Here is Microsoft's definition: "The Windows Telephony services are provided as a WOSA (Windows Open Services Architecture) component. It consists of both an application programming interface (API) used by applications and a service provider interface (SPI) implemented by service providers.

The focus of the API is to provide "personal telephony" to the Windows platform. Telephony services break down into Simple Telephony services and Full Telephony services. Simple Telephony allows telephony-enabled applications to be easily created from within these applications without these apps needing to become aware of the details of the Full Telephony services. Word processors, spreadsheets, data bases, personal information managers can easily be extended to take advantage of this.

Complete call control is only possible through the use of the Full Telephony services. Applications access the Full Telephony API services using a first-party call control model. This means that the application controls telephone calls as if it is an endpoint of the call. The application can make calls, be notified about inbound calls, answer inbound calls, invoke switch features such as hold, transfer, conference, pickup, park, etc., detect and generate DTMF for signaling with remote equipment. An app can also use the API to monitor call-related activities occurring in the system.

The fact the API presents a first-party call control model does not restrict its use to only first-party telephony environments. The Windows Telephony API can be meaningfully used for third-party call control.

The API provides an abstraction of telephony services that is independent of the underlying telephone network and the configuration used to connect the PC to the switch and phone set. The API provides independent abstractions of the PC connections to the switch or network and the phone set. The connection may be realized in a variety of arrangements including pure client based wired or wireless connections, or client/server configurations using some sort of local area network.

The Telephony API by itself is not concerned with providing access to the information exchanged over a call. Rather, the call control provided by the API is orthogonal to the information stream management. The Telephony API can work in conjunction with other Windows services such as the Windows multimedia wave audio, MCI, or fax APIs to provide access to the information on a call. This guarantees maximum interoperability with existing audio or fax applications.

The Telephony API defines three levels of service. The lowest level of service is called Basic Telephony and provides a guaranteed set of functions that corresponds to "Plain Old Telephone Service" (POTS - only make calls and receive calls). The next service level is Supplementary Telephone Service providing advanced switch features such as hold, transfer, etc. All supplementary services are optional. Finally, there is the Extended Telephony level. This API level provides numerous and well-defined API extension mechanisms that enable application developers to access service provider-specific functions not directly defined by the Telephony API.

Windows Toolkits Windows toolkits are libraries of code that implement the graphical user interface objects that every software application uses. The toolkits save time by eliminating the need for software developers to re-implement the same code repeatedly for each application. Toolkits also have the benefit of consistent user interface implementation across all applications that use the toolkit. See also WINDOWS.

Wink A signal sent between two telecommunications devices as part of a hand-shaking protocol. It is a momentary interruption in SF (Single Frequency) tone, indicating that the distant central office is ready to receive the digits that have just been dialed. In telephone switching systems, a single supervisory pulse. On a digital connection such as a T-1 circuit, a wink is signaled by a brief change in the A and B signaling bits. On an analog line, a wink is signaled by a change in polarity (electrical + and -) on the line.

Wink Operation A timed off-hook signal normally of 140 milliseconds, which indicates the availability of an incoming register for receiving digital information from the calling office. A control system for phone systems using address signaling.

Wink Pulsing Recurring pulses of a type where the off-pulse is very short with respect to the on-pulse, e.g., on key telephone instruments, the hold position (condition) of a line is often indicated by wink pulsing the associated lamp at 120 impulses per minute, 94 percent make, 6 percent break (470 ms on, 30 ms off).

Wink Release On most modern central offices when the person or device at the other end hangs up, your local central office will send you a single frequency tone. That tone is called wink release. Such a tone can be used to alert a data device that the device at the other end has hung up. (Remember it can't tell by just listening — like you and me.) When a data device hears a wink release, it usually takes it as a signal to hang up also.

Wink Signal A short interruption of current to a busy lamp causing it to flicker. Indicates there is a line on hold.

Wink Start Short duration off hook signal. See WINK OPERATION.

WINS Windows Internet Name Service. A name resolution service that resolves Windows networking computer names to IP addresses in a routed environment. A WINS server, which is a Windows NT Server computer, handles name registrations, queries, and releases.

Winsock 2 Winsock stands for Windows Sockets. Winsocks are standard APIs between Microsoft Windows (3.1, 95 and NT) application software and TCP/IP protocol software (also called a protocol stack). Winsock 2 is a network programming interface at the transport level in the ISO reference model. It is being defined by an open, industry wide workgroup, called the Winsock Forum. Microsoft currently includes a Winsock 1.1 interface in all versions of ITS operating systems from Windows 3.11 onward, and has committed to supply version 2 of Winsock (when its definition has been finalized) as well. Virtually all vendors of TCP/IP protocol stacks supply a Winsock 1.1 programming interface for their stacks, and a large number of TCP/IP applications are written to the Winsock 1.1 standard. The Winsock 2 API maintains backwards compatibility with Winsock 1.1 which is based on the familiar Berkeley "sockets" paradigm for establishing connections and sending and receiving data over those connections. Winsock version 2 significantly extends the interface by providing applications with transport independence across a wide range of communication transports such as TCP/IP, NetBEUI or SPX/IPX local area networks, analog and digital telephone networks, wirEless networks such as cellular and CDPD, and ATM. In addition, Winsock 2 will include mechanisms for applications to negotiate quality of service with a network, thus improving its use for multimedia applications. Stack vendors who wish to make their transport protocols accessible via Winsock 2 need only implement the Winsock 2 service provider interface (SPI), which is itself very socket-like. The Winsock Forum's stated objective is to finalize the Winsock 2 specifications during the second quarter of '95, and have the needed SDKs available (for both 16 and 32 bit operating environments) during the 2nd half of 95. Key Contacts: Martin Hall, Stardust Technologies, 408-438-6643 (Leader of the Winsock Forum) Dave Treadwell, Microsoft, davidtr@microsoft.com Dan Ohlemacher, Intel Architecture Labs, 503-264-5914 dan_ohlemacher@ccm.jf.intel.com As regards the current status on defining version 2, draft copies of the proposed API and SPI specs are available via anon ftp:
ftp://aurora.intel.com/pub/winsock2/gen_api_ext /* API spec */
ftp://aurora.intel.com/pub/winsock2/oper_framework /* SPI spec */
A series of email reflectors are established and are the primary means of conducting forum business. To get information on these, send an email to majordomo@mailbag.intel.com Leave the subject blank, and put the following in the body: lists help. You will be mailed the names of all the mailing lists and instructions on how to subscribe.
Several Web pages also exist:
http://sunsite.unc.edu:80/winsock/
http://www.microsoft.com/pages/developer/winsock/
http://www.stardust.com/

Wintel A combination of the words Windows and Intel. Wintel refers to PCs that run Microsoft Windows and use Intel microprocessors. Wintel PCs are by far the biggest-selling PCs, amounting to around 80% of all PCs sold. Observers of the PC industry refer to the "Wintel standard" to refer the phenomenon of such high market dominance by one type of PC.

Wire Center The location where the telephone company terminates subscriber outside cable plant (i.e. their local lines) with the necessary testing facilities to maintain them. Usually the same location as a class 5 central office. A wire center might have one or several class 5 central offices, also called public exchanges or simply switches. A customer could get telephone service from one, several or all of these switches without paying extra. They would all be his local switch.

Wire Center Serving Area That area of an exchange served by a single wire center.

Wire Concentrator A conduit; a pipe within which a large number of individual wires are routed through.

Wire Pair Two separate conductors traveling the same route, serving as a communications channel.

Wire Printer A matrix printer which uses a set of wire hammers to strike the page through a carbon ribbon, generating the matrix characters.

Wire Running Tools Tools that help you run wire in and around a building. The most common form of wire running tools that help you fish wire through hollow drywalls.

Wire Speed The rate at which bits are transmitted over a cable. Ethernet's wire speed is 10 Million bits per second. Wire speed is not transmission speed. That depends on many factors, including how many devices are transmitting simultaneously on the same cable. See MULTIPLEX.

Wire Stripper A tool which takes the insulation off a wire without hurting the inside metal conductor.

Wire Tap The attaching to a phone line of a piece of equipment whose job is to record all conversations on that phone line. Wire taps are illegal. Law enforcement agencies use them, but must receive authorization from a court to apply the tap. Such authorizations are given if the law enforcement agency argues that applying the tap will prevent crime or help bring a suspected criminal to justice. Wire taps are not authorized lightly. See also TRAP and TRACE.

Wire Telephony The transmission of speech over wires.

Wired For Capacity The wired-for capacity represents the upper limit of capacity for a particular configuration. To bring to a phone system to its "wired for capacity," all that's necessary is to fill the empty slots in the system's metal shelving (its cage) with the appropriate printed circuit boards. "Wired-for Capacity" is a marginally useful term, giving little indication of the type of printed circuit boards — trunk, line, special electronic line, special circuit, etc. — that can be installed. And many PBXs allow only their printed circuit boards to go into assigned slots. Your PBX cabinet might, for example, have plenty of empty space for extra printed circuit boards, but it may not have any more space for boards which service electronic phones. Thus, it is effectively maxed out.

Wired Logic A required logic function implemented in hardware, not software.

Wireless Without wires. An English and Australian word for radio and, now in the U.S., a phone system that operates locally without wires. Cellular is wireless in the strictest sense of the term. But "wireless" has come to mean wire-less systems that work within a building. See WIRELESS ACCESS CONTROLLER.

Wireless Access Controller The first component in an in-building wireless phone systems is the wireless access controller. It does many things. It provides access to the host network, be it a host PBX or the public switched telephone network (including Centrex). The access controller also manages the picocellular infrastructure of the wireless system through connections to the radio base stations. In the case of a Northern Telecom wireless business systems, base stations are connected to the controller via 144-kilobit-per-second (kbit/s) digital links that offer 2B+D interface connectivity. This digital connectivity (two 64-kbit/s channels for voice and data, and one 16-kbit/s channel for signaling information) provides the high-speed signaling capability needed by the controller to offer advanced business services and to manage mobility across several base stations. These digital links also make it possible to enhance radio system capacity by having the controller synchronize all base stations.

The controller software structures are designed so that untethered personal directory numbers and physical ports (specific interface circuits wired to a particular location) are dynamically associated at every call and at every hand-off to another base station. This dynamic assignment makes it possible for the same personal or group directory number to be used for a variety of wireless and wireline terminals, irrespective of location. The controller also handles user registration, roaming, and hand-off.

Roaming which is the capability to redirect incoming calls to the appropriate base station, is accomplished through a combination of radio protocols, system software, and databases. The databases make it possible to locate portable terminals, through various broadcasting or polling schemes, without incurring excessive search delays.

Hand-off, on the other hand, is the capability needed in order to cope with the fact that a user will continuously move from one location, and hence one cell, to another, while communicating. As this happens, the link must be maintained in a manner transparent to the end user, always maintaining communications with the strongest base station signal in the neighborhood of the portable terminal. The controller monitors the radio signal strength of the portable and, when the signal weakens, switches it to a base station with a stronger signal. It then switches the communications link from the former base station to the new one and signals the terminal to begin radio communication on the new channel. Interference could be caused, for example, by other portable terminals in the same cell or an adjacent cell, or by external influences, such as nearby traffic or people moving partitions in an office. In such cases, the base station redirects the call rapidly to a less noisy channel in the same cell or an adjacent cell.

Wireless Cable An oxymoron which means that TV signals are broadcast by microwave to antennas on customers' homes. The former name for wireless cable was MMDS, short for Multichannel Multipoint Distribution Service.

Wireless Digital Standards See DIGITAL WIRELESS STANDARDS.

Wireless LANS The conventional local area network (LAN) uses wires or optical fiber as a common carrier medium. However, other possibilities exist. Low microwave frequencies (lower than about 10 GHz) can provide data rates as high 10 Mbit/s. Millimetric waves at around 60 GHz could support several 10 Mbit/s channels, while infra-red beams could support even greater data throughputs. The area covered by such a scheme would be restricted by the low allowable power radiation. The data rates of such systems tend to be restricted by walls, by interference and by multipath propagation problems that arise due to reflections within the building. Because of the wide bandwidth available, channeling can easily be provided by using spread spectrum methods and code division multiple access (CDMA), a technique that significantly improves the system security.

Wireless Messaging Technology allowing the exchange of electronic messages without plugging into a wired land-based phone line. Two wireless messaging types are available: one-way, based on existing radio paging channels; and two-way, based on either radio-packet technology or cellular technology. Some people include in-room infra-red links in the term "wireless messaging." Some of the PDAs use wireless links.

Wireless Packet Switching Unlike existing cellular networks, wireless packet-switched networks are designed specifically for data communications. Packet switching breaks messages into packets and sends these packets individually over the network. Here's how a message is sent over the RAM Mobile Data Wireless Networks, one of the packet radio networks in operation today:
1) After you've written a message and turned on your modem, you enter a send command in your e-mail software.
2) The modem breaks your message into packets. A typical packet has a message space for as much as 512 bytes (about 100 words). Longer messages are divided into 512 byte sections.

3) The modem then sends each packet separately over the RAM packet radio network. Each packet includes the sender's and receiver's addresses.
4) The network routes the message to the recipient.
5) The recipient's packet radio modem reassembles the individual packets into a single message.
6) Your recipient can then read the message.
A packet radio network, typically uses a hierarchical architecture to route messages. At the lowest level, base stations exchange wireless messages with nearby mobile computers. Base stations can route messages to other users who are within its service area, or the local switch to read recipients who are in other areas, on LANs, or on public e-mail services. The local switch can either route the message to a different base station or to a regional switch. Users of these packet radio networks can typically send messages anywhere in the network — regardless of the physical distance — for the same rate per message.

Wireless Service Provider WASP. A carrier authorized to provide wireless communications exchange services (for example, cellular carriers and paging services carriers).

Wireless Switching Center WSC. A switching system used to terminate wireless stations for purposes of interconnection to each other and to trunks interfacing with the Public Switched Telephone Network (PSTN) and other networks.

Wireline 1. Another name for a telephone company that uses cables, not radio.
2. Cellular licenses received from the FCC with initial association to telephone company. Also referred to as B-Block. See WIRELINE CELLULAR CARRIER.

Wireline Cellular Carrier Also called the Block B carrier. Under the FCC's initial cellular licensing procedures, the Block B carrier is the local telephone company. The FCC reserved one of the two systems in every market for the local telephone — or wireline company. Wireline or Block B systems operate on the frequencies 869 to 894 Megahertz. See NON-WIRELINE CELLULAR COMPANY.

Wiretapping To listen in clandestinely to someone else's conversation. Other than scrambling, there is no known method to protect your telephone call against wiretapping, no matter what equipment you buy from companies advertising their wares nationally. Wiretapping can be accomplished without physical connection to a phone line, though technically this would be called "bugging." For all intents and purposes you should consider your telephone conversations as public and treat your conversations as such.

Wiring Closet Termination point for customer premises wiring, offering access to service personnel. Generally serves a specific area, with cross-connected multiple wiring closets.

Wiring Concentrator A wiring concentrator is an FDDI node that provides additional attachment points for stations that are not attached directly to the dual ring, or for other concentrators in a tree structure. The concentrator is the focal point of Digital's Dual Ring of Trees topology.

Wiring Density Refers to the number of wires that may be terminated on a connecting block in a given area. A high density block may terminate twice as many wires as a low density block, while a low density block may provide better wire management since fewer wires are being dressed into and out of the connecting block.

Wiring Grid The overall architecture of building wiring.

Witting In on a secret.

WITS Wireless Interface Telephone System.

WIZOP A Chief Sysop (System Operator). See SYSOP.

WLAN Wireless Local Area Network. A LAN without wires. There are major benefits, the biggest being the ability to move items on the LAN around fast, easily and cheaply.

Woo Woo Tone A tone on a phone line indicating the number is unavailable. Also the words to a neat Jeffrey Osborne song, as in "Will you woo woo with me?"

Word A collection of bits the computer recognizes as a basic information unit and uses in its operation. Usually defined by the number of bits contained in it, e.g. 8, 16 or 32 bits. Using DOS, the IBM PC (and compatibles) defines a word as eight bits. Here's another explanation: A group of characters capable of being processed simultaneously in the processor and treated by computer circuits as an entity.

Word Length The number of bits in a data character without parity, start or stop bits.

Words Per Minute WPM. The speed of printing, typing or communications. 100 WPM is 600 characters per minute (six characters per average word) or 10 characters per second. In ASCII, asynchronous transmission at this rate is also 100 or 110 bits per second, depending on the number of stop bits.

Work Location Wiring Subsystem The part of a premises distribution system that includes the equipment and extension cords from the information outlet to the terminal device connection.

Work Order A term used in the secondary telecom equipment business. Internal document used by a remarketer specifying: 1. Work to be performed; 2. Machine or item on which work is to be performed; 3. Required completion date; 4. Cost of work; 5. Customer purchase order number and/or other pertinent billing information. This document is used internally to: 1. Implement required work; 2. Monitor progress; and 3. Issue final billing. A work order is implemented once a written request has been received authorizing the work to be performed.

Word Spotting In speech recognition over the phone, word spotting means looking for a particular phrase or word in spoken text and ignoring everything else. For example, if the word to spot was "brown," then it wouldn't matter if you said "I want the brown one," or "how about something in brown?" In short, word spotting is the process whereby specific words are recognized under specific speaking conditions (i.e. natural, unconstrained speech). It can also refer to the ability to ignore extraneous sounds during continuous word recognition.

Work Station In this dictionary I spell it as one word WORKSTATION. See WORKSTATION.

Workflow The way work moves around an organization. It follows a path. That path is called workflow. Here's a more technical way of defining workflow: The automation of standard procedures (e.g. records management in personnel operations) by imposing a set of sequential rules on the procedure. Each task, when finished, automatically initiates the next logical step in the process until the entire procedure is completed.

Workflow Management The electronic management of work processes such as forms processing (e.g. for insurance policy acceptances, college admissions, etc.) or project management using a computer network and electronic messaging as the foundation. See WORKFLOW.

Workforce Management According to Jim Gordon of TCS in Nashville, Call center workforce management is the art and science of having the right number of people...agents...at the right times, in their seats, to answer an accurately forecasted volume of incoming calls at the service level you desire.

Workgroup A fancy new word for a department, except that the members of the workgroup may belong to different departments. The idea is that members of the workgroup work with themselves, so they'd be perfect candidates to buy electronic mail packages that could send messages between themselves and other software packages that would allow them to share their collective wisdoms and schedule their meeting times. Typically members of the workgroup would be on the same local area network and share the same telephone system. See WORKGROUP TELEPHONY.

Workgroup Computing An approach to the supply of computer services whereby access to computer power and information is organized on a workgroup by workgroup basis. Such systems normally consist of computers of varying capabilities connected to a local area network. See WORKGROUP.

Workgroup Manager An assistant network supervisor with rights to create and delete bindery objects (such as users, groups, or print queues) and to manage user accounts. A Workgroup Manager has supervisory privileges over a part of the bindery. When several groups share a file server, Workgroup Managers can provide autonomous control over their own users and data.

Workgroup Telephony See TELEPHONY WORKGROUPS.

Workload A call center term. The total duration of all calls in a given period (half hour or quarter hour), not counting any time spent in queue. This figure is equal to the number of calls times the average handle time per call.

Workstation In the telecom industry, a workstation is a computer and a telephone on a desk and both attached to a telecom outlet on the wall. The computer industry tends to refer to workstations as high-speed personal computers, such as Sun workstations, which are used for high-powered processing tasks like CAD/CAM, engineering, etc. A common PC — like the one you find on my desk — is not usually considered a workstation. The term workstation is vague.

World Numbering Zone One of eight geographic areas used to assign a unique telephone address to each telephone subscriber.

World Wide Web Also called WEB or W3. The World Wide Web is the universe of accessible information available on many computers spread through the world and attached to that gigantic computer network called the Internet. The Web has a body of software, a set of protocols and a set of defined conventions for getting at the information on the Web. The Web uses hypertext and multimedia techniques to make the web easy for anyone to roam, browse and contribute to. The Web makes publishing information (i.e. making that information public) as easy as creating a "home page" and posting it on a server somewhere in the Internet. Pick up any Web access software (e.g. Netscape), connect yourself to the Internet (through one of many dial-up, for-money, Internet access providers or one of the many free terminals in Universities) and you can discover an amazing diversity of information on the Web. From weather to stock reports to information on how to build nuclear bombs to the best tennis tips, it can be posted on the Web for all to read. See HOME PAGE, HTML and INTERNET.

World Zone 1 The area of the World Numbering Plan which is identified with the single-digit country code "1" and includes the territories of the United States and Canada, and the following Caribbean countries: Antigua, Bahamas, Barbados, Bermuda, British Virgin Islands, Cayman Islands, Dominican Republic, Granada, Jamaica, Montserrat, Puerto Rico, St. Kitts, St. Lucia, St. Vincent, Virgin Islands.

Worldwideweb WWW. An easy but powerful global information system, based on a combination of information retrieval and hypertext techniques.

WORM 1. Write Once, Read Many times. Refers to the new type of optical disks (similar to compact discs) which can be written to only once, but read many times. In other words, once the data is written, it cannot be erased. WORM disks typically hold around 600 megabytes. See also ERASABLE OPTICAL DRIVE.
2. A computer program which replicates itself. The Internet worm was perhaps the most famous; it successfully (and accidentally) duplicated itself on many of the systems across the Internet.

WORN Write Once, Read Never (A joke).

Worst Hour Of The Year That hour of the year during which the median noise over any radio path is at a maximum. This hour is considered to coincide with the hour during which the greatest transmission loss occurs.

WOSA Windows Open Services Architecture. According to Microsoft, WOSA provides a single system level interface for connecting front-end applications with back-end services. Windows Telephony, announced in May 1993, is part of WOSA. According to Microsoft, application developers and users needn't worry about conversing with numerous services, each with its own protocols and interfaces, because making these connections is the business of the operating system, not of individual applications. WOSA provides an extensible framework in which Windows based applications can seamlessly access information and network resources in a distributed computing environment. WOSA accomplishes this feat by making a common set of APIs available to all applications. WOSA's idea is to act like two diplomats speaking through an interpreter. A front-end application and back-end service needn't speak each other's languages to communicate as long as they both know how to talk to the WOSA interface (e.g. Windows Telephony). As a result, WOSA allows application developers, MIS managers, and vendors of back-end services to mix and match applications and services to build enterprise solutions that shield programmers and users from the underlying complexity of the system.

This is how WOSA works: WOSA defines an abstraction layer to heterogeneous computing resources through the WOSA set of APIs. Initially, this set of APIs will include support for services such as database access, messaging (MAPI), file sharing, and

printing. Because this set of APIs is extensible, new services and their corresponding APIs can be added as needed.
WOSA uses a Windows dynamic-link library (DLL) that allows software components to be linked at run time. In this way, applications are able to connect to services dynamically. An application needs to know only the definition of the interface, not its implementation. WOSA defines a system level DLL to provide common procedures that service providers would otherwise have to implement. In addition, the system DLL can support functions that operate across multiple service implementations. Applications call system APIs to access services that have been standardized in the system. The code that supports the system APIs routes those calls to the appropriate service provider and provides procedures and functions that are used in common by all providers. The primary benefit of WOSA is its ability to provide users of Windows with relatively seamlessly connections to enterprise computing environments. Other WOSA benefits, according to Microsoft include:
* Easy upgrade paths. * Protection of software investment. * More cost-effective software solutions. * Flexible integration of multiple-vendor components. * Short development cycle for solutions. * Extensibility to include future services and implementations. See also ODBC, MAPI, and TAPI.
WPM See WORDS PER MINUTE.
WRAM Windows Random Access Memory. Similar to VRAM, but with added logic to accelerate common video functions such as bit-block transfers and pattern fills. See also VRAM.
Wrap 1. In data communications, to place your diagnostic and test equipment around parts of a network so you can monitor their use (i.e. do network diagnostics on them). You are, in essence, wrapping your products around theirs.
2. To make a connection between a flexible wire and a hard tag by tightly wrapping the cable around the tag. There are automatic wire wrapping tools available for this job.
3. Redundancy measure in IBM Token Ring LANS. Trunk cabling used in Token Ring TCUs contains two data paths: a main and backup normally unused). If the trunk cable is faulty, the physical disconnection of the connector at the TCU causes the signal from the main path to wrap around on to the backup path, thus maintaining the loop. The term wrap is now used on FDDI networks. If a failure occurs on one of the FDDI rings, the stations on each side of the failure reconfigure. The two rings then are combined into a single ring topology that allows all functioning stations to remain interconnected.
Wrap-Up Between-call work state that an ACD agent enters after releasing a caller. It's the time necessary to complete the transaction that just occurred on the phone. In wrap-up, the agent's ACD phone is removed from the hunting sequence. After wrap-up is completed, it is returned to the hunting sequence and is ready to take the next call.
Wrap-Up Data Ad hoc data gathered by an agent in the ACD system following a call.
Wrap-Up Time A call center term. The time an employee spends completing a transaction after the call has been disconnected. Sometimes it's a few seconds. Sometimes it can be minutes. Depends on what the caller wants.
Wrapping In token-ring networks, the process of bypassing cable faults without changing the logical order of the ring by using relays and additional wire circuits.
Write To record information on a storage device, usually disk or tape.
Write Head A magnetic head capable of writing only. You find write heads on everything from tape recorders to computers.
Write Protect Using various hardware and software techniques to prohibit the computer from recording (writing) on storage medium, like a floppy or hard disk. You can write protect a 5 1/4 diskette by simply covering the little notch with a small metal tag. The idea of "Write Protect" is to stop someone (including yourself) from changing your precious data or program. You can't write protect a hard disk easily. The easiest way to stop someone changing a file is to use the program ATTRIB.EXE. See ATTRIBUTES.
Write Protection A scheme for protecting a diskette from accidental erasure. 5 1/4" diskettes have a notch which must be uncovered to allow data on the diskette to be modified. 3 1/2" diskettes have small window with a plastic tab which must be slid into place to cover the window to allow data on the diskette to be modified. See WRITE PROTECT and ATTRIBUTES.
Write Protection Label A removable label, the presence or absence of which on a diskette prevents writing on the diskette.
WRT With respect to.
WSC Wireless Switching Center. A switching center designed for wireless communications services, typically fixed wireless services including data and voice.
WTAC World Telecommunications Advisory Council. WTAC is comprised of telecommunications leaders from the private and public sectors and from every region of the world. WTAC gives advice to the ITU — the International Telecommunications Union. The WTAC held its first meeting in Geneva, Switzerland, in April, 1992. In February 1993, it published a small booklet called "Telecommunications Visions of the Future."
WTNG WaiTiNG.
WWW 1. World Wide Web; a hypertext-based system for finding and accessing resources on the Internet network. See WORLD WIDE WEB and INTERNET.
2. Also referred to as World Wide Wait, which according to some frustrated observers, is the real meaning of WWW.
WYPIWYF Acronym for "What You Print Is What You Fax," also "The Way You Print Is the Way You Fax." Coined by Intel to describe its one-step pop-up menu that makes sending faxes from the PC as easy as sending a document to a printer.
WYSIWYG (pronounced Whiz-i-wig) What You See Is What You Get. A word processing term meaning what you see on your computer screen is what you will see printed on paper. The exact typeface, the correct size, the right layout, etc. Some word processors do WYSIWYG. Others don't. You usually need a screen with graphics to get the full effect.
WZ1 World Zone One. The part of the earth covered by what used to be called The North American Numbering Plan. It includes the U.S., Canada, Alaska, Hawaii, and the Caribbean islands, but does not include Mexico or Cuba.

NEWTON'S TELECOM DICTIONARY

X An abbreviation for the word "cross," as in crossbar. 5XB would be the abbreviation for a No. 5 Crossbar.

X Band 7 GHz and 8 GHz. Used by military satellites.

X Recommendations The ITU-T documents that describe data communication network standards. Well-known ones include: X.25 Packet Switching standard, X.400 Message Handling System, and X.500 Directory Services.

X Series Recommendations drawn up by the ITU-T to establish communications interfaces for users' Data Terminal Equipment (DTE) and Data Circuit Terminating Equipment (DCE). They govern the attachment of data terminals to public data networks (PDNs) and the Public Switched Telephone Network (PSTN). In short, a set of rules for interfacing terminals to networks.

X Windows X Windows is officially called X Window, or The X Window System, or X. Sometimes it is spelled with an hyphen, e.g. X-Window. It is used primarily with UNIX systems, but not exclusively. X was originally designed to be an industry-standard network windowing system, and because UNIX is widely used, X has become the windowing standard for UNIX.

X-10 Protocol A protocol found in home automation. You can use X-10 to control conforming black boxes and devices. The protocol is a command signal that rides on the AC 60 cycle sine power curve. The signal is a series of 120 kHz pulses sent on the "zero crossing" of each cycle. The signal is in a binary fashion, transmitting the letter code, unit code and command for a device. All receivers monitor the power line waiting for a command to respond. The limit of number of unique codes available is 256. This is derived by having 16 letter codes and 16 unit codes, hence 16x16=256 "addresses." More than one device can share an address. If you decide that every time you select "A1" on, you want all the front lights to come on, then you can place all the receivers to the same address. The system is easy to use and very flexible.

X-Axis Horizontal axis on a graph or chart.

X-Base A term used to describe any database application capable of generating custom programs with dBASE-compatible code.

X-Off/X-On A flow control protocol for asynchronous serial transmission. Flow control is a method of adjusting information flow. For example, in transmitting between a computer and a printer, the computer sends the information to be printed at 9600 baud. That's several times faster than the printer can print. The printer, however, has a small memory. The computer dumps to the memory, called a buffer, at 9600 baud. When it fills up, the printer signals the computer that it is full and please stop sending. When the buffer is ready to receive again, the printer (which also has a small computer in it) sends a signal to the desktop computer (the one doing the printing) to please start sending again. X-OFF means turn the transmitter off (xmit in Ham radio terms). It is the ASCII character Control-S. X-ON means turn the transmitter on. It is the ASCII character Control-Q. You can use these characters with many microcomputer functions. For example, if you do DIR in MS-DOS and you want to stop the fast rush of files, then type Control-S.

X-Dimension Of Recorded Spot In facsimile, the center-to-center distance between two recorded spots measured in the direction of the recorded line. This term applies to facsimile equipment that responds to a constant density in the subject copy by yielding a succession of discrete recorded spots.

X-Dimension Of Scanning Spot In facsimile, the center-to-center distance between two scanning spots measured in the direction of the scanning line on the subject copy. The numerical value of this term will depend upon the type of system used.

X-Open An international consortium of computer vendors working to create an internationally supported vendor-independent Common Applications Environment based on industry standards.

X-Series Recommendations Set of data telecommunications protocols and interfaces defined by the ITU-T.

X-Windows The UNIX equivalent of Windows. What Windows is to MS-DOS, X-Windows is to Unix. A network-based windowing system that provides a program interface for graphic window displays. X-Windows permits graphics produced on one networked station to be displayed on another. Almost all UNIX graphical interfaces, including Motif and OpenLook, are based on X-Windows. X-Windows is a networked window system developed and specified by the MIT X Consortium. Members of the X Consortium include IBM, DEC, Hewlett-Packard and Sun Microsystems. Sun Microsystems has been contracted by the MIT X Consortium to implement PEX (PHIGS Extensions to X), which will be the standard networking protocol for sending PHIGS (Programmers Hierarchical Graphics System) graphics commands through X-Windows. Some people spell it X-Windows and some spell it X Windows.

X-Y A specific variety of electromechanical switch. Does the same things as a Stronger step-by-step switch but in a horizontal plane. It's so called because it's a two motion switch with horizontal and vertical movements. The first pulse sends the switch horizontally to the right place, then the next pulse sends it vertically up to the right place and so on, until it has switched the call through. One of the most reliable switches ever produced. Unfortunately, it's slow, space-consuming and unable to be programmed with many new customer pleasing features.

X.1 A ITU-T specification that defines classes of service in a packet-switched network, such as virtual-circuit, datagram, and fast-packet services.

X.110 International routing principles and routing plan for PDNs.

X.121 International Numbering Plan for public data networks. X.121 defines the numbering system used by data devices operating in the packet mode. X.121 is used by ITU-T X.25 packet-switched networks and has been proposed by several computer vendors as the future universal addressing scheme.

X.121 Address A standard O/R (Originator/Recipient) attribute that allows Telex terminals to be identified in the context of

store-and forward communications.
X.130 Call setup and clear down times for international connection to synchronous PDNs.
X.132 Grade of service over international connections to PDNs.
X.150 DTE and DCE test loops in public data networks.
X.2 International user services and facilities over public switched telephone networks.
X.20 Asynchronous communications interface definitions between data terminal equipment (DTE) and data circuit terminating equipment (DCE) for start-stop transmission services on public switched telephone networks.
X.20 bis Used on public data networks of data terminal equipment (DTE) that is designed for interfacing to asynchronous duplex V-series modems.
X.200 A series of CCITT (International Telegraph and Telephone Consultative Committee) recommendations that defines the type of service offered by specific layers in the OSI (Open Systems Interconnection) Model, and defines the protocol to be used at those layers. CCITT is now the ITU.
X.21 Interface between data terminal equipment (DTE) and data circuit-equipment (DCE) for synchronous operation on public switched telephone networks.
X.21 bis Used on public switched telephone networks of data terminal equipment (DTE) that is designed for interfacing to synchronous V-series modems.
X.21 TSS Specification for Layer 1 interface used in the X.25 packet-switching protocol and in certain types of circuit-switched data transmissions.
X.224 A ITU-T standard associated with the transport layer of the Open Systems Interconnect (OSI) architecture used in networks employing circuit-switched techniques.
X.225 A ITU-T standard associated with the session layer of the Open Systems Interconnect (OSI) architecture used in networks employing circuit-switched techniques.
X.226 A ITU-T standard associated with the Open Systems Interconnect (OSI) architecture that defines specific presentation layer services used with circuit-switched network services.
X.24 List of definitions for interchange circuits between data terminal equipment (DTE) and data circuit terminating equipment (DCE) on public switched telephone networks.
X.25 From its beginning as an international standards recommendation from ITU-T, the term X.25 has come to represent a common reference point by which mainframe computers, word processors, mini-computers, VDUs, microcomputers and a wide variety of specialized terminal equipment from many manufacturers can be made to work together over a type of data communications network called a packet switched network. On a packet switched data network (private or public), the data to be transmitted is cut up into blocks. Each block has a header with the network address of the sender and that of the destination. As the block enters the network, the number of bits in the block are put through some mathematical functions (an algorithm) to produce a check sum.
The check sum is attached as a "trailer" to the packet as it enters the network. Packets may travel different routes through the network. But, ultimately, the packets are routed by the network to the node where the destination computer or terminal is located. At the destination, the packet is disassembled. The bits are put through the same algorithm, and if the digits computed are the same as the ones attached as the trailer, there are no detected errors. An ACK, or acknowledgement, is then sent to the transmitting end. If the check sum does not match, a NAK, or Negative Acknowledgement is sent back, and the packet is retransmitted. In this manner, high speed, low error rate information can be transmitted around the country using shared telecommunications circuits on public or private data networks.
X.25 is the protocol providing devices with direct connection to a packet switched network. These devices are typically larger computers, mainframes, minicomputers, etc. Word processors, personal computers, workstations, dumb terminals, etc. do not support the X.25 packet switching protocols unless they are connected to the network via PADs — Packet Assembler/Disassemblers. A PAD converts between the protocol used by the smaller device and the X.25 protocol. This conversion is performed on both outgoing (from the network) and incoming data (to the network), so the transmission looks transparent to the terminal. (See TRANSPARENT.) There's a very good book on X.25 called X.25 Explained, Protocols for packet switching networks by R. J. Deasington of IBM. It is available from Telecom Library on 1-800-LIBRARY, or 1-212-691-8215.
X.25 Network Any network that implements the internationally accepted ITU-T standard governing the operation of packet-switching networks. The X.25 standard describes a switched communications service where call setup times are relatively fast. The standard also defines how data streams are to be assembled into packets, controlled, routed, and protected as they cross the network.
X.28 DTE/DCE interface for start-stop-mode data terminal equipment accessing the packet assembly/disassembly facility (PAD) in a public switched telephone networks situated in the same country.
X.29 Procedures for the exchange of control information (handshaking) and user data between a packet assembly/disassembly facility (PAD) and a packet mode DTE or another PAD.
X.3 ITU-T recommendation describing the operation of a Packet Assembly/Disassembly (PAD) device or facility in a public data network. X.3 defines a set of 18 parameters that regulate basic functions performed by a PAD to control an asynchronous terminal. The setting of these parameters governs such characteristics as terminal speed, terminal display, flow control, break handling and data forwarding conditions, and so on.
X.30 Support of X.20 bis, X.21 and X.21 bis DTEs by an ISDN.
X.31 Support of packet mode DTEs by an ISDN.
X.32 Interface between Data Terminal Equipment (DTE) and Data Circuit Terminating Equipment (DCE) operating in a packet mode and accessing a packet switched public data network via a public switched telephone network or a circuit switched public data network. X.32 describes the functional and procedural aspects of the DTE/DCE interface for DTEs accessing a packet

switched public data network via a public switched network.

X.38 ITU-T recommendation for the access of Group 3 facsimile equipment to the Facsimile Packet Assembly/Disassembly (FPAD) facility in public data networks situated in the same country.

X.39 ITU-T recommendation for the exchange of control information and user data between a Facsimile Packet Assembly/Disassembly (FPAD) facility and a packet mode Data Terminal Equipment (DTE) or another pad, for international networking.

X.3T9.3 The ANSI committee responsible for the creation and perpetuation of Fiber Channel standards.

X.3T9.5 The ANSI committee responsible for the creation and perpetuation of FDDI Standards.

X.4 International Alphabet No.5 for character oriented data.

X.400 X.400 is an international standard which enables disparate electronic mail systems to exchange messages. Although each e-mail system may operate internally with its own, proprietary set of protocols, the X.400 protocol acts as a translating software making communication between the electronic mail systems possible. The result is that users can now reach beyond people on their same e-mail system to the universe of users of interconnected systems. One problem with e-mail sent between X.400 networks is that the sender's name is not sent. (I kid you not.) This was one element of the protocol the committees forgot! If your message crosses an X.400 network, remember to sign your name. The X.400 standard itself is an overview which is broken down under subsequent numbers:

X.402 Overall Architecture
X.403 Conformance testing
X.407 Abstract service definition conventions
X.408 Encoded information type conversion rules
X.411 Message transfer system
X.413 Message store
X.419 Protocol specifications

X.420 Interpersonal messaging system. An IPM format specification using the X.400 transfer protocol. In addition to text, it also allows CAD/CAM, graphics, Fax, and other electronic information.

X.435 An EDI (Electronic Data Interchange) format specification based on the X.400 transfer protocol. It can also allow for CAD/CAM, graphics, Fax, and other electronic information to accompany an EDI interchange.

X.440 A VM (Voice Messaging) format specification, using the X.400 transfer protocol. In addition to voice, it can also contain CAD/CAM, graphics, fax, and other electronic information.

X.445 The X.445 standard, or APS (Asynchronous Protocol Specification), lets X.400 clients and servers exchange all types of digital data over public telephone networks rather than over X.25 leased lines, which are required today. Among those backing the spec are Intel, AT&T, Microsoft Corp., Lotus and Isocor. X445 is an extension of the X.400 standard. X.400 provides an option to other messaging backbones such as System Network Architectural Distribution Services and SMTP. It supports multimedia data traffic including text, binary files, E-mail, voice images, and sound.

In its December 19, 1994 issue, PC Week said that the X.445 standard, which earlier this month gained final approval from the International Telecommunications Union, should ease messaging dramatically by allowing users to exchange X.400 data traffic over standard telephone networks.

X.50 Fundamental parameters of multiplexing scheme for the international interface between synchronous data networks.

X.500 The ITU-T international standard designation for a directory standard that permits applications such as electronic mail to access information which can either be central or distributed. The X.500 standard for directory services provides the means to consolidate e-mail directory information through central servers situated at strategic points throughout the network. These X.500 servers then exchange directory information so each server can keep all its local mail directory information current. With X.500, any e-mail user, whether on OpenVMS, Macintosh, DOS, or UNIX workstations, can be listed in a central directory that can be accessed using an X.500-compatible user agent.

X.509 A cryptography term. The part of the ITU-T X.500 Recommendation which deals with Authentication Frameworks for Directories. Within X.509 is a specification for a certificate which binds an entity's distinguished name to its public key through the use of a digital signature. Also contains the distinguished name of the certificate issuer.

X.51 Fundamental parameters of multiplexing scheme for the international interface between synchronous data networks using 10-bit envelop structure.

X.51 bis Fundamental parameters of a 48Kbit/s transmission scheme for the international interface between synchronous data networks using a 10-bit envelope structure.

X.58 Fundamental parameters of multiplexing scheme for the international interface between synchronous data networks using a 10-bit envelope structure.

X.60 Common channel signaling for circuit switched data applications.

X.61 Signaling system no. 7 — data user part.

X.70 Terminal and transit control signaling for asynchronous services on international circuits between anisochronous data networks.

X.71 Decentralized terminal and transit control signaling on international circuits between synchronous data networks.

X.75 An international standard for linking X.25 packet switched networks. X.75 defines the connection between public networks, i.e. for a gateway between X.25 networks. See X.25.

X.80 Interworking of inter-exchange signals for circuit switched data services.

X.92 Hypothetical reference connections for synchronous PDNs in a packet switched network.

X.95 ITU-T specification dealing with a number of internal packet-switched network parameters such as packet size limitations and service restrictions.

X.96 Call progress signals in PDNs.
X/Open A group of computer manufacturers that promotes the development of portable applications based on UNIX. They publish a document called the X/Open Portability Guide.
X12 ANSI (American national Standards Institute) standard that is the dominant EDI (Electronic Data Interchange) standard in the U.S. today; designed to support cross-industry exchange of business transactions. Standard specifies the vocabulary (dictionary) and format for electronic business transactions.
X3.15 Bit sequencing of ASCII in serial-by-bit data transmission.
X3.16 Character structure and character parity sense for serial-by-data communications in ASCII.
X3.36 Synchronous high speed data signaling rates between data terminal equipment and data circuit terminating equipment.
X3.41 Code extension techniques for use with 7-bit coded character set of ASCII.
X3.44 Determination of the performance of data communications systems.
X3.79 Determination of performance of data communications systems that use bit oriented control procedures.
X3.92 Data encryption algorithm.
X.Windows A networked GUI developed at the Massachusetts Institute of Technology (MIT) as part of Project Athena. Based on a client/server architecture it displays information from multiple networked hosts on a single workstation.
XA Transaction management protocol.
XA-SMDS Exchange Access SMDS. An access service provided by a local exchange carrier to an interexchange carrier. It enables the delivery of a customer's data over local and long distance SMDS networks.
XAPIA X.400 Application Program Interface Association. Microsoft has a new concept. It's called an Enterprise Messaging Server (EMS). The idea is to allow users to transparently access the messaging engine from within desktop applications to route messages, share files, or retrieve reference data. According to Microsoft, corporate developers will be able to add capabilities using Visual Basic and access EMS by writing either to the X.400 Application Program Interface Association's (XAPIA's) Common Mail Calls (CMC) or to Microsoft's Messaging API (MAPI). (ftp://nemo.ncsl.nist.gov/pub/olw/dssig/xapla/) See MAPI.
Xbar Crossbar.
XC Cross connect.
XDP Ron Stadler of Panasonic dreamed this one up. It stands for eXtra Device Port. It's an analog RJ-11 equipped port on the back of a Panasonic digital telephone, which is driven by Panasonic's Digital Super Hybrid switch. The XDP is an extension line completely separate to your digital voice line. You can be speaking on the phone while receiving or sending a fax or while sending or receiving data. Or plug a cordless phone or answering machine into the XTP.
XDSL A generic — X is the generic — name for Asymmetrical Digital Subscriber Line equipments and services. See ADSL for a much fuller explanation.
Xenix Microsoft trade name for a 16-bit microcomputer operating system derived from AT&T Bell Labs' UNIX.
Xerographic Recording Recording by action of a light spot on an electrically charged photoconductive insulating surface where the latent image is developed with a resinous powder.
Xerox Network Services XNS. A multilayer protocol system developed by Xerox and adopted, at least in part, by Novell and other vendors. XNS is one of the many distributed-file-system protocols that allow network stations to use other computers files and peripherals as if they were local. XNS is used by some companies on Ethernet LANs. In local area networking technology, special communications protocol used between networks. XNS/ITP functions at the 3rd and 4th layer of the Open Systems Interconnection (OSI) model. Similar to transmission control protocol/internet protocol (TCP/IP).
XFR TransFeR.
XGA eXtended Graphics Array. A new IBM level of video graphics which has a screen resolution of 1,024 dots horizontally by 768 vertically, yielding 786,432 possible bits of information on one screen, more than two and a half times what is possible with VGA. See also MONITOR.
XID Frame A High-Level Data Link Control (HDLC) transmission frame used to transfer operational parameters between two or more stations.
XIP eXecute-In-Place. Refers to specification for directly executing code from a PCMCIA Card without first having to load it into system memory.
XIWT Cross-Industry Working Team. About 28 companies (and growing) whose work centers on the Clinton-Gore administration's document "National Information Infrastructure — Agenda for Action." The document spells out policy initiatives required to achieve the benefits of widespread, convenient and affordable access to existing and future information resources. In short, the Information SuperHighway.
XJACK A registered trademark of modem manufacturer, MegaHertz, for one of the most innovative ideas in PCMCIA modems. XJACK is the world's tiniest female RJ-11. It's about a quarter of an inch wide and the most convenient way of attaching your laptop's modem to the phone network.
XLIU X.25/X.75/X.75' Link Interface Unit.
XMA eXtended Memory specificAtion. Interface that lets DOS programs cooperatively use extended memory in 80286 and higher computers. One such driver is Microsoft's HIMEM.SYS, which manages extended memory and HMA (high memory area), a 64k block just above 1Mb.
Xmit Transmit.
XModem Also called "Christiansen Protocol". An error-correcting file transfer, data transmission protocol created by Ward Christiansen of Chicago for transmitting files between PCs. A file might be anything — a letter, an article, a sales call report, a Lotus 1-2-3 spreadsheet. The XModem protocol sends information in 128 bytes blocks of data. Some sums (check sums) are done on each block and the result is sent along with the block. If the result does not check out at the other end, the computer at the other end sends a request (a NAK — Negative AcKnowledgement) to re-transmit that block once again. If the block checks

out, the computer sends an ACK — an Acknowledgement. In this way, relatively error-free transmission can be accomplished. XModem was first used by computer hobbyists and then by business users of PCs. If you're buying a telecommunications software program for your PC — IBM, Radio Shack, Compaq, Apple, etc. — it's a good idea to buy a program with XModem. It's among the most common data communications protocols. But it's not the fastest, just the most common. AT&T Mail supports XModem protocol. So does TELECONNECT Magazine's own E-mail InfoBoard system (212-989-4675). MCI Mail does not support XModem protocol. We don't know why. There are many variations of XModem including XModem 1K (which uses blocks of 1,025 bytes), Modem7, YModem, Y-Modem-G, and ZModem. Most common communications software packages only support (i.e. will handle) the original version of XModem (checksum) and the newer CRC variation. A study in Byte Magazine (March, 1989) showed ZModem to be a far more efficient file transfer protocol than XModem, YModem, or W/XModem. The author now tends to use ZModem more commonly. It is supported by most on-line services, such as CompuServe, etc. See also DATA COMPRESSION PROTOCOLS, ERROR CONTROLS PROTOCOLS, FILE TRANSFER PROTOCOL, YMODEM and ZMODEM.

XModem-1K Xmodem-1K is an error-correcting file transfer, data transmission protocol for transmitting files between PCs. It is essentially Xmodem CRC with 1K (1024 byte) packets. On some systems and bulletin boards it may also be referred to as Ymodem.

XModem-CRC Cyclic Redundancy Checking is added to XMODEM frames for increased reliability of errors detection. See XMODEM.

XMP X/Open Management Protocol; an API and software interface specified in the Open Software Foundation's Distributed Management Environment.

XMS An acronym for eXtended Memory Specification. To run this standard, your system must have 350K of extended memory. XMS creates the HMA (High Memory Area), then governs access to and the allocation of the remainder of extended memory.

XNMS Trademark for MICOM's IBM PC-based packet data network (PDN) network management system software products.

XNS Xerox Network System. The lan architecture developed at the Xerox Palo Alto Research Center (Parc). It is a five-layer architecture of protocols and was the foundation of the OSI seven-layer model. It has been adopted in part by Novell and other vendors. XNS is one of the many distributed-file-system protocols that allow network stations to use other computers files and peripherals as if they were local. XNS is used by some companies on Ethernet LANs.

XO Crystal Oscillator.

XON/XOFF XON/XOFF are standard ASCII control characters used to tell an intelligent device to stop or resume transmitting data. In most systems typing <Ctrl>-S sends the XOFF character, i.e. to stop transmitting. Some devices understand <Ctrl>-Q as XON, i.e. start transmitting again. Others interpret the pressing of any key after <Ctrl>-S as XON.

XOpen A group of computer manufacturers that promotes the development of portable applications based on UNIX. They publish a document called the X/Open Portability Guide. X/Open is the correct spelling.

XPAD An eXternal Packet Assembler/Disassembler.

XPC A set of protocols developed by British Telecom Tymnet to allow asynchronous terminals to connect to an X.25 packet-switched network.

XPM Extended Peripheral Module.

XRB Transmit Reference Burst.

XSG X.25 Service Group.

XT Abbreviation for crosstalk.

XTEL API The Sun Solaris Teleservices Application Programming Interface. See TELESERVICES for a long explanation.

XTELS The Sun Solaris Teleservices protocol. The XTELS protocol is essentially the XTEL provider protocol with extensions to support multiple clients and multiple XTEL providers. See TELESERVICES for a long explanation.

XTELTool A window-based tool used to configure providers in the Sun Solaris Teleservices platform. Basically a graphical user interface. See TELESERVICES for a long explanation.

XTEND EXTEND Telecommunications Integration Platform, a series of network software utilities and calls that allow a PBX to be integrated and directed by workstations on a local area network. XTEND is an offering of XTEND Communications, Inc. New York City.

XWindows See X-WINDOWS.

Y-Dimension Of Recorded Spot In facsimile, the center-to-center distance between two recorded spots measured perpendicular to the recorded line.

Y-Dimension Of Scanning Spot In facsimile, the center-to-center distance between two scanning spots measured perpendicular to the scanning line on the subject copy. The numerical value of this term will depend upon the type of system used.

Y/C A two channel video channel. One is for color (chrominance) and the other for black and white (luminance).

Yagi Antenna A type of directional antenna.

Yahoo Huge directory of all kinds of places on the Web. Apparently Yahoo stands for "Yet Another Hierarchical Officious Oracle." See SEARCH ENGINES for a list of other indexes to what's on the Web. http://akebono.stanford.edu/yahoo/

Yellow Alarm A T-1 alarm signal sent back toward the source of a failed transmit circuit in a DS-1 2-way transmission path. A yellow sends 0's (zeros) in bit two of all time slots. See also T-1.

Yellow Cable A coaxial cable used in 10Base-5 networks. It is also referred to as "Thick" coax. This was the first cable used on many early LANs.

Yellow Pages A directory of telephone numbers classified by type of business. It was printed on yellow paper throughout most of the twentieth century until it was obsoleted in the late 1990s by dial up yellow page directories operated by voice processing systems and in the early 21st century by electronic directories delivered on disposable laser disks. As a concession to history, the laser disks are now painted bright yellow. Actually, yellow pages remain one of the phone companies' most lucrative sources of revenues. Advertising rates are not cheap. There is now competition. There are many "Yellow Pages" directories, since AT&T never trademarked the term "Yellow Pages." Some "yellow page" directories are better value than others. And some are more legitimate than others. Some actually never get printed or, if they are printed, are not printed in great quantity and are not distributed as widely as their sales literature implies. Many businesses have been suckered into paying money for listings and advertisements in directories that never appeared. This fictitious directory scam also has happened with "telex" and "fax" directories. This "scam" is fraud by mail and is heavily stomped upon by the US Postal Service. As a result, many fake directories (especially the telex ones) are "published" abroad.

Yes A program which Novell created to certify products — hardware and software — that work with its products. It's a rigorous program. It takes significant work to be "Yes" certified by Novell. Novell has Yes programs for NetWare and UnixWare. It's Novell's equivalent of the Good Housekeeping Seal of Approval.

YMMV Your Mileage May Vary. A response usually given when the answer is not precise and depends on the user's own circumstances. YMMV is an acronym used in electronic mail on the Internet to save words or to be hip, or whatever.

YModem A faster transfer variation of XMODEM. In YMODEM, XMODEM's 128-byte block grew to YMODEM's 1024 bytes (1 kilobyte). YMODEM combines the 1K block and the 128-byte block modems into the same protocols. YMODEM, or 1K as it is known, became the thrifty way to send files (i.e. it saved on phone time). Enhancements were added, such as auto-fallback to 128-byte blocks if too many errors were encountered (because of bad phone lines, etc.) See XMODEM for a much larger explanation of file transfer using X, Y and ZMODEM protocols.

YModem-G Ymodem-g is a variant of Ymodem. It is designed to be used with modems that support error control. This protocol does not provide software error correction or recovery, but expects the modem to provide it. It is a streaming protocol that sends and receives 1K packets in a continuous stream until told to stop. It does not wait for positive acknowledgement after each block is sent, but rather sends blocks in rapid succession. If any block is unsuccessfully transferred, the entire transfer is canceled. See also ZMODEM, which we prefer.

Yurt A Mongolian circular shed. Some companies are making Yurt-like sheds for installation in the back yards of telecommuters — especially those telecommuters who don't have enough room inside for a home office.

YUV A color encoding scheme for natural pictures in which luminance and chrominance are separate. The human eye is less sensitive to color variations than to intensity variations. YUV allows the encoding of luminance (Y) information at full bandwidth and chrominance (UV) information at half bandwidth.

YUV9 The color encoding scheme used in Indeo Video Technology. The YUV9 format stores information in 4x4 pixel blocks. Sixteen bytes of luminance are stored for every one byte of chrominance. For example, a 640x480 image will have 307,200 bytes of luminance and 19,200 bytes of chrominance.

Z

Z Abbreviation for Zulu time. See GREENWICH MEAN TIME.
Zap To eradicate all or part of a program or database, sometimes by lightning, sometimes intentionally.
ZBTSI Zero Byte Time Slot Interchange. A technique used with the T carrier extended superframe format (ESF) in which an area in the ESF frame carries information about the location of all-zero bytes (eight consecutive "0"s) within the data stream.
Zen A Japanese sect of Buddhism that stresses attaining enlightenment through intuition rather than by studying scripture.
Zener Diode A particular type of semiconductor which acts as a normal rectifier until the voltage applied to it reaches a certain point. At this point — at the zener voltage or the avalanche voltage — the zener diode becomes conducting.
Zero Beat Reception Also called "homodyne" reception. A method of reception using a radio frequency current of the proper magnitude and phase relation so that the voltage impressed on the detector will be of the same nature as that of the wave. An old radio term.
Zero Bit The high-order bit in a byte or a word.
Zero Byte Time Slot Interchange ZBTSI. A method of coding in which a variable address code is exchanged for any zero octet. The address information describes where, in the serial bit stream, zero octets originally occurred. It is a five-step process where data enters a buffer, zero octets are identified and removed, the nonzero bytes move to fill in the gaps, the first gap is identified, and a transparent flag bit is set in front of the message to indicate that one or more bytes originally contained zeros. See ZBTSI.
Zero Code Suppression The insertion of a "one" bit to prevent the transmission of eight or more consecutive "zero" bits. Used primarily with digital T1 and related telephone-company facilities which require a minimum "ones density" keep the individual sub channels of a multiplexed, high-speed facility active. Several different schemes are currently employed to accomplish this. Proposals for a standard are being evaluated by the ITU-T. See also ZERO SUPPRESSION.
Zero Fill See ZEROFILL.
Zero Frequency The frequency (wavelength) at which the attenuation of the lightguide is at a minimum.
Zero Power Modem A modem that takes its power from the phone line and therefore needs no battery or external power. Such modems are often limited in their speed and capabilities.
Zero Slot LAN A Local Area Network (LAN) that uses a PC's serial port to transmit and receive data. It doesn't require a network interface card to be installed in a slot in the PC, thus the name "zero-slot" LAN. RS-232 LANs usually use standard RS-232 or phone cable to link PCs. Software does the rest of the work. Due to the slow speed of serial communications on a PC, RS-232 LANs are usually restricted to speeds of around 19.2K bits per second. What they lose in speed, however, they make up in low price.
Zero Stuffing Get a cup of coffee right now. Synchronous data transmission is done by sending what IBM and AT&T call Frames, and what everyone else calls Packets. A frame starts off by sending a bit pattern of 01111110 (notice the six 1's in a row). Synchronous transmission is for sending a bit stream, which means that the bits may (but probably do not) have any relation to the transmission of characters. This is especially true when sending digitized voice. As the bits pass to the receiver, they go through a shift register. When the flag signifying the end of a frame goes by, the last 16 bits in the shift register are the check digits.
The receiver computes the check digits based on the data bits that have gone by. As the sender sent the data, it computed the check digit, sent it after the end of the frame, and then sent the flag. If the receiver computes the same check digit that the sender sent, then one can be reasonably assured the data came through without error. But that's not what I came to talk to you about. I came to talk about Zero Stuffing. The problem is that somewhere in the bit stream, there is the possibility of there being six 1 bits in a row. To the receiving computer, six 1's means a flag. Therefore the sending computer, if it "sees" six 1 bits, will send five 1 bits, and stuff a zero in the bit stream.
In fact, if it sees even five 1 bits, it will stuff a zero anyway, so there will be no ambiguity. The rule is, "If there are five ones in a row and it is NOT the end of a frame, stuff a zero into the bit stream." This way the receiver will know that this is in no way the end of the frame yet. Now if the receiver sees six 1's in a row, it knows without a doubt that it IS at the end of a frame, and should proceed with the error checking.
Zero Suppression The elimination of nonsignificant zeros from a numeral. Zero suppression is the replacement of leading zeros in a number with blanks so that when the number appears, the leading zeros are gone. The data becomes more readable. For example, the number 00023 would be displayed on the monitor or printed as 23.
Zero Test Level Point A level point used as a reference in determining loss in circuits. Analogous to using sea level when defining altitude. Written as 0 TLP.
Zero Transmission Level Point ZTLP. In telephony, a reference point for measuring the signal power gain and losses of telecommunications circuit, at which a zero dBm signal level is applied.
Zero Transmission Level Reference Point A point in a circuit to which all relative transmission levels are referenced. The transmission level at the transmitting switchboard is frequently taken as the zero transmission level reference point.
Zero Usage Customer An MCI definition. An MCI customer who has not placed a call over the network, even though he/she is an active customer. Sometimes used interchangeably, but incorrectly, with the term "no usage customer."
Zerofill 1. To fill unused storage locations with the character "0."

2. Here's definition from GammaLink, a fax board maker: A traditional fax device is mechanical. It must reset its printer and advance the pages as it prints each scan line it receives. If the receiving machine's printing capability is slower than the transmitting machine's data sending capability, the transmitting machine adds "fill bits" (also called Zero Fill) to pad out the span of send time, giving the slower machine the additional time it needs to reset prior to receiving the next scan line.

ZIF Zero Insertion Force. Intel makes a bunch of math co-processor chips which are used with their 80XXX range of microprocessors. ZIF is a special device which is typically soldered to the motherboard. You place an 80387 chip on this device, move the handle down, it grabs the chip and pulls the chip down, seating it electrically. When you want to remove the chip, you simply lift the handle and up the chip comes. The device was invented by Intel because so many people were apparently breaking the legs on their math co-processor chips each time they removed them. Apparently the problem was most prevalent in the computer rental business.

Zinc Spark Gap A spark gap having zinc as the electrode.

Zip It all started years ago with a very popular PC program called PKZIP. You could run this program on a file on your PC and bingo PKZIP would reduce the size of the file by as much as 90%. You could then transmit the zipped file over a phone line, saving 90% of the time and 90% of the cost. At the other end the recepient would run another file called PKUNZIP and bingo your file would be returned to its original size. You get the program from PKWARE, Inc. Glendale, WI 414-352-3670.

Zip Code The specific five-digit code assigned to each post office to facilitate mail delivery. The Postal Service is beginning to use nine-digit zip codes. The additional four digits will determine the precise letter carrier routing from the post office into the street. Some post offices now optically read zip codes. Allegedly this speeds up delivery.

Zip Tone Short burst of dial tone to an ACD agent headset indicating a call is being connected to the agent console.

Zipped See ZIP.

ZModem ZMODEM is an error-correcting file transfer, data transmission protocol for transmitting files between PCs. A file might be anything — a letter, an article, a sales call report, a Lotus 1-2-3 spreadsheet. Always use ZMODEM if you can. It's the best and fastest data transmission protocol to use. This is not my sole advice. Virtually every writer in data communications recommends it. Here's an explanation, beginning with XMODEM, an older, more common and less efficient protocol. Both XMODEM and YMODEM transmit, then receive, then transmit. The handshake (ACK or NAK) happens when the sender isn't sending. ZMODEM adds full duplex-transmission to the transfer protocol. ZMODEM does not depend on any ACK signals from the host computer. It keeps sending unless it receives a NAK, at which time it falls back to the failed block and starts to retransmit at that point. ZMODEM was written by Chuck Forsberg. According to PC Magazine (April 30, 1991) ZMODEM is the first choice of most bulletin boards. ZMODEM, according to PC Magazine, features relatively low overhead and significant reliability and speed. ZMODEM dynamically adjusts it packet size depending on line conditions and uses a very reliable 32-bit CRC error check. It has a unique file recovery feature. Let's say ZMODEM aborts a transfer because of a bad line (or whatever), it can start up again from the point it aborted the transfer. Other file transfer protocols have to start all over again. ZMODEM's ability to continue is a major benefit. ZMODEM in some communications program is a little more automated than other protocols. For example, ZMODEM will start itself when the other end gives a signal — thus saving a keystroke or two and speeding things up. See FILE TRANSFER PROTOCOL, XMODEM and YMODEM.

Zoomed Video ZV. A technology that allows certain streams of digital information to write directly to a laptop's screen, bypassing the CPU and its bus (ISA, PCI, EISA, etc.). Zoomed video can show full 30 frames per second movies to a laptop screen. Zoomed video used MPEG-2. To show NTSC on a laptop, you need a special ZV (Zoomed Video) PCMCIA card. It feeds NTSC video directly to the screen. Zoomed Video technology allegedly will bring full laptop screen video conferencing to laptops.

Zone 1. A telephony definition: One of a series of specified areas, beyond the base rate area of an exchange. Service is furnished in zones at rates in addition to base rates.

2. A LAN definition. Part of a local area network, typically defined by a router. A router will let you get into one part of someone else's network. They define what you are able to get access to. You might get to that router by an external telecommunications circuit — dial up, ISDN, Switched 56, T-1 etc.

Zone Bits 1. One or two leftmost bits in a commonly used system of six bits for each character.
2. Any bit in a group of bit positions that are used to indicate a specific class of items, i.e., numbers, letters, commands.

Zone Method A ceiling distribution method in which ceiling space is divided into sections or zones. Cables are then run to the center of each zone to serve the information outlets nearby. See also CEILING DISTRIBUTION SYSTEMS.

Zone Of Silence Skip zone.

Zone Paging Ability to page a specific department or area in or out of a building. "Page John in the Accounting Department." Zone paging is useful for finding people who wander, as most of us do.

Zulu Time Coordinated Universal Time.

ZV See ZOOMED VIDEO.

ZZF Zentralamt fur Zulassungen im Fernmeldewessen (Approval Authority — Germany).

Centronics Parallel Interface

RS-449 Interface

RS-232 Interface

IEEE 488 Interface

IBM PC Color Monitor Interface

IBM PC Monochrome Monitor Interface

IBM PS/2 Keyboard Interface

IBM PC Keyboard Interface

RJ-11-4

RJ-11-6

RJ-45

RJ-45 KEYED

RJ-11-6 MODIFIED

PUBLISHED BY FLATIRON PUBLISHING

Flatiron Publishing publishes books and magazines and organizes trade conferences on computer telephony, telecommunications, networking and voice processing. It also distributes the books of other publishers, making it the "central source" for all the above materials. Call or write for you FREE catalog.

236 Killer Voice Processing Applications
ATM Users' Guide
Client Server Computer Telephony
Complete Traffic Engineering Handbook
Customer Service Over the Phone
Frames, Packets and Cells in Broadband Networking
The Guide to Frame Relay
The Guide to SONET
The Guide to T-1 Networking
Local & Long Distance Telephone Billing Practices
Moore's Imaging Dictionary
The MVIP Book
Newton's Telecom Dictionary
PC-Based Voice Processing
SCSA
Speech Recognition
Telephony for Computer Professionals
VideoConferencing: The Whole Picture

Quantity Purchases

If you wish to purchase this book, or any others, in quantity, please contact:

Christine Kern, Manager
Flatiron Publishing, Inc.
12 West 21 Street
New York, NY 10010
212-691-8215 or 1-800-LIBRARY
facsimile orders: 212-691-1191

Flatiron Publishing, Inc.
12 West 21 Street
New York, NY 10010
212-691-8215 1-800-LIBRARY